ENCYCLOPEDIA OF

FOOD SCIENCE AND TECHNOLOGY

VOLUME 4

WILEY ENCYCLOPEDIA OF FOOD SCIENCE AND TECHNOLOGY, Second Edition

Editor-in-Chief
Frederick J. Francis
University of Massachusetts, Amherst

Associate Editors

Christine M. Bruhn
Center for Consumer Research

Pavinee Chinachoti
University of Massachusetts, Amherst

Fergus M. Clydesdale
University of Massachusetts, Amherst

Michael P. Doyle
University of Georgia

Kristen McNutt
Consumer Choices, Inc.

Carl K. Winter
University of California, Davis

Editorial Staff
Publisher: **Jacqueline I. Kroschwitz**
Associate Editor: **Glenn Collins**
Managing Editor: **John Sollami**
Editorial Assistants: **Susan O'Driscoll, Hugh Kelly**

ENCYCLOPEDIA OF

FOOD SCIENCE AND TECHNOLOGY
Second Edition

VOLUME 4

Frederick J. Francis
University of Massachusetts
Amherst, Massachusetts

A Wiley-Interscience Publication
John Wiley & Sons, Inc.
New York / Chichester / Weinheim / Brisbane / Singapore / Toronto

Published simultaneously in Canada.

For ordering and customer service, call 1-800-CALL-WILEY.

Library of Congress Cataloging-in-Publication Data:

Wiley encyclopedia of food science and technology.—2nd ed. / [edited by] Frederick J. Francis.
    p. cm.
Rev. ed. of: Encyclopedia of food science and technology / Y.H. Hui, editor-in-chief. c1992.
Includes bibliographical references.
ISBN 0-471-19285-6 (set : cloth : alk. paper).—ISBN 0-471-19255-4 (v. 1 : cloth : alk. paper).—ISBN 0-471-19256-2 (v. 2 : cloth : alk. paper).—ISBN 0-471-19257-0 (v. 3 : cloth : alk. paper).—ISBN 0-471-19258-9 (v. 4 : cloth : alk. paper)
 1. Food industry and trade Encyclopedias.   I. Francis, F. J. (Frederick John), 1921-   . II. Encyclopedia of food science and technology.
TP368.2.E62   2000
664′.003—dc21   99-29003
                                                          CIP

Printed in the United States of America.

10 9 8 7 6 5 4 3 2 1

S

SALAD OILS. See FATS AND OILS: PROPERTIES, PROCESSING TECHNOLOGY, AND COMMERICAL SHORTENINGS.

SALMONELLA. See FOODBORNE DISEASES.

SANITIZERS. See DISINFECTANTS.

SAUSAGES

Sausages are a diverse group of foods made from ground or comminuted meats, salt and spices. They originated during prehistoric times when our ancestors discovered that addition of salt and drying would delay spoilage. Adding various spices improved palatability and stuffing them into intestines enhanced convenience. Homer mentions sausages in the Odyssey and the word, sausage, come from Latin meaning salted or meat preserved by salting. Over the centuries virtually every culture developed distinctive sausages depending on their live stock, spices, climate, and culture. The European origin of most of our sausages is indicated by the city names associated with them: Bologna, Genoa, Braunschweig, and Thuringen. Dried sausages originated in southern Europe where warm climates required a more spoilage-resistant sausage. Northern Europeans, having a cooler climate, created cooked and summer sausages. Americans contributed Lebanon bologna and the popularization of the hot dog.

Sausages are an excellent source of high quality proteins. In addition, they contribute iron, zinc, and B vitamins, particularly folic acid, B_6 and B_{12}. Controversy has existed over the addition of nitrite and possible subsequent formation of nitrosamines, sodium content and fat levels, especially saturated fats.

Americans manufactured 6.0 billion pounds of sausages in 1987, 10% of the total amount of all red meats. Frankfurters amounted to 1.5 billion pounds, bologna 640 million, uncooked cured sausages 24 million, dried and semi-dried 445 million, liver sausages 96 million, loaf products 1.3 billion, and other sausage and loaf products 1.0 billion pounds. Poultry based luncheon meats totaled 151 million pounds and poultry frankfurters were 280 million pounds.

Classification of sausages is not clear cut because of the many variations in materials, spices and processes used. Meat is ground or finely chopped (comminuted) and mixed with salt and spices. It may have curing salts added and may then be stuffed, smoked, cooked, fermented, or dried. The United States Department of Agriculture classifies sausages as fresh; uncooked-smoked; cooked-smoked; cooked; dry/semi-dry; and luncheon meats, loaves, and jellied products.

Fresh sausages are made from uncured (no nitrate or nitrite added) meats, usually pork. They are ground and mixed with salt and spices, then stuffed into chubs, natural casings or manufactured casings. They have a limited shelf-life even when refrigerated and must be cooked before serving.

Uncooked-smoked sausages (country-style sausage and kielbasa) are usually cured, require refrigeration for preservation and are cooked before serving.

Cooked-smoked sausages (frankfurters and bologna) are cured and cooked in casings inside a smokehouse. When properly refrigerated they have a long storage life and do not require final cooking by the consumer.

Cooked sausages (liver sausage and Braunschweiger) may be cured and are usually comminuted, seasoned, stuffed, and cooked. They are generally consumed cold.

Dry/semi-dry sausages utilize bacteria to ferment sugars into lactic acid. This increases the sausages' acidity and contributes to their characteristic flavor and to their bacterial preservation. They are then dried to varying extents. These sausages do not need cooking before serving. Semi-dried sausages (summer sausages, cervelat, thuringer) need refrigeration for optimal storage. Dry sausages (hard salami, Genoa salami and pepperoni) need little or no refrigeration. Lebanon bologna is fermented and smoked but not appreciably dried.

The category of luncheon meats encompasses a wide variety of popular products. Typically a mixture of chopped meats with extenders and other ingredients is processed into a loaf form. Pimento loaf, olive loaf, and honey loaf are examples. Also included in this group are sandwich spreads, which are cooked, soft mixtures packaged in chubs. Jellied products, such as jellied tongue, souse and head cheese, consist of chunks of cooked and cured meats held in a loaf form by gelatin.

INGREDIENTS

Meats

Beef and pork are the primary meats, but goat and sheep are used by several cultures and fish sausages are consumed in the Orient. Poultry luncheon meats and frankfurters are increasing in popularity. Skeletal muscle has the highest binding and water holding properties compared to other meats and non-meat extenders. Other meats (lips, tripe, cardiac muscle), variety meats (tongue, livers, pork stomachs, etc) and mechanically deboned meats must be labelled if used. Pork trimmings and backfat are important ingredients, because too little fat results in a dry, unpalatable product. Maximum levels of fat for different sausages are regulated by the U.S. Department of Agriculture; for example, frankfurters must have less than 30% fat.

Salt

Salt is necessary for extracting the meat proteins to bind fat and water and to form the sausages' texture. Salt also contributes to the sausages' flavor and retards microbiological growth. Most sausages contain 2.5 to 3.0% salt, although fresh sausages contain only about 0.75%. To reduce the sodium content, potassium chloride may be substituted for salt (sodium chloride). The practical limit to the amount of potassium chloride that can be added is half of the total salt because of the bitter taste of the potassium. Addition of phosphates is another way to improve the binding and water holding of the proteins. Pyrophosphates, tripolyphosphate, and metaphosphates may be added to a combined maximum of 0.5%.

Curing Salts

Originally part of natural salts, sodium and potassium nitrate and nitrite causes formation of the cured meat color and flavors. The nitrate is converted by naturally occurring bacteria to nitrite, which in turn forms nitric oxide. This binds to the heme iron of myoglobin to form the pink color of cured meats. Nitrite also contributes to the flavor of cured meats and retards rancidity. Nitrite inhibits the outgrowth of *Clostridium botulinum* spores; growth of these bacteria can produce a potentially fatal toxin. Nitrite addition is strictly limited because of the potential for formation of mutagenic nitrosamines. Use of sodium ascorbate or erythorbate prevents formation of nitrosamines by competing with secondary amines for nitrite, speeds the curing reactions, and stabilizes the cured meat color.

Spices

The flavor of a particular sausage is strongly dependent on the spices added; every sausage maker has a special mixture. Sage gives the flavor to fresh breakfast sausages; white pepper, mustard, coriander and nutmeg flavor frankfurters; garlic is added to Italian sausages; and cayenne pepper, paprika and garlic flavor pepperoni.

Sugars

Sucrose (common sugar), dextrose, corn syrups, or sugar derivatives may be added to many sausages. About 1% is necessary in fermented sausages for growth of the lactic acid bacteria. Sugars contribute to the flavors by their sweetness, participation in browning flavors, and masking (mellowing) the saltiness.

Water

Often added as ice to chill the meats, addition of water permits more efficient chopping and mixing, aids in dissolving the salts, and improves the fluidity of the mixture for proper filling of the casings. The texture of sausages is greatly enhanced by having the proper amount of moisture.

Additives

Acidulents, such as glucono-delta-lactone, are occasionally added to increase the acidity rapidly. This accelerates the curing reaction in quickly processed products such as frankfurters, or controls the growth of undesirable microorganisms in fermented sausages until fermentation occurs. Soy proteins, milk proteins (casein and whey), yeast protein, flours, and starch are permitted in limited amounts in some sausages where they improve emulsion stability, fat and water binding, flavor, cooking yield, and textural characteristics. Antioxidants to retard rancidity and mold inhibitors are added to some products. All of these additives must be declared on the label if they are used.

PROCESS

Grinding and Comminuting

The meat for fresh and fermented sausages is ground through a grinder and then mixed with salt and any other ingredients. Some loaf and luncheon meat products use chunks of meat bound together by the extracted meat proteins that coat the surface after mixing with salt and water.

Frankfurters and bolognas are comminuted or finely chopped with the salt and ice to extract/solubilize the proteins. Then the fat and remaining ingredients are added and chopped further to form a smooth batter.

Casings

The sausage mixture or batter is stuffed into casings of various types, sizes and shapes for further processing, storage, and consumer convenience. Originally natural or animal casings made from stomachs, intestines and bladders of cattle, hogs and sheep were used, and are still associated with many traditional, high quality sausages. They are in limited supply and highly perishable. They are less uniform in size, more fragile and require more care during stuffing than manufactured casings. The latter are either cellulose from cotton linters (cotton by-products) or collagen from the corium layer of beef hides. Cellulose casings are available in many sizes, strengths, and properties. Small types are used to cook frankfurters in a smokehouse and then removed before packaging. Large fibrous cellulose casings, which remain on luncheon meats and fermented and dried sausages, are removed by the consumer. Small collagen casings can be consumed and are used for breakfast and other uncooked sausages.

Stuffing and Linking

The sausage mixtures and frankfurter batters can be pumped through stuffers by either impeller or piston pumps. The casings are placed over a stuffing horn and the sausage extruded into the casing. Links are formed by twisting the casing or by using string ties or metal clasps.

Fermenting and Drying

Originally the bacteria for fermented meats were natural contaminants of the meats, equipment and facilities. Small amounts of successfully fermented products were often added back to new batches to start the fermentation. Most manufacturers now add cultures of selected bacteria and a fermentable sugar and then hold the mixture at 70–82°F and 80% relative humidity for 2–3 days to ensure a desirable fermentation. The bacteria produce lactic acid which gives the sausages their tangy flavor and reduces the pH to 4.6–5.4, aiding preservation. The cured color also forms at this time. The sausages then go into the drying room where they are held at 45–55°F and 69–72% relative humidity. Semi-dry sausages are dried for 10 to 21 days and contain 50% moisture. Dry sausages are dried for 60 to 120 days and contain only 35% moisture.

Smoking and Cooking

Smoke is no longer as important as a preservative, but it contributes to flavor, color, and microbial preservation. Controlled burning of hardwood chips produces a mixture of organic acids, carbonyls, and phenols. The acids cause surface coagulation of the proteins forming a skin on the product. Carbonyls contribute a brownish color and phenols plus some carbonyls add the smoke flavors. Many continuous and large volume processes use an aerosol of liquid smoke which is more consistent and easier to use. It may also be added directly as a flavoring or applied to the surface. In addition, liquid smoke does not contain any of the carcinogenic polycyclic compounds that may be formed if the burning temperatures are too high.

Frankfurters are smoked during the beginning of a cooking sequence. The temperature is raised in steps with humidity control and good circulation until the internal temperature reaches 155–160°F. The frankfurters are then showered with cold water to cool.

Packaging

Most sausages are sold in the casings they were stuffed into. Frankfurters, however, are stripped of the cellulose casing after cooking in the smokehouse. Many sausages are vacuum packaged to extend shelf life by preventing water loss, retarding oxidation, and slowing microbial growth.

Developments

Sausages continue to have great popularity because of their appealing flavor and convenience. The increasing variety available in the retail stores reflect consumer interest in convenience and specialty foods. Salt levels have been reduced although many sausages still contain high amounts of sodium. Research is underway to reduce or replace the animal fat while retaining the expected flavor and textural feel of the products. Meat inspection and quality assurance programs are under revision as the inspection agencies include more microbial and residue testing with the traditional visual inspection. Hazard analysis-critical control point (HACCP) procedures are particularly important in controlling pathogenic microorganisms such as *Salmonella, Staphylococci* and *Listeria*.

GENERAL REFERENCES

H. W. Ockerman, *Sausage and Processed Meat Formulations*, AVI Publishing Co., Inc., Westport, Conn., 1989.

A. M. Pearson and F. W. Tauber, *Processed Meats*, 2nd ed., AVI Publishing Co., Inc., Westport, Conn., 1984.

R. E. Rust, "Sausage Products," in J. F. Price and B. S. Schweigert, eds., *The Science of Meat and Meat Products*, 3rd ed., Food & Nutrition Press, Inc., Westport, Conn., 1987.

U.S.D.A., *Agriculture Statistics*, U.S. Government Printing Office, Washington, D.C., 1988.

R. C. WHITING
A. J. MILLER
United States Department of Agriculture
Philadelphia, Pennsylvania

SEAFOOD: FLAVORS AND QUALITY

BACKGROUND

With few but important exceptions, the majority of seafoods are relatively mild in flavor intensity when compared to foods prepared from beef, poultry, pork, and lamb. Indeed, American and European consumer preferences for seafoods seem to favor varieties that have mild flavor profiles, for example, various white-fleshed species such as soles, cods, snappers, and rock fishes (1). Because these flavors are so subtle, they can be overwhelmed by very small quantities of exogenous or endogenous flavors due to contamination or spoilage, such as fuel oils, rancidity, ammoniacal, acetic acid, and earthy flavors. In addition, because of this subtlety and low intensity, the presence of flavor enhancers and modifiers, such as monosodium glutamate or 5'-ribonucleotides (disodium 5'-inosinates[5'-IMP] and 5'-guanylates[5'-GMP]), can exert enormous and profound influences on the perception of quality of a seafood. Added to this complex interplay is the creation of new flavors that can occur during processing that are identified with the final product, for example, canned salmon flavor, that may or may not have been present in the original seafood.

For the purposes of this review, some definitions and discussion of flavor and quality are in order. We will use the description of flavor put forward by Hall (2) and Chang (3). Hall (2) defined flavor as the "sum of those characteristics of any material taken in the mouth, perceived principally by the senses of taste and smell, and also by the general pain and tactile receptors in mouth, as reviewed and interpreted by the brain." The reader is particularly directed to a recent review by Kinsella (4) for a biochemical/biophysical discussion of flavor and the role of binding of the flavor compounds to physiological sites. Although Chang (3) and Hall (2) identify three components to flavor (taste, odor, and "pain-tactile"), we will limit our discussion of seafood flavors to only taste and odor. The pain-tactile responses of pungency, heat, cooling, or other nonspecific or trigeminal neural responses are not *directly* associated with commonly consumed seafoods themselves but are usually associated with external materials added to a finished seafood, such as condiments, peppers, herbs, and spices.

When a seafood, or any food, for that matter, is consumed, a series of conscious and subconscious responses to flavor and texture stimuli are integrated into a single unified perception. Because of this facility of the brain to integrate many stimuli into a single unifying concept, the definition of seafood quality becomes elusive and difficult. Typically, discussions of quality usually end with the statement, "but I'll know it when I see it." Therefore, any attempt to discuss seafood quality objectively must include a compilation of various measurable physical, chemical, and sensory attributes that have been correlated with predetermined quality assessments.

The underlying assumption of all quality assessment or assurance is that if a series of standards are met upon analysis, the product under evaluation is then considered to be of a certain level of quality. Unfortunately, these standards defining levels of quality (eg, excellent, good, bad, poor, decomposed, or passable) for measured attributes have been difficult to develop. Part of the reason is that the human perception of quality in seafoods is multifactorial, comprising a large number of attributes that are integrated in a *judgmental* response. This judgmental response, particularly in the evaluation of sensory attributes such as flavor and texture, can be strongly influenced or biased by personal, cultural, and societal factors. For example, some popular fermented fish/vegetable sauces would be assessed as spoiled or rotten by some palates yet be considered highly desirable by others. In short, what is "spoilage" to one is "fermentation" to another. Because of these cultural differences, it is sometimes difficult to arrive at a precise and consistent international set of standards, particularly for attributes that are the result of sensory judgment.

Today a great reliance is still placed on subjective assessments of quality in seafoods, such as presence of decomposition or quality grades. Inspectors and graders, in both government agencies and the seafood industry, are trained to various degrees in sensory techniques to conduct these evaluations. In many cases, potentially large sums of money and the health and safety of many consumers may ride on the nose, eyes, and hands of a few seafood inspectors (in some cases, only one inspector). Clearly, it would be desirable to minimize reliance on these subjective methods with all their inherent drawbacks, namely, fatigue, boredom, and sensory overload, and use chemically and physically measurable indicators of seafood quality.

Therefore, due to the elusive and complex nature of quality as it applies to seafood, the emphasis of this review will be on specific identifiable flavor/odors and their relationship to quality insofar as information or standards may exist. Our approach in this chapter will be to first identify those components that can be identified as specific flavors/odors that characterize specific seafoods. We will then discuss various modifiers and enhancers that influence the integration of the basic flavor components. Finally we will discuss various efforts to use seafood flavors to classify fish species. We feel that this last section is necessary because of the large number of marine species that complicate the seafood market. Recently, various efforts have been made to simplify the marketing and retailing of seafoods due to the overwhelming array of species that can confront the consumer. These efforts range from species groupings derived from statistical analysis of trained sensory panels to categories developed by consumer groups or knowledgeable seafood-industry personnel. Therefore we have divided the review into three sections:

1. Chemical compounds responsible for specific flavors found in seafoods.
2. Compounds responsible for modifying and/or influencing seafood flavors.
3. Attempts to use seafood flavors and texture attributes to classify fish.

Though the major portion of this chapter will involve a discussion of seafoods, that is, fishery products taken from the marine environment, we will include, of necessity, many products that derive from freshwater sources due to the enormous growth of aquaculture (eg, catfish, trout, and salmon) and the introduction of anadromous marine species (Coho salmon, *Oncorhynchus kisutch*; pink salmon, *Oncorhynchus gorbuscha*; chinook salmon, *Oncorhynchus tshawytscha*) into freshwater systems such as the Great Lakes. In fact, some of the more serious flavor problems arise in aquaculture systems due to feed components and environmental sources such as algae. In addition, the tremendous growth of marine aquaculture in the past 10 years has introduced similar problems to seawater-raised species.

Clearly, significant use of sensory-evaluation techniques are employed in seafood flavor and odor research. It is not our intent to review basic sensory methodologies and associated techniques and utilities; to that end the reader is encouraged to examine several excellent texts by O'Mahony (5), Jellinek (6), Warren and Walradt (7), and Boudreau (8). The book by O'Mahony (5) is particularly recommended for its statistical methods and procedures used in sensory evaluation. For developing sensory panels and programs, the reader is especially directed to the series of monographs published by the American Society for Testing and Materials (ASTM) for particularly useful training techniques and requirements of sensory panels (9–15). Recently, an excellent monograph on "Flavor and Lipid Chemistry of Seafoods", the result of an American Chemical Society symposium, has just appeared (16).

CHEMICAL COMPOUNDS RESPONSIBLE FOR FLAVORS FOUND IN SEAFOODS

Quality Flavor Attributes

Chemical compounds responsible for seafood flavors can be of either an endogenous or exogenous origin. Endogenous compounds are those that are present in fish and shellfish as the result of their inherent metabolism, whereas exogenous compounds arise from the diet, from immediate contact with the aquatic environment, for example, skin absorption and adsorption, or from both. The common endogenous compounds usually found in fish give rise to the characteristics that distinguish one fish species from another, such as salmon from sole, and are responsible for our perception of seafood quality attributes such as fresh, stale, and rancid. On the other hand, exogenous compounds are present accidentally in the fish, commonly arising from the environment, and are usually considered taints or contaminants.

Up until the 1970s high-resolution gas chromatography (HRGC) and mass spectrometry (MS) were available as expensive research tools only to well-financed institutions or corporations. With the advent of new technologies in engineering, manufacturing, and computers, HRGC and MS came within the reach of many research laboratories, and as a result came a rapid increase in flavor and odor research. Before the use of these new techniques, research into odors and flavors associated with fishery products were relegated largely to "scratch-and-sniff" techniques. Although these early techniques may appear crude and unsophisticated by today's standards, they were in their limited way remarkably productive and accurate. Many of these early studies were based on the abilities of very perceptive sensory analysts who could detect various odors and flavors present in a food and then search for simple chemical compounds that duplicated the flavor or odor. Usually for these studies the chemical compounds were selected on their probability of being present in the food, perhaps due to oxidative breakdown of fats and oils, spoilage, or other possible biochemical changes. These techniques (so-called "P and P," or postulate and prove, and "I and I," or isolate and identify; see Forss [17], p. 191) were similar to those used in the fragrance industry in the formulation of duplicate perfumes.

To overcome some of these past empirical approaches, Stansby and Jellinek (18) attempted to place fish odors and flavors research on a more rational but still descriptive basis. They classified nine types of flavors or odors: A, amine; B, burnt; F, freshwater-fish; G, green; I, iodine; N, natural species; O, pure oxidation types; P, putrid; and S, sweet. These descriptive terms are still in use today in the search for the responsible chemical compounds in seafood flavor and odor research. During this study, Stansby and Jellinek also noted that the green flavor of fish oils was similar to that of cis-3-hexen-1-ol and cis-3-hexenal, the latter found in the reversion of soybean oil (19). Other odors identified in their studies was that of green cucumber, similar to that of hex-2-enal and trans-2-cis-6-nonadienal (2,6-nonadien-1-al) found by Forss et al. (20).

As will be seen from the preceding brief historical perspective in research of flavors and odors in fishery products, these early studies were remarkably accurate in identifying specific flavor compounds.

In terms of quality, what is a fresh fish? In a redundant way, a fresh fish has a fresh-fish flavor or odor. What compound or compounds are responsible for the fresh-fish aroma? At the same time, poor-quality fish are said to have "fishy" odors and flavors. This section will cover the compounds responsible for fresh-fish flavor/odor, "fishy" flavors, rancidity/scorch flavors, and other miscellaneous flavors.

Fresh-Fish Flavor/Odor. After catch, a series of metabolic changes begin in the tissues of fish and shellfish (21,22). Part of the catabolic changes is the formation of odors indicative of decomposition. In general, the odor of a fillet removed immediately from a landed fish has a "fresh" odor; with time this odor disappears and "sweet" (per-

fumelike) odors begin to appear (23). Eventually, depending on holding conditions and length of time, odors such as fruity, sour, and putrid develop (24–26). However, the disappearance of the fresh-fish aroma is usually the first sign of a quality change in the fish and has been a subject of considerable interest to flavor/odor investigators (27).

Josephson et al. (28–30) found a variety of odorous compounds (Table 1), but three families of compounds were major contributors to the characterization of Lake Michigan whitefish (*Coregonus clupeaformis*): C_6 aldehydes and alcohols contributed green and plantlike aromas (eg, 2-hexenal and 1-hexanol), C_9 aldehydes and alcohols contributed cucumber and melon aromas (eg, 2-nonenal, 2,6-nonadienal, 6-nonen-1-ol, and 3,6-nonadien-1-ol), and the C_8 ketones and alcohols provided a characteristic mushroom aroma (eg, 1-octen-3-one, 1,5-octadien-3-one, 1-octen-3-ol, 1,5-octadien-3-ol, 2-octen-1-ol, and 2,5-octadien-1-ol). In this work, they found that increasing amounts of corresponding alcohols (1,5-octadien-3-ol and 3,6-nonadien-1-ol) and a concomitant decrease in carbonyls (1,5-octadien-3-one and 3,6-nonadienal) was associated with the change in aroma from that of very fresh whitefish to aromas that were characterized as "suppressed, flat, and sweet melon-like." Josephson et al. (27) found that greater losses of fresh-fish carbonyl compounds were observed in fish held under air (2°C, 7 days) than under a carbon dioxide atmosphere.

In a study comparing freshwater species (whitefish [*C. clupeaformis*], smelt [*Osmerus mordax*], black crappie [*Pomoxis nigromaculatus*], bluegill [*Lepomis macrochirus*], muskellunge [*Esox masquinongy*], perch [*Perca flavescens*], northern pike [*E. lucius*], ciscoe [*Coregonus artedii*], walleye pike [*Stizostedion vitreum*], sauger [*Stizostedion canadense*], rainbow trout [*Oncorhynchus mykiss*], and emerald shiners [*Notropis atherinoides*]) and saltwater species (ocean perch [*Sebastes marinus*], cod [*Gadus morhua*], petrale sole [*Eopsetta jordani*], and haddock [*Melanogramus aeglefinus*]), Josephson et al. (29) found a common occurrence of hexanal, 1-octen-3-ol, 1,5-octadien-3-ol, and 2,5-octadien-1-ol, whereas freshwater species contained in addition 1-octen-3-one and 1,5-octadien-3-one. Interestingly, Whitfield et al. (36) found that (Z)-1,5-octadien-3-ol and 1-octen-3-ol were involved in the metallic off-flavor of deep sea prawn (*Hymenopenaeus sibogae*) and sand lobster (*Ibacus peronii*). Also Berra et al. (37) found that (E,Z)-2,6-nonadienal was responsible for the cucumber aroma of Australian grayling (*Prototroctes maraena*).

It is postulated that the origin of these fresh-fish flavor/odor compounds is the enzymes and nonenzymic oxidation of polyunsaturated lipid fractions in the flesh (28,29,38). Josephson et al. (39) found that the formation of carbonyls and alcohols that characterize the fresh-fish aroma of emerald shiners (*N. atherinoides*) were almost completely inhibited when the fish were sacrificed and then immediately exposed to acetylsalicylic acid (aspirin) or stannous chloride, both specific inhibitors of fatty acid cyclooxygenase and lipoxygenase, respectively. They concluded that these observations implicated the involvement of enzymic conversions of ω-6 and ω-3 fatty acids to the development of volatile aroma compounds in fresh fish. Josephson et al. (39) found that the highest concentrations of volatiles from

Table 1. Volatile Compounds Identified in Fresh Whitefish (*C. clupeaformis*)

Peak no.	Compounds	I_E	Estimated concentration[a], ppb	GC effluent odor quality (packed column)	Odor threshold[b], ppb	Identification means
1	2,3-Butandione	3.16	1.7	Butterlike	8.6	MS,Rt,odor
2XO[c]	Hexanal	4.49	20.5 ± 9.2	Green	4.5	MS,Rt,odor
3	1-Penten-3-ol	5.08	9.5 ± 2.1		400	MS,Rt
4	3-Heptanone	5.20	4.0	Earthy		MS,Rt
5	2-Methyl-2-pentenal	5.24	1.95 ± 0.19			MS,Rt
6	2-Heptanone	5.40	3.2 ± 3.1		300	MS,Rt
7	Dodecane	5.50	0.6 ± 0.1			MS,Rt
8	Heptanal	5.51	1.8 ± 0.6		3.0	MS,Rt
9	3-Methyl-1-butanol	5.61	Internal standard (1.25 ppm)			MS,Rt
10O	(E)-2-Hexenal	5.83	1.4 ± 0.6		17	MS,RT
11	1-Pentanol	6.03	1.6 ± 0.2		120	MS,RT
12X	3-Octanone	6.18	2.0 ± 1.1		50	MS,Rt
13	Ethenylbenzene	6.26	1.4			MS,Rt
14O	Trimethylbenzene	6.40	3.4 ± 0.2			MS
15	Acetoin	6.46	Not measured			MS,Rt
16	2-Octanone	6.52	1.0 ± 0.1	Peaks 16 and 17-green, earthy, aldehyde	150	MS,Rt
17	Octanal	6.57	0.6 ± 0.1		0.7	MS,Rt
18X	1-Octen-3-one	6.57	2.6 ± 1.5	Boiled mushrooms	0.09	MS,Rt,odor
19	Tridecane	6.71	0.8 ± 0.2			MS,Rt
20	2,3-Octanedione	6.89	2.0 ± 1.6			MS,Rt
21	1-Hexanol	7.14	1.6 ± 0.8		4870	MS,Rt
22X	1,5-Octadien-3-one	7.42	1.3 ± 0.3	Geranium leaves	0.001	MS,Rt,odor
23X	3-Octanol	7.54	1.0 ± 0.6		18	MS,Rt
24O	Nonanal	7.62	3.2 ± 1.1	Planty, aldehyde	1.0	MS,Rt,odor
25	Tetradecane	7.73	0.8 ± 0.2			MS,Rt
26X	(Z)-2-Octenal	7.91	1.4 ± 0.3	Peaks 26 and 27-fatty, heavy, green	3.0	MS,Rt
27	Acetic acid	7.95	4.5 ± 0.6		7.0	MS,Rt
28X	1-Octen-3-ol	8.12	18.6 ± 4.1	Raw mushrooms	10	MS,Rt,odor
29X	1,5-Octadien-3-ol	8.45	24.8 ± 2.0	Earthy, mushroom	10	MS,Rt,odor
30XO	Benzaldehyde	8.69	1.0	Peaks 30, 31, and 32-slight cucumber, green, vinelike	0.44	MS,Rt
31O	Decanal	8.69	2.6 ± 0.4		0.1	MS,Rt
32	Pentadecane	8.75	2.4 ± 0.2			MS,Rt
33O	(E)-2-Nonenal	9.00	5.8 ± 1.5	Cardboardlike, cucumber	0.08	MS,Rt,odor
34X	1-Octanol	9.11	0.5		480	MS,Rt
35	2,3-Butanediol	9.15	6.4 ± 2.6			MS,Rt
36O	(E,Z)-2,6-Nonadienal	9.43	17.2 ± 0.1	Cucumber rind, peeling	0.01	MS,Rt,odor
37X	2-Octen-1-ol	9.72	6.3 ± 0.4	Peaks 37 and 38-green, musty	40	MS,Rt,odor
38	Hexadecane	9.74	1.4 ± 0.2			MS,Rt
39O	1-Nonanol	10.19	0.8			MS,Rt
40X	2,5-Octadien-1-ol	10.34	3.8 ± 1.5	Peaks 40 and 41-fresh fish undertone	MS,Rt	
41O	6-Nonen-1-ol	10.41	4.8 ± 0.8			MS,Rt
42	Heptadecane	10.67	2.0 ± 1.0			MS,Rt
43O	3,6-Nonadien-1-ol	11.05	5.7 ± 0.7	Clean cucumber	10	MS,Rt
44	Naphthalene	11.14	1.3 ± 0.2			MS,Rt
45	Octadecane	11.62	1.6 ± 0.7			MS,Rt
46	Nexanoic acid	11.82	1.1			
47	Unknown	12.38	12.4 ± 4.1	Green, cucumberlike		43(100), 71(60), 41(35), 56(32), 55(30), 83(21), 39(14), 89(12), 98(10)

Source: Reprinted from Ref. 28 with permission from *J. Agric. Food Chem.* Copyright 1983 American Chemical Society. Data also from Refs. 31–35.
[a]Based on duplicate determinations; μg/L slime-water extract.
[b]Threshold in water.
[c]X = previously identified in mushrooms. O = previously identified in cucumber/melon.

enzymic conversions appear to be associated with the slime and the surfaces of the fish. In their model, polyunsaturated fatty acids (eg, eicosapentaenoic acid; $C_{20;5'}$ ω-3) undergo enzymic oxidation to fresh-fish aromas by 12-lipoxygenase or 15-lipoxygenase at the skin-water interface because these enzymically derived volatiles appear to occur at higher concentrations in the skin than in fish muscle. They speculated that the enzymic oxidation and cleavage of the C_{20} polyunsaturated fatty acids may be involved with the induction of cellular repair because of the close involvement of the mucus or slime layer. In addition, these volatile compounds present in the slime may also provide signals for feeding, schooling, or fright responses.

Research on enzyme-mediated oxidations of fatty acids has been pursued in recent years (40–43) German and Kinsella (44) demonstrated the ability of lipoxygenase from trout tissue to catalyze the oxidation of polyunsaturated fatty acids (PUFA) typically associated with fish (arachidonic, eicosapentaenoic, and docosahexaenoic acid) to acyl peroxides. Hsieh and Kinsella (45), working with gill preparations from rainbow trout (*O. mykiss*), focused on the properties of enzyme-initiated lipid oxidation in tissues, specifically lipoxygenase, to produce specific oxidative flavor compounds in fish tissues. The major volatile compounds generated from the PUFA were 1-octen-3-ol, 2-octenal, 2-nonenal, 2-nonadienal, 1,5-octadien-3-ol, and 2,5-octadien-1-ol. They concluded that these compounds may be important in imparting fresh aroma to fish and also in causing off-flavors. Following the proposal of Josephson et al[34], Hsieh et al[38] demonstrated the presence of 12-lipoxygenase activity in fish skin, which may play a role in generating off-flavors. Some fish have little of the lipoxygenase activity in the skin, according to Hsieh et al. (43). For example, white bass (*Morone chrysops*), rock bass (*Ambloplites rupestris*), sheepshead (presumably Archosargus probatocephalus), and bluegill (*L. macrochirus*) did not have detectable activities in the skin but did have the enzyme present in the gill tissue. With the exception of rainbow trout, most species examined by Hsieh et al. (43) had higher lipoxygenase activity in the gills than in the skin. These studies point out the need for proper postharvest processing, such as low-temperature storage, and perhaps most importantly, the avoidance of damage to skin and muscle tissue, such as bruising, cutting, and punctures. Hsieh and Kinsella (45) also found that antioxidants such as butylated *tert*-hydroxyanisole (BHA), *tert*-butylated hydroxytoluene (BHT), esculetin (6,7-dihydroxycoumarin), and esculin (esculetin 6-glucoside) could inhibit lipoxygenase activity. Esculetin was found to be a potent inhibitor of lipoxygenase activity, whereas BHT had negligible inhibitory effect. The use of these compounds could effectively control lipid oxidation and slow the deterioration of fish after landing.

Earlier workers noted that the aroma of freshly harvested fish had some recognizable components, described in such terms as green, cucumber, and leafy. In turn, various compounds were suggested and identified that might logically be present to play a role in this aroma. Perhaps the most significant finding in the series of papers by Josephson et al. (27–30,39) is that the fresh-fish aroma is composed of a variety of chemical compounds that are produced by a dynamic system involving a complex of lipids, enzymic and nonenzymic reactions, and physiological compartments in the fish. What role these compounds play in the life of the fish is not yet clear, but Josephson et al. (39) and Josephson and Lindsay (30) suggest, not unreasonably, that these compounds may play some biologically active role, such as trauma recovery or defense.

It is well known that during cooking and storage of cooked foods, significant flavor changes occur. In some cases, new and favorable flavors are developed, such as browning flavors during baking or boiling. However, after a food is cooked, degradative changes can also occur that yield undesirable flavors. Although the baking industry has expended a large effort on staling and its prevention, little work has been done toward similar research on seafood products. Recently, Josephson and Lindsay (46) reported on some lipid changes in chicken and fish. Briefly, they found that 2,4-alkadecadienals were formed from the autoxidation of polyunsaturated fatty acids. Though these compounds are thought to contribute desirable aromas to various foods, they are also found in foods with undesirable flavors such as warmed-over, stale, and oxidized. In model systems, Josephson and Lindsay (46) demonstrated that 2,4-decadienal was degraded to 2-octenal and ethanal by a water-mediated alpha/beta double-bond hydration, retro-aldo condensation reaction series. The resulting 2-octenal was further degraded to hexanal and ethanal. Working with battered, deep-fried Alaskan pollock fillets, they found that immediately after frying and after 3 days of refrigerated storage (6°C), the 2,4-decadienals were higher than in fried chicken. In both these foods, the concentration of the 2,4-decadienals declined after refrigerated storage with concomitant increases in 2-octenal. A major difference was observed between chicken and fish in hexenal levels. In the chicken, an initially large hexenal concentration tripled during storage, but hexenal levels declined in fish during storage. Josephson and Lindsay (46) suggest that the staling flavors that are formed during refrigerated storage may arise from retro-aldol-related reactions. Kubota and Kobayashi (47) reported two ketones in the aroma concentrate of cooked shrimp (*Euphausia pacifica*, *E. superba*, and *Sergia lucens*) and reported the structures as tentatively (5Z, 8Z, 11Z)- and (5E, 8Z, 11Z)-5,8,11-tetradecatrien-2-ol. Recently, in more detail, Kobayashi et al. (48) synthesized a series of tetradecatriens, isomers of 5,8,11-tetradecatriene-2-one, determined their structures, and related their sensory aroma characteristics to the chemical structures. The descriptions of the isomers of 5,8,11-tetradecatrien-2-one are listed in Table 2. Compounds I and II were the naturally occurring isomers reported earlier by Kubota and Kobayachi (47).

Though alcohols and ketones are indeed responsible for the variety of odors and flavors noted in seafoods, other compounds containing nitrogen and sulfur are also involved in fresh, cooked, and fermented products. In a similar vein, Tanchotikul and Hsieh (49) examined the volatile flavor components in crayfish (*Procambarus clarkii*) waste. Using dynamic headspace sampling with gas chromatography (GC)/MS techniques, they identified 117 compounds; of these compounds, 106 were positively identified and the other 11 were tentatively identified in the head-

Table 2. Aroma Characteristics of Isomers of 5,8,11-Tetradecatrien-2-One

Isomer	Aroma characteristic
I (5Z, 8Z, 11Z)	Shrimp, crab, shellfish
II (5E, 8Z, 11Z)	Shrimp, crab, sea cucumber
III (5Z, 8E, 11Z)	Fruity, oily, seafood
IV (5Z, 8Z, 11E)	Fruity, oily, milk
V (5E, 8E, 11Z)	Short-legged clam, seafood
VI (5E, 8Z, 11E)	Short-legged, clam, oily
VII (5Z, 8E, 11E)	Fruity, cucumber, white-meat fish
VIII (5E, 8E, 11E)	Oily, fishy, dry bonito

space of crayfish waste. Interestingly, they found a large amount of phenol, which appeared to give the sample a medicinal aroma. In addition to a variety of compounds resulting from oxidation of polyunsaturated fatty acids, Tanchotikul and Hsieh (49) also identified nitrogen-containing compounds, pyrazines and pyrroles. 2,5-dimethylpyrazine was the most abundant nitrogen-containing compound they detected.

Recent work by Vejaphan et al. (50) with cooked crayfish (*P. clarkii*) tail meat revealed 70 volatile compounds in many different chemical classes. These authors identified green, grassy, woody, buttery, metallic, or sulfurlike aromas. Although ketones were the major components found in crayfish waste and in tail meat, significant sulfur-containing compounds were also found. Buttery et al. (51) identified two straight-chain sulfur compounds, dimethyl disulfide and dimethyl trisulfide, that gave the odor of cooked cabbage and spoilage. Indeed, these compounds are found as flavor/odor components in cabbage, broccoli, and cauliflower (51). In addition, the organic sulfides (dimethyl sulfide, dimethyl disulfide, and dimethyl trisulfide) are also involved with anaerobic putrefactive decomposition of many foods (26). Clearly, these compounds would not be expected to contribute desirable aromas to cooked crayfish.

Defect Flavor Attributes

Fishy Flavors and Odors Whereas the fresh-fish flavor is a desirable component of high-quality fish, "fishy" flavors and odors represent the other end of the quality spectrum, namely, poor or low quality. The fishy odors and flavors discussed here are in contrast to decomposition or spoilage aromas that involve bacterial and catabolic changes in fish that would be declared unfit for human consumption. In general, a fishy odor or flavor in either the raw or the cooked product marks a seafood as low quality. However, there are marked, and perhaps unusual, exceptions to this rule. For example, the authors of this article and Jellinek (6, p. 76) have observed sensory panelists (American and German) who were raised in environments where fresh fish was not available reject freshly caught fish as being too "bland." Clearly, the perception of fishy flavors and odors is somewhat culturally biased. Nevertheless, in the broad mainstream of seafood quality, the presence of fishy flavors and odors marks a seafood of low quality.

In 1962, a pioneer in fish flavors and odors, M. E. Stansby (52,53), commented, "With few exceptions . . . the relation of fishy odors and flavors to specific chemical substances in fish has been largely neglected in recent years." At this time, Stansby (53) pursued earlier work by Davies and Gill (54) and Davies (55) relating fishy odors in fish to nitrogenous compounds (ie, trimethylamine) and fat in the flesh. Besides basic organoleptic work on the odor of trimethylamine (TMA), Stansby (52,53) related the interaction of TMA with fish oils in attempting to define the odor that most people identify as fishy. Stansby concluded that probably no one compound was responsible for the fishy odor and that TMA, fish oil, and air were necessary for the development of fishy odors.

TMA is undoubtedly responsible for the major portion of the fishy odors noted in most cooked samples of marine fish. TMA usually arises from the reduction of TMA oxide (TMAO) in fish tissues (56). This reduction proceeds quite easily, due to either enzymic reactions or simple redox reactions utilizing ferrous ion (Fe^{2+}) complexed in some way in the tissue (57). With heating at 80°C, EDTA-complexed FE^{2+}, can quantitatively reduce TMAO to TMA within 5 min and serves as the basis of a method for TMAO/TMA analysis (58). TMAO occurs primarily in the marine fishes, particularly the gadoids (cods, haddock, hakes, and pollock), whereas in freshwater fishes the TMAO content is low to nondetectable. Even in species that contain large amounts of TMAO (ie, the gadoids), the variation in muscle concentration of TMAO is highly variable (57). TMAO can also break down to formaldehyde and dimethylamine. This latter reaction is thought responsible for toughening found in some fish fillets during frozen storage due to cross-linking reactions of the protein and formaldehyde (59). The odor of dimethylamine (bp 7°C) is similar to that of TMA. The origin and role of TMAO in fish is not well understood at present, and the interested reader is directed to the review by Hebard et al. (57) for a more complete discussion concerning origin and occurrences of TMAO.

TMAO itself has a moderately intense odor but it is not fishy; its odor is similar to that of stale air released from an automobile tire. On the other hand, TMA (bp 4°C) is ammoniacal in high concentration, but as the concentration is lowered a fishy note appears. Nevertheless, although diluted TMA has an odor that is somewhat fishy, it is not quite identical to that observed in deteriorating fish (52,53). At least in fish with high levels of TMAO, TMA is probably responsible for most of the undesirable fishy aromas (60,61) but other compounds may also be involved, such as acetic, butyric, and valeric acids (61). TMA or TMAO may not be completely necessary for the development of a fishy aroma. For example, one of the authors (J. C. W.) noted that when freeze-dried bovine serum albumin was coated with a molecularly distilled fish oil fraction and stored at −20°C for several weeks under nitrogen in a ground glass stoppered flask, a distinctly fishy aroma was observed in the headspace of the container. It was assumed that oxidative free radical reactions interacted with the proteinaceous material to yield unknown odorous compounds.

Clearly the production of the fishy aroma is complex, involving a multiplicity of compounds and reactions, perhaps similar to those described earlier in the production of fresh-fish aromas and flavors, that can produce a myriad of compounds (eg, acids, alcohols, esters, aldehydes, and ketones). These compounds in turn mix with TMA to yield the distinct fishy odors that may vary from one fish species to another.

Rancidity In plant oils and fats, two kinds of rancidity are noted (62): oxidative and hydrolytic. Oxidative rancidity is caused by the presence of oxygen and its reaction with the unsaturated lipid fraction, producing typical off-flavors and odors. Hydrolytic rancidity is the result of the hydrolysis of triglycerides, producing free fatty acids. These free fatty acids, particularly capric, lauric, and myristic acids, have very strong off-flavors. While hydrolysis of fish lipid triglycerides occurs, the resulting fatty acids have much higher molecular weights (ca 300 daltons) with concomitant lower flavor and odor intensities (63).

By and large the development of rancidity in a seafood is noted by flavor more than odor. However, in some frozen fatty species of fish, definite rancid odors can be detected quite easily. Nevertheless, for most low- to moderate-fat species, rancidity is usually noticed first by taste when the product is eaten, particularly in the fattier tissues. Many of the odors and flavors *and* reaction mechanisms that we described earlier that play a role in the fresh-fish flavor also play a role in rancidity. This is not surprising in that many of the fresh-fish flavors are developed through hydroperoxidation reactions of the polyunsaturated lipids present in seafoods. It is not our intent here to go into details of the reactions involved in the oxidation of fats and oils. The interested reader is directed to several excellent texts by Hamilton (64), Hardy (65), and Sargent et al. (66).

The flavors we associate with rancidity arise from the oxidation of lipids and are very complex in nature, similar to those responsible for fresh-fish flavor. These reactions of oxygen and the double bonds in seafood lipids can be catalyzed by trace metals, such as Fe^{2+} complexed in the heme molecule (67). A large portion of the work on rancid flavors and odors has been conducted on plant lipids, using linolenic or linoleic acids as model compounds that contain 3 and 2 *cis*-double bonds, respectively. Marine fish, on the other hand, contain not only both linolenic ($C_{18:3}$) and linoleic ($C_{18:2}$) acids but also other highly unsaturated fatty acids containing 4, 5, and 6 *cis*-double bonds (arachidonic—$C_{20:4}$, eicosapentaenoic—$C_{20:5}$, and docosahexaenoic acids—$C_{22:6}$, respectively). Furthermore, these C_{20} unsaturated acids are highly labile to both enzymic and nonenzymic oxidation (68–70). If one considers the production of compounds from the oxidation of C_{18} fatty acids complex, the complexity of products from the C_{20} and C_{22} unsaturates must be increased by at least an order of magnitude. Therefore, much of the work on the production of odorous and flavor compounds in model systems has understandably been done using linolenate ($C_{18:3}$) and linoleate systems ($C_{18:2}$).

In a model system Ullrich and Grosch (71) identified the most intense odor compounds in autoxidized methyl linoleate at room temperature using flavor units (FU) and flavor dilution factors (FD). After a reaction time of 48 h, 20 mol percent of the methyl linolenate was converted into hydroperoxides. Ullrich and Grosch (71) found that *trans,cis*-2,6-nonadienal followed by 1,*cis*-5-octadien-3-one, *trans,cis*-3,5-octadien-2-one and *cis*-3-hexanal had the highest flavor units. However, after 102 h, 1,*cis*-5-octadien-3-one was by far the most important odor compound, which was followed by *cis*-3-hexenal and *trans,cis*-2,6-nonadienal. 1,*cis*-5-octadien-3-one has a potent geranium-like note that gives a metallic off-flavor (72). Ullrich and Grosch (71) found that the odor was so potent that no peak could be detected for this compound in the gas chromatographic chart until enrichment procedures were employed. They postulate that 1,*cis*-5-octadien-3-one is the oxidation product of the 1,*cis*-5-octadien-3-ol.

Hsieh et al. (73) identified volatile compounds in undeodorized menhaden fish (*Brevoortia tyrants*) oil. They used a dynamic headspace sampler and stripped the volatiles at 65°C onto a Tenax TA sorbent cartridge held at ambient temperature. Desorbed compounds were separated and identified using either gas chromatography, mass spectrometry, or both. In their method, the low-molecular-weight alkanes (C_1 to C_7) were not detected due to insufficient cold trapping. Though the major portion of the volatiles found were alkenals and alkadienals, the largest component was heptan-3-one at 8530 ppb (sickly sweet odor) followed by butanoic acid (8110 ppb, odor of dirty socks). The volatile short-chain fatty acids (C_2–C_6) gave very intense and objectionable sweaty odors. The origin of these low-molecular-weight acids (ie, C_2–C_6) was not discussed, but they could arise from hydrolytic cleavage of short-chain triglycerides or from the oxidation of short-chain aldehydes arising from the hydroperoxidation of unsaturated fatty acids. In addition, they found significant amounts of decadienals, decatrienals, and nonatrienals that have oxidized oil odors and are characteristic of crude menhaden oil.

Tainting Flavors Unlike their terrestrial counterparts, fish are intimately associated with their environment. Given this situation, whatever is present in that aquatic environment has more than ample opportunity to enter the living aquatic organism. In this way, fish and shellfish are particularly prone to tainting, or the acquisition of environmental odors and flavors. For example, oil spills into the marine environment near aquaculture pens or fish raised in closed-pond aquaculture will sometimes produce earthy, muddy, or musty odors or flavors in the fish (74). Some of these flavors can be acquired through the food, but most originate from microorganisms (75) living in the water, soil, or pond detritus. These muddy or tainting flavors can present economic problems to fish growers. For example, sometimes commercially reared catfish (*Ictalurus punctatus*) will develop such an intense musty odor or flavor that the fish cannot be marketed. A lexicon of flavor descriptors has been developed for pond-raised catfish by Johnsen et al. (76). Fish reared in closed-pond cultures are susceptible to acquiring odorous compounds when these ponds become eutrophic, usually caused by high feeding rates, temperatures, and stocking densities and low aeration. Though tainting occurs in small ponds, stagnant water, or both, tainted fish have also been reported from lakes and salt water. Yurkowski and Tabachek (77,78) reported that the muddy flavor in rainbow trout (*O. mykiss*) and other species, taken from some western Canadian lakes, was due to geosmin and methylisoborneol. In saltwater studies, Persson (79) correlated the muddy flavor of bream (*Abramis brama*) with an increase in the biomass of *Oscillatoria agardhii* in the Gulf of Finland. Clearly, the tainting of seafood products can also occur due to the presence of human-made pollution, such as petroleum hydrocarbons from oil spills or the discharge of other odorous chemicals as by-products of industrial production. Although we believe that the human-made taints can and do seriously impact seafood quality, a detailed discussion of this particular complex problem is beyond the intent of this chapter. Therefore, we have limited our discussion of seafood flavors and quality to those compounds arising from natural environmental origins and leave a discussion of the contribution of human-made contaminants to others.

Several heterocyclic compounds have been reported to be involved in muddy or earthy odors: 2-methylisoborneol (80,81), 2-methylenebornane, and 2-methyl-2-bornene (82). Examination of the commercial catfish ponds revealed 2-methylisoborneol in both water and mud. Martin et al. (81) report that they found the highest levels of 2-methylborneol in catfish flesh (284 ng/g), whereas mud and water were much lower, 14.9 and 10.06 ng/g, respectively. The range for the 2-methylisoborneol was 5.0 to 284 ng/g in the fish flesh, whereas water ranged from a low of 0.1 ng/g to a high of 13.32 and mud ranged from a low of 1.0 ng/g to 14.9 ng/g. They determined a bioconcentration factor (concentration in water/concentration in fish) for 2-methylisoborneol of 28.1 ± 14.0 ($\pm$ std dev) for the channel catfish. An unsaturated aliphatic compound, 1,10-*trans*-dimethyl-*trans*-(9)-decalol or geosmin (83), has been isolated from muddy-tasting catfish. These compounds, the 2-methylisoborneol and geosmin, are the products of the metabolism of certain species of cyanobacteria and actinomyces (75,84). Gerber (75) proposed the name geosmin for 1,10-*trans*-dimethyl-*trans*-(9)-decalol based on the Greek words *ge*, meaning earth, and *osme*, odor. Geosmin and methylisoborneol have been reported to be produced by *Microbispora rosea* 3758, *Nocardia* spp., and *Streptomyces* spp. (75,85). These organisms are readily found in water and soils (86), so it is not surprising that geosmin and methylisoborneol have been found in other foods besides fish, for example, soil-grown vegetables such as beets (87,88). Blue-green algae are also responsible for the production of various compounds that are the cause of off-flavors in water (89,90).

Lovell and Sackey (91) were able to induce earthy-musty flavors in channel catfish by culturing the fish in tanks containing a geosmin-producing blue-green alga, *Symploca muscorum*; however, they did not identify the chemical compounds responsible for the off-flavor. Lovell et al. (83) attempted with mixed success to correlate sensory findings of earthy-musty flavors in catfish with geosmin content. In their study, 10 of 13 fish were judged to have an earthy-musty taint and were found by instrumental analysis to contain geosmin. Using an inverse scoring system (highest intensity of earthy-musty flavor receives a 0; no off-flavor receives a score of 10), they found that one fish that contained over 200 µg/kg geosmin had a score of 2.7, whereas a fish containing 15.6 µg/kg was given a score of 1.7. Clearly, other compounds, such as the bornanes, may also be present in these fish that may contribute to the scoring of flavor quality.

Sulfur-containing compounds, such as dimethyl disulfide and dimethyl trisulfide, are also responsible for tainting. However, of these two compounds, dimethyl disulfide is somewhat of a two-edged sword in that depending on concentration, it can play both beneficial and detrimental roles. For example, at low concentrations, it is responsible for the typical odor of fresh oysters, but at higher concentrations it has been reported to be responsible for a petroleum refinery odor in Nova Scotia mackerel (*Scomber scombrus*) (92), a petroleum odor in canned salmon (93), and a blackberry odor in Labrador cod (94). In Nova Scotia mackerel, levels of 0.2 to 0.3 µg/g of dimethyl sulfide were found in tail muscle whereas stomach contents had levels as high as 17 µg/g. The high levels of dimethyl sulfide were thought to arise from the mackerel's consumption of large quantities of the pteropod *Spiratella retroversa*. The metabolic precursor was probably dimethyl-β-propiothetin because large quantities were also found in the stomach contents.

Another sulfur compound, dimethyl trisulfide ($CH_3SCH_2SSCH_3$), was implicated in the off-flavor of some Australian red prawns (*Hymenopenaeusi sibogae*) (95). The prawns exhibited a strong disagreeable cooked-onion flavor and aroma. Close inspection of the lot revealed that some of the prawns were damaged, the flesh was discolored, and the shellfish were slimy. Initial speculation was that the onionlike odor was due to microbial spoilage, possibly as a result of poor handling. Whitfield et al. (95) analyzed the product and found that dimethyl trisulfide was responsible for the off-flavor. In their studies, they collected approximately 10 µg of dimethyl trisulfide from 100 g of shrimp, giving an estimated level of 100 µg/kg in the original product. The odor threshold for dimethyl trisulfide is very low; Buttery et al. (51) reported it to be 0.01 µg kg^{-1}, or 0.01 ppb in water. Interestingly, adding dimethyl trisulfide to minced samples of fresh normal-flavored shrimp at levels of 1 to 10 µg kg^{-1} followed by cooking for 3 min at 100°C gave a product with a distinctive and unpleasant onion flavor; however, when it was added at 100 µg kg^{-1}, the material became inedible. Because the shrimp were heavily damaged and had high indole levels (50 µg kg^{-1}), they speculated that two pseudomonads, *Pseudomonas putrifaciens* and *P. perolens*, might be possible causative agents. Both *P. putrifaciens* and *P. perolens* have been reported to produce dimethyl trisulfide when inoculated onto sterile fish muscle and incubated at 0°C (96,97).

During the early stages of spoilage, chilled fish muscle will develop a musty, potatolike odor (98,99). Castell et al. (100) determined the causative organism to be *P. perolens*. Miller et al. (96) found a variety of odorous compounds in spoiling rockfish (*Sebastes melanops*) muscle that had been inoculated with *P. perolens* ATCC 10757 and allowed to incubate at 5, 15, and 25°C. The following compounds were detected and identified: methyl, dimethyl disulfide, dimethyl trisulfide, 3-methyl-1-butanol, butanone, and 2-methoxy-3-isopropylpyrazine. Also tentatively identified were 1-penten-3-ol and 2-methoxy-3-isopropylpyrazine. The musty, potatolike odor was primarily caused by 2-methoxy-3-isopropylpyrazine (96). In a slightly later work, again using sterile *S. melanops* muscle tissue, Miller et al. (97) found that *P. putrifaciens, P. fluorescens*, and an *Achromobacter* species (strain H15) produced methyl mercaptan, dimethyl disulfide, dimethyl trisulfide, 3-methyl-1-butanol, and trimethylamine. The bacterial strains used in these studies were isolated from the environment. The *P. fluorescens* was isolated from ocean perch whereas the *P. putrifaciens* strain H6 and the *Achromobacter* strain H15 were isolated from Dungeness crab. Only the *Achromobacter* species did not produce dimethyl trisulfide. The major sulfur-containing compounds produced by *P. fluorescens* were methyl mercaptan and dimethyl disulfide.

Other sulfur-containing compounds, such as heterocyclic sulfur compounds, have been associated with meaty aromas in meat products and may also be important in crustaceans (101–103). Kubota et al. (101) found that the cooked-odor concentrate from boiled antarctic krill (*Euphausia superba* Dana) contained 11 sulfur-containing compounds. Significant odorous sulfur compounds were 3,5-dimethyl-1,2,4-trihiolane (onionlike to garlicky) depending on replacement of the methyl groups with ethyl groups. Dimethylthiolanes have been identified in the volatiles of cooked beef, beef broth, cooked mutton, and cooked chicken. As in the case with the dimethyl disulfide, in concentrated form they have a sulfide odor, but upon dilution they yield a meaty aroma. 3-ethyl-5-methyl-1,2,4-trithiolane and 3,5-diethyl-1,2,4-thiolane have been identified in the flavor compounds of onion. In a later work, Kubota et al. (102) studied the volatile components of roasted shrimp (*Pleoticus muelleri*). In this work, they found 77 compounds, including isovaleric acid, alkyl pyrazines, isovaleramide, ketones, and some sulfur-containing compounds. In roasted shrimp, 3-methylthiopropanol was present at the highest level, but in boiled shrimp, dimethyl disulfide was present whereas the 3-methylthiopropanol was absent.

In another study that examined both cooked and fermented shrimp (*Acetes japonicus*) Choi et al. (103) found that sulfur compounds and pyrazines seemed to be important contributors to the cooked odor of the fermented product. In their work, shrimp were allowed to ferment in 20% salt at 20°C for 3 months, while the shrimp were kept frozen (−25°C) until used. Volatiles from both the raw and fermented shrimp were collected using a combined steam distillation-extraction (water–diethyl ether = 1000:100 mL) for 4 h. The main components of the volatile concentrate made from the nonfermented shrimp were thialdine (dihydro-2,4,6-trimethyl-4H-1,3,5-dithiazine, 41.4%), isovaleraldehyde + ethanol (22.5%) and pyrazines (6%), whereas in the fermented shrimp the main odor constituents were the pyrazines (32%), isovaleraldehyde + ethanol (26.5%), and furfuryl alcohol (25.1%). Indole was low (0.3%) in the nonfermented shrimp and present only in trace amounts in the fermented product. The presence of furfuryl alcohol in the fermented product would probably account for its nutty flavor notes. Choi et al. (103) speculate that the origin of sulfur and nitrogen compounds in the fermented shrimp arise from free amino acids released during the fermentation process, yielding smaller molecular compounds that in turn condense to form products

such as the pyrazines, thialdines, trithiolanes, and thiazoles.

Rancidity. Perhaps the most common defect in all foods, and specifically in seafoods, that consumers recognize most readily is rancidity. In both formal studies and informal interviews, rancidity is commonly reported by both trained and untrained panelists. Unfortunately, although it can be easily identified organoleptically, its chemical measurement and correlation with sensory observation has been somewhat more difficult. Attempts at measuring the components of oxidative rancidity and correlating them with sensory findings has proved difficult because of the compounds' transient nature, particularly in a complex food matrix. Many of the compounds responsible for rancidity, such as malonaldehyde, are very reactive and are presented with a variety of substrates in foods. Numerous methods have been proposed and used to measure the degree of rancidity of foods, and the reader is directed to Rossell (62), particularly pages 25 to 34. The oxidation of oils, particularly in a food matrix, is a very dynamic system, producing many highly labile compounds that in turn can react with each other or with other components in the food. All of these yield odorous or flavor compounds that are characteristic of rancidity.

The presence of highly unsaturated fatty acids (eg, $C_{22:6}$ and $C_{20:5}$) in seafoods makes for one of the most complex systems for study. In the early stages of oxidative rancidity, peroxides are formed at the double bonds in the unsaturated lipids, yielding compounds that are odorless and flavorless. Through further oxidation, these compounds are eventually converted to aldehydes and ketones, many of which have undesirable flavors, such as fishy or rancid (104). For the measurement of the early stages of oxidation, the peroxide value (PV) is useful, whereas for intermediate stages, the thiobarbituric acid value (TBA) method, which is claimed to be specific for malonaldehyde, is most useful. Many of these oxidation products, namely, the aldehydes and ketones, are volatile and highly reactive. In addition, they are present in a very complex food matrix with many available and similarly reactive functional groups (eg, amino groups, —NH_2). Therefore, as rancidity proceeds, the TBA values will increase and then reach a maximum and begin to decline as the malonaldehyde is consumed in the food matrix. For a discussion of these methods, the reader is directed to Rossell (62, pp. 25 and 29). Official methods for PV (105,106) exist, but methods for estimating malonaldehyde using TBA are still largely empirical. In our laboratory the method of Lemon (107) based on the method of Vyncke (108), developed at the Canadian Fisheries laboratory at Halifax, Nova Scotia, has been found useful. Researchers must be aware that peroxides, and particularly malonaldehyde, can increase and then decline in food samples. Consequently, some degree of knowledge about the history of the food sample is necessary to ensure meaningful interpretation of PV and/or TBA values.

COMPOUNDS RESPONSIBLE FOR MODIFYING SEAFOOD FLAVORS AND TASTES

Typically when seafoods are described by either trained or untrained sensory panels, a set of common descriptors are usually found, such as "sweetness," "saltiness," "briny," and "salty-briny" (109,110). The cause of these tastes has been variously ascribed to the presence of sodium chloride or carbohydrates in the fish flesh. However, recent work indicates that it may not be necessary to rely on sodium chloride and simple sugars to explain some of these taste phenomena. For example, the peptide N-L-α-aspartyl-L-phenylalanine 1-methyl ester, or aspartame as it is commonly known, is now widely used as an artificial sweetener because it is about 160 times sweeter than sucrose. It has also been recognized that peptides and nucleotides can be responsible for bitter tastes. Furthermore, Tada et al. (111) reported L-ornithyltaurine as a salty-tasting peptide. More recently, Kawasaki et al. (112) and Seki et al. (113) reported further studies on salty peptides, such as L-ornithyl-β-alanine and a series of synthesized peptides and amino acid derivatives. Some of these compounds appear to have the ability to potentiate the salty taste. Though some controversy appears to surround the early work on salty peptides, recent papers by Kawasaki et al. (112) and Seki et al. (113) confirm that peptides and amino acid derivatives can have a salty taste or potentiate the salty perception. Clearly, seeking to profile foods by merely measuring the quantities of sodium chloride, simple carbohydrates, and other low-molecular-weight compounds ought to be reassessed. The saltiness and sweetness that are commonly attributed to fresh fish may possibly have their origins in some of these peptides.

Umami

An important taste sensation associated with monosodium glutamate (MSG) is described as "meaty" or "mouthfilling." Many who are aficionados of Asian cuisines are perhaps very familiar with this flavor. Besides MSG, other compounds, such as the 5'-nucleotides, also have the same flavor and in addition have some capability of potentiating the taste of MSG (114). A recent computer search of the literature by the authors indicates that up to two-thirds of the reports on the taste of MSG and 5'-nucleotides originated in either Japan or Asia. Because of the Japanese keen interest and large body of work on this flavor, we will use here the Japanese term *umami* to describe the taste sensation of MSG and the 5'-nucleotides (5'-ribonucleotides and their derivatives such as disodium 5'-inosinate [IMP] and 5'-guanylate [GMP]). For more information about the derivation and description of umami, the reader is directed to the review by Yamaguchi (114,115).

Although MSG contributes the basic taste sensation of umami to a food, the 5'-nucleotides may play just as important a role. Some of the nucleotides have the umami flavor, but their ability to potentiate this flavor may be more important. The 5'-nucleotides are naturally present in many foodstuffs and have probably been used for their flavor-enhancing abilities long before their role was understood. For example, Table 3 lists some values for the nucleotides in common foods. It can be seen from this data that the use of meat and poultry broths or soups prepared with shiitake mushrooms would have a potentiating effect, that is, one similar to that of the commercially produced Ribotide® (Takeda Chemical Industries, Ltd.), if used with a food high in glutamic acid. The prevalence of these broths in the cooking traditions of various cultures, particularly Chinese and French, may be due to the presence of these naturally occurring enhancers. In fact, fermented fish sauces, which are common throughout Asia, are prepared from fish in the sardine family, a group of fish high in 5'-IMP. In Thailand, these fish sauces are prepared by layering salt and fish, then permitting autolysis to occur slowly, sometimes over a 2-year period at tropical temperatures (27°C–35°C). After the fermentation process is finished, the liquefied fish portion is drawn off and filtered if necessary. The finished sauce will contain a variety of amino acids, nucleotides, salt, and other components. A high-quality fish sauce will be clear, almost odorless, and moderate to dark brown in color and be used as a condiment with various foods; typically pieces of food are dipped in the sauce before being eaten. In some cases, the fish sauce may be added as an ingredient to soups or cooking broths. In Japan, shaved or small pieces of dried bonito are a common ingredient added to miso soups; the bonito's very high 5'-IMP content obviously contributes to the miso profile.

In addition to IMP and other 5'-nucleotides, peptides may also be responsible for flavor enhancement. Fujimaki et al. (117) isolated pleasant brothy tastes from the hydrolysis of fish protein concentrate using commercial peptidase enzymes. Molsin® (obtained from Seishin Seiyaku Co.), Rapidase® (obtained from Takeda Chemical Co.), and Pronase® (obtained from Kaken Kagaku Co.) enzymes gave the strongest brothy flavor. The release of brothy taste enhancers appeared enzyme dependent in that pepsin yielded little of the enhancing flavor. Noguchi et al. (118) found that peptides from fish protein hydrolysates having flavor activity qualitatively similar to MSG had high molar ratios of glutamic acid residues. Arai et al. (119) suggest that highly acidic (hydrophilic) L-glutamyl oligopeptides possibly possess a brothy taste and contribute to the favorable taste of food protein hydrolysates. Yamasaki and Maekawa (120) isolated what they term a "delicious" peptide from the gravy of beef meat having a proposed structure of H-Lys-Gly-Asp-Glu-Glu-Ser-Leu-Ala-OH. Later Yamasaki and Maekawa (121) synthesized this peptide and compared the synthetic compound with that isolated from beef and found that the sensory properties were the same. During this work they examined several precursor peptides and found that removal of the Lys-Gly group, yielding Asp-Glu-Glu-Ser-Leu-Ala, created a sour-tasting peptide, whereas the tripeptide Ser-Leu-Ala produced a bitter-tasting peptide. The peptide Glu-Ser-Leu-Ala was both sour and astringent. More recently, Tamura et al. (122) studied the interactions of the "delicious" peptide's amino acid groups. They confirmed that the umami and salty taste were produced by a combination of the basic and acidic groups of the peptide. They proposed that the cationic and anionic groups must be located next to each other to produce the salty and umami taste.

In careful determinations, Yamaguchi et al. (123) reported a threshold value for the umami at 0.012 g/100 mL or 6.25×10^{-4} M. Under the conditions of their testing, the MSG threshold was higher than that of quinine sulfate or tartaric acid and lower than sucrose, but similar to that

SEAFOOD: FLAVORS AND QUALITY   2081

Table 3. Nucleotide Content of Some Foodstuffs

Food	Nucleotides, mg %					
	5'-CMP	5'-UMP	5'-IMP	5'-GMP	5'-AMP	ADP + ATP
Sardine			192.6		6.6	15.4
Bonito			285.2		7.6	15.8
Salmon			154.5	0.0	6.9	
Perch			124.9	0.0	8.4	
Sea bream			214.8			30.0
Codfish			43.8		23.9	295.6
Swordfish			19.9	0.0	3.1	
Rainbow trout			117.0		14.6	223.4
Trout			187.0		4.2	27.7
Oyster	+	31	0	0	21	53
Crab			0		10.1	372.5
Prawn (Shrimp)					11.5	701.5
Beef	1.0	1.6	106.9	2.2	6.6	17.4
Pork	1.9	1.6	122.2	2.5	7.6	12.2
Chicken	2.3	3.6	212	3.6	5.2	30.1
Asparagus		72			27	77
Tomato	0.5	2.2			10.4	
Green beans	0.8	2.6			1.8	
Cucumber	±	0.6			0.5	
Onion	±	0.5	±		0.8	
Mushroom (dry)[a]	114.2	135.2		156.5	131.6	

Note: Abbreviations used: 5'-CMP, 5'-cytidylic acid; 5'-UMP, 5'-uridylic acid; 5'-IMP, 5'-inosinic acid; 5'-GMP, 5'-guanylic acid; 5'-AMP, 5'-Adenylic acid; ADP, Adenosine diphosphate; ATP, Adenosine triphosphate.
[a]Hot water extract of dried *shiitake*.
Source: Ref. 116.

of sodium chloride. MSG appears to interact with the four tastes. In this experiment, the thresholds for the four tastes were measured again in a 5-mM solution of MSG or IMP. The thresholds for sucrose and sodium chloride were unaffected by either of the umami substances, but IMP raised the threshold of tartaric acid from 0.94 mg % to 30 mg % (nearly a 30-fold increase), whereas MSG raised the tartaric acid threshold to 1.9 mg % (approximately 2-fold increase). MSG had no apparent effect on the quinine sulfate threshold (0.049 mg %), but IMP raised the threshold to 0.4 mg % (about an 8-fold increase). Perhaps the most important aspect of the umami flavor is the synergistic relationship between MSG and IMP. Yamaguchi (114) found that when mixtures of MSG and IMP containing 25 mg % MSG and 25 mg % IMP were rated by a sensory panel, the flavor intensity was equivalent to 780 mg % MSG alone, a 16-fold increase in apparent flavor. This synergistic effect was not affected by the four taste substances (sucrose, sodium chloride, tartaric acid, and quinine sulfate). He did find that the basic amino acids, histidine and arginine, could suppress the synergistic effect of IMP; however, when the pH was adjusted to between 5 and 6.5, the synergism was recovered.

At a more practical level, Yamaguchi (114), in extensive taste tests with untrained sensory personnel, examined the effect of umami substances on food flavors. These tests employed from 180 to 300 people in panels that varied from 25 to 50 persons. The results of these tests indicated that MSG has no effect on the aroma of food but does increase the total taste intensity. MSG enhances certain flavor characteristics of foods, such as continuity, mouth fullness, im-

pact, mildness, and thickness. Specific food flavors are also increased, such as meatiness. Probably most importantly, MSG increases the palatability of foods.

The presence of 5'-nucleotides, particularly IMP, may significantly influence our perception of the flavor, and hence the quality, of seafoods. Some of these nucleotides arise from the metabolism of adenosine-5'-triphosphate (ATP):

$$ATP \rightarrow ADP \rightarrow AMP \rightarrow IMP\ (HxR) \rightarrow hypoxanthine\ (Hx)$$
$$\rightarrow xanthine\ (Xa) \rightarrow uric\ acid\ (UA)$$

In resting muscle tissue, ATP is the predominant nucleotide. ATP undergoes a series of enzymatic dephosphorylation steps, forming adenosine-5'-diphosphate (ADP) and finally adenosine-5'-monophosphate (AMP). At this stage, deamination of AMP by AMP deaminase yields inosine-5'-monophosphate (124). These reactions proceed very rapidly, even during slow freezing of fish muscle (125). In trawl-caught fish, most ATP is depleted during capture and is converted to IMP within a day or two if the fish is stored on ice (126). IMP is then dephosphorylated by 5'-nucleotidase to form inosine (HxR); however, this reaction is slow compared to the formation of IMP. Inosine is cleaved by muscle ribosidase to form hypoxanthine and ribose; however, this reaction appears to be quite variable and depends on the fish species. For example, in lemon sole, petrale sole, and redfish very little HxR is seen (125,127,128). In other species, the further metabolism of inosine is much slower than the dephosphorylation of IMP, causing inosine to accumulate. In haddock, plaice, and pacific salmon inosine accumulates and then very slowly breaks down to hypoxanthine (125,129,130). The further oxidation of hypoxanthine to xanthine and uric acid is also slow, probably due to microbial metabolism (123). A measure of the various ATP metabolites has been adopted as an indicator of freshness or quality of seafoods—the K value. The K value is defined as:

$$K = \frac{[Hx] + [HxR]}{[HxR] + [Hx] + [ATP] + [ADP] + [AMP] + [IMP]} \times 100$$

where the concentrations of the various nucleotides in the fish muscle are presented in brackets. In some cases, because the tissue concentrations of ATP, ADP, and AMP are so low, the equation is shortened to:

$$K_1 = \frac{[HxR] + [Hx]}{[Hxr] + [Hx] + [IMP]} \times 100$$

The K value has been supported by the Japanese government in an experimental project since 1982 aimed at improving the distribution of fish by controlling its freshness (131). Apparently, in this project, the K values of fish were measured and then shown on packaging labels for the benefit of consumers. Though the K value appears to be accepted in Japan, it has not been widely adopted in the United States, where freshness or quality of seafoods is still determined by empirical observation based on experience. Unfortunately, determining K values requires relatively expensive laboratory equipment and procedures that are beyond the financial resources of many small processors and retailers, which may account for its current lack of widespread use. Nevertheless, the K value does appear to be a potentially useful tool that could find application in determining seafood quality in Western countries if the chemical analyses could be simplified, automated, or both. It is beyond the scope of this chapter to detail the use of K values in determining fish freshness; therefore, the reader is directed to the chapter by Ehira and Uchiyama (131).

Though it is clear that freshness of fish depends on the mixture of nucleotides, the levels of IMP in fish tissue appear to be directly related to acceptability. Bremmer et al. (132) found that flavor acceptability scores were positively correlated with IMP levels in four species of North West Shelf Australian tropical fish. The study involved long-spined sea bream (*Argyrops spinifer*), painted sweetlip (*Diagramma pictum*), snapper (*Lutjanus vittus*), and threadfin bream (*Nemipterus furcosus*). The fish were maintained on ice for up to 24 days. As expected, IMP levels declined in almost a linear fashion while sensory panelists' scores also decreased. In two species, long-spined sea bream and threadfin bream, hypoxanthine, a catabolite of IMP, was low and did not exhibit much change. The snapper and painted sweetlip, on the other hand, showed dramatic increases in hypoxanthine and would correlate negatively with flavor acceptability. Initial bacterial loads on the fish samples were low, typically of the order of 10^2 to 10^3 cfu/cm^2 and were identified to be *Moraxella*, *Vibrios*,

Aeromonas, and *Flavobacterium*; at the end of 23 days of storage the predominant species were *Pseudomonas*, *Moraxella*, and *Aeromonas* and were present at levels of 10^6 cfu/cm^2, with a high of 10^7. Nevertheless, Bremner et al. (129) concluded that acceptability or shelf life was more related to IMP loss rather than bacterial spoilage. In temperate waters off Alaska, Greene and Bernatt-Byrne (133) found that both inosine and 5'-inosine monophosphate were positively correlated with overall desirability of both Pacific cod (*Gadus macrocephalus*) and pollock (*Theragra chalcogramma*). On the other hand, hypoxanthine was negatively correlated with flavor and desirability similar to the findings of Bremner et al. (132).

Over the past decade, there has been considerable interest in the taste of shellfish in Japan, probably due to the development and manufacture of crustacean and molluskan surimi analogs. Konosu (134) reviewed the role of amino acids, low-molecular-weight organic acids, and nucleotides on shellfish tastes. In work on synthetically reconstituted extracts that paralleled extracts from abalone (*Haliotis gigantea discus*), Konosu and Maeda (135) found that glycine and glycine betaine contributed to the sweetness and umami of abalone. As expected, the major contributor to umami in these extracts was glutamic acid. The presence of 5'-AMP also contributed to the umami taste. Glycogen, which was present in high concentration in the extract (7.4%), appeared to contribute a body effect on the taste, although glycogen itself was tasteless. Konosu (134) concluded that the taste characteristic of abalone is due to umami produced by glutamic acid and AMP; the sweetness is derived from glycine and glycine betaine. Glycogen produces a harmonizing and smoothing of the taste. Konosu claimed that AMP is accumulated in marine invertebrates, in contrast to fish where IMP accumulates, because the muscle lacks AMP aminohydrolase, or if it is present, its activity is low.

Free amino acids apparently also play a role in the desirability of squid. Like Konosu and Maeda's (135) earlier work with abalone, Endo et al. (136) found in squid (Hirakensaki-Ika, *Logigo chinensis*; Kensaki-Ika, *L. Kensaki*; Aori-Ika, *Sepioteusthis lessoniana*; and Ko-Ika *S. esculenta*) that high levels of free amino acids (glycine, alanine, and proline) were correlated with palatability. The role of glutamic acid and umami in the palatability of these squid was unclear in that it was high in one species (*Ommastrephes sloani pacificus*, Surume-Ika) considered inferior in taste, yet in another preferred species (*S. lessoniana*, Aori-Ika), glutamic acid was low. In two species that were considered inferior in taste (Sode-Ika, *T. rhombus* and Surume-Ika, *O. sloani pacificus*), TMA oxide content was about twice that found in the desirable species.

Reddy et al. (137) found that freeze-dried wash water from clam-processing plants could be added to clam dips in place of clam meat and juice without any significant changes in overall acceptability, taste, odor, texture, and appearance. Major amino acid components in these wash water samples were glycine, glutamic acid, alanine, arginine, serine, and in some cases aspartic acid. In their work, Reddy et al. (137) determined gas chromatographic volatile profiles but did not identify individual components that may be responsible for the clam flavor profile in these freeze-dried extracts.

In processing the 5'-nucleotide can have a significant role in potentiating flavor. Hashida et al. (138) studied the application of 5'-nucleotides to canned seafoods. In canned short-necked clams (*Venerupis semidecusata*), oyster (*Ostrea giges*), and red crab (*Acanthodes armatus*) the flavor of the seafood was enhanced at the following levels: 0.04 or 0.08% Ribotide® (a 1:1 mixture of 5'-IMP · Na$_2$ and 5'-GMP · Na$_2$, Takeda Chemical Industries Ltd.) for short-necked clam and red crab meat. The 5'-nucleotides appeared to be somewhat stable to heating and storage. Canned red crab meat retained 77 to 88% of added nucleotides during storage for 28 days, and canned short-necked clams retained 52 to 61% after 189 days.

ATTEMPTS TO USE SEAFOOD FLAVOR AND TEXTURE ATTRIBUTES TO CLASSIFY FISH

Due to a greatly expanded U.S. fishery, the American seafood marketplace contains a variety of new commercially harvested fish with little name familiarity, which has created a minor identity crisis in the industry and for the consumer. Although the National Marine Fisheries Service of the U.S. Department of Commerce (NMFS), the American Fisheries Society, and the International Congress of Zoology maintain scientific names of glossaries of common fish names, these are generally unsuited for use in market identification. To make matters worse, common names for fish species used in the United States contain little or no useful information for the consumer (139). As an example of the problem, some fish have many common names (as many as 5) from as many different locations, and many popular commercial fish have several common names, such king salmon which is also known as chinook, blackmouth, and spring salmon. Also, commonly used names in one region of the United States may be misleading in another, for example, the fish known as "snapper" along the Pacific coast is a species of rockfish (*Sebastes* spp.), not the acceptable *Lutjanidae* spp. from the Gulf of Mexico and Caribbean that the Food and Drug Administration has officially designated as snapper. In other cases, names may be unattractive or pejorative or may reflect ethnic slurs or epithets (eg, ratfish, hagfish, wolffish, jewfish, etc). As a result, consumers can learn little about the edibility characteristics of a product from its name, thus a large variety of seafoods remain effectively unavailable to them. Faced with a bewildering array of common names, it is not surprising that consumers confine their purchases to a few familiar items rather than deal with the unknown.

Because the American consumer was perceived as resisting the purchase of new seafoods that appeared in the marketplace because of a lack of information about what they taste like, NMFS launched a research project to determine the feasibility of developing a national nomenclature system that would group fish species according to their edibility characteristics (flavor and texture). It was felt that both processors and retailers could employ such a system to group together fish that share the same edibility characteristics so that consumers, both new and experienced, would be able to consistently and knowledgeably select their seafood. The system could work much like our classification of wines in the United States (ie, Chablis, Burgundy, claret, etc). Each of these wine classifications can contain one or more specific grape varietals, for example, claret can be made from Cabernet sauvignon, Merlot, or both.

It is a common human trait to classify and group that which interests us. For example, we readily group foods based on ethnic origin (eg, Italian, Mexican, Chinese, Japanese); we even break down the cuisines of a single country (eg, French/Provençal, Chinese/Szechuan, American/Southern, Mexican/"Tex-Mex"). The classification of seafoods is no exception. It is quite common to hear a buyer in the fish market ask, "What is this fish like for eating?" Questions of this type are not at all unreasonable considering the large number of fish species that crowd both Western and Asian marketplaces (more than 1,000 worldwide, 500 in the U.S. alone). Inevitably, the answers to such questions will involve one or more flavor attributes (eg, sweet, mild, strong, etc). Clearly, the flavor of a fish species is one of the most important criteria that the consumer exercises when he or she purchases seafood.

It should be clear that the various chemical compounds that impart characteristic flavors and odors to a particular species of fish are many and complex. Despite the efforts of scientists, certainly as detailed in this chapter, the relationships of specific chemical compounds and observed flavors and odors are yet to be completely understood. To date, the marketplace must ultimately rely on the consumer to determine what a fish "should be." However, to be "knowledgeable" about fish has, in recent years, become increasingly more difficult because of a large increase in the number of fish species and products finding their way into the marketplace (136). Demands on the consumer's discriminatory abilities will be increased as even more fish species are introduced into the marketplace, both in the United States and elsewhere. This increase is due to a variety of political, economic, social, and health reasons. Economic factors are perhaps the easiest to identify. In the past, fishermen, particularly those of developed countries, would keep only the most desirable fish, such as halibut, salmon, and cod, in the catch and return the less-desirable species, the so-called "by-catch," to the sea. However, with increased global demand for fishery products as well as improved (and sometimes costlier) fishing technology and processing methods, many species of fish in the "by-catch" heretofore not always utilized are now harvested and processed into edible products.

To help resolve these problems of nomenclature, NMFS, in its role of providing technical and marketing assistance to the fishery industry, organized and coordinated a program to attempt to clarify existing marketing names and provide improved procedures for establishing a uniform seafood nomenclature. As part of this project, NMFS funded a study that eventually involved elements of the U.S. Army Research Laboratories at Natick, Massachusetts (NLABS); the University of Massachusetts; Arthur D. Little Co.; and NMFS laboratories in Gloucester, Massachusetts; Charleston, South Carolina; and Seattle, Washington. As envisioned by NMFS and its collaborators, the system would be based on similarities in edibility characteristics among the various fish species produced by U.S. commercial fisheries. The system would enable consumers and the fishing industry to make informed choices among species (familiar and unfamiliar) by providing comparative sensory data consumers need to select a desired texture, flavor, and so on, of fish (140,141).

Although sensory analyses are commonly used in fisheries research, quality control, and inspection (142), it was recognized at the beginning that there was no comprehensive set of sensory terms available that was meaningful and useful *to consumers* for making comparisons between different species (141). In addition, because sensory comparisons were to be a part of this project, there was a need for a reliable scaling method that permitted the development of quantifiable numeric sensory profiles that could be statistically manipulated into valid groups. This work included research in methods of cooking; sensory panel techniques; and the development of sensory attributes to describe fish samples, including the development of descriptive terms and selection of a psychophysical method for scaling the magnitude of sensory attributes. Ultimately, a complex protocol was developed for classifying fish based on edibility characteristics, including texture, flavor, and chemical compositional attributes. Although this edibility profile involved both flavor and texture attributes, for purposes of this chapter only the work with flavor will be discussed. More detail will be given to this work because we feel that the sensory methods combined with statistical methods, such as cluster analysis, and the descriptors developed for this project have useful application to other areas of seafood technology.

Sensory Panels

Because of the consumer-oriented nature of this project, a comprehensive list of sensory attributes that could be applied to a wide range of edible fish species (141) was developed. This was achieved through the use of both extensive consumer and trained-taste panel assessments. The descriptive sensory analysis of aroma, flavor-by-mouth, and aftertaste attributes of cooked fish muscle was performed by the Natick flavor profile panel using the Arthur D. Little Flavor Profile Method (143,144). From these panels and related activities, a list of descriptive terms that was applicable to the description of the edibility characteristics of fish was generated through the use of statistical methods that were able to identify attributes that best discriminated among the various species of fish. Concomitantly, the researchers had to select a reliable scaling method for the tasters to use to describe the magnitude of the sensory attributes. The method selected was based on a 7-point category scale of intensity where 1 = slight, 4 = moderate, and 7 = extreme. Complete details of the procedures used to develop the scalar method can be found in Kapsalis (140), Cardello et al. (145), and Sawyer et al. (146). As the result of extensive testing and winnowing, a list of flavor descriptors emerged (Table 4).

The resulting system was tested by applying the method to a model system of 17 species of North Atlantic fish. Among the species of fish tested were Atlantic whiting

Table 4. Sixteen Flavor Attributes Identified by Consumers and Trained Panelists as Important to the Characterization of Fish

Flavor attributes

Overall flavor intensity[a]
Delicate or fresh fish
Heavy or gamey fish[a]
Fish (old fish)
Sweet
Briny, salty[a]
Sour[a]
Seaweed
Bitter
Fish oil[a]
Buttery
Nutty
Musty
Ammonia
Metallic
Shellfish[a]

[a]These flavor attributes were found to be significant discriminators on the basis of stepwise discriminant analysis.

(*Merluccius bilinearis*), Atlantic mackerel (*Scomber scombrus*), monkfish (goosefish) (*Lophius americanus*), Atlantic pollock (*Pollachius virens*), and others. Once completed, the results of the texture and flavor panel analyses were compiled as mean texture attributes and consensus ratings for flavor attributes of the 17 species of fish (139). From the compiled data, it was possible to compare and contrast the various species using a composite profile such as that shown in Figure 1 (137), where weakfish and cusk are contrasted. The model was further analyzed to identify groups of fish with similar characteristics using cluster analysis (147). The tree diagram in Figure 2 (140) shows the results of cluster analysis of the combined sensory data. Although limited in scope by the low number of fish evaluated, the purpose of the tree is to show the order of combination of the species. Pollock and tilefish, located opposite number 1 on the left side of the diagram, are paired

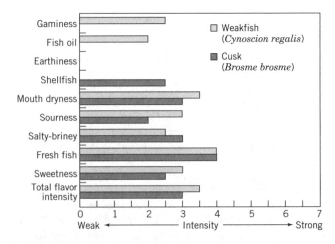

Figure 1. Composite (flavor/texture) sensory profile for weakfish (*Cynoscion regalis*) and cusk (*Brosme brosme*).

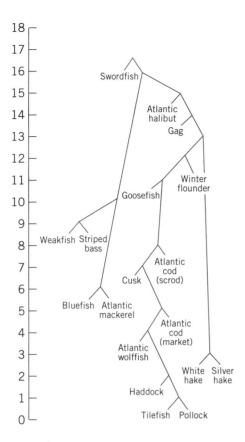

Figure 2. Tree diagram depicting the results of the cluster analysis of combined (flavor/texture) sensory data for 17 species of North Atlantic fish.

together because of their strong sensory similarities. Haddock, appearing opposite step 2, was next most similar to those paired in step 1 whereas white hake and whiting, grouped together, were next in line of sensory similarity (step 3). Hake and whiting were shown together because their profiles were more alike than either tilefish, pollock, or haddock. The amalgamation of all the species was continued until the cluster diagram was completed. The last few species (halibut, grouper, and swordfish) represent fish with the least similar sensory attributes and are weak linkages in the branching system. Figure 2 was divided into two major groups of fish. The first group consists of fish with low fat content, flavor intensity, and white flesh (tilefish, pollock, haddock, wolffish, cod, etc). The second major group includes species such as bluefish, mackerel, weakfish, and striped bass. This group is characterized by high fat content, high flavor intensity, and dark flesh. The third group was represented by swordfish. Within the two major groups of white flesh fish are two subgroups. These are represented by tilefish, pollock, haddock, wolffish, cod, cusk, monkfish, and flounder on one branch and hake and whiting on the adjacent branch.

After the 17 fish species were arranged by cluster analysis according to their characteristic sensory attributes, the original sensory data matrix was used to determine the "reasons" that the various species of fish were grouped together through the use of multidimensional unfolding

(148). With this technique, it was possible to visualize the relationships among stimulus objects (individual fish species) and sensory variables (the various flavor or texture attributes) by constructing a statistical map. The resulting map (Figure 3) has embedded within it both the sensory profile attributes as well as the 17 species of fish that were evaluated. Fish species in close proximity to one another are perceptually similar in texture, flavor, and appearance. The proximity of any fish species to any attribute point is an index of the perceived amount of that attribute in that fish species. For a more complete description of these methods and interpretation of the results, the reader is directed to the final report by NLABS summarizing the nomenclature project (140). For a more detailed explanation of the application and uses of cluster analysis, the text by Romesburg (147) is recommended.

After the initial phase of the work was completed, personnel at NLABS trained teams from three NMFS laboratories (Gloucester, Charleston, and Seattle) in their profiling methods. The NMFS laboratories began profiling commercially harvested fish species from their respective regions. Although the NMFS nomenclature program called for a comprehensive study of edibility characteristics, including chemical composition, instrumental color and texture measurements, and evaluation of sensory texture characteristics, we will deal with only the flavor attributes.

For the purposes of edibility profiling, the best-quality fish were used, usually 1 to 3 days postharvest; however, in some cases frozen fish fillets were used because they represented typical commercial product. Nevertheless, all samples used were carefully identified as to species, location, and time (date) of catch. Open consensus sensory panels (with 8–12 members) were run as described by Kapsalis (140) and Prell et al. (110). After scoring each sample, scores were tabulated in common and the results discussed to clarify individual profiles. Where there was disagreement, panelists were encouraged by the panel leader to reach a consensus. Sometimes this was difficult to achieve because of real organoleptic differences between samples of the same species. Each fish sample was replicated a minimum of 3 times during the course of this work. Therefore, the sensory profile of each species described here is a composite of data determined by the taste panel after discussion and agreement. All the combined flavor data describing each species of fish evaluated by NMFS laboratories were subjected to cluster analysis using the hierarchical method employing the BMDP-P2M "Cluster Analysis of Cases" computer program (149).

In all, 74 combined species of fish were analyzed. Results of the initial sensory evaluations are presented in the cluster analysis diagram in Figure 4 and were based on raw, unweighted data. Numbers appearing at the bottom of the diagram indicate the strength of relationships (amalgamations) based on similarities. Low numbers, or early amalgamation, indicate strong similarities between fish species and their flavor characteristics, whereas larger numbers represent fish that were quite dissimilar. From this cluster diagram, it is clear that complex relationships between fish species and their respective flavor profiles were found. Because of the large number of fish species represented in Figure 4 and the complex nature of their

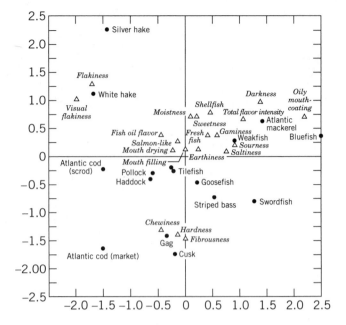

Figure 3. Multidimensional unfolding of analysis (two-dimensional) of relationships among 17 species of North Atlantic fish and estimates of combined sensory (flavor/texture) attributes.

flavor profile relationships, it became obvious that it was necessary to further simplify the cluster diagram in order to provide a far less complicated basis for describing general flavor characteristics, particularly for marketing purposes.

A less-complex cluster diagram or tree could be achieved by analyzing the same group of fish but using only four of the most significant sensory terms from Table 5 (total flavor intensity, sweetness, salty, and sourness). Results of that analysis are shown in Figure 5. Using only these four flavor attributes, the cluster analysis yielded three distinct major groups of fish. Each group, in turn, was composed of a series of subgroups of fish with increasing similarities in sensory attributes. Associations of species are different from those in Figure 4 in that only four attributes were used. We will discuss the three groups that appear in Figure 5 to give a feel for what cluster analysis can yield in its application to flavor profiling. In the discussion of each of these three groups, we will make use of the other flavor attributes, not used in the cluster analysis, as an aid in describing each group.

In Figure 5, the overall flavor profile of fish represented in group 1 was generally characterized by a moderate to moderately low total flavor intensity (TFI); that is, early impact of combined flavors, with a predominant sweet and salty-briny flavor. The intensity of a persistent, but very low, sour flavor was also observed in group 1 fish. Other flavor notes encountered in group 1 fish were gamey (as in fresh-cooked venison and opposed to beef); earthy, fish oil, and mouth drying (astringency), particularly in widow rockfish (*Sebastes entomales*). The first three flavor attributes were considered low to very low, but the sensation of mouth drying was more pronounced. Chinook salmon (*O. tshawytscha*), on the other hand, has a very low earthy flavor note but was characterized by the ever-present fish oil flavor associated with the more oily species. Mouth drying in chinook salmon had a moderately low impact.

Pen-raised (in salt water) Atlantic salmon (*Salmo salar*) tended to have slightly more exaggerated minor flavor notes as compared to the king salmon and a more pronounced earthy flavor. The flavor notes fish oil and earthy were also more pronounced (low intensity) in the Atlantic salmon than in the chinook salmon. The effect of mouth drying at the point of swallowing was moderately low. A low-intensity shellfish (lobster, clam), earthy, and fish oil flavor note characterized longspine porgy (*Stenotomus caprinus*), a species with a moderate TFI. However, the mouth drying flavor attribute was found to be very low in this species. The most predominant flavor characteristic of king mackerel (*Scomberomus cavalla*) was a moderately low gamey note followed by a low-intensity earthy flavor and mouth-drying sensation. Although mackerel was considered to be a somewhat flavorful species, in this grouping it falls in the moderate group, due in part to the presence of the tunas and sharks with their high sour notes.

Although defined by a moderate TFI, the flavor profile of spot (*Leiostomus xanthurus*) was most influenced by a slightly low salty flavor, a low gamey note, and a sweet afterflavor. Mouth drying was very minimal in spot. Residual aftertaste of earthy and fish oil were very low intensity. Monkfish (*L. americanus*), or goosefish, had a moderately low overall TFI with a slight salty, sweet flavor and near threshold sour note. The most notable sensory characteristic of this fish was its moderate shellfish flavor note, which is responsible for the fish being referred to as "poor man's lobster." The fish has a persistent shellfish aftertaste while mouth drying is low.

Unlike group 1 fish, which have a higher overall TFI, fish in group 2 are generally characterized by a low TFI and are typically described as mild in flavor. These fish have moderately low to low salty-briny and/or sweet flavor notes, and sourness was generally noticeable as a very low to just barely detectable flavor note. A typical species in

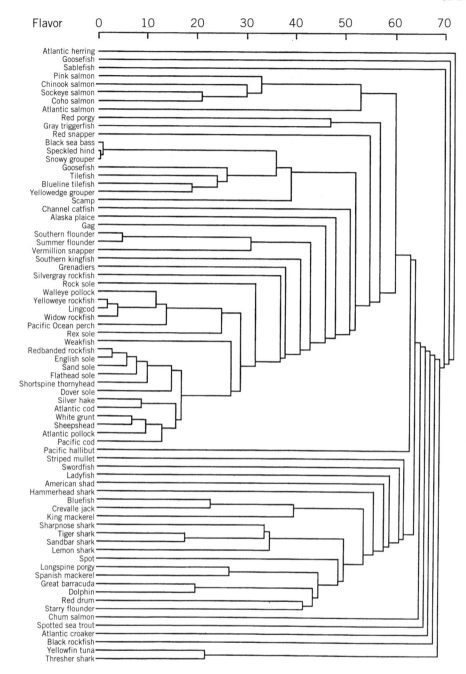

Figure 4. Expanded cluster analysis diagram (tree) showing relationships (amalgamations) between 74 species and their individual flavor profiles.

this group is English sole (*Parophrys vetulus girard*), which has a low TFI characterized by a somewhat sweet-salty flavor. It also has a very light or threshold gamey-shellfish flavor and is sometimes known to possess a very low earthy note. A bitter or iodine aftertaste was the mouth-drying characteristic in these samples. Similarly, Dover sole (*Microstomus pacificus*) was also a very mild flavored fish with a low TFI with a characteristic slight salty-sweet flavor and a hint of sour and threshold levels of earthy-gamey flavors. The effect of mouth drying was also low and sometimes accompanied by a low iodine and/or bitter aftertaste. Atlantic cod (*Gadus morhua*), also characterized by a low TFI, often had a slight earthy flavor and mouth-drying note and was sometimes characterized

by a low shellfish flavor note or a slight sour aftertaste. All of these attributes were found at low intensity levels when encountered. With a slightly stronger TFI (low) than the aforementioned fish, summer flounder (*Paralichthys dentatus*) and southern flounder (*Paralichthys lethostigma*) were relatively mild-flavored fish characterized by a low salty-briny and sour flavor with a very low level of shellfish, sweet, or sometimes earthy flavor notes observed. A fish oil flavor note was sometimes detected in the dark flesh but at or near threshold levels. Mouth drying was observed but at a low intensity. Scamp grouper (*Myteroperca phenox*) was also found to be a species with a low TFI, with a mild salty-sweet flavor and a very low sour note. Scamp was characterized by a slight shellfish flavor

Table 5. Sensory Terms for the Description of Fish Flavor

Total flavor intensity

The initial or early total impact of flavors

Salty-brine	A combination of the taste sensations of sodium chloride and the other salt compounds found in ocean water.
Sour	The taste sensation produced by acids. The taste of vinegar or lemon are typical examples.
Shellfish	The flavor associated with any cooked shellfish, such as lobster, clam, or scallop.
Gamey fish	The flavor associated with the heavy, gamey characteristics of some cooked fish such as Atlantic mackerel, as opposed to a delicate flavor such as sole. Analogous to the relationship of the heavy, gamey characteristics of fresh cooked venison compared to fresh cooked beef, or duck to chicken.
Fish oil	The flavor associated with fish oil, such as found in mackerel and canned sardines or cod liver oil.
Sweet	The basic taste sensation of which the taste of sucrose is typical.
Earthy	The flavor associated with slightly undercooked boiled potato, soil, or muddy fish.
Mouth drying	The sensation of dry skin surfaces of the oral cavity; dry feeling in the mouth after swallowing; astringency.

Other flavor notes that may be encountered

Bitter	The taste sensation of which the taste of a solution of caffeine, quinine, or iodine are typical examples.
Metallic	The taste sensation suggesting the use of slightly oxidized metal, such as tin or iron.
Nutty-buttery	The flavor associated with the rich, full flavor of chopped nuts, such as pecans and warm melted butter.
Canned salmon	The flavor associated with the salmon character in canned salmon.
Old fish	The flavor associated with cooked fish, with an off-note related to trimethylamine.
Stale fish	The flavor associated with cooked fish that is getting "off," but is not yet old.

note and a very low earthy and fish oil aftertaste. Mouth drying was very low in intensity.

Two species of fish in group 2, black rockfish (*Sebastes melanops*) from the Pacific Ocean and spot (*L. xanthurus*) from the Atlantic Ocean, were noted for the mildly heightened TFI as compared to other fish species represented in group 2. Both species were considered to be low salty-sweet flavored with very low sour notes. These fish were also characterized for their slight gamey and earthy flavor notes. Black rockfish was also found to have a very slight fish oil flavor associated with its dark flesh. This species

had a moderately low mouth-drying aftertaste whereas spot had a very low level of mouth drying.

As a family, fish in group 3 tended to have a much higher TFI profile than fish from the preceding groups. In this analysis, group 3 represents all of the shark species tested interspersed with several nonshark species, including tuna and herring. The least flavorful in this group were the starry flounder (*Platichthys stellatus*) and sandbar shark (*Carcharhinus plumbeus*). These two different species were characterized by a moderately low TFI and very low sweetness; however, this was accentuated by low salty and/or sour flavors that were more likely to be associated with sharks. Mouth drying or astringency was low but definitely present. Afterflavors of gamey, earthy, and fish oil were often present but were found at very low concentrations. Somewhat stronger-flavored scalloped hammerhead shark (*Spherna lewini*), ladyfish (*Elops sarus*), and lemon shark (*Negaprion brevirostris*) were characterized by a moderately low TFI that was somewhat influenced by a low salty-sour flavor. The effect of mouth drying was slight, and the gamey and earthy notes were predominate afterflavors but at low intensities. Perhaps the most flavorful species in group 3 were yellowfin tuna (*Thunnus albacores*) and thresher shark (*Alopias verlpinus*). These moderately flavored fish had a relatively strong TFI that was characterized by a moderate to moderately strong initial sour flavor. Usually equated with early impact of sour was a salty-briny note, which was found at low levels of flavor intensity. At the same level of intensity was a gamey and/or meaty flavor that was itself somewhat modified by a very slight sweet flavor. An earthy note was also part of the overall flavor profile but at very low levels. The impact of mouth drying was moderately low, and the aftertaste was sour-salty with some gamey note.

It is essentially from the results of research by other workers studying the chemistry of flavors that we are beginning to understand that those characteristics that distinguish one fish from another, such as salmon from sole, are attributed to chemical compounds that are endogenous in origin. We also know that exogenous flavor compounds from the environment can accidentally manifest themselves in fish, but these are generally considered as taints. Chemical compounds responsible for flavor attributes such as sweet, salty, or salty-briny, flavor characteristics common to the soles, are usually identified with the presence of sodium chloride (salt) or carbohydrates. However, the same flavors can also be caused by other chemical components such as peptides (aspartame) or nucleotides (IMP), chemicals known for their ability to impart or potentiate salty or sweet flavors in fish. Peptides are also known to influence bitter, sour, and acidic flavors common to tunas and sharks. Low pH, caused by the accumulation of lactic acid in fish muscle resulting from the breakdown of glycogen, can also produce sour flavors in fish.

Other flavor notes such as clamlike, shellfishlike, buttery, grassy, metallic, or sulfurlike, are the result of chemicals known as organic sulfides. One especially, dimethylsulfide (DMS), has been identified as producing the clam or shellfish flavor in certain gadoids and salmon when present in very small concentrations. Organic sulfides, which by their very nature can be complex, have also been

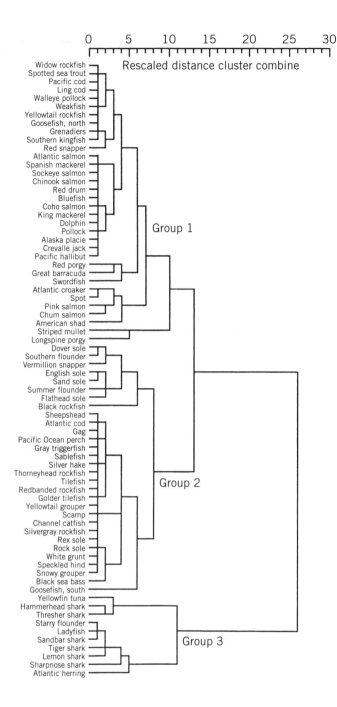

Figure 5. Grouping diagram (modified tree) of 74 species by cluster analysis showing their relationships (amalgamations) based on four significant flavor attributes.

identified with the meaty flavors prominent in some species of tuna and shark. Although organic sulfides occur naturally in fish, natural feeds such as *Spiratella heliciana*, a plankton that contains dimethyl-β-propiothetin that can be converted to DMS, can also contribute to different flavors in fish.

In addition to the various flavors attributed to organic sulfides, alcohols, ketones, and phenols have also been identified as producing their own set of characteristic flavor notes in fish. Phenols are known to cause medicinal flavors and aromas in shrimp and crayfish and possibly are responsible for the metallic-bitter or iodoformic odors and flavors sometimes seen in flatfish, such as Dover or English sole. These flavors and odors are also found in natural feeds, such as algaes and seaweeds.

Fishy odors and flavors, in contradistinction to the desirable fresh-fish flavor, predominantly found in cooked dark-muscled fish like herring or mackerel and sometimes in gadoids (cods, haddocks, hakes, and pollock), are generally caused by the presence of TMA. TMA is easily formed by the breakdown of trimethylamine oxide, which can be accumulated through the diet. The combination of lipid oxidation with TMA appears to contribute to the overall fishy odor and flavor.

The fish oil flavor, identified in the NMFS nomenclature work, was associated with the initial oxidation of fish lipids (triglycerides and fatty acids) and is best described as resembling a light cod liver oil similar to the oils used in packing sardines, and also containing grassy notes (odor of fresh-mown grass) as in fresh king or coho salmon, or a nutty-buttery or light creamy flavor characteristic of the mild-flavored black cod (sablefish).

Because of their close relationship to the environment, fish frequently reflect this connection with off-flavors or taints. For example, descriptors such as earthy, muddy, or musty are often used to describe the flavor of catfish or trout that are raised in artificial environments. Although these flavor notes are often attributed to artificial feeds, the flavors can also be caused by microorganisms living in the water, soil, or pond detritus where the fish are raised. In the marine environment, muddy flavor found in bream were identified as coming from the blue-green algae *Oscillatoria agardhii*. These off-flavors are also thought to be produced by heterocyclic organic compounds such as 1,10-*trans*-dimethyl-*trans*-(9)-decolol or geosmin and 2-methyl-lisoborneal and are the metabolic products of certain species of cyanobacteria and actinomyces.

It is easy to see that edibility profiling combined with the clustering technique provides useful tools for establishing a rational basis for the development of fish edibility groups that can be conveniently used in marketing fishery products. Results from the nomenclature studies conducted by NMFS laboratories show that, although they are complex, fish can indeed be classified into groups according to their inherent flavor characteristics. These studies also suggest that flavor profiling can be used to classify new or commercially underutilized species of fish into edibility groups as they enter the market. Because we limited ourselves to the subject of flavor profiling in this section, only a small portion of the NMFS nomenclature work was described here. However, more species were evaluated but were not included in this discussion because the minimum replicate requirements were not met. Considerably more work remains to be done to complete the edibility profiling.

FUTURE WORK

As we updated this chapter, it became clear that efforts devoted to identifying the chemical components of seafood flavors appeared to have reached a plateau in the late 1980s and very early 1990s, based at least in part on the number of recent publications. At least part of the reason for this may lie in the fact that in the decade since the first edition chapter was written, some very significant changes have occurred within the seafood industry related to the outstanding growth and importance of value-added convenience food products. Part of the cause of these changes have been cultural, such as the increased numbers of families with both spouses working; political, for example, passage of the 1976 Magnuson Fishery Conservation and Management Act (MFCMA) (150); and environmental/economic, for example, overfished, high-valued species such as salmon.

Demand for value-added convenience foods in both North America and Europe is evidenced by the increased popularity of fast-food restaurants worldwide and sales of ready-to-eat and prepared entrées, particularly those that minimize and simplify both home and institutional food preparation—that is, "heat and eat" foods. To lower costs and reduce time, institutional users (restaurants, hospitals, school lunch programs) rely less on in-house food preparation, preferring to utilize centralized production of whole fresh or frozen entrées. Many of these entrées involve new combinations of food forms. Convenience foods have been a staple in supermarkets, but now warehouse/club stores, which may have exclusively serviced the restaurant trade (both fast-food and white-tablecloth types), are now marketing these same items for home consumption. To service all these market segments, the food industry has created new and improved, value-added, complex-formulated food forms. To maintain market niches, the seafood industry has followed suit. New seafood product forms include the use of analogue simulated traditional forms (eg, crabmeat and shrimp), flavored breadings, sauces, and stuffings applied to traditional seafood forms, such as steaks and fillets.

Clearly, interest in seafood flavors and flavor components is now complicated by the presence of other food components, such as extracts, spices, herbs, induced cooking flavors (eg, browning of fried and baked breadings). In such complex cases, the inherent mild flavors of most fish and shellfish are completely lost or submerged in the final presentation.

Passage of MFCMA effectively extended U.S. jurisdiction over the waters of all coastal states and U.S. marine territories. The effect of this legislation was to establish a fisheries conservation zone that enabled the United States to greatly expand its commercial fisheries, most notably in Alaska, where an estimated 10 billion lb of fish and shellfish were available to U.S. fishers. Although much of this fishery pre-existed before the act, namely, cod, halibut, salmon, and crab, a biomass in excess of 5 billion lb of groundfish consisting of walleye pollock, assorted rockfish (*Sebastes* spp.), soles, and flounders was opened up to the U.S. seafood industry when foreign entities were excluded from the newly formed economic zones. At the same time, extension of the Pacific coast's economic zones expanded a fishery for Pacific whiting (200,000 MT) (151), a white fish, facilitating the growth of a new industry in the eastern Pacific. Passage of this act led to one of the most significant developments in the seafood processing industry: the appearance of surimi-based analogue products, primarily artificial crab and shrimp forms. Produced from very mild flavored pollock or whiting, surimi (160 MT in 1997) (155) is finely minced, deboned, washed fish flesh containing stabilizers to which crab or shrimp meats or flavorings are added to make simulated products. These can be used in seafood salads, casseroles, hors d'oeuvres, and other dishes. Here the role of extractable flavorings and the fates of flavor components is of major importance and will need future study.

Stresses on our marine fishery resources, such as overfishing, environmental factors, and economics, have led to the creation of new fisheries enterprises and modified traditional ones. For example, aquaculture and fish culturing are now major contributors to the seafood marketplace. In the past 10 years, there has been outstanding growth in the production of farmed salmon. The largest producer of farmed salmon (500,000 MT) (152) is Norway. Norway competes with Chile, which is the fastest-growing producer of farmed salmon, as well as with Scotland and Canada for sustainable markets. A considerable portion of the salmon consumed in the United States (855 million lb) (152), however, still comes from the capture fisheries. But within the last 10 years, farmed salmon produced in the United States and imported has successfully taken a significant market share. Although salmon, catfish, and trout have been successfully farmed in the United States, other species are now being investigated for possible cultivation. Convenience, value added, and institutional processors are major forces behind this growth. They in turn are being driven by the need for quality, portion and cost control, and year-round availability.

Fish culture raises some interesting questions concerning flavor research. It has been known for over 30 years that cultured fish, such as catfish and salmon, can take on both physical and flavor characteristics of their diets, for example, astaxanthin and canthaxanthin are used to achieve the pink or reddish color expected with salmon. Cultured trout or salmon fed diets high in soybean meal sometimes accrued a "beany" flavor. We have also recognized that a similar phenomenon occurs in wild fish, such as salmon, where unusual flavor notes, or taints, are observed. We have discussed some of these in this chapter. Fish can, at least from a flavor standpoint, become what they eat. This raises the potential of including desired and selective flavor components in the diet of cultured fish, which might include herbs, spices, and other extractables, to achieve new, flavorful seafood products. If not the inclusion of specific flavor components, perhaps alteration of the lipid content could also achieve similar, subtle flavor changes.

CONCLUSIONS

The consumer, acting as the ultimate sensory evaluator, is the final judge of the quality of *any* food. In this capacity, seafood flavor plays a central role in the evaluation of seafood quality. Because of this intimate relationship between quality and flavor, a clear understanding of the components that constitute flavor are significant and important. In this chapter, we have attempted to point out that consumers use seafood flavor for two broad decision-making functions: to classify fish species to determine their use and to determine acceptability. In a broad sense, the consumer asks first, "*Will* I like it?" followed by "*Do* I like it?" The answer to these questions determines repeat purchases of the item, on which rests the future of the seafood industry. Clearly, flavor science and technology has matured during the past decade, primarily due to the advent of modern instrumentation capable of observing and measuring compounds in the nanomole to femtomole range. The combined use of sensitive mass spectroscopy and high-resolution gas and liquid chromatography have helped our understanding of the flavor components associated with varying quality levels of seafoods. In addition, the combined use of sophisticated statistical tools along with sensory scoring techniques has enabled food technologists to put classification of fish species on a rational basis.

It would appear from a brief perusal of this chapter that the data generated over these last two decades on flavor compounds has diverged from the work of the sensory specialist. What then for the future? It is our belief that the dichotomy between flavor and sensory data is more apparent than real. New capabilities (both instrumental and chemical) will continue to be developed and applied to flavor research. In the last decade, instrumentation such as GC–mass spectroscopy, once the domain only of sophisticated academic research laboratories, has become increasingly available because it has decreased remarkably in price while becoming far easier to use and maintain. As with most current technology, this trend is expected to continue, and these methods will find their way into everyday routine Q&A procedures. In addition, the phenomenal and dramatic decrease in the costs of computation has placed extremely powerful analytical and statistical programs on virtually every desk in most laboratories. All of this taken together should yield even more data and information on chemical compounds responsible for flavor. However, even with all of this technical firepower, the role of the human judge or sensory specialist will not be replaced but will become more important. It will be the sensory judges and consumers who will in the final analysis have to integrate the myriad of data observations into a cohesive role for flavor in quality.

BIBLIOGRAPHY

1. M. Hamilton and R. K. Bennett, "Consumer Preferences for Fresh White Fish Species," *J. Consumer Studies and Home Economics* **8**, 243–249 (1984).

2. R. J. Hall, "Food Flavors: Benefits and Problems," *Food Technol.* **22**, 1388–1392 (1968).

3. S. S. Chang, "Food Flavors: A Scientific Status Summary by the Institute of Food Technologists' Expert Panel on Food Safety & Nutrition," *Food Technol.* **43**, 99–106 (1989).

4. J. E. Kinsella, "Flavor Perception and Binding," *Inform* **1**, 215–226 (1990).

5. M. O'Mahony, *Sensory Evaluation of Food—Statistical Methods and Procedures*, Marcel Dekker, New York, 1986.

6. G. Jellinek, *Sensory Evaluation of Food—Theory and Practice*, Ellis Horwood, Chichester, United Kingdom, 1985.

7. C. B. Warren and J. P. Walradt, eds., *Computers in Flavor and Fragrance Research*, ACS Symposium Series 261, American Chemical Society, Washington, D.C., 1984.

8. J. C. Boudreau, ed., *Food Taste Chemistry*, ACS Symposium Series 115, American Chemical Society, Washington, D.C., 1979.

9. *Guidelines for the Selection and Training of Sensory Panel Members*, ASTM Special Technical Publication 758, American Society for Testing and Materials, Philadelphia, Pa., 1981.

10. *ASTM Manual on Consumer Sensory Evaluation*, ASTM Special Technical Publication 682, American Society for Testing and Materials, Philadelphia, Pa., 1979.

11. *Compilation of Odor and Taste Threshold Values Data*, ASTM Publication DS 48A, American Society for Testing and Materials, Philadelphia, Pa., 1978.

12. *Correlating Sensory Objective Measurements*, ASTM Special Technical Publication 594, American Society for Testing and Materials, Philadelphia, Pa., 1976.

13. *Sensory Evaluation of Appearance of Materials*, ASTM Special Technical Publication 545, American Society for Testing and Materials, Philadelphia, Pa., 1973.

14. *Manual on Sensory Testing Methods*, ASTM Special Technical Publication 434, American Society for Testing and Materials, Philadelphia, Pa., 1968.

15. *Basic Principles of Sensory Evaluation*, ASTM Special Technical Publication 433, American Society for Testing and Materials, Philadelphia, Pa., 1968.

16. F. Shahidi and K. R. Cadwallader, eds., *Flavor and Lipid Chemistry of Seafoods*, ACS Symposium Series 674, American Chemical Society, Washington, D.C., 1997.

17. D. A. Forss, "Odor and Flavor Compounds from Lipids," in R. T. Holman, ed., *Progress in the Chemistry of Fats and Oils and Other Lipids*, Vol. 13, Pergamon Press, Oxford, United Kingdom, 1972, pp. 177–258.

18. M. E. Stansby and G. Jellinek, "Flavor and Odor Characteristics of Fishery Products with Particular Reference to Early Oxidative Changes in Menhaden Oil," in R. Kreuzer, ed., *The Technology of Fish Utilization*, The Fishing News (Books) Ltd., London, 1965, pp. 171–176.

19. G. Hoffmann, "3-*cis*-Hexenal, The 'Green' Reversion Flavor of Soybean Oil," *J. Amer. Oil Chem. Soc.* **38**, 1–3 (1961).

20. D. A. Forss et al., "The Flavor of Cucumbers," *J. Food Sci.* **27**, 90–93 (1962).

21. R. M. Love, *The Food Fishes—Their Intrinsic Variation and Practical Implications*, Van Nostrand Reinhold, New York, 1988.

22. L. F. Jacober and A. G. Rand, "Biochemical Evaluation of Seafood," in R. E. Martin et al., eds., *Chemistry and Biochemistry of Marine Food Products*, AVI Publishing Co., Westport, Conn., 1982, pp. 347–365.

23. C. H. Castell and R. E. Triggs, "Spoilage of Haddock in the Trawlers at Sea: The Measurement of Spoilage and Standards of Quality," *J. Fish. Res. Board Canada* **12**, 329–341 (1955).

24. J. M. Shewan et al., "The Development of a Numerical Scoring System for the Sensory Assessment of the Spoilage of Wet White Fish Stored in Ice," *J. Sci. Food Agric.* **4**, 283–298 (1953).

25. B. G. Shaw and J. M. Shewan, "Psychrophilic Spoilage Bacteria of Fish," *J. Appl. Bacteriol.* **31**, 89–96 (1968).

26. A. Kamiya and Y. Ose, "Study of Odorous Compounds Produced by Putrefaction of Foods. V. Fatty acids, Sulphur Compounds and Amines," *J. Chromatogr.* **292**, 383–391 (1984).

27. D. B. Josephson, R. C. Lindsay, and G. Olafsdottir, "Measurement of Volatile Aroma Constituents as a Means for Following Sensory Deterioration of Fresh Fish and Fishery Products," in D. E. Kramer and J. Liston, eds., *Seafood Quality Determination*, Elsevier Science Publishers, Amsterdam, The Netherlands, 1986, pp. 27–47.

28. D. B. Josephson, R. C. Lindsay, and D. A. Stuiber, "Identification of Compounds Characterizing the Aroma of Fresh Whitefish (*Coregonus clupeaformis*)," *J. Agric. Food Chem.* **31**, 326–330 (1983).

29. D. B. Johsephson, R. C. Lindsay, and D. A. Stuiber, "Variations in Occurrences of Enzymatically Derived Volatile Aroma Compounds in Salt- and Freshwater Fish," *J. Agric. Food Chem.* **32**, 1344–1347 (1984).

30. D. B. Josephson and R. C. Lindsay, "Enzymic Generation of Volatile Aroma Compounds from Fresh Fish," in T. H. Parliment and R. Croteau, eds., *Biogeneration of Aromas*, ACS Symposium Series 317, American Chemical Society, Washington, D.C., 1986, pp. 201–219.

31. F. A. Frazzalari, ed., *Compilation of Odor and Taste Threshold Value Data*, American Society for Testing and Materials, Philadelphia, Pa., 1978, pp. 79–80, 84, 116–117, 120.

32. H. Pyysalo and M. Suihko, *Lebensm.-Wiss. Technol.* **9**, 371 (1976).

33. P. A. T. Swoboda and K. E. Peers, *J. Sci. Food Agr.* **28**, 1019 (1977).

34. F. B. Whitfield et al., *Aust. J. Chem.* **35**, 373 (1982).

35. R. G. Buttery, in R. Teranishi, R. A. Flath, and H. Sugisawa, eds., *Flavor Research—Recent Advances*, Marcel Dekker, New York, 1981, pp. 180–184.

36. F. B. Whitfield et al., "Oct-1-en-3-ol and (5Z)-Octa-1,5-dien-3-ol, Compounds Important in the Flavour of Prawns and Sand-Lobsters," *Austr. J. Chem.* **35**, 373–383 (1982).

37. T. M. Berra, J. F. Smith, and J. D. Morrison, "Probable Identification of the Cucumber Odor of the Australian Grayling *Prototroctes maraena*," *Trans. Amer. Fisheries Soc.* **111**, 78–82 (1982).

38. C. Karahadian and R. C. Lindsay, "Role of Oxidative Processes in the Formation and Stability of Fish Flavors," in R. Teranishi, R. G. Buttery, and F. Shahidi, eds., *Flavor Chemistry Trends and Developments* ACS Symposium Series 388, American Chemical Society, Washington D.C., 1989, pp. 60–75.

39. D. B. Josephson, R. C. Lindsay, and D. A. Stuiber, "Biogenesis of Lipid Derived Volatile Compounds in the Emerald Shiner (*Notropis atherinodes*)," *J. Agric. Food Chem.* **32**, 1347–1352 (1984).

40. J. Kanner and J. E. Kinsella, "Lipid Deterioration Initiated by Phagocytic Cells in Muscle Foods: β-Carotene Destruction by a Myeloperoxidase-Hydrogen Peroxide-Halide System," *J. Agric. Food Chem.* **31**, 370–376 (1983).

41. J. B. German and J. E. Kinsella, "Lipid Oxidation in Fish Tissue. Enzymatic Initiation via Lipoxygenase," *J. Agric. Food Chem.* **24**, 680–683 (1985).

42. J. Kanner, J. B. German, and J. E. Kinsella, "Initiation of Lipid Peroxidation in Biological Systems," *CRC Crit. Rev. Food Sci. Nutrition* **25**, 317–364 (1987).

43. R. J. Hsieh, J. B. German, and J. E. Kinsella, "Lipoxygenase in Fish Tissue: Some Properties and the 12-Lipoxygenase from Trout Gill," *J. Agric. Food Chem.* **36**, 680–685 (1988).

44. J. B. German and J. E. Kinsella, "Hydroperoxide Metabolism in Trout Gill Tissue: Effect of Glutathione on Lipoxygenase Products Generated from Arachidonic Acid and Docosahexaenoic Acid," *Biochim. Biophys. Acta* **879**, 378–387 (1986).

45. R. J. Hsieh and J. E. Kinsella, "Lipoxygenase Generation of Specific Volatile Flavor Carbonyl Compounds in Fish Tissues," *J. Agric. Food Chem.* **37**, 279–286 (1989).

46. D. B. Josephson and R. C. Lindsay, "Retro-Aldol Related Degradations of 2,4-Decadienal in the Development of Staling Flavors in Fried Foods," *J. Food Sci.* **52**, 1186–1190, 1218 (1987).

47. K. Kubota and A. Kobayashi, "Identification of Unknown Methyl Ketones in Volatile Flavor Components from Cooked Small Shrimp," *J. Agric. Food Chem.* **36**, 121–123 (1988).

48. A. Kobayachi et al., "Syntheses and Sensory Characterization of 5,8,11-Tetradecatrien-2-One Isomers," *J. Agric. Food Chem.* **37**, 151–154 (1989).

49. U. Tanchotikul and T. C. Y. Hsieh, "Volatile Flavor Components in Crayfish Waste," *J. Food Sci.* **54**, 1515–1520 (1989).

50. W. Vejaphan, T. C. Y. Hsieh, and S. S. Williams, "Volatile Flavor Components from Boiled Crayfish (*Procambarus clarkii*) Tail Meat," *J. Food Sci.* **53**, 1666–1670 (1988).

51. R. G. Buttery et al., "Additional Volatile Components of Cabbage, Broccoli, and Cauliflower," *J. Agric. Food Chem.* **24**, 829–832 (1976).

52. M. E. Stansby, "Odors and Flavors," in M. E. Stansby, ed., *Fish Oils, Their Chemistry, Technology, Stability, Nutritional Properties, and Uses*, AVI Publishing Co., Westport, Conn., 1967, pp. 171–180.

53. M. E. Stansby, "Speculations on Fishy Odors and Flavors," *Food Technol.* **16**, 28–32 (1962).

54. W. L. Davies and E. Gill, "Investigations on Fishy Flavors," *Chem. and Ind.* (London) **55**, 141T (1936).

55. W. L. Davies, "Fishiness as a Flavor and Taint," *Flavours* **2**, 18 (1939).

56. J. M. Regenstein and M. A. Schlosser, in R. E. Martin et al., eds., *Chemistry and Biochemistry of Marine Food Products*, AVI Publishing Co., Westport, Conn., 1982, pp. 137–148.

57. C. E. Hebard, G. J. Flick, and R. E. Martin, in R. E. Martin et al., eds., *Chemistry and Biochemistry of Marine Food Products*, AVI Publishing Co., Westport, Conn., 1982, pp. 149–304.

58. J. C. Wekell and H. J. Barnett, "A New Method for the Analysis of Trimethylamine Oxide using Ferrous Sulfate and EDTA," *J. Food Sci.* **56**, 132–135, 138 (1991).

59. J. Spinelli and B. Koury, "Nonenzymic Formation of Dimethylamine in Dried Fishery Products," *J. Agric. Food Chem.* **27**, 1104–1108 (1979).

60. N. R. Jones, "Fish Flavors," in H. W. Schultz, E. A. Day, and L. M. Libbey, eds., *Symposium on Food: The Chemistry and Physiology of Flavors*, AVI Publishing Co., Westport, Conn., 1967, pp. 267–295.

61. T. Kikuchi, S. Wada, and H. Suzuki, "Significance of Volatile Bases and Volatile Acids in the Development of Off-flavor of Fish Meat," *J. Jpn. Soc. Food. Nutrition* **29**, 147–152 (1976).

62. J. B. Rossell, "Measurement of Rancidity," in J. C. Allen and R. J. Hamilton, eds., *Rancidity in Foods*, Applied Science Publishers, London, 1983, pp. 21–45.

63. C. Karahadian and R. C. Lindsay, "Evaluation of Compounds Contributing Characterizing Fishy Flavors in Fish Oils," *J. Am. Oil Chemists Soc.* **66**, 953–960 (1989).

64. R. J. Hamilton, "The Chemistry of Rancidity in Foods," in J. C. Allen and R. J. Hamilton, eds., *Rancidity in Foods*, Applied Science Publishers, London, 1983, pp. 1–20.

65. R. Hardy, "Fish Lipids Part 2," in J. J. Connell, ed., *Advances in Fish Science and Technology*, Fishing News Books, Surrey, United Kingdom, 1980, pp. 103–111.

66. J. Sargent, R. J. Henderson, and D. R. Tocher, "The Lipids," in J. E. Halver, ed., *Fish Nutrition*, 2nd ed., Academic Press, San Diego, Calif., 1989, pp. 154–218.

67. T. Roubal, *Lipid Peroxidation Damage to Biological Materials*, Ph.D. Dissertation, University of California, Davis, Calif., 1965.

68. A. Khayat and D. Schwall, "Lipid Oxidation in Seafood," *Food Technol.* **37**, 130–140 (1983).

69. H. O. Hultin, R. E. McDonald, and S. D. Kelleher, "Lipid Oxidation in Fish Muscle Microsomes," in R. E. Martin et al., eds., *Chemistry and Biochemistry of Marine Food Products*, AVI Publishing Co., Westport, Conn., 1982, pp. 1–11.

70. C. H. Castell, "Metal Catalyzed Lipid Oxidation and Changes of Proteins in Fish," *J. Am. Oil Chemists Soc.* **48**, 645–649 (1971).

71. F. Ullrich and W. Grosch, "Identification of the Most Intense Odor Compounds Formed During Autoxidation of Methyl Linolenate at Room Temperature," *J. Am. Oil Chemists Soc.* **65**, 1313–1317 (1988).

72. P. A. T. Swoboda and K. E. Peers, "Metallic Odour caused by Vinyl Ketones Formed in the Oxidation of Butterfat. The Identification of Octa-1,*cis*-5-dien-3-One," *J. Sci. Food. Agric.* **28**, 1019–1024 (1977).

73. T. C. Y. Hsieh et al., "Characterization of Volatile Components of Menhaden Fish (*Brevoortia tyrannus*) Oil," *J. Am. Oil Chemists Soc.* **66**, 114–117 (1989).

74. N. N. Gerber, "Geosmin, An Earthy-smelling Substance Isolated from *Actinomycetes*," *Biotechnol. Bioeng.* **9**, 321–327 (1967).

75. N. N. Gerber, "Volatile Substances from *Actinomycetes*: Their Role in the Odor of Pollution of Water," *Water Sci. Tech.* **15**, 115–125 (1983).

76. P. B. Johnsen, G. V. Civille, and J. R. Vercellotti, "A Lexicon of Pond-raised Catfish Flavor Descriptors," *J. Sensory Studies* **2**, 85–91 (1987).

77. M. Yurkowski and J.-A. L. Tabachek, "Identification, Analysis, and Removal Ofgeosmin from Muddy Flavoured Trout," *J. Fish. Res. Board Can.* **31**, 1851–1858 (1974).

78. M. Yurkowski and J.-A. L. Tabachek, "Geosmin and 2-Methylisoborneol Implicated as a Cause of Muddy Odor and Flavor in Commercial Fish from Cedar Lake, Manitoba," *Can. J. Fish. Aquat. Sci.* **37**, 1449–1450 (1980).

79. P. E. Persson, "The Source of Muddy Odor in Bream (*Abramis brama*) from the Porvoo Sea Area (Gulf of Finland)," *J. Fish. Res. Board Can.* **36**, 883–890 (1979).

80. N. Sugiura, O. Yagi, and R. Suda, "Musty Odor from Bluegreen Alga, *Phormidium tenue* in Lake Kasumigaura," *Environ. Tech. Lett.* **7**, 77–86 (1986).

81. J. F. Martin et al., "Analysis of 2-Methylisoborneol in Water, Mud, and Channel Catfish (*Ictalurus punctatus*) from Commercial Culture Ponds in Mississippi," *Can. J. Aquat. Sci.* **44**, 909–912 (1987).

82. J. F. Martin, T. H. Fisher, and L. W. Bennett, "Musty Odor in Chronically Off-Flavored Channel Catfish: Isolation of 2-Methyleneboronane and 2-Methyl-2-Bornene," *J. Agric. Food Chem.* **36**, 1257–1260 (1988).

83. R. T. Lovell et al., "Geosmin and Musty-Muddy Flavors in Pond-Raised Channel Catfish," *Trans. Amer. Fisheries Soc.* **115**, 485–489 (1986).

84. N. N. Gerber and H. A. Lechevalier, "Production of Geosmin in Fermentors and Extraction with Ion-Exchange Resin," *Appl. Microbiol.* **34**, 857–858 (1977).

85. K. Sivonen, "Factors Influencing Odour Production by *Actinomycetes*," *Hydrobiologia* **86**, 165–170 (1982).

86. R. G. Buttery and J. A. Garibaldi, "Geosmin and Methylsioborneol in Garden Soil," *J. Agric. Food Chem.* **24**, 1246–1247 (1976).

87. T. E. Acree et al., "Geosmin, the Earthy Component of Table Beet Odor," *J. Agric. Food Chem.* **24**, 419–430 (1976).

88. K. E. Murray, P. A. Bannister, and R. G. Buttery, "Geosmin: An Important Volatile Constituent of Beetroot (*Beta vulgaris*)," *Chem. Ind.* (London), 973 (1975).

89. G. P. Slater and V. C. Blok, "Volatile Compounds of the Cyanophyceae—A Review," *Water Sci. Techn.* **15**, 181–190 (1983).

90. G. Izaguirre et al., "Production of 2-Methylisoborneol by Two Benthic Cyanophyta," *Water Sci. Tech.* **15**, 211–220 (1983).

91. R. T. Lovell and L. A. Sackey, "Absorption of Musty Flavor by Channel Catfish Held in Monospecies Cultures of Blue-

Green Algae," *Trans. Amer. Fisheries Soc.* **102**, 774–777 (1973).

92. R. G. Ackman, J. Hingley, and K. T. McKay, "Dimethyul Sulfide as an Odor Component in Nova Scotia Fall Mackerel," *J. Fish. Res. Board Can.* **29**, 1085–1088 (1972).

93. T. Motohiro, "Studies on the Petroleum Odour in Canned Chum Salmon," referenced in S. Ikeda in J. J. Connell, ed., *Advances in Fish Science and Technology*, Fishery News Books Ltd., Surrey, United Kingdom, 1980.

94. J. C. Spios and R. G. Ackman, "Association of Dimethyl Sulphide with the 'Blackberry Odour' in Cod from the Labrador Area," *J. Fish. Res. Board Can.* **21**, 423–425 (1964).

95. F. B. Whitfield, D. J. Freeman, and P. A. Bannister, "Dimethyl Trisulphide: An Important Off-flavour Component in the Royal Red Prawn (*Hymenopenaeus sibogae*)," *Chem. Ind.* (London), 692–693 (1981).

96. A. Miller, III, et al., "Volatile Compounds Produced in Sterile Fish Muscle (*Sebastes melanops*) by *Pseudomonas perolens*," *Appl. Microbiol.* **25**, 257–261 (1973).

97. A. Miller, III, et al., "Volatile Compounds Produced in Sterile Fish Muscle (*Sebastes melanops*) by *Pseudomonas purtrefaciens, Pseudomonas fluorescens*, and an *Achromobacter* Species," *Appl. Microbiol.* **26**, 18–21 (1973).

98. C. H. Castell and M. F. Greenough, "The Action of *Pseudomonas* on Fish Muscle. I. Organisms Responsible for Odours Produced During Incipient Spoilage of Chilled Fish Muscle," *J. Fish. Res. Board Can.* **14**, 617–625 (1957).

99. C. H. Castell and M. F. Greenough, "The Action of *Pseudomonas* on Fish Muscle. IV. Relation Between Substrate Composition and the Development of Odours by *Pseudomonas fragi*," *J. Fish. Res. Board Can.* **16**, 21–31 (1959).

100. C. H. Castell, M. F. Greenough, and N. L. Jenkin, "The Action of *Pseudomonas* on Fish Muscle. II. Musty and Potatoe-like Odours," *J. Fish. Res. Board Can.* **14**, 775–782 (1957).

101. K. Kubota, A. Kobayashi, and T. Ymanishi, "Some Sulfur Containing Compounds in Cooked Odor Concentrate from Boiled Antarctic Krills (*Euphausia superba* Dana)," *Agric. Biol. Chem.* **44**, 2677–2682 (1980).

102. K. Kubota, H. Shijimaya, and A. Kobayashi, "Volatile Components of Roasted Shrimp," *Agric. Biol. Chem.* **50**, 2867–2873 (1986).

103. S. H. Choi, A. Kobayashi, and T. Yamanishi, "Odor of Cooked Small Shrimp, *Acetes japonica* Kishinouye: Difference Between Raw Material and Fermented Product," *Agric. Biol. Chem.* **47**, 337–342 (1983).

104. H. H. Huss, "Fresh Fish—Quality and Quality Changes," FAO Fisheries Series No. 29, Food and Agriculture Organization of the United Nations, Danish International Development Agency, Rome, Italy, 1988.

105. "Peroxide Value of Oils and Fats Titration Method Final Action," American Oil Chemists Society Method 965.33, in K. Helrich, ed., *Official Methods of Analysis of the Association of Official Analytical Chemists*, 15th ed., Vol. 2, Association of Analytical Chemists, Inc., Arlington, Va., 1990.

106. *British Standard 684*, British Standards Institution, London, 1976.

107. D. W. Lemon, "An Improved TBA Test for Rancidity," New Series Circular No. 51, Canadian Fisheries and Marine Service Halifax, Nova Scotia, 1975.

108. W. Vyncke, "Evaluation of the Direct Thobarbituric Acid Extraction Method for Determining Oxidative Rancidity in Mackerel (*Scomber scrombus* L.)," *Fette, Seifen, Anstrichm.* **77**, 239–240 (1975).

109. F. M. Sawyer, "Sensory Methodology for Estimating Quality Attributes of Seafoods," in D. E. Kramer and J. Liston, eds., *Seafood Quality Determination*, Elsevier Science Publishers, Amsterdam, The Netherlands, 1987, pp. 89–97.

110. P. A. Prell and F. M. Sawyer, "Flavor Profiles of 17 species of North Atlantic Fish," *J. Food Sci.* **53**, 1036–1042 (1988).

111. M. Tada, I. Shinoda, and H. Okai, "L-Ornithyltaurine, A New Salty Peptide," *J. Agric. Food Chem.* **32**, 992–996 (1984).

112. Y. Kawasaki et al., "Glycine Methyl Ester Hydrochloride as the Simplest Examples of Salty Peptides and Their Derivatives," *Agric. Biol. Chem.* **52**, 2679–2681 (1988).

113. T. Seki et al., "Further Study on the Salty Peptide Ornithyl-β-Alanine. Some Effects of pH and Additive Ions on the Saltiness," *J. Agric. Food Chem.* **38**, 25–29 (1990).

114. S. Yamaguchi, "The Umami Taste." in J. C. Boudreau, ed., *Food Taste Chemistry*, ACS Symposium Series 115, American Chemical Society, Washington D.C., 1979, pp. 33–51.

115. S. Yamaguchi and K. Ninomiya, "Umami and Palatability," abstract for 216th American Chemical Society Meeting, Boston, Mass., August 23–27, 1998.

116. *Ribotide, A Flavor Enhancer*, Takeda Chemical Industries, Ltd.

117. M. Fujimaki et al., "Taste Peptide Fractionation from a Fish Protein Hydrolysate," *Agric. Biol. Chem.* **37**, 2891–2898 (1973).

118. M. Noguchi et al., "Isolation and Identification of Acidic Oligopeptides Occurring in a Flavor Potentiating Fraction from a Fish Protein Hydrolysate," *J. Agric. Food Chem.* **23**, 49–53 (1975).

119. S. Arai et al., "Tastes of L-Glutamyl Oligopeptides in Relation to Their Chromatographic Properties," *Agric. Biol. Chem.* **37**, 151–156 (1973).

120. Y. Yamsaki and K. Maekawa, "A Peptide with Delicious Taste," *Agric. Biol. Chem.* **42**, 1761–1765 (1978).

121. Y. Yamasaki and K. Maekawa, "Synthesis of a Peptide with Delicious Taste," *Agric. Biol. Chem.* **44**, 93–97 (1980).

122. M. Tamura et al., "The Relationship Between Taste and Primary Structure of 'Delicious Peptide' (Lys-Gly-Asp-Glu-Glu-Ser-Leu-Ala) from Beef Soup," *Agric. Biol. Chem.* **53**, 319–325 (1989).

123. S. Yamaguchi et al., "Measurement of the Relative Taste Intensity of some L-α-Amino Acids and 5'-Nucleotides," *J. Food Sci.* **36**, 846–849 (1971).

124. J. M. Kennish and D. E. Kramer, "A Review of High-Pressure Liquid Chromatographic Methods for Measuring Nucleotide Degradation in Fish Muscle," in D. E. Kramer and J. Liston, eds., *Seafood Quality Determination*, Elsevier Science Publishing, New York, 1987.

125. T. Saito and K. Arai, "Studies on the Organic Phosphates in Muscle of Aquatic Animals. V: Changes in Muscular Nucleotides of Carp During Freezing and Storage," *Bull. Jpn. Soc. Sci. Fish.* **23**, 265–268 (1957).

126. B.-O. Kassemsarn et al., "Nucleotide Degradation in the Muscle of Iced Haddock (*Gadus aeglefinus*), Lemon Sole (*Pleuronectes microcephalus*), and Plaice (*Pleuronectes platessa*)," *J. Food Sci.* **28**, 28–37 (1963).

127. J. Spinelli, M. Eklund, and D. Miyauchi, "Measurement of Hypoxanthine in Fish as a Method of Assessing Freshness," *J. Fish. Res. Board Can.* **29**, 710–714 (1964).

128. D. I. Fraser, S. G. Simpson, and W. J. Dyer, "Very Rapid Accumulation of Hypoxanthine in the Muscle of Redfish Stored in Ice," *J. Food Sci.* **25**, 817–821 (1968).

129. D. E. Kramer et al., data presented at the 4th International Congress on Engineering and Food, Edmonton, Alberta, Canada, July, 1985.

130. V. M. Creelman and N. Tomlinson, "Inosine in the Muscle of Pacific Salmon Stored in Ice," *J. Fish. Res. Board Can.* **17**, 449–451 (1960).

131. S. Ehira and H. Uchiyama, "Determination of Fish Freshness Using the K Value and Comments on Some Other Biochemical Changes in Relation to Freshness," in D. E. Kramer and J. Liston, eds., *Seafood Quality Determination*, Elsevier Science Publishing, New York, 1987, pp. 185–207.

132. H. A. Bremner et al., "Nucleotide Catabolism: Influence on the Storage Life of Tropical Species of Fish from the North West Shelf of Australia," *J. Food Sci.* **53**, 6–11 (1988).

133. D. H. Greene and E. I. Bernatt-Byrne, "Adenosine Triphosphate Catabolites as Flavor Compounds and Freshness Indicators in Pacific Cod (*Gadus macrocephalus*) and Pollock (*Theragra chalcogramma*)," *J. Food Sci.* **55**, 257–258 (1990).

134. S. Konosu, "The Taste of Fish and Shellfish," in J. C. Boudreau, ed., *Food Taste Chemistry* ACS Symposium Series 115, American Chemical Society, Washington D.C., 1979, pp. 185–203.

135. S. Konosu and Y. Maeda, "Muscle Extracts of Aquatic Animals. IV. Distribution of Nitrogenous Constituents in the Muscle Extracts of an Abalone, *Haliotis gigantea discus* Reeve," *Bull. Jpn. Soc. Sci. Fish.* **27**, 251–254 (1961).

136. K. Endo, M. Hujita, and W. Simidu, "Studies on Muscle of Aquatic Animals. Free Amino Acids, Trimethylamine oxide, and Betaine in Squids," *Bull. Jpn. Soc. Sci. Fish.* **28**, 833–836 (1962).

137. N. R. Reddy et al., "Characterization and Utilization of Dehydrated Wash Waters from Clam Processing Plants as Flavoring Agents," *J. Food. Sci.* **54**, 55–59, 182 (1989).

138. W. Hashida, T. Mouri, and I. Shiga, "Application of 5'-Ribonucleotides to Canned Seafoods," *Food Technol.* **22**, 1436–1441 (1968).

139. R. E. Martin, W. H. Doyle, and J. R. Brooker, "Toward and Improved Seafood Nomenclature System," *Mar. Fish. Rev.* **45**, 1–20 (1983).

140. J. G. Kapsalis, "Consumer and Instrumental Edibility Measures for Grouping of Fish Species," Final Report, National Technical Information Service, U.S. Dept. of Commerce, Springfield, Va., 1980.

141. A. V. Cardello et al., "Sensory Methodology for the Classification of Fish According to Edibility Characteristics," *Lebensm. Wiss. und Technol.* **16**, 190–194 (1983).

142. F. J. King et al., "Consumer and Instrumental Edibility Measures for Grouping Fish Species," in J. J. Connell, ed., *Advances in Fish Science and Technology*, Fishing News (Books), Ltd., London, 1980.

143. S. E. Cairncross and L. Sjostrom, "Flavor Profiles—New Approach to Flavor Problems," *Food Technol.* **4**, 308–311 (1950).

144. J. F. Caul, "The Profile Method of Flavor Analysis," E. M. Mrak and G. F. Stewart, eds., in *Advances in Food Research*, Academic Press, New York, 1957.

145. A. V. Cardello et al., "Sensory Evaluation of the Texture and Appearance of 17 Species of North Atlantic Fish," *J. Food Sci.* **47**, 1818–1823 (1982).

146. F. M. Sawyer, A. V. Cardello, and P. A. Prell, "Consumer Evaluation of the Sensory Properties of Fish," *J. Food Sci.* **53**, 12–18, 24 (1988).

147. H. C. Romesburg, *Cluster Analysis for Researchers*, Van Nostrand Reinhold, New York, 1984.

148. C. J. Coombs, *A Theory of Data*, Wiley, New York, 1964.

149. L. Engelman, "Cluster Analysis of Cases," in W. J. Dixon and M. B. Brown, eds., *Biomedical Computer Programs, P-Services*, University of California Press, Berkeley, Calif., 1977.

150. National Marine Fisheries Service, *Our Living Oceans. The Economic Status of U.S. Fisheries*, Technical Memorandum. NMFS F/SPO-22, U.S. Department of Commerce, National Oceanic and Atmospheric Administration, 1996.

151. National Marine Fisheries Service, *Fisheries of the United States, 1996*, U.S. Department of Commerce, National Oceanic and Atmospheric Administration, Washington, D.C., 1997.

152. H. M. Johnson et al., *1998 Annual Report on the United States Seafood Industry*, 6th ed., H. M. Johnson and Associates, Bellevue, Wash., 1998.

JOHN C. WEKELL
HAROLD J. BARNETT
National Oceanic and Atmospheric Administration
Seattle, Washington

See also SENSORY SCIENCE: PRINCIPLES AND APPLICATIONS.

SEAFOOD: SENSORY EVALUATION AND FRESHNESS

The term fresh seafood refers to a concept, not a distinct object or a specified actuality. Therefore it has many different definitions, such as (1) seafood that has never been frozen, cooked, cured, or otherwise preserved (1,2); (2) seafood that has the characteristics of being newly harvested and is not the opposite of stale (3); (3) a seafood that is the opposite of stale (4); (4) seafood that arrived at the store at a particular time (5); and (5) raw product that has not progressed beyond a certain degree of microbiological or chemical degradation (6). Depending on which definition is used, it may be very difficult to objectively determine if a specific seafood is actually "fresh." However, when freshness is defined in terms of the sensory characteristics (appearance, flavor, odor, and/or texture) of the specific seafood being evaluated, it is very possible to objectively determine the freshness of that seafood. A seafood of optimum freshness would be one that possesses the characteristics concerning appearance, flavor, odor, and/or texture that are normally associated with that particular seafood product or species, that is, it is caught at the best time of year; caught in the best location; caught by the best method; and handled, processed, prepared, and served in the best manner.

Seafood freshness is considered an extremely important factor in determining overall quality of a particular seafood item (7–9). Depending on the particular seafood item being purchased (eg, chilled, frozen, or canned), the buyer may or may not be able to readily determine the freshness of that seafood item. However, each user determines, either consciously or unconsciously, the freshness (characteristics regarding appearance, flavor, odor, and/or texture) of each seafood item as it is consumed (9). Thus, the degree to which the freshness of a seafood item meets the user's expectations will greatly affect whether the user will purchase that seafood item again (9–12).

INDIRECT DETERMINATION OF SEAFOOD FRESHNESS

Seafood freshness may be indirectly determined by chemically analyzing a sample to determine the concentration of a specific chemical(s) within the sample or by measuring the magnitude of one or more physical parameters (eg, color, distance, force) of a sample. This observed concentration of a chemical or magnitude of a physical parameter is used to indirectly measure (predict) the level of a specific sensory attribute, which allows for the immediate determination of the freshness once the predicted level of the sensory attribute is compared to that specified in the appropriate product standards. Because the measurement of chemical concentrations and physical magnitudes are often very precise (exact), these types of methods have often been considered to be objective methods and therefore superior (because they are less variable) to methods involving sensory evaluation (13–15), which have often been referred to as subjective methods. It is very important to realize that just because the results of these objective methods of determining seafood freshness may be less variable does not necessarily mean the results are thereby more accurate (16). The tremendous practical importance of understanding the difference between results that are precise and results that are accurate is clearly demonstrated by this example: "The archer or marksman may group shots close together, thus being precise, but the shots may be at the rim of the target instead of at the bull's-eye, thus being inaccurate."

Depending on the circumstances, objective chemical and physical methods may be very useful, but only if scientifically sound studies have definitely revealed that a close relationship between the results of the specific chemical or physical method and objective sensory evaluation methods does indeed exist. The results proving such a close relationship apply only to the conditions (method of catching and handling, time of season during which the seafood was harvested, species being examined, specified chemical/physical method, specified sensory evaluation method, etc) under which the investigation was conducted. Altering only one of these conditions may alter the relationship to such a degree that the close relationship may no longer apply, causing the estimated level of the sensory attribute to be of questionable value. Also during these investigations, the sensory evaluations must have been conducted using scientifically sound procedures. Whenever a chemical or physical method has met these requirements, it can be extremely useful. For additional information concerning chemical and/or physical methods of determining seafood freshness, see references 17–23.

DIRECT DETERMINATION OF SEAFOOD FRESHNESS

In addition to its use in confirming the validity of a specific chemical or physical method of indirectly determining sea-

food freshness, sensory evaluation procedures are frequently used to directly measure seafood freshness.

The results of a sensory evaluation depend on the type of sensory test used. In general, the many different sensory tests may be grouped into a few categories, each with a common purpose. These categories are discriminative or difference tests (which are used to determine whether an overall sensory difference exists between samples or how a particular sensory attribute differs between samples), descriptive tests (which attempt to identify sensory characteristics concerning appearance, odor, flavor, texture, or sound of a food product and to quantify them), and affective tests (which are used to determine which sample is preferred and how acceptable a particular sample is) (24–26). Although affective tests are subjective, both discriminative and descriptive (including grading) methods are objective (24–29). Thus, arbitrarily classifying all sensory methods as subjective is now widely recognized as wrong. Grading (a descriptive test) is the major sensory evaluation method used to determine seafood freshness.

Freshness grading is a technique that directly determines the level of one or more specific sensory attributes. This is achieved by having a small number of graders (who have been highly trained) who individually use their appropriate senses (sight, smell, taste, and/or touch) and a written grade standard (consisting of words that clearly define specific sensory properties) to sort the seafood being evaluated into clearly defined categories (grades) (3,27,30–32). Although this type of sensory evaluation is used in a wide variety of food industries, these assessors are usually trained to evaluate only one class of product, such as coffee, dairy products, seafood, tea, or wine (31).

Regardless of the seafood being evaluated, technologically sound freshness grading directly measures freshness by quantifying the characteristics of appropriate sensory attributes. This is achieved by using a single type of sensory test (ie, a structured category scale, which is commonly called a grade standard) and a small number of highly specialized graders who have been trained to function as analytical instruments (31,32).

Just because freshness grading is an objective procedure does not necessarily mean the result will be accurate (ie, hit the bull's-eye). Whenever freshness grading is being conducted, extreme care must be taken to ensure that all graders always function as analytical instruments (21,33). It is extremely difficult to ensure the accuracy of the graders' assessments without first clearly stating, in writing, all of the reasons why the seafood freshness of a particular product is being graded (29). The importance of each factor that affects the objectivity of the assessments can then be readily established by comparing it with these reasons.

Ensuring Graders Function as Analytical Instruments

Ensuring that graders function as analytical instruments can best be achieved by properly screening all graders, properly training all graders, reducing the effects of the immediate surroundings, minimizing the effects of psychological factors, and monitoring the accuracy of the assigned grades.

Screening primarily involves testing the potential grader's basic ability to assess each type of attribute (appearance, odor, flavor, and/or texture) that the grader will likely be required to evaluate by subjecting the person to a broad range of triangle tests, which vary only in appearance, aroma, flavor, or texture (34). This will determine if the potential grader is actually capable of accurately individually assessing appearance, odor, flavor, and texture. However, screening also involves assessing an individual's health, general attitudes, availability, and ability to communicate (32,34).

Because it depends on the specific situation, the extent of training that is actually necessary is best established by critically examining the written reasons for determining freshness. This is very important because if the extent of training is not closely related to the duties expected, a person may receive costly training for a specific grading duty that he or she may never perform, or conversely, the grader may be asked to conduct a specific grading duty for which he or she has not been adequately trained. For example, extensive training is necessary whenever graders are required to assess specific off-odors and off-flavors, commonly referred to as taints (35).

Training involves teaching the potential grader (1) the general principles and practices of sensory evaluation; (2) to readily identify and accurately assess each sensory characteristic defined in each of the different product standards that he or she will be expected to use; (3) to have a long-term memory so that standards that have not been recently used can be readily and accurately implemented; (4) to grade a large number of samples at a single session, and (5) to evaluate his or her performance (35,36).

The grader's learning of general principles and practices usually involves formal lectures and demonstrations concerning topics such as the operation of each of the different senses (sight, smell, taste, touch); how the perception of color (and other types of appearances), flavor, odor, and texture is affected by a wide range of variables; the different types (affective, discriminative, descriptive) of sensory assessment; and technologically valid testing procedures (31,35). The importance of graders possessing a sound knowledge of the principles and practices of sensory evaluation must not be underestimated.

The assessment area may greatly affect the grader's ability to function as an analytical instrument. Any detrimental effects of the immediate surroundings may be minimized by (1) ensuring that all of the surfaces (booths, cabinets, ceilings, countertops, doors, floors, tables, walls, etc) are of a neutral color; (2) ensuring that there is no distraction by noise (including talking); (3) ensuring that the assessment area is both well ventilated and cleaned using only non-odor-generating cleaners, thereby preventing/eliminating foreign odors; (4) ensuring that the samples are not prepared but are only evaluated within the assessment area; and (5) ensuring that the assessment area is illuminated to give an intensity of approximately 1,000 lx/m^2 and a color temperature of 5,000 to 5,500 K (29,33,35). Whenever the entire assessment area is not properly illuminated, a properly illuminated grading cabinet must be used.

The negative effect of psychological factors (Table 1) must not be ignored but must always be seriously considered. For example, the grader(s) who assess the seafood

Table 1. Psychological Factors Influencing Sensory Measurements

Factor	Explanation
Expectation error	Any information panelists receive about the test can influence the lts
Stimulus error	The desire to be right may cause the panelists' judgment to be influenced by irrelevant characteristics of the samples.
Logical error	Can cause a panelist to assign ratings to particular characteristics because they appear to be logically associated with other characteristics.
Leniency error	Occurs when panelists rate products based on their feelings about the researcher, thereby ignoring product differences.
Halo effect	Tendency of the rating of one factor to influence the rating of another factor when panelists are asked to evaluate odor, texture, color, and taste simultaneously.
Suggestion effect	Occurs when the response of one panelist is influenced by reactions of other panelists.
Order effect	Tendency of panelists to score the second sample (of a set of samples) higher or lower than expected.
Contrast effect	Two products that are markedly different cause panelists to exaggerate the difference in their scores.
Convergence error	Occurs when a large difference between two products masks small differences between other samples in the test.
Proximity error	When a set of samples is being rated on several characteristics, panelists usually rate as more similar the characteristics that follow one another (in close proximity) on the ballot sheet than those that are either far apart or read alone.
Central tendency error	Occurs when panelists score samples in the mid-range of a scale to avoid extremes.
Motivation	An interested panelist is always more efficient and reliable.

Source: Ref. 37.

must not have been involved with actually obtaining the seafood (30). This is because knowledge concerning where the seafood was caught, how the seafood was handled, and the length of time the seafood was stored can cause the grader to have specific expectations concerning the results, and the grader would no longer be acting as an analytical device (25). In addition, (1) graders should not be involved with the preparation of the sample; (2) each sample should be labeled with a three-digit random number; and (3) the samples should be presented in a random order, ensuring that the order in which the samples are graded by the grader is balanced so that each different product appears in a given position an equal number of times.

Although grading different products in a balanced manner sounds very simple, it can be extremely difficult to achieve, particularly when a large number of different products are to be assessed at one time. For example, when the number of different products being evaluated at any one time is increased from 1 to 2 to 3 to 4 to 5, the required number of different combinations necessary to achieve a balanced presentation increases from 1 to 2 to 6 to 24 to 120, respectively. Thus, when more than three different products are evaluated at the same time, the large number of combinations necessary to achieve a balanced presentation becomes very cumbersome.

Regardless of the extent to which the graders have been screened and trained and the effects of the surroundings, and the psychological factors have been reduced, the grader must accurately implement the grade standards. The necessity of monitoring the assessments of graders who have been screened and trained to operate as analytical instruments is similar to the necessity of periodically checking the accuracy of a pH meter (or other instruments used for chemical analysis) even though it had previously been calibrated (38–40). Monitoring the accuracy of assigned grades may be achieved by having the grader check his or her consistency, comparing the grader's assessments with those of other graders, and conducting formal examinations (30). A grader's performance may also be monitored by having the grader, without his or her knowledge, assess samples with sensory characteristics of a known level.

Defined Grade Standards

In addition to having graders use their senses (sight, smell, taste, and/or touch) as analytical instruments to sort the seafood into specified categories, the second extremely essential part of seafood freshness grading is the defined grade standard that describes each category. Even if the effects of both immediate surroundings and psychological factors are minimized and the graders are both well screened and well trained, the assigned grades will probably not be valid unless the grade standards clearly allow the grader to objectively assess the samples.

The traditional grade standard used to grade intact (ie, not filleted, split, or steaked) gutted fish consists of specified sensory characteristics of three to six different criteria (eg, visual characteristics of the eyes, gills, outer slime, peritoneum, and skin and odorous characteristics of the gills) associated with the fish (41). Depending on the specific grade standard, each criterion of each sample being assessed is individually categorized as being one of two grades (accept or reject), three grades (A, B, or C), or four grades (A, B, C, or D), and so on, whichever is specified in the grade standard being used. Each grade description, for each of the three to six criteria, consists of three to six terms. Only when the grade for each criterion has been determined can the final grade (based on the grades assigned each of the different criteria) be assigned. However, the sensory characteristics of a criterion being assessed may not agree with all of the three to six terms used to describe the specific grade of a specific criterion. Whenever this lack of agreement occurs, it often causes the grader to be confused and therefore increases the time it takes to decide which grade to assign, decreasing the objectivity of this type of grading. Consequently, methods that avoid this serious disadvantage were eventually developed.

One method of increasing both the speed and the objectivity of a seafood freshness system was to increase both the number of grades within a specific criterion and the number of different criteria to be assessed (42). Rather than having four different grades for each of six different criteria (41), six different grades for each of 11 criteria were used (42). These changes reduce the number of terms required to describe each grade of each criterion and thereby (1) increase the chances that a sample's sensory characteristics agree with all of the terms used to describe a specific grade of a specific criterion; (2); decrease confusion, (3) increase objectivity, and (4) increase speed of assessment. However, depending on both the system and the sample being assessed, the increased number of grades for each criterion may increase the probability that an individual sample's sensory characteristics (of one or more specific criteria) agree with the terms of more than one specific grade of that specific criterion. This would also cause the grader to be confused, resulting in decreased speed and reduced objectivity.

An alternate method of increasing both the speed and the objectivity of the traditional freshness grading system is to greatly increase the number of criteria but to also decrease the number of grades within each criterion. This method was first used by Australian scientists (43). Eighteen different criteria (appearance of surface, skin, scales, slime, stiffness, clarity of eyes, shape of eyes, iris, blood of eyes, color of gills, mucus of gills, smell, belly discoloration, firmness, condition of vent, odor of vent, nature of stains in belly cavity, and color of blood) were used, but only two to four grades (demerit points) were used within each criterion (43). Once the sensory characteristic of a criterion was determined, it was immediately assigned a demerit point ranging from 0 to 3. Every description of each demerit point was very brief, usually involving only one or two words, and, if possible, was very precise. Thus, while assessing each criterion, a grader was exposed to minimum confusion. Another unique feature of this system was that the final freshness grade of a sample was based not on the average of the different grades assigned the different criteria, but on the total number of demerit points assigned the sample assessed (43). Individuals using this system have been reported to assess the freshness of intact fish both rapidly and objectively (43–46). Graders may readily use a pencil and paper to record their assessments of each sample, but programming this grading system into a handheld computer helps ensure that all questions have been answered and decrease the time required to assess a fish (47). Other important aspects of this system initially reported by Branch and Vail (43) are that (1) no undue emphasis is placed on a single feature, and the sample cannot be rejected on the basis of a single criterion; (2) minor differences in any one criterion being assessed do not unduly influence the total score; and (3) the combination of the different criteria and different demerit scores gives a total possible score of a reasonable magnitude (47).

Although the grading system initially reported by Branch and Vail (43) is both objective and nondestructive, it is extremely rapid, requiring only 5 minutes to grade 10 fish (45). Since its development this system has been reported to be a successful method of objectively determining the freshness grade of intact anchovy, Atlantic cod, herring, hoki, plaice, redfish, saithe, sardine, and whiting (44–

46,48,49). In addition to being a useful method of grading intact fish, this procedure has also been reported to be useful for grading fillets (44). Thirteen different criteria (fillet color, bloodstains, clotting, skin color, texture, gaping, bruising, autolysis discoloration, parasite infestation, other discoloration or contaminations, ease of filleting, ease of skinning, and wetness) with two to four demerit points for each criterion was used (44).

It is extremely important to stress that whenever one intends to use this system for a new species, preliminary studies must be conducted to ensure that all the criteria and their corresponding defined characteristics incorporated in the grade standard are appropriate and will actually be used (45,46,48,49).

CONCLUSION

Sensory evaluation (eg, grading) of seafood is an extremely useful rapid and objective method of both directly determining seafood freshness and verifying the accuracy of chemical and/or physical methods used to indirectly measure seafood freshness. However, if sensory grading is being used to objectively determine seafood freshness, it is extremely important that all graders be appropriately screened, trained, and monitored; all effects of the immediate surroundings be minimized; the effects of psychological factors be minimized; and an appropriate grade standard (one that actually allows the graders to function as analytical instruments) be used. Otherwise the procedure may not be objective, and the results may not hit the bull's-eye (ie, be accurate).

BIBLIOGRAPHY

1. C. Perkins, ed., *The Advanced Seafood Handbook*, Seafood Business Magazine, Rockland, Maine, 1992.
2. Canada Department of Fisheries and Oceans, *Fish Inspection Regulations Amended February 25, 1992*, Government of Canada, Ottawa, Canada, 1992.
3. J. J. Waterman, *Composition and Quality of Fish: A Dictionary*, Torry Advisory Note No. 87, U.K. Ministry of Agriculture Fisheries and Food, Torry Research Station, Her Majesty's Stationery Office, Edinburgh, Scotland, 1982.
4. I. Dore, *Fish and Shellfish Quality Assessment*, Van Nostrand–Reinhold, New York, 1991.
5. J. M. Regenstein and C. E. Regenstein, *Introduction to Fish Technology*, Van Nostrand–Reinhold, New York, 1991.
6. F. W. Wheaton and T. B. Lawson, *Processing Aquatic Food Products*, Wiley, New York, 1985.
7. N. K. Sorsensen, "Physical and Instrumental Methods for Assessing Seafood Quality," in H. H. Huss, M. Jakobsen, and J. Liston, eds., *Quality Assurance in the Fish Industry*, Elsevier Science, Amsterdam, The Netherlands, 1992, pp. 321–332.
8. M. Sakaguchi and A. Koike, "Freshness Assessment of Fish Fillets Using the Torrymeter and K-Values," in H. H. Huss, M. Jakobsen, and J. Liston, eds., *Quality Assurance in the Fish Industry*, Elsevier Science, Amsterdam, The Netherlands, 1992, pp. 333–338.
9. P. F. Howgate, "Quality Assessment and Quality Control," in A. Aitken et al., eds., *Fish Handling and Processing*, Her Majesty's Stationery Office, Edinburgh, Scotland, 1982, pp. 177–186.
10. L. K. Freeman, "The Quest for Quality," *Food Engineering* 62, 62–71 (1990).
11. R. Martin, "Contaminants in Relation to Quality of Seafood," *Food Technol.* 42, 104–108 (1988).
12. K. McNutt, "Consumer Attitudes and the Quality Control Function," *Food Technol.* 42, 97–98 (1988).
13. T. A. Gill, "Objective Analysis of Seafood Quality," *Food Rev. Int.* 6, 681–714 (1990).
14. T. A. Gill, "Biochemical and Chemical Indices of Seafood Quality," in H. H. Huss, M. Jakobsen, and J. Liston, eds., *Quality Assurance in the Fish Industry*, Elsevier Science, Amsterdam, The Netherlands, 1992, pp. 377–388.
15. L. F. Pivarnik et al., "Freshness Assessment of Six New England Fish Species Using the Torrymeter," *J. Food Sci.* 55, 79–82 (1990).
16. F. Garfield, "Sampling and Sample Analysis," *Quality Assurance Principles for Analytical Laboratories*, Association of Official Analytical Chemists, Arlington, Va. 85 1991, pp. 64–85.
17. J. R. Botta, "Chemical Methods of Evaluating Freshness Quality," in J. R. Botta, *Evaluation of Seafood Freshness Quality*, VCH, New York, 1995, pp. 9–33.
18. J. R. Botta, "Physical Methods of Evaluating Freshness Quality," in *Evaluation of Seafood Freshness Quality*, VCH, New York, 1995, pp. 35–63.
19. F. R. Jack, A. Paterson, and J. R. Piggott, "Perceived Texture: Direct and Indirect Methods for Use in Product Development," *International Journal of Food Science and Technology* 30, 1–12 (1995).
20. J. Oehlenschlager, "Evaluation of Some Well Established and Some Underrated Indices for the Determination of Freshness and/or Spoilage in Ice Stored Wet Fish," in H. H. Huss, M. Jakobsen, and J. Liston, eds., *Quality Assurance in the Fish Industry*, Elsevier Science, Amsterdam, The Netherlands, 1992, pp. 339–350.
21. H. H. Huss, ed., *Quality and Quality Changes in Fresh Fish*, Fisheries Technical Paper No. 348, Food and Agriculture Organization (FAO) of the United Nations, Rome, 1995.
22. G. Olafsdottir et al., "Methods to Evaluate Fish Freshness in Research and Industry," *Trends Food Sci. Technol.* 8, 258–265 (1997).
23. C. N. G. Scotter, "Non-Destructive Spectroscopic Techniques for the Measurement of Food Quality," *Trends Food Sci. Technol.* 8, 285–292 (1997).
24. Institute of Food Technologists, "Sensory Evaluation Guide for Testing Food and Beverage Products," *Food Technol.* 35, 50–59 (1981).
25. M. Meilgaard, G. V. Civille, and B. T. Carr, *Sensory Evaluation Techniques*, 2nd ed., CRC Press, Boca Raton, Fla., 1991.
26. H. Stone and J. Sidel, *Sensory Evaluation Practices*, Academic Press, New York, 1985.
27. J. J. Connell and P. F. Howgate. "Fish and Fish Products," in S. M. Herschdoerfer, ed., *Quality Control in the Food Industry*, Vol. 2, Academic Press, London, 1986, pp. 347–405.
28. G. Jellinek, *Sensory Evaluation of Food: Theory and Practice*, Ellis Horwood, Chichester, U.K., 1993.
29. E. Larmond, "Sensory Evaluation Can Be Objective," in J. K. Kapsalis ed., *Objective Methods in Food Quality Assessment*, CRC Press, Boca Raton, Fla., 1986, pp. 3–14.

30. K. Straus, "Seafood Standards," in K. Straus, ed., *The Seafood Handbook: Seafood Standards*, Seafood Business Magazine, Rockland, Maine, 1991, pp. 7–17.

31. R. K. York, "Canadian Fish Products: Fish Inspection and Sensory Evaluation," *Canadian Institute of Food Science and Technology Journal* 22, AT441–AT444 (1989).

32. International Standards Organization, *Sensory Analysis: General Guidance for the Selection, Training, and Monitoring of Assessors, Part II: Experts, Draft International Standards*, International Standards Organization, Geneva, Switzerland, 1991.

33. J. J. Connell, *Control of Fish Quality*, 3rd ed. Fishing News Books, London, 1990.

34. American Society of Testing and Materials, *Guidelines for the Selection and Training of Sensory Panel Members*, American Society of Testing and Materials Special Technical Publication No. 758, Philadelphia, Pa., 1981.

35. P. Howgate, *Review of Inspection Procedures (Sensoric Evaluation) for Fish and Shellfish*, Joint FAO/WHO Food Standards Programme, Codex Alimentarius Commission, Committee on Fish and Fishery Products, Bergen, Norway, 1992.

36. International Standards Organization, *International Standard: Sensory Analysis—Vocabulary*, ISO Reference No. ISO 5492, International Standards Organization, Geneva, Switzerland, 1992.

37. L. M. Poste et al., *Laboratory Methods for Sensory Analysis of Food* Publication No. 1864E, Government of Canada, Research Branch, Department of Agriculture, Ottawa, Ontario, 1991.

38. J. K. Taylor, "Validation of Analytical Methods," *Anal. Chem.* 55, 600A–605A (1983).

39. J. K. Taylor, "Role of Collaborative Studies in Evaluation of Analytical Methods," *J. Assoc. Off. Anal. Chem.* 69, 398–400 (1986).

40. P. J. Wagstaffe, "Errors in Analytical Methods: Use of Intercomparisons to Locate Sources of Error and to Improve Accuracy in Food Analysis," *Analysis* 17, 455–459 (1989).

41. P. Howgate, A. Johnston, and A. Whittle, eds., *Multilingual Guide to EC Freshness Grades for Fishery Products*, Torry Research Station, Food Safety Directorate, Ministry of Agriculture Fisheries and Food Aberdeen, Scotland, 1992.

42. H. M. Lupin et al., "Storage of Chilled Patagonian Hake (*Merluccius hubbsi*)," *J. Food Technol.* 15, 285–300 (1980).

43. A. C. Branch and A. M. A. Vail, "Bring Fish Inspection into the Computer Age," *Food Technology in Australia* 37, 352–355 (1985).

44. H. A. Brenner, "A Convenient Easy-to-Use System for Estimating the Quality of Chilled Seafoods," *Fish Processing Bulletin* 7, 59–70 (1985).

45. E. Larsen et al., "Development of a Method for Quality Assessment of Fish for Human Consumption Based on Sensory Evaluation," in H. H. Huss, M. Jakobsen, and J. Liston, eds., *Quality Assurance in the Fish Industry*, Elsevier Science, Amsterdam, The Netherlands, 1992, pp. 351–358.

46. J. Nielsen, et al., "Development of Methods for Quality Index of Fresh Fish," *FAR Meet.*, Noordwijkkerhort, The Netherlands, 1992.

47. H. A. Brenner, J. Olley, and A. M. A. Vail, "Estimating Time-Temperature Effects by a Rapid Systemic Sensory Method," in D. E. Kramer and J. Liston, eds., *Seafood Quality Determination*, Elsevier Science, Amsterdam, The Netherlands, 1987, pp. 413–435.

48. E. Martinsdottir and A. Atnason, "Redfish, Kapittel 4, Sluttraport," in *Nordic Industrial Fund, Quality Standards for Fish, Final Report Phase II*, 1992, pp. 21–35.

49. S. Jonsdottir, "Kvalitetsnomer pa sild, Slutrapport," in *Nordic Industrial Fund, Quality Standards for Fish: Final Report Phase II*, 1992, pp. 36–69.

J. R. BOTTA
Moncton Food Quality Investigations, Inc.
Moncton, New Brunswick
Canada

SEAWEED AQUACULTURE: TAIWAN

INTRODUCTION

Aquaculture of seaweed in Taiwan started in 1961 with the cultivation of *Gracilaria*. Presently one species of *Gracilaria, G. tenuistipitata* Chang et Xia var. *liui* Chang et Xia (Fig. 1); two species of *Porphyra, P. dentata* Kjellman (Fig. 2) and *P. haitanensis* T. J. Chang et Zhang Baofu; and two species of brown algae, *Undaria undarioides* (Yendo) Okamura and *Laminaria japonica* Areschong are cultivated. But only *Gracilaria* and *Porphyra* are cultivated on a commercial scale and *Undaria* and *Laminaria* are still in the

Figure 1. Habit of *Gracilaria tenuistipitata* var. *liui* from Anping (scale = 1 cm).

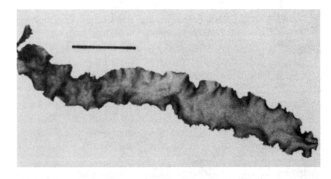

Figure 2. Habit of *Porphyra dentata* from Penghu (scale = 1 cm).

experimental stages. The actual methods of cultivating *Gracilaria* and *Prophyra* practiced in Taiwan are described as follows.

CULTIVATION OF *GRACILARIA*

Gracilaria is a red algal genus, widely distributed in many temperate to tropical regions and many species of this genus are cultivated for manufacturing an industrial colloid, agar, in many parts of the world (1–3). Species of *Gracilaria* are popularly known in Taiwan as "Long-Shi-Tsai" (meaning vegetable of dragon's beard). Originally it was commercially cultivated in ponds for extracting agar but now it has been cultivated chiefly for feeding small abalone, *Haliotis diversiolar* Lischke. There are thirteen species of *Gracilaria* reported from the coast of Taiwan and its offshore islands (4). Among them, *G. tenuistipitata* var. *liui* (Fig. 1) (5) is the main species cultivated, but sometimes plants of *G. gigas* Harvey and *G. lichenoides* (L.) Harvey can also be found among the cultivated species in the same pond (1).

In Taiwan, cultivation of *Gracilaria* started in early 1960 (6). Then, the raw material for agar extraction had decreased significantly due to overexploitation of natural agarophyte, *Gelidium* resources. Thus, in order to meet the increasing demand for agarophyte from the agar industry, the Taiwan Fisheries Bureau supported the Chilou Fishermen's Association in carrying out experimental cultivation of *Gracilaria* in Hsinta-Kang Bay (Kauhsiung Hsien) in 1961. The initial method of cultivation was by inserting branches of the seaweed into ropes and growing them in shallow coastal water. Later, in the same year (6), another experiment was done at Si-Kun-Shen (Tainan Hsien) growing the seaweeds by scattering them on the bottom of shallow, brackish water ponds in which milkfish, *Chanos chanos* Forskal, were cultivated. Subsequent experiments proved that the latter method of cultivation was better than the former one.

In 1970s, with the assistance from the Joint Commission on Rural Reconstruction, the methods of cultivation were improved greatly. Therefore, during this period both the cultural area and the production of the seaweed had increased. According to the Taiwan Fisheries Year Book, the total culture area in 1972 was about 111 ha and had increased to more than 240 ha in 1980. The Anping area (Tainan City) (Fig. 3) alone reached about 200 ha and the annual production increased from about 3,200 tons of fresh seaweed in 1972 to about 9,600 tons in 1980. In the mid 1980s, the area of cultivation decreased significantly to less than 200 ha due to the fact that many *Gracilaria* ponds were utilized to cultivate grass shrimps, *Penaeus monodon* Fabricius. But from 1988, prevalent shrimp diseases caused many former *Gracilaria* ponds to return to growing seaweed again. At the same time, the number of small abalone farms has increased greatly, and the increase in the numbers of abalone farms has brought about a great demand for fresh *Gracilaria* material. Thus, having been stimulated by the increasing demands for fresh seaweed, the number of *Gracilaria* ponds was also increased. Now the total area of cultivation is about 350 ha and it produced about 9,310 tons of fresh seaweed in 1989.

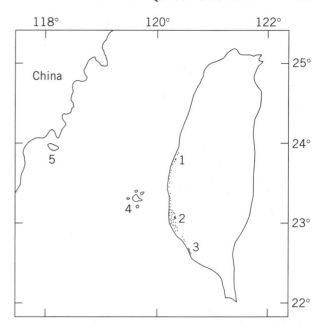

Figure 3. Map of Taiwan showing principal areas of seaweed cultivation. 1, Chanhua Hsien; 2, Anping, Tainan City; 3, Pington Hsien; 4, Penghu; 5, Kinmen. 1–3, *Gracilaria* cultivation. 4, *Porphyra dentata*; 5, *P. haitanensis*.

All *Gracilaria* ponds are located on the west coast of Taiwan, covering the coastal area from Chan-hua to Pington Hsien (counties) (Fig. 3), mainly in the Anping area (Tainan City) which occupies about 250 ha. The farming method employed is a polyculture system. That is, plants of *Gracilaria* are scattered randomly on the bottom of ponds in which milkfish, *Chanos chanos* are cultivated. Sometimes, grass shrimps, *Penaeus monodon* and crabs, *Scylla serrata* Forskal are also cultivated together to increase the income of the pond. Presently almost all seaweed produced is used for feeding small abalone culture. Only a small portion of the seaweed which is harvested from late autumn to winter goes to the agar industry. Sketches of the commercial cultivation of *Gracilaria* are presented as follows:

The Pond

These were formerly used in cultivation of milkfish or grass shrimp. The best sites for *Gracilaria* cultivation are those where no strong wind prevails and where both sea and fresh water can be obtained easily. A place with sufficient tidal difference to facilitate exchange of water is preferable to ones with little tidal difference. Most of the ponds are rectangular and are quite different in size. Smaller ponds are preferable to larger ones, because the former are easier to manage when shrimps and/or crabs are additionally cultivated. Each pond has an inlet and an outlet to facilitate the exchange of water.

Ponds are generally about 50 to 80 cm deep with a flat bottom of clay loam, sand silt loam, sandy loam or loamy sand (7). Seaweed growing in the pond with a sandy bottom can be buried in the sand during windy days. Therefore, the size of the pond should be smaller and the water should be deeper to minimize the effects of wind. Since seaweed in larger ponds (larger than one hectare) generally pile up at one corner of the pond by the wind, installation of one or two rows of wind breaks made of bamboo pieces perpendicular to the direction of the wind in the center of the pond is needed to minimize the damage which might be caused by the wind.

Seeding

At present *Gracilaria tenuistipitata* var. *liui* (Fig. 1) is the species most extensively cultivated.

The procedures of cultivation are as follows: (*1*) Drying the pond in the sun for several days. At the same time the bottom of the pond is flattened. These tasks are usually finished during early spring. Once the cultivation is started and goes well, the seaweed can be grown continuously in the same pond for many years without drying the pond again. (*2*) Introduction of water. Depending on the season, about 30 to 60 cm deep of water is introduced. (*3*) Seeding. It is usually done from March to May. Cultured strains are used as planting stock. Depending on the fertility of the ponds, about 5,000 to 12,000 kg/ha of fresh material are added by tearing them randomly into pieces and scattered evenly into the pond. (*4*) Introduction of milkfish to control epiphytes. Generally about 1,000 fish (about 15 cm long) per hectare are introduced before the appearance of epiphytes.

Pond Management

The depth of the pond water (the distance between the top of the seaweed and the water surface) is usually kept at 30 to 40 cm. But during the summer time, the water level is elevated to 50 to 60 cm high to prevent damage from rise in water temperature. While in the winter, the water level is lowered about 20 to 30 cm to receive more sunlight and to raise water temperature for the seaweed.

Frequent change of water is needed to keep a proper range of salinity and to provide nutrients for the growth of the seaweed. One change of water every two or three days is the usual practice. Only one half to two-thirds of the water is changed. The optimum salinity of the pond water is about 15 to 25 ppt (7). In some extreme cases the salinity can be as low as 4 ppt during the rainy season or as high as 50 ppt in the winter. Under such extreme conditions, even though they may not cause death of the seaweed in one or two days, the pond water should be changed more often.

To accelerate the growth of the seaweed, about 3 kg/ha of urea or 120 to 180 kg/ha of fermented manure of pigs or chickens are applied every two or three days at the time of the introduction of new water. The amount of fertilizer application are based on the color of the thalli of *Gracilaria*. If they are yellowish brown, it means they need fertilizer. When they are dark reddish-brown, it means the seaweed are healthy (8).

The main pests of *Gracilaria* are species of *Enteromorpha*, *Chaetomorpha*, and *Ectocarpus*, with the former two being the most harmful and most difficult to control. At present, control of epiphytes with milkfish seems quite satisfactory. However, milkfish may also eat *Gracilaria* after

the epiphytes are gone. When this is the case, the larger fish are removed and smaller ones are introduced again. During winter because of lower grazing power of the fish, epiphytes, especially plants of the species *Ectocarpus*, may prevail. Fortunately plants of *Ectocarpus* are small and their influence on the seaweed is relatively small. Besides epiphytes, *Tilapia mossambica* Peters may become a pest of *Gracilaria*. *Tilapia* are widely distributed in southern Taiwan and get into *Gracilaria* ponds very easily during the changing of water. Because of their rapid propagation, the population of *Tilapia* usually becomes very large in a short time and causes great damage to seaweed because of their strong grazing ability and their digging holes into the bottom of the pond. Therefore, efforts should be made to catch as many *Tilapia* every two or three days to keep their population down.

Depending on the season, the quantity of seaweeds seeded, and the fertility of the pond, seaweeds are harvested once every 16 to 20 days during summer, or 35–45 days during the winter. In general, (in fertile ponds) about one third to one half of the seaweed is harvested each time. Harvesting is done with a scoop net which collects the seaweed in rows and washes them with pond water. The cleaned seaweed is brought to the bank of the pond with a raft and is put into nylon bags. Each bag contains about 60 kg of fresh seaweed. They are shipped to abalone farms in trucks. If the seaweed is going to agar factories, it is sun-dried on bamboo screens or plastic sheets. About 6 kg of fresh plants yield 1 kg of dry seaweed. After harvesting, the plants remaining in the pond are torn into pieces and are seeded again. Generally about 16,000 to 50,000 kg/yr/ha of fresh *Gracilaria* are produced.

In conclusion, the cultivation of *Gracilaria* in Taiwan is quite successful with vegetative propagation in brackish-water ponds with milkfish. It seems that the present technique is adequate and the production is profitable to the fisherman. Previously the seaweed produced was used as raw material for the domestic agar industry and production was quite stable. But since 1988, the increased demand for fresh seaweed for the small abalone, *Haliotis diversiolor* mariculture has changed the situation greatly. Now almost all of the seaweed produced is used for abalone cultivation, and the *Gracilaria* needed for agar production has to be imported.

AQUACULTURE OF *PORPHYRA*

Plants of the genus *Porphyra* are known as "Tsu-tsai" in China, "nori" in Japan and "purple larve" in the west. They are highly appreciated as food by many oriental (Asian) people, and are cultivated on a large scale in China mainland, Japan, and Korea (9).

Every year Taiwan has to import a large amount of "Tsu-tsai" for human consumption. In 1980 about USD 2.0 million dollars worth of "Tsu-tsai" was imported, but this increased to about USD 3.0 million in 1987.

Cultivation of *Porphyra* started in 1970, in the Penghu Islands (Fig. 3). Due to a lack of basic studies on the biology of the cultivated species, during the first several years the whole process and method of the cultivation took place through much trial and error. Later on, (the biology of) some local species of *Porphyra*, such as *P. angusta* Ueda and *P. dentata* Kjellman, were studied both in the laboratory and in the field. Based on the results of these studies, the methods of cultivation were improved and gave very good result. Encouraged by the successful cultivation in Penghu (Islands), the cultivation of this red alga was tried in Kinmen (Fig. 3) in 1978 with a local species, *P. haitanensis*. The prospect of developing the *Porphyra* industry in this country appears to be very promising. Now *P. dentata* and *P. haitanensis* are the main species cultivated in Penghu and Kinmen, respectively.

CULTIVATION OF *CONCHOCELIS*

Cultivation of *Porphyra* begins with the culture of filamentous sporophytic phase, or *Conchocelis*. The culture of *Conchocelis* begins at the end of February to the middle of March depending on the water temperature. During this time the water temperature (17–22°C) becomes suitable for the germination of carpospores and the growth of the conchocelis plants. The flat upper parts of oyster shells are used as substrate for the carpospores.

The attachment of carpospores to the shells is initiated by sprinkling carpospore suspension over the shells placed side by side on the bottom of small shallow tanks which are 40 × 61 × 15 cm in size, containing seawater about 10 cm deep. The suspension of carpospores is prepared by filtering the suspension of pulverized mature plants of *Porphyra* crushed in a grinder. After the completion of seeding carpospores on shells, the tank is placed on a woody shelf under 1000 to 1500 lux light intensity for one week to allow the carpospores to germinate on the shells. The water then is changed. The carpospores germinate into conchocelis plants and burrow into the pearl layer. The conchocelis are cultured in the tank until the end of October for seeding conchospoes on the net. During the cultivation of conchocelis, the water in the shell-tank should be changed once every two to three weeks, and the shells should be cleaned with a brush every one to two months depending on the growth of epiphytes on the shells. Once the conchocelis grow into the shell, the growing conditions of the conchocelis and epiphytes can be controlled by changing light intensity. During summer, special attention should also be paid to the temperature of the nursery room to avoid undue rise in temperature of the tank water to affect on the growth of the conchocelis. If the room temperature rises to higher than 30°C, seawater is used to lower the temperature by spraying it on the floor, at the same time, the light intensity of the room is lowered to about 400 lux or less.

Another method of growing conchocelis is seeding the conchocelis fragments on shells. Seeding conchocelis fragments on the shells starts during the middle of May through the middle of June, by grinding several conchocelis colonies which are growing in a flask and kept in the laboratory. The colonies are cut in a sterile grinder with 500-mL filtered seawater for about 30 seconds. Then conchocelis-fragments solution is diluted with about 30 times of filtered seawater and spread thinly on the shells to allow the conchocelis to grow into it.

Seeding Conchospores on Nets and Cultivation of Leafy *Porphyra*

Conditions for conchospore formation and liberation of these *Porphyra* species are still not well understood. Therefore, seeding conchospores on cultivation nets is still (at the mercy of nature) and the annual production fluctuates greatly. It usually starts in the middle of October at Penghu, or from the end of October through early November at Kinmen.

The attachment of conchospores on the net is accomplished by putting the mature conchocelis-bearing shells into troughs of bamboo which form the frames of the net rafts, or into plastic bags opened at top which are attached beneath the net. About 75 to 100 shells are needed for each net. At Penghu, the floating raft type (9) is employed in the cultivation of leafy plants. The net rafts are floated with anchor ropes fixed in the middle part of the intertidal zone. The net rafts are made of bamboo. They are rectangular in shape, about 2.0 × 9.0 m in size and with six bamboo legs of about one meter long. The cultivation nets are made of synthetic strings. They are rectangular in shape and measures 1.5 × 8.0 m (Penghu) or 2 × 20 m (Kinmen) in size. In general, used old nets are preferable to new ones, because the surface of new strings are too slippery for the conchospores to attach. Two nets which are stacked one upon another are fastened on a net raft.

About 40 days after seeding, the seedings of *Porphyra* plants may reach 15 to 30 cm long and can be harvested. They are cultivated until March of the next year and can then be harvested 3 or 4 times, producing about 18 to 30 kg of fresh seaweed per raft (2 nets).

In Taiwan, fishermen usually cultivate *Porphyra* as profitable side job during winter when they cannot go out fishing. Since Mainland China produces large quantity of very cheap dried "Tsu-tsai" the fishermen in Taiwan are gradually losing their interest to cultivate *Porphyra* and the area of cultivation in Taiwan is decreasing.

BIBLIOGRAPHY

1. Y. M. Chiang, "Cultivation of *Gracilaria* (Rhodophyta, Gigartinales) in Taiwain," in T. Levering, ed., *Proceedings 10th International Seaweed Symposium*, 1981, pp. 569–574.

2. M. D. Hanisak, B. E. Lapointe, and J. H. Ryther, "Cultivation of Unattached Macroalgae in Florida," *J. Phycol.* 21, 15 (1985).

3. A. Pizarro and H. Barrales, "Field Assessment of Two Methods for Planting the Agar-Containing Seaweed, *Gracilaria*, in Northern Chile," *Aquaculture* 59, 31–43 (1986).

4. H. N. Yang and Y. M. Chiang, "Taxonomical Study on the *Gracilaria* of Taiwan," *J. Fish. Soc. Taiwan* 9, 55–71 (1982).

5. Y. M. Chiang and J. L. Lin, "Nitrate Uptake by Nitrogen-Starved Plants of the Red Alga *Gracilaria tenuistipitata* var. *liui*, *Jp. J. Phycol (Sorui)* 37, 187–193 (1989).

6. W. H. Kuan, "*Gracilaria* Cultivation," *China Fish. Monthly* 3, 3–9, 1962. (in Chinese).

7. S. P. Huang, "Effects of Salinity on the Growth and Uptake of Nitrogen of *Gracilaria verrucosa* and *Gelidium japonicum*," M. S. Thesis, National Taiwan University, 1980 (in Chinese), 6 pp.

8. B. E. Lapointe and J. H. Ryther, "Some Aspects of the Growth and Yield of *Gracilaria tikvahiae* in Culture," *Aquaculture* **15**, 185–193 (1979).

9. C. K. Tseng, "Commercial Cultivation," in C. S. Lobban and M. J. Wynne, eds., *The Biology of Seaweeds*, University of California Press, Berkeley and Los Angeles, 1981.

YOUNG-MENG CHIANG
National Taiwan University
Taipei, Taiwan

SENSORS AND FOOD PROCESSING OPERATIONS

Sensors link the real world with the electronic and computer world. The effectiveness of a process monitoring and control system is greatly dependent on the accuracy, repeatability and the rate of response of the various sensors employed to provide the input parameters (eg, values of temperature, moisture, composition, flow, pH, etc.). Unavailability of the right sensors (1) is one reason why the U.S. food industry has not benefitted as much from computer-based monitoring and control systems that have dramatically improved efficiency, energy savings and labor savings in most other manufacturing sectors (2–4). Automated process monitoring and control can also help to meet some of the unique process quality control requirements of the food industry such as the rigid standards of identity and regulatory requirements of the United States Food and Drug Administration. Need for automation is also felt as more batch operations in food industry are replaced with continuous processing. Development of better sensing systems are an integral part of such automation.

The unique and unsatisfied sensing needs of the food industry are often related to measuring composition of materials, usually in the solid state. One study (5) in the food industry revealed that almost 40% of the needs were for moisture sensors, 23% were for fat sensors, 14% were for protein sensors, 6% were for solids content sensors, and less than 3% were for acidity sensors. The remaining 14% were for a number of different sensors. The urgent need for the composition sensors had blocked out requests for temperature, pressure, flow and level sensors although it is known that there are unsatisfied needs in these areas. Composition sensing involves high specificity and sensitivity in the assay of food ingredients, trace compounds, additives, contaminants, toxins, microbial state, and extent of oxidative reactions such as fat rancidity. It is the need for sensitivity to very specific chemical, biochemical, or biological components that makes each composition sensing process in the food industry truly unique and difficult.

Some of the sensing techniques having limited success in rapid composition sensing include spectroscopic, microwave, ultrasonic, and fiber-optic methods (6). Of these, the near infrared spectroscopic technique has been the most successful in analyzing food components, but the process frequently requires sample preparation and is generally off-line and more expensive. Microwave, ultrasonic, and fiber-optic sensing techniques are primarily unable to provide the high specificity and sensitivity need for numerous food components. Off-line laboratory (wet chemistry) procedures are available for most measurement needs in the food industry. Typically, samples are removed from the process, taken to a laboratory, and answers received minutes or hours later, as opposed to providing immediate results to allow the utilization of computer controlled processes. For example, several techniques exist for moisture measurement. These include oven-dry, Karl-Fischer, electrical conductance, electrical capacitance, radio frequency absorption, microwave absorption and infrared absorption. Of these, only the microwave absorption method is reasonably rapid and can be used on-line. Off-line techniques similarly exist for measurements of fat, protein, etc, but their continuous on-line sensing is limited by the lack of suitable sensors.

Chemical and biological sensors can provide some of the needed specificity and sensitivity for rapid sensing of various components in the food industry. They minimize chances of 'mistaken identity' and produce faster results with accuracy comparable to conventional laboratory techniques. They are generally small, easy to use and can be made highly sensitive. For example, some biosensors can be made sensitive down to parts-per-billion and parts-per-trillion detection level (7). Information on composition sensors, which is a more immediate necessity in food processing, is harder to locate (8) compared to sensing techniques for physical parameters (eg, temperature, flow). This is due to the very specific nature of chemical and biological reactions exploited in these sensors and the resulting limited range of use. Also, due to the interdisciplinary nature of such sensing processes, the little information that is available and has possible relevance to food applications (9,10) is scattered over literature sources spanning several scientific disciplines. This study attempts to identify such novel and existing composition sensing techniques from a great variety of literature sources, categorize the techniques, and discuss their potential for rapid monitoring of food industry processes.

CHEMICAL SENSORS

The three main types of chemical sensors—ion selective electrodes, chemically sensitive field effect transistors, and metal oxide gas sensors—will be discussed.

Ion-Selective Electrodes

Ion-selective electrodes (ISEs) are electrochemical devices (11) whose voltage output, at virtually zero current, is directly related to the concentration of some ionic species around the electrode. They have been in use since the early 1930s and are discussed here since they form the basis of many advanced chemical and biological sensors. The ISEs have a membrane that selectively allows only certain ions to pass through. This selective migration of ions from a higher to a lower concentration results in an uneven charge distribution across the membrane, with a consequential building up of a potential across the membrane in a direction which opposes further movement of the ions, and an equilibrium is reached. At equilibrium, the resulting potential would exactly balance the net ionic flow in each direction, and no further net flow of charge would occur. In a system in which no current can flow, the total number of ions required to develop this opposing potential is extremely small compared to usual solution concentrations. This potential difference developed across the membrane is of the same form as that described for redox electrodes and may be written as:

$$E = E^0 + \frac{RT}{nF} \ln A$$

where E is the observed potential; E^0 is a constant potential which varies with the choice of reference electrodes; R is the gas constant; T is the temperature; n is the integral value of the charge on the measured species, A; F is the Faraday constant; and A is the thermodynamic activity of the measured species.

The properties and performance of the electrodes depend on the composition and choice of membrane material. The most well-known ISE is the glass electrode used for pH measurements (Fig. 1). A typical glass electrode consists of a glass tube sealed in one end with a very thin glass membrane and filled with a reference electrolyte. Electrical contact with the inner solution is normally made with a metal wire coated with Ag/AgCl. The glass electrode is immersed in the analyte together with a reference electrode. Hydrogen ions build up a pH dependent voltage across the glass membrane and this voltage is measured by connecting the pH electrode and the reference electrode to a very high input resistance voltmeter.

An example of a food application is the measurement of the sodium content of foods (12). This study described a glass electrode with special composition and having selectivity and sensitivity for sodium ions. The operation was performed in a range where hydrogen ion activity was negligible compared to sodium ion activity. Sodium glass-membrane electrodes contained 11% Na_2O, 18% Al_2O_3 and 71% SiO_2 and could detect sodium ion concentrations as low as 0.23 ppm. Varying percentages of these three oxides caused a desired selectivity for other monovalent ions such as potassium. The electrode was successfully used to find sodium in meat samples. Use of this electrode required less sample preparation than other analytical methods but was still an off-line method for solids since they needed sample

preparation. Other reported ion-selective electrodes include measuring sodium and potassium in practically all types of foods (13,14) and measuring nitrite content of animal feed (15).

Practical use of ion-selective electrodes in foods have not been very successful due to an incomplete understanding of the underlying principles and due to the poor performance of some of the older instruments. They are normally not suitable for aggressive environments and sample preparation is often required with these electrodes. However, they are still faster and less expensive than more traditional wet chemistry techniques and have potential for future on-line use in food and agricultural processing, particularly in composition sensing of fluids.

Chemically Sensitive Field Effect Transistor (CHEMFET)

Chemical parameters can be measured with semiconductor devices. The ion-selective electrodes can be combined with a transistor to develop an ion-selective field effect transistor (ISFET). The relationship between ISEs and ISFETs is shown in Figure 2 (16), in which the conventional ISE and a reference electrode 'R' are first shown connected to the insulated gate field effect transistor 2(IGFET) input of a voltmeter (Fig. 2a). An IGFET is commonly called MOSFET (Metal-oxide-semiconductor field effect transistor). It has very high input impedance and is used in ultra-high-input impedance amplifiers as needed in the potentiometric electrode measurement to minimize the flow of current. The integration process is achieved by attaching the ion-selective membrane 'M' directly to the gate of the input transistor (Fig. 2b,c). The resulting device is called an ion-selective field effect transistor (ISFET). The ISFET (or CHEMFET) chip is completely insulated except for the gate whose surface complexation with the specific electrolyte in solution determines the ionic selectivity of the chip.

Food application of CHEMFETs include on-line measurement of pH. Due to the chemical selectivity of the ion-selective membrane, these sensors can potentially be made to respond to other food components (liquid and gaseous) and therefore measure their concentration on-line. Compared to ion-selective electrodes, CHEMFETs offer several advantages including flexibility due to small size, a fast response, and the requirement of less complex labor-intensive calibration and preconditioning procedures (18). At present, they are used as research tools in physiology

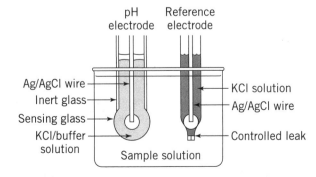

Figure 1. Schematic representation of a pH sensing electrode with a pH electrode and a reference electrode. *Source:* Ref. 11.

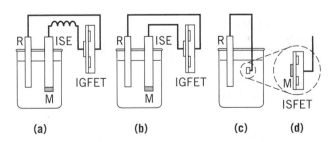

Figure 2. Evolution of ISFET (ion-selective field effect transistor, (**c**) and (**d**)) from ISE (ion-selective electrode) and IGFET (insulated gate field effect transistor) in (**a**) and (**b**). *Source:* Ref. 17.

and medicine, surface adsorption studies, gas detection, and the study of stochastic processes at various interfaces. The compatibility of the ion-selective membranes and semiconductor materials used in the construction of CHEMFETs is the major practical problem standing in the way of wide usage of these devices. By deliberate choice both the electrochemical element (the membrane) and the electronic preamplifier (FET chip) are exposed to a very hostile environment, the electrolyte solution. In such a situation, the requirements that are placed on the encapsulation materials are much more stringent than those required for ISEs. Once the encapsulation and membrane attachment problems are solved, design of sensor packages that will include data acquisition as well as data-processing elements will be feasible (16).

Using an operational amplifier, Perez-Olmos (19) improved the sensitivity of ISE for measuring potassium and calcium in wines. Knee and Srivastava (20) adapted an ISE to measure calcium in apple fruit tissue; calcium is a critical mineral affecting postharvest handling and quality. A fully automated battery-operated computerized field-based ISE method for monitoring fluoride in water was reported by Bond et al. (21).

Metal Oxide Gas Sensors

The most widely used sensors for combustible gases today are based on semiconducting metal oxides. Semiconducting gas sensors are usually based on the surface properties of the oxides of tin or zinc (SnO_2 or ZnO). It is generally agreed that the surface conductivity of semiconductors can be markedly changed by the adsorption and subsequent reaction of gases with already-adsorbed atmospheric oxygen. This implies that for an n-type material (where electrical conduction is associated with electrons, as opposed to holes) such as SnO_2 or ZnO, the concentration of electrons available for conduction can be changed by either oxidation or reduction processes. At elevated temperatures, atmospheric oxygen is adsorbed and it accepts electrons to become O_2^-, O^-, or O^{2-}. If a reducing gas is then also adsorbed, it may either simply donate electrons and become a positively charged species; or it may react with oxygen, thus releasing bound electrons. In either case, electrons become available for conduction and the resistance of the surface layers decreases drastically. For an oxidizing gas, the converse mechanisms will operate and the resistance will rise. This conductivity versus gas concentration response is exploited in the metal oxide gas sensors.

An example of a possible food industry use of a gas sensor can be seen in the work of Mandenius and Mattiasson (22) for the on-line monitoring of ethanol during fermentation processes. A sensor (Fig. 3) similar to the one used by them is described by Watson (23). The sensor consisted of a small tubular ceramic former having an interdigitated metallization pattern on the outer surface, upon which was deposited the active materials, which was largely tin oxide plus catalytic dopants, notably palladium. A heater filament was put inside the tube, while the gold alloy sensor leads were bonded to annuli on the outer surface at the ends. These annuli were in fact part of the interdigitated

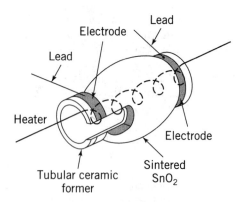

Figure 3. Configuration of a semiconductor gas sensor. *Source:* Refs. 23 and 24.

metallization deposited on the outer surface coated with the sintered, but porous, tin oxide mix. In the work of Mandenius and Mattiasson (22), ethanol, which is a reducing gas, was allowed to flow over the tin oxide coated surface of the sensor. The gas was absorbed onto the sensor surface and produced a marked decrease in the surface electrical resistance. By using a continuous dilution of the gas flow streams, the lower limit of detection of ethanol was extended by their work to operate within the concentration ranges of importance in biotechnological processes. A catalytic gas sensor device combined with a carrier gas flow control was recently developed into an integrated ethanol-sensing system (25). This permitted online monitoring of ethanol during the fermentation operation in a brewery. Gas detectors can be potentially made sensitive to other gases or volatiles present in biotechnological processes (for example, butanol, formate, acetate, and formaldehyde). For example, using metal oxide semiconductor gas sensors, it was possible to analyze wine vapors (26) or coffee aroma (27).

The advantages of this form of sensor include small size, convenient operation from low voltage power supplies, very high sensitivity in most applications and the need for only simple associated electronics. The disadvantages include a continuous power drain for sensor heating, sensitivity to ambient temperature and humidity and the presence of long-term drift. The first generation of gas sensors has suffered from poor selectivity, but many improvements are taking place, and the above example illustrates their potential on-line uses for food and agricultural processing.

BIOSENSORS

A biosensor comprises a biologically sensitive material immobilized in intimate contact with a suitable transducing system which converts the biochemical signal into a quantifiable and processible electrical signal (28,29). The biologically sensitive material is typically an enzyme, multienzyme system, antibody, membrane component, organelle, bacterial or other cell, or whole slices of mammalian or plant tissues. To produce an electrical signal, the biological interacting system is placed in close proximity to a suitable transducer (Fig. 4). When biological molecules

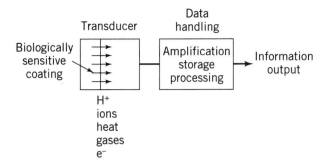

Figure 4. A generalized biosensor configuration. The biologically sensitive coating comprising an antibody, receptor protein or biocatalyst, such as an enzyme, organelle, whole cell or tissue slice, generates a physico-chemical change in response to the analyte which is converted into an electrical signal by the transducer, amplified and subsequently processed and outputted. *Source:* Ref. 3.

interact specifically and reversibly, a change takes place in one or more physico-chemical parameters associated with the interaction, such as a change in proton concentration, release or uptake of gases (O_2, CO_2, NH_3), specific ions (NH_4^+, monovalent cations, anions), heat, optical density, non-specific ions or electron transfer which if generated in close proximity to a suitable transducer, may be converted into an electrical signal. Specific interactions with the biological material can be exploited in the biosensors for sensing specific food ingredients as well as trace compounds, additives or contaminants, toxins, and marker chemicals indicating the microbial state of the food (28,31,32).

Intimate contact between the biosystem and the transducer is quite important. This is achieved by immobilization of the biological material (biocatalyst) at the device surface, generally by one of several methods. First, by chemically cross-linking the biocatalyst with an inert, generally proteinaceous, material with a bifunctional reagent to form intermolecular bonds between the catalyst and the inert protein. An alternative and commonly employed procedure, physically restrains the biocatalyst at the transducer surface by entrapment in polymer matrices such as polyacrylamide or agarose, or by retention with a polymer membrane comprising cellophane, cellulose acetate/nitrate, poly(vinyl alcohol) or polyurethane. Finally, a preferred procedure in some cases is to covalently attach the biologically sensitive system directly to the surface of the transducer and thereby achieve intimate contact without incurring the diffusional limitations, sometimes observed with membrane or matrix entrapped systems (30).

The transducer is an electrical device which responds to the products of the biocatalytic process and outputs the response in a form which can subsequently be electrically amplified and displayed. The design of the transducer should accommodate the following features: it must be highly specific for the analyte of interest and respond in an appropriate concentration range; it should have a moderately fast response time, typically 1–60 s; it must be amenable to miniaturization and should ideally compensate internally for adverse environmental effects such as temperature dependency, drift, etc.

Three major types of biosensors currently available are biosensors based on either potentiometric or amperometric transducers (33,34) and those combined with optical fibers (35).

Potentiometric Biosensors

Potentiometric devices operate under equilibrium conditions and measure the accumulation of charge at the electrode surface brought about by some selective process. The best known potentiometric biosensor is based on the ion-selective electrode (ISE) discussed earlier, where an immobilized enzyme is coated over the electrode, making it a potentiometric enzyme electrode. The specificity and sensitivity of various enzymes in catalyzing different reactions are exploited here and the substrates or products are measured as they are produced or consumed at the electrode surface. The glass pH electrode was the first ISE to be used with enzymatic reactions that proceed with the consumption or production of hydrogen ions, although other electrodes have subsequently been utilized. For example, the enzyme urease catalyzes the hydrolysis of urea and may be exploited in the determination of urea by immobilization around a pH, NH_4^+, HCO_3^- or NH_3-gas electrode (30).

The integration of an ion-selective membrane with a solid state FET results in the ion-selective field effect transistor (ISFET) as described under Chemical Sensors. It is possible to immobilize enzymes directly over the gate of an ISFET. The resulting device would be called an enzymatically sensitive field effect transistor (ENFET). A major advantage of the ENFET over an equivalent enzyme electrode is its small size. Other advantages of ENFETs are similar to those of ISFETs described earlier. Caras et al. (36–38) reported ENFETs sensitive to glucose and penicillin. In their glucose ENFET (Fig. 5), they had immobilized glucose oxidase enzyme by covalently linking it to an open, polyacrylamide gel matrix and placing it on the gate of a pH-sensitive transistor. The figure shows two pH-sensitive ISFETs on a chip which are identical except one of them has the glucose oxidase enzyme added to its gel while the other does not have any enzyme added. The two ISFETs

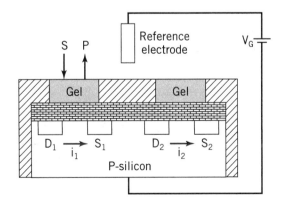

Figure 5. Schematic diagram of the ENFET chip sensitive to glucose. One of the gels has glucose oxidase enzyme for glucose sensing. D and S are transistor drain and source, respectively. V_G is the applied gate voltage and I is the drain-to-source current. *Source:* Ref. 38.

were put side-by-side so that the non-specific variations in the transistor outputs which were caused by variations in the ambient temperature, pH, and common noise could be eliminated by differential measurement. Such devices are under development and have future potential for on-line monitoring of glucose and other components in various food processes (39).

Amperometric Biosensors

Amperometric electrodes measure the flux of electroactive species. An example of an amperometric biosensor for food applications is a microbial sensor for detection of fish freshness (40). During storage of fish, larger molecular weight compounds such as proteins and glycogen are gradually degraded into smaller molecular weight compounds, which can be utilized more readily by microorganisms. The sensor (Fig. 6) was prepared by immobilizing spoilage causing bacteria, *A. putrefaciens*, on a membrane filter. The filter was fixed at the tip of an oxygen electrode and covered with a cellulose acetate membrane. The extent of the assimilation of organic substances (resulting from deterioration of fish during storage) by these immobilized microorganisms can be determined from the respiratory activity of the microorganisms by directly measuring the consumption of oxygen in the oxygen electrode. When extract from a stored fish was allowed to flow over the membranes, oxygen consumption due to the increased respiratory activity of the microorganisms caused a decrease in dissolved oxygen around the microbial membrane and consequently brought a marked decrease in the output current of the oxygen electrode. The current decrease between the initial and the minimum current was used as the measure of fish freshness. The output of the sensor, like many other sensors, is influenced by the number of immobilized living cells, pH, temperature, etc. The fish freshness sensor is one of the few biosensors developed for specific applications to the food industry. A xanthine oxidase-immobilized and carbon-based screen-printed electrode was recently developed for determining fish freshness (41). Amperometric measurements of uric acid and hypoxanthine in fish muscle exudates were effective in calculating K value, the index of fish freshness based on the levels of uric acid and hypoxanthine.

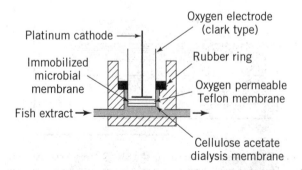

Figure 6. Schematic diagram of a microbial sensor (an amperometric enzyme electrode) system for fish freshness determination. *Source:* Ref. 40.

Another example of an amperometric enzyme electrode is a sensor for rapid in-line determination of lactose and other milk components (42). The sensor was prepared by immobilizing glucose oxidase and β-galactosidase on a nylon membrane and placing it on the platinum electrode surface. A cellulose acetate membrane was placed between the electrode surface and immobilized enzyme to eliminate electroactive compounds such as ascorbic acid. The layer of immobilized enzymes were covered with a cellulose acetate dialysis membrane in order to prevent microbial attack and leaching of the enzymes. The platinum electrode surface can measure hydrogen peroxide concentration amperometrically. Lactose from milk entered through the cellulose acetate dialysis membrane and first changed to glucose by the action of β-galactosidase. The glucose formed reacted with glucose oxidase and formed hydrogen peroxide at the platinum electrode which was reflected by a change in current in the electrode. This method of lactose determination in milk samples does not require any preliminary sample treatment, is inexpensive and also very simple and quick compared to existing titrimetric methods. It is likely to be useful for obtaining rapid analytical data on milk in dairy farms and can be used in-line. More recent applications of this technology include the measurement of aspartame in diet beverages (43), nitrate in drinking water (44), glutamate in seasonings (45), and malate in grape musts and wines (46).

Fiber-Optic Biosensors

Optical fibers are being used for numerous sensing applications (35). The fibers are rugged, more resistant to corrosion than other sensors and immune to electromagnetic interference. The transparent fibers of glass are generally used to guide a light signal to the point of measurement where the light signal changes its parameters in response to physical, chemical, and biological changes at the point of measurement. This modified light signal transmitted back along the fibers is analyzed to derive information about the physical, chemical, and biological changes. Common fiber-optic-based biochemical sensing techniques involve the use of enzymes or substrates immobilized on the tip of an optical fiber for substrates or enzyme activity measurement respectively. Trettank et al. (47) described an optical fiber sensor capable of continuously monitoring glucose. In this feasibility study, glucose oxidase enzyme was physically immobilized in a sensing layer at the end of a fiber optic light guide (Fig. 7). The enzyme catalyzed the oxidation of glucose to give gluconic acid, which in turn, lowered the pH in the microenvironment of the sensor. By following the changes in the fluorescence of a pH sensitive dye incorporated in the sensing layer, the enzymatic reaction could be monitored. The optical isolation layer prevented ambient light or intrinsic sample fluorescence from interfering. The pH sensitive dye used was 1-hydroxypyrene-3,6,8-trisulfonate (HPTS) which had an almost linear decrease in bright green fluorescence over the pH 7-6 range. The food industry, with a need for glucose sensing for fermentation monitoring and control, can benefit from such developments. A highly specific and sensitive fluorometric fiber-optic biosensor immobilized with glucose-

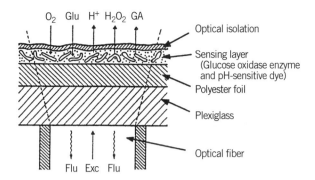

O_2 Glu H^+ H_2O_2 GA

Optical isolation

Sensing layer
(Glucose oxidase enzyme
and pH-sensitive dye)

Polyester foil

Plexiglass

Optical fiber

Flu Exc Flu

Figure 7. Cross-section through the sensing layer of the fiber-optic glucose sensor. The directions of the exciting light (Exc) and fluorescence (Flu) are also shown. *Source:* Ref. 47.

fructose-oxidoreductase from *Zymomonas mobilis* was later developed for the dual analysis of glucose and fructose (48).

Another type of fiber-optic-based biological sensor involves bioluminescence, where light is emitted during highly sensitive and specific enzymatic reactions. For example, a bio-luminiscent system (47) for the determination of NADH was produced using the bacterial luciferase and NAD(P)H:FMN oxidoreductase coimmobilized on a polyamide membrane. The bioluminiscent system was maintained in close contact with one end of a optical-fiber bundle. The light intensity obtained during reactions was analyzed using a photomultiplier tube which is a simpler instrumentation as compared to the need for a light source and monochromators in many other types of fiber-optic-based biosensors. The maximum light intensity correlated linearly with the concentration of NADH in this study. Such systems are quite new but hold future promises for composition sensing. An extremely sensitive dual-enzyme fiber-optic biosensor immobilized with glutamate dehydrogenase and glutamate–pyruvate transaminase was reported for measuring glutamate based on reduced NAD luminescence (49). A fiber-optic biosensor immobilized with xanthine oxidase and peroxidase was subsequently reported for assaying the quality of seafood products (50). Krug et al. (51) developed a fiber-optic system to measure total and free cholesterol. Immobilized cholesterol esterase cleaved cholesterol esters, while cholesterol oxidase converted cholesterol to cholest-4-ene-3-one and hydrogen peroxide. Hydrogen peroxide formed a colored complex with a dye that was measured by fiber-optic instruments.

A variety of fiber-optic biosensors have been reported capable of detecting bacterial and mold toxins including *Clostridium botulinum* toxin A (52), Staphylococcal enterotoxin B (53), *Eschericha coli* lipopolysaccharide (54), and the mycotoxin fumonisin B1 (55).

Advantages of biosensors are that they do not require highly technically qualified users in order to generate precise results, they are inexpensive, they generate results in a short period of time thus enabling better control of processes or better information for users, and that they can be used in remote locations. Estimates of possible markets of biosensors in the 1990s for food and other industrial

process monitoring have been put at $59 million. Novel biosensors are currently under development at several institutions around the world. Although much of this development is geared towards biomedical needs, food processing can significantly benefit from them due to the similar types of measurements involved. Future research related to biosensors will focus on immobilization techniques, the matching of biosensitive detection substances to analytes, proper transducing mechanisms, proper packaging of the sensor for use in hostile environments, and the optimization of the sensor response.

SUMMARY

The food industry has unique and unsatisfied needs for rapid, on-line sensing of process parameters, particularly the composition of various food components. Chemical and biological sensing techniques have been identified that can provide the specificity for a particular reaction and can be fast, accurate, inexpensive, and easier-to-use when compared to standard laboratory tests. The chemical sensors explored include semiconductor gas sensor, ion selective field effect transistors, and ion selective electrodes. Also included are enzyme electrodes, enzyme field effect transistors, and fiber-optic biosensors. These future sensing techniques have potential for on-line monitoring and control of food and agricultural processing operations.

ACKNOWLEDGMENTS

This research was supported through Hatch project 408, USDA. Source of this article is A. K. Datta, "Novel Chemical and Biological Sensors for Monitoring and Control of Food Processing Operations," *J. Food Eng.* **12** (1990). Copyright 1990. Reproduced with permission, Elsevier Science Publishers, England.

BIBLIOGRAPHY

1. H. C. J. Frost, The Food Industry's Unsatisfied Needs for Sensing Systems, *Prepared Foods*, Gorman Publishing Co., Chicago, July 1984.

2. J. P. Clark, K. J. Valentas, and D. B. Lund, Report of the Workshop Session on Food Process Engineering/Automation, *Food Technology* **39**, 21R–22R (1985).

3. G. W. Isaacs, Research Needs in Sensor Technology for Agriculture, Experiment Station Committee on Organization and Policy, 1986.

4. L. D. Pederson, W. W. Rose, H. Redsun, and S. H. Bogosian, Assessment of Sensors Used in the Food Industry Interim Report, National Food Processors Association, Dublin, Calif., 1989.

5. H. C. J. Frost, "Unsatisfied Needs for Continuous Sensing Systems. Instrumentation in the Food and Pharmaceutical Industries," *Proceedings of the Process Automation Symposium sponsored by the Instrument Society of America*, Instrument Society of America, N.C., 3–4 April 1985.

6. A. K. Datta, Sensor Needs and Their Development for On-Line Control of Food Processing Operations, *Paper no. 87-6530*, American Society of Agricultural Engineers, 1987.

7. D. Best, "Biosensors Revolutionize Quality Control," *Prepared Foods* **10**, 182–186 (1987).

8. L. Gould, Sensors in Food Processing: Part Two, *Sensors* **6**, 75–84 (1989).

9. E. Kress-Rogers, "Instrumentation in the Food Industry I: Chemical, Biochemical and Immunological Determinants," *Journal of Physics E: Scientific Instrumentation* **19**, 13–21 (1986).

10. E. Kress-Rogers, Instrumentation in the Food Industry. II: Physical Determinants in Quality and Process Control *Journal of Physics E: Scientific Instrumentation* **19**, 105–109 (1986).

11. M. S. Frant, "Ion-Selective Electrodes, in M. Grayson, ed., *Kirk-Othmer Encyclopedia of Chemical Technology*, Vol. 13, Wiley-Interscience, New York, 1978.

12. W. F. Averill, Ion-Sensitive Electrode System Measures Sodium Content of Foods, *Food Technology* **87**, 44–2 (1983).

13. E. Rabe, Zur natrium-und kaliumbestimmung mit ionensensitiven elektroden. (Determination of Sodium and Potassium by Ion-Sensitive Electrodes.) *Zeitscheiift fur Lebensmittel-Untersuchung und-Forschung* **76**, 270–274 (1983).

14. Y. Yamamoto et al., "A Volatile Amine Sensor Based on the Amperometric Ion Selective Electrode," *Buneski-Kagaku* **38**, 589–595 (1989).

15. H. J. Issaq, G. M. Muschik, and N. H. Risser, "Determination of Nitrite in Animal Feed by Spectrophotometry or Potentiometry with a Nitrite-Selective Electrode," *Analytica Chimica Acta* **154**, 335–339 (1983).

16. J. Janata, Chemically Sensitive Field Effect Transistors, in J. Janata and R. J. Huber, eds., *Solid State Chemical Sensors*, Academic Press, Inc., New York, 1985.

17. J. Janata and R. J. Huber, Ion-Selective Field Effect Transistors, *Ion-Selective Electrode Review* **1**, 31–79 (1979).

18. S. Moire, "Chemfet Sensors in Industrial Process Control," *Sensors* **5**, 39–44 (1988).

19. R. Perez-Olmos et al., "Construction and Evaluation of Ion-Selective Electrodes for Potassium and Calcium with a Summing Operational Amplifier, Application to Wine Analysis," *Journal of Analytical Chemistry* **360**, 659–663 (1998).

20. M. Knee and P. Srivastava, "Binding of Calcium by Cell Walls and Estimation of Calcium in Apple Fruit Tissue with an Ion Selective Electrode," *Postharvest Biology and Technology* **5**, 19–27 (1994).

21. A. M. Bond et al., "Development and Application of a Fully Automated Battery-Operated Computerized Field-Based Fluoride Monitor," *Analytica Chimica Acta* **237**, 345–352 (1990).

22. C. F. Mandenius and Mattiasson, Improved Membrane Gas Sensor Systems for On-Line Analysis of Ethanol and Other Volatile Organic Compounds in Fermentation Media," *European Journal of Applied Microbiology and Biotechnology* **18**, 197–200 (1983).

23. J. Watson, "The Tin Oxide Gas Sensor and its Applications," *Sensors and Actuators* **5**, 29–42 (1984).

24. U.S. Pat 3,631,436 (to Figaro Gas Sensor).

25. G. D. Austin et al., "A Gas-Sensor-Based On-Line Ethanol Meter for Breweries," *Journal of the American Society of Brewing Chemists* **54**, 212–215 (1996).

26. C. di Natale et al., "Metal Oxide Semiconductor Gas Sensors Array as a Tool for the Analysis of Wine Vapors, Current and Future Trends," *Proceedings of Euro Food Chem VIII*, Vol. 1, Vienna, Austria, 1995, pp. 131–134.

27. T. Aishima, "Aroma Discrimination by Pattern Recognition Analysis of Responses from Semiconductor Gas Sensor Array," *Journal of Agricultural and Food Chemistry* **39**, 752–756 (1991).

28. S. S. Deshpande and R. M. Rocco, "Biosensors and Their Potential Use in Food Quality Control," *Food Technology* **48**, 146–150 (1994).

29. D. L. Wise, *Applied Biosensors*, Butterworth Scientific Ltd., U.K., 1989.

30. C. R. Lowe, An Introduction to the Concept and Technology of Biosensors, *Biosensors* **1**, 3–16 (1985).

31. E. Kress-Rogers, and E. J. D'Costa, Biosensors for the Food Industry. *Analytical Proceedings* **23**, 149–151 (1986).

32. G. Wagner and G. G. Guilbault, *Food Biosensor Analysis*, Marcel Dekker, New York, 1993.

33. M. K. Y. Ho and G. A. Rechnitz, "An Introduction to Biosensors," in R. M. Nakamura, Y. Kasahara, and G. A. Rechnitz, eds., *Immunochemical Assays and Biosensor Technology for the 1990s*, American Society of Microbiologists, Washington, D.C., 1992, pp. 275–290.

34. A. N. Reshetilov, "Models of Biosensors Based on Principles of Potentiometric and Amperometric Transducers: Use in Medicine, Biotechnology, and Environmental Monitoring (Review)," *Applied Biochemistry and Microbiology* **32**, 72–85 (1996).

35. G. Yazbak, "Fiberoptic Sensors Solve Measurement Problems," *Food Technology* **45**, 76–78 (1991).

36. S. D. Caras and J. Janata, "pH-Based Enzyme Potentiometric Sensors, Part 3, Penicillin-Sensitive Field Effect Transistor," *Analytical Chemistry* **57**, 1924–1925 (1985).

37. S. D. Caras, J. Janata, D. Saupe, and K. Schmidt, "pH-Based Enzyme Potentiometric Sensors, Part I, Theory," *Analytical Chemistry* **57**, 1917–1920 (1985).

38. S. D. Caras, D. Petelenz, and J. Janata, "pH-Based Enzyme Potentiometric Sensors, Part 2, Glucose-Sensitive Field Effect Transistor," *Analytical Chemistry* **57**, 1920–1923 (1985).

39. R. Ulber et al., "ISFET Biosensors, Applications in Analysis and Biological Process Control," *Lebensmittel & Biotechnologie* **13**, 187–189 (1996).

40. E. Watanabe, A. Nagumo, M. Hoshi, S. Konagaya, and M. Tanana, "Microbial Sensors for Detection of Fish Freshness," *Journal of Food Science* **52**, 592–595 (1987).

41. M. A. Carole, G. Volpe, and M. Mascini, "Amperometric Detection of Uric Acid and Hypoxanthine with Xanthine Oxidase and Carbon Based Screen-Printed Electrode, Application for Fish Freshness Determination," *Talanta* **44**, 2151–2159 (1997).

42. M. Mascini, D. Moscone, G. Palleschi, and R. Pilloton, "In-Line Determination of Metabolites and Milk Components with Electrochemical Biosensors," *Analytica Chimica Acta* **213**, 101–111 (1988).

43. R. L. Villarta, A. A. Suleiman, and G. G. Guilbault, "Amperometric Enzyme Electrode for the Determination of Aspartame in Diet Food," *Microchemical Journal* **48**, 60–64 (1993).

44. S. A. Glazier, E. R. Campbell, and W. H. Campbell, "Construction and Characterization of Nitrate Reductase-Based Amperometric Electrode and Nitrate Assay of Fertilizers and Drinking Water," *Analytical Chemistry* **70**, 1511–1515 (1998).

45. R. L. Villarta, D. D. Cunningham, and G. G. Guilbault, "Amperometric Enzyme Electrodes for the Determination of L-Glutamate," *Talanta* **38**, 49–55 (1991).

46. M. C. Messia et al., "A Bienzyme Electrode for Malate," *Analytical Chemistry* **68**, 360–365 (1996).

47. W. Trettank, M. J. P. Leiner, and O. S. Wolfbeis, Fiber-Optic Glucose Sensor with a pH Optrode as the Transducer," *Biosensors* **4**, 15–26 (1988).

48. L. Sei-Jin et al., "A Fluoremetric Fiber-Optic Biosensor for Dual Analysis of Glucose and Fructose Using Glucose-Fructose-Oxidoreductase from *Zymomonas mobilis*," *Journal of Biotechnology* **36**, 39–44 (1994).

49. A. J. Wang and M. A. Arnold, "Dual-Enzyme Fiber-Optic Biosensor for Glutamate Based on Reduced Nicotinamide Adenine Dinucleotide Luminescence," *Analytical Chemistry* **64**, 1051–1055 (1992).

50. C. Coppersmith, P. Pivarnik, and A. G. Rand, "Fiber-Optic Chemiluminescent Biosensor for the Analysis of Quality Status in Seafood Products," IFT Annual Meeting, 1996, Abstract p. 132.

51. A. Krug et al., "Colorimetric Determination of Free and Total Cholesterol by Flow Injection Analysis with a Fiber Optic Detector," *Enzyme and Microbial Technology* **14**, 313–317 (1992).

52. R. A. Ogert et al., "Detection of *Clostridium botulinum* Toxin A Using a Fiber Optic-Based Biosensor," *Analytical Biochemistry* **205**, 306–312 (1992).

53. L. A. Tempelman et al., "Quantitating Staphylococcal Enterotoxin B in Diverse Media Using a Portable Fiber-Optic Biosensor," *Analytical Biochemistry* **233**, 50–57 (1996).

54. E. A. James, K. Scmeltzer, and F. S. Ligler, "Detection of Endotoxin Using an Evanescent Wave Fiber-Optic Biosensor," *Applied Biochemistry and Biotechnology* **60**, 189–202 (1996).

55. V. S. Thompson and C. M. Maragos, "Fiber-Optic Immunosensor for the Detection of Fumonisin B1," *Journal of Agricultural and Food Chemistry* **44**, 1041–1046 (1996).

GENERAL REFERENCES

W. J. Albery, P. N. Bartlett, A. E. G. Cass, and K. W. Sim, "Amperometric Enzyme Electrodes. Part IV. An Enzyme Electrode for Ethanol," *Journal of Electroanalytical Chemistry* **218**, 127–134 (1987).

R. C. Hochberg, "Fiber-Optic Sensors," *IEEE Transactions on Instrumentation and Measurement* **IM-35**, 447–450 (1986).

C. Nylander, Chemical and Biological Sensors, *Journal of Physics E: Scientific Instrumentation* **18**, 736–749 (1985).

F. W. Scheller and R. Renneberg, Glucose-Eliminating Enzyme Electrode for Direct Sucrose Determination of Glucose-Containing Samples, *Anal. Chim. Acta* **152**, 265–269 (1983).

SENSORY SCIENCE: PRINCIPLES AND APPLICATIONS

INTRODUCTION

Over the past fifty years sensory analysis of foods has grown from an informal "taste test" performed by bench top chemists and product developers to a science comprising basic tenets, accepted methods, and defined statistical analyses. The scientific literature of sensory analysis comprises investigations of sensory processes (the way we respond to physical stimuli), methodological studies of "taste testing" (how to assess reactions to food products in a test situation), and statistical experimental design (how to set up systematically varied formulations of products). The actual physiology and psychophysics of the sensory systems involved in sensory analysis (viz, taste, smell, touch, appearance) are dealt with elsewhere in this volume. Data about the physical characteristics of food belong properly to the domain of food science. Sensory analysis is the nexus where these different disciplines join.

THE THREE AREAS OF SENSORY ANALYSIS

The easiest approach to sensory analysis for both theory and practice divides it into three major subject areas:

1. *Descriptive analysis*. The language of food and drink, and how we describe our perceptions.
2. *Intensity measurement*. Perception of the sensory characteristics of foods on a scale, which shows how we process physical stimulus magnitudes to generate sensory responses.
3. *Hedonic measurement*. Our likes/dislikes.

Each of these three topics has been addressed by numerous investigations in a large scientific literature stretching back centuries, but with much of that literature appearing in the ever burgeoning science of the 1950s and later (1,2). In the main this article deals with both conventional and new techniques available to the sensory analyst.

DESCRIPTIVE ANALYSIS

At the heart of sensory analysis lies the complex field of language or descriptive analysis. While at first blush it may seem easy to describe a food, individuals familiar with a specific food use many terms to describe the fine nuances. Some of these terms are idiosyncratic, but many are standard terms easily understood. There is no single descriptive language of general and universal application, but rather attempts to capture on paper and numerically the elusive character of different foods (3).

Since ancient times researchers have tried to create standardized lists of terms by which to summarize (and classify) sensory perceptions. Researchers using any of these lists soon find that the set of terms in any list is unbalanced and necessarily incomplete, comprising too many unusable terms for the specific product being studied, but too few relevant terms to capture the key "note" or attributes. Human beings use a stretchable "rubber bag" of terms to describe each product. The number of terms grows or shrinks to fit the specific product, but the list for any product may be sure to contain terms that would never be found in a general list.

People process a constant amount of "information" when they describe products. When forced to describe a wide range of qualitatively different products during a single test panelists overlook or jettison the nuances and focus on the more general terms that differentiate this broad array of qualitatively different stimuli. For products which vary only slightly from each other (eg, different samples of strawberry aroma) panelists employ the more rarely used terms, focusing on the minor differences and highlighting them with these terms.

Fixed Descriptor Systems Versus User Developed Systems For Flavor

Tables 1–3 shows lists of descriptors for smell (4). With a rating scale the panelist rates each odor or flavor on a limited set of characteristics. The investigator may compare two or more samples across the profile of the descriptors to determine their similarity (or difference). Academic researchers, mainly, use the ready-made lists, often to profile the sensory characteristics of pure odorous stimuli (so-called "model systems.) Table 4 shows some of the profiles obtained by this method.

Applied researchers interested in specific foods and their nuances prefer to develop their own lists customized to each product in order to capture the appropriate characteristics. A general list of descriptors selected to apply (potentially) to all flavors would not contain many of these unique terms. In order to develop a list comprising these "relevant" terms researchers follow a protocol designed to elicit many candidate terms, and then cull the list to select specific terms which truly apply to the specific product. One protocol comprises the three steps laid out below:

Initial List Development. Panelists assess the products, and either individually record all the terms that they feel would apply, or in a group discuss the product (in which case the terms would emerge from the discussion). By requiring panelists to list the terms in isolation (without interacting with other panelists) the researcher need not worry about the panelist being influenced by opinions of others. In contrast, by eliciting the terms in a "focus group" (eg, 4–12 panelists participating in a structured discussion) the researcher may discover that the panelists are "sparked" to search for other terms. The interaction among the panelists inspires an extra measure of creativity from which emerge new terms that would ordinarily not come to the fore.

Culling the List to the Final Set. Once the full set of possible terms is collected, panelists or the researcher cull the list to select those descriptors that are meaningful (rather than being idiosyncratic to one panelist). The final list is often developed by polling the panelists separately to see which terms in the reduced list apply, and retaining only those terms which receive at least a certain minimum number of votes. Panelists may discuss each term in a follow-up group, and decide by group consensus whether or not each term belongs in the list.

Scale Development. The final step consists of erecting scales for each term as well as selecting reference samples to define that term (or at least explicating the term in a paragraph). This third step ensures that the panelists comprehend the meaning of the term, and have available a reference sample to represent the characteristic. Without references or explication, panelists may have different ideas about what a descriptor term means. This final step ensures that all panelists share the same concept of the attribute (6).

Three systems have been developed commercially using the methods described above. Historically, the first one was the A.D. Little Inc. "Flavor Profile" (7,8). The Flavor Profile method demands intense discussion by panelists of the flavor aspects of a product, followed by the development of standards. Panelists use an easily understood intensity scale with defined levels. In addition to assessing the flavor notes, panelists also evaluate the blend of the attributes and the order in which the attributes appear. The Flavor Profile method gained wide acceptance and support from industry because it partially solved the pressing problem of measuring "flavor." Requiring extensive training and being fairly expensive, the Flavor Profile technique has generally been limited to industry, rather than academia.

More recently, sensory analysts have had the benefit of other methods, such as the QDA technique (Quantitative Descriptive Analysis; (9,10)) and Spectrum (11). These two systems use better scales, making their data more amenable to statistical analyses. QDA, Spectrum and the Flavor Profile all operate according to the same principles, however. Furthermore, the consultants at Arthur D. Little Inc. have presented a variant of the Flavor Profile technique more appropriate for statistical analysis. They call the approach PAA (Profile Attribute Analysis), which is an offshoot of the Flavor Profile. PAA is developed in the same way (viz, by group discussion), but has scale properties appropriate for a wide range of statistical procedures (8).

Flavor, in contrast to appearance and texture, has enjoyed the primary attention of researchers involved in developing descriptive language because there is no immediate language of flavor. Most flavor words are either too general without easy definition (eg, "aromatic," "harsh"), or too specific and refer to an object (eg, "winy" or "goaty"). There are no easily understood terms that are general, as there are for taste (eg, sweet, salty, sour, bitter) and texture (eg, hard, soft, liquid). Descriptive analysis thus provides the necessary language to capture the sensory nuances of flavor.

Descriptive Analysis For The Other Senses

Texture has received attention by researchers, who patterned their efforts after the efforts expended for flavor. Following the development of the Flavor Profile, Szczesniak and her colleagues at the General Foods Corporation created the Texture Profile (12). Unlike the Flavor Profile which was developed specifically for each food product, the Texture Profile comprised a standard set of descriptor terms, along with different reference samples to represent each scale term. Table 5 shows the relevant scales and standard reference stimuli to accompany each scale point.

The texture profile has been expanded to include cosmetics as well (13). Since, however, the textural characteristics of cosmetics differ (because we invoke different mechanical forces when we apply cosmetics to the skin) the descriptive terms differ. Table 6 shows how the Texture Profile has been modified to accommodate a cosmetic product that is applied by rubbing, rather than a food product that is chewed and ingested.

The Contribution Of Statistics—Multivariate Analysis

Recognizing that there are many descriptors that we can use for product evaluation, psychologists and statisticians

Table 1. Classification Systems for Odorants

Zwaardemaker (1895): 30 subclasses	Linnaeus (1758): 7 classes	Henning (1915): 6 classes	Croker and Henderson (1927): 4 classes	Amoore (1952): 7 classes	Schuts (1964): 9 classes	Wright and Michaela (1964): 8 classes	Harper and co-workers (1968): 44 classes	Misc. add.
1 Fruity						Hexyl-acetate	Fruity	
2 Waxy							Soapy	
3 Ethereal				Ethereal	Etherish		Etherish, solvent	
4 Camphor				Camphor			Camphor, mothballs	
5 Clove	Aromatic						Aromatic	
6 Cinnamon		Spicy				Spice	Spicy	
7 Aniseed						Benzo-thiazole		
8 Minty				Minty			Minty	
9 Thyme								
10 Rosy								
11 Citrus		Fruity				Citral	Citrus	
12 Almond					Spicy		Almond	
13 Jasmine		Flowery		Floral			Floral	
14 Orange blossom	Fragrant		Fragrant		Fragrant		Fragrant	
15 Lily								
16 Violet								
17 Vanilla					Sweet		Vanilla, sweet	
18 Amber							Animal	
19 Musky	Ambrosial			Musky			Musk	
20 Leek	Alliaceous						Garlic	
21 Fishy							Ammonia	Fishy
22 Bromine								
23 Burnt		Burnt	Burnt		Burnt	Affective	Burnt	
24 Phenolic							Carbolic	
25 Caproic	Hircine		Caprylic				Sweety	
26 Cat-urine								
27 Narcotic	Repulsive							
28 Bed-bug								
29 Carrion	Nauseous						Sickly	
30 Fecal							Fecal	
31		Resinous				Resinous	Resinous; paint	
32		Foul		Putrid	Sulfurous	Unpleasant	Putrid sulfurous	
33			Acid				Acid	
34					Oily		Oily	
35					Raneid		Raneid	
36					Metallic		Metallic	
37							Meaty	
38							Moldy	
39							Grassy	
40							Bloody	
41							Cooked-vegetable	
42								Sandal
43								Watery
44								Urinous
(Nonolfactory)				(Pungent)		(Trigeminal)	(Pungent and 5 others)	

Source: Ref. 4.

have questioned the "independence" of these descriptors. Are these descriptive terms truly independent of each other? Or, do many of the terms overlap with each other, so that the full array of many terms simply reflects different combinations of the same and limited number of "primaries."

Color science provides us with a model showing how the many colors of the rainbow can be expressed as combinations of three distinct primaries, whether these primaries be the familiar red, blue, and yellow, or more esoteric combinations of wavelengths whose combination in various ways regenerates all colors.

In the late 1930s statisticians, especially those involved with psychological issues, began to recognize that they could reduce data from many psychological tests to just a few primary dimensions using the statistical procedure of factor analysis. According to factor analytical theory, the many tests of mental abilities, as an example, all measure just a few simple "primary" aspects of intelligence. Each test comprises a unique and specifiable combination of

Table 2. Examples of Attributes for a Variety

Eucalyptus	Strawberrylike
Buttery	Stale
Like burnt paper	Corklike
Cologne	Lavender
Caraway	Cat-urinelike
Orange (fruit)	Barklike, birchlike
Household gas	Roselike
Peanut butter	Celery
Violets	Burnt candle
Tea-leaflike	Mushroomlike
Wet wool, wet dog	Pineapple (fruit)
Chalky	Fresh cigarette smoke
Leatherlike	Nutty (walnut, etc)
Pear (fruit)	Fried fat
Stale tobacco smoke	Wet paperlike
Raw cucumber-like	Coffeelike
Raw potatolike	Peach (fruit)
Mouselike	Laurel leaves
Pepperlike	Scorched milk
Bananalike	Sewer odor
Burnt rubberlike	Sooty
Geranium leaves	Crushed weeds
Urinelike	Rubbery (new rubber)
Beery (beerlike)	Bakery (fresh bread)
Cedarwoodlike	Oak wood, cognaclike
Coconutlike	Grapefruit
Ropelike	Grapejuicelike
Seminal, spermlike	Eggy (fresh eggs)
Like cleaning fluid (Carbone)	Bitter
Cardboardlike	Cadaverous, like dead animal
Lemon (fruit)	Maple (as in syrup)
Dirty linenlike	Seasoning (for meat)
Kippery (smoked fish)	Apple (fruit)
Caramel	Soup
Sauerkrautlike	Grainy (as grain)
Crushed grass	Raisins
Chocolate	Hay
Molasses	Kerosene

Table 3. Fragrance Classifications

Class	Subclass
1. Citrus	Classic
	Modern
2. Green	Fresh
	Balsam
	Floral
3. Floral	Fresh
	Floral
	Heady
	Sweet
4. Aldehydic	Fresh
	Floral
	Sweet
	Dry
5. Oriental	Spicy
	Sweet
6. Chypre	Fresh
	Floral-animal
	Sweet
7. Fougere	Classic
	Modern
8. Woody	Dry
	Warm
9. Tobacco/leather	
10. Musk	

these primaries. Factor analysis uses the correlations (or covariances) of the tests with each other to estimate the number of these "primaries" and their respective contributions to each test. The factor analysis approach *does not name the primaries*. Rather it demonstrated the existence of the primaries, and shows both the number of these primaries, and the correlation of each test with each primary. It is left up to the ingenuity of the investigator to name the statistically uncovered primaries.

Factor analysis lends itself to analyzing the perception of odors, tastes and textures. The panelist profiles the stimuli (eg, pure odorants) on a list of characteristics using one or another scale. The investigator then processes the data, to uncover the underlying "factors" or "attributes." Table 7 shows results for perfumes (14). Table 8 and 9 show factor analysis of texture. Table 8 shows results for 79 food names each rated on 40 texture attributes (15). Table 9 shows results for 50 nonfood stimuli, each profiled on 20 texture attributes (16).

Beginning in the late 1950s and early 1960s with the work of two psychologists, Clyde Coombs at the University of Michigan (17) and Roger Shepard at Bell Laboratories (18) alternate ways to derive these "fundamental dimensions" were developed. They developed statistical procedures known as "proximities analysis." Rather than instructing panelists to rate the samples on predefined scales, panelists rated (or ranked) the perceived "distance" or dissimilarity of pairs of stimuli. Either the actual ratings of "distance" or the rank order of dissimilarity between pairs were processed to yield the basic set of dimensions. The analysis produced a geometrical map in which the stimuli could be placed. Distances between pairs of stimuli in this geometrical map correspond to subjective dissimilarity between the pairs of stimuli.

Overall multivariate techniques nicely display dissimilarities between different stimuli. The data plots show which stimuli are similar and which are different, as well as an idea of the nature of dimensions. As a method for uncovering basic sensory dimensions, however the statistical procedures have not lived up to their promise. They have found use primarily as devices for data reduction of many attributes to a simple set.

Descriptor Systems for Sensory Analysis—Their Outlook

Historically, descriptor systems began as a method to learn how the senses work. The Linnean approach (classification as a way to understand the physiology of the senses), gave way to description as a way to understand products. Today we have available many systems, appropriate either when we desire a "ready made" descriptor system, or when we need to target a descriptor system to a specific product.

We can best understand the future of descriptive analysis from the problems which it helps to solve.

Table 4. Examples of Attribute Profiles for a Variety of Odorants

Experienced panelist		Inexperienced panelist	
(A) Musk xylol (solid)[a]			
Fragrant	2.6 (13/15)	Light	2.3 (12/20)
Sweet	1.8 (12/15)	Fragrant	2.0 (13/20)
Floral	1.3 (12/15)	Sweet	1.9 (11/20)
Musklike	1.0 (6/15)	Floral	1.8 (11/20)
Dry, powdery	0.7 (7/15)	Dry, powdery	1.2 (8/20)
		Aromatic	1.0 (7/20)
(B) Naphthalene (1% in DNP)[a,b]			
Like mothballs	4.1 (15/15)	Like mothballs	4.3 (20/20)
Camphorlike	1.7 (9/15)	Camphorlike	2.4 (13/20)
Cool, cooling	0.7 (8/15)	Cool, cooling	1.6 (11/20)
Etherish, anaesthetic	0.6 (8/15)	Heavy	1.5 (10/20)
		Aromatic	1.1 (11/20)
		Sickly	0.8 (9/20)
(C) Octanol (0.5% in DNP)[a,b]			
Soapy	2.3 (9/15)	Soapy	2.4 (12/20)
		Light	1.8 (13/20)
		Cool, cooling	1.6 (11/20)
		Sweet	1.4 (10/20)
		Oily, fatty	1.3 (9/20)
		Sickly	1.9 (11/20)
		Fragrant	1.0 (7/20)
(D) Phenylacetic Acid (1.5% in DNP)[a,b]			
Sickly	1.3 (9/15)	Sickly	1.7 (14/20)
Sharp, pungent, acid	1.1 (10/15)	Sweaty	1.7 (13/20)
Floral	1.1 (7/15)	Putrid, foul decayed	1.4 (13/20)
Sweaty	0.9 (8/15)	Heavy	1.4 (10/20)
Sweet	0.9 (8/15)	Sharp, pungent acid	1.4 (9/20)
Fragrant	0.9 (7/15)	Warm	1.3 (10/20)
Sour, acid, vinegar	0.8 (7/15)	Sour, acid, vinegar	1.3 (9/20)
		Animal	1.2 (8/20)
		Sweet	1.1 (9/20)
		Rancid	1.1 (9/20)
		Fecal (like dung)	1.0 (8/20)
(E) Phenylethanol (0.5% in DNP)[a,b]			
Floral	2.7 (11/15)	Floral	3.1 (17/20)
Fragrant	2.2 (10/15)	Fragrant	3.0 (19/20)
Sweet	2.0 (13/15)	Sweet	2.4 (17/20)
		Light	2.2 (12/20)
		Herbal, green, cut, grass, etc	1.4 (11/20)
		Aromatic	1.3 (12/20)
		Cool, cooling	1.2 (12/20)
(F) Nitrobenzene (10% in DNP)[a,b]			
Almondlike	3.1 (13/15)	Almondlike	3.7 (17/20)
Aromatic	1.2 (9/15)	Sweet	1.9 (13/20)
Petro, chemical solvent	0.9 (8/15)	Heavy	1.8 (10/20)
Vanillalike	0.9 (7/15)	Oily, fatty	1.6 (11/20)
Etherish, anaesthetic	0.7 (7/15)	Sickly	1.5 (14/20)
		Fragrant	1.5 (12/20)
		Aromatic	1.1 (8/20)
		Like mothballs	1.0 (8/20)

Source: Ref. 5.
[a]First number = mean rating; ratio = number in panel using that term.
[b]DNP = dinitrophthalene.

Table 5. The General Foods Texture Profile System for Foods

Standardized rating scales for mechanical properties with generalized standards (intensity of parameter increases downward)

Hardness	Fracturability	Chewiness
Cream cheese	Corn muffin	Rye bread
Velveeta cheese	Egg Jumbos	Frankfurter
Frankfurters	Graham crackers	Large gumdrops
Cheddar cheese	Melba toast	Well-done round
Giant stuffed olives	Bordeaux cookies	steak
Cocktail peanuts	Ginger snaps	Nut chews
Shelled almonds	Treacle brittle	Tootsie Rolls

Adhesiveness	Viscosity
Hydrogenated shortening	Water
Cheese Whiz	Light cream
Cream cheese	Evaporated milk
Marshmallow topping	Maple syrup
Peanut butter	Chocolate syrup
	Cool'n Creamy
	pudding
	Condensed milk

Geometrical properties related to particle size and shape		Geometrical properties related to particle size and orientation	
Property	Example	Property	Example
Powdery	Confectioner's sugar	Flaky	Flaky pastry
Chalky	Tooth powder	Fibrous	Breast of chicken
Grainy	Cooked Cream of Wheat	Pulpy	Orange sections
		Cellular	Apples, cake
Gritty	Pears	Aerated	Whipped cream
Course	Cooked oatmeal	Puffy	Rice pudding
Lumpy	Cottage cheese	Crystalline	Granulated sugar
Beady	Cooked tapioca		

Source: Ref. 12.

Quality Control (eg, plant to plant variation, modifications of an ingredient or a process by new suppliers or new machinery). Descriptive analysis develops a sensory "fingerprint" of the product, which can then be compared to standards on file. Often it is vital that descriptive analysis record the sensory characteristics of those products which defy description by chemical or physical analysis. It is difficult, if not impossible, to assess some products (eg, coffee), simply from the tracings of the gas chromatograph, (which records several hundred ingredient components in the product, each at its own concentration). Sensory analysis, using an experienced evaluator (in contrast to instrumental analysis) can more efficiently classify the sample as "same" or "different" from a "gold" standard reference. To the degree that quality assurance becomes increasingly important in the food industry we may expect to see increasing use of descriptive analysis to maintain sensory identity. The descriptive analysis report will be of the sort generated by the Flavor Profile technique, or the QDA technique, ie, designed to capture the sensory nuances of the particular product.

New Product Development. Product developers searching for new creations use descriptive analysis to help them achieve their goal. For instance, in the creation of a new cola beverage, descriptive analysis may highlight specific flavor notes such as winy, pruny, etc, which differentiate one cola from another. Descriptive analysis can point out the need for the specific characteristic to be present. Descriptive analysis proves even more valuable when it shows the product developer that he has, indeed, incorporated the desired note into the product. In a parallel fashion, descriptive analysis proves useful when it shows that the new product possesses an undefined "off-note." Identifying and labeling that note helps the product developer to find the cause and remove the offending compound, flavor or process.

Relating Sensory Characteristics to Physical Measures. Much scientific interest has focused on the relation between what we subjectively perceive and the objective physical (or chemical) properties of products. Chemists analyze the aroma of food products into hundreds of constituents by using gas chromatography and mass spectrometry. A trained panelist sniffing the "effluent" or separated chemicals at the exit port of the chromatograph (after the components have been separated) can describe what each component smells like. This dual analysis, combining objective and subjective measures, is often done by analytical chemists to characterize aromas.

Texture researchers correlate well defined physical characteristics that they measure instrumentally (eg, force applied, deformation obtained) with the sensory attributes. The "Instrumental Texture Profile" proposed by Bourne (19) as an extension of Szczesniak's subjective texture profile (12), exemplifies this subjective-objective type of correlation. Corresponding to each of Szczesniak's terms, Bourne recommends a well-defined instrumental measure. Szczesniak also has provided "Instrumental Analogs" of texture perceptions, using the General Foods Texturometer.

MEASUREMENT OF PERCEIVED INTENSITY

Intensity measurement (quantification) comprises the second major thrust of sensory analysis. Quantification takes many forms, including determination of threshold (the lowest level of a stimulus that one can either detect or recognize), assessment of discrimination ability (the smallest physical change between two samples that is just noticeable), and scaling of suprathreshold intensity (how strong does the stimulus seem).

Is It Possible to Measure Perceived Intensity Validly?

Scientists agree that they can measure "objective" characteristics such as temperature and mass using well-accepted scientific procedures. It is more problematic for sensations. We cannot point to a sensory perception in the way that we can point to a physical stimulus. Measurement of "private sensory experience" does not, intuitively, appear to be "public" or open to validation and replication in the same way that we understand scientific measurement to be.

Table 6. The Texture Profile System Modified for Use in Skin Care Product Evaluation

Stage of evaluation parameters	Skin care product attributes and definition	Texture profile
Pick-up: product removed from container, product poured or squeezed from bottle onto fingertips, or products lifted from jar with forefinger	Thickness: perceived denseness of product; evaluated as force required to squeeze between thumb and forefinger, rated as thin-medium-thick or Consistency perceived structure of product; evaluated as resistance to deformation and difficulty of lifting from container; rated as light-medium-heavy	Viscosity (for lotions) or Hardness, cohesiveness
Rubout (application): spread of product over and into skin with fingertips using gentle circular motion at a rate of two rubs per second for a specified period of time, depending on the product	Spreadability: ease of moving product from point of application over rest of face; evaluated as resistance to pressure; rated or described as: "slips"—very easy to spread "glides"—moderately easy "drags"—difficult to spread Absorbency: rate at which product perceived as absorbed into skin; evaluated by noting changes in character or product and in amount of product remaining (tactile and visual) and by changes in skin surfaces; rated as slow-moderate-fast	Viscosity, cohesiveness, springiness, gumminess, adhesiveness Other characteristics—oil and water content of product
After-feel (and appearance): evaluation of skin surface with fingertips, visually and kinesthetically immediately after product application and possibly at varying intervals thereafter	After feel: type and intensity of product residue left on skin; changes in skin feel; product residue described by type ie, film (oily or greasy), coating (waxy or dry), flaky or powdery particles, the amount of such residue, identified as slight-moderate-large Skin feel described as dry (taut, pulled tight), moist (supple, pliant), oily (dirty, clogged) Other sensations also noted and identified where applicable (ie, clean, stimulated, irritated, etc)	Other characteristics—oil and water content Geometrical characteristic—gritty, powdery, etc

Source: Ref. 13.

Despite these roadblocks, however, researchers have developed an array of reliable and thus far valid techniques to measure perception (20). Each measurement technique is grounded in specific assumptions. Each technique yields numerical scale values for perception which permit decisions to be made—whether pertaining to the perceiver (eg, the panelist can or cannot perceive the stimulus, two stimuli are or are not "different," the stimulus is strong or weak), or to the product (eg, the product developer should increase or decrease the sensory characteristic of an ingredient, by a specific amount).

Measurement of Thresholds

The most fundamental and intuitively obvious measurement is the threshold, defined as the lowest physical intensity of a stimulus that a panelist can correctly detect. Occasionally other definitions may apply (eg, recognition of stimulus quality rather than detection). Threshold measurement requires the least intellectual effort on the part of the panelist. The panelist simply responds that either he detects or he does not detect the stimulus. There are no other intellectual demands. By varying the physical intensity of the stimulus and determining how many panelists detect the stimulus (or recognize its quality—a different task, but one requiring the same response) the investigator determines the level of the stimulus that is just detectable.

The physical intensity at which the percentage detecting correctly equals 50% is defined as the threshold.

The early psychology literature is replete with reports on the measurement of threshold. In practice and prior to the widespread use of subjective intensity scales, food scientists assumed (albeit wrongly) that the threshold corresponded to relative intensity. For instance, artificial sweeteners such as saccharin possess a threshold concentration hundreds of times lower than the threshold concentration of sugar. Researchers misinterpreted this concentration ratio to mean that saccharin was several hundred times sweeter than sugar. Thresholds only indicate the physical intensity of a stimulus which generates a fixed sensory intensity—namely "just perceptible." Thresholds do not, and cannot provide a direct measure of subjective intensity, however.

Discrimination Testing and Sensory Measurement

A second milestone in sensory measurement was reached when investigators quantified the ability of panelists to discriminate. Discrimination testing traces its beginning to research conducted more than one hundred fifty years ago. E. H. Weber (21), a German physiologist, reported that stimuli which are just noticeably different from each other often lie in a constant ratio. That is, if the ratio is 1.05, then it requires an approximately 5% increase in a stim-

Table 7. Results of a Factor Analysis of Perfumes

Odorants tested

McCall's scheme of perfumes, which classifies perfumes
  according to user groups

Example

Group 1 = single florals (eg, White Rose, Jasmin, Heliotropin)
Group 2 = floral bouquet (eg, Arpège, Paris, Amour Amour)
Group 3 = modern blends (eg, Chanel No. 5, l'Aimant)
Group 4 = oriental blends (eg, Emeraude, Shalimar, Tabu,
  l'Origan)
Group 5 = woodsy-mossy-leafy (eg, Soir de Paris, Mitsuoko,
  Cypre)
Group 6 = spicy bouquet (eg, Capri)
Group 7 = fruity bouquet (eg, Fleurs de Rocailles)

Subjects

25 women

Scaling

Each subject rated each perfume on a series of 25 bipolar scales
  (ie, semantic differential scales).

Results

Four dimension extracted but not named:

1. Chanel No. 5 vs White Rose (on opposite ends)
2. l'Origan vs Aspège (on opposite ends)
3. Aspège vs Cypre (on opposite ends)
4. Not specified

Source: Ref. 14.

Table 8. Results of a Factor Analysis of Texture Food Names

Food names used in questionnaire

79 names profiled and 40 attributes

1. Hard	15. Creamy	28. Rustling
2. Soft	16. Flaky	29. Saku (Japanese)
3. Juicy	17. Fibrous	30. Coarse
4. Chewy	18. Thick	31. Sticky
5. Not chewy	19. Light	32. Slushy
6. Warm	20. Heavy	33. Mushy
7. Cold	21. Slippery	34. Sticky (other term)
8. Oily	22. Crunchy	35. Fluffy
9. Greasy	23. Melting	36. Hoku (Japanese)
10. Viscous	24. Brittle	37. Smooth
11. Sticky	25. Slippery	38. Lumpy
12. Moist	26. Crunchy	39. Crisp
13. Elastic	(other term)	40. Sticky
14. Gummy	27. Sprinkling	(other term)

Results

8 factors:	Hard-soft	Flaky
	Cold-warm	Heavy
	Oily-juicy	Viscous
	Elastic	Smooth

Source: Ref. 15.

Table 9. Results of a Factor Analysis of Texture–nonfoods

Samples tested

Felt, line, handkerchief, tricot, nylon, broadcloth, silk, acetate,
  satin, etc (50 different samples)

Scales

20 bipolar scales; panelists rated each sample on every one of
  the bipolar scales, each of which contained 7 points

Subjects

25 untrained female students

Results

Four factors extracted:

1. Coldness (attributes loading high included painful, stiff,
   nonplastic, cold, coarse, heavy, massive, substantial)
2. Wet and smooth (attributes loading high included were,
   heavy, substantial, smooth, massive)
3. Hardness (attributes loading high included sharp, stiff, hard,
   plastic)
4. Opposition of tactual impression of viscosity, elasticity, and
   plasticity from visual aesthetic impression of luster or gloss,
   smoothness

Source: Ref. 16.

ulus intensity before panelists will recognize the change
and report that the second stimulus seems stronger.

It took the genius of another German physiologist and
physicist, G. T. Fechner (22), to recognize that discrimination ability could become the foundation for a scale of
subjective sensory magnitude. Fechner conjectured that to
create a sensory scale one should add successive physical
intensities that are just discriminable from each other.
Each stimulus increment, (corresponding to one just noticeable difference), would equal one psychological unit of
perceived intensity. One would then cumulate just noticeable differences to develop a true sensory scale. Mathematically sensory intensity relates to physical intensity by
a simple function, a logarithmic function, written as:

$$\text{Sensory Intensity} = A[\log(\text{Physical Intensity})] + B$$

Researchers erected psychological scales of sensory intensity using the cumbersome discrimination testing (23).
As a method for applied sensory analysis, however, discrimination tests found more use in the practical realm,
with problems such as deciding whether or not two stimuli
were perceptibly different from each other. Applied researchers were not particularly interested in erecting a
sensory scale based on discrimination testing. They, however, did need discrimination tests to compare products
which differed in terms of ingredients, processing conditions, storage time, etc. The analysis of differences were
far more statistical however, invoking procedures such as
the T test for difference of means or percentages, or the
analysis of variance (ANOVA) to assess differences among
many samples.

Direct Scales of Sensory Magnitude

More productive uses of scaling have been made with "direct scaling," in which the investigator instructs the panelist to rate the stimulus intensity using a numbering system. The intensity scale may comprise graded categories (eg, from weak to strong, on a set of categories ranging from as few as 3, up to as many as 100 categories or more), or require the panelist to place a mark on a line of fixed length to denote intensity (with one end of the line denoting weak, and the opposite end denoting strong), or even require the panelist to assign numbers so that the ratios of the numbers assigned reflect the ratios of perceived intensity (the method of magnitude estimation (24). The investigator need not use numbers nor line length as the scale measure. The panelists can adjust the intensity of a stimulus such as white noise so that the intensity of the adjusted stimulus matches the perceived intensity of the stimulus being measured (25,26).

Direct scaling methods (in contrast to threshold and discrimination tests) enable the panelist to behave as a true measuring instrument. Data can be collected from many panelists relatively quickly and painlessly. The investigator straightforwardly treats the data by conventional statistical procedures, including computing the average and measures of variability (standard deviation, variance), determining differences among samples ("t" tests, analysis of variance), and modeling the relation between perceived intensity and physical measures by regression analysis (curve fitting).

In practical terms, scaling procedures provide a great deal of useful information. Consider the data shown in Table 10, which shows average consumer ratings of five cereals, on a variety of attributes. (Some of the attributes in Table 10 are sensory attributes, some are liking attributes, whereas some are "image" attributes). Panelists used a 0–100 scale. When panelists rate each product on all of the attributes the investigator generates a "report card" for both marketed products and test prototypes.

Profiles of products on many attributes enable the researcher to compare the full profile of one product to the full profile of another and assess the degree of difference. Graphical display of data using the so-called spider plots, (Fig. 1) shows the difference between products. The "spider plot" is not analytical—it does not show the user anything beyond the fact that the two samples being plotted overlap, or do not. (One gets this same information from tables such as Table 10). The Spider plots, however, make the data come alive and increase its impact.

HEDONICS—MEASUREMENT OF LIKING

Acceptance measurement, the third branch of sensory analysis, may well be, the most vital aspect. We select and eat foods on the basis of their visual appeal and palatability. A large and growing body of scientific literature has been published on the measurement of acceptance. Appropriate measures of liking or purchase intent, and the assessment of possible consumer boredom with a product are critical issues when applying hedonic tests in the world of commercial research.

The Search for the Liking Scale

Buried in the archives of scientific literature are numerous attempts to create scales which measure acceptance, and techniques beyond scaling which probe other aspects of hedonics. Table 11 shows a variety of scales, ranging from classification (like/dislike) up to refinements which go beyond liking to probe other activities such as the expected effort one would make to eat the food (eg, FACT scale) (27,28). Still other efforts have focused on the assessment of the boredom factor in food—how time itself and repeated consumptions of the food item modify acceptance (29).

Directional Liking Scales

Although researchers use many different liking scales, other scales also incorporate aspects of liking. The "directional scale," often used to guide product development is a prime example. Directional scales combine intensity with liking, as shown by the following statement:

> "Considering the amount of visible spice in this product, is the amount of spice: Too little, Just right, Too much? (Choose the appropriate answer.)"

Directional questions assume that the panelist knows the "optimal level" or "ideal level" of the attribute. When evaluating the product the panelist also estimates deviations from that optimum. Directional scales appeal to product developers because they appear, superficially, to provide accurate direction. Ostensibly the developer need only know the degree to which the product under-delivers or over-delivers an attribute. Table 12 shows data obtained from directional scales for a complex product—spaghetti and meatballs rated on a variety of directional attributes.

In many cases the panelists cannot accurately gauge the direction of change, however. For attributes which have a strong emotional positive or negative connotation (eg, chocolate intensity, coffee bitterness, etc) the product never has enough or always has more than one desires. As a result chocolate products are often rated as not having enough "chocolate" on the directional scales, whereas coffee is too often rated as having too much bitterness, even when it has too little bitter taste to make it palatable. Panelists do not know their ideal points for these emotion laden attributes. They find it hard to provide direction to product developers, who consequently continue to interpret literally the panelist's desire for more of the attribute, (eg, add more chocolate), never quite reaching the goal of "just right." Furthermore, all too often consumers may request several incompatible changes in the product, making the data less usable. Despite these inadequacies and failings, however, directional scales are widely used in product development to guide consumer driven reformulation. They are easy to use, they incorporate hedonics, and they do provide some guidance.

EXPERIMENTAL DESIGN AND SENSORY PRODUCT OPTIMIZATION

Sensory analysis has achieved its major commercial impact by providing reliable tests results from panelists.

Table 10. Ratings of Five Cereals on Attributes Comparison of Attribute Profiles—High versus Low Scorers[a]

	High-scoring cereals			Low scoring cereals	
	100% Natural	Cheerios	Honey Nut Cheerios	100% Bran	Grape Nuts
Overall					
Purchase interest	72	66	63	15	28
Liking—dry	78	62	68	31	24
Liking—w/milk	77	68	66	31	27
Appearance					
Like appearance—dry	65	65	65	14	26
Like appearance—wet	68	63	63	22	23
Size of pieces	28	37	37	12	4
Thickness	36	47	47	18	20
Darkness	43	28	28	75	45
Familiar looking	54	95	95	28	46
Flavor					
Like flavor—dry	74	57	65	36	25
Like flavor—w/milk	76	61	67	36	31
Like wheat flavor	43	24	34	25	31
Intensity—dry	75	46	60	50	56
Nutty—dry	82	33	36	34	45
Intensity—w/milk	74	52	64	47	57
Nutty—w/milk	84	25	37	34	47
Sweet	66	13	80	33	17
Spicy	46	15	32	18	15
Texture					
Like texture—dry	72	61	72	29	23
Like texture—milk	74	60	65	22	28
Crunchy—dry	83	60	75	49	91
Amt. milk absorbed	28	39	28	79	57
Quickly absorbs milk	23	36	32	75	59
Holds together	83	61	69	38	76
Crunchy—milk	86	46	66	28	97
Crispy—milk	87	40	59	15	81
Image					
Approp. as dry snack	80	59	70	19	12
Nutritious	76	60	50	51	65
Quality	83	71	77	52	66
Satisfy as snack	81	55	70	11	14
Calorie	69	37	73	39	44
Unique	59	23	38	55	51
Tire of it (boredom)	33	46	44	73	60
Kids would like	65	67	81	9	7
Teens would like	70	59	70	13	20

[a]Percent of panelists who rate the product as definitely or probably would buy on a 5-point buy scale (1 = definitely not buy to 5 = definitely would buy.) Remaining scales = 0–100.

When coupled with experimental design and modeling (by regression analysis), sensory analysis becomes a powerful tool to guide product development.

Experimental Design

An alternative approach beyond directional scales to develop products using consumers, consists of systematically varying the formula ingredients in a way which allows one to assess the effects of each ingredient, and the interactions among ingredients on acceptance and attribute perceptions.

Experimental designs are properly the province of statisticians (30). Sensory analysts and product developers have used experimental designs with great success to understand consumer reactions to test prototypes comprising known ingredients and processes. Product categories amenable to these designs range from simple food systems such as a fruit-flavored beverage, to complex systems such as pizza, apple pie, and sausage. In all cases the assessment

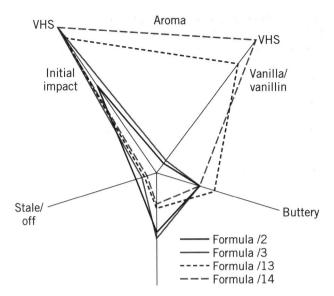

Figure 1. Example of quantitative descriptor analysis "spider plot" of the aroma of a sample. Scale: 0 at center to 30–40 at edge. *Source:* Ref. 10.

Table 11. Verbal Descriptors for Hedonic Scales

Number of scale points descriptors[a]

2	Dislike, unfamiliar
3	Acceptable, dislike, not tried
3	Like a lot, dislike, do not know
4	Well liked, indifferent, disliked, seldom if ever used
5	Like very much, like moderately, neutral, dislike moderately, dislike very much
5	Very good, good, moderate, dislike, tolerate
5	Very good, good, moderately well, tolerate, dislike
6	Very good, good, moderate, tolerate, dislike, never tried
9	Like extremely, like very much, like moderately, like slightly, neither like nor dislike, dislike slightly, dislike moderately, dislike very much, dislike extremely

Food action (fact) rating scale[b]

Eat every opportunity
Eat very often
Frequently eat
Eat now and then
Eat if available
Do not like—eat on occasion
Hardly every eat
Eat if no other choice
Eat if forced

Source: [a]Ref. 27 and [b]Ref. 28.

Table 12. Example of Attribute Profile Ratings Using Different Attributes Liking, Image, and Directional, Product: Spaghetti and Meatballs[a]

Attribute		Product[b]	
		#1	#2
Overall			
Like	Overall liking of product	55	63
Appearance			
Like	Appearance	63	75
Like	Appearance of meatballs	63	66
Dir	Darkness of sauce	15	9
Dir	Size of portion	−8	0
Overall taste			
Like	Overall taste of product	55	12
Dir	Strength of afterstate	15	4
Meatballs			
Like	Taste of meatballs	57	53
Like	Texture of meatballs	63	50
Dir	Amount of meatballs	−3	−1
Dir	Size of meatballs	−5	−2
Dir	Moistness of meatballs	0	−11
Dir	Softness vs firmness of meatballs	1	13
Sauce (taste)			
Like	Overall flavor of sauce	53	62
Dir	Strength of tomato flavor	11	−1
Dir	Spiciness of sauce	7	−3
Dir	Sweetness of sauce	−4	−1
Dir	Amount of seasoning	7	−5
Dir	Amount of salt	6	2
Sauce (texture, amount)			
Like	Overall texture of sauce	65	65
Dir	Amount of sauce	18	−1
Dir	Thickness of sauce	7	2
Dir	Smoothness vs chunkiness of sauce	0	2
Pasta			
Dir	Amount of pasta	−11	3
Dir	Softness vs firmness of pasta	−14	−5

[a]Explanation of ratings: Panelists rated each product on a variety of directional questions (0 = much too little, 50 = on target, 100 = much too much).
Data for directional (DIR) questions shown, after 50 was subtracted from each rating. Positive numbers = over-delivery on the attribute. Negative numbers = under-delivery on the attribute. Like = liking rating (0 = hate, 100 = love).
[b]Base size: 112.

of systematically varied alternatives has educated the researcher and provided concrete direction for product modification.

A Case History Using Sensory Analysis—Tomato Sauce

The easiest way to understand experimental design and its link with sensory analysis is with a case history. This case history concerns tomato sauce. The issue was to de-

velop a new formulation to take advantage of a new processing technology.

Phase 1. Formula Selection and Design

Phase 1 comprises the selection of appropriate variables, and the recommended array of formula combinations. R&D selected 6 variables to investigate. Different experimental designs can be used to investigate the 6 variables, depending upon the expected relation between formula variables and consumer ratings. If one expects the attributes rated by consumers to change only linearly with ingredients, with no interaction among the variables, as few as 12 prototypes need to be developed using the Plackett Burman screening design. Table 13 (part A) shows this design (31). If one expects curvature in the data, ie, the attributes (whether sensory or liking) do not follow a simple straight line versus physical changes, then one may wish to fit a quadratic curve, necessitating a quadratic design. If one expects curvature and interactions among the variables than the central composite screening design shown in Table 13 (part B) is appropriate (32). When there are simply too many prototypes to create, but one wishes to use the central composite design (because it captures curvature in the response surface and allows for interaction between pairs of ingredients) one would select the reduced design such as that shown in Table 13 (part C). R&D selected the design, shown in Table 10 (part C), encompassing 29 prototypes.

Along with the experimentally designed products, researchers test in-market competitors which provide "anchors" for the category. Panelists do not know which samples belong to the experimental design and which are actual in-markets products. All products are tested "blind," and in a randomized order. The only clues come from the panelists' sensory perceptions.

Attribute Selection for the Questionnaire

Attributes define the characteristics to which consumers will attend when evaluating the sample. It is important to select appropriate attributes, but the choice of attributes in an experimental design is not as critical as would be the case were the products to simply represent unconnected "rifle shots."

The key attribute is usually the overall criterion of acceptance, whether this be overall liking, purchase intent, or some integrative measure such as "high quality." The key attribute measures the single response that is of real interest to the investigator, and usually corresponds to the characteristic of the product that is being optimized.

The questionnaire should also deal with relevant sensory characteristics. Optimization methods can uncover formulations which are highly acceptable and simultaneously possess specific, required sensory characteristics (eg, to support an advertising position). The questionnaire can encompass image attributes as well (eg, "refreshing" for a beverage, "caloric" for a snack, etc). Image attributes are more complex cognitive characteristics.

Often there are other data available for the experimentally varied products. Additional data may include objective physical measurements, cost of goods, storage stability, ratings by experts which describe qualitative nuances, and evaluations by a quality control panel. If available, these other data sets can be added to the data file. The augmented attribute set appears in Table 14.

In this study data was available from all of these sources. Generally, however, studies involving experimental design and optimization are of a more modest scale. The

Table 13. Plackett Burman Screen Design

Product	A	B	C	D	E	F
	A. Six Variables in 12 Prototypes[a]					
101	1	1	0	1	1	1
102	0	1	1	0	1	1
103	1	0	1	1	0	1
104	0	1	0	1	1	0
105	0	0	1	0	1	1
106	0	0	0	1	0	1
107	1	0	0	0	1	0
108	1	1	0	0	0	1
109	1	1	1	0	0	0
110	0	1	1	1	0	0
111	1	0	1	1	1	0
112	0	0	0	0	0	0
	B. Six Variables—Full Replicate[b]					
301	1	1	1	1	1	1
302	1	1	1	1	1	−1
303	1	1	1	1	−1	1
304	1	1	1	1	−1	−1
305	1	1	1	−1	1	1
306	1	1	1	−1	1	−1
307	1	1	1	−1	−1	1
308	1	1	1	−1	−1	−1
309	1	1	−1	1	1	1
310	1	1	−1	1	1	−1
311	1	1	−1	1	−1	1
312	1	1	−1	1	−1	−1
313	1	1	−1	−1	1	1
314	1	1	−1	−1	1	−1
315	1	1	−1	−1	−1	1
316	1	1	−1	−1	−1	−1
317	1	−1	1	1	1	1
318	1	−1	1	1	1	−1
319	1	−1	1	1	−1	1
320	1	−1	1	1	−1	−1
321	1	−1	1	−1	1	1
322	1	−1	1	−1	1	−1
323	1	−1	1	−1	−1	1
324	1	−1	1	−1	−1	−1
325	1	−1	−1	1	1	1
326	1	−1	−1	1	1	−1
327	1	−1	−1	1	−1	1
328	1	−1	−1	1	−1	−1
329	1	−1	−1	−1	1	1
330	1	−1	−1	−1	1	−1
331	1	−1	−1	−1	−1	1
332	1	−1	−1	−1	−1	−1
333	−1	1	1	1	1	1
334	−1	1	1	1	1	−1
335	−1	1	1	1	−1	1
336	−1	1	1	1	−1	−1
337	−1	1	1	−1	1	1
338	−1	1	1	−1	1	−1
339	−1	1	1	−1	−1	1
340	−1	1	1	−1	−1	−1
341	−1	1	−1	1	1	1
342	−1	1	−1	1	1	−1
343	−1	1	−1	1	−1	1
344	−1	1	−1	1	−1	−1
345	−1	1	−1	−1	1	1
346	−1	1	−1	−1	1	−1
347	−1	1	−1	−1	−1	1
348	−1	1	−1	−1	−1	−1

Table 13. Plackett Burman Screen Design (*continued*)

Product	A	B	C	D	E	F
349	−1	−1	1	1	1	1
350	−1	−1	1	1	−1	1
351	−1	−1	1	1	−1	1
352	−1	−1	1	1	−1	−1
353	−1	−1	1	−1	1	1
354	−1	−1	1	−1	1	−1
355	−1	−1	1	−1	−1	−1
356	−1	−1	1	−1	−1	−1
357	−1	−1	−1	1	1	1
358	−1	−1	−1	1	1	−1
359	−1	−1	−1	1	−1	1
360	−1	−1	−1	1	−1	−1
361	−1	−1	−1	−1	1	1
362	−1	−1	−1	−1	1	−1
363	−1	−1	−1	−1	−1	1
364	−1	−1	−1	−1	−1	−1
365	1	0	0	0	0	0
366	−1	0	0	0	0	0
367	0	1	0	0	0	0
368	0	−1	0	0	0	0
369	0	0	1	0	0	0
370	0	0	−1	0	0	0
371	0	0	0	1	0	0
372	0	0	0	−1	0	0
373	0	0	0	0	1	0
374	0	0	0	0	−1	0
375	0	0	0	0	0	1
376	0	0	0	0	0	−1
377	0	0	0	0	0	0

C. Six Variables—Central Composite, Quarter Replicate[c]

101	1	1	1	1	1	1
102	1	1	1	−1	1	−1
103	1	1	−1	1	−1	−1
104	1	1	−1	−1	−1	1
105	1	−1	1	1	−1	−1
106	1	−1	1	−1	−1	1
107	1	−1	−1	1	1	1
108	1	−1	−1	−1	1	−1
109	−1	1	1	1	−1	1
110	−1	1	1	−1	−1	−1
111	−1	1	−1	1	1	−1
112	−1	1	−1	−1	1	1
113	−1	−1	1	1	1	−1
114	−1	−1	1	−1	1	1
115	−1	−1	−1	1	−1	1
116	−1	−1	−1	−1	−1	−1
117	1	0	0	0	0	0
118	−1	0	0	0	0	0
119	0	1	0	0	0	0
120	0	−1	0	0	0	0
121	0	0	1	0	0	0
122	0	0	−1	0	0	0
123	0	0	0	1	0	0
124	0	0	0	−1	0	0
125	0	0	0	0	1	0
126	0	0	0	0	−1	0
127	0	0	0	0	0	1
128	0	0	0	0	0	−1
129	0	0	0	0	0	0

[a]1 = high; 0 = low.
[b]1 = high; 0 = medium; −1 = low.
[c]1 = high; 0 = medium; −1 = low, E in this section = ABC, F = BCD.

Table 14. Attributes Used in the Sauce Study

Independent Variable	Quality Control[b]
Ingredient A	QC/Coarse
Ingredient B	QC/Visible-spice
Ingredient C	QC/Pieces
Ingredient D	QC/Thick
Ingredient E	QC/Flavor
Cook-process	QC/Bitter
Cost-of-Goods	QC/Burnt
	QC/Metallic
Consumer-Data[a]	*Instrumental-Readings[c]*
C/Like/Overall	
C/Like/Segment-A	I/Pieces-surface
C/Like/Segment-B	I/Color
C/Like/Segment-C	I/NaCl
C/Like/Appearance	I/Acid
C/Like/Aroma	I/Consistency
C/Like/Flavor	
C/Like/Texture	*Expert-R&D-Panel[d]*
C/Image/Fresh	
C/Image/Authentic	E/Particulates
C/Image/Sophisticated	E/Burnt-sweet
C/Sen/Darkness	E/Onion-flavor
C/Sen/Aroma	E/Pepper-flavor
C/Sen/Taste	E/Sweet-taste
C/Sen/Sweet	E/Sour-taste
C/Sen/Spicy	E/Salt-taste
C/Sen/Aftertaste	E/Bitter-taste
C/Sen/Chunky	E/Throat-burn
C/Sen/Grainy	

[a]C = Consumer rating.
Like = liking.
Sen = sensory.
Segment = sensory preference segments (showing different patterns of liking).
[b]QC = Quality Control Panel.
[c]I = Instrumental measure.
[d]E = Expert panel (QDA Method).

studies involve fewer variables (often no more than 2–3), and may use only one panel (eg, consumers from a research guidance panel). However the data becomes more variable once the data set is augmented to comprise consumer and instrumental data.

Panel Selection

Depending upon when the research is undertaken (early versus late in the development cycle) the researcher will work with different consumer panels. In the earliest development stages researchers work with an in-house group of consumers or experts. Often, however, it is necessary to obtain ratings from users of different and competing products in the category. It is also instructive to run the study in different geographical areas (or even in different countries). The researcher hires field services in the different markets and "tailors" a consumer panel with specific demographic and usage characteristics. Fewer consumers participate in these guidance tests than in subsequent large scale market tests. This study tested products among 120 female consumers, 40 consumers in 3 markets, (in order to ensure representativeness). Optimization studies are usually of this magnitude, with panel sizes ranging from 50 to 200 participants.

Panels for optimization studies usually comprise users of the category, occasionally (and more specifically) users of the product being optimized and/or users of competitive brands. Here half the panel comprised consumers who used the manufacturer's product most often, and the other half comprised consumers who used competitor products. By incorporating different groups of consumers in the same panel, (all of whom test the same array of products), the investigator can compare ratings assigned to the same products by the two groups. The comparison can be used to develop optimal products which satisfy one target group, or both groups simultaneously.

Activities During the Evaluation

Depending upon the product being tested, panelists may either participate in a supervised session in a central location, or test the products at home, without supervision.

Panelists followed the protocol presented in Table 15. A central location setup makes it easy to monitor the quality of the data as the panelists assign their ratings. Close supervision of the evaluation ensures that panelists remain alert and motivated. Neither the interviewers nor the panelists "know the correct answer." Panelists must, perforce, answer honestly because there are no other cues to aid them other than their sensory perceptions. The ongoing question and answer dialogue between the panelist and the interviewer (after each product is rated) maintains motivation. For home use tests (appropriate when panelists prepare the product themselves, or use the product as an ingredient), other procedures are necessary. These procedures include giving the panelist one or two products to use, instructing the panelist to rate each product just after eating it, and then requiring the panelist to return to the test site where the panelist is questioned about the ratings that he or she assigned. Upon returning to the test site the panelist receives the next set of products, and follows the same protocol. Control in a home use test is maintained by

Table 15. Protocol for the Sauce Study

Panelists prerecruited to participate, by a telephone call.
Panelists are screened to be appropriate (viz, category users, interested in the concept which an interviewer reads to them).
Panelists show up, prepared for a 4-hour session.
   Panelists show up, in groups of 15–20 (randomly recruited, so that the panelists do not know each other).
Panelists are oriented in scaling by a short practice exercise.
   Panelists rate water and cracker on sensory and liking attributes.
Panelists try the first product, rating it on all attributes.
   Products are randomized to reduce order bias. Products are rated on attributes as they appear (viz, appearance first, then aroma, then texture). Liking ratings and sensory ratings are interdigitated.
Ratings checked by an attending interviewer on a panelist-by-panelist basis, to ensure panelist comprehension.
Panelist waits 15 minutes and goes to the next product.
After panelist finishes, the panelist is paid and dismissed.
Each panelist rates 11 products (randomized from the full set).

the careful allocation of products to panelists, along with an orientation session at the start of the study (before actual home use). Motivation is maintained by payment for the successful completion of the test and by the ongoing question and answer interchanges between panelists and the interviewers.

Initial Data Analysis—Product × Attribute Matrix

The initial data comprises a matrix of mean ratings of each product by each attribute. Table 16 shows part of this matrix. Statisticians may prefer to adjust the means in Table 16 to account for the fact that each panelist evaluated only a randomized subset of products. These "adjusted means" rather than the actual means, would appear in Table 16.

Table 16 alone teaches a lot about the products because the researcher can compare products, on the different attributes. Even without an experimental design several instructive analyses (correlation, regression, and histogram plotting) highlight key dimensions of the category.

Questions investigators can answer with Table 16 (in its entirety) include:

1. For a specific attribute do the products differ a great deal or relatively little? What is the range of scores? By looking at the highest and lowest means across products on the attribute, and graphing the distribution of the means within that range the investigator sees the "spread" of products. If the raw data is also available, then a 1 way analysis of variance (ANOVA) provides a measure of discrimination, (viz, the F ratio). The F value measures the "signal to noise" ratio, or variability due to product differences versus variability due to random error (eg, panelist differences).

2. What is the distribution of the mean liking ratings across products? If competitor products score in the 50s and 60s (on a 0–100 point scale), whereas most of the prototypes (from the experimental design) score in the 30s and 40s, then the optimized product will probably not score higher than the competitors. The only way that one could develop a superb product would be if the optimal formulation lies in a region of ingredient values not tested in the original experimental design. (In that case the optimal solution would be suspect).

3. Which liking attributes covary with overall liking? This analysis reveals important attributes. Figure 2 shows a schematic plot of overall liking (ordinate) versus attribute liking (abscissa). The correlation between the two attributes describes the strength of a *linear relation* between attribute liking and overall liking, but does not show the slope. A straight line shows the quantitative nature of the relation. The steeper the slope of the straight line the more small changes in attribute liking covary with large changes in overall liking (and therefore the more important the attribute).

4. Which sensory attributes covary with overall liking? As a formula variable or sensory attribute changes, liking also changes. Often this relation exhibits an

Table 16. Partial Data Base for Sauces

Product	110	111	112	113	114	115	116	117
			Independent variables[a]					
I/Ingred-A	3	2	2	1	1	1	2	3
I/Ingred-B	1	2	3	1	1	1	2	3
I/Ingred-C	1	2	2	1	3	3	2	1
I/Ingred-D	3	3	2	3	3	1	2	1
I/Ingred-E	1	2	2	1	1	1	1	3
I/Cook-process	1	2	2	3	0	3	2	3
O/Cost-of-Goods	116	138	132	116	141	132	119	123
			Consumer-data[b]					
C/Like/Overall	44	63	67	57	56	49	42	61
C/Like/Segment-A	52	61	78	63	48	57	53	56
C/Like/Segment-B	39	65	73	53	58	50	42	63
C/Like/Segment-C	47	60	51	58	56	43	37	59
C/Like/Appearance	48	61	66	55	66	60	56	59
C/Image/Sophisticated	46	47	47	49	52	43	43	56
C/Sen/Darkness	47	57	71	51	59	55	53	69
C/Sen/Aroma	39	44	49	42	46	45	43	51
C/Sen/Taste	51	65	64	57	58	44	54	72
C/Sen/Sweet	33	48	73	39	39	43	38	48
C/Sen/Spicy	42	55	52	52	46	36	36	43
C/Sen/Afterstate	40	44	32	39	38	32	40	41
C/Sen/Thick	53	72	70	55	76	55	51	59
			Quality-Control[c]					
QC/Coarse	56	77	74	68	68	63	68	81
QC/Visible-spice	39	52	48	44	42	47	30	68
QC/Pieces	25	32	39	24	48	50	33	37
QC/Thick	62	82	73	67	85	64	71	70
QC/Flavor	59	63	59	61	68	52	72	65
QC/Bitter	13	12	14	13	20	17	13	23
QC/Burnt	55	56	53	54	54	50	58	53
QC/Metallic	41	27	28	36	32	40	45	34
			Instrument-reading					
I/Pieces-surface	26	30	33	24	38	51	30	31
I/Color	46	52	63	44	53	51	47	64
I/NaCl	42	42	35	41	38	23	29	38
I/Acid	53	41	29	42	46	32	43	46
I/Consistency	57	26	39	53	27	59	46	56
			Expert-R&D-QDA[d]					
E/Particulates	35	47	49	41	52	49	50	34
E/Burnt-sweet	17	32	28	16	23	12	14	23
E/Salt-taste	60	54	49	57	56	28	53	55
E/Bitter-taste	1	8	5	5	6	1	1	7
E/Throat-burn	19	20	21	23	17	12	16	21

[a] I = instrumental, O = objective, 3 = high 2 = medium 1 = low.
[b] C = consumer.
[d] QC = quality control.
[d] E = expert panel.

inverted U shape (32,33). One can plot the relation between sensory attribute level (abscissa) and liking (ordinate). The data will scatter around the curve but the underlying data should follow a pattern that can be extracted. Fitting a parabola [liking = A + B(Sensory Level) + C(Sensory Level^2)] to the data, uncovers this relation. The fitted curve shows the *probable* nature of the relation, as Figure 3 illustrates (schematically).

Developing Relations Between Variables (Modeling)

Modeling creates a mathematical relation between the systematically varied independent variables and dependent

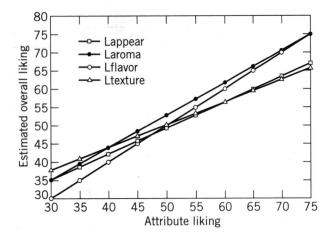

Figure 2. Relation between attribute liking and overall liking. The steeper the slope of the straight line the more important is the attribute.

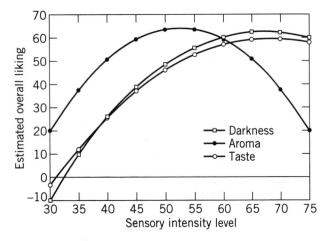

Figure 3. Schematic relation between sensory attribute level and overall liking. The relation usually follows an inverted U shaped curve. Large areas under the curve (subtended by the curve) correspond to important attributes.

variables. Dependent variables can be sensory attributes, acceptance ratings (all from consumers), or physical measures (including cost of goods), etc.

Relations between variables are expressed by equations. Researchers can use one of many available equations, depending upon the expected relation between the independent variable and the dependent variable. Some of these equations are

The Linear Equation. Changes in the dependent variable correspond to *linear* changes in the independent variable. The linear equation often emerges when we vary one ingredient (eg, sucrose) in a simple food system (eg, sweetened carbonated beverage). Over the limited concentration range consumer ratings of perceived sweetness versus sweetener concentration follow a straight line. Although a straight line might not describe the "true, underlying" relation between the two variables (sucrose level versus per-

ceived sweetness) it is an adequate and useful representation.

We express the relation by the simple linear equation:

$$\text{Dependent Variable} = A + B(\text{Independent Variable})$$

Depending upon how closely the data points fall to the line, the equation may fit the data well or poorly. Of critical importance is the value of slope, B, which shows how unit changes in the independent variable correspond to changes in the dependent variable. When B is high there is great sensitivity to the independent variable. When B is low there is little sensitivity. Large changes in the independent variable correspond to small changes in the dependent variable.

Quadratic Relation. As the independent variable increases the rating first increases, peaks and then drops down. The parabolic or quadratic equation is a good summary of how liking varies with physical level (33). The parameters of the quadratic equation are unique to each product. Parabolic equations allow for increases, peaks, and then decreases (the classical inverted U curve), or simply increase and flatten (or the corresponding decrease and flattening). The equation is expressed as:

$$\text{Dependent Variable} = A + B(\text{Independent}) + C(\text{Independent})^2$$

The Plane (for Two or More Independent Variables, Jointly Predicting the Dependent Variable). The plane generalizes the line. We can envision it easily as a flat sheet in 3 dimensions. The slope of the plane in any dimension (corresponding to a single independent variable) is given by the coefficient for that independent variable. The plane equation does not account for interactions among the variables. The plane assumes that each variable acts entirely independently to determine the dependent variable. We write the equation of the plane as a multiple linear equation:

$$\text{Dependent} = A + B(\text{Independent 1}) + C(\text{Independent 2}). \; \dots$$

In the absence of theory about the relation between independent variables and dependent variable it is prudent to use the multiple linear equation. The equation makes no assumptions other than there exists a simple linear relation between the dependent variable and each of the independent variables.

Quadratic Surface. Like the plane the quadratic surface also relates several independent variables simultaneously to the dependent variable. We assume that there exists a parabolic or curvilinear relation between the independent variables and the dependent variable (eg, for overall liking). However, for a quadratic or parabolic surface not only do we have linear terms, but also square terms.

If the equation comprises only linear terms and square terms, then we assume that the variables act independently of each other to determine the value of the dependent variable. The square terms allow each independent variable to act in a nonlinear fashion. There may be a flattening of the surface, so that as an independent variable increases the dependent variable can first increase, but then flatten out, and then perhaps even decrease. If there are no interactions between the independent variables, then we write the equation as:

$$\text{Dependent} = A + B(\text{Independent 1}) + C(\text{Independent 1})^2$$
$$+ D(\text{Independent 2}) + E(\text{Independent 2})^2$$
$$+ \dots.$$

Like the plane equation, the parabolic or quadratic equation can be extended to accommodate many independent variables simultaneously. If there exists interaction between pairs of independent variables, we can write the equation as above, but also include a multiplicative term [(Independent 1) × (Independent 2)] to account for interactions (here between the first two independent variables). Many such pairwise interactions may exist. The equation can accommodate them.

There are many other equations that an investigator can use in order to model the relations between independent variables and dependent variables. For product testing, however, we generally do not really know the true form of the relation (viz, how variables interact with each other). It is prudent to use the most parsimonious equation. Most researchers opt for the linear equation, and add square terms and pairwise interactive (multiplicative) terms only when specifically required, usually for one of two reasons:

1. *Theory.* We know from experimentation that liking does not continue to increase linearly with physical level *ad infinitum*, but eventually peaks, plateaus, and then drops. Curvature requires the quadratic term in the equation.

2. *Parsimony.* The quadratic equation is relatively parsimonious, allowing both curvature and interactions between variables. Other equations might fit the data better, but would demand more terms.

Whether the investigator fits a linear or quadratic model to the data, the utility of the model is the same—it predicts responses to combinations of independent variables, even when the combinations were not actually tested but lie within the range tested.

Tables 17 and 18 show the equations developed for the products (darkness, liking). The equations are simply mathematical statements. They lack substantive meaning for the product developer, until the developer substitutes meaningful values for the independent variables and estimates the attribute ratings.

Optimizing a Response by Finding the Highest Value

A major benefit of modeling is the ability to discover the particular combination of physical variables which correspond to a desired condition, such as the highest (optimal) rating on an attribute (eg, liking). Table 19 shows several optimizations, including overall optimization of liking (column A), three optimizations for liking subject to imposed cost constraints (columns B,C,D), and optimization of lik-

Table 17. Model (Equation) Relating Formula and Process Variables to Liking[a]

Variable[b]	Coefficient	Std error	T value[c]	P value[d]
Constant	16.5	11.9	1.3	0.1
A	−2.3	4.5	−0.5	0.6
B	6.1	4.5	1.3	0.1
C	−0.5	4.5	−0.1	0.9
D	15.1	4.6	3.2	0.0
E	15.5	4.5	3.4	0.0
F	−2.6	2.0	−1.2	0.2
A*A	0.2	1.1	−0.2	0.8
B*B	−0.5	1.1	−0.4	0.6
C*C	0.9	1.1	0.8	0.3
D*D	−2.9	1.1	−2.6	0.0
E*E	−3.3	1.1	−3.0	0.0
F*F	0.6	0.5	1.3	0.1

Analysis of variance

Source	Sum-of-squares	DF	Mean-square	F-ratio	P
Regression	2376.813	12	198.068	6.836	0.000
Residual	898.187	31	28.974		

[a]Multiple R: .85 Squared multiple R: .73
Adjusted squared multiple R: .62 Standard error of estimate: 5.3
[b]A–E = Ingredients A to E, respectively; F = Cook process.
[c]T Value = T test for significance of the coefficient. (T values in excess of +1 or −1 denote significant coefficients—vis, those significantly different from 0).
[d]P value = probability that the coefficient is really "0". (Low P value means that the coefficient is significantly different from 0).

Table 18. Model Relating Formula and Process Variables to Perceived Darkness[a]

Variable[b]	Coefficient	STD error	T[b]	P[b]
Constant	36.7	7.1	5.1	0.0
A	−0.6	2.7	−0.2	0.0
B	3.0	2.7	1.1	0.2
C	−0.5	2.7	−0.1	0.8
D	9.0	2.8	3.2	0.0
E	−0.2	2.7	−0.1	0.9
F	0.1	1.2	0.1	0.8
A*A	−0.1	0.6	−0.2	0.8
B*B	−0.0	0.6	−0.0	0.9
C*C	0.7	0.6	1.0	0.2
D*D	−1.9	0.6	−2.7	0.0
E*E	1.7	0.6	2.5	0.0
F*F	0.0	0.3	0.1	0.8

Analysis of variance

Source	Sum-of-squares	DF	Mean-square	F-ratio	P
Regression	2717.953	12	226.496	21.499	0.000
Residual	326.593	31	10.535		

[a]Multiple R: .94 Squared multiple R: .89
Adjusted squared multiple R: .85 Standard error of estimate: 3.2
[b]See footnotes in Table 17 for explanation.

Table 19. Sequence of Optimal Formulations (Maximal Liking) Under Reduced Cost of Goods and High Spiciness

	No constraints	Cost constraints			Spiciness constraint
	A^a	B^b	C^c	D^d	E^e
I/Ingred-A	1.0	1.0	1.0	1.0	1.0
I/Ingred-B	3.0	3.0	3.0	3.0	3.0
I/Ingred-C	3.0	2.3	1.4	1.0	3.0
I/Ingred-D	2.5	2.4	2.5	2.1	2.5
I/Ingred-E	2.3	2.1	2.3	2.0	2.7
I/Cook Process	3.0	3.0	3.0	3.0	3.0
O/Cost-of goods	151	140	130	120	154
		Consumer data			
C/Like/Overall	71	68	65	64	70
C/Like/Appearance	74	69	64	60	74
C/Like/Aroma	63	59	57	55	64
C/Like/Flavor	71	68	65	64	70
C/Like/Texture	76	72	65	60	77
C/Image/Fresh	68	64	62	61	67
C/Image/Authentic	63	59	59	54	64
C/Image/Sophisticated	55	52	51	49	58
C/Sen/Darkness	70	66	66	63	74
C/Sen/Aroma	54	50	50	48	57
C/Sen/Taste	67	65	65	62	70
C/Sen/Sweet	56	56	55	54	56
C/Sen/Spicy	61	58	58	55	65
C/Sen/Aftertaste	35	36	37	35	38
C/Sen/Thick	79	72	65	58	82
C/Sen/Chunky	60	53	44	38	61
C/Sen/Grainy	30	29	28	24	33
		Quality control			
QC/Coarse	82	77	76	72	86
QC/Visible-Spices	59	53	54	50	66
QC/Pieces	50	43	32	28	50
QC/Thick	83	77	73	67	86
QC/Flavor	64	62	59	58	64
QC/Bitter	16	14	15	13	20
QC/Burnt	53	52	52	52	54
QC/Metallic	21	23	25	25	25
		Instrument reading			
I/Pieces-Surface	47	40	30	25	48
I/Color	62	60	60	57	66
I/NaCl	37	37	38	37	38
I/Acid	34	36	37	36	35
I/Consistency	28	36	41	48	24
		Expert R&D			
E/Particulates	61	54	43	35	63
E/Burnt-sweet	29	27	26	23	31
E/Onion-flavor	30	29	31	29	32
E/Pepper-flavor	12	11	11	10	12
E/Sweet-taste	33	32	31	30	33
E/Sour-taste	52	54	55	56	51
E/Salt-taste	48	48	50	51	48
E/Bitter-taste	9	8	9	6	12
E/Throat-burn	21	21	22	21	21

[a] A = No constraints.
[b] B = Cost of goods less than 140.
[c] C = Cost of goods less than 130.
[d] D = Cost of goods less than 120.
[e] E = Perceived spiciness greater than 65.

ing for a product that must be perceived as highly spiced (column E). Optimization yields the likely formulations, which are then used to estimate the likely response profile on all of the attributes. If the model includes attributes derived from physical measures, then one can calculate the profile of physical measures corresponding to the optimal formulation. This link between physical measures and ratings can be used to develop a quality control system (see below).

Fitting a Predesignated "Goal" Profile

The models allow one to prescribe a desired goal profile of subjective attributes, and then discover the particular combination of physical variables which generates that desired goal profile. Rather than optimizing a single response (eg, overall liking), the optimizer system varies the independent variables until the expected response profile matches as closely as possible the predesignated goal profile. (The closeness of fit can be expressed either in absolute terms, or in percentage terms. Percentage for "closeness of fit" is preferable when the units of measurement for the response attributes differ from attribute to attribute). Table 20 shows the solution, for two goal profiles. The first profile comes from consumers who evaluated a competitor product that the researcher wishes to "match" using the set of ingredients. The second goal profile was generated at the plant by instrumental probes which recorded the characteristics of each batch as it was produced. The objective was to estimate consumer reactions to this batch, using the instruments as quality probes.

Profile fitting enables the investigator to estimate the likely profile for one data set, given the profile for another data set. Each target profile constitutes a goal. It is possible to:

1. *Relate Difference Sources of Data*. Given a target profile from one source (eg, instrumental measures) the investigator can estimate the corresponding attribute profile which would emerge from another source of data for the same product (eg, consumer sensory ratings).

2. *Program a set of "probes" (in-line) in a processing plant to respond to the formulation as consumers would*. Given the profile of responses to a batch obtained from the probe, the quality assurance engineer can estimate the likely formulation that would have yielded the profile, and then from that formulation estimate the likely consumer ratings.

Sensitivity Analysis Around a Point

Estimates of responses to a single combination of variables do not tell the whole story. Often the formulation projected by the optimizer corresponds to the absolutely highest point of what may be a very wide plane of virtually equally acceptable optima. Furthermore, changes in one physical variable may not change ratings, whereas changes in another variable may dramatically change ratings.

Sensitivity analysis estimates effects due to one variable, with all other variables held constant. For changes in one independent variable the investigator estimates the attribute profile using the equations.

Table 21 shows the sensitivity analysis. The table shows how the dependent variable covaries with the independent variable.

Using the Product Model to Assure Quality

The product model relates independent attributes (under research or production control) to dependent variables, be

Table 20. Example of Goal Fitting

	Consumer goals	Goal	Instrumental goals	Goal
Ingredient				
I/Ingred-A	1.0		2.5	
I/Ingred-B	1.9		1.3	
I/Ingred-C	2.1		2.8	
I/Ingred-D	1.0		I.0	
I/Ingred-E	1.6		1.0	
I/Cook-process	3.0		3.0	
O/Cost-of-goods	125		139	
Consumer data				
C/Like/Overall	56		54	
C/Image/Fresh	54		56	
C/Image/Authentic	46		49	
C/Image/Sophisticated	46	Goal	52	
C/Sen/Darkness	56	50	55	
C/Sen/Aroma	45	60	45	
C/Sen/Taste	57	50	60	
C/Sen/Sweet	45	60	39	
C/Sen/Spicy	50	50	49	
C/Sen/Aftertaste	36	60	40	
C/Sen/Thick	50	50	66	
C/Sen/Chunky	42	60	49	
C/Sen/Grainy	24	50	26	
Quality control				
QC/Coarse	62		73	
QC/Visible-Spices	45		45	
QC/Pieces	38		43	
QC/Thick	58		78	
QC/Flavor	56		64	
QC/Metallic	29		32	
Instrument reading				Goal
I/Pieces-surface	37		40	40
I/Color	52		49	45
I/NaCl	37		39	30
I/Acid	40		45	55
I/Consistency	58		40	40
Expert R&D panel				
E/Particulates	40		50	
E/Burnt-sweet	19		22	
E/Sweet-taste	27		26	
E/Sour-taste	57		58	
E/Salt-taste	48		55	
E/Throat-Burn	18		18	

Table 21. Sensitivity Analysis (From midpoint)

	1.0	1.3	1.7	2.0	2.3	2.7	3.0
Ingredient A							
O/Cost-of-goods	132	132	132	132	132	132	133
C/Like/Overall	61	61	60	60	59	59	58
C/Sen/Darkness	61	61	60	60	59	59	58
C/Sen/Aroma	48	47	47	47	47	47	47
C/Sen/Taste	63	63	64	64	65	66	66
C/Sen/Sweet	47	47	47	46	46	45	44
C/Sen/Spicy	55	55	56	56	56	57	57
C/Sen/Aftertaste	38	38	39	40	41	42	42
C/Sen/Thick	64	64	64	64	64	63	63
C/Sen/Chunky	46	45	45	45	45	45	45
C/Sen/Grainy	27	27	27	27	28	28	28
Ingredient B							
O/Cost-of-Goods	131	132	132	132	132	133	133
C/Like/Overall	55	57	58	60	61	62	63
C/Sen/Darkness	57	58	59	60	61	62	63
C/Sen/Aroma	46	46	47	47	47	47	48
C/Sen/Taste	63	63	64	64	65	65	65
C/Sen/Sweet	40	42	44	46	49	52	55
C/Sen/Spicy	54	55	55	56	56	56	56
C/Sen/Aftertaste	41	41	41	40	39	39	38
C/Sen/Thick	60	61	63	64	65	65	66
C/Sen/Chunky	44	44	45	45	45	46	47
C/Sen/Grainy	27	27	27	27	27	27	27
Ingredient C							
O/Cost-of-Goods	118	123	128	132	137	142	146
C/Like/Overall	57	58	58	60	61	62	64
C/Sen/Darkness	58	59	59	60	61	62	63
C/Sen/Aroma	45	46	46	47	48	48	49
C/Sen/Taste	62	63	64	64	65	65	65
C/Sen/Sweet	45	45	46	46	47	47	47
C/Sen/Spicy	54	55	55	56	56	57	58
C/Sen/Aftertaste	39	40	40	40	40	39	39
C/Sen/Thick	54	57	61	64	67	70	72
C/Sen/Chunky	34	38	41	45	48	52	55
C/Sen/Grainy	24	25	27	27	28	28	28
Ingredient D							
O/Cost-of-Goods	127	128	130	132	134	136	138
C/Like/Overall	53	56	58	60	60	60	60
C/Sen/Darkness	57	58	59	60	60	60	59
C/Sen/Aroma	45	46	46	47	47	47	47
C/Sen/Taste	61	62	64	64	65	64	63
C/Sen/Sweet	45	46	46	46	47	46	46
C/Sen/Spicy	53	54	55	56	56	56	55
C/Sen/Afterstate	39	39	40	40	40	40	40
C/Sen/Thick	51	56	61	64	66	66	66
C/Sen/Chunky	40	42	44	45	45	45	44
C/Sen/Grainy	25	26	27	27	28	28	28
Ingredient E							
O/Cost-of-Goods	126	128	130	132	134	136	138
C/Like/Overall	54	57	58	60	60	59	58
C/Sen/Darkness	55	56	58	60	62	65	68
C/Sen/Aroma	43	44	45	47	49	51	54
C/Sen/Taste	59	61	62	64	67	69	71
C/Sen/Sweet	44	45	46	46	46	46	45
C/Sen/Spicy	48	50	53	56	59	62	65
C/Sen/Aftertaste	38	38	39	40	42	44	46

Table 21. Sensitivity Analysis (From midpoint) *(continued)*

	1.0	1.3	1.7	2.0	2.3	2.7	3.0
C/Sen/Thick	58	60	62	64	66	67	69
C/Sen/Chunky	41	42	44	45	46	47	48
C/Sen/Grainy	25	25	26	27	29	31	34
Cool process							
O/Cost-of-Goods	133	32	132	132	132	132	132
C/Like/Overall	60	60	60	60	60	60	60
C/Sen/Darkness	60	60	60	60	60	60	60
C/Sen/Aroma	46	47	47	47	47	47	47
C/Sen/Taste	64	64	64	64	64	64	64
C/Sen/Sweet	46	46	46	46	47	47	47
C/Sen/Spicy	56	56	56	56	56	56	57
C/Sen/Aftertaste	40	40	40	40	40	40	40
C/Sen/Thick	65	64	64	64	64	64	64
C/Sen/Chunky	46	46	45	45	45	45	46
C/Sen/Grainy	27	27	27	27	28	28	28

Note: selected attriutes shown.
All ingredients held constant at the midpoint of the design (each variable held at "2"). Then each independent variable (ingredients A–E, cooking process was separately increased in small steps from a low of 1 to a high of 3. The numbers in the table show changes in attributes, including cost of goods and consumer ratings.

these consumer ratings, expert panel ratings, etc. The product model plays a role in assuring quality, because it can be used in two distinct ways to control production:

1. *Process Control.* At the most basic level sensitivity analysis reveals how process or ingredient variations affect liking. Once the desired formulation has been located within the grid of alternative formulations, and the range of independent variables set (high to low for each independent variable), the investigator computes the sensitivity curves for each independent variable, holding the other independent variables fixed (at the prescribed level for production). The changes in the sensory ratings show how the independent variable affects the sensory character of the product (viz, sensory attributes). The change in liking indicates how changes in the independent variable affect acceptance.

   Quality control tables based upon sensitivity analyses highlight those independent variables which produce noticeable sensory changes, and the degree to which those changes generate acceptance changes in their wake. The manufacturer should maintain tighter control over the key critical variables (perhaps at greater cost) and maintain looser control (at lower cost) over the less important variables (where departure from the optimal or production specifications do not reduce acceptance nor affect sensory integrity).

   The approach is *consumer driven*, because it is based upon the reactions of consumers to actual variations of the product, rather than hypothesis of what production variables might be important. It might well turn out that some variables play little or no role in determining acceptance, and can either be reduced to save money, or ignored by quality assurance over a wide range of levels. Conversely, more attention would then be paid to those important variables which influence acceptance.

2. *Batch Analysis.* The inter-relation among formula variables, consumer responses, instrumental variables, expert panel responses, and quality control panel responses assures quality at the production level on a batch-to-batch basis. Each batch is measured, either by experts, by the quality control panel, or by instruments. The measurements generate a goal profile. The goal profile determines a corresponding set of estimated ingredients or process variables which would generate that profile. Once the levels of the independent variables are estimated the model estimates the corresponding consumer sensory profile and/or the likely acceptance rating. As each batch emerges from the production line quality assurance can calculate the expected difference from the reference or "gold standard" on a sensory or acceptance basis. At the time of the measurement, plant personnel decide whether the product is sufficiently similar to a target standard (or reasonably acceptable) to warrant shipment, or whether the batch must be reworked (or even discarded).

AN OVERVIEW

Sensory analysis in the last decade of the Twentieth Century has matured considerably since its informal beginnings almost ninety years before. At its inception the field comprised individuals who had little experience in evaluation, who were practitioners of other disciplines, and who made sensory analysis a professional avocation. Techniques developed haltingly as the field developed, with various home grown approaches proposed, implemented, the best culled for further use, and the remainder discarded. The archival early literature of sensory analysis reveals these discarded byways.

Today, thanks to the continued importance of sensory satisfaction placed on products by the food processor and the need to remain competitive in a marketplace of intensely competing brands, manufacturers have refined their test methods. By combining contributions from academia, statistical procedures, psychological and psychophysical test methods, and applying the tests procedures in a business environment which needs accurate and actionable feedback, practitioners have advanced the field considerably. What was once an informal bioassay of a product to determine palatability has become, over the past years, a computer-based technology which quantitatively guides product development to improve consumer acceptance.

Sensory analysis comprises many different methods. Its strength coming from a body of diverse approaches is also its inherent weakness. There is no single set of approaches deemed "appropriate, correct and true" in sensory analysis, but rather different approaches to the same problem, proffered by practitioners. Sensory analysis is a science in development, and a technology in practice. The slow de-

velopment towards a science is witnessed by the recent emergence of journals in the U.S., and the U.K devoted to sensory analysis as a science. These journals are belated witnesses to that recognition, and to the evolving nature of the field. Whether it takes a year or another lifetime, however, the field will become a true science. The wealth of methods presented to the users in product development, and the literature now available certainly today qualify sensory analysis as a dynamic, evolving, intellectually stimulating and business-relevant technology which has a vital place in product development and marketing. Sensory analysis has earned a key role in food science, and continues to be worthy of that role.

BIBLIOGRAPHY

1. M. A. Amerine, R. M. Pangborn, and E. B. Roessler, *Principles of Sensory Evaluation of Food*, New York, Academic Press, New York, 1965.

2. R. M. Pangborn and I. Trabue, "Bibliography of the Sense of Taste," in M. R. Kare and O. Maller, eds., *The Chemical Senses and Nutrition*, Johns Hopkins Press, Baltimore, 1967.

3. R. Harper, E. C. Bate-Smith, and D. G. Land, *Odour Description and Odour Classification*, American Elsevier, New York, 1968.

4. J. E. Amoore, "A Plan to Identify Most of the Primary Odors," in C. Pfaffmann, ed., *Olfaction and Taste IV*, Rockefeller University Press, New York, 1969, pp. 158–171.

5. R. Harper, E. C. Bate-Smith, and N. M. Griffiths, "Odour qualities: A glossary of usage," *British Journal Of Psychology* **59**, 231–252 (1968).

6. H. Sokolow, "Qualitative Methods for Language Development, in H. Moskowitz, ed., *Applied Sensory Analysis of Foods*, Vol I, CRC Press, Boca Raton, Fla., 1988, pp. 3–20.

7. J. F. Caul, "The Profile Method of Flavor Analysis," *Advances in Food Research*, 1957 pp. 1–40.

8. A. J. Neilson, V. B. Ferguson, and D. A. Kendall, "Profile methods: Flavor profile and profile attribute analysis," in H. R. Moskowitz, ed., *Applied Sensory Analysis of Foods*, Vol I, CRC Press, Boca Raton, Fla., 1988, pp. 22–41.

9. H. Stone, J. L. Sidel, S. Oliver, A. Woolsey, and R. Singleton, "Sensory Evaluation by Quantitative Descriptive Analysis," *Food Technology* **28**, 24–34 (1974).

10. K. L. Zook and J. H. Pearce, "Quantitative Descriptive Analysis," in H. R. Moskowitz, ed., *Applied Sensory Analysis of Foods*, Vol I, CRC Press, Boca Raton, Fla., 1988, pp. 44–71.

11. M. Meilgaard, G. V. Civille, and B. T. Carr, *Sensory Evaluation Techniques*, CRC Press, Boca Raton, Fla., 1987.

12. A. S. Szczesniak, M. A. Brandt, and H. H. Friedman, "Development of standard rating scales for mechanical parameters of texture and correlation between the objective and sensory methods of texture evaluation," *Journal of Food Science* **28**, 397–403 (1963).

13. N. Schwartz, "Adaptation of the Sensory Texture Profile to Skin Care Products," *Journal of Texture Studies* **6**, 33 (1975).

14. M. Yoshida, "Studies in the Psychometric Classification of Odor," *Japanese Psychological Research* **6**, 111–115 (1964).

15. M. Yoshida, "Dimensions of Tactile Perceptions," *Japanese Psychological Research* **10**, 157–173 (1968).

16. S. Yoshikawa, S. Nishimaru, T. Tashiro, and M. Yoshida, "Collection and Classification of Words for Description of Food Texture, I, II, III," *Journal of Texture Studies* **1**, 437–463 (1970).

17. C. Coombs, *A Theory of Data*, John Wiley & Sons, Inc., New York, 1964.

18. R. N. Shepard, "The Analysis of Proximities: Multidimensional Scaling with an Unknown Distance Function," *Psychometrika* **27**, 219–246 (1962).

19. M. Bourne, personal communication, 1975.

20. G. A. Gescheider, *Psychophysics: Method and Theory*, Lawrence Erlbaum Associates, Hillsdale, N.J. 1976.

21. E. H. Weber, *De Pulsu Resorptime, Auditor, et Tache. Annotations Anaomatical et Physiological*, Leipzig, Koehler, 1834, (cited by E. G. Boring, *Sensation and Perception In; The History of Experimental Psychology*, New York, Appleton-Century Crofts, New York, 1942.

22. G. T. Fechner, *Elemente der Psychophysik*, Breitkopf und Hartel, Leipzig, 1860.

23. F. Lemberger, "Psychophysiche Untersuchungen uber den Geschmack von Zucker und Saccharin," *Pfleuger's Archiv fur die Gesamte Physiologie* **123**, 293–311 (1908).

24. S. S. Stevens, "On the Brightness of Lights and the Loudness of Sounds," *Science* **118**, 576 (1953).

25. B. Bond and S. S. Stevens, "Cross Modality Matches of Brightness and Loudness by 5 Year Olds," *Perception & Psychophysics* **6**, 337–339 (1969).

26. H. R. Moskowitz, "Intensity Scales for Pure Tastes and Taste Mixtures," *Perception & Psychophysics* **9**, 51–56 (1971).

27. H. L. Meiselman, "Scales for Measuring Food Preference," in M. S. Petersen and A. H. Johnson, eds., *Encyclopedis of Food Science*, Avi, Westport, Conn., 1978, pp. 675–678.

28. H. G. Schutz, "Food Action Rating Scale for Measuring Food Acceptance," *Journal of Food Science* **30**, 365–374 (1965).

29. J. L. Balintfy, P. Sinha, H. R. Moskowitz, and J. G. Rogozenski, *The Time Dependence of Food Preferences*, Food Product Development, Nov. 1975.

30. G. E. Box, W. G. Hunter, and J. S. Hunter, *Statistics For Experimenters*, John Wiley & Sons, Inc., New York, 1978.

31. R. L. Plackett and J. D. Burman, "The Design of Optimum Multifactorial Experiments," *Biometrika* **33**, 305–325 (1946).

32. J. G. Beebe-Center, *The Psychology of Pleasantness And Unpleasantness*, Van Nostrand Reinhold, New York, 1932.

33. H. R. Moskowitz, "Relative Importance of Perceptual Factors to Consumer Acceptance: Linear Versus Quadratic Analysis," *Journal of Food Science* **46**, 244–248 (1981).

HOWARD R. MOSKOWITZ
Moskowitz Jacobs Inc.
White Plains, New York

SENSORY SCIENCE: STANDARDIZATION AND INSTRUMENTATION

Whenever the sensory properties of a substance are at issue, the sine qua non is that sensory evaluation is supreme. No chemical, physical, or instrumental test can substitute for sensory judgment under all circumstances. Humans and other animals are the only ones who can make decisions about acceptability or other sensory judgments. Notwithstanding that, there have been efforts as long as chemical and sensory measurements have coexisted for technologists and others to attempt to substitute chemical or physical measurements for sensory ones. Some reasons are obvious. Others are more illusory than factual. Among the obvious ones is the fact that chemical and mechanical tests usually can be conducted at any time whereas the "expert" tea taster, fish smeller, or the sensory panel—if the firm is more in the modern mode—are often available only during the daytime shift. Among the reasons more illusory than factual is the conception that chemical tests are more objective and precise; they often are, but that is not necessarily better. As in target shooting, there are differences between precision and accuracy. The archer or marksman may group shots close together, thus being precise, but the shots may be at the rim of the target instead of at the bull's-eye. Although a chemical test may be very precise in itself, it may miss the target simply because it is measuring an attribute not particularly well correlated with sensory quality. The purpose of this article is to describe how correlations may be established, if they exist, and how to use those that do exist most effectively as substitutes for, or complements to, sensory evaluation. The statement made above that no instrumental test can substitute for sensory evaluation under all circumstances is true, but fortunately there are many instances where instrumental measurements, once shown to be correlated with sensory judgment, can be used beneficially as replacements for sensory evaluation provided from time to time the instrumental test is restandardized or validated against sensory evaluation. In spite of some of the limitations just mentioned, the use of instrumental tests to supplement sensory evaluation for quality-control purposes is already a well-established industrial practice. In basic research, such tests are even more valuable. If we ever hope to understand sensory perception, most certainly knowledge of the role played by specific compounds, functional groups, and various forms of interaction among compounds must be ascertained. The interaction referred to here is not the chemical reaction between compounds but the perceptual interaction in the brain where there are often additive, masking, or synergistic effects. While correlations themselves are not adequate, they are a step toward establishing cause-and-effect relations.

EMERGING AND COALESCING TECHNOLOGIES

A happy set of circumstances has made sensory–instrumental correlations both easier to detect and more useful than was true 45 yr ago. Hereafter the term instrumental will be used in a generic sense to encompass all forms of chemical, mechanical, physical, spectral, or auditory tests where there is some type of physical measurement being obtained to match against sensory evaluation. In other instances, the particular form of physical measurement will be named where that is necessary to provide accuracy. That is so for the first development to be described. The detection and isolation of chemical compounds became easier once gas chromatography was established (1). Prior to that time, analytical chemists often had to spend weeks isolating and identifying even a single substance. Inasmuch as almost every food contains hundreds of compounds, if not thousands, and most of these have sensory import, there was little prospect of being able to establish good correlations when only a few compounds in a given kind of food could be identified and quantified. Fortunately, sometime before gas chromatography enabled chemistry to go "multicomponent," food and other sensory substances began to be looked at in the same way.

Prior to 1949, sensory judgments were generally quite simple. Expert tea tasters or wine judges rendered a composite judgment that this tea or that wine met their company's buying specifications, perhaps with a few comments as to particularly pertinent qualities. Generally, they kept the basis for their judgments to themselves. In fact, their status as an expert depended on the imperiousness of their judgments. In some industries, sensory panels came into use and worked in parallel with experts. In other industries, the first sensory analyses were those provided by sensory panels. In either case, the panels also rendered simple judgments. At the most they examined the material for odor, taste, texture, or color or for such specific attributes as sweetness, bitterness, saltiness, and sourness. Even the latter attributes are often not as specific as is desirable, because different substances can have different nuances of bitterness. The same is true of the other character notes.

The development that changed things in 1949 was the development of the flavor profile method (FPM) (2). The method required panelists to attempt to pick out from the melange of sensory notes comprising odor or flavor the particular odor or taste character notes that were there. FPM set the feet of sensory scientists on the right path (3), but it was inadequate in some respects. The responses sought were only semiquantitative. Furthermore, one of the objectives of the method was to get the panelists to arrive at a consensus as to which sensory notes were present, the order of their perception, and their intensities. Semiquantitative data and the merging into a consensus differences expressed by the individual panelists made difficult the establishment of sound correlations between the sensory–instrumental measurements. The deficiency of strict quantitative data to match against the quantitative data of instrumental methods changed in 1974 when quantitative descriptive analysis (QDA) was devised (4). QDA requires that the panelists make their judgments independently, and the rather crude scoring system of the FPM was replaced with more carefully defined scoring systems. Other differences between the FPM and QDA procedures as well as the quantitative sensory profiling (QSP) methods that have been developed more recently have been described (3). The FPM and QDA procedures were singular developments for the influence they had on sensory analysis.

EXAMINATION OF MULTICOMPONENT DATA SETS

In 1966 there was a seminal development (5,6), which was the use of statistical multivariate analysis (MVA) to seek out from all the data from gas chromatography those peaks most effective for classifying samples into the sensory category to which they belonged. Earlier efforts had been made

to find the peaks most closely correlated with sensory quality, but they were often futile or only partially successful because simple, univariate statistical methods, such as ξ^2, were employed rather than MVA procedures. The new method used discriminant analysis (6). Almost immediately other researchers began to follow this lead; sometimes they used discriminant analysis, but quite logically they began to examine other forms of MVA for their applicability to sensory–instrumental analysis. The result is that today several multivariate methods play a role in examining and correlating sensory and instrumental data.

The flux that enabled multicomponent sensory data to be effectively melded with multicomponent instrumental data was the development of computers. Some of MVA procedures had existed since early in the twentieth century, but they were not widely used because calculation was too onerous and tedious a task when done by hand or by the hand-cranked calculating machines then extant. In the 1960s and the 1970s the critical components to effective correlation of sensory and instrumental data fused. The development of gas chromatography in 1952 (1), the application of discriminant analysis to multicomponent data in 1966 (5,6), and the refining of multicomponent sensory evaluation in 1974 (4) were all encompassed by advances in computer technology during the same period, and subsequent to that time, and set the stage for these individual achievements to coalesce into a major thoroughfare of application. Their coalescing had an influence much like the confluence of the Ohio, the Missouri, and other rivers with the Mississippi, enabling it to bear a heavier traffic load than it could if other rivers did not add to its majesty.

KINDS OF MULTIVARIATE METHODS

There are at least 14 kinds of MVA procedures used to examine sensory–instrumental data. Some are of chief benefit for exploratory examination of data to gleam from large data sets certain components or relations for more detailed examination or later use. Others are employed to learn whether sample differences exist. Among the procedures are

Principal component analysis
Cluster analysis
Multidimensional scaling
Discriminant analysis
Multiple regression analysis
Partial least squares analysis
Response surface analysis
Procrustes analysis
Canonical analysis
Correspondence analysis
Multivariate analysis of variance
Fuzzy logic (mathematics)
Neural network
Factor analysis

Of the above procedures, principal component analysis (PCA), cluster analysis (CA), and factor analysis (FA) are known as methods of internal analysis. They are used to study the relations of variables within the same data set, ie, a set of sensory measurements or a set of instrumental measurements, but not sensory compared with instrumental measurements. Discriminant analysis (DA) and multiple regression (MRA) perform just the opposite functions. DA is a procedure for assigning products, or some other entity, to a class based on measurements made on those products, with prior knowledge that classes do exist. For example, yellow and red globe onions and Vidalia (Grano–Granex) onions were analyzed gas chromatographically to learn if they could be so classified (7). That species differences existed was, of course, known in advance. MRA works somewhat analogously to DA except it applies to things that progress in some manner rather than things in discrete classes, as species of onions are. An illustration would be predicting the intensity of vanilla flavor in ice cream when different levels of vanilla are added. DA and MRA are the most useful of the MVA methods for predicting the identity or state of a sample. They are useful because once a discriminant function or a multiple regression equation has been calculated, the identity or state of a new sample can be predicted by merely inserting its measurement values in the equation. The other methods listed above are more commonly used to study relations between sensory and instrumental measurements. Some, such as canonical analysis or partial least squares regression can be used for prediction purposes, but generally they are used to examine relations among variables rather than to pinpoint the identity of a single sample.

PRELIMINARY OR EXPLORATORY USES

While not always confined to being exploratory methods, PCA, CA, and FA are often used for that purpose. Of the three, PCA is probably the one most used to examine data to learn how the data should be subsequently examined to secure the greatest amount of value from them. Basically, PCA is a procedure for examining the variance of data to learn which variables go together and which others belong to a different group. The procedure thus performs two functions. It establishes whether there is correlation among some of the variables (sensory scores and instrumental measurements); then it attempts to reduce the dimensionality of the data, ie, groups together in a component those things that are correlated with each other.

Variance, or dispersion, should perhaps be defined. It is a measure of the degree to which measurement values should in theory be the same rather than differ slightly, as they usually do. The variation represents error, not in the sense that the measurement is wrong but in the sense that measurements made on the same sample are usually not identical. Among causes of error, generally minor, are periodic inconsistencies in the performance of the instrument. Slightly more common are variations in results caused by the operator of the instrument not being entirely consistent in his or her handling of the instrument or samples. The most prevalent cause of variance is the inherent differences that characterize all biological materials—and a very high proportion of specimens examined for sensory

qualities are biological in nature: nearly all our foods and beverages are. Other materials such a perfumes may be compounded from differences substances, but many of the most desirable ingredients are extracted from natural substances. Such things as air or water pollutants, although not necessarily biologically derived, likewise vary according to the way the wind blows and how an industrial plant is being operated. Error should not, therefore, be looked on as "wrongness" but rather "rightness" in the sense that all things exhibit some variation in their properties and performance, and the error term is merely reflecting that.

CALCULATION AND FUNCTIONING OF PCA

PCA operates by searching for the variables accounting for the greatest amount of variance. The items under investigation (samples, people, and descriptive terms), correlated and contributing the greatest amount of variance in ratio to the total, are grouped into the first principal component. The second component is derived in the same way. It is a measure of the variance remaining after the first component has been extracted and accounts for the next greatest amount of variance. The process continues in that fashion until all the variance has been accounted for. In practice, the investigator is rarely interested in all the components. Usually, the first few account for so much of the existing total variance that the other components can be ignored because they are not supplying enough additional information to justify being retained for further use or evaluation. The consequence of PCA is that the volume and complexity of the data set is reduced. The investigator may give up 5–15% of the information contained in the whole mass of data. In turn, there will be fewer entities to work with in the future, maybe three components, instead of the original number of samples (people, descriptive terms, or whatever was being subjected to PCA). Not only may data reduction be involved but PCA is useful to learn whether products thought to be different are in fact different. A common problem is to judge whether products sold under different brand names are indeed different sensorially. The same producer may, for example, market the same product under two or more different labels.

Not quite in line with the illustration above is another type of problem (8). Nine brands of Scotch whisky and one of Irish whiskey were analyzed. Upon applying PCA it was learned that some of the brands were so close together in their sensory qualities they should be treated as one set of samples, not two or three different samples. A cooperative research project was conducted between the food chemistry and technology department of the University of Helsinki and the University of Georgia (9). The Finns had selected five kinds of Finnish sour rye bread, which they considered to be representative of different types, for sensorial and chemical examination. The chemical data permitted the five breads to be separated except for breads no. 1 and 4. The sensory data did the same except breads no. 1 and 2 could not be separated. Putting together the two kinds of data permitted sample resolution of the five kinds of bread, but a pictorial representation of just how close together the five kinds were was desired. Figure 1 illus-

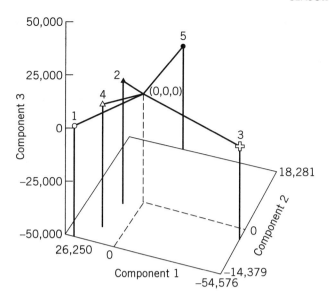

Figure 1. Spatial relations of five types of Finnish wholemeal sour rye bread in three-dimensional space based on principal component analysis. Data set reduced from responses by 20 panelists for 12 attributes, replicated four times. Numerals designate the five kinds of bread.

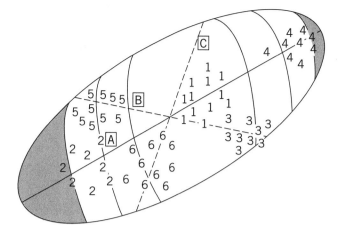

Figure 2. Depiction of basis for calculation of principal components. Regression curve through measurement types 2, 1, and 4 establishes first component, regression curve through measurement types 5 and 3 establishes second component, curve through measurement type 6 illustrates formation of a third component, the regression curve for each being orthogonal to the others.

trates the positions of the five types of bread, one to the other, in three-dimensional space. Other ways of providing evidence that the samples differed will be described later. PCA was used here as a preliminary check on the validity of the assumption that the five kinds of bread were different.

Perhaps the theory underlying PCA should have been given first, but it is hoped that the reader will find the explanation of what PCA is about more understandable having first seen some of the results of its application. In the case of the Finnish sour rye bread, the data mass was reduced from judgments by 20 panelists for 12 sensory characteristics and several chemical measurements to five components, ie, the five coordinates of the five kinds of bread. Figure 2 depicts what is going on. The kinds of measurements represented by the numbers 1, 2, and 4 are correlated with each other because they lie along the same, and the longest, axis. They account for the greatest amount of variance; thus they represent the first component. Measurement values 3 and 5 are along the next longest axis, and they account for the next greatest amount of variance, once the variance of the first component has been removed. They thus constitute the second component. Measurement value 6 is off by itself. It, therefore, constitutes the third component. Although the three axes are not at right angles to each other, one of the goals of PCA is to derive components that are orthogonal, ie, at right angles, to each other. The drawing was purposely made not to be ideal in terms of orthogonality, because in practice the axis of each component is not always exactly at a right angle to the other axes. The drawing depicts the way PCA was originally calculated. The problem of calculating principal components was solved by fitting a line to the longest axis of the ellipsoid of objects such that the sum of the squared residual

distances was at a minimum for highly correlated objects (10). The same was then done for the next longest axis and those next most highly correlated. The reason the components are orthogonal to each other is that once the first component has been extracted, subsequent ones extracted are not correlated, ie, are orthogonal, because the variance of the first is no longer in play to allow correlation to exist. To carry on PCA calculations today, Hotelling's (11) procedure is generally used. It depends on transposing and inverting the variance–covariance matrix, or the correlation matrix, to accomplish the same purpose. The procedure selects the variables accounting for the greatest proportion of the variance, it arrays them and then rotates them in such a way that they are in the order of decreasing variance and are orthogonal.

Figure 3 provides an illustration of the use of PCA to reduce data dimensionality. This study was mentioned above (7). Forty-five gc peaks were useful to resolve product differences, but only 7 peaks were needed to be 100% correct in classifying 47 samples of the three different kinds of onions. For many, a pictorial display is more convincing than numerical data. To take advantage of the information contained in the 7 peaks, yet still be able to show that the onions did indeed differ, PCA was applied to the 7 peaks to reduce them to three components. The graph shows quite clearly that the three kinds of onions occupy three distinct locations in space.

A more useful purpose of PCA than merely to reduce a number of variables to a lesser number of components to permit visual display is the use of it to classify panelists undergoing training for possible membership on a sensory panel. Reference was made above to biological differences often being the cause of errors in measurement. Humans are no exception. Every one is an individual, responding in different ways when called on to be a sensory panelist. That is true even after training. Some of the variation can be overcome by training panelists to respond nearly alike

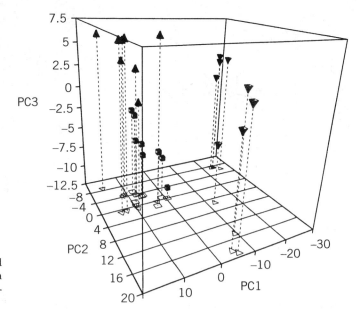

Figure 3. Principal component analysis applied to yellow and red globe onions and to Vidalia onions based on reducing seven gas chromatographic peaks to three components, four replications each.

to the same stimulus, but it can never be fully overcome. In deciding whether to admit an individual to membership on a sensory panel, it must be considered whether the individual will be in reasonable concert with other members of the panel, or if the individual will respond so individually as to be a statistical outlier. Figure 4 shows PCA applied to the results of 11 panelists when they examined a baked bean product for the intensity of 14 attributes (12). Panelist 9 is clearly an outlier. Panelists 1, 2, and 3 do not

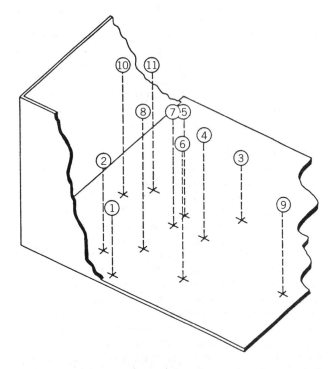

Figure 4. Interspatial relations of 11 panelists based on principal component analysis, eight baked bean products evaluated for 14 attributes, four replications.

agree with the others as closely as is desirable. If panelist 9 is to be retained as a panelist, he will have to receive additional training in an effort to bring his responses into line with the rest of the panel. The sensory leader might decide that panelists 1, 2 and 3 might likewise improve in performance with additional guidance and training. One of the advantages sensory technologists have today is that PCA can be readily applied to the responses of panelists, and if there is a program to graph the results, a visual display can be created so that panelists can see how far they are from the rest. Knowing that, often panelists can readjust their mental scaling processes to bring their responses into line with the rest of the panel. For any of the illustrations above, the basic information is supplied as numerical output. The investigator generally just examines it to make decisions. Naturally the numerical output is the only way of examining the output beyond the third component. Graphic illustrations were given here merely to make clear the kinds of information obtainable from PCA.

In reducing the dimensionality of the data, some information is given up, but usually not much. In return for the loss of information, relations among the entities being evaluated, whatever they are, are made clearer. The entities can be products as in the first illustration, gas chromatographic peaks in the third, the responses of people in the fourth, or any pertinent category such as descriptive terms, geographical location, or demographic information.

CLUSTER ANALYSIS

Cluster analysis (CA) is another data-reduction method. It, too, can be used to examine the responses of panelists to learn whether panel members are responding alike. CA can take several different forms. Whereas PCA depends on examination of variance, CA is basically a measure of proximity. It can be calculated on the basis of distance, correlation, or the effect they have on variance by lumping together or separating other individual elements or tentative clusters. There are several forms of depicting proximity. There are single linkage, centroid linkage, average linkage, density linkage, and minimum variance. One of the most useful is the agglomerative hierarchical clustering technique, of which Ward's minimum variance method is one. Ward's method uses the square of the Euclidean distance as the proximity measure. Some publications refer to the representation as a tree; others call it an icicle. Whether distance, correlation, or variance, the basic purpose of CA is the same. Things having characteristics in common form a cluster; things dissimilar to the first have to go into a second cluster, etc. Having characteristics in common does not necessarily mean that the objects in a cluster are related. Sensory leaders hope that their panelists will form a single cluster; but just being grouped together does not mean the panelists are related. They merely happen to be together because they respond alike; only rarely are panel members kin. Table 1 shows cluster analysis applied to the 20 panelists who examined the five kinds of Finnish sour rye bread for 12 attributes (13). Tables 2 and 3 show the kinds of output obtained by different clustering processes. There are some differences because hierarchial clustering was used for the icicle formation (Table 2) and the VARCLUS routine of the SAS package (14) was used for Table 3. Either one of the clustering processes serves the purpose of the sensory analyst, who is generally interested in only the major groups formed. Here there are a few differences in the assignment of panelists, but they reveal essentially the same information. Both processes said that panelists 7 and 12 responded least like any of the others. From other information (15) every one of the 20 panelists was an accomplished sensory judge who could reproduce his or her assignment of scores. The only flaw was that the panel did not operate as one unit. The panelists of each of the clusters arrived at the same decision, three of the kinds of bread were sensorially different, but they didn't reach the same conclusion by the same pathway.

Sometimes CA is used to learn if resolution of product differences can be stepped up. By applying CA to a group

Table 1. Ward's Minimum Variance Cluster Analysis Applied to the Scale Values of 20 Panelists Evaluating 12 Attributes of Finnish Wholemeal Sour Rye Bread

i—Partial R^2	Panelists
	1 13 3 12 7 5 19 8 15 9 2 6 17 20 4 18 10 11 14 16
0.35	XXXXXXXXXXXXXXXXXXXXXXXXXXXXXXXXXXXXXXXXXXXXXXXXXXXXXXXXXXXXXXXXXXXXXXXXXXX
	XXXXXXXXXXXXXXXXXXXXXXXXXXXXXXXXXXXXXXXXX XXXXXXXXXXXXXXXXXXXXXXXXXXXXXXXXX
	XXXXXXXXXXXXXXXXXXXXXXXXXXXXXXXXXXXXXXXXX XXXXXXXXXXXXXXXXXXXXXXXXXXXXXXXXX
	XXXXXXXXXXXXXXXXXXXXXXXXXXXXXXXXXXXXXXXXX XXXXXXXXXXXXXXXXXXXXXXXXXXXXXXXXX
0.3	XXXXXXXXXXXXXXXXXXXXXXXXXXXXXXXXXXXXXXXXX XXXXXXXXXXXXXXXXXXXXXXXXXXXXXXXXX
	XXXXXXXXXXXXXXXXXXXXXXXXXXXXXXXXXXXXXXXXX XXXXXXXXXXXXXXXXXXXXXXXXXXXXXXXXX
	XXXXXXXXXXXXXXXXXXXXXXXXXXXXXXXXXXXXXXXXX XXXXXXXXXXXXXXXXXXXXXXXXXXXXXXXXX
	XXXXXXXXXXXXXXXXXXXXXXXXXXXXXXXXXXXXXXXXX XXXXXXXXXXXXXXXXXXXXXXXXXXXXXXXXX
	XXXXXXXXXXXXXXXXXXXXXXXXXXXXXXXXXXXXXXXXX XXXXXXXXXXX XXXXXXXXXXXXXXXXXXXX
	XXXXXXXXXXXXXXXXXXXXXXXXXXXXXXXXXXXXXXXXX XXXXXXXXXXX XXXXXXXXXXXXXXXXXXXX
0.25	XXXXXXXXXXXXXXXXXXXXXXXXXXXXXXXXXXXXXXXXX XXXXXXXXXXX XXXXXXXXXXXXXXXXXXXX
	XXXXXXXXXXXXXXXXXXXXXXXXXXXXXXXXXXXXXXXXX XXXXXXXXXXX XXXXXXXXXXXXXXXXXXXX
	XXXXXXXXXXXXXXXXXXXXXXXXXXXXXXXXXXXXXXXXX XXXXXXXXXXX XXXXXXXXXXXXXXXXXXXX
	XXXXXXXXXXXXXXXXXXXXXXXXXXXXXXXXXXXXXXXXX XXXXXXXXXXX XXXXXXXXXXXXXXXXXXXX
	XXXXXXXXXXXXXXXXXXXXXXXXXXXXXXXXXXXXXXXXX XXXXXXXXXXX XXXXXXXXXXXXXXXXXXXX
0.2	XXXXXXXXXXXXXXXXXXXXXXXXXXXXXXXXXXXXXXXXX XXXXXXXXXXX XXXXXXXXXXXXXXXXXXXX
	XXXXXXXXXXXXXXXXXXXXXXXXXXXXXXXXXXXXXXXXX XXXXXXXXXXX XXXXXXXXXXXXXXXXXXXX
	XXXXXXXXXXXXXXXXXXXXXXXXXXXXXXXXXXXXXXXXX XXXXXXXXXXX XXXXXXXXXXXXXXXXXXXX
	XXXXXXXXXXXXXXXXXXXXXXXXXXXXXXXXXXXXXXXXX XXXXXXXXXXX XXXXXXXXXXXXXXXXXXXX
0.15	XXXXXXXXXXXXXXXXXXXXXXXXXXXXXXXXXXXXXXXXX XXXXXXXXXXX XXXXXXXXXXXXXXXXXXXX
	XXXXXXXXXXXXXXXXXXXXXXXXXXXXXXXXXXXXXXXXX XXXXXXXXXXX XXXXXXXXXXXXXXXXXXXX
	XXXXXXXXXXXXXXXXXXXXXXXXXXXXXXXXXXXXXXXXX XXXXXXXXXXX XXXXXXXXXXXXXXXXXXXX
	XXXXXXXXXXXXXXXXXXXXXXXXXXXXXXXXXXXXXXXXX XXXXXXXXXXX XXXXXXXXXXXXXXXXXXXX
	XXXXXXXXXXXXXXXXXXXXXXXXXXXXXXXXXXXXXXXXX XXXXXXXXXXX XXXXXXXXXXXXXXXXXXXX
0.1	XXXXXXXXXXXXXXXXXXXXXXXXXXXXXXXXXXXXXXXXX XXXXXXXXXXX XXXXXXXXXXXXXXXXX
	XXXXXXXXXXXXXXXXXXXXXXXXX XXXXXXXXX XXXXXXXXXXX XXXXXXXXXXXXXXXXX
	XXXXXXXXXXXXXXXXXXXXXXXXX XXXXXXXXX XXXXXXXXXXX XXXXXXXXXXXXXXXXX
	XXXXXXXXXXXXXXXXXXXXXXXXX XXXXXXXXX XXXXXXXXXXX XXXXXXXXXXXXXXXXX
	XXXXXXXXXXXXXXXXXXXXXXXXX XXXXXXXXX XXXXXXXXXXX XXXXXXXXXXXXXXXXX
0.05	XXXXXXXXXXXXXXXXXXXXXXXXX XXXXXXXXX XXXXXXX XXXXXXXXXXXXXXXXX
	XXXXXXXXXXXXXXXXXXXXXXXXX XXXXXXXXX XXXXXXX XXXX XXXXXXXXX
	XXXXX XXXXXXXX XXXXXXXXX XXX XXXX XXXXX
	XXXXX XXXXXXXX XXXXXX XXXX XXXXX
	XXX

Table 2. Tabular Output of Ward's Minimum Variance Cluster Analysis[a]

Variables	Eigenvalues of the covariance matrix			
	Eigenvalue	Difference	Proportion	Cumulative
1	796.987	313.406	0.429565	0.42956
2	485.580	254.313	0.260643	0.69021
3	229.267	132.981	0.123572	0.81378
4	96.286	22.873	0.051897	0.86568
5	73.413	25.975	0.039569	0.90525
6	47.438	9.109	0.025568	0.93081
7	38.329	7.307	0.020659	0.95147
8	31.023	6.098	0.016721	0.98163
9	24.925	8.704	0.013434	0.98163
10	16.221	5.445	0.008743	0.99037
11	10.776	3.687	0.005808	0.99618
12	7.089		0.003821	1.00000

Number of clusters	Clusters joined		Frequency of new cluster	Semipartial R^2	R^2
19	3	12	2	0.010001	0.989999
18	11	14	2	0.011672	0.978327
17	7	CL19	3	0.014082	0.964245
16	4	18	2	0.014111	0.950134
15	8	15	2	0.014587	0.935547
14	1	13	2	0.016947	0.918600
13	2	6	2	0.019158	0.899442
12	CL15	9	3	0.024987	0.874454
11	10	CL18	3	0.025403	0.849052
10	CL17	5	4	0.027946	0.821106
9	CL14	CL10	6	0.029373	0.792732
8	CL13	17	3	0.030646	0.761086
7	CL16	CL11	5	0.034871	0.726215
6	CL8	20	4	0.041325	0.685890
5	CL9	19	7	0.054941	0.684890
4	CL7	16	6	0.060708	0.569241
3	CL5	CL12	10	0.062688	0.506553
2	CL6	CL4	10	0.189603	0.316950
1	CL3	CL2	20	0.316950	0.000000

[a]Root-mean-square total-sample standard deviation = 12.4343; Root-mean-square distance between observations = 60.9153.

of panelists, it was possible to demonstrate more conclusively that three cheese products were different (16). Before the panel was split into three clusters, classification of 180 samples by discriminant analysis into the product category to which they belonged was 76% correct. When DA was applied to each of the clusters separately, the success level was raised to 90%. Either PCA or CA can be used for the purpose of determining whether samples differ. Table 4 shows Ward's minimum variance cluster analysis applied to five grades of tea to learn how much alike or different they are. Tea no. 5 was a "fine" tea selected by an expert tea buyer. Tea no. 1 was the poorest as judged by the tea buyer. Note that the CA tells the investigator that the panelists can most readily tell apart teas no. 1 and 5. The fact that teas nos. 2, 3 and 4 did not break off from each other as readily indicates that the panelists had more difficulty in distinguishing among them.

MULTIDIMENSIONAL SCALING

Multidimensional scaling (MDS) will be explained at this point chiefly because it, too, is a form of proximity mea-

surement. It is a method for finding a configuration of points for individuals from information about the distance between the points. The usual metric is similarity or dissimilarity judgments, but correlation matrices, attribute ratings, or Euclidean distances can be used. The $r \times c$ matrix is generally not one of $n \times p$ but rather $n \times n$. It has been stated that the strategy in MDS is to look for the solution with the smallest dimensionality in which the differences between the reconstructed distances and the original proximities are acceptably small (17). Smaller stress values indicate more satisfactory fit. Various programs for the analysis of proximity data have been described (18). It has been pointed out that MDS and correspondence analysis have some features in common with PCA (19). For MDS, if $n > p$, then a PCA should generally be preferred because it is easier to find the eigenvectors of a $p \times p$ matrix than the larger $n \times n$ matrix. The theory and methods of MDS have been published (20).

DISCRIMINANT ANALYSIS

FA and correspondence analysis will be discussed later. Although PCA and CA can be used to depict product differ-

Table 3. Varclus (Oblique Principal Component) Clustering of 20 Panelists Based on Their Correlation Matrices

| | Variable (panelists) | R^2 with | | |
		Own cluster	Next closest	$1\text{-}R^2$ ratio
Cluster 1	2	0.6955	0.4487	0.5524
	4	0.6727	0.5217	0.6842
	10	0.4667	0.2598	0.7205
	11	0.5618	0.3385	0.6625
	14	0.5137	0.2311	0.6325
	16	0.6652	0.3906	0.5495
	17	0.6051	0.3610	0.6180
	18	0.6990	0.3981	0.5000
	20	0.7621	0.3548	0.3687
Cluster 2	1	0.4873	0.2060	0.6458
	3	0.4340	0.2197	0.7253
	5	0.5326	0.3257	0.6931
	6	0.6405	0.5085	0.7314
	8	0.4476	0.2818	0.7691
	9	0.5211	0.2681	0.6544
	13	0.5267	0.3455	0.7231
	15	0.5462	0.3316	0.6788
	19	0.5076	0.2401	0.6480
Cluster 3	7	0.6645	0.1541	0.3967
	12	0.6645	0.1898	0.4142

| | | Cluster summary for 3 clusters | | | |
Cluster	Members	Cluster variation	Variation explained	Proportion explained	Second eigenvalue
1	9	9.00000	5.64174	0.6269	0.71773
2	9	9.00000	4.64371	0.5160	0.91497
3	2	2.00000	1.32892	0.6645	0.67108

Total variation explain = 11.6144   Proportion = 0.580718

Source: Refs. 13 and 14.

ences, DA and MRA serve that purpose more admirably. DA is a procedure that depends for its effectiveness on transposing the measurement values in such a way that the linear distance between samples is maximized. There are four kinds of DA: stepwise (SDA), linear or multiple (LDA), canonical, and nearest-neighbor. Normally SDA is used to winnow from the measurements those that are not effective as discriminators, then ferret from among those retained the ones that are most effective. In actuality, computer programs search for the most effective discriminator among the measurements, eg, gas chromatographic peaks, liquid chromatographic peaks, or mechanical measurements. Having selected the best single discriminator, the program then looks for a second measurement to go with the first to form a pair of measurements for discrimination. The process goes on in that fashion until enough measurements have been selected to permit all the samples to be assigned to the proper class, or else the process fails because it runs out of discriminating variables before full resolution of product differences is obtained. Table 5 shows how the process works schematically and Table 6 lists the kind of output obtained. The data are for the classification of a certain industrial produced in four branch plants. The objective was to learn whether the products could be assigned by chemical measurement to the plant where each was actually produced. Sometimes sample differences cannot be resolved by instrumental measurements. Generally where there are many measurements available such as gas

chromatographic or liquid chromatographic analysis yields, there is no difficulty in finding sufficient discriminating variables to permit the samples to be correctly classified. When sensory scores for various attributes (such as crumbliness, graininess, sweetness, a sulfidic odor, and a grassy note) are the kinds of quantitative values available, there is generally less success in attaining 100% classification. Sometimes 100% success can be achieved, but usually the maximum is approximately 85%.

LINEAR DISCRIMINANT ANALYSIS

SDA is really a screening operation. The investigator wants to find out which measurements are useful to resolve product differences. Then either the discriminant or the classification coefficients are utilized to form a discriminant function. The form of the equation is

$$Z = \lambda_1 X_1 + \lambda_2 X_2 + \lambda_3 X_3 + \lambda_4 X_4 - C$$

where the λ's are weighting values, the X's are the measurement values, C is a correction factor, and Z is the weighted mean. The magnitude of the weighted mean determines the class to which the sample is assigned. The correction term is provided (it is a part of the computer output) to enable original (raw) measurement values to be used. As a part of the DA calculation, the measurement

Table 4. Ward's Minimum Variance Cluster Analysis Applied to Five Grades of Tea, Grade Five Being a High-Quality Tea, Grade One Being the Tea Lowest in Quality

```
                           Product
                 1      2      4      3      5
Semipartial R²  0.7 +
                    |
                    |
                    |
                    |
                    |
                    | XXXXXXXXXXXXXXXXXXXXXXXXXXX
                0.6 + XXXXXXXXXXXXXXXXXXXX        X
                    | XXXXXXXXXXXXXXXXXXXX        X
                    | XXXXXXXXXXXXXXXXXXXX        X
                    | XXXXXXXXXXXXXXXXXXXX        X
                    | XXXXXXXXXXXXXXXXXXXX        X
                    | XXXXXXXXXXXXXXXXXXXX        X
                    | XXXXXXXXXXXXXXXXXXXX        X
                0.5 + XXXXXXXXXXXXXXXXXXXX        X
                    | XXXXXXXXXXXXXXXXXXXX        X
                    | XXXXXXXXXXXXXXXXXXXX        X
                    | XXXXXXXXXXXXXXXXXXXX        X
                    | XXXXXXXXXXXXXXXXXXXX        X
                    | XXXXXXXXXXXXXXXXXXXX        X
                0.4 + XXXXXXXXXXXXXXXXXXXX        X
                    | XXXXXXXXXXXXXXXXXXXX        X
                    | XXXXXXXXXXXXXXXXXXXX        X
                    | XXXXXXXXXXXXXXXXXXXX        X
                    | XXXXXXXXXXXXXXXXXXXX        X
                    | XXXXXXXXXXXXXXXXXXXX        X
                    | XXXXXXXXXXXXXXXXXXXX        X
                0.3 + XXXXXXXXXXXXXXXXXXXX        X
                    | X         XXXXXXXXXXXXXXX   X
                    | X         XXXXXXXXXXXXXXX   X
                    | X         XXXXXXXXXXXXXXX   X
                    | X         XXXXXXXXXXXXXXX   X
                    | X         XXXXXXXXXXXXXXX   X
                    | X         XXXXXXXXXXXXXXX   X
                0.2 + X         XXXXXXXXXXXXXXX   X
                    | X         XXXXXXXXXXXXXXX   X
                    | X         XXXXXXXXXXXXXXX   X
                    | X         XXXXXXXXXXXXXXX   X
                    | X         XXXXXXXXXXXXXXX   X
                    | X         XXXXXXXXXXXXXXX   X
                0.1 + X         XXXXXXXXXXXXXXX   X
                    | X         XXXXXXXXXXXXXXX   X
                    | X         XXXXXXXXXX        X
                    | X         XXXXXXXXXX        X
                    | X         XXXXXXXXXX        X
                    | X         XXXXXXXXXX        X
                    | X                          X
                  0 + X                          X
```

Table 5. Illustration of Selecting and Combining Variables to Improve Percent Success in Classifying Products[a]

Step	Gc peak(s) selected	Percent correct jackknifed classification
1	4	63.1
2	4 + 36	68.4
3	4 + 36 + 26	75.9
4	4 + 36 + 26 + 25	88.6
5	4 + 36 + 26 + 25 + 2	95.8
6	4 + 36 + 26 + 25 + 2 + 33	95.8
7	4 + 36 + 26 + 25 + 2 + 33 + 17	100.0

[a]Three onion varieties: yellow and red globe onions and Vidalis (Grano-Granex) onions.

Table 6. Order of Selection and Deletion of Variables to Classify Product Produced in Four Branch Plants[a]

Step	Gc peak Enter	Gc peak Delete	F-value	U-statistic	Multivariate F-value
1	20		361.90	0.0153	361.90
2	11		134.93	0.0006	219.15
3	10		16.70	0.0001	126.72
4	6		32.02	0.0000	101.73
5	42		24.95	0.0000	87.97
6	3		32.30	0.0000	76.93
7	41		7.35	0.0000	69.08
8	12		8.59	0.0000	62.55
9	18		12.97	0.0000	57.66
10		3			52.54
11	2		3.51	0.0000	49.79

[a]F-value = univariate F-value for that peak; U-statistic is a measure of the error, ie, the degree to which the classification is approaching 100% correctness, values rounded off to fourth decimal; multivariate F-value = F-value for values as combined.

values are standardized. The correction constant adjusts for that fact so that the raw, ie, nonstandardized, values can be inserted in the equation.

SDA was used to learn whether instrumental measurements could be used as a purchasing specification for an herb bought from five different countries (21). It was demonstrated that the chemical means was quite effective at distinguishing the same kind of an herb grown in five different geographical countries of the world. At intervals, the chemical means need to be revalidated by matching the instrumental output with sensory judgments in case growers have made some change in their growing practices or in the way the product is stored or shipped following harvesting, or for other reasons. However, if the equation is established with sufficient replication of measurements and samples to start with, it is usually quite robust.

The opposite approach has also been used (22). Chemical analysis has been used in an attempt to standardize more closely batches of distillers' spirits. It was decided that sensory evaluation coupled with LDA was a more suitable way to maintain a high level of quality control. The procedure is a more direct means of ascertaining sensory quality than going through chemical analysis, which in turn must be correlated with sensory quality. It is less suitable for high-speed lines or for plants that run 16 or 24 h/day. A disadvantage to using sensory panelists is that if a panelist leaves the panel, the SDA process and development of a LDA equation should be recalculated to be sure the replacement panelist has not caused a change in the weighting values. Recalculation, however, is not a time-consuming operation. In fact, if the original panelists were selected carefully to be sure they were in good agreement in their responses and if records of performance have been kept (as they should be), replacing a panelist is no more a touchy subject than it is for replacement of any panelist. A comparison of the new panelist's performance with the prior history of the panel should tell the sensory analyst whether the panel mean is likely to be influenced.

MULTIPLE REGRESSION ANALYSIS

MRA is used similarly to LDA. The chief difference is that MRA is appropriate for applications where the dependent variable is along a continuum and the measurement values are continua, whereas LDA is suitable for assigning a sample to a disjoint set of categories, ie, classes. In the typical regression application, flavor intensity or something that progressively changes is predicted from several measurement values that themselves can be any value along continua. The item being predicted is called the dependent value; the predictors are the independent values. A multiple regression equation takes the form:

$$Y = a + b_1X_1 + b_2X_2 + b_3X_3 + b_4X_4 + e$$

where Y is the value of the dependent variable; a is the intercept of the regression line; the b's are the slopes, ie, the weighting values attached to each predicting variable; e is residual error; and the X's are the measurement values. Stepwise regression analysis (SRA) can be applied just as can SDA. The difference is that an estimate is given for the intercept and for the b values (the weighting value) for regression of each of the independent values as they are added to the total number of independent regressors included in the equation.

Premises to Use of SDA or MRA

The MVA procedures above are risky ventures unless relations among the variables are based on a firm foundation of replication of measurement and the initial set of samples is truly representative of those likely to be encountered in the future. If there is not adequate sampling and replication of measurements, there is the risk that relations that appear to exist may be spurious. The usual statistical risk is set at a probability level of 0.05. When there are 7 or 10 or 20 measurement values, each one of which will fluctuate some from determination to determination, there is no practical way to calculate the total risk involved in making a wrong decision. The only ways to overcome the uncertainty are to carry on determinations long enough in time, to evaluate a large number of samples, and to choose the sources of the samples carefully to be fairly sure of having selected a fair representation of nature's variability. Although on first use an equation may appear to be a good predictor of the class to which a product belongs or the degree of some quality it has, it should be looked at with suspicion. Some industrial firms that have replaced routine sensory evaluation with instrumental analysis conducted sensory and instrumental analyses

side by side for five years to be sure that the first indications of effectiveness were not just a case of good luck at the first try. That kind of effort is not wasted. As data accumulates, the predicting equation can be refined to make it even more applicable and robust than it probably was originally. It has been pointed out that the process of validating an equation by accumulating more and more data is much like the building up of a delta (23). Eventually through accretion, the shifting sands of random happenings become firm earth on which one can build with confidence. The caveat is to be cautious at first, no matter how good an equation seems to be, until it has proved itself on repeated occasions.

There are some features of MVA that the user should be aware of. Novices to the use of SDA and SRA are sometimes disconcerted because the process of selecting good discriminators appears to be overlooking some that are good. The step process bypasses a measurement that by itself is moderately effective as a discriminator in favor of one less of a discriminator by itself, because the computer program is looking for a discriminator not correlated with one already selected. If discriminating measurements are correlated, use of both of them adds little to the discrimination. All that is being provided is redundant information. Table 7 provides an illustration of bypassing measurements that have a high F-value, ie, are effective discriminators, in favor of measurements that are not correlated to one already selected. A noncorrelated variable is adding more to the discrimination needed than would one merely providing redundant information. Note from Table 7 that often a term not appearing to be particularly discriminating by itself is selected ahead of terms appearing to be better discriminators because the one selected in conjunction with the terms already selected adds more to discrimination than its betters would have. Sometimes a

Table 7. Illustration of Order Variables Selected to Avoid Covariation[a]

Gc peak	F-value	Order selected	Gc peak	F-value	Order selected
12	175.8	1	16	25.4	
39	139.2		6	23.2	4
5	96.8		32	22.3	
38	94.9	2	21	19.5	
46	70.7		4	18.6	
9	69.2		26	17.4	
48	67.4		34	14.4	
22	60.9	5	43	12.4	10
36	60.3	9	37	11.8	
28	56.2		17	11.4	
30	49.7	8	45	9.7	
2	37.2		8	9.0	
33	36.7		11	7.7	
25	34.3		7	6.8	
29	30.9	6	44	6.7	
14	30.3		7	5.2	3
10	30.0		3	4.2	7
31	26.9				

[a]F-value specified to enter or delete variable, <4.0. F-value is univariable F-value.

good term, once selected, may be later eliminated (Table 6) because in combination with the set of terms then at play, the term doing the replacing steps up the discriminating power more than the term selected earlier did.

Table 8 shows for SRA the same sort of thing as described above. At the fourth step, the computer had to go over possible combinations again and again to find the combination that yielded the highest coefficient of determination. In this instance the gain in effectiveness of prediction is so slight that most investigators would decide to go with a three- or four-term equation rather than one having a larger number of terms.

In some cases, investigators may not select the most efficient LDA or MRA equation because while there is one best equation there usually is a whole tier of second-best equations. A user of LDA or MRA may decide that one of the tests among the combination yielding the most efficient equation is too lengthy or too expensive to run, that a slight sacrifice in efficiency is merited in exchange for a more speedy or less costly test. In one example, an attempt was made to determine which compounds in oil of onion were the essential ones to an oil of onion flavor so that a synthetic product could be formulated (24). The most effective equation generated did not include compounds that the literature and the experience of the investigators indicated were compounds likely to be essential to onion flavor. The ones not selected were colinearly related (see the discussion above concerning correlated versus orthogonal terms) to the ones selected; thus both would not be included in the same equation. The researchers, therefore, forced into the equation the gas chromatographic peaks they thought were critical to authenticity of onion flavor. This goes back to the statement made in the opening paragraph that a chemical test may be precise but miss the target because it is not measuring a key sensory quality. Computer programs are blind. They will select the most efficient equation. Sensory technologists and chemists sometimes must

interpose their judgments into the process to obtain a formula that fits the task. In fact, in every case the analyst should examine the variables selected to be sure to have an equation that preferably is related to sensory quality.

There is a parallel to the matter of accepting a secondary equation. Sometimes an equation extending over the whole range of samples fails to discriminate well somewhere along the range. By calculating an equation for the part of the range giving trouble, the second equation permits the samples to then be readily separated. An illustration of that procedure to make product resolution tractable has been given (25). Being able to do that is one of the benefits of having analyzed sufficient samples to know where the problems are and having the information at hand to overcome the problem.

PARTIAL LEAST SQUARES REGRESSION

A procedure that has come to the fore in recent years is partial least squares regression analysis (PLS). The calculation and functioning have been described (26). Assume one has two blocks of variables: sensory and instrumental. PSC searches for a few latent orthogonal variables (factors) to describe the interrelationships between the two blocks of variables (27). It extracts from the regressor matrix (ie, the instrumental block), those factors relevant to prediction of the sensory parameters of the food. The procedure may be looked upon as a combination of PCA and multiple linear regression performed simultaneously on the latent variables. Several research publications are available that detail applications in sensory-instrumental correlation (26,27).

RESPONSE SURFACE METHODOLOGY

The response surface method (RSM) will be mentioned at this point, too, because it is another of the regression methods. The basic idea underlying RSM is to find the combination of several ingredients that lead to optimal sensory quality. An elementary description of the procedures for setting up a response surface trial and of analyzing the results have been given (28). Other treatises have been published (29,30). Sometimes the resulting configuration is not in the form of a mountain peak. The area of optimal performance may be a ridge. At other times a saddle formation is evident. RSM is highly effective in permitting reduction of the number of trials that must be carried out; unfortunately, for sensory purposes its application is generally limited to a few variables only, because otherwise the amount of sensory testing would be beyond the capability of most laboratories to handle the enormous number of samples that would have to be evaluated at any one session. Figures 5, 6, and 7 illustrate some of the kinds of response surfaces that can result (31).

Table 8. R^2 Improvement and F-Value Decline as Variables are Added, Stepwise Regression

Step	Variable added	Variable deleted	R^2	Regression F-value
1	GC5		0.954	458.60
2	GC16		0.969	326.36
3	GC9		0.975	256.28
4	GC38		0.978	209.93
4	GC33	GC9	0.979	244.66
4	GC12	GC16	0.979	244.88
4	GC7	GC5	0.981	247.92
4	GC26	GC33	0.983	276.43
4	GC30	GC38	0.984	296.00
5	GC44		0.986	245.02
6	GC9		0.987	213.13
20[a]			0.994	25.11

Note: Product H with various amounts of ingredients added; dependent variable is intensity of flavor.

[a]Beyond 20 variables, the SAS stepwise regression program became quite costly (52 variables available here). Normally, a limitation on the number of steps would be specified. Actually, Step 3 was the last step where all the variables entered had probabilities < 0.05.

PROCRUSTES ANALYSIS

Like PLS, generalized Procrustes analysis (GPA) has come to play a major role in the relating of one set of variables

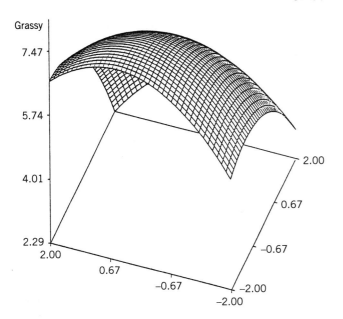

Figure 5. Illustration of response surface showing a peak. The vertical axis shows sensory evaluation for a grassy note in tea as influenced by varying amounts of linalool (abscissa) and epicatechin (ordinate).

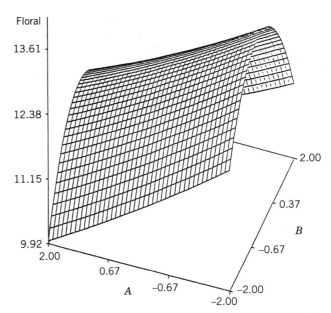

Figure 7. Illustration of ridge formation, response surface analysis. The vertical axis shows the degree of a floral note imparted to tea after being treated with varying amounts of 1-penten-3-ol (abscissa) and *cis*-3-hexen-1-ol (ordinate).

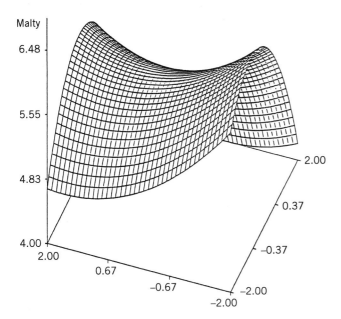

Figure 6. Illustration of a saddle formation, response surface analysis. The vertical axis shows degree of maltiness imparted to tea on addition of varying amounts of epicatechin (abscissa) and 1-penten-3-ol (ordinate).

to another. Dijksterhuis (32) has explained the principles and techniques underlying GPA. GPA has especial application when consumers are allowed to use "free-choice profiling" to describe their perceptions of the qualities of a set of foods (33,34). Trained panelists are normally asked to rate the intensities of an established set of terms. There are good reasons for not training consumers to use a re-

stricted set of terms. In free-choice profiling, the panelists are free to use whatever terms seem appropriate to them to describe the sensations they are perceiving. A consequence is that the configuration of terms and intensities applied to the sensation almost always differ for each panelist. GPA involves rescaling of the intensity values and rotation of the configurations to bring them as closely as possible into alignment with each other. The investigator thereby secures a consensus configuration of the terms and of the relationships of the products to each other. The process derives its name from the mythical innkeeper Procrustes, who made all his guests exactly fit the beds. If the guest were longer than the bed, Procrustes cut off the guest's legs at the appropriate place, if the guest were smaller than the bed, Procrustes used a rack to stretch the unwary guest. Kaye (35) commented that "in a procrustean approach to a theory, the scientist stretches and pulverizes any data until it [they] fits the preconceived ideas." That is not a valid criticism applied to free-choice profiling. The investigator might have a preconceived notion as to the pattern of the consensus configuration, but free-choice profiling hardly favors much in the way of a firm idea of the pattern to be evolved. Others, in only two publications that will be cited (36,37), have commented upon or reported comparisons between GPA and other multivariate forms of analysis.

MULTIVARIATE ANALYSIS OF VARIANCE

Multivariate analysis of variance (MANOVA) is another of the MVA procedures of use in analyzing multicomponent data. It is akin to analysis of variance (ANOVA). In ANOVA, the total sum of squared deviations about the grand mean, SS (total), is partitioned into a sum of squares due to one or more sources and a residual sum of squares. In MANOVA, the p-dimensional multivariate analysis is based on the same design, there are pSS (total)'s to partition, one for each component measured. In addition, there are the measures of covariance between pairs of components presented as sums of products. MANOVA is concerned with partitioning of both of these measures, the variance and covariance, collected in a matrix of sums of squares and of products. The result is that MANOVA yields a global estimate as to whether there are any significant differences among the different variables or their correlations. MANOVA can be separated into ANOVAs yielding the same information. The number of ANOVA calculations required is $p(p + 1)/2$. To illustrate, if there are seven dimensions, 28 ANOVAs would yield the same information as the one MANOVA.

The effect of the above is that ANOVA for one type of measurement acts as if the others did not exist whereas MANOVA takes all of the different kinds of measurements made into account in one operation. The global aspect means that if the multivariate F-value is not significant, there is not a significant difference anywhere among the variables. If the multivariate F-value is significant, there is statistical significance somewhere. For most purposes, the analyst then must apply other tests to ascertain where significance lies.

A use of MANOVA that has value in sensory analysis is to ascertain just how effective a judge has been at performing several tasks. When a prospective sensory judge has been trained to assign scores to several different attributes of the sensory material, there is a problem at the end of the training period to decide just how effective the judge should be for the various tests. Must each judge be significant for each one of the attributes examined, or must the sensory leader be willing to accept the fact that not every judge can be statistically significant for each of the tasks. The usual way is to look at all the probability values that go with the F-test (ANOVA) for each of the tasks, then make a judgment based on a composite opinion of overall performance. MANOVA puts a little objectivity into the process. The thing to do is to calculate the multivariate F-value; from it, some of the tediousness of examining a whole matrix of F-values or probability values for each judge can be eliminated. If a judge exhibits a multivariate F that has a high degree of probability, it is almost certainly possible to eliminate that judge as a suitable candidate for appointment to the sensory panel. If the probability is very low, the judge should be retained. Only for judges intermediate in performance need the sensory leader go back to the ANOVA matrix to make the final decision. Perhaps the judge is a poor performer for only one attribute in which case the leader might decide to accept the judge in spite of a questionable multivariate F-value. If the ANOVA documents that performance has been spotty across the board, the sensory leader more likely will reject the judge. Illustrations of the joint use of the frequency of significant probability according to ANOVA calculations and by MANOVA calculation have been given (15,38,39). MANOVA's merit is chiefly one of telling the analyst to stop at a particular point, because significance doesn't exist, or to go on, because significance does exist, but other methods of analysis must be applied to determine where or how often.

FUZZY LOGIC

Fuzzy logic will barely be mentioned, for its use in the sensory field is not as common as the other methods described here. The theory behind fuzzy logic is that uncertain phenomena can be treated mathematically to offset ambiguous or conflicting information to yield instead a yes-or-no answer to allow the best product, choice of ingredients, or process method to be selected (40).

NEURAL NETWORKS

Neural networks (NN) involves making decisions by learning the results of past performance. In that regard, NN is not much different from most of the methods discussed previously where experimentation is used to yield a data set from which, in turn, a predicting equation may be derived for future application. Although, as was pointed out earlier, instrumental approaches to relating sensory-to-instrumental data are often blind in the sense that the chemical/physical predictor(s) chosen may not be at all related to sensory quality, that risk is even greater for NN. NN consists of three parts: input (measurement) variables, hidden interrelations, and output variables (41,42). The output values are the result of examination of the thousands of permutations arising from even a moderate number of different measurement values. The process is spoken of as "training." By means of 10,000 or 100,000 iterations—or even more—interrelations among the hidden values are repeatedly evaluated until a set is found that correlates with the known differences in the test samples, Based on evaluation of numerous test samples, output values are eventually found that allow categorization of an unknown to possible classes, ie, the best product, a product matching the traditional product, one suffering from flaws such as off-flavor, or some one of the sensory notes being so dominant as to be objectionable. Once the necessary output value(s) have been obtained, NN may then be used for routine evaluation such as in quality control.

CANONICAL ANALYSIS

Canonical analysis is the counterpart of simple correlation analysis. In simple correlation two things increase jointly, decrease jointly, or one increases as the other decreases. Whatever they do, they stay in step with each other. Canonical correlation works the same way. The difference is that in canonical correlation sets of variables are correlated with each other instead of individual items. Many pairs of correlates may exist among the different variables. The intent is to find the set with maximum linear correlation. The canonical variates are derived in essentially the same manner as principal components. In PCA the intent is to account for as much of the variance as possible within one set. In canonical correlation, the intent is to account for the maximum amount of correlation between sets. The analysis process searches out the first canonical variate from the first set and the first from the second set that are maximally correlated with each other. It then does the same for the second set and all that follow. Like simple correlation, the two first correlates may be plotted against each other; so too may the second set, and any others that exist. Like PCA, only the first few correlates may be of interest. Correlation may be between sets of sensory measurements, sets of instrumental measurements, or more commonly between sensory and instrumental measurements. Canonical analysis has some similarity to MRA. In MRA the dependent variable is singular whereas the independent variables are multiple; in canonical correlation, both sets are multiple.

CANONICAL DISCRIMINANT ANALYSIS

A form of analysis that is particularly valuable is canonical DA. Table 9 shows canonical DA applied to the five kinds of rye bread referred to earlier. The values above the diagonal give the interspatial distance. The values below the line give the multivariate F-value. All are significant. If some are significant and some not significant, it is more informative in that case to give the probability values that go with the F-values. When all are significant, differences between products are easier to discern from the F-values themselves. In determining interspatial relations, canonical DA is preferable to PCA because canonical analysis allows the setting of statistical limits.

CORRESPONDENCE ANALYSIS

Correspondence analysis is a procedure for the analysis of contingency tables. The cells of the table contain frequencies or concentrations. They are given unit weight by dividing by the grand sum. The contingency table is essentially a two-way classification where neither the rows nor columns need be regarded as the variables and measurements. Both are commonly called elements. To permit plotting of both sets of elements on one plot, a fully symmetrical treatment of the data is employed. That can be done by calculating a χ^2 distance as the similarity measure for both rows and columns. Once that is done, PCA and prin-

cipal coordinate analysis, another name for MDS, are applied to yield a factor analysis in terms of the χ^2 profile measurements. A short description of correspondence analysis with two simple illustrations has been presented (43). There have been extensive publications on correspondence analysis (44); it is a method of analysis particularly favored in France. A book has also been published on the subject (45).

FACTOR ANALYSIS

Factor analysis (FA) has been left to the last because it is a method about which an unusual amount of controversy swirls. Some statisticians look on all the internal analysis methods with some suspicion on the grounds that the results are high provincial simply because only one data set is involved. Others frown on FA particularly because of indeterminacy in factor solutions (19). There is no doubt that different solutions can be obtained according to the particular method of factor analysis employed. Notwithstanding such criticisms, FA is a procedure that is used more and more in sensory–instrumental analysis. FA is different from PCA in that PCA is concerned with variance only. FA requires a statistical model and is concerned with explaining the covariance structure as well as variance. Unlike PCA where there is no particular assumption about possible underlying structure among the variables, FA is based on the faith that the observed correlations are mainly the result of some underlying regularity in the data. The various factoring techniques initially extract common factors, but generally they are uninterpretable. To enhance interpretability, there are several models for rotation and determination of the number of factors. The ideal situation is to have a variable highly loaded on one factor and scarcely loaded on the others.

There are three kinds of matrices involved in FA. One is the factor pattern. It provides the weights to estimate variables from the factors. A second one is the factor score. That matrix provides weights to estimate factors from the variables. The third kind of a matrix is the factor structure. It gives the correlations between the factors and the variables. The pattern matrix delineates more clearly the groupings or clustering of variables than the structure matrix.

If two rotations give rise to different relationships, the two interpretations should not be looked on as conflicting. They are two different ways of looking at the same thing, ie, they are two different points of view in the common-factor space. The basic impetus to employing any rotational method is the same: somehow to achieve simpler and theoretically more meaningful factor patterns. Orthogonal factors are mathematically simpler to handle; oblique factors are empirically more realistic.

Although FA is frowned on by some statisticians, some of their reasons are highly suspect because provinciality does not apply to many sensory–instrumental data sets. Most of the applications of FA in the past dealt with such things as determining people's opinion or some similar data where there was no opportunity to replicate responses. Sensory panel results are different. Often there

Table 9. Canonical Discriminant Analysis Applied to Five Types of Finnish Wholemeal Sour Rye Bread

Products	Products				
	1	2	3	4	5
1		1.11	4.50	1.11	1.80
2	3.91		4.23	1.16	1.60
3	59.67	52.21		4.25	3.31
4	3.73	4.00	55.17		1.50
5	9.17	7.56	33.50	6.81	

Note: Values above the diagonal line show the Mahalanobis distances between products; values below the diagonal line list the multivariate F-values (probability < 0.0001). Evaluation was by 20 panelists for 12 attributes in four replications.

is replication; thus the analyst has the opportunity to determine whether the panelist is responding randomly or consistently for the same sample given at different times. The objection to FA that it is likely to be provincial may still apply to the factoring of consumer responses, because in most instances there is no replication. If, however, the survey is large enough, that objection loses some of its validity.

Whether or not replication is involved, there is one phase of FA that has been overlooked by most investigators. As is true for all methods of analysis where a mean is calculated, the following question should be asked. Is the mean merely the average of a group of disparate responses or do the responses tend to agree with each other? To illustrate, if a particular perfume receives scores of 7, 7, 6, 8, 8, 6, and 6 out of a possible score of 10, or an average score of 6.86, and another perfume receives scores of 10, 10, 4, 5, 10, 6, and 3, or an average score of 6.86, the conclusion should be entirely different although the means are the same. The first perfume is likely to have general acceptance whereas the second will appeal to some and be scorned by others. In the same fashion, rather than just generate a factor pattern for the whole sensory panel and assume that it represents the pattern for all members of the panel, the patterns yielded by the different panelists should be examined to learn if they agree in general or the pattern for the panel is merely the mean for a group of disparate patterns. Rarely, have factor patterns been examined in that light. Early studies began to look at the homogeneity of the factor patterns of a panel in that manner. One study visually examined the eigenvalues to learn if they were similar (46). Another study applied statistical tests to learn objectively how well they agree (38). Later other studies (13,47,48) refined the process even more with the result that today factor patterns should be examined as most other sensory–instrumental data are, to learn if the individual patterns are quite homogeneous or if they are heterogeneous. If they are heterogeneous, then, of course, considerably less credence should be put into their value and representativeness of how things really are. Figure 8 shows the extent of agreement between a panel of 18 and a subset of 8 from the same panel. Often it is difficult to compare the pattern for an individual member versus the panel because of lack of degrees of freedom to carry on the analysis for the individual's responses. To apply some of the FA methods there must be at least limited pooling of responses to provide sufficient degrees of freedom to enable a comparison to be made. Figure 9, which is a little easier to interpret than Figure 8, shows one comparison (45). While a little harder to visualize the agreement, Figure 8 is more realistic of the pattern that exists. Note that the cube standing for total flavor fits in factor 1 or 2 as does sour flavor. That is as it should be. Pointed out in the beginning is the fact that an attribute may be partly in one factor and partly in another, because FA involves both variance and covariance. The same sort of standing in one factor with one foot and in another with the other foot applies to chemical data as well. A chemical may have more than one type of functional group; consequently its covariance may be shared by two different factors reflecting the two kinds of properties caused by different functional

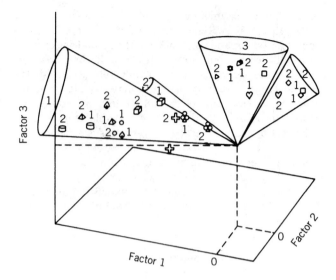

Figure 8. Comparison of factor patterns exhibited by a panel of 18 and a subset consisting of 8 of the panel members. The numbers in the opening of each cone stand for the factor pattern, the symbols within a cone represent the different attributes evaluated. pyramid = rye flavor; spade = rye aroma; circle = total aroma; cube = total flavor; cross = sour aroma; club symbol = sour flavor; triangle = sweet aroma; flower = sweet flavor; heart = saltiness; diamond = bitter flavor; square = flour flavor; cylinder = desirability.

groups. A major use both of PCA and FA is to discern redundancies in the choice of descriptive terms. Table 10 shows a factor that was derived in one study. Panelists evaluated canned and frozen green beans for 27 attributes. The seven terms comprising Factor 4 reveal that the panelists had been put to a lot of unnecessary work. In theory, any one of the 7 terms could substitute for all the others. In practice, a sensory technologists might not want to go that far and might still want to use three or four different terms in case some panelists are more adept at distinguishing among products based on one descriptor whereas a different one is more meaningful to other panelists. It can readily be understood that crispness and tenderness are the complements of each other, but it is not as clear that slimy, juicy, and soggy represent nearly the same characteristic.

WHITHER SENSORY–INSTRUMENTAL CORRELATION?

The tools to carry on sensory–instrumental analysis are pretty much in place. Some industrial firms have made good use of the tools available by working out for themselves beneficial applications involving the complementary uses of sensory and instrumental analysis. There has been little full substitution of instrumental methods for sensory ones because chemical and other analytical methods themselves are not complete. There are instances of as many as 400 or more peaks being discernible on gas chromatograms, but that number often falls short of the number of compounds comprising the sensory character of a food product; thus no chemical analysis is probably ever really

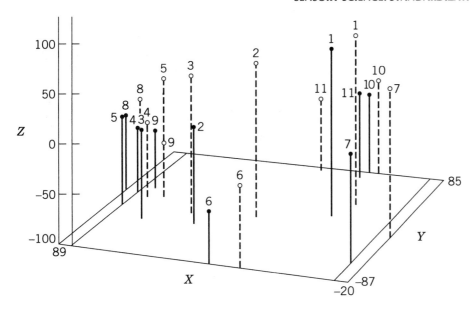

Figure 9. Comparison of factor patterns: ○, panel results pooled; ●, two of the panelists.

Table 10. Texture Factor Pattern for Canned, Frozen, and Fresh Green Beans as Judged by 21 Panelists, Four Replications (Factor Pattern Derived from among 21 Attributes Evaluated)

Term	Correlation with factor
Coarse	0.57
Fibrous	0.57
Crisp	0.53
Slimy	−0.53
Juicy	−0.53
Soggy	−0.63
Tender	−0.66

exhaustive of all the compounds in a food. More important, no chemical analysis reveals the exact pattern of the chemicals contained within a food or most other sensory materials. Not every compound is detected to the same extent by gas chromatography or liquid chromatography. Thus the exact ratios between compounds is often not known. In concentrating materials for analysis so as to bring the concentrations within the range of the instrument, the ratios among compounds are initially changed because of effects of vapor pressure or solubility in the extracting medium. The interrelations of factors affecting texture present even greater difficulties. Characteristics such as gumminess and the hardness of embedded particles need not be related to each other, thus it is not possible to speak of total texture in the same way as it is possible to speak of total flavor. There are some difficulties in measuring color, but on the whole color can be more reliably expressed physically than is true for the other sense modalities. Relations between acoustical properties and sensory characteristics have been extensively studied (49). The point of mentioning each of the senses is that a food or a perfume or the odor of a barnyard is usually judged by more than one sense. That is true even for the seemingly simple response to the odor of a barnyard. It is thought that the sensation

perceived is odor, but actually generally more than odor is involved. Some of the volatile substances reaching the nose are compounds possessing trigeminal characteristics causing a stinging sensation or even pain. Considering all that is not known, it is remarkable that sensory–instrumental applications have been as useful as they are.

To make correlation methods even better, it is necessary to know more about the sensory receptors and how the brain processes information. Humans and other animals probably utilize a process somewhat like PCA. Hundreds of sensory signals are detected and these sensory signals must be combined in various ways to form patterns of varying complexity; in other words, data reduction is carried out. Like PCA components, some of the patterns may encompass signals from only a dozen or so sensory notes, other may encompass several dozen. The point is that humans and other animals are responding to several patterns representing different kinds of sensations and patterns of varying size, then somehow this semiprocessed information is ultimately taken and put together as a composite sensation that tells the mind that this food is delectable, that food is almost without taste, and that food has a trace of cinnamon in it. In the brain a whole series of complex interaction effects have taken place. They are not the chemical interactions a chemist thinks of, a reaction between this physical substance and that. They are chimera, not amenable to understanding even if all the kinds of chemical reactions that could occur among the hundreds of chemicals comprising flavor were known, because they are perceptive reactions not physical ones.

Before sensory–instrumental correlation can reach the ultimate in applicability, knowledge of sensing, brain processing of information, and all the other sciences (physiology, psychology, biochemistry, and anatomy) that impinge on the reception, the processing, and the analyzing of sensory information must be greatly expanded. At present, statistical methods are adequate. There must be enlistment of a greater number of statisticians who have an interest in sensory research and are willing to work around some of the limitations that sensory analysis imposes. That which is ideal is often not practical in sensory analysis, such as the evaluation of a large number of samples all at one session. For the time being, present-day computers can handle most of the data sets presented to them, but the day will come when computing demands will be far greater than they are now. If perceptual interactions are begun to be understood, far more will be demanded of computing capacity and complexity than presently exists. Chemical measurement of food components must be made to be "true" to avoid giving a distorted picture of the relations among the hundreds of compounds comprising foods and other natural substances of sensory importance. The characteristics of compounds that make them be sensory substances must be learned. There is fair knowledge of the functional grouping that must exist for substances to be sweet, the properties of chemicals that make them be sour or salty are known with even more certainty, but by and large it is not really known why this compound has the property of being a sensory substance and that one does not. It is necessary know as much about the laws that govern sensory properties as does a synthetic chemist who is seeking to synthesize a new compound. From knowledge of the properties of chemicals, their functional groupings, and possible pathways to foster reactions otherwise particularly difficult to make occur, the chemist can predict in advance the likelihood of success in synthesizing a new compound. Chemists, once they know the identity of a compound, have at their fingertips, or at least readily ascertainable, a whole series of other information about properties of that substance. Sensory properties are not hard facts like vapor pressure and molecular weight. They must be described in terms of perception. Their description is mutable according to who is doing the describing. The description should be an immutable property, not merely a perception. Notwithstanding as important as are all these other considerations, there is nothing to stop the far wider use of multivariate analysis of sensory–instrumental relations. There must be a will on the part of practitioners to stay at the task for some time to avoid having their results be parochial because of acquiring insufficient data. Once sufficient data has been secured, to be sure the results represent real relations, sensory–instrumental correlations can be of great practical importance for present industrial operations such as ensuring ingredients purchased meet the company's buying specifications or the finished product meets its quality specifications for marketing. Of greater importance, correlations so determined can serve as a base for the advancement of fundamental knowledge. From such correlations will ultimately arise an understanding of cause-and-effect relations. Still more important, they in turn will provide a good share of the base of knowledge required to advance comprehension of the detection of sensory signals, their analysis and data processing within the brain.

BIBLIOGRAPHY

1. A. T. James, A. J. F. Martin and G. H. Smith, "Gas Liquid Partition Chromatography. The Separation and Microestimation of Ammonia and Methylamines," *Biochemistry Journal* **52**, 238–247 (1952).

2. S. E. Cairncross and L. B. Sjostrom, "Flavor Profiles—A New Approach to Flavour Problems. Annual Meeting, Institute of Food Technologists (1949)." *Food Technology* **4**, 308–311 (1950).

3. J. Powers, "Descriptive Methods of Analysis," in J. R. Piggott, ed., *Sensory Analysis of Foods*, 2nd ed., Elsevier Applied Science Publishers, Barking, UK, 1988.

4. H. Stone, J. L. Sidel, S. Oliver, A. Woolsey, and R. C. Singleton, "Sensory Evaluation by Quantitative Descriptive Analysis," *Food Technology* **28**, 24–34 (1974).

5. J. J. Powers and E. S. Keith, "Evaluation of the Flavour of Foods from Gas Chromatographic Data," in *Abstracts of Papers*, Second International Congress of Food Science and Technology, Warsaw, 1966.

6. J. J. Powers and E. S. Keith, "Stepwise Discriminant Analysis of Gas Chromatographic Data," *Journal of Food Science* **33**, 207–213 (1968).

7. H. W. Dodo, "Sensory and Chemical Attributes of Three Varieties of Onions (*Allium cepa*)," Master of Science thesis, University of Georgia, Athens, 1985.

8. J. R. Piggott and S. P. Jardine, "Descriptive Sensory Analysis of Whisky Flavour," *Journal of the Inst. Brew.* **85**, 82–85 (1979).

9. U. Hellemann, J. J. Powers, H. Salovaara, K. Shinholser, and M. Ellila, "Relation of Sensory Sourness to Chemically Analyzed Acidity in Wholemeal Sour Rye Bread," *Journal of Sensory Studies* **3**, 95–111 (1988).

10. K. Pearson, "On Lines and Planes of Closest Fit to Systems of Points in Space," *Phil. Mag.* **6**, 559–572 (1901).

11. H. Hotelling, "Analysis of a Complex of Statistical Variables into Principal Components," *Journal of Educational Psychology* **24**, 417–441, 498–520 (1933).

12. J. J. Powers, S. Cenciarelli, and K. Shinholser, "El uso de programas estadisticos generales en la evaluacion de los resultados sensoriales," *Rev. Agroquim. Technol. Aliment.* **24**, 469–484 (1984).

13. K. Shinholser, U. Hellerman, H. Salvaara, M. Ellila, and J. J. Powers, "Factor Patterns Yielded by Subsets of Panelists Examining Finnish Sour Rye Bread," *Journal of Sensory Studies* **2**, 119–213 (1987).

14. *SAS User's Guide: Statistics*, Version 5 ed., SAS Institute Inc., Cary, N.C., 1985.

15. J. J. Powers, "Uses of Multivariate Methods in Screening and Training Sensory Panelists," *Food Technology* **42**, 123–127 (1988).

16. I. K. Hwang Choi, "A Case Study of the Use of Standard Methods to Analyze Multivariate Data," Ph.D. dissertation, University of Georgia, Athens, 1982.

17. S. L. Bieber and D. V. Smith, "Multivariate Analysis of Sensory Data: A Comparison of Methods," *Chemical Senses* **11**, 19–47 (1986).

18. *SPSS (Statistical Package for the Social Sciences)*, 2nd ed., SPSS, Inc., Chicago, Ill., 1986.

19. C. Chatfield and A. J. Collins, *Introduction to Multivariate Analysis*, Chapman and Hall, London, 1980.

20. S. S. Schiffman, M. L. Reynolds, and F. W. Young, *Introduction to Multidimensional Scaling, Theory, Methods and Applications*, Academic Press, Inc., Orlando, Fla., 1981.

21. M. Gillette, "Applications of Descriptive Analysis," *Journal of Food Protection* **47**, 403–409 (1984).

22. G. D. Wilkin, M. A. Weber, and E. A. Lafferty, "Appraisal of Industrial Continuous Still Products," in J. R. Piggott, ed., *Flavour of Distilled Beverages*, Ellis Horwood Ltd., Chichester, UK, 1983.

23. J. J. Powers and G. O. Ware, "Discriminant Analysis," in J. R. Piggott, ed., *Statistical Procedures in Food Research*, Elsevier Applied Science Publishers Ltd., Barking, UK, 1986.

24. W. G. Galetto and A. Bednarczyk, "Relative Flavour Contribution of Individual Volatile Components of the Oil of Onion (*Allium cepa*)," *Journal of Food Science* **40**, 1165–1167 (1976).

25. J. J. Powers, "Techniques of Analysis of Flavours: Integration of Sensory and Instrumental Methods," in I. D. Morton and A. J. MacLeod, eds., *Food Flavors, Part A. Introduction*, Elsevier, Amsterdam, The Netherlands, 1982.

26. M. Martens and H. Martens, "Partial Least Squares Regression," in J. R. Piggott, ed., *Statistical Procedures in Food Research*, Elsevier Applied Science Publishers Ltd., Barking, UK, 1986.

27. M. Martens and Eeke van der Burg, "Relating Sensory and Instrumental Data from Vegetables Using Different Multivariate Techniques," in J. Adda, ed., *Progress in Flavour Research*, Elsevier, Amsterdam, The Netherlands, 1985, pp. 131–148.

28. M. E. Wells, "Response Surface Methodology and Subjective Data," in J. J. Powers and H. R. Moskowitz, eds., *Correlating Sensory Objective Measurements–New Methods for Answering Old Problems*, Publication STP 594, American Society for Testing and Materials, Philadelphia, 1976.

29. G. E. P. Box, W. G. Hunter, and J. S. Hunter, *Statistics for Experimenters—An Introduction to Design, Data Analysis and Model Building*, John Wiley & Sons, Inc., New York, 1978.

30. L. Vuataz, "Response Surface Methods," in Ref. 23.

31. D. R. Godwin, "Relationships between Sensory Response and Chemical Composition of Tea," Ph.D. dissertation, University of Georgia, Athens, 1984.

32. G. B. Dijksterhuis, *Multivariate Data Analysis in Sensory and Consumer Science*, Food & Nutrition Press, Inc., Trumbull, Conn., 1997.

33. G. M. Arnold and A. A. Williams, "The Use of Generalised Procrustes Techniques in Sensory Analysis," in J. Adda, ed., *Progress in Flavour Research*, Elsevier, Amsterdam, The Netherlands, 1985, pp. 35–50.

34. H. T. Lawless and H. Heymann, *Sensory Evaluation of Food*, Chapman & Hall, New York, 1998.

35. B. Kaye, *Chaos & Complexity*, VCH Publishers, New York, 1993.

36. E. A. Hunter and D. D. Muir, "A Comparison of Two Multivariate Methods for the Analysis of Sensory Profile Data," *Journal of Sensory Studies* **10**, 89–104 (1995).

37. R. Popper, H. Heymann and Frank Rossi, "Three Multivariate Approaches to Relating Consumer to Descriptive Data," in A. M. Munoz, ed., *Relating Consumer, Descriptive, and Laboratory Data*, Manual 30, ASTM, West Conshohocken, Pa., 1997, pp. 39–61.

38. J. J. Powers, K. Shinholser, and D. R. Godwin, "Evaluating Assessors' Performance and Panel Homogeneity Using Univariate and Multivariate Statistical Analysis," in J. Adds, ed., *Progress in Flavour Research 1984*, Elsevier, Amsterdam, The Netherlands, 1985.

39. J. J. Powers, "Utilizing General Statistical Programs to Evaluate Sensory Data," *Food Technology* **38**, 74–84 (1984).

40. Q. Zhang and J. B. Litchfield, "Applying Fuzzy Mathematics to Product Development and Comparison," *Food Technology* **45**, 108–118 (1991).

41. C. N. Thai and R. L. Shewfelt, "Modeling Sensory Color Quality of Tomato and Peach: Neural Networks and Statistical Regressionk," *Transactions of the ASAE* **34**, 950–955 (1991).

42. G. Peters et al., "Linear Regression, Neural Network and Industion Analysis to Determine Harvesting and Processing Effects on Surimi Quality," *Journal of Food Science* **61**, 876–880 (1996).

43. D. L. Hoffman and G. R. Franke, "Correspondence Analysis: Graphical Representation of Categorical Data in Marketing Research," *Journal of Marketing Research* **23**, 213–227 (1986).

44. J. P. Benzecri et al., *L'Analyse des donnees, Vol 11. L'Analyse des correspondances*, Dunod, Paris, 1973.

45. M. J. Greenacre, *Theory and Application of Correspondence Analysis*, Academic Press, Inc., Orlando, Fla., 1984.

46. J. M. Harries, D. N. Rhodes, and B. B. Chrystall, "Meat Texture. 1. Subjective Assessment of the Texture of Cooked Beef," *Journal of Text. Studies* **3**, 101–114 (1972).

47. E. I. Fischman, K. J. Shinholser, and J. J. Powers, "Examining Methods to Test Factor Patterns for Concordance," *Journal of Food Science* **52**, 448–454 (1987).

48. J. J. Powers, K. Shinholser, and D. Brett, "Testing Factor Patterns for Agreement," *Proceedings Tecno Alimentaria, Barcelona* (1987).

49. Z. M. Vickers, "Sensory, Acoustical, and Force-Deformation Measurements of Potato Chip Crispness," *Journal of Food Science* **52**, 138–140 (1987).

GENERAL REFERENCES

P. L. Bernstein, *Against the Gods, The Remarkable Story of Risk*, John Wiley and Sons, Inc., New York, 1996.

BMDP Biomedical Computer Programs, P-series, *University of California Press*, 1979.

J. Caul, "The Profile Method of Flavor Analysis," *Advances in Food Research* **7**, 1–40 (1957).

A. Dravnieks, L. Keith, B. K. Krotoszynski, and J. Shah, "Vaginal Odors: GLC Assay Method for Evaluating Odor Changes." *Journal of Pharmaceutical Science* **63**, 36–40 (1974).

M. C. Gacula, Jr., *Descriptive Sensory Analysis in Practice*, Food and Nutrition Press, Inc., Trumbull, Conn., 1997.

P. E. Green, *Analyzing Multivariate Data*, Dryden Press, Hinsdale, Ill., 1978.

T. Persson and E. von Sydow, "A Quality Comparison of Frozen and Refrigerated Cooked Sliced Beef. 2 Relationships between Gas Chromatographic Data and Flavor Scores," *Journal of Food Science* **37**, 234–239 (1972).

J. J. Powers, "Applying Multivariate Statistical Methods to Enhance Information Obtainable from Flavor Analysis Results," in David B. Min and Thomas H. Smouse, eds., *Flavor Chemistry of Lipid Foods*, American Oil Chemists' Society, Champaign, Ill., 1989.

B. G. Tabachnick and L. S. Fidell, *Using Multivariate Statistics*, Harper & Row, New York, 1983.

L. Vuataz, "Some Points of Methodology in Multidimensional Data Analysis as Applied to Sensory Evaluation," *Nestlé Research News 1976/77*, Nestec Ltd., Vevey, Switzerland, 1977.

JOHN J. POWERS
University of Georgia
Athens, Georgia

SHELLFISH

This article concerns three major groups of shellfish: oysters, scallops, and shrimps. When discussing these popular groups of seafoods, the government, academics, and industry are interested in harvesting, processing, sanitation, and economic considerations. Limited by space, this article will discuss the quality control of processing oysters, scallops, and shrimps with some emphasis on definitions and standards, general processing, sanitary practices, and sizing. It is obvious that the thrust of this article is most applicable to the United States.

OYSTERS

For many years packers of canned oysters in the Gulf of Mexico area of the United States have labeled their output with a declaration of the drained weight of oysters in the containers. Packers in other areas have marketed canned oysters with a declaration of the total weight of the contents of the container. Under present-day practice consumers generally do not discard the liquid packing medium, but use it as a part of the food. Compliance with the label declaration of quantity of contents requirement will be met by an accurate declaration of the total weight of the contents of the can.

General Definitions

Oysters, raw oysters, and shucked oysters, are the class of foods each of which is obtained by shucking shell oysters and preparing them in a prescribed manner. If water or salt water containing <0.75% salt is used in any vessel into which the oysters are shucked, the combined volume of oysters and liquid should not be less than four times the volume of the water or salt water. Any liquid accumulated with the oysters is removed. The oysters are washed, by blowing or otherwise, in water, salt water, or both. The total time that the oysters are in contact with the water or salt water after leaving the shucker, including the time of washing, rinsing, and any other contact with water or salt water is not more than 30 min. In computing the time of contact with water or salt water, the length of time that oysters are in contact with water or salt water that is agitated by blowing or otherwise, should be calculated at twice its actual length. Any period of time that oysters are in contact with salt water containing not less than 0.75% salt before contact with oysters should not be included in computing the time. Before packing into containers for shipment, or other delivery for consumption, the oysters are thoroughly drained and are packed without any added substance.

Shell oysters means live oysters of any of the species *Ostrea virginica*, *Ostrea gigas*, or *Ostrea lurida*, in the shell, which, after removal from their beds, have not been floated or otherwise held under conditions that result in the addition of water. The oysters are drained on a strainer or skimmer that has an area of not less than 300 in.²/gal of oysters, drained, and has perforations of at least 0.25 in. in diameter and not more than 1.25 in. apart, or perforations of equivalent areas and distribution. The oysters are distributed evenly over the draining surface of the skimmer and drained for not less than 5 min.

Extra Large Oysters

Extra large oysters, oysters counts (or plants), extra large raw oysters, raw oysters counts (or plants), extra large shucked oysters, and shucked oysters counts (or plants) are of the species *O. virginica*. One gallon of this size oyster contains not more than 160 oysters and a quart of the smallest of these oysters contains not more than 44 oysters.

Large Oysters

Large oysters, oysters extra selects, large raw oysters, raw oysters extra selects, large shucked oysters, and shucked oysters extra selects are of the species *O. virginica* and 1 gal contains more than 160 oysters but not more than 210 oysters. A quart of the smallest of these oysters contains not more than 58 oysters, and a quart of the largest oysters contains more than 36 oysters.

Medium Oysters

Medium oysters, oysters selected, medium raw oysters, raw oysters selects, medium shucked oysters, and shucked oysters selects are of the species *O. virginica*. One gallon contains more than 210 oysters but not more than 300 oysters; a quart of the smallest oysters contains not more than 83 oysters, and a quart of the largest oysters contains more than 46 oysters.

Small Oysters

Small oysters, oysters standards, small raw oysters, raw oysters standards, small shucked oysters, and shucked oysters standards are of the species *O. virginica*; 1 gal contains more than 300 oysters but not more than 500 oysters. A quart of the smallest oysters contains not more than 138 oysters, and a quart of the largest oysters contains more than 68 oysters.

Very Small Oysters

Very small oysters, very small raw oysters, and very small shucked oysters are of the species *O. virginica*; 1 gal contains more than 500 oysters and a quart of the largest oysters contains more than 112 oysters.

Olympia Oysters

Olympia oysters, raw Olympia oysters, and shucked Olympia oysters are of the species *O. lurida*.

Large Pacific Oysters

Large Pacific oysters, large raw Pacific oysters, and large shucked Pacific oysters are of the species *O. gigas*, and 1 gal contains not more than 64 oysters. The largest oyster in the container is not more than twice the weight of the smallest oyster therein.

Medium Pacific Oysters

Medium Pacific oysters, medium raw Pacific oysters, and medium shucked Pacific oysters are of the species *O. gigas*, and 1 gal contains more than 64 oysters and not more than 96 oysters. The largest oyster in the container is not more than twice the weight of the smallest oyster therein.

Small Pacific Oysters

Small Pacific oysters, small raw Pacific oysters, and small shucked Pacific oysters are of the species *O. gigas*. One gallon contains more than 96 oysters and not more than 144 oysters; the largest oyster in the container is not more than twice the weight of the smallest oyster therein.

Extra Small Pacific Oysters

Extra small Pacific oysters, extra small raw Pacific oysters, and extra small shucked Pacific oysters are of the species *O. gigas*, and 1 gal contains more than 144 oysters. The largest oyster in the container is not more than twice the weight of the smallest oyster therein.

Canned Oysters

Canned oysters contain one or any mixture of two or all of the forms of oysters described above in a packing medium of water, the watery liquid draining from oysters before or during processing, or a mixture of such liquid and water. The food may be seasoned with salt. It is sealed in containers and so processed by heat as to prevent spoilage. The forms of oysters referred to are prepared from oysters that have been removed from their shells and washed and that may be steamed while in the shell or steamed or blanched or both after removal therefrom, and are as follows:

1. Whole oysters with such broken pieces of oysters as normally occur in removing oysters from their shells, washing, and packing.
2. Pieces of oysters obtained by segregating pieces of oysters broken in shucking, washing, or packing whole oysters.
3. Cut oysters obtained by cutting whole oysters.

The name of the food is oysters or cove oysters if the species is *O. virginica*; oysters or Pacific oysters if the species is *O. gigas*; and oysters or Olympia oysters if the species is *O. lurida*.

The standard of fill of container for canned oysters is a fill such that the drained weight of oysters taken from each container is not less than 59% of the water capacity of the container. If canned oysters fall below the standard of fill of container, the label should bear the general statement of substandard fill: "A can of this size should contain _ oz. of oysters. This can contains only _ oz.," the blanks being filled in with the applicable figures.

Abbreviated Inspections

Microbiological considerations are of prime importance in any shellfish gathering and processing plant. Time-temperature abuses enter into most problems with the products. However, the high value of these products has made economic violations even more profitable to the unethical operator. The FDA conducts both abbreviated and comprehensive inspections of oyster processing plants.

When conducting an abbreviated inspection, the FDA uses the following critical factors:

1. Check for evidence of contamination from the presence of cats, dogs, birds, or vermin in the plant.
2. Check results of any testing conducted on incoming oysters including filth, decomposition, pesticides, or bacteria.
3. Check for possible incorporation of excessive fresh water through prolonged contact with water or by insufficient drainage.
4. Be alert to misbranding of oysters by size.
5. Determine if employee sanitation practices preclude adding contamination (clean dress and proper use of 100 ppm chlorine equivalent hand sanitizers).
6. Determine if equipment is washed and sanitized about every two hours.
7. Check for time-temperature abuses that may cause rapid bacterial growth.
8. Weigh 20 retail packages for net weight declaration, and check accuracy of labeling statements at this time.

Comprehensive Inspection

During a comprehensive inspection of an oyster processing plant, an FDA inspector pays particular attention to the points listed below.

1. Check insanitation in chucking plants and decomposition of oysters, by conventional means. Pay particular attention to frozen oysters rolled in batter ready to cook. When dealing with economic violations, emphasis should be placed on consumer-size packages, pints, and quarts, with less attention to larger-size institutional packs.
2. The abuses most often encountered in the marketing of shucked oysters are
   a. The incorporation of excessive fresh water into the bodies of the oysters by keeping them in prolonged contact with fresh water.
   b. Increasing the output of oysters, usually by insufficient drainage before packing.
   c. Short volume.
   d. Misbranding as to size. Actions on misbranding as to size should be based only on significant evidence of financial gain.
3. Cover in detail the amount of water in shucking pails, the total length of time of contact with water, length of blowing time, length of time in contact with fresh water when not being blown, length of draining time, description of skimmer and draining process, and a count of the oysters in one or more sizes. De-

termine the source of oysters and whether they are all from the same locality.

SCALLOPS

The scallop industry encompasses three primary species:

1. Sea scallops.
2. Bay scallops.
3. Calico scallops.

The processing of sea scallops is accomplished on board the vessel actually harvesting the product. Boats that process sea scallops remain at sea from 3 to 12 days depending on area and catch. In most cases, the calico scallops are harvested daily and processed at shore processing plants rather than on board the vessel. The trend, however, is toward on board processing for this species also. Bay scallops pose a unique problem in that they may be processed in a commercial plant or at home.

Comprehensive Inspection

During a comprehensive inspection of scallop processors, the following will be covered.

1. Raw materials, determine
   a. Geographical area where the scallops are harvested.
   b. Type of scallops harvested and processed by common or species name.
   c. How scallops are handled between harvesting and processing.
2. Processing
   a. Observe in detail the scallop processing operation. Make a flow plan.
   b. Check shucking and evisceration process and see if this process is physically separated from the packaging and other operations.
   c. Determine source of water used in the scallop washing and rinsing operations. If treated by the processor, determine nature and extent of treatment.
   d. See if equipment used in processing operation is of proper construction and design.
   e. Check firm's equipment cleaning and sanitizing operation.
   f. Determine time and temperature of processing operation. Check
      i. How long between harvest and chuckling and the temperature of the scallops.
      ii. How long scallops are held at ambient air temperature and determine the ambient temperature.
      iii. How long between shucking and rinsing and the temperature of the scallops.
      iv. After being iced, how long before scallops reach an internal temperature below 40°F.

g. Check finished product packaging.
h. Determine source of ice used in icing operation and if bagged ice is used, source and type of bag, condition of bags, and conditions storage.
i. Check finished product storage facilities and condition.
j. Obtain any coding system used and key.
k. Check on the use of any food additives to determine if used at allowable levels.
3. Sanitation
   a. See if building or vessel is free from rodent or insect activity.
   b. Check that toilets and handwashing facilities provided are properly located and maintained.
   c. Determine strength and type of hand sanitizing solutions used and the sanitizer's location.
   d. Note any employee practices that could lead to the contamination of the scallops with filth or bacteria.
   e. See if water and ice used in the process is from an approved source and list source.
   f. Determine method of shell and waste material disposal.
   g. Evaluate the firm's operation for compliance with FDA's Human Foods (Sanitation) Good Manufacturing Practice Regulations.
   h. Document any insanitary conditions noted that could lead to the contamination of this firm's products with filth and/or bacteria.
   i. Obtain distribution for the collection of official samples.

Frozen Raw Scallops

Frozen raw scallops are clean, wholesome, adequately drained, whole or cut adductor muscles of the scallop of the regular commercial species. The portion of the scallop used should be only the adductor muscle eye, which controls the shell movement. Scallops should be washed, drained, packed, and frozen in accordance with good manufacturing practices and are maintained at temperatures necessary for the preservation of the product. Only scallops of a single species should be used within a lot.

Solid pack scallops are frozen together into a solid mass that may be glazed or not glazed. Individually quick-frozen pack (IQF) scallops are individually quick frozen. Individual scallops can be separated without thawing. Again, they may or may not be glazed.

Scallops are of two types: adductor muscle and adductor muscle with catch (gristle or sweet meat) portion removed. Good flavor and odor (essential requirements for a U.S. grade A product) means that the product has the typical flavor and odor of the species and is free from bitterness, staleness, and off-flavor and off-odors of any kind. Reasonably good flavor and odor (minimum requirements for a U.S. Grade B product) means the product is lacking in good flavor and odor but is free from objectionable off-flavors and off-odors of any kind. Dehydration refers to the loss of moisture from the scallops surface during frozen storage.

Small degree of dehydration is color-masking but can be easily scraped off. Large degree of dehydration is deep, color-masking, and requires a knife or other instrument to scrape it off.

Undesirable small pieces are pieces that will pass through the openings in a 0.75-in. sieve for larger-size scallops. For the smaller by scallops, undesirable pieces are pieces of scallops that do not have the general conformation of the other scallops. The total weight of these pieces within a sample unit will be obtained. These pieces should not be used for determining the weight ratio.

Color refers to reasonably uniform color characteristics of the species used within an individual container. Only noticeable variation in color from the predominating color of the scallops in the container is considered. Medium gray to black colored scallops are not to be graded.

Extraneous materials are pieces or fragments of undesirable material that are naturally present in or on the scallops and that should be removed during processing. An instance of minor extraneous material includes each occurrence of intestines, seaweed, etc, and each aggregate of sand and grit up to 0.5-in.^{2} and located on the scallop surface. An instance of major extraneous material includes each instance of shell or aggregate of embedded sand or other extraneous embedded material that affects the appearance or eating quality of the product. Texture refers to the firmness, tenderness, and moistness of the cooked scallop meat that is characteristic of the species.

Frozen Raw Breaded Scallops and Frozen Fried Scallops

Frozen raw breaded or fried scallops are

1. Prepared from wholesome, clean, adequately drained, whole or cut adductor muscles of the scallop of the regular commercial species, or scallop units cut from a block of frozen scallops that are coated with wholesome batter and breading.
2. Packaged and frozen according to good commercial practice and maintained at temperatures necessary for preservation.
3. Composed of a minimum of 50% by weight of scallop meat.

The styles of frozen raw breaded scallops and frozen fried scallops include

1. *Random Pack.* Scallops in a package are reasonably uniform in weight or shape. The weight or shape of individual scallops are not specified.
2. *Uniform Pack.* Scallops in a package consist of uniformly shaped pieces that are of specified weight or range of weights.

The scallops are of two types: adductor muscle and adductor muscle with catch (gristle or sweet meat) portion removed.

Condition of the package refers to freedom from packaging defects and the presence in the package of oil, loose breading, or frost. Ease of separation refers to the difficulty of separating scallops that are frozen together after the frying operation and during freezing.

Continuity refers to the completeness of the coating of the product in the cooked state. Lack of continuity is exemplified by breaks, ridges, or lumps of breading. Each 1/16-in.^{2} area of any break, ridge, or lump of breading is considered an instance of lack of continuity. Individual breaks, ridges, or lumps of breading measuring less than 1/16 in.^{2} are not considered objectionable. Deduction points are based on the percentage of the scallops within the package that contain small or large instances of lack of continuity.

Workmanship defects refer to the degree of freedom from doubled and misshaped scallops and extraneous material. The defects of doubled and misshaped scallops are determined by examining the frozen product, while the defects of extraneous materials are determined by examining the product in the cooked state. Doubled scallops are two or more scallops that are joined together during the breading or frying operations. Misshaped scallops are elongated, flattened, mashed, or damaged scallop meats.

Extraneous materials are pieces or fragments of undesirable material that are naturally present in or on the scallops and that should be removed during processing. Examples of minor extraneous material include intestines, seaweed, and each aggregate of sand and grit within an area of 0.5 in.^{2} Examples of major extraneous material include shell and aggregate of embedded sand or other extraneous embedded material that affects the appearance or eating quality of the product.

Character refers to the texture of the scallop meat and of the coating and the presence of gristle in the cooked state. Gristle is the tough elastic tissue usually attached to the scallop meat. Each instance of gristle is an occurrence. Texture refers to the firmness, tenderness, and moistness of the cooked scallop meat and to the crispness and tenderness of the coating of the cooked product. The texture of the scallop meat may be classified as a degree of mushiness, toughness, and fibrousness. The texture of the coating may be classified as a degree of pastiness, toughness, dryness, mushiness, or oiliness.

SHRIMP

Canned Wet Pack Shrimp in Transparent or Nontransparent Containers

Canned wet pack shrimp is the food consisting of the processed meat of peeled shrimp; free of heads; and, to the extent practicable under good manufacturing practice, free of shells, legs, and antennae. One or any combination of species may be canned, prepared in one of the styles specified, in sufficient water or other suitable aqueous packing medium to fill the interstices and permit proper processing in accordance with good manufacturing practice.

Canned shrimp may contain one or more of the optional ingredients specified. It is packed in hermetically sealed transparent or nontransparent containers and so processed by heat as to prevent spoilage.

The species of shrimp that may be used in the food are of the families Penaeidae, Pandalidae, Crangonidae, and

Palaemonidae. Canned shrimp is prepared in one of the following styles:

1. Shrimp with readily visible dark vein (dorsal tract, back vein, or sand vein).
2. Deveined shrimp containing not less than 95% by weight of shrimp prepared, by removing the dark vein from the first five segments by deliberate cutting action.
3. Shrimp, other than "deveined" as described, containing not less than 95% by weight of shrimp with no readily visible dark vein within the first five segments.
4. Broken shrimp, consisting of less than four segments and otherwise conforming to one of the styles described.

The following safe and suitable optional ingredients may be used:

1. Salt.
2. Lemon juice.
3. Organic acids.
4. Nutritive carbohydrate sweeteners.
5. Spices or spice oils or spice extracts.
6. Flavorings.
7. Sodium bisulfite.
8. Calcium disodium ethylenediaminetetraacetate (EDTA), complying with prescribed regulations.

The name of the food is shrimp, or shrimps. The word *prawns* may appear on the label in parentheses immediately after the word *shrimp*, or shrimps, if the shrimp are of large or extra large size. When the food is of the style described above, the words *cleaned, cleaned (deveined), deveined*, or *contains no dark veins* may appear. The name of the food should include a declaration of any flavoring that characterizes the food and the term spiced if spice characterizes the food. Each of the ingredients used should be declared on the label as required.

Frozen Raw Breaded Shrimp: Definitions

Frozen raw breaded shrimp is the food prepared by coating one of the optional forms of shrimp with safe and suitable batter and breading ingredients. The food is frozen. The food tests not less than 50% of shrimp material. The term shrimp means the tail portion of properly prepared shrimp of commercial species. Except for composite units, each shrimp unit is individually coated. The optional forms of shrimp are

1. *Fantail or Butterfly.* Prepared by splitting the shrimp; the shrimp are peeled, except that tail fins remain attached and the shell segment immediately adjacent to the tail fins may be left attached.
2. *Butterfly, Tail Off.* Prepared by splitting the shrimp; tail fins and all shell segments are removed.

3. *Round.* Round shrimp, not split; the shrimp are peeled, except that tail fins remain attached and the shell segment immediately adjacent to the tail fins may be left attached.
4. *Round, Tail Off.* Round shrimp, not split; tail fins and all shell segments are removed.
5. *Pieces.* Each unit consists of a piece or a part of a shrimp; tail fins and all shell segments are removed.
6. *Composite Units.* Each unit consists of two or more whole shrimp or pieces of shrimp, or both, formed and pressed into composite units prior to coating; tail fins and all shell segments are removed; large composite units, prior to coating, may be cut into smaller units.

The batter and breading ingredients referred to are the fluid constituents and the solid constituents of the coating around the shrimp. These ingredients consist of suitable substances that may or may not be food additives. Chemical preservatives that are suitable are

1. Ascorbic acid, which may be used in a quantity sufficient to retard development of dark spots on the shrimp.
2. Antioxidant preservatives may be used to retard development of rancidity of the fat content of the food, in amounts within the limits prescribed.

The label should name the food as follows:

1. Breaded fantail shrimp (the word *butterfly* may be used in lieu of fantail in the name).
2. Breaded butterfly shrimp, tail off.
3. Breaded round shrimp.
4. Breaded round shrimp, tail.
5. Breaded shrimp pieces.
6. Composite units.

If the composite units are in a shape similar to that of breaded fish sticks the name is breaded shrimp sticks; if they are in the shape of meat cutlets, the name is breaded shrimp cutlets. If prepared in a shape other than that of sticks or cutlets, the name is breaded shrimp _, the blank to be filled in with the word or phrase that accurately describes the shape, but that is not misleading.

The word *prawns* may be added in parentheses immediately after the word *shrimp* in the name of the food if the shrimp are of large size; for example, "fantail breaded shrimp (prawns)." If the shrimp are from a single geographical area, the adjectival designation of that area may appear as part of the name; for example, "breaded Alaska shrimp sticks."

Frozen Raw Lightly Breaded Shrimp: Definitions

Frozen raw lightly breaded shrimp differs from the regularly breaded ones in that it contains not less than 65% of shrimp material and that the word *lightly* immediately precedes the words *breaded shrimp* on the label.

Shrimp: General Processing Quality and Standards

These standards should apply to clean, wholesome shrimp of the regular processed commercial species that are fresh or frozen, raw or cooked. Such shrimp are processed and maintained in accordance with good commercial practice at temperatures necessary for the preservation of the product. Types of fresh shrimp include

1. Frozen individually (IQF) glazed or unglazed.
2. Frozen solid pack, glazed or unglazed.

The styles of the shrimp can be as follows:

1. *Raw.* Uncoagulated protein.
2. *Parboiled.* Heated for a period of time such that the surface of the product reaches a temperature adequate to coagulate the protein.
3. *Cooked.* Heated for a period of time such that the surface of the product reaches a temperature adequate to coagulate the protein.
4. *Cooked.* Heated for a period of time such that the thermal center of the shrimp reaches a temperature adequate to coagulate the protein.

The market forms of shrimp include

1. Heads on (head, shell, tail fins on).
2. Headless (only head removed; shell, tail fins on).
3. Peeled, round, tail off (all shell and tail fins removed with segments unslit and vein not removed).
4. Peeled, round, tail off (all shell and tail fins removed with segments unslit and vein not removed).
5. Peeled and deveined, round, tail off (all shell and tail fins removed with segments shallowly slit to last segment and vein removed to last segment).
6. Peeled and deveined, round, tail on (all shell removed except last shell segment, and tail fins, with segments shallowly slit to last segment and vein removed to last segment).
7. Peeled and deveined, fantail or butterfly tail off (all shell and tail fin removed with segments deeply slit to last segment and vein removed to last segment).
8. Peeled and deveined, fantail or butterfly, tail on (all shell removed except last shell segment and tail fins, with segments deeply slit to last segment and vein removed to last segment).
9. Peeled and deveined, western (all shell removed except last shell segment and tail fins, with segments split completely to last segment and vein removed to last segment).
10. Shell on pieces (head removed, shell and tail fins when existing not removed).
11. Peeled and deveined, round pieces (all shell removed with segments shallowly slit except last segment when existing; vein removed except last segment removed with segments deeply slit except last segment when existing; vein removed except last segment when existing).
12. Peeled undeveined pieces (all shell removed).

Evaluation of Flavor and Odor. Sensory evaluation of flavor and odor on each of the sample units should be carried out only by those trained to do so. For raw odor evaluation, frozen shrimp should be thawed. Fresh or thawed shrimp are broken and the broken flesh is held close to the nose immediately to detect off-odor. Cooked style shrimp should be evaluated for flavor and odor as is, thawed if frozen. Sensory evaluation of cooked samples should be completed as soon as practical after cooking while the samples are still warm.

Good flavor and odor means that the product has the normal, pleasant flavor and odor characteristic of freshly caught shrimp that is free from off-flavors and odors of any kind. A natural odor or flavor reminiscent of iodoform is acceptable unless excessive. Reasonably good flavor and odor means that the product may be somewhat lacking in good flavor and odor characteristic of freshly caught shrimp but is free from objectionable off-flavors and off-odors of any kind. Minimum acceptable flavor and odor means that the raw product and the cooked product have a moderate storage-induced odor (for the raw product) and flavor and odor (for the raw product) and flavor and odor (for the cooked product), but the product is reasonably free from any objectionable off-flavors and off-odors that may be indicative of spoilage or decomposition.

Evaluation of Physical Characteristics and Defects. Each sample unit should be evaluated as to physical characteristics and defects in accordance with the following definitions and methods of analysis. Dehydration refers to a general desiccation of the shrimp flesh that is noticeable after the shell and glaze are removed. It is evaluated by noting any detectable change from the normal, characteristic bright appearance of freshly caught, properly iced or properly processed shrimp. The degree of dehydration is as follows:

1. Slight dehydration means scarcely noticeable dessication of the shrimp flesh that will not affect the desirability or eating quality of the shrimp.
2. Moderate dehydration means conspicuous desiccation of the shrimp flesh that will not seriously affect the desirability or eating quality of the shrimp.
3. Severe dehydration means conspicuous dessication that will seriously affect the desirability or eating quality of the shrimp.

Deterioration is evaluated by noting any detectable change in the normal good odor of freshly caught, properly iced or properly processed shrimp.

1. Slight deterioration means that overall the sample unit lacks the normal characteristic pleasant odor of freshly caught, properly iced or properly processed shrimp; the desirability or eating quality of the shrimp is not affected.

2. Moderate deterioration means that overall the sample unit has scarcely noticeable odors of prolonged storage off-odors that materially affect the desirability or eating quality of the shrimp.

3. Severe deterioration means that overall the sample unit has definite odors of prolonged storage or spoilage odors that seriously affect the desirability or eating quality of the shrimp.

Fresh or thawed shrimp (glaze removed) should be visually examined for the presence of broken or damaged shrimp. Broken or damaged shrimp, identified as follows, are grouped together and evaluated by noting the percent by weight of such broken or damaged shrimp in the total net weight of the sample unit.

1. Broken means a shrimp having a break in the flesh greater than one-third of the thickness of the shrimp measured where the break occurs.

2. Damaged means a shrimp that is crushed or mutilated so as to materially affect its appearance or usability.

Frozen Raw Breaded Shrimp: Processing Quality and Standards

Frozen raw breaded shrimp are whole, clean, wholesome, headless, and peeled shrimp that have been deveined, where applicable, of the regular commercial species, coated with a wholesome, suitable batter or breading. Whole shrimp consist of five or more segments of unmutilated shrimp flesh. They are prepared and frozen in accordance with good manufacturing practice and are maintained at temperatures necessary for the preservation of the product.

Frozen raw breaded shrimp should contain not less than 50% by weight of shrimp material. The styles of shrimp are as follows:

1. Regular breaded shrimp are frozen raw breaded shrimp containing a minimum of 50% of shrimp material.

2. Lightly breaded shrimp are frozen raw breaded shrimp containing a minimum of 65% of shrimp material.

Types of breaded shrimp are breaded fantail shrimp and breaded round shrimp. Breaded fantail shrimp includes the following:

1. Split (butterfly) shrimp with the tail fin and the shell segment immediately adjacent to the tail fin.

2. Split (butterfly) shrimp with the tail fin but free of all shell segments.

3. Split (butterfly) shrimp without attached tail fin or shell segments.

Breaded round shrimp includes the following:

1. Round shrimp with the tail fin and the shell segment immediately adjacent to the tail fin.

2. Round shrimp with the tail fin but free of all shell segments.

3. Round shrimp without attached tail fin or shell segments.

Hygienic processing of frozen raw breaded shrimp includes processing and maintenance in accordance with the applicable requirements of the good manufacturing practice regulations. Good flavor and odor means that the cooked product has flavor and odor characteristics of freshly caught or well-refrigerated shrimp and the breading is free from staleness and off-flavors and off-odors of any kind. Iodoform is not to be considered in evaluating the product for flavor and odor. Reasonably good flavor and odor means that the cooked product may be somewhat lacking in the good flavor and odor characteristics of freshly caught or well-refrigerated shrimp but is free from objectionable off-flavors and objectionable off-odors of any kind.

Dehydration refers to the occurrence of whitish areas on the exposed ends of the shrimp (due to the drying of the affected area) and to a generally desiccated appearance of the meat after the breading is removed. Deterioration refers to any detectable change from the normal good quality of freshly caught shrimp. It is evaluated by noting in the thawed product deviations from the normal odor and appearance of freshly caught shrimp.

Extraneous material consists of nonedible material such as sticks, seaweed, shrimp thorax, or other objects that may be accidentally present in the package. Halo means an easily recognized fringe of excess batter and breading extending beyond the shrimp flesh and adhering around the perimeter or flat edges of a split (butterfly) breaded shrimp. Balling up means the adherence of lumps of the breading material to the surface of the breaded coating, causing the coating to appear rough, uneven, and lumpy. Holidays means voids in the breaded coating as evidenced by bare or naked spots. Damaged frozen raw breaded shrimp means frozen raw breaded shrimp that have been separated into two or more parts or that have been crushed or otherwise mutilated to the extent that their appearance is materially affected. Black spot means any blackened area that is markedly apparent on the flesh of the shrimp. Sand vein means any black or dark sand vein that has not been removed, except for that portion under the shell segment adjacent to the tail fin when present.

Breading of shrimp has long posed a problem from an economic standpoint. In addition, the time–temperature abuses present a great potential for food poisoning organisms. The growing scarcity and consequential high value of the raw material make the breading standards even more important.

Abbreviated Inspection

During an abbreviation inspection of a shrimp processing plant, an FDA inspector is advised of the following:

1. Check for the presence of cats, dogs, birds, or vermin in the plant.

2. Review testing of incoming shrimp. Check results of tests for decomposition, bacterial load, pesticides, and other possible adulterants.

3. Evaluate operation for compliance with standards for raw breaded shrimp.

4. Watch for any time—temperature abuses in the handling of seafood.

5. Determine that employee hygienic practices are satisfactory, eg, clean dress, washing of hands, and use of 100 ppm chlorine equivalent hand sanitizers.

6. Note any equipment defects that cause seafood to lodge, decompose, then dislodge into the pack.

7. Observe breading operations for suspected excesses, lack of coolant to keep batter mix below 50°F in an open system and below 40°F in a closed system.

8. Check labeling declarations while weighing 20 retail units against the net weight statement on four different products or sizes.

Comprehensive Inspection

During a comprehensive inspection, the FDA inspector will note the following.

Raw Materials: Receipt and Storage. The inspector will determine if:

1. Shrimp and other raw materials are inspected on receipt for decomposition, microbial load, pesticides, and filth.

2. Raw materials susceptible to microbial contamination are received under a suppliers guarantee.

3. Raw material specifications exist and only wholesome raw materials are accepted into active inventory; determine disposition of rejected raw materials.

4. Shrimp receiving and storage facilities are physically adequate.

5. Frozen shrimp are stored at −18°C (0°F) or below.

6. Fresh or partially processed shrimp are iced or otherwise refrigerated to maintain a temperature of 4°C (40°F) or below until they are ready to be processed.

7. Decomposed shrimp are being processed.

   a. Examine shrimp as received and again after sorting, for decomposition; classify as passable (class 1), decomposed (class 2), or advanced decomposition (class 3); less-experienced inspectional personnel should submit some of class 2 and class 3 shrimp for confirmation by the laboratory.

   b. Prompt handling and adequate sorting is necessary to prevent decomposition; check times and temperatures.

   c. Where decomposed shrimp are going into canned or cooked-peeled shrimp, collect investigational samples of the finished pack; give attention to disposition of loads showing a high percentage of decomposition that cannot be adequately sorted,

and to disposition of reject shrimp; make certain that bait shrimp is denatured.

8. Fresh raw shrimp are washed and chilled to 4°C (40°F) or below within 2 h of receipt; frozen shrimp should be held at −18°C (0°F) or below; determine if they are examined organoleptically when received.

9. Peeled and deveined shrimp are promptly chilled to 4°C (40°F) or below.

Plant Sanitation. The inspector will determine if:

1. The water (ice) is
   a. From an approved source.
   b. Disinfected and contains residual chlorine.
   c. Sampled and analyzed for contamination.
   d. Handled in a sanitary manner.

2. Drainage facilities are adequate to accommodate all seepage and wash water.

3. The plant has readily cleanable floors that are sloped and equipped with trap drains.

4. The plant is free of the presence of vermin, dogs, cats, or birds.

5. The screening and fly control are adequate.

6. Offal, debris, refuse is placed in covered containers and removed at least daily or continuously.

7. Adequate handwashing and sanitizing facilities are located in processing area and are easily accessible to the preparation, peeling, and subsequent processing operations.

8. Signs are posted directing employees handling shrimp and other raw materials to wash and sanitize their hands after each absence from the workstation.

9. Employees actually wash and sanitize their hands as necessary (before starting work, after absences from the workstation, when hands become soiled, etc).

10. Hand-sanitizing solutions are maintained at 100 ppm available chlorine or the equivalent and are used.

11. Persons handling food or food contact surfaces wear clean outer garments, maintain a high degree of personal cleanliness, and conform to good hygienic practices.

12. Management prevents any person known to be affected with boils, sores, infected wounds, or other sources of microbiological contamination from working in any capacity in which there is a reasonable probability of contaminating the food.

13. The product is processed to prevent contamination by exposure to areas involved in earlier processing steps, refuse, or other objectionable conditions or areas.

14. Food contact surfaces are constructed of metal or other readily cleanable materials.

15. Seams are smoothly bonded to prevent accumulation of shrimp, shrimp material or other debris.

16. Each freezer and cold storage compartment used for raw materials, in process or finished product is fitted with required temperature indicating devices.

17. Unenclosed batter application equipment is flushed and sanitized at least every 4 h during plant operations and all batter application equipment is cleaned and sanitized at the end of and the beginning of the days operation.

18. Breading application equipment and utensils are thoroughly cleaned and sanitized at the end of the days operations.

19. Utensils used in processing and product contact surfaces of equipment are thoroughly cleaned and sanitized at least every 4 h during operation.

20. All utensils and product contact surfaces excluding breading application equipment and utensils are rinsed and sanitized before beginning the days operation.

21. Containers used to convey or store food are handled in a manner to preclude direct or indirect contamination of the contents.

22. The nesting of containers is prohibited.

Processing. The inspector will determine if:

1. Raw frozen shrimp are defrosted at recommended temperatures: air defrosting at 7°C (45°F) or below or in running water at 21°C (70°F) or below in less than 2 h.

2. Fresh raw shrimp are washed in clean potable water and chilled to 4°C (40°F) or below.

3. Fresh shrimp are adequately washed, culled, and inspected.

4. Every lot of shrimp that has been partially processed in another plant, including frozen shrimp, is inspected for wholesomeness and cleanliness.

5. Shrimp entering thaw tank are free from exterior packaging material and substantially free of liner material.

6. On removal from thaw tank, shrimp are washed with a vigorous potable water spray.

7. Shrimp are removed from thaw tank within 30 min after they are thawed.

8. During the grading, sizing, or peeling operation the:
   a. Equipment is cleaned and sanitized before use.
   b. Water is maintained at proper chemical strength and temperature.
   c. Raw materials are protected from contamination.

9. Sanitary drainage is provided to remove liquid waste from peeling tables.

10. Firm prohibits the practice of salvaging shrimp (ie, repacking the accumulated hulls and shells for missed shrimp or shrimp pieces).

11. Peeled and deveined shrimp are promptly chilled to 4°C (40°F) or below.

12. Peeled shrimp are transported from peeling machines or tables immediately, or, if containerized, within 20 min.

13. Peeled shrimp containers, if applicable, are cleaned and sanitized as often as necessary, but in no case less frequently than every 3 h.

14. When a peeler is absent from his duty post, his container is cleaned and sanitized prior to resuming peeling.

15. Peeled shrimp that are transported from one building to another are properly iced or refrigerated, covered, and protected.

16. Shrimp are handled minimally and protected from contamination.

17. Shrimp that drop off processing line are discarded or reclaimed.

18. Shrimp are washed with a low-velocity spray or in unrecirculated flowing water at 10°C (50°F) or below just prior to the initial batter or breading application, whichever comes first, except in cases where a predust application is included in the process.

19. Removal of batter or breading mixes or other dry ingredients from multiwalled bags is accomplished in an acceptable manner.

20. Batter in enclosed equipment is insured a temperature of not more than 4°C (40°F) and disposed of at the end of each work day, but in no circumstances less often than every 12 h.

21. Batter in an unenclosed system is maintained at or below 10°C (50°F) and disposed of at least every 4 h and at the end of the day's operation.

22. Breading reused during a day's operation is sifted thru a 0.5-in. or smaller mesh screen.

23. Breading remaining in the breading application equipment at the end of a day's operation is reused within 20 h and is sifted as above and stored in a freezer in a covered sanitary manner.

24. Hand batter pans are cleaned, sanitized, and rinsed between each filling with batter or breading.

Finished Product Process and Quality Assurance. The inspector will determine if:

1. Processing and handling of finished product is
   a. Performed in a sanitary manner.
   b. Protected from contamination.
   c. Arranged to facilitate rapid freezing.

2. Manual manipulation of breaded shrimp is kept to a minimum.

3. Aggregate processing time, excluding the time required for thawing frozen material, is less than 2 h, exclusive of iced or refrigerated storage time.

4. Breaded shrimp are placed into freezer within 30 min of packaging.

5. Breaded shrimp are frozen in a plate or blast freezer at −29°C (−20°F) or below.

6. Storage freezer is maintained at or below −18°C (0°F).

7. In-line, environmental, and finished product samples are analyzed and evaluated at least weekly for microbial conditions; review these analytical records if available.

8. Firm has established microbiological specifications for the final product; if so, review and report these specifications.

9. Firm withholds from distribution lots that do not meet their established microbiological standards.

10. Finished product is handled and stored in a manner that precludes contamination.

11. Labels bear a cautionary statement to keep product frozen.

12. Packaged finished product bears a label code; report interpretation or breakdown of firm's coding.

GENERAL REFERENCES

Code of Federal Regulations, Titles 21 and 50.

Y. H. Hui, *United States Food Laws and Regulations*, 2 vols., Wiley-Interscience, New York, 1986.

Y. H. HUI
American Food and Nutrition Center
Cutten, California

SNACK FOOD TECHNOLOGY

Snacks can be considered as tasty, savory, or sweet foods eaten at nonmeal occasions. Indeed, *Snack World's* 1987 Consumer Attitude Survey indicated that salted snacks are most often eaten while watching television, at parties, or between meals (1). With the trend today toward grazing, the boundaries of snack foods are being broadened. In such a view it is possible to consider the influx of finger foods and microwaveable items such as dough enrobed meat and cheese, pizza, and hot dogs as snacks. *Snack Food* magazine (2) includes the following categories: candy, cookies–crackers, frozen pizza, and snack cakes–pies, along with such traditional salted items as potato chips, tortilla chips, nuts, pretzels, popcorn, granola items, dried fruit, extruded snacks, and meat snacks in its annual review of the snack food market. While a strong case can be made for including all of these items, only the technology for potato chips, tortilla chips, popcorn, pretzels, extruded items, and meat snacks will be considered in this overview due to space constraints.

BUSINESS AND TRENDS

Of the salted snacks, potato chips have the greatest share of the market, followed by corn and tortilla chips, nuts, dried fruit, popcorn, extruded snacks, meat snacks, pretzels, and granola. Estimated sales (2) for representative items are found in Table 1. The strong growth of the popcorn sector is directly attributable to microwave popcorn

sales. Tortilla and corn chips have shown sales growth in each year from 1978 to 1989. Total sales have more than doubled in that period. During the mid-1980s granola product sales climbed to $439 million but have been declining since. Currently, dried fruit snack sales are increasing, presumably due to their wholesome image. There also has been a sharp increase in imported items and healthful products. Rice cakes have had a strong market impact because they are low in salt, fat, and calories. In addition, there has been a proliferation of miscellaneous items including flat breads, plantain and apple chips, bagel crisps, and sourdough bread crisps.

POTATO CHIPS

Potato chip manufacture is much more involved than the mere frying of thinly sliced potatoes. Potatoes are living organisms and as such have respiration and energy requirements. Because it is impossible to provide a year-round supply of potatoes to the chipper, potatoes must be stored. During storage, the plant's energy requirements are met by conversion of starch to sugars through enzymatic action. After a short curing period that allows cuts and bruises to heal, chipping potatoes are placed in well-ventilated cold storage (95% RH and 50–55°F) (3). However, during storage there is an accumulation of the reducing sugars, glucose, and fructose, which lead to unacceptably dark chips (4). Stored potatoes must be reconditioned to reconvert the reducing sugars back to starch through enzymatic action. Reconditioning is carried out by holding potatoes at 65–70°F until frying trials produce chips of the desired color, usually in one to four weeks (3). Potatoes with higher specific gravities are desired because they have increased chip yield and reduced oil consumption (4).

The block flow diagram of Figure 1 outlines a large commercial chipping process. After dumping, the potatoes are conveyed to a washer–destoner. Often, the conveying system is a water flume where a large centrifugal pump moves potatoes and water around the plant. The turbulence of the water helps remove dirt and stones, which are separated by gravity and are removed by a conveyor. Other types of destoner include an elevating screw and a bubble unit in which jets of air or water tumble the potatoes and dislodge

Table 1. Estimated Sales by Manufacturers (Millions of Dollars)

Snack food	1988	1987	1986
Potato chips	2,890	2,850	2,651
Corn and tortilla chips	1,401	1,324	1,283
Snack nut meats	1,098	1,160	1,160
Dried fruit snacks	587	508	418
Popcorn	554	458	365
Extruded snacks	472	467	415
Meat snacks	424	400	325
Pretzels	303	293	271
Granola snacks	298	359	439

Source: Ref. 2.

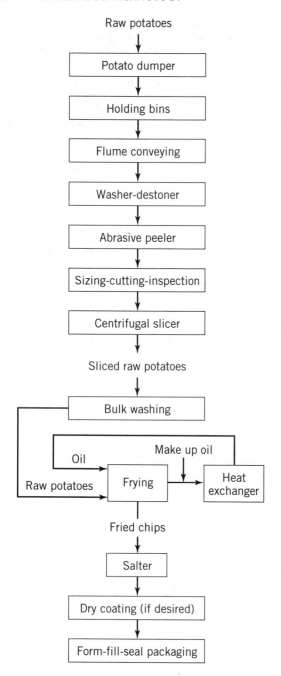

Figure 1. Commercial potato chip process flow diagram.

debris. Once washed, the potatoes are abrasively peeled. The peeler contains long cylindrical rollers covered with a rough material. The units are designed to provide uniform tumbling of potatoes against the rollers. Steam and chemical (lye) peelers are generally not used with chips because such methods lead to unsightly rings, while chemicals can carry over to the oil. Peeled potatoes can be sized, if necessary. Most North American manufacturers use rotary centrifugal slicers. Potatoes are fed into the center of a rotating bowl where centrifugal action and carefully designed impellers force the potatoes against knives mounted in the stationary outer housing. A large variety

of knives are available including straight, corrugated, wave, and julienne types. To maintain efficient cutting, blades must be replaced frequently. Raw slice thickness is about 1/16 in. Slice thickness must be adjusted to compensate for such variables as potato variety, physical condition, specific gravity, consumer preference, and ease of processing. Sliced potatoes are washed and sprayed with either cold or hot water to remove surface starch and sugars. After draining, they are fried.

Modern potato fryers incorporate a variety of devices to insure long oil life and uniformity in frying conditions. Hot oil, 360–380°F, is pumped into the front end of the fryer where the washed slices enter. The turbulent action of water escaping help separate the slices. The velocity of the oil and the action of conveying paddles helps move the chips downstream. To achieve final frying to less than 1.5% moisture, the last section of the fryer contains a hold-down conveyor to uniformly remove the last traces of water from the buoyant chips. During frying the oil temperature drops by about 40°F. External heat exchangers are used to bring the oil back to the desired inlet temperature (5). After frying, chips are salted and optionally flavored before they are packaged. Packaging for chips should include moisture, oxygen, and light barriers.

KETTLE CHIPS

A recent market trend has been the emergence of crunchy and crispy, kettle type, homestyle, or hand cooked chips. Kettle chips achieve their desired crispy texture and stronger taste through batch frying. Raw, unwashed sliced potatoes are added to a batch or kettle of hot oil, which causes the oil temperature to drop. The chips fry as the oil temperature recovers. Frying times in the kettle are almost three times longer than in the continuous fryer (nine minutes vs three). It is believed that some large producers have simulated kettle fry conditions in the large continuous units.

TORTILLA CHIPS AND CORN CHIPS

Modern methods for the production of corn snacks such as tortilla chips and corn chips have origins in the ancient Aztec process of nixtamalization or alkaline cooking and steeping of corn (6). In nixtamalization whole corn is cooked in excess water containing lime. A mixture of white and yellow corn is used because corn chips made from 100% yellow corn have an unacceptable flavor profile probably resulting from carotenoid degradation (7). After cooking, the corn is allowed to steep overnight. During the process, some starch is gelatinized, the pericarp or hull disintegrates, and the endosperm hydrates and softens. After washing, the cooked corn or the nixtamal is stone ground into masa (6,7).

A modern tortilla chip process is outlined in Figure 2 and a more detailed account follows. Corn is cooked in excess water and lime (1% of corn weight) at temperatures above 200°F. Water-to-corn weight ratios range from 1.5 to 2.0:1. Cooking times vary depending on end use of the masa but are typically in the 10–30 min range. After cook-

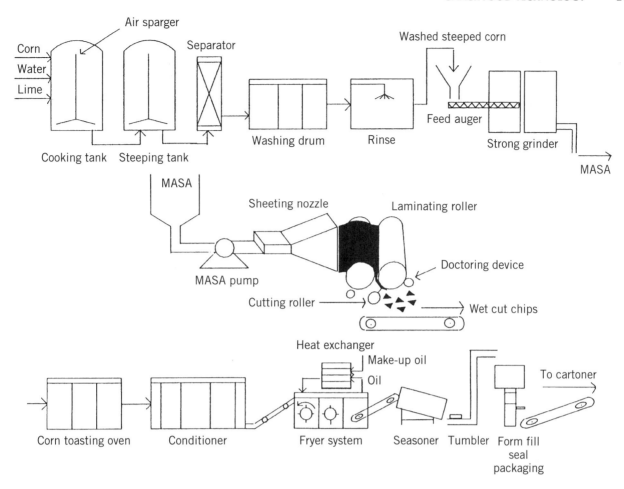

Figure 2. Tortilla chip process outline.

ing, the corn is quenched to below 150°F to stop starch gelatinization. The cooked corn is then held in water or steeped from 8 to 24 h. At this point the moisture content has reached 45–50%. After steeping, the corn is pumped to a washing drum where lime and loose hulls are removed. The nixtamal is ground to masa in a stone grinding mill. Stone grinding is preferred because it provides the range of particle sizes needed for optimum frying performance and texture. Water added to the mill to facilitate grinding brings moisture of the ground corn or masa to 50–54%. The ground masa is immediately pumped to a spreading device that deposits a 0.5–1-in.-thick sheet onto sizing rollers. The rollers extrude the masa into an extremely thin sheet (less than 1/16 in), which is cut into the desired shape with roller cutters. The wet, newly cut chips, which lie on the belt in a monolayer are baked at extremely high temperatures (600–850°F) for 10–20 s. During the baking some moisture is driven off and a dry skin is formed around a moist interior. To prepare the baked chips for frying, the chips pass through a conditioning chamber where the moisture of the chip equilibrates and the chip cools. If moisture equilibration did not take place, the chips would expand and blister in the fryer. Conditioning equipment varies widely and consequently equilibration times range

from 15 s to 15 min. In the conditioner the baked chips loose an additional 2% water and enter the fryer at 35–38% (7).

Tortilla chip fryers resemble potato chip fryers. They usually contain paddles to keep the chips moving downstream. The last section of the fryer contains a hold-down belt, which submerges the buoyant chips and ensures that all the chips will contain less than 1.5% water. Due to the lower water content of the tortilla chips, frying times are shorter than for potato chips, about 1 min at 350–370°F. The fat content of fried tortilla chips is about 25%. Chips are seasoned and coated while hot, then packaged. The process for fried corn chips is very similar to the tortilla chip process. Longer cook times and an increased water-to-corn ratio produces a somewhat softer corn. After stone grinding, the resulting masa contains 50–52% water and has a particle size range somewhat more coarse than the tortilla masa. The wet masa is extruded and cut into pieces that fall into very hot frying oil, about 400–410°F. The coarse structure of the masa ensures that pathways exist for the water to escape. Otherwise an expanded blistered chip would result. The fried chips are more friable than the tortilla chips and contain about 35% fat.

POPCORN

Why does popcorn pop when heated at atmospheric pressure and why do other cereal grains such as wheat, barley, rice, and dent corn fail to do so? The answer lies in the structure of the popcorn kernel itself. Popcorn kernels contain a strong pericarp or hull and a high percentage of translucent endosperm cells where densely packed starch granules are located. Most other grains contain opaque cells that have many intergranular spaces. During popping, the hull serves as a pressure vessel that allows the moisture held in the kernel to turn into superheated steam. Eventually the hull can no longer withstand the internal pressure and it fails. Rupture occurs at temperatures in the range 350–375°F (8). These temperatures correspond to saturated steam pressures inside the kernel of 135–185 psi. Microscopic studies have shown that after popping, cell walls are generally intact (8,9). It is believed that the tight packing of the translucent cells helps the escaping steam to create the foamed network. Opaque starch contains air spaces that provide channels for the escaping steam; consequently it does not foam or expand.

Popcorn quality, as perceived by the consumer, is strongly related to its expanded volume. Highly expanded popcorn is desired for its tenderness and "melt-in-the-mouth" characteristics. Kernel moisture and kernel test weight (bulk density or number of kernels per 10 g) both strongly influence the amount of expansion. Maximum expansion of popcorn typically occurs in the 12.5–15% total moisture range (8). However, each hybrid has its own optimum moisture content. The level of expansion at moistures below 10% is poor (8). Among hybrids, those with the highest test weights have the greatest expanded volumes (10). The percentage of totally popped kernels is strongly related to both popping temperature and moisture content. In one study, when the temperature of cooking oil was raised from 345 to 350°F the percentage of totally popped kernels rose from 70 to 94%. Similarly reduction in moisture content from 11.1 to 9.9% caused a reduction from 100% popping to 82% (8).

Commercial Popping

Today, most large commercial poppers use continuous dry or hot air poppers. Continuous poppers generally consist of a rotating drum with helical flights for conveying the popcorn. Hot air (410–430°F) is blown in from the feed end and heats the tumbling kernels, which are metered into the drum from above. Unpopped kernels, small pieces and hulls are separated with a screen, then the popped kernels are cooled and coated if desired. Because fresh popcorn is extremely hygroscopic, it must be immediately packaged in a moisture barrier, eg, foil, to preserve freshness. Coatings range from mixtures of oil, salt, seasonings, and coloring to candy caramel and nuts. Caramel corn is made by combining fresh popcorn with caramel, a molten mixture of sugar, glucose, butter, sodium bicarbonate, and flavor in a specialized agitated vessel. After coating, the caramel corn is cooled and separated into the appropriate cluster size before packaging.

Microwave Popcorn

Through the late 1980s microwave popcorn has been one of the fastest growing grocery product categories in the United States. By the end of 1987 annual microwave popcorn sales were estimated at $351 million, up from $124 million in 1985 and $246 million in 1986 (2). Most microwave popcorn is packed in an expandable bag that incorporates design features that focus or concentrate the microwave energy on a mixture of solid fat, popcorn and salt. Popcorn-to-fat ratios are approximately 3:1. The bags are generally designed to expand to allow room for the popped corn (11). Early commercial microwave popcorn products were refrigerated or frozen to provide fat and moisture stability. Newer packaging materials such as polyester-lined kraft paper provide about six months shelf stability for the fat-covered kernels and eliminate the need for freezing or refrigeration. The introduction of susceptors (metallized films that are good absorbers of microwave energy) to the microwave popcorn package has improved poppability by providing a localized source of intense heat (400–450°F) beneath the kernel–oil mixture.

EXTRUDED SNACKS

Extruded snack products fall into two broad categories: directly expanded or puffed, and pellets (dense half-products). Extrusion literally means forcing out. Many food formers such as hydraulic presses and double rollers are called extruders. However, the extruders used in snack food processing resemble those of the plastic industry. Rotating screws enclosed in a barrel convey a cereal dough and force it through a die. Products are cut to size with rotating knives. Both single and twin-screw extruders are used. Twin-screw extruders are more complex, more flexible, more easily controlled and operate over a wider range of moisture conditions. However, they are more costly. Single-screw extruders still are used extensively, particularly in the manufacture of expanded corn snacks and pellets.

Direct Expansion

Directly expanded snacks are made by extruding cereal grains and flours at rather low moisture levels (12–16%) and forcing them through a die. Corn grits are the most common ingredient. Rice flour, wheat, potato starch flour, and tapioca all expand well. Due to its high native fat content, oat flour does not provide the expanded puffy texture desired in a snack base. In puffing extrusion friction and shear combine to heat and plasticize the grains or flours into a pressurized mass of molten gelatinized dough. The pressure at the die end of the extruder is much greater than the saturated pressure of steam at the temperature where the dough is extruded, about 350°F. Thus the water becomes superheated. Consequently when the dough is forced through the die into the atmosphere, water and steam are released explosively and cause the gelatinized starch to expand into a foamed structure that sets up on cooling. The extruders used for puffed snacks range from extremely short length single-screw collet extruders that have a length-to-diameter ratio (L/D) of only 2–3 to 1 to long single- and twin-screw models of 20:30 L/D (12). Expanded puffs contain 5–9% water and must be dried or baked to below 2% moisture to achieve the desired crispness. Once dried, puffs are coated with oil and flavors. Cheese puffs are an example of a baked, coated expanded snack.

Fried Expanded

In a variation of direct expansion, corn grits of 15–18% water are extruded in a collet type extruder that contains a die or auxiliary rolling device modified to restrict the expansion of corn grit extrudate. The extrudate, in the form of irregular rods, is deep fried to remove residual water, impart a fried flavor, and slightly expand the extrudate. After frying the rods are coated, usually with cheese flavoring.

Pellets

The use of extruders to manufacture pellets is actually a new version of an old process. A number of Asian cultures make a starch paste of about 50% water containing flavoring such as shrimp. The paste is molded, gelatinized with steam, sliced, dried, and fried to a puff. Extruded pellets are made in much the same way. A blend of flours and starches that have good puffing characteristics, such as corn and potato, is combined with water and flavor (if desired) and extrusion cooked to gelatinize the starch. Moisture content ranges from 25 to 35%. The cooked mass must be cooled before it is formed to avoid expansion and produce a dense, glassy pellet. If a twin-screw extruder is used the dough can be cooled by venting steam and by external heat transfer before it is forced through a pasta-type die and knife assembly. In another version, two single-screw extruders are used (13). The first cooks and gelatinizes. As the dough exits the first extruder, water evaporates, and cools the dough mass that has foamed. The second extruder recompresses the mass and forces it through a shaping die. Pellet shapes are virtually unlimited. Shapes include grills, wagon wheels, spirals, shells, tubes, and playing card suits among others. Freshly extruded pellets are tacky, therefore they must be predried with air to prevent sticking in the final dryer. Final drying takes place in a rotating drum dryer equipped with temperature and humidity control. Pellets of about 10% moisture emerge after 6–8 h of drying (14). The shelf-stable pellets are extremely versatile and can be fried in only 10–15 s, puffed in hot air without any fat and can even be puffed in a microwave. Fried pellets contain 20–25% fat.

PRETZELS

Pretzels are a baked item made from a very stiff dough containing flour, shortening, malt, yeast, salt and sodium bicarbonate. Intricate machinery is used to make the classic twist pretzels from cut lengths of dough. Today many twist pretzels are extruded. Multiple sets of individually controlled augers force the dough through a die where the exiting dough is cut by wires. The pretzel receives its characteristic glaze and dark brown color as the result of a caustic dip. Before baking the raw, shaped pretzels are conveyed through a hot, 200°F, sodium hydroxide bath (0.5–1.25%) for about 10–15 s (15). The dipped pretzels are immediately sprinkled with enough salt so that about 2% remains on the finished product. Baking ovens vary considerably. In the United States many companies use a two-pass oven. In the first pass pretzels are baked at 425–475°F for about 5 min. At the end of the first pass the pretzels, now at 15% water, are removed from the baking belt with a doctor blade. A second belt returns the pretzels under the baking zone to dry them down to approximately 2% water.

NUTS

In the United States, peanuts are the major snack nut, commanding over a 50% share. Other major snack nuts include cashews, almonds, pistachios, macadamias, and sunflower seeds. Walnuts, pecans, and filberts are viewed primarily as in shell or baking commodities—not snacks. Most snack nuts are blanched (skin removed) and roasted to develop a nutty flavor and crunchy texture. Nuts are either roasted dry in air or in oil, in units that resemble fryers. Oil-roasted nuts are salted when the nuts are still warm to promote adhesion. A recent market trend has been honey roasting, where nuts are covered with a honey–water solution and coated with a sugar–starch mixture prior to roasting (16). Due to their high fat level nuts are extremely susceptible to oxidative rancidity. Consequently many snack nuts are vacuum packed.

MEAT-BASED SNACKS

Meat snack sales have grown steadily through the 1980s and now account for roughly 5% of the salted snack market. Major constituents of the segment are beef sticks, beef jerky, and popped pork skins.

Popped Pork Skins

Popped pork skins are produced from diced, rendered green pork skins. After a hot brine dip, skins are drained, cooled and diced into 0.5–1-in. pieces. Diced skins must be rendered, to remove fat and moisture. Skins are rendered by heating at 230–240°F in oil or lard that contains antioxidants and antifoaming agents (17). After 4 h or so of rendering, the diced pellets rise and are removed. Following cooling and draining the pellets are fried at 400–425°F to create a puffed product.

Jerky

Jerky probably has its origins with native Americans who simply stripped or jerked the muscle from game animals and dried it slowly in the wind or sun or over smoky fires. This has evolved to a process where beef is marinated, optionally smoked, and carefully dried. Restructured jerkys, made from ground meat, are gaining in popularity owing to ease of manufacture. Ground meat is mixed with curing aids, seasonings such as pepper and garlic, sugar, salt, dextrose, sodium nitrite and sodium erythorbate and extruded or formed into thin strips. Formed strips, which may be frozen are cured and dried for approximately 12 h at 150°F (18).

Beef Sticks

Beef sticks and other related meat stick products such as snack-size pepperoni are classified as fermented sausages. However, some meat snack sticks are acidulated, often with glucose-δ-lactone to achieve the desired pH and sharp taste. Chilled meat (32–34°F) is ground or chopped with a silent cutter into coarse pieces. Salt, spices, flavorings, curing salts and a fermentable carbohydrate such as dextrose is blended in along with a lactic acid starter culture, or A_w acidulant. In preparing fermented sausages it is crucial to control the amount of carbohydrate and the distribution of the starter culture (19). Good distribution ensures uniformity, whereas the amount of dextrose dictates the final pH. After stuffing, the sausages ferment at temperatures that range from 65 to 120°F but typically 90 to 110°F depending on the culture. Sausages can be stuffed into strippable cellulose casings or coextruded into reconstituted collagen. The end of fermentation is signaled by achievement of the final desired pH, eg, 4.8. Following fermentation, which may include smoking, the sticks are dried. Drying temperatures vary considerably. However, sausages that contain pork must reach an internal temperature of 137°F to destroy trichinae parasites (19).

PACKAGING

Snack packaging must answer to many challenges. Snacks are generally high in fat, very low in moisture, and often fragile. Both oxygen and light barriers are needed to combat rancidity. Potato chips in particular are subject to light damage. Moisture barriers are needed to prevent sogginess and to maintain crispness. Metallized films that protect against these elements have become the packaging material of choice. The opacity of the metallized film also offers the ability to hide crumbs from the consumer. Polyester and polypropylene laminates are common. Polypropylene is preferred when it is necessary to protect the bag or pouch from punctures from sharp snack pieces such as tortilla chips. Pouches or bags made on form–fill–seal machines have become the primary package for fragile, salted snacks due to their low relative cost, their ability to display attractive graphics and their ability to cushion. In the most modern plants, computer controlled, multiple head scaling units dose products into pouches with a high level of accuracy.

TRENDS

Although snacks are an indulgent item, they have not been immune from consumer demands for less fat, more fiber, low salt, and no cholesterol. Nutritional snacks such as rice cakes, dried fruit, and sunflower seeds have established strong market niches. No salt potato chips and pretzels are on the market, while the resurgence of popcorn may be due to its high fiber content. It is anticipated that more snacks will incorporate fiber. There is ample opportunity for introducing fiber into fabricated chips, extruded pellets, and tortilla chips. Technology is available for producing lower fat potato chips. It is expected that consumer demand will soon establish a segment for lower-calorie fried snacks despite the superior taste and texture of the full-fat products. Over the last few years flavors have become more sophisticated and spicier. Cajun, Mexican, cream cheese and dill, seafood, Italian (herbs and cheese), Indian (curry), and even hot pepper flavored snacks have all been introduced. Shape and texture are also becoming increasingly important. The introduction of extruded pellet technology has dramatically opened up the possibilities of snack shapes. Kettle potato chips have established a crunchier, tastier potato chip category. Rippled, ridged, corrugated, and undulated cuts offer diversity in texture. At the same time other snack segments are moving toward simplicity. Such products include white popcorn and uncoated tortilla chips. Sweet tastes have intruded into the salted snack market. Honey coating, introduced with nuts, has helped launch sweet coated popcorn and potato chips. Despite all the innovations, it is expected that traditional fried, salted potato chips will continue to lead the salted snack market.

BIBLIOGRAPHY

1. "1987 Snack Food Association Consumer Attitude Report," *Snack World* **44**, 1–35 (Sept. 1987).

2. "20th Annual State of the Snack Food Industries Report," *Snack Food* **77**, M1–M40 (June 1988).

3. R. E. Hardenburg, A. E. Watanda, and C. Y. Weng, *The Commercial Storage of Fruits and Vegetables, and Florist and Nursery Stocks*, USDA Agriculture Handbook no. **66**, U.S. Government Printing Office, Washington, D.C., 1986.

4. O. Smith, "Potato Chips," in W. F. Talburt and O. Smith, eds., *Potato Processing*, 3rd ed., AVI Publishing Co., Inc., Westport, Conn., 1975.

5. "*Potato Chip Processing*," Heat and Control, Inc., So. San Francisco, 1986.

6. M. S. Bedolla and L. W. Rooney, "Cooking Maize for Masa Production," *Cereal Foods World* **27**, 219–221 (May 1982).

7. M. H. Gomez, L. W. Rooney, R. D. Waniska, and R. L. Pflugfelder, "Dry Corn Masa Flours for Tortilla and Snack Food Production," *Cereal Foods World* **32**, 372–377 (May 1987).

8. R. C. Hoseney, K. Zeleenak, and A. Abdelrahman, "Mechanism of Popcorn Popping," *Journal of Cereal Science* **1**, 43–52 (Jan. 1983).

9. R. M. Reeve and H. G. Walker, "The Microscopic Structure of Popped Cereals," *Cereal Chemistry* **46**, 227–241 (May 1969).

10. C. G. Haugh, R. M. Lien, R. E. Hanes, and R. B. Ashman, "Physical Properties of Popcorn," *Trans. ASAE* **19**, 168–171, 176 (Jan.–Feb. 1976).

11. U.S. Pat. 4,450,180 (May 22, 1984), J. D. Watkins (to Golden Valley Foods, Inc.).

12. G. Toft, "Snack Foods: Continuous Processing Techniques," *Cereal Foods World* **24**, 142–143 (Apr. 1979).

13. O. B. Smith, "Extrusion—Cooked Snacks in a Fast Growing Market," *Cereal Science Today* **19**, 312–316 (Aug. 1974).

14. M. Byrne, "The Shape of Things to Come," *Food Manufacture*, 59–61 (Oct. 1982).

15. S. A. Matz, *Snack Food Technology*, 2nd ed., AVI Publishing Company, Inc., Westport, Conn., 1984, pp. 180–182.

16. U.S. Pat. 4,161,545 (July 17, 1979), W. M. Green and M. W. Hoover.

17. S. A. Matz, *Snack Food Technology*, 2nd ed., AVI Publishing Company, Inc., Westport, Conn., 1984, pp. 130–133.

18. U.S. Pat. 4,239,785 (Dec. 16, 1980), E. N. Roth.

19. L. Long, S. L. Komarik, and D. K. Tressler, *Food Products Formulary, Vol. I, Meats, Poultry, Fish, Shellfish*, 2nd ed., AVI Publishing Company, Inc., Westport, Conn., 1982.

GENERAL REFERENCES

M. Byrne, "The Boom in Snack Foods," *Food Eng. Int.* **18**, 49–50, 52, 54, 57–58 (1993).

J. I. Duffy, ed., *Snack Food Technology Recent Developments*, Noyes Data Corporation, Park Ridge, N.J., 1981.

B. Gebhardt, "Oils and Fats in Snack Foods," in Y. H. Hui, ed., *Bailey's Industrial Oil and Fat Products*, Vol. 3, 5th ed., John Wiley & Sons, New York, 1996, pp. 409–427.

R. C. E. Guy, "Snack Foods Product Design II, Snack Food Formulation," *Food Ingred. Analysis Int.*, 25–26 (1997).

J. M. Harper, *Extrusion of Foods*, Vols. 1 and 2, CRC Press, Inc., Boca Raton, Fla., 1981.

M. Hilliam, "Trends in Snack Foods," *Food Processing* **66**, 14–15 (1997).

R. Holding, "Coextruded Three Dimensional Snack Foods," *Proc. Grain Processing Asia* **95**, 56–57 (1995).

P. Hollingsworth, "Snack Foods," *Food Technol.* **49**, 58, 60, 62 (1995).

D. Howling, "Convenience and Snack Foods," *Eur. Food and Drink Rev.*, 69, 71 (1991).

K. J. Lee, "Utilization of Korean Maizes in Production of Alkaline Processed Snack Foods," *J. Food Sci. Nutrition* **2**, 11–16 (1997).

S. A. Matz, *Snack Food Technology*, 2nd. ed., AVI Publishing Co., Westport, Conn., 1984.

T. M. Midden, "Impingement Air Baking for Snack Foods," *Cereal Foods World* **40**, 532–535 (1995).

M. N. Riaz, "Technology of Producing Snack Foods by Extrusion," *Tech. Bull. Amer. Inst. Baking Res. Dept.* **19**, 1–8 (1997).

T. P. Shukla, "Future Snacks and Snack Food Technology," *Cereal Foods World* **39**, 704–705 (1994).

W. F. Talburt and O. Smith, *Potato Processing*, 3rd. ed., AVI Publishing Co., Westport, Conn., 1975.

P. Tettweiler, "Snack Foods Worldwide," *Food Technol.* **45**, 58, 60, 62 (1991).

ROBERT E. ALTOMARE
D. MARK KETTUNEN
CHARLES J. CANTE
General Foods USA
Tarrytown, New York

SOLID-STATE FERMENTATION AND VALUE-ADDED UTILIZATION OF FRUIT AND VEGETABLE PROCESSING BY-PRODUCTS

SOLID-STATE FERMENTATION

Definition

Solid-state fermentation, or solid-substrate fermentation (SSF), while difficult to define precisely, is generally referred to as the process in which microbial growth and product formation occur on the surface of solid materials in the absence or near-absence of free water, and the substrate contains certain moisture that exists in absorbed form within the solid matrix (1,2).

Solid-state fermentation deals with the utilization of water-insoluble materials for microbial growth and metabolic activities. It is different from surface culture, which uses either a solid or liquid substrate, and refers primarily to the mode of growth (3). It is also distinguished from submerged liquid fermentation/culture by the fact that microbial growth and product formation occur at or near the surfaces of solid materials with low moisture contents. Microbial activities cease at a certain low level of moisture content, and this establishes the lower limit at which solid-state fermentation can take place (3). The upper limit for solid-state fermentation is a function of absorbency of the medium which varies with the substrate material type (3).

Solid-state fermentations are not as well as characterized on a fundamental scientific or engineering basis as are the submerged liquid cultures that have been used almost exclusively in the West for the industrial production of microbial metabolites (2). They are, however, widely used in the orient for thousands of years, and traditional methods used in food processing have been modernized and extended to nontraditional products (1,2). Today solid-state fermentation is increasingly gaining interest in the world for development of value-added products from a variety of cheap materials, and for bioremediation of agricultural and industrial wastes.

History

Solid-state fermentation has been used long before the underlying microbiological or biochemical processes involved were understood. The use of naturally occurring microorganisms in the preparation of foods such as bread and cheese, or directly as food such as mushrooms, dates back many centuries, and these are some examples of traditional solid-state fermentation systems (4). As early as 2600 B.C., Egyptians were making bread by methods essentially similar to those of today (4). In Asia, cheese had been prepared as food for several hundred years before the birth of Christ (4). The preparation of koji for soy sauce and miso production in Japan and Southeast Asia goes back as far as 1000 years ago and probably 3000 years ago in China (4,5). Preservation of fish, meat, and other animal products by solid-state fermentation goes back about 2500 years (4). Vinegar was produced by solid-state fermentation from fruit pomace in the eighteenth century (5). The production of gallic acid is another early example of solid-state fermentation, and its discovery was made in the eighteenth century (4). Solid-state composting was used for sewage treatment in the late nineteenth century (4,5). The production of fungal and other microbial enzymes by solid-state fermentation started during the early twentieth century. The new fermentation introduced from 1920 to 1940 was the production of gluconic acid, citric acid, and enzymes, as well as the development of rotary drum fermenter (4,5). Between 1940 and 1950, the fermentation industry developed very rapidly, and the first clinically useful antibiotic, penicillin, was produced by both solid-state fermentation and submerged culture method (4). During the decade from 1950 to 1960, steroid transformation by fungal spores, which was produced by solid-state fermentation, was developed (4). From 1960 to 1980, many important new solid-state fermentation processes were developed, which include the production of mycotoxins and the treatment and reuse of animal, plant, and domestic wastes (4,5). Since 1980 the developments of solid-state fermentation have been made in every aspect: expansion of microbial types, utilization of wastes as cheaper substrates, improvement of fermenter design, and discovery of new products. The history and recent developments in solid-state fermentation have been reviewed in detail by several authors (3–5).

Microbial Types

Although a wide range of microorganisms are able to grow on solid substrates, a relatively few genera and species are employed in the main commercial systems. Filamentous microorganisms are most widely used in solid-state fermentations. The ability of such microorganisms to colonize substrates by apical growth and penetration gives them a considerable ecological advantage over nonmotile bacteria and yeasts, which are less able to multiply and colonize on low-moisture substrates (6). Among the filamentous fungi, three classes have gained the most practical importance in SSF: the phycomycetes such as the genera *Mucor*, and *Rhizopus*, the ascomycetes with the genera *Aspergillus* and *Penicillium*, and the basidiomycetes, especially the white rot fungi such as edible mushrooms (3).

Bacteria and yeasts usually grow on solid substrates at the 40 to 70% moisture levels and can play important roles in some solid-state fermentations (3,6). In composting, thermophilic bacteria grow predominantly when the temperature exceeds 60°C, while ensiling processes are predominated by lactic acid bacteria (7). In food fermentations, the best-known yeast genera such as *Saccharomyces*, *Candida*, and bacterial genera such as *Lactobacillus* and *Bacillus*, have established their commercial roles (6).

Very few actinomycetes were used in solid-state fermentation. Jermini and Demain (8) first developed a solid-state fermentation system for the production of cephalosporin antibiotics by *Streptomyces clavuligerus* grown on barley, and approximately 300 μg of cephalosporins per gram of substrate was produced under the optimal conditions for seven days. Sircar et al. (9) recently optimized a solid-state fermentation medium for the production of clavulanic acid antibiotics by *Streptomyces clavuligerus* grown on a medium consisted of wheat rawa, soya flour, dipotassium hydrogen phosphate, and sunflower oil cake.

Substrates

Both natural and synthetic substances can be used in solid-state fermentation. The main substrates for solid-state fermentation are insoluble in water, and water is absorbed into substrate particles that can be used by microorganisms for growth and metabolic activity. Bacteria and yeasts grow on the surface of the substrate while fungal mycelia penetrate into the particles of the substrate (5). The substrates used are, with the exception of synthetic media, usually cheap agricultural raw materials and its by-products (10).

The most widely used substrates for solid-state fermentation in practice are mainly materials of plant origin, which include starchy materials such as grains, rice, corn, roots, tubers and legumes, and cellulosic, lignin, proteins, and lipid materials (6). These organic materials in nature are mixtures of polymeric compounds in structure, which act as a source of carbon, nitrogen, and other nutrients as well as providing anchorage for the microorganisms (5). Recent interest in solid-state fermentation has put more weight on cheaper substrates, including various agricultural or agroindustrial by-products. Agricultural and food processing wastes such as wheat bran, cassava, sugar beet pulp, bagasse, citrus peel, corn cob, banana waste, sawdust, wheat and rice straw, and fruit pomace are the most commonly used substrates for solid-state fermentation. Such by-products are usually lignocellulosic and pectin-rich wastes in nature and have been used in SSF to produce many value-added products.

Products

Several products are generated from solid-state fermentation. Some products are directly used as foods. For example, many kinds of higher fungi, edible mushrooms, have been cultivated and used as human food for centuries in China and Japan. They usually grow on agricultural residues, such as wheat straw, wood sawdust, grasses, horse manure, cotton seed crust, and fruit pomaces (11,12). In ancient time, the bread was made through natural solid-state fermentation by indigenous microorganisms. Today people use the pure culture of baker's yeast to produce bread (13). Roquefort and Camembert represent the two important types of mold-fermented cheese. *Penicillium roqueforti* is the blue mold of Roquefort cheese while *Penicillium camemberti* is the white mold of Camembert (14). *Penicillium* species and *Mucor* species are often used in fermented sausage production (15,16).

Among many oriental fermented products, soy sauce is probably the most well known in the diet of Western countries (17). Soy sauce is made from soybeans and wheat flour fermented by molds. The main microorganisms include *Aspergillus oryzae* (koji) and yeasts and lactic acid bacteria (18). Tempeh, a vegetarian meat analogue and source of vitamin B-12 (generally lacking in vegetarian diets), is yet another product made in the orient by solid-state fermentation (19,20).

Solid-state fermentation can be used not only as a tool for nutritional enrichment but also as a means of reducing toxins in raw substrates (21). It was reported that the tempeh mold *Rhizopus oligosporus* could decrease the aflatoxin content of peanut presscake by 70% during fermentation (22). A significant reduction in antinutritional and toxic components in breads and other plant-derived foods by solid-state fermentation was observed (23). Biomass production is another important aspect of bioconversion from wastes to value-added feed or food through solid-state processes (24). A recent study shows that carotenoid could be produced from cornmeal by solid-state fermentation using *Penicillium* species, with a maximum yield of 5.26 mg of carotenoid per 100 g of substrate (25).

Solid-state fermentation has been used for producing several industrially important products today. Production of many kinds of important organic acids has been successfully achieved via solid-state fermentation. Such organic acids include citric acid (26), lactic acid, acetic acid, gluconic acid, glutamic acid (27), linolenic acid (28), and other fatty acids (29), and phenolics such as ferulic acid (30). In addition, Hesseltine (31) used rice as the substrate to produce aflatoxin by *Aspergillus flavus* and *Aspergillus parasiticus* via solid-state fermentation.

The best potential application of solid-state fermentation, however, is the production of various industrial enzymes such as amylases, proteases, cellulases, pectinases, rennet, invertase, lactase, peroxidase, and so on. The production of cellulase, α-amylase and β-glucosidase from agricultural by-products by solid-state fermentation was investigated by Zheng et al. (32,33). Glucoamylases were produced by solid-state fermentation from cassava starch using *Rhizopus* species (34) or corn flour and wheat bran (35). A thermostable α-amylase was produced by thermophilic *Bacillus coagulans* in solid-state fermentation (36). A thermostable α-L-arabinofuranosidase was produced by solid-state fermentation with *Thermoascus aurantiacus* on sugar beet pulp (37). α-Galactosidase was formed by *Aspergillus niger* on wheat and rice bran–based solid-state medium (38). A novel carbohydrase complex was produced from solid-state fermentation by the aerobic fungus *Penicillium capsulatum* (39). A protease was produced from wheat bran by *Rhizopus oryzae*, with a maximum enzyme activity of 341 units per gram of wheat bran (40). A lipase was produced by solid-state fermentation using gingelly oil cake as the substrate from *A. niger*, and the enzyme activity of 363.6 units per gram of dry substrate was obtained under optimal conditions (41).

Numerous products were produced by solid-state fermentation from fruit processing wastes, and they will be discussed later in this article.

Factors Affecting SSF

Solid-state fermentations are in general carried out in fermenters of simple construction and operation and without the range of control units found in liquid fermentation systems. Laboratory studies are usually carried out in conical flasks, beakers, Rowx bottles, jars, or glass incubators. For a pilot-plant or large-scale SSF, however, the design of bioreactors in batch or continuous mode has been empirical in nature (42). For a solid-state fermentation with a selected microorganism, the success of the process will depend on culture condition with the controlled parameters including mainly nutrient supplements, moisture content, temperature, pH, aeration, and agitation.

Nutritional factors usually limit the growth of fungi. In solid-state fermentation, this limitation is much more severe due to the limited diffusion rate of the substrate and the limited access of the fungus to the substrate. An important indicator of nutritional regulation of growth in solid-state fermentation is the C/N ratio. Optimal C/N ratio varies in a wide range from 10 to 100 or higher in vari-

ous SSF processes, but the availability of C and N can be more important than the ratio (3). In most SSF systems, the C source comes from the natural starchy or cellulosic materials while the N source is added. The most commonly used synthetic N sources are NH_4Cl, $(NH_4)SO_4$, NH_4NO_3 and urea, while some organic N sources such as soybean cake and yeast extract are often used in solid-state fermentation.

The moisture level of the substrate will have a determining influence on the success of the overall process. A better way of expressing moisture content is water activity (a_w), which gives the availability of water for growth of microorganisms. In general, the types of the microorganisms that can grow in SSF systems are determined by the water activity. Most bacteria grow at higher a_w values while filamentous fungi and some yeasts can grow at lower a_w level. The microorganisms, which can grow and are capable of carrying out their metabolic activities at lower a_w level, are suitable for SSF processes. Water activity level in SSF is governed by the nature of substrate, the type of end product, and the requirement of microorganisms. A high a_w level results in decreased porosity or intracellular spaces, lower oxygen diffusion, decreased gas exchange, enhanced formation of aerial mycelium, decreased substrate degradation and an increased risk of bacterial contamination (43). In contrast, low a_w will lead to decreased substrate swelling and decreased microbial growth (43).

Temperature is another critical parameter that can affect SSF processes, since every microorganism has its own optimal temperatures for growth and for metabolism, which are often different. During solid-state fermentation, a large quantity of metabolic heat is generated and this is directly related to the metabolic activities of the microorganism (43,44). It is not difficult to control the temperature for solid-state fermentation in the laboratory; in large-scale SSF, however, due to high substrate concentration, microbial heat generation is much greater per unit volume than in liquid fermentation, and therefore, constant heat removal for temperature control is a major problem (43,44).

Although pH is one of the critical factors, the monitoring and control of pH during solid-state fermentation is not usually attempted. It is difficult to monitor and control pH in the solid-state fermentation, since pH electrodes able to measure the pH of the moist solids, in the absence of free water, are not available (43). On the other hand, good buffering capacity of some of the substrates used in SSF helps in eliminating the need for pH control during fermentation. This advantage is, therefore, exploited in the initial adjustment of the pH of the solids in the range of 4.4 to 5.0 during moistening by using water at the desired pH level (43). Another approach in counteracting the acidification of the fermenting mass is the use of urea as the nitrogen source rather than ammonium salts (44).

For most solid-state fermentations, oxygen is essential and is usually achieved by free air exchange or forcing sterile air under pressure through the fermenting mass (43). The mechanism of O_2 transfer from the circulating air into the aerobic microorganisms is unclear, but most probably it occurs via the O_2 dissolved in the water film around the particle surface (43,44). Aeration not only provides oxygen but also simultaneously removes carbon dioxide, other volatile metabolites and heat from the fermenter (44).

Agitation is not employed in many solid-state fermentation processes; however, it is usually an essential part of periodically or continuously agitated aerobic solid-state fermentations (44). Agitation of the fermenting mass can ensure homogeneity with respect to temperature and gaseous environment and promote heat and gas transfers. It also helps for the uniform distribution of nutrients during the course of fermentation (43,44).

Comparison to Liquid Fermentation

Solid-state fermentation has numerous advantages over the conventional stirred or aerated liquid fermentation both on a laboratory and large scale. First of all, the medium is relatively simple (4). For instance, cereal grain, legumes, or other plant and animal products, various agricultural and food processing wastes, are all potential substrates of solid-state fermentation. Usually the only other medium component required in SSF is water, although occasionally other nutrients such as nitrogen sources or minerals may be added (4). Other advantages of solid-state fermentation over liquid fermentation include higher substrate concentration, less probability of contamination, superior productivity, improved product recovery, reduced energy and water requirements and wastewater output, and lower capital investment (4,45).

Despite solid-state fermentation being both economically and environmentally attractive, their biotechnological exploitation has been rather limited (5). Compared with liquid fermentation, the major problems associated with solid-state fermentation include the limitations on microorganisms, medium heterogeneity, heat and mass transfer control, growth measurement and monitoring, and scale-up problems (4,44,45).

On the whole, SSF may be the exclusive tool, especially when demand for the product is limited, as is the case for some industrial enzymes. Furthermore, SSF has demonstrated a great potential in bioconversion of agricultural and industrial wastes into useful value-added products. In addition to obtaining a variety of value-added products, SSF system can also be used in environmental remediation of agricultural by-products.

FRUIT AND VEGETABLE PROCESSING WASTES

Apple Pomace

Apple pomace is the residual left after juice extraction and represents about 25% of the original fruit. It is rich in carbohydrate but low in protein and fat contents. The freshly pressed apple pomace has a low pH ranging from 3.1 to 3.8, with a bulk density of 935 kg/m^3 (46,47). Since it is produced wet with high moisture content, it is susceptible to rapid growth of microorganisms. The major composition of apple pomace is shown in Table 1.

It is estimated that nearly 36 million tons of apples are produced annually in the United States, and approximately 45% of which is used for processing purposes, with a primary by-product of apple pomace, which results from

Table 1. Chemical Composition of Apple Pomace

Constituent	Fresh pomace (%)	Dried pomace (%)
Moisture	66.4–78.2	11.0–12.5
Carbohydrates	9.5–22.0	—
Nitrogen-free extract	—	54.8–59.3
Pectin	1.5–2.5	15.0–18.0
Crude fiber	4.3–10.5	15.3–20.6
Proteins	1.03–1.82	4.45–5.67
Fat (ether extract)	0.82–1.43	3.75–4.65
Ash	0.56–2.27	2.11–3.50
Potassium (as K$_2$O)	0.2–0.6	—
Phosphorus (as P$_2$O$_5$)	0.4–0.7	—

Source: Ref. 46.

processing for juice, cider, applesauce, or slices (46). More than 500 apple-processing plants in the United States generate a total of about 1.3 million metric tons of apple pomace each year, and the annual disposal fees for apple pomace alone exceed $10 million in the United States (47,48). The high level of organic contents in apple pomace will result in environmental pollution if they are disposed directly to the environment.

Other Fruit and Vegetable Processing Wastes

The food processing industry in the United States generates large amount of various by-products as wastes each year, a large portion of which comes from fruit and vegetable processing industries. It is estimated that approximately 9 million tons of solid residuals are generated annually from the fruit and vegetable processing industry in the United States (49). In addition to apple pomace, other major fruit and vegetable processing wastes come from citrus, grape, cherries, berries, banana, olive, peach, pear, pineapple, kiwifruit, apricot, sweet corn, tomato, potato, asparagus, beans, peas, carrots, beet, pumpkin, squash, spinach, and sauerkraut processing industries (49–56). The primary sources of solid residues resulting from fruit and vegetable processing operations include sorting, cutting, slicing, peeling, pulping, and pressing (49).

Traditionally about 79% of the fruit and vegetable processing wastes is used for animal feed, and the remaining 21% is handled as waste (49). The solid wastes contain mainly soluble sugar and other hydrolyzable materials with small amounts of crude fiber. Disposal of such wastes may present an added cost to processors, and direct disposal to soil or in a landfill may not continue to be acceptable. Thus, exploration of potential value-added uses for pomace is highly attractive.

UTILIZATION OF FRUIT AND VEGETABLE PROCESSING WASTES VIA SOLID-STATE FERMENTATION

Fertilizer

Although direct disposal of fruit processing wastes in landfills has become environmentally unacceptable, some fruit processing wastes can be composted under anaerobic condition, and then used as fertilizer in landfills, because such wastes were readily degraded under anaerobic digestion conditions (57,58). Solid-state fermentation can also be used for composting of fruit and vegetable processing wastes such as apple waste (59) and tomato pomace (60). Composted apple pomace and other fruit and vegetable processing wastes can be used in nursery potting mixes and as field soil amendments (61). Composted grape pomace was used as an organic fertilizer in vineyards for growing grapes (62).

Animal Feed and Single-Cell Protein

Apple pomace was directly used as a feed for cows and sheep, but its value as an animal feed is less than $7 per ton (47). Apple pomace is a poor animal feed supplement because of its low protein content. The nutritional value of apple pomace was improved by solid-state fermentation with a food yeast, *Candida utilis* (62). Yeast fermentation resulted in a 2.5-fold increase in protein, 3.4-fold increase in niacin, 10-fold increase in pantothenic acid, 1.5-fold increase in riboflavin, and 1.2-fold increase in thiamine (62). An improved stock feed was produced from apple pomace by solid-state fermentation with *Kloeckera apiculate* and *C. utilis* (63,64).

A solid-state fermentation of orange processing wastes with *Aspergillus niger* and *Rhizopus* species enriched the protein content by 300%, and the fermented product could be readily sold as animal feed (65). Banana wastes were also used for protein enrichment and protein and biomass production by solid-state fermentation using *A. niger* (66), yeast *Pichia spartinae* (67), and *Saccharomyces uvarum* (68).

As the world population increases, there is a greater demand for new sources of protein for formulating new types of food. For this reason, there is a great interest in the use of single-cell protein as a source of protein for animals or humans. Apple pomace is a potential substrate for producing single-cell protein. The microorganisms that can be used for single-cell protein production include yeast, filamentous fungi, bacteria, and algae. It is well recognized that these microorganisms have the ability to produce protein and other essential nutrients (49). Yeasts have been studied extensively and are already widely used as a source of protein for animals and humans. The production of "food yeast," *C. utilis*, has long been accepted for use in foods by the regulatory authorities. *C. utilis* has high nutritive value and is able to grow rapidly in high yields utilizing a variety of carbon and nitrogen sources, with high tolerance to low pH. A general review on the production of feed and food yeasts from various plant materials has been published (69). A single-cell protein of *C. utilis* was successfully produced from apple pomace (70).

Mushrooms

Mushroom production is one of the few large-scale commercial applications of microbial technology profitable for bioconversion of cellulosic waste materials to valuable foods (11). Apple pomace was found to be a good substrate for the cultivation of edible mushrooms. Various oyster mushrooms (*Pleurotus* species) were growing very well on apple pomace with a biological efficiency ranging between 30 and 40% (71). The biological efficiency was a measurement of the conversion rate, a ratio of the fruit body yield to the weight of the cultivation medium. In another experiment, a mixture of apple pomace and sawdust was used as a substrate for production of shiitake (*Lentinula edodes*) and oyster mushroom (*Pleurotus ostreatus*) on synthetic logs (72). Apple pomace is complex and has readily usable carbohydrates that complement the nutritional properties of sawdust. As a result, the mushroom mycelia grew faster and more densely in logs containing apple pomace than in sawdust alone (73).

Ethanol

In general, ethanol is produced from a liquid substrate by a culture of *Saccharomyces cerevisiae*. Because of its physical nature, apple pomace is not readily amenable to submerged yeast fermentation (74). A solid-state fermentation system for the production of ethanol from apple pomace with a strain of *S. cerevisiae* was devised (74–76). The yield of ethanol varied from about 29 g to more than 40 g per kilogram of apple pomace fermented at 30°C in 24 h, depending on the samples fermented; the fermentation efficiency of this process was approximately 89% (76). The spent pomace resulting from the separation of alcohol should be a better animal feed supplement, because its protein content was enriched by the yeast biomass. A novel process for concomitant production of ethanol and animal feed from apple pomace by solid-state fermentation was developed (77). Ethanol production from other fruit processing wastes such as corn fiber (78,79), carob (80,81), and banana waste (68) has also been reported.

Gupta et al. (80) further studied the effect of nutrition variables on solid-state alcoholic fermentation of apple pomace by several strains of yeasts and improved the fermentation efficiency by adding various phosphates, nitrogen sources, or trace elements. Ngadi and Correia (82) studied the effect of factors such as moisture and bioreactor mixing speed on the ethanol production from apple pomace. They even proposed a mathematical model to describe the kinetics of solid-state ethanol fermentation from apple pomace and suggested a logistic function of cell growth and ethanol production during solid-state fermentation (83).

Biogas

Fruit and vegetable processing wastes were readily degraded under anaerobic digestion conditions (57,58,84). Jewell and Cummings (47) investigated the biodegradability of apple pomace using long-term digestion experiments and by graphic analysis. A total of 12 steady-state loading conditions were examined, and kinetic analysis indicated that a hydraulic retention time of 45 days would result in nearly 90% conversion of the biodegradable organics to biogas, and the methane content of the biogas was 60% (47). Under these conditions, the net energy yield from pomace would range from about 0.7 to 1.7 million kcal per wet metric ton, with dry matter contents varying between 20 and 40% of the wet weight, and the energy has a value of between $12 and $30 per wet metric ton of apple pomace (47).

Orange processing waste was also used to produce biogas by solid-state fermentation. Srilatha et al. (85) reported that pretreatment of orange processing waste by solid-state fermentation using selected strains of *Sporotrichum, Aspergillus, Fusarium*, and *Penicillium* improved overall productivity of biogas and methane.

Citric Acid and Other Organic Acids

Citric acid, a tricarboxylic acid, has a wide range of applications in the food, pharmaceutical, and beverage industries as an acidifying and flavor-enhancing agent. The use of apple pomace as a substrate for microbial production of citric acid under solid-state fermentation conditions has been reported (86,87). In the experiment, 40 g of apple pomace was introduced into 500-mL Erlenmeyer flasks and inoculated with A. niger spore suspension, and then incubated at 30°C for five days under stationary conditions. Methanol was added to the substrate before fermentation to enhance the yield of citric acid. Under the optimal fermentation conditions, this process yielded up to 90 g of citric acid per kilogram of wet pomace fermented, with a conversion rate of more than 88% based on the amount of sugar consumed (46). A process for leaching citric acid from apple pomace fermented with A. niger was also devised (88).

Pineapple waste could be a better substrate for citric acid production in solid-state fermentation than apple pomace (89), and A. niger was used in the process (90,91). The highest citric acid yield achieved on pineapple waste in four days was 161 g per kilogram of dried pineapple waste of 70% moisture content, and in the presence of 3% methanol, and the conversion rate was 62.4% based on the sugar consumed (89).

Kiwifruit peel is a by-product resulting from the manufacture of kiwifruit into nectars or slices and represents nearly 10 to 16% of the weight of the original fruit, depending on the peeling method used (52). Like apple or pineapple processing wastes, kiwifruit peel waste can be a good substrate for citric acid production by solid-state fermentation, and about 100 g citric acid per kilogram of kiwifruit peel was produced by A. niger in the presence of 2% methanol at 30°C in four days. The yield was more than 60% based on the amount of fermented sugar consumed (52).

Corn wastes were used as a substrate for citric acid production by A. niger with a maximum yield of 250 g/kg dry matter of corn cobs after 72 h of growth at 30°C (92). Carob kibble, a waste material from carob pod fruit was also an attractive substrate for citric acid production by solid-state fermentation (93,94). Orange processing wastes were also used as a substrate to produce citric acid by A. niger (95). During citric acid fermentation, the presence of methanol was a critical factor in increasing citric acid yield, and this appears to be a general phenomenon with strains of A. niger (52,95).

Other organic acids were also produced by solid-state fermentation from fruit processing wastes. Grape pomace could also be used for citric acid production by solid-state fermentation (96). The production of tartaric acid from grape pomace was also attempted (97), while banana wastes were targeted as a substrate for lactic acid production (98).

Butanol

The acetone-butanol fermentation is a well-known process that is being looked at with renewed interest recently, owing to the increasing value of oil products. The classical substrates used for butanol production by fermentation are a variety of starchy materials and molasses, which are of relatively high cost. Apple pomace was used as the substrate for butanol production by solid-state fermentation with the strains of *Clostridium acetobutylicum* and *C. butylicum*, and the yields of butanol from apple pomace were between 1.9 and 2.2% of fresh apple pomace (99).

Volatile Compounds

During solid-state fermentation with edible fungus *Rhizopus oryzae* on apple pomace and cassava bagasse, substantial volatile compounds were produced in association with the fungal growth (100). The major volatile metabolites produced by R. oryzae from apple pomace were identified as acetaldehyde, 1-propanol, ethyl acetate, ethyl propinioate and 3-methyl butanol (100). Ethyl acetate and ethyl propionate probably originated from the esterification between ethanol and acetic acid and propionic acid, respectively, which are the products of carbohydrate metabolism (100).

Enzymes

Pectic enzymes derived from *Aspergillus* species have been widely used to increase juice yields and clarify juices. The commercial enzyme preparations are produced by cultivating the mold in a synthetic medium under submerged fermentation conditions. Recently apple pomace has been reported to be an attractive raw material for the production of pectinases by *Aspergillus foetidus* in solid-state cultures (101). To obtain maximum enzyme yields, it is essential to supplement apple pomace with a sufficient amount of an organic nitrogen constituent such as corn steep, yeast extract, or peptone (102). Berovic and Ostroversnik (103) developed a solid-state bioprocess for production of pectolytic enzymes from apple pomace and the process parameters such as inoculation, influence of mixing, aeration, temperature, and moisture content on pectolytic enzymes production were studied. Under the optimal conditions, maximal amounts of 15 g of polygalacturonase and 200 mg of pectinesterase per kilogram of solid medium were obtained (103). Polygalacturonase is one of the most important pectinases and is widely used in juice processing industries (104). Apple pomace was used as the substrate of solid-state fermentation to produce polygalacturonase by A. niger, and the highest yield of 25,000 units per kilogram of fermented apple pomace was achieved (105). In addition to apple pomace, citrus processing wastes were also alternative substrates for pectinase production in solid-state fermentation (106,107).

Amylases are enzymes catalyzing the hydrolysis of α-1,4-glucosidic linkages of polysaccharides and have wide applications in the food and beverage processing indus-

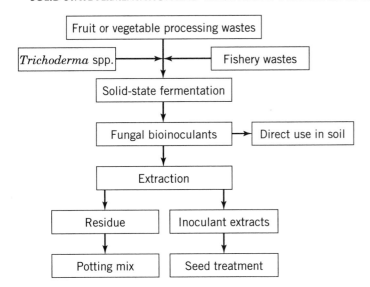

Figure 1. A value-added approach to utilization of fruit and vegetable processing wastes.

tries. Banana waste was a good substrate for α-amylase production by *Bacillus subtilis* under solid-state fermentation (108).

β-Glucosidase (β-D-glucoside glucohydrolase) can catalyze the hydrolysis of glycosidic linkages in aryl and alkyl β-D-glucosides as well as glycosides containing only carbohydrate residues (109,110), and it has been used to increase the concentration of aroma volatile compounds from wine and fruit juice through enzymatic hydrolysis of nonvolatile precursors (111,112). A recent study by Hang and Woodams (113) shows that apple pomace is a potential substrate for production of β-glucosidase by solid-state fermentation, and the highest yield of 900 units of the enzyme per kilogram of fermented apple pomace was obtained by using *A. foetidus*.

β-Fructofuranosidase catalyzes the enzymatic hydrolysis of a fructofuranoside to an alcohol and D-fructose. This enzyme is used commercially in the conversion of sucrose to glucose and fructose and in the manufacture of chocolate-coated soft cream candies (114). Apple pomace was used as the substrate for β-fructofuranosidase production by solid-state fermentation using *Aspergillus* species with the highest yield of 2700 units per kilogram of fermented apple pomace (115).

Lipases (triacylglycerol acylhydrolases) catalyze the hydrolysis of triglycerides to glycerol and free fatty acid at an oil–water interface. This enzyme was produced by solid-state fermentation of olive cake and sugar cane bagasse (116).

A multienzyme complex containing pectinase, cellulase, and xylanase enzymes was produced by a coculture of six fungal isolates grown on orange peels as the sole carbon source (117). This multienzyme preparation also contains insignificant levels of amylase and lipase activities, suggesting potential uses in the extraction of the major components such as starches and lipids form plant materials (117). A cellulase-free xylanase was produced by solid-state fermentation from *Thermomyces lanuginosus* grown on corncobs (118).

A New Value-Added Approach to Utilization of Fruit and Vegetable Processing Wastes

Recently, Zheng and Shetty (119,120) proposed a novel strategy for better utilization of apple and cranberry processing wastes. In this process, a solid-state bioconversion of such wastes into various fungal bioinoculants, which have multiple applications in agricultural and environmental industries, was suggested. For example, *Trichoderma* inoculants produced from apple pomace significantly enhanced the seedling vigor of peas germinated in potting soil (Z. Zheng and K. Shetty, unpublished data, 1999). Commercial seed treatments are being viewed as a means to substantially increase the value of the seed and to improve plant growth and productivity. As a result of enhanced seedling vigor, the plant growth and productivity may be significantly improved. The seed vigor-enhancing compounds could also be extracted with water from fermented fruit pomace, and the residue remaining after extraction is being targeted for use as potting soil mix for several plant species. Therefore, the results of this research have showed value-added application potential for agricultural and environmental industries. A schematic diagram of the new value-added approach for utilization of food processing wastes is shown in Fig. 1.

BIBLIOGRAPHY

1. E. Cannel and M. Moo-Young, "Solid State Fermentation Systems," *Process Biochem.* **4**, 2–7 (1980).

2. R. E. Mudgett, "Solid-State Fermentations," in A. L. Demain and N. A. Solomon, eds., *Manual of Industrial Microbiology and Biotechnology*, American Society for Microbiology, Washington, D.C., 1986, pp. 66–83.

3. M. Moo-Young, A. R. Moreira, and R. P. Tengerdy, "Principle of Solid-Substrate Fermentation," in J. E. Smith, D. R. Berry, and B. Kristiansen, eds., *The Filamentous Fungi*, vol. 4, *Fungal Technology*, E. Arnold, London, 1983, pp. 117–144.

4. K. E. Aidoo, R. Hendry, and B. J. B. Wood, "Solid Substrate Fermentation," *Adv. Appl. Microbiol.* **28**, 201–237 (1982).

5. A. Pandey, "Recent Progress Developments in Solid-State Fermentation," *Process Biochem.* **27**, 109–117 (1992).

6. J. E. Smith and K. E. Aidoo, "Growth of Fungi on Solid Substrates," in D. R. Berry, ed., *Physiology of Industrial Fungi*, Blackwell, Oxford, England, 1988, pp. 249–269.

7. D. A. Mitchell, "Microbial Basis of Processes," in H. W. Doelle, D. A. Mitchell, and C. E. Rolz, eds., *Solid Substrate Cultivation*, Elsevier, Essex, England, 1992.

8. M. F. G. Jermini and A. L. Demain, "Solid State Fermentation for Cephalosporin Production by *Streptomyces clavuligerus* and *Cephalosporium acremonium*," *Experientia* **45**, 1061–1065 (1989).

9. A. Sircar, P. Sridhar, and P. K. Das, "Optimization of Solid State Medium for the Production of Clavulanic Acid by *Streptomyces clavuligerus, Process Biochem.* **33**, 283–289 (1998).

10. J. P. Smits et al., "Solid-State Fermentation—A Mini Review," *Agro Food Industry Hi-Technol.* **9**, 29–36 (1998).

11. E. Moyson and H. Verachtert, "Growth of Higher Fungi on Wheat Straw and Their Impact on the Digestibility of the Substrate," *Appl. Microbiol. Biotechnol.* **36**, 421–424 (1991).

12. S. Ohga and Y. Kitamoto, "Future of Mushroom Production and Biotechnology," *Food Rev. Int.* **13**, 461–469 (1997).

13. G. Reed, "Production of Bakers' Yeast," in G. Reed, ed., *Prescott and Dunn's Industrial Microbiology*, 4th ed., AVI, Westport, Conn., 1982, pp. 593–633.

14. C. S. Pederson, *Microbiology of Food Fermentations*, 2nd ed., AVI, Westport, Conn., 1979.

15. M. D. Selgas et al., "Potential Technological Interest of a *Mucor* Strain to Be Used in Dry Fermented Sausage Production," *Food Res. Int.* **28**, 77–82 (1995).

16. F. K. Lucke, "Fermented Meat Products," *Food Res. Int.* **27**, 299–307 (1994).

17. H. L. Wang and C. W. Hesseltine, "Oriental Fermented Foods," in G. Reed, ed., *Prescott & Dunn's Industrial Microbiology*, 4th ed., AVI, Westport, Conn., 1982, pp. 492–538.

18. B. J. Wood, "Technology Transfer and Indigenous Fermented Foods," *Food Res. Int.* **27**, 269–280 (1994).

19. G. Campbell-Platt, "Fermented Foods—A World Perspective," *Food Res. Int.* **27**, 253–257 (1994).

20. S. Keuth and B. Bisping, "Vitamin B12 Production by *Citrobacter freundii* or *Klebsiella pneumoniae* during Tempeh Fermentation and Proof of Enterotoxin Absence by PCR," *Appl. Environ. Microbiol.* **60**, 1495–1499 (1994).

21. K. H. Steinkraus, "Nutritional Significance of Fermented Foods," *Food Res. Int.* **27**, 259–267 (1994).

22. A. G. Van-Veen, D. C. W. Graham, and K. H. Steinkraus, "Fermented Peanut Presscake," *Cereal Sci. Today* **13**, 96–99 (1968).

23. N. R. Reddy, "Reduction in Antinutritional and Toxic Compounds in Plant Foods by Fermentation," *Food Res. Int.* **27**, 281–290 (1994).

24. A. Noomhorm, S. Ilangantileke, and M. B. Bautista, "Factors in the Protein Enrichment of Cassava by Solid State Fermentation," *J. Sci. Food Agri.* **58**, 117–123 (1992).

25. J. R. Han, "Sclerotia Growth and Carotenoid Production of *Penicillium* sp. PT95 during Solid State Fermentation of Corn Meal," *Biotechnol. Lett.* **20**, 1063–1065 (1998).

26. M. Gutierrez-Rojas et al. "Heat Transfer in Citric Acid Production by Solid State Fermentation," *Process Biochem.* **4**, 363–369 (1996).

27. K. M. Nampoothiri and A. Pandey, "Solid State Fermentation for L-Glutamic Acid Production Using *Brevibacterium* sp.," *Biotechnol. Lett.* **18**, 199–204 (1996).

28. E. V. Emelyanova, "γ-Linolenic Acid Production by *Cunninghamella japonica* in Solid State Fermentation," *Process Biochem.* **31**, 431–434 (1996).

29. S. Argelier, J. P. Delgenes, and R. Moletta, "Design of Acidogenic Reactors for the Anaerobic Treatment of the Organic Fraction of Solid Food Waste," *Bioprocess Eng.* **18**, 309–315 (1998).

30. C. B. Faulds and G. Williamson, "Release of Ferulic Acid from Wheat Bran by a Ferulic Acid Esterase (FAE-III) from *Aspergillus niger*," *Appl. Microbiol. Biotechnol.* **43**, 1082–1087 (1995).

31. C. W. Hesseltine, "Biotechnology Report: Solid State Fermentation," *Biotechnol. Bioeng.* **14**, 517–532 (1972).

32. Z. Zheng, F. B. Elegado, and Y. Fujio, "Production and Some Properties of Cellulase from *Rhizopus japonicus* IFO5318," *Annu. Rep. ICBiotech* **16**, 223–232 (1993).

33. Z. Zheng et al., "Screening of Cellulase-Rich Microbial Strains and the Conditions for Enzyme Production," *J. Microbiol.* **16**, 35–38 (1996).

34. C. R. Soccol et al. "Breeding and Growth of *Rhizopus* in Raw Cassava by Solid State Fermentation," *Appl. Microbiol. Biotechnol.* **41**, 330–336 (1994).

35. A. Pandey, P. Selvakumar, and L. Ashakumary, "Performance of a Column Bioreactor for Glucoamylase Synthesis by *Aspergillus niger* in SSF," *Process Biochem.* **31**, 43–46 (1996).

36. K. R. Babu and T. Satyanarayana, "α-Amylase Production by Thermophilic *Bacillus coagulans* in Solid State Fermentation," *Process Biochem.* **30**, 305–309 (1995).

37. N. Roche, C. Desgranges, and A. Durand, "Study on the Solid-State Production of a Thermostable α-L-Arabinofuranosidase of *Thermoascus aurantiacus* on Sugar Beet Pulp," *J. Biotechnol.* **38**, 43–50 (1994).

38. R. I. Somiari and E. Balogh, "Production of an Extracellular Glycosidase of *Aspergillus niger* Suitable for Removal of Oligosaccharides from Cowpea Meal," *Enzyme Microbiol. Technol.* **17**, 311–316 (1995).

39. I. C. Connelly et al., "Novel Carbohydrase 'Complex' from Solid-State Cultures of the Aerobic Fungus *Penicillium capsulatum*," *Enzyme Microbiol. Technol.* **13**, 470–477 (1991).

40. R. Tunga, R. Banerjee, and B. C. Bhattacharyya, "Optimizing Some Factors Affecting Protease Production under Solid State Fermentation," *Bioprocess Bioeng.* **19**, 187–190 (1998).

41. N. R. Kamini, J. G. S. Mala, and R. Puvanakrishnan, "Lipase Production from *Aspergillus niger* by Solid-State Fermentation Using Gingelly Oil Cake," *Process Biochem.* **33**, 505–511 (1998).

42. A. Pandey, "Aspects of Fermenter Design for Solid-State Fermentations," *Process Biochem.* **26**, 355–361 (1991).

43. B. K. Lonsane et al., "Engineering Aspects of Solid State Fermentation," *Enzyme Microbiol. Technol.* **7**, 258–265 (1985).

44. B. K. Lonsane et al., "Scale-up Strategies for Solid-State Fermentation Systems," *Process Biochem.* **27**, 259–273 (1992).

45. P. E. Cook, "Fermented Foods as Biotechnological Resources," *Food Res. Int.* **27**, 309–316 (1994).

46. Y. D. Hang, "Production of Fuels and Chemicals from Apple Pomace," *Food Technol.* **41**, 115–117 (1987).

47. W. J. Jewell and R. J. Cummings, "Apple Pomace Energy and Solids Recovery," *J. Food Sci.* **48**, 407–410 (1984).

48. K. J. Carson, J. L. Collins, and M. P. Penfield, "Unrefined, Dried Apple Pomace as a Potential Food Ingredient," *J. Food Sci.* **59**, 1213–1215 (1994).

49. Y. D. Hang, "Production of Single-Cell Protein from Food Processing Wastes," in J. H. Green and A. Kramer, ed., *Food Processing Management*, AVI, Westport, Conn., 1979, pp. 442–455.

50. R. H. Walter and R. M. Sherman, "Fuel Value of Grape and Apple Processing Wastes," *J. Agric. Food Chem.* **24**, 1244–1245 (1976).

51. H. Al-Wandawi, M. Abdul-Rahman, and K. Al-Shaikhly, "Tomato Processing Wastes as Essential Raw Materials Source," *J. Agric. Food Chem.* **33**, 804–807 (1985).

52. Y. D. Hang, B. S. Luh, and E. E. Woodams, "Microbial Production of Citric Acid by Solid State Fermentation of Kiwifruit Peel," *J. Food Sci.* **52**, 226–227 (1987).

53. S. N. Onyeneho and N. S. Hettiarachchy, "Antioxidant Activity, Fatty Acids and Phenolic Acids Compositions of Potato Peels," *J. Sci. Food Agric.* **62**, 345–350 (1993).

54. F. Carvalheiro, J. C. Roseiro, and M. T. A. Collaco, "Biological Conversion of Tomato Pomace by Pure and Mixed Fungal Cultures, *Process Biochem.* **29**, 601–605 (1994).

55. W. P. Weiss, D. L. Frobose, and M. E. Koch, "Wet Tomato Pomace Ensiled with Corn Plants for Dairy Cows," *J. Dairy Sci.* **80**, 2896–2900 (1997).

56. G. A. Fenton and M. J. Kennedy, "Rapid Dry Weight Determination of Kiwifruit Pomace and Apple Pomace Using an Infrared Drying Technique," *N. Z. J. Crop Hortic Sci.* **26**, 35–38 (1998).

57. W. Knol, M. M. Van der Most, and J. de Waart, "Biogas Production by Anaerobic Digestion of Fruit and Vegetable Wastes," *J. Food Sci. Agric.* **29**, 822–827 (1978).

58. A. G. Lane, "Methanol, Anaerobic Digestion of Fruit and Vegetable Pressing Wastes," *Food Technol. Australia* **31**, 201–206 (1979).

59. S. F. Barrington, K. Elmoueddeb, and B. Porter, "Improving Small-Scale Composting of Apple Waste," *Can. Agr. Eng.* **39**, 9–16 (1997).

60. C. Chong, "Experiences with the Utilization of Wastes in Nursery Potting Mixes and as Field Soil Amendments," *Can. J. Plant Sci.* **79**, 139–148 (1999).

61. G. Logsdon, "Pomace Is a Grape Resource," *Biocycle* **33**, 40–41 (1992).

62. Y. D. Hang, "Improvement of the Nutritional Value of Apple Pomace by Fermentation," *Nutr. Rep. Int.* **38**, 207–211 (1988).

63. H. Rahmat, R. A. Hodge, and G. J. Manderson, "Solid-Substrate Fermentation of *Kloeckera apiculate* and *Candida utilis* on Apple Pomace to Produce an Improved Stock-Feed," *World J. Microbiol. Biotechnol.* **11**, 168–170 (1995).

64. V. K. Joshi and D. K. Sandhu, "Preparation and Evaluation of an Animal Feed Byproduct Produced by Solid-State Fermentation of Apple Pomace," *Bioresource Technol.* **56**, 251–255 (1996).

65. T. J. B. Menezes et al., "Protein Enrichment of Citrus Wastes by Solid Substrate Fermentation," *Process Biochem.* **24**, 167–171 (1989).

66. J. Baldensperger et al., "Solid State Fermentation of Banana Wastes," *Biotechnol. Lett.* **7**, 743–748 (1985).

67. S. L. Chung and S. P. Meyer, "Bioprotein from Banana Wastes," *Dev. Ind. Microbiol.* **20**, 723–731 (1979).

68. O. Enwefa, "Biomass Production from Banana Skins," *Appl. Microbiol. Biotechnol.* **36**, 283–284 (1991).

69. G. Reed and H. J. Peppler, *Yeast Technology*, AVI, Westport, Conn., 1973.

70. P. J. Fellows and J. T. Worgan, "Growth of *Saccharomycopsis fibuliger* and *Candida utilis* in Mixed Culture on Apple Processing Wastes," *Enzyme Microbiol. Technol.* **9**, 434–437 (1987).

71. R. C. Upadhyay and H. S. Sohi, "Apple Pomace—A Good Substrate for the Cultivation of Edible Mushrooms," *Curr. Sci.* **57**, 1189–1190 (1988).

72. J. J. Worrall and C. S. Yang, "Shiitake and Oyster Mushroom Production on Apple Pomace and Sawdust," *HortScience* **27**, 1131–1133 (1992).

73. E. Gabler et al., "Feasibility of Ethanol Production from Cheese Whey and Fruit Pomace in New York," in *Proceedings of Feed and Fuel from Ethanol Production Symposium*, NRAES Report no. 17, Cornell University, Ithaca, N.Y., 1982, p. 13.

74. Y. D. Hang, "Production of Alcohol from Apple Pomace," *Appl. Environ. Microbiol.* **42**, 1128–1129 (1981).

75. Y. D. Hang et al., "Production of Alcohol from Apple Pomace," *Appl. Environ. Microbiol.* **42**, 1128–1129 (1981).

76. Y. D. Hang, "A Solid State Fermentation System for Production of Ethanol from Apple Pomace," *J. Food Sci.* **47**, 1851–1852 (1982).

77. D. K. Sandhu and V. K. Joshi, "Solid State Fermentation of Apple Pomace for Concomitant Production of Ethanol and Animal Feed," *J. Sci. Ind. Res. India* **56**, 86–90 (1997).

78. R. J. Bothast and B. C. Saha, "Ethanol Production from Agricultural Biomass Substrates," *Adv. Appl. Microbiol.* **44**, 261–286 (1997).

79. B. C. Saha, B. S. Dien, and R. J. Bothast, "Fuel Ethanol Production from Corn Fiber-Current Status and Technical Prospects," *Appl. Biochem. Biotechnol.* **70**, 115–125 (1998).

80. L. K. Gupta, G. Pathak, and R. P. Tiwari, "Effect of Nutrition Variations on Solid State Alcoholic Fermentation of Apple Pomace by Yeasts," *J. Sci. Food Agric.* **50**, 55–62 (1990).

81. T. Roukas, "Solid-State Fermentation of Carob Pods for Ethanol Production," *Appl. Microbiol. Biotechnol.* **41**, 296–301 (1994).

82. M. O. Ngadi and L. R. Correia, "Solid State Ethanol Fermentation of Apple Pomace as Affected by Moisture and Bioreactor Mixing Speed," *J. Food Sci.* **57**, 667–670 (1992).

83. M. O. Ngadi and L. R. Correia, "Kinetics of Solid-State Ethanol Fermentation from Apple Pomace," *J. Food Eng.* **17**, 97–116 (1992).

84. R. Borja and C. J. Banks, "Kinetic Study of Anaerobic Digestion of Fruit Processing Wastewater in Immobilized-Cell Bioreactors," *Biotechnol. Appl. Biochem.* **20**, 79–92 (1994).

85. H. R. Srilatha et al., "Fungal Pretreatment of Orange Processing Waste by Solid-State Fermentation for Improved Production of Methane," *Process Biochem.* **30**, 327–331 (1995).

86. Y. D. Hang and E. E. Woodams, "Apple Pomace: A Potential Substrate for Citric Acid Production by *Aspergillus niger*," *Biotechnol. Lett.* **6**, 763–764 (1984).

87. Y. D. Hang and E. E. Woodams, "Solid State Fermentation of Apple Pomace for Citric Acid Production," *J. Appl. Microbiol. Biotechnol.* **2**, 281–285 (1986).

88. Y. D. Hang and E. E. Woodams, "A Process for Leaching Citric Acid from Apple Pomace Fermented with *Aspergillus niger* in Solid-State Culture," *Mircen J. Appl. Microbiol.* **5**, 379–382 (1989).

89. C. T. Tran and D. A. Mitchell, "Pineapple Waste—A Novel Substrate for Citric Acid Production by Solid State Fermentation," *Biotechnol. Lett.* **17**, 1107–1110 (1995).

90. V. DeLima, T. Stamford, and A. Salgueiro, "Citric Acid Production from Pineapple Waste by Solid State Fermentation Using *Aspergillus niger*," *Arq. Biol. Technol.* **38**, 773–783 (1995).

91. C. T. Tran, L. I. Sly, and D. A. Mitchell, "Selection of a Strain of *Aspergillus* for the Production of Citric Acid from Pineapple Waste in Solid-State Fermentation," *World J. Microbiol. Biotechnol.* **14**, 399–404 (1998).

92. Y. D. Hang and E. E. Woodams, "Production of Citric Acid from Corncobs by *Aspergillus niger*," *Bioresource Technol.* **65**, 251–253 (1998).

93. T. Roukas, "Carob Pod: A New Substrate for Citric Acid Production by *Aspergillus niger*," *Appl. Biochem. Biotechnol.* **74**, 43–53 (1998).

94. T. Roukas, "Citric Acid Production from Carob Pod by Solid-State Fermentation," *Enzyme Microbiol. Technol.* **24**, 54–59 (1999).

95. G. Aravantinos-Zafiris et al., "Fermentation of Orange Processing Wastes for Citric Acid Production," *J. Sci. Food Agric.* **65**, 117–120 (1994).

96. Y. D. Hang, "Recovery of Food Ingredients from Grape Pomace," *Process Biochem.* **23**, 2–4 (1988).

97. C. Nurgel and A. Canbas, "Production of Tartaric Acid from Pomace of Some *Anatolian* Grape Cultivars," *Am. J. Enol. Vitic.* **49**, 95–99 (1998).

98. A. Lopez-Baca and J. Gomez, "Fermentation Pattern of Whole Banana Waste Liquor with Four Inocula," *J. Sci. Food Agric.* **60**, 85–89 (1992).

99. C. E. Voget, C. F. Mignone, and R. J. Ertola, "Butanol Production from Apple Pomace," *Biotechnol. Lett.* **7**, 43–46 (1985).

100. A. Bramorski et al., "Production of Volatile Compounds by the Edible *Rhizopus oryzae* during Solid State Cultivation on Tropical Agro-Industrial Substrates," *Biotechnol. Lett.* **20**, 359–362 (1998).

101. R. A. Hours, C. E. Voget, and R. J. Ertola, "Apple Pomace as Raw-Material for Pectinases Production in Solid-State Fermentation," *Biol. Wastes* **23**, 221–228 (1988).

102. R. A. Hours, C. E. Voget, and R. J. Ertola, "Some Factors Affecting Pectinase Production from Apple Pomace in Solid-State Cultures," *Biol. Wastes* **24**, 147–157 (1988).

103. M. Berovic and H. Ostroversnik, "Production of *Aspergillus niger* Pectolytic Enzymes by Solid State Bioprocessing of Apple Pomace," *J. Biotechnol.* **53**, 47–53 (1997).

104. F. M. Rombouts and W. Pilnik, "Enzymes in Fruit and Vegetable Juice Technology," *Process Biochem.* **13**, 9–13 (1978).

105. Y. D. Hang and E. E. Woodams, "Production of Fungal Polygalacturonase from Apple Pomace," *Food Sci. Technol.* **27**, 194–196 (1994).

106. C. G. Garzon and R. A. Hours, "Citrus Waste—An Alternative Substrate for Pectinase Production in Solid-State Culture," *Bioresource Technol.* **39**, 93–95 (1992).

107. H. E. Hart et al., "Orange Finisher Pulp as a Substrate for Polygalacturonase Production by *Rhizopus oryzae*," *J. Food Sci.* **56**, 480–483 (1991).

108. C. Krishna and M. Chandrasekaran, "Banana Waste as Substrate for α-Amylase Production by *Bacillus subtilis* (CBTK 106) under Solid-State Fermentation," *Appl. Microbiol. Biotechnol.* **46**, 106–111 (1996).

109. J. Woodward, "Fungal and Other β-D-Glucosidase—Their Properties and Applications," *Enzyme Microb. Technol.* **4**, 73–79 (1982).

110. T. R. Yan, Y. H. Lin, and C. L. Lin, "Purification and Characterization of an Extracellular β-Glucosidase II with High Hydrolysis and Transglucosylation Activities from *Aspergillus niger*," *J. Agric. Food Chem.* **46**, 431–437 (1998).

111. P. J. Williams, "Hydrolytic Flavor Release in Fruit and Wines through Hydrolysis of Nonvolatile Precursors," in T. Acree and R. Teranishi, eds., *Flavor Science: Sensible Principles and Techniques*, American Chemical Society, Washington, D.C., 1993.

112. Y. Z. Gunata et al., "Hydrolysis of Grape Monoterpenyl β-Glucosides by Various β-Glucosidases," *J. Agric. Food Chem.* **38**, 1232–1236 (1990).

113. Y. D. Hang and E. E. Woodams, "Apple Pomace: A Potential Substrate for Production of β-Glucosidase by *Aspergillus foetidus*," *Food Sci. Technol.* **27**, 587–589 (1994).

114. G. Reed, *Enzymes in Food Processing*, Academic, New York, 1966.

115. Y. D. Hang and E. E. Woodams, "β-Fructofuranosidase Production by *Aspergillus* Species from Apple Pomace," *Food Sci. Technol.* **28**, 340–342 (1995).

116. J. Cordova, M. Nemmaoui, and M. Ismaili-Alaoui, "Lipase Production by Solid State Fermentation of Olive Cake and Sugar Cane Bagasse," *J. Mol. Catal. B-Enzyme* **5**, 75–78 (1998).

117. A. M. S. Ismail, "Utilization of Orange Peels for the Preparation of Multienzyme Complexes by Some Fungal Strains," *Process Biochem.* **31**, 645–650 (1996).

118. H. Purkarthofer, M. Sinner, and W. Steiner, "Cellulase-Free Xylanase from *Thermomyces lanuginosus*: Optimization of Production in Submerged and Solid-State Culture," *Enzyme Microbiol. Technol.* **15**, 677–682 (1993).

119. Z. Zheng and K. Shetty, "Solid-State Production of Beneficial Fungi on Apple Processing Wastes Using Glucosamine as the Indicator of Growth," *J. Agric. Food Chem.* **46**, 783–787 (1998).

120. Z. Zheng and K. Shetty, "Cranberry Processing Waste for Production of Solid State Fungal Inoculants Production," *Process Biochem.* **33**, 323–329 (1998).

ZUOXING ZHENG
KALIDAS SHETTY
University of Massachusetts
Amherst, Massachusetts

SOY FOODS, FERMENTED

Fermented soy foods are an important part of the human diet in Asia. The fermented soy foods that originated many thousands of years ago in the Asian countries are often referred to as indigenous fermented foods (1,2). Some of the fermented soy foods are now popular in the West (3). The increased palatability of fermented soy foods is due to desirable changes in soybean properties, including texture and organoleptic characteristics (flavor, aroma, and appearance or consistency). Elimination of beany flavors, improvement in digestibility, enhanced keeping quality of the product, improved safety (absence of toxins and partial and/or complete elimination of antinutritional factors), improved nutrition, and reduced cooking time (3) are results of fermentation of soybeans. Fermentation makes the organoleptic characteristics of soybean more attractive to the consumer than the raw soybean.

Of the great number of fermented soybeans consumed in Asia, only a few have been studied in the West; these are described in Table 1 (3,4). Some of these products are used primarily as flavoring agents that add some protein or amino acid to generally bland and low-protein diets. Others, such as tempe and natto, are served as staples. The origin of some fermented and nonfermented soy foods is presented in Table 2. The majority of soy foods, such as soy milk, soybean sprout, tofu, fermented tofu, soy sauce, and miso, originated in China. Natto is a soy food of Japanese origin, whereas tempeh originated in Indonesia (5).

SOY SAUCE

History

The recorded use of soy sauce dates back 2,700 years in China. The prototypes of soy sauce are believed to have been introduced from China to Japan more than 1,300

Table 1. Asian Fermented Soyfoods

Foods	Local names	Microorganisms involved	Substrate(s)	Use and nature of product
Soy sauce	Chiang-Yu (Chinese) Inyu (Chinese) Shoyu (Japanese) Ketjap (Indonesian) Kecap (Indonesian) Kanjang (Korean) Tayo (Philippine) See-ieu (Thai)	*Aspergillus* *Pediococcus* *Torulopsis* *Saccharomyces*	Whole soybean Defatted soyflake Wheat	Seasoning, a flavoring agent Dark reddish brown liquid, salty taste suggesting the meat extract
Miso	Chiang (Chinese) Tou-chiang (Chinese) Wei-cheng (Chinese) Miso (Japanese) Doenjang (Korean) Tauco (Indonesian) Tao-chieo (Thai) Soybean paste	*Aspergillus* *Pediococcus* *Streptococcus* *Saccharomyces* *Torulopsis*	Whole soybean Rice Barley	Flavoring agent, soup base Light yellow to dark reddish brown paste, smooth or chunky, salty and strongly flavored, resembling soy sauce
Sufu	Sufu, Furu, Fuju, Toufuru, Tao-yu (Chinese) Fu-nyu (Japanese) Tohuri (Philippine) Chinese cheese Soybean cheese Fermented tofu	*Actinomucor* *Mucor* *Rhizopus* *Aspergillus* *Monascus*	Soybean curd (firm tofu)	Condiment Creamy cheese-type cubes, salty, served with or without further cooking
Natto	Natto (Japanese) Tau-nou or Thua-nao (Thai) Kenima (Nepalese)	*Bacillus natto* *Bacillus subtilis*	Whole soybean	Served with cooked rice as a main dish or snack Cooked beans bound together and covered with viscous, sticky substance produced by the bacteria, ammonium odor, musty flavor
Tempe	Tempe kedelee (Indonesian) Tenpei (Japanese)	*Rhizopus oligosporus*	Whole soybean	Cooked and served as main dish or snack Cooked soft soybeans bound together by mycelia as cake, a clean fresh and yeasty odor
Taosi	Tou-shih (Chinese) Tao-si (Philippine) Tao-cheo (Malaysian) Taotjo (Indonesian) Hamanatto (Japanese)	*Aspergillus* *Streptococcus* *Pediococcus*	Whole soybean Wheat flour	Condiment, seasoning Nearly black soft bean, salty flavor resembling soy sauce
Meitauza	Meitauza (Chinese)	*Mucor* *Actinomucor*	Soybean presscake	Snack food
Meju	Meju (Korean)	*Aspergillus* *Rhizopus*	Black soybean	Seasoning agent

Source: Refs. 3 and 4.

Table 2. The Origin of Soyfoods

Soyfoods	Country of origin	Date of origin
Soy milk	China	164 B.C.
Soybean sprout	China	
Tofu	China	164 B.C.
Fermented tofu	China	Before 1400 A.D.
Soy sauce	China	700 B.C.
Miso	China	722 B.C.
Natto	Japan	Before 1286 A.D.
Tempe	Indonesia	Before 1750 A.D.

Source: Ref. 5.

years ago. The history of soy sauce in China and Japan is summarized in Table 3 (5,8).

Country of Consumption

Soy sauce is produced and consumed mostly in Asian countries. Owing to progress in world trade, soy sauce is now consumed almost worldwide. In 1978, soy sauce was exported from Hong Kong, the Korean Republic, Singapore, and Japan to a total of 97 countries (7,8).

Preparation

Different countries have different methods of preparing soy sauce. Even in Japan, there are five kinds of soy sauce: koikuchi, usukuchi, tamari, saishikomi, and shiro. The preparation of koikuchi soy sauce is outlined in Figure 1. The process includes (*1*) the treatment of raw materials, (*2*) koji making, (*3*) mash production and aging, (*4*) pressing, and (*5*) refining and pasteurization (8).

Treatment of Raw Materials. Cleaned soybeans or soybean meal is moistened (add 120–135% water) and cooked with steam under pressure. Wheat kernels are roasted at 170 to 180°C for several minutes and coarsely crushed into four or five pieces.

Koji Making. Equal parts of the resultant cooked soybeans or defatted soybean meal and roasted cracked wheat are mixed together and inoculated with 1 to 2% wt/wt of

Table 3. History of Soy Sauce

China	Japan
Shu-Ching (700 s.c.)[a]	
Chu[c]	
Chiang (meat from fish, bird, or meat)	
Chi-Min-Yao-Shu (532–549 A.D.)[a]	Manyo-shu (350–759)[a]
Chu (made from crushed wheat, or wheat flour made into balls or cakes, or cooked rice)	Koji (same as chu)
Chiang (made from soybeans or wheat)	Hishio (same as chiang, made from fish, meat, or soybean)
Shi and Shi-tche	Koma-hishio and miso
Tang dynasty (618–906)	Taiho-law (701)[a]
Ben-chao-gong mu (1590)[a]	Soybean-hishio, miso, kuki same as shi), taremiso, usudare, misodamari
Chiang-yu	Ekirinbon-setsuyoshu (1598)[a]
Tao-yu	Shoyu (same Chinese charactors as chiang-yu)
	Honcho-shokukan
	Shoyu, miso, tamari
	Industrial production of Koikuchi-shoyu in Noda (1561) and Chochi (1616), that of usukuchi shoyu in Sendai (1645); export of shoyu from Nagasaki, Japan (1668)

Source: Ref. 6.

[a]Names of old references

Note: Chu: mold-cultured cereals; chiang: a mixture of chu, proteinous foodstuffs, and salt; shi: mold-cultured soybeans with or without salt; shi-tche: the saltwater extract of shi: chiang-yu: the liquid separated from chiang; tao-yu: the liquid separated from soybean chiang.

the seed koji or a pure culture of *Aspergillus oryzae* or *A. sojae*. This mixture, containing 45% water, is spread on larger perforated stainless steel or wooden trays to a depth of 30 to 40 cm and incubated in a room at 25 to 30°C with humidity control for 2 to 3 days. During this period, the temperature, moisture, and aeration are controlled to al-

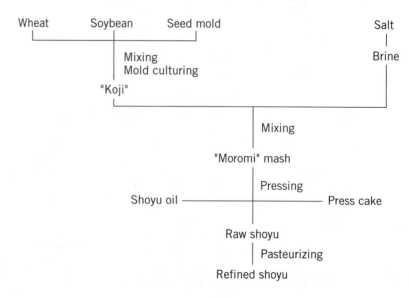

Figure 1. Preparation of Japanese fermented koikuchi shoyu. *Source:* Ref. 8.

low the seed koji to grow on the mixture, to prevent development of *Mucor* spp. or bacteria, and to enhance production of proteolytic and amylolytic enzymes. A rather high moisture level (30–32%) is necessary for good mycelial growth and enzyme activity. The resulting end product (clear yellow to yellowish green in color), is called koji, which is a mixture of hydrolytic enzymes and the substrate.

Mash (Moromi) Production and Aging. The koji is mixed with salt water, which has a 22 to 23% salt content and a volume 120 to 130% that of the raw materials. The mash or moromi is transferred to deep fermentation tanks. Approximately 5- to 10-kL wooden kegs or 10- to 20-kL concrete tanks for soy sauce fermentation are now being replaced by resin-coated iron tanks of 50 to 300 kL. The moromi is held for 4 to 8 months, depending on its temperature, with occasional agitation with compressed air to mix the dissolving contents uniformly and to promote the microbial growth. During the fermentation period, the enzymes from koji mold hydrolyze most of the protein to amino acids and low molecular weight peptides. Approximately 20% of the starch is consumed by the mold during koji cultivation, but almost all the remaining starch is converted into simple sugars. More than half this is fermented to lactic acid and alcohol by lactobacilli and yeasts, respectively. The initial pH value drops from 6.5–7.0 to 4.7–4.9. The lactic acid fermentation carried out in the beginning stage is gradually replaced by yeast fermentation. Pure culture of *Pediococcus halophylus* and *Saccharomyces rouxii* are sometimes added to the mash. The salt concentration of mash remains at 17 to 18% (weight per volume) after 1 or 2 months. The high concentration of mash effectively limits the growth to only a few desirable types of microorganisms.

Pressing. Once the aging of mash is completed, the liquid (raw shoyu) is separated from the mash by a hydraulic filter press in commercial operations or by a simple mechanical press on a domestic level. Sometimes, fresh salt water is added to the press cake, and a second fermentation is allowed to proceed for 1 to 2 months before a second pressed shoyu is produced, which is of lower quality than the first one. The final moisture content of the press cake is less than 30%. The press cake is called shoya cake, which is used as animal feed.

Refining and Pasteurization. Raw shoyu is separated into three layers (sediment at the bottom, clear shoyu in the middle, and an oil layer at the top) on standing in tanks. The oil layer is called shoyu oil and is removed by decantation. The middle layer, clear shoyu, is pasteurized at 80 to 85°C in a kettle or in a heat exchanger to kill vegetative microbial cells, denature enzymes, and coagulate proteins. Sometimes a filter aid is added to the pasteurized shoyu to enhance clarification. The clear shoyu is filtered, bottled, and marketed. Preservatives such as benzoic acid or propyl- or butyl p-hydroxy benzoate are sometimes added to the filtered shoyu during pasteurization. Steps for the industrial preparation of shoyu in Japan are mechanized, which prevents contamination by undesirable microorganisms.

The preparation of usukuchi, tamari, and shiro shoyu are almost the same as those of koikuchi shoyu, except procedures for usukuchi are directed at getting lighter color and aroma in the final product. Tamari shoyu uses soybean with a small amount of wheat (20:3) and dryer koji. Tamari mash cannot be agitated with compressed air. The liquid is obtained by dripping, not by pressing. Shiro shoyu is made mostly from wheat with less soybean, the volume of which is 10 to 20% that of wheat. Saishikomi is made from enzymatically degraded soybeans and wheat instead of the usual salt water.

Soy Sauce Produced in Other Oriental Countries

The fermented soy sauce industrially produced in Korea is similar to the Japanese koikuchi type. Almost the same amount of soy sauce is prepared at home as is industrially produced. The homemade soy sauce is prepared by a traditional method in which cooked soybeans are mashed and made into small balls, then subjected to natural inoculation of *Aspergillus* and *Rhizopus* molds, a process taking several months in winter. When spring comes these mold-cultured materials are extracted with salt water. The liquid part is boiled and fermented under the sun to make soy sauce. The residue of extraction is mixed with salt and stored to make miso, sometimes along with red pepper (6). Of Taiwan soy sauce, 5 to 10% is estimated to be inyu, which is made only from soybeans. The remarkable characteristics of inyu are that it is prepared from black soybeans instead of yellow soybeans and that the black bean koji is washed with water before it is mixed with salt water to make mash. Fermented soy sauce similar to inyu and tamari is still being produced in the southern part of China, and it seems to be the prototype of the soy sauce prepared only from soybeans. In Japan, tamari mash is usually fermented in wooden kegs, but the inyu in Taiwan, Singapore, and the southern part of China is fermented in big china pots placed under sunlight.

However, most of the soy sauce made in Peking and Shanghai today is prepared differently. The koji is prepared by using a large-scale method to culture *A. oryzae* with a mixture of steamed soybeans and wheat or wheat bran (6:4), and the koji is mixed with salt water to make hard mash, the moisture content of which is about 80% and the salt concentration about 6 to 7%. This hard and low-salt mash is kept at 45 to 50°C for about 3 weeks for enzymatic digestion. The digested mash is extracted with hot salt water and then with plain hot water. The residue without salt is good for animal feed. There is no alcoholic fermentation of mash or processing of mash as in Japanese shoyu manufacture.

The soy sauce produced in Indonesia is called kecap. The soybean koji is put in a wooden tank containing a salt solution of 20% and left to soak for 3 to 4 weeks at room temperature. The mash is boiled in a certain volume of water and then filtered. The filtrate liquor is cooked in water, two to three times the amount of liquor, and mixed with caramel and other spices (9).

New Development

Recent research and technology advances can be summarized as follows (8,10):

1. Greater use of defatted soybean instead of whole beans

2. Increase in protein digestibility of raw materials from 65 to 90% as the result of improved cooking methods of soybeans and wheat; the selection and mutation of starter molds; improved conditions for culturing molds or making koji; and the control of mash in terms of the temperature, pH, types and behavior of lactobacilli and yeasts; and the chemical components

3. Reduction of the time for koji cultivation from 72 to 48 hours

4. Decrease of fermentation period of mash from 1 to 3 years to about 6 months

5. The use of pure cultured starters of lactobacilli and yeasts

6. Mechanization of the equipment and expansion of the production scale

7. Improvement of quality and reduction of cost

Composition

The typical composition of the five types of shoyu are presented in Table 4. Both koikuchi and usukuchi shoyu contain appreciable amounts of total nitrogen in the range of 1.2 to 1.6%, of which 80 to 90% are lower peptides, peptones, and amino acids. Ammonia makes up the remaining nitrogen. Lactic, pyroglutamic, acetic, succinic, and formic acids are present in the koikuchi and usukachi shoyu. Shoyu contains neutral sugars such as glucose, galactose, arabinose, and xylose. A total of 271 flavor components have been found in Japanese shoyu in which 4-hydroxy-2(or 5)-ethyl-5(or 2)-methyl-3(2H)-furanone (HEMF) possesses a shoyulike sweet flavor (11).

Organisms Used

The microorganisms used in koji are *A. oryzae* and *A. sojae*. The molds produce a great variety of enzymes including α-amylase, proteases (acid, neutral, and alkaline), nucleases, sulfatases, phosphatases, transglycosidases, peptidases, acylase, ribonucleo-depolymerases, mononucleotide phosphatase, adenyl-deaminases, and purine nucleosidases (12).

The predominant active microbes in shoyu mash are salt-tolerant lactobacilli and yeasts such as *Zygosaccharomyces rouxii* and *Candida (Torulopsis) versatillis* or *C. etchellsii*. The major lactobacilli are *P. soyae* or *P. halophylus*. Good results have also been obtained by adding pure cultured lactobacilli to the new mash. The initial pH value of mash, 6.5 to 7.0, gradually decreases as the raw materials are degraded and lactic acid fermentation proceeds, and at around pH 5.5, yeast fermentation takes the place of lactic acid fermentation. To accelerate the alcoholic fermentation and to shorten its development time, pure cultured yeasts, *Z. rouxii* or *Torulopsis*, are sometimes added to the shoyu mash when its pH value reaches about 5.3, usually 3 to 4 weeks after the mash making.

Safety

Soy sauce mold mostly belongs to the *A. oryzae*, *A. sojae*, and *A. tamarii*, whereas all the aflatoxin-producing molds belong to *A. parasiticus* and *A. flavus*. Although fluorescent compounds are produced by *Aspergillus* molds with Rf values resembling those of aflatoxins, there is no hazard in human consumption of soy sauce (1,7,9,11).

Future Development

1. Soy sauce–like seasonings can be prepared from a mixture of plant proteins and starches other than soybeans and wheat.

2. An effort to find better strains that can degrade plant tissue is needed.

3. A much greater reduction in fermentation time is necessary for economic reasons.

4. Supplementing the enzyme systems would help reduce both the fermentation period and the viscosity of mash.

5. The heat-coagulant substances produced by pasteurization still pose a problem of refining.

6. It would be interesting to produce flavor compounds biochemically.

7. To increase the stability of fermented soy sauce is a difficult but important problem (7).

MISO

History

There is evidence to suggest that miso fermentation predates the production of soy sauce (10,12,13). Soy sauce was first called tamari miso. Tamari literally means a liquid drip and is the name given to the liquid that drips to the bottom of a miso keg while miso is aging. Thus, tamari, as the original form of soy sauce in Japan, was made from miso (14).

Countries of Consumption

Miso is consumed in China, Taiwan, Japan, the Philippines, Indonesia, and Korea. In these countries, miso is called by different names (Table 1).

Table 4. Typical Composition of Different Types of Shoyu

Types	Salt % (w/v)	Total nitrogen % (w/v)	Formyl nitrogen % (w/v)	Reducing sugars % (w/v)	Alcohol % (w/v)	pH
Koikuchi	16.95	1.58	0.92	2.99	2.45	4.79
Usukuchi	19.95	1.21	0.82	3.88	2.87	4.79
Tamari	16.85	2.35	1.14	6.98	0.68	4.85
Shiro	17.86	0.51	0.28	17.21	0.11	4.74
Saishikomi	14.30	2.35	1.17	9.43	1.33	4.84

Source: Ref. 8.

Preparation

There are many varieties of miso that differ from one another depending on the manufacturing process. Rice miso is the most popular type. The manufacturing method of salty rice miso is presented in Figure 2 (12,14). Miso fermentation is about the same as that of soy sauce. Unlike soy sauce, however, miso is a paste product and is mainly consumed as a miso soup. Most recently, there has been new development in the process of the brine fermentation of miso by an automatic system (10).

In Indonesia a similar product is called tauco. It uses soybean as raw material. The fermented product is cooked with palm sugar with two times the amount of salt used in the brine fermentation.

Composition

The approximate composition of various types of miso is presented in Table 5. The moisture content ranges from 42.6 to 45.7%. Sweet and barley miso have higher amounts of carbohydrates compared to others. This is due to a short fermentation period (12,14).

Organisms Used

The principal and indispensable microorganisms in miso fermentation are koji molds, salt-tolerant yeasts, and lactic acid bacteria. The koji molds are *A. oryzae* and *A. sojae*.

The bacteria *Streptococcus faecalis* and *S. faecium* are the dominant weak salt-tolerant lactic acid bacteria. *P. halophilus* and *P. acidi lactici* are the most dominant lactic acid types. They play an important role in the fermentation of miso. These organisms can survive in 23% brine. Because the acid tolerance of these bacteria is comparatively weak, they cannot survive below pH 5.0. The role of these bacteria in miso fermentation is to produce organic acids from carbohydrates, which lowers the pH of the fermenting miso and accelerates the growth of salt-tolerant yeasts. *Z. rouxii*, *T. versatilis*, and *T. etchellsii* are the salt-tolerant yeasts. *Z. rouxii* grows in a high-salt medium (18% sodium chloride) of pH 4.0 to 5.0 and is the main yeast involved in the fermentation of miso. The yeasts produce ethanol from sugars and phenolic compounds such as 4-ethylguaiacol, 4-ethylphenol, and 2-phenylethanol. These compounds give a characteristic flavor to the fermented miso. *Z. rouxii* var halomembranis nov var, *Pichia* spp., and *Hansenula* spp. are the aerobic film yeasts, which often produce films on substrate surfaces, resulting in off-flavor of the products.

Safety

No aflatoxin-producing mold strains were found in mold strains collected from miso factories. Similarly, no aflatoxin was detected in kojis from miso factories (12,14).

Future Development

Miso production in Japan decreased from 590,000 tons in 1973 to 572,000 tons in 1983. The decrease may be due to

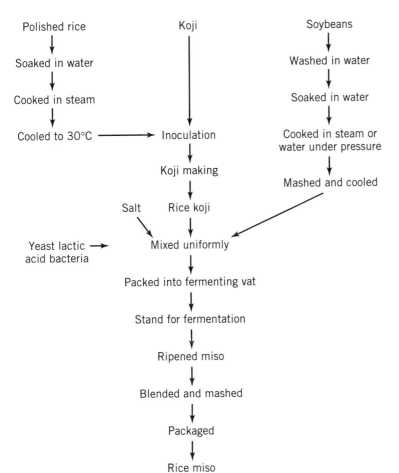

Figure 2. Flow chart of the manufacture of salty rice miso. *Source:* Ref. 15.

Table 5. Chemical Composition of Various Types of Miso

Component	Sweet miso	Salty light-yellow miso	Salty red miso	Barley miso	Soybean miso
Moisture (%)	42.6	45.4	45.7	44.0	44.9
Protein (%)	9.7	12.5	13.1	9.7	17.2
Fat (%)	3.0	6.0	5.5	4.3	10.5
Carbohydrate (%)	36.7	19.4	19.1	28.3	11.3
Crude fiber (%)	1.2	2.5	2.0	1.7	3.2
Ash (%)	6.8	14.2	14.6	12.0	12.9
Salt (%)	6.1	12.4	13.0	10.7	10.9
Calcium (mg/100 g)	80.0	100.0	130.0	80.0	150.0
Phosphorus (mg/100 g)	130.0	170.0	200.0	120.0	250.0
Iron (mg/100 g)	3.4	4.0	4.3	3.0	6.8
Thiamin (mg/100 g)	0.05	0.03	0.03	0.04	0.04
Riboflavin (mg/100 g)	0.1	0.1	0.1	0.1	0.12
Niacin (mg/100 g)	1.5	1.5	1.5	1.5	1.2
Vitamin B_{12} (μg/100 g)	—	0.17	—	—	—

Source: Ref. 15.

the high sodium chloride content. If the salt level could be lowered, some of the problems related to sodium would be lessened and consumption probably would increase. Therefore, further studies should be conducted to find new flavoring materials to replace salt.

SUFU

History

Manufacture of tofu (unfermented soybean curd) began during the era of the Han dynasty. The Pen Ts'ao, or Chinese Materia, of 1596, compiled by Li Shi-Chin, indicated that tofu was invented by Liu An (179 B.C. to 122 S.C.), king of Wainan. It is not known when sufu production began (16). Sufu is produced and consumed in China and Taiwan. It is used directly as a condiment or is cooked with vegetables or meats. Because of its creamy texture, it is suitable for use as a cracker spread or as an ingredient for dips and dressings in the Western world (16,17).

Preparation

The preparation of sufu consists of five steps: (1) preparation of tofu, (2) preparation of pehtze (freshly prepared tofu grown with the mold) (3) salting, (4) processing, and (5) aging. A flow sheet for preparation of sufu is shown in Figure 3 (16,18,19).

Preparation of Tofu. Soybeans are washed, soaked overnight in water, and ground with water to a milky slurry. A water-to-bean ratio of 10:1 is common. The ground slurry is boiled or steamed and filtered to separate soy milk. A coagulant such as calcium sulfate, magnesium sulfate, or sea salt (about 2.5–3.5% of dry weight of soybean) is added to the hot soy milk. Generally, 20% higher weight of coagulant is used to produce tofu for sufu preparation than that used for regular tofu. After the addition of coagulant to the soy milk, the mixture is agitated vigorously to break the curd into small pieces. The agitated mixture is set aside for 10 min to complete the coagulation process. The coagulated material is referred to as tofu. Tofu is separated from the whey by filtering and pressing the coagulated mixture. Pressing is done in a cloth-lined box to form firm

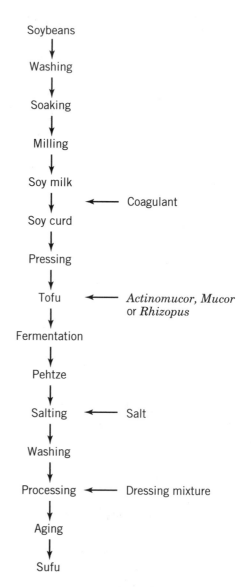

Figure 3. Flow sheet for preparation of sufu. *Source:* Ref. 18.

blocks of desired size. The firm blocks of tofu are cut into small cubes (2 × 4 × 4 cm) weighing approximately 18 g. Because adept tofu workers are becoming scarce, the automatic tofu machine was developed in recent years. This machine saves a lot of labor and unifies the quality of tofu.

Preparation of Pehtze. Pehtze is tofu overgrown with grayish hairlike mycelium of molds of the genus *Actinomucor* spp., *Mucor* spp., or *Rhizopus* spp. To avoid contamination and bacterial spoilage, tofu cubes are exposed to bright sunlight for several hours before inoculation with mold. Instead of being sun dried, tofu cubes can be dried in a hot-air oven at 100°C for about 15 min (17). Before hot-air drying, tofu cubes may be immersed in an acidic saline solution (prepared by dissolving 6 g NaCl and 2.5 g citric acid in 100 mL water) (20). The hot tofu cubes are cooled to 20°C. A pure culture of molds is inoculated onto the surface of the tofu cubes. The inoculated tofu cubes are put in a tray for incubation. The room temperature is maintained at 10 to 20°C. After 3 to 7 days, white fungal mycelium can be seen on the surface of tofu cubes, and at that time they are taken out and immediately salted.

Salting. Pehtze is transferred to large earthenware jars (having a volume of 7 hL) for further curing. It is put in layers in the jars. Each layer of pehtze is sprinkled with a layer of salt and kept for 3 to 4 days. During this period, the pehtze absorbs most of the salt. The salted pehtze is removed from the jars, washed with water, and transferred to other jars for further processing.

Processing and Aging. Pehtze is processed and aged in earthenware jars containing a dressing mixture. The most commonly used dressing mixture consists of 2 kg salt, 1 kg soy sauce mash, 0.6 kg red koji, 0.6 kg jaggery, and 6 kg water. Additional essence can be added to the dressing mixture to give a special flavor. For aging, alternate layers of pehtze and dressing mixture are packed into the jar to 80% of its volume, then the jar is filled with brine containing 12 to 20% salt. The mouth of the jar is wrapped with sheath leaves of bamboo shoots and sealed with clay. The sealed jars are aged for 1 to 3 months or longer for further fermentation before packing into small containers for marketing (16,19). The salt in the dressing mixture imparts a salty taste to the sufu and retards the growth of contaminating microorganisms and mold. Shelf life is a major problem with sufu. Alcohol in the dressing mixture helps to preserve sufu; however, it gives a harsh flavor to the sufu. Therefore it is important to find a method for preservation of fermented pehtze and sufu without the use of alcohol in the dressing mixture.

Composition

The chemical composition of fresh and dried tofu, pehtze, and sufu is presented in Table 6. Enzymes such as invertase, emulsin, trypsin-like protease, pepsin-like protease, oxidase, and catalase are found in pehtze. These enzymes are produced by the mold during fermentation and act on their respective substrates. During the first 10 days of aging, the mold enzymes act on substrates such as proteins

Table 6. Composition of Tofu, Pehtze, and Sufu

Component (%)	Tofu Fresh	Tofu Dried	Pehtze Fresh	Pehtze Dried	Sufu Fresh	Sufu Dried
Moisture	75.8	—	70.0	—	59.7	—
Protein	16.0	66.0	17.9	59.7	15.9	39.4
Fat	7.2	29.7	9.8	32.8	20.3	50.4
Carbohydrate	0.1	0.4	0.5	1.7	0.0	0.0
Fiber	0.0	0.0	0.4	1.3	1.1	3.7
Ash	0.9	3.9	1.4	4.5	3.0	7.4

Source: Ref. 18.

and lipids to produce various hydrolyzed products. These hydrolyzed products provide the principal constituents for development of mild characteristic flavors of sufu.

Organisms Used

In the preparation of pehtze, a variety of fungi are involved. They are *Rhizopus chinensis* var. chungyuen, *Actinomucor elegans*, *Mucor hiemalis*, *M. silvaticus*, and *M. praini*. The fungi, namely *R. chinensis* var chungyuen, are most commonly present in commercial sufu. In the dressing mixture, the fungi *Monascus anka* or *M. purpureus* is present in the red koji, and the microorganisms used for the processing of soy sauce are present in the soy mash.

Safety

When Sprague-Dawley mice were fed 30 g sufu per kg body weight, the mice did not die, and there were no toxic or growth effects. If properly fermented, sufu and its related products are not hazardous to health. The microorganisms involved in sufu fermentation are not known to produce harmful toxins (16).

Future Development

Sufu is a creamy, cheeselike product that has a mild flavor. It can be used in the same way as cheese. If further improvements in taste, flavor, and texture are made, it may become more widely accepted. Although a pure culture method for preparing sufu has been developed, future studies are needed to develop uniform high-quality products and well-defined economical processes to manufacture sufu. Nutritional quality evaluation of sufu needs further investigation (16).

NATTO

History

The earliest record for the term natto appeared in 1068, but there was no description of the preparation method. Natto, during early periods of use, was prepared simply by wrapping cooked soybeans with rice-straw bundles and fermenting the wrapped soybeans at ambient temperature (21,22). In the natural fermentation of soybeans, mold usually dominates, but natto is one of the few products in which bacteria predominate during fermentation. *Bacillus natto*, identified as *B. subtilis*, is the organism responsible for natto fermentation. Consequently, natto possesses the characteristic odor and persistent musty flavor of this organism and is also covered with viscous, sticky polymers produced by *B. subtilis*. Because of its characteristic odor, flavor, and slimy appearance, natto, even though it is well known in Japan, is not as popular nor as widely consumed as miso (23). A similar product, kenima, is known in Nepal, Sikkim, and the Darjeeling districts of India. No yeasts or filamentous fungi were recovered from the samples analyzed from Darjeeling, but two rod-shaped, acid-producing bacteria, present at levels of 10^6 to 10^7/g of wet weight were recovered. Uncooked kenima was unappealing to the Western taste, but when deep-fried and salted, it had a pleasant, nutlike flavor. Thai thua-mao and Nigerian dawadawa are also alkaline-fermented foods closely related to Japanese natto and Indian kenima (22).

Preparation

The manufacture of natto is very simple, as shown in Figure 4. The spores of *B. natto* are sprayed on the cooked soybeans. The resultant soybeans are put into a porous plastic container with a cover and packed. The packed container is put into a fermentation room and kept for 14 to 18 h. The cooked whole soybean grains are kept in the original shape, and the surface of each cooked grain is covered with a very viscous, stringy substance, which is produced by *B. natto* (10).

Composition

There is no change in fat and fiber contents of soybeans during a 24-h period of fermentation, but carbohydrate almost totally disappears. A great increase in water-soluble and ammonia nitrogen is noted during fermentation as well as during storage. The amino acid composition remains the same. Boiling markedly decreases the thiamine level of soybeans, but fermentation by *B. natto* enhances the thiamine content of natto approximately to the same level of soybeans. Riboflavin in natto greatly exceeds that in soybeans. Vitamin B_{12} in natto was found to be higher than in soybeans (23).

TEMPEH

History

Even though tempeh has been produced and consumed in Indonesia for centuries, there are no written records of its origin. The central and east Java areas of Indonesia are still the major tempeh producers. For that reason, several scientists thought that tempeh probably originated in these areas. Besides these areas, it is produced and consumed in other parts of Indonesia (1,9,24).

Countries of Consumption

Tempeh is relatively unknown in surrounding countries such as Thailand, China, and Japan, where soybeans form an important part of the diet. Tempeh is produced in small quantities and consumed by immigrants from Indonesia in Malaysia, Surinam, Canada, and the Netherlands. It is becoming a popular food for many vegetarians in the United States. Tempeh experienced a sudden rise in popularity in the United States in 1975, although it was studied extensively and known in the United States as early as 1950 (25). In 1971, soybeans and the fermented food tempeh were introduced into the Republic of Zambia to improve the nutritional status of that country. Tempeh was introduced recently into Sri Lanka through trainees who received special training in production and processing technologies of tempeh in Indonesia.

Preparation

The essential steps for traditional preparation of tempeh are presented in Figure 5. At first, the soybeans are soaked and cooked for 30 min in boiling water to loosen the soybean seed coats. The seed coats of cooked soybeans are hand removed or rubbed with feet to loosen the seed coats and washed with water to separate beans from seed coats. The dehulled beans are again soaked overnight to be hydrated and to allow bacterial acid fermentation. The soaked, dehulled beans are cooked again, drained, cooled, and inoculated with a starter culture or an inoculum from

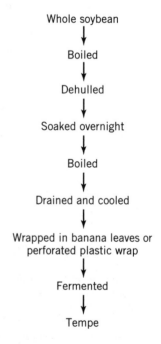

Whole soybean
↓
Boiled
↓
Dehulled
↓
Soaked overnight
↓
Boiled
↓
Drained and cooled
↓
Wrapped in banana leaves or
perforated plastic wrap
↓
Fermented
↓
Tempe

Figure 5. Flow sheet: Indonesian household tempe process. *Source:* Ref. 24.

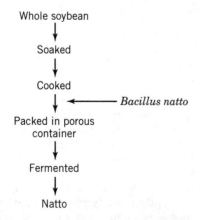

Whole soybean
↓
Soaked
↓
Cooked
↓ ← *Bacillus natto*
Packed in porous
container
↓
Fermented
↓
Natto

Figure 4. Manufacturing process of natto. *Source:* Ref. 10.

Table 7. Composition of Nutrients Per 100 g of Tempe

Tempe type	Food energy (cal)	Moisture (%)	Protein (%)	Fat (%)	Carbohydrate (%)	Fiber (%)	Ash (%)
Soy tempe fresh	157	60.4	19.5	7.5	9.9	1.4	1.3
Soy tempe, sun-dried		8.9	43.1	18.0	26.2	3.8	
Soy tempe, freeze-dried		1.9	46.2	23.4	25.8	2.7	2.7
Soy tempe, dry basis		0.0	54.6	14.1	27.9	3.1	3.5
Soy tempe, deep fried		50.0	23.0	18.0	8.0	2.0	1.0

Source: Ref. 24.

a previous batch, wrapped in banana leaves or perforated plastic bags, and incubated for up to 48 h at room temperature (1).

The tempeh should be harvested as soon as the bean cotyledons have been completely overgrown with mycelium and knitted into a compact cake. Freshly made tempe can be stored for a day or two at room temperature without changing many of its qualities or flavor characteristics. Traditionally in Indonesia, tempeh is consumed on the day it is made. If the fresh tempeh is stored for longer periods (more than 2 days) at room temperature, it becomes unsuitable for consumption because of off-flavors and odors produced by excessive proteolysis or by contaminating bacteria and other microorganisms.

Composition

Tempeh is an excellent source of proteins, vitamins, and minerals. It is often used in the diet of diabetics because of the low utilizable carbohydrate content (1,25). During a 72-h fermentation of tempeh, total soluble solids and soluble nitrogen increase, respectively, from 13 to 28% and 0.5 to 2.5%, while the total nitrogen remains fairly constant. The nutrient composition of tempeh is presented in Table 7 (24).

Organisms Used

At least four species of *Rhizopus* (*R. Oligosporus saito*, *R. oryzae*, *R. stolonifer (Ehrenbex Fries) Lind*, and *R. arrhizus* (Fisher) could be used in the preparation of tempeh. In addition to the mold *R. oligosporus*, numerous bacteria (both spore-forming and non-spore-forming types) and yeasts are reported to exist in tempeh. The presence of bacteria in the tempeh fermentation causes the off-odors and flavors of tempeh.

Safety

It is reported that the mold species responsible for tempeh fermentation do not produce aflatoxin. There have been no reports of toxin found in foods prepared from soybean. The mold required for tempeh fermentation is reported to protect tempeh and tempeh-like foods against aflatoxin production and aflatoxin-producing molds. The tempeh mold *R. oligosporus* prevents accumulation of aflatoxin in tempeh by hydrolyzing it (9).

Future Development

Tempeh has great potential as a key protein source in developing countries owing to its low cost and easy preparation. In the future, tempeh may play a major role in sat-

isfying the protein deficiency problems in underdeveloped countries. During tempeh preparation, two major problems are usually encountered: large energy requirement and losses of solids during preparation. There is a need for research on how to reduce energy input and losses of solids.

TAOSI

Taosi is a product made by fermenting whole soybeans with strains of *A. oryzae*. The product is known as toushih by the Chinese, taotjo by the East Indians, and hamanatto by the Japanese (23). In the United States, such fermented beans are often referred to as black beans because of their

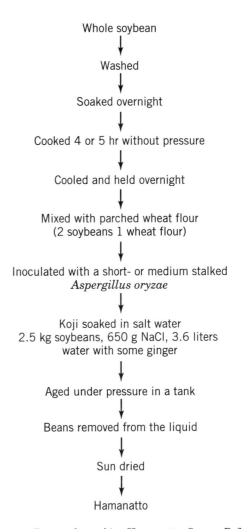

Whole soybean
↓
Washed
↓
Soaked overnight
↓
Cooked 4 or 5 hr without pressure
↓
Cooled and held overnight
↓
Mixed with parched wheat flour
(2 soybeans 1 wheat flour)
↓
Inoculated with a short- or medium stalked
Aspergillus oryzae
↓
Koji soaked in salt water
2.5 kg soybeans, 650 g NaCl, 3.6 liters
water with some ginger
↓
Aged under pressure in a tank
↓
Beans removed from the liquid
↓
Sun dried
↓
Hamanatto

Figure 6. Process for making Hamanatto. *Source:* Ref. 23.

color. In China, the product is described as "shi" in the old literature and Chi-Min-Yao-Shu (Table 3) (15). It is mainly consumed as a side dish. The methods of preparing soybeans for fermentation and the composition of the brine may vary from country to country, but the essential features are similar. Soybeans are soaked and steamed until soft, drained, cooled, mixed with parched wheat flour, and then inoculated with a strain of *A. oryzae*. After incubation, the beans are packed with the desired amount of salt, spices, wine, and water and aged for several weeks or months. The finished products are blackish, they have a salty taste, and their flavor resembles that of shoyu. However, they may differ in salt and moisture contents. Hamanatto is rather soft, having a high moisture content. Toushih has a much lower moisture content than that of hamanatto and therefore is not as soft as hamanatto. Taotjo tends to have a sweet taste because sugar is often added to the brine (23).

A typical process for making hamanatto in Japan is outlined in Figure 6. The finished product has a salt content of 13% and a moisture content of 38%. The fermented beans can be used as an appetizer to be consumed with bland foods, such as rice gruel, or they can be cooked with vegetables, meats, and seafoods as a flavoring agent (23).

BIBLIOGRAPHY

1. K. H. Steinkraus, ed., *Handbook of Indigenous Fermented Foods*, 2nd ed., Marcel Dekker, New York, 1996.

2. C. W. Hesseltine and H. L. Wang, "Fermented Foods," *Chem. and Industry* **12**, 393 (1979).

3. N. R. Reddy et al., "Legume-based Fermented Foods: Their Preparation and Nutritional Quality," *CRC Crit. Rev. Food Sci. Nutrition* **17**, 335 (1982).

4. N. R. Reddy, M. D. Pierson, and D. K. Salunkhe, eds., *Legume-Based Fermented Foods*, CRC Press, Boca Raton, Fla., 1986.

5. S. Chen, "Soy Foods in the Far East and USA," *The 1st European Soy Foods Workshop*, Amsterdam, Netherlands, September, 1984, pp. 227–228.

6. H. L. Wang, "Tofu and Tempe as Potential Protein Sources in the Western Diet," *J. Am. Oil Chemists Soc.* **61**, 528–534 (1984).

7. T. Yokotsuka, "Soy Sauce Biochemistry," in C. O. Chichester, E. M. Mrak, and B. S. Schweigert, eds., *Advances in Food Research*, Vol. 30, Academic Press, Orlando, Fla., 1986, pp. 196–329.

8. D. Fukushima, "Industrialization of Fermented Soy Sauce Production Centering Around Japanese Shoyu," in K. H. Steinkraus, ed., Industrialization of Indigenous Fermented Foods, Marcel Dekker, New York, 1989, pp. 1–88.

9. F. G. Winarno, "Traditional Fermented Soy Foods," in F. G. Winarno, ed., *International Soy Foods Symposium*, Yogyakarta, Indonesia, September, 1986.

10. D. Fukushima, "New Development in the Process of Traditional Soy Foods in Japan," in F. G. Winarno, ed., *International Soy Foods Symposium*, Yogyakarta, Indonesia, September, 1986.

11. N. Nunomura and M. Sasaki, "Soy Sauce," in N. R. Reddy, M. D. Pierson and D. K. Salunkhe, eds., *Legume-Based Fermented Foods*, CRC Press, Boca Raton, Fla., 1986, pp. 5–46.

12. H. Ebino, "Industrialization of Japanese Miso Fermentation," in *Industrialization of Indigenous Fermented Foods*, in K. H. Steinkraus, ed., Marcel Dekker, New York. 1989, pp. 89–126.

13. W. Shurtleff and A. Aoyagi, *The Book of Miso*, Autumn Press, Kanagawa-ken, Japan, 1976.

14. S. H. Abiose, M. C. Allan, and B. J. B. Wood, "Microbiology and Biochemistry of Miso Fermentation," in A. I. Laskin, ed., *Advances in Applied Microbiology*, Vol. 28, Academic Press, New York, 1982, pp. 239–265.

15. H. Ebine, "Miso," in N. R. Reddy, M. D. Pierson, and D. K. Salunkhe, eds., *Legume-Based Fermented Foods*, CRC Press, Boca Raton, Fla., 1986, pp. 47–68.

16. Y.-C. Su, "Sufu," in N. R. Reddy, M. D. Pierson and D. K. Salunkhe, eds., *Legume-Based Fermented Foods*, CRC Press, Boca Raton, Fla., 1986, pp. 69–83.

17. H. L. Wang and C. W. Hesseltine, "Mold-Modified Foods," in H. J. Peppler and D. Perlman, eds., *Microbial Technology*, Vol. 2, 2nd ed., Academic Press, New York, 1979.

18. Y. C. Su, "Traditional Fermented Foods in Taiwan," *Proc. Oriental Fermented Foods*, Food Industry Research and Development Institute, Hsinchu, Taiwan, **15** (1980).

19. N. S. Wai, "Investigation of the Various Processes Used in Preparing Chinese Cheese by the Fermentation of Soybean Curd with *Mucor* and Other Fungi," Final Technical Report, UR-A6-(40)-1, U.S. Department of Agriculture, Washington, D.C., 1968.

20. W. Shurtleff and A. Aoyagi, *The Book of Tofu*, Ten Speed Press, Berkeley, Calif., 1983.

21. T. Ohta, "Natto," in N. R. Reddy, M. D. Pierson, and D. K. Salunkhe, eds., *Legume-Based Fermented Foods*, CRC Press, Boca Raton, Fla., 1986, pp. 85–93.

22. K. H. Steinkraus, "African Alkaline Fermented Foods and Their Relation to Similar Fermented Foods in Other Parts of the World," in A. Westby and P. J. A. Reilly, eds., *Traditional African Foods Quality & Nutrition*, International Foundation for Science, Stockholm, Sweden, 1991, pp. 87–92.

23. C. W. Hesseltine and H. L. Wang, "Fermented Soybean Food Products," in A. K. Smith and S. J. Circle, eds., *Soybeans: Chemistry and Technology*, AVI Publishing Co., Westport, Conn., 1972, pp. 389–419.

24. F. G. Winarno and N. R. Reddy, "Tempe," in N. R. Reddy, M. D. Pierson, and D. K. Salunkhe, eds., *Legume-Based Fermented Foods*, CRC Press, Boca Raton, Fla., 1986, pp. 95–117.

25. W. Shurtleff and A. Aoyagi, *The Book of Tempeh: A Super Soy Food from Indonesia*, Harper & Row Publishers, New York, 1979.

KEITH H. STEINKRAUS
Cornell University
Ithaca, New York

WEN-LIAN CHEN
Food Industry Research and Development Institute
Hsinchu, Taiwan

SOYBEANS AND SOYBEAN PROCESSING

The four principal oilseed crops grown in the United States are soybeans, cottonseed, peanuts, and sunflowers. Some are consumed directly as foods, and all serve as sources of edible oils. After removal of these oils, the resulting meals are rich in proteins and are used mainly for animal feeds. Soybeans belong to the family Leguminosae (legume) and the genus and species of *Glycine max Merrill*. Soybeans are principally produced in the United States, Brazil, People's Republic of China, and Argentina. They are used for edible oil, animal feed, food, edible proteins and industrial oil.

Soybeans are now the principal oilseed crop in the United States. They are believed to have been domesticated in the eastern half of northern China around the eleventh century B.C. or earlier. They were later introduced and established in Japan and other parts of Asia, brought to Europe, and were introduced in the United States in the late eighteenth century or early nineteenth century. Soybeans became established as an oilseed crop in the United States in the late 1920s and attained major commercial and importance during World War II.

PROPERTIES AND COMPOSITIONS

A common feature of the structure of soybean seed is storage of the bulk of the protein and oil in distinct membrane-bound, subcellular organelles called protein bodies and lipid bodies, (spherosomes) as illustrated in Figure 1. Although not shown in Figure 1, the protein bodies contain inclusions referred to as globoids (>0.1 μm) that are storage sites for phytate and cations such as potassium, magnesium, and calcium. During germination, the contents of the storage organelles are mobilized and utilized by the growing seedling. Soybeans grow on erect, bushy annual plants, 6.3–2.0 m high with hairy stems and trifoliolate leaves. The flowers are white or purple or combinations of white and purple. Growing season varies with latitude (120–130 days in central Illinois). Seeds are produced in pods, usually containing three almost spherical to oval seeds weighing 0.1–0.2 g. Commercial varieties have a yellow seed coat plus two cotyledons, plumule, and hypocotyl-radicle. Cotyledons contain protein and lipid bodies.

On a moisture-free basis, soybeans contain approximately 8% hulls, 20% oil, 43% protein ($N \times 6.25$), 5% ash, and 52% protein in dehulled, and defatted meal (variable depending on efficiency of dehulling and oil extraction, variety of seed, and climatic conditions during growth). The proteins found in soybeans are complex mixtures consisting of four characteristic fractions with molecular weights of ca 8,000–700,000, as illustrated by the ultracentrifuge pattern (Fig. 2). The 7 S and 11 S fractions usually predominate. The 7 S and 11 S fractions are considered to be storage proteins and are located in the protein bodies. The principal portion of soybean 7 S fraction, β-conglycinin, consists of at least seven isomers resulting from various combinations of three subunits (α, α', and β) $\alpha\beta$, $\alpha'\beta2$, $\alpha\alpha'\beta$, $\alpha2\beta$, $\alpha'\alpha2$, $\alpha3$, and $\beta3$. Based on the sizes of the subunits, the β-conglycinin isomers have molecular weights between 126,000 and 171,000.

The 11 S molecule, also called glycinin, is more complex than β-conglycinin. It consists of ca 12 polypeptides, half of which have acidic isoelectric points (molecular weight ca 37,000–44,000) and half of which have basic isoelectric points (molecular weight ca 17,000–22,000); six different acidic and five different basic polypeptides have been separated and identified. Some acidic and basic subunits are linked in nonrandom pairs through disulfide bonds. They are apparently synthesized as single-chain proteins that are subsequently modified by proteolysis to form the acidic and basic polypeptide chains.

In addition to the storage proteins, soybeans contain a variety of minor proteins, including trypsin inhibitors, hemagglutinins, and enzymes (eg, urease and lipoxygenase). Amino acid compositions of soybeans are given in Table 1 along with the essential amino acid pattern for a high-quality protein that meets human requirements as established by the Food and Nutrition Board of the National

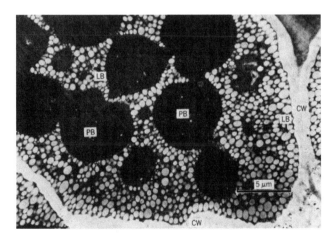

Figure 1. Transmission electron micrograph of a section of a mature, hydrated soybean cotyledon. Protein bodies (PB), lipid bodies (LB), and cell wall (CW) are identified. *Source:* Ref. 1.

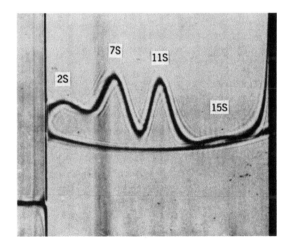

Figure 2. Ultracentrifugal pattern for the water-extractable proteins of defatted soybean meal in pH 7.6, 0.5 ionic strength buffer. Numbers across the top of the pattern are sedimentation coefficients in Svedberg units (S). Molecular weight ranges for the fractions are 2 S: 8,000–50,000; 7 S: 100,000–180,000; 11 S: 300,000–350,000; and 15 S: 700,000 *Source:* Ref. 2.

Table 1. Ammonia and Amino Acid Composition of Defatted Oilseed Meals, g/16 g Nitrogen

Amino acid	Soybean[a]	FNB Pattern[b]
Lysine	6.4	5.1
Histidine	2.6	1.7
Ammonia	1.9	
Arginine	7.3	
Aspartic acid	11.8	
Threonine	3.9	3.5
Serine	5.5	
Glutamic acid	18.6	
Proline	5.5	
Glycine	4.3	
Alanine	4.3	
Valine	4.6	4.8
Cystine	1.4	
Methionine	1.1	2.6
Isoleucine	4.6	4.2
Leucine	7.8	7.0
Tyrosine	3.8	
Phenylalanine	5.0	7.3
Tryptophan	1.4	1.1

[a]Means for seven varieties (3).

[b]Essential amino acid pattern (Food and Nutrition Board pattern) for high-quality protein for humans. An essential amino acid cannot be synthesized by an organism at a sufficiently rapid rate to meet metabolic needs (4).

Research Council. Soybeans meet or exceed the reference pattern except for valine and the sulfur amino acids, which are only slightly low.

Representative fatty acid compositions of the unprocessed triglyceride oils found in soybeans are given in Table 2. Although the oleic and linoleic acid content of soybean oil are high, soybean oil is distinguished from the others by a content of 4–10% linolenic acid, and hence is called a linolenic acid oil.

In addition to the triglycerides, soybeans also contain phosphatides that consist mainly of phosphatidylcholines, phosphatidylethanolamines, phosphatidylinositols, and phosphatidylserines. Generally, ca one-half the phosphatides are extracted from soybeans with hexane.

Table 2. Fatty Acid Composition of Unprocessed Oilseed Oils, Percentage

Carboxylic acid	Soybean
Saturated fatty acids	
10:0	
12:0	0.10
14:0	0.16
16:0	10.7
18:0	3.87
20:0	0.22
22:0	
24:0	
Unsaturated fatty acids	
16:1	0.29
18:1	22.8
18:2	50.8
18:3	6.76
20:1	

Source: Ref. 5.

Sterols are present in concentrations of 0.2–0.4% in soybean oil in the following compositions: campesterol, 20%; stigmasterol, 20%; β-sitosterol, 53%; and δ^5 avenasterol, 3%. The sterols exist in the seeds in four forms: free, esterified, nonacylated glucosides, and acylated glucosides. Soybeans contain a total of 0.16% of these sterol forms in the ratio of ca 3:1:2:2.

Soybeans contain two types of carbohydrates: soluble monosaccharides and oligosaccharides and insoluble polysaccharides. The contents of soluble sugars and total carbohydrates in the defatted oilseed meals are given in Table 3; values for intact seeds are different because of their oil and seed-coat contents. Sucrose, raffinose, and stachyose are the principal soluble sugars present. The polysaccharide (not including crude fiber) content is roughly equal to the total carbohydrates minus total soluble sugars, and it ranges from 11 to 19% of the flours.

Soybeans contain minor components that affect the defatted seeds, especially when used for feed and food. Soybean seeds contain 1.0–1.5% phytic acid. The isoflavone glucosides genistin, daidzin, and glycetein-7-β-glucoside, plus small amounts of the corresponding aglycones, are constituents of soybeans. Isoflavone contents range from 0.047 to 6.36% (Fig. 3). Soybeans also contain saponins. Recent results indicate that saponin content of soybeans is 5.6% and that of defatted soy flour 2.2–2.5%; these values are ca tenfold higher than those obtained in earlier investigations.

Table 3. Carbohydrate Contents of Defatted Oilseed Flours, Percentage

Constituent	Soybean
Soluble Sugars	
Glucose	trace
Sucrose	7.80
Trehalose	
Raffinose	1.25
Stachyose	6.30
Total	*15.35*
Total carbohydrate[a]	*31.0*

Source: Ref. 6.

[a]Obtained by different: 100 − (protein + oil + ash + crude fiber) = nitrogen-free extract.

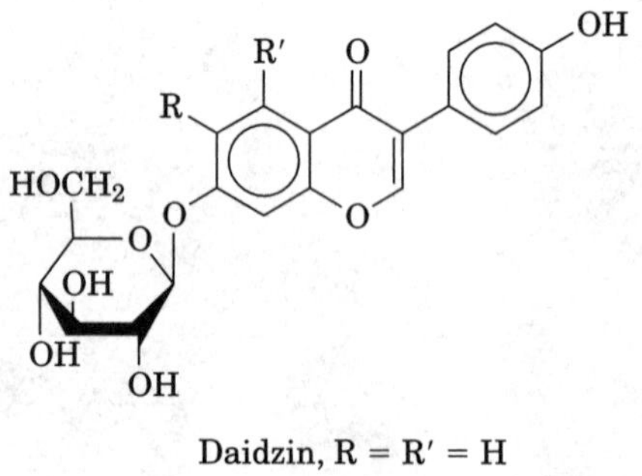

Daidzin, R = R′ = H
Genistin, R = H, R′ = OH
Glycetein-7β-glucoside, R = OCH₃, R′ = H

Figure 3. Daidzen, Genisten and glycetein-7β-glucoside.

HARVESTING AND STORAGE

The U.S. soybean crop normally is harvested in September or October. Ideal moisture content for harvesting is 13%, and the crop can be successfully stored at this moisture content until the following summer. Soybeans at <12% moisture can be stored for 2 yr or more with no significant deterioration. Beans with moisture above safe storage limits are dried or placed in aerated bins for gradual moisture reduction.

Soybeans are trucked from the farm to country elevators. From there they are moved by truck or rail to processing plants, subterminal elevators, or terminal elevators. From the terminal elevators, the beans are shipped by rail or barge to export elevators or processors. Soybeans are stored in concrete silos 6–12 m in diameter with heights of > 46 m. The silos are often arranged in multiple rows, and the resulting interstitial silo areas are likewise used for storage. In bulk storage, seasonal temperature changes cause variations in temperature between the different portions of the grain mass (eg, in the winter, soybeans next to the outer walls are colder than those in the center of the silo). Such temperature differences initiate air currents that transfer moisture from warm to cold portions of the seed mass. Thus bulk soybeans originally at safe moisture concentrations may, after storage, have localized regions of higher moisture that cause growth of microorganisms, which in turn can lead to heating. Under these conditions, the beans turn black and may even ignite. Such seasonal moisture transfer also occurs with other oilseeds. In commercial practice the temperature is carefully monitored and, when it rises, the beans are either remixed or processed. Aflatoxin contamination is not a problem as it is with cottonseed and peanuts. Although fungi invade soybeans stored at high moisture and temperatures. *Aspergillus flavus* does not grow well on soybeans and aflatoxins are negligible.

PROCESSING

Virtually all soybeans processed in the United States are solvent extracted (Fig. 4). Beans arriving at the plant are cleaned and dried if necessary before storage. When the beans move from storage to processing, they are cleaned further, cracked to loosen the seed coat or hulls, dehulled, and then conditioned to 10–11% moisture. The conditioned meats are flaked and extracted with hexane to remove the oil. Hexane and oil in the miscella are separated by evaporation and the hexane is recovered. Residual hexane in the flakes is removed by steam treatment in a desolventizer-toaster. The heat treatment inactivates antinutritional factors (trypsin inhibitors) in the raw flakes and increases protein digestibility. A metric ton of soybeans yields ca 180 kg oil and 790 kg meal.

NUTRITION AND TOXICITY

Oil

Because of their high linoleic acid contents, unhydrogenated and partially hydrogenated soybean oils are good sources of this essential fatty acid. Soybean oil is the principal vegetable oil consumed 4.66 × 10^6 t in 1985), and ca 75% is partially hydrogenated to impart high temperature stability to cooking oil, extend shelf life, and improve flavor stability and physical and plastic properties. Linoleic and linolenic acid contents of soybean oil are reduced by hydrogenation, but more important from a nutritional viewpoint are the migration of double bonds up and down the carbon chain and the conversion of cis to trans isomers, ie, positional and geometrical isomerization. Although long-term studies with rats and short-term tests with humans have failed to reveal toxic effects on ingestion of partially hydrogenated soybean oil, this complex problem is under active investigation.

Heating and oxidation of fats, especially under severe conditions, results in the formation of a variety of compounds including hydrocarbons, cyclic hydrocarbons, alcohols, cyclic dimeric acids, and polymeric fatty acids. Some of these compounds are toxic, but the present consensus is that an oil such as soybean oil is safe and non-toxic when used under normal cooking conditions.

Proteins and Meals

Nutritional properties of the oilseed protein meals and their derived products are determined by the amino acid compositions, content of biologically active proteins, and various nonprotein constituents found in the defatted meals. Phytic acid, which is common to soybean meals, is believed to interfere with dietary absorption of minerals such as zinc, calcium, and iron. Numerous studies have demonstrated that methionine is the first limiting amino acid in soybean protein, ie, it is in greatest deficit for meeting the nutritional requirements of a given species. Although it is common practice to add synthetic methionine to broiler feed to compensate for this deficiency, there is growing evidence that this is not necessary when soy proteins are fed to humans, with the possible exception of infants. The presence of trypsin inhibitors in soybeans is well documented, and when ingested, their primary physiological effect is to enlarge the pancreas. They are largely inactivated by moist heat, and there are no documented cases where ingestion of soybean proteins by humans has affected the pancreas. Long-term effects of ingestion of soy products by rats are under study.

PRODUCTS AND USES

Oil

Most crude oil obtained from oilseeds is processed further and converted into edible products. Only a small fraction of the total oils is utilized for industrial (nonedible) purposes. For edible purposes, oilseed oils are processed into salad and cooking oils, shortenings, and margarines. These products are prepared by a series of steps as shown for soybean oil in Figure 5.

Degumming removes the phosphatides and gums, which are refined into commercial lecithin or returned to the defatted flakes just before the solvent removal and toasting step. Next, free fatty acids, color bodies, and metallic prooxidants are removed with aqueous alkali. Some processors omit the water-degumming step and remove the phosphatides and free fatty acids with alkali in a single

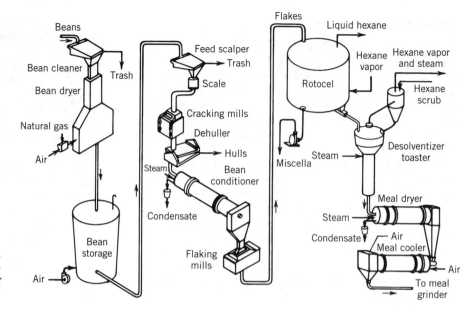

Figure 4. Schematic outline for processing soybean into oil and meal by solvent extraction. *Source:* Courtesy of Dravo Corp.

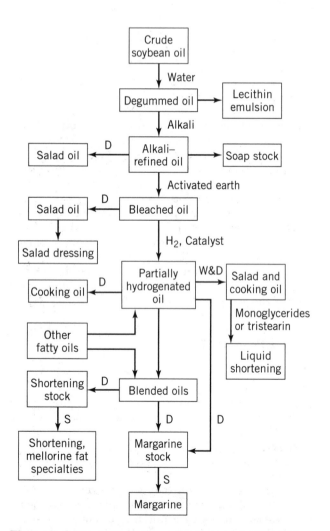

Figure 5. Schematic outline for manufacture of edible soybean oil products. D, deodorization; W, winterization; S, solidification. *Source:* Courtesy of the American Soybean Association and the American Oil Chemists' Society.

operation. High, vacuum steam distillation in the deodorization step removes undesirable flavors to yield a product suited for salad oil. Partial hydrogenation, under conditions where linolenate is selectively hydrogenated, results in an oil with greater stability to oxidation and flavor deterioration. After winterization (cooling and removal of solids that crystallize in the cold), the product is suited for use as salad and cooking oils. Alternatively, soybean oil can be hydrogenated under selective or nonselective conditions to increase its melting point and produce hardened fats. Such a partially hydrogenated soybean oil, by itself or in a blend with other vegetable oils or animal fats, is used for shortening and margarine. Blends of soybean or other oils of varying melting point ranges are utilized to obtain desired physical characteristics, eg, mouthfeel and plastic melting ranges, and the least expensive formulation.

Polyunsaturated fatty acids in vegetable oils are subject to oxidative deterioration. Linolenic esters in soybean oil are particularly sensitive to oxidation; even a slight degree of oxidation, commonly referred to as flavor reversion, results in undesirable flavors, eg, beany, grassy, painty, or fishy. Oxidation is controlled by the exclusion of metal contaminants (iron and copper), addition of metal inactivators (citric acid), minimum-exposure to air, protection from light, and selective hydrogenation to decrease the linolenate content to ca 3%.

Nonfood Uses

Vegetable oils are utilized in a variety of nonedible applications, but only 2–3% of U.S. soybean oil production is used for such products. Soybean oil can be converted into alkyd resins for protective coatings, plasticizers, dimer acids, surfactants, and a number of other products.

Protein Products

Most of the meal obtained in processing of soybeans is used as protein supplements in animal feeds. Only in the last two decades have appreciable amounts been converted into products for human consumption, and these have been almost exclusively derived from defatted soybean flakes.

Feeds

Because of their high content of protein, protein meals are essential ingredients of poultry and livestock feeds. Proximate compositions are shown in Table 4. In contrast, dehulled soybean meal is low in crude fiber and high in protein. Although limiting in methionine, soybean meal is high in lysine is a key ingredient for blending with corn in formulating feeds for nonruminats, eg, poultry and swine. The two proteins complement each other; soy supplies the lysine and corn the methionine necessary to provide a balanced ration at relatively low cost.

Edible Products

At present, only defatted soybean flakes are converted into edible-grade products. Defatted soybean flakes for edible purposes are prepared essentially as outlined in Figure 4, except that more attention is paid to sanitation than in processing for feed use, and the solvent is removed in a vapor desolventizer—deodorizer or flash desolventizer to permit preparation of flakes ranging from raw to fully cooked. Desolventizer–toasters are used to prepare fully cooked (toasted) flakes to give maximum nutritive value. Degree of cooking is determined by estimating the amount of water-soluble protein remaining after a moist heat treatment with the protein dispersibility index (PDI) or the nitrogen solubility index (NSI). A raw, uncooked flake has a PDI or NSI of ca 90, whereas a fully cooked flake has values of 5–15.

Defatted soybean flakes give three classes of products differing in minimum protein content (expressed on a dry basis) flours and grits (50% protein); protein concentrates (70% protein); and protein isolates (90% protein). Typical analyses are shown in Table 5. Flours and grits are made by grinding and sieving flakes. Concentrates are prepared by extracting and removing the soluble sugars from defatted flakes by leaching with dilute acid at pH 4.5 or leaching with aqueous ethanol. Isolated soy proteins are obtained by extracting the soluble proteins with water at pH 8–9, precipitating at pH 4.5, centrifuging the resulting protein curd, washing, redispersing in water (with or without adjusting the pH to ca 7), and finally spray drying (Fig. 6). Flours and concentrates are also specially processed into textured products that are used as meat extenders and substitutes. Typical composition of soybean protein products and their uses are summarized in Table 5.

Oilseed proteins are used in foods at concentrations of 1–2 to nearly 100%. Because of their high protein contents, textured soy flours and concentrates serve as meat substitutes. At low concentrations, the proteins are added primarily for their functional properties, eg, emulsification,

Table 5. Typical Compositions of Soybean Protein Products and Their Uses,[a] wt %

Constituents	Defatted flours and grits[b]	Protein Concentrates[c]	Isolated[d]
Protein[e]	56.0	72.0	96.0
Fat	1.0	1.0	0.1
Fiber	3.5	4.5	0.1
Ash	6.0	5.0	3.5
Soluble carbohydrates	14.0	2.5	0
Insoluble carbohydrates	19.5	15.0	0.3

[a]Analytical values on a moisture-free basis (8).
[b]For baked goods, ground and processed meats, breakfast cereals, diet foods, infant foods, meat substitutes, confections, and milk replacers.
[c]For ground and processed meats and infant foods.
[d]For processed meats, meat substitutes, infant foods, and dairy substitutes.
[e]$N \times 6.25$.

fat absorption, water absorption, texture, dough formation, adhesion, cohesion, elasticity, film formation, and aeration. The use of some oilseed proteins in foods is limited by flavor, color, and flatus problems. Raw soybeans, for example, taste grassy, beany, and bitter. Even after processing, residues of these flavors may limit the amounts of soybean proteins that can be added to a given food. Flatus production by defatted soy flours has been attributed to raffinose and stachyose. These sugars are removed by processing the defatted flours into concentrates and isolates.

Industrial Products

In the United States, only soybean protein isolates are used for industrial applications. They are available commercially in unhydrolyzed types are made by the process outlined in Figure 6; the acid-precipitated curd is washed, dried, and then ground. For the hydrolyzed types, the acid-precipitate curd is suspended in alkali and heated to dissociate the proteins into subunits and partially hydrolyze the polypeptide chains. The reaction is terminated by acidifying to a pH of ca 4.5 to precipitate the modified protein, which is dried. The chemically modified isolates are made by proprietary processes. Isolates are employed primarily as adhesives for clays used in coating of paper and paperboard to render surfaces suitable for printing.

Food Products

Soybeans seeds are consumed as such or are processed into edible products. Small amounts of soybeans are roasted and salted for snacks. Nut substitutes for baked products and confections are also manufactured from soybeans. Larger amounts are used in Oriental foods, some of which are becoming increasingly popular in the United States.

Soy Milk. In the traditional Chinese process, soybeans are soaked in water, ground into a slurry, cooked, and filtered to remove the insoluble cell wall and hull fractions. A number of process modifications have been made since the early 1960s, including a heat treatment before or during grinding to inactivate the enzyme lipoxygenase and thus prevent formation of grassy and beany flavors. Markets for the blander products made by the new processes

Table 4. Proximate Compositions of Soybean Meals, wt %

Meal	Process	Dry matter	Crude protein	Crude fat	Crude fiber	Ash
			Soybean			
With hulls	Solvent	89.6	44.0	0.5	7.0	6.0
Dehulled	Solvent	89.3	47.5	0.5	3.0	6.0

Source: Ref. 7.

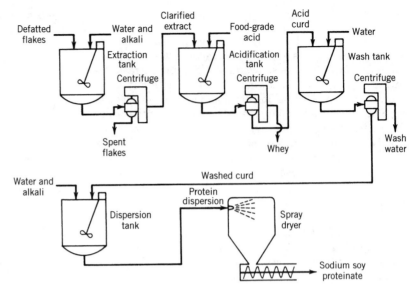

Figure 6. Schematic outline for manufacture of soybean protein isolates.

have developed rapidly in Japan since the late 1970s. Some of the products are sweetened and flavored; fermented (yogurtlike) soy milks are also available in Japan and other parts of Asia.

Tofu. Tofu is prepared by adding a coagulant such as calcium sulfate to soy milk to precipitate the protein and oil into a gelatinous curd. The curd is then separated from the soluble portion (whey), pressed, and washed to yield a market-ready product that is a traditional food in Japan and throughout Asia. It has become popular in the United States since the late 1970s and was produced in 1981 by more than 150 shops.

Miso. Miso is a pastelike food that resembles peanut butter in consistency and is made by fermenting cooked soybeans and salt with or without a cereal such as rice or barley. It is used as a base for soups and is consumed in Japan, China, Indochina, Indonesia, and the East Indies. In 1990, it was produced by ca 20 small establishments in the United States.

Tempeh. Cooked dehulled beans are inoculated with the mold *Rhizopus oligosporus* and allowed to ferment for 24 h. The mold mycellium binds the soybean cotyledons together. When sliced and deep fried in fat, the product is crisp and golden brown. Tempeh, a traditional food of Indonesia, was made by more than 30 concerns in the United States in 1981.

Soy Sauce. This condiment, well known to U.S. consumers, is made by fermentation or acid hydrolysis. In the fermentation process, cooked soybeans or defatted soybean meal are mixed with roasted wheat and the mixture is blended with a pure culture of *Aspergillus oryzae*. Brine is added and the mixture is allowed to ferment 6–8 months. The product is then filtered and pasteurized. In the acid hydrolysis process, defatted soybean flour is refluxed with hydrochloric acid until the proteins are hydrolyzed. The hydrolysate is then filtered, neutralized, and bottled.

BIBLIOGRAPHY

Adapted from W. J. Wolfe, "Soybeans and Other Oilseeds." *ECT*, 3rd ed., Vol. 21, p. 407. Refer to the original document for specific reference citation.

1. *Nippon Shokuhin Kogyo Gakkai-shi* **15**, 290 (1968).
2. *Journal of Agricultural and Food Chemistry* **18**, 969 (1970).
3. *Journal of the American Chemical Society* **45**, 876 (1968).
4. *RDA*, 10th ed., FNB/NAS, 1990.
5. *Journal of the American Dietetic Association* **68**, 224 (1976).
6. *Journal of the American Chemical Society* **54**, 150 (1977).
7. *Feedstuffs* **53**, 25 (1981).
8. *Journal of the American Oil Chemists' Society* **51**, 67A (1974).

GENERAL REFERENCES

R. L. Anderson and W. J. Wolf, "Compositional Changes in Trypsin Inhibitors, Phytic acid, Saponins and Isoflavones Related to Soybean Processing," *J. Nutrition* **125**, 5815–5885 (1995).

D. R. Erickson, *Practical Handbook of Soybean Processing and Utilization*, American Oil Chemists' Press, Champaign, Ill., 1995.

W. J. Huei and P. A. Murphy, "Mass Balance Study of Isoflavones During Soybean Processing," *J. Ag. and Food Chem.* **44**, 2377–2383 (1996)

S. H. Lence, D. J. Hayes, and W. H. Meyers, "Futures Markets and Marketing Firms: The U.S. Soybean-Processing Industry," *Amer. J. Ag. Econ.* **74**, 716–725 (1992).

H. Miyamura, Y. Takenaka, and T. Takenaka, "Fibrinolytic Activity of Akara Fermented by *Bacillus subtilis*. II. The Utility of Akara, A By-Product of the Soybean Processing Industry," *J. Jap. Soc. for Food Sci. and Technol.* **45**, 100–107 (1998).

Z. L. Nikolov et al., "Potential Applications for Supercritical Carbon Dioxide Separations in Soybean Processing," *Developments in Food Sci.* **29**, 595–616 (1992).

G. A. Sulebele, "Soybean Processing Industry, An Update," *Indian Food Industry* **10**, 23–26 (1991).

L. Thomas and A. Hohn, "Beneficial Use of Enzymes in Soybean Processing," *Food Marketing and Technol.* **11**, 14, 16, 18 (1997).

Y. H. HUI
American Food and Nutrition Center
Cutten, California

SPICES AND HERBS: NATURAL EXTRACTIVES

The importance of herbs and spices in flavor and color applications cannot be overstated, but there are shortcomings as well. The natural extractives of spices and herbs are designed to overcome many of the difficulties inherent in the use of whole or ground plant material in food and flavor applications.

The aroma and flavor produced by the use of plant material varies, depending not only on the variety and source of the raw material, but also on the growing conditions and weather patterns of a given year. This variability makes the consistent application in food products very difficult. In addition, flavor components in plant tissue are unstable, leading to a decrease in quality due to volatility losses, oxidation, isomerization, hydrolysis, and other chemical changes. Plant tissue can also support molds and other types of undesirable microbial contamination.

Spices and herbs require storage under regulated conditions, such as refrigeration, that will minimize deterioration with time. Given the large bulk volume, this can be an expensive proposition. Extractives of spices and herbs are highly concentrated, which simplifies storage conditions and greatly decreases the amount of additive needed in product formulations. They are more stable in terms of flavor strength and have lower microbiological activity.

The biggest advantage of extractives versus dried spices, however, is the increased control of the flavoring agents. A more consistent flavoring agent is available due to the quality control of the extraction process. A broader range of flavor effects is made possible by blending or further processing of the extractives. Natural extractives are divided into essential oils, oleoresins, and products made from these.

ESSENTIAL OILS

Essential oils can be defined as the volatile materials present in plants. These are complex mixtures of flavor and fragrance substances usually obtained by distillation. U.S. import statistics are published periodically (1).

Manufacturing Process

Essential oils, due to their relatively high boiling points, are usually obtained by steam distillation, by codistillation with water, or by the combined use of steam and water (2,3). In steam distillation the plant material (whole or ground) is packed into a still, and the essential oil is removed by passing steam through the plant material. The steam pressure is most commonly atmospheric, although both higher and lower pressures are used.

Hydrodistillation, or direct-heating distillation, is the process in which plant material is soaked in the boiler while heat is applied to codistill essential oil with water (4). The plant material is supported above boiling water. This method is used for collecting the volatile oil of green plant material.

On cooling, the raw distillate separates into aqueous layer and an organic layer. The organic layer may require further water removal by centrifugation or by the use of a drying agent. The essential oil can be purified and refined to obtain the desired flavor and aroma characteristics and to remove undesired flavor components. Vacuum distillation, liquid/liquid extraction, and adsorption are the most commonly used refinement techniques.

Vacuum distillation provides for the separation of the components with less thermal degradation than redistillation at atmospheric pressure. At reduced pressures the essential oil is redistilled or fractionated to remove undesired compounds and to concentrate desired compounds. Several types of products are obtained depending on the distillation conditions used:

1. Rectified oils are redistilled with rejection of the last distillation fraction (single rectified) or the first and last fractions (double rectified). Usually only a small percentage of the raw essential oil containing water, color, off-aromas, and other undesirable components are rejected.

2. Terpeneless oils can be produced if efficient separation conditions can be achieved so that the terpene fraction can be removed from the higher boiling fractions that contain the more valuable flavor compounds.

Liquid/liquid extraction is a process using differences in polarity and solubility to separate oxygenated compounds from terpenes. A mixture of the essential oil and solvent is agitated and allowed to settle into two liquid layers that are then separated. The more polar compounds will be concentrated in the polar solvent and recovered after solvent removal. In cold-solvent extraction of citrus oils, for example, the oxygenated compounds are concentrated into aqueous alcohol. The use of a second, nonpolar solvent to dilute the terpenes may improve the separation and selectivity of the extraction.

Adsorption collects and concentrates selected compounds according to their affinities to a solid, for example, silica gel. The essential oil, with or without the use of a solvent, is passed through a solid material that selectively adsorbs some components. These compounds are then stripped from the solid phase by a suitable solvent and concentrated by removal of the solvent.

When the essential oil has been enriched in the ingredients that are responsible for the characteristic flavor, it is commonly referred to as a folded oil. Although this term originally referred to concentration by distillation, it is currently used without specification of the method used to increase the flavor strength.

Physical Characteristics

Physical characteristics, such as refractive index, optical rotation, specific gravity, among others, of essential oils are tabulated in various listed references (5–7). It is important to remember that the physical characteristics do not necessarily determine quality, flavor, or natural origin, although they can be a useful part of quality control.

Chemical Composition

Essential oils are complex mixtures of compounds; some essential oils have been separated into hundreds of different components by modern techniques (6,8,9). Although some essential oils have only one main constituent, it is not necessary that the main component determine the characteristic aroma. In some cases a minor component can be the most important for defining an aroma, for example, dill ether in dill seed oil. In general, the minor components can be expected to determine the quality of the aroma, or at least the completeness of its sensory impact (10,11). The constituents of essential oils can be grouped into three major categories:

1. Hydrocarbons, usually olefinic, are compounds whose structures are made up of isoprene units. Terpenes, sesquiterpenes, and diterpenes are in this group. They have a low contribution to aroma and flavor strength even though they may constitute the major percentage of the oil. The unsaturated nature of these compounds can lead to deterioration due to oxidation and polymerization.

2. Oxygenated hydrocarbons, which can be subcategorized as alcohols, esters, aldehydes, ethers, ketones, phenols, and so on, have much lower sensory thresholds than hydrocarbons. In some cases an individual oxygenated compound gives the characteristic aroma of a spice or herb, and its level in the essential oil can be correlated to flavor strength.

3. Other compounds containing sulfur or nitrogen can be important for the flavor of essential oils or can impart off-flavors at extremely low levels.

Gas chromatography (GC) is by far the most widely used technique in the study of the components of essential oils (12–14). These compounds are separated by boiling point and by interactions with the stationary-phase coating of the GC columns. The improved resolution possible with capillary columns compared to packed columns has greatly improved separations. One of the advantages of GC is the flexibility and selectivity possible in the detection of the compounds as they are eluted from the column. A flame ionization detector can provide universal and sensitive detection of all compounds in an essential oil. However, a detector specific for sulfur- or oxygen-containing compounds can also be used. Using a gas chromatograph coupled to a mass spectrometer or a Fourier transform infrared spectrometer can provide compound identification as well as quantification. The use of a computerized library search greatly simplifies the identification of previously characterized compounds. The gas chromatograph is not a suitable instrument for all essential oils; thermally labile compounds can decompose during the assay.

OLEORESINS

Oleoresins are the solvent extractables of the botanical raw material after solvent removal. They contain the volatile and nonvolatile active ingredients of the raw materials and have been an item of commerce for more than 50 years (15). As with spices and herbs, numerous oleoresins are available to the flavor-food chemist to achieve the desired impact. The usage of oleoresins parallels the volume of raw commodities as listed in U.S. government import statistics (16). Major oleoresins are black pepper, capsicum, celery, ginger, nutmeg, paprika, and turmeric. Minor oleoresins are basil, bay, cinnamon, clove, coriander, mace, marjoram, oregano, rosemary, sage, and thyme. A more complete listing of the products can be found in Title 21 *Code of Federal Regulations* (CFR), part 182, section 182.20, and part 184 (17).

MANUFACTURING PROCESS

Individual manufactures differ in the process details used for production, but all have in common the following stages:

grinding > extraction > desolventization
> standardization

The grinding operation is employed to make the extractives available for dissolution into the solvent. Grinding also increases the surface area available for extraction. Many types of grinding equipment are used, depending on the raw material and the extraction process. This information is not readily available from manufacturers because it is one of the most critical steps in the process and is considered proprietary by most manufacturers. What is important to remember is that the grinding must have minimum deleterious effect on the aroma and flavor attributes of the raw material but must be effective in insuring exhaustive extraction of the components contributing to the essence. Because heat is generated in most grinding processes, cryogenic techniques are often employed to retain volatile fractions.

The extraction process can be subdivided into two types: single stage and double stage. The choice for the individual manufacturer depends on the solvent used to perform the extraction and solvent removal capability. The permitted solvents for extraction are listed in 21 *CFR*, part 173, and include acetone, ethylene dichloride, hexane, isopropyl alcohol, methyl alcohol, methylene chloride, and trichloroethylene (17). Use of chlorinated solvents is being reevaluated within the industry and will be further discussed under the section on regulatory issues.

In single-stage extraction, the ground product is transferred to the extraction vessel and washed with solvent to remove the essence, and the exhausted material (marc) is desolventized to recover entrained solvent. The extraction vessel may be batch type, where all of the extraction functions occur in a steam-jacketed pot containing a basket. The ground material is loaded into a basket, sealed, and washed with solvent. The oleoresin is then desolventized; the marc is removed and desolventized separately. Alternatively, this process may be done on a continuous basis, with the ground material introduced via an air lock to the extractor and conveyed through solvent wash stages; the marc passes through another air lock for desolventization.

Double-stage extraction utilizes the batch-type extractor. The difference is that the volatile fraction is removed by steam distillation prior to solvent extraction of the resinous fraction. These two fractions are then recombined in the standardization stage. This procedure is most often used with the manufacture of oleoresin black pepper; the advantage of this process is that the resinous fraction can be desolventized under more rigorous conditions, because the preservation of the volatile fraction is not an issue. The disadvantage is that the process limits the extraction solvent to the chlorinated hydrocarbons to produce commercially acceptable product, because residual water from the first stage inhibits extraction of the resinous fraction (18).

The desolventization process reduces the extraction solvent in the oleoresin to permissible levels, as defined in the applicable regulations (17). Various methods are employed, dependent on the physical characteristics of the oleoresin. Two basic strategies are used. The first, where preservation of the volatile fraction is not required and viscosity of the product is low, is called wiped-film evaporation. The oleoresin-laden solvent, or miscella, is pumped through a column and spread by an impeller along the inside surface. The column is heated via a steam jacket and partial vacuum is applied to aid in solvent removal. The miscella is introduced at the top of the column, with the thickness of the film controlled by addition rate, speed of the impeller, viscosity, and temperature. The evaporated solvent flows countercurrent to the oleoresin and is recovered for reuse. Oleoresins that are processed in this manner include the capsicums and turmeric.

The second method, where retention of the volatile fraction is desired or the viscosity is high, is called batch vacuum distillation. The miscella is loaded into a steam-jacketed vacuum vessel, which is fitted with a fractionation column. This equipment is normally called a still. Steam is applied to heat the vessel to the atmospheric boiling point of the solvent. The solvent vapors are refluxed to return the volatile fraction back to the product. After the majority of the solvent is removed from the product (which is indicated by a temperature rise in the still), vacuum is applied to further reduce the solvent content to acceptable levels. Although care is taken to exclude the volatile fraction from the distillate, it is necessary with some products to fractionate the distillate to recover some of the volatiles. These light fractions are reserved for blending in the standardization step.

Standardization is necessary to re-create the flavor and aroma profile of the raw material while providing the customer with a consistent product. Edible oils, such as soybean or cottonseed, are used to standardize flavor intensity. Emulsifiers are also incorporated with some products at this stage to make for easier application. Standardization occurs on a batch basis, combining the extractives into a mixing vessel. If the manufacturer uses standardization agents, then they must be clearly stated on the label. The ingredients are sequentially combined based on decreasing viscosity, with the recovered volatile fractions being added last. The blending may be achieved by use of a high shear stirrer, colloidal mill, or homogenizer. Mixing times are product dependent. Samples are drawn for testing to assure conformance to requirements of the final mixture before packaging.

Physical Characteristics

Oleoresins are liquids; viscosities vary from that of vegetable oil, as with capsicum and celery, to that of a semisolid paste, as with nutmeg. The oleoresins are concentrated essences of the raw material with an increase of flavor strength in the range of 6-fold for celery to greater than 40-fold for cinnamon. This is due to the varying amount of extractables within each raw material and to the fact that the aroma and flavor components on extraction from the plant matrix are more readily incorporated into the end product. Along with the information provided by the manufacturer, many of the products have criteria established in the *Food Chemical Codex* (19) to aid the user in developing meaningful specifications.

Chemical Composition

Unlike the essential oils, where only the volatile notes are available for aroma and flavor effect, oleoresins contain the volatile fraction, active compounds and fixed compounds. The volatile fraction is very similar in composition to the previously discussed essential oils. The active compounds are defined, for the purpose of this article, as the nonvolatile constituents perceived predominantly in the mouth and not by the nasal membranes. To give a few examples, these compounds are the pungent alkaloid group related to piperine in black pepper and white pepper, the amide group related to capsaicin in capsicums, and the keto-alcohols related to gingerol in ginger. The fixed compounds are defined as the nonvolatile, nonpungent compounds. Examples of these are coloring compounds, such as chlorophyll found in herbs, cartenoids found in capsicums, and curcuminoids found in turmeric. The fixed compounds also include various saturated and unsaturated fats, which lend balance to the flavor and aroma profiles of the oleoresins, making them more representative of the starting raw material. The extraction process does not yield the carbohydrate or protein fraction of the raw material, for the most part, and therefore these fractions are not an important component of the oleoresin.

FORMS AND APPLICATIONS

The common form of both essential oils and oleoresins is an oil-dispersible liquid. Because most foods contain a high amount of water, homogenous distribution is difficult. Concentrated products are required in small dosages to achieve the desired effect, making it doubly difficult to ensure consistency. To overcome these difficulties, various means have been developed to incorporate the extractives into commercial products:

1. Dispersed products (dry-soluble seasonings) are prepared by dispersing the essential oils and oleoresins on an edible carrier: salt, dextrose, or flour. Limited in concentration by the ability of the carrier to remain flowable, these blends can incorporate up to 10% liquids.

2. Solubilized products are blended with polysorbate esters, such as polysorbate 80; these products are clearly dispersible in deionized water. Such formulations may contain up to 50% extractives.

3. Emulsions are blended into water with gum arabic or other permitted emulsifying agents; these products have a limited shelf life.

4. Encapsulation (spray drying) forms products in which essential oils or oleoresins are homogenized into water with vegetable gums or food starches, then atomized into a drying chamber. The resultant dry powder may contain up to 20% liquids, with the encapsulation providing some increased shelf life stability in comparison to dispersed products.

5. Water-dispersible liquids are also available. Since 1980, advances in emulsifier technology have led to products that are dispersible in water without the use of polysorbate esters. These products contain about 50% extractives and are much more shelf stable than many of the other products listed in this section. They are easily incorporated into the water phase of the finished product. Some manufacturer formulations are also soluble in the oil-fat phase of a product, evenly distributing the extractives in both phases.

Decisions must be made by the product formulator as to which form of the extractives is most suitable for the application. The standard essential oils and oleoresins have been successfully used in oil-fat-based foods, such as salad dressings, without special incorporation techniques.

Traditionally the dispersed products have had the widest application in food systems, being mixed in as dry ingredients in early processing. Most formulated meats are flavored in this manner, along with bakery and breading goods and seasonings for snack foods. Pickling processors use the oleoresins solubilized with polysorbate esters to flavor the brine, which migrates to the product during retort and storage. Emulsions are not widely used, although they do have limited applications in emulsified meats such as bologna and hot dogs. Encapsulated products are used in dry mixes (soups, salad dressings, drinks, etc) where extended shelf life is required.

Water-dispersible liquids can replace most of the aforementioned application needs, though only to a limited degree in pickling and dry-mix applications. Most extractive producers provide technical support based on their applications research and experience with a variety of processes. This can be a valuable asset to the flavor formulator.

NUTRITIONAL INFORMATION

It is not possible to generalize the nutritional value of essential oils or oleoresins. Nutritional information is available from the manufacturer for individual products. In foods, the weight percentage of these added flavor ingredients does not usually make a significant contribution to overall nutrition. Essential oils, because they are both ether soluble and combustible, assay as fat. Oleoresins contain the natural vegetable oils extracted with aroma and flavor components, predominantly unsaturated fats normally found in vegetable sources. Carriers, emulsifiers,

and diluents also need to be considered in nutritional data in that they often make up more than 50% of commercial forms. Specific components may assay as protein, such as piperine in black pepper, although they are not types of protein. Mineral and vitamin content is not usually significant, with the exception of vitamins A and C in some products.

Herbs and spices contain many compounds with proven antioxidant, anti-inflammatory, and antimutagenic activity. Nutraceuticals is a term used to describe the beneficial components in spices and herbs as part of a diet or as food supplements (19).

METHODS OF ANALYSIS

The most important criteria for evaluation of extractives are the flavor and aroma profiles. Organoleptic evaluation is the best means by which to judge these flavor additives. Even though it is difficult and expensive to do on a routine basis in production operation, it is necessary. As a pre-screening tool, physical measurements can reduce the number of organoleptic evaluations somewhat. It is critical that the analytical methods to be employed are agreed on by both customer and supplier, because different methods give different results, although in theory they are measuring the same attributes. Control criteria (20) and methods of analysis (21,22) have been published.

REGULATORY STATUS

In the United States, Title 21 of the *CFR* governs usage of spice extractives in foods and drugs. Under part 182.20 and part 184 are listed the essential oils and oleoresins that are considered to be generally recognized as safe (GRAS) and their requirements. Oleoresins used as vegetable colors are listed under separate headings of part 73. Under part 173, the maximum residue for the solvent used in extraction is as follows:

Solvent	Level (ppm)
Acetone	30
Ethylene dichloride	30
Isopropyl alcohol	50
Methyl alcohol	50
Methylene chloride	30
Hexane	25
Trichloroethylene	30

Some countries do not permit the use of extracts made with chlorinated solvents. This must be taken into account when finished goods are to be exported. Because the list of areas in the world that enforce this ban is growing, any listing here would be obsolete before publication. The standardizing agents used in the extracts are considered incidental additives. The labels of finished foods in which these products are used need only state "natural flavors" or "natural flavoring." It is best to check with legal counsel to confirm labeling status.

BIBLIOGRAPHY

1. United States Department of Agriculture, *U.S. Essential Oil Trade*, USDA Foreign Agricultural Service, Washington, D.C., 1980–1989.

2. K. Bauer and D. Garbe, *Common Fragrance and Flavor Materials*, VCH Verlagsgesellschaft, Germany, 1985.

3. H. B. Heath, *Flavor Technology: Profiles, Products, Applications*, AVI Publishing Co., Inc., Westport, Conn., 1978.

4. E. F. K. Denny, "Hydro-distillation of Oils from Aromatic Herbs," *Perfumer and Flavorist* **14**, 57–63 (1989).

5. G. Fenaroli, *Fenaroli's Handbook of Flavor Ingredients*, T. E. Furia and N. Bellanca, eds. and trans., The Chemical Rubber Co., Cleveland, Ohio, 1971.

6. B. M. Lawrence, ed., *Essential Oils (1976–1978), (1979–1980), (1981–1987), (1989–1991), (1992–1994)*, Allured Publishing Co., Carol Stream, Ill.

7. *EOA Book of Standards and Specifications*, Essential Oil Association of USA, Inc., New York, 1979.

8. S. R. Srinivas, *Atlas of Essential Oils*, published by the author, New York, 1986.

9. E. Guenther, *The Essential Oils*, Vols. 1–6, Van Nostrand Reinhold Co., Inc., New York, 1949.

10. F. A. Fazzalari, ed., *Compilation of Odor and Taste Threshold Values Data*, American Society for Testing and Materials, Philadelphia, Pa., 1978.

11. A. Dravnieks, *Atlas for Odor Character Profiles*, American Society for Testing Materials, Philadelphia, Pa., 1985.

12. W. Jennings and T. Shibamoto, *Qualitative Analysis of Flavor and Fragrance Volatiles by Glass Capillary Gas Chromatography*, Academic Press, Inc., Orlando, Fla., 1980.

13. Y. Masada, *Analysis of Essential Oils by Gas Chromatography and Mass Spectroscopy*, John Wiley & Sons, Inc., New York, 1976.

14. P. Sandra and C. Bicchi, eds., *Capillary Gas Chromatography in Essential Oil Analysis*, Heutig, New York, 1987.

15. H. B. Heath, *Source Book of Flavors*, AVI Publishing Co., Inc., Westport, Conn., 1981, p. 212.

16. United States Department of Agriculture, *U.S. Spice Trade*, USDA Foreign Agricultural Service, Washington, D.C.

17. Office of the Federal Register National Archives and Records Administration, *Code of Federal Regulations, Title 21 Food and Drugs*, U.S. Government Printing Office, Washington, D.C., 1997.

18. V. S. Govindarajan, "Pepper—Chemistry, Technology, and Quality Evaluation," *CRC Critical Review of Food Science and Nutrition*, 115–225, 1977.

19. I. Goldberg, ed., *Functional Foods: Designer Foods, Pharmafoods, Nutraceuticals*, Chapman and Hall, New York, 1994.

20. Committee on Codex Specifications, *Food Chemical Codex*, 4th. ed., National Academy Press, Washington, D.C., 1996.

21. Association of Official Analytical Chemists, *Method of Analysis of the AOAC*, 16th ed., AOAC, Arlington, Va., 1995.

22. American Spice Trade Association, *Official Analytical Methods of the American Spice Trade Association*, 4th ed., ASTA, Englewood Cliffs, N.J., 1997.

James A. Guzinski
Ted M. Lupina
Kalsec, Inc.
Kalamazoo, Michigan

SPICES AND SEASONINGS

Spices are the whole or ground seed, fruit, bark, or root of a plant. Examples are cumin, allspice, cinnamon, and ginger. Herbs are the leafy parts of annual or perennial shrubs or plants. Oregano, marjoram, and basil would fit this classification. Seasonings, on the other hand, are blends of spices with other functional and nonfunctional ingredients. Seasonings are usually blended with other food items to produce a finished food product. Spices and seasonings will be discussed separately.

SPICES

The use of all spices and herbs has increased dramatically over the years. According to the United States Department of Agriculture (USDA), imports of spices in 1997 was more than 638 million lb (1).

Definition

The Food & Drug Administration (FDA) has defined *spice* as any "aromatic vegetable substance in the whole, broken or ground form, except for those substances which have been traditionally regarded as foods, such as onions, garlic, and celery; whose significant function in food is seasoning rather than nutritional; that is true to name; and from which no portion of any volatile oil or other flavoring principle has been removed" (2). The Code of Federal Regulations (CFR) goes on to define the specific spices and herbs as "spices." In addition, paprika, saffron, and turmeric are identified by the FDA also as colors and must be declared as such on ingredient labels (2). See Table 1 for a listing of the common spices.

Table 1. Spice Classification

Spices	
Allspice	Dill seed
Anise	Fennel seed
Anise, star	Fenugreek
Caraway seed	Ginger
Cardamom	Mustard seed
Celery seed	Nutmeg
Cinnamon	Mace
Cloves	Pepper, black & white
Coriander	Pepper, red
Cumin seed	

Herbs	
Basil	Rosemary
Bay leaves	Sage
Dill weed	Savory
Marjoram	Tarragon
Oregano	Thyme

Colors	
Paprika	Turmeric
Saffron	

Economic Factors

Because spices are agricultural commodities, various political and climatic conditions can cause prices to rise and fall due to supply and demand. If a war or other political situation occurs, often the spice cannot be exported. For example, during the 1979 Iran hostage crisis, cumin was difficult to obtain and prices rose substantially due to U.S. trade restrictions. Cumin was often shipped to Pakistan and France from Iran, before shipping to the United States.

If growing conditions are poor in countries where a spice is usually imported from, prices may increase dramatically, or the spice may be unavailable completely. In 1988 and 1989, Dundicut chillies, commonly used to make red pepper, were unavailable from Pakistan due to a drought. Beginning in 1992, the quality of sage imported from the former Yugoslavia decreased dramatically due to internal ethnic conflicts and the resulting division of the country. In addition, in recent years, black pepper prices have risen about 50%. This is mostly due to poor growing conditions over a multiyear period. In addition, low commodity prices prior to the increase did not give farmers an incentive to grow pepper.

When purchasing spices from overseas, either direct from source or through brokers in New York, it is important to keep abreast of supply problems and not restrict specifications to a particular country of origin unless absolutely necessary.

Cleaning and Production

Spices are commonly brought into the United States through New York brokers and sold to spice processors who clean and sometimes grind the product. Large spice companies import direct from processors overseas. The spices are then sold to other food processors, packaged and sold to consumers or food service establishments, or blended with other products to produce seasonings.

The United States has cleanliness specifications that all imported spices must meet. These are the FDA Defect Action Levels (DALs) (3). An excellent discussion of how spices are legally imported and passed by FDA can be found in Spices and Seasonings, A Food Technology Handbook (4). The most commonly used cleanliness guidelines are the American Spice Trade Association (ASTA) Cleanliness Specifications (5). Both the DALs and the ASTA specifications are designed to be the minimum measure of cleanliness and to realize that spices, being agricultural commodities, cannot be free of foreign material and do require additional cleaning by processors. Table 2 shows the current ASTA specifications. Processors receive spices in bales or often burlap bags and are responsible for cleaning and grinding (if required) to meet their internal specifications or those set by their customer. Reputable spice processors first check the product for cleanliness in the whole form by sampling a representative number of containers and physically checking for extraneous material such as live and dead insects, stones, excreta, hair, metal, and glass. If the product is found to have these items present, it must be cleaned. Infestations with insects are treated with methyl bromide or heat to kill them, then removed as described next. Metal is generally removed by magnets, which should be present on all cleaning and grinding equipment. See Figure 1 for a flowchart of spice processing.

Although a variety of equipment is used to clean spices, depending on the foreign matter present and the spice being cleaned, three basic types of separation techniques are used: air separators, gravity separators, and centrifugal separators. Air separators use the movement of air to separate material according to the velocity of air required to remove the foreign material. Centrifugal separators use centrifugal force to separate product, and gravity separators separate particles by differences in weight or specific gravity of the product. Most equipment uses a combination of these three methods to clean spices. The most commonly used cleaning equipment is the vacuum gravity separator, or air table. This piece of equipment uses two separation techniques: air separation and gravity separation (Fig. 2).

In a vacuum gravity separator, product is metered onto a porous deck where air is forced through, suspending the product. The air and deck motion causes the particles to separate into layers of different densities. The heavy material moves uphill and the light material moves downhill. This piece of equipment can be adjusted to separate product into many different density fractions. More information on cleaning spices is available from ASTA (6).

After cleaning, the product is typically checked again for cleanliness. If the product passes, it is checked for quality attributes and either sold whole or ground. Grinding can be done to any specific size by using screens to separate the product into specific fractions.

Spice Quality

Chemical and Physical. Quality attributes for spices are usually present on a specification from the processor. These include volatile oil, moisture, heavy and light filth, ash, and acid-insoluble ash. Product specific tests include Scoville heat units for red peppers, ASTA color units for paprika, and curcumin content for turmeric. ASTA has published the Official Analytical Methods of the American Spice Trade Association (7). The most common methods are outlined in the following:

Volatile oil is the major aroma and flavor component of all spices. It will volatilize at low temperatures, and most spices contain up to 20 or more different components in the volatile oil. A spice generally contains 0.5 to 3.0% volatile oil, and typical levels are higher in seed spices than in herbs. Whole spices have a higher volatile oil level than ground spices. When a spice is ground, it loses its volatile oil content quickly due to the crushing of the protective cell structure present in the whole spice. In addition, the heat produced by grinding can vaporize the volatile oil during processing. Volatile oil content is an important quality parameter in spices. The higher the volatile oil content, the more flavor the spice has. Volatile oil is lost with age, and it can vary between different crop years, origins, and even within lots. In the industry, a minimum level is usually set that the spice must adhere to. Using liquid nitrogen to cool the mill head during grinding can help retain spice volatiles.

Moisture levels also indicate spice quality. A high moisture content indicates improper storage or inadequate dry-

Table 2. ASTA Cleanliness Specifications

Name of spice seed, or herb	Whole insects, dead by count[k]	Excreta, mammalian by mg/lb)	Excreta, other (by mg/lb)	Mold (% by wgt)	Insect defiled/infested (% by wgt)	Extraneous/foreign matter (% by wgt)
Allspice	2	5	5.0	2.00	1.00	0.50
Anise	4	3	5.0	1.00	1.00	1.00
Sweet Basil	2	1	2.0	1.00	1.00	0.50[l]
Caraway	4	3	10.0	1.00	1.00	0.50
Cardamom	4	3	1.0	1.00	1.00	0.50
Cassia	2	1	1.0	5.00	2.50	0.50
Cinnamon	2	1	2.0	1.00	1.00	0.50
Celery Seed	4	3	3.0	1.00	1.00	0.50
Chillies	4	1	8.0	3.00	2.50	0.50
Cloves[a]	4	5	8.0	1.00	1.00	1.00[a]
Coriander	4	3	10.0	1.00	1.00	0.50
Cumin Seed	4	3	5.0	1.00	1.00	0.50
Dill Seed	4	3	2.0	1.00	1.00	0.50
Fennel Seed	SF[e]	SF[e]	SF[e]	1.00	1.00	0.50
Ginger	4	3	3.0	SF[f]	SF[f]	1.00
Laurel Leaves[b]	2	1	10.0	2.00	2.50	0.50
Mace	4	3	1.0	2.00	1.00	0.50
Marjoram	3	1	10.0	1.00	1.00	1.00[l]
Nutmeg (Broken)	4	5	1.0	SF[g]	SF[g]	0.50
Nutmeg (Whole)	4	0	0.0	SF[h]	SF[h]	0.00
Oregano[c]	3	1	10.0	1.00	1.00	1.00[l]
Black Pepper	2	1	5.0	SF[i]	SF[i]	1.00
White Pepper[d]	2	1	1.0	SF[j]	SF[j]	0.50
Poppy Seed	2	3	3.0	1.00	1.00	0.50
Rosemary Leaves	2	1	4.0	1.00	1.00	0.50[l]
Sage[b]	2	1	4.0	1.00	1.00	0.50
Savory	2	1	10.0	1.00	1.00	0.50[l]
Sesame Seed	4	5	10.0	1.00	1.00	0.50
Sesame Seed, Hulled	4	5	1.0	1.00	1.00	0.50
Tarragon	2	1	1.0	1.00	1.00	0.50[l]
Thyme	4	1	5.0	1.00	1.00	0.50[l]
Turmeric	3	5	5.0	3.00	2.50	0.50

[a]Clove stems: A 5% allowance by weight for unattached clove stems over and above the tolerance for Other Extraneous Matter is permitted.

[b]Laurel Leaves and Sage: "Stems" will be reported separately for economic purposes and will not represent a pass/fail criteria.

[c]Oregano: Analysis for presence of Sumac shall not be mandatory if samples are marked "Product of Mexico."

[d]White Pepper: "Percent Black Pepper" will be reported separately for economic purposes and will not represent a pass/fail criteria.

[e]Fennel Seed: In the case of Fennel Seed, if more than 20% of the subsamples contain any rodent, other excreta or whole insects, or an average of 3 mg/lb of mammalian excreta, the lot must be reconditioned.

[f]Ginger: More than 3% moldy pieces and/or insect infested pieces by weight.

[g]Broken Nutmeg: More than 5% mold/insect defiled combined by weight.

[h]Whole Nutmeg: More than 10% insect infested and/or moldy pieces, with a maximum of 5% insect defiled pieces by count.

[i]Black Pepper: 1% moldy and/or infested pieces by weight.

[j]White Pepper: 1% moldy and/or infested pieces by weight.

[k]Whole Insects, Dead: Cannot exceed the limits shown.

[l]Extraneous Matter: Includes other plant material, eg foreign leaves

ing. High moisture contents can cause mold or microbiological growth. Testing for moisture can be done by two methods: vacuum oven drying or toluene distillation. Vacuum oven drying generally gives a higher result because some of the volatile oils may be driven off as well as the moisture. Therefore, this method is generally only recommended for capsicums and dehydrated vegetables where volatile oil is not an important parameter.

Heavy and light filth indicates the amount of foreign material present in the spice. Metal shavings, rocks, insect fragments, and glass will all be found by these methods. Filth testing is especially important in ground spices because these items may be difficult to identify by other methods. The spice to be analyzed is placed in a dense organic solvent such as carbon tetrachloride. The heavy foreign matter will sink and the spice will float. By physically separating these two fractions, heavy and light filth can be determined.

Total ash is the residue left after ignition of the spice. Acid-insoluble ash is the residue left after treating the total ash with hydrochloric acid. These tests also determine the amount of foreign material present. Rocks, metal, and other foreign items will remain behind as residue. Ash is a measure of metal, sand, and minerals naturally occurring in the spice. Acid-insoluble ash is a measure of only metal and sand present.

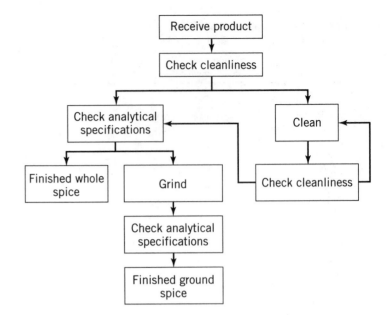

Figure 1. Flowchart of spice processing.

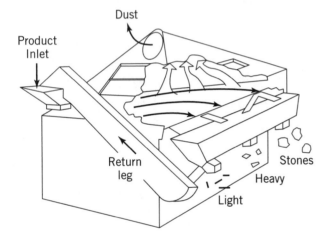

Figure 2. Vacuum gravity separator (air table).

Heat units indicate the relative heat of red peppers. The higher the number, the hotter the red pepper. The Scoville test is a threshold taste test of a red pepper ethanol extraction using five trained panelists. Scoville test has been the most common method for determining the heat level in red peppers, although variance between labs can be very high. ASTA has recently dropped the approval for the Scoville method. A more accurate and ASTA-approved method is high-pressure liquid chromatography (HPLC), which is an analytical test instead of an organoleptic method. The correlation between Scoville and HPLC methods is questionable.

Extractable color of paprika is determined by an extraction with acetone followed by a spectrophotometric test. This method results in the ASTA color of paprika, such as 100 ASTA paprika. These units do not necessarily correlate with surface color. Curcumin content of turmeric is determined by a similar method.

Microbiological. Spices can contain very high levels of bacteria. Most come from third world countries where sanitary conditions are poor. Often spices are laid out in fields, sides of roads, or banks of rivers to dry and are subject to a high degree of contamination. It is not uncommon to have total plate counts on black pepper up to 40 million per gram. Even onion powder grown in the United States can have counts as high as 1 million per gram. The highest counts are usually found on allspice, black pepper, caraway, celery seed, cumin, paprika, and onion powder. Typical microbiological tests performed on natural spices are total plate count, yeast, mold, total coliforms, *Escherichia coli*, and *Salmonella*.

Often spices are treated with sterilizing agents to reduce bacterial loads. The two most common treatments are ethylene oxide and irradiation. Some spice suppliers also use steam to sterilize spices. Both ethylene oxide and irradiation have disadvantages; however, microbial counts can be significantly reduced by using one of these methods.

Ethylene oxide is a gas that is used to reduce microbial loads in spices. It can cause about a 90% reduction in microbial counts. This chemical has been found to be a carcinogen and may be physically retained on the spice after the process has been completed. Chlorohydrins that are toxic can also form. The FDA has a 50 ppm residual tolerance. Currently the FDA and Environmental Protection Agency (EPA) are debating its continued approved use.

Irradiation is an ionizing radiation that is used to kill microorganisms in spices. It is a very effective treatment, providing almost 100% kill of bacteria (8). However, certain consumer groups are very vocal about their opposition to this treatment due to its misguided association with radiation. Spices can be irradiated with levels up to 3 Mrad. FDA regulations state that an irradiated spice must be labeled as "treated with/by irradiation" and the irradiation logo prominently displayed. If a food contains irradiated ingredients, the finished product is not required to be labeled as irradiated (9). With the increasing *E. coli* incidence in meats, this may be an ideal time to influence consumer perception about this very effective treatment. FDA is currently considering alternate labeling regulations.

Description of Whole Spices

Allspice, Pimenta dioica L. (formerly *Pimenta officinalis*) is the dried, unripe berry of a shiny leafed evergreen tree indigenous to the Western Hemisphere. The berries are reddish brown, round and smooth, and about 1 cm in diameter. The name allspice comes from the flavor description: a combination of cloves, cinnamon, and nutmeg. Another, less common term for allspice is pimento. Allspice is imported from Jamaica, Guatemala, Mexico, and Honduras. In 1997, the United States imported about 2.3 million lb of allspice (1).

Anise seed, Pimpinella anisum L., is the seed of a small annual herb of the parsley family. The grayish-brown oval seeds are about 0.5 cm long. Anise seed has a strong licorice-like flavor and is grown primarily in Spain, Egypt, and Turkey. The United States imported about 2.6 million lb in 1997 (1).

Basil, Ocimum basilicum L., is the leaf of an annual herb of the mint family native to India and the Persian countries. The leaves are greenish gray and about 5 cm long and 2 cm wide before drying. When purchased whole, basil is actually dried pieces of leaves. Primary sources are France, Egypt, and the United States. Domestic basil is much more expensive than the imported varieties because it is mechanically harvested and dried. Basil has seen phenomenal growth in the last 30 years. Only 40,000 lb were imported into the United States in 1964 compared with 6.1 million lb in 1997! (1).

Bay leaves or laurel leaves, *Laurus nobilis* L., are large light-green elliptical leaves, up to 8 cm in length and 3 to 4 cm wide. The flavor is fragrant and sweet with a bitter note and is used most commonly in soups and stews as well as pickling spices. Principal countries exporting to the United States are Turkey and Greece. There is also a domestic bay leaf that is packed for retail sale, *Laurus californica*, which does has a slightly different flavor and is longer and narrower than its imported counterpart.

Capsicum peppers, Capsicum frutescens L., include red peppers, chili peppers, and sweet bell peppers. Red pepper, also known as cayenne, is ground chili pods, generally the small elongated and hottest varieties, which are blended to produce a product with standard heat units. Heat levels, expressed as either Scoville heat units or HPLC heat units, in whole pods vary from crop to crop. The hottest varieties are grown in Africa, China, and India. Chili pepper is the ground product of larger, milder peppers, primarily grown in California, New Mexico, and Mexico. Variances in color are primarily due to a caramelization process that processors employ and pepper variety.

Caraway seed, Carum carvi L., is the fruit of a biennial plant of the parsley family. Each seed is a half of the fruit and is about 0.75 cm long, curved, with ridges lengthwise. Caraway is native to Europe, Asia, and North Africa, although it is primarily exported now from The Netherlands. The flavor is warm and acrid, similar to dill seed. The United States imported about 6.9 million lb in 1997 (1).

Cardamon, Elettaria cardamomum Maton, is a large perennial belonging to the ginger family. The spice is im-ported whole in oval, greenish pods, 1 to 2 cm long, which contain 15 to 20 brownish-black seeds. These seeds produce the sweet, aromatic flavor of cardamon. Most cardamon is grown in India, Guatemala, and Ceylon. It is a minor spice, with the United States only importing about 0.5 million lb a year (1).

Celery seed, Apium graveolens L., is the dried fruit of a biennial herb native to southern Europe called smallage. It is different from Pascal celery, which is eaten as a vegetable in the United States. The seed is oval, brown, and very small, about 0.25 cm in length. It has a warm, bitter flavor. It is now primarily imported from India.

Cinnamon. There are many species of cinnamon or cassia. The type most commonly imported into the United States is not cinnamon at all, but rather cassia (*Cinnamomum cassia* Presl—Chinese cassia or *Cinnamomum burmannii* Blume, Indonesian or Korintji cassia). Cinnamon (*Cinnamomum verum* Presl—Ceylon or Seychelles cinnamon) is lighter in color and milder than the two varieties of cassia previously mentioned. It is most commonly exported to Great Britain and other countries. Both cinnamon and cassia are the bark of evergreen trees, which is peeled off and allowed to curl into cinnamon sticks. Another, less common type is Saigon cinnamon, *Cinnamomum loureirii*, which recently is being imported from Vietnam again. Before the Vietnam War, this was a much more common supply for cinnamon.

Cloves (*Eugenia caryophyllata*, Thunb.) are the dried, unopened flower buds of an evergreen tree of the myrtle family. Cloves are about 1 cm long and resemble a rounded top nail. The flavor is strong and aromatic. Cloves are grown in many parts of the world; however, the chief exporters are Madagascar and Brazil.

Coriander seed (*Coriandrum sativum*) is the dried ripe fruit of an annual herb native to southern Europe and the Mediterranean region. Coriander is a round light brown seed about 0.5 cm in diameter with vertical ridges. The leaves of the coriander plant are known as cilantro, which is used extensively in Latin American, Thai, and Vietnamese cuisine. The flavor of coriander seed is mild and fragrant. Most coriander is currently imported from Canada and Morocco.

Cumin, Cuminium cyminum L., is a small annual herb of the parsley family. The dried fruit, commonly referred to as the seed, is oval, about 0.5 cm long and similar in appearance to caraway. Cumin is used extensively in Mexican and Indian foods and has a strong earthy and bitter flavor. Cumin is grown and exported from the Middle East, India, and Turkey.

Dill, Anethum graveolens L., is an annual herb related to the other spices of the parsley family. Dill seed and dill weed are from the same plant. The seeds are oval, tan, and about 0.5 cm long. The flavor is reminiscent of caraway. Dill is grown in Egypt, India, and the United States.

Fennel seed, Foeniculum vulgare Mill., is the dried fruit of a perennial plant native to southern Europe. The seed is oval and curved, about 0.75 cm long, and greenish in color. The flavor is similar to anise, but less sweet. The consumption of fennel in the United States has increased significantly in the last 20 years, primarily due to its use in pizza topping and Italian sausage. Fennel is imported primarily from India and Egypt, about 7.5 million lb in 1997 (1).

Fenugreek (*Trigonella foenum-graecum* L.) is a little-used spice in the United States. Its flavor is maplelike and bitter. The spice itself is a hard, tan pod containing 10 to 20 seeds. It is primarily used to make imitation maple flavor and curry powders in the United States. Fenugreek is imported from India and Morocco.

Ginger, Zingiber officinale, is a perennial herb with thick, tuberous roots or rhizomes. These tubers are often described as shaped like deer antlers and are tan. To cultivate, the rhizomes are dug up, washed, dried, and often peeled. The flavor is pungent and spicy. Ginger is available fresh, crystallized, and dried. It is imported from China, India, and Jamaica. More than 29.5 million lb (including fresh) were imported in 1997 (1).

Marjoram (*Marjorana hortensis*, Moench) is the leaves of a perennial of the mint family and is a close relative of oregano. Its leaves are green-gray and about 1.25 cm long. The flavor is aromatic, camphoraceous, and bitter. Marjoram is native to western Asia and the Mediterranean region. Principal sources today are France and Egypt.

Mustard seed (*Brassica hirta*—white or yellow mustard and *Brassica juncea*—Oriental or brown mustard) is small in size, about 3 mm in diameter. Color ranges from a light tan (*B. hirta*) to a reddish brown (*B. juncea*). Both types of mustard contain the enzyme myrosinase, which reacts with glycosidic compounds (sinalbin in *B. hirta* and sinigrin in *B. juncea*) in the presence of moisture to release its characteristic pungency (10). Acidic liquids do not trigger this reaction, due to the lowered pH inactivating the myrosinase, but will preserve the pungency if added after it first develops in water. *B. juncea*, upon the enzymatic reaction, produces the compound allylisothiocyanate, which gives a powerful aroma, much like horseradish (10).

Mustard seeds are available in three primary forms: whole mustard, often used in pickling spices; ground mustard, which is used mainly in the sausage industry; and mustard flour. Ground mustard can be treated with heat to deactivate the myrosinase enzyme. This type of ground mustard is used in fresh sausages because the active enzyme can react with the protein in the meat. Mustard flour is the ground seed after the husk or bran is removed. Mustard flour is generally milled to a very fine flour and is a blend of yellow and Oriental mustard seeds. Mustard flour is used primarily in the salad dressing and sauce industries. The principal producing countries of mustard are Canada and the northern plains of the United States.

Nutmeg and *mace* (*Myristica fragrans*, Houtt) are both from an evergreen tree native to the islands of the East Indian archipelago. The tree produces an apricot-like fruit containing a large brown seed about 4 cm long and 2 cm wide. This is the nutmeg seed. Mace is a lacy, netlike orange covering over the seed. The flavor of both is very similar: sweet and warm. Nutmeg is generally milder and sweeter while mace has a sharper flavor. Nutmeg and mace are primarily grown in Indonesia, Grenada, and Trinidad. Nutmeg is the product used most often in the United States. Import volumes from 1997 show it 14 times the volume of mace (1).

Oregano. Two types of oregano are commonly used in the United States. *Origanum vulgare* L. is a perennial of the mint family and is indigenous to the Mediterranean region. Its leaves are dark green, hairy, and about 1.5 cm long. The flavor is bitter with a green note. It is grown primarily in Greece and Turkey. Mexican oregano, *Lippia graveolens* has a distinctly different flavor. Its leaves are larger than the Mediterranean variety. Mexican oregano is commonly available on the West Coast and used most often in Mexican, Southwest, and Tex Mex recipes.

Parsley. Petroselinum sativum L. and *Petroselinum crispum* are two varieties of parsley. The former is the curly leaf variety grown primarily in California and the most common type dried today. The flakes are available in a variety of sizes and bright green in color. The flavor is very mild.

Paprika, Capsicum annum L., is the ground dried fruit pods of a small bushy plant. The pods are orange to bright red and used primarily for their coloring properties. Most paprika is sweet and mild and is grown primarily in California, Spain, Morocco and Hungary.

Pepper, black and white. Piper nigrum L. is the berry of an evergreen-climbing vine native to the coast of southwestern India. The berries are shriveled, dark brown to black, and about 0.75 to 1.0 cm in diameter. Black and white pepper both come from the same plant. White pepper is a ripe peppercorn with the dark hull removed. Black pepper is the unripe berry, which is simply dried. Green peppercorns, growing in popularity recently, are the immature berry of the same vine and are often freeze-dried. Primary sources of black pepper are the Malabar Coast of India (often called Tellicherry if the berry is large enough), the Lampung district of Sumatra in Indonesia, and Brazil. Vietnam has recently emerged as a viable source for black pepper. White pepper is usually imported from the Muntok area of Indonesia. Black pepper is the largest volume spice imported into the United States. In 1997, 112 million lb were imported (1)!

Rosemary, Rosmarinus officinalis L., is an evergreen shrub of the mint family. The leaves are green, narrow, and about 2 cm long, resembling pine needles. It is currently exported from Albania, France, and Spain.

Saffron, Crocus sativus L., is the most expensive spice, costing an average wholesale price of about $200/lb in 1997. It is the dried stigmas of the flower of a purple crocus. The cost is due to the hand harvesting of the three delicate stigmas on each flower. It takes about 250,000 stigmas to produce one pound of saffron (11). Saffron has a bitter flavor but is most prized for its bright yellow color. Spain is the primary producer.

Sage, Salvia officinalis L., is an evergreen shrub of the mint family. The leaves are silver gray to green and velvety to the touch, ranging from 5 to 7 cm long. Sage is primarily used in poultry dishes and sausage, especially fresh pork sausage. In 1997, 4.4 million lb were imported into the United States (1), primarily from Albania and Turkey.

Savory, Satureja hortensis L., is an annual herb of the mint family. It has narrow, dark green leaves about 0.5 to 1.0 cm long. Savory is used very little in the United States. The flavor is sharp, aromatic, and resinous. The major producing countries are Albania and France.

Tarragon, Artemisia dracunculus L., has grown in popularity in recent years. It is a small perennial plant of the sunflower family. Its leaves are large (about 5 cm long), narrow, and dark green. The flavor is fragrant and bittersweet with a licorice-like undertone. It is grown in California and also imported from France.

Thyme, Thymus vulgaris L., is a perennial shrub native to the Mediterranean region. The leaves are narrow, grayish green, and only about 0.5 cm long. The flavor is aromatic and pungent, often used in poultry seasonings and Creole dishes. Thyme is imported primarily from Spain and France.

Turmeric, Curcuma longa L., is a perennial herb of the ginger family. Its thick underground rhizome is boiled, cleaned, dried, and ground into a powder. It has a musty, earthy flavor and is mainly utilized for its bright yellow color. Turmeric is used in curry powders and prepared mustard. There are two types grown in India, Alleppy and Madras. Alleppy is most commonly imported into the United States and has a higher curcumin content than the Madras type.

Other Spice Products

Extractives of spices are the oil-soluble flavor extract of the natural spice. Two types are available: essential oils and oleoresins. Essential oils are the steam-distilled fraction of the oil from the spice. Spice oleoresins are the solvent extract of the spice, containing the essential oil as well as other nonvolatile extractives.

Oleoresins provide the user with a more rounded, closer duplication of the natural spice flavor than essential oils. For example, oleoresin black pepper contains not only its volatile oil, but also piperine, which is the compound that gives it the black pepper bite, or pungency. Black pepper essential oil contains the volatile flavor components only, with no piperine present. Other nonvolatile components present in spice oleoresins include heat components (capsaicin-red pepper), fixatives (from the fatty oils from the seeds of celery, anise, fennel, etc), antioxidants (rosemary and sage), and pigments (paprika and turmeric) (12).

Both essential oils and oleoresins are very concentrated forms of the spice flavor, with usage levels in finished products usually less than 0.01%. Using extractives will standardize the flavor of the finished product compared with natural spices, which can vary from years, origins, or lots. Spice extractives are free from microbial contamination and enzyme activity found in cinnamon and black pepper (12). Extractives also provide flavor with no particulates present in a finished food product. Shipping and storage are easier with extractives, and shelf life is longer than natural spices. The flavor of extractives, however, is not as complex as in their natural spice counterparts.

Many extractives are also available in water-soluble versions from suppliers for use in products where water is the aqueous phase. Most commonly available products contain emulsifiers to make them water soluble.

Spice extractives can be difficult to use in food processing because they are very concentrated. Oleoresin black pepper, for example, is very viscous and hard to pour. Soluble spices were developed by the spice industry to alleviate these problems. Soluble spices are spice extractives plated on a salt or dextrose carrier. They are free-flowing powders, often formulated to approximate a one-to-one replacement to the ground spice. Other products that have been developed for manufacturers include spray-dried extractives and spray-dried products mixed with a carrier to again approximate a one-to-one replacement ratio.

SEASONINGS

Seasoning blends can be defined as dry mix products containing spices; other flavoring agents; and functional ingredients such as salt, sugars, and starches that enhance or provide flavor to a food item. A seasoning blend can be spices only, spices with starches, salt, hydrolyzed vegetable proteins, monosodium glutamate, and flavors such as a gravy mix, or even a blend that has no spices present at all, such as a cheese snack seasoning.

The spice industry has defined analytical testing procedures and standards due to the efforts of ASTA. The seasoning industry does not. It does require experienced, creative food technologists with a well-rounded knowledge of food chemistry, food processing, and laws governing labeling and legal limits of ingredients. The technologist must know how ingredients react in the finished food system, how the finished product is cooked or processed to select the proper ingredients, the intended target market, and finally have a well-developed palate to produce the wide variety of seasonings desired by food processors and consumers.

Seasoning blends are used by food processors for a variety of reasons, primarily for convenience. Seasoning blends can be purchased in batch-sized units, thus reducing ingredient inventories. In addition, fewer quality control personnel are needed for raw material inspection. The responsibility for batching each ingredient separately is eliminated, thus reducing the possibility of error. Consumers purchase seasoning blends for many of the same reasons, most commonly convenience and flavor.

Food Industry

Seasonings are used in the food industry in almost every product category, most commonly in those industries where dry ingredients must be blended with wet ingredients or topically applied. Processed meats, snacks, sauces, gravies, dips, soups, and salad dressing manufacturers are the major users of seasonings. When you think of all the items that are composed of these basic products, the possibilities are endless. Frozen entrees, pizza, boxed side dish items, and retorted sauces and entrees are just a few examples.

Table 3 lists some of the more common meat, snack, and gravy products and the basic spices and other ingredients associated with them. The meat items highlight the spices, although other functional and flavor imparting ingredients (salt, dextrose, sugar, sodium erythorbate) are usually included in the formula. In the snack and gravy area, spices as well as other ingredients are listed because the spices themselves do not produce all the flavors associated with the seasonings. To formulate these products, the flavors

Table 3. Common Seasoning Ingredients

Meat seasonings	
Wiener and bologna	*Italian sausage*
Coriander	Fennel
Nutmeg	Anise
Allspice	Black pepper
Ginger	Red pepper
Clove	Garlic
Black pepper	
Paprika	*Pork sausage*
	Sage
Corned beef	Black pepper
	Red pepper
Garlic	
Allspice	
Clove	
Bay	
Cassia	

Snack seasonings	
Barbecue	*Cheese*
Sugar	Cheese solids
Salt	Whey and other dairy
Citric acid	ingredients
Chili pepper	Dehydrated shortening
Clove	Buttermilk
Allspice	Salt
Smoke	Onion
Yeast (Torula)	Flavorings
Red and black pepper	Yeast (Torula)
Tomato	Salt
Garlic	
Onion	

Sour cream and onion	
Sour cream solids	
Buttermilk	
Whey and other dairy	
ingredients	
Artificial flavors	
Citric acid	
Onion	
Parsley	
Salt	

Gravies	
Thickeners (starches and	Celery seed
flours)	Black pepper
Hydrolyzed vegetable proteins	Onion
Meat flavors	Turmeric (chicken gravy)
Dehydrated meat or meat	Caramel color (brown gravy)
stock	Salt
Shortening or nondairy	
creamers	
Milk solids	

must be balanced to achieve the exact flavor profile desired. This is where the knowledge of an experienced seasoning food technologist is important. There are literally hundreds of variations of each type of flavor.

Ethnic food products are always popular. Everyone is looking for the next "hot" trend. Table 4 lists five basic ethnic flavors and the spices needed to create them. By first formulating a base product, such as a marinade, with the proper amount of salt, hydrolyzed vegetable protein, sugar, gum or starch, phosphate, and so on, the spices listed in Table 4 can then be added and balanced to produce the desired flavor of the finished product. It should be noted that seasonings not only use the natural spice for flavor but also oleoresins and essential oils to produce a product with the exact profile desired.

Specifications

When purchasing seasonings, the specification should adequately reflect the important attributes of the blend. Generally, flavor profile, appearance, granulation, salt, and color are included, as well as other tests such as pH if acid is present; cold, hot, or retort viscosity where applicable; or heat units if red pepper is present in appreciable amounts. If the food item is very sensitive to microbial loads, such as a sour cream dip, then microbiological spec

Table 4. Common Ethnic Seasonings

Italian	*Asian*
Oregano	Ginger
Basil	Garlic
Marjoram	Sesame
Garlic	Red pepper
Fennel	5 spice (cinnamon, anise, ginger,
Anise	fennel, cloves)
Red pepper	
	Indian
Mexican	Cumin
Chili pepper	Coriander
Cumin	Fenugreek
Oregano	Cardamom
Garlic	Turmeric
Onion	Red pepper
Red pepper	Black pepper
Paprika	Garlic
Cinnamon	Onion
Red pepper	
	Caribbean
Cajun	Red pepper
Red pepper	Allspice
White pepper	Ginger
Black pepper	Nutmeg
Thyme	Thyme
Oregano	Bay leaves
Garlic	Onion
Onion	Garlic
Paprika	Chili pepper
Sassafras	Cumin
	Coriander
	Turmeric

ifications should be considered. When writing or reviewing specifications, it is essential to know what the important parameters are that the seasoning blend is contributing to the finished food.

FUTURE TRENDS

With the increasing demand for high-quality and unique food items, the need for the seasoning company is more evident than ever before. Although many seasoning companies began by servicing the needs of processed meat manufacturers, the trend for reduced red meat consumption, as well as competition within the industry, has limited growth in this area. Conversely, snacks, specialty sauces, marinades, glazes, and seasonings are on the rise. These types of products hold the future for seasoning manufacturers.

In addition, leveraged buyouts and consequent cutbacks in research and development staffs should cause a higher demand for outside product development work. Finally, small start-up companies often require help with new products. Seasoning companies can offer their services in new product development. The new millennium promises a bright future for the seasoning industry.

BIBLIOGRAPHY

1. United States Department of Agriculture, *United States: Imports of Specified Condiments, Seasonings and Flavoring Materials*, U.S. Government Printing Office, Washington, D.C., 1998.
2. *CFR* 21:101.22 & 182.10, U.S. Government Printing Office, Washington, D.C., 1996.
3. Food and Drug Administration, Center for Food Safety and Applied Nutrition, *The Food Defect Action Levels* (HFS-565), U.S. Government Printing Office, Washington, D.C., 1997.
4. D. R. Tainter and A. T. Grenis, *Spices and Seasonings; A Food Technology Handbook*, VCH-Wiley, New York, 1993.
5. *ASTA Cleanliness Specifications for Spices, Seeds, and Herbs*, American Spice Trade Association, Englewood Cliffs, N.J., 1997.
6. *Cleaning and Reconditioning of Spices, Seeds, and Herbs*, American Spice Trade Association, Englewood Cliffs, N.J.
7. *Official Analytical Methods of the American Spice Trade Association*, American Spice Trade Association, Englewood Cliffs, N.J., 1997.
8. M. Vajdi and R. R. Pereira, "Comparative Effects of Ethylene Oxide, Gamma Irradiation and Microwave Treatments on Selected Spices," *J. Food Sci.* **38**, 893–895 (1973).
9. *CFR* 21:179.26, U.S. Government Printing Office, Washington, D.C., 1996.
10. *Mustard Flour*, R. T. French Co., Rochester, N.Y., 1988.
11. J. D. Dziezak, "Spices," *Food Technol.* **43**, 102–116 (1989).
12. G. D. Deline, "Modern Spice Alternatives," *Cereal Foods World* **30**, 697–700 (1985).

GENERAL REFERENCES

K. T. Farrell, *Spices, Condiments and Seasonings*, 2nd ed., Van Nostrand Reinhold, New York, 1990.
J. Giese, "Spices and Seasoning Blends: A Taste for All Seasons," *Food Technol.* **48**, 88–98 (1994).
S. Hegenbart, "Spices and Seasoning Blends," *Food Product Design* **3**, 43–64 (1993).
E. L. Ortiz, *The Encyclopedia of Herbs, Spices, and Flavorings*, Dorling Kindersley, New York, 1992.
J. W. Purseglove et al., *Spices*, Vols. 1 and 2, Longman, Essex, England, 1987.
E. W. Underriner and I. R. Hume, *Handbook of Industrial Seasonings*, Chapman and Hall, London, 1993.

DONNA R. TAINTER
Tone Brothers, Inc.
Ankeny, Iowa

STARCH

Next to water, starch is the most abundant constituent in the human diet. Starch is abundantly available, is inexpensive, is a desirable source of energy, and occurs in the form of granules (see the section "Granular Nature, Cooking Characteristics, and Solution and Gel Properties"). Starch occurs naturally in most plant tissues, including roots and tubers, cereal grains, vegetables, and fruits. Large amounts of starches are consumed as components of wheat, corn (maize), and rye flours and as components of potatoes and whole kernels of corn, rice, and barley. Starch composition and properties vary with the plant sources from which they are obtained. Each starch from each source is unique in terms of its behavior and characteristics. Starches may be isolated and added to food products as ingredients. Often, they are modified after isolation and added during the preparation process in modified form.

Native and modified starches serve a variety of roles in food products. They are added primarily to modify texture and consistency. Principally they serve to bind water, to thicken, and to form soft, spoonable gels, all controllable properties. Starches generally must be cooked in order to realize their physicochemical properties and to impart their functionalities. Heating starch granules in the presence of water causes the granules to swell, lose their crystallinity, and in some cases, break apart (see the section "Granular Nature, Cooking Characteristics, and Solution and Gel Properties"). When a starch is cooked in excess water, the resulting dispersion is called a paste.

Corn (maize) is the principal source of commercial starch in the United States (1–3). Lesser amounts of potato, tapioca (cassava, manioc), sorghum (milo), wheat, rice, and arrowroot starches are used in the United States (1), but a starch such as one from potato, wheat, or cassava may be the major starch in other countries. Minor amounts of sago, barley, sweet potato, mung bean, and rye starches find their way into commerce elsewhere.

MOLECULAR COMPONENTS

Although commercial starch granules, especially those from cereal grains, contain small amounts of protein, lipids (especially phosphoglycerides and free fatty acids), and other components, the principal components are amylose and/or amylopectin (Tables 1 and 2) (1,4).

Amylose is an essentially linear polysaccharide composed of (1→4)-linked α-D-glucopyranosyl units (see the article CARBOHYDRATES: CLASSIFICATION, CHEMISTRY, LABELING). Its degree of polymerization (DP) is 1,000 to 16,000 (MW 160,000–2,650,000), depending on the source and method of preparation. Amylose can have several conformations. In the solid state, it probably exists most often as a left-handed, sixfold helix. In solution, it seems to be a loosely wound and extended helix that behaves as a random coil. Because of its helical structure, amylose is able to complex with hydrophobic molecules. In foods, amylose often complexes with mono- and diglycerides and/or free fatty acids or their salts. Complexed amylose molecules retrograde less effectively (see the section "Retrogradation"). Hence, molecules complexed with hydrocarbon chains provide greater stability to foods, so emulsifiers and other surfactants are added to baked goods to retard staling, which at least in part results from retrogradation.

Amylopectin has a branch-on-branch structure. Amylopectin molecules are composed of chains of (1→4)-linked α-D-glucopyranosyl units; branches are formed by joining these chains with α-D-(1→6) linkages. The average chain length is 20 to 30 units, although branch points are not equally spaced. The currently accepted architecture of an amylopectin molecule is that of a long main chain to which are attached clusters of branches, some of which are themselves branched (Fig. 1). This model has been termed the cluster model. The molecular weight of amylopectin has been measured as $5 \times 10^7 - 4 \times 10^8$ (DP $3 \times 10^5 - 2.5 \times 10^6$), depending on the source and method of preparation. Potato starch amylopectin occurs as a natural phosphate ester.

Corn and rice cultivars with altered polysaccharide composition have been developed. Normal corn starch is composed of approximately 28% of the linear polysaccharide amylose and approximately 72% of the branched polysaccharide amylopectin. Those starches that result from a mutation that makes them composed of essentially 100%

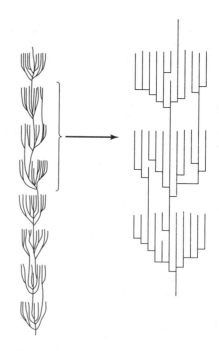

Figure 1. Structures of segments of an amylopectin molecule.

amylopectin are designated waxy starches. Waxy maize (waxy corn) starch is an important food starch. Other all-amylopectin starches have been produced but are not commercially available. High-amylose corn starches containing approximately 55% and 70% amylose are in commercial production and find some food use. Starches from other special corn cultivars are both available and under development.

The physical properties of amylose and amylopectin (Table 2) are reflected in starches. For example, high-amylose starches are difficult to gelatinize (because of the extra energy needed to dissociate and hydrate the aggregates of amylose); form firm, opaque gels; and can be used

Table 1. General Properties of Some Native Starch Granules and Their Pastes

	Common corn	Waxy maize	High-amylose corn	Potato	Tapioca
Amylose content, %	28	<2	50–70	21	17
Cooking temperature, °C	62–78	63–72	160–170	58–65	52–65
Relative viscosity	Medium	Medium high	Very low[a]	Very high	High
Paste rheology (body)	Short	Long	Short	Very long	Long
Paste clarity	Opaque	Slightly turbid	Opaque	Translucent	Translucent
Relative shear stability	Stable	Unstable	Stable	Unstable	Unstable
Tendency to gel	Strong	Weak	Very strong	Weak	Medium

[a]Under ordinary cooking conditions (100°C).

Table 2. General Properties of the Two Starch Polysaccharides (Unmodified)

Property	Amylose	Amylopectin
Solution stability	Unstable (precipitates)	Stable
Gels	Soft, reversible, flowable gels become irreversibly stiffer with time; undergo syneresis	Solutions do not gel
Solubility	Difficult soluble; high-temperature solubility only	Soluble
Complex formation	Complexes with iodine, lipids, and various polar organic molecules	Does not complex

to make strong, tough films. Their solutions and gels undergo retrogradation rapidly. Waxy maize starches, even when unmodified, gelatinize easily and yield viscous, almost transparent solutions that do not gel.

In general, the properties of a starch paste or gel are a function of the amounts of amylose and amylopectin dispersed or solubilized, the amounts of insoluble swollen granules and granule fragments, and interactions between components. Amylose increases gel strength; amylopectin decreases gel strength and viscosity.

GRANULAR NATURE, COOKING CHARACTERISTICS, AND SOLUTION AND GEL PROPERTIES

All green plants package and store carbohydrate (D-glucose) in the form of discrete particles of starch called granules (1,5). In granule form, starch is insoluble in cold water and only slightly hydrated. The sizes and shapes of granules are specific for the plant of origin and can be identified microscopically (1,6). Diameters (in micrometers) of some commercial starches are in the following ranges: rice, 3 to 8; corn, 5 to 25; tapioca, 5 to 35; potato, 15 to 100.

Occurrence in granule form makes starch unique. Starch granules can be dispersed in water, producing low-viscosity slurries that can be easily mixed and pumped. Starch granules can be reacted while slurried. They can be isolated/recovered, either with or without reaction, by filtration or centrifugation and resuspended for cooking. The thickening power of granular starch is realized only when the slurry is cooked.

Starch granules reversibly absorb water and swell slightly but remain as granules until an aqueous suspension (slurry) is heated. When dry corn starch (normal moisture content 10–12%) is placed in cold water, the moisture content of its granules increases to approximately 30%, and the average granule size of normal yellow dent corn starch increases by ca 9% and of waxy maize starch approximately 23%.

When heated in water, starch granules gelatinize. Gelatinization is the collapse (disruption) of molecular orders within granules resulting in irreversible changes in properties such as granule swelling, native crystallite melting, loss of birefringence, and leaching of soluble components (primarily amylose) (7). (Some amylose leaching can occur at temperatures below the gelatinization temperature.) The temperature of initial gelatinization and the range over which gelatinization occurs depends on the method used to determine it and is governed by the starch type and concentration and heterogeneities within the granule population under observation (Table 1).

Pasting is the phenomenon following gelatinization when a starch slurry containing excess water is heated (7). It involves further granule swelling, additional leaching of soluble components, and eventually, especially with the application of shear, total disruption of granules, resulting in molecules and aggregates of molecules in dispersion or solution. In most if not all cases, some granule remnants remain.

The cooking behavior of starches and the viscosity of the resulting pastes is most often followed with an instrument

with which a starch suspension is heated at a designated rate to a designated temperature while continuously measuring and recording the viscosity. The resulting paste is held at the designated temperature (usually 95°C) for a designated time and then cooled. The resulting curve reveals the pasting temperature, rate of viscosity development, peak viscosity, rate and extent of viscosity breakdown, and rate and extent of viscosity development during paste cooling (Fig. 2).

The property of forming thick pastes or gels is the basis of most starch uses. The extent of starch gelatinization and pasting is the principal factor controlling texture, other product properties such as storageability, and digestibility. In some baked goods, many starch granules remain ungelatinized (as much as 90% in pie crust and some cookies that are high in fat and low in water content). In the processing and preparation of most other food products containing starch, which usually occur in the presence of excess water, starch granules swell beyond the reversible point. On heating a starch slurry (usually 2–5% starch), granules absorb water until almost all water is within highly swollen granules, forcing the granules to press against one another, filling the container with a highly viscous paste or gel. These highly swollen granules are fragile. Stirring then effects granule rupture and disintegration and a decrease in viscosity.

Starches from different sources have different cooking characteristics (Table 1). Tuber and root (potato and tapioca) starches gelatinize more easily than do cereal starches and produce more viscous pastes that easily and rapidly lose viscosity on application of shear. These pastes are rather clear and slow to gel and have a bland flavor. Ordinary corn starch produces an opaque, cohesive gel that undergoes syneresis and has a slight cereal flavor. The properties of waxy maize starch pastes are in between those of potato starch and corn starch. Rice and wheat starches also produce opaque pastes. All starches can be overcooked to give stringy pastes; they can also be undercooked.

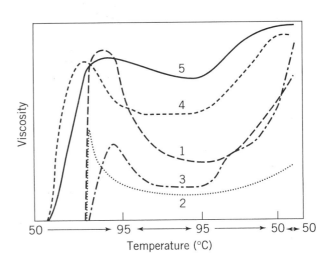

Figure 2. Generalized curves showing the cooking behaviors of selected starches; 1, potato starch; 2, waxy maize starch; 3, common corn starch; 4, stabilized common corn starch; 5, moderately crosslinked and stabilized common corn starch.

The viscosity obtained by cooking a suspension of starch is determined by starch type, type and amount of modification, solids concentration, pH, amount of agitation during heating, rate of heating, maximum temperature reached, time held at that temperature, agitation during holding, and the presence of other ingredients. High sugar concentrations decrease the rate of gelatinization and lower the peak viscosity and gel strength. Emulsifiers such as monoacylglycerols and lipids that complex with amylose inhibit granule swelling, increase both the gelatinization temperature and the temperature needed to produce maximum viscosity, and decrease both the temperature needed for gel formation and gel strength.

Pregelatinized Starches

Starches can be cooked and the paste dried in a way that destroys granular structure and allows the product to be redispersed (dissolved) in water at temperatures below the gelatinization temperature. Various types of these pregelatinized (instant) starches that do not require cooking to achieve thickening are produced. Some will produce smooth dispersions; others produce pulpy or grainy dispersions and find use in fruit drinks and tomato products. Pregelatinized starches are often used in dry mixes because they disperse readily, even when mixed with other ingredients such as sugar.

Cold Water-Swellable Starches

By heating common corn starch in media containing a limited amount of water, granular products that swell extensively in cold water can be made. Cold water-swellable starches can be used in instant products of various kinds. When they are dispersed in sugar syrups with rapid stirring and the mixture is poured into molds, a rigid gel that can be sliced is formed.

Retrogradation

Starch molecules in an unordered state (in solution, in a dispersion, or in gelatinized granules) will undergo a process called retrogradation. Retrogradation (setback) is the reassociation of starch polymer molecules. It occurs when molecules that have become disordered during cooking begin to reassociate in an ordered structure (8). In the initial phases of retrogradation, linear segments of two or more starch chains may form a simple juncture point that may develop into more extensively ordered regions. Ultimately, under favorable conditions, a crystalline order appears. The result is gelation or precipitation. Generally, extensive retrogradation is undesirable. Various approaches are used to retard retrogradation; one is the complexation described earlier (see the section "Molecular Components").

Derivatization to produce stabilized starch is also used to reduce the tendency for retrogradation (see the section "Stabilized Starch").

MODIFIED FOOD STARCH

In the native form, starches do not have the properties desired by food processors. For example, when aqueous slurries of native starches are heated, the starch granules swell, then rupture—causing the viscosity to increase rapidly, then fall. Cooling of the cook produces weak-bodied, cohesive pastes or gels. Modification can correct such a defect. According to the U.S. *Code of Federal Regulations* (21 *CFR* 172.892), modified food starch (1,9) may be prepared using the following treatments: acid modification with hydrochloric and/or sulfuric acids; bleaching with hydrogen peroxide, peracetic acid, potassium permanganate, or sodium hypochlorite; oxidation with sodium hypochlorite; esterification with acetic anhydride, adipic–acetic mixed anhydride, monosodium orthophosphate, 1-octenylsuccinic anhydride, phosphoryl chloride, sodium tripolyphosphate and/or sodium trimetaphosphate, or succinic anhydride; etherification with propylene oxide; or various combinations of the preceding. Also allowed under the regulations are bleaching with ammonium persulfate or sodium chlorite, acetylation using vinyl acetate, and etherification with acrolein or epichlorohydrin (crosslinking); but these reactions are not currently used to produce food starches. Limitations of modifications are given in the regulations. Most U.S. modified food starch is prepared from waxy maize, common corn, or potato starch. In the United States, the ingredient label of the product containing one of these food starches must designate "modified food starch" or "food starch modified" as one of the ingredients but does not need to indicate the source of the starch or the modification(s).

In general, derivatization increases solution and gel clarity, reduces the tendency to gel and/or crystallize, improves water binding, increases freeze-thaw stability, reduces the gelatinization temperature, and increases the peak viscosity. Combinations of derivatizations and other modifications are used to obtain desired properties for specific applications. The useful properties of starches that can be modified and controlled by various treatments are their property of forming a suspension of insoluble granules, without thickening, until the slurry is heated; their ability to thicken aqueous systems and to form a paste on heating; their ability to form semisolid gels on cooling of pastes; their ability to form strong, adhesive films and coatings and to act as a binder; the ability of their pastes to disperse and suspend fats, oils, and particulate matter; and the ability of their pastes and gels to provide important textural qualities to prepared foods.

Crosslinked Starch

Crosslinking is the most important modificaton applied to food starches. It is used to control texture and provide heat, acid, and shear tolerance and, thus, to provide better control over end-product quality and greater flexibility in dealing with formulations and processing and storage conditions. Diphosphate esters, which can be introduced by reaction of starch with either phosphoryl chloride or sodium trimetaphosphate (Fig. 3), are the most common crosslinks. Crosslinking of the polymer chains of starch granules is also done by making an adipic acid diester.

Crosslinking strengthens the granule. A small amount of crosslinking (eg, treatment with only 0.0025% sodium trimetaphosphate) greatly reduces both the rate and the

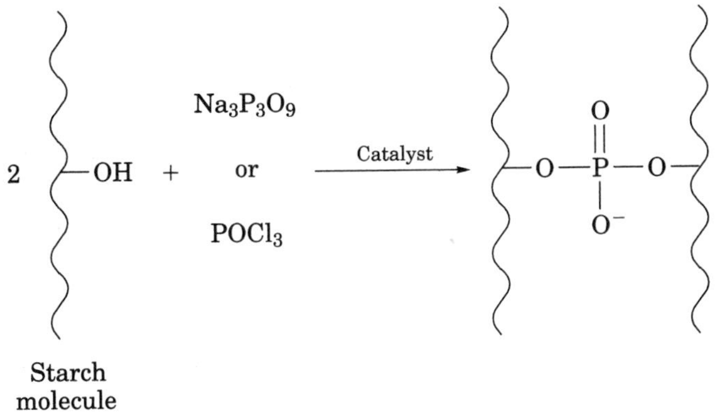

Figure 3. Crosslinking of molecular chains of starch granules with phosphate diester linkages.

degree of granule swelling and the sensitivity of starch pastes to processing conditions (high temperature; extended cooking times; high shear during mixing, milling, homogenization, or pumping; low pH). Pastes of crosslinked starches are more viscous, more stable, shorter-textured (more pseudoplastic), and generally heavier-bodied than those of native starch from which they were prepared. Crosslinked waxy maize starch products are especially popular in the food industry. Starches with a range of properties are prepared by varying the kind and degree of crosslinking (Table 3). In general, when choosing a starch for a particular application, one should select a starch that is crosslinked sufficiently to enable it to withstand the processing conditions to which it will be subjected and still give optimum viscosity in the final product. Crosslinked starches can be pregelatinized.

Stabilized Starch

Pastes of ordinary starch will gel, and the gels will generally be cohesive, long-textured, and prone to syneresis— all undesirable characteristics. Pastes of underivatized waxy maize starch, which has no amylose, are less inclined to retrograde or gel; however, in the refrigerator or freezer, they will eventually increase in opacity and become chunky. This is especially true if processing or cooking conditions effect cleavage of chains, producing linear fragments.

To improve textural aesthetics and/or other properties, other starch derivatives are prepared (Table 4). Few of the hydroxyl groups of modified food starches are derivatized, in other words, ether or ester groups are present in small amounts. Most modified food starches contain, on average, one substituent group per every 5 to 10 α-D-glucopyranosyl units, a DS (degree of substitution, the average number of hydroxyl groups derivatized per monomeric unit of the polymer, of 0.1–0.2). These small degrees of derivatization change dramatically the properties of the starch and increase its range of application.

The derivatives most commonly used to prepare a stabilized starch are the hydroxypropyl ether, the acetate ester, and the monophosphate ester. Addition of these groups reduces the ability of starch molecules to form junction zones (intermolecular associations).

Crosslinked Stabilized Starches

Often, it is desired to improve both processing and cooking tolerances and shelf and storage stability, so many food starches are derivatized with two reagents to introduce both crosslinking and non-crosslinking substituent groups.

Addition of Hydrophobic Groups

Other derivatives are made to add specific functional properties. For example, nonwetting hydrophobic starches are used as release agents for dusting on dough sheets or as processing aids. Other starch products with hydrophobic groups are used as emulsifiers and emulsion stabilizers.

Acid-Modified Starches

Acid-modified (acid-thinned, thin-boiling) starches are prepared by treating a suspension of native or derivatized

Table 3. General Reasons for Crosslinking Starches

Changes in granule properties

Modification of cooking characteristics
Inhibition of swelling

Changes in paste properties

Reduction of cohesiveness
Inhibition of gel formation
Improvement of acid stability
Improvement of heat stability
Improvement of shear tolerance

Table 4. General Reasons for Making Stabilized Starches

Changes in granule properties

Reduction in energy required to cook

Changes in paste properties

Increased stability
Increased freeze-thaw stability
Enhancement of clarity
Increased sheen
Inhibition of gel formation
Reduction of gel syneresis
Increased viscosity
Improved interactions with other substances
Improvement in stabilizing properties

starch with dilute mineral acid at temperatures below the gelatinization temperature for varying periods. When it is determined that a product that gives the desired viscosity after cooking has been produced, the acid is neutralized and the starch is recovered by centrifugation or filtration, washed, and dried. In this process, a small amount of glycosidic bond hydrolysis occurs, resulting in products that produce much less viscosity. A concurrent weakening of the granule structure occurs. The result is that there is more granule disintegration when acid-modified starches are heated in water; and although they have reduced viscosity-imparting power, they form gels with increased strength and improved clarity.

These so-called thin-boiling starches are used when strong gel strength is desired, such as in gum candies (eg, jelly beans, orange slices, and spearmint leaves) and in processed cheese loaves. Where especially strong or fast-setting gels are desired, high-amylose starches are used. Other of these low-viscosity products, which allow higher-concentration pastes to be formed, are used as film formers and adhesives in products such as pan-coated nuts and candies.

Bleaching

Absolute whiteness, particularly of corn starch, requires bleaching. The bleach most commonly used is sodium hypochlorite. During the bleaching operation, the starch is oxidized. Small amounts of carboxyl and carbonyl groups are introduced and some glycosidic bonds are cleaved. The result is decreased pasting temperature, thickening power, and tendency to retrograde.

USES

Uses of starches and starch products in foods are extensive. In the granular form (dry powder), starches are used as anticaking agents and diluents and for dusting and molding. When nearly fully cooked to a state approaching a molecular dispersion, starches are used to provide viscosity (thickening), impart texture, suspend solids, and form films. Pastes are also employed as binders, processing aids, protective colloids, and encapsulating agents. As swollen granules (produced by heating slurries of cross-linked granules), starches are used to provide viscosity, impart texture, and suspend solids and as binders and processing aids. Use of a high-amylose starch will make breaded products crispier. Table 5 lists important properties exhibited by starch in both the dry and cooked states and a few examples of products containing a starch or modified food starch.

DIGESTION

Essentially, only cooked starch can be digested effectively by humans. Amylases are the enzymes that catalyze the hydrolysis of the glycosidic bonds of the polysaccharides of starch (10). α-Amylases are endo-enzymes, that is, enzymes that catalyze the hydrolysis of internal bonds of starch polysaccharides. Although saliva contains an α-

Table 5. Important Properties of Specific Food Starch Products

Property	Product Examples
Adhesion	Breaded products, pan-coated nuts and candies
Anticaking (flowing aid)	Baking powder, powdered sugar, salt
Antistaling	Baked goods
Binding	Extruded foods, meat products
Clouding	Beverages, cream fillings
Crispiness	Breaded products
Dusting	Bread, chewing gum, rolls
Foam strengthening	Confectioneries
Gelling	Gum candies, processed cheeses, puddings, spoonable dressings
Glazing	Nuts
Hydration	Dry mixes
Moisture retention	Breadings
Molding	Gumdrops
Oil absorption	Peanut butter
Pulpiness	Fruit drinks, tomato products
Shaping	Meat products, pet foods
Stabilizing emulsions and suspensions	Beverages, salad dressings
Thickening	Baby foods, cream-style corn, gravies, pie fillings, sauces, soups, yogurt

amylase, very little starch hydrolysis occurs in the mouth because of the short dwell time.

Almost all starch digestion and absorption takes place in the small intestine. The pancreatic juice secreted into the small intestine contains another α-amylase. This enzyme effects a rapid reduction in molecular weight of the starch polysaccharides, producing starch oligosaccharides (maltooligosaccharides), primarily of six and seven α-D-glucopyranosyl units (Fig. 4). The α-amylase then acts more slowly on these oligosaccharides to reduce them to smaller fragments (maltose and maltotriose). The enzyme acts even more slowly on maltotriose; it does not catalyze the hydrolysis of the glycosidic bond of the disaccharide

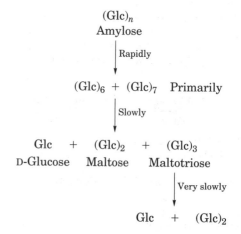

Figure 4. Action of α-amylase on amylose. Glc = D-glucose or α-D-glucosyl unit.

maltose. α-Amylases catalyze the hydrolysis of α-D-(1→4) linkages only, never α-D-(1→6) linkages. Therefore, the products of the action of pancreatic α-amylase on amylose and amylopectin are both linear and branched maltooligosaccharides.

Other enzymes are then needed to catalyze the hydrolysis of the maltooligosaccharides. Complete hydrolysis to D-glucose is required because only monosaccharides can be absorbed. Disaccharidases are located on the surface of cells lining the inner surface of the small intestine. Maltase catalyzes the hydrolysis of maltose, 4-O-(α-D-glucopyranosyl)-D-glucose (see the article CARBOHYDRATES: CLASSIFICATION, CHEMISTRY, LABELING), to D-glucose. Isomaltase catalyzes the hydrolysis of isomaltose, 6-O-(α-D-glucopyranosyl)-D-glucose, to D-glucose. Both enzymes act on higher oligosaccharides but more slowly than they do on the disaccharides.

Starch that resists and escapes digestion by endogenous enzymes of the upper gastrointestinal tract and is thus available to the large intestine as a fermentable substrate is called resistant starch. Resistant starch has physiological benefits, primary among which may be a decreased incidence of colorectal cancer and ulcerative colitis. The resistant starch content of foods can be manipulated by choice of the botanical source of starch, processing, cooking, and storage conditions.

STARCH-RELATED PRODUCTS

Starches can be converted into various products of depolymerization, namely, D-glucose (dextrose), dextrins, maltodextrins, syrup solids, and syrups, including high-fructose syrups (see the articles CARBOHYDRATES: CLASSIFICATION, CHEMISTRY, LABELING and SWEETENERS: NUTRITIVE) (1,11).

BIBLIOGRAPHY

1. R. L. Whistler, J. N. BeMiller, and E. F. Paschall, eds., *Starch: Chemistry and Technology*, 2nd ed., Academic Press, Orlando, Fla., 1984.

2. J. B. May, "Wet Milling: Process and Products," in S. A. Watson and P. E. Ramsad, eds., *Corn: Chemistry and Technology*, American Association of Cereal Chemists, St. Paul, Minn., 1987, pp. 377–397.

3. F. T. Orthoefer, "Corn Starch Modification and Uses," in S. A. Watson and P. E. Ramstad, eds., *Corn: Chemistry and Technology*, American Association of Cereal Chemists, St. Paul, Minn., 1987, pp. 479–499.

4. A. Guilbot and C. Mercier, "Starch," in G. O. Aspinall, ed., *The Polysaccharides*, Vol. 3, Academic Press, Orlando, Fla., 1985, pp. 209–282.

5. L. F. Hood, "Current Concepts of Starch Structure," in D. R. Lineback and G. E. Inglett, eds., *Food Carbohydrates*, AVI Publishing Co., Westport, Conn., 1982, pp. 217–236.

6. J. Seidemann, *Starke-Atlas*, Paul Parey, Berlin, 1966.

7. W. A. Atwell et al., "The Terminology and Methodology Associated with Basic Starch Phenomena," *Cereal Foods World* **33**, 306–311 (1988).

8. V. J. Morris, "Starch Gelation and Retrogradation," *Trends in Food Sci. Technol.* **1**, 2–6 (1990).

9. P. S. Smith, "Starch Derivatives and Their Use in Foods," in D. R. Lineback and G. E. Inglett, eds., *Food Carbohydrates*, AVI Publishing Co., Westport, Conn., 1982, pp. 237–269.

10. The Amylase Research Society of Japan, eds., *Handbook of Amylases and Related Enzymes*, Pergamon Press, Oxford, 1988.

11. G. M. A. van Beynum and J. A. Roels, eds., *Starch Conversion Technology*, Marcel Dekker, New York, 1985.

J. N. BeMiller
Purdue University
West Lafayette, Indiana

See also CARBOHYDRATES: CLASSIFICATION, CHEMISTRY, LABELING; CARBOHYDRATES: FUNCTIONALITY AND PHYSIOLOGICAL SIGNIFICANCE.

SUGAR: SUBSTITUTES, BULK, REDUCED CALORIE

Absorption of polyols is slower than absorption of their carbohydrate analogs, allowing a portion to reach the large intestine where metabolism by the colonic bacteria yields short-chain fatty acids and reduced calories to the host. Measuring the caloric value of a polyol in humans is a difficult task, and regulatory agencies may reach different conclusions despite having similar databases from which to render an opinion. The Life Sciences Research Office Expert Panel of the Federation of American Societies for Experimental Biology prepared a report titled "The Evaluation of the Energy of Certain Sugar Alcohols Used as Food ingredients" in June 1994. The Expert Panel estimated the net energy of sugar alcohols from 1.6 to 3.3 kcal/g. The European Union promulgated a Nutritional Labeling Directive stating that all polyols have a caloric value of 2.4 cal per gram.

Polyols are resistant to metabolism by oral bacteria that break down sugars and starches to release organic acids, leading to erosion of tooth enamel and eventually to dental cavities. Polyols are noncariogenic. Polyols are useful alternatives to sugars and facilitate formulation of sweet, but not cariogenic snacks.

Polyols are slowly and poorly absorbed, compared with simple sugars, and do not raise blood glucose levels following consumption. Compared with sucrose or starch, both of which raise blood glucose levels, polyols do not. The major polyols are described next.

Hydrogenated Starch Hydrolysates

Hydrogenated starch hydrolysates (HSHs) including hydrogenated glucose syrups, maltitol syrups, and sorbitol syrups, are a family of products found in a wide variety of foods. They serve a number of functional roles, including use as bulk sweeteners, viscosity or bodying agents, humectants, crystallization modifiers, cryoprotectants, and rehydration aids.

HSHs are produced by the partial hydrolysis of corn, wheat, or potato starch and subsequent hydrogenation of the hydrolysate at high temperature under pressure. The end product is composed of sorbitol, maltitol, and higher hydrogenated saccharides. By varying the manufacturing conditions and by controlling the extent of hydrolysis, one can tailor the chemical species produced. The term HSH is applied to describe the broad groups of polyols that contain substantial quantities of hydrogenated oligo- and polysaccharides in addition to any monomeric or dimeric polyols.

Because hydrolysis and subsequent hydrogenation does not yield a single compound but, rather, a variety of compounds, manufacturers have adopted a practice of using the prevalent polyol to name the product. For example, polyols containing sorbitol as the majority component are called sorbitol syrups; those with maltitol as the majority are maltitol syrups or maltitol solutions. Polyols without a single polyol at 50% or more of the total continue to be referred to by the general term HSH.

HSHs are outstanding humectants that do not crystallize, enabling the production of sugar-free confections with the same cooking and handling systems used to produce sugar candies. These products are used extensively in confections, baked goods, a broad range of other foods, dentifrices, and mouthwashes.

HSHs provide 40 to 90% of the sweetness of sugar, texture, and bulk to a variety of sugarless products.

Isomalt

Isomalt is derived from sucrose in a two-step process. In step one, the normal linkage of 1 (glucose) to 5 (fructose) is changed to a 1 to 6 linkage. In step two, hydrogenation converts the aldehyde to one of two alcohols, either gluco-mannitol or gluco-sorbitol in approximately equal proportions as shown in Figure 1.

The use of isomalt in place of sucrose retains the volume and textural qualities of sucrose. Because isomalt can be heated without losing its sweetness or being broken down, it is used in products that are boiled, baked, or subjected to high temperatures.

Isomalt absorbs very little water. Therefore, products made with isomalt tend not to be as sticky as products made with sucrose. Extended shelf life is observed with isomalt-containing foods.

Isomalt's sweetening power depends on its concentration, temperature, and the form of the product in which it is used. When used alone, it contributes 45 to 65% of the sweetness that would result from the same amount of sucrose. For nutrition labeling purposes, isomalt has a caloric density of 2.0 cal per gram, one-half that of sucrose. Isomalt's lower caloric value is partly due to the fact that intestinal enzymes are not able to easily hydrolyze its more stable disaccharide bond. Therefore, less is absorbed from the small intestine into the blood, and absorption takes place at a slower rate than with sucrose.

Lactitol

Lactitol is manufactured by reducing the glucose half of the disaccharide lactose (2).

The resulting compound has two more hydrogen atoms than does lactose. Unlike lactose, lactitol is not hydrolyzed by lactase enzyme. It is neither hydrolyzed nor absorbed in the small intestine. Lactitol is metabolized by bacteria in the large intestine, where it is converted into carbon dioxide, short-chain fatty acids, and a small amount of hydrogen.

The organic acids are further metabolized resulting in a calorie contribution of 2 cal per gram. The Scientific Committee for Food of the European Union has provided a Nutrition Labeling Directive stating that all sugar alcohols, including lactitol, have a caloric value of 2.4 kcal/g. Canada has assigned a 2.6 kcal/g value to lactitol.

As a replacement for sucrose in formulated foods, lactitol does not induce an increase in blood glucose or insulin levels. Control of blood glucose, lipids, and weight are the three major goals in diabetes management. Foods using lactitol to replace sugar can be used by people with diabetes, giving them a wider variety of low-calorie and sugar-free choices.

Lactitol is not metabolized by oral bacteria that break down sugars and starches to release acids that may lead to cavities.

Due to its stability, solubility, and similar taste to sucrose, lactitol can be used in a variety of low-calorie, low-fat and/or sugar-free foods such as ice cream, chocolate, hard and soft candies, baked goods, sugar-reduced preserves, chewing gum, and sugar substitutes.

Sorbitol

Sorbitol is manufactured from glucose by high-pressure hydrogenation or by electrolytic reduction, which results in the addition of two hydrogens per molecule. The chemical structure is shown as follows (2):

$$
\begin{array}{c}
CH_2OH \\
| \\
HC-OH \\
| \\
HO-CH \\
| \\
HC-OH \\
| \\
HC-OH \\
| \\
CH_2OH
\end{array}
$$

Sorbitol is about 60% as sweet as sucrose with fewer calories, 2.6 cal per gram in the United States, and 2.4 cal per gram in the European Union. Sorbitol is a natural constituent of a wide variety of fruits and berries.

Sorbitol is used as a humectant in many types of products for protection against loss of moisture content. The moisture-stabilizing and textural properties of sorbitol are used in the production of confectionery, baked goods, and chocolate where products tend to become dry or harden. Its moisture-stabilizing action protects these products from drying and maintains their initial freshness during storage.

Sorbitol is stable and chemically unreactive. Sorbitol is resistant to metabolism by oral bacteria that break down sugars and starches to release acids. Sorbitol can with-

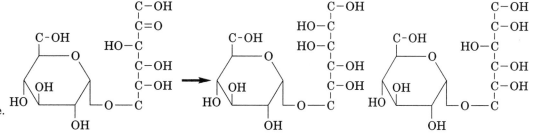

Figure 1. How isomalt is produced. *Source:* Ref. 1.

stand high temperatures and does not participate in browning reactions.

Sorbitol functions well in many food products such as chewing gums, candies, frozen desserts, cookies, cakes, icings, and fillings as well as oral care products including toothpaste and mouthwash.

Xylitol

Xylitol is a five-carbon sugar alcohol prepared by the reduction of xylose. The chemical structures of xylitol and xylose are shown as follows (2):

$$CH_2OH$$
$$HC-OH$$
$$HO-CH$$
$$HC-OH$$
$$CH_2OH$$

α-D-Xylose

Xylitol occurs naturally in many fruits and vegetables and is produced in the liver during metabolism of D-glucose. Produced commercially from plants such as birch and other hardwood trees and fibrous vegetation, xylitol has the same sweetness and bulk as sucrose with reduced calories (about 2.4 cal/g in the United States and European Union). Xylitol quickly dissolves and produces a cooling sensation in the mouth.

Xylitol is a caries fighter. In clinical tests, the consumption of xylitol-containing foods between meals was associated with significantly reduced new caries formation, even when participants were already practicing good oral hygiene. Xylitol inhibits the growth of *S. mutans*, the primary bacterium associated with dental caries. Use of xylitol-containing foods was correlated with a significant decrease in plaque accumulation. In addition, the sweetness and cooling effect of xylitol-sweetened products such as mints or chewing gum creates an increase in salivary flow. The combination of demonstrated increased salivary flow, retardation of plaque growth, and caries prevention qualifies xylitol as a good-for-your-teeth food ingredient.

Mannitol

Mannitol is synthesized from hydrogenation of invert sugar, monosaccharides, and sucrose. It is used in the food industry as an anticaking and free-flow agent, flavoring agent, lubricant and release agent, stabilizer, and thickener. Its chemical structure is shown as follows (2):

$$CH_2OH$$
$$HO-CH$$
$$HO-CH$$
$$HC-OH$$
$$HC-OH$$
$$CH_2OH$$

The caloric density of mannitol is 1.6 cal per gram in the United States.

Erythritol

Erythritol is a four-carbon sugar alcohol with the following chemical structure (2):

$$CH_2OH$$
$$HC-OH$$
$$HC-OH$$
$$CH_2OH$$

Erythritol has the lowest caloric density of the polyols— 0.2 cal per gram in the United States.

Polyols as a group provide the bulk of sugar, without as many calories as sugar. Since the compounds as a class are poorly absorbed, substantial percentages my enter the large intestine and undergo metabolism by the colonic bacteria. This can cause adverse gastrointestinal events. Gastrointestinal events are usually mild and temporary. These events are controlled by either reducing the polyol intake, and/or by gradually increasing the intake to achieve the desired level over time. This provides an opportunity for the colonic bacteria to adapt to the presence of the polyol in the large intestine.

Compared with sucrose, polyols as a class are lower in calories, have little effect on blood glucose levels, require moderate to no insulin for metabolism, are not cariogenic, replace the bulk of sugar on a one-to-one basis, and have valuable food processing properties. For sugar-reduced and -free food products, the polyols are central to food formulation.

Polydextrose

Polydextrose is a randomly bonded polymer of dextrose containing minor amounts of bound sorbitol and citric acid. It was designed to replace the body and texture lost when sugar or carbohydrates were removed from a traditional food or beverage. Important in its commercialization and utilization is its reduced calorie content, deemed as 1 kcal/g by the U.S. Food and Drug Administration.

Polydextrose is made from food-grade starting materials by vacuum polycondensation. Dextrose is thermally polymerized using sorbitol as a plasticizer and citric acid as a catalyst. The resulting product is a slightly acidic, water-soluble polymer. Through further processing, vari-

ous grades are produced with bland, desirable features. Polydextrose is not sweet.

Polydextrose is used in confectionery to achieve calorie and sugar reduction in hard and chewy candy; to replace sucrose in chocolate; to replace the bulk, creaminess, smoothness, and mouth-feel of sugar and fat in frozen desserts; to provide viscosity without gumminess in cultured dairy products; to replace sugars and some fat in baked applications; and to replace sugars and build solids in fruit spreads and filling applications.

Food ingredient manufacturers have responded to the desire for reduced calorie/fat nutrition claims with a diverse array of products that help food processors reduce the level of fat in formulated foods. These include lipid or lipidlike materials including salatrim and olestra, microparticulated proteins (Simplesse), and carbohydrate-based ingredients. The carbohydrate-based food ingredients are reviewed here.

Fiber

Fiber sources from cereal products—oat, wheat, corn and combinations of the three—are available. Through a series of purification steps, manufacturers can provide products that are light in color, essentially bland in flavor, and neutral in odor. Typically these products have the ability to absorb water, from 2 to more than 6 g of water per gram of fiber. The hydrated fiber products are frequently used in place of fats and oils on a gram for gram basis.

Similar products are available based on starch technology. Combinations of starch and water are able to replace fats and oils in some low-heat applications, ice cream, and processed meats. Limitations in this technology are the inability to carry flavors, and the inability to cook and fry with these reduced-fat products.

Hydrocolloids

Hydrocolloids have also found favor in replacing fat and oils in formulated food products. M. Glicksman classified hydrocolloids into a number of major classes including (1) exudates (gum Arabic), (2) extracts from seed weed (carrageenans) and land plants (pectin) and animals (gelatin), (3) flours from seeds (guar) and cereals (starches), (4) biosynthesis or fermentation (xanthan gum), and (5) cellulose and chemically modified cellulose derivatives.

Food hydrocolloids provide a wide array of technological benefits in formulated food products. In calorie-reduced foods, reduced in fats and/or sugars, these products restore the eating qualities associated with full-fat and full-sugar formulations. In an era of full nutritional disclosure, the bulking agents can facilitate the reduction or removal of fats and oils while achieving products that are acceptable to calorie- and fat-conscious consumers.

Hydrocolloids can perform many and varied functions, including as adhesives, binding agents, bodying agents, crystallization inhibitors, clarifying agents, clouding agents, fibers, emulsifiers, encapsulating agents, film formers, flocculating agents, foam stabilizers, gelling agents, moldings, flavor emulsifiers, suspending agents, swelling agents, syneresis inhibitors, thickening agents, whipping agents, and more.

Exudates. Exudates from various plants are among the oldest gums available to humans. The gums are exuded in teardrop or flake shapes that are normally harvested by hand, brought to a central location for sorting, and ground into powder for final sale. For food applications, it is common to spray-dry the gums to reduce the bacterial content and yield a clean, white powder. The four commercial gums include arabic, ghatti, karaya, and traagacanth.

Gum arabic is exuded from the Acacia trees where it forms in scars or wounds on the tree. Gum arabic is a complex, highly branched, globular molecule with low viscosity, and extremely high solubility in water, up to 55%. The stabilizing and emulsifying properties of gum arabic are highly prized characteristics. The other three gums are much more viscous in aqueous solutions. Gum arabic is used for many important functions, including for flavor fixation, for prevention of sugar crystallization and to keep fatty components distributed in confectionery products, as an emulsifier in flavor emulsion concentrates, for preparation of glazes and toppings in baked applications, and to prepare coatings for vitamin and mineral supplements.

Extracts. Extract examples include carrageenan, pectin, and gelatin. Carrageenan is isolated from certain members of the class Rhodophyceae (red seaweeds). It is a hydrocolloid consisting mainly of the potassium, sodium and magnesium, calcium, and ammonium sulfate esters of galactose, and 3,6-anhydrogalactose copolymers. Carrageenans are soluble in hot water and hot milk. Carrageenan reacts with the casein fraction of milk and is very effective in stabilizing milk-based products. Carrageenans are used in the formulation of milk and milk-based beverages, ice cream products, milk puddings, dessert gels, meat analogs, salad dressings, and more.

Pectin is the designation for a group of valuable polysaccharides extracted from edible plant material and used extensively as gelling agents and stabilizers. Pectic substances are abundant in fruits and vegetables and to a large extent are responsible for firmness and form retention of their tissue. Pectin and pectic substances are heteropolysaccharides mainly consisting of galacturonic acid and galacturonic acid methyl ester residues. Pectin is obtained by aqueous extraction of citrus peels and apple pomance.

Pectin solutions show relatively low viscosity compared with other plant hydrocolloids; hence, pectin has limited use as a thickener. Pectin is primarily used as a gelling agent to impart texture to jams, jellies, and preserves. Other applications include bakery fillings and glazings, yogurt fruit preparations, fruit beverages and sauces, fruit jellies and jelly centers, and dairy products.

Gelatin is extracted from cowhides, pigskins, and animal bones. Gelatin is prepared by either alkaline or acid treatment of collagenous tissue and is a form of hydrolyzed, denatured collagen; it is well known for its ability to form food gels. Collagen has an unusual amino acid distribution, containing repeating triplets of -(Gly-X-Y)-where a large percentage of X and Y are proline or hydroxyproline residues. *In vivo* collagen self-assembles to give a triple helical structure; the structure is stabilized by the formation of chemical cross-links. Above about 40°C, cross-linked collagen unwinds and forms a random, denatured configuration. On cooling, such denatured collagen gels. Gelation is thought to occur through development of junction zones with partial reformation of the collagen triple helix structure.

Gelatin can be used as a thickening agent at low concentrations, and a gelling agent at high concentrations. Gelatin is unique in the spectrum of gelling agents in that it can form aqueous gels with water at any pH and without the need for any other additives. Gelatin gels are thermally reversible and can be melted and reset by heating and cooling.

Flours from Seeds and Cereals. Flours containing gums such as guar and starch are separated by mechanical means from the plant seed or cereal. The guar plant, *Cyamposis tetragonolobus*, is a member of the legume family. The endosperm of guar seed is an important hydrocolloid. The guar molecule is a straight-chain galactomannan with galactose on every other mannose unit. Guar gums form a colloidal dispersion to yield a highly viscous system. Guar gum solutions of 1% concentrations or higher are thixotropic with thixotropy decreasing below 1%. In comparison with other hydrocolloids, guar gum at equivalent solids in water has the highest viscosity.

Guar gum modifies the behavior of water in food systems in a highly efficient manner. It reduces and minimizes friction in food products, thereby aiding processing and palatability of foods. Guar viscosity aids in the control of crystal size in saturated sugar solutions. Guar imparts smoothness to ice cream by promoting small ice crystals. Guar yields a homogeneous finished texture to cottage cheese. The addition of guar to cold-packed cheese eliminates syneresis and results in more uniform texture and flavors.

Biosynthesis or Fermentation. Fermentation or biosynthesis is the route of production of xanthan gum. Microorganisms that produce extracellular polysaccharides are widely distributed in marine and land environments. *Xanthomonas campestris*, a naturally occurring bacterium originally isolated from the rutabaga plant, is grown in fermentation vats to produce xanthan gum, a high-molecular-weight polysaccharide gum. The gum is extracted with isopropyl alcohol, dried, and milled. It contains D-glucose and D-mannose as the dominant hexose units, along with D-glucuronic acid, and is prepared as the sodium, potassium, or calcium salt. The polymer backbone consists of 1,4-linked β-D-glucose and is, therefore, identical to that of cellulose. At the 3-position of alternate glucose monomer units is a trisaccharide side chain containing a glucuronic acid residue between two mannose units.

Xanthan gum is completely soluble in hot or cold water. Low concentrations of xanthan gum exhibit high viscosity. Solutions of xanthan gum at 1% or higher concentration appear almost gel-like at rest, yet these solutions pour readily and have low resistance to mixing and pumping.

Xanthan gum has found application in bakery fillings and icings; in beverages to build body, clouding agent, and to suspend insoluble ingredients; in confectionery in processing starch jelly candies and in xylitol-coated chewing gum; in dairy products where it performs as a stabilizer;

in dairy substitutes as a stabilizer; and in pourable dressings as an emulsion stabilizer.

Powdered Cellulose and Cellulose Derivatives. Powdered cellulose is a polymer composed of carbon, hydrogen, and oxygen. Chemically, it is a chain of glucose units linked in a 4-(β-D-glucosido)-D-glucose, not to be confused with α-1,4-glucan, which is common in starch. The manufacture of powdered cellulose begins with wood pulp that undergoes a series of bleaching steps and drying. Following drying, the pure white, virgin cellulose is cut to the desired fiber length by cutters or by a ball mill. The lengths are varied to achieve specific applications in the food industry. Due to its nature, powdered cellulose is more than 99% fiber and considered to have zero calories.

Powdered cellulose has a number of interesting properties, including (1) a greater affinity for water than fat, (2) hydrogen bonding that restricts displacement of water by fat, (3) no nonenzyme browning, and (4) increased pliability of powdered cellulose–containing foods.

Powdered cellulose has a number of food applications, including (1) increasing the fiber content of formulated foods, (2) reducing the caloric content of the food by displacement of fats, and (3) increasing water retention capacity and viscosity.

Being a white, flavorless, and odorless powder enhances the use of powdered cellulose as a noncaloric bulking agent in food products. Low-level inclusion of powdered cellulose in fried food formulations can result in reduced fat pickup during frying. This is attributed to the hydrophilic nature of powdered cellulose and its molecular structure, which permits the formation of significant amounts of additional hydrogen bonds that require more energy to break during the frying process.

Other uses for powdered cellulose include binding and thickening (with gums and stabilizers), anticaking, antisticking extrusion aid, enhancing the volume of baked goods, as a texturizing agent, and more.

There are a variety of chemically modified cellulose products, including carboxymethylcellulose, methylcellulose, hydroxypropylcellulose, and hydroxypropylmethylcellulose with food applications.

There are a variety of reasons a food scientist may wish to reduce the content of simple sugars and or fat in food products. Among the reasons are to reduce calories, reduce simple sugars, reduce fat, attain nutrition labeling advantages, attract consumers who seek diet or dietetic foods, and more. In the case of sugar-reduced foods, these are normally formulated with a high-intensity sweetener and certain bulking agents that fill the space occupied by sugar(s).

BIBLIOGRAPHY

1. K. McNutt and Sentko, "Sugar Replacers, A Growing Group of Sweeteners in the United States," *Nutrition Today* **31**, 255–261 (1996).

2. S. Budavari, ed., *The Merck Index*, 11th ed., Merck & Co., Inc., Rahway, N.J., 1989, pp. 1375, 1591, 3614, 5633, 9996.

ROBERT HARKINS
North Brunswick, New Jersey

SUGAR: SUCROSE

Sugar (sucrose) is a common food ingredient, often used to satisfy the innate human desire for sweetness (see the article SWEETENERS: NUTRITIVE) (1–3). In most of its forms, sugar is a very pure food ingredient. It is also a versatile ingredient and, as a digestible carbohydrate, a macronutrient.

STRUCTURE

The chemical name of sucrose is α-D-glucopyranosyl β-D-fructofuranoside or β-D-fructofuranosyl α-D-glucopyranoside (4). In a sucrose molecule, the two monosaccharide units are linked reducing end to reducing end, that is, anomeric carbon atom to anomeric carbon atom, by a glycosidic linkage. Because both the potential aldehydo group of the D-glucosyl unit and the potential keto group of the D-fructosyl unit are covalently bound in a mutual glycosidic bond, sucrose has no reducing end, as do essentially all other oligo- and polysaccharides (with few exceptions), so it is classified as a nonreducing carbohydrate.

SOURCES

Sucrose is produced by all green plants, but only sugar cane and sugar beets are used to provide commercial quantities. (Maple syrup, fruits, honey, etc provide a relatively insignificant amount of additional sucrose to the human diet.)

Sugar cane (12–18 months old) is brought to the mill and crushed as soon as possible to reduce exposure to microorganisms, which hydrolyze sugar by means of the enzyme invertase as a first step in using it as a carbon and energy source. Some organisms also convert a portion of the sugar to dextran, a soluble polysaccharide that thickens the sugar solution and causes clogging of filters and other mechanical problems.

To obtain sugar from the cane, it is crushed between rolls (5). A spray of water helps remove the juice. The juice, which is approximately 16% sucrose, is made slightly alkaline by addition of lime (calcium hydroxide) to prevent hydrolysis of the acid-labile glycosidic linkage, and the mixture is heated to coagulate proteins. A heavy scum or cake containing a variety of impurities results. The mixture is filtered and concentrated under reduced pressure at a carefully controlled temperature to approximately 50% solids. When crystals of about 300-μm diameter develop, they are removed by centrifugation and washed. The mother liquor is further concentrated to obtain another crop of crystals (6). Concentration and crystallization is continued until impurities build up to the point where the remaining sucrose will not crystallize. Usually this occurs after two or three crops of crystals are obtained. The final mother liquor, termed black strap molasses, is dark, black, heavy, bitter-flavored, and high in ash.

The light brown product, raw sugar, is shipped, usually across an ocean, to a refining mill for purification. First, the crystals are washed to remove the film of molasses from the surfaces of the crystals. The washed raw sugar is dissolved, and the solution is treated with lime and carbon dioxide or lime and phosphates and filtered to produce a clear, golden solution, which is decolorized by passing it through a column of activated carbon. The colorless solution is then evaporated to supersaturation. By controlling the rate of evaporation, crystal size can be controlled. Crystals are collected, washed, and dried.

Sugar beets contain approximately 76% water, 16% sucrose, 5.5% insoluble pulp, and 2.5% other solubles (7). Sugar is obtained by countercurrent extraction of V-shaped slices of beets called cosettes. Heating weakens cell walls and enhances diffusion of sugar from cells. Approximately 98% of the sugar is extracted in this operation. The extract contains approximately 12% sucrose by weight and 2% soluble impurities. Among these impurities are a trisaccharide, raffinose, which has a single D-galactopyranosyl unit attached to sucrose, and a tetrasaccharide, stachyose, which contains an addition D-galactosyl unit attached to the first galactosyl unit. Lime and carbon dioxide are added for clarification of the extract and removal of some soluble impurities, just as they are in the cane sugar purification process. Following centrifugation and filtration to remove precipitated impurities, the clarified extract (approximately 13% dissolved solids) is treated again with carbon dioxide to remove residual calcium ions. Sulfur dioxide is added to minimize color formation during subsequent processing steps. Then, the solution is evaporated to 60 to 65% concentration for crystallization as in the cane sugar process, and as in the cane sugar process, several crops of crystals are obtained by successive repetitions of the concentration and crystallization steps. About 83% of the sugar present in fresh sugar beets is obtained as pure, white sugar.

FORMS

Refined sugar from either source is pure crystalline sucrose. Although refined sugar is available in a variety of forms, each is at least 99.8% sucrose, with some forms being 99.96% sucrose. Crystalline refined sugar is produced in a variety of crystal size distributions (granulations) (8). There is no standard definition of granulation grades, so grades with the same designation from different manufacturers may not have the same mesh-size distribution. From the largest to the smallest crystal sizes, the most common grades are as follows. Coarse granulated sugar, generally the highest purity crystalline product with the lowest color, is used in fondants and other formulations where a colorless product is desired. Sanding granulated sugar, with a crystal size that normally ranges between U.S. 20 and U.S. 40 mesh, is used mainly as sprinkle on

baked goods and as sanding on starch gum confectioneries. Extra Fine (X-Fine) or Fine sugar normally has a crystal size that ranges between U.S. 20 and U.S. 100 mesh. It is the largest volume industrial granulated sugar because it is best for bulk-handling without being susceptible to caking. Fruit granulated sugar normally has a crystal size that ranges between U.S. 40 and U.S. 100 mesh, making it slightly finer than Extra Fine. It is used mainly in dry mixes such as gelatin dessert, pudding, and drink mixes. Bakers Special sugar, with a size distribution normally between U.S. 50 and U.S. 140 mesh, is used for sugaring of cookies and in doughs. Because it dissolves more rapidly than the other granulated forms, it is often used to sweeten bar drinks. Powdered confectioners' sugar is produced by grinding or milling granulated sugar with starch, which is added to prevent caking. It is used for dusting and to form wet fondants. Other grades and granulated forms with other names are also marketed.

Some food or beverage processors, rather than dissolve crystalline sucrose, prefer to purchase sugar in solution. Liquid sucrose is at least 99.5% sucrose at a concentration of at least 67%. It can be used whenever a dissolved granulated sugar product might be used. Amber liquid sugar, which is darker in color and higher in ash than is liquid sucrose, is used where color and minor inorganic impurities can be tolerated. In liquid invert sugar, about half of the sugar has been inverted, that is, converted into equimolar amounts of D-glucose and D-fructose (see the article SWEETENERS: NUTRITIVE), making liquid invert sugar an approximately equimolar mixture of sucrose, D-glucose, and D-fructose. It is sweeter than liquid sucrose and will not crystallize.

Brown sugars are used to impart molasses flavor to cookies, candies, and other products. Brown sugar is composed of very fine sugar crystals in a thin film of syrup. Brown sugar is produced in two different ways. Soft brown sugars (light/golden and dark) are crystallized directly from dark syrups of cane sugar selected for color and flavor during the refining process. In an alternative method, cane or beet sugar crystals are coated with a proper cane syrup or molasses.

Less-refined sugars, available under a variety of names, are essentially first, second, or third crops of crystals. Dried cane juice is prepared by vacuum evaporation of filtered cane juice to a dry solid that is ground. Sugarcane syrups are prepared by evaporation of unfiltered sugarcane juice.

Molasses is both the syrup (concentrated juice) that remains after cane sugar has been recovered by crystallization and the product of raw cane sugar manufacture. It is 40 to 60% sucrose, D-glucose, and D-fructose. It is available in several grades based on color and flavors and can be used to flavor food products. It may be unsulfured or lightly sulfured. Most molasses is used in animal feed supplements.

SWEETNESS

The relative sweetness of nutritive sweeteners is a function of concentration, pH, and temperature (see the article

SWEETENERS: NUTRITIVE). Nevertheless, relative sweetnesses of the common nutritive sweeteners can be roughly determined (Table 1). The clean, pure sweet taste of sucrose, without aftertaste or unpleasant secondary reactions, is the standard by which all other sweeteners are judged (9,10). Sucrose also balances other flavor components (11).

OTHER PROPERTIES

Crystalline sucrose, in addition to being sweet tasting, is colorless and odorless (12,13). Dry crystalline sucrose is stable up to its melting point. At temperatures of approximately 160 to 186°C (320–367°F), sucrose crystals melt and the molecules decompose with formation of a family of dark-brown carmelization products.

The glycosidic bond joining the two sugar units is essentially a high-energy bond and unstable; its cleavage is the first step in the thermal decompositions mentioned in the preceding paragraph. The bond is also very susceptible to acid-catalyzed hydrolysis into an equimolar mixture of D-glucose and D-fructose. This process is called inversion and the product invert or invert sugar. The products of hydrolysis, both of which are reducing sugars, will react with amino acids, peptides, proteins, and ammonium ions in typical Maillard reactions, producing other caramels and resulting in the brown colors, aromas, and flavors associated with cooking and baking.

In addition to acid-catalyzed hydrolysis, the extent of which is a function of time, temperature, and pH, sucrose undergoes enzyme-catalyzed hydrolysis to the same products. Yeast and bacterial hydrolytic enzymes are called invertases; the enzyme of the human digestive tract, which operates by a slightly different mechanism, is called a sucrase.

Its high degree of water solubility is a key property of sucrose (14). As a result, several methods have been developed for determining its concentration in solution, primary among which are measurement of specific gravity and refractive index. Equations and tables have been developed to determine its saturation concentration at various temperatures; and scales have been developed to report concentrations, the primary one being the Brix scale, in which the percentage by weight of sucrose in a water solution is expressed as degrees Brix.

Table 1. Approximate Relative Sweetnesses of Common Nutritive Sweeteners

Sweetener	Approximate relative sweetness
D-Fructose (crystalline, initial)	140
55% High-fructose corn syrup	120
Invert sugar syrups	105
Xylitol	105
Sucrose	100
D-Glucose	75
Sorbitol (D-glucitol)	50
Maltose	40
D-Mannitol	40
Lactose	25

Also important to sugar producers and food processors are the specific heat and heat of crystallization of sucrose, and again equations and tables have been developed to determine each as functions of temperature and concentration. Other tables relate sucrose concentration to the freezing point (depressed) and boiling point (elevated) of its solutions. Surface tension increases and the vapor pressure of water and water activity decreases with increasing sugar concentration (12,13).

USES

Sucrose is a macronutrient food ingredient that provides a variety of functionalities in processed food products. It always provides sweetness and may be used to provide bulk, body, and texture. It may also function as a moisturizing agent, dispersing agent, stabilizer, fermentation substrate, flavor carrier, browning agent, decorative agent, and preservative. Important characteristics of sucrose related to its use in processed foods are (1) its solubility in water; (2) its ease of crystallization upon evaporation of water; (3) its function as a precursor to caramel and Maillard reaction colors, flavors, and aromas; (4) its stability and non-hygroscopic nature, which allows it to be used as a diluent, dispersant, and bulking agent in dry mixes; (5) its being a readily fermentable substrate for yeasts and other specific microorganisms; and (6) its ability to decrease water activity appreciably and thereby to provide protection from chemical and/or microbiological deterioration and extend shelf life (15).

The largest amount of sucrose is used in bakery and breakfast cereal products. In bakery products, in addition to providing sweetness and being a source of fermentable carbohydrate, sucrose may provide one or more of the following functionalities: crumb tenderization, moisture retention, texture improvement, formation of crust color, shelf-life extension, and an aid to whipping and creaming (16). It is used in both crystalline and liquid forms.

In yeast-leavened products, the amount of sugar used can range from approximately 1% to approximately 11% (based on flour) in breads and can be as much as 20% in Danish sweet products. In these products, sucrose is used as a source of fermentable carbohydrate and to enhance flavor, improve crust color and toasting properties, improve crumb texture and moistness, and extend shelf life. The principal contributors to bread flavor are fermentation and crust browning, both of which involve sugar. Browning is the result of both carmelization (thermal decomposition) and the Maillard reaction.

Two processes are involved in producing crumb texture: gluten hydration and starch gelatinization. With increasing amounts of sugar in bread doughs, less and less water is available for gluten hydration, the rate of gluten development is reduced, and the tenderness of the bread crumb increases. Sucrose also inhibits starch gelatinization in a concentration-dependent manner. A delay of starch gelatinization is important in cake making, but its role in bread making is unclear. In some cake formulations, sugar acts as a creaming and whipping aid. It also modifies crumb texture, aids in crust color formation, and retains moisture. In cookie manufacture, sucrose increases the temperature at which starch gelatinizes, prevents hydration of wheat flour proteins, and retards gluten formation. Sucrose also serves as a creaming aid in mixing cookie dough, provides flavor and browning, increases cookie volume and spread of the dough, imparts crispness through crystallization, and produces the characteristic surface cracking pattern of some cookies. Substitution with, or addition of a sufficient amount of, a corn syrup to prevent sucrose recrystallization will produce a cookie with a soft (rather than a crisp) texture and prevent surface cracking. In icings, frostings, and glazes, sucrose provides sweetness, browning, stability, and a pleasing appearance and balances flavor.

Sucrose is blended with other ingredients and added to grain during cooking, prior to flaking, or in the dough stage of ready-to-eat (RTE) breakfast cereals to contribute binding, flavor, browning, and texture (18). It is also a potentiator of other flavors. However, by far the major use of sucrose in RTE cereals is for frosting or sugar coating.

Confectionery production consumes the second greatest amount of sucrose (17). The production of hard candies depends on the solubility of sucrose in water and the fact that, as water is boiled off a sugar solution, the boiling point increases. Therefore, the boiling temperature can be used as a determinant of sugar concentration; for example, when the boiling point of a sugar solution is 93°C (200°F), the concentration of sucrose is approximately 82% by weight. In hard candies, sucrose is combined with glucose, a corn syrup, and/or a maltodextrin to achieve the proper consistency. Soft candies consist of very small crystals of sucrose embedded in a matrix of the other ingredients. Very small crystals of sucrose are also a major ingredient in chocolate, which requires a dry sweetener with low hygroscopicity. Cocrystallization of citric acid with sucrose provides a nonhygroscopic sanding material.

Sucrose is also used in the preparation of other food products. Gelation of pectin in nondietetic jams, jellies, preserves, and marmalades requires a pH in the range 3.0 to 3.4 and soluble solids of approximately 65%, which is most often met with sucrose, making it the principal ingredient (19). Because of its high concentration, it also functions as preservative. Sucrose is used in very few carbonated beverages in the United States, but it is the sweetener used in most "non-diet" carbonated beverages in the rest of the world, dry beverage mixes, and liqueurs (20). In ice cream and other frozen dairy products, sucrose provides a clean, sweet taste and contributes to the creamy texture and body (21). Sucrose is also used in canned fruit, condiments (salad dressings, tomato catsup, mayonnaise, barbecue and steak sauces, etc), cured meats and sausage products, and many other products such as puddings, pie fillings, and granola bars (22).

PHYSIOLOGICAL EFFECTS

While several specific cause-and-effect relationships between consumption of sucrose and pathological conditions have been suggested, they have not been substantiated by research (23). Only the link between sucrose consumption and the incidence of dental caries has been documented.

BIBLIOGRAPHY

1. M. A. Godshall, "Use of Sucrose as a Sweetener in Foods," *Cereal Foods World* **35**, 384–389 (1990).
2. N. L. Pennington and C. W. Baker, eds., *Sugar, a User's Guide to Sucrose*, Van Nostrand Reinhold, New York, 1990.
3. M. Mathlouthi and P. Reiser, eds., *Sucrose: Properties and Applications*, Blackie Academic & Professional, London, 1995.
4. A. B. Rizzuto, "Cane Sugar Refining," in N. L. Pennington and C. W. Baker, eds., *Sugar, a User's Guide to Sucrose*, Van Nostrand Reinhold, New York, 1990, pp. 22–25.
5. S. Perez, "The Structure of Sucrose in the Crystal and in Solution," in M. Mathlouthi and P. Reiser, eds., *Sucrose: Properties and Applications*, Blackie Academic & Professional, London, 1995, pp. 11–32.
6. G. Vaccari and G. Mantovani, "Sucrose Crystallisation," in M. Mathlouthi and P. Reiser, eds., *Sucrose: Properties and Applications*, Blackie Academic & Professional, London, 1995, pp. 33–74.
7. S. E. Bichsel, "Beet Sugar Production," in N. L. Pennington and C. W. Baker, eds., *Sugar, A User's Guide to Sucrose*, Van Nostrand Reinhold, New York, 1990, pp. 26–35.
8. J. F. Dowling, "Sugar Products," in N. L. Pennington and C. W. Baker, eds., *Sugar, A User's Guide to Sucrose*, Van Nostrand Reinhold, New York, 1990, pp. 36–45.
9. R. L. Knecht, "The Flavor of Sucrose in Foods," in N. L. Pennington and C. W. Baker, eds., *Sugar, A User's Guide to Sucrose*, Van Nostrand Reinhold, New York, 1990, pp. 66–70.
10. I. Ramirez, "The Sweetness of Sugar," in N. L. Pennington and C. W. Baker, eds., *Sugar, A User's Guide to Sucrose*, Van Nostrand Reinhold, New York, 1990, pp. 71–81.
11. M. A. Godshall, "Role of Sucrose in Retention of Aroma and Enhancing Flavor of Foods," in M. Mathlouthi and P. Reiser, eds., *Sucrose: Properties and Applications*, Blackie Academic & Professional, London, 1995, pp. 248–263.
12. R. L. Knecht, "Properties of Sugar," in N. L. Pennington and C. W. Baker, eds., *Sugar, A User's Guide to Sucrose*, Van Nostrand Reinhold, New York, 1990, pp. 46–65.
13. P. Reiser, G. G. Birch, and M. Mathlouthi, "Physical Properties," in M. Mathlouthi and P. Reiser, eds., *Sucrose: Properties and Applications*, Blackie Academic & Professional, London, 1995, pp. 186–222.
14. Z. Bubnik and P. Kadlec, "Sucrose Solubility," in M. Mathlouthi and P. Reiser, eds., *Sucrose: Properties and Applications*, Blackie Academic & Professional, London, 1995, pp. 101–125.
15. A. L. Raoult-Wack, G. Rios, and S. Guilbert, "Sucrose and Osmotic Dehydration," in M. Mathlouthi and P. Reiser, eds., *Sucrose: Properties and Applications*, Blackie Academic & Professional, London, 1995, pp. 279–290.
16. J. G. Ponte, Jr., "Sugar in Bakery Foods," in N. L. Pennington and C. W. Baker, eds., *Sugar, A User's Guide to Sucrose*, Van Nostrand Reinhold, New York, 1990, pp. 130–151.
17. C. M. Barnett, "Sugar in Confectionery," in N. L. Pennington and C. W. Baker, eds., *Sugar, A User's Guide to Sucrose*, Van Nostrand Reinhold, New York, 1990, pp. 103–129.
18. C. E. Walker, "Sugar in Ready-to-Eat Breakfast Cereals," in N. L. Pennington and C. W. Baker, eds., *Sugar, A User's Guide to Sucrose*, Van Nostrand Reinhold, New York, 1990, pp. 182–197.
19. E. E. Meschter, "Sugar in Preserves and Jellies," in N. L. Pennington and C. W. Baker, eds., *Sugar, A User's Guide to Sucrose*, Van Nostrand Reinhold, New York, 1990, pp. 212–227.

20. G. J. Marov and J. F. Dowling, "Sugar in Beverages," in N. L. Pennington and C. W. Baker, eds., *Sugar, A User's Guide to Sucrose*, Van Nostrand Reinhold, New York, 1990, pp. 198–211.

21. D. E. Smith, "Sugar in Dairy Products," in N. L. Pennington and C. W. Baker, eds., *Sugar, A User's Guide to Sucrose*, Van Nostrand Reinhold, New York, 1990, pp. 152–164.

22. N. D. Pintauro, "Sugar in Processed Foods," in N. L. Pennington and C. W. Baker, eds., *Sugar, A User's Guide to Sucrose*, Van Nostrand Reinhold, New York, 1990, pp. 165–181.

23. T. Cayle, "Sugar in the Body," in N. L. Pennington and C. W. Baker, eds., *Sugar, A User's Guide to Sucrose*, Van Nostrand Reinhold, New York, 1990, pp. 82–102.

J. N. BeMiller
Purdue University
West Lafayette, Indiana

See also CARBOHYDRATES: CLASSIFICATION, CHEMISTRY, LABELING; CARBOHYDRATES: FUNCTIONALITY AND PHYSIOLOGICAL SIGNIFICANCE.

SULFITES AND FOOD

Sulfites, or alternatively, sulfiting agents, refers to sulfur dioxide (SO_2) and some inorganic sulfites that will release SO_2 under conditions of use. Sulfites have been widely used in foods for various purposes, including control of enzymatic and/or nonenzymatic browning; inhibition of microbial growth; and as antioxidants and reducing and bleaching agents. Typical applications of sulfites in foods may be seen in such cases as preservation of fresh fruits and vegetables, prevention of discoloration of dehydrated fruits and vegetables, inhibition of undesirable microorganisms in wine brewing, and bleaching of food starches.

Since 1959 sulfites have been classified by the Food and Drug Administration of the United States (USFDA) as GRAS (generally recognized as safe) substances when they are used in foods according to the Code of Federal Regulations (CFR). However, the numerous reports of sulfite-induced asthma in some cases of sulfite sensitivity had prompted USFDA to commission an expert panel to reexamine the GRAS status of sulfites in 1984. Consequently, on August 8, 1986 USFDA banned the use of sulfites in fruits and vegetables intended to be served or sold raw to consumers and on January 9, 1987 announced that any foods containing a detectable level of residual sulfites (ie ≥10 ppm SO_2 based on the currently official method), either directly from addition during processing or indirectly from other ingredients, must be labeled with sulfite content disclosed.

According to CFR, six sulfites are currently listed as GRAS that may be used in foods, except for meats and other foods recognized as a main source of thiamine and for fruits and vegetables intended to be served and sold raw to consumers: sulfur dioxide (SO_2), sodium bisulfite ($NaHSO_3$), sodium metabisulfite ($Na_2S_2O_5$), potassium metabisulfite ($K_2S_2O_5$), sodium sulfite (Na_2SO_3), and potassium bisulfite ($KHSO_3$). Potassium sulfite (K_2SO_3) and sulfurous acid (H_2SO_3), which are not listed as GRAS, are specifically allowed for use only in caramel processing.

The species of sulfites approved for food use may vary with country. For instance, the officially approved sulfites in Taiwan ROC (Republic of China) for use in foods are potassium sulfite (K_2SO_3), sodium sulfite (Na_2SO_3), sodium bisulfite ($NaHSO_3$), sodium hydrosulfite ($Na_2S_2O_4$), potassium metabisulfite ($K_2S_2O_5$), sodium metabisulfite ($Na_2S_2O_5$), and anhydrous sodium sulfite (1).

CHEMICAL REACTIONS OF SULFITES IN FOOD SYSTEM

Basic Chemistry of Sulfites

In aqueous solution, sulfites may exist in various forms as shown in the following formulas depending on the pH value of the solution (2,3):

$$H_2O + SO_2 \rightleftharpoons H_2SO_3 \quad (SO_2 \text{ dissolves readily in } H_2O)$$

$$H_2SO_3 \rightleftharpoons H^+ + HSO_3^-, \quad pK_1 \simeq 2$$

$$HSO_3^- \rightleftharpoons H^+ + SO_3^{2-}, \quad pK_2 \simeq 7$$

From the pK values of sulfurous acid (H_2SO_3) and the pH value of the solution, one can calculate the approximate percentage of various species of sulfites (see Table 1).

Since most food systems have a pH of 3–7, the predominant form of sulfites is HSO_3^-. Sulfurous acid and SO_2 gas may exist only in high-acid foods with pH values less than 4.

Reactions of Sulfites with Food Components

Sulfites react readily with many food components such as aldehydes, ketones, reducing sugars, proteins, and amino acids and form various combined organic sulfites (4). Some of the reactions are desirable and some are undesirable.

Reactions between sulfites and carbonyls such as aldehydes and ketones will primarily form hydroxysulfonates which are very stable at pH 1–8 (5). This reaction enables sulfites to prevent the formation of enzymatic and nonenzymatic browning by binding the carbonyl intermediates of browning reactions (6–8). In general, all carbonyl hydroxysulfonates have good stability, with α,β-unsaturated carbonyl hydroxysulfonates as the most stable (7).

Reducing sugars do not readily react with sulfites as carbonyls. To form sugar hydroxysulfonates, a considerable excess in molar equivalent of reducing sugar is required (4). Sugar hydroxysulfonates are much less stable than carbonyl hydroxysulfonates (2,3). At lower pH, sugar hydroxysulfonates have better stability (2).

Table 1. Percentage of Four Sulfite Species

pH	Species of sulfites			
	SO_2	H_2SO_3	HSO_3^-	SO_3^{2-}
2	37	31.5	31.5	
3	6	8.5	85	
4	0.5	1	98	
5			99	1
6			91	9
7			50	50

Sulfites can break down the disulfide bonds of proteins and amino acids and form thiol (R-SH) and S-sulfonate (R-SSO_3^-) (9). Tertiary amines and Schiff's bases produced from browning reactions also react with sulfites and form amine bisulfites (3). Methionine will be oxidized and converted to methionine sulfoxide by sulfites through free-radical reaction (10).

In addition to these reactions, sulfites react with many vitamins, including B_1, B_{12}, C, and K, resulting in loss of vitamin activity (4).

APPLICATION OF SULFITES IN THE FOOD INDUSTRY

By making use of the reactions with various food components, sulfites can be applied to the food industry for one or more of the following purposes.

Inhibition and Control of Microorganisms

Sulfites are often used in wine brewing to prevent undesirable bacterial fermentation. Table grapes and fruits intended for jam and juice are also often preserved by sulfites to prevent decay (11,12).

The inhibitory action of sulfites is more effective to acetic acid bacteria, lactic acid bacteria, and various molds and is less effective in yeasts (3). The antimicrobial activity of sulfites is generally in the following order. Gram-negative bacteria > Gram-positive bacteria > molds > yeasts (13). The mechanisms of the antimicrobial action of the sulfites is not well understood; however, it is believed that undissociated sulfurous acid is the active form possessing antimicrobial activity (14). With this regard, the pH of the food system should be low enough (≤ 4) to have a significantly inhibitory effect. The combined sulfites do not have antimicrobial action.

Inhibition of Enzymatic and/or Nonenzymatic Browning

Sulfites can effectively prevent or minimize nonenzymatic browning by forming stable hydroxysulfonates with carbonyls and reducing sugars as mentioned in the previous section. In this respect, sulfites are widely used in wine making and in dehydrated fruits and vegetables to prevent the discoloration of the finished products (3,7,8). Sulfites can also inhibit some oxidative enzymes such as polyphenoloxidase, ascorbate oxidase, lipogenase, and peroxidase (6) and therefore can retard the enzymatic browning resulting from polyphenoloxidase. In addition, sulfites can form stable hydroxysulfonates with browning intermediates such as quinones, and prevent further reaction to form browning pigment (6). Thus, sulfites are commonly used in fresh vegetables for salad bars (this application has been banned by USFDA), peeled and sliced potatoes, apple dice and other fruits used in bakery products, fresh mushroom for processing, and table grapes and fresh shrimp, where enzymatic browning presents a serious problem.

Use as Antioxidants and Reducing Agents

As an antioxidant, sulfites can be used in citrus juices (12), beer (13), and peas (6) to prevent the oxidative off-flavor. As an reducing agent, sulfites are widely employed as dough conditioners in the baking industry to break down the disulfide bonds of the gluten fraction and facilitate the processing of biscuits, crackers, and cookies and obtain the desired texture properties of the finished products (15).

TREATMENT AND RESIDUAL LEVELS OF SULFITES IN FOODS

It is difficult to establish criteria for the treatment and residual levels of sulfites used in foods because of the complexity of foods, the wide variations in individual daily intake of sulfite-treated foods, and the dependence of the residual level of sulfites on food composition, processing methods, and storage conditions (4). In spite of this, it is believed that the residual amount of sulfites at the point of consumption is very limited provided foods are treated under normal conditions (2,3).

Currently the treatment and residual level of sulfites in most foods is not strictly limited in the United States. The residual levels or treatment levels are regulated only for those food products that are commonly consumed. These include glucose syrup, dextrose monohydrate, and grape wine, and the maximum residual levels are 40, 20, and 350 ppm (SO_2), respectively. For food starch bleaching the treatment of SO_2 has been established at $\leq 0.05\%$ (4). Regulations in this regard may vary with countries. For instance, the official regulation on maximum residual levels of sulfites in Taiwan, ROC for various foods is as follows (all calculated as SO_2); dehydrated fruits and vegetables and gelatin, 500 ppm; candies, 300 ppm; shrimps and shellfish, 100 ppm; fruit wines, 25 ppm and other processed foods, 30 ppm; and in beverages other than fruit juices, wheat flour, and flour products, the use of sulfites is prohibited (1).

After treatment, sulfites may react with various food components forming combine sulfites, may be oxidized into sulfates (SO_4^{2-}), liberated as SO_2 gas, or may still exist in free inorganic sulfites depending on food composition, pH, processing conditions, and storage conditions. Combined sulfites predominate in most food systems.

BIBLIOGRAPHY

1. Department of Health of Taiwan ROC, *Regulations on Scope and Application Level of Food Additives* (in Chinese), Department of Health of ROC, 1993, pp. 18-4-1–18-4-6.

2. L. F. Green, "Sulfur Dioxide and Food Preservation (a Review)," *Food Chemistry* **1**, 103–104 (1976).

3. M. A. Joslyn and J. B. S. Braverman, "The Chemistry and Technology of the Pretreatment and Preservation of Fruits and Vegetable Products with Sulfur Dioxide and Sulfites," *Advances in Food Research* **5**, 97–160 (1954).

4. S. L. Taylor, N. A. Higley, and R. K. Bush, "Sulfites in Foods," *Advances in Food Research* **30**, 2–64 (1986).

5. L. F. Burroughs and A. H. Sparks, "Sulfite-Binding Power of Wines and Ciders," *Journal of the Science of Food and Agriculture* **24**, 187–198 (1973).

6. D. R. Haisman, "The Effect of Sulfur Dioxide on Oxidizing Enzyme Systems in Plant Tissues," *Journal of the Science of Food and Agriculture* **25**, 803–810 (1974).

7. D. J. McWeeny, M. E. Knowles, and J. F. Hearne, "The Chemistry of Non-enzymatic Browning in Foods and Its Control by Sulfites," *Journal of the Science of Food and Agriculture* **25**, 735–746 (1974).

8. B. L. Wedzicha, O. Lamikanva, J. C. Herrera, and S. Panahi, "Recent Developments in the Understanding of the Chemistry of Sulphur(IV) Oxospecies in Dehydrated Vegetables," *Food Chemistry* **15**, 141–155 (1984).

9. G. W. Means and R. E. Feeney, *Chemical Modification of Proteins*, Holden-Day, San Francisco, 1971, p. 152.

10. S. F. Yang, "Sulfoxide Formation from Methionine or Its Sulfide Analogs during Aerobic Oxidation of Sulfite," *Biochemistry* **9**, 5008–5014 (1970).

11. K. E. Nelson and M. Ahmedullah, "Packaging and Decay Control Systems for Storage and Transit of Table Grapes for Export," *American Journal of Enology and Viticulture* **27**, 74–79 (1976).

12. A. C. Roberts and D. McWeeny, "The Uses of Sulfur Dioxide in the Food Industry (a Review)," *Journal of Food Technology* **7**, 221–238 (1972).

13. R. S. Applebaum, "Sulfites and Food (a Review)," *The Manufacturing Confectioner*, 45–55 (November 1985).

14. J. G. Carr et al. *Journal of Applied Bacteriology* **40**, 201–212 (1976).

15. P. Wade, "Action of Sodium Metabisulfite on the Properties of Hard Sweet Biscuit Dough," *Journal of the Science of Food and Agriculture* **23**, 333–336 (1972).

PING YANG CHANG
Food Industry Research and Development Institute
Hsinchu, Taiwan

See also FOOD ADDITIVES.

SUPERCRITICAL FLUID TECHNOLOGY

Supercritical fluid (SCF) technology has many potential applications in extraction and fractionation processes as an alternative to conventional separation processes. The chemical, pharmaceutical, polymer, and food industries are actively engaged in developing and evaluating this technology.

The ability of a SCF to dissolve low-vapor-pressure materials was first reported in 1879 (1). Solubility experiments carried out in high-pressure glass cells showed that several inorganic salts such as cobalt chloride, potassium iodide, potassium bromide, and ferric chloride could be dissolved or precipitated solely by changes in pressure on ethanol above its critical point. Increasing the pressure of the system caused the solutes to dissolve; decreasing the pressure caused the dissolved materials to nucleate and precipitate. An overview of the background, early findings, and historical development of the technology up to the mid-1980s has been presented (2).

Intensive industrial interest in the application of supercritical fluid extraction (SCFE) used in food and biological products resulted in numerous patents, and a number of journal articles have been written to introduce or to review the subject (3–7).

Some of the motivations to employ SCF technology as a viable separation technique are (1) tightening governmental regulations on solvent residues, and pollution control; (2) consumer concerns over the use of chemical solvents in food manufacture; (3) increased demand for higher quality products that traditional processing techniques cannot meet; and (4) the increased cost of energy. SCFE can enhance product recovery, improve product quality and safety, and minimize energy requirements. There are many potential applications, especially in the food and pharmaceutical fields. Today, SCFE is applied on a large scale in the decaffeination of coffee (8,9), and in the recovery of hop extracts used in beer brewing (10). Both of these applications employ SC CO_2 to replace conventional liquid solvents such as methylene chloride, which is under toxicological investigation by the U.S. Food and Drug Administration (FDA) and faces increasing restrictions on its use. In addition, SCF chromatography is being developed as an analysis tool (11).

FUNDAMENTALS OF SUPERCRITICAL FLUID EXTRACTION

Definitions and Properties

The critical temperature T_c of a substance is the temperature above which it cannot be liquefied, no matter how high the pressure. There is a corresponding pressure on the P-T diagram called the critical pressure, P_c (Fig. 1). The region on the P-T diagram above both T_c and P_c is called the supercritical region. In this region, the compressed gas is called a SCF, having characteristics of both gases and liquids. Fluids under SC conditions exhibit enhanced dissolving power and have transport properties that favor high extraction capabilities (Table 1). The SCF has a density similar to that of a liquid and functions like a liquid solvent, but it diffuses easily like a gas because its viscosity is low (9).

Thermodynamic Principles

The high density of a SCF allows it to dissolve relatively large quantities of organic compounds that normally have

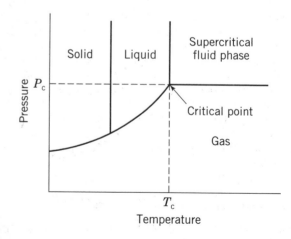

Figure 1. Pressure-temperature phase diagram of a pure substance.

Table 1. Ranges of Some Physical Properties Associated with Different Fluid States

	Density, kg/m^3	Diffusivity, m^2/s	Viscosity, MPa.s
	Gas		
$P = 101.3$ kPa $T = 288$–303 K	0.6–2	$(0.1$–$0.4) \times 10^{-4}$	$(1$–$3) \times 10^{-4}$
	Supercritical		
$T = T_cP = P_c$	200–500	0.7×10^{-7}	$(1$–$3) \times 10^{-4}$
$T = T_cP = 4P_c$	400–900	0.2×10^{-7}	$(3$–$9) \times 10^{-4}$
	Liquid		
$T = 288$–303 K	600–1600	$(0.22) \times 10^{-9}$	$(0.2$–$3) \times 10^{-2}$

Source: Ref. 11.

low solubility in the ordinary gaseous state of the same fluid. However, the solubility of compounds in a SCF also depends on solute and solvent properties. The dissolved compounds can be recovered from the fluid by decreasing the pressure or increasing the temperature, both of which reduce fluid density and allow it to be separated from the dissolved compound.

The reason SCF technology has been slow to find commercial applications has been due in part to the lack of available thermodynamic data to develop large-scale operations. Thermodynamic principles that can be used for quantitative and mathematical descriptions of solvent power of the SCF are models of phase equilibria and equations of state. For a process that involves the dissolution of a solute or a mixture of solutes from a solid matrix or a liquid mixture at constant temperature and pressure, the equilibrium conditions for the system would be given by

$$\hat{f}_i^{food}(T, P, x_i) = \hat{f}_i^{SCF}(T, P, y_i) \qquad (1)$$

where

$$\hat{f}_i^{food} = \text{food system component fugacity}$$

$$\hat{f}_i^{SCF} = \text{SCF component fugacity}$$

T is the temperature, P is the pressure, x_i is the mole fraction of solute in food, and y_i is the mole fraction of solute in SCF phase.

Earlier studies have considered the food as an inert solid containing the soluble material. Then, the problem was treated similarly to the calculation of water activity (12). Later, it was confirmed that the solid matrix interacts with the soluble material; therefore, it is not appropriate to consider it as inert. For example, the solubility of pure caffeine in CO_2 can be 200 times larger than its solubility in CO_2 when dissolved in coffee beans (13). Therefore, for complex materials such as black pepper, clove buds, ginger rhizomes, and so on, an empirical treatment is advisable, since currently there is no equation available to describe their thermodynamic behavior. If, however, the system involves extraction or concentration of solutes present in a

liquid phase, equation 1 can be used together with any appropriate cubic equation of state (EOS), such as Peng–Robinson (PR) or Soave–Redlich–Kwong (SRK). A simplified thermodynamic description of the calculation of solubility of a food system in CO_2 follows (14,15). The solute can be a mixture of substances like orange essential oil, a mixture of fatty acids or any similar system. For constant temperature and pressure, equation 1 is written for each component of the mixture, plus one for CO_2. Therefore, for a mixture with the following composition:

$$Z^* = (Z_1^*, Z_2^*, \ldots, Z_{n_c-1}^*) \qquad (2)$$

n_c equations similar to equation 1 should be written, where Z^* is the composition vector, Z_i^* is the molar fraction of component in the mixture, and $(n_c - 1)$ is the number of substances from the original mixture that will be treated with CO_2.

The data required for SCF plant scale-up is the solubility of the mixture (food system) with the composition represented by equation 2 in the SC solvent for a given temperature and pressure. This is accomplished by combining equation 1 with an appropriate EOS capable of calculating the fugacity for a given temperature, pressure, and composition. From thermodynamics we have (14,16):

$$\ln(\hat{\varphi}_i) = \ln\left(\frac{\hat{f}_i}{x_iP}\right) = \frac{1}{RT}\int_{\underline{V}=\infty}^{\underline{V}}\left[\frac{RT}{\underline{V}} - P\right]d\underline{V} - \ln(Z) + (Z - 1) \qquad (3)$$

where $\hat{\varphi}_i$ is the fugacity coefficient of component i, $\hat{f}_i$ is the fugacity of component, i, T is the temperature, P is the pressure, x_i is the mole fraction of component i, $\underline{V}$ is the specific volume of the mixture, and $Z = P\underline{V}/RT$ is the compressibility factor.

To evaluate the fugacity coefficients for all substances that form the mixture, the PR EOS for a mixture can be used:

$$P = \frac{RT}{\underline{V} - b} - \frac{a(T)}{\underline{V}(\underline{V} + b) + b(\underline{V} - b)} \qquad (4)$$

Using the compressibility factor Z, equation 4 can be written as (17):

$$Z^3 + \alpha'Z^2 + \beta'Z + \gamma' = 0 \qquad (5)$$

where $Z = P\underline{V}/RT$, $\alpha' = (-1 + B)$, $\beta' = (A - 3B^2 - 2B)$, $\gamma' = (-AB + B^2 + B^3)$, $A = a(T)P/(RT)^2$, and $B = bP/RT$. In general, the parameter $a(T)$ is referred to as the attractive parameter and b as the volumetric parameter, and they are given by:

$$a(T) = \sum_{i,j} x_ix_ja_{i,j} \quad a_{i,j} = a_{j,i} = \sqrt{a_{i,i}a_{j,j}}\,(1 - k_{a_{i,j}}) \qquad (6)$$

$$b = \sum_{i,j} x_ix_jb_{i,j} \quad b_{i,j} = b_{j,i} = \left[\frac{b_{i,i} + b_{j,j}}{2}\right](1 - k_{b_{i,j}}) \qquad (7)$$

where a_{ij} is the attractive parameter for binary i-j, b_{ij} is the volumetric parameter for the binary i-j, a_{ii} is the attractive parameter for component i, b_{ii} is the volumetric parameter for component i, $k_{a_{i,j}}$ is the attractive binary interaction parameter, and $k_{b_{i,j}}$ is the volumetric binary interaction parameter. The pure components' attractive and volumetric parameters are calculated using the critical properties of the solute. For instance (17):

$$a_{ii}(T) = 0.45724R^2T_c^2/P_c\alpha(T) \qquad (8)$$

$$b_{ii} = 0.07780RT_c/P_c \qquad (9)$$

$$\sqrt{\alpha(T)} = 1 + \kappa(1 - \sqrt{T/T_c}) \qquad (10)$$

and

$$\kappa = 0.37464 + 1.54226\omega - 0.226992\omega^2 \qquad (11)$$

where w is the acentric factor.

To solve equation 1, which is a nonlinear problem, a nonlinear trial-and-error solution is required. The objective function to be minimized is defined accordingly with the available experimental data. The following information is needed: (1) the system temperature and pressure, (2) the composition vector Z^*, and (3) the matrices of binary attractive and volumetric interaction parameters given by equations 6 and 7. This should include binary systems such as component-i/CO_2 and component-i/component-j. Then, there will be n_c binary parameters of the kind component i-CO_2, plus $(n_c)^2/2$ parameters of the kind component-i/component-j. Binary liquid-SC and liquid-vapor equilibrium data are required to evaluate these parameters. Currently, scarce information is available in the literature for systems involving foods; therefore, some difficulties can be anticipated to solve the preceding equations. Finally, one should be aware that to solve equation 1, equation 4 is solved twice, once for each phase in equilibrium.

The procedure just described was used to evaluate the solubility of orange essential oil in CO_2 (15,16). The problem was treated as a dewpoint problem; therefore, the oil solubility in CO_2 was considered to be the amount of carbon dioxide required to dissolve a given oil sample. Defining α as the essential oil fraction in the final SC mixture (essential oil and CO_2), the composition vector in the CO_2 rich phase (SCF) when the oil phase is completely dissolved is given by:

$$y_{i+1} = \alpha Z_i^* \quad i = 1, (n_c - 1) \qquad (12)$$

The solubility of the essential oil in the SC phase will be:

$$S_b = \frac{\sum_{i=2}^{n_c} m_i}{m_{CO_2}} = \frac{\sum_{i=1}^{n_c-1} y_{i+1}M_{i+1}}{y_{CO_2}M_{CO_2}} = \frac{\alpha \sum_{i=1}^{n_c-1} Z_i^* M_i}{(1 - \alpha)M_{CO_2}} \qquad (13)$$

where S_b is the solubility of the oil (mg of oil/g of CO_2), m_i are the mass fractions of essential oil components, M_i are the molecular mass of essential oil components, and y_i are molar fractions of essential oil components, respectively. Figure 2 shows the experimental and calculated solubilities of the orange oil (oil-phase) in CO_2 made using both

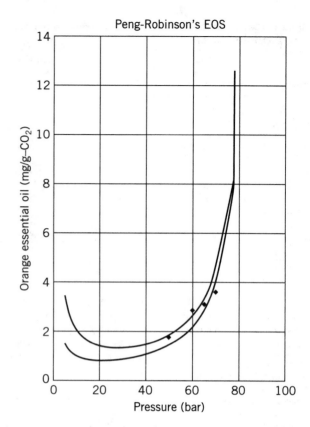

Figure 2. Predicted solubility of orange essential oil in CO_2 according to PR and SRK equations of state. *Source:* Experimental solubility determined by Santana (Ref. 19) at 308.15 K.—Calculated using orange essential oil molar composition determined by Marques (Ref. 18);—Calculated using orange essential oil molar composition determined by Santana (Ref. 19).

PR and SRK equations of state. The calculations were done using the oil with the composition given in Table 2 (18) and the composition given in Table 3 (19).

Experimental data on the physical properties of essential oil components is scarce in the literature. In addition, many of these substances are thermally degraded before T_c can be measured. Therefore, T_c, P_c, and acentric factor (w) used in equations 8 and 9 were predicted. The choice of methods to be used in these calculations depends on the available experimental data on pure components. The binary parameters used are given in Table 4 (16). For binary systems (orange essential oil component/CO_2 or essential oil component/essential oil component) where experimental data is missing, interaction parameters were set equal to zero. This choice does not imply that there is no interaction or that the interaction are negligible between these binary systems, but this simplifying assumption was required to make the problem solvable.

Although the orange essential oil used by Marques (18) and Santana (19) have different compositions, their calculated solubilities did not differ much from the experimental value measured by Santana (19), as can be observed in Figure 2. The PR and SRK equations gave similar solubility values. In this case, the calculated solubilities were not sensitive to the initial oil composition. The

Table 2. Molar Composition of Orange Essential Oil Determined by Marques (18) and Pure Component Properties

Component	Z_i^*	T_{eb} (K)	T_c (K)	P_c (bar)	w
Ethanol	0.0886	351.65[a]	513.92[d]	61.48[d]	0.6452[d]
Linalool	0.00586	472.15[a]	635.99[c]	25.82[b]	0.7617[f]
α-Terpineol	0.00055	493.15[a]	675.59[c]	29.50[b]	0.7133[f]
trans-2, Hexenal	0.00012	419.65[a]	615.15[c]	35.94[b]	0.4199[f]
Octanal	0.00361	444.15[a]	620.10[e]	27.35[b]	0.5558[f]
Nonanal	0.00053	464.15[a]	637.67[e]	24.80[b]	0.6053[f]
Decanal	0.00287	481.65[a]	651.94[e]	22.59[b]	0.6536[f]
Dodecanal	0.00027	484.94[b]	675.98[b]	18.97[b]	0.7585[f]
Citronelal	0.00050	480.65[a]	663.86[e]	24.05[b]	0.5570[f]
Neral	0.00051	502.15[c]	699.97	25.25[b]	0.7174[f]
Geranial	0.00125	502.15[a]	699.97[c]	25.25[b]	0.7174[f]
β-Simensel	0.00012	592.58[b]	782.72[b]	17.64[b]	0.6853[f]
α-Simensel	0.00015	600.06[b]	794.25[b]	17.79[b]	0.6749[f]
Ethyl butyrate	0.00100	369.15[a]	571.00[d]	30.60[b]	0.4190[d]
α-Pinene	0.00421	429.35[a]	630.87[b]	28.90[b]	0.3242[f]
Δ-3-Carene	0.00102	440.15[b]	646.74[b]	28.90[b]	0.3242[f]
β-Mircene	0.01620	440.15[a]	642.32[e]	28.08[b]	0.3425[f]
Valencene	0.00342	564.52[b]	780.55[b]	18.97[b]	0.4324[f]
d-Limonene	0.94895	451.15[a]	661.11[e]	27.56[b]	0.3170[f]

[a]T_{eb} from Ref. 20.
[b]P_c calculated using Ref. 18.
[c]Ref. 21.
[d]Ref. 22.
[e]T_c calculated using Ref. 18 and T_{eb} from Ref. 20.
[f]Lee-Kesler method to calculate acentric factor, Ref. 23.

Table 3. Molar Composition of Orange Essential Oil Determined by Santana (19) and Pure Components Properties

Component	Z_i^*	T_{eb} (K)	T_c (K)	P_c (bar)	w
Linalool	0.00727	472.15[a]	635.99[a]	25.82[b]	0.7617[f]
α-Pinene	0.00610	429.35[a]	630.87[e]	28.90[b]	0.3242[f]
β-Mircene[2]	0.00726	440.15[a]	642.32[e]	28.08[b]	0.3425[f]
Sabinene	0.00177	437.15	640.12	29.35	0.3547
Limonene	0.97760	451.15[a]	661.11[e]	27.56[b]	0.3170[f]

[a]T_{eb} from Ref. 20.
[b]P_c calculated using Ref. 19.
[c]Ref. 21.
[d]Ref. 22.
[e]T_c calculated using Ref. 19 and T_{eb} from Ref. 20.
[f]Lee-Kesler method used to calculate acentric factor (23).

Table 4. Binary Interaction Parameters Available for the System: Orange Oil/CO₂.

	PR		SRK	
	$k_{aij}10^2$	$k_{bij}10^2$	$k_{aij}10^2$	$k_{bij}10^2$
CO₂/Ethanol	9.048	−1.414	8.543	−1.412
CO₂/Linalool	4.281	−3.156	4.363	−3.249
CO₂/α-Pinene	9.482	−2.820	10.28	−2.805
CO₂/d-Linomene	10.15	1.960	9.921	−1.415

Source: Ref. 16.

PR and SRK equations gave similar solubility values. This is expected since, in general, these mixtures are formed from the same family of substances.

Different substances have different critical temperatures and pressures, important in selecting the appropriate fluid and operating conditions for an application (Table 5). If the critical pressure is too high, processing will be more difficult and costly because equipment must be designed for high pressures. If the critical temperature is too high, heat-sensitive materials such as many food constituents will be altered or destroyed.

Carbon dioxide is the solvent of choice for food applications for many reasons (25). It is (1) inert and noncorrosive in dry environments, (2) nonflammable and nonexplosive, (3) abundant and inexpensive, (4) nontoxic and accepted as a harmless ingredient of foods and beverages, (5) easy to separate the solute from the solvent, and (6) has desirable physical properties such as low T_c (31°C) and low P_c (7.39 MPa), low viscosity, low surface tension, and high diffusivity. When pressure is decreased, CO_2 will separate out from the solute as gas. CO_2 can also protect foods from oxygen and limit oxidation reactions. The solubilities of some substances in liquid CO_2 are given in Table 6. This can be extrapolated to the SC CO_2 extraction. Since the CO_2 molecule is nonpolar, at high densities it is a good solvent of lipophilic organic substances and esters, ethers, and lactones. In general, hydroxyl and carboxyl groups reduce the solubility of a substance in SC CO_2. Polar or charged materials such as sugars and amino acids are not soluble. Large molecules such as cellulose and proteins are also insoluble. The solubilities of substances such as waxes, resins, and pigments increase with an increase in the density of SC CO_2. Solubility is also affected by the intermolecular forces of the solute molecules, which enable the CO_2 molecules to surround and hold them in solution (5). In general, as the number of carbon—carbon double

Table 5. Critical Property Data for Some Supercritical Solvents

Substance	Critical temperature[a] (K)	Critical pressure[b] (MPa)	Critical density[b] (kg/m³)
Methane	190.6	4.60	162
Ethylene	282.4	5.03	218
Chlorotrifluoromethane	302.0	3.92	579
Carbon dioxide	304.2	7.38	468
Ethane	305.4	4.88	203
Propylene	365.0	4.62	233
Propane	369.8	4.42	217
Ammonia	405.6	11.30	235
Diethyl ether	467.7	3.64	265
n-Pentane	469.6	3.37	237
Acetone	508.1	4.70	278
Methanol	512.6	8.09	272
Benzene	562.6	4.89	302
Toluene	591.7	4.11	292
Pyridine	620.0	5.63	312
Water	647.3	22.00	322

[a]Ref. 24.
[b]Ref. 4.

Table 6. Solubilities of Some Compounds in liquid CO₂

Fully miscible	Partially miscible (% by wt)		Low solubility
H₂S, SO₂	Water	0.1	β-carotene
Limonene	Iodine	0.2	Urea
Benzene	Oleic acid	2.0	Glycine (not soluble)
Acetic acid	Lactic acid	0.5	Citric acid
Ethyl alcohol	Glycerol	0.05	Ascorbic acid
Benzaldehyde	Aldol	11.0	Sucrose (not soluble)
	Orange oil (308, 15K)[a]		
	5.0 MPa	0.17	
	6.0 MPa	0.28	
	6.5 MPa	0.31	
	7.0 MPa	0.36	

Source: Ref. 26.
[a]Ref. 19.

bonds increases, solubility decreases. Substances belonging to the same chemical class are dissolved in the SC phase in order of their increasing boiling points. The use of cosolvents or entrainers such as water or methanol enhances the solubility of substances normally not miscible with SC CO_2.

PROCESS SYSTEMS AND OPERATION CHARACTERISTICS

A simplified diagram of a process that uses a SCF as a pressure-dependent, variable-power solvent for extraction is shown in Figure 3. A typical system consists of an extraction vessel at high pressure, expansion valves, separator vessels, and a compressor or pump. The feed enters the extraction vessel and contacts the SC CO_2. The pressure and temperature are controlled. Since its density is high, the SCF dissolves some components and separates them from the feed. The rest is obtained as residue. The SC phase containing dissolved materials flows through an expansion valve where the pressure is reduced. Since its density is reduced, the SCF deposits some dissolved components into the separation vessel as product 1. The re-

maining solutes in the SC phase are separated from CO_2 in the last separator and collected as product 2. The CO_2 is now at a low pressure, and carries no dissolved material. It is recycled by going through the pump, and reentering the extraction vessel. Therefore, multiple stages can be used to fractionate different compounds from the original feed. However, fractionation of extracts is possible only if the constituents exhibit large differences in vapor pressure.

As an alternative to fractionation by pressure reduction, the process can also be operated at constant pressure with temperature changes in the SCF to cause separation. Such a system is shown in Figure 4.

Operation Characteristics

The conditions of pressure and temperature for extraction and fractionation applications using CO_2 vary greatly with the materials to be extracted. Typical ranges of pressure and temperature for food applications are given in Figure 5. To evaluate the technical and economic feasibility for SCF extraction of a specific material, a pilot-plant study is carried out to determine factors such as solubility and selectivity properties, yields, quality, and operation condi-

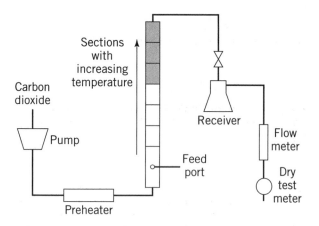

Figure 4. Column-type separation by supercritical fluids.

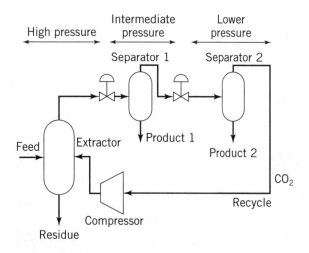

Figure 3. Typical process flow diagram of SCFE.

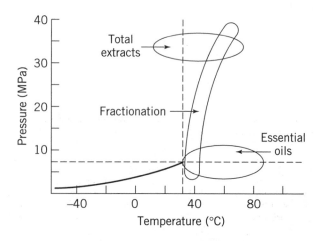

Figure 5. Typical applications of SCFE.

tions. Analysis of SCF operations applicable to the food industry, design and economic considerations, and optimization of SCFE operations have been discussed in the literature. The products of SCFE processes show different characteristics as compared with those obtained by traditional vacuum distillation processes. In general, the product quality of SCFE processes may or may not be superior to those produced by conventional methods.

COMMERCIAL APPLICATIONS

Coffee

In the decaffeination process of coffee, green coffee beans are treated with CO_2. The coffee beans are not roasted prior to the extraction. Thus, the inherent moisture in the beans acts as a chemical agent to free the caffeine from its bound form (2). Presoaked green coffee beans are treated with CO_2 at 16 to 22 MPa. The CO_2 is continuously recycled until the caffeine diffuses out of the beans into the SC CO_2 phase. The caffeine is washed out with water at 70 to 90°C. Caffeine content in the bean can be decreased from the initial value in the range of 0.7 to 3% to a value as low as 0.02%.

Hop Extraction

The use of hop extracts by the brewing industry is a recent, commercial application of SC extractions. The soft resin constituents of hops that contain the flavor components, humulones and lupulones, are conventionally extracted with dichloromethane. This results in a pasty, dark-green to black-green mass. In addition, the dichloromethane must be removed by evaporation after the resins have been extracted. In contrast, SC CO_2 extraction of hops between 35 and 80°C and 80 to 300 atm results in an "olive-green, pasty extract with an intense aroma of hops" (10).

POTENTIAL APPLICATIONS

Extraction of Flavors

As compared with conventional solvent extraction, SCFE could be a potential alternative to extract and fractionate flavors and fragrances from fermentation-induced and natural products (8,27). SC CO_2 has been used to extract the following materials: essential oils and piperine from black pepper, alkaloids from chili pepper, eugenol from cloves, cinnamic aldehyde from cinnamon, limonene and carvone from caraway, sesamin from sesame seeds, vanillin from vanilla pods, menthol and menthone from peppermint leaves, and geraniol and citronella from roses. Supercritical CO_2 was used for concentration of aroma and flavor compounds in citrus oils (28). Citrus oils, like many other essential oils, consist of mixtures of hydrocarbons of the terpene and sesquiterpene groups, oxygenated compounds, and nonvolatile residues. The compounds that provide much of the characteristic flavor are mostly oxygenated compounds that account for less than 4% of the orange oil. Thus, the concentration requirement is high and the amount of yield is relatively low.

EDIBLE OILS AND FATTY ACIDS

There is intense interest in the extraction and fractionation by SC CO_2 of vegetable, animal, and seed oils used for human consumption. In conventional extraction, hexane residue in the material leftover from extraction is routinely detected in ppm levels. Concerns about possible carcinogenicity, hazards, and cost associated with hexane have prompted researchers to develop alternative methods. To design a SCFE system, two types of information are necessary: equilibrium distribution of the oil between the SC phase and the plant matrix, and mass-transfer rate of the oil from the oil source to the solvent. Soybean oil, canola seed oil, palm oil, wheat germ oil, mustard oil, corn oil, and cottonseed oil have been extracted with SC CO_2. The effects of moisture content and particle size of the seeds, CO_2 flow rate, chamber dimensions, and other factors on the oil yield have been investigated (29). For soybeans, oxidative stability of SC CO_2 extracted oil was lower than conventional hexane extracts, since phosphatides that are natural antioxidants were not extracted with SC CO_2. Extraction of freeze-dried phytoplankton by SC CO_2 at 40°C and between 17 and 32 MPa resulted in an orange-red oil rich in omega-3 fatty acids, without extracting chlorophyll (30). Conventional separation or concentration of fatty acids, which have low vapor pressures and normally high boiling points, requires vacuum distillation at high operating temperatures. Some thermal degradation is unavoidable in this process. In contrast, it has been shown that the separation of fatty acid ethyl esters differing by two carbon atoms is possible by using SC CO_2 under mild conditions (31). Cholesterol is readily soluble in SC CO_2. There are intense efforts to remove it from foods from animal origin, without damaging their quality (32). For all of its potential benefits, there are challenges for the food scientist and engineer to implement the SCF technology. The lack of accumulated process and scale-up experience, the existence of many patents in the field, and high capital and maintenance costs are some of these challenges.

BIBLIOGRAPHY

1. J. B. Hannay and J. Hogarth, "On the Solubility of Solids in Gases," *Proc. R. Soc.* London **29**, 324–326 (1879).

2. M. McHugh and V. J. Krukonis, *Supercritical Fluid Extraction—Principles and Practice*, Butterworth, Boston, Mass., 1986.

3. G. M. Schneider, E. Stahl, and G. Wilke, eds., *Extraction with Supercritical Gasses*, Verlag Chemie, Deerfield Beach, Fla., 1980.

4. D. F. Williams, "Extraction with Supercritical Gases," *Chem. Eng. Sci.* **36**, 1769–1788 (1981).

5. M. E. Paulaitis et al., "Supercritical Fluid Extraction," *Rev. Chem. Eng.* **1**, 179–250 (1983).

6. S. S. H. Rizvi et al., "Supercritical Fluid Extraction: Fundamental Principles and Modeling Methods," *Food Technol.* **40**, 55–65 (1986).

7. S. S. H. Rizvi et al., "Supercritical Fluid Extraction: Operating Principles and Food Applications," *Food Technol.* **40**, 57–64 (1986).

8. R. A. Novak and R. J. Robey, "Supercritical Fluid Extraction of Flavoring Materials: Design and Economics," paper presented at the AIChE 1988 Annual Meeting in Washington, D.C., November 27–December 2, 1988.

9. K. Zosel, "Separation with Supercritical Gases: Practical Applications," in G. M. Schneider, E. Stahl, and G. Wilke, eds., *Extraction with Supercritical Gases*, Verlag Chemie, Deerfield Beach, Fla., 1980.

10. P. Hubert and O. G. Vitzthum, "Fluid Extraction of Hops, Spices, and Tobacco with Supercritical Gases," in G. M. Schneider, E. Stahl, and G. Wilke, eds., *Extraction with Supercritical Gases*, Verlag Chemie, Deerfield Beach Fla., 1980.

11. L. G. Randall, "The Present Status of Dense (Supercritical) Gas Extraction and Dense Gas Chromatography: Impetus for DGC/MS Development," *Separation Sci. Technol.* **17**, 1–118 (1982).

12. M. A., A. Meireles and Z. Nikolov, "Extraction and Fractionation of Essential Oils with Liquid Carbon Dioxide," in G. Charalambous, ed., *Herbs, Spices, and Edible Fungi*, Elsevier, Amsterdam, The Netherlands, 1994, pp. 171–199.

13. G. Brunner, *Gas Extraction: An Introduction to Fundamentals of Supercritical Fluids and the Applications to Separation Process*, Steinkopff, Darmstadt, Germany, 1994.

14. F. A. Cabral and M. A. A. Meireles, "The Solubility and the Phase Equilibria of Essential Oil with CO_2 Calculated Using a Cubic Equation of State," in G. Charalambous, ed., *Food Flavors: Generation, Analysis and Processing Influence*, Elsevier, Amsterdam, The Netherlands, 1995, pp. 331–354.

15. L. Cardozo-Filho, F. Wolff, and M. A. A. Meireles, "High Pressure Phase Equilibrium: Prediction of Essential Oil Solubility," *Ciência e Tecnologia de Alimentos* **17**, 485–488 (1997).

16. L. Cardozo-Filho et al., "The Generalized Maximum Likelihood Method Applied to High Pressure Phase Equilibrium," *Ciência e Tecnologia de Alimentos* **17**, 481–484 (1997).

17. S. I. Sandler, *Chemical Engineering Thermodynamics*, 2nd ed., Wiley, New York, 1989.

18. D. S. Marques, "Terpene Reduction in Orange Essential Oil by Preparative Chromatography and Supercritical Carbon Dioxide Extraction," Master's Thesis, Faculty of Food Engineering, State University of Campinas, São Paulo, Brazil, 1997, 331 pp.

19. H. B. Santana, "Determination of the Solubility of Essential Oil in CO_2," Master's Thesis, Faculty of Food Engineering, State University of Campinas, São Paulo, Brazil, 1996, 119 pp.

20. R. C. Weast and M. J. Astle, *CRC—Handbook of Data on Organic Compounds*, Vols. 1 and 2, CRC Press, Boca Raton, Fla., 1992.

21. R. H. Perry and C. H. Chilton, *Chemical Engineers' Handbook*, 5th ed., McGraw-Hill, New York, 1973.

22. R. P. Danner and T. E. Daubert, *Data Compilation Tables of Properties of Pure Compounds*, DIPPR, The Pennsylvania State University, University Park, 1984.

23. R. C. Reid, J. M. Prausnitz, and B. E. Pouling, *The Properties of Gases and Liquids*, 4th ed., McGraw-Hill, New York, 1987.

24. R. C. Reid, J. M. Prausnitz, and T. P. Sherwood, *The Properties of Gases and Liquids*, McGraw-Hill, New York, 1977.

25. H. Broyle, "CO_2 as a Solvent Its Properties and Application," *Chem. Industry (London)* **19**, 385–390 (1982).

26. W. G. Schultz and J. M. Randall, "Liquid Carbon Dioxide for Selective Aroma Extraction," *Food Technol.* **24**, 1282–1287 (1970).

27. V. J. Krukonis, "Supercritical Fluid Extraction in Flavors Applications," in D. D. Bills and C. J. Mussinan, eds., *Characterization and Measurement of Flavor Compounds*, ACS Symposium Series, 289, American Chemical Society, Washington, D.C., 1985, pp. 154–175.

28. F. Temelli, C. S. Chen, and R. J. Braddock, "Supercritical Fluid Extraction in Citrus Oil Processing," *Food Technol.* **42**, 145–150 (1988).

29. J. P. Friedrich, G. R. List, and A. J. Heaking, "Petroleum-Free Extraction of Oil from Soybeans with Supercritical CO_2," *J. Am. Oil Chem. Soc.* **59**, 288 (1982).

30. J. T. Polak et al., "SC CO_2 Extraction of Lipids from Algae," in K. P. Johnston and J. M. L. Penninger, eds., *Supercritical Fluid Science and Technology*, ACS Symposium Series, 406, American Chemical Society, Washington, D.C., 1989, pp. 449–467.

31. K. Arai and S. Saito, "Fractionation of Fatty Acids and Their Derivatives by Extractive Crystallization Using Supercritical Gas as Solvent," paper presented during the 1986 Annual AIChE meeting in Miami Beach, Fla., November 2–7, 1986.

32. J. M. Wong and K. P. Johnston, "Solubilization of Biomolecules in CO_2 Based Supercritical Fluids," *Biotechnol. Prog.* **2**, 29 (1986).

M. O. BALABAN
University of Florida
Gainesville, Florida

M. A. A. MEIRELES
State University of Campinas—UNICAMP
São Paulo, Brazil

SURFACE (INTERFACIAL) TENSION

Surface tension can be referred to as free energy per unit area or force per unit length. Customary units are either ergs per square centimeter or dynes per centimeter. In SI (International System of Units) units, they are joules per square meter or newtons per meter (J/m^2, N/m).

The mathematical notation used for surface tension is usually γ, and in performing dimensional analysis, γ is represented by M/t^2, where M is mass and t is time. The methods available for the measurement of surface tension include the capillary rise method, the capillary wave method, the ring method, and the tensiometer method (1).

Surface energy (tension) at the boundary of two phases, such as solid–liquid in skim milk, liquid–liquid in salad dressing, gas–liquid in meringue, gas–solid in foam candy, and solid–gas in smoke, is an important parameter for studying the manufacturing and the shelf-life stability of food systems (2). Furthermore, many chemical and physical processes can occur at the boundary; for example, dissolution and crystallization, heterogeneous catalysis, and phenomena related to the colloidal state. Surface energy at the boundary often determines whether the process will occur.

London, van der Waals, or cohesive forces act among the molecules of all substances irrespective of their state of aggregation. Such forces are significant only when the molecules are rather close to each other, say, only several nanometers apart. For polar compounds, such as water, hydrogen bonding is also a significant cohesive force. In a single bulk phase, intermolecular cohesive forces are balanced. This is because molecules in the interior of a phase are attracted equally in all directions to the other molecules in their vicinity. However, in either solids, liquids, or gases, if the atoms, ions, or molecules exist in the interface, they are exposed to the action of unbalanced forces due to both phases. In other words, those at the interface are not surrounded completely by other entities of the same type or same physical state. Thus, the surface molecules or atoms are in an energy state different from that of those in the bulk phase. This additional energy, generally called surface energy, imparts to the surface region distinct features that are unique to the region. Surface tension acts parallel to the substance's surface, opposing any attempt to expand the surface area. A component of surface energy causes forces to act normally at the phase boundary, resulting in an inward attraction, which, in turn, tends to reduce the number of molecules at the interface and to reduce the interfacial area to a minimum. A liquid, for example, has a tendency to obtain a spherical shape, the shape with the smallest ratio of surface to volume, if it is not contained. On the other hand, if the liquid is contained, it does not alter to the spherical shape because gravity imposes an added requirement. The surface must be uppermost and parallel to the plane of the earth, as it takes the shape of its container. Work or energy is required to increase the area of a surface or of an interface. This requires molecules to be moved from the body of the substance to the surface (interface). The total surface energy required to expand a surface by 1 cm^2 is the sum of the work required to overcome the surface tension and the amount of heat that must be supplied to maintain the expanding surface at a constant temperature. If the surface of water is at 20°C, the thermal energy required is 47.7 ergs/cm^2, and since surface tension, $\gamma = 72.8$ ergs/cm^2, the total energy is 120.5 ergs/cm^2 (2). With an increase in temperature, the kinetic energy of the molecules increases and the cohesive forces among them decreases, which leads to a decreased surface tension. The influence of temperature on the surface tension of water and vegetable oils is shown in Table 1 (3). The surface tensions shown in Table 1 are those between the liquid (water or oil) phase and the gaseous phase (air). For making stable food emulsions, the interfacial tensions between edible oils and water, as shown in Table 2 (4,5), are important factors. An important consequence of surface tension (and its existence) is that there is a pressure difference on the two sides of a surface. This can be visualized by considering a gas-filled balloon stretched by the internal pressure of the gas. It experiences an elastic force

Table 1. Influence of Temperature on the Surface Tension (dyne/cm) of Water and Vegetable Oil

Temperature, °C	Water	Cottonseed oil	Coconut oil	Olive oil
0	72.6			
20	72.8	35.4	33.4	33.0
30	71.2			
50	67.9			
80	62.6	31.3	28.4	
100	58.9			
130		27.5	24.0	

Source: Ref. 3.

Table 2. Interfacial Tension (dyne/cm) Between Water and Various Oils

Purified Oil	25°C	75°C
Triolein	14.6	13.5
1,3-Dioleo-2-palmitin	14.5	12.3
Peanut (screw press)	18.1	
Peanut (solvent, extracted)	18.5	
Cottonseed (screw press)	14.9	
Olive	17.6	
Coconut	12.8	

Source: Refs. 4 and 5.

parallel to the surface that counteracts the stretching; otherwise the balloon would pop. The internal pressure of a balloon must be larger than the external pressure since the gas is not only counteracting the external pressure but also stretching the balloon. This force is analogous to the surface tension of a liquid, which arises due to the rubber molecules' attraction or the liquid's molecular attraction with each other.

In general, the smaller the interfacial tension, the less energy is required to create an emulsion between the phases. As a rule of thumb, the interfacial tension between water and edible oil should be below 10 dyne/cm to facilitate emulsification of the two phases (2). To meet this requirement, surface-active agents, or surfactants, are often added to reduce the interfacial tensions between the two phases. Surfactants in small quantities are known to reduce interfacial tensions. However, there are other solutes, such as inorganic salts and compounds with a large number of hydroxyl groups, such as sugars, when added to water, the surface tension of the solution increases slightly as the concentration of the solute increases. The molecular-level explanations for such distinctive phenomena are the following. Surface-active chemicals exhibit positive adsorption phenomena that lead to a higher concentration of the added chemicals at the surface (or interface) than in the bulk. For sugars and salts that show a negative adsorption phenomenon in water, the concentration in the bulk is higher than that in the surface (or interface). Not only are their effects in opposite directions, but the effects of surface-active agents on the surface (or interfacial) tension are much more drastic than those caused by sugars and salts. Consequently, small amounts of surfactants are normally used. At the molecular level, surfactant molecules are oriented at the surface (or interface) by pointing their polar (hydrophilic) groups toward the aqueous phase and their apolar (hydrophobic) groups toward the gaseous (nonaqueous) phase. In this process, the surface free energy is minimized. The surfactant molecules will form a monolayer or a film on the surface until the surface is completely covered and will form micelles with organized structures in the bulk solution as the concentration of surfactant exceeds the critical micelle concentration, which is in the range 0.004 to 0.15 mol/L for many surfactants (2).

Surfactants are either natural or synthetic, nonionic, or ionic. Food-grade surfactants are generally nonionic and are preferably natural. Soaps and detergents act as surfactants in order to lower the surface tension, which helps to remove oily dirt particles from solid surfaces (including skin). This enables these products to do what they do best—clean! Many proteins are good surfactants. Protein molecules can be adsorbed on the interface to form either a monolayer or multiple layers, depending on the individual protein. This is related to the fact that different proteins have different abilities in reducing the surface tension of water. Table 3 shows the equilibrium surface tension of four different protein solutions at different concentrations.

Similarly, Table 4 shows the influence of several milk proteins on the interfacial tension between water and butter oil. The values of surface and interfacial tension shown in Table 3 and 4 are both equilibrium values that were obtained after a transient period of surface (or interfacial) tension reduction after the addition of the protein. An equilibrium is reached when the surface tension measured stays at a constant value over an arbitrarily selected time period, such as 30 min.

Table 3. The Equilibrium Surface Tension of Protein Solutions at Various Concentrations

Protein	Protein concentration, %w/v	Equilibrium surface tension[a]
Ovalbumin	8.0	41.9
	4.0	40.3
	2.0	41.7
	1.0	44.0
Soybean protein	3.0	41.2
	2.0	41.7
	1.0	42.2
	0.3	42.5
	0.1	42.9
Bovine serum albumin	2.0	49.4
	1.5	49.1
	1.0	48.6
	0.5	49.7
Casein	3.0	48.6
	2.0	42.2
	1.0	46.5
	0.6	48.4
	0.3	47.0
	0.1	47.6

Source: Ref. 6.
[a]Surface tension values (dyne/cm) attained and maintained constant for more than 30 min.

Table 4. Influence of Milk Proteins on the Interfacial Tension Between Water and Butter Oil at 40°C

Protein	Concentration, %	Interfacial tension, dyne/cm
None	—	19.2
Euglobulin	0.2	18
	0.6	18
β-Lactoglobulin	0.2	14
	0.6	14
α-Lactalbumin	0.2	11
	0.6	11
Interface protein	0.2	11
	0.6	9

Source: Ref. 7.

When protein molecules are adsorbed onto the interface, they may change tertiary structure to suit the new environment, such as specific orientation and film formation to obey the law of thermodynamics. Specifically this results in the reduction of surface energy, which makes the achieved thermodynamic state more stable. Compounded by the fact that proteins are slow-diffusing molecules, the kinetics of reducing interfacial tension by the addition of protein is not a fast one. According to Ghosh and Bull (8), the surface tension of 0.06% egg albumin solution (pH 4.9, 30°C) dropped from 62 to 48 dyn/cm within a period of 55 min. Graham and Phillips (9) also reported that the process of reduction of surface tension due to the addition of relatively low protein concentrations can take between 10 and 60 min. Kitabatake and Doi (6) reported that the reduction of surface tension due to added protein follows first-order kinetics. They also found that the formability of the protein solution correlated better with the first-order rate constant of the surface tension reduction process than with the equilibrium surface tension of the solution. In other words, formability involves the capacity to entrap air on whipping, which is promoted by proteins that will cause a fast reduction in the surface tension of water.

There are structural similarities between foams and emulsions; surface tension and interfacial tension are important physical parameters to be dealt with in food processing for the former and the latter, respectively.

BIBLIOGRAPHY

1. A. W. Adamson, *Physical Chemistry of Surfaces*, 4th ed., John Wiley and Sons, New York, 1982.

2. W. D. Powrie and M. A. Tung, "Food Dispersions," in O. R. Fennema, ed., *Food Dispersions in Principles of Food Science Part I: Food Chemistry*, Marcel Dekker, New York, 1976, pp. 185–194.

3. A. Halpern, "The Surface Tension of Oils," *J. Physical Colloid Chem.* **53**, 895–897 (1949).

4. R. R. Benerito, W. S. Singleton, and R. O. Feuge, "Surface and Interfacial Tensions of Synthetic Glycerides of Known Composition and Configuration," *J. Physical Chem.* **58**, 831–834 (1954).

5. W. S. Singleton and R. R. Benerito, "Surface Phenomena of Fats for Parenteral Nutrition," *J. Amer. Oil Chemists' Soc.* **32**, 23–36 (1955).

6. N. Kitabatake and E. Doi, "Surface Tension and Foamability of Protein and Surfactant Solutions," *J. Food Sci.* **53**, 1542–1545 (1988).

7. R. H. Jackson and M. Pallanasch, "Influence of Milk Proteins on Interfacial Tension Between Butter Oil and Various Aqueous Phases," *J. Agricultural and Food Chem.* **9**, 424–427 (1961).

8. S. Ghosh and H. B. Bull, "Adsorbed Films of Bovine Serum Albumin: Tensions at Air-Water Surfaces and Paraffin Water Interfaces," *Biochemica et Biophysica Acta* **66**, 150–157 (1963).

9. D. E. Graham and M. C. Phillips, "Proteins at Liquid Interfaces I, Kinetics of Adsorption and Surface Denaturation," *J. Colloid Interface Sci.* **70**, 403–414 (1979).

Shaw Wang
Rutgers University
Piscataway, New Jersey

SURIMI: SCIENCE AND TECHNOLOGY

Surimi is a Japanese term for mechanically deboned fish mince that has been washed with water and mixed with cryoprotectants for a long frozen shelf life. It is used as an intermediate product for a variety of shellfish analog products, such as crab legs, shrimp, and lobster. Minced fish, on the other hand, is a mechanically separated flesh that has not been washed and does not have good freeze storability. Washing not only removes fat and undesirable matter, such as blood, pigments, and odorous substances but, more important, increases the concentration of myofibrillar proteins (primarily actomyosin) through removal of water-soluble sarcoplasmic proteins. As a result, washing improves gel strength and elasticity, essential properties for surimi-based products.

Unlike soy protein, surimi, because of its high concentration of myofibrillar proteins, produces an elastic and chewy texture that can be made to resemble that of shellfish. Because it has this unique property, surimi has been used extensively in Japan for many centuries in a variety of traditional as well as new fabricated products. Surimi technology has led to the development of commercially acceptable shellfish analogs that were not successful in the U.S. market when soy protein was used. It appears that surimi has great potential as a functional protein ingredient that can be substituted for a variety of traditional animal and vegetable proteins. The potential of surimi in developing new products is not limited to shellfish analogs; it has already been realized in developing products based on surimi–meat blends. The virtually unlimited resources of underutilized fish species will ensure a sufficient supply of surimi at a reasonable cost to meet the need for base material for surimi-based products.

The development of surimi technology in the United States began in the early 1980s, utilizing Alaska pollock (*Theragra chalcogramma*). Within a short period, the U.S. surimi industry established a strong commercial base with technology development assisted by the government and university research laboratories. As a result, U.S. surimi production from Alaska pollock has grown from about 4,000 t in 1984 to an estimated 160,000 t in 1997, employing 19 factory ships and four shore plants. In nine years, the U.S. consumption of surimi analog products rose from 2,700 t (6 million lb) in 1980 to an estimated 80,000 t (176 million lb) in 1996 (1). The world surimi production topped 540,000 t (2) in 1992. As the Alaska pollock resource weakens, Southern blue whiting (*Micromesistius australis*) caught off Argentine and Pacific whiting (*Merluccius productus*) off the U.S. West Coast became the next largest sources for surimi production.

The interest of the U.S. industry in these products stems from the following: the surimi market continues to grow worldwide; new products with high profit margins can be developed; nontraditional fish species can be processed at a profit; and the nonseafood food companies can enter the manufacturing of surimi seafood products without having to become involved in fish processing. This article focuses on the historical background of surimi technology, and the important physical and chemical principles behind the process for manufacturing surimi and surimi-based seafood products.

Traditionally, Japanese surimi was freshly prepared from fresh fish and immediately processed into kamaboko products. Kamaboko is a generic term that includes a variety of products prepared from surimi and is distinguished from texturized shellfish meat analogs, which are a new breed of surimi-based products. The technique for making kamaboko products from minced and washed fish evolved around A.D. 1100, when Japanese fishermen discovered that they could keep the product longer if washed minced fish was mixed with salt, ground, and steam cooked or broiled (Fig. 1). Although all products made from surimi are generally called kamaboko, strictly speaking, kamaboko is one that is mounted on a wood plate and steamed or broiled. There are other narrowly defined products such as chikuwa, which is broiled, and tempura, which is fried.

The traditional surimi production was run on a day-to-day basis, depending on the supply of fresh fish. Consequently, the surimi industry could not expand to any great extent and remained a small-scale operation. However, in 1959, the surimi industry took a new turn when a group of scientists at the Hokkaido Fisheries Laboratories discovered a technique to stabilize frozen surimi (3,4). This discovery was made from an incidental finding of a cryoprotectant that kept the surimi from freeze denaturation during frozen storage. This technique enabled Japanese manufacturers to stockpile surimi. Previously, most of the surimi was produced on shore, but subsequently about half of the surimi has been produced on processing ships as a result of an intensive joint effort by government and industry to mechanize on-board surimi production. Subsequently, frozen surimi production increased from 32,000 t in 1965 to 380,000 t in 1975, together with a record production of 1.1 million t of kamaboko (5,6).

THE SURIMI PROCESS

A typical surimi process on a pilot-plant scale is shown in Figure 2. The fish is headed, gutted, and cleaned in a washing tank. The washed fish is then put through a belt-drum-type meat separator that separates the flesh from the bone and skin. The fillets processed by a high-speed filleter are often favored over headed and gutted fish for surimi of superior quality. The diameter of the drum perforations ranges from 4 to 7 mm and is selected in accordance with the size and freshness of the fish (7). For the production of high-quality surimi with limited washing, particularly in the case of on-board processing, the backbone is mechanically removed, or the fillet is used. Filleting is done mechanically using a filleting machine at a fast speed.

Basically, surimi is produced by repeatedly washing mechanically separated fish flesh with chilled water (10°C) until it becomes odorless and colorless or, technically, until most of the water-soluble proteins are removed. The water temperature does not need to be kept unnecessarily low. It should be determined by the type of fish (species), specifically the thermostability of fish protein; for example, 10°C for Alaska pollock (a cold-water fish) and 15°C for red hake (*Urophycis chuss*, a temperate-water fish) (8). On the basis

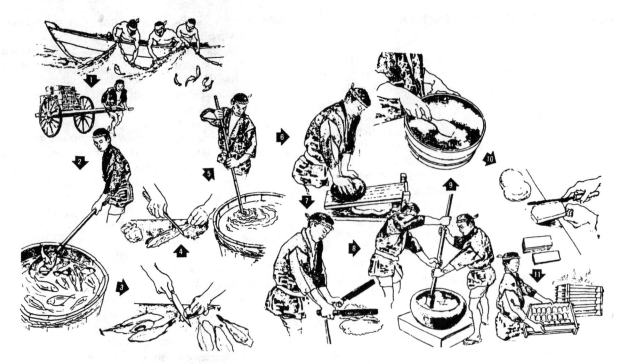

Figure 1. An old surimi and kamaboko making process; (**1**) fish harvest, (**2**) gutting and cleaning, (**3**) filleting, (**4**) mincing, (**5**) washing, (**6**) dewatering, (**7**) chopping, (**8**) stone grinding with salt and spices, (**9**) straining, (**10**) shaping of kamaboko, and (**11**) steaming of kamaboko. *Source:* Courtesy of Suzuhiro Kamaboko Kogyo Co., Ltd., Odawara, Japan.

of the relationship between species and thermostability of actomyosin ATPase (9), it can be assumed that warm-water fish can tolerate a higher water temperature than cold-water fish without a reduction in protein functionality.

In a washing process, the number of washings, the volume of water, and the washing time will vary with the fish species, the initial condition of the fish (freshness), the type of washing unit, the ratio of water to meat weight, and the desired quality of surimi (10). Generally, in the shore plant process three washings are recommended at a 4:1 ratio of water to meat, while in the factory ship process two washings are done at two to three parts water to one part meat because of fresh water shortage. Based on 5 min for each washing and two to three washing cycles, 15 to 20 min for the entire leaching process is appropriate for a commercial surimi manufacturing operation.

In a commercial process, the washing is done continuously, with mechanical agitation in a series of washing tanks and a rotary screen rinser. During repeated washings with continuous agitation, much of the water-soluble proteins are removed, along with undesirable substances and enzymes, causing the level of actomyosin to increase. The level of functional actomyosin is a measure of the gel-forming ability of the surimi. This explains why surimi gives a more elastic texture than unwashed minced fish meat. For the last washing, a 0.1 to 0.3% NaCl solution is used to ease the removal of the water. The washing is followed by refining with the aid of a refiner that removes connective tissue, black skin, bone, and scale. The white flesh passed through the refiner's perforations is collected and referred to as first grade, while the portion rejected is

put through the refiner again to recover the additional refined flesh, which is darker and less functional than the first grade. The second-time refined mince is, therefore, referred to as second grade.

The refined flesh is transferred to a screw press, which removes excess water. At this point, the flesh should be white, odorless, and residue free and have 82 to 84% moisture. Using a silent cutter, cryoprotectants sucrose, sorbitol, and sodium tripolyphosphate are mixed into the dewatered flesh at levels of 4, 4 to 5, and 0.2 to 0.3%, respectively. The levels and ratio of sucrose and sorbitol can be adjusted depending on the type of product to be made and the sweetness desired. Some surimi are prepared only with sorbitol, while some are produced only with sucrose but at a level no more than 5% because of sweetness. The principal steps of the rotary screen surimi manufacturing process are summarized in Figure 2.

The material balance of surimi manufacturing is shown in Figure 3. Most of losses occur during washing in the form of water-soluble proteins and secondarily from refining in which connective tissue is removed. Fine particles of water-insoluble myofibrillar proteins are also lost through the screen during draining. A variety of surimi are made from more than 60 different fish species. Each species requires slightly different processing techniques. Primary species include Alaska pollock, Southern blue whiting, Pacific whiting, and treadfin bream (*Nemipterus tambuloides*) in Thailand, which are processed in commercial volume for the production of frozen surimi.

Pacific whiting is caught off the U.S. West Coast, and its estimated production has exceeded 30,000 t in 1997.

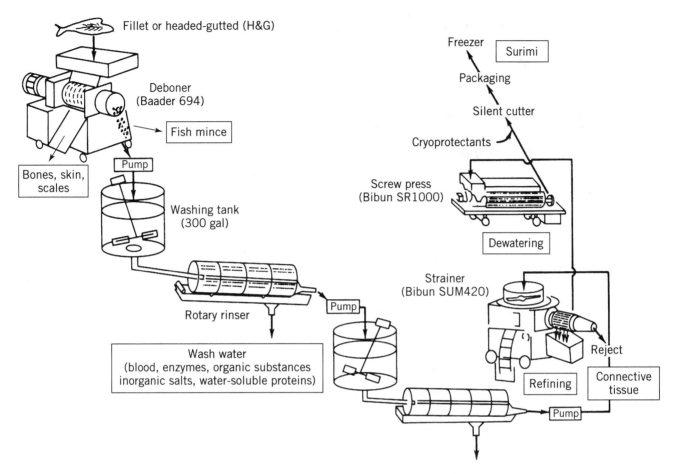

Figure 2. Flowchart of a pilot-plant surimi manufacturing process. *Source:* Courtesy of the Department of Food Science and Nutrition, University of Rhode Island.

Due to a strong parasitic proteolytic activity, Pacific whiting requires a special process that includes holding fish at below 5°C and the use of enzyme inhibitor such as beef plasma protein (1%) or egg white (2%) in mince prior to freezing. Currently, the surimi produced from Alaska pollock and Southern blue whiting make up the majority of the commercial production and command superior quality in terms of color, odor, and gel-forming ability.

The factory ship–processed surimi always commands high quality because of the freshness of the fish used and the prior removal of the backbone and belly flap. This leaves a minimum amount of blood and cavity residues with no autolysis (proteolytic degradation) in deboned fish mince, requiring less water in washing. Due to limited water availability, it is an absolute necessity for factory ships to have minimum washing using the freshest fish mince with minimum organ residues, whereas shore plants process boat-delivered fish, which could be one to five days old, depending on the length of the fishing trip. Thus, sometimes these shore-processed fish could lack freshness and produce a low-grade surimi. In the shore process, fish is often deboned without prior removal of the backbone and belly flap. Fish that is less fresh and has inadequate deboning requires more extensive washing to remove proteolyzed products, blood, and other undesirable organ resi-

dues and fluids. The preceding explains why only two washing cycles are employed in the ship process, compared with three to four cycles in the shore process. The modern factory ship (Fig. 4) has the capability of processing surimi, as well as by-products such as fish meal and fish oil. The only limitation of the ship process is the water supply, which must be generated from seawater by desalinization.

Cryoprotectants were originally incorporated into the dewatered meat by a kneader or stone grinder; however, this incorporation is now accomplished by a high-speed cutter, which is more effective and faster than the kneader. Caution must be taken not to allow the temperature of the mix to exceed 10°C, above which protein functionality could be damaged.

PROCESS WASTEWATER

Surimi making requires a large amount of fresh water during the washing process; this ranges from 10 to 20 times the weight of the deboned meat, depending on the species, condition of fish, and extent of cleaning required. Currently, all the water used during the surimi process is discharged as wastewater. This is of concern because the wastewater contains an average of 3.4 g of protein per liter

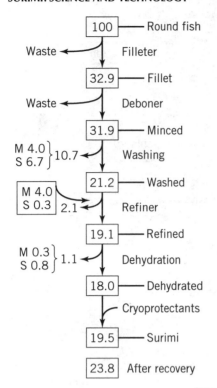

Figure 3. Material balance of surimi manufacturing process. M, myofibrillar proteins; S, sarcoplasmic proteins. *Source:* Ref. 11.

(0.34%). About 80% of the protein present in the wastewater is water soluble. The total protein lost accounts for approximately 30% of the deboned meat weight (12) and varies from plant to plant, depending on the amount of water used and the number of washing cycles employed. The wastewater generated from the shore plants must be treated to reduce its biological oxygen demand (BOD) level before being discharged.

PROTEIN RECOVERY

During washing of minced meat, a considerable amount of water-soluble and some insoluble proteins are lost, as much as 33% (on a minced meat weight basis, Fig. 3), of which water-insoluble myofibrillar proteins in fine particles unretained by a rotary screen can be recovered by a decanter centrifuge or a settling tank with a conical bottom (10). Both have been successfully used to improve the yield in recent years. Some efforts were made without commercial success to recover water-soluble proteins employing an ultrafiltration (UF) membrane system.

GEL QUALITY EVALUATION

Surimi is highly concentrated with myofibrillar proteins, primarily actomyosin. The myofibrillar proteins are solubilized by salt during chopping, which is an initial step of the manufacturing shellfish analog products. The solubilized protein sol, or paste, gels on heating. The gel-forming

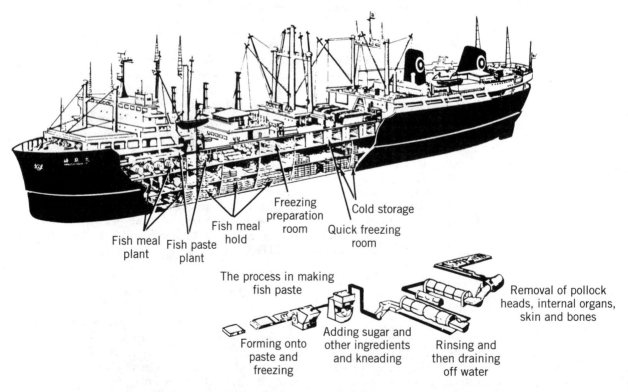

Figure 4. Surimi factory ship, the 21,000-ton Mineshima Maru. *Source:* Courtesy of Hippon Suisan Kaisha Co., Ltd., Tokyo, Japan.

ability, measured by the water-binding capacity of the comminuted tissue (protein solubilization) and gel strength, is determined by the level of functional actomyosin. The level of functional actomyosin, measured as extractable actomyosin or ATPase activity, increases with an increase in the number of washing cycles and decreases as the freshness of the fish decreases; such a decrease in extractable actomyosin is caused by tissue autolysis, which is proteolytic breakdown of myofibrillar protein by catheptic protease. The proteolytic breakdown increases with extended storage or during the spawning season. The quality of surimi during frozen storage is affected by storage temperature, storage period, the level of remaining moisture, the type and level of cryoprotectants used, and remaining inorganic salts (10,13).

The quality of surimi is graded on the basis of its gel-forming ability and the chemical and visual conditions of raw surimi. A new version of the Japanese grading system for frozen surimi was developed initially by Tokai Regional Fisheries Research Laboratory (1980) and adopted by various individual companies with some modifications for its internal standards. The modified standard procedures of Japan Fisheries Ltd. (14) and some of the procedures from the Tokai Regional Fisheries Laboratory are summarized in Table 1.

In the United States, on the other hand, methods for specifying properties of raw surimi instead of a grading system were proposed by the U.S. surimi technical committee under the direction of the National Fisheries Institute (15). Additional testing procedures for surimi (16) and fiberized analog products (17) were also proposed. They were intended not only to supplement the existing Japanese grading system but also to make procedures more practical for the United States, where the system is intended for specification rather than grading. In this way, a suitable type of surimi, which meets the gel-forming requirements of specific formulations, can be selected.

USE OF CRYOPROTECTANTS

Freezing of surimi become commercially possible after the discovery of the cryoprotective role of sucrose, which prevents muscle protein, particularly actomyosin, from denaturation during frozen storage. Initially, 8% sucrose was used, but it made the surimi too sweet and caused a color change during frozen storage. Subsequently, the level of sucrose was reduced to 4%, and 4% sorbitol was added to compensate for this reduction (3). The increased ratio of sorbitol to sucrose resulted in an increase in the gel-forming ability (compressive force), suggesting that sorbitol is more effective cryoprotectant than sucrose. The most desirable sweetness was attained at a 6:2 ratio of sorbitol to sucrose (18). The cryoprotective effect of these sugars was greatly enhanced by adding polyphosphate at a level of 0.2 to 0.3%.

Prevention of protein denaturation by sugars can be explained by their ability to increase the surface tension of water (19) and the amount of bound water, resulting in increased hydration of protein molecules. This prevents withdrawal of water molecules from the protein, thus stabilizing the protein (20).

CHANGES DURING FROZEN STORAGE

The gel-forming ability of surimi made from fresh (one- to two-day-old) fish in good condition does not change significantly for up to one year when held at a constant temperature below −20°C (21,22). However, when the surimi is stored at higher temperatures, the gel-forming ability gradually decreases; this is attributed to a decrease in extractable actomyosin subsequent to freeze denaturation of proteins (22). Temperature fluctuation during short-term transportation does not significantly reduce surimi quality. However, extended periods of temperature fluctuation (more than three weeks) do cause significant loss in quality (23).

GEL FORMATION

Myofibrillar proteins, primarily actomyosin, in surimi need to be solubilized by salt during chopping to form an extrudable paste (sol) essential for fabrication. The surimi paste must be extruded right after chopping so that it can be shaped or formed before it sets to a gel. The extruded paste is subjected to partial heat setting for fiberization, followed by cooking. With or without heat, the surimi sol undergoes gelation where randomly arranged peptide chains irreversibly transform to an ordered state with the aid of both hydrogen and hydrophobic bonds (5,24). This irreversible transition from sol to gel is called *setting*. Hydrogen bonding is primarily involved in setting at low temperature, while hydrophobic bonding dominates in subsequent gel setting at high temperature (25). In a commercial practice, a partial heat-setting is done at temperatures of 40 to 50°C before a final cooking at 80 to 90°C. The textural quality of the finished product is highly dependent upon the way the partial heat setting is carried out. A proper partial heat setting with a right temperature–time relationship is required for development of a desirable texture. Gel-setting behavior and gel strength vary from species to species (26).

Gel setting (swari) takes place at temperatures up to 50°C without cooking and occurs even below room temperature. However, when a comminuted fish (paste) passes through a heating zone of 60 to 70°C, part of the gel structure is altered such that the gel becomes soft. This phenomenon is called softening (modori). Such softening is believed to be associated with proteolytic activity of alkaline protease, because this enzyme is present in many fish species and has an optimum activity at 60 to 70°C (27–29). The degree of gel softening is significantly reduced after washing the fish mince. This further confirms the involvement of protease in gel softening and suggests that gel softening is specific to the type and amount of water-soluble proteins (sarcoplasmic) in the fish tissue. This theory has been challenged by a series of reports in which softening of the texture still occurred after protease was inhibited (30,31). This led to a theory of temperature-dependent gel-setting behavior. At a fast heating rate, a tight, cohesive network with a large number of small aggregates is formed, whereas at a slow heating rate, as in the case of 60 to 70°C heating, a loose network with a small number of large aggregates is formed (32).

Table 1. Japanese Grading System for Quality of Frozen Surimi

Item	Procedure
	Chemical and visual conditions
Moisture	5 g of half-thawed surimi is dried either in the oven at 105°C or in their moisture balance, and the moisture loss determined
pH	10 g of thawed surimi is homogenized with 90 mL of distilled water, and pH is measured
Impurities	The presence of impurities is scored by counting the number of black skin and bone residues in a 10 cm × 10 cm field, which is prepared by spreading 10 g of surimi in 1 mm thickness
Whiteness	The whiteness of surimi gel is measured by a Hunter color meter (model C-1) with red filter (620 mm)
	Physical properties
Expressible drip	50 g of thawed surimi is placed in a cylinder (35 mm in diameter × 150 mm long) having a base with a 3 mm mesh and pressed at a 500-g load for 5–10 min and at an additional 500-g load for 20 min; expressible drip (%) = (drip weight/sample weight) × 100
Viscosity	143 g of thawed surimi is homogenized for 8 min at 8–10°C with 857 mL of 3.5% NaCl solution in an antifoaming mixer having cooling capability; after 40 min at room temperature, the viscosity is measured using a Brookfield viscometer (Tokyo Instruments, Model C) at 10°C at 4 rps
	Gel-forming ability
Gel preparation	A gel is prepared using 2–5 kg of thawed surimi and 3% salt by grinding in a kneader for 30 min or chopping in a silent cutter for 15 min; during chopping, the temperature of the surimi paste must be kept below 10°C to preserve the functionality of actomyosin; potato starch is added according to the grade: none to AA grade, 3% to A grade, and 5% to B grade; the resulting paste is stuffed into a 30-mm-diameter polyvinylidene chloride casing, linked in 25-cm lengths, and immediately heated in a water bath at 90°C for 40 min; it is then cooled in cold water and left at room temperature for 18–48 h until tested
Gel strength	The gel strength is measured by either a rheometer or an Okada gelometer using a cylindrical specimen (30 mm in diameter × 25 mm long) and a 5-mm-diameter plunger with a round end, gel strength (g-cm) = breaking force × deformation
Folding test	Grade is determined according to the extent of crack when a 33-mm-thick slice of gel is folded between thumb and index finger AA = No crack occurs after folding twice A = Crack occurs after folding twice, but no crack occurs after folding once B = Crack occurs gradually after folding once C = Crack occurs immediately after folding once D = Breakable by finger press without folding
Sensory score test	Firmness is scored by the bite test: Extremely firm = 10 Very firm = 8 Moderately firm = 6 Slightly soft = 4 Very soft = 2 Extremely soft = 1

The texture of gels prepared from a given surimi system is affected by the moisture content of the surimi, the levels of added salt and polyphosphates, the extent of actomyosin solubilization (chopping time), pH, and heating schedule. Generally, the gel weakens as the moisture content increases. The salt concentration required for gel formation ranges from a minimum of 2% to a maximum of 3% of the weight of surimi, depending on the process and saltiness requirements; the average level used in the commercial process is 2.5%, where the formula contains other ingredients in addition to surimi. When only surimi and water are used in the formula, the gel strength of Alaska pollock surimi reaches a maximum at 0.26 M NaCl (1.5%) on a gel weight basis and gradually decreased as a salting-out effect occurred on further increases in salt concentration (8). Solubilization of actomyosin increases with chopping time and reaches a maximum. The chopping time for maximum solubilization is determined by the amount of surimi used and the size and speed of chopper. Beyond this maximum chopping time, the mechanical shear damage and temperature rise occur along with protein–protein interaction, resulting in a decrease in gel-forming ability (33,34). Therefore, chopping should be stopped when a sufficient solubilization is achieved, while keeping the temperature of the batter below the temperature that protein can tolerate. The present Japanese practice calls for chopping at 10°C as a maximum tolerable temperature for Alaska pollock. The maximum allowable chopping temperature thus depends on the type of fish species. Temperate species such as red hake and croaker can tolerate a higher temperature than cold-water species such as Alaska pollock (8). Thermostability of actomyosin is closely related to the body temperature of the fish and the temperature of the water where the fish lives. The lower the water temperature, the more unstable the fish muscle actomyosin is (9,35).

Elasticity and resilience of surimi gel increase with an increase in the concentration of actomyosin, but decrease with an increase in the concentration of water-soluble (sarcoplasmic) proteins. The presence of water-soluble proteins retards gel setting by interfering with the actomyosin cross-linking process. A mechanism was proposed whereby water-soluble proteins bind actomyosin, making it less available for the cross-linking process (36,37).

The strength of surimi gel is affected by the pH of the comminuted surimi paste. The optimum pH varies with the fish species from which surimi is prepared, as well as with formulation, because ingredients interact with one another. A pH around 7 with a moderate salt concentration gives maximum gel strength. A pH below 7 results in a sharper decrease in the gel strength than above 7 (5). The pH dependency of gel formation is related to the dependence of water-binding ability of myofibrillar protein on pH (38).

PRODUCTION OF SURIMI-BASED PRODUCTS

Surimi-based products are prepared by extruding the surimi paste into various shapes that resemble shellfish meat such as crab, lobster, scallop, or shrimp. The closer the simulation desired, the greater the sophistication of the extrusion technique that is required. The products may be divided into four major categories according to their fabrication and structural features: molded, fiberized, composite-molded, and emulsified (Fig. 5).

Molded Products

These products are produced by molding the chopped surimi paste into the desired shape and allowing it to set and form an elastic gel. Molding may be done by either a single extrusion or a coextrusion. For the former, the paste is extruded through a single opening of the nozzle without concurrent texturization. For the latter, the paste is extruded through a nozzle having many separate openings such that strings of extrudate are laid over one another during forming. Coextrusion, therefore, gives a meatlike texture, whereas the single extrusion results in a uniform and rather rubbery mouthfeel.

Fiberized Products

As shown in Figure 6, these products are produced by extruding the paste into a thin sheet through a rectangular nozzle having a narrow opening (1–2 mm). The extruded sheet is then partially heat set and cut into strips of desired width by a cutter, similar to a noodle cutter, having a clearance that allows only partial cutting (<4/5 of the thickness), so that a sheet of strips results. The surimi used in this process should be of high grade so that the paste remains sufficiently cohesive and elastic while it is stretched, cut, and pulled during a fabrication process. The greatest pulling tension occurs between the cutter and wrapper. Strip width is determined by the type of finished product desired. Fine strips are preferred for the fibrous

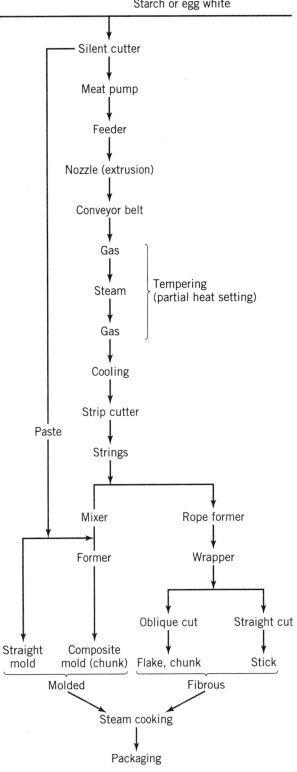

Figure 5. Flowchart of process for fibrous, flake, chunk, and composite-mold-type products. *Source:* Courtesy of the Institute of Food Technologists.

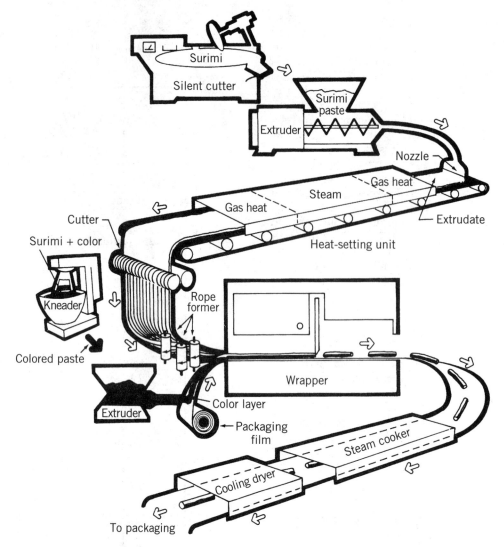

Figure 6. Schematic of small-scale crab leg processing line (capacity 1 t/day). The color layer is transferred from the film to the meat surface during steam heating. *Source:* Courtesy of the Institute of Food Technologists.

crab leg product, whereas wider strips are more suitable for the crab flakes and chunks as well as for scallop analogs.

The resulting sheet of strips is folded into a rope (a bundle of fibers) by a simple narrowing device called a rope former. The rope is then colored, wrapped, and cut into a desired length by a wrapping machine. Coloring is done by applying a thin layer of colored paste onto a wrapping film. The film of colored paste sticks onto the surface of the rope as wrapping and cooking progress. Coloring is also done by directly applying a thin colored extrudate onto the rope without wrapping film. The crab leg product is produced by a straight cut, whereas the flake and chunk types are formed by an oblique cut. During the folding process, the texture of the finished products can be further altered by manipulating the adhesion between the folded layers. Recently, a fiberization by the twin-screw extrusion was developed in Japan. In this process, protein salt-solubilization is followed by texturization of surimi pro-

tein, both done in an extruder. Although the texturization by extrusion is found to be efficient and less labor intensive, the process has not been universally accepted due to inferior texture and flavor retention compared to the conventional fiberization method.

Composite-Molded Products

For these products, the strings or shreds of desired length are mixed with surimi paste and extruded into a desired shape. Strings or shreds are produced either by the method just described or by shredding a block of surimi gel into thin rectangular pieces (<1 mm thick). Texture can be manipulated by adjusting the mixing ratio of strings and shreds and surimi paste (binder). This type of product gives a better bite than the strictly molded variety, which often tends to be rubbery and uniform in texture. Composite-molded products are found in stick form and sold mixed with fiberized products. Most typical products under this category that can be found in the market are shrimp and lobster analogs. Another type of composite-molded product called *fish ham* is prepared by mixing the dice of cured tuna and pork into the fish paste before extrusion.

Emulsified Products

To make this type of product, surimi is treated in a manner similar to that used when meat is processed for emulsion products. The level of fat added is usually less than 10%, and the type of fat used is not limited to animal fat. In fact, vegetable oil is often added, because unlike mammal and bird meat, fish meat readily produces a stable emulsion with oil (39). The wiener-type products have been developed and successfully marketed in Japan. Sausage-type products can be produced by a method similar to that used for composite-molded products, as illustrated in Figure 7.

EFFECT OF INGREDIENTS ON TEXTURE

The sequence of incorporation of ingredients affects the textural quality of the final product and varies with the type of formulation. In general, salts and a portion of ice-chilled water (one-third) are added first, then the mixture is chopped for the first third of the chopping period to allow

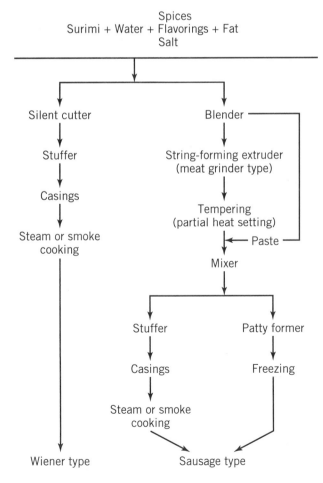

Figure 7. Flowchart of process for surimi-based emulsion-type products. *Source:* Courtesy of the Institute of Food Technologists.

the solubilization of myofibrillar protein. Next, for the improvement of texture and water binding, starch and egg white (raw, pasteurized, and frozen) are added with the remaining water, and the mixture is chopped for the remainder of the chopping period. Flavor is added during the last third of the chopping period for maximum delivery.

Starch

Currently, corn or wheat starch is used at around 6% to improve the textural properties of surimi gel, or at higher levels for cost reduction. Starch must be uniformly dispersed and gelatinized to strengthen the gel. Starch participates in gel formation as a dispersed phase, whereas protein does so as a continuous phase. Starch increases gel strength and elasticity through composite reinforcement and water binding. Gel-strengthening ability varies greatly from starch to starch. The gel-strengthening ability of starch is affected by its water-binding capacity during gelatinization and the viscosity of the gelatinized starch (40). The greater the water-binding capacity and viscosity of the starch, the greater is its gel-strengthening ability.

In general, strong, elastic gels are produced from unmodified high amylopectin starches, such as potato and waxy corn starches, which are less prone to retrogradation at room temperature. In addition to the water-binding capacity and rheological properties of starch, its retrogradation behavior and interaction with protein during gel network formation are also important factors that should be considered in studying the role of starch in surimi gel formation.

Despite the ability of potato starch and waxy maize to produce strong, elastic gels, the gels prepared with these starches have poor freeze–thaw stability and become rubbery and rigid with extended frozen storage. Addition of unretrogradable modified starch (hydroxypropylated type) greatly improves freeze–thaw stability, but it weakens gel strength in terms of elasticity and firmness. It is, therefore, advisable to use a proper mix of unmodified and modified starch to produce a desirable balance between gel strength and freeze–thaw stability. Starch modification through cross-linking in addition to hydroxypropylation has been a commercial practice to overcome gel-weakening problem associated with use of starch that is only hydroxypropylated.

Proteins

Addition of nonmuscle proteins tends to reduce gel strength and rubberiness. Nonmuscle proteins include egg white, milk protein, soy protein, and wheat gluten. The appropriate reduction in such textural properties leads to texture modifications and often results in an improvement of sensory quality. The gel-texture-modifying effect of nonfish proteins is closely related to their water-binding properties (41,42). Egg white is most commonly used in surimi-based products. Egg white is added in a raw form (<10%) on a surimi weight basis to enhance gel strength. When dry egg white is used, an appropriate amount of water should be added to adjust the moisture content in the finished product. Another important function of egg white is its ability to make the product whiter and glossier.

Hydrocolloids (Gums)

Gums being considered for use in surimi-based seafoods include carrageenan and alginate (extracted from seaweeds) and curdlan and xanthan gum (microbial extract). Among these ι-carrageenan gum has shown a gel strength improvement effect when used in either a dry or a cold hydrated form as low as at 0.2% (dry powder) (43,44). The gel-strengthening ability of gums highly depends on their firmness and elastic properties (45).

Cross-linking Agents

Microbially produced transglutaminase is being applied in surimi to increase the gel strength. The gel-strengthening mechanism involves cross-linking of myosins through the formation of nonsulfide covalent bonds between the carboxyl group of glutamic acid and ε-amino group of lysine during a cold setting, which requires several hours (Seki and Kimura). Therefore, it may not be useful for a high-speed production of crab analog, but suitable for some surimi-based products that require a cold setting as part of the preparation step.

Sodium ascorbate, not ascorbic acid, increases surimi gel strength with maximum at a 0.2% level (46). Its gel-strengthening effect is believed to be from cross-linking through oxidation of sulfhydryl (-SH) groups in proteins. Despite its gel-strengthening effect, ascorbate causes the surimi gel to undergo freeze syneresis with increased expressible moisture and rubbery texture development.

IMPORTANT PROCESSING CONSIDERATIONS

There are a number of important steps and considerations in making surimi-based seafood products:

1. Use of high-quality surimi is essential in making fiberized products that require a highly elastic and resilient texture. Surimi of good quality can be readily recognizable, as its paste becomes tacky, glossy, and translucent on chopping with salt and is extruded smoothly. In contrast, surimi of poor quality produces a dull, opaque, and less tacky paste that breaks easily when extruded.

2. Selection of ingredients is determined by the formulation needs, primarily gel strength and freeze-thaw stability requirements. Type and level to be used should be carefully determined so that a desirable texture can be obtained in the finished products.

3. Water may be added so that the moisture level remains at 78 to 80%, as a maximum, in the paste. The amount of water to be added depends on the quality of the surimi, particularly its water-holding capacity. In a given surimi system, the amount of water to be added must be determined on the basis of the moisture content of surimi, the amount of starch and other water-absorbing ingredients added, and the desired texture, particularly the firmness and elasticity of the finished products.

4. Chopping must be sufficient for maximum solubilization of actomyosin, and carried out below the temperature that fish actomyosin tolerates; for example, 10°C for Alaska pollock.

5. Setting of the paste extrudate should occur slowly and to a moderate degree without surface drying. The setting time is generally controlled by the length and width of the conveyor belt. The longer the setting takes, the better the product becomes. If setting is achieved in a short time at a high temperature, the extrudate becomes less manageable for fiberization, and the texture of finished products may be less elastic and resilient. The flavor may also be thermally altered by the elevated temperature. An optimum temperature–time relationship must be determined to obtain a desirable setting condition.

6. In selecting formulation, several considerations must be made as to what the product requirements are, what form of product will be manufactured (fiberized, molded, or composite-molded), what kind of texture is required (soft-moist, rubbery, fibrous, etc), and what form of product will be marketed (refrigerated or frozen). If it is to be distributed frozen, the expected time span on the market shelf and the kind of freeze–thaw stability needed for that time span must be considered. Other considerations are the type of formulation (starch, protein, starch–protein combination, low sugar, and salt), the nutritional equivalency, and the level of ingredients to be used (low or high). These are all interrelated and the optimization of formulation would vary with each ingredient group, namely, starch, protein, or starch–protein.

BIBLIOGRAPHY

1. National Marine Fisheries Service, *Fishery Market News Report*, National Marine Fisheries Service, Washington, D.C., 1990.

2. L. de Franssu, "The World Surimi Industry Prospects for Europe," *FAO/Globefish Research Programme*, Vol. 18, FAO, Rome, 1991.

3. Jpn. Pat. 306,857 (1961), K. Nishiya.

4. K. Tamoto et al., "Studies on Freezing of Surimi and Its Application. IV. On the Effect of Sugar upon the Keeping Quality of Frozen Alaska Pollock Meat," *Bull. Hokkaido Regional Fish Res. Lab.* **23**, 50–60 (1961).

5. M. Okada, "History of Surimi Technology in Japan," in T. C. Lanier and C. M. Lee, eds., *Surimi Technology*, Marcel Dekker, New York, 1992, pp. 3–22.

6. J. J. Matsumoto, "Minced Fish Technology and Its Potential for Developing Countries," in *Proceedings on Fish Utilization Technology and Marketing*, Vol. 18, Indo-Pacific Fishery Commission, Bangkok, 1978, sec. HI, p. 267.

7. K. Toyoda et al., "The Surimi Manufacturing Process," in T. C. Lanier and C. M. Lee, eds., *Surimi Technology*, Marcel Dekker, New York, 1992, pp. 79–112.

8. M. Douglas-Schwarz and C. M. Lee, "Comparison of the Thermostability of Red Hake and Alaska Pollock Surimi during Processing," *J. Food Sci.* **53**, 1347–1351 (1988).

9. K. Arai, K. Kawamura, and C. Hayashi, "The Relative Thermostabilities of the Actomyosin-ATPase from the Dorsal Muscles of Various Fish Species," *Bull. Jpn. Soc. Sci. Fish* **39**, 1077–1085 (1973).

10. C. M. Lee, "Surimi Processing from Lean Fish," in F. Shahidi and J. R. Botta, eds., *Seafoods: Chemistry, Processing Technology and Quality*, Blackie Academic & Professional, London, 1994, pp. 263–287.

11. C. M. Lee, "Surimi Manufacturing and Fabrication of Surimi-Based Products," *Food Technol.* **40**(3), 115–124 (1986).

12. H. Watanabe et al., "An Estimation of the Amount of Protein Lost in the Effluent from Frozen Surimi Manufacture," *Bull. Jpn. Soc. Sci. Fish* **48**, 869–871 (1982).

13. T. Tamoto, "Effect of Leaching on the Freezing Denaturation of Surimi Proteins," *New Food Industry* **13**, 61–69 (1971).

14. Nippon Suisan Kaisha, Ltd., *Standard Procedure for Quality Evaluation of Frozen Surimi*, Nippon Suisan Kaisha, Ltd., Tokyo, 1980.

15. National Fisheries Institute, *A Manual of Standard Methods for Measuring and Specifying the Properties of Surimi*, National Fisheries Institute, Washington, D.C., 1991.

16. C. M. Lee, "Surimi Process Technology," *Food Technol.* **38**(11), 69–80 (1984).

17. K. S. Yoon and C. M. Lee, "Effect of Powdered Cellulose on the Texture and Freeze–Thaw Stability of Surimi-Based Shellfish Analog Products," *J. Food Sci.* **55**, 87–91 (1990).

18. P. Akengo and C. M. Lee, "Assessment of Cryoprotective Effectiveness of Various Combinations of Sucrose and Sorbitol and the Manner of Incorporation," paper presented at the 30th Atlantic Fisheries Technological Conference, Boston, Mass., August 25–29, 1985.

19. T. Arakawa and S. N. Timasheff, "Stabilization of Protein Structure by Sugars," *Biochemistry* **21**, 6536–6544 (1982).

20. J. J. Matsumoto and S. F. Noguchi, "Cryostabilization of Protein in Surimi," in T. C. Lanier and C. M. Lee, eds., *Surimi Technology*, Marcel Dekker, New York, 1992, pp. 357–388.

21. K. Iwata et al., "Influences of Storage Temperatures on Quality of Frozen Alaska Pollock Surimi," *Reigto (Refrigeration)* **43**, 1145–1150 (1968).

22. K. Iwata et al., "Study of the Quality of Frozen Stored Alaska Pollock Surimi. I. The Influence of Freshness of the Material and Changes in Storage Temperature," *Bull. Jpn. Soc. Sci. Fish* **37**, 626–633 (1971).

23. M. Okada et al., "Influences of Transportation on the Quality of Frozen Alaska Pollock Surimi," *Reito (Refrigeration)* **43**, 1270–1277 (1968).

24. E. Niwa and M. Miyake, "Physico-Chemical Behavior of Fish Meat Proteins. I. Behavior of Polypeptide Chains of Proteins during Setting of Fish Meat Paste," *Bull. Jpn. Soc. Sci. Fish* **37**, 877–883 (1971).

25. E. Niwa, Y. Matsubara, and I. Hamada, "Hydrogen and Other Polar Bondings in Fish Flesh Gel and Setting Gel," *Bull. Jpn. Soc. Sci. Fish* **48**, 667–670 (1982).

26. Y. Shimizu, R. Machida, and S. Takenami, "Species Variations in the Gel-Forming Characteristics of Fish Meat Paste," *Bull. Jpn. Soc. Sci. Fish* **47**, 95–104 (1981).

27. Y. Makinodan and S. Ikeda, "Studies on Fish Muscle Protease. IV. Relationship between Himodori of Kamaboko and Muscle Protease," *Bull. Jpn. Soc. Sci. Fish* **37**, 518–523 (1971).

28. C. S. Cheng, D. D. Hammann, and N. B. Webb, "Effect of Thermal Processing on Minced Fish Gel Texture," *J. Food Sci.* **44**, 1080–1086 (1979).

29. T. C. Lanier et al., "Effects of Alkaline Protease in Minced Fish on Texture of Heat-Processed Gels," *J. Food Sci.* **46**, 1643–1645 (1981).

30. K. Iwata, K. Kobashi, and J. Hase, "Studies on Muscle Alkaline Protease. H. Sime Enzymatic Properties of Carp Muscular Alkaline Protease," *Bull. Jpn. Soc. Sci. Fish* **40**, 189–200 (1974).

31. C. M. Lee, "Differentiation between Enzymatic and Thermal Effects on Textural Modification in Fish Gels," paper presented at the 43rd Annual Meeting of the Institute of Food Technologists, New Orleans, La., June 19–22, 1983.

32. E. Niwa, "Functional Aspect of Surimi," in R. E. Martin and R. L. Collete, eds., *Proceedings of the International Symposium on Engineered Seafood-Including Surimi*, National Fisheries Institute, Washington, D.C., 1985, pp. 136–142.

33. J. C. Deng, R. T. Toledo, and D. A. Lillard, "The Effect of Temperature and pH on Protein–Protein Interactions in Actomyosin Solutions," *J. Food Sci.* **41**, 273–277 (1976).

34. C. M. Lee and R. T. Toledo, "Factors Affecting Textural Characteristics of Cooked Comminuted Fish Muscle," *J. Food Sci.* **41**, 391–397 (1976).

35. I. Kimura, T. Murozuka, and K. Arai, "Comparative Studies on Biochemical Properties of Myosins from Frozen Muscle of Marine Fishes," *Bull. Jpn. Soc. Sci. Fish* **43**, 315–321 (1977).

36. M. Okada, "Effect of Washing on the Jelly Forming Ability of Fish Meat," *Bull. Jpn. Soc. Sci. Fish* **30**, 255–261 (1964).

37. Y. Shimizu and F. Nishioka, "Interactions between Horse Mackerel Actomyosin and Sarcoplasmic Proteins during Heat Coagulation," *Bull. Jpn. Soc. Sci. Fish* **40**, 231–234 (1974).

38. R. Hamm, "Functional Properties of the Myofibrillar System and Their Measurements," in P. J. Bechtel, ed., *Muscle as Food*, Academic, Orlando, Fla., 1986, pp. 135–199.

39. C. M. Lee and A. Abdollahi, "Effect of Physical Properties of Plastic Fat on Structure and Material Properties of Fish Protein Gels," *J. Food Sci.* **40**, 1755–1759 (1981).

40. J. M. Kim and C. M. Lee, "Effect of Starch on Textural Properties of Surimi Gel," *J. Food Sci.* **52**, 722–725 (1987).

41. K. H. Chung and C. M. Lee, "Relationship between Physio-Chemical Properties of Nonfish Protein and Textural Properties of Protein-Incorporated Surimi Gel," *J. Food Sci.* **55**, 972–975 (1990).

42. C. M. Lee and K. H. Chung, "The Role of Hydrodynamic Properties of Biopolymers in Texture Strengthening/Modification and Freeze–Thaw Stabilizing of Surimi Gel," in M. N. Voigt and R. Botta, eds., *Advances in Fisheries, Technology and Biotechnology for Increased Profitability*, Technomic Publishing, Lancaster, Pa., 1990, pp. 397–412.

43. C. W. Bullens et al., "The Function of Carrageenan-Based Stabilizers to Improve Quality of Fabricated Seafood Products," in M. N. Voigt and R. Botta, eds., *Advances in Fisheries, Technology and Biotechnology for Increased Profitability*, Technomic Publishing, Lancaster, Pa., 1990, pp. 313–323.

44. I. Filipi and C. M. Lee, "Preactivated Iota-carrageenan and Its Rheological Effects in Composite Surimi Gel," *Lebensm.-Wiss. u.-Technol.* **31**, 129–137 (1998).

45. C. M. Lee, "Relationships of Rheological and Hydrodynamic Properties of Biopolymeric Ingredients to the Composite Surimi Gel Texture," in T. Yano, R. Matsuno and K. Nakamura, eds., *Developments in Food Engineering—Part 1*, Blackie Academic & Professional, London, 1994, pp. 99–101.

46. H. G. Lee et al., "Sodium Ascorbate Affects Surimi Gel-Forming Properties," *J. Food Sci.* **57**, 1343–1347 (1992).

CHONG M. LEE
University of Rhode Island
Kingston, Rhode Island

SWEETENERS: NUTRITIVE

Sweeteners are used to satisfy the inborn human desire for sweet taste. They also make sour and bitter foods more palatable. And the functionalities non-high-intensity sweeteners impart to prepared foods is often as important as, or more important than, their taste characteristics. Each sweetener, either natural or synthetic, has specific properties and therefore specific applications and limitations. For this reason and because the demand and hence the market for sweeteners is so great, many different sweeteners have been developed, are available, and are used.

Those sweeteners that provide calories to humans are categorized as nutritive sweeteners (1); those that do not are nonnutritive sweeteners (see the article SWEETENERS: NONNUTRITIVE). Nutritive sweeteners are covered in this article with one exception. Peptide sweeteners such as aspartame (APM) are classified as nutritive by the U.S. Food and Drug Administration because they, like other peptides and proteins, contribute approximately the same number of calories per gram as do carbohydrates. However, because they are high-intensity sweeteners, they need only be used in small amounts and therefore contribute an insignificant number of calories to the products in which they are used; so in this work, they are considered as nonnutritive sweeteners.

Table 1 lists the nutritive sweeteners available for food products. Relative amounts used depend on the country and/or region and are a function of the food type, such as bakery product, beverage, or canned fruit. In the remainder of this article, the term sweetener refers to the sweeteners in Table 1, that is, nutritive sweeteners as defined previously.

Table 1. Non-High-Intensity, Nutritive Sweeteners

Naturally occurring	
Sucrose	Maple syrup
Honey	

Derived from starch	
Glucose (corn) syrups	Glucose (dextrose)
High-fructosee syrups	Fructose (levulose)
Corn syrup solids	Reduced corn syrups
Maltodextrins	Trehalose

Polyhydric alcohols	
Sorbitol	Reduced corn syrups
Mannitol	Erythritol
Xylitol	

Derived from sucrose	
Invent sugar	

Byproduct of sucrose refining	
Molasses	

Derived from inulin	
Fructose	Inulin syrups

In the United States, sucrose (ca 45%), glucose (corn) syrups (ca 13%), high-fructose syrups (HFCS) (ca 40%), and glucose (dextrose) (ca 2.5%) account for 99% of total sweetener use (2). About 40 billion lb $(20 \times 10^9 \, \text{kg})$ of these products are produced and used in beverages (ca 36%), pharmaceuticals and miscellaneous products (ca 17%), confections (ca 12.5%), bakery products (ca 10%), breakfast cereals (ca 8.5%), dairy products (ca 8%), and preserved fruits and vegetables (ca 8%) (2). (About two-thirds of the sugar used to make alcoholic beverages and bakery products is not actually consumed by humans but used up in fermentation.)

Liquid sweeteners are usually analyzed for and graded in terms of color, clarity, concentration/solids content, pH, sulfur dioxide content, presence of other dissolved solids, and viscosity. If they are produced by hydrolysis of starch, that is, if they are glucose (corn) syrups, the dextrose equivalency (DE) and saccharide distribution is also determined and reported. Crystalline sweeteners are graded by particle size. They are dissolved and analyzed as the liquids are.

In baked goods, sweeteners provide sweetness, texture, and humectancy (1). Providing sweetness is most important in sweet breads, sweet rolls, doughnuts, cakes, and cookies. Sweeteners provide the fermentable substrate for yeast-leavened products. Sweeteners also reduce/retard starch gelatinization (completely in some cases) (see the article STARCH) and gluten development. Moisture content/retention is key to texture (soft and moist vs crisp, for example). The primary determinant of texture is the glass transition temperatures of the ingredients (especially starch, sugar, or both), and the primary determinant of the glass transition temperature is the moisture content. By raising the sugar concentration, the glass transition temperature of wheat starch granules is raised and the degree of starch gelatinization at baking temperatures is reduced. The degree to which this occurs is also a function of the type of sweetener used.

About 55% of the sweetener used to make bakery products is sucrose. Sucrose provides sweetness, bulk, flavor enhancement, texture/crumb structure (by influencing starch gelatinization and gluten development), shelf life, and the fermentable substrate for yeast-leavened products. HFCS, which accounts for ca 25% of the total sweetener used in bakery products, can often replace sucrose, except in certain cookies where crispness results from low moisture content and the glassy state of sucrose. Glucose (corn) syrups (ca 11%) provide moisture retention, freeze-thaw stability, and dough viscosity and control sugar crystallization. Glucose (dextrose) is the most efficient sweetener in terms of promoting browning, that is, crust color development.

Sucrose is also the most important ingredient in confections, in which it is usually present in either a crystalline or glassy state or a mixture of the two states. Crystals are often an important contributor to texture, and thus crystal size is important. Glucose (corn) syrup is often used to prevent crystallization so that the product is a glassy solid. Carbohydrate sweeteners also function as humectants.

Other applications of sweeteners are in the preparation of jams, jellies, preserves, and marmalades; in beverages; in frozen desserts; in the preservation of canned fruits; and in the preparation of breakfast cereals.

SUCROSE

Sucrose may be obtained from different sources, primarily from sugarcane and sugar beet, and is available in a variety of forms. For example, maple syrup is a solution of sucrose (58–66%) and less than 10% of other constituents. Because of its overall dominance worldwide as a sweetener, sucrose is presented separately (see the article SUGAR: SUCROSE).

FRUCTOSE

D-Fructose (levulose) is available in both crystalline and solution forms and is the sweetest (on a weight basis) and most soluble of the natural sugars (1–3). A 5 to 15% aqueous solution of D-fructose at room temperature is about 1.15 to 1.25 times as sweet as an equal concentration of sucrose. (Actual relative sweetness is a function of temperature, pH, and concentration.) In its pure crystalline form, β-D-fructopyranose (see Fig. 1), fructose has 1.8 times the sweetness of sucrose on an equal-weight basis; but few, if any, foods are prepared with crystalline fructose as the sole sweetener. Fructose dissolves rapidly and has a rapidly forming and intense sweet taste. Its sweetness

perception dissipates before that of sucrose peaks. This makes fructose an ideal sweetener for products such as sorbets in which lingering sweetness is not desired. Fructose intensifies perceived flavor and acid tastes, that is, both flavor and acid levels can be reduced when products such as puddings, dessert gels, pie fillings, and beverages are formulated with fructose. A blend of sucrose and fructose is sweeter than the sweetness of either sugar used alone (3).

As both fructose itself and sucrose–fructose blends are sweeter than sucrose alone, less of them is needed for the same degree of sweetness. Fructose and high-fructose syrups are used, therefore, in reduced-calorie foods and beverages, and because fructose is sweeter than sucrose, use of crystalline fructose, usually in combination with granulated sugar, in sweetened, dry-mix fruit drinks reduces package size and weight and shipping costs.

Because of its water-binding capacity, which is greater than that of other sugars, fructose can be used as a humectant in baked goods (where it reduces staling) and in jams, jellies, pet foods, and certain confections. Crystalline fructose can also reduce the water activity of intermediate-moisture foods such as granola bars and fruit products and thus can act as a preservative, but because of its hygroscopic nature, it requires special handling and storage conditions (3). In addition, higher-solids solutions can be prepared with fructose than with sucrose (83% vs 71% at 30°C [86°F], 90% vs 76% at 60°C [140°F]). When used with food starches, fructose provides more rapid development of vis-

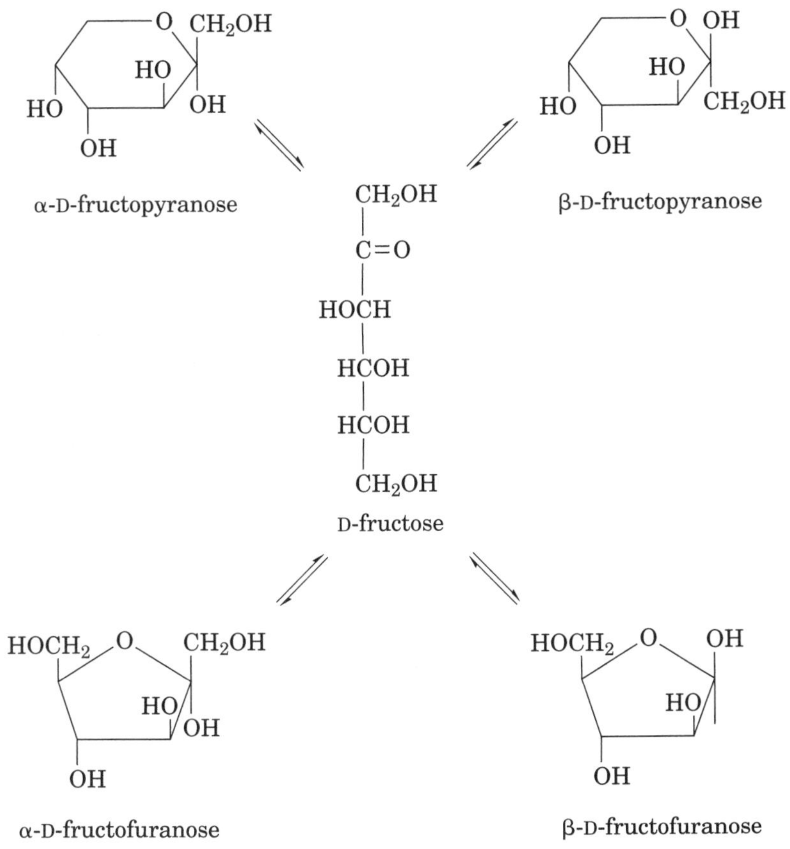

Figure 1.

cosity and increased final gel strength, as compared to sucrose on an equal-weight basis.

Fructose is less glucogenic than is glucose or sucrose, so it has been recommended as a replacement for those sweeteners in the diets of diabetic and obese persons. However, fructose is more lipogenic than is glucose or starches and can cause increases in blood pressure, uric acid, and lactic acid in susceptible segments of the population (4).

In Europe small amounts of inulin syrup, which is primarily fructose with some fructooligosaccharides, is made and used. The primary source of the inulin is chicory. Jerusalem artichoke is an alternative source.

GLUCOSE

D-glucose (dextrose) occurs naturally in low concentrations in many fruits and other plant tissues (Table 2). Glucose used in processed food formulations is produced by enzyme-catalyzed hydrolysis of starch, usually to produce dextrose syrups of DE 95 or DE 99; see the section "Glucose (Corn) Syrups." Crystalline products of DE 93 to DE 99 are available in various granulations. Solutions of glucose (2–50%) are 0.50 to 0.90 times as sweet as those of equal concentrations of sucrose. Its negative heat of solution effects a cool sensation that enhances fruit flavors. It promotes browning. Glucose is available commercially as crystalline anhydrous dextrose and dextrose monohydrate (ca 8.5% water) and in solution, microcrystalline, and agglomerated forms.

Glucose is used in a wide variety of products, particularly in the confectionery industry in jellies, gums, marshmallows, pressed products, panned products, chocolate, and compound coatings; as a bulking agent for high-intensity sweeteners; as a coating for dried fruits such as raisins, dates, apricots, and pears; as a binder in tabletted confections; and as a carrier for liquid flavors and oils. It is used to improve product density in some cakes and cookies.

GLUCOSE (CORN) SYRUPS

Syrups are produced from starch by enzyme-catalyzed hydrolysis or a combination of acid-catalyzed and enzyme-catalyzed hydrolyses (see the section on "Fructose-Glucose Syrups," also see the article CARBOHYDRATES: CLASSIFICATION, CHEMISTRY, LABELING). In the United States, where these syrups are prepared from corn starch, they are known as corn syrups. In Europe, where they may be made from potato or wheat starch, in addition to corn starch, they are known as glucose syrups.

Glucose syrups are described primarily by the term dextrose equivalence (DE), which is defined as the percent reducing sugar calculated as glucose on a dry-weight basis and is an indicator of the degree of hydrolysis. Glucose has a DE value of 100 and starch a DE value of 0. Corn syrups (also labeled corn sweetener) with DE values of from ca 24 to at least 80 are available, those with DE values of 42 and 62 being the most common types. All are colorless, viscous liquids of 74 to 85% solids. Those with the lower DE values have the highest-average-molecular-weight saccharides and those with the higher DE values contain the lowest-average-molecular-weight saccharides (glucose, maltose, etc). Low-DE syrups are the least hygroscopic and impart less stickiness but bind water, preventing its loss. High-DE syrups are the most hygroscopic and the sweetest. A DE 24 syrup contains ca 5% glucose, 6% maltose, 11% maltotriose, and 78% oligosaccharides with degrees of polymerization greater than 3. A DE 80 syrup contains ca 56% glucose.

Corn syrups provide sweetening and viscosity, control humectancy and water activity, give surface glaze, and impart desirable texture to a wide variety of foods, beverages, confectionery, and bakery products, including canned and frozen desserts, meat products, peanut butter, pickles, salad dressings, table syrups, and pet foods.

Syrups can also be described by their saccharide composition (6,7). High-maltose syrups can contain 28 to 65% maltose and little glucose. They provide reduced browning and hygroscopicity when compared with a glucose syrup of equivalent DE. Maltose, or malt sugar, is a disaccharide. It is created in yeast doughs by the action of enzymes on the starch or flour. Barley malt, in which barley starch has been broken down to a high percentage of maltose during germination of the grain, is used as an adjunct in brewing to enhance the color and flavor of beer. Products containing maltose produced by malting or by the action of yeast have a distinctive flavor.

FRUCTOSE–GLUCOSE SYRUPS

Fructose and glucose occur together in HFCS (known outside the United States as high fructose [HF] and isoglucose syrups), invert sugar, honey, and certain fruits and vegetables. HFCS is made from starch by the action of combinations of enzymes (8,9). First, a mixture of α-amylase and glucoamylase, and often a debranching enzyme (pullulan-

Table 2. Sweetener Composition of Some Fruits

	Apple	Grape	Peach	Strawberry
Glucose	2.03 (1.40–2.35)	7.18 (3.98–9.05)	1.03 (0.72–1.50)	2.17 (1.90–2.33)
Fructose	5.74 (4.80–6.40)	7.44 (3.86–9.30)	1.23 (0.86–1.80)	2.30 (2.13–2.40)
Sucrose	2.55 (0.54–2.78)	0.43 (0.18–1.61)	5.73 (4.50–6.80)	1.00 (0.08–1.45)
Sorbitol	0.51 (0.51–0.58)	0.20	0.89 (0.50–0.90)	0.32
Xylitol				0.28

Note: Percent of fresh weight: mean (range).
Source: Ref. 5.

ase or isoamylase [amylo-1,6-glucosidase]) is used to convert native or acid-thinned starch to a solution that is as close to 100% D-glucose as can be obtained (6,7,10). A heat-stable α-amylase is generally used for the initial breakdown (liquefaction) of a native starch paste, which is performed at 80 to 95°C (185–205°F). The resulting glucose solution can be concentrated to a syrup and is the source of crystalline glucose (6,7).

The glucose solution is then treated with an immobilized enzyme commonly known as glucose isomerase (6,7) to convert a portion of the glucose into fructose, that is, to achieve thermodynamic equilibrium. An HFCS containing 42% fructose, 53% glucose, and 5% maltooligosaccharides is prepared in this way. Using an industrial-scale chromatographic separation, fructose pure enough for crystallization (ca 90%) is obtained. By blending a 42% fructose solution with a 90% fructose solution, a 55% fructose (41% glucose and 5% maltooligosaccharides) solution is obtained. In the U.S. soft drink industry, 55% fructose syrup, designated HFCS-55 (HF-55), is the main nutritive sweetener used; some 42% fructose syrups are also used. HFCS-55 has a sweetness ca 1.2 times that of sucrose; therefore, less HFCS, as compared to sucrose, is required to give equivalent sweetness, resulting in products with fewer calories. HFCS is used in many other food, confectionery, and beverage products, including soft cookies, where its humectancy and noncrystallizability are important; other baked goods; salad dressings; pickle products; catsup; table syrups; canned and frozen fruits; desserts; soft chewy candies; and granola bars. HFCS is easy to blend, store, and ship.

INVERT SUGAR

True invert sugar is a 50:50 mixture of glucose and fructose produced by cleavage of the glycosidic linkage of sucrose (see the article SUGAR: SUCROSE) with dilute acid or the enzyme invertase. It too is sweeter than sucrose. Invert sugar and HFCS-50 (HF-50) are identical in composition and properties. Invert sugar acts as a humectant and a water binder and promotes nonenzymic browning. However, although an invert sugar syrup composed predominately of invert sugar can be obtained, more commonly used is a sugar solution with only a portion of the sucrose hydrolyzed. Invert sugar is used in the confectionery industry to control or prevent crystallization of sucrose. In the presence of the proper amount of invert sugar, only small crystals of sucrose that are below the perceptual threshold of the mouth form.

Inversion, a term used to denote the hydrolysis of sucrose to glucose plus fructose, often occurs during processing. Deliberate postprocessing conversion is used, for example, in producing chocolate-covered cherries, where invertase in the fondant acts on sucrose after the product is enrobed with chocolate, turning the fondant into a liquid syrup that has a reduced tendency to undergo fermentation. (Invert sugar is also available in a creamy fondant form.) Inadvertent inversion caused by heating too long or at too high a temperature or in the presence of acid can negatively effect stickiness, graininess, color formation, and shelf life.

Invert sugar can be used in creme centers, fudge, marshmallows, caramels, chewy candies, nougats, and jellies. It can be added to the ingredient blend but is often formed during candy manufacture.

HONEY AND MOLASSES

Honey and molasses are used primarily to impart their flavors. Honey contains about 38% fructose and 31% glucose. In addition to providing its unique flavor and humectancy, honey can enhance spicy or fruity flavors in such products as gum and fruit jelly candies and mint creams. It is a more costly ingredient than HFCS, sucrose, or invert sugar.

Molasses, a by-product of cane sugar refining, also contains a high percentage of invert sugar. Molasses is available in several grades. Lighter-colored molasses are generally used in foods and confectioneries requiring long cooking times. Darker-colored molasses are generally added at the end of processing to products requiring its robust flavor. Both honey and molasses are available in liquid and dry powder forms.

CORN SYRUP SOLIDS

Corn syrup solids, those products in the approximate range of DE 20 to DE 36, are produced by less breakdown of starch than occurs during the manufacture of corn syrups. They have less reducing power and hence a lesser tendency to brown, only slight sweetness, and low to moderate hygroscopicity. They are used as binders and crystallization inhibitors; to impart a chewy texture and improved mouthfeel and flavor; and to provide easy digestibility and control of sweetness in such products as infant formulas, chocolate drinks, coffee whiteners, whipped toppings, imitation sour creams, frozen foods, prepared meats, and granola bars. In frozen novelties, corn syrup solids provide viscosity, an increased freezing point, controlled water crystallization, more body, slower meltdown, better stability, and a smoother texture.

MALTODEXTRINS

Maltodextrins have DE values from ca 5 to ca 15 and are even less sweet than corn syrup solids. They find application because of their quick dispersion, rapid solubility, film-forming ability, low hygroscopicity, crystallization-inhibitor property, ability to impart body, low browning potential, low osmolality, resistance to caking, flavor-carrying ability, good binding power, fat dispersibility, textural effects, and ability to provide clean flavor in food products such as baked goods (cream-type fillings, glazes, frostings), powdered mixes (soups, salad dressings, sauces, flavors, spice blends), frozen foods and novelties, meat spreads, sauces (white, cheese, pizza), salad dressings, imitation cheeses and sour creams, confectioneries (chewy candies, hard candies, pan coatings, nut and snack coatings), sports beverages, and infant formulas. A key to their use in powdered mixes is their ability to be spray-dried and to produce agglomerated products. Because of their high solubility and low hygroscopicity and their ability to form complexes with some hydrophobic substances, maltodextrins are often used as carriers, bulking agents, or both. Special crystalline maltodextrins find use as fat replacers.

LACTOSE

Lactose (see the article LACTOSE) is only slightly sweet and is used where only low sweetness is desired, for its aroma- and flavor-binding properties, or for a combination of these qualities.

TREHALOSE

Trehalose, a naturally occurring disaccharide of D-glucosyl units that can be produced from starch enzymically, has ca 45% of the sweetness of sucrose and low hygroscopicity. It is noncariogenic and does not contribute to browning. It is reported to be the most effective sugar in protecting proteins against changes effected by drying or freezing and in reducing the rate of starch retrogradation. It is approved for use as a food ingredient in Japan.

SUGAR ALCOHOLS

Alditols (also known as sugar alcohols, polyhydric alcohols, and polyols) with a sweet taste include glucitol (sorbitol), mannitol, xylitol, erythritol, and hydrogenated corn syrups, especially high-DE and high-maltose syrups.

Sorbitol, the most widely used alditol, is one-half to two-thirds as sweet as sucrose, depending on concentration and temperature. It is available in crystalline form and as a 70% solution. Sorbitol can be used as a humectant. It is used in special dietary foods, breath mints, cough drops, hard and soft candies, and chewing gum. Its negative heat of solution provides a cooling effect. Excessive consumption (>50–80 g/day) may have a laxative effect.

Mannitol is also about two-thirds as sweet as sucrose. It is used in sugarless chewing gum as a dusting agent because of its low solubility and low hygroscopicity. It is also used to a limited extent as a bulking agent in powdered foods.

Xylitol has about the same sweetness as sucrose. Commercial xylitol is produced by reduction of D-xylose obtained from corncobs, sugarcane bagasse, birch wood, or a combination of these. Of all nutritive sweeteners, xylitol has the largest negative heat of solution and provides the greatest cooling effect. Xylitol is used in sugarless chewing gum and mints, other confectioneries, dietetic foods, and foods for diabetics.

Erythritol is the newest alditol to be introduced. Like other alditols, it is classified as a nutritive sweetener because it does have some caloric value, although it is less than 5% that of sucrose. The sweetness of a 10% solution is 50 to 60% that of sucrose at the same concentration. Crystalline erythritol has a strong cooling effect. When

used with high-intensity sweeteners, the sweet taste is improved. It is proposed for use in chewing gum as a softener that extends shelf life.

Sorbitol, mannitol, xylitol, and erythritol all have good heat stability, resistance to thermal and nonenzymic browning, and a low tendency to undergo fermentation with the production of acids and are considered to be noncariogenic. There is evidence that xylitol may actually be effective in reducing the number of new dental caries (11). All sugar alcohols have about the same caloric content as sucrose but are absorbed more slowly from the digestive tract and do not raise postprandial blood sugar and insulin levels; thus they are suitable for the diets of diabetics.

Chemical reduction (hydrogenation) of syrups, especially high-maltose syrups, produces products available as viscous 75% solutions with a sweetness about three-fourths that of sucrose. These syrups, like regular glucose syrups, do not crystallize, prevent crystallization of other ingredients, and impart hygroscopicity. They can be used to prepare sugarless chewing gum and hard, chewy, fondant, and jellied candies. Crystalline maltitol is produced and used in Japan.

Lactitol, the reduced form of lactose, is about 40% as sweet as sucrose; has a clean, sweet taste; is produced in the Netherlands; and is approved for food use in the United Kingdom as a sweetener for dietetic foods and foods for diabetics.

BIBLIOGRAPHY

1. R. J. Alexander, *Sweeteners: Nutritive*, Eagen Press, St. Paul, Minn., 1998.

2. R. J. Alexander, "Carbohydrate Sweeteners," *Cereal Foods World* **42**, 420 (1997).

3. "Crystalline Fructose: A Breakthrough in Corn Sweetener Process Technology," *Food Technol.* **41**, 66, 67, 72 (1987).

4. J. Hallfrisch, "Metabolic Effects of Dietary Fructose," *FASEB J.* **4**, 2652–2660 (1990).

5. H. Scherz and F. Senser, eds., *Food Composition and Nutrition Tables*, 5th ed., CRC Press, Boca Raton, Fla., 1994.

6. N. E. Loyd and W. J. Nelson, "Glucose- and Fructose-containing Sweeteners from Starch," in R. L. Whistler, J. N. BeMiller, and E. F. Paschall, eds., *Starch: Chemistry and Technology*, 2nd ed., Academic Press, Orlando, Fla., 1984, pp. 611–660.

7. G. M. A. van Beynum and J. A. Roels, eds., *Starch Conversion Technology*, Marcel Dekker, New York, 1985.

8. R. Bernetti, "From Corn Syrup to Fructose," *Cereal Foods World* **35**, 390 (1990).

9. J. P. Casey, "High Fructose Corn Syrup: A Case History of Innovation," *Starch* **29**, 196 (1977).

10. Amylose Research Society of Japan, *Handbook of Amylases and Related Enzymes*, Pergamon Press, Oxford, 1988.

11. A. Bar, "Caries Prevention with Xylitol," *World Rev. Nutrition and Diet* **55**, 183–209 (1988).

J. N. BeMiller
Purdue University
West Lafayette, Indiana

SWEETENERS: NONNUTRITIVE

The sense of sweet taste has directed people, as well as many types of animals, insects and probably even microorganisms, to nutritive substances since early in time. Although it is almost certain that the correlation of sweet taste with nutrition was noted earlier, it appears the first written record of this recognition was made by Aristotle in 330 B.C. (1). He speculated that the taste system could be subdivided into the basic taste qualities of sweet, bitter, sour, salty, astringent, pungent, and harsh and that it was the function of this sensory system to direct organisms to sources of nutrition. In modern times, at least for most of the western hemisphere, adequate nourishment has not been an issue. On the contrary, 20th-century westerners generally tend to consume excess calories. Obesity is very common and people are increasingly developing illnesses that are at least partly attributed to excess calorie consumption (eg, cardiovascular disease, diabetes, and cancer), although decreased physical activity is also a factor.

Consumers are very interested in low calorie food and beverage products. Such products require low or zero calorie replacements for full-calorie ingredients. Significant contributors to the caloric content of many foods are sucrose and other nutritive carbohydrate sweeteners. The discovery of the potent sweet taste of saccharin by Fahlberg in 1878 made possible, for the first time, the formulation of zero- or low-calorie equivalents to full calorie food products (2). Since 1878, chemists have synthesized, or identified from nature, thousands of novel, potently sweet organic compounds. Many of these sweeteners are substantially better than saccharin in taste quality. Saccharin's sweetness is limited by the presence of bitter and metallic off tastes.

The sweetener discovery of greatest commercial significance is that of aspartame, a peptide-type sweetener that was serendipitously discovered in 1965 by Schlatter, a chemist at G.D. Searle & Company (3). Aspartame's sweetness is not limited by off-tastes and as such allowed, for the first time, formulation of many low-calorie food products with minimal compromise in taste. Because aspartame so closely mimics sucrose taste, it is not surprising that aspartame has revolutionized the low-calorie food and beverage industry. It must be pointed out that aspartame, because of its natural amino acid content, is a nutritive rather than nonnutritive sweetener. Aspartame is metabolized to provide energy proportional to the amount ingested. However, since about 200 g of sucrose may be replaced by 1 g of aspartame, it functions, in effect, as a nonnutritive sweetener.

Aspartame, saccharin, and other sweeteners that are substantially more potent than sucrose are generally referred to as high-potency sweeteners. These sweeteners are often erroneously called intense or high-intensity sweeteners. This latter terminology is confusing and therefore not recommended because no sweet substances are known to exhibit a higher intensity of sweetness than is possible with sucrose. Occasionally, high-potency sweeteners are referred to as artificial sweeteners. This term is also inaccurate because sweet taste from any source is real and not artificial. It is recommended that sweeteners be referred to as nutritive, nonnutritive, or high-potency, as accurately describes the particular substance. This review will focus on nonnutritive and low-calorie sweeteners, among which sweetness potencies vary from less than 1 to more than 10,000 times that of sucrose.

In some food systems, such as confections, frozen desserts, and baked goods, acceptable products cannot be prepared by simply replacing sugar with high-potency sweeteners. In these systems, sucrose and other nutritive carbohydrate sweeteners provide functionality beyond sweet taste. In confections, sucrose may provide nearly 100% of the product bulk, and in baked goods, the contribution to product bulk is important but somewhat less so. In some baked goods, sucrose is a substrate for yeast because the action of yeast is essential for development of the airy texture. Sucrose also provides, following hydrolysis, the necessary reducing sugar needed for Maillard browning of baked goods. In frozen desserts, sucrose performs the key function of freezing-point depression. Without sucrose or other sugars, frozen desserts are not malleable. Thus, for many foods, it is not possible to prepare low-calorie alternatives with high-potency sweeteners. Necessary for the formulation of such food products are nonnutritive sweeteners that are one-for-one substitutions for sucrose. These sweeteners are often referred to as sugar macronutrient substitutes or sugar MNSs.

This article begins with a discussion of the properties required for commercial viability of nonnutritive sweeteners. The nonnutritive sweeteners of broadest use in the food industry today are saccharin salts, cyclamate salts, aspartame, and acesulfame-K. Two sweeteners recently approved in many countries and that may experience significant use in the future are sucralose and alitame. These six sweeteners are discussed in detail. Five nonnutritive sweeteners that have experienced limited application in foods are neohesperidin dihydrochalcone, thaumatin, glycyrrhizin, the stevia sweeteners (stevioside, rebaudioside A, etc) and Lo Han Kuo sweetener (mogroside). Short summaries of these sweeteners are given. These 11 sweeteners are all high-potency sweeteners. Five sugar MNSs that are being used in foods are the polyols maltitol, lactitol, isomalt, fructooligosaccharide sweetener, and erythritol. These sweeteners are also discussed. Finally, the article concludes with a summary of the most significant recent developments in the field of nonnutritive sweeteners.

REQUIREMENTS FOR COMMERCIAL VIABILITY OF NONNUTRITIVE SWEETENERS

Taste Quality

The number of food products sweetened with nonnutritive sweeteners has increased dramatically since the 1960s. The most successful of these products are beverages, especially carbonated soft drinks (CSDs). Good tasting zero- or near-zero-calorie CSDs are now available as alternatives to the sucrose and high-fructose corn syrup sweetened products that have approximately 150 calories per 12-oz serving. This commercial success only occurred, however, once good-tasting noncalorie sweeteners approaching the quality of sucrose were identified. Consumers have never shown any significant willingness to sacrifice taste quality in their choices of food products. Helgren of Abbott Laboratories made a key discovery in 1957, that greatly facilitated the 1960s burst of growth in low-calorie foods (4). He found that the combination of sodium saccharin with sodium or calcium cyclamate, in a ratio such that each sweetener contributed equal sweetness to the mixture, resulted in a product with improved taste quality relative to either sweetener taken separately. This blend was one of approximately 10 parts cyclamate salt to 1 part of the saccharin salt, since the latter is roughly 10 times more potent than the former. Cyclamate and especially saccharin salts exhibit off-tastes that are often described as bitter and metallic. As a result of these off-tastes, neither saccharin- nor cyclamate-sweetened products approach the taste quality of sugar-sweetened products. Helgren found, however, that these off-tastes are substantially ameliorated in the 10:1 cyclamate/saccharin blend. In fact, the improvement was so significant that, for the first time, zero- and low-calorie alternatives to many food and beverage products were possible without major compromise in taste.

The taste quality of a sweetener is really only meaningful in the context of a food product, and the taste quality of a food product is best assessed by consumer studies. However, less resource-intensive techniques are often used to predict taste quality. Flavor profile analysis (FPA) is such a technique (5). In FPA, pioneered at the Arthur D. Little Company in the 1940s, expert sensory panels are trained to break down complex and multiple-flavor-attribute systems and to rate component intensities. The FPA technique has also been used to assess the taste attributes of sweeteners. As examples, the flavor profiles of saccharin and cyclamate are illustrated in Figure 1. These data, as well as most other quantitative sensory data provided in this review, were generated by an expert sensory panel of The NutraSweet Company. The sensory studies were conducted according to the methodology described in the work by Carr et al. (6).

Some nonnutritive sweeteners are quite similar to sucrose in flavor profile. One might expect that such sweeteners would make possible diet products equivalent in taste to sucrose-sweetened products. However, this is not the case. One factor contributing to the difference in taste between the nonnutritive-sweetener- and sucrose-sweetened products is sweetener temporal profile (7). DuBois et al. have comprehensively studied this effect in the flavonoid glycoside class of sweeteners, the best-known member of which is neohesperidin dihydrochalcone (8). In the case of this sweetener, 9 s are required to reach the sweetness intensity maximum for a solution *iso*-sweet with 10% sucrose. This time has been defined as the sweet-taste appearance time (AT). For comparison, 10% sucrose exhibits an AT of 4 s. In addition to a delayed AT, neohesperidin dihydrochalcone shows a prolonged sweet aftertaste. This effect has been defined as the time required for the perceived sweetness intensity to decline from a 10% sucrose equivalent to the low, but greater than threshold, 2% sucrose equivalent. This time has been defined as the extinction time (ET). Neohesperidin dihydrochalcone exhibits an ET of 40 s in comparison to 14 s for *iso*-sweet 10% sucrose.

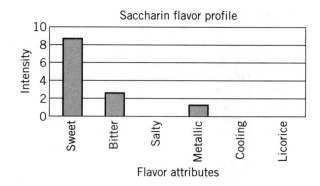

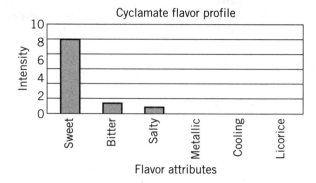

Figure 1. FPA of sodium saccharin (384 ppm) and sodium cyclamate (5930 ppm) in water.

These temporal effects combine to cause neohesperidin dihydrochalcone to be a substantially less acceptable sweetener than sucrose in most food systems. Thus, it is not sufficient to know only a sweetener's flavor profile to predict viability. The temporal profile must also be known. Ultimately, of course, the taste quality of a sweetener may be appreciated only after consumer acceptability studies in the food category of interest.

Safety

In the United States, the use of sweeteners is regulated by the 1958 Food Additives Amendment to the Food, Drug and Cosmetic Act of 1938. This legislation and its effects on the regulation of sweeteners and other food additives has been reviewed (9,10). By this act, saccharin and cyclamates were exempted as generally recognized as safe (GRAS) food ingredients. Not all sweeteners are included on the GRAS list, however, which in its original form, listed 675 food ingredients. Surprisingly, even sucrose is not included. However, the omission of sucrose and many other obviously safe food ingredients prompted the following official FDA comment:

It is impractical to list all substances that are generally recognized as safe for their intended use. However, by way of illustration, the Commissioner regards such common food ingredients as salt, pepper, sugar, vinegar, baking powder and monosodium glutamates as safe for their intended use. (11)

In order to achieve GRAS status, a sweetener or any food ingredient may be generally recognized as safe among

... experts qualified by scientific training and experience to evaluate the safety of substances directly or indirectly added to food. The basis of such views may be either (1) scientific procedures or (2) in the case of a substance used in food prior to January 1, 1958, through experience based on common use in food. (12)

For the cases of substances not included on the GRAS list, two tracks toward approval for use in foods are defined. First, and somewhat simpler where applicable, is the GRAS Affirmation Process. To qualify for this track, it must be either (1) demonstrated that the substance was in common use in food in the United States prior to 1958 or (2) be based on expert judgment of safety demonstrated by published safety studies. If the available safety data are not sufficient to support the requested uses or increase in projected exposure levels, the FDA may affirm GRAS status but limit levels of use until further safety data are established. Examples of sweeteners for which GRAS affirmation has been requested are discussed in the latter part of this article.

The second track toward approval of a sweetener for use in foods is the food additive petition (FAP) process. The FAP process requires extensive safety studies in test animals as well as studies in humans. One objective of these studies is the determination of the highest dose that may be given without adverse effects. This dose is termed the no observed adverse effect level (NOAEL). The NOAEL, in the most sensitive animal species evaluated, is then used by the FDA to regulate the level at which the food additive may be used in foods. This level, termed the acceptable daily intake (ADI), is defined as 0.01 of the NOAEL in the most sensitive animal species. These NOAEL and ADI exposure levels are given in milligrams per kilogram of body weight. Thus, as an example, if the NOAEL of a proposed new sweetener is determined to be 500 mg/kg in the most sensitive animal species evaluated, an ADI of 5 mg/kg would be allowed. The question then naturally arises as to the meaning of a 5 mg/kg ADI allowance. This level is then employed by the FDA to determine the food categories and levels in those categories in which the new sweetener may be used. The objective of this exercise is to ensure that the 5 mg/kg exposure level is not exceeded on a chronic basis. In order to do this, 90th percentile, 14-day average food category consumption data are employed. Approval of the new sweetener may then be granted for use in food categories where, in the aggregate, 90th percentile, 14-day average consumption data on these categories does not exceed the ADI. Examples of sweeteners that have successfully undergone, or are currently undergoing the FAP Process, are discussed in this article. Marshall and Pollard have comprehensively reviewed the category and level approvals that have been granted in the United States and abroad for many of the sweeteners discussed herein (13). Broulik reviewed the regulatory issues related to new sweetener development in the United States in 1996 (14).

Although individual countries assume responsibility for the regulation of food additives within their boundaries, there has been an attempt at international standardization of food additive regulation. Vettorazzi reviewed this process in 1989 (15). In 1956, the Food and Agriculture Organization of the United Nations (FAO) and the World Health Organization (WHO) established the Joint FAO/WHO Expert Committee on Food Additives (JECFA). The fundamental objective of JECFA is the establishment of ADIs for food additives following the assembly and interpretation of all relevant biological and toxicological data. It is important to recognize that ADIs provide a wide margin of safety. According to JECFA, "an ADI provides a sufficiently large safety margin to ensure that there need be no undue concern about occasionally exceeding it provided the average intake over longer periods of time does not exceed it" (16). Over its lifetime, more than 700 intentional food additives have been evaluated by JECFA.

Solubility

Many sweeteners are insufficiently water soluble to be of general utility. Commonly, sweetness intensity levels at least equivalent to 10% sucrose are required. In some systems (eg, frozen desserts), soluble sweetener levels are required that match the sweetness of 15 to 20% sucrose. In addition, for many food systems, manufacturers may require that sweeteners dissolve sufficiently rapidly so as to not interfere with current manufacturing processes. A particularly relevant illustration of requirements imposed by manufacturing processes is that for CSD products. In this case, it is necessary that a concentrate solution of the sweetener–flavor system complex be rapidly attainable. Thus, high solubility and rapid dissolution rates are very desirable properties for nonnutritive sweeteners.

Stability

To be commercially viable, a sweetener must be stable to the intended conditions of use. Degradation may be from hydrolytic, pyrolytic, or photochemical processes depending on the food application. Stability is required for two reasons. First, the sweetener must not degrade such that the level of sweetness of the food product would be substantially reduced during the product lifetime. Also, the sweetener must not break down to cause any off taste. The second reason for the stability requirement relates to sweeteners that are defined as food additives. By the FAP process, degradation products must be shown to be safe. Currently, for most organic compounds, if exposure to the degradation product may reach or exceed 0.0063 mg/kg, then the requisite safety assessment studies would be equal to those mandated for the sweetener itself (17). As such, a sweetener's stability is a very important factor when assessing its viability for development as a new food additive.

Cost

Nonnutritive sweeteners are always compared to sucrose, the consumer's standard. With sucrose in plentiful supply, presently at 36 cents/lb, potential replacements must either be cost-competitive or present sufficient advantages to justify a premium (18). The effective cost to the food product manufacturer of an alternative sweetener is equivalent to the wholesale price divided by the potency of the alternate product in delivering the desired property. In most cases, the desired property is sweet taste intensity. Thus, the effective cost or cost per sucrose equivalent (CSE) is the quotient of wholesale price and sweetness potency (P). P is generally expressed as a multiple of the P of sucrose, which we define to be 1. Although sometimes P is expressed on a molar basis, most commonly P is expressed on a weight basis (P_w). It is important to recognize that although P_w is not constant for high-potency sweeteners, it often is for sugars and other polyol sweeteners of low potency (19). P_w for a high-potency sweetener varies according to the sucrose reference concentration. Concentration/response (C/R) functions demonstrating saccharin's and cyclamate's nonlinear dependencies of P_w on sucrose reference concentration are illustrated in Figure 2. DuBois et al. demonstrated that the law of mass action hyperbolic function $R = R_m C/(k_d + C)$ provides a good fit for high-potency sweetener C/R function data and that the simple linear function $R = P_w C + b$ provides a good fit for most sugar and polyol sweetener C/R function data (20). When sweetener C/R data are fit to these equations, the constant terms provide very useful information. If a sweetener's C/R data are well modeled by $R = P_w C + b$, then P_w for the sweetener is a constant and is determined to be the slope of the line. On the other hand, if a sweetener's C/R data are well modeled by $R = R_m C/(k_d + C)$, then P_w is not a constant and is dependent on sucrose reference concentration. It is important to note, however, that the parameters R_m (maximal response) and k_d (apparent sweetener/receptor dissociation constant) in this equation

provide useful information on the properties of the sweetener. R_m is the maximal response, in sucrose equivalents, possible for that sweetener. Thus, for sodium saccharin and sodium cyclamate, the equations given in Figure 2 show that the maximum sweetness responses are equivalent to 10.1 and 15.2% sucrose, respectively. The R_m value for saccharin is rather low and indicates that it would be difficult to formulate a food product based on saccharin as the sole sweetener. k_d also provides useful information. It is the sweetener concentration in milligrams per liter that gives 50% of the maximal response. Thus, for sodium saccharin and sodium cyclamate, the equations given in Figure 2 show that 115 and 3140 mg/L are equivalent to 5.0 and 7.6% sucrose, respectively. These values are useful in assessment of the potential of a sweetener as a sole sweetener or as a component of a blend system.

In addition to the reference concentration effect on P, substantial effects can also be found that are due to food systems and temperature. As an example of the food system effect, it has been reported that in water, aspartame is 133 times more potent than sucrose (10% reference), whereas in CSDs, it is 180 times sucrose at the same reference concentration level (21). As an example of the temperature effect, it has been reported, and is generally accepted, that fructose is more potent than sucrose (22). At elevated temperatures, however, sucrose is more potent than fructose. Thus, calculation of sweetener CSE requires specification of the application to ensure that the relevant value of P_w is used.

Carbohydrate sweeteners (glucose, maltose, etc) generally are less potent than sucrose. As such, CSE values for carbohydrate sweeteners usually exceed that of sucrose. During the past century, many high-potency noncarbohydrate sweeteners have been discovered. As a consequence, very low CSE values are often realized for these sweeteners. For illustration, consider sodium saccharin, which currently is available in the United States for a wholesale price of $3.90/lb (18) and exhibits P_w (8) in water of 180 times that of sucrose. The CSE of sodium saccharin, used in water at an 8% sucrose sweetness equivalent level, would therefore be $3.90/180 = 2.2 cents/lb. This contrasts quite sharply with sucrose, which is presently available for a wholesale price of 36 cents/lb (18). Clearly, such cost factors attainable with high-potency sweeteners offer food manufacturers substantial increased profitability in the low-calorie sector of product categories. Although sweetener cost is certainly an important consideration for the food manufacturer, it is discussed here only for the cases of saccharin and cyclamates, which, as commodity chemicals, have relatively stable prices.

HIGH-POTENCY NONNUTRITIVE SWEETENERS OF MAJOR COMMERCIAL SIGNIFICANCE

The four nonnutritive sweeteners saccharin, cyclamate, aspartame, and acesulfame-K are used extensively today. In addition, sucralose and alitame are newcomers that may see significant use in the future. These six sweeteners are discussed in detail in this section. These sweeteners are approved for general use in the United States except

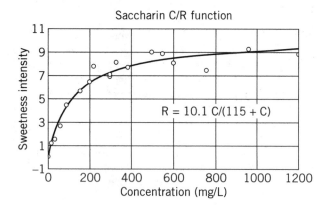

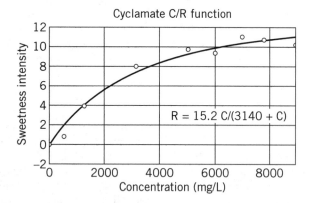

Figure 2. C/R functions for sodium saccharin and sodium cyclamate in water.

for alitame. The chemical composition, a history of discovery and development, and discussions of taste quality, safety, solubility, and stability issues are presented for each.

Saccharin

Saccharin (**1**) is the first sweet-tasting organic compound identified that is significantly more potent than sucrose. It was discovered in 1878 by Fahlberg while he was working in the Remsen's laboratory at Johns Hopkins University (2). Saccharin is exemplary of the *N*-sulfonyl amide structural class of sweeteners that is distinguished by the common -$CONHSO_2$- substructural unit.

(**1**)

After its discovery, saccharin soon found its way into many food applications. It was particularly popular among persons with diabetes as well as obese persons. Saccharin use was limited, however, by a taste quality substantially inferior to that of sucrose. Its sweet taste is accompanied by significant bitter and metallic taste attributes. Interestingly, it has been found that the population is heterogeneous in this respect, approximately one-third is hypersensitive to these off tastes, and the remaining two-thirds is moderately sensitive to insensitive (23–26). Saccharin's FPA data and C/R function are illustrated in Figures 1 and 2, respectively, and are also given in Table 1. As has been generally discussed earlier, sweetness potency is strongly dependent on the concentration of sucrose used as reference. For saccharin, the falloff in sweetness potency as a function of increasing sucrose reference concentration is particularly dramatic. Thus, from the equation in Figure 2, saccharin P_w values of 710, 180, and 9 are calculated relative to sucrose references of 2, 8, and 10%, respectively. The abrupt drop in sweetness potency shown here for sac-

charin is due to its low R_m of 10.1. Persons hypersensitive to saccharin bitterness do not experience an R_m that even approaches 10, whereas individuals insensitive to this off-taste experience a much higher R_m. However, the average result for the expert sensory panel used for the development of the data in this study is 10.1. Although the reason for a low R_m for some panelists is not known, it seems likely that it may be due to mixture suppression effects. This phenomenon of suppression of one taste attribute (eg, sweetness) by another (eg, bitter or sour) is well known (19). The temporal profile of saccharin is essentially identical to that of sucrose. ATs of 4 and ETs of 14 s, respectively, have been determined for both saccharin and sucrose (7). Thus, with the exception of a marginal flavor profile, which is particularly objectionable for some subjects, the taste quality of saccharin mimics that of sucrose sufficiently to be useful in food applications.

Questions concerning the safety of saccharin date back to the early 1900s. A staunch defender of saccharin safety was President Theodore Roosevelt. In response to those who questioned its safety, Roosevelt stated, "Anybody who says saccharin is injurious to health is an idiot" (27). Despite such high-level support, however, saccharin has continued to come under fire. The chronologies of safety studies and regulatory agency actions concerning saccharin have been comprehensively reviewed (28,29). The principal events in this chronology have had many ups and downs. In 1958, saccharin was listed as one of the 675 substances on the original GRAS list. In 1972, however, the FDA retracted the GRAS status of saccharin based on concern over results in a long-term rat feeding study conducted by the Wisconsin Alumni Research Foundation. Then, in 1977, as a result of a Canadian multigeneration rat study in which bladder tumors were found in the second-generation animals, the FDA announced its intention to ban saccharin. The FDA held that it had no choice other than to ban saccharin because of the Delaney Clause of the Food Additive Amendment to the Food, Drug and Cosmetic Act. This law requires that if any food additive, at any dose, in any animal species, is found to cause cancer, its further use is to be outlawed. Acting FDA Commissioner Gardner stated,

We have no evidence that saccharin has ever caused cancer in human beings. But we do now have clear evidence that the

Table 1. FPA and C/R Function Data on Major High-Potency Nonnutritive Sweeteners

Sweetener	C/R function data		FPA data					
	R_m	k_d	Sweet	Bitter	Salty	Metallic	Cooling	Licorice
Saccharin[a]	10.1	115	8.6	2.6	0.0	1.2	0.0	0.0
Cyclamate[b]	15.2	3240	10.0	1.2	0.6	0.0	0.0	0.0
Aspartame[c]	16.0	562	8.4	0.0	0.0	0.0	0.0	0.0
Acesulfame-K[d]	11.6	472	7.7	2.8	0.0	1.3	0.0	0.0
Sucralose[e]	14.7	142	8.7	0.0	0.0	0.0	0.0	0.0
Alitame[f]	14.6	28.1	8.4	0.0	0.0	0.0	0.0	0.0

[a]Sodium salt; 384 mg/L for FPA.
[b]Sodium salt; 5930 mg/L for FPA.
[c]560 mg/L for FPA.
[d]974 mg/L for FPA.
[e]172 mg/L for FPA.
[f]29 mg/L for FPA.

safety of saccharin does not meet the standards for food additives established by Congress.

Acting in response to strong consumer and industry pressure to ensure the continued availability of saccharin, Congress placed a moratorium on the FDA ban pending additional research. This moratorium on the saccharin ban has since been extended several times, and the one presently in place is set to expire on May 1, 2002. During this period of moratorium, the regulatory climate has evolved considerably such that, at the present time, a policy that considers both risks and benefits is now considered as more appropriate than the absolute statements of the Delaney Clause. In addition, risk is now recognized to be more of a function of dose than an all-or-none phenomenon as was previously considered. Mechanism of action is also recognized as an important factor to consider. Since the 1970s, research has been conducted in effort to elucidate the mechanism whereby saccharin initiates bladder cancer in rats. Interestingly, in one important study, evidence was obtained that the effect, which could only be demonstrated in rats, is due to the sodium component of sodium saccharin and not due to saccharin, the sweetener (30). In this and related studies, sodium ascorbate (vitamin C), sodium chloride (salt), and other sodium salts were shown to promote similar effects on the rat bladder. Most scientists think that the FDA should withdraw its requirement of a cancer warning label for food products containing saccharin in the very near future.

Temporary U.S. regulatory restrictions regarding the use of saccharin have been defined (31). For example, in beverages, saccharin may be used up to a level of 12 mg/oz (406 ppm). JECFA has established 5 mg/kg as an ADI for saccharin and stated that the rodent bladder tumors caused by sodium saccharin are not relevant to humans.

The physical properties of saccharin are ideal for use as a nonnutritive sweetener. First, it is extremely stable toward all the conditions to which it may be exposed in food applications. In studies conducted at the Sherwin Williams Company, it was determined that at pH 3.3 and 120°C, 18 and 69% loss occurred at 27 and 219 h, respectively (M. L. Mitchell, unpublished results, 1989). The exclusive product of this hydrolysis is 2-sulfobenzoic acid, which is also present as a low-level contaminant in commercial samples of saccharin. At pH 7.0 and 120°C, even greater stability was found with no significant degradation after 27 h and only 6% loss after 219 h. At pH 9.0 and 120°C, stability was also high with 2 and 12% losses, respectively, found after 27 and 219 h. The sole degradation product formed under neutral and alkaline pH conditions is 2-sulfonamidobenzoic acid, which is also a low-level contaminant in commercial samples of saccharin. As a consequence of this high stability, neither loss of sweetness during food product lifetime nor degradation product safety is a significant concern for saccharin. Second, saccharin is highly water soluble. As the sodium salt, 1200 g is soluble in 1 L of water, whereas in the acid form, a solution of 3.4 g/L may be obtained (32). If we consider saccharin P_w (8) to be 180, as an example for applications where an 8% sucrose level of sweetness intensity is required, it can be calculated that saccharin is either 2700

or 8 times as soluble as necessary, depending on whether the sodium salt or free-acid form is employed.

At the present time, no sweetener matches saccharin in terms of cost. As discussed earlier in the general discussion on sweetener cost, the current price of $3.90/lb (18) coupled with $P_w(8) = 180$ combine to result in a saccharin CSE of 2.2 cents/lb. Thus, saccharin is a very inexpensive sweetener.

Cyclamates

Although many sweet-tasting organic compounds were found during the 59 years between the discovery of saccharin and 1937, none achieved significant usage in foods. Then, in 1937, Sveda, a graduate student in Audrieth's laboratory at the University of Illinois, discovered that metal salts of cyclohexylsulfamic acid are sweet (33). These salts have become known as cyclamate salts or, more commonly, cyclamate. Cyclamate, in the acid form, has the chemical structure **2** and was the first of the sulfamate structural class of sweeteners to be discovered. Sweeteners of this class exhibit the —NH-SO₃—moiety as the common structural feature. Cyclamate is generally used in foods as either the sodium or calcium salt. Since 1937, many sweet analogues of **2** have been synthesized, although none have been developed for use in foods.

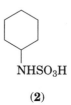

NHSO₃H

(2)

Sensory panel studies on cyclamate salts demonstrate them to better reproduce the taste quality of sucrose than does saccharin. Nonetheless, a significant bitter taste attribute is noted for cyclamate. The flavor profile of sodium cyclamate is illustrated in Figure 1. Interestingly, a salty attribute is also present in the sodium cyclamate profile. This factor is almost certainly due to the high sodium ion concentration present in intensely sweet solutions. The C/R function for sodium cyclamate is illustrated in Figure 2. As is true for all high-potency sweeteners, the potency of sodium cyclamate is dependent on sucrose reference concentration. Thus, $P_w(2) = 42$, $P_w(8) = 23$, and $P_w(10) = 17$ are calculated from the equation given in Figure 2. The temporal profile of sodium cyclamate is very similar to that of sucrose. ATs of 4 and ETs of 14s have been determined for both substances (7). Thus, although sodium cyclamate exhibits a flavor profile with significant bitter and salty notes, it exhibits a sucrose-like temporal profile and, in the aggregate, mimics the taste of sucrose sufficiently to be generally useful in food applications.

The history of the evaluation of cyclamate safety has been reviewed comprehensively (34,35). Cyclamates were first approved in 1951 for use as drugs for use by people with diabetes and others who had to restrict their sugar intake. In 1958, they were listed by the FDA on the original GRAS list. As was discussed earlier, the use of cyclamate sweeteners experienced explosive growth during the 1960s after it was found that a 10:1 mixture of cyclamate and saccharin exhibits taste quality superior to either sweetener individually and approaches that of sucrose. In 1969, the FDA became concerned about cyclamate safety. At that time, the FDA was advised of a rat study in which the commonly used 10:1 cyclamate/saccharin mixture was shown to induce bladder tumors. These results were interpreted by the FDA to indicate that cyclamate salts (ie, not saccharin salts) are bladder carcinogens in rats. As a consequence, cyclamates were immediately removed from the GRAS list and, in 1970, were banned from use. At the same time, however, cyclamates have remained on the market in more than 50 countries.

Since the 1970 FDA action, Abbott Laboratories and the Calorie Control Council have maintained continuous efforts to have cyclamates reapproved. As a result of this effort, it is now generally accepted that cyclamates are not carcinogens (34). The principal remaining areas of concern are related to the biological activity of cyclohexylamine, the major cyclamate metabolite. In early studies, hypertensive activity and testicular atrophy effects were noted for this metabolite. However, in recent studies, no adverse hypertensive or reproductive effects have been observed in humans. Nonetheless, these weak biological effects for cyclohexylamine may limit the ADI to be granted by the FDA in the event that it is reapproved. The ADI established by JECFA for cyclamate is 11.0 mg/kg. In the European community, cyclamate levels in CSDs are restricted to a maximum of 400 ppm (36). From the equation given in Figure 2, it can be calculated that 400 ppm cyclamate is equivalent to 1.7% sucrose in sweetness intensity. Thus, although cyclamates may be reapproved in the United States, restrictions may limit their application to sweetener blend systems.

Cyclamates exhibit excellent solubility and stability characteristics for use in essentially all imaginable applications. The sodium and calcium salt forms are both commercially available. Although the acid form is sufficiently water soluble (13.3 g/100 mL), its high acidity results in preference for the very soluble sodium (20 g/100 mL) or calcium (25 g/100 mL) salts (37). Since $P_w(8) = 23$, as an example for applications where an 8% sucrose equivalent level of sweetness is desired, it can be calculated that sodium cyclamate is 570 times as soluble as necessary. Hydrolytic degradation of cyclamate salts proceeds to yield cyclohexylamine and inorganic sulfate. As a consequence of the adverse biological activity of cyclohexylamine, FDA scientists conducted a comprehensive evaluation of cyclohexylamine levels in a spectrum of food products (38). Cyclohexylamine was found in the majority of these products albeit at low levels. Interestingly, even in the most acidic samples (cola CSDs), cyclohexylamine levels increased insignificantly during 4 months of ambient-temperature storage. Cyclohexylamine in food products appears to be substantially derived as a cyclamate sweetener contaminant. Data have been reported on the hydrolysis of cyclamate under extreme conditions (39). After 1 h at 100°C, 13.7, 8.1, 0.98, 0.10, 0.52, and 0.58% of cyclamate sweetener is lost at pH values of 0.9, 1.6, 2.5, 4.5, 5.3, and 6.5, respectively. In summary, cyclamate sweeteners are quite stable. No significant loss of sweetness or generation of unsafe degradation products is expected in any common applications.

Cyclamates are economical sweeteners. On the world market, sodium cyclamate is presently available for $1.04/lb (40). On a CSE basis, this is equivalent to 4.5 cents/lb if one employs $P_w(8) = 23$ times that of sucrose as the relevant sweetness potency. Thus, although sodium cyclamate is approximately two times more expensive than saccharin, it still represents a very substantial economy over sucrose.

Aspartame

L-Aspartyl-L-phenylalanine methyl ester (3) is an example of one of several novel structural types of sweet-tasting organic compounds identified in the 1960s. It is known under the generic name aspartame, is often abbreviated as APM, and is very well known to consumers as Nutra-Sweet®, a brand name of Monsanto. As was true for all the sweeteners described herein that are not of botanical origin, the sweetness of APM was discovered by accident. The discovery was made by Schlatter while working under the direction of Mazur of G.D. Searle & Company (3). APM was prepared by Schlatter as an intermediate in a drug discovery program aimed at finding new ulcer treatments. Interestingly, this compound had been prepared some years earlier by chemists at ICI in Great Britain. Its sweet taste had not been noted, however.

(3)

Aspartame was the first sweet-tasting member of a structural class of sweeteners generally known as the dipeptides. Peptides are oligomers of α-amino acids. Aspartame is simply derived from the two natural amino acids L-aspartic acid and L-phenylalanine. It is ironic that even though substantially more than 1000 sweet-tasting analogues of aspartame have been prepared since 1965, none are more advantageous than aspartame after consideration of all the properties requisite for commercial viability.

The phenomenal commercial success of aspartame is easily explained by its exceptional sucrose mimicry. As suggested by its flavor profile summarized in Table 1, it is essentially indistinguishable from sucrose; no nonsweet taste attributes are observed. Interestingly, however, the temporal profile of aspartame is different from sucrose. Aspartame exhibits both a slightly delayed AT [5 s (APM) vs. 4 s (sucrose)] and a protracted ET [19 s (APM) vs. 14 s (sucrose)] (7). These differences are not sufficient, however, to adversely affect acceptability in food products. An interesting and useful manner in which aspartame has an advantage over other high-potency sweeteners is that it enhances other flavor attributes. This effect was noted early on and has been systematically studied by Baldwin and Korschgen (41) and also by Wiseman and McDaniel (42). It has been hypothesized that this unique advantageous property of aspartame is due to its atypical temporal profile (43). More specifically, aspartame's flavor enhancement properties are suggested to be consequences of its slightly delayed AT relative to that of sucrose. Thus, greater temporal resolution of sweetness (taste) and flavor (olfaction) signals to the brain occurs for APM than is the case for sucrose. This increased resolution of neural olfactory and taste signals for APM over sucrose may result in the flavor intensity to be judged higher in the better-resolved aspartame-sweetened system. This effect is then interpreted as a flavor enhancement effect of aspartame.

The sweetness potency of aspartame is again dependent on the sucrose reference concentration, although somewhat less so than for the other sweeteners discussed in this section. Thus, from the APM C/R function given in Table 1, it is calculated that $P_w(2) = 250$, $P_w(8) = 140$, and $P_w(10) = 107$. Disparities between these potencies and literature potencies quoted earlier are a consequence of differences in methodology.

The safety of aspartame has been as extensively studied as any food additive. Most of the principal metabolism, preclinical, and clinical studies have been reviewed (44–46). On the basis of evaluation of all the safety assessment studies, aspartame was approved by the FDA in 1981 with an ADI of 20 mg/kg. This ADI is based on a NOAEL in preclinical studies of greater than 2,000–4,000 mg/kg. In 1983, based on clinical studies, the ADI for aspartame was raised by the FDA to 50 mg/kg. The ADI established by JECFA for aspartame is 40 mg/kg. No effects were noted in the clinical studies at doses many times greater than approved for human consumption. Consideration of aspartame metabolism explains why this is the case. On ingestion, aspartame is completely broken down to the two natural amino acid building blocks and methanol. The safety of the amino acids L-aspartic acid and L-phenylalanine is, of course, no surprise since normal dietary protein (from meat, milk, eggs, etc) provides substantially greater quantities of these nutrients. Methanol also is not a safety concern when exposure occurs as a metabolite of aspartame. Beverages formulated with aspartame to a sweetness level matching 10% sucrose contain the equivalent of 50 to 60 mg/L of methanol; this is substantially less than the average of 140 mg/L content for fruit juices. Thus, the safety of aspartame should be no surprise to anyone. Consumption of aspartame does not result in exposure of the internal body tissues to any novel substances. Natural subunit assembly and metabolic subunit disassembly is a unique high-potency sweetener concept presently exemplified only by aspartame. Despite the safety of aspartame, however, products containing it are required to carry an informational statement for people with a rare genetic disease known as phenylketonuria (PKU). Approximately 1 of 15,000 people has this disorder, which involves an inability to metabolize phenylalanine. Unchecked, PKU results in mental retardation. In the United States, this disorder is detected at birth, if present. The normal treatment is a phenylalanine-restricted diet from infancy through childhood and sometimes into adulthood. It should be emphasized, however, that phenylalanine consumption is only restricted, not eliminated, since it is an essential amino acid and, as such, is necessary for life. Persons with PKU must carefully monitor their consumption of phenylalanine from all dietary sources, including aspartame.

Aspartame exhibits sufficient solubility for all food applications. At ambient temperature in water, a solubility of approximately 1.0% can be attained at pH 4 and also at neutral pH; solubility reaches a minimum at pH 5.5, the isoelectric point (47). Thus, using $P_w(8) = 140$, it can be calculated that aspartame is greater than 70 times more soluble than necessary to provide an 8% sucrose level of sweetness intensity.

If aspartame has a drawback, it is hydrolytic stability. At 25°C (77°F), its hydrolytic half-lives in aqueous buffer are 116, 260, 242, 82, and 2 days for pHs of 3, 4, 5, 6 and 7, respectively (21). Clearly, maximum utility would be expected in the pH range 3 to 5. Fortuitously, this is a representative range for most food systems. It is particularly fortuitous that none of the three principal degradation products, L-aspartyl-L-phenylalanine (AP), 3-benzyl-6-carboxymethyl-2,5-diketopiperazine (DKP), and β-L-aspartyl-L phenylalanine methyl ester (β-APM) exhibit either off-tastes in aged food products or safety problems. All the degradation products have been tested and demonstrated to be safe. Interestingly, β-L-aspartyl-L-phenylalanine (β-AP), the product of ester hydrolysis of β-APM, has been demonstrated to be naturally produced on metabolism of dietary protein (48). In summary, aspartame is sufficiently stable for use in acidic food products. On the other hand, the neutral pH range encountered in baked goods is problematic. An encapsulated form of aspartame has been developed, however, to address this limitation (49).

Acesulfame

The discovery of acesulfame was accidental, just as was the case for saccharin, cyclamate, and aspartame. In 1967, Clauss and Jensen, while working in the laboratories of Hoechst AG on the reactions of fluorosulfonylisocyanate with acetylenes, obtained the product 5,6-dimethyl-1,2,3-oxathiazin-4(3H)-one-2,2-dioxide and noted it to be sweet (50). As a consequence of the 1969 ban on cyclamates in the United States, Hoechst initiated a systematic program to optimize the properties found in this compound. In the end, it was determined that the preferred product candidate was 6-methyl-1,2,3-oxathiazin-4(3H)-one-2,2-dioxide, which has since been given the generic name acesulfame and has chemical structure 4. The potassium salt of acesulfame is known as acesulfame-K, is often referred to as ACE-K, and has been given the brand name Sunette®. Acesulfame is structurally related to saccharin in that both exhibit the essential-for-activity N-sulfonyl amide structural motif (ie, -CONHSO$_2$-) and thus are both members of the N-sulfonyl amide sweetener class. The chemistry, safety assessment, sensory properties, and applications of acesulfame-K have been reviewed (51).

(4)

The structural analogy of acesulfame with saccharin is not the only area of commonality. ACE-K exhibits a flavor profile with bitter and metallic attributes that are equivalent to or stronger than those of saccharin. The FPA data for ACE-K are summarized in Table 1. Interestingly, it has been found that the population is heterogeneous with respect to its sensitivity to the bitter and metallic off-tastes of acesulfame-K, as has also been noted for saccharin. Particularly striking in this study is the finding that the same individuals who are sensitive to saccharin off-taste characteristics are also sensitive to those of ACE-K. As a consequence, ACE-K, like saccharin, has minimal utility as a sole sweetener in food applications. Although quantitative temporal profile data on acesulfame-K are not available, it appears, qualitatively, to be quite similar to sucrose in this dimension. In summary, it is expected that ACE-K will find a niche in food applications as an alternative to saccharin. Thus, its principal application should be as a component of blends with aspartame and other sweeteners with sucroselike flavor profiles.

C/R function data on acesulfame-K are provided in Table 1. The sweetness potency of ACE-K drops off with increasing sucrose reference concentration just as is the case for saccharin, cyclamate, and aspartame. The rate of this decrease is rapid, very much as is the case for saccharin. Thus, from the C/R function for ACE-K given in Table 1, $P_w(2) = 204$, $P_w(8) = 76$, and $P_w(10) = 34$ are calculated.

ACE-K received its first approval in the United States for tabletop use in 1988 and finally received approval for all food categories in 1998. An ADI of 15 mg/kg was established by the FDA after review of the safety assessment studies submitted by Hoechst. An ADI of 15 mg/kg has also been established by JECFA. The safety studies conducted on ACE-K have been reviewed (52).

The solubility of acesulfame-K is quite high, with 27 g dissolving in 100 mL of water (53). If, for illustration, the $P_w(8) = 76$ value of acesulfame-K is employed, it can be calculated that acesulfame-K is more than 250 times more soluble than necessary to match the sweetness intensity of 8% sucrose.

The hydrolytic stability of acesulfame-K is substantially less than that of saccharin. For example, under conditions designed to model a cola CSD application, 15% is lost in 1 year at room temperature and 25% is lost in 3 months at 40°C (104°F) (54). Products that have been identified from ACE-K degradation include acetoacetamide-N-sulfonic acid, acetoacetamide, acetoacetic acid, and acetone. In summary, it can be said that although ACE-K is not completely stable toward hydrolytic breakdown, its slight instability does not create any problems (eg, loss of sweetness) from an application perspective.

Sucralose

Until the discovery of saccharin, nearly all known sweet substances were carbohydrate based. In 1976, in a collaborative program between Hough's laboratory at Queen Elizabeth College in the University of London and scientists at Tate & Lyle, it was discovered that certain halodeoxysucrose derivatives are potently sweet (55). This collaboration led to the selection of a trichlorinated derivative of sucrose that they named sucralose (5) as a product candidate (56). The halodeoxysucroses, of which sucralose is the best known example, are the first carbohydrate-based sweeteners that are substantially more potent than the simple sugars. The commercialization of 5 was pursued by a joint venture of Tate & Lyle and McNeil Specialty Products Company, a Johnson & Johnson subsidiary. Sucralose is known under the brand name of Splenda®. The properties of sucralose were reviewed in 1996 (57).

(5)

FPA data on sucralose are provided in Table 1. As can be seen by inspection of these data, sucralose exhibits a clean sweet taste equivalent to that of sucrose. No off-tastes are observed. Unfortunately, temporal profile data on sucralose are not available, and so it is not possible to quantitatively-compare it to sucrose in this dimension. In general, however, sucralose's sweetness appears to linger somewhat similar to the behavior of aspartame.

C/R function data on sucralose are provided in Table 1. Its sweetness potency drops off with increasing sucrose reference concentration just as is the case for saccharin, cyclamate, aspartame, and acesulfame-K. Thus, from the C/R function given in Table 1, $P_w(2) = 910$, $P_w(8) = 470$, and $P_w(10) = 330$ are calculated for sucralose.

In 1998, the FDA granted approval for the general use of sucralose in foods and beverages. An ADI of 5 mg/kg was established by the FDA after review of the safety assessment studies submitted. In contrast, an ADI of 15 mg/kg has been recommended by JECFA. Sucralose slowly hydrolyzes in acidic food and beverage products to the monosaccharides 4-chloro-4-deoxy-galactose and 1,6-dichloro-dideoxy-fructose, and for this reason, it was required that safety studies be carried out on sucralose and the two degradation products. The solubility of sucralose in water is very high and is not significantly affected by pH. Sucralose is commercially available as a 25% (w/v) solution. If, for illustration, the $P_w(8) = 470$ value of sucralose is employed, it can be calculated that sucralose is >1400 times more soluble than necessary to match the sweetness intensity of 8% sucrose.

The hydrolytic stability of sucralose is high, though not as high as that of saccharin. As already mentioned, it does slowly hydrolyze to the two chlorinated monosaccharides 4-chloro-4-deoxy-galactose and 1,6-dichloro-dideoxy-fructose in acidic food systems. However, the loss of sucralose over 1 year is <5%, and so there is no significant loss of sweetness. In the solid form, sucralose does break down with loss of hydrochloric acid, causing the solid product to caramelize. This problem has substantially been addressed by reducing the particle size of the solid material to less than 12 μM. The microparticulated material has adequate shelf life for common food and beverage manufacturing operations. In summary, it can be said that although sucralose is not completely stable, its slight instability does not create any problems (eg, loss of sweetness) from an application perspective.

Alitame

Alitame is the product of an intensive research program carried out by Hendrick and his coworkers at Pfizer Central Research during the 1970s (58). The program was aimed at the identification of a stable, cost-effective analogue of aspartame that, of course, still retains the quality taste of aspartame. Alitame was selected from a group of hundreds of aspartame analogues, which were synthesized by the Pfizer chemists. Alitame has recently been reviewed (59).

(6)

FPA data on alitame are provided in Table 1. As can be seen by inspection of these data, alitame exhibits a clean, sweet taste comparable to sucrose taste. No off-tastes are observed. Unfortunately, temporal profile data on alitame are not available, and so it is not possible to quantitatively compare it to sucrose in this dimension. In general, however, alitame's sweetness appears to linger somewhat, much as is the case for aspartame.

C/R function data on alitame are provided in Table 1. Its sweetness potency drops off with increasing sucrose reference concentration just as is the case for saccharin, cyclamate, aspartame, acesulfame-K, and sucralose. Thus, from the C/R function given in Table 1, $P_w(2) = 4440$, $P_w(8) = 2350$, and $P_w(10) = 1640$ are calculated for alitame.

In 1986, Pfizer submitted an FAP to the U.S. FDA requesting broad approval for use in foods and beverages. At about the same time, approval was also requested in many other countries. At this point, approval for use in the United States has not yet been granted. However, broad clearance for use has been obtained in several countries, including Australia, Chili, China, Columbia, Indonesia, Mexico, and New Zealand. An ADI for alitame of 1 mg/kg was recommended by JECFA after review of all the safety assessment studies conducted. Alitame slowly breaks down in acidic food and beverage products to L-β-aspartyl-N-(2,2,4,4-tetramethyl-3-thietanyl)-D-alanine amide and N-(2,2,4,4-tetramethyl-3-thietanyl-D-alanine amide, and for this reason, it was required that safety studies be carried out on alitame and these two degradation products.

Alitame's solubility in water is high (131g/L w/v) and is only slightly sensitive to pH. If, for illustration, the $P_w(8)$ = 2350 value of alitame is employed, it can be calculated that alitame is >3800 times as soluble as necessary to match the sweetness intensity of 8% sucrose.

The stability of alitame is substantially better than that of aspartame, though not as high as that of saccharin. Its hydrolytic stability is very sensitive to pH. In the pH range of 2 to 4, alitame is 2 to 3 times more stable than aspartame, while in the pH range of 5 to 8, it is >1000 times more stable. As already mentioned, it does slowly break down to L-β-aspartyl-N-(2,2,4,4-tetramethyl-3-thietanyl)-D-alanine amide and N-(2,2,4,4-tetramethyl-3-thietanyl)-D-alanine amide in acidic food systems. However, the loss of alitame is insufficient to significantly affect the shelf life of most products. Alitame is limited, however, in that it does break down in acidic systems, by an unknown pathway, to produce trace levels of a sulfurous breakdown product or products. This limitation is particularly evident in cola beverages. In summary, it can be said that alitame is sufficiently stable for most food applications but is not an alternative for acidic products that require a long shelf life.

Summary

At the present time, there are four high-potency nonnutritive sweeteners that have achieved widespread use in the food industry, and there are two newcomers that are expected to see significant use in the future. Comparative data on these sweeteners are given in Table 2.

OTHER HIGH-POTENCY NONNUTRITIVE SWEETENERS OF COMMERCIAL SIGNIFICANCE

In this section, the five nonnutritive high-potency sweeteners neohesperidin dihydrochalcone, stevia sweeteners, glycyrrhizin, thaumatin, and Lo Han Kuo sweetener are discussed. In view of the limited approvals for use of these sweeteners in the United States, discussions of them will be abbreviated. Comparative data on these sweeteners are given in Table 3.

Neohesperidin Dihydrochalcone

As was the case for the high-potency sweeteners mentioned above, neohesperidin dihydrochalcone (NDC) also was discovered in a serendipitous manner. Horowitz and Gentili, of the USDA laboratory of Pasadena, California,

Table 2. Summary of High-Potency Nonnutritive Sweeteners of Major Commercial Significance

Sweetener	Year of discovery	Taste quality	Safety concern	Solubility	Hydrolytic stability	Cost[a]	U.S. regulatory status	Other properties
Saccharin	1878	Fair; bitter and metallic off-taste	None	Excellent	Excellent	2.2	Food additive permitted on an interim basis	
Cyclamates	1937	Good; weak, bitter and salty off-taste	Metabolite associated with testicular atrophy and hypertension	Excellent	Excellent	4.5	Not approved	
Aspartame	1965	Excellent	None	Good	Fair		ADI = 50 mg/kg	Flavor enhancement
Acesulfame-K	1967	Fair, bitter and metallic off-taste	None	Excellent	Good		ADI = 15 mg/kg	
Sucralose	1976	Excellent	None	Excellent	Good		ADI = 5 mg/kg	
Alitame	1979	Excellent	None	Excellent	Good		Not approved	

[a]Cost in cents/lb on a cost per sucrose equivalent basis.

as part of an investigation on the bitter tastes of citrus flavonoid glycosides, made a surprising observation (60). Based on a model of bitter taste structure–activity relationships they were developing in the early 1960s, they predicted that naringin dihydrochalcone would be bitter like naringin, the bitter constituent of grapefruit rinds. Naringin dihydrochalcone was then synthesized, and the USDA investigators were astonished to find it to be sweet. Subsequent studies, initiated in attempt to optimize the properties of sweeteners of this type, resulted in the identification of NDC (7) as the preferred product candidate. Neohesperidin, the flavanone glycoside precursor of 7, occurs naturally in the Seville orange. NDC and other related sweeteners have been reviewed (61).

(7)

NDC, until later discoveries in the 1970s, was one of the most potent sweeteners known. At near threshold levels of sweetness intensity, Horowitz and Gentili reported NDC to be 7 times more potent than saccharin on a weight basis. Problematical, however, are the presence of bitter, cooling, and licorice taste attributes (8). In addition to a flavor profile quite different from that of sucrose, NDC exhibits a substantially delayed and protracted temporal profile rela-

tive to that of sucrose. An AT = 9 s (sucrose, 4 s) and an ET = 40 s (sucrose, 14 s) have been measured (7). In food systems, the complex flavor and atypical temporal profiles of NDC combine to provide a very unnatural taste. Thus, the utility of NDC in food systems is marginal unless employed as a minor component in combination with other more sucroselike sweeteners.

NDC is not approved in the United States for use as a sweetener, and no FAP for such use has been filed. The Flavor and Extract Manufacturers' Association of the United States (FEMA) has, however, approved it as GRAS. By this approval, it may be used as a flavor modifier within the range of 1 to 5 ppm in a variety of foods and beverages (62). At these low concentrations, NDC has been reported to enhance flavors, to enhance mouth-feel, and to make many nonnutritive sweeteners taste more sugarlike. These properties of NDC have recently been reviewed (63).

Stevia Sweeteners

In Paraguay is found a small shrub known as *Stevia rebaudiana* (Bertoni), which is distinguished by its intensely sweet leaves. Investigations began to identify the sweet component of these leaves in the early part of this century. These efforts have been reviewed (64). The first significant work was by the French researchers Bridel and Lavieille, who isolated a crystalline sweet product. They determined this substance to be 300 times as potent as sucrose and to be a glycoside composed of three glucose moieties and a diterpenoid aglycone. This diterpenoid glycoside is now known as stevioside. The structure determination of stevioside as 8a was completed in the 1950s and early 1960s by a combination of the efforts of Mosettig and Wood. Since that time, Tanaka et al. at Hiroshima University in Japan have isolated and characterized at least eight additional sweet glycosidic analogues of 8a that are present in the

Table 3. FPA and C/R Function Data on Other Nonnutritive Sweeteners

	C/R function data		FPA Data					
Sweetener	R_m	k_d	Sweet	Bitter	Salty	Metallic	Cooling	Licorice
Neohesperidin dihydrochalcone[a]	9.8	53.2	7.2	0.5	0.0	0.0	3.3	2.3
Stevioside[b]	9.9	406	7.0	3.6	0.0	0.0	0.0	0.0
Rebaudioside A[c]	10.0	200	8.5	2.6	0.0	0.0	0.0	1.3
Glycyrrhizin[d]	7.8	240	7.2	1.6	1.0	0.0	1.2	1.3
Thaumatin[e]	10.1	3.6	8.7	1.2	0.0	0.0	1.8	0.0

[a] 160 mg/L for FPA.
[b] 1000 mg/L for FPA.
[c] 400 mg/L for FPA.
[d] 300 mg/L for FPA.
[e] 35 mg/L for FPA.

plant source. The most important of these is rebaudioside A (**8b**) because of its somewhat improved flavor profile. Also of particular significance are the oligomeric glucose conjugates of **8a**, which have been described by Tanaka and associates. These compounds are obtained on reaction of stevioside with starch and glucosyl transferase enzymes. This work was reviewed in 1991 (65). All these natural and enzyme-modified diterpenoid glycosides are approved for use in Japan. The plant-derived glycosides are also approved in South Korea, the Peoples Republic of China, Brazil, and Paraguay. Stevia sweeteners have been comprehensively reviewed (66,67).

(**8a**) R=H
(**8b**) R=β-D-Glu

Of the natural stevia sweeteners, rebaudioside A is preferred because of its improved flavor profile relative to stevioside. Even rebaudioside A, however, exhibits substantial bitter and cooling taste attributes. In general, the stevia sweeteners exhibit temporal profiles that are sucroselike in AT (stevioside AT = 4 s; sucrose AT = 4 s) but not in ET (stevioside ET = 22 s; sucrose ET = 14 s) (7).

Thus, as a consequence of flavor and temporal profiles that are rather unlike those of sucrose, rebaudioside A is not likely to find broad utility in food applications unless as a component of a blend.

At the present time, no stevia sweeteners are approved for use in the United States as either food additives or flavors. Furthermore, it is not expected that any will be approved since it is apparent, for a combination of performance, economic, and competitive reasons, that the comprehensive studies requisite for a FAP are not being undertaken. At the same time, however, stevia sweeteners are being used in the United States as dietary supplements. As such, they may be sold in the form of the plant leaves, but they may not be added to foods where their principal use is to provide sweetness.

Glycyrrhizin

Glycyrrhizin, a mixture of calcium, magnesium, and potassium salts of glycyrrhizic acid, occurs at a level of 6 to 14% in the roots of the European and Central Asian shrub *Glycyrrhiza glabra* Linn. (Fabaceae). The crude extract of the plant is well known as licorice. Glycyrrhizin is obtained from licorice extract by sulfuric acid precipitation. Ammoniated glycyrrhizin (AG) is obtained by dissolution of glycyrrhizin in aqueous ammonia, concentration, and recrystallization from alcohol. AG, which contains more than one molar equivalent of ammonia, may be converted to the monoammonium salt known as monoammonium glycyrrhizinate (MAG) by careful recrystallization. Glycyrrhizic acid is a triterpenoid glycoside of structure **9**. The original structural work leading to this structural assignment was completed by Lythgoe and Trippett in 1950 (68). MAG has been reported to have $P_w(10) = 33$ times that of sucrose and to exhibit significant bitter, licorice, and cooling flavor attributes (69). The temporal profile of MAG is not at all sucroselike. A very slow AT of 16 s (sucrose AT = 4 s) and a very prolonged ET of 69 s (sucrose ET = 14 s) have been reported (7). As a consequence of these strong deviations from sucroselike sweetness, MAG and the related licorice-derived products have only very limited utility in food applications. These limited applications have been reviewed (70).

(9)

In 1985, the glycyrrhizic-acid-derived sweeteners were confirmed in the United States as GRAS (71). However, the approval is specific with regard to their use as flavoring agents, flavor enhancers, and surfactants. Specifically excluded from this approval is as a component of sugar substitutes. As a consequence of this limited approval, the applications of glycyrrhizin-type sweeteners are extremely limited. Limiting utility also is the lack of sucroselike taste quality.

Thaumatin

Thaumatin, currently marketed under the trademark Talin®, is a protein that occurs naturally in the fruit of the West African plant *Thaumatococcus daniellii*. This protein is a single chain of 207 amino acids with eight disulfide bridges and a molecular weight of 22,209. van der Wel and colleagues at Unilever carried out most of the classical structure determination work on thaumatin. Their work has been reviewed (72). More recently, X-ray crystallographic studies by Sung-Hou Kim at the University of California, Berkeley, have completed the structure determination (73).

On a molar basis, thaumatin is one of the most potent sweet substances ever found. A molar basis potency of 100,000 times that of sucrose has been reported relative to a threshold concentration of sucrose. On a weight basis, however, and relative to a 10% sucrose reference, thaumatin's potency is only 1,600 times that of sucrose. In flavor profile studies, thaumatin is, in many respects, similar to the glycyrrhizin sweeteners. Significant licorice and cooling taste attributes are observed. Particularly problematical is a very atypical, temporal profile relative to sucrose. The sweetness AT is significantly delayed and the sweetness ET, very protracted. As a consequence of these strong deviations from sucrose taste, thaumatin has only limited utility.

Thaumatin was approved in the United States as GRAS by FEMA in 1984 and has approvals in many other countries as well (74). FEMA's approval was extended in 1996 to include recombinant thaumatin (62). These approvals are limited to the use of thaumatin as a flavor modifier,

and it may now be used in a broad range of food and beverage products where it provides this functionality. It may not be used, however, where its function is as a sweetener. In view of the unnatural sweetness of thaumatin, this limitation on its approval is not a hindrance for the development of quality low-calorie food and beverage products.

Lo Han Kuo Sweetener

The fruit of the Chinese plant known as Lo Han Kuo (systematic name, *Tlàdiantha grosvenorii* [Swingle] C. Jeffrey [Cucurbitaceae]) are very sweet. In the early 1980s, Takemoto et al. demonstrated the major sweet principle of this fruit to be a triterpenoid glycoside that they named mogroside V (**10**) (75). This sweetener is present at a level of approximately 1% of the weight of the dried fruits. Mogroside V has been reported to exhibit $P_w(5.1) = 256$ (76). No quantitative information is available on its FPA, C/R function, or temporal profile. However, it is noteworthy that efforts at Proctor & Gamble (77), as well as other companies, are underway to commercialize this sweetener in the form of a plant extract. This extract is on the market in the United States and is being used in beverages and as a tabletop sweetener.

(10)

SUGAR MACRONUTRIENT SUBSTITUTES OF COMMERCIAL SIGNIFICANCE

In this section, the five sweeteners maltitol, lactitol, isomalt, fructooligosaccharide sweetener, and erythritol are discussed. Each substance attempts to deliver all the functionality of sucrose; thus, they are sugar macronutrient substitutes (MNSs). All these sweeteners are approved for use following the GRAS affirmation process rather than the FAP process. The bioavailable calorie content of the sugar MNSs is a subject of substantial debate. In the United States, specific bioavailable calorie contents have been assigned to each sugar MNS, whereas in the European Union, a value of 2.4 cal/g has been assigned for all sugar MNSs. Comparative data on these sweeteners are presented in Table 4.

Maltitol

Maltitol is a disaccharide alcohol having structure **11**. Maltitol is prepared commercially by hydrogenation of maltose, which in turn is obtained from starch by a combination of enzymatic hydrolysis and chromatography. It is generally agreed to have a sweetness potency of approximately 0.9 times that of sucrose and is relatively invariant

Table 4. FPA and C/R Function Data on Sucrose, Glucose, and Sugar MNS Sweeteners

Sweetener	C/R function data		FPA data					
	P_w	b	Sweet	Bitter	Salty	Metallic	Cooling	Licorice
Sucrose[a]	0.98	0.63	10.0	0.0	0.0	0.0	0.0	0.0
Glucose[b]	0.52	0.52						
Maltitol[c]	0.46	1.33	8.0	0.0	0.0	0.0	0.0	0.0
Lactitol[d]	0.40	−0.58	9.6	0.0	0.0	0.0	0.0	0.0
Isomalt[e]	0.42	−0.02						
FO sweetener[f]	0.27	−0.50	2.3	0.0	0.0	0.0	0.0	0.0

[a]10.0% (w/v) for FPA.
[b]10.0% (w/v) for FPA.
[c]10.0% (w/v) for FPA.
[d]25.5% (w/v) for FPA.
[e]10.0% (w/v) for FPA.
[f]8.0% (w/v) for FPA.

with respect to sucrose reference concentration and application. Pure maltitol was first described as a sweetener for reduced-calorie foods by Mitsuhashi and coworkers of Hyashibara Company in 1973 (78). Crystalline maltitol is known principally under the Hyashibara brand name Malbit®. Maltitol is currently approved for food applications in Denmark, Norway, Finland, the United Kingdom, Switzerland, Sweden, Belgium, France, Austria, and Italy. A GRAS affirmation petition was filed with the FDA by Towa Chemical Industry Company in 1986. Commercial development of maltitol as a food ingredient in the United States awaits action on this petition. The historical development, physical properties, safety assessment studies, metabolism, and food applications studies on maltitol have been comprehensively reviewed and will not be discussed further here (79,80). Particularly relevant to the present discussion, however, are bioavailable calorie content and laxative effects. Substantial controversy has developed over the bioavailable calorie issue, and it is unlikely that an absolute number of calories per gram will ever be agreed on. Arguments on this subject have been summarized by Secard and LeBot (81). As a result of a consideration of the rates of cleavage and absorption in all regions of the gastrointestinal tract, Ziesenitz and Siebert suggest that a value of 2.4 cal/g may be approximately correct (82). The generally accepted bioavailable calorie content in the United States, however, is 3.0 cal/g. Thus, for maltitol, only an approximate 25% reduction in calories may be possible relative to fully nutritive carbohydrate sweeteners. Ziesenitz and Siebert also indicate the intestinal discomfort (laxation, flatulence, etc) experienced from maltitol to be relatively mild.

Lactitol

Lactitol is a disaccharide alcohol of structure **12**. It is obtained commercially by the hydrogenation of lactose (milk sugar), which is available in quantity as a by-product of the dairy industry. Lactitol exhibits a sweetness potency that varies only slightly with sucrose reference concentration; $P_w(2) = 0.30$ and $P_w(8) = 0.39$ have been reported in a general review on this sweetener (83). Lactitol has been commercially developed by CCA Biochem of The Netherlands and is known under the brand name Lacty®. Lactitol was affirmed as GRAS in the United States in 1993. In an attempt to understand the bioavailable calorie content of lactitol, studies have been conducted on the rate of lactitol cleavage in the small intestine. In comparison to lactose and isomaltose, the cleavage rate was sharply reduced. Thus, lactitol passes largely unchanged into the large intestine. Anaerobic microbial fermentation at this point converts it into organic acids and gases. The organic acids are, in part, absorbed by the host as an energy source. In general, after review of the results of many animal and human studies, it is accepted that lactitol has a bioavailable calorie content of approximately 2.0 cal/g. According to Ziesenitz and Siebert, the intestinal discomfort to be expected from lactitol is high (82).

(11)

(12)

Isomalt

Isomalt is approximately a 1:1 mixture of the two disaccharide alcohols 1-O-α-D-glucopyranosyl-mannitol (**13a**) and 1-O-α-D-glucopyranosyl-sorbitol (**13b**). The development and properties of isomalt have been reviewed (84–87). Isomalt is best known under the trade name Palatinit® of Palatinit Sussungsmittel GmbH, a wholly owned subsidiary of Sudzucker AG. Isomalt was discovered at Suddeutsche Zucker in the early 1950s. In the late 1970s, Bayer AG joined with Suddeutsche Zucker in the development of isomalt in the joint venture Palatinit Sussungsmittel GmbH. Isomalt is obtained from sucrose in a two-step process in which it is first enzymatically rearranged to produce the reducing disaccharide isomaltulose, which is then hydrogenated to give the **13a/13b** mixture known generically as isomalt. The sweetness potency of isomalt is moderately dependent on sucrose reference concentration; $P_w(1) = 0.28$ and $P_w(10) = 0.45$ have been reported. Isomalt is generally approved for food usage in a number of countries, including the United Kingdom, Switzerland, France, Israel, The Netherlands, South Africa, Austria, Australia, Denmark, Finland, Sweden, West Germany, Italy, and Norway. In addition, specific category approvals have been given in a number of other countries. In the United States, isomalt was affirmed as GRAS in 1990 following the publication of the last of four scientific papers detailing the results of safety assessment studies on isomalt (W. Irwin, personal communication, 1990). As is the case for all sweeteners of this type, the bioavailable energy content for isomalt is controversial. Although some studies suggest isomalt to be fully caloric, it is generally accepted that 2.0 cal/g is approximately correct. Thus, isomalt use may permit only a 50% reduction of caloric intake when used as a substitute for fully nutritive sweeteners. Ziesenitz and Siebert have indicated the intestinal discomfort experienced as a consequence of isomalt consumption to be relatively mild (82).

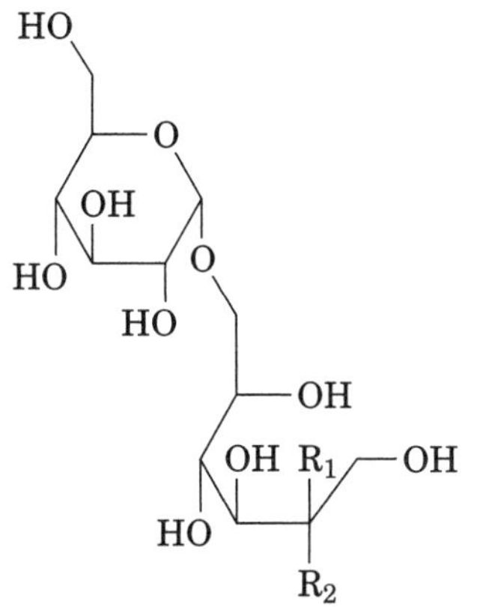

Fructooligosaccharide Sweetener

Fructooligosaccharide sweetener (FO sweetener) is a mixture of fructooligosaccharides of the type found naturally in onions, asparagus, Jerusalem artichoke tubers, and wheat. Specifically, it is largely a mixture of kestose (**14a**), nystose (**14b**), and 1-β-fructofuranosyl-nytose (**14c**). Typically, the ratio of these sugars in FO sweetener is 28:60:12. Meiji Seika Company of Japan has conducted most of the development work of FO sweetener, to which they have given the brand name Neosugar®. The sweetness potency of FO sweetener has been reported to be 0.4 to 0.6 times that of sucrose, depending on reference concentration. FO sweetener is prepared from sucrose by treatment with the microbial enzyme fructosyl transferase. To date, FO sweetener is approved for food use only in Japan. It has been reported that the glycolytic breakdown of FO sweetener in the small intestine is slight and that FO sweetener is therefore a nonnutritive sweetener (88). In other studies, however, it has been demonstrated that FO sweetener is extensively fermented in the large intestine (82). Studies specifically conducted on nystose suggest a very substantial calorie provision to the host following anaerobic microbial fermentation in the large intestine (81). The allocation of a precise value for the bioavailable calories of FO sweetener is difficult; a value of at least 2.0 cal/g is probably realistic, however. The intestinal discomfort and laxation effects promoted by FO sweetener were indicated to be high by Ziesenitz and Siebert (82). This report is consistent with the results of a clinical study by Stone-Dorshow and Levitt (89).

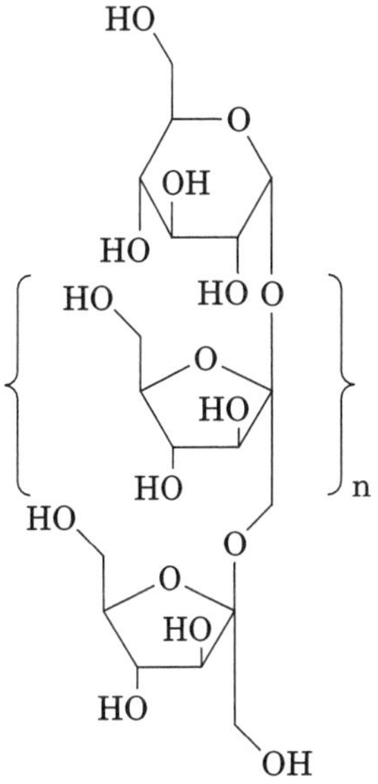

(**13a**) R₁ = OH, R₂ = H
(**13b**) R₁ = H, R₂ = OH

(**14a**) n = 1
(**14b**) n = 2
(**14c**) n = 3

Erythritol

Erythritol is a four-carbon polyol of structure **15**. It is present naturally in a wide range of microorganisms, plants, and animals and, as such, is present in many foods. This natural form of erythritol is the symmetrical isomer and is more properly referred to as *meso*-erythritol. Erythritol has been commercially developed by at least two companies, Mitsubishi in Japan and Eridania Beghin-Say in Europe. In the United States, Mitsubishi Chemical Company and Cargill have formed a partnership known as M&C Sweeteners for the purpose of commercializing erythritol. Erythritol's properties and food applications have recently been reviewed (90). The sweetness potency of erythritol has been reported to be 0.6 to 0.7 times that of sucrose. Unfortunately, flavor and temporal profile data, as have been reported for most of the other sweeteners discussed in this review, are not available for erythritol. However, it is the general consensus that erythritol exhibits a clean sweet taste with a sugarlike temporal profile. Erythritol has fair solubility in water (37% at 25°C) and is very stable. It is prepared on yeast fermentation of glucose. It is easily isolated in a purity of >99% on recrystallization from the filtered fermentation broth. At this time, erythritol has already experienced significant use in foods and beverages in Japan and is expected to see use in many other countries. It was affirmed as GRAS in the United States in 1997. Of all the sugar MNSs that have been commercially developed, erythritol is the most promising because it is lowest in bioavailable calories and it is also the lowest in intestinal discomfort. It has been conservatively estimated that the bioavailable calorie content of erythritol is 0.2 cal/g. This estimate assumes that the portion of a normal dose (ie, 20–50 g in a single intake) that is not absorbed is completely fermented to short-chain fatty acids by the anaerobic bacteria of the colon. In fact, however, erythritol is very poorly fermented by these bacteria (G. DuBois and G. Stark, unpublished results, 1988), and therefore the actual bioavailable calorie content of erythritol is zero or near zero. The minimal intestinal discomfort of erythritol follows from the fact that it is readily absorbed from the small intestine, taken up into the bloodstream, and excreted in the urine. Thus, most of the fermentative and osmotic effects of a large quantity of sugar MNS in the colon are avoided.

(**15**)

Summary

Only the high-potency sweeteners (saccharin, cyclamates, acesulfame-K, aspartame, sucralose, alitame, etc) allow the food technologist the opportunity to formulate noncaloric products. Even this opportunity, however, is limited to a few product categories (eg, CSDs) where the product bulking and other nonsweet taste properties of sucrose are not required. Other food product categories require properties of sucrose in addition to sweet taste (eg, bulking and freezing-point depression), and these can be realized only when employing sugar MNSs (eg, maltitol or isomalt), perhaps in combination with a high-potency sweetener. Most of these one-for-one sugar substitutes are substantially caloric, however, and therefore only reduced calorie food products may be formulated in product categories requiring their use. In addition, most sugar MNSs may have laxative effects. Thus, at this point in time, fully satisfactory zero- or low-calorie food products may be formulated only in product categories where sweet taste is the sole property of sucrose required. Other food product categories will be significantly impacted by the drive for zero- or low-calorie products only on the identification of a new ingredient. To meet this need, this new substance must deliver all or most of the properties of sucrose and, at the same time, be both truly noncaloric and without laxative effects. A strong need exists for such a food ingredient.

NEW DEVELOPMENTS IN NONNUTRITIVE SWEETENERS

Significant progress has been made in the identification of novel high-potency nonnutritive sweeteners. Most of the effort aimed at this objective has been focused on the dipeptide structural class of sweeteners in follow-up of the 1965 discovery of aspartame, the first dipeptide sweetener. The search for analogues of aspartame was intensified by the 1975 report by Fujino and coworkers of Takeda Chemical Industries that dipeptide ester **16** exhibits a clean sweet taste, with $P_w(8.3) = 33,200$ times that of sucrose (91). This represents more than a 200-fold potency increase over aspartame. Problematical, however, was that **16** is much less stable than aspartame. Nonetheless, **16**, along with aspartame, are exciting lead compounds that stimulated research in at least eight laboratories, including the following:

1. *Pfizer Pharmaceutical Company*: Hendrick and coworkers discovered alitame (**6**), as has already been described (58).

2. *Proctor & Gamble Company*: Rizzi and coworkers discovered dipeptide ester **17** (92). They reported **17** to exhibit $P_w(8-10) = 16,450$ and to be substantially more stable than aspartame. Problematical for **17**, however, is a delayed AT and a prolonged ET.

3. *Shanghai Institute of Organic Chemistry*: Zheng and coworkers discovered dipeptide ester **18** (93). This compound was said to be stable and to exhibit $P_w(10) = 2000$.

4. *General Foods Corporation*: In 1988, Roy et al. also discovered dipeptide ester **18** as well as analogue **19**. They reported these compounds to be stable and to exhibit $P_w(8.6) = 2300$ and $P_w(8.5) = 1133$ times that of sucrose, respectively (94).

5. *Takasago Corporation*: Takasago chemists disclosed their discovery of the same compounds (ie, **18** and **19**) as were discovered by the General Foods chemists (95).

6. *Coca-Cola Company*: Iacobucci and coworkers reported dipeptide ester **20** (P_w = 1930) (96) in 1988 and dipeptide amide **21** (P_w = 2500) (97) in 1994.

7. *University of California at San Diego*: Goodman and coworkers discovered **22** as the preferred product candidate (98,99). Sweetener **22** was reported to be 800 to 1000 times more potent than sucrose and also to be much more stable than aspartame.

8. *Université Claude Bernard*: Nofre and Tinti discovered aspartame derivative **23**, which they named neotame and which they report exhibits $P_w(10)$ = 10,000 (100). This sweetener is reported to be equivalent to aspartame in sweetness quality, although it is not more stable than aspartame in acidic products. This new sweetener has been commercially developed by Monsanto. A FAP for tabletop use was filed on neotame in December 1997, and a second petition for general use was filed in January 1999.

A rational approach to engineering safety into a high-potency nonnutritive sweetener is to construct it from building blocks that are common components of the diet. Aspartame is exemplary of such a rationale. Clearly, a protein sweetener would also be consistent with this idea because of the natural amino acid subunit composition. This logic led to a program on the synthesis of thaumatin variants at International Genetic Engineering, Inc. (Ingene) (101). The objective was to identify analogues having a more sucroselike taste. Although the preparation of improved thaumatin analogues has been reported, it appears that the level of improvement is insufficient to justify commercialization. A similar program was undertaken by Kim at the University of California at Berkeley based on another sweet protein known as monellin (102,103). Monellin is a two-chain protein having a molecular weight of 10,700; it is roughly equivalent in sweetness potency to thaumatin and is similarly disadvantaged by an atypical temporal profile. Guided by X-ray crystallographic data, the Kim group designed and genetically engineered an equally potent, and much more stable, single-chain analogue of monellin. This sweetener has also been expressed in plants, including lettuce and tomatoes. Several other sweet proteins have been identified in nature, including the mabinlins (104), curculin (105), and brazzein (106). All these proteins exhibit a sweetness slow to develop that lingers relative to that of sucrose. Nonetheless, studies aimed at the identification of analogues with a more sugarlike taste may be a fruitful area for future research.

Progress has also been made in the identification of novel sugar MNSs. One type of carbohydrate-based nonnutritive sweetener that has been pursued is based on the L- or so-called left-handed sugars (107). It is argued that these substances, being enantiomerically related to the natural D- or right-handed sugars, will be nonmetabolizable and therefore nonnutritive. However, evidence has been presented that one of these sweeteners, L-glucose, is fermented by human fecal bacteria (82). Thus, L-glucose, and perhaps all the L-sugars, are partially caloric and may cause significant intestinal discomfort as well. Of greater

(**16**) R_1 = H; $R_2(R_3)$ = COOCH$_3$; $R_3(R_2)$ = H; R_4 = FnOCO

(**17**) R_1 = H; $R_2(R_3)$ = [structure]; $R_3(R_2)$ = H; R_4 = FnOCO

(**18**) R_1 = H; R_2 = CH$_3$; R_3 = H; R_4 = FnOCO

(**19**) R_1 = H; R_2 = CH$_3$; R_3 = CH$_3$; R_4 = FnOCO

FnOCO = [structure]

(**20**) R_1 = H; $R_2(R_3)$ = COOCH$_3$; $R_3(R_2)$ = H; R_4 = [structure]

(**21**) R_1 = H; R_2 = CH$_2$CH$_3$; R_3 = CONH-(S)-CH(CH$_2$CH$_3$)Ph

(**22**) R_1 = H; R_2 = CH$_3$; R_3 = H; R_4 = [structure]

(**23**) R_1 = (CH$_2$)$_2$C(CH$_3$)$_3$; R_2 = COOCH$_3$; R_3 = H; R_4 = CH$_2$Ph

potential interest is D-tagatose (**24**), a hexose sugar formed from lactose in heated milk and is claimed by Zehner and associates to be noncaloric (108,109). MD Food Ingredients has developed a commercial process to prepare **24** from lactose and is initiating a GRAS affirmation petition on it for general use in foods and beverages. D-Tagatose is claimed to exhibit a clean sweet taste, with $P_w = 0.5$. In rigorous studies, D-tagatose was estimated by MD Food Ingredient scientists to exhibit a bioavailable calorie content of 2.0 Cal/g.

(**24**)

A breakthrough in the design of novel sugar MNSs was described by DuBois et al. in 1992 (110). They identified a wide variety of cleanly sweet sugar MNSs. This work focused on the design of replacements for sucrose and, therefore, to reproduce the colligative properties of sucrose (freezing point depression, etc), novel molecules with molecular weight comparable to that of sucrose were designed. In addition, this work was based on recognition of the fact that neutral, polyhydroxylic organic compounds of low molecular weight, in general, are sweet. Thus, in the attempt to reproduce the taste of sucrose, novel molecules with the same ratio of C atoms to OH groups as exhibited by sucrose (ie, C/OH = 1.5) were identified as targets for synthesis and evaluation. One of the preferred sugar MNS sweeteners identified in this work is polyol **25**. This sweetener exhibits a clean sweet taste and a P_w of 0.46 which is independent of sucrose reference concentration. This polyol is very stable and quite water soluble. Sugar MNS **25** is easily prepared in quantitative yield in one step on reaction of gluconolactone with 1-amino-1-deoxy-sorbitol. The latter starting material is commercially available and is easily prepared by reductive amination of glucose with ammonia. Of particular interest is the fact that **25** is not fermented at all by human fecal bacteria and therefore has an apparent bioavailable calorie content of 0 Cal/g.

(**25**)

At about the same time as the work on **25** and related compounds was described by DuBois et al., Mazur and as-

sociates at Proctor & Gamble described another series of sugar MNSs that also appear to have 0 Cal/g bioavailable calorie content (111). They referred to these new sugar MNSs as the 5-C-(hydroxymethyl)hexoses. The preferred compound in the Mazur series of compounds is **26**, which is prepared in three steps from lactose. Sugar MNS **26** is claimed to exhibit a clean sweetness, a P_w of 0.5, and it is not fermented at all by human fecal bacteria. Further, it is claimed to function well as a sugar replacement in a broad range of food products.

(**26**)

CONCLUSION AND FUTURE DEVELOPMENT

In summary, the nonnutritive sweetener market in the United States is dominated by aspartame and saccharin. In the near future, it is reasonable to expect that the recently approved acesulfame-K and sucralose, as well as cyclamates and alitame, which are not yet approved, may be competitors in this market. In the longer term, it is likely that blends of these sweeteners will be the most favored nonnutritive sweetener systems. Sweetener cost, as a consequence of the competition among these sweeteners as well as the substantial sweetness synergies realized in sweetener blends, is thus likely to be only a minor component of food product manufacturing cost. This statement is applicable, however, only to food categories in which the sucrose function may be fulfilled by a high-potency sweetener alone. Categories that require the type of functionality delivered by a sugar MNS such as maltitol, isomalt, lactitol, fructooligosaccharide sweetener, and erythritol are very unlikely ever to be favored by low-sweetener cost. The colligative properties of sucrose that are requisite for such applications necessitate one-for-one replacement. These materials have high manufacturing cost, relative to sucrose. In addition, the laxative effects when consumed at high levels are a disadvantage. A strong need exists for a sugar MNS that is truly noncaloric and nonlaxative.

BIBLIOGRAPHY

1. G. R. T. Ross, ed., and trans., "Aristotle," in *Aristotle: De-Sensu and de Memoria*, Cambridge University Press, Cambridge, U.K., 1906, pp. 41–99 (originally published ca. 330 B.C.).

2. C. Fahlberg and I. Remsen, *Berichte* **12**, 469–473 (1879).

3. R. H. Mazur, J. M. Schlatter, and A. H. Goldkamp, *J. Am. Chem. Soc.* **91**, 2684 (1969).

4. U.S. Pat. 2,803,551 (August 20, 1957), F. J. Helgren (to Abbott Laboratories).

5. M. Meilgaard, G. Vance Civille, and B. T. Carr, *Sensory Evaluation Techniques*, Vol. II, CRC Press, Boca Raton, Fla., 1987, pp. 5–6.

6. B. T. Carr et al., in C.-T. Ho and C. H. Manley, eds., *Flavor Measurement*, Marcel Dekker, New York, 1993, pp. 219–237.

7. G. E. DuBois and J. F. Lee, *Chem. Senses* **7**, 237–246 (1983).

8. G. E. DuBois, G. A. Crosby, and R. A. Stephenson, *J. Med. Chem.* **24**, 408–428 (1981).

9. R. J. Ronk, in J. H. Shaw and G. G. Roussos, eds., *Sweeteners and Dental Caries*, Information Retrieval, Inc., Washington, D.C., 1978, pp. 131–144.

10. H. W. Schultz, *Food Law Handbook*, AVI Publishing Company, Westport, Conn., 1981.

11. *Code of Federal Regulations*, Title 21 (Food and Drugs), Section 182.1, U.S. Government Printing Office, Washington, D.C., 1988, pp. 385–386.

12. *Code of Federal Regulations*, Title 21 (Food and Drugs), Section 570.30, p. 611 U.S. Government Printing Office, Washington, D.C., 1989, p. 611.

13. J. P. Marshall and J. A. Pollard, *Food Legislation Surveys No. 1: Sweeteners—International Legislation*, 4th ed., Leatherhead Food, R. A., United Kingdom (1987).

14. F. J. Broulik, in T. H. Grenby, ed., *Advances in Sweeteners*, Blackie Academic & Professional, London, 1996, pp. 35–55.

15. G. Vettorazzi, in R. D. Middlekauf and P. Shubik, eds., *International Food Regulation Handbook*, Marcel Dekker, New York, 1989.

16. "Toxicological Evaluation of Certain Food Additives with a Review of General Principles and of Specifications," World Health Organization Technical Report Series No. 539, Geneva, Switzerland, 1974, p. 11.

17. *Toxicological Principles for the Safety Assessment of Direct Food Additives and Color Additives Used in Food*, U.S. Food and Drug Administration, Bureau of Foods, 1982, pp. 1–19.

18. *Chemical Marketing Reporter*, Vol. 256, Schnell Publishing Company, New York, 1999.

19. H. R. Moskowitz, ed., *Product Testing and Sensory Evaluation of Foods*, Food and Nutrition Press, Westport, Conn., 1983.

20. G. E. DuBois et al., in D. E. Walters, F. T. Orthoefer, and G. E. DuBois, eds., *Sweeteners: Discovery, Molecular Design, and Chemoreception*, American Chemical Society, Washington, D. C., 1991, pp. 261–276.

21. B. E. Homler, in L. D. SteginK and L. J. Filer Jr., eds., *Aspartame: Physiology and Biochemistry*, Marcel Dekker, New York, 1984.

22. L. Hyvonen and P. Koivistoinen, in G. G. Birch and K. J. Parker, eds., *Nutritive Sweeteners*, Applied Science Publishers, Englewood, N.J., 1982, pp. 135–137.

23. F. J. Helgren, M. F. Lynch, and F. J. Kirchmeyer, *J. Am. Pharm. Assoc. (Sci. Ed.)* **44**, 353–355 (1955).

24. L. M. Bartoshuk, *Science* **205**, 934–935 (1979).

25. J. F. Gent and L. M. Bartoshuk, *Chem. Senses* **7**, 265–272 (1983).

26. S. Pecore and B. T. Carr, "Sweeteners: Carbohydrate and Low Calorie, International Conference," paper presented at The Agricultural and Food Chemistry Division of the American Chemical Society Symposium, Los Angeles, Calif., September 22–25, 1988.

27. E. M. Whelan, *The Conference Board Magazine* **16**, 54 (1977).

28. G. J. Walter and M. L. Mitchell, in L. O'Brien Nabors and R. C. Gelardi, eds., *Alternative Sweeteners*, Marcel Dekker, New York, 1986, pp. 15–41.

29. A. G. Renwick, *Comments Toxicol.* **3**, 289–305 (1989).

30. S. M. Cohen et al., *Cancer Res.* **58**, 2557–2561 (1998).

31. *Code of Federal Regulations*, Title 21 (Food and Drugs), Section 180.37, U.S. Government Printing Office, Washington, D.C., 1988, pp. 378–379.

32. *The Merck Index*, 10th ed., Merck & Co., Rahway, N.J., 1983.

33. L. F. Audrieth and M. Sveda, *J. Org. Chem.* **9**, 89 (1944).

34. B. A. Bopp, R. C. Sonders, and J. W. Kesterson, *CRC Crit. Rev. Toxicol.* **16**, 213–306 (1986).

35. L. O'Brien-Nabors and W. T. Miller, *Comments Toxicol.* **314**, 307–315 (1989).

36. *Food Chemical News*, November 13, 1989, p. 19.

37. K. M. Beck, in T. E. Furia, eds., *CRC Handbook of Food Additives*, Vol. 11, 2nd ed., CRC Press, Boca Raton, Fla., 1980, p. 125.

38. T. Fazio, J. W. Howard, and E. O. Haenni, *J. Assoc. Off. Anal. Chem.* **53**, 1120–1128 (1970).

39. "Cyclamates," Hearings before Subcommittee No. 2 of the Committee on the Judiciary House of Representatives, Serial No. 2, Ninety-Second Congress, Washington, D.C., September 29, 30 and October 6, 1971.

40. Zhong Hua Fang BA (Hong Kong, China), quotation for sodium cyclamate: $2.30/kg (April 28, 1999).

41. R. E. Baldwin and B. M. Korschgen, *J. Food Sci.* **44**, 938–939 (1979).

42. J. J. Wiseman and M. R. McDaniel, paper presented at the Institute of Food Technology Meeting, Chicago, Ill. (1989).

43. G. DuBois, "Sweeteners, Nonnutritive," in Y. H. Hui, ed., *Encyclopedia of Food Science and Technology*, John Wiley & Sons, New York, 1991.

44. L. D. Stegink and L. J. Filer, Jr., eds., *Aspartame, Physiology and Biochemistry*, Marcel Dekker, New York, 1984, pp. 29–199, 289–653.

45. H. H. Butchko and F. N. Kotsonis, *Comments Toxicol.* **3**, 253–278 (1989).

46. C. Tschanz et al., eds., *The Clinical Evaluation of a Food Additive: Assessment of Aspartame*, CRC Press, Boca Raton, Fla., 1996.

47. *NutraSweet Technical Applications Manual*, The NutraSweet Company, Deerfield, Ill., 1987, p. 5.

48. E. G. Burton et al., *J. Nutr.* **119**, 713–721 (1989).

49. U.S. Pat. 4,704,288 (November 3, 1987), J. H. Tsau and J. G. Young (to The NutraSweet Company).

50. K. Clauss and Harald Jensen, *Angew. Chem.* (International Edition in English) **12**, 869–876 (1973).

51. D. G. Mayer and F. H. Kemper, eds., *Acesulfame-K*, Marcel Dekker, New York, 1991.

52. G.-W. von Rymon Lipinski and D. Maver, *Comments Toxicol.* **3**, 279–287 (1989).

53. G.-W. von Rymon Lipinski and B. E. Huddart, *Chem. Ind.* **6**, 427–432 (1983).

54. *United States Food Additive Petition*, Hoechst Celanese Corporation, Somerville, N.J.

55. L. Hough and S. P. Phadnis, *Nature* (London) **263**, 800 (1976).

56. U.S. Pat. 4,435,440 (March 6, 1984), L. Hough (to Tate & Lyle Public Limited Company, Reading, England).

57. M. R. Jenner, in T. H. Grenby, ed., *Advances in Sweeteners*, Blackie Academic and Professional, London, 1996, pp. 253–262.

58. U.S. Pat. 4,411,925 (October 25, 1983), T. M. Brennan and M. E. Hendrick (to Pfizer, Inc.).

59. M. E. Hendrick, H. L. Mitchell, and P. R. Murray, in T. H. Grenby, ed., *Advances in Sweeteners*, Blackie Academic and Professional, London, 1996, pp. 226–239.

60. R. M. Horowitz and B. Gentili, *J. Agric. Food Chem.* **17**, 696–700 (1969).

61. R. M. Horowitz and B. Gentili, in *Product Testing and Sensory Evaluation of Foods*, Food and Nutrition Press, Westport, Conn., 1983, pp. 135–153.

62. R. L. Smith et al., *Food Technol.* **10**, 72–81 (1996).

63. M. G. Lindley, in T. H. Grenby, ed., *Advances in Sweeteners*, Blackie Academic & Professional, London, 1996, pp. 240–252.

64. G. A. Crosby and R. E. Wingard, Jr., in C. A. M. Hough, K. J. Parker, and A. J. Vlitos, eds., *Developments in Sweeteners-1*, Applied Science Publishers, New York, 1979, pp. 137–140.

65. S.-H. Kim and G. E. DuBois, in S. Marie and J. R. Piggott, eds., *Handbook of Sweeteners*, Blackie, Glasgow, 1991, pp. 116–185.

66. K. C. Phillips, in T. H. Grenby, ed., *Developments in Sweeteners-3*, Elsevier Applied Science, New York, 1987, pp. 1–43.

67. B. Crammer and R. Ikan, in T. H. Grenby, ed., *Developments in Sweeteners-3*, Elsevier Applied Science, New York, 1987, pp. 45–64.

68. B. Lythgoe and S. Trippett, *J. Chem. Soc.*, 1983–1990 (1950).

69. G. E. DuBois, in H. J. Hess, ed., *Annual Reports in Medicinal Chemistry*, Vol. 17, Academic Press, New York, 1982, pp. 323–332.

70. L. O'Brien Nabors and G. E. Inglett, in *Product Testing and Sensory Evaluation of Foods*, Food and Nutrition Press, Westport, Conn., 1983, pp. 311–313.

71. *Fed. Regist.* **50**, 21043–21045 (1985).

72. J. D. Higginbotham, in *Product Testing and Sensory Evaluation of Foods*, Food and Nutrition Press, Westport, Conn., 1983, pp. 103–134.

73. S.-H. Kim, A. Devos, and C. Ogata, *Trends Biol. Sci.* **13**, 13–15 (1988).

74. B. L. Oser, R. A. Ford, and B. K. Bernard, *Food Technol.* **38**, 66–89 (1984).

75. T. Takemoto et al., *Yakugaku Zasshi* **103**, 1167–1173 (1983).

76. A. D. Kinghorn and D. D. Soejarto, *CRC Crit. Rev. Plant Sci.* **4**, 79–120 (1986).

77. M. J. Mohlenkamp, oral presentation, *International Symposium on Sweeteners*, Jerusalem, Israel, July 14–20, 1996.

78. U.S. Pat. 3,741,776 (June 26, 1973), M. Mitsuhashi, M. Hirao, and K. Sugimoto (to Hyashibara Company, Okayama, Japan).

79. I. Fabry, in T. H. Grenby, ed., *Developments in Sweeteners-3*, Elsevier Applied Science, New York, 1987, pp. 83–108.

80. M. Heume and A. Rapaille, in T. H. Grenby, ed., *Advances in Sweeteners*, Blackie Academic & Professional, London, 1996, pp. 85–108.

81. P. J. Secard and Y. LeBot, paper presented at the Agricultural and Food Chemistry Division of the American Chemical Society Symposium, *Sweeteners: Carbohydrate and Low Calorie*, International Conference, Los Angeles, Calif., September 22–25, 1988.

82. S. C. Ziesenitz and G. Siebert, in T. H. Grenby, ed., *Developments in Sweeteners-3*, Elsevier Applied Science, New York, 1987, pp. 83–108.

83. C. H. Den Uyl, in T. H. Grenby, ed., *Developments in Sweeteners-3*, Elsevier Applied Science, New York, 1987, pp. 83–108.

84. W. E. Irwin, paper presented at The Agricultural and Food Chemistry Division of theamerican Chemical Society Symposium, *Sweeteners: Carbohydrate and Low Calorie*, International Conference, Los Angeles, Calif., September 22–25, 1988.

85. P. J. Strater, in *Product Testing and Sensory Evaluation of Foods*, Food and Nutrition Press, Westport, Conn., 1983, pp. 217–244.

86. S. C. Ziesenitz, in T. H. Grenby, ed., *Advances in Sweeteners*, Blackie Academic and Professional, London, 1996, pp. 109–133.

87. I. Willibald-Ettle and H. Schiweck, in T. H. Grenby, ed., *Advances in Sweeteners*, Blackie Academic and Professional, London, 1996, pp. 134–149.

88. T. Oku, T. Tokunaga, and N. Hosoya, *J. Nutr.* **114**, 1574–1581 (1984).

89. T. Stone-Dorshow and M. D. Levitt, *Am. J. Clin. Nutr.* **46**, 61–65 (1987).

90. J. Goossens and M. Gonze, in T. H. Grenby, ed., *Advances in Sweeteners*, Blackie Academic and Professional, London, 1996, pp. 150–186.

91. M. Fujino et al., *Chem. Pharm. Bull.* **24**, 2112–2117 (1976).

92. U.S. Pat. 4,692,513 (September 8, 1987), R. B. Blum et al. (to The Proctor & Gamble Company).

93. G. Z. Zeng and S. T. Wei, *Molecular Recognition of Taste*, Science Publisher, Beijing, China, 1984

94. U.S. Pat. 4,766,246 (August 23, 1988), P. R. Zanno, R. E. Barnett, and G. M. Roy (to General Foods Corporation).

95. Jpn. Pat. 61291596 (December 22, 1986), A. Nagakura et al. (to Takasago Perfumery Co., Ltd., Japan).

96. U.S. Pat. 4,788,069 (November 29, 1988), G. A. Iacobucci, J. G. Sweeny, and J. G. King III (to The Coca-Cola Company).

97. U.S. Pat. 5,286,509 (February 15, 1994), L. L. D'Angelo and J. G. Sweeny (to The Coca-Cola Company).

98. U.S. Pat. 4,571,345 (February 18, 1986), M. S. Verlander, W. D. Fuller, and M. Goodman (to Cumberland Packing Corp.).

99. W. D. Fuller, M. Goodman, and M. S. Verlander, *J. Am. Chem. Soc.* **107**, 5821–5822 (1985).

100. U.S. Pat. 5,480,668 (January 1, 1996), C. Nofre and J.-M. Tinti (to Universite Claude Bernard).

101. International Pat. Application W085/01746 (April 25, 1985), M. Grunstein, S. Holienberg, R. Koduri, and T. Date (to Beatrice Companies, Inc.).

102. International Pat. Application W088/10265 (December 29, 1988), S.-H. Kim and J. M. Cho (to Lucky Biotech Corp., Emeryville, Calif.).

103. S.-H. Kim et al., *Prot. Eng.* **2**, 571–575 (1989).

104. X. Liu et al., *Eur. J. Biochem.* **211**, 281–287 (1993).

105. H. Yamashita, S. Theerasilp, and Y. Kurihara, *Tenth International Symposium on Olfaction and Taste*, Oslo, Norway, 1989, p. 77.

106. U.S. Pat. 5,346,998 (September 13, 1994), B. G. Hellekant and D. Ming (to Wisconsin Alumni Research Foundation).

107. U.S. Pat. 4,262,032 (April 14, 1981), G. V. Levin (to Biospherics, Inc., Rockville, Md.).

108. G. V. Levin et al., *Am. J. Clin. Nutrition* **62**, 11615–11685 (1995).

109. U.S. Pat. 4,786,722 (November 22, 1988), L. R. Zehner (to Biospherics, Inc.).

110. G. E. DuBois et al., *J. Chem. Soc. Chem. Commun.*, 1604–1605 (1992).

111. A. W. Mazur and M. J. Mohlenkamp, in M. Yalpani, ed., *New Technologies for Healthy Foods & Nutraceuticals*, ATL Press, Mt. Prospect, Ill., 1997, pp. 124–142.

GRANT E. DUBOIS
The Coca-Cola Company
Atlanta, Georgia

SYRUPS (STARCH SWEETENERS AND OTHER SYRUPS)

Pure crystalline glucose (dextrose) and glucose-containing syrups are hydrolysis products of starch and can be referred to as starch-based sweeteners or starch sugars. Crystalline maltose and high maltose content syrups are also produced by starch hydrolysis. Starch-based sweeteners are produced almost exclusively from corn starch in the United States, and the syrups are usually called corn syrups. Elsewhere, other starch sources are also used, such as wheat, rice, potato, and tapioca, to name a few. Although the commercial products are analytically distinguishable, they have in common the general methods of preparation and many properties and applications.

Starch-based sweeteners display increasing sweetness with increasing degree of conversion until complete conversion to D-glucose is achieved. Additional sweetness can be gained by isomerizing the glucose molecule to fructose. Sweetness is impacted by concentration as well as the content and stereochemical configuration of the individual saccharides. When tasted by different taste panels at approximately 10 wt % solids concentration, the various starch sweeteners exhibit a range of sweetness relative to sucrose. This range is illustrated in Figure 1. The degree of conversion can be quantified by the dextrose equivalent (DE) of the products. From the lesser converted maltodextrins through the corn syrups to pure glucose, sweetness increases from about zero to 60 to 85% that of sucrose. Then, as glucose is isomerized to fructose, sweetness continues to rise, up to a value of 105 to 135% that of sucrose when tasting pure fructose solutions.

Dextrose is the common or commercial name for D-glucose, the monosaccharide liberated by complete hydrolysis of starch. Maltose is a disaccharide that is produced by the action of β-amylase enzyme on liquefied starch. Corn syrups (starch hydrolysates having a DE value of 20 or greater) are clear, colorless, viscous liquids prepared by hydrolysis of starch to solutions of dextrose, maltose, and higher molecular weight saccharides. Maltodextrins have a lower degree of hydrolysis than corn syrups, with DE values less than 20. High-fructose corn syrups (HFCS) are products containing fructose produced by enzymatic isom-

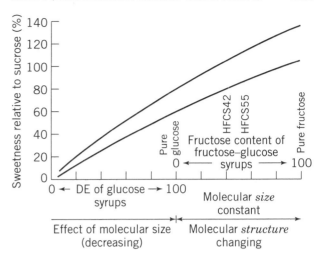

Figure 1. Range of sweetness of starch sugars relative to sucrose. All data at about 10 wt % solids water solutions. *Source:* Ref. 1, reproduced with permission.

erization of dextrose. Starch sugar production and pricing data for a number of countries around the world are shown in Table 1. It should be noted that accurate production data are sometimes difficult to obtain, prices are usually published list prices, and quantity discounts are normal. Production and price data shown in this article should be considered as approximations only.

Maple syrup is a nutritive sweetener produced by evaporating maple tree sap to a concentrated solution of carbohydrate (nearly all sucrose). Molasses is a syrup produced as a by-product during sugar purification. Because of differing sucrose recoveries and the presence of gums, minerals, and nitrogenous materials, molasses composition can vary. Sorgo syrup is made from the juice extracted from the canes of the sorghum plant. Palatable syrups can also be made from artichokes by hydrolyzing the inulin contained in aqueous extracts to form fructose-rich syrups. Today, inulin-based fructose is produced commercially, but the primary source is reported to be chicory.

The per capita consumption of the major nutritive sweeteners is shown in Figure 2. Before the development of HFCS in the late 1960s, starch sugars comprised less than one-fifth of the total; sucrose was the sweetener of choice. As HFCS rapidly captured market share, primarily in the beverage industry, and consumption of dextrose, corn syrups, and maltodextrins continued to grow, starch sugars increased their market share. Around 1985, they captured more than half of the total nutritive sweetener market. Today, they hold about 60% of the nutritive sweetener market in the United States. The growth of starch sugars in various market segments is shown in Figure 3. The growth of the beverage segment and its dominant position today can be clearly seen.

DEXTROSE

Dextrose (D-glucose, corn sugar, starch sugar, blood sugar, grape sugar) is by far the most abundant sugar in nature. It occurs either in the free state (monosaccharide form) or chemically linked with other sugar moieties. In the free

Table 1. Production and Pricing of Starch-Based Sweeteners by Country

Country	Glucose syrups	Maltose syrups	Crystalline glucose Monohydrate	Anhydrous	Total sugar	Fructose-glucose syrup 42%	55%
United States							
Production	3200		645	25	68	2300	5000
Price, US$/kg ds	0.30		0.62	1.25	0.88	0.41	0.45
Europe (EU15)							
Production	2200[a,b]		320[a,b]		45	330[a,b]	
Price, US$/kg ds	0.73[c]		0.96[a]		0.88	0.69[c]	
Canada							
Production	118[d]						300[e]
People's Republic of China							
Production	250	100	250[f]		50	100[e]	
Price, RMB[g]/kg ds	4	4	6		3	6[h]	
Japan							
Production		440[a,i]	35[d]	39[d]	44[d]	845[a,e]	
Price, yen/kg/ds	34[a]	46[a]	95[a]	126[a]	113[d]	48[a]	54[a]
Latin America[d]							
Production	285		9			36	155
Price, US$/kg ds	0.30		2.86			0.17	0.40
Korea							
Production	200[j]		33[d,k]	3[d,k]	55[d,k]	250[e,l]	
Price, won/kg ds	384[l]		420[l]	420[l]	390[m]	390[l]	
South Africa							
Production		175[i]	3				
Price, US$/kg ds	0.43[n]		0.68				
Australia[o]							
Production	57					4.3	
Malaysia[o]							
Production	3.8		2.4				
Price, US$/kg ds	0.57		0.59				

Note: all production values in 10^3 t of dry substance-per year; 1996 values used except where noted.
Source: Ref. 2, reproduced with permission.
[a] 1995 data.
[b] 90+% of the volume shown is sold within the EU15; less than 10% is sold in other European countries.
[c] 1983 data.
[d] 1986 data.
[e] Combined production of 42% db and 55% db fructose-bearing syrups shown.
[f] Most crystalline (pure) dextrose is monohydrate; some is anhydrous. Main uses are medical in nature.
[g] At the time of the preparation of this chart, US$1.00 = 8.5 RMB.
[h] Price for 42% db fructose syrup shown; only one known producer of 55% fructose syrup in the China.
[i] Combined production data for glucose and maltose syrups.
[j] 1992 data.
[k] Totals for the production of these three products in 1992 virtually unchanged from 1986, individual product data not available in 1992 so 1986 values are shown.
[i] 1994 data.
[m] 1991 data.
[n] 1996 list price for 43 DE glucose syrup.
[o] 1987 data.

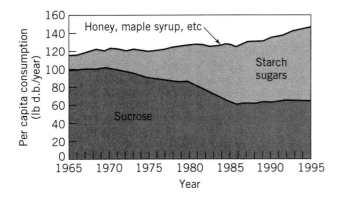

Figure 2. Per capita U.S. sweetener consumption by type. *Source:* Ref. 1, reproduced with permission.

state, it occurs in substantial quantities in honey, fruits, and berries. As a polymer of anhydrodextrose units, it occurs in starch, cellulose, and glycogen. Sucrose is a disaccharide of dextrose and fructose. Commercial production of dextrose by hydrolysis of starch yields white crystalline sugars that are either anhydrous ($C_6H_{12}O_6$) or hydrated by a single water molecule ($C_6H_{12}O_6 \cdot H_2O$). Dextrose monohydrate, with its one molecule of water of crystallization per molecule of sugar, separates from concentrated solutions at <50°C (122°F). Anhydrous D-glucose does not contain water of crystallization and separates at 50 to 115°C (122 to 239°F). Another anhydrous form, β-D-glucose, separates if crystallization is carried out at temperatures >110 to 115°C (230 to 239°F).

In one of the first attempts to prepare commercial dextrose, grapes were used as the starting material (3). It is generally conceded that Kirchoff's work in 1811 was the forerunner of the starch hydrolysate industry (4). It was first reported in 1815 that acid conversion of starch to sugar was the result of hydrolysis of the starch rather than dehydration, and that the starch sugar was identical with

grape sugar (5). It was not until 1842, however, that starch hydrolysis was first practiced commercially in the United States. Crystalline dextrose became a main industrial product when a commercially feasible crystallization process was patented in 1923 (6). This patent was one of the rare instances of a crystal structure being the subject of a patent claim. Fifty metric tons of monohydrate dextrose were sold in the United States in 1923. Today, the annual production is more than 500,000 metric tonnes, a 10,000-fold increase.

Properties

Physical properties of the three crystalline forms of dextrose are listed in Table 2. In solution, dextrose exists in both the α and β forms. When α-dextrose dissolves in water, its optical rotation diminishes gradually as a result of mutarotation until, after a prolonged time, an equilibrium value is reached (see arrows, Table 2). At this point, about 62% of the dextrose is present in the β form. This equilibrium value is not significantly changed over a wide range of temperatures and concentrations. The same equilibrium value apparently exists in anhydrous melts as well as in glassy materials. Pure crystalline β-dextrose is quite sensitive to moisture, and it changes to the more stable α-form if exposed to high humidity. At 25°C (77°F), α-dextrose monohydrate dissolves fairly rapidly, yielding a solution containing about 30 wt % dextrose. Very slowly thereafter, further quantities of dextrose dissolve until a saturated solution containing ca 50 wt % dextrose is obtained. The first phase of the dissolving process results from the limited solubility of α-dextrose. The slow subsequent dissolution is caused by the transformation of part of the dissolved α-dextrose to the more soluble β form. When saturation is finally reached, a mixture of α- and β-dextrose in solution is in equilibrium with solid α-dextrose monohydrate. At 25°C (77°F), anhydrous α-dextrose dissolves rapidly and beyond the limit of solubility of α-dextrose monohydrate. Since the monohydrate is the stable form at

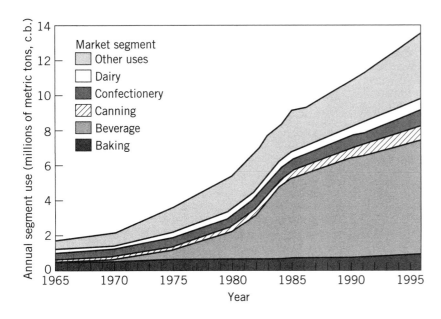

Figure 3. Applications of starch sugars in the United States, 1965–1996. *Source:* Ref. 1, reproduced with permission.

Table 2. Physical Properties of D-Glucose

Property	α-D-Glucose	α-D-Glucose hydrate	β-D-Glucose
Molecular formula	$C_6H_{12}O_6$	$C_6H_{12}O_6 \cdot H_2O$	$C_6H_{12}O_6$
Melting point (°C)	146	83	150
Solubility (at 25°C), g/100-g solution	$62 \rightarrow 30.2 \rightarrow 51.2^a$	$30.2 \rightarrow 51.2^{a,b}$	$72 \rightarrow 51.2^a$
Optical rotation, $[\alpha]_D$	$112.2 \rightarrow 52.7^a$	$112.2 \rightarrow 52.7^{a,b}$	$18.7 \rightarrow 52.7^a$
Heat of solution (at 25°C), J/g^c	$+59.4$	$+105.1$	$+26.0$

[a]Equilibrium value.
[b]Anhydrous basis.
[c]To convert joules to calories, divide by 4.184.

this temperature, crystallization of the hydrate occurs to its limit of solubility and the pattern then follows that of the monohydrate as described earlier. The rate of attainment of equilibrium is increased by heating or in the presence of acids or bases. Data for the solubility of the equilibrium mixtures, as interpolated from previous data (7), are listed in Table 3.

Dextrose in solution or in solid form exists in the pyranose structural conformation. In solution, a small amount of the open-chain aldehyde form exists in equilibrium with the two cyclic forms (Fig. 4) and is responsible for the reducing properties of dextrose (8).

Dextrose exhibits the reactions of an aldehyde, a primary alcohol, a secondary alcohol, and a polyhydric alcohol. In acid solution, either after standing for a prolonged time or after heating, dextrose undergoes polycondensation, i.e., dehydration. This reaction yields a mixture of di- and oligosaccharides, most of which are the disaccharides gentiobiose and isomaltose. In acid solution and at high temperature, dehydration leads to formation of 5-hydroxymethylfurfural, which is a water-soluble, high-boiling, and relatively unstable compound. Polymerization of 5-hydroxymethylfurfural yields dark-colored compounds, and is an intermediate in the discoloration of sugar solutions (9). Dextrose decomposition under these conditions also yields levulinic and formic acids.

In mildly alkaline solution, the principal reaction of dextrose is partial transformation, i.e., isomerization, to fructose and other ketoses. D-Mannose, saccharinic acids, and other decomposition products are formed to a lesser extent. Highly alkaline solutions, particularly in the presence of atmospheric oxygen, can form a complex mixture of products of decomposition and rearrangement. Mild oxidation in slightly alkaline solution gives D-gluconic acid in quantitative yield. More vigorous oxidation with nitric acid yields glucaric acid, tartaric acid, oxalic acid, and other compounds resulting from fragmentation of the dex-

trose molecule. Alkaline Fehling's solution is reduced by dextrose with roughly 5 atoms of copper reduced per molecule of dextrose. Electrolytic reduction or catalytic hydrogenation of dextrose is practiced commercially to manufacture sorbitol.

When dextrose is heated with methanol containing a small amount of anhydrous hydrogen chloride, α-methyl-D-glucoside is obtained in good yield and can be isolated by crystallization. Similar reactions occur with higher alcohols, but the reaction products are more difficult to isolate by crystallization. Dextrose reacts with acid anhydrides in the presence of basic catalysts, yielding esters. The complete reaction gives the pentaacetylated derivative.

The reaction of dextrose with a nitrogen-containing compound, for example, amino acids or proteins, yields a series of intermediates that form pigments of varied molecular weight (Maillard reaction). The type of pigments produced is dependent on reaction conditions such as pH, temperature, and concentration of reactants (10).

Dextrose is the common intermediary metabolite in carbohydrate metabolism because other utilizable monosaccharides are converted to dextrose before they are further metabolized. Starch, glycogen, and the common monosaccharides are hydrolyzed enzymatically in the alimentary canal. Dextrose is normally absorbed into the portal-vein blood, by which it is transported first to the liver and then circulated to all parts of the body. Before dextrose or any other monosaccharide can be utilized metabolically, it must be phosphorylated and enter the glycolytic cycle. Other monosaccharides, such as galactose and fructose, eventually are converted to glucose-6-phosphate and metabolized like dextrose.

Manufacture

Until 1960, a commercial high dextrose content syrup was produced by acid-catalyzed hydrolysis of starch at elevated

Table 3. Solubility of Dextrose in Water

Temperature (°C)	Dextrose in solution (wt %)	Temperature (°C)	Dextrose in solution (wt %)
0	34.9	30	54.6
5	38.0	40	61.8
10	41.2	50	70.9
15	44.5	60	74.8
20	47.8	70	78.2
25	51.3	80	81.3

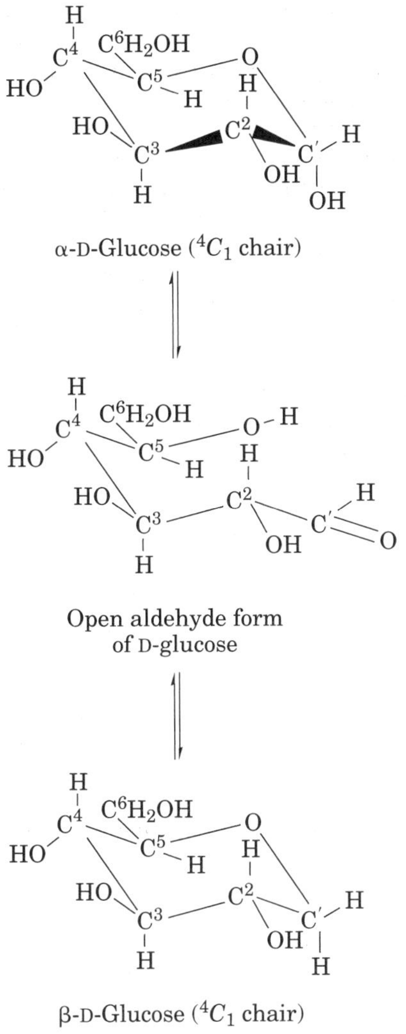

α-D-Glucose ($^4C_1$ chair)

Open aldehyde form
of D-glucose

β-D-Glucose ($^4C_1$ chair)

Figure 4. Structures of D-glucose.

temperatures and pressures. Dextrose content of the starch hydrolysate was limited to about 86% dry basis (db) as a result of the formation of degradation products (11). Dextrose content was increased by partial or complete replacement of acid with one or more enzymes. These processes are referred to as acid–enzyme (A–E) or enzyme–enzyme (E–E), depending on whether initial starch hydrolysis (the starch thinning or liquefaction step) is conducted with acid or a bacterial α-amylase. In either case, subsequent conversion to dextrose (saccharification) is achieved with glucoamylase, a fungal enzyme that releases dextrose from the nonreducing end of starch polymers by successive hydrolysis of α-1,4 and 1,6 glucosidic linkages.

By 1960, the A–E process was in general use, and hydrolysate dextrose content was increased to 92 to 94% db (12). A typical process involved thinning a 30 to 35 wt % starch slurry with acid (HCl or H_2SO_4) at a temperature and pressure necessary to achieve a DE level of 10 to 20. DE is a measure of the reducing-sugar content of a starch hydrolysate calculated as dextrose and expressed as a percentage of the total dry substance. Formation of acid-

reversion products during thinning limited the dextrose yield attained during saccharification. Higher dextrose content was achieved by the E–E process that was developed in the 1960s. This process used a heat-resistant bacterial α-amylase for starch thinning. Lower temperature and nearly neutral liquefaction conditions limited side reactions and resulted in hydrolysate dextrose contents of 95 to 97% db (11,13). Initial processes were based on α-amylase derived from the bacterium *Bacillus subtilis*. In a typical process, 30 to 40 wt % starch is thinned at 85°C (185°F) for 1 h followed by a short heat treatment at 120 to 140°C (248 to 284°F). Then, a second enzyme addition is made, and reaction at 85°C (185°F) is continued to complete the liquefaction step (Fig. 5). The resulting 10 to 15 DE hydrolysate is then saccharified with glucoamylase. The high temperature heat treatment is required to solubilize insoluble starch particles that are believed to be amylose–fatty acid complexes formed during initial liquefaction (14). Since the high temperature inactivates the enzyme, a second enzyme addition is necessary to continue hydrolysis before the complex reforms.

Elimination of the heat treatment step was made possible by the commercialization of α-amylases that are extremely thermostable and capable of efficient starch hydrolysis at a temperature above 100°C (212°F). Enzymes derived from *Bacillus stearothermophilus* (15) and *Bacillus licheniformis* (16) are used for this purpose (Fig. 6). Starch slurry is thinned continuously with a steam-injection heater at 30 to 40 wt % solids, pH 6 to 6.5, and 103 to 107°C (217 to 225°F) for 6 to 10 min. A 1 to 2 h hold at about 95°C (203°F) completes the thinning and yields a 10 to 15 DE

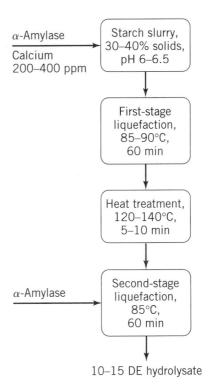

Figure 5. Simplified process flowsheet. Starch liquefaction using *Bacillus subtilis* α-amylase.

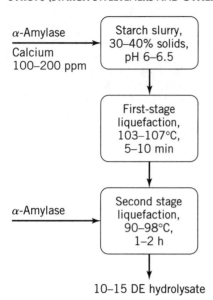

Figure 6. Simplified process flowsheet. Starch liquefaction using *Bacillus stearothermophilus* or *Bacillus licheniformis* α-amylase.

hydrolysate for saccharification. The high initial temperature is sufficient to disrupt the amylose–lipid complex. Concurrently, the thermostable enzyme hydrolyzes the starch to a point where reassociation of the complex cannot occur. The thermostable α-amylases are able to survive the initial liquefaction with little loss in activity since the half-life of α-amylases derived from *B. stearothermophilus* and *B. licheniformis* is about 2 h (15) and 1 h (16), respectively, at 105°C (221°F).

Regardless of the type of thinning utilized, dextrose is produced by the action of glucoamylase during saccharification (17). Glucoamylase is produced from strains of *Aspergillus niger* in submerged fermentation. A broth is obtained containing two or more glucoamylase isozymes, α-amylase, transglucosidase, and other enzymes, eg, protease and cellulase. The α-amylase assists in saccharification (11); however, transglucosidase catalyzes the formation of isomaltose and, therefore, must be removed before use. Removal of the enzyme is accomplished by adsorption on a clay mineral or by other techniques. Alternatively, selection for an *Aspergillus* mutant that does not produce transglucosidase eliminates the need for a removal step (18). Glucoamylase is also produced from a *Rhizopus* organism in Japan by surface fermentation; however, enzyme properties are somewhat inferior to those of *Aspergillus*.

Saccharification of liquefied starch hydrolysate is conducted by batch or continuous processes in large agitated reactors that are often several million liters in size. The reaction is conducted at 58 to 61°C (136 to 142°F), pH 4 to 4.5 with a glucoamylase dosage that is sufficient to produce the maximum dextrose yield in 1 to 4 days. At the normal solids level of 30 to 35 wt %, maximum hydrolysate dextrose content is about 95 to 96% db. During saccharification, isomaltose and maltose are produced by the reverse reaction (reversion) of glucoamylase, in which two dextrose

molecules are combined. Consequently, if the reaction is extended beyond the time needed to achieve maximum dextrose content, the dextrose level decreases as a result of excessive formation of these reversion products.

Dextrose content can be increased by conducting saccharification at lower solids. As solids level is reduced, the forward reaction is favored and a higher dextrose content is achieved. At a solids level of 10 to 12 wt %, a dextrose content of 98 to 99% db can be attained (19). However, operation at low solids is not commercially viable because of increased evaporation cost, the need for larger saccharification tanks, and the risk of microbial infection.

Dextrose level can also be increased by using enzymes that enhance the action of glucoamylase during saccharification. Enzymes available for this purpose are pullulanase and *Bacillus megaterium* amylase. Pullulanase is specific for the hydrolysis of α-1,6 glucosidic linkages and, when used in combination with glucoamylase, increases the rate and extent of dextrose production (20). A *B. megaterium* enzyme has been commercialized that combines hydrolysis and transferase activity to assist glucoamylase in increasing dextrose content (21). Either enzyme is effective in increasing hydrolysate dextrose content by 0.5 to 1.0% db.

After saccharification, the high dextrose content hydrolysate is clarified by centrifugation or filtration (usually either precoat or membrane) to remove insolubles. Concentration (usually by evaporation) to lower the water content and refining to reduce color and ionic contaminants are the next steps. Refining can be done with activated carbon (either powdered or granular), or with ion exchange resins (decolorizing and/or demineralizing resins), either singly or in combination. A simplified flowsheet for the manufacture of high dextrose content products is shown in Figure 7.

Commercial dextrose products include crystalline monohydrate or anhydrous dextrose and liquid dextrose. In addition, the hydrolysate can be refined and evaporated to a high DE corn syrup or dried to a solid product. Compositions are listed in Table 4.

Crystalline dextrose products can be manufactured using either batch or continuous crystallization technology. Sites for crystal growth can be provided either by the addition of seed crystals or by spontaneous nucleation. Subsequent crystal growth is encouraged by maintaining a controlled degree of supersaturation. Control of the entire process is critical to ensure that the resulting crystals are of a size and shape that will readily separate from the remaining mother liquor in perforate bowl centrifuges to provide a dry product of high purity.

The predominant technique for the production of monohydrate dextrose uses batch cooling-type crystallizers, although continuous crystallization using either evaporation (23) or cooling (24) to maintain supersaturation have been described. A process employing a continuous precrystallizer supplying batch crystallizers has been used to increase productivity (25).

In the dominant process, clarified, refined, and concentrated liquor having a dextrose content above about 80% db is mixed with a substantial bed of seed crystals (typically 20 to 30 wt %) left from the previous batch in the

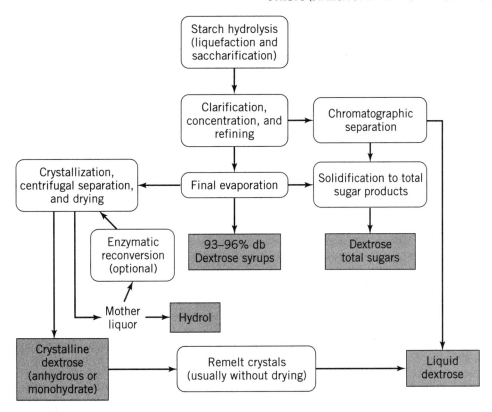

Figure 7. Simplified process flowsheet for the manufacture of high dextrose content products.

Table 4. Saccharide composition of dextrose, high-fructose corn syrups, corn syrups, and maltodextrins (% db)

		Constituent							
Sweetener type	DE	Fructose	Dextrose	DP-2	DP-3	DP-4	DP-5	DP-6	DP-7+
Dextrose	>99.5	0	>99.5	<0.25	←————— DP-3+ totals <0.1 ————→				
High DE corn syrup	97	0	95	<4	<1	←——— DP-4+ totals <1 ———→			
HFCS42		43	52	←———————— DP-2+ totals 5 ————————→					
HFCS55		55	40	←———————— DP-2+ totals 5 ————————→					
HFCS90		90	7	←———————— DP-2+ totals 3 ————————→					
Acid-converted corn syrup	25	0	2.5	7	8.5	6	6	13	57
	33	0	9	10	12	9	7	6	47
	43	0	19	14	12	10	8	7	30
Acid–enzyme corn syrup	43	0	6	45	12	3	2	2	30
Enzyme–enzyme corn syrup	28	0	3	11	14	←——— DP-4+ totals 72 ———→			
	43	0	2	52	15	1	1	1	28
High-maltose corn syrup	50	0	4	65	15	←——— DP-4+ totals 16 ———→			
	53	0	3	75	13	←——— DP-4+ totals 9 ———→			
High-conversion corn syrup	70	0	46	29	7	5	3	2	8
Maltodextrin	1	0	0.3	0.1	0.2	←——— DP-4+ totals 99.4 ———→			
	5	0	0.9	0.9	1	1.1	1.3	1.4	93.4
	10	0	0.5	2.5	4.2	3.4	3.6	5.4	81.4
	15	0	1.4	4.2	6.2	4.4	4.4	9.2	70.2
	18	0	1.6	4.7	6.5	4.6	4.6	10	68

Source: Refs. 2, 22, and Manufacturer's Product Data Sheets.
Notes: DP is the degree of polymerization (DP-2 represents disaccharides, DP-3, trisaccharides, etc). Values are typical for the syrup types shown but will vary depending on the process conditions used during manufacture. Values are expressed as weight percent dry basis.

crystallizer. The common form of crystallizer is a horizontal cylindrical tank fitted with a slowly turning agitator and a means of cooling the contents by indirect heat transfer. The initial temperature of the mixture is kept below 50°C (122°F) to ensure that crystals grow in the monohydrate form. The mass is slowly cooled, usually over a period of days, until about 50 to 60 wt % of the dry substance has crystallized out as monohydrate crystals of a form suitable for separation and washing.

The values for concentration, initial and final temperature and rate of cooling vary depending primarily on the purity of the supply material. Monohydrate dextrose crystals can be successfully grown in liquors of purity ranging from pure dextrose to syrups containing less than 60% db dextrose. Dextrose crystallizes to form monohydrate α-dextrose although the existence of a hydrated β-dextrose has been reported (26).

The resulting magma in the crystallizers is passed into perforated screen centrifuge baskets where it is spun to remove as much of the mother liquor as practical. Then a spray of water is applied to the rapidly rotating cake to wash off residual mother liquor. After spinning an additional period to remove as much liquid as practical, the cake is plowed out of the basket and fed to a dryer to reduce the moisture content to about 8.5 to 9.0%, slightly below the theoretical monohydrate moisture content.

Monohydrate crystallization can be done in several sequential steps starting with high purity hydrolysate to maximize the production of higher value crystalline product and to minimize the amount of mother liquor (usually sold as a lower value process co-product called Hydrol) that contains the polysaccharides that result from incomplete starch hydrolysis. Or, a portion of the mother liquor can be recycled and mixed with the incoming hydrolysate to provide an equivalent yield of crystals, usually about 75 to 80% db of the incoming hydrolysate. Such processes still result in 20 to 25 wt % of the hydrolysate supplied to the crystallization process being sold as Hydrol. Enzymatic reconversion of the mother liquor has been reported to allow nearly complete elimination of the production of lower value Hydrol (27).

The anhydrous form of dextrose is more fastidious than is the monohydrate and does not form easily separated crystals when crystallized from liquors below about 88 to 90% db dextrose content. Although it can be produced by batch crystallization, the predominant method uses vacuum evaporative crystallizers operated either continuously or in a batch mode. And because the starting material requires a minimum purity of about 95% db to produce purgable crystals, the dominant production process begins with a supply produced by remelting monohydrate crystals, resulting in the production of a very high purity product that meets the requirements of the *United States Pharmacopeia*. When producing USP-grade dextrose, the supply liquor can be passed through an ultrafilter membrane filter to reject pyrogens and make the final product acceptable for injection into the human body. Anhydrous α-dextrose is produced when the crystallization temperature is between about 55 and 100°C (about 131 and 212°F), preferably about 65°C (149°F) to minimize color development during the 6 to 8 h crystallization cycle required to

produce a magma containing about 50% of the dry substance on the crystalline form. Crystals are separated, washed, and dried much like the monohydrate variety, although centrifuge cycles are usually shorter. Anhydrous β-dextrose crystallizes at temperatures above about 100°C (212°F), although the β-form will continue to crystallize at lower temperatures if all α-form crystals have been excluded from the crystallizer. The β-form is the most soluble and rapidly dissolves in water.

Clarified, refined high dextrose content hydrolysate can be evaporated to about 71% ds and sold as a liquid product. Or, it can be solidified to either a monohydrate or anhydrous total sugar product (containing all the polysaccharides that result from incomplete starch hydrolysis) using a variety of processes (28). If a higher purity product is desired, chromatographic separation technology can be used to separate the high dextrose content syrup into a very pure (>99.5% db dextrose) product and a raffinate stream that contains nearly all the polysaccharides that result from incomplete starch hydrolysis. The raffinate stream can either be concentrated and sold as a lower value (Hydrol-type) product or be enzymatically reconverted. The very pure dextrose syrup can be sold as a liquid product or converted to a very pure total sugar. Liquid dextrose can also be produced by remelting crystalline dextrose.

Dry dextrose products can be shipped in vapor-resistant bags usually containing 50 or 100 pounds, or the approximately equivalent metric sizes of 20, 25, and 40 kg. Reusable containers holding about 1 metric ton of material are used for customers who do not require bulk delivery but who do not want the bother and expense of bag opening and disposal. Bulk transport via truck or railcar (jumbo railcars can hold as much as 82 metric tons or 180,000 lb) is also used. Solid products should be stored below 40°C (104°F), and rapid temperature changes and extremes should be avoided to minimize product caking.

Although most liquid products are shipped in bulk trucks and railcars (the largest railcars holding as much as 90 metric tons), some products are sold in 200-L (55 gallon) drums and 1000-L (250-gallon) disposable containers. Most small customers are supplied the dry equivalent (where available) of the desired syrup until their needs grow large enough to justify the installation of facilities for bulk truck delivery.

Liquid product storage facilities must be carefully designed to minimize product oxidation and microbial problems, especially in the headspace of storage tanks. Typical storage conditions of common liquid starch sweeteners are shown in Table 5. Storage temperatures selected are a compromise between a low temperature that would minimize product color development with time and the higher temperatures needed to either reduce the viscosity of the lower DE products to allow pumping or to prevent formation of crystals in the higher DE syrups and HFCS.

Economic Aspects

The price of starch is a major factor in the cost of production of starch sugars such as crystalline dextrose. Equipment investment, notably for crystallizers, is another im-

Table 5. Typical Storage Conditions of Liquid Starch Sweeteners

Syrup type	Solids content (wt %)	Temp. (°C)	Factor-controlling storage conditions
42 DE acid syrup	80	38	Product viscosity
42 DE acid syrup	83	49	Product viscosity
60 DE acid–enzyme syrup	82	32	Product viscosity
60 DE acid–enzyme syrup	84	41	Product viscosity
95 DE high-dextrose syrup	71	49	Dextrose crystallization
HFCS42	71	32	Dextrose crystallization
HFCS55	77	27	Dextrose crystallization
HFCS90	80	16	Fructose crystallization
Liquid dextrose	71	52	Dextrose crystallization

portant factor. Since Hydrol usually sells for a small fraction of the price of dextrose, optimizing process yield also has a meaningful impact on profits. Nearly eliminating the concomitant production of Hydrol by enzymatic mother liquor reconversion has been shown to provide a double-digit profit increase (27).

The price of dextrose monohydrate since 1975 is given in Table 6. Although demand, processing capacity, and production costs no doubt exert some effect on pricing, dextrose price has averaged about 95% of sucrose price on a dry basis for the period 1975 to 1996. Compared to dextrose monohydrate, anhydrous dextrose sells at a premium; prices in early 1989 were $1.12/kg (50 cents/lb) for anhydrous dextrose and $1.21/kg (55 cents/lb) for USP anhydrous dextrose.

Analysis and Specifications

The United States and international specifications for crystalline dextrose are given in References 29 to 32, re-

Table 6. Wholesale List Prices for Some Starch Sugars in the United States ($/kg db)

Year	Dextrose	HFCS42	HFCS55	Corn syrup
1975	0.46	0.50	NA	0.34
1976	0.34	0.31	NA	0.25
1977	0.31	0.27	NA	0.18
1978	0.36	0.27	NA	0.19
1979	0.38	0.29	NA	0.22
1980	0.64	0.52	NA	0.32
1981	0.65	0.47	0.52	0.35
1982	0.60	0.31	0.41	0.30
1983	0.58	0.41	0.46	0.28
1984	0.58	0.44	0.50	0.28
1985	0.53	0.39	0.44	0.24
1986	0.52	0.40	0.44	0.23
1987	0.50	0.36	0.38	0.22
1988	0.56	0.36	0.41	0.26
1989	0.56	0.42	0.47	0.30
1990	0.54	0.43	0.48	0.31
1991	0.54	0.46	0.51	0.34
1992	0.54	0.46	0.51	0.33
1993	0.54	0.41	0.46	0.29
1994	0.57	0.44	0.49	0.34
1995	0.56	0.37	0.42	0.32
1996	0.56	0.41	0.45	0.36

Source: Data from USDA Economic Research Service publications.

spectively. High-quality anhydrous dextrose produced for the pharmaceutical industry is prepared in accordance with specifications given in Reference 33. Typical product analyses include reducing sugar value, solution color, solution clarity, ash, particle size distribution, and moisture. High-performance liquid chromatography (HPLC) is used to measure the concentration of dextrose and other saccharides in solution.

Uses

The main use of dextrose is in the food-processing industry, where it is of value for its physical, chemical, and nutritive properties. Distribution of dextrose to various industries in the United States is shown in Table 7. In the baking industry, it serves primarily as a fermentable sugar that contributes to crust color. In the beverage industry, it is used as a source of fermentables in low-calorie beer and as a sweetener in beverage powders. In the canning industry, it supplies sweetness, body, and osmotic pressure and also contributes to better natural color retention in certain products. In the confectionery industry, it is used to supply sweetness and softness control and to regulate crystallization. Dextrose is also used in tableted products; flavor is often enhanced by the cooling effect obtained when dextrose hydrate dissolves in the mouth. In frozen desserts, it prevents oversweetness and improves flavor.

In many cases, dextrose is used in conjunction with sucrose. Although dextrose by itself is somewhat less sweet than sucrose, the combination with sucrose may be as sweet as pure sucrose at the same concentration. Dextrose is also used in the pharmaceutical industry for intravenous feeding as well as tableting and other formulations. In fermentation, dextrose is a raw material for biochemical synthesis, and, chemically, it is a raw material for sorbitol, mannitol, and methyl glucoside production. Dextrose in the form of unrefined hydrolysate is used as a raw material in yeast fermentations for production of alcohol that is used as an octane enhancer in gasoline.

HIGH-FRUCTOSE CORN SYRUPS

High-fructose corn syrups (HFCS, isosyrup, isoglucose) are concentrated solutions containing primarily fructose and dextrose with lesser quantities of higher molecular weight saccharides. HFCS is produced by partial enzymatic isomerization of dextrose hydrolysates followed by refining and

Table 7. Utilization of Starch Sugars by Market Segment in the United States

Year	Baking	Beverage	Canning	Confectionery	Dairy	Total[a]
Dextrose, thousands of metric ton/year[b]						
1965	178	8	21	37	6	468
1970	174	8	23	52	7	547
1975	157	18	15	62	7	561
1980	51	66	4	55	2	513
1985	56	81	2	58	1	479
1988	71	88	3	66	1	555
1996	—	—	—	—	—	700
HFCS (all types), thousands of metric ton/year[c]						
1970	18	39	9	0.5	5	99
1975	129	279	64	4	36	715
1980	365	1,039	235	15	140	2,659
1985	428	4,246	288	39	213	6372
1988	410	5,097	394	60	253	7288
1996	—	—	—	—	—	9560
Corn syrup (all types), thousands of metric ton/year[c]						
1965	209	36	104	410	159	1211
1970	222	93	98	439	213	1449
1975	315	205	174	416	278	2278
1980	196	384	126	446	241	2201
1985	151	377	150	496	290	2469
1988	166	400	150	456	238	2677
1996	—	—	—	—	—	3295

Note: Dash indicates data on individual market segments not available. 1996 data are author's personal estimates of U.S. production.
[a]Includes other market segments/applications not listed in table.
[b]Monohydrate basis (8.5–9.0% moisture).
[c]Commercial basis.

concentration. Commercial syrup products contain 42, 55, or 90% fructose on a dry-weight basis (db). Crystalline fructose is also commercially available in large quantities.

In nature, fructose (levulose, fruit sugar) is the main sugar in many fruits and vegetables. Honey contains about 50 wt % fructose on a dry-weight basis. Sucrose is composed of one unit each of fructose and dextrose combined to form the disaccharide. Fructose exists in polymeric form as inulin in plants such as Jerusalem artichokes, chicory, dahlias, and dandelions and is liberated by treatment with acid or enzyme (34).

Fructose was first isolated in 1847 (35). In 1874, it was recognized that fructose had advantages over sucrose as a sweetener for diabetics (36). Following the discovery of the alkaline conversion of dextrose to fructose in 1895 (37), a considerable number of investigations were conducted in an attempt to develop a commercial process (38). However, because of problems associated with color, off-flavor, degradation products, and low fructose yield, a process was never commercialized. Enzymatic conversion of dextrose to fructose with glucose isomerase was first reported in 1957 and patented in 1960 (39). Research in this area continued for several years in Japan, resulting in commercial production in 1966 (40) and a U.S. patent in 1971 (41). Japanese technology was licensed by Standard Brands and production was initiated in the United States in 1967 by a

batch process. A 15 wt % fructose syrup was manufactured initially, followed in 1968 by a 42% db fructose product. In 1972, a continuous system was initiated using an immobilized enzyme process (42,43).

Properties

Fructose, a ketohexose monosaccharide, crystallizes as β-D-fructopyranose and has a molecular weight of 180 and a melting point of 102 to 104°C (216 to 219°F). In solution, fructose undergoes rapid mutarotation to a mixture of α-D- and β-D-fructopyranose and α-D- and β-D-fructofuranose (Fig. 8), during which time specific rotation changes from $-132°$ to $-96°$ (45). At equilibrium, about 60 to 70% of the fructose is in the β-pyranose form, and 20 to 30% is in the β-furanose form. The remainder exists as α-D-fructose. Relative amounts of each depend on temperature and pH. At equilibrium, the solubility of fructose in water is 80 wt % at 25°C (77°F). The crystalline form of fructose is the sweetest of the tautomeric forms, 1.8 times sweeter than sucrose. Because the crystalline form is the sweetest form, relative sweetness may change when the fructose dissolves in a high moisture or liquid food product. Regardless of the tautomeric form, fructose is sweeter than sucrose and thus can be used in reduced quantities to attain equivalent sweetness. A discussion of the metabolism of fructose is given in Reference 46.

HFCS is sweeter than conventional corn syrups because of the presence of fructose, although intensity of sweetness results from many factors, eg, temperature, pH, and concentration. In general, 42% db HFCS (HFCS42) has the same sweetness as sucrose, 55% db HFCS (HFCS55) is as sweet as medium invert and 2 to 5% sweeter than sucrose, and 90% db HFCS (HFCS90) is 15 to 30% sweeter than sucrose (47). Other properties of HFCS include high solubility, which reduces the possibility of crystallization during shipment; humectant properties allowing for increased shelf life of bakery products; easier decomposition of fructose during baking, resulting in improved color and flavor; and high osmotic pressure for containment of microbial growth.

Manufacture

HFCS42 is produced commercially by continuous isomerization of clarified and refined dextrose hydrolysate with immobilized glucose isomerase. Enriched syrup containing 90% db fructose is prepared by chromatographic separation and blended with HFCS42 to obtain HFCS55 (Fig. 9).

During the initial development of glucose isomerase, it became obvious that batch reactions would not be commercially feasible because of several factors. The long residence time required with soluble enzyme caused the production of color, off-flavors, and by-products, eg, psicose, which is a nonmetabolizable sugar formed under alkaline conditions. In addition, enzyme cost was high, and it became necessary to develop a process that allowed reuse of enzyme. Consequently, many different methods of immobilization were examined in an attempt to develop a continuous reaction system. As a result of these studies, two types of immobilization processes were developed for commercial use: whole-cell and soluble-enzyme processes.

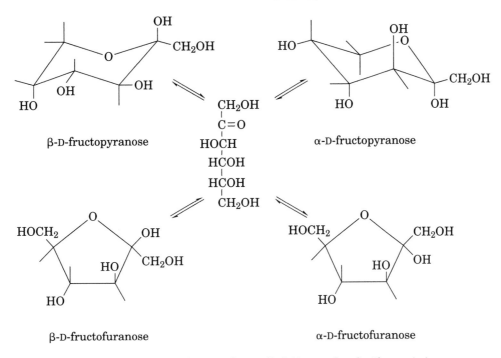

β-D-fructopyranose

α-D-fructopyranose

β-D-fructofuranose

α-D-fructofuranose

Figure 8. Structures of fructose. *Source:* Ref. 44, reproduced with permission.

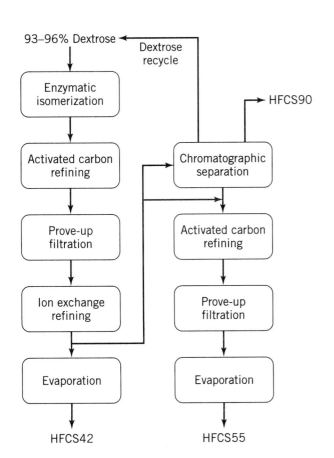

Figure 9. Simplified process flowsheet for fructose production.

In the whole-cell process, the microbial cell containing intracellular enzyme is recovered from the fermentation broth and treated to preserve enzymatic activity and maintain particle integrity. In one process (48), culture broth is centrifuged, and the concentrated cells are disrupted by homogenization and cross-linked with glutaraldehyde. The material is then diluted and flocculated with a cationic flocculant, and a final product is prepared by extrusion, drying, and sieving. Another method (49) involves fixing the cells in gelatin and cross-linking with glutaraldehyde, followed by washing, particle size classification, and drying.

In the soluble-enzyme process, enzyme is separated from the cells and purified before immobilization. Since glucose isomerase is an intracellular enzyme, the cell wall is first disrupted by physical (sonification, homogenization) or enzymatic methods. The solubilized glucose isomerase is then separated by centrifugation and/or filtration and concentrated by ultrafiltration. In one method of immobilization (50), binding takes place by simply contacting the soluble enzyme with a particle containing a combination of DEAE cellulose, titanium dioxide, and polystyrene. Other supports for glucose isomerase include anion-exchange resin, anion-exchange cellulose, porous ceramics, and porous alumina.

Glucose isomerase enzymes have been isolated from a variety of microbial sources (51,52). However, only a few bacterial organisms have been used to produce glucose isomerase for commercial use. These include *Streptomyces olivochromogenes, S. murinus, S. rubiginosus, Bacillus coagulans, Actinoplanes missouriensis,* and *Microbacterium arborescens.* Enzyme properties vary depending on the source, but all are similar in terms of operational pH and temperature.

HFCS is produced from dextrose hydrolysate that has been clarified, refined (including demineralization), and evaporated to 40 to 50 wt % dry substance (ds), as shown in Figure 9. Magnesium is added as a cofactor to maintain isomerase stability and to prevent inhibition by trace amounts of residual calcium. If the calcium level is below 1 ppm, a magnesium level (added as $MgSO_4$) of 0.4 mM is sufficient, although a higher calcium level requires additional magnesium. Hydrolysate is passed through a fixed bed of immobilized isomerase at a controlled flow rate. Process conditions depend on the particular enzyme system used but are generally 55 to 65°C (131 to 149°F), pH 7.5 to 8.5, and an initial reaction time of 1 h or less.

Maximum fructose content at equilibrium is 50 to 55% db; however, residence time is adjusted to attain 42 to 45% db fructose since a greatly increased reaction time is required to attain higher levels. The enzyme can be used for as long as several months, and compensation for loss in activity during operation is made by regulating the residence time, ie, flow rate through the column. Enzyme reactors are operated in parallel or in series until activity is reduced to about 10% of the initial level. Isomerized hydrolysate containing about 42% db fructose is adjusted to pH 4 to 5, refined to remove color and salts using resins either alone or in combination with activated carbon, and concentrated by evaporation to about 71 wt % solids. Product shipment is by tank trucks or by rail. Since the dextrose content in the HFCS product is more than 50% db, storage at 32 °C (90°F) is required to prevent crystallization. If crystallization occurs, the syrup is heated to 38°C (100°F) to dissolve crystals before unloading.

Products containing higher levels of fructose are produced by chromatographic separation of HFCS42 and were first commercially available in limited quantities in 1976. Enrichment is accomplished by passing HFCS42 through a column of absorbent (usually a strong-acid cation-exchange resin) containing calcium or other cation groups acting as a counter ion. Fructose is retained to a greater degree than dextrose or oligosaccharides, and therefore a stream containing most of the nonfructose is collected first in a raffinate stream, followed by elution of a fructose-enriched extract with water. Separation can be achieved using batchwise, semicontinuous, or continuous operation; commercial operations favor continuous processes. The continuous procedure involves a simulated moving bed in which feed and desorbent enter the column at different points while fructose and raffinate streams are withdrawn at others. Points of entry and withdrawal are changed periodically to correspond to flow through the column, and hence separation efficiency is maximized. Typical operation involves the addition of HFCS42 at a dry substance content of 50 to 60% to a column at 50 to 70°C (122 to 158°F). Dextrose-rich raffinate is recycled to isomerization columns for additional production of HFCS42. Enriched HFCS is blended with HFCS42 to produce a product containing 55% db fructose. A solids level of 77% ds is suitable for shipment at 27°C (81°F) with minimal dextrose crystallization problems. In addition, enriched HFCS at about 90% db fructose is evaporated to 80% ds for shipment as an essentially noncrystallizable syrup at normal room temperature.

Fructose of >99% purity is produced by crystallization to the β-D-fructopyranose form. In one process, chromatographic separation of 42% HFCS is used to produce a fraction containing 97% fructose. The material is evaporated to 70% solids, and 50% of the fructose is recovered by crystallization in 80 to 100 h (53).

Economic Aspects

Published U.S. wholesale prices of 42 and 55% HFCS since 1980 are listed in Table 6. The 1989 price for 90% fructose was about $1/kg (45 cents/lb) on a commercial basis. The price of HFCS has not been intimately tied to the price of sucrose for a number of years. Supply and demand seem to be the most important factors influencing price.

During the early 1970s, the price of refined sugar increased dramatically, resulting in a concomitant increase in the price of HFCS and the construction of new production facilities. When sugar prices fell during 1975 to 1978, HFCS prices also dropped, falling to a level of about 30 to 35% below sugar rather than the 15 to 20% discount that had existed previously. Excess production capacity also lowered HFCS prices, delaying the start-up of some new plants. However, during the late 1970s through the early 1980s, demand for HFCS increased significantly in the soft-drink industry, and the discount to sucrose decreased, especially for the sweeter HFCS55 product.

As can be seen from Table 7, the new sweetener also took market share from existing starch sugars in some segments, especially the more expensive dextrose. Dextrose consumption in the United States actually fell for a time before rebounding in the late 1980s.

Because of the increased demand and higher sucrose prices, processing capacity was again increased throughout the industry. From 1980 to 1985, per capita consumption of HFCS increased from 8.7 to 20.5 kg (19 to 45 lb) due primarily to 100% substitution for sugar in many soft drinks. Because of this near maximum penetration of the soft drink market, HFCS growth slowed to 2 to 4% per year through the end of the 1990s, and prices relative to sucrose fell again.

Then, in 1991, consumption accelerated to a level of 5 to 6% annual growth, and prices firmed until the industry once more made substantial capacity increases that caused prices relative to sucrose to fall. This was particularly painful in 1996 when corn prices increased dramatically, averaging nearly $4.00/bushel ($160/metric ton) for the year. Prices in 1997 reached lows not seen in 20 years but have begun to firm as demand continues to grow by about 1 billion pounds per year. HFCS prices in the United States are expected to continue to be influenced primarily by factory utilization and to a lesser degree by sugar prices and the cost of corn.

Analysis and Specifications

International (including U.S.) specifications for HFCS are the same as those for corn syrup. Fructose content is determined by high-performance liquid chromatography; dry substance, by refractive index; and color, by spectroscopy.

High-fructose corn syrup is used as a partial or complete replacement for sucrose or invert sugar in food applications to provide sweetness, flavor enhancement, fermentables, or humectant properties. It is used in soft drinks, baking, canned fruit, dairy products, and confections (Table 7). HFCS is used in combination with sucrose as well as other corn sweeteners. The main application of HFCS55 is in soft drinks. HFCS containing 90 wt % fructose is used in low calorie or specialty foods because of its high sweetness and, therefore, reduced usage level and lower caloric value. Other uses are as a liquid tabletop sweetener or as a honey-flavored product for use in baked goods and confections. Crystalline fructose is essentially pure and is used at a level that provides sweetness at a lower caloric level than that in other sweeteners. Uses include health foods and medicines.

CORN SYRUPS AND MALTODEXTRINS

Corn syrups (glucose syrup, starch syrup) are concentrated solutions of partially hydrolysed starch (having a DE value greater than 20) that contain dextrose, maltose, and higher molecular weight saccharides. Corn syrups are generally sold in the form of viscous liquid products, although solid forms are also available, especially in the lower DE ranges. They vary in physical properties (eg, viscosity, humectancy, hygroscopicity, sweetness, and fermentability), primarily as a result of their individual saccharide distributions. Maltodextrins are starch hydrolysates having a DE value less than 20.

Properties

Corn syrups are defined on the basis of reducing-sugar content as having a DE of 20 or greater. They are classified into four types: Type I syrups have DE values from 20 up to 38; type II, from 38 up to 58; type III, from 58 up to 73, and type IV, 73 and above. Syrups may also be classified as acid-conversion, acid-enzyme-conversion, high-maltose, etc. However, as shown in Table 4, the most adequate characterization is with respect to their content of individual saccharides. In many cases, it is the content of individual saccharides or groups of saccharides that determines the characteristics of syrups. Functional properties of corn syrups include fermentability, viscosity, humectancy and hygroscopicity, sweetness, colligative properties, and browning reactions.

Fermentability of corn syrups by yeast is important in certain food applications, eg, baking and brewing. The fermentable sugars present in corn syrup are dextrose, maltose, and maltotriose. Fermentability of maltose or maltotriose depends on the specific fermentation process and organism. In general, greater fermentability is obtained at the higher DE levels.

Viscosity of corn syrup is a function of DE value, temperature, and solids concentration. Viscosity decreases with increasing DE and temperature and increases with increasing concentration. For a 43-DE corn syrup, viscosity at 1.42 sp gr (43°Bé) is 56,000 mPa · s (= cP) at 27°C (81°F), 14,500 mPa · s at 38°C (100°F), and 4900 mPa · s at 49°C (120°F). Corresponding values for a 55-DE syrup at the same density and temperatures are 31,500 mPa · s, 8500 mPa · s, and 2900 mPa · s.

The hygroscopic and humectant properties of corn syrups are of great importance in many applications. Depending on the type of syrup and on the specific conditions of temperature and humidity, the products may either resist or facilitate moisture loss or moisture absorption. The ability to attract moisture or retard its loss increases with increasing DE value. Prevention of moisture pickup is more characteristic of syrups with low DE values.

Sweetness is primarily a function of the levels of dextrose and maltose present and therefore is related to DE (see Fig. 1). Other properties that increase with increasing DE value are flavor enhancement, flavor transfer, freezing-point depression, and osmotic pressure. Properties that increase with decreasing DE value are bodying contribution, cohesiveness, foam stabilization, and prevention of sugar crystallization.

Manufacture

Acid hydrolysis of starch is conducted by batch or continuous processes. Batchwise conversion is carried out in large cookers or converters that are usually built of manganese bronze and have capacities of about 10 m³ (2650 gallons). A suspension of starch at 35 to 40 wt % solids is placed in the converter. Hydrochloric acid is added to provide a concentration of 0.015 to 0.02 N, and the converter is steam heated until a temperature of 140 to 160°C (284 to 320°F) is reached. The mixture is held at this temperature for a period of time, usually 15 to 20 min, to produce the desired degree of hydrolysis.

Improved process control and, therefore, better product uniformity, is achieved by continuous processes using indirect heating. Acidified starch slurry is pumped at a constant rate through a series of heat exchangers at reaction conditions similar to those used in batch operations. In either process, hydrolysate is neutralized to pH 4 to 5.5 by addition of soda ash, clarified by centrifugation or filtration, refined (using activated carbon and/or ion exchange resins) to remove color and acid degradation products, and concentrated to 77 to 85 wt % solids. Sulfur dioxide used to be added during evaporation to some grades of syrup to reduce color development, but improved refining techniques have all but eliminated this practice.

Syrups produced by straight acid hydrolysis do not exceed 55 DE because of the formation of difficult-to-remove color and flavor by-products created at higher DE levels. Some syrup compositions are given in Table 4. Syrups are also produced by A–E or E–E processes. Starch is first hydrolyzed by acid, as described above, or by enzyme, as described for dextrose, followed by saccharification to the desired composition with one or more enzymes.

Maltose syrup, for example, is prepared from about 20 DE, partially hydrolyzed starch by saccharification with a maltose-producing enzyme at ca 55°C (131°F) and pH 5. Maltogenic enzymes, eg, β-amylase obtained from barley, wheat, or soybeans or fungal α-amylase derived from *Aspergillus oryzae*, are used to produce a hydrolysate containing about 45 to 60 wt % maltose (see Table 4). Higher levels of maltose (up to 80 to 90 wt %) are produced by saccharification with a combination of enzymes, including a maltogenic enzyme and a debranching enzyme, eg, pullulanase.

High-maltose syrups can be separated chromatographically to produce a 95 to 97% db maltose stream from which pure maltose can be crystallized. The sweetness of maltose is 30 to 40% that of sucrose. Maltose exists as both α- and β-forms, and aqueous solutions equilibrate at 42% α- and 58% β-form. At 20°C (68°F), crystalline α-anhydrous maltose will dissolve to about 64% ds, whereas β-monohydrate crystals dissolve to initially form a 27% ds solution. At equilibrium, a 42% α- and 58% β-form solution has a solubility of about 62% ds at 20°C (68°F) (50).

High-conversion syrups of 60 to 70 DE containing high levels of dextrose and maltose are also produced (Table 4). These types of syrups are produced in the same manner as maltose syrup, except a combination of glucoamylase and maltogenic enzyme is used. Maltose and high-conversion syrups are refined by the same procedures as described earlier and shipped in rail cars and tank trucks. Syrups are generally heated to about 30 to 50°C (86 to 122°F) to facilitate unloading.

Some syrups, particularly those with a limited extent of hydrolysis, are reduced to dry form by spray-drying or roll-drying. These products are commonly called dry maltodextrins (DE less than 20), corn syrup solids, and total sugars (for the highly converted products containing predominantly one saccharide). They are usually shipped in moisture-resistant bags.

Products of <20 DE are referred to as maltodextrins or hydrolyzed cereal solids (Table 4). Maltodextrins are usually prepared from regular or waxy corn starch, although some recently introduced products use high-amylose starch to enhance the retrogradation of the linear polymers to form maltodextrins resistant to digestion. Maltodextrins are produced by enzyme or acid hydrolysis processes to products of 5 to 20 DE. They are clarified, refined, and either spray-dried to a moisture content of 3 to 5 wt % and shipped in 100-lb (45.4-kg) bags or sold as a syrup of about 75 wt % solids. These products are bland-tasting, and the solid forms are free-flowing and nonhygroscopic.

Economic Aspects

Prices of corn syrup in the United States since 1975 are listed in Table 6. In the early 1970s, competition depressed prices to a very low level (12.4$/cwt, equal to $273/metric ton, in 1972). Pricing recovery followed when sucrose prices increased, and production capacity was reduced because of the manufacture of other competitive products. Recent prices have been about 50 to 60% that of sucrose.

Analysis and Specifications

International (including U.S.) specifications for corn syrups and dried corn syrups are given in References 55 to 57. A list of maltodextrins and their physical properties for U.S. and non-U.S. producers is available in Reference 58.

Corn syrups are usually sold with a specification of the Baumé measurement, and these can be related to the solids content. The most common value is sp gr 1.42 (43°Bé), corresponding to 80.3 wt % solids for a 42-DE acid-converted syrup. Solids contents for such a syrup at sp gr 1.39, 1.41, 1.44, and 1 45 (41, 42, 44, 45°Bé) are 76.3, 78.3, 80.3, 82.3, and 84.3 wt %, respectively. Higher DE syrups have slightly higher solids contents at the same densities. Saccharide composition is determined by high-

performance liquid chromatography. Other important analyses include color, iron, and pH (although with the development of totally demineralized syrups, pH measurement has much less meaning).

Uses

Principal uses of corn syrups are shown in Table 7. Corn syrups are often used in combination with sucrose, dextrose, or HFCS. The specific type of syrups used depends on the properties desired in the final product. Properties listed above are important to varying degrees in the various products, and changes in formulation can affect the choice of corn syrup required to supply the most desirable properties. In many cases, corn syrups are used as supplementary sweeteners, with sucrose remaining the primary one.

In the confectionery industry, corn syrups are used extensively in nearly every type of confection, ranging from hard candy to marshmallows. In hard candies, which are essentially solid solutions of nearly pure carbohydrates, corn syrup contributes resistance to heat discoloration, prevents sucrose crystallization, and controls hygroscopicity, viscosity, texture, and sweetness. Maltose syrups, high-conversion syrups, and acid-converted syrups (36 and 42 DE) are all used for this application.

In the canning and preserving industries, corn syrups are used to prevent crystallization of sucrose, provide body, accentuate true fruit flavors, and improve color and texture. In the beverage industry, the predominant use is in the beer and malt-liquor areas. High-conversion syrups are used to replace dry-cereal adjuncts, provide fermentable sugars, enhance flavor, and provide body. These syrups contain controlled amounts of dextrose and maltose for proper fermentation.

Corn syrups used in baking are generally of the highly converted type. They are incorporated into cakes, cookies, icings, and fillings to increase the amount of moisture retained, retard crystal growth of other sugars, enhance tenderness, and increase shelf life. In yeast-raised goods, fermentability is of importance, and therefore only high DE syrups are used.

Corn syrups used in ice cream and frozen desserts are generally 36- or 42-DE acid-converted syrups. The syrup serves primarily to provide maximum flexibility in adjusting flavor, texture, body, and smoothness; it also aids in grain control and in the modification of melt-down and shrinkage characteristics of the frozen product.

Syrups of 25 to 30 DE are used as spray-drying aids in products such as coffee whiteners. High-conversion syrup, maltose syrup, and 42-DE syrup are used in jams and jellies. Additional uses of corn syrup include applications in table syrups, baby food, meat packing, breakfast foods, salad dressing, pickles, dehydrated powdered foods, medicinal syrups, textile furnishings, adhesives, and numerous other products and processes.

MAPLE SYRUP

Maple syrup is prepared by evaporating sap from the maple tree to a concentrated solution containing predominantly sucrose. Its characteristic flavor and color are formed during evaporation. Maple syrup is produced from the sap of several varieties of mature maple trees, eg, the sugar maple (*Acer saccharum*) and black maple (*Acer nigrum*). Main producing areas include the northeastern and upper midwestern United States and eastern Canada.

Manufacture and Uses

Collection of sap is made sometime between late fall and midspring, depending on weather; the best time is when the temperature is about 7°C (45°F) during the day and below freezing at night. Sap generally contains 2 to 3 wt % solids, of which about 96% is sucrose and the remainder other carbohydrates, organic acids, ash, protein, and ligninlike materials. A taphole is drilled into the tree, the hole is sanitized with a germicidal pellet, and a spout is driven in. Sap is collected in a bucket or bag or alternatively in plastic tubes directed to a centralized collection tank by gravity or vacuum.

Evaporation is conducted at atmospheric pressure until a boiling point of 104°C (219°F) is reached to produce a syrup meeting federal specifications of at least 66 wt % solids (59). Control of syrup concentration is critical. A syrup concentrated to only 65 wt % of solids has a thin taste, and one concentrated to more than 67 wt % crystallizes when cooled. A refractometer or hydrometer is used to measure concentration. Final specific gravity should be 1.35 (37.75°Bé) at 15.6°C (60°F). Flavor and color develop during evaporation as a result of reactions occurring between sugar and other components.

If change in the sucrose content is minimized during evaporation, a light-colored syrup of delicate maple flavor is produced. Darker syrups usually suffer from greater sucrose inversion and the production of caramelized off-flavors. Color is the principal grading factor, and the USDA has developed glass color filter standards referred to as light, medium, and dark amber. Syrup is clarified, graded as to color, flavor, and density, and finally packaged in small containers for retail sale as table syrup. Typically, the product contains 88 to 99% db sucrose and 0 to 12% db invert sugar. Maple syrup is also used in candy manufacture by blending with sucrose.

Other applications include addition to cookies, cakes, ice cream, baked beans, baked ham, and baked apples. Maple sugar is prepared by evaporating sap to a high solids content, ie, a boiling temperature of 116 to 121°C (241 to 250°F), and then allowing the supersaturated solution to crystallize or solidify during cooling. Maple cream or maple butter is made by stirring a supersaturated solution while cooling rapidly to produce a product of creamy texture.

Production and Economics

Maple syrup production in the United States averaged 4.9 million L (1.3 million gal) per year between 1989 and 1998. Canada produced about three times this amount. Production in the United States was as high as 6.4 million L in 1992 and 6.1 million L in 1996. Prices have increased from $6.00 per liter in 1992 to $7.00 per liter in 1998 (60).

MOLASSES

Molasses, another type of syrup, is a by-product of the sugar industry; it is the mother liquor remaining after crystallization and removal of sucrose from the juices of sugarcane or sugar beet and is used in a variety of food and nonfood applications. Molasses was first produced from sugarcane in China and India centuries ago, later spreading into Europe and Africa. It was introduced as the by-product of cane sugar production into Hispaniola (island Española in modern West Indies) by Columbus in 1493. During colonial times, molasses was very important to the American colonies for the production of rum. In 1733, the British Parliament passed the Molasses Act to tax importation of molasses from foreign countries. This attempt to restrict trade was ignored by the colonies and was, in part, responsible for the American Revolution.

Manufacture

Raw sugar is produced from sugarcane by a process that involves extraction of the sugar in water, treatment to remove impurities, concentration, and several crystallizations. After the first crystallization and removal of first sugar, the mother liquor is called first molasses. First molasses is recrystallized to obtain a second lower-quality sucrose (second sugar) and a second molasses. After a third crystallization, the third molasses contains considerable nonsucrose material, and additional recovery of sucrose is not economically feasible. The third molasses is sold as blackstrap, final, or cane molasses. Raw sugar obtained from this process is mixed with water to dissolve residual molasses and is separated by centrifugation. This process is called affination, and the syrup is referred to as affination liquor. The sugar is dissolved in water, treated to remove color and impurities, and subjected to several crystallizations to obtain refined sugar. The mother liquor from the final crystallization is combined with affination liquor and crystallized to produce a dark sugar (remelts) that is recycled to raw sugar. The remaining mother liquor, called refiners molasses, is similar to final molasses but usually of better quality.

In beet sugar manufacture, the beet juice does not contain reducing sugars (fructose and glucose) that are present in cane juice but may contain raffinose. Because of the absence of reducing sugars, the sucrose level in beet molasses is not reduced to the same extent as with cane.

Final molasses from beet contains about 60 wt % sucrose (dry basis) compared to 30 wt % sucrose (dry basis) in cane molasses. Treatment of diluted beet molasses with calcium oxide precipitates sucrose as tricalcium sucrate (Steffen process), which is recycled to the incoming hot beet juice. During recycling, raffinose accumulates in the final molasses and retards crystallization if not removed. Therefore, a portion of the final molasses is periodically removed and called discard molasses. High-test molasses (invert molasses) is produced from cane sugar when sucrose manufacture is restricted because of overproduction. The cane sugar at about 55 wt % solids is enzymatically converted to invert syrup to prevent crystallization and is evaporated to a syrup. The product is used in the same applications as blackstrap molasses.

Nearly 8% of the beet sugar production forecast for fiscal 1996/1997 is sucrose recovered from molasses. First employed in 1989, molasses desugarization using ion exchange technology is performed at eight facilities across the United States (61). After recovery of about 85% of the sucrose contained in the molasses feedstock, the process can also produce two co-products, CMSB (consolidated molasses solids from beets) that contains 18 to 21% sucrose and betaine, used as a growth stimulant for poultry and fish.

Molasses from other sources include citrus and corn sugar (Hydrol) molasses. Citrus molasses is produced from citrus waste and contains 60 to 75% sugars. Corn sugar molasses, commonly called Hydrol, is the mother liquor remaining after dextrose crystallization and contains a minimum of 43% reducing sugars expressed as dextrose. Molasses is shipped in drums, barrels, tank trucks, tank cars, barges, and ocean-going vessels. Because of high viscosity, molasses must be heated in some situations to facilitate pumping. However, prolonged heating must be avoided to prevent carmelization.

Composition

Molasses composition depends on several factors, eg, locality, variety, soil, climate, and processing. Cane molasses is generally at pH 5.5 to 6.5 and contains 30 to 40 wt % sucrose and 15 to 20 wt % reducing sugars. Beet molasses is about 7.5 to 8.6 pH, and contains about 50 to 60 wt % sucrose, a trace of reducing sugars, and 0.5 to 2.0 wt % raffinose. Cane molasses contains less ash, less nitrogenous material, but considerably more vitamins than beet molasses. Composition of selected molasses products is listed in Table 8.

Uses

The primary use of molasses is in animal feed. Molasses provides a carbohydrate source, salts, protein, vitamins, and palatability and may be used directly or mixed with other feeds.

The carbohydrate content of 24.6 L (6.5 gallons) of blackstrap molasses is considered to be equal to 0.035 m^3 (one bushel) of corn as measured by the energy produced from 0.035 m^3 of corn and the amount of molasses required to produce the same amount of energy. When molasses is less expensive than corn, sales increase; when the reverse is true, sales decrease. Relative corn and molasses costs are given in Table 9 for 1981 to 1986.

The second main use for molasses is in fermentation processes as an inexpensive source of carbohydrate. Molasses is the basic raw material for rum production and is also used for production of yeast and citric acid.

Food applications utilize first and second molasses in baking (bread, cakes, cookies) for flavor. Molasses is also used in curing of tobacco and meats, in confections such as toffees and caramels, and in baked beans and glazes. Distribution of molasses to various industries is shown in Table 10.

Production and Economics

Molasses production along with imports and exports in the United States are shown in Table 11. Principal molasses producers are listed in Table 12. Molasses prices since 1980 are given in Table 13. Because of declining price, desugaring processes have been developed using ion exchange technology to recover about 85% of the sucrose content of molasses as refined sugar. When desugaring techniques are used, more than 90% of the sucrose contained in the beet feedstock is recovered as beet sugar. About 8% of the beet sugar production forecast for fiscal 1996/1997 is sucrose recovered from molasses.

SORGHUM SYRUPS

Sorghum syrup, also known as sorgo, is made from the juice of the sorghum plant that is related to the sugar cane. The finished syrup flavor is similar to that of cane syrup but has a sharper tang. Some sorghum syrups exhibit a strong, distinctive flavor. For this reason, blends of sorghum and corn syrup are found to be preferred by some (62).

Although sorghum has been grown for grain and animal feed for many centuries, syrup production is a relatively recent use that began in the United States shortly before the Civil War. In 1859, nearly 7 million gallons (26 million L) were produced, increasing to 28 million gallons by 1879 (63). By 1938, production had decreased to about 15 million gallons annually. Today, only 1 to 2 million gallons are produced each year. Much of this sorghum syrup is produced by small-scale producers for local consumption. Because of the fragmented nature of the market, it is difficult to obtain an accurate measure of the amount of sorghum syrup produced.

Although many variations exist on the process for sorghum syrup production, typically the sorghum plants are topped and the leaves are removed from the canes. The juice is then pressed from the canes, usually in a roller mill essentially like the mills used for crushing sugar cane. A ton of sorghum yields enough juice to produce between 10 and 20 gallons (38 and 76 L) of finished syrup, depending on the growing and crushing conditions as well as the sorghum variety grown. Sweet sorghum varieties yield more syrup than forage varieties.

Table 8. Molasses Composition

Molasses Type	Solids (%)	Total sugars as invert (% db)	Crude protein (% db)	Total ash (% db)
Cane (Louisiana)	80.8	59.5	3.0	7.2
Cane (refiner's)	75.4	55.9	2.1	8.6
High test (Cuba)	80.4		0.7	1.4
Beet (Wisconsin)	78.6	52.7	11.4	9.3
Corn	74.9	50.3	0.4	8.9
Citrus	71.4	42.4	4.7	4.8

Table 9. Price Comparisons between Molasses and No. 2 Yellow Corn

	Minneapolis, Minnesota			Chicago, Illinois			Stockton, California		
Year	Corn, 0.035 m^3 1 bushel	Molasses 24.6 L	Difference	Corn, 0.035 m^3 1 bushel	Molasses 24.6 L	Difference	Corn, 0.035 m^3 1 bushel	Molasses 24.6 L	Difference
1981	293.2	422.7	− 129.5	315.8	420.0	− 104.2	393.1	339.1	+ 54.0
1982	240.0	269.2	− 29.2	250.8	269.1	− 18.3	341.0	230.0	+ 111.0
1983	308.1	292.3	+ 15.8	317.3	294.6	+ 22.7	397.0	267.6	+ 129.4
1984	309.7	307.9	+ 1.8	324.0	307.5	+ 16.5	390.3	273.5	+ 116.8
1985	254.1	276.8	− 22.7	267.3	271.2	− 3.9	340.5	237.5	+ 103.0
1986	192.8	349.8	− 157.0	208.2	348.0	− 139.8	289.5	279.2	+ 10.3

Note: Prices are shown in cents.

Table 10. Estimated Utilization of Molasses in the United States, 1000 Metric Ton

Year	Distilled spirits	Yeast and citric acid	Pharmaceuticals and edible molasses	Sub total	Mixed feeds direct feeding, and silage	Total utilization
1981	31.4	400.4	101.8	533.6	2,496.4	3,030.0
1982	75.1	435.1	82.5	592.7	2,291.4	2,884.1
1983	266.3	489.9	102.2	858.4	2,334.0	3,192.4
1984	372.8	505.9	37.3	916.0	2,305.2	3,221.2
1985	372.8	532.5	34.6	939.0	2,455.9	3,395.8
1986	410.0	484.6	59.1	953.7	2,014.0	2,967.7

Table 11. U.S. Molasses Production, Imports, and Exports

	Production					
Year	Mainland cane	Hawaii cane	Domestic beet	Refiners blackstrap	Imports	Exports
1985	502	168	625	97	1186	97
1986	520	195	729	91	945	275
1987	517	197	758	79	710	237
1988	527	127	NA	NA	729	210

Note: Data are expressed as 10^6 L.

Table 12. World Production of Industrial Molasses, 1000 Metric Ton

Country	1984/1985	1985/1986	1986/1987	1987/1988
Brazil	4,800	3,986	2,587	4,000
USSR	3,200	2,900	3,100	3,300
India	2,500	2,850	3,100	3,280
China	1,130	1,909	1,814	1,850
Cuba	2,000	2,000	1,750	1,750
United States[a]	1,725	1,716	1,710	1,700
Mexico	1,400	1,608	1,590	1,630
Thailand	1,356	1,194	1,205	1,300
France	1,165	1,126	1,013	1,000
Argentina	702	854	950	950
Indonesia	720	840	950	930
South Africa	852	796	735	770
Poland	708	636	675	700
Australia	736	658	693	680
FRG	640	751	715	600
Philippines	744	551	518	580
Others	10,173	9,970	10,661	10,675
Totals	*34,551*	*34,345*	*33,766*	*35,695*

[a]Does not include Puerto Rico or Hawaii.

Table 13. Molasses Price, US$/Ton

Year	Blackstrap[a]	Beet[b]
1980	106.37	105.38
1985	55.39	NA
1986	76.68	88.49
1987	55.64	66.44
1988	71.26	80.36
1997 (Nov.)	57.50	NA

[a]New Orleans.
[b]Colorado.

The juice contains starch and is sometimes treated with a hydrolyzing enzyme to convert the starch to soluble sugars. Treatment with invertase can be used to invert the sucrose into glucose and fructose to reduce the tendency of the finished syrup to crystallize. The juice is clarified and boiled in an atmospheric evaporating pan similar to a maple syrup concentrator. The traditional sorghum pan is a shallow rectangular pan 2-1/2 to 4 feet (3/4 to 1-1/4 m) wide and 12 to 18 feet (3-3/4 to 5-1/2 m) in length divided internally by staggered strips of metal to impart a zigzag flow to the syrup. Heat can be supplied by direct fire beneath the pan or from a steam jacket. The syrup flow is regulated so that product discharges at the desired solids level. The syrup is concentrated to about 76 to 80% solids and is then cooled and transferred to cans, bottles, or jars for sale.

INULIN-BASED PRODUCTS

Inulin is a naturally occurring polymer usually consisting of 17 to 30 anhydrofructose units and a terminal anhydroglucose unit that replaces starch as the food reserve in many plants of the family Compositae, such as Jerusalem artichoke, chicory, salsify, cardoon elecampane, burdock, dahlia, and ti plants. The polymer chain length varies during the growing season and consequently, the glucose–fructose ratio of products of inulin hydrolysis is not constant (64). Although the Jerusalem artichoke is reported to produce tuber yields as high as 100 ton/hectare (65), present commercial sources of inulin are primarily obtained from chicory (66).

A number of processes have been described for the production of crystalline fructose and fructose-rich syrups from inulin-containing plant material (34,67–69). Most commercial crystalline fructose and fructose-rich syrups are produced by isomerization of glucose obtained from starch hydrolysis and enriched by chromatographic separation. Although some fructose is being produced from inulin, much of the inulin being recovered today is partially enzymatically hydrolyzed to make a range of nondigestible oligofructose ingredients that are finding food use as fat mimetics.

ACKNOWLEDGMENTS

The author acknowledges that a portion of the information contained in this article was the work of Ronald E. Hebeda, previously published as the chapter entitled "Syrups" that appeared in *Encyclopedia of Food Science and Technology*, 3rd ed., John Wiley & Sons, New York, 1992, pp. 2490–2504.

BIBLIOGRAPHY

1. F. W. Schenck, "Worldwide Starch Sugars—Past and Future," *Corn Annual*, Corn Refiners Assoc., Inc., Washington, D.C., 1997, pp. 9–12.

2. F. W. Schenck, "Glucose and Glucose-containing Syrups," in *Ullman's Encyclopedia of Industrial Chemistry*, 6th ed., Wiley-VCH, New York.

3. H. Wichelhaus, *Der Starkezucker*, Akademische Verlagsgesellschaft, Leipzig, Germany, 1913.

4. G. S. C. Kirchoff, *Memoirs of the Imperial Academy of Science at St. Petersbourg* **4**, 27 (1811).

5. T. de Saussere, *Ann. Phys.* **49**, 129 (1815).

6. U.S. Pat. 1,471,347 (1923), W. B. Newkirk (to Corn Products Refining Company).

7. R. F. Jackson and C. G. Silsbee, *Natl. Bureau Standards (U.S.) Sci. Pap.* **437**, 715 (1922).

8. S. M. Cantor and Q. P. Peniston, *J. Am. Chem. Soc.* **62**, 2113 (1940).

9. B. Singh, G. R. Dean, and S. M. Cantor, *J. Am. Chem. Soc.* **70**, 517 (1948).

10. D. J. McWeeny, "The Role of Carbohydrate in Non-Enzyme Browning," in G. G. Birch and L. F. Green, eds., *Molecular Structure and Function of Food Carbohydrate*, Wiley, New York, 1973, p. 21.

11. E. R. Kooi and F. C. Armbruster, "Production and Use of Dextrose," in R. L. Whistler and E. F. Paschall, eds., *Starch Chemistry and Technology*, Vol. II, Academic Press, New York, 1967, p. 562.

12. A. D. Woolhouse, *Starch as a Source of Carbohydrate Sweeteners*, Report No. CD 2237, Department of Scientific and Industrial Research, New Zealand, 1976, p. 20.

13. R. V. MacAllister, "Nutritive Sweeteners Made from Starch," in R. S. Tipson and D. Horton, eds., *Advances in Carbohydrate Chemistry and Biochemistry*, Vol. 36, Academic Press, New York, 1979, p. 40.

14. R. E. Hebeda and H. W. Leach, "The Nature of Insoluble Starch Particles in Liquefied Corn-Starch Hydrolysates," *Cereal Chem.* **51**, 272–278 (1974).

15. W. E. Henderson and W. M. Teague, "A Kinetic Model of *Bacillus stearothermophilus* α-Amylase under Process Conditions," *Starch / Stärke* **40**, 412–418 (1988).

16. P. Rosendal, B. H. Nielsen, and N. K. Lange, "Stability of Bacterial α-Amylase in the Starch Liquefaction Process," *Starch / Stärke* **31**, 368–872 (1979).

17. B. C. Saha and J. G. Zeikus, "Microbial Glucoamylases: Biochemical and Biotechnological Features," *Starch / Stärke* **41**, 57–64 (1989).

18. B. E. Norman, "The Application of Polysaccharide Degrading Enzymes in the Starch Industry," in R. C. W. Berkeley, G. W. Gooday, and D. C. Ellwood, eds., *Microbial Polysaccharides and Polysaccharases*, Academic Press, New York, 1979; p. 359.

19. U.S. Pat. 4,017,363 (April 12, 1977), W. H. McMullen and R. Andino (to Novo Industri A/S).

20. U.S. Pat. 4,560,651 (December 24, 1985), G. C. Nielsen et al., (to Novo Industri A/S).

21. R. E. Hebeda, C. R, Styrlund, and W. M. Tezgue, "Benefits of *Bacillus megaterium* Amylase in Dextrose Production," *Starch / Stärke* **40**, 33–36 (1988).

22. R. E. Hebeda, "Syrups," in M. Grayson, ed., *Kirk-Othmer Encyclopedia of Chemical Technology*, 3rd ed., Vol. 22, John Wiley and Sons, New York, 1983.

23. U.S. Pat. 3,257,665 (1966), L. R. Idaszak (to Corn Products Company).

24. U.S. Pat. 4,620,880 (1986), G. Bodele et al., (to Roquette Freres).

25. U.S. Pat. 4,357,172 (1982), L. W. Edwards (to CPC International).

26. G. R. Dean, "An Unstable Crystalline Phase in the D-Glucose-Water System," *Carbohydr. Res.* **34**, 315–322 (1974).

27. A. F. M. Mahbubar Rahman and F. W. Schenck, "Reconversion Improves Monohydrate Dextrose Economics," paper 13 presented at the 1997 AACC Annual Meeting, San Diego, Calif., October 13, 1997.

28. F. W. Schenck, "Solid Starch Hydrolysates," *Cereal Foods World* **41**, 388–390 (1996).

29. *Code of Federal Regulations*, 21 CFR 168.110 and 21 CFR 168.111, U.S. Government Printing Office, Washington, D.C.

30. Codex Alimentarius Commission, *Dextrose Monohydrate*, CAC Vol. III, 1st ed., Codex Standard 8-1981, FAO/WHO, Rome, 1981.

31. Codex Alimentarius Commission, *Dextrose Anhydrous*, CAC Vol. III, 1st ed., Codex Standard 7-1981, FAO/WHO, Rome, 1981.

32. Codex Alimentarius Commission, *Powdered Dextrose (Icing Dextrose)*, CAC Vol. III, 1st ed., Codex Standard 54-1981, FAO/WHO, Rome, 1931.

33. *The United States Pharmacopoeia*, 21st revision (USP XXI NF XVI), United States Pharmacopoeial Convention, Inc., Rockville, Md., 1985, p. 300.

34. M. Kierstan, "Production of Fructose Syrups from Inulin," *Process Biochem.* **2** (1980).

35. T. Doty, "Fructose: The Rationale for Traditional and Modern Uses," in P. Koivistoinen and L. Hyvonen, eds., *Carbohydrate Sweeteners in Foods and Nutrition*, Academic Press, London, 1980, p. 259.

36. C. Morris, "America's Gold," *Food Eng.* **95** (1980).

37. C. A. L. DeBruyn and W. A. van Eckenstein, *Recueil des Travaux Chimiques des Pays-Bas* **14**, 203 (1895).

38. M. Seidman, "New Technological Developments in D-Fructose Production," in G. G. Birch and R. S. Shallenberger, eds., *Developments in Food Carbohydrate*, Vol. 1, Applied Science Publishers, London, 1977, p. 19.

39. U.S. Pat. 2,950,228 (August 23, 1960), R. O. Marshall (to Corn Products Co.).

40. Y. Takasaki, "Studies on Sugar-Isomerizing Enzyme Production and Utilization of Glucose Isomerase from *Streptomyces sp.*", *Agric. Biol. Chem.* **30**, 1247 (1966).

41. U.S. Pat. 3,616,221 (October 26, 1971), Y. Takasaki and O. Tanabe (to Agency of Industrial Science and Technology, Tokyo).

42. U.S. Pat. 3,694,314 (September 26, 1972), N. E. Lloyd, L. T. Lewis, R. M. Logan, and D. N. Patel (to Standard Brands, Inc.)

43. U.S. Pat. 3,788,945 (January 29, 1974), K. N. Thompson, R. A. Johnson, and N. E. Lloyd (to Standard Brands, Inc.).

44. L. M. Hanover, "Crystalline Fructose: Production, Properties, and Applications," in F. W. Schenck and R. E. Hebeda, eds., *Starch Hydrolysis Products*, Wiley-VCH, New York, 1992, pp. 201–231.

45. R. V. MacAllister and E. K. Wardrip, "Fructose and High Fructose Corn Syrups," in M. S. Peterson and A. H. Johnson, eds., *Encyclopedia of Food Science*, AVI Publishing Co., Westport, Conn., 1978, p. 330.

46. G. L. S. Pawan, "Fructose," in G. G. Birch and L. F. Green, eds., *Molecular Structure and Function of Food Carbohydrate*, Wiley, New York, 1973, p. 65.

47. *Food Product Development* **39** (April 1978).

48. O. B. Jorgensen et al., "A New Immobilized Glucose Isomerase with High Productivity Produced by a Strain of *Streptomyces murinus*," *Starch/Stärke* **40**, 307–313 (1988).

49. J. V. Hupkes, "Practical Process Conditions for the Use of Immobilized Glucose Isomerase," *Starch/Stärke* **30**, 24–28 (1978).

50. R. L. Antrim and A. L. Auterinen, "A New Regenerable Immobilized Glucose Isomerase," *Starch/Stärke* **38**, 132–137 (1986).

51. R. van Tilburg, "Enzymatic Isomerization of Corn Starch Based Glucose Syrups," in G. M. A. Van Beynum and J. A. Roels, eds., *Starch Conversion Technology*, Marcel Dekker, New York, 1985.

52. Y. Takasaki, "Data on Individual Related Enzymes" in *The Amylase Research Society of Japan*, Pergamon Press, New York, 1988.

53. C. E. Morris, "First Crystalline-Fructose Plant in U.S.," *Food Eng.* **53**, 70–71 (1981).

54. M. Okada and T. Nakakuki, "Oligosaccharides: Production, Properties, and Applications," in F. W. Schenck and R. E. Hebeda, eds., *Starch Hydrolysis Products*, VCH, New York, 1992, pp. 351–352.

55. *Code of Federal Regulations*, 21 CFR 168.120 and 21 CFR 168.121, U.S. Government Printing Office, Washington, D.C.

56. Codex Alimentarius Commission, *Glucose Syrup*, CAC Vol. III, 1st ed., Codex Standard 9-1981, FAO/WHO, Rome, 1981.

57. Codex Alimentarius Commission, *Dried Glucose Syrup*, CAC Vol. III, 1st ed., Codex Standard 10-1981, FAO/WHO, Rome.

58. R. J. Alexander, "Maltodextrins: Production, Properties, and Applications," in F. W. Schenck and R. E. Hebeda, eds., *Starch Hydrolysis Products*, VCH, New York, 1992, Tables 8.1–8.6, pp. 241–248.

59. *Code of Federal Regulations*, 21 CFR 168.140, U.S. Government Printing Office, Washington, D.C.

60. *Sugar and Sweetener Situation and Outlook Report*, ERS, U.S. Department of Agriculture, Washington, D.C., 1999.

61. *Sugar and Sweetener Situation and Outlook Report*, ERS, U.S. Department of Agriculture, Washington, D.C., 1996.

62. J. L. Collins, I. C. Yachouh, and I. E. McCarty, "Quality of Sorghum-Corn Syrup Blends," *Tennessee Farm Home Sci.*, 10–13 (1980).

63. L. R. Wilhelm and I. E. McCarty, *An Evaluation of Sorghum Syrup Processing Operations in Tennessee*, University of Tennessee Ag. Exp. Station Research Report 85-01 (1995).

64. J. S. Bacon, *Biochem. J.* **51**, 208 (1952).

65. M. Parameswaran, "Jerusalem Artichoke," *Food Australia* **46**, 473–475 (1994).

66. P. Coussement, *Food Ingredients and Analysis International* **18**, 25–27 (1996).

67. J. W. Eichinger et al., *Ind. Eng. Chem.* **24**, 41 (1932).

68. P. V. Golovin, N. A. Bryukhanova, and A. T. Fridman, *J. Sugar Ind. USSR* **3**, 140 (1929).

69. F. A. Dykins et al., *Ind. Eng. Chem.* **25**, 937 (1933).

GENERAL REFERENCES

G. G. Birch, L. F. Green, and C. B. Coulson, eds., *Glucose Syrups and Related Carbohydrates*, Elsevier, Amsterdam, The Netherlands, 1970.

R. T. Foulds, Jr., "Maple Sugar," in A. H. Johnson end M. S. Peterson, eds., *Encyclopedia of Food Technology*, AVI Publishing Co., Westport, Conn., 1974.

R. E. Hebeda, "Corn Sweeteners," in S. A. Watson and P. E. Ramstad, eds., *Corn: Chemistry and Technology*, American Association of Cereal Chemists, St. Paul, Minn., 1987.

S. H. Hemmingsen and B. E. Norman, "Enzyme Technology in the Manufacture of Sugars from Cereals," in G. E. Inglett and L. Munck, eds., *Cereals for Food and Beverages*, Academic Press, New York, 1980.

D. Howling, "The General Science and Technology of Glucose Syrups," in G. G. Birch and K. J. Parker, eds., *Sugar: Science and Technology*, Applied Science Publishers, London, 1979.

A. Lachmann, "Molasses," in A. H. Johnson and M. S. Peterson, eds., *Encyclopedia of Food Technology*, AVI Publishing Co., Westport, Conn., 1974.

N. E. Lloyd and W. J. Nelson, "Glucose and Fructose Containing Sweeteners from Starch," in R. L. Whistler, J. N. BeMiller, and E. F. Paschall, eds., *Starch: Chemistry and Technology*, 2nd ed., Academic Press, New York, 1984.

R. V. MacAllister, "Manufacture of High Fructose Corn Syrup Using Immobilized Glucose Isomerase," in W. H. Pitcher, Jr., ed., *Immobilized Enzymes for Food Processing*, CRC Press, Boca Raton, Fla., 1980.

R. V. MacAllister, E. K. Wardrip, and B. J. Schnyder, "Modified Starches, Corn Syrups Containing Glucose and Maltose, Corn Syrups Containing Glucose and Fructose, and Crystalline Dextrose," in G. Reed, ed., *Enzymes in Food Processing*, 2nd ed., Academic Press, New York, 1975.

R. A. McGinnis, ed., *Beet-Sugar Technology*, 3rd ed., Beet gar Development Foundation, Fort Collins, Colo., 1982.

G. P. Meade and J. C. P. Chen, *Cane Sugar Handbook*, 10th ed., John Wiley & Sons, New York, 1977, pp. 359–377.

Molasses Market News, Market Summary 1966, Agricultural Marketing Service, USDA, Denver, Colo., 1966.

H. M. Pancoast and W. R. Junk, eds., *Handbook of Sugars*, 2nd ed., AVI Publishing Co., Westport, Conn., 1980.

F. W. Schenck and R. E. Hebeda, eds., *Starch Hydrolysis Products*, VCH, New York, 1992.

J. W. White, Jr. and J. C. Underwood, "Maple Syrup and Honey," in G. E. Inglett, ed., *Symposium: Sweeteners*, AVI Publishing Co., Westport, Conn., 1974.

C. O. Willits and C. H. Hills, eds., *Maple Syrup Producers Manual*, Agricultural Handbook No. 134, Agricultural Research Service, USDA, Washington, D.C., 1976.

"Enzyme Technology," in L. B. Wingard, Jr., E. Katchalski-Katzir, and L. Goldstein, eds., *Applied Biochemistry and Bioengineering*, Vol. 2, Academic Press, New York, 1979.

G. M. A. Van Beynum and J. A. Roels, eds., *Starch Conversion Technology*, Marcel Dekker, New York, 1985.

FRED W. SCHENCK
Sun City, Arizona

T

TANNINS

BACKGROUND

Tannins have been defined as water-soluble polyphenolic compounds with molecular weights between 500 and 3000 that have the ability to precipitate alkaloids, gelatin, and other proteins (1). The usefulness of such a definition stems from the original interest in these substances in the tanning process, that is, the conversion of hides to leather, and is enhanced by the recent recognition of other important chemical and physical properties. The tanning of hides has been practiced since around 1500 B.C. The process involves the cross-linking of adjacent collagen chains with polyphenol attachment at several points. The molecular weight limitations apply because a minimum molecular size is required to engage protein chains; too large a molecule cannot easily diffuse into the collagen fibers (2). The plant polyphenols that exhibit the property of tannage also interact with other proteins, peroxides, and many other biological molecules. This has broad implications for fields as diverse as plant evolution, animal nutrition, other aspects of animal physiology, food technology, and, possibly, human therapeutic applications. The significance of the plant polyphenols referred to as tannins goes far beyond their role in the production of leather.

Tannin molecules are complex. Not all have been completely characterized or even recognized. Basically, there are two types: proanthocyanidins, often referred to as condensed; and glucose polyesters of gallic or hexahydroxydiphenic acids, also known as hydrolyzable. The polyphenols are secondary metabolites and not integral parts of the primary growth, energy producing, or reproductive processes of the plant.

Examples of the two major types of tannins are shown in Figure 1.

PROANTHOCYANIDINS

The proanthocyanidins are condensation products of flavanols, chiefly the catechins. Acid hydrolysis under stringent conditions ruptures carbon–carbon bonds connecting flavanol moities. Anthocyanidins and catechins are the final products. Alkaline hydrolysis results in the formation of quinone intermediates that may then be converted to anthocyanidins (3). The acid hydrolysis reaction is demonstrated in Figure 2.

The proanthocyanidins are almost invariably associated with their corresponding flavanols (catechins) from which they are derived. During flavanol biosynthesis a portion of the cation intermediate becomes attached to an already synthesized flavanol molecule to form a dimer rather than undergoing enzymic reduction to the monomer. Trimers and higher oligomers (condensation products of flavanols or other tannin units) are formed in an analogous fashion (4). The proportion of the various species produced is largely a function of the availability of the necessary reducing enzyme.

It has also been suggested that quinone, rather than cation, intermediates participate in proanthocyanidin formation (3). The quinones involved also react with proteins in a process that has been called quinone tanning. The reactions are irreversible with the formation of new covalent bonds. Insects exploit quinone tanning to the ultimate extent. Their bodies are covered with a hard exoskeleton consisting of quinone–protein adducts that provide protection from desiccation and predators and that account for their biological success (5).

Formation of a dimeric condensation product by the cation mechanism is illustrated in Figure 3. Procyanidins are the most frequently encountered group of the proanthocyanidins. They are derived from the simple catechins (R = H). Prodelphinidins are not as widespread. They are derived from the gallocatechins (R = OH).

Biosynthesis of the parent flavanol proceeds through a complex pathway starting with the conversion of phenylalanine to cinnamic acid. Incorporation of acetate units leads to the formation of chalcones. Ring closure results in flavonoid structures from which catechins are derived (4).

The procyanidins in foods occur in soluble free forms and in insoluble combinations with polysaccharides. Such adducts are the result of covalent bonds formed during the biosynthesis of the procyanidins. Dimeric procyanidins, barely within the molecular size range of tannins, and trimeric forms are widely distributed in berries and in the leaves of many fruit plants at levels of 50 to 500 mg/100 g fresh weight (6). They are often accompanied by dimeric prodelphinidins, which predominate in strawberries. Fruits and grains are also rich in higher oligomers of catechins. As molecular size increases, the solubility of the proanthocyanidins decreases. Varieties of sorghum, one of the most widely consumed of all grains by humans, differ in tannin levels by a factor of 100. Some fodder varieties contain as much as 6% tannin (7).

Proanthocyanidins and their parent flavanols that occur in human food undergo changes during preparation and storage that often lead to the formation of highly colored polymeric material. Enzymic browning occurs when susceptible polyphenols come in contact with plant oxidases in the presence of atmospheric oxygen as a result of cell disruption caused by spoilage, insect infestation, or other mechanical damage. This effect is exploited in the manufacture of black tea. The cell structure of fresh green tea leaf is disrupted mechanically to promote contact between the abundant cellular flavanol supply and oxidizing enzymes. Proanthocyanidins constitute one fraction of the oxidized catechin group and are responsible for some of the desirable organoleptic properties of black tea. Procyanidins are also important constituents of cocoa. They undergo many changes during the processing of the bean that impact on flavor and color (6). Gallotannins are also present and constitute over 4% of the dried fermented cotyledons.

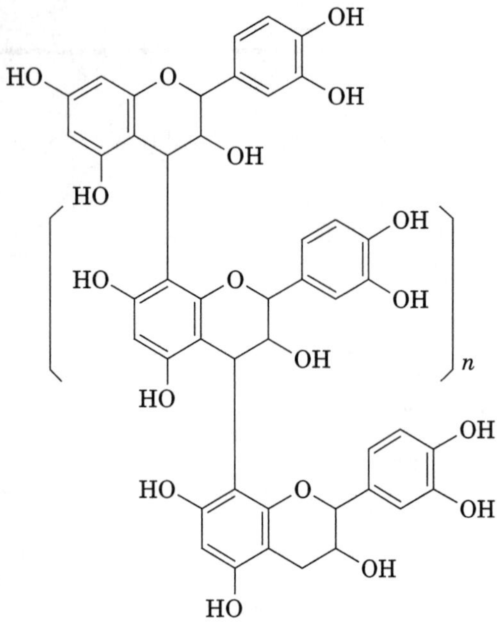

Proanthocyanidin

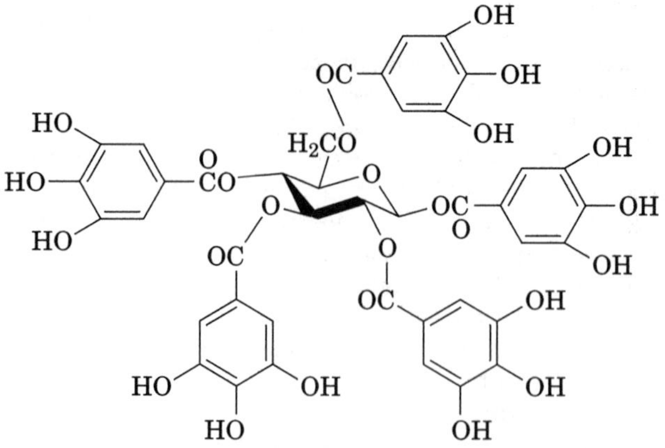

Pentagalloyl glucose

Figure 1. Tannins.

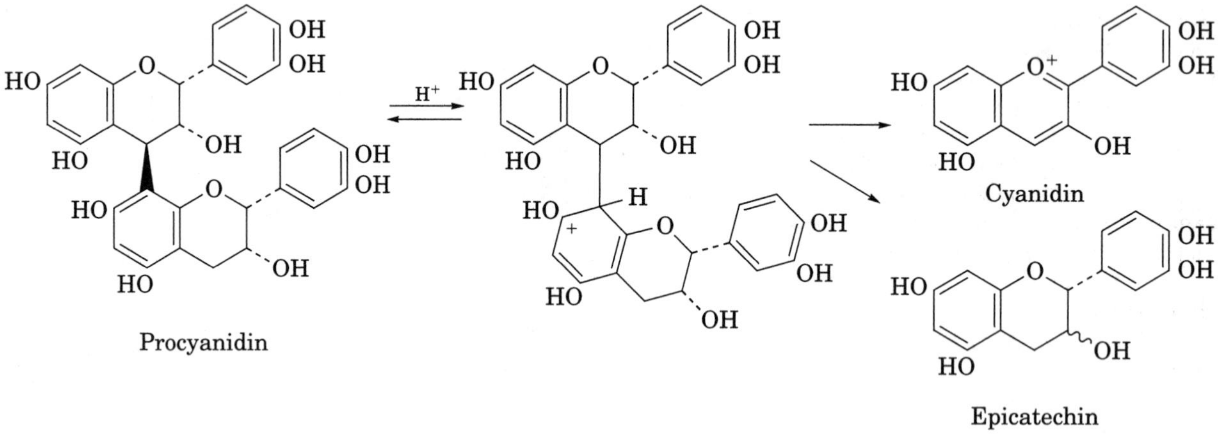

Procyanidin

Cyanidin

Epicatechin

Figure 2. Acid hydrolysis of proanthocyanidins.

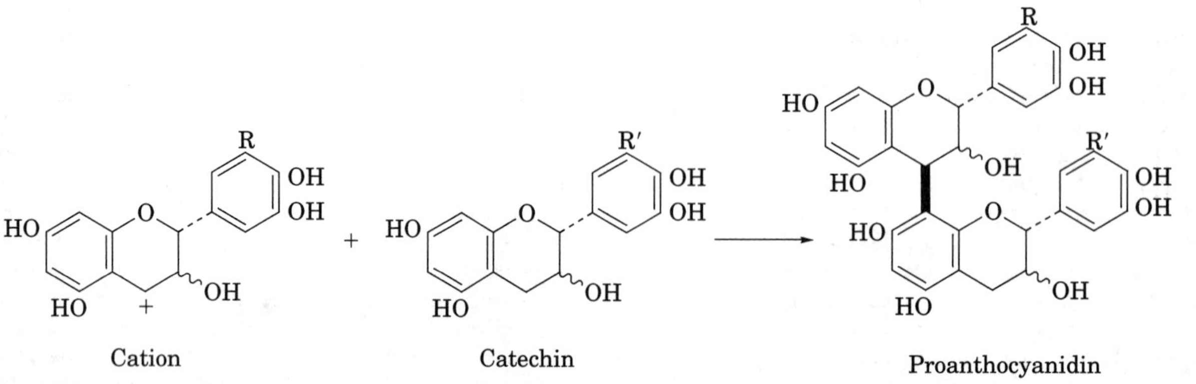

Cation   Catechin   Proanthocyanidin

Figure 3. Formation of proanthocyanidins.

In other food products enzymic browning may be disadvantageous. It can be prevented by enzyme inactivation (often blanching), by the use of antioxidants, or by the rigorous exclusion of oxygen. Wine technology is largely an exercise in tannin control (6). White wine production is managed so as to exclude the polyphenol-rich grape skins and seeds from the juice in order to minimize enzymic browning. Procyanidins are present at about 10 mg/L and may be desirable at this low level. Red wine is produced so as to include phenolic components in the fermenting juice. Final procyanidin content is in the range of 40 to 65 mg/L. Color changes that occur during the aging of red wine include the diminution of the original color (primarily from anthocyanins) and the formation of new, more stable, oligomeric pigments (8). Procyanidin dimers in beer originate from the malt and are troublesome haze formers. The tannins involved are primarily dimers of catechin and epicatechin (9).

Fruit, vegetables, grain, and many beverages are sources of condensed tannins in the diet as shown by a few examples in Table 1.

POLYESTERS

Most of the hydrolyzable tannins are essentially polygallates of glucose. Hydroxyl groups of the gallate segments can be further esterified with gallic acid to build up large complex molecules. Gallotannins occur in chestnut, oak, sumac, and in several plant galls. The Peruvian plant, tara, is an important commercial source. The gallotannins of commerce are known as tannic acid. They contain 5 to 10 esterified gallic acid residues. Tannic acid is used as a mordant in the dyeing of cotton, as a clarifying agent for wine and beer, and as a food flavorant at low concentrations. It is approved for use in the United States as a GRAS additive as shown in Table 2.

When galloyl groups of glucose esters are appropriately oriented, oxidative coupling may take place during plant metabolism to form ellagitannins that contain hexahydroxydiphenic acid groups. Hydrolysis yields ellagic acid as well as gallic acid (10). The ellagitannins are rigid molecules in contrast to the gallotannins, which exhibit a great degree of flexibility. The structural relationships among the ellagitannis, casuaricitin, and pentagalloyl glucose that are depicted in Figure 4. provide an indication of the biosynthetic pathway for the formation of ellagitannins.

Many ellagitannins yield additional phenolic compounds on hydrolysis such as chebulic, dehydrodigallic,

Table 1. Proanthocyanidin Content of Foods

Food	Proanthocyanidins (mg/100 g)
Strawberries	110–150
Raspberries	100–140
Black currants	350
Persimmons	800–1700
White wine	1–2
Red wine	4–7
Sorghum	100–1000
Barley	20
Tea (black, leaf)	15,000
Cocoa (powder)	4500

Table 2. Approved Use Levels for Tannic Acid

Product	Percentage of product as used
Baked goods	0.01
Alcoholic beverages	0.015
Nonalcoholic beverages, beverage bases, gelatins, puddings, fillings	0.005
Frozen desserts and soft candy	0.04
Hard candy and cough drops	0.013
Meat products	0.001

and flavogallonic acids. These substances are also formed by modification of galloyl groups attached to glucose. In some ellagitannins, glucose is in its open-chain state. The ellagitannins occur in the leaf, bark, and heartwood of many plant species.

The complexity and large molecular size of some of the ellagitannins is demonstrated by the isolation of a series of newly found substances in the flower buds of plants of the *Camellia* sp (11). One of these, *Camelliin B*, while only a dimer, exhibits a complex macroring structure as shown in Figure 5. Several other macroring dimeric ellagitannins have been isolated from other plants and are part of this newly recognized class of tannins.

As of 1993 more than 400 discrete ellagitannins had been reported (12,13). Taxonomic studies have shown that each of these tannins is associated with specific plant species that follow particular paths of plant evolution (14).

Gallotannins and ellagitannins are not present in food at the high levels observed for the proanthocyanidins. Although *tannic acid* has often been referred to as a component of tea and coffee, none is present. Gallic acid is formed in tea, however, by hydrolysis of gallated flavonoids.

ASTRINGENCY

The reaction of tannins with protein is responsible for their astringency—a property of considerable importance in human and animal feeding. It can be described as the complex of sensations due to shrinking, drawing, or puckering of the epithelium in the buccal cavity. Astringency is not one of the basic taste sensations, as its effect is noticed away from the site of the primary taste receptors. It is a tactile sensation arising from reduced lubrication of the tongue and soft palate (15). The primary reaction that leads to astringency is the precipitation of proteins and mucopolysaccharides in saliva. The astringency of tannins is their most noticeable organoleptic property and is correlated with their protein complexation ability. The simple glucose–gallic acid esters bind more strongly to protein than do the proanthocyanidins because of conformational restraints imposed by the interflavan bonds of the latter group. The gallotannins are, accordingly, more astringent than the proanthocyanidins. It has been shown that proteins that are rich in the amino acid proline, and that possess open, flexible structures have the highest affinity for polyphenols. The presence of these proteins in human saliva is protective against the entrance of excessive quantities of polyphenols into the digestive tract. There is evidence that the prevalence of tannins in the human diet tends to increase the formation of the proline-rich proteins (16).

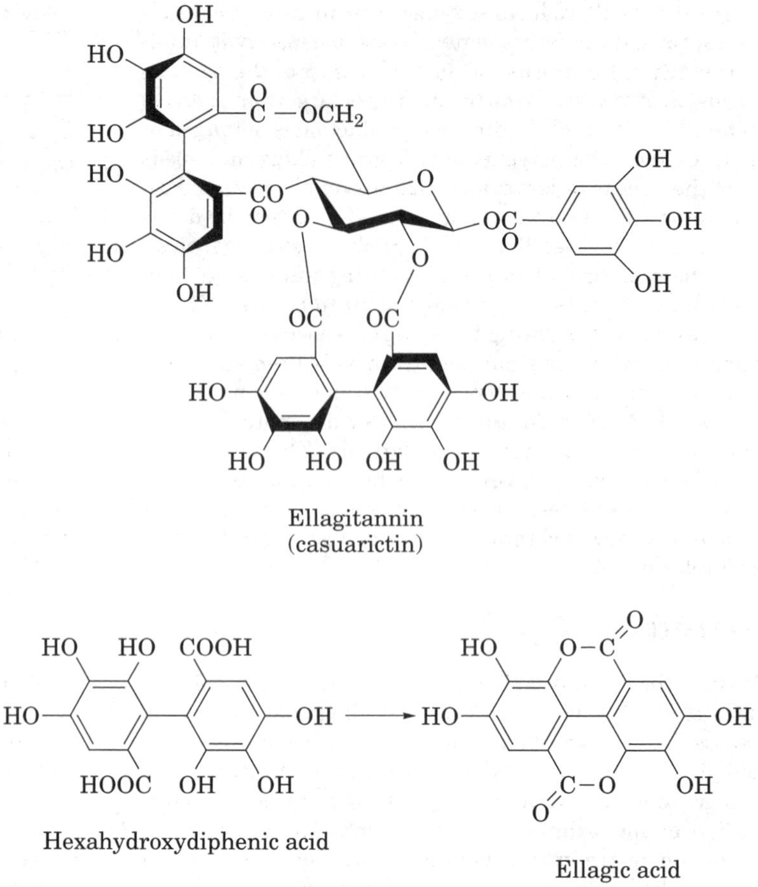

Ellagitannin
(casuarictin)

Hexahydroxydiphenic acid

Ellagic acid

Figure 4. Ellagitannins and component relationships.

The structural feature of a tannin responsible for the phenomenon of astringency has been stated as "the accumulation within a molecule of moderate size, of a substantial number of phenolic groups, many of which are associated with 1,2-dihydroxy or 1,2,3-trihydroxy orientation within a phenyl ring" (16).

The human diet basically includes two groups of tannins: the ellagic acid esters of glucose (ellagitannins) found in the persimmon, pomegranate, and chestnut; and flavanol-derived substances such as condensed catechin oligomers occurring in many fruits, and compounds produced by the enzymic oxidation of the flavanols accompanied by condensation reactions. These are found in wine, beer, and black tea.

The more water-soluble tannins, such as the ellagitannins, are less efficient in precipitating protein than the polygalloyl glucose esters. Molecular size is also important because of its effect on the orientation of the di/trihydroxylphenyl groups with respect to the proline-rich proteins. Astringency decreases because of steric hindrance as the number of epicatechin units in an oligomer exceeds six or seven. The loss of astringency that occurs during fruit ripening may be caused by further polymerization or by complexation with increased amounts of sugar.

ANALYSIS

The isolation, determination of structure, and quantitation of tannins have been exhaustively investigated because of

concerns with their possible physiological effects and because of their interest as pharmaceutical agents. The separation of closely related oligomers has been accomplished with droplet countercurrent and centrifugal partition chromatography (17). By use of these techniques, large molecules, such as nobotanin K, consisting of 4 glucose and 20 gallic acid groups have been isolated in the pure state. Accurate molecular mass determination of these nonvolatile substances has been perfected with the use of fast atomic bombardment (FAB) and mass spectrometry (MS). Nuclear magnetic resonance (NMR) spectroscopy has also provided details of the intricate molecular structures of large complex species. The use of fast atom bombardment mass spectrometry (FABMS) has allowed for the accurate structure determination of large tannin molecules—up to molecular weight of 3200 (18). The most significant methods for the quantitation of tannins are based on measurements of protein precipitation, since their most important properties are attributed to this effect. Procedures based on interaction with bovine serum albumen measure both tannins and protein in the precipitate (19). As little of 5 μg of tannic acid can be detected.

TANNIN–PROTEIN INTERACTIONS

The primary mode for tannin–protein interaction is hydrogen bond formation (20). It is largely a surface phenomenon. Bonding occurs between the amide carbonyl groups of

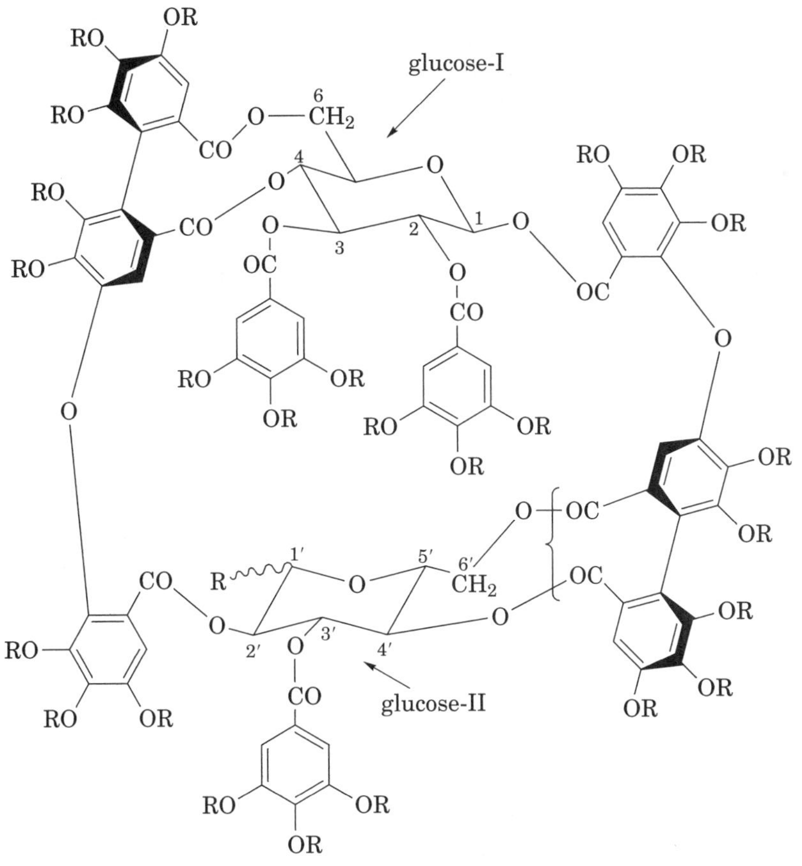

R=H, R′=OH

Figure 5. Camelliin B.

the protein and the hydroxyl groups of the polyphenol. The polyfunctional tannin molecule is able to bind at many sites on the surface of the protein molecule, thus increasing its hydrophobic nature and promoting precipitation. Inhibition of enzymic activity is often a result of tannin–protein complexation. At high protein concentrations the interaction results in cross-linking. There is also evidence for some hydrophobic bonding.

The formation of the tannin–protein complex is pH dependent and reaches a maximum at the isoelectric point of the protein. Most complex formation is reversed at pH >9. Solvents, such as acetone, that tend to form hydrogen bonds competitively inhibit the interaction. Polyphenol molecules with a high degree of flexibility are most reactive. Proanthocyanidins are more rigid than the gallotannins and are less active complexing agents on a molar basis.

Tannin–protein complexation is favored as the polyphenol solubility is decreased. Proteins with a high proline or hydroxyproline content, such as gelatine, have a greater affinity for polyphenols as these amino acids impart flexibility to the protein molecule and form strong hydrogen bonds. The tendency of a protein to bind to a tannin may be determined by the degree to which the protein competitively prevents inhibition of a specific enzymic activity by the tannin.

Complexation and precipitation of caffeine by tannins has been recognized as the most important cause of "cream" formation when infusions of black tea are allowed to cool. The enzyme tannase catalyzes the cleavage of the gallate linkage in tannin molecules and has been used to solubilize caffeine–polyphenol adducts during preparation of instant tea (see the article TEA). It may be induced in large quantities in some strains of *Aspergillus* molds. Caffeine–polyphenol systems have been used as models for the investigation of tannin–protein interactions because there is a resemblance between the molecular structures of caffeine and peptides.

The tanning of animal hides with vegetable extracts accounts for only a small proportion of the leather now produced in the United States. The use of chromium salts has supplanted the traditional process for most applications. Vegetable tannage depends on packing the amorphous regions of collagen fibers with large quantities of tannins so as to bring about extensive cross-linking. The collagen may adsorb up to 50% of its weight of tannins. Tanning renders the mass impermeable to bacterial attack, decreases water penetration, and increases tensile strength. Vegetable-tanned leather has superior qualities of resilience. The extracts used are derived from oak galls, chestnuts, and the bark and heartwood of trees such as quebracho, gambier, and mimosa. Wastewater from tanneries is highly contaminated. Its purification before discharge is made difficult by the inhibitory action of residual tannins on the microbiological organisms normally utilized for anaerobic fermentation (21).

PHYSIOLOGICAL EFFECTS

Dietary tannin decreases the growth rate of many mammals and the net energy available to domestic grazers (22). For ruminants there is a compensating effect since complexation of dietary protein diminishes its microbial degradation in the rumen. Egg production by hens is decreased. The antinutritional effects probably result from the formation of less digestible complexes.

The presence of tannins in plants imparts some degree of protection against insects and grazing animals because of their negative nutritional impact. It is suggested that this may be the evolutionary basis for their presence (20).

Conversely, some insects and higher animals have developed compensatory mechanisms that enable them to consume tannins with no antinutritional effects.

The widespread consumption of *Sorghum bicolor* in tropical semiarid zones provides large human populations with their major source of calories and protein. The nutritional consequences of high sorghum diets have been studied extensively and much specie variation has been observed (22). The rat goes through an adaptation period during which salivary glands are stimulated to produce proline-rich proteins. These compete effectively for tannins and inhibit dietary protein complexation. In other animal species, different salivary proteins are produced that have similar inhibitory effects. The hamster has no compensatory mechanism and soon succumbs. Several techniques have been utilized to minimize the antinutritional effects of the tannins in high-sorghum human diets. Cooking is not effective. Dehulling is carried out in many cultures with considerable benefit since tannin is concentrated in the layer just below the seed coat. Alkaline treatment of

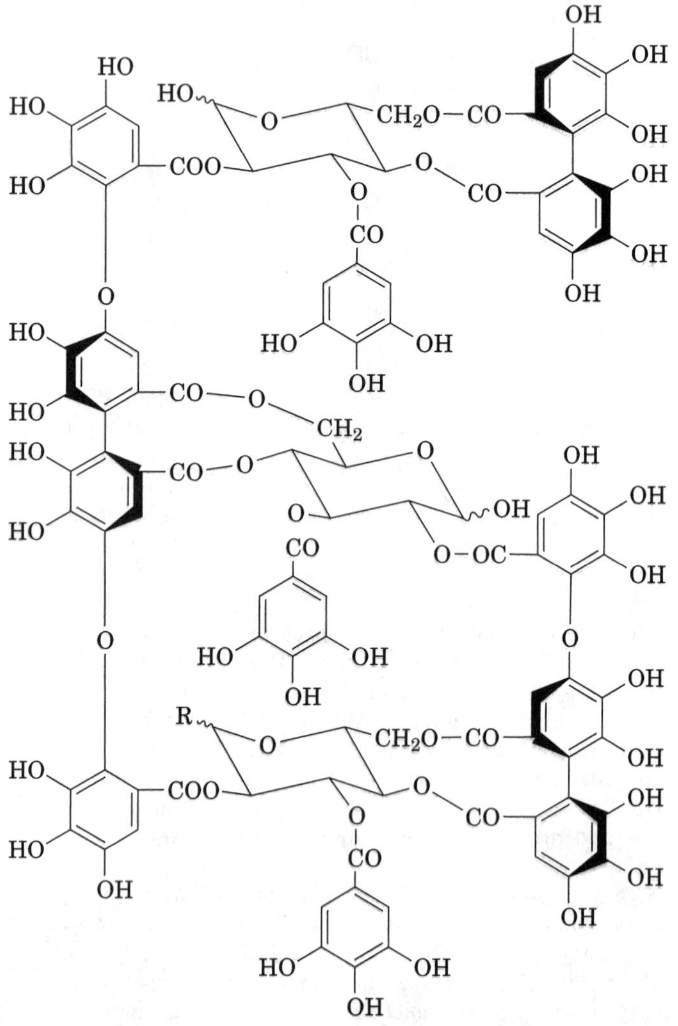

Oenothein A: R = OH
Woodfordin D: R = (α)-OG

Figure 6. Oenthein A and Woodfordin D.

the grain and soaking in dilute formaldehyde are also effective procedures for improving its nutritional value.

The polyphenolic content of the tannins imparts antioxidant activity to biological systems. The ability of condensed tannins from several plant sources to scavenge oxygen-containing free radicals has been demonstrated (23). Exploitation of this effect to produce beneficial consequences for humans has been suggested since peroxidation has been implicated in tumerogenesis. Some dimeric ellagitannins exhibit antitumor activity in mice. Structure and activity have been correlated (23). It has been suggested that possible mechanisms for pharmacological activity of tannins may depend on several tannin propensities: complexation with transition-metal ions; free-radical quenching; complexation with proteins and peptides (25).

Inhibition of tumor growth in rodents by the administration of some oligomeric hydrolyzable tannins after intraperitoneal inoculation of tumor cells has been observed. Examples of these active ellagitannins are oenothein and woodfordin, shown in Figure 6. This effect is an example of host-mediated antitumor activity as interleukin-1 is induced in the treated animals (26,27).

Life expectancy of stroke-prone hypertensive rats has been extended by feeding persimmon tannin. Grape seed procyanidin exhibits anti-inflammatory activity in experimental animals. (28).

Some selected purified tannins have been tested and shown to be nonmutagenic by the Ames Test (29).

BIBLIOGRAPHY

1. E. Haslam, "Twenty-Second Procter Memorial Lecture; Vegetable Tannins—Renaissance and Reappraisal," *J. Soc. Leather Technol. Chem.* **72**, 45–64 (1988).

2. T. White, *The Chemistry of Vegetable Tannins, Society of Leather Trades Chemists*, Croydon, London, 1956, p. 7.

3. R. W. Hemingway and L. Y. Foo, "Condensed Tannins: Quinone Methide Intermediates in Procyanidin Synthesis," *J. Chem. Soc. Commun.* **18**, 1035–1036 (1983).

4. E. Haslam, "Symmetry and Promiscuity in Procyanidin Biochemistry," *Phytochemistry* **16**, 1625–1640 (1977).

5. M. Sugumaran, "Quinone Methide Sclerotization: A Revised Mechanism for β-Sclerotization of Insect Cuticle," *Bio-Org. Chem.* **15**, 194–211 (1987).

6. W. S. Pierpoint, "Flavonoids in the Human Diet," in V. Cody, E. Middleton Jr. and J. B. Harborne, eds., *Progress in Clinical and Biological Research*, Vol. 213, Alan R. Liss, New York, 1986, pp. 125–140.

7. R. Gujer, D. P. Magnolato, and R. Self, "Proanthocyanidins and Polymeric Flavonoids from Sorghum," in V. Cody, E. Middleton Jr., and J. B. Harborne, eds., *Progress in Clinical and Biological Research*, Vol. 213, Alan R. Liss, New York, 1986, pp. 159–166.

8. T. C. Somers and E. Verette, "Phenolic Composition of Natural Wine Types," in H. F. Linkskens and J. F. Jackson, eds., *Modern Methods of Plant Analysis, New Series Wine Analysis*, Vol. 6, Springer-Verlag, Berlin, 1988, pp. 219–257.

9. K. J. Siebert, A. Carrasco, and P. Y. Lynn, "Formation of Protein-Polyphenol Haze in Beverages," *J. Agric. Food Chem.* **44**, 1997–2005 (1996).

10. E. Haslam, "Vegetable Tannins," in T. Swain, J. B. Harborne, and C. V. Van Sumere, eds., *Recent Advances in Phytochemistry*, vol. 12, *Biochemistry of Plant Phenolics*, Plenum Press, New York, 1979, pp. 475–523.

11. T. Yoshida et al., "Tannins of Theaceous Plants II. Camelliians A and B. Two New Dimeric Hydrolyzable Tannins from Flower Buds of Camellia-japonica and Camellia-sasanqua THUNB," *Chem. Pharm. Bull.* **38**, 2681–2686 (1990).

12. E. Haslam, "Polyphenols—Phytochemical Chameleons," *Proc. Phytochem. Soc. Eur.* **34**, 219 (1993).

13. T. Okuda, T. Yoshida, and T. Hatano, "Hydrolyzable Tanins and Related Polyphenols," *Prog. Chem. Org. Nat. Prod.* **66**, 1–117 (1995).

14. T. Okuda, T. Yoshida, and T. Hatano, "Classification of Oligomeric Hydrolyzable Tannins and Specificity of Their Occurrence in Plants," *Phytochemistry* **32**, 507–521 (1993).

15. P. A. S. Breslin et al., *Chem. Senses* **18**, 405–407 (1993).

16. M. N. Clifford, "Astringency," *Proc. Phytochem. Soc. Eur.* **41**, 87–107 (1997).

17. T. Okuda, T. Yoshida, and T. Hatano, "New Methods of Analyzing Tannins," *J. Natural Products* **52**, 1–31 (1989).

18. T. Okuda, T. Yoshida, and T. Hatano, "Hydrolyzable Tannins and Related Polyphenols," *Prog. Chem. Org. Nat. Prod.* **66**, 62 (1995), pp. 2–117.

19. H. P. S. Makkar, "Protein Precipitation Methods for Quantitation of Tannins: A Review," *J. Agr. Food Chem.* **37**, 1197–1202 (1989).

20. C. M. Spencer et al., "Polyphenol Complexation—Some Thoughts and Observations," *Phytochemistry* **27**, 2397–2409 (1988).

21. H. S. Bajwa and C. F. Forster, "The Inhibition of Anaerobic Processes by Vegetable Tanning Agents," *Environ. Technol. Lett.* **9**, 1245–1256 (1988).

22. L. G. Butler et al., "Dietary Effects of Caffeine," in V. Cody, E. Middleton Jr., and J. B. Harborne, eds., *Progress in Clinical and Biological Research*, Vol. 213, Alan R. Liss, New York, 1986, pp. 141–157.

23. S. Uchida et al., "Radical Scavenging Action of Condensed Tannins," *Neurosciences (Kobe, Japan)* **14**, 243–245 (1988).

24. K. Miyamoto et al., "Relationship between the Structures and the Antitumor Activities of Tannins," *Chem. Pharm. Bull.* **35**, 814–822 (1987).

25. E. Haslam, "Natural Polyphenols (Vegetable Tannins) as Drugs: Possible Modes of Action," *J. Nat. Prod.* **59**, 205–215 (1996).

26. K. T. Miyamoto et al., "Antitumor Activity and Interleukin-1 Induction by Tannins," *Anticancer Res.* **31**, 37–42 (1993).

27. T. Yoshida et al., "Woodfordin D and Oenothien A, Trimeric Hydrolyzable Tannins of Macro-Ring Structure with Antitumor Activity," *Chem. Pharm. Bull.* **39**, 1157–1162 (1991).

28. M. Gábor, "Anti-allergic and Anti-inflammatory Properties of Flavonoids," in V. Cody, E. Middleton Jr., and J. B. Harborne, eds., *Progress in Clinical and Biological Research*, Vol. 213, Alan R. Liss, New York, 1986, pp. 471–480.

29. C. L. Yu and B. Swaminathan, "Mutagenicity of Proanthocyanidins," *Food Chem. Toxicity* **25**, 135–139 (1987).

Harold N. Graham
New York City, New York

See also TEA.

TASTE AND ODOR. See SENSORY SCIENCE:
  PRINCIPLES AND APPLICATIONS; SENSORY SCIENCE:
  STANDARDIZATION AND INSTRUMENTATION.

TEA

The term tea refers to the plant *Camellia sinensis*, the dried, processed leaf manufactured from it, extracts derived from the leaf, and beverages prepared from leaf or extract of this species. Tea is the most widely consumed of all beverages aside from water. Annual worldwide per capita consumption, although geographically far from uniform, is 40 to 50 L (based on total known tea production of 2.667 million tons (1) and a beverage yield of ca 100 L/kg). Apart from its extensive usage, it is of interest to the food technologist as well as to the biochemist, organic chemist, and physiologist because of the dependence of the final product characteristics on the unusual chemical composition of the fresh leaf and because of the complex series of biochemical and organochemical reactions that occur during processing. There has also been increasing interest in its health benefits.

Two major varieties of *C. sinensis* are recognized: *assamica*, a large-leaved (15–20 cm) plant, and *sinensis*, a smaller-leaved (5–12 cm) plant. Intervarietal hybrids are common. There are also hybrids with a related species, *Camellia irrawadiensis*. Tea originated in southeast Asia in an area that includes parts of China and India, and probably Myanmar, Laos, and Vietnam.

The first authenticated consumption of tea dates from the fourth century A.D. in China, although legend dates it much earlier. Its use spread to Japan, where cultivation started in the eighth century. Tea from China reached Europe in the sixteenth century, and by the end of the seventeenth century it had become a popular beverage served in London teahouses. Tea was discovered growing wild in Assam Province of India in 1823 and cultivation followed rapidly. The plant was introduced into the Dutch East Indies (Indonesia) and then into Ceylon (Sri Lanka) after the destruction through disease of that country's coffee plantings in the 1870s. Cultivation of tea on a large scale subsequently spread to other parts of Asia, to Georgia in the former USSR, and to Africa, South America, and several tropical islands. Tea production was carried out on a very small scale in South Carolina around the beginning of the twentieth century; at its peak, ca 10,000 lb/yr was produced (2). Commercial operation was discontinued in 1915 but resumed in 1987 (3).

AGRICULTURE

Tea grows best in tropical and subtropical areas where adequate rainfall (200 cm), good drainage, and slightly acid soils prevail. *Sinensis* variety is the more cold resistant. In hot tropical areas, tea quality is improved by planting at high altitudes, as practiced in India and Sri Lanka, where tea is cultivated at elevations up to 2,000 m. The tea plant is maintained by pruning as a low shrub (1–1.5 m) at a density of 5,000 to 10,000 plants/ha. Tea is now primarily propagated from leaf cuttings rather than from seed. The establishment of desirable clones is accomplished by selection on the basis of beverage quality, yield, pest and frost resistance, and other criteria dictated by local conditions and marketing considerations. Yields have been greatly increased and exceed 6,000 kg/ha in some clonal fields, although national averages are very much lower (1). Tissue culture of tea is under investigation as a method for producing and rapidly replicating new clonal material. Plants have been regenerated and set out in soil (4). Tea can remain productive for many decades; some 100-year-old plantings are still being harvested. The use of irrigation, fertilizers, mulches, and pesticides has been extensively researched in many tea-growing areas and the results reported in the journals of the various tea research institutes (5–8). Pesticides banned in the United States are not used (9).

New tea growth (flush) is harvested at intervals of 6 to 12 days depending on climatic conditions. In cooler areas such as North India, Japan, and Georgia there is a dormancy period. Plucking is a hand operation on most estates. Varying degrees of mechanization are being applied in some countries where terrain and labor costs are appropriate. These range from handheld cutters powered by small back-packed gasoline engines to large self-propelled harvesters that straddle the tea row. Removing the apical bud and two leaves below it ("two and a bud") constitutes the ideal plucking practice, but it is not always realized, and never with mechanical harvesting. Tea estates range in size from small peasant holdings of 1 ha or less to large establishments of up to 1,000 ha. The larger estates usually include housing, schools, hospitals, and other facilities for the workers employed in this labor-intensive activity. Labor utilization is often at the rate of 3 persons/ha. Plucking may account for half of the total requirement.

FRESH LEAF COMPOSITION

Tea leaf contains all of the enzymes, metabolic intermediates, and structural elements normally associated with photosynthesis and plant growth. In addition, the tea plant contains a large quantity of substances that are responsible for its unique properties as a beverage. Chief among these are polyphenols (mostly flavonoids) and caffeine. A representative analysis of young shoots from fresh *assamica* leaf is depicted in Table 1.

Flavanols

The most striking chemical characteristic of tea flush is its very high content of flavanols. Although amounts vary, depending on all of the factors that influence plant metabolism, such as light, rainfall, temperature, nutrient availability, leaf age, and genetic makeup, a small number of flavanols alone may constitute more than 20% of the dry matter. The most important group of the tea flavanols are the catechins. These include the free catechins as well as monogallate and digallate esters and a single caffeic acid ester (12). Free catechins and the monogallate esters predominate, and their reaction products are the best known (Fig. 1).

Table 1. Composition of Tea Flush (% Dry Weight)

Flavanols	25
Flavonols and flavonol glycosides	3
Flavone C-glycosides	0.2
Polyphenolic acids and depsides	5
Other polyphenols	3
Caffeine	3
Theobromine	0.2
Amino acids	4
Organic acids	0.5
Monosaccharides	4
Polysaccharides	14
Protein	15
Cellulose	7
Lignin	6
Lipids, waxes, sterols	3.5
Chlorophyll and other pigments	0.5
Ash	5
Volatiles	<0.1

Source: Refs. 10 and 11.

The tea flavanols are water-soluble, colorless, astringent substances that oxidize rapidly, especially in an alkaline environment. They occur in the cytoplasmic vacuoles of the leaf cells. They form complexes with many other substances (13). Epigallocatechin gallate constitutes a high proportion of fresh-leaf weight. Tea beverage quality is correlated with fresh-leaf flavanol content, especially with the gallated compounds that decrease in quantity as the leaf ages on the growing plant. They are the most important components in the black-tea manufacturing process.

Catechins are notable antioxidants, a property also exhibited by their oxidation products present in black tea and by other flavonoid compounds (14). This is demonstrated by their ability to scavenge free radicals generated in many *in vitro* systems. The physiological significance of this property will be discussed later in this article.

Other Polyphenols

Other polyphenols include gallic acid (**7**) and quinic acid; flavonols, such as kaemferol (**11**), quercitin (**12**), myricetin (**13**), and their glycosides; a group of chalcan–flavan dimers; proanthocyanidins; a group of flavone-*C*-glycosides (**10**); and a group of bisflavanols also known as theasinensins (**12**). Depsides present include chlorogenic acid (**8**), *p*-coumarylquinic acid (**9**), and theogallin (3-galloylquinic acid) (**10**), which is unique to tea (Fig. 2). It occurs at relatively high concentrations (1% of dry weight) and is said to correlate with quality. Most of the polyphenols participate in the important reactions that occur during manufacture.

Caffeine and Other Xanthines

Tea flush contains 2 to 4.5% caffeine and much smaller amounts of theobromine. Theophylline may be present but in small quantities. Caffeine is responsible for the mild stimulatory properties of tea beverage and is generally

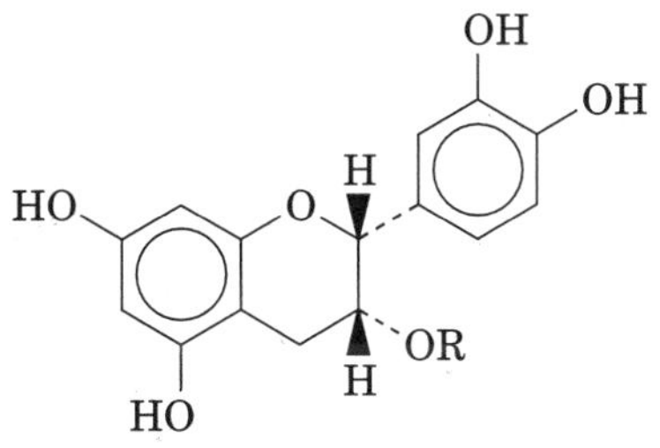

(**1**) (−)-epicatechin;
    R = H; 1–3% of dry wt
(**2**) (−)-epicatechin gallate;
    R = 3,4,5-trihydroxybenzoyl;
    3–6% of dry wt

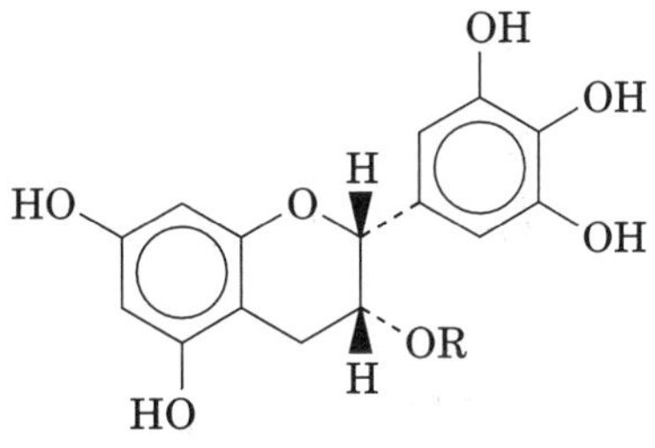

(**3**) (−)-epigallocatechin;
    R = H; 3–6% of dry wt
(**4**) (−)-epigallocatechin gallate;
    R = 3,4,5-trihydroxybenzoyl;
    9–13% of dry wt

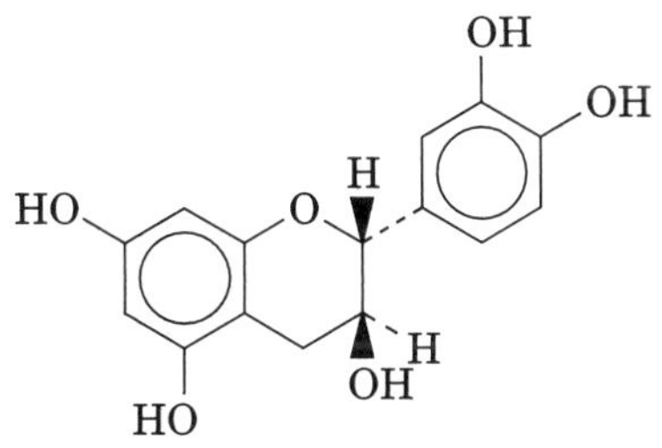

(**5**) (+)-catechin; 1–2% of dry wt

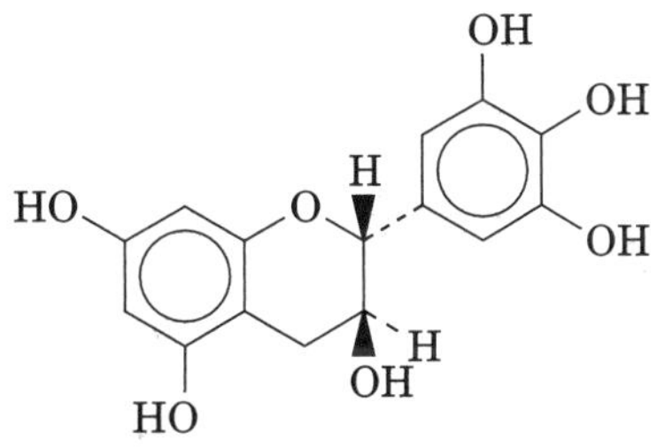

(**6**) (+)-gallocatechin; 3–4% of dry wt

Figure 1. The flavanols occurring in tea.

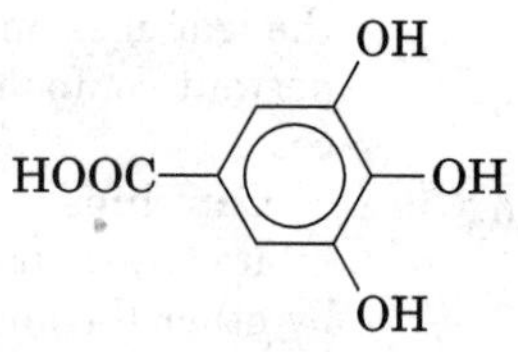

(7) Gallic acid

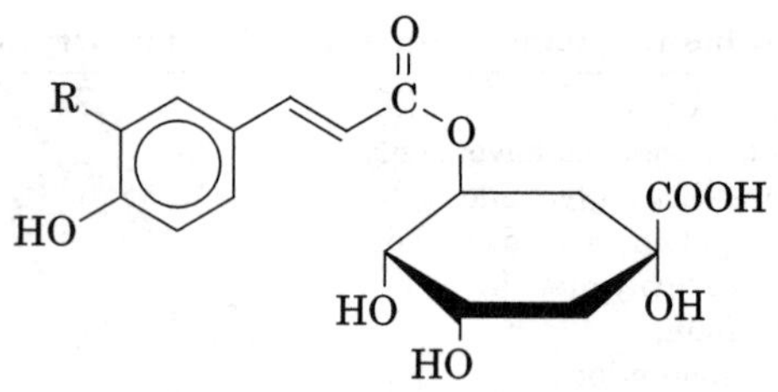

(8) Chlorogenic acid; R = OH
(9) p-coumarylquinic acid; R = H

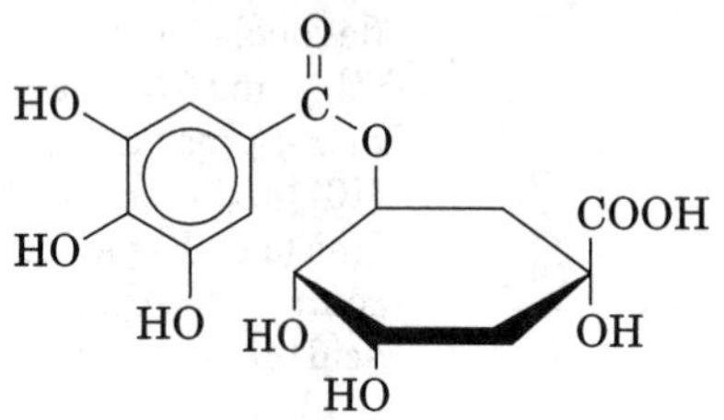

(10) Theogallin

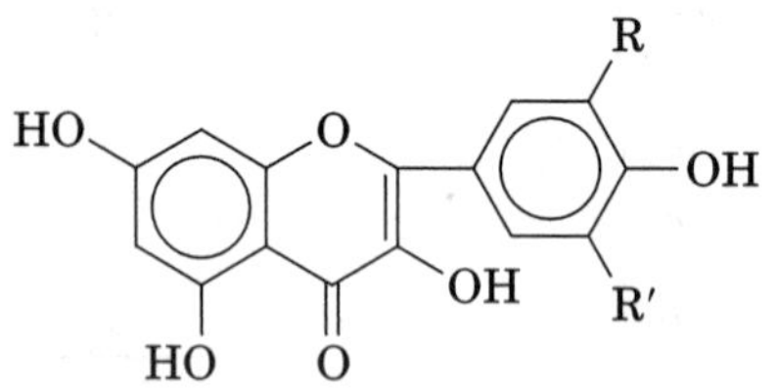

(11) Kaempferol; R = R' = H
(12) Quercitin; R = OH, R' = H
(13) Myricetin; R = R' = OH

Figure 2. Other polyphenols occurring in tea.

$$HO_2CCH(NH_2)CH_2CH_2CONHC_2H$$

Figure 3. Theanine (14).

present at a level of 30 to 45 mg/cup. Its tendency to complex with polyphenols affects the organoleptic properties of tea beverage and the technology of instant-tea manufacture. The properties of caffeine and tea decaffeination processes are not discussed in this article.

Amino Acids

In addition to the normal plant leaf amino acids, theanine (5-N-ethylglutamine) (14) (Fig. 3) is unique to tea and correlates with green-tea quality (15). Several of the other amino acids in the leaf take part in aroma formation during manufacture.

Carotenoids

Carotenoid compounds constitute only 0.03 to 0.06% of tea flush but are important in aroma generation during manufacture. They include β-carotene, lutein, zeaxanthine, and violaxanthine (Fig. 4) (16).

Volatiles

More than 50 volatile compounds have been reported in fresh leaf, although some of these may be artifacts of the isolation procedures (17). Aroma components, their origin, and their significance will be discussed in the manufacturing section as it is during leaf processing that most aroma character is developed.

Lipids, Waxes, and Pigments

The presence of unsaturated fatty acids is also of significance to aroma generation, as will be shown. Linoleic and linolenic acids are present at a level of ca 1.2% in fresh leaf

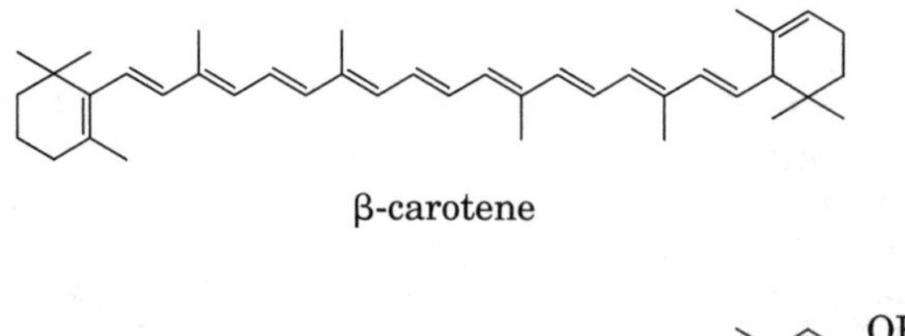

β-carotene

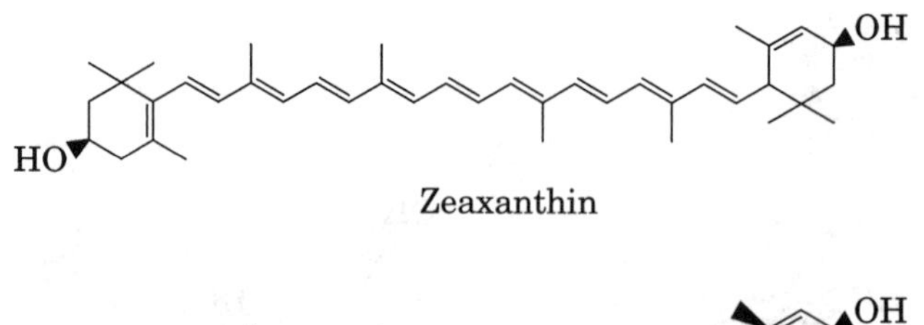

Zeaxanthin

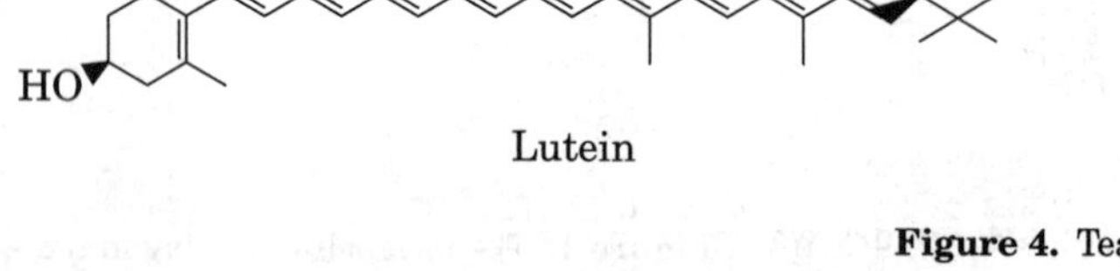

Lutein

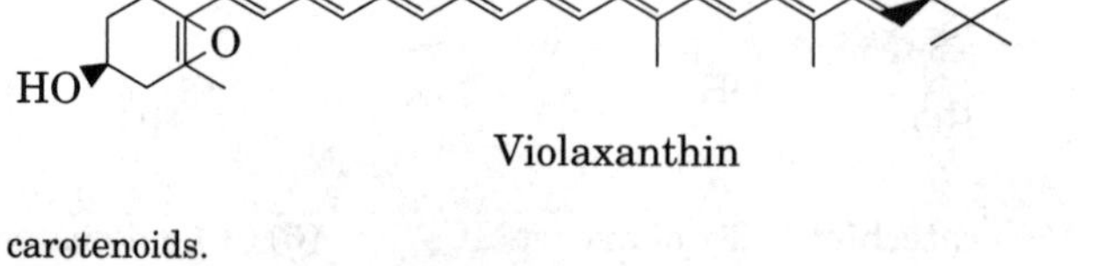

Violaxanthin

Figure 4. Tea carotenoids.

(18). Other hydrophobic compounds such as plant waxes and sterols may be sufficiently extracted to affect beverage appearance (19). Chlorophyll is of significance as a possible precursor of some dark-colored pigments in manufactured tea.

Enzymes

Tea leaf polyphenol oxidase that catalyzes the aerobic oxidation of the catechins during manufacture of black tea is of prime importance. It contains copper (0.32%) and exists as a mixture of several isoenzymes. Its main component has a molecular weight of 142,000 (20). Peroxidase also takes part in the polyphenol oxidation processes (21).

The complex enzyme systems involved in the synthesis of the flavanols and related tea leaf components have been described in two reviews (22,23). Alcohol dehydrogenase (24) and leucine α-ketoglutarate transaminase (25) contribute to the development of aroma components during manufacture. Glycosidases catalyze the breakdown of several glucosides to yield other important aroma components (26).

Minerals

In addition to the common plant minerals, the tea plant is rich in potassium and fluoride and also accumulates aluminum if soil conditions so favor (27).

MANUFACTURE

The manufacturing process converts freshly harvested leaf to products of commerce. Black tea is the most widely consumed form of tea. It is produced by promoting the aerobic oxidation of fresh leaf catechins in reactions catalyzed by tea polyphenol oxidase. Green tea, consumed mostly in China, Japan, the Middle East, and North Africa, is processed so as to prevent the oxidation of catechins. Oolong tea is partially oxidized. It is manufactured and consumed primarily in China and Taiwan.

Black Tea

The most significant change that occurs during the manufacture of black tea is the conversion of the colorless catechins to a complex mixture of yellow-orange to red-brown substances accompanied by the development of a large number of volatile compounds. The process was historically referred to as fermentation, but no microbiological processes take place. The changes result in a dark-colored leaf that produces an amber-colored beverage, less astringent than that derived from fresh leaf. The product possesses an exceedingly complex flavor profile.

Catechin Oxidation

The first step in catechin transformation is oxidation to highly reactive quinones (Fig. 5) (28). All catechins undergo this reaction, but the gallocatechins (3) and (4) are preferentially oxidized. Quinone formation is the primary driving force for the transformation of catechins to black-tea components.

Theaflavins

During fermentation, quinones derived from a simple catechin (1), (2), or (5) react with quinones derived from gallocatechins (3), (4), or (6) to form compounds with a seven-membered benztropolone ring known as theaflavins (Fig. 6). The predicted theaflavins derivable from each of the possible quinone combinations have been isolated (29,30). In solution, theaflavins are bright orange-red in color and provide "brightness," a desirable quality, to the beverage. (Many of the named attributes of tea arise from the tea taster's lexicon and are difficult to define objectively (31).

The distinctively colored theaflavins are determined spectrophotometrically in a tea brew (32). Total theaflavin concentration in black tea does not exceed 2.5% and is sometimes as low as 0.3% (33). Prolongation of the oxidation period and high fermentation temperatures decrease theaflavin content (34). Theaflavins are oxidized by epicatechin quinone, formed late in the oxidation process because of its high oxidation potential (35). Low oxygen tension also results in decreased theaflavin formation (36). At most, only 10% of the original catechin content of tea flush is accounted for as theaflavins in black tea.

Theaflavic Acids

Theaflavic acids are formed by the reaction between quinones derived from (1), (2), or (5) and gallic acid quinone (Fig. 7). It is postulated that gallic acid is formed by degallation during processing. Although it is not directly oxidized by tea polyphenol oxidase, it is converted to a quinone by the catechin quinones that possess a higher oxidation potential, such as that of epicatechin (37,38). The theaflavic acids are bright red, acidic substances present in black tea only in very small quantities because of their reactivity (39). Theaflagallins originate through a similar route but involve the gallocatechins (3), (4), or (6). They are also present in very small quantities (40).

Bisflavanols

Coupling of quinones derived from (3) and (4) produces a series of colorless substances known as bisflavanols (35). They are highly reactive and occur only at low levels in black tea (Fig. 8). The bisflavanols are also known as theasinensins. There have been some reports of small amounts of theasinensins in fresh leaf (12).

Thearubigins

In the course of black-tea formation, ca 10 to 15% of the leaf catechin fraction remains unchanged and ca 10% is accounted for by theaflavin, theaflavic acid, and bisflavanol formation. Seventy five to 80% of the original catechin fraction is converted to a complex mixture, incompletely defined and largely intractable to resolution. This red-brown fraction was named thearubigen when first described in 1957 (42). Investigations of its chemistry over a period of four decades have resulted in conflicting data and conclusions, attesting to the complexity of the mixture. Molecular weight determinations have indicated a range of 700 to 40,000 (43), but this must be considered in context with the known increase in apparent molecular weight

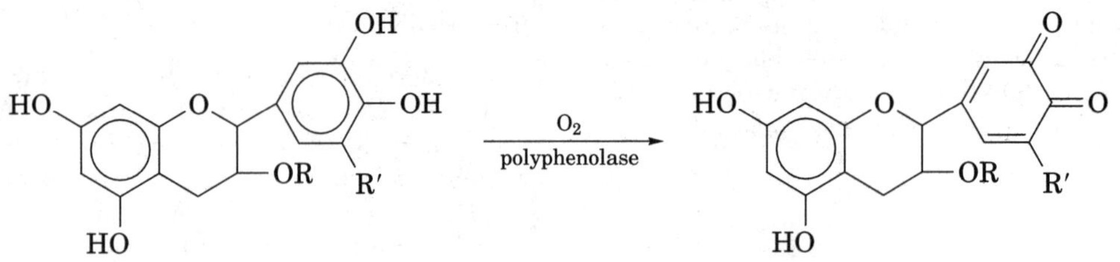

R = H or 3,4,5-trihydroxybenzoyl
R' = H or OH

Figure 5. Enzymic oxidation of catechins.

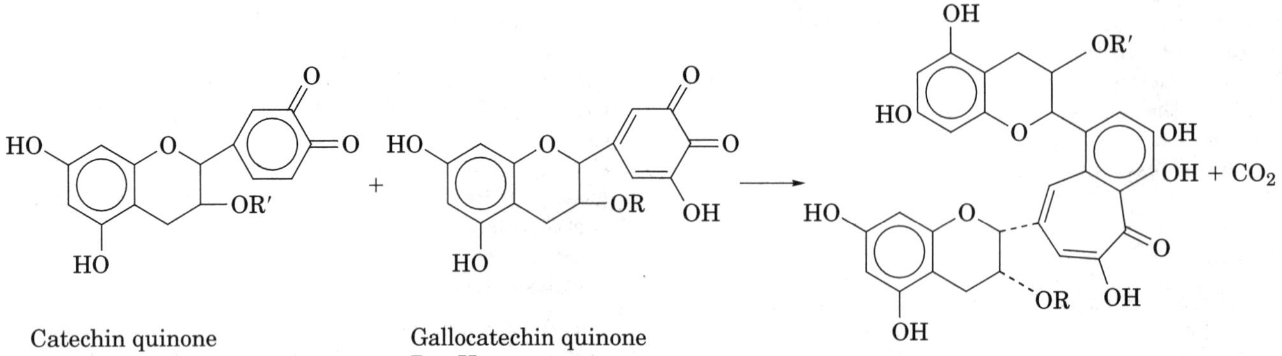

Catechin quinone
R' = H
Catechin gallate quinone
R' = 3,4,5-trihydroxybenzoyl

Gallocatechin quinone
R = H
Gallocatechin gallate quinone
R = 3,4,5-trihydroxybenzoyl

(**15**) Theaflavin; R= R' = H
(**16**) Theaflavin gallate A; R= 3,4,5-
trihydroxybenzoyl, R' = H
(**17**) Theaflavin gallate B; R= H
R' = 3,4,5-trihydroxybenzoyl
(**18**) Theaflavin digallate; R= R' = 3,4,5-
trihydroxybenzoyl

Figure 6. Formation of theaflavins.

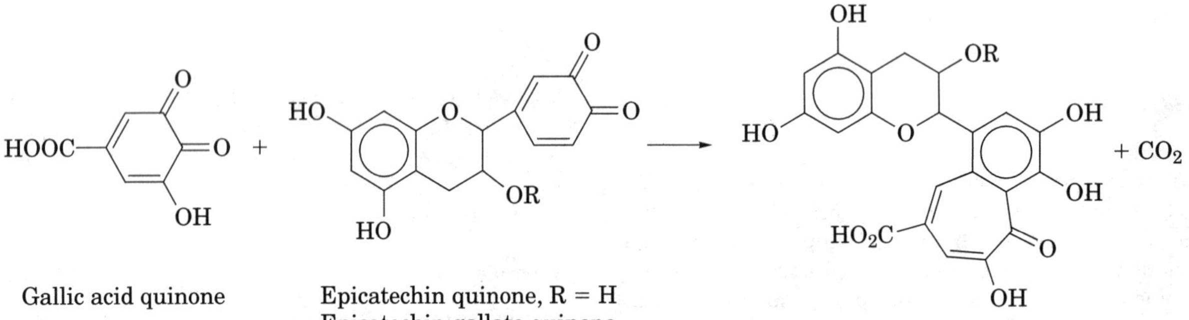

Gallic acid quinone

Epicatechin quinone, R = H
Epicatechin gallate quinone
R = 3,4,5-trihydroxybenzoyl

(**19**) Epitheaflavic acid; R = H
(**20**) Epitheaflavic acid-3'-gallate;
R = 3,4,5-trihydroxybenzoyl

Figure 7. Formation of theaflavic acids.

that occurs when tea infusions are allowed to stand. Polyphenolic compounds in this molecular weight range are active complexing agents and are therefore likely to be present in combination with other substances. Flavonol glycosides have been shown to constitute one group of compounds that appear as part of the thearubigin fraction as originally designated, but, as colorless compounds, they are not usefully regarded as such. New high-performance liquid chromatography (HPLC) techniques have allowed for the isolation of several other components of this mass.

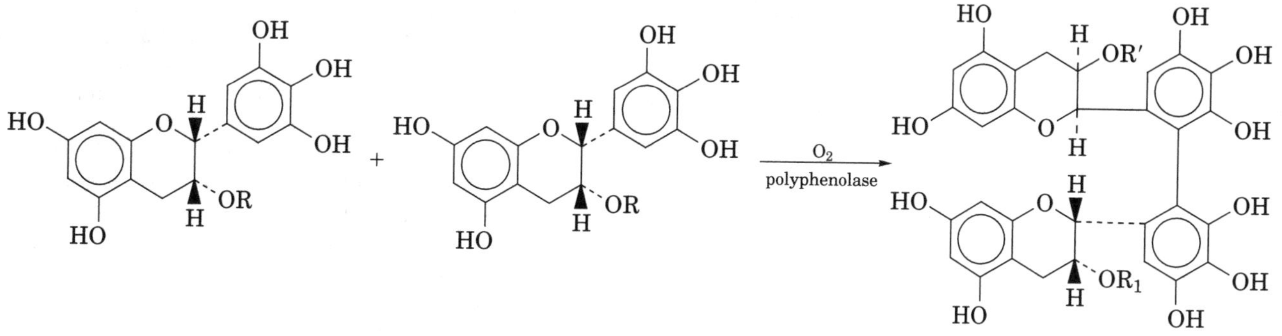

(21) Bisflavanol A; R = R′ = 3,4,5-
    trihydroxybenzoyl
(22) Bisflavanol B; R = 3,4,5-tri-
    hydroxybenzoyl; R′ = H
(23) Bisflavanol C; R = R′ = H

Figure 8. Formation of bisflavanols.

A fraction designated as theafulvin has been separated but not resolved. It accounts for only a small proportion of the thearubigin fraction. It is a buff-colored material containing some proanthocyanidin groups, but most of its mass is not completely characterized (44). Theacitrins have also been isolated. They are red substances, similar to the theaflavins but possibly containing additional carboxylic acid groups. They are also present in very small quantities (45,46). Separation techniques have not yet provided methodology for the resolution of the major portion of the thearubigin mixture.

Degradative treatments of the unresolved material have been helpful in piecing together more information. Acid hydrolysis results in the formation of cyanidin and delphinidin (Fig. 9), suggesting proanthocyanidin structures (47). Other degradative procedures have provided evidence for the presence of heterogeneous polymers of flavanols and flavanol gallates (48).

Decreases in theaflavins and bisflavanols during fermentation suggest that these compounds are further oxidized and subsequently incorporated into the thearubigin complex. It is also likely that catechin quinones oxidize (8), (10), and glycosides of (11), (12), and (13), leading to incorporation of their oxidation products as well (29). In ad-

dition, minor amounts of hydrogen peroxide generated during fermentation may be activated by the peroxidase enzyme and bring about further oxidative condensations that augment the thearubigin fraction (49). Model system studies in which single catechins and known catechin mixtures are enzymatically oxidized *in vitro* have shown some promise for further structure elucidation (50).

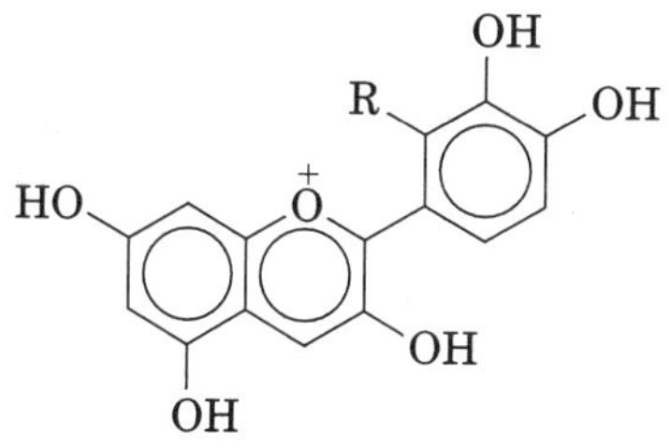

This fraction constitutes the largest group of compounds in black tea (up to 20% of dry weight) and contributes significantly to color, strength, and mouth-feel of the beverage, but is its least well defined. It has been suggested that thearubigin may not be a useful term for the chemical characterization of this tea fraction because of its heterogeneity and the more recently observed facts that some unchanged fresh-leaf components are included (13). It does refer, however, to a group of substances that have significance to the tea taster with regard to beverage organoleptic properties.

BLACK TEA

Black-tea manufacture is carried out in several stages known as withering, rolling, fermenting, firing, and sorting. An understanding of the changes that take place during these processes has been gradually accumulated since the 1930s.

Withering

Plucked leaf arrives at the estate factory within a few hours of its harvest. It is handled carefully to prevent premature bruising and to promote dissipation of heat generated during continued respiration. Moisture level is reduced from 75–80% to 55–65% without heat buildup, usually in troughs designed for effective airflow through leaf beds up to 30 cm thick. Moisture reduction converts the turgid leaf to a flaccid material that is easily handled without excessive fracture. The process takes place over a period of 6 to 18 h, depending on factory equipment and

(24) Cyanidin; R = H
(25) Delphinidin; R = OH

Figure 9. Anthocyanidins from tea thearubigins.

weather conditions. Important chemical changes occur during withering: cell membrane permeability is increased, which allows for more efficient disruption of cell structure during the next stage of manufacture; amino acid and organic acid levels increase; and polyphenol oxidase activity is enhanced. These changes have been referred to as chemical wither. Degree of physical wither (moisture loss) is estimated by leaf texture or by checking the weight loss of an isolated leaf batch. The withering process affects final product value (51).

Rolling

Rolling serves to establish proper conditions for the enzymic oxidation of the flavanols by atmospheric oxygen. This is accomplished by disrupting cell structure to effect enzyme–substrate contact. The leaf mass must also be maintained in a physical state conducive to oxygen diffusion. Rolling processes are designed to meet these requirements. The conventional roller consists of a circular platform, 1 to 1.3 m in diameter, equipped with battens. Above this, a smaller circular sleeve moves eccentrically over the surface of the platform. Withered leaf is charged into the sleeve and a pressure cap is lowered into it. The rates of movement and the pressure applied are chosen in accordance with leaf properties and atmospheric conditions. Intermittently, sufficiently rolled leaf (dhool) is removed by sieving and the remainder is rerolled. Leaf juices become well distributed on the surface of the cut, twisted leaf. Teas manufactured by this process are known as orthodox. They include almost all Ceylon teas.

More modern equipment has come into use in most of the other tea-growing areas. The McTear Rotorvane consists of a cylinder 20 or 37.5 cm in diameter, lined with vanes, through which a vaned rotorshaft propels the withered leaf. Constriction at the end applies back pressure and increases the work done on the leaf. Cutter blades are sometimes added to the shaft. From the Rotorvane, tea is usually passed through a CTC (crush, tear, curl) machine, which consists of a pair of ridged cylindrical rollers revolving at different speeds. There is a narrow, adjustable clearance between the rollers. The Rotorvane-CTC combination provides efficient and uniform maceration and is well suited to the establishment of a continuous manufacturing process. The macerated leaf is fluffed up by passage through a high-speed Rotorvane, with no back pressure, to eliminate matting and facilitate air diffusion. It is desirable to maintain leaf temperature below 35°C during maceration to prevent flavor deterioration. The Legg Cutter, originally designed for processing tobacco, is used in India to process unwithered or lightly withered leaf. The Lawrie Tea Processor (LTP) is essentially a hammer mill. It is used in Africa for lightly withered leaf.

The final physical state of the leaf is dependent on the maceration conditions employed and affects the course of the oxidation reactions. This, in turn, affects the chemical composition and some aspects of the quality of the final product, especially the theaflavin-to-thearubigin ratio (52).

Fermentation

The term fermentation arose from the erroneous concept of black-tea production as a microbial process (53). Around 1901 the conversion of green tea to black tea became recognized as an oxidative process initiated by tea enzymes (54). The process actually starts at the onset of maceration and is allowed to continue under ambient conditions in most instances, but leaf temperatures are preferably controlled so as not to exceed 25 to 30°C. Lower temperatures (15–25°C) improve flavor (34). Temperature control and air diffusion are facilitated by spreading the macerated leaf in layers 5 to 8 cm thick on racked trays in a fermentation room maintained at high humidity and at the lowest feasible temperature. Depending on the nature of the leaf, maceration technique, ambient temperature, and the style of tea desired, the fermentation time ranges from 45 min to 3 h, although the shorter process times are more common. Completion of oxidation is usually judged by the development of the characteristic coppery color and the fermented tea aroma. More highly controlled fermentation systems have been described. They depend on the time-controlled conveyance of rolled leaf on mesh belts through which there is forced air circulation. In a few factories the fermentation belt is part of a closed system that allows for control of temperature, aeration, and humidity (55).

Firing

Fermentation is terminated by firing in large ovens with hot forced-air circulation. In most driers, loaded trays traverse the drying zone countercurrent to the airflow. Incoming air temperature is ca 90°C and emergent air is ca 50°C. Time-temperature profiles are controlled to effect drying in 18 to 20 min with little energy wastage. Fluid-bed driers are sometimes utilized (56). During the firing process leaf moisture is reduced to 2 to 3%. Proper drying control is crucial to product quality. Many organochemical processes are accelerated during this period, as are the enzymic reactions before total thermal inactivation occurs. Incomplete enzyme inactivation can cause accelerated deterioration during storage. One noticeable effect of firing is the color change brought about by the conversion of chlorophyll to pheophytins and pheophorbide, which are dark brown and black, respectively. Small amounts of caffeine are lost through sublimation. A high proportion of the aroma of black tea is generated during the firing process.

Grading and Packaging

Fired tea is graded by the use of a series of oscillating screens. The grades most frequently produced, in descending particle size order, are orange pekoe (OP), pekoe (P), broken orange pekoe (BOP), broken orange pekoe fannings (BOPF), fannings, and dust. Fine dust, debris, and some fiber are removed by winnowing. Stalky materials tend to retain moisture and may be removed from the dryer tea fractions by use of an electrostatic sorter. It is essential to protect tea from heat, moisture, and light to insure reasonable shelf life. Color, taste, and aroma deteriorate if poor storage conditions prevail. Theaflavin decrease is marked (57). Aroma quality also deteriorates. Multilayered natural Kraft bags lined with aluminum foil have come into general use in shipment.

Blending

Most branded black tea is blended by the packer. This is essential to insure constancy of taste, appearance, and, to the greatest extent possible, cost. Seasonal, climatic, and market changes would make a constant product impossible to achieve unless the tea blender had access to many different teas for use in a single blend. This allows for necessary substitutions without departure from product image. The special talents and experience of the tea taster are still required. Instrumental analysis is not yet adequate to define and control all of the attributes important to the consumer.

Aroma Generation

Black-tea aroma is a complex system in which well over 500 components have been identified (58). The continuing improvement of analytical techniques results in the proliferation of known tea volatiles, for example, one research paper reported the identification of 133 new components, all of which are present in quantities <10 μg/kg of beverage (59). Aroma ingredients may originate either as components of the fresh leaf, as products of enzymic or organochemical activity during the manufacturing processes, or as products of the high-temperature firing step. Almost all aroma components develop during postharvest processes. It is probable that only a few dozen are significant to tea aroma quality.

Aldehydes are generated in fresh leaf by the oxidative breakdown of unsaturated fatty acids catalyzed by lipoxygenases. The short-chain unsaturated aldehydes are generally deleterious to tea flavor. Leaf alcohol dehydrogenase systems catalyze their partial reduction to alcohols that exhibit a characteristic green leaf flavor not desirable at high levels (60). The more volatile of these compounds are partially lost during the firing stage. Aldehydes are also formed during black-tea manufacture by a Strecker-type synthesis resulting from the reaction of catechin quinones with amino acids. Phenylacetaldehyde is formed from phenylalanine in this manner (61).

Terpene alcohols, phenylethyl alcohol, methyl salicylate, and other important aroma components are generated by enzymic hydrolysis of their glycosides (62). Their reported presence in fresh leaf may be partially the result of initiating hydrolysis during analytical procedures. These compounds contribute significantly to the character of tea aroma. Several aromatic compounds present in fine teas are formed by the oxidation of the tea carotenoids by catechin quinones. These include cis-jasmone, theaspirone, β-ionone, and dihydroactinodiolide (61). Carotenoids are also thermally decomposed to produce these and other aroma components during firing.

Other thermal reactions are responsible for the development of a large group of cyclic aroma components. These include pyrazines, pyrroles, pyridines, and quinolines (63). Their significance is not known. Some compounds that impart the most desirable flavor characteristics to black tea are shown in Table 2. Tea tasters use the term quality to refer to tea with desirable aroma character. Black-tea aroma also contains the group of short-chain (5–7 carbon atoms) aldehydes and alcohols that arise from lipid oxi-

Table 2. Desirable Black-Tea Aroma Components

Benzyl alcohol
Dihydroactinodiolide
Geraniol
cis-3-hexenyl hexanoate
β-ionone
cis-jasmone
Linalool
Linalool oxides
Methyl salicylate
Nerolidol
Phenylacetaldehyde
2-phenylethanol
1-terpineol
Theaspirone

dation. An abundance of these substances is undesirable. Aroma quality is determined by the relative quantities of the two groups of components (58,64).

BLACK TEA BEVERAGE

Preparation and Properties

Black-tea beverage is characterized by its amber color and distinctive aroma. In the United States it is generally prepared from tea bags containing 2.27 g of tea (200 servings/lb). Tea bag paper must be sufficiently permeable to provide rapid infusion but also retain small leaf particles. It should not impart taste to the beverage, it must be consistent in weight and thickness to insure machinability, and it must have adequate dry and wet strength. Abaca fiber obtained from a plant grown in Ecuador and the Philippines is commonly included. Tea bagging machines are able to produce up to 1,000 bags/min.

Infusing tea bags with boiling water in a 6- to 8-oz cup for 3 min provides a beverage containing 0.25 to 0.35% solids. (The use of packaged tea in the United States has declined greatly and now accounts for only approximately 3% of retail sales). Beverage concentration is highly dependent on extraction time, temperature, and water-to-tea ratio. Caffeine is extracted proportionally with the other soluble components. Excessively hard water produces a murky beverage. Black-tea beverage color is one of the useful indicators of value in that it reflects the quality of the cultivar, the success of the manufacturing procedure, and the control of storage conditions. Desirable tea beverage color requires the use of fresh leaf that contains a sufficient level of polyphenols and oxidizing enzyme held under conditions favorable to achieving adequate oxidation, and proper theaflavin-to-thearubigin ratios. Inferior aroma may result from suboptimal manufacturing or storage conditions. Brewed tea beverage is relatively unstable. Storage for more than 30 to 60 min results in noticeable flavor deterioration. Theaflavin levels decrease during long-term holding at elevated temperatures (65).

When black-tea beverages are allowed to cool, cloudiness slowly develops, and eventually a precipitate becomes visible. The tea taster refers to this as cream and considers it a normal characteristic of most acceptable teas. The composition of cream and its significance to instant tea manufacture will be discussed under that topic.

Composition

The composition of a characteristic black-tea beverage is shown in Table 3.

GREEN TEA

The primary objective in green-tea manufacture is to prevent catechin oxidation to the greatest extent possible while imparting an attractive, twisted appearance to the leaf and developing the characteristic green tea aroma. There are many styles of green tea. In Japan sencha is the most extensively drunk. Gyokuro, grown under shade, is the most highly esteemed. Gunpowder tea is a pelleted green tea consumed in China.

Enzyme Inactivation

The leaf enzymes may be inactivated either by steaming, most often carried out in Japan, or by panning, the procedure of choice in China. The pan, however, has been superseded by 5-m-long, open, rotating cylinders that are heated during the passage of the tea over a 7- to 10-min period. Catechin oxidation is prevented and the leaf retains its green color and almost all of its original polyphenol content.

Rolling

The primary function of the rolling process in green-tea manufacture is to impart the desirable twisted appearance to the leaf. Lighter equipment is used than that required for black-tea production as small-leaved varieties are gen-

erally utilized, and there is no need for cell disruption and oxygen permeability.

Firing and Aroma Development

Green teas are often fired in horizontal rotating drums. The chemical changes that take place are primarily responsible for the development of aroma (67).

Beverage Composition

The composition of a representative green tea beverage is shown in Table 4.

Approximately 300 components of green-tea aroma have been identified. The most important and a representative aroma composition is shown in Table 5.

OOLONG TEA

Oolong teas are only partially oxidized and retain a considerable amount of the original polyphenolic material so

Table 3. Composition of Black-Tea Beverage (Weight % of Extract Solids)

Thearubigins	36.0
Theaflavins	2.6
Theaflavic acids	Trace
Bisflavanols	Trace
Catechins	11.2
Flavonol glycosides	Trace
Gallic acid	1.1
Chlorogenic acid	0.2
Caffeine	7.6
Theobromine	0.7
Oxalic acid	1.5
Other aliphatic acids	1.2
Lipids and waxes	4.8
Sugars	6.9
Polysaccharides	4.2
Pectin	0.2
Peptides	6.0
Theanine	3.5
Other amino acids	3.0
Potassium	4.8
Other minerals	4.7
Aroma	<0.1

Source: Ref. 66.

Table 4. Composition of Green-Tea Beverage (Weight % Solids)

Epigallocatechin gallate	20.3
Epicatechin gallate	5.2
Epigallocatechin	8.4
Epicatechin	2.0
Flavonols	2.2
Caffeine	7.4
Theanine	4.7
Glutamic acid	0.5
Aspartic acid	0.5
Argenine	0.7
Other amino acids	0.8
Sugars	6.7
Alcohol precipitatable material	12.2
Potassium	4.0
Volatiles	<0.1

Source: Ref. 68.

Table 5. Significant Green-Tea Aroma Components (% Total in a Representative Sample)

Geraniol	17.9
Linalool oxides	16.0
Linalool	9.5
Nerolidol	8.8
cis-jasmone	7.5
2,6,6,trimethyl-2-hydroxycyclohexane-1-one	7.0
β-ionone	5.5
Benzyl alcohol	4.7
cis-3-hexenyl hexanoate	3.5
5,6-epoxy-β-ionone	2.7
1-pentene-3-al	2.7
1-terpineol	2.2
cis-3-hexene-1-ol	2.0
Acetylpyrrole	1.8
2-phenylethanol	1.3
cis-pentene-1-ol	1.1
Pentane-1-ol	0.7

Source: Ref. 69.

there is little or no cream formation from beverage-strength extracts.

Manufacture is most often a cottage industry involving a series of withering, gentle rolling, and drying steps what vary greatly from facility to facility. Sun drying is often utilized as the first step. Appearance of leaf is considered an important aspect of quality. Hand labor is often utilized.

Composition of Oolong Tea

The color of oolong tea is intermediate between that of green and black teas, and to some extent its composition reflects a similar relationship. Residual catechin levels vary with the intensity and duration of the oxidation processes. There is little or no theaflavin or thearubigin. It is interesting to note, however, that there are also substances present that have not been found in either green or black teas. These may represent intermediates on the oxidative pathways that would lead to theaflavins or thearubigins on further oxidation or end products only formed under the gentle oxidation conditions prevailing because of less-severe cell disruption. These conditions result in lower quinone concentrations that may favor different oxidative pathways.

Components of oolong teas include several theasinensins and their oxidation products, known as oolongtheanins and oolonghomobisflavans (methylene-bridged bisflavanols) (70,71).

WORLD PRODUCTION AND TEA TRADE

Total world tea production has increased over the last few decades largely in proportion to population growth, but not uniformly. Increases have been greatest where investment and technology have been encouraged, such as in Kenya and north India, and declining where political and economic stability are minimal, such as in Sri Lanka and Georgia. Table 6 is a compilation of tea-production data for the major producing countries in 1997.

Tea consumption throughout the world is far from uniform. In Europe it ranges from ca 3 kg per capita annually in the United Kingdom and Ireland to less than 0.1 kg in Italy. Similar variations occur in the other continents. U.S. consumption is 0.36 kg; in Canada it is 0.68 kg. Tea consumption has been slowly declining in the traditional Anglo-Saxon markets where the popularity of coffee and soft drinks has been increasing. It has increased markedly in India as the economic situation has improved.

Importation of tea into the United States has remained fairly stable on a per capita basis over the period from 1938 to 1998. Product origin has changed considerably. China, Indonesia, Argentina, Sri Lanka, Kenya, and India are now the major suppliers to the U.S. tea market (1). Tea is imported into the United States without duty. Standards established by the International Standards Organization (ISO) and Codex are intended to function as guidelines for teas in international trade. These include extractable solids (32% minimum), ash (4–8%), and crude fiber (16.5% maximum). The real price of tea has tended to decline with only infrequent increases occasioned by temporary shortages. This has caused problems for producing countries,

Table 6. World Tea Production in 1997 (1000 t dry leaf)

India	811
China	613
Sri Lanka	277
Indonesia	131
Turkey	110
Japan	91
Iran	60
Bangladesh	53
Other Asia	73
Total Asia	*2,219*
Kenya	221
Malawi	44
Tanzania	22
Uganda	21
Zimbabwe	17
Other Africa	36
Total Africa	*361*
Argentina	55
Other S. America	14
Total S. America	*69*
Other	18
Total world	*2,667 (ca 600 is green tea)*

Source: Ref. 1.

many of which depend on tea as a major source of foreign currency. Efforts to control tea prices through international agreements among the producing countries have never been successful.

The consumption pattern has also undergone large changes. Iced tea use in the United States has grown from a very small proportion of the total to approximately 80%. This factor has contributed greatly to the importance of convenience products based on cold-water-soluble instant teas and ready-to-drink (RTD) beverages.

INSTANT TEA

Instant-tea powder is generally prepared by the aqueous extraction of black tea followed by concentration and drying. It is also possible to prepare instant tea from green tea or from oxidized leaf prior to the drying step. As most tea consumed in the United States is drunk as a cold beverage, the cold-water insolubility of the important cream fraction introduces technological problems in manufacture. A product prepared simply by extraction with hot water, concentration, and drying will not rehydrate completely without applied heat or the use of additives. Cold-water extraction does not recover sufficient solids to meet organoleptic or economic criteria. Most instant-tea technology is concerned with ways to resolve this problem. Tea cream, the fraction that precipitates when a black-tea solution is cooled, constitutes 15 to 30% of the original extract solids. The quantity varies with the type of tea, the extract concentration, and the temperature to which the extract is cooled. Beverage color, caffeine content, and taste are disproportionately concentrated in the cream fraction. Tea cream is a complex of the oxidized polyphenolic material with caffeine along with other occluded substances. The aggregates that develop on standing include other components of tea extract, such as proteinaceous material, lipids, and waxes. Freshly precipitated cream can be resolubilized by heating. The first instant teas that appeared on the U.S. market were prepared by discarding the cream after cooling the hot extract. The cream problem was solved in many ways as indicated by a copious patent literature. Most approaches are based on the disruption of the caffeine–tannin complexes.

Manufacture

Tea is extracted with hot water in a countercurrent mode, either batchwise in kettles or in columns. Continuous extraction systems have been described (72). Aroma fractions are collected by vacuum stripping and condensation or by separation and preservation of the first column effluents. They are retained for later addition to the product stream. Vacuum concentration to the desired solids level for adequate cream separation may then be carried out. Prolonged storage of extracts at elevated temperatures is avoided to prevent oxidative and microbiological deterioration. Tea cream is solubilized or avoided by several different procedures. Treatment of cream with the enzyme tannase results in the selective hydrolysis of gallate linkages, thereby increasing solubility (73). Similarly, it is possible to treat black tea with tannase to produce a coldwater-extractable product (74). The addition of unoxidized (green tea) catechins to black-tea extract prevents the formation of insoluble substances by competitive complexation (75).

Several oxidative procedures for the conversion of green-tea extracts to black tea have been described. These include the use of oxygen, ozone, or hydrogen peroxide (76–78). The reactions are carried out in an alkaline medium so that oxidation and hydrolysis occur simultaneously. These oxidized green-tea extracts are useful in instant-tea manufacture because color and solubility control are easily effected by their addition to the black-tea extract stream.

If cream has been solubilized apart from the remainder of the extract, the fractions are recombined and the total extract is brought to the concentration desired for drying. Vacuum or freeze concentration may be utilized. Concentration by reverse osmosis has also been described as useful for flavor conservation (79). Retained aroma fractions, often concentrated, are added to the drier feed.

Most instant tea is spray-dried. Pectins extracted from tea or other sources have been used to control final product density (80). Inflation of particles with carbon dioxide during spray-drying controls density and wall thickness. Both parameters affect product solubility. Instant tea may also be freeze-dried. The treatment of spent tea leaf with a β-glucanase or cellulase to hydrolyze leaf components adds to the yield of extract solids and decreases the disposal load. The additional use of proteases enhances this effect. The extraction of fermented green leaf, omitting the firing stage, is an attractive process for use in tea-growing countries to conserve energy. Minor amounts of instant tea are prepared for hot beverage consumption. Processing is simpler and taste is more authentic.

Product Characteristics

Instant-tea powders are usually manufactured with a density of ca 0.1 g/mL. Products are soluble in cold tap water and produce a clear solution with an authentic tea color and taste. Instant-tea mixes that contain sugar or a noncaloric sweetener and various fruit flavors are widely used. Canned iced tea beverages containing sweeteners and flavors are also based on instant tea. Decaffeinated instant tea has also come into use. Most instant tea is manufactured and consumed in the United States, where it accounts for ca 15% of total tea usage. Production in 1998 was ca 2,500 t. Kenya, India, and Sri Lanka together manufacture ca 5,000 t, more than half of which is exported to the United States (1).

READY-TO-DRINK (RTD) TEA

RTD teas represent the fastest-growing mode of tea consumption in the United States. In 1997 ca 1 billion L was produced. They are usually presented as a sweetened, flavored, slightly acidified beverage. Taste stability and maintenance of clarity under refrigeration pose major technical problems in their manufacture. The addition of high methoxy pectin to an acidified, flavored tea extract at normal beverage strength (ca 0.3%) results in a clear, coldwater-stable product (81). Gum arabic provides the same benefits (82).

PHYSIOLOGICAL EFFECTS

Soon after the introduction of tea into England, extreme positive and negative effects were attributed to it. A London teahouse advertisement stated that "Tea removeth lassitude, vanquisheth heavy dreams, easeth the frame, and strengtheneth the memory. It overcometh superfluous sleep, and prevents sleepiness in general, so that without trouble whole nights may be passed in study." Though some doctors extolled its curative powers, others attacked tea for causing "palsies, impotence, leanness, sterility, and nervousness" (83).

Tea is now considered to be a healthful beverage. It is a useful source of fluoride and potassium. Its vitamin content is too low to be significant except for heavy users (5 cups per day), who may obtain 5 to 10% of the RDA for riboflavin, niacin, folic acid, and pantothenic acid.

Catechins and their oxidation products have been shown to be effective antioxidants. They exhibit significant quenching of active oxygen species such as hydroxyl and peroxyl radicals as well as more stable free radicals. In lipophilic systems they inhibit the free radical–mediated oxidation of low-density lipoproteins (LDL). Active oxygen radicals are known to play a role in mutagenesis and chemical carcinogenesis. Oxidized LDL is believed to promote atherosclerotic lesions. Tea has therefore been studied for its possible usefulness in preventing these developments. It has exhibited protective effects on laboratory animals treated with mutagenic, carcinogenic, or atherogenic agents.

Prior administration of green-tea extracts inhibits the chromosome aberrations normally induced by aflatoxin B_1

in rat bone marrow cells (84). Green-tea polyphenol consumption exhibits protective effects for rodents treated with carcinogens (85). Topical application of green-tea polyphenols is protective against the initiation of skin tumors in mice treated with 3-methylcholanthrene (86). Though most of these studies have been conducted with green tea or its components, black tea also exhibits protective effects in biological systems (87). Extension of these studies to humans (primarily with epidemiological approaches) has not produced definitive results, partly due to the confounding factors inherent in studies of this type. Epidemiological studies carried out in different parts of the world are difficult to compare and evaluate because of dietary differences. There is often a lack of quantification concerning the tea solids consumed. In a few such investigations, however, some evidence for its anticarcinogenic effect for humans has been obtained (88).

Green tea and green-tea polyphenols lower serum cholesterol levels of rats maintained on a cholesterogenic diet (89). A strong negative correlation has been observed between green-tea consumption and the onset of stroke in a large Japanese population (90).

Tea may be inhibitory to the absorption of iron. Borderline iron deficiencies may be accentuated by high levels of intake (91). A similar effect occurs with thiamine (92). Reports linking tea consumption with esophageal cancer have been based on faulty epidemiological studies: In some areas where the condition is rampant and where tea consumption is high, such as in parts of China, dietary contamination with carcinogenic molds and nitrosamines is widespread (93); in an area of Iran, beverages are commonly consumed scaldingly hot—a practice in itself conducive to esophageal cancer.

Confusion of herbal teas with *Camellia sinensis* and tannic acid with tea components has led to further misunderstanding concerning the effects of tea consumption.

ACKNOWLEDGMENTS

The author is grateful for the considerate and expert assistance provided by Terris Binder and Diane Cassatly of the Lipton Library and Information Services.

BIBLIOGRAPHY

1. *Annual Bulletin of Tea Statistics*, International Tea Committee, London, 1998.
2. G. F. Mitchell, *The Cultivation and Manufacture of Tea in the United States*, Bulletin No. 234, USDA Bureau of Plant Industry, Washington, D.C., 1912.
3. S. F. McLester, *Tea and Coffee Trade Journal* 169, 78–81 (1997).
4. P. V. Arulpragasam, R. Latiff, and P. Seneviratne, "Studies on the Tissue Culture of Tea," *Sri Lanka Journal of Tea Science* 1, 20–23 (1988).
5. *Annual Reports*, Tea Research Institute of East Africa, Kericho, Kenya.
6. *Tea Quarterly*, Journal of the Tea Research Institute of Sri Lanka, Talawakelle, Sri Lanka.
7. *Two and a Bud*, Tocklai Experimental Station, Jorhat, Assam, India.
8. *Annual Reports*, United Planters Association of South India, Cinchona P.O., Coimbatore District, South India.
9. B. Banerjee, "Revised List of Approved Pesticides," *Two and a Bud* 27, 80–84 (1980).
10. D. A. Balentine, M. E. Harbowy, and H. N. Graham, "Tea: The Plant and Its Manufacture, Chemistry, and Consumption of the Beverage," in G. Spiller, ed., *Caffeine*, CRC Press, New York, 1998, pp. 35–72.
11. U. H. Engelhardt, A. Finger, and S. Kuhr, "Determination of Flavone C-glycosides in Tea," *Z. Lebensm.-Unters.-Forsch.* 197, 239–244 (1993).
12. F. Hashimoto, G. Nonaka, and I. Nishioka, "Tannins and Related Compounds: LXXVII," *Chem. Pharm. Bull.* 37, 77–85 (1989).
13. M. E. Harbowy and D. A. Balentine, "Tea Chemistry," *Critical Reviews in Plant Science* 16, 415–478 (1997).
14. S. A. Wiseman, D. A. Balentine, and B. Frei, "Antioxidants in Tea," *Critical Rev. Food Sci. Nutr.* 37, 705–718 (1997).
15. R. A. Cartwright, E. A. H. Roberts, and D. J. Wood, "Theanine, an Amino Acid N-Ethylamide, Present in Tea," *J. Sci. Food Agric.* 5, 597–599 (1954).
16. S. Venkatakrishna, B. R. Premachandra, and H. R. Cama, "Distribution of Carotenoid Pigments in Tea Leaves," *Tea Quarterly* 47, 28–31 (1977).
17. H. Yingfang et al., "Study on the Chemical Constituents of the Volatile Oils from the Fresh Leaves of *Camellia sinensis*," *Acta Botanica Sinica* 24, 440–450 (1982).
18. A. C. Liyanage, M. J. de Silva, and A. Ekanayake, "Analysis of Major Fatty Acids in Tea," *Sri Lanka Journal of Tea Science* 57, 46–49 (1988).
19. R. Seshadri and N. Dhanaraj, "New Hydrophobic Lipid Interactions in Tea Cream," *J. Sci. Food Agric.* 45, 79–86 (1988).
20. S. R. Whitaker, "Polyphenol Oxidase," *Food Sci. Technol.* 61, 543–556 (1994).
21. A. Finger, "*In vitro* Studies on the Effect of Polyphenol Oxidase and Peroxidase on the Formation of Polyphenolic Black Tea Constituents," *J. Sci. Food Agric.* 66, 293–305 (1994).
22. D. A. Balentine, "Tea," in *Kirk-Othmer Encyclopedia of Chemical Technology*, 4th edition, John Wiley & Sons, New York, 1997, pp. 747–768.
23. W. Heller and G. Forkmann, "Biosynthesis of Flavonoids," in J. B. Harborne, ed., *The Flavonoids: Advances in Research since 1986*, Chapman & Hall, London, 1994, pp. 495–535.
24. A. Hatanaka and T. Harada, "Formation of *cis*-3-Hexenal, *trans*-2-Hexenal, and *cis*-3-Hexenol in Macerated *Thea sinensis* Leaves," *Phytochemistry* 12, 2341–2346 (1973).
25. R. L. Wickremasinghe, "The Mechanism of Action of Climatic Factors in the Biogenesis of Tea Flavor," *Phytochemistry* 13, 2057–2063 (1974).
26. K. Morita et al., "Aglycone Constituents in Fresh Tea Leaves Cultivated for Green and Black Tea," *Biosci. Biotechnol. Biochem.* 58, 687–690 (1994).
27. T. Nagata, M. Hayatsu, and N. Kosuge, "Aluminum Kinetics in the Tea Pland Using ^{27}Al and ^{19}F NMR," *Phytochemistry* 32, 771–775 (1993).
28. G. W. Sanderson, "The Chemistry of Tea and Tea Manufacturing," in V. C. Runeckles, ed., *Structural and Functional Aspects of Phytochemistry*, Academic Press, Orlando, Fla., 1972, pp. 247–316.
29. Y. Takino, "Reddish Orange Pigments of Black Tea: Structure and Oxidative Formation from Catechins," *Japanese Agric. Res. Quarterly* 12, 94 (1978).
30. D. J. Cattell and H. E. Nursten, "Separation of Thearubigens on Sephadex LH-20," *Phytochemistry* 16, 1269–1272 (1977).

31. International Organizations for Standards, *Black Tea Vocabulary*, International Standards Organization, London, 1982.

32. D. L. Whitehead and C. M. J. Temple, "Rapid Method for Measuring Thearubigins and Theaflavins in Black Tea Using C_{18} Sorbent Cartridges," *J. Sci. Food Agric.* **58**, 149–152 (1992).

33. E. A. H. Roberts and R. F. Smith, "The Phenolic Substances of Manufactured Tea: IX, The Spectrophotometric Evaluation of Tea Liquors," *J. Sci. Food Agric.* **14**, 689–700 (1963).

34. J. B. Cloughley, "The Effect of Fermentation Temperature on the Quality Parameters and Price Evaluation of Central African Black Teas," *J. Sci. Food Agric.* **31**, 911–919 (1980).

35. K. L. Bajaj et al., "Effects of (−)-Epicatechin on Oxidation of Theaflavins by Polyphenol Oxidase from Tea Leaves," *Agric. Biol. Chem.* **51**, 1767–1772 (1987).

36. A. Robertson, "Effects of Physical and Chemical Conditions on the *in vitro* Oxidation of Tea Leaf Catechins," *Phytochemistry* **22**, 889–896 (1983).

37. J. E. Berkowitz, P. Coggon, and G. W. Sanderson, "Formation of Epitheaflavic Acid and Its Transformation to Thearubigens during Tea Fermentation," *Phytochemistry* **10**, 2271–2278 (1971).

38. D. T. Coxon, A. Holmes, and W. D. Ollis, "Theaflavic and Epitheaflavic Acids," *Tetrahedron Lett.* **60**, 5247–5250 (1970).

39. T. Bryce et al., "Three New Theaflavins from Black Tea," *Tetrahedron Lett.* **72**, 463–466 (1972).

40. G. Nonaka, F. Hashimuto, and I. Nishioka, "Tannins and Related Compounds: XXXVI, Isolation and Structures of Theaflagallins, New Red Pigments from Black Tea," *Chem. Pharm. Bull.* **34**, 61–65 (1986).

41. G. Nonaka, O. Kawahara, and I. Nishioka, "Tannins and Related Compounds: XV, A New Class of Dimeric Flavan-3-ol Gallates, Theasinensins A and B, and Proanthocyanidin Gallates from Green Tea Leaf," *Chem. Pharm. Bull.* **31**, 3906–3914 (1983).

42. E. A. H. Roberts, R. A. Cartwright, and M. Oldschool, "Fractionation and Paper Chromatography of Water Soluble Substances from Manufactured Tea," *J. Sci. Food Agric.* **8**, 72–80 (1957).

43. D. J. Millin and D. W. Rustidge, "Tea Manufacture," *Process Biochemistry* **2**, 9–13 (1967).

44. R. G. Bailey, H. E. Nursten, and I. McDowell, "Isolation and Analysis of a Polymeric Thearubigin Fraction from Tea," *J. Sci. Food Agric.* **59**, 365–375 (1992).

45. A. L. Davis et al., "A Polyphenolic Pigment from Black Tea," *Phytochemistry* **46**, 1397–1402 (1997).

46. C. Powell et al., "Fractionation of Thearubigins: Isolation of Theacitrins," *Symp. Recent Advances Chem. Biochem. Tea*, School of Biological Sciences, University of Surrey, Guilford, Surrey, U.K., March 24, 1994.

47. A. G. Brown et al., "The Identification of the Thearubigens as Polymeric Proanthocyanidins," *Phytochemistry* **8**, 2333–2340 (1969).

48. T. Ozawa et al., "Elucidation of the Partial Structure of Thearubigins from Black Tea by Chemical Degradation," *Biosci. Biotechnol. Biochem.* **60**, 2023–2027 (1996).

49. S. C. Opie, M. N. Clifford, and A. Robertston, "Formation of Black Tea Thearubigins by *in vitro* Enzymic Oxidation of Flavan-3-ols," *Symp. Recent Advances Chem. Biochem. Tea*, School of Biological Sciences, University of Surrey, Surrey, England, Month day(s), 1994.

50. M. A. Dix et al., "Fermentation of Tea in Aqueous Suspension: Influence of Tea Peroxidase," *J. Sci. Food Agric.* **32**, 920–932 (1981).

51. K. L. Tomlins and A. Mashingaidze, "Influence of Withering, including Leaf Handling, on the Manufacturing and Quality of Black Teas: A Review," *Food Chem.* **60**, 573–580 (1997).

52. P. O. Owuor and C. O. Othieno, "Effects of Maceration on the Chemical Composition and Quality of Clonal Black Teas," *J. Sci. Food Agric.* **49**, 87–94 (1989).

53. *The Tea Cyclopedia*, Whittingham, London, 1882.

54. W. H. Ukers, *All about Tea*, Tea and Coffee Trade Journal, New York, 1935.

55. Brit. Pat. 1,274,002 (May 10, 1972), A. C. K. Krishnaswami and C. Hariprasad.

56. Brit. Pat. 1,484,540 (September 1, 1977), D. Kirtisinghe and D. P. Ranasinghe (to Tea Research Institute).

57. J. B. Cloughley, "Storage Deterioration in Central African Tea: Changes in Chemical Composition, Sensory Characteristics, and Price Evaluation," *J. Sci. Food Agric.* **32**, 1213–1223 (1981).

58. J. M. Robinson and P. O. Owuor, "Tea Aroma," in K. C. Wilson and M. N. Clifford, eds., *Tea: Cultivation to Consumption*, Chapman & Hall, London, 1992, pp. 603–647.

59. W. Mick and P. Schreier, "Additional Volatiles of Black Tea Aroma," *J. Agric. Food Chem.* **32**, 924–929 (1984).

60. P. O. Owuor, "Flavor of Black Tea: A Review," *Tea* **7**, 29–42 (1986).

61. G. W. Sanderson and H. N. Graham, "On the Formation of Black Tea Aroma" *J. Agric. Food Chem.* **21**, 576–584 (1973).

62. N. Fischer, S. Nitz, and F. Drawert, "Freie und Gebundene Aromastoffe in Grünem und Schwarzem Tee," *Z. Lebensm.-Unters.-Forschung* **185**, 195–201 (1987).

63. O. G. Vitzhum, P. Werkhoff, and P. Hurbert, "New Volatile Constituents of Black Tea Aroma," *J. Agric. Food Chem.* **23**, 999–1003 (1975).

64. M. A. Bokuchava and N. I. Skobeleva, "The Flavor of Beverages," in I. D. Morton and A. J. Mcleod, eds., *Food Flavours: Part B*, Elsevier Science, Amsterdam, The Netherlands, 1986, pp. 49–84.

65. G. R. Roberts and C. C. Rajasingham, "Changes in the Polyphenolic Constituents of Tea Liquors during Storage," *Tea Quarterly* **50**, 71–73 (1981).

66. G. W. Sanderson et al., "Contribution of Polyphenolic Compounds to the Taste of Tea," in G. Charalambous and I. Katz, eds., *Sulfur and Nitrogen Compounds in Food Flavors*, ACS Symposium, Vol. **26**, American Chemical Society, Washington, D.C., 1976, pp. 14–46.

67. T. Anan, "Changes of Chemical Compounds during Green Tea Manufacturing," *Japanese Agric. Res. Quart.* **22**, 195–199 (1988).

68. M. Nakagawa, "Relationship between Chemical Constituents and the Taste of Green Tea," *Tea Res. J. (Japan)* **40**, 1 (1973).

69. M. Kosuge, H. Aisaka, and T. Yamanishi, "Flavor Constituents of Chinese and Japanese Pan-Fried Green Teas," *Eiyot Shokuryo* **34**, 545–549 (1981).

70. F. Hashimoto, G. Nonaka, and I. Nishioka, "Tannins and Related Compounds: LXIX, Isolation and Structure Elucidation of B,B′-Linked Bisflavanols, Theasinensins D-G and Oolongtheanin from Oolong Tea (2)," *Chem. Pharm. Bull.* **36**, 1676–1684 (1988).

71. F. Hashimoto, G. Nonaka, and I. Mishioka, "Tannins and Related Compounds: XC, 8-C-Ascorbyl(−)-Epigallocatechin 3-O-Gallate and Novel Dimeric Flavan-3-ols, Oolonghomobisflavans A and B from Oolong Tea (3)," *Chem. Pharm. Bull.* **37**, 3255–3263 (1989).

72. U.S. Pat. 3,992,983 (November 23, 1976), R. J. Gasser and S. N. Watercutter (to Socit d'Assistance Technique pour Produits Nestle SA).

73. Brit. Pat. 1,380,135 (January 8, 1975), P. Coggon, H. N. Graham, and G. W. Sanderson (to Unilever, Ltd.).

74. U.S. Pat. 4,051,254 (September 27, 1977), G. W. Sanderson, A. C. Hoefler, and H. N. Graham (to Thomas J. Lipton, Inc.).

75. U.S. Pat. 4,680,193 (July 14, 1987), T. L. Lunder and C. M. Nielsen (to Nestec SA).

76. U.S. Pat. 3,484,246 (December 16, 1969), T. R. Moore, Jr., H. N. Graham, and M. Gurkin (to Thomas J. Lipton, Inc.).

77. U.S. Pat. 3,484,247 (December 16, 1969), H. N. Graham, V. V. Studer, and M. Gurkin (to Thomas J. Lipton, Inc.).

78. U.S. Pat. 3,484,248 (December 16, 1969), H. N. Graham, V. V. Studer, and M. Gurkin (to Thomas J. Lipton, Inc.).

79. Eur. Pat. 267660 (May 18, 1988), B. Hoogstad (to Unilever NV).

80. U.S. Pat. 3,666,484 (May 30, 1972), M. Gurkin, G. W. Sanderson, and H. N. Graham (to Thomas J. Lipton, Inc.).

81. U.S. Pat. 5,529,796 (June 25, 1996), D. A. Balentine, S. A. Gobbo, and J. W. Tobin (to Thomas J. Lipton, Inc.).

82. Eur. Pat. 675,683 (October 11 1995), D. A. Balentine and J. W. Tobin (to Unilever PLC).

83. J. F. Greden, "The Tea Controversy in America," *JAMA* **236**, 63–66 (1976).

84. Y. Ito, S. Ohnishi, and K. Fujie, "Chromosome Aberrations Induced by Aflotoxin B-1 in Rat Bone Marrow Cells *in vivo* and Their Suppression by Green Tea," *Mutat. Res.* **222**, 253–261 (1989).

85. S. K. Katiyar and H. Mukhtar, "Tea Consumption and Cancer," in A. P. Simopoulos, ed., *Metabolic Consequences of Changing Dietary Patterns. World Rev. Nutr. Diet*, Vol. 79, Karger, Basel, Switzerland, 1996, pp. 154–184.

86. Z. Y. Wang et al., "Protection against Polycyclic Aromatic Hydrocarbon-Induced Skin Tumor Initiation in Mice by Green Tea Polyphenols," *Carcinogenesis* **10**, 411–415 (1989).

87. Z. Y. Wang et al., "Inhibition of N-Nitrosobenzylamine-induced Esophageal Tumorigenesis in Rats by Green and Black Tea," *Carcinogenesis* **16**, 2143–2148 (1995).

88. Y. T. Gao et al., "Reduced Risk of Esophageal Cancer Associated with Green Tea Consumption," *J. Natl. Cancer Inst.* **86**, 855 (1994).

89. H. Matsuda et al., "Effects of Crude Drugs on Experimental Hypercholesterolemia: I, Tea and Its Active Principles," *Journal of Ethnopharmacology* **17**, 213–224 (1986).

90. Y. Sato et al., "Possible Contribution of Green Tea Drinking Habits to the Prevention of Stroke," *Tohoku J. Experimental Med.* **157**, 337–343 (1989).

91. C. S. Farkas, "Survey of Tea Intake by Northern Canadian Indians with Emphasis on the Implications for Iron Status," *J. Can. Diet Assoc.* **4**, 318 (1980).

92. R. B. H. Wills and K. J. McBrien, "Antithiamin Activity of Tea Fractions," *Food Chem.* **6**, 111–114 (1980).

93. "Mold Linked to Esophageal Cancer," *Science News* **130** (1986).

HAROLD N. GRAHAM
New York, New York

THERMAL PROCESSING OF FOOD

Thermal processing of canned foods has been one of the most widely used methods of food preservation during this century. In its broadest sense, it refers to the technology of using heat sterilization to preserve ready-to-eat foods and other biological products so that they can remain safe and wholesome under long-term extended storage at room temperature without chemical additives or preservatives. Foods preserved in this manner have become so commonplace in the human diet that the health of the world population now depends in great measure on the safety and wholesomeness of these foods. Thermal processing consists of heating food containers in pressurized retorts at specified temperatures for prescribed lengths of time. These process times are calculated on the basis of achieving sufficient bacterial inactivation (lethality) in each container to comply with public health standards and to ensure that the probability of spoilage will be less than some minimum. Associated with each thermal process are some degradation of heat-sensitive vitamins and other undesirable quality factors. Because of these quality and safety factors, great care is taken in the calculation of these process times and in the control of time and temperature during processing to avoid either underprocessing or overprocessing. The heat transfer considerations that govern the temperature profiles achieved within the container of food are critical factors in the determination of time and temperature requirements for sterilization. The sections that follow explain the distinction between sterilization and pasteurization, provide a brief history describing the evolution of canning technology leading to the current state of the art, a description of the various types of commercial equipment systems operating today in modern food processing factories, a review of the scientific principles that must be understood in order to establish these processes and specify necessary process conditions, and a look at future trends and perspectives for this important food preservation technology.

STERILIZATION VERSUS PASTEURIZATION

Thermal processing covers the broad area of food preservation technology in which heat treatments are used to inactivate microorganisms to accomplish either commercial sterilization or pasteurization. Sterilization processes are used with canning to preserve the safety and wholesomeness of ready-to-eat foods over long terms of extended storage at normal room temperature (nonrefrigerated) without additives or preservatives, and pasteurization processes are used to extend the refrigerated storage life of fresh foods. Although both processes make use of heat treatments for the purpose of inactivating microorganisms, they differ widely with respect to the classification or type of microorganisms targeted, and thus the range of temperatures that must be used and the type of equipment systems capable of achieving such temperatures.

Pasteurization

Pasteurization is a relatively mild heat treatment given to foods with the purpose of destroying selected vegetative microbial species (especially the pathogens) and inactivating the enzymes. Because the process does not eliminate all the vegetative microbial population and almost none of the spore formers, pasteurized foods must be contained and stored under conditions of refrigeration with chemical additives or modified atmosphere packaging, which minimize microbial growth. Depending on the type of product, the shelf life of pasteurized foods could range from several days (milk) to several months (fruit juices). Because only mild heat treatment is involved, the sensory characteristics and nutritive value of the food are minimally affected. The severity of the heat treatment and the length of storage depends on the nature of the product, pH conditions, the resistance of the target microorganism or enzyme, the sensitivity of the product, and the method of heating. Some of these are summarized in Table 1 (1).

Most pasteurization operations involving liquids (milk, milk products, beer, fruit juices, liquid egg, etc) are carried out in continuous heat exchangers. The product temperature is quickly raised to the pasteurization levels in the first heat exchanger, held for the required length of time in the holding tube, and quickly cooled in a second heat exchanger. For viscous fluids, a swept surface heat exchanger is often used to promote faster heat transfer and to prevent surface fouling problems. In-package pasteurization is similar to conventional thermal processing of foods except that it is carried out at lower temperatures. The thermal processing of high acid foods (natural or acidified) is also sometimes termed pasteurization to indicate that relatively milder heat treatment is involved (generally carried out at boiling water temperatures).

Sterilization

Sterilization implies the destruction of all viable microorganisms and is not the appropriate word to be used for thermal processing of foods, because these foods are far from being sterile in the medical sense of the word. The success of thermal processing does not lie in destroying all viable microorganisms but in the fact that together with the nature of the food (pH), environment (vacuum), hermetic packaging, and storage temperature, the given heat process prevents the growth of microorganisms of spoilage and public health concern. In essence, it represents a thermal process in which foods are exposed to a high-enough temperature for a sufficiently long time to render them commercially sterile. The process takes into account the heat resistance of the spore formers in addition to their growth sensitivity to oxygen, pH, and temperature. The presence of vacuum in cans prevents the growth of most aerobic microorganisms, and if the storage temperature is kept below 25°C, the heat-resistant thermophiles pose little or no problem. From the public health perspective, the most important microorganism in low-acid (pH > 4.5) foods is *Clostridium botulinum*, a heat-resistant, spore-forming, anaerobic pathogen that, if it survives processing, can potentially grow and produce the deadly botulism toxin in foods. Because *C. botulinum* and most spore formers do not grow at pH < 4.5 (acid and medium-acid foods), the thermal processing criterion for these foods is the destruction of heat-resistant yeasts and molds, vegetative microorganisms, or enzymes. The temperature and pH requirements of some common spoilage microorganisms are summarized in Table 2 (2). Because spore formers generally have high heat resistance, the low-acid foods that support their growth are processed at elevated temperatures (115–125°C), whereas acid foods need only to be brought to 80–90°C for adequate inactivation of enzymes or destruction of vegetative cells, yeasts, and molds.

STERILIZATION EQUIPMENT SYSTEMS

Historical Perspectives

The practice of canning as a method of food preservation originated in the early 1800s in France when Emperor Napoleon Bonaparte offered a prize of 12,000 francs to anyone who could develop better and more diversified foods to feed his troops on their military campaigns. A man named Nicholas Appert won the prize for successfully preserving, for the first time, a variety of perishable food products by heat-processing them in glass jars and bottles. At the time of Appert's discovery, the reasons for food spoilage were not known; process times and temperatures were selected by trial and error. Pasteur discovered the existence of microscopic organisms many years later in 1860.

From Appert's work, plus the invention of the metal container and the pressure cooker or retort, evolved the present-day thermal processing technology. The development of metal and glass containers capable of withstanding more than 1 atm of added internal pressure was a ma-

Table 1. Pasteurization Objectives and Conditions for Selected Foods

Food	Purpose	Typical processing conditions
Fruit juice	Inactivation of enzymes (pectinesterase and polygalacturonase)	65°C for 30 min; 77°C for 1 min; 88°C for 15 s
Beer	Destruction of spoilage microorganisms (wild yeasts, *Lactobacillus* sp.), and residual yeasts (*Saccharomyces* sp.)	65–68°C for 20 min (in bottle); 72–75°C for 1–4 min at 900–1,000 kPa
Milk	Destruction of pathogens: *Brucella aboritis, Mycobacterium tuberculosis, Coxiella burnettii*	63°C for 30 min; 71.5°C for 15 s
Liquid egg	Destruction of pathogens *Salmonella seftenburg*	64.4°C for 2.5 min; 60°C for 3.5 min
Ice cream	Destruction of pathogens	65°C for 30 min; 71°C for 10 min; 80°C for 15 s

Source: Ref. 1.

Table 2. Spore-Forming Bacteria Important in Spoilage of Food

	Acidity of food	
Approximate temperature (°C) range for vigorous growth	Acid 3.7 < pH < 4.5	Low acid pH > 4.5
Thermophilic (55–35°)	B. coagulans	C. thermosaccharolyticum C. nigrificans Bacillus stearothermophilus
Mesophilic (40–10°)	C. butyricum C. pasteurianum B. mascerans B. polymyxa	C. botulinum, A and B C. sporagenes B. licheniformis B. subtilis
Psychrophilic (35–<5°)		C. botulinum, E

Source: Ref. 2.

jor breakthrough. This allowed a concurrent development of devices permitting the exposure of the filled and sealed containers to steam pressures above atmospheric, namely, processing temperatures of 120°C instead of only 100°C (boiling water). Because the thermal inactivation of food spoilage organisms is a function of both time and temperature, the higher temperature under pressure allowed a very considerable reduction in the time necessary to ensure product sterility. Equally important was the marked improvement of the canned food quality. Commercial equipment systems in use today for heat sterilization of canned foods are described in the following sections.

Retort Processing

Batch Systems. There are two fundamentally different process methods by which canning is accomplished in the food industry. These two methods are known as retort processing and aseptic processing. In retort processing, foods to be sterilized are first filled and hermetically sealed in cans, jars, or other retortable containers. Then, they are heated in their containers using hot steam or water under pressure so that heat penetrates the product from the can wall inward, and both product and can wall become sterilized together. In aseptic processing, a liquid food is first sterilized outside the container by pumping it through heat exchangers that deliver very rapid heating and cooling rates. Then, the cool sterile product is filled and sealed in a separately sterilized package under a sterile environment at room temperature. Thus, retort processing can be thought of as in-container or in-can sterilization, and aseptic processing can be thought of as out-of-container sterilization.

In retort processing (in-can sterilization), the food to be sterilized is first filled and hermetically sealed in rigid, flexible, or semirigid containers such as metal cans, glass jars, retort pouches, or plastic bowls or trays that are placed within large steam retorts, sometimes called autoclaves. These are pressure vessels that work like giant pressure cookers, as shown in Figure 1. Once the retorts are full of containers to be sterilized, the retort doors are closed tightly and the air is replaced by hot steam under pressure to achieve temperatures above the atmospheric boiling point of water. A common retort temperature for sterilizing canned foods is 121°C (250°F), at approximately 1 atm of added internal pressure. After the containers have

Figure 1. Empty batch retort with doors ajar showing interior rails for entry and exit of wheeled crates used in loading and unloading operations. *Source:* Courtesy FMC Food Process Systems Division, Madera, Calif.

been exposed to the sterilizing temperature for sufficient time to achieve the desired level of sterilization, the steam is shut off, and cooling water is introduced to cool the containers and reduce the pressure, thus ending the process. Once the retort pressure has returned to atmosphere, the doors can be opened, and the processed containers are removed for labeling, case-packing, and warehousing to await distribution to the marketplace.

Although the unloading and reloading operations are labor intensive, a well-managed cook room can operate with surprising efficiency. The cook room is the room or area within a food canning factory in which the retorts are located (Fig. 2). Some cook rooms are known to have more than 100 retorts operating at full production. Although each retort is a batch cook operation, the cook room as a whole operates as a continuous production system in that filled and sealed unsterilized cans enter the cook room continuously from the filling line operations, and fully processed sterilized cans leave the cook room continuously enroute to subsequent case packing and warehousing. Within the cook room, teams of factory workers move from retort to retort to carry out loading and unloading operations, and retort operators are responsible for a given number or bank of retorts. These operators carefully monitor the

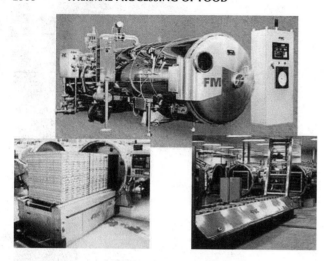

Figure 2. Exterior view of operating batch retort (above), and commercial system of batch retorts showing automated loading/unloading operations in a typical food processing plant cook room (below). *Source:* Courtesy FMC Food Process Systems Division, Madera, Calif.

operation of each retort to make sure the scheduled process is delivered for each batch. For convection-heating products that benefit from mechanical agitation during processing, agitating batch retorts are available that accomplish axial rotation of the cans or end-over-end agitation by rotating the entire crate during processing operations (3,4).

Continuous Retort Systems. Continuous retort operations require some means by which cans are automatically and continuously moved from atmospheric conditions into a pressurized steam environment, held or conveyed through that environment for the specified process time, and then returned to atmospheric conditions for further handling operations. The most well-known commercially available systems that accomplish these requirements are the crateless retort, the continuous rotary cooker, and the hydrostatic sterilizer (3,4).

Crateless Retorts. A crateless retort system is, in a sense, an automatic cook room in that the system is made up of a series of individual retorts, each operating in a batch mode, with loading, unloading, and process scheduling operations all carried out automatically without the use of crates. An individual crateless retort is shown schematically in Figure 3. When ready to load, the top hatch opens automatically, and cans fed from an incoming conveyor literally fall into the retort, which is filled with hot water to cushion the fall. Once fully charged, the hatch is closed while the incoming conveyor diverts the flow of cans to another retort that is ready for loading. Steam entering from the top displaces the cushion water out the bottom. When the cushion water has been fully displaced, all valves are closed and processing begins. At the end of the process time, the retort is refilled with warm water and the bottom hatch, which lies beneath the water level in the discharge cooling canal, is opened to let the cans gently fall onto the moving discharge conveyor in the cooling canal.

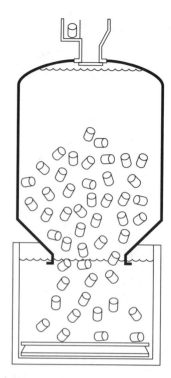

Figure 3. Operating schematic of a crateless retort showing "free fall" discharge of sterilized food containers onto submerged cooling canal exit conveyor.

After all cans are discharged, the bottom hatch is reclosed, and the retort is ready to begin a new cycle. A commercial system of crateless retorts would consist of several such retorts in a row sharing a common infeed and discharge conveyor system to achieve continuous operation of any design capacity.

Continuous Rotary Cookers. The continuous rotary pressure sterilizer or "cooker" is a horizontal rotating retort through which the cans are conveyed while they rotate about their own axis through a spiral path on a revolving reel mechanism. Residence time through the sterilizer is controlled by the rotating speed of the reel, which can be adjusted to accomplish the specified process time. This, in turn, sets the line speed for the entire system. Cans are transferred from an incoming can conveyor through a synchronized feeding device to a rotary pressure sealed transfer valve, which indexes the cans into the sterilizer while preventing the escape of steam and loss of pressure (much like a revolving door). Once cans have entered the sterilizer, they travel in the annular space between the reel and the shell. They are held between spines on the reel and a helical or spiral track welded to the interior shell wall. In this way, the cans are carried by the reel around the inner circumference of the shell imparting a rotation about their own axes, while the spiral track in the shell directs the cans forward along the length of the sterilizer by one can length for each revolution of the reel. At the end of the sterilizer, cans are ejected from the reel into another rotary valve and into the next shell for either additional cooking or cooling (Figs. 4 and 5).

Most common systems require at least three shells in series to accomplish controlled cooling through both a pres-

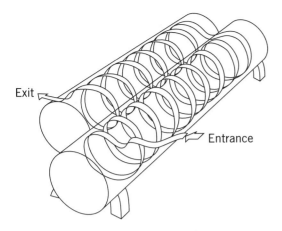

Figure 4. Schematic of helical path traveled by food containers moving through a series of rotary cooker/cooler shells in a continuous rotary sterilizer system.

STERILMATIC

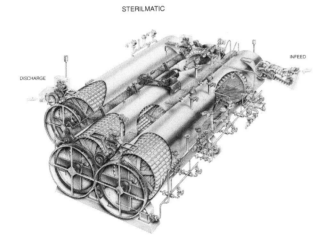

Figure 5. Continuous rotary sterilizer system with cutaway views showing rotary pressure-seal valves and internal helical spline and real conveying mechanism. *Source:* Courtesy FMC Food Process Systems Division, Madera, Calif.

sure cool shell and an atmospheric cool shell following the cooker or sterilizer. For cold-fill products that require controlled preheating, as many as five shells may be required in order to deliver an atmospheric preheat, pressure preheat, pressure cook, pressure cool, and atmospheric cool. By nature of its design and principal of operation, a continuous rotary sterilizer system is manufactured to accommodate a specified can size and cannot be easily adapted to other sizes. For this reason, it is not uncommon to see several systems in operation in one food canning plant, each system dedicated to a different can size filling line.

Hydrostatic Sterilizers. These systems are so named because steam pressure is controlled hydrostatically by the height of a leg of water. Because of the height of water leg required, these sterilizers are usually installed outdoors adjacent to a canning plant. They are self-contained structures with the external appearance of a rectangular tower as shown in Figure 6. They are basically made up of four

Figure 6. Exterior view of continuous hydrostatic sterilizer. *Source:* Courtesy FMC Food Process Systems Division, Madera, Calif.

chambers: a hydrostatic bring-up leg, a sterilizing steam dome, a hydrostatic bring-down leg, and a cooling section.

The principal of operation for a hydrostatic sterilizer can be explained with reference to the schematic in Figure 7. Containers are conveyed through the sterilizer on carriers connected to a continuous chain link mechanism that provides positive line speed control. This provides residence time control to achieve specified process time in the steam dome. Carriers are loaded automatically from incoming can conveyors and travel to the top of the sterilizer

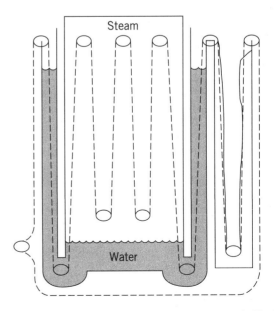

Figure 7. Schematic of continuous conveyor path traveled by food containers through inlet water leg, sterilizing steam dome, outlet water leg, and cooling section of a continuous hydrostatic retort.

where they enter the bring-up water leg. They travel downward through this leg as they encounter progressively hotter water. As they enter the bottom of the steam dome, the water temperature will be in equilibrium with steam temperature at the water steam interface. In the steam dome, the cans are exposed to the specified process or "retort" temperature controlled by the hydrostatic pressure for the prescribed process time controlled by the carrier line speed. When cans exit the steam dome, they again pass through the water steam interface at the bottom and travel upward through the bring-down leg as they encounter progressively cooler water until they exit at the top. Cans are then sprayed with cooling water as the carriers travel down the outside of the sterilizer on their return to the discharge conveyor station.

Aseptic Processing

Aseptic canning systems have rapidly developed in recent years primarily to allow for the marketing of shelf-stable foods in novel or more economical packaging systems that cannot withstand normal retort processing conditions. The primary goal in earlier development work on aseptic canning systems was to use the high temperature–short time (HTST) benefit of aseptic processing to minimize quality losses that occur in the slow heating of foods processed in conventional retorting systems. In either case, aseptic canning circumvents the need for retort operations by sterilizing the product outside of the container through heat exchanger systems before it is filled aseptically into separately sterilized containers or packaging systems.

Heat Exchangers. The benefits of HTST processing have been known for a number of years and have led to the rapid development of aseptic canning systems wherever possible. These methods generally apply only to fluid products that can be pumped through heat exchangers capable of applying ultra-high temperature–short time (UHT) heating conditions to the product before it is filled and sealed aseptically. The general types of heat exchangers commonly used with aseptic canning systems fall into the two basic categories of direct and indirect heating. In direct heating, the product is brought into direct contact with live steam through either steam injection or steam infusion heaters followed by holding and flash cooling under controlled pressure (Fig. 8). In indirect heating, the product contacts the heated metal surfaces of a heat exchanger, which separate the product from direct contact with the heat exchange medium. Either plate, tubular, or swept-surface heat exchangers are most often used for this purpose (Fig. 9). The residence time experienced by the product as it flows through an insulated holding tube or holding section between the heating and cooling heat exchangers accomplishes the necessary process time for delivering the specified sterilizing value or lethality, and is controlled by flow rate.

Aseptic Processing Systems. Among the first commercially successful aseptic canning systems is the Dole system, illustrated schematically in Figure 10. The system was designed to aseptically fill conventional steel cans, and

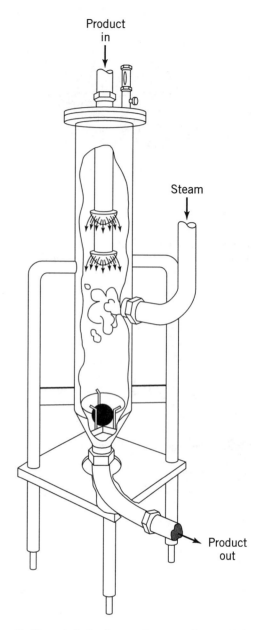

Figure 8. Steam-infusion heat exchangers. *Source:* Ref. 5, courtesy of Crepaco, Inc.

made use of superheated steam chambers to sterilize empty can bodies and covers as they were slowly conveyed to the filling chamber. The filling chamber was also maintained sterile by superheated steam under positive pressure and received cool sterile product from the heat exchangers in the product sterilizing subsystem. The entire system was presterilized before operation by passing superheated steam through the can tunnel, cover and closing chamber, and filling chamber for a prescribed start-up program of specified times and temperatures. The product sterilizing line was presterilized by passing pressurized hot water through the cooling heat exchanger (with coolant turned off), product filling line, and filler heads. This start-up procedure had to be repeated every time a compromise in sterility occurred at any system component. Careful

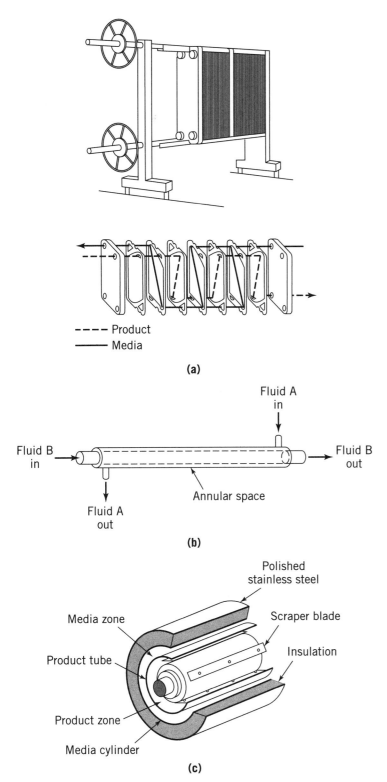

---- Product
——— Media

(a)

Fluid A
in

Fluid B
in
Fluid B
out

Annular space

Fluid A
out

(b)

Polished
stainless steel

Media zone

Scraper blade

Product tube

Insulation

Product zone

Media cylinder

(c)

Figure 9. (**a**) Plate heat exchangers with product flow schematic, (**b**) schematic of tubular heat exchangers, and (**c**) cutaway section of swept surface heat exchangers. *Source:* Ref. 5, courtesy of Cherry-Burrell.

monitoring and control by skillful and highly trained operators is a must for such an intricately orchestrated system.

Although the Dole system had been the mainstay for aseptic filling into metal cans, subsequent regulatory approval for the use of chemical sterilants such as hydrogen peroxide to sterilize the surfaces of various paper, plastics, and laminated packaging materials opened the door to a wide array of commercially available aseptic filling systems that produce shelf-stable liquid foods in a variety of gable-topped, brick-packed, and other novel package configurations. Filling machines designed for these packaging

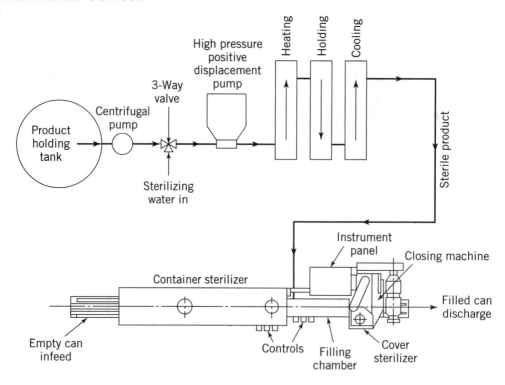

Figure 10. Dole Aseptic Canning Line. *Source:* Ref. 4, courtesy of CTI Publications, Inc.

systems are usually based on the use of form-fill-seal operations. In these machines, the packaging material is fed from either precut blanks or directly from roll stock, passed through a chemical sterilant bath or spray treatment, formed into the final package shape while being filled with cool sterile product from the product sterilizing system, and then sealed and discharged, all within a controlled aseptic environment.

Another important commercial application of aseptic processing technology is in the storage and handling of large bulk quantities of sterilized food ingredients, such as tomato paste, fruit purees, and other liquid food concentrates that need to be purchased by food processors or institutional end users for use as ingredients in further processed prepared foods. The containers for such applications can range in size from the classic 55-gallon steel drum to railroad tank cars or stationary silo storage tanks. Specially designed aseptic transfer valves and related handling systems make it possible to transfer sterile product from one such container to another without compromising sterility (Fig. 11).

Pasteurization Processes Systems

Pasteurization can be carried out by either in-container or out-of-container processes. The main difference from sterilization is that the lower temperatures used for pasteurization do not require the need for operating under pressure. Thus, the equipment systems needed for pasteurization are much simpler in design and easier to operate and maintain.

Normally, liquid foods with delicate heat-sensitive quality attributes, such as milk and fruit juices, are pasteur-ized out-of-container using HTST heat exchangers to pasteurize with minimum quality degradation before filling in clean packages. These HTST pasteurization systems are similar to the aseptic process systems used in sterilization except that they operate at lower temperatures and at atmospheric pressure, and they do not require rigid aseptic filling conditions. Some liquid dairy products, such as dairy cream and coffee whitener, are given a sterilization heat treatment by operating the heat exchanger under pressure to achieve sterilizing temperatures, but are filled into conventional sanitary cartons without aseptic filling systems. Such products are marketed as ultrapasteurized with markedly longer storage life under refrigeration.

Less heat-sensitive foods as well as most nonliquid foods are pasteurized in-container much like the retort process for sterilization, except that an open tank of hot or near-boiling water is sufficient, and there is no requirement to use pressure vessels like retorts or autoclaves. A third method of pasteurization, known as hot fill, makes use of the high pasteurizing temperature reached by the product in a batch tank or mixing kettle as part of the product preparation. The clean empty containers are filled with the hot product and sealed. They are held upright for a few minutes to transfer sufficient heat to the container walls and bottom, and then they are inverted for an additional few minutes to complete pasteurization of the container lid and seal area using heat transferred from the still hot product. Most canned fruits, fruit preserves, and acidified (pickled) products are pasteurized in this way.

Note that the food examples given for the hot fill method of pasteurization are nonrefrigerated foods that enjoy long-term storage at room temperature without the use of sterilization heat treatments. That is because they are high-

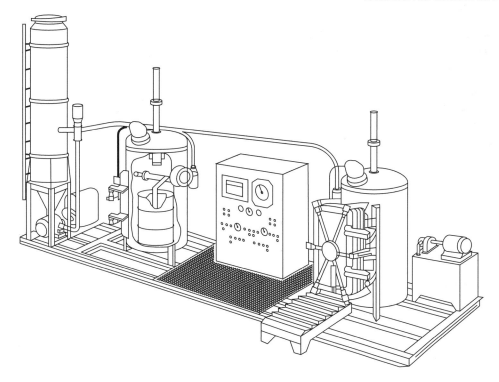

Figure 11. Aseptic filling system for 55-gallon drums. *Source:* Ref. 4, courtesy of Cherry-Burrell Corporation and CTI Publications.

acid foods (pH < 4.5) that cannot support the growth of heat-resistant spore forming pathogens. High-acid foods are subject to spoilage principally by yeasts and molds, which have low heat resistance and can be inactivated by pasteurization heat treatments alone. These are technically canned foods, but are essentially processed by the use of pasteurization technology. That is why it is important to distinguish between high-acid and low-acid canned foods in the context of thermal processing.

SCIENTIFIC PRINCIPLES OF THERMAL PROCESSING

Important Interrelationships

An understanding of two distinct bodies of knowledge is required to appreciate the basic principles involved in thermal process calculation. The first of these is an understanding of the thermal inactivation kinetics (heat resistance) of food-spoilage-causing organisms. The second body of knowledge is an understanding of the heat transfer considerations that govern the temperature profiles achieved within the food container during the process, commonly referred to in the canning industry as heat penetration.

Figure 12 conceptually illustrates the interdependence between the thermal inactivation kinetics of bacterial spores and the heat transfer considerations in the food product. Thermal inactivation of bacteria generally follows first-order kinetics and can be described by a logarithmic reduction in the population or concentration of bacterial spores with time for any given lethal temperature, as shown in the upper family of curves in Figure 12. These

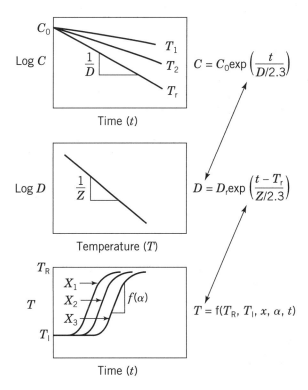

Figure 12. Time and temperature dependence of the thermal inactivation kinetics of bacterial spores in the thermal processing of canned foods. *Source:* Ref. 3, reprinted with permission, copyright 1992 by Marcel Dekker, Inc., New York.

are known as survivor curves. The decimal reduction time, D, is expressed as the time in minutes to achieve 1 log cycle of reduction in concentration, C. As suggested by the family of curves shown, D is temperature dependent and varies logarithmically with temperature, as shown in the second graph. This is known as a thermal death time (TDT) curve and is essentially a straight line over the range of temperatures used in food sterilization. The slope of the line that describes this relationship is expressed as the temperature difference, Z, required for the line to traverse 1 log cycle (10-fold change in D). The temperature in the food product, in turn, is a function of the retort temperature (T_R), initial product temperature (T_I), location within the container (x), thermal diffusivity of the product (α), and time (t), as shown by the heat penetration curves at the bottom of Figure 12. Thus, the concentration of viable bacterial spores during thermal processing decreases as a function of the inactivation kinetics, which are a function of temperature. The temperature, in turn, is a function of the heat transfer considerations, involving time, space, thermal properties of the product, and initial and boundary conditions of the process.

Microbiological Considerations

Heat Resistance. The heat resistance of microorganisms varies considerably. At any given temperature, this is generally expressed as a decimal reduction time (D-value), which is the heating time required to reduce the number of microorganisms by 90% (or to one-tenth of the initial). The temperature sensitivity of these D-values is expressed in terms of a Z-value, which represents the temperature range that results in a 10-fold change in the D-values. These two values can be realized as negative reciprocal slopes of logarithm of surviving microbial numbers vs time (D-value or survivor curve) and logarithm of D-values vs temperature (Z-value curve) as described by the functional expressions in Figure 12. Some typical D- and Z-values for selected microorganisms are given in Table 3.

As mentioned earlier, the Z-value curve, which describes the temperature dependency of the D-value, is often referred to as the TDT curve. It forms the basis upon which thermal process times and temperatures are determined and is shown in more detail in Figure 13. Once the TDT curve has been established for a given microorganism in a given food product, it can be used to calculate the time

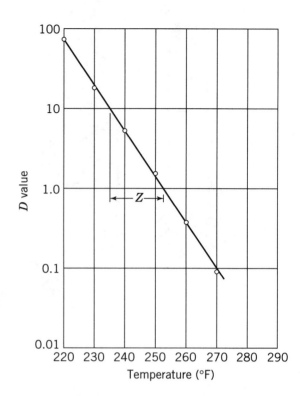

Figure 13. Thermal death time (TDT) curve showing temperature dependency of D value (decimal reduction time of microorganisms) given by temperature change (Z) required for tenfold change in D value. *Source:* Ref. 3, reprinted with permission, copyright 1992 by Marcel Dekker, Inc., New York.

and temperature requirements for any idealized thermal process. For example, assume a process is required that will achieve a 6-log-cycle reduction in the population of bacterial spores whose kinetics are described by the TDT curve in Figure 13, and that a temperature of 235°F has been chosen for the process. The TDT curve shows that the D-value at 235°F is 10 min. This means that 10 min will be required for each log cycle reduction in population at that temperature. If a 6-log-cycle reduction is required, a total of 60 min is required for the process. If a temperature of 270°F had been chosen for the process, the D-value at that temperature is approximately 0.1 min, and only 0.6 min (or 36 s) would be required at that temperature to accomplish the same 6-log-cycle reduction.

Sterilization F-Value (Lethality). The example process calculations carried out using the TDT curve in Figure 13 showed clearly how two widely different processes (60 min at 235°F and 0.6 min at 270°F) were equivalent with respect to their ability to achieve a 6-log-cycle reduction for that organism. Therefore, for a given Z-value, the specification of any one point on the line is sufficient to specify the sterilizing value of any process combination of time and temperature on that line. The reference point that has been adopted for this purpose is the time in minutes at the reference temperature of 250°F (121°C), or the sterilizing F-value (lethality) for the process. Since the F-value is expressed in minutes at 250°F (121°C), the unit of lethality

Table 3. Thermal Resistance of Spore Formers Used as a Basis for Thermal Processing

Microorganism	D_{250}-value, min	z-value, °C
B. stearothermophilus	4.0	7.0
B. subtilis	0.48–0.76	7.4–13.0
B. cereus	0.0065	9.7
B. megaterium	0.04	8.8
C. perfringens		10.0
C. sporogenes	0.15	13.0
C. sorogenes (PA 3679)	0.48–1.4	10.6
C. botulinum	0.21	9.9
C. thermosaccharolyticum	3.0–4.0	8.9–12.2

is 1 min at 250°F (121°C). Thus, if a process is assigned an F-value of 6, it means that the integrated lethality achieved by whatever time–temperature history employed by the process must be equivalent to the lethality achieved by 6 min of exposure to 250°F.

To illustrate, the example process calculation using the TDT curve in Figure 13 will be repeated by specifying the F-value for the required process. Recall from that example that the process was required to accomplish a 6-log-cycle reduction in spore population. All that is required to specify the F-value is to determine how many minutes at 250°F will be required to achieve that level of log-cycle reduction. The D_{250}-value is used for this purpose, since it represents the number of minutes at 250°F to accomplish 1 log-cycle reduction. Thus, the F-value is equal to D_{250} multiplied by the number of log cycles required in population reduction, or

$$F = D_{250}(\log a - \log b) \qquad (1)$$

where a is the initial number of viable spores and b is the final number of viable spores (or survivors).

In this example, $D_{250} = 1.16$ min as taken from the TDT curve in Figure 13, and $(\log a - \log b) = 6$. Thus, $F = 1.16(6) = 7$ min, and the sterilizing value for this process has been specified as $F = 7$ min. This is normally the way in which a thermal process is specified for subsequent calculation of a process time at some other temperature. In this way, information regarding specific microorganisms or numbers of log cycles reduction can be replaced by the F value as a process specification. Note also that this F-value serves as the reference point to specify the equivalent process curve discussed earlier. By plotting a point at 7 min on the vertical line passing through 250°F in Figure 13 and drawing a curve with a slope of $1/z$ parallel to the TDT curve through this point, the line will pass through the two equivalent process points that were calculated earlier (60 min at 235°F and 0.6 min at 270°F). Alternatively, the equation of this straight line can be used to calculate the process time (t) at some other constant temperature (T) when F is specified:

$$F = 10^{[(T-250)/Z]}t \qquad (2)$$

Equation 3 becomes important in the general case when the product temperature varies with time during a process, and the F-value delivered by the process must be integrated mathematically, such as at the center of a container of solid food.

$$dF = \int_0^t 10^{[(T-250)/Z]}dt \qquad (3)$$

Specification of Process Lethality. Establishing the lethality (F-value to be specified for a low-acid canned food) is undoubtedly one of the most critical responsibilities taken on by a food scientist or engineer acting on behalf of a food company in the role of a competent thermal processing authority. The steps normally taken for this purpose are outlined here.

There are two types of bacterial populations of concern in canned food sterilization. First is the population of organisms of public health significance. In low-acid foods with pH above 4.5, the chief organism of concern is *Clostridium botulinum*. A safe level of survival probability that has been accepted for this organism is 10^{-12}, or one survivor in 10^{12} cans processed. This is known as the 12-D concept for a botulinum cook. Since the highest D_{250} value known for this organism in foods is 0.21 min, the minimum process sterilizing value for a botulinum cook assuming an initial spore load of one organism per can is

$$F = 0.21 \times 12 = 2.52 \text{ min}$$

Essentially, all low-acid foods are processed far beyond the minimum botulinum cook in order to deal with spoilage-causing bacteria of much greater heat resistance. For these organisms, acceptable levels of spoilage probability are usually dictated by economic considerations. Most food companies accept a spoilage probability of 10^{-5} from mesophilic spore-formers (organisms that grow and spoil food at room temperature). The organism most frequently used to characterize this classification of food spoilage is a strain of *Clostridium sporogenese* known as PA 3679 with a maximum D_{250} value of 1.00. Thus, a minimum process sterilizing value for a mesophilic spoilage cook assuming an initial spore load of one spore per can is

$$F = 1.00 \times 5 = 5.00 \text{ min}$$

Where thermophilic spoilage is a problem, more severe processes may be necessary because of the high heat resistance of thermophilic spores. Fortunately, most thermophiles do not grow readily at room temperature; they require incubation at unusually high storage temperatures (110–130°F) in order to cause food spoilage. Generally, foods that show no more than 1% spoilage (spoilage probability of 10^{-2}) upon incubation after processing will show less than the minimum 10^{-5} spoilage probability in normal commerce. Therefore, when thermophilic spoilage is a concern, the target value for the final number of survivors is usually taken as 10^{-2}, and the initial spore load needs to be determined through microbiological analysis since contamination from these organisms varies greatly. For a situation with an initial thermophilic spore load of 100 spores per can and an average D_{250} value of 4.00, the process sterilizing value required would be

$$F = 4.00(\log 100 - \log 0.01)$$

$$F = 4.00(4) = 16 \text{ min}$$

These procedural steps are guidelines for average conditions; they often need to be adjusted up or down in view of the types of contaminating bacteria that may be present, the initial level of contamination or bioburden of the most resistant types, the spoilage risk accepted, and the nature of the food product.

Heat Transfer Considerations

In traditional thermal processing of canned foods, containers are placed in steam retorts that apply heat to the outside wall. The product temperature cannot respond instantaneously, but will gradually rise in an effort to approach the temperature at the wall followed by a gradual fall in response to cooling at the wall. In this situation, the sterilizing value delivered by the process will be the integrated result of the time–temperature profile experienced at the slowest heating point of the container. This profile shape will depend in large part upon the mode of heat transfer experienced by the product.

Modes of Heat Transfer

Conduction-Heating. Solid-packed foods in which there is essentially no product movement within the container, even when agitated, heat largely by conduction heat transfer. Because of the lack of product movement and the low thermal diffusivity of most foods, these products heat very slowly and exhibit a nonuniform temperature distribution during heating and cooling, which is caused by the temperature gradient set up between the can wall and geometric center. For conduction-heating products, the geometric center is the slowest heating point in the container. Therefore, process calculations are based on the temperature history experienced by the product at the can center. Solid-packed foods such as canned fish and meats, baby foods, pet foods, pumpkin, and squash fall into this category. These foods are usually processed in still cook or continuous hydrostatic retorts that provide no mechanical agitation.

Convection-Heating. Thin-bodied liquid products packed in cans, such as soups, sauces, and gravies, will heat by either natural or forced convection heat transfer, depending upon use of mechanical agitation during processing. In a still cook retort that provides no agitation, product movement will still occur within the container because of natural convective currents induced by density differences between the warmer liquid near the hot can wall and the cooler liquid near the can center. The rate of heat transfer in nearly all convection heating products can be increased substantially by inducing forced convection through mechanical agitation. For this reason, most convection-heating foods are processed in agitating retorts designed to provide either axial or end-over-end can rotation. Normally, end-over-end rotation is preferred and can be provided in batch retorts; continuous rotary retorts can provide only limited axial rotation. Unlike conduction heating products, because of product movement in forced convection-heating products, the temperature distribution throughout the product is reasonably uniform under mechanical agitation. In natural convection, the slowest heating point is somewhat below the geometric center and should be located experimentally in each new case.

Broken-Heating. Broken-heating canned food products exhibit a break between the two modes of heat transfer; they will heat part of the time by convection and part of the time by conduction. The more common of these foods are those that initially heat by convection, then, because of starch gelatinization or other thickening agent activity, they set-up or thicken and proceed to heat by conduction. Less common are products that begin heating first by conduction, then for the remainder of the period heat by convection. Generally, these are products with solid pieces in

a liquid brine that settle and pack into the lower two-thirds or so of the container when placed in the retort. After some time of heating, when convective currents become sufficiently strong, the solid pieces are lifted and disperse to begin moving with the liquid phase.

Heat Penetration Measurement. The primary objective of heat penetration measurements is to obtain an accurate recording of the product temperature at the can cold spot over time while the container is being heated under a controlled set of retort processing conditions. This is normally accomplished through the use of copper constantan thermocouples inserted through the can wall so as to have the junction located at the can geometric center. Thermocouple lead wires pass through a packing gland in the wall of the retort for connection to an appropriate data acquisition system in the case of a still cook retort. For agitating retorts, the thermocouple lead wires are connected to a rotating shaft for electrical signal pick up from the rotating armature outside the retort.

The precise temperature–time profile experienced by the product at the can center will depend on the physical and thermal properties of the product, size and shape of the container, and retort operating conditions. Therefore, it is imperative that test cans of product used in heat penetration tests be truly representative of the commercial product with respect to ingredient formulation, fill weight, head space, and can size and that the laboratory or pilot plant retort being used is capable of accurately simulating the operating conditions that will be experienced during commercial processing on the production-scale retort systems intended for the product. If this is not possible, then heat penetration tests should be carried out using the actual production retort during scheduled breaks in production operations.

Heat Penetration Curves. During a heat penetration test, both the retort temperature history and product temperature history at the can center are measured and recorded over time. A typical test process will include venting of the retort with live steam to remove all atmospheric air and then closing of vents to bring the retort up to operating pressure and temperature. This is the point at which process time begins, and the retort temperature is held constant over this period of time. At the end of the prescribed process time, steam is shut off, and cooling water is introduced under overriding air pressure to prevent sudden pressure drop in the retort. This begins the cooling phase of the process, which ends when the retort pressure returns to atmosphere and the product temperature in the can has reached a safe low level for removal from the retort. A typical temperature–time plot of these data is shown in Figure 14, which illustrates the degree to which the product center temperature can lag behind the retort temperature during both heating and cooling. The product center temperature history can be taken directly from this plot to perform a process calculation by numerical integration of equation 3; this will be discussed in further detail later.

A heat balance between the heat absorbed by the product and the heat transferred across the can wall from the

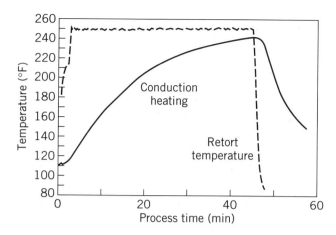

Figure 14. Center temperature profile in cylindrical container of conduction-heating food product in response to constant retort temperature process during heating and cooling.

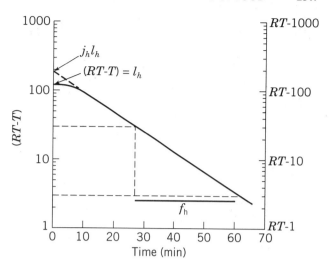

Figure 15. Semilogarithmic heat penetration curve showing predominent log-linear decay of difference between retort and product temperature over process heating time.

steam retort could be expressed as follows for an element of food volume facing the can wall of surface area A and thickness L:

$$\rho L A C_p \frac{dT}{dt} = \frac{k}{L} A(T_r - T) \tag{4}$$

where T is product temperature, T_r is retort temperature, and ρ, C_p, and k are density, specific heat, and thermal conductivity of the product, respectively. Because of the high surface heat transfer coefficient of condensing steam at the can wall and high thermal conductivity of the metal can, overall surface resistance to heat transfer can be assumed negligible in contrast to the product's resistance to heat transfer. After rearranging terms, expression 4 can be written in the form of an ordinary differential equation:

$$\frac{dT}{dt} = \left(\frac{k}{\rho C_p}\right) L(T_r - T) \tag{5}$$

By letting the thermal diffusivity (α) represent the combination of thermal and physical properties ($k/\rho C_p$) and letting T_o represent the initial product temperature, the solution to equation 5 becomes

$$\frac{T_r - T}{T_r - T_o} = \exp\left(\frac{\alpha}{L^2} t\right) \tag{6}$$

Thus, the product center temperature can be seen to be an exponential function of time; a semilog plot of the temperature difference $[T_r - T]$ against time would produce a straight line sloping downward having a slope related to the product's thermal diffusivity and can dimensions (Fig. 15). Because the heat penetration rate factor, f_h, is the reciprocal slope of the heat penetration curve, it is related to the product's thermal diffusivity and container dimensions. For a finite cylinder, the following relationship can be used to obtain the thermal diffusivity from a heat penetration curve:

$$\alpha = \frac{2.303}{\left[f_h\left(\dfrac{2.4048^2}{R^2} + \dfrac{\pi^2}{L^2}\right)\right]} \tag{7}$$

where R = can radius, L = one-half can height, and f_h is the heating rate factor in minutes. This relationship is also useful to determine the heating rate factor, f_h, for the same product in a different sized container, since the thermal diffusivity is a combination of physical properties that characterize the product and its ingredient formulation and remains unaffected by different container sizes.

COMPUTER SIMULATION

Numerical Models

Another purpose for obtaining the thermal diffusivity of products from a heat penetration curve is to make use of numerical computer models capable of simulating the heat transfer in canned foods. One of the primary advantages of these models is that once the thermal diffusivity has been determined, the model can be used to predict the product temperature history at any specified location within the can for any set of processing conditions and container size. With the use of such models, it is unnecessary to carry out repeated heat penetration tests in the laboratory or pilot plant in order to determine the heat penetration curve for a different retort temperature or can size. A second advantage of even greater importance is that the retort temperature need not be held constant, but can vary in any prescribed manner throughout the process, and the model will predict the correct product temperature history at the can center. Use of these models has been invaluable for simulating the process conditions experienced in continuous sterilizer systems, in which cans pass from one chamber to another experiencing a changing retort temperature at the can wall as they pass through the system. Another important application of these models is in the rapid evaluation of an unscheduled process deviation, such as when an unexpected drop in retort temperature occurs during the course of the process. The model can quickly predict the product center temperature history in response to such a deviation and calculate the delivered F-value for comparison with the sterilizing value specified for the product (6–8).

The first published numerical computer model for simulating the thermal processing of canned foods made use of a numerical solution by finite differences of the two-dimensional partial differential equation (equation 8) that describes conduction heat transfer in a finite cylinder (9). Temperature in the food product is a function of the retort temperature (T_R), initial product temperature (T_I), location within the container (x), thermal diffusivity of the product (α), and time (t) in the case of a conduction-heating food. In practice, α is obtained from the slope of the heat penetration curve (f_h) and is readily known. Because heat is applied only at the can surface, temperatures will rise first only in regions near the can walls, and temperatures near the can center will begin to respond only after a considerable time. Mathematically, the temperature (T) is a distributed parameter in that at any point in time (t) during heating, the temperature takes on a different value with location in the can (r,y), in any one location, the temperature changes with time as heat gradually penetrates the product in accordance with the thermal diffusivity (α).

Equation 8 can be written in the form of finite differences for numerical solution by digital computer, as shown in equation 9. The finite differences are discrete increments of time and space defined as small fractions of process time and container height and radius (Δt, Δh, and Δr, respectively). As a framework for computer iterations, the cylindrical container is imagined to be subdivided into volume elements that appear as layers of concentric rings having rectangular cross-sections, as illustrated in Figure 16 for the upper half of the container. Temperatures nodes are assigned at the corners of each volume element on a vertical plane as shown in Figure 17, where i and j are used to denote the sequence of radial and vertical volume elements, respectively. By assigning appropriate boundary and initial conditions to all the temperature nodes (interior nodes set at initial product temperature and surface nodes

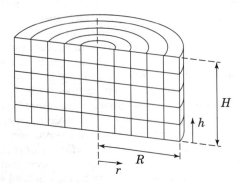

Figure 16. Subdivision of a cylindrical container for application of finite differences for numerical solution of heat conduction equation in a finite cylinder. *Source:* Ref. 3, reprinted with permission, copyright 1992 by Marcel Dekker, Inc., New York.

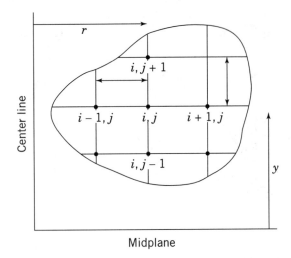

Figure 17. Labeling of grid nodes in matrix of volume elements on a vertical plane for application of finite differences. *Source:* Ref. 3, reprinted with permission, copyright 1992 by Marcel Dekker, Inc., New York.

set at retort temperature), the new temperature reached at each node can be calculated after a short time interval (Δt) that would be consistent with the thermal diffusivity of the product obtained from heat penetration data (f_h). This new temperature distribution is then taken to replace the initial one, and the procedure is repeated to calculate the temperature distribution after another time interval. In this way, the temperature at any point in the container at any instant in time is obtained. At the end of process time, when steam is shut off and cooling water is admitted to the retort, the cooling process is simulated by simply changing the boundary conditions from retort temperature to cooling temperature at the surface nodes and continuing with the computer iterations described earlier.

$$\frac{dT}{dt} = \alpha\left[\frac{d^2T}{dr^2} + \frac{1}{r}\frac{dT}{dr} + \frac{d^2T}{dy^2}\right] \quad (8)$$

$$T_{ij}^{(t+\Delta t)} = T_{ij}^{(t)} + \frac{\alpha\Delta t}{\Delta r^2}[T_{i-1j} - 2T_{ij} + T_{i+1j}]^{(t)}$$
$$+ \frac{\alpha\Delta t}{2r\Delta r}[T_{i-1j} - T_{i+1j}]^{(t)}$$
$$+ \frac{\alpha\Delta t}{\Delta h^2}[T_{ij-1} - 2T_{ij} + T_{ij+1}]^{(t)} \quad (9)$$

In this way, the temperature at the can center can be calculated after each time interval to produce a heat penetration curve upon which the process lethality or F-value can be calculated. When the numerical computer model is used to calculated the process time required at a given retort temperature to achieve a specified lethality, the computer follows a programmed search routine of assumed process times that quickly converges on the precise time at which cooling should begin in order to achieve the specified F-value. Thus, the model can be used to determine the process time required for any given set of constant or variable retort temperature conditions. The next section describes the method for calculating the process time required at any given retort temperature to deliver a specified process sterilizing value or lethality.

Process Calculation

The numerical integration of equation 3 is the most versatile method of process calculation because it is universally applicable to essentially any type of thermal processing situation. It makes direct use of the product temperature history at the slowest heating point of the container obtained from a heat penetration test or predicted by a computer model for calculating the process sterilizing value delivered by a given temperature–time history. This method is particularly useful in taking maximum advantage of computer-based data acquisition systems used in connection with heat penetration tests. Such systems are capable of reading temperature signals received directly from thermocouples monitoring both retort and product center temperature and processing these signals through the computer. Both retort temperature and product center temperature are plotted against time without any data transformation. This allows the operator to see what has actually happened throughout the duration of the process test. As the data are being read by the computer, additional programming instructions call for calculation of the incremental process sterilizing value at each time interval between temperature readings and summing these over time as the process is under way (numerical integration of equation 3). As a result, the accumulated sterilizing F-value is known at any time during the process and can be plotted on the graph along with the temperature histories to show the final value reached at the end of the process. An example of the computer printout from such a heat penetration test is shown in Figure 18.

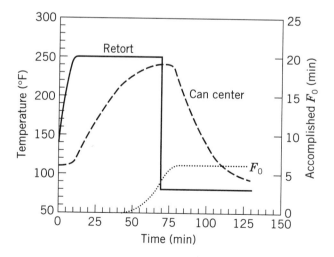

Figure 18. Computer-generated plot of measured retort temperature and calculated center temperature and accomplished F_0 for a given thermal process. *Source:* Ref. 3, reprinted with permission, copyright 1992 by Marcel Dekker, Inc., New York.

Process Optimization

The principle objective of thermal process optimization is to maximize product quality, minimize undesirable changes, minimize cost, and maximize profits. At all times, a minimal process must be maintained to exclude the danger from microorganisms of public health and spoilage concern. Five elements common to all optimization problems are performance or objective function (nutrients, texture, and sensory characteristics), decision variables (retort temperature and process time), constraints (practical limits for temperatures and required minimal lethality), mathematical model (analytical, finite differences, and finite element), and optimization technique (search, response surface, and linear or nonlinear programming).

Optimization theory makes use of the different temperature sensitivity of microbial and quality factor destruction rates. Microorganisms have lower decimal reduction time (less resistant to heat) and a lower Z-value (more sensitive to temperature) than most quality factors. Hence, a higher temperature will result in preferential destruction of microorganisms over the quality factor. Especially applied to liquid product either in a batch in-container mode or in continuous aseptic systems, the higher temperature with shorter time offers a great potential for quality optimization. However, for conduction heating foods, one of the major limitations is the slower heating. All higher temperatures do not necessarily favor the best quality retention because they also expose the product nearer the surface to more severe temperature than the product at the center, which might result in diminished overall quality. Using a finite differences computer simulation program, it has been demonstrated that the optimal process temperature is around 250°F (121°C) for maximized thiamin retention. In fact, processing at temperatures beyond 265°F was shown to result in poorer thiamine retention than processing at 240°F (9).

Product quality optimization can also be accomplished by promoting faster and more uniform heating in the product by other means, such as optimized container geometry in the form of appropriate height-to-diameter ratios, alternate packaging materials and shave such as the thin-profile retort pouch or semirigid container, or an agitated process for foods that are normally conduction heated but can flow under the influence of agitation (6).

On-Line Computer Control

Traditional control of thermal process operations has consisted of maintaining specified operating conditions that have been predetermined from product and process development research, such as the process calculations for the time and temperature of a batch cook. Sometimes unexpected changes can occur during the course of the process operation or at some point upstream in a processing sequence such that prespecified processing conditions are no longer valid or appropriate, and off-specification product that is produced must be either reprocessed or destroyed at appreciable economic loss. These types of situations can be of critical importance in food processing operations because the physical process variables that can be measured and controlled are often only indicators of complex biochemical reactions that are required to take place under the specified process conditions.

Because of the important emphasis placed on the public safety of canned foods, processors operate in strict compliance with the Food and Drug Administration's Low-Acid Canned Food Regulations. Among other things, these regulations require strict documentation and record-keeping of all critical control points in the processing of each retort load or batch of canned product. Particular emphasis is placed on product batches that experience an unscheduled process deviation, such as when a drop in retort temperature occurs during the course of the process that may result from loss of steam pressure. In such a case, the product will not have received the established scheduled process and must be either destroyed, fully reprocessed, or set aside for evaluation by a competent processing authority. If the product is judged to be safe, then batch records must contain documentation showing how that judgement was reached. If judged unsafe, then the product must be fully reprocessed or destroyed. Such practices are costly.

In recent years, food engineers knowledgeable in the use of engineering mathematics and scientific principles of heat transfer have developed computer models (described earlier in this section) capable of simulating thermal processing of conduction-heated canned foods. These models make use of numerical solutions to mathematical heat transfer equations capable of predicting accurately the internal product cold spot temperature in response to any dynamic temperature experienced by the retort during the process. The accomplished lethality (F_o) for any thermal process is easily calculated by numerical integration of the predicted cold spot temperature over time, as explained previously. Thus, if the cold spot temperature can be accurately predicted over time, so can accumulated process lethality.

Computer-based intelligent on-line control systems make use of these models as part of the decision-making software in a computer-based on-line control system. Instead of specifying the retort temperature as a constant boundary condition, the actual retort temperature is read directly from sensors located in the retort and is continually updated with each iteration of the numerical solution. Using only the measured retort temperature as input to the control system, the model operates as a subroutine calculating the internal product cold spot temperature at small time intervals for computer iteration in carrying out the numerical solution to the heat conduction equation by finite differences. At the same time, the model also calculates the accomplishing process lethality associated with cold spot temperature in real time as the process is under way. At each time step, the subroutine simulates the additional lethality that will be contributed by the cooling phase if cooling were to begin at that time. In this way, the control system decision of when to end heating and begin cooling is withheld until the model has determined that

final target process lethality will be reached at the end of cooling (10).

By programming the control logic to continue heating until the accumulated lethality has reached some designated target value, the process will always end with the desired level of sterilization (F_o) regardless of an unscheduled process temperature deviation. At the end of the process, complete documentation of measured retort temperature history, calculated center temperature history, and accomplished sterilization (F_o) can be generated in compliance with regulatory record-keeping requirements. Such documents are shown in Figure 19 for a normal process and for the same intended process with an unexpected deviation.

FUTURE TRENDS

Thermal processing has been in use as a predominant method of food preservation since the middle of the nineteenth century. People throughout the world have become

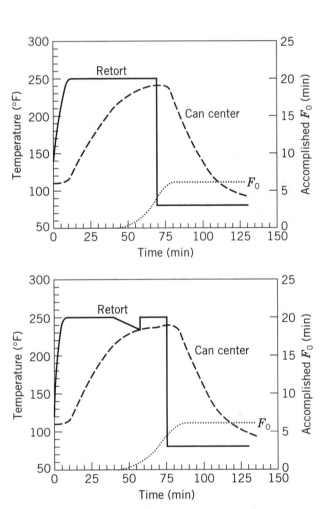

Figure 19. Output documentation of computer-based on-line control system showing scheduled heating time of 68 minutes for normal process (above), and heating time extended automatically to 76 minutes in compensation for unscheduled temporary loss of retort temperature (process deviation) (below).

quite familiar with canned foods packed in traditional metal cans and glass jars. Perhaps less apparent is the important role that this technology has had and continues to have in promoting and sustaining the health and well-being of populations throughout the world. Although the major portion of canned foods produced throughout history has been used to help feed the consuming public, this technology has also played a very strategic role in major world events. The famous C-rations that supported military troops through the World Wars I and II were canned food rations. These rations remained safe and wholesome for consumption after long periods of storage and handling under highly stressful and abusive conditions.

Developments in new packaging materials and retort systems have brought about a host of innovative new canned food products that are often not recognized as being canned foods. Most canned food rations for today's military troops are packaged in flexible retort pouches; they are convenient, comfortable (soft), and lightweight when being carried on maneuvers. For feeding large numbers of military troops in field kitchens, canned foods come as fully prepared meals in large institutional-sized rectangular steam-table trays ready to heat and serve.

The increasing popularity of the microwave oven has created a demand for canned foods in microwavable containers that are ready to "pop and zap." This has led to the increasing success of complete prepared meals in attractive lunch bowls or dinner trays that can be placed in the microwave oven and be taken directly to the dinner table. Such products are hardly recognizable as traditional canned foods, but they are.

Perhaps least recognizable, however, is the important role that thermal processing technology has played in the pharmaceutical and health care industry. Large quantities and varieties of sterile solutions are required daily for surgical and patient care procedures. Sterile saline solutions, irrigation solutions, intravenous solutions with dextrose or glucose, and dialysate solutions, along with a host of other large-volume parenteral solutions in glass, plastic, flexible and semirigid containers, are sterilized in retort systems using the technology of canning for food preservation. Such products, of course, are neither thought of nor considered to be canned foods, but they are in fact a very important use of thermal processing technology throughout the world.

BIBLIOGRAPHY

1. P. Fellows, *Food Processing Technology: Principles and Practices*, Ellis Horwood Ltd., Chichester, United Kingdom, 1988.
2. D. B. Lund, "Heat Processing," in M. Karel, O. R. Fennema, and D. B. Lund, eds., *Principles of Food Science, Part II: Physical Principles of Food Preservation*, Marcel Dekker, Inc., New York, 1975.
3. A. A. Teixeira, "Thermal Process Calculations," in D. R. Heldman and D. B. Lund, eds., *Food Engineering Handbook*, Marcel Dekker, New York, 1992, pp. 563–620.
4. A. Lopez, *A Complete Course in Canning and Related Processes*, CTI Publications, Baltimore, Md., 1987.
5. R. P. Singh and D. R. Heldman, *Introduction to Food Engineering*, Academic Press, San Diego, Calif., 1984.
6. A. A. Teixeira, G. E. Zinsmeister, and J. W. Zahradnik, "Computer Simulation of a Variable Retort Control and Container Geometry as a Possible Means of Improving Thiamine Retention in Thermally Processed Foods," *J. Food Sci.* **40**, 656–659 (1975).
7. A. A. Teixeira and J. E. Manson, "Computer Control of Batch Retort Operations with On-line Correction of Process Deviations," *Food Technol.* **36**, 85–90 (1982).
8. A. K. Datta, A. A. Teixeira and J. E. Manson, "Computer-based Retort Control Logic for On-line Correction of Process Deviations," *J. Food Sci.* **51**, 480–483, 507 (1986).
9. A. A. Teixeira et al., "Computer Optimization of Nutrient Retention in Thermal Processing of Conduction-heated Foods," *Food Technol.* **23**, 137–142 (1969).
10. A. A. Teixeira and C. F. Shoemaker, *Computerized Food Processing Operations*, Van Nostrand Reinhold, New York, 1989.

A. A. TEIXEIRA
University of Florida
Gainesville, Florida

See also CANNING: REGULATORY AND SAFETY CONSIDERATIONS.

THERMAL STERILIZATION OF CANNED LIQUID FOODS

Sterilization of food involves the destruction of microorganisms carried in it so that the food can be stored longer and be safe to consume. Thermal sterilization is the most often used and most time-tested method of sterilization. It involves use of heat to destroy the harmful microorganisms. But heat also affects nutrients and other desirable food properties adversely. Design of sterilization processes involves selection and implementation of heating temperature and duration to minimize the destruction of desirable food components while maintaining the required level of destruction of the harmful microorganisms (1,2).

Due to the finite size of any food material, time-temperature histories will always vary spatially, the exact variation being a function of the food properties, the design of the machinery, and the properties of container (batch) or tube (continuous). Only by analyzing all of these factors, which requires detailed heat transfer studies, would the required time-temperature history be known. Sterilization of liquids is further complicated because it is not the time-temperature history of locations but that of the individual moving liquid particles that determine the extent of sterilization, the latter being difficult to observe experimentally or to calculate theoretically.

The sterilization literature spans more than 100 years. Most studies are product, process, or equipment specific. The studies generally do not explain the underlying physics in sufficient detail to be able to make generalized conclusions about a large category of problems. This article synthesizes the engineering studies of liquid food sterilization to provide insight into the underlying physics of the process. After introducing the general concepts involving heating of fluids, three broad categories of liquid food sterilization systems are discussed in detail. These are continuous heating, agitated containerized heating, and unagitated containerized heating. Included under each category are a brief description of the processing equipment, methods of heating, flow and temperature profiles, sterilization values, effect of non-Newtonian liquid behavior, and effect of presence of particulates.

QUANTITATIVE MEASURE OF STERILIZATION

The rates of reactions in foods, such as the destruction of microorganisms or nutrients, are generally first-order reactions described by

$$-\frac{dc}{dt} = k_T c \tag{1}$$

where c is the concentration of any component at time t and k_T is the rate constant at temperature T. The temperature dependency of the rate constant is given by the well-known Arrhenius law

$$k_T = k_0 e^{-E_a/RT} \tag{2}$$

where k_0 is a constant called frequency factor, E_a is the activation energy, and R is the gas constant. The temperature T normally varies in a heating process. To obtain the final concentration in changing temperature, equation 1 is integrated between initial concentration c_i and final concentration c at time t to get

$$\ln \frac{c_i}{c} = k_0 \int_0^t e^{-E_a/RT} dt \tag{3}$$

In the food literature, equation 3 is used in a different form. Instead of referring to a final concentration, an equivalent heating time of F_0 is used that gives the same final concentration when temperature T is constant at a reference value. For a reference temperature of T_R, if the rate constant is k_{T_R},

$$F_0 = \frac{\ln(c_i/c)}{k_{T_R}} = \frac{k_0 \int_0^t e^{-E_a/RT} dt}{k_0 e^{-E_a/RT_R}}$$
$$= \int_0^t e^{[(E_a/R)(1/T_R - 1/T)]} dt \tag{4}$$

When the process temperature T stays close to T_R, equation 4 can be simplified by noting $TT_R \approx T_R^2$. A new parameter Z is defined in terms of the activation energy E_a to describe the temperature dependence of reaction as

$$Z = \frac{2.303 R T_R^2}{E_a} \tag{5}$$

Using this simplification and the definition of parameter Z, equation 4 can be rewritten as

$$F_0 = \int_0^t 10^{(T-T_R)/Z} dt \tag{6}$$

This is the familiar equation in food sterilization that provides the equivalent heating time F_0 at a reference temperature T_R for a process whose actual temperature T varies with time t.

In any finite mass of liquid (or any material), the temperature history $T(t)$ varies spatially throughout the mass. Thus F_0 as given by equation 6 will vary spatially at any given time. Thus in reality, there is always a distribution of F_0 values due to spatial variation of temperature history. For bacterial destruction the lowest value of F_0 is taken as the measure of the extent of sterilization. The average F_0 for nutrients, which is sometimes referred to as the cook value, provides a measure of the nutritional quality of the processed food material.

The complete distribution of quality parameters such as sterilization and nutrient retention is generally quite difficult to obtain due to the complex variations of temperature T with position and time in most processing situations of practical interest. Under these circumstances, design of the heating process is aimed at minimizing the spatial variation of temperature, for example, by introducing turbulence. For many practical calculations, where spatial variation is ignored, mean temperature is used in equation 6.

SPECIAL CONSIDERATIONS FOR LIQUIDS

Several factors uniquely complicate the study of liquid sterilization as compared to solids, which are now discussed.

Distribution of Residence Time and Thermal Time

Chemical and biological reactions such as destructions of microorganisms and nutrients associated with sterilization refer to elements in the fluid and solid food. For pure conduction heating in solids, the elements stay in a fixed location. Thus sterilization of material particles becomes synonymous with sterilization at various locations. To obtain the extent of these reactions in liquids that are often flowing, material elements would need to be followed conceptually. Generally speaking, for every liquid particle the temperature history as well as the duration of heating would be different.

The durations of heating are studied using the concepts of residence-time distributions (RTDs). The time it takes a molecule to pass through the sterilization vessel is called its residence time. In continuous operations, all liquid particles do not remain in the equipment for the same time. Particles near the walls (boundary layer) or in dead space, for example, move more slowly than particles farther away. Such behavior of liquids result in a distribution of residence times for various liquid elements. From equation 6 the extent of sterilization would vary for the various liquid elements. It is therefore important to have an idea about the RTD to insure that the liquid residing for the shortest time receives adequate sterilization. Figure 1 shows an RTD where the area between any two residence times is

Residence-time distribution

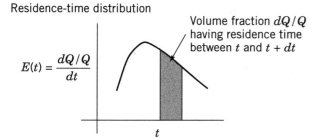

$$E(t) = \frac{dQ/Q}{dt}$$

Volume fraction dQ/Q having residence time between t and $t + dt$

Thermal-time distribution

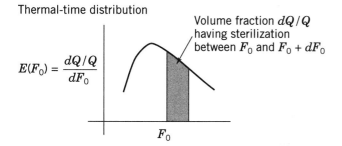

$$E(F_0) = \frac{dQ/Q}{dF_0}$$

Volume fraction dQ/Q having sterilization between F_0 and $F_0 + dF_0$

Figure 1. Representations of residence-time distributions and thermal-time distributions of fluids during a sterilization process. The symbol Q refers to either the volumetric flow rate or the total volume of liquid being sterilized.

the volume (or mass) fraction of liquid that resides between those two times. RTDs are described in detail in various sources (3–5).

But RTDs have only a duration of heating and not a temperature history and therefore cannot provide information on sterilization distribution in general. In analogy to RTDs, thermal-time distributions (TTDs) have been defined (6) that include the effect of temperature history. Figure 1 shows a typical TTD. The area under the curve between two values of sterilization is the fraction of liquid volume between those sterilization values. Unlike RTDs, TTDs have complete information on the temperature history in addition to duration of heating. Three parameters that can characterize the distribution are the lowest value, range, and average values. Average values are of interest in estimates of nutrient retention. For sterilization, the lowest value is of interest. The lowest value in the distribution corresponds to the particle whose time-temperature history combines (as given by equation 6) to be the lowest F_0 value. This least value of sterilization of a particle is difficult to obtain experimentally. In practice, least value of sterilizations obtained at various locations in the flow domain is used as the least sterilization for the process.

It is important to note the conceptual distinction between the time-temperature histories of location as opposed to a liquid particle (7). Time-temperature history at a location in a moving liquid would not correspond to a single liquid particle in general. Instead it would correspond to all particles passing through that point over time. Depending on the nature of the flow, point sterilization can be different from particle sterilization. From the food safety point of view, point sterilization is inherently safe because it assumes a particle is stagnant at a location

where temperature always stays the least. All physical particles stay only some time at the slowest heating point, and no particle is expected to stay all the time in the slowest heating point. Thus all particles in the system obtain a sterilization more than or equal to the sterilization calculated based on the temperature at the slowest heating location. It becomes difficult to refer to slowest heating location when no one location stays the coldest over the entire period, as in some complex recirculating flows described later. In practice, the region over which the coldest point moves over time is called the slowest heating zone. The time-temperature history at any location in the slowest-heating zone is considered to provide the least value of sterilization.

Rates of Convective Heat Transfer

In heating of liquids, whether as continuous flow in a heat exchanger or as batch processing in a can, it is useful to think of the heat transfer process as comprising several thermal resistances. These thermal resistances are related to the commonly used heat transfer coefficients that describe a heating process involving liquids, that is, in flowing media. Thus, the total thermal resistance between a fluid food and the heating medium is made up of internal thermal resistance between the fluid food and the wall of the heat exchanger or the container holding the fluid, resistance of the wall of the heat exchanger or the container, and the thermal resistance between the outside wall of the heat exchanger or the container and the heating medium such as steam. Symbolically, we can write (8)

$$\underbrace{1/U}_{\substack{\text{total}\\\text{resistance}}} = \underbrace{1/h_i}_{\substack{\text{resistance of}\\\text{liquid food}\\\text{inside the}\\\text{container}}} + \underbrace{\Delta x/k}_{\substack{\text{resistance of}\\\text{container wall}}} + \underbrace{1/h_o}_{\substack{\text{resistance of}\\\text{heating fluid}\\\text{outside the}\\\text{container}}} \quad (7)$$

Here U is the overall heat transfer coefficient between the fluid food and the heating medium, h_i is the heat transfer coefficient between the fluid food and the container or exchanger wall, k and Δx are the thermal conductivities and thicknesses, respectively, of the container or exchanger wall, and h_o is the heat transfer coefficient between the container or exchanger outside wall and the heating medium. The internal convective heat transfer coefficient, h_i, is generally small and the most significant one. The purpose of agitation is to significantly enhance it. The contribution due to wall conductivity and thickness can be ignored for metal cans with large thermal conductivity but can be significant with nonmetallic materials such as pouches. The external convective heat transfer coefficient is very large for steam condensation, and therefore its contribution can be ignored for steam heating. For heating with other liquids, the external heat transfer resistance can be significant.

The overall heat transfer coefficient U can be used to make an energy balance to calculate the change in food temperature. For a continuously flowing liquid, this energy balance can be written as

$$\dot{m}c_p dT = UdA(T_m - T) \qquad (8)$$

where dA is a small area of the heat exchanger surface over which heat transfer takes place and during which time the temperature of the liquid changes by an amount dT. The mass flow rate of the liquid being heated is $\dot{m}$, its specific heat is c_p, and T_m is the temperature of the heating medium. For batch heating of a liquid of mass m in a can, the energy balance can be written as

$$mc_p dT = UA(T_m - T)dt \qquad (9)$$

where A is typically the surface area of the can and m is the total mass of liquid that changes its temperature by an amount dT during the time dt. Integrating equation 8 over the complete area of the heat exchanger or integrating equation 9 over the total duration of heating gives the time-temperature history $T(t)$ of the liquid to be used with equation 6 to obtain the net sterilization of the liquid. This is particularly useful in complex heat transfer situations where spatial variation in temperature is either small or quite difficult to obtain.

Newtonian and Non-Newtonian Liquids

The relationship between shear stress and shear rate of a large number of liquids can be represented as

$$\sigma = \sigma_0 + K\left(-\frac{dv}{dz}\right)^n \qquad (10)$$

where σ is the shear stress, σ_0 is the shear stress needed to initiate flow, dv/dz is the velocity gradient of shear rate, K is called the consistency coefficient, and n is called the flow behavior index. When σ_0 is zero and n is equal to one, the liquid is called Newtonian and K becomes the viscosity of the liquid. For all other values of σ_0 and n, the liquid is described as non-Newtonian. Both Newtonian and non-Newtonian behavior are common to liquid foods (9). Among non-Newtonian liquid foods, a more common type is a pseudoplastic liquid for which σ_0 is zero and n is less than one. Many of the studies on non-Newtonian liquids referred to in this article are pseudoplastic type.

Laminar and Turbulent Flow

Turbulent flow is generally desired. Mechanical agitation is one way of getting turbulent flow as in a scraped-surface heat exchanger or in an agitating retort. In continuous flow, tube dimensions, flow rates, and liquid properties are adjusted to get a Reynolds number in the turbulent range. There are times, however, when turbulent flow is uneconomical to achieve, as in the case of many non-Newtonian liquids. These liquids exhibit high apparent viscosity, and pumping pressures required to obtain fully turbulent conditions are not economic for production rates of interest (10). Likewise, extent of agitation required for containerized non-Newtonian liquid to attain turbulent flow may not always be feasible for practical industrial applications.

Presence of Solid Particles in the Liquid

When solid particles are present in the liquid, a very complex heat-transfer system results. The particles in the liquid would need to be sterilized besides the liquid itself. Additional conductive resistance for the inside of the solid food and convective resistance between solid food surface and the liquid carrying it are added to the three resistances on the right side of equation 7. The thermal resistance represented by the surface heat transfer coefficient between the solid particles and the fluid carrying it is difficult to obtain experimentally or estimate from theoretical considerations. Additionally, the interior of the solid particles heat by a slow conduction process. The combined effect is that the sterilization of the food is limited by the sterilization of the particles.

AGITATED BATCH (IN-CONTAINER) HEATING OF LIQUIDS

Processing Equipment and Methods of Heating

Mechanical agitation is frequently used to enhance heat transfer by increasing the surface heat transfer coefficient between the container surface and the liquid in it. Processing time can be markedly shortened this way, generally resulting in improvement of quality. Quality also may improve because there is less chance for food to cook onto the can walls because the can contents are in motion. The critical factors that decide the effective agitation are the headspace or the amount of airspace in the can, consistency of the liquid, and speed and mode of agitation (11).

Several modes of agitation are used in the industry, some of which are shown in Figure 2. In retorts known as sterilmatic type (13), the can rotates part of the time about its own axis and part of the time about an axis parallel to its own, as it travels through the retort. As the can rolls, the headspace travels along the cylindrical contours of the can, resulting in the agitation of the contents. In a spin cooker the can is axially rotated (Fig. 2) at speeds as high as 500 rpm as it travels through the length of the retort.

In the other major type of agitation called end-over-end (EOE) agitation, the can is rotated around an axis perpen-

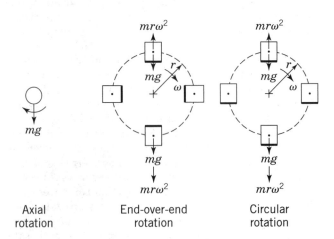

Axial rotation   End-over-end rotation   Circular rotation

Figure 2. Some common types of can rotation. *Source:* Ref. 12.

dicular to its own (Fig. 2). The axis of rotation is also located externally to the container. A positive movement of the headspace bubble is created throughout the rotating cycle. Because the headspace is moving along the length of the can and reverses direction every half a rotation, excellent mixing of the contents is achieved. In another rotating method of agitation (12) the cans are rotated in a circular path (Fig. 2) in such a way that the can orientation remains fixed. Agitation forces similar to those in EOE agitation are produced in this system.

The heating medium generally used is steam. Heating in direct flame with agitation of the cans is also used (14,15). The flame temperatures can be close to 2,000°C, and this high temperature is the primary contributor to rapid heat transfer.

Flow and Temperature Profiles

The complex nature of the agitation provided by the equipment discussed previously has generally defied any theoretical analysis to obtain detailed temperature and flow profiles. The temperature and flow patterns are affected by the headspace, fill of the container, solid-to-liquid ratio when solid particles are present, consistency of the liquid, and speed and type of agitation (11). There is only one theoretical study of liquid flow and heat transfer in a container that incorporates some agitations of the container similar to the sterilmatic-type retort discussed previously (16). The effects of container rotation about its own axis were considered to have the dominant effect on heat transfer and liquid flow over simultaneous rotation about an axis parallel to its own. Thus only container axial rotation was considered. Faster heating rates were observed in presence of rotation (centrifugally driven flows).

Experimental studies have measured temperature at several locations and correlated the heat transfer coefficient with other parameters. Nusselt-Prandtl correlations for obtaining heat transfer coefficient in axially rotated cans in a steam retort using water and silicone oil as model liquids have been provided (17). The temperature distribution in direct-flame heating of axially rotating cans was very uniform (15). Transient temperatures in an axially rotating can heated in a steam retort was shown to be quite uniform (Fig. 3). Temperature measurements in an end-over-end rotating can were also similar (Fig. 4) and showed uniformity, which further improved when the direction of rotation was reversed at intervals. It has been noted that the heat transfer coefficient was much higher in EOE as compared to axial rotation (18). It was also noted that the mere presence of a minimal-size headspace in EOE rotation markedly increases the heat transfer coefficient, although its contribution becomes progressively smaller (Fig. 5).

Due to the complexities of a theoretical or an experimental study, and the presence of fairly uniform temperatures inside the agitated container, detailed spatial variations of temperature and velocities are often bypassed. Instead, the energy balance in equation 9 is based on T being the mean fluid temperature. Using this formulation, overall heat transfer coefficient, U, is made available from experiment often in the form of standard Nusselt-Prandtl

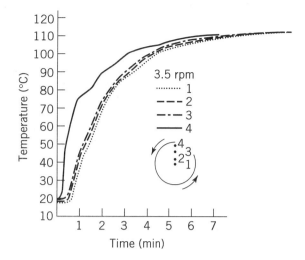

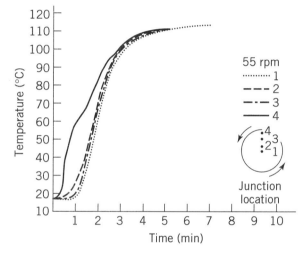

Figure 3. Temperature distribution at radial locations in an axially rotating horizontal can. *Source:* Ref. 18.

number correlations. The coefficient U depends on a number of processing parameters. For example, in EOE agitation, increasing the retort temperature increased the overall heat transfer coefficient, U, as shown in Figure 6, probably by lowering the viscosity of the fluid. Rotational speed also increased the heat transfer coefficient (Fig. 7), due to enhanced mixing. Figure 7 also shows the expected reduction in overall heat transfer coefficient for a more viscous fluid (oil).

The solution to equation 9 when starting from a uniform initial temperature of T_0 is

$$\frac{T - T_m}{T_0 - T_m} = e^{-(UA/mc_p)t} \tag{11}$$

which can also be written in the form

$$\log\left(\frac{T - T_m}{j(T_0 - T_m)}\right) = -t/f \tag{12}$$

where $j = 1$ and $f = mc_p/2.303UA$. Thus a plot of $\log((T - T_m)/(T_o - T_m))$ versus t would be a straight line charac-

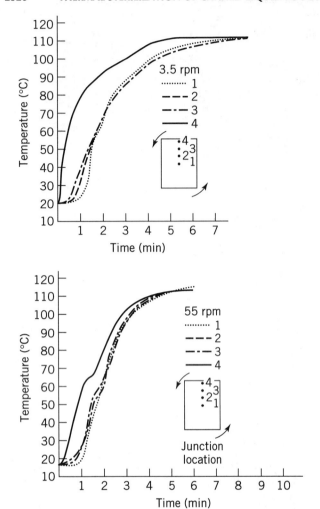

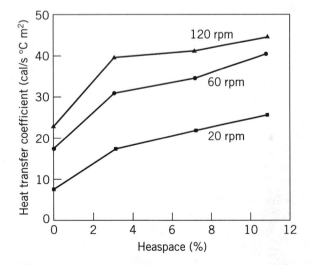

Figure 4. Temperature distribution at axial locations in an end-over-end agitated can. *Source:* Ref. 18.

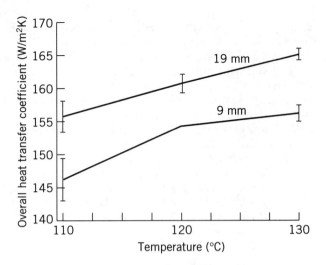

Figure 6. Overall heat transfer coefficient during end-over-end rotation at two (9 and 19 cm) radii of rotation at a rotational speed of 15 rpm as influenced by the retort (heating medium) temperature. *Source:* Ref. 20.

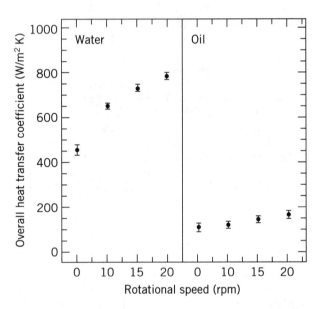

Figure 7. Overall heat transfer coefficient between a canned liquid and the heating medium during end-over-end rotation as influenced by the rotational speed of the can and the liquid viscosity. *Source:* Ref. 21.

terized by the slope $-1/f$ and intercept j. This enables the f and j values to be obtained experimentally from measured $T(t)$ data. These f and j values are used to calculate sterilization time for a process as discussed in books such as Ball and Olson (1).

Sterilization Values and Distribution

Because the detailed temperature and flow patterns have not been solved or experimentally obtained for the various complex heating situations, knowledge of distribution of sterilization (TTD) inside the agitated containers is not

Figure 5. Inside surface heat transfer coefficient of an end-over-end agitated can, as influenced by headspace. *Source:* Ref. 19.

available. Due to the presence of agitation, liquid would generally have a smaller spatial variation of temperature within the container and therefore a smaller spread in sterilization values. The precise cold point in the can is generally unavailable for specific heating situations due to unavailability of detailed spatial variation in temperature. In axially rotating cans under flame sterilization (15), the cold points were on the axis near the ends of the can, although temperatures at other locations varied very little except near the wall. The center of the can is generally assumed to be the cold point for agitated heating, although for small variations in temperature within the container this is somewhat irrelevant.

Effect of Non-Newtonian Liquid Behavior

Several non-Newtonian liquids in the steritort have been studied, and correlations similar to Nusselt-Prandtl number correlations for Newtonian liquids have been provided. Several other studies have also included non-Newtonian liquids, and these studies have been summarized (11). Studies in nonfood applications indicate that the convective heat transfer characteristics of non-Newtonian liquids are generally similar to those for Newtonian liquids (22,23)

Effect of Presence of Particulates

Several authors have studied agitated canned liquids in the presence of particulates, although rigorous engineering studies providing detailed temperature and velocity profiles in the liquid in presence of particulates are generally lacking (24,25). The additional critical parameter due to the presence of particles is, of course, the convective surface heat transfer coefficient between the fluid and the particles. It was noted that for agitations in a steritort, this surface heat transfer coefficient depended on the ratio of particle to container diameter, rotational speed of steritort, and liquid properties (26). In axial rotation of horizontal cylindrical containers, liquid to particle heat transfer coefficient for a physically restrained particle was found to be nearly invariant at $160 \pm 30 \text{W/m}^2\text{K}$ (27).

In EOE agitation, the overall heat transfer coefficient is shown (Fig. 8) to increase initially as particle concentration is increased from a single particle. This is attributed to the additional particles aiding the fluid mixing as compared to a single particle. After some concentration value, further increase in concentration reduces the overall heat transfer coefficient. Such decrease was also observed by others (26,27) and is attributed to the decreased ability of the fluid to move inside the can (28). The overall heat transfer coefficient decreased with increase in particle size (28), as shown in Figure 9. This trend is supported by reference 29 but is opposite that reported by reference 26. Spherical-shaped particles lead to a higher overall heat transfer coefficient, followed by cylindrical shape, with the cubic shape producing the smallest heat transfer coefficient (30).

NONAGITATED BATCH (IN-CONTAINER) HEATING OF LIQUIDS

Processing Equipment and Methods of Heating

Nonagitating or still retorts are used when agitators are to be avoided to keep the product or package integrity or

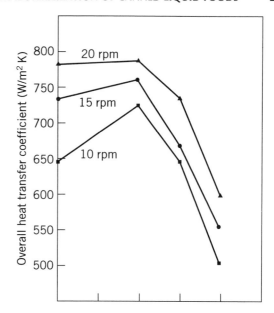

Figure 8. Overall heat transfer coefficient between solid food particles in a canned liquid and the heating medium as influenced by particle concentration during end-over-end rotation. *Source: Ref. 28.*

for economical reasons in small production volume. Due to slow movement of the liquid from natural convection, heating times required for sterilization are somewhat long. Steam is generally the heating medium, however, hot water is also used.

Flow and Temperature Profiles

In natural convection heat transfer, the liquid moves only due to the buoyancy induced by the change in temperature. This couples the flow and temperature fields and makes their computation a challenging task, as can be seen for pasteurization of beer in bottles (31) and sterilization of canned Newtonian (32) and non-Newtonian (33,34) fluids. Figures 10 and 11 show the typical flow patterns and temperature profiles in such nonagitated heating starting from a fluid at rest at a uniform temperature and raising its walls to the retort temperature. The liquid next to the sidewall, the top wall, and the bottom wall receives heat from the hot walls. As the liquid is heated, it expands and thus gets lighter. Liquid farther from the sidewall is still at a much lower temperature. The buoyancy force created by the liquids at different temperatures forces the hot liquid, next to the wall, upward. A sharp drop in temperature occurs from the wall to the core, creating a boundary layer. The buoyancy force resulting from this large temperature difference contributes to the largest velocities near the wall. The hot liquid going up is interrupted by the top wall and then travels radially toward the core. The core liquid, being heavier, moves downward and toward the wall. Thus a recirculating flow is created with a boundary layer at the sidewall, core flow at the centerline, and a mixing region at the top. Because the liquid is going up only through a small cross-sectional area (the thickness of the boundary

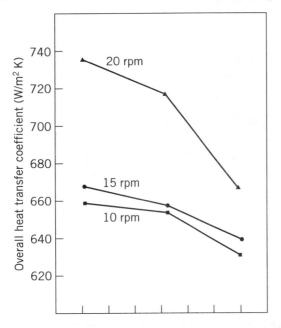

Figure 9. Overall heat transfer coefficient between solid food particles in a canned liquid and the heating medium as influenced by particle size during end-over-end rotation. *Source:* Ref. 28.

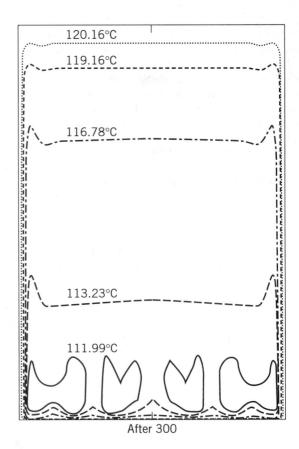

After 300

Figure 11. Computed transient temperature contours in natural convection heating of liquid in a cylindrical can heated from all sides with the wall temperature set at 121°C. *Source:* Ref. 32.

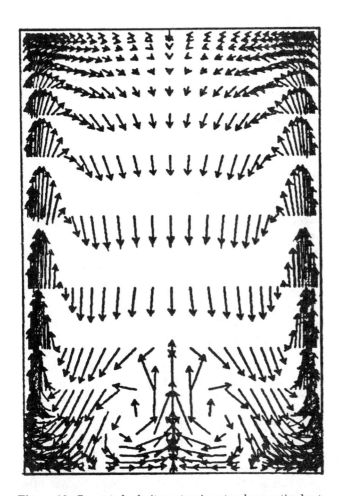

Figure 10. Computed velocity vectors in natural convection heating of liquid in a cylindrical can heated from all sides. *Source:* Ref. 32.

layer) while its downward movement is over a much larger area at the core, the downward velocity is much smaller.

Temperature values along the axis show that due to deposition of hot liquid, the top stays consistently at a higher temperature. However, at the very bottom, because the bottom wall was heated, the temperature is higher again. The cold point, therefore, exists somewhere in between and was found not to stay at the same location over time. These cold points were seen to migrate within the bottom 15% of can height with no particular pattern of migration (35). Similar location of cold points were experimentally observed (36–38).

Experimental studies of flow patterns during natural convection heating of a fluid food in a container have been performed. These are mostly qualitative visual observations, except in (37), where velocities were measured using a laser–Doppler velocimeter. Temperature profiles were measured in these studies and were often represented as the slope $-1/f$ and intercept j of semi–log time-temperature line as shown in equation 12. It has been shown (39) that use of such f and j values to represent time-temperature data for natural convection heating has been merely an extension of the formulas for conduction (1) and forced convection (equation 12) heating and is empirical

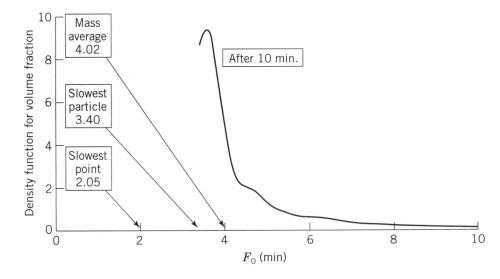

Figure 12. Computed distribution of sterilization (thermal-time-distribution) in liquid for the conditions in Figures 10 and 11.

without any physical or mathematical justification. Nevertheless, the f and j values measured from experiment are used to calculate sterilization time, as described in Ball and Olson (1).

Sterilization Values and Distribution

A typical transient distribution of sterilization (TTD described previously) in a cylindrical container (35) is shown in Figure 12. This was calculated by implicitly following the liquid elements (7) throughout the duration of heating. The volume (or mass) average sterilization lies within the range of the distribution, as expected. However, the sterilization calculated based on the temperature at the slowest heating point lies completely outside the distribution. This is because all physical fluid particles stay only some time in the slowest heating zone, and no particle stays all the time in the slowest heating zone. Thus all particles in the system obtain a sterilization more than what is calculated based on the slowest heating zone. Because sterilization at the slowest heating zone is easily measured by placing a thermocouple, this is the temperature used in practice to calculate sterilization. The slowest point value in Figure 12 demonstrates that use of such slowest heating thermocouple data provides additional overprocessing (and safety) beyond the true least sterilization of a fluid.

Effect of Non-Newtonian Liquid Behavior

For non-Newtonian (pseudoplastic) liquid with temperature-dependent viscosity, the recirculation pattern, radial and axial profile, and location of the slowest heating points were found (36) to be qualitatively similar to that of Newtonian liquid (35) discussed previously. As could be expected for the higher effective viscosity of pseudoplastic liquids, the maximum axial velocity near the wall was lower, being of the order of 10^{-4} m/s as opposed to 10^{-2} m/s for Newtonian liquid. The non-Newtonian nature itself was not expected to change the physics considerably because at the low shear rates involved in a naturally

convective flow, the liquid behaves close to a Newtonian liquid.

Starch gelatinization has been included (34,40) in unagitated heating of a canned 3.5% cornstarch dispersion. The thermorheological behavior of the starch dispersion was described in terms of three phases: (1) a pregelatinization phase where the magnitudes of apparent viscosity were assumed to be those of the continuous phase, (2) a gelatinization phase where the magnitudes of apparent viscosity increased, and (3) a postgelatinization phase where the magnitudes of apparent viscosity decreased. Although the spatial velocity and temperature profiles were qualitatively similar to the previous studies of unagitated heating (32), they (34,40) showed for the first time how the well-known broken heating curves (temperature-time curves with drastic changes in slope) develop during heating of starch-containing products, due to changes in the product apparent viscosity with temperature. The complex thermorheological behavior of the starch dispersion made it difficult to develop prediction equations for temperature-time histories using the commonly used f and j parameters.

Effect of Presence of Particulates

The complexities of natural convection heating are considerably amplified when solid particles are present in the liquid food, for which little information is available. If the overall heat transfer coefficient between the solid particles inside the container and the heating medium is available, equations for conduction heating can be used to estimate the particle temperatures and insure sterility. Dimensionless Nusselt-Prandtl type correlations for the overall heat transfer coefficients between mushroom-shaped particles and water in a still retort have been provided (41), although there was considerable scatter in the data. Presence of spherical glass particles reduced natural convection in viscous liquid such as silicone, whereas the particles had very little effect on convection in thin liquid such as water (37). Smaller particles caused greater reduction in the convective flow.

BIBLIOGRAPHY

1. C. O. Ball and F. C. W. Olson, *Sterilization in Food Technology*, McGraw-Hill, New York, 1957.

2. C. R. Stumbo, *Thermobacteriology in Food Processing*, Academic Press, Orlando, Fla., 1965.

3. J. M. Smith, *Engineering Kinetics*, McGraw-Hill, New York, 1981.

4. A. M. Scalzo et al., "Residence Time of Egg Products in Holding Tubes of Egg Pasteurizers," *Food Technol.* **23**, 678–681 (1969).

5. M. A. Rao and M. A. Loncin, "Residence Time Distribution and Its Role in Continuous Pasteurization, Part 1," *Lebensm.-Wiss. Technol.* **7**, 5–13 (1974).

6. E. B. Nauman, "Nonisothermal Reactors: Theory and Applications of Thermal Time Distributions," *Chem. Eng. Sci.* **32**, 359–367 (1977).

7. A. K. Datta, "Integrated Thermokinetic Modeling of Processed Liquid Food Quality," in W. E. L. Spiess and H. Schubert, eds., *Engineering and Food*, Vol. 1, *Physical Properties and Process Control*, Elsevier Applied Science, New York, 1989, pp. 838–847.

8. F. P. Incropera and D. P. Dewitt, *Introduction to Heat Transfer*, Wiley, New York, 1984.

9. M. A. Rao and S. S. H. Rizvi, *Engineering Properties of Foods*, Marcel Dekker, New York, 1986.

10. A. H. P. Skelland, *Non-Newtonian Fluid Flow and Heat Transfer*, Wiley, New York, 1967.

11. M. A. Rao and R. C. Anantheswaran, "Convective Heat Transfer to Fluid Foods in Cans," *Advances in Food Research* **32**, 39–84 (1988).

12. P. Parchomchuk, "A Simplified Method for Agitation Processing of Canned Foods," *J. Food Sci.* **42**, 265–268 (1977).

13. A. Lopez, *A Complete Course in Canning: Book 1, Basic Information on Canning*, Canning Trade, Baltimore, Md., 1981.

14. D. J. Casimir, "Flame Sterilization," *CSIRO Food Research Quarterly* **35**, 34–39 (1975).

15. R. L. Merson et al., "Temperature Distributions and Liquid Side Heat Transfer Coefficients in Model Liquid Foods in Cans Undergoing Flame Sterilization Heating," *Journal of Food Process Engineering* **4**, 85–98 (1981).

16. F. Ladiende, "Studies on Thermal Convection in Self-Gravitating and Rotating Horizontal Cylinders in a Vertical External Gravity Field," Ph.D. Dissertation, Cornell University, Ithaca, N.Y., 1988.

17. C. L. Soule and R. L. Merson, "Heat Transfer Coefficients to Liquids in Axially Rotated Cans," *Journal of Food Process Engineering* **8**, 33–46 (1985).

18. S. Hotani and T. Mihori, "Some Thermal Engineering Aspects of the Rotation Method in Sterilization," in T. Motohiro and K. Hayakawa, eds., *Heat Sterilization of Food*, Koseisha-Koseikaku, Tokyo, 1983.

19. D. Naveh and I. J. Kopelman, "Effects of Some Processing Parameters on the Heat Transfer Coefficients in a Rotating Autoclave," *Journal of Food Processing and Preservation* **4**, 67–77 (1980).

20. S. S. Sablani and H. S. Ramaswamy, "Particle Heat Transfer Coefficients under Various Retort Operating Conditions with End-over-End Rotation," *Journal of Food Process Engineering* **19**, 403–424 (1996).

21. H. S. Ramaswamy and S. S. Sablani, "Particle Motion and Heat Transfer in Cans during End-over-End Rotation: Influence of Physical Parameters and Rotational Speed," *Journal of Food Process Engineering* **21**, 105–127 (1997).

22. E. M. Parmentier, "A Study of Thermal Convection in Non-Newtonian Fluids," *J. Fluid Mech.* **84**, 1–11 (1978).

23. S. F. Liang and A. Acrivos, "Experiments on Buoyancy Driven Convection in Non-Newtonian Fluid," *Rheol. Acta* **9**, 447–454 (1970).

24. N. G. Stoforos and R. L. Merson, "Estimating Heat Transfer Coefficients in Liquid/Particulate Canned Foods Using Only Liquid Temperature Data," *J. Food Sci.* **55**, 478–483 (1990).

25. A. N. Lekwauwa and K. Hayakawa, "Computerized Model for the Prediction of Thermal Responses of Packaged Solid Liquid Food Mixture Undergoing Thermal Processes," *J. Food Sci.* **51**, 1042–1049 (1986).

26. M. K. Lenz and D. B. Lund, "The Lethality-Fourier Number Method: Heating Rate Variations and Lethality Confidence Intervals for Forced-Convection Heated Foods in Containers," *Journal of Food Process Engineering* **2**, 227–271 (1978).

27. M. F. Deniston, B. H. Hassan, and R. L. Merson, "Heat Transfer Coefficients to Liquids with Food Particulates in Axially Rotating Cans," *J. Food Sci.* **52**, 962–966 (1987).

28. S. S. Sablani and H. S. Ramaswamy, "Heat Transfer to Particles in Cans with End-over-End Rotation: Influence of Particle Size and Concentration (% v/v), *Journal of Food Process Engineering* **20**, 265–283 (1997).

29. B. Hassan, "Heat Transfer Coefficients for Particles in Liquid in Axially Rotating Cans," Ph.D. Dissertation, University of California, Davis, 1984.

30. H. S. Ramaswamy and S. S. Sablani, "Particle Shape Influence on Heat Transfer in Cans Containing Liquid Particle Mixtures Subjected to End-over-End Rotation," *Lebensm.-Wiss. Technol.* **30**, 525–535 (1997).

31. M. S. Engelman and R. L. Sani, "Finite-Element Simulation of an In-Package Pasteurization Process," *Numerical Heat Transfer* **6**, 41–54 (1983).

32. A. K. Datta and A. Teixeira, "Numerical Modeling of Natural Convection Heating in Canned Liquid Foods," *Transactions of the ASAE* **30**, 1542–1551 (1987).

33. A. Kumar, M. Bhattacharya, and J. Blaylock, "Numerical Simulation of Natural Convection Heating of Canned Thick Viscous Liquid Products," *J. Food Sci.* **55**, 1403–1410 (1990).

34. W. H. Yang and M. A. Rao, "Numerical Study of Parameters Affecting Broken Heating Curve," *Journal of Food Engineering* **37**, 43–61 (1998).

35. A. K. Datta and A. Teixeira, "Numerically Predicted Transient Temperature and Velocity Profiles during Natural Convection Heating of Canned Liquid Foods," *J. Food Sci.* **53**, 191–195 (1988).

36. J. L. Blaisdell and W. J. Harper, "Location of Slowest Cooling Zone and Rate Cooling of Homogenized Milk in Various Containers," *Transactions of ASAE* **13**, 433–435 (1970).

37. J. Hiddink, *Natural Convection Heating of Liquids, with Reference to Sterilization of Canned Foods*, Agricultural Research Reports No. 839, Center for Agricultural Publishing and Documentation, Wageningen, The Netherlands, 1975.

38. L. G. Zechman and I. J. Pflug, "Location of the Slowest Heating Zone for Natural-Convection-Heating Fluids in Metal Containers," *J. Food Sci.* **54**, 205–209 (1989).

39. A. K. Datta, "On the Theoretical Basis of the Asymptomatic Semilogarithmic Heat Penetration Curves Used in Food Processing," *Journal of Food Engineering* **12**, 177–190 (1990).

40. W. H. Yang and M. A. Rao, "Transient Natural Convection Heat Transfer to Starch Dispersion in a Cylindrical Container: Numerical Solution and Experiment," *Journal of Food Engineering* **36**, 395–415 (1998).

41. S. K. Sastry, "Convective Heat Transfer Coefficients for Canned Mushrooms Processed in Still Retorts," *1984 Winter Meet. ASAE*, New Orleans, La., December 11–14, 1984.

ASHIM K. DATTA
Cornell University
Ithaca, New York

THERMODYNAMICS

Thermodynamics is the science of the dynamics of heat and the quantitative relationship between heat and other forms of energy. It is the basis for:

1. Analyzing and studying the transformation of energy from one form to another.
2. The use and availability of energy to do useful work.
3. The stability and equilibrium associated with chemical substances.

Thermodynamics relationships pertain to processes in which matter changes state or in equilibria in which matter remains fixed. The application of thermodynamic relationships are most frequently used in food processing operations that involve transferring heat and changing the state of materials.

Processing changes of state take place during heating and cooling, expanding and contracting, vaporizing and condensing, and when chemical reactions result from reactions in a food system. Since all forms of energy tend to change into heat, the quantitative measurement of temperature is a most important controlling factor in studying and controlling food systems.

There must always be a driving force or energy potential difference between two systems before a useful relationship can be realized. The end point of any process is reached when mechanical, physical, thermal, or chemical equilibrium is reached or when the two systems are disengaged. These concepts are embodied in the Laws of Thermodynamics, the first law being the basis of the science.

LAWS OF THERMODYNAMICS

There are four laws of thermodynamics which are developed into basic equations describing property relationships, energy balances, entropy balances, and mass balances. An excellent way of stating the four laws of thermodynamics can be demonstrated by visualizing three bodies, A, B, and C (1). A and C are identical rigid hot bodies at the same temperature and body B is a cold body below the temperature of A and C. In discussing the laws and the applications, especially in food processing, it is useful to visualize the transfer of heat between these bodies.

Zeroth Law

When bodies A and C are placed in contact with each other, no change will occur. This is a relatively recent law, added to the original three laws, to emphasize the fact that when two bodies or systems are at the same state of energy, no useful work can be obtained by putting the two systems in contact with each other. From this, it follows that the zeroth law also states that two systems in thermal equilibrium with a third are in thermal equilibrium with each other.

First Law

If A is placed in contact with B, there will be an increase in the internal energy of B equaling a decrease in the internal energy of A. The first law as originally stated is that the sum of all the energies in an isolated system is constant or that energy may be transformed from one form into another, but it cannot be created or destroyed.

This law is particularly important as applied to food processing in which the basis for operations is the transfer of mass and energy. Food processing involves heating, cooling, changing state (freezing and evaporating), adding chemicals that may cause internal energy changes, and irradiation.

Second Law

After A has been in contact with B, the initial conditions will never be restored without outside influence. The cooling of A and heating of B constitute a spontaneous, irreversible process resulting in loss of ability to do work. Simply stated the second law is that all systems tend to approach a state of equilibrium.

Third Law

It is impossible by any procedure, no matter how idealized, to reduce any system to the absolute zero of temperature in a finite number of operations. As in the case of the zeroth law, the third law has little practical application in determining energy relationships involving food processing. It is primarily applied to the physics of very low temperatures.

USING THERMODYNAMIC RELATIONSHIPS IN FOOD PROCESSING

Prior to discussing specific food processing operations, it is necessary to define the world or system in which the process takes place. That is, boundaries must be defined as a means of isolating the process from the surrounding environment. There is usually some interchange between a system and the surroundings but they do not affect the principal relationships within the system. For example, if a food is being heated in a retort the heat balance must account for the heat interchange between the source and the food plus allow for a heat loss through the insulated or lagged walls of the retort.

The most important thermodynamic applications reflected in food processing involve: (a) transferring mass within the processing environment or between phases in a process, (b) transferring heat to heat or cool a food and, (c) transferring heat to change the phase of a component in a food processing system.

Basic Relationships

In considering units of measurement, it is necessary to define base values from which one can compare changes. For example, when one is standing on a ladder, the height of 3 meters would be compared to the base value of zero, the floor. Otherwise 3 meters would have no meaning. Likewise, the base for measuring temperature, and thus being able to determine heat energy changes in a body, is absolute zero.

The heat content or enthalpy of a food and the change of this parameter is basic to processing. It is defined as:

$$H = E + pv \tag{1}$$

where E is the internal energy, p is the absolute pressure, and v is the volume. The changes in enthalpy and not the absolute values are important in food processing. Hence,

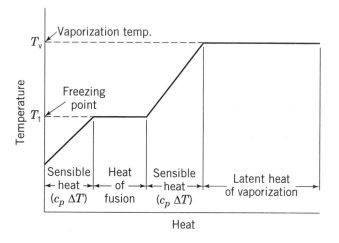

Figure 1. Heat stages when heating a food from a frozen solid to a vapor at constant pressure.

Table 1. Physical Constants of Water and Ice

Molecular weight	18.01534
Phase transition properties	
Melting point at 101.3 kPa (1 atm)	0.000°C
Boiling point at 101.3 kPa (1 atm)	100.000°C
Critical temperature	374.15°C
Triple point	0.0099°C and 610.4 kPa (4.579 mm Hg)
Heat of fusion at 0°C	6.012 kJ (1.436 kcal)/mol
Heat of vaporization at 100°C	40.63 kJ (9.705 kcal)/mol
Heat of sublimation at 0°C	50.91 kJ (12.16 kcal)/mol

Source: Refs. 5 and 6.

at constant pressure, which is the case in most food processing:

$$\Delta H = \Delta E + p\Delta v \tag{2}$$

Often in a food process, there is no pv work so that the change in internal energy is equal to the change in enthalpy.

The specific heat of a food is the change in heat content per unit mass per unit temperature change. At constant volume

$$c_{v\ \text{Avg}} = \frac{Q_v}{T_2 - T_1} = \frac{E_2 - E_1}{T_2 - T_1} \tag{3}$$

and at constant pressure

$$c_{p\ \text{Avg.}} = \frac{Q_p}{T_2 - T_1} = \frac{H_2 - H_1}{H_2 - H_1} \tag{4}$$

In actuality, specific heat is not constant over significant temperature and pressure ranges and each of the above equations must be considered in integral form for precise definitions. However, over most ranges of temperature and pressure at which food process and auxiliary food operations take place, the specific heats can be considered constant and the average value is satisfactory.

Entropy (S) expresses the second law in precise mathematical terms. The first law is concerned with the accounting of the many kinds of energy that are involved in a given process whereas the second law is concerned with how available the energy of this system is for doing useful work. The concept of entropy is hard to visualize since it is not defined in terms which one can equate to normal physical entities such as a mass of material or a given amount of heat. It is expressed as follows:

$$dS = \frac{dq}{T} \tag{5}$$

The value of entropy is always positive and is actually a measure of the decline of the universe.

The total energy balance of a fluid flowing at steady state in a closed system (eg, food flowing through a pipe or chamber) accounts for the various changes that effect the thermodynamic state of the system between two given locations. An energy balance between the two sections of the fluid which is flowing is simplified for most food processing conditions in which the food is incompressible and the specific volume is essentially constant:

$$x_1 + E_1 + p_1v_1 + V_1^2/2g + W$$
$$= x_2 + E_2 + p_2v_2 + V_2^2/2g + F \tag{6}$$

x is the potential energy due to height above reference plane, E is the internal energy, pv is the pv work done by the system, $V^2/2g$ is the kinetic energy of the flowing fluid, W is the work done by the system, and F is the friction loss due to contact with walls of the vessel. The total change in

Table 2. Useful Properties of Water

			Properties of saturated water			
T (°C)	ρ (kg/m³)	C_p (J/kg·K)	k (W/m·K)	μ (Pa·s)	α (m²s)	Pr
0	1002	4218	0.552	17.9×10^{-4}	1.31×10^{-7}	13.06
20	1001	4182	0.597	10.1	1.43	7.02
40	995	4178	0.628	6.55	1.51	4.34
60	985	4184	0.651	4.71	1.55	3.02
80	974	4196	0.668	3.55	1.64	2.22
100	960	4216	0.680	2.82	1.68	1.74
120	945	4250	0.685	2.33	1.71	1.45
140	928	4283	0.684	1.99	1.72	1.24
160	910	4342	0.680	1.73	1.73	1.10
180	889	4417	0.675	1.54	1.72	1.00
200	867	4505	0.665	1.39	1.71	0.94
220	842	4610	0.652	1.26	1.68	0.89
240	816	4756	0.635	1.17	1.64	0.88
260	786	4949	0.611	1.08	1.58	0.87
280	753	5208	0.580	1.02	1.48	0.91
300	714	5728	0.540	0.96	1.32	1.02

Source: From Ref. 7. Reprinted with permission.

heat content of the system or the amount of heat transferred is, therefore:

$$Q = \Delta PE + \Delta E + \Delta pv + \Delta KE - W - F \quad (7)$$

ΔPE is the change in potential energy, $x_2 - x_1$, ΔE is the change in internal energy, $E_2 - E_1$, Δpv is the change in pv work, $p_2 v_2 - p_1 v_1$, and ΔKE is the change in kinetic energy, $V_2^2/2g - V_1^2/2g$.

In considering food processes, there are many simplifications to the rigorous thermodynamic relationships. This is possible since most of the fluids are being heated or cooled and there are negligible changes in specific volume, and kinetic energy and no shaft work is being done by the flowing product. Therefore, since $\Delta E + \Delta pv = \Delta H$ = enthalpy, the work required for compressing a gas or the heat change in cooling or heating is equal to the enthalpy and equation 7 becomes:

$$W = Q = \Delta H \quad (8)$$

Heating or Cooling Foods

The amount of heat required to raise or lower the temperature of a food is dependent on the specific heat. As previously discussed, over the temperature ranges encountered in food processing, the specific heat (c_p) can be considered constant and the heat added or lost (called sensible heat) is

$$Q = c_p(T - T) \quad (9)$$

When a change in phase takes place, there is an additional heat input required called the "latent" heat. Hence, the latent heat of vaporization or the latent heat of fusion must be considered when a product is being evaporated or frozen. Although certain foods (eg, pure fats) have independent latent heat values for phase change, the latent heat of most foods is associated with water. When foods

are dehydrated or frozen, water is normally the major component that determines the latent heat required for the phase change. In fact, water is the sole consideration in dehydration processes since that is the only component that is removed by a change of state.

Figure 1 is a generalized diagram showing the stages of heat change taking place as a product is heated and goes through phases changes as follows:

1. Heating in frozen stage to freezing point, $Q_1 = c_p \Delta T$ (c_p is that for the frozen product).
2. Changing from frozen to solid or liquid food, $Q_2 =$ Latent heat of fusion.
3. Heating liquid or solid food, $Q_3 = c_p \Delta T$. (c_p is that for the nonfrozen product).
4. Changing to vapor phase, $Q =$ Latent heat or vaporization

$$Q_T = Q_1 + Q_2 + Q_3 + Q_4$$

Values for thermodynamic properties of foods and other substances are available in many scientific handbooks. These give the thermal properties as related to temperature, pressure, and volume. Most frequently used are Tables giving the basic properties of water (Tables 1, Table 2) and saturated steam (Table 3). The base or reference point for published thermodynamic properties have been established by international agreements as $E = 0$ and entropy at 0.01°C.

NOMENCLATURE

c specific heat of substance, J/(kg)(K) or Btu/(lb)(F).

c_p specific heat at constant pressure, J/(kg)(K) or Btu/(lb)(F).

c_v specific heat at constant volume, J/(kg)(K) or Btu/(lb)(F).

Table 3. Properties of Saturated Steam: Temperature Table (SI Units)

Temp. (°C) T	Press. (kPa) P	Specific volume m³/kg		Internal energy kJ/kg			Enthalpy kJ/kg			Entropy kJ/kg · K		
		Sat. liquid v_f	Sat. vapor v_g	Sat. liquid u_f	Evap. u_{fg}	Sat. vapor u_g	Sat. liquid h_f	Evap. h_{fg}	Sat. vapor h_g	Sat. liquid s_f	Evap. s_{fg}	Sat. vapor s_g
0.01	0.6113	0.001 000	206.14	0.00	2375.3	2375.3	0.01	2501.3	2501.4	0.0000	9.1562	9.1562
5	0.8721	0.001 000	147.12	20.97	2361.3	2382.3	20.98	2489.6	2510.6	0.0761	8.9496	9.0257
10	1.2276	0.001 000	106.38	42.00	2347.2	2389.2	42.01	2477.7	2519.8	0.1510	8.7498	8.9008
15	1.7051	0.001 001	77.93	62.99	2333.1	2396.1	62.99	2465.9	2528.9	0.2245	8.5569	8.7814
20	2.339	0.001 002	57.79	83.95	2319.0	2402.9	83.96	2454.1	2538.1	0.2966	8.3706	8.6672
25	3.169	0.001 003	43.36	104.88	2304.9	2409.8	104.89	2442.3	2547.2	0.3674	8.1905	8.5580
30	4.246	0.001 004	32.89	125.78	2290.8	2416.6	125.79	2430.5	2556.3	0.4369	8.0164	8.4533
35	5.628	0.001 006	25.22	146.67	2276.7	2423.4	146.68	2418.6	2565.3	0.5053	7.8478	8.3531
40	7.384	0.001 008	19.52	167.56	2262.6	2430.1	167.57	2406.7	2574.3	0.5725	7.6845	8.2570
45	9.593	0.001 010	15.26	188.44	2248.4	2436.8	188.45	2394.8	2583.2	0.6387	7.5261	8.1648
50	12.349	0.001 012	12.03	209.32	2234.2	2443.5	209.33	2382.7	2592.1	0.7038	7.3725	8.0763
55	15.758	0.001 015	9.568	230.21	2219.9	2450.1	230.23	2370.7	2600.9	0.7679	7.2234	7.9913
60	19.940	0.001 017	7.671	251.11	2205.5	2456.6	251.13	2358.5	2609.6	0.8312	7.0784	7.9096
65	25.03	0.001 020	6.197	272.02	2191.1	2463.1	272.06	2346.2	2618.3	0.8935	6.9375	7.8310
70	31.19	0.001 023	5.042	292.95	2176.6	2469.6	292.98	2333.8	2626.8	0.9549	6.8004	7.7553
75	38.58	0.001 026	4.131	313.90	2162.0	2475.9	313.93	2321.4	2635.3	1.0155	6.6669	7.6824
80	47.39	0.001 029	3.407	334.86	2147.4	2482.2	334.91	2308.8	2643.7	1.0753	6.5369	7.6122
85	57.83	0.001 033	2.828	355.84	2132.6	2488.4	355.90	2296.0	2651.9	1.1343	6.4102	7.5445
90	70.14	0.001 036	2.361	376.85	2117.7	2494.5	376.92	2283.2	2660.1	1.1925	6.2866	7.4791
95	84.55	0.001 040	1.982	397.88	2102.7	2500.6	397.96	2270.2	2668.1	1.2500	6.1659	7.4159
	MPa											
100	0.101 35	0.001 044	1.6729	418.94	2087.6	2506.5	419.04	2257.0	2676.1	1.3069	6.0480	7.3549
105	0.120 82	0.001 048	1.4194	440.02	2072.3	2512.4	440.15	2243.7	2683.8	1.3630	5.9328	7.2958
110	0.143 27	0.001 052	1.2102	461.14	2057.0	2518.1	461.30	2230.2	2691.5	1.4185	5.8202	7.2387
115	0.169 06	0.001 056	1.0366	482.30	2041.4	2523.7	482.48	2216.5	2699.0	1.4734	5.7100	7.1833
120	0.198 53	0.001 060	0.8919	503.50	2025.8	2529.3	503.71	2202.6	2706.3	1.5276	5.6020	7.1296
125	0.2321	0.001 065	0.7706	524.74	2009.9	2534.6	524.99	2188.5	2713.5	1.5813	5.4962	7.0775
130	0.2701	0.001 070	0.6685	546.02	1993.9	2539.9	546.31	2174.2	2720.5	1.6344	5.3925	7.0269
135	0.3130	0.001 075	0.5822	567.35	1977.7	2545.0	567.69	2159.6	2727.3	1.6870	5.2907	6.9777
140	0.3613	0.001 080	0.5089	588.74	1961.3	2550.0	589.13	2144.7	2733.9	1.7391	5.1908	6.9299
145	0.4154	0.001 085	0.4463	610.18	1944.7	2554.9	610.68	2129.6	2740.3	1.7907	5.0926	6.8833
150	0.4758	0.001 091	0.3928	631.68	1927.9	2559.5	632.20	2114.3	2746.5	1.8418	4.9960	6.8379
155	0.5431	0.001 096	0.3468	653.24	1910.8	2564.1	653.84	2098.6	2752.4	1.8925	4.9010	6.7935
160	0.6178	0.001 102	0.3071	674.87	1893.5	2568.4	675.55	2082.6	2758.1	1.9427	4.8075	6.7502
165	0.7005	0.001 108	0.2727	696.56	1876.0	2572.5	697.34	2066.2	2763.5	1.9925	4.7153	6.7078
170	0.7917	0.001 114	0.2428	718.33	1858.1	2576.5	719.21	2049.5	2768.73	2.0419	4.6244	6.6663
175	0.8920	0.001 121	0.2168	740.17	1840.0	2580.2	741.17	2032.4	2773.6	2.0909	4.5347	6.6256
180	1.0021	0.001 127	0.194 05	762.09	1821.6	2583.7	763.22	2015.0	2778.2	2.1396	4.4461	6.5857
185	1.1227	0.001 134	0.174 09	784.10	1802.9	2587.0	785.37	1997.1	2782.4	2.1879	4.3586	6.5465
190	1.2544	0.001 141	0.156 54	806.19	1783.8	2590.0	807.62	1978.8	2786.4	2.2359	4.2720	6.5079
195	1.3978	0.001 149	0.141 05	828.37	1764.4	2592.8	829.98	1960.0	2790.0	2.2835	4.1863	6.4698
200	1.5538	0.001 157	0.127 36	850.65	1744.7	2595.3	852.45	1940.7	2793.2	2.3309	4.1014	6.4323
205	1.7230	0.001 164	0.115 21	873.04	1724.5	2597.5	875.04	1921.0	2796.0	2.3780	4.0172	6.3952
210	1.9062	0.001 173	0.104 41	895.53	1703.9	2599.5	897.76	1900.7	2798.5	2.4248	3.9337	6.3585
215	2.104	0.001 181	0.094 79	918.14	1682.9	2601.1	920.62	1879.9	2800.5	2.4714	3.8507	6.3221
220	2.318	0.001 190	0.086 19	940.87	1661.5	2602.4	943.62	1858.5	2802.1	2.5178	3.7683	6.2861
225	2.548	0.001 199	0.078 49	963.73	1639.6	2603.3	966.78	1836.5	2803.3	2.5639	3.6863	6.2503
230	2.795	0.001 209	0.071 58	986.74	1617.2	2603.9	990.12	1813.8	2804.0	2.6099	3.6047	6.2146
235	3.060	0.001 219	0.065 37	1009.89	1594.2	2604.1	1013.62	1790.5	2804.2	2.6558	3.5233	6.1791
240	3.344	0.001 229	0.059 76	1033.21	1570.8	2604.0	1037.32	1766.5	2803.8	2.7015	3.4422	6.1437
245	3.648	0.001 240	0.054 71	1056.71	1546.7	2603.4	1061.23	1741.7	2803.0	2.7472	3.3612	6.1083
250	3.973	0.001 251	0.050 13	1080.39	1522.0	2602.4	1085.36	1716.2	2801.5	2.7927	3.2802	6.0730
255	4.319	0.001 263	0.045 98	1104.28	1496.7	2600.9	1109.73	1689.8	2799.5	2.8383	3.1992	6.0375
260	4.688	0.001 276	0.042 21	1128.39	1470.6	2599.0	1134.37	1662.5	2796.9	2.8838	3.1181	6.0019
265	5.081	0.001 289	0.038 77	1152.74	1443.9	2596.6	1159.28	1634.4	2793.6	2.9294	3.0368	5.9662
270	5.499	0.001 302	0.035 64	1177.36	1416.3	2593.7	1184.51	1605.2	2789.7	2.9751	2.9551	5.9301
275	5.942	0.001 317	0.032 79	1202.25	1387.9	2590.2	1210.07	1574.9	2785.0	3.0208	2.8730	5.8938
280	6.412	0.001 332	0.030 17	1227.46	1358.7	2586.1	1235.99	1543.6	2779.6	3.0668	2.7903	5.8571
285	6.909	0.001 348	0.027 77	1253.00	1328.4	2581.4	1262.31	1511.0	2773.3	3.1130	2.7070	5.8199

Table 3. Properties of Saturated Steam: Temperature Table (SI Units) (*continued*)

Temp. (°C) T	Press. (kPa) P	Specific volume m³/kg		Internal energy kJ/kg			Enthalpy kJ/kg			Entropy kJ/kg · K		
		Sat. liquid v_f	Sat. vapor v_g	Sat. liquid u_f	Evap. u_{fg}	Sat. vapor u_g	Sat. liquid h_f	Evap. h_{fg}	Sat. vapor h_g	Sat. liquid s_f	Evap. s_{fg}	Sat. vapor s_g
	MPa											
290	7.436	0.001 366	0.025 57	1278.92	1297.1	2576.0	1289.07	1477.1	2766.2	3.1594	2.6227	5.7821
295	7.993	0.001 384	0.023 54	1305.2	1264.7	2569.9	1316.3	1441.8	2758.1	3.2062	2.5375	5.7437
300	8.581	0.001 404	0.021 67	1332.0	1231.0	2563.0	1344.0	1404.9	2749.0	3.2534	2.4511	5.7045
305	9.202	0.001 425	0.019 948	1359.3	1195.9	2555.2	1372.4	1366.4	2738.7	3.3010	2.3633	5.6643
310	9.856	0.001 447	0.018 350	1387.1	1159.4	2546.4	1401.3	1326.0	2727.3	3.3493	2.2737	5.6230
315	10.547	0.001 472	0.016 867	1415.5	1121.1	2536.6	1431.0	1283.5	2714.5	3.3982	2.1821	5.5804
320	11.274	0.001 499	0.015 488	1444.6	1080.9	2525.5	1461.5	1238.6	2700.1	3.4480	2.0882	5.5362
330	12.845	0.001 561	0.012 996	1505.3	993.7	2498.9	1525.3	1140.6	2665.9	3.5507	1.8909	5.4417
340	14.586	0.001 638	0.010 797	1570.3	894.3	2464.6	1594.2	1027.9	2622.0	3.6594	1.6763	5.3357
350	16.513	0.001 740	0.008 813	1641.9	776.6	2418.4	1670.6	893.4	2563.9	3.7777	1.4335	5.2112
360	18.651	0.001 893	0.006 945	1725.2	626.3	2351.5	1760.5	720.5	2481.0	3.9147	1.1379	5.0526
370	21.03	0.002 213	0.004 925	1844.0	384.5	2228.5	1896.5	441.6	2332.1	4.1106	0.6865	4.7971
374.14	22.09	0.003 155	0.003 155	2029.6	0	2029.6	2099.3	0	2099.3	4.4298	0	4.4298

Source: Adapted from Ref. 8.

F	friction work, J/kg or Btu/lb.
H	enthalpy of product, kJ/kg or Btu/lb.
KE	kinetic energy, J/kg or Btu/lb.
m	mass, kg or lb.
p	pressure, Pa (N/m) or lb/in. (psi).
PE	potential energy, J/kg or Btu/lb.
Q	quantity of heat transferred, J/kg or Btu/lb.
T	temperature, °C or °F.
E	internal energy, J/kg or Btu/lb.
v	volume, m or ft or specific volume/unit mass, m/kg or ft/lb.
W	work, J/kg or Btu/lb.
Δ	change in a property (eg, temperature, enthalpy).

BIBLIOGRAPHY

1. J. J. Martin, *Thermodynamics Program, A.I.Ch.E. Teaching Aids*, American Institute of Chemical Engineers, New York, 1965.

2. E. L. Watson and J. C. Harper, *Elements of Food Engineering*, Van Nostrand Reinhold Company, New York, 1986.

3. D. R. Heldman, *Food Process Engineering*, AVI Publishing Company, Inc., Westport, Conn., 1977.

4. G. M. Pigott, compilation of classroom lecture notes and unit operations laboratory experiments over 28 years of teaching Food Engineering, University of Washington, Seattle, 1989.

5. F. Franks, ed., *Water—A Comprehensive Treatise*, 6 vols., Plenum Press, New York, 1972–1979.

6. R. C. Weast and M. J. Astle, eds., *CRC Handbook of Chemistry and Physics*, CRC Press, Boca Raton, Fla., 1981.

7. J. A. Duffle and W. A. Beckman, *Solar Engineering of Thermal Processes*, Wiley-Interscience, New York, 1980.

8. J. H. Keenan, F. G. Keyes, P. G. Hill, and J. G. Moore, *Steam Tables*, John Wiley & Sons, Inc., New York, 1969; Reprinted in G. J. Van Wylen and R. E. Sonntag, *Fundamentals of Classical Thermodynamics*, John Wiley & Sons, Inc., New York, 1978.

GENERAL REFERENCES

H. Cabezas, Jr., M. B. Kabiri, and D. C. Szlag, "Statistical Thermodynamics of Phase Separation and Ion Partitioning in Aqueous Two-Phase Systems," *Bioseparation* **1**, 227–233 (1990).

M. Certel and M. E. Ertugay, "The Thermodynamics of Water Activity in Foods," *Gida* **21**, 193–199 (1996).

A. Cesaro, "The Role of Conformation on the Thermodynamics and Rheology of Aqueous Solutions of Carbohydrate Polymers," *J. Food Eng.* **22**, 27–42 (1994).

G. Grasso, "Modellistic Description of Food Materials and Their Technologies, II: The Macrostructural Approach," *Industrie Alimentari* **35**, 923–932 (1996).

C. Kapseu et al., "Thermodynamics of Solid Formation in Cottonseed Oil," *J. American Oil Chemists' Society* **68**, 237–240 (1991).

H. Kumagai et al., "Application of Solution Thermodynamics to the Water Sorption Isotherms of Food Materials," *Bioscience Biotechnol. and Biochem.* **58**, 475–481 (1994).

H. Kumugai, A. Mizuno, and T. Yano, "Analysis of Water Sorption Isotherms of Superabsorbent Polymers by Solution Thermodynamics," *Bioscience Biotechnol. and Biochem.* **61**, 936–941 (1997).

M. Rutgers, K. D. Van, and H. V. Westerhoff, "Control and Thermodynamics of Microbial Growth Rational Tools for Bioengineering," *CRC Critical Rev. in Biotechnol.* **11**, 367–395 (1991).

V. A. Taran, O. G. Fedorov, and A. I. Pokatilov, "Mathematical Formulae for Determining the Dessication Thermodynamics of Chilled Foods by Means of Energy Entropy," *Kholodil'naya Tekhnika* **12**, 21–24 (1988).

GEORGE M. PIGOTT
University of Washington
Seattle, Washington

TOXICANTS, NATURAL

In addition to the many well-known major components (protein, fat, carbohydrate, and fiber) and trace nutrients (vitamins, minerals, and nonessential compounds), our food contains thousands of naturally present toxic compounds. Although these chemicals are in every meal we eat and are present in much greater quantities than residues of synthetic chemicals such as PCBs and pesticides, they have traditionally received relatively little attention compared to these well-known human-made chemicals. Furthermore, the mechanisms of action of natural toxins—metabolic activation, interaction with critical cellular macromolecules—are no different from those of synthetic toxins. In short, our bodies handle toxins similarly regardless of their origin. Many of the toxins presented here are known to cause or strongly suspected of causing cancer in laboratory animals and are therefore potentially carcinogenic in people. The human health risk posed by individual natural toxins varies considerably due to a variety of factors such as dose, inherent potency, variety of the diet, and presence of detoxifying factors. In any event, natural toxins pose a far greater health risk than do synthetic chemicals in our foods, despite the popular notion that "natural is good."

An issue of major concern is the relationship between diet and various human diseases, including cancer. As the single most important cancer variable to be identified from epidemiology studies, diet contributes to approximately 35% of the variation in cancer rates among individuals and populations in the United States. Other factors such as food additives, genetic predisposition, industrial pollution, and pesticide contamination play comparatively minor roles in human cancer rates. Excluding smoking, diet-related factors account for over 50% of all remaining cancer deaths in this country, or 150,000 food-related cancer deaths per year in the United States alone.

Natural toxicants are defined here as naturally occurring substances in plants or other food products that exert undesirable or unhealthy effects when they are consumed. Although not intended to be an exhaustive, all-inclusive discussion of natural toxins, this section will center on those that are particularly well studied and characterized or are of contemporary interest in food toxicology. The focus of this chapter will be not on bacterial toxins, but on three major categories of food toxicants: (1) toxins from plant or plant-derived foods, (2) mold-produced toxins (or mycotoxins), and (3) toxic substances created during cooking or other processing of the food (or induced toxins). An additional yet important category of induced toxicants is endogenous toxic substances that may appear inadvertently in genetically manipulated plant materials that result from efforts to alter plant quality. This latter category will also not be considered here.

Our food also contains natural chemicals that can counteract and thereby prevent the adverse effects of many natural and synthetic toxins. Though much more work on these chemopreventives is needed, the data suggesting that some plant foods can actually reduce the incidence of certain types of cancer thus far are very encouraging. Hundreds of animal and epidemiological studies have identified several foods or specific compounds that offer protection against the carcinogenic effects of a wide variety of natural and synthetic chemicals. A few compounds have been shown to actually reverse the carcinogenic process in animals. As might be imagined, the field of chemoprevention is one of the most exciting areas in nutritional toxicology and cancer research. The reader is encouraged to consult a review on food chemopreventives.

NATURAL PLANT TOXINS IN FOODS

The following is a survey of some of the most well studied and characterized plant toxins.

Allyl Isothiocyanates

Allyl isothiocyanates are a group of major naturally occurring compounds that confer the pungent flavor to foods such as mustard and horseradish, where it is present at about 50 to 100 ppm. These compounds are in *Brassica* vegetables such as broccoli and cabbage, and in cassava and other tropical staple foods, but at much lower concentrations. Normal dietary exposure to isothiocyanate-containing foods releases milligram amounts of isothiocyanates. Nominal processing steps (chopping, rinsing, milling) renders the food safe when wash water is discarded. In high doses, isothiocyanates are carcinogenic in rats but nonmutagenic in bacteria. Isothiocyanates do not occur in foods per se but occur as glucosinolate conjugates that are hydrolyzed when the plant releases enzymes as it is disturbed, such as during chopping, processing, or ingestion (Fig. 1). The major concern with isothiocyanates is their goitrogenic properties in that they inhibit binding of iodine in the thyroid gland. Because iodine is required for the formation of the critical thyroid hormones thyroxine (T_4) and triiodothyronine (T_3), isothiocyanate-induced hyperthyroidism (goiter) mimics iodine deficiency. Hyperthyroidism is a physiological response as the thyroid attempts to compensate for reductions in both T_4 and T_3 production.

Endemic goiter is seen in geographical areas like India and Africa where consumption of poorly processed foods is coincident with iodine deficiency. Like many food toxins, allyl isothiocyanates are "double-edged swords" in that they have been shown to be chemopreventive in certain animal-testing protocols.

Canavanine

Despite their reputation as being the ultimate health food, alfalfa sprouts contain up to 15,000 ppm canavanine, an arginine analogue that can substitute for this amino acid in cellular proteins, thereby altering their function. Canavanine is also produced in other legumes such as the jack bean. Canavanine inhibits nitric oxide synthetase and induces heat-shock proteins in human cells *in vitro* (1). By virtue of its antimetabolic action, canavanine is under current consideration as an antitumor drug in combination with other antimetabolites such as 5-fluorouracil, but it has not yet been tested for carcinogenicity. Canavanine may cause autoimmune disorders such as lupus erythematosus in people (2). Primates fed alfalfa sprouts develop a severe toxic syndrome resembling human lupus.

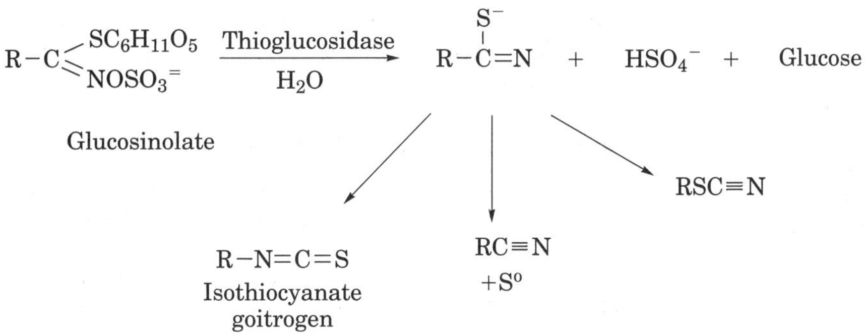

Figure 1. Mechanism of formation of isothiocyanate goitrogen from glucosinolates.

Cyanogenic Glycosides

Seeds from apples, apricots, cherries, peaches, pears, plums, and quinces as well as almonds, sorghum, lima beans, cassava, corn, yams, chickpeas, cashew nuts, and kirsch contain compounds that are toxic due to their release of free hydrogen cyanide, which occurs when the plant tissue is disturbed as during chopping, processing, or ingestion. These conditions initiate the hydrolysis of the glycoside by the action of β-glucuronidases and other enzymes naturally present in the plant tissue and in the intestinal lumen. Although acid also initiates this process, it doesn't appear to occur in the digestive tract to any great extent, despite the acid environment in the stomach. Hydrolysis by β-glucuronidases produce the sugar and a cyanohydrin, the latter spontaneously or enzymatically degrading to form free hydrogen cyanide (Fig. 2). There are several such cyanogenic glycosides, of which linamarin, amygdalin, and dhurrin are examples (Fig. 2). In the 1970s, amygdalin, as laetrile, was a fad remedy touted as a cure and/or preventive for cancer and other ailments. Underground "clinics" briefly flourished where patients were given large quantities of amygdalin or amygdalin-rich seeds and nuts.

Cyanide is one of the most acutely toxic chemicals. It binds to and inactivates heme enzymes, specifically mitochondrial cytochrome aa_3 oxidase, resulting in an acute, life-threatening anoxia. The two-step therapy is initiated with sodium nitrite, which induces methemoglobinemia, permitting the release of cyanide from heme proteins, followed by sodium thiosulfate, which acts as a substrate for rhodanese, an endogenous hepatic enzyme that catalyzes the conversion of free cyanide to the less toxic thiocyanate.

Cases of acute human poisoning from the cyanide released from certain varieties of lima beans, cassava, and bitter almonds are a regular occurrence (3). Due to the importance of cassava as a subsistence crop in Africa and South America, cyanogenic glycosides in that food probably represent the greatest health risk. High-cyanide varieties of cassava, distinguished by their bitter taste, may contain over 600 ppm cyanide on a dry weight basis, whereas "sweet" varieties contain significantly less. Processing steps such as sun drying, soaking, boiling, and fer-

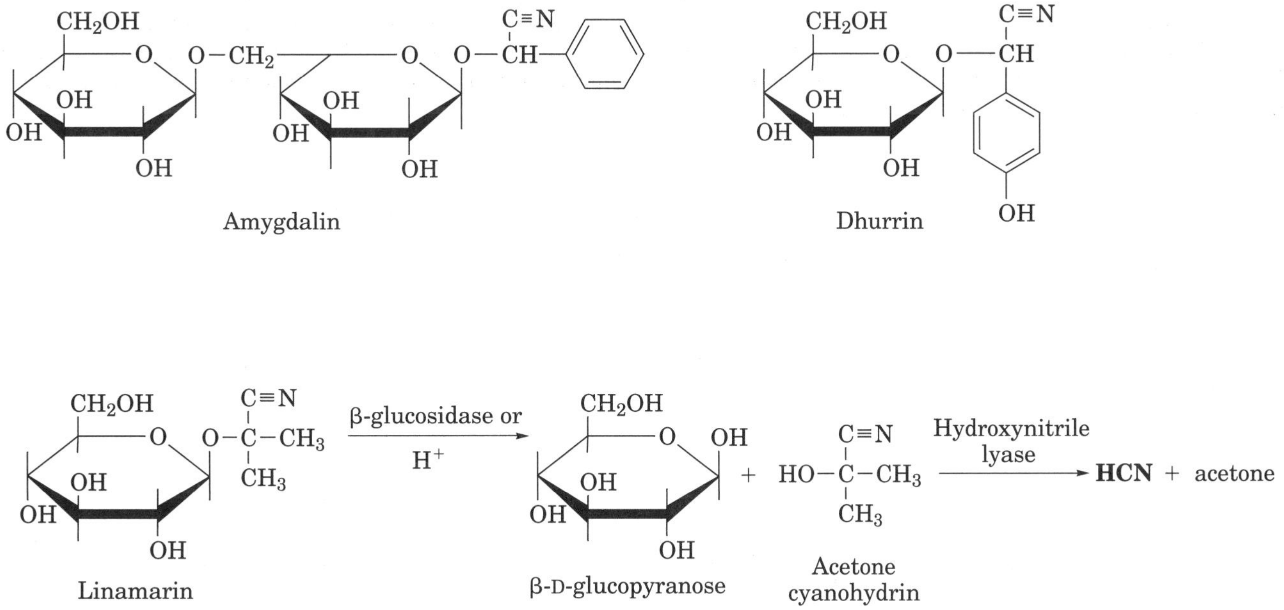

Figure 2. Natural cyanogenic glycosides and mechanism of formation of hydrogen cyanide.

menting render the food safe by eliminating most of the cyanide (4). In addition to regular cases of human deaths, cyanogenic glycosides in cassava may be responsible for birth defects, endemic goiter, and konzo, an upper myelo- pathic motor neuron disease endemic to East Africa (5). Cyanogenic glycosides have also been implicated as a caus- ative agent of diabetes. The risk associated with cyanide poisoning due to cassava is negligible in the United States because cassava-based foods (such as tapioca pudding) are highly processed and rarely eaten here.

Hydrazines and Other Toxins in Edible Mushrooms

Commercial mushrooms, such as the cultivated mushroom (*Agaricus bisporus*), the shiitake mushroom (*Cortinellus shiitake*), and the false morel (*Gyromitra esculenta*), all contain substantial amounts of compounds in the hydra- zine family (Fig. 3). Many hydrazines are potent liver toxins and animal carcinogens. Commonly found in con- centrations of 500 ppm in mushrooms, *N*-methyl-*N* for- mylhydrazine is a lung carcinogen in mice and is also car- cinogenic in hamsters. People eating a 100-gram serving of mushrooms (therefore ingesting 50 mg *N*-methyl-*N* formylhydrazine) get nearly the same dose (on a per- kilogram-body-weight basis) that will cause cancer in mice on sustained daily exposure.

Shiitake and *Agaricus* mushrooms contain up to 3000 ppm agaritine, a metabolic product of which (a diazonium derivative) is a potent carcinogen and a mutagen. The ma- jor carcinogenic hydrazine in the false morel, gyromitrin (acetaldehyde-*N*-methyl-*N*-formylhydrazone), is also pres- ent in similar concentrations. Other carcinogenic hydra- zines include *p*-hydrazinobenzoate (present in *A. bisporus* at 10 ppm) and 4-(hydroxymethyl) benzenediazoate (HMBD), the latter shown to induce DNA strand breaks, presumably through a carbon-centered free-radical inter- mediate, a possible mechanism of the carcinogenic action of hydrazines in general (6). Another carcinogenic hydra- zine, methylhydrazine, is present in smaller concentra- tions (14 ppm).

Whole mushrooms have been shown in numerous stud- ies to cause cancer in laboratory animals, but whether they are a significant cause of cancer in people is uncertain.

Rats fed a diet of whole *A. bisporus* mushrooms (30% of total diet) did not have a significant increase in tumors compared to control animals.

Toxic Substances in Spices and Flavoring Agents

Safrole, estragole, myristicin, *β*-asarone, piperine, and iso- safrole (Fig. 4) are closely related alkenylbenzenes found in many spices, essential oils, and herbs. They are also present, in much lower levels, in parsnips, parsley, and ses- ame seeds. All are weak to moderate rodent hepatocarcin- ogens.

Oil of sassafras (*Sassafras albidum*), which was once used to flavor root beer, contains approximately 85% saf- role. The oil has been banned as a flavor additive since 1960 but Safrole is also a minor, natural component of nut- meg, mace, star anise, cinnamon, and black pepper. Safrole is also found in sassafras tea. Sassafras bark is an ingre- dient in filé powder used to make gumbo, a spicy Cajun stew. A procarcinogen, safrole is activated by endogenous enzymes to a DNA-reactive intermediate that forms co- valent adducts with guanine *in vitro* (7). Estragole, a re- lated aromatic flavor agent, is found in tarragon, basil, and fennel and is likewise a weak carcinogen.

Isosafrole, a component of ylang-ylang (*Cananga odor- ata*) oil, a flavorant and scent, is carcinogenic in mice. Many of these alkenylbenzenes interact with cytochrome P-450 (CYP)–mediated metabolism. For example, both iso- safrole and safrole are powerful inducers of CYP 1A en- zymes. Safrole and isosafrole also inhibit CYP 2E1 en- zymes and in so doing protect against carbon tetrachloride liver toxicity in mice. *β*-asarone is a major component of oil of calamus (derived from the *Acorus calamus* root, which is a folk remedy for indigestion) and was once used to flavor vermouth and bitters. It is an intestinal carcinogen in rats.

Myristicin is a major flavor component of nutmeg, which is derived from the dried seed of the *Myristica fra- grans* tree. The world's principal commercial supply of nut- meg is grown in the Malay peninsula. Approximately 2% of nutmeg is myristicin, which is present in the steam- distilled volatile oil produced from the dried seeds. Mace, a closely related spice, is derived from the outer coating of the seed. Myristicin is also found in black pepper, parsley,

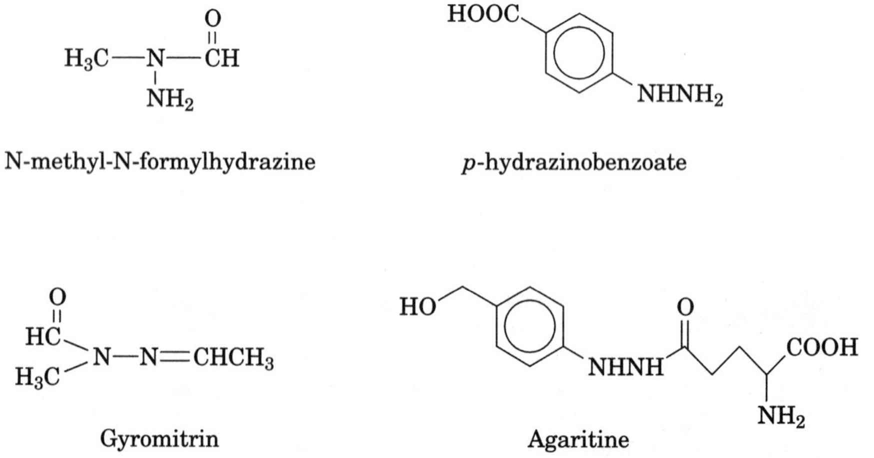

Figure 3. Representative hydrazines in mushrooms.

N-methyl-N-formylhydrazine

p-hydrazinobenzoate

Gyromitrin

Agaritine

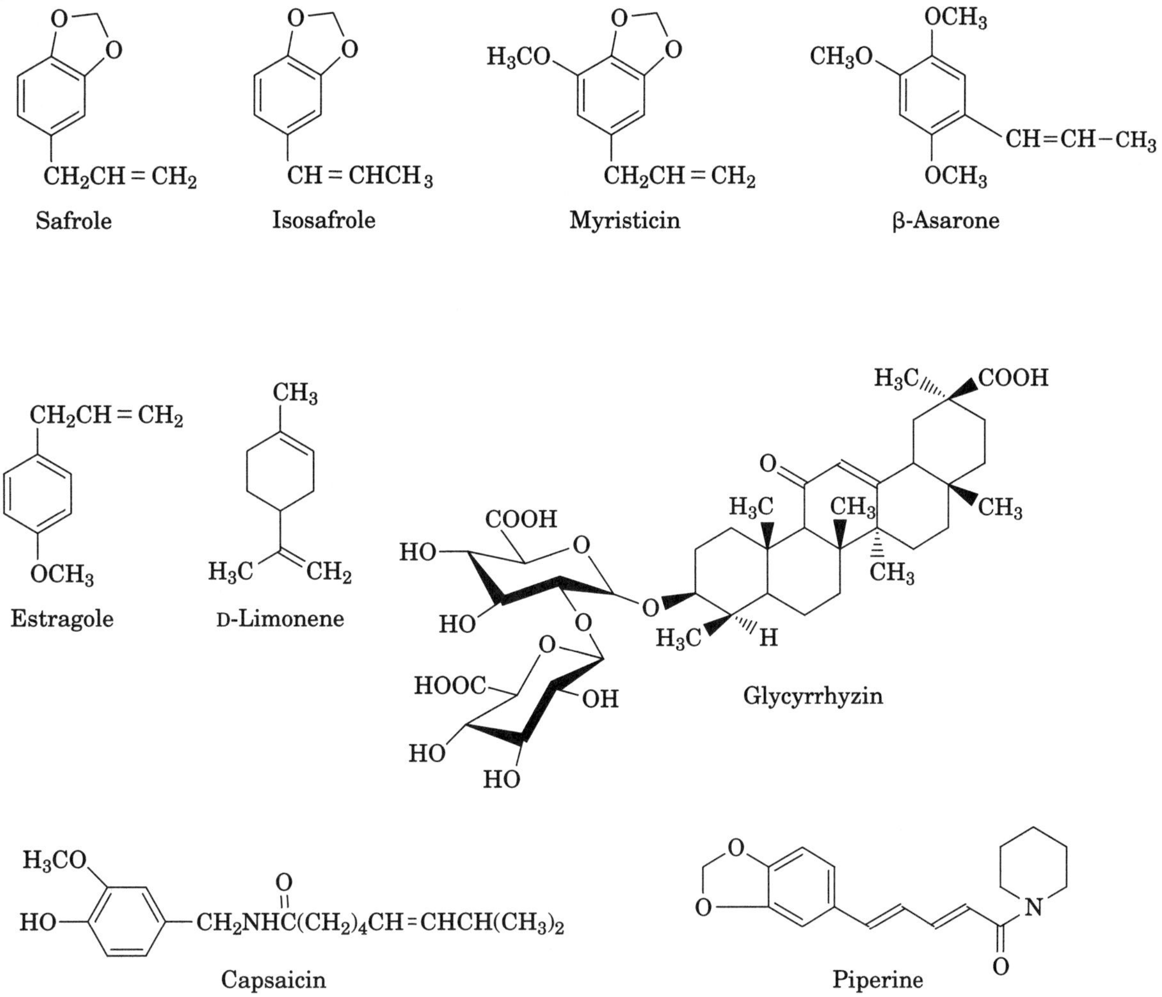

Figure 4. Toxins in spices and flavoring agents.

celery, dill, and carrots. Though not thought to be carcinogenic, large amounts of nutmeg, equivalent to two whole nutmeg seeds (ca 15 g), are intoxicating and allegedly hallucinogenic, and large doses cause undesirable side effects such as tachycardia, flushed skin, and dry mouth. Because pure myristicin is not as hallucinogenic as nutmeg, other components in nutmeg are believed to contribute to its reported psychoactive properties.

Piperine, an alkaloid present in high concentrations (10%) in black pepper (*Piper nigrum* and other spp.), is largely responsible for the pungent "bite" of this condiment. Powdered *P. cubeba* berries are added to cigarettes and smoked as a remedy for throat irritation, and oil derived from these berries is added to some throat lozenges. Reports of the cancer-causing ability of this compound are conflicting. Extracts of black pepper caused cancer in mice at several sites in skin painting tests, whereas orally injected piperine did not (8). Furthermore, piperine is not mutagenic in a number of *in vitro* screening assays. Under appropriate conditions, however, piperine is chemically converted to potentially carcinogenic intermediates. In the presence of nitrite, piperine is nitrosated to form highly mutagenic nitrosamine intermediates *in vitro* that may

have potential carcinogenic activity. Like the related alkenylbenzenes, piperine also affects CYP expression and activity. Piperine specifically inhibits CYP 2E1 while specifically inducing the expression and activities of CYP 1A and 2B (9).

The ingredient responsible for the pungency of red and yellow chili peppers (*Capsicum frutescens, C. conoides*, and *C. annum*) is capsaicin, which represents up to 0.5% of the weight of the fruit. Aerosol capsaicin sprays are popular dog repellents for mail carriers. Topical creams containing capsaicin (0.025%) are commercially available as analgesics. Although its pain-relieving qualities are debatable, capsaicin has been shown to cause a local depletion of substance P, an endogenous neuropeptide known to transmit pain impulses. Thus, even though the physiological conditions causing pain may persist, capsaicin prevents pain impulses from reaching the brain.

Capsaicin may be a weak carcinogen. It is a bacterial mutagen in the Ames test and causes benign digestive tract adenomas in mice given lifelong dietary exposure at 0.03125%. Intraperitoneal injections of capsaicin causes the formation of sister-chromatid exchanges and micronuclei in mice. Sister-chromatid exchanges and micronuclei are genotoxic endpoints presumably associated with cancer risk. One possible toxic mechanism is that CYP 2E1 converts capsaicin to an active phenoxy radical intermediate that has the potential for alkylating tissue macromolecules such as DNA and protein (10).

Glycyrrhizin is a saponin-like glycoside found in licorice (*Glycyrrhiza glabra*). Licorice is one of the oldest traditional medicines; it has been used as an expectorant, a flavoring agent (also used to mask the bitter taste of medicines), and a demulcent. Reportedly, licorice was used by the Sumerians as far back as 4000 B.C., and pieces of licorice root were found in King Tut's tomb. The one caveat to the many benefits of licorice is that ingesting large amounts of it promotes hypertension.

Glycyrrhizin is thought to be responsible for the hypertensive properties of licorice. The mechanism is through inhibition of the enzyme 11β-hydroxysteroid dehydrogenase, which acts as a protective modulator by metabolizing receptor-active glucocorticoids such as cortisol to 11-keto derivatives (eg, cortisone) that are not receptor agonists. Inhibition of this enzyme mimics aldosterone excess, which results in severe sodium retention, loss of potassium (hypokalemia), and hypertension (11). Licorice binges reportedly have been responsible for fatal episodes of acute hypertension in people. Consequently, people with heart problems or hypertension should avoid licorice; as little as 100 to 200 g/day can cause persistent, heightened mineralocorticoid activity. Recently, a healthy Salt Lake City man was hospitalized with congestive heart failure and pulmonary edema for eating 21/2 lb of licorice in 3 days.

A major component of citrus oils (orange, grapefruit, lemon, and others) is d-limonene, which is also found in much lower amounts in other fruits and vegetables. Citrus peel oil can contain as much as 95% d-limonene. Citrus oils or pure d-limonene is the major constituent in flavoring agents and/or fragrances in perfumes and soaps and in a variety of foods such as ice cream, soft drinks, baked goods, gelatin, chewing gum, and puddings. It is also the active ingredient in "natural" citrus-based degreasing solvents and in insect repellents. d-limonene is specifically toxic to the kidneys of male rats. d-limonene binds specifically but reversibly to $\alpha_{2\mu}$-globulin, which is the major low-molecular-weight protein produced by the renal proximal tubules and hence excreted in the urine of the male rat. Female rats excrete much less $\alpha_{2\mu}$-globulin. Accordingly, male rats that do not excrete $\alpha_{2\mu}$-globulin (NBR strain) do not exhibit nephrotoxicity following d-limonene treatment.

Some studies indicate that d-limonene causes renal tumors in rodents. When administered orally, d-limonene induced renal adenomas and carcinomas in male rats but not in mice. Oral d-limonene was also shown to significantly promote the development of N-nitrosoethylhydroxyethylamine-induced renal tumors in male rats. However, the toxicity and carcinogenicity of d-limonene appears to be absolutely species- and gender-specific due to the specific binding of this natural compound with $\alpha_{2\mu}$-globulin. Because humans do not excrete $\alpha_{2\mu}$-globulin, d-limonene is not thought to be harmful to people.

Pyrrolizidine Alkaloids

Pyrrolizidine alkaloids (PAs) are common plant toxins produced by more than 200 species of flowering plants, from genera such as *Senecio, Crotalaria,* and *Cynoglossum*. They are often present at very high concentrations—as much as 5% of the plant's dry weight. PA-containing plants pose significant health hazards to people who consume some kinds of "natural" herbal teas and traditional folk remedies and those who eat grain-based foods contaminated with PA-containing plant parts. Some PAs have been investigated in clinical trials for their anticancer potential.

These chemicals are acutely and chronically hepatotoxic, and many are carcinogenic, mutagenic, and teratogenic. PAs are derivatives of a necine base like retronecine, otonecine, or heliotriine esterified to various necic acid substituents (Fig. 5). PAs are activated by CYPs primarily of the 3A4 subfamily to reactive bifunctional pyrrolic electrophiles that form covalent cross-links to a variety of cellular nucleophiles, such as DNA and proteins. Pyrrolic intermediates then reportedly from electrophilic carbonium ions at atoms 7 and 9 and cross-link cellular nucleophiles at these positions. Cytochrome P450s also convert PAs to the less toxic and more easily excreted N-oxides that do not interact with cellular constituents (Fig. 6). Accordingly, animals that metabolize PAs to produce proportionally more N-oxides (such as sheep) appear to be relatively resistant to the toxic effects of PAs compared to animals that produce more of the pyrrole (rats and horses). Other hydrolysis reactions may also occur that decrease the toxicity of PAs.

DNA cross-links are probably a critical event in PA bioactivity in that the cytotoxic, antimitotic, and megalocytic activity of PAs closely corresponds with the formation of cross-links *in vitro*. PAs form both DNA interstrand and DNA–protein cross-links in equal amounts *in vitro* (12). Structure-activity studies have revealed that the presence of a continuous macrocyclic diester and α,β-unsaturation are important structural determinants for DNA cross-link formation (13). Thus PAs like senecionine are more potent cross-linkers than monocrotaline, which is more potent than open diesters such as latifoline and heliosupine. Of those examined, the simple necine retronecine is the least active cross-linker. The pattern of proteins cross-linked by PAs is similar to those cross-linked by other bifunctional compounds, such as *cis*platinum and mitomycin C. Actin was recently found to be the major protein involved in the PA-induced cross-links in mammalian cells (14).

Petasitenine is found in *Petasites japonicus*, a medicinal herb used as an expectorant and cough suppressant. The flower stalks of this herb are used as a food or herbal remedy. When incorporated into the diet, dried stalks are hepatocarcinogenic to rats. Purified petasitenine is also hepatocarcinogenic in rats as well as mutagenic in bacteria.

Coltsfoot (*Tussilago farfara*) is a common herb that has been used for centuries in Europe and Asia as a medicine for coughs and bronchitis. (*Tussilago* is the ancient Roman term for "cough suppress.") The plant contains the PA senkirkine at concentrations as high as 150 ppm, as well as high concentrations of senecionine, another very hepatotoxic (and carcinogenic) PA. Again, both the dried buds of coltsfoot (when ground and mixed in the diet) and purified senkirkine or senecionine cause liver tumors in rats, and both are bacterial mutagens.

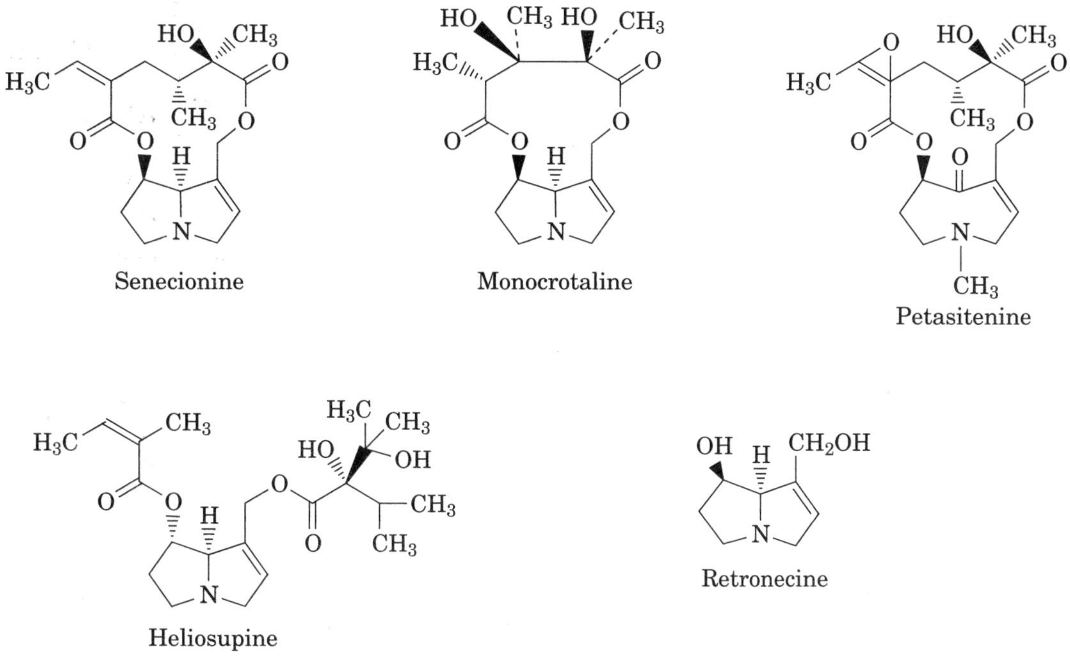

Figure 5. Five representative pyrrolizidine alkaloids.

Human intoxication by PA-containing plants is well recognized and reported in the medical literature and is endemic in Jamaica, India, and parts of Africa. Liver cirrhosis, veno-occlusive disease, and liver cancer are linked to consumption of PA-containing plants. Hispanic and Native American populations in the western and southwestern United States are at high risk for PA intoxications due to their traditional widespread use of herbs, occasional lack of confidence in traditional medicine, and, more commonly, lack of access to medical care.

Comfrey (*Symphytum officinale*) is a nearly universal herb commonly sold not only in health food stores and by herbalists but also in supermarkets. Since ancient Greek and Roman times, both leaves and roots have been used to make teas and compress pastes to treat a variety of external and internal diseases, such as wounds, skin disorders, and respiratory diseases. Numerous vegetarian recipes call for comfrey leaves in the preparation of soufflés, salads, and breads. Comfrey leaves and roots contain up to 0.29% of PAs such as intermedine, lycopsamine, symphytine, and others. Diets containing powder from dried leaves and roots cause liver tumors in rats. Additionally, these pure PAs are animal carcinogens and bacterial mutagens. Several cases are cited in the medical literature of comfrey-related intoxications in people. The well-known toxicity and carcinogenicity of comfrey is such a significant cause for concern that the governments of Australia, Canada, Great Britain, and Germany either restrict comfrey's availability or have banned its sale entirely. The U.S. Food and Drug Administration (FDA) has not yet acted to restrict the sale of PA-containing foods.

Substances in Bracken Fern

Bracken fern (*Pteridium aquilinum, P. esculentum*, and others) is widely used as greens or in salads in places like New Zealand, Australia, Canada, the United States, and especially Japan. It is also a forage plant for sheep and cattle. Veterinarians first noticed severe toxicity—bladder cancer, bone marrow depression, severe leukemia, thrombocytopenia, and a hemorrhagic syndrome—in livestock grazing on this plant. Bracken is a potent bladder, lung, and intestinal carcinogen in rodents. Lactating cows fed bracken fern produced milk that was carcinogenic to rats, showing that human exposure may also occur through cow's milk. Human consumption of bracken fern has been linked to an increased incidence of esophageal cancer in Japan. Ptaquiloside, potent norsesqiterpenoid glucoside (Fig. 7), often present at up to 1.3% dry weight, is thought to be the major carcinogen in bracken. Like other natural carcinogens, ptaquiloside alkylates DNA at codon 61 in the *Ha*-ras oncogene (15).

Solanum Alkaloids

Members of the family Solanaceae contain a variety of toxic steroidal glycoalkaloids. Potatoes (*Solanum tuberosum*) are an important food staple in many parts of the world, and under certain conditions they produce a variety of glycoalkaloids. Concentrations of solanum alkaloids vary considerably between species and strain of potatoes, but more important factors include wound damage, fungal or bacterial infection, exposure to light, and whether the potatoes have sprouted. The major glycoalkaloids are α-solanine and α-chaconine (Fig. 8), which may be present at concentrations over 100 ppm. Like physostigmine, solanine and chaconine are potent inhibitors of the enzyme acetylcholine esterase. Higher amounts of α-solanine and α-chaconine are present in the potato greens (tops). Healthy potatoes contain negligible amounts of these toxins.

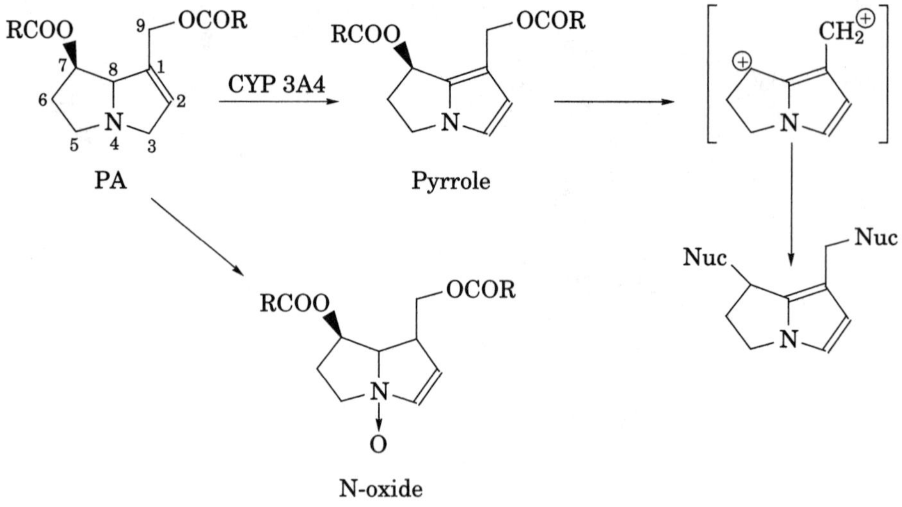

Figure 6. Metabolic formation of pyrroles and N-oxides from pyrrolizidine alkaloids. Nuc = cellular nucleophile such as glutathione, DNA, or protein.

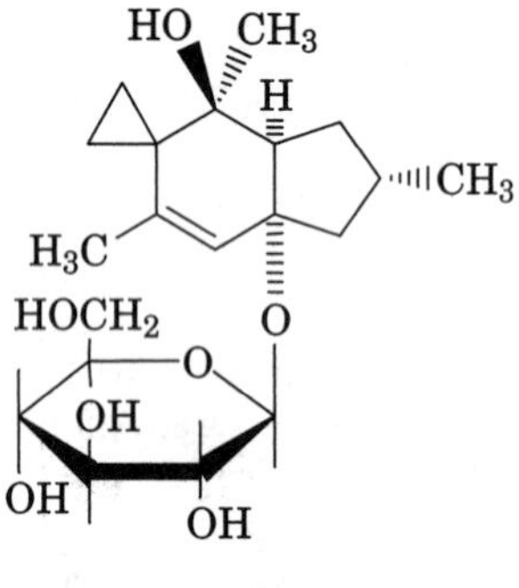

Ptaquiloside

Figure 7. Chemical structure of the bracken carcinogen ptaquiloside.

Episodes of human poisoning by green potatoes have been documented. Fatalities from solanum alkaloid are rare, but gastrointestinal symptoms—gastric pain, weakness, nausea, vomiting, and labored breathing—that are consistent with acetylcholinesterase inhibition are more common. These symptoms have been duplicated in clinical trials with human volunteers. Studies have indicated that the acetylcholinesterase inhibitory activity of solanine is probably insufficient to cause these toxic effects, which are probably due to the combined toxicity of solanine with other cholinesterae inhibitors in the potato, such as chaconine.

Most cases of human poisoning and deaths have occurred in Europe, but they are occasionally seen in the Western Hemisphere. Livestock fed damaged potatoes or peel, greens, or trim are occasionally poisoned. A small number of studies in which animals are fed toxic doses of blighted potatoes or pure glycoalkaloid have indicated that these compounds may have weak teratogenic activity. For example, solanine and chaconine, and their aglycone derivative solanidine, induced craniofacial malformations (exencephaly, encephalocele, and anophthalmia) in Syrian hamsters (16). In that study, solanidine was a much stronger teratogen than solanine and chaconine, which

were classified as weakly teratogenic. The teratogenic and embryotoxic effects of solanine and chaconine appear to be synergistic (17).

Caffeic Acid and Chlorogenic Acid

High concentrations (often over 1,500 ppm) of caffeic acid and its quinic acid conjugate, chlorogenic acid (Fig. 9), occur in an extremely wide range of vegetables (eg, lettuce, potatoes, radishes, and celery), fruits (eg, grapes, berries, eggplant, and tomatoes), and seasonings (eg, thyme, basil, anise, caraway, rosemary, tarragon, marjoram, sage, and dill). Coffee is especially rich in caffeic and chlorogenic acid, as well as other compounds, such as caffeine. A cup of coffee contains about 190 mg of chlorogenic acid. Other minor conjugates of caffeic acid also exist.

Chlorogenic acid is hydrolyzed in the gastrointestinal tract to caffeic and quinic acids. In people, caffeic acid is metabolized to the o-methylated derivatives ferulic, dihydroferulic, and vanillic acids and meta-hydroxyphenyl derivatives, which are excreted in the urine. Caffeic acid inhibits 5-lipoxygenase, which is a key enzyme in the biosynthesis of various eicosanoids such as leukotrienes and thromboxanes. These eicosanoids are mediators of a wide variety of physiological and disease states and are involved in immunoregulation, asthma, inflammation, and platelet aggregation. At high doses (2% in the diet), caffeic acid causes forestomach squamous cell papillomas and carcinomas in both sexes of rats and mice, renal tubular cell hyperplasia in male rats and female mice, and alveolar type II cell tumors in male mice (18). Chlorogenic acid is a bacteria mutagen but has not been tested for carcinogenicity.

Coumarin and Furocoumarin

Coumarin (Fig. 9) is a natural anticoagulant found in a variety of plant foods such as cabbage, radish, and spinach and in flavoring agents such as lavender and sweet woodruff (*Asperula odorata*), the latter an essential herb in making German May wine. Coumarin is found in herb teas based on tonka beans (*Dipteryx odorata*) and sweet clover

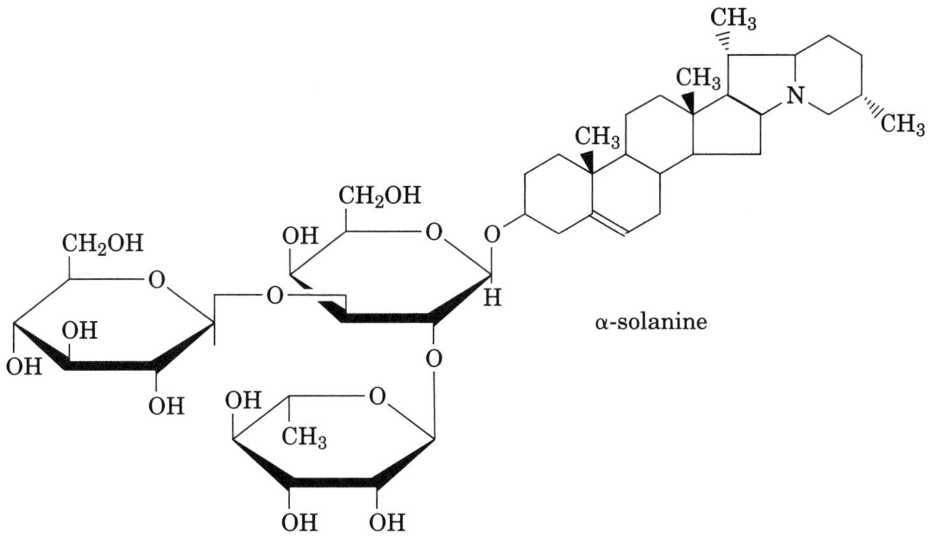

α-solanine

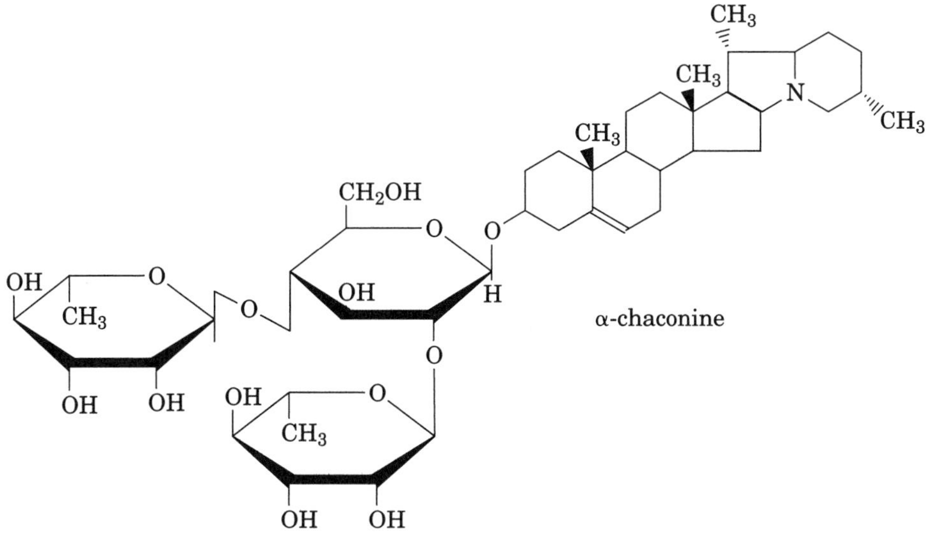

α-chaconine

Figure 8. Acetylcholinesterase inhibitors in potatoes—α-solanine and α-chaconine.

(*Melilotus albus* and *M. officinalis*) called melilot. (In fact, the name coumarin originates from *coumarou*, the Carribean name for tonka beans.) Purified coumarin was once used as a food additive but was banned by the FDA after it was discovered that high doses caused liver damage in test animals. Coumarin has also been reported to cause bile duct carcinomas in rats. The anticoagulant action of coumarin is based on its interference with the action of vitamin K in the synthesis of clotting factors II, VII, IX, and X. The anticoagulant nature of coumarin was discovered when cows grazing on sweet clover developed bruising and internal bleeding. Shortly after it was found that rodents (which were used to bioassay for the then-unknown toxic principle in clover) were extremely susceptible to clover, the newly isolated 3-(α-acetonylbenzyl)-4-hydroxycoumarin, known as warfarin (a named derived from the acronym for Wisconsin Alumni Research Fund), was developed as a rodent bait. Known as coumadin, warfarin is also used in human medicines as a blood-thinning agent and to prevent the formation of blood clots.

Psoralens are a group of phototoxic furocoumarins widespread in a number of plant families such as Apiaceae (formerly Umbelliferae—celery and parsnips), Rutaceae (eg, bergamot, limes, cloves), and Moraceae (eg, figs). Celery contains 100 ppb psoralens, whereas parsnips contain approximately 40 ppm. When activated by sunlight, psoralens are mutagenic, presumably due to their ability to form interstrand and protein cross-links with DNA. Many members of this chemical family are carcinogenic as well, including 5-methoxypsoralen and 8-methoxypsoralen (also called methoxsalen and xanthotoxin, respectively, Fig. 9). Methoxsalen, in addition to forming DNA cross-links, causes a specific mutation in the tumor suppressor gene p53. Dietary exposure to psoralens is probably not a significant health risk; however, the margin of safety is thought to be narrow. Human volunteers who ingested 300 g of celery roots (with a total phototoxic furocoumarin content of 28 ppm) experienced no skin reactions after UVA exposure, and the blood levels of psoralen, methoxsalen, and 5-methoxypsoralen were below the analytical detection limit.

MYCOTOXINS

Mycotoxins are a diverse group of mold-produced chemicals that elicit a wide range of toxic responses in animals

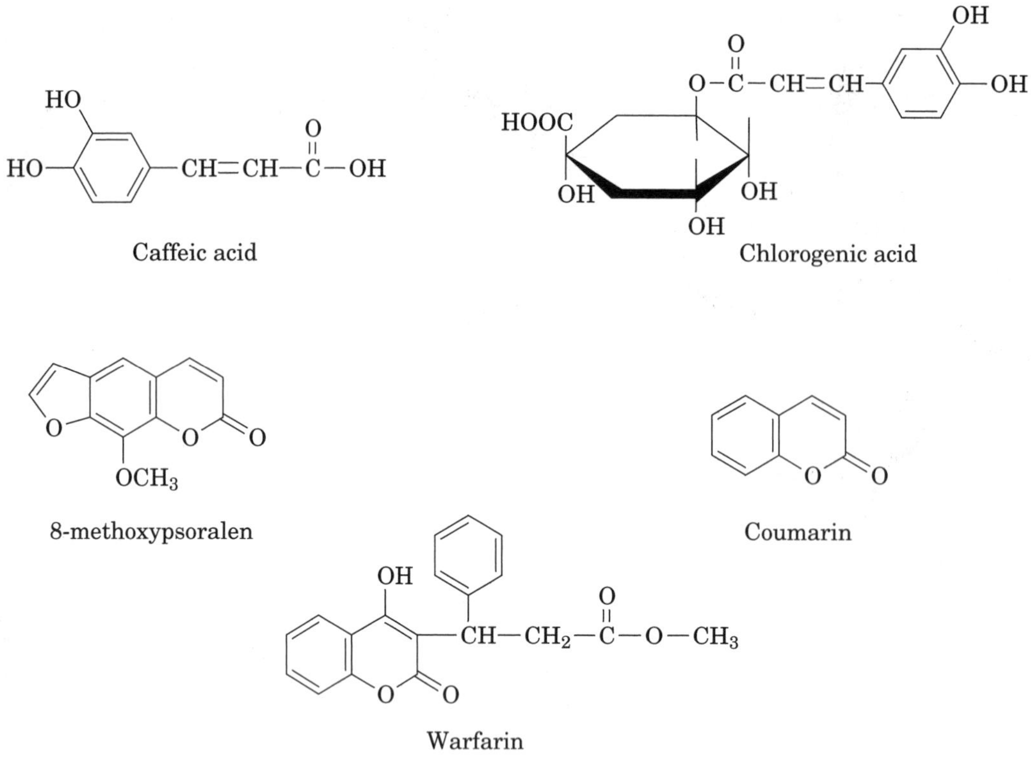

Figure 9. Compounds in coffee (chlorogenic and caffeic acid) and coumarins.

and humans. Pre- or postharvest contamination of various food crops by mycotoxigenic fungi is a common problem; approximately 25% of the world's food supply is contaminated by mycotoxins annually. The severity of mycotoxin contamination of agricultural commodities varies from year to year, depending on factors such as excessive moisture in the field and in storage, temperature extremes, humidity, drought, variations in harvesting practices, and insect infestation. Although the actual resultant economic loss to agriculture is difficult to determine with accuracy, it is considerable.

Many mycotoxins have been implicated in outbreaks of human diseases. Some are potent animal and presumed human carcinogens. In domestic animals, such as dairy cattle, swine, and poultry, mycotoxin contamination is known to reduce growth efficiency, lower feed conversion, lower reproductive rates, impair resistance to infectious diseases, reduce vaccination efficacy, and induce pathologic damage to the liver and other organs. For these reasons, mycotoxins pose a major threat to public and animal health. A few classes of mycotoxins that are problems in foods will be considered here.

Aflatoxin B₁

Aflatoxin B$_1$ (AFB$_1$) represents a group of potent mycotoxins produced by strains of the filamentous fungus *Aspergillus flavus* and *A. parasiticus*. Of the *Aspergillus* mycotoxins, AFB$_1$ has generated the greatest concern and has stimulated the most research because of its extreme toxicity and its widespread occurrence in staple foods and

feeds (such as corn, peanuts, and cottonseed). For these reasons, the U.S. FDA regulates AFB$_1$ in foods. The current action level, the concentration above which the commodity is condemned and discarded, is 20 ppb of total aflatoxins. This regulatory value was established in the 1960s, in large part from the analytical detection limits at that time. The action level for the major AFB$_1$ metabolite present in milk and milk products, aflatoxin M$_1$ (AFM$_1$), is 0.5 ppb in fluid milk. Other regulatory guidelines for AFB$_1$ include 20 ppb in corn for dairy cows, 300 ppb in corn for finishing beef cattle and swine, and 100 ppb for breeding stock. Permissible AFB$_1$ concentrations in cottonseed for beef cattle, swine, and poultry is 300 ppb. These regulatory concentrations generally preclude detectable AFB$_1$ in the various products from these animals. Worldwide, established tolerances for AFB$_1$ in animal feeds range from 10 to 600 ppb.

Prevention of *Aspergillus* infection in foods and feeds is the most desirable method of reducing contamination. Despite the best agricultural practices, however, AFB$_1$ contamination is mostly unavoidable. Thus, several methods of reducing postharvest product contamination have been developed. Some of these methods involve early identification and segregation of grossly contaminated kernels of corn or peanuts, or electronic devices to identify and reject grains that exhibit fluorescence due to AFB$_1$. Ammoniation results in nearly complete elimination of aflatoxins and associated toxicity in commodities, and it is suitable for treating large batches of product (19). This method has not yet been approved by the FDA for interstate shipments, but it is in use in some states. Another experimental strategy is the use of inorganic adsorbent feed additives that prevent absorption of mycotoxins in animals. Hydrated sodium calcium aluminosilicate (HSCAS), an FDA-approved anticaking agent, significantly reduces AFB_1 bioavailability as well as many of its specific toxic effects in pigs (20).

The many toxic effects of AFB_1 are initiated by its conversion, principally by hepatic and extrahepatic microsomal CYPs, to a variety of metabolites (Fig. 10). The reputed carcinogenic intermediate is the AFB_1-8,9-epoxide. Presumably because of its extreme reactivity (it has a half-life of approximately 0.5 s), the AFB_1-8,9-epoxide has been isolated only indirectly from biological systems as adducts of glutathione (GSH), DNA bases, or other macromolecules. Most other metabolic products are less toxic than parent AFB_1, the most prevalent of which is AFM_1, so named for its appearance in the milk of dairy cows that consume AFB_1-contaminated feeds. Other detoxified metabolites produced from the CYP oxidation of AFB_1 include aflatoxin Q_1 (AFQ_1) and aflatoxin P_1 (Fig. 10). Soluble NADPH-dependent cytosolic enzymes reduce AFB_1 to produce aflatoxicol (AFL), which is nearly as toxic, mutagenic, and carcinogenic as the parent compound. Thus AFL is not considered a detoxified metabolite. Because the reduction is reversible, AFL is postulated to represent a storage form of AFB_1.

The most critical detoxification AFB_1 route is via glutathione S-transferase (GST) using GSH as cofactor. Species whose GST has a high affinity for the AFB_1-8,9-epoxide are generally protected from the toxic and carcinogenic effects of AFB_1, regardless of how well the epoxide is formed by CYPs. The AFB_1-8,9-epoxide is a substrate for GST, producing some form of nontoxic AFB_1-GSH adduct that is often the simple sulfhydryl derivative of AFB_1-GSH. Aflatoxin B_1 may also be detoxified via conjugation with sulfates and glucuronic acid. The AFB_1-8,9-epoxide may also be catalytically (by epoxide hydrolase) or spontaneously hydrolyzed to the AFB_1-8,9-dihydrodiol. A soluble AFB_1 aldo-keto reductase (AFAR) has also been isolated from liver and extrahepatic tissues from human, rat, and other species that has strong affinity for reducing the AFB_1-diol, thereby contributing to epoxide detoxification (21).

The electrophilic and highly reactive AFB_1-8,9-epoxide is reportedly responsible for the carcinogenic and mutagenic action of AFB_1. This intermediate binds to cellular nucleophiles, such as DNA. Activated AFB_1 binds exclusively to guanyl residues, and the AFB_1-N^7-guanine (AFB_1-N^7 Gua) adduct is the most predominant (Fig. 10). Additional adducts have been isolated, of which the "ring-opened" derivative of AFB_1-N^7-Gua, the formamidopyrimidine, or AFB_1-FAPyr, is the most common. In hepatic DNA from livers of rats injected with AFB_1, approximately 80% of the adducts present are AFB_1-N^7-Gua, whereas the AFB_1-FAPyr comprises approximately 7% (22). The formation of these adducts is the presumed first step in the development of heritable mutations from which tumors may arise. Repair of these genetic lesions occurs in living cells enzymatically or spontaneously, and the removed adduct is excreted in the urine. The ring-opened adduct appears to be more resistant to DNA repair enzymes. In rats treated with a single dose of AFB_1, the AFB_1-N^7-Gua was rapidly removed with an apparent half-life of 7.5 h, whereas other adducts, such as FAPyr, were removed much more slowly (23).

Aflatoxin B_1 is a potent acute toxin that primarily targets the liver. The primary lesions include hemorrhagic necrosis, fatty infiltration, and bile duct proliferation. In pigs, guinea pigs, and dogs, these effects are found most commonly in the centrilobular region, whereas in ducklings and rats, the periportal region is the site of action. Vertebrate species show considerable variation in susceptibility to the acute effects of AFB_1, and no species appears to be totally resistant. Poultry, rainbow trout, rats, and monkeys are particularly sensitive to the acute effects of AFB_1. Mice and hamsters are much less sensitive (24).

Aflatoxin B_1 is carcinogenic in a wide variety of animals. As is the case following acute exposures, the major target organ is the liver, although tumors in other organs result from long-term dietary exposure to AFB_1. Aflatoxin B_1 at 0.4 ppb fed over a 14-month period resulted in a 14% incidence of hepatocellular carcinomas in rainbow trout, the most sensitive animal species known to the carcinogenic effects of this mycotoxin. By contrast, only a 5% tumor incidence was observed during a similar time frame in Fischer rats exposed to 5 ppb.

Epidemiological data indicate that at least in sub-Saharan Africa and Southeast Asia, where AFB_1 contamination in foods is considerable, dietary AFB_1 is an important risk factor for human hepatocellular carcinoma (HCC). In these geographical areas there is a linear relationship between levels of AFB_1 contamination of food and the incidence of HCC. A factor that complicates epidemiology is that the incidence of hepatitis B virus infection, which is another reputed factor for HCC in humans, is also high in these regions. Specific biomarkers of human exposure to AFB_1, such as adducts of DNA and serum albumin, have proven to be valuable tools in the study of the role of dietary AFB_1 in human cancer. Measurement of these biomarkers in samples of blood or urine has allowed a direct determination of actual AFB_1 exposure in populations, which is an improvement over performing random dietary analysis for AFB_1 and imprecise dietary recall surveys.

Although the majority of interest in possible health effects has correctly focused on dietary exposure to AFB_1, workers in food and grain production, harvest, transport, and processing industries are also exposed to considerable amounts of airborne, respirable AFB_1-contaminated grain dusts. For example, airborne dust sampled in a corn-processing plant contained 107 ng/m^3 AFB_1, and the daily occupational exposure to this toxin was estimated to be between 40 and 856 ng (25). In another survey, concentrations of AFB_1 in smaller, more easily retained airborne grain particles were found to contain more AFB_1 than did larger grain particles; AFB_1 in particles under 7 μm were as high as 1,814 ppb, whereas particles in the size range of 7 to 11 μm had an average content of 695 ppb (26).

Some studies indicate that inhalation exposure to AFB_1 may result in adverse health effects to those exposed. Aflatoxin-contaminated peanut dusts have been associated with liver and lung cancer in Dutch peanut-processing workers who were continually exposed to be-

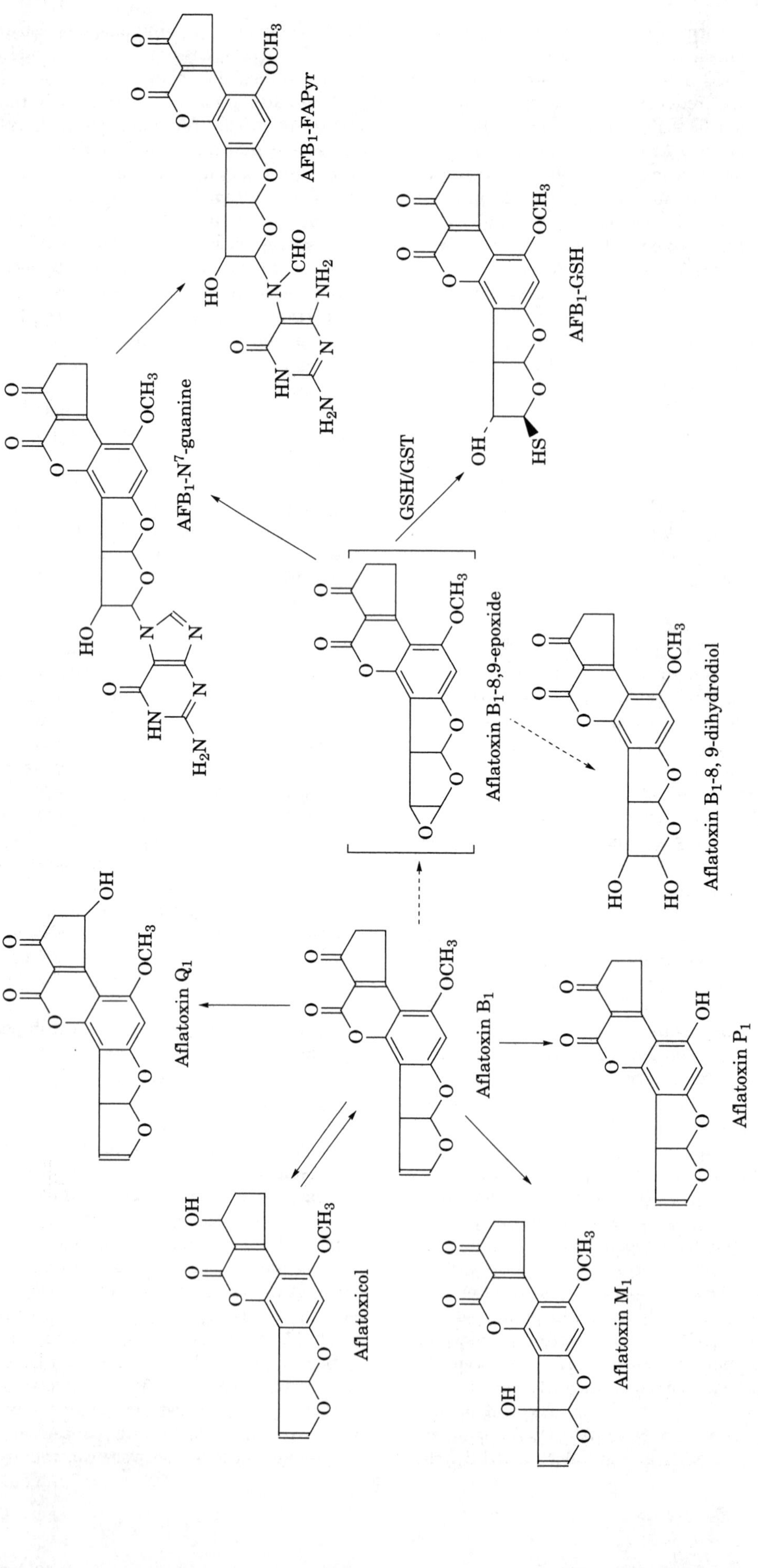

Figure 10. Major metabolites and fate of aflatoxin B_1 in animals.

tween 0.04 and 2.5 μg AFB$_1$ per week, compared to unexposed cohorts. In laboratory studies, AFB$_1$ has been shown to be converted to stable metabolites and activated to mutagenic and DNA-binding intermediates by cells of the mammalian respiratory epithelium. In cultured airway epithelium from a variety of mammalian species, AFB$_1$ forms AFB$_1$-N^7-Gua and FAPyr in patterns similar to those found in hepatic systems. Compared to liver, human airway tissue has comparatively little AFB$_1$-metabolizing ability, however (27).

Tricothecenes

Another economically and toxicologically significant class of mycotoxins are the tricothecenes, which are a group of more than 150 structurally related compounds produced by several genera of fungi, the most common of which are *Fusarium sporotrichioides* and *F. graminearum*. Tricothecenes are common contaminants of grains such as corn, wheat, and barley. In the United States, corn from the upper Midwest is especially affected. Low temperatures, high moisture, and humidity appear to increase toxin production. Tricothecenes possess a sesquiterpenoids tetracyclic 12,13-epoxy-tricothecane skeleton. Examples of important tricothecenes include T-2 toxin, nivalenol, and deoxynivalenol (Fig. 11). T-2 toxin was the first tricothecene discovered in grains, but of the tricothecenes, the incidence of deoxynivalenol (DEN; vomitoxin) is more widespread worldwide.

The most acutely toxic tricothecene is verrucarin J, which has a LD$_{50}$ of 0.5 mg/kg (iv) in the mouse. In mice, T-2 and nivalenol are nearly equitoxic; LD$_{50}$ values are 5.2 and 4.1 mg/kg (Ip), respectively. The *in vivo* toxicity of T-2 and nivalenol are approximately 10 times that of deoxynivalenol. In a variety of *in vitro* systems, T-2 toxin consistently has been the most toxic, followed by nivalenol and then deoxynivalenol (28).

Acute tricothecene toxicity is characterized by vomiting, diarrhea, and inflammation. Dermal irritation, feed refusal, abortion, and hematological sequelae such as anemia and leukopenia also are common. In cattle, dietary T-2 toxin at 0.64 ppm for 20 days results in death with bloody feces, enteritis, and abomasal and ruminal ulcers. In poultry, tricothecenes at levels as low as 5 ppm cause oral necrosis and reduced body weight gains. Decreased egg production and shell quality result when poultry are fed a diet of 20 ppm T-2. Tricothecenes also are toxic when administered dermally and cause symptoms similar to those seen when these mycotoxins are administered orally.

Several widespread cases of human tricothecene poisoning have occurred. Known as *alimentary toxic aleukia*, the disease follows a multistage pathogenesis. Shortly after ingestion of contaminated cereal grains, initial symptoms include a burning sensation in the mouth, tongue, throat, esophagus, and stomach and gastrointestinal disturbances. Severe hematologic effects, such as leukopenia and granulopenia, may then follow, which may progress to white cell counts as low as 100/μL of blood. Continuous exposure to trichothecenes results in rashes on the skin and elsewhere that may progress to severe necrotic lesions.

Tricothecenes have widespread adverse effects on synthesis of proteins and other macormolecules and on membrane and immune functions. Tricothecenes inhibit all steps in protein synthesis (initiation, elongation, and termination). Protein synthesis inhibition likely is a result of binding of the toxin to ribosomes. The binding of T-2 to ribosomes is specific and saturable (0.3 nM), and the stoichiometry is 1 toxin molecule bound per ribosome (29).

The amphipathic nature of T-2 has led some to believe that it could be incorporated into the lipid or protein moieties of cellular plasma membranes, thereby interfering with membrane function. Myoblasts treated with T-2 toxin had reduced uptake of calcium, glucose, leucine, and tyrosine within 10 min of exposure, effects apparently independent of protein synthesis inhibition.

The toxin significantly alters several immune parameters, and the major effects appear to be associated with the cellular immune response. Specifically, effects such as an inhibition of the mitogen response, reductions of protein, DNA and interleukin-2 synthesis in concanavalin A–treated normal spleen cells, thymic atrophy, and reductions in plaque-forming spleen cells have been demonstrated. In animals, decreases in resistance to many challenges that are cellular immune dependent have been observed, such as to bacterial, fungal, and mycobacterial infections and skin grafts.

Zearalenone

Zearalenone (ZEN) is a phenolic resorcyclic acid lactone (Fig. 11) produced by strains of *Fusarium*, primarily by *F. graminearum* and *F. sporotrichiodes*. It is a natural contaminant of corn, wheat, barley, oats, sorghum, and hay. As with most mycotoxins, significant ZEN production is promoted by high humidity and low temperatures, common conditions in the upper Midwest during autumn harvest. Zearalenone is a nearly universal corn contaminant and is often found in the same samples with tricothecenes. Despite its dissimilarity to steroid compounds, ZEN produces potent hyperestrogenic responses in susceptible animals. Thus the toxicity of this compound is unique among known mycotoxins.

The animal most affected by ZEN is swine, but other animals, such as cattle, poultry, and laboratory animals, are affected to a lesser degree. Symptoms of ZEN poisoning include uterine enlargement and swollen vulva and mammae. In pigs, symptoms of hyperestrogenism generally appear when contamination of ZEN in corn exceeds 1 ppm, but it can occur as low as 0.1 ppm. In ewes, ZEN exposure results in decreased ovulation rate and cycle length and an increase in duration of estrus. Young male pigs exposed to ZEN undergo symptoms of feminization, such as enlarged nipples and testicular atrophy.

Zearalenone causes significant adverse reproductive effects in animals. For example, dairy heifers exposed to ZEN have reduced conception rates; in rats, ZEN (10 mg/kg in the diet) causes decreased growth rate, food intake, fertility, fetal resorptions, stillbirths, abortion and fetal bone malformations. ZEN reduces egg production in hens. In newborn female mice, ZEN (1 μg daily for 5 days) results in significant ovary-dependent reproductive tract

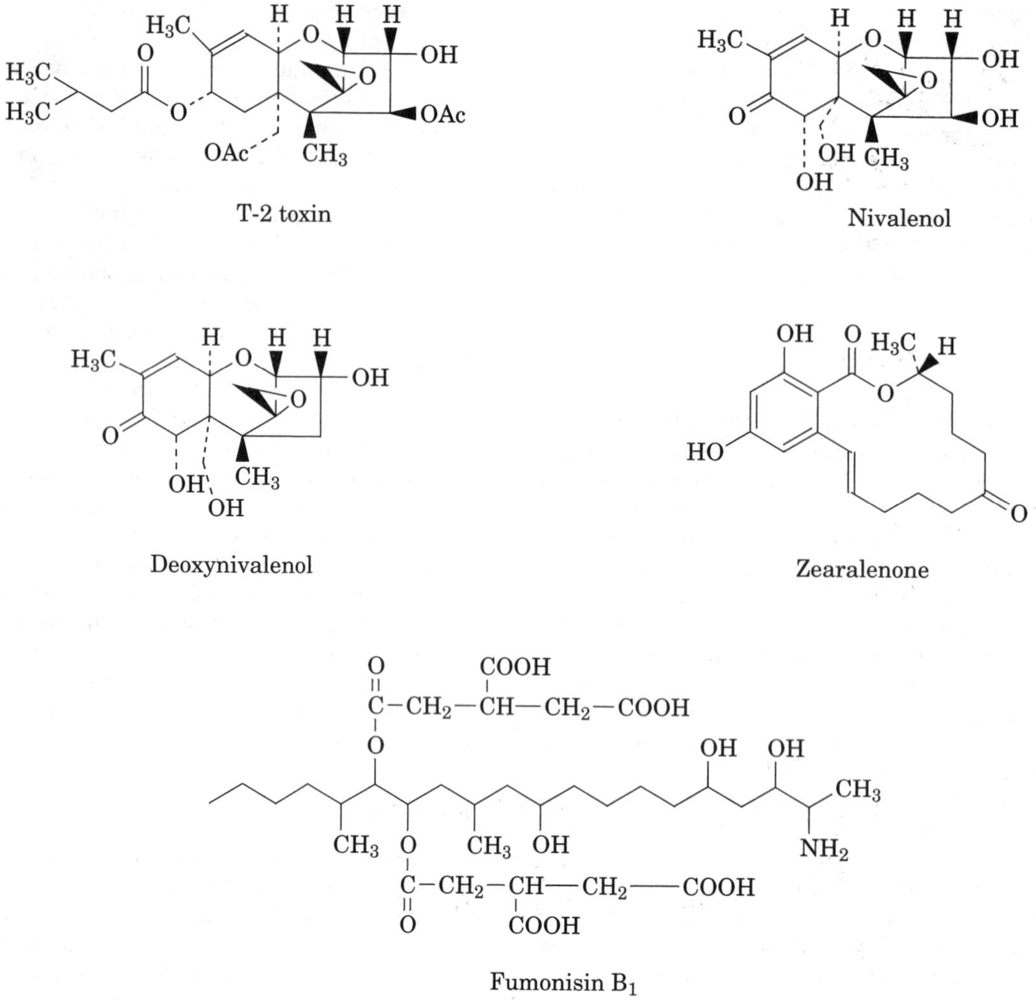

Figure 11. Representative *Fusarium* mycotoxins.

alterations at 8 month posttreatment; the majority of treated animals lacked corpus lutea and uterine glands and exhibited squamous metaplasia of the uterine luminal epithelium. Ovarectomized, ZEN-treated animals showed none of these ovary-dependent alterations.

The mechanism of the estrogenic effect of ZEN appears to be mediated via binding of ZEN or metabolite(s) to the cytoplasmic estrogen receptor. In rat uterine tissue, *trans*- and *cis*-ZEN and two ZEN derivatives compete with 17β-estradiol for binding with the cytosolic receptor, but with a lesser affinity that that of estradiol. Accordingly, ZEN appears to have a greater affinity for estrogen receptors from animals that are more susceptible to the estrogenic effects of the mycotoxin. The affinity of ZEN to uterine and oviduct estrogen receptors follows the pattern pig, rat, and then chicken.

Fumonisins

Fumonisins are a group of recently discovered mycotoxins that have been associated with toxicity and mortality in horses and pigs following ingestion of *Fusaria*-contaminated corn-based feeds. Fumonisin B₁ (FB₁) is the most

toxic representative of the fumonisins (Fig. 11). One such animal disease is the neurotoxic syndrome Equine Leukoencephalomalacia (ELEM), which is characterized by facial paralysis, nervousness, lameness, ataxia, and an inability to eat or drink. The principal pathologic lesions include severe cerebral edema, focal malacia, and liquefaction of cerebral white matter. The onset of such severe symptoms can be as short as a few hours. ELEM syndrome often is epizootic and nearly always is associated with ingestion of moldy corn.

Several studies have positively linked fumonisins to ELEM. A majority of horses fed a corn-based feed contaminated with 37 to 122 ppm FB₁ developed fatal ELEM (30). Hepatic involvement is often coincident with central nervous involvement in horses and swine. Intravenous or dietary administration of FB₁ causes both pulmonary edema and hydrothorax in swine (31). In swine, lower doses of FB₁ result in a slowly progressive hepatic necrosis, whereas higher doses result in acute pulmonary edema coincident with hepatic toxicity. That fumonisins are potent inhibitors of sphingosine biosynthesis in cultured hepatocytes has been postulated to account for both the hepatotoxic and central nervous system effects of this toxin.

Besides the neurotoxic and hepatotoxic effects, FB_1 is a potent rodent carcinogen. Initial studies showed that a diet containing culture material inoculated with *F. moniliforme* was carcinogenic in rats. Dietary FB_1 not only initiates γ-glutamyltranspeptidase-positive (γ-GT+) hepatic foci, but also promotes those induced by compounds such as diethylnitrosamine (32). Long-term dietary FB_1 (50 ppm) induces a high incidence (66%) of HCC along with metastases to the heart, lungs, or kidneys (33). Symptoms of hepatic involvement such as macronodular cirrhosis and cholangiofibrosis are also observed. Later studies have led to the conclusion that FB_1 is probably only a modest initiator of liver tumors, because γ-GT$^+$ foci and hepatocellular nodules were observed only after prolonged feeding of very high (0.1%) FB_1 in rats. It was postulated that the carcinogenic effect of FB_1 likely involves promotion and the selection of initiated hepatocytes, events that occur during the postinitiation phase of hepatocarcinogenesis. Additionally, FB_1 carcinogenicity does not appear to involve interaction with DNA. Neither FB_1 nor FB_2 elicits unscheduled DNA repair in primary rat hepatocyte cultures treated either *in vitro* or *in vivo* (34). It is therefore possible that the carcinogenic activity of fumonisins is mediated via epigenetic mechanisms, such as is the case of peroxisome proliferators.

Research on the health effects of fumonisins is relatively modest, and information on the mechanism of fumonisin toxicity or on the mechanisms underlying species sensitivity is only beginning to be compiled. Obviously, more information is needed to provide a fuller understanding of the extent of the adverse effects of fumonisins on human and animal health. Because corn is a staple in many parts of the world, and because *Fusarium* contamination of corn is nearly universal, it is likely that fumonisins may be involved in human toxicoses and other health effects. *Fusarium*-contaminated corn has been epidemiologically associated with human esophageal cancer in some regions of South Africa.

Ergot Alkaloids

Documented since the Middle Ages, the fungus *Claviceps purpurea* has caused periodic outbreaks of mycotoxicoses in many parts of the world. This mold, which grows in grasses and cereal products, is especially prevalent in over-wintered rye and wheat. Growth of the fungus takes the form of hard, black-purple sclerotia (or ergots) that germinate in the spring. *Claviceps* produces several ergot alkaloids, which are derivatives of lysergic acid (ergotamine; Fig. 12). The major effect of these compounds is vasoconstriction, resulting in a reduction of blood flow and subsequent gangrene in the feet, legs, and hands. Symptoms of ergot poisoning include a burning sensation in the arms and legs, gangrene, abortion, vomiting, diarrhea, weakness, and sometimes hallucinations.

Ochratoxins

Ochratoxins are produced by *Aspergillus ochraceus* and selected *Penicillium* species. Ochratoxin A (Fig. 12), which is the most toxicologically significant of the ochratoxins, is produced in a variety of cereal grains (wheat, barley, oats, corn), dry beans, peanuts, and cheese. Ochratoxin A is a potent nephrotoxin in nearly all animals studied and is a probable etiologic agent for Balkan endemic nephropathy (BEN), a fatal chronic renal disease. The syndrome, which is associated with nephritis and associated urinary tumors, is especially prevalent in Bulgaria, Romania, and Yugoslavia. Pigs exposed to ochratoxin A experimentally present a nephropathy strikingly similar to BEN.

Surveys have repeatedly shown that a significant percentage of food supplies from several Balkan countries are contaminated with ochratoxin. In addition to the aforementioned cereal and other products, ochratoxin is often found in animal products such as sausage, bacon, and ham.

HETEROCYCLIC AMINES IN COOKED FOODS

It has been known for years that heating food produces several classes of toxic compounds. For example, polycyclic aromatic hydrocarbons, the same compounds present in chimney soot, are also produced by charbroiling meat, chicken, and fish. More recently, however, a new class of heat-produced compounds, the heterocyclic amines (HCA), have been discovered and characterized. Like polycyclic aromatic hydrocarbons, HCAs are produced by pyrrolysis reactions that occur under ordinary cooking conditions. HCAs have been found in meat, fish, and poultry. Components in raw food that react to form HCAs include creatinine, sugars, and amino acids. Both creatinine and amino acids appear to be rate limiting, but the role of sugars in formation of HCAs is uncertain (35). Cooking practices involving higher temperatures, such as flame broiling or panfrying, produce higher amounts of these toxicants, which are stable in food under most storage situations. In addition to temperature, cooking time and type of food are factors important in HCA formation.

More than 20 HCAs have been isolated from cooked foods, nearly all of which are powerful mutagens, and in most cases animal carcinogens. In fact, HCAs are among the most potent bacterial mutagens known. Some of the most common and intensely studied include IQ (2-amino-3-methylimidazo[4,5-*f*]quinoline), methyl IQ (MeIQ; 2-amino-3,4-dimethylimidazo[4,5-*f*]quinoline), methyl IQx (MeIQx; 2-amino-3,8-dimethylimidazo[4,5-*f*]quinoxaline), PhIP (2-amino-1-methyl-6-phenylimidazo[4,5-*b*]pyridine), Trp-P-1 (3-amino-1,4-dimethyl-5*H*-pyrido[4,3-*b*]indole), and Glu-P-1 (2-amino-6-methyldipyridol[1,2-*a*:3′,2′-*d*]imidazole) (Fig. 13). Heterocyclic amines are well absorbed from the intestine and, like many other natural carcinogens, require hepatic metabolic activation for toxic action. Cytochrome P450 1A2 has been identified as an important isozyme that catalyses the N-hydroxylation of the exocyclic amino groups of HCAs. Reactive electrophilic intermediates are produced by subsequent ester formation with either acetic acid, sulfuric acid, or proline (Fig. 14). The mutagenic and carcinogenic action of HCAs are thought to occur via binding of reactive nitreuniom ion intermediates with the C-8 atom of guanine in DNA. This resultant bulky adduct causes frameshift mutations by small base deletions.

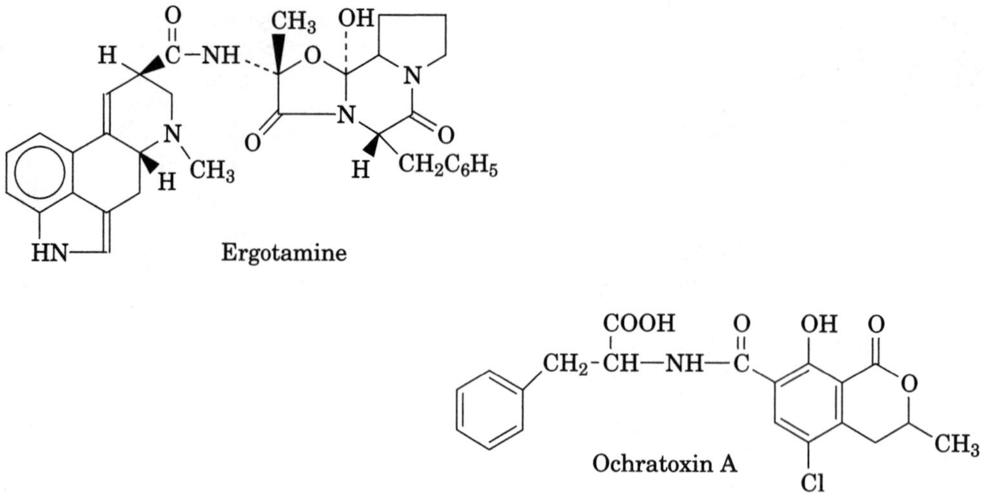

Figure 12. Chemical structures of mycotoxins ergotamine and ochratoxin A.

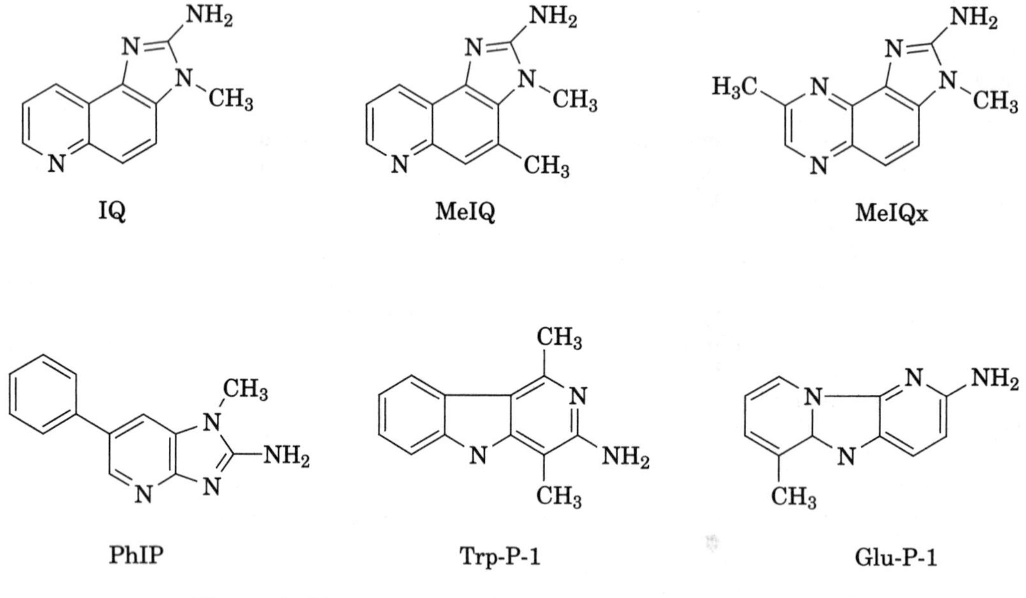

Figure 13. Major carcinogenic heterocyclic amines in cooked foods.

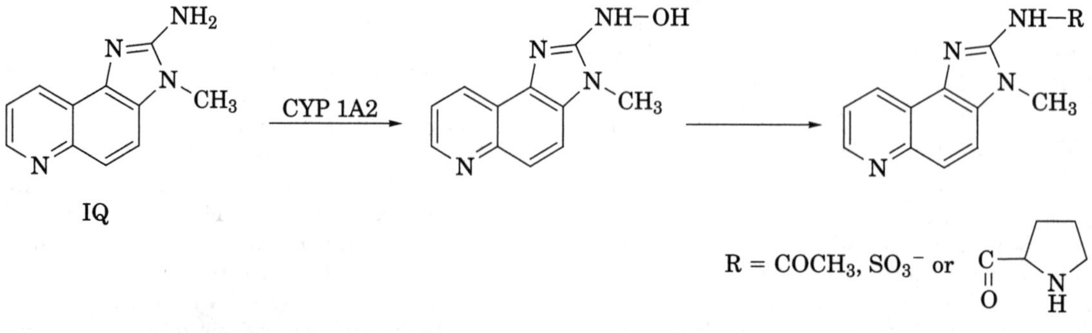

$$R = COCH_3, SO_3^- \text{ or }$$

Figure 14. Mechanism of metabolic activation of IQ.

As animal carcinogens, heterocyclic amines are not as potent as would be expected from their impressive mutagenic activity. However, HCAs are still potent rodent carcinogens, causing tumors in a wide variety of tissues, mainly the liver, blood vessels, intestines, and forestomach. The most abundant HCA in cooked food, PhIP, causes tumors primarily of the large intestines, mammary gland, and lymphoid tissue (36).

Although the intake of HCAs is unavoidable, the average amount consumed in people varies dramatically. In the general population, the average intake of HCAs has been calculated to be between 0.4 and 16 μg/day (37). Urinary HCAs have been isolated in all healthy volunteers who eat normal diets. Whether HCAs are in fact carcinogenic in people at the nominal exposure levels is uncertain and is difficult to ascertain. Amounts of individual HCAs in the human diet are probably insufficient to cause cancer per se. Differences in diet, meat intake, and cooking methods can result in 50,000 to 100,000-fold differences in HCA exposure among individuals (38). Animal studies seem to indicate that HCAs act additively, or perhaps synergistically. Therefore, mixtures of HCAs, which would be present in most cooked foods, may be associated with human cancer risk.

ACKNOWLEDGMENTS
The author wishes to acknowledge support from grant ES04813 from the National Institute of Environmental Health Sciences, NIH; from National Research Initiative—National Competitive Program grants 970-3081 and 980-3754 from the USDA; and from the Utah Agricultural Experiment Station, Utah State University, Logan, Utah, 84322-4810, where this is designated as document number 7089.

BIBLIOGRAPHY

1. E. Mattei et al., "Stress Response, Survival and Enhancement of Heat Sensitivity in a Human Melanoma Cell Line Treated with L-Canavanine," *Anticancer Res.* **12**, 757–762 (1992).
2. A. Montanaro and E. J. Bardana, Jr., "Dietary Amino Acid-Induced Systemic Lupus Erythematosus," *Rheum. Dis. Clin. North Amer.* **17**, 323–332 (1991).
3. A. Akintonwa and O. L. Tunwashe, "Fatal Cyanide Poisoning from Cassava-Based Meal," *Hum. Exp. Toxicol.* **11**, 47–49 (1992).
4. O. C. Kemdirim, O. A. Chukwu, and S. C. Achinewhu, "Effect of Traditional Processing of Cassava on the Cyanide Content of Gari and Cassava Flour," *Plant Foods Hum. Nutr.* **48**, 335–339 (1995).
5. T. Tylleskar et al., "Cassava Cyanogens and Konzo, an Upper Motoneuron Disease Found in Africa," *Lancet* **339**, 208–211 (1992); published erratum appears in *Lancet* **339**, 440 (1992)
6. K. Hiramoto et al., "DNA Strand Breaking by the Carbon-Centered Radical Generated from 4-(Hydroxymethyl) Benzenediazonium Salt, a Carcinogen in Mushroom *Agaricus bisporus*," *Chem. Biol. Interact.* **94**, 21–36 (1995).
7. G. Luo and T. M. Guenthner, "Covalent Binding to DNA *In vitro* of 2',3'-Oxides Derived from Allylbenzene Analogs," *Drug Metab. Dispos.* **24**, 1020–1027 (1996).
8. H. Wrba et al., "Carcinogenicity Testing of Some Constituents of Black Pepper (*Piper nigrum*)," *Exp. Toxicol. Pathol.* **44**, 61–65 (1992).
9. M. H. Kang et al., "Piperine Effects on the Expression of P4502E1, P4502B and P4501A in Rat," *Xenobiotica* **24**, 1195–1204 (1994).
10. Y. J. Surh and S. S. Lee, "Capsaicin, A Double-Edged Sword: Toxicity, Metabolism, and Chemopreventive Potential," *Life Sci.* **56**, 1845–1855 (1995).
11. B. R. Walker and C. R. Edwards, "Licorice-Induced Hypertension and Syndromes of Apparent Mineralocorticoid Excess," *Endocrinol. Metab. Clin. North Am.* **23**, 359–377 (1994).
12. J. R. Hincks et al., "DNA Cross-linking in Mammalian Cells by Pyrrolizidine Alkaloids: Structure–Activity Relationships," *Toxicol. Appl. Pharmacol.* **111**, 90–98 (1991).
13. H. Y. Kim et al., "Structural Influences on Pyrrolizidine Alkaloid-Induced Cytopathology," *Toxicol. Appl. Pharmacol.* **122**, 61–69 (1993).
14. R. A. Coulombe, Jr., G. L. Drew, and F. R. Stermitz, "Pyrrolizidine Alkaloids Cross-link DNA with Actin," *Toxicol. Appl. Pharmacol.* **54**, 198–202 (1999).
15. A. S. Prakash et al., "Mechanism of Bracken Fern Carcinogenesis: Evidence for H-*ras* Activation via Initial Adenine Alkylation by Ptaquiloside," *Nat. Toxins* **4**, 221–227 (1996).
16. W. Gaffield and R. F. Keeler, "Induction of Terata in Hamsters by Solanidane Alkaloids Derived from *Solanum tuberosum*," *Chem. Res. Toxicol.* **9**, 426–433 (1996).
17. J. R. Rayburn, M. Friedman, and J. A. Bantle, "Synergistic Interaction of Glycoalkaloids α-Chaconine and α-solanine on Developmental Toxicity in *Xenopus* Embryos," *Food. Chem. Toxicol.* **33**, 1013–1019 (1995).
18. A. Hagiwara et al., "Forestomach and Kidney Carcinogenicity of Caffeic Acid in F344 Rats and C57BL/6N × C3H/HeN F1 Mice," *Cancer Res.* **51**, 5655–5660 (1991).
19. D. L. Park et al., "Review of the Decontamination of Aflatoxins by Ammoniation: Current Status and Regulation," *J. Assoc. Off. Anal. Chem.* **71**, 685–703 (1988).
20. R. W. Beaver et al., "Distribution of Aflatoxins in Tissues of Growing Pigs Fed an Aflatoxin-Contaminated Diet Amended With a High Affinity Aluminosilicate Sorbent," *Vet. Human Toxicol.* **32**, 16–18 (1990).
21. E. M. Ellis and J. D. Hayes, "Substrate Specificity of an Aflatoxin-Metabolizing Aldehyde Reductase," *Biochem. J.* **312**, 535–541 (1995).
22. J. M. Essigmann et al., "Metabolic Activation of Aflatoxin B$_1$: Patterns of DNA Adduct Formation, Removal, and Excretion in Relation to Carcinogenesis," *Drug Metab. Rev.* **13**, 581–602 (1982).
23. R. G. Croy and G. N. Wogan, "Temporal Patterns of Covalent DNA Adducts in Rat Liver after Single and Multiple Doses of Aflatoxin B$_1$," *Cancer Res.* **41**, 197–203 (1981).
24. D. S. P. Patterson, "Metabolism as a Factor in Determining the Toxic Action of the Aflatoxins in Different Animal Species," *Food Cosmet. Toxicol.* **11**, 287–294 (1973).
25. W. R. Burg, O. L. Shotwell, and B. E. Saltzman, "Measurements of Airborne Aflatoxins during the Handling of Contaminated Corn," *Amer. Indust. Hyg. Assoc.* **42**, 1–11 (1981).
26. W. G. Sorenson, J. P. Simpson and M. J. Peach, III, "Aflatoxin in Respirable Corn Dust Particles," *J. Toxicol. Environ. Health* **7**, 669–672 (1981).
27. J. D. Kelly et al., "Aflatoxin B$_1$ Activation in Human Lung," *Toxicol. Appl. Pharmacol.* **144**, 88–95 (1997).
28. Y. Ueno, "Toxicological Features of T-2 Toxin and Related Trichothecenes," *Fund. Appl. Toxicol.* **4**, S124–S132 (1984).
29. J. L. Middlebrook and D. L. Leatherman, "Specific Association of T-2 Toxins with Mammalian Cells," *Biochem. Pharmacol.* **38**, 3093–3102 (1989).

30. T. M. Wilson and P. F. Ross, "Fumonisin Mycotoxins and Equine Leukoencephalmalacia," *J. Amer. Vet. Med. Assoc.* **198**, 1104 (1991).

31. P. F. Ross et al., "Production of Fumonisins by *Fusarium moniliforme* and *Fusarium proliferatum* Isolated Associated with Equine Leukoencephalomalacia and a Pulmonary Edema Syndrome in Swine," *Appl. Environ. Microbiol.* **56**, 3225–3226 (1990).

32. W. C. A. Gelderblom et al., "Fumonisins: Isolation, Chemical Characterization and Biological Effects," *Mycopathologia* **117**, 11–16 (1992).

33. W. C. A. Gelderblom et al., "Toxicity and Carcinogenicity of the *Fusarium moniliforme* Metabolite, Fumonisin B₁, in Rats," *Carcinogenesis* **12**, 1247–1251 (1991).

34. W. C. A. Gelderblom et al., "The Cancer-Initiating Potential of the Fumonisin B Mycotoxins," *Carcinogenesis* **13**, 433–437 (1992).

35. F. T. Hatch, M. G. Knize, and J. S. Felton, "Quantitative Structure–Activity Relationships of Heterocyclic Amine Mutagens Formed During the Cooking of Food," *Environ. Molecular Mutagenesis* **17**, 4–19 (1991).

36. T. Sugimura, "Overview of Carcinogenic Heterocyclic Amines," *Mutation Res.* **376**, 211–219 (1997).

37. K. Wakabayashi et al., "Food-Derived Mutagens and Carcinogens," *Cancer Res.* **52**, 2092S–2098S (1992).

38. J. S. Felton et al., "Health Risks of Heterocyclic Amines," *Mutation Res.* **376**, 37–41 (1997).

ROGER A. COULOMBE, JR.
Utah State University
Logan, Utah

See also MYCOTOXINS.

TOXICOLOGY AND RISK ASSESSMENT

Risk assessment is the process used to determine the probability, type, and magnitude of human toxic effects anticipated from exposure to specific levels of chemicals. All chemicals have the potential to produce adverse effects on health under some conditions of exposure. According to a sixteenth-century principle expressed by Paracelsus (ca 1493–1541), it is the right dose that differentiates a poison and a remedy. There is, however, a great variation of potency among chemicals in regard to their ability to cause adverse health effects; some chemicals may cause toxicological effects at very low doses, whereas others require much higher doses before effects are produced. All our food is composed of chemicals, and there are numerous types of chemicals that have been subject to risk assessment and/or regulation (Table 1).

The U.S. Food and Drug Administration (FDA) manages risks from food additives and veterinary drugs that may persist as residues in foods of animal origin (milk, meat, and eggs). The FDA has published comprehensive guidelines for safety assessment of direct food and color additives (1). The U.S. Environmental Protection Agency (EPA) is responsible for the control of several chemicals that may contaminate food, including pesticides and water pollutants. The EPA has developed a variety of guidelines for the toxicological testing of pesticides. At the international level, the regulatory authorities include different

Table 1. Food Chemicals Considered in Risk Analysis

Food additives

Preservatives
Colors
Flavors
Stabilizers
Supplements

Biogenic compounds

Plant toxins or phytotoxins
Bacterial toxins
Enterotoxins
Mycotoxins
Marine toxins (eg, algal toxins responsible for paralytic shellfish poisoning and diarrhetic shellfish poisoning)

Residues

Processing aids (catalysts, filtration aids, antifoaming agents, extraction solvents, etc)
Pesticides
Animal feed additives
Veterinary or therapeutic drugs (medicines, antibiotics, hormones)
Environmental contaminants (heavy metals, industrial chemicals such as polychlorinated biphenyls and polycyclic aromatic hydrocarbons)

Contaminants

Processing contaminants (eg, mutagenic heterocyclic amines)
Food packaging migrants (plastics, waxes, inks, etc); also called indirect food additive

Others

Adulterants, chemicals used for tampering

committees of the Food and Agriculture Organization of the World Health Organization (FAO/WHO), such as the Joint FAO/WHO Expert Committee on Food Additives, the Joint FAO/WHO Meeting on Pesticide Residues, and the Codex Alimentarius.

TOXICITY: TYPES AND TARGETS

Based on the duration of exposure to chemicals, toxicity is often classified as acute (involving a single dose), chronic (generally involving exposure over a lifetime), and subchronic (repeated exposures).

Different organs can be damaged by chemicals, including the liver, kidney, skin, brain, lungs, stomach, spleen, intestines, bladder, eyes, blood, and blood vessels. Different systems can also be affected, including the cardiovascular, nervous, immune, and reproductive systems, as well as the developing embryo or fetus. Although some toxicological effects can be reversed, others are irreversible. Carcinogens are chemicals capable of producing tumors, teratogens cause birth defects (2).

Toxicity Testing of Food Chemicals

In early times, trial-and-error methods were required to allow the distinction between safe and unsafe foods. In fact, early risk assessors worked for ancient pharaohs as food tasters. Currently, the preferred method for predicting potential health risks is human epidemiology, which involves statistical treatment of data collected from human populations. Epidemiological studies have been used to correlate human cancers with various factors. It is estimated that one-third of cancers can be attributed to diet (3) and that dietary changes can bring a substantial decrease in the rate of certain cancers (4).

Despite its usefulness in predicting risks from human exposure, epidemiology is limited by several factors, such as ethical considerations that prohibit human studies, the difficulty in measuring health effects with low probabilities of occurrence, inaccuracy due to recall bias, problems in identifying control groups that have not been exposed to the chemical under study, and the impossibility of obtaining human data on newly developed chemicals prior to their release.

As the preferred substitute for epidemiological data, animal toxicology studies are commonly performed in a variety of animal species, and the results of such studies are extrapolated to allow risk assessors to predict potential human health risks.

The major types of toxicology studies commonly conducted are whole animal studies and *in vitro* studies.

Whole Animal Studies. For practical reasons, most whole animal toxicology studies are conducted in rodents (mice and rats), although tests in other nonrodent species such as dogs and nonhuman primates are also commonly performed.

The most comprehensive toxicology test performed is the lifetime animal bioassay using rodent species administered daily doses of the test material over their normal life span. Typical dosing schemes involve a high dose (the maximum tolerated dose, or MTD), a lower dose (often one-fourth to one-half of the MTD), and a control (zero) dose. (5). In testing many food chemicals, high doses are often used to maximize the potential to identify toxicological effects; many of these doses are several orders of magnitude greater than the typical human intake.

In vitro Studies. Commonly, a number of *in vitro* studies are also performed that may utilize specific components of mammalian systems (for example, liver enzymes obtained from whole animals) or using other living systems such as bacterial colonies. *In vitro* studies are frequently employed to identify the potential metabolic pathways that a specific chemical is subject to and to screen for various forms of toxicity such as mutations and endocrine disruption. Results from initial *in vitro* studies may provide preliminary information for chemical and pharmaceutical manufacturers that allow them to identify suitable chemical candidates they may wish to study in expanded and costlier whole animal studies. Some *in vitro* studies also allow risk assessors to determine the similarities and differences between test animals and humans in their abilities to metabolize the chemicals under investigation. Specific tests have been developed to determine the DNA-damaging properties of certain chemicals and DNA-repair tests (6). It is anticipated that increasingly sophisticated methods,

with major inputs from molecular biology, will play a more extensive and important role in risk assessment (7).

RISK ASSESSMENT

In 1983, the National Research Council (NRC) of the National Academy of Sciences published the first U.S. federal guidelines for performing chemical risk assessments (8). The NRC identified four major components of the risk assessment process: hazard identification, dose-response assessment, exposure assessment, and risk characterization.

Hazard Identification

Hazard identification is the determination of whether a particular chemical is or is not causally linked to specific health effects. This first step involves a critical evaluation of all the available information on the chemical under study, including epidemiological (human) data and experimental (animal) toxicity data, conditions of exposure under which the effects may be produced, and studies identifying possible biochemical mechanisms of toxic action. The information may have been developed from acute, subchronic, or chronic toxicology studies and might include a variety of harmful effects, such as cancer, allergic reactions, birth defects, developmental defects, reproductive abnormalities, neurotoxicity, immunotoxicity, or toxicity to the liver, kidney, or lung.

A critical component of hazard identification is the determination of whether a chemical does or does not cause cancer. This distinction is important because risk assessment practices use different criteria for carcinogens (cancer-causing chemicals) and noncarcinogens. Cancer studies usually are performed using the chronic lifetime rodent bioassays and frequently make use of the MTD. The MTD is usually determined following the results of 90-day toxicity studies and is roughly described as the highest dose that does not alter the test animal's longevity or well-being because of noncancer effects (9). However, it has been argued that many chemicals may induce cancer at the MTD through biological mechanisms that do not occur at lower dose. Such mechanisms include increased cell proliferation rates in response to high-dose toxicity (10). Exposure at lower doses, where these mechanisms are not active, would not result in cancer.

Dose-Response Assessment

Once a specific toxicological hazard has been identified, it is possible to determine the relation between the level of exposure and the probability of occurrence of the adverse effects in question. The procedures used to establish this dose-response relationship are governed by the type of hazard; noncarcinogenic and carcinogenic hazards are typically treated differently.

There are two general models used to determine the dose-response relationships: (1) the threshold model and (2) the nonthreshold model.

Threshold Model. This model is adopted for noncarcinogenic hazards and assumes that toxic effects will not be observed until a minimum, or threshold, dose is reached.

To estimate the threshold, toxicology studies generally try to identify two dose levels: one above the threshold at which effects are seen (known as the lowest observed effect level, or LOEL) and one presumably below the threshold at which no effects are seen (known as the no observed effect level, or NOEL). The degree to which the NOEL and LOEL estimates approximate the threshold cannot be determined because of statistical and biological limitations as well as limitations in the number of dose levels used in toxicology studies. As a prudent measure, the NOEL is generally used as a conservative estimate of the threshold (11).

To determine acceptable levels of human exposure following identification of the NOEL value from animal studies, uncertainty factors (often called safety factors) are also incorporated into the process. Such factors presumably account for differences in animal to human extrapolation and consider biological variability among human individuals.

The NOEL values are derived from toxicology studies involving small homogeneous groups of animals and, therefore, may not represent appropriate thresholds for large and nonhomogeneous human populations. The choice of uncertainty factors is governed by the availability of human data; the nature, severity, and chronicity of the effect; the quality of animal toxicology data; and the need to accommodate human response variability for sensitive subgroups. Overall uncertainty factors may range from 1 to 10,000 (12).

The most common uncertainty factors is 100, which is rationalized as a 10-fold uncertainty factor for species variation (assuming humans are 10 times more sensitive than the animals studied) multiplied by another 10-fold uncertainty factor for human variation (assuming some humans are 10 times more sensitive than the average humans). It has been recommended that up to an additional 10-fold uncertainty factor be incorporated to provide greater protection for infants and children from possible effects of pesticide residue exposure if data relating the relative sensitivities of infants and children to the sensitivities of adults are not available (12).

An acceptable daily intake (ADI) is calculated by dividing the NOEL by the selected uncertainty factor. The ADI is defined as an estimate of the amount of a food chemical (such as an additive or pesticide) that can be ingested by humans on a daily basis over a lifetime without anticipated harm. The EPA has recently replaced the term ADI with an analogous term, the toxicity reference dose, or RfD, thereby removing the inference of acceptability, which may carry the connotation of a nonscientific, value judgment (2). For metal contaminants with cumulative properties (such as mercury or lead), the concept of provisional tolerable weekly intake (PTWI) was established by the WHO/FAO (13). For chemicals that are essential nutrients but cause toxicological effects when consumed at high levels, the WHO (14) established the concept of provisional maximum tolerable daily intakes (PMTDI).

As an alternative to the use of the NOEL approach in the dose-response assessment, the concept of the benchmark dose has been proposed (15). This approach is intended to reduce the uncertainty related to the NOEL determination. The benchmark dose is defined as the lower confidence limit for the effective dose that produces a specific increase in the incidence over control (background) levels. The benchmark dose provides a consistent basis for calculating the RfD, considers the dose-response model, and uses all available experimental data in contrast to the NOEL approach, which ignores the shape of the dose-response curve. The benchmark dose approach can also be applied to the risk assessment of carcinogens.

Nonthreshold Model. This model is used for assessment of the dose-response relationship for carcinogenic hazards. It is assumed that any level of exposure to a carcinogen may provide some probability for the development of cancer (although the probability is still related to the dose), which implies that a NOEL does not exist for carcinogens. Mechanistically, it is assumed that many carcinogens act as mutagens that cause direct damage to the genes; it has been proposed, in what is often called the one-hit model of carcinogenesis, that exposure to a single molecule of a carcinogen could ultimately lead to a mutation that could develop into cancer.

As mentioned previously, the study of cancer-causing chemicals usually involves long-term rodent (eg, rat and mouse) assays. In an effort to maximize the chance of detecting cancer in the animal studies, special strains of animals that may be more susceptible to developing cancer are often used; this practice raises questions about the validity of extrapolating such results to humans (16). Moreover, although cancer itself requires tumors to invade other tissues, benign (noninvasive) tumors are also usually considered evidence of potential carcinogenicity.

The notion that carcinogens lack toxicity thresholds is controversial. Chemicals may, in fact, use several biological mechanisms to cause cancer (17). Genotoxic chemicals, which cause mutations, may indeed lack threshold doses. In other cases, such as the induction of thyroid tumors (18) or the induction of tumors resulting from increased cell proliferation (10), it is argued that such carcinogenic effects are exerted through mechanisms consistent with a threshold hypothesis. Currently, however, most carcinogenicity risk assessment models still rely on the nonthreshold assumption.

Typical human exposure to animal carcinogens is often several thousand times lower than doses that produce tumors in experimental studies. Toxicology testing protocols for carcinogens usually involve the administration of massive levels of chemicals (at the MTD) to animals throughout their lifetimes. Calculation of carcinogenic risks therefore requires the results of high-dose animal studies to be extrapolated to predict human risks at low exposures. A number of mathematical models have been developed for the dose-response assessment of carcinogens; each yield a value known as the cancer potency factor (19). Cancer potency factors may vary widely depending on the choice of the model and its assumptions. The most commonly used model is the linearized multistage model that assumes a cell, which may be a target for a carcinogenic chemical, goes through a specific number of different stages and that the probability of a carcinogenic "hit" on the cell, which leads to the development of cancer, is stage-specific. At the lowest level of exposure, the relationship between exposure level and excess cancer is linear. Also commonly performed are statistical corrections that express cancer risks on the basis of the upper 95% confidence interval of the slope of the dose-response curve, adding an additional element of conservatism to the risks (20). Upper-confidence-level cancer risks may be orders of magnitude greater than the best estimates obtained using the mathematical models.

Exposure Assessment

Before risks from exposure to chemicals in food can be properly assessed, the information derived from the hazard identification and dose-response assessment processes must be combined with an estimate of the likely amount of human exposure, or dose, before the levels of risk may be assessed. To obtain reliable information on dietary exposure to chemicals in food, it is necessary to obtain information on the frequency of occurrence and levels of the chemicals in food and the amounts and types of various food items that are consumed. Human exposure to chemicals in the diet is typically expressed as the product of the concentration of the chemical in various foods and the amounts of the foods consumed. Estimation of both factors requires several assumptions and involves considerable uncertainty.

The exposure assessment depends on the type of food chemical under study. With food additives, the concentration of a chemical in food may be well known and relatively constant. For incidental chemicals in foods (eg, pesticide residues, hormones, antibiotics, and environmental contaminants), however, the concentration may vary dramatically from sample to sample, making an accurate estimate of the actual level of concentration difficult to obtain. The choice of assumptions used to predict the concentration levels may also be related to the availability of reliable monitoring data.

Several techniques are available to estimate pesticide residue levels in foods. The highly theoretical and conservative assumption (worst case) is that all pesticides are present at a predetermined level, typically at the maximum allowable level throughout the whole production system, including the raw food commodity, processed forms of the food, and the food in table-ready form. More complex data-intensive approaches are based upon actual measurements of residue levels at the time the food is ready to be consumed. Also of use are a variety of intermediate techniques that consider factors such as residue results from field monitoring studies, effects of postharvest factors on residue levels, and incorporation of actual pesticide-use data (20). Results from the use of the various techniques may differ dramatically. Archibald and Winter (21), using realistic pesticide residue data, reported exposure levels ranging from hundreds to tens of thousands of times lower than those estimated using much more conservative assumptions.

Obtaining useful and accurate food consumption data represents another scientific challenge. Most food consumption data are collected from national food consumption surveys. Typically, such surveys rely on eight general methods that are used to assess food consumption: food

disappearance data (correcting food production and import data for food exports, waste, storage, and nonhuman food use), household disappearance data, dietary histories, dietary frequencies, 24-hour recalls, food records, weighted intakes, and duplicate portions (22). The method used depends upon the purposes of the study and availability of resources. For dietary risk assessment purposes, the most common food consumption estimates are derived from the results of U.S. Department of Agriculture nationwide food consumption surveys based on three-day dietary records describing the types and amounts of foods consumed, both at home and away from home. Such surveys involve tens of thousands of individuals in the 48 continental states and Hawaii, Alaska, and Puerto Rico. The amount of each food item consumed and the individual's weight are specified. Additional information concerning demographic and socioeconomic background, age, gender, ethnicity, and geographic location are tabulated. The accuracy of such food consumption estimates is directly related to the accuracy of those surveyed in identifying the amounts and types of foods they actually consumed during the survey period.

A critical step in the exposure assessment phase is to identify the target audience exposed to the food chemical under study. It is widely accepted that dietary chemical exposures of different population subgroups may vary dramatically due to differences in food consumption patterns. One of the limitations of data from nationwide food consumption surveys is the ability to accurately identify the food consumption patterns of population subgroups for which the sample size is limited in the overall data set.

During the past decade, the assessment of food chemical risks to infants (up to 1 y) and children (1–12 y) has been a key issue in risk assessment. Infants and children represent a unique subpopulation group, and concerns are common not only because of their small body size, but also because of their physiological and developmental differences. Infants and children eat fewer foods than adults, but consume more food on a per-body-weight basis. Hence, their exposure to pesticide residues is often greater than that for adults (12). Other vulnerable population subgroups are represented by the elderly and people with existing health conditions.

The main focus of chemical food safety evaluation has been on pesticides. In the case of acute (short-term) risk assessments, the diets of the most highly exposed individuals—those representing the upper 90th, 95th, or 99th percentiles (high-level consumers)—are often considered rather than median consumers, and chemical concentrations are often considered at highest detected levels rather than at median levels.

Traditional approaches to estimate food chemical exposure have relied upon "point estimates" derived by assuming a particular concentration level of the chemical and a specific food consumption estimate. In reality, both chemical levels and food consumption patterns may vary dramatically among different food samples and individual consumers and are more appropriately presented as distributions. Through the process of statistical convolution, it is possible to combine the distributions of chemical levels and food consumption patterns to develop an exposure distribution rather than simply a point estimate. Through such a process, for example, it is possible to identify what percentage of a population (or subpopulation) is exposed to a specific amount of a chemical (such as the RfD) on a given day or in a short time period. Such a distributional analysis approach is also more amenable to the assessment of exposure to multiple chemicals that may have similar mechanisms of toxic action (12).

Risk Characterization

The final stage of the risk assessment process involves describing the nature, and often the magnitude, of risk and optimally includes statements reflecting the uncertainty in the final risk estimates. The core of the risk characterization stage is a comparison between estimated human exposures with levels of toxicological significance. Risk may be expressed as a probability, but optimally, risk characterization should include qualitative evaluation in addition to single numerical depictions of risk or ranges of numerical depictions. Such qualitative factors include the strength of the evidence that a chemical produces the particular effect from which the risk was estimated. The numerous uncertainties and assumptions inherent in the risk assessment process should also be discussed (23). In fact, the analysis of uncertainty in the risk estimate is an essential component of risk characterization and may provide valuable information for risk managers.

For noncarcinogens, risk characterization typically relates the estimated exposure to the RfD or ADI. It is critical to understand that the RfD or ADI is not a toxicity threshold level in itself that provides a cutoff between "safe" and "unsafe" human exposures. Exposure at the RfD or ADI presents a very low risk, although "very low" is undefined (2). Qualitatively, risk increases at levels above the RfD or ADI, with greater exposure resulting in greater potential risk.

For carcinogens, estimated cancer risks are obtained by multiplying exposure estimates by cancer potency factors. This practice often results in the reporting of cancer risks indicating, for example, a specific number of cancers (above normal cancer rates) per 1 million persons exposed. The actual numbers presented usually represent the upper bounds of cancer risks calculated using conservative treatment of various uncertainties (such as nonthreshold behavior) and, as such, probably do not accurately reflect true cancer rates (24). Although fairly arbitrary, a cancer risk of 1 excess cancer per million persons exposed typically represents a level of risk considered to be "negligible" (19).

BIBLIOGRAPHY

1. *Toxicological Principles for the Safety Assessment of Direct Food Additives and Color Additives Used in Food*, Redbook II, U.S. Food and Drug Administration, Washington, D.C., 1993.

2. J. V. Rodricks, *Calculated Risks: The Toxicity and Human Health Risks of Chemicals in Our Environment*, Cambridge Univ. Press, New York, 1992.

3. B. N. Ames and L. S. Gold, "The Causes and Prevention of Cancer: Gaining Perspective," *Environ. Health Persp.* **105** (Suppl. 4), 865–873 (1997).

4. R. Doll, "The Lessons of Life: Keynote Address to the Nutrition and Cancer Conference," *Cancer Res.* **52**, 2024s–2029s (1992).

5. M. W. Pariza, "Risk Assessment," *Crit. Rev. Food Sci. Nutr.* **31**, 205–209 (1992).

6. L. A. P. Hoogenboom and H. A. Kuiper, "The Use of *In Vitro* Models for Assessing the Presence and Safety of Residues of Xenobiotics in Food," *Trends Food Sci. Technol.* **8**, 157–166 (1997).

7. D. R. Tennant, "Food Chemicals and Risk Analysis," in D. R. Tennant, ed., *Food Chemical Risk Analysis*, Blackie Academic & Professional, London, 1997, pp. 3–18.

8. *Risk Assessment in the Federal Government: Managing the Process*, National Research Council, National Academy Press, Washington, D.C., 1983.

9. *Issues in Risk Assessment*, National Research Council, National Academy Press, Washington, D.C., 1993.

10. B. N. Ames and L. S. Gold, "Too Many Rodent Carcinogens: Mitogenesis Increases Mutagenesis," *Science* **249**, 970–971 (1990).

11. Environ, *Elements of Toxicology and Chemical Risk Assessment*, Environ Corp., Washington, D.C., 1986.

12. *Pesticides in the Diets of Infants and Children*, National Research Council, National Academy Press, Washington, D.C., 1993.

13. WHO/FAO, *Evaluation of Certain Food Additives and the Contaminants Mercury, Lead and Cadmium*, 16th Report, WHO Technical Report, Series No. 505, Geneva, Switzerland, 1972.

14. WHO, *Evaluation of Certain Food Additives and Contaminants*, 26th Report, WHO Technical Report, Series No. 683, 1982.

15. K. S. Crump, "A New Method for Determining Allowable Daily Intakes," *Fundam. Appl. Toxicol.* **4**, 854–871 (1984).

16. P. H. Abelson, "Health Risk Assessment," *Reg. Toxicol. Pharmacol.* **17**, 219–223 (1993).

17. G. B. Gori, "Cancer Risk Assessment: The Science That Is Not," *Reg. Toxicol. Pharmacol.* **16**, 10–20 (1992).

18. O. E. Paynter et al., "Goitrogens and Thyroid Follicular Cell Neoplasia: Evidence for a Threshold Process," *Reg. Toxicol. Pharmacol.* **8**, 102–119 (1988).

19. *Regulating Pesticides in Food: The Delaney Paradox*, National Research Council, National Academy Press, Washington, D.C., 1987.

20. C. K. Winter, "Dietary Pesticide Risk Assessment," *Rev. Environ. Contam. Toxicol.* **127**, 23–67 (1992).

21. S. O. Archibald and C. K. Winter, "Pesticides in Our Foods: Assessing the Risks," in C. K. Winter, J. N. Seiber, and C. F. Nuckton, eds., *Chemicals in the Human Food Chain*, Van Nostrand Reinhold, New York, 1990, pp. 1–50.

22. J. A. T. Pennington, "Methods for Obtaining Food Consumption Information," in I. Macdonald, ed., *Monitoring Dietary Intakes*, Springer-Verlag, New York, 1991, pp. 3–8.

23. F. D. Hoerger, "Presentation of Risk Assessments," *Risk Anal.* **10**, 359–361 (1990).

24. C. K. Winter, "Lawmakers Should Recognize Uncertainties in Risk Assessment," *Calif. Agric.* **48**, 21–29 (1994).

ELISABETH L. GARCIA
CARL K. WINTER
University of California
Davis, California

See also FOOD SAFETY AND RISK COMMUNICATION; FOOD SAFETY AND RISK MANAGEMENT; FOOD TOXICOLOGY.

U

ULTRAFILTRATION AND REVERSE OSMOSIS

BACKGROUND INFORMATION

Ultrafiltration (UF) and reverse osmosis (RO) are unit operations in which water or some solutes in a solution are selectively removed through a semipermeable membrane. Both processes are similar in that the driving force for the transport of water and solutes across the membrane is the pressure applied to the feed solution. However, the membrane and pressure used and the separation achieved are quite different. Membranes used in both UF and RO have a very thin skin layer or an asymmetric structure formed by phase-inversion processes. In RO, the thin skin layer consists of a very dense film; thus, virtually all suspended matter and most dissolved species are retained by the membrane, and only water passes through the membrane. The operating pressure is in the order of 700 kPa or higher to overcome the high osmotic pressure of the low molecular weight solutes. UF membranes have a thin skin layer that is microporous and, hence, they separate species based on molecular size. In addition to water, some dissolved low molecular weight species can pass through the membrane. The driving force for the separation is the pressure difference across the membrane, typically in the range of 70 to 700 kPa. Another important difference between UF and RO is the effect of the chemical nature of the membrane polymer. In UF, the chemical nature of the membrane polymer has only a small effect on the separation process since separation is based on a sieving mechanism. By contrast, RO separations are based on preferential sorption or solution/diffusion phenomena and are greatly influenced by the chemical nature of the polymer membrane.

The advantages of UF and RO, comparing to other separation or dewatering processes, have been summarized by Porter and Nelson (1). First, they do not require a phase change in solvent to effect the separation or dewatering processes. Second, UF and RO can be conducted at ambient or other selected temperatures. These can result in considerable savings in processing energy, elimination of thermal or oxidation degradation problems associated with evaporation and collapse of gels, breaking of emulsions, and mechanical damages associated with freezing. For UF processes, in addition, changes in ionic strength and pH that occur during other methods of separation or dewatering are also avoided because low molecular weight acids, bases, or salts are not retained and are freely permeable to the UF membranes.

The main limitations of UF and RO are (1) fouling of the membranes, which reduces the operating time between periods of cleaning; (2) the maximum concentration is limited to 30% total solids; (3) higher capital costs than evaporation; and (4) variation in the product flow rate when changes occur in the concentration of feed solution (2).

The history of UF and RO has been reviewed by Sourirajan (3), Matsuura and Sourirajan (4), and Cheryan (5),

and the costs of their operations has been reviewed by Cheryan (5).

MEMBRANES AND MEMBRANE MODULES

UF and RO membranes are usually asymmetric, which means the pores change in size from one surface of the membrane to the other. They also differ from microporous membranes, which have pores of uniform size throughout the membrane. The two techniques have been developed to manufacture asymmetric membranes: one utilizes the phase-inversion process (6) and the other forms a composite structure by depositing a very thin homogeneous solid polymer film on a microporous membrane (7). The phase-inversion process consists of three steps: (1) a polymer is dissolved in an appropriate solvent to form a solution containing 10 to 30 wt % polymer; (2) the solution is cast into a film about 100 to 500 μm thickness; and (3) the film is quenched in a nonsolvent, which for most polymers is typically water or an aqueous solution (3,8,9). Variations of the phase-inversion process such as wet, dry, or melt spinning processes have been reviewed (5). More than 90 different homopolymers, copolymers, and blends have been investigated for the manufacture of the phase-inversion membranes (10). Cellulose acetate, polysulfone, and polyamide membranes are the most common types used commercially.

The methods of forming composite membranes have been reviewed (5,11). These methods can be briefly summarized as follows: (1) casting of the barrier layer separately, which is followed by lamination to the support film; (2) dip-coating of the support film in a polymer solution and drying, or in a reactive monomer or prepolymer solution followed by curing with heat or radiation; (3) gas-phase deposition of the barrier layer from a glow-discharge plasma; and (4) interfacial polymerization of reactive monomers on the surface of the support film.

Mineral or ceramic membranes are another class of membranes that are lately receiving increased attention. They are formed by the deposition of inorganic solutes, such as zirconium oxide, onto reusable porous carbon tubes or porous metal tubes. Because they are made of inorganic materials, they have little or none of the disadvantages associated with polymeric membranes. They possess a high degree of resistance to chemical and abrasion degradation and tolerate a wide range of pH and temperature (5).

For the successful application of UF and RO, the design of the membrane module and the layout of the system in which the module is installed are as important as the selection of the proper membranes (9). There are four important membrane module designs used in both UF and RO: tubular, plate-and-frame, spiral-wound, and capillary or hollow-fiber.

The tubular membrane module is shown in Figure 1 and consists of a membrane cast on the inside of a porous fi-

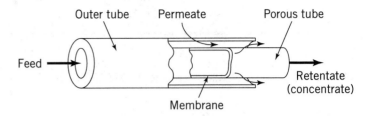

Figure 1. A tubular membrane module.

berglass reinforced plastic tube or a synthetic fiber tube and then inserted into a supporting, perforated stainless steel tube. These tubes are usually 12.5 mm (0.5 in.) to 25 mm (1 in.) in internal diameter and 0.6 to 6.4 m (2 to 20 ft) in length. A number of these tubes are then placed inside of a PVC or stainless steel sleeve or shroud to form a module. This type of membrane module provides a good control of the concentration polarization and membrane fouling. The system can usually tolerate large quantities of suspended solid matter in the feed and can be cleaned by foam swabs without dismantling the equipment. However, it has a lower membrane surface per module volume than other types of membrane module design.

A typical plate-and-frame membrane module is shown in Figure 2. It is similar to the conventional filter press. The membranes, porous support plates, and spacers with curved ribs forming the feed flow channels are clamped together and stacked between two end plates in a vertical or horizontal frame. Permeate is led out through fine, stainless steel tubes from the edge of the plates. All plate-and-frame membrane modules provide a larger membrane surface per unit volume than tubular membrane modules, but the control of concentration polarization is more difficult. Handling of solutions with suspended solids may cause plugging of the flow channel.

Figure 3 shows a spiral-wound membrane module. Two flat membrane sheets, separated by a porous membrane support, are placed together with their active sides facing away from each other. The three sides are then sealed together and the fourth side is attached to a perforated center stainless tube. A cover leaf and a spacer screen are then

placed on each side of the membrane and the whole assembly is rolled around the center tube and placed inside a PVC or stainless steel housing fitted with manifolds to form a spiral-wound module. The feed is pumped into one end of the module and flows in parallel to the axis of the module. The permeate flows radially through both the membrane and porous membrane support and finally into the perforations of the center tube. The retentate or concentrate then flows out from the other end of the membrane module. The membrane surface per unit volume is higher than the plate-and-frame type, but the control of concentration polarization effects is more difficult. Severe membrane fouling may occur even with moderate concentrations of suspended solid materials. Thus, the spiral-wound membrane module is usually used in RO applications, and its use in UF is limited (9).

The hollow-fiber geometry is a newer approach to the membrane module design (Fig. 4). The membrane is a self-supporting tube with the active surface on the inside of the tube. Each hollow fiber has a fairly uniform bore and is bonded on each end in an epoxy tube sheet and is housed in a clear polysulfone or translucent PVC shell to form a membrane cartridge. The feed enters the inside of a group of hollow-fiber tubes, and the permeate passes through the tube and enters the shell side of the chamber for collection. Hollow-fiber membrane modules provide the largest membrane surface per unit volume of all membrane module design. However, the small fiber tubes are highly susceptible to fouling and plugging. Pretreatment or prefiltering of the

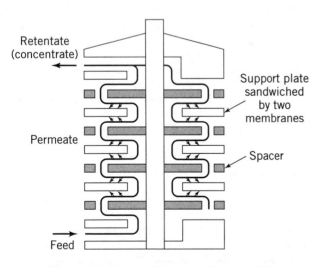

Figure 2. A plate-and-frame membrane module.

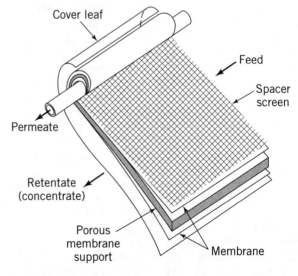

Figure 3. A spiral-wound membrane module.

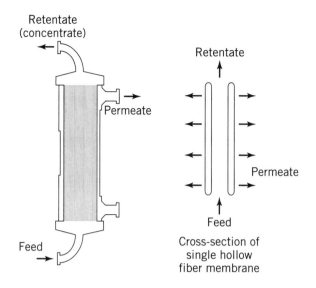

Figure 4. A hollow-fiber membrane cartridge.

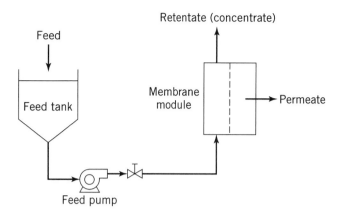

Figure 5. A schematic diagram of single-pass UF or RO system.

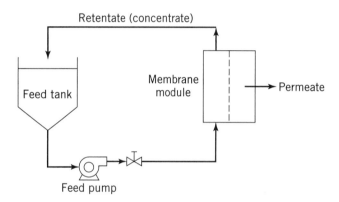

Figure 6. A schematic diagram of batch UF or RO system.

feed is often required and, furthermore, the flow rate of the feed is limited because it is proportional to the inlet pressure, which is limited to 1.8 atm (25 psig) maximum operating pressure.

PROCESS DESCRIPTION

UF and RO can be conducted either as a single-pass, batch, feed-and-bleed, or multistage recycle system (5). In single-pass systems, the feed solution is brought to contact with the membrane module only once (Fig. 5). This limits the attainable concentration of the retentate. Batch systems return the retentate continuously to the feed tank for recycling through the membrane module (Fig. 6). Because the permeate is collected in another tank, the retentate is concentrated continuously until the desirable retentate concentration is reached. Feed-and-bleed systems are essentially a combination of the batch and single-pass operations (Fig. 7). Initially, the retentate is totally recycled similar to a batch system until the final retentate concentration is reached. At this point, a portion of the retentate

is bled from the system, and fresh feed is added to the recycle loop. The quantity of feed into the loop is controlled at a rate equal to the permeate flowing out of the loop plus the amount of the retentate being taken out of the system. Multistage operations are achieved by linking two or more feed-and-bleed stages together (Fig. 8). The stages are operated in series with respect to the retentate/feed flow but in parallel for permeate flow. Each stage operates at a constant feed concentration, which increases from the first to

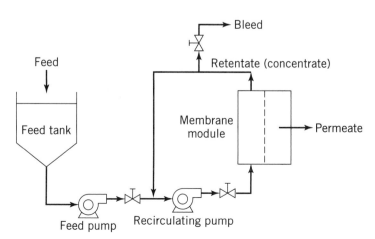

Figure 7. A schematic diagram of feed-and-bleed UF or RO system.

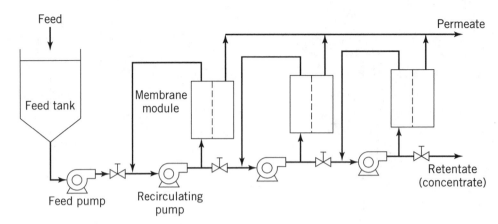

Figure 8. A schematic diagram of multistage recycle UF or RO system (only three stages are shown).

the last stage. The final stage operates at the desired retentate concentration. The advantage of the multistage system is that the retentate can be concentrated to high levels without the long residence time that occurs in a batch system, which can lead to potential microbial problems (12).

EFFECTS OF OPERATING PARAMETERS

The performance of a UF system is quantified by permeate flux. There are four important operating parameters that affect the flux: pressure, temperature, feed concentration, and flow rate. The pressure-flux behavior in typical UF processes is pressure dependent at low pressures and pressure independent at high pressures (5); increasing pressure increases permeate flux when the pressure is low, but permeate flux becomes independent of pressure when the pressure is high. Both the resistance model and the osmotic pressure model describe correctly the observed pressure-flux behavior of the UF processes (13). In the resistance model, the increase in pressure increases the flux and compacts the boundary layer. The compaction of the boundary layer increases its hydraulic resistance and thus opposes further increases in flux. On the other hand, the osmotic model does not consider the hydraulic resistance of the boundary layer. The increase in pressure increases the concentration of solute at the membrane surface and, therefore, the osmotic pressure. The increase in the osmotic pressure opposes the increase in flux. The mathematical expressions for the resistance model and the osmotic pressure model are shown in the following equations (5,13). The resistance model:

$$J = \frac{\Delta P}{r_m + r_b} = \frac{\Delta P}{r_m + \varphi \Delta P}$$

The osmotic pressure model:

$$J = \frac{\Delta P - \Delta \pi}{r_m}$$

where J is the permeate flux, ΔP is the pressure difference, r_m is the membrane resistance, r_b is the resistance due to

the boundary layer and gel-polarized layer, φ is the proportional coefficient between r_b and ΔP, and $\Delta \pi$ is the osmotic pressure.

Higher temperatures generally favors the increase in flux of the UF processes. This is because increasing temperature reduces the solution viscosity. Cheryan (5) suggested that it is best to operate UF at the highest possible temperature that is within the limits of the feed solution and the membrane. Increasing feed concentration increases the viscosity and density of the solution, but reduces the solute diffusivity; the net results are increased concentration polarization and increased thickness of the boundary layer. Observed experimental results, in general, show that the flux decreases exponentially with increasing feed concentration (5). On the other hand, increasing flow rate or turbulence tends to remove the accumulated solids near the membrane surface and reduce concentration polarization near the membrane surface and the thickness of the boundary layer. Therefore, the permeate flux increases.

In addition to permeate flux, the performance of a reverse osmosis (RO) system is also quantified by solute rejection. Similar to a UF system, pressure, temperature, feed concentration, and flow rate are four major influencing processing parameters. The effect of pressure on permeate flux and solute rejection is complex. Although permeate flux increases with pressure, the osmotic pressure at the membrane surface also increases because of increased solute concentration. The latter tends to reduce the increase of permeate flux. Because the solute flux is not driven by pressure, it is not affected by increasing pressure. Therefore, increasing pressure increases solute rejection. However, increased concentration at the membrane surface indirectly increases the solute flux, which tends to moderate the increase in solute rejection.

Temperature increases both permeate flux and solute flux. This is because increasing temperature reduces viscosity of the permeate and increases diffusivity and solubility of the solute in the membrane (13). Because the increase in solute flux with temperature is usually faster than the increase in permeate flux, solute rejection decreases with increasing temperature.

Increasing feed concentration reduces permeate flux by increasing osmotic pressure at the membrane surface because of increased solute concentration. Because solute flux is concentration driven, increased solute concentration or concentration polarization at the membrane surface increases solute flux. Both increased solute flux and reduced permeate flux cause rapid reduction in solute rejection.

The effect of flow rate on permeate flux and solute rejection is similar to that of pressure (13). Increasing flow rate reduces concentration polarization near the membrane surface and the thickness of the boundary layer. Thus, permeate flux increases, causing the osmotic pressure to increase; this tends to reduce or moderate the permeate flux. The initial reduction in the concentration polarization caused by increase in flow rate decreases solute flux and hence increases solute rejection. However, this is also reduced or moderated by the osmotic pressure increases since solute flux is concentration driven.

APPLICATIONS IN FOOD INDUSTRIES

Dairy Products

UF and RO are finding increased applications in the dairy products. Milk and skim milk are concentrated for the manufacture of soft and hard cheeses (14–16), dried whole milk and skim milk (17), ice cream (18), milk beverages (19), condensed milk (20), and yogurt (14,21,22). Proteins recovered from sweet and acid wheys are by-products from the manufacture of the various kinds of cheeses. Lactose can be separated from wheys for uses in alcohol or beer production (23).

Cereal and Oilseeds

For cereal and oilseeds, UF and RO are used for (1) recovery of proteins from ground coconut meat (24), cottonseed and soy flour (25–27), and rapeseed meal (28); (2) removal of undesirable components such as raffinose, stachyose, and phytic acid from soy milk (29); (3) treatment and processing of cottonseed and soy whey (30,31); (4) protein modification (32); and (5) degumming of edible oils (33).

Agricultural Raw Materials and By-products

For agricultural raw materials and by-products, notable applications for UF and RO are (1) recovery of meat proteins from animal blood and meat waste (1); (2) recovery of useful by-products from potato, sweet potato, and wheat starch processes (34–36); and (3) removal of glucose while concentrating egg white (37–39).

Others

UF and RO have found many other applications. Examples include concentration fractionation, or purification of alfalfa juice (40), apple juice (41–44), beet juice (45,46), cane juice (47), citrus juices (48), maple sap (49), onion juice (50), orange juice (51), passion fruit juice (52), perilla anthocyanins (53), strawberry juice (54), tomato juices (55), and vegetable juices (56). UF and RO are also finding many applications in biotechnology (5,57) and removal of toxic components from rapeseed meal extracts (58). The discussion is beyond the scope of this article, however.

BIBLIOGRAPHY

1. M. C. Porter and L. Nelson, "Ultrafiltration in the Chemical, Food Processing, Pharmaceutical, and Medical Industries," in N. N. Li, ed., *Recent Developments in Separation Science*, Vol. 2, CRC Press, Boca Raton, Fla., 1972, pp. 227–267.

2. P. Fellows, "Membrane Concentration," in P. Fellows, ed., *Food Processing Technology*, Ellis Horwood, Chichester, London, 1988, pp. 148–158.

3. S. Sourirajan, *Reverse Osmosis*, Academic Press, New York, 1970.

4. T. Matsuura and S. Sourirajan, *Advances in Reverse Osmosis and Ultrafiltration*, National Research Council Canada Publ., Ottawa, Canada, 1989.

5. M. Cheryan, *Ultrafiltration and Microfiltration Handbook*, Technomic Publ. Co., Lancaster, Pa., 1997.

6. R. E. Kesting, *Synthetic Polymeric Membranes*, McGraw-Hill, New York, 1971.

7. R. L. Riley, G. R. Hightower, and C. R. Lyons, "Preparation, Morphology, and Transport Properties of Composite Reverse Osmosis Membranes for Seawater Desalination," in H. K. Longsdale and H. E. Podall, eds., *Reverse Osmosis Membrane Research*, Plenum Press, New York, 1972, pp. 437–456.

8. T. Matsuura, *Synthetic Membranes and Membrane Separation Processes*, CRC Press, Boca Raton, Fla., 1994.

9. H. Strathmann, "Membrane Separation Processes," *J. Membrane Sci.* **9**, 121–189 (1981).

10. D. R. Lloyd and T. B. Meluch, "Selection and Evolution of Membrane Materials for Liquid Separations," in D. R. Lloyd, ed., *Materials Science of Synthetic Membranes*, American Chemical Society, Washington, D.C., 1985, pp. 47–79.

11. J. E. Cadotte, "Evolution of Composite Reverse Osmosis Membranes," in D. R. Lloyd, ed., *Materials Science of Synthetic Membranes*, American Chemical Society, Washington, D.C., 1985, pp. 273–294.

12. F. V. Kosikowski, "Membrane Separations in Food Processing," in W. C. McGregor, ed., *Membrane Separations in Biotechnology*, Marcel Dekker, New York, 1986, pp. 201–254.

13. J. D. Mannapperuma, "Design and Performance Evaluation of Membrane Systems," in K. J. Valentas, E. Rotstein, and R. P. Singh, eds., *Handbook of Food Engineering Practice*, CRC Press, Boca Raton, Fla., 1997, pp. 167–209.

14. H. R. Chapman et al., "Use of Milk Concentrated by Ultrafiltration for Making Hard Cheese, Soft Cheese and Yogurt," *J. Soc. Dairy Technol.* **27**, 151–155 (1974).

15. F. L. Davies, P. A. Shankar, and H. M. Underwood, "The Use of Milk Concentrated by Reverse Osmosis for the Manufacture of Yogurt," *J. Soc. Dairy Technol.* **30**, 23–28 (1977).

16. D. J. Phillips, "Application of Membrane Processing to Cheese Manufacture," in R. W. Field and J. A. Howell, eds., *Process Engineering in the Food Industry*, Elsevier Applied Science, London, 1989, pp. 249–258.

17. J. Abbot et al., "Application of Reverse Osmosis to the Manufacture of Dried Whole Milk and Skim Milk," *J. Dairy Res.* **46**, 663–672 (1979).

18. A. Bundgaard, "Hyperfiltration of Skim Milk for Ice Cream Manufacture," *Dairy Ind.* **39**, 119–122 (1974).

19. F. V. Kosikowski, "Low Lactose Yogurts and Milk Beverages by Ultrafiltration," *J. Dairy Sci.* **62**, 41–46 (1979).

20. D. Serpa Alverez, M. Bennasar, and B. De la Fuente, "Application of Ultrafiltration to the Manufacture of Sweetened Condensed Milk," *Le Lait* **59**, 376–386 (1979).

21. G. Morgensen, "Production and Properties of Yogurt and Ymer Made from Ultrafiltered Milk," *Desalination* **35**, 213–222 (1980).

22. P. H. Ferguson, "Membrane Processing in the Food and Dairy Industries," in R. W. Field and J. A. Howell, eds., *Process Engineering in the Food Industry*, Elsevier Applied Science, London, 1989, pp. 237–247.

23. T. I. Hedrick, "Reverse Osmosis and Ultrafiltration in the Food Industry: Review," *Drying Technol.* **2**, 329–352 (1984).

24. P. Chakraborty, "Functional Properties of Coconut Protein Isolate by Ultrafiltration," in M. Le Maguer and P. Jelen, eds., *Food Engineering and Process Applications*, Vol. 2, Elsevier Applied Science, London, 1986, pp. 305–315.

25. J. T. Lawhon et al., "Production of Protein Isolates and Concentrates from Oilseed Flour Extracts Using Industrial Ultrafiltration and Reverse Osmosis Systems," *J. Food Sci.* **42**, 389–394 (1977).

26. O. Omosaiye and M. Cheryan, "Ultrafiltration of Soybean Water Extracts: Processing Characteristics and Yields," *J. Food Sci.* **44**, 1027–1031 (1979).

27. N. Devereux and M. Hoare, "Membrane Recovery of Soluble and of Precipitated Soya Proteins," in M. Le Maguer and P. Jelen, eds., *Food Engineering and Process Applications*, Vol. 2, Elsevier Applied Science, London, 1986.

28. Y. Tzeng, L. L. Diosady, and L. J. Rubin, "Preparation of Rapeseed Protein Isolate by Sodium Hexametaphosphate Extraction, Ultrafiltration, Diafiltration, and Ion-Exchange," *J. Food Sci.* **53**, 1537–1541 (1988).

29. H. G. Ang, W. L. Kwik, and C. K. Lee, "Ultrafiltration Studies of Foods: Part 1—The Removal of Undesirable Components in Soymilk and the Effects on the Quality of the Spray-Dried Powder," *Food Chem.* **20**, 183–199 (1986).

30. E. S. K. Chian and M. A. Aschauer, "Effect of Freezing Soybean Whey upon the Performance of Ultrafiltration Process," in G. F. Bennett, ed., *Industrial Waste Water Treatment*, AIChE Symp. Ser., No. 144, 1974, pp. 163–169.

31. J. T. Lawhon, et al., "Fractionation and Recovery of Cottonseed Whey Constitutes by Ultrafiltration and Reverse Osmosis," *Cereal Chem.* **52**, 34–41 (1975).

32. S. D. Cunningham, C. M. Cater, and K. F. Mattil, "Cottonseed Protein Modification in an Ultrafiltration Cell," *J. Food Sci.* **43**, 1477–1480 (1978).

33. C. D. O'Donnell, "Lipid Science: Refining Knowledge of Fats & Oils," *Prepared Foods* **167**, 87–90 (1998).

34. E. O. Strolle et al., "Recovering Useful Byproducts from Potato Starch Factory Waste Effluents," *Food Technol.* **34**, 90–95 (1980).

35. F. Meuser and F. Köbler, "Membrane Filtration of Process Water from Potato and Wheat Starch Plants," in C. Cantarelli and C. Peri, eds., *Progress in Food Engineering*, Forster-Verlag AG, Küsnacht, Switzerland, 1983, pp. 335–340.

36. B. H. Chiang and W. D. Pan, "Ultrafiltration and Reverse Osmosis of the Water from Sweet Potato Starch Process," *J. Food Sci.* **51**, 971–974 (1986).

37. E. Lowe et al., "Egg White Concentrated by Reverse Osmosis," *Food Technol.* **23**, 45–54 (1969).

38. R. E. Payne, C. G. Hill Jr., and C. H. Amundson, "Concentration of Egg White by Ultrafiltration," *J. Milk Food Technol.* **36**, 359–363 (1973).

39. L. S. Tsai, K. Ijichi, and M. W. Harris, "Concentration of Egg White by Ultrafiltration," *J. Food Protect.* **40**, 449–455 (1977).

40. B. E. Knuckles et al., "Pilot Scale Ultrafiltration of Clarified Alfalfa Juice," *J. Food Sci.* **45**, 730–734, 739 (1980).

41. F. Chou, R. C. Wiley, and D. V. Schlimme, "Reverse Osmosis and Flavor Retention in Apple Juice Concentration," *J. Food Sci.* **56**, 484–487 (1991).

42. M. J. Sheu and R. C. Wiley, "Preconcentration of Apple Juice by Reverse Osmosis," *J. Food Sci.* **48**, 422–429 (1983).

43. O. I. Padilla and M. R. McLellan, "Molecular Weight Cut-off of Ultrafiltration Membranes and the Quality and Stability of Apple Juice," *J. Food Sci.* **54**, 1250–1254 (1989).

44. V. Gokmen, Z. Bornemman, and H. H. Nijhuis, "Improved Ultrafiltration for Color Reduction and Stabilization of Apple Juice," *J. Food Sci.* **63**, 504–507 (1998).

45. S. Landi et al., "Purification of Beet Raw Juice by Means of Ultrafiltration Membranes," *Z. Zuckerind* **24**, 585–591 (1974).

46. A. Bayindirli, F. Yildiz, and M. Özilgen. "Modeling of Sequential Batch Ultrafiltration of Red Beet Extract," *J. Food Sci.* **53**, 1418–1422 (1988).

47. R. F. Madsen, "Application of Ultrafiltration and Reverse Osmosis to Cane Juice," *Int. Sugar J.* **75**, 163–167 (1973).

48. R. J. Braddock, S. Nikdel, and S. Nagy, "Composition of Some Organic and Inorganic Compounds in Reverse Osmosis-Concentrated Citrus Juices," *J. Food Sci.* **53**, 508–512 (1988).

49. C. O. Willits, J. C. Underwood, and U. Merten, "Concentration by Reverse Osmosis of Maple Sap," *Food Technol.* **21**, 24–26 (1967).

50. J. S. Nuss and D. E. Guyer, "Concentration of Onion Juice Volatiles by Reverse Osmosis and Its Effects on Supercritical CO_2 Extraction," *J. Food Process Eng.* **20**, 125–139 (1997).

51. B. G. Medina and A. Garcia, "Concentration of Orange Juice by Reverse Osmosis," *J. Food Process Eng.* **10**, 217–230 (1988).

52. Z. R. Yu and B. H. Chiang, "Passion Fruit Juice Concentration by Ultrafiltration and Evaporation," *J. Food Sci.* **51**, 1501–1505 (1986).

53. M. Y. Chung, L. S. Hwang, and B. H. Chiang. "Concentration of *Perilla Anthocyanins* by Ultrafiltration," *J. Food Sci.* **51**, 1494–1497, 1510 (1986).

54. S. Rwabahizi and R. E. Wrolstad, "Effects of Mold Contamination and Ultrafiltration on the Color Stability of Strawberry Juice and Concentrate," *J. Food Sci.* **53**, 857–861, 872 (1988).

55. D. Pepper, A. C. J. Orchard, and A. J. Merry, "Concentration of Tomato Juices and Other Fruit Juices by Reverse Osmosis," in R. W. Field and J. A. Howell, eds., *Process Engineering in the Food Industry*, Elsevier Applied Science, London, 1989, pp. 281–290.

56. S. S. Köseoglu, J. T. Lawhon, and E. W. Lusas "Vegetable Juices Produced with Membrane Technology," *Food Technol.* **45**, 124–130 (1991).

57. W. C. McGregor, *Membrane Separations in Biotechnology*, Marcel Dekker, New York, 1986.

58. M. J. Lewis and T. J. A. Finnigan, "Removal of Toxic Components Using Ultrafiltration," in R. W. Field and J. A. Howell, eds., *Process Engineering in the Food Industry*, Elsevier Applied Science, London, 1989, pp. 291–306.

Fu-hung Hsieh
University of Missouri
Columbia, Missouri

UNITED STATES: ARMED FORCES FOOD RESEARCH AND DEVELOPMENT

Food and foodservice equipment research and development for the Army, Navy Air Force, Marine Corps and the Defense Logistics Agency is conducted at the U.S. Army Natick Research, Development and Engineering Center in Natick, Massachusetts. Since 1970, Natick has served as the executive agent for the Department of Defense Food Program, conducting research, development and engineering to meet the unique feeding needs of the Armed Forces through a combination of in-house and contractual efforts.

Research and Development conducted under the DOD Food Program can generally be divided into two categories: Technology Base and Military Service Requirements. Technology Base efforts encompass research in areas of known or anticipated technology gaps that impede solution of current problems or development of future systems to meet military feeding needs. Military Service Requirements are recognized, validated needs identified by one or more of the Services and generally address known current deficiencies or needs contemplated in the next three to seven years.

The DOD Food and Nutrition Research and Engineering Board (FNREB), consisting of representatives of each of the four military Services and other involved organizations, functions as a board of directors, providing direction to Natick in formulation of the DOD Food Program. The FNREB meets annually to prioritize the projects, including initiation of effort on new requirements identified by the Services during the preceding year and revalidation of the need for continuation of ongoing work.

Each Service maintains a full-time representative at Natick as a member of the Joint Technical Staff (JTS). These representatives provide military guidance to the Natick scientists relative to their Services' projects and provide liaison between Natick personnel and the military community. They assist in matters such as organizing field tests and provide a communication link between the item developer (Natick) and the customers, the military planners and men and women in the field, aboard ships and on aircraft. The JTS members represent the FNREB on day-to-day actions and participate fully in all aspects of the DOD Food Program formulation and execution.

The annual program, formulated and prioritized as described above, is executed in three Natick elements, the Food Engineering Directorate, the Soldier Science Directorate and the Advanced Systems Concepts Directorate, whose specific functions are summarized below.

FOOD ENGINEERING DIRECTORATE

The Food Engineering Directorate (FED) is responsible for design, development and evaluation of military rations and foodservice equipment, to fulfill the requirements of the Army, Navy, Air Force, and Marine Corps, as well as special requirements to support the Defense Logistics Agency. Response to identified needs of the above Services involves long-range research, item development and engineering support to procurement agencies for food and foodservice equipment during the entire life of the item. The end products of FED's effort are Technical Data Packages (specification, drawings, etc), which are used to procure items from commercial suppliers.

The Food Engineering Directorate is divided into three product divisions and a support division. The product-oriented divisions are the Technology Acquisition Division, the Food Technology Division, and the Food Equipment and Systems Division.

The Technology Acquisition Division (TAD) is structured and staffed to conduct basic research and exploratory development involving new technologies for food preservation and processing, and for foodservice equipment and operations. The TAD's function is to explore, assess, adapt and refine new materials, equipment and operational concepts as basis for establishing new technological opportunities for exploitation to meet near- and long-term military feeding requirements. As with all FED Divisions, the emphasis on R&D conducted by TAD is on military-unique requirements, for which there are no civilian counterparts.

Exploratory and feasibility studies are conducted on new foods and new processing techniques needed to provide optimally configured and formulated field rations, which can be assembled from a limited number of basic components, to meet specific needs, such as high altitude or low temperature operating environments, for future military scenarios. Objectives of studies on foods and processes include, in addition to optimizing nutrient content, extending shelf life, improving product functionality and reducing overall food costs through means such as automating processing technologies and developing methods to objectively monitor quality and serviceability of food products in storage.

Research and engineering studies on food service equipment encompass diverse areas, including combustion technology, energy conversion technologies, robotics, electronics, material science and sanitation technologies. The goal of these studies is to assure that new and emerging technologies are applied to military food service equipment needs, and to develop new technology as necessary to meet anticipated unique military needs for feeding military personnel on the "high-tech" battlefields of the future.

The Food Technology Division (FTD) conducts exploratory development, advanced development and engineering programs to meet current and future military feeding needs. Division responsibilities include development of new or improved foods, food processes and operational rations in response to specific Military Service Requirements. Disciplines include food technology, packaging technology, and physical sciences. Product development and evaluation are focused primarily on operational rations for field feeding and include design of rations to meet designated nutritional content, acceptance by troops, serviceability, storage life, and convenience of use under field conditions. Commercial items are evaluated for suitability and are utilized wherever possible, or modified, if feasible, to meet any special requirements resulting from the unique military environment. This Division is responsible for preparation of technical requirements for approximately 700 procurement documents and for providing engineering and technical support to the Defense Personnel

Support Center, the agency responsible for purchasing foods and operational rations used by all military services.

The Food Equipment and Systems Division (FESD) is responsible for development, testing and fielding of new or improved foodservice equipment and systems in support of Military Services Requirements. Its mission includes evaluation of commercial foodservice equipment and methods, including foreign military equipment, for applicability to military requirements. Advanced and engineering development tasks are accomplished through in-house design and contracted design and fabrication of prototype equipment. Development of foodservice equipment encompasses all aspects of evaluation, including laboratory and field testing, to assure that all requirements are met.

In addition to specific equipment development and testing, this Division is responsible for managing the development of selected systems throughout the acquisition cycle. Systems engineering activities involve integration of components, ie, food, packaging, preparation and serving equipment, into a total feeding system. Engineering support is provided to procuring activities throughout the life cycle of the hardware system.

ADVANCED SYSTEMS CONCEPTS DIRECTORATE

The Advanced Systems Concepts Directorate (ASCD) combines the functions of advanced systems development, long-range planning, and interfacing with military customers. With respect to foodservice systems, ASCD works closely with each of the uniformed Services through the DOD Food and Nutrition Research, Development and Engineering Program. From its creation in 1970, as the Operations Research and Systems Analysis Office, ASCD has evolved into an organization employing roughly 20 individuals working in the foodservice concept development and long-range planning arena. Typically, there are 10 to 15 projects being pursued in ASCD's portion of the food program. These projects place ASCD in the forefront of developing the most advanced feeding system concepts for land, sea, and air military missions.

ASCD applies a systems approach, which uniquely combines appropriate statistical and mathematical modeling techniques with field verification when possible. This approach provides analyses of the benefits and characteristics of the existing system and alternatives, and validates the performance of the selected candidate system with field experiments. The success of this approach has been demonstrated in the development and rapid adoption of a new foodservice system concept for aircraft carriers, the development and fielding of a new concept for combat field feeding in the Army, and the development and deployment of a new feeding concept for Air Force Ground-Launched Cruise Missiles.

Current systems analysis development projects include a foodservice concept for the proposed Peacekeeper Rail Garrison system. Peacekeeper units must operate without assigned foodservice personnel, without resupply, and, despite extreme space constraints, must still provide high quality food for the crews. The problems of providing qual-

ity foodservice to the Navy's antisubmarine warfare squadrons is the subject of another project wherein the current system cannot provide the flexibility required by air crews engaged in long surveillance and combat missions. Also for the Navy, ASCD is developing a new concept for disposal of plastic waste products. Under international treaty, by 1993 the Navy was to cease disposing of plastics at sea. With the large volume of plastic packaging generated by aircraft carriers, which are essentially floating cities containing over 5,000 individuals, and with industry's trend toward increasing use of plastics packaging material, this is a crucial project addressing a highly visible problem.

In addition to projects involving long-term improvements to current systems, ASCD is involved in the definition and evaluation of "Next Generation" (1990s) and "Future Systems" (2000 and beyond), recognizing future threats and warfighting concepts. One example is a project to explore concepts for undersea subsistence storage and resupply. Another project is looking at the value of concepts involving mobile, automated food production systems in terms of futuristic battlefield threats and warfighting concepts.

Working closely with representatives from the Food Engineering and Soldier Science Directorates, ASCD also conducts a proactive program to obtain feedback from users in the field. For this purpose ASCD representatives makes a large number of visits to Army installations worldwide. These visits familiarize General Officers and their staffs with Natick's products as well as providing opportunities for collecting survey data from soldiers using those products in field exercises.

SOLDIER SCIENCE DIRECTORATE

The Soldier Science Directorate has the mission of science and technology leadership for all of Natick's products and systems in support of the individual soldier. The Directorate, a multidisciplinary research organization, draws from the behavioral, biological, physical, chemical, and engineering sciences to maintain a decided edge in technology in order to avoid surprise and to ensure the timely transfer of information to Natick's product Directorates, other Army research and development activities, industry, and academia. The SSD thrusts in food science and technology are aimed at providing soldiers with acceptable, nutritionally complete and environmentally stable meals for the whole spectrum of operational scenarios for the battlefield of the future.

The Directorate is organized into three divisions known as the Behavioral Sciences Division, the Biological Sciences Division, and the Physical Sciences Division. Approximately half of the Directorate's 129 permanent civilian and 17 military personnel have doctorate or master's degrees.

The Behavioral Sciences Division provides human factors support to all projects and programs at Natick. Its Food Systems Human Factors Branch has responsibility for supporting ration design and development. Modern marketing research techniques are being introduced into the behavioral research efforts in order to develop sensory and behavioral rules, criteria, and limits for optimizing the acceptance of novel rations and to engineer into these rations the desirable sensory characteristics that will ensure their acceptance and consumption in the targeted field environment. Methods are being developed to attenuate or eliminate the effects of jet lag on individuals by modifying diet or feeding patterns prior to, during, or after time zone changes. Artificial intelligence technology is being implemented to aid in configuring ration components that optimally meet the nutrition and load-carrying requirements for a given mission. An expert system is being developed to aid soldiers who have to forage for food in unfamiliar and hostile environments.

The Biotechnology Branch of the Biological Sciences Division focuses on the development of new food packaging based on biopolymers. These materials, which provide long-storage stability due to low oxygen permeability, can be microwaved and are readily biodegradable. Efforts are also underway to develop a survival kit with the capability to convert environmental substrates into edible materials through the action of enzymes and/or microorganisms.

The Biohazard Assessment and Control Branch of the Biological Sciences Division studies the nutritional adequacy and stability of rations and the microbiological quality of military rations stored for varying lengths of time under a wide range of temperature and humidity levels. The research focus is on developing preservative systems that minimize oxidative events causing membrane damage during processing or storage. Other projects are directed toward developing a nonanimal assay to aid in the selection of fats for use in energy-dense military foods, improved thermal processing techniques using combinations of pH, water activity, and preservatives to prevent microbial growth in current and future combat ration components, and water extraction, recovery, and generation concepts for enhanced ability to supply potable drinking water under combat feeding conditions.

The primary contribution of the Physical Sciences Division to rations research is in the development of new or improved techniques for materials testing. Among other important contributions, the Physical Sciences Division has developed more sensitive as well as more precise methodology for measuring the amount of oxygen permeation of packaging materials. This division also possesses the capability to provide analytical support plus textural engineering via electron microscopy characterization of membranous materials used for encapsulating and packaging rations.

The Divisions within the Soldier Science Directorate do not pursue their scientific and technological objectives in a vacuum. Rather, they work closely with each other, other product directorates at Natick, and outside agencies. The Directorate maintains agreements with other federal agencies, academia, and industry for the pursuit of solutions to common perplexing scientific and engineering problems. The focus of all these efforts is to ensure that new rations, materials, and packaging processes are consistent with improving the total performance, health, and safety of today's and tomorrow's soldier, whether operating under peacetime training or extremely hostile combat conditions.

GERALD L. SCHULZ
D. PAUL LEITCH
STANLEY HOLGATE
U.S. Army Natick Research, Development and Engineering
    Center
Natick, Massachusetts

UNITED STATES DEPARTMENT OF AGRICULTURE

In the United States, the principal in-house research agency of the U.S. Department of Agriculture (USDA) is the Agriculture Research Service (ARS). It is one of the four component agencies of the Research, Education, and Economic (REE) mission area.

This article provides an overview of research activities conducted by the USDA in cooperation with universities in the country. The information has been adapted from its website, *http://www.ars.usda.gov*.

MISSION OF AGRICULTURE RESEARCH SERVICE

ARS conducts research to develop and transfer solutions to agricultural problems of high national priority and provides information access and dissemination to:

- Ensure high-quality, safe food and other agricultural products
- Assess the nutritional needs of Americans
- Sustain a competitive agricultural economy
- Enhance the natural resource base and the environment
- Provide economic opportunities for rural citizens, communities, and society as a whole

GOALS OF AGRICULTURE RESEARCH

ARS goals are:

1. To enhance the production, value, and safety of foods and other products derived from animals that have a major impact on the American economy, world markets, and the U.S. balance of trade
2. To enhance the quality of the environment through a better understanding of and building on agriculture's complex links with soil, air, and biotic resources
3. To enhance the production, value, and safety of foods and other products derived from plants that have a major impact on the American economy, world markets, and the U.S. balance of trade.

The remaining part of this article will discuss the details of these goals as expressed by the USDA national programs under the following three research areas:

1. Animal production, product value, and safety (APPVS)
2. Natural resources and sustainable agricultural systems (NRSAS)
3. Crop production, product value, and safety (CPPVS)

ANIMAL PRODUCTION, PRODUCT VALUE, AND SAFETY

Multidisciplinary research conducted by ARS in these areas is built on the following national programs discussed in the sections that follow.

Animal Genome, Germ Plasm, Reproduction, and Development

Program Rationale. Production of foods derived from animals has a major impact on the U.S. agricultural economy. Annual cash receipts of $93 billion from livestock and poultry products account for about 50% of all agricultural products. The main components of this program are designed to identify and use livestock and poultry with appropriate genotypes that will have a major impact on the quality of animal products used for food, international competitiveness, and efficiency of production.

Program Components

Animal Germ Plasm. A major goal is to identify, facilitate, acquire, and characterize potentially useful germ plasm and selection to improve efficiency of production and to retain genetic diversity of animals for future generations. Another major goal is to store gametes or embryos on a long-term basis (eg, 25 years or more). The research effort emphasizes germ plasm cryopreservation and vitrification to enhance the feasibility and efficiency under which desired germ plasm would be stored. The program includes research on the sexing of male and female sperm so as to provide sexed gametes and embryos for storage. It also includes research on stem cells, primordial cells, cell lines, and DNA as they relate to the overall objective of the national program.

Animal Genome. A major goal is to develop genomic maps and associated DNA markers to improve accuracy of selection, increase frequency of desirable genes in populations, and characterize valuable germ plasm populations. For each livestock species genetic maps of polymorphic loci must be developed with sufficient resolution to permit location, definition, and use of genes affecting economically important traits. Related program components include analyses of the fine structure of candidate genes and gene families, definition of the genetic basis of quantitative trait loci to be used to implement genetic marker–assisted selection, and development of new experimental technologies for utilization of genome information.

Reproductive Efficiency. The overall goal is to improve reproductive efficiency of livestock, poultry, and aquaculture species. Research will focus on improving reproductive performance of animals through genetics, nutrition, and health management and on management of environmental factors such as temperature and humidity. Research advances and new biotechnologies will be developed to reduce losses due to reproduction problems in all species, and to maximize reproductive output of animals that produce high-quality products.

National Animal Germ Plasm Database. The overall goal is to develop a fully operational national animal germ plasm database system that will be linked with a central and satellite database that is now part of the USDA-ARS Genetic Resources Information Network (GRIN) System at Beltsville, Maryland. Access is readily available to the public through the Internet at *http://www.ars-grin.gov/nag*. An important component of the program is to expand the database to contain descriptors and characteristics of the world's animal genetic resources and DNA sequences or mapping information on relevant animal genomes. An additional program component is development of cooperative arrangements with outside groups such as the National Association of Animal Breeders (NAAB) and the American Livestock Breeds Conservancy to fully develop the database and to establish, for each species, national species committees representing industry, scientists, and commodities.

Animal Gene Bank/Repository. The purpose of the animal gene bank/repository is to develop long-term storage space for sperm, embryos, oocytes, stem cells, cell lines, and DNA from designated genotypes covering a wide range of domestic animal species and aquaculture. An important component of this program is to establish a national committee on cryopreservation to work with species committees. This national committee will establish priorities for genetic material to be maintained in repositories and will define long-term storage requirements for this material.

Animal-Production Systems

Program Rationale Producers are challenged with integrating knowledge from diverse disciplines into production practices suitable for their individual operation. Research on food animal–production systems assesses the interactions between nutrition, genetics, reproduction, physiology, microbiology, immunology, and molecular biology and related effects on animal health, productivity, and the environment. Improvements in the production efficiency of individual components are often realized in the total system and ensure a continuing supply of economical nutrient-dense products for the consumer. Research leading to discoveries and applications in production efficiency, sustainability, animal and environmental well-being, and high-quality products are imperative if animal agriculture is to be economically viable.

Program Components

Animal Nutrition. Nutrition is the single most costly component in modern animal production. Suboptimal nutrition is a significant factor in the failure to realize genetic potential for production and increased susceptibility to disease. Economically optimizing nutrient supply and use is imperative for improving growth and reproduction and maximizing overall production efficiency. Research is needed in the following areas:

1. Chemical composition and availability of nutrients in feedstuffs

2. Nutritional requirements of grazing and nongrazing animals

3. More efficient use of nutrients

4. Specific attention to improving the use of environmentally sensitive nutrients such as phosphorus, nitrogen, copper, and potassium

5. Increasing understanding of nutrient partitioning between biological functions (eg, reproduction, growth, and lactation)

Integrated Animal Systems Research. Production output and efficiency of the whole animal represent the biological integration of nutritional, physiological, and genetic components. Factors that affect an individual component influence animal production, either partially or totally, through their effects on other biological components. It is imperative that research be conducted that develops our knowledge of how animals integrate nutrition, endocrinology, immunology, and genetic factors to optimize efficiency of nutrient use, reproduction efficiency, and product quantity and quality. With this knowledge, one can understand the animal as a production unit in developing new approaches and decision aids applicable to improving the sustainability of animal-production systems.

Integrated Information for Animal-Production Systems. The overwhelming amount of information about animal-production efficiency is difficult to use without the aid of computer-based technology. This technology application is needed to improve management decisions and strategies that will yield the greatest economic returns. New-generation computer models are needed for evaluating production options. Decision aids are needed to integrate the components of animal production in modular forms compatible with farming systems programs. These decision aids must be useful to farmers and producers.

Animal Diseases

Program Rationale. Animal disease is the single greatest hindrance to efficient livestock and poultry production, with economic losses estimated to be 17% of production costs in the developed world and more than twice this figure in the developing world. Rapid diagnostic tests, novel genetic vaccines, immune modulatory strategies, disease-resistant genes, and increased biosecurity measures are needed to prevent or control outbreaks and the spread of animal diseases and to protect people from zoonotic diseases. Protecting the livestock industry from animal diseases ensures competitiveness of the U.S. animal agriculture in the global marketplace and promotes a sustainable and profitable production system in the United States.

Program Components

Pathogen Detection. New generations of rapid, sensitive, and specific serological and molecular diagnostic tests are needed to:

- Detect newly emerged disease pathogens
- Identify new variants of known disease pathogens
- Control or eradicate zoonotic diseases
- Control diseases that impact production trade

Epidemiology of Disease. Epidemiological surveillance can better define the economic impact of livestock diseases, enabling animal producers to better understand the ecology of emerging disease and natural transmission cycles. This is needed to develop strategies to prevent disease. The emergence or introduction of an exotic disease into the United States could rapidly escalate into an epidemic due to lack of resistance in host animals, absence of vaccines or effective drugs, and limited resources to effectively manage the spread of these pathogens.

Mechanism of Disease. A better understanding of the molecular and cellular mechanisms responsible for disease requires studies of disease agents, the host animal's response, and the environment. The genetic basis for pathogenicity differences among selected viral, bacterial, mycotic, and parasitic agents affecting livestock is needed to define the molecular factors that promote the emergence of highly virulent pathogens and to provide an understanding of how virulence correlates with pathologic manifestation of disease. The interaction of host-specific and environmental factors such as the role of air quality and incidence of respiratory disease is needed to develop immune modulatory strategies for improved vaccine efficiency. Mechanisms controlling vector–pathogen–host interactions will be investigated. Molecular tools will aid in determining the method and frequency of pathogen persistence in recovered and subclinically infected animals.

Genetic Resistance to Disease. The genetic basis, when present, for disease resistance to parasitic diseases; arthropod-vectored viral diseases; oncogenic viral diseases, such as Marek's disease; bacterial diseases, such as Johne's disease and scrapie will be investigated. Host immunological and physiological responses will be related to genes and genetic markers. Immunogenetic studies will lead to the identification of genes of pathologic significance that can be used to develop novel non-immunity-based disease-control strategies.

Disease Prevention/Control through Vaccines and Novel Strategies. To prevent animal disease, the biology and ecology of pathogens must be understood and their weaknesses exploited. The knowledge of the variation in genetic sequences of genes from pathogens could explain the basis for some vaccine failures as well as provide a rational approach to developing improved vaccines. Modification of existing vaccines to enhance effectiveness and safety will be adequate in some cases. Development and testing of new subunit, whole virus, viral-vectored, deletion mutants, naked DNA, and peptide vaccines for the prevention of important livestock diseases will lead to a new generation of more-effective vaccines. The evaluation of new vaccination schemes and immune modulators to improve herd protection against selected diseases will provide better disease protection. Identification of environmental factors that contribute to disease and the design of improved management practices to control diseases are an integral part of disease-control programs.

Arthropod Pests of Animals and Humans

Program Rationale Animal pests and pathogens are a serious threat to the U.S. agricultural economy, to food safety, and to human health. Losses caused by pests and pathogens amount to hundreds of millions of dollars annually and comprise productivity losses and damage to the quality of animal products. These problems restrain the exportation of animals and animal products from the United States and lead to higher prices in the domestic market. Arthropods that transmit human disease pose risks to the health of U.S. military troops deployed overseas and to the global human population. Pesticides and drugs used for the control of pests and pathogens pose a distinct and identifiable threat to animal and human health. Research is needed to develop safe, efficacious, environmentally benign, and user-friendly methods for the detection, surveillance, and control of pests and for the development of new and effective vaccines and immunomodulators, biologically based pest- and pathogen-control strategies, and the development and use of pest- and pathogen-resistant livestock.

Program Components

Ticks and Mites (Cattle Fever Ticks, Lyme Ticks, and Mange Ticks). Research focus is on the development and application of technologies for detection, surveillance, monitoring, pesticide-resistance management, and biologically and ecologically based area-wide control.

Blood-Sucking and Filth Flies. Research thrust is on the development, integration, application of technologies for detection, surveillance, monitoring, pesticide-resistance management, and biologically and ecologically based control.

Athropod-Borne Animal Pathogens. Research thrust is on the development and application of technologies for detection, surveillance, and control of pathogens on methods to enhance production, quality, and safety of animals and animal products; and on the development of a database to support research on pests and pathogens of veterinary and medical importance.

Arthropod-Borne Diseases. Research thrust is on the development of information technology leading to intervention strategies to prevent or reduce arthropod-borne pathogen infections in animals and humans and to reduce the risk of U.S. animal livestock from domestic, exotic, and emerging arthropod-borne animal pathogens.

Repellent for Personal Protection from Biting Insects and Arthropod-Borne Diseases. Research thrust is on the development of new personal-protection technology for use by U.S. military troops and the American public and on the development of alternatives to tropical repellents, such as deet, and permethrin, which is a toxicant used on clothing.

Household Pests. Research focus is on development, integration, and application of tools for detection, surveillance, monitoring, pesticide-resistance management, and biologically and ecologically based control.

Structural Pests. Research focus is on the development, integration, and application of tools for the detection, surveillance, monitoring, pesticide-resistance management, and biologically and ecologically based technologies for termite control.

Imported Fire Ants. Research thrust is on development and application of areawide fire ant–management technologies.

Screwworm Eradication. Research focus is on development and application of improved technologies for sterile fly production and development and application of improved technologies for detection, surveillance, monitoring, and eradication of flies.

Animal Well-Being and Stress-Control Systems

Program Rationale. Measures of well-being are needed to give producers and consumers the information they need to evaluate management practices and determine which techniques best assess the well-being of animals used for food production. Development of scientific measures of well-being and an enhanced ability to interpret such measures are crucial to the evaluation of current agricultural practices and development of improved alternatives. Stress caused by social and environmental stressors needs to be understood to limit negative impacts on production efficiency and well-being. Animal well-being research will benefit animals, producers, and ultimately consumers by reducing animal health-care costs and by improving food-production efficiencies. Lack of sensitivity to animal-welfare issues may be used in domestic marketing and as an artificial trade of animal products in world markets.

Program Components

Scientific Measures of Well-Being. Measures of well-being of food-producing animals are needed to make scientific assessments. These measures must be scientifically sound and relevant. The measurements will integrate behavioral, physiological, and productivity parameters of economic importance.

Adaptation and Adaptedness. Most food animals have been domesticated for thousands of years. Selection under intensive management conditions has occurred only recently and is oriented primarily toward the improvement of production traits. Research in this area will determine the roles that genetics and environment play in well-being. Research information on adaptedness will serve as a basis for modifying management practices. Genetic research will be evaluated to improve animal fitness and determine the basis of adaptation to environmental stressors such as heat and cold. Marker-assisted selection techniques will be explored.

Social Behavior and Spacing. With the intensification of animal agriculture and the greater number of animals at each location or in production units, a major question is whether the intensive management adversely affects an animal's well-being. Research will be conducted to provide a scientific basis for understanding the social behavior of food animals and how the quality and quantity of space influences behavior. Research to show consequences of the change in patterns of social interactions and space utilization will require an integrated research approach.

Cognition and Motivation. The mental state, fear, frustration, suffering, pleasure, and boredom of animals are major concerns of the public; however, there is currently little scientific information that can be used as a basis for addressing these concerns. Research is needed to learn how animals perceive and process sensory information from the environment and what animals learn.

Evaluate Practices and Systems to Improve Well-Being.
Management practices such as transportation and slaughter, and special agricultural practices such as beak trimming, deboning, branding, tail docking, and castration, are important and necessary elements of animal management in current production systems. These practices affect the well-being of animals. Research will address evaluation of the current practices and alternative practices concerning potential pain, stress or discomfort, and production efficiency. Alternative environment systems and current management practices will be evaluated for their effect on farm animal well-being and overall goals to improve animal comfort, well-being, and production efficiency. Research to improve both production efficiency and animal well-being will be conducted.

Bioenergetic Criteria for Environmental Management. Adverse environmental conditions cause livestock and poultry losses, decreased production efficiency, and decreased animal well-being. Available technology needs to be adopted for proactively managing environmental stressors. Research to develop decision-support tools is needed to help producers deal with environmental stressors, provide protective measures, recognize livestock and poultry in distress, and take appropriate management action.

Aquaculture

Program Rationale. With increasing seafood demand, declining capture fisheries, and a fisheries trade deficit exceeding $4 billion annually, aquaculture is poised to become a major U.S. growth industry in the twenty-first century. The confirmed growth and competitiveness of U.S. aquaculture will be related to the resources invested in research and technology development. A strong research and development program for U.S. agriculture, led by ARS would offer significant benefits to both producers and consumers by enhancing production efficiency, profitability, and quality of aquaculture products and systems.

Program Components

Genetic Improvement. There has been limited genetic improvement of aquaculture, so there are major opportunities for improvement through traditional animal breeding, broodstock development, germ plasm preservation, molecular genetics, and allied technologies. Research will address improvement in growth rates, feed efficiency, survival, disease resistance, fecundity, yield, and product quality; genetic characterization and gene mapping; and conservation utilization of important aquatic germ plasm.

Integrated Aquatic Animal Health Management. Despite progress in aquatic animals, significant losses to disease still occur. Research will address improvement of survival vigor and well-being of cultivated aquatic animal stocks through integrated aquatic and health research; improved technologies and practices, such as population health management; and development of health-management products, including vaccines and therapeutics disease detection/diagnostic techniques.

Reproduction and Early Development. There are major opportunities for improved reproduction and early development of cultivated aquatic organisms. Research will ad-

vance year-round maturation and production of gametes and fry of economically valuable species; development of specialized brood stock; and improvement of systems to enhance reproduction efficiency.

Growth, Development, and Nutrition. There are substantial opportunities to improve growth, development, and nutrition of cultivated aquatic organisms. Research will advance improving survival, growth rates, feed conversion, environmental tolerances, and feed formulations and feeding strategies to reduce dependence on marine fish–based protein aquaculture diets.

Aquaculture Production Systems. There are opportunities to improve the performance of aquaculture production systems through development and application of innovative early approaches and technologies. Research will address development and successful application to aquaculture of new technologies as well as relevant existing technologies and engineering presently employed in other sectors of the economy.

Sustainability and Environmental Compatibility of Aquaculture. The goal is to protect and conserve the nation's water resources and natural environments by conducting research and technology transfer to improve the sustainability and environmental compatibility of production systems. Of primary concern is the protection and conservation of the nation's resources.

Quality, Safety, and Variety of Aquaculture Products to Consumers. The goal is to improve the quality, safety, and variety of aquaculture products through research and technology transfer. Research will address improvement of safety, freshness, flavor, texture and nutritional characteristics, and shelf life of cultivated fish and shellfish; and development of value-added products and processes.

Information and Technology Transfer. Aquaculture in the United States can benefit greatly from improvements in technology-transfer programs, information dissemination, and access information and technology generated and adopted in foreign countries. Programs that address international information-retrieval networks and partnerships improve access to important technology and development of appropriate databases linked to electronic systems to enhance information exchange.

HUMAN NUTRITION REQUIREMENTS, FOOD COMPOSITION, AND INTAKE

Program Rationale

A viable agricultural enterprise demands that producers target consumers' needs for desirable, safe, and nutritious foods.

Human nutrition science has moved from a focus on the prevention of nutrient deficiencies to an emphasis on health maintenance and reduced risks of chronic disease. Diet affects growth, development, aging, and the ability to enjoy life to its fullest. Dietary intake has been linked to risks for development of a variety of common chronic diseases. The annual economic impact of cardiovascular disease in the United States exceeds $8 billion; obesity, $86 billion; osteoporosis, $10 billion; cancer, $104 billion; and cataracts, $4 billion. The American Cancer Society estimated in 1996 that one-third of the 500,000 cancer deaths annually in the United States are due to dietary factors.

Changing population dynamics, lifestyle habits, food technologies, animal- and crop-production capabilities, and the globalization of the market for foods all demand revisions in targeting food production, distribution, and marketing strategies to assure optimization of health through nutrition. Nutrition research design and capabilities are increasingly needed to address the changing dynamics and needs of the populations to be fed and the dynamics of the food supply.

ARS is an ideal location for human nutrition research because of the proximity to research on basic agriculture. A nutrition research program within the USDA focused on dietary intake and its health significance will offer significant benefits to producers, consumers, and industry by strengthening the rational basis for agricultural and public health programs and policy.

Program Components
Changes in Needs for Nutrients Throughout the Life Cycle. One needs to understand the nutrient requirements of infants, children, adolescents, and adults to understand the effects of nutrition at different points in the life cycle on the need for nutrients later in life and the prevention of chronic disease.

What Do Americans Eat and What Are the Trends for Changes in Diet? Research is needed on the nature of the diet of individuals and populations to understand their particular risk for disease at various stages of life, as a basis for modifying food and agriculture policy and practice, and as a strategy in marketing.

Composition of the Diet. There is a need for more extensive information about key foods, which are major contributors of the U.S. population's intake of nutrients of public health significance. A national nutrient databank is needed to accurately reflect the dynamic diet of the United States, with representative data on a variety of foods, including effects of biotechnologies and ethnic, restaurant-prepared, and carry-home foods. Nutritionally significant components of food must be identified, followed by the development of analytical methods to quantify those nutrients.

Function and Metabolism of Nutrients for Cognitive and Physical Behavior. Information is required to provide an understanding of how various dietary components act to improve learning and physical performance potential.

Definition of Marginal Deficiencies or Borderline Deficiencies. Frank nutritional deficiencies are rare in the United States, but there is abundant evidence that subclinical deficiencies have profound negative effects on health, intellectual development, growth, and disease resistance. Research is needed to define borderline deficiencies and to update indicators of deficiency.

Bioavailability of Nutrients in Foods. A need exists to understand how nutrients interact and how the effects of agricultural practices including postharvest handling and cooking affect nutrient content and bioavailability. It is important in this context that nutrition research methods be adequate to meet the growing demands and opportunities

for genetic engineering in plant and animal production strategies.

What Are Effective Intervention Strategies? Simply knowing the nutritional requirements for humans will not ensure adequate nutrition in the population for a variety of reasons, including poor education, low income, and inadequate access to food, among others. Research on biological outcomes is needed to identify effective ways of communicating food and nutrition knowledge to individuals and populations to elicit changes in food intake and to design effective food-assistance programs.

Nutritional Needs of a Diverse Population. Research is needed to provide an understanding of the nutritional needs of diverse individuals that differ based on gender, race, genetic backgrounds, environment, behavior, and lifestyle.

Food Safety
Program Rationale. The U.S. food supply is among the world's safest and most wholesome. However, significant food-safety problems can cause either human illness or economic losses and threaten the international competitiveness of American agriculture. The recent food safety initiative (FSI) has highlighted food safety, and in particular, the control of food-borne pathogens, as an important concern of the entire federal government. Reduction in the potential health risks to consumers from human pathogens in food is the most important food-safety goal.

Program Components
Microbial Pathogens. Determining how to reduce microbial pathogens in food products throughout food operations from farm to fork is the most urgent food-safety problem today. A major goal of this program is to develop tests that are precise, reliable, and rapid enough to detect contamination in all foods before they enter into commerce. Equally important is the development of effective, reliable, and cost-effective methods to control or eliminate pathogens in and on food-producing animals throughout production and processing. Additionally, recently recognized pathogen problems to be addressed include:

- The presence of pathogens in fruits and vegetables
- The presence and persistence of pathogens in animal waste
- Pathogen resistance to traditional processing techniques and to drugs
- The need for development of pathogen growth and survival models to support the risk-assessment process

Chemical Residues. The objective of the chemical residue program is to reduce the risks of chemical residues from animal drugs, food additives, herbicides, and pesticides and environmental contaminants that are potentially present in foods. A major goal is to develop reliable, effective, accurate, user-friendly, cost-effective residue-detection methodology that requires minimal amounts of organic solvents to detect these residues. The program also increases knowledge of the adsorption, distribution, metabolism, and excretion and toxicity of certain chemicals and environmental contaminants in food-producing animals so that animal producers gain the knowledge and the means to control their residues in foods.

Mycotoxins. The presence and potential for the presence of mycotoxins in crops not only is a direct food-safety problem but also threatens the competitiveness of U.S. agriculture in the world market. Major goals are to control aflatoxin in peanuts, corn, cottonseed, tree nuts, and figs; fumonisins in corn; and deoxynivalenol in wheat and barley through an understanding of the biology of plant–fungus interactions and toxin production in the field. Specific approaches include combined altered agronomic practices, chemical and biological control, improved plant resistance to mycotoxins through conventional plant breeding and transgenic approaches, and reduction of insect damage leading to fungal infections. This program also develops methods to measure mycotoxins in important crops.

Toxic Plants. This program component seeks to minimize exposure of animals and humans to natural toxins from poisonous plants. The research focuses on:

- Identifying toxic principles in plants to which animals might be exposed (from range or pasture)
- Determining toxic manifestations of these plants in animals
- Developing appropriate technology to reduce losses in livestock
- Developing appropriate technology to control elevated cadmium in sunflower seeds and wheat that concern some foreign markets
- Using biotechnology to reduce the presence of solanaceous alkaloids in new varieties of potatoes

NATURAL RESOURCES AND SUSTAINABLE AGRICULTURAL SYSTEMS

ARS seeks to enhance the quality of the environment through seeking a better understanding of and building on agriculture's complex links with soil, water, air, and biotic resources. Multidisciplinary research is conducted to solve problems arising from the interaction between agriculture and the environment. New practices and technologies will be developed to conserve the nation's natural resource base and balance production efficiency and environmental quality. ARS collaborates with foreign research entities to address global environmental quality problems. This research area is built on the national programs discussed in the sections that follow.

Water Quality and Management

Program Rationale. Life depends on water. Safe drinking water is mandated by the 1996 amendments to the Safe Drinking Water Act; clean water in lakes, rivers, streams, and aquifers is mandated by the 1972 Clean Water Act; and safe, dependable, and abundant food and fiber depends on safeguarding our nation's soil and water resources. Agricultural production systems that sustain and protect our national water resources while satisfying our needs for clean water and safe, dependable food and fiber supply are paramount national issues. Agriculture, however, is identified as a major nonpoint source of water contamination. Water quality can be degraded by agricultural chemicals and crop nutrients as well as by salts, toxic trace elements, and waterborne pathogens. Water and soils can be degraded by application of waste products and waters from municipal and industrial sources as well as from agricultural processing plants and agricultural animal-production practices. Innovations and improvements in American agriculture are needed to sustain our vital interests in a safe, abundant, and reliable food and fiber supply while enhancing our nation's water quality and sustaining our precious water resources.

In February 1998, the President of the United States released a clean water action plan that contained 111 action items addressing land and environmental issues. ARS targets the following national programs:

Program Components
Land Area Issues: Watershed Management on Agricultural Lands. By fostering research on weather and climate characterization, hydrological processes and watershed characteristics, and watershed management, this component focuses on the interaction between the environment and agricultural production on irrigated and rain-fed croplands and grazing lands at the field, farm, and watershed scales. For instance, research exploring the natural variability of weather is a critical component of determining the reliability of water supplies, the economic risks associated with farming and ranching operations, as well as the frequency and severity of natural hazards. Research on the interaction of water with soil and vegetation will help assess natural and human-made impacts on the environment and the nation's natural resource base. Research on watershed management will promote more effective use of precipitation in food and fiber production, better allocation and assessment of irrigation water resources, and other advances in integrated ecosystem management. Land use policy, management, and stewardship reflect responsible consideration of the ability of rain-fed crop production and grazing land practices to provide high-quality water in sufficient amounts to fulfill on- and off-site requirements. Research in advancing strategies for water management is needed to promote and safeguard the amount and quality of these water supplies. Research will quantify the impacts of rain-fed and grazing lands tactics and determine how they affect the hydrologic cycle.

Land Resource Issues: Irrigation and Drainage Management. Existing and future water resources for irrigation are projected to decline even further, which further emphasizes the need to improve irrigation and drainage practices to enhance water quality and sustain American food production for strategic national economic and social benefits. Innovative irrigation and drainage techniques and management are required that can improve water quality, reduce soil erosion, conserve water, and reduce energy requirements while enhancing and sustaining crop production and water use efficiency. Advanced technologies, like precision agriculture, site-specific management, remote sensing, and decision-support systems, are needed to address the water quantity and quality needs in irrigated agriculture.

Water Resource Issues: Water Quality Management on Agricultural Lands. The decline in the expansion of agriculture can be attributed in part to the high costs of developing new water resources. Research is needed to develop technologies for use by growers and by water-management agencies and regulatory agencies at all levels of government. Research must focus on improving the efficiency and safety of structures used to store and regulate water flows. With these needed improvements, rural and urban communities will be better protected from the worst ravages of floods, loss of productive irrigated and rain-fed croplands from stream bank failures, and landscape degradation from concentrated flow erosion. In addition, the development and implementation of more effective water-use strategies for irrigated and dryland farming, including water measurement and conservation, wetland and riparian area management, and water reuse are necessary priorities in order to enhance crop production and improve its efficiency, increase biodiversity in the environment, and conserve water on agricultural lands in an economic and socially responsible manner.

Water Resource Issues: Water-Quality Management on Agricultural Lands. New and improved strategies are needed to reduce water contamination from agricultural lands. Improved technologies are required for the management of wastewaters and for transferring specific farming-management systems from one geographic area to another. Field practices will be developed to reduce impacts of pesticides, nutrients, sediments, salts, toxic trace elements, and bacterial contaminants in surface waters and groundwaters. New strategies will be developed that use management systems evaluation areas to determine the water quality and other environmental benefits of alternative farming systems at field, farm, watershed, and river basin scales on irrigated and rain-fed croplands and grazing lands. Water-resource management to improve water quality will be evaluated on regional ecosystems in the context of its diversity and risk management in sustainable agricultural systems.

Water Resource Issues: Water Resource Databases, Decision-Support Systems, and Cross-Cutting Technologies. The development and validation of improved water resource–management technologies requires long-term databases. Models and decision-support systems are needed to develop optimal strategies for managing water resources in irrigated and dryland agriculture to:

- Establish the benefits of precision agriculture and integrated farming technologies
- Resolve conflict among competing water demands when the supplies are limited
- Determine the socioeconomic and environmental impact of proposed water, nutrient, and pesticide management programs and policies

Science-based results and technologies will be transferred to various user groups and policy makers and integrated into other ARS national programs.

Soil Assessment and Management

Program Rationale. The thin layer of soil at the surface of the earth functions as the central natural resource to sustain life. Soil management determines plant productivity, which in turn sustains animal productivity. Soils also remove impurities to benefit water and air quality. It is imperative that a balance be reached between the short-term use of the soil and the long-term sustainability of this critical resource. Protective, preservation, and management practices need to be developed to overcome limitations to productivity while maintaining or enhancing environmental quality. Tools need to be developed to assess the overall quality of a soil to determine the effectiveness and sustainability of soil-management practices. The overall goal of this national program is to develop and transfer science-based knowledge regarding the physical, chemical, and biological properties and processes within soil resources to ensure sustainable and economically viable food, feed, and fiber production throughout the nation and world.

Program Components

Resource Conservation. Protection of our soil resource is a key component of this program. Control of water, wind, and tillage-induced erosion needs to be developed to stop and reverse soil degradation (eg, loss of organic matter, soil compaction, infertility, and poor water retention). An improved understanding is needed of how soil influences greenhouse gases through carbon sequestration. Techniques are needed to measure soil salinity and develop remediation strategies at field and watershed scales. A greater understanding of the impact of nonagricultural activities on soil productivity is needed. Although individual soil properties can be measured, the capability to assess soil quality or health for a range of uses and functions is not actual yet. Soil quality assessment will encompass studies of soil properties, processes, and indicators; measurement tests and protocols; and interpretive tools and decision-support aids.

Water Use Efficiency. Improving soil water infiltration, storage, and use by crops has economic and environmental benefits. Research to increase soil infiltration and reduce runoff and sediment transport of agricultural chemicals to surface waters is one focus of this component. Improving soil physical properties to improve soil water storage and use by crops will also be a focus of this program component (eg, tillage and residue management systems and biological systems to improve crop rooting). Coordinating research to improve our fundamental understanding of soil crusting and sealing processes will result in the development of methods to reduce the impact of these processes on runoff and infiltration. Other research will focus on the development of techniques to evaluate the impact of irrigation practices on soil and other ecosystem resources. Especially critical in this area will be research to effectively use wastewater for irrigation without degrading soil physical, chemical, or biological properties and processes.

Nutrient-Use Efficiency. Efficient crop nutrient use results in increased profitability and reduced potential for negatively impacting the environment. Research to improve our understanding of how tillage (primary and secondary), crop rotation, cover crops, and residue management affect nutrient cycling and fertilization requirements will result in improved management systems and decision aids. Coordinating research to improve nutrient (both inorganic and organic) input, application, and utilization efficiencies will also be conducted, including site-specific and precision nutrient management. Synchronizing nutrient supply in soil with crop need and developing quick, reliable tests for nutrient availability in soil and organic matter, manures, wastes and amendments, and cover crops will also be addressed. Another focus of this program component is research to improve our understanding of how soil organic matter fractions influence soil structure, water relations, and nutrient cycling. Quantification of how nutrient-use practices coordinate to soil-management practices to affect environment quality will also be addressed in this program. Research to evaluate effects of grazing and livestock systems on nutrient cycling and other indicators of soil quality is also critical to this program component.

Resource-Based Profitability. Soil-management assessments and practices that sustain the soil resource must be economically viable or they will not be adopted. This research component will identify and develop management systems that improve profitability and competitiveness by increasing productivity (output per unit input). In addition, research will develop quantitative economic assessments at various scales for impacts of soil-management practices both on-farm and off-farm. These results can be used by action agencies to increase acceptance of sustainable practices. Finally, research to identify and develop production systems that are less dependent on nonrenewable resources will be conducted.

Air Quality

Program Rationale. The health of humans, plants, and animals, as well as other aspects of the environment, are affected by dusts, odors, and other volatilized substances released to the atmosphere by agricultural activities. Excess levels of ozone damage plants. The air-quality national program conducts research on both these issues:

- To understand contaminant releases and their impacts, and how to minimize them
- How ozone affects plants, and how to reduce the impact

In many areas, dusts are intimately related to wind erosion, the effects of which are deleterious to soils and to downwind environments.

Program Components

Dust Emission-PM-10. These are dusts that are regulated by the Environmental Protection Agency (EPA) because they pass through the human nasal air-filtering system and lodge in the lungs. This research program seeks to elucidate the biological, physical, and chemical mechanisms by which 10-μm and smaller particles are generated, how they are transported and suspended in the air, and

their patterns of movement. As understanding of these mechanisms improves, advances are made in technology that can be used on farms to reduce or prevent the emission of dusts to the atmosphere. Research is also conducted on identifying geographical areas that significantly emit dusts, so land managers and policy makers can facilitate application of controls where they are most needed. As resources become available, the air-quality national program will be expanded to consider dust emissions of such agricultural operations as tillage and harvest.

Effect of Ozone on Crop Systems. High levels of ozone in the atmosphere result in reduction of crop yields by as much as 20%. ARS research assesses the effects of ozone on crop production and plant health and develops techniques for mitigating the problems. Carbon dioxide is intimately associated with ozone problems, and the effects of this gap on crop physiology and productivity are also under study.

Emission of Odors from Agricultural Operations. Most animal-production enterprises and some crop operations result in unpleasant odors that are carried downwind and impact neighboring households and communities. Basic research is needed to understand the biological and chemical processes that produce odors and emit them to the atmosphere. This will permit the development of mitigating measures to be applied at the emitting source.

Emission of Volatilized Pesticides by Agriculture. Volatilized pesticides are a complex of very fine liquid particles and gaseous phase compounds, or they may be attached to particulates from other agricultural or nonagricultural sources. The research objectives are:

- To understand the chemical and biological mechanisms that influence pesticide volatilization and transport
- To understand transport processes
- To understand the impacts of deposited volatiles
- To develop means to reduce volatile emissions

Wind Erosion. High winds on exposed soils give rise to dust storms that are locally devastating in terms of denuding productive fields of their soils and depositing large amounts of soil sediment in the immediate vicinity of the storm. A large storm may also result in emitting into the atmosphere large amounts of PM-10 that can travel for thousands of miles. Research objectives include:

- Understanding the biological, physical, and chemical mechanisms of wind erosion
- Developing effective control measures for application on farm fields

Develop Wind-Erosion and Air-Quality Simulation Models and Aids for Agricultural Decision Making. Simulation models are based on scientific knowledge developed as a result of research discussed in the air quality problem areas. The aim of simulation models is to integrate and express science knowledge to facilitate the selection of the best management practices. One may test various combinations of control measures (such as roughness of a land surface in relation to wind breaks, or surface area versus depth of waste lagoon) to find the optimum emission-controlling combinations for any given farm or farm component. This capability allows the land manager to select practices with a high degree of confidence so that they will have the desired impact on the farm's emissions or on lessening excess ozone impacts.

Global Change

Program Rationale. The term global change is a shortened form of the phrase global environmental change. It refers to large-scale changes in the earth's biological, geological, and atmospheric systems. Agriculture is vulnerable to these environmental changes. Successfully adapting to change depends on a quantitative and predictive understanding of the ways that agriculture and food supplies may be affected by climate change, rising atmospheric carbon dioxide and ozone levels, and increased UV-B radiation. Title XXIV of the 1990 farm bill directed the Secretary of Agriculture to "study the effects of global climate change on agriculture," including "the effects of simultaneous increases in temperature and carbon dioxide on crops of economic importance; the effects of more frequent or more severe weather events on such crops; the effects of potential changes in hydrological regimes on current crop yields," and other possible impacts.

Program Components

Climate Change Effects on Agriculture. Research conducted under this component focuses on the potential impacts of a changing climate on crop and livestock production. Both direct and indirect effects are considered.

Direct effects include positive and negative influences of rising temperatures, altered precipitation amounts and patterns, increased cloudiness, and similar climatic changes on:

- Geographical areas of crop and forage production
- Geographical distribution of weeds, insects, and diseases
- Plant physiology and crop yields
- Animal performance

Implications of an increase in the frequency of extreme weather conditions and climatic events, as predicted by some climate models, are also considered here. Important is the possibility of increases in the occurrence level of various stresses and the interaction of additional stresses induced by climate change with existing ones. Many such interactions can be envisioned in that determining the significance of those with potential to intensify adverse or positive effects of climate change is an important goal.

Agriculture's Role in Greenhouse Gas Emissions and Sequestration. Changes in agricultural land uses and increased use of production practices that conserve natural resources may be reducing greenhouse gas emissions (or removing them from the atmosphere and sequestering them in vegetation and soils). Increased productivity of vegetation due to the "fertilization effect" of rising atmospheric carbon dioxide levels also may be removing atmospheric carbon and storing it in soils as organic matter. Limited studies show soil carbon increased after conversion of cropland from conventional to limited or no-tillage systems, reduction or elimination of fallow periods, the use of cover crops, and plantings of perennial grasses on cropland under the conservation reserve program. However, effects of these new and larger sinks on net greenhouse gas emissions are not well quantified and have not been related to the spatial distributions of such practices in ways that allow the magnitude of carbon sequestering to be compared with estimates of agricultural and other sources of atmospheric carbon.

Effects of Rising Atmospheric Carbon Dioxide Levels on Plants and Ecosystems. Many questions remain concerning the reality or magnitude of the carbon dioxide fertilization effect on vegetation. Until these are resolved by additional experimentation, a wide range of assumptions must be accommodated in predictive models, and model output will remain very general. Will increases in growth and productivity due to rising carbon dioxide eventually decline, because limiting availability of water and nutrients precludes continued responsiveness of crop stands and plant communities? To what extent do the favorable responses of plants to elevated carbon dioxide, observed under controlled conditions, also occur in the real world? Do the carbon dioxide–induced increases in growth and production observed at a doubling of current carbon dioxide levels extend to a tripling, or beyond? ARS conducts research on the effects of tropospheric ozone on plant physiology and crop yield as a part of the air-quality national program because ozone is a pollutant and ozone damage to plants is most appropriately considered an air-quality problem.

UV-B Radiation. Fundamental questions about UV-B effects on plants and animal health also remain unresolved. No one understands how plants cope with current background levels of UV-B, so it is not possible to predict the extent to which crops and native vegetation can adapt to higher levels or repair more extensive damage to DNA and chlorophyll. Seedlings of annual species were used in most of the research conducted to date, so little is known about long-term effects or responses of perennial or woody species. Of special interest are the interactions of current and increased levels of UV-B with other plant stresses as global change effects. For instance, deleterious effects of UV-B are eliminated when plants are grown in carbon dioxide concentrations expected in the next century. The question of how to experimentally control UV-B radiation is an issue. Source lamps and filters may not provide the proper spectral quality, and radiation levels (dosages) calculated from the available action spectrum may not be appropriate or accurate.

Grazing-Land Management

Program Rationale. More than half of the earth's land surface is grazed. In the United States, rangelands, pastures, and hay lands together make up about 55% of the total land surface. More than 85% of the 307 million acres of publicly owned lands in the western United States are grazed. These lands, the plants that grow on them, and the domestic and wild animals that graze them all contribute to the environmental, economic, and social well-being of our nation. The estimated annual value of U.S. hay production alone is $11 billion, third in value behind corn and soybeans and exceeding the value of wheat, vegetables, cotton, and noncitrus fruits. Livestock numbers vary with time, but grazing lands are usually home to more than 100 million cattle and 8 million sheep, supporting a livestock industry that annually contributes $70 billion in farm sales to the U.S. economy. Grazing lands also function as watersheds. Maintaining adequate supply of clean water for irrigated agriculture, industrial uses, and urban areas is critical to society, and much of our water supply originates as rainfall or snowmelt on rangeland watersheds. A strong ARS research effort will be required to maintain the economic stability and environmental acceptability of the use of grazing lands for animal agriculture.

Program Components

Forage Germ Plasm. Limitations in the availability of adapted, nutritious, highly productive forages continue to hamper the production potential of the nation's grazing lands. New forage, pasture, and rangeland varieties with higher nutritive quality and resistance to diseases, insects, and weeds are needed to maximize the utilization efficiency of the nation's grazing lands.

Forage Management. Management of timely forage availability in hay fields, in pastures, and on rangelands to maximize seasonal distribution, yield, and quality is one of the greatest limitations in enhancing the productivity of the nation's grazing lands. New forage-management technologies and more efficient approaches to forage management are needed to help managers maximize economic efficiencies and facilitate the integration of grazing lands into livestock-production systems.

Grazing Management. Environmentally sound management of grazing animals on the nation's grazing lands has become a public concern. Grazing-management systems are needed that help grazing-lands managers maximize economic efficiencies in grazing-animal production while minimizing or avoiding negative impacts on the environment.

Ecosystem Processes. Increased knowledge of the natural processes (competition, fire, herbivory, carbon and nutrient cycling, water use, energy capture and flow, and vegetation change) that control productivity and promote stability of these grazing lands is required to develop better approaches to their management. A thorough understanding of the basic biology of grazing lands is needed to provide land managers with the best information for managing pastures and, particularly, rangelands.

Grazing-Lands Databases, Simulation Models, and Decision-Support Systems. Management decision making by the grazing lands (particularly rangelands) landowners and managers would be greatly enhanced by the availability of decision-support systems. Databases that provide necessary inputs to decision-support tools are sorely needed.

Conservation of Grazing Lands as a Natural Resource. Conservation practices on grazing lands are critical to the long-term preservation of the earth's soil, wildlife, biodiversity, and air and water quality. Efforts are needed to provide the best information on the management of grazing lands to enhance the environment as well as provide a feed source for livestock.

Manure and By-Product Utilization

Program Rationale. Our nation's farms, cities, and industries generate over 1 billion tons of by-products each year. Animal production in large confinement units, often concentrated on relatively small land areas, has exacerbated the already difficult problem of how best to manage manure. Odors and other emissions to the atmosphere from confinement units as well as other by-product sources are of increasing concern. Accumulation of large amounts of animal, municipal, and industrial by-products at the production site or inappropriate subsequent disposal of the by-products can harm soil, water, and air quality. Many by-products, however, also contain essential nutrients (nitrogen, phosphorus, etc) that can help meet crop requirements if properly applied to soil. Organic by-products can serve as soil conditioners and increase soil organic matter levels. Research on the development of systems to optimally integrate animal manure and other products into sustainable agricultural practices is indispensable to agriculture today. Manures and by-products can be agricultural resources rather than wastes if they are appropriately managed. The goal of the manure and by-product utilization national program is to develop agronomic management practices that effectively utilize manure and other by-products and are protective of the total environment (soil, water, air, etc) and public and animal health.

Program Components

Nutrient Management. Movement of nutrients in excess amounts from manure and other by-products to water and air can cause significant environmental problems. These nutrient losses to the environment can occur during manure handling and storage and during and after field application. Nitrogen and phosphorus from manure and other sources have been associated with algal bloom and accelerated eutrophication of lakes and streams. Proposed regulations and congressional legislation will limit application of manure nitrogen and phosphorus to levels that will protect water quality. Historically, manure applications have been based on meeting crop nitrogen needs, but recently phosphorus has become a greater concern in many areas. A variety of research areas will need to be addressed to protect soil, water, and air from excess nutrients in manure and other by-products, including:

- A more efficient use of nutrients in animal feed
- Improved manure handling, storage, and treatment technologies
- Improved tests for nutrients in manure and soil treated with manure
- Development of soil threshold nutrient levels that protect water quality
- Development of methods to identify areas in a watershed that are most susceptible to nutrient losses
- Improved methods for precise application of manure

- Development of integrated animal and cropping systems to effectively use and recycle nutrients

Atmospheric Emissions. Odor-causing volatile organic compounds and greenhouse gases are generated at most animal-production facilities. These were of less concern when the volume of manure generated at a farm or production site was less. Because volatile organic compounds are often immediately irritating, they are the primary concern of the general populace at many locations. Furthermore, they are now more frequently perceived as real threats to air quality and therefore are subject to air-quality regulations. Although they can be generated at any place in the process, it is in the handling and storage system (eg, lagoons) that the bulk of the odors and trace gases are emitted. Improved management practices, new treatment technologies, or alternative production systems are needed to solve these atmospheric emission problems. This will require an increased understanding of the processes involved in the generation and transport of volatile organic compounds and greenhouse gases.

Pathogens. Contamination of water, food, and air by pathogens potentially present in manure and other by-products is an increasing concern. Most of these outbreaks were from more virulent strains of pathogens such as *Escherichia coli* O157:H7, *Salmonella*, *Campylobacter jejuni*, *Listeria monocytogens*, *Cryptosporidium parvum*, *Cyclospora*, and *Yersinia enterocolitica*. In addition, viruses remain a concern, and *Pfiesteria* outbreaks on the east coast of the United States may have been related to animal waste mismanagement. The survival, transport, and ultimate fate of pathogens from manure and other by-products need to be determined. The effects of manure management on other harmful organisms like *Pfiesteria* also need to be determined. The entire production process needs investigation to ascertain where the greatest risks occur and to develop the most effective controls so as to protect public and animal health.

Integrated Farming Systems

Program Rationale. Because of narrow profit margins and due to their fixed geographical locations, most farms and ranches have only limited flexibility to respond to changing factors in their environments. Every farm or ranch is a complex system of interacting components residing in a natural and socioeconomic environment. A high degree of management skill is required of modern producers. The primary products of the integrated farming systems national program are information and tools for use by farmers and by consultants in farm planning and decision making within the complex socioeconomic-physical-technical environment of farm or farms for which they are responsible. The overall problem to be addressed is provision of science-based information and its effective retrieval, management, and analysis for agricultural management decision making. This includes historic, recent, and real-time information and data. The goal is to help farmers retrieve this information, understand it, and apply it to the development and effective management of whole farm systems within the context of their ecosystems and communities. Effective response to issues concerning farms and ranches requires that partnerships be established in which research is conducted jointly with farmers and with participation from community stakeholders.

Program Components

Interactions among Components of Whole Farm Systems. A collection of components selected on the basis of their individual descriptions may form a farming system, but the operation of the whole system greatly depends on how the components interact—whether they are mutually compatible and supportive or whether one or more of them conflicts with the effectiveness of others. Considerable knowledge is available regarding many individual agricultural components. An essential aspect of the systems program is to understand interactions among and between the components and then to effectively apply this understanding. The integrated farming systems national program focuses on projects that foster understanding and managing of interactions among system components and how these components in turn act together within the larger process or system. On-farm research will often be used to ascertain system functions.

System Impacts. The integrated farm system is viewed as a component of a larger system, such as an ecological or watershed system. Phenomena such as eroded sediments, pathways of pesticides, and speed and sites of breakdown of pesticides become potentially important to ecological and watershed systems beyond the farm. The integrated farming systems national program will address this issue with the aim of assisting farmers and consultants to manage their environmental impacts. In a similar manner, with the assistance of farmers, consultants, economists, and social scientists, issues related to farm systems and economic viability and rural communities will be assessed. There is a critical need for knowledge of how management practices influence system processes and interactions.

Development of Integrated Farming Systems. At meetings with farmers and ranchers, a common refrain is that data and information are the primary need, and that farmers and ranchers will develop their own farm or ranch systems, given full information regarding components and their interactions. However, there may be situations in which it is appropriate for the agricultural research community to design and propose management strategies for whole or partial systems for use in agriculture. For example, there appears to be a special need for information on the transition from farming with high inputs to farming with reduced nonrenewable inputs, such as organic farming. The entire farming system needs to be evaluated to provide optimal entry points for farmers to make such a transition. The adoption of scientific management practices, whether in a precision or a sustainable agriculture context or both, is fostered by an understanding of how the site-specific practices affect the entire operation.

Decision-Support Systems. Integrated farming systems is a national program whose primary organizational function is to integrate knowledge and information developed in ARS and elsewhere in world agricultural research for direct application in decision making at farm (and field) level. The target audiences are farmers, ranchers, and their primary enterprise consultants, but benefits will also accrue to policymakers, financial and other service providers, and regulators. Possible outcomes include Internet systems, agricultural-management gaming programs, individual computer-based decision aids, printed guidelines, sustainable agriculture decision support, precision agriculture, and other products. These products will take multiple forms so that they will be accessible by farmers and others with a wide variety of educational backgrounds, skills, economic positions, and types of agricultural enterprises. New delivery technologies will be sought, but methods that have worked in the past, such as field days, county farmer meetings, and on-farm consultations and demonstrations, will continue to be used.

CROP PRODUCTION, PRODUCT VALUE, AND SAFETY

ARS seeks to enhance the production, value, and safety of foods and other products derived from plants that have a major impact on the American economy, world markets, and the U.S. balance of trade. The ARS scientific program in CPPVS consists of multidisciplinary research to solve problems of high national priority that threaten the security, safety, and productivity of U.S. agriculture. In the area of human nutrition, foods from plants are a major contributor of vital nutrients and fiber in the diets of Americans. ARS collaborates with public, private, academic, and foreign research entities to increase crop production and improve product quality and safety. Through these efforts, the United States can preserve its preeminent role as food provider to the world and overcome artificial trade barriers in world markets. Research in this area is based on the national programs discussed in the sections that follow.

Plant Microbial and Insect Germ Plasm, Conservation, and Development

Program Rationale. Genetic diversity is the basis of a sustainable agriculture. Habitat destruction, emerging pests and disease, changes in climate, and human-made pollution affecting soil and water are threatening the occurrence and distribution of traditional agricultural systems and natural plant communities on a global scale. The disappearance of natural plant habitats has a major impact on agriculture and industrial production. Genetic resources of plants play a major role in crop improvement for the introduction of new genes to increase resistance to pests and diseases and improve both productivity and crop quality. Microbial resources are essential for cycling the elements essential for long-term sustainability and productivity. New technologies are needed to develop more rapid and efficient methods of identifying useful properties of genes for manipulating genetic material. New approaches in the characterization and classification of genetic material and in the application of new biotechnologies will support the development of new crops and more efficient use of microbes and beneficial insects. The national program in genetic resources encompasses conservation, enhancement, and the development of new approaches for breeding and genetic improvement utilizing molecular technologies.

Program Components
Germ Plasm Acquisition, Maintenance, and Management.
Higher plant, microbial, and insect germ plasm should be acquired and safeguarded through the expansion of gene banks or *in situ* preserves for long-term accessibility. The genetic makeup of acquired germ plasm should be characterized to insure broad-spectrum genetic variability while minimizing redundancy.

Germ Plasm Enhancement and Manipulation. Sources of useful and important genes should be identified and incorporated into crop germ plasm. The genetic bases of crop, microbial, and beneficial insect gene pools should be broadened through genetic improvement to combat potential losses due to pests, diseases, and stresses imposed by the natural environment. The long-term economic value of genetic material for use by plant breeders and other scientists will be increased.

Genomics and Database Management. Genomic databases are an essential component in the management of new information that is generated at an increasingly rapid rate. Genomic and DNA sequence analysis should be emphasized, and the development of software to manage DNA sequence information should be continuously updated. Software to store, process, and organize the data generated by these activities should also be further developed. New DNA probes and primers will be needed to facilitate germ plasm characterization. Infrastructure will be developed that provides collaborative links through computer networks among ARS laboratories, universities, the private sector, and other government research facilities.

Germ Plasm Resources Information Network. GRIN is the official database of the National Plant Germplasm System (NPGS) and is currently maintained on a minicomputer at Beltsville, Maryland. Data in GRIN are available to any plant scientist or researcher worldwide. The functions of GRIN are:

- To act as a repository for all information of interest about plant germ plasm maintained by the NPGS
- To unify the NPGS with regard to data standards, crop location, and movement of germplasm
- To allow fast access to the most current data available for all users of the germ plasm and its accompanying information
- To facilitate and track the distribution of germplasm
- To provide germ plasm maintenance sites with a system of inventory management.

Improving Plant Biological and Molecular Processes

Program Rationale. Products of American agriculture are increasingly marketed to consumers around the globe. Increasing population and improving economic status of other countries, as well as changing environmental and other safeguards domestically, have set American farmers on a course of rapid change. Better food safety and security, crop protection, crop yield, and crop quality require advances at the frontiers of agricultural sciences. Yesterday's scientific advances support the rapid pace of change today; today's research breakthroughs will support tomorrow's innovations. This program is organized around fundamental long-term research needed for breakthrough advances in crop production, protection, product value, and safety to meet the changing needs of society and the global marketplace.

Program Components

Increased Crop Productivity and Stress Tolerance. Research will provide fundamental scientific knowledge required for continued increases in crop productivity and production efficiency to keep abreast with the growing demand for food. Basic information will be developed to increase the efficiency by which inputs are used by crops and to reduce unnecessary depletion of resources, extra expense in crop production, and contamination of the agroecosystem with excess nutrients or water not used by the crop. Innovative approaches will be developed to increase crop productive capacity in the future while safeguarding the environment. Mechanisms of crop responses to environmental stress will be identified and the processes improved to minimize yield losses and yield variability. New knowledge of plant gene organization and function will fuel more effective genetic improvement of crops.

Improved Crop Protection. Research focuses on providing fundamental plant knowledge to control pests (primarily insects, weeds, and nematodes) and pathogens in a more environmentally friendly manner than presently available. The relationships among host plants, pests and pathogens, and beneficial organisms will be characterized and expanded into new pest-management technology that will be sustainable over the long term. The processes to be investigated include sensing of pests and pathogens by the host plant, triggering of plant defense reactions, signaling processes within or between individual plants or between species that contribute to plant health, transfer and expression of resistance genes, and other fundamental plant processes that determine plant interactions with pests and pathogens. The research is expected to provide new approaches to crop protection that will be more effective, less costly, more targeted to the pest or pathogen species, and less risky to the environment.

Enhanced Crop Product Value, Quality, and Safety. Research will provide basic plant information to increase the diversity of products from crop plants, to improve crop product quality, and to prevent occurrence of contaminants (from either microorganisms or from plants themselves) that make plants and plant products unsuitable for consumption. Research within this program will lead to development of new crops and crop products. In addition, it will delineate the properties of crop products and their roles in product utilization as well as innovative ways of measuring quality and safety determinants. This advance in knowledge will be the basis for identifying limitations to achieving best product quality and developing novel ways to overcome those limitations.

Plant Disease

Program Rationale. Since the beginning of agriculture, diseases have reduced the yield, quality, and value of plant products in agricultural, landscape, and forest settings. Despite considerable success over the years in managing them, plant diseases still cause more than $9 billion in annual losses in the United States alone. This is largely because:

1. Current control strategies are not 100% effective.
2. Pathogens evolve and overcome once-effective management tactics.
3. Exotic pathogens are introduced.

Thus an ongoing research program is needed to devise effective management strategies to keep up with the changing disease situation.

Identification and Classification of Pathogens. Effective disease control usually depends on rapid and accurate identification of the pathogens involved so that appropriate control measures may be taken. Accurate identification of pathogens is also critical for making sound decisions regarding quarantines of imported and exported plant material commodities. Knowing how pathogens are related to each other can be helpful in suggesting possible control strategies.

Biological Control. Biological control is a strategy that uses beneficial microorganisms to block pathogens from causing disease. In addition to finding such organisms and, in some cases, genetically improving them, it is necessary to learn how they are influenced by factors such as weather, soil type, conventional pesticide application, and crop variety to make biological control as effective and reliable as possible. Also needed are commercially feasible methods to mass produce, store, and apply biological control agents.

Cultural Control. Changing the way a crop is managed (soil preparation, time of planting, irrigation regime, storage conditions, etc) often has a significant impact on disease severity. Knowledge of how such practices affect pathogen survival and disease progression can be used to devise cultural systems that allow acceptable crop production yet are less favorable for disease development.

Pathogen Biology, Genetics, Population Dynamics, Spread, and Relationship with Hosts and Vectors. Research on the processes that take place during disease development can uncover vulnerable steps in the life cycles of pathogens where control measures can be successfully used. Similarly, understanding how pathogens move from plant to plant in the field or within harvested commodities in storage, how they survive in the absence of host material, or how they are affected by their environment can suggest possible control strategies. Knowledge of how and where pathogens survive on plant parts and seeds is essential for developing methods to reduce the domestic and international spread of disease.

Host Plant Resistance to Disease. Enhancing genetic resistance to disease in plants is sometimes the only disease-management option and is often the most cost-effective, environmentally benign one. This approach depends on research to identify and characterize genes for resistance in crop species themselves; in plants closely related to them; and in unrelated, alien species. Incorporating such genes into commercially acceptable varieties can be accomplished through conventional plant breeding and genetic engineering.

Crop and Commodity Pest Biology, Control, and Quarantine

Program Rationale. Economic loss of food and fiber due to damage caused by insect, mite, and weed pests costs U.S. consumers and producers billions of dollars every year. Besides direct losses, pests may further reduce the quality and value of products, increase the costs of production, restrict U.S. products from foreign markets, damage environmental areas, and place native species at risk. Technology developed through the crop and commodity pest biology, control, and quarantine national program provides the pest-control methods used to devise effective and sustainable integrated pest management (IPM) programs that support agricultural-production systems for a diversity of situations throughout the country. The overall philosophy of this national program is to develop and help implement sustainable approaches to managing pests through a combination of biological, cultural, mechanical, and chemical tactics that reduce pest populations to acceptable levels while minimizing economic loss, impact on human health, and environmental risk.

Program Components

Identification and Classification of Insects, Mites, Microbes, and Plants. Accurate taxonomic identification and classification of organisms is the first step in providing information on the biology and control of known and new pests; in determining the geographic origin of introduced pest species; and in the location, collection, and importation of effective biological control agents for IPM programs. ARS scientists will maintain specimen and germ plasm collections of agriculturally important insects, mites, microbes, and plants, conduct taxonomic assessments of important groups of organisms; develop identification keys; and support private and governmental action agencies in characterizing critical pest problems.

Investigation of Pest Biology. Inadequate knowledge of pest biology, movement, genetics, physiology, biochemistry, organismal biology, reproduction, and population ecology currently limits our ability to develop and implement biologically sound and environmentally compatible pest-control technologies. Continued ARS research on insect, mite, microbe, and plant biology and on important individual pest species is necessary to provide the scientific knowledge base to create new and effective pest-control alternatives that are safer for humans, nontarget organisms, and the overall environment.

Understanding Pest/Host Plant Interactions and Economic Impact. ARS scientists are discovering how insect, mite, and weed pests interact with valuable crops and commodities to cause losses and are determining the biological mechanisms that allow pests to cause this damage. The economic assessment of pest attack and the associated costs of pest control provide further vital information on pest-control alternatives and action thresholds for IPM programs. This research will lead to new cost-effective methods of pest control such as pest-resistant crops, biological control agents, or other environmentally friendly technologies.

Investigation of Pest Exclusion and Quarantine Treatment Procedures. Development of pest detection, exclusion, and quarantine treatment technologies is important both in keeping new pests from invading the United States and in treating our crops and commodities so they can be shipped to and sold in foreign markets. In this era of expanded free trade and global shipping, the movement, introduction, and establishment of exotic pests is an ever-increasing threat that will be addressed through research that supports action and regulatory agencies such USDA Animal and Plant Health Inspection Service.

Development of New and Improved Pest-Control Technologies. Research and development of new pest-control technologies, particularly biologically based methods such as host plant resistance, biological control, semiochemical treatment methods (eg, pheromones, mating disruption), and cultural strategies will be expanded. Additionally, pest and natural enemy sampling will be improved, pesticide formulation technology will be made more efficient, new precision agriculture methods will be evaluated, and the impact of these technologies on both target and nontarget organisms will be evaluated and optimized. The new technology resulting from this research will provide the tools to protect ourselves and future generations from the continuous pressure of pest damage that threatens our food and fiber supply each and every year.

Integration of Component Technologies for IPM Systems. IPM is the best method known to control pest problems; however, research on combining different pest-control technologies into effective, economical, and sustainable IPM systems needs to be explicitly conducted or integration often fails. Although integration of pest-control tactics is the primary responsibility of other ARS national programs, consideration is given in this national program to the planning of holistic IPM systems early in the development of each component pest-control technology.

Integrated Crop Production and Protection Systems

Program Rationale. Production capacity, production efficiency, and crop protection are major pillars supporting national crop productivity. A high priority is the development and subsequent transfer to customers of efficient crop production and sustainable cropping systems. Overall challenges are:

- To substantially increase the knowledge base of and sustainable technology for crop production and cropping systems
- To improve the delivery of technologies generated
- To promote the use of these systems.

Meeting these goals will require bringing emerging technological capabilities together to support immediate and long-range strategies aimed at future crop production, protection, and food-safety challenges for small, medium, and large farms. It will also require substantive collaborations with the customers in program planning, research, and evaluation. The program closely ties in the whole farm–management strategies of the integrated farming systems national program and with the components of the crop protection and quarantine national program.

Program Components

Crop Management and Production Efficiency. The goal is to develop new cutting-edge crop-production principles and practices and provide the necessary agronomic principles, technologies, and approaches associated with the efficient production of crops. Research undertaken in this component will (1) help fill the knowledge gap needed for applying appropriate principles and practices under differing ecological and climatic conditions, and (2) furnish the agronomic principles needed for crop production.

Bees and Pollination. The goal is to develop new technologies leading to sustainable strategies for using bees as pollinators to increase crop production. At the same time these technologies will help maintain the profitability for providers of bees that are used in the pollination and honey-production industry and will help U.S. farms to remain competitive in the global marketplace. The discovery of two parasitic mites of honeybees in the 1980s, the invasion of the Africanized honeybee in 1990, and diseases of both Apis and non-Apis bees is seriously impacting the availability of pollinators. Technology for the rapid detection and control of parasitic mites and other diseases of pollinators will enhance pollination efficiency and honey production.

Agroengineering, Agrochemical, and Related Technology. The goal is to develop more effective production, harvesting, and tillage equipment; sensor technology; and pesticide-application methodologies and other tools for crop production. The component also includes developing appropriate practices for the use in and incorporation into cropping systems of products generated through biotechnology.

Models and Other Decision Aids. The goal is to provide user-friendly crop-production models and decision aids for (1) assessing alternative biological, economic, and related production and management practices; (2) maximizing processing; and (3) increasing energy-use efficiency.

New Uses, Quality, and Marketability of Plant and Animal Products

Program Rationale. American agriculture faces increasing, intense competition in the global marketplace. Worldwide, agricultural production has increased faster than demand in many areas, resulting in current commodity surpluses, low prices, and unreliable profitability. Recent shifts in U.S. farm policy to remove price supports emphasize the need for American agriculture to move beyond production of ever-larger quantities of ever-cheaper commodities. In response, farmers and ranchers must be able to produce higher-quality products that can be differentiated from lower-value commodities, commodities and co-products must be converted into useful value-added food and nonfood products, and products must be protected from contamination or loss of quality after harvest to ensure marketability. Research must also be responsive to consumer demands for high-quality, safe products; government and consumer pressure to provide products that are "environmentally friendly" or that are produced using processes that are "friendly" to people, animals, and the environment; and the need for a sustainable and profitable

agricultural-production system. In this program, research will develop knowledge and technology for crop and animal product quality measurement and maintenance or enhancement during processing and marketing, commodity and co-product processing into value-added materials, and new specialty products from crops and animals.

Program Components

Intrinsic Product Quality. Fundamental research will clarify the roles of product composition, molecular structure, and physical state in determining end-use quality and functionality. Genetic improvement of crops and animals for quality attributes will lead to the best possible product quality at the farm gate. Applied research will develop new processes to maintain or enhance product quality during harvest, storage, transport, and marketing.

Pest and Disease Control. The goal is to provide information that will alleviate commodity trade barriers and quality and economic losses attributed to microbial or insect infestation. The major thrust of this research is to prevent spoilage or contamination by managing or eliminating postharvest pests and pathogens, especially those resulting in formation of mycotoxins. This effort includes both basic and applied research to improve pest and disease resistance, new treatments to eliminate pests and pathogens (especially treatments that do not rely on synthetic chemical pesticides), and biological control of pests and pathogens. The focus of this work is on preservation of harvested farm products. Research is specifically aimed at overcoming quarantine barriers to export or ensuring that food safety is found in different programs.

Product Handling and Grading. Efficient technologies and improved or new equipment will be developed to preserve the quality of agricultural commodities during harvest, transport, and storage. Research will establish basic needs for material handling that preserves quality characteristics, and processes and equipment will be developed that meet the requirements. Research will also develop new technology for product grading to provide rapid, accurate, and reproducible information on quality. The research in this component of the program will lead to both expanded and better-documented value of agricultural products. The improved quality and the improved grading will make agricultural products from the United States more competitive and marketable domestically and overseas.

New Processes, New Uses, and Value-Added Products. The goal is to develop new uses of agricultural products and co-products. Innovative new processes will be created and existing ones adapted for the extraction and purification or the manufacture of superior products from agricultural commodities. Application of these innovative technologies will expand the range and value of agricultural products and reduce the cost of their production, making processed goods from the United States more competitive. Sources of natural products will be identified for use as nutriceuticals, pharmaceuticals, biopesticides, or other high-value uses. This program will identify alternative sources and create technology leading to an expanded, diverse range of value-added food and nonfood products from commodities and undervalued by-products of agriculture. New crops will be "mined" for valuable materials.

Bioenergy and Energy Alternatives

Program Rationale. The last decade has been characterized by huge U.S. trade deficits. In fact, petroleum import for transportation purposes alone was $50 billion in 1996. America's dependency on foreign oil (now at 50% and rising) is not only an economic issue but is one of national security as well, particularly in times of global unrest. These factors, coupled with environmental concerns regarding the use of fossil fuels and production of CO_2, fostered the expansion of the fuel ethanol industry. The capacity of the U.S. industry now exceeds 1.5 billion gallons per year. The industry has become an important partner with American agriculture and the USDA estimates that 17,000 jobs are created for every billion gallons of ethanol produced. Similarly, a nascent biodiesel industry has been developing in recent years.

Current use of ethanol and biodiesel as fuel additives or alternative fuels (eg, E-85, a good, neat biodiesel) depends on many factors, including political actions, tax policies, agricultural practices, regulatory issues, and international economic trends. The relatively high cost of ethanol production and the very high cost of biodiesel production, however, remain as important constraints on their use. Removing technical constraints is the key to a viable biofuel industry of the future.

A likely market for alternative energies is within agriculture itself. Low-cost alternative fuels can be used to power farm tractors and small agricultural production and processing facilities within rural communities. Wind and solar energy, as well as biofuels, may also be utilized to supply or supplement electrical energy for water pumping, small-scale irrigation systems, and other farmstead needs.

Program Components

Ethanol. Advances in enzymology; microbiology; and chemical, biochemical, and process engineering are required to underpin this effort. Technologies are needed to reduce the cost of producing ethanol from cornstarch (95% of production). In addition, the conversion of biomass, a vastly more abundant feedstock, is an economic necessity. This program's first biomass target is corn fiber, due to its ready availability at ethanol plants and extremely low value. Immediate integration of corn fiber to ethanol at the plant is feasible and would result in at least a 10% production increase. These research findings will be applicable to other forms of biomass as well, ranging from agricultural and forestry wastes to fast-growing crops that could be grown solely for energy production. Research will focus on, but not be limited to, developing new enzymic processes for saccharification and improved microorganisms to ferment the multiple sugars found in biomass. To improve process economics, value-added co-products will be developed from current by-products, and separation technologies will be improved for ethanol as well as co-product recovery.

Biodiesel. Vegetable oils and animal fats and their derivatives (biodiesel) are attractive as alternative fuels, ex-

tenders, and additives for compression ignition (diesel) engines. Opportunities for biodiesel include off-road markets such as underground mines, marine applications, mass transit (subways, trains), and stationary power generation. However, research is needed to improve cold start-up and operability, to identify and reduce harmful exhaust emissions (eg, nitrogen oxides), to develop a rapid and low-cost fuel-quality test, and to reduce feedstock and formulation costs.

Energy Alternatives for Rural Practices. Autonomous water-pumping systems are needed to supply water to livestock in semiarid areas where precipitation is sufficient for forage production but is not adequate for watering livestock. Renewable energy systems are well suited to this application because they are independent of environmentally hazardous, expensive distribution systems. Remote renewable energy systems are needed to supply additional electric power to rural areas that are undersupplied by overload rural electric distribution systems. On-farm electric-generating systems powered by a combination of wind, solar, and biofuels will be developed.

Methyl Bromide Alternatives

Program Rationale. Methyl bromide (MB) is a highly efficacious soil fumigant for more than 100 crops and foreign ornamental nurseries. It is also used to kill arthropod pests in postharvest commodities fumigate structures. The U.S. EPA published a listing of MB as an ozone depletor, banning the production and importation of MB in January 2001. The development of alternatives to MB will have a major impact on the U.S. agricultural industry and alternatives to its use must be found.

Program Components

Preplant Soil Fumigation Alternatives. The loss of methyl bromide for soil fumigation will result in serious disease, pest, and weed problems. These problems will likely be highly variable and will depend on crop, soil type, environment, and cropping conditions. Therefore, separate research efforts will be required to develop multiple management strategies. Approaches will include host plant resistance, biological control, cultural practice modifications, alternative chemicals, and combinations of all of these.

Postharvest Commodity Treatment (Including Structural). The loss of MB in 2001 will have significant negative impacts on U.S. domestic and international trade. The time frame is very short given that treatments will have to be developed commodity by commodity, and many years may be required to gain appropriate new commodity treatment by a trading partner country. Approaches will include heat; controlled atmospheres; irradiation; chemicals, including other fumigants; exclusion; as well as exotic pest-eradication technology.

UNITED STATES FOOD MARKETING SYSTEM.

See UNITED STATES FOOD MARKETING SYSTEM in the Supplement section.

V

VANILLA EXTRACT

HISTORY OF VANILLA

Vanilla, the world's most prized flavor, is one of the most valuable treasures the Europeans brought back from the New World. When Cortez landed his army in Eastern Mexico in 1519, he formed an alliance with the local Indians, who then helped lead the Spanish troops against Montezuma and his Aztec empire (Fig. 1). Montezuma royally welcomed Cortez with a vanilla-cocoa brew. This drink, chocolatl, was concocted from cocoa beans, ground corn, honey and Tlilxochitl (vanilla pods). Cortez took Montezuma's life and his treasures, including the secret of vanilla, which was brought back to the Old World. For several hundred years thereafter, cured vanilla beans were imported from Mexico to Europe for the production of vanilla flavor and perfume. Although the Spaniards had this supply of cured beans, they did not have the complete secret of how to grow and cure vanilla. It was over a hundred years before the Europeans could successfully cultivate the vanilla plant in greenhouses. Propagation through cuttings was somewhat successful in the early 1700s, although the plants seldom flowered and never produced fruits.

Plants were started in tropical regions, including Indonesia, also with no fruit. Eventually, in 1836, a botanist noted that the flowers needed to be individually hand pollinated in order to fruit. In Mexico, this may have been done naturally by bees, hummingbirds and/or a species of leaf-cutting ant. There is no experimental proof that any insect is actually effective in pollinating the vanilla flowers (1,2).

By the mid 1800s improvements in the human hand-pollination techniques had been developed which led to successful vanilla plantations in Madagascar, Reunion, Mauritius, the Seychelles Islands, Tahiti, the Comoros Islands, Ceylon, Java, the Philippines, and parts of Africa. By the 19th century more vanilla was being grown in Madagascar and tropical Asia than in Mexico, breaking the monopoly enjoyed by Mexico for over 200 years. Mexico continued to be a major producer of vanilla beans until the mid-1900s.

HORTICULTURE

Vanilla is the common term for the alcoholic extract of the vanilla bean. The vanilla bean is actually the fruit of a thick, tropical orchid vine. Of the 35,000 or more species in the orchid family, *Orchidacae*, the vanilla orchids produce the only edible fruit. There are over 50 vanilla orchid species, of which only two are of commercial use. *Vanilla planifolia* Andrews (also known as *Vanilla fragrans* (Salisbury) Ames) is the species responsible for 99% of the vanilla imported into the United States (Fig. 2). The other species, *Vanilla tahitensis* (Tahitian Vanilla) grows on the French Pacific Islands and is visually quite different. The tahitensis pods are shorter, have a thicker skin, less seeds, and are much broader than the planifolia beans. Tahitian vanilla beans are primarily exported to France and Europe, although roughly nine tons are imported into the United States.

Hernan Cortez was the first white man to taste vanilla, in Mexico, in 1520.

Figure 1. The Spanish troops led by Cortez conquered Montezuma and stole the secret of vanilla. *Source:* Photograph courtesy of McCormick/Schilling Spices.

Figure 2. The vanilla orchid. *Source:* Photograph courtesy of McCormick/Schilling Spices.

Vanilla pompona is frequently cited as a third commercial species of vanilla orchid used in perfumes. Today, this species is rarely seen and is mostly a curiosity. Visually, the pods resemble small bananas (2).

Vanilla planifolia is indigenous to southeastern Mexico, the West Indies, Central America and the northern part of South America. Vanilla vines will grow between 25° north and south of the equator; in hot, moist tropical climates; in a 50/50 mixture of sun and shade; from sea level to 2,000 ft. altitude; in areas with frequent moderate rainfall and no extended droughts or high winds; and with gentle slopes for drainage. For commercial production of vanilla beans, it is optimum to have the rainfall evenly distributed throughout 10 months of the year followed by a two-month dry spell to check vegetative growth and spur flower formation. These conditions describe a typical tropical island climate. Today, virtually all vanilla beans are grown on islands such as Madagascar and the Indonesian Islands, where temperatures range between 70–90°F, with 80–100 in. of rain per year.

Vanilla planifolia has smooth, succulent bright green leaves and aerial roots which cling to some type of support. If left untended, vanilla vines will grow 75 ft to the tree tops. On vanilla plantations they are pruned or bent downward to keep the flowers and beans in reach of the workers for pollination and harvest. The pruning and bending also seems to increase flowering.

Commercially, vanilla is propagated entirely by means of 3 ft cuttings, 8 to 12 nodes in length. Longer cuttings usually bring the vines into early production. Vanilla is very difficult to grow from seeds (1–3).

The vines are cultivated from cuttings on trellises and supports, but most frequently on living trees. The species of tree depends upon the geographical area. Small trees with low branches make the best supports, and those with smaller leaves will allow an appropriate amount of sunlight through. Quick growing banana trees are commonly used to provide shade as well (4). Even when man-made

supports are used, trees are still planted for partial shading of the vines. The character and growth of the vines depends greatly on the amount of shade and exposure (1).

Healthy vines produce up to 1000 satiny-yellow orchid blossoms, of which only 5–30% will be hand-pollinated depending upon the age and health of the plant. The orchids flower in the morning and die by sundown if not fertilized. Although the flowering period extends over 2 months, each blossom has only this small time-window in which to be fertilized. An expert can pollinate 1000 blossoms a day. If too many blossoms bear fruit, the vines will be weakened for subsequent production. Roughly 100 beans are the maximum number that should be expected from a cultivated plant.

After pollination, the pods ripen in seven to nine months. The vanilla beans are green, long, smooth and slender, somewhat resembling a large green bean filled with thousands of tiny seeds (Fig. 3). As the beans mature they lengthen to up to 12 in. (by 6 months) and gradually turn yellow in the 9th month. The beans are usually harvested before they fully ripen, when only the blossom-end tips are pale yellow (Fig. 4). The best quality vanilla extract comes from blossom-end-yellow beans rather than less ripened green beans (5,6).

At this time, the beans are anywhere from 5–10 in. long and about one-inch in circumference. Longer beans are more desirable as they bring in more poundage per bean. If the beans are picked too early, their flavor precursor content and their quality will be low. They will also become too woody in flavor character and will be more prone to molding. If they are allowed to fully ripen prior to harvest (blossom-end-brown) they split. Splits are very susceptible to molding. Although splits are often considered inferior quality and bring a lower price, if not moldy their vanillin content and resulting extracts can be superior to whole beans (7,8). This may be due to their fully ripe nature.

A very important consideration in the ripening of the vanilla fruit is the formation of the vanillin precursor, Glucovanillin, and most importantly the enzyme B-glucosidase. During the curing process glucovanillin is en-

Figure 3. Unripe, green, three month old Madagascar vanilla beans. *Source:* Photograph courtesy of Benjamin H. Kaestner III, McCormick/Schilling Spices.

Figure 4. Vanilla beans on the vine showing mature (blossom-end-yellow) split (blossom-end-brown and split) and unripe green pods. *Source:* Photograph courtesy of Benjamin H. Kaestner III, McCormick/Schilling Spices.

zymatically converted by the glucosidase to glucose and the aldehyde vanillin, the single most characteristic component of vanilla extract flavor. The longer a bean ripens before harvest the more concentrated the glucosidase is, and thus the more concentrated the vanillin and other important flavor precursors are after curing (9). Higher vanillin beans are considered higher quality and command a higher market value.

Privately funded bioengineering research on vanilla has progressed slowly over the past 5 years. Research has focused on improving the resilience of the plant and on manipulating its cells to mass produce natural vanilla flavors in the lab. No patents have been issued yet, nor has the technology been commercialized (10).

CURING

The vanilla orchid and the mature vanilla bean have no aroma; it is the curing process that develops the characteristic flavor of vanilla beans. Vanilla beans left on the vine will cure naturally, but the pod splits, loses contents, and ultimately decreases in flavor and value.

The Mexican Indians developed the original, very labor-intensive process for curing green vanilla beans. The

French made slight modifications to this original process, and this Bourbon process is generally practiced in the Madagascar and Comoros Islands today. Other modifications are practiced in Indonesia. No geographical source of vanilla beans strictly employs one method of curing and many parts of the process are interchanged (mixed and matched). The curing operation is not a regulated, large-scale, sophisticated procedure. It is crude, nonhygienic, subject to personal modifications, and is practiced by individual small farmers up to larger scale curer-exporters. Any given shipment of cured beans may represent the composite curing operations of dozens of individual producers. During the curing process, each bean is individually handled and inspected at least a dozen times (2,11).

In 1952, 1952, and in 1972, McCormick & Company, Inc. patented processes for rapid, controlled curing of vanilla beans (12–14). However, none of these modern methods are commercially practiced in full today. Portions of the process, modified to use whole beans, are being commercially applied in some parts of Java and Bali (2).

All curing methods involve four basic phases:

1. *Wilting* or *killing* of the beans, which stops the natural respiratory metabolism and vegetative life of the pod (Figs. 5 and 6). The specific technique used to kill the beans will affect subsequent vanillin contents (6,15,16). After wilting, the beans may have begun to turn chocolate brown in color.

2. *Sweating* the wilted beans until they are flexible in the hand and can be easily wrapped around the finger. This step involves a fairly rapid dehydration and slow fermentation. The characteristic flavor compounds develop here. Key enzymatic and nonenzymatic reactions occur during this phase forming sugars, phenols, quinones, pigments, vanillin and other aromatic compounds. After sweating, the beans are deep chocolate brown.

3. *Drying* of the sweated beans very slowly at low temperature to 20–25% finished moisture (Fig. 7). Beans should still be flexible; over-drying, or too rapid drying, reduces flavor quality and value.

4. *Conditioning* of the dried beans in closed boxes for a few months, where they finish the development of their characteristic fragrance. Unless moldy, beans can be kept indefinitely in this state.

Generally about five to seven pounds of green vanilla beans yields one pound of finished product, or about one hundred twenty pounds of finished beans per acre of vanilla vines.

The basic steps in the curing process are summarized, by method, Table 1 with illustration in Figure 8. Many bean-curers employ a combination of methods and/or their own customized approach. Thus, there is a certain amount of variation in quality even within one growing region in any given year (2).

When properly cured, vanilla beans should resemble long, very thin cigarillos; supple, very dark brown, a raisin-like texture and a somewhat oily sheen (Figs. 8–10). High quality beans may have white crystals of vanillin clinging

Figure 5. Killing of the beans in Madagascar. *Source:* Photograph courtesy of Benjamin H. Kaestner III, McCormick/Schilling Spices.

Figure 6. Killed beans are poured into sweating chambers in Madagascar. *Source:* Photograph courtesy of Benjamin H. Kaestner III, McCormick/Schilling Spices.

to the outside, but this is rarely seen today. The beans should be free of mold and insects (2,7,8).

Beans should be less than 25–30% moisture or a greenish-white mold will develop. If mold does develop, it is removed by washing with alcohol in mild cases, or by cutting off the affected portion if severe. The odor of mold from a few beans can permeate an entire box. This reduces the quality and the value of the beans.

After curing, the vanilla beans can be divided into the following four quality grades by appearance (the numbers of, and names given to, these grades differ according to source).

1. *Whole beans, no defects*; oily sheen, smooth exterior, moist, aromatic, very dark brown.
2. *Whole beans, some defects*; rough exterior, somewhat reddish color, spotted, dry.
3. *Splits*; whole split beans.
4. *Cuts*; beans chopped into 1–2 in. lengths, may include very small whole beans.

The sorted beans are grouped by quality grade, packed loosely or in bundles of 50–100 in tin, wooden or cardboard boxes, sealed and shipped to their destination (Fig. 11).

Figure 7. Vanilla beans sundrying in Madagascar. *Source:* Photograph courtesy of Benjamin H. Kaestner III, McCormick/Schilling Spices.

The bulk of global vanilla bean trading is handled by only a handful of importer-dealers. Internationally, all beans are traded in the cured form. Locally, many vanilla farmers sell their crops of green beans to regional curing operations, but green beans are not traded internationally (2).

Table 1. Vanilla Curing: Four Typical Methods

Step	Mexican[a]	Bourbon[b]	McCormick[c]	Java[d]	Miscellaneous[e]
Wilting or killing	Stored in shed for a few days until the beans shrivel; then *Sun/traditional* Green beans spread on wool blankets or straw mats in the direct sun until too hot to hold (ca 1 h). During the dry season; Nov.–May, or *Oven* (generally used today) Beans rolled in wet blankets, placed in humid oven or room at 60°C for 36–48 h. Indirect heat from firewood.	Beans placed in straw or metal baskets or in blankets, immersed in hot (150–190°F) water for 2–3 minutes, drained.	Beans are chopped into 1/2 in. lengths	Process extremely variable between regions Immerse in boiling water, or Nothing—no killing	Freezing for 3–4 h Ethylene gas 100 ppm for 8–16 h Scratch bean with pin (surface wounding) Place beans in ashes of fire (Guiana method)
Sweating	*Sun/traditional* Spread on blankets or straw mats and exposed daily to sunshine for 1–2 h, then placed the remainder of each day inside blanket and/or inside a shelter. Repeated daily for several weeks until proper aroma and texture are obtained. *Oven/current procedure* Sweating continues in the closed warm room/oven until brown. Beans in blankets may be transferred into chests.	Hot, drained beans packed into chests, or boxes covered with tarp, retained temperature about 122°F. Removed after several days when brown. Then placed in sun as in Mexico, but on elevated platforms, for 6–8 days.	Chopped beans placed into perforated trays, trays are stacked in curing tank, cured at 140°F for about 72 h, until beans are no longer green.	McCormick process/ whole beans, or Nothing—no sweating, or Sun-sweating as in traditional Mexico	
Drying	Sweated beans spread on grass mats or blankets in the sun every day for a few hours for 2 months or longer; then dried to finish inside a shelter.	Same as in Mexico, but often on elevated platforms.	Dried in rotary or fluidized dryer with forced air at 140°F until moisture content is 35–40%.	McCormick process, or In tobacco curing sheds	
Conditioning	Beans are sorted; placed in closed boxes for several months.	Placed in closed tin boxes or chests for several months.	Remove from rotary drier, spread 4 in. deep in perforated conditioner. Blow room temp/ humidity air through conditioner until moisture is 20–25%. Place in loosely covered container for 3 months (during shipping).	No conditioning in some cases, or McCormick process, or Placed in closed boxes for several months.	
Packaging	Bundles of 50–130 beans in paper or metal boxes.	Bundled 50–100 beans/ bundle in tin or cardboard boxes.	Metal or cardboard box, not hermetically sealed.	Highly variable, whatever the farmer has on hand.	
Total time	5–6 months	5–8 months	3 months	2 weeks to 2 months	

Source: [a]Refs. 2, 4, 16, and 17.
[b]Refs. 2, 4, and 17.
[c]Refs. 12–14.
[d]Ref. 2.
[e]Ref. 5.

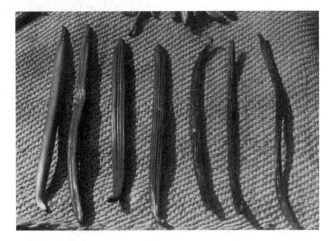

Figure 8. Vanilla bean at various stages of curing; green, killed, sweated, dried, and conditioned (left to right). *Source:* Photograph courtesy of Benjamin H. Kaestner III, McCormick/Schilling Spices.

Figure 9. Cross section of cured vanilla beans showing vanillin, resins, and seeds. *Source:* Photograph courtesy of McCormick/Schilling Spices.

Figure 10. Cured vanilla beans from Madagascar, Comoros, and Java; whole splits and cuts. *Source:* Photograph courtesy of McCormick/Schilling Spices.

Moisture and vanillin contents vary from crop year to year, but are fairly stable within a given period. Table 2 illustrates the typical ranges in 1988–1990.

EXTRACTION

Regulatory

Vanilla extract is the only flavoring material with a U.S. FDA standard of identity. It is included in the *Code of Federal Regulations* (21-CFR-169). The standard was developed and promulgated concurrent, and in close relationship, with the ice cream standard (21-CFR-135.110). This coordination was necessary since vanilla extract and related flavorings are ice cream's most widely used flavorants; and the labeling of ice cream is dependent on the type of flavoring used. Category I (21-CFR-135) vanilla ice cream contains only pure vanilla components and no arti-

Figure 11. Sorting of vanilla beans in Madagascar. *Source:* Photograph courtesy of Benjamin H. Kaestner III, McCormick/Schilling Spices.

Table 2. Moisture and Vanillin Contents Ranges in 1988–1990

Vanilla bean origin	Moisture, bean (%)	Vanillin, one-fold extract (% weight/volume)
Madagascar, whole	13–30	0.09–0.26
Comoros, whole	16–26	0.14–0.25
Indonesian, whole superior quality	12–30	0.09–0.25
Indonesian, whole average quality	10–19	0.04–0.13
Indonesian, whole poor quality	9–11	0.04–0.06
Mexican	17–23	0.14–0.17
Java cuts	8–12	0.02–0.09
Bourbon cuts	11–16	0.06–0.13

ficial flavors. This product can be labeled "Vanilla Ice Cream." Category II (21-CFR-135) vanilla ice cream can be flavored with up to one ounce of synthetic vanillin per unit (defined below) of vanilla extract. This natural and artificial product, where the natural is the characterizing and predominant contributor to the flavor, must be labelled "Vanilla Flavored Ice Cream." Finally, Category III (21-CFR-135) ice cream contains predominantly or exclusively an artificial vanilla flavoring which includes primarily synthetic vanillin. This product must be labelled "Artificially Flavored" or "Artificial Vanilla."

Both the ice cream standard and the vanilla standard nomenclatures rely heavily on the definition of a unit of vanilla constituent. This term is defined by the Vanilla Standard 21-CFR-169.3. The types of vanilla beans are identified as "the properly cured and dried fruit pods of *Vanilla planifolia* Andrews and of *Vanilla Tahitensis* Moore." But the term properly cured and dried is not defined. The quantity of beans necessary to make a unit weight of vanilla beans is also identified. This last part has led to confusion in that the definition of quantity was set at 13.35 oz of 25% moisture vanilla beans. This is 10 oz of dry weight solids. If the beans contain greater than 25% moisture, logic states (as does the standard) that the quantity of beans in a unit be based on 10 oz of dry weight solids, although the wording refers again to the 25% moisture vanilla beans. However, when the moisture content of the beans used is below 25% moisture, the standard requires 13.35 oz regardless of the dry weight solids. No guidance is given concerning when the moisture content should be measured, when packed, when purchased, when received, or when used. There has been much discussion within the industry recently concerning the relationship between moisture content of the beans and the flavor of the extract. A positive correlation (eg, increase in moisture accompanies an increase or improvement in flavor) would be necessary to require an adjustment in the quantity and quality of bean usage in the extraction. The intent of the regulation was to provide a consistent product to the consumer, and therefore, an analytically valid dry weight standard would be more appropriate and defendable. This debate will probably go on for several more years. Until resolution, the government is evaluating and approving

appropriate industrial permits as requested by vanilla extract producers.

Because the moisture content is so important to policing the standards (and will be regardless of which refinement prevails), the standard goes on to define how moisture should be analyzed—azeotropic distillation with toluene/benzene. This, also, leads to a concern in that most, if not all, extraction companies and analytical labs no longer permit the use of benzene in their facilities. Therefore, the moisture analysis needs to be revised as the entire standard is being overhauled. Currently, the Technical Committee of the Flavor Extract Manufacturer's Association of the United States (FEMA) has undertaken this task. In addition, alternative methods to the wet technique should be investigated and incorporated as appropriate. These include near infrared (NIR), electroconductivity, nuclear magnetic resonance (MMR), etc.

The remainder of the standard is involved with describing in general terms how the extract is made and what other ingredients can be used. The standard also defines other products related to pure vanilla extract—what constitutes them and how they can be labeled:

- Concentrated vanilla extract
- Vanilla flavoring
- Concentrated vanilla flavoring
- Vanilla powder
- Vanilla–vanillin extract
- Vanilla–vanillin flavoring
- Vanilla–vanillin powder

The more significant statements in the standard, which should be noted, require that the finished extract have not less than 35% ethyl alcohol and contain no less than one unit of vanilla beans per gallon.

As a federal regulation, the FDA standard is enforceable by agencies of the Executive Branch. Because this product uses potable alcohol subject to tax controls, the Bureau of Alcohol, Tobacco and Firearms (BATF) has been in the forefront of assuring compliance to the standard. In addition, the Flavor Extract Manufacturing Association of the United States (FEMA) has a self-policing policy aimed at maintaining quality and integrity of vanilla extracts and natural vanilla flavors.

As a result of this unique standard, the manufacture of vanilla extract is tightly controlled and requires a means for assuring compliance via analytically supported criteria.

Manufacture

As a result of the restrictions imposed by the Vanilla Standard, there is a limited number of methods to produce this regulated product. All methods are required (1) to extract the total rapid and odorous principles extractable by 35% alcohol and (2) to use no less than a unit quantity of vanilla beans in a gallon of extract. Therefore, the manufacturing variations involve how the alcohol and vanilla beans are brought into contact, the temperature during the extraction, the percent alcohol in the miscella (extracting solution) and the length of time the vanilla beans and solvent are in contact.

The temperature is usually between ambient and 60°C and the alcohol concentration between 40% and 60% (vol/vol). The extraction time can be days, weeks, or months. Economics inhibit the longer extraction times and encourage higher temperatures and other innovations to accelerate extraction of the desired vanilla constituents. The alcohol concentration is controlled in order to encourage efficient extraction of the desirable constituents, while inhibiting the non-polar lipids and waxes which are soluble in the higher proof alcohol solvents. Some extractors prefer those processes which encourage rapid, indiscriminating conditions followed by settling tanks, filters, and/or centrifuges to remove constituents not soluble at ambient temperature and in the final 35% alcohol solution.

Maceration (soaking) and percolation are the two methods for contacting vanilla beans and solvents. Soaking involves mixing the chopped beans with the solvent and allowing these to set and extract for a given time period. The first extract is removed and the beans reextracted. This is repeated until the desired level of extraction is achieved. The combined solutions produce a complete and full-bodied extract.

However, soaking has been replaced almost entirely by percolation (Fig. 12). As the name implies, percolation involves a continuous circulation of the aqueous ethanol through the chopped vanilla beans. Unlike coffee percolation, the heat applied, if any, is not enough to cause the solvent circulation. The solvent is pumped mechanically through the beds of vanilla beans. The temperatures can vary between ambient and 60°C and the drip percolation can be hours to days.

The entire extraction process can be divided into four subroutines (1) sample preparation, (2) extraction, (3) adjustment and/or concentration (where a concentrate is desired), and (4) solvent recovery. Sample preparation involves chopping or shredding the vanilla beans. The extraction process has also been detailed previously. Concentration involves removing all or part of the extraction solvent. This is called folding and is usually accomplished by solvent distillation under vacuum. A one-fold (1X) extract contains one unit of vanilla beans per gallon. The

Figure 12. Vanilla extract percolators. *Source:* Photograph courtesy of McCormick/Schilling Spices.

extract removed is usually over one-fold and over 35% alcohol. This permits the adjustment to the optimum flavor and cost for a minimum standard product as well as compliance with the standard and customer specifications. This also allows for the preparation of a concentrate with lower energy use. Some higher fold vanilla extracts can be made by adjusting the ratio of beans to alcohol, and by using counter-current extraction methods. Extracts of four (4X) fold or less are attainable in this manner. Higher folds are best made by producing an oleoresin through total solvent removal, followed by dilution to the desired 5X, 10X or 20X. One-fold extracts are sold retail; the concentrated or multifolded extracts are favored by industrial users.

The solvent recovery step involves the recovery of residual solvent from the exhausted or spent vanilla bean material. Economics and regulation demand this for two reasons—the most obvious being the ability to reuse the costly solvent for further extraction. In addition, the U.S. Bureau of Alcohol, Tobacco & Firearms (BATF) stipulates that the spent beans be free of recoverable alcohol. The alcohol is often recovered from the spent material through distillation.

More modern methods of extraction and concentration have been used. These include supercritical carbon dioxide extraction and reverse osmosis for concentration. These methods produce useful products for industrial flavoring, but they either fail to fit the regulatory requirements for the standardized product, and/or are very expensive. These specialized products differ in solubility, flavor profile and appearance, but add to the list of natural vanilla flavorings available to food and beverage manufacturers.

Finally, vanilla extract, like a fine wine, improves on aging! Storage in oak barrels (new or previously used for whiskey) is a traditional method for vanilla. Modern technology can accelerate the process (19,20).

FLAVOR

Chemistry

The chemistry and analysis of vanilla has been studied extensively for two main reasons; first, in order to understand what causes this popular delicate unique flavor which also enhances many other flavors, and secondly, to protect the integrity and therefore the quality of this product from unscrupulous manufacturers and suppliers who are willing to adulterate vanilla or even present a completely synthetic imitation product as natural. A complete review of past endeavors is beyond the limits of this review, so a brief historic background and summary of the more current results of applications that have been developed from these many efforts and discoveries over the past 100 years are presented.

Flavor Chemistry

Vanillin is the most abundant of the vanilla flavor constituents. However, since its discovery by Golby in 1858 (21), there has been much controversy as to its contribution to the overall flavor impact and quality of vanilla extract. Klimes and Lamparsky (22), who at the time utilized very innovative nanogram techniques, identified 169 components, most below 1 ppm, in vanilla beans. Other than to state the general importance of the trace constituents to the overall sensation of vanilla flavour, no specifics on individual constituent contributions to flavor were identified. This has been the general trend in vanilla research: identify the unique constituents but ignore their ultimate sensory contributions. This intentional withholding of information is a result of proprietary and costly research undertaken by various companies whose livelihood and future is tied in part to understanding and exploiting vanilla flavor.

However, Galetto and Hoffman (23) synthesized the methyl and ethyl ethers of p-hydroxybenzyl alcohol and vanillyl alcohol found in vanilla (Compounds I, II, III & IV in Fig. 13). They described them as imparting "sweet, vanilla-like flavor notes with creamy, coconut secondary flavor which could contribute to the 'character' of vanilla." Schulte-Elte and co-workers (24) stereoselectively synthesized two diastereoisomers of vitisporanes (Compound V, Fig. 13) which had been found to be "important aroma components of vanilla." They report the odors of each to be "unmistakably different." The cis-compound is fresher and more intense, is reminiscent of green chrysanthemum and has a flowery-fruity wine note. The trans has the heavy scent of exotic flowers with an earthy-woody undertone and a note of dry wines as in marc. Feyertag and Hutchins

(25) isolated and identified a trimethylpiperidonene with two possible isomers (Compound VI A or B, Fig. 13) which they report as having a camphoraceous minty odor and a burning taste.

Walbaum (26) identified the oxidative series of anisyl compounds—alcohol, aldehyde and acid in Tahitian vanilla (Compounds VII, VIII, IX in Fig. 13). Gnadinger (20) confirmed that the alcohol was the major component. These could account for the unique floral aroma of the extract from Tahitian beans. Purseglove and co-workers (27) have done a commendable job of comparing the constituents of three species of vanilla as reported in the literature. They observed that "differences in volatile-component composition between species are probably quantitative rather than qualitative. . . . "

As of 1989, we have reviewed 98 articles and patents on vanilla which identify about 190 constituents in vanilla. Riley and Kleyn (28), in 1989, report that over 250 volatile flavor components had been identified, but did not specify which.

Despite all this research, the key constituents responsible for vanilla's characteristic flavor remain undefined. The general belief in the industry is that 1 oz of synthetic vanillin is equivalent in flavor strength, not character, to one gallon of vanilla extract. However, 1 gal of pure extract contains approximately 7 g of vanillin, or only one-quarter of an ounce. Therefore, these other undefined constituents

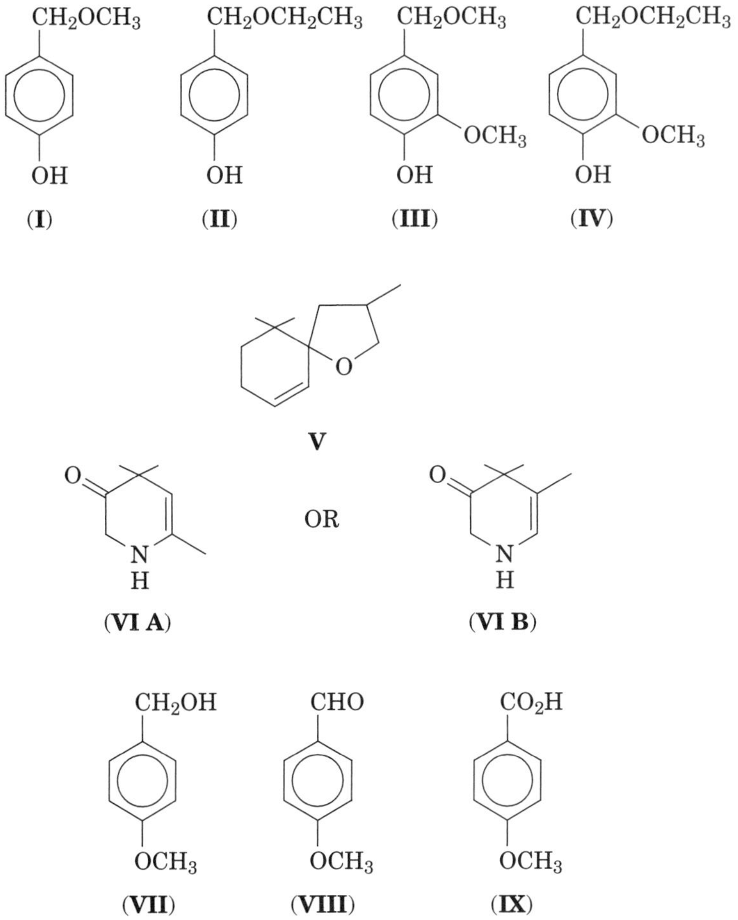

Figure 13. Vanilla flavor compounds.

provide the remaining 75% of the flavor strength and character (18). The vanilla standard implies agreement with this ratio in the definition of vanilla–vanillin extract (21-CFR-169.180).

Analytical Methodology

Methods to analyze vanilla have evolved over the years, guided by two forces—regulation of the vanilla standard and specification of the industrial user. Before adoption of the FDA Standard of Identity, manufacture and "enforcement" were guided by the *U.S. Pharmacopoeia* and expectations of the customer. The late nineteenth century and early twentieth saw reliance on lead numbers (Wichmann), vanillin content, and the determination of coloring content (Marsh Test) and related resins content. These analyses could detect the lack of vanilla bean extractives, the use of prune juice, caramel, wine or artificial colors and the addition of coumarin (29–31). The lead number which by the 1930s was used as a measure of quality was investigated in detail by Purcell (32). He separated the individual fractions contributing to the lead number and demonstrated a very small percentage of the constituents contributed to vanilla flavor. Broderick (31) confirmed and expanded on Purcell's efforts in an attempt to establish analytically supported quality measurements. He supported the use of the related color and resin analyses. However, lead number is still used as an analytical index, primarily as an indicator of geographic origin.

With the promulgation of the vanilla standard in 1962, new authentication methods became available, as did more sophisticated adulteration techniques. The use of St. John's Wort, cherry bark extracts and the addition of organic acids to by-pass existing detection methods required the application of paper chromatographic techniques (33,34) and gas chromatography (35,36) to detect the adulteration.

Each of the new methods reduced the number of adulterated products, but also challenged the "creative" manufacturers.

Many of these techniques and others have been upgraded and refined and are contained in the Official Methods of Analysis (1990), published by the Association of Official Analytical Chemists (AOAC), Section 36. Flavors (37):

973.23 Alcohol in Flavors-Gas Chromatographic Method

920.127 Specific Gravity of Vanilla Extract-Pycnometer Method

920.128 Alcohol in Vanilla Extract

920.129 Glycerol in Vanilla Extract

947.09 Propylene Glycol in Vanilla Extract

946.10 Vanillin in Vanilla Extract—Ultraviolet Screening Method

966.12 Vanillin in Vanilla Extract—Ultraviolet Spectrophotometric Method

966.13 Vanillin in Ethyl Vanillin in Vanilla Extract—Paper Chromatographic Method

920.130 Coumarin in Vanilla Extract—Photometric Method

955.31 Vanillin, Ethyl Vanillin, and Coumarin in Vanilla Extract—Chromatographic Separation Method

968.15 Lead Number (Wichmann) of Vanilla Extract—Titrimetric Method

920.01 Solids (Total) in Vanilla Extract

920.131 Ash of Vanilla Extract

926.09 Vanilla Resins in Vanilla Extract—Quantitative Method

960.36 Vanilla Resins in Vanilla Extract—Paper Chromatographic Qualitative Test

920.132 Methanol in Vanilla Extract—Titrimetric Method

920.133 Color (Insoluble in Amyl Alcohol) in Vanilla Extract—Colorimetric Method

960.37 Plant Material (Foreign) in Vanilla Extract—Paper Chromatographic Method

Table 3. Vanillin Carbon SIRA

Vanillin source	$\delta^{13}C$ 0/00 PDB
Madagascar	−20.4
Java	−18.7
Mexico	−20.3
Tahiti	−16.8
Lignin	−27.0
Eugenol	−30.8
Guaiacol	−32.7

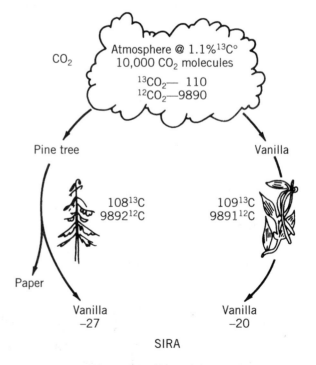

Figure 14. Biosynthesis of vanillin.

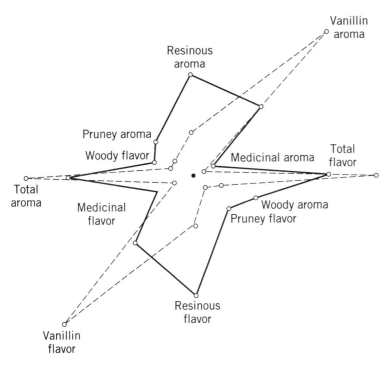

Figure 15. Vanilla aroma and flavor sensory profiles. Products evaluated at 0.5% in water. Intensity of flavor and aroma characteristics (eg, resinous, pruney) increases with distance from center point.

——— Natural bourbon vanilla extract (single-fold)
– – – – Typical imitation vanilla extract (single-fold)

964.11 Flavoring Additives in Vanilla Extract—Thin Layer Chromatographic Method

963.16 Organic Acids in Vanilla Extract—Paper Chromatographic Sorting Method

969.22 Organic Acids in Vanilla Extract—Derivative Gas Chromatographic Method

971.12 Nonvanillin Vanilla Volatiles in Vanilla Extract—Direct Gas Chromatographic Method

These methods, and the technique for moisture analyses elaborated in the standard of identity can be used for quality control of manufacturing and customer specification. They are also useful for identifying gross, but not the newer more sophisticated adulterations, of vanilla products.

High pressure liquid chromatography was used to authenticate vanilla product by several workers. Herrmann and Stocki (38) utilized a reverse-phase gradient system and reported the quantitation of 4-hydroxybenzyl alcohol, vanillyl alcohol, 4-hydroxybenzoic acid, 4-hydroxybenzaldehyde, and vanillin.

Guarino and Brown (39) developed an HPLC method which separated p-hydroxybenzoic acid, p-hydroxybenzaldehyde, vanillic acid, vanillin and ethyl vanillin. The method was evaluated by six industrial laboratories and the authors recommended a collaboration of this method as an alternative method for the quantitation of vanillin in extracts.

Archer (40) developed an HPLC procedure with phenoxyacetic acid as an internal standard. He quantitated p-hydroxybenzoic acid, vanillic acid, p-hydroxybenzaldehyde and vanillin in a series of oleoresins, vanilla extracts and vanilla essences. From this data, he suggested a range of ratios for authentic essences.

In 1975, Martin and co-workers (41) proposed the use of identification ratios for determining authentic products. He demonstrated that from the analyses of vanillin, potassium, inorganic phosphates, and nitrogen and the interrelationship of their ratios, the BATF could not only determine adulteration, but also the strength of an extract (units of vanilla beans per gallon). Martin and co-workers (42) elaborated further on this by including the analyses of the amino acids: alanine, valine, glycine, isoleucine, leucine, proline, threonine, serine, methionine, hydroxyproline, phenylalanine, aspartic acid, glutamic acid, tyrosine, lysine, histidine, arginine and tryptophan.

Martin's analyses for the 1975 vanilla crop found the lead number for Madagascar, Mexican and Comoros extracts to be approximately 1.0, whereas Java extracts were around 1.5 and Tahitian about 0.8.

Bricout (43), and later, Hoffman and Salb (44) investigated the high resolution mass spectrophotometric method based on the analyses of the stable isotopes of carbon (C-12 and C-13). Vanillin from vanilla bean origin as a result of the biosynthesis, has a greater C-13 concentration, and therefore, less negative C-13/C-12 ratio than vanillin from synthetic sources—eugenol/clove oil, guaiacol or lignin (Table 3, Fig. 14). However, this was circumvented through the use of synthetic vanillin doped with C-13 labeled material. In an attempt to stay at least even with the fraudulent vendors, additional techniques were developed to detect each carbon position which could be labelled (45–47). In addition, several workers, including a group at the University of Georgia (CAIS) funded by FEMA, are looking at

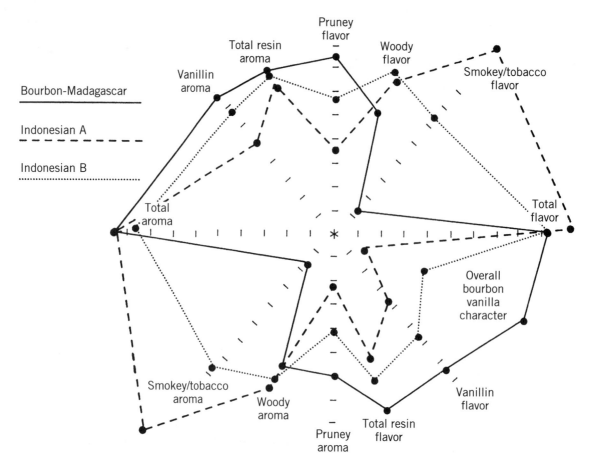

Figure 16. Vanilla aroma and flavor profiles for Bourbon and Indonesian pure vanilla extracts; tested at 1.0% in room temp. spring water, Apr. 25, 1990.

Table 4. Flavor Characteristics of Pure Vanilla Extract

Flavor character	Bourbon beans	Java beans	Tahitian beans
Vanillin	Slight	Very slight	Slight
Resinous/leathery	Slight-moderate	Moderate	Very slight
Woody	Slight	Moderate	Very slight
Pruney	Slight	Very slight	Very slight
Fruity	Very slight	None	Moderate
Chocolate	Very slight	Very slight	None
Smokey–tobacco	None	Moderate	None
Bourbon–rummy	Moderate	Very slight	Slight
Sweet–floral	Very slight	None	Moderate

Note: These descriptions reflect the norm. Flavor quality does very between lots, years, and manufacturers, depending upon a variety of factors (see text).

other isotopic techniques—C-14 (radiocarbon), D/H and 0-16/0-18, each merely a stop gap; none absolute (48).

A battery of tests is possibly the only solution to eliminating most of the fraud. Martin and co-workers (49) demonstrated this technique when they coupled their vanillin/potassium ratios with the stable carbon isotopic method. Fayet and co-workers (50), coupled stable isotope ratio analyses of both vanillin and *p*-hydroxybenzaldehyde and

HPLC analyses of vanillin, *p*-hydroxybenzaldehyde, *p*-hydroxybenzoic and vanillic acids.

Toulemonde (51) investigated the application of high resolution deuterium NMR for the detection of adulterated vanilla products and postulated that the aldehydic deuterium should be investigated further. Maubert and co-workers (52) in France developed site-specific isotopic fractionation-nuclear magnetic resonance or SNIF-NMR. This technique uses multivariant analysis of the deuterium concentration of each site on the vanillin molecule in order to differentiate synthetic and natural materials.

Sensory

If beauty is in the eye of the beholder, then flavor quality is in the mouth of the consumer! The flavor of vanilla extracts does vary considerably, depending primarily upon country of origin, but also upon crop-year, curing techniques used, storage conditions, extraction method and age of the vanilla extract itself. It is difficult to state what is the best quality vanilla, as taste is largely a matter of personal opinion. Most experts generally consider the flavor of a high vanillin ($\geq 0.20\%$), high moisture ($\geq 20\%$) bourbon bean to deliver the best quality extract. The lower the vanillin content, the lower the flavor quality of the bean will be, not just because of the vanillin itself, but also due to

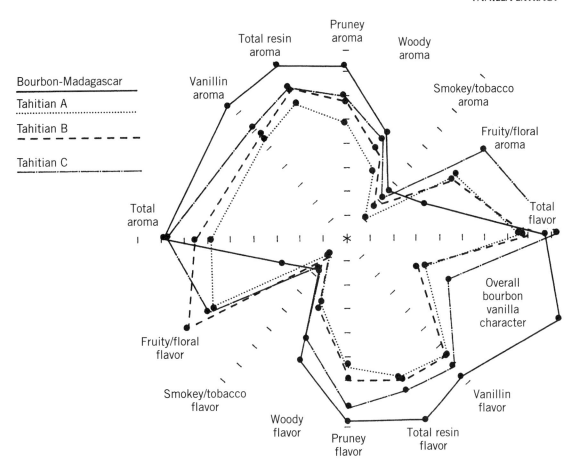

Figure 17. Vanilla aroma and flavor profiles for Bourbon and Tahitian pure vanilla extracts; tested at 1.0% in room temp. spring water, May 23, 1990.

the other flavor notes which develop along with vanillin during curing. These flavor notes include characteristic woody, pruney, resinous, leathery, floral and fruity aromatics. These flavor characteristics do not develop and/or are lost, in over-dried (low moisture) beans.

With undue emphasis, vanillin content is generally used as the prime indicator of flavor quality of beans and extracts. As a result, this focus on vanillin content has distracted many in the vanilla business from due consideration of the remaining aromatics. Collectively, these other flavor notes such as pruney, woody, floral, fruity and rummy are even more important than vanillin character alone. This point is illustrated quite clearly in Mexican vanilla bean extract which is traditionally quite low in vanillin yet superior in flavor quality. On the other hand, most imitation vanilla extracts are very high in vanillin content, but lacking in the remaining full-bodied character notes (Fig. 15). Vanillin alone is probably responsible for no more than 25% of the character of quality vanilla extract (18).

Over the past 30 years, the best quality vanilla beans have come most consistently from the islands of Madagascar, Comoros and Reunion ("Bourbon" beans). The quality of beans from the Indonesian islands has been generally low, but that of some lots has improved significantly over

the past 2–3 years. The quality of Java beans has typically been low for several reasons: (1) the beans are picked too green for optimal formation of glucovanillin, beta-glucosidase and other flavor precursors, and/or (2) the beans pick up a typical smokey-tobacco flavor during curing, and/or (3) the beans are shipped without the conditioning phase. The quality of beans from certain growers/processors in Indonesia is improving due to the implementation of parts of the quick cure (McCormick) process (Table 1), allowing beans to ripen on the vine and to improved handling in general. However, the majority of beans coming from Indonesia today are still lower in vanillin content and exhibit an undesirable smokey flavor (Fig. 16).

Tahitian vanilla has a very different flavor profile in comparison to *vanilla planifolia* (Table 4). It is very floral, fruity/cherry and perfumey, and less resinous/leathery than *vanilla planifolia*. Tahitian vanilla extract is not easily found in the United States because production volumes of the beans are relatively small and the flavor is not substitutable for *vanilla planifolia* (Fig. 17). Mexican vanilla beans are also rarely used in the United States due to high price and low production volumes. They produce a relatively good quality woody/resinous flavored extract (Fig. 18) although lower than Bourbon in vanillin content (0.14–

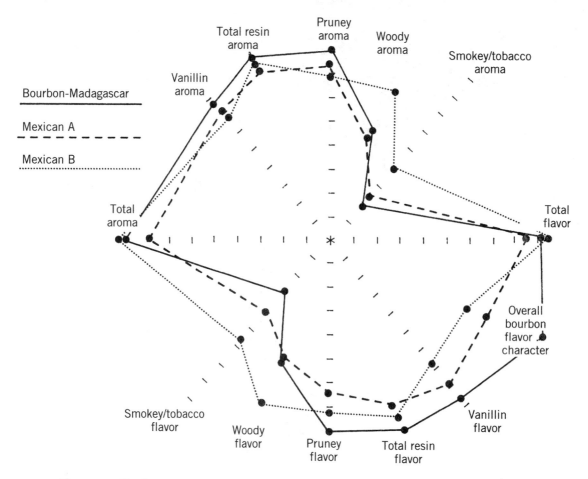

Figure 18. Vanilla aroma and flavor for Bourbon and Mexican pure vanilla extracts; tested at 1.0% in room temp. spring water, June 20, 1990.

0.17). These fine quality Mexican extracts should not be confused with the inexpensive, often coumarin-containing imitation "extracts" from Mexico which are popular with U.S. tourists.

Descriptive flavor profiles (Figs. 16–18) illustrate the differences in flavor notes and flavor balance between vanilla extract samples made from beans of different origin.

The flavor of vanilla extracts improves with aging. A minimum of 60 days aging of a finished extract is advisable to develop the resinous, bourbon character fully. The extract continues to improve over a several year period until it more resembles a delightful liquor than vanilla extract. Although aging improves a good vanilla extract, it does not compensate for poor starting material (poor beans). Likewise, aging of good beans improves their quality, but aging does not compensate for or improve poor quality beans.

THE VANILLA MARKET

The bulk of global vanilla trade is in the form of whole or cut cured beans. Further processing into vanilla extract is generally done in the consuming countries. The bean itself is used in Europe to flavor home recipes while in North

America the pure extract of vanilla is widely used by retail customers. In Japan, vanilla remains virtually undiscovered. Imitation and artificial vanilla flavors (vanillin-based) are heavily used in the North American and European industrial markets.

Madagascar normally produces at least 70% of the world supply of cured vanilla beans (53). The United States is the largest consumer of vanilla bean exports, importing more than 50% of total worldwide production, followed by France at 15–20% and West Germany at 10–15% (53). Over the past decade, the proportion of vanilla beans used by the United States that are from Madagascar has dropped consistently from 60–70% of vanilla bean imports to 38% in 1989, while imports of Indonesian beans have increased from less than 20% to almost 50% in 1989. Less than 10% of United States beans comes from the Comoros and Reunion islands and negligible amounts from other sources (Table 5).

Historically, wide fluctuations in supply and thus, price, have characterized the vanilla bean market. Over the past 30 years, U.S. bean prices have ranged from $2/lb up to $90/lb (2,54). Traditionally, the Tahitian beans are the most expensive; the Java beans the least expensive in price, 1990–1991 U.S. bulk prices for cured beans were (2):

Table 5. Vanilla Import Figures (metric tons)

Country of origin	1980[a]	1981[b]	1982[b]	1983[b]	1984[b]	1985[b]	1986[c]	1987[c]	1988[c]	1989[c]	1990[c]
Madagascar	205	438	580	651	603	462	672	820	578	421	398
Indonesia	90	123	103	240	192	184	306	443	433	526	482
Comoros and Reunion	31	64	99	59	34	87	49	97	185	107	59
Malaysia			56	12							
France	7	4	23	2							
Mexico						2.78	1.58	11	10	16	1.6
Other Pacific Islands (Tonga)						6.72	14	10	7.68	28	2.6
French Pacific Islands						1.04	2.28	4.48	2.59	8.67	3.8
French Indian Ocean Areas							0.75	1.22			5.0
Costa Rica								0.16			
Argentina							2.02				
Pacific Islands including Caroline							0.05				
Belgium									2.0		
New Caledonia									0.51	0.34	
Total	343	640	884	977	841	743	1048	1387	1219	1107	975

[a]Data from the General Agreement on Tariffs & Trade (GATT), *Spices, A Survey of the World Market*, 1982.
[b]Data from the General Agreement on Tariffs & Trade (GATT), *Spices, A Survey of the World Market*, 1986.
[c]Data from the U.S. Department of Commerce, Bureau of Trade Statistics, 1987–1989.

Java $8/lb–24/lb (depending upon flavor quality)

Tonga $28.00–30.00/lb

Comoros 35.50/lb

Madagascar $37.50/lb

Tahitian $55.00/lb

In order to stabilize vanilla prices, the Vanilla Alliance was formed with Madagascar, the Comoros and Reunion (Bourbon Islands) as members. This organization regulates prices and inventories of Bourbon beans. The pricing of Bourbon vanilla is set by the producing countries with little or no negotiation possible. In Indonesia and other pro-

Table 6. Vanilla Extract Usage

Market/application	Usage level[a]	% of supply used in this application
Food service		5
Retail		20
Industrial		75
Ice cream	1 oz/10 gal mix (10×)	30
	2.5–3.0 oz/10 gal mix (5×)	
	8–12 oz/10 gal mix (1×)	
	1 oz/10 gal mix (1×)	
Soft beverages		17
Cola	0.05% (1×)	
Cream soda	0.1% (1×)	
Alcoholic beverages	0.25% (1×)	11
Yogurt	1 oz/10 gal mix (10×)	10
	2.5–3.0 oz/10 gal mix (5×)	
	14 oz/10 gal mix (1×)	
	1 oz/10 gal mix (1×)	
Others		7
Cakes	0.15–0.2% (5×)	
	1.0% (1×)	
Candy, cream	0.5% (1×)	
Cookies	0.5% (1×)	
Cream sauces	0.1% (1×)	
Frosting	0.1% (10×)	
	0.5% (1×)	
Pudding	1.0% (1×)	
	0.5% (1×)	
Cereal	0.25% (10×)	
	2.0% (1×)	
Tobacco products	0.04% (1×)	

Source: McCormick & Co., Inc. Industrial Marketing Estimates, 1988.
[a]10× = 10 fold concentration, 5× = 5 fold concentration, etc.

ducing countries, vanilla trade is still conducted through traditional spice trading channels (shippers, brokers, dealers, merchants).

Roughly 75% of the 3.2 million gallons of U.S. vanilla extract is distributed through industrial channels, 20% through retail and 5% through Food Service (Table 6). Industrially, pure vanilla extract's primary use is in dairy desserts, particularly ice cream (40%). Soft beverages account for 23% of the industrial market and alcoholic beverages for 15%. The retail consumer of vanilla extract uses it primarily for baking of cookies and cakes.

Ice cream and ice cream novelties are clearly the largest category of vanilla extract usage in the United States. Category I and II ice creams rely in total or in part on pure vanilla extract from a legal as well as a marketing standpoint (see Regulatory section). About one-third of all ice cream is vanilla flavored. Additionally, vanilla is found as an ingredient in other flavors such as chocolate, strawberry, fudge-swirl, butter pecan, etc.

The soft beverage industry uses vanilla extract in cola, cream, pepper-type, root beer and other sodas.

Natural flavors such as vanilla extract are required by BATF regulations for spirits and specialty liquors. Vanilla can be used to round-out harsh flavor profiles in whiskey products. The alcoholic beverage industry generally uses the more concentrated forms of vanilla extract; 3-fold, 5-fold and 10-fold.

Cereals, bakery products, confectionery and tobacco products also use vanilla extract, although they use more artificials or imitations than natural. The yogurt industry, which is rapidly growing, is another major user of vanilla.

Table 6 summarizes major markets and usages for pure vanilla extract in 1988–1989. Vanilla has many other nontraditional uses, such as its use as a paint additive to improve the aroma of a freshly painted room and for grafting an orphaned lamb to an ewe by dabbing the ewe's nose and the lamb's rump with vanilla!

BIBLIOGRAPHY

This article is used with permission. Copyright 1990, McCormick & Company, Inc.

1. N. Childers, H. Cibes, and E. Hernandez-Medina, "Vanilla—The Orchid of Commerce," in C. Withner, ed., *The Orchids–A Scientific Survey*, Ronald Press Company, New York, 1959, pp. 477–508.

2. B. Kaestner, Director of Spice Procurement, McCormick & Co., Inc., Personal interview, April 1990.

3. L. Knudson, "Germination of Seeds of Vanilla," *American Orchid Society Bulletin* 19, 241–247 (1950).

4. D. Correll, "Vanilla—It's Botany, History, Cultivation and Economic Import," *Economic Botany* 7, 291–358 (1953).

5. F. Arana, *Vanilla Curing*, Circular #25, Federal Experiment Station, Washington, D.C., May 1945.

6. M. Jones and G. Vicente, "Criteria for Testing Vanilla in Relation to Killing and Curing Methods," *J. Agr. Res.* 78, 425–450 (1949).

7. M. Anzalone, Specifications Supervisor, McCormick & Co., Inc., personal interview, 1990.

8. K. Lentz, Laboratory Technician, McCormick & Co., Inc., personal interview, 1990.

9. F. Arana and A. Kevorkian, "Relation of Moisture Content to Quality of Vanilla Beans," *The Journal of Agriculture at the University of Puerto Rico* 27, 105–114 (July 1943).

10. "Biotech Vanilla's Promise Still Remains Down the Road," *Chemical Marketing Reports*, 36 (March 1990).

11. J. Sullivan, retired Director of Technical Services, McCormick & Co., Inc., personal interview, 1990.

12. U.S. Pat. 2,621,127, (Dec. 9, 1952), L. Towt (to McCormick & Co., Inc.).

13. U.S. Pat. 2,835,591 (May 20, 1958), R. Graves, R. Hall, and A. Karas (to McCormick & Co., Inc.).

14. U.S. Pat. 3,663,238 (May 16, 1972), A. Karas et al., (to McCormick & Co., Inc.).

15. F. Arana, *Vanilla Curing and Its Chemistry*, Bulletin #4, Federal Experiment Station of the USDA, Mayaguez, Puerto Rico, 1944.

16. F. Arana, "Action of a Beta-Glucosidase in the Curing of Vanilla" *Food Research* 8, 343–351. (July–Aug. 1943).

17. A. Karas, retired Principal Scientist, McCormick & Co., Inc., personal interview, 1990.

18. E. Merwin, retired Principal Scientist, McCormick & Co., Inc., personal interview, 1990.

19. J. Broderick, "The Chemistry of Vanilla," *Coffee Tea Ind. Flavor Field* 78, 74 (1955).

20. C. Gnadinger, "Identification of Sources of Vanilla Extracts," *Ind. Eng. Chem.* 17, 303 (1927).

21. P. M. Golby, "Research on the Principal Odor Constituents in Vanilla" *J. Pharm. Chem.* 34, 401 (1858).

22. I. Klimes and D. Lamparskey, "Vanilla Volatiles—A Comprehensive Analysis," *Int. Flavours Food Additives* 7, 272–291 (1976).

23. W. Galetto and P. Hoffman, "Some Benzyl Ethers Present in the Extract of Vanilla (*Vanilla plainfolia*)," *J. Agr. Food Chem.* 26, 195–197 (1978).

24. K. Shulte-Elte, "Vitispiranes, Important Constituents of Vanilla Aroma," *Helvetica Chim. Acta* 61, 1125–1133 (1978).

25. E. Feyertag and R. Hutchins, "Isolation and Tentative Identification of a New Constituent of Bourbon Vanilla Bean Extract," *J. Food Chem.* 7, 311–315 (1981).

26. H. Walbaum, "Das Vorkommen von Anis-alkohol, und Anisaldehyd in den Fruchten der Tahitivanille," *Wallachs Fastschrift*, 649–653 (*Ber. Schimmel*, 140–141 Oct. 1909).

27. J. W. Purseglove, E. G. Brown, C. L. Green, and S. R. J. Robbins, in *Spices*, Vol. 2, *Vanilla*, Longman, Inc., New York, 1981.

28. K. Riley and D. Kleyn, "Fundamental Principles of Vanilla/Vanilla Extract Processing and Methods of Detecting Adulteration in Vanilla Extracts," *Food Technol.* 43, 64 (1989).

29. A. Wenton and E. Berry, *The Chemical Composition of Authentic Vanilla Extracts, Together With Analytical Methods and Limits of Constants*, U.S. Department of Agriculture, Bureau of Chemistry, Bulletin No. 147, 146–159 (1912).

30. W. Denis, "The Detection of Prune Juice and Caramel in Vanilla Flavoring Extracts," *The Journal of Industrial and Engineering Chemistry* III, 254–255 (1911).

31. J. Broderick, Jr., "Vanilla Color," *The Flavor Field* (Feb. 1954); J. Broderick, Jr., "Lead Number and Vanilla Quality", *The Flavor Field* (March 1954); J. Broderick, Jr., "The Relationship of Vanilla Resin to Vanilla Color," *The Flavor Field* (May 1954).

32. C. Purcell, "More On The Lead Number," *The Spice Mill Flavors*, 694–695 (1934).

33. W. Stahl, W. Voelker, "Analysis of Vanilla Extracts II. Preparation and Reproduction of Two Dimensional Fluorescence Chromatograms," *J. AOAC* **43**, 606–610 (1960).

34. W. Stahl et al., "Analysis of Van. Extracts III. Use of Paper Chromatography in the Analysis of Pure and Adulterated Van. Ext." *J. AOAC* **44**, 549 (1961).

35. J. Fitelson and G. Bowden, "Determination of Organic Acids in Vanilla Extracts by GLC Separation of Trimethylsilylated Derivatives," *J. AOAC* **51**, 1224 (1968).

36. K. Schoen, "Organic Acids in Vanilla by Gas-Liquid Chromatography: Interpretation of Analytical Data," *J. AOAC* **55**, 1150 (1972).

37. K. Helrich, *Official Methods of Analysis 1990*, 36, 15th ed. *Flavors*, AOAC, Arlington, Va., 1990.

38. A. Hermann and M. Stockli, "Rapid Control of Vanilla-Containing Products Using High Performance Liquid Chromatography," *J. Chromatog.* **246**, 313–316 (1982).

39. P. Guarino and S. Brown, "Liquid Chromatographic Determination of Vanillin and Related Flavor Compounds in Vanilla Extract: Cooperative Study," *J. AOAC* **68**, 1198 (1985).

40. A. Archer, "Analysis of Vanilla Essences by High Performance Liquid Chromatography, *J. Chromatog.* **462**, 461–466 (1989).

41. G. Martin et al., "Determining the Authenticity of Vanilla Extracts," *Food Technol.* **29**, 54 (1975).

42. G. Martin et al., "Determining the Authenticity of Vanilla Extracts," *J. Food Science* **42**, 1580–1586 (1977).

43. J. Bricout, J. C. Fontes, and L. Merlivat, "Detection of Synthetic Vanillin in Vanilla Extracts by Isotopic Analysis," *J. AOAC* **57**, 713 (1974).

44. P. Hoffman and M. Salb, "Isolation and Stable Isotope Ratio Analysis of Vanillin," *J. Agr. Food Chem.* **27**, 352 (1979).

45. D. Krueger and H. Krueger, "Carbon Isotopes in Vanillin and the Detection of Falsified "Natural" Vanillin," *J. Agr. Food Chem.* **31**, 1265–1268 (1983).

46. D. Krueger and H. Krueger, "Detection of Fraudulent Vanillin Labeled with ^{13}C in the Carbonyl Carbon," *J. Agr. Food Chem.* **33**, 323:25 (1985).

47. J. Bricout et al., "New Possibilities for the Analysis of Stable Carbon Isotopes in Monitoring Vanilla Quality," *Ann. Falsif. Expert. Chem. Toxicol.* **74**, 691 (1981).

48. R. Culp and J. Noakes, "Identification of Isotopically Manipulated Cinnamic Aldehyde and Benzaldehyde," *J. Agr. Food Chem.* **38**, 1249–1255 (May 1990).

49. G. Martin, F. Alfonso, D. Figert and J. Burggraff, "Stable Isotope Ratio Determination of the Origin of Vanillin in Vanilla Extracts and Its Relationship to Vanillin/Potassium Ratios," *J. AOAC* **64**, 1149–1153 (1981).

50. B. Feyet et al., "New Analytical Specifications In The Study of Vanilla Beans," *Analysis* **15**, 217–226 (1987).

51. B. Toulemonde, "Differentiation Between Natural and Synthetic Vanillin Samples Using ^{2}H-NMR," *Helvetica Chim. Acta* **66**, 2342 (1983).

52. C. Maubert et al., "Determination of the Origin of Vanillin By Multivariate Analysis of the Site-specific Natural Isotope Fractionation Factors of Hydrogen," *Analysis* **16**, 434–439 (1988).

53. N. Anand and A. Smith, *The Market for Vanilla*, Report of the Tropical Development and Research Institute, London, England, G198, 1986, V + 33p.

54. R. Bracecklein, Team Leader Extract Manufacture, McCormick & Co., Inc., personal interview, 1990.

GENERAL REFERENCES

Ref. 27 is a good general reference.

A. K. Balls and F. E. Arana, "The Curing of Vanilla," *Industrial and Engineering Chemistry* **33**, 1073–1075 (1941).

M. Mariani, A. Scotti, and E. Colombo, "Vanilla: From the Origin to Its Analysis," in Italian and English, *Quintessenza* **17**, 3–25 (1982).

P. Rain, *Vanilla Cookbook*, Celestial Arts Publisher, Berkeley, Calif. 1986.

F. Rosengarten, Jr., *The Book of Spices*. Pyramid Communications, Inc., New York, 1973.

R. Sax, "All About Vanilla," *Bon Appetit*, 132–138 (March 1986).

Standard #5565, International Standards Organization. *Vanilla (Vanilla Fragrans, Salisbury Ames) Specification*, 1982-12-01, Ref. #ISO 5565-1982(E), 1982.

R. Theodose, "Traditional Methods of Vanilla Preparation and Their Improvement," *Trop. Sci.* **15**, 47–57 (1973).

MARIANNE H. GELLETTE
PATRICK G. HOFFMAN
McCormick & Co., Inc.
Hunt Valley, Maryland

See also FLAVOR CHEMISTRY.

VEGETABLE DEHYDRATION

Dehydration involves the removal of the water normally present in the tissues by evaporation. Vegetable dehydration has one or both of the following purposes: (*1*) to preserve the perishable raw commodity against deterioration, and (*2*) to reduce the cost of packaging, handling, storing, and transporting the material by converting it to a dry solid, thus reducing its weight and also, usually but not necessarily, its volume.

Essential to the dehydration process is the application of heat to vaporize water, and some means of removing water vapor after its separation from the plant tissue. The removal of water involves mass transfer, and the application of heat in some manner also involves heat transfer. Three basic methods of heat transfer are used in dryers in varying degrees of prominence and combinations. These are convection, conduction, and radiation. Dehydration terms are described in the article FRUIT DEHYDRATION.

Dehydrated vegetables are an important and steadily growing portion of the food industry. Vegetables commonly dehydrated include potatoes, beets, cabbage, carrots, parsley, chili peppers, horseradish, onion, garlic, bell peppers, green beans, turnips, parsnips, sweet corn, mushrooms and celery.

The U.S. vegetable-dehydration industry may be divided into four groups:

1. Onions and garlic, almost exclusively in California

2. Potatoes, primarily in Idaho and also in Washington, Oregon, and a few other states

3. All other vegetable products, almost exclusively in California

4. Freeze-drying of certain high-value crops, mainly in California and Oregon

Dehydrated vegetables are used in dry soup mixes, canned soups and sauces, frozen entrées, processed meats, baby foods, dairy products, snack foods, and seasoning blends.

PREDRYING TREATMENTS

Almost all vegetable crops are dehydrated fresh from the field, although some items such as green beans may be dehydrated from the frozen state. Harvesting, washing, sorting, grading, peeling, and cutting operations are essentially identical to those described for canned or frozen foods, and the raw material is handled by a similar method.

Some exceptions to standard operating conditions are that most vegetables are blanched before drying; beets are not peeled—all dehydrated beets are sold as dried ground powder; and parsley is dried in a specially designed hot-air tower for a short time to discharge to a more conventional dryer for final moisture equalization.

Enzyme Inactivation

Vegetables contain catalase, peroxidase, polyphenoloxidase (PPO) and many other enzymes. At the time of cutting and peeling, enzymatic reactions are accelerated and discoloration will often occur. Enzymes can be inactivated by heating or acidification to low pH values. Inactivation by heat treatment (blanching) is usually done by immersing the vegetables in hot water at 95 to 100°C for a few minutes, or exposing it to steam. Vegetables have also been successfully blanched by hot air, infrared, or microwave radiation. In addition to enzyme inactivation, blanching serves to remove the gases between the tissues, soften tissues, decrease the amount of microorganisms, and cleanse the vegetables.

Commercially continuous blanchers are favored over the batch type, which involve a 2- to 10-min exposure to live steam. A conveyor or an exhausting box is suitable for the continuous blanching process. The effectiveness of blanching is controlled by measuring the activity of catalase in cabbage and PPO in other vegetables, and a guaiacol test is used as an index for the activity of peroxidase. If fresh vegetables are not properly blanched, enzymatic deterioration of the dehydrated products may take place during processing or storage or after reconstitution. Flavor changes resulting from lipid oxidation may take place during storage. Browning reactions may occur if the dried product is stored at temperatures higher than 200°C and when the moisture content is higher than 5%.

Sulfuring

Vegetables undergo browning due to enzymatic or nonenzymatic reactions that occur during processing and storage. Until recently the addition of sodium sulfite or metabisulfite to cut vegetables before dehydration has been a standard practice to control both enzymatic and nonenzy-

matic browning. Sulfites act as a PPO enzyme inhibitor and also react with intermediates to prevent pigment formation. Also, sulfiting agents control the growth of microorganisms, act as bleaching agents, and provide other technical functions. Sodium sulfite, sodium bisulfite, sodium metabisulfite, potassium bisulfite, and potassium metabisulfite salts are generally recognized as safe (GRAS) compounds for use in foods by the FDA. For small-scale operations, blanched vegetables may be dipped for a specified time in a sulfite salt solution. Spraying, however, is the most practical method. Cabbage, potatoes, and carrots are routinely sulfite treated before dehydration. The highest sulfite levels are required for cabbage, between 750 and 1500 ppm. Limits of 200 to 500 ppm are imposed on potatoes and carrots. Soaking in 500 to 1000 ppm sodium bisulfate solution for 5 to 10 min before slicing retained good color of mushrooms. The residual sulfur dioxide after such treatment was less than 10 ppm in the dry mushrooms.

Sulfites have been associated with allergylike reactions in some asthmatics, prompting limit of their use. Because of these reactions, and the legal requirement to declare on the label of packaged food that the product has been treated with sulfites, food processors are seeking alternative methods for producing high-quality dehydrated vegetables. However, sulfites are multifunctional agents. The search for alternatives has yielded compounds that are effective substitutes for only one or two of the functionalities obtained with sulfites. Perhaps the best alternative to sulfiting agents is the use of ascorbic acid or erythorbic acid as a browning inhibitor. Blends of citric/ascorbic or citric/erythorbic acids are also effective and somewhat less expensive. These compounds can be applied to prepared vegetables by immersion in a water solution containing 1% citric and 0.5 to 1% ascorbic or erythorbic acid for 30 to 60 s. The penetration of these browning inhibitors may be enhanced by vacuum infiltration of treatment solutions, which also removes air from within the product's void spaces.

DRYING METHODS

Selection of a drying method for a given food product is determined by quality requirements, raw material characteristics, and economic factors. In general, dryers are suitable for either batch or continuous operation. Batch-operated equipment usually is related to small production runs or to operations requiring great flexibility.

As most vegetables are dehydrated to under 5% (low moisture) level, sun drying with vegetables has only limited application, except for a small quantity of home-dried mushrooms and limited use of solar dryers for carrots, celery, and the like. Most dehydrated vegetables are processed by the forced air–drying method. A disadvantage of air-dried vegetables is their low rate of rehydration. The process of rehydration after drying is not a simple reversal of the drying mechanism. Some of the changes produced by drying are irreversible; also, the swelling of outside layers as water is reabsorbed puts severe stress on the softened outside layers. Irreversible changes of the colloidal components of vegetable tissue occur if the material is held for a period of time at high temperature, even if the exposure is insufficient to produce scorching. However, depending on the particular use of dehydrated vegetables, slow rehydration may not be of critical importance. Large quantities of carrots, onions, garlic, potatoes, and other vegetables are forced air–dried. For rapid rehydration and for highly heat-sensitive vegetables (chives, mushrooms, etc), the freeze-drying process may be justified.

For liquid, purees, or the flowable suspension form of vegetable material, spray drying is the most important dehydration method. Among vegetable substances, concentrated tomato puree is commercially spray-dried. Drum drying, a high-temperature, short-time method, is an inexpensive way to dry pureed foods. Drum drying is successfully used for drying mashed potatoes and tomato concentrates.

The common kinds of vegetable-drying equipment are cabinet, tunnel, continuous conveyor belt, pneumatic, belt trough, fluidized bed, bin, drum, spray, and freeze dryers.

Cabinet Drying

A cabinet dryer is essentially a small batch tray dryer. The trays are made of stainless angle iron and stainless net or apertured plate with a height of about 5 cm. The dryer usually loads more than 10 trays vertically. The heat source usually is a steam coil or electrical coil through which the air from a fan passes across the trays loaded with the raw vegetables. Cabinet dryers are usually held at a constant temperature, though humidity may decrease during the process of drying. Airflow may be across or through the trays with or without recirculation. Cabinet dryers are flexible as to type and size of operation, although production may be lower and labor requirements greater than for a tunnel dehydrator.

Tunnel Drying

A tunnel dryer is basically a group of truck-and-tray dryers operated in a programmed series so as to be semicontinuous. Truckloads of freshly prepared material are moved at intervals into one end of the long, closely fit enclosure; the string of trucks is periodically advanced a step; and the dried truckloads are removed at the other end of the tunnel. The hot drying air is supplied to the tunnel in any of several ways: counterflow, concurrent or parallel flow, center exhaust, multistage, and compartment arrangements.

Continuous Conveyor Drying

The continuous conveyor dryer consists of an endless belt that carries the material to be dried through a tunnel of warm circulating air. The speed of the conveyor is variable to suit both the product and the heat conditions. Most dryers have several sections, allowing different flow rates, humidities, and temperatures to be set to each section and the rotation of the product when it moves from one section belt to the next. This drying method has the advantage of essentially automatic operation, which minimizes labor requirements.

These dryers are being used extensively in vegetable dehydration plants for such crops as beets, carrots, onions,

garlic, and potato pieces. The conveyor dryer is best adapted to the large-scale drying of a single commodity for the whole operating season; it is not well suited to operations in which the raw material or drying conditions are changed frequently.

Pneumatic Drying

The method of drying finely divided solids while they are being conveyed in a hot airstream has been used for the first stage or rough drying of granulated cooked potatoes in the manufacture of potato granules. The partial drying desired takes place in a few seconds for finely granulated wet potato, whose initial moisture content generally has been reduced to around 35 to 40% by an admixture of fresh mash with recycled dry product. Most of the material is finer than 60 mesh, but a small portion may be 20 mesh or coarser. The moist granulated material is usually sucked into the conveying pipeline at a low-pressure throat. Air velocity in the pipeline is just enough to overcome the free-fall velocity of the coarser particles to keep them suspended while avoiding unnecessary abrasion damage to the delicate potato cells. The pipe is sufficiently long to give the necessary retention time at that air velocity.

The operating section is generally terminated by a diverting section that helps to separate the light, completely dry particles from the heavier, still somewhat moist ones and gives the latter more time in the hot air. The conveying riser ends with a cyclone collector where the granular solid is separated from the outgoing airstream.

Belt-Trough Drying

Belt-through drying is a type of continuous trough-flow drying that can be used for a wide range of vegetables that have a firm texture in piece form. It consists of a wire-mesh belt running on sprockets mounted on three driveshafts. During part of its path of travel, the belt forms a trough whose bottom rests on an inclined, flat-surfaced air gate. The material to be dried is continuously cascaded in this trough while a current of hot air blows up through the grade and bed of material. High operating temperatures result in high drying rates without causing appreciable heat damage to vegetable pieces because the constant agitation keeps any individual piece from being exposed to the very hot dry air for more than a moment. Then each piece is surrounded by air at much lower temperatures for a longer time before again moving into the zone of intense drying.

Fluidized-Bed Drying

In fluidized-bed drying, heated air is blown up through the food particles with just enough force to suspend the particles in a gentle motion. Semidry particles such as potato granules gradually migrate through the apparatus until they are discharged dry. Heated air is introduced through a porous plate that supports the bed of granules. The moist air is exhausted at the top. This process is continuous, and the depth of the bed and other means can regulate the amount of time that the particles remain in the dryer. This type of drying can be used to dehydrate peas and other vegetable particles also.

Bin Drying

The bin drying method is widely used as a finishing step in the dehydration of piece-form vegetables when they constitute a deep bed of granular, nonsticky material. Bin dryers are used to complete the drying operation after most of the moisture has been removed by other drying techniques. Typically, they reduce the moisture content of a partially dried, 10 to 15% moisture cut vegetable to 3 to 6% or even lower in the case of onion slices and possibly cabbage shreds.

A characteristic unit consists of a metal or wooden box equipped with an air inlet at the bottom and a wire mesh deck or false bottom with an air supply duct below it, arranged so that warm dry air can be passed through the nearly dry product piled on the deck.

Drum Drying

Drum dryers are designed to handle solutions, vegetable purees, and pastes. Drum drying processes are used successfully in the production of powdered tomato, pea and bean soups, potato flour, tomato juice flakes, mashed potato flakes, and sweet potato flakes.

A drum dryer consists of one or two hollow drums, which are fitted so that a heating medium, usually steam, or a heat-transfer liquid can be circulated through them. The drums are mounted to rotate about the symmetrical axis and are customarily driven with a variable-speed mechanism. A feeding device applies a thin, uniform layer of the material to be dried on the hot drum surface. A knife is fitted to the drum at an appropriate location. The material is dried as the heated drum rotates toward the blade, which scrapes the thin layer of dry material from the drum surface.

Spray Drying

In spray drying, the fluid to be dried is dispersed into a stream of heated air. The dry particles are separated from the air and collected, and the moist, cold air is exhausted. The process is used to dehydrate vegetable juices. Materials that are easily damaged by heat or oxidation can be handled in a properly designed and operated spray dryer.

Freeze Drying

Freeze dehydration consists of freezing the product and then evaporating the water in the form of ice (by sublimation) directly into the vapor phase without liquid interim. Water vapor is removed from the frozen food, which is achieved by reducing the pressure in a sealed drying chamber by a vacuum pump to 50 to 1000 mm mercury. This pressure holds the vegetables below their eutectic temperature, facilitates rapid evaporation of water molecules from the frozen state, and condenses the vapor either within the chamber or in a separate chamber. Heat energy applied to the frozen product causes the water molecules to break loose and migrate to the block of ice on the condenser, where their energy is removed by cooling. Heat may be applied to the frozen product by conduction or radiation, or both.

Freeze dehydration produces one of the highest-quality food products obtainable by any drying method. The porous, nonshrunken structure of the freeze-dried vegetable facilitates rapid and nearly complete rehydration when water is added. The low processing temperatures and the rapid transition minimizes the extent of various degradative reactions associated with other drying methods and also help reduce the loss of volatile flavor compounds. However, freeze drying is an expensive way to evaporate water, and in addition freeze-dried vegetables must be protected from deterioration by adequate packaging.

PACKAGING AND STORAGE OF DEHYDRATED VEGETABLES

Oxygen concentration, moisture content, temperature, exposure to light, and the length of storage all affect deterioration of dehydrated vegetables. The relative importance of these parameters vary from commodity to commodity. For instance, carrots and other high-carotene-containing tissues have been found to be very sensitive to the presence of oxygen.

Because freeze-dried foods differ significantly in physical and chemical properties from food dried by conventional methods, they must be stored differently than other dehydrated products. Moisture absorption and changes in quality due to oxidation are the most important problems encountered in the storage of freeze-dried vegetables.

Packaging Materials

Cartons, cans, glass containers, and pouches can be used to package dehydrated vegetables. At the time of packaging, dried vegetables should not be above 4% moisture, and in some cases a maximum of 3% is preferable. Most dried vegetables are packed in paper cartons lined with some moisture-proof material, preferably in high-density polyethylene bags that are heat sealed and placed in the cartons.

To destroy insects, dehydrated vegetables may be fumigated with methyl bromide as they come off the drying trays, while they are still hot and before insects have had the opportunity to lay eggs on them. Two serious objections to the use of cartons for packing dehydrated vegetables are their susceptibility to later insect infestation and to penetration of moisture.

The ideal container for dehydrated vegetables is the key-top can such as is used for canned meats and for coffee. If the cans are reinforced to prevent collapse under vacuum, or if the cans are tightly filled so that the dried product will support the walls, the filled cans may be sealed under high vacuum. The vacuum protects against oxidative changes and will prevent insect infestation. Friction-top cans of the type used for paint are usually insectproof and reasonably moistureproof. They have been used very successfully for dried vegetables. Powdered vegetables should be packed only in airtight tin or glass containers to prevent not only insect infestation but also the absorption of moisture and subsequent caking.

Dehydrated vegetables can also be packed in 5-gal cans of rectangular shape. Glass jars are moisture- and insect-proof. Some can be sealed under vacuum, and they are preferable to other types for dehydrated foods. One great advantage of glass is the visibility of the products; another is the resealability of most glass jars. However, the greater weight and fragility of these containers are disadvantages. Plastic pouches, laminated pouches, and aluminum–film combination pouches may be used for packaging dehydrated vegetables.

Freeze- or vacuum-dried foods must be kept in airtight containers made from a material that is impermeable to oxygen and light. Transparent packaging materials are usually unsuitable for this purpose. Because freeze-dried products are fragile, the packaging materials should protect them from mechanical damage.

In-Package Desiccants, Anticaking Agents, and Compression

Use of in-package desiccants to bring moisture content to 2% or lower should permit storage of dehydrated vegetables for an extended time. Addition of fumed silica at a concentration of 1 to 2% or calcium stearate at 0.1 to 0.2% is recommended in powdered vegetables to prevent agglomerating or caking of spray-dried tomato powder.

Dried vegetables may be compressed into dense bricks or cylinders. These can be wrapped in any suitable manner, for example, in paper with an outer aluminum foil wrap, in cellophane or plastic, or in cartons or cans. If compressed to high density, the products are resistant to insect attack and represent a substantial saving of storage space. Compression technology utilizes hydraulic press to reduce product volumes to as little as 1/16th of the original volume. Carrots, green beans, spinach, cabbage, and celery are suitable for utilizing this process.

QUALITY CONTROL

Important laboratory tests specific for dehydrated food are moisture, sulfite contents, particle size distribution, enzyme activity determinations, and behavior on rehydration.

Moisture

Moisture may be determined in about 10 min with an analytical balance, which uses an infrared lamp and a calibrated dial to indicate weight loss in powdered foods. The Karl Fischer titration method is recommended for moisture determinations on low-moisture products.

Sulfur Dioxide

Sulfur dioxide residue can be determined by the Monier-Williams method, which has a detection limit of 50 ppm. The ion-chromatographic technique is accurate to 1 ppm, and it has a short analysis time of 25 min.

Particle Size Distribution

Particle size distribution, an important characteristic of powdered products in particular, is determined by a screen analysis. In this procedure, the product is shaken for a specified time (usually 3–5 min) through a series of standard sieves and the residue on each sieve is determined.

Enzymes

The enzyme peroxidase and catalase have been used as indicators of blanching adequacy before dehydration. Some vegetables such as green peas may be overblanched when peroxidase is used as an indicator, and lipoxygenase has been proposed as a new indicator to determine the effectiveness of blanching. It takes less than half the heat to destroy lipoxygenase as it does to destroy peroxidase.

Rehydration Test For the rehydration test, a 50-g sample from tray, bin, or final package is placed in a beaker or jar. For such products as peas, potatoes, corn, and sweet potatoes about 200 mL of cold water is added; for other vegetables, 300 to 500 mL depending on the drying ratio. If the pieces float, a wire screen cut to fit the beaker will keep them submerged. After standing 12 h at room temperature the samples are drained for 3 to 4 min on a screen and again weighed. After experience with different lots, a minimum drained weight for each vegetable can be established and used as an indication of rehydration value.

DEHYDRATION PROCEDURES FOR SELECTED COMMODITIES

Potatoes

In the United States about 20% of the processing potatoes (about 2 million t in 1997) are dehydrated. To achieve maximum use and efficiency of manufacturing facilities, the processor attempts to operate the dehydration plant year-round. Thus it is necessary for large quantities of potatoes to be placed in storage at the end of each harvest. A high relative humidity must be maintained to prevent dehydration of stored potatoes, which is associated with as much as 10% loss of carbohydrates.

From storage or the field, the potatoes are directed to a water flume or other transport system by high-pressure water hoses. A metering wheel feeds the potatoes into the process system. The potatoes are washed by passing them through a rotary drum or cylindrical washer where the potatoes are scrubbed either with brushes or merely by tumbling them together. Water sprays remove additional foreign material and soil particles. Following the washer the potatoes pass over a short draining belt, which permits internal recirculation of the wash water. An inspection of the potatoes is made on the drainage belt, and the undesirable whole potatoes are removed.

Four systems are used to remove potato peel: abrasive peeling, steam peeling, lye (caustic) or wet lye peeling, and dry caustic peeling. The peel loss, including trimming, can result in a 15 to 30% loss of the potatoes processed. The trimming process should be considered part of the peel removal. In this process the presence of eyes, blemishes, and remaining peel are often detected electronically, which directs the imperfect potato to the trim table, where the imperfections are manually cut out of the potato. After being peeled, the potato is sliced or diced, depending on the desired end product. In any cutting process a number of potato cells are ruptured, releasing a considerable amount of starch. Many processors are installing hydrocyclones to remove the starch from the wash water in the form of slurry.

After the potato is peeled and sliced, the pieces are blanched to deactivate the enzymes, to remove surface air, to partially cook the potato, and to remove excess sugars from the potato pieces. The potato slices or dices are dehydrated as individual pieces in an atmospheric recirculating air tunnel or conveyor dryer. If granules or flakes are to be processed, then blanched potato pieces are mashed and conditioned before drying as flakes on a drum dryer or as granules in a fluid bed dryer.

Onions and Garlic

The harvest period is from mid-May to November for onions and from mid-June until the beginning of October for garlic. Topped and screened bulbs are transported to the plant, where they are screened to remove dirt and loose stems and then gently conveyed to sorting bins. These bins are fitted with fans that circulate warm air through the onions (from bottom to top) in order to dry remaining tops and stems, the outer layer of skin, and the bulbs themselves to prevent microbial spoilage.

The dehydration of onions involves cleaning them with both wet and dry methods. Dry cleaning is used to remove dried tops, some loose skins, and dirt. Thus machinery usually consists of a series of vibrating screens, parallel rollers, or air aspirators or a combination of all three. Dried tops are usually pinched off by a series of rollers. Wet cleaning is usually done by a series of dip-and-soak tanks and high-pressure water sprays. Hand trimming is sometimes employed to further remove tops, defective parts of bulbs, or other undesirable blemishes.

Garlic normally contains more dirt than onions do, and dry cleaning is essential. This involves a series of rollers or screens, which both screen the dirt and aspirate the skins from the bulk. A cracking device is included to break the bulbs into individual cloves. It is the cloves that are eventually sliced and dried into a finished product. The cloves are inspected visually (culls, trash, and foreign debris removed) and conveyed through a riffle washer, the purpose of which is to wash the individual cloves and separate them from small rocks and any adhering dirt. They then enter a flotation tank, where they are immersed in water. The good cloves sink, whereas defective (dry) clove and loose skin float and are simmered off as solid waste.

Onions may be size graded as a part of the dry-cleaning operation, or they may be sorted by size and quality after wet washing. In some cases the larger onions are separated from the main stream and used primarily in the production of larger onion pieces (ie, large sliced, sliced, and large chopped portion). These bulbs are normally cored by rotating circular saws, which remove the root and root crown from the bulb, and the dry outer scales are removed by revolving high-pressure washers or by a flame peeler.

Following grading, sorting, and washing, the onion bulbs or garlic cloves are conveyed to specially designed machines that slice the whole bulbs and cloves into thin layers. A belt or vibratory conveyor then transfers these layers to a continuous-belt dryer. All onions and garlic dehydrated in the United States are now dried on continuous-belt conveyors. They are usually long and multistaged with baffled chambers that blow heated and sometimes desiccated air from over and under the bed depth of the raw slices. Residence time in this type of dryer is usually 19 to 20 h, resulting in a product that has a finished moisture content of no greater than 4.25% for onions and 6.0% for garlic. Alternatively, the onion and garlic slices may be taken from the final stage of drying at a slightly higher moisture and reduced to the desired moisture content by bin drying. After dehydration, the dried slices are usually screened, milled, aspirated, separated, and ground in various mechanical combinations to achieve the final desired piece size.

Dehydrated onions are sold in a variety of sizes. Classification includes sliced, chopped, minced, granulated, and powdered. Care is taken during dehydration to insure minimal powder production because of its low value. Pieces are separated by screening or air classification. Screening, grinding, and packaging operations are carried out in special dehumidified rooms because of the hygroscopic nature of the dried onions. Air in these rooms is kept below 30% RH.

Mushrooms

Mushrooms are one of the important vegetables (in terms of value) to be preserved by dehydration. *Agaricus bisporus* is the most frequently used for drying. The process of commercial drying begins with trimming the stalks that are contaminated with soil and hard to wash, followed by soaking whole mushrooms in 0.05 to 0.1% sodium sulfite for 5 to 10 min before they are cut into slices to retard discoloration. The residual sulfur dioxide of dried mushrooms is less than 10 ppm. The mushroom is sliced longitudinally with a slicer machine to a thickness of 4 mm. Two drying methods are applied to mushrooms air drying and freeze drying. Air drying involves two stages: the first is drying at 40 to 45°C for 13 h and then at 70°C for 2 h, both with air velocity of outlet of about 70 m/min. Drying at a higher initial temperature could cause browning of the product. In freeze drying two stages are involved also. The first is to freeze the mushrooms at −34°C and then dehydrate with a freeze dryer until the moisture content of the products is reduced to 2 to 3%. The temperature may be raised gradually to about 60 to 70°C in 10 to 12 h. The drying ratio of mushrooms is 11 to 11.75 and the yield is about 8.5 to 9.0%. The dried product should be stored in airtight containers at 20°C.

Peppers

The principal dried peppers are hot red peppers for cayenne and sometimes pimientos for paprika. Ancho is the pepper that is partially or completely dried and used as the principal flavor source in chili powders. Peppers are harvested by hand. At the plant they are washed, inspected, and sorted for size and color. They are then cored and sliced and the seeds removed by screening and water sprays. Stems and core material are also removed. Sulfiting to a final sulfite level of 1000 to 2500 ppm is applied by spraying. Pimientos are not sulfited, but due to their very thick and tough skin, pimientos must be peeled before dehydration. This is accomplished in the same manner as for canning, which may involve flame, hot oil, or lye peeling.

The initial temperature in dehydrators can be as high as 76.6°C with a finishing temperature of 63°C or below. Moisture is removed so that a final level of 5% is accomplished in sweat or finishing bins. The dried product is used to flavor and improve the appearance of various processed foods. Some sliced peppers are partially dried and mixed with salt for preservation for ultimate use in various processed products. Grinding of sweet peppers for paprika and of hot peppers for cayenne pepper is carried out immediately at a very low moisture content because dried peppers rapidly absorb moisture and become tough.

Bell peppers are handpicked and delivered to the plant in large wooden bins. Bell peppers change color and solids content as they mature on the plant. The first portion of the harvest results in the prime green bell pepper. As the season grows longer, peppers remaining on the vine begin to develop yellow and yellow-orange stripes and blotches. Reaching full maturity, they develop an intense red color. Typically, then, bell peppers are run as three distinct products: green, mixed, and red cut into dice, slice, and so on. Dehydrated bell peppers are used in soup and gravy mixes, salad dressings, dip mixes, and spreads.

The peppers are initially graded according to size to facilitate cracking of the pods and later core separation. Grading is usually done by parallel rollers, the smaller vegetables dropping to another processing line. The peppers are then usually washed in immersion-type or spray-type washers and conveyed to a popper or cracker. This can be a device consisting of two closely spaced belts moving in the same direction (one over the other to pull the peppers in and crack them), or it may be two revolving wheels between which the peppers fall and are popped. Separation of pods and cores is usually accomplished by means of flotation. The very light and buoyant core floats whereas the flesh of the pepper is denser and sinks.

The cores are skimmed and go to solid waste. Inspection for defects and hand sorting of remaining attached cores is then accomplished before a final wash to remove the remaining bits of core and seeds. The peppers are diced, sliced, or cut into the desired piece size and sprayed with a sodium bisulfite solution to give a final sulfur dioxide content of 1000 to 2500 ppm. Most bell pepper products are dried on a continuous-belt dryer. The dice are removed from the belt at about 10% moisture content, and the final moisture content of 5% is achieved in bin drying. Packing is usually done in bulk (drums), although the peppers may be repacked into smaller packages at a later date.

Other Vegetable Crops

Other vegetables commonly dehydrated include beets, cabbage, carrots, parsley, horseradish, turnips, parsnips, and celery. Additionally, other vegetables such as asparagus, tomatoes, green beans, spinach, and green onion tops may be dehydrated on commercial demand. These items are commonly used as ingredients for various food specialties, baby food, and dry soup mix.

Almost all crops are dehydrated fresh from the field, although some items such as green beans may be dehydrated from the frozen state. As examples, the dehydration process for celery and carrots are described in detail.

Dehydrated celery is commercially available as 1/8-in. stalk dice, stalk and leaf flakes, stalk and leaf granules, and powder. Celery is delivered to the plant in bulk fresh from the field. The butt and leaf ends are trimmed by mechanical circular saws. Following trimming the inner yellow stalks are separated by hand and typically discarded. The celery stalks are then washed in immersion-type or spray-type washers and conveyed directly to slicers or dicers. The cut fractions are sprayed with sodium bisulfite solution (to preserve color) and fed in a steady stream to continuous-belt dryers. The dried product may be further dried in forced-air bin dryers or may be packed directly into bulk containers.

Carrots are size graded, inspected, washed, trimmed, and peeled almost exactly as for canning or freezing. After final inspection and wash, they are conveyed to a dicer or slicer. Diced carrots are sold in various commercial sizes, cross-cut slices made at right angles to the axis of the carrot, and so on. Immediately after cutting the carrot pieces are blanched, usually in a steam blancher, by being spread on a continuous stainless-steel mesh belt at a loading of about 4 lb/sq ft and heating in flowing steam for 6 to 8 min and spraying with sodium bisulfite solution of about 1% concentration. Blanching is critical, and both catalase and peroxidase enzymes must be inactivated. The carrot pieces are usually dried in conveyor or tunnel-type dehydrators; the initial temperature can be 46 to 93°C and the finished temperature 96°C. Four percent moisture is the desirable endpoint for adequate storage and retention of carotene. The dehydrated pieces are packed in airtight containers and filled with inert gas such as carbon dioxide or nitrogen. One hundred kilos of raw carrots will yield approximately 9 kg of low-moisture carrots (11:1 drying ratio).

GENERAL REFERENCES

Anonymous, "New Insight into Blanching Vegetables," *Food Eng.* **55**, 161 (1984).

C. Andres, "Compressed Dehydrated Cabbage has 'Fresh' Color and Texture," *Food Process.* **37**, 93 (1977).

J. G. Brennan, *Food Dehydration, A Dictionary and Guide*, Technomic Publishing Co., Lancaster, Pa., 1994.

T. P. Labuza, S. R. Tannenbaum, and M. Karel, "Water Content and Stability of Low Moisture and Intermediate Moisture Foods," *Food Technol.* **24**, 543–550 (1970).

H. S. Lambrecht, "Sulfite Substitutes for the Prevention of Enzymatic Browning in Foods," C. Y. Lee and J. R. Whitaker, eds., *American Chemical Society Symposium Series 600*, Washington, D.C., 1995, pp. 313–323.

S. B. Lin, "Drying and Freeze-Drying of Vegetables," D. S. Smith et al., eds., *Processing Vegetables, Science and Technology*, Technomic Publishing Co, Lancaster, Pa., 1997.

G. M. Sapers, "Browning of Foods: Control by Sulfites, Antioxidants, and Other Means," *Food Technol.* **47**, 75–84 (1993).

L. P. Somogyi and B. S. Luh, "Vegetable Dehydration," in B. S. Luh and G. J. Woodroof, eds., *Commercial Vegetable Processing*, 2nd ed., Van Nostrand Reinhold, New York, 1988.

W. F. Talburt and O. Smith, *Potato Processing*, Van Nostrand Reinhold, New York, 1987.

S. L. Taylor and R. K. Bush, "Sulfites as Food Ingredients," *Food Technol.* **40**, 47–52 (1986).

W. B. Van Arsdel, M. J. Copley, and A. L. Morgan, *Food Dehydration*, 2nd ed., Vols. 1 and 2, AVI Publishing Co., Westport, Conn., 1973.

Laszlo P. Somogyi
Consulting Food Scientist
Kensington, California

VEGETABLE PROCESSING

The vegetable processing industry in the United States has made technological progress during the past decades. More computers are being used in the handling, processing, and marketing of processed vegetables (1). Increasing recognition of the importance of vegetables to human nutrition has stimulated greater consumer interest and the food industry in improving quality of processed vegetables.

Vegetable processing has greatly expanded the farm produce markets through conversion of perishable produce into various stable forms that can be stored and shipped to distant markets. Canning, freezing, dehydration, pickling, and freeze drying are the basic methods of preserving vegetables. The ability to lengthen the period of vegetables' availability in a preserved form that retains their nutritive value and palatability improves human health, adds variety to the human diet, and reduces the time for preparation. Equally important is that these gains have been achieved at low cost due to mechanization and automation of the processing methods. This article updates the recent developments in vegetable processing.

GENERAL PRINCIPLES

The preservation of vegetables in a stable form that can be stored and shipped to distant markets is a primary objective. The large quantity of canned vegetables preserved every year helps growers as well as consumers (2,3). Frozen vegetables can maintain better color, texture, and flavor after processing. Dehydrated vegetables, such as beans, corn, and soy beans, are also very important items.

Pickling is a good process for preserving cucumbers, olives, and cabbages. The principles of preserving vegetables by canning depends on killing the microorganisms in sealed containers (4). Frozen vegetables are preserved by freezing at $-18°C$ or lower and storage of the product in containers to prevent moisture loss. Dehydrated vegetables are preserved by decreasing the moisture content so that no microorganisms can grow. Storage temperature (usually at 20°C or lower) and packaging materials are very important to the shelf life of canned vegetables. In the pickling of vegetables, lactic acid fermentation lowers the pH of the vegetable to 3.0 to 4.0 so that the products can be easily canned or dehydrated.

MICROBIAL SPOILAGE OF FOODS

There are several methods of preventing spoilage of foods. The most common method is to destroy the microorganisms in the food and prevent recontamination by microorganisms from the outside of the container. Canned vegetables are sterilized by heating the prepared product in vacuum-sealed containers in a pressure cooker. Due to the high pH value of most vegetables, usually above 4.5, it is necessary to heat the container under pressure at 116 to 121°C so that the heat-resistant bacterial spores are killed. The pH of fruits is usually below 4.5, and the vegetative non-spore-forming microbes present in fruits are readily killed in boiling water at 100°C (212°F). The details for the operation of the pressure cooker in vegetable canning were described by Luh and Kean (4). In general, the safety of canned vegetables on the market is under the strict control of the Food and Drug Administration, the Environmental Protection Agency, and the health authority in each state.

The second method of preventing food spoilage is to alter the environment so as to prevent or retard the growth of undesirable microorganisms. Frozen vegetables are preserved by preheating the prepared vegetable with steam or in boiling water sufficient to inactivate the undesirable enzymes, followed by cooling, packaging, and freezing. Rapid freezing of the packaged foods by low-temperature air blast is very important to prevent undesirable textural changes.

Dehydration is probably the oldest method used in food preservation. Removal of water from foods is usually accomplished by dehydration in a tunnel with hot air of low relative humidity, or by sun drying. It is very important to control the dehydration process so that undesirable oxidation catalyzed by polyphenolase, peroxidase, and lipoxygenase enzymes are prevented.

Preparation and handling of foods before canning, freezing, and dehydration may greatly affect the quality of the preserved foods. Packaging and postprocessing storage conditions are very important to the keeping quality of canned, frozen, and dehydrated foods. Vegetables are dehydrated for protection against spoilage by microorganisms by controlling the moisture content and water activity in the preserved foods. Large quantities of soybeans, corn, rice, and wheat are preserved by dehydration each year for export. Ray (5) published a book entitled *Fundamental Food Microbiology* that contains detailed information on food microbiology.

CHEMICAL AND BIOLOGICAL CHANGES

The most common chemical and biological change in foods that causes degradation in sensory quality is oxidation of unsaturated oil and fats and flavoring compounds. These oxidation reactions occur much more rapidly when catalyzed by lipase, lipoxygenase, and polyphenoloxidase when the tissues are damaged by mechanical impact or by physiological changes during postharvest storage. It is extremely important to destroy the enzymes by blanching with steam or in boiling water to inactivate the undesirable enzymatic deterioration during the canning and freezing processes (Table 1). Removal of molecular oxygen from a container of foods also retards oxidation.

A common form of texture breakdown in processed foods is related to the breakdown of the pectic materials in the

Table 1. Blanching Time for Vegetable Freezing

Vegetables	Min. at 100°C	Vegetables	Min. at 100°C
Artichokes	3–10	Kale	2
Asparagus	2–4	Mustard greens	2
Beans, green or wax	3	Okra	3–4
Beans, lima	2–4	Peas, black-eyed	2
Broccoli	3	Peas, green	1-1/2
Brussels sprouts	3–5	Peppers, green & red	2–3
Carrots (cut)	2	Rhubarb	1
Cauliflower	3	Spinach	1-1/2–2
Celery	3	Squash, summer	3
Collards	3	Turnip greens	2
Corn (cut)	4	Turnips	2
Corn on the cob	7–11		

Source: Ref. 6.

cell wall caused by pectic enzymes. Softening of pickled cucumbers and olives can be attributed to degradation of cell walls by pectin esterase and polygalacturase from microorganisms or by excessive heating in the presence of acid. In addition to quality attributes, such as flavor and color, the nutritive values of the processed foods must be maintained as much as possible. The use of sulfur dioxide in dried-fruit processing has been a common practice in preventing oxidation of ascorbic acid. Another advantage in the application of sulfur dioxide at 1,000- to 2,000-ppm levels in dehydrated vegetables is to prevent the Maillard type of browning reaction. Because of the allergic effect (irritation with respiration) that sulfur dioxide produces in some people, recent trends have been toward using other antioxidants to accomplish the maintenance of dehydrated vegetables (7–9). The ideal in developing a new processing technique or a new food product would involve preventing growth of microorganisms through control of moisture and water activity.

Freezing

Microorganisms will not grow in the frozen state. The key to holding the fresh flavor of frozen foods was not freezing per se, but rather proper inactivation of flavor-deteriorating enzymes by blanching before freezing (Table 1). The most common method is to test for any residual enzyme activity of peroxidase, catalase, or lipoxygenase in the blanched foods. Recent research interests of the frozen food industry includes searching for better blanching conditions to stop undesirable biochemical changes during processing and improving storage stability of frozen foods (6,10,11).

Dehydration

Removal of water from vegetables is accomplished primarily by dehydration in a tunnel with hot air of low relative humidity, or by sun drying (12). Microorganisms will not grow if the moisture content of the foods is reduced to a safety zone that varies with the chemical components of the food. The principle is to control the water activity of the final product, and also to retain the sensory quality and storage stability of the dehydrated product. The processor

must control the moisture content and water activity (a_w) of the dehydrated product. Most foods will not spoil if the a_w is at 0.60 or lower. The limitation is dependent on the chemical components of the foods. For better keeping quality, dehydrated onions must be at 3 to 4% moisture content. Water activity is a measure of the free moisture in a product. It is determined by dividing water vapor pressure of the substance by the vapor pressure of pure water at the same temperature. The dehydrated vegetables should be packaged in containers that will prevent moisture, air, and light from entering the package. Use of sulfur dioxide at proper levels in dehydrated vegetables will inhibit the browning reaction, thus lengthening the shelf life of the product. Several methods of dehydration are used commercially, namely, forced hot air, drum drying, spray drying, vacuum drying, and freeze-drying (3,12). Use of continuous-belt convey and hot-air dryers has increased in recent years. Addition of sodium sulfite or metabisulfite to cut vegetables before dehydration has been a standard practice. In addition to improving storage stability of the dehydrated product, the presence of a small amount of sulfite in a blanched cut vegetable makes it possible to increase the drying temperature and thus shorten the drying time, increasing the production capacity of the dehydrator (12,13). More recently, use of sulfite in food dehydration has received considerable scrutiny throughout the food industry (8,9,12). Processors are looking for alternative methods for producing high-quality dehydrated vegetables. Lambrecht (7) reported that the family of erythorbates, erythorbic acid and sodium erythorbate, are sterioisomers of the ascorbates and function in a similar manner to that of antioxidants. These compounds are reducing agents and are preferentially oxidized in foods, thus preventing or minimizing oxidative flavor and color deterioration. Erythorbates can prevent enzymatic browning in many products such as fruits, vegetables, and beverages. It is a prime requirement in any vegetable-dehydration process that no opportunity exist for the development of bacterial toxins by *Clostridium* spp. and other causative organisms that may be present. Such toxin formation is most likely to occur when a moist product, usually with some soil contamination, is given a heat treatment followed by prolonged holding under moderately warm conditions without access to air. Thorough washing and proper blanching are required to obtain a satisfactorily low level of microorganisms. The second requirement is the absence of pathogenic bacteria such as *Salmonella* and *Staphylococcus aureus*. The third requirement is maintenance of reasonably low general bacteriological content so that no undesirable odor or flavor develops in processing. The quality of the finished dehydrated product is reflected not only in its texture, flavor, and color but also in its ability to rehydrate readily. Storage temperature is very important to the keeping quality of dehydrated vegetables. Even though dehydrated vegetables will not spoil at room temperature, better-quality retention can be achieved by proper packaging and storage at temperatures lower than 20°C.

Pickling

Pickled food may be defined as a product to which an edible acid, for example, lactic or acetic acid in the form of vinegar, has been added. A fermented food is a food in which the acid is produced from the sugars in foods by the fermentation of lactic acid bacteria (14).

FREEZE DRYING

The freeze-drying process consists of removing moisture from foods in the frozen state by sublimation under high vacuum. The low temperature used in the process inhibits undesirable chemical and biochemical reactions and minimizes loss of aromatic compounds. The dried product is light in weight and can be stored in airtight containers for long periods without refrigeration.

The important steps in making freeze-dried vegetables include raw material selection, grading, size separation, washing, removal of undesirable parts, sorting, slicing, blanching, freezing, freeze drying, inspection, screening, packaging, and marketing.

Freeze drying is applicable to vegetables of high intrinsic value such as asparagus, mushroom, and parsley. For bell peppers and pimientos, a common commercial practice is to remove part of the water in the prepared raw produce in a hot-air dehydrator, followed by freeze drying. This two-stage method greatly reduces the cost of production without significant loss of quality in the finished product.

HARVESTING, HANDLING, AND STORING

Vegetables reach the peak of quality at a definite time, depending on the varietal characteristics, time of planting, location, temperature, soil type, available water, adequate fertilizer application, and cultural practices (15).

By the application of a formula that must be worked out for each variety, a succession of plantings of the same variety may be made at intervals to assure a continuous supply of raw produce during the season. The processing may be further extended by the use of early, midseason, and late varieties.

The growers as well as the processors have to know how to harvest their crops within a narrow time limit. This can be coordinated by the fieldpersons of the processors with the growers. In the United States, slightly more than half of the vegetable crops are processed. Beets, corn, cucumbers, green peas, potatoes, lima beans, potatoes, tomatoes, and several other crops are mechanically harvested.

Bulk bins are used for holding field-run vegetables until they can be transported to the processing plant and for holding precooled vegetables in refrigeration until processing is completed.

Each bulk bin can hold 25 to 27 field boxes of vegetables. When filled, the bins weigh about 1,000 lb. The bins are emptied by mechanical dumpers either onto inspection/grader belts or into washers.

PREPARING VEGETABLES FOR PROCESSING

The operations involved in processing vary somewhat with the type of vegetable and the method to be used, for example, canning, freezing, dehydration, freeze drying, or

pickling. The general preparation procedures for vegetable freezing may be summarized as follows:

1. *Harvest at the optimum maturity stage before any portion becomes fibrous and tough.* Objective testing methods should be applied before harvest. The decision must be made by experienced field workers as to when and how to harvest. Vegetables at the peak of quality are highly perishable and should be harvested, transported, and processed promptly after harvest.

2. *Sorting and grading.* This operation removes diseased, insect-infested, and trash materials. This may involve a roller grader, air blower, rod shaker, or any mechanical device followed by sorting on conveyor belts.

3. *Washing.* Rinsing with water will remove dirt, insects, and small trash. Food-grade detergents may be used when washing vegetables taken from the soil such as potatoes, red beets, mechanically harvested tomatoes, and some leafy vegetables.

4. *Preparation such as peeling, shelling, trimming, cutting, and dicing.* Lye peeling of potatoes and mechanically harvested tomatoes has been a common practice. Due to greater restrictions on waste disposal, peeling waste that contains sodium hydroxide will pose a difficult problem with regard to disposal. There is a trend to return to steam peeling for mechanically harvested tomatoes.

5. *Blanching.* This is very important step to inactivate the enzymes present in the vegetables. The blanching process inactivates the enzymes that will cause discoloration and the formation of off-flavor and off-aroma during freezing storage and reduces the number of microorganisms. This process makes the product easier to pack into the container. The blanching process can also remove harsh flavors in collards, snap beans, and spinach for freezing preservation. The blanching time required for freezing some vegetables is presented in Table 1.

Test for Adequacy of Blanching

For the canning and freezing of vegetables, it is very important to have an adequate blanching procedure to avoid development of off-flavor and off-aroma in the product during postprocessing storage. Several enzymes, such as catalase, peroxidase, polyphenoloxidase, and lipoxygenase, are present in vegetable tissues. These enzymes should be properly controlled by blanching in order to produce a stable frozen vegetable. A simple method, but not quantitative in nature, is to add a drop of 0.5 to 3.0% hydrogen peroxide to the freshly cut surface of the blanched vegetable. The appearance of pinkish-orange color indicates the presence of peroxidase, whereas a perfusion of bubbles indicates the presence of oxidases.

Robinson (16) reviewed the literature on peroxidases in fruit and vegetables. Thermal inactivation of peroxidases in horseradish and other vegetables has been studied in detail. Although thermostable enzymes may represent only a small proportion of the total enzymatic activity in fresh foods, for blanched vegetables their effects on the quality may be quite substantial.

Peroxidase Assay

Peroxidase activity in green asparagus extract can be measured spectrophotometrically at 470 nm. The assay mixture contains 0.225 mL 0.3% H_2O_2, 0.225 mL 1% guaiacol in water, and 2.4 mL 0.2 M sodium acetate (pH 4.5). The reaction is initiated at 30°C after addition of 0.01 to 0.02 mL enzyme solution. A Perkin-Elmer spectrophotometer equipped with thermostatical cell compartment and recorder is used to monitor absorbance change with time at 470 nm. Enzyme activity was calculated from the linear portion of the curve. One unit of peroxidase activity is defined as 1 optical density/min at 470 nm.

The peroxidase paper test developed by the USDA Western Regional Research Laboratory, Albany, California, can be used by the frozen vegetable–processing industry for any residual peroxidase activity in the blanched vegetables.

Lipoxygenase Assay

Most assays for lipoxygenase employ the substrate, linoleic acid, with measurement of the disappearance of dissolved oxygen or by the increase in absorbance at 234 nm. In one assay, the bleaching of indigo carmine by the hydroperoxide reaction product was monitored. Lipoxygenase assay has been used by the industry with an intention of reducing the blanching time for preparing frozen vegetables.

PROCESSING

Canning

Important vegetables being preserved by canning are asparagus, beans, beets, carrots, corn, hominy, mushrooms, okra, olives, peas, pimientos, potatoes, spinach, whole tomatoes, tomato paste, and various other types of tomato products (4). Tin cans and glass bottles are used as containers for processed vegetables. The quantities of some selected canned vegetables in the United States (1977–1990) are presented in Table 2. Corn, beans, and olives were separated on a moving screen with different mesh sizes, or over differently spaced rollers. Separation into groups according to degree of ripeness or perfection of shape may also be done on conveyor belts. Green peas and lima beans are frequently separated into more or less mature portions by flotation in a salt solution, the operation being performed continuously in automatic equipment. Trimming, if necessary, is done by hand operators well trained in locating and removing blemishes. In many cases, vegetables are sliced, diced, halved, or peeled, usually by machines specially designed for each type of produce. In each of these steps, the raw vegetable is continuously inspected, and a final inspection is made to pick out mashed or broken pieces as well as any foreign matter that may have passed the cleaning, washing, and trimming operations.

Blanching

The prepared vegetables are immersed in hot water or exposed to live steam for a proper amount of time in an operation known as blanching. The process serves to inactivate enzymes and thus arrest changes in flavor, and to soften the product so that more may be filled into the container. Proper blanching removes the occluded gases in the tissue and reduces strain on the seam of the can during processing. Hot-water blanching causes loss of some nutrients and formation of large volumes of liquid waste. Microwave and hot-air blanching may also be used to minimize wastewater problems. The processor has to select the best blanching method for the particular type of vegetable. Lin and Schynes (18) studied the influence of blanching on the texture and color of 11 kinds of sterilized vegetables. They observed that low-temperature-long-time (LTLT) blanching significantly increased the firmness of seven kinds of canned vegetables, including carrot, endive, beetroot, leek, green bean, onion, and white cabbage. Among these, the texture of the canned carrot, endive, leek, and green bean was improved to a great extent. Analysis of variance revealed that LTLT treatment in solution containing calcium and zinc ions significantly affected the texture and color retention of sterilized green beans and endive. The reason for the improvement may be related to the activation of pectin–esterase enzyme during the blanching process, which enhances the firming action of demethylated pectin by calcium ions.

Filling

Metal or glass containers are conveyed to the point where they are filled with the prepared vegetables. Before being filled, the containers are cleaned by air blast.

Most filling is done by machine, although some vegetables such as asparagus may be put into containers by hand. The container is filled with the solid product and then usually topped with a liquid, which can be juice, as for whole peeled tomatoes; brine; or water that contains ingredients to improve the sensory quality of the processed foods. The choice of ingredients includes sucrose, high-fructose corn syrup, salt, organic acids, spices, and thickening agents. Sucrose is added to enhance the flavor of the product, to blend with other flavor notes, and to suppress the effects of undesirable compounds. A slightly sweet taste is desirable in some vegetables such as corn, green peas, red beets, and peeled tomatoes. Sucrose acts as a seasoning agent rather than as a sweetener in vegetable canning.

Brines are added to the cans during canning. If sucrose is incorporated into the brine, an interaction between the two ingredients may take place that has a beneficial effect on the flavor acceptance of the canned product. Canned and frozen vegetables are regulated by USDA and FDA standards (Table 3) that relate to the produce itself and all dry and liquid ingredients incorporated into the brine.

Vacuum in the Container

The purpose of keeping a vacuum in canned foods is to remove air so that the pressure inside the container follow-

Table 2. Quantity (United States pack—24/303 basis) of Selected Vegetables Canned from 1977 to 1990

Year	Asparagus	Beans, lima	Beans, snap	Beets (× 1,000 cases)	Carrots	Corn, sweet	Peas, green	Tomatoes[a]
1977	3,705	2,657	54,494	11,348	5,973	56,300	30,238	54,124
1978	3,382	3,356	57,121	12,834	6,609	57,907	25,269	49,241
1979	2,819	3,061	66,281	14,990	6,298	60,022	36,492	52,896
1980	2,535	2,833	59,689	11,322	5,084	50,574	30,056	53,096
1981	2,844	2,602	52,808	9,555	4,639	57,949	27,296	51,937
1982	2,727	2,275	48,832	11,034	6,222	60,522	24,790	42,968
1983	2,549	2,338	46,506	9,035	4,958	51,595	20,827	37,597
1984	2,928	2,316	52,089	9,292	4,832	56,789	23,312	43,170
1985	3,097	1,966	55,503	9,386	4,787	55,729	29,306	37,485
1986	3,154	1,738	48,939	9,042	4,321	55,090	20,919	38,438
1987	3,361	1,326	56,081	12,127	5,249	59,117	23,057	44,634
1988	3,846	1,267	46,610	8,519	4,505	50,831	12,494	50,626
1989	3,060	1,586	60,500	9,175	N.A.	65,437	24,336	N.A.
1990	N.A.	N.A.	58,750	N.A.	N.A.	62,216	24,298	N.A.
Average	3,077	2,255	63,684	1,089	5,342	57,148	25,192	46,351
%	1.5	1.1	31.2	0.5	2.6	28.0	12.3	22.7

Source: Ref. 17.

[a] Excludes soups and baby foods.

ing heat treatment and cooling will be less than atmospheric. The vacuum helps to reduce strain on containers during heat treatment and cooling and minimizes the level of oxygen remaining in the headspace.

Vacuum in the container can be obtained by heating the filled container through a steam blancher for 5 to 7 min to 80 to 90°C, followed by sealing the container, while it is still hot. The alternative is to seal the can in a mechanical vacuum seamer at room temperature. Another method is to replace the air in the headspace with steam before double seaming. The steam in the headspace condenses after heat processing and thereby creates a vacuum.

The vacuum in the canned, cooled container will vary with the type of container, the depth of the headspace, and the type of product. Quality-assurance personnel should always check the vacuum, headspace, and drained weight according to USDA regulations.

Sealing

To seal the lids onto the flange of the metal can, a double seam is created by interlocking the curl of the lid and the flange of the can. Various types of closing machines (double seamers) can create the vacuum in the headspace either mechanically in a vacuum-sealing chamber or by steam flow before the sealing.

Glass containers are sealed under vacuum created mechanically or by steam flow. The bottles are sealed with a close-fitting cover of tin-plate or a threaded or lug cap. Sealing speeds of 800 to 1,000 bottles/min have been achieved for baby foods.

Heat Processing

Heat sterilization of the canned vegetables either in metal cans or glass jars applied after hermetic (airtight) sealing is called the *process*. During the process, microorganisms that can cause food spoilage are destroyed by heat in either stationary or rotary pressure cookers. The temperature

and time of heat processing vary with the nature of the product and the size of the container (4,19). In Table 4, the time–temperature requirement for canning selected vegetables is presented. The calculated sterilizing values for heat processing some vegetable in commercial practices are presented in Table 5.

Acid products with pH values below 4.5 are readily preserved at the temperature of boiling water. The containers holding these products are processed in atmospheric steam or hot-water cookers. Vegetables and other low-acid foods require higher temperatures for sterilization in steam-tight pressure cookers (retorts or continuous pressure cookers) usually controlled by automatic devices. The size of the container is an important factor in determining the correct combination of time and temperature in processing. Obviously, heat will penetrate to the center of a small can more quickly than to the center of a large one. Because it is vital for canning processes to be exact for every size of container as well as for every type of product, precise information about the rate of heat penetration is required (Table 5).

Steam is usually used to sterilize canned foods. Improved methods of heat processing are being used extensively. These include continuous-agitating pressure cookers and hydrostatic cookers in which the necessary temperature is maintained by the pressure of the water column. Another new process is flame sterilization, in which the cans are exposed to direct flames while rotating rapidly. After being held for the time necessary to ensure sterilization, the cans are cooled by means of water sprays.

Cooling

Containers are cooled quickly to prevent overcooking. This may be done by adding water to the cooker under air pressure or by conveying containers from the cooker to a rotary cooler equipped with a cold-water spray. Cooling water should be treated with 1 to 2 ppm free chlorine to avoid recontamination of the canned foods.

Table 3. Processed Vegetables for which U.S. Standards of Grades Have Been Established

Asparagus, canned (2541)	Peas (field and black-eyed), canned, (1641)
Asparagus, frozen (381)	Peas (field and black-eyed), frozen (1661)
Beans, canned baked (6461)	
Beans, canned dried (411)	Peppers, frozen sweet (3001)
Beans, green and wax, frozen (2321)	Pickles (1681)
Beans, canned lima (471)	Pimientos, canned (2681)
Beans, frozen lima (501)	Pork and beans, canned (6441)
Beans, frozen speckled butter (lima) (5241)	Potatoes, frozen French-fried (2391)
Beets, canned (521)	Potatoes, hash brown (6401)
Broccoli, frozen (631)	Potatoes, canned white (1811)
Brussels sprouts, frozen (651)	Pumpkin (squash), canned (2741)
Carrots, canned (671)	Sauerkraut, canned (2951)
Carrots, frozen (701)	Sauerkraut, bulk (3451)
Cauliflower, frozen (721)	Spinach, canned (1901)
Chili sauce, canned (2191)	Spinach, frozen (1921)
Corn, canned cream-style (851)	Squash, frozen cooked (1941)
Corn on the cob, frozen (931)	Squash, canned summer-type (3581)
Corn, canned whole-kernel (881)	Squash, frozen summer-type (1961)
Corn, frozen whole-kernel (911)	Succotash, canned (6001)
Hominy, canned (3281)	Succotash, frozen (2011)
Leafy greens (other than spinach), frozen (1371)	Sweet potatoes, canned (2041)
Mushrooms, canned (1481)	Sweet potatoes, frozen (5001)
Okra, canned (3331)	Tomatoes, canned (5161)
Okra, frozen (1511)	Tomato catsup (2101)
Okra and tomatoes (tomato and okra), canned (3421)	Tomato juice, canned (3621)
Olives, canned green (5441)	Tomato juice, concentrated (5201)
Olives, canned ripe (3751)	Tomato paste, canned (5041)
Onion rings, frozen breaded (4061)	Tomato puree (tomato pulp), canned (5081)
Onions, canned (3041)	Tomato sauce, canned (2371)
Peas, canned (2281)	Turnip greens with turnips, frozen (3731)
Peas, frozen (3551)	Vegetables, frozen mixed (2131)
Peas and carrots, canned (6201)	
Peas and carrots, frozen (2501)	

Note: Numbers in parentheses refer to the section in the Code of Federal Regulations (7 CFR 52) in which the standard is published.

Labeling and Casing

After the cooking, cooling, and drying operations, the containers are ready for labeling. Labeling machines apply glue and labels in one high-speed operation. The labeled cans or jars are conveyed to devices that pack them into shipping cartons.

Aseptic Processing and Packaging

In aseptic processing, the problems of chemical and quality changes resulting from the slow heat penetration inherent in an in-container process are avoided by sterilizing and cooling the food separately from the container. Presterilized containers are filled with the sterilized and cooled product and the containers are sealed in a sterile atmo-

sphere with a sterile cover. Because aseptic processing is a continuous operation, the behavior of one part of the system can affect the overall performance of the entire system. As a result, numerous critical factors are associated with aseptic processing and packaging, often requiring automated control systems. The processor must consider the sterilization of the product, the processing equipment, sterilization of the packaging materials, and maintenance of sterile conditions throughout the aseptic system (21). Aseptically canned products are superior to retorted products in color, aroma, flavor, and nutrient retention.

Reuter (22) published a book entitled *Aseptic Packaging of Foods*. Aseptic processing of liquid with small particles (a few millimeters in size) and liquids with larger particles (15–25 mm in size) were treated differently. In liquids with small particles, plate and tubular heat exchangers are used for low-viscosity liquids in heating and cooling, whereas scraped-surface heat exchangers are used for high-viscosity liquids. There is a well-known process in which the carrier liquid of high viscosity is separately ultraheat treated by continuous method in a scraped-surface heat exchanger and the solid product is first sterilized in a discontinuous aseptic heater by direct steam injection. After sterilization, the two components are reunited aseptically, mixed, and then aseptically packaged. Because of the complicated process, this method still experiences only limited usage by the industry. Aseptic processing of tomato paste in bulk (50-gal or 300-gal containers) is practiced in commercial processing because of the improvement of the sensory quality of the product compared with those made by conventional heating and cooling methods.

SOME ILLUSTRATIVE SAMPLES OF VEGETABLE PROCESSING

Beet Canning

The annual pack of canned beets (*Beta vulgaris* L.) is 10 to 12 million cases in the United States (17). Beets for canning should have a uniform deep red color. Detroit Dark Red is the most important cultivar for canning. The beet-canning season starts in August and ends in the early part of November. The mechanical harvester travels along the row digging, cleaning, topping, and discharging beets into a truck ready for transport to the cannery. If storage is necessary, beets for canning should be stored in slatted bins in a cool, well-ventilated warehouse. Under proper conditions beets may be stored for several months.

Canning Process. After delivery to the cannery, beets are washed thoroughly in a rotary-drum spray washer. Beets are usually graded into three size classes by a Magnuson Shufflo beet sizer or other rod-and-reel sizer. They are peeled by a two-stage procedure in which they are first heated sufficiently to loosen the skin without overcooking the flesh and are then peeled in an abrasion peeler. The peeled beets are inspected and trimmed. Smaller beets, 46 mm or less in diameter, are canned whole; the medium sizes, between 46 and 70 mm, are sliced; and the large sizes, over 70 mm, are diced or cut into shoestring strips. The dices and julienne cuts are washed under sprays to

Table 4. Time–Temperature Requirement for Canning Selected Vegetables

Vegetables	Can sizes	Fill weight (oz)	Initial tempreature (°F)	Retort time (min) 240°F	Retort time (min) 250°F
Asparagus cuts, in brine	300 × 409	11.0	120	27	17
Beans, green/waxy, whole/cuts, in brine	303 × 406	12.5	120	21	12
Beans, lima, succulent, in brine	303 × 406	11.0	120	46	23
Beets, whole/cuts, in brine	303 × 406	10.9	140	31	19
Carrots, whole/cuts, in brine	303 × 406	11.6	140	30	20
Carrots and peas, in brine	303 × 406	11.5	140	37	18
Corn, whole kernel, in brine	303 × 406	12.0	140	46	22
Mushrooms, sliced, in brine	300 × 400	9.0	140	—	31
Peas, green, succulent, in brine	303 × 406	11.5	140	38	19
Potatoes, white, whole, in brine	303 × 406	11.6	140	34	22
Spinach, in brine	303 × 406	12.6	140	71	55

Source: Ref. 20.

Table 5. Calculated Sterilizing Values (F$_0$) for Some Commercial Processes

Product	Can sizes	Approximate: calculated sterilizing value (F$_0$)
Asparagus	All	2–4
Green beans, brine packed	No. 2	3.5
Green beans, brine packed	No. 10	6
Chicken, boned	All	6–8
Corn, whole kernel, brine packed	No. 2	9
Corn, whole kernel, brine packed	No. 10	15
Cream-style corn	No. 2	5–6
Cream-style corn	No. 10	2.3
Dog food	No. 2	12
Dog food	No. 10	6
Mackerel in brine	301 × 411	2.9–3.6
Meat loaf	No. 2	6
Peas, brine packed	No. 2	7
Peas, brine packed	No. 10	11
Sausage, Vienna, in brine	Various	5
Chili corn carne	Various	6

Source: Courtesy of American Can Co.

remove fines. Beets are canned in enamel-lined cans to prevent bleaching of the color. Peeled beets should be handled in stainless-steel equipment. In some plants, beets are washed, heated in a high-pressure steam peeler for a short time, and then released to atmospheric pressure. The skins are loosened in this process and are removed in a rotary-drum washer under heavy sprays of water. They are then inspected, sorted, and trimmed.

Filling and Heat Processing. Sliced, diced, or julienne-cut beets are packed into cans by machine; whole beets are packed by hand or a hand-pack filler. A light brine is added with a small amount of added sucrose. The amount of sweetener added, together with the natural sugar contents of the beets, should result in a level of about 7% sugar in the beet pack. After brining, the cans are exhausted in steam, sealed, and heat processed in a retort. The cans are sealed by the steam-flow method. Beets in no. 303 cans are usually processed in a continuous-agitating Food Machinery and Chemical Corporation (FMC) retort at 115°C for 22 min and cooled in a two-stage continuous water cooler. No. 10 cans (1-gal can) are processed for 33 min at 115°C in the same type of retort and then cooled in water. Sliced beets packed in light syrup are processed at 115°C for 35 min in no. 303 jars and for 40 min in no. 2-1/2 jars.

Green Beans

Two types of green beans (*Phaseolus vulgaris* L.) are cultivated: bush and climbing, or pole, types. Most commercial cultivars are of the bush type because they can be machine harvested. The Blue Lake cultivar is preferred for canning or freezing.

Canning Process. After arrival at the factory, green beans are conveyed to size graders. These consist of revolving cylinders with slots of various diameters through which the beans fall onto conveyors that carry them to the shippers. The beans are separated into the following sizes: no. 0, 4.8 mm or less; no. 1, 4.8 to 5.8 mm; no. 2, 5.8 to 7.3 mm; no. 3, 7.3 to 8.3 mm; no. 4, 8.3 to 9.5 mm; no. 5, 9.5 to 10.7 mm; and no. 6, over 10.7 mm.

Snipping. After size grading, the beans are carried through snippers consisting of metal cylinders having narrow slots. As the cylinder revolves, the ends of the beans are caught in the slots, and fixed knives lying snugly against the lower surface of the cylinder cut off the tips and stems that have been caught in the slots.

Inspection. The snipped beans then pass over inspection belts, where defective beans are removed. Those that escape snipping are returned to the snipping machine again.

Cutting. Smaller beans (no. 1, 2, and 3) are canned whole. The larger ones are cut crosswise by machine into lengths of 25 to 28 mm for regular-cut beans. Some beans of smaller size are cut lengthwise after blanching; these are called French-cut beans. Cross-cut beans are cut before blanching. The cut beans are screened to remove fine fragments and short pieces, which are canned as a special small-cut pack.

Blanching. The cut beans and whole no. 1, 2, and 3 sizes are blanched for 1.5 to 2 min, usually in hot water at 82°C to give better-filled cans. Blanched whole beans may be returned to French-style cutters and cut into shoestring-shaped pieces.

Filling and Heat Processing. Cut blanched beans are packed to volume mechanically. A plunger beyond the filler automatically levels the fill of each can. The cans are then filled with hot water, and dry salt is added by a dispenser. The cans are steam exhausted for about 5 min and are sealed hot or are steam flow sealed. The 303, no. 1 tall and no. 10 cans are the most popular. The processing time for no. 2 (307 × 409) and smaller cans with initial temperature at 49°C are 22 min at 115°C and 12 min at 121°C; for no. 10 cans, 27 min at 115°C, and 15 min at 121°C. For French-cut green beans, the processing time for no. 2 cans (307 × 409) is 26 min at 115°C; for no. 10 cans (603 × 700) 36 min with the initial temperature at 49°C (4).

Frozen Green Beans. In 1988, the total pack of frozen green beans was 235,986,000 lb. The preparation procedures and varietal characteristics of green beans for freezing are similar to those for canned beans, except for the last step in the terminal end of the preservation (6,20).

Corn

Sweet corn (*Zea mays* L.) is one of the common vegetables being preserved by canning, freezing, and dehydration. In 1988, the pack of frozen cut corn was 423,399,000 lb and that of frozen corn on the cob was 345,869,000 lb. The important corn-canning states are Illinois, Minnesota, Wisconsin, Indiana, Iowa, Maine, Maryland, New York, Washington, and Michigan. The corn cultivars suitable for canning are Country Gentleman, Crosob, Golden Bantam, Golden Cream, Potters Excelsior, Stowell's Evergreen (white cultivars), and Yellow Evergreen. Golden Bantam was more popular before the hybrid cultivars came in general use. Yellow cultivars have large kernels that make an attractive pack. For canning, the corn should be young and tender and the ears well formed. Nearly all commercially grown corn is now from hybrid cultivars. These give increased yield; better uniformity; and excellent flavor, color, and canning quality. Examples of single crosses are Golden Cross Bantam, Ioana, and Golden Hybrid 2409.

Styles of Canned Corn. Canned corn is packed mostly as *Maine style* or *Maryland style*. Maine style is obtained by cutting through the kernels, scraping the remaining portions of the kernels from the cobs, and mixing the scrapings with the cut kernels. Water flavored with salt and sugar is added to give the desired consistency. Starch is also usually added. The canned product is creamier in texture and is the more popular of the two styles. It is also called *cream-style* corn. Maryland-style, or whole kernel, corn consists of the whole kernels cut from the cob and canned in brine; the cobs are not scraped. A third style, *double-cut* corn, is

made from overmature corn or from large kernels. The corn is cut from the cob, and then the kernels are cut a second time with special knives to produce a cream-corn effect. *Brineless whole kernel corn*, vacuum sealed in cans, is now being favorably received.

Harvesting. Corn may be at best maturity for canning about 15 to 19 days after the appearance of the silk. The fieldperson uses the thumbnail test by noting the pressure necessary to break the hull of the kernel with the thumbnail. Experienced persons can judge maturity quite well based on the thumbnail test plus assessment of the appearance and feel of the ears. A pressure tester can also be used to test the toughness of the corn and hence its fitness for canning. The Succulometer measures the volume of liquid expressible from a weighed sample of kernels; other tests used are sugar content, starch content, tenderometer readings, and percentage of alcohol-insoluble matter. The FMC harvester cuts the cornstalks near ground level and elevates them to a mechanism that cuts the ears off the stalks. An additional machine separates the ears from the leaves and stalk. A corn harvester can harvest 10 to 12 acres/day.

Husking and Silking. Conveying systems are arranged so that the corn is handled mechanically throughout the entire canning process. The husking machine has a pair of rapidly revolving rubber or milled steel rolls that catch the husks and remove them. Modern huskers are equipped with two sets of rolls and are known as double huskers. The rolls in a husking machine are spaced so that an ear cannot pass between them; friction of the rolls against the ear loosens the husk, which is caught between the rolls and torn off. The rolls are slightly wider apart at the upper end, where the larger portion of the husks is removed. The rolls are mounted so that they can spread apart sufficiently to allow the husks to pass between them but are then returned instantly to their former position by powerful springs. They are kept wet with a spray of water during operation to prevent fouling. Fixed knives cut off both the butt and the silk ends. A hooklike device loosens and opens the husk on the ear before it reaches the rolls. The rolls then easily catch the husk and are apt to do a cleaner job of husking and removing the silks. The capacity of most single-husking machines is rated at 60 to 80 ears/min and of double machines at 120 to 160 ears/min (approximately 2 t/h). In some canneries, the husked ears are passed through a silking machine equipped with revolving brushes and rolls. The ears are carried forward through the brushes, which remove nearly all the silk and deliver the silked ears to the washing machine. Silking of the machine can be omitted if the work of the mechanical huskers and the trimmers is well done.

Sorting and Washing. In the first sorting, improperly husked ears are returned to the husking department, where the culls and trash are removed. In the second sorting, the ears may be sorted into A and B grades, the A grade being used for "fancy" and "choice" canned corn and the B grade for "standard" quality. The more mature ears are sorted out for cream-style corn. There are two satisfactory methods of washing husked ears: (1) by means of a silker-brusher washer, and (2) by means of a revolving cylinder or conical reel. The corn is passed under a heavy spray of water.

Canning Whole-Kernel Corn. In making whole-kernel corn, some canners blanch the husked ears in water near the boiling point to coagulate the juice before cutting. The ears are fed, small end first, through curved knives that accommodate themselves by means of springs to the size of the ear. The cut corn is inspected on a slowly moving belt and the unfit material discarded.

Cut whole-kernel corn is next passed through a shallow tank filled with water in which a paddle stirrer moves slowly. The kernels sink, and the floating trash is removed automatically. The kernels are then conveyed to the silking and washing reels (cob reels). As the reels revolve, the kernels fall through the screen to the product-collecting pan. From this point, they are pumped to a dewatering screen, and the dewatered corn is delivered to the filling machine.

Whole-kernel corn is heated by steam in a tubular heater on its way to the can filler and is filled hot. The cans are lined on the inside with C-enamel (corn enamel), which contains zinc oxide to prevent discoloration caused by formation of black FeS. The FMC 15-station filler, a gravity-type filling and brining machine, is commonly used. It is equipped with 15 telescopic product-measuring pockets mounted on a revolving turret. The pockets measure the exact required amount of product and discharge it through filling tunnels into cans. Cutoff plates automatically open and close the measuring pockets. The pockets rotate in succession under a supply hopper to receive their load. The hopper rotates and is equipped with an offset stud that agitates the corn and thus prevents bridging, ensuring uniform fill pockets and cans. Below the hopper is a special drain that drains away any water entering the hopper with the corn. The cans travel on supporting tracks that can be vibrated to settle the product in the cans. Filled cans travel from the filler to the briner. Hollow cylindrical measuring tubes within the brine tank separate a column of brine and fill it into the cans. Up to 350 cans/min can be filled and brined. The brines contains 2.0 to 2.7% salt and 6 to 8% sucrose.

The brine may contain as little as 1.5% sugar depending on the grade and the processor's standards. Recently, the amount of salt in the canned product may be reduced by 50 to 70% to suit the needs of the public. Cans are filled with brine at 46 to 49°C. Corn has also been canned in butter and cheese sauce to improve the flavor acceptance. To each no. 2 can, 383 to 400 g of whole-kernel corn is added. The filled cans pass beneath a topper, which is a leveling plunger that automatically gives the proper headspace in the can. In many plants, the cans are sealed in a steam-flow (or steam-vac) sealer. In some plants they are well exhausted in a steam-exhaust box and sealed hot. The process time for canned whole-kernel corn in no. 2 cans (303 × 409) is 46 min at 115°C or 22 min at 121°C, provided that the initial temperature is at 60°C. For no. 10 cans (603 × 700) the process time is 81 min at 115°C and 43 min at 121°C.

Frozen Corn. The harvest and preparation procedures for frozen corn are similar to those for canned corn, except for the last terminal step in freezing. Yellow corn cultivars such as Golden Cross Bantam are usually preferred for freezing. Sweet corn loses its quality rapidly after harvest and should be frozen within a few hours of harvest. Blanched whole-kernel corn is produced in a number of ways: (1) the corn may be completely blanched (7–11 min) on the cob before cutting; (2) it may be partially blanched on the cob to set the milk, then cut and blanched again; or (3) it may be cut and subsequently blanched. This last practice results in loss of flavor, lower yield, and microbiological problems. The split blanch is recommended to keep the bacterial counts at a minimum. After the corn is cut, the kernels must be cleaned to remove bits of kernels; husk; silk; light fiber; and dry, immature kernels. Both froth-washing and brine-flotation graders are excellent for this purpose. The corn kernels are removed at the sink or discharge end while the trash floats off the top discharge or is picked up by a rotating reel (6).

Whole-kernel corn is usually individually quick-frozen and either packaged directly into cartons and polyethylene bags or bulk stored. A fluidized freezing process can be used for both cut corn or cobbed corn.

Green Peas

The U.S. pack of canned peas was 24,298,000 cases (24 × no. 303 cans) in 1990 (Table 2). The pack of frozen green peas was 358,779,000 lb in 1988. There appears to be a gradual shifting of canned peas to frozen peas in the past few years. Wisconsin, New York, Minnesota, Illinois, Washington, and Oregon are important pea-producing states. Peas (*Pisum sativum* L.) are grown as a field crop by tractor cultivation. The most important cultivar is Alaska; the most widely used wrinkle-seeded cultivars are Perfection, Horsford Market Garden, Advancer, and Admiral. Alaska is a favorite in the eastern United States because it is a more reliable producer than most wrinkled cultivars. Popular cultivars in the Midwest are Early Perfection, Prince of Wales, and Green Giant s-537. In Oregon and Washington, Perfection is the principal cultivar used for canning. The words *smooth* and *wrinkled* apply to the ripe, dry peas; before drying, all are smooth.

Green Pea Canning. Green peas are harvested and vined at vining stations in the field, thereby avoiding long haulage of the vines. The correct time for harvesting is determined by the appearance of the pods, which should be swollen and well filled with young tender peas. Most canners use the tenderometer to determine maturity and grade of peas.

The shelled peas are collected at the viners in bins of 45.5-kg capacity. Normally no more than 4 h elapse between vining in the field and canning of the peas. The shelled peas are emptied onto a conveyor that moves them to a broad vibrating screen. The peas drop through the screen, but leaves, clods, pods, and other trash pass over the end of the screen. When peas are vined at the cannery, the shelled peas go directly from the viner over the scalper screen. After screening, the peas are cleaned in a clipper cleaner, which operates on the same principle as a grain-cleaning mill. A strong blast of air from a fan blows light trash away, and screens remove coarser waste material. The peas fall through the screen and are conveyed into stainless-steel tanks of water to which has been added a special soap in the form of a paste and a highly refined mineral oil. The separation tank is in the form of two cones, one inverted and the other upright, with a cylindrical portion between the cones. A special pump violently aerates and circulates the liquid, converting it into a frothy mixture. As the peas drop to the bottom of the lower cone, they are picked up by a pump and delivered to a squirrel-cage spray washer in which any adhering trash is thoroughly removed. The trash flows to the surface and overflows from the upper cone. Cleaned, shelled peas are size graded in a slowly revolving screen cylinder or through a series of nested or parallel screens; in either case the screens have holes of varying diameters.

Blanching. Hot-water blanching is done in a pea blancher of special design for 3 min at 88 to 93°C. If the peas are fairly mature, the blanching time may be increased to 4 to 5 min; young tender peas are blanched for 1.5 min. Water for blanching should be soft because the calcium in hard water will harden the peas. A continuous fluidized-bed steam-blanching process for peas can be used. Steam blanching for over 30 s prepares peas adequately for canning. However, steam blanching does not have as great a cleansing and plumping action as water blanching.

Quality Grading and Sorting. Only the largest peas of any cultivar need to be quality graded. After the blanched peas are cooled, the larger ones are fed into a brine of 38 to 40° Salometer (9.5–10.0% salt). The younger tender peas are lower in density than the riper peas and will float in brine of this concentration. The more mature grade B peas will sink; they are removed from the bottom of the brine separator continuously and are pumped or flumed to the next operation. The lighter grade A peas are carried over into the overflow and conveyed to the next stage. Both classes are thoroughly washed immediately after brine separation to remove the adhering brine.

Filling and Heat Processing. The sorted peas are fed by gravity to cone-shaped hoppers on the can-filling machine. The peas fall into the cups of the automatic rotary can filler. Each cup delivers exactly the desired volume of peas into a can. The cans are carried to the briner. The cans for peas are usually enameled with C-enamel to prevent blackening of the tinplate due to FeS formation. The brine for canned peas usually contains sugar as well as salt. The composition of the brine differs considerably according to the preference of the distributors and canners and the maturity of the peas. As less-mature peas are sweeter, they need less sugar in the brine. The brine is added at 71°C by adjustable cups. Filled cans are sealed under steam flow by the steam-vac procedure. The recommended heat-process time for peas in no. 2 cans (303 × 409) is 31 min at 115°C, or 15 min at 121°C; for no. 10 cans, 48 min at 115°C or 21 min at 121°C, provided that the initial tem-perature is at 60°C. Canned peas should be cooled rapidly following sterilization to avoid excessive softening and the formation of a cloudy brine through gelatinization of starch.

Frozen Green Peas. The procedures for freezing green peas are similar to those for canned peas except for the step of freezing preservation (6).

TRENDS IN VEGETABLE-PROCESSING DEVELOPMENT

1. Studies on plant genetics of vegetables is becoming a very important topic. The objectives are to improve the varietal characteristics of the vegetable seeds regarding yield, flavor, quality, texture, resistance to pathogens, and adaptability to stress. Tailoring genes for crop improvement and genetic engineering of plants are attracting more horticultural scientists to pursue further studies on vegetable improvement.

2. Value-added vegetable products have been of great research interest to various research institutes, university and agricultural experiment stations, and the biotechnology industries. Products such as vegetable juices, sauces, soups, and pickled vegetables also attract more attention within industrial research programs. The success of the vegetable industry depends on quality control and sanitation departments to keep up the high levels of quality (23) and nutrition demanded by the consumer as well as the government. The ultimate goal is to have processed vegetables with good storage life, quality retention, nutritive value, and safety for the consumer. Use of computers for process control and record is now a must for modern vegetable processors.

BIBLIOGRAPHY

1. A. Przybyla, "Computers in Food Processing," in B. S. Luh and J. G. Woodroof, eds., *Commercial Vegetable Processing*, 2nd ed., Van Nostrand Reinhold, New York, 1988, pp. 741–764.

2. J. M. Connor, *Food Processing—An Industrial Powerhouse in Transition*, D. C. Heath and Company, Lexington, Mass., 1988.

3. D. S. Smith et al., "Processing Vegetables," in D. S. Smith, eds., *Science and Technology*, Technomic Publishing Co., Lancaster, Pa., 1997.

4. B. S. Luh and C. E. Kean, "Canning of Vegetables," in B. S. Luh and J. G. Woodroof, eds., *Commercial Vegetable Processing*, 2nd ed., Van Nostrand Reinhold, New York, 1988, pp. 195–285.

5. B. Ray, *Fundamental Food Microbiology*, CRC Press, Boca Raton, Fla., 1996.

6. B. S. Luh and M. C. Lorenzo, "Freezing of Vegetables," in B. S. Luh and J. G. Woodroof, eds., *Commercial Vegetable Processing*, 2nd ed., Van Nostrand Reinhold, New York, 1988, pp. 343–386.

7. H. S. Lambrecht, "Sulfite Substitutes for the Prevention of Enzymatic Browning in Foods," in C. Y. Lee and J. R. Whitaker, eds., *Enzymatic Browning and Its Prevention*, pp. 313–323, ACS Symposium Series 600, American Chemical Society, Washington, D.C., 1995, pp. 313–323.

8. S. J. Pintauro, S. L. Taylor and C. O. Chichester, A review of the safety of sulfites as food additives, The Nutrition Foundation Status Report, The Nutrition Foundation, Inc., Washington, DC, 1983.

9. S. L. Taylor and R. K. Bush, "Sulfites as Food Ingredients," *Food Technol.* **40**, 47–52 (1986).

10. D. P. Judge, *The Almanac of the Canning, Freezing, Preserving Industries*, 78th ed., Edward E. Judge and Sons, Inc., Westminster, Md., 1995.

11. C. P. Mallett, *Frozen Food Technology*, Blackie Academic and Professional, London, 1993.

12. L. P. Somogyi and B. S. Luh, "Vegetable Dehydration," in B. S. Luh and J. G. Woodroof, eds., *Commercial Vegetable Processing*, 2nd ed., Van Nostrand Reinhold, New York, 1988, pp. 387–473.

13. W. F. Talburt and O. Smith, *Potato Processing*, 4th ed., Van Nostrand Reinhold, New York, 1987.

14. B. S. Luh and J. G. Woodroof, *Commercial Vegetable Processing*, 2nd ed., Van Nostrand Reinhold, New York, 1988.

15. J. G. Woodroof, "Harvesting, Handling and Storing Vegetables," in B. S. Luh and J. G. Woodroof, eds., *Commercial Vegetable Processing*, 2nd ed., Van Nostrand Reinhold, New York, 1988, pp. 135–174.

16. D. S. Robinson, "Peroxidases and Their Significance in Fruits and Vegetables," in P. F. Fox, ed., *Food Enzymology*, Vol. 1, Elsevier Applied Science, London, 1991, pp. 399–426.

17. *Agricultural Statistics*, U.S. Department of Agriculture, Washington, D.C., 1992.

18. Z. M. Lin and E. Schyens, "Influence of Blanching Treatments on the Texture and Color of Sterilized Vegetables," in *Symp. Second International Conference Food Science and Technology*, Wuxi Institute of Light Industry, Wuxi, China, 1994.

19. D. Downing, *A Complete Course in Canning and Related Processes*, 13th ed., Book III, *Processing Procedures for Canned Food Products*, The Canning Trade Inc., Timonium, Md., 1996.

20. A. Lopez, "Canning of Vegetables," in *A Complete Course in Canning*, Book III, 12th ed., The Canning Trade Inc., Baltimore, Md., 1987.

21. *Canned Foods—Principles of Thermal Process Control, Acidification and Container Closure Evaluation*, 5th ed., The Food Processors Institute, Washington, D.C., 1988.

22. H. Reuter, *Aseptic Packaging of Food*, Technomic Publishing Co., Inc., Lancaster, Pa., 1988.

23. H. Patino and B. S. Luh, "Quality Control," in B. S. Luh and J. G. Woodroof, eds., *Commercial Vegetable Processing*, 2nd ed., Van Nostrand Reinhold, New York, 1988, pp. 561–609.

B. S. Luh
University of California
Davis, California

VEGETABLE PRODUCTION

The focus of this article is on the commercial production of vegetables for sale. Commercial producers range from those growers that operate on a very limited scale to those with thousands of hectares in several locations. Current data on production and consumption are provided and the contributions of vegetables to human health discussed. The utilization of genetic resources in concert with various environmental situations to optimize production is considered along with environmentally sensitive pest-management strategies. Harvest and postharvest handling practices that provide for timely delivery of high-quality, nutritious products to consumers are included.

DEFINITION

Vegetables are so diverse (Fig. 1) and their use so ubiquitous that it is difficult to contrive a definition that will not have exceptions. One authority (1) proclaims in desperation, "The terms are impossible of close definition because the plants that fall within their scopes are so various. The best definition is an enumeration of the plants." One attempt at defining vegetable has been proposed by Maynard (2): " . . . a vegetable is a herbaceous plant, some parts of which may be eaten. Plant parts usually eaten, and representative vegetables, include roots (beet, carrot, turnip); stems (asparagus, kohlrabi); tubers (potato); bulbs (leek, onion); leaf petioles (celery, rhubarb); entire leaves (cabbage, lettuce, spinach); flower parts (broccoli, cauliflower); immature fruit (cucumber, snowpea, summer squash); and mature fruit (cantaloupe, tomato, winter squash)." A more complete listing of vegetables with designated parts used for food is shown in Table 1. The foregoing definition, although quite encompassing, excludes one important vegetable—mushroom—which is not herbaceous. The pileus (cap) and stipe (stalk) are the edible portions of the common mushroom.

CLASSIFICATION OF VEGETABLES

Vegetables, because of their diversity, can be classified in many ways for the convenience of the grower, crop adviser, shipper, or marketer. One consideration for classification, based on plant part consumed, has already been illustrated in Table 1. Another classification scheme (Table 2) is based on familial relationships that focus on characteristics of the flower. The appearance of the plant may be

Figure 1. Vegetable display at Hunt's Point Wholesale Market in New York City.

Table 1. Vegetable Plant Parts Used as Food

Below-ground structures		Above-ground structures		
Bulb	*Root*	*Leaf*	*Leaf petiole*	*Immature fruit*
Garlic	Beet	Cabbage	Celery	Cucumber
Leek	Carrot	Chard	Rhubarb	Eggplant
Onion	Celeriac	Chicory		Okra
Shallot	Horseradish	Chinese cabbage	*Stem*	Snap bean
	Jicama	Chive		Snowpea
Corm	Parsnip	Collards	Asparagus	Summer squash
	Rutabaga	Endive	Kohlrabi	
Taro	Sweet potato	Kale		
	Turnip	Lettuce	*Flower bud*	*Mature fruit*
Rhizome		Mustard	Broccoli	Cantaloupe
		Parsley	Cauliflower	Chayote
Ginger		Spinach	Globe artichoke	Pepper
				Pumpkin
Tuber			*Immature seed*	Tomato
				Watermelon
Jerusalem artichoke			Lima bean	Winter squash
Potato			Pea	
Yam			Sweet corn	

drastically altered by environmental extremes, but the structure of the flower remains unchanged, making it a constant factor for classification purposes. The principal vegetables of commerce in developed temperate regions are found in a relative handful of plant families. The most important of these are Solanaceae (potato, tomato, pepper), Brassicaceae (cabbage, Chinese cabbage, broccoli), Cucurbitaceae (cucumber, squash, watermelon), Alliaceae (onion, garlic), and Fabaceae (peas, beans).

Even though commercial vegetable production is realized in a relatively small number of plant families, the potential is enormous. Kays and Silva Dias (3) list an astounding 389 species in 69 families as vegetables (not necessarily on a commercial basis) cultivated in the world. In China alone, Chen, Zhu, Fu, and Wang (4) report that 177 species in 65 families may be grown as vegetables, but only about 20% of these are now grown commercially. Clearly, the potential for diversity in the diet is available if demand can be generated and production practices developed. Aside from their familial relationships, related vegetables may have similar growth habits that make for common cultural practices, for example, the vining crops cantaloupe, squash, and watermelon of the gourd family have similar spatial requirements. Similarly, tomato, pepper, and eggplant in the nightshade family grow best in a long, warm, frost-free season. Hazards to crop production such as plant diseases and insects may find common hosts within a family, for example, similar pests attack a number of vegetables within the mustard family.

PRODUCTION STATISTICS

United States

The current status of the commercially important vegetables in the United States (5) is shown in Table 3. The area (1000 ha) harvested, production per ha (t) and in total (1000 t), and value per ha (U.S. dollars) and in total

(million U.S. dollars) is shown for each vegetable. Eggplant, escarole/endive, and brussels sprouts are grown on the smallest area whereas dry bean, potato, and sweet corn are grown on the largest area. Production per hectare is lowest for dry pea, asparagus, and lima bean and highest for celery, tomato, and onion. Total production varies on the low side with escarole/endive, brussels sprouts, and eggplant to potato, tomato, and sweet corn on the high side. Crop value per hectare is low for dry pea, dry bean, and lima bean and high for celery, bell pepper, and artichoke. On a total value basis beet, escarole/endive, and eggplant are lowest while potato, tomato, and lettuce are highest. Overall, vegetables were harvested from almost 3 million ha that produced over 56 million t and had a farm value exceeding $13 billion U.S. in 1997. Accordingly, the U.S. vegetable industry has a very great economic impact on the country.

The localization of the U.S. industry (5) is illustrated in Figure 2. By any standard, California is the most important state for vegetable production for fresh and processing use. California is favored by a variety of climatic niches, from cool coastal areas to warm interior valleys, and changing seasons related to the extreme south–north situation of the state. For example, the Mediterranean climate necessary for successful artichoke production is found in coastal areas south of San Francisco whereas the long, warm season that favors tomato production is found in the vast Central Valley.

Florida, Georgia, Arizona, and Texas are major contributors to the fresh-produce industry. Production in these states occurs primarily in the "off-season" fall, winter, and spring months when local production is not possible in the northeastern and Midwest population centers.

For processing, Wisconsin, Minnesota, Washington, and Oregon are important producing states. High yields, essential for profitability of processing crops, are favored by the long days of summer in these northern states.

Table 2. Common Vegetable Family Groupings

Family	Common name	Vegetables
Aizoaceae	Carpetweed family	New Zealand spinach
Alliaceae	Onion family	Leek, shallot, onion, garlic, chive
Apiaceae	Carrot family	Celery, carrot, parsnip, parsley
Araceae	Arum family	Taro
Asteraceae	Sunflower family	Endive, chicory, cardoon, globe artichoke, lettuce, Jerusalem artichoke, salsify
Brassicaceae	Mustard family	Horseradish, mustard, kale, rutabaga, collards, cauliflower, cabbage, brussels sprouts, kohirabi, broccoli, Chinese cabbage, turnip, radish
Cactaceae	Cactus family	Prickly pear
Chenopodiaceae	Goosefoot family	Chard, garden beet, spinach
Convolvulaceae	Bindweed family	Sweet potato
Cucurbitaceae	Gourd family	Watermelon, cantaloupe, cucumber, pumpkin, gourd, squash, chayote
Dioscoreaceae	Yam family	Yam
Fabaceae	Pea family	Jicama, lima bean, French bean, pea, gram, southern pea
Liliaceae	Lily family	Asparagus
Malvaceae	Mallow family	Okra
Polygonaceae	Buckwheat family	Rhubarb
Poaceae	Grass family	Sweet corn
Solanaceae	Nightshade family	Pepper, tomato, eggplant, sweet pepino, potato
Zingiberaceae	Ginger family	Ginger

Table 3. U.S. Vegetable Industry—Area, Production, and Value (1997)

Vegetable	Area harvested (1000 ha)	Production		Value	
		Ha (t)	Total (1,000 t)	Ha (USD)	Total (million USD)
Artichoke[a]	3.68	10.66	39.24	18,375	67.62
Asparagus[b]	29.76	3.02	89.77	6,089	181.22
Bean, dry[b]	787.32	5.95	132.22	745	586.45
Bean, lima[b]	22.96	3.31	76.00	1,740	39.96
Bean, snap[b]	111.51	7.51	836.88	2,566	286.13
Beet[c]	3.02	36.70	110.84	2,695	8.14
Broccoli[b]	54.31	14.46	785.40	9,124	495.52
Brussels sprouts[a]	1.70	16.81	28.58	15,765	26.80
Cabbage[b]	34.53	40.81	1,409.24	8,329	287.59
Carrot[b]	48.43	41.80	2,024.29	9,883	478.64
Cauliflower[b]	19.18	15.33	294.06	10,321	197.96
Celery[a]	10.90	75.16	819.28	25,087	273.45
Corn, sweet[b]	278.03	14.53	4,039.60	2324	646.12
Cucumber[b]	64.60	16.39	1,058.63	5128	331.24
Eggplant[a]	1.05	31.89	33.48	16,724	17.56
Escarole/Endive[a]	1.26	20.02	25.22	10,413	13.12
Garlic[a]	14.97	16.82	251.74	17,470	261.52
Lettuce[a]	112.37	35.14	3,949.07	14,271	1,603.63
Melons					
Cantaloupe[a]	46.04	23.21	1,068.48	9,076	417.86
Honeydew[a]	11.90	22.09	262.86	9,192	109.39
Watermelon[a]	74.71	24.73	1,847.66	4,139	309.23
Mushroom[b]	–	–	352.61	–	730.28
Onion[a]	62.70	46.22	2,897.69	10,342	648.44
Pea, dry[b]	148.68	2.06	306.59	289	43.02
Pea, green[b]	108.82	3.97	431.77	1,259	137.00
Pepper, bell[a]	26.57	28.63	760.81	18,916	502.60
Potato[b]	536.83	38.86	20,857.01	4,855	2,604.19
Spinach[b]	13.98	15.71	219.62	5,323	74.41
Sweet potato[b]	33.74	18.17	612.77	6,327	213.49
Tomato[b]	165.41	65.06	10,762.03	11,198	1,852.19
Total	*2,828.96*		*56,383.44*		*13,444.77*

[a]Includes product for the fresh market only.
[b]Includes product for the fresh and processing markets.
[c]Includes product for the processing market only.

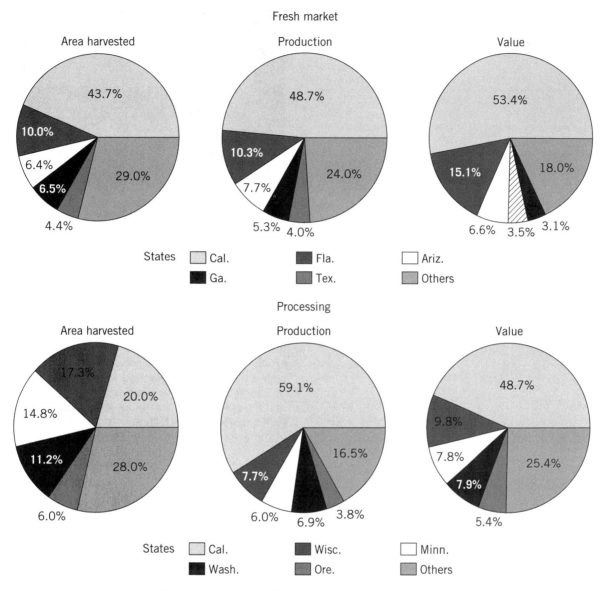

Figure 2. Important U.S. vegetable-producing states, 1997.

World

The area harvested (1,000 ha), yield (kg/ha), and total production (1,000 t) of the world's most important vegetables (6) are shown in Table 4. In terms of area harvested, dry bean, potato, and cassava are the most important crops, whereas artichoke, green bean, and cauliflower are relatively unimportant. Highest yields are obtained from tomato, cabbage, and carrot. The dry legumes—bean, chickpea, and lentil—produce the lowest yields per hectare because of their low moisture content. Total world production is highest for potato, sweet potato, and cassava—the root crops that provide massive amounts of energy and can be stored for extended periods to provide foodstuff when other foods are unavailable. Vegetables entering into commerce on a global basis are harvested from over 118 million ha and produce more than 1 billion t. Because much of the population in developing countries is engaged in sub-

sistence farming where farm produce, including vegetables, does not enter into traditional commerce, it is clear that total worldwide production is much greater than that reported.

The ranking of the primary countries in production (t) of vegetables (6) is shown in Table 5. This listing is related to the size of the country, its population, presence of a favorable climate, and ethnic culinary habits of the population. Even a casual examination of Table 5 will reveal the overwhelming importance of China (Fig. 3) as a vegetable-producing country. Of 25 vegetables or vegetable groups listed, China ranks first in 15. India, the next most frequent listing, appears four times as the leading producer. According to data developed by the Chinese Ministry of Agricultural Statistics and quoted by Chen and Liu (7), the 21 major vegetables in China were grown on 4,365,000 ha, which produced more than 142 million t of vegetables in 1991.

Table 4. World Vegetable Production—Principal Vegetables (1996)

Vegetable	Area harvested (1,000 ha)	Yield (kg/ha)	Production (1000 t)
Artichoke	106	10,836	1,150
Cabbage	1,974	23,632	46,656
Cantaloupe and other melons	994	16,283	16,190
Carrot	770	21,396	16,477
Cassava	16,322	9,983	162,942
Cauliflower	681	18,681	12,725
Chickpea	12,009	742	8,908
Chilies and pepper, green	1271	11,064	14,068
Cucumber and gherkin	1424	16,182	23,051
Dry bean	27,470	679	18,639
Dry broad bean	2,355	1,499	3,531
Dry pea	6,515	1,680	10,945
Eggplant	722	16,589	11,981
Garlic	986	10,549	10,401
Green bean	491	7,376	3,620
Green pea	806	6,467	5,214
Lentil	3,389	832	2,819
Onion, dry	2204	16,174	35,644
Potato	18,353	16,065	294,834
Pumpkin, squash, and gourd	768	12,789	9,822
Sweet potato	9,156	14,662	134,244
Taro	1,016	5,650	5,739
Tomato	3,094	27,435	84,873
Watermelon	2,393	16,601	39,725
Yam	3,173	10,435	33,110
Total	*118,442*	*–*	*1,007,308*

Figure 3. A retail vegetable display in Shanghai, China.

VEGETABLE CONSUMPTION

Estimated per capita consumption of commercially produced vegetables in the United States in 1997 (8) is shown in Table 6. Total consumption varies from 0.2 kg each for artichoke, eggplant, and dry pea to 64.7 kg for potato. Tomato at 41.7 kg is the second most popular vegetable in the United States. Substantial quantities of sweet corn and lettuce are also consumed. The large number of distinct vegetables for which data are available is indicative of the diversity available to American consumers. In addition, the "other vegetable" category accounts for several additional vegetables. Total U.S. per capita consumption

Table 5. World Vegetable Production—Leading Countries (1996)

Vegetable	Rank 1	2	3	4	5
Artichoke	Italy	Spain	Argentina	France	United States
Cabbage	China	Russian Fed.	India	Korea Rep.	Japan
Cantaloupe and other melons	China	Turkey	Iran	United States	Spain
Carrot	China	United States	Russian Fed.	Poland	United Kingdom
Cassava	Nigeria	Brazil	Zaire	Thailand	Indonesia
Cauliflower	India	China	France	Italy	United Kingdom
Chickpea	India	Turkey	Pakistan	Iran	Australia
Chilies and peppers, green	China	Turkey	Nigeria	Mexico	Spain
Cucumber and gherkin	China	Iran	Turkey	United States	Japan
Dry bean	India	Brazil	China	Mexico	United States
Dry broad bean	China	Egypt	Ethiopia	Morocco	Italy
Dry pea	France	Ukraine	Canada	China	Russian Fed.
Eggplant	China	Turkey	Japan	Egypt	Italy
Garlic	China	Korea Rep.	India	United States	Spain
Green bean	China	Turkey	Indonesia	Spain	Italy
Green pea	United States	China	France	United Kingdom	India
Lentil	India	Turkey	Canada	Bangladesh	Syria
Onion, dry	China	India	United States	Turkey	Japan
Potato	China	Russian Fed.	United States	Poland	Ukraine
Pumpkin, squash, and gourd	China	Ukraine	Turkey	Mexico	Egypt
Sweet potato	China	Indonesia	Uganda	Vietnam	Rwanda
Taro	Ghana	China	Nigeria	Ivory Coast	Japan
Tomato	China	United States	Turkey	Italy	Egypt
Watermelon	China	Turkey	Iran	United States	Korea Rep.
Yam	Nigeria	Ivory Coast	Ghana	Benin	Togo

Table 6. Per Capita Annual Consumption of Commercially Produced Vegetables—United States (1997)

Vegetable	Fresh	Canned	Frozen	Total
Artichoke, all	–	–	–	0.2
Asparagus	0.3	0.1	0.1	0.5
Bean, dry, all	–	–	–	3.5
Bean, snap	0.7	1.8	0.9	3.3
Broccoli	2.3	–	1.1	3.4
Cabbage	4.8	0.6	–	5.4
Cantaloupe	5.5	–	–	5.5
Carrot	5.8	0.6	1.2	7.6
Cauliflower	0.7	–	0.2	1.0
Celery	3.0	–	–	2.7
Corn, sweet	3.7	4.8	4.3	12.8
Cucumber	2.8	2.3	–	5.1
Eggplant, all	–	–	–	0.2
Garlic, all	–	–	–	0.9
Honeydew melon	1.2	–	–	1.2
Lettuce, head	10.9	–	–	10.9
Lettuce, leaf and romaine	2.9	–	–	2.9
Mushroom	1.0	0.8	–	1.8
Onion[a]	8.6	–	–	8.6
Pea, green	–	0.7	1.0	1.6
Pea & lentil, dry, all	–	–	–	0.2
Pepper, bell	3.2	–	–	3.2
Pepper, chile	–	2.8	–	2.8
Potato[b]	21.8	–	42.9	64.7
Spinach	0.3	0.3	0.3	0.8
Sweet potato, all	–	–	–	2.1
Tomato	8.6	33.2	–	41.7
Watermelon	7.7	–	–	7.7
Other vegetables, all	–	–	–	4.8
Total	*95.8*	*48.0*	*9.1*	*207.1*

[a]Includes fresh and processed onion.
[b]Includes all processed potato.

Table 7. U.S. Per Capita Annual Use of Commercially Produced Vegetables, 1970–1998

Year	Fresh[a]	Canned	Frozen[b]	Total
1970	83.1	44.4	34.1	161.6
1975	78.6	42.9	38.5	160.0
1980	79.2	45.1	36.2	160.5
1985	86.1	43.4	43.2	172.7
1990	91.1	48.7	44.3	184.1
1995	96.3	48.3	58.8	203.4
1998	103.6	47.1	53.5	204.2

[a]Includes fresh and processed sweet potato and mushroom.
[b]Includes canned and frozen potato.

ons (watermelon, cantaloupe, honeydew) 47%, lettuce 37%, onion 68%, bell pepper 322%, tomato 24%, potato 17%, and mushrooms 307%. Only a few vegetables recorded less consumption during the period, notably celery, green peas, and sweet potato (8).

What factors have fueled the increases in vegetable consumption? The industry-financed "Five a Day" program has focused attention on the health benefits of vegetable and fruit consumption. This and other educational programs have promoted vegetables in general and even specific constituents such as antioxidants, lycopene, and beta-carotene for their general and specific health benefits. The trend toward increased vegetable consumption is expected to continue as consumers gain a greater understanding of the role of vegetables in human health.

VEGETABLE IMPROVEMENT

Vegetables cultivated by the very first farmers were similar to those that had been collected in the wild by their immediate predecessors. Improvement came slowly as the result of careful observation by the best farmers or by acquisition from neighbors via trade transactions. Surely the need for food during famine necessitated consumption of valuable seeds or other propagating material, so the improvement process was temporarily set back. Early horticulturists would prize and select for such qualities as ability to thrive under stressful growing conditions, good keeping quality, and good taste. The development of the earliest varieties occurred over time by careful selection and isolation of germplasm. The seed trade probably had its beginning in barter but gradually became a distinct enterprise toward the end of the eighteenth century. According to Nonnecke (9), David Landreth of Philadelphia established the first commercial seed business in the United States in 1784. The area around Philadelphia became the center of the vegetable seed business in the early nineteenth century with the establishment of several other firms including Bernard McMahon. Large-scale commercial bean and pea seed production was initiated in New York state by the Keeney Seed Company shortly thereafter.

As the country developed westward, it was found that the low humidity and low rainfall there during the growing

of over 200 kg is impressive. For comparison, per capita consumption of Chinese cabbage, cucumber, tomato, cabbage, eggplant, kidney bean, celery, sweet pepper, and Chinese chive was 195 kg in Beijing, China, in 1995, according to Chen and Liu (7).

The data in Table 6 present a snapshot of U.S. per capita vegetable consumption in 1997. But what have been the trends in vegetable consumption in recent times? Per capita consumption of fresh, canned, frozen, and total vegetables (8) from 1970 to the present at 5-year intervals is shown in Table 7. Over the period fresh vegetable consumption increased from about 83 to 104 kg, canned vegetable consumption from about 44 to 47 kg, frozen vegetable consumption from 34 to 54 kg, and total consumption from 162 to 204 kg. Accordingly, there have been major increases in fresh and frozen vegetable consumption, but not much change in the consumption of canned vegetables over the period 1970 through 1998. Much of the increase in frozen vegetable consumption, however, can be attributed to the increased popularity of potato products, notably french fries.

There were major increases in consumption of certain vegetables during the period. For example, broccoli consumption increased 500%, carrot consumption 75%, mel-

season resulted in greatly improved quality, including disease-free seed. Much of the open-pollinated and some of the hybrid vegetable seed production is still centered in the western United States.

Continued improvement in vegetable varieties took place during the nineteenth and early twentieth centuries by careful selection. As knowledge of the new science of genetics became more commonplace, its power was utilized to make controlled and planned improvements in the performance of vegetable varieties. Because these newly developed varieties were all true breeding, the successes obtained by one company could quickly be utilized by competitors.

All of this changed beginning in 1934 when Glen Smith at Purdue University developed the first widely accepted vegetable hybrid, Golden Cross Bantam sweet corn. Hybrids, because they are not true breeding, offer the opportunity for exclusivity by the originator and, if successful, a very substantial financial incentive. The grower also benefits from hybrids by use of seed that produces a uniform crop with useful characters, such as disease resistance from both parents combined into the single hybrid, albeit at higher seeds costs. From this early beginning, hybrids have become the norm in many of the important vegetables—beet, broccoli, cabbage, carrot, sweet corn, cucumber, eggplant, melons, onion, pepper, spinach, and tomato. The promise of exclusivity with hybrids has contributed to the development of major national and international firms devoted to vegetable seed production and sales.

Currently, the use of molecular genetics is being exploited in vegetable improvement. Using these techniques, foreign genes can be introduced into vegetable germplasm. Despite some public concern, products such as virus-resistant summer squash, beetle- and virus-resistant potato, and worm-resistant sweet corn are now entering the marketplace. The possibilities for vegetable improvement using these techniques in concert with the knowledge of the trained horticulturist appear to be very great.

VEGETABLE GROWTH REQUIREMENTS

Vegetable plants, like all plants, thrive and produce successful crops when factors required for growth are present in suitable proportions. These factors are enumerated and discussed briefly.

Light

The ambient light conditions are employed for field production of vegetables. Field production is timed for favorable temperatures that coincide with summertimelike conditions and ample light for maximum growth and yield. Protected culture, on the other hand, occurs during the off-season, and crops produced in greenhouses, plastichouses, or tunnels may respond favorably to artificial light.

Water

Plants require water to maintain turgidity and replace that lost through transpiration to the surrounding environment. The water molecule is cleaved in photosynthesis to free H^+, which combines with CO_2 to produce simple sugars that can then be transformed into the myriad of compounds produced by plants. So crop yields are dependent on an adequate water supply provided through either rainfall or irrigation. Because of their high value, virtually all vegetable production in developed countries is irrigated. The recent introduction of microirrigation (Fig. 4), as a system for delivery of water directly to the crop in thin, polyethylene tubes, has enhanced the use of irrigation in developing countries where water for irrigation is often in critically short supply. The frequency and intensity of irrigation is dependent on the soil's water-holding capacity; prevailing weather, especially temperature, wind, and relative humidity; and characteristics of the plant (11). Those crops with shallow root systems (Table 8) require more frequent, lower volume irrigations than those with deep root systems.

Essential Elements

In addition to the organic elements carbon, hydrogen, and oxygen provided from carbon dioxide from the atmosphere and from water, certain inorganic elements are necessary for growth of all plants. The macronutrients nitrogen, phosphorous, and potassium are required in highest concentrations. The secondary elements calcium, magnesium, and sulfur are also necessary in substantial quantity, but in addition they affect soil pH, which has a profound effect on availability of other elements. Iron, boron, manganese, zinc, copper, and molybdenum are required in minute concentrations by plants and collectively are termed micronutrients.

Medium

The traditional medium for field and protected culture of vegetables has been native soil, often amended with organic matter and inorganic fertilizer and sometimes treated with limestone or sulfur compounds for pH adjustment. This is still the situation for field culture but is rarely so for protected culture. Various materials, including peat, perlite, rockwool, straw bales, and liquid cultures, have replaced soil as a growing medium in protected culture. These materials provide superior physical characteristics and are generally free of pathogenic organisms. Essential elements are provided on a scheduled basis when soil substitutes are used for vegetable culture.

Carbon Dioxide

Carbon dioxide (CO_2) is present in the atmosphere at an average concentration of 0.03%. There is no practical way yet of altering the ambient CO_2 concentrations under field conditions. Experimental results have been inconclusive when CO_2-enriched water has been introduced into low tunnels in the field. There has been considerable success in growth and yield enhancement, however, when CO_2 concentrations are elevated to about 0.10% in protected culture. This practice is universal in areas like western Europe, where ventilation of structures is not required on a continuous basis and the additional CO_2 is trapped in the structure.

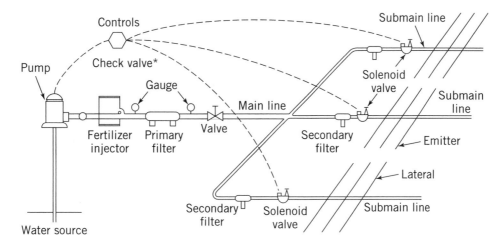

*A backflow preventer or vacuum breaker is required in some areas.

Figure 4. Microirrigation system components. *Source:* Ref. 10, reprinted with permission.

Table 8. Characteristic Maximum Rooting Depths of Various Vegetables

Shallow (45–60 cm)	Moderately deep (90–120 cm)	Deep (>120 cm)
Broccoli	Bean, bush	Artichoke
Brussels sprouts	Bean, pole	Asparagus
Cabbage	Beet	Bean, lima
Cauliflower	Cantaloupe	Parsnip
Celery	Carrot	Pumpkin
Chinese cabbage	Chard	Squash, winter
Corn	Cucumber	Sweet potato
Endive	Eggplant	Tomato
Garlic	Mustard	Watermelon
Leek	Pea	
Lettuce	Pepper	
Onion	Rutabaga	
Parsley	Squash, summer	
Potato	Turnip	
Radish		
Spinach		

Source: Ref. 10, reprinted with permission.

Oxygen

Postharvest life of certain vegetables is enhanced by use of controlled-atmosphere storage whereby O_2 concentration is lowered with concomitant increases in CO_2 and H_2, which reduces respiration rates.

Temperature

Each vegetable has a specific minimum, maximum, and optimum temperature for growth at a particular life cycle stage. These cardinal temperatures may be quite different as the plant passes through seed germination, establishment, vegetative growth, reproductive growth, and maturation. Consequently, a general scheme (Table 9) giving a range of temperatures usually suffices (11). The frost sensitivity of vegetables is of special interest in that it deter-

mines time of planting in areas subject to frosts. For example, one would feel quite comfortable planting peas any place in New England in April, but it would be extremely unwise to plant tomatoes in the field at that time. Peas are extremely frost tolerant; in my own experience peas 8 cm high were totally unaffected by a late-season snowfall of about the same depth. On the other hand, tomato will be irreparably damaged at temperatures slightly below 0°C.

PROPAGATION OF VEGETABLES

Most vegetables are propagated by seed, either by planting directly in the field or transplanting seedlings that were started from seeds in protected culture (Fig. 5) some weeks previously. The latter procedure is more conservative of costly seeds, provides a more hospitable environment for fragile seedlings, and lengthens the growing season in areas having a restricted period of favorable weather. Vegetative propagation (Table 10) is necessary for some vegetables that do not breed true from seeds or do not readily produce seeds. Various organs or organ sections may be used for propagation, depending on the crop. Although seeds may be used to propagate globe artichoke and potato, vegetative organs are by far the most common in commercial uses. Vegetative propagation is used only when sexual (seed) propagation is not possible because, at least in temperate areas, the expense of storage from one season to the next is extremely great. Vegetative planting stock may be stored *in situ* in tropical areas.

VEGETABLE-PRODUCTION SYSTEMS

Methods employed for commercial vegetable production vary with geographical location, the crop being grown, season of production, and sometimes by the grower and resources available for production. Consequently, it is possible to discuss production systems not in specific terms but only in generalities.

Table 9. Temperature Demands and Frost Sensitivities of Vegetables

	Frost sensitivity		Frost sensitivity
Hot: Optimum range 25–27°C; growth range 18–35°C			
Okra	+	Cantaloupe	+
Roselle	+	Hot pepper	+
Watermelon	+	Sweet potato	+
Warm: Optimum range 20–25°C; growth range 12–35°C			
Cucumber	+	Sweet corn	+
Eggplant	+	Tomato	+
Sweet pepper	+	Bean	+
Pumpkin, squash	+	Purslane	u
New Zealand spinach	u		
Cool-hot: Optimum range 20–25°C; growth range 7–30°C			
Taro	u	Chicory	−
Globe artichoke	u	Bok choy	−
Onion, shallot	−	Scorzonera	−
Leek	−	Chives	−
Garlic	−		
Cool-warm: Optimum range 18–25°C; growth range 7–25°C			
Pea	−	Red beet	(−)
Broad bean	−	Swiss chard	(−)
Cauliflower	−	Spinach	−
Broccoli	−	Lettuce	(+)
Cabbage	−	Endive	−
Kohlrabi	−	Carrot	−
Kale	−	Celery, celeriac	(+)
Brussels sprout	−	Parsnip	−
Turnip	−	Potato	+
Rutabaga	−	Lamb's lettuce	−
Chinese cabbage	−	Rhubarb	−
Parsley	−	Asparagus	(+)
Fennel	(+)	Horseradish	−
Dill	(+)	Garden cress	(+)
Radish	(+)		

Note: + indicates sensitive to weak frost; − indicates relatively insensitive; ( ) indicates uncertain; u indicates undetermined.

FIELD PRODUCTION OF VEGETABLES

Traditional Practices

Vegetables are high-value crops with little or no tolerance for blemished products in the marketplace. Accordingly, vegetable growers have used the best agricultural practices of the day, including such things as large additions of composted manure to the soil, use of appropriate crop rotations, use of cover crops in the off-season to prevent soil erosion and to trap residual fertilizers, and early disposal of crop residue to assist in pest control. In addition, successful vegetable growers routinely have provided inputs such as quality seeds, abundant fertilizer, pesticides, irrigation, and weed control together with extensive use of

Figure 5. Production of vegetable transplants for field setting has become a specialized segment of the vegetable industry.

Table 10. Vegetative Propagation of Vegetables

Vegetable	Propagule
Cassava	Stem cuttings
Ginger	Rhizome divisions
Globe artichoke	Divisions, suckers, seeds
Horseradish	Rhizome divisions
Jerusalem artichoke	Tubers
Potato	Tubers, seeds
Rhubarb	Divisions
Sweet potato	Slips from roots, vine cuttings
Taro	Corms, cormels
Yam	Tubers, crowns, sets

labor and a high level of management for profitable yields of quality produce. Advanced growers use south-facing slopes to produce earlier crops, planted (Fig. 6) or constructed windbreaks to restrict damaging winds early in the season, and various techniques for frost prevention.

Enhanced Inputs

Research is continually developing new technologies for crop production. To remain competitive and enhance effi-

Figure 6. A sugarcane windbreak creates a favorable microclimate for pepper production in south Florida.

ciencies, the most advanced growers adopt these technologies as soon as available. For some growers use of polyethylene mulch with microirrigation/fertigation is new technology even though it has been commonplace for some time in other production areas. Integrated crop management (ICM) is a concept whereby all production and pest-management decisions are considered before invoking a single variable. Best management practices for judicious use of fertilizers is an integral part of ICM, as is precision farming that provides for chemical applications where needed rather than farmwide. Enhanced germplasm fits particularly well into the ICM concept. Together these practices provide for sustainability of the vegetable enterprise.

Organic Production

Some consumers desire products that are produced without chemicals, even though growers use extreme care in chemical use and governmental agencies monitor pesticide residues. This has led to an alternative cultural system known as organic farming. Organic farming is a production system that avoids or excludes the use of synthetically compounded fertilizers, pesticides, or growth regulators. It relies on crop rotations, mechanical cultivation, and biological pest control to maintain soil productivity; to supply nutrients; and to control insects, diseases, and weeds. Specific aspects of organic farming are defined in several state laws. National organic-farming guidelines are being developed. Organic vegetable production represents a very small but rapidly growing part of the industry.

Protected Culture

Vegetable production in glasshouses, plastichouses (Fig. 7), and high tunnels (Fig. 8) is the most rapidly growing segment of the entire vegetable industry. Nearly complete control of the growing environment is achieved in modern growing structures. When combined with superior genetic material, yields are manyfold those of field-grown vegetables. For example, tomato yields in the Netherlands from protected culture are over 7 times greater than in the United States, mostly from field culture. Products derived

Figure 8. High tunnels for vegetable production in southern Spain.

from protected culture such as tomato, pepper, eggplant, cucumber, and lettuce are sold to discriminating, affluent buyers around the world.

VEGETABLE HARVEST

Vegetables grown for fresh market have a high value and a low tolerance for blemishes, so they are mostly hand harvested (12). Selection, sorting, grading, trimming if necessary, and packing usually are done in the field for some vegetables such as lettuce, cabbage, broccoli, and endive. Other vegetables like asparagus, tomato, and snap beans usually are brought to a central packing area where these tasks are performed. Some vegetables destined for the fresh market are usually harvested mechanically, including carrot, potato, brussels sprouts, and sweet potato. Snap beans for fresh market are hand harvested when prices are high and machine harvested (Fig. 9) when prices are low.

Whenever possible, vegetables destined for processing are harvested mechanically to reduce costs and provide the volume of product necessary for efficient plant operation. Virtually all of the beet, carrot, pea, bean, sweet corn, spin-

Figure 7. A vast expanse of plastichouses for vegetable production cover the landscape in southern Spain.

Figure 9. Machine harvest of snap beans in Miami, Dade County, Florida.

Table 11. Status of Hand versus Mechanical Harvest of Vegetables

Area hand harvested (%)	Vegetable			
76–100	Artichoke	Asparagus	Broccoli[a]	Cabbage
	Cauliflower	Celery	Cucumber[a]	Lettuce
	Green onion	Collards	Cress	Dandelion
	Eggplant	Endive	Escarole	Fennel
	Kale	Kohlrabi	Mushroom[a]	Okra
	Pepper	Rapini	Rhubarb[a]	Romaine
	Sorrel	Squash	Watercress	Cassava
	Celeriac	Ginger	Parsley root	Parsnip
	Rutabaga	Salsify	Turnip	Taro[a]
	Jerusalem artichoke			
51–75	Sweet potato	Mustard greens	Parsley	Swiss chard
	Turnip greeens			
26–50	Dry onion[a]	Pumpkin[a]	Tomato[a]	
0–25	Beet[a]	Carrot	Potato[a]	Lima bean[a]
	Snap bean[a]	Sweet corn[a]	Spinach[a]	Horseradish[a]
	Pea[a]	Garlic	Brussels sprouts[a]	Malanga
	Boniato	Radish		

[a]More than 50% of the crop is processed.
Source: Ref. 13, reprinted with permission.

ach, and potato crops for processing are harvested by mechanical means. A summary of the status of hand versus mechanical vegetable harvest is shown in Table 11.

POSTHARVEST HANDLING

Vegetables are perishable commodities. Maximum postharvest life varies from a few days to several months depending on the vegetable. Harvesting at the correct maturity, careful handling at harvest and during packing and shipping, good sanitation, timely and proper cooling (Fig. 10), shipping and storing at optimal temperature and relative humidity, and freedom from exposure to ethylene gas for some commodities will help to assure maximum postharvest life. The importance of rapid cooling and maintenance of recommended temperature throughout the postharvest period to the consumer cannot be overemphasized

as it is the most important factor in determining life of the product.

Shipping containers provide a measure of volume or weight and must be designed to protect the product. Their shape should conform to standard pallets (Fig. 11) for convenient vehicle loading and unloading. Vegetable-shipping

Figure 11. Pepper cartons designed to utilize the full pallet area are used in south Georgia.

Figure 10. Forced-air cooling provides rapid cooling of beans and other products.

containers may be constructed of burlap, paper, fiberboard, wood, or plastic. Because of the mass of materials at receiving points, there is increasing emphasis on materials that are recyclable or reusable.

PROJECTIONS FOR THE FUTURE

What are the expectations for the commercial vegetable industry as we enter the twenty-first century? Some thoughts on this question follow: Vegetables will continue to be consumed in greater quantities; there will be an ever-increasing variety of vegetables from which consumers can choose; organically grown vegetables will command a greater proportion of the market; the vegetable industry will be global rather than national as trade barriers are loosened; there will be fewer new pesticides available to growers and increased monitoring of pesticide usage; the use of genetically enhanced vegetable varieties will become commonplace; and the number of vegetable enterprises will decrease, but those remaining will be larger.

BIBLIOGRAPHY

Reprinted from the Florida Agricultural Experiment Station Journal Series No. R-06636.

1. The Staff of The Liberty Hyde Bailey Hortorum, *Hortus Third*, MacMillan Publishing Co., New York, 1976.
2. D. N. Maynard, "Vegetable," in S. J. Foderaro, ed., *Academic American Encyclopedia*, Vol. 19, Aretê Publishing Co., Princeton, N.J., 1980, pp. 531–534.
3. S. J. Kays and J. C. Silva Dias, *Cultivated Vegetables of the World*, Exon Press, Athens, Ga., 1996.
4. H. Chen et al., "Diversity of Edible Vegetables in China," in V. E. Rubatzky, H. Chen, and J. Y. Peron, eds., *Proc. Third Int. Symp. on Diversification of Vegetable Crops*, Acta Horticulturae No. 467, International Society for Horticultural Science, Leuven, Belgium, 1998.
5. Anonymous, *Vegetables, 1997 Summary*, Vg 1-2(98), U.S. Department of Agriculture Washington, D.C., 1998.
6. Anonymous, *FAO Production Yearbook 1996*, FAO Statistics Series No. 135, Food and Agriculture Organization of the United Nations, Rome, 1997.
7. D. Chen and M. Liu. "An Overlook at Vegetable Production, Consumption, Distribution and its Progress in China," in V. E. Rubatzky, H. Chen, and J. Y. Peron, eds. *Proc. Third Int. Symp. Diversification of Vegetable Crops*, Acta Horticulturae No. 467, International Society for Horticultural Science, Leuven, Belgium, 1998.
8. G. Lucier, *Vegetables and Specialties. Situation and Outlook*, VGS-275, U.S. Department of Agriculture, Washington, D.C., 1998.
9. I. L. Nonnecke, *Vegetable Production*, Van Nostrand Reinhold, New York, 1989.
10. D. N. Maynard and G. J. Hochmuth, *Knott's Handbook for Vegetable Growers*, 4th ed., John Wiley & Sons, 1997.
11. H. Krug, "Environmental Influences on Development, Growth and Yield," in H. C. Wein, ed., *The Physiology of Vegetable Crops*, CAB International, New York, 1997.
12. J. F. Thompson, "Harvesting Systems," in A. A. Kader, ed., *Postharvest Technology of Horticultural Crops*, 2nd ed. University of California, Oakland, Calif., 1992.

GENERAL REFERENCES

J. L. Brewster, *Onions and Other Vegetable Alliums*, CAB International, New York, 1994.

A. A. Kader, ed., *Postharvest Technology of Horticultural Crops*, University of California, Oakland, Calif., 1992.

D. N. Maynard and G. J. Hochmuth, *Knott's Handbook for Vegetable Growers*, 4th ed., John Wiley & Sons, New York, 1997.

R. W. Robinson and D. S. Decker-Walters, *Cucurbits*, CAB International, New York, 1997.

V. E. Rubatzky and M. Yamaguchi, *World Vegetables, Principles, Production, and Nutritive Value*, 2nd ed., Chapman and Hall, New York, 1997.

E. J. Ryder, *Lettuce, Endive, and Chicory*, CAB International, New York, 1999.

H. C. Wein, ed., *The Physiology of Vegetable Crops*, CAB International, New York, 1997.

R. Wills et al., *Postharvest, An Introduction to the Physiology & Handling of Fruit, Vegetables, & Ornamentals*, 4th ed., CAB International, New York, 1998.

DONALD N. MAYNARD
University of Florida
Bradenton, Florida

VEGETABLES, PICKLING

Vegetables may be preserved by fermentation, direct acidification, or a combination of these along with other processing conditions and additives to yield products that are referred to as pickles. Pasteurization and refrigeration are used to ensure stability of certain products. Organic acids and salt (NaCl) are primary preservatives for most types of pickles. Lactic acid is produced naturally in fermented products. Acetic acid (or vinegar) is the usual acid added to pasteurized, nonfermented (fresh-pack) pickles. Acetic acid also is added to many products made from fermented (salt-stock) cucumbers. Other preservatives such as sodium benzoate, potassium sorbate, and sulfur dioxide may be added to finished products. Although the term *pickles* in the United States generally refers to pickled cucumbers, the term is used herein in a broader sense to refer to all vegetables that are preserved by fermentation or direct acidification. Cucumbers, cabbage, olives, and peppers account for the largest volume of vegetables and fruits commercially pickled. Lesser quantities of onions, tomatoes, cauliflower, carrots, melon rinds, okra, artichokes, beans, and other produce also are pickled.

Although this article emphasizes pickling of cucumbers and cabbage, which are major vegetable commodities commercially preserved and consumed in the United States and Europe, and kimchi, which is a mixture of fermented vegetables consumed in Korea, most vegetables have been preserved by pickling, either commercially or in the home.

HISTORY

The preservation of vegetables by fermentation is thought to have originated before recorded history, and the technology was developed by trial and error. Pederson (1) presumed that early humans observed that when vegetables were flavored with salt or brine and packed tightly in a vessel, they changed in character but remained appetizing and nutritious. Pederson concluded that the Chinese were the first to preserve vegetables in this manner and assumed that fermentation in salt brines occurred first and that dry salting came later. The Chinese have been credited with introducing fermented vegetables into Europe.

Pasteurization, as a means of preserving acidified vegetables, was introduced in the United States in the 1940s. Before then, all pickled vegetables were preserved commercially by fermentation. Pasteurization resulted in a doubling of the per capita consumption of pickled vegetables by the 1950s. Refrigerated pickles were introduced into U.S. commerce in the 1960s and have been an increasing segment of the industry since.

PROCESSING

Processing methods for the preservation of cucumbers, cabbage, and kimchi will serve to illustrate the principles involved in vegetable pickling.

Cucumbers

Various authors have attributed the origin of the cucumber (*Cucumis sativus* L.) to Africa, China, India, or the Near East (2). Later domestication occurred throughout Europe, and cucumbers are now grown throughout the world using field or greenhouse culture, but with various characteristics, depending upon region.

Cucumbers are bred either for fresh market or processing (pickling). The fresh market varieties possess a relatively tough skin, which serves to extend their storage life as fresh produce. Pickling varieties, however, possess a thin, relatively tender skin. Pickling cucumbers are harvested in a relatively immature stage, before the seeds mature and before the seed area becomes soft and starts to liquify.

Cucumbers are harvested by hand or mechanical means, depending upon availability of labor, land size and conformation, and other factors. Cucumbers are a seasonal crop grown in various geographical regions and shipped to the processor. Great changes have occurred in the United States during the past 20 years as to the origin of the fruit that a processor receives. Although once a mainly regional and seasonal enterprise, some large processors now receive fresh cucumbers nearly the entire year. The fruit is grown from Mexico to Canada and shipped fresh to processors according to their demands. Cucumbers grown near the processor may be processed within 24 hours. Cucumbers are shipped under refrigeration if grown at distant locations from the processor. The demand for a year-round supply of fresh cucumbers varies according to the types of products that the processors manufacture. Brined cucumbers, being more stable, are transported intercontinentally.

Pickling cucumbers are preserved by three basic methods: fermentation (40% of overall production), pasteurization (40%), and refrigeration (20%), as outlined in Figure

1 (3). After pasteurization of fresh cucumber pickles was introduced into the U.S. industry in the 1940s, the consumption of pickles increased significantly because of the milder acid flavor and more uniform quality of the pasteurized product. The process involves heating properly acidified cucumbers to an internal temperature of 74°C and holding for 15 min (4). Some processors deviate from this standard process, depending upon their products and experiences. Fermented cucumbers also may be pasteurized to increase shelf stability, but at lower temperatures and times. Refrigerated pickles are preserved by addition of low concentrations of vinegar and a chemical preservative (eg, sodium benzoate), in addition to refrigeration at 1 to 5°C. Microbial growth in these products is not desired. Nonacidified, refrigerated pickles, originally popular among certain ethnic groups, also are marketed in some metropolitan areas. These products may or may not be allowed to undergo fermentation before refrigeration. After packaging, these nonacidified pickles undergo a slow lactic acid fermentation while under refrigeration.

Cabbage

The modern, hard-head cultivars of cabbage (*Brassica oleracea*) are reported to have descended from wild, nonheading *Brassias* originating in the eastern Mediterranean and in Asia Minor (5). Cabbage is grown for the fresh market and for the production of sauerkraut. For use in the production of sauerkraut, it is desired that the cabbage heads be large (typically 8–12 lb) and compact (ie, dense), have a minimum of green outer leaves, and possess desirable flavor, color, and textural properties when converted into sauerkraut.

Fresh cabbage for sauerkraut is harvested mechanically or by hand mostly from August to November. The cabbage is transported to the processor, where it is graded, cored, trimmed, shredded, and salted. The waste from the coring and trimming operations typically is returned to the field, where it is plowed into the soil. This waste constitutes about 30% of the weight of the fresh cabbage.

After shredding (about 1-mm thick), the cabbage is conveyed by belt, where salt is added, to the fermentation tanks. The tanks typically hold 20 to 180 tons of shredded cabbage. Most tanks today are constructed of reinforced concrete, but some wooden tanks remain. The tank is uniformly filled, heaped to extend slightly above the top of the tank, and loosely covered with plastic sheeting. After about 24 h, brine generated and located at the bottom of the tank is allowed to drain from the tank to allow the top of the cabbage to settle below the top of the tank. Then, the cabbage is manually distributed to create a slightly concave surface. Plastic sheeting then is placed on the surface, and water is added on top to weight it down and to provide an anaerobic seal. Gas generated during fermentation escapes by forcing its way between the tank wall and the cabbage, or through a plastic tube placed between the cabbage and the cover.

In the United States, the cabbage is allowed to remain in the tanks until at least 1% lactic acid is formed (about 30 days minimum, depending upon the temperature), and is stored beyond this time and until such time as needed

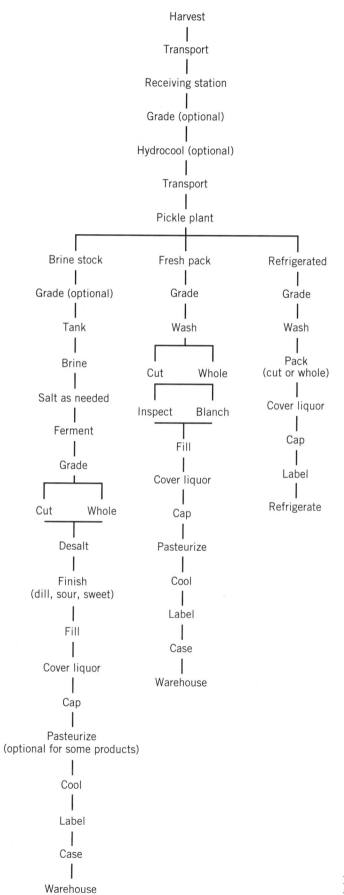

Figure 1. Flow diagrams for three methods of cucumber processing. *Source:* Ref. 3.

for processing. Thus, the sauerkraut tanks are used for fermentation as well as storage. Although extended periods of holding in the tanks can result in excess acid formation and waste generation due to the need to wash excess acid from the product before canning, this disadvantage is offset by the economic advantages of bulk storage. Also, most U.S. sauerkraut companies specialize in this product only and desire the option of further processing of sauerkraut throughout the year so as to distribute labor and equipment needs. This U.S. procedure differs significantly from that of many European manufacturers who process the sauerkraut into finished products when it reaches the desired level of titratable acidity (calculated as lactic acid, typically 1%). Thus, European manufacturers have greater control over product uniformity, but lose the economic advantages of bulk storage. Also, the European method results in less waste generation since less excess acid is produced.

The sauerkraut is removed pneumatically, by mechanical fork, or by hand from the tanks as needed for processing during the year. The sauerkraut may be packaged in cans, glass, or plastic containers. When packaged in glass or plastic, the product is not heated. Rather, sodium benzoate (0.1%, w/w) and potassium metabisulfite are added as preservatives, and the product is held under refrigeration (5°C).

Canned sauerkraut is preserved by pasteurization without the addition of preservatives. Heating is performed either by steam injection into a thermal screw or by a sauerkraut juice immersion-type cooker. The product is heated to 74 to 82°C for about 3 min and hot-filled into the cans. After closure, the cans are immediately cooled to less than 32°C.

Kimchi

Kimchi is a general term applied to a Korean product made by the lactic acid fermentation of salted vegetables (dry salted or brined) with or without secondary ingredients. Unless otherwise specified, however, kimchi generally means a product made from Chinese cabbage as the primary vegetable with secondary ingredients. Kimchis made from vegetables other than Chinese cabbage as the primary vegetable are specified with a qualifying word, eg, radish kimchi, cucumber kimchi, and green onion kimchi. Both solids and liquids of kimchis are consumed.

Kimchi has been a major table condiment in Korea for about 200 years. Kimchi provided a spicy and flavorful adjunct to a rather narrow choice of foods for Koreans in difficult times, complementing the bland taste of cooked rice.

Winter kimchi has been most traditional and is made between mid-November and early December, depending on the climate of the particular year. It is consumed until the following spring. The fresh ingredients are tightly filled into large earthen jars with earthen lids. The jars are partially buried (80–90% of container depth) under ground and are covered with bundles of rice straw for protection from direct sunlight and the climate. This procedure is still practiced in much of the rural areas and some households in urban areas. The subsurface temperature of the earth during the winter season is fit for slow but excellent fer-

mentation of kimchi and also for the subsequent storage. Overacidification and development of yeasty flavor are the two most important problems in the long-term storage of kimchi. However, as the temperature goes up late in the following spring, the quality of winter kimchi is reduced.

Today, lesser amounts of winter kimchi are made, and Koreans now enjoy freshly fermented kimchi anytime of the year owing to year-round availability of fresh vegetables, general availability of household refrigerators, and distribution of commercial products through efficient cold distribution systems. Also, it is common for Korean families to make small batches of kimchi as needed and to store it under refrigeration for short periods to extend the fresh quality.

Kimchis can be divided into two types simply depending on the way the primary vegetable is salted. Primary vegetables are salted (either by dry salting or brining) and then the liquid is drained off completely before blending with secondary ingredients in the preparation of general kimchis. However, sufficient brine is poured over the packed ingredients to cover whole cucumber fruits or radish roots. Pickles are usually prepared without secondary ingredients.

FERMENTATION AND MICROBIOLOGY

Most vegetable fermentations are the result of growth by naturally occurring lactic acid bacteria present on the raw vegetable and the concentration of salt present, which dictates the types and extent of growth by these bacteria. The concentration of salt used varies among commodities (Table 1) and depends upon the flavor and textural properties desired. Cucumbers tend to soften (enzymatically) more if brined below about 5% NaCl. Fermentation of vegetables occurs in various stages (Table 2). The principal species of lactic acid bacteria and their metabolic end products are summarized in Table 3.

Glucose and fructose are the principle sugars of cucumbers and cabbage. Cabbage also contains low levels of sucrose. The metabolic pathways involved in fermentation of these sugars to end products are illustrated in Figure 2.

Leuconostoc mesenteroides is an important species for initiating the fermentation of sauerkraut and kimchi because of its production of acetic as well as lactic acid and nonacidic end products (ethanol, mannitol), which results in a more flavorful product. However, this species is rather acid sensitive and soon is overtaken by *Lactobacillus plan-*

Table 1. Brining Procedures for Vegetables

Method of salting	Concentration (%, w/v) of salt		Vegetable
	Fermentation	Storage	
Dry salting	1.5–2.5	1.5–2.5	Cabbage
Dry salting	3	3	Kimchi
Brine solution	5–8	8–16	Cucumbers

Note: The concentrations of salt indicated generally are used for the vegetables listed. Wide variations exist in the salt concentration used for peppers, onions, and cauliflower.

Table 2. Stages of Microbial Activities During the Natural Fermentation of Vegetables

Stage	Prevalent microorganisms (conditions)
Initiation of fermentation	Various Gram-positive and Gram-negative bacteria
Primary fermentation	Lactic acid bacteria, yeasts (sufficient acid has been produced to inhibit most bacteria)
Secondary fermentation	Fermentative yeasts (when residual sugars remain and lactic acid bacteria have been inhibited by low pH)
	Spoilage bacteria (degradation of lactic acid when pH is too high and/or salt/acid concentration is too low, eg, propionic acid bacteria, clostridia)
Postfermentation	Open tanks: surface growth of oxidative yeasts, molds, and bacteria
	Anaerobic tanks: none (provided the pH is sufficiently low and salt or acid concentrations are sufficiently high)

Source: Ref. 6.

Table 3. Lactic Acid-Producing Bacteria Involved in Vegetable Fermentations

Genus and species	Fermentation type[a]	Main product (molar ratio)	Configuration of lactate
Streptococcus faecalis	Homofermentative	Lactate	L(+)
Streptococcus lactis	Homofermentative	Lactate	L(+)
Leuconostoc mesenteroides	Heterofermentative	Lactate/acetate/CO_2 (1:1:1)	D(−)
Pediococcus pentosaceus	Homofermentative	lactate	DL, L(+)
Lactobacillus brevis	Heterofermentative	Lactate/acetate/CO_2 (1:1:1)	DL
Lactobacillus plantarum	Homofermentative	Lactate	D(−), L(+), DL
	Heterofermentative[b]	Lactate/acetate (1:1)	D(−), L(+), DL
Lactobacillus bavaricus	Homofermentative	Lactate	L(+)

Source: Ref. 7.
[a]With respect to hexose fermentation.
[b]Heterofermentative with respect to pentoses (facultatively heterofermentative).

tarum, which produces mainly lactic acid and can result in a harsh, too-acidic product. However, *L. mesenteroides* (a heterofermentor) produces CO_2 (a gas) that can cause bloater damage (gas pockets) in fermented cucumbers. For this reason, *L. plantarum* or *Pediococcus pentosaceus* (homofermentor of hexoses, ie, produces only lactic acid and no CO_2) is preferred for cucumber fermentations. Fortuitously, the texture of cabbage is retained at relatively low NaCl concentrations (1.5–2.5%), which is conducive to growth by *L. mesenteroides*. The higher salt concentration required for texture retention of fermented cucumber (5–8%) prevents growth by *L. mesenteroides*, but not *L. plantarum*, which is a preferred species to dominate cucumber fermentation because of its acid tolerance and homofermentative metabolism.

NUTRITIVE VALUE

The chief nutritive value of vegetables is in their content of vitamins, which may include important amounts of β-carotene, ascorbic acid, and folic acid, and less important amounts of riboflavin and some of the other B vitamins. Vitamin B_{12} is not present in vegetables. Vegetables used for fermentation are an important source of fiber, but contain low levels of protein and calories.

The effect of brining and fermentation on the nutritive value of vegetables has been reviewed (8–10); however, relatively little information is available on the subject. Lactic acid bacteria conventionally important in vegetable fermentations are nutritionally fastidious and would not be expected to increase essential nutrients during fermentation.

Although microbial action may alter the nutrient content of vegetables during fermentation, other factors may significantly influence the retention of nutrients during storage and further processing. If the vegetables are brined at high levels of salt, large losses in nutrients will result when the vegetables are subsequently desalted before use. From a nutritive standpoint, therefore, it would be preferable to brine vegetables at a low level of salt. Unfortunately, softening and other spoilage problems dictate the use of higher levels of salt in some vegetables than would otherwise be desirable. There is nearly complete loss of ascorbic acid from salt-stock cucumbers, which are stored at 8 to 16% salt, when the cucumbers are desalted to 2 to 4% salt for use in finished products. Fellers (9) reported 86, 82, and 28% losses for vitamin C (ascorbic acid), thiamin, and carotene, respectively, in desalted salt-stock cucumbers; but, in genuine dills, which are not desalted, 33 to 60% vitamin C retention was found. Sauerkraut requires only 2 to 3% salt for preservation and, therefore, is not desalted before use. Thus, vitamin C is largely retained during storage of sauerkraut, but may be diminished during later processing, depending upon exposure to air and the extent of heating.

Ascorbic acid and thiamin are stabilized by acid and exclusion of oxygen. Because fermented vegetables are normally held under such conditions, good retention of these vitamins would be expected. Current nutritional labels on commercial sauerkraut show a usual ascorbic acid content of 50% of the U.S. RDA per 100 g serving.

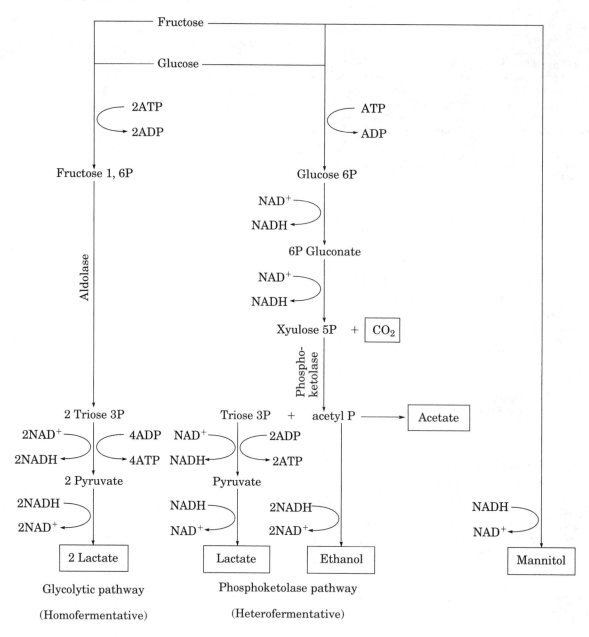

Figure 2. Major pathways for sugar fermentation by lactic acid bacteria. *Source:* Ref. 12.

SHELF LIFE AND BULK STORAGE

The shelf life of pasteurized pickles typically is 12 to 18 months, depending on storage temperature (22°C or below preferred). The shelf life of refrigerated pickles varies from a few weeks (nonacidified type) to several months (acidified type, with preservatives). The shelf life of sauerkraut depends largely on the integrity of the can. With enamel-lined cans, the shelf life typically is 18 to 30 months. Sauerkraut packaged in glass or plastic bags usually contains preservatives such as potassium bisulfite (for maintaining a light color) and sodium benzoate or potassium sorbate (as microbial inhibitors), which results in a shelf life of 8 to 12 months.

Fermented cucumbers held at 8% NaCl or higher before being processed into finished products can be held for a year or longer. The storage life is extended at higher salt concentrations. Sauerkraut is held at relatively low salt concentration (1.5–2.5%) in bulk storage, and typically is held for up to a year (United States) or only for a month or so (Europe).

SAFETY

There are no authenticated reports of pathogenic microorganisms associated with standard commercial pickle products prepared under good manufacturing practices of acid, salt, and sugar content (and combinations thereof) from brined, salted, and pickled vegetable brine-stock, including cucumbers. The Commissioner of the FDA stated that "No instances of illness as the result of contamination of commercially processed fermented foods with *Clostridium botulinum* have been reported in the United States" (11). Essentially the same pattern of consumer safety applies to fresh-pack (pasteurized) pickle products. The pasteurization process calls for the packed product to be acidified at the outset with a sufficient amount of food-grade organic acid (eg, vinegar, acetic acid, or lactic acid) to result in an equilibrated brine product pH of 4.0 or below (preferably 3.8). Vinegar (acetic acid) is usually the acidulant of industry choice for cucumber pickle products. The basic pasteurization procedure, with acidified product heated to an internal temperature of 74°C and held for 15 min, has been used successfully by industry since about 1940. Insufficient acidification of pasteurized pickles can result in butyric-acid-type spoilage, possibly involving public health concerns.

STARTER CULTURES

Currently, most vegetable fermentations rely on the natural microflora, although use of starter cultures has been suggested by various authors. Various reasons have been proposed to explain the lack of commercial use of cultures, including economics, lack of sufficiently unique and valuable properties to justify their use, and the fact that vegetables will undergo a natural lactic fermentation under proper environmental conditions (12). However, recent research with cucumbers, sauerkraut, and olives indicate that use of special cultures for vegetable fermentations may find application in the near future. Examples of novel properties of lactic acid bacteria cultures that make them potentially useful in commercial vegetable fermentations:

- *Inability to degrade malic acid to lactic acid and CO_2.* A culture of *L. plantarum* was developed by mutation and selection for use in cucumber fermentation. Because this culture does not produce CO_2 from malic acid, the natural acid present in cucumbers, it is being evaluated for use to ferment whole cucumbers to avoid the need for purging to prevent bloater damage.

- *Inability to produce biogenic amines.* Biogenic amines such as histamine are formed by decarboxylation of amino acids. Selection of cultures lacking this ability has been considered.

- *Exclusive l(+)-lactic acid formation.* Health concerns have been raised about the production of D(-)-lactic acid (10). Starter cultures that produce only L(+)-lactic acid have been developed.

- *Bacteriocin production.* Cultures of lactic acid bacteria that produce nisin and other bacteriocins have been isolated and may prove useful in inhibiting the growth of undesired bacteria in vegetable fermentations.

- *Bacteriophage resistance.* Bacteriophage sensitivity of starter cultures for use in preparing fermented dairy products is a serious concern. Although bacteriophage problems are not currently apparent with fermented vegetables, the problem may become significant when the use of starter cultures becomes more prominent.

BIBLIOGRAPHY

1. C. S. Pederson, *Microbiology of Food Fermentations*, 2nd ed., AVI Westport, Conn., 1979.
2. C. H. Miller and T. C. Wehner, "Cucumbers," in N. A. M. Eskin, ed., *Quality and Preservation of Vegetables*, CRC Press, Boca Raton, Fla., 1989, pp. 245–264.
3. H. P. Fleming and W. R. Moore, "Pickling," in I. A. Wolff, ed., *CRC Handbook of Processing and Utilization in Agriculture*, Vol. II, CRC Press, Boca Raton, Fla., 1983, pp. 397–463.
4. R. J. Monroe et al., "Influence of Various Acidities and Pasteurizing Temperatures on the Keeping Quality of Fresh-Pack Dill Pickles," *Food Technol.* **23**, 71–77 (1969).
5. M. H. Dickson and D. H. Wallace, "Cabbage Breeding," in M. A. Bassett, ed., *Breeding Vegetable Crops*, AVI, Westport, Conn., 1986, pp. 395–432.
6. H. P. Fleming, K. H. Kyung, and F. Breidt, "Vegetable Fermentations," in H. J. Rehm and G. Reed, eds., *Biotechnology*, Vol. 9, VCH, New York, 1995, pp. 629–661.
7. O. Kandler, "Carbohydrate Metabolism in Lactic Acid Bacteria," *Antonie van Leeuwenhoek* **49**, 209–224 (1983).
8. I. D. Jones and J. L. Etchells, "Nutritive Value of Brined and Fermented Vegetables," *Am. J. Public Health* **34**, 711–718 (1944).
9. C. R. Fellers, *Effects of Fermentation on Food Nutrients*, in R. S. Harris and H. Von Loesecke, eds., AVI, Westport, Conn., 1960, p. 161.
10. R. F. McFeeters, "Effects of Fermentation on the Nutritional Properties of Food," in E. Karmas and R. S. Harris, eds., *Nutritional Evaluation of Food Processing*, 3rd ed., Van Nostrand, New York, 1988, pp. 423–446.
11. FDA, "Acidified Foods and Low-Acid Canned Foods in Hermetically Sealed Containers," *U.S. Food Drug Admin.*, Fed. Reg. **44**, 16204, 1979.
12. H. P. Fleming, R. F. McFeeters, and M. A. Daeschel, "The Lactobacilli, Pediococci, and Leuconostocs: Vegetable Products," in S. E. Gilliland, ed., *Bacterial Starter Cultures for Foods*, CRC Press, Boca Raton, Fla., 1985, pp. 97–118.

GENERAL REFERENCES

S. Davidson et al., *Human Nutrition and Dietetics*, Churchill Livingstone Inc., New York, 1979.

J. L. Etchells, T. A. Bell, and W. R. Moore, Jr., "Refrigerated Dill Pickles—Questions and Answers," *Pickle Pak Science* **5**, 1–20 (1976).

H. P. Fleming and R. F. McFeeters, "Use of Microbial Cultures: Vegetable Products," *Food Technol.* **35**, 84–88 (1981).

H. P. Fleming, R. F. McFeeters, and M. A. Daeschel, "Fermented and Acidified Vegetables," in C. Vanderzant and D. F. Splittstoesser, eds., *Compendium of Methods for the Microbiological Examination of Foods*, 3rd ed., American Public Health Association, Washington, D.C., 1992, pp. 929–952.

H. O. Lee and I. W. Yang, "Studies on the Packaging and Preservation of Kimchi," *J. Korean Agricultural Chemical Soc.* **13**, 207–210 (1975).

J. R. Stamer, "Lactic Acid Fermentation of Cabbage and Cucumbers," in H. J. Rehm and G. Reed, eds., *Biotechnology*, Vol. 5, Verlag Chemie, Weinheim, Germany, 1983, pp. 365–378.

J. R. Stamer and B. O. Stoyla, "Stability of Sauerkraut Packaged in Plastic Bags," *J. Food Protection* **41**, 525–529 (1978).

HENRY FLEMING
USDA/ARS
Raleigh, North Carolina

VEGETARIAN AND VEGETARIAN-AWARE EATING TRENDS

Today, more Americans are willing to make meatless choices, at least occasionally, than were willing to do so in the past. Vegetarian menus are more acceptable among shoppers today, and they are seen as a healthful and smart way of eating. Americans are not giving up meat altogether, but they are no longer as loyal to meat on the menu. They are seeking solutions for preventing disease and maintaining daily good health through vegetarian food choices.

Two percent of shoppers say they always maintain a vegetarian diet, and 26% say they usually (7%) or sometimes (19%) do so. This 28% of shoppers drives demand for vegetarian products on grocery shelves, in restaurants, and at school cafeterias.

Growth in the market for vegetarian products in recent years has been largely fueled by vegetarian-aware shoppers; the number of true vegetarian shoppers has remained relatively stable. Not interested in avoiding meat altogether; vegeterian-aware shoppers eat meatless meals more often for health reasons; they seek out fruits, vegetables, and whole grains and adopt meatless burgers, soy hot dogs, and other vegetarian choices.

WHAT VEGETARIAN MEANS

As a dietary habit or lifestyle, vegetarianism includes a spectrum of dietary choices:

- Vegans do not eat any animal-derived products.
- Ovo-vegetarians do not eat meats or dairy products, but do eat eggs.
- Lacto-ovo-vegetarians do not eat meats, but do eat eggs and dairy products.
- Pesco-vegetarians do not eat meats, but do eat eggs, dairy products, and fish.
- Semi-vegetarians do not eat red meats, but do eat eggs, dairy products, fish, and poultry.
- Vegetarian-aware shoppers do not avoid meats, dairy products, or eggs altogether, but are trying to eat meatless meals more often.

A BRIEF HISTORY

Vegetarianism has existed for centuries, often as a tenant of eastern religions and philosophies. In the United States, vegetarianism has Christian roots. Reverend William Metcalfe introduced vegetarianism to the United States in 1817. In 1863, the Seventh-Day Adventist church was founded, and members adopted eating guidelines prohibiting meat, tobacco, alcohol, tea, and coffee consumption. Around the turn of the century, Dr. John Harvey Kellogg, of cereal fame, founded a sanitarium in Battle Creek, Michigan, that advocated vegetarianism and healthy living. During the 1960s, vegetarianism became tied to the counter-culture as environmental issues fueled new interest in vegetarian eating habits. In 1971, Frances Moore

Lappé harshly examined the meat industry and the resources necessary to support it in her book, *Diet for a Small Planet*. An immensely popular book, *Diet for a Small Planet* increased awareness of the environmental reasons for becoming a vegetarian among the American public.

NUTRITIONALLY MISUNDERSTOOD

Until recently, the health benefits of vegetarian diets have been largely misunderstood. Just 25 years ago, vegetarian diets were predicted to give rise to osteoporosis, anemia, and protein deficiencies. Vegetarian eating habits are now nutritionally acceptable and are generally recognized as healthy. Although vegetarian once meant counter-culture, vegetarian today is viewed by mainstream consumers as a smart way to eat healthfully. Male and female vegetarians have significantly less occurrence of cancer, heart disease, and obesity.

Once the domain of health food or other specialty brands, major brands such as Green Giant and Stouffer's now offer vegetarian frozen burgers and dinners. Two icons of American culture, Walt Disney World and Denny's, have vegetarian menu options in their restaurants.

VEGETARIAN-AWARE SHOPPERS

Vegetarian-aware shoppers are interested in eating less meat, primarily because of their interest in eating less fat and eating more vegetables, fruits, whole grains, and beans.

This term was coined by HealthFocus, a market research and consulting firm based in Des Moines. HealthFocus has conducted their National Study of Public Attitudes and Actions Toward Shopping and Eating every 2 years since 1990. The national survey is conducted in two stages: telephone recruitment and a mailed questionnaire. More than 300 questions are asked. More than 2000 people participated in 1996 and 1994, and more than 1000 people participated in 1992 and 1990.

The survey objectives are to

- Identify current issues in consumer health and nutrition attitudes and actions, such as avoiding red meats and eating more vegetables.
- Assess trends in consumer priorities regarding eating habits and nutrition issues, from price and convenience to better nutrition.
- Examine where consumers are headed in their behavior toward health and diet.
- Determine what nutritional issues will be most important to shoppers in the near future.

MEAT ON THE MENU

It important to recognize that vegetarian and vegetarian-aware shoppers are not giving up meat altogether; rather, they are viewing meat as a less prominent aspect of their meals. Some shoppers say, "I'm a vegetarian; I only eat chicken and never beef or pork." This is not really being a true vegetarian.

According to the 1996 HealthFocus Study, both vegetarian shoppers and vegetarian-aware shoppers still eat meat on occasion, although vegetarian shoppers are far less likely to eat meat. Fish and seafood are the most widely used meats among vegetarian and vegetarian-aware shoppers, followed by poultry.

Almost one-third (31%) of vegetarian shoppers always or usually avoid dairy products and ingredients; 79% always or usually avoid foods with red meat. Still, only 78% of shoppers who say they always maintain a vegetarian diet eat meatless meals twice a week or more often (Fig. 1).

MARKET SIZE

According to the HealthFocus Survey, almost 30% of American shoppers, shopping for 25.7 million households, are choosing vegetarian or meatless meals. Since 1990, the number of vegetarian shoppers has remained relatively stable, with about 2% of all shoppers always maintaining a vegetarian diet.

The number of shoppers making vegetarian-aware choices is increasing. In 1996, 26% of shoppers maintain a vegetarian diet usually (7%) or sometimes (19%), up from 24% of shoppers in 1994. The number of shoppers maintaining a vegetarian diet at least on occasion increased by 6 percentage points, from 48% in 1994 to 54% in 1996 (Table 1).

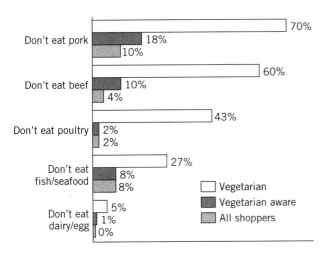

Figure 1. Meat and dairy consumption as reported by respondents.

Table 1. How Many Vegetarians Are There?

	1994 (%)	1996 (%)	No. of households (1996)
Vegans	0.5	0.5	460,000
Vegetarians	1.5	1.5	1,380,000
Vegetarian aware	24	26	23,920,000
Total	*26*	*28*	*25,760,000*

DEMOGRAPHICS OF VEGETARIAN AND VEGETARIAN-AWARE SHOPPERS (TABLE 2)

Men are somewhat more likely to be vegetarian shoppers, and women are more likely to be vegetarian-aware shoppers: 3% of men always maintain a vegetarian diet, compared to 2% of women. Twenty-six percent of women usually or sometimes maintain a vegetarian diet, compared to 21% of men.

Vegetarian shoppers tend to be younger: 21% are 18–29 years old, compared to 12% of shoppers overall in this age group. They are also more likely to have young children at home: 42% have children younger than 2 years of age, and 42% have children 3 to 6 years of age.

In contrast, vegetarian-aware shoppers are older: 26% are 50–64 years old, compared to 23% of all shoppers in this age group; 22% are over age 65, compared to 17% of shoppers overall. Vegetarian-aware shoppers have older children at home: 47% have teenagers, compared to 44% of all shoppers and 42% of vegetarian shoppers.

Shoppers making vegetarian and vegetarian-aware choices are somewhat better educated than other shoppers. Forty percent of vegetarian shoppers and 36% of vegetarian-aware shoppers have at least a college degree; 32% of all shoppers have college degrees.

Vegetarian-aware shoppers live in higher income households than vegetarian shoppers, likely because of their age and experience in the workplace. Vegetarian shoppers have a median household income of $29,000, and vegetarian-aware shoppers have a median household income of $37,800.

Geographically, vegetarian and vegetarian-aware shoppers live in highest concentrations in the midwest region of the United States. Vegetarian shoppers are most likely to live in the North Central (33%), Pacific (21%), Mid-Atlantic (14%), and New England (12%) regions. Vegetarian-aware shoppers are most likely to live in the North Central region (27%).

Compared to the general population, vegetarian shoppers are found in lower concentrations in the South Central (9%) and Mountain (2%) regions. Vegetarian-aware shoppers are found in higher concentrations in the South Atlantic (18%) and South Central (19%) regions.

DIETARY PRIORITIES

The most significant aspect of the vegetarian and vegetarian-aware consumer market is their strong com-

Table 2. Selected Demographics of Vegetarian Shoppers

	Vegetarian	Vegetarian aware	All shoppers
Men	23%	16%	19%
Women	77%	84%	81%
Adult (18–39 years)	42%	29%	35%
Mid-Life (40–55 years)	33%	36%	35%
Prime (56–64 years)	16%	13%	12%
Senior (65+ years)	9%	22%	18%

mitment to healthy eating habits. Healthy eating is more important to vegetarian and vegetarian-aware shoppers, compared to shoppers overall. Their attitudes are extreme and their behavior is more consistent, compared to other shoppers (Fig. 2).

Fifty percent of vegetarian shoppers and 24% of vegetarian-aware shoppers strongly agree that their diet is very important to them, compared to 16% of all shoppers. Fifty-eight percent of vegetarian shoppers and 32% of vegetarian-aware shoppers strongly agree that they think about the healthfulness or nutritional value of what they eat, compared to just 24% of all shoppers.

Thirty-six percent of vegetarian shoppers and 17% of vegetarian-aware shoppers always select foods for healthy reasons. Forty-eight percent of vegetarian and 36% of vegetarian-aware shoppers always read labels on food packages, compared to 27% of all shoppers.

HEART-HEALTHY, LOWFAT, AND LOW-CALORIE DIETS

Eighty percent of vegetarian shoppers and 67% of vegetarian-aware shoppers always or usually maintain a heart-healthy diet, compared to about half of all shoppers who place similar importance on heart-healthy eating habits. Seventy-four percent of vegetarian shoppers and 65% of vegetarian-aware shoppers always or usually maintain a lowfat diet (Fig. 3).

Vegetarian and vegetarian-aware shoppers also put a higher priority on maintaining a low-calorie diet. Sixty percent of vegetarians and 43% of vegetarian-aware shoppers always maintain a low-calorie diet, compared to 26% of shoppers overall.

This emphasis on health, rather than on religious or environmental issues, has meant that many longstanding vegetarian brands and products have had to reformulate and reposition themselves to address heart-healthy, lowfat, low-calorie diets. For many consumers, the simplest way to cut back on fat is to cut back on meats. These choices are preventive measures that are motivated by interest in better future health.

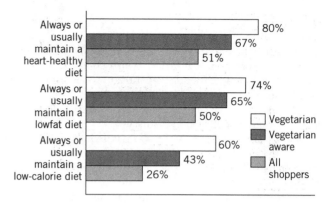

Figure 3. Importance of lowfat eating habits among respondents.

HEALTH ISSUES

Ensuring future good health and managing weight are the primary reasons Vegetarian and Vegetarian Aware Shoppers make healthy choices (Fig. 4). Ensuring future good health is the primary reason for choosing healthy foods for 67% of vegetarian-aware shoppers, 65% of all shoppers, and 56% of vegetarian shoppers.

Although philosophical or spiritual reasons are strong motivators for those strictly vegetarian (18%), the secondary reason vegetarian shoppers (24%) choose healthy foods is to lose weight. Only 10% of vegetarian-aware shoppers and 13% of all shoppers select healthy foods to lose weight.

The secondary reason vegetarian-aware shoppers (16%) choose healthy foods is to control existing health problems.

HEALTH CONCERNS

Vegetarian and vegetarian-aware shoppers have more extreme concerns about future health problems, compared to

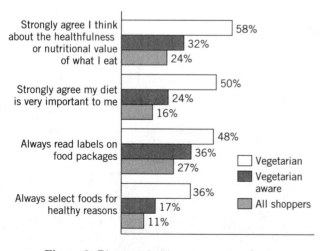

Figure 2. Dietary priorities among respondents.

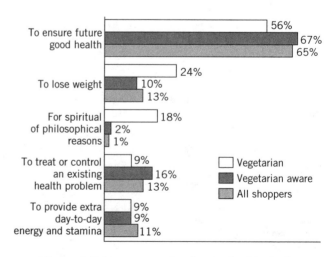

Figure 4. Primary reason for choosing healthy foods.

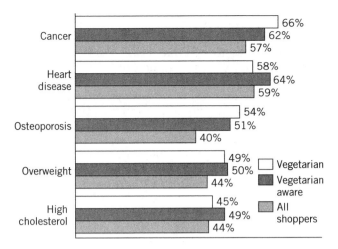

Figure 5. Health issues respondents are extremely or very concerned about.

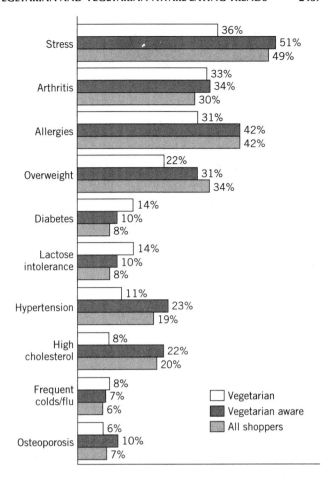

Figure 6. Examples of health problems personally affecting respondents.

shoppers overall. Cancer tops the list of health concerns, followed by heart disease, osteoporosis, being overweight, and having high cholesterol (Fig. 5).

Vegetarians have the highest cancer concerns. Sixty-six percent of vegetarians are extremely or very concerned about cancer, compared to 62% of vegetarian-aware shoppers and 57% of shoppers overall. They also express high concerns about osteoporosis and their weight.

Vegetarian-aware shoppers are especially concerned about heart disease and high cholesterol, more so than vegetarians are. Sixty-four percent of vegetarian-aware shoppers are extremely or very concerned about heart disease, compared to 58% of vegetarians and 59% of shoppers overall. Similarly, 49% of vegetarian-aware shoppers are extremely or very concerned about high cholesterol, compared to 45% of vegetarians and 44% of shoppers overall.

HEALTH PROBLEMS

The comparatively low levels of concern about heart disease and high cholesterol among vegetarian shoppers is probably in part a reflection of their better health status (Fig. 6). Sixty-five percent of vegetarian shoppers describe their health as excellent or very good, compared to 53% of vegetarian-aware shoppers and 53% of shoppers overall.

Vegetarian shoppers are significantly less likely to be personally affected by health problems than are other shoppers. Three exceptions are lactose intolerance, diabetes, and arthritis, where vegetarian choices may be helping to manage the health problem once it occurs or is diagnosed. Younger and more attentive to heart-healthy eating habits, vegetarian shoppers are relatively less likely to report being overweight, having high cholesterol levels or hypertension, or having allergies or stress, compared to vegetarian-aware shoppers and shoppers overall.

Vegetarian-aware shoppers are somewhat more likely than all shoppers to report hypertension or high cholesterol levels, stress, diabetes, arthritis and osteoporosis, but they are less likely than all shoppers to report being overweight. They are equally likely to report allergies.

Stress, arthritis, and allergies are the most common health problems reported by vegetarian shoppers. Thirty-six percent of vegetarian shoppers are affected by stress, 33% have arthritis, and 31% have allergies. Osteoporosis, frequent colds and flus, and high cholesterol levels are the least widespread.

The most common health problems among vegetarian-aware shoppers are stress, allergies, and arthritis. Fifty-one percent of vegetarian-aware shoppers have stress, 42% report allergies, and 34% have arthritis. Frequent colds and flus, diabetes, lactose intolerance, and osteoporosis are their least common health problems.

SHOPPING HABITS

Just as vegetarian and vegetarian-aware shoppers differ in their nutrition attitudes and health priorities, they have different shopping habits (Fig. 7). Both shop at natural

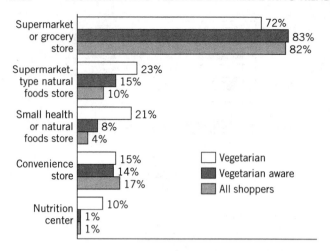

Figure 7. Shopping habits of respondents. Percent of those shopping one or more times per week at various stores.

foods and health food stores more often than shoppers overall, but vegetarian shoppers are much more likely than vegetarian-aware shoppers to go often and to go to nutrition centers. Vegetarian shoppers go to the supermarket less often: 72% visit their supermarket weekly, compared to 83% of vegetarian-aware shoppers and 82% of shoppers overall.

Twenty-three percent of vegetarian shoppers go to a supermarket-type natural foods store weekly, and 21% go to a small health or natural foods store as often. Fifteen percent of vegetarian-aware shoppers go to a supermarket-type natural foods store weekly, and 8% go to a small health or natural foods store. This compares to 10% of all shoppers who go to a supermarket-type natural foods store weekly, and 4% who go to a small health or natural foods store.

Ten percent of vegetarian shoppers frequent nutrition centers (stores specializing in vitamins and supplements) weekly; 1% of vegetarian-aware shoppers and 1% of shoppers overall shop at nutrition centers at least once a week.

Surprisingly, vegetarian shoppers are about as likely as shoppers are overall to shop at a convenience store at least once a week. Fifteen percent of vegetarian shoppers, 14% of vegetarian-aware shoppers, and 17% of all shoppers make the convenience store stop weekly.

PACKAGE INFORMATION

When shopping, vegetarian shoppers are label readers: 79% always or usually read labels on packages. Seventy-six percent of vegetarian-aware shoppers and 66% of all shoppers do the same.

The freshness date is the most important part of the package labeling, followed by the nutrition fact box and ingredient statement. Vegetarian shoppers give extreme importance to this label information. Seventy percent consider freshness dating extremely important, 60% consider the nutrition fact box extremely important, and the same number say the ingredient statement is extremely important.

Freshness dating is extremely important to 60% of vegetarian-aware shoppers and to 58% of shoppers overall. Forty-six percent of vegetarian-aware shoppers and 36% of shoppers overall consider the nutrition fact box extremely important. Forty-three percent of vegetarian-aware shoppers and 33% of shoppers overall consider the ingredient statement extremely important.

Other important label information includes content claims and health claims, a customer service or 800 number, and recipes. Thirty-four percent of vegetarian shoppers consider a customer service number on the label extremely important, compared to 22% of vegetarian-aware shoppers and 17% of all shoppers. Twenty-six percent of vegetarian shoppers like recipes on the package and consider them extremely important. Half as many vegetarian-aware shoppers give recipes this level of importance, and only 11% of shoppers overall do so.

LABEL CLAIMS

Vegetarian shoppers are significantly more interested in and influenced by health claims and content claims than are vegetarian-aware shoppers or shoppers overall. Content claims on labels are extremely important to 43% of vegetarian shoppers and 31% of vegetarian-aware shoppers, compared to only 25% of shoppers overall. Health claims are important to 37% of vegetarian shoppers and 24% of vegetarian-aware shoppers, compared to only 19% of shoppers overall.

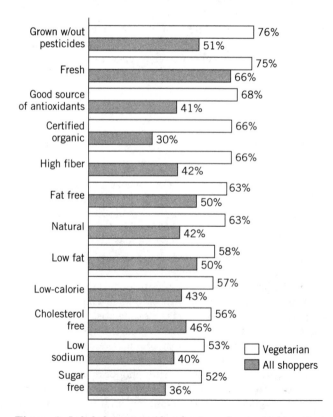

Figure 8. Label claims considered extremely or very important to vegetarian shoppers.

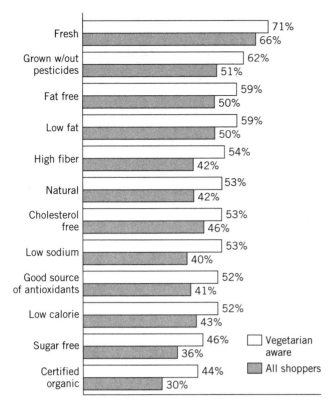

Figure 9. Label claims considered extremely or very important to vegetarian-aware shoppers.

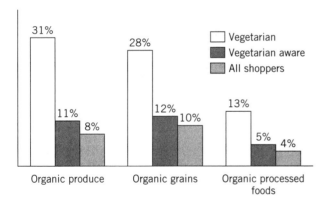

Figure 10. Percent of respondents using organic foods twice a week or more often.

The differing health priorities among vegetarian and vegetarian-aware shoppers are apparent in their label interests (Figs. 8 and 9). Vegetarian shoppers give the most importance to labels reading "grown without pesticides," "fresh," "good source of antioxidants," and "certified organic." Vegetarian-aware shoppers have more mainstream interests, giving the most importance to labels reading "fresh," "grown without pesticides," "fat free," and "low fat."

NATURAL AND ORGANIC

Just as vegetarian and vegetarian-aware shoppers are more likely to shop at health and natural foods stores, they are more likely to look for brands and to support companies that share their social, environmental, and other values. Seventy-nine percent of vegetarian shoppers strongly agree or agree that environmental issues have had an impact on their purchases, compared to 64% of vegetarian-aware shoppers and 51% of shoppers overall. Vegetarian (73%) and vegetarian-aware shoppers (61%) prefer to buy products from companies that support their social, community, or environmental interests.

This social and environmental sense of responsibility is reflected at the checkout counter in organic foods purchases. Thirty-one percent of vegetarian shoppers use organic produce twice a week or more, compared to only 11% of vegetarian-aware shoppers and 8% of all shoppers. Twenty-eight percent of vegetarian shoppers use organic grains twice a week or more, compared to 12% of vegetarian-aware shoppers and 10% of all shoppers. Organic processed foods are used twice a week or more by 13% of vegetarian shoppers, 5% of vegetarian-aware shoppers and 4% of all shoppers.

A GROWING MARKET

The use of vegetarian products and meatless meal options is growing and is expected to continue to grow, driven by consumer interest in healthier eating and scientific evidence supporting the importance of more produce and grains in the diet and less meat. Fifty-four percent of all shoppers in 1996 report they eat a vegetarian diet at least on occasion, up from 48% just 2 years previously. Nine percent of shoppers call themselves vegetarian, although fewer than 2% practice strict vegetarian eating habits including no meat, no eggs, and no dairy products.

In 1996, 40% of all shoppers reported eating meatless meals such as pastas or salads weekly, which is up from 13% in 1990.

Meat substitutes are also becoming more popular with American shoppers. Fifteen percent of all shoppers use these products weekly, up from 9% in 1994 and 7% in 1992.

The weekly use of soy foods such as tempeh or tofu has also increased, from 3% of shoppers in 1992 and 4% in 1994 to 5% in 1996.

Clearly, use of vegetarian products and meatless meal options is growing and will continue to grow. Market growth will be driven by consumer interest in healthier eating and fueled by scientific evidence supporting the importance of more produce and grains in the diet and less meat.

BIBLIOGRAPHY

All data are taken from the 1996 HealthFocus Survey, HealthFocus, Inc., Des Moines, Iowa.

LINDA GILBERT
HealthFocus, Inc.
Des Moines, Iowa

VITAMINS

VITAMINS: SURVEY

Vitamins are specific organic compounds that are essential for normal metabolism. Many participate as cofactors or coenzymes in mammalian biochemical reactions. The common thread for the diverse chemical structures of the vitamins is that they are micronutrients. Micronutrients are compounds that are required in only small amounts and are not synthesized by humans, either at all or, at least, in sufficient quantity for metabolic needs. Vitamins are obtained from the diet or as synthetic preparations used in food fortification or supplements.

The term vitamin is derived from *vitamine*, a word created in 1911 (1) when it was believed that the antiberiberi factor (thiamine) was an amine essential to the maintenance of life (*vita*). Later discoveries showed that not all vitamins are amines. However, a form of the name has remained.

The history of the vitamins can be divided into five major, overlapping periods (2).

PREHISTORY TO CA 1900

In this period, the empirical healing of certain diseases by foods was established. Examples (3) were the treatment of night blindness (vitamin A deficiency) with liver in many cultures over centuries, of beriberi (vitamin B_1 deficiency) by use of unpolished rice by the Japanese navy, of scurvy (vitamin C deficiency) by citrus fruits in the British navy or pine needle extracts by North American natives, and pellagra (niacin deficiency) by a dietary shift away from corn-based foods in many countries. Other, nondietary empirical treatments involved, eg, exposure of children in northern latitudes to sunlight to cure rickets (vitamin D deficiency) (4).

1880–1905

The ability to induce disease states in animals by manipulation of the diet was established in this period. The classical work by Eijkman (5), in which a beriberi-like condition was induced in chickens fed on polished rice, was significant. These findings led to the concept by Hopkins that small amounts of accessory growth factors are necessary for survival and growth.

1900–1972

During this 70 year period, all 13 of the substances now recognized as vitamins were discovered and isolated in pure form. Structure elucidation for each vitamin was completed, as was its total synthesis (Table 1).

1933–PRESENT

The first commercial synthesis of a vitamin occurred in 1933 when the Reichstein approach was employed to manufacture vitamin C (6). All 13 vitamins are available in commercial quantities, and their biological functions have largely been established (7). A list of Nobel prize winners associated with vitamin research is given in Table 2.

1955–PRESENT

The possibility that vitamins might have physiological functions beyond the prevention of deficiency diseases was first recognized in 1955 with the finding (8) that niacin can affect serum cholesterol levels in humans. An explosion of research (9–11) in the intervening years has been aimed at establishing optimal vitamin levels and anticipating the health consequences.

STRUCTURE

Common names, chemical structure, and synonyms for the vitamins are given in Table 3. The names given to the vi-

Table 1. History of the Vitamins

Vitamin	Discovery	Isolation	Chemical structure	Synthesis
A	1909	1931	1931	1947
D	1918	1932	1936	1959
E	1922	1936	1938	1938
K	1929	1939	1939	1939
B_1	1897	1926	1936	1936
B_2	1920	1933	1935	1935
Niacin	1936	1935	1937	1867
B_6	1934	1938	1938	1939
B_{12}	1926	1948	1956	1972
Folic acid	1941	1941	1946	1946
Pantothenic acid	1931	1938	1940	1940
Biotin	1931	1935	1942	1943
C	1912	1928	1933	1933

Source: Ref. 4.

Table 2. Nobel Prizes for Vitamin Research

Year	Recipient	Field	Citation
1928	Adolf Windaus	Chemistry	Research into constitution of steroids and connection with vitamins
1929	Christiaan Eijkman	Medicine, physiology	Discovery of antineuretic vitamins
	Sir Frederick G. Hopkins	Medicine, physiology	Discovery of growth-stimulating vitamin
1934	George R. Minot, William P. Murphy, George H. Whipple	Medicine, physiology	Discoveries concerning liver therapy against anemias
1937	Sir Walter N. Haworth	Chemistry	Research into constitution of carbohydrates and vitamin C
	Paul Karrer	Chemistry	Research into constitution of carotenoids, flavins, and vitamins A and B_2
	Albert Szent-Györgyi	Medicine, physiology	Discoveries in connection with biological combustion processes, with special reference to vitamin C and catalysis of fumaric acid
1938	Richard Kuhn	Chemistry	Work on carotenoids and vitamins
1943	Carl Peter Henrik Dam	Medicine, physiology	Discovery of vitamin K
	Edward A. Doisy	Medicine, physiology	Discovery of chemical nature of vitamin K
1953	Fritz A. Lipmann	Medicine, physiology	Discovery of coenzyme A and its importance for intermediary metabolism
1964	Konrad E. Bloch, Feodor Lynen	Medicine, physiology	Discoveries concerning mechanism and regulation of cholesterol and fatty acid metabolism
	Dorothy C. Hodgkin	Chemistry	Structural determination of vitamin B_{12}
1967	Ragnar A. Granit	Medicine, physiology	Research which illuminated electrical properties of vision by studying wavelength discrimination by eye
	Halden K. Hartine	Medicine, physiology	Research on mechanisms of sight
	George Wald	Medicine, physiology	Research on chemical processes that allow pigments in eye retina to convert light into vision

Source: Ref. 2.

tamins and/or their letter designations do not follow a logical pattern and are of historical significance only. Despite this, the nomenclature is in common use.

Vitamins are classified by their solubility characteristics into fat-soluble and water-soluble groups. The fat-soluble vitamins A, E, D_3 and K result from the isoprenoid biosynthetic pathway. Vitamin A is derived by enzymic cleavage of the symmetrical C_{40} beta-carotene, also known as pro-vitamin A. Vitamins E and K result from condensations of phytyldiphosphate (C_{20}) with aromatic components derived from shikimic acid. Vitamin D results from photochemical ring opening of 7-dehydrocholesterol, itself derived from squalene (C_{30}).

The water-soluble vitamins, B_1, B_2, B_6, B_{12}, niacin, folic acid, pantothenic acid, biotin, and C, have a much less common biosynthetic heritage. Furthermore, multiple biosynthetic paths exist for several of the vitamins. Thus, vitamins B_1, B_2 and folic acid are derived from purines, with components coming from carbohydrates. Portions of vitamins B_1 and folic acid also originate in amino acids. One biosynthesis of niacin and that of vitamin B_6 are derived from glycerin. Acetyl-CoA serves as common precursor, in extremely complex processes, to biotin (via pimelic acid) and vitamin B_{12} (through 5-aminolevulinic acid). Pantothenic acid is constructed from glyoxylic acid and beta-alanine, itself derived from several pathways. Vitamin C, a carbohydrate, is biosynthesized from glucose or galactose (4,12).

VITAMERS

Although the 13 substances described in Table 3 are commonly recognized as the vitamins, significant complexities exist because of the following factors: (1) A number of similar chemical structures can exhibit, at least qualitatively, the biological activity of the vitamin; (2) The biological activity can vary by organism type, eg, in microorganisms vs humans; (3) The physiologically active form of the vitamin may be other than that indicated; for example, several water-soluble vitamins are active in co-enzyme form; (4) A metabolite may be the active form of the vitamin, as in the case of 1-α-25-dihydroxy-vitamin D_3; (5) Biochemical interconversion of several forms of the vitamins can occur. The best studied example is pyridoxine, pyridoxal, pyridoxamine, and their 5'-phosphates; (6) Synthetic commercial forms, often manufactured and sold because of their increased stability, are convertible in the target organism to the vitamin. Esters of α-tocopherol and retinol are the most prominent examples. These compounds are known as vitamers. Table 4 presents a partial list of these vitamers. Specifically segmented are the major commercial forms of the vitamin.

PROPERTIES

Because of their diverse structures, there are few common threads to vitamin chemical properties aside from their fat or water solubility. Of general concern in all applications, however, is vitamin stability. Table 5 provides generic information regarding stability under several conditions. Levels of stability vary greatly and are impacted by acid or base strength, light intensity, etc.

The relative instability of many of the vitamins has led to the creation of derivatives and stabilized product forms capable of withstanding the rigors of food, animal feed, and pharmaceutical process conditions. Depending on the vitamin and application, stabilized forms may involve spray

Table 3. Vitamin Names and Structures

Vitamin	Structure	CAS Registry Number	Common name and synonyms
A		[68-26-8]	Retinol, axerophthol
D		[520-91-2]	Vitamin D_3, cholecalciferol, antirachitic vitamin
E		[59-02-9]	α-Tocopherol, (RRR)-α-tocopherol, antisterility vitamin
K		[84-80-0]	Vitamin K_1, phylloquinome, coagulation vitamin, antihemorrhagic vitamin, phytonadione
B_1		[67-03-8]	Thiamine, aneurin
B_2		[83-88-5]	Riboflavin, vitamin G, lactoflavin, hepatoflavin, ovoflavin, verdoflavin
Niacin		[59-67-6]	Vitamin B_3, nicotinic acid
B_6		[58-56-0]	Pyridoxine, pyridoxol, antiacrodynia factor

Table 3. Vitamin Names and Structures (*continued*)

Vitamin	Structure	CAS Registry Number	Common name and synonyms
B$_{12}$		[*68-19-9*]	Cobalamin, cyanocobalamin
Folic acid		[*59-30-3*]	Folacin, pteroylglutamic acid, antianemia factor, fermentation, *L. casei* factor
Pantothenic acid		[*79-83-4*]	Chick antidermatitis factor
Biotin		[*58-85-5*]	*d*-Biotin, vitamin H, anti-egg-white-injury factor, bios II, coenzyme R
C		[*520-91-2*]	Ascorbic acid, antiscorbutic vitamin

Table 4. Vitamers

Vitamin	Vitamers	Major commercial forms
A	Retinal (vitamin A aldehyde), retinoic acid (vitamin A acid), retinol cis-isomers, β-carotene, other carotenoids	Vitamin A acetate, vitamin A propionate, vitamin A palmitate, vitamin A acid, β-carotene, β-apo-8'-carotenal
D	Ergocalciferol (vitamin D_2), 1-α-25-dihydroxycholecalciferol (calcitriol)	Cholecalciferol (vitamin D_3), ergocalciferol (vitamin D_2)
E	(RRR)-β-Tocopherol, (RRR)-γ-tocopherol, (RRR)-δ-tocopherol, 2R-α-tocotrienol, 2R-β-tocotrienol, 2R-γ-tocotrienol, 2R-δ-tocotrienol, all-rac-α-tocopherol	(RRR)-α-Tocopherol, (RRR)-α-tocopheryl acetate, all-rac-α-tocopherol, all-rac-α-tocopheryl acetate, all-rac-α-tocopheryl succinate
K	Vitamin K_2 (menaquinone), vitamin K_3 (menadione), 2-(R,S)-vitamin K_1 (trans)	2-(R,S)-Vitamin K_1 (cis, trans-mixture), menadione sodium bisulfite, menadione dimethylpyrimidinol bisulfite
B_1	Thiamine diphosphate (cocarboxylase), thiamine orthophosphate	Thiamine chloride, thiamine mononitrate, thiamine diphosphate, thiamine disulfide
B_2	Riboflavin mononucleotide (FMN), riboflavin dinucleotide (FAD)	Riboflavin, riboflavin sodium phosphate
Niacin	Niacinamide (nicotinamide), nicotinamide adenine dinucleotide (NAD, coenzyme I), nicotinamide adenine dinucleotide phosphate (NADP, coenzyme II)	Niacin, niacinamide
B_6	Pyridoxine-5'-phosphate, pyridoxal, pyridoxal-5'-phosphate (codecarboxylase), pyridoxamine, pyridoxamine-5'-phosphate	Pyridoxine hydrochloride, pyridoxal-5'-phosphate
B_{12}	Methylcobalamin, adenosyl-cobalamin (coenzyme B_{12}), hydroxocobalamin, aqua-cobalamin chloride	Cyanocobalamin, hydroxocobalamin
Folic acid	Tetrahydrofolic acid, N-formyl-tetrahydrofolic acid, pteroic acid, pteroyltriglutamic acid, pteroylhexaglutamic acid	Folic acid
Pantothenic acid	Coenzyme A, pantethein	Calcium pantothenate, d-panthenol, d,l-panthenol
Biotin	Desthiobiotin	d-Biotin
C	Dehydroascorbic acid	Ascorbic acid, ascorbyl palmitate, sodium ascorbate, calcium ascorbate, ascorbyl-2-polyphosphate

Table 5. Vitamin Stability

Vitamin	Oxygen	Light	100°C	Acids[a]	Bases[a]
A	U	U	S	U	S
D	U	U	U	U	U
E	U	U	S	S	S
K	S	U	S	S	U
B_1	U	U	U	S	U
B_2	S	U	U	S	U
Niacin	S	S	S	S	S
B_6	S	U	U	S	S
B_{12}	U	U	U	S	S
Folic acid	U	U	U	U	S
Pantohenic acid	S	S	U	U	U
Biotin	S	S	U	S	S
C	U	U	U	S	U

Note: S = stable; U = unstable.
[a]In absence of oxygen.

drying in a suitable matrix (eg, gelatin), encapsulation (beadletting), or coating with fats or waxes (qv).

ANALYSIS

Analytical methods for the vitamins involve physical and chemical, chromatographic, microbiological, and biological methods. The method of choice for a particular vitamin depends on the substance being measured, as well as its concentration. For example, detection of biotin, folic acid, and vitamin B_{12} is accurately achieved by high pressure liquid chromatography (hplc) or gas—liquid chromatography (glc). However, chromatographic methods often are not sufficiently sensitive for the low levels at which these vitamins usually are found. Microbiological methods, though slower, more expensive, and usually less accurate, are then employed for these low level assays. Biological assays, although extremely significant in the history of the vitamins, are rarely used in the 1990s.

An excellent overview of vitamin analytics is available (13). Specifications for the vitamins are available for food use (14) and pharmaceutical use (15).

PHYSIOLOGICAL FUNCTION

Physiological functions as well as clinical symptoms that occur in humans deficient in specific vitamins are given in Table 6. It is becoming more authenticated that vitamins have additional potential health benefits when administered, via the diet or by supplementation, at levels above those required for obviating deficiency. Although for most vitamins the optimal levels are not yet established, some

Table 6. Vitamin Functions, Clinical Deficiency Symptoms, and Potential Health Benefits

Vitamin	Function	Clinical deficiency symptoms	Potential health benefits
A	Needed for normal vision, reproduction, and maintenance of healthy skin, mucous membranes, bones, red blood cells, cell differentiation, and immune function	Night blindness, xerophthalmia, loss of appetite, increased susceptibility to infections, skin disorders, poor growth, defective reproduction	Regression of precancerous lesions, reduces measle-associated morbidity in children
D	Enhances calcium absorption, regulates calcium and phosphorus metabolism, promotes mineralization of bones, plays role in cell differentiation	Insufficient bone mineralization, skeletal deformities (rickets and osteomalacia)	Protection against certain types of cancer, reduces risk of osteoporosis, psoriasis (topical)
E	Lipid-soluble antioxidant, prevents lipid oxidation of membranes, needed for healthy blood cells and tissues, blocks nitrosamine formation, protects PUFAs from autoxidation, important for normal immune function	Neuromuscular disorders, red blood cell rupture (both uncommon)	Reduces risk of chronic disease (cardiovascular, precancerous lesions, cancer), immunoenhancement, protection from exercise-induced muscle injury, improves metabolic control, reduces risk of complications in diabetes
K	Essential for normal blood clotting, needed for formation and maintenance of healthy bones, essential for formation of osteocalcin (bone protein)	Reduced blood clotting, increased risk for brittle bones	Reduces risk of osteoporosis
B_1	Helps convert carbohydrates from food into energy, required by muscle, nervous system, and brain	Mental confusion, loss of appetite, muscular weakness and paralysis, heart failure (beriberi)	Important role in energy metabolism
B_2	Involved in carbohydrate, protein and fat metabolism	Oral lesions, cracks in corners of mouth, dermatitis	Important role in energy metabolism
Niacin	Involved in carbohydrate, protein, and fat metabolism; at pharmacologic levels used to lower blood cholesterol	Mouth soreness, diarrhea, abdominal pain, skin and nervous system disorders (pellegra)	Reduces risk of atherosclerosis
B_6	Active in protein metabolism, required for bone health, involved in homocysteine metabolism	Dermatitis, anemia, nervous system disorders	Reduces risk of cardiovascular disease (also B_{12} and folic acid), immunoenhancement, reduces risk of osteoporosis, protection against carpal tunnel syndrome
Folic acid	Involved in amino acid interconversions, needed for red blood cell formation and DNA and RNA synthesis, protects against neural tube birth defects, involved in homocysteine metabolism	Anemia, nervous system disorders, gastrointestinal lesions	Reduces risk of cardiovascular disease (also B_6 and B_{12}), certain cancers, and precancerous lesions; birth defect prevention
Pantothenic acid	Active in carbohydrate, protein, and fat metabolism	Abnormal sensations in legs and feet (rare)	
Biotin	Involved in protein, fat, and carbohydrate metabolism	Dermatitis, hair loss	Helps develop healthy nails
C	Helps form and maintain collagen; important for healthy cartilage, bones, teeth, and gums; needed for wound healing; enhances nonheme iron absorption; water-soluble antioxidant protects cells from free-radical damage; blocks nitrosamine formation; increases body's resistance to infection	Swollen or bleeding gums, joint pain, slow wound healing, lowered resistance to infection, scurvy	Protection against chronic diseases (cancer, cardiovascular, eye disease), reduces symptoms of common cold, periodontal disease protection, improves fertility in men, reduces risk of osteoporosis

Table 7. Recommended Dietary Allowances

Category	Age	Vitamin A (µg RE)[a]	Vitamin D (µg)[b]	Vitamin E (mg α-TE)[c]	Vitamin K (µg)	Thiamin (mg)	Riboflavin (mg)	Niacin (mg)	Vitamin B_6 (mg)	Vitamin B_{12} (µg)	Folate (µg)	Vitamin C (mg)
Infants	0.0–0.5	375	7.5	3	5	0.3	0.4	5	0.3	0.3	25	30
	0.5–1.0	375	10	4	10	0.4	0.5	6	0.6	0.5	35	35
Children	1–3	400	10	6	15	0.7	0.8	9	1.0	0.7	50	40
	4–6	500	10	7	20	0.9	1.1	12	1.1	1.0	75	45
	7–10	700	10	7	30	1.0	1.2	13	1.4	1.4	100	45
Males	11–14	1000	10	10	45	1.3	1.5	17	1.7	2.0	150	50
	15–18	1000	10	10	65	1.5	1.8	20	2.0	2.0	200	60
	19–24	1000	10	10	70	1.5	1.7	19	2.0	2.0	200	60
	25–50	1000	5	10	80	1.5	1.7	19	2.0	2.0	200	60
	51+	1000	5	10	80	1.2	1.4	15	2.0	2.0	200	60
Females	11–14	800	10	8	45	1.1	1.3	15	1.4	2.0	150	50
	15–18	800	10	8	55	1.1	1.3	15	1.5	2.0	180	60
	19–24	800	10	8	60	1.1	1.3	15	1.6	2.0	180	60
	25–50	800	5	8	65	1.1	1.3	15	1.6	2.0	180	60
	51+	800	5	8	65	1.0	1.2	13	1.6	2.0	180	60
Pregnant lactating		800	10	10	65	1.5	1.6	17	2.2	2.2	400	70
1st six months		1000	10	12	65	1.6	1.8	20	2.1	2.6	280	95
2nd six months		1200	10	11	65	1.6	1.7	20	2.1	2.6	260	90

Source: Ref. 18.

[a]RE = retinol equivalents; 1 RE = 1 µg retinol or 6 µg β-carotene.
[b]As cholecalciferol; 10 µg cholecalciferol = 400 IU vitamin D.
[c]TE = α-Tocopherol equivalents; 1 α-TE = 1 mg (RRR)-α-tocopherol.

of the potential health benefits to be derived from vitamins are indicated (16). In one case, the level of scientific proof is such that the U.S. FDA has allowed "a health claim that women who are capable of becoming pregnant and who consume adequate amounts of folate daily during their childbearing years may reduce their risk of having a pregnancy affected by spina bifida or other neural tube defects" (17).

The Recommended Dietary Allowances (RDA) of the National Academy of Sciences are intended as daily intake guidelines for preventing deficiencies and maintaining reasonable reserves among most healthy people. These recommendations, revised every 5 to 10 years, are made by a committee of the U.S. Food and Nutrition Board and are based on scientific information that is available at the time of the review. This includes studies on subjects on deficient diets, nutrient balance studies, measures of tissue saturation, normal intakes of the nutrient, and extrapolation from animal data. The most recent RDAs are given in Table 7 (18). The 1989 RDA has a special requirement of 100 mg of vitamin C for cigarette smokers. Otherwise, there is no recognition of special needs for specific groups such as those using certain drugs chronically, people with disease, or the elderly (eg, over 65 years). Neither does it address the needs for the prevention of any of the chronic diseases.

Because there is less information upon which to base dietary allowances for biotin and pantothenic acid, ranges of intake are provided, as in Table 8.

Along with increasing evidence of health benefits from consumption of vitamins at levels much higher than RDA recommendations comes concern over potential toxicity. This topic has been reviewed (19). Like all chemical substances, a toxic level does exist for each vitamin. Traditionally it has been assumed that all water-soluble vitamins are safe at any level of intake and all fat-soluble vitamins are toxic, especially at intakes more than 10 times the recommended allowances. These assumptions are now known to be incorrect. Very high doses of some water-soluble vitamins, especially niacin and vitamin B_6, are associated with adverse effects. In contrast, evidence indicates that some fat-soluble micronutrients, especially vitamin E, are safe at doses many times higher than recommended levels of intake. Chronic intakes above the RDA for vitamins A and D especially are to be avoided, however.

PRODUCTION

Preparation of the vitamins in commercial quantities can involve isolation, chemical synthesis, fermentation, and mixed processes, including chemical and fermentation steps. The choice of process is economic, dictated by the need to obtain materials meeting specifications at the lowest cost. Current process technologies (ca 1997) employed for each vitamin are indicated in Table 9.

More than one process is available for some of the vitamins. Further, manufacturers have developed variants of the classical syntheses during optimization. Whereas some of this information is available, as described in the individual sections on vitamins, much is closely held as trade secrets. Judging from the more recent patent literature, the assessment can be made that vitamin production technologies are in general mature. However, the economic value of these products drives continuing research aimed at breakthrough processes. Annual production of vitamins varies greatly, from ca 10 metric tons of vitamin B_{12} to ca 50,000 metric tons of vitamin C.

APPLICATIONS

It is generally assumed that adequate vitamin levels in humans can be obtained through a balanced diet. However, ongoing studies continue to indicate that the majority of the U.S. population is not receiving even the RDA through diet. Supplementary vitamins are thus provided for fortification of foods (20) and as oral or parenteral dosage forms.

The use of vitamins in humans consumes ca 40% of vitamins made worldwide. The majority of the vitamins, particularly in countries outside the United States, are used in animal husbandry. It is well established (21) that vitamins are critical to animal productivity, especially under confined, rapid growth conditions. Newer information (22) has shown that vitamin E added to cattle feed has the additional effect of significantly prolonging beef shelf life in stores. Additional applications of vitamins exist. A small but growing market segment involves cosmetics (qv) (23). The use of the chemical properties of the vitamins, particularly as antioxidants (qv) in foods and, more recently, in plastics (vitamin E (24)), has emerged as a growing trend.

ECONOMIC ASPECTS

Vitamins are sold for direct application to foods and animal feeds. In addition, they are further processed into nutritional supplements. This last market is particularly significant in the United States. In many other countries, vitamins are regulated as drugs, leading to a much lower supplement usage.

It can be estimated that approximately $3,000,000,000 of vitamins were sold in 1996. Market growth is slightly

Table 8. Estimated Safe and Adequate Daily Dietary Intake

Category	Age	Biotin (μg)	Pantothenic acid (mg)
Infants	0–0.5	10	2
	0.5–1	15	3
Children and adolescents	1–3	20	3
	4–6	25	3–4
	7–10	30	4–5
	11+	30–100	4–7
Adults		30–100	4–7

Source: Ref. 18.

Table 9. Vitamin Production Processes

Isolation	Chemical synthesis	Fermentation	Synthesis/fermentation
(Vitamin A)	Vitamin A	Vitamin B_2	Vitamin B_2
Vitamin E[a]	Vitamin D	Vitamin B_{12}	Vitamin C
(Vitamin D)	Vitamin E[b]		
(Vitamin K)	Vitamin K		
	Vitamin B_1		
	Niacin		
	Vitamin B_6		
	Folic acid		
	Pantothenic acid		
	Biotin		

Note: Parentheses indicate minor processes primarily of historical significance.
[a](RRR)-α-Tocopherol.
[b]All-rac-α-tocopherol.

higher than population growth, but varies widely by individual vitamin, geographical area, and/or application. The largest vitamin manufacturer is Hoffmann-La Roche. Other significant producers include BASF, Takeda, Eisai, and Rhône-Poulenc. Additional vitamins are produced in China, Russia and India.

VITAMIN-LIKE COMPOUNDS

Although it is a general consensus that these 13 substances represent the known vitamins, other compounds upon which there is less agreement often enter the vitamin discussion (25). Included in this group are choline, myo-inositol (Bios I), vitamin F (essential fatty acids including linoleic acid, γ-linolenic acid, and arachidonic acid), α-lipoic acid, ubiquinones, plastoquinone, vitamin P (bioflavonoids), vitamin L (o-aminobenzoic acid), vitamin T (mixture including folic acid and vitamin B_{12}), vitamin U (L-methioninylmethylsulfonium chloride), orotic acid (vitamin B_{13}), and pyrroloquinoline quinone.

ACKNOWLEDGMENTS

The Vitamins articles have been reprinted from the *Kirk-Othmer Encyclopedia of Chemical Technology*, Vol. 25, John Wiley & Sons, New York, 1998.

BIBLIOGRAPHY

1. C. Funk, *J. Physiol. (London)* **43**, 395 (1911).

2. *Vitamins Basics*, F. Hoffmann-La Roche, Ltd., Basel, Switzerland, 1994, p. i.

3. K. Reese, *Today's Chem. Work* **59** (Oct. 1995).

4. O. Isler and G. Brubacher, *Vitamine I: Fettlösliche Vitamine*, Georg Thieme Verlag, Stuttgart, Germany, 1982.

5. C. Eijkman, *Geweskund Tijdsche Ned. Indie* **30**, 295 (1896).

6. T. Reichstein and A. Grüssner, *Helv. Chim. Acta* **17**, 311 (1934).

7. L. J. Machlin, ed., *Handbook of Vitamins*, 2nd ed., Marcel Dekker, New York, 1991.

8. R. Altschul, A. Hoffer, and J. D. Stephen, *Arch. Biochem. Biophys.* **54**, 558 (1955).

9. H. E. Sauberlich and L. J. Machlin, eds., *Annal. NY Acad. Sci.* **669** (1992).

10. A. Bendich and C. E. Butterworth, Jr., *Micronutrients in Health and Disease Prevention*, Marcel Dekker, New York, 1991.

11. S. K. Gaby, A. Bendich, V. N. Singh, and L. J. Machlin, *Vitamin Intake and Health: A Scientific Review*, Marcel Dekker, New York, 1991.

12. O. Isler, G. Brubacher, S. Ghisla, and B. Kräutler, *Vitamine II: Wasserlösliche Vitamine*, Georg Thiem Verlag, Stuttgart, Germany, 1988.

13. J. Augustin, in *Encyclopedia of Food Technology*, Vol. 4, John Wiley & Sons, Inc., New York, 1991, p. 2697.

14. Food and Nutrition Board, National Research Council, *Food Chemicals Codex*, 3rd ed., National Academy Press, Washington, D.C., 1981.

15. *The United States Pharmacopeia XXIII* (USP XXIII-NF XVIII), United States Pharmacopeial Convention, Inc., Rockville, Md., 1995.

16. V. Singh, personal communication, Hoffmann-La Roche Inc., Nutley, N.J., 1996.

17. *Fed. Reg.* **59**, 433 (Jan. 4, 1994).

18. Food and Nutrition Board, National Research Council, *Recommended Dietary Allowances*, 10th ed., National Academy Press, Washington, D.C., 1989.

19. *Vitamin Nutrition Information Service Backgrounder*, Vol. 5 (No. 1), Hoffmann-La Roche Inc., Nutley, N.J., 1994.

20. J. Giese, *Food Technol.*, 110 (May 1995).

21. *Vitamin Compendium*, F. Hoffmann-La Roche & Co., Ltd., Basel, Switzerland, 1976.

22. J. A. Sherbeck, D. M. Wulf, J. B. Morgan, J. D. Tatum, G. C. Smith, and S. N. Williams, *J. Food Sci.* **60**, 250 (1995).

23. D. Djerassi, *The Role of Vitamins in Skin Care*, Hoffmann-La Roche Inc., Nutley, N.J., 1993.

24. S. F. Laermer, S. S. Young, and P. F. Zambetti, *Converting Mag.*, 80 (Feb. 1996).

25. M. M. Cody, in Ref. 7, p. 565.

JOHN W. SCOTT
Hoffmann-La Roche, Inc.

VITAMINS: ASCORBIC ACID

Ascorbic acid [50-81-1] (1) (Fig. 1) is the name recognized by the IUPAC-IUB Commission on Biochemical Nomenclature for Vitamin C (1). Other names are: L-ascorbic acid, L-xyloascorbic acid, and L-*threo*-hex-2-enoic acid γ-lactone. The name implies the vitamin's antiscorbutic properties, eg, the prevention and treatment of scurvy. L-Ascorbic acid is widely distributed in plants and animals. The pure vitamin, $C_6H_8O_6$, mol wt 176.13, is a water-soluble, strongly reducing, optically active (chiral) white crystalline substance.

L-Ascorbic acid biosynthesis in plants and animals as well as the chemical synthesis starts from D-glucose. The vitamin and its main derivatives, sodium ascorbate, calcium ascorbate, and ascorbyl palmitate, are officially recognized by regulatory agencies and included in compendia such as the *United States Pharmacopeia/National Formulary (USP/NF)* and the *Food Chemicals Codex (FCC)*.

The most significant chemical characteristic of L-ascorbic acid (1) is its oxidation to dehydro-L-ascorbic acid (L-*threo*-2,3-hexodiulosonic acid γ-lactone) (3) (Fig. 1). Vitamin C is a redox system containing at least three substances: L-ascorbic acid, monodehydro-L-ascorbic acid, and dehydro-L-ascorbic acid. Dehydro-L-ascorbic acid and the intermediate product of the oxidation, the monodehydro-L-ascorbic acid free radical (2), have antiscorbutic activity equal to L-ascorbic acid.

The reversible oxidation of L-ascorbic acid to dehydro-L-ascorbic acid is the basis for its known physiological activities, stabilities, and technical applications (2). The importance of vitamin C in nutrition and the maintenance of good health is well documented. Over 22,000 references relating only to L-ascorbic acid have appeared since 1966.

L-Ascorbic acid was isolated first by Albert Szent-Györgyi in 1930. However, its early history is associated with the etiology, treatment, and prevention of scurvy (3–5). Scurvy is one of the oldest diseases known to humankind. It affected many people in ancient times in Egypt, Greece, and Rome and influenced the course of history. Outbreaks of the disease resulting from inadequate

amounts of vitamin C in rations spontaneously ended many military campaigns and long ocean voyages (6). During the Middle Ages, scurvy was endemic in northern Europe, occurring mostly in the winter season when fresh fruits and vegetables were unavailable. By the end of the seventeenth century, scurvy became a severe problem among sailors on long voyages. It was treated as a venereal disease, with disastrous results.

The first clues to the treatment of scurvy occurred in 1535–1536 when Jacques Cartier, on advice from Newfoundland Indians, fed his crew an extract from spruce tree needles to cure an epidemic. Various physicians were recommending the use of citrus fruits to cure scurvy in the mid-sixteenth century. Two hundred years later, in 1753, it was proved by Dr. James Lind, in his famous clinical experiment, that scurvy was associated with diet and caused by lack of fresh vegetables. He also demonstrated that oranges and lemons were the most effective cure against this disease. In 1753, in "*A Treatise on the Scurvy*," Lind published his results and recommendations (7). Forty-two years later, in 1795, the British Navy included lemon juice in seamen's diets, resulting in the familiar nickname "limeys" for British seamen. Evidence has shown that even with undefined scorbutic symptoms, vitamin C levels can be low, and can cause marked diminution in resistance to infections and slow healing of wounds.

Research leading to the discovery of vitamin C began in 1907 when it was observed by Axel Holst and Theodor Fröhlich that guinea pigs were as susceptible to scurvy as humans and that the disease could be produced experimentally in these animals (8). These findings led to the development of an assay for the biological determination of antiscorbutic activity of food products (9).

Between 1910 and 1921, the vitamin was obtained in almost pure form from lemons and some of the physical and chemical properties were determined (10). It was discovered that vitamin C is easily destroyed by oxidation and best protected by reducing agents, and that 2,6-dichlorophenolindophenol is reduced by solutions of the vitamin. Subsequent studies showed that the dichlorophenolindophenol test measured only the vitamin and not dehydroas-

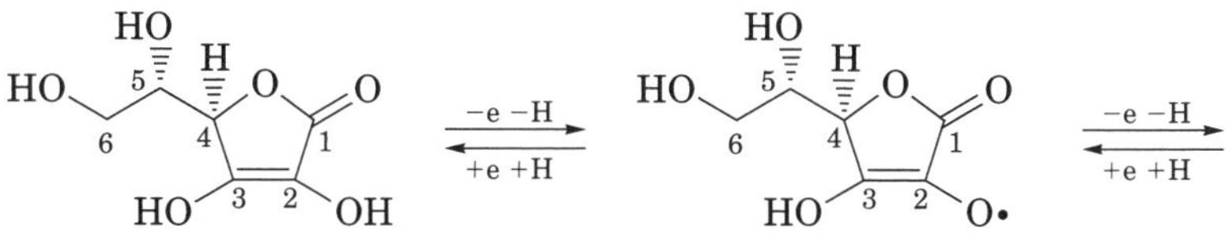

(**1**) L-Ascorbic acid   (**2**) Monodehydro-L-ascorbic acid (free radical)

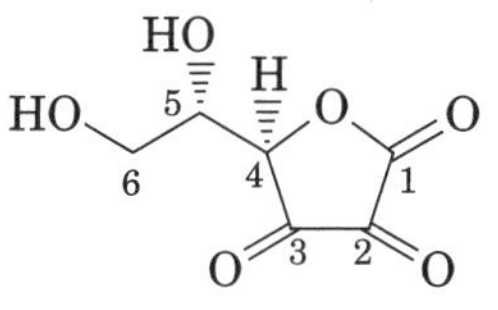

(**3**) Dehydro-L-ascorbic acid

Figure 1. Vitamin C redox system.

corbic acid, which also has antiscorbutic activity. The isolation of vitamin C was first accomplished by Szent-Györgyi in 1928 from cabbage, paprika, and the adrenal glands of animals (11). The relationship between vitamin C and the antiscorbutic factor was established in 1932 by Szent-Györgyi and, at the same time, by King and Waugh (12). Also in 1932, the chemical structure of ascorbic acid was determined independently by Haworth and Hirst (13). One year later, in 1933, Reichstein synthesized D-ascorbic acid and L-ascorbic acid (14). L-Ascorbic acid was synthesized from D-glucose (4) because the chiral centers of C-2 and C-3 were in the correct configuration to become C-4 and C-5, respectively, of L-ascorbic acid (15).

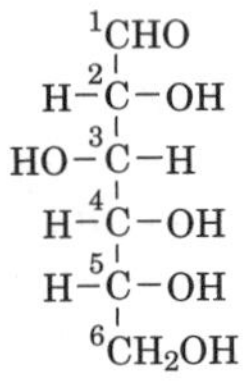

(5) 3-oxo-L-sorbosone

(1)

(6)

(7)

(4) D-Glucose

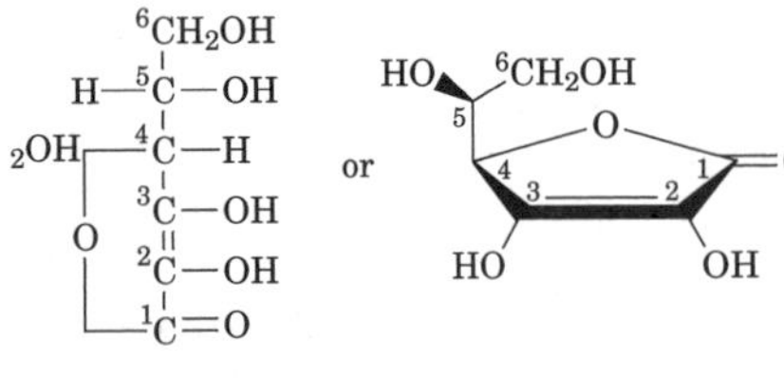

(1) L-Ascorbic acid

This synthesis was the first step towards industrial vitamin production, which began in 1936. The synthetic product was shown to have the same biological activity as the natural substance. It is reversibly oxidized in the body to dehydro-L-ascorbic acid (3) (L-*threo*-2,3-hexodiulosonic acid γ-lactone), a potent antiscorbutic agent with full vitamin activity. In 1937, Haworth and Szent-Györgyi received the Nobel Prize for their work on vitamin C.

STRUCTURE DETERMINATION

Albert Szent-Györgyi demonstrated that "hexuronic acid," $C_6H_8O_6$, which was first isolated from sources such as cabbage juice, was identical to vitamin C (L-ascorbic acid). Ultraviolet absorption studies carried out in the mid-1930s led to the proposal that L-ascorbic acid was 3-oxo-L-sorbosone (5), which exists in a variety of tautomeric forms, including (6), (7), and L-ascorbic acid (1). The chemical structure of L-ascorbic acid was determined independently by Haworth and Hirst using x-ray crystallographic techniques (13). Soon afterwards, additional data supporting the structure 1 was published by other authors (16–18). Excellent review articles summarizing the early work on vitamin C have appeared (19,20). Years later, with improved x-ray techniques (21,22) and neutron-diffraction methods (23), the structure of L-ascorbic acid was refined (24,25). The five-membered ring containing the enediol group is almost planar. The conformation of the side chain in the crystal is as shown by structure 1. The chiral center at position C-4 has the R- or D-configuration, whereas C-5 has the S- or L-configuration.

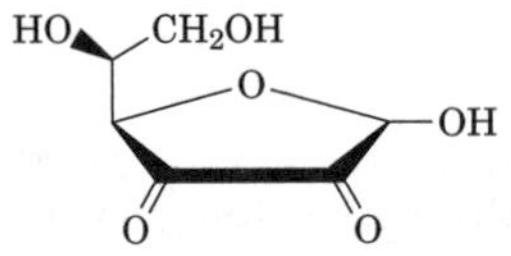

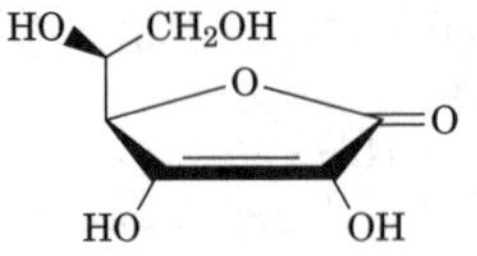

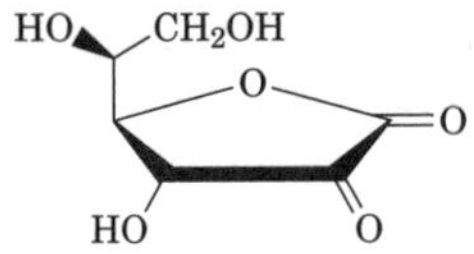

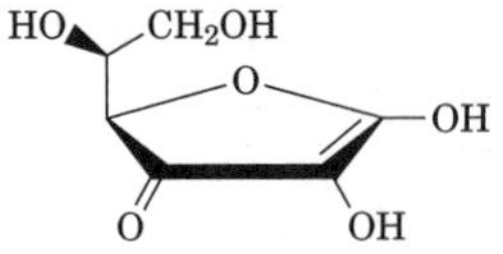

As a result of having two chiral centers, four stereoisomers of ascorbic acid are possible (Table 1) (Fig. 2). Besides L-ascorbic acid (Activity = 1), only D-araboascorbic acid (erythorbic acid (9)) shows vitamin C activity (Activity = 0.025–0.05). The L-ascorbic acid structure (1) in solution and the solid state are almost identical. Ascorbic acid crystallizes in the space group $P2_1$ with four molecules in the unit cell. The crystal data are summarized in Table 2.

Crystalline dehydro-L-ascorbic acid (3) is reported to exist as the dimer (11) (26). In water, it is present as the hemiacetal monomer (12).

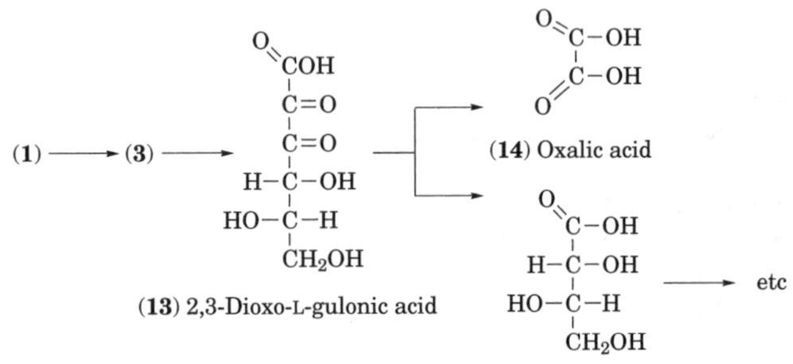

(13) 2,3-Dioxo-L-gulonic acid

(14) Oxalic acid

(15) L-Threonic acid

PROPERTIES

Physical Properties

Table 3 contains a summary of the physical properties of L-ascorbic acid. Properties relating to the structure of vitamin C have been reviewed and summarized (32). Stabilization of the molecule is a consequence of delocalization of the π-electrons over the conjugated enediol system. The highly acidic nature of the H-atom on C-3 has been confirmed by neutron diffraction studies (23).

Chemical Properties

The most significant chemical property of L-ascorbic acid is its reversible oxidation to dehydro-L-ascorbic acid. Dehydro-L-ascorbic acid has been prepared by uv irradiation and by oxidation with air and charcoal, halogens, ferric chloride, hydrogen peroxide, 2,6-dichlorophenolindophenol, neutral potassium permanganate, selenium oxide, and many other compounds. Dehydro-L-ascorbic acid has been reduced to L-ascorbic acid by hydrogen iodide, hydro-

Table 1. Isomeric Ascorbic Acids

Substance	Structure	Mp (°C)	$[\alpha]_D$ (H$_2$O) (°)	Activity
L-Ascorbic acid	(1)	192	+24	1
D-Xyloascorbic acid (D-ascorbic acid, D-*threo*-hex-2-enonic acid γ-lactone)	(8)	192	−23	0
D-Araboascorbic acid (erythorbic acid, D-*erythro*-hex-2-enonic acid γ-lactone)	(9)	174	−18.5	0.025–0.05
L-Araboascorbic acid (L-*erythro*-hex-2-enonic acid γ-lactone)	(10)	170	+17	0

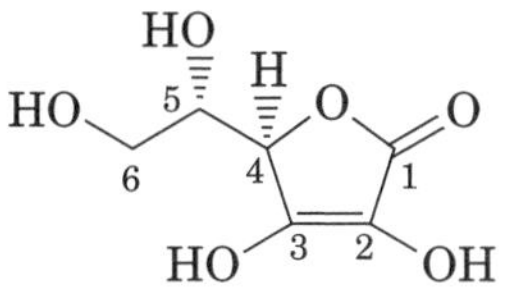

(**1**) L-Ascorbic acid

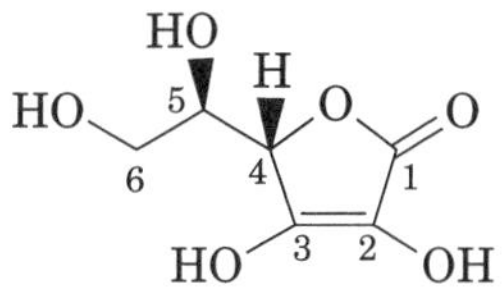

(**8**) D-Ascorbic acid

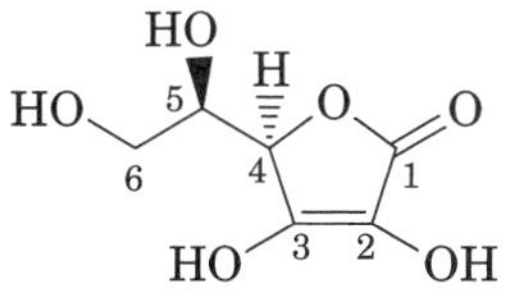

(**9**) D-Araboascorbic acid (erythorbic acid)

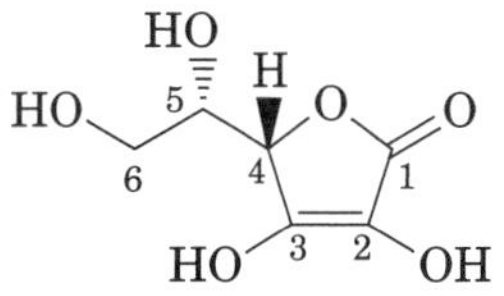

(**10**) L-Araboascorbic acid

Figure 2. Isomers of ascorbic acid.

Table 2. X-Ray Crystal Data of L-Ascorbic Acid

Space group	$P2_1$
a	1.7299 nm
b	0.6353 nm
c	0.6411 nm
β	102° 11′
V	0.68859 nm³
Z	4
d_{calcd}	1.699 g/cm³
μ(CuK$_a$)	13.9 cm^{-1}

Source: Ref. 24.

gen sulfide, 1,4-dithiothreitol (1,4-dimercapto-2,3-butane-diol), and the like (33).

The degradation of L-ascorbic acid in aqueous solution depends on several factors, eg, pH, temperature, presence of oxygen, or metals. Comprehensive reviews of degradation reactions and mechanisms have been published (34–37). In aqueous solution, L-ascorbic acid is more sensitive in the presence of oxygen to bases than to acids. The pH range with the highest stability is between 4 and 6. L-Ascorbic acid is sensitive to heat. In the presence of oxygen and heat, it is oxidized at a rate proportional to the temperature rise. On standing, dehydro-L-ascorbic acid (**3**), the initial oxidation product, undergoes irreversible hydrolysis to 2,3-dioxo-L-gulonic acid (*threo*-2,3-hexodiulosonic acid (**13**)), which is further oxidized to oxalic acid (**14**) and L-threonic acid ((R-(R*,S*)-2,3,4-trihydroxybutanoic acid)) (**15**) (Fig. 3).

In acidic solution, the degradation results in the formation of furfural, furfuryl alcohol, 2-furoic acid, 3-hydroxyfurfural, furoin, 2-methyl-3,8-dihydroxychroman, ethylglyoxal, and several condensation products (36). Many metals, especially copper, catalyze the oxidation of L-ascorbic acid. Oxalic acid and copper form a chelate complex which prevents the ascorbic acid-copper-complex formation and therefore oxalic acid inhibits effectively the oxidation of L-ascorbic acid. L-Ascorbic acid can also be stabilized with metaphosphoric acid, amino acids, 8-hydroxyquinoline, glycols, sugars, and trichloracetic acid (38). Another catalytic reaction which accounts for loss of L-ascorbic acid occurs with enzymes, eg, L-ascorbic acid oxidase, a copper protein-containing enzyme.

Stability

Ascorbic acid, a white crystalline compound, is very soluble in water and has a sharp, acidic taste. In solution, the vitamin oxidizes on exposure to air, light, and elevated temperatures. Solutions of ascorbic acid turn yellowish, followed by development of a tan color. Ascorbic acid is stable to air when dry but gradually darkens on exposure to light.

SYNTHESIS

Chemical Synthesis

The first synthesis of ascorbic acid was reported in 1933 by Reichstein and co-workers (14,39–42) (Fig. 4). Similar, independent reports published by Haworth and co-workers followed shortly after this work (13,43–45). L-Xylose (**16**)

Table 3. Physical Properties of L-Ascorbic Acid

Property	Characteristic(s)	Reference
Appearance	White, odorless, crystalline solid with a sharp acidic taste	
Formula; mol. wt.	$C_6H_8O_6$; 176.13	
Crystalline form	Monoclinic; usually plates, sometimes needles	
Mp °C	190–192 (dec)	27
Density, g/cm^3	1.65	27
Optical rotation	$[\alpha]^{25}$ + 20.5° to +21.5° (c = 1 in water) $[\alpha]^{23}$ + 48 (c = 1 in methanol)	27
pH		27
5 mg/mL	3	
50 mg/mL	2	
pK_1	4.17	27
pK_2	11.57	27
Redox potential	First stage: E + 0.166 V (pH 4)	27
Solubility, g/mL		27
water	0.33	
95 wt. % ethanol	0.033	
absolute ethanol	0.02	
glycerol USP	0.01	
propylene glycol	0.05	
ether	Insoluble	
chloroform	Insoluble	
benzene	Insoluble	
petroleum ether	Insoluble	
oils	Insoluble	
fats	Insoluble	
fat solvents	Insoluble	

Spectral properties

UV	pH 2: E_{max} (1%, 1 cm) 695 at 245 nm (nondissociated form)	28
	pH 6.4: E_{max} (1%, 1 cm) 940 at 265 nm (monodissociated form)	
IR (KBr)	Characteristic wavelengths, cm^{-1}	29
	3455, 3405, 3155 ν OH groups	
	2570 associated OH groups	
	1770, 1670 carbonyl lactone	
	1254 C–O–C lactone	
	1057 δ OH groups	
nmr[a]	^{1}H nmr (D$_2$O)	30
	δ 4.97 (d, 1 H, $J_{4.5}$ = 2 Hz, H-4), 4.10	
	(ddd, 1 H, $J_{5.6}$ = 5 and 7 Hz, $J_{4.5}$ = 2 Hz, H-5), 3.78 (m, 2 H, C-6)	
	^{13}C nmr (D$_2$O)	31
	δ 174.0 (C-1), 118.8 (C-2), 156.3 (C-3)	
	77.1 (d, $J_{c\text{-}4, H\text{-}4}$ = 158 Hz, C-4)	
	69.9 (d, $J_{c\text{-}5, H\text{-}5}$ = 145.0 Hz, C-5),	
	63.2 (t, $J_{C\text{-}6, H\text{-}6}$ = 145.0Hz, C-6)	

[a]d = Doublet; ddd = doublet of doublet of doublet; m = multiplet; and t = triplet.

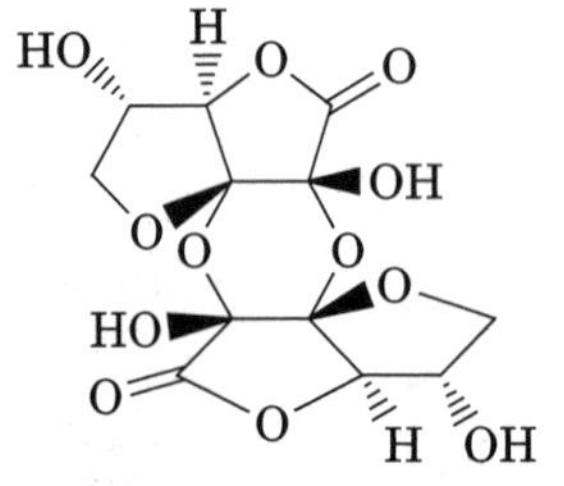

Figure 3. Degradation of L-ascorbic acid.

(**11**) Dehydro-L-ascorbic acid dimer

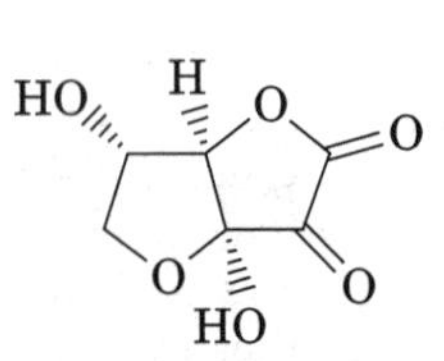

(**12**) Dehydro-L-ascorbic acid hemiacetal

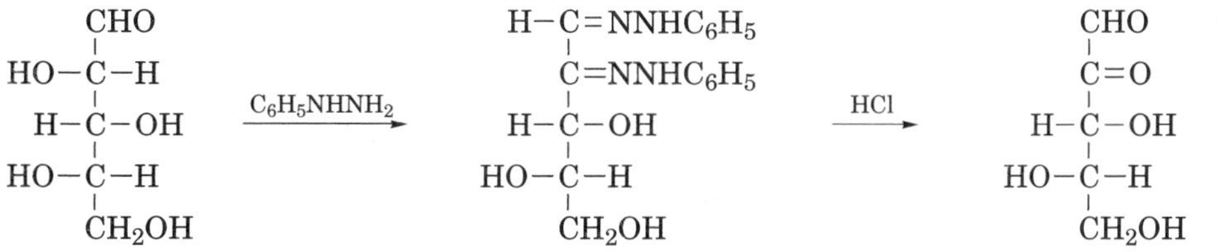

CHO
HO—C—H
H—C—OH
HO—C—H
CH$_2$OH

(**16**) L-Xylose

$\xrightarrow{C_6H_5NHNH_2}$

H—C=NNHC$_6$H$_5$
C=NNHC$_6$H$_5$
H—C—OH
HO—C—H
CH$_2$OH

(**17**) L-Xylose osazone

$\xrightarrow{HCl}$

CHO
C=O
H—C—OH
HO—C—H
CH$_2$OH

(**18**) L-Xylosone

$\xrightarrow{HCN}$

CN
HO—C—OH
C=O
H—C—OH
HO—C—H
CH$_2$OH

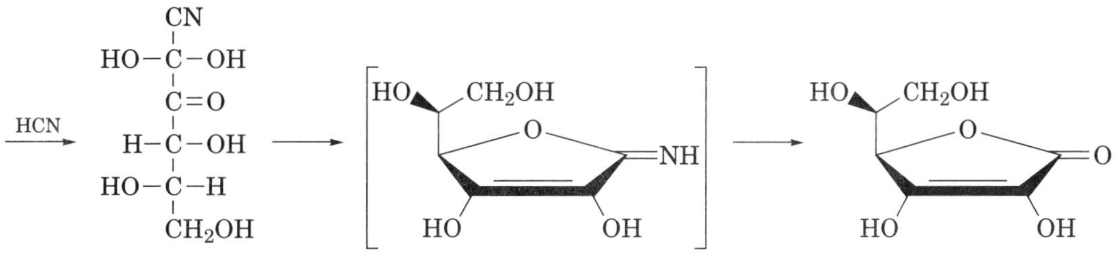

(**19**) L-Xylonitrile

Cycloimine of L-ascorbic acid

(**1**) L-Ascorbic acid

Figure 4. First syntheses of L-ascorbic acid.

was converted by way of its osazone (**17**) into L-xylosone (**18**), which reacted with hydrogen cyanide forming L-xylonitrile (**19**). L-Xylonitrile cyclized under mild conditions to the cycloimine of L-ascorbic acid. Hydrolysis of the cycloimine yielded L-ascorbic acid. The yield for the conversion of L-xylosone to L-ascorbic acid was ca 40%.

The L-xylosone pathway to L-ascorbic acid was never commercialized because L-xylosone was not readily available and was too expensive to prepare. The route, however, was valuable for L-ascorbic acid structure determination and for the preparation of derivatives.

Most current industrial vitamin C production is based on the efficient second synthesis developed by Reichstein and Grüssner in 1934 (15). Various attempts to develop a superior, more economical L-ascorbic acid process have been reported since 1934. These approaches, which have met with little success, are summarized in Crawford's comprehensive review (46). Currently, all chemical syntheses of vitamin C involve modifications of the Reichstein and Grüssner approach (Fig. 5). In the first step, D-glucose (**4**) is catalytically (Ni-catalyst) hydrogenated to D-sorbitol (**20**). Oxidation to L-sorbose (**21**) occurs microbiologically with *Acetobacter suboxydans*. The isolated L-sorbose is reacted with acetone and sulfuric acid to yield 2,3:4,6 diacetone-L-sorbose, (DAS) (2,3:4,6-bis-O-isopropylidene-α-L-sorbofuranose) (**22**). The remaining unprotected primary hydroxyl group of DAS is oxidized, either catalytically (O$_2$, Pd), electrochemically, or by sodium hypochlorite with nickel chloride as catalyst to the corresponding carboxylic acid. The resulting 2,3:4,6-diacetone-2-keto-L-gulonic acid (DAG) (2,3:4,6-bis-O-isopropylidene-2-oxo-L-gulonic acid) (**23**) is treated with hydrogen chloride in an inert solvent system. Under these conditions, acetone removal occurs followed by consecutive lactonization and enolization to form L-ascorbic acid (**1**).

Fermentation

Much time and effort has been spent in undertaking to find fermentation processes for vitamin C (47). One such approach is now practiced on an industrial scale, primarily in China. It is not certain, however, whether these processes will ultimately supplant the optimized Reichstein synthesis. One important problem is the instability of ascorbic acid in water in the presence of oxygen; it is thus highly unlikely that direct fermentation to ascorbic acid will be economically viable. The successful approaches to date involve fermentative preparation of an intermediate, which is then converted chemically to ascorbic acid.

L-Sorbose to 2-KGA Fermentation. In China, a variant of the Reichstein-Grüssner synthesis has been developed on an industrial scale (see Fig. 5). L-Sorbose is oxidized directly to 2-ketogulonic acid, (2-KGA) (**24**) in a mixed culture fermentation step (48). Acid-catalyzed lactonization and enolization of 2-KGA produces L-ascorbic acid (**1**).

A Chinese publication (47) with 17 references reviews the use of genetically engineered microorganisms for the production of L-ascorbic acid and its precursor, 2-KGA (49). For example, a 2-keto-L-gulonic acid fermentation process from sorbose has been published with reported yields over 80% (50).

D-Glucose to 2-KGA Fermentation. A different fermentative route to 2-KGA proceeds via 2,5-diketo-L-gluconic acid (51,52). In a two-stage fermentation (Shionogi-Process), D-glucose (**4**) is oxidized to 2,5-diketo-D-gluconic acid (2,5-DKGA) (**25**) with *Erwinia sp.*, followed by stereospecific reduction at C-5 by *Corynebacter sp.*, forming 2-ketogulonic acid (**24**) (Fig. 6). The 2-KGA is rearranged upon treatment with acid to give ascorbic acid. A produc-

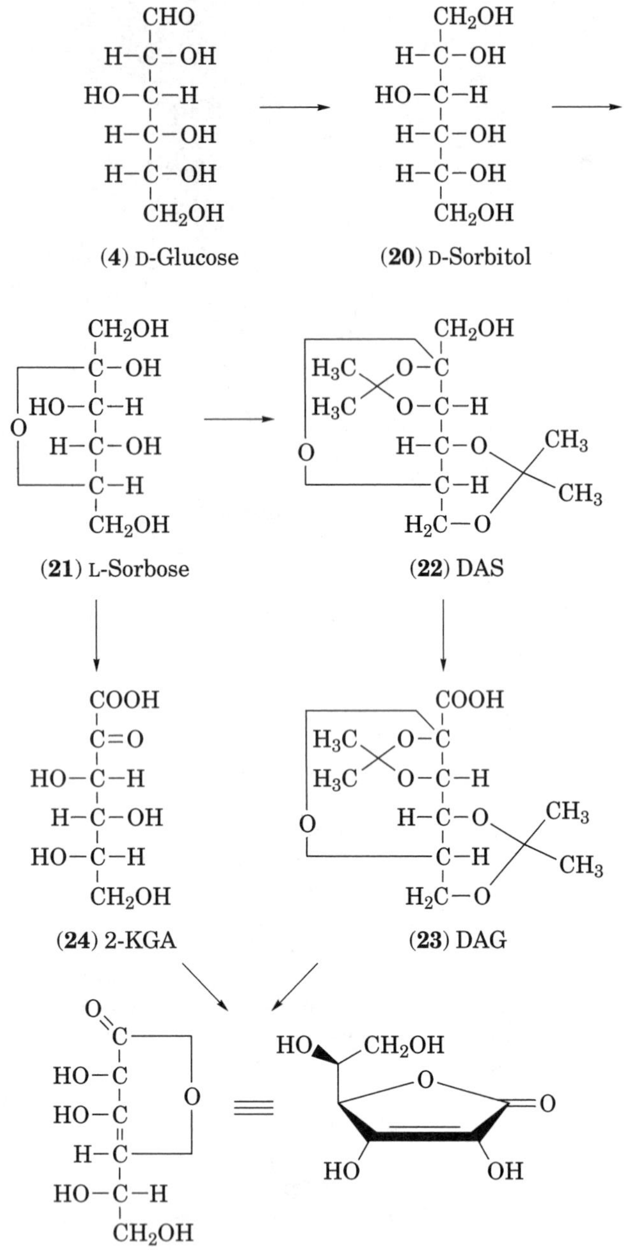

Figure 5 structures

CHO
H–C–OH
HO–C–H
H–C–OH
H–C–OH
CH₂OH

(**4**) D-Glucose

CH₂OH
H–C–OH
HO–C–H
H–C–OH
H–C–OH
CH₂OH

(**20**) D-Sorbitol

CH₂OH
C–OH
HO–C–H
H–C–OH
C–H
CH₂OH

(**21**) L-Sorbose

CH₂OH
H₃C–O–C
H₃C–O–C–H
H–C–O–CH₃
C–H
H₂C–O–CH₃

(**22**) DAS

COOH
C=O
HO–C–H
H–C–OH
HO–C–H
CH₂OH

(**24**) 2-KGA

COOH
H₃C–O–C
H₃C–O–C–H
H–C–O–CH₃
C–H
H₂C–O–CH₃

(**23**) DAG

O=C
HO–C
HO–C
H–C
HO–C–H
CH₂OH
≡
HO CH₂OH ... O ... =O ... HO ... OH

(**1**) L-Ascorbic acid

Figure 5. Syntheses of L-ascorbic acid.

tion of ascorbic acid by this route is reportedly being developed, ca 1997. An analogous process using a recombinant strain of *Erwinia citreus* was developed at Biogen (53).

D-Glucose to L-Ascorbic Acid Fermentation. The direct heterotrophic fermentation of D-glucose to L-ascorbic acid with algae is disclosed in a European Patent Application (54). The overall yield of L-ascorbic acid is 4%.

INDUSTRIAL TECHNOLOGY

Vitamin C was the first vitamin to be manufactured by chemical synthesis on an industrial scale. Major suppliers of vitamin C are Hoffmann-La Roche, BASF, Takeda, E. Merck, and various companies in China. Additional production occurs in Eastern Europe and India.

Reichstein and Grüssner's second L-ascorbic acid synthesis became the basis for the industrial vitamin C production. Many chemical and technical modifications have improved the efficiency of each step, enabling this multi-step synthesis to remain the principal, most economical process up to 1997 (46). L-Ascorbic acid is produced in large, integrated, automated facilities, involving both continuous and batch operations. The process steps are outlined in Figure 7. Procedures require ca 1.7-kg L-sorbose/kg of L-ascorbic acid with ca 66% overall yield in 1977 (55). Since 1977, further continuous improvement of each vitamin C production step has taken place. Today's overall ascorbic acid yield from L-sorbose is ca 75%. In the mid-1930s, the overall yield from L-sorbose was ca 30%.

The catalytic hydrogenation of D-glucose to D-sorbitol is carried out at elevated temperature and pressure with hydrogen in the presence of nickel catalysts, in both batch and continuous operations, with >97% yield (56,57). The cathodic reduction of D-glucose to L-sorbitol has been practiced (58). D-Mannitol is a by-product (59).

Sterile aqueous D-sorbitol solutions are fermented with *Acetobacter suboxydans* in the presence of large amounts of air to complete the microbiological oxidation. The L-sorbose is isolated by crystallization, filtration, and drying. Various methods for the fermentation of D-sorbitol have been reviewed (60). *Acetobacter suboxydans* is the organism of choice as it gives L-sorbose in >90% yield (61). Large-scale fermentations can be carried out in either batch or continuous modes. In either case, sterility is im-

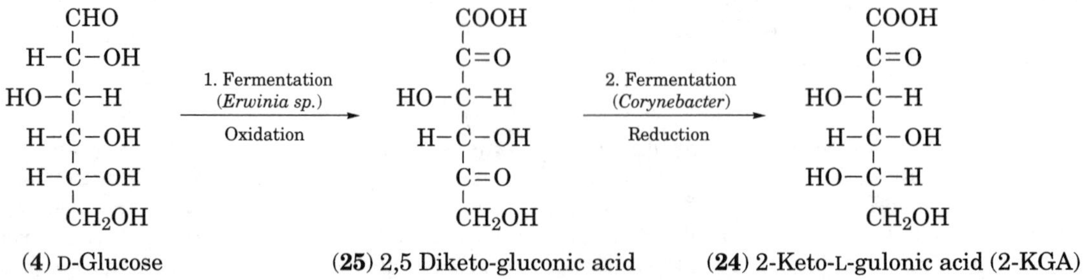

Figure 6 structures

CHO
H–C–OH
HO–C–H
H–C–OH
H–C–OH
CH₂OH

(**4**) D-Glucose

1. Fermentation
(*Erwinia sp.*)
Oxidation

COOH
C=O
HO–C–H
H–C–OH
C=O
CH₂OH

(**25**) 2,5 Diketo-gluconic acid

2. Fermentation
(*Corynebacter*)
Reduction

COOH
C=O
HO–C–H
H–C–OH
HO–C–H
CH₂OH

(**24**) 2-Keto-L-gulonic acid (2-KGA)

Figure 6. D-Glucose to 2-KGA fermentation.

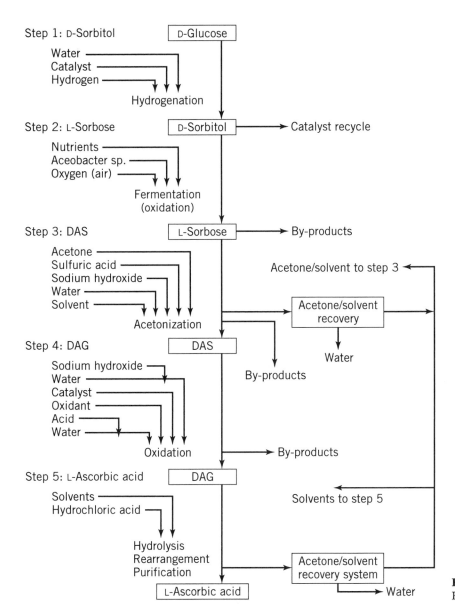

Step 1: D-Sorbitol

Step 2: L-Sorbose

Step 3: DAS

Step 4: DAG

Step 5: L-Ascorbic acid

Figure 7. L-Ascorbic acid manufactured by Reichstein and Grüssner Synthesis.

portant to prevent contamination, with subsequent loss of product.

In the third step, L-sorbose is reacted with acetone and excess sulfuric acid at low temperatures. The sorbose dissolves on conversion into the 2,3-mono-O-isopropylidene-L-sorbose (2,3 monoacetone-L-sorbose) (MAS), and 2,3:4,6-bis-O-isopropylidene-α-L-sorbofuranose (2,3:4,6-diacetone-L-sorbose) (DAS). The equilibrium mixture consists of ca 65% DAS and 35% MAS. The sulfuric acid acts as a catalyst and a dehydrating agent. The reaction mixture is worked up by dilution, neutralization, and extraction to separate the DAS from the MAS. MAS is recycled and the mother liquors are distilled to recover acetone and solvents. The original Reichstein and Grüssner process has been optimized over decades by other workers, giving ca 85% yield (55). Other acidic catalysts, besides sulfuric acid, have been investigated, eg, zinc chloride–phosphoric acid, p-toluenesulfonic acid, copper(II) sulfate, and cationic exchange resins. Ferric chloride and perchloric acid are ac-

tive catalysts that give >90% yields of DAS with azeotropic removal of the water (62). Other methods, including use of acetone dimethylacetal as a water removal reagent, have been reported (63). The use of other ketonic and aldehydic protective agents for L-sorbose, eg, cyclohexanone and its derivatives, formaldehyde, benzaldehyde, etc, has also been described (15,64).

2,3:4,6-Diacetone-L-sorbose (DAS) is oxidized at elevated temperatures in dilute sodium hydroxide in the presence of a catalyst (nickel chloride for bleach or palladium on carbon for air) or by electrolytic methods. After completion of the reaction, the mixture is worked up by acidification to 2,3:4,6-bis-O-isopropylidene-2-oxo-L-gulonic acid (2,3:4,6-diacetone-2-keto-L-gulonic acid) (DAG), which is isolated through filtration, washing, and drying. With sodium hypochlorite/nickel chloride, the reported DAG yields are >90% (65). The oxidation with air has been reported, and a practical process was developed with palladium–carbon or platinum–carbon as catalyst (66,67).

The electrolytic oxidation with nickel salts as the catalyst has also been published (68).

DAG is treated with ethanol and hydrochloric acid in the presence of inert solvent, eg, chlorinated solvents, hydrocarbons, ketones, etc. The L-ascorbic acid precipitates from the mixture as it forms, minimizing its decomposition (69). Crude L-ascorbic acid is isolated through filtration and purified by recrystallization from water. The pure L-ascorbic acid is isolated, washed with ethanol, and dried. The mother liquor from the recrystallization step is treated in the usual manner to recover the L-ascorbic acid and ethanol contained in it. The crude L-ascorbic acid mother liquor contains solvents and acetone liberated in the DAG hydrolysis. The solvents are recovered by fractional distillation and recycled. Many solvent systems have been reported for the acid-catalyzed conversion of DAG to L-ascorbic acid (46). Rearrangement solvent systems are used which contain only the necessary amount of water required to give >80% yields of high purity crude L-ascorbic acid (70).

The DAG conversion to L-ascorbic acid also can occur by a base-catalyzed mechanism. Methyl 2-oxo-L-gulonate (methyl DAG) is converted, on treatment with sodium methoxide, to sodium-L-ascorbate, which is then acidified to L-ascorbic acid. Various solvent systems have been evaluated and reported (46).

PACKAGING

L-Ascorbic acid is screened or pulverized into a variety of particle sizes. It is usually packaged in 25-kg and 50-kg quantities in standard, polyethylene-lined containers, eg, fiber drums, corrugated boxes, etc. The recommended storage conditions are low humidity and temperatures of ≤23°C.

ENVIRONMENTAL ISSUES

The environmental concerns of an ascorbic acid manufacturing facility are those typical of a chemical processing plant. Its operating design must be patterned to conform to environmental protection regulations. Measures must be taken to contain solvents and to keep emissions within official guidelines. Special condensers, continuous instrumental monitoring, and emergency containment and cleanup systems are required. Wastewater-treatment facilities have to be provided to remove by-product organics and inorganics from effluent streams before disposal. The extent of these treatment facilities depends upon the location of the plant and the local tolerances. Usually, there are secondary treatment facilities for organic removal; at some plant sites, additional treatment may be required to remove inorganic salts.

ECONOMIC ASPECTS

Strong growth in demand during the period 1985–95 from existing and new applications led to firm pricing and expansion of world capacity. Total world demand in 1995 was estimated to be approximately 60,000 metric tons. Balance of capacity and demand at that time resulted in a selling price of $16.00/kg. More recently (ca 1997), prices have dropped to less than $10/kg.

Continuing demand from food, pharmaceutical, animal nutrition, and cosmetic applications will result in growth rates of 3–4% annually throughout the remainder of the 1990s.

SPECIFICATIONS

Specifications for ascorbic acid, sodium ascorbate, calcium ascorbate, and ascorbyl palmitate are found in the *United States Pharmacopeia / National Formulary* (71) and the *Food Chemicals Codex* (72). The official assay for all four compounds is the iodimetric titration with 0.1 *N* iodine solution and starch as the indicator.

ANALYTICAL METHODS

Many different analytical methods have been developed to determine L-ascorbic acid in feed, biological, and pharmaceutical samples. An excellent review article describes the methodology for finding the proper L-ascorbic acid assay method (73). Comprehensive reviews of all analytical methods, including the extraction of ascorbic acid from foods and biological samples, have been published (74). Ascorbic acid has been determined by a variety of methods, including uv absorption; redox and derivatization reactions; electrochemical and enzymatic oxidation reactions; chromatographic, eg, hplc methods; and biological methods with animals. The guinea pig, one of the few animal species requiring ascorbic acid, is used in the bioassay for ascorbic acid activity. The various methods practiced have been described in detail (75).

Because of the time and expense involved, biological assays are used primarily for research purposes. The first chemical method for assaying L-ascorbic acid was the titration with 2,6-dichlorophenolindophenol solution (76). This method is not applicable in the presence of a variety of interfering substances, eg, reduced metal ions, sulfites, tannins, or colored dyes. This 2,6-dichlorophenolindophenol method and other chemical and physiochemical methods are based on the reducing character of L-ascorbic acid (77). Colorimetric reactions with metal ions as well as other redox systems, eg, potassium hexacyanoferrate(III), methylene blue, chloramine, etc, have been used for the assay, but they are unspecific because of interferences from a large number of reducing substances contained in foods and natural products (78). These methods have been used extensively in fish research (79). A specific photometric method for the assay of vitamin C in biological samples is based on the oxidation of ascorbic acid to dehydroascorbic acid with 2,4-dinitrophenylhydrazine (80). In the microfluorometric method, ascorbic acid is oxidized to dehydroascorbic acid in the presence of charcoal. The oxidized form is reacted with o-phenylenediamine to produce a fluorescent compound that is detected with an excitation maximum of ca 350 nm and an emission maximum of ca 430 nm (81).

Another method that determines both ascorbic acid and dehydroascorbic acid first reduced the dehydroascorbic acid to ascorbic acid and then retains the ascorbic acid on an anionic Sephadex column (82). The ascorbic acid is oxidized on the column to dehydroascorbic acid by *p*-benzoquinone, which simultaneously elutes the dehydroascorbic acid. The dehydroascorbic acid is reacted with 4-nitro-1,2-phenylenediamine and absorbance of the resulting yellow solution produced is measured at 375 nm.

Chromatographic methods, notably hplc, are available for the simultaneous determination of ascorbic acid as well as dehydroascorbic acid. Some of these methods result in the separation of ascorbic acid from its isomers, eg, erythorbic acid and oxidation products such as diketogulonic acid. Detection has been by fluorescence, uv absorption, or electrochemical methods (83–85). Polarographic methods have been used because of their accuracy and their ease of operation. Ion exclusion (86) and ion suppression (87) chromatography methods have recently been reported. Other methods for ascorbic acid determination include enzymatic, spectroscopic, paper, thin layer, and gas chromatographic methods. Excellent reviews of these methods have been published (73,88,89).

USES

L-Ascorbic acid is used as a micronutrient additive in pharmaceutical, food, feed, and beverage products, as well as in cosmetic applications. The over-the-counter (OTC) vitamin market is strong, growing in demand, and vitamin C is available in the form of pills and tablets to supplement the daily diet to maintain peak physical performance.

Industrial uses of L-ascorbic acid relate to its antioxidant and reducing properties. It is used as an antioxidant in the commercial preparation of beer, fruit juices, cereals, and canned and frozen foods, etc.

A proposal was made to use L-ascorbic acid as an antioxidant replacement for sulfites in the food industries (90,91). Ascorbic acid's antioxidant property also inhibits nitrosamine formation in cured meat. The addition of ascorbic acid in flour improves baking qualities of dough and appearance of baked goods (92). Vitamin C also prevents discoloration of food during cooking and storage. Its fatty acid esters, eg, L-ascorbyl palmitate, are used to stabilize fats and oils (90). Ascorbic acid and its more stable derivatives L-ascorbyl-2-sulfate, ascorbyl 2-monophosphate, etc, are added to fish feed to improve feed utilization and decrease the rate of infection (93–95). The biological and pharmacological activities of L-ascorbyl-2-sulfate have been reviewed (96,97). Ascorbic acid is also used in agriculture as an abscission agent for fruit, in photography as a developing agent, in metallurgy as a reducing agent, in the polymer industry as a catalyst, in cosmetic formulations, in the manufacture of inks, in explosives, and in a variety of other applications (92,98).

In applications where vitamin C activity is unimportant, often D-erythorbic acid (D-araboascorbic acid) can also be used, providing the same antioxidant and reducing properties as L-ascorbic acid.

DERIVATIVES

Ascorbic acid has a variety of reactive positions that can be used to synthesize derivatives (99). Various derivatives and analogues have been prepared in attempts to find substances with increased biological activity (100,101).

Only L-ascorbic acid and its salts and C-6-substituted esters have full vitamin activity; sodium L-ascorbate, calcium L-ascorbate, and L-ascorbyl palmitate are commercially significant. L-Ascorbic acid 2-sulfate is bioavailable to fish. 6-Chloro-6-deoxy-L-ascorbic acid was prepared in 1977 and its activity, compared to L-ascorbic acid, is 4/5. Derivatives, eg, 6-deoxy-L-ascorbic acid, L-ascorbic acid 3-*O*-methylether, and 2-amino-2-deoxy-L-ascorbic acid have been prepared; their respective activities compared to L-ascorbic acid are 1/3, 1/25–1/50, and 0. Many more derivatives with and without biological activity have been synthesized (102,103). The highest vitamin C activity correlates with the enediol lactone group, D-configuration for the C-4 hydrogen group, at least a two-carbon substituent on C-4, and L-configuration for the C-5 hydroxyl group. The primary C-6 hydroxyl group has minor impact on the biological activity.

Methods for the preparation of L-ascorbic acids having isotopic C, H, O in various positions have been described and reviewed (104,105). Labeled L-ascorbic acid has played an important role in the elucidation of the metabolic pathway of L-ascorbic acid in plants and animals.

BIOSYNTHESIS

In all plants and most animals, L-ascorbic acid is produced from D-glucose (**4**) and D-galactose (**26**). Ascorbic acid biosynthesis in animals starts with D-glucose (**4**). In plants, where the biosynthesis is more complicated, there are two postulated biosynthetic pathways for the conversion of D-glucose or D-galactose to ascorbic acid.

$$
\begin{array}{cc}
\text{CHO} & \text{CHO} \\
\text{H}-\text{C}-\text{OH} & \text{H}-\text{C}-\text{OH} \\
\text{HO}-\text{C}-\text{H} & \text{HO}-\text{C}-\text{H} \\
\text{H}-\text{C}-\text{OH} & \text{HO}-\text{C}-\text{H} \\
\text{H}-\text{C}-\text{OH} & \text{H}-\text{C}-\text{OH} \\
\text{CH}_2\text{OH} & \text{CH}_2\text{OH} \\
(\mathbf{4})\ \text{D-Glucose} & (\mathbf{26})\ \text{D-Galactose}
\end{array}
$$

Biosynthesis in Animals

Amphibians and reptiles carry the enzyme L-gulono-γ-lactone oxidase which can transform a sugar-like glucose or galactose into ascorbic acid in the kidneys. In mammals (106) and birds (107) this enzyme system has been transferred from the kidneys to the liver. Humans, other primates, guinea pigs, fruit bats, and monkeys from the top of the evolutionary tree, as well as insects and invertebrates from the bottom end of the evolutionary tree, cannot synthesize L-ascorbic acid. Thus, they must consume vitamin C from exogenous sources to survive (108). Animals that are able to produce ascorbic acid do so by the glucuronic acid pathway in the liver or kidneys. The reactions involved in rats are illustrated in Figure 8. In this pathway, the D-glucose chain remains intact, and the C-1 and C-6 of D-glucose become C-6 and C-1, respectively, of L-ascorbic acid as the sequence of carbon-chain numbering is inverted. By measuring the incorporation of ^{14}C from D-glucose-1-^{14}C into urinary L-ascorbic acid, it was determined that the label is found at the C-6 position of L-ascorbic acid (109).

In animals, the glucuronic pathway (Fig. 9) is an important route for glucose utilization leading to the formation of glucuronides and mucopolysaccharides. D-glucose is first phosphorylated to D-glucose-6-phosphate, then isomerized to D-glucose-1-phosphate, which reacts with uridine-triphosphate (UTP) to form UDP-glucose. UDP-glucose is oxidized and hydrolyzed to D-glucuronic acid. D-Glucuronic acid is reduced to L-gulonic acid. L-Gulonic acid lactonizes and forms L-gulono-γ-lactone. Finally, oxidation of this intermediate is carried out by L-gulono-γ-lactone oxidase, the essential oxidizing enzyme, leading to L-ascorbic acid.

D-Galactose may also serve as a precursor of vitamin C because it can be converted to D-glucose. The pathway is important in the metabolism of sugars under normal and diseased conditions and in regulating physiological functions (110). The biosynthesis of L-ascorbic acid is inhibited by deficiencies of certain vitamins, eg, vitamin A, vitamin E, and biotin, but is stimulated by certain drugs, eg, barbiturates, chlorobutanol, aminopyrine, and antipyrine, and by carcinogens, eg, 3-methylcholanthrene and 3,4-benzpyrene (111–113). It has been proposed that the excretion of D-glucaric acid can be used to diagnose both the exposure to the body for foreign substances and the drug metabolic capacity of the liver (114).

Biosynthesis in Plants

As in animals, L-ascorbic acid is also the product of hexose phosphate metabolism in plants, but its biosynthesis is more complicated. There are two biosynthetic pathways for the conversion of D-glucose or D-galactose to L-ascorbic acid in plants (115). The main pathway is postulated as involving retention of configuration by oxidation at C-1 to yield D-gluconic-acid (27). Lactonization to yield D-glucono-γ-lactone (28) is followed by oxidation at C-2 or C-3 and epimerization at C-5 to afford L-2-oxogulono-γ-lactone (29) (Fig. 10). The other biosynthetic pathway from D-glucose and D-galactose may be postulated similarly to Figure 8 to account for configurational inversion (116). The main precursor of L-ascorbic acid is L-galactono-γ-lactone, rather than L-gulono-γ-lactone, which is less active (116,117), and is thought to be epimerized to L-galactono-γ-lactone prior to oxidation to L-ascorbic acid (Fig. 11). Little is known about the functions of ascorbic acid in plants. It is involved in cellular respiration and may contribute to plant growth.

SOURCES OF VITAMIN C

Fruits and vegetables that are good sources of vitamin C include peppers, greens, broccoli, cabbage, spinach, potatoes, tomatoes, strawberries, and citrus products. Meats, fish, poultry, eggs, and dairy products contain small

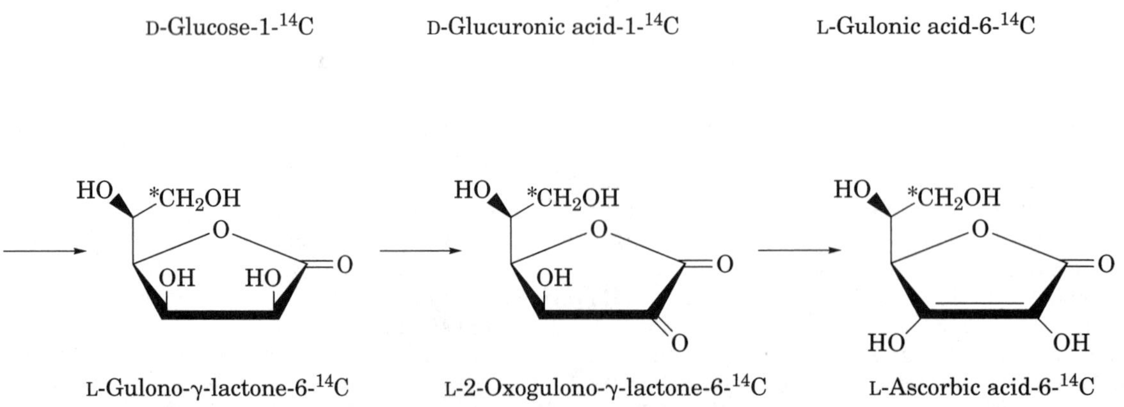

Figure 8. Pathway for the biosynthesis of L-ascorbic acid in rats using C-1-labeled D-glucose; * indicates position of [14]C.

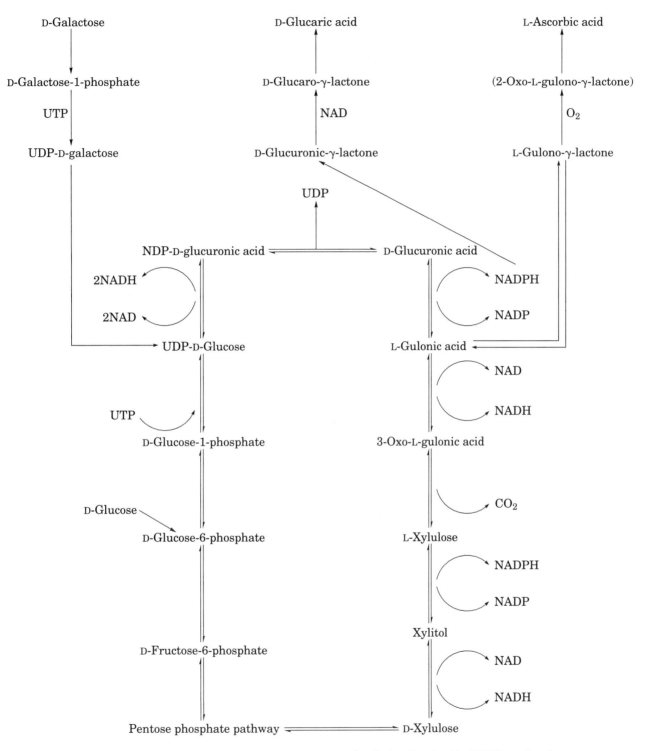

Figure 9. Glucuronic acid pathway. NAD = nicotinamide-adenine dinucleotide; NADH = reduced nicotinamide-adenine dinucleotide; NADP = nicotinamide-adenine dinucleotide phosphate; NADPH = reduced nicotinamide-adenine dinucleotide phosphate; NDP = nucleoside diphosphate; UDP = uridine diphosphate; and UTP = uridine triphosphate.

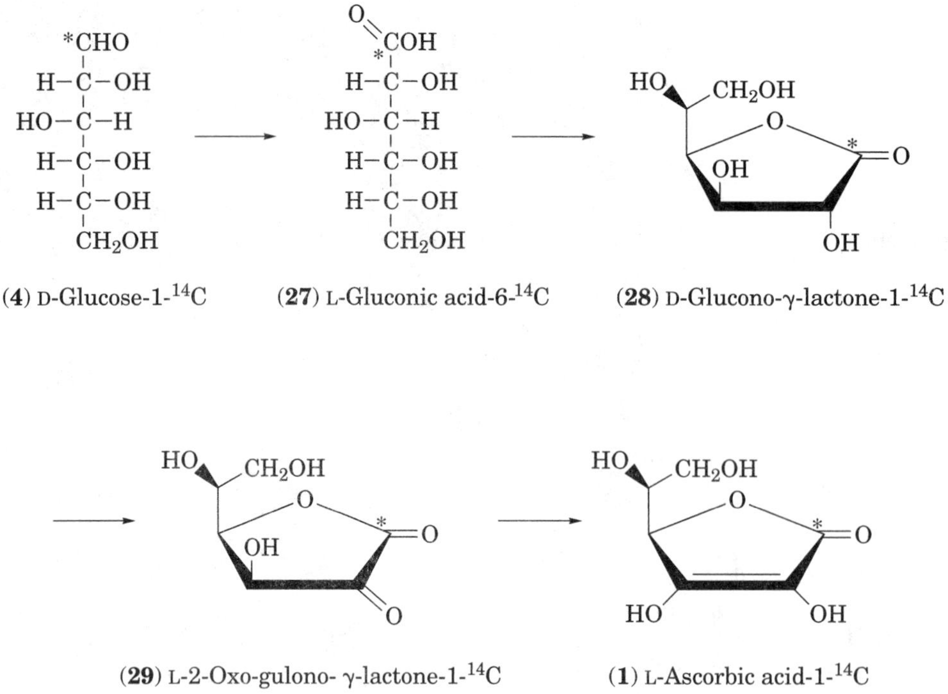

(**4**) D-Glucose-1-^{14}C   (**27**) L-Gluconic acid-6-^{14}C   (**28**) D-Glucono-γ-lactone-1-^{14}C

(**29**) L-2-Oxo-gulono- γ-lactone-1-^{14}C   (**1**) L-Ascorbic acid-1-^{14}C

Figure 10. Suggested pathway for the biosynthesis of L-ascorbic acid (with retention of configuration) in higher plants based on D-glucose-1-^{14}C experiments; * indicates position of the label.

amounts and grain contains none (118–121). The vitamin C content of some representative foods is listed in Table 4. Potatoes and cabbage have traditionally been the most important sources of vitamin C for the majority of people in the Western World during the winter season.

The contribution of any fruit and vegetable to the vitamin C content of the diet varies depending on climate, soil, and freshness. In actuality, there is a 30% coefficient of variability for ascorbic acid in fresh food products (122). In the case of field-grown spinach, for example, transportation and storage losses have been reported to be as high as 90% (122). During storage after harvesting, the vitamin C content of fruit and vegetables will decrease depending on time and temperature of storage. Ascorbic acid readily oxidizes both enzymatically and chemically on exposure to oxygen. Based on its water solubility, losses are caused through leaching during washing and blanching operations. Heat sterilization of canned foods inactivates completely the enzyme ascorbic acid oxidase which destroys vitamin C. Freezing inactivates the enzymes and is likewise a good method for preserving the vitamin C content of foodstuffs.

Restoration means adding to the food the amount of vitamin C which has been lost during food processing. The term enrichment or fortification is used when a nutrient is added to a food in an amount in excess of that naturally present. Fortification of foods with vitamin C is widely practiced. Technologically, it is important to avoid oxygen, copper, iron, and heat as much as possible to minimize the overages necessary above label claim to ensure compliance

throughout the shelf life of the product (123). In canned foods, breakfast drinks, and cereal products (106), overages of 25–50% are commonly used.

PHYSIOLOGY AND BIOCHEMISTRY

Biochemical Functions

Ascorbic acid has various biochemical functions, involving, for example, collagen synthesis, immune function, drug metabolism, folate metabolism, cholesterol catabolism, iron metabolism, and carnitine biosynthesis. Clear-cut evidence for its biochemical role is available only with respect to collagen biosynthesis (hydroxylation of proline and lysine). In addition, ascorbic acid can act as a reducing agent and as an effective antioxidant. Ascorbic acid also interferes with nitrosamine formation by reacting directly with nitrites, and consequently may potentially reduce cancer risk.

Enzymatic Reactions

The polypeptide collagens are the main component of skin and connective tissue, the organic substance of bones and teeth, and the ground substance between cells. The role ascorbic acid plays in collagen formation has been reviewed (124). The synthesis of collagens involves enzymatic hydroxylations of proline and of lysine. The former produces a stable extracellular matrix and the latter is needed for glycosylation and formation of cross-linkages in

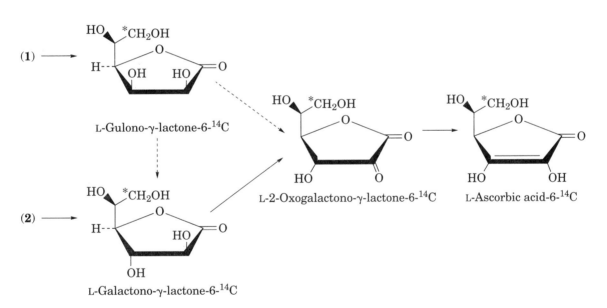

Figure 11. Proposed pathway for the biosynthesis of L-ascorbic acid (with inversion of configuration) in plants using C-1-labeled D-glucose or C-1-labeled D-galactose; * indicates position of the ^{14}C.

the fibers. Vitamin C's role in collagen formation is of importance in wound healing. Intake of 8–50 times the RDA level of 60 mg ascorbic acid per day before and after surgery increases the rate of wound healing considerably (125–127). Many of the clinical signs of scurvy are attributed to defects in collagen synthesis.

Ascorbic acid is involved in carnitine biosynthesis. Carnitine (γ-amino-β-hydroxybutyric acid, trimethylbetaine)

(**30**) is a component of heart muscle, skeletal tissue, liver and other tissues. It is involved in the transport of fatty acids into mitochondria, where they are oxidized to provide energy for the cell and animal. It is synthesized in animals from lysine and methionine by two hydroxylases, both containing ferrous iron and L-ascorbic acid. Ascorbic acid donates electrons to the enzymes involved in the metabolism of L-tyrosine, cholesterol, and histamine (**128**).

Table 4. Content of L-Ascorbic Acid in Representative Foods

Food substances	L-Ascorbic acid (mg/100 g)
Meat, fish, and milk	
Beef, pork, fish	≤2
Liver, kidney	10–40
Cow's milk	1–2
Vegetables	
Asparagus	15–30
Brussel sprouts, broccoli	90–150
Cabbage	30–60
Carrots	9
Cauliflower	60–80
Kale	120–180
Leek	15–30
Onion	10–30
Peas, beans	10–30
Parsley	170
Peppers	125–200
Potatoes	10–30
Spinach	50–90
Tomatoes	20–33
Fruit	
Apples	10–30
Bananas	10
Grapefruit	40
Guava	300
Hawthorne berries	160–800
Oranges, lemons	50
Peaches	7–14
Pineapples	17
Rose hips	1000
Strawberries	40–90

$$(CH_3)_3N^+ \underset{\text{OH}}{\wedge} \overset{\text{O}}{\underset{||}{C}O^-}$$

(**30**) Carnitine

L-Tyrosine metabolism and catecholamine biosynthesis occur largely in the brain, central nervous tissue, and endocrine system, which have large pools of L-ascorbic acid (128). Catecholamine, a neurotransmitter, is the precursor in the formation of dopamine, which is converted to noradrenaline and adrenaline. The precise role of ascorbic acid has not been completely understood. Ascorbic acid has important biochemical functions with various hydroxylase enzymes in steroid, drug, and lipid metabolism. The cytochrome P-450 oxidase catalyzes the conversion of cholesterol to bile acids and the detoxification process of aromatic drugs and other xenobiotics, eg, carcinogens, pollutants, and pesticides, in the body (129). The effects of L-ascorbic acid on histamine metabolism related to scurvy and anaphylactic shock have been investigated (130). Another cellular reaction involving ascorbic acid is the conversion of folate to tetrahydrofolate. Ascorbic acid has many biochemical functions which affect the immune system of the body (131).

Antioxidant Activity

Ascorbic acid serves as an antioxidant to protect intracellular and extracellular components from free-radical damage. It scavenges free radicals and forms the less reactive ascorbyl radical. The ascorbyl radical can be either reduced to ascorbic acid or oxidized to dehydroascorbic acid (132). Vitamin E is the major fat-soluble antioxidant involved in protecting cells from free-radical impact. During the process of fatty acid oxidation, tocopherol (vitamin E) forms the tocopheroxyl radical. Ascorbic acid has been proven to protect membrane and other hydrophobic compartments from such damage by regenerating the antioxidant form of vitamin E (131). These free-radical scavenging reactions of vitamin C are important in protecting the intracellular and extracellular structure of the lung by quenching free radicals generated by smoke, ozone, and singlet oxygen. The possibility that free-radical damage is also involved in the pathogenesis of HIV and that antioxidants may reduce infection is an area of intense interest (134).

Inhibition of Nitrosamine Formation

Nitrites can react with secondary amines and *N*-substituted amides under the acidic conditions of the stomach to form *N*-nitrosamines and *N*-nitrosamides. These compounds are collectively called *N*-nitroso compounds. There is strong circumstantial evidence that *in vivo N*-nitroso compounds production contributes to the etiology of cancer of the stomach (135,136), esophagus (136,137), and nasopharynx (136,138). Ascorbic acid consumption is negatively correlated with the incidence of these cancers, due to ascorbic acid inhibition of *in vivo N*-nitroso compound formation (139). The concentration of *N*-nitroso compounds formed in the stomach depends on the nitrate and nitrite intake. Nitrite is part of the preserving process for cured meats. Cigarette smoke contains high levels of nitrite.

Iron Absorption

A very important effect of ascorbic acid is the enhancement of absorption of nonheme iron from foods. Ascorbic acid also enhances the reduction of ferric iron to ferrous iron. This is important both in increasing iron absorption and in its function in many hydroxylation reactions (140,141). In addition, ascorbic acid is involved in iron metabolism. It serves to transfer iron to the liver and to incorporate it into ferritin. Ascorbic acid also forms soluble chelate complexes with iron (142–145). It seems ascorbic acid has no effect on high iron levels found in people with iron overload (146). It is well known, in fact, that ascorbic acid in the presence of iron can exhibit either prooxidant or antioxidant effects, depending on the concentration used (147). The combination of citric acid and ascorbic acid may enhance the iron load in aging populations. Iron overload may be the most important common etiologic factor in the development of heart disease, cancer, diabetes, osteoporosis, arthritis, and possibly other disorders. The synergistic combination of citric acid and ascorbic acid needs further study, particularly because the iron overload produced may be correctable (147).

Deficiency

Scurvy is a vitamin C-specific disease. It is characterized by anemia and alteration of protein metabolism; weakening of collagenous structures in bone, teeth, and connective tissues; swollen, bleeding gums with loss of teeth; fatigue and lethargy; rheumatic pains in the legs and degeneration of the muscles, skin lesions, and capillary weakness, massive hematomas in the thighs; and hemorrhages in many organs, including the eyes. Small (10–60 mg/d) quantities of L-ascorbic acid are sufficient to reverse the trend of both subclinical and clinical scurvy and alleviate their symptoms. Plasma and leukocyte ascorbic acid levels are the most reliable markers of vitamin C intake. Leukocyte levels are more reliable, less sensitive to recent vitamin C intake, and better reflect tissue ascorbate, but require relatively large amounts of blood. Normal leukocyte vitamin C levels are 20–40 $\mu g/10^8$ cells (148). Plasma concentrations of ascorbic acid less than 0.2 mg/dL and leukocyte concentrations below 2 $\mu g/10^8$ cells are seen in scurvy (149). Plasma ascorbic acid levels of 0.5 mg/dL are considered to prevent deficiency symptoms. Normal plasma levels are 0.8–1.4 mg/dL.

Absorption, Transport, and Excretion

The vitamin is absorbed through the mouth, the stomach, and predominantly through the distal portion of the small intestine, and hence, penetrates into the bloodstream. Ascorbic acid is widely distributed to the cells of the body and is mainly present in the white blood cells (leukocytes). The ascorbic acid concentration in these cells is about 150 times its concentration in the plasma (150,151). Dehydroascorbic acid is the main form in the red blood cells (erythrocytes). White blood cells are involved in the destruction of bacteria.

Up to 80% of oral doses of ascorbic acid are absorbed in humans with intakes of less than 0.2 g of vitamin C. Absorption of pharmacological doses ranging from 0.2 g to 12 g results in an inverse relationship, with less than 20% absorption at the higher doses. A single oral dose of 3 g has been reported to approach the absorptive capacity (tissue saturation) of the human intestine. Higher blood levels can be attained by providing multiple divided vitamin C doses per day.

The adrenal glands and pituitary glands have the highest tissue concentration of ascorbic acid. The brain, liver, and spleen, however, represent the largest contribution to the body pool. Plasma and leukocyte ascorbic acid levels decrease with increasing age (152). Elderly people require higher ascorbic acid intakes than children to reach the same plasma and tissue concentration (153).

Ascorbic acid is very soluble in water and mainly excreted in the urine. No ascorbic acid is excreted during vitamin C deficiency. A minimum amount is lost in the feces, even after intake of gram dosages (154).

Mobilization and Metabolism

The total ascorbic acid body pool in healthy adults has been estimated to be approximately 1.5 g, which increases to 2.3–2.8 g with intakes of 200 mg/d (151–158). Depletion of the body pool to 600 mg initiates physiological changes, and signs of clinical scurvy are reported when the body pool falls below 300 mg (149). Approximately 3–4% of the body pool turns over daily, representing 40–60 mg/d of metabolized, or consumed, vitamin C. Smokers have a higher metabolic turnover rate of vitamin C (approximately 100 mg/d) and a lower body pool than nonsmokers, unless compensated through increased daily intakes of vitamin C (159). The metabolism of ascorbic acid varies among different species.

In rats and guinea pigs, respiratory carbon dioxide is the major oxidation product of vitamin C (160). Given to humans in physiological doses, approximately 10% of a dose will be recovered in the urine, and urinary oxalic acid is the predominant metabolite. The formation of urinary metabolites is limited to 30–50 mg/d. Intake of up to 10 g/d ascorbic acid does not result in a considerable increase of urinary metabolites. The vitamin is excreted largely unmetabolized (161). In humans, the metabolites are dehydroascorbic acid, oxalate, 2,3-diketo-L-gulonic acid and derivatives of L-threose and L-threonic acid. L-Ascorbate-2-sulfate has been found in human urine as well (162). The feces are not a significant excretory route unless doses over 1 g are given.

The half-life of ascorbic acid is inversely related to the daily intake and is 13–40 d in humans and 3 d in guinea pigs, which is consistent with the longer time for humans to develop scurvy.

Requirements

The level of ascorbic acid intake required is dependent in part upon the body's handling of the nutrient. The recommendations for the daily intake of vitamin C in various countries range from 30 to 100 mg/d. There is an extensive lack of knowledge about the biochemical and physiological functions of vitamin C. Although as little as 10 mg/d of ascorbic acid can prevent clinical scurvy, this intake is insufficient to maintain an adequate body pool of the vitamin for peak physical and mental health. The RDA (Recommended Dietary Allowances) for vitamin C in the United States is 60 mg for men and women to maintain the body pool (Table 5). Vitamin C levels are higher for pregnant and lactating women to account for losses to the fetus and to breast milk.

The most recent RDA has included a vitamin C recommendation of 100 mg/day for cigarette smokers. An increasing number of investigators have concluded that the current RDA for vitamin C may not be adequate for elderly individuals. Plasma vitamin C level is generally accepted as an indicator of vitamin C status.

A recent study recommended that the current RDA be increased from 60 mg/d to 200 mg. The researchers indicated, however, that vitamin C daily doses above 400 mg have no value (163).

Toxicity

The acceptable daily allowance, which may be ingested without any risk of harm, is 1050 mg for a 70-kg healthy person (20). There is also no evidence in the literature that

Table 5. The 1989 RDA and US RDA for Vitamin C

Group	Age (yr)	Vitamin C (mg)
	RDA	
Infants	0–0.5	30
	0.5–1.0	35
Children	1–3	40
	4–10	45
Males	11–14	50
	15–51+	60
Females	11–14	50
	15–51+	60
pregnant		70
lactating		
1st 6 mo		95
2nd 6 mo		90
Cigarette smokers		100
	U.S. RDA	
Infants under 13 mo		35
Children under 4 yr		40
Adults and children over 4 yr		60
Pregnant or lactating women		60

ingestion of up to 10 g vitamin C per day constitutes a serious health risk for humans.

BIBLIOGRAPHY

1. *Biochem. Biophys. Acta* **107**, 4 (1965).

2. W. N. Haworth, *Chem. Ind.* (London) **52**, 482 (1993); W. N. Haworth and E. L. Hirst, *Ergeb. Vitamin-u: Hormonforsch.* **2**, 160 (1939).

3. R. E. Hodges and E. M. Baker, in R. S. Goodhart and M. E. Shils, eds., *Modern Nutrition in Health and Disease*, 5th ed., Lea & Febiger, Philadelphia, Pa., 1973, p. 245; U. Wintermeyer, H. Lahann, and R. Vogel, *Vitamin C: Entdeckung, Identifizierung und Synthese—Heutige Bedeutung in Medizin und Lebensmitteltechnologie*, Deutscher Apotheker Verlag, Stuttgart, Germany, 1981.

4. L. Stone, *The Healing Factor—Vitamin C Against Disease*, Grosset & Dunlap, Inc., New York, 1972, p. 7.

5. I. M. Sharman in G. G. Birch and K. J. Parker, eds., *Vitamin C: Recent Aspects of its Physiological and Technological Importance*, John Wiley & Sons, Inc., New York, 1974, p. 1.

6. A. F. Hess, *Scurvy, Past and Present*, J. B. Lippincott Co., Philadelphia, Pa., 1920.

7. J. Lind, *A Treatise on the Scurvy*, London, 1753 (republished by Edinburgh University Press, Edinburgh, Scotland, 1953).

8. A. Holst and T. Fröhlich, *J. Hyg.* **7**, 634 (1907).

9. H. Chick and E. M. Hume, *J. R. Army Med. Corps* **29**, 121 (1917); *J. Biol. Chem.* **39**, 203 (1919).

10. A. Harden and S. S. Zilva, *Biochem. J.* **12**, 259 (1918); S. S. Zilva and F. M. Wells, *Proc. R. Soc. London Ser. B* **90**, 505 (1917–1919); S. S. Zilva, *Biochem. J.* **21**, 689 (1927).

11. A. Szent-Györgyi, *Biochem. J.* **22**, 1387 (1928); J. L. Svirbely and A. Szent-Györgyi, *Biochem. J.* **26**, 865 (1932).

12. C. G. King and W. A. Waugh, *Science* **75**, 357 (1932).

13. W. N. Haworth and E. L. Hirst, *J. Soc. Chem. Ind.* **52**, 645–646 (1933).

14. T. Reichstein, A. Grüssner, and R. Oppenhauer, *Helv. Chim. Acta* **16**, 1019–1033 (1933).

15. T. Reichstein and A. Grüssner, *Helv. Chim. Acta* **17**, 311 (1934).

16. P. Karrer, K. Schöpp, and F. Zehner, *Helv. Chim. Acta* **16**, 1161–1163 (1933).

17. F. Michael and K. Kraft, Hoppe-Seyler's *Z. Physiol. Chem.* **222**, 235–249 (1933).

18. H. von Euler and C. Martius, *Ann.* **505**, 73–87 (1933).

19. E. L. Hirst, *Fortsch. Chem. Org. Naturst.* **2**, 132–159 (1939).

20. *Wld. Hlth. Org. Techn. Rep. Ser.* 1972, No. 505; *Wld. Hlth. Org. Techn. Rep. Ser.* 1974, No. 539.

21. Armour Research Foundation, *Anal. Chem.* **20**, 986–987 (1948).

22. C. von Planta, *Helv. Chim. Acta* **44**, 1444–1446 (1961).

23. J. Hvoslef, *Acta Crystallogr. Sect. B* **24**, 1431–1440 (1968).

24. J. Hvoslef, *Acta Chem. Scand.* **18**, 841–842 (1964).

25. J. Hvoslef, *Acta Crystallogr. Sect. B* **24**, 23–35 (1968).

26. W. Müller-Mulot, Hoppe-Seyler's *Z. Physiol. Chem.* **351**, 52,56 (1970); J. Hvoslef, *Acta Chem. Scand.* **24**, 2238 (1970); *Acta Crystallogr.* **B28**, 916 (1972); J. Hvoslef and B. Pedersen, *Acta Chem. Scand.* **B33**, 503 (1979); *Carbohydr. Res.* **92**, 9 (1981).

27. R. J. Kutsky, *Handbook of Vitamins and Hormones*, Van Nostrand Reinhold Co., Inc., New York, 1973, p. 71; *The Merck Index*, 11th ed., Merck & Co., Inc., Rahway, N.J., 1989, p. 130.

28. J. S. Koevendel, *Nature* **180**, 434 (1957); Y. Ogata and Y. Kosugi, *Tetrahedron* **26**, 4711 (1970).

29. Sadtler Standard Spectra, *Spectrum No. 5424*, Sadtler Research Laboratory, Inc., Philadelphia, Pa., 1972.

30. High Resolution NMR Spectra Catalog, Vol. 2, *Spectrum No. 464*, Varian Assoc., Palo Alto, Calif., 1963.

31. V. F. Johnson and W. C. Jankowski, Carbon-13 NMR Spectra, *Spectrum No. 171*, Wiley-Interscience, New York, 1972; S. Berger, *Tetrahedron* **33**, 1587 (1977); T. Ogama, J. Uzawa, and M. Matsui, *Carbohydr. Res.* **59**, C32 (1977); T. Radford, J. G. Sweeny, and G. A. Jacobucci, *J. Org. Chem.* **44**, 658 (1979).

32. B. M. Tolbert et al., *Ann. N.Y. Acad. Sci.* **258**, 44 (1975); B. M. Tolbert, *Int. J. Vit. Nutr. Res. Suppl.* **19**, 127 (1979).

33. B. Tolbert and J. B. Ward, in *Ascorbic Acid, Chemistry, Metabolism and Uses, Adv. Chem. Sci.* **200**, 101–123 (1982); M. Ohmori, M. Takage, *Agr. Biol. Chem.* **42**, 173–174 (1978).

34. F. E. Huelin, *Food Res.* **18**, 633 (1953).

35. J. Campbell and W. G. Tubb, *Can. J. Res.* **28E**, 19–32 (1950).

36. K. Mikova and J. Davidele, *Chem. Listy* **68**, 715 (1974).

37. F. E. Huelin, *Food Res.* **18**, 633 (1953); J. Campbell and W. G. Tubb, *Can. J. Res.* **28E**, 19 (1947).

38. D. W. Bradley, G. Emery, and J. E. Maynard, *Clin. Chim. Acta* **4**, 47–52 (1973).

39. T. Reichstein, A. Grüssner, and R. Oppenauer, *Helv. Chim. Acta* **16**, 561–565 (1933).

40. T. Reichstein, A. Grüssner, and R. Oppenauer, *Nature (London)* **132**, 280 (1932).

41. T. Reichstein, A. Grüssner, and R. Oppenauer, *Helv. Chim. Acta* **17**, 510–520 (1934).

42. U.S. Pat. 2,056,126 (Sept. 29, 1936), T. Reichstein (to Hoffmann-La Roche Inc.).

43. W. N. Haworth, *Chem. Ind. (London)* **52**, 482–485 (1933).

44. R. G. Ault et al., *J. Chem. Soc.* 1419–1423 (1933).

45. W. N. Haworth and E. L. Hirst, *Helv. Chim. Acta* **17**, 520–523 (1934).

46. T. C. Crawford and S. A. Crawford, *Adv. Carbohydr. Chem. Biochem.* **37**, 79 (1980).

47. M. Kulhanek, *Adv. Appl. Microbiol.* **12**, 11–33 (1970).

48. G. Xin, *Weishengwuxue Tongbao* **14**, 73 (1987).

49. Eur. Pat. Appl. 278 447 (Prior. Feb. 2, 1987) W. Ning, Z. Tao, C. Wang, S. Wang, Z. Yan, and G. Yin (to Institute of Microbiology Academia Sinica).

50. Eur. Pat. Appl. 221 707 (Prior. Oct. 22, 1985) I. Nogami, H. Shirafuji, M. Oka, and T. Yamaguchi (to Takeda Chemical Industries, Ltd.).

51. T. Sonoyama et al., *Appl. Environ. Microbiol.* **43**, 1064 (1982).

52. S. Anderson et al., *Science* **230**, 144–149 (1985).

53. J. F. Grindley, M. A. Payton, H. van do Pol, and K. G. Hardy, *Appl. Environ. Microbiol.* **54**, 1770 (1988).

54. Eur. Pat. Appl. No. 207 763 (1985), T. J. Skatrud and R. J. Huss (to Bio-Technical Resources).

55. I. T. Strukov and N. A. Kopylova, *Farmatsiya* **10**, 8 (1974); *Drug Cosmet. Ind.* **40** (Aug. 1977).

56. W. Ipatiew, *Ber.* **45**, 3218 (1912).

57. L. W. Cover, R. Connor, and H. Adkins, *J. Am. Chem. Soc.* **54**, 1651 (1932).

58. A. T. Kuhn, *J. Appl. Electrochem.* **11**, 261 (1981).

59. W. E. Cake, *J. Am. Chem. Soc.* **44**, 859 (1922).

60. M. Kulhanek, *Adv. Appl. Microbiol.* **12**, 11 (1970).

61. P. A. Wells, J. J. Stubbs, L. B. Lockwood, and E. T. Roe, *Ind. Eng. Chem.* **29**, 1385 (1937); E. J. Fulmer and L. A. Underkofler, Iowa State Coll. *J. Sci.* **21**, 251 (1974).

62. U.S. Pat. 3,622,560 (Nov. 23, 1971), N. Halder, M. J. O'Leary, and N. C. Hindley; U.S. Pat. 3,598,804 (Aug. 10, 1971), N. Halder, N. C. Hindley, G. M. Jaffe, M. J. O'Leary, and P. Weinert; U.S. Pat. 3,607,862 (Sept. 21, 1971), G. M. Jaffe, W. Szkrybalo, and P. Weinert (all to Hoffmann-La Roche Inc.).

63. R. S. Glass, S. Kwoh, and E. P. Oliveto, *Carbohydr. Res.* **26**, 181 (1973).

64. U.S. Pat. 2,039,929 (May 5, 1936), T. Reichstein, U.S. Pat. 2,301,811 (Nov. 10, 1942), T. Reichstein (both to Hoffmann-La Roche Inc.); V. F. Kazimirova, *J. Gen. Chem. (USSR)* **24**, 1559 (1955).

65. J. W. Weijlard, *J. Am. Chem. Soc.* **67**, 1031 (1945); U.S. Pat. 2,367,251 (Jan. 16, 1945), J. Weijlard and J. B. Ziegler (to Merck and Co.); K. Nakagawa, R. Konaka, and T. Nakata, *J. Org. Chem.* **27**, 1597 (1962).

66. L. O. Shnaidman and I. N. Kushehinskaya, *Tr. Vses. Nauchno-Issled. Vitam. Inst.* **8**, 13 (1961).

67. U.S. Pat. 3,832,355 (Aug. 27, 1972), G. M. Jaffe and E. Pleven (to Hoffmann-La Roche Inc.).

68. U.S. Pat. 3,453, 191 (July 1, 1969), G. J. Fröhlich, A. J. Kratavil, and E. Zrike (to Hoffmann-La Roche Inc.); Eur. Pats. 40331 (Nov. 25, 1981) and 40709 (Dec. 2, 1981), R. Whittmann, W. Wintermeyer, and J. Butska (to Merck GMBH); P. M. Robertson, P. Berg, H. Reimann, K. Schleich, and P. Seiler, *Chimia* **36**, 305 (1982).

69. P. P. Regna and B. P. Caldwell, *J. Am. Chem. Soc.* **66**, 246 (1944).

70. F. Elger, *Festschr. Emil Barell*, 229 (1936); Brit. Pat. 428,815 (May 20, 1935); and Brit. Pat. 466,548 (May 31, 1937), T. Reichstein (to Hoffmann-La Roche, Inc.).

71. *The United States Pharmacopeia / National Formulary (USP XXIII / NFXVIII)*, United States Pharmacopeial Convention, Inc., Rockville, Md., 1995, pp. 130, 245, 1414, 2215.

72. Food and Nutrition Board, National Research Council, *Food Chemicals Codex*, 3rd ed., National Academy Press, Washington, D.C., 1981, pp. 27, 28, 277–278.

73. L. A. Pachla and D. L. Reynolds, *J. Assoc. Off. Anal. Chem.* **68**, 1–12 (1985).

74. J. R. Cooke and R. E. D. Moxon, in *Vitamin C (Ascorbic Acid)*, J. N. Counseil and D. H. Hornig, eds., Applied Science Publishers, London, 1981, pp. 167–198; *Adv. Chem. Ser.* **200**, 199 (1982); R. C. Rose and D. L. Nahrwold, *Anal. Biochem.* **123**, 389 (1982).

75. C. I. Bliss and P. György, in P. György, ed., *Vitamin Methods*, Vol. II, Academic Press, New York, 1951, pp. 244–261.

76. J. Tillmans, *Z. Unters. Lebensmittel* **54**, 33 (1927).

77. R. Strohecker, *Wiss. Veroff. Dtsch. Ges. Ernahr.* **14**, 157 (1965).

78. T. Tono and S. Fujita, *Agric. Biol. Chem.* **45**, 2947 (1981).

79. K. Dabrowski et al., *J. Aquaculture* **124**, 169–192 (1994).

80. J. H. Roe and C. A. Kuether, *J. Biol. Chem.* **147**, 399–407 (1943).

81. M. J. Deutsch and C. E. Weeks, *J. Assoc. Off. Anal. Chem.* **48**, 1248–1256 (1965).

82. G. Brubacher, W. Müller-Mulot, and D. A. T. Southgate, *Methods for the Determination of Vitamins in Food Recommended by Cost 91*, Elsevier Applied Science, New York, 1985, Chapt. 7.

83. J. Geigert, D. S. Hirano, and S. L. Neidleman, *J. Chromatogr.* **206**, 396–399 (1981).

84. S. Garcia-Castineiras, V. D. Bonnet, R. Figueroa, and M. Miranda, *J. Liq. Chromatogr.* **4**, 1619–1640 (1981).

85. W. Lee, P. Hamernyik, M. Hutchinson, V. A. Raisys, and R. F. Labbe, *Clin. Chem.* **28**, 2165–2169 (1982).

86. H. J. Kim, *J. Assoc. Official Anal. Chem.* **72**, 681–686 (1989).

87. L. L. Lloyd, F. Warner, J. F. Kennedy, and C. A. White, *J. Chromatogr.* **437**, 453–456 (1988).

88. O. Pelletier, in J. Augustin, B. P. Klein, D. A. Becker, and P. B. Venugopal, eds., *Methods of Vitamin Assay*, 4th ed., John Wiley & Sons, Inc., New York, 1985, Chapt. 12.

89. M. H. Bui-Nguyen, in A. P. DeLeenheer, eds., *Modern Chromatographic Analysis of the Vitamins*, Marcel Dekker, New York, 1985, Chapt. 5.

90. B. Borenstein, *Food Tech.* **41**, 98–99 (1987).

91. M. Liao and P. A. Seib, *Food Tech.* **41**, 104–107 (1987).

92. J. C. Bauernfeind, in *Ascorbic Acid: Chemistry, Metabolism, and Uses*, P. A. Seib and B. M. Tolbert, eds., American Chemical Society, Washington, D.C., 1982, pp. 395–497.

93. V. S. Durve and R. T. Lovell, *Can. J. Fish Aquat. Sci.* **39**, 948–951 (1982).

94. M. Matusiewicz, K. Dabrowski, L. Volker, and K. Matusiewicz, *J. Nutr.* **125**, 3055–3061 (1995).

95. R. T. Lovell and G. O. Elnaggar, *Proceedings of the Third International Symposium on Feeding and Nutrition in Fish*, Toba, Japan, (Aug. 28–Sept. 1, 1989), pp. 159–165.

96. G.-O. Toa, *Sheng Lik'o Hseuh Chin Chan* **12**, 26 (1981).

97. K. Dabrowski, M. Matusiewicz, and J. H. Blom, *Aquaculture* **124**, 169–190 (1994).

98. U.S. Pat. No. 4,728,376 (Mar. 1, 1988), E. F. Kurtz (to Golden Powder of Texas, Inc.).

99. O. Isler, G. Brubacher, S. Ghisla, and B. Kräutler, *Vitamin II*, Georg Thieme Verlag, Stuttgart, Germany, 1988, pp. 396–444.

100. F. Smith, *Adv. Carbohydr. Chem. Biochem.* **2**, 79 (1946).

101. J. Stanek, M. Cerny, J. Kocourek, and J. Pacak, *The Monosaccharides*, Academic Press, Inc., New York, 1963, p. 735.

102. G. Mohn, in W. Dirscherl and W. Friedrich, eds., *Fermente, Hormone, Vitamine, und die Beziehungen dieser Wirkstoffe sueinander*, 3rd ed., Vol. 3, Pt. 1 (Vitamine), Thieme, Stuttgart, Germany, 1982, pp. 910–913.

103. R. S. Harris, *The Vitamins*, 2nd ed., Vol. 1, Academic Press, New York, London, 1967, 383–385.

104. M. Williams and F. A. Loewus, *Carbohydr. Res.* **63**, 149 (1978).

105. S. L. von Schuching and A. F. Abt, *Methods Enzymol.* **18**, 1 (1970).

106. I. B. Chatterjee, *Science* **82**, 1271–1272 (1973).

107. I. B. Chatterjee, A. K. Majumder, B. K. Nandi, and N. Subramanian, in *Second Conference on Vitamin C*, Vol. 258, C. G. King and J. J. Burns, eds., New York Academy of Sciences, New York, 1975, pp. 24–47.

108. J. J. Burns, *Am. J. Med.* **26**, 740 (1959).

109. F. A. Isherwood, Y. T. Chen, and L. W. Mapson, *Biochem. J.* **56**, 11 (1954); J. J. Burns and C. Evans, *J. Biol. Chem.* **23**, 897 (1956).

110. G. J. Dutton, ed., *Glucuronic Acid-Free and Combined: Chemistry, Biochemistry, Pharmacology and Medicine*, Academic Press, Inc., New York 1966; B. L. Horecker in G. Ritzel and G. Brubacher, eds., *Monosaccharides and Polyols in Nutrition, Therapy and Dietetics*, Hans Huber, Bern, Switzerland, 1967, p. 1.

111. A. H. Conney, G. A. Bray, C. E. Evans, and J. J. Burns, *Ann. N.Y. Acad. Sci.* **92**, 115 (1961).

112. J. J. Burns, C. Evans, and N. Trousof, *J. Biol. Chem.* **227**, 785–794 (1957).

113. F. Horio and A. Yoshida, *J. Nutr.* **112**, 416–425 (1982).

114. W. R. F. Notten, Stichting Studentenpers, University of Nijmegen, Nijmegen, The Netherlands, 1975; W. R. F. Notten and P. T. Henderson, *Int. Arch. Occup. Environ. Health* **38**, 197, 209 (1977).

115. F. A. Loewus and M. M. Baig, *Methods Enzymol.* **18**, 22 (1970); F. A. Loewus, G. Wagner, and J. C. Yang, *Ann. N.Y. Acad. Sci.* **258**, 7 (1975); F. A. Loewus in Preiss, ed., *Biochemistry of Plants*, Vol. 3, Academic Press, Inc., New York, 1980, p. 77; C. H. Foyer and B. Halliwell, *Phytochemistry* **16**, 1347 (1977).

116. F. A. Isherwood, Y. T. Chen, and L. W. Mapson, *Biochem. J.* **56**, 1 (1954).

117. M. M. Baig, S. Kelley, and F. A. Loewus, *Plant. Physiol.* **46**, 277 (1970).

118. Committee on Dietary Allowances, Food and Nutrition Board, *Recommended Dietary Allowances*, 10th ed., National Academy of Sciences, Washington, D.C., 1989, pp. 115–124.

119. W. H. Sebrell and R. S. Harris, eds., *The Vitamins, Chemistry, Physiology, Pathology, Methods*, 2nd ed., Vol. 1, Academic Press, Inc., New York, 1967, p. 305.

120. J. Marks, *A Guide to the Vitamins, Their Role in Health and Disease*, Medical and Technical Publishing Co., Lancaster, U.K., 1975.

121. S. Nobile and J. M. Woodhill, *Vitamin C: The Mysterious Redox-System—A Trigger of Life*, MTP Press, Boston, Mass., 1981, p. 38.

122. P. A. La Chance and M. C. Fisher, in R. Karmas and R. S. Harris, eds., *Nutritional Evaluation of Food Processing*, 3rd ed., Van Nostrand Publishing Co., New York, 1988, p. 506.

123. B. Borenstein, in T. E. Furis, ed., *Handbook of Food Additives*, 2nd ed., CRC Press, Cleveland, Ohio, 1972, pp. 91, 101.

124. B. Peterkofsky and S. Udenfriend, *Proc. Nat. Acad. Sci.* **53**, 335 (1965); M. J. Barnes and E. Kodicek, *Vitam. Horm.* **30**, 1 (1972).

125. A. H. Hunt, *Br. J. Surg.* **28**, 436–461 (1941).

126. S. P. Shukla, *Experientia* **25**, 704 (1969).

127. T. T. Irvin, D. K. Chattopadhyay, and A. Smythe, *Surg. Gynec. Obstet.* **147**, 49–55 (1978).

128. C. J. Bates, in J. N. Counsell and D. H. Hornig, eds., *Vitamin C (Ascorbic Acid)*, Applied Science Publishers, London, 1981.

129. V. G. Zannoni, J. I. Brodfuehrer, R. C. Smart, and R. L. Susick, in J. J. Burns, J. M. Rivers, and L. J. Machlin, eds., *Third Conference on Vitamin C*, N.Y. Academy of Sciences, New York, 1987.

130. W. Dawson and G. B. West, *Br. J. Pharmacol.* **24**, 75 (1965).

131. A. Bendich, *Food Tech.* **41**, 112–114 (1987).

132. A. Bendich, L. J. Machlin, O. Scandurra, G. W. Burton, and D. D. M. Wayner, *Adv. in Free Radical Biol. Med.* **2**, 419–444 (1986).

133. R. E. Beyer, *J. Bioenerget. Biomembr.* **26**, 349–358 1994.

134. R. J. Jariwalla, S. Harakch, in R. R. Watson, ed., *Nutrition and AIDS*, CRC Press, Boca Raton, Fla., 1994, pp. 117–139.

135. S. S. Mirvish, *J. Natl. Cancer Inst.* **71**, 629–647 (1983).

136. P. N. Magee, *Cancer Surv.* **8**, 207–239 (1989).

137. V. M. Craddock, *Eur. J. Cancer Prev.* **1**, 89–103 (1992).

138. Y. Zeng and co-workers, in *Cancer Epidemiol. Biomarkers & Prev.* **2**, 195–200 (1993).

139. S. S. Mirvish, *Cancer Res.* (Suppl.), 1948–1951 (Apr. 1, 1994).

140. L. Hallberg, in *Vitamin C (Ascorbic Acid)*, J. N. Counsell and D. H. Hornig, eds., Applied Science Publishers, London, 1981, pp. 49–61.

141. L. Hallberg, M. Brune, and L. Rossander-Hulthen, in J. J. Burns, J. M. Rivers, and L. J. Machlin, eds., *Third Conference on Vitamin C*, Vol. 498, New York Academy of Sciences, New York, 1987, pp. 324–332.

142. E. R. Monsen and J. F. Page, *J. Agric. Food Chem.* **26**, 223–226 (1978).

143. N. W. Solomons and F. E. Viteri, in P. A. Seib and B. M. Tolbert, eds., *Ascorbic Acid: Chemistry, Metabolism, and Uses*, American Chemical Society, Washington, D.C., 1982, pp. 551–569.

144. C. M. Plug, D. Decker, and A. Bult, *Pharm. Weekblad* **16**, 245–248 (1984).

145. J. J. M. Marx and J. Stiekema, *Eur. J. Clin. Pharmacol.* **23**, 335–338 (1982).

146. A. Bendich in L. Packer and J. Fuchs, eds., *Vitamin C in Health and Disease*, J. Marcel Dekker Inc., NY, 1997, pp. 411–416.

147. R. D. Crawford, *Biochem. Molec. Med.* **54**, 1–11 (1995).

148. C. J. Schorah, in T. G. Tayler, ed., *The Importance of Vitamins to Health*, MTP Press, Lancaster, U.K., 1979, pp. 61–72.

149. R. E. Hodges, E. M. Baker, J. Hood, H. E. Sauberlich, and S. C. March, *Am. J. Clin. Nutr.* **22**, 535–548 (1969).

150. R. Anderson, in Ref. 128, pp. 249–272.

151. A. Castelli, G. E. Martorana, E. Meucci, and G. Bonetti, *Acta Vitaminol. Enzymol.* **4**, 189–196 (1982).

152. C. J. Schorah et al., *Am. J. Clin. Nutr.* **34**, 871–876 (1981).
153. P. J. Garry, J. S. Goodwin, W. C. Hunt, and B. A. Gilbert, *Am. J. Clin. Nutr.* **36**, 332–339 (1982).
154. D. Hornig, J. P. Vuileumier, and D. Hartmann, *Intern. J. Vitamin Nutr. Res.* **50**, 309–314 (1980).
155. E. M. Baker, J. C. Saari, and B. M. Tolbert, *Am. J. Clin. Nutr.* **19**, 371–378 (1966).
156. E. M. Baker, R. E. Hodges, J. Hood, H. E. Sauberlich, and S. C. March, *Am. J. Clin. Nutr.* **22**, 549–558 (1969).
157. E. M. Baker, R. E. Hodges, J. Hood, H. E. Sauberlich, and E. M. Baker, *Am. J. Clin. Nutr.* **24**, 444–454 (1971).
158. R. E. Hodges, J. Hood, J. Canham, H. E. Sauberlich, and E. M. Baker, *Am. J. Clin. Nutr.* **24**, 432–443 (1971).
159. A. B. Kallner, D. Hartmann, and D. Hornig, *Am. J. Clin. Nutr.* **34**, 1347–1355 (1981).
160. D. Hornig and D. Hartmann, in P. A. Seib and P. M. Tolbert, eds., *Ascorbic Acid: Chemistry, Metabolism and Uses*, American Chemical Society, Washington, D.C., 1982, pp. 293–316.
161. A. Kallner, D. Hartmann, and D. Hornig, *Am. J. Clin. Nutr.* **32**, 530–539 (1979).
162. E. M. Baker, D. C. Hammer, S. C. March, B. M. Tolbert, and J. E. Canham, *Science* **173**, 826–827 (1971).
163. M. Levine et al., *Proc. Natl. Acad. Sci. USA* **93**, 3704 (1996).

GENERAL REFERENCES

Committee on Dietary Allowances, Food and Nutrition Board, *Recommended Dietary Allowances*, 10th ed., National Academy of Sciences, Washington, D.C., 1989, pp. 115–124.

D. H. Hornig, U. Moser, and B. E. Glatthaar, in M. E. Shils and V. R. Young, eds., *Modern Nutrition in Health and Diseases*, 7th ed., Lea & Febiger, Philadelphia, 1988, pp. 417–435.

R. A. Jacob, J. H. Skala, and S. T. Omave, *Am. J. Clin. Nutrit.* **46**, 818–826 (1987).

R. Marcus and A. M. Coulston, in A. G. Gilman, L. S. Goodman, T. W. Rall, and F. Murari, eds., *Goodman and Gilman's Pharmacological Basis of Therapeutics*, 7th ed., Macmillan Publishing Co., New York, 1985, pp. 1151–1573.

U. Moser and A. Bendich, in L. J. Machlin, ed., *Handbook of Vitamins*, 2nd ed., Marcel Dekker, New York, 1991, pp. 195–232.

R. L. Pike and M. L. Brown, *Nutrition: An Integrated Approach*, 3rd ed., Macmillan Publishing Co., New York, 1986, pp. 131–136.

H. E. Sauberlich, in M. L. Brown, ed., *Present Knowledge in Nutrition*, 6th ed., International Life Sciences Institute, Nutrition Foundation, Washington, D.C., 1990, pp. 132–141.

VOLKER KUELLMER
Hoffmann-La Roche, Inc.

VITAMINS: BIOTIN

Biotin, [3aS-(3aα, 4β, 6aα)]-hexahydro-2-oxo-1*H*-thieno-[3,4-*d*]imidazole-4-pentanoic acid [58-85-5] (vitamin H, vitamin B_8, bios IIB, and coenzyme R) (**1**) is a water-soluble B complex vitamin. The name coenzyme R was coined during work on a protective factor, from the liver, for egg white injury. This protective factor was also called factor S, factor W or vitamin B_W (1–4). Biotin is a complex molecule having three stereocenters. There are eight stereoisomers of biotin; only the naturally occurring one is active in metabolism. The richest sources of biotin are yeast, liver, kidney, egg yolks, pancreas, and milk (5–7). The highest content of biotin in cow's milk occurs early in lactation. Plant materials, such as nuts, seeds, cereals such as oats and bulgar wheat, pollen, molasses, rice, soybeans, mushrooms, fresh vegetables such as cauliflower, split peas, cow peas, and legumes, and some fruits, are also good sources. In addition, small amounts of biotin are found in most fish, eg, mackerel, salmon, and sardines.

(**1**)

ISOLATION

In 1936, biotin was isolated from egg yolks (8), in 1939 from beef liver (9), and in 1942 from milk concentrates (10). Biotin-producing microorganisms exist in the large bowel but the extent and significance of this internal synthesis is unknown.

BIOCHEMICAL FUNCTION

Biotin forms part of several enzyme systems and is necessary for normal growth and body function. Biotin functions as a cofactor for enzymes involved in carbon dioxide fixation and transfer. These reactions are important in the metabolism of carbohydrates, fats, and proteins, as well as promotion of the synthesis and formation of nicotinic acid, fatty acids, glycogen, and amino acids (5–7). Biotin is absorbed unchanged in the upper part of the small intestine and distributed to all tissues. Highest concentrations are found in the liver and kidneys. Little information is available on the transport and storage of biotin in humans or animals. A biotin level in urine of approximately 160 nmol/24 h or 70 nmol/L, and a circulating level in blood, plasma, or serum of approximately 1500 pmol/L seems to indicate an adequate supply of biotin for humans. However, reported levels for biotin in the blood and urine vary widely and are not a reliable indicator of nutritional status.

NUTRITIONAL REQUIREMENTS

Since exact requirements for biotin are uncertain owing to incomplete knowledge regarding biotin availability from food and a lack of definitive studies concerning biotin requirements, the United States National Research Council has established a safe and adequate daily dietary level of intake for biotin rather than a recommended dietary allowance (RDA). The recommended daily intake of biotin in the United States for all persons ages seven years and older is

30–100 μg/d. In France and South Africa, a recommended daily intake for adults of up to 300 μg/d has been established, whereas in Singapore an intake of up to 400 μg/d is recommended. Diets consisting of a daily biotin intake of 28–42 μg/d are considered adequate. A level of 60 μg/d is sufficient for patients under long-term total parenteral nutrition (fed intravenously). An infant's daily intake ranges from 15–20 μg/d and is acquired mostly through human milk containing 3–20 μg/L, or formulas fortified with biotin. An adequate intake level of 10–30 μ/d for infants and young children is recommended. The safe and adequate levels for daily dietary intake of biotin are listed in Table 1 (11). No side effects have been reported with oral doses of biotin as great as 40 mg or parenteral doses of 5–10 mg/d in infants. No toxicity of biotin has been found (5–7).

PHYSIOLOGICAL SIGNIFICANCE

Animal

Dietary biotin deficiencies are extremely rare, perhaps because of the biosynthesis of biotin by intestinal microorganisms. However, biotin deficiency can be easily induced in most animals by ingestion of large amounts of raw egg white, which contain the biotin binding protein avidin. Avidin has a high affinity for biotin and binds with the ureido group to form a complex that is resistant to digestive enzymes. Biotin deficiency in animals causes a decrease in growth rate, loss of weight, alopecia, scaly dermatitis, hyperkeratosis, achromatrichia, and transverse fissures across the soft sole and cracks in the hard horn of the sole and claw wall. Minute amounts of biotin are known to be adequate to support body functions; therefore, biotin requirements for animals may be covered by the natural content of the feed and by the intestinal biosynthesis of biotin. However, biotin-responsive disease conditions not caused by primary biotin deficiency have been observed. One such condition is the fatty liver and kidney syndrome (FLKS). This syndrome has caused heavy economic losses in commercial broiler flocks. FLKS was found to be the result of a suboptimal biotin content in diet, coupled with other nutritional and environmental factors. Although the symptoms of FLKS are not those of classic biotin deficiency, they can be eliminated by biotin supplements. Supplementation has also been found to reduce the incidence and severity of claw lesions in pigs and weak hoof horn in horses (5–7).

Human

Biotin deficiencies in humans are extremely rare. The symptoms of deficiency are anorexia, fatigue, nausea, vomiting, hyperesthesia, glossitis, pallor, mental depression, dry scaly dermatitis, alopecia, and localized parasthesias. Both seborrheic dermatitis and Leiner's disease could be the signs of a biotin deficiency in infants (12). Several studies (5–7) have indicated that an erythematous exfoliative dermatitis is the first clinical sign of a biotin deficiency. Infants under six months of age and people who eat large quantities of raw egg whites are probably the groups most susceptible to biotin deficiency. Another susceptible group would be people who lack biotinidase. Biotinidase is the only enzyme capable of catalyzing the cleavage of biocytin (biotinyl-ϵ-lysine), the bound form of biotin. It has also been postulated that there may be a connection between biotin and the etiology of the sudden infant death syndrome (SIDS), which is a common cause of death in the first year of life. Unless biotin is included in their alimentation, patients receiving long-term parenteral nutrition are susceptible to biotin deficiency. All symptoms can be reversed and conditions corrected with biotin treatment. For adults, biotin dosage levels of 150–300 μg per day would be effective treatment. Finally, no drugs have been found to cause a potential biotin deficiency through short- or long-term use. However, low circulating biotin levels have been observed in heavy smokers, alcoholics, and patients under prolonged treatment with anticonvulsant drugs (5–7).

CHEMICAL AND PHYSICAL PROPERTIES

The empirical formula for biotin, $C_{10}H_{16}N_2O_3S$, was established in 1941 (13). The full structure of biotin was elucidated in 1942 by two independent groups (14–16). The first total chemical synthesis of biotin was achieved by Harris in 1945; this work confirmed the structure of biotin (17). The configuration of d-biotin was definitively established in 1956 by x-ray crystallographic analysis (18,19). The physical properties of d-biotin are found in Table 2. Chemically pure biotin is stable to air and heat. Biotin is also stable for months in mildly acidic and alkaline solutions; however, alkaline solutions, particularly above pH 9, are the least stable. Although biotin is not affected by reducing agents, it is incompatible with formaldehyde, chloramine T, oxidizing agents such as hydrogen peroxide and potassium permanganate, and strong acids such as nitrous acid (20). In most foodstuffs, biotin is bound to proteins, from which it is released in the intestine by protein hydrolysis and the enzyme biotinidase. Biotin is a highly stable, water-soluble vitamin that is resistant to most processing procedures, long-term storage, and normal cooking heat. Most of the biotin losses during cooking are the result of leaching into the cooking water. On the other hand, canning and food processing cause a moderate reduction in biotin content, owing to decomposition.

CHEMICAL SYNTHESIS

Original Synthesis

The first attempted synthesis of d-biotin in 1945 afforded racemic biotin (Fig. 1). In this synthetic pathway, L-

Table 1. 1989 Estimated Safe and Adequate Daily Dietary Intake for Biotin

Group	Age (yr)	Biotin (μg)
Infants	0–0.5	10
	0.5–1.0	15
Children	1–3	20
	4–6	25
	7–10	30
Adolescents and adults	11 +	30–100

Table 2. Physical Properties of d-Biotin

Property	Characteristic
Appearance	Fine long needles
Color	White
Molecular weight	244.31
Molecular formula	$C_{10}H_{16}N_2O_3S$
Elemental analysis, wt %	
carbon	49.16
hydrogen	6.60
nitrogen	11.47
sulfur	13.12
Melting point, °C	232–233
α^{21}, specific optical rotation, degrees	$+89-91^a$
Dissociation constant, pK_A	6.3×10^{-6}
Isoelectric point, pH	3.5
Solubility, mg/mL	
H_2O, RT	$\sim 0.22^b$
95% alcohol, RT	~ 0.80
common organic solvents	Insoluble

$^a c = 1$ in $0.1\,N$ NaOH.
^{b}Higher in dil alkali.

cysteine [52-90-4] (**2**) was converted to the methyl ester [5472-74-2] (**3**). An intramolecular Dieckmann condensation, during which stereochemical integrity was lost, was followed by decarboxylation to afford the thiophanone [57752-72-4] (**4**). Aldol condensation of the thiophanone with the aldehyde ester [6026-86-4] (**5**) afforded the conjugated thiophanone [85269-47-2] (**6**). The aldol condensation allowed for attachment of the properly functionalized C-5 side chain in one step, a weakness in many subsequent syntheses. The introduction of the second nitrogen was achieved by converting the keto group to the oxime [85269-48-3] (**7**), followed by zinc reduction to the thiophene [85269-49-4] (**8**). Stereoselective catalytic hydrogenation of the thiophene double bond gave mainly the thiophane [78763-60-7] (**9**), which was easily transformed to biotin. The nonselective catalytic hydrogenation and the epimerization that occurred during the conversion of (**3**) to (**4**) were the major factors leading to formation of racemic biotin (17).

Original Asymmetric Synthesis

The efficient introduction of the three stereocenters in the all-cis configuration was first accomplished in the elegant synthesis developed by Sternbach (21–23) in 1949. This process, the Hoffmann-La Roche industrial synthesis of biotin, is still (ca 1997) the basis of industrial preparations. An improved version of the original Sternbach synthesis is shown in Figure 2 (24). The fixed cis position of the carboxyl groups of the imidazolidine dicarboxylic acid (cycloacid) (**13**) and the stereochemistry of the chiral carbons 4 and 5 of the imidazolidine ring, throughout the synthetic scheme, are established by the starting material, fumaric acid [110-17-8] (**10**). Fumaric acid is brominated to dibromosuccinic acid [608-36-6] (**11**) followed by reaction with benzylamine to give dibenzylaminosuccinic acid [55645-40-4] (**12**). Treatment with phosgene forms the imidazolone ring of cycloacid [51591-75-4] (**13**). Cycloacid is dehy-

drated with acetic anhydride to form cycloanhydride [56688-83-6] (**14**). Opening of cycloanhydride with cyclohexanol forms the racemic monoester (**15**). The mixture is resolved by salt formation with (+)-ephedrine [299-42-3] (**16**) in high yield. The desired d-ephedrine salt undergoes acid cleavage, followed by selective reduction of the ester functionality with lithium borohydride to form, after acidification, the d-lactone [56688-82-5] (**17**). The undesired l-ephedrine salt is recycled via the meso-cycloacid to cycloanhydride. The d-lactone, which possesses the desired configuration at two of the stereocenters, is converted to the d-thiolactone [56688-83-6] (**18**) using potassium thioacetate. The introduction of the C-5 side chain is accomplished by reacting the thiolactone with the Grignard reagent derived from 3-methoxypropyl chloride, followed by dehydration to give the thiophene [85611-62-7] (**19**). Stereoselective hydrogenation establishes the third stereocenter and affords the thiophane [33607-59-9] (**20**). The thiophane undergoes acid-catalyzed cyclization to form the key intermediate, l-thiophanium bromide [60209-10-1] (**21**). The last two carbons of the biotin side chain are added by reaction of thiophanium bromide with sodium dimethylmalonate to form the diester [8554-84-3] (**22**). Hydrolysis of the ester groups, decarboxylation and debenzylation using strong acid forms homochiral, pure d-biotin. Although a number of significant modifications to the Sternbach synthesis have appeared over the last 45 years, it is still the basis of today's industrial preparations.

Synthetic Drawbacks

The major drawbacks in the Sternbach-Goldberg synthesis are the resolution/recycling of the intermediate that leads to d-lactone and the multiple manipulations required to add the five-carbon side chain. This sequence is inefficient, bringing with it a net loss of methanol, hydrogen bromide, carbon dioxide, and water that were once part of the molecule. In the resolution of the intermediate that leads to d-lactone, 50% of the product is the undesirable isomer, although this material is converted to the cycloacid for recycling. This is inherently inefficient and limits single-run production capacity by at least 50%. Recycling the undesired isomer also requires additional labor.

Industrial Synthetic Improvements

One significant modification of the Sternbach process is the result of work by Sumitomo chemists in 1975, in which the optical resolution–reduction sequence is replaced with a more efficient asymmetric conversion of the meso-cycloacid (**13**) to the optically pure d-lactone (**17**) (Fig. 3) (25). The cycloacid is reacted with the optically active dihydroxyamine [2964-48-9] (**23**) to quantitatively yield the chiral imide [85317-83-5] (**24**). Diastereoselective reduction of the pro-R-carbonyl using sodium borohydride affords the optically pure hydroxyamide [85317-84-6] (**25**) after recrystallization. Acid hydrolysis of the amide then yields the desired d-lactone (**17**). A similar approach uses chiral alcohols to form diastereomic half-esters stereoselectivity. These are reduced and directly converted to d-lactone (26). In both approaches, the desired diastereomeric half-amide

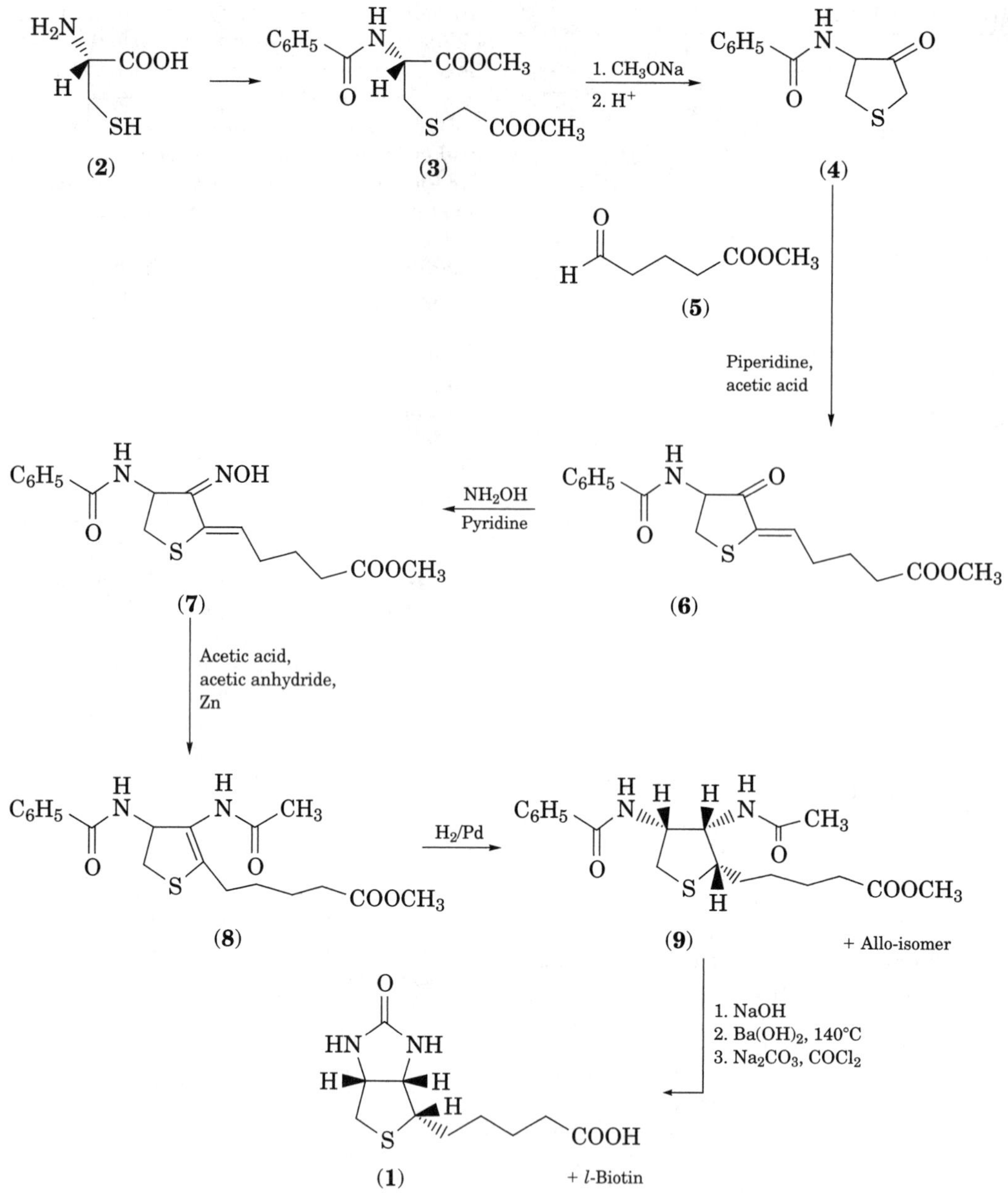

Figure 1. Synthetic pathway for racemic biotin.

or half-ester is formed in excess, thus avoiding the costly resolution step required in the Sternbach synthesis.

Another modification of the Sternbach method involves the direct attachment of the C-5 side chain to thiolactone or a functionalized thiolactone intermediate. One such method, possibly utilized by Sumitomo and reported several times by Lonza (27–29), involves the treatment of the thiolactone (14) with the five-carbon Wittig reagent (26) to give the olefin (27). Hydrogenation of the thiophene (27) followed by acid hydrolysis completes the synthesis of *d*-biotin. Another method that directly attaches the five-carbon side chain is patented by Lonza (Fig. 4). This method involves a Wittig reaction of a phosphonium salt derived from the thiolactone (14) and a C-5 aldehyde (30). The thiolactone (14) is reduced to the thiophanol (28),

which is converted to the Wittig salt (29) with Ph_3PH^+ BF_3^-. The ylide (29), in the presence of base, undergoes condensation with the aldehyde ester (30) to form the thiophene (27), which in turn is catalytically hydrogenated to *d*-biotin (30).

A final variation of the Sternbach method was reported by Lonza (27–29,31). This method (Fig. 5) not only involves direct attachment of the C-5 side chain to the lactone [118609-09-9] (34), it also introduces the key stereocenters early in the synthetic pathway via a chiral mono-protected imidazole intermediate [116291-87-3] (32). Catalytic diastereoselective hydrogenation of the furoimidazole (32) affords the *d*-lactone (34) as the major product, in 54% yield. The undesired diastereomeric *l*-lactone (33) can be easily separated by chromatography or crystallization. In the

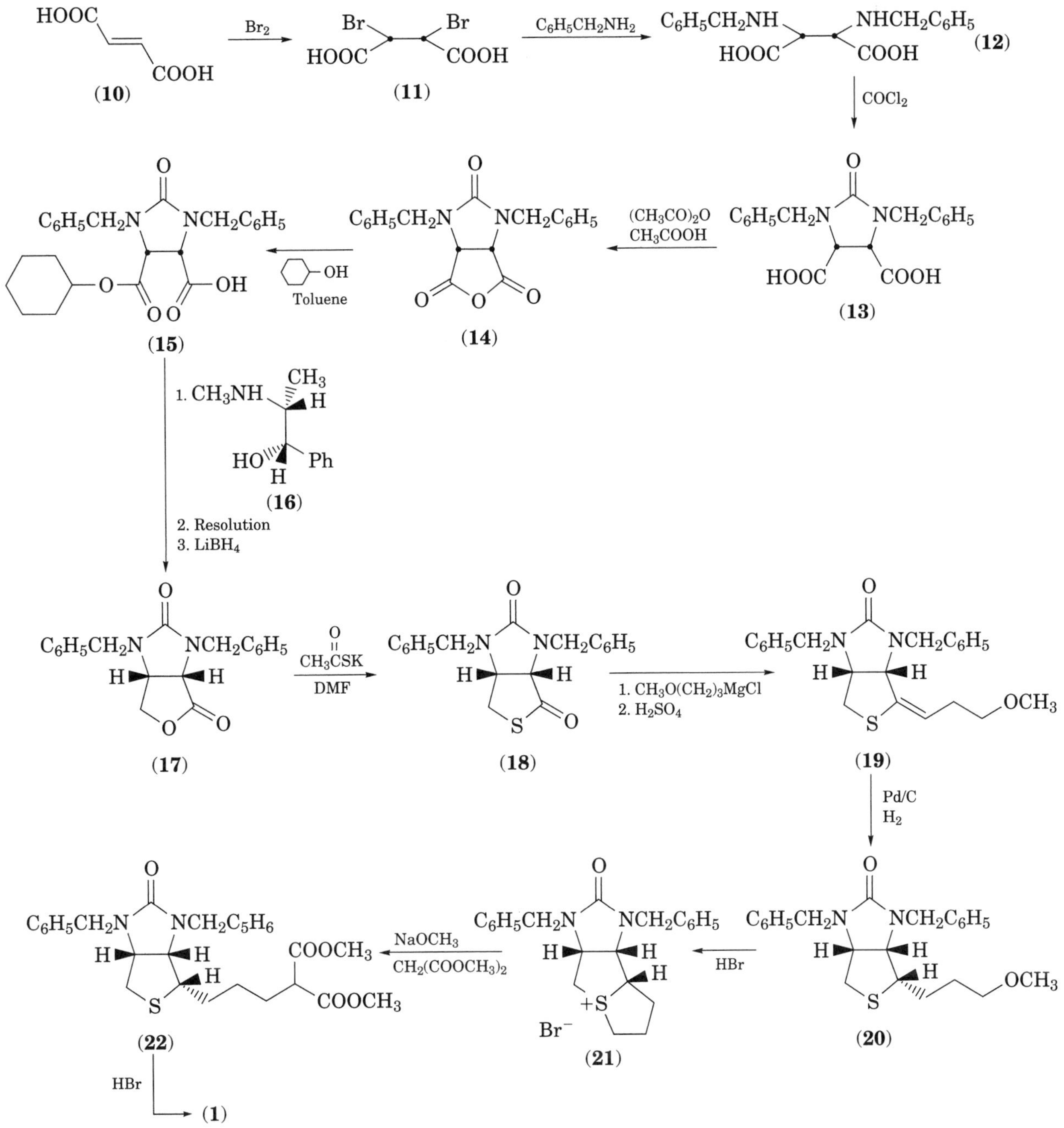

Figure 2. Synthetic pathway for *d*-biotin (Sternbach synthesis).

step following separation, the *d*-lactone is converted to the thiolactone [*118609-15-7*] (**35**) using a thiocarboxylic acid salt. The thiolactone undergoes a Grignard or Wittig reaction to incorporate the C-5 side chain. The thiophene [*118609-16-8*] (**36**) is then catalytically hydrogenated. This process hydrogenates the carbon–carbon double bond and also removes the chiral protecting group on the nitrogen. This process takes advantage of a short synthetic pathway and the use of a relatively inexpensive starting material. A recent modification of this Lonza process uses a chiral-substituted ferrocenyldisphosphine with a rhodium cata-

lyst to stereoselectively reduce the enamine (**32**) to the *d*-lactone (**34**) in 99% yield. This improvement eliminates the need to separate the undesired *l*-lactone by chromatography or crystallization and increases the yield of the desired lactone from 54 to 99%, making the process extremely efficient (32,33).

Novel Synthetic Methods

More recently, several novel syntheses of *d*-biotin, starting from a variety of chiral starting materials, have been de-

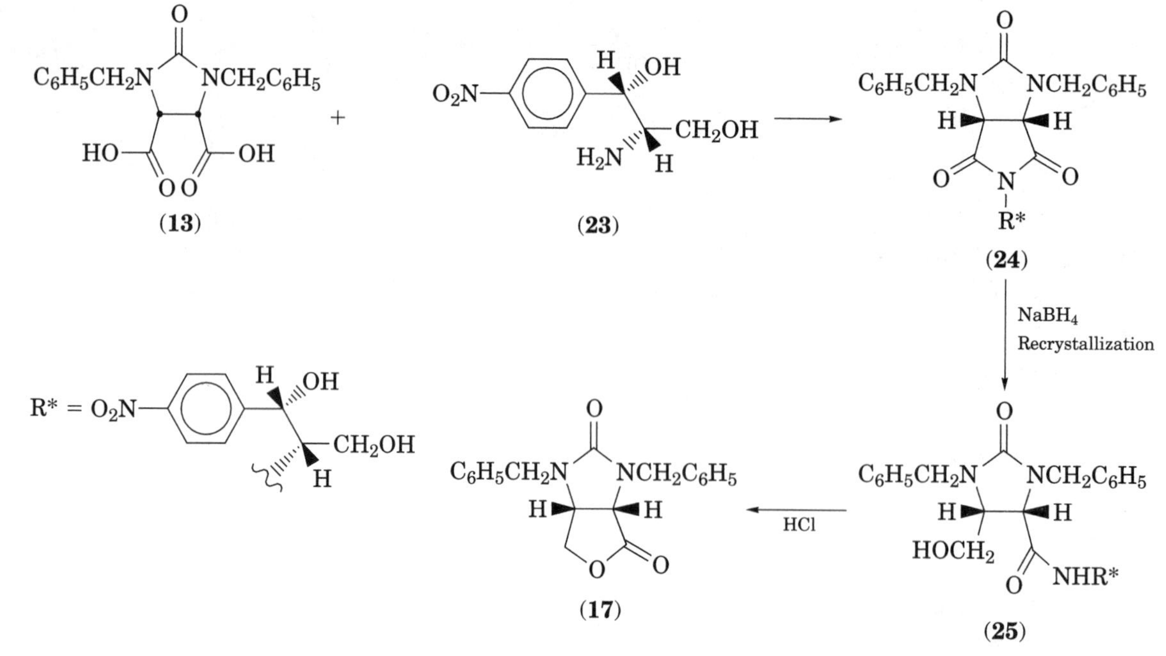

Figure 3. Synthetic pathway for d-biotin (Sumitomo synthesis).

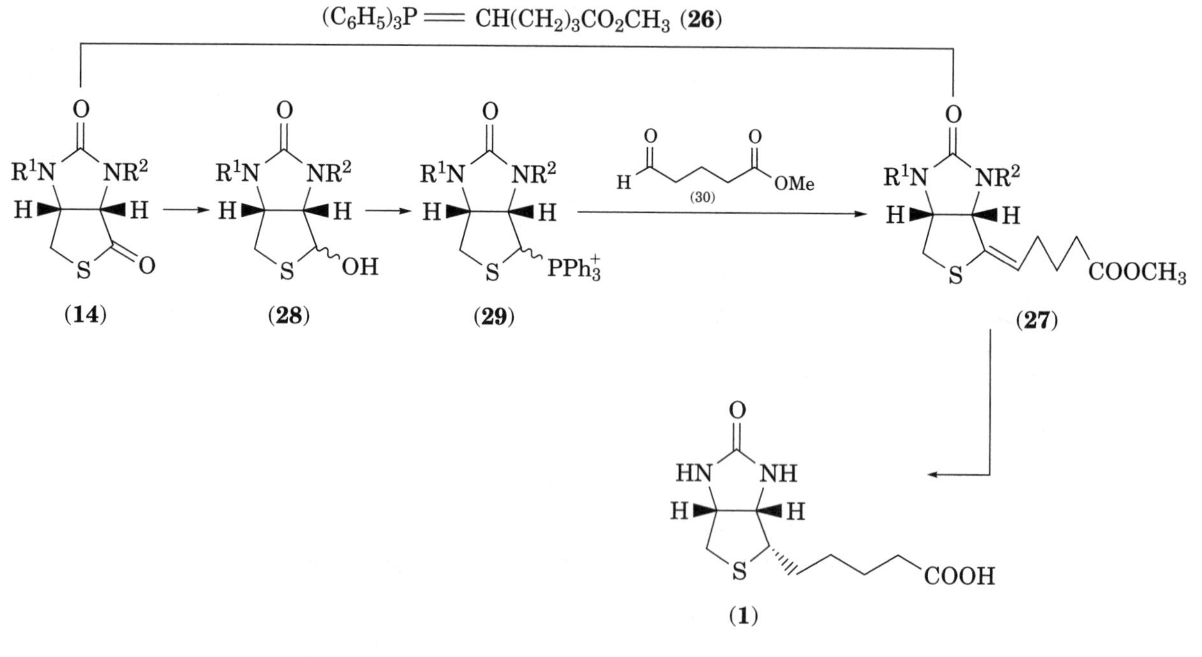

Figure 4. Synthetic pathways for d-biotin (Lonza syntheses).

veloped. Seven of these synthetic pathways start from L-cysteine (34–41), two from L-cysteine (42,43), two from D-arabinose (44,45), and one from ribitol (46). Each of these methods has at least one of the following drawbacks: multiple steps with overall low yields; low yield steps associated with cyclization reactions; safety issues associated with reagents such as metal azides, methyl iodide, diazomethane, organoazide reagents, etc; environmental issues

with reagents and by products such as triphenyl phosphine oxide and stannane salts; and costs of reagents such as 1,3-dicyclohexylcarbodiimide and trialkyltin hydride. Only one synthetic pathway starts from an achiral starting material, 2,5-dihydrothiophene-1,1-dioxide; this route requires a stereochemical resolution step (47–49). The drawbacks of this pathway are the commercially unavailable starting material and low conversions in several steps.

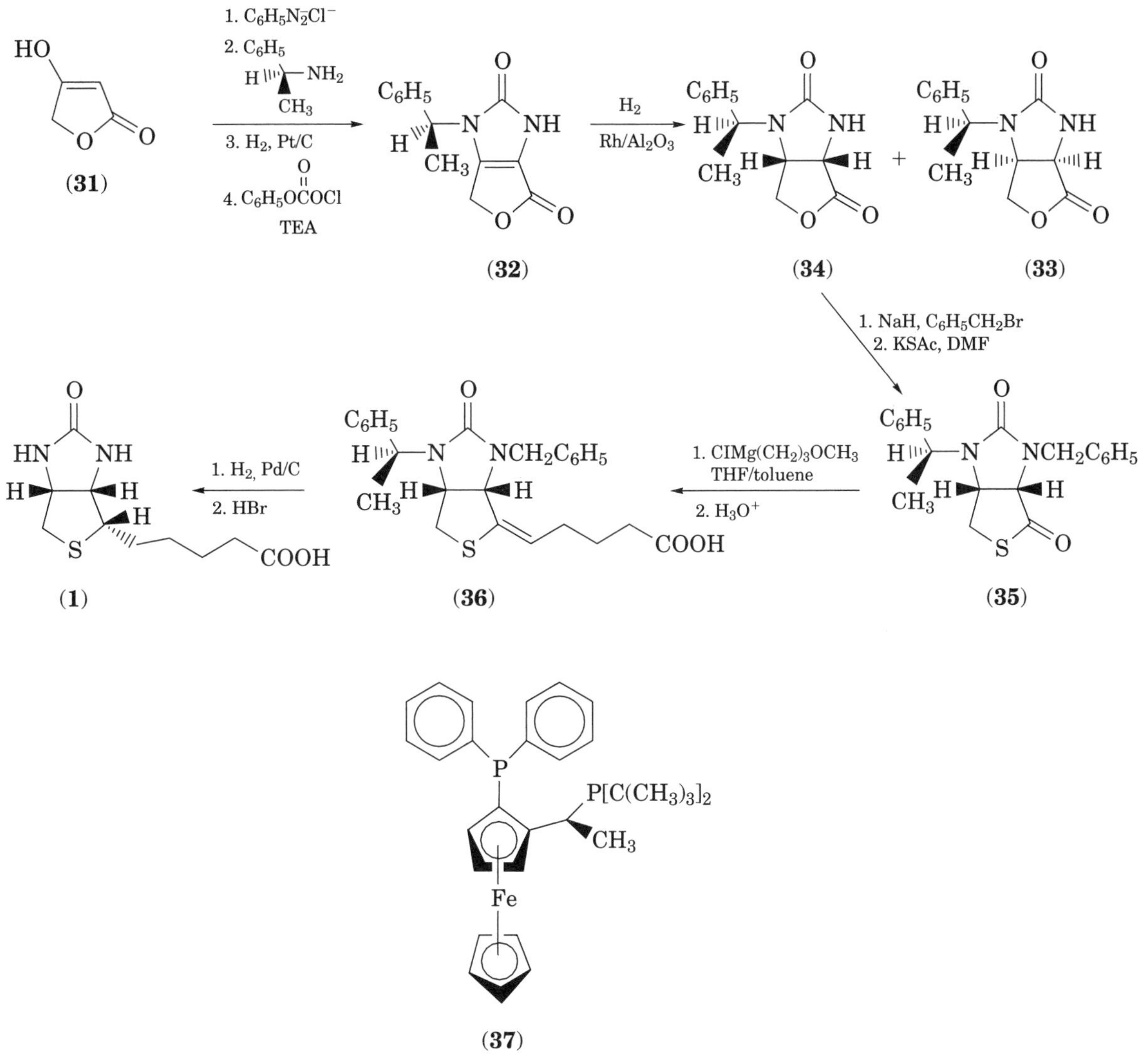

Figure 5. Synthetic pathway for *d*-biotin (Lonza synthesis). An improved process uses the chiral ferrocenyldisphosphine (**37**) to introduce stereospecificity during the hydrogenation of lactone (**32**).

BIOSYNTHESIS

Biotin is produced by a multistep pathway in a variety of fungi, bacteria, and plants (50–56). The established pathway (50,56) in *E. coli* is shown in Figure 6. However, little is known about the initial steps that lead to pimelyl-CoA or of the mechanism of the transformation of desthiobiotin to biotin. Pimelic acid is believed to be the natural precursor of biotin for some microorganisms (51).

Evidence that pimelic acid is a biotin precursor has been found in *Bacillus* species by feeding pimelic acid and observing the concomitant increase in biotin titers. Also, if labeled pimelic acid is used, labeled biotin is formed. In the biosynthetic pathway, pimelic acid [111-16-0] (**38**) is converted into pimelyl-CoA (**39**) by the enzyme pimelyl-CoA-synthetase in the presence of ATP and Mg^{2+} at 32°C and pH 7–8 (57). On the other hand, *E. coli* does not seem to rely on pimelic acid as a starting material for biotin synthesis. *E. coli* seems to form pimelyl-CoA by a pathway

similar to that of fatty acid and polyketide synthesis (58). Pimelyl-CoA (**39**) is transformed into 7-keto-8-aminopelargonic acid [4707-58-8] (7-KAPA) (**40**) by reaction with L-alanine. The reaction requires pyridoxal-5′-phosphate (PLP) as a co-reactant and is catalyzed by the enzyme KAPA-synthetase (59). 7-Keto-8-amino-pelargonic acid (**40**) is converted into 7,8-diamino-pelargonic acid [157120-40-6] (DAPA) (**41**) by the enzyme DAPA-aminotransferase, which uses *S*-adenosyl-L-methionine (SAM) instead of a simple amino acid as the amino group donor (60,61). 7,8-Diaminopelargonic acid (**41**) is converted by the enzyme DTB-synthetase to desthiobiotin [533-48-2] (**42**). The optimal biological production of desthiobiotin requires ATP, Mg^{2+} and bicarbonate at 50°C and pH 7–8 (62–66). Finally desthiobiotin (**42**) is converted to biotin, supposedly by the enzyme biotin synthetase, which is encoded by the Bio B gene (56).

The conversion of desthiobiotin to biotin occurs in growing as well as resting cells in which the internal source of

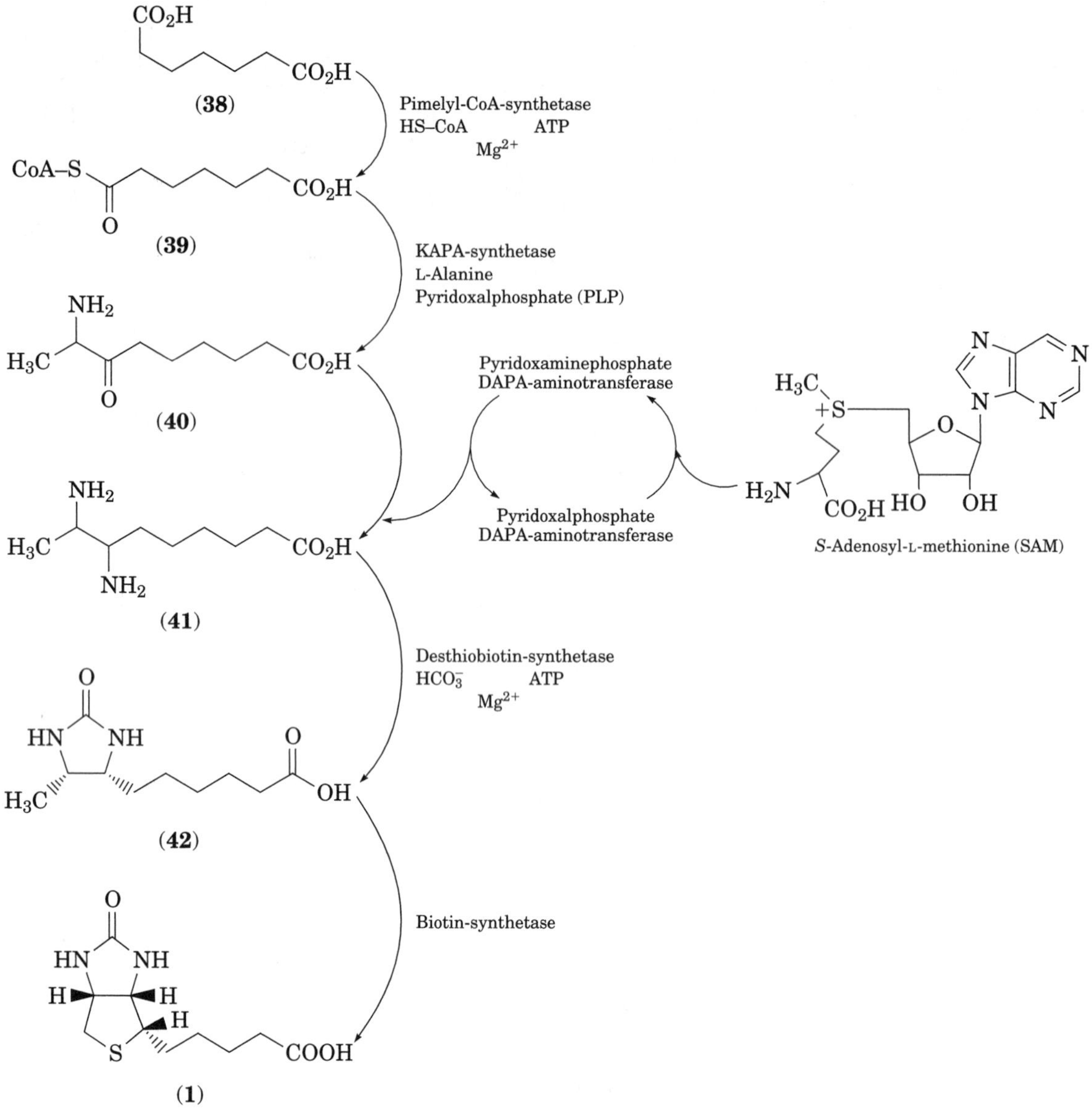

Figure 6. Biosynthetic pathway of *d*-biotin.

sulfur is believed to be cysteine or methionine. The introduction of sulfur into desthiobiotin has been investigated by several groups and it is proposed that the sulfur is introduced at C-4 of desthiobiotin with retention of configuration (67–69). The biosynthesis of biotin has led several groups to explore the feasibility of synthesizing biotin on a commercial scale by microorganisms. To date, no commercial scale total synthesis of biotin by fermentation is known. However, several microorganisms and mutant strains have been evaluated (Table 3). One of the problems associated with biotin biosynthesis via fermentation is that total biotin production decreases with increasing biotin concentrations in the fermentation broth. In fact, biotin strongly inhibits all steps of the biosynthesis except the synthesis of pimelyl-CoA. This inhibition by biotin leads to the isolation of several biotin vitamers. A vitamer is a compound, structurally similar to a vitamin, that ex-

hibits varying degrees of vitamin activity. The most important biotin vitamer is desthiobiotin. Inhibition of biotin biosynthesis is the subject of much research.

ANALYTICAL METHODS

Biological Assay

Various analytical methods, including microbiological, biological, chemical, enzymatic and chromatographic assays, have been used to determine biotin levels in food, feed, and body fluids (79–81). The biological assay of biotin is conducted with rats or chicks made biotin-deficient by special diets containing either raw egg whites or avidin. The assay is based on growth response curves since test-animal weight gain is proportional to the logarithm of the biotin dose. A biological assay of biotin is advantageous because

Table 3. Organisms Used for Biotin Fermentation

Organism name	Titer (mg/L)	References
Cornebacteria flavum		70,71
Brevibacterium flavum	0.5[a]	70,72
Serratia marcescens		73–78
	600	73
	500	74
	120	75,78
	83	77
Escherichia coli	2	79,80
Bacillus sphaericus		81–85
	70	81
	20	82
	365	83
Rhodotorula rubra		86
Sporobolomyces roseus		86
Yarrowia lipolytica		86
Candida shehatae		87
Rhizopus delemar	0.6	88,89

no pretreatment of the sample is necessary and the assay measures the amount of biotin available to the test animal. However, biological assays are time-consuming (four weeks to actually run the assay and six to seven weeks to prepare the biotin-deficient animals), costly to run, and require a large number of chicks or rats for accurate results.

Microbiological Assay

An alternative to the biological assay is the microbiological assay, which takes less time, costs less, requires less space, and has greater sensitivity than the biological assay. The typical microorganisms used for a microbiological assay of biotin are *Lactobacillus casei* ATCC 7469, *Neurospora crassa*, *Ochromonas danica*, *Saccharomyces cerevisiae* ATCC 7745, and *Lactobacillus plantarum* ATCC 8014. *Lactobacillus plantarum* ATCC 8014 is the test organism employed by most laboratories for biotin assay. Since biotin occurs in both the free and bound forms in nature and *L. plantarum* ATCC 8014 responds only to free biotin, all available biotin must be converted to free biotin prior to the microbiological assay. Bound biotin is usually converted to free biotin by acid hydrolysis with sulfuric acid or by digestion with papain. Hydrochloric acid should not be used in place of sulfuric acid for the acid hydrolysis because it may inactivate biotin. Lipids that stimulate the growth of *L. plantarum* interfere with the assay and have to be removed by filtration of the acid extracts or by preliminary ether extraction prior to the assay. A microbiological assay involves extraction, addition of graded levels of standard and sample to the assay tubes, addition of the medium, sterilization of the assay tubes, inoculation with *L. plantarum* and incubation. After the cell growth stops, the biotin content of the test material is determined by measuring the growth response of the organism (eg, colorimetrically, spectrophotometrically, or titrimetrically with sodium hydroxide) as compared to cell growth with known biotin concentration standards. Microbiological assay of biotin using *L. plantarum* is capable of detecting as little as 0.05 ng of biotin/mL of test solution.

Isotope Dilution Assay

An isotope dilution assay for biotin, based on the high affinity of avidin for the ureido group of biotin, compares the binding of radioactive biotin and nonradioactive biotin with avidin. This method is sensitive to a level of 1–10 ng biotin (82–84), and the radiotracers typically used are [^{14}C]biotin (83), [^{3}H]biotin (84,85) or an ^{125}I-labeled biotin derivative (86). A variation of this approach uses ^{125}I-labeled avidin (87) for the assay.

High-Performance Liquid Chromatography Analysis

The analysis of biotin has also been achieved by high performance liquid chromatography (hplc). Biotin has been analyzed in B-complex tablets, vitamin premixes, and multi-vitamin–multimineral preparations by reverse-phase, high performance liquid chromatographic methods (hplc) using a C^{18} column and uv detection at either 230 nm or 200 nm (88–91). Although this method can detect biotin in vitamin premixes, it is not sensitive enough to determine typical biotin levels in food or feed. Another method, one having greater sensitivity, is a reverse-phase ion-interaction reagent hplc method, with a biotin detection limit of 4 μg. An hplc method with even greater sensitivity involves derivatizing biotin prior to chromatographic separation. This technique can be used for either uv-absorbing derivatives, such as the bromoacetophenone ester (92), or fluorescent derivatives, such as methyl methoxycoumarin ester (92) and anthryldiazomethane ester (93). The biotin detection limit for the bromoacetophenone ester derivative at uv 254 nm was 50 ng. The detection limit for the methyl methoxycoumarin ester at excitation wavelength 360 nm and emission wavelength 410 nm was 5 ng; whereas, the limit of the anthryldiazomethane ester at excitation wavelength at 365 nm and emission wavelength 412 nm was 0.1–10 ng.

Gas Chromatography Analysis

From a sensitivity standpoint, a comparable technique is a gas chromatographic (gc) technique using flame ionization detection. This method has been used to quantify the trimethylsilyl ester derivative of biotin in agricultural premixes and pharmaceutical injectable preparations at detection limits of approximately 0.3 μg (94,95).

SPECIFICATIONS AND PRODUCT FORMS

According to the *Food Chemicals Codex* (96), the biologically active, food-grade form of biotin must have an assay of 97.5%, a melting point in the range of 229–232°C, and a specific rotation at 25°C in 0.1 N NaOH in the range of +89–93°. It must also contain less than 3 ppm of arsenic and less than 10 ppm of heavy metals, eg, lead, mercury, and copper. Finally, it must be able to be quantitatively sieved through U.S. Standard Sieves No. 80 using a mechanical shaker. Biotin is listed as GRAS in the *Code of Federal Regulations* (97,98) for use as a nutrient or dietary supplement (21 CFR 182.5159). The specifications for *d*-biotin for pharmaceutical applications are similar to those for food and are listed in the *United States Pharmacopeia*

(99). No specifications for the optically inactive, *d,l*-biotin are given for either food or pharmaceutical applications.

ECONOMIC ASPECTS

The biotin market is divided between agricultural and human use, with ~90% of biotin used in the animal health care market and ~10% for the human nutritional market. The major producers of biotin are Hoffmann-La Roche, Lonza, E. Merck-Darmstadt, Rhône-Poulenc, Sumitomo Pharmaceutical, E. Sung, and Tanabe Seiyaku (100). Worldwide production of biotin in 1994 was approximately 60 metric tons. The list price for pure biotin in 1995 was ~$7.00/g; whereas, the list price for technical feed-grade biotin was ~$5.50/g. Biotin is used in various pharmaceutical, food, and special dietary products, including multivitamin preparations in liquid, tablet, capsule, or powder forms. One of the commercially available products of *d*-biotin is Britrit-1, which is a 1% biotin trituration used in food premixes.

BIBLIOGRAPHY

1. P. György and F.-W. Zilliken, in R. Amman and W. Dirscherl, eds., *Fermente, Hormone, Vitamine*, Vol. 31, 3rd ed., Georg Thieme Verlag, Stuttgart, Germany, 1974, p. 766.

2. O. Isler, G. Bracher, S. Ghisla, and B. Kräutler, *Vitamine II. Wasserlösliche Vitamine*, Georg Thieme Verlag, Stuttgart, Germany, 1988, p. 231.

3. P. N. Achuta Murthy and S. P. Mistry, *Prog. Food Nutr. Sci.* **2**, 405 (1977).

4. F. A. Robinson, *The Vitamin Co-Factors of Enzyme Systems*, Pergamon Press, Oxford, U.K., 1966, p. 497.

5. *Encyclopedia of Food Science and Technology*, Pts. 1–8, Vol. 4, John Wiley & Sons, Inc., New York, 1991, p. 2764.

6. J. P. Bonjour, in L. Machlin, ed., *Handbook of Vitamins*, 2nd ed., Marcel Dekker Inc., New York, 1991, p. 393.

7. J. Gallagher, *Good Health with Vitamins and Minerals: A Complete Guide to a Lifetime of Safe and Effective Use*, Summit Books, New York, 1990, p. 67.

8. F. Kögl and B. Tönnis, *Z. Physiol. Chem.* **242**, 43 (1936).

9. P. György, R. Kuhn, and E. Lederer, *J. Biol. Chem.* **131**, 745 (1939).

10. D. B. Melville, K. Hofmann, E. Hague, and V. du Vigneaud, *J. Biol. Chem.* **142**, 615 (1942).

11. Food and Nutritional Board, National Research Council, *Recommended Dietary Allowances*, 10th ed., National Academy Press, Washington, D.C., 1989.

12. J. P. Bonjour, *Int. J. Vit. Nutr. Res.* **47**, 107 (1977).

13. V. du Vigneaud, K. Hofmann, D. B. Melville, and J. R. Rachele, *J. Biol. Chem.* **140**, 763 (1941).

14. V. du Vigneaud et al., *J. Biol. Chem.* **146**, 475 (1942).

15. D. B. Melville, A. W. Moyer, K. Hofmann, and V. du Vigneaud, *J. Biol. Chem.* **146**, 487 (1942).

16. P. György and B. W. Langer, Jr., in W. H. Sebrell, Jr., R. S. Harris, eds., *The Vitamins*, 2nd ed., Academic Press, New York, 1968, p. 261.

17. S. A. Harris et al., *JACS* **67**, 2096 (1945).

18. W. Traub, *Nature* **178**, 649 (1956).

19. J. Trotter and Y. A. Hamilton, *Biochemistry* **5**, 713 (1966).

20. S. Budavari, ed., *The Merck Index: An Encyclopedia of Chemicals, Drugs, and Biologicals*, 11th ed., Merck & Co., Inc., Rahway, N.J., 1989, p. 192.

21. U.S. Pat. 2,489,232 (Nov. 22, 1949), M. W. Goldberg and L. H. Sternbach (to Hoffmann-La Roche Inc.).

22. U.S. Pat. 2,489,235 (Nov. 22, 1949), M. W. Goldberg and L. H. Sternbach (to Hoffmann-La Roche Inc.).

23. U.S. Pat. 2,489,238 (Nov. 22, 1949), M. W. Goldberg and L. H. Sternbach (to Hoffmann-La Roche Inc.).

24. M. Gerecke, J. P. Zimmermann, W. Aschwanden, *Helv. Chim. Acta* **53**, 991 (1970).

25. U.S. Pat. 3,876,656 (Apr. 8, 1975), A. Hisao, A. Yasuhiko, O. Shigeru, and S. Hiroyuki (to Sumitomo Chemical Co., Ltd.); Pat. Appl. E.P.O. 044,158 (Jan. 20, 1982), A. Hisao, A. Yasuhiko, O. Shigeru, and S. Hiroyuki (to Sumitomo Chemical Co., Ltd.).

26. Eur. Pat. Appl. EP 161,580 (Nov. 21, 1985), C. Wehrli and H. Pauling (to F. Hoffmann-La Roche AG).

27. Eur. Pat. Appl. EP 273270-A1 (July 6, 1988), J. McGarrity and L. Tenud (to Lonza AG).

28. Eur. Pat. Appl. EP 270076-A1 (Feb. 12, 1986), M. Eyer, R. Fuchs, D. Laffan, J. F. McGarrity, T. Meul, and L. Tenud (to Lonza AG).

29. U.S. Pat. 4,851,540 (Dec. 1, 1987), J. F. McGarrity, L. Tenud, and T. Meul (to Lonza AG).

30. Eur. Pat. Appl. EP 387,747 (Sept. 19, 1990), M. Eyer (to Lonza AG).

31. U.S. Pat. 4,873,339 (Nov. 30, 1988), J. F. McGarrity, L. Tenud, and T. Meul (to Lonza AG).

32. Eur. Pat. 624587 (Nov. 17, 1994), J. McGarrity, F. Spindler, R. Fuchs, and M. Eyer (to Lonza AG).

33. PCT Int. Appl. WO 94 24,137 (Oct. 27, 1994), M. Eyer and R. E. Merrill (to Lonza AG).

34. U.S. Pat. 4,837,402 (Jan. 6, 1988), E. Poetsch and M. Casutt (to Merck Patent GmbH).

35. E. Poetsch and M. Casutt, *Chimia* **41**, 148 (1987).

36. Ger. Offen. DE 3,926,690 (Feb. 14, 1991), M. Casutt, E. Poetsch, and W. N. Spekamp (to Merck Patent GmbH).

37. F. D. Deroose and P. J. DeClercq, *Tetrahedron Lett.* **34**, 4365 (1993).

38. F. D. Deroose and P. J. DeClercq, *Tetrahedron Lett.* **35**, 2615 (1994).

39. F. D. Deroose and P. J. DeClercq, *J. Org. Chem.* **60**, 321 (1995).

40. T. Fujisawa, M. Nagai, Y. Koike, and M. Shimizu, *J. Org. Chem.* **59**, 5865 (1994).

41. H. L. Lee, E. G. Baggiolini, and M. R. Uskokovic, *Tetrahedron* **43**, 4887 (1987).

42. E. J. Corey and M. M. Mehrotra, *Tetrahedron Lett.* **29**, 57 (1988).

43. T. Ravindranathan, S. V. Hiremath, D. R. Reddy, and R. B. Tejwani, *Synth. Commun.* **18**, 1855 (1988).

44. Jpn. Kokai Tokkyo Koho, JP 61254590 (Nov. 12, 1986), T. Kono, Y. Shimakawa, M. Takahashi, T. Horisaki, and S. Masuda (to Teikoku Chemical Industry Co., Ltd.).

45. N. A. Hughes, K.-M. Kuhajda, and D. A. Miljkovic, *Carbohydr. Res.* **257**, 299 (1994).

46. D. A. Miljkovic and S. Velimirovic, *J. Serb. Chem. Soc.* **53**, 37 (1988).

47. H. A. Bates, L. Smilowitz, and J. Lin, *J. Org. Chem.* **50**, 899 (1985).

48. H. A. Bates, L. Smilowitz, and S. B. Rosenblum, *J. Chem. Soc. Chem. Commun.*, 353 (1985).

49. H. A. Bates and S. B. Rosenblum, *J. Org. Chem.* **51**, 3447 (1986).

50. M. Eisenberg, in F. Neidhardt, ed., *Escherichia coli and Samonella typhimzerium Cellular and Molecular Biology*, American Society for Microbiology, New York, 1987, p. 544.

51. R. Gloeckler et al., *Gene* **87**, 63 (1990).

52. Y. Izumi and K. Ogata, *Adv. Appl. Microbiol.* **22**, 145 (1977).

53. P. Baldet, C. Alban, S. Axiotis, and R. Douce, *Arch. Biochem. Biophys.* **303**, 67 (1993).

54. T. Schneider, R. Dinkins, K. Robinson, J. Shellhammer, and D. Meinke, *Dev. Biol.* **131**, 161 (1989).

55. J. Shellhammer and D. Meinke, *Plant Physiol.* **93**, 1162 (1990).

56. M. A. Eisenberg, *Ann. N.Y. Acad. Sci.* **447**, 335 (1985).

57. O. Ploux, P. Soularue, A. Marquet, R. Gloeckler, and Y. Lemoine, *Biochem. J.* **287**, 685 (1992).

58. I. Sanyal, S.-L. Lee, and D. H. Flint, *JACS* **116**, 2637 (1994).

59. M. A. Eisenberg and C. Star, *J. Bacteriol.* **96**, 1291 (1968).

60. G. L. Stoner and M. A. Eisenberg, *J. Biol. Chem.* **250**, 4029 (1975).

61. C. H. Pai, *J. Bacteriol.* **105**, 793 (1971).

62. K. Krell and M. A. Eisenberg, *J. Biol. Chem.* **245**, 6558 (1970).

63. P. Cheeseman and C. H. Pai, *J. Bacteriol.* **104**, 726 (1970).

64. O. Ifuku, J. Kishimoto, S. Haze, M. Yanagi, and S. Fukushima, *Biosci. Biotechnol. Biochem.* **56**, 1780 (1992).

65. R. L. Baxter, A. J. Ramsey, L. A. McIver, and H. C. Baxter, *J. Chem. Soc., Chem. Commun.*, 559 (1994).

66. R. L. Baxter and H. C. Baxter, *J. Chem. Soc., Chem. Commun.*, 759 (1994).

67. F. Frappier, M. Jouany, A. Marquet, A. Olesker, and J.-C. Tabet, *J. Org. Chem.* **47**, 2257 (1982).

68. D. A. Trainer, R. J. Parry, and A. Gutterman, *JACS* **102**, 1467 (1980).

69. L. Even, D. Florentin, and A. Marquet, *Bull. Soc. Chim. Fr.* **127**, 758 (1990).

70. Jpn. Kokai Tokkyo Koho, JP06339371, (Dec. 13, 1994), Y. Yoneda, T. Abe, N. Hara, I. Ohsawa, and A. Fujisawa (to Nippon Zeon Co., Ltd.).

71. I. Ohsawa et al., *J. Ferment. Bioen.* **73**, 121 (1992).

72. Eur. Pat. Appl. EP 316229 (May 17, 1989), S. Haze, O. Ifuku, and J. Kishimoto (to Shiseido Co., Ltd.).

73. Jpn. Kokai Tokkyo Koho, JP 58060996 (Apr. 11, 1983), (to Nippon Zeon Co., Ltd.).

74. Eur. Pat. Appl. EP 375525 (June 27, 1990), R. Gloeckler, D. Speck, J. Sabatie, S. Brown, and Y. Lemoine (to Transgene S. A., Fr.).

75. B. M. Pearson, D. A. MacKenzie, and M. H. J. Keenan, *Lett. Appl. Microbiol.* **2**, 25 (1986).

76. J. C. Du Preez, J. L. F. Kock, A. M. T. Monteiro, and B. A. Prior, *FEMS Microbiol. Lett.* **28**, 271 (1985).

77. Otkrytiya, Izobret., *Prom. Obraztsy, Tovarnye Znaki* (38), 93 (1983); Sov. Pat. SU 1047956 (Oct. 15, 1983), L. I. Vorob'eva, E. V. Shchelokova, and E. S. Naumova (to Moscow State University) (in Russian).

78. V. N. Maksimov, E. V. Shchelokova, and L. I. Vorob'eva, *Prikl. Biokhim. Mikrobiol.* **19**, 353 (1983).

79. I. D. Lumley and P. R. Lawrance, *J. Micronutr. Anal.* **7**, 301 (1991).

80. J. Scheiner, J. Augustin, B. P. Klein, D. A. Becker, and P. B. Venugopal, eds., *Methods of Vitamin Assay*, 4th ed., John Wiley & Sons, Inc., New York, 1985, p. 535.

81. F. Frappier and M. Gaudry, in A. P. DeLeenheer, W. E. Lambert, and M. G. M. De Ruyter, eds., *Modern Chromatographic Analysis of the Vitamins*, Marcel Dekker, New York, 1985, p. 482.

82. Dakshinamurti, A. D. Landman, L. Ramamurti, and R. J. Constable, *Anal. Biochem.* **61**, 225 (1974).

83. R. L. Hood, *J. Sci. Food Agric.* **26**, 1847 (1975).

84. R. Rettenmaier, *Anal. Chim. Acta* **113**, 107 (1980).

85. T. Suormala et al., *Clin. Chim. Acta* **177**, 253 (1988).

86. E. Livaniou, G. R. Evangeliatos, and D. S. Ithakissions, *Clin. Chem.* **33**, 1983 (1987).

87. D. M. Mock and D. B. DuBois, *Anal. Biochem.* **155**, 272 (1986).

88. T. S. Hudson, S. Subramanian, and R. J. Allen, *J. Assoc. Official Anal. Chemists* **67**, 995 (1984).

89. S. L. Crivelli, P. F. Quirk, D. J. Steible, and S. P. Assenza, *Pharm. Res.* **4**, 261 (1987).

90. A. Rizzolo, C. Baldo, and A. Polesello, *J. Chromatogr.* **553**, 187 (1991).

91. M. C. Gennaro, *J. Chromatogr. Sci.* **29**, 410 (1991).

92. P. L. Desbene, S. Coustal, and F. Frappier, *Anal. Biochem.* **128**, 359 (1983).

93. K. Hayakawa and J. Oizumi, *J. Chromatogr.* **413**, 247 (1987).

94. V. Viswanathan, F. P. Mahn, V. S. Venturella, and B. Z. Senkowski, *J. Pharm. Sci.* **59**, 400 (1970).

95. H. Janecke and H. Voege, *Naturwissenschaften* **55**, 447 (1968).

96. Food and Nutrition Board, National Research Council, *Food Chemicals Codex*, 3rd ed., National Academy Press, Washington, D.C., 1981, p. 38.

97. *Evaluation of the Health Aspects of Biotin as a Food Ingredient*, Life Sciences Research Office, Federation of American Societies for Experimental Biology, Bethesda, Md., 1978.

98. *Code of Federal Regulations, Title 21, Food and Drugs, Parts 100–199 rev.*, Office of the Federal Register, General Services Administration, U.S. Government Printing Office, Washington, D.C., 1977.

99. *The United States Pharmacopeia XXIII (USP XXIII-NFXVIII)*, United States Pharmacopeial Convention, Inc., Rockville, Md., 1995, p. 206.

100. J. D. Greer and B. Rhomberg, *CEH Marketing Research Report, Animals: Chemical Inputs for Nutrition and Health—Overview*, Chemical Economics Handbook-SRI International, Palo Alto, CA 1995, 201.8001A-B, Aug. 22, 1996.

ROBERT A. OUTTEN
Hoffmann-La Roche, Inc.

VITAMINS: FOLIC ACID

Folic acid [*59-30-3*] (**1**) belongs to the group of B-vitamins. The term folate is used to designate all members of the family of compounds based on the *N*-[(6-pteridinyl)methyl]-*p*-aminobenzoic acid skeleton conjugated with one or more L-glutamic acid units. In 1930, a dietary factor in yeast and crude liver extract was found to cure megaloblastic anemia in pregnant women (1). Purified folate was isolated by using two bioassay procedures, the microbiological growth assay and the assay for the antianemic factor for chickens (2–4). The growth factor from yeast and the antianemic factor (vitamin B_c) were later shown to be different entities. A crystalline compound was isolated from spinach, which was called folic acid (5,6). It was later shown that several of the above-mentioned factors belonged to the nutritionally and chemically related family of pteroylglutamic acid compounds. The structure of pteroylglutamic acid was elucidated in 1946 (7). The metabolically active forms of folic acid have a reduced pteridine ring and several glutamic acids residues. A detailed account on the discovery and early development of folic acid is available (8).

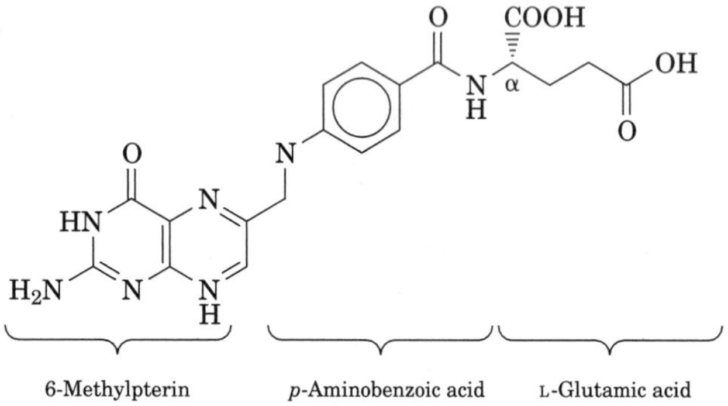

(**1**)

OCCURRENCE, SOURCE, AND BIOAVAILABILITY

Good food sources of folate are liver; fresh, dark green, leafy vegetables; beans; wheat germ; and yeasts (qv) (Table 1). Folic acid is synthesized only by microorganisms and plants (10–14). Most dietary folates exist in the polyglutamate form, which are converted to the more readily bioavailable monoglutamate form in the small intestine by the jejunal brush border folate conjugase. Certain foods such as cabbage and legumes contain conjugase inhibitors, which can decrease folate absorption.

The total folate content of food varies, based on the method of preparation and length of storage (15–17). Different forms of folates occur in nature and the stability or bioavailability of each form varies. Most folates in food are easily oxidizable and therefore are susceptible to oxidation under aerobic conditions during storage and processing. Folic acid (commercial form) has superior bioavailability because it is more readily absorbed when compared to the tri- or heptaconjugates. The factors affecting the bioavailability of food folates are not well understood, but seem to

Table 1. Select Contributors of Folate in the U.S. Diet

Ranking	Description	Total folate (%)
1	Orange juice	9.7
2	White bread, rolls, crackers	8.6
3	Pinto, navy, other dried beans	7.1
4	Green salad	6.8
5	Cold cereals	5.0
6	Eggs	4.6
9	Liver	3.1
23	Hamburger	1.2
25	Spinach	1.0
30	Green beans	0.8
34	Broccoli	0.7

Source: Ref. 9.

include iron and vitamin *C* (qv) status. Deficiencies of both of these nutrients in humans is associated with impaired utilization of dietary folate. Improved research techniques such as use of radiolabeled folates has provided a powerful tool for determining bioavailability (18). Under fasting conditions, folic acid is almost completely absorbed, whereas only 50–80% of folyl polyglutamate is absorbed, as determined by urinary excretion measurements (19).

CHEMICAL AND PHYSICAL PROPERTIES

L-Folic acid (**1**) contains three subunits: 6-methylpterin, *p*-aminobenzoic acid, and L-glutamic acid. The *Chemical Abstracts* name is *N*-[4-[{(2-amino-1,4-dihydro-4-oxo-6-pteridinyl)methyl}amino]benzoyl]-L-glutamic acid.

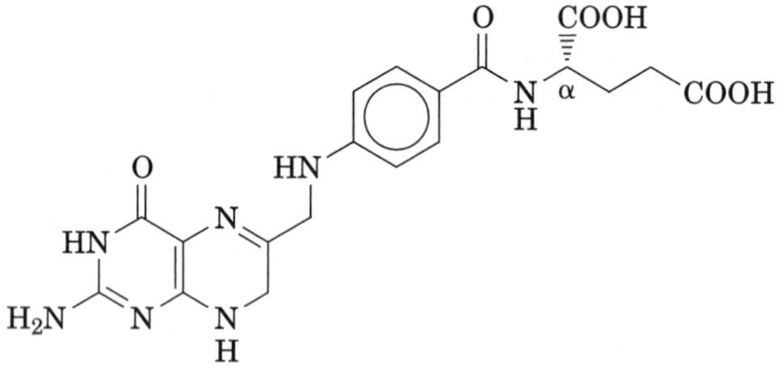

(**2**)

Enzymatic reduction of folic acid leads to the 7,8-dihydrofolic acid (H_2 folate) (**2**), a key substance in biosynthesis. Further reduction, catalyzed by the enzyme dihydrofolic acid reductase, provides (6*S*)-5,6,7,8-tetrahydrofolic acid (H_4 folate) (**3**). The H_4 folate (**3**) is the key biological intermediate for the formation of other folates (**4–8**) (Table 2).

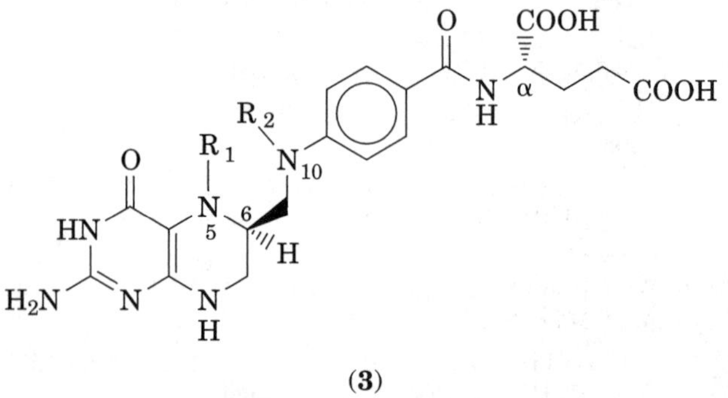

(**3**)

Table 2. H₄ Folate Cofactors

Structure	R_1	R_2	Nomenclautre	Configuration
(3)	H	H	Tetrahydrofolic acid	6S, α(S)
(4)	CH₃	H	5-Methyltetrahydrofolic acid	6S, α(S)
(5)	CH₂		5,10-Methylenetetrahydrofolic acid	6R, α(S)
(6)	HC=O	H	5-Formyltetrahydrofolic acid	6S, α(S)
(7)	H	HC=0	10-Formyltetrahydrofolic acid	6R, α(S)
(8)	HC=NH	H	5-Formiminotetrahydrofolic acid	6S, α(S)
(9)	CH⁺		5,10-Methenyltetrahydrofolic acid	6R, α(S)

Nomenclature and symbols for folic acid assigned based on recommendation published by IUPAC–IUB joint commission on Biochemical Nomenclature 1986 (20,21); see structure (3).

Folic acid (1) is found as yellow, thin platelets which char above 250°C. The uv spectrum of L-folic acid at pH 13 shows absorptions at $\lambda = 256$ nm ($\epsilon = 30{,}000$), 282 nm ($\epsilon = 26{,}000$), and 365 nm ($\epsilon = 9800$). Folic acid has a specific rotation of $[\alpha]_D^{27} = +19.9°$ ($c = 1, 0.1\ N$ NaOH). Solutions of folic acid are stable at room temperature and in the absence of light. It is slightly soluble in aqueous alkali hydroxides and carbonates but is insoluble in cold water, acetone, and chloroform. Table 3 lists some physical properties of selected folic acid derivatives.

SYNTHESIS

The first L-folic acid synthesis was based on the concept of a three-component, one-pot reaction (7,22). Triamino-4(3H)-pyrimidinone [1004-45-7] (10) was reacted simultaneously with C_3-dibromo aldehyde [5221-17-0] (11) and p-amino-benzoyl-L-glutamic acid [4271-30-1] (12) to yield folic acid (1).

All known commercial syntheses are based on this approach with improvements in preparations of the three components (23). Shortly after the first synthesis, similar methods were published employing other C_3-halo compounds, such as 1,1,3-tribromo-2-propanone, 2,2,3-tribromopropanal (24), 2,2,3-trichloropropanal, and 1,1,3-trichloro-2-propanone (23).

Yields were improved to ≥37% by the addition of sodium sulfite to the reaction mixture. Apart from the sulfite, the C_3-component unit has the greatest influence on the yield of folic acid. The use of nickel(II) chloride as an additive has been claimed to give higher yields (25).

The required triamino-4(3H)-pyrimidinone is prepared in three steps starting from guanidine [50-01-1] (14) (26). Condensation with methylcyanoacetate [105-34-0] (15) under basic conditions, followed by nitrosation of the intermediate [56-06-4] (16), gives 2,6-diamino-5-nitrosopyrimidinone [2387-48-6] (17). Chemical reduction using sodium sulfite (27) or catalytic hydrogenation using Raney nickel (28) furnishes (10).

p-Aminobenzoyl-L-glutamic acid (12) is obtained by condensation of p-nitrobenzoyl chloride [122-04-3] (18) with L-glutamic acid [56-86-0] (19) under Schotten-Baumann conditions. This is followed by reduction of the nitro group with either sodium hydrogen sulfide (29) or by electrochemical methods (30).

1,1,3-Trichloroacetone [921-03-9] (13a) is prepared by chlorination of acetone. The reaction is nonselective and the required compound is isolated by distillation. The selectivity has been improved by catalyzing the reaction with iodine (31).

Table 3. Physical Properties of Folic Acid Derivatives

Compound	Structure number	CAS Registry Number	uv λ_{max} (nm)	ϵ	Stability	Molecular formula (mol wt)
Folic acid Pteroylglutamic acid PteGlu	(1)	[59-30-3]	(pH 13) 256	30,000	Unstable in alkaline and acidic solutions	$C_{19}H_{19}N_7O_6$ (441.41)
Dihydrofolic acid H_2 Folate	(2)	[4033-27-6]	(pH 7.2) 282	28,600	Highly air-sensitive	$C_{19}H_{21}N_7O_6$ (443.42)
Tetrahydrofolic acid H_4 Folate	(3)	[135-16-0]	(pH 7.8) 296	28,000	Sensitive to oxygen	$C_{19}H_{23}N_7O_6$ (445.44)
5-Methyltetrahydrofolic acid 5-CH_3-H_4 Folate	(4)	[134-35-0]	(pH 7) 290	32,000	Stable to oxygen	$C_{20}H_{25}N_7O_6$ (459.46)
5,10-Methylenehydrofolic acid 5,10-CH_2-H_4 Folate	(5)	[3432-99-3]	(pH 7.2) 294	32,000	Sensitive to hydrolysis	$C_{20}H_{23}N_7O_6$ (457.45)
10-Formyltetrahydrofolic acid 10-CHO-H_4 Folate	(6)	[2800-34-2]	(pH 7.5) 260	17,000	Quite unstable	$C_{20}H_{23}N_7O_7$ (473.45)
5-Formyltetrahydrofolic acid (6R,S)-5-CHO-H_4 Folate	(7)	[58-05-9]	(pH 13) 282	32,600	Most stable	$C_{20}H_{23}N_7O_7$ (473.45)
(6S)-5-CHO-H_4 Folate		[68538-85-2]				$C_{20}H_{23}N_7O_7$ (473.45)
5-Formiminotetrahydrofolic acid 5-NHCH-H_4 Folate	(8)	[2311-81]	(pH 7) 285	35,400	Hydrolyzed in aqueous solution	$C_{20}H_{24}N_8O_6$ (472.46)
5,10-Methyltetrahydrofolic acid 5,10-CH^+-H_4 Folate	(9)	[65981-89-7]	(pH 1) 352	23,900		$C_{20}H_{22}N_7O_6^+$ (cation)

Alternative Approaches for Synthesis of L-Folic Acid

L-Folic acid (1) has been prepared in two steps by condensing 6-bromomethylpterin (20) with p-aminobenzoyl-L-glutamic acid (12) in 80% yield (32). Dissolved folic acid further reacts easily with one more equivalent of 6-bromomethylpterin to form the undesired dialkylated aminobenzoyl-L-glutamic acid.

In spite of the good yields of L-folic acid obtained in this reaction, all of the published methods for the synthesis of 6-bromomethylpterin (20) are multistep procedures with low overall yields (33–36). For example, the route starting from 2,4,5,6-tetraaminopyrimidine [5392-28-9] (21) gave 6-bromomethylpterin (20) in three steps with an overall yield of only 18% (33,35,36). This synthesis is not economical because the intermediate 6-bromomethyl-2,4-diamino-4-pterin (22) has to be deaminated in an additional step to form 6-bromomethylpterin (20).

Another viable method for the synthesis of L-folic acid (1) starts from 6-formylpterin (23). The diester of L-glutamic acid (24) is condensed with 6-formylpterin (23). Reduction of the Schiff base with sodium borohydride is followed by hydrolysis to yield L-folic acid (37).

A cost-efficient synthesis of folic acid via Schiff base formation is feasible only if 6-formylpterin (23) is readily available. This compound is prepared by the reaction of 2-bromomalondialdehyde dimethylacetal [59453-00-8] (25) with triaminopyrimidinone (10), followed by acetylation and cleavage of the acetal to give compound (23) in 51% overall yield (38).

(10) + (25) → (23)

A second approach for the synthesis of 6-formylpterin (23) involves the condensation of triaminopyrimidinone (10) with 5-deoxy-L-arabinose (26). The key diol is obtained in four steps starting from compound (10). Cleavage of the diol side chain is achieved either with periodate (39) or with lead(IV) (40) to furnish 6-formylpterin (23) in 45% overall yield.

A third approach for the synthesis of 6-formylpterin (23) starts from iminodipropionitrile [2869-25-2] (27). The intermediate pyrazine (28) is also prepared starting from chloropyruvaldehyde oxime (41,42). The required formylpterin (23) is obtained in three steps in 71% yield, starting from the intermediate pyrazine (28). A few other routes for the synthesis of 6-formylpterin (23) are described in the literature. All are multistep procedures with only moderate overall yields (43–45).

A new variant of the three-component, one-pot synthesis of L-folic acid has been reported by Hoffmann-La Roche Inc.

One equivalent of 2-hydroxymalondialdehyde [497-15-4] (29) is condensed with two equivalents of p-aminobenzoyl-L-glutamic acid (12). The intermediate dimine (30) is treated with one equivalent of triaminopyrimidinone (10) to obtain L-folic acid in 84% yield (46).

Fermentation

Development of an economically viable production process for folic acid either by genetically engineered microorganisms or by extraction from natural sources is not yet feasible.

LABELED COMPOUNDS

Radiolabeled folate provides a powerful tool for folate bioavailability studies in animals and for diagnostic procedures in humans. Deuteration at the 3- and 5-positions of the central benzene ring of folic acid (31) was accomplished by catalytic debromination (47,48) or acid-catalyzed exchange reaction (49). Alternatively, deuterium-labeled folic acid (32) was prepared by condensing pteroic acid with commercially available labeled glutamic acid (50).

DERIVATIVES AND ANALOGUES

The metabolically active H$_4$ folate cofactors (see Table 2) are prepared synthetically as follows. 7,8-Dihydrofolic acid

(2) and 5,6,7,8-tetrahydrofolic acid (3) are prepared via catalytic hydrogenation of folic acid under controlled reaction conditions (51,52). Optical rotation of (6S,R)-H$_4$ folate (3) is $[\alpha]_D^{27} = +14.9$ (0.1 N NaOH) and the natural (6S)-H$_4$ folate (3) is $[\alpha]_D^{27} = -16.9°$ (0.1 N NaOH).

5-Methyltetrahydrofolic acid (5-CH$_3$-H$_4$ folate) (4) is involved in methionine biosynthesis. Condensation of formaldehyde with H$_4$ folate (3), followed by the reduction of the intermediate 5,10-CH$_2$-H$_4$ folate (5) with sodium borohydride gave 5-CH$_3$-H$_4$ folate (4) (53). 5,10-Methylenetetrahydrofolic acid (5,10-CH$_2$-H$_4$ folate) (5) is a coenzyme in thymidylate biosynthesis; the natural (6R)-stereoisomer is prepared by enzymatic reduction of H$_2$ folate (2), followed by condensation with formaldehyde (54).

Formylation of H$_4$ folate (3) or hydrolysis of 5,10-CH$^+$-H$_4$ folate (9) gives (6R,S)-5-formyltetrahydrofolic acid (6) (5-HCO-H$_4$ folate) (55). On the other hand, (6S)-5-HCO-H$_4$ folate is obtained by selective crystallization in the form of its calcium salt from the diastereomeric mixture of (6S,R)-5-HCO-H$_4$ folate (56). 10-Formyltetrahydrofolic acid (7) is a coenzyme in purine synthesis which is synthesized by hydrolysis of 5,10-CH$^+$-H$_4$ folate (9) or by hydrogenation of 10-CHO-folate (57).

Folic acid analogues containing amino acids other than glutamate, and also folate covalently bound to a protein for the purposes of antibody production, have been prepared (58–60). Methotrexate is an analogue of folic acid that is widely used in cancer chemotherapy (61). Other analogues such as trimethoprim and pyrimethamine are used in the treatment of malaria and protozoal diseases (62). These analogues bind extremely tightly to dihydrofolate reductase.

BIOSYNTHESIS

Folic acid is synthesized both in microorganisms and in plants. Guanosine-5-triphosphate (GTP) (33), p-aminobenzoic acid (PABA), and L-glutamic acid are the precursors. Reviews are available for details (63,64). The sequence of reactions responsible for the enzymatic conversion of GTP to 7,8-dihydrofolic acid (2) is shown (see Scheme 1. In E. coli, GTP cyclohydrolase catalyzes the conversion of GTP (33) into 7,8-dihydroneopterin triphosphate (34) via a three-step sequence. Hydrolysis of the triphosphate group of (34) is achieved by a nonspecific pyrophosphatase to afford dihydroneopterin (35) (65). The free alcohol (36) is obtained by the removal of residual phosphate by an unknown phosphomonoesterase. The dihydroneopterin undergoes a retro-aldol reaction with the elimination of a hydroxy acetaldehyde moiety. Addition of a pyrophosphate group affords hydroxymethyl-7,8-dihydropterin pyrophosphate (37). Dihydropteroate synthase catalyzes the condensation of hydroxymethyl-7,8-dihydropteroate pyrophosphate with PABA to furnish 7,8-dihydropteroate (38). Finally, L-glutamic acid is condensed with 7,8-dihydropteroate in the presence of dihydrofolate synthetase.

ANALYTICAL METHODS

Analysis of folic acid is difficult because most natural folates exist in the polyglutamate form and there is variation in the oxidation state of the single-carbon substituent. Determination of the individual folate vitamers is complicated; assay simplification is achieved by determining folate in the monoglutamyl or diglutamyl forms after enzymatic deconjugation. Methods for determining folic acid in food and feed include biological, microbiological, chemical, chromatographic, and radiometric assays. The microbiological assay using Lactobacillus casei is the official method of the Association of Official Analytical Chemists (AOAC). Folyl polyglutamates react very slowly with this organism compared to mono- and diglutamates. As a result, it is required to hydrolyze the polyglutamate chain using γ-glutamylhydrolase prior to microbiological analysis (66). The monoglutamate in the food and feed extract is separated using anion-exchange column chromatography, followed by differential assay of the separated fraction by a microbiological assay.

A radioassay procedure has been developed to determine folic acid in erythrocyte and blood samples. The method is based on competitive protein binding between radiolabeled and unlabeled folate compounds for folic acid binding protein. A very sensitive, nonisotopic microtitration plate, folate-binding protein assay (FBPA) was developed to measure down to the 6 pg level of folyl monoglutamate and vitamin-active folate (67). A hplc method is useful for analysis of high potency premix samples containing folic acid, along with other water-soluble vitamins. Ion-pair reverse-phase columns using uv detection at 280 nm or post-column derivatization techniques have been employed to determine free folic acid. Details on hplc applications are available (68–70).

DEFICIENCY

Folic acid is a precursor of several important enzyme cofactors required for the synthesis of nucleic acids (qv) and the metabolism of certain amino acids. Folic acid deficiency results in an inability to produce deoxyribonucleic acid (DNA), ribonucleic acid (RNA), and certain proteins (qv). Megaloblastic anemia is a common symptom of folate deficiency owing to rapid red blood cell turnover and the high metabolic requirement of hematopoietic tissue. One of the clinical signs of acute folate deficiency includes a red and painful tongue. Vitamin B$_{12}$ and folate share a common metabolic pathway, the methionine synthase reaction. Therefore a differential diagnosis is required to measure folic acid deficiency because both folic acid and vitamin B$_{12}$ deficiency cause megaloblastic anemia. Serum and red blood cell levels of folate are measured to confirm and diagnose the deficiency. Serum folate levels less than 3 ng/mL are diagnostic of a deficiency. Some forms of dietary folic acid are more poorly absorbed by the elderly than by younger individuals. However, vitamin supplements containing the monoglutamate form are well absorbed in all age groups (71).

Folate antagonists (eg, methotrexate and certain antiepileptics) are used in treatment for various diseases, but their administration can lead to a functional folate deficiency. Folate utilization can be impaired by a depletion of

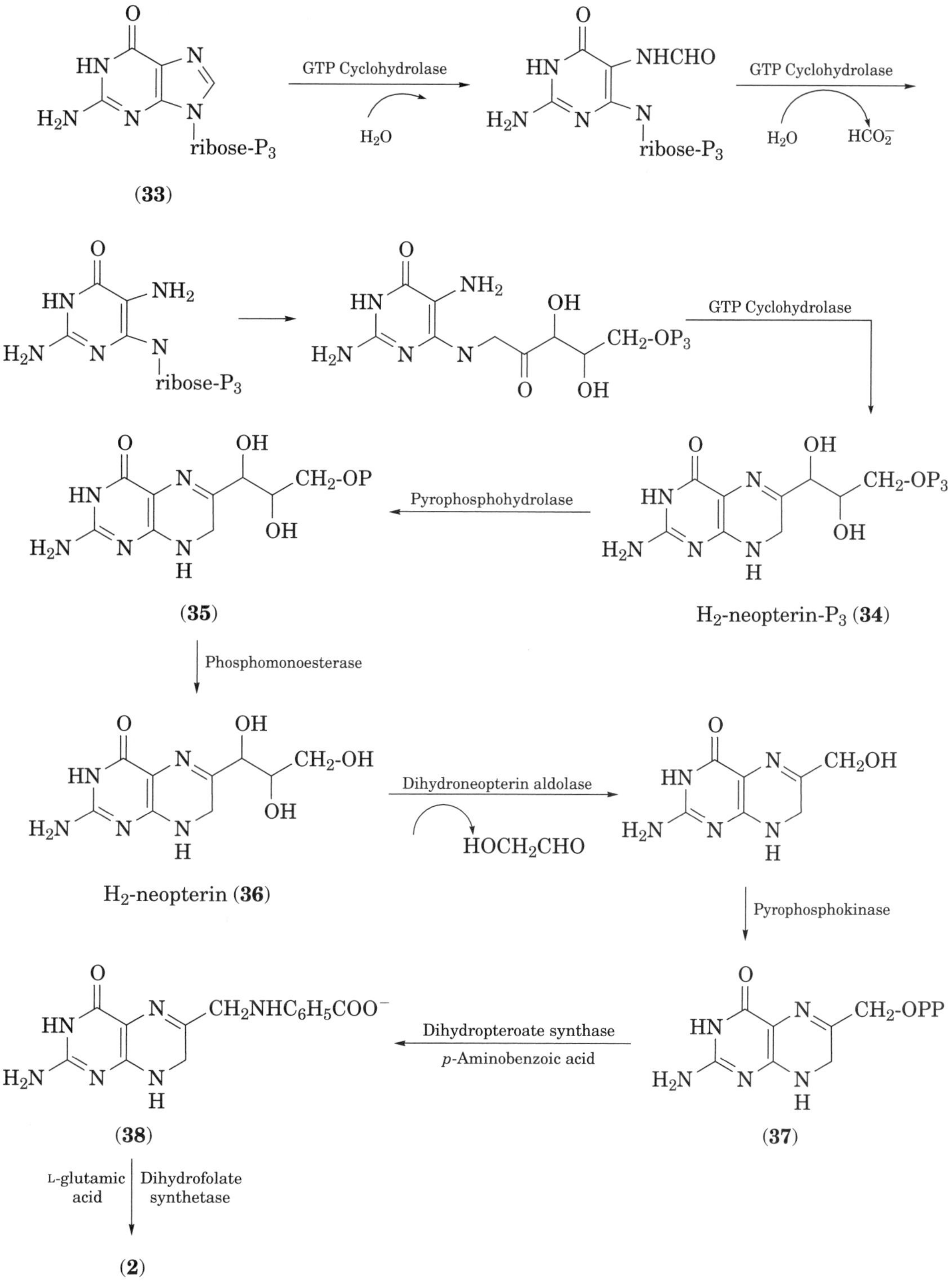

Scheme 1.

zinc. In humans, the intestinal brush border folate conjugase is a zinc metalloenzyme (72). One study indicates that the substantial consumption of alcohol, when combined with an inadequate intake of folate and methionine, may increase the risk of colon cancer (73). Based on this study, it is recommended to avoid excess alcohol consumption and increase folate intake to lower the risk of colon cancer.

Incomplete closure of neural tube during the embryo development in humans can lead to spina bifida. The condition is characterized by an opening in the spinal cord and results in physical disability in a child. Incomplete closure of the skull produces anencephaly. These and similar conditions are collectively called neutral tube defects (NTD). Each year in the United States approximately 2500 infants are born with spina bifida and anencephaly and an estimated 1500 fetuses affected by these birth defects are spontaneously aborted (74). It has been shown that folic acid given at 400 µg/day prevents the recurrence of NTD and that doses of 800 µg/day prevent both the occurrence and recurrence of NTD in the majority of cases. Published studies also indicate that folic acid supplementation taken six weeks before conception may reduce the risk of neural tube defects by at least 50% (74,75). Folic acid may be more effective in reducing neural tube defect incidence than conjugated food folate because free folic acid is more readily absorbed (76). A cost-benefit analysis, based on the U.S. population, of preventable neural tube defects indicates that folic acid fortification of grains in the United States may yield a substantial economic benefit (77). The U.S. Food and Drug Administration (FDA) has amended the standards of identity for several enriched grain products. The agency is requiring that these products be fortified with folic acid at levels ranging from 0.95 to 3.09 mg per kg of product (78). This is the first B-vitamin fortification requirement since 1943 when the U.S. government mandated fortification of flour with niacin, thiamin, and riboflavin. Incidence of cleft lip/cleft palate has been reported in animal studies due to folic acid deficiency. Poor folic acid status has been associated with megaloblastic changes in the cells of the uterine, cervix (79), and intestinal epithelium (80).

Homocysteine arises from dietary methionine. High levels of homocysteine (hyperhomocysteinemia) are a risk factor for occlusive vascular diseases including atherosclerosis and thrombosis (81–84). In a controlled study, serum folate concentrations of ≤9.2 nmol/L were linked with elevated levels of plasma homocysteine. Elevated homocysteine levels have been associated also with ischemic stroke (9). The mechanism by which high levels of homocysteine produce vascular damage are, as of yet, not completely understood. Interaction of homocysteine with platelets or endothelial cells has been proposed as a possible mechanism. Clinically, homocysteine levels can be lowered by administration of vitamin B_6, vitamin B_{12}, and folic acid.

REQUIREMENTS

The amount of folic acid required for daily intake is estimated based on the minimum amount required to maintain a certain level of serum folate. The recommended dietary allowance (RDA) for folic acid accounts for daily losses and makes allowances for variation in individual needs and bioavailability from food sources (85). The U.S. recommended daily allowance for adults is 400 µg and for pregnant women is 800 µg (Table 4).

ANIMAL NUTRITION

To obtain optimal performance of farm animals, folic acid supplementation is required (86) and as is the case with most of the vitamins, the majority of worldwide consumption is as feed supplements. The folic acid requirement for chickens and pigs is about 0.2–0.5 mg of folic acid/kg diet

Table 4. RDA and U.S. RDA for Folic Acid

Group	Age	Folic acid (µg)
RDA		
Infants	0–0.5	25
	0.5–1.0	35
Children	1–3	50
	4–6	75
	7–10	100
Males	11–14	150
	>15–51	200
Females	11–14	150
	15–51	180
Pregnant, lactating		400
1st six months		280
2nd mix months		260
U.S. RDA		
Infants <13 months		100
Children <4 yr		200
Adults and children >4 yr		400
Pregnant or lactating women		800

Source: Ref. 21.

and 0.3 mg/kg diet, respectively. Increased amounts, 0.5–1.0 mg/kg feed for chickens and 0.5–2.0 mg/kg for swine, are recommended under commercial production conditions (87). The degree of intestinal folic acid synthesis and the utilization by the animal dictates the folic acid requirements for monogastric species. Also, the self-synthesis of folacin is dependent on dietary composition (88).

Folacin requirements are related to the type and level of production. The more rapid the growth or production rates, the greater the need for folacin owing to its role in DNA synthesis. In poultry, the requirement for egg hatchability is higher than for production (88). In swine, folic acid supplementation has been shown to increase fertility and growth rates (89).

METABOLISM

The principal function of the folate coenzyme is to carry one-carbon units. Dietary folylpolyglutamate is hydrolyzed to the monoglutamate form by folyl polyglutamate hydrolyases (conjugases) prior to transport across the intestinal mucosa. Intestinal folate absorption occurs in the jejunum. A slightly acidic pH of 6 is optimum for intestinal absorption and transportation into the blood stream (90). It was shown by the competitive inhibition method (91) that a single protein seems to be responsible for transportation of the monoglutamate. Transportation may occur by an anion-exchange mechanism (92). Folate transportation may be partially dependent on sodium ions in the cell medium (93).

Three forms of folate appear to be transported in the blood: folic acid, folate loosely bound to low affinity binder serum proteins (such as albumin, α-macroglobulin, and transferrin), and folate bound to high affinity protein binders. Approximately 5% of total serum folate is being transported by high affinity protein binders but the function of these proteins is not well understood (94,95). Folic acid is stored in folyl polyglutamate form. The liver and other tissues (mitochondria) convert folic acid and methyltetrahydrofolate into folylpolyglutamate by employing polyglutamate synthetase. The total folate pool in adult humans is estimated to be around 5–10 mg, with half of this in the liver; folylpolyglutamate synthetase activity is highest in liver.

The metabolism of folic acid involves reduction of the pterin ring to different forms of tetrahydrofolylglutamate. The reduction is catalyzed by dihydrofolate reductase and NADPH functions as a hydrogen donor. The metabolic roles of the folate coenzymes are to serve as acceptors or donors of one-carbon units in a variety of reactions. These one-carbon units exist in different oxidation states and include methanol, formaldehyde, and formate. The resulting tetrahydrofolylglutamate is an enzyme cofactor in amino acid metabolism and in the biosynthesis of purine and pyrimidines (10,96). The one-carbon unit is attached at either the N-5 or N-10 position. The activated one-carbon unit of 5,10-methylene-H_4 folate (5) is a substrate of T-synthase, an important enzyme of growing cells. 5-10-Methylene-H_4 folate (5) is reduced to 5-methyl-H_4 folate (4) and is used in methionine biosynthesis. Alternatively, it can be oxidized to 10-formyl-H_4 folate (7) for use in the purine biosynthetic pathway.

TOXICITY

Folic acid is safe, even at levels of daily oral supplementation up to 5–10 mg (97). Gastrointestinal upset and an altered sleep pattern have been reported at 15 mg/day (98). A high intake of folic acid can mask the clinical signs of pernicious anemia which results from vitamin B_{12} deficiency and recurrence of epilepsy in epileptics treated with drugs with antifolate activity (99). The acute toxicity (LD_{50}) is approximately 500 and 600 mg per kg body weight for rats and mice, respectively (100).

USES

L-Folic acid is available as a crystalline dihydrate containing 8% water. Approximately 80% of the commercial production is consumed for feed enrichment in animal nutrition. Folic acid is being offered by the pharmaceutical industry for therapeutic and prophylactic use. Pharmacological doses of folic acid are commonly used as a rescue dose during cancer chemotherapy, in women using oral contraceptives, and alcoholics. Several studies have provided evidence that multivitamins or folic acid (0.8–4 mg/day) supplementation prevent the majority of neural tube defects (101).

ECONOMIC ASPECTS

The world production of synthetic L-folic acid 1996 was estimated at 400 metric tons per year. The total market is expected to grow with increasing recognition of need, especially during pregnancy and lactation. The principal producers of folic acid are Hoffmann-La Roche, Takeda, Sumika Fine Chemical (previously Yodogawa Pharmaceuticals), Kongo, and three Chinese companies, Changzhou Pharmaceuticals, Changshu Hugang Pharmaceuticals, and Zhejiang Jiangnan Pharmaceuticals. Smaller quantities are also produced by companies in India, China, and Russia. The 1996 sale price varied between $50 to $130/kg.

CONCLUSIONS

All known commercial syntheses of folic acid are based on the three-component process developed in the late 1940s. Industrial production of folic acid by genetically engineered microorganisms or extraction from natural sources is not yet economically viable. The mechanism governing folate turnover and excretion is still poorly understood. Improved research techniques will aid in the development of a better understanding of factors affecting intestinal absorption and *in vivo* kinetics in human beings. Folic acid and multivitamin supplement use are associated with a decreased occurrence of neural tube defects. Abnormalities in homocysteine metabolism are observed in many women who have given birth to children with neural tube defects and in individuals with cardiovascular disease. Folic acid is likely to be involved in overcoming this abnormality. The exact mechanism by which folic acid prevents these diseases is a current active area of research.

BIBLIOGRAPHY

1. L. Wills, P. W. Clutterbuck, and P. D. F. Evans, *Biochem. J.* **31**, 2136 (1937).
2. A. G. Hogan and E. M. Parrott, *J. Biol. Chem.* **132**, 507 (1940).
3. E. E. Snell and W. H. Peterson, *J. Bacteriol.* **39**, 273 (1940).
4. H. K. Mitchell, E. E. Snell, and R. J. Williams, *J. Am. Chem. Soc.* **63**, 2284 (1941).
5. S. B. Binkley, O. D. Bird, E. S. Bloom, R. A. Brown, D. G. Calkins, C. J. Campbell, A. D. Emmett, and J. J. Pfiffner, *Science* **100**, 36 (1944).
6. J. J. Pfiffner, D. G. Calkins, E. S. Bloom, and L. B. O'Dell, *J. Am. Chem. Soc.* **68**, 1392 (1946).
7. R. B. Angier et al., *Science* **103**, 667 (1946).
8. A. D. Welch, *Perspect. Biol. Med.* **27**, 64 (1983).
9. W. H. Giles, S. J. Kittner, R. F. Anda, J. B. Croft, and M. L. Casper, *Stroke* **26**, 1166 (1995).
10. W. Friedrich, *Handbuch der vitamine*, Urban & Schwarzenberg, Baltimore, Md., 1987, p. 398.
11. B. Botticher and R. Kluthe, in K. Pietrzik, ed., *Folsäure-Mangel*, W. Zuckschwerdt Verlag, Germany, 1987, p. 15.
12. T. Brody, in L. J. Machlin, ed., *Handbook of Vitamins*, Marcel Dekker Inc., New York, 1991, p. 453.
13. L. B. Bailey, in L. B. Bailey, ed., *Folate in Health and Disease*, Marcel Dekker Inc., New York, 1995, p. 123.
14. K. H. Bässler, E. Grühn, D. Loew, and K. Pietrzik, *Vitamin-Lexikon*, G. Fischer Verlag, New York, 1992, p. 127.
15. J. Leichter, A. F. Landymore, and C. L. Krumdieck, *Am. J. Clin. Nutr.* **32**, 92 (1979).

16. S. P. Rothenberg, M. P. Iqbal, and M. Da Costa, *Anal. Biochem.* **103**, 152 (1980).

17. J. F. Gregory, *Adv. Food Nutr. Res.* **33**, 1 (1989).

18. J. F. Gregory, *Food Technol.* **42**, 230 (1988).

19. H. E. Sauberlich, M. J. Kretsch, J. H. Skala, H. L. Johnson, and P. C. Taylor, *Am. J. Clin. Nutr.* **46**, 1016 (1987); J. F. Gregory, in L. B. Bailey, ed., *Folate in Health and Disease*, Marcel Dekker Inc., 1995, p. 195.

20. IUPAC-IUB Joint commission on biochemical nomenclature (JCBN), *Eur. J. Biochem.* **168**, 251 (1987); *Pure Appl. Chem.* **59**, 834 (1987).

21. D. Bhatia, in *Encyclopedia of Food Science and Technology*, John Wiley & Sons, Inc., New York, 1991, p. 2770.

22. C. W. Waller et al., *J. Am. Chem. Soc.* **70**, 19 (1948).

23. F. Weygand and V. Schmied-Kowarzik, *Chem. Ber.* **82**, 333 (1949).

24. S. Uyeo, S. Mizukami, T. Kubota, and S. Takagi, *J. Am. Chem. Soc.* **72**, 5339 (1950).

25. Jpn. Pat. 7405995 (1974), K. Ito, H. Fukushima, and K. Nakagawa (to Nisshin Flour Milling Co., Ltd.).

26. W. Traube, *Berliner Berichte* **33**, 1371 (1900).

27. Ger. Pat. 3403468 A1 (Aug. 8, 1985), H. Blum and G. Dreesmann (to Federal Republic of Germany).

28. Eur. Pat. 444266 A1 (Sept. 4, 1991), A. Hunds and W. Rogler (to Huls AG).

29. Jpn. Pat. 59204158 (1974) (to Daicel Chem Ind., KK).

30. Ger. Pat. 3419817 A1 (Dec. 6, 1984), K. Yoshida, T. Niinobe, and T. Baba (to Takeda Chemical Industries Ltd.).

31. Ger. Pat. 3605484A (Aug. 8, 1987), W. Deinhammer and H. Petersen (to Walker-Chemie GmbH); Eur. Pat. 394968 A (Oct. 31, 1990), B. D. Dombek and T. T. Wenzel (to Union Carbide Chemical).

32. J. R. Piper, G. S. McCaleb, and J. A. Montgomery, *J. Heterocycl. Chem.* **24**, 279 (1987).

33. C. M. Baugh and E. Shaw, *J. Org. Chem.* **29**, 3610 (1964).

34. Ger. Pat. 2741383 (Feb. 22, 1979), E. Catalucci (to Lonza AG).

35. J. R. Piper and J. A. Montgomery, *J. Org. Chem.* **42**, 208 (1977).

36. J. R. Piper and J. A. Montgomery, *J. Heterocycl. Chem.* **11**, 279 (1974); U.S. Pat. 4,077,957A (Mar. 7, 1978), J. A. Montgomery and J. R. Piper (to U.S. Sec. Dept. Health); U.S. Pat. 4,079,056 (Mar. 13, 1978), J. A. Montgomery and J. R. Piper (to U.S. Sec. Dept. Health).

37. J. H. Bieri and M. Viscontini, *Helv. Chim. Acta* **56**, 2905 (1973); K. Khalifa, P. K. Sengupta, J. H. Bieri, and M. Viscontini, *Helv. Chim. Acta* **59**, 242 (1976).

38. M. Sletzinger, D. Reinhold, J. Grier, M. Beachem, and M. Tishler, *J. Am. Chem. Soc.* **77**, 6365 (1955).

39. M. Viscontini and J. H. Bieri, *Helv. Chim. Acta* **54**, 2291 (1971).

40. Swiss Pat. 255409 (1949) (to Hoffmann-La Roche Inc.).

41. Ger. Pat. 3242193 A1 (May 17, 1984), F. Brunnmueller and M. Kroener (to BASF AG).

42. E. C. Taylor and T. Kobayashi, *J. Org. Chem.* **38**, 2817 (1973); E. C. Taylor, R. N. Henrie, and R. C. Portnoy, *J. Org. Chem.* **43**, 736 (1978).

43. B. Schirks, J. H. Bieri, and M. Viscontini, *Helv. Chim. Acta* **68**, 1639 (1985).

44. E. C. Taylor and K. Lenard, *Liebigs. Ann. Chem.* **726**, 100 (1969); E. C. Taylor and D. J. Dumas, *J. Org. Chem.* **46**, 1394 (1981).

45. Ger. Pat. 3242195 A1 (May 17, 1984), F. Brunnmueller and M. Kroener (to BASF AG); Eur. Pat. 175263 A2 (Mar. 26, 1986), H. Leninger, W. Littmann, J. Paust, and W. Trautmann (to BASF AG); Eur. Pat. 175264 A2 (Mar. 26, 1986), H. Leininger, W. Littmann, and J. Paust (to BASF AG).

46. Eur. Pat. Appl. 608 693 A2 (Aug. 3, 1994), C. Wehrli (to Hoffmann-La Roche AG).

47. J. F. Gregory and J. P. Toth, *J. Labelled Comp. Radiopharm.* **25**, 1349 (1988).

48. J. F. Gregory, *J. Agric. Food Chem.* **38**, 1073 (1990).

49. D. L. Hachey, L. Palladino, J. A. Blair, I. H. Rosenberg, and P. D. Klein, *J. Labelled Comp. Radiopharm.* **14**, 479 (1978).

50. J. F. Gregory and J. P. Toth, *Anal. Biochem.* **170**, 94 (1988).

51. P. A. Charlton, D. W. Young, B. Birdsall, J. Feeney, and G. C. K. Roberts, *J. Chem. Soc. Perkin Trans. I*, 1349 (1985).

52. C. M. Tatum, M. G. Fernald, and J. P. Schimel, *Anal. Biochem.* **103**, 255 (1980).

53. J. A. Blair and K. J. Saunders, *Anal. Biochem.* **34**, 376 (1970).

54. C. Zarow, A. M. Pellino, and P. V. Danenberg, *Prep. Biochem.* **12**, 381 (1983).

55. E. Khalifa, A. N. Ganguly, J. H. Bieri, and M. Viscontini, *Helv. Chim. Acta* **63**, 2554 (1980).

56. D. B. Cosulich, J. M. Smith, and H. P. Broquist, *J. Am. Chem. Soc.* **74**, 4215 (1952).

57. J. E. Baggott and C. L. Krumdieck, *Biochemistry* **18**, 1036 (1979).

58. C. M. Baugh, J. Stevens, and C. Krumdieck, *Biochim. Biophys. Acta* **212**, 116 (1970).

59. B. L. Hutchings et al., *J. Biol. Chem.* **170**, 323 (1947).

60. J. D. Cook, D. J. Cichowicz, S. George, A. Lawler, and B. Shane, *Biochemistry* **26**, 530 (1987).

61. M. C. Li, R. Hertz, and D. M. Bergenstal, *New Engl. J. Med.* **259**, 66 (1958).

62. J. Burchanall and G. Hitchings, *Mol. Pharmacol.* **1**, 216 (1965).

63. G. M. Brown and H. Williamson, *Adv. Enzymol. Relat. Areas Mol. Biol.* **53**, 345 (1982).

64. G. M. Brown and H. Williamson, in F. C. Neidhart, ed., *Escherichia Coli and Salmonella Typhmurium*, Vol. 1, American Society for Microbiology, Washington, D.C., 1987, p. 521.

65. Y. Suzuki and G. M. Brown, *J. Biol. Chem.* **249**, 2405 (1974).

66. J. Kas and J. Cerna, *Methods Enzymol.* **66E**, 443 (1980).

67. P. M. Finglas, R. M. Faulks, and M. R. A. Morgan, *J. Micronutrient Anal.* **4**, 295 (1988).

68. C. L. Krumdieck, T. Tamura, and I. Eto, *Vitamins Horm.* **40**, 45 (1983).

69. G. Brubacher, W. Müller-Mulot, and D. A. T. Southgate, *Methods for Determination of Vitamins in Food*, Elsevier Applied Science Publisher, New York, 1985, p. 158.

70. D. B. McCormick and L. D. Wright, ed., *Methods Enzymol.* **66E**, 429 (1980).

71. H. Baker, S. P. Jaslow, and O. Frank, *J. Am. Geriatr. Soc.* **26**, 218 (1978).

72. S. A. Anderson and J. M. Talbot, *FDA Technical Report* FDA/RF-82/13, Washington, D.C., 1981.

73. E. Giovannucci, E. B. Rimm, A. Ascherio, M. J. Stampfer, G. A. Colditz, and W. C. Willett, *J. Natl. Cancer Inst.* **87**, 265 (1995).

74. *MMWR Morb Mortal Wkly Rep.* **44**, 716 (1995).

75. A. Milunsky, J. Herschel, S. S. Jick, C. L. Bruell, D. S. Maclaughlin, K. J. Rothman, and W. Willett, *J. Am. Med. Assoc.* **262**, 2847 (1989).

76. C. Bower and J. F. Stanley, *Med. J. Austral.* **150**, 613 (1989).

77. P. S. Romano, N. J. Waitzman, R. M. Scheffler, and R. D. Pi, *Am. J. Public Health* **85**, 667 (1995).

78. B. F. Satchell, *FDA Technical Report*, FDA/61RF-8781, Washington, D.C., 1996.

79. N. Whitehead, F. Reyner, and J. Lindenbaum, *J. Am. Med. Assoc.* **226**, 1421 (1973).

80. A. Bianchi, D. W. Chipman, A. Dreskin, and N. S. Rosensweig, *New Engl. J. Med.* **282**, 859 (1970).

81. K. S. McCully, *Am. J. Pathol.* **56**, 111 (1969).

82. C. S. Berwanger, J. Y. Jeremy, and G. Stansby, *Br. J. Surg.* **82**, 726 (1995).

83. J. Selhub, P. F. Jacques, A. G. Bostom, R. B. D'Agostino, P. W. Wilson, A. J. Belanger, D. H. O'Leary, P. A. Wolf, E. J. Schaefer, and I. H. Rosenberg, *N. Engl. J. Med.* **332**, 286 (1995).

84. J. L. Mills, J. M. McPartlin, P. N. Kirke, Y. J. Lee, M. R. Conley, D. G. Weir, and J. M. Scott, *Lancet* **345**, 149 (1995).

85. A. F. Subar, G. Block, and L. D. James, *Am. J. Clin. Nutr.* **50**, 508 (1989).

86. *Nutrient Requirements of Poultry*, 9th. ed., National Research Council, National Academy Press, Washington, D.C., 1994; *Nutrient Requirements of Swine*, 9th ed., National Research Council, National Academy Press, Washington, D.C., 1988.

87. L. R. McDowell, *Vitamins in Animal Nutrition; Comparative Aspects to Human Nutrition*, Academic Press, Inc., San Diego, CA, 1989, pp. 298–325.

88. *Vitamin Nutrition for Poultry*, Dept. of Animal Health and Nutrition, Hoffmann-La Roche, Inc., Nutley, N.J., 1991.

89. *Vitamin Nutrition for Swine*, Dept. of Animal Health and Nutrition, Hoffmann-La Roche, Inc., Nutley, N.J., 1991.

90. J. Zimmerman, Z. Gihula, J. Selhub, and I. H. Rosenberg, *Int. J. Vit. Nutr. Res.* **59**, 151 (1989).

91. J. Zimmerman, J. Selhub, and I. H. Rosenberg, *Am. J. Clin. Nutr.* **46**, 518 (1987).

92. C. H. Young, F. M. Sirtnak, and M. Dembo, *J. Membrane Biol.* **79**, 285 (1984).

93. D. W. Horne, W. Y. Briggs, and C. Wagner, *J. Biol. Chem.* **253**, 3529 (1978).

94. G. B. Henderson, *Ann. Rev. Nutr.* **10**, 319 (1990).

95. C. Wagner, *Ann. Rev. Nutr.* **2**, 229 (1982).

96. C. Wagner, in L. B. Bailey, ed., *Folate in Health and Disease*, Marcel Dekker, Inc., New York, 1995, p. 23.

97. C. E. Butterworth and T. Tamura, *Am. J. Clin. Nutr.* **50**, 353 (1989).

98. R. Hunter, J. Barnes, H. F. Oakeley, and D. M. Matthews, *Lancet* **1**, 61 (1970).

99. *New Engl. J. Med.* **237**, 713 (1947).

100. A. Hanck, in *Spektrum Vitamine* **42**, 81 (1986).

101. A. F. Czeisal and J. Dudas, *New Engl. J. Med.* **327**, 1832 (1992); MRC Vitamin Study Research Group, *Lancet* **338**, 131 (1991).

GENERAL REFERENCES

J. F. Gregory, in L. B. Bailey, ed., *Folate in Health and Disease*, Marcel Dekker Inc., New York, 1995, p. 195.

T. Brody, in L. J. Machlin, ed., *Handbook of Vitamins*, Marcel Dekker Inc., New York, 1991, p. 453.

O. Isler and G. Brubacher, in O. Isler, G. Brubacher, S. Ghisla, and B. Kräutler, eds., *Vitamine II*, Georg Thieme Verlag, New York, 1982, p. 264.

S. K. Gaby and A. Bendich, in S. K. Gaby, A. Bendich, V. N. Singh, and L. J. Machlin, eds., *Vitamin Intake and Health*, Marcel Dekker Inc., New York, 1991, p. 175.

THIMMA R. RAWALPALLY
Hoffmann-La Roche, Inc.

VITAMINS: NIACIN, NICOTINAMIDE, AND NICOTINIC ACID

3-Pyridine carboxamide [*98-92-0*] (nicotinamide) (**1**) and 3-pyridine carboxylic acid [*59-67-6*] (nicotinic acid) (**2**) have a rich history and their early significance stems not from their importance as a vitamin but rather as products derived from the oxidation of nicotine. In 1867, Huber prepared nicotinic acid from the potassium dichromate oxidation of nicotine. Many years later, Engler prepared nicotinamide. Workers at the turn of the twentieth century isolated nicotinic acid from several natural sources. In 1894, Suzuki isolated nicotinic acid from rice bran, and in 1912 Funk isolated the same substance from yeast (1).

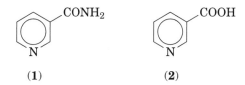

(**1**)   (**2**)

In 1913, Goldberger demonstrated that pellagra was due to a dietary deficiency. Pellagra had been earlier described by Thiery, who had coined the term *mal de la rosa* for this disease. Several decades later, Elvehjem and co-workers isolated nicotinamide from a liver extract and identified it as a pellagra-preventing factor (1).

There is considerable confusion regarding the nomenclature of these simple substances. The term niacin is a generic descriptor for pyridine 3-carboxylic acid and derivatives that exhibit the biological activity of nicotinic acid (1). However, niacin and vitamin PP (pellagra preventing) are frequently used interchangeably with nicotinic acid. Vitamin B_3 is often used as a designation for nicotinamide (1). The following is a list of common names for these substances.

Nicotinamide	Nicotinic acid
Aminicotin	Akotin
Benicot	Apelagrin
Delonin amide	Daskil
NAM	Efacin
Niacinamide	Linic
Niavit PP	Niacin
Nicasir	Nicacid
Nicosan 2	Nicangin
Nicovit	Nicolar
Pelmine	Pellagrin
Pelonin amide	Pelonin
Vitamin B	SR 4390
Vitamin B_3	

The biological importance of these compounds stems from their use as cofactors. Both nicotinamide and nicotinic acid are building blocks for coenzyme I (Co I), nicotinamide—adenine dinucleotide (NAD) (3) and coenzyme II (Co II), nicotinamide—adenine dinucleotide phosphate (NADP) (4) (2).

(3)   (4)

CHEMICAL AND PHYSICAL PROPERTIES

Nicotinamide is a colorless, crystalline solid. It is very soluble in water (1 g is soluble in 1 mL of water) and in 95% ethanol (1 g is soluble in 1.5 mL of solvent). The compound is soluble in butanol, amyl alcohol, ethylene glycol, acetone, and chloroform, but is only slightly soluble in ether or benzene. Physical properties are listed in Table 1.

Nicotinic acid is an amphoteric solid with needle-shape crystals. It is less soluble than nicotinamide and its poor solubility in diethyl ether can be used as a basis to separate these compounds. Because nicotinamide has some solubility in ether, extraction of aqueous solutions of the acid and the amide with ether allows for selective extraction of the amide into the organic phase. Table 2 lists the physical properties of this vitamin.

The ring nitrogen of both the amide and the carboxylic acid can be quaternized and oxidized. *N*-methylnicotinate (trigonelline) (5) is an important component of green coffee beans and is converted to nicotinic acid during the roasting process. The reactivity of the 3-substituent parallels standard functional groups. Acid chlorides, esters, amides, and anhydrides have been prepared from nicotinic acid. Both the corresponding aldehyde and alcohol are available from the acid with a variety of reducing agents. Nicotinamide can be converted to nicotinic acid esters, the nitrile and acylamidines by routine methods.

Table 1. Physical Properties of Nicotinamide

Property	Value
Molecular weight	122.12
Melting point, °C	
Stable modification	129–132
Unstable modifications	
I	105
II	110
III	111
IV	113
V	116
Boiling point, °C (0.067 Pa)	150–160
Sublimation range, °C	80–100
True dissociation constants in water, at 20°C	
K_{b1}	2.24×10^{-11}
K_{b2}	3.16×10^{-14}
Specific heat, kJ/(kg·K)^a	
Solid, 55°C	1.30
65°C	1.34
75°C	1.39
Liquid, 135°C	2.18
Heat of solution in water, kJ/kg^a	-148
Heat of fusion, kJ/kg^a	381
Density of melt, at 150°C, g/cm^3	1.19

^{a}To convert J to cal, multiply by 4.184.

Table 2. Physical Properties of Nicotinic Acid

Property	Value
Molecular weight	123.11
Melting point, °C	236–237
Sublimation range	≥150
Density of crystals, g/cm^3	1.473
Dissociation constants in water, at 25°C	
K_a	1.50×10^{-5}
K_b	1.04×10^{-12}
Isoelectric point in water, at 25°C, pH	3.42
pH of saturated aqueous solution	2.7
Solubility, g/L	
Water	
0°C	8.6
38°C	24.7
100°C	97.6
Ethanol, 96%	
0°C	5.7
78°C	76.0
Methanol	
0°C	63.0
62°C	345.0

(5)

SYNTHESIS

Key intermediates in the industrial preparation of both nicotinamide and nicotinic acid are alkyl pyridines (Fig. 1).

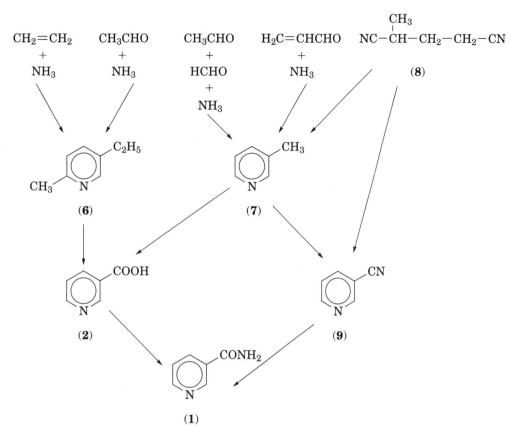

Figure 1. Alkyl pyridines.

2-Methyl-5-ethylpyridine (**6**) is prepared in a liquid-phase process from acetaldehyde. Also, a synthesis starting from ethylene has been reported. Alternatively, 3-methylpyridine (**7**) can be used as starting material for the synthesis of nicotinamide and nicotinic acid and it is derived industrially from acetaldehyde, formaldehyde (qv), and ammonia. Pyridine is the principal product from this route and 3-methylpyridine is obtained as a by-product. Despite this and largely due to the large amount of pyridine produced by this technology, the majority of the 3-methylpyridine feedstock is prepared in this fashion.

3-Methylpyridine can also be prepared by the condensation of acrolein and ammonia. Yields of 40–50% are obtained with pyridine as a by-product. Higher yields have been claimed when both propionaldehyde and acrolein have been used. A recent U.S. patent claims better selectivity if the cyclization is carried out in the presence of zeolites (3).

In an alternative approach, 2-methylglutaronitrile (**8**) is hydrogenated and cyclized to give high yields of 3-methylpyridine. The feedstock for this process is produced as a by-product of the production of adiponitrile. Oxidative cyclization of 2-methylglutaronitrile to nicotinonitrile (**9**) has been described (4).

The alkyl pyridines (**6**) and (**7**) can be transformed either to nicotinic acid or nicotinonitrile. In the case of nicotinic acid, these transformations can occur by either chemical or biological means. From an industrial stand-

point, the majority of nicotinic acid is produced by the nitric acid oxidation of 2-methyl-5-ethylpyridine. Although not of industrial significance, the air oxidation has also been reported. Isocinchomeronic acid (**10**) (Fig. 2) is formed as an intermediate.

Although an inherently more efficient process, the direct chemical oxidation of 3-methylpyridine does not have the same commercial significance as the oxidation of 2-methyl-5-ethylpyridine. Liquid-phase oxidation procedures are typically used (5). A Japanese patent describes a procedure that uses no solvent and avoids the use of acetic acid (6). In this procedure, 3-methylpyridine is combined with cobalt acetate, manganese acetate and aqueous hydrobromic acid in an autoclave. The mixture is pressurized to 101.3 kPa (100 atm) with air and allowed to react at 210°C. At a 32% conversion of the picoline, 19% of the acid was obtained. Electrochemical methods have also been described (7).

In more recent work, several groups have reported on fermentative approaches to nicotinic acids from 3-picoline. *Rhodococcus* (8), *Acinetobactorr* (9), and *Pseudomonas* (10) have found utility in this application.

Nicotinonitrile is produced by ammoxidation of alkylpyridines (11–24). A wide variety of different catalysts have been developed for this application. For example, a recent patent describes a process in which 3-methylpyridine is reacted over a molybdenum catalyst supported on silica gel. The catalyst ($PV_4Mo_{12}O_x$) was prepared from

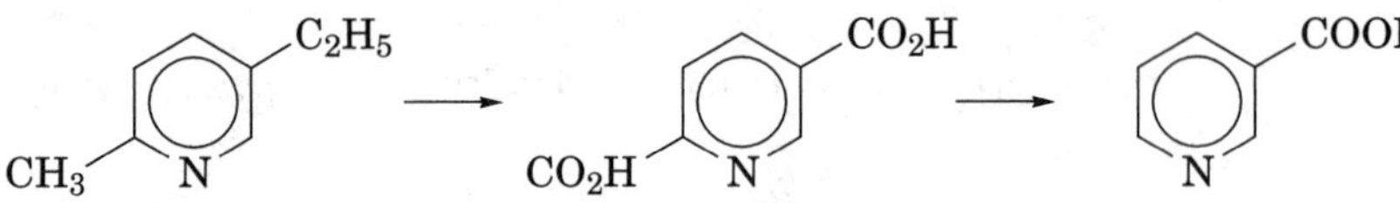

Figure 2. Formation of isocinchomeronic acid (**10**).

NH₄VO₃, H₃PO₄, and (NH₄)₅Mo₇O₂₄. Reaction at 380°C at a residence time of 2.5 seconds gave 95% of nicotinonitrile at a 99% conversion (16).

Conversion of the nitrile to the amide has been achieved by both chemical and biological means. Several patents have described the use of modified Raney nickel catalysts in this application (25,26). Also, alkali metal perborates have demonstrated their utility (27). Typically, the hydrolysis is conducted in the presence of sodium hydroxide (28–31). Owing to the fact that the rate of hydrolysis of the nitrile to the amide is fast as compared to the hydrolysis of the amide to the acid, good yields of the amide are obtained. Other catalysts such as magnesium oxide (32), ammonia (28,29,33), and manganese dioxide (34) have also been employed.

Since the late 1980s, there has been considerable activity in the development of fermentative approaches for the preparation of the amide from the nitrile. Organisms such as *Achromobacter* (35), *Agrobacterium* (36), *Streptomyces* (37), *Rhodococcus* (38,39), and *Cornebacterium* (40) have been described. Purified enzymes in either free (41) or immobilized form (42,43) have also been used in this application.

Nicotinic acid has also been produced by microbial means from the nitrile. *Rhodococcus* (44,45) has frequently been used in this regard. Interestingly, irradiation of a *Corynebacterium* suspension during the fermentation led to higher yields of nicotinic acid (46).

BIOSYNTHESIS

A significant pathway for the formation of nicotinic acid mononucleotides begins with tryptophan (**11**). In the first step, cleavage of the indole ring is catalyzed by the enzyme L-tryptophan-2,3-dioxygenase. This step is rate determining unless it is hormonally induced by glucocortocoids and even tryptophan itself. The resulting *N*'-formylkynurenine (**12**) is deformylated to the free aniline derivative, kynurenine (**13**). Hydroxylation to yield (**14**) is followed by oxidative removal of alanine to form 3-hydroxyanthranilic acid (**15**). Oxidative cleavage of the aromatic ring to the semialdehyde (**16**) is followed by dehydration to the important metabolite, quinolinic acid (**17**). Quinolinic acid is decarboxylated with concomitant alkylation at nitrogen to yield nicotinic acid mononucleotide (**18**) (Fig. 3).

The result of this biosynthesis is that the product is nicotinic acid mononucleotide rather than free nicotinic acid. Ingested nicotinic acid is converted to nicotinic acid mononucleotide which, in turn, is converted to nicotinic acid adenine dinucleotide. Nicotinic acid adenine dinucleotide is then converted to nicotinamide adenine dinucleotide. If excess nicotinic acid is ingested, it is metabolized into a series

of detoxification products (Fig. 4). Physiological metabolites include *N*-methylnicotinamide (**19**) and *N*-methyl-6-pyridone-2-carboxamide (**24**) (1).

Nicotinamide is incorporated into NAD and nicotinamide is the primary circulating form of the vitamin. NAD has two degradative routes: by pyrophosphatase to form AMP and nicotinamide mononucleotide and by hydrolysis to yield nicotinamide adenosine diphosphate ribose.

ANALYTICAL METHODS

Both nicotinic acid and nicotinamide have been assayed by chemical and biological methods. Owing to the fact that niacin is found in many different forms in nature, it is important to indicate the specific analyte in question. For example, if biological assay procedures are used, it is necessary to indicate whether the analysis is to determine the quantity of nicotinic acid or if niacin activity is the desired result of the analysis. If nicotinic acid is desired, then a method specific for nicotinic acid should be used. If quantitation of niacin activity is the desired outcome, then all compounds (bound and unbound) which behave like niacin will assay biologically for this substance (1).

The König reaction (Fig. 5) has been used to determine the amount of nicotinic acid and niacinamide. In this procedure, quaternization of the pyridine nucleus by cyanogen bromide is followed by ring opening to generate the putative dialdehyde intermediate. Reaction of this compound with an appropriate base, such as *p*-methylaminophenol sulfate (47) or sulfanilic acid (48), generates a dye. The concentration of this dye is determined colorimetrically.

In the case of nicotinamide, the color yield is often low. This problem can be circumvented by either hydrolysis to nicotinic acid or by conversion of the amide to a fluorescent compound. Treatment of nicotinamide with methyl iodide yields the quaternary ammonium salt, *N*-methyl nicotinamide (**5**). Reaction of this compound with acetophenone yields a fluorescent adduct (49). Other carbonyl compounds have also been used (50–54).

For more specific analysis, chromatographic methods have been developed. Using reverse-phase columns and uv detection, hplc methods have been applied to the analysis of nicotinic acid and nicotinamide in biological fluids such as blood and urine and in foods such as coffee and meat. Derivatization techniques have also been employed to improve sensitivity (55). For example, the reaction of nicotinic amide with DCCI (*N*'-dicyclohexyl-*O*-methoxycoumarin-4-yl)methyl isourea to yield the fluorescent coumarin ester has been reported (56). After separation on a reversed-phase column, detection limits of 10 pmol for nicotinic acid have been reported (57).

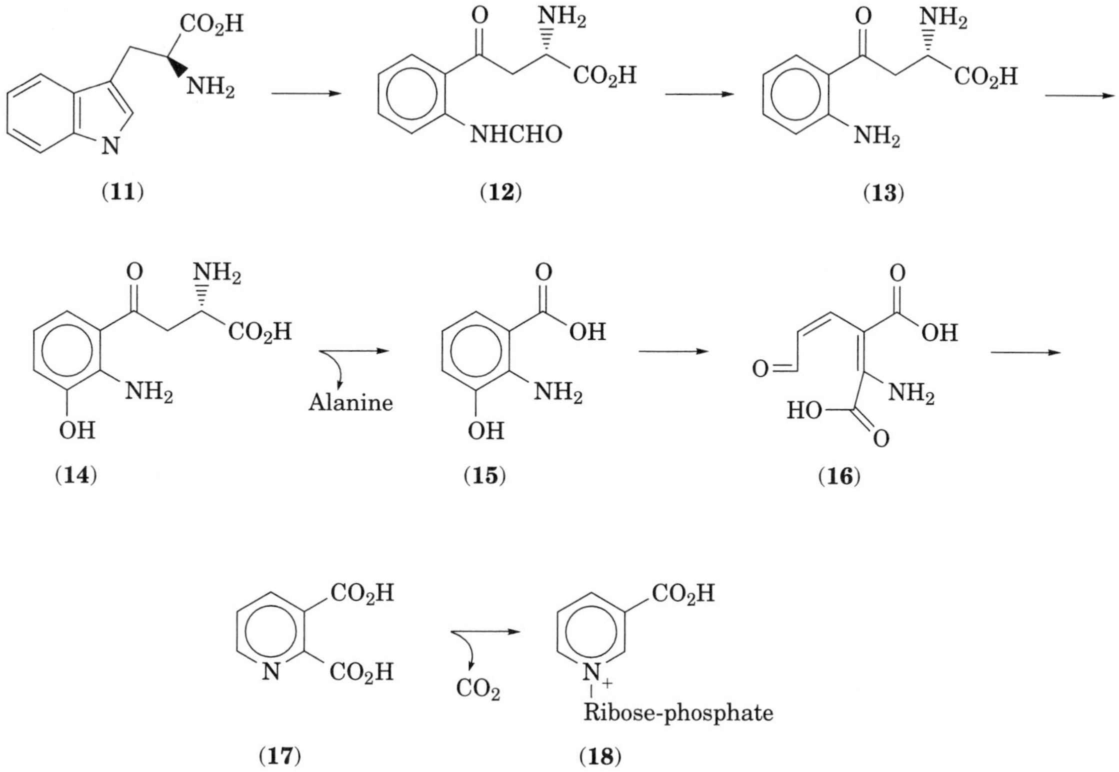

Figure 3. Formation of nicotinic acid mononucleotides.

Owing to poor volatility, derivatization of nicotinic acid and nicotinamide are important techniques in the gc analysis of these substances. For example, a gc procedure has been reported for nicotinamide using a flame ionization detector at detection limits of ~0.2 μg (58). The nonvolatile amide was converted to the nitrile by reaction with heptafluorobutryic anhydride (56). For a related molecule, quinolinic acid, fmol detection limits were claimed for a gc procedure using either packed or capillary columns after derivatization to its hexafluoroisopropyl ester (58).

As with many of the vitamins, biological assays have an important historical role and are widely used. For example, microbiological assays use *Lactobacillus plantarum* ATCC No. 8014 (57,59) or *Lactobacillus arabinosus* (60). These methods are appropriate for both nicotinamide and nicotinic acid. Selective detection of nicotinic acid is possible if *Leuconostoc mesenteroides* ATCC No. 9135 is used as the test organism (61). The use of microbiological assays have been reviewed (62).

Bioassays procedures have been developed in species such as chicks which have been fed a niacin-deficient diet. Due to the fact that, for example, tryptophan is a biological precursor of niacin, niacin can be produced from other sources (55). As a result, the tryptophan content of the diet has to be monitored carefully for accurate results.

Specifications for niacin and niacinamide for food use are given in the *Food Chemicals Codex* (63) and for pharmaceutical use in the *United States Pharmacopeia* (64). The *Codex* also gives specifications for niacinamide ascorbate.

OCCURRENCE

Nicotinamide and nicotinic acid occur in nature almost exclusively in the bound form. In plants, nicotinic acid is prevalent whereas in animals nicotinamide is the predominant form. This nicotinamide is exclusively in the form of NAD and NADP.

Nicotinic acid is found in plants associated with both peptides and polysaccharides. For example in wheat bran, two forms are described: a peptide with a molecular weight of approximately 12,000 and a carbohydrate complex that is called niacytin. Polysaccharides isolated from wheat bran have been found to contain 1.05% nicotinic acid in bound form. Hydrolysis yielded a fragment identified as β-3-O-nicotinoyl-D-glucose (**25**.

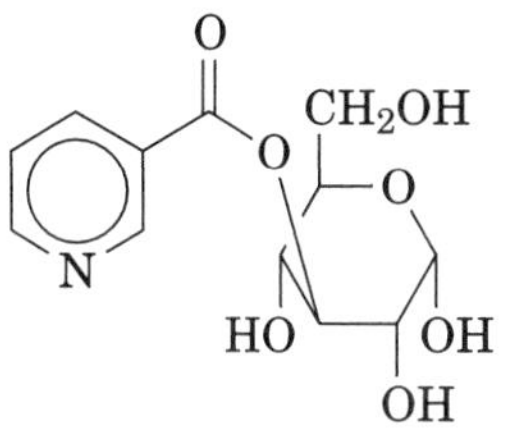

(**25**)

From a bioavailability standpoint, the fact that a significant amount of nicotinic acid is in a bound form has important biological consequences. Poor bioavailability

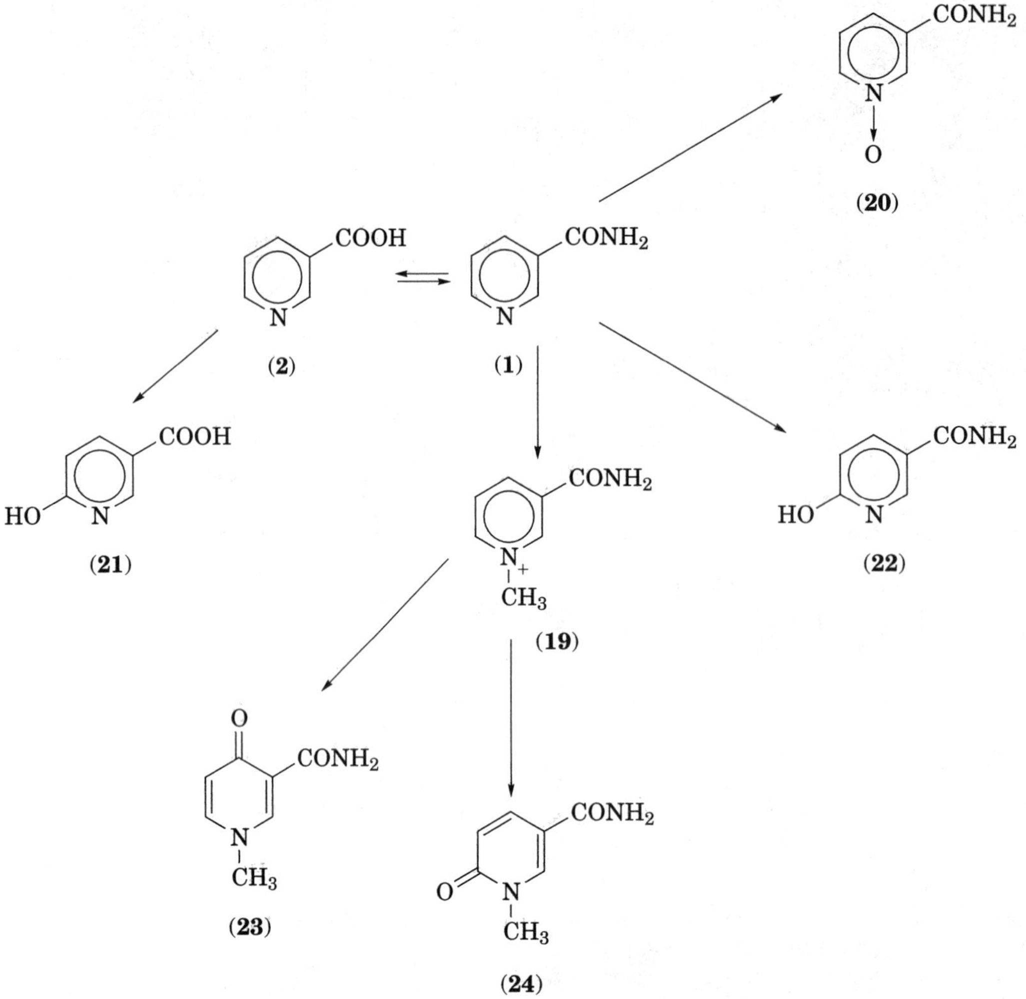

Figure 4. Detoxification products as the result of digestion of excess nicotinic acid.

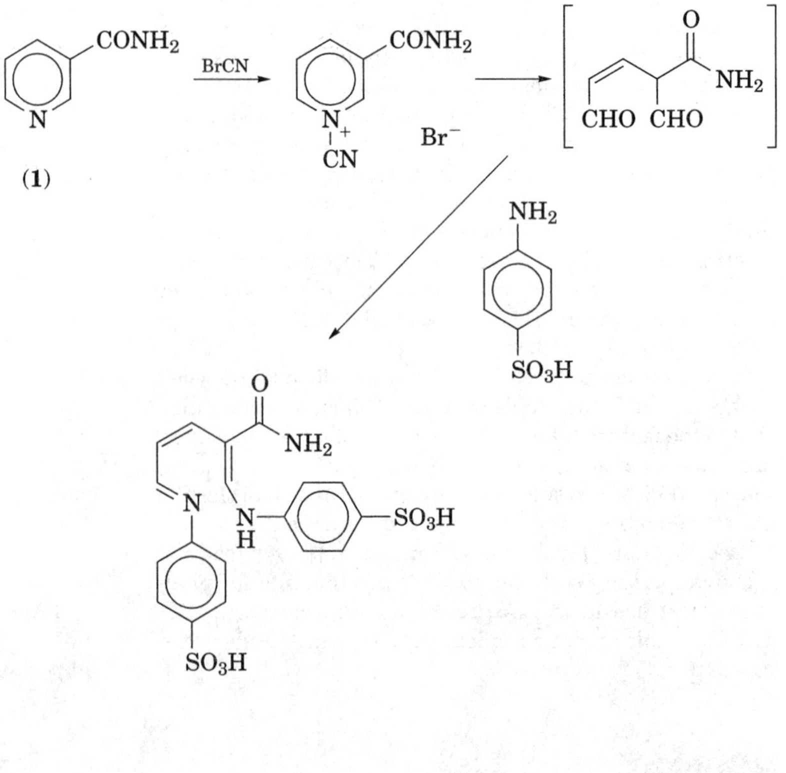

Figure 5. The König reaction.

stems from the fact that the ester linkage is resistance to digestive enzymes. In the case of corn, this condition can be alleviated if corn is pretreated with alkali. This food preparation method is frequently practiced in Mexico for the preparation of tortillas.

Nicotinamide and nicotinic acid are prevalent in many common foodstuffs and are especially concentrated in brewer's yeast, wheat germ and liver. In this regard, tryptophan is considered a provitamin and is assigned a niacin equivalent of 1/60. The following lists the vitamin B_3 content of many common foodstuffs and in Table 3, values of vitamin B_3 content are compared to niacin potential from tryptophan.

Corn, yellow	Soybean meal
Corn, gluten feed	Rapeseed-extract meal
Rye	Cottonseed cake
Oats	Palm kernel meal
Milo	Groundnut cake
Barley	Sunflower cake
Wheat	Blood meal
Wheat bran	Meat and bone meal
Tapioca flour	Fish meal
Potatoes	Skim milk, dried
	Torula yeast

Table 3. Nicotinic Acid and Nicotinamide (Vitamin B_3) Content of Foodstuffs (mg/kg)

Foodstuff	Vitamin B_3	Potential from tryptophan
Cow milk, fresh whole summer	0.8	7.8
Cheese, cream cheese	0.8	7.4
Eggs, whole, raw	0.7	36.1
Butter, salted	Trace	Trace
Beef, lean, average, raw	52	43
Beef liver, fresh	178	
Pork, lean, average, raw	62	38
Pork liver, fresh	118	
Corn, whole	12	
Corn, flour	Trace	1
Breakfast cereals, all-bran	490	32
Cornflakes		
Fortified	210	9
Unfortified	6	
Wheat flour		
Whole meal	56	25
White, 72% for bread		
Fortified	20	23
Unfortified	7	
Brown, 85%		
Fortified	42	26
Unfortified	12	
Bread, white	14	16
Macaroni, boiled	3	9
Rice, polished, boiled	3	5
Rice, bran	366–437	
Soybean flour, full fat	20	86
Spaghetti, boiled	3	9
Potatoes, raw	12	5
Yeast. Baker's dry	257	

Source: Ref. 2.

Coffee represents a fertile source of nictonic acid and the average cup of coffee contains between 1–2 mg. The amount of nicotinic acid depends on the roasting conditions (Table 4). During the roasting process, trigonelline (5), a plentiful component in green beans is demethylated to nictonic acid (65). The ratio of trigonelline (5) to nicotinic acid and nicotinamide is defined as the roasting number.

BIOCHEMICAL FUNCTION

NAD and NADP are required as redox coenzymes by a large number of enzymes and in particular dehydrogenases (Fig. 6). NAD^+ is utilized in the catabolic oxidations of carbohydrates, proteins, and fats, whereas $NADPH_2$ is the coenzyme for anabolic reactions and is used in fats and steroid biosynthesis. $NADP^+$ is also used in the catabolism of carbohydrates (2).

In addition to their *in vivo* function, these cofactors have important *in vitro* functions as co-factors in enzymatic synthesis. Because these cofactors are too expensive to be used in equimolar amounts, many methods have been developed for their regeneration. These include chemical (66), biological (67) and electrochemical (68) methods. Enzymatic regeneration has found particular utility in this application (69).

DEFICIENCY

A deficiency of niacin manifests itself in the disease pellagra. The symptoms of pellagra include dermatosis, dementia, and diarrhea. The dermatosis is expressed as lesions. These lesions are most frequently found on the hands, wrists, elbows, face and neck, knees, and the feet. At the onset of the disease, the affected areas resemble sunburn, progressing to keratosis and scaling of pigmented skin. Other symptoms include insomnia, loss of appetite, weight and strength loss, soreness of the tongue and mouth, indigestion, diarrhea, abdominal pain, burning sensations, vertigo, headache, numbness, nervousness, distractability, apprehension, mental confusion, and forgetfulness (1).

A deficiency of niacin also affects the nervous system. Numbness is initially observed and later, paralysis, particularly in the extremities is common. Severe cases are characterized by tremor and a spastic or ataxic gait and are frequently associated with peripheral neuritis. Left untreated, severe thought disorders can ensue (1).

Table 4. Vitamin B_3 Content in Roasting Coffee

Coffee beans	Vitamin B_3 (mg/kg)
Green	20
Medium roasted	80–150
Dark roasted	≤500

Source: Ref. 2.

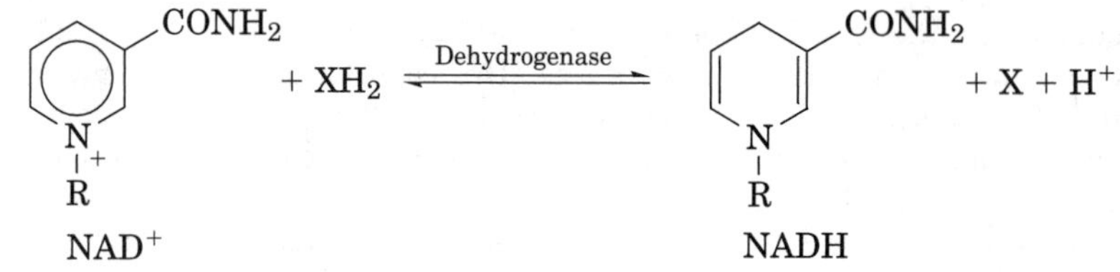

Figure 6. Coenzymes required for dehydrogenase. **(3)**

Table 5. RDA and U.S. RDA for Niacin

Category	Niacin (mg NE)
RDA	
Infants	
0.0–0.5 yr	5
0.5–1.0 yr	6
Children	
1–3 yr	9
4–6 yr	12
7–10 yr	13
Males	
11–14 yr	17
15–18 yr	20
19–50 yr	19
>51 yr	15
Females	
11–50 yr	15
>51 yr	13
pregnant	17
lactating	20
U.S. RDA	
Infants and children <4 yr	9
Adults and children >4 yr	20
Pregnant or lactating women	20

REQUIREMENTS

The RDA for niacin is based on the concept that niacin coenzymes participate in respiratory enzyme function and 6.6 niacin equivalents (NE) are needed per intake of 239 kJ (1000 kcal). One NE is equivalent to 1 mg of niacin. Signs of niacin deficiency have been observed when less than 4.9 NE/239 kJ or less than 8.8 NE per day were consumed. Dietary tryptophan is a rich source of niacin and the average diet in the United States contains 500–1000 mg of tryptophan. In addition, the average diet contains approximately 8–17 mg of niacin. In total, these two quan-

tities total 16–34 NE daily. Table 5 lists the RDA and U.S. RDA for niacin (69).

SAFETY

Despite structural similarities, the pharmacological consequences of excesses of these substances are quite different. Due to the interest in the effects of nicotinic acid on atherosclerosis, and in particular its use based on its ability to lower serum cholesterol, the toxicity of large doses of nicotinic acid has been evaluated. For example, in a study designed to assess its ability to lower serum cholesterol, only 28% of the patients remained in the study after receiving a large initial dose of 4 g of nicotinic acid due to intolerance at these large doses (70).

Nicotinamide can also be toxic to cells at concentrations that increase the NAD levels above normal. Individuals consuming nicotinamide at levels of 3 g/d for extended periods (3–36 months) have experienced various side effects such as heartburn, nausea, headaches, hives, fatigue, sore throat, dry hair, and tautness of the face (1).

USES

Both nicotinic acid and nicotinamide have been used in the enrichment of bread, flour, and other grain-derived products. Animal feed is routinely supplemented with nicotinic acid and nicotinamide. Nicotinamide is also used in multivitamin preparations. Nicotinic acid is rarely used in this application. The amide and carboxylic acid have been used as a brightener in electroplating baths and as stabilizer for pigmentation in cured meats.

DERIVATIVES

Nicotinyl alcohol (3-pyridinylcarbinol, 3-pyridinemethanol) (27) has use as an antilipemic and peripheral vasodilator. It is available from either the reductions of nicotinic acid esters or preferably, the reduction of the ni-

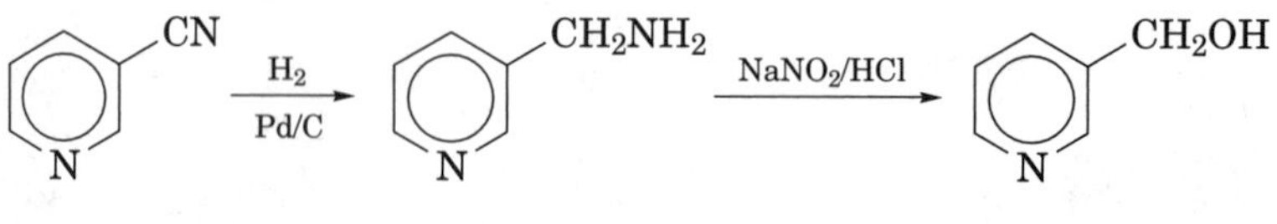

Figure 7. Formation of nicotinyl alcohol (**27**).

trile to the amine followed by diazotation and nucleophilic displacement. It is frequently administered in the form of the tartrate (Fig. 7). Nicotinic acid is frequently used as a salt in conjunction with basic drugs such as the peripheral vasodilator xanthinol niacinate (**28**). Nicotinic acid and its derivatives have widespread use as antihyperlipidemic agents and peripheral vasodilators (1).

(**28**)

ECONOMIC ASPECTS

U.S. manufacturers of niacin and niacinamide include Nepera, Inc. and Reilly Industries, Inc. U.S. suppliers include BASF Corporation, Hoffmann-La Roche Inc., and Rhône-Poulenc. Western European producers and suppliers include Degussa, Rhône-Poulenc, BASF, Hoffmann-La Roche, and Lonza (71). In 1995, the prices for niacin and nicotinamide were $9.75/kg and $9.25/kg, respectively (72,73).

BIBLIOGRAPHY

1. J. van Eys, in L. Machlin, ed., *Handbook of Vitamins*, Marcel Dekker, New York, 1991, pp. 311–340.

2. D. Bhatia, *Encyclopedia of Food Science and Technology*, John Wiley & Sons, Inc., New York, 1991, pp. 2276–2783.

3. U.S. Pat. 5,395,940 (Mar. 7, 1995), P. J. Angevine, C. T. Chu, and T. C. Potter (to Mobil Oil Corp.).

4. U.S. Pat. 4,721,789 (Dec. 6, 1988), R. Dicosimo, J. D. Burrington, and R. K. Grasselli (to Standard Oil Co.).

5. Jpn. Pat. 94-26603 (Feb. 24, 1994), Y. Asamidori, I. Hashiba, and S. Takigawa (to Nissan Chemical Ind Ltd. Japan).

6. World Pat. 9305022 A1 (Mar. 18, 1993), M. Hatanaka and N. Tanaka (to Nissan Chemical Industries, Ltd., Japan).

7. U.S. Pat. 4,482,439 A (Nov. 13, 1984), J. E. Toomey, Jr. (to Reilly Tar and Chemical Corp.).

8. Jpn. Pat. 93-193212 (July 8, 1993), T. Ishikawa, K. Maeda, Y. Mukohara, and T. Kojima (to Nippon Soda Co., Japan).

9. Jpn. Pat. 90-80698 (Mar. 30, 1990), F. C. Fujita (to Nitto Chemical Industry Co., Ltd.).

10. Eur. Pat. 442430 A2 (Aug. 21, 1991), A. Kiener (to Lonza, AG).

11. Eur. Pat. 684074 A1 (Nov. 29, 1995), Y. K. Lee, C. H. Shin, T. S. Chang, D. H. Cho, and D. K. Lee (to Korea Research Institute of Chemical Technology, South Korea).

12. Eur. Pat. 525367 A1 (Feb. 3, 1993), Y. Onda, K. Tsukahara, and N. Takahashi (to Mitsubishi Gas Chemical Co.).

13. U.S. Pat. 5,028,713 (July 2, 1991), R. Dicosimo, J. D. Burrington, and R. K. Grasseli (to Standard Oil Co.).

14. U.S. Pat. 4,876,348 (Oct. 24, 1989), R. Dicosimo, J. D. Burrington, and D. D. Suresh (to Standard Oil Co.).

15. Eur. Pat. 339680 A2 (Nov. 2, 1989), M. Saito, K. Tsukahara, K. Yamada, and H. Imai (to Mitsubishi Gas Chemical Co., Inc.).

16. Int. Pat. 163169A (Aug. 20, 1988), B. V. Suvorov and coworkers (to the Institute of Chemical Sciences, Academy of Sciences, Kazakh SSR).

17. Ger. Pat. 241903 A1 (Jan. 7, 1987), B. Luecke and co-workers (to Akademie der Wissenschaften der DDR).

18. Eur. Pat. 253360 A2 (Jan. 20, 1988), S. Shimizu and coworkers (to Koei Chemical Co., Ltd., Japan).

19. Jpn. Pat. 86-154741 (July 1, 1986), N. Abe, M. Dojo, and A. Iguchi (to Koei Chemical Industry Co., Ltd.).

20. Jpn. Pat. 85-169554 (July 30, 1985), H. Sato and K. Hirose (to Sumitomo Chemical Co., Ltd., Japan).

21. U.S. Pat. 4,603,207 A (July 29, 1986), R. Dicosimo, J. D. Burrington, and D. D. Suresh (to Standard Oil Co., Ohio).

22. Eur. Pat. 96-100676 (Jan. 18, 1996), L. V. Hippel, A. Neher, and D. Arntz (to Degussa Aktiengesellschaft, Germany).

23. World Pat. 9532055 (Nov. 30, 1995), D. C. Sembaer, F. A. Ivanovskaya, and E. M. Guseinov (to Lonza Ltd, Switzerland).

24. World Pat. 9532054 (Nov. 30, 1995), B. V. Suvorov, L. A. Stepanova, B. N. Anatoljevna, J. R. Chuck, and D. Pianzola (to Lonza Ltd, Switzerland).

25. Jpn. Pat. 93-206579 (Aug. 20, 1993), H. Hirayama (to Showa Denko KK, Japan).

26. Eur. Pat. 85-306670 (Sept. 19, 1985), S. Asano and J. Kitagawa (to Mitsui Toatsu Chemicals, Inc.).

27. World Pat. 9009988 A1 (Sept. 7, 1990), A. McKillop and D. Kemp (to Interlox Chemicals, Ltd.).

28. U.S. Pat. 2,471,518 (May 31, 1949), B. F. Duesel and H. H. Friedman (to Pyridium Co.).

29. U.S. Pat. 4,721,789 (Dec. 6, 1988), R. Dicosimo, J. D. Burrington, and R. K. Grasselli (to Standard Oil Co.).

30. Brit. Pat. Appl. 11357/55 (Aug. 28, 1942), B. F. Duesel and H. H. Friedman (to Pyridium Co.).

31. Ger. Offen. Appl. 2,517,054 (Apr. 17, 1975), H. Beschke, H. Friedrich, K.-P. Muller, and G. Schreyer (to Degussa Co.).

32. C. B. Rosas and G. B. Smith, *Chem. Eng. Sci.* **35**, 330 (1975).

33. Brit. Pat. 777,517 (Mar. 28, 1956), E. J. Gasson and D. J. Hadley (to The Distillers Co., Ltd.).

34. U.S. Pat. 4,008,241 (Feb. 15, 1977), A. P. Gelbein and coworkers (to the Lummus Co.).

35. Ger. Pat. 2,1313,8134 (Oct. 19, 1972), L. R. Haefele (to R. J. Reynolds Tobacco Co.).

36. Jpn. Pat. 93-18645 (Feb. 5, 1993), A. Hatamori, M. Shimada, and K. Ishiwatari (to Mitsui Toatsu Chemicals).

37. Eur. Pat. 568072Y A2 (Feb. 27, 1991), Y. Takashima, K. Kumagai, and S. Mitsuda (to Sumitomo Chemicals Co.).

38. Jpn. Pat. 90-80697 (Mar. 30, 1990), F. To and C. Fujita (to Nitto Chemical Industry Co., Ltd.).

39. Jpn. Pat. 89-81137 (Mar. 31, 1989), H. Yamada, T. Nagasawa, and H. Shimizu (to Nitto Chemical Industry Co., Ltd.).

40. Eur. Pat. 204555 A2 (Dec. 10, 1986), K. Kawakami, T. Tanabe, and O. Nagano (to Asahi Chemical Industry Co., Ltd.).

41. Jpn. Pat. 82-91408 (May 31, 1982) (to Nitto Chemical Industry Co., Ltd.).

42. Eur. Pat. 579907 A1 (Jan. 26, 1994), K. Yamada, M. Ochiai, Y. Yotsumoto, Y. Morimoto, and Y. Teranishi (to Mitsubishi Kawei Corp., Japan).

43. J. Eyal and M. Charles, *Ann. N.Y. Acad. Sci.* **589**, 678 (1990).

44. Jpn. Pat. 86-109401 (May 15, 1986), K. Kawakami and T. Tanabe (to Technology Research Association for New Application Development for Light-Weight Fractions, Japan).

45. Eur. Pat. 91-102938 (Feb. 27, 1991), H. Yamada, T. Nagasawa, and T. Nakamura (to Nitto Chemical Co., Ltd., Japan).

46. C. D. Mathew, T. Nagasawa, and M. Kobayashi, *Appl. Environ. Microbiol.* **54**, 1030–1032 (1988).

47. Eur. Pat. 86-100305 (Jan. 11, 1986), K. Enomoto, Y. Sato, Y. Nakashima, A. Fujiwara, and T. Doi (to Nitto Chemical Industry Co., Ltd., Japan, Mitsubishi Rayon Co., Ltd.).

48. F. E. Friedemann and E. J. Frazier, *Arch. Biochem.* **26**, 361 (1950).

49. *Official Methods of Analysis*, 8th ed., W. Horwitz, ed., Association of Official Agricultural Chemists, Washington, D.C., 1955, p. 826.

50. B. R. Clark, in D. B. McCormick and L. D. Wright, eds., *Methods in Enzymology*, Vol. 66, Academic Press, Inc., New York, 1980, p. 5.

51. J. W. Huff and W. A. Perlzweig, *J. Biol. Chem.* **167**, 157 (1947).

52. N. Levitas, J. Robinson, F. Rosen, J. W. Huff, and W. A. Perlzweig, *J. Biol. Chem.* **167**, 169 (1947).

53. H. B. Burdch et al., *J. Biol. Chem.* **212**, 897 (1955).

54. K. H. Carpenter and E. Kodicek, *Biochem. J.* **46**, 421 (1950).

55. G. A. Goldsmith and O. N. Miller, in P. György and W. N. Pearson, eds., *The Vitamins*, 2nd ed., Vol. VII, Academic Press, Inc., New York, 1967, p. 137.

56. Y. Tsuruta, K. Kohashi, S. Ishida, and Y. Ohkura, *J. Chromatogr.* **309**, 309 (1984).

57. K. Shibata and T. Shimono, in A. D. P. De Leenheer, W. E. Lambert, and H. J. Nelis, eds., *Modern Chromatographic Analysis of Vitamins*, 2nd ed., Marcel Dekker, Inc., New York, 1992, pp. 285–317.

58. A. Tanaka, M. Iijima, Y. Kikuchi, Y. Hoshino, and N. Nose, *J. Chromatogr.* **466**, 307 (1989).

59. M. P. Heyes and S. P. Markey, *Anal. Biochem.* **174**, 349 (1988).

60. *Official Methods of Analysis*, Association of Official Analytical Chemists, Washington, D.C., 1980, pp. 759–761, 763–764.

61. *Methods of Vitamin Assay*, Interscience Publishers, New York, 1966, pp. 172–176.

62. B. C. Johnson, *J. Biol. Chem.* **159**, 227 (1945).

63. M. R. S. Fox and G. M. Briggs, *J. Nutr.* **72**, 243 (1960).

64. Food and Nutrition Board, National Research Council, *Food Chemicals Codex*, 3rd ed., National Academy Press, Washington, D.C., 1981, pp. 205–206.

65. *The United States Pharmacopeia XXII* (USP XXII-NF XVII), United States Pharmacopeial Convention, Inc., Rockville, Md., 1990, pp. 943–944.

66. H. Tagachi, M. Sakaguchi, and Y. Shimabayashi, *Agri. Bio. Chem.* **49**, 3467–3471 (1985).

67. R. Ruppert, S. Hermann, and E. Steckhan, *J. Chem. Soc. Chem. Comm.*, 1150–1151 (1988).

68. Y. Morita, H. Furui, H. Yoshii, and R. Matsuno, *J. Ferment. Bioeng.* **77**, 13–16 (1994).

69. C. H. Wong and G. Whitesides, *Enzymes in Organic Chemistry*, Pergamon Press, Elsevier Science Ltd., Oxford, U.K., 1994.

70. Committee on Dietary Allowances, Food and Nutrition Board, *Recommended Dietary Allowances*, 10th ed., National Academy of Sciences, Washington, D.C., 1989, pp. 137–142.

71. H. K. Schock, in W. L. Holmes, L. H. Carlson, and R. Paoletts, eds., *Drugs Affecting Lipid Metabolism*, Plenum Press, New York, 1969, p. 405.

72. *Chemical Economics Handbook*, SRI International, Menlo Park, Calif., July 1995.

73. *Chem. Mark. Rep.* (Mar. 18, 1996).

GENERAL REFERENCES

Refs. 2, 3, and 57 are good general references.

O. Isler, G. Brubacher, S. Ghisla, and B. Kräutler, "Die Niacin-Gruppe (Vitamin PP)" in *Vitamine II*, Georg Thieme Verlag Stuttgart, New York, 1988, pp. 160–192.

SUSAN D. VAN ARNUM
Westfield, New Jersey

VITAMINS: PANTOTHENIC ACID

(*R*)-Pantothenic acid [79-83-4] (**1**) is a member of the B-complex vitamins. It is a water-soluble vitamin and has been designated as vitamin B_5. It was discovered by Williams and co-workers in 1933 by extracting a growth factor for yeast from a wide range of biological tissues (1). Later the growth factor was identified as pantothenic acid (**1**). Independently two groups, Woolley and co-workers (2) in 1939 and Jukes (3) between 1939 and 1941, identified the liver filtrate factor in rats and antidermatitis factor in chicks. These factors later proved to be identical with pantothenic acid. The total synthesis of (*R*)-pantothenic acid was achieved in 1940 by three groups: Stiller and co-workers (4), Reichstein and Grüssner (5), and Kuhn and Wieland (6).

(**1**)

After a full structural elucidation of coenzyme A in 1953 by Baddiley, it became evident that pantothenic acid is one of the components of coenzyme A (**2**) (7).

(**2a**, X = NH; coenzyme A)
(**2b**, X = S; thio analogue)

Biosynthesis of coenzyme A (CoA) in mammalian cells incorporates pantothenic acid. Coenzyme A, an acyl group carrier, is a cofactor for various enzymatic reactions and serves as either a hydrogen donor or an acceptor. Pantothenic acid is also a structural component of acyl carrier protein (ACP). ACP is an essential component of the fatty acid synthetase complex, and is therefore required for fatty acid synthesis. Free pantothenic acid is isolated from liver, and is a pale yellow, viscous, and hygroscopic oil.

OCCURRENCE, SOURCE, AND BIOAVAILABILITY

Good food sources of pantothenic acid include yeast, chicken, beef, potatoes, oat cereal, vegetables, legumes, and whole grain. Pantothenic acid in foods and feedstuffs is fairly stable to ordinary means of cooking and storage, but appreciable losses have been reported as a result of canning and storage of some foods. Pantothenic acid is synthesized only by plants and microorganisms. Estimation of the dietary intake of pantothenic acid is difficult because the acid exists in the free form and is also incorporated in coenzyme A and fatty acid synthetase. Total pantothenic acid content present in the food is estimated after releasing the bound acid enzymatically (8,9).

Relatively little is known about the bioavailability of pantothenic acid in human beings, and only approximately 50% of pantothenic acid present in the diet is actually absorbed (10). Liver, adrenal glands, kidneys, brain, and testes contain high concentrations of pantothenic acid. In healthy adults, the total amount of pantothenic acid present in whole blood is estimated to be 1 mg/L. A significant (2–7 mg/d) difference is observed between different age-group individuals with respect to pantothenic acid intake and urinary excretion, indicating differences in the rate of metabolism of pantothenic acid.

PHYSICAL AND CHEMICAL PROPERTIES

(R)-Pantothenic acid (1) contains two subunits, (R)-pantoic acid and β-alanine. The chemical abstract name is N-(2,4-dihydroxy-3,3-dimethyl-1-oxobutyl)-β-alanine (11). Only (R)-pantothenic acid is biologically active. Pantothenic acid is unstable under alkaline or acidic conditions, but is stable under neutral conditions. Pantothenic acid is extremely hygroscopic and there are stability problems associated with the sodium salt of pantothenic acid. The major commercial source of this vitamin is thus the stable calcium salt (3) (calcium pantothenate).

(3)

Panthenol (4) is the reduced form of pantothenic acid and is the pure form most commonly used. The alcohol is more easily absorbed and is converted into the acid *in vivo*

(12). Both panthenol and pantyl ether are used in hair care products.

(4, R = H)
(5, R = C$_2$H$_5$)

The biologically active R- or d-pantothenic acid can be obtained upon hydrolysis of coenzyme A with a combination of two enzymes, alkaline phosphatase and pantotheinase (13) (Fig. 1). The phosphatase catalyzes the selective cleavage of the phosphate bond in coenzyme A to afford adenosin-3'5'-diphosphate (6) and 4-phosphopantetheine (7). The latter substance is dephosphorylated enzymatically to yield pantetheine (8), which is rapidly converted by pantotheinase to pantothenic acid (1). Table 1 lists some physical properties of pantothenic acid and its derivatives.

SYNTHESIS

Currently (ca 1997) pantothenic acid is produced mainly by chemical methods. Initial efforts in this area are summarized in Reference 14. Several groups are actively involved in developing syntheses of pantothenic acid or its precursor, (R)-pantolactone (9) by microbial methods.

(R)-Calcium pantothenate (3) is prepared by condensing (R)-pantolactone (9) with β-alanine (10) in the presence of base, followed by treatment of the sodium salt (11) with calcium hydroxide.

An alternative procedure for the preparation of (R)-calcium pantothenate (3) is to condense (R)-pantolactone (9) with the preformed calcium salt (12) of β-alanine (15).

Similarly, panthenol (4) and pantyl ether (5) are prepared by condensing 3-aminopropanol (13) and 3-ethoxypropylamine (14) with (R)-pantolactone (16,17).

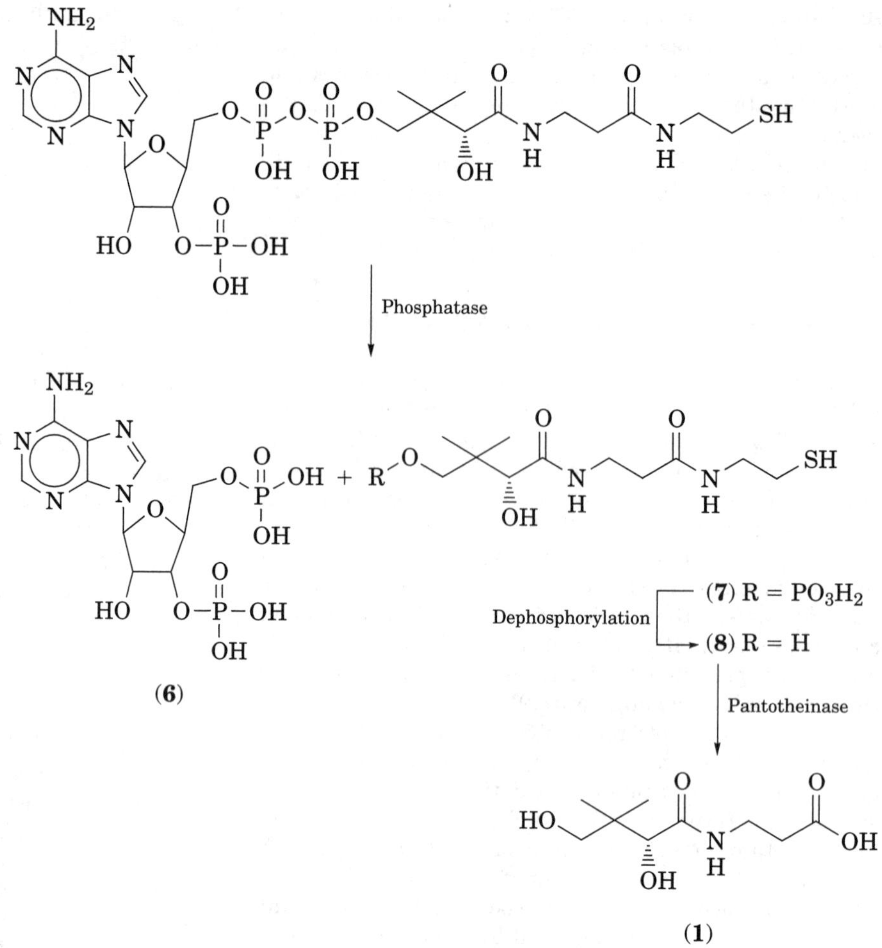

Figure 1. Enzymatic hydrolysis of coenzyme A (**2**).

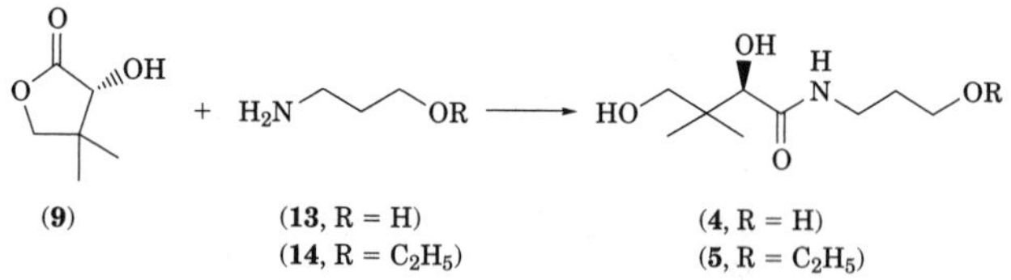

Racemic pantolactone is prepared easily by reacting isobutyraldehyde (**15**) with formaldehyde in the presence of a base to yield the intermediate hydroxyaldehyde (**16**). Hydrogen cyanide addition affords the hydroxy cyanohydrin (**17**). Acid-catalyzed hydrolysis and cyclization of the cyanohydrin (**17**) gives (*R,S*)-pantolactone (**18**) in 90% yield (18).

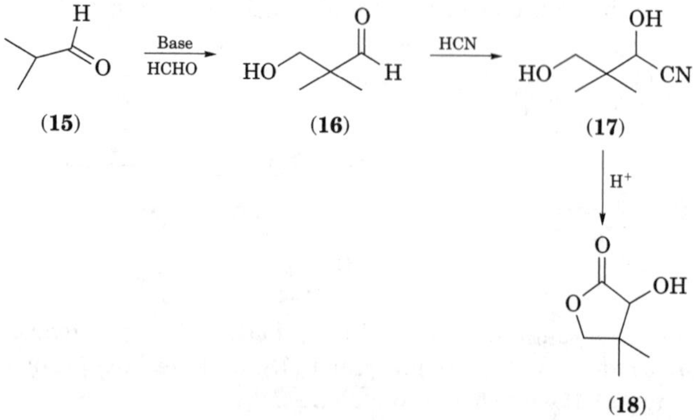

(*R*)-Pantolactone (**9**) is prepared in either of two ways: by resolution of the (*R,S*)-pantolactone mixture (**18**), or by stereoselective reduction of ketopantolactone (**19**) by chemical or microbial methods (19).

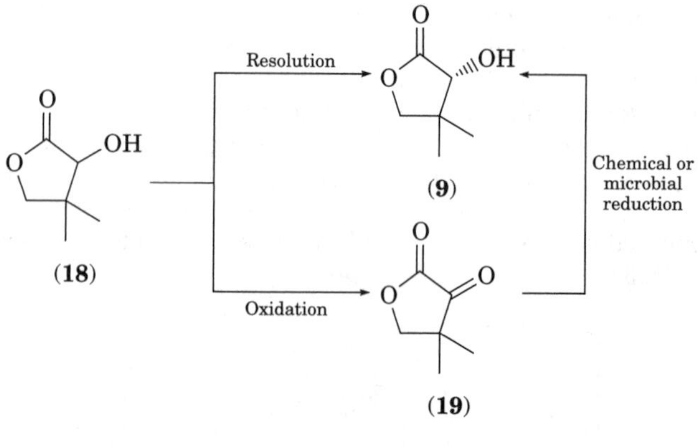

Chemical resolution of pantoic acid, readily available from (*R,S*)-pantolactone (**18**), is the most efficient method currently being practiced on an industrial scale. A wide variety of chiral resolving agents are being used, such as (+)-2-aminomethylpinane by BASF (20), 2-benzylamino-1-phenylethanol by Fuji (21), chiral aminoalcohols by Alps and Sumitomo (22,23), (1*R*)-3-endoaminoborneol by Hoff-

Table 1. Physical Properties of Pantothenic Acid and Derivatives

Compound	CAS registry number	α_D^{20a}	bp/mp (°C)	Molecular formula	Molecular weight
(R)-Pantothenic acid	[79-83-4]			$C_9H_{17}NO_5$	219.24
(S)-Pantothenic acid	[37138-77-5]			$C_9H_{17}NO_5$	219.24
(R)-Pantothenic acid calcium salt	[137-08-6]	+28.2°	195°C	$(C_9H_{16}NO_5)_2Ca$	476.5
(R)-Pantothenic acid sodium salt	[876-81-2]	+27.7°	171–178°C	$C_9H_{16}NO_5Na$	241.2
(R)-Panthenol	[81-13-0]	+29.5°	120°C[b]	$C_9H_{19}NO_4$	205.25
(S)-Panthenol	[74561-18-5]	−29.5°	120°C[b]	$C_9H_{19}NO_4$	205.25
(R,S)-Panthenol	[16485-10-2]	0	66–69°C	$C_9H_{19}NO_4$	205.25
(R,S)-Ethylpanthenol	[667-84-5]	0	210°C	$C_{11}H_{23}NO_4$	233.1
(R)-Ethylpanthenol	[667-83-4]			$C_{11}H_{23}NO_4$	233.1

[a]C = 5 in H_2O for all values other than 0.
[b]At 0.02 mm Hg.

mann-La Roche (24), and 1-ethylbenzylamine by Daicel (25). Other approaches include optical resolution by inclusion crystallization (26) and spontaneous resolution of (R,S)-pantolactone by seeding (27). (R,S)-Pantolactone can be resolved directly using chiral phase chromatography (28), or after derivatization using optically active acids or isocyanates (29). The sodium salt of pantoic acid is enantioselectively protonated in the presence of (1S)-(+)-10-camphorsulfonic acid and then undergoes spontaneous cyclization to (R)-pantolactone (9) (30).

Ketopantolactone (19) is conveniently prepared by oxidation of (R,S)-pantolactone (18). Various oxidizing agents have been patented for the oxidation of pantolactone, such as MnO_2 (31), DMSO–Ac_2O (32), and hypohalites (33). An improved yield of ketopantolactone (19) via electrolytic oxidation of pantolactone with an aqueous solution containing an alkali metal salt was reported (34). Ketopantolactone (19) has been prepared in good yield via cyclocondensation of the 2-keto-3-methylbutyrate (20) with formaldehyde (35).

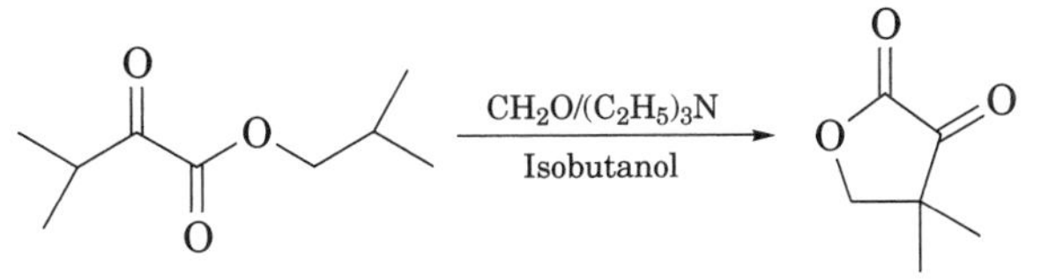

Table 2. Asymmetric Hydrogenation of Ketopantolactone (19)

Catalyst/chiral ligand	Isolated yield (%)	ee (%)	Ref.
[Rh(bppm) (1,5-Hexadiene)Cl]	93	56	37
[Rh(bppm) (COD)Cl]	93	86.7	38
Dimethyl-(R)-malate	93	>99	39
Di-(μ-carboxylato)bis(aminophosphine-phosphinite) dirhodium complexes	100	91	36

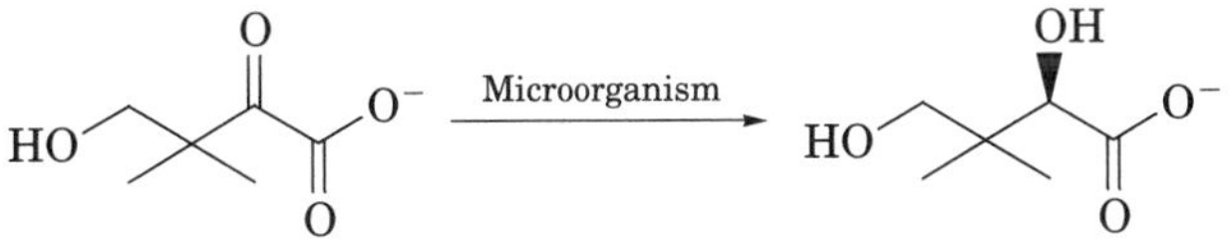

microorganisms (40–44). Reduction with *Proteus vulgaris* gave (R)-pantoate (22) in high (460 mmol) concentration, excellent selectivity (97% ee), and 98.5% yield with acceptable productivity numbers. Productivity number (PN) is derived by dividing the concentration in mmol of product formed by the dry weight in kg of the cells, multiplied by the time in hours required for the conversion. Higher PN is favored for production (44) (Table 3).

(20)   (19)

Asymmetric hydrogenation of ketopantolactone (19) in the presence of chiral dirhodium complexes gave (R)-pantolactone (9) in high yield and excellent selectivity (36) (Table 2).

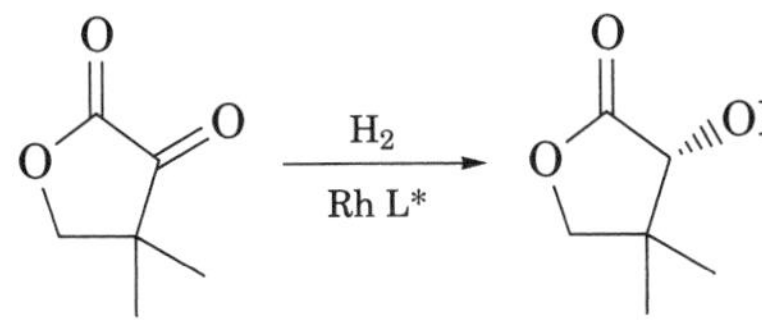

(19)   (9)

The microbial reduction of ketopantoate (21) to (R)-pantoate (22) has been investigated, employing one or two

(21)   (22)

Reduction of ketopantolactone (19) to (R)-pantolactone (9) also was evaluated using microbes (45–52) (Table 4).

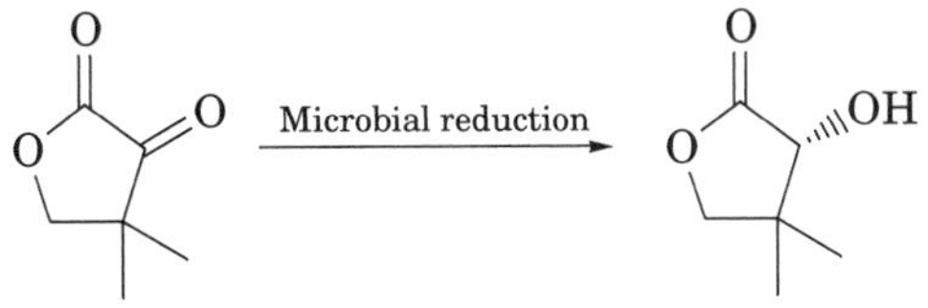

(19)   (9)

These microbial methods have afforded excellent ee values and yields (40–52). However, owing to low, ie, 30–130, productivity numbers these processes have not yet been commercialized.

Table 3. Microbial Reduction of Ketopantoate (21) to (R)-Pantoate (22)

Microorganism	Conversion yield (%)	ee (%)	Final product concentration (mM)	PN[a]	Reference
Rhodotorula erythropolis	90.5	94.4	140		40
Agrobacterium sp. S-242	90	>98	910	120	41
Nocardia asteroides/Agrobacterium radiobacter	89	82.8	620	130	42
Candida macedonienis-AK4 4588	97.2	98	80		43
Proteus vulgaris	98.5	>97	460	740	44

[a]Productivity number (PN) = product in mmol/dry wt of cells in kg x time, h.

In a novel approach, enantiomerically enriched (*R*)-pantolactone (**9**) is obtained in a enzymatic two-step process starting from racemic pantolactone.

In a first step, *Nocardia asteroides* selectively oxidizes only (*S*)-pantolactone to ketopantolactone (**19**), whereas the (*R*)-pantolactone remains unaffected (47). The accumulated ketopantolactone is stereospecifically reduced to (*R*)-pantolactone in a second step with *Candida parapsilosis* (product concentration 72 g/L, 90% molar yield and 100% ee) (48). Racemic pantolactone can also be converted to (*R*)-pantolactone by one single microbe, ie, *Rhodococcus erythropolis*, by enantioselective oxidation to (*S*)-pantolactone and subsequent stereospecific reduction in 90% yield and 94% ee (product concentration 18 g/L) (40).

Although not of industrial importance, several asymmetric syntheses of (*R*)-pantolactone (**9**) have been developed. Stereoselective abstraction of the *si*-proton of the achiral 1,3-propanediol derivative (**23**) by *sec*-butyllithium-(−)-sparteine, followed by carboxylation and hydrolysis, results in (*R*)-pantolactone in 80% yield and 95% ee (53).

(*R*)-Pantolactone is also prepared in a sequence involving Claisen rearrangement of the chiral glycolate (**24**), although with poor enantioselectivity (54).

By employing Sharpless epoxidation as a key step, a multistep chemical synthesis of (*R*)-pantolactone has also been reported (55).

Enantioselective addition of hydrogen cyanide to hydroxypivaldehyde (**25**), catalyzed by (*R*)-oxynitrilase, afforded (*R*)-cyanohydrin (**26**) in good optical yield. Acid-catalyzed hydrolysis followed by cyclization resulted in (*R*)-pantolactone in 98% ee and 95% yield after one recrystallization (56).

LABELED COMPOUNDS

Various labeled degradation products of pantothenic acid and coenzyme A are known. Both (1-^{14}C)-sodium pantothenate (**27**) and (*R*)-(1-^{14}C)-panthenol are

known and are commercially available from New England Nuclear Corporation (NEN, Boston, Mass.). The syntheses of ^{14}C-phosphopantotheine (57) and ^{14}C-labeled coenzyme A (58) have also been reported.

DERIVATIVES AND ANALOGUES

Methyl and diacetyl derivatives of pantothenic acid are prepared via methylation and acetylation of pantothenic acid (59). Pantethein (**8**) is one of the biochemical degradation products derived from coenzyme A. Synthetically, it can be prepared in several different ways, starting from either methyl pantothenate (**28**) or (*R*)-

Table 4. Microbial Reduction of Ketopantolactone (19) to (R)-Pantolactone (9)

Microorganism	Conversion yield (%)	ee (%)	Final product concentration (mM)	PN	Reference
Saccharomyces cerevisiae	>90	>98	40	30	45
Byssochamys fulva	>90	>96			46
Nocardiaasteroides / candida parapsilosis	>90	>98	390	130	47
Candida parapsilosis	>90	94–98	700		48
Rhodotorulaminuta	80.3	99.7	380		49
Saccharomyces cerevisiae	39	93	20		50
Zygosaccharo-myces	96.8	90			51
Baker's yeast/β-cyclodextrin	39	93			52

pantolactone. Pantethein (8) is prepared either by condensing cysteamine (29) with methyl pantothenate (28) (60) or by coupling (R)-pantolactone (9) with 3-amino-N-(2-mercaptoethyl)propanamide (30) (61). Phosphopantetheine analogue (31) was prepared starting from (R)-pantothenic acid. Enzymatic condensation of this compound with adenosin-3,5-diphosphate (6) gave CoA analogue (2b) (62).

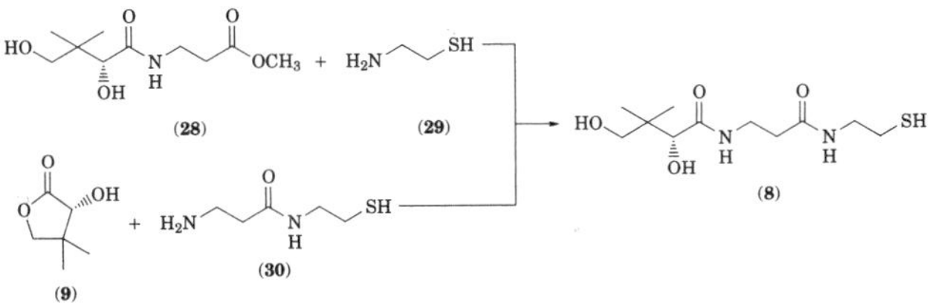

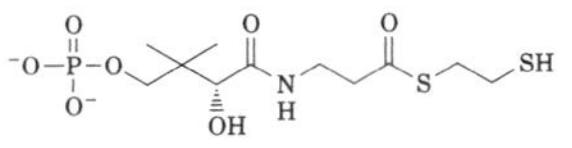

BIOSYNTHESIS

The metabolically active form of pantothenic acid is coenzyme A. Pantothenic acid is produced only by microorganisms, starting from (R)-pantoate (22) and β-alanine. (R)-Pantoate is synthesized by a set of enzymatic reactions, as follows (63,64):

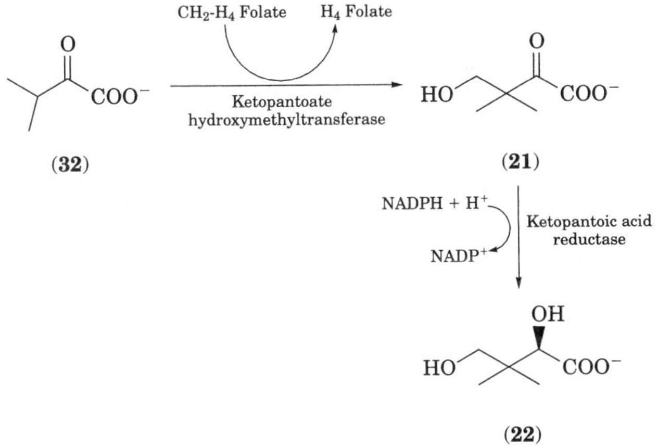

The conversion of α-ketoisovalerate (32) to ketopantoate (21) is catalyzed by ketopantonate hydroxymethyl-

transferase and a cofactor tetrahydrofolate (65). Further reduction of ketopantoate (21) to (R)-pantoate (22) is catalyzed by ketopantoic acid reductase (66).

Aspartic acid decarboxylase catalyzes the decarboxylation of asparatic acid to yield β-alanine (10), a precursor for the biosynthesis of pantothenic acid (67). Finally, (R)-pantothenic acid is obtained by coupling β-alanine (10) with (R)-pantoate (22) in the presence of pantothenate synthetase:

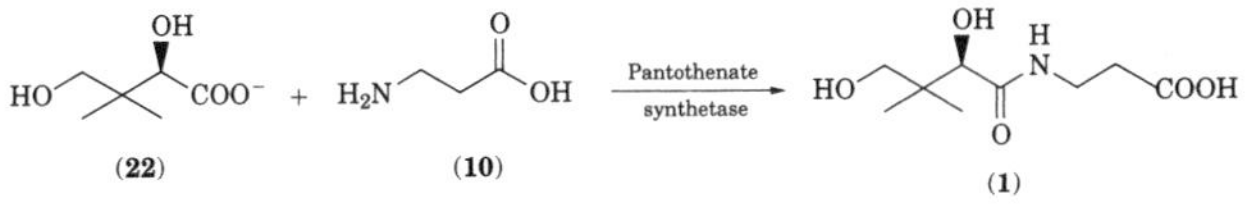

Various pathways have been proposed for the conversion of pantothenic acid to coenzyme A (68,69). The currently accepted pathway involves the following sequence; pantothenate to 4-phosphopantothenic acid (33), 4-phosphopantothenyl cysteine (34), 4-phosphopantetheine (7), dephosphocoenzyme A (35), and coenzyme A (2). Phosphorylation of pantothenic acid to 4-phosphopantothenic acid is catalyzed by pantothenic kinase in the rate-limiting step in the overall synthesis of coenzyme A (70).

ANALYTICAL METHODS

Chemical, physical, animal, microbiological, and biochemical assays have been used to determine pantothenic acid. Pantothenic acid in foods is found in both the free form and as the vitamin moiety of coenzyme A and phosphopantetheine (see Scheme 1). For most assays, the incorporated pantothenic acid has to be liberated enzymatically. Usually, a combination of pantotheinase and alkaline phosphatase is used to liberate the bound pantothenic acid. The official method for pantothenic acid of the Association of Official Analytical Chemists (AOAC) is the microbiological assay that uses *L. Plantarium* (ATCC 8014) as the test organism (71). Samples are extracted at 121°C at pH 5.6–5.7, proteins are precipitated at pH 4.5, and the resulting clear extracts are adjusted to pH 6.8 prior to assay. This procedure is only suitable to determine calcium pantothenate or other free forms of pantothenic acid.

A radioimmunoassay method has been developed by Wyse and co-workers (72). It is a sensitive method for the determination of small amounts of pantothenic acid in biological fluids. The assay is based on the binding of an en-

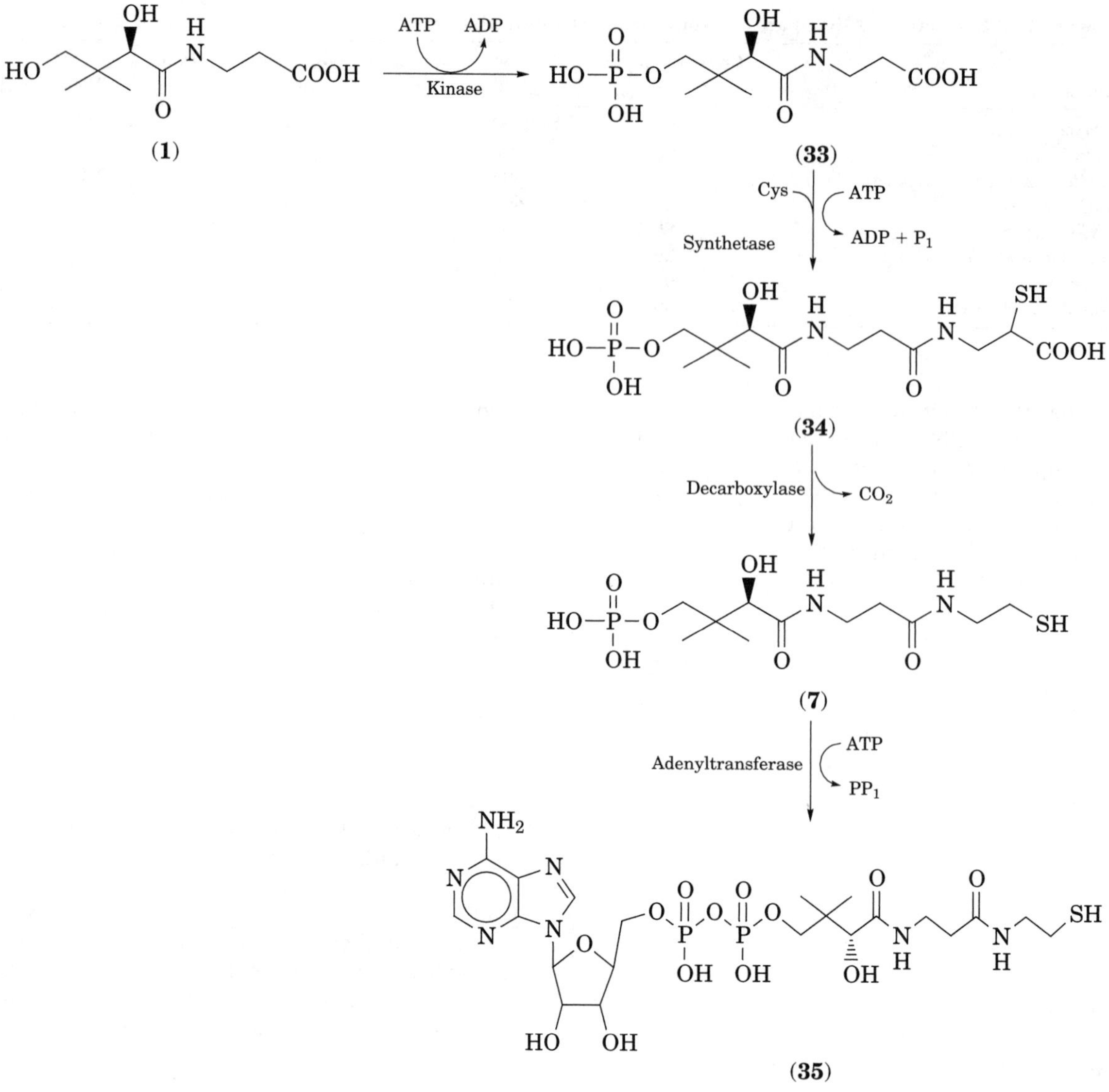

Scheme 1.

zyme specific for pantothenic acid. Hemolyzed blood samples are subjected to a double-enzyme treatment with 5 IU of bovine intestinal alkaline phosphatase and 0.1 IU of pantetheinase. After hydrolysis, the clear supernatant extract is analyzed for pantothenate by a radioimmunoassay method. In the case of pharmaceutical preparations, vitamin concentrations are determined by using spectrophotometric and fluorometric methods (73). For spectrophotometric methods, detection is done at 358 nm and for fluorescence, excitation is done at 350 nm and collection at 450 nm. Food pantothenate assays involve hydrolysis of the food by 25% hydrochloric acid in a 95–100°C water bath, extraction of the pantoyl lactone into dichloromethane and analysis by gas chromatography with flame ionization detection (74). Chiral polysiloxane fused silicon capillary columns (XE-60-L-Val-(S)- or (R)-α-phenylethylamide) have been developed for separation of the (R)- and (S)-pantothenic acid mixture (75). Recently, clinical separation and simultaneous determination of (R)- and (S)-pantothenic acids in rat plasma using gc–ms has been described (76). Deproteinization of the plasma samples is done by eluting the plasma sample through an anion-exchange resin using 1 M sodium chloride solution. Pantothenic acid is extracted quantitatively with ethyl acetate after acidifying the basic aqueous solution. After esterification of the carboxylic group, each derivative is identified using gs–ms.

DEFICIENCY

A deficiency of pantothenic acid in humans has not been detected because it is widely available in food (77). In the case of prisoners of war in the 1940s, a burning feet syndrome, which was attributed to pantothenic acid deficiency as a result of severe malnutrition, was observed (78). Pantothenic acid deficiency in humans is induced experimentally by feeding a diet deficient in pantothenic acid along with a pantothenic acid antagonist such as ω-methylpantothenic acid. This results in clinical malaise, fatigue, headache, sleep disturbance, nausea, vomiting, and cardiovascular instability. Impaired responses to insulin, histamine, and ACTH (stress hormone) have also been observed. A wide variety of abnormalities such as retarded growth, impaired fertility, gastrointestinal lesions, and adrenal necroses are reported when rats are fed a pantothenic acid-deficient diet (79,80). Chicks on a pantothenic acid-deficient diet develop lesions at the corners of the mouth, swollen eyelids, hemorrhagic cracks on the feet, and listlessness (81).

ABSORPTION

Pantothenic acid occurs in most foods and feedstuffs as CoA and acyl-carrier protein. The utilization of the vitamin depends upon the hydrolytic digestion of these protein complexes to release the free vitamin. Coenzyme A is hydrolyzed in the intestinal lumen before absorption into the cell by a sodium ion-dependent, passive mechanism (82,83). High concentrations of pantothenic acid are found in the heart and kidney, compared to other organs such as the intestine and liver (84). Pantothenic acid utilization in the formation of acetyl CoA is impaired in alcoholics because the ethanol metabolite acetaldehyde inhibits the conversion (85). Erythrocytes, which carry most of the vitamin in the blood, carry it predominantly in the form of CoA. Blood and urinary levels of pantothenic acid are lower for oral contraceptive users than for nonusers (86,87). Absorption is also inhibited in the presence of known antagonists, such as ω-methylpantothenic acid, and to some extent with (S)-pantothenic acid. Another factor which influences vitamin absorption is a high fat diet. This condition significantly lowers the utilization of pantothenic acid in the formation of coenzyme A in the liver because of changes in lipid metabolism (88). On the other hand, high levels of protein in the diet may promote better utilization of this vitamin in the synthesis of coenzyme A (89).

NUTRITIONAL REQUIREMENTS

Pantothenic acid is widely distributed in food and because of the lack of conclusive evidence regarding quantitative needs, a recommended dietary allowance (RDA) for pantothenic acid has not been established. In 1989, the Food and Nutrition Board of the United States National Research Council suggested a safe intake of 4–7 mg/d for adults. The provisional allowance for infants is 2–3 mg daily (90).

FUNCTIONS

The most important functions of pantothenic acid are its incorporation in coenzyme A and acyl carrier protein (ACP). Both CoA and ACP/4-phosphopantetheine function metabolically as carriers of acyl groups. Coenzyme A forms high-energy thioester bonds with carboxylic acids. The most important coenzyme is acetyl CoA. Acetic acid is produced during the metabolism of fatty acids, amino acids, or carbohydrates. The active acetate group of acetyl CoA can enter the Krebs cycle and is used in the synthesis of fatty acids or cholesterol. ACP is a component of the fatty acid synthase multienzyme complex. This complex catalyzes several reactions of fatty acid synthesis (condensation and reduction). The nature of the fatty acid synthase complex varies considerably among different species (91).

TOXICITY

Pantothenic acid toxicity has not been reported in humans. Massive doses (10 g/d) in humans have produced mild intestinal distress and diarrhea. Acute toxicity was observed in case of mice and rats by using calcium pantothenate at fairly large doses (92).

USES

The bulk of the industrial supply of the calcium salt of (R)-pantothenic acid is used in food and feed enrichment. Food enrichment includes breakfast cereals, beverages, dietetic, and baby foods. Animal feed is fortified with calcium-(R)-pantothenate which functions as a growth factor.

Panthenol is used in skin care products. When applied topically to alleviate itching, it helps to keep the skin moist and supple, stimulates cell growth, and accelerates wound healing by increasing the fibroblast content of scar tissue (93,94). It is also used as a moisturizer and conditioner in hair care products. It is believed to protect hair against chemical and mechanical damage caused by perming, coloring, and shampooing (95).

ECONOMIC ASPECTS

The major producers of the calcium salt of pantothenic acid and panthenol are Hoffmann-La Roche, Daiichi, BASF, and Alps. Racemic panthenol is used mainly in hair care products, whereas (R)-panthenol is exclusively used in top-of-the-line, more expensive skin and hair care products. The current (1997) price of (R,S)-panthenol varies between $12 and $18 per kg, that of D-panthenol varies between $30 and $45 per kg, and that of D-calcium pantothenate (Calpan) varies between $22 and $30 per kg.

FUTURE PROSPECTS

Despite the progress made in the stereoselective synthesis of (R)-pantothenic acid since the mid-1980s, the commercial chemical synthesis still involves resolution of racemic pantolactone. Recent (1997) synthetic efforts have been directed toward developing a method for enantioselective synthesis of (R)-pantolactone by either chemical or microbial reduction of ketopantolactone. Microbial reduction of ketopantolactone is a promising area for future research.

BIBLIOGRAPHY

1. R. J. Williams, C. M. Lyman, G. H. Goodyear, J. H. Truesdail, and H. Holaday, *J. Am. Chem. Soc.* **55**, 2912 (1933).

2. D. W. Woolley, H. A. Waisman, and C. A. Elvehjem, *J. Biol. Chem.* **129**, 673 (1939).

3. T. H. Jukes, *J. Am. Chem. Soc.* **61**, 975 (1939).

4. E. T. Stiller, S. A. Harris, J. Finkelstein, J. C. Keresztesy, and K. Folkers, *J. Am. Chem. Soc.* **62**, 1785 (1940).

5. T. Reichstein and A. Grüssner, *Helv. Chem. Acta* **23**, 650 (1940).

6. R. Kuhn and Th. Wieland, *Chem. Ber.* **73**, 962 (1940).

7. J. Baddiley and E. M. Thain, *J. Chem. Soc.* **3421** (1951); J. Baddiley and E. M. Thain, *J. Chem. Soc.*, 1610 (1953).

8. M. L. Orr, *Home Economic Research Report* No. 36, U.S. Department of Agriculture, Washington, D.C., 1969.

9. J. H. Walsh, B. W. Wyse, and R. G. Hansen, *J. Am. Diet. Assoc.* **78**, 140 (1981).

10. J. B. Tarr, T. Tamura, and E. L. R. Stokstad, *Am. J. Clin. Nutr.* **34**, 1328 (1981).

11. IUPAC-IUB Joint Commission on Biochemical Nomenclature (JCBN), *Pure Appl. Chem.* **59**, 834 (1987).

12. J. Marks, *The Vitamins in Health and Disease*, Little, Brown & Co., Boston, MA, 1968, pp. 121–125.

13. D. Novelli, N. O. Kaplan, and F. Lipman, *Fed. Proc.* **9**, 209 (1950).

14. O. Isler, G. Brubacher, S. Ghisla, and B. Kräutler, *Vitamine II*, Georg Thieme Verlag, New York, 1988, p. 309.

15. E. H. Wilson, J. Weiglard, and M. Tischler, *J. Am. Chem. Soc.* **76**, 5177 (1954); Ger. Pat. DE 2919515 (Nov. 22, 1979), D. Bartoldus and E. Broger (to Hoffmann-La Roche Inc.).

16. O. Schnider in *Festschrift Emil Barell*, Hoffmann-La Roche, Basel, Switzerland, 1946, p. 85; E. E. Snell and W. Shive, *J. Biol. Chem.* **158**, 551 (1945); G. S. Kozlowa, T. I. Erokhina, and V. I. Gunar, *Pharm. Chem. J.* **11**, 505 (1977).

17. J. Baddiley and A. P. Mathias, *J. Chem. Soc.*, 2803 (1954).

18. Ger. Pat. DE 2758883 (July 5, 1979), H. Distler and W. Goetze (to BASF); E. Glaser, *Herstellung Monatsch. Chem.* **25**, 46 (1904); Jpn. Pat. 55062080 (May 10, 1980) (to UBE Industries Ltd.).

19. S. Shimizu and H. Yamada, in T. O. Baldwin, F. M. Raushel, and A. I. Scott, eds., *Chemical Aspects of Enzyme Biotechnology*, Plenum Press, New York, 1990, p. 151.

20. Ger. Pat. DE 2404305 (Oct. 27, 1983), S. Pfohl and co-workers (to BASF AG).

21. Jpn. Pat. 02240073, (Sept. 25, 1990) S. Takeda, M. Takeda, K. Suzuki, and M. Yuya (to Fuji Chemical Industrial Co., Ltd.).

22. Ger. Pat. DE 2558508, (Sept. 7, 1976) A. Kinugasa, T. Okuda, M. Goto, and M. Saito (to Alps Pharmaceutical Industrial Co., Ltd.).

23. Jpn. Pat. 8115240 (Aug. 10, 1982) (to Sumitomo Chemical Co., Ltd.).

24. C. Fizet, *Helv. Chim. Acta.* **69**, 404 (1986).

25. Jpn. Pat. 8735340, (Aug. 23, 1988) H. Nohira, S. Yoshida, M. Nohira, K. Sato (to Daicel Chemical Industries Ltd.).

26. F. Toda, A. Sato, K. Tanaka, and T. C. W. Mak, *Chem. Lett.*, 873 (1989).

27. I. Masahiro, *Res. Inst. Yakugaku Zasshi* **96**, 71 (1976); Jpn. Pat. 47022220, (June 22, 1972) S. Nabeta, Y. Nakabe, and M. Nagaki (to Daiichi Seiyaku Co., Ltd.).

28. N. Oi, T. Doi, H. Kitahara, and Y. Inda, *J. Chromatogr.* **208**, 404 (1981); W. A. König and U. Sturm, *J. Chromatogr.* **328**, 357 (1985); E. Francotte and D. Lohmann, *Helv. Chim. Acta.* **70**, 1569 (1987); D. W. Armstrong, W. Li, C. Chang, and J. Pitha, *Anal. Chem.* **62**, 914 (1990); W.-Y. Li, H. L. Jin, and D. W. Armstrong, *J. Chromatogr.* **509**, 303 (1990).

29. T. Arai, H. Matsuda, and H. Oizumi, *J. Chromatogr.* **474**, 405 (1989); A. Takasu and K. Ohya, *J. Chromatogr.* **389**, 251 (1987).

30. K. Fuji, M. Node, M. Murata, S. Terada, and K. Hashimoto, *Tetrahedron Lett.* **27**, 5381 (1986).

31. Jpn. Pat. 04095087 A2, (Mar. 27, 1992) N. Kuroda and K. Kashiwa (to Takeda Yakuhin Kogyo K. K.).

32. Jpn. Pat. 04095086 A2, (Mar. 27, 1992) N. Kuroda and K. Kashiwa (to Takeda Yakuhin Kogyo K. K.).

33. Jpn. Pat. 04095084 A2, (Mar. 27, 1992) N. Kuroda and K. Kashiwa (to Takeda Yakuhin Kogyo K. K.).

34. Jpn. Pat. 01208488 A2, (Aug. 22, 1989) H. Sato, S. Takeda, and M. Yuya (to Fuji Chemical Industrial Co., Ltd.).

35. DE 3229026 A1, (Aug. 4, 1982) K. Halbritter and M. Eggersdorfer (to BASF AG) (Aug. 4, 1982).

36. J.-F. Carpentier, F. Agbossou, and A. Mortreux, *Tetrahedron Asymm.* **6**, 39 (1995).

37. K. Achiwa, T. Kogure, and I. Ojima, *Tetrahedron. Lett.* **18**, 4431 (1977).

38. T. Ojima, T. Kogure, and Y. Yoda, *Org. Synth.* **63**, 18 (1985).

39. I. Ojima, T. Kogure, and T. Terasaki, *J. Org. Chem.* **43**, 3444 (1978).

40. S. Shimizu, S. Hattori, H. Hata, and H. Yamada, *Appl. Environ. Microbiol.* **53**, 519 (1987).

41. M. Kataoka, S. Shimizu, and H. Yamada, *Agric. Biol. Chem.* **54**, 177 (1990).

42. M. Kataoka, S. Shimizu, and H. Yamada, *Recl. Trav. Chim. Pays-Bas.* **110**, 155 (1991).

43. M. Kataoka, S. Shimizu, Y. Doi, and H. Yamada, *Appl. Environ. Microbiol.* **56**, 3595 (1990).

44. R. Eck and H. Simon, *Tetrahedron Asymm.* **5**, 1419 (1994).

45. R. Kuhn and Th. Wieland, *Chem. Ber.* **75**, 121 (1942).

46. R. P. Lanzilotta, D. G. Bradley, and K. M. McDonald, *Appl. Microbiol.* **27**, 130 (1974).

47. S. Shimizu, S. Hattori, H. Hata, and H. Yamada, *Enzyme Microb. Technol.* **9**, 411 (1987).

48. H. Yamada and S. Shimizu in D. S. Clark, D. Estell, and J. Dordick, eds., *Enzyme Engineering*, Vol. 11, The New York Academy of Sciences, New York, 1992, p. 374.

49. S. Shimizu, H. Yamada, H. Hata, T. Morishita, S. Akutsu, and M. Kawamura, *Agric. Biol. Chem.* **51**, 289 (1987).

50. K. Nakamura, S.-I. Kondo, Y. Kawai, and A. Ohno, *Tetrahedron Asymm.* **4**, 1253 (1993).

51. Jpn. Pat. 05227987 A2, (Sept. 7, 1993) K. Sakamoto, H. Yamada, and A. Shimizu (to Fuji Yakuhin Kogyo KK).

52. Jpn. Pat. 06311889 A2, (Apr. 30, 1993) J. Ohno, K. Nakamura, K. Yasushi and K. Shinichi (to Kokai Tokkyo Koho); K. Nakamura, K. Shin-ichi, and O. Atsuyoshi, *Bioorganic. Med. Chem.* **2**, 433 (1994).

53. M. Paetow, H. Ahrens, and D. Hoppe, *Tetrahedron Lett.* **33**, 5323 (1992).

54. J. Kallmerten and T. Gould, *J. Org. Chem.* **51**, 1152 (1986).

55. A. V. Rama Rao, S. Mahender Rao, and G. V. M. Sharma, *Tetrahedron Lett.* **35**, 5735 (1994).

56. F. Effenberger, J. Eichhorn, and J. Roos, *Tetrahedron Asymm.* **6**, 271 (1995).

57. C. J. Chesterton, P. H. W. Butterworth, and J. W. Porter, *Meth. Enzymol.* **18A**, 371 (1970).

58. C. J. Chesterton, P. H. W. Butterworth, and J. W. Porter, *Meth. Enzymol.* **18A**, 364 (1970).

59. U.S. Pat. 2441949 (May 25, 1948) S. H. Babcock (to the University of California, Berkeley, Calif.).

60. E. E. Snell et al., *J. Am. Chem. Soc.* **72**, 5349 (1950).

61. J. Baddiley and E. M. Thain, *J. Chem. Soc.*, 800 (1952).

62. D. P. Martin and D. G. Drueckhammer, *J. Am. Chem. Soc.* **114**, 7288 (1992).

63. D. S. Vallari and C. O. Rock, *J. Bacteriol.* **162**, 1156 (1985).

64. G. M. Brown and J. M. Williamson, *Adv. Enzymol.* **53**, 345 (1982).

65. J. H. Teller, S. G. Powers, and E. E. Snell, *J. Biol. Chem.* **251**, 3780 (1976).

66. S. Shimuzu, M. K. Qtaoka, M. C. M. Chunag, and H. Yamandu, *J. Biol. Chem.* **263**, 1207 (1988).

67. J. M. Williamson and G. M. Brown, *J. Biol. Chem.* **254**, 8074 (1979).

68. D. G. Novelli, *Physiol. Rev.* **33**, 523 (1953).

69. Y. Shigeta and M. Shichiri, *J. Vitamin.* **12**, 186 (1966).

70. Y. Abiko, S. Ashida, and M. Shimuzu, *Biochim. Biophys. Acta* **268**, 364 (1972).

71. D. Bhatia, in *Encyclopedia of Food Science and Technology*, John Wiley & Sons, Inc., New York, 1991, p. 2783; B. W. Wyse, in J. Augustin, B. P. Klein, D. A. Becker, and P. B. Venugopal, eds., *Methods of Vitamin Assay*, 4th ed., John Wiley & Sons, Inc., New York, 1985, Chapt. 16, p. 2783.

72. B. W. Wyse, C. Withwer, and R. G. Hansen, *Clin. Chem.* **25**, 108 (1979).

73. R. B. Roy and A. Buccafuri, *J. Assoc. Off. Anal. Chem.* **61**, 720 (1978).

74. J. Davidek, J. Velisek, J. Cerna, and T. Davidek, *J. of Micronutr. Anal.* **1**, 39 (1985).

75. W. A. König and U. Sturm, *J. Chromatogr.* **328**, 357 (1985).

76. K. Banno, S. Horimoto, and M. Matsuoka, *J. Chromatogr.* **564**, 1 (1991).

77. R. E. Hodges, M. A. Ohlson, and W. B. Bean, *J. Clin. Invest.* **37**, 1642 (1958).

78. M. Glusman, *Am. J. Med.* **3**, 211 (1947).

79. E. Kazuko, N. Kubato, T. Nishigaki, and M. Kikutani, *Chem. Pharm. Bull.* **23**, 1 (1975).

80. M. M. Nelson, F. Van Nouheys, and H. M. Evans, *J. Nutr.* **34**, 189 (1947).

81. D. M. Hegsted, J. J. Oelson, R. C. Mills, C. A. Elvehjem, and E. B. Hart, *J. Nutr.* **20**, 599 (1940).

82. G. D. Lopaschuk, M. Michalak, and H. Tsang, *J. Biol. Chem.* **262**, 3615 (1987).

83. K. Shibata, C. J. Gross, and L. M. Henderson, *J. Nutr.* **113**, 2107 (1983).

84. D. K. Reibel, B. W. Wyse, D. A. Berkich, and J. R. Neely, *Am. J. Physiol.* **240**, H 606 (1981).

85. B. R. Eissenstat, B. W. Wyse, and R. G. Hansen, *Am. J. Clin. Nutr.* **44**, 931 (1986).

86. W. B. Bean and R. E. Hodges, *Proc. Soc. Exp. Biol. Med.* **86**, 693 (1954).

87. P. C. Fry, H. M. Fox, and H. G. Tao, *J. Nutr. Sci. Vitaminol.* **22**, 339 (1976).

88. Y. Furukawa and S. Kimura, *J. Vitamin.* **18**, 213 (1972).

89. R. W. Luecke, J. A. Hoefer, and F. Thorp, Jr., *J. Anim. Sci.* **11**, 138 (1952).

90. Committee on Dietary Allowances, Food and Nutrition Board, *Recommended Dietary Allowances*, 10th ed., National Academy Press, Washington, D.C., 1989.

91. G. F. Combs, *The Vitamins*, Academic Press, San Diego, Calif., 1992, Chapt. 15, p. 345.

92. K. Unna and J. C. Greslin, *J. Pharmacol. Exp. Ther.* **73**, 85 (1941).

93. J. F. Grenier, M. Aprahamian, C. Genot, and A. Dentinger, *Acta Vita. et Enzy.* **4**, 81 (1982).

94. M. Aprahamian, A. Dentinger, C. Stoch-Damge, J. C. Kouassi, and J. F. Grenier, *Am. J. Clin. Nutr.* **41**, 578 (1985).

95. F. Vaxman et al., *Eur. Surg. Res.* **27**, 158 (1995).

GENERAL REFERENCES

H. M. Fox in L. J. Machlin, eds., *Handbook of Vitamins*, Marcel Dekker Inc., New York, 1991, pp. 429–451.

O. Isler, G. Brubacher, S. Ghisla, and B. Kräutler, *Vitamine II*, Georg Thieme Verlag, New York, 1982, pp. 309–339.

S. Shimizu and Y. Yamada, in T. O. Baldwin, F. M. Raushel, and A. I. Scott, eds., *Chemical Aspects of Enzyme Biotechnology*, Plenum Press, New York, 1990, p. 151.

B. W. Wyse, in J. Augustin, B. P. Klein, D. A. Becker, and P. B. Venugopal, eds., *Methods of Vitamin Assay*, 4th ed., John Wiley & Sons, Inc., New York, 1985, Chapt. 16.

G. F. Combs, *The Vitamins*, Academic Press, San Diego, Calif., 1992, Chapt. 15, pp. 345 ff.

THIMMA R. RAWALPALLY
Hoffmann-La Roche, Inc.

VITAMINS: PYRIDOXINE (B₆)

Pyridoxine [65-23-6] (vitamin B₆) is the recommended IUPAC-IUB designation for a group of closely related, biologically interconvertible 3-hydroxy-2-methylpyridines which exhibit the biological activities of pyridoxine in rats (1). This group includes pyridoxine [65-23-6] (**1**), pyridoxal [66-72-8] (**2**), and pyridoxamine [85-87-0] (**3**) and their respective 5′-phosphates (**4–6**). In common usage, though, the term vitamin B₆ most often refers to pyridoxine hydrochloride [58-56-0] (**7**), the most important commercial form (Fig. 1). Contrary to previous proposals and explanations, pyridoxine is not to be used as a general name for this group of vitamers. Pyridoxine is the official name of the specific substance 5-hydroxy-6-methyl-3,4-bis(hydroxymethyl)pyridine (**1**) which usually comprises only a fraction of the total amount of vitamin B₆ in natural sources. The older term pyridoxol is seldom used.

Vitamin B₆ is widely distributed in small amounts in the tissues and fluids of nearly all living substances. Bound to enzymes as the active form, pyridoxal phosphate (**2**), it acts as a coenzyme for a number of biochemical conversions of amino acids important to general cellular energy supply, to growth, and to specific organ functions. Deficiency in animals leads to a variety of symptoms ranging from mild (dermatitis, loss of appetite, irritability, and muscular weakness) to serious (anemia, weight loss, and

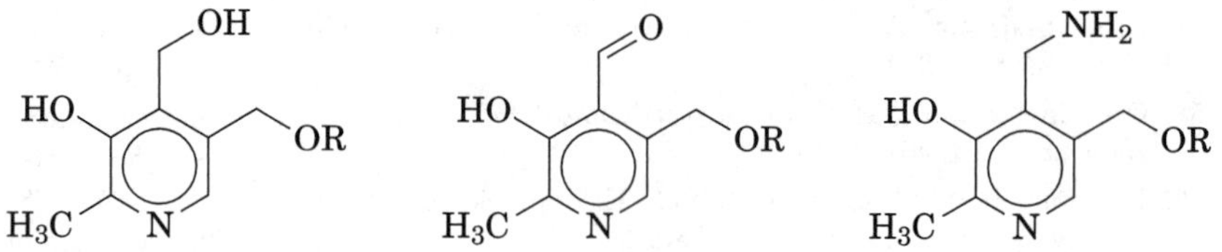

R = H   (**1**)
R = OPO₃H₂   (**4**)
R = H HCl salt   (**7**)

(**2**)   (**3**)

(**5**)   (**6**)

Figure 1. Forms of vitamin B_6.

nerve degradation leading to convulsive seizures) and even death. Vitamin B_6 requirements depend on protein intake and other factors. Meats, cereals, beans, and nuts are rich sources. Bioavailability and stability vary with the food sources and methods of preparation and storage. As a preventive measure, pyridoxine hydrochloride, the most stable form, is used in food, feed, and nutrient supplements. Pyridoxine is also used therapeutically for treatment of deficiency conditions and other diseases. No natural sources are rich enough, nor is biosynthetic understanding sufficiently advanced, to allow cost-effective production from biological sources. All commercially available pyridoxine hydrochloride is manufactured by chemical processes practiced at fine chemical scale.

Vitamin B_6 was found in the 1940s during the unraveling of the mysteries of the nitrogen-base B vitamin complex. A diet-induced pellagra-like condition in rats was shown to be cured by a water-soluble base, later differentiated from the known vitamins B_1 and B_2 and named vitamin B_6 by Gyorgy in 1934. Several research groups undertook isolation and characterization of this substance and five different laboratories announced this achievement in 1938. The chemical structure of pyridoxine, the form first isolated, was correctly identified and reported along with two independent syntheses in 1939 (2,3). Snell showed microorganisms that could not use pyridoxine were satisfied with the natural vitamin B_6 extracts. This led to the discovery and synthesis of the vitamers pyridoxal (**2**) and pyridoxamine (**3**) in 1944. Subsequent clinical and biochemical investigations have shown the vitamin B_6 complex has therapeutic and prophylactic uses for treatment of a number of diseases. Commercial production of pyridoxine hydrochloride began in the 1940s. Production rose rapidly to more than 3000 metric tons per year worldwide.

CHEMICAL AND PHYSICAL PROPERTIES

Physical and Chemical Characterization

Ultraviolet absorptions of the B_6 vitamers vary with pH and the substituent at position 4. As hydroxypyridines, the B_6 vitamers show strong fluorescence (excitations at 310–330 nm, emissions at 365–425 nm, depending on structure and pH) which are useful in their detection (4,5). Ir and ¹H, ¹³C, and ¹⁵N nmr show the expected effects of N-protonation, hydrogen bonding, and exchange of the acidic phenolic proton (6–8). In mass spectrometry (ms), normal

electron impact ionization is sufficient to observe the molecular ion and an intense peak for a quinone—methide fragment from loss of water at the 4-hydroxymethyl position (9). Thermospray hplc/ms at picogram sample levels showed particularly strong molecular ions for pyridoxine and derivatives (10).

Reactions and Stability

The B_6 vitamers show weakly basic and acidic properties (4) and undergo reactions expected of their functionalities (11,12). Acidity constants are reported in Table 1. Pyridoxine gives the characteristic reactions of phenols such as color tests, ether formation, diazo coupling, etc. Ketalization or acetalization of pyridoxine gives the thermodynamically more stable six-membered ring products involving the 3- and 4-positions. Esterification gives first mainly the 4,5-diester, then the triester with excess acylating agent. Pyridoxal and pyridoxamine are phosphorylated on the 5′-position with reagents such as phosphorus pentoxide in phosphoric acid or metaphosphoric acid. Pyridoxine requires prior protection of substituents at positions 3- and 4- as a ketal to allow selectivity. In biological systems, glycosylation occurs at the 5′-hydroxyl.

The B_6 vitamers are labile. Pyridoxine is the most stable and pyridoxal the least. The reactions of the substituents at the 4-carbon dominate. On mild oxidation, pyridoxine is converted to pyridoxal (**2**), which is in equilibrium with its hemiketal [*17281-92-4*] (**8**) (13). Air oxidation gives pyridoxic acid [*82-82-6*] (**9**), an intensely blue fluorescent substance of value for quantitation (4). Heating pyridoxine with ammonia converts it to pyridoxamine (**3**) and with alcohols to the 4-ethers (13). Pyridoxine, pyridoxal, and pyridoxamine are relatively stable in acid when cold or heated at 100°C for short periods. Prolonged heating of pyridoxine in hydrochloric acid gives the 4,5-bis-chloromethyl analogue, the 4-chloromethyl group of which is preferentially hydrolyzed. Heating pyridoxine in neutral solution in the absence of good nucleophiles gives dimeric products (14). Some of these reactions are readily rationalized by an intermediate quinone—methide (**10**) (Fig. 2).

Pyridoxal and its 5′-phosphate undergo typical aldehyde reactions, such as aldol condensations, bisulfite addition, etc (4,11,12). Both form imines, assisted by the electronegative 2-methylpyridine ring and the acidic 3-hydroxyl. Extensive kinetic and thermodynamic data on imine formation in model systems are available (4). The epsilon-amino group of a lysine unit is usually involved in the active sites of enzymes but other groups can bind pyr-

Table 1. Properties of Pyridoxine, Pyridoxamine, Pyridoxal, and Derivatives

Substance	Structure number	CAS registry number	Formula	Mol wt	Form	Mp in water (°C)	Solubility (g/100 mL)	pK Values
Pyridoxine	(1)	[65-23-6]	C$_8$H$_{11}$NO$_3$	169.18	Needles	160	Soluble	5.0, 9.0
Pyridoxine HCl	(7)	[58-56-0]	C$_8$H$_{12}$ClNO$_3$	205.64	Platelets	204–206 dec	22	
Pyridoxine-5′-PO$_4$	(4)	[447-05-2]	C$_8$H$_{12}$NO$_6$P	249.16	Needles	212–213 dec	Soluble	5.0, 9.4
Pyridoxamine	(3)	[85-87-0]	C$_8$H$_{12}$N$_2$O$_2$	168.20	Crystals	193	Soluble	3.4, 8.1, 10.5
Pyridoxamine di-HCl		[524-36-7]	C$_8$H$_{14}$Cl$_2$N$_2$O$_2$	241.12	Platelets	226–227 dec	50	
Pyridoxamine-5′-PO$_4$	(6)	[529-96-4]	C$_8$H$_{13}$N$_2$O$_5$P	248.18			Soluble	3.5, 5.8, 8.6, 10.8
Pyridoxal	(2)	[66-72-8]	C$_8$H$_9$NO$_3$	167.16			50	4.2, 8.7, 13
Pyridoxal HCl		[65-22-5]	C$_8$H$_9$NO$_3$	203.63	Rhombic	173 dec	50	
Pyridoxal-5′-PO$_4$	(5)	[54-47-7]	C$_8$H$_{10}$NO$_6$P	247.14			Soluble	3.5, 8.4

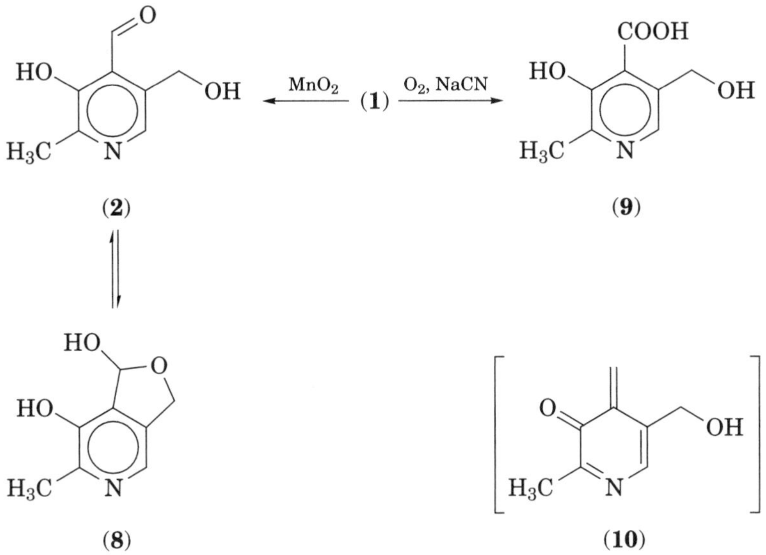

Figure 2. Reactions of pyridoxal (2).

idoxal during thermal processing and storage, reducing bioavailability (4,15). Hydroxylamines and hydrazine derivatives, such as isoniazid, react and cause enzyme inhibition (4). Imine formation with pyridoxal is used as a tool for selective modification of proteins in biochemical studies (4).

Reactions of imines and their tautomers are the basis of the important biochemistry of pyridoxal 5′-phosphate (5), the coenzyme form for over 100 enzymatic reactions. Reaction with an epsilon-amino group of a lysyl residue and ionic interaction of the 5′-phosphate binds pyridoxal to the enzyme in the resting state. Displacement of lysine by a substrate amino group initiates a sequence of facile equilibrations of the aldimine, ketimine, and enamine forms, along with elimination and addition reactions which rationalize the various conversions and products. Known reactions involving the alpha-, beta-, and gamma-carbons of the amino acid and their substituents include decarboxylations, transaminations, racemizations, aldol and retroaldol reactions, eliminations, and hydration reactions (4) (Fig. 3). Many of these transformations have been modeled *in vitro* in chemical studies. Some of this chemistry can cause covalent bonding of a substrate to other reactive residues in the active site, resulting in irreversible "suicide inactivation" of the enzyme. This is a mechanism of action

of some naturally occurring toxins and a protocol in rational drug design (4).

Pyridoxine hydrochloride is stable if kept dry or in acidic solution. Its aqueous solutions may be heated for 30 minutes at 120°C without decomposition (16). Losses have been reported for mixtures of the vitamers stored in solution at room temperature or below unless they are kept acidic (17). Decomposition in foods on processing and storage depends mainly on pH, moisture, and temperature. First-order kinetics are often followed and stabilities can be estimated using Arrhenius calculations (4,18). All B$_6$ vitamers are susceptible to decomposition by near-uv light to give reddish brown materials, especially in solution at neutral or higher pH. Pyridoxal and pyridoxamine decompose most rapidly, even in acid (18,19). Gamma irradiation of food can also cause losses (20). Hydroxyl radical formed from ascorbate gives the 6-hydroxyl derivative of pyridoxine. This process may occur during the storage of fruits and vegetables containing high ascorbate levels (21,22).

Salts and Derivatives

Generally the B$_6$ vitamers are high melting crystalline solids that are very soluble in water and insoluble in most other solvents. Properties of the common forms are listed

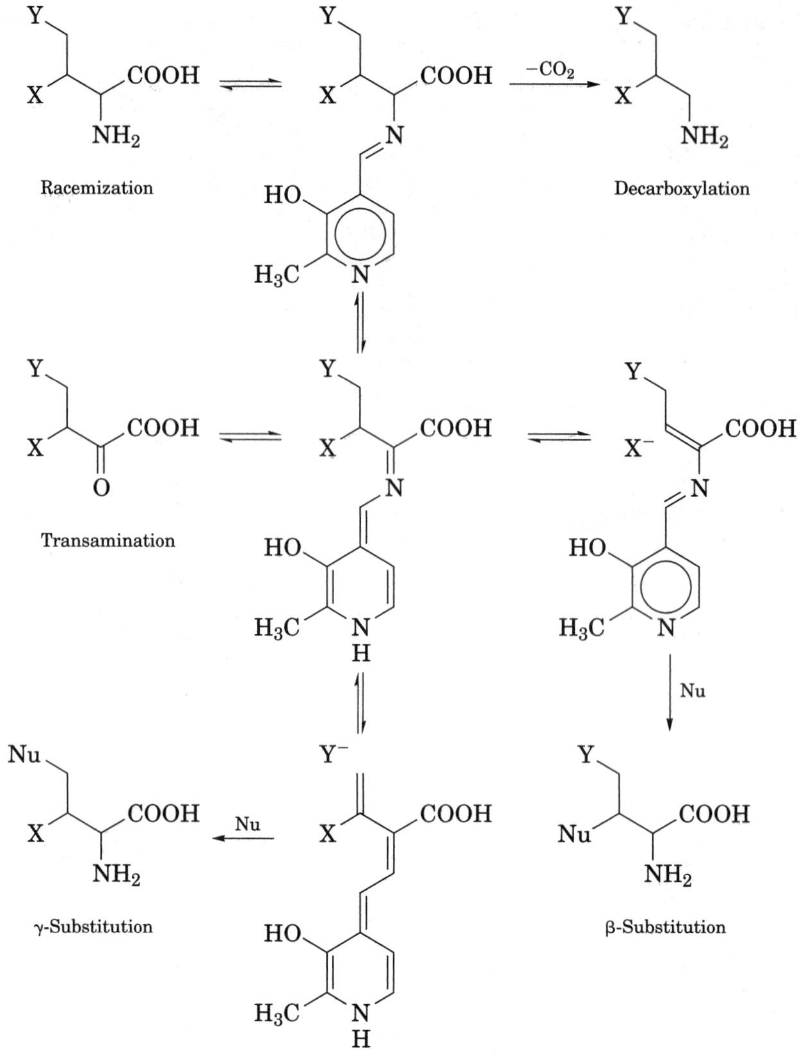

Figure 3. Reactions of imines and their tautomers.

in Table 1. The only commercially important form of vitamin B$_6$ is pyridoxine hydrochloride (**7**). This odorless crystalline solid is composed of colorless platelets melting at 204–206°C (with decomposition). In bulk, it appears white and has a density of ~0.4 kg/L. It is very soluble in water (ca 0.22 kg/L at 20°C), soluble in propylene glycol, slightly soluble in acetone and alcohol (ca 0.014 kg/L), and insoluble in most lipophilic solvents. A 10% water solution shows a pH of 3.2. Both the hydrochloride and corresponding free base sublime without decomposition (16).

OCCURRENCE

Vitamin B$_6$ occurs in the tissues and fluids of virtually all living organisms, attesting to its essential role in amino acid biochemistry. Pyridoxal phosphate bound to protein is the main form in animals. Pyridoxine, pyridoxamine, and their phosphates are the usual forms in plants (23,24). Significant percentages in plants occur as pyridoxine glycosides, which are substantially less bioavailable (23,25,26). Nearly all foods in a typical human diet contain detectable amounts of the B$_6$ vitamers. Relatively rich sources are

edible yeast, meats (especially salmon and calf liver), whole grain cereals, legumes, and nuts (qv). Cheeses with extensive mold growth are also good sources. Broccoli, cauliflower, spinach, corn, potatoes, bananas, and raisins are good sources, but vegetables and fruits in general are not (24) (Table 2).

Table 2. Average Vitamin B$_6$ Contents of Raw Foods, mg/ 100 g^a

Food	Vitamin B$_6$	Food	Vitamin B$_6$
Beef liver	0.84	Brown rice	0.55
Beef	0.33	White rice	0.17
Chicken	0.33–0.68	Whole wheat flour	0.34
Pork	0.35	White flour	0.06
Fish	0.17–0.43	Beans	0.56–0.81
Eggs	0.11	Peanuts	0.40
Whole milk	0.04	Corn	0.25
Bananas	0.51	Potatoes	0.25
Raisins	0.24	Vegetables, green	0.15–0.28
Other fruits	0.02–0.07	Tomatoes	0.10

^{a}By microbial assay.

As vitamin B$_6$ is mainly located in the germ and aleurone layer in cereal grains; polishing for the production of flour removes a substantial portion. White bread is therefore a poor source unless fortified. Some nonedible yeasts contain up to 38 mg/100 g dry weight vitamin B$_6$, the highest level of the natural sources (4,27). As a rule, these amounts are too low for cost-effective isolation.

BIOCHEMICAL AND PHYSIOLOGICAL FUNCTIONS

Because vitamin B$_6$ is widely distributed in nature, severe deficiency symptoms are seldom observed in humans with adequate diets. However, there is little doubt vitamin B$_6$ is essential in human and animal nutrition. As enzyme cofactors, pyridoxal 5'-phosphate and pyridoxamine 5'-phosphate are required for important biotransformations of amino acids (4). Decarboxylations of amino acids give rise to a number of important amines, among them the vasodilator histamine, the vasoconstrictor and neurotransmitter serotonin, and the major neurotransmitters gamma-aminobutyric acid, epinephrine, norepinephrine, tyramine, and dopamine. Transamination interconverts specific pairs of amino acids and their corresponding ketoacids and amino groups and aldehydes, key functions in amino acid biosynthesis and catabolism. Racemization provides D-amino acids, constituents of the cell walls of bacteria. Other important transformations requiring pyridoxal include bioconversions of tryptophan (involved in the biosynthesis of the vitamin niacin), serine transhydroxymethylation (a key step in one carbon metabolism leading to nucleic acid synthesis), glycogen phosphorylation (in liver and muscle-maintaining glucose supplies), biosynthesis of delta-aminolevulinic acid (a precursor to heme synthesis), and key steps in the metabolism of sulfur-containing amino acids. Other roles of pyridoxal and its phosphate include binding to hemoglobin (affecting oxygen binding affinity), effects on the conversion of linoleic acid to arachidonic acid, possible effects on cholesterol and fatty acid metabolism, and modulation of steroid action. As a result, vitamin B$_6$ levels affect gluconeogenisis, nervous system activity, red cell formation and metabolism, immune function, and hormone functions (4,23).

A large variety of biochemical lesions are found in deficiency (23). Depressed antibody, nucleic acid, and protein synthesis are characteristic. Consequences of vitamin B$_6$ deficiency in humans and many animals include scaling dermatitis (especially around the eyes, nose, and mouth), anemia, loss of appetite, poor growth or weight loss, muscular weakness, central nervous system changes such as irritability or depression, kidney stones, and abnormalities seen by electroencephalography. In severe cases, nerve degradation leading to convulsive seizures and death can occur. Possible vitamin B$_6$ deficiency must be assessed; estimation of dietary intake is not sufficient. An older method measures the level of pyridoxal-dependent metabolite xanthurenic acid in the urine after ingestion of a large dose of tryptophan. Improved methods measure pyridoxal phosphate in plasma or pyridoxic acid excretion in urine by sensitive chromatographic procedures (23).

Humans cannot biosynthesize vitamin B$_6$ and must obtain it from dietary sources. Ruminant animals derive some benefit from microfloral supply in the stomach. All animals interconvert the six major forms (1–6). Thus, requirements of the enzyme cofactor forms can be maintained by ingestion of pyridoxine only. The bioavailability of vitamin B$_6$ in foods is still an active research area with some unresolved issues (4,15,23). Reported levels of availability vary widely. Earlier studies suggest processing, storage, and preparation could lead to losses of up to 80% of vitamin B$_6$, depending on conditions. Irreversible reductive binding to lysyl residues is reported to give inactivity or even antivitamin effects (15,28). One study estimates bioavailability of vitamin B$_6$ from foods representing the average North American human diet at ca 70–80% (29).

In humans, vitamin B$_6$ in phosphorylated forms is mostly unavailable until hydrolyzed, then passively absorbed in the small intestine. The free vitamins are interconverted and supplied to the circulation by the liver, then enter cells by simple diffusion and are phosphorylated. A typical adult pool of pyridoxal is estimated to be about ≥25 mg, most of it bound to enzymes in muscle (23). Small amounts of all forms are excreted in the urine. Most is oxidized in the liver to the main urinary metabolite 4-pyridoxic acid (9), the corresponding diacid, and ring-cleaved fragments (4,30).

Human requirements for vitamin B$_6$ parallel protein metabolism, with increased intake of protein requiring higher intake of the vitamin. As little as 0.010–0.015 mg of vitamin B$_6$ per gram of protein prevents deficiency signs with adequate protein intake. Slightly higher intake maintains acceptable metabolite excretion and plasma vitamin levels. The Recommended Daily Allowance (RDA) for vitamin B$_6$ assumes consumption of twice the recommended amount of protein (Table 3) (31). Allowances for pregnant and lactating women reflect increased protein needs. Low levels in breast milk place infants at risk for deficiency (4,23).

The typical U.S. daily diet contains 1.1–3.6 mg of vitamin B$_6$, most coming from meats and vegetables. Poor diets may provide less than half of these amounts and less than the RDA. Some populations require higher amounts: persons with high protein intakes, pregnant and lactating women, users of oral contraceptives, alcoholics, users of drugs which interfere with vitamin B$_6$ function, and those afflicted with some diseases. Several reviews have examined the relationship of vitamin B$_6$ and specific diseases in more detail (4,23).

MANUFACTURE

Pyridoxal (2) can be prepared by oxidation of pyridoxine (1) (13). Pyridoxamine (3) can be made by reduction of pyridoxal oxime (32). Only pyridoxine (1), the most stable of

Table 3. U.S. RDA for Vitamin B$_6$

Age	RDA (mg)
Infants and children <4 yr	0.7
Adults and children >4 yr	2.0
Pregnant or lactating women	2.5

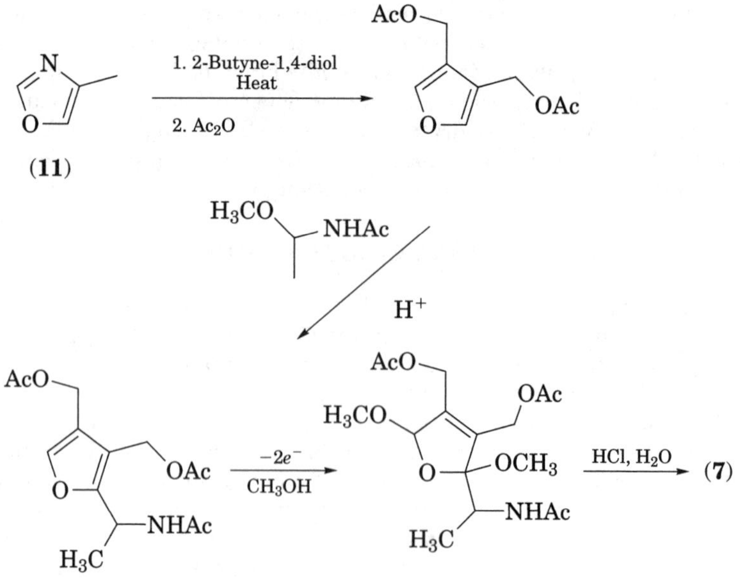

Figure 4. Furan route to pyridoxine.

the three, is manufactured. The earliest syntheses, by degradation of isoquinoline rings, are only of historical interest. The first syntheses to be applied on a large scale relied on classical condensation reactions. Many variations were explored during the 1940s and 1950s and some formed the basis of early production. Such routes required numerous steps giving low overall yields, in part due to difficulties of obtaining the correct oxidation levels of the 4- and 5-substituents. These methods have been reviewed (11,12). In the mid-1950s, electrooxidation of furan derivatives was shown to provide direct precursors to hydroxypyridines (33). In later improvements, cycloaddition of readily available 2-butyne-1,4-diol to 4-methyloxazole [693-93-6] (**11**) followed by amidoalkylation conveniently provided the furans (34,35). The overall yields of these linear sequences are modest (Fig. 4).

The current paradigm for B$_6$ syntheses came from the first report in 1957 of a synthesis of pyridines by cycloaddition reactions of oxazoles (36) (Fig. 5). This was adapted for production of pyridoxine shortly thereafter. Intensive research by Ajinomoto, BASF, Daiichi, Merck, Roche,

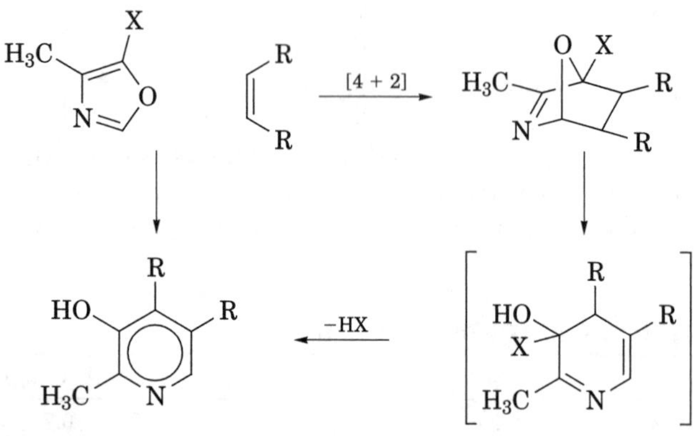

Takeda, and other companies has resulted in numerous publications and patents describing variations. These routes are convergent, shorter, and of reasonably high throughput.

Substitution on the azadiene and alkene affect reactivity and only certain pairs react efficiently. Electrophilic alkenes and oxazoles bearing donor groups generally allow faster reactions at lower temperatures (37,38). Steric effects play a lesser role. Earlier variations with 4-methyloxazole (**11**) and maleic acid derivatives require the presence of oxidants such as hydrogen peroxide or nitroarenes to aromatize the cycloadducts in order to obtain good yields (11). Subsequent costly reduction steps are also needed to correct the oxidation level of the product side chains. One advantage is the use of leaving groups on one of the cycloaddition partners to facilitate aromatization without loss of the product 3-hydroxyl group. These superfluous groups can be added at the 5-position of the oxazoles or on the alkene.

Oxazoles preferred in practice bear 5-ethoxy- or 5-cyanosubstituents. Other alkoxy (39–42), trimethylsilyloxy (43), alkythio (44), amino (45), and carboalkoxy (46) substituents are reported. Processes using 5-ethoxyoxazoles (**12**), (**13**), and (**14**) have been published (38,47–60). Extra carbons carried by the Takeda, Chinese, and Ajinomoto intermediates are lost as waste by decarboxylation. A process with 5-cyano-4-methyloxazole (**15**) has been disclosed by Roche (59). Reactions of these azadienes with butenediol derivatives, the usual alkene partner, take place at higher temperatures (115–180°C) and require longer reaction times (10–20 h) than those with maleate partners. The ethoxy compound (**12**) reacts under milder conditions and the adduct is isolable. Mild acid converts it to the product and ethanol. With the higher temperatures required of the cyano compound [1003-52-7] (**15**), the intermediate cycloadduct is converted directly to the product by elimination of waste hydrogen cyanide. Often the reactions are run with neat liquid reagents having an excess of alkene as solvent. Polar solvents such as sulfolane and N-methyl-

Figure 5. Synthesis of pyridines by cycloaddition reactions of oxazoles.

pyrrolidone are claimed to be superior for reactions of the ethoxy compound with butenediol (53). Organic acids, phenols, maleic acid derivatives, and inorganic bases are suggested as catalysts (51,52,54,59,61,62) (Fig. 6).

As the alkene partners, (Z)-2-butene-1,4-diol acetals, ketals, esters or ethers (11), and a polymer-bound acetal (63) are described. Butenediol may be used if polar solvents are employed (54). Current manufacturers generally use the cyclic acetal [5417-35-6] (16) of butenediol with isobutyraldehyde, a liquid of conveniently high boiling point that readily allows the product to be deprotected with no waste (Fig. 7). Butenediol is readily and cheaply available from the reaction of acetylene and formaldehyde to 2-butyne-1,4-diol, then catalytic hydrogenation. In an unusual variant of this cycloaddition approach, a leaving group on the alkene facilitates aromatization. Cyclic sulfone [41409-84-1] (17) paired with 4-methyloxazole (11) effectively gives the pyridine nucleus (64,65) taking advantage of the increased reactivity of an electrophilic alkene while maintaining the correct oxidation level of the side chains. Electrophilic dihydrofuran [332-77-4] (18) reacts

efficiently with excess ethoxymethyloxazole (12), also at lower temperatures (66).

Ethoxyoxazole (12) is constructed from N-formyl alanine esters (67) and ethoxyoxazole acid (13) from N-oxalyl alanine esters (55–58). Alanine is readily available from China by chlorination of propionic acid, then ammonation. Both ethoxyoxazole (12) and the acid (14) are derived from maleic anhydride by N-formyl aspartate esters (52,68). Cyanooxazole (15) is derived from ketene by ethyl chloroacetoacetate and formamide (69) or, alternatively, from diketene (70). All routes require dehydration of amides as a critical step. Traditionally these dehydrations have been accomplished with phosphorus pentoxide, phosphoryl chloride, or phosgene; however, more environmentally acceptable procedures use acetic anhydride (71), cyanuryl chloride (72), or gas-phase catalytic dehydration (73). In some cases, the oxazoles are generated in situ in the cycloaddition reactions from N-formyl compounds by the isonitriles under dehydrating conditions (45,74,75). Other cycloaddition routes include reactions of 1,2,4-triazines with butynediol or enamine derivatives give the pyridoxine nu-

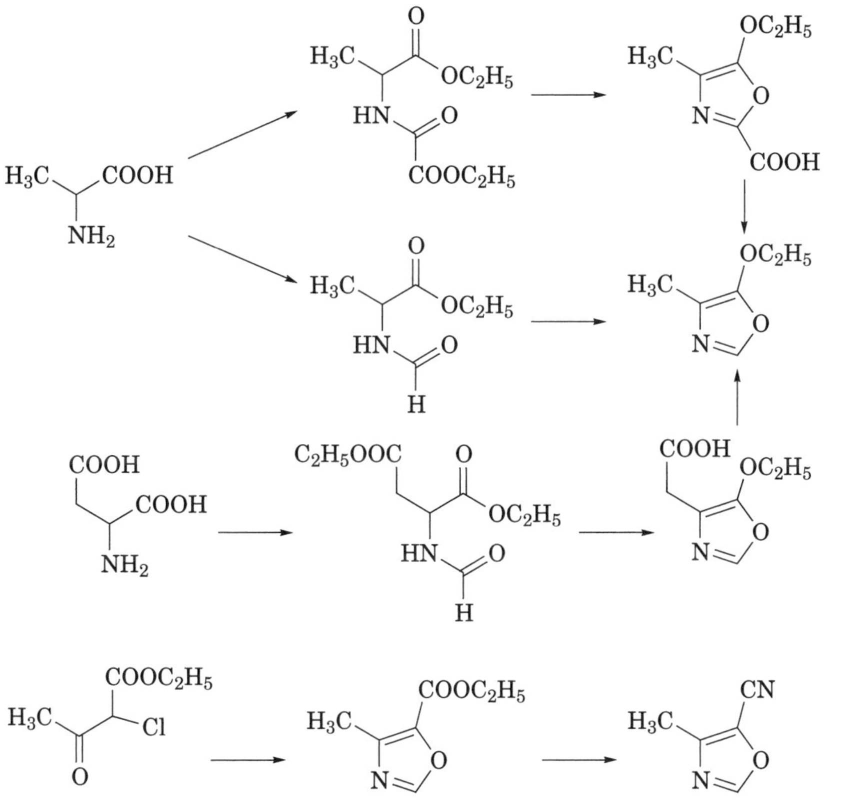

(13)

(12)

(14)

(15) **Figure 6.** Process chemistry for oxazoles.

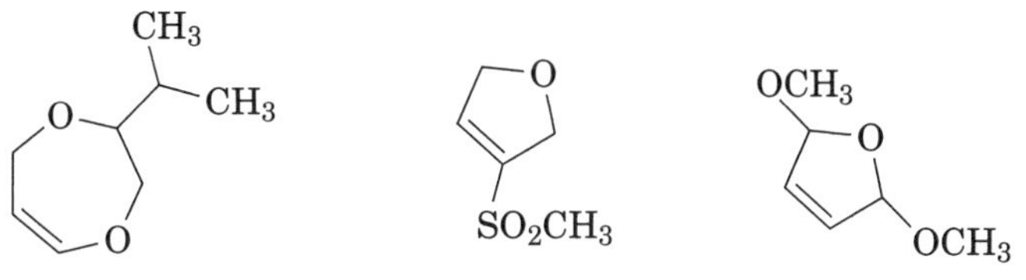

(16)   (17)   (18)

Figure 7. Dienophiles.

cleus (76,77), or a cobalt-catalyzed [2 + 2 + 2] cycloaddition of acetylenic ethers with acetonitrile (78,79) (Fig. 8).

ECONOMIC ASPECTS

Significant producers include Daiichi, Hoffmann-La Roche, Takeda, and several factories in mainland China. Takeda and Daiichi practice processes based on alanine by ethoxyoxazole (12), at least one Chinese producer from alanine by oxazole acid (13), and Roche from ketene via cyanooxazole (15). Production is practiced at many hundreds of metric tons annually by batchwise or semicontinuous operation in automated glass and stainless steel equipment typical of fine chemical manufacture. World production in 1995 was estimated at about 3000 MT with sales prices in the United States in the range of $22–$30 per kilogram.

PRODUCT SPECIFICATIONS AND TESTING

Nutrients and diet supplements without claims of therapeutic effects are considered foods, and are thus regulated by the U.S. Food and Drug Administration. These are further subject to specific food regulations. Specifications for pyridoxine hydrochloride (7) for foods are given in the *Food Chemicals Codex* (80) and for pharmaceuticals in the *U.S. Pharmacopeia* (81). General test methods have been summarized (82).

SAFETY AND HANDLING

Pyridoxine hydrochloride is typically packaged in standard 20–25 kg polylined fiber cartons or drums. Smaller packing sizes are also available. Pyridoxine hydrochloride is kept under normal dry storage conditions not to exceed 25°C and protected from light. It is listed in the TSCA inventory and material safety data sheets are available from suppliers. Protection against dust exposure to the eyes, skin, or lungs (IOEL of 2.0 mg/m^3 time weighted average) and fire and dust explosion are indicated. NPFA ratings for health, fire, and reactivity are 1, 2, and 1. Neither national nor international transportation is regulated. The U.S. EPA has not established reportable quantities for environmental releases. The acute LD$_{50}$ (rat oral) of 7750 mg/kg classifies this material as practically nontoxic orally (83), Oral doses of 100–300 mg/d (50–150 times the RDA) have been used in therapeutic studies without adverse effects.

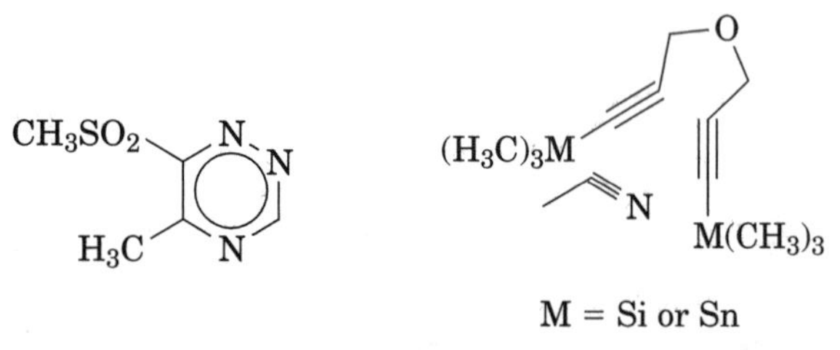

Excess amounts are readily cleared by the kidney. More recent studies show that long-term consumption of very large doses can lead to sensory nerve damage. Supplements of more than 200 mg/d should be taken only under medical supervision. Intravenous administration is not associated with toxicity (84).

ANALYTICAL METHODS

Determining vitamin B$_6$ in foods and tissues is complicated by the number of forms, typical low levels, varying degrees of ionization, lability to heat, light, and base, interconversion, and binding to proteins and glucose. Colorimetric, fluorometric, microbiological, animal, and chromatographic methods are used and have been reviewed in depth (5,11,85,86). Few are fully suitable for simultaneous analysis of all six bioactive forms at the typical low levels. Colorimetric assays involve derivatization, are relatively insensitive, and used only for assay of high potency samples such as premixes. Microbiological assays have previously been the most commonly used, especially for foods. The yeast *Saccharomyces uvarum* (ATCC 9080, also known as *S. carlsbergensis* 4228) responds well to pyridoxine and pyridoxal, less to pyridoxamine, and not at all to the phosphate forms. Hydrolysis is first necessary; by the official AOAC method, the matrix is autoclaved with hydrochloric acid (87). Such methods are slow and overestimate bioavailability. Animal assays in vitamin B$_6$-deficient chicks or rats are time consuming and expensive but have been useful in determining bioavailability of derivatives and related materials.

Increasingly, chromatographic methods are used for their ease, accuracy, and ability to determine all forms simultaneously. Gas-liquid chromatography is hampered by the polarity of the compounds unless silyl or acyl derivatives are used (5,88). The comprehensive method of choice generally is high pressure liquid chromatography (HPLC) (5). Separations by ion exchange or reverse-phase absorption with ion-pair reagents and isocratic or gradient mobile phases are common. Direct uv detection is less sensitive but useful for premixes and vitamin mixtures. Post-column derivatization and fluorescence detection allows measurement to nanogram levels. High performance capillary zone electrophoresis (cze) coupled with fluorescence detection allows measurement to below picogram levels (89). Gc/ms, hplc/ms, and cze/ms hold promise for measurement and identification of femtomole levels of vitamers and metabolites (90–92). Immunoassay (qv) method are being developed (4).

BIOSYNTHESIS

Almost all microorganisms synthesize B$_6$ vitamers intracellularly at levels of 0.05–0.3 mg/g dry weight of cells and excrete at levels of 0.05–0.5 mg/L. Biosynthesis in *E. coli* is known to be controlled by both feedback inhibition and repression. Some organisms can overproduce, for example *B. subtilis*, up to 2.0–5.0 mg/L, *Flavobaterium sp* 238-7 up to 20 mg/L, and yeast *Pichia guilliermondi* up to 25 mg/L (4,93). The mechanisms of vitamin B$_6$ biosynthesis are

M = Si or Sn

Figure 8. Other cycloaddition routes.

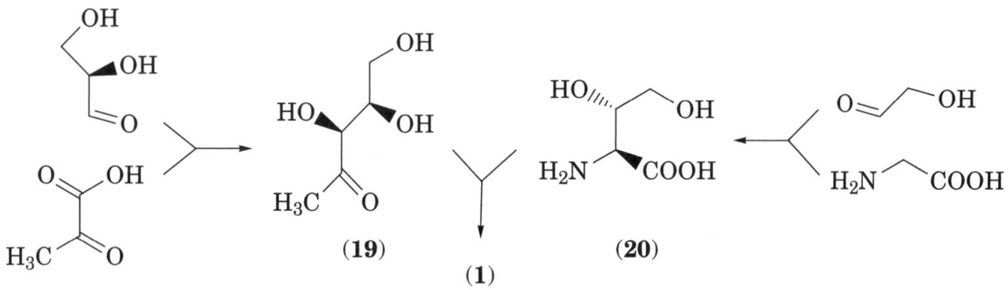

Figure 9. Biosynthetic pathway to pyridoxine.

partly understood. Extensive isotopic label feeding studies with *E. coli* mutants show all carbons are derived from glucose and the nitrogen from glycine. Union of the intermediates 1-deoxy-D-xylulose [*16709-34-9*] (**19**) and 4-hydroxy-L-threonine [*21768-45-6*] (**20**) gives pyridoxine in several steps, only two of which require enzymatic catalysis (Fig. 9) (94). In aerobic bacteria, pyridoxal phosphate is made as needed by the action of oxidase or dehydrogenase enzymes on pyridoxine phosphate or pyridoxamine phosphate (4,93). The pathways in *Flavobacterium* sp 238-7 (95) and in *Saccharomyces cerevisiae* (96) may differ from that in *E. coli* in the early steps. In microorganisms, some genes and related enzymes for biosynthesis have been identified (94,97–99). Significant additional progress will be necessary before cost-effective production of vitamin B₆ by bioprocesses is possible.

ANALOGUES AND ANTAGONISTS

Over 250 analogues of the B₆ vitamers have been reported (11,100). Nearly all have low vitamin B₆ activity and some show antagonism. Among these are the 4-deshydroxy analogue, pyridoxine 4-ethers, and 4-amino-5-hydroxymethyl-2-methylpyrimidine, a biosynthetic precursor to thiamine. Structurally unrelated antagonists include drugs such as isoniazid, cycloserine, and penicillamine, which are known to bind to pyridoxal enzyme active sites (4).

USES

The primary uses of pyridoxine hydrochloride are in multivitamin supplement tablets and for fortification of human food and animal feed, especially for poultry and pigs. Most breakfast cereals and infant formulas in the United States are supplemented. Lesser amounts are used therapeutically to correct deficiencies or to treat specific disorders. Pyridoxine hydrochloride has been used experimentally to treat a variety of conditions with varying degrees of effectiveness (4,23). Pyridoxine hydrochloride is readily incorporated into premixes and foods.

BIBLIOGRAPHY

1. Commission on Biological Nomenclature, *Biochem. J.* **137**, 417 (1974).
2. S. A. Harris and K. Folkers, *J. Am. Chem. Soc.* **61**, 1245 (1939).
3. R. Kuhn et al., *Naturwissenschaften* **27**, 469 (1939).
4. D. Dolphin, R. Poulson, and O. Avramovic, eds., *Coenzymes and Cofactors*, Vols. 1A & 1B, John Wiley & Sons, Inc., New York, 1986.
5. J. Ubbink, in A. P. De Leenheer, W. Lambert, and H. Nelis, eds., *Modern Chromatographic Analysis of Vitamins*, 2nd ed., Marcel Dekker, New York, 1992, Chapt. 10.
6. W. Korytnyk and R. P. Singh, *J. Am. Chem. Soc.* **85**, 2813 (1963).
7. I. D. Spenser et al., *J. Biol. Chem.* **262**, 7463 (1987).
8. A. Lycka and J. Cizmarik, *Pharmazie* **45**, 371 (1990).
9. W. Korytnyk et al., *J. Am. Chem. Soc.* **88**, 1233 (1966).
10. J. Iida and T. Murata, *Anal. Sci.* **6**, 273 (1990).
11. O. Isler, G. Brubacher, S. Ghisla, and B. Kraeutler, eds., *Vitamine II*, Verlag, New York, 1981, pp. 193–227.
12. J. M. Osbond, *Vitamins and Hormones* **22**, 387 (1964).
13. K. Folkers et al., *J. Am. Chem. Soc.* **73**, 3430 (1951) and cites therein.
14. S. A. Harris, *J. Am. Chem. Soc.* **63**, 3363 (1941).
15. *Nutr. Rev.* **36**, 346 (1978).
16. *The Merck Index*, 11th ed., Merck & Co., Rahway, N.J., 1989, p. 1269.
17. G. S. Shephard and D. Labadarios, *Clin. Chim. Acta* **60**, 307 (1981).
18. B. Saida and J. Warthesen, *J. Agric. Food Chem.* **31**, 876 (1983).
19. C. Y. W. Ang, *J. Assoc. Official Anal. Chem.* **62**, 1170 (1979).
20. A. R. Forester and I. G. Davidson, *Radiat. Phys. Chem.* **36**, 403 (1990).
21. K. Tadera et al., *Agric. Biol. Chem.* **52**, 2359 (1988).
22. M. L. Hu et al., *Chem. Biol. Interact.* **97**, 63 (1995).
23. J. E. Leklem, in L. J. Machlin, ed., *Handbook of the Vitamins*, 2nd ed., Chapt. 6, Marcel Dekker, New York, 1991.
24. M. L. Orr, *Pantothenic Acid, Vitamin B6 and Vitamin B12 in Foods*, Home Economics Report No. 36, U.S. Dept. of Agriculture, Washington, D.C., 1969.
25. J. F. Gregory and D. B. Sartain, *J. Agr. Food Chem.* **39**, 899 (1991).
26. K. Tadera and Y. Naka, *Agr. Biol. Chem.* **55**, 563 (1991).
27. K. D. Lunan and C. A. West, *Arch. Biochem. Biophys.* **101**, 262 (1963).

28. J. Gregory et al., *J. Food Sci.* **51**, 1345 (1986).

29. J. B. Tarr, T. Tamura, and E. L. R. Stokstad, *Am. J. Clin. Nutr.* **34**, 1328 (1981).

30. A. H. Merrill and M. J. Henderson, *Ann. N.Y. Acad. Sci.* **585**, 110 (1990).

31. Committee on Dietary Allowances, Food and Nutrition Board, *Recommended Daily Allowances*, 10th ed., National Academy of Sciences, Washington, D.C., 1980, pp. 142–150.

32. P. Karrer et al., *Helv. Chim. Acta* **32**, 1004 (1948).

33. N. Clauson-Kaas et al., *Acta Chem. Scand.* **9**, 1,9,14,17,23 (1955).

34. Ger. Pat. 1,935,009 (Jan. 14, 1971), F. Graf and H. Konig (to BASF).

35. Y. Matsumura et al., *Chem. Lett.*, 1121 (1981).

36. M. Y. Karpeiskii and V. L. Flornet'ev, *Russ. Chem. Rev.* **38**, 540 (1969).

37. I. Turchi and M. J. S. Dewar, *Chem. Rev.* **75**, 388 (1975).

38. G. Y. Kondrat'eva et al., *Dokl. Akad. Nauk SSR* **142**, 593 (1962).

39. T. Naito et al., *Chem. Pharm. Bull.* **13**, 873 (1965).

40. Z. I. Itov et al., *Khim. Farm. Zh.* **13**, 59 (1980).

41. E. E. Harris et al., *Tetrahedron* **23**, 943 (1967).

42. Ger. Pat. 3,029,231 (Nov. 3, 1982), H. R. Mueller (to Merck).

43. H. Takagaki et al., *Chem. Lett.*, 183 (1979).

44. Jpn. Pat. 69-23,088 (Oct. 1, 1969), T. Naito, K. Ueno, and T. Miki (to Daiichi Seiyaku Co. Ltd.).

45. Jpn. Pat. 81-128,761 (Oct. 8, 1981) (to Daiichi Seiyaku Co, Ltd.).

46. H. Murakami and M. Iwanami, *Bull. Chem. Soc. Jpn.* **41**, 726 (1968).

47. E. E. Harris et al., *J. Org. Chem.* **27**, 2705 (1962).

48. U.S. Pat. 3,227,721 (May 24, 1965), E. E. Harris, K. Pfister, and R. A. Firestone (to Merck & Co.).

49. Jpn. Pat. 73-30,636 (Sept. 21, 1973), Y. Morita, S. Onishi, and T. Yamagami (to Daiichi Seiyaku Co.).

50. Jpn. Pat. 70-39,259 (Dec. 10, 1970), T. Naito, Y. Morito, and U. Toshihiro (to Daiichi Seiyaku Co.).

51. Jpn. Pat. 72-39,115 (Oct. 3, 1972), T. Miki and T. Matsuo (to Takeda Chemical Industries, Ltd.).

52. Jpn. Pat. 81-99,461 (Aug. 10, 1981) (to Takeda Chemical Industries, Ltd.).

53. T. Miki and T. Matsuo, *Yakugaku Zasshi* **87**, 323 (1967).

54. U.S. Pat. 3,796,720 (Mar. 12, 1974), H. Miki and H. Saikawa (to Takeda Chemical Industries).

55. J. Maeda et al., *Bull. Chem. Soc. Jpn.* **42**, 1435 (1969).

56. Fr. Pat. 1,533,817 (July 19, 1968) (to Ajinomoto Co.).

57. H. Zhou et al., *Zhongguo Yiyao Gongye Zazhi* **25**, 385 (1994).

58. Chin. Pat. 86-101,512 (July 7, 1986), H. Zhou (to Shanghai Industrial Res. Inst.).

59. U.S. Pat. 3,250,778 (Nov. 11, 1962), W. Kimel and W. Leimgruber (to Hoffmann-La Roche Inc.).

60. M. Dumic et al., *Prax. Vet.* **37**, 81 (1989).

61. D. Szlopmek-Nesteruk and co-workers, *Przem. Chem.* **54**, 238 (1975).

62. U.S. Pat. 3,822,274 (July 2, 1974), E. E. Harris and R. Currie (to Merck & Co.).

63. H. Ritter and R. Sperber, *Macromolecules* **27**, 5919 (1994).

64. W. Boell and H. Koenig, *Ann. Chem.*, 1657 (1979).

65. U.S. Pat. 3,876,649 (Apr. 8, 1975), W. Boell and H. Koenig (to BASF).

66. T. Naito et al., *Tetr. Lett.*, 5767 (1968).

67. P. Karrer and C. Granacher, *Helv. Chem. Acta* **7**, 763 (1924).

68. Jpn. Pat. 68-30,286 (Dec. 26, 1968), T. Kitazawa, T. Matsuyama, K. Kunimitsu, and K. Masuya (to Fujisawa Pharmaceutical Co.).

69. T. Rinderspacher and B. Prijs, *Helv. Chim. Acta* **43**, 1522 (1960).

70. U.S. Pat. 4,026,901 (June 6, 1978), D. L. Coffen (to Hoffmann-La Roche Inc.).

71. U.S. Pat. 4,011,234 (Mar. 8, 1977), H. Hoffmann-Paquotte (to Hoffmann-La Roche Ltd.).

72. Eur. Pat. 612736 (Aug. 31, 1994), W. Bonrath, R. Karge, and H. Pauling (to Hoffmann-La Roche Ltd.).

73. Eur. Pat. 492,233 (July 1, 1992), P. Noesberger (to Hoffmann-La Roche Ltd.).

74. S. Shimada and M. Oki, *Chem. Pharm. Bull.* **32**, 38 (1984).

75. Jpn. Pat. 79-20,493 (July 23, 1979), S. Ohnishi and Y. Morita (to Daiichi Seiyaku Co.).

76. M. Oki and S. Shimada, *Chem. Pharm. Bull.* **35**, 4705 (1987).

77. E. C. Taylor and J. Macor, *J. Org. Chem.* **54**, 1249 (1989).

78. R. E. Geiger et al., *Helv. Chim. Acta* **67**, 1274 (1984).

79. C. A. Parnell and K. P. C. Vollhardt, *Tetr.* **41**, 5791 (1985).

80. Food and Nutrition Board, National Research Council, *Food Chemicals Codex* 3rd ed., National Academy Press, Washington, D.C., 1981, p. 260.

81. United States Pharmacopieal Convention, Inc., *The United States Pharmacopeia XXII* (USP XXII-NF XVII), Rockeville, Md., 1990, p. 1194.

82. H. Y. Aboul-Enein and M. A. Loutfy, in K. Florey, ed., *Analytical Profiles of Drug Substances*, Vol. 13, Academic Press, New York, 1984, p. 447.

83. *Pyridoxine Hydrochloride*, MSDS 11076, Hoffmann-LaRoche Inc., Nutley, N.J., 1997.

84. M. Cohen and A. Bendich, *Toxicol. Lett.* **34**, 129 (1986).

85. M. M. Polansky, in J. Augustin, B. P. Klein, D. A. Decker, and P. B. Venugopal, eds., *Methods of Vitamin Assay*, 4th ed., Association of Vitamin Chemists, John Wiley & Sons, Inc., New York, 1985, Chapt. 17.

86. J. F. Gregory, *J. Food Comp. Anal.* 1, 105 (1988).

87. K. Helrich, ed., *Official Methods of Analysis of the Association of Official Analytical Chemists*, 15th ed., Association of Official Analytical Chemists, Inc., Arlington, Va., 1990, p. 961.15.

88. S. P. Colburn and J. D. Mahuren, *J. Biol. Chem.* **262**, 2642 (1987).

89. Y. Kurosu et al., *Bitamin* **66**, 91 (1992).

90. R. D. Smith et al., *Anal. Chem.* **60**, 436 (1988).

91. R. Benyon et al., *J. Chrom.* **581**, 179 (1992).

92. J. Iida and T. Murata, *Anal. Sci.* **6**, 273 (1990).

93. Y. Tani, in E. Vandamme, ed., *Biotechnology of Vitamins, Pigments and Growth Factors*, Elsevier, New York, 1988, Chapt. 13.

94. R. E. Hill et al., *J. Am. Chem. Soc.* **117**, 1661 (1995).

95. M. Morisaki, *Kurume Igakkai Zasshi* **53**, 649 (1990).

96. K. Tazuya et al., *Biochem. Biophys. Acta* **1244**, 113 (1995).

97. M. Winkler et al., *J. Bacteriol.* **177**, 2804 (1995) and cites therein.

98. M. Winkler et al., *FEMS Microbiol. Lett.* **135**, 275 (1996).

99. T. Copeland et al., *J. Bacteriol.* **174**, 1544 (1992).

100. W. Korytnik and M. Ikawa, in S. P. Colowick and N. O. Kaplan, eds., *Methods in Enzymology*, Vol. 18A, Academic Press, New York, pp. 524–566.

DAVID BURDICK
Hoffmann-La Roche, Inc.

VITAMINS: RIBOFLAVIN (B$_2$)

Riboflavin [83-88-5] (vitamin B$_2$, vitamin G, lactoflavin, ovoflavin, lyochrome, hepatoflavin, uroflavin) has the chemical name 1-deoxy-1-(3,4-dihydro-7,8-dimethyl-2,4-dioxobenzo[*g*]pteridin-10(2*H*)-yl)-D-ribitol, 7,8-dimethyl-10-D-ribitylisoalloxazine, C$_{17}$H$_{20}$N$_4$O$_6$ (**1**), mol wt 376.37.

(**1**)

In 1933, R. Kuhn and his co-workers first isolated riboflavin from eggs in a pure, crystalline state (1), named it ovoflavin, and determined its function as a vitamin (2). At the same time, impure crystalline preparations of riboflavin were isolated from whey and named lyochrome and, later, lactoflavin. Soon thereafter, P. Karrer and his co-workers isolated riboflavin from a wide variety of animal organs and vegetable sources and named it hepatoflavin (3). Ovoflavin from egg, lactoflavin from milk, and hepatoflavin from liver were all subsequently identified as riboflavin. The discovery of the yellow enzyme by Warburg and Christian in 1932 and their description of lumiflavin (4), a photochemical degradation product of riboflavin, were of great use for the elucidation of the chemical structure of riboflavin by Kuhn and his co-workers (5). The structure was confirmed in 1935 by the synthesis by Karrer and his co-workers (6), and Kuhn and his co-workers (7).

For therapeutic use, riboflavin is produced by chemical synthesis, whereas concentrates for poultry and livestock feeds are manufactured by fermentation using microorganisms such as *Ashbya gossypii* and *Eremothecium ashbyii*, which have the capacity to synthesize large quantities of riboflavin.

In the free form, riboflavin occurs in the retina of the eye, in whey, and in urine. Principally, however, riboflavin fulfills its metabolic function in a complex form. In general, riboflavin is converted into flavin mononucleotide (FMN, riboflavin-5′-phosphate) and flavin-adenine dinucleotide (FAD) (**2**), which serve as the prosthetic groups (coenzymes), ie, they combine with specific proteins (apoenzymes) to form flavoenzymes, in a series of oxidation–reduction catalysts widely distributed in nature. In several riboflavin coenzymes, the apoenzyme is covalently attached to $C - 8\alpha$ through a linkage to the nitrogen of a histidine imidazolyl group, to the sulfur of a cysteine residue, or to the oxygen of a tyrosine moiety. Riboflavin is not a nucleotide, since it is derived from D-ribitol rather than D-ribose, and therefore FMN and FAD are not truly nucleotides; yet this designation has been accepted overwhelmingly and continues to be used.

(**2**)

As a coenzyme component in tissue oxidation–reduction and respiration, riboflavin is distributed in some degree in virtually all naturally occurring foods. Liver, heart, kidney, milk, eggs, lean meats, malted barley, and fresh leafy vegetables are particularly good sources of riboflavin (see Table 1). It does not seem to have long stability in food products (8).

Riboflavin is widely used in the pharmaceutical, food-enrichment, and feed-supplement industries. Riboflavin USP is administered orally in tablets or by injection as an aqueous solution, which may contain nicotinamide or other solubilizers. As a supplement to animal feeds, riboflavin is usually added at concentrations of 2–8 mg/kg, depending on the species and age of the animal.

PROPERTIES

Riboflavin forms fine yellow to orange-yellow needles with a bitter taste from 2 *N* acetic acid, alcohol, water, or pyridine. It melts with decomposition at 278–279°C (darkens at ca 240°C). The solubility of riboflavin in water is 10–13 mg/100 mL at 25–27.5°C, and in absolute ethanol 4.5 mg/100 mL at 27.5°C; it is slightly soluble in amyl alcohol, cyclohexanol, benzyl alcohol, amyl acetate, and phenol, but insoluble in ether, chloroform, acetone, and benzene. It is very soluble in dilute alkali, but these solutions are unstable. Various polymorphic crystalline forms of riboflavin exhibit variations in physical properties. In aqueous nicotinamide solution at pH 5, solubility increases from 0.1 to

Table 1. Riboflavin Content of Various Foods

Food	mg/100 g
Fruits	
Apple, raw	0.01
Banana, raw	0.04
Citrus, grapefruit, orange	0.03–0.04
Strawberry	0.03
Vegetables	
Broccoli, raw	0.27
Cabbage, raw	0.05
Fresh green peas	0.14
Mushroom	0.57
Parsley	0.24
Potato, raw	0.03
Sweet corn	0.14
Sweet potato, raw	0.05
Tomato, raw	0.03
Meat	
Beef muscle	0.16–0.32
Pork muscle	0.19–0.33
Chicken muscle	0.10–0.31
Liver, beef, pork	3.00–3.60
Fish	
Cod, haddock, raw	0.17
Salmon, raw	0.17
Salmon, canned	0.12
Tuna, canned	0.13
Whitefish, herring, halibut	0.17–0.29
Grain	
Corn, entire	0.10
Wheat, entire	0.10
Wheat, germ	0.6
Rice, entire	0.06
Rye, entire	0.20
Cereal products	
Refined	
Bread	0.07–0.10
Cereal	0.10–0.15
Soda cracker	0.02–0.10
Whole grain and enriched	
Bread	0.12–0.25
Cereal	0.20–1.25
Dairy products	
Cheese	0.33–0.68
Eggs	0.48
Milk	0.15–0.18

Note: Averages from several sources, often encompassing a wide range of analytical results; should be regarded as working estimates which vary with geography, season, and preparative method.

2.5% as the nicotinamide concentration increases from 5 to 50% (9).

In aqueous solution, riboflavin has absorption at ca 220–225, 226, 371, 444 and 475 nm. Neutral aqueous solutions of riboflavin have a greenish-yellow color and an intense yellowish green fluorescence with a maximum at ca 530 nm and a quantum yield of $\Phi_f = 0.25$ at pH 2.6 (10). Fluorescence disappears upon the addition of acid or alkali. The fluorescence is used in quantitative determinations. The optical activity of riboflavin in neutral and acid solutions is $[\alpha]_D^{20} = +56.5 - 59.5°$ (0.5%, dil HCl). In an alkaline solution, it depends upon the concentration, eg, $[\alpha]_D^{25} = -112 - 122°$ (50 mg in 2 mL 0.1 N alcoholic NaOH diluted to 10 mL with water). Borate-containing solutions are strongly dextrorotatory, because borate complexes with the ribityl side chain of riboflavin; $[\alpha]_D^{20} = +340°$ (pH 12).

Photochemical decomposition of riboflavin in neutral or acid solution gives lumichrome (3), 7,8-dimethylalloxazine, which was synthesized and characterized by Karrer and his co-workers in 1934 (11). In alkaline solution, the irradiation product is lumiflavin (4), 7,8,10-trimethylisoalloxazine; its uv–vis absorption spectrum resembles that of riboflavin. It was prepared and characterized in 1933 (5). Another photodecomposition product of riboflavin is 7,8-dimethyl-10-formylmethylisoalloxazine (12).

(3)   (4)

Riboflavin is stable against acids, air, and common oxidizing agents such as bromine and nitrous acid (except chromic acid, KMnO$_4$, and potassium persulfate). Upon reduction by conventional agents such as sodium dithionite, Na$_2$S$_2$O$_4$, zinc in acidic solution, or catalytically activated hydrogen, riboflavin readily takes up two hydrogen atoms to form the almost colorless 1,5-dihydroriboflavin (5) (Fig. 1), which is reoxidized by shaking with air. This oxidation–reduction system has considerable stability, a normal potential of -0.208 V (referred to as the normal hydrogen electrode), and is probably responsible for the physiological functions of riboflavin. The flavins are reduced to dihydroflavins through intermediate semiquinone radicals (13,14), which have been directly observed by electron spin resonance (esr) (15–18) and electron double resonance (endor) (19).

Riboflavin forms a deep-red silver salt (1). The strong bathochromic shift of the spectra of riboflavin analogues occurring by interaction with Ag$^+$ can also be obtained with Cu$^+$ and Hg^{2+} complexes (20). These complexes contain the flavin and the metal ligand anion in a ratio of 1:1. Their color is the result of a charge transfer between the metal and flavin (21). The chelates with Fe(II/III), Mo(V/VI), Cu(I/II), and Ag(I/II) belong to this group; the last two are stable in the presence of water. Another group of metal complexes, radical chelates, are formed with Mn(II), Fe(II),

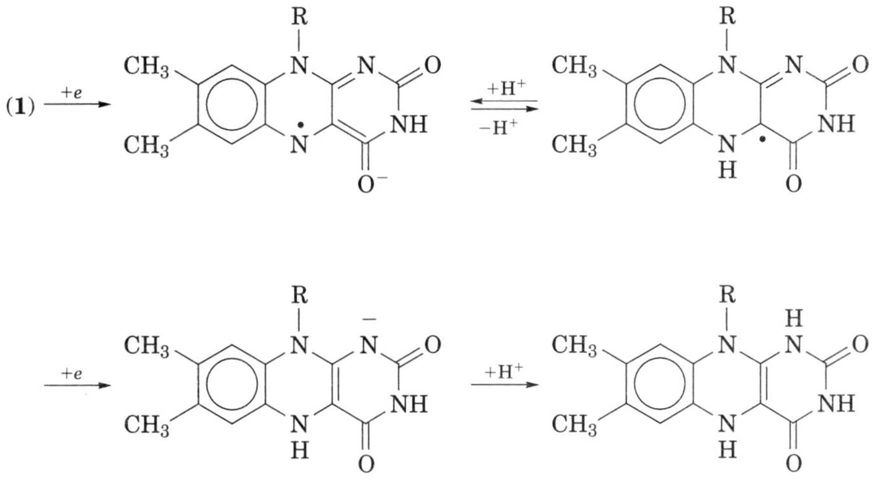

R = D-Ribityl

Figure 1. Formation of dihydroriboflavin.

Co(II), Ni(II), Zn(II), and Cd(II); in these cases, the radical character of the ligand is conserved (22).

CHEMICAL SYNTHESIS

In 1935, Karrer (6) and Kuhn (7) each proved independently that riboflavin was 7,8-dimethyl-10-D-ribitylisoalloxazine by total synthesis (see Fig. 2). These syntheses are essentially the same and involve a condensation of 6-D-ribitylamino-3,4-xylidine (6) with alloxan (7) in acid solution. Boric acid as a catalyst increases the yield considerably (23). The intermediate (6) was prepared by a condensation of 6-nitro-3,4-xylidine (8) with D-ribose (9), followed by catalytic reduction of the riboside (10). The yield based on D-ribose was increased (24) by using N-D-ribityl-3,4-xylidine (11), which was prepared by the condensation of 3,4-xylidine (12) with D-ribose (9), followed by catalytic reduction. The reduced product (11) was coupled with p-nitrophenyldiazonium salt to give 4,5-dimethyl-2-p-nitrophenylazo-1-D-ribitylaminobenzene (13), which was reduced to (6) and treated with alloxan (7) (25) to give riboflavin. Replacement of (7) by 5,5-dichlorobarbituric acid (26), 5,5-dibromobarbituric acid, or 5-bromobarbituric acid (27) in the above syntheses yields riboflavin.

More conveniently, compound (13) was directly condensed with barbituric acid (14) in acetic acid (28) or in the presence of an acid catalyst in an organic solvent (29). The same azo dye intermediate (13) and alloxantin give riboflavin in the presence of palladium on charcoal in alcoholic hydrochloric acid under nitrogen. This reaction may involve the reduction of the azo group to the o-phenylenediamine by the alloxantin, which is dehydrogenated to alloxan (30).

Although it is not suitable for large-scale manufacture, the synthesis of riboflavin from lumazine derivatives is interesting in connection with the biosynthesis of riboflavin (see Fig. 3). Thus, 5-amino-6-D-ribitylaminouracil (15) was condensed with a dimeric or trimeric aldol of biacetyl to give riboflavin (1) through the formation of intermediary

6,7-dimethyl-8-D-ribityllumazine (16) (31). A variation of the above synthesis involves the condensation of monomeric biacetyl and preformed (16) prepared by the condensation of (15) with biacetyl (32). The condensation of (15) with 4,5-dimethyl-1,2-benzoquinone (17) is another pathway to riboflavin, although in low yield (33).

Later, a completely different and more convenient synthesis of riboflavin and analogues was developed (34). It consists of the nitrosative cyclization of 6-(N-D-ribityl-3,4-xylidino)uracil (18), obtained from the condensation of N-D-ribityl-3,4-xylidine (11) and 6-chlorouracil (19), with excess sodium nitrite in acetic acid, or the cyclization of (18) with potassium nitrate in acetic in the presence of sulfuric acid, to give riboflavin-5-oxide (20) in high yield. Reduction with sodium dithionite gives (1). In another synthesis, 5-nitro-6-(N-D-ribityl-3,4-xylidino) uracil (21), prepared in situ from the condensation of 6-chloro-5-nitrouracil (22) with N-D-ribityl-3,4-xylidine (11), was hydrogenated over palladium on charcoal in acetic acid. The filtrate included 5-amino-6-(N-D-ribityl-3,4-xylidino)uracil (23) and was maintained at room temperature to precipitate (1) by autoxidation (35). These two pathways are suitable for the preparation of riboflavin analogues possessing several substituents (Fig. 4).

The chemistry of flavins, including several synthetic methods for the preparation of N-D-ribityl-3,4-xylidine (11) is reviewed in Reference 36.

MICROBIAL SYNTHESIS

Biosynthetic Mechanism

Riboflavin is produced by many microorganisms, including Ashbya gossypii, Asperigillus sp, Eremothecium ashbyii, Candida yeasts, Debaryomyces yeasts, Hansenula yeasts, Pichia yeasts, Azotobactor sp, Clostridium sp, and Bacillus sp.

These organisms have been used frequently in the elucidation of the biosynthetic pathway (37,38). The mecha-

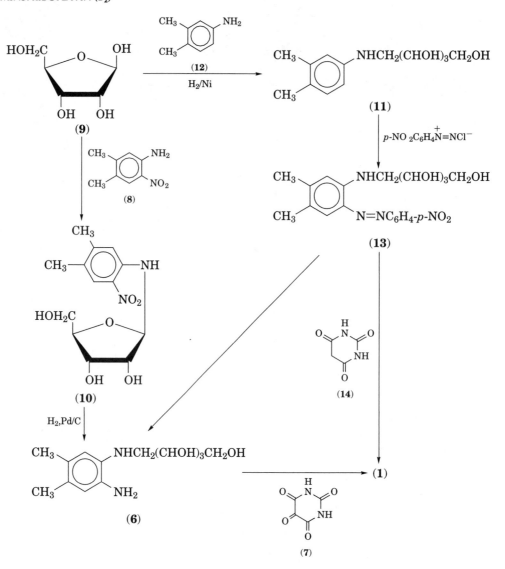

Figure 2. Syntheses of riboflavin.

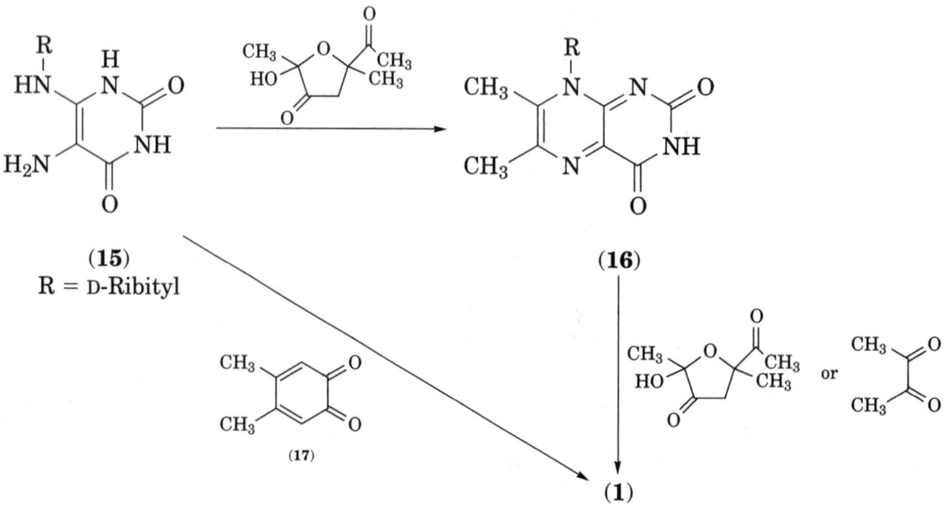

Figure 3. Syntheses of riboflavin from a lumazine.

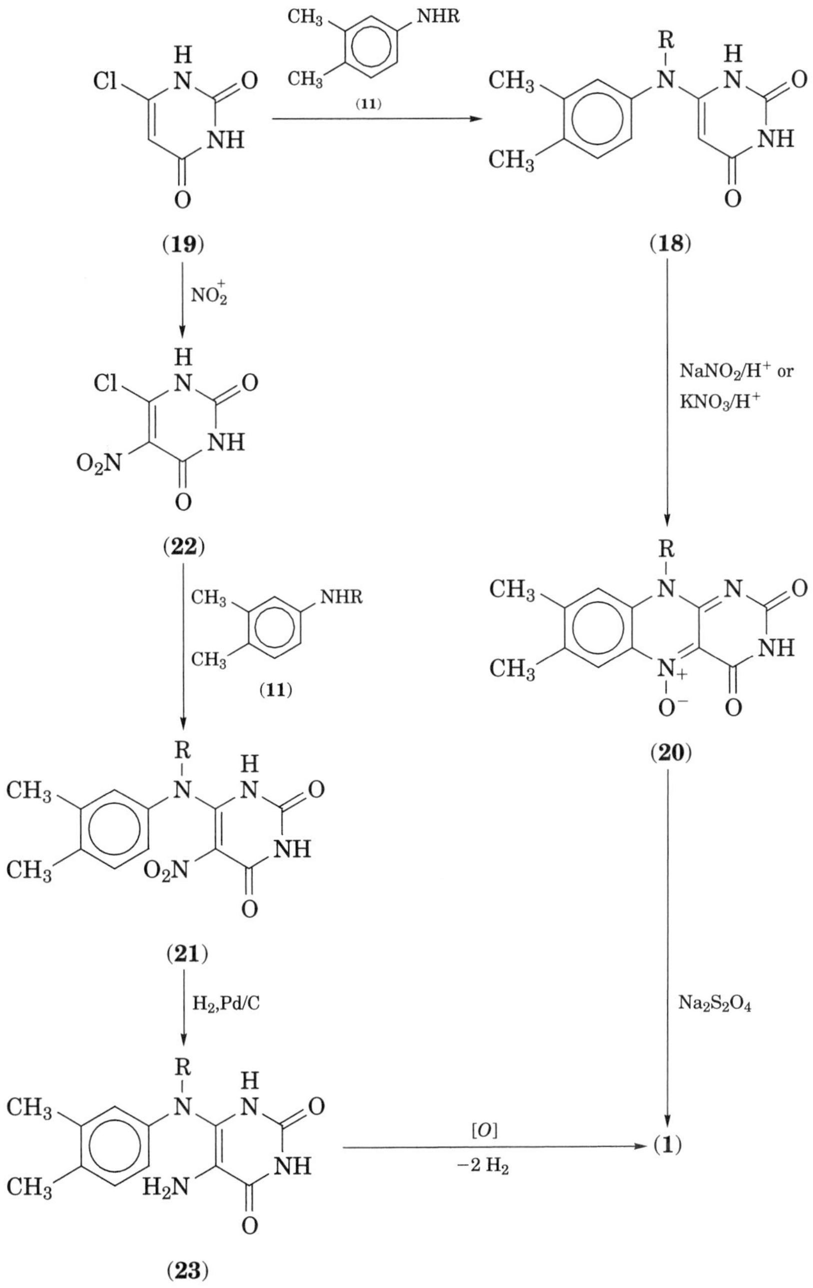

R = D-Ribityl

Figure 4. Alternative syntheses of riboflavin.

nism of riboflavin biosynthesis has formally been deduced from data derived from several experiments involving a variety of organisms (Fig. 5). Included are conversion of a purine such as guanosine triphosphate (GTP) to 6,7-dimethyl-8-D-ribityllumazine (**16**) (39), and the conversion of (**16**) to (**1**). This concept of the biochemical formation of riboflavin was verified *in vitro* under nonenzymatic conditions (40).

Fermentative Manufacture

Throughout the years, riboflavin yields obtained by fermentation have been improved to the point of commercial feasibility. Most of the riboflavin thus produced is consumed in the form of crude concentrates for the enrichment of animal feeds. Riboflavin was first produced by fermentation in 1940 from the residue of butanol–acetone fermentation. Several methods were developed for large-scale production (41). A suitable carbohydrate–containing mash is prepared and sterilized, and the pH adjusted to 6–7. The mash is buffered with calcium carbonate, inoculated with *Clostridium acetobutylicum*, and incubated at 37–40°C for 2–3 d. The yield is ca 70 mg riboflavin/L (42).

Most varieties of *Candida* yeasts produce substantial amounts of riboflavin when glucose is the carbon source. Particularly, *Candida guilliermondia* and *Candida flaveri*

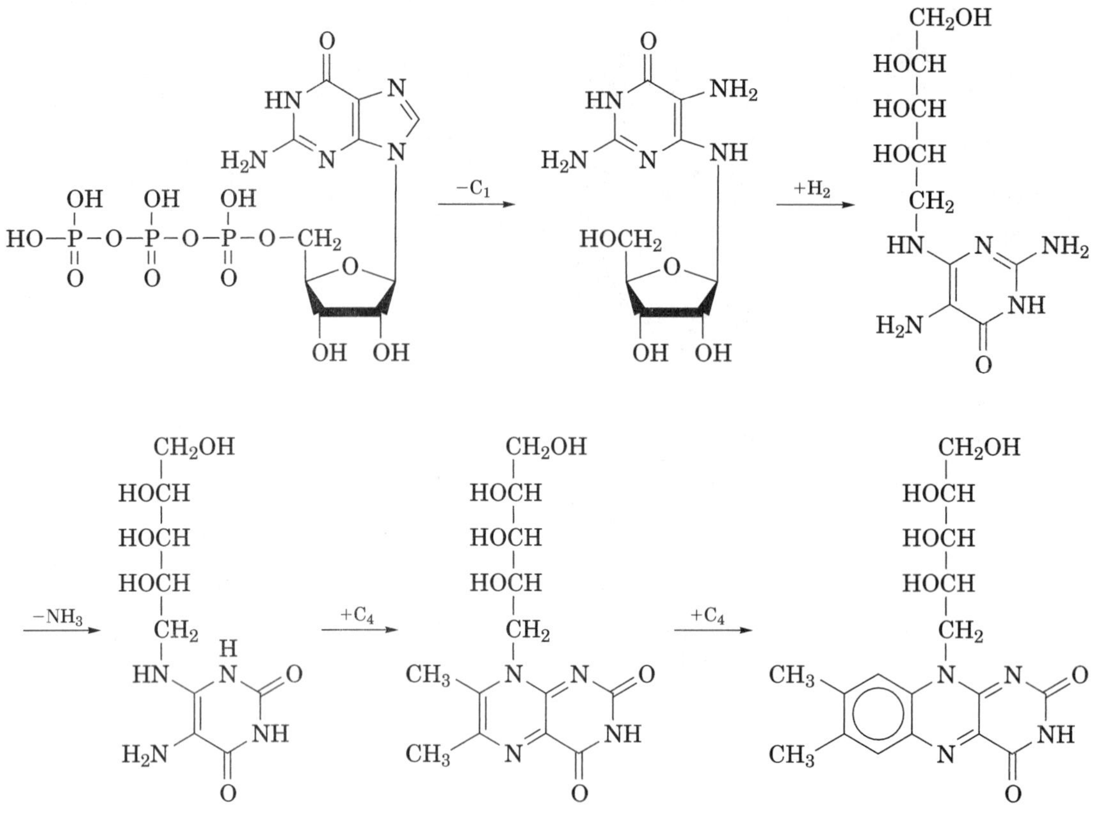

Figure 5. Biosynthesis pathway to riboflavin.

produce high yields on a simple synthetic medium of low cost. Some modifications employing several *Candida* yeasts have been patented; eg, *C. intermedia var A*, a newly isolated microorganism assimilating lactose and ethanol, gave yields of 49.2 mg riboflavin/L from ethanol in the presence of biotin (43). *Candida T-3*, assimilating methanol, produces riboflavin by this method from 50 g methanol (44). *Candida* bacteria have the advantage of extremely low iron tolerance. Chelating agents, such as 2,2'-dipyridyl, are recommended to control the iron content (45).

Most of the commercial riboflavin production by aerobic fermentation is obtained by biosynthesis with the yeast-like fungus *Eremothecium ashbyii*. Many variations for the production of riboflavin by *E. ashbyii* have been patented. Employing *E. ashbyii* grown on a yeast medium, riboflavin production on an industrial scale is said to have reached 11.6 g/kg from dried powder (46). Riboflavin is also obtained by fermentation with *E. ashbyii* preserved on Difco-treated millet seeds. The culture, kept for 7–8 d at 32°C on the medium containing crude collagen, corn extracts, unrefined plant oil, glucose, KH_3PO_4, trace elements, and H_2O at pH 7–8, gave 4.5–5.0 g/L (47). In another procedure, *E. ashbyii* was cultivated on a culture medium containing sources of assimilable nitrogen, essential minerals, and growth factors, along with sources of assimilable carbohydrates, unsaturated fatty acids, saccharides, and amino acid or their salts. Incubation at 29°C for 6 d on a

rotary shaker with aeration gave average yields of 3.8 g/L (48). *Eremothecium ashbyii* grown in a culture medium containing the oil cake obtained after extraction of lipids from the biomass grown on hydrocarbons gives an even better yield (49).

In operations similar to the *E. ashbyii* procedures, the closely related fungus *Ashbya gossypii* gave similar yields. Thus, a yield of 7.3 g/L was obtained with a lyophilized culture in a medium containing fat, leather glue, and corn extracts (50), and 6.420 g/L with bone or hide fat, alone or in a mixture with other plant or animal fats as the carbon source (51). The yield from immersed cultures of *A. gossypii* was increased to 6.93–7.20 g/L by use of waste fats or technical cod-liver oil (52).

Riboflavin is also made by aerobic culturing of *Pichia guilliermondii* on a medium containing n-C_{10}–C_{15} paraffins in a yield of 280.5 mg/L (53). A process employing *Pichia* yeasts, such as *P. miso, P. miso* Mogi, or *P. mogii*, in a medium containing a hydrocarbon as the carbon source, has been patented (54).

Processes employing *Torulopsis xylinus* (55), *Hansenula polymorpha* (56), *Brevibacterium ammoniagenes* (57), *Achromobactor butrii* (58), *Micrococcus lactis* (59), *Streptomyces testaceus* (60), and others have also been patented. These procedures yield, at most, several hundred milligrams of riboflavin per liter.

Manufacturing procedures of riboflavin have also appeared using *Saccharomyces* bacteria, eg, fermentation with a purine-independent *S.* reverse mutant (61) and with *S. cerevisiae* NH-268 (62) produced 2.79 g/L and 4.9 g/L, respectively.

Further efficient fermentative methods for manufacture of riboflavin have been patented; one is culturing *C. famata* by restricting the carbon source uptake rate, thereby restricting growth in a linear manner by restriction of a micronutrient. By this method, productivity was increased to >0.17 g riboflavin/L/h (63). The other method, using *Bacillus subtilis* AJ 12644 low in guanosine monophosphate hydrolase activity, yielded crude riboflavin 0.9 g/L/3 days, when cultured in a medium including soy protein, salts, and amino acids (64).

In recent (ca 1997) years, fermentative manufacturing methods using recombinant microorganisms have been developed. The mutant clone GA18Y8-6#2, prepared by fusion and mutagenation of *C. flaveri* mutants A22 and GA18, produced riboflavin 7.0–7.5 g/L/6 d, when cultivated in 4B medium with a supplement of $FeCl_3$, yeast extracts, peptone, and malt extracts (65). Riboflavin overproducing bacteria prepared by expression of the cloned *rib* operon of *Bacillus subtilis* showed increases in riboflavin manufacture of up to a hundredfold (up to 0.7 g riboflavin/L/48 h). The *rib* operon was cloned as series of overlapping genes using an oligonucleotide derived from the amino acid sequence of the riboflavin synthase β subunit (66). Culturing recombinant *Corynebacterium ammoniagenes* KY13313 harboring gene for at least guanidine triphosphate cyclohydrolase and riboflavin synthase produced riboflavin ~30-fold higher than that with the controlled bacteria (67).

INDUSTRIAL ASPECTS

For the industrial production of riboflavin as pharmaceuticals, the traditional methodology comprising the direct condensation of (**13**) with (**14**) in an acidic medium with continuous optimization of the reaction conditions is still used (28). A great part of riboflavin manufactured by fermentative methods is used for feeds in the form of concentrates. The present world demand of riboflavin may be about 2500 t per year. Of this amount, 60%, 25%, and 15% are used for feeds, pharmaceuticals, and foodstuffs, respectively. The main producers are Hoffmann-La Roche, BASF, Merck & Co., and others.

ANALYTICAL METHODS

Riboflavin can be assayed by chemical, enzymatic, and microbiological methods. The most commonly used chemical method is fluorometry, which involves the measurement of intense yellow-green fluorescence with a maximum at 565 nm in neutral aqueous solutions. The fluorometric determinations of flavins can be carried out by measuring the intensity of either the natural fluorescence of flavins or the fluorescence of lumiflavin formed by the irradiation of flavin in alkaline solution (68). The later development of a laser-fluorescence technique has extended the limits of detection for riboflavin by two orders of magnitude (69,70).

Polarography is applied in the presence of other vitamins, eg, in multivitamin tablets, without separation. The polarography of flavins is reviewed in Reference 71.

The microbial assay is based on the growth of *Lactobacillus casei* in the natural (72) or modified form. The lactic acid formed is titrated or, preferably, the turbidity measured photometrically. In a more sensitive assay, *Leuconostoc mesenteroides* is employed as the assay organism (73). It is 50 times more sensitive than *L. casei* for assaying riboflavin and its analogues (0.1 ng/mL vs 20 ng/mL for *L. casei*). A very useful method for measuring total riboflavin in body fluids and tissues is based on the riboflavin requirement of the protozoan cliate *Tetrahymena pyriformis*, which is sensitive and specific for riboflavin. This method can be applied to large-scale nutrition studies.

Although riboflavin can be assayed more readily by chemical or microbiological methods than by animal methods, the latter are preferred for nutritional studies and as the basis of other techniques. Such assays depend upon a growth response; the rat or chick is the preferred experimental animal. This method is particularly useful for assaying riboflavin derivatives, since the substituents frequently reduce or eliminate the biological activity.

An enzymatic method for assessing riboflavin deficiency in humans has been developed (74). It is based on the fact that NADPH-dependent glutathione reductase of red cells reflects riboflavin fluctuations.

High pressure liquid chromatography (hplc) has been extensively used for the riboflavin determinations. This method is usually automated and more rapid and sensitive than the microbial method. It has been used in combination with fluorometric detection for the riboflavin assay in foods (75), meat and meat products (76), and enriched and fortified foods (77), as well as in a simple assay for animal tissues (78). A rapid and efficient reversed-phase hplc method is described for the quantitative separation of flavin coenzymes and their structural analogues such as 5-deazaflavin [19342-73-5] (**24**) and 8-hydroxy-7-demethyl-5-deazariboflavin [37333-48-5] (**25**) (F_{420} chromophore) (79). Comprehensive reviews of the analytical methods for riboflavin are given in References 80 and 81.

(**24**)   (**25**)

R = D-Ribityl

BIOLOGICAL FUNCTION

In biological systems, riboflavin functions almost exclusively in the form of flavo-proteins, in which the FMN or FAD is generally bound as prosthetic group or coenzyme to specific proteins. These enzymes catalyze oxidation-reduction reactions (Table 2). The flavin group of the oxidized coenzyme is reduced chemically or enzymatically to 1,5-dihydroflavin coenzyme, probably in two one-electron steps, each involving the addition of a single electron. Stable semiquinone radicals are formed as intermediates, be-

Table 2. Some Reactions Catalyzed by Flavoproteins

Enzyme	Electron donor	Product	Coenzyme and other components	Electron acceptor
D-Amino acid oxidase	D-Amino acids	α-Keto acids + NH_3	2FAD	$O_2 \rightarrow H_2O_2$
L-Amino acid oxidase (liver)	L-Amino acids	α-Keto acids + NH_3	2FAD	$O_2 \rightarrow H_2O_2$
L-Amino acids oxidase (kidney)	L-Amino acids	α-Keto acids + NH_3	2FMN	$O_2 \rightarrow H_2O_2$
L(+)-Lactate dehydrogenase (yeast)	Lactate	Pyruvate	1FMN; 1heme (cyt b_5)	Respiratory chain
Glycolate oxidase	Glycolate	Glyoxylate	FMN	$O_2 \rightarrow H_2O_2$
NADH-Cytochrome c reductase	NADH	NAD^+	FMN; 2Mo, NHI	Cytochrome c_{ox} respiratory chain
NADH-Cytochrome b_5 reductase	NADH	NAD^+	FAD; Fe	Cytochrome b_5
Aldehyde oxidase (liver)	Aldehydes	Carboxylic acids	FAD; Fe, Mo	Respiratory chain
α-Glycerol phosphate dehydrogenase	Glycerol 3-phosphate	Dihydroxyacetone phosphate	FAD; Fe	Respiratory chain
Succinate dehydrogenase	Succinate	Fumarate	FAD; Fe, NHI	Respiratory chain
Acyl-CoA(C_6–C_{12}) dehydrogenase	Acyl-CoA	Enoyl-CoA	FAD	Electron-transferring flavoprotein
Nitrate reductase	NADPH	$NADP^+$	FAD; Mo, Fe	Nitrate
Nitrite reductase	NADPH	$NADP^+$	FAD; Mo, Fe	Nitrite
Xanthine oxidase	Xanthine	Uric acid	FAD; Mo, Fe	O_2
Lipoate dehydrogenase	Reduced lipoic acid	Oxidized lipoic acid	2FAD	NAD^+
Dehydroorotate dehydrogenase	Dihydroorotic acid	Orotic acid	2FMN; 2FAD, 4Fe	Undetermined

Source: Ref. 82.

cause the unpaired electron is highly delocalized by the conjugated isoalloxazine structure.

In contrast to the nicotinamide nucleotide dehydrogenases, the prosthetic groups FMN and FAD are firmly associated with the proteins, and the flavin groups are usually only separated from the apoenzyme (protein) by acid treatment in water. However, in several covalently bound flavoproteins, the enzyme and flavin coenzymes are covalently affixed. In these cases, the flavin groups are isolated after the proteolytic digestion of the flavoproteins.

Many flavoproteins react directly with molecular oxygen to produce hydrogen peroxide. Some flavoproteins, such as the flavin-containing monooxygenase, give water instead of hydrogen peroxide. In these cases, one atom of oxygen is introduced into a substrate to undergo hydroxylation, whereas the other oxygen atom is released as water. Several flavoproteins include metal complexes where these reactions take place. A number of reviews of the preparation, properties, and mechanism of action of these enzymes have been published (83,84).

DEFICIENCY, REQUIREMENTS, AND TOXICITY

Riboflavin is essential for mammalian cells. A lack of riboflavin in the human diet causes characterized deficiency syndromes, such as sore throat, hyperemia, cheilosis, angular stomatitis, glossitis (magenta tongue), a generalized seborrheic dermatitis, scrotal and vulval skin changes, and a normocytic anemia. Because riboflavin is essential to the functioning of vitamins B_6 and niacin, some symptoms attributed to riboflavin deficiency are actually due to the failure of systems requiring these other nutrients to operate effectively (85).

The 1989 Recommended Dietary allowances (RDA) of the Food and Nutrition Board (86) are 0.6 mg riboflavin per 239 kJ (1000 kcal) for essentially healthy people of all ages. This leads to the ranging from 0.4 mg/day for early infants to 1.8 mg/day for young males. For elderly people and others whose daily calorie intake may be less than 478 kJ (2000 kcal), a minimum of 1.2 mg/day is recommended. During pregnancy, an additional riboflavin intake of 0.3 mg/day is recommended in view of the increased tissue synthesis for both fetal and maternal development. For the lactating woman, the requirement is assumed to increase by an amount at least equal to that excreted in milk, which has a mean riboflavin content of 35 μg/100 mL. At an average milk production of 750 mL/day and 600 mL/d during the first and second 6 months of lactation, riboflavin secretion is 0.26 mg/d and 0.21 mg/d, respectively. Because the utilization of the additional riboflavin for milk production is assumed to be 70%, and the coefficient of variation of milk production is 12.5%, an additional daily intake of 0.5 mg is recommended for the first 6 mo of lactation and 0.4 mg thereafter.

Riboflavin is essentially nontoxic. The LD_{50} values in mice and rats by intraperitoneal injection are 340 mg/kg (87) and 560 mg/kg (88), respectively. The oral administration of 10 g/kg to rats or 2 g/kg to dogs showed no toxic effects (89).

DERIVATIVES

Riboflavin-5′-Phosphate

Riboflavin-5′-phosphate [146-17-8] (vitamin B_2 phosphate, flavin mononucleotide, FMN, cytoflav), $C_{17}H_{21}N_4O_9P$, mol wt 456.35, is a microcrystalline yellow solid, mp 195°C, $[\alpha]_D^{28} = +44.5°$ (2% soln in conc HCl) with biological and enzymatic activity. It is prepared by phosphorylation of riboflavin with chlorophosphoric acid (90), pyrophosphoric acid (91), metaphosphoric acid (92), or catechol cyclic phosphate (93). It is soluble in water to the extent of 3 g/100

Table 3. Covalently Bound Flavins

Prosthetic group (R = D-Ribityl)	CAS Registry Number	Formula	Enzyme
 8 α-(N^1-Histidyl)riboflavin	[58525-92-1]	C$_{23}$H$_{27}$C$_7$O$_8$	Thiamine dehydrogenase β-Cyclopiazonate oxidocyclase L-Gulono-γ-lactone oxidase L-Galactonolactone oxidase Cholesterol oxidase
 8 α-(N^3-Histidyl)riboflavin	[37854-44-7]	C$_{23}$H$_{27}$N$_7$O$_8$	Succinate dehydrogenase Fumarate reductase Salcosine dehydrogenase Salcosine oxidase Dimethylglycine dehydrogenase D-Hydroxynicotine oxidase Choline oxidase D-Gluconate dehydrogenase
 8 α-S-Cysteinylriboflavin	[35836-22-7]	C$_{20}$H$_{25}$N$_5$O$_8$S	Monoamine oxidase A Monoamine oxidase B *Chlorobium* Cytochrome C$_{552}$ *Chlomatium* cytochrome C$_{553}$
 8 α-O-Tyrosylriboflavin	[38065-74-6]	C$_{26}$H$_{29}$N$_5$O$_9$	*p*-Cresol methylhydroxylase
 6-S-Cysteinylriboflavin	[73647-60-6]	C$_{20}$H$_{25}$N$_5$O$_8$S	Trimethylamine Dehydrogenase Dimethylamine Dehydrogenase

mL at 25°C as the sodium salt but tends to gel. Because of the high sensitivity of FMN to uv, it must be preserved in dark, tight containers.

Flavin mononucleotide was first isolated from the yellow enzyme in yeast by Warburg and Christian in 1932 (4). The yellow enzyme was split into the protein and the yellow prosthetic group (coenzyme) by dialysis under acidic conditions. Flavin mononucleotide was isolated as its crystalline calcium salt and shown to be riboflavin-5′-phosphate; its structure was confirmed by chemical synthesis by Kuhn and Rudy (94). It is commercially available as the monosodium salt dihydrate [6184-17-4], with a water solubility of more than 200 times that of riboflavin. It has wide application in multivitamin and B-complex solutions, where it does not require the solubilizers needed for riboflavin.

Riboflavin-5′-Adenosine Diphosphate

Riboflavin-5′-adenosine diphosphate [146-14-5] (flavin–adenine dinucleotide, FAD), $C_{27}H_{33}N_9O_{15}P_2$ (2), mol wt 785.56, was first isolated in 1938 from the D-amino acid oxidase as its prosthetic group (95), where it was postulated to be flavin–adenine dinucleotide. The structure was established by the first synthesis in Todd's laboratory (96); the monosilver salt of FMN was condensed with 2′,3′-isopropylidene-adenosine-5′-benzylphosphorchloridate, followed by removal of protective groups. It was also synthesized directly from FMN and adenosine-5′-monophosphate (AMP) with di-p-tolylcarbodiimide as the condensation agent (97). Another direct synthesis was achieved by dehydration between FMN and AMP with trifluoroacetic acid anhydride (98). A 40% yield was obtained by condensation of adenosine-5′-phosphoramidate and FMN using a mixture of pyridine and o-chlorophenol as the solvent (99). Condensation of AMP with FMN in ethoxyacetylene gives a 10–15% yield; by-products such as riboflavin-4′,5′-cyclic phosphate are avoided (100). An efficient procedure was developed, which transforms the free acid of both nucleotides into their tri-n-butyl-ammonium salts and uses a reactive carbodiimide for the coupling reaction. The yield of FAD can be as high as 70% (101). In addition to D-amino acid oxidase, FAD is the prosthetic group for the other flavoproteins including glucose oxidase, glycine oxidase, fumarate hydrogenase, histaminase, and xanthine oxidase.

Covalently Bound Flavins

The FAD prosthetic group in mammalian succinate dehydrogenase was found to be covalently affixed to protein at the 8 α-position through the linkage of 3-position of histidine (102,103). Since then, several covalently bound riboflavins (104,105) have been found successively from the enzymes listed in Table 3. The biosynthetic mechanism, however, has not been clarified.

6-Hydroxyriboflavin

This compound [86120-61-8] (26) was isolated as a green coenzyme of the NADH dehydrogenase from Peptostreptococcus elsdenii and also from glycolate oxidase of porcine liver. It is not fluorescent, and its structure was established by synthesis (106). The 5′-monophosphate serves as a cofactor for glycolate oxidase from pig liver.

8-Nor-8-hydroxyriboflavin

A prosthetic group of red color has been isolated from NADH dehydrogenase of the electron-transferring flavoprotein of Peptostreptococcus elsdenii. Its structure [52134-62-0] (27) has been established as the FAD derivative of 8-hydroxy-7-methylisoalloxazine. Proof has been obtained by the synthesis of 8-hydroxy-7-methylisoalloxazine models and stepwise degradation of the naturally occurring compound (107).

Roseoflavin

Roseflavin [51093-55-1], $C_{18}H_{23}N_5O_6$, (28), mol wt 405.41, mp 274–297°C, $[\alpha]_D = -320°$ (0.1M NaOH), was isolated from a culture medium of Streptomyces davawensis as dark, reddish-brown fine needles (from ethanol); the 8-methyl group of riboflavin is substituted by a dimethylamino group. This structure was confirmed by the synthesis. Roseflavin shows antimicrobial activity against gram-positive bacteria (108).

R = D-Ribityl

5-Deazariboflavin

In 5-deazariboflavin (24), the N-5 of riboflavin is replaced by CH; it serves as cofactor for several flavin-catalyzed reactions (109). It was first synthesized in 1970 (110); improved synthetic processes were reported later (111).

A low potential electron carrier, the fluorescent factor F_{420} [37333-48-5, 64885-97-8] (29) (it absorbs maximally at 420 nm), possessing a 5-deazaflavin moiety, was isolated from methane-producing bacteria (112). F_{420} is an obligate intermediate for passage of an electron from H_2 to $NADP^-$ to generate NADPH. The structure of F_{420} was proposed as an 8-hydroxy-7-demethyl-5-deazaflavin derivative (25) (113) and confirmed as compound (29) by the total synthesis (114).

BIBLIOGRAPHY

1. R. Kuhn, P. György, and T. Wagner-Jauregg, Ber. Dtsch. Chem. Ges. 66, 576 (1933).

2. Ref. 1, 66, 317 (1933).

3. P. Karrer and K. Schöpp, Helv. Chem. Acta 17, 735, 771 (1934).

4. O. Warburg and W. Christian, *Biochem. Z.* **254**, 438 (1932); **266**, 377 (1933).

5. R. Kuhn, H. Ruby, and T. Wagner-Jauregg, *Ber. Dtsch. Chem. Ges.* **66**, 1950 (1933).

6. P. Karrer, K. Schöpp, and F. Benz, *Helv. Chim. Acta* **18**, 426, 522 (1935).

7. R. Kuhn, K. Reinemund, H. Kaltschmitt, K. Stöbele, and H. Trischmann, *Naturwissenschaften* **23**, 260 (1935).

8. E. A. Woodcock and J. J. Warthesen, *J. Food Sci.* **47**, 545 (Apr. 1982).

9. U.S. Pat. 2,407,412 (Sept. 10, 1946), D. V. Frost (to Abbott Laboratories).

10. A. W. Varnes, R. B. Dodson, and E. L. Wehry, *J. Am. Chem. Soc.* **94**, 946 (1972).

11. P. Karrer, H. Salmon, K. Schöpp, E. Schlittler, and H. Fritsche, *Helv. Chim. Acta* **17**, 1010 (1934).

12. E. C. Smith and D. Metzler, *J. Am. Chem. Soc.* **85**, 3285 (1963).

13. R. Kuhn and T. Wagner-Jauregg, *Ber. Dtsch. Chem. Ges.* **67**, 361 (1934).

14. L. Michaelis, M. P. Schubert, and C. V. Smythe, *J. Biol. Chem.* **116**, 587 (1936).

15. A. Ehrenberg, *Acta Chem. Scand.* **11**, 205 (1957).

16. B. Commoner and B. Lippincott, *Proc. Natl. Acad. Sci. U.S.A.* **44**, 1110 (1958).

17. F. Müller, P. Hemmerich, and A. Ehrenberg in H. Kamin, ed., *Flavins and Flavoproteins*, University Park Press, Baltimore, Md., 1971, p. 107.

18. F. Müller et al., *Eur. J. Biochem.* **116**, 17 (1981).

19. H. Kurreck et al., *J. Am. Chem. Soc.* **106**, 737 (1984).

20. P. Hemmerich, *Experientia* **16**, 534 (1969).

21. P. Bamberg and P. Hemmerich, *Helv. Chim. Acta* **44**, 1001 (1961); K. H. Dubley, A. Ehrenberg, P. Hemmerich, and F. Müller, *Helv. Chim. Acta* **47**, 1354 (1964).

22. A. Ehrenberg and P. Hemmerich, in T. P. Singer, ed., *Biological Oxidations,* Wiley-Interscience, New York, 1968, p. 722.

23. R. Kuhn and F. Weygand, *Ber. Dtsch. Chem. Ges.* **68**, 1282 (1935).

24. P. Karrer and H. F. Meerwein, *Helv. Chim. Acta* **18**, 1130 (1935).

25. Ref. 24, **19**, 264 (1936).

26. M. Tishler, J. W. Wellman, and K. Ladenburg, *J. Am. Chem. Sco.* **67**, 2165 (1945).

27. T. Matsukawa and K. Shirakawa, *Yakugaku Zasshi* **69**, 208 (1949).

28. M. Tishler, K. Pfister, R. D. Babson, K. Ladenburg, and A. J. Fleming, *J. Am. Chem. Soc.* **69**, 1487 (1947).

29. U.S. Pat. 4,673,742 (June 16, 1987), J. Grimmer and H. C. Horn (to BASF AF); Ger. Pat. 3,542,837 (June 11, 1987), J. Grimmer (to BASF AG).

30. F. Bergel, A. Cohen, and J. W. Haworth, *J. Chem. Soc.*, 165 (1945).

31. R. M. Cresswell and H. C. S. Wood, *J. Chem. Soc.*, 4768 (1960).

32. Jpn. Pat. 10,031 (Nov. 13, 1959), A. Masuda (to Takeda Pharmaceutical Industries).

33. J. Davol and D. D. Evans, *J. Chem. Soc.*, 5041 (1960); see also R. M. Cresswell, T. Neilson, and H. C. S. Wood, *J. Chem. Soc.*, 477 (1961).

34. F. Yoneda, Y. Sakuma, M. Ichiba, and K. Shinomura, *J. Am. Chem. Soc.* **98**, 830 (1976).

35. F. Yoneda, Y. Sakuma, and K. Shinozuka, *J. Chem. Soc. Perkin Trans. 1*, 348 (1978).

36. T. Wagner-Jauregg in W. H. Sebrell, Jr., and R. E. Harris, eds., *The Vitamins*, Vol. V, Academic Press, Inc., New York, 1972, p. 19; J. P. Lambooy in R. C. Elderfield, ed., *Heterocyclic Compounds*, Vol. 9, John Wiley & Sons, Inc., New York, 1967, p. 118.

37. G. W. E. Plaut in M. Florkin and E. H. Stotz, eds., *Comprehensive Biochemistry*, Vol. 21, Elsevier Publishing Co., Amsterdam and New York, 1971, p. 11.

38. A. Bacher in F. Müller, ed., *Chemistry and Biochemistry of Flavoenzymes*, Vol. 1, CRC Press, Boca Raton, Fla., 1991, p. 215.

39. M. Mitsuda, K. Nakajima, and T. Nadamoto, *J. Nutr. Sci. Vitaminol.* **22**, 477 (1976); **23**, 71 (1977).

40. T. Rowan and H. C. S. Wood, *J. Chem. Soc.*, 452 (1968); T. Paterson and H. C. S. Wood, *J. Chem. Soc. Perkin Trans. 1*, 1051 (1972).

41. U.S. Pat. 2,202,161 (May 28, 1940), C. S. Miner (to Commercial Solvents Corp.).

42. U.S. Pat. 2,369,680 (Feb. 20, 1945), R. E. Maeda, H. L. Polland, and N. E. Rodgers (to Western Condensing Co.); U.S. Pat. 2,449,144 (Sept. 14, 1948), N. E. Rogers, H. L. Polland, and R. E. Maeda (to Western Condensing Co.).

43. Jpn. Pat. 73 19,958 (June 18, 1973), S. Sugawara and K. Sato (to Nippon Beet Sugar Manufacturing Co.).

44. Jpn. Pat. 76 19,187 (Feb. 16, 1976), Y. Ichida, H. Abe, and A. Aoike (to Kuraray Co.).

45. U.S. Pat. 2,425,280 (Aug. 5, 1947), R. Hickey (to Commercial Solvents Corp.).

46. Ger. Pat. 1,936,238 (Jan. 28, 1971), E. M. Dikanskaya and A. A. Balabanova (to All-Union Scientific Research Institute of Protein Biosynthesis).

47. Ger. Pat. 2,028,355 (Jan. 14, 1971), I. Nitelea and co-workers (to Romania Antibiotic Plant).

48. Ger. Pat. 1,767,260 (Aug. 19, 1976), G. M. Miescher (to Commercial Solvents Corp.).

49. USSR Pat. 194,261 (June 15, 1974), E. M. Dikanskaya (to All-Union Scientific-Research Institute of Protein Biosynthesis).

50. Ger. Pat. 2,453,827 (May 15, 1975), T. Slave and co-workers (to Institutul de Cercetari Chimico Farmaceutice).

51. Pol. Pat. 66,611 (Mar. 15, 1973), T. Szczesniak and co-workers (to Instytut Przemyslu Fermentacyjnego).

52. Pol. Pat. 76,481 (Mar. 10, 1975), T. Szezesniak and co-workers (to Instytut Przemyslu Farmaceutycznego).

53. Ger. Pat. 2,037,905 (Feb. 3, 1972), T. Kamikubo and N. Hiroshima (to Kanegafuchi Chemical Industry Co.).

54. U.S. Pat. 3,433,707 (Mar. 18, 1969), T. Matsubayashi and Y. Suzuki (to Dai Nippon Suger Manufacturing Co. and Nitto Physico-Chemical Research Institute).

55. Jpn. Pats. 73 96,790 and 73 96 791 (Dec. 10, 1973), T. Fukukawa, T. Matsuyoshi, and J. Hiratsuka (to Mitsui Petrochemical Industries).

56. Jpn. Pat. 79 80,495 (June 27, 1979), S. Uragami (to Mitsubishi Gas Chemical Co.).

57. Jpn. Pat. 77 110,897 (Sept. 17, 1977), K. Nakayama, K. Araki, and S. Shimojo (to Kyowa Hakko Kogyo Co.).

58. Jpn. Pat. 77 54,094 (May 2, 1977), I. Chibata and co-workers (to Tanabe Seiyaku Co.).

59. USSR Pat. 511,742 (Sept. 25, 1976), T. E. Popova and co-workers (to All-Union Scientific-Research Institute of Protein Biosynthesis).

60. Jpn. Pat. 75 116,690 (Sept. 12, 1975), H. Umezawa and co-workers (to Sanraku-Ocean Co.).

61. Jpn. Kokai Tokkyo Koho Jpn. Pat. 87 25,996 (Feb. 3, 1987), A. Matsuyama and co-workers (to Daicel Chemical Industries, Ltd.).

62. Jpn. Kokai Tokkyo Koho Jpn. Pat. 88 112,996 (May 18, 1988), A. Matsuyama and co-workers (to Daicel Chemical Industries, Ltd.).

63. PCA Int. Appl. WO 9201,060 (Jan. 23, 1992), R. B. Bailey, G. W. Lauderdale, D. L. Heefner, C. A. Weaver, M. J. Yarus, L. A. Burdzinski, and A. Boyte (to Coors Biotechnology, Inc.).

64. Eur. Pat. Appl. EP 531,708 (Mar. 17, 1993), N. Usui, Y. Yamamoto, and T. Nakamatu (to Ajinomoto Co.).

65. Eur. Pat. Appl. EP 231,605 (Aug. 12, 1987), D. L. Heefner, M. Yarus, A. Boyts, and L. Burdzinski (to Adolph Coors Co.).

66. Eur. Pat. Appl. EP 405,370 (Jan. 2, 1991), D. B. Perkins, J. K. Pero, and A. Sloma (to Hoffmann-La Roche, F., Co.).

67. Eur. Pat. Appl. EP 604,060 (Jan. 29, 1994), S. Koizumi, Y. Yonetani, and S. Teshiba (to Kyowa Hakko Kogyo Co.).

68. J. Koziol, in D. B. McCormick and L. D. Wright, eds., Methods in Enzymology, Vol. 43, Academic Press, Inc., New York, 1971, p. 253.

69. J. H. Richardson in D. B. McCormick and L. D. Wright, eds., Methods in Enzymology, Vol. 66, Academic Press, Inc., New York. 1980, p. 416.

70. N. Ishibashi, T. Ogawa, T. Imasaka, and M. Kunitake, Anal. Chem. 51, 2096 (1979).

71. E. Knobloch in Ref. 63, p. 305.

72. E. E. Snell and F. M. Strong, Ind. Eng. Chem. Anal. Ed., 11, 346 (1939).

73. H. A. Kornberg, R. S. Langdon, and V. H. Cheldelin, Anal. Chem. 20, 81 (1948).

74. J. A. Tillotson and E. M. Baker, Am. J. Clin. Nutr. 25, 425 (1972).

75. P. J. Richardson, D. J. Favell, G. C. Gidley, and A. D. Jones, Proc. Anal. Div. Chem. Soc. 15, 53 (1978).

76. C. Y. Wang and F. A. Moseley, J. Agric. Food Chem. 28, 483 (1980).

77. J. F. Kamman, T. P. Labuza, and J. J. Warthesen, J. Food Sci. 45, 1497 (1980).

78. K. Yagi and M. Sato, Biochem. Int. 2, 327 (1981).

79. D. R. Light, C. Walsh, and M. A. Marletta, Anal. Biochem. 109, 87 (1980).

80. W. N. Pearson, in P. György and W. N. Pearson, eds., The Vitamins, Vol. VII, Academic Press, Inc., New York, 1967, p. 99.

81. H. Baker and O. Frank in R. S. Rivilin, ed., Riboflavin, Plenum Press, New York, 1975, p. 49.

82. E. E. Conn, P. K. Stumpf, G. Bruening, and R. H. Doi, Outlines of Biochemistry, 5th ed., John Wiley & Sons, Inc., New York, 1987, p. 207.

83. K. M. Horowitt and L. A. Wittig, in W. H. Sebrell, Jr., and R. E. Harris, eds., The Vitamins, Vol. V, Academic Press, Inc., New York, 1972, p. 53; C. A. Hamilton, Prog. Bioorg. Chem., 1, 83 (1971); T. C. Bruice, Prog. Bioorg. Chem. 4, 1 (1976); C. Walsh, Annu, Rev. Biochem. 47, 881 (1978); V. Massey and P. Hemmerich in P. D. Boyer, ed., The Enzymes, Vol. 12, Academic Press, Inc., New York, 1976, p. 191.

84. F. Müller, ed., Chemistry and Biochemistry of Flavoenzymes, Vol. 1, 1991; Vol. 2, 1991; and Vol. 3, 1992 CRC Press, Boca Raton.

85. D. B. McCormick, in M. E. Sils and V. R. Young, eds., Modern Nutrition in Health and Disease, Lea and Febiger, Philadelphia, Pa., 1988, p. 362.

86. Recommended Dietary Allowances, 10th ed., Food and Nutrition Board, National Research Council, National Academy Press, Washington, D.C., 1989, p. 132.

87. R. Kuhn and P. Boulanger, Hoppe-Seyler's Z. Physiol. Chem. 241, 233 (1936).

88. K. Unna and J. G. Greslin, J. Pharmacol. Exp. Ther. 76, 75 (1942).

89. V. Demole, Z. Vitaminforsch. 7, 138 (1938).

90. U.S. Pats. 2,610,178 and 2,610,179 (Sept. 9, 1952), L. A. Flexser and W. G. Farkas (to Hoffmann-La Roche Inc.).

91. U.S. Pat. 2,535,385 (Dec. 26, 1950), P. J. Breivogel (to White Laboratories).

92. M. Viscontini, C. Ebnother, and P. Karrer, Helv. Chim. Acta 35, 457 (1952); M. Viscontini and co-workers, Helv. Chem. Acta 38, 15 (1955).

93. T. Ukita and K. Nagasawa, Chem. Pharm. Bull. 7, 465 (1959).

94. R. Kuhn and H. Rudy, Ber. Dtsch. Chem. Ges. 68, 353 (1935).

95. O. Warburg and W. Christian, Biochem. Z. 296, 294 (1938); 298, 150 (1938); O. Warburg, W. Christian, and A. Griese, Biochem. Z. 297, 417 (1938).

96. S. M. H. Christie, G. W. Kenner, and A. R. Todd, Nature (London) 170, 924 (1952); J. Chem. Soc., 46 (1954).

97. F. M. Huennekens and G. L. Kilgour, J. Am. Chem. Soc. 77, 6716 (1955).

98. C. DeLuca and N. O. Kaplan, J. Biol. Chem. 223, 569 (1956).

99. J. G. Moffatt and H. G. Khorana, J. Am. Chem. Soc. 80, 3756 (1958).

100. H. Wassermann and D. Cohen, Chem. Eng. News, 47 (1962).

101. F. Cramer and H. Neunhöffer, Ber. Dtsch. Chem. Ges. 95, 1664 (1962).

102. E. B. Kearney and T. P. Singer, Biochem. Biophys. Acta 17, 596 (1955).

103. T. Y. Wang, C. L. Tsuo, and Y. L. Wang, Sci. Sin. 5, 73 (1956).

104. T. P. Singer and D. E. Edmondson in Ref. 64, p. 253 and references cited therein.

105. D. E. Edmondson and R. De Francesco in Ref. 37, p. 73.

106. S. G. Mayhew, C. D. Whitefield, S. Ghisla; and M. Schuman-Jörns, Eur. J. Biochem. 44, 579 (1974).

107. S. Ghisla and S. G. Mayhew, Eur. J. Biochem. 63, 373 (1976).

108. S. Otani, M. Takatsu, M. Nakano, S. Kasai, R. Miura, and K. Matsui, J. Antibiot. 27, 88 (1974); S. Kasai, R. Miura, and K. Matsui, Bull. Chem. Soc. Jpn. 48, 2877 (1975).

109. P. Hemmerich, V. Massey, and H. Fenner, FEBS Lett. 84, 5 (1977).

110. D. E. O'Brien, L. T. Weinstock, and C. C. Cheng, J. Heterocycl. Chem. 7, 99 (1970).

111. F. Yoneda in Ref. 64, p. 267; F. Yoneda and B. Kokel in Ref. 37, p. 121.

112. D. Eirich, G. D. Vogels, and R. S. Wolfe, Biochemistry 17, 4583 (1978).

113. W. T. Ashton, R. D. Brown, F. Jacobson, and C. Walsh, J. Am. Chem. Soc. 101, 4419 (1979).

114. T. Kimachi, M. Kawase, S. Matsuki, K. Tanaka, and F. Yoneda, J. Chem. Soc., Perkin Trans. 1, 253 (1990).

FUMIO YONEDA
Fujimoto Pharmaceutical Corporation

VITAMINS: THIAMINE (B₁)

Thiamine [59-43-8] is the official IUPAC-IUB name for 3-(4-amino-2-methyl-5-pyrimidinyl) methyl-5-(-2-hydroxyethyl)-4-methylthiazolium chloride, $C_{12}H_{17}$-N_4OSCl (1). The names thiamine, thiamin (used in many official and commercial documents), aneurine, and the older term vitamin B₁ are also casually used. These names usually refer to the chloride hydrochloride [67-03-8] (2), the common biological and commercial form.

(1)

(2)

(3)

(4, n = 1)
(5, n = 2)
(6, n = 3)

Thiamine is found in varying, low levels as its salts and phosphate esters in the tissues of practically all life forms, where its pyrophosphate [154-87-0] (5), known as cocarboxylase, plays essential roles in carbohydrate metabolism. Plants and most microorganisms biosynthesize thiamine, but animals are incapable of doing so, require it in their diets in amounts varying with their carbohydrate use, and excrete the excess. Thiamine content of foods varies and can be partially lost during storage or processing, as it is one of the less stable of the vitamins. Thiamine deficiencies can lead to a range of effects, from the malaise and fatigue of a lesser deficiency to the serious disorders of gross deficiency observed for many animal species, such as beriberi, where weight loss, heart disease, and serious neurological degeneration can lead to death. As preventive measures, thiamine is used to enrich foods and feeds and as an ingredient of dietary supplements. Thiamine is also used therapeutically for treatment of specific deficiency conditions. Thiamine was one of the first of the vitamins to be manufactured for commerce. The common commercially available forms of thiamine, the chloride hydrochloride (2) and the mononitrate [532-43-4] (3), are manufactured by chemical processes operated at fine chemical scales. At present, natural sources and bioprocesses are not cost-competitive for bulk production.

The history of thiamine is linked with the disease beriberi, which once caused a great toll of suffering and death in parts of the world where milled or polished rice is a staple of the diet (1). Beriberi was recognized in China as early as 2600 B.C., but the first cure was only demonstrated in 1885. Takaki, then surgeon general of the Japanese navy, eradicated beriberi in the fleet by adding more protein to the sailors' diet of mainly refined rice to achieve a more Western-style diet. By 1901 the Dutch physicians Eijkman and Grijn showed hens kept on a diet of dehusked rice developed a disease similar to human beriberi which could be reversed or prevented by adding rice bran and other nitrogenous materials to their feed. They suggested beriberi was caused not by pathogens or toxins but by the lack of a vitally important food constituent, which was a radical idea at the time. They also showed the preventive factor was leached by water and destroyed by heat. Searches for a specific water-soluble nitrogenous substance led the English chemist Funk in 1911 to coin the term vitamine (vital amine, an amine essential for life), popularizing the concept of deficiency diseases. In 1926, a small, pure crystalline sample of thiamine hydrochloride was painstakingly first isolated from rice bran extracts by Dutch chemists Jansen and Donath. After eight years of effort, in 1932 the German chemist Windaus and his coworkers obtained pure thiamine from yeast extracts and established the correct empirical formula. In 1935–1936, following similarly prolonged isolation efforts, the American chemists Williams and Cline and German chemist Grewe independently proposed the correct structure based on degradation studies. The name thiamine was first used by Williams. Shortly thereafter, the Williams group confirmed the structure by rational synthesis (2), followed closely by two different syntheses by other workers (3–5). Whereas in 1933 Williams had succeeded in isolating only gram quantities from metric tons of rice polishings, a highly enriched source, in 1937 chemists at Merck and Hoffmann-La Roche developed a production level of about 100 kg within a year based on two different, relatively involved chemical syntheses. Demand and production levels have risen steeply since then to their present estimated level of about 3300 t/yr worldwide.

PHYSICAL AND CHEMICAL PROPERTIES

Salt Formation

As a weakly basic pyrimidine and a thiazolium cation, thiamine forms both mono- and dipositive salts, eg, the two commercial forms.

Thiamine chloride hydrochloride [67-03-8], (thiamine hydrochloride) $C_{12}H_{18}N_4OSCl_2$ (2), crystallizes as colorless monoclinic needles, mp 248–250°C (with decomposition), which in bulk appear white and have an approximate density of 0.4 kg/L. Several polymorphic crystal forms have been reported. The salt has a characteristic thiazole meat-like odor and a slightly bitter taste. On exposure to air of average humidity, the hydrochloride (2) can adsorb up to one mole of water (more typically slightly less, to about 4% by weight), which may be removed by heating to 100°C or by vacuum drying. It is very soluble in water (over 1 kg/L at 25°C), soluble in glycerol (0.056 kg/L), propylene glycol, and methanol, sparingly soluble in 95% ethanol (0.01 kg/L), and practically insoluble in less polar organic solvents. In 1–5% solutions in water it shows a pH of 3–3.5 (6,7).

Thiamine mononitrate [532-43-4] (3), $C_{12}H_{17}N_5O_4S$, is an apparently white, colorless crystalline solid with a typical odor, melting point of ca 196–200°C (dec.) and an approximate bulk density of 0.5 kg/L. It is much less soluble in water than the hydrochloride (0.027 kg/L at 25°C, 0.030 kg/L at 100°C) and practically nonhygroscopic. Dilute solutions in water show a pH of 6.5–7 (6,8).

Numerous other salts have been reported in the literature, including some which are insoluble in water.

Physical Chemical Characterization

Thiamine, its derivatives, and its degradation products have been fully characterized by spectroscopic methods (9,10). The ultraviolet spectrum of thiamine shows pH-dependent maxima (11). ^1H, ^{13}C, and ^{15}N nuclear magnetic resonance spectra show protonation occurs at the 1-nitrogen, and not the 4-amino position (12–14). The ^1H spectrum in D_2O shows no resonance for the thiazole 2-hydrogen, as this is acidic and readily exchanged via formation of the thiazole ylid (13) an important intermediate in the biochemical functions of thiamine. Recent work has revised the pK_a values for the two ionization reactions to 4.8 and 18 respectively (9,10,15). The mass spectrum of thiamine hydrochloride shows no molecular ion under standard electron impact ionization conditions, but fast atom bombardment and chemical ionization allow observation of both an intense peak for the parent cation and its major fragmentation ion, the pyrimidinylmethyl cation (16).

Reactions and Stability

Thiamine hydrochloride is stable as a solid if kept dry. Heating to 100°C for 24 h does not diminish its potency. In solution, its stability depends heavily on conditions of pH, temperature, and oxygen. Aqueous solutions below pH 5 are stable to oxygen, heating, and even autoclaving (11). Heating in water to 140°C under pressure gives decomposition to 4-amino-5-hydroxymethyl-2-methyl-pyrimidine [73-67-6] (7) and 4-methyl-5-(2-hydroxyethyl)thiazole [137-00-8] (8) (17). Heating with 20% hydrochloric acid hydrolyzes the 4-amino group to yield oxythiamine [582-36-5] (9), an antagonist of thiamine (18). Above pH 5, aqueous thiamine is destroyed relatively rapidly by boiling. At pH 7 destruction occurs even at room temperature. Concentrated solutions of thiamine in alcohol when neutralized degrade rapidly at room temperature to liberate thiazole (8) and form oligomers of the pyrimidine where the 5-methylene is linked to N-1 of another pyrimidine unit (10,14).

(7)   (8)   (9)

At pH 8 or above, even at room temperature, water solutions of thiamine turn yellow, then fade as a series of reactions ensues (9,10,19). Basification allows intramolecular attack of the 4-amino function on the thiazole ring to generate dihydrothiochrome [80483-97-2] (10), which eliminates thiolate to generate the salt of the yellow form, 5,6-dihydropyrimido(4,5d)pyrimidine [84825-03-6] (11), as a kinetic product (9,10,19) (Fig. 1). The sequence is reversible. More slowly, under the same conditions, a competing hydrolysis of the thiazolium ring via the pseudobase generates another product, known as thiamine thiol or the

thiol form [554-45-0] (12) as the thermodynamic product. Also present in the system is a small amount of thiamine ylid [84812-92-0] (13), formed by deprotonation of C-2. The products dihydrothiochrome (10), yellow form (11), and ylid (13) undergo significant and useful chemistry.

The yellow form (11) on acidification is converted to the more stable thiol form (12). On oxidation, typically with alkaline ferricyanide, yellow form (11) is irreversibly converted to thiochrome [299-35-4] (14), a yellow crystalline compound found naturally in yeast but with no thiamine activity. In solution, thiochrome exhibits an intense blue fluorescence, a property used for the quantitative determination of thiamine.

The thiol form (12) undergoes reactions mainly via its N-formyl or ene–thiol groups. Heating an aqueous solution of the thiol form (Fig. 2) effects hydrolysis to the diamine 4-amino-5-aminomethyl-2-methyl-pyrimidine [95-02-3] (16), 5-hydroxy-3-mercaptopentan-2-one [15678-01-0] (17), and formic acid (20). Neutralization of a solution of the thiolate anion with carbon dioxide gives a fat-soluble basic material [21682-72-4, 35922-43-1] (20), presumably via dihydrothiochrome (10) (21). Acylation of the thiolate occurs on both sulfur and oxygen to give mono- or diacyl thiamines, some of which are interesting fat-soluble depot forms of thiamine.

The thiol form (12) is susceptible to oxidation (see Fig. 2). Iodine treatment regenerates thiamine in good yield. Heating an aqueous solution at pH 8 in air gives rise to thiamine disulfide [67-16-3] (21), thiochrome (14), and other products (22). The disulfide is readily reduced to thiamine in vivo and is as biologically active. Other mixed disulfides, of interest as fat-soluble forms, are formed from thiamine, possibly via oxidative coupling to the thiol form (12).

Whereas a claim of isolation of the thiamine ylid (13) has been the subject of controversy (15,23), kinetic and product studies and molecular orbital calculations support the formation and reactivity of a thiamine ylid as an unstable intermediate (15,24–26). Ylid (13) is accepted as an intermediate in explanations of the enzymatic and nonenzymatic reactions of thiamine. Among these reactions are the enzymatic oxidative decarboxylation of pyruvic and 2-ketoglutaric acids, the formation of acetoin, the reversible alpha-ketol transfer reactions catalyzed by transketolase, and the nonenzymatic acyloin condensation. According to the thiazolium ylid mechanism, ylid (13) reacts reversibly with carbonyl reagents, facilitating aldol–retroaldol and decarboxylation reactions (Fig. 3), via the acyl anion equivalent thiazolium ylids (23), the so-called active aldehydes, which are the key intermediates (27). In the specific case of oxidative decarboxylation of pyruvic acid, for example, ylid (23) transfers the acyl group to lipoic acid. Other reactions involving thiamine similarly involve nucleophilic additions of a thiazolium ylid to a suitable acceptor. Syntheses of a thiamine–pyruvate adduct (22) and a protonated form of its decarboxylation product (23) (28), as well as model studies with thiazolium ions (29), further support the thiazolium ylid mechanism. There is still current debate and investigation of the mechanistic details of the involvement of the thiamine ylid (13) in biological systems, including the role of the associated metal ions (15,30).

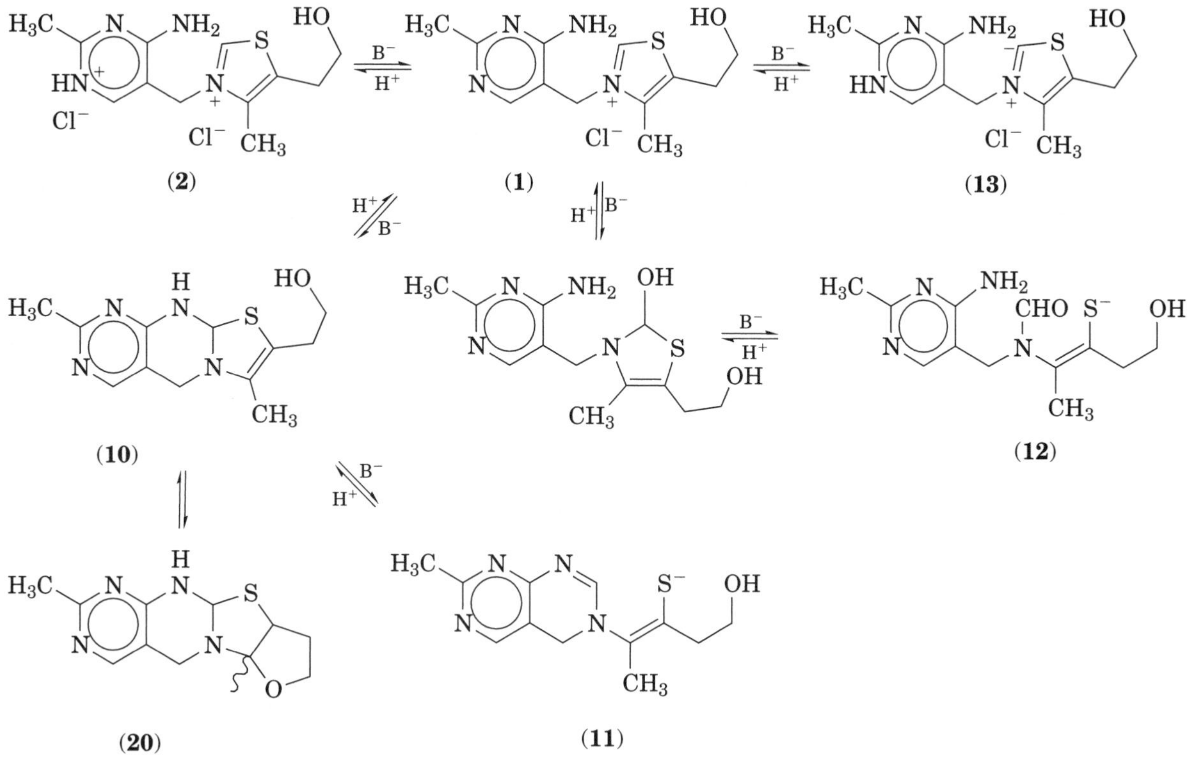

Figure 1. Structural changes of thiamine with pH.

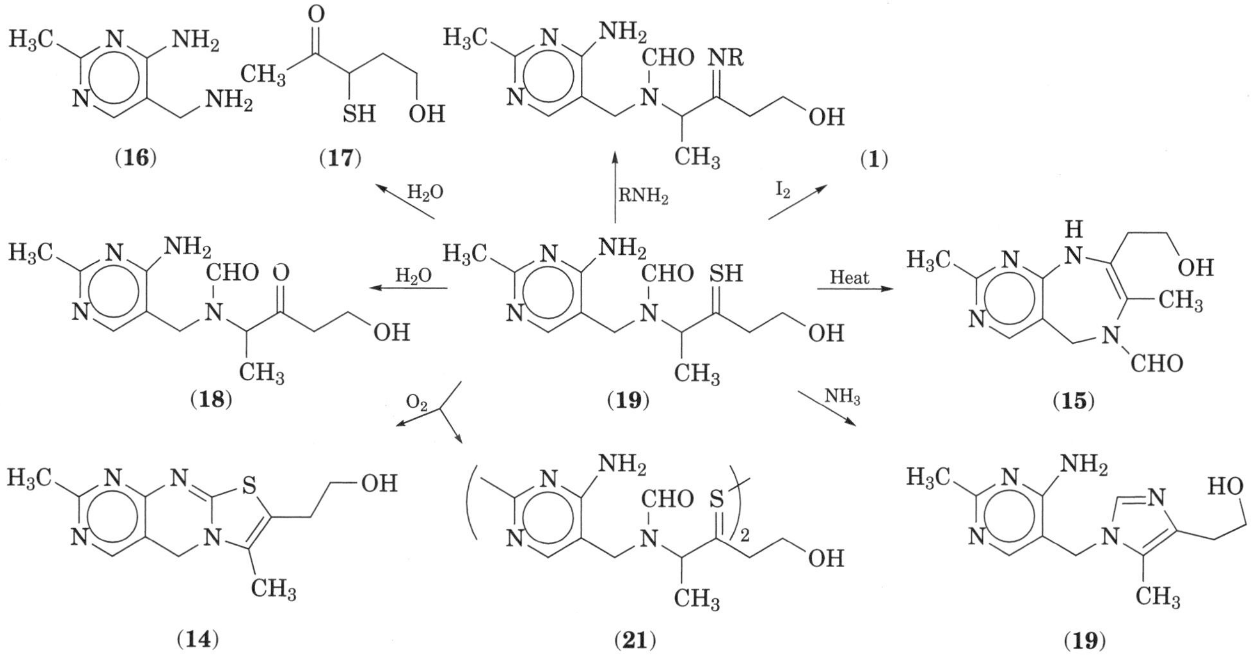

Figure 2. Reactions of the thiol form (**12**).

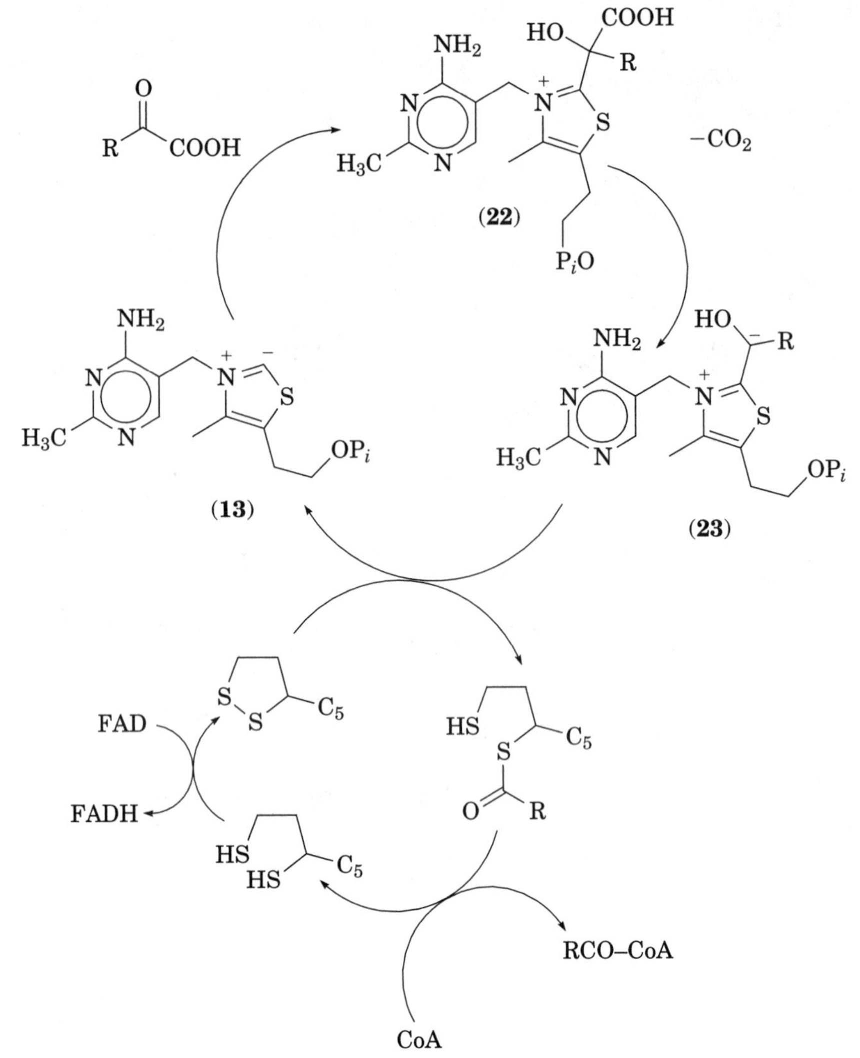

Figure 3. Oxidative decarboxylation of alpha-ketoacids.

Thiamine is susceptible to reaction with various nucleophilic species. During the race for the isolation of thiamine, it was found by chance during attempted stabilization of extracts that thiamine is readily ruptured by sulfite treatment. This occurs slowly at pH 3 and rapidly at pH 5 and above, to give 4-amino-2-methyl-5-pyrimidine-methane-sulfonic acid [108084-76-0] (24) (Fig. 4), with liberation of the thiazole moiety (8) (2,10). Other nucleophilic groups such as pyridines, phenolates, anilines, and azide react similarly in the required presence of sulfite. An addition–elimination mechanism involving sulfite attack on the pyrimidine ring has been elucidated (9,10). Similar reactions are observed in the degradation of thiamine by thiaminases, enzymes found in certain foods and bacteria, including some found in the human intestine (1,10,31). Thiaminase I (EC 2.5.1.2), found in shellfish, ferns, some vegetables and some bacteria, promotes the replacement of the thiazole by other organic bases, including purines. Thiaminase II (EC 3.5.99.2), found mostly in bacteria, catalyzes the cleavage of thiamine into (7), the biosynthetic precursor of thiamine and an antagonist of pyridoxine (vitamin B₆). A thiol group at the active site is strongly im-

plicated (11). These enzymes can have strong effects on animals as they promote thiamine deficiency (32). Fortunately, they are thermolabile, and hence a problem mostly in uncooked foods or feeds.

Thermostable substances which inactivate thiamine have been found in a large number of plants and in some animal tissues. Among these are polyphenols such as caffeic acid, tannic acid, hydroxylated derivatives of tyrosine, and some flavonoids (1). Thiamine is susceptible to destruction by x-rays, gamma rays, and ultraviolet light to generate diamine (16) by cleavage of the thiazole ring (1). Photochemical and thermal degradation of thiamine gives rise to numerous sulfur-containing heterocycles, some with meaty or bread aromas of interest to the flavor industry (33).

Operations in preparing or preserving food and feeds can sometimes lead to significant losses of thiamine value through leaching, chemical destruction (from high heat, high energy irradiation, or especially alkaline oxidizing conditions), and specific chemical interactions with thiamine-destroying substances. Such losses range in amounts from a few percent to as high as 85%. In many

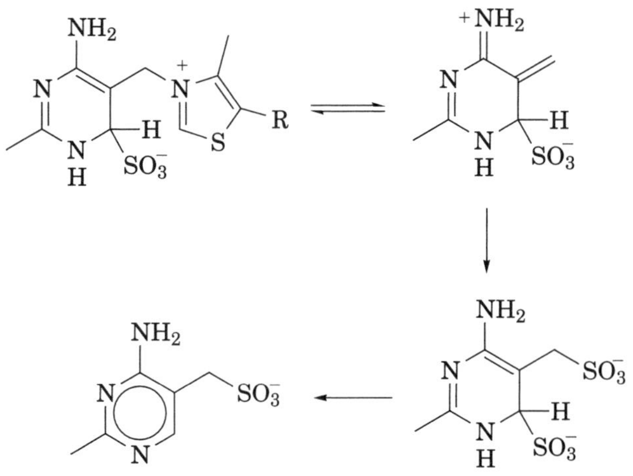

(24)

Figure 4. Mechanism of nucleophilic degradation.

cases, the destruction of thiamine shows apparent first-order kinetics, and Arrhenius calculations can be used to estimate losses (1,34).

Thiamine forms the expected derivatives of the thiazole alcohol function, such as carboxylic and phosphate esters. Few reactions at the pyrimidine 4-amino function have been reported. Most of the usual conditions used for formation of amides, for example, lead to destruction of the thiazolium ring.

NATURAL OCCURRENCE

Thiamine is widespread in nature, although generally in only relatively minute quantities (1,35). In microorganisms it is found mainly intracellularly, although minute amounts are lost to the natural environment upon cell lysis. In higher plants the most abundant form is free thiamine along with lesser amounts of the phosphate esters (4–6). Within the plant, thiamine is unevenly distributed, its amounts and location depending on the life stage. Seeds and uptake from the soil supply the plant until biosynthesis in the leaves begins. Thiamine is then transported from the leaves to the roots where it exhibits hormone-like effects on their growth. Later thiamine is again concentrated in the seeds, especially in the germ and in the pericarpal layers surrounding the starchy areas of the seed (36). As the germ and bran is often removed during processing for aesthetic reasons, highly refined rice or wheat, for example, can have significantly lowered contents of thiamine, and can be a cause of thiamine deficiency when used as staple foods.

In the tissues of animals, most thiamine is found as its phosphorylated esters (4–6) and is predominantly bound to enzymes as the pyrophosphate (5), the active coenzyme form. As expected for a factor involved in carbohydrate metabolism, the highest concentrations are generally found in organs with high activity, such as the heart, kidney, liver, and brain. In humans this typically amounts to 1–8 μg/g of wet tissue, with lesser amounts in the skeletal mus-

cles (35). A typical healthy human body may contain about 30 mg of thiamine in all forms, about 40–50% of this being in the muscles owing to their bulk. Almost no excess is stored. Normal human blood contains about 90 ng/mL, mostly in the red cells and leukocytes. A value below 40 ng/mL is considered indicative of a possible deficiency. Amounts and proportions in the tissues of other animal species vary widely (31,35).

Good natural human dietary sources of thiamine are unrefined cereal grains, organ meats, pork, legumes, and nuts (Table 1). Oils, fats, and highly refined foods are essentially devoid of thiamine (3,36). Although thiamine is widespread in foods, some can be lost in food preparation and storage. As a result, dietary intake can vary significantly. In most developed countries, foods (typically white rice and white flour) and feeds are supplemented with thiamine and its use in vitamin tablets as dietary supplements is common. Enriched grains, cereals, and baked products contribute about 30–45% of the recommended daily allowance (RDA) for the adult diet in the United States.

BIOCHEMICAL AND PHYSIOLOGICAL FUNCTIONS

Thiamine serves essential functions and its deficiency causes particularly deleterious effects on an organism's energy status and, in higher organisms, its nerve functions. In living systems, the only established biochemically active form is the pyrophosphate (cocarboxylase) (5), which plays a vital role in intermediate metabolism as a cofactor for some important enzymatic reactions. Dehydrogenase enzymes require cocarboxylase (5) for oxidative decarboxylation of 2-ketoacids, notably pyruvate in glycolysis, 2-oxoglutarate in the citric acid cycle, and other ketoacids from amino acid decarboxylation (1). Transketolase enzymes require cocarboxylase (5) for the reversible transfer of alpha-ketols in ketose–aldose transformations important in the production of pentoses for RNA and DNA synthesis (1). Thiamine triphosphate (6) occurs in unusually higher concentrations in nerve tissues and the brain and may play an essential role in the stimulation of peripheral nerves (35).

Table 1. Average Thiamine Contents of Foods

Food	Thiamine (μg/100 g)
Wheat germ	2050
Dried brewer's yeast	1820
Soybeans	1300
Pork	600–950
Dried beans and peas	680
Dried milk whey	500
Nuts	300–560
Brown rice	300
Beef liver	300
Potatoes	170
Fish	50–90
Eggs	70
Vegetables (fruit, leaf, stem, root)	60–70
Whole milk	30–70
White rice	50

In humans, thiamine is both actively and passively absorbed to a limited level in the intestines, is transported as the free vitamin, is then taken up in actively metabolizing tissues, and is converted to the phosphate esters via ubiquitous thiamine kinases. During thiamine deficiency all tissues stores are readily mobilized. Because depletion of thiamine levels in erythrocytes parallels that of other tissues, erythrocyte thiamine levels are used to quantitate severity of the deficiency. As deficiency progresses, thiamine becomes indetectable in the urine, the primary excretory route for this vitamin and its metabolites. Six major metabolites, of more than 20 total, have been characterized from human urine, including thiamine fragments (**7,8**), and the corresponding carboxylic acids (1,37,38).

The classic pathology resulting from severe thiamine deficiency in humans is called beriberi. Similar conditions have been described for many other animals. Beriberi develops primarily from inadequate intake of thiamine or from ingestion of food containing antithiamine factors and is somewhat rare in developed areas of the world. Less severe thiamine deficiency is more common and is characterized by anorexia and mental disturbances, such as irritability, inattention, memory defects, depression, and insomnia. If a lesser deficiency is left untreated, one of several clinical forms of beriberi develops, the symptoms of which include mental changes, peripheral neuritis, paresthesias, muscle cramps, edema, muscular atrophy, and cardiac failure. The most commonly encountered type of thiamine deficiency in Western countries is associated with alcohol abuse. This is generally thought to result from high intake of empty calories and low intake of nutritionally adequate foods. Other factors which influence thiamine status include general level of muscular activity; dietary practices such as high intake of refined carbohydrates, tea or coffee, or raw seafood; reduced thiamine intake as a result of disease, or parasites or drugs lowering food intake and utilization or thiamine absorption; pregnancy and lactation; heavy smoking; advanced age; genetic background; and stress. In other animals, climate and intestinal microflora also become important, and toxic effects can occur from changes in the microflora (39). Thiamine status has been monitored by blood thiamine levels, thiamine and metabolite levels in the urine, blood pyruvate and lactate levels, and blood transketolase activity (1,37,38).

Thiamine requirements vary and, with a lack of significant storage capability, a constant intake is needed or deficiency can occur relatively quickly. Human recommended daily allowances (RDAs) in the United States are based on calorie intake at the level of 0.50 mg/4184 kJ (1000 kcal) for healthy individuals (Table 2). As little as 0.15–0.20 mg/4184 kJ will prevent deficiency signs but 0.35–0.40 mg/4184 kJ are required to maintain near normal urinary excretion levels and associated enzyme activities. Pregnant and lactating women require higher levels of supplementation. Other countries have set different recommended levels (1,37,38).

MANUFACTURE

Chemistry

Isolation of thiamine from natural sources (rice bran, yeast extracts, or wheat germ) is only of historical interest. Production by bioprocesses is not cost-effective at present. All of the thiamine produced worldwide is manufactured by chemical processes operated at moderately large scale. Two major synthetic routes have been used: alkylation of a preformed thiazole, or construction of the thiazolium salt from a pyrimidine carrying the ultimate thiazole nitrogen (9,40). The latter approach is now generally preferred for manufacturing.

The first approach parallels the known biosynthetic pathway where the alkylating agent is the pyrophosphate ester of alcohol (**7**). Typical of this approach is Williams' first convergent synthesis, which is the basis for the first industrial method developed by Merck & Co. (2,5,41). Synthetic 4-amino-5-bromomethyl-2-methylpyrimidine [*2908-71-6*] (**26**) and thiazole (**8**), or its *O*-acetate, were condensed and then the bromide was replaced by use of silver salts. Numerous variants were published or patented in the 1930s and 1940s. Such methods generally gave only moderate yields, used higher temperatures in polar solvents, required tedious purification of the colored products and are no longer used.

In the second general method, the amine (**16**), known as Grewe diamine, is the paradigm intermediate onto which the thiazolium ring is constructed. Differences occur only in the raw materials and methods used for the manufacture of the diamine. The same end of the linear sequence is universally practiced by all large manufacturers.

The synthesis which forms the basis of production at Hoffmann-La Roche (Fig. 5) proceeds via the pyrimidinenitrile [*698-29-3*] (**27**) made from malononitrile, trimethylorthoformate, ammonia, and acetonitrile (42,43). High pressure catalytic reduction of the nitrile furnishes diamine (**16**). The overall sequence is short, highly efficient, and generates almost no waste. However, malononitrile is a relatively expensive and hazardous three-carbon source.

Other syntheses produce Grewe diamine using inexpensive acrylonitrile and alkyl formates as raw materials. Such routes deliver the pyrimidine bridge carbon at the correct aminomethyl oxidation level without a need for reduction. Thus, in the older method of Shionogi, base-catalyzed acylation of 3-methoxypropanenitrile followed by enolate alkylation and ring closure with two equivalents

Table 2. U.S. RDAs for Thiamine

Population segment	RDA (mg)
Infants and children <4 yr	0.7
Adults and children >4 yr	1.5
Pregnant or lactating women	1.7

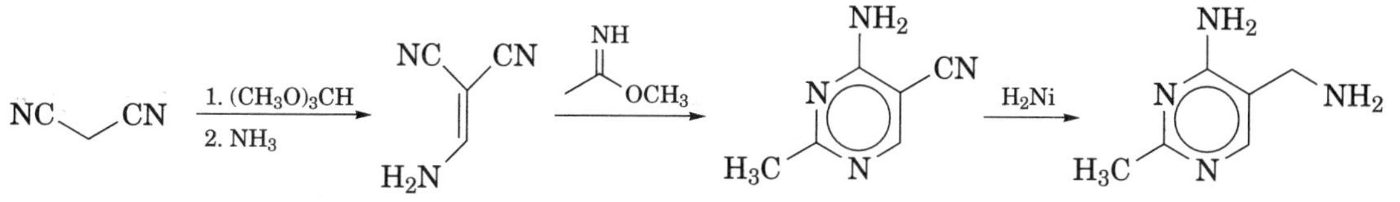

Figure 5. Roche process.

of acetamidine gives the acetyl derivative [23676-63-3] (**28**) of Grewe diamine (Fig. 6) (44–46). Alternatively, in a newer method by BASF, acylation of 3-formamidopropanenitrile, followed by ring closure with acetamidine, gives diamine as its formamide [1886-34-6] (**29**) (Fig. 7) (47–50). Both the Shionogi and BASF methods are simple technologies based on inexpensive raw materials, but both generate significant levels of salts and organic carbon as wastes.

In the past decade Takeda and Ube, in a joint venture have developed a new process for diamine, also based on acrylonitrile and alkyl formate, but carrying the pyrimidine bridge carbon at the carbonyl level (Fig. 8) (50–55). Metal-catalyzed oxidation of acrylonitrile in methanol generates 3,3-dimethoxypropanenitrile [57597-62-3] (**30**) which is acylated with methoxide/carbon monoxide under pressure. Unlike other methods in which the intermediate enolate is alkylated, in this process the enolate is acetalized and the acetal thermally converted to the acetal enol ether [87466-78-2] (**31**). Condensation with acetamidine provides the remaining carbons. After hydrolysis of the acetal [16057-06-0] (**32**), reductive amination at high pres-

sure with a specialized catalyst provides diamine (**16**) in very high overall yield. The process is operated in a new, large-scale, automated, continuous, technically complex plant in Japan. Advantages include use of inexpensive acrylonitrile and carbon monoxide, avoidance of the costs of an alkylating agent, and consumption of only one equivalent of acetamidine.

In a much different approach based on cyanamide, acrylonitrile, and acetonitrile, cyanoacetamidine [56563-07-6] (**33**) is cyanoethylated and the condensation product [56563-10-1] (**34**) is dehydrogenated and hydrogenated directly to Grewe diamine in the presence of Raney cobalt and ammonia (56).

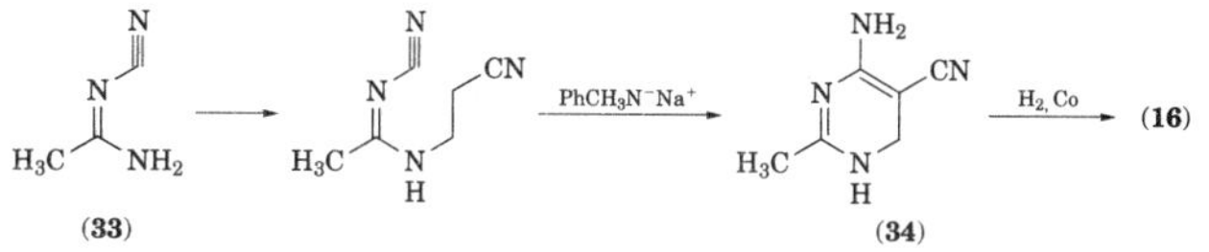

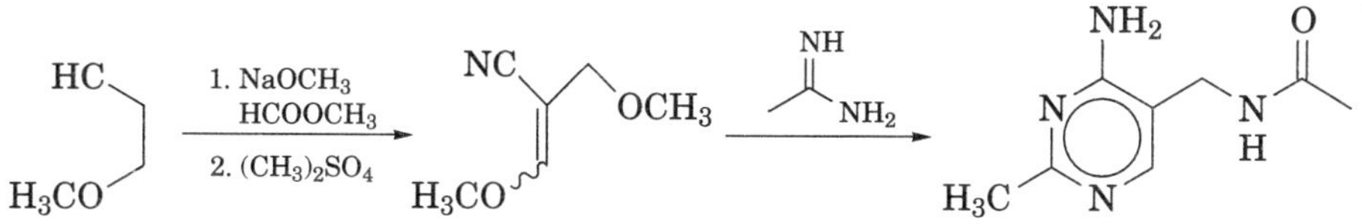

Figure 6. Shionogi process.

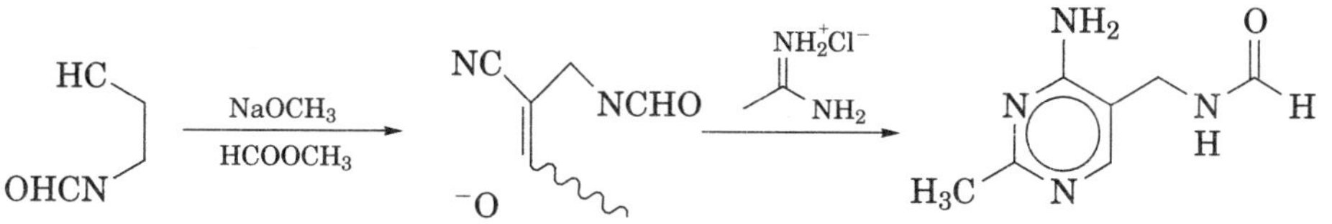

Figure 7. BASF process.

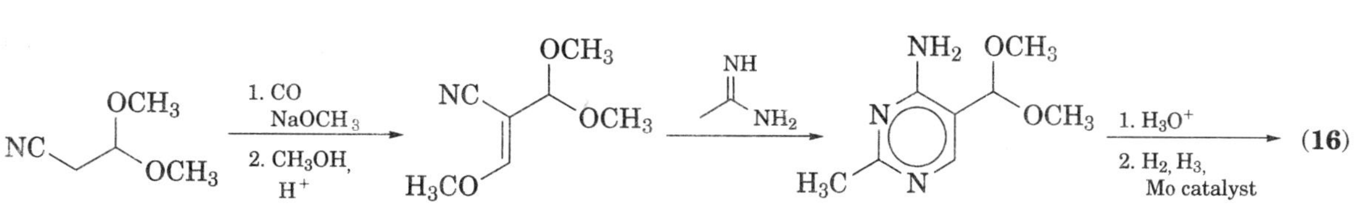

Figure 8. Takeda–Ube process.

Methods for synthesis of the thiazolium ring also have matured technically based on the cost, throughput, and waste disposal issues of production. In earlier syntheses, the 2-carbon and the sulfur atom were supplied as potassium dithioformate, made from chloroform and potassium sulfide. N-(4-Amino-2-methyl-5-pyrimidinyl)-methylthioformamide [31375-20-9] (35) served as a partner to 5-acetoxy-3-chloropentan-2-one [119867-66-2] (36) in a classical Hantzch thiazole synthesis (3,57). Such routes suffered from similar limitations to the quaternization routes mentioned before, but were used in commercial production until the discovery of the paradigm intermediate known as thiothiamine [299-35-4] (37). In the 1940s and 1950s, chemists at Tanabe and Takeda showed that the 2-carbon and the sulfur atom can be supplied very efficiently via carbon disulfide, the extra sulfur atom being readily removed by oxidation in nearly quantitative yield (58,59). Typically, an aqueous solution of diamine and alkali is treated in succession with carbon disulfide to form the dithiocarbamate [2882-49-7] (38), then with chloroketone (36), then acid to form the relatively insoluble, thiothiamine (37). Oxidation with hydrogen peroxide forms thiamine sulfate, which is converted by ion exchange (qv) to a solution of the hydrochloride (2) which is concentrated, crystallized, and dried. The much less soluble nitrate (3) is precipitated from aqueous solution with alkali metal nitrate. The advantages of this approach, ie, a cheap, easily handled sulfur source, very high overall yield and throughput, excellent product color and purity, and the use of water as solvent, have made it the preferred method of manufacture worldwide. Disadvantages include use of chlorinated intermediates, salt load from the sulfate waste stream, and malodorous aqueous waste which must be well contained and treated. This chemistry is practiced by Takeda and Hoffmann-La Roche in automated factories using standard glass and stainless steel fine chemical processing equipment operated continuously at many hundreds of metric tons annually. Production in the several smaller Chinese factories is probably basically similar.

In two unique, convergent approaches to the thiazole ring, other formate synthons are used for the 2-carbon. In one, reaction of formaldehyde with diamine (16) and thiol (17) gives dihydrothiamine [959-18-2] (39), which is oxidized to thiamine (60). In another, reaction of diamine (16) with orthoformate gives intermediate dihydropyrimidine [31375-19-6] (40), to which thiol (17) is added. Acidic rearrangement gives thiamine in high yield (60).

Although numerous other materials have been proposed as carbon sources for the thiazole ring, all manufacturing of thiamine is believed to use 3-chloro-5-acetoxypentan-2-one (36) or the corresponding alcohol [13045-13-1] (41) as intermediates. These are made by chlorination of acetylbutyrolactone, the latter from inexpensive butyrolactone and methyl acetate, generating chlorinated wastes.

Worldwide production of thiamine was estimated at 3300 t in 1995. The principal suppliers were Hoffmann-La Roche, Takeda, and several Chinese factories. Prices in the United States were in the range of $20–$28/kg in 1995.

Product Specifications and Testing

Nutrients and diet supplements without claims are considered foods, and thus are regulated by the U.S. Food and Drug Administration and are further subject to specific food regulations. Specifications for the hydrochloride (2) and the mononitrate (3) for foods are given in the *Food Chemicals Codex* (62) and for pharmaceuticals in the *U.S. Pharmacopeia* (63). General test methods have been summarized (64).

Safety and Handling

The hydrochloride (2) and the nitrate (3) are typically packaged in standard 20–25-kg foil or poly-lined cardboard boxes. Smaller packaging sizes are also available. Both hydrochloride (2) and nitrate (3) are required to be kept under normal cool, dry storage conditions, not to exceed 70°C and protected from light. Both are listed in the TSCA Inventory and material safety data sheets (MSDS) are available from suppliers. Precautions against dust exposure to the eyes, skin, or lungs (IOEL of 3.0 mg/m³ time weighted average) and against fire and dust explosion are indicated. The NPFA ratings for health, fire, and reactivity for the hydrochloride (2) are 1, 2, and 1, respectively, and for the nitrate (3) they are 1, 3, and 1. Neither national nor international transportation is regulated. The U.S. EPA has not established reportable quantities for environmental releases (7,8). The acute LD₅₀ (rat. oral) of 12,300 mg/kg for the hydrochloride (2) and 15,900 for the nitrate (3) qualify these materials as relatively harmless orally. There is no evidence of toxicity from thiamine taken orally by humans, even when doses of 500 mg, over 300 times the RDA, are taken daily for a month. Excess thiamine is easily cleared by the kidney (7,8,65). Parenteral doses of the hydrochloride (subcutaneous, intramuscular, or intravenous) have generally been well tolerated up to 100–500 mg with very few toxic effects. Some cases of sensitization or anaphylactic shock on repeated injections have been reported (1).

ANALYTICAL METHODS

Fluorometric, chromatographic, microbiological, and animal assays have been used for thiamine and its derivatives (66). The most widely used and officially sanctioned method has been the fluorometric assay (67), although high performance liquid chromatography has been increasingly employed (68). In natural materials, thiamine is often present as its phosphate esters and is protein-bound, therefore procedures to free it are necessary steps in most assays. Determination of thiamine by the fluorometric method involves acid extraction, phosphatase hydrolysis, ion exchange to remove interferences, and oxidation with ferricyanide or other reagents to thiochrome, whose fluorescence (excitation 365 nm/emission 435 nm) is measured and compared with a standard (67). Various hplc methods have been developed for quantitation of thiamine and its phosphates in biological matrices to picomole to femtomole levels. Separations are made with reversed-phase, ion-exchange, and specialized straightphase packings and detection with uv, post-column derivatization—fluorometry, or electrochemical techniques. Alternatively, precolumn derivatization and measurement of thiochrome has been used (68). High performance capillary electrophoresis has been applied (69). Gas chromatography has been used to determine thiamine to picomole level as the thiazole portion following cleavage of thiamine with sulfite (71). Microbiological assays are simple, inexpensive, and sensitive (5–50 ng thiamine). The results agree well with fluorometric methods but can take longer to achieve results. Microorganisms that have been most widely used include *Lactobaccillus fermenti* (ATCC 9338) and *Lactobacillus viridescens* (ATCC 12706), partly because they respond only to intact thiamine and not to its precursors (31). Bioassays in rats based on growth or cure of polyneurititis symptoms are time consuming and expensive but have been historically valuable in determining thiamine availability in foods and thiamine activity of related compounds (31).

FORMS, DERIVATIVES, AND ANALOGUES

The hydrochloride and mononitrate are the only commercial forms approved in the United States. The mono-, di-, and triphosphate esters (4–6) are colorless, water-soluble, organic-insoluble, high melting solids found naturally. Although they have been synthesized (9), they have not been produced commercially on large scale. In Japan and Europe, other forms have been approved for human use. Thiamine disulfide (21) is somewhat more soluble than thiamine in organic media. This material and its more fat-soluble *O,O*-dibutyrate and *O,O*-dibenzoate [2667-89-2] (42) have been used therapeutically for treatment of thiamine deficiency (70). Treatment of thiamine with extracts

of garlic or other *Allium* species converts it to a lipid-soluble disulfide derivative which is a very physiologically active depot form of thiamine. Thiamine allyl disulfide [554-44-9] (allithiamine) (43), was the first of several thiamine alkyl disulfides to be used therapeutically for thiamine deficiency, including prosultiamine [59-58-5] (44) and fursultiamine [804-30-8] (45). Because of their greater lipid solubility, they are absorbed and retained more strongly in the body. *In vivo* they are converted back to thiamine. They are of interest almost exclusively in Japan and are not yet approved in the United States (70,71). Sugar derivatives (qv) are claimed to have better taste and no odor (73). In applications where water solubility is detrimental, such as fish feeds with high thiaminase activity, formulations of the hydrochloride or nitrate with a waxy coating are effective.

(42)

(43), Allyl
(44), Propyl
(45), Tetrahydrofurfuryl

Numerous analogues of thiamine have been synthesized by structural modifications of the pyrimidine or thiazole rings or the bridging atoms (1,9,10). None has been found to exceed the biological activity of thiamine but some act as antagonists. Substitution at the 2-methyl, 4-amino, or 6-position of carbon for ring nitrogens, or of the methylene bridge, results in loss of activity. In the thiazolium ring, the sulfur atom, hydrogen at the 2-position, and the 2-hydroxyethyl group are essential for activity. The pyridine analogue, pyrithiamine [534-64-5] (46) is the most potent thiamine antagonist known. Some thiamine antagonists were found at low levels to selectively inhibit thiamine uptake by *Coccidia*. A modified thiamine-like molecule, called Amprolium [121-25-5] (47) was once claimed to be useful as a coccidiostat for poultry (73).

(46)   (47)

BIOSYNTHESIS

Higher plants, most bacteria, and some fungi biosynthesize thiamine. Humans or most other animals cannot, although some of their gut microflora can. Many microbial species are self-sufficient, others can synthesize thiamine if one or both of immediate precursors are available, and still others require the complete substance. Few produce much of an excess over their own requirements. Biosynthesis of thiamine in microorganisms has been reviewed (9,74–76). Media modifications and mutagenesis have achieved small increases of thiamine levels with microorganisms, the highest levels of a few mg/L being found with yeasts (qv) (77). A broader search among many yeast species has reported a *Saccharomyces cerevisiae* strain which accumulates up to 200 mg/L in the culture broth (78).

The pathways for thiamine biosynthesis have been elucidated only partly. Thiamine pyrophosphate is made universally from the precursors 4-amino-5-hydroxymethyl-2-methylpyrimidine pyrophosphate [841-01-0] (48) and 4-methyl-5-(2-hydroxyethyl)thiazole phosphate [3269-79-2] (49), but there appear to be different pathways in the earlier steps. In bacteria, the early steps of the pyrimidine biosynthesis are same as those of purine nucleotide biosynthesis, 5-Aminoimidazole ribotide [41535-66-4] (AIR) (50) appears to be the sole and last common intermediate; ultimately the elements are supplied by glycine, formate, and ribose. AIR is rearranged in a complex manner to the pyrimidine by an as-yet undetermined mechanism. In yeasts, the pathway to the pyrimidine is less well understood and may be different (74–83) (Fig. 9).

In the biosynthesis of the thiazole, cysteine is the common sulfur donor. In yeasts, the C-2 and N may be supplied by glycine, and the remaining carbons by D-ribulose-5-phosphate [108321-99-9] (51). In anaerobic bacteria, the C-2 and N may be recruited from tyrosine and the carbons from D-1-deoxyxylulose [16709-34-5], (52), whereas in aerobic bacteria the C-2 and N may be derived from glycine, as in yeasts (74–76,83–86) (see Fig. 9).

Biosynthesis of pyrophosphate (5) from pyrimidine phosphate (48) and thiazole phosphate (49) depends on the

activity of five enzymes, four of them kinases (87). In yeasts and many other organisms, including humans, pyrophosphate (5) can be obtained from exogenous thiamine in a single step catalyzed by thiamine pyrophosphokinase (88).

A number of the genes involved in the biosynthesis of thiamine in *E. coli* (89–92), *Rhizobium meliloti* (93), *B. subtilis* (94), and *Schizzosaccharomyces pombe* (95,96) have been mapped, cloned, sequenced, and associated with biosynthetic functions. Thiamine biosynthesis is tightly controlled by feedback and repression mechanisms limiting overproduction (97,98). A cost-effective bioprocess for production of thiamine will require significant additional progress.

USES

Most of the thiamine sold worldwide is used for dietary supplements. Primary market areas include the following applications: addition to feed formulations, eg, poultry, pigs, cattle, and fish fortification of refined foods, eg, flours, rice, and cereal products; and incorporation into multivitamins. Small amounts are used in medicine to treat deficiency diseases and other conditions, in agriculture as an additive to fertilizers (qv), and in foods as flavorings. Generally for dry formulations, the less soluble, nonhygroscopic nitrate is preferred. Only the hydrochloride can be used for intravenous purposes. Coated thiamine is used where flavor is a factor.

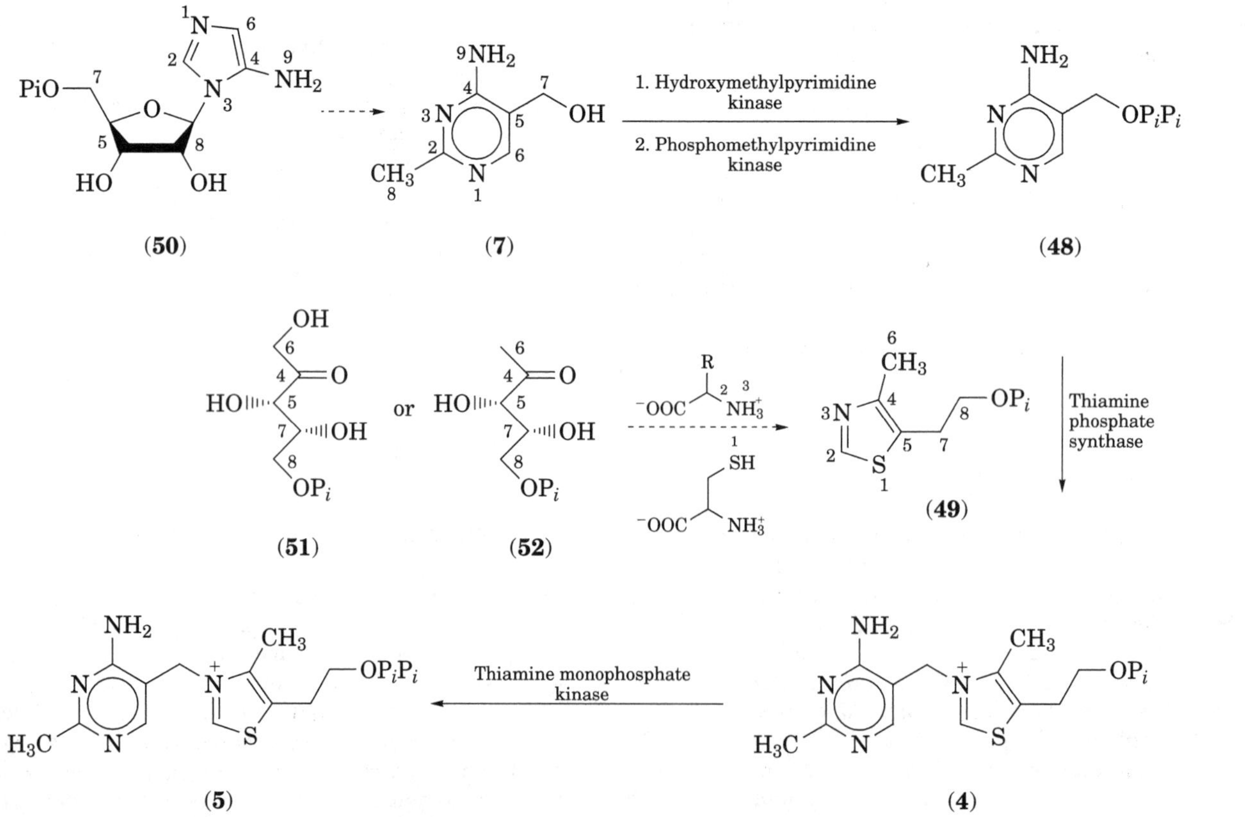

Figure 9. Biosynthesis of thiamine.

BIBLIOGRAPHY

1. C. J. Gubler, in L. J. Machlin, ed., *Handbook of Vitamins*, 2nd ed., Marcel Dekker, New York, 1991, pp. 233–280.

2. R. R. Williams and J. K. Cline, *J. Am. Chem. Soc.* **58**, 1504 (1936); *ibid.* **59**, 1052 (1937).

3. A. R. Todd and F. Bergel, *J. Chem. Soc.*, 364 (1936).

4. A. R. Todd and F. Bergel, *J. Chem. Soc.*, 26 (1938).

5. H. Andersag and K. Westphal, *Ber. Dtsch. Chem. Ges.* **70**, 2035 (1937).

6. *The Merck Index*, 11th ed., Merck & Co., Rahway, N.J., 1989, p. 1464.

7. *Thiamine Hydrochloride*, MSDS 3834, Hoffmann-La Roche Inc., Nutley, N.J.

8. *Thiamine Mononitrate*, MSDS 3835, Hoffmann-La Roche Inc., Nutley, N.J.

9. O. Isler, G. Brubacher, S. Ghisla, and B. Kraeutler, *Vitamine II*, Georg Thieme Verlag, New York, 1988, pp. 17–49.

10. J. A. Zoltewicz and G. Uray, *Bioorganic Chem.* **22**, 1–28 (1994) and references therein.

11. J. M. E. H. Chahine and J. E. Dubois, *J. Am. Chem. Soc.* **105**, 2335 (1983).

12. J. D. Roberts et al., *J. Am. Chem. Soc.* **99**, 6423 (1977).

13. A. Lycka and J. Cizmarik, *Pharmazie* **45**, 371 (1990).

14. J. M. E. H. Chahine, *J. Chem. Soc., Perkin Trans. II*, 505 (1990).

15. R. Kluger, *Chem. Rev.* **87**, 863 (1987).

16. R. D. Sedgwick et al., *J. Chem. Soc., Chem. Commun.*, 325 (1987).

17. J. J. Windheuser and T. Higuchi, *J. Pharm. Sci.* **51**, 3545 (1962).

18. T. Matsukawa and S. Yurugi, *Yakugaku Zasshi* **71**, 827 (1951).

19. J. Herrmann et al., *J. Chem. Soc., Perkin Trans. II*, 463 (1995).

20. A. Watanabe and Y. Asahi, *Yakugaku Zasshi* **77**, 153 (1957).

21. A. Takamizawa et al., *Chem. Pharm. Bull.* **19**, 759 (1971).

22. T. Matsukawa and S. Yurugi. *Yakugaku Zasshi* **72**, 1599 (1952).

23. H. Sugimoto et al., *J. Org. Chem.* **55**, 46 (1990).

24. J. Fournier, *Bull. Soc. Chim. Fr.*, 854 (1988).

25. J. M. E. H. Chahine and J. E. Dubois, *J. Chem. Soc., Perkin Trans. II*, 25 (1989).

26. Y.-T. Chen and F. Jordan, *J. Org. Chem.* **56**, 5029 (1988).

27. R. Breslow, *J. Am. Chem. Soc.* **80**, 3719 (1958).

28. J. T. Stiners and M. W. Washabaugh, *Bioorg. Chem.* **18**, 425 (1990).

29. R. Breslow, *Pure Appl. Chem.* **62**, 1859–1866 (1990).

30. M. Louloudi and N. Hadjiliadis, *Coord. Chem. Rev.* **135/136**, 429–468 (1994).

31. W. H. Sebrell and R. S. Harris, eds., *The Vitamins*, 2nd ed., Vol. V, Academic Press, Inc., New York, 1972, p. 98.

32. W. C. Evans, in P. L. Munson, J. Glover, E. Diezfalusy, and R. E. Olson, eds., *Vitamins and Hormones*, Vol. 33, Academic Press, Inc., New York, 1975, pp. 467–504.

33. P. Werkhoff et al., *ACS Symp. Ser.* **564**, 199–223 (1994).

34. L. Mauri et al., *Int. J. Food Sci. Technol.* **24**, 1–9 (1989).

35. M. P. Lamden, in W. H. Sebrell, Jr. and R. S. Harris, eds., *The Vitamins*, 2nd ed., Vol. V, Academic Press, Inc., New York, 1972, pp. 114–120.

36. A. Mozafar, *Plant Soil* **167**, 305–311 (1994).

37. R. E. Davis and G. C. Icke, *Adv. Clin. Chem.* **23**, 93 (1983).

38. P. M. Finglas, *Int. J. Vit. Nutr. Res.* **63**, 270 (1993).

39. J. Harmeyer and U. Kollenkirchen, *Nutr. Res. Rev.* **2**, 201–225 (1989).

40. D. Brown et al., *The Pyrimidines*, 2nd ed., John Wiley & Sons, Inc., New York, 1994.

41. U.S. Pat. 2,235,862 (Mar. 25, 1940), O. Zima (to Merck & Co.).

42. R. Grewe, *Z. Physiol. Chem.* **242**, 89 (1936).

43. Fr. Pat. 831,110 (Aug. 23, 1938), (to Hoffmann-La Roche).

44. A. Takimizawa and K. Hirai, *Chem. Pharm. Bull.* **12**, 393 (1964).

45. H. Miromoto et al., *Chem. Ber.* **106**, 893 (1973).

46. K. Tokuyama et al., *Bull. Chem. Soc. Jap.* **46**, 253 (1973).

47. U.S. Pat. 4,226,799 (Oct. 16, 1978), W. Bewert and W. Littmann (to BASF).

48. Ger. Pat. 3,511,273 (Mar. 28, 1985), W. Ernst and J. Paust (to BASF).

49. Ger. Pat. 3,520,982 (Dec. 18, 1986), H. Kiefer and W. Bewert (to BASF).

50. K. Nishihara et al., *Kagaku Kogaku* **55**, 433 (1991).

51. Eur. Pat. 55,108 (June 30, 1982), K. Matsui et al. (to Ube Industries).

52. Eur. Pat. 279,556 (Aug. 24, 1988), K. Nishihara et al. (to Ube Industries).

53. U.S. Pat. 4,536,577 (Aug. 20, 1985), H. Yoshida and S. Niida (to Ube Industries).

54. Ger. Pat. 3,303,789 (Aug. 11, 1983), H. Yoshida et al. (to Ube Industries).

55. Jpn. Pat. 6,307,869 (July 10, 1988). Y. Miyashiro et al. (to Takeda Chemical Industries).

56. A. Edenhoffer et al., *Helv. Chim. Acta* **58**, 1230 (1975).

57. T. Matsukawa et al., *J. Pharm. Soc. Japan* **63**, 216 (1943).

58. K. Washimi, *J. Pharm. Soc. Japan.* **66**, 62 (1946).

59. T. Matsukawa and S. Hojiro, *J. Pharm. Soc. Japan* **69**, 550 (1949); *ibid.* **70**, 28 (1950); **71**, 667, 720, 1215 (1951).

60. H. Hirano and Y. Yonemitsu, *J. Pharm. Soc. Japan* **76**, 1332 (1956).

61. G. Moine et al., *Helv. Chim. Acta* **73**, 1300 (1990).

62. Food and Nutrition Board, National Research Council, *Food Chemicals Codex*, 3rd ed., National Academy Press, Washington, D.C., 1981, pp. 324–325.

63. *The United States Pharmacopeia XXII* (USP XXII-NF XVII) United States Pharmacopeial Convention, Inc., Rockville, Md., 1990, pp. 1356, 1357–1358.

64. F. J. Al-Shammary et al., *Anal. Profiles Drug Subst.* **18**, 414 (1989).

65. H. E. Samerlich and L. J. Machlin, *Ann. NY Acad. Sci.* **699**, 1 (1992).

66. W. C. Ellefson, in J. Augustine, B. P. Klein, D. A. Becker, and P. B. Venugopal, eds., *Methods of Vitamin Assay*, 4th ed., John Wiley & Sons, Inc., New York, 1985, pp. 351–352.

67. K. Helrich, ed., *Official Methods of Analysis of the Association of Official Analytical Chemists*, 15th ed., Association of Official Analytical Chemists, Inc., Arlington, Va., 1990, pp. 942.23, 953.17, 957.17, and 986.27.

68. T. Kawasaki, in A. P. Leenheer, W. E. Lambert, and M. G. M. DeRuyter, eds., *Modern Chromatographic Analysis of the Vitamins*, Marcel Dekker, New York, 1992, pp. 319–354.

69. U. Jegle, *J. Chrom.* **652**, 495–501 (1993).

70. C. Kawasaki, in R. Hamb, I. Wool, and J. Lomaine, eds., *Vitamins and Hormones*, Vol. 21, Academic Press, Orlando, Fla., 1963, pp. 69–111.

71. R. E. Echols et al., *J. Chrom.* **347**, 89–97 (1985).

72. C. E. Mueller et al., *Int. J. Pharm.* **57**, 41–47 (1989).

73. Y. Susuki and K. Uchida, *Biosci. Biotech. Biochem.* **58**, 1273 (1994).

74. W. H. Ott et al., *Poult. Sci.* **44**, 920 (1965).

75. A. Iwashima in E. Vandamme, ed., *Biotechnology of Vitamins, Pigments and Growth Factors*, Elsevier, Amsterdam, the Netherlands, 1988.

76. D. W. Young, *Nat. Prod. Rep.* **3**, 395 (1986).

77. G. M. Brown and J. M. Williamson, in F. C. Neidhardt, ed., *Escherichia coli and salmonella typhimurium*, American Society for Microbiology, Washington, D.C., 1987, p. 521.

78. A. Silhamkeva, *J. Inst. Brew.* **191**, 78 (1985).

79. G. G. Stewart et al., *Can. J. Microbiol.* **38**, 1156 (1992).

80. T. Kozlok and I. D. Spenser, *J. Am. Chem. Soc.* **109**, 4968 (1987).

81. D. M. Downs and L. Peterson, *J. Bacteriol.* **176**, 4858 (1994).

82. N. J. Leonard et al., *Proc. Natl. Acad. Sci.* **85**, 7174 (1988).

83. B. Estramareix et al., *Biochem. Biophys. Acta* **1035**, 154 (1990).

84. K. Tazuya et al., *Biochem. Mol. Biol. Int.* **30**, 893 (1993).

85. K. Tazuya et al., *Biochem. Int.* **10**, 689 (1985).

86. R. H. White and F. B. Rudolph, *Biochem. Biophys. Acta* **542**, 340 (1978).

87. K. Tazuya et al., *Biochem. Int.* **14**, 153 (1987).

88. Y. Kawasaki, *J. Bacteriol.* **175**, 5153 (1993).

89. A. Iwashima et al., *J. Biol. Chem.* **268**, 17440 (1993).

90. T. Nohno, Y. Kasai, and T. Saito, *J. Bacteriol.* **170**, 4097 (1988).

91. N. Imamura and H. Nakayama, *J. Bacteriol.* **151**, 708 (1982).

92. J. Ryals et al., *J. Bacteriol.* **151**, 899 (1982).

93. T. P. Begley et al., *J. Am. Chem. Soc.* **117**, 2351 (1995).

94. H. Brennen et al., *J. Bacteriol.* **167**, 66 (1986).

95. J. A. Hoch et al., in A. L. Sonenshein, J. A. Hoch, and R. Losick, eds., *Bacillus subtilis*, American Society for Microbiology, Washington, D.C., 1993, pp. 425.

96. K. G. Maundrell et al., *Yeast* **10**, 1075 (1994).

97. A. Zurlinden and M. E. Schweingruber, *J. Bacteriol.* **176**, 6631 (1994).

98. A. Iwashima et al., *J. Bacteriol.* **172**, 6145 (1990); *ibid.* **174**, 4701 (1992).

99. A. Schweingruber et al., *Genetics* **130**, 445 (1992).

DAVID BURDICK
Hoffmann-La Roche, Inc.

VITAMINS: VITAMIN A

The curative role of the juice of cooked liver extracts in the treatment of night blindness was first recognized and practiced by the Egyptians in ancient times (1). Several millenniums later, in the early part of the twentieth century, many groups identified a lipid-soluble factor found in milk, butter, and egg yolks which was therapeutic for this condition. In 1913, McCollum and Davis coined the term fat-soluble A and subsequently, ascribed the growth-promoting effects of liver extracts to this material (2,3). Later, Osborne and Mendel found that cod liver oil contained an ingredient that was essential for growth promotion in rats (4).

The structure of vitamin A [11103-57-4] and some of the important derivatives are shown in Figure 1. The parent structure is all-*trans*-retinol [68-26-8] and its IUPAC name is (all-*E*)-3,7-dimethyl-9-(2,6,6-trimethyl-1-cyclohexen-1-yl)-2,4,6,8-nonatetraen-1-ol (**1**). The numbering system for vitamin A derivatives parallels the system used for the carotenoids. In older literature, vitamin A compounds are named as derivatives of trimethyl cyclohexene and the side chain is named as a substituent. For retinoic acid derivatives, the carboxyl group is denoted as C-1 and the trimethyl cyclohexane ring as a substituent on C-9. The structures of vitamin A and β-carotene were elucidated by Karrer in 1930 and several derivatives of the vitamin were prepared by this group (5,6). In 1935, Wald isolated a substance found in the visual pigments of the eye and was able to show that this material was identical with Karrer's retinaldehyde [116-31-4] (**5**) (7).

Vitamin A$_2$ [79-80-1] (**6**) is structurally similar to vitamin A$_1$ [68-26-8] and is also found in fish oils. This compound is important biologically for fish and other lower animals. Interestingly, tadpoles require vitamin A$_2$ but after metamorphosis require vitamin A$_1$ (8).

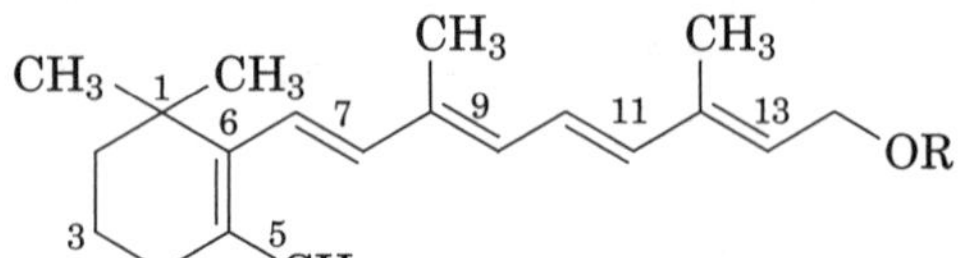

(**1**, R = H)
(**2**, R = CH$_3$CO)
(**3**, R = CH$_3$(CH$_2$)$_{14}$CO)
(**4**, R = CH$_3$CH$_2$CO)

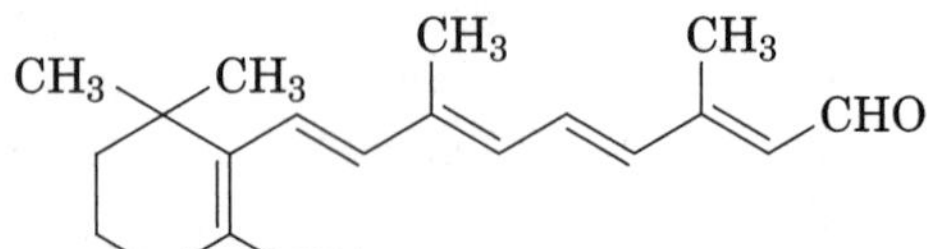

(**5**)

Figure 1. Vitamin A and derivatives: retinol (**1**), retinyl acetate [127-47-9] (**2**), retinyl palmitate [79-81-2] (**3**), and retinyl propionate [7069-42-3] (**4**).

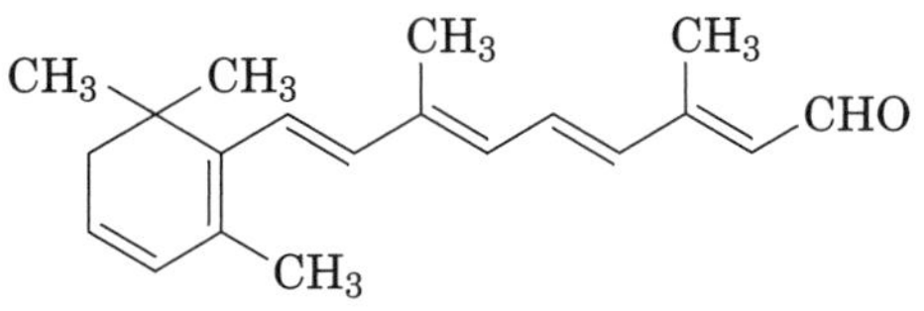

(6)

Vitamin A constitutes the most significant sector of the commercial retinoid market and is used primarily in the feed area. In the pharmaceutical area, there are several important therapeutic dermatologic agents which structurally resemble vitamin A and they are depicted in Figure 2. The carotenoids as provitamin A compounds also represent an important commercial class of compounds with β-carotene [7235-40-7] (10) occupying the central role (Fig. 3) (9).

CHEMICAL AND PHYSICAL PROPERTIES

Because of the presence of an extended polyene chain, the chemical and physical properties of the retinoids and carotenoids are dominated by this feature. Vitamin A and related substances are yellow compounds which are unstable in the presence of oxygen and light. This decay can be accelerated by heat and trace metals. Retinol is stable to base but is subject to acid-catalyzed dehydration in the presence of dilute acids to yield anhydrovitamin A [1224-18-8] (16). Retrovitamin A [16729-22-9] (17) is obtained by treatment of retinol in the presence of concentrated hy-

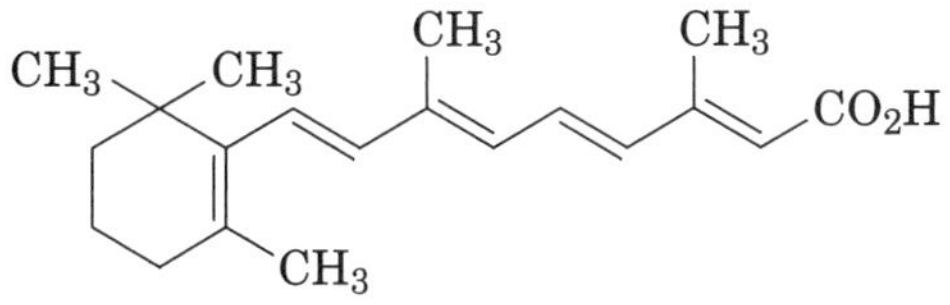

(7)

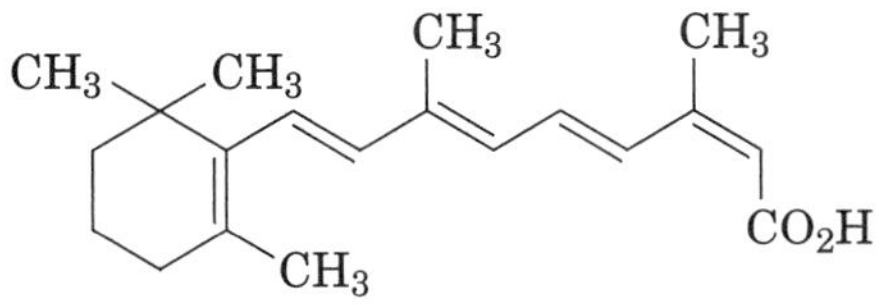

(8)

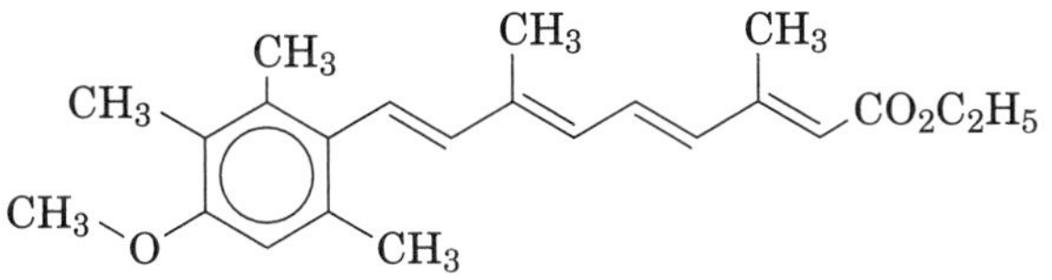

(9)

Figure 2. Commercially important retinoids: retinoic acid [302-79-4] (Tretinoin) (7), 13-Z-retinoic acid [4759-48-2] (Isotretinoin) (8), and etretinate [54350-48-0] (9).

drobromic acid. In the case of retinoic acid and retinal, re-isomerization is possible after conversion to appropriate derivatives such as the acid chloride or the hydroquinone adduct. Table 1 lists the physical properties of β-carotene [7235-40-7] and vitamin A.

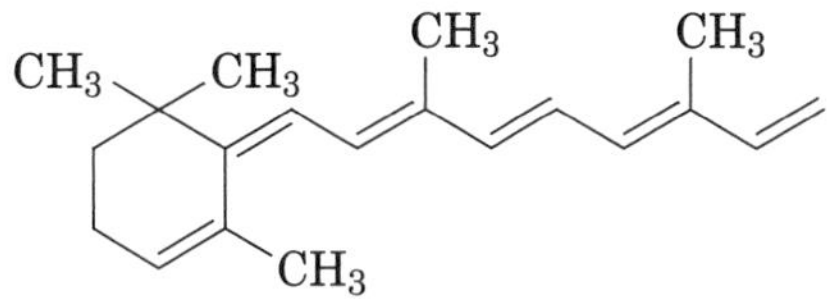

(16)

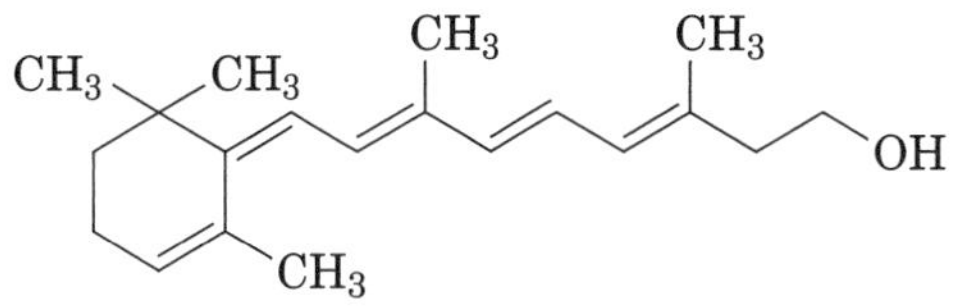

(17)

Of the 16 possible geometric isomers of vitamin A, only the all-trans form has full vitamin A activity and these compounds are commonly named according to the older, nonsystematic nomenclature (Table 2). From a biological standpoint, 11-cis-retinal plays a critical role in vision (vide infra). As earlier described, 13-cis-retinoic acid is important as a dermatological agent. Recently, 9-cis-retinoic acid has also been shown to be an important biological isomer and has been identified as a novel endogenous hormone in mammalian tissues (11). The other 13 isomers have no biological or commercial significance. Isolation, characterization, and synthesis of these isomers have been reported (12).

The most conspicuous physical feature of the retinoids and of the carotenoids is the uv spectrum. In the case of the carotenoids, this coloration property is one reason for their commercial significance. In general, there are several factors which influence the position of λ_{max}, the intensity of the absorbance, and the degree of fine structure (13). These include the length of the polyene chain, the number of cis double bonds, and end group functionality. For example, all-trans-retinal (5) has an absorption maximum at 368 nm and a molecular extinction coefficient at 48,000, whereas the 7,9,11,13-cis isomer has values of 308 nm and 15,500. Both hypsochromic shifts and hypochromic effects are observed when the stereochemistry of the double bond is changed from trans to cis. External factors such as solvent, temperature, and molecular environments also influence the uv spectra. A striking example of the latter phenomenon is the observed significant bathochromic shift (ca 150 nm) during the association of a carotenoid with a lipoprotein. From a molecular standpoint, the origins of this effect are not completely understood and this remains an area of active research (14,15).

Other spectroscopic methods such as infrared (ir), ^{1}H and ^{13}C nuclear magnetic resonance (nmr), circular dichroism (CD), and mass spectrometry (MS) are invaluable tools for identification and structure elucidation. Nmr spectroscopy allows for geometric assignment of the carbon–carbon double bonds, as well as relative stereo-

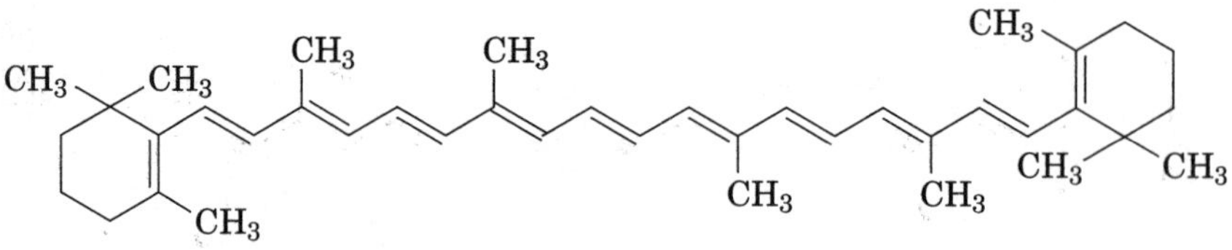

(10)

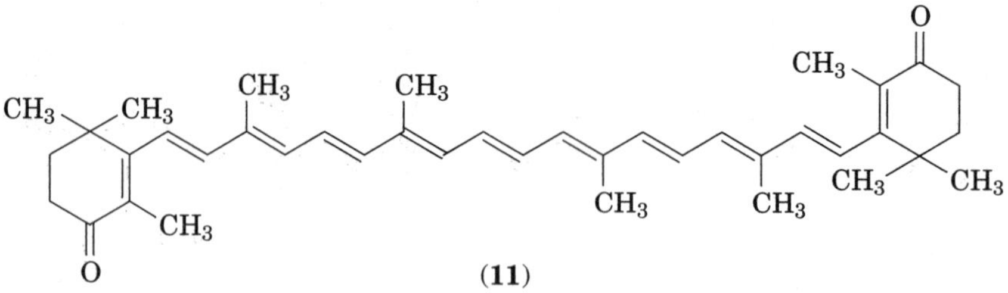

(11)

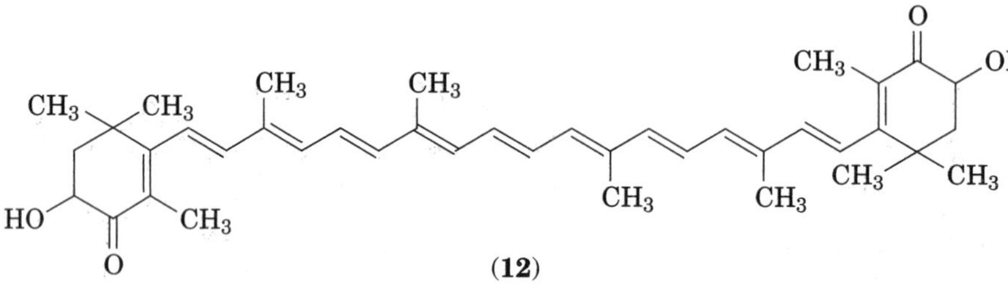

(12)

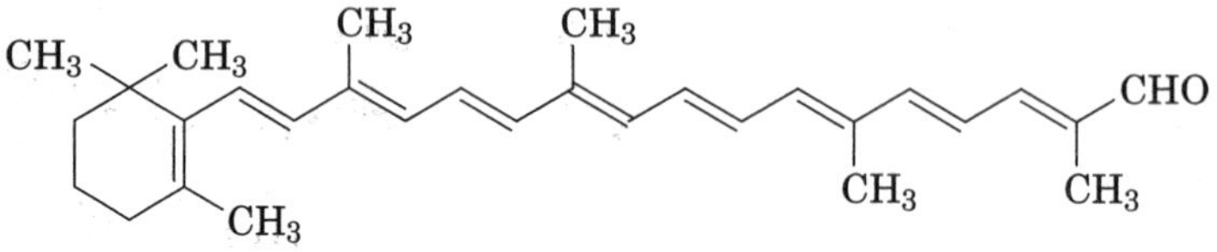

(13)

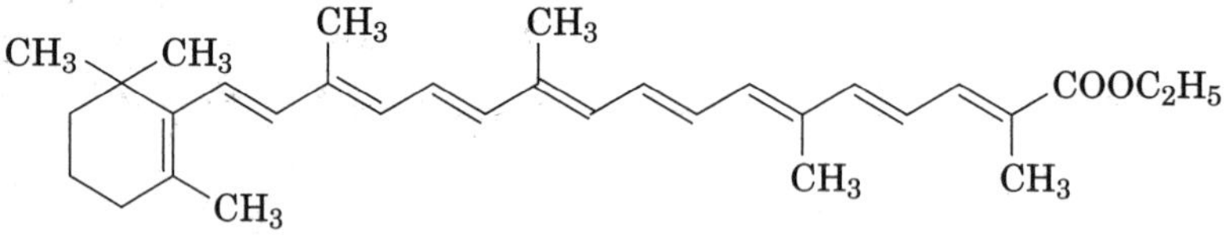

(14)

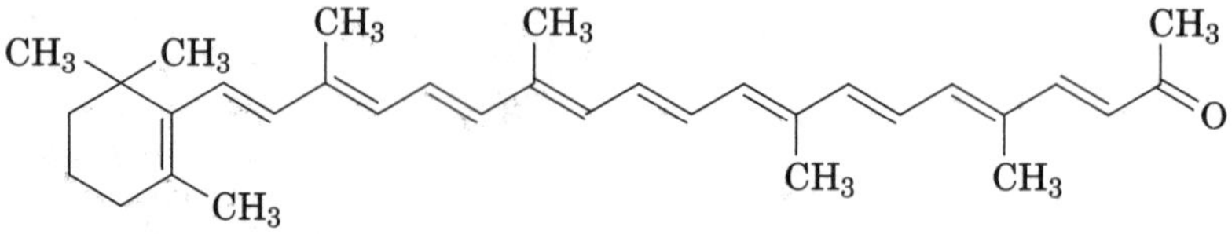

Figure 3. Commercially important carotenoids: β-carotene (10), canthaxanthin [514-78-3] (11), astaxanthin [472-61-7] (12), β-apo-8'-carotenal [1107-26-2] (13), β-apo-8'-carotenoic acid ethyl ester [1109-11-1] (14), and citranaxanthin [3604-90-8] (15).

(15)

chemistry of ring substituents. These spectroscopic methods coupled with traditional chemical derivatization techniques provide the framework by which new carotenoids are identified and characterized (16,17).

SYNTHESIS

Vitamin A acetate [11098-51-4] (2) is the commercially significant form of the vitamin and is mainly produced by

Hoffmann-La Roche, BASF, and Rhône-Poulenc (Fig. 4). All of these processes have β-ionone (18) as their key intermediate and in this regard are based on work performed in the 1940s (18,19). Their differences lie in methodology to this key intermediate as well as in the methodology to elaborate the side chain. A review of the early work is available (20).

In the process practiced by Hoffmann-La Roche, a C_{13} + C_1 + C_6 strategy is employed. In this approach, β-ionone

Table 1. Properties of β-Carotene and Vitamin A Derivatives

Property	Retinol[a]	Retinyl acetate[a]	Retinyl palmitate[a]	Retinyl propionate[b]	β-Carotene[a]
Appearance	Crystalline	Crystalline	Crystalline or amorphous	Oil	Crystalline
Color	Yellow	Yellow	Yellow	Light yellow	Dark red–dark purple
Odor	Faint, hay-like	Faint, hay-like	Faint, hay-like	Slight	Faint, hay-like
Molecular weight	286.46	328.50	534.88	342.50	536.85
Molecular formula	$C_{20}H_{30}O$	$C_{22}H_{32}O_2$	$C_{36}H_{60}O_2$	$C_{23}H_{34}O_2$	$C_{40}H_{56}$
Melting point, °C	63–64	57–59	28–29		180–182
Solubility, g/100 mL					
Water	Insoluble	Insoluble	Insoluble	Insoluble	Insoluble
Ethanol	Soluble	Soluble	Soluble	Soluble	Slightly soluble
Isopropanol	Soluble	Soluble	Soluble		Slightly soluble
Chloroform	Soluble	Soluble	Soluble		Soluble
Acetone	Soluble	Soluble	Soluble		Slightly soluble
Fats, oils	750	750	750		0.05–0.08
Spectrophotometric properties, nm, max	375^c	326^c	325^c		$497, 466^d$
Fluorescence					
Excitation, max, nm	325	325	325		
Emission, max, nm	470	470	470		

Source: [a]Ref. 8 and [b]Ref. 10.
[c]Isopropanol.
[d]Chloroform.

Table 2. Physical Properties of Stereoisomers of Vitamin A

Isomer	Alcohol mp (°C)	λ_{max} (nm)	E_{max}	Vitamin A potency (%)	Aldehyde mp (°C)	λ_{max} (nm)	E_{max}	Vitamin A potency (%)
All-*trans*-Vitamin A	62–64	325	1,832	100	57/65	381	1,530	91
13-*cis*-Vitamin A	58–60	328	1,686	75	77	375	1,250	93
11-*cis*-Vitamin A	Oil	319	1,220	23	63.5–64.7	376.5	878	48
11,13-di-*cis*-Vitamin A	86–88	311	1,024	15	oil	373	700	31
9-*cis*-Vitamin A	81.5–82.5	323	1,477	22	64	373	1,270	19
9,13-di-*cis*-Vitamin A	58–59	324	1,379	24	49/85	368	1,140	17

(**18**) is subjected to a Darzen's condensation to yield the C_{14} aldehyde [116-31-4] (**19**) (see Fig. 4). Construction of the side chain is completed by a metal acetylide coupling reaction with compound (**20**). Acetylation, partial reduction of the triple bond, and acid-catalyzed elimination of water completes the synthesis. BASF and Rhône-Poulenc use a different scheme and extend the side chain of β-ionone in an initial step via a Grignard reaction with a metal acetylide. Semihydrogenation yields vinyl β-ionol (**21**) and it is at this point that the approaches diverge. In the Rhône-Poulenc $C_{15} + C_5$ synthesis, the carbon terminus of vinyl β-ionol is activated by conversion to the sulfone (**22**). In a second step, the anion of the sulfone is reacted with C_5-chloroacetate (**23**) to yield vitamin A acetate. BASF utilizes a Wittig olefination and first prepares the phosphonium salt of the C_{15} unit. Reaction of the salt (**24**) with the C_5 aldehyde (**25**) leads to vitamin A acetate (21).

Vitamin A palmitate [*79-81-2*] (**3**), a commercially important form of the vitamin, is produced from vitamin A acetate (**2**) via a transesterification reaction with methyl palmitate. Enzymatic preparation of the palmitate from the acetate has also been described (22).

In addition to differences in their methodology to extend the carbon chain, these manufacturers differ in their syntheses of β-ionone. β-Ionone is commercially prepared via an acid-catalyzed rearrangement of pseudoionone (**26**). This intermediate is manufactured on an industrial scale from either citral (**27**) or dehydro-linalool (**28**) (21) (Fig. 5).

Citral is prepared starting from isobutene and formaldehyde to yield the important C_5 intermediate 3-methylbut-3-enol (**29**). Pd-catalyzed isomerization affords 3-methylbut-2-enol (**30**). The second C_5 unit of citral is derived from oxidation of (**30**) to yield 3-methylbut-2-enal (**31**). Coupling of these two fragments produces the dienol ether (**32**) and this is followed by an elegant double Cope rearrangement (21) (Fig. 6).

The synthesis of dehydro-linalool (**28**) relies on the basic chemicals acetone and acetylene. Addition of a metal acetylide to acetone yields methylbutynol (**33**). Semihydrogenation affords the alkene (**34**) which is reacted with *i*-propenyl-methyl ether. A Cope rearrangement of the adduct yields methylheptenone (**35**). Addition of a second mole of metal acetylide to dehydro-linalool (**28**) is followed

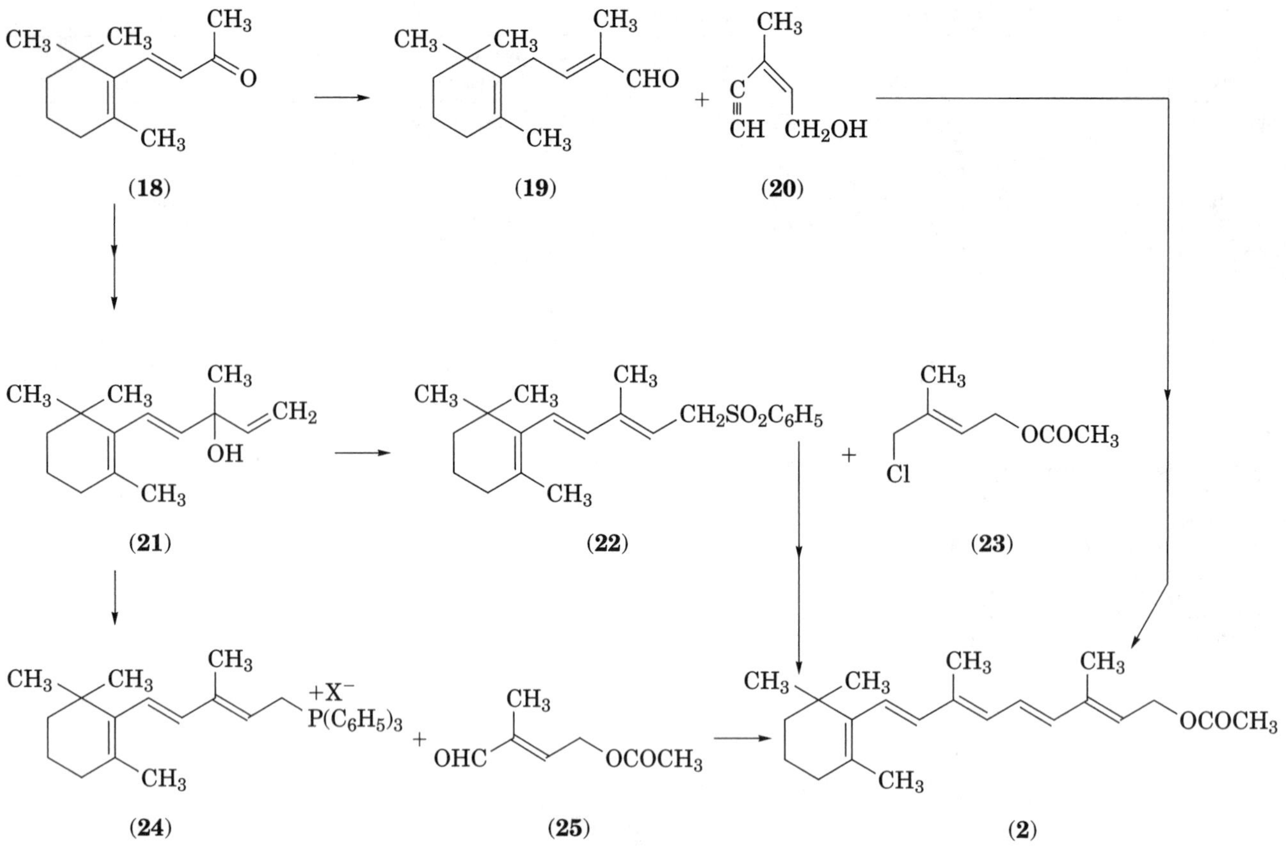

Figure 4. Commercial synthesis of vitamin A acetate (**2**).

by a second Cope rearrangement to yield pseudoionone (**26**) (9,21) (Fig. 7).

In other work, sulfone chemistry plays an integral part of the syntheses of both β-carotene and vitamin A by workers at Kuraray. In this approach, the anion of C_{10} β-cyclogeranyl sulfone (**36**) is condensed with the C_{10} aldehyde (**37**). The resulting β-hydroxy sulfone (**38**) is treated with dihydropyran followed by a double elimination to yield vitamin A acetate. Alternatively, the β-hydroxy sulfone (**38**) can be converted to the δ-halo sulfone (**39**) and a similar double elimination scheme is employed (23,24) (Fig. 8).

Work at Rhône-Poulenc has involved a different approach to retinal and is based on the palladium-catalyzed rearrangement of the mixed carbonate (**41**) to the allenyl enal (**42**). Isomerization of the allene (**42**) to the polyene (**43**) completes the construction of the carbon framework. Acid-catalyzed isomerization yields retinal (**5**). A decided advantage of this route is that no by-products such as triphenylphosphine oxide or sodium phenylsulfinate are formed. However, significant yield improvements would be necessary for this process to compete with the current commercial syntheses (25–27) (Fig. 9).

In contrast to the similarities seen in the majority of the industrial syntheses of vitamin A, substantial diversity is observed in the preparations of the carotenoids. Owing to the fact that all-trans stereochemistry is required in the

final product and that many olefination reactions yield the cis product, the ease of isomerization at a given locus on the polyene chain has influenced the choice of building blocks. In addition, technological strengths in the construction of these building blocks as well as synergies with a core olefination strength have played an important role in the choice of synthesis.

Hoffmann-La Roche has produced β-carotene since the 1950s and has relied on core knowledge of vitamin A chemistry for the synthesis of this target. In this approach, a five-carbon homologation of C_{14} vitamin A aldehyde (**19**) is accomplished by successive acetalizations and enol ether condensations to prepare the C_{19} aldehyde (**46**). Metal acetylide coupling with two molecules of aldehyde (**46**) completes construction of the C_{40} carbon framework. Selective reduction of the internal triple bond of (**47**) is followed by dehydration and thermal isomerization to yield β-carotene (**21**) (Fig. 10).

In the BASF synthesis, a Wittig reaction between two moles of C_{15} phosphonium salt (vitamin A intermediate (**24**) and C_{10} dialdehyde (**48**) is the important synthetic step (9,28,29). Thermal isomerization affords all trans-β-carotene (Fig. 11). In an alternative preparation by Roche, vitamin A process streams can be used and in this scheme, retinol is carefully oxidized to retinal, and a second portion is converted to the C_{20} phosphonium salt (**49**). These two halves are united using standard Wittig chemistry (8) (Fig. 12).

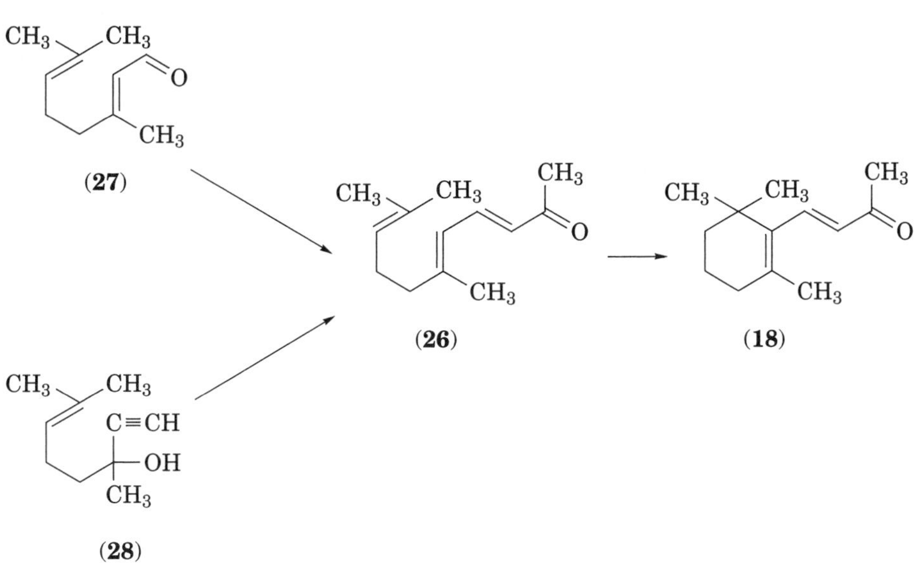

Figure 5. Industrial approaches to β-ionone (18).

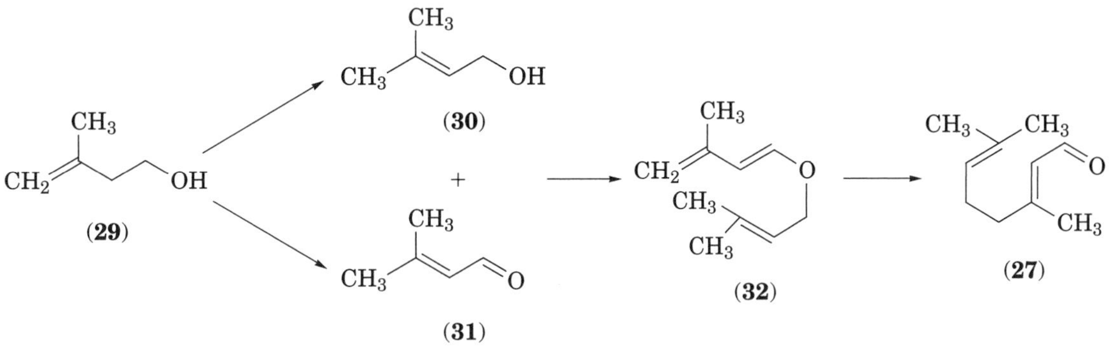

Figure 6. Synthesis of citral (27).

Other approaches to direct C_{20} couplings have been reported (9,30–35). Based on their knowledge of sulfone chemistry, Rhône-Poulenc has patented many syntheses of β-carotene which use this olefination chemistry (36–41). Horner-Emmons chemistry has also been employed for this purpose (42). The synthetic approaches to the carotenoids have been reviewed (43).

Compounds labeled ^{14}C, ^{13}C, ^{3}H, or ^{2}H are extremely important in understanding the absorption, distribution, metabolism, and excretion of these materials in biological systems. The preparation of these materials has been reviewed (44,45) and has been the subject of recent investigations (46).

BIOSYNTHESIS

In nature, vitamin A aldehyde is produced by the oxidative cleavage of β-carotene by 15,15′-β-carotene dioxygenase. Alternatively, retinal is produced by oxidative cleavage of β-carotene to β-apo-8′-carotenal followed by cleavage at the 15,15′-double bond to vitamin A aldehyde (47). Carotenoid biosynthesis and fermentation have been extensively studied both in academic as well as in industrial laboratories.

On the commercial side, the focus of these investigations has been to increase fermentation titers by both classical and recombinant means.

The carotenoid skeleton is assembled in a primary step by the coupling of two moles of geranylgeranyl pyrophosphate (52) by an enzyme which is encoded by the crt B (carotenogenic) gene. The resulting prephytoene pyrophosphate (53) is further transformed to phytoene (54) possibly by products also derived from the crt B gene. Phytoene is a common intermediate in carotenoid biosynthesis. For example, in *E. uredovora* phytoene is converted to lycopene (55) in sequential dehydration steps. These reactions are catalyzed by an enzyme called phytoene desaturase which is encoded on the gene cluster crt I. Interestingly, only *cis*-phytoene and *trans*-lycopene are detected and it is hypothesized that *trans*-phytoene is formed in a nonenzymatic step. By further biosynthetic transformations, β-carotene (10) is produced from lycopene and zeaxanthin (56) from β-carotene (Fig. 13) (48).

The majority of industrial research describes classical approaches to yield improvement (49). However, there has been some work using genetically modified organisms. In the case of these recombinant organisms, the carotenoid biosynthetic gene cluster has been expressed in non-

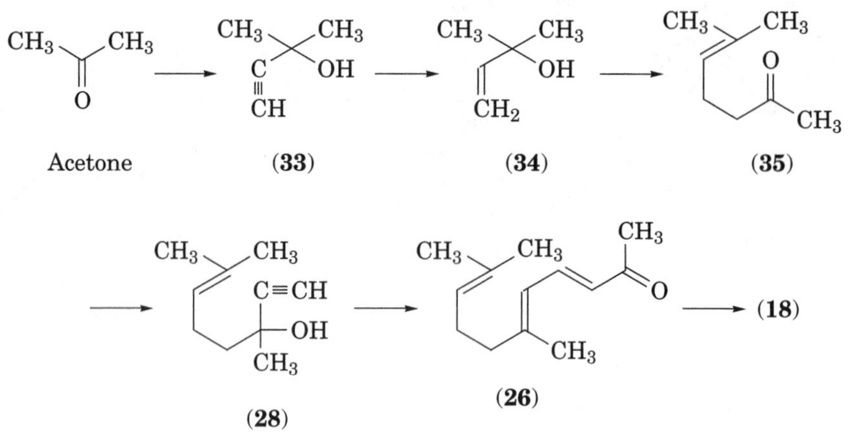

Figure 7. Acetone-based synthesis of β-ionone.

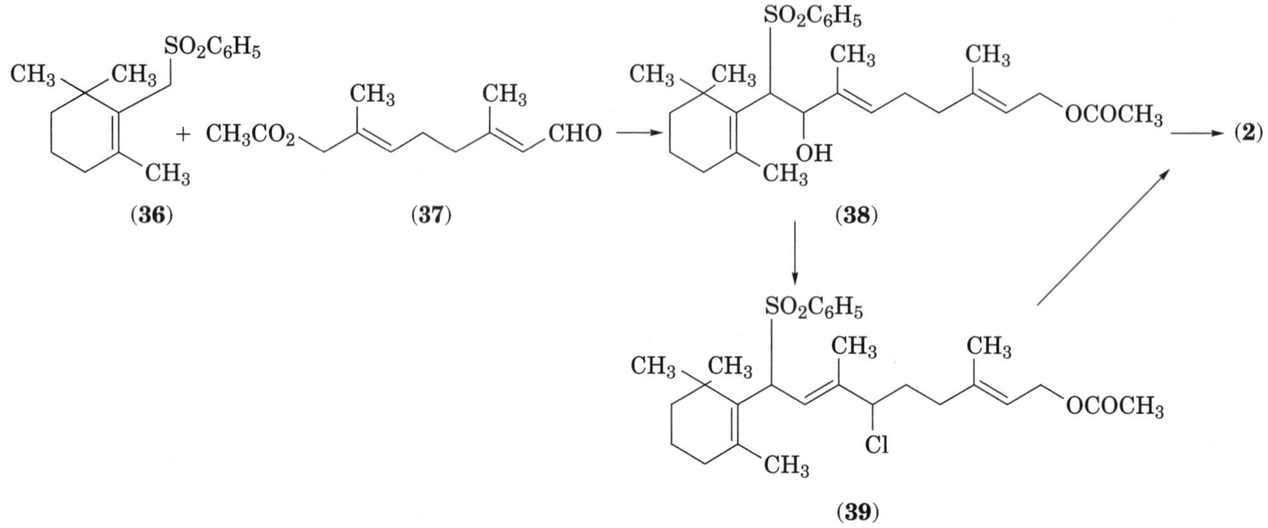

Figure 8. Kuraray synthesis.

carotegenic species such as *E. coli* (50) and *S. cerevisiae* (51).

ANALYTICAL METHODS

Biological, spectroscopic, and chromatographic methods have been used to assay vitamin A and the carotenoids. Biological methods have traditionally been based on the growth response of vitamin A-deficient rats. The utility and shortcomings of this test have been reviewed (52,53). This test has found applicability for analogues of retinol (54,55). Carotenoids which function as provitamin A precursors can also be assayed by this test (56).

Spectroscopic methods such as uv and fluorescence have relied on the polyene chromophore of vitamin A as a basis for analysis. Indirectly, the classical Carr-Price colorimetric test also exploits this feature and measures the amount of a transient blue complex at 620 nm which is formed when vitamin A is dehydrated in the presence of Lewis acids. For uv measurements of retinol, retinyl acetate, and retinyl palmitate, analysis is done at 325 nm. More sen-

sitive measurements can be obtained by fluorescence. Excitation is done at 325 nm and emission at 470 nm. Although useful, all of these methods suffer from the fact that the method is not specific and any compound which has spectral characteristics similar to vitamin A will assay like the vitamin (57).

More specific methods involve chromatographic separation of the retinoids and carotenoids followed by an appropriate detection method. This subject has been reviewed (57). Typically, hplc techniques are used and are coupled with detection by uv. For the retinoids, fluorescent detection is possible and picogram quantities of retinol in plasma have been measured (58–62). These techniques are particularly powerful for the separation of isomers. Owing to the thermal lability of these compounds, gc methods have also been used but to a lesser extent. Recently, the utility of cool-on-column injection methods for these materials has been demonstrated (63).

Owing to the light and air sensitivity of the carotenoids and retinoids, sample handling is a critical issue. It is recommended to conduct extraction of these materials with peroxide-free solvents, to store biological samples at

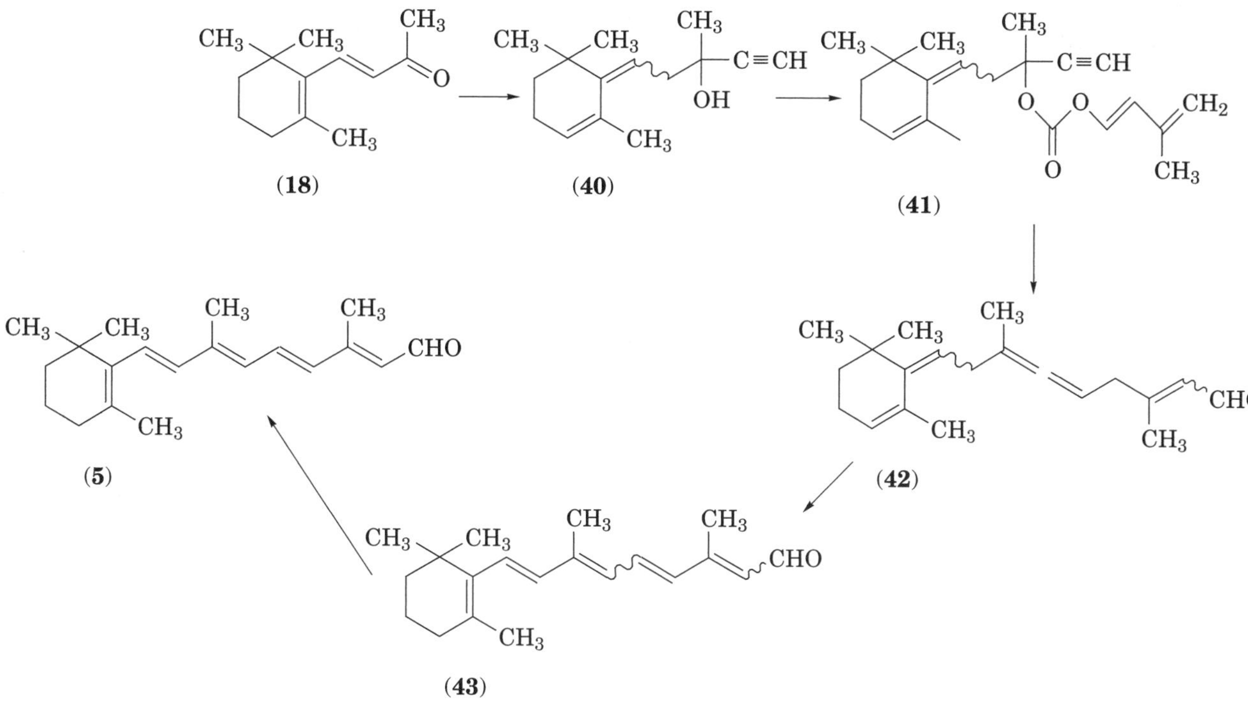

Figure 9. Allene synthesis of vitamin A aldehyde.

−70°C under argon and in the dark, to perform the analysis under yellow light, and to use reference compounds of high purity (57).

In the *United States Pharmacopeia* (USP), vitamin A content is determined by the ratio of the corrected absorbance to the observed absorbance and is not to be less than 0.85 at 325 nm. Total vitamin A content is to be 95% of label claim. For β-carotene, the assay is performed using uv spectroscopy and is determined by the absorbance at 455 nm. The range of the assay should be from 96.0 to 101.0% (64).

OCCURRENCE

Rich sources of vitamin A include dairy products such as milk, cheese, butter, and ice cream. Eggs as well as internal organs such as the liver, kidney, and heart also represent good sources. In addition, fish such as herring, sardines, and tuna, and in particular the liver oil from certain marine organisms, are excellent sources. Because the vitamin A in these food products is derived from dietary carotenoids, vitamin A content can vary considerably. Variation of vitamin A content in food can also result from food processing and in particular, oxidation processes (8).

Fertile sources of carotenoids include carrots and leafy green vegetables such as spinach. Tomatoes contain significant amounts of the red carotenoid, lycopene. Although lycopene has no vitamin A activity, it is a particularly efficient antioxidant. Oxidation of carotenoids to biologically inactive xanthophylls represents an important degradation pathway for these compounds (56).

BIOLOGICAL FUNCTIONS

In humans, vitamin A has important functions in vision, growth, and tissue differentiation. Its role in vision is fairly well understood and involves initial absorption of all-*trans*-retinol by the pigment epithelium cells of the eye. This event is followed by isomerization to 11-*cis*-retinol by all-*trans*-11-*cis*-retinol isomerase. After conversion of 11-*cis*-retinol to 11-*cis*-retinal, reaction with opsin forms rhodopsin. Absorption of light by this pigment generates the visual signal by cis/trans isomerization. After release of *trans*-retinal from the pigment, *trans*-retinal is rapidly reduced to all-*trans*-retinol and the cycle repeats itself. Consequently, vitamin A is not consumed in the process. One feature of this system is that the basic chemistry is common to many photosensitive systems from such diverse species as halophilic bacteria to higher organisms such as vertebrates (65).

As previously described, the biological assay for vitamin A was based on the growth response of rats fed a vitamin A-deficient diet. Vitamin A is necessary for normal bone growth and remodeling and is required for the activity of epiphyseal cartilage. Retinoic acid is the active form of vitamin A for this function and it has been observed that ingestion of retinoic acid restores growth but cannot maintain vision (8). In animals cycled on retinoic acid, ie, fed a retinoic acid-containing diet then deprived of a retinoic acid-containing diet, these animals become sensitive to the removal of retinoic acid. This condition is characterized by a loss in appetite followed by a depression in growth. As a result, loss of appetite is one of the first symptoms of a vitamin A-deficient animal (8).

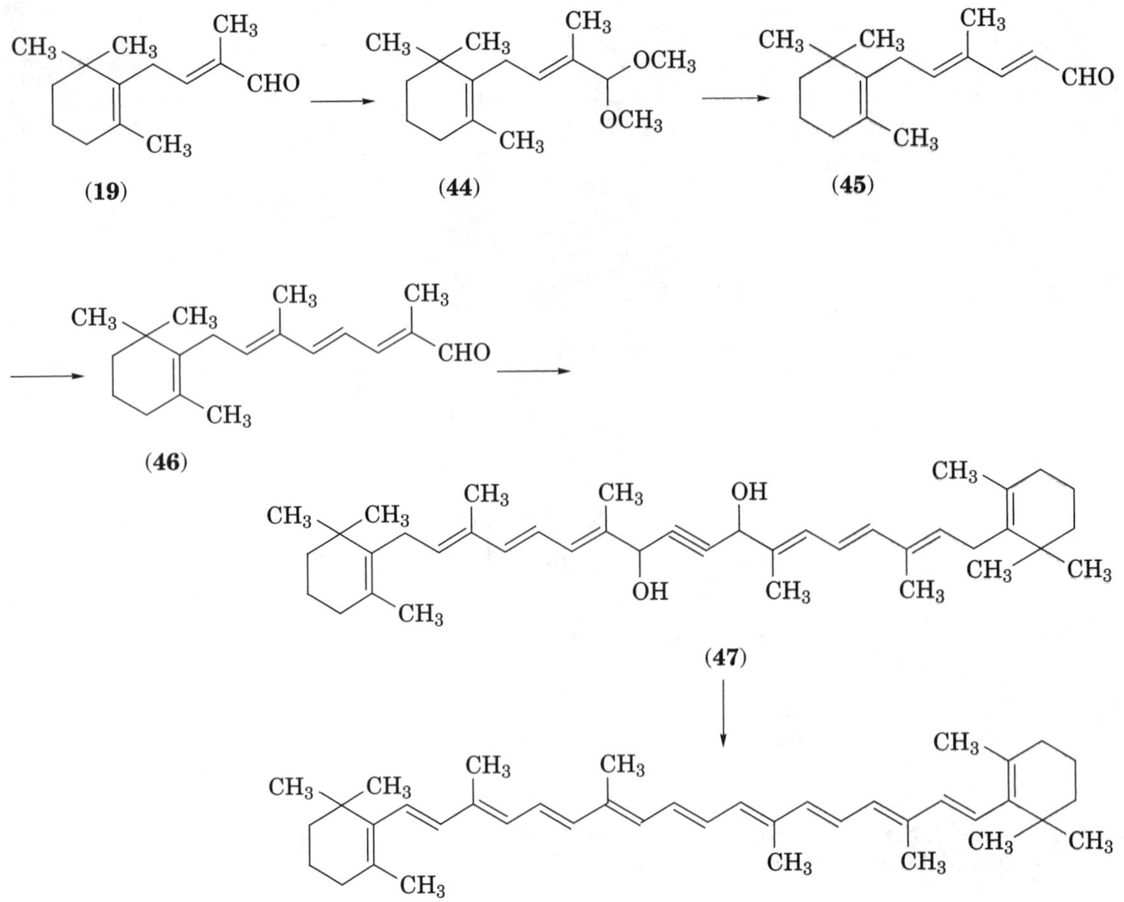

Figure 10. Hoffmann-La Roche synthesis of β-carotene.

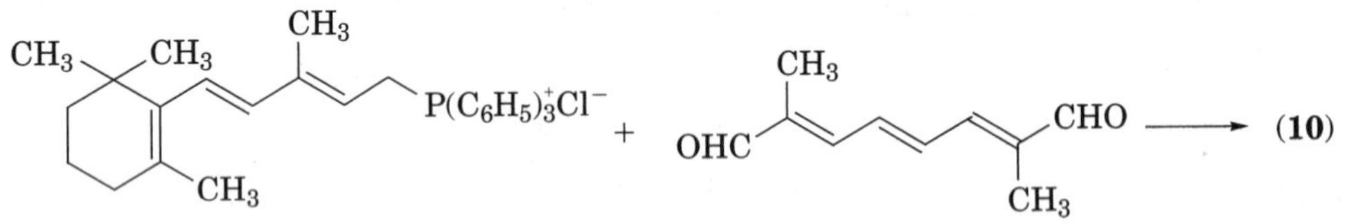

Figure 11. BASF synthesis of β-carotene.

The specific role of vitamin A in tissue differentiation has been an active area of research. The current thinking, developed in 1979, involves initial delivery of retinol by *holo*-RBP (retinol-binding protein) to the cell cytosol (66). Retinol is then ultimately oxidized to retinoic acid and binds to a specific cellular retinoid-binding protein and is transported to the nucleus. Retinoic acid is then transferred to a nuclear retinoic acid receptor (RAR) which enhances the expression of a specific region of the genome. Transcription occurs and new proteins appear during the retinoic acid-induced differentiation of cells (56).

In contrast to vitamin A, the carotenoids are important in both plant and animal kingdoms. In plants, carotenoids are associated with photosynthetic structures. The func-

tion of these pigments is twofold: (1) the pigments serve as acceptors and act as energy-transfer agents and allow for excitation energy to be transferred from the carotenoid to the porphyrin; (2) the pigments protect the organism from light-induced photooxidative damage. In this regard, protection is achieved by two mechanisms: quenching of the excited triplet state of chlorophyll (itself a producer of singlet oxygen), and quenching of singlet oxygen. In both cases, a triplet excited state of the carotenoid, which decays back to the ground state by a radiationless conversion, is produced (67).

Many carotenoids function in humans as vitamin A precursors; however, not all carotenoids have provitamin A activity (Table 3). Of the biologically active carotenoids, β-

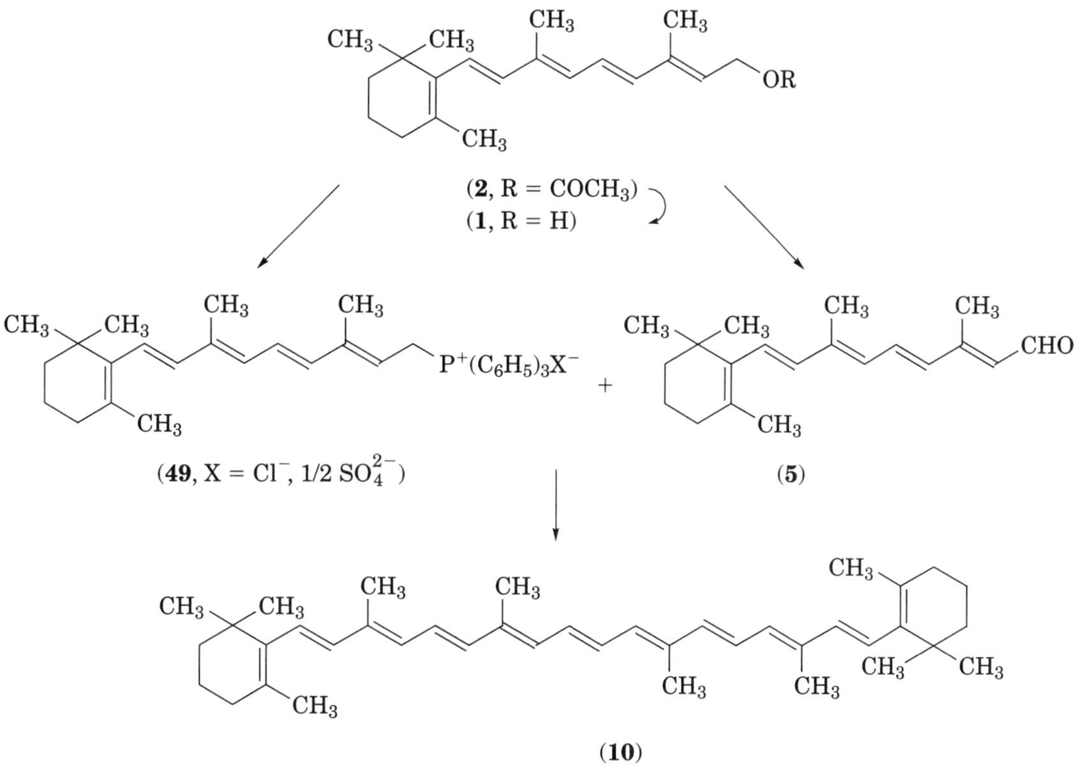

Figure 12. Alternative Roche synthesis of β-carotene.

carotene has the greatest activity. Despite the fact that theoretically one molecule of β-carotene is a biological source of two molecules of vitamin A, this relationship is not observed and 6 μg of β-carotene is equivalent to 1 μg of vitamin A. Although β-carotene and vitamin A have complementary activities, they cannot totally replace each other. Because the conversion of β-carotene to vitamin A is highly regulated, toxic quantities of vitamin A cannot accumulate and β-carotene can be considered as a safe form of vitamin A (8).

Owing to the presence of an extended polyene chain, all carotenoids are effective single-oxygen quenchers and antioxidants. In addition, they can stimulate the immune response and, as a result, may protect against certain forms of cancer. These separate functions, ie, vitamin A equivalents or antioxidants, have allowed for carotenoids to be classified according to whether they are nutritionally or biologically active. For example, β-carotene is both nutritionally and biologically active whereas 14′-β-apocarotenal is nutritionally active but biologically inactive. Lycopene is nutritionally inactive but biologically active, whereas phytoene, the natural precursor to β-carotene, is both nutritionally and biologically inactive (56).

REQUIREMENTS

Animals cannot synthesize vitamin A-active compounds and necessary quantities are obtained by ingestion of vitamin A or by consumption of appropriate provitamin A compounds such as β-carotene. Carotenoids are manufac-

tured exclusively by plants and photosynthetic bacteria. Until the discovery of vitamin A in the purple bacterium *Halobacterium halobium* in the 1970s, vitamin A was thought to be confined to only the animal kingdom (56). Table 4 lists RDA and U.S. RDA for vitamin A (67).

DEFICIENCY

In humans, vitamin A deficiency manifests itself in the following ways: night blindness, xerophthalmia, Bitot's spots, and corneal involvement and ulceration. Changes in the skin have also been observed. Although vitamin A deficiency is seen in adults, the condition is particularly harmful in the very young. Often, this results from malnutrition (56).

On a vitamin A-deficient diet, mucus-secreting tissues become keratinized. This condition tends to occur in the trachea, the skin, the salivary glands, the cornea, and the testes. When this occurs in the cornea, it can be followed by blindness. Vitamin A deficiency is the principal cause of blindness in the very young. This problem is particularly acute in the third world (8).

SAFETY

Vitamin A toxicity can be categorized as either acute or chronic. Acute toxicity results from extremely high doses (≥500,000 IU of vitamin A). In children, approximately half of that amount causes problems. Hypervitaminosis A is characterized by headache, blurred vision, loss of coor-

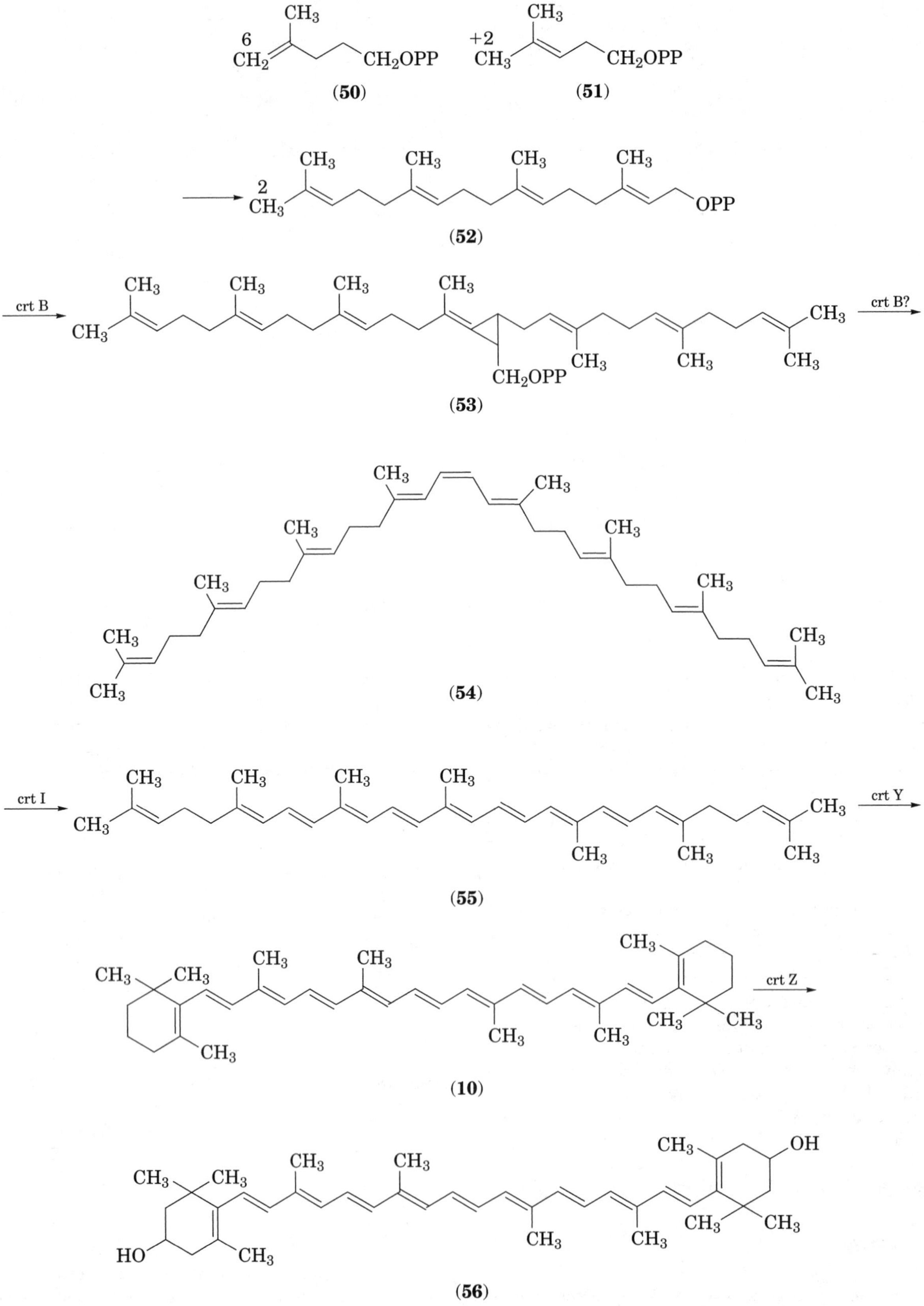

Figure 13. Biosynthetic transformation of β-carotene.

Table 3. Provitamin A Activity of Selected Carotenoids

Provitamin A activity	No provitamin A activity
β-Carotene	Astaxanthin
α-Carotene	Canthaxanthin
γ-Carotene	Lutein
β-Cryptoxanthin	Lycopene
β-Zeacarotene	Zeaxanthin

Source: Ref. 8.

Table 4. Recommended Daily Allowances of Vitamin A

Sex/age group	1989 RDA (RE)[a]	U.S. RDA (IU)[b]
Infants <12 months	375	1,500
Children		
1–3 yr	400	2,500
4–6 yr	500	5,000
7–10 yr	700	5,000
Males ≥ 11 yr	1,000	5,000
Females ≥11 yr	800	5,000
Additional during pregnancy		+ 3,000
Additional during lactation		+ 3,000

Source: Ref. 8.
[a]RE = retinol equivalents (1 RE = 1 μg of all-*trans*-retinol).
[b]IU = international units (1 IU = 0.30 μg of all-*trans*-retinol).

dination, nausea, and peeling and itchy skin. Chronic vitamin A toxicity occurs in adults with long-term intakes of ≥50,000 IU/d. Symptoms include dry hair, hair loss, weakness, headache, bone thickening, enlarged liver and spleen, anemia, abnormal menstrual periods, stiffness, and joint pain. Most of these symptoms are reversible. In animals, extremely high doses of vitamin are teratogenic (56). On the other hand, the carotenoids are generally nontoxic and there have been only a few isolated cases of problems associated with a large daily intake. Tanning pills which contain large doses of canthaxanthin were shown to cause canthaxanthin retinopathy in patients with eyrthropoietic porphyria (68,70).

USES

Vitamin A is generally used in feeds, foods, and pharmaceutical applications; however, the principal use of carotenoids is as a colorant. β-Carotene, for example, is used to color fat products such as margarine, shortening, butter, cheese, egg nog, and ice cream. It is also used in baked goods, pasta products, juices, and beverages. It imparts a natural yellow to orange shade. Conversely, if an orange to reddish orange color is desired, the carotenoid of choice is β-apo-8'-carotenal [1107-26-2]. This carotenoid is used to color juices, fruit drinks, soups, jams, jellies, and gelatin (qv). Carotenoids are also added to feed to color poultry, fish, and of egg yolks. In order to guard against oxidative damage during feed processing, these compounds are protected in a gelatin matrix as beadlets.

ECONOMIC ASPECTS

Vitamin A is manufactured by Hoffmann-La Roche (Switzerland), BASF (Germany), and Rhône-Poulenc (France), as well as by some smaller suppliers in India, China, and Russia. The worldwide production is estimated to be 2500 to 3000 metric tons. About three-quarters of this production is for animal feed; the remainder is for food fortification and pharmaceuticals (qv). The main trade names of feed products are Rovimix, Lutavit, and Microvit. Prices depend on application forms and are approximately $60–$70/10^9 IU retinol (1995); ie, $200–$233/10^9 RE. One IU is equivalent to 0.300 μg of all-*trans*-retinol and 1 RE is equivalent to 1 μg of all-*trans*-retinol.

BIBLIOGRAPHY

1. G. Wolf, *Am. J. Clin. Nutr.* **31**, 290 (1978).
2. E. V. McCollum and M. Davis, *J. Biol. Chem.* **15**, 167 (1913).
3. E. V. McCollum and M. Davis, *J. Biol. Chem.* **23**, 291 (1915).
4. T. B. Osborne and L. B. Mendel, *J. Biol. Chem.* **37**, 187 (1923).
5. P. Karrer, A. Helfensten, H. Wehrli, and A. Wettstein, *Helv. Chim. Acta* **13**, 1084 (1930).
6. P. Karrer, R. Morf, and K. Schopp, *Helv. Chim. Acta* **14**, 1431 (1931).
7. G. Wald, *Nature (London)* **134**, 65 (1934).
8. D. Bhatia, in *Encyclopedia of Food Science and Technology*, John Wiley & Sons, Inc., New York, 1991, pp. 2701–2712.
9. J. Paust, *Pure Appl. Chem.* **63**, 45 (1991).
10. Technical data, Hoffmann-La Roche, Inc., Nutley, N.J., 1996.
11. R. A. Heyman, D. J. Mangelsdorf, J. A. Dyck, R. B. Stein, G. Eichelle, R. M. Evans, and C. Thaller, *Cell* **68**, 397 (1992).
12. R. S. H. Liu and A. E. Asato, *Tetrahedron* **40**, 1931 (1984); Y. Bennani, *J. Org. Chem.* **61**, 3542 (1996); A. E. Asato, A. Kini, M. Denney, and R. S. H. Liu, *J. Am. Chem. Soc.* **105**, 2923 (1983); G. Wald, P. K. Brown, R. Hubbard, and W. Oroshnik, *Natl. Acad. Sci. USA* **41**, 438 (1955); G. Wald, P. K. Brown, R. Hubbard, and W. Oroshnik, *Natl. Acad. Sci. USA* **42**, 578 (1956); K. Ramamurthy and R. Liu, *Tetrahedron* **31**, 201 (1975); A. E. Asato and R. S. H. Liu, *J. Am. Chem. Soc.* **97**, 4128 (1975); R. S. H. Liu and A. E. Asato, *Methods. Enzymol.* **88**, 506 (1982); A. Kini, H. Matsumoto, and R. S. H. Liu, *J. Am. Chem. Soc.* **107**, 5078 (1979); A. Kini, H. Matsumoto, and R. S. H. Liu, *Biorg. Chem.* **9**, 406 (1980); G. G. Ladson, S. C. Cary, and W. H. Okamura, *J. Am. Chem. Soc.* **102**, 6355 (1980).
13. G. Britton, in G. Britton, S. Liaaen-Jensen, and H. Pfander, eds., *Carotenoids, Vol. 1B: Spectroscopy*, Birkhauseer, Basel, Switzerland, 1995, pp. 13–62.
14. G. Britton, G. Armitt, S. Y. M. Lau, A. K. Patel, and C. S. Shone, in G. Britton and T. W. Goodwin, eds., *Carotenoid Chemistry and Biochemistry*, Pergamon Press, Oxford, U.K., 1984, pp. 237–251.
15. P. Zagalsky, in Ref. 13, *Vol. 1A: Isolation and Analysis*, pp. 287–294.
16. S. Liaaen-Jensen, in Ref. 13, pp. 343–354.
17. C. H. Eugster, in Ref. 15, pp. 343–354.
18. O. Isler, W. Huber, A. Ronoco, and M. Kofler, *Helv. Chim. Acta* **30**, 1911 (1942).
19. O. Isler, A. Ronco, W. Guex, N. C. Hindley, W. Huber, K. Dialer, and M. Kofler, *Helv. Chim. Acta* **32**, 489 (1949).

20. O. Isler, *Pure Appl. Chem.* **51**, 447 (1979).
21. C. Mercier and P. Chabardes, *Pure Appl. Chem.* **66**, 1509 (1994).
22. Jpn. Pat. 86-93840 (1986), Y. Inada (to Bihaman, Hishara).
23. J. Otera, H. Misana, T. Mandai, T. Onishi, S. Suzuki, and Y. Fujita, *Chem. Lett.*, 1883 (1985).
24. J. Otera, H. Misawa, T. Onishi, S. Suzuki, and Y. Fujita, *J. Org. Chem.* **51**, 3834 (1986).
25. H. Bienayme, *Tetrahedron Lett.* **35**, 7387 (1994).
26. Ref. 25, p. 7383.
27. H. Bienayme, *Bull. Soc. Chim. Fr.* **132**, 696 (1995).
28. H. Pommer and P. C. Thieme, *Top. Curr. Chem.* **109**, 165 (1983).
29. A. Nurrenbach, J. Paust, H. Pommer, J. Scheider, and B. Schultz, *Liebigs Ann. Chem.*, 1146 (1972).
30. H. J. Bestman, C. Kisielowski, and N. Distler, *Angew. Chem.* **88**, 297 (1976).
31. Ger. Pat. 2,551,914 (1975), H. J. Bestmann and N. Distler (to Hoechst AG).
32. Ger. Pat. 2,702,633 (1977), B. Schulz, J. Paust, and J. M. Schneider (to BASF AG).
33. Jpn. Pat. 76-108106 (1979), M. Mukaiyama and A. Ishida (to Japan); A. Ishida, T. Mukaiyama, *Chem. Lett.*, 1127–1130 (1972).
34. Ger. Pat. 2,641,075 (1977), J. McMurray (to University of California).
35. Ger. Pat. 2,515,011 (1976), J. McMurray (to University of California).
36. Ger. Pat. 2,224,606 (1971), P. Chabardes and J. Marc (to Rhône-Poulenc).
37. Ger. Pat. 2,264,501 (1971), P. Chabardes and M. Julia (to Rhône-Poulenc).
38. Ger. Pat. 2,305,217 (1972), P. Chabardes, M. Julia, and A. Menet (to Rhône-Poulenc).
39. Ger. Pat. 2,355,898 (1972), P. Chabardes, M. Julia, and A. Menet (to Rhône-Poulenc).
40. Ger. Pat. 2,305,267 (1972), P. Chabardes, M. Julia, and A. Menet (to Rhône-Poulenc).
41. Ger. Pat. 2,202,689 (1971), M. Julia (to Rhône-Poulenc).
42. U.S. Pat. 4,916,250 (Apr. 10, 1990), J. H. Babler (to University of Loyola); J. H. Babler and S. A. Schlidt, *Tetrahedron Lett.* **33**, 7697–7700 (1992).
43. K. Bernhard and H. Mayer, *Pure Appl. Chem.* **63**, 35–44 (1991); E. Widmer, *Pure Appl. Chem.* **57**, 741–752 (1985).
44. O. Isler, R. Rüegg, U. Schwieter, and J. Wursch, *Vitam. Horm. (NY)* **18**, 295 (1960).
45. H. H. Kaegi, in M. B. Sporn, A. R. Roberts, and D. S. Goodman, eds., *The Retinoids*, Vol. I, Academic Press, Orlando, Fla., 1984, pp. 147–174.
46. M. F. Boehm et al., *J. Med. Chem.* **37**, 408–414 (1994); I. M. Mishina, E. N. Karnaukhova, V. I. Shvets, and M. G. Vladimirova, *Bioteknologiya*, 13–17 (1993); A. A. Liebman, W. Burger, S. C. Choudhry, and J. Cupano, *Methods Enzymol.* **213**, 42–49 (1992); M. Groesbeek and J. Lugtenberg, *Photochem. Photobiol.* **56**, 903–908 (1992); R. W. Curley, Jr., M. A. Blust, and K. A. Humphries, *J. Labelled Compd. Radiopharm.* **29**, 1331–1335 (1991); E. M. M. Van den Berg, A. Van der Bent, and J. Lugtenburg, *Recl. Trav. Chim. Pays-Bas* **109**, 160–167 (1990); H. Akita, *Chem. Pharm. Bull.* **31**, 1796–1799 (1983); S. C. Welch and J. M. Gruber, *J. Org. Chem.* **47**, 385–389 (1982).
47. J. Ganguly, *Biochemistry of Vitamin A*, CRC Press, Boca Raton, Fla., 1989.
48. N. Misawa, M. Nakagawa, K. Kobayashi, S. Yamano, Y. Izawa, K. Nakamura, and K. Harahima, *J. Bacteriol.* **172**, 6704 (1990).
49. World Pat. 9406918 (1994), A. J. van Ooyen (to Gist-Brocades, Inc.); Eur. Pat. 635576 A1 950125 (1995), T. Kiyota, M. Takaki, A. Tsubokura, and H. Yoneda (to Nippon Oil KK); Eur. Pat. 454024 A 911030 (1991), W. D. Prevatt and R. E. Torregrossa (to Phillips Petroleum Co.); World Pat. 9103571 (1991), D. L. Gierhart (to Applied Food BioTechnology, Inc.).
50. Eur. Pat. 107493 900420 (1990), N. Misawa, K. Kobayashi, and K. Nakamura (to Kirin Brewery Co, Ltd.).
51. World Pat. 9113078 A1 91095 (1991), R. L. Ausich, F. L. Brinkhaus, I. Mukharji, J. H. Proffitt, J. G. Yarger, and B. Yen (to Amoco Corp.).
52. N. D. Embree, S. R. Ames, R. W. Lehman, and P. L. Harris, *Meth. Biochem. Anal.* **4**, 43 (1957).
53. P. L. Harris, *Vitam. Horm.* **18**, 341 (1960).
54. P. Tosukhowong and J. A. Olson, *Biochem. Biophys. Acta* **529**, 438 (1978).
55. A. B. Barua and J. A. Olson, *Biochem. J.* **244**, 231 (1987).
56. J. A. Olson, in L. J. Machlin, ed., *The Handbook of Vitamins*, Marcel Dekker, New York, 1991, pp. 1–57.
57. H. C. Furr, A. B. Barua, and J. A. Olson, in A. P. De Leenheer, W. F. Lambert, and H. J. Nelis, eds., *Modern Chromatographic Analysis of the Vitamins*, Marcel Dekker, New York, 1992, pp. 1–71.
58. R. Schindler, A. Kloop, C. Gorny, and W. Feldheim, *Int. J. Vit. Nutr. Res.* **55**, 25 (1985).
59. A. T. Rhys Williams, *J. Chromatogr.* **341**, 198 (1985).
60. A. J. Speek, C. Wongkham, N. Limratana, W. Saowakontha, and W. H. Schreurs, *Curr. Eye. Res.* **5**, 841 (1986).
61. C. A. Collins and C. K. Chow, *J. Chromatogr.* **317**, 349 (1984).
62. H. K. Biesalski and H. Weiser, *J. Clin. Chem. Clin. Biochem.* **27**, 65 (1989).
63. C. R. Smidt, A. D. Jones, and A. J. Clifford, *J. Chromatogr.* **21**, 434 (1988).
64. *The United States Pharmacopeia XXII* (USP XXII-NF XVII), United States Pharmacopeial Convention, Inc., Rockville, Md., 1990, pp. 156–157, 1451, 1550–1551.
65. C. D. B. Bridges, in Ref. 45, pp. 126–176.
66. F. Chytil and D. Ong, *Fed. Proc.* **38**, 2510 (1979).
67. P. Mathis and C. C. Schenck. in Ref. 14, pp. 339–351.
68. Committee on Dietary Allowances, Food and Nutrition Board, *Recommended Dietary Allowances*, 10th ed., National Academy Press, Washington, D.C., 1989.
69. U. Weber, G. Goerz, and R. Hennekes, *Klin. Monatsbl. Augenheilkd.* **186**, 351 (1985).
70. B. Daicker, K. Schiedt, J. J. Adnet, and P. Bermond, *Graefe's Arch. Clin. Exp. Ophthalmol.* **225**, 189 (1987).

GENERAL REFERENCES

J. Ganguly, *Biochemistry of Vitamin A*, CRC Press, Boca Raton, Fla., 1989.
A. P. De Leenheer, W. F. Lambert, and H. J. Nelis, eds., *Retinoids and Carotenoids* in *Modern Chromatographic Analysis of the Vitamins*, Marcel Dekker, New York, 1992.

M. A. Livrea and G. Vidali, eds., *Retinoids: From Basic Science to Clinical Applications*, Birkhauser Verlag, Germany, 1994.

L. J. Machlin, ed., *The Handbook of Vitamins*, Marcel Dekker, New York, 1991.

M. B. Sporn, A. B. Roberts, and D. S. Goodman, eds., *The Retinoids*, Vols. 1 and 2, Academic Press, Inc., Orlando, Fla., 1994.

SUSAN D. VAN ARNUM
Westfield, New Jersey

See also CAROTENOID PIGMENTS.

VITAMINS: VITAMIN B$_{12}$

Vitamin B$_{12}$ [68-19-9] (1,2) is the generic name for a closely related group of substances of microbial origin. Although the last of the vitamins to be characterized, its history is a long one, dating from 1824 when Combe (3) proposed the relationship of pernicious anemia, a disease characterized by defective (megoblastic) red blood cell formation, to disorders of the digestive system. Additional study of pernicious anemia, in particular the work of Addison (4), continued for over 100 years before Minot and Murphy (5) reported that a diet containing large quantities of raw liver restored the normal level of red blood cells in patients with pernicious anemia. This clinical breakthrough was based on the findings of Whipple and Robscheit-Robbins (6) that liver was of benefit in regeneration of blood in anemic dogs. For this work, Whipple, Minot, and Murphy were awarded the Nobel prize in medicine and physiology in 1934.

In 1929, Castle (7) tied the work of Combe and Addison with that of Whipple, Minot, and Murphy by proposing that both an extrinsic factor and an intrinsic factor are involved in the control of pernicious anemia. The extrinsic factor, from food, is vitamin B$_{12}$. The intrinsic factor is a specific B$_{12}$-binding protein secreted by the stomach. This protein is required for vitamin B$_{12}$ absorption.

Work for the next 20 years focused on purification of the extrinsic factor from liver. The work was slow and tedious, because fractionation was guided by tests on pernicious anemia patients. The discovery (8) that *Lactobacillus lactis* Dorner requires liver extracts for growth greatly accelerated the isolation process. In 1948, groups at Merck (9) in the United States and Glaxo (10) in England reported isolation of vitamin B$_{12}$ (cyanocobalamin) as a crystalline, red pigment. The clinical efficacy of this material in the treatment of pernicious anemia was rapidly established (11).

Parallel to the activities in the treatment of pernicious anemia were observations in the 1930s that most farm animals had a requirement for an unknown factor beyond the vitamins then known. The lack of this factor became apparent, eg, when chicks or pigs fed a diet with only vegetable protein evidenced slow growth rate and high mortality. It became apparent that the required factor, termed animal protein factor, was present in animal sources such as meat and tissue extracts, milk, whey, and cow manure. Subsequent to its isolation, it was rapidly shown that vitamin B$_{12}$ is the same as animal protein factor.

After its separation from liver extracts, vitamin B$_{12}$ was isolated also from cultures of *Streptomyces aureofaciens* (12). All vitamin B$_{12}$ sold commercially is produced by microbial fermentation.

STRUCTURE

The structure of the first isolated vitamin B$_{12}$, cyanocobalamin [68-19-9] (1a) is known to occur only sporadically, at best, in biological systems. Its isolation was an artifact resulting, probably, from use of charcoal containing cyanide in the purification process.

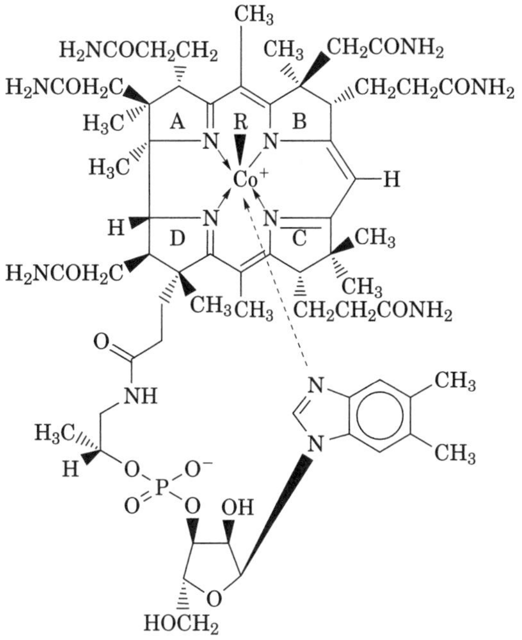

(1a, R = CN)
(1b, R = HO OH)

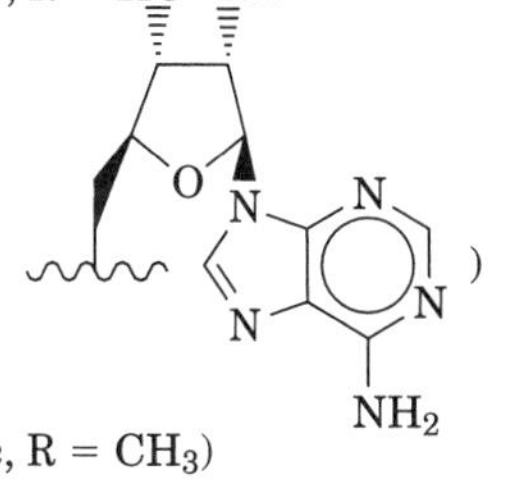

(1c, R = CH$_3$)
(1d, R = OH)
(1e, R = NO$_2$)
(1f, R = OH$_2$Cl)

It is recognized that there are several important forms of vitamin B$_{12}$. The active coenzyme forms are adenosylcobalamin [13870-90-1] (coenzyme B$_{12}$ (1b)) and methylcobalamin [13422-55-4] (1c). These, along with hydroxocobalamin [13422-51-0] (vitamin B$_{12a}$ (1d)), are the forms found in humans and other animals. Other forms of interest are nitrocobalamin (vitamin B$_{12c}$ (1e)) and aquacobalamine chloride [13422-52-1] (vitamin B$_{12b}$ (1f)). The primary commercial form is cyanocobalamin, due to its ease of isolation and purification as well as its stability.

The IUPAC-IUB Commission on Biochemical Nomenclature (13) recommends that the term vitamin B$_{12}$ be used as the generic descriptor for all corrinoids exhibiting qualitatively the biological activity of cyanocobalamin. However, because of its commercial importance, cyanocobalamin is used interchangeably with vitamin B$_{12}$ herein.

Determination of the structure of vitamin B$_{12}$ was a slow, tedious process with the tools available in the 1940s. Despite extensive information gathered by degradation studies, the structures of cyanocobalamin and its coenzyme forms were only established beginning in 1955 by x-ray crystallography (14–16). The value of this work in the study of natural products was recognized by the award to Hodgkin of the Nobel prize in chemistry in 1964.

Vitamin B$_{12}$ belongs to the class of molecules known as corrins. The core corrin structure (2) consists of four linked, partially saturated pyrrole rings. Corrin is a truncated form (no CH$_2$ between rings A and D) of the more common porphyrin skeleton. The corrin ring in vitamin B$_{12}$ is substituted in a highly regio- and stereospecific manner with eight methyl groups, three acetic acid chains, and four propionic acid chains, as in cobyrinic acid (3). The six conjugated double bonds of the corrin system give octahedral complexes with a number of metals. However, only complexes with cobalt exhibit vitamin B$_{12}$ activity.

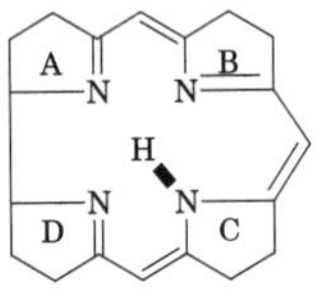

(2)

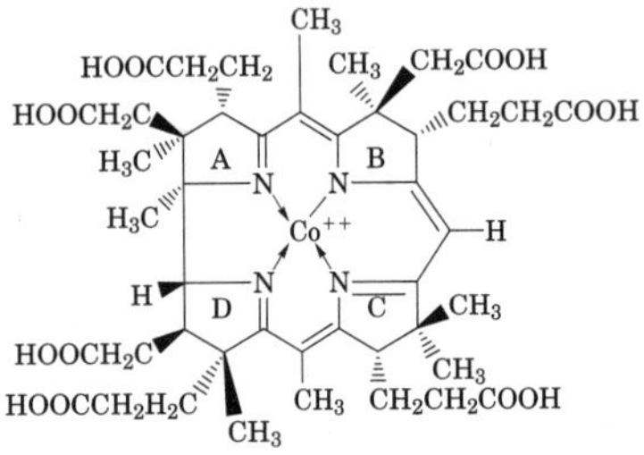

(3)

Cobrynic acid in which the propionic acid is amidated with 1-amino-2-propanol is known as cobinic acid, whereas the compound in which all other acids are primary amides is cobinamide (4).

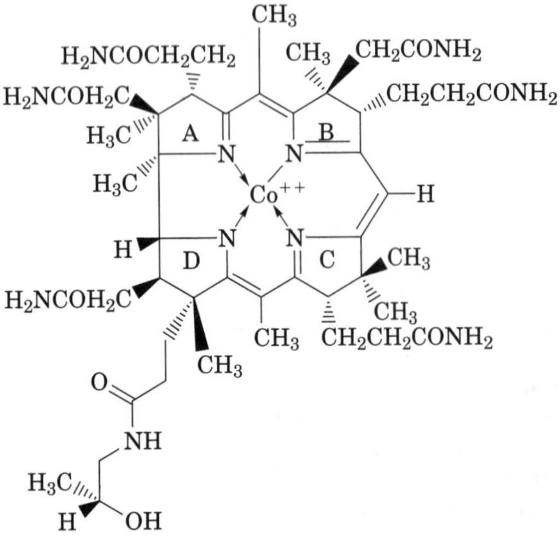

(4)

Cobinic acid and cobinamide, which are linked to ribose 3-phosphate, are known as cobamic acid and cobamide (5), respectively.

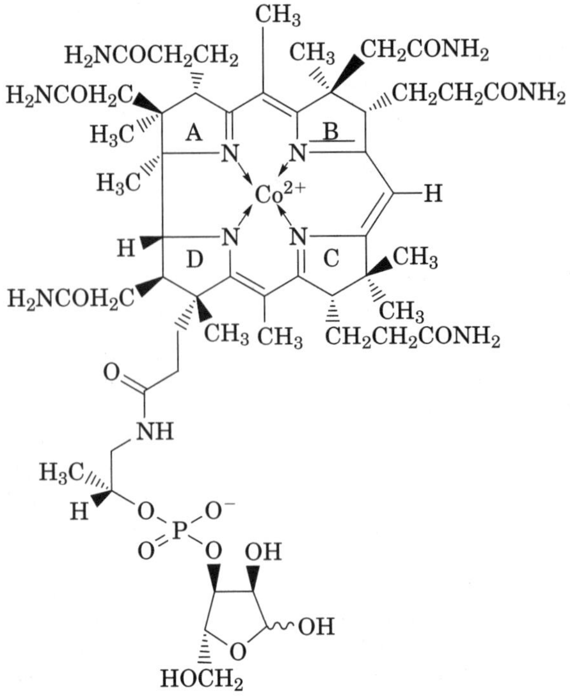

(5)

The cobalt corrin complex is octahedral. The four nitrogens of the corrin ring occupy the four equatorial ligand positions in a virtually planar arrangement. The nucleotide formed from cobamide and 5,6-dimethylbenzimidazole provides a fifth ligand in the alpha, axial position of vitamin B$_{12}$. The group occupying the sixth, beta-position can vary substantially, as noted in structure (1), and dictates the compound name. Thus, when the ligand is cyanide, the compound name is cyanocobalamin; when methyl, methylcobalamin, etc.

Vitamin B$_{12}$ exists as a neutral complex. The beta-ligand and one nitrogen atom in the corrin ring each contribute a negative charge to Co(III). The nucleotide phosphate contributes the final negative charge. The structure of vitamin B$_{12}$ is highly organized and compact, as dictated by charge neutralization and the ligand sphere of the cobalt atom.

Many analogues of vitamin B$_{12}$ are known. These occur either naturally or are obtained by synthesis, degradation of the vitamin, or directed biosynthesis. Most of these compounds have little or no biological activity in animals. However, many have significant microbiological activity. Among the analogues found are those in which the 5,6-dimethylbenzimidazole is replaced by another (or no) base. These include 2-methyl adenine (Factor A), no base (Factor B), guanine (Factor C), 5-hydroxybenzimidazole (Factor III), 5-methoxybenzimidazole (Factor III$_m$) and adenine (pseudovitamin B$_{12}$) (17–20).

OCCURRENCE

For many years, it was thought that the occurrence of vitamin B$_{12}$ was limited to animal tissues and bacteria, with all of the material originating from bacterial sources. More recently, the presence of vitamin B$_{12}$ and/or vitamin B$_{12}$-like activity at low levels in plants has been recognized (21–24). Nonetheless, the largest portion of the vitamin B$_{12}$ requirement is met by consumption of animal tissues and from microorganisms in the animal's digestive tract. Herbivorous animals satisfy their vitamin B$_{12}$ needs by absorption of material produced by rumenal or intestinal flora. In humans and carnivorous/omnivorous animals, intestinal production of vitamin B$_{12}$ occurs but at an insufficient level. As a result, vitamin B$_{12}$ levels are maintained by consumption of foods rich in vitamin B$_{12}$ such as liver, heart, and kidney. Egg yolk and some fish and shellfish are also good sources. Vitamin B$_{12}$ supplements, both oral and parenteral, are also available. Dietary sources of foodstuffs are provided in Table 1 (25–27).

BIOCHEMICAL FUNCTIONS

Methylcobalamin and adenosylcobalamin are the two coenzyme forms of vitamin B$_{12}$ in animals and humans. Each is involved in the catalysis of a specific transformation. In humans, it appears that there are only two enzymes requiring vitamin B$_{12}$ as an essential coenzyme although many other, particularly bacterial enzyme systems also require a vitamin B$_{12}$ coenzyme (2,28).

Adenosylcobalamin (coenzyme B$_{12}$) is required in a number of rearrangement reactions; that occurring in humans is the methylmalonyl-CoA mutase-mediated conversion of (R)-methylmalonyl-CoA (6) to succinyl-CoA (7) (eq. 1). The mechanism of this reaction is poorly understood,

although probably free radical in nature (29). The reaction is involved in the catabolism of valine and isoleucine. In bacterial systems, adenosylcobalamin drives many 1,2-migrations of the type exemplified by Scheme 1 (30).

Methylcobalamin is involved in a critically important physiological transformation, namely the methylation of homocysteine (8) to methionine (9) (Scheme 2) catalyzed by N^5-methyltetrahydrofolate homocysteine methyltransferase. The reaction sequence involves transfer of a methyl group first from N^5-methyltetrahydrofolate to cobalamin (yielding methylcobalamin) and thence to homocysteine. Once again, the intimate details of the reaction are not well known (31). Demethylation of tetrahydrofolate to tetrahydrofolic acid is a step in the formation of thymidine phosphate, in turn required for DNA synthesis. In the absence of the enzyme, excess RNA builds up in red blood cells.

Homocysteine has been identified as an independent risk factor for atherosclerosis (32) and thus metabolic control over homocysteine levels has major health implications.

Deficiency

Macrocytic anemia, megaloblastic anemia, and neurological symptoms characterize vitamin B$_{12}$ deficiency. Alterations in hematopoiesis occur because of the high requirement for vitamin B$_{12}$ for normal DNA replication necessary to sustain the rapid turnover of the erythrocytes. Abnormal DNA replication secondary to vitamin B$_{12}$ deficiency produces a defect in the nuclear maturational process of committed hematopoietic stem cells. As a result, the erythrocytes are either morphologically abnormal or die during development.

Neurological symptoms result from demyelination of the spinal cord and are potentially irreversible. The symptoms and signs characteristic of a vitamin B$_{12}$ deficiency include paresthesis of the hands and feet, decreased deep-tendon reflexes, unsteadiness, and potential psychiatric problems such as moodiness, hallucinations, delusions, and psychosis. Neuropsychiatric disorders sometimes develop independently of the anemia, particularly in elderly patients. Visual loss may develop as a result of optic atrophy.

Clinical manifestation of vitamin B$_{12}$ deficiency is usually a result of absence of the gastric absorptive (intrinsic)

Table 1. Dietary Sources of Vitamin B$_{12}$

>50 µg/100 g	5–50 µg/100 g	<5 µg/100 g
Liver	Liver	Beef (lean)
Lamb	Rabbit	Lamb
Beef	Chicken	Pork
Calf	Kidney	Chicken
Pork	Rabbit	Egg (whole)
Kidney	Beef	Cheese
Lamb	Fish	American
Brain	Sardines	Swiss
Beef	Salmon	Milk (cow)
	Berring	Fish
	Heart	Cod
	Beef	Flounder
	Rabbit	Haddock
	Chicken	Sole
	Egg yolk	Halibut
	Clams	Swordfish
	Oysters	Tuna
	Crabs	Mackerel
		Lobster
		Scallop
		Shrimp
		Green vegetables

$$\text{HOOC}-\underset{\underset{\textbf{(6)}}{\overset{|}{\underset{\text{H}}{\diagup}}\;\overset{}{\underset{\text{CH}_3}{}}}{\text{C}}-\text{COSCoA} \longrightarrow \text{HOOCCH}_2\text{CH}_2\text{COSCoA}$$

$$\textbf{(7)}$$

Scheme 1.

$$\text{HSCH}_2\text{CH}_2\underset{\underset{\textbf{(8)}}{\overset{}{\underset{\text{H}}{\diagup}\;\underset{\text{NH}_2}{}}}}{\text{CCOOH}} \longrightarrow \text{CH}_3\text{SCH}_2\text{CH}_2\underset{\underset{\textbf{(9)}}{\overset{}{\underset{\text{H}}{\diagup}\;\underset{\text{NH}_2}{}}}}{\text{CCOOH}}$$

Scheme 2.

factor. Dietary deficiency of vitamin B$_{12}$ is uncommon and may take 20 to 30 years to develop, even in healthy adults who follow a strict vegetarian regimen. An effective entero-hepatic recycling of the vitamin plus small amounts from bacterial sources and other contaminants greatly mini-mizes the risk of a complete dietary deficiency. Individuals who have a defect in vitamin B$_{12}$ absorption, however, may develop a deficiency within three to seven years.

Dietary deficiency in the absence of absorption defects can be effectively reversed with oral supplementation of 1 μg of vitamin B$_{12}$ daily. If deficiency is related to a defect in vitamin absorption, daily doses of 1 μg administered subcutaneously or intramuscularly are effective (33). How-ever, a single intramuscular dose of 100 μg of cobalamin once per month is adequate in patients with chronic gastric or ileal damage. Larger doses are generally rapidly cleared from the plasma into the urine and are not effective unless the patient demonstrates poor vitamin retention.

Requirement

A daily intake of 1 μg should cover the daily loss of vitamin and maintain an adequate body pool. The RDA (34), how-ever, has been established at 2 μg/day to cover metabolic variation among individuals and to ensure normal serum concentrations and adequate pool sizes (Table 2).

Smaller pool sizes with normal serum B$_{12}$ levels may be maintained with dietary intakes below 1 μg. However, more substantial pool sizes are considered advantageous as protection against the development of pernicious ane-mia, which may occur in advanced age; achlorhydria be-comes more common after age 60, resulting in compro-mised absorption of vitamin B$_{12}$.

In general, maternal stores of vitamin B$_{12}$ are consid-ered adequate to meet the demands of pregnancy.

Safety

No toxicity has been associated with acute or chronic in-takes of vitamin B$_{12}$ in doses of 100 and 1 mg, respectively. Vitamin B$_{12}$ absorption is both limited and affected by vi-tamin status. Therefore, absorption is reduced with im-proved status, lessening the risk of toxicity.

Cobalamin should be administered parenterally by the intramuscular or subcutaneous route. Isolated cases of anaphylaxis have been reported with intravenous admin-istration.

METABOLISM

Absorption

An absorption mechanism capable of handling between 1.5 to 3.0 μg of vitamin B$_{12}$ is responsible for most of the in-testinal absorption. This mechanism includes the binding of vitamin B$_{12}$ to a specific transport protein (intrinsic fac-tor) and the presence of specific membrane-bound recep-tors on the ileal cell surface in the small intestine. In ad-dition, there is a second mechanism for the absorption of vitamin B$_{12}$ by diffusion. This mechanism can provide a physiologically significant source of the vitamin when de-livered in pharmacological dosages.

Food vitamin B$_{12}$ appears to bind to a salivary transport protein referred to as the R-protein, R-binder, or haptocor-rin. In the stomach, R-protein and the intrinsic factor com-petitively bind the vitamin. Release from the R-protein oc-curs in the small intestine by the action of pancreatic proteases, leading to specific binding to the intrinsic factor. The resultant complex is transported to the ileum where it is bound to a cell surface receptor and enters the intes-tinal cell. The vitamin B$_{12}$ is then freed from the intrinsic factor and bound to transcobalamin II in the enterocyte. The resulting complex enters the portal circulation.

Transport

Transcobalamin II delivers the absorbed vitamin B$_{12}$ to cells and is the primary plasma vitamin B$_{12}$-binding trans-port protein. It is found in plasma, spinal fluid, semen, and extracellular fluid. Many cells, including the bone marrow, reticulocytes, and the placenta, contain surface receptor sites for the transcobalamin II-cobalamin complex.

Other plasma vitamin B$_{12}$ proteins, transcobalamines I and III, appear to have primarily a storage function and only a lesser role in transport.

Tissue Uptake and Storage

Cell surface receptors take up the transcobalamin II-cobalamin complex which is internalized into endosomes. The complex is dissociated and the transcobalamin II re-leased. The mechanism by which cobalamin leaves the en-dosome is uncertain.

The liver is the principal site of vitamin B$_{12}$ storage, containing between 50 and 99% of the total body pool. The average total body pool is estimated to range between 2.0 and 5.0 mg. Higher storage values reported probably re-flect noncobalamin analogues as well as cobalamin. The main storage form of vitamin B$_{12}$ appears to be coenzyme B$_{12}$. The primary circulating form in the plasma appears to be methylcobalamin.

Table 2. The 1989 RDA for Vitamin B$_{12}$

Group	Age (yrs)	Vitamin B$_{12}$ (μg)
RDA		
Infants	0–0.5	0.3
	0.5–1.0	0.5
Children	1–3	0.7
	4–6	1.0
	7–10	1.4
Males	11–51+	2.0
Females	11–51+	2.0
	Pregnant	2.0
	Lactating	2.6
U.S. RDI (daily value)		
Infants and children <4 yr		3.0
Adults and children >4 yr		6.0
Pregnant or lactating women		8.0

Metabolism and Mobilization

On entry of vitamin B$_{12}$ into the cell, considerable metabolism of the vitamin takes place. Co(III)cobalamin is reduced to Co(I)cobalamin, which is either methylated to form methylcobalamin or converted to adenosylcobalamin (coenzyme B$_{12}$). The methylation requires methyl tetrahydrofolate.

Approximately 0.05 to 0.2% of vitamin B$_{12}$ stores are turned over daily, amounting to 0.5–8.0 μg, depending on the body pool size. The half-life of the body pool is estimated to be between 480 and 1360 days with a daily loss of vitamin B$_{12}$ of about 1 μg. Consequently, the daily minimum requirement for vitamin B$_{12}$ is 1 μg. Three micrograms (3.0 μg) of vitamin B$_{12}$ are excreted in the bile each day, but an efficient enterohepatic circulation salvages the vitamin from the bile and other intestinal secretions. This effective recycling of the vitamin contributes to the long half-life. Absence of the intrinsic factor interrupts the enterohepatic circulation. Vitamin B$_{12}$ is not catabolized by the body and is, therefore, excreted unchanged. About one-half of the vitamin is excreted in the urine and the other half in the bile.

PROPERTIES

Table 3 lists some of the physical and chemical properties of vitamin B$_{12}$ (1,35). Crystalline vitamin B$_{12}$ is stable in air and is not affected by moisture. The anhydrous compound, however, is very hygroscopic, and when exposed to moist air may absorb about 12% of water.

Aqueous solutions of vitamin B$_{12}$ at pH 4.0 to 7.0 show no decomposition during extended storage at 25°C. For optimum stability at elevated temperatures, solutions should be adjusted to pH 4.0 to 4.5. Aqueous solutions in this pH range may be autoclaved for 20 minutes at 120°C without significant decomposition.

Redox Reactions

Critical to the function of cobalamins as enzyme co-factors is the ability of the cobalt atom to exist in the Co(III) and Co(I) oxidation states. Chemically, Co(II) forms are also known. Each oxidation state has different ligand-accepting abilities. The chemical (1a, 1d–1f) and coenzyme (1b, 1c) forms of cobalamin are trivalent. These compounds are readily reduced to Co(II) and Co(I)cobalamins. The redox potentials for the aquacobalamin to Co(II)cobalamin and Co(II)cobalamin to Co(I)cobalamin couples are −0.04 and −0.85 V, respectively (36). Chemical reduction of aquacobalamin to Co(II)cobalamin is effected by thiols or carbon monoxide. Cobalt(II)cobalamin contains a single unpaired electron in the $3d_{z2}$ orbital of the cobalt ion; one of the two axial positions is unoccupied and thus this material is five-coordinate. Chemically, Co(I)cobalamin is obtained with strong reducing agents, eg, NaBH$_4$ or zinc and NH$_4$Cl. It is a powerful reducing agent, and reacts rapidly with oxygen and reduces protons to hydrogen. It is therefore unstable in aqueous acid solution, less so in neutral or basic aqueous solution. It is frequently used for the preparation of organocobalamins, eg, adenosylcobalamin and methylcobalamin. Co(I)cobalamin contains two electrons in the $3d_{z2}$ orbital. It has been postulated that the coordination number for the cobalt is four and that both axial ligands are vacant (37).

Exchange of Axial Ligands

Many ligand-exchange reactions involve groups in which the coordination to the metal is through nitrogen (NH$_3$, N$_3^-$), oxygen (H$_2$O, OH$^-$), sulfur (SH$^-$, SO$_3^{-2}$), halogen, or carbon (CN$^-$, CH$_3^-$). Important reactions involve displacement of the heterocyclic base from the alpha-coordination position by a solvent, usually water. This displacement occurs in acidic solution and results from the protonation of the heterocyclic base. The protonation is associated with a characteristic change in the spectrum. The pK_a for the base-on/base-off equilibrium depends on the nature of the beta-ligand. Displacement of H$_2$O, adenosyl, or methyl from cobalt by cyanide has also been studied. In the presence of cyanide ion, aquacobalamin and adenosylcobalamin are converted to cyanocobalamin. In contrast, methylcobalamin and other alkyl corrinoids are stable in the presence of 0.1 M cyanide in the dark (37,38). The equilibrium, aquacobalamin ⇌ hydroxocobalamin + H$^-$ (pK_a = 6.9–7.8), is not a pure ligand exchange but only a ligand modification. The cobalt-bound water is acidic and reversibly loses a proton at neutral pH.

Chemical Reactions

A wealth of information exists on the chemistry of vitamin B$_{12}$ (1,39,40). Much of this chemistry was established during the studies leading to structure determination, partial synthesis, and the total synthesis of vitamin B$_{12}$. Vitamin B$_{12}$ is slowly decomposed by ultraviolet or strong visible light. By controlled irradiation with visible light, cyanide may be selectively liberated from cyanocobalamin without destruction of the cobalamin structure, but long exposure to light results in complete inactivation of the vitamin. Vitamin B$_{12}$ is also inactivated by treatment with strong acids or bases.

One development involves the use of vitamin B$_{12}$ to catalyze chemical, in addition to biochemical processes. Vi-

Table 3. Physical and Chemical Properties of Vitamin B$_{12}$ (Cyanocobalamin)

Property	Characteristic
Appearance	Crystalline
Color	Dark red
Molecular weight	1355.42
Empirical formula	C$_{63}$H$_{88}$CoN$_{14}$O$_{14}$P
Melting point, C°	Darkens 210–220; does not melt <300
Specific rotation	$\alpha_{656}^{23} = -59 \pm 9°$ (diluted aqueous)
UV Absorption	278,361,550 nm (water)
pH (aq)^a	Neutral
Solubility %	
H$_2$O	1.25
alcohol	2.03
acetone	Insoluble
ether	Insoluble

Note: Properties given for the anhydrous compound.

^{a}Vitamin B$_{12}$ exhibits maximum stability between pH 4.5 and 5.0.

tamin B$_{12}$ derivatives and B$_{12}$ model compounds (41,42) catalyze the electrochemical reduction of alkyl halides and formation of C-C bonds (43,44), as well as the zinc–acetic acid-promoted reduction of nitriles (45), alpha, beta-unsaturated nitriles (46), alpha, beta-unsaturated carbonyl derivatives and esters (47,48), and olefins (49). It is assumed that these reactions proceed through intermediates containing a Co-C bond which is then reductively cleaved.

ANALYSIS

Specifications

Cyanocobalamin is the commercial form of vitamin B$_{12}$. It is sold under the following trade names (35):

Anacobin	Ducobee
Antipernicin	Duodecibin
Bedoce	Embiol
Bedodeka	Emociclina
Bedoz	Eritrone
Behepan	Erycytol
Berubi	Erythrotin
Berubigen	Euhaemon
Betalin-12	Fresmin
Betolvex	Hemo-B-Doze
Bevatine-12	Hemomin
Bevidox	Hepagon
Bexii	Hepavis
Bexil	Hepcovite
Biocobalamine	Hydroxamin
Biocres	Hydroxobase
Bitevan	Macrabin
B-Telve	Megabion (Indian)
B-Twelv	Megalovel
Byladoce	Milbedoce
Claretin-12	Millevit
Cobalin	Nagravon
Cobamin	Normocytin
Cobamine	Peraemon
Cobione	Pernaevit
Covit	Pernipur
Crystamin	Plecyamin
Cycobemin	Poyamin
Cycolamin	Redamina
Cykobeminet	Redisol
Cytacon	Rhodacryst
Cytamen	Rubesol
Cytobion	Rubivitan
Distivit (B$_{12}$ peptide)	Rubramin
Dobetin	Rubripca
Docemine	Rubrocitol
Docibin	Sytobex
Docigram	Vitalt
Docivit	Vibisone
Dodecabee	Virubra
Dodecabee	Virubra
Dodecavite	Vitarubin
Dodex	Vita-Rubra
	Vitral

Specifications are found in the *Codex* for food use (50) and in the USP (51) for pharmaceutical use.

Analytical Methodology

Vitamin B$_{12}$ can be determined by microbiological, radio-isotope dilution, spectrophotometric, chemical, or biological methods employing animals (52–54). Microbiological assays involve the extraction and stabilization of vitamin B$_{12}$ from the food, feed, or pharmaceutical matrix prior to assay. The official method of the AOAC (55) accomplishes this by autoclaving samples in a phosphate–citric acid buffer containing metabisulfite, whereas the British Analytical Methods Committee (56) recommends extraction with aqueous cyanide solution at pH 4.6–5.0 in a boiling water bath. The AOAC extraction procedure was found to yield somewhat higher results than the British Analytical Methods Committee procedure (57).

The AOAC assay is based on the graded growth of the bacterium *Lactobacillus leichmannii* ATCC 7830 in a medium containing all required growth factors but vitamin B$_{12}$. Results are obtained by determining the transmittance of the sample and standard tubes after 16–24 h incubation at 30–40°C or by titrating the acid produced after 72 h incubation. The British Analytical Methods Committee assay employs the protozoan *Ochromonas malhamensis* ATCC 11532 in an assay based on the graded growth of the protozoan. The assay incubation period varies from 3–6 days and degree of growth is measured turbidimetrically. Although the *L. leichmannii* assay is claimed to be less specific than the *O. malhamensis* assay it has several advantages over the latter. The AOAC procedure has a shorter incubation period and the assay is set up in test tubes, whereas the *O. malhamensis* assay is set up in 25-mL micro-Fernbach flasks, the assay medium is less complex, and samples whose turbidity or color would interfere in a turbidimetric assay can be analyzed by the titrimetric assay. Derivatives of deoxynucleic acid that stimulate the growth of *L. leichmannii* in the absence of vitamin B$_{12}$ can be measured after destroying the vitamin B$_{12}$ by autoclaving the sample at pH 11–12 and determining the residual activity. The bacteria *Lactobacillus lactis* Dorner ATCC 8000 and *Escherichia coli* ATCC 10799 and 14169 as well as the protozoan *Euglena gracilis* ATCC 12716 have been used also for the determination of vitamin B$_{12}$.

Radioisotope dilution assays are based on the principle of competition between radioactive labeled (^{57}Co) vitamin B$_{12}$ and cobalamins extracted from matrices for binding sites on the intrinsic factor (a glycoprotein). Binding is in proportion to the concentration of the radioactive and non-radioactive B$_{12}$ with the concentration of intrinsic factor as the limiting factor. Free cobalamins are separated from those bound on the intrinsic factor by absorption onto treated charcoal and the amount of free-labeled vitamin B$_{12}$ is determined. Vitamin B$_{12}$ content of the sample is determined from a standard curve. Results obtained by a radioisotopic dilution method are very similar to those obtained by the AOAC microbiological assay (58).

Spectrophotometric determination at 550 nm is relatively insensitive and is useful for the determination of vitamin B$_{12}$ in high potency products such as premixes. Thin-layer chromatography and open-column chromatography have been applied to both the direct assay of cobalamins and to the fractionation and removal of interfering substances from sample extracts prior to microbiological or radioassay. Atomic absorption spectrophotometry of cobalt has been proposed for the determination of vitamin B$_{12}$ in dry feeds. Chemical methods based on the estimation of cyanide or the presence of 5,6-dimethylbenzimidazole in the vitamin B$_{12}$ molecule have not been widely used.

Various aspects of the chromatography of vitamin B$_{12}$ and related corrinoids have been reviewed (59). A high performance liquid chromatographic (hplc) method is reported to require a sample containing 20–100 μg cyanocobalamin and is suitable for premixes, raw material, and pharmaceutical products (60).

Bioassays are based on the growth response of vitamin-depleted rats or chicks to graded amounts of vitamin B$_{12}$ added in the diet. These assays are not specific for vitamin B$_{12}$ because factors, other than vitamin B$_{12}$ present in biological materials, produce a growth response. Because coenzyme B$_{12}$, a primary form of natural vitamin B$_{12}$, is light sensitive, assays should be carried out in subdued light.

SYNTHESIS

The achievements in total synthesis of organic compounds in recent years are perhaps nowhere better illustrated than in the synthesis of vitamin B$_{12}$ by the groups of Woodward (61–63) and Eschenmoser (64–69) in a collaborative effort (70,71). The work has been reviewed (1,72,73). The real value of the synthesis lies in the synthetic methodology developed in the course of the work. During the synthesis of the A–D fragment, the Woodward-Hoffmann rules for the conservation of orbital symmetry (74) were developed. For this work, Hoffmann shared the Nobel prize in chemistry in 1981 with Fukui. Woodward was awarded the Nobel prize in chemistry in 1965 for the synthesis and structure elucidation of natural products, including the work leading up to the synthesis of vitamin B$_{12}$. Thus, the history of vitamin B$_{12}$ is intimately connected with the awarding of four Nobel prizes to date.

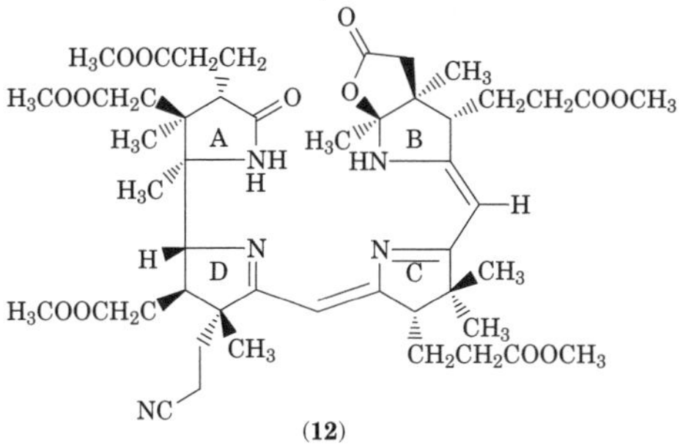

(12)

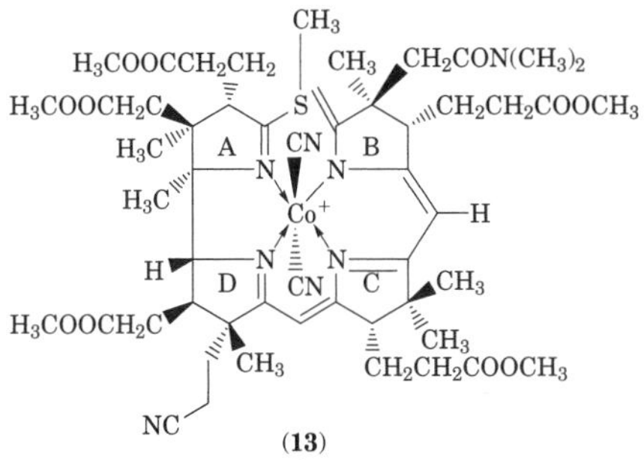

(13)

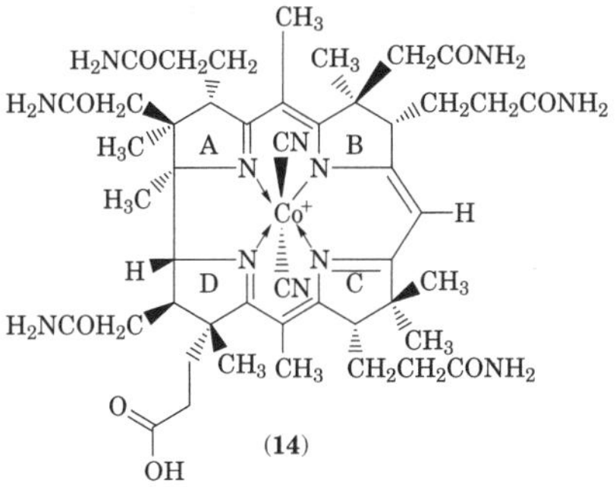

(14)

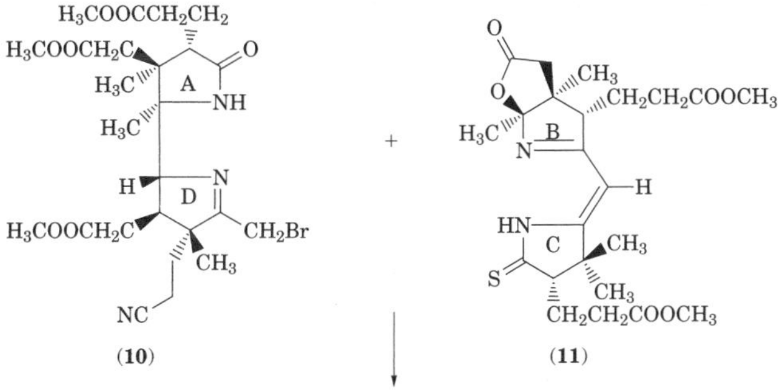

The core of the first synthesis of vitamin B$_{12}$ involved condensation of the A–D ring fragment (10) with the B–C fragment (11). The former compound was obtained by the Harvard group in ca 35 steps, including a classical resolution, from 3-methoxyaniline and camphor.

The synthesis of the second fragment (11) by the group at the ETH in Zurich occurred in 18 steps from camphorquinone (ring C) and trans-3-methyl-4-oxopentenoic acid (ring D). Base-catalyzed alkylation of (11) with (10) yielded a thioiminoether which underwent sulfide contraction upon treatment with acid to give the tetrapyrrole (12). Manipulation of substituents and introduction of the cobalt atom gave complex (13). The stereo-organizing template effect of the complex allowed base-catalyzed cyclization to complete the corrin ring system. Functional group exchange and addition of the final methyl group then yielded cobyric acid (14). This material had previously been con-

(10)   (11)

verted to vitamin B$_{12}$ (75) and thus its obtention by total synthesis completed the synthesis.

H$_3$COOCCH$_2$CH$_2$
H$_3$COOCH$_2$C
H$_3$C   A
H$_3$C   CN   NH   S

(15)

O
CH$_3$ CH$_2$CH$_2$COOCH$_3$
H$_3$C   B
HN   S

(16)

N
H$_3$COOCH$_2$C   D   CH$_2$Br
CH$_3$
NC

(18)

HN
CH$_3$
O   C   CH$_3$
CH$_2$CH$_2$COOCH$_3$

(17)

H$_3$COOCCH$_2$CH$_2$ CH$_3$ CH$_2$CON(CH$_3$)$_2$
H$_3$COOCH$_2$C   CH$_2$CH$_2$COOCH$_3$
H$_3$C   A   N   N   B
Cd
H$_3$COOCH$_2$C   D   N   N   C   CH$_3$
CH$_3$ CH$_3$
NC   CH$_2$CH$_2$COOCH$_3$

(19)

H$_3$COOCCH$_2$CH$_2$ CH$_3$ CH$_2$CON(CH$_3$)$_2$
H$_3$COOCH$_2$C   CH$_2$CH$_2$COOCH$_3$
H$_3$C   A   N   N   B
H$_3$C
H$_3$COOCH$_2$C   D   N   HN   C   CH$_3$
CH$_3$ CH$_3$
NC   CH$_2$CH$_2$COOCH$_3$

(20)   ⟶ (14)

A second synthesis of cobyric acid (14) involves photochemical ring closure of an A–D secocorrinoid. Thus, the Diels-Alder reaction between butadiene and *trans*-3-methyl-4-oxopentenoic acid was used as starting point for all four ring A–D synthons (15–18). These were combined in the order B + C → BC + D → BCD + A → ABCD. The resultant cadmium complex (19) was photocyclized in buffered acetic acid to give the metal-free corrinoid (20). A number of steps were involved in converting this material to cobyric acid (14).

The total syntheses have yielded cobyric acid and thence cyanocobalamin. Routes to other cobalamins, eg, methylcobalamin and adenosylcobalamin, are known (76–79). One approach to such compounds involves the oxidative addition of the appropriate alkyl halide (eg, CH$_3$I to give methylcobalamin) or tosylate (eg, 5′-*p*-tosyladenosine to yield adenosylcobalamine) to cobalt(I)alamine.

The complexity of the vitamin B$_{12}$ molecule makes it extremely unlikely that total synthesis will ever be employed for preparation of commercial quantities.

BIOSYNTHESIS

The study of the biosynthesis of vitamin B$_{12}$ is a saga whose resolution, due primarily to Battersby (80–83) and Scott (84,85), required an effort on the same magnitude as the total synthesis. It was only when recent molecular biology tools became available to complement enzymology, isotopic labeling, chemical synthesis, and spectroscopy that solution of this problem became possible.

The biosynthesis of vitamin B$_{12}$ in the species *Pseudomonas dentrificans* has received the most attention due to its importance as a commercial source of the vitamin. The 22 genes involved have been identified and the sequence of the resulting proteins described (83). The sequence is initiated from 5-aminolevulinic acid (21) which is biosynthesized from succinyl-CoA (7) and glycine. Condensation of two molecules of 5-aminolevulinic acid yields porphobilinogen (22) which is tetramerized as a linear, head-to-tail arrangement to give hydroxymethylbilane (23). Ring closure and rearrangement (formally on exchange of acetate and propionate substituents in ring C) gives uroporphyrinogen III (uro'gen III) (24). Uro'gen III is a key biosynthetic branchpoint, leading not only to vitamin B$_{12}$ but also to chlorophyll and heme.

The sequence to vitamin B$_{12}$ proceeds by the introduction of the eight methyl groups, punctuated by ring contraction to the corrin, an oxidation and reduction, and a methyl migration. Thus, the first three methylations lead stereospecifically to precorrin-1 (25) and thence by precorrin-2 (26) to precorrin-3A (precorrin-3) (27). Oxidation, ring contraction to the corrin, and introduction of the fourth methyl group give precorrin-4 (28). Further methylation leads to precorrin-5 (29). Loss of the acetyl group created by ring contraction as acetate occurs with methylation to give precorrin-6A (precorrin-6x) (30) which is reduced to precorrin 6B (precorrin-6y) (31). Decarboxylation of the ring C acetic acid subsequent to addition of the last two methyl groups (precorrin-8x) (32) and methyl migration gives hydrogenobyrinic acid (33).

Introduction of the cobalt atom into the corrin ring is preceeded by conversion of hydrogenobyrinic acid to the diamide (34). The resultant cobalt(II) complex (35) is reduced to the cobalt(I) complex (36) prior to adenosylation to adenosylcobyrinic acid a,c-diamide (37). Four of the six remaining carboxylic acids are converted to primary amides (adenosylcobyric acid) (38) and the other amidated with (R)-1-amino-2-propanol to provide adenosylcobinamide (39). Completion of the nucleotide loop involves conversion to the monophosphate followed by reaction with guanosyl triphosphate to give diphosphate (40). Reaction with α-ribazole 5′-phosphate, derived biosynthetically in several steps from riboflavin, and dephosphorylation completes the synthesis.

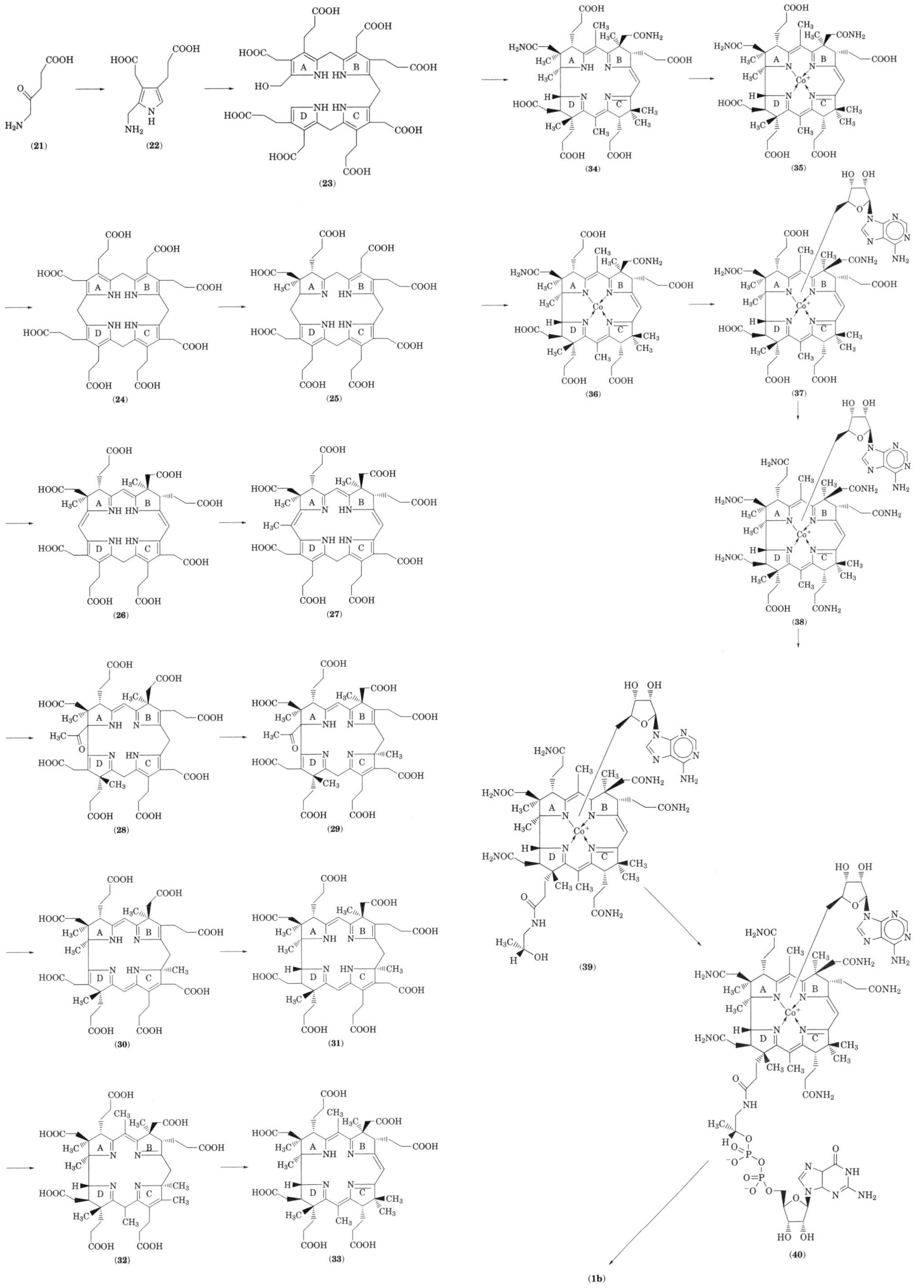

(21)

(22)

(23)

(34)

(35)

(24)

(25)

(36)

(37)

(38)

(26)

(27)

(28)

(29)

(39)

(30)

(31)

(40)

(32)

(33)

(1b)

The biosynthetic sequence for other aerobic bacteria appears, where known, to be similar to that in *Pseudomonas dentrificans* although the genes involved, and thus the enzymes, exhibit differences.

In anaerobic (or more correctly, almost anaerobic or microaerophilic) bacteria such as *Propionibacterium shermanii*, a fundamental difference occurs in that the cobalt atom is introduced at a much earlier stage, possibly to precorrin-2 or precorrin-3A.

MANUFACTURE

As noted above, all vitamin B₁₂ is produced by microbial fermentation. A partial list of microorganisms which synthesize vitamin B₁₂ under appropriate conditions follows. Most strains, in their wild state, produce less than 10 mg/L vitamin B₁₂, although a few approach 40 mg/L. The organisms are both aerobes and anaerobes. The carbon requirements in the fermentations are satisfied from sources as wide ranging as hydrocarbons, methanol, and glucose.

Arthrobacter hyalinus	*Propionibacterium*
Bacillus megaterium	*arabinosum*
Butyribacterium rettgeri	*Propionibacterium*
Clostridium sticklandii	*freudenreichii*
Clostridium tetanomorphum	*Propionibacterium*
Clostridium thermoaceticum	*pentosaceum*
Corynebacterium and	*Propionibacterium peterssoni*
Rhodopseudomonas	*Propionibacterium shermanii*
Crithidia fasciculata	*Propionibacterium technicum*
Methanobacterium	*Propionibacterium vannielli*
arbophilicum	*Protaminobacter ruber*
Methanobacterium formicicum	*Pseudomonas denitrificans*
Methanobacterium	*Rhizobium meliloti*
ruminantium	*Rhodopseudomonas capsulata*
Methanobacterium	*Rhodopseudomonas*
thermoautotrophicum	*spheroides*
Methanosarcina barkeri	*Strigomonas oncopelti*
Micromonospora purpurea	*Streptomyces aureofaciens*
Nocardia gardneri	*Streptomyces griseus*
Nocardia rugosa	*Streptomyces olivaceus*

Commercial Production

Vitamin B₁₂, as cyanocobalamin, is produced by several companies. The market is dominated, however, by two French firms, Rhône-Poulenc and, to a lesser extent, Roussel-Uclaf. Smaller amounts are produced in Japan by Nippon Petrochemical, in Hungary by Medimpex-Richter, and by minor producers in several other countries. Earlier manufacturers, particularly Merck (U.S.) and Glaxo (U.K.), have exited the market. Although estimates vary, it appears that ca 10,000 kg/yr of vitamin B₁₂ is produced (1).

The process employed by Rhône-Poulenc for production of vitamin B₁₂ has not been revealed. However, from a variety of sources (83,86) it can be inferred that a *Pseudomonas dentrificans* producing over 200 mg/L is employed. The high production is the result of classical mutation as well as (possibly) genetic engineering.

A fermentation such as that of *Pseudomonas dentrificans* typically requires 3–6 days. A submerged culture is employed with glucose, cornsteep liquor and/or yeast extract, and a cobalt source (nitrate or chloride). Other minerals may be required for optimal growth. pH control at 6–7 is usually required and is achieved by ammonium or calcium salts. Under most conditions, adequate 5,6-dimethylbenzimidazole is produced in the fermentation. However, in some circumstances, supplementation may be required.

The fermentation product, which is primarily adenosylcobalamin, is retained to the largest extent within the cell. Centrifugation of the broth yields a sludge which, when dispersed in a minimum of water, water–alcohol, or water–acetone and heated, releases the vitamin into the solution. Addition of cyanide converts the cobalamins into cyanocobalamin which is then extracted from the filtered solution. Many procedures have been reported for this step, including adsorption on charcoal (87), bentonite (88), ion-exchange resins (89–91), or aluminum oxide (92). For elution, water, water–alcohol, organic bases, or hydrochloric acid are used. Chromatography on aluminum oxide and crystallization from methanol–acetone or water–acetone complete the process (93,94).

Market Forms

Vitamin B₁₂ is sold almost exclusively as cyanocobalamin. Approximately one-third of the material is for the human pharmaceutical market whereas two-thirds is used in the animal feed market, primarily for poultry and swine. Modest growth in both markets has occurred in the period 1980–1995 and this trend is expected to continue.

In the human market, oral and parenteral dosage forms are prepared from the crystal. However, because of the extremely high potency, more dilute (0.1–10%) forms are available. These include dilutions with mannitol, triturations on dicalcium phosphate or resins, and spray-dried forms. Prices for these forms are driven by that of the crystal, which in early 1996 was ca $9.50/gram (95). Prices for the vitamin have risen during the first half of the 1990s. However, little growth in price beyond inflation is anticipated.

For animal feed use, vitamin B₁₂ is usually provided in a diluted form on a carrier such as calcium carbonate and/or rice hulls. An earlier practice of using a spray-dried fermentation biomass in this application appears to be no longer used.

BIBLIOGRAPHY

1. B. Kräutler, in O. Isler, G. Brubacher, S. Ghisla, and B. Kräutler, eds., *Vitamin II: Wasserlösliche Vitamine*, Georg Thieme Verlag, New York, 1988, pp. 340–388.
2. L. Ellenbogen and B. A. Cooper, in L. J. Machlin, ed., *Handbook of Vitamins*, 2nd ed., Marcel Dekker, Inc., New York, 1991, pp. 491–536.
3. J. S. Combe, *Trans. Med. Chirurg. Soc (Edinburg)*, **1**, 194 (1824).
4. T. Addison, *On the Constitutional and Local Effects of Disease of the Suprarenal Capsules*, Samuel Highley, London, 1855, p. 2.
5. G. R. Minot and W. P. Murphy, *JAMA* **91**, 923 (1926).
6. G. H. Whipple, C. W. Hooper, and F. S. Robscheit, *Am. J. Physiol.* **53**, 236 (1920).

7. W. B. Castle, *Am. J. Med. Sci.* **178**, 748 (1929).

8. M. S. Shorb and G. M. Briggs, *J. Biol. Chem.* **176**, 1463 (1948).

9. E. L. Rickes, N. G. Brink, F. R. Koniuszy, T. R. Wood, and K. Folkers, *Science* **107**, 396 (1948).

10. E. L. Smith, *Nature (London)* **161**, 638 (1948).

11. R. West, *Science* **107**, 398 (1948).

12. J. V. Pierce, A. C. Page, E. L. R. Stokstad, and T. H. Jukes, *J. Am. Chem. Soc.* **71**, 2952 (1949).

13. "The Nomenclature of Corrinoids (1973 Recommendations)," *Biochemistry* **13**, 1555 (1974).

14. D. C. Hodgkin and co-workers, *Nature (London)* **176**, 325 (1995).

15. P. G. Lenhert and D. C. Hodgkin, *Nature (London)* **102**, 937 (1961).

16. D. C. Hodgkin, *Science* **150**, 979 (1965).

17. J. W. G. Porter, in H. C. Heinrich, ed., *Vitamin B$_{12}$ and Intrinsic Factor 1*, Europäisches Symposion, Hamburg, 1956, F. Enke, Stuttgart, 1957, p. 43.

18. M. E. Coates and S. K. Kon, in Ref. 17, p. 72.

19. H. C. Heinrich, W. Friedrich, E. Gabbe, S. P. Manjrekar, and M. Staak, in *Abstracts of the 5th International Congress on Nutrition*, Washington, D.C., 1960, p. 62.

20. E. L. Smith, *Vitamin B$_{12}$*, Methuen, London, 1965.

21. L. Fries, *Physiol. Plantarum* **15**, 566 (1962).

22. G. G. Laties and C. Hoelle, *Phytochemistry* **6**, 49 (1967).

23. A. P. Petrosyan, L. A. Abramyan, and M. B. Sarkisyan, *Vopr. Mikrobiol.* **4**, 181 (1969).

24. J. M. Poston, *Science* **195**, 301 (1977).

25. L. J. Bogert, G. M. Briggs, and D. H. Calloway, *Nutrition and Physical Fitness*, W. B. Saunders, Philadelphia, Pa., 1973.

26. W. Friedrich, *Vitamin B$_{12}$ und Verwandte Corrinoide*, Georg Thiem, Stuttgart, Germany, 1975, p. 170.

27. J. Marks, *A Guide to the Vitamins, Their Role in Health and Disease*, Medical and Technical Publishing Co., Lancaster, U.K., 1975, p. 118.

28. J. P. Glusker, in G. Litwack, ed., *Vitamins and Hormones*, Vol. 50, Academic Press, Inc., New York, 1995, pp. 1–76.

29. Y. Zhao, P. Such, and J. Rétey, *Angew, Chem. Int. Ed. Engl.* **31**, 215 (1992); M. He and P. Dowd, *J. Am. Chem. Soc.* **118**, 711 (1996).

30. T. C. Stadtman, *Science* **102**, 859 (1971).

31. R. G. Mathews, in R. L. Blakely and S. J. Benkovic, eds., *Folates and Pteridins*, Vol. 1, John Wiley & Sons, Inc., New York, 1984, p. 497.

32. C. S. Berwanger, J. Y. Jeremy, and G. Stansby, *Brit. J. Surgery* **82**, 726 (1995).

33. V. Herbert and N. Coleman, in M. E. Shils and V. R. Young, eds., *Modern Nutrition in Health and Disease*, 7th ed., Lea and Febiger, Philadelphia, Pa., 1988, pp. 388–416.

34. Food and Nutrition Board, National Research Council, *Recommended Dietary Allowances*, 10th ed., National Academy Press, Washington, D.C., 1989.

35. *The Merck Index*, 11th ed., Merck and Co., Rahway, N.J., 1989, pp. 9921–9922.

36. H. P. C. Hogenkamp, *Am. J. Clin. Nutr.* **33**, 1 (1980).

37. H. P. C. Hogenkamp, in B. M. Babior, ed., *Cobalamin-Biochemistry and Pathophysiology*, John Wiley & Sons, Inc., New York, 1975, p. 21.

38. B. M. Babior and J. S. Krouwer, *CRC Crit. Rev. Biochem.* **6**, 35 (1979).

39. R. Bonnett, in D. Dolphin, ed., *B$_{12}$*, John Wiley & Sons, Inc., New York, 1982, p. 201; J. Halpern, *ibid.*, p. 501; B. T. Golding, *ibid.*, p. 543.

40. E. L. Smith, *Vitamin B$_{12}$*, Methuen, London, 1965.

41. G. N. Schrauzer, *Acc. Chem. Res.* **1**, 97 (1968).

42. D. Dodd and M. D. Johnson, *J. Organomet. Chem.* **52**, 1 (1973).

43. G. Rytz, L. Walder, and R. Scheffold, in B. Zagalak and W. Friedrich, eds., *Vitamin B$_{12}$, Proceedings of the Third European Symposium on Vitamin B$_{12}$, Zurich, Switzerland, 1979*, Walter de Gruyter, Berlin, 1979, p. 173.

44. R. Scheffold, M. Dike, S. Dike, T. Herold, and L. Walder, *J. Am. Chem. Soc.* **102**, 3642 (1980).

45. A. Fischli, *Helv. Chim. Acta* **61**, 2560, 3028 (1978).

46. A. Fischli, *Helv. Chim. Acta* **62**, 882 (1979).

47. A. Fischli and D. Süss, *Helv. Chim. Acta* **62**, 48, 2361 (1979).

48. A. Fischli and J. J. Daly, *Helv. Chim. Acta* **63**, 1628 (1980).

49. A. Fischli and P. M. Müller, *Helv. Chim. Acta* **63**, 529, 1619 (1980).

50. Food and Nutrition Board, National Research Council, *Food Chemicals Codex*, 3rd ed., National Academy Press, Washington, D.C., 1981, p. 343.

51. *The United States Pharmacopeia XXIII (USP XXIII–NF XVIII)*, United States Pharmacopeial Convention, Inc., Rockville, Md., 1995, pp. 1719–1721.

52. H. B. Chin, J. Augustin, B. P. Klein, D. A. Becker, and P. B. Venugopal, eds., *Methods of Vitamin Assay*, 4th ed., John Wiley & Sons, Inc., New York, 1985, Chapt. 19.

53. H. L. Rosenthal, in W. H. Sebrell, Jr. and R. S. Harris, eds., *The Vitamins*, 2nd ed., Vol. II, Academic Press, Inc., New York, 1968, pp. 145–170.

54. H. R. Skeggs, in P. György and W. N. Pearson, eds., *The Vitamins*, 2nd ed., Vol. VII, Academic Press, Inc., New York, 1968, pp. 277–293; H. Baker and O. Frank *ibid.*, pp. 293–301.

55. K. Helrich, ed., *Official Methods of Analysis of the Association of Official Analytical Chemists*, 15th ed., Association of Official Analytical Chemists, Inc., Arlington, Va., 1990, pp. 952.20, 986.23.

56. Analytical Methods Committee, *Analyst* **81**, 132 (1956).

57. H. L. Newmark, J. Scheiner, M. Marcus, and M. Prabhudesai, *Am. J. Clin. Nutr.* **29**, 645 (1976).

58. P. J. Casey, K. R. Speckman, F. J. Ebert, and W. E. Hobbs, *J. Assoc. Off. Anal. Chem.* **65**, 85 (1982).

59. J. Lindemans and J. Abels, in A. P. De Leenheer, W. E. Lambert, and M. G. M. De Ruyter, eds., *Modern Chromatographic Analysis of the Vitamins*, Marcel Dekker, New York, 1985, Chapt. 12.

60. T. S. Hudson, S. Subramanian, and R. J. Allen, *J. Assoc. Off. Anal. Chem.* **67**, 996 (1984).

61. R. B. Woodward, *Pure Appl. Chem.* **17**, 519 (1968).

62. Ref. 61, **25**, 283 (1971).

63. Ref. 61, **33**, 145 (1973).

64. A. Eschenmoser, *Q. Rev. Chem. Soc. London* **24**, 366 (1970).

65. A. Eschenmoser, *Pure Appl. Chem. Suppl.* **2**, 69 (1971).

66. A. Eschenmoser, *Naturwissenschaften* **61**, 513 (1974).

67. A. Eschenmoser, *Chem. Soc. Rev.* **5**, 377 (1976).

68. A. Eschenmoser and C. E. Wintner, *Science* **196**, 1410 (1977).

69. A. Eschenmoser, in Ref. 43, p. 89.

70. J. H. Kreiger, *Chem. Eng. News*, 16 (1973).

71. T. H. Maugh, *Science* **179**, 266 (1973).

72. A. H. Jackson and K. M. Smith, in J. ApSimon, ed., *The Total Synthesis of Natural Products*, Vol. 1, John Wiley & Sons, Inc., New York, 1983, p. 232; *Ibid.*, Vol. 6, 1984, p. 258.

73. N. Anand, J. Bindra, and S. Ranganathan, *Art in Organic Synthesis*, 2nd ed., John Wiley & Sons, Inc., New York, 1988, p. 375.

74. R. B. Woodward, *Chem. Soc. Spec. Publ.* **21**, 217 (1967).

75. W. Friedrich, G. Gross, K. Bernhauer, and P. Zeller, *Helv. Chim. Acta* **43**, 704 (1960).

76. E. L. Smith, L. Mervyn, A. W. Johnson, and N. Shaw, *Nature (London)* **194**, 1175 (1962).

77. A. W. Johnson, L. Mervyn, N. Shaw, and E. L. Smith, *J. Chem. Soc.*, 4146 (1963).

78. K. Bernhauer and O. Müller, *Biochem., Z.* **336**, 102 (1962).

79. D. Autissier, P. Barthelemy, and L. Penasse, *Bull. Soc. Chim. Fr. Pt. II*, 192 (1980).

80. A. R. Battersby, *Acc. Chem. Res.* **26**, 15 (1993).

81. A. R. Battersby, *Pure Appl. Chem.* **65**, 1113 (1993).

82. A. R. Battersby, *Science* **264**, 1551 (1994).

83. F. Blanche et al., *Angew. Chem. Int. Ed. Engl.* **34**, 383 (1995).

84. A. I. Scott, *Angew. Chem. Int. Ed. Eng.* **32**, 1223 (1993).

85. A. I. Scott, *Tetrahedron* **50**, 13315 (1994).

86. J. Crouzet, B. Cameron, F. Blanche, D. Thibaut, L. Debussche, in R. H. Baltz, G. D. Hegeman, and P. L. Skatrud, eds., *Industrial Microorganisms: Basic and Applied Molecular Genetics*, American Society for Microbiology, Washington, D.C., 1993, p. 195.

87. U.S. Pat. 2,505,053 (Apr. 25, 1950), F. A. Kuehl and L. Chaiet (to Merck & Co., Inc.).

88. U.S. Pat. 2,626,888 (Jan. 27, 1953), S. Kutosh, G. B. Hughey, and R. Malcolmson (to Merck & Co., Inc.).

89. U.S. Pat. 2,628,186 (Feb. 10, 1953), W. Shive (to Research Corp.).

90. Ger. Pat. 953,643 (June 14, 1952), H. M. Shafer and A. J. Holland (to Merck & Co., Inc.).

91. H. Vogelmann and F. Wagner, *J. Chromatogr.* **76**, 359 (1973).

92. Ger. Pat. 1,037,066 (Feb. 12, 1959), K. Bernhauer and W. Friedrich (to Aschaffenburger Zellstoffwerke).

93. U.S. Pat. 2,563,794 (Aug. 7, 1951), E. L. Rickes and T. R. Wood (to Merck & Co., Inc.).

94. U.S. Pat. 2,582,589 (Jan. 15, 1952), H. H. Friecke (to Abbott Laboratories).

95. *Chem. Mark. Rep.* (Feb. 5, 1996).

JOHN W. SCOTT
Hoffmann-La Roche, Inc.

VITAMINS: VITAMIN D

Vitamin D [1406-12-2] is a material which is formed in the skin of animals upon irradiation by sunlight and serves as a precursor for metabolites which control the animal's calcium homeostasis and act in other hormonal functions. A deficiency of vitamin D can cause rickets, as well as other disease states. This tendency can be a problem wherever animals, including humans, especially infants and children, receive an inadequate amount of sunshine. The latter phenomenon became prevalent with the advent of the industrial revolution, and efforts to cure rickets resulted in the development of commercial sources of vitamin D for supplementation of the diet of livestock, pets, and humans.

Research conducted during and subsequent to the 1970s revealed that vitamin D is better defined as those natural or synthetic substances that are converted by animals into metabolites that control calcium and phosphorus homeostasis and act in a variety of other hormonal-like functions.

Vitamin D_2 and vitamin D_3 are the two economically important forms of vitamin D. The other D vitamins have relatively little biological activity and are only of historical interest. Vitamin D_2 (ergocalciferol; ercalciol), (5Z,7E,22E)-(3S)-9,10-*seco*-5,7,10(19)-22-ergostatraene-3-ol (**2**), is active in humans and other mammals, although recently (ca 1997) it has been shown to be less active than vitamin D_3 in cattle, swine, and horses. It is relatively inactive in poultry. It is prepared by the uv irradiation of ergosterol (provitamin D_2), (24-methylcholesta-5,7,22-triene-3B-ol) (**1**), a plant sterol.

Vitamin D_3 (cholecalciferol; calciol), (5Z,7E)-(3S)-9,10-seco-5,7-10(19) cholestatriene-3-ol (**4**), is the naturally occurring active material found in all animals. It is produced in the skin by the irradiation of stored 7-dehydrocholesterol (provitamin D_3), cholesta-5,7-diene-3B-ol (**3**).

NOMENCLATURE

The Vitamin D compounds are steroidal materials and thus are named according to the IUPAC-IUB rules for nomenclature (1) (Table 1). Vitamin D_1 [520-91-2] is a mixture of vitamin D_2 and lumisterol.

The common name vitamin D is used throughout the pharmaceutical industry for simplicity. The trivial name calciferol has also been used extensively with the prefix ergo- and chole-, which indicate vitamin D_2 (**2**) and vitamin D_3 (**4**), respectively. Vitamin D_2 was originally named calciferol in 1931 by Angus and his co-workers (2). Historically, a number of substances were referred to as vitamin D and were distinguished from one another by a subscript numeral, eg, vitamin D_2, vitamin D_3, etc.

Table 1. Vitamin D Substances

Common name	CAS registry number	Provitamin	Trivial name	IUPAC-IUB name
Vitamin D$_2$	[50-14-6]	Ergosterol	Ergocalciferol	9,10-seco-5,7,10(19),22-Ergostatetraen-3β-ol
Vitamin D$_3$	[67-97-0]	7-Dehydrocholesterol	Cholecalciferol	9,10-seco-5,7,10(19),Cholestatrien-3β-ol
Vitamin D$_4$	[511-28-4]	22,23-Dihydroergosterol		24-Methyl-9,10-seco-5,7,10(19)-cholestatrien-3β-ol
Vitamin D$_5$	[71761-06-3]	7-Dehydrositosterol	Sitocalciferol	2,4-Ethyl-9,10-seco-5,7,10(19)-cholestatrien-3β-ol
Vitamin D$_6$	[481-19-6]	7-Dehydrostigmasterol		24-Ethyl-9,10-seco-5,7,10(19)-22-ergo-statetraen-3β-ol
Vitamin D$_7$	[20304-51-2]	7-Dehydrocampesterol		2,4-Methyl-9,10-seco-5,7,10(19)-cholestatrien-3β-ol

The vitamins D are 9,10-secosteroids, that is, steroid molecules with an opened 9,10 bond of the B-ring. The relationship between the provitamin steroid (perhydro-1,2-cyclopentanophenanthrene ring system) and the 9,10-secosteroid nucleus is shown in structures (5) and (6), cholestane and 9,10-secocholestane (calcitane), respectively.

(5)   (6)

In 1981, the IUPAC-IUB Joint Commission on Biochemical Nomenclature proposed that there be a set of trivial names for the important vitamin D compounds, including calciol [67-97-0] for vitamin D, calcidiol [19356-17-3] for 25-hydroxy-vitamin D$_3$, and calcitriol [32222-06-3] for 1α,25-dihydroxy-vitamin D$_3$. This nomenclature has met with varying degrees of acceptance, as has the proposal to use calcine [69662-75-5] (deoxy-vitamin D$_2$) and ercalcine [68323-40-0] (deoxy-vitamin D$_3$) to name the triene hydrocarbon structure for 9,10-seco-cholesta-5,7,10(19)-triene and 9,10-seco-ergosta-5,7,10(19),22-tetraene, respectively. In systematic nomenclature, calcitane can be used for the basic 27-carbon skeleton instead of 9,10-secocholestane.

ISOLATION AND STRUCTURE DETERMINATION

The isolation and structure elucidation of vitamin D is closely related to the efforts to understand and cure rickets and related bone diseases. The advent of the use of soft coal, the migration of people to cities, and the tendency of people and animals to spend less time in sunshine caused a decline in the ability of populations to synthesize sufficient quantities of vitamin D$_3$. This led to the increased incidence of rickets, beginning around the mid-1600s (3).

The bone disease rickets in children, and a similar condition in adult known as osteomalacia, is characterized by the body's inability to calcify the collagen matrix of growing bone, resulting in wide epiphyseal plates and large areas of uncalcified bone called osteoid. The resultant lack of rigidity of bones leads to the ends becoming twisted and bent, particularly in long bones. The ribs develop a bumpy and uneven texture known as rosary ribs and the legs become bowed. Also, the cranium becomes soft and misshapen. In adults, no long-bone growth occurs, but new bone which is being continually remodeled, activates cells to resorb bone, followed by osteoblast-mediated bone-growth replacement (4,5). Metabolites of vitamin D act in the body to modulate these activities and maintain strong bone structure.

Although there is evidence that rickets was manifest in humans as early as 800 B.C., it was not until 1645 that it was first described (6). The progress of finding a cure for rickets was relatively slow until the late 1800s, when a sufficient number of scientific developments began to allow workers to unravel the difficult puzzle of vitamin D metabolism. Tarret in 1889 had isolated ergosterol from ergot of rye and demonstrated that it was different from cholesterol. The similarity of the structures, however, led to difficulty in the structural elucidation of the vitamin D molecule. Mellanby (7) demonstrated the lack of a dietary component could be used to develop rickets and was able to raise D-deficient animals. This allowed research aimed at finding the antirachitic factor to be carried out.

In 1919, Huldschinski (8) realized that uv light cured rickets and impacted on its etiology. The uv light and cod liver oil were found to be useful in the treatment of the disease, and irradiation of food produced the same effect as irradiation of the animal. The link between irradiation and plant materials led to the conclusion that ergosterol was an antirachitic substance, and an extensive effort was made to characterize the chemistry of irradiated ergosterol.

Windaus in 1933 derived the skeletal structure of vitamin D$_2$. He found the elemental formula to be $C_{28}H_{44}O$. The side chain was unequivocally characterized by x-ray crystallography later, in 1948. Reindel and Kipphan (9) in 1932 ozonized the side chain and elucidated the C_{22} double bond. The reaction to form a Diels Alder adduct with maleic anhydride and the typical diene uv spectrum led to the conclusion there was a diene in the B ring. In 1934 Fernholz and Chakravorty (10) determined the provitamin had a 3β-hydroxy group because the molecule could be precipitated with digitonin, a reaction characteristic of the 3β-ol functionality in steroids.

Perbenzoic acid gave a doubly unsaturated triol monobenzoate. Only two hydroxyl groups could be acetylated, and one was tertiary. The saturated triol reacted with lead tetracetate to give an α glycol. When reacted with chromic acid, it gave a hydroxy lactone. From these observations,

Windaus and Grundmann (11) described the correct structure for ergosterol (1).

The observation that the uv spectrum of provitamin D changed with uv irradiation and also produced antirachitic activity led to the conclusion that vitamin D was derived from the provitamin. Windaus found the vitamin D_2 formula $C_{28}H_{44}O$ to be isomeric with the provitamins.

Bunal in 1932 used x-ray crystallography to demonstrate that many of the preparations described in the literature were mixtures and that vitamin D_2 was one crystal structure.

Calciferol, when hydrogenated catalytically, took up 4 moles of hydrogen and gave a compound with the empirical formula $C_{28}H_{52}O$. Sodium in ethanol reduction gave a dihydroproduct which reacted with 3 moles of perbenzoic acid, thus demonstrating the derivative to have three double bonds.

Ozonolysis of vitamin D_2 gave 2,3-dimethylbutanol, showing the side chain to contain the 22 double bond and a C_{24} methyl group.

Vitamin D_2 reacted with maleic anhydride to give a mono Diels-Alder adduct, which hydrolyzed to yield a dicarboxylic acid. Acetylation of the alcohols, esterification of carboxylic acids, and hydrogenation gave a compound which, when ozonized, gave a saturated ketone, $C_{19}H_{34}O$. This molecule contains the C and D ring and side chain from vitamin D_2, indicating that the B ring must have been opened by photolytic cleavage of the C_9–C_{10} bond (12). The vitamin D_2 molecule was thus shown to be a tricyclic material with four double bonds.

Windaus and Boch (13) isolated and characterized 7-dehydrocholesterol in 1937 from pig skin. They showed, further, that vitamin D_3 could be generated from the provitamin by uv irradiation.

Irradiated ergosterol was found not to be as antirachitic in the chick as in the rat, whereas the chick could be protected by direct irradiation. The provitamin in cholesterol was shown not to be ergosterol. Rygh (14) in 1935 found that 1 rat unit of cod liver oil was 100 times more potent in chicks than 1 rat unit of vitamin D_2. Brockmann (15) in 1936, prepared the pure crystalline 3,5-dinitrobenzoate derivative of vitamin D_3 obtained from tuna liver oil (subsequently demonstrated in other species like halibut and blue fin tuna). This material was shown to be identical to the dinitrobenzoate derivative of vitamin D_3 obtained by Windaus from the irradiation of 7-dehydrocholesterol. Thus, the chemical identity of vitamin D_2 and D_3 were well established. A more detailed description of the events leading to the structure elucidation of the vitamin Ds can be found in Reference 6.

OCCURRENCE

The provitamins, precursors of the vitamin Ds, are distributed widely in nature, whereas the vitamins themselves are less prevalent. The amounts of provitamins D_2 and D_3 in various plants and animals are listed in Table 2.

Fish-liver oil, liver, milk, and eggs are good natural sources of the D_3 vitamin. Most milk sold in the United States is fortified with manufactured vitamin D. Fish oil is the only commercial source of natural vitamin D_3, and

the content of the vitamin varies according to species as well as geographically, ie, Atlantic cod contain 100 IU/g where IU (International Unit) = 0.025 μg of vitamin D_3, whereas oriental tuna (*Percomorpli*) contain 45,000 IU/g of oil.

Vitamin D_3 rarely occurs in plants. However, *Solanum glaucophyllum*, *Solanum malacoxylon*, *Cestrum diurnum*, and *Trinetum flavescens* have been shown to contain water-soluble glycosides of vitamin D analogues with $1\alpha,25$-dihydroxy-vitamin D activity (16–22). The vitamin D content in various plant and animal materials is shown in Table 3. Vitamin D_3 occurs naturally in all animals (24).

CHEMICAL AND PHYSICAL PROPERTIES

Provitamin

The chemistry of the D vitamins is intimately involved with that of their precursors, the provitamins. The manufacture of the vitamins and their derivatives usually involves the synthesis of the provitamins, from which the vitamin is then generated by uv irradiation. The chemical and physical properties of the provitamins are discussed below, followed by the properties of the vitamins.

3β-Hydroxy steroids which contain the 5,7-diene system and can be activated with uv light to produce vitamin D compounds are called provitamins. The two most important provitamins are ergosterol (1) and 7-dehydrocholesterol (3). They are produced in plants and animals, respectively, and 7-dehydrocholesterol is produced synthetically on a commercial scale. Small amounts of hydroxylated derivatives of the provitamins have been synthesized in efforts to prepare the metabolites of vitamin D, but these products do not occur naturally. The provitamins do not possess physiological activities, with the exception that provitamin D_3 is found in the skin of animals and acts as a precursor to vitamin D_3, and synthetic dihydroxalated analogues of pro- and previtamins have been found to have selective activity towards nuclear and nonnuclear receptors (24–27).

Provitamin D_2

Ergosterol is isolated exclusively from plant sources. The commercial product is ca 90–100% pure and often contains up to 5 wt % of 5,6-dihydroergosterol. Usually, the isolation of provitamin D_2 from natural sources involves the isolation of the total sterol content, followed by the separation of the provitamin from the other sterols. The isolation of the sterol fraction involves extraction of the total fat component, its saponification, and then reextraction of the unsaponifiable portion with an ether. The sterols are in the unsaponifiable portion. Another method is the saponification of the total material, followed by isolation of the nonsaponifiable fraction. Separation of the sterols from the unsaponifiable fraction is done by crystallization from a suitable solvent, eg, acetone or alcohol. Ethylene dichloride, alone or mixed with methanol, has been used commercially for recrystallization. In the case of yeasts, it is particularly difficult to remove the ergosterol by simple extraction, thereby obtaining only ca 25% recovery. Industri-

Table 2. Occurrence of the Provitamins D in Selected Plants and Animals, Parts per Thousand of Total Sterol

Source	Amount	Source	Amount
Cottonseed oil	28	Gallstones, man	0.25
Rye grass	15	*Mytilus edulis*, sea mussel	100
Wheat germ oil	10	*Modiolus demissus*, ribbed mussel	370
Carrot	1.7	*Ostrea virginica*, oyster	80
Cabbage	0.5	*Ostrea edulis*, oyster	34
Skin, pig	46	*Asterias rubens*, starfish	3.8
Skin, chicken feet	25	Common sponges	20
Skin, rat	19	Common coral	10
Skin, mouse	9	*Aspergillus niger*, mold	1,000
Skin, calf	7	*Cortinellus shiitake*, mushroom	1,000
Skin, human infant	1.5	*Claviceps purpurea*, ergot	9,000
Skin, human adult	4.2	*Saccharomyces cerevisiae*, yeast	800
Liver, Japanese tuna	11	*Penicillium puberculum*, mold	1,000
Liver, Atlantic cod	4.4	*Fucus vesiculosus*, alga, seaweed	0.8
Liver, shark	1.0	*Tubifex* sp, waterworm	210
Liver, halibut	0.6	*Bumbricus terrestris*, earthworm	170
Liver, tuna	1.0	*Tenebrio molitor*, mealworm	120
Eggs, Chinese duck	60	*Gyronomus*, sp, goat	61
Eggs, cod (roe)	5.5	*Cancer pagurus*, common crab	15
Eggs, hen	1.6	*Dephnia* sp, water flea	7.5
Wool fat, sheep	3.9	*Musca domestica*, housefly	7.0
Milk, cow	2.3	*Crangon vulgaris*, shrimp	3.8
Pancreas, beef	1.8	*Homarus vulgaris*, lobster	2.5
Spinal cord, beef	1.2	*Helix pomatia*, edible snail	97
Blood serum, cow	0.5	*Arion empiricorum*, slug, red road snail	220
Herring oil	0.5	*Littorina littorea*, periwinkle	170
Heart, calf	0.32	*Sepia* sp, cuttlefish	12

Table 3. Distribution of Vitamin D Activity

Sample	Amount[a]
Phytoplankton	0
Sargassum (a gulfweed)	Some activity
Clover hay	
Sun-cured	Slight
Dark-cured	0
Mushrooms (*Agaricus campestris*)	0.21 IU/g
Milk (unfortified)	
Winter (bovine)	5.3 IU/L
Summer (bovine)	53 IU/L
Milk (human)	63 IU/L
Milk colostrum (human)	315–635 IU/L
Egg yolk	150–400 IU/g
Butter	4–8 IU/g
Fish-liver oils	50–45,000 IU/g

Source: Ref. 23.

[a]IU (International Unit) = 0.025 μg of vitamin D_3.

ally, therefore, the ergosterol is obtained by preliminary digestion with hot alkalies or with amines (28–33). Variations of the isolation procedure have been developed. For example, after saponification, the fatty acids may be precipitated as calcium salts, which tend to absorb the sterols. The latter are then recovered from the dried precipitate by solvent extraction.

Provitamin D_3

Provitamin D_3 is made from cholesterol, and its commercial production begins with the isolation of cholesterol from

one of its natural sources. Cholesterol occurs in many animals, and is generally extracted from wool grease obtained by washing wool after it is sheared from sheep. This grease is a mixture of fatty-acid esters, which contain ca 15 wt % cholesterol. The alcohol fraction is obtained after saponification, and the cholesterol is separated, usually by complexation with zinc chloride, followed by decomplexation and crystallization. Cholesterol can also be extracted from the spinal cords and brains of animals, especially cattle, and from fish oils.

Cholesterol (**7**) is converted to 7-dehydrocholesterol (**3**) (see Fig. 1). This process usually involves the Ziegler allylic bromination of the 7 position followed by dehydrobromination (34). Esterification of the cholesterol (**8**) is necessary to prevent oxidation of the 3β-alcohol by the brominating agent. Allylic bromination may be accomplished with a variety of brominating agents, eg, *N*-bromosuccinimide, *N*-bromophthalimide, or preferably 5,5-dimethyl-1,3-dibromohydantoin (35). Bromine in carbon disulfide can be used if the free-radical bromination is photocatalyzed (36). A mixture of 7α-(**9**) and 7β-bromo cholesteryl esters (**10**) is obtained and treated with an appropriate base to dehydrohalogenate the molecule and give the 7-dehydrocholesteryl ester (**11**) (37,38). Proper conditions for this reaction are necessary to generate a high yield of the desired 7-dehydro product instead of the undesired cholesta-4,6-dien-3β-ol ester (**12**), which is formed as a by-product. Various reagents can be used to perform the dehydrohalogenation. Trimethyl phosphite or pyridine bases, particularly trimethylpyridine, have been used; the symmetrical collidine is the reagent of choice (35,38,39). *t*-Butylammonium fluo-

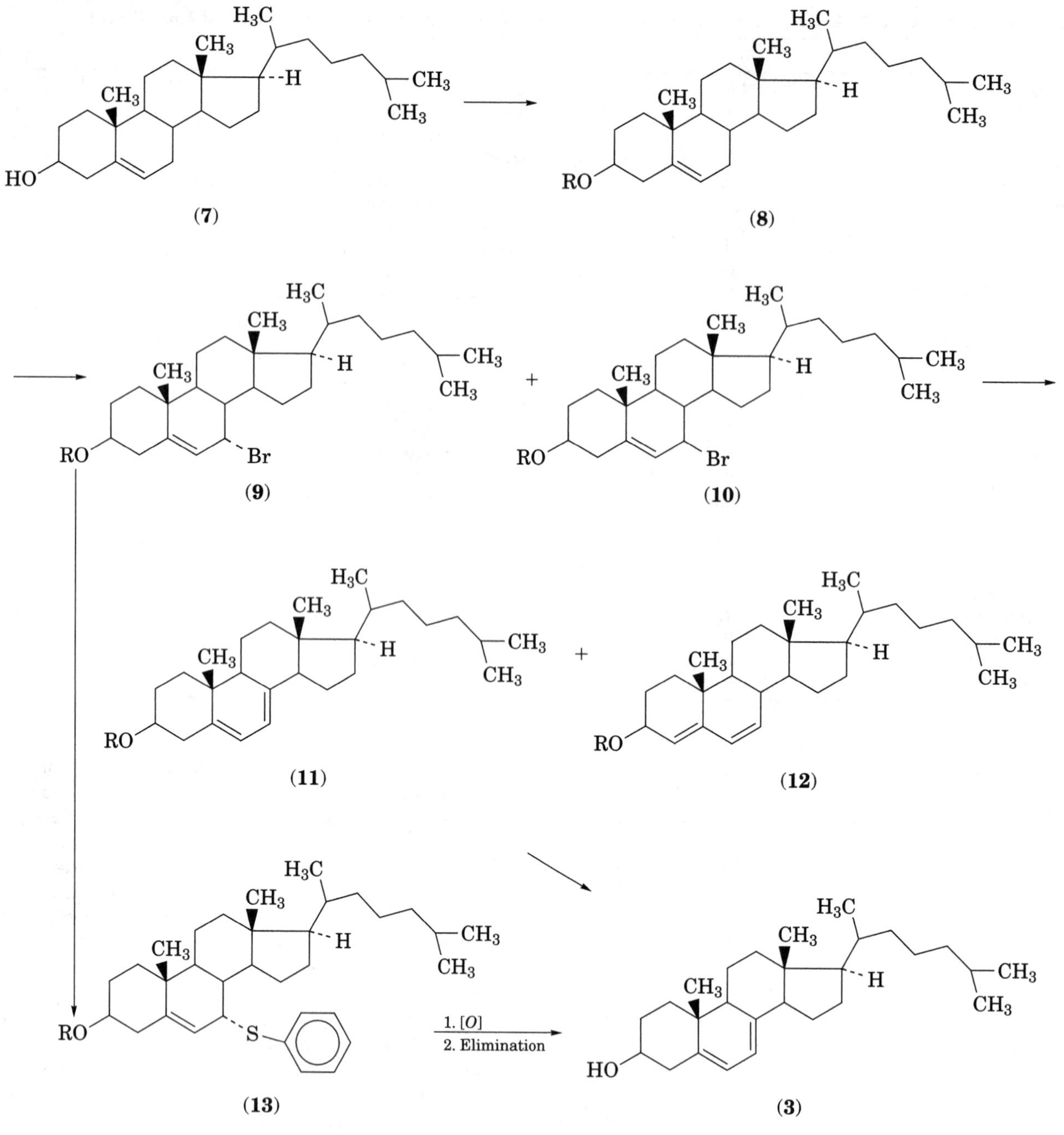

Figure 1. Conversion of cholesterol to 7-dehydrocholesterol.

ride has also been used to improve the yield of high quality 5,7-diene (40).

The 7α-bromo steroid (**9**) can also be treated with sodium phenyl selenolate (41). The resultant 7β-phenyl selenide (**13**) can be oxidized and the corresponding phenyl selenoxide eliminated to form the 7-dehydrocholesteryl ester (**11**).

7-Dehydrocholesterol has also been made from cholesterol by the Windaus procedure (Fig. 2); the 3,7-dibenzoate (**16**) is obtained (via (**14**) and (**15**) by oxidation and reduction), which undergoes thermal elimination to give the 7-dehydrocholesteryl benzoate (**11**) (42–44). However, the

yields are substantially lower than those achieved by the bromination–dehydrobromination method.

7-Tosylhydrazone and 7-phenyl sulfoxide groups have also been introduced into cholesterol and eliminated to prepare the 5,7-diene (45,46). The method of choice is the allylic bromination–dehydrobromination procedures, and the commercial yields in converting cholesterol to 7-dehydrocholesterol are in the range of 35–50%.

Vitamin D

The irradiation of the provitamins to produce vitamin D as well as several isomeric substances was first studied with

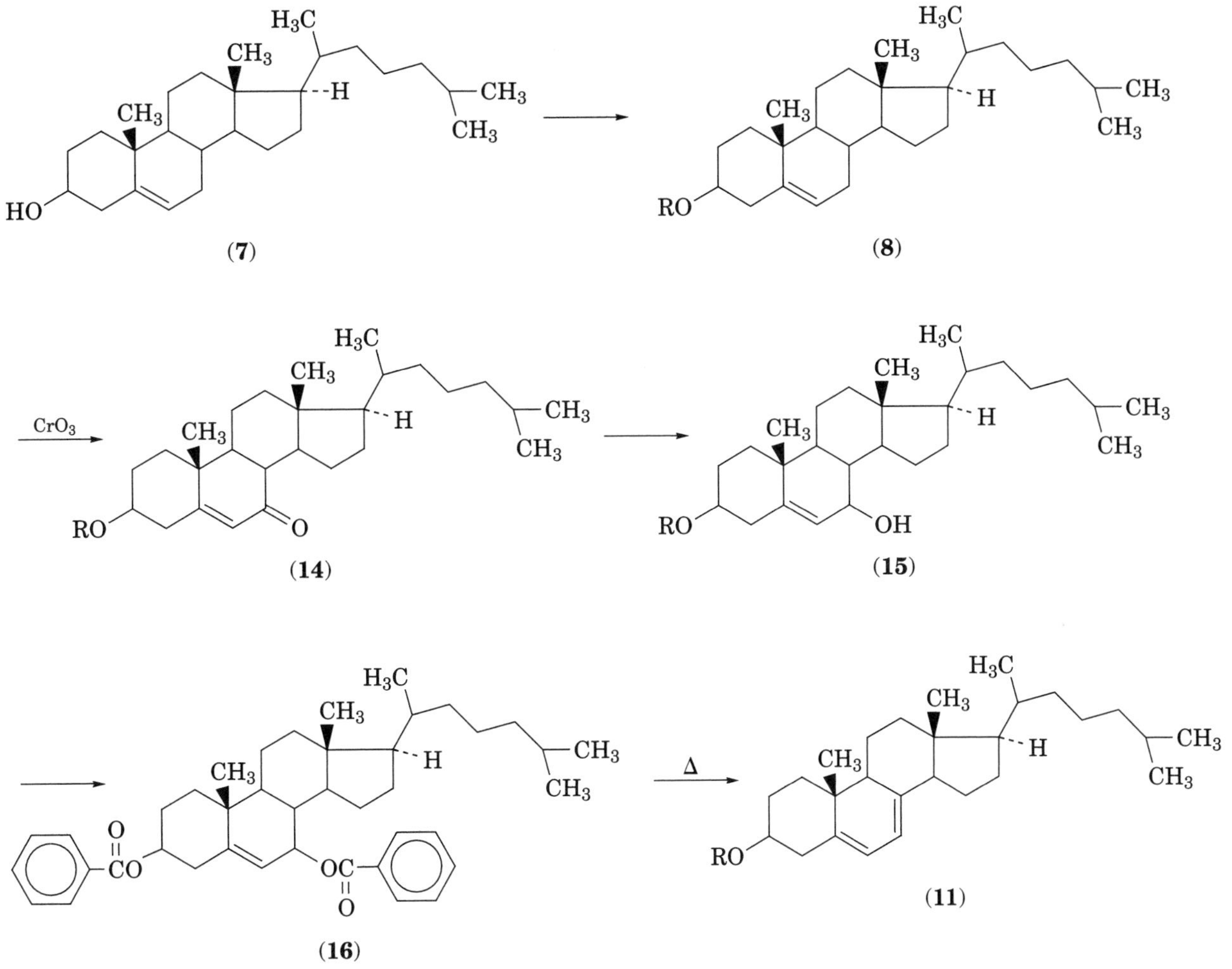

Figure 2. The Windaus procedure. *Source:* Ref. 42.

ergosterol. The chemistry is identical for the vitamin D_3 series and yields analogous isomers. In 1932, a scheme for the irradiation of ergosterol leading to vitamin D_2 was proposed (45). Twenty years later, the mechanism of the irradiation of the provitamins to vitamin D and its photoisomers was further elucidated (46,47). More recently, Jacobs (48) has reviewed the photochemistry. A number of products associated with the irradiation process are shown in Figure 3. The geometry and electronic characteristics of these molecules have been well established by x-ray crystallographic analysis and valence force-field calculations (50–52). The irradiation process, which converts 7-dehydrocholesterol to vitamin D also occurs in the skin of animals if sufficient sunlight is available. The photochemical and thermal isomerizations occur during the generation of vitamin D *in vivo* as well as during its synthetic photochemical preparation (24,53,54). The initial step involves ring opening of the B-ring of the sterol by ultraviolet activation of the conjugated diene. The absorbance of uv energy activates the molecule, and the $\pi \rightarrow \pi^*$ excitation (absorption, 250–310 nm; $\lambda_{max} = 291$ nm, $\epsilon = 12,000$) results in the opening of the 9,10 bond and the formation of the (Z)-hexadiene, previtamin D_3 (R) [1173-13-3] (17) or

previtamin D_2 (R′) [21307-015-1] (18). The uv irradiation of 7-dehydrocholesterol or ergosterol results in the steady diminution in concentration of the provitamin, initially giving rise to predominantly previtamin D. The pre- levels reach a maximum as the provitamin level drops below ca 10%. The concentration of the previtamin then falls as it is converted to tachysterol and lumisterol, which increase in concentration with continued irradiation (see Fig. 4). Temperature, frequency of light, time of irradiation, and concentration of substrate all affect the ratio of products. Previtamin D undergoes thermal equilibration to vitamin D (2) or (4), *cis*-vitamin D_2 [50-14-6] and *cis*-vitamin D_3 [67-97-0], respectively. The conversion of previtamin D (18) or (17) at temperatures of $\leq 80°C$ by thermal isomerization to give the cis vitamin (ergocalciferol) (2) or cholecalciferol (4) involves an equilibrium, as indicated in Table 4 (55,56).

The equilibrium composition is normally ca 80% vitamin D and 20% previtamin D. This reaction is an intramolecular {1–7} H sigmatropic shift and occurs through a rigid cyclic transition state (57).

Additionally, the $\pi \rightarrow \pi^*$ excitation of previtamin D can result in ring closure back to the provitamin (1) [57-87-4] or (3) [434-16-2] or to lumisterol$_2$ [474-69-1] (19) or lumi-

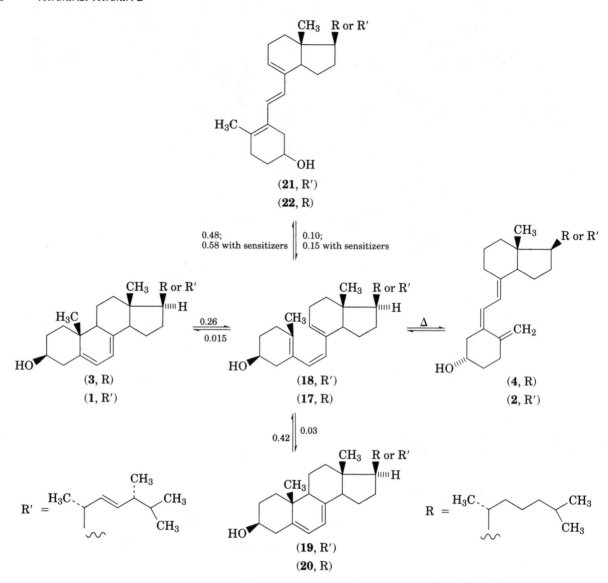

Figure 3. Photochemical and thermal isomerization products of vitamin D manufacture (49). The quantum yields of the reactions are listed beside the arrows for the given reactions.

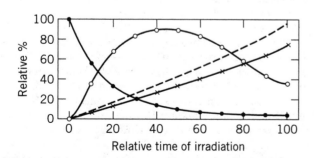

Figure 4. Time course for uv-irradiation of 7-dehydrocholesterol (●): (○), previtamin D; (×), lumisterol; (—), tachysterol.

sterol$_3$ [5226-01-7] (**20**), which have the 9β,10α configuration. This excitation can also exhibit (Z) ⇌ (E) photoisomerization to the 6,7-(E)-isomer, tachysterol$_2$ [115-61-7] (**21**) or tachysterol$_3$ [63902-44-3] (**22**) (58,52).

Other photoinduced cyclization reactions can occur by conrotatory bond formation to give the 9β,10β-antiisomers, isopyrocalciferol$_2$ [474-70-4] (**23**) or isopyrocalciferol$_3$ [10346-44-8] (**24**) (Fig. 5), whereas thermal cyclization at >100°C leads to the two 9,10-syn isomers, (9α,10α)-pyrocalciferol (**27**) [128-27-8] or (**28**) [10346-43-7] by a disrotatory bond formation mechanism (47). Ultraviolet overirradiation leads to photopyro- (**29**) [41411-05-6] or (30) and photoisopyrocalciferols (**25**) [26241-65-6] or (**26**) [85354-28-5], respectively and to the formation of suprasterols of the type shown in structure (**37**) (59,60) (Fig. 6). Prolonged irradiation of the mixture of isomers can also lead to toxisterols of the type shown in structures (**38**) and (**39**), where R = D$_2$ and R' = D$_3$ side chains (see Fig. 3)

Table 4. Inverconversion Time of Previtamin D₃ and Vitamin D₃ at 20°C, and 40°C

Formation (%)	Vitamin D₃ from previtamin D₃			Previtamin D₃ from vitamin D₃	
	Days at −20°C	Days at 20°C	Hours at 40°C	Days at 20°C	Hours at 40°C
2	27	0.2	0.4	2.2	3.7
5	68	0.4	1.1	8.2	11.2
7	96	0.5	1.5		18.6
10	140	0.8	2.3		43.3
20	269	1.6	4.8		
30	474	2.6	7.8		
40	681	3.8	11.3		
50	926	5.2	15.6		
60	1230	7.0	21.2		
70	1628	9.5	29.2		
80	2204	13.3	43.4		
90	2910	23.3			

Source: Ref. 55.

(61–63). More than 20 members of this type of substance have been identified. There is little evidence that these materials are toxic; however, their nomenclature leads to some misunderstanding. These compounds generally show little if any biological activity and have not been found *in vivo* (15,63–66). Normal irradiation conditions for the production of vitamin D include sufficiently low temperatures so that these products do not form, and they are not found in normal commercial samples of D₃ resins.

The irradiation of calciferol in the presence of iodine leads to the formation of 5,6-*trans*- vitamin D₂ [14449-19-5] (**31**) or D₃ [22350-41-0] (**32**) (67,68). 5,6-*trans*-Vitamin D as well as vitamin D (**2**) or (**4**) can be converted to isovitamin D by treatment with mineral or Lewis acids. Isocalciferol (**35**) [469-05-6] or (**36**) [42607-12-5] also forms upon heating of 5,6-*trans*-vitamin D. Isotachysterol (**33**) [469-06-7] or (**34**) [22350-43-2] forms from isocalciferol or vitamin D upon treatment with acid, and its production appears to be the result of sequential formation of *trans*- and isocalciferol from calciferol. These reactions are the basis of the antimony trichloride test for vitamin D (69–72).

Commercially, the irradiation of the 5,7-diene provitamin to make vitamin D must be performed under conditions that optimize the production of the previtamin while avoiding the development of the unwanted isomers. The optimization is achieved by controlling the extent of irradiation, as well as the wavelength of the light source. The best frequency for the irradiation to form previtamin is 295 nm (64–66). The unwanted conversion of previtamin to tachysterol is favored when 254 nm light is used. Sensitized irradiation, eg, with fluorenone, has been used to favor the reverse, triplet-state conversion of tachysterol to previtamin D (73,74).

The molecular extinction coefficients (at various wavelengths) of the four main components of the irradiation are shown in Table 5. The absorption of light above 300 nm is favored by tachysterol. A yield of 83% of the previtamin at 95% conversion of 7-dehydrocholesterol can be obtained by

irradiation first at 254 nm, followed by reirradiation at 350 nm with a yttrium aluminum garnet (YAG) laser to convert tachysterol to previtamin D. A similar approach with laser irradiation at 248 nm (KrF) and 337 nm (N₂) has also been described (76).

The irradiation of the provitamin has been achieved using the acetate and benzoate esters, although the free alcohol form of the provitamin is usually used (77).

The physical properties of the provitamins and vitamins D₂ and D₃ are listed in Table 6. The values are listed for the pure substances. The D vitamins are fat-soluble and, as such, are hydrophobic.

SHIPPING AND HANDLING

Vitamin D and its products are sensitive to uv light, heat, air, and mineral acids. Its sensitivity to these conditions is exaggerated by the presence of heavy-metal ions, eg, iron. Care should be taken to store and ship vitamin D and its various product forms so that exposure to these conditions is minimized. Pharmaceutical grade vitamin D₃ is supplied in the pure crystalline form. This is used in vitamin D as well as multivitamin preparations. Commercial sources of feed-grade vitamin D are usually the vitamin D₃ resin stabilized by spray- or roll-drying a starch or gelatin suspension of the vitamin. These products should be stored in a cool, dry place in opaque, hermetically-sealed containers under nitrogen. Vitamin D is generally recognized as safe when used in accordance with good manufacturing or feeding practices (79).

Shipping vitamin D in crystalline or resin form should be done in containers marked appropriately to indicate the material is toxic by DOT standards. Its proper DOT labeling is DOT Hazard Class 6.1, poisonous. Waste material should be burned or placed in an appropriate landfill.

The provitamins are also unstable to heat and light. They generally should be stored in a dark, cool, dry place. The provitamin is more stable if shipped with 10–15 wt % methanol rather than in a dry form.

ANALYTICAL AND TEST METHODS

The development of reliable uv analysis permitted the dependable detection and assay of the provitamins and vitamins. Prior to this, the Lieberman-Bouchard chemical test was used, but the color reaction gave many false positives and was relatively inaccurate.

In 1949 the World Health Organization adopted the biological activity of 1 mg of an oil solution containing 0.025 μg of crystalline D₃ as the analytical standard for vitamin D₃. This standard was discontinued in 1972. USP uses crystalline cholecalciferol as a standard (80). Samples of reference standard may be purchased from U.S. Pharmacopeial Convention, Inc., Reference Standards Order Department, 12601, Twinbrook Parkway. Rockville, Maryland 20852. One international unit of vitamin D activity is that activity demonstrated by 0.025 μg of pure crystalline *cis*-vitamin D₃. One gram of vitamin D₃ is equivalent to 40

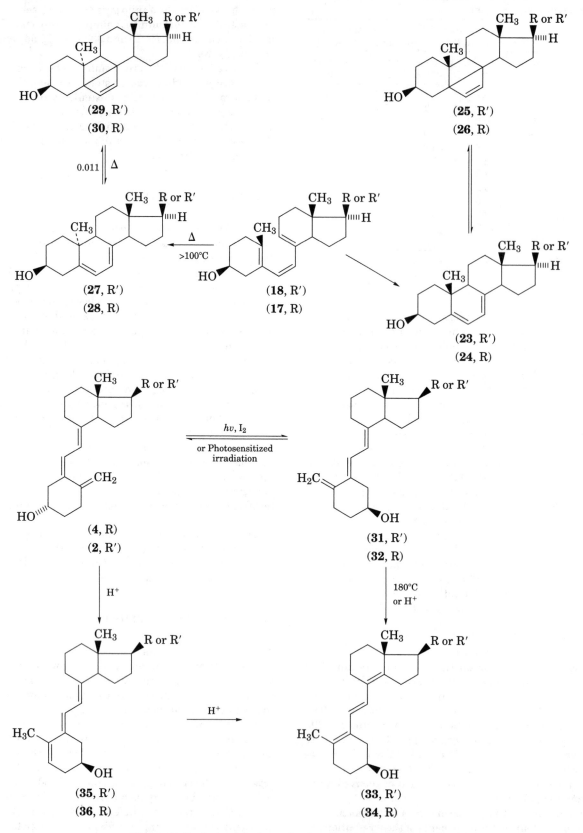

Figure 5. Products of the isomerization of pre- and *cis*-vitamin D.

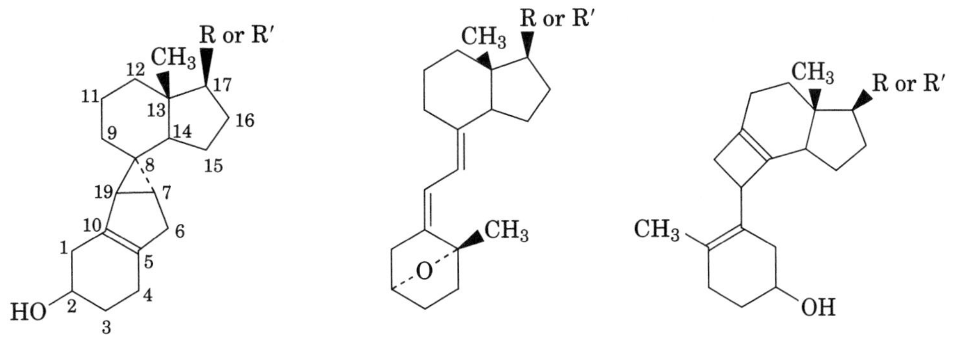

Figure 6. Products of overirradition of vitamin D: suprasterol II (**37**), toxisterol-E (**38**), and toxisterol E_1 (**39**).

Table 5. Molecular Extinction Coefficients of Irradiation Products

λ (nm)	7-Dehydrocholesterol	Previtamin D_3	Tachysterol$_3$	Lumisterol$_3$
254	4,500	725	11,450	4,130
300	1,250	930	11,250	1,320
330	25	105	2,940	30
340	20	40	242	25
350	10	25	100	20

Source: Ref. 75.

$\times$ 10^6 IU or USP units. The international chick unit (ICU) is identical to the USP unit.

USP also issues vitamin D_3 capsules for AOAC determination in rats and an oil solution for the vitamin D_3 AOAC determination in chicks. Historically, the following units (shown with their approximate international unit equivalence) have been used but are currently abandoned: 1 clinical unit = 12–17 IU; 1 biological unit = 0.125 IU; 1 protection unit = 0.125 IU; 1 Laquer unit = 0.14 IU; 1 Poulson unit = 0.2 IU; 1 Steenbach unit = 3 IU. The MRC, ICU, and Coward units all approximated the international unit and are also no longer in common use.

The standard chemical and biological methods of analysis are those accepted by the *United States Pharmacopeia XXIII* as well as the ones accepted by the AOAC in 1995 (81–84). The USP method involves saponification of the sample (dry concentrate, premix, powder, capsule, tablet, or aqueous suspension) with aqueous alcoholic KOH; solvent extraction; solvent removal; chromatographic separation of vitamin D from extraneous ingredients; and colormetric determination with antimony trichloride and comparison with a solution of USP cholecalciferol reference standard.

The AOAC (978.42) recognizes a similar procedure, except that the unsaponifiable material is treated with maleic anhydride to remove the trans-isomer which may possibly be present (83). The antimony trichloride colorimetric assay is performed on the trans-isomer-free material. This procedure cannot be used to distinguish certain inactive isomers, eg, isotachysterol; if present, these are included in the result, giving rise to a falsely high analysis. A test

must therefore be performed to check for the presence of isotachysterol.

Preferably, high pressure liquid chromatography (hplc) is used to separate the active pre- and cis-isomers of vitamin D_3 from other isomers and allows their analysis by comparison with the chromatograph of a sample of pure reference cis-vitamin D_3, which is equilibrated to a mixture of pre- and cis-isomers (82,84,85). This method is more sensitive and provides information on isomer distribution as well as the active pre- and cis-isomer content of a vitamin D sample. It is applicable to most forms of vitamin D, including the more dilute formulations, ie, multivitamin preparations containing at least 1 IU/g (AOAC Methods 979.24; 980.26; 981.17; 982.29; 985.27) (82). The practical problem of isolation of the vitamin material from interfering and extraneous components is the limiting factor in the assay of low level formulations.

A number of methods have been developed for the chromatographic separation of vitamin D and related substances and can be found in Table 7.

Biological Assay

The USP and AOAC recognize a biological method for the determination of vitamin D. The rat line test, however, is slow, expensive and not as accurate as the chemical or chromatographic methods. Rachitic rats are fed diets containing the vitamin D sample. This test measures bone growth on the proximal end of the tibia or distal end of the ulna, which is visualized by staining with silver nitrate. This test is not applicable to products offered for poultry

Table 6. Physical Properties of Provitamins and Vitamins D$_2$ and D$_3$

Properties	7-Dehydrocholesterol	Ergosterol	Vitamin D$_2$ (ergocalciferol)	Vitamin D$_3$ (cholecalciferol)
		Substance		
Melting point, °C	150–151	165	115–118	84–85
Color and form	Solvated plates from ether–methanol	Hydrated plates from alcohol; needles from acetone	Colorless prisms from acetone	Fine colorless needles from dilute acetone
CAS Registry Number	[434-16-2]	[57-87-4]	[50-14-6]	[67-97-0]
Optical rotation (α_D^{20}),°				
Acetone			82.6	83.3
Ethanol			103	105–112
Chloroform	−113.6	−135	52	51.9
Ether			91.2	
Petroleum ether			33.3	
Benzene	−127.1			
Coefficient of rotation per °C in alcohol			0.515	
UV max, nm	282	281.5	264.5	264.5
Specific absorption, E_{max}, at 1% conc	308		458.9 ± 7.5	473.2 ± 7.8
Potency[a] IU/g			40 × 10^6	40 × 10^6
Biological activity			in mammals	in mammals and birds
Chicken efficacy, %			8–10[b]	100
Solubility, g/100 mL				
Acetone at 7°C			7	
Acetone at 26°C			25	
Absolute ethanol	sl sol		28	
At 26°C		0.15		
At Δ°C		2.2		
Ethyl acetate at 26°C			31	
Water	Insol	Insol	Insol	Insol

[a]The international standard for vitamin D is an oil solution of activated 7-dehydrocholesterol (3). The IU is the biological activity of 0.025 μg of pure cholecalciferol.

[b]Studies have claimed an efficacy as high as 10% (77).

feeding. The AOAC recognizes another procedure which measures the vitamin D sample activity in increasing bone ash of growing chicks compared to the activity of a USP cholecalciferol reference standard. It too is slow, expensive, and gives variable results.

Physical Methods

Vitamins D$_2$ and D$_3$ exhibit uv absorption curves that have a maximum at 264 nm and an E_{max} (absorbance) of 450–490 at 1% concentration (Table 8). The various isomers of vitamin D exhibit characteristically different uv absorption curves. Mixtures of the isomers are difficult to distinguish. However, when chromatographically separated by hplc, the peaks can be identified by stop-flow techniques based on uv absorption scanning or by photodiodearray spectroscopy. The combination of elution time and characteristic uv absorption curves can be used to identify the isomers present in a sample of vitamin D.

Infrared and nmr spectroscopy have been used to help distinguish between vitamins D$_2$ and D$_3$ (87–89). X-ray crystallographic techniques are used to determine the vitamin D structure, and gas chromatography also is a method for assaying vitamin D (49,90–95).

Provitamin D

The molecular extinction coefficient of 7-dehydrocholesterol at 282 nm is 11,300 and is used as a measure of 7-

dehydro isomer content of the provitamin (96,97). High pressure liquid chromatography can also be used to analyze the provitamins. There are a variety of chemicals that show characteristic colors when reacted with the provitamins. Some of these are listed in Table 9.

The extremely low levels of vitamin D and its metabolites in biological systems make it very difficult to assay these products by traditional methods. Calcium-binding protein is not found in the intestinal mucosa of vitamin D-deficient animals. It is synthesized only in response to the presence of a material with vitamin D activity. Thus, using antiserum specific to intestinal calcium-binding protein, a radioimmunodiffusion assay (98) conducted on homogenates of intestinal mucosa of chicks fed the test material for 7–10 d allows assay of the material for vitamin D activity down to 1–200 IU. The development of this technique has allowed research to be conducted since the mid-1970s to elucidate the vitamin D metabolism and biochemistry and is now used routinely to assay vitamin D-active materials in clinical as well as research samples (6,40,51,53,55).

SYNTHESIS

Manufacture

Most of the vitamin D produced in the world is made by the photochemical conversion of 7-dehydrocholesterol. Er-

Table 7. Methods for Chromatographic Separation of Vitamin D and Related Steroid

Method	Comments
Paper chromatography	
Quinoline-impregnated paper	
Reversed-phase paper	
Column chromatography	
Alumina	
Floridin	
Celite	Useful for multivitamin tablets
Silicic acid	Useful for metabolites
Factice	Can resolve vitamin D_2 and vitamin D_3
Sephadex LH-20	Highly useful for metabolites
High-pressure liquid chromatography	
Zorbax-Sil support	Separates 24(R), 25-dihydroxy-vitamin D_3 from 24(S), 25-dihydroxy-vitamin D_3
	Useful for metabolites
	Assay of 25-hydroxy-vitamin D_3
ODS[a]-Permaphase	Useful for vitamin D metabolites
Silica gel	Commercial vitamin D_3 assay
Thin-layer chromatography	
Silicic acid	Effective for irradiation mixtures
Silica gel G	Two-dimensional
Gas chromatography	
	Separates trimethylsilyl ethers of vitamin D_3
	Cyclizes vitamin D_3
	Useful for 25-hydroxy-vitamin D_3
	Useful for multivitamin tablets
	Useful for vitamin D_2 in milk

Source: Ref. 86.
[a]ODS = octadecyl (C_{18}) silane.

Table 8. Approximate UV Absorbance of 7-Dehydrocholesterol, Pre- and *cis*-Vitamin D

Absorbance at μm	(E 1% 1 cm)		
	7-Dehydrocholesterol	Previtamin D	*cis*-Vitamin D
230	35	190	250
235	40	170	280
240	50	210	325
245	60	235	360
250	85	250	400
255	120	260	430
260	150	270	460
265	200	265	470
270	270	250	450
273	282		
275	250	220	420
277	240		
280	290	180	340
282	293		
285	240	150	300
290	155	100	180
295	170	70	120
300	70	50	80

gosterol is not used as extensively as it once was, because it offers no real price advantage and, upon irradiation, it gives vitamin D_2, which has been shown to be less active in many species. The pig, chicken, cow, and horse have been shown to discriminate against vitamin D_2 (99). Irradiation of 7-dehydrocholesterol or ergosterol is carried out by dissolving the steroid in an appropriate solvent, eg, peroxide-free diethyl ether. Solvents such as methanol, cyclohexane, and dioxane have also been used. The cooled solution is pumped through uv-transparent quartz reactors which permit the light from high pressure mercury lamps to impinge upon the solution (Fig. 7). The solution is recycled until the desired degree of irradiation has been achieved. This results in a mixture of unreacted 7-dehydrosterol, previtamin D, vitamin D, and irradiation by-products.

Among the light sources used for irradiation are carbon arcs, metal-corded carbon rod, magnesium arcs, and mercury-vapor lamps; the high pressure mercury lamp is most widely used. Higher yields and more favorable isomer distribution can be achieved if the frequency of light is kept at 275–300 nm. Several arrangements of the components of the simple reactor shown are possible and have been used. An important feature is a cooling jacket which controls the high temperature (ca 800°C) of the mercury-vapor lamps. Water solutions for cooling may contain salts for screening frequencies of light. Light below 275 nm can be filtered by aromatic compounds, as well as by a 5-wt % lead acetate or other inorganic salt solution. Glass filters can also be used as screens for frequencies which are outside the chemical filter ranges. Photosensitizers, eg, eosin, erythrosin, dibromodinitrofluorescein, and others, have been suggested to limit light frequencies and improve the isomer distribution.

The vitamin D resin is stabilized against oxidation by the addition of ≤1 wt % butylated hydroxyanisole or butylated hydroxytoluene.

The solvent is then evaporated, and the unconverted sterol is recovered by precipitation from an appropriate solvent, eg, alcohol. The recovered sterol is reused in subsequent irradiations. The solvent is then evaporated to yield vitamin D resin. The resin is a pale yellow-to-amber oil that flows freely when hot and becomes a brittle glass when cold; the activity of commercial resin is 20–30 × 10⁶ IU/g. The resin is formulated without further purification for use in animal feeds. Vitamin D can be crystallized to give the USP product from a mixture of hydrocarbon solvent and aliphatic nitrile, eg, benzene and acetonitrile, or from methyl formate (100,101). Chemical complexation has also been used for purification.

In 1938, it was estimated that 7.5 × 10¹³ quanta of light were required to convert ergosterol to 1 USP unit of vitamin D_2 (102). The value was later determined to be 9.3 × 10¹³ quanta.

A flow diagram for the D_3 manufacturing process is shown in Figure 8. First, ether solution containing 7-

Table 9. Chemical Test Methods for Provitamin D

Name of reaction	Components	Results	Interpretation
Revised Salkowski reaction	$CHCl_3 + H_2SO_4$ (conc)	Deep red acid layer	Differentiates from sterols lacking conjugated diene (red color in $CHCl_3$ layer) acid gives green fluorescence
Lieberman-Burchard reaction	$CHCl_3$; acetic acid—H_2SO_4 added dropwise	Red color develops and changes to blue-violet to green	Can be quantitative; acts similarly, but red color lasts longer
Tortelli-Jaffé reaction	Acetic acid + 2 wt % Br_2 in $CHCl_3$	Green	Sterols with ditertiary double bonds; vitamin D and compounds that give similar bonds upon isomerization or reaction
Rosenheim reaction	$CHCl_3$ + trichloroacetic acid in H_2O	Red color develops and changes to light blue	
Rosenheim reaction	$CHCl_3$ + lead tetraacetate in CH_3COOH is added; then trichloroacetic acid is added	Green fluorescence	Not given by esters of provitamin D; can be used to distinguish between provitamin and provitamin ester; sensitive to 0.1 μg and is quantitative
Chloral hydrate	Mixture of crystalline provitamins and chloral hydrate heated slowly; melts at 50°C	Color develops and changes red to green to deep blue	Other sterols, eg, cholesterol, do not react to give color
Antimony trichloride reaction	$CHCl_3 + SbCl_3$	Red color	
Chugaev reaction	Glacial acetic acid plus acetyl chloride and zinc chloride heated to boiling	Eosin-red greenish yellow fluorescence	1:800,000 sensitivity

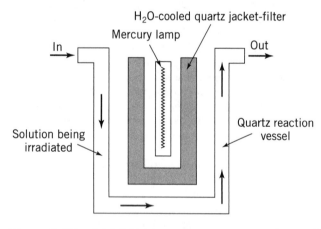

Figure 7. Ultraviolet-light-transparent quartz reactor for sterol irradiation.

dehydrocholesterol is recirculated through a quartz uv reactor, and the ether is distilled off. Methanol is added to the 7-dehydrocholesterol–vitamin D_3 mixture, and the remaining ether is azeotroped. The resulting solution is transferred to a crystallizer, and the 7-dehydrocholesterol is crystallized and recovered by filtration. The methanol is distilled, and the vitamin D_3 resin is heated to isomerize the pre-vitamin D to the cis-vitamin D isomer. Vitamin D_3 resin is then packaged for shipment.

Total Synthesis

Poor yields encountered during the manufacture of vitamin D stimulated early attempts to synthesize vitamin D.

In 1959 Inhoffen synthesized vitamin D_3 from 3-methyl-2-(2-carboxyethyl)-2-cyclohexenone (**40**), using the Wittig reaction extensively (103).

(40)

5,6-*trans*-Vitamin D_3 was prepared, followed by photochemical isomerization into vitamin D_3. Direct preparation of the previtamin and vitamins D is described in References 104 and 105.

The discovery that vitamin D_3 was metabolized to biologically active derivatives led to a significant effort to prepare 25-hydroxy vitamin D_3 and, subsequently, the 1α-hydroxy and 1,25 dihydroxy derivatives. Initial attempts centered around modification of steroidal precursors, which were then converted to the D derivatives by conventional means.

Chemical syntheses of 25-hydroxy-vitamin D_3 were achieved by several groups of researchers (106–112). Many of these syntheses depended on the availability of the precursor 25-oxo-27-norcholesterol. Grignard reaction followed by introduction of the 7-dehydro function and irradiation allows 25-hydroxy-vitamin D_3 to be isolated (109). Fucosterol (stigmasta-5,24(28)-dien-3β-ol) as well as bile acids, pregnenolone, and desmosterol have also been used

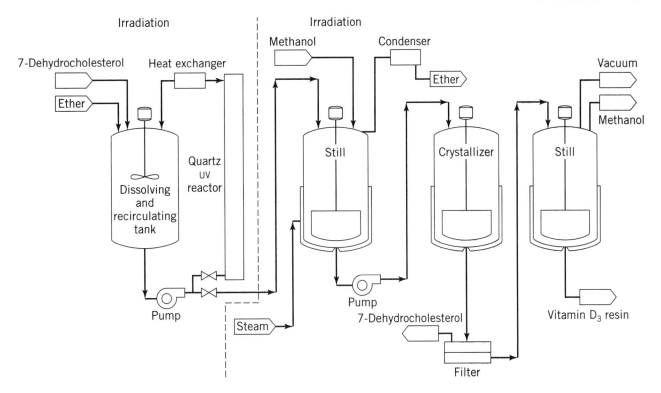

Figure 8. Vitamin D irradiation process.

as starting materials for the synthesis of 25-hydroxy intermediates. 24,25-Dihydroxy-vitamin D$_3$ [40013-87-4] was isolated and chemically characterized in 1972 (113). The first synthesis of this important D$_3$ catabolite was of a racemic mixture, and it was followed by the stereospecific syntheses of the two epimers (114–123). The biosynthesized material migrates exclusively with the synthetic 24(R) epimer.

The yield of the first chemical synthesis of 1,25-dihydroxy-vitamin D$_3$ was <0.005% (124). A key intermediate compound was 1,25-dihydroxy-cholesterol (109,125–130).

Subsequent synthesis of Vitamin D metabolites involved oxidative degradation of the vitamin D molecule to obtain the C- and D-ring portion with the intact side chain. Recombination of this molecule with an appropriate structure containing the A-ring was then carried out by a Wittig-type condensation.

The chemical syntheses of 1,24(R),25-trihydroxy-vitamin D$_3$ [56142-94-0] and 1,24(S),25-trihydroxy-vitamin D [56142-95-1] were reported (131,132) in 1975. The chemical synthesis of 25,26-dihydroxy-vitamin D$_3$ [29261-12-9] has also been described, and it has been determined that the biologically occurring epimer is 25(R),26-dihydroxy-vitamin D$_3$ (117,133–135). The 23,25-dihydroxy-24-oxo metabolite has been isolated (136) as well. 1α-Hydroxycalcitroic acid (1-hydroxy-24-nor-9,10-secochola-5,7-10(19)-trien-23-oic acid) [71204-89-2], 25-hydroxy-26,23-lactone vitamin D$_3$ [71203-34-6], and homocalcitroic acid (9,10-secohola-5,7,10(19)-trien-24-oic acid) have also been reported, and the biological activity of these metabolites has been studied (137). The synthetic chem-

istry associated with these and other metabolites of vitamin D$_2$ and vitamin D$_3$ is described in References 6, 40, and 138, and more recently in References 16, 51, 55, 139–141.

The discovery that vitamin D metabolites play a much larger biochemical role than just maintaining calcium homeostasis has stimulated a number of groups around the world to develop more economical chemical syntheses for the vitamin D metabolites and analogues, which might be useful in studying and treating D$_3$-related diseases and conditions. Many of these methods are reviewed in References 139 and 140.

The most useful synthetic routes include the following pathways to the vitamin D structures and their derivatives:

Photochemical ring opening of 7-dehydrocholesterol derivatives which have ring A or the side chain modified (142,143).

A phosphine oxide of type (**41**) can be coupled with Grundman's ketone (**42**) to produce the D$_3$ skeleton (105,144–151).

(**41**, X = a protective group)   (**42**, R' = cholesterol side chain or derivative)

Synthetic dienynes like (43) are semihydrogenated to form previtamin D and are then rearranged to the D structure (152–155).

Vinyl allenes (44) are rearranged with heat or metal catalysis and photosensitized isomerization to produce the vitamin D triene (156–160).

(43, R = H, OH)   (44)

R' = cholesterol side chain or derivative

Intermolecular cross-coupling of (45) with a synthetic alkyne structure like (46) leads to the D₃ skeleton (161,162).

(45, R = cholesterol side chain or derivative)   (46, X = protective group)

Direct modification of vitamin D₂ or D₃ or its metabolites to give derivatives by a variety of synthetic methods has also been used extensively (163–171).

Whereas these preparations are extremely useful for obtaining sufficient quantities of the D₃ metabolites and analogues for study and as possible therapeutic treatments, their cost is high compared to the cost of manufacture of vitamin D₃. For this reason, the metabolites are unlikely to be successful in replacing vitamin D₃ as an ingredient in animal or human nutrition.

BIOCHEMISTRY

Biochemistry

Vitamin D is introduced into the bloodstream either from the skin after natural synthesis by the irradiation of 7-dehydrocholesterol stored in the epidermis (172) or by ingestion and absorption of vitamin D_2 or vitamin D_3 through the gut wall (40). Between 60 and 80% of the vitamin introduced in the blood is taken up by the liver, where cholecalciferol is transferred from chylomicrons to a

vitamin D-binding protein (DBP), an α-globulin specific for vitamin D and its metabolites but one which does not bind with previtamin D in the skin (173). Cholecalciferol is hydroxylated in the liver at the C-25 position (51,141,174). This hydroxylation occurs in the endoplasmic reticulum and requires NADPH, a flavoprotein, cytochrome P-450, Mg^{2+}, and O_2 (175). 25-Hydroxylation also occurs in intestinal homogenates of chicks (176), but does not appear to occur outside the liver in mammals (177). 25-Hydroxycholecalciferol is the main circulating form of vitamin D_3, and its normal concentration level is 15–47 ng/mL (178).

25-Hydroxy vitamin D pools in the blood and is transported on DBP to the kidney, where further hydroxylation takes place at C-1 or C-24 in response to calcium levels. 1-Hydroxylation occurs primarily in the kidney mitochondria and is catalyzed by a mixed-function monooxygenase with a specific cytochrome P-450 (52,179,180). 1α- and 24-Hydroxylation of 25-hydroxycholecalciferol has also been shown to take place in the placenta of pregnant mammals and in bone cells, as well as in the epidermis. Low phosphate levels also stimulate 1,25-dihydroxycholecalciferol production, which in turn stimulates intestinal calcium as well as phosphorus absorption. It also mobilizes these minerals from bone and decreases their kidney excretion. Together with PTH, calcitriol also stimulates renal reabsorption of the calcium and phosphorus by the proximal tubules (51,141,181–183).

Vitamin D receptors have been identified in intestine, bone, and kidney. Receptors have also been shown, to a lesser degree, to be present in the pituitary, brain, skin, reproductive organs, and cells of the immune system, suggesting vitamin D may play a broader role in controlling cellular proliferation and differentiation (184). The nuclear receptor (genomic) for 1,25-dihydroxy vitamin D_3 has been shown to be a member of the super-family of transactivating regulators of gene transcription similar to the receptors of other steroid hormones (185). Nongenomic activation by 1,25-dihydroxy vitamin D_3 on the cellular and subcellular level has been observed and is believed to be responsible for calcium transport in tissue known as transcaltachia (186,187).

Further side-chain oxidation of vitamin D_3 metabolites may be necessary for phosphate transport (188,189). 24,25-Dihydroxycholecalciferol is produced by kidney mitochondria, but can also be produced elsewhere (190,191). $1\alpha,24,25$-Trihydroxy-vitamin D_3 stimulates intestinal calcium mobilization to the extent of ca 60% of the $1\alpha,25$ activity and has 10% of the $1\alpha,25$ activity in causing bone resorption (192). It is less active in chicks and is excreted.

More than 20 other, naturally occurring metabolites of vitamin D have been isolated and characterized, and many derivatives have been synthesized. Their function is the subject of continuing research (16,51,141,162).

Although it is being found that vitamin D metabolites play a role in many different biological functions, metabolism primarily occurs to maintain the calcium homeostasis of the body. When calcium serum levels fall below the normal range, $1\alpha,25$-dihydroxy-vitamin D_3 is made; when calcium levels are at or above this level, 24,25-dihydroxycholecalciferol is made, and 1α-hydroxylase activity is discontinued. The calcium homeostasis mechanism involves a hypocalcemic stimulus, which induces the secretion of parathyroid hormone. This causes phosphate diuresis in the kidney, which stimulates the 1α-hydroxylase activity and causes the hydroxylation of 25-hydroxy-vitamin D to 1α,25-dihydroxycholecalciferol. Parathyroid hormone and 1,25-dihydroxycholecalciferol act at the bone site cooperatively to stimulate calcium mobilization from the bone. Calcium blood levels are also influenced by the effects of the metabolite on intestinal absorption and renal resorption.

Interaction of vitamin D and its metabolites with sex hormones has been demonstrated, particularly in birds in which the egg-laying functions combine calcium needs and reproductive activity. The metabolites of vitamin D behave as hormones. As such, they play an active role in the endocrine system, along with other hormones, to maintain the various body functions. Several biological influences of metabolites of vitamin D have been studied, including effects related to cancer (193–197), skin diseases (198–201), immunomodulatory effects (202,203), and Alzheimer's disease (204–206) (Fig. 9).

The metabolism of vitamin D_2 follows a pathway similar to that described for Vitamin D_3.

Vitamin D_3 Deficiency

Vitamin D_3 deficiency is uncommon in normal adults. However, when it does occur, it can be serious, particularly in pregnant women. Some vitamin D_3 deficiency can occur because of a large reduction of fat intake, which decreases D_3 absorption. Strict vegetarians also risk reduced vitamin D_3 intake. Premature infants and elderly people who are exposed to minimal sunlight and consume little vitamin D_3 also have a reduced capacity to metabolize D_3 and can develop vitamin D_3 deficiency.

Clinical stresses which interfere with vitamin D_3 metabolism, can result in calcium deficiency leading to osteomalacia and osteoporosis (secondary vitamin D deficiency). These stresses include: intestinal malabsorption (lack of bile salts); stomach bypass surgery; obstructive jaundice; alcoholism; liver or kidney failure decreasing hydroxylation of vitamin D_3 to active forms; inborn error of metabolism; and use of anticonverdiants that may lead to increased D_3 requirement.

People who experience the exclusion of sunlight (living in northern climates; cultures where apparel limits sunlight), as well as infants and children that are confined to bed or limited outdoor activity because of weather or illness, also can show a tendency towards vitamin D_3 deficiency.

Vitamin D deficiency in animals may be caused by the fact that the vitamin is not available to the livestock. Modern animal husbandry subjects animals to total confinement with little or no exposure to sunlight. This mandates that they be given vitamin D-fortified diets. The vitamin is sensitive to oxidation, heat, light, and minerals, and significant losses may occur in the fortified feed unless the product is adequately protected. Mycotoxins in feeds also interfere with utilization of vitamin D in feeds (207–209).

Symptoms of D_3 deficiency in animals include poor appetite, stunted growth, and weight loss; increased inci-

dence of irritability and convulsions (tetany); some growth abnormalities; decreased egg production in poultry with reduced hatchability and thin eggshell quality; and birth of weak, dead, or deformed offspring in other animals.

For a more detailed description of symptoms, see References 4,5,210, and 211.

Dietary Requirements

Vitamin D is essential for growth and maintenance of good health. According to the National Research Council (NRC), the vitamin D requirement for optimum health is ca 400 IU/d in humans, regardless of age. This amount of vitamin D gives ample protection from rickets, provided a sufficient amount of the other essential nutrients, including calcium and phosphorus, is supplied (212). Recommended NRC amounts of vitamin D per kilogram of feed for various species are as follows: starting and growing chicks, 200 IU; laying and breeding hens, 500 IU; turkeys, 1100 IU; ducks, 200 IU; quail, 480–900 IU; geese, 200 IU; and swine 125–220 IU. Calves require 600 IU per 100 kg of body weight (213–215). Practical feeding levels are somewhat higher, in order to assure delivery of the vitamin to the animals.

The present (ca 1997) maximum safe level of vitamin D_3 for long-term feeding in most species is four to ten times the NRC dietary requirements. Short-term (<60 d), most species can tolerate 100 times their apparent dietary requirements (210).

Animals exposed to sunlight for extended lengths of time do not require substantial dietary vitamin D. Current livestock management practices place an emphasis on high productivity, and most feed manufacturers recommend vitamin D supplementation of diets. Recommendations for practical levels of vitamin D in feeds for various animals, as recommended by feed manufacturers, are listed in Table 10.

Historically, rickets prevention or cure was used to evaluate adequate vitamin D_3 nutrient levels. More recently, in the absence of uv light, Edwards (216) found different vitamin D_3 levels were required for the optimization of the various effects of vitamin D_3 in poultry, ie, 275 IU/kg for growth, 503 IU/kg for bone ash, 552 IU/kg for blood plasma calcium, and 904 IU/kg for rickets prevention.

Toxicity

Vitamin D toxicity was known as early as the year 1429 (217). Accidental toxicity has been reported in monkeys, dogs, horses, pigs, chinchillas, and humans, and particularly in cattle when extremely high doses of vitamin D have been used to treat milk fever.

Vitamin D intoxication causes 25-hydroxy vitamin D_3 blood levels to go from a normal of 30–50 ng/mL to 200–400 ng/mL. At this high level, the metabolite can compete with 1α-25-dihydroxy vitamin D_3 for receptors in the intestine and bone and induce effects usually attributed to the dihydroxy vitamin D_3. Thus, 25-hydroxy vitamin D_3 is believed to be the critical factor in vitamin D intoxication. Vitamin D_2 is metabolized slower than vitamin D_3 and thus appears to be less toxic (218).

The overall effect in most animals is to stimulate intestinal absorption of calcium with a concomitant increase in

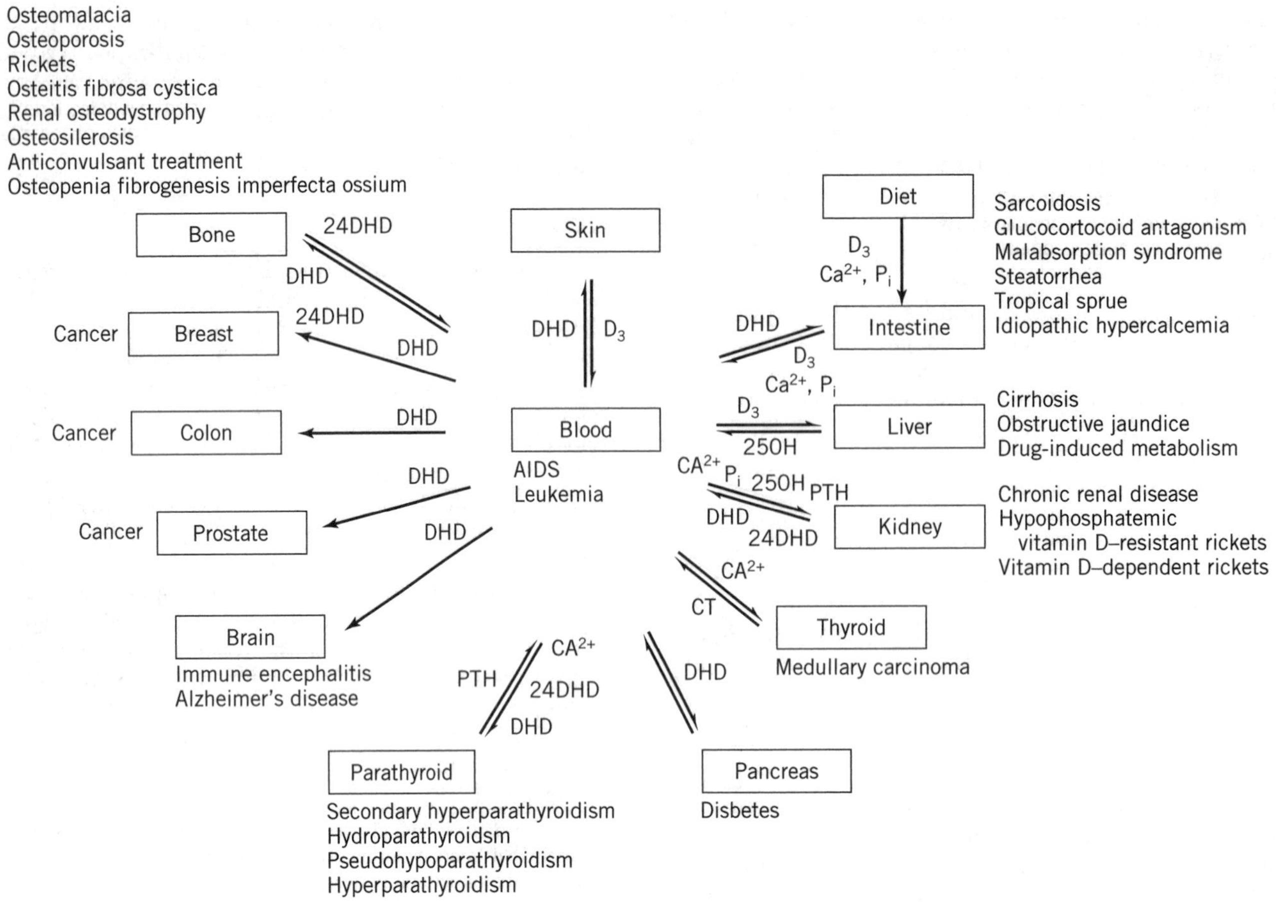

Osteomalacia
Osteoporosis
Rickets
Osteitis fibrosa cystica
Renal osteodystrophy
Osteosilerosis
Anticonvulsant treatment
Osteopenia fibrogenesis imperfecta ossium

Figure 9. Human disease states in which vitamin D has been implicated. D_3 = vitamin D_3; DHD = 1a,25-dihydroxyvitamin D_3; 24DHD = 24(R),25-dihydroxyvitamin D_3; 25OH = 25-hydroxyvitamin D_3; PTH = parathyroid hormone; CT = calcitonin; P_i = inorganic phosphorus.

serum calcium and a reduction in parathyroid hormone (PTH). Modest hypercalcemia allows the glomerular filtration rate to remain stable and hypercalciuria to occur because of increased filtered load of calcium and reduction of tubular resorption of calcium with reduced PTH. However, with further increases in serum calcium, the glomerular filtration rate decreases, resulting in an even more rapid increase in serum calcium and the subsequent fall in urinary calcium.

Polyuria, along with vomiting can cause extracellular fluid to be reduced, contributing to further renal function disruption. Renal dysfunction then, becomes the major contributor to loss of calcium homeostasis, and the resulting hypercalcemia during vitamin D intoxication. Loss of appetite, gastrointestinal disturbances, head and joint pain, and muscle weakness are typical clinical symptoms. Death from hypervitaminosis D is usually caused by renal failure. Cardiovascular and kidney mineralization results. Respiratory tract mineralization has also been observed, as well as calcification in the salivary gland. The severity of effects of toxic levels of vitamin D depend on the dose, functional state of the kidneys, composition of the diet, as well as differences in toxicology of the various species (210).

Vitamin D withdrawal is an obvious treatment for D toxicity (219). However, because of the 5–7 d half-life of plasma vitamin D and 20–30 d half-life of 25-hydroxy vitamin D, it may not be immediately successful. A prompt reduction in dietary calcium is also indicated to reduce hypercalcemia. Sodium phytate can aid in reducing intestinal calcium transport. Calcitonin glucagon and glucocorticoid therapy have also been reported to reduce serum calcium resulting from D intoxication (210).

Vitamin D_3 is nonirritating to rabbit eye and dermal tests. It has an oral LD_{50} of 35–42 mg/kg in rats; 42 mg/kg in mice (136 mg/kg IP), and 4–80 mg/kg in dogs (10 mg/kg IP and 5 mg/kg IM and IV). It is nontoxic on inhalation.

The metabolites of vitamin D are usually more toxic than the vitamin because the feedback mechanisms that regulate vitamin D concentrations are circumvented. 25-Hydroxycholecalciferol has a one-hundredfold increase in toxicity over vitamin D_3 when fed to chicks (220) and 1α,25-dihydroxy vitamin D_3 is several times more toxic than the 25-hydroxy analogue. Vitamin D_2 seems to have less toxicity than vitamin D_3, a circumstance which is believed to be caused by the more efficient elimination of 25-hydroxy and the 1α,25-dihydroxy vitamin D_2 from the ani-

Table 10. Practical Feeding Levels of Vitamin D, 10^6 IU/t

Animal	Amount
Poultry	
Chickens	
Broilers	2–4
Replacement birds	1–3
Layers	1–3
Breeding hens	2–4
Turkeys	
Starting	3–5
Growing	2–4
Breeding	3–5
Ducks	
Market	1–3
Breeding	2–4
Swine	
Prestarter (to 10 kg)	2–3
Starter (10–35 kg)	1–2
Growing-finishing (35 kg to market)	1–2
Gestation	1–2
Lactation	2–3
Boars	2–3
Fish	
Trout	1–2
Catfish	1–2
Dairy cattle	
Calf starter	1–2
Calf milk replacer	2–4
Replacement heifers	1–2
Dry cows	1–2
Lactating cows	2–4
Bulls	2–4
Beef cattle	
Calf starter	1–2
Replacement heifers	1–2
Feedlot	2–3
Dry pregnant cows	1–2
Lactating cows	1–2
Bulls	1–2
Sheep	
Fattening lambs	2–3
Breeding	1–2
Other	
Dogs	5–1
Cats	1–2
Horses	1–2

Table 11. Estimated Safe Upper Dietary Levels

Species	NRC recommended dietary daily requirement (IU/kg)	Exposure, multiple of daily requirement Long-term (<60 d)	Exposure, multiple of daily requirement Short-term (>60 d)
Chicken	200	200	14
Quail	1,200	100	4
Turkey	900	100	3.9
Cow	300	83	3.9
Catfish	1,000		20
Trout	1,800		555
Horse	400		5.5
Sheep	275	91	8
Pig	220	150	10

Source: Ref. 210.

shown to be related to vitamin D. These can involve a lack of the vitamin, deficient synthesis of the metabolites from the vitamin, deficient control mechanisms, or defective organ receptors. The control of calcium and phosphorus is essential in the maintenance of normal cellular biochemistry, eg, muscle contraction, nerve conduction, and enzyme function. The vitamin D metabolites also have a function in cell proliferation. They interact with other factors and receptors to regulate gene transcription.

In the treatment of diseases where the metabolites are not being delivered to the system, synthetic metabolites or active analogues have been successfully administered. Vitamin D_3 metabolites have been successfully used for treatment of milk fever in cattle, turkey leg weakness, plaque psoriasis, and osteoporosis and renal osteodystrophy in humans. Many of these clinical studies are outlined in References 6, 16, 40, 51, and 141. The vitamin D receptor complex is a member of the gene superfamily of transcriptional activators, and 1,25 dihydroxy vitamin D is thus supportive of selective cell differentiation. In addition to mineral homeostasis mediated in the intestine, kidney, and bone, the metabolite acts on the immune system, β-cells of the pancreas (insulin secretion), cerebellum, and hypothalamus.

Vitamin D metabolites may therefore play an active role in diseases related to these functions, ie, leukemia, cancer (breast, colon, prostate), and autoimmune diseases (AIDS, immune encephalitis, and diabetes) (51,141,193–197,202,203).

ECONOMIC ASPECTS

Vitamin D is available in a variety of forms. Cod-liver oil and percomorph-liver oil historically were good sources of vitamin D. Recent cost increases of these materials have caused a decline in their market position. Cod-liver oil sold for \$0.40–0.45/L in 1970 and as high as \$1.45/L in 1979 and \$3.43/L in 1996. The prices of the cod-liver oils and of vitamin D_2 and vitamin D_3 from 1955 to 1995 are shown in Table 12.

Vitamin D_3 is produced as a resin (20–30 × 10^6 IU/g) which sold in 1980 for ca \$1.02/10^6 IU. The 1979 prices of

mals. Estimated safe upper dietary levels are given in Table 11.

Disease States

Rickets is the most common disease associated with vitamin D deficiency. Many other disease states have been

Table 12. Price History of Vitamin D Materials

	Source	Vitamin D	
Year	Cod-liver oil, \$/L	Ergocalciferol (crystalline vitamin D_2), \$/g	Cholecalciferol D_3 (\$/g)[a]
1955	1.51	0.60	0.045
1960	1.40	0.54	0.045
1965	1.71	0.48	0.045
1970	1.70	0.48	0.045
1975	5.50	0.63	0.045
1980	8.00	0.63	0.045
1985	7.25	[b]	24[c]
1990	13.00	[b]	24[c]
1995	13.00	[b]	24[c]

Source: Ref. 221.
[a]850,000 IU/g unless otherwise indicated.
[b]D_2 costs were not quoted after 1983.
[c]40×10^6 IU/g.

vitamin D were $24.25/10^9$ IU (\$9.70/kg of product at 400,000 IU/g) (222). In the United States in 1979, 3.6×10^6 worth of vitamin D was sold. The 1996 price for vitamin D_3 products was \$13.50–14.00/g for resin and \$24.00/g (223) for crystalline material. Vitamin D_3 feed-grade formulated products sell for approximately $0.03–0.04/10^6$ units of activity.

Estimates of world demand in 1979 were as high as 1300×10^{12} IU of vitamin D_3. This was divided into thirds for Europe, the United States, and the rest of the world, respectively. Of this demand, 90% was estimated for animal-feed fortification and 100% for food and pharmaceutical uses. It is estimated that the demand will be $1500–1600 \times 10^{12}$ IU in 1997 for animal usage and 100×10^{12} IU for human use. The United States will require approximately 500 TU (1 Trillion Units = 25 kg *cis*-vitamin D_3 or 17 t of resin) for animal use and 30 TU (approximately 1 t of crystalline *cis*-vitamin D_3) for human use. This represents approximately 50 t of vitamin D_3 resin/yr for animal use worldwide and about 2.5 t of crystalline vitamin D_3 for human use. A substantial proportion of the vitamin D_3 is imported, and with all uses included, it is estimated that 80–90% of the sales are of vitamin D_3.

The vitamin D_2 volume is estimated at only 1200 kg/yr, exclusive of Chinese sources.

USES

Most of the vitamin D sold is synthetic. Vitamin D_2 as a concentrate or in microcrystalline forms is used in many pharmaceutical preparations, although vitamin D_3 is preferred by many manufacturers and consumers. Vitamin D_2 in the form of irradiated yeast has been used as a feed supplement for cattle, swine, and dogs, but its use has declined in favor of Vitamin D_3. In the United States, swine usage accounts for 13% and poultry for 41% of vitamin D consumption in animal agriculture. The beef and dairy industries account for 44% of vitamin D consumption, of which 22% is by dairy calves. The remaining 1% of total consumption of D vitamins is largely in prepared pet foods. European usage in 1994 was estimated at: 58% cattle, 19% swine, 20% poultry, and 3% other. Crystalline vitamin D_3

is used for medicinal preparations and formulations in the pharmaceutical industry, as well as for the fortification of fresh and evaporated milk and nonfat dry milks. Preparations based on the use of the resin are less expensive than those using crystalline vitamin D. Essentially all milk produced in the United States is fortified with vitamin D_3. Cereals and margarine are also fortified with vitamin D_3. The vitamin must be diluted to a proper dosage form, and many supplement preparations are marketed, eg, tonics, drops, capsules, and tablets; oil-based injectables are also available. Combination formulations are used widely, particularly with vitamin A for humans and with vitamins A and E for animals. Preemulsified products are used for increased bioavailability. Water-miscible formulations have also been developed (224). Solutions of vitamin D in oil or in oil-on-dry carriers, eg, corn or flour, are used in animal feeds. These diets contain high levels of minerals, and the vitamin D formulation is not stable unless it is protected. Several stable forms are patented and sold commercially; they include beadlets or powders of dry suspensions in gelatin, carbohydrates, wax, and cellulose derivatives (89,225–228).

Animal uses employ resin in various stabilized forms at levels of 200,000; 400,000; 500,000; and 1×10^6 units of D_3 per gram. Combination products containing A and D are also available, with 650,000 units of vitamin A and 325,000 units vitamin D per gram of product being the most common dosage form.

Approximately 200 kg/yr of Vitamin D_3 formulations are also marketed as rat poisons. The metabolites of vitamin D_3 and synthetic derivatives are being used or developed for treatment of osteoporosis, skin psoriasis, and other diseases in humans. 1α-Hydroxy vitamin D_3 is being used for milk fever in cows, and 25-hydroxy vitamin D_3 has been proposed for eggshell thickness in poultry and is being marketed as an animal dietary nutritional supplement.

BIBLIOGRAPHY

1. P. Karlson et al., IUPAC-IUB Joint Commission on Biochemical Nomenclature. "Nomenclature of Vitamin D Recommendations 1981," *Eur. J. Biochem.* **124**, 223 (1982).

2. T. C. Angus et al., *Proc. R. Soc. London, Ser. B* **105**, 340 (1931).

3. W. F. Loomis, *Sci. Am.* **223**, 77 (1970).

4. H. M. Frost, *Bone Dynamics in Osteoporosis and Osteomalacia*, Surgery Monograph Series, Charles C. Thomas, Publisher, Springfield, Ill., 1966.

5. H. F. DeLuca in R. B. Alfin-Slater and D. Kratchevsky, eds., *Nutrition and the Adult: Micronutrients*, Plenum Press, New York, 1980, p. 207.

6. A. W. Norman, *Vitamin D—The Calcium Homeostatic Steroid Hormone*, Academic Press, Inc., New York, 1979, p. 3.

7. E. Mellanby, *J. Physiol. (London)* **52**, L 111 (1919); *Lancet* **196**, 407 (1919).

8. K. Huldschinski, *Dtsch. Med. Wochenschr.* **45**, 712 (1919).

9. F. Reindel and H. J. Kipphan, *Justus Liebigs Ann. Chem.* **493**, 181 (1932).

10. E. Fernholz and P. N. Chakravorty, *Ber. Deutsch. Chem. Ger. A.* **67**, 2021 (1934).

11. A. Windaus and H. Grundmann, *Justus Liebigs Ann. Chem.* **524**, 295 (1936).

12. Ref. 6, pp. 39–48.

13. A. Windaus and F. Bock, *Z. Physiol. Chem. Hoppe-Seglers* **245**, 168 (1937).

14. O. Rygh, *Nature (London)* **136**, 396 (1935).

15. H. Brockmann et al., *Z. Physiol. Chem. Hoppe-Seglers* **241**, 104 (1936); **245**, 96 (1937).

16. A. W. Norman, K. Schaefer, H.-G. Grigoleit, D. V. Herath, eds., *Vitamin D. Chemical, Biochemical and Clinical Update, Proceedings of the Sixth Workshop on Vitamin D*, Merano, Italy, March, 1985, Walter de Gruyter, Berlin, 1985, p. 55.

17. R. H. Wasserman, J. D. Henion, M. R. Haussler, and T. A. McCain, *Science* **194**, 853 (1976).

18. M. Peterlik et al., *Biochem. Biophys. Res. Commun.* **70**, 797 (1976).

19. J. P. Simonit, K. M. L. Morris, and J. C. Collins, *J. Endocrinol.* **68**, 18 (1976).

20. H. Zucker and W. A. Rambeck, *Centralbl. Veterinaermed.* **28**, 436 (1981).

21. W. A. Rambeck et al., *Z. Pflanzenphysiol.* **104**, 9 (1981).

22. R. H. Wasserman et al., *Nutr. Rev.* **33**, 1 (1975); *J. Nutr.* **106**, 457 (1976); *Biochem. Biophys. Res. Commun.* **62**, 85 (1975).

23. Ref. 6, p. 49.

24. M. F. Hollick in Ref. 16, p. 219.

25. A. W. Norman et al., *J. Biol. Chem.*, 13811–13819 (1994).

26. M. C. Dormanen et al., *Biochem. Biophys. Res. Comm.* **201**, 394–401 (1994).

27. I. Nemere et al., *J. Biol. Chem.* **269**, 23750–23756 (1994).

28. U.S. Pat. 2,395,115 (Feb. 19, 1946), K. J. Goering (to Anheuser-Busch, Inc.).

29. U.S. Pat. 2,874,171 (Feb. 17, 1959), H. A. Nelson (to Upjohn Co.).

30. U.S. Pat. 3,006,932 (Oct. 31, 1961), J. Green, S. A. Price, and E. E. Edwin (to Vitamins, Ltd.).

31. U.S. Pat. 2,794,035 (May 28, 1957), O. Hummel (to Zellstaff-Fabrik Waldhof).

32. U.S. Pat. 2,865,934 (Dec. 23, 1958), R. A. Fisher (to Biofermentation Corp.).

33. Ger. Pat. 1,252,674 (Oct. 26, 1967), K. Petzoldt, K. Klieslich, and H. J. Koch (to Schering A.G.).

34. K. Ziegler, *Ann.* **551**, 80 (1942).

35. S. Bernstein, L. J. Benovi, L. Dorfman, K. S. Sax, and Y. Subbarow, *J. Org. Chem.* 14, 433 (1949); U.S. Pat. 2,498,390 (Feb. 21, 1950), S. Bernstein and K. J. Sax (to American Cyanamid Co.).

36. U.S. Pat. 2,446,091 (May 4, 1968), J. Van Der Vliet and W. Stevens (to Hartford National Bank and Trust Co.).

37. H. Schaltagger, *Helv. Chim. Acta* **33**, 2101 (1950).

38. D. N. A. Holwerda, Doctoral Dissertation, Imperial University Leydon, 1970.

39. F. Hunziker and F. X. Mullner, *Helm. Chim. Acta* **41**, 70 (1958).

40. A. W. Norman, K. Schaefer, D. V. Herrath, and H. G. Grigoleit, eds., *Vitamin D: Chemical, Biochemical, and Clinical Endocrinology of Calcium Metabolism, Proceedings of the Fifth Workshop on Vitamin D*, Williamsburg, Va., Feb. 1982, Walter de Gruyter, Berlin, 1982, p. 1133.

41. W. G. Salmond, M. A. Serta, A. M. Cain, and M. C. Sobala, *Tetrahedron Lett.* **20**, 1683 (1977).

42. U.S. Pat. 2,098,984 (Nov. 16, 1937), A. Windaus and F. Schenck (to Winthrop Chem. Co., Inc.).

43. L. E. Fieser, *J. Am. Chem. Soc.* **75**, 4394 (1953).

44. U.S. Pat. 2,505,646 (Apr. 25, 1950), W. E. Meuly (to E. I. du Pont de Nemours & Co., Inc.).

45. A. Windaus, F. VonWeder, A. Luttringhaus, and E. Fernholz, *Ann.* **499**, 188 (1932).

46. L. Velluz, G. Amiard, and B. Goffinet, *Bull. Soc. Chim. France* **22**, 1341 (1955).

47. E. Havinga, R. J. DeKoch, and M. Rappoldt, *Tetrahedron* **11**, 276 (1960).

48. H. J. C. Jacobs, *Pure & Appl. Chem.* **67**, 63 (1995).

49. J. G. Bell and A. A. Christie, *Analyst* **99**, 385 (1974).

50. P. B. Braun, J. Hornstra, C. Knobles, E. W. M. Rutten, and C. Romer, *Acta Crystallogr.* **B29**, 463 (1973).

51. E. Havinga, *Experimentia* **29**, 1181 (1973).

52. A. W. Norman, R. Bouillon, M. Thomasset, eds., *Vitamin D. A Pluripotent Steroid Hormone: Structural Studies, Molecular Endocrinology and Clinical Applications. Proceedings of the Ninth Workshop on Vitamin D*, Orlando, Florida, May 1994, Walter de Gruyter, Berlin, 1994, p. 89.

53. J. I. Pedersen, J. G. Ghazarian, N. R. Orme-Johnson, and H. F. DeLuca, *J. Biol. Chem.* **251**, 3933 (1976).

54. A. W. Norman et al., eds., *Vitamin D; Basic Research and its Clinical Application, Proceedings of the Fourth Workshop on Vitamin D*, Berlin, Feb. 1979, Walter de Gruyter, Berlin, 1979, p. 173.

55. K. H. Hanewald, M. P. Rappoldt, and J. R. Roborgh, *Rec. Trav. Chim. Pays-Bas* **80**, 1003 (1961).

56. A. W. Norman, K. Schefer, H.-G. Grigoleit, D. Herrath, eds., *Vitamin D, Molecular, Cellular and Clinical Endocrinology. Proceedings of the Seventh Workshop on Vitamin D*, April, 1988, Walter de Gruyter, Berlin, 1988, p. 83.

57. A. Verloop, A. L. Koevoet, and E. Havinga, *Rec. Trav. Chim. Pays-Bas* **76**, 689 (1957).

58. A. L. Hoevoet, A. Verloop, and E. Havinga, *Rec. Trav. Chim.* **74**, 788 (1955).

59. W. G. Dauben et al., *J. Am. Chem. Soc.* **80**, 4117 (1958).

60. W. H. Okamura, M. L. Hammond, A. J. C. Jacobs, and J. Von Thiegil, *Tetrahedron Lett.* **52**, 4807 (1976).

61. E. Havinga, *Chimia* **30**, 27 (1976); D. H. R. Barton and co-workers, *Chem. Comm.*, 65 (1976).

62. F. Boosman, H. J. C. Jacobs, E. Havinga, and A. Van den Gen, *Tetrahedron Lett.* **7**, 427 (1975).

63. A. W. Norman et al., eds., *Vitamin D Biochemical, Chemical and Clinical Aspects Related to Calcium Metabolism; Proceedings of the Third Workshop on Vitamin D*, Asilomar, Pacific Grove, Calif., Jan. 1977, Walter de Gruyter, Berlin, 1977, p. 15.

64. S. Kobayshi and M. Yasumura, *J. Nutr. Sci. Vitaminol.* **19**, 123 (1973); S. Kobayashi et al., *J. Nutr. Sci. Vitaminol.* **26**, 545 (1980).

65. D. H. R. Barton, R. H. Heese, M. M. Pecket, and E. Rizzardo, *J. Am. Chem. Soc.* **95**, 2748 (1973).

66. K. Pfoertner and J. D. Weber, *Helv. Chim. Acta* **55**, 921, 937 (1972).

67. A. Verloop, A. L. Koevoet, R. Van Moorselase, and E. Havinga, *Rec. Trav. Chim. Pays-Bas* **78**, 1004 (1959).

68. A. Verloop, A. L. Koevoet, and E. Havinga, *Rec. Trav. Chim. Pays-Bas* **74**, 1125 (1955).

69. T. Kobagashi, *J. Vitaminol. (Jpn).* **13**, 255 (1967).

70. H. H. Inhoffen, G. Quinkert, J. H. Hess, and H. M. Erdman, *Chem. Ber.* **89**, 2273 (1956).

71. C. H. Nields, W. C. Russell, and A. Zimmerli, *J. Biol. Chem.* **136**, 73 (1940).

72. J. B. Wilkie, S. W. Jones, and O. L. Kline, *J. Amer. Pharm. Assoc. Sci. Ed.* **47**, 395 (1958).

73. S. C. Eyley and D. H. Williams, *J. Chem. Soc. Chem. Comm.*, 858 (1975).

74. E. C. Snoeren, M. R. Daha, J. Lugtenburg, and E. Havinga, *Rec. Trav. Chim. Pays-Bas* **89**, 261 (1970).

75. W. G. Dauben and R. B. Phillips, *J. Am. Chem. Soc.* **104**, 355 (1982).

76. V. Malatesta, C. Willis, and P. A. Hackett, *J. Am. Chem. Soc.* **103**, 6781 (1981).

77. U.S. Pat. 3,661,939 (May 9, 1972), M. Toyoda, Y. Tawara (to Nisshin Flour Milling Co., Ltd.).

78. P. S. Chen and H. B. Bosman, *J. Nutr.* **83**, 133 (1964).

79. Code of Federal Regulations, Title 21-Food and Drugs, Part 582.5923, p. 541, April 1, 1995.

80. *Chron. WHO* **3**, 747 (1965).

81. *The United States Pharmacopeia 23 (USP 23-NF 18)*, The United States Pharmacopeial Convention, Inc., Rockville, Md., 1995.

82. P. Cunniff, ed., *Official Methods of Analysis of the International Association of Official Analytical Chemists*, 16th ed., Washington, D.C., 1995.

83. F. Mulder, *J. Assoc. Off. Anal. Chem.* **60**, 989 (1977).

84. H. Hofsass, N. J. Alicino, A. L. Hirsch, L. Amieka, and L. D. Smith, *J. Assoc. Off. Anal. Chem.* **61**, 774 (1978).

85. F. Mulder et al., *J. Assoc. Off. Anal. Chem.* **64**, 61 (1981); **65**, 1228 (1982); **68**, 822 (1985).

86. Ref. 6, p. 64.

87. J. Carol, *J. Pharm. Sci.* **50**, 451 (1961).

88. W. W. Morris, Jr., J. B. Wilkie, S. W. Jones, and L. Friedman, *Anal. Chem.* **34**, 381 (1962).

89. R. Strohecker and H. M. Henning, *Vitamin Assay-Tested Methods*, CRC Press, Cleveland, Ohio, 1966, p. 281.

90. U.S. Pats. 2,777,797 and 2,777,798 (Jan. 15, 1977), M. Hochberg and M. L. MacMillan (to Nopco Chemical Co.).

91. J. R. Evans, *Clin. Chim. Acta* **42**, 167 (1972).

92. T. K. Murray, K. C. Day, and E. Kodicek, *Biochem. J.* **98**, 29P (1966).

93. L. V. Aviolo and S. W. Lee, *Anal. Biochem.* **16**, 193 (1966).

94. D. Sklan, P. Budowski, and M. Katz, *Anal. Biochem.* **56**, 606 (1973).

95. D. O. Edlund and F. A. Filippini, *J. Assoc. Anal. Chem.* **56**, 1374 (1973).

96. T. G. Hogness, A. E. Sidwell, Jr., and F. P. Zscheile, Jr., *J. Biol. Chem.* **120**, 239 (1937).

97. W. Huber, G. W. Ewing, and J. Kriger, *J. Am. Chem. Soc.* **67**, 609 (1945).

98. R. H. Wasserman et al., *Vitam. Horm. (N.Y.)* **32**, 299 (1974).

99. R. L. Horst et al., Ref. 55, p. 93.

100. U.S. Pat. 3,334,118 (Aug. 1, 1967), K. Schaff, S. Schmuhler, and H. C. Klein (to Nopco Chemical Co.).

101. U.S. Pat. 3,665,020 (May 23, 1972), R. Marbet (to Hoffmann-LaRoche, Inc.).

102. R. S. Harris, J. W. M. Bumker, and L. M. Moser, *J. Am. Chem. Soc.* **60**, 2579 (1938).

103. H. H. Inhoffen et al., *Chem. Ber.* **91**, 2309 (1958); H. H. Inhoffen et al., *Chem. Ber.* **92**, 1564 (1959).

104. T. M. Dawson, J. Dixon, P. S. Littlewood, B. Lythgoe, and A. K. Saksena, *J. Chem. Soc. C*, 2960 (1971).

105. B. Lythgoe, M. E. N. Nambudiry, and J. Tideswell, *Tetrahedron Lett.* **31**, 3685 (1977).

106. J. A. Campbell, D. M. Squires, and J. C. Babcock, *Steroids* **13**, 567 (1969).

107. S. J. Halkes and N. P. VanVliet, *Rec. Trav. Chim. Pays-Bas* **88**, (1969).

108. J. W. Blunt and H. F. DeLuca, *Biochemistry* **8**, 671 (1969).

109. M. Morisaki, J. Rubio-Lightbourn, and N. Ikekawa, *Chem. Pharm. Bull.* **21**, 457 (1973).

110. J. J. Partridge, S. Faber, and M. R. Uskokovic, *Helv. Chim. Acta* **57**, 764 (1974).

111. U.S. Pat. 4,172,076 (Apr. 4, 1977), A. Hirsch and J. Pikl (to Diamond Shamrock, subsequently assigned to A. L. Labs, Jan. 1, 1982).

112. U.S. Pat. 4,226,770 (Oct. 7, 1980), E. Kaiser.

113. M. H. Holick et al., *Biochemistry* **11**, 4251 (1972).

114. Y. Tanaka, H. Frank, H. F. DeLuca, N. Koizumi, and N. Ikekawa, *Biochemistry* **14**, 3293 (1975).

115. Y. Tanaka, H. F. DeLuca, N. Ikekawa, M. Morisaki, and N. Koizumi, *Arch. Biochem. Biophys.* **170**, 620 (1975).

116. H.-Y. Lam, H. K. Schnoes, H. F. DeLuca, and T. C. Chen, *Biochemistry* **12**, 4851 (1973).

117. M. Seki, J. Rubio-Lightbourn, J. Morisaki, and N. Ikekawa, *Chem. Pharm. Bull.* **21**, 2783 (1973).

118. J. Redel, P. Bell, F. Delbarre, and E. Kodicek, *C. R. Acad. Sci.* **278**, 529 (1974).

119. J. Redel et al., *J. Steroid Biochem.* **6**, 117 (1975).

120. N. Ikekawa, M. Morisaki, N. Koizumi, Y. Kato, and T. Takeshita, *Chem. Pharm. Bull.* **23**, 695 (1975).

121. G. Milhaud, M.-L. Labat, J. Redel, *C. R. Acad. Sci.* **279**, 827 (1974).

122. M. Seki, N. Koizumi, M. Morisaki, and N. Ikekawa, *Tetrahedron Lett.*, 15 (1975).

123. J. J. Partridge, V. Toome, and M. R. Uskokovic, *J. Am. Chem. Soc.* **98**, 3739 (1976).

124. E. J. Semmler, M. F. Holick, H. K. Schnoes, and H. F. De-Luca, *Tetrahedron Lett.*, 4147 (1972).

125. K. Ochi et al., *J. Chem. Soc. Perkin I*, 165 (1979).

126. T. A. Narwid, J. F. Blunt, J. A. Iacobelli, and M. R. Uskokovic, *Helv. Chim. Acta* **57**, 781 (1974).

127. J. Rubio-Lightbourn, M. Morisaki, and N. Ikekawa, *Chem. Pharm. Bull.* **21**, 1854 (1973).

128. M. Morisaki, K. Bannai, and N. Ikekawa, *Chem. Pharm. Bull.* **21**, 1853 (1973).

129. M. Morisaki, J. Rubio-Lightbourn, N. Ikekawa, and T. Takeshita, *Chem. Pharm. Bull.* **21**, 2568 (1973).

130. H. E. Paaren, D. E. Hamer, H. K. Schnoes, and H. F. DeLuca, *Proc. Natl. Acad. Sci. U.S.A.* **75**, 2080 (1978).

131. N. Ikekawa et al., *Biochem. Biophys. Res. Commun.* **62**, 485 (1975).

132. J. J. Partridge, S. J. Shivey, E. G. Baggiolini, B. Hennessy, and M. Uskokovic, in A. W. Norman and co-eds., *Vitamin D: Biochemical, Chemical and Clinical Aspects Related to Calcium Metabolism*, Walter de Gruyter, Berlin, 1977, p. 47.

133. H.-Y. Lam, H. K. Schnoes, and H. F. DeLuca, *Steroids* **25**, 247 (1975).

134. J. Redel, P. Bell, F. Delbarre, and E. Kodicek, *C. R. Acad. Sci.* **276**, 2907 (1973).

135. J. Redel et al., *Steroids* **24**, 463 (1974); J. Redel, N. Bazely, Y. Tanaka, and H. F. DeLuca, *FEBS Lett.* **94**, 228 (1978).

136. A. W. Norman et al., *Biochemistry* **22**, 1798 (1983).

137. H. F. DeLuca and H. K. Schnoes, in A. W. Norman and co-eds., *Vitamin D: Recent Basic Advances and Their Clinical Application*, Walter de Gruyter, Elmsford, N.Y., 1979.

138. D. E. M. Lawson, *Vitamin D*, Academic Press, Inc., New York, 1978.

139. H. Dai and G. H. Posner, *Synthesis*, 1383 (1994).

140. W. H. Okamura and C. D. Zhu, *Chemical Reviews* **95**, 1877 (1995).

141. A. W. Norman, R. Bouillon, and M. Thomasset, eds., *Vitamin D, Structure, Function Analysis and Clinical Application, Proceedings of the Eighth Workshop on Vitamin D*, Paris, July 5, 1991, Walter de Gruyter, Berlin, 1991.

142. D. H. R. Barton, R. H. Hesse, M. M. Pechet, and E. Rizzardo, *J. Chem. Soc., Chem. Commun.*, 203–204 (1974).

143. D. H. R. Barton, R. H. Hesse, M. Pechet, and E. Rizzardo, *J. Am. Chem. Soc.* **95**, 2748 (1973).

144. B. Lythgoe et al., *Tetrahedron Lett.*, 3863–3866 (1975).

145. B. Lythgoe et al., *J. Chem. Soc., Perkin Trans. 1*, 590–595 (1978).

146. B. Lythgoe, T. A. Moran, M. E. N. Nambudiry, and S. Ruston, *J. Chem. Soc., Perkin Trans. 1*, 2386–2390 (1976).

147. B. Lythgoe et al., *J. Chem. Soc., Perkin Trans. 1*, 387–395 (1978).

148. J. V. Frosch, I. T. Harrison, B. Lythgoe, A. K. Saksena, *J. Chem. Soc., Perkin Trans. 1*, 2005–2009 (1974).

149. E. G. Baggiolini et al., *J. Org. Chem.* **51**, 3098–3108 (1986).

150. J. DeSchrijver and R. J. De Clercq, *Tetrahedron Lett.* **34**, 4369 (1993).

151. G. H. Posner and C. M. Kinter, *J. Org. Chem.* **55**, 3967–3969 (1990).

152. L. Castedo, J. L. Mascarenas, A. Mourino, and L. A. Sarandeses, *Tetrahedron Lett.* **29**, 1203–1206 (1988).

153. L. Castedo, A. Mourino, and L. A. Sarandeses, *Tetrahedron Lett.* **29**, 1203–1206 (1988).

154. S. A. Barrack, R. A. Gibbs, and W. J. Okamura, *J. Org. Chem.* **53**, 1790–1796 (1988).

155. J. M. Aurrecoechea and W. H. Okamura, *Tetrahedron Lett.* **28**, 4947–4950 (1987).

156. W. H. Okamura, J. M. Aurrecoechea, R. A. Gibbs, and A. W. Norman, *J. Org. Chem.* **54**, 4072–4083 (1989).

157. E. M. VanAlstyne, A. W. Norman, and W. H. Okamura, *J. Am. Chem. Soc.* **116**, 6207–6216 (1994).

158. M. L. Hammond et al., *J. Am. Chem. Soc.* **100**, 4907 (1978).

159. P. Condran et al., *J. Am. Chem. Soc.* **102**, 6259 (1980).

160. W. H. Okamura et al., in Ref. 53, pp. 5–12.

161. B. M. Trost, J. Dumas, and M. Villa, *J. Am. Chem. Soc.* **114**, 9836–9845 (1992).

162. K. Nagasawa, Y. Zako, H. Ishihara, and I. Shimizu, *Tetrahedron Lett.* **32**, 4937–4940 (1991).

163. M. Sheves and Y. Mazur, *J. Am. Chem. Soc.* **97**, 6249–6250 (1975).

164. H. E. Paaren, H. F. DeLuca, and H. K. Schnoes, *J. Org. Chem.* **45**, 3253–3258 (1980).

165. M. Kabat et al., *Tetrahedron Lett.* **32**, 2343–2346 (1991).

166. S. R. Wilson, A. M. Venkatesan, C. E. Augelli-Szafran, and A. Yasmin, *Tetrahedron Lett.* **32**, 2339–2342 (1991).

167. D. R. Andrews et al., *J. Org. Chem.* **51**, 1635–1637 (1986).

168. H. H. Inhoffen, G. Quinkert, and S. Schutz, *Chem. Ber.* **90**, 1283–1286 (1957).

169. H. H. Inhoffen, G. Quinkert, H.-J. Hess, and H. Hirschfeld, *Chem. Ber.* **90**, 2544–2553 (1957).

170. D. R. Andrews, D. H. R. Barton, R. H. Hesse, and M. M. Pechet, *J. Org. Chem.* **51**, 4819–4828 (1986).

171. L. Vanmaele, P. J. De Clercq, and M. Vandewalle, *Tetrahedron Lett.* **41**, 141–144 (1985).

172. M. F. Holick, *Am. J. Clin. Nutr.* **60**, 619–630 (1994).

173. M. F. Holick et al., *Endocrinology* **135**, 655 (1994).

174. G. Ponchon, A. L. Kennan, and H. F. DeLuca, *J. Clin. Invest.* **48**, 1273 (1969).

175. T. C. Madhok and H. F. DeLuca, *Biochem. J.* **184**, 491 (1979); E. Axen, T. Bergman, and K. Wikvall, *J. Steroid Biochem. Mol. Biol.* **51**, 97 (1994).

176. S. A. Holick, M. F. Holick, T. E. Tavela, H. K. Schnoes, and H. F. DeLuca, *J. Biol. Chem.* **251**, 397 (1976).

177. E. B. Olson, Jr., J. C. Knutson, M. H. Bhattacharyya, and H. F. DeLuca, *J. Clin. Invest.* **57**, 1213 (1976).

178. J. A. Eisman, R. M. Shepard, and H. F. DeLuca, *Anal. Biochem.* **80**, 198 (1977).

179. D. R. Fraser and E. Kodicek, *Nature (London)* **228**, 764 (1970).

180. J. G. Ghazarian, C. R. Jefcoate, J. C. Knutson, W. H. Orme-Johnson, and H. F. DeLuca, *J. Biol. Chem.* **249**, 3026 (1974).

181. L. T. Boyle, L. Miravet, R. W. Gray, M. F. Holick, and H. F. DeLuca, *Endocrinology* **90**, 605 (1972).

182. T. C. Chen, L. Castillo, M. Korycka-Dahl, and H. F. DeLuca, *J. Nutr.* **104**, 1056 (1974).

183. H. F. DeLuca et al., in C. Hansch, P. G. Sammes, and J. B. Taylor, eds., *Comprehensive Medicinal Chemistry*, Vol. 3, Pergamon Press, Oxford, U.K., 1991, p. 1129.

184. A. W. Norman et al., *J. Steroid Biochem. Mol. Biol.* **41**, 231 (1992).

185. K. E. Lowe, A. C. Maiyor, and B. W. Norman, *Crit. Rev. Eukar. Gene. Exp.* **2**, 65–109 (1992).

186. D. T. Baran, *J. Cell. Biochem.* **270**, 303 (1994).

187. D. W. Beno et al., *J. Biol. Chem.* **270**, 3542 (1995).

188. R. Kumar, D. Harnden, and H. F. DeLuca, *Biochemistry* **15**, 2420 (1976).

189. D. Harnden, R. Kumar, M. F. Holick, and H. F. DeLuca, *Science* **193**, 493 (1976).

190. M. F. Holick et al., *Biochemistry* **11**, 4251 (1972).

191. Y. Tanaka, L. Castillo, H. F. DeLuca, and N. Ikekawa, *J. Biol. Chem.* **252**, 1421 (1977).

192. M. F. Holick et al., *J. Biol. Chem.* **248**, 6691 (1973).

193. J. N. M. Heersche and J. A. Kanis, eds., *Bone and Mineral Research*, Vol. 8, Elsevier Science B.V., Amsterdam, the Netherlands, 1994, p. 45.

194. M. Gross et al., *J. Bone Miner. Res.* **1**, 457 (1986).

195. T. V. Wijngaarden et al., *Cancer Res.* **54**, 5711 (1994).

196. R. J. Skowronski, D. M. Pechl, and D. Feldman, *Endocrinology* **136**, 20 (1995).

197. M. Inaba et al., *Blood* **82**, 53 (1993).

198. J. A. MacLaughlin et al., *Proc. Natl. Acad. Sci.* **82**, 5409 (1985).

199. S. Morimoto et al., *Calcif. Tissue Int.* **39**, 209 (1986).

200. S. Morimoto et al., *Calcif. Tissue Int.* **38**, 119 (1986).

201. K. J. Kragballe, *J. Cell. Biochem.* **49**, 46 (1992).

202. J. M. Lemire, *J. Cell. Biochem.* **49**, 26 (1992).

203. E. P. Amento, *Steroids* **49**, 55 (1987).

204. M. S. Saporito et al., *Brain Res.* **633**, 189 (1994).

205. M. S. Saporito et al., *Exp. Neurol.* **123**, 295 (1993).

206. S. Carswell, *Exp. Neurol.* **124**, 36 (1993).

207. H. Kohler et al., *Zentralbl. Veterinaermed. Reihe B.* **25**, 89 (1978).

208. F. H. Bird, *Poult. Sci.* **57**, 1293 (1978).

209. B. Jedek et al., *Zentralbl. Veterinaermed. Reihe B.* **25**, 29 (1978).

210. *Vitamin Tolerances of Animals*, National Research Council, National Academy Press, Washington, D.C., 1987, p. 10.

211. Y. H. Hull, *Encyclopedia of Food Science and Technology*, Vol. 4, John Wiley & Sons, Inc., New York, 1991, Pts. 1–8, p. 2713.

212. *Recommended Dietary Allowances*, 7th ed., National Academy of Science, National Research Council, Washington, D.C., 1968, p. 1964.

213. *Nutrient Requirements of Poultry*, 9th ed., National Academy of Science, National Research Council, Washington, D.C., 1994, p. 15.

214. *Nutritient Requirements of Swine*, 9th ed., National Academy of Science, National Research Council, Washington, D.C., 1988, p. 35.

215. *Nutritional Requirements of Dairy Cattle*, 3rd ed., National Academy of Science, National Research Council, Washington, D.C., 1966, p. 1349.

216. H. Edwards, "Factors Influencing Leg Disorders in Broilers," p. 21 in (Proceedings of the Maryland Nutrition Conference, March 23, 1995. University of Maryland, College Park, Maryland).

217. W. Putscher, *Z. Kinderheilkd.* **48**, 269 (1929).

218. D. O. Harrington and E. H. Page, *J. Am. Vet. Med. Assoc.* **182**, 1358 (1983).

219. K. Diem and C. Lentner, eds., *Scientific Tables*, Ciba Geigy, Ltd., Basel, Switzerland, 1971, p. 464.

220. R. L. Morrissey et al., *J. Nutr.* **107**, 1027 (1977).

221. *Oil, Paint and Drug Reporter* **167**, No. 2, p. 23 (1955) and subsequent issues.

222. *Chemical Economics Handbook*, Stanford Research Institute, Menlo Park, Calif., 1980.

223. *Chem. Mktg. Rep.* **251**, No. 5, p. 39 (1997).

224. U.S. Pat. 2,417,299 (Mar. 11, 1947), L. Freedman and E. Green (to U.S. Vitamin Corp.).

225. U.S. Pat. 2,702,262 (Feb. 15, 1955), Burley and A. E. Timrech (to Charles Pfizer and Co., Inc.).

226. U.S. Pat. 2,827,452 (Mar. 18, 1978), H. Schlenk, D. M. Sand, and J. A. Tillotson (to University of Minnesota).

227. U.S. Pat. 3,067,104 (Dec. 4, 1962), M. Hochberg and C. Ely (to Nopco Chemical Co.).

228. U.S. Pat. 3,143,475 (Aug. 4, 1964), A. Koff and R. F. Widmer (to Hoffmann-LaRoche, Inc.).

GENERAL REFERENCES

Refs. 5, 6, 16, 51, 62, and 141 are also general references.

C. E. Bells, in W. H. Sebrell and R. S. Harris, eds., *The Vitamins*, 1st ed., Vol. 2, Academic Press, New York, 1954, p. 132.

G. F. Combs and H. F. DeLuca, *The Vitamins: Fundamental Aspects in Nutrition and Health*, Academic Press, Inc., 1992, p. 151.

E. D. Collins and A. W. Norman, in L. J. Macklin, ed., *Handbook of Vitamins*, 2nd ed., Marcel Dekker, New York, 1991, Chapt. 2, p. 59.

H. F. DeLuca in R. B. Olfin-Slater and D. Kritchevsky, eds., *Nutrition and the Adult: Micronutrients*, Plenum Press, New York, 1986, p. 205.

W. Friedrich, *Vitamins*, Walter de Gruyter, New York, 1988, p. 141.

A. W. Norman, in M. Brown, ed., *Present Knowledge in Nutrition*, 6th ed., International Life Sciences Institute–Nutrition Foundation, Washington, D.C., 1990, Chapt. 12, p. 108.

ARNOLD L. HIRSCH
Alpharma, Inc.

VITAMINS: VITAMIN E

Vitamin E was first described in 1922 and the name was originally applied to a material found in vegetable oils. This material was found to be essential for fertility in rats. It was not until the early 1980s that symptoms of vitamin E deficiency in humans were recognized. Early work on the natural distribution, isolation, and identification can be attributed to Evans, Burr, and Emerson (University of California) and Mattill and Olcott (University of Iowa). Subsequently a group of substances (Fig. 1), which fall into either the family of tocopherols or tocotrienols, were found to act like vitamin E (1–4). The structure of α-tocopherol was determined by degradation studies in 1938 (5).

CHEMICAL, BIOLOGICAL, AND PHYSICAL PROPERTIES

The structures (see Fig. 1) of all vitamin E compounds are characterized by a 6-chromanol ring with a C_{16} side chain. As a result of three asymmetric carbons, there are eight possible optical isomers. However, it is the configuration around the 2-position on the ring which determines, to the largest extent, biological activity (6). Natural vitamin E is characterized by all R stereochemistry. The tocopherols are distinguished by a saturated phytol side chain and the to-

cotrienols are characterized by an unsaturated side chain. The isomers within the two families differ by the extent and position of the methylation on the ring. Although tocopherols and tocotrienols are chemically related, they have varying degrees of vitamin E activity depending on their structure and stereochemistry (Table 1).

The biological activity of various vitamin E forms was established by the fetal resorption assay in rats and is assumed to be applicable to humans. The results of some human studies may indicate that the ratio of 1.36 underestimates the biological activity of the *RRR* form relative to the *all-rac* form of α-tocopheryl acetate (10–12).

α-Tocopherol and α-tocopheryl acetate are viscous oils. The fat solubility of α-tocopherol and α-tocopheryl acetate is characteristic of the family of compounds. Both are readily soluble in ethanol, chloroform, and acetone, but are insoluble in water. Other physical properties of α-tocopherols and their acetate esters are given in Table 2.

Although they are generally stable to heat and alkali, and acids in the absence of oxygen, free tocopherols and tocotrienols can be oxidized by atmospheric oxygen in the presence of light, unsaturated fatty acids, heat, or metal ions. The reported oxidation products of α-tocopherol are shown in Figure 2. These products can be readily synthesized by oxidizing α-tocopherol with such reagents as nitric acid, silver nitrate, ferric chloride, or benzoyl peroxide. As

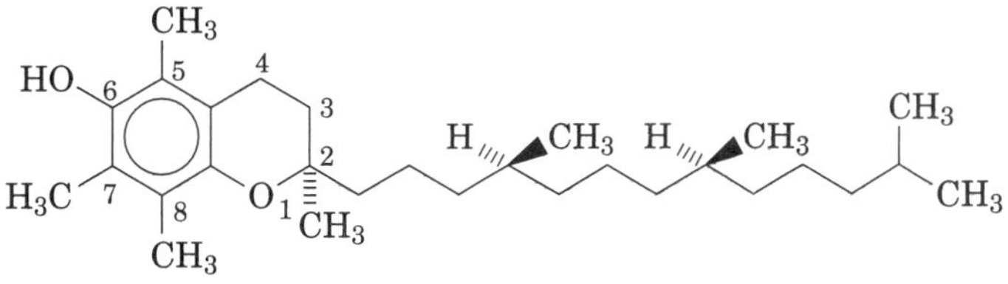

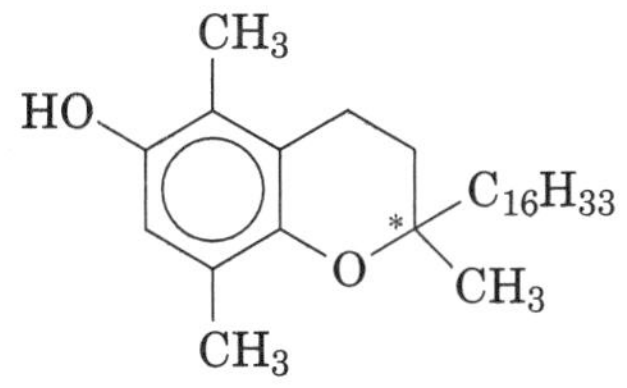

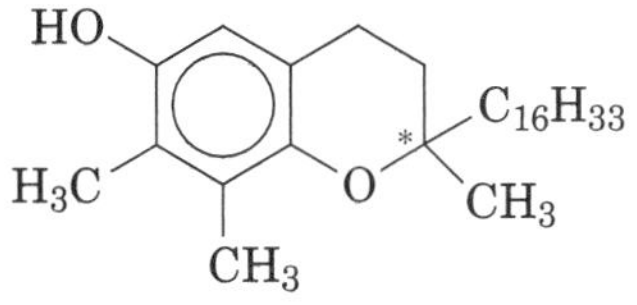

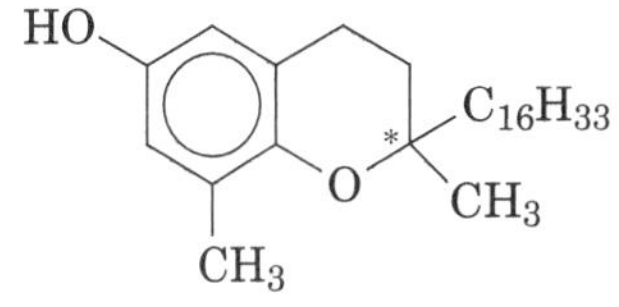

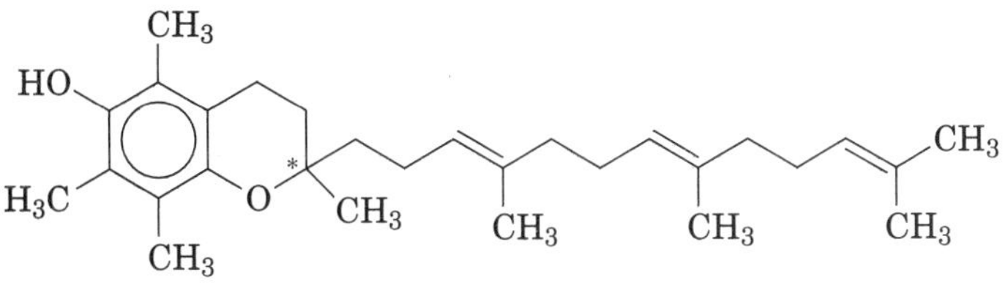

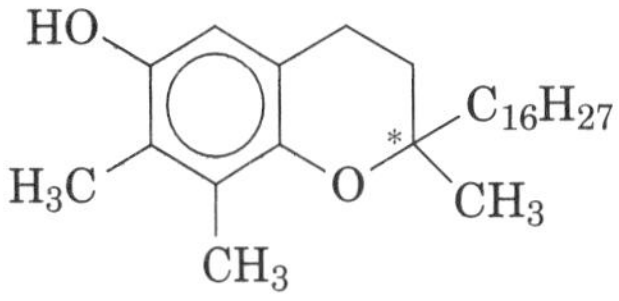

Figure 1. The four naturally occurring tocopherols (α-tocopherol, *RRR* [59-02-9]/*all-rac* [2074-53-5] (**1**); β-tocopherol [148-03-8] (**2**); γ-tocopherol [54-28-4] (**3**); δ-tocopherol [119-13-1] (**4**)), α-tocotrienol [1721-51-3] (**5**), and β-tocotrienol [14101-61-2] (**6**) where asterisks denote asymmetric centers and the absolute configuration of *RRR*-α-tocopherol is shown.

Table 1. Relative Activity of Vitamin E Forms

Substance	Vitamin E activity
all-rac-α-Tocopheryl acetate	1.00
RRR-α-Tocopheryl acetate	1.36
all-rac-α-Tocopheryl acid succinate	0.89
RRR-α-Tocopheryl acid succinate	1.21
all-rac-α-Tocopherol	1.10
RRR-α-Tocopherol	1.49
β-Tocopherol	0.50
γ-Tocopherol	0.35
α-Tocotrienol	0.30
γ-Tocotrienol	0.10

Source: Refs. 7–9.

a result of the ease of oxidation, significant amounts of vitamin E (tocopherols) may be lost during food processing (qv). The severity of the processing, ie, cooking, canning, and baking, determines the amount of tocopherol remaining in foods and edible oils. Storage conditions can also impact on the losses of vitamin E in foods and animal feeds. The stability of α-tocopherol is significantly enhanced by esterification. Although the acetate ester does not have any inherent antioxidant activity, it is bioactive and is the commercially important form because of its greater stability.

NATURAL OCCURRENCES AND SOURCES

Unesterified tocopherols are found in a variety of foods: however, concentration and isomer distribution of tocopherols vary greatly with source. Typically, meat, fish, and dairy contain <40 mg/100 g of total tocopherols. Almost all (>75%) of this is α-tocopherol for most sources in this group. The variation in the content of meat and dairy products can be related to the content of the food ingested by the animal. A strong seasonal variation can also be observed. Vegetable oils contain significant levels of γ-, β-, and δ-tocopherol, along with α-tocopherol (Table 3).

Vegetable oils, typically soybean, are important feedstocks for the commercial production of the *RRR* forms of vitamin E.

SYNTHESIS

Although all four tocopherols have been synthesized as their *all-rac* forms, the commercially significant form of tocopherol is *all-rac-*α-tocopheryl acetate. The commercial processes in use are based on the work reported by several groups in 1938 (15–17). These processes utilize a Friedel-Crafts-type condensation of 2,3,5-trimethylhydroquinone with either phytol (16), a phytyl halide (7,16,17), or phytadiene (7). The principal synthesis (Fig. 3) in current commercial use involves condensation of 2,3,5-trimethylhydroquinone (**13**) with synthetic isophytol (**14**) in an inert solvent, such as benzene or hexane, with an acid catalyst, such as zinc chloride, boron trifluoride, or orthoboric acid/oxalic acid (7,8,18) to give the *all-rac*-α-tocopherol (**15a**). Free tocopherol is protected as its acetate ester (**15b**) by reaction with acetic anhydride. Purification of tocopheryl acetate is readily accomplished by high vacuum molecular distillation and rectification (<1 mm Hg) to achieve the required USP standard.

Trimethylhydroquinone can be synthesized from various isomeric trimethylphenols such as 2,4,6- or 2,3,6-trimethylphenol (Fig. 4). When starting with 2,3,6-trimethylphenol (**16**), this can be accomplished by catalytic oxidation using oxygen and cupric halide catalysts in a solvent such as water-alcohol, 2-methoxyethanol, acetone, or *N,N*-dimethylformamide (19,20). The resulting benzoquinone (**17**) can then be hydrogenated catalytically using Pd or Pt in an inert solvent, such as methanol, isobutanol, or toluene (21) to yield 2,3,5-trimethylhydroquinone (**13**). When 2,4,6-trimethylphenol (**18**) is used, 4-hydroxy-2,4,6-trimethyl-2,5-cyclohexadien-1-one (**19**) is obtained by oxidation with sodium hypochlorite (22) or oxygen and base (23). This compound can then be rearranged by base to trimethylhydroquinone (22,23). The trimethylphenols can be obtained by methylation of phenol.

The isophytol side chain can be synthesized from pseudoionone (Fig. 5) using chemistry similar to that used in the vitamin A synthesis (9). Hydrogenation of pseudoionone (**20**) yields hexahydropseudoionone (**21**) which can be reacted with a metal acetylide to give the acetylenic alcohol (**22**). Rearrangement of the adduct of (**22**) with isopropenylmethyl ether yields, initially, the allenic ketone (**23**) which is further transformed to the C_{18}-ketone (**24**). After reduction of (**24**), the saturated ketone (**25**) is treated with a second mole of metal acetylide. The acetylenic alcohol (**26**) formed is then partially hydrogenated to give isophytol (**14**).

Although the eight stereoisomers of α-tocopherol have been synthesized (24), these preparations are at present

Table 2. Physical Properties of Tocopherols

Property	*all-rac*-α-Tocopherol	(*RRR*)-α-Tocopherol	*all-rac*-α-Tocopheryl Acetate	*RRR*α-Tocopheryl Acetate
CAS Registry Number	[2074-53-5]	[59-02-9]	[7695-91-2]	[58-95-7]
Color	← colorless to pale yellow →			
Form	← viscous oil →			
Boiling point, °C	200–220 (0.1 mm)		224 (0.3 mm)	
sp gr$_{25}^{25}$,	0.947–0.958	0.950–0.964	0.950–0.964	
n_D^{20}	1.5030–1.5070	1.4940–1.4985	1.4940–1.4985	
Molecular weight	430.69	430.69	472.73	472.73
UV absorption				
maxima, nm	292–294	292–294	285.5	285.5
$E_{1\,cm}^{1\%}$ (ethanol)	71–76	71–76	40–44	40–44

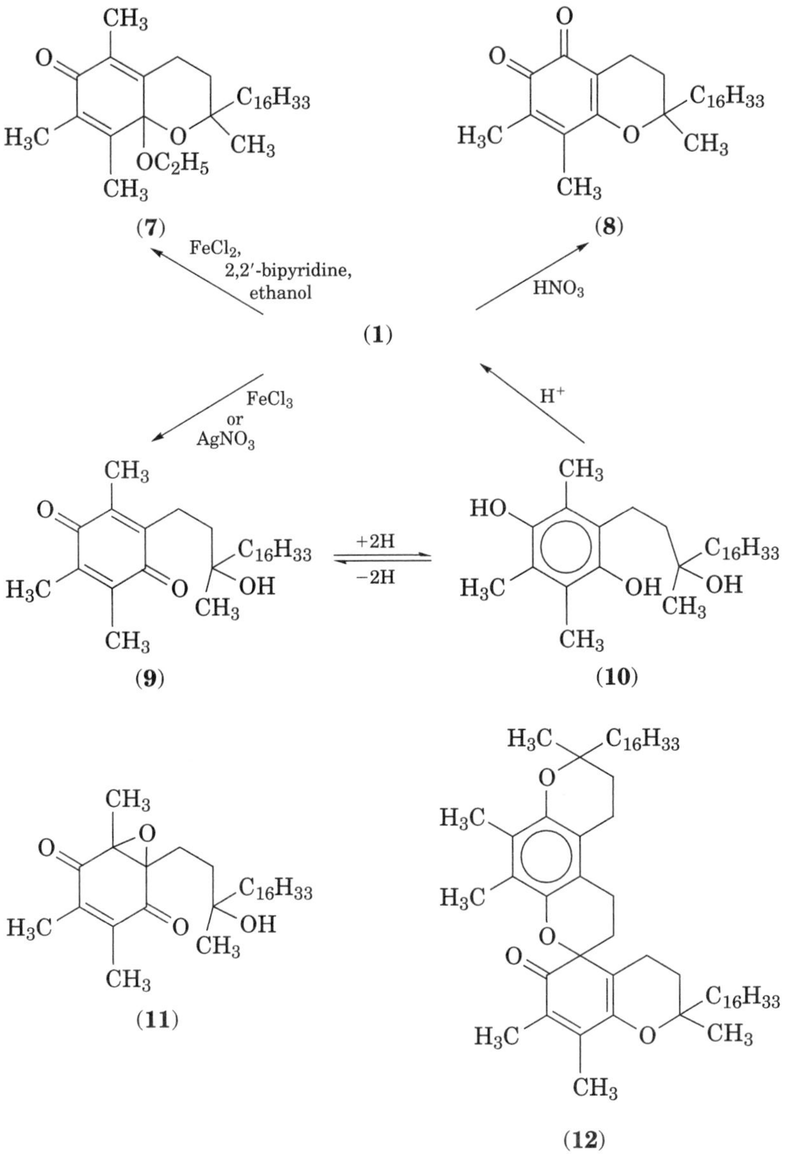

Figure 2. Oxidation products of α-tocopherol: tocopherethoxide [511-72-8] (**7**), tocored [17111-16-9] (**8**), α-tocoquinone [7559-04-8] (**9**), α-tocohydroquinone [14745-36-9] (**10**), α-tocoquinone-2,3-oxide [35499-91-3] (**11**), and α-tocopherol dimer [1604-73-5] (**12**).

Table 3. Vitamin E Content of Common Oils, mg/100 g

Oil source	Total tocopherols	α-Tocopherol	γ-Tocopherol	β-Tocopherol	δ-Tocopherol
Soybean	14	8–10	59		26
Corn	24	10–16	60	5	2
Safflower	41	40	17		1
Canola	23	19	43		
Cottonseed	48	39–44	39		
Sunflower	49	49	5		0.8
Coconut	1	0.5–1			0.6
Palm	29	20–26	32		7
Olive	10	5–10			
Peanut		13	22		2
Rapeseed		18	17		1
Wheatgerm		133	26	71	27

Source: Refs. 13 and 14.

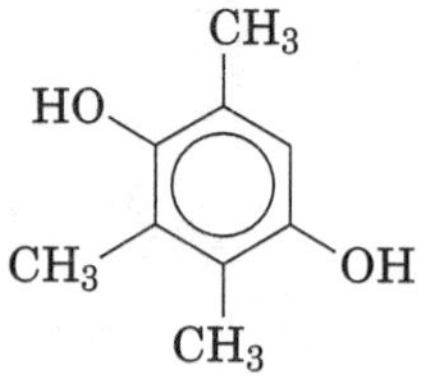

Figure 3. Synthesis via trimethylhydroquinone [700-13-0] (TMHQ) (13) and isophytol [505-32-8] (14) of α-tocopherol (15a) and α-tocopherol acetate (15b).

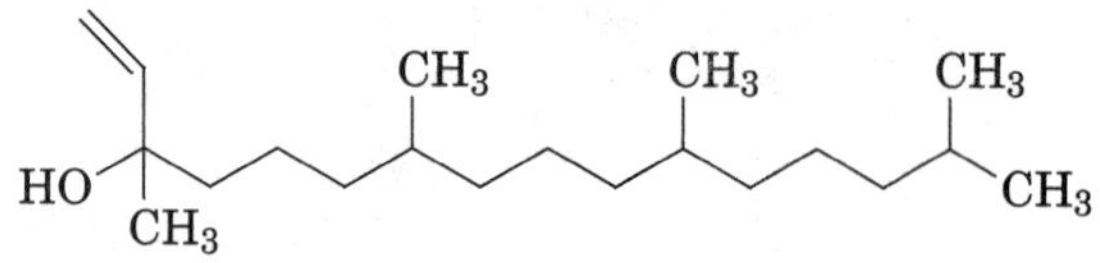

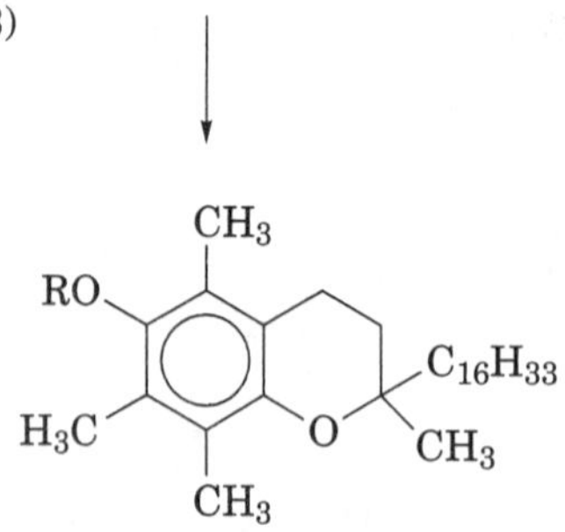

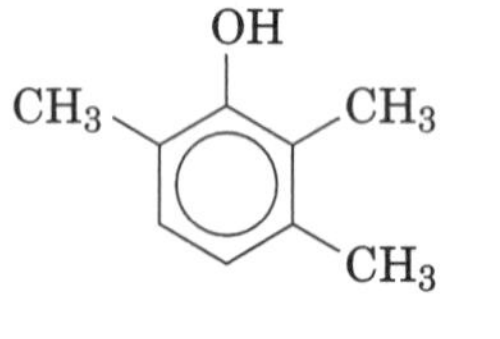

Figure 4. Starting materials and intermediates in the TMHQ syntheses: 2,3,6-trimethylphenol [2416-94-6] (16), 2,3,5-trimethylquinone [935-92-2] (17), 2,4,6-trimethylphenol [527-60-6] (18), and 4-hydroxy-2,4,6-trimethyl-2,5-cyclohexadien-1-one [28750-52-9] (19).

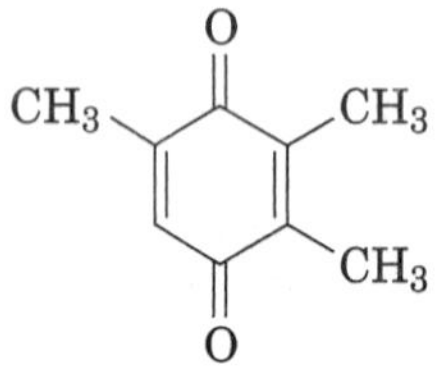

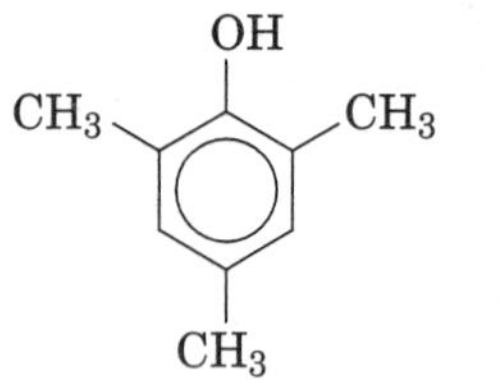

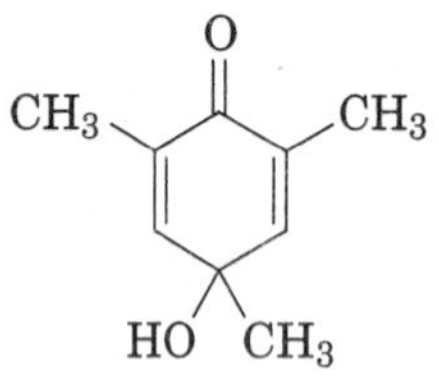

only of academic interest. However, *RRR*-α-tocopherol and the other natural forms of vitamin E can be isolated from deodorizer distillates produced as by-products of vegetable oil processing. This requires a series of steps built around key molecular distillations which are required to increase the purity of the *RRR*-α-tocopherol and mixed tocopherol fractions for use in commercial preparations. A typical process would involve a distillation of alkali-treated soybean oil at high vacuum (<1 mm Hg) in a continuous molecular still. This minimizes thermal degradation. The distillate, which contains α-, γ-, and δ-tocopherols, is then cooled to

low temperature (~ −10°C) in a solvent to remove impurities such as sterols. The other impurities, such as fatty acids, can be removed by saponification. Further molecular distillation is required to produce a fraction containing high levels (≥60%) of tocopherols. The sterols and fatty acids can be sold.

The yield of the more active *RRR*-α-tocopherol can be improved by selective methylation of the other tocopherol isomers or by hydrogenation of α-tocotrienol (25,26). Methylation can be accomplished by several processes, such as simultaneous haloalkylation and reduction with an aldehyde and a hydrogen halide in the presence of stannous chloride (27), aminoalkylation with ammonia or amines and an aldehyde such as paraformaldehyde followed by catalytic reduction (28), or via formylation with formaldehyde followed by catalytic reduction (29).

The *all-rac* forms of β-, γ-, and δ-tocopherols can be synthesized using the same condensation reaction as used for *all-rac*-α-tocopherol. To synthesize *all-rac*-β-tocopherol, 2,5-dimethylhydroquinone instead of trimethylhydroquinone is condensed with isophytol. For *all-rac*-γ- and δ-tocopherol, 2,3-dimethylhydroquinone and methylhydroquinone are used, respectively.

Although apparently not commercially important, fermentation (qv) processes have been reported for the production of tocopherols. *Aspergillus niger* (30), *Lactobacter* (31), *Euglena gracilis Z.* (32,33), and *Mycobacterium* (34) have been shown to produce (*RRR*)-α-tocopherol. In the case of *Euglena*, titers of 140–180 mg/L have been reported.

PHYSIOLOGICAL EFFECTS AND REQUIREMENTS

The symptoms of vitamin E deficiency in animals are numerous and vary from species to species (13). Although the deficiency of the vitamin can affect different tissue types such as reproductive, gastrointestinal, vascular, neural,

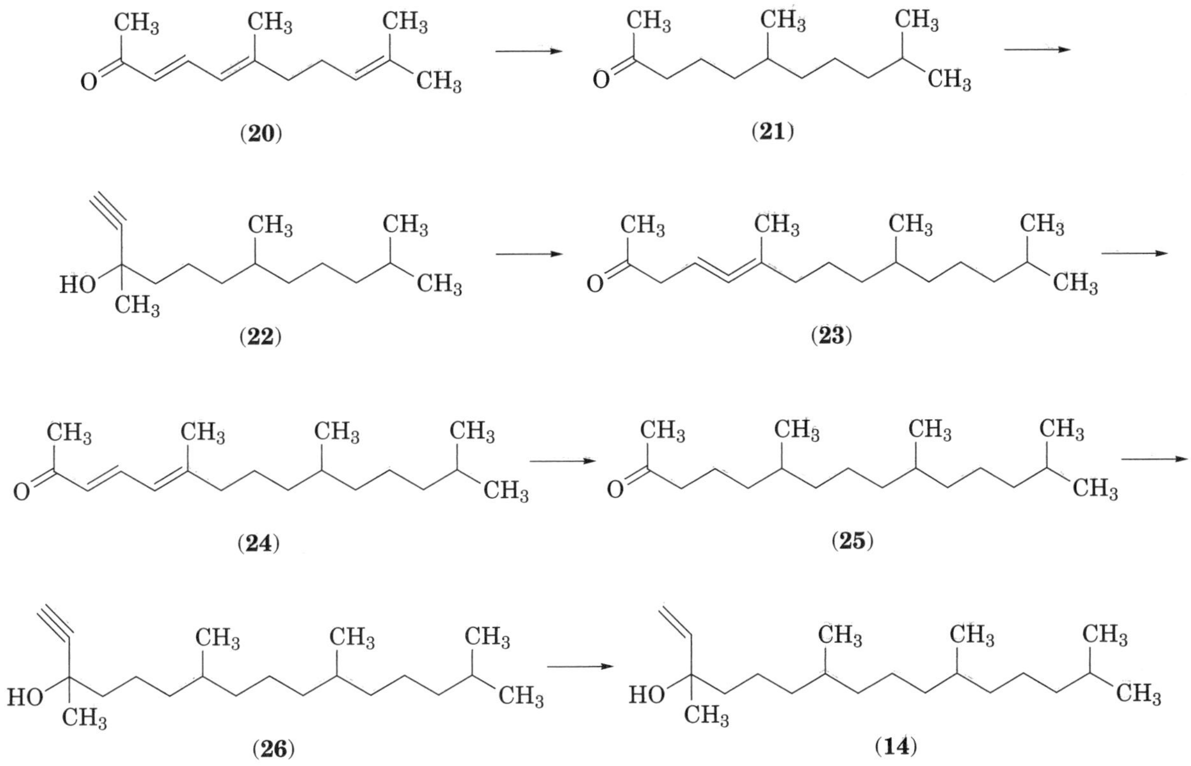

Figure 5. Synthesis of isophytol (**14**): pseudoionone [*141-10-6*] (**20**), hexahydropseudoionone [*1604-34-8*] (**21**), C$_{15}$-acetylenic alcohol [*1604-35-9*] (**22**), C$_{18}$-allenic ketone [*16647-10-2*] (**23**), C$_{18}$-diene ketone [*1604-32-6*] (**24**), C$_{18}$-saturated ketone [*16825-16-4*] (**25**), and C$_{20}$-acetylenic alcohol [*29171-23-1*] (**26**).

hepatic, and optic in a variety of species such as pigs, rats, mice, dogs, cats, chickens, turkeys, monkeys, and sheep, it is generally found that necrotizing myopathy is relatively common to most species. In humans, vitamin E deficiency can result from poor fat absorption in adults and children. Infants, especially those with low birth weights, typically have a vitamin E deficiency which can easily be corrected by supplements. This deficiency can lead to symptoms such as hemolytic anemia, reduction in red blood cell lifetimes, retinopathy, and neuromuscular disorders.

Vitamin E can also act as an antioxidant (qv) in animals and humans alone or in combination with vitamin C (qv). Both are good free-radical scavengers with the vitamin C acting to preserve the levels of vitamin E (35). Vitamin E in turn can preserve the levels of vitamin A in animals (13). It has been shown that vitamin E reduces the incidence of cardiovascular disease (36–39). This most likely results from the antioxidant property of the vitamin which inhibits the oxidation of low density lipoproteins (LDLs) (40–42). The formation of the oxidized LDLs is considered important in decreasing the incidence of cardiovascular disease (43).

The recommended daily allowance for vitamin E ranges from 10 International Units (1 IU = 1 mg *all-rac-α*-tocopheryl acetate) for children under 4 years of age to 30 IU for adults and children over 4 years of age. This should be adequate to prevent vitamin E deficiency in humans. High levels enhance immune responses in both animals

and humans. Requirements for animals vary from 3 USP units/kg diet for hamsters to 70 IU/kg diet for cats (13). The complete metabolism of vitamin E in animals or humans is not known. The primary excreted breakdown products of α-tocopherol in the body are gluconurides of tocopheronic acid (**27**) (Fig. 6). These are derived from the primary metabolite α-tocopheryl quinone (**9**) (see Fig. 2) (44,45).

Vitamin E is considered nontoxic at levels up to 3200 mg/day. It has not been found to be teratogenic, mutagenic, or carcinogenic at doses below 1 g/kg of body weight. Vitamin E can heighten the effect of vitamin K deficiency (coagulation defect) or anticoagulation therapy. A recent study, however, indicates that it may be the oxidation product α-tocoquinone that causes the effect (46).

ECONOMIC ASPECTS

World production in the late 1990s of both natural and synthetic forms of vitamin E is estimated at 22,000 metric tons and growth is expected to keep pace with increasing need. The 1993 U.S. production was 14,096 metric tons (47) with an additional 1080 metric tons from imports. The principal U.S. producers of the natural form are Eastman Chemical Company, Archer Daniels Midland Company, and Henkel, and of synthetic vitamin E, Hoffmann-La Roche and BASF. International producers include Hoffmann-La Roche, BASF, Eisai, and Rhône-Poulenc.

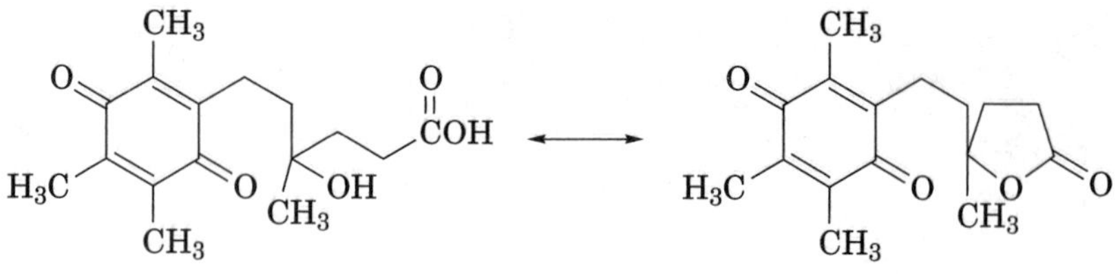

Figure 6. α-Tocopheronic acid [1948-76-1] (27) and its lactone, α-tocopheronolactone [3121-68-4] (28).

(27)   (28)

The price of synthetic vitamin E decreased steadily until about 1990 when prices began to increase. The price of the natural form, although higher than the synthetic form, has approximately paralleled the synthetic form (Table 4).

ANALYTICAL METHODS AND SPECIFICATIONS

Specifications and standards for various vitamin E forms and preparations for use in pharmaceutical applications are given in the *United States Pharmacopeia* (52). All products should contain not less than 96.0% or more than 102.0% of the appropriate form. The products must be labeled to indicate both the chemical and stereochemical forms contained in the product.

Label claims for tocopherol levels in preparations can be based on milligrams or International Units. Only the *RRR* or *all-rac-α*-tocopherol and its esters can be claimed. No vitamin E activity can be claimed for the tocotrienols and the non-α-tocopherols. International Units are also used in some reference books and compendia, eg, *Food Chemicals Codex* (40,53), which is of particular importance for specifications for food fortification.

All formulations of vitamin E must show low acidity, and contain not more than 0.004% heavy metals (reported as Pb) and not more than 10 ppm Pb. Formulations which contain (*RRR*)-α-tocopherol must have a specific rotation of +24° for the oxidation product with alkaline potassium ferricyanide.

Analysis of vitamin E can be done by a variety of methods depending on the form and level present and the preparation in which the form is found (54). For pure or highly concentrated forms, this is accomplished by reaction with Emmerie-Engel reagent (2,2′-bipyridine (α,α′-dipyridyl) and ferric chloride) to give a red color. This color results from the combination of bipyridine with ferrous ions formed from the reduction of the ferric ions by the tocopherol and is directly proportional to the amount of tocopherol present. Analysis of α-tocopherol in forms which contain low levels, such as vegetable oils, foods, or feeds, is difficult because the colorimetric method is nonspecific and significant sample preparation is involved to remove

interferences such as other tocopherols, tocotrienols, and β-carotene. The AOAC Official Methods of Analysis describes a packed column gas chromatographic method for the analysis of tocopherol isomers in mixed concentrates (41). This method separates α- and δ-tocopherols as discrete peaks, but β- and γ-tocopherols elute as a combined peak.

Numerous high pressure liquid chromatographic techniques have been reported for specific sample forms: vegetable oils (55,56), animal feeds (57,58), sera (59,60), plasma (61,62), foods (63,64), and tissues (63). Some of the methods require a saponification step to remove fats, to release tocopherols from cells, and/or to free tocopherols from their esters. All require an extraction step to remove the tocopherols from the sample matrix. The methods include both normal and reverse-phase hplc with either uv absorbance or fluorescence detection. Application of supercritical fluid (qv) chromatography has been reported for analysis of tocopherols in marine oils (65).

Bioassay Method

A modification of the Evans resorption–gestation method can be used to determine α-tocopherol activity in supplements. Although the method is time-consuming, when carefully performed, it can produce reasonably accurate results. The method requires raising female rats on a vitamin E-free diet and mating them with normal males. Unless a vitamin E supplement containing more than 0.3–1.0 mg (depends on methodology) of α-tocopherol is administered in the first 10–12 days of pregnancy, the embryos die and are reabsorbed without apparent harm to the mother. If the test supplement contains greater than the threshold dose of α-tocopherol, the pregnancy proceeds normally.

APPLICATIONS AND PRODUCT FORMS

Both α-tocopherol and its esters are constituents of multivitamin and single-dose nutrient capsules or liquid dietary supplements. Supplements for human use range from a few milligrams in multivitamin preparations to 500–

Table 4. Bulk Prices of Pharmaceutical-Grade α-Tocopheryl Acetate, $/kg

Product	1948	1951	1954	1960	1970	1982	1991	1994	1995
(*RRR*)-form	250	250	185	122	68	68	57	59	59
all rac-form	750	350	136	90	50	27	23	34	34

Source: Refs. 48–51.

1000 mg in single-dose supplements. Specialty items, such as ointments, creams, salves, and suppositories containing vitamin E provide other outlets for α-tocopherol. Tocopherols have significant application in cosmetics (qv) and dermatology. Tocopherols can be used in topical cream or oil forms to treat chronic skin diseases (66), reduce scarring in wound healing (67), reduce inflammation (68), and protect against uv radiation (69). Tocopherols are also incorporated into cosmetic formulations to reduce nitrosamine and nitrosamide formation from amines and amides also present in the formulation (70).

Animal feeds consume approximately 40% of the commercial production of α-tocopherol acetate. Although tocopherol is normally found in feed, the poultry, beef and dairy cattle, lamb, and swine industries use vitamin E in supplements and concentrates to replace tocopherol lost during feed processing and storage. The acetate ester is added to the feeds as either a dry, granular, free-flowing, nondusting powder containing 44,000–276,000 units per kilogram (20,000–125,000 units per pound) or as a high potency oil concentrate (23–100% α-tocopheryl acetate).

Foods are not typically fortified with vitamin E because of the lack of signs of deficiency in the general population. Fortification does occur for infant formulas as a supplement and cereals to replace processing losses. Tocopherols are finding additional applications as antioxidants in foods as the less expensive butylated hydroxytoluene (BHT) and butylated hydroxyanisole (BHA) used are being prohibited in more and more countries. Foods such as lards and citrus oils which do not contain appreciable levels of natural tocopherols can be protected with α-tocopherol at 0.02% levels (weight based on lipid phase) (9). The meat from turkeys and chickens is less prone to rancidity when refrigerated or frozen if the animal's diet contains vitamin E. The meat from pigs fed vitamin E supplements also has better storage (71) and color (72) stability than meat from pigs with normal diets. It has also been reported that veal has better storage stability (73) and beef has better color stability (74) when vitamin E supplements are used in the animal feeds. Tocopherol can be used in bacon and similar foods to prevent the formation of nitrosamines. The traditional antioxidants, such as BHT and phosphites, used in high density polyethylene (HDPE) and polypropylene (PP), can be replaced by *all-rac*-α-tocopherol (75–77).

BIBLIOGRAPHY

1. J. F. Pennock, F. W. Hemming, and J. D. Kerr, *Biophys. Res. Commun.* **17**, 542 (1964).

2. International Union of Nutritional Sciences, *Nutr. Abstr. Rev.* **48**, 831 (1978).

3. R. H. Bunnell, J. Keating, A. Quaresimo, and G. K. Parman, *Am. J. Clin. Nutr.* **17**, 1 (1965).

4. J. G. Bieri and R. P. Evarts, *J. Am. Diet. Assoc.* **62**, 147 (1973).

5. E. Fernholz, *J. Am. Chem. Soc.* **60**, 1741 (1938).

6. H. Weiser and M. Vecchi, *Int. J. Vitam. Nutr. Res.* **52**, 351–370 (1982).

7. S. Kasparak, in L. J. Machlin, ed., *Vitamin E: A Comprehensive Treatise*, Marcel Dekker, Inc., New York, 1980, pp. 7–65.

8. S. Kijima, in M. Mino, H. Nakamura, A. I. Diplock, and H. J. Kayden, eds., *Vitamin E: Its Usefulness in Health and Curing Diseases*, Japan Societies Press, Tokyo, Japan, and S. Karger AG, Basel, Switzerland, 1993, p. 7.

9. D. Bhatia et al., "Vitamins," in *Encyclopedia of Food Technology*, Vol. 4, parts 1–8, John Wiley & Sons, Inc., New York, 1991.

10. G. W. Burton and K. U. Ingold, in L. Packer and G. Fuchs, eds., *Vitamin E in Health and Disease*, Marcel Dekker, Inc., New York, 1993, pp. 329–344.

11. R. V. Acuff, S. S. Thedford, N. N. Hidiroglou, A. M. Papas, and T. Odom, *Am. J. Clin. Nutr.* **60**, 397–402 (1994).

12. H. J. Kayden and M. G. Traber, *T. Lipid Res.* **34**, 343–358 (1993).

13. L. J. Machlin, in L. Machlin, ed., *Handbook of Vitamins*, Marcel Dekker, Inc., New York, 1990, pp. 99–144.

14. H. T. Slover, *Lipids* **6**, 291–296 (1971).

15. P. Karrer, H. Fritzsche, B. H. Ringier, and H. Solomon, *Helv. Chim. Acta* **21**, 520 (1938).

16. F. Bergel, A. Jacob, A. R. Todd, and T. S. Work, *Nature* **142**, 36 (1938).

17. L. I. Smith, H. E. Ungnade, and W. W. Prichard, *Science (Washington, D.C.)* **88**, 37 (1938).

18. Ger. Pat. 92-4208477 (1993), P. Grafen, H. Kiefer, and H. Jaedicke (to BASF AG).

19. U.S. Pat. 5,041,572 (Aug. 20, 1991), U. Hearcher, B. Jessel, B. Bockstiegel, P. Grafen, and H. Laas (to BASF AG).

20. Ger. Pat. 2221624 (Nov. 1972), W. Brenner (to Hoffmann-LaRoche AG).

21. Eur. Pat. Appl. 87-115070 (1988), T. Yui and A. Ito (to Mitsubishi Gas Co. Inc.).

22. Jpn. Pat. 62,263,136(86-104088) (1987), T. Kunitomi and H. Tamai (to Kuraray Co. Ltd.).

23. Y. Ichikawa, Y. Yamanaka, N. Suzuki, T. Naruchi, O. Kobayashi, and H. Tsuruta, *Ind. Eng. Chem. Prod. Res. Dev.* **18**, 373–375 (1979).

24. N. Cohen, C. G. Scott, C. Neukom, R. J. Lopresti, G. Weber, and G. Saucy, *Helv. Chim Acta* **64**, 1158 (1981).

25. J. F. Pennock, F. W. Hemming, and J. D. Kerr, *Biochem. Biophys. Res. Commun.* **17**, 542 (1964).

26. H. May, P. Schindl, R. Ruegg, and O. Isler, *Helv. Chim. Acta* **46**, 963 (1963).

27. U.S. Pat. 2,486,539 (Nov. 1, 1949), L. Weisler (to Distillation Products).

28. U.S. Pat. 2,519,863 (Aug. 22, 1950), L. Weisler (to Eastman Kodak Co.).

29. U.S. Pat. 2,592,628 (Apr. 15, 1952), L. Weisler (to Eastman Kodak Co.).

30. Y. Kawai, M. Otaka, M. Kakio, Y. Oeda, N. Inoue, and H. Shinano, *Hokkaido Daigaku Suisangakubu Kenkyo Iho* **45**, 26–31 (1994).

31. H. Ariga, M. Okazaki, and M. Yamada, *Rakuna Kagaku, Shokuhin no Kenkyu* **41** (1992).

32. Y. Tani and S. Osuka, *Agric. Biol. Chem.* **53**, 2313–2318 (1989).

33. Y. Tani and H. Tsumura, *Agric. Biol. Chem.* **53**, 302–315 (1989).

34. Jpn. Pat. 57,163,490(82,163,490) (Oct. 1982), (to Shiseido Co., Ltd.).

35. E. R. Pacht, H. Kaseki, J. R. Mohammed, D. G. Cornwell, and W. B. Davis, *J. Clin. Invest.* **77**, 789 (1986).

36. N. G. Stephens, A. Parsons, P. M. Schofield, F. Kelly, K. Cheeseman, M. J. Mitchinson, and M. J. Brown, *Lancet* **347**, 781–786 (1996).

37. M. J. Stampfer, C. H. Hennekens, J. E. Manson, G. A. Colditz, B. Rosner, and W. C. Willett, *N. Eng. J. Med.* **328**, 1444–1449 (1993).

38. E. B. Rimm, M. J. Stampfer, A. Ascherio, E. Giovannucci, G. A. Colditz, and W. C. Willett, *N. Eng. J. Med.* **328**, 1450–1456 (1993).

39. K. G. Losonczy, T. B. Harris, and R. J. Havlik, *Am. J. Clin. Nutr.* **64**, 190–196 (1996).

40. H. M. G. Princen, G. Poppel, C. Vogelzang, R. Buytenhek, and F. J. Kok, *Arterioscler. Thromb.* **12**, 554–562 (1992).

41. H. Esterbauer, M. Dieber-Rotheneder, M. Striegel, and G. Waeg, *Am. J. Clin. Nutr.* **53**, 314s–321s (1991).

42. I. Jialal, C. J. Fuller, and B. A. Huet, *Arterioscler. Thromb. Vasc. Biol.* **15**, 190–198 (1995).

43. D. Steinberg, *Circulation* **84**, 1420–1425 (1991).

44. C. Drevon, in Ref. 4, pp. 75–76.

45. C. K. Chow, *World Rev. Nutr. Diet.* **45**, 133–166 (1985).

46. P. Dowd and Z. B. Zheng, *Proc. Natl. Acad. Sci. USA* **92**, 8171–8175 (1995).

47. Synthetic Organic Chemicals, *United States Production and Sales, 1993*, USITC Publication 2810, U.S. Government Printing Office, Washington, D.C., 1994, pp. 3–191.

48. *Chem. Mark. Rep.* **221**, 47 (1982).

49. *Chem. Mark. Rep.* **240**, 32 (1991).

50. *Chem. Mark. Rep.* **245**, 31 (1994).

51. *Chem. Mark. Rep.* **247**, 45 (1995).

52. *The United States Pharmacopeia XXIII (USP XXIII-NF XVIII)*, The United States Pharmacopeia Convention, Inc., Rockville, Md., 1995, p. 1631.

53. Food and Nutrition Board, National Research Council, *Food Chemicals Codex*, 3rd ed., National Academy Press, Washington, D.C., 1981, p. 330.

54. K. Helrich, ed., *Official Methods of Analysis of the Association of Official Analytical Chemists*, 15th ed., Association of Official Analytical Chemists, Inc., Arlington, Va., 1990, pp. 1070–1079.

55. K. Abe, Y. Yuguchi, and G. Katsui, *J. Nutr. Sci. Vitaminol.* **21**, 183–188 (1975).

56. E. Guzman-Contreras and F. C. Strong III, *J. Agric. Food Chem.* **30**, 1109–1112 (1982).

57. H. Cohan and M. R. LaPointe, *J. Assoc. Off. Anal. Chem.* **63**, 1254–1257 (1980).

58. C. H. McMurray and W. J. Blanchflower, *J. Chrom.* **176**, 488–492 (1979).

59. L. Jansson, B. Nilsson, and R. Lingren, *J. Chrom.* **181**, 242–247 (1980).

60. W. J. Driskell, J. Neese, C. C. Bryant, and M. M. Bashor, *J. Chrom.* **231**, 439–444 (1982).

61. C. H. McMurray and W. J. Blanchflower, *J. Chrom.* **178**, 525–531 (1979).

62. J. Lehman and H. Martin, *Clin. Chem.* **28**, 1784–1787 (1982).

63. J. N. Thompson and G. Hatina, *J. Liq. Chrom.* **2**, 327–344 (1979).

64. U. Manz and K. Philipp, *Int. J. Vit. Nutr. Res.* **51**, 342–348 (1981).

65. A. Staby, C. Borch-Jensen, S. Balchen, and J. Mollerup, *Chromatographia* **39**, 697–705 (1994).

66. W. Nikolowski, in *Vitamins* (3) Editones (Roche), F. Hoffmann-LaRoche & Co. Ltd., Basle, Switzerland, (1973).

67. M. Ehrlich, H. Traver, and T. Hunt, *Annal. Surg.* **175** (Feb. 1972), pp. 31–36.

68. M. Kamimura, *J. Vitaminol.* **18**, 201–209 (1972).

69. W. A. Pryor, *Free Radicals in Biology*, Academic Press, Inc., New York, 1976, Chapt. 1.

70. W. Mergens and E. DeRiter, *Cosmetic Technol.* (Jan. 1980).

71. F. J. Monahan, D. J. Buckley, P. A. Morrissey, P. B. Lynch, and J. I. Gray, *Food Sci. Nutr.* **42F**, 203 (1990).

72. A. Asgher, J. I. Gray, A. M. Booser, E. A. Gomaa, M. M. Abonzied, and E. R. Miller, *J. Sci. Food Agric.* **57**, 31 (1991).

73. R. Ellis, W. I. Kimato, J. Bitman, and L. F. Edmonson, *J. Am. Chem. Soc.* **51**, 4 (1974).

74. F. B. Shorland, J. O. Igene, A. M. Pearson, J. W. McGuffy, and A. E. Aldridge, *J. Agric. Food Chem.* **29**, 863 (1981).

75. D. Burdick, S. Laermer, S. Young, and P. Zambetti, "A New Primary Antioxidant System for Polyolefins," presented at *Additives '95*, Clearwater, Fla., Feb. 22–24, 1995.

76. S. F. Laermer and P. F. Zambetti, *J. Plast. Film Sheet.* **8**, 228–248 (1992).

77. Y. C. Ho, K. L. Yam, S. S. Young, and P. F. Zambetti, *J. Plast. Film Sheet.* **10**, 194–212 (1994).

GENERAL REFERENCES

L. J. Machlin, ed., *Vitamin E: A Comprehensive Treatise*, Marcel Dekker, Inc., New York, 1980.

ROBERT CASANI
Hoffmann-La Roche, Inc.

VITAMINS: VITAMIN K

Vitamin K represents a class of substances which contain the 2-methyl-1,4-naphthoquinone moiety and are characterized by their antihemorraghic properties. Vitamin K was first discovered by Dam in 1929 during experiments with chicks on a lipid-deficient diet (1). Dam coined the term *Koagulations vitamin* to describe this new substance (2,3). In the late 1930s, several groups identified the causative agent responsible for the antihemorraghic properties of alfalfa (4–7). Structural determination of vitamin K_1 (**1**) was the result of work from several laboratories (8,9). Stereochemical assignment came much later from Mayer (10). Additional studies isolated and characterized related materials. Vitamin K_2 (**2**) was first found in decaying fish meal (9); it is a crystalline compound and is distinguished by its repeating isoprenyl units. Vitamin K_3 [58-27-5] (**3**) is the simplest form of the vitamin. The chemical name and common names for some important forms of the vitamin follow.

Name	Vitamin K_1
CAS Registry Number	[84-80-0]
Chemical name	1,4-Naphthalenedione, 2-methyl-3-(3,7,11,15-tetramethyl-2-hexadecenyl)-, [R-[R*,R*-(E)]]-
Common name	Phylloquinone, phytomenadione, phytonadione
Name	Vitamin $K_{2(20)}$
CAS Registry Number	[863-61-6]
Chemical name	1,4-Naphthalenedione, 2-methyl-3-(3,7,11,15-tetramethyl-2,6,10,14-hexadecatetraenyl)-, (E,E,E)-
Common name	Menaquinone 4, menaquinone K4, MK-4
Name	Vitamin K_3
CAS Registry Number	[58-27-5]
Chemical name	1,4-Naphthalenedione, 2-methyl
Common name	Menadione; menaquinone-0; synkay

(1)

(2)

(3)

The K vitamins are named after the original vitamin and the number of carbon atoms in the side chain. Using this convention, vitamin $K_{2(20)}$ is so named because it contains 20 carbon atoms in the chain. In the biological literature, vitamin K_2 is frequently referred to as menaquinone and is further designated by the number of isoprene units in the side chain. For example, vitamin $K_{2(20)}$ is also called menaquinone-4 or MK-4. Vitamin K_3 is also referred to as menadione or MK-0.

Vitamin K is typically found in green, leafy vegetables such as cabbage, broccoli, and spinach at levels of 95–200 μg/100 g of fresh vegetables. Cauliflower at a level of 136 μg/100 g also represents an excellent source of dietary vitamin K. Additionally, animal sources such as liver and eggs provide good sources of vitamin K (11).

CHEMICAL/PHYSICAL PROPERTIES

Vitamin K compounds are yellow solids or viscous liquids. The natural form of vitamin K_1 is a single diastereoisomer with $2'(E)$, $7'(R)$, $11'(R)$ stereochemistry. The predominant commercial form of vitamin K_1 is the racemate and a $2'(E)$/(Z) mixture. Table 1 lists some physical and spectral properties of vitamin K_1.

Vitamin K_1 is insoluble in water and is soluble in 70% alcohol, chloroform, petroleum ether, benzene, and hexane. Vitamin K_1 is stable in air but should be protected from light. Although unstable in alkali, the vitamin is stable in acidic medium. Its facile decomposition in basic solution forms the basis of the Dam-Karrer color test.

Early structural determination lends insight into the chemical reactivity of vitamin K_1. Catalytic hydrogenation of vitamin K_1 consumes four moles of hydrogen and affords a colorless compound. Because complete hydrogenation of a 1,4-naphthoquinone structure consumes three molecules of hydrogen, consumption of the fourth mole indicates unsaturation in the side chain. Reductive acetylation of vitamin K_1 affords the diacetate of dihydrovitamin K_1. This behavior further supports the conclusions that vitamin K_1 contains a naphthoquinone group (13).

In addition to its reactivity toward reducing agents, vitamin K_1 is also reactive toward oxidizing agents. Chromic acid oxidation of vitamin K_1 gives rise to two principal products, 2-methyl-1,4-naphthoquinone-3-acetic acid and phthalic acid. By comparison with known compounds, it has been concluded that the benzene ring is unsubstituted and the substitution pattern on the naphthoquinone ring is 2,3-dialkyl (8,14–16).

ANALYTICAL METHODS

The classical method for the determination of vitamin K is based on the clotting time of a vitamin K-deficient chick. It is relatively easy to produce a hemorraghic state in chicks (17). Vitamin K-deficient rats have also been used for this assay (18). Owing to the development of modern chromatographic techniques, this method of analysis has been supplanted by other methodology.

There are several comprehensive reviews of analytical methods for vitamin K (19,20). Owing to the presence of a naphthoquinone nucleus, the majority of analytical methods use this structural feature as a basis for analysis. Several identity tests such as its reaction with sodium bisulfite or its uv spectrum exploit this characteristic. Although not specific, titrimetric, polarographic, and potentiometric methods have also been used (20).

Chromatographic methods including thin-layer, hplc, and gc methods have been developed. In addition to developments in the types of columns and eluents for hplc applications, a significant amount of work has been done in the kinds of detection methods for the vitamin. These detection methods include direct detection by uv, fluorescence after post-column reduction of the quinone to the hydroquinone, and electrochemical detection. Quantitative gc methods have been developed for the vitamin but have found limited applications. However, gc methods coupled

Table 1. Physical and Spectral Properties of Vitamin K₁

Item	Vitamin K$_1$	Vitamin K$_{2(30)}$	Vitamin K$_{2(35)}$	Vitamin K$_3$
Color	Yellow, Viscous oil	Yellow crystals	Yellow crystals	Bright yellow crystals
Melting point, °C	−20	50	54	105–107
Molecular weight	450.68	580.9	649.02	172.17
Molecular formula	$C_{31}H_{46}O_2$	$C_{41}H_{56}O_2$	$C_{46}H_{64}O_2$	$C_{11}H_8O_2$
Spectrophotometric data, λ_{max}, nm	242, 248, 260, 269, 325	243, 248, 261, 270, 325	243, 248, 261, 270, 325	244
ϵ, petroleum ether	396, 419, 383, 387, 68	304, 320, 290, 292, 53	278, 195, 266, 267, 48	1150 (Hexane)

Source: Ref. 12.

with highly sensitive detection methods such as gc/ms do represent a powerful analytical tool (20).

As described in the USP, phytonadione is a mixture of the cis- and trans-isomers of vitamin K₁. This mixture should not contain more than 103% and not less than 97.0% of total vitamin K content. The amount of the cis-isomer is also specified and is not to exceed 21%. In addition to the pure substance, the USP also describes methods for the analysis of parental as well as tableted forms of the vitamin (21).

SYNTHESIS

The first synthesis of vitamin K₁ was reported by several workers in the late 1930s and the synthetic approaches have been reviewed (22). Vitamin K₁ was prepared by the reaction of menadione with phytyl bromide in the presence of zinc (23).

(3)

Vitamin K₁ has been synthesized from the condensation of the monosodium salt of menadiol with phytyl bromide (24).

In methodology developed by Fieser, unprotected menadiol (5) was condensed with natural phytol using oxalic acid as catalyst (25). Oxidation of the hydroquinone to the naphthoquinone yielded vitamin K₁. A similar approach has been reported (26). The commercial synthesis of vitamin K₁ is largely based on the above with some important improvements from Roche (27) and Merck (28).

(5)

A significant improvement in this technology was the discovery that BF₃-etherate could be substituted as a condensation catalyst in place of oxalic acid (qv). Owing to the fact that oxalate esters of isophytol did not form, this substitution reduced the formation of undesired phytadiene by-products. A second notable advance was the use of 1-acetyl-2-methyl-1,4-naphthoquinone as starting material for the condensation. In addition to phytadiene, a significant by-product is 2-methyl-2-phytyl-2,3-dihydro-1,4-naphthoquinone. The use of the monoprotected compound eliminates the formation of this impurity. Moreover, in the earlier process, the sensitivity of the unprotected compound toward air oxidation necessitated a reduction before purification by base extraction of the crude condensation mixture. The acetylated compound is not labile toward air oxidation. As a result, the reduction step is not warranted and its inherent complications are avoided.

As practiced by Hoffmann-La Roche, the commercial synthesis of vitamin K₁ is outlined in Figure 1. Oxidation of 2-methylnaphthalene (4) yields menadione (3). Catalytic reduction to the naphthohydroquinone (5) is followed by reaction with a benzoating reagent to yield the bis-benzoate (6). Selective deprotection yields the less hindered benzoate (7). Condensation of isophytol (8) (see VITAMINS: VITAMIN E) with (7) under acid-catalyzed conditions yields the coupled product (9). Saponification followed by an air oxidation yields vitamin K₁ (1) (29).

Although the industrial synthesis of vitamin K₁ remains largely unchanged from its early beginnings, significant effort has been devoted to improvements in the condensation step, the oxidation of dihydrovitamin to vitamin K₁, and in economical approaches to vitamin K₃ (*vide infra*). Also, several chemical and biochemical alternatives to vitamin K₁ have been developed.

A Japanese patent has claimed improvements in the direct condensation of menadione with phytyl chloride in the presence of a reducing metal such as zinc or iron powder (30). Tin chloride has been reported to be a useful catalyst for this condensation (31,32).

In a patent assigned to Mitsubishi, air oxidation is carried out in the presence of copper salts to avoid the formation of complicating impurities in the oxidation of di-hydrovitamin K_1 to vitamin K_1 (33). In other work, high yields of vitamin K_1 were obtained by performing the oxidation in an alkali medium (34). High purity vitamin K_1 can also be obtained by an oxidation in dimethyl sulfoxide (35).

In a novel approach to vitamin K_1, Hoffmann-La Roche has exploited the potential acidity at C-3 as a means to attach the side chain of vitamin K_1 (36). Menadione was reacted with cyclopentadiene to yield the Diels-Alder adduct. The adduct is treated with base and alkylated at C-3 with phytyl chloride. A retro Diels-Alder reaction yields vitamin K_1. Process improvements in this basic methodology have been claimed by Japanese workers (37).

Although not of industrial importance, many organo-metallic approaches have been developed (38). A one-pot synthesis of vitamin K_1 has been described and is based on the anionic [4 + 2] cycloaddition of three-substituted isobenzofuranones to 1-phytyl-1-(phenylsulfonyl)propene. Owing to the rather mild chemical conditions, the (E)-stereochemistry is retained (39).

Although the predominant commercial form of vitamin K_1 is the racemate, natural (2'(E), 7'(R), 11'(R))-vitamin K_1 is accessible either from a natural source or from condensation with natural phytol. In the curative blood clotting test, these compounds have been shown to have nearly equivalent biopotencies. The synthesis and spectral properties of all four stereoisomers of (E)-vitamin K_1 has been described and their biopotencies determined. Using natural vitamin K_1 as a standard, the relative activities were found to be 1.0, 0.93, 1.19, and 0.99 for 2'(E), 7'(R), 11'(R);

2'(E), 7'(R), 11'(S'); 2'(E), 7'(S), 11'(S); and 2'(E), 7'(S), 11'(R), respectively. These differences were not considered significant and it was concluded that these compounds have identical activities (40).

Aside from chemical methods, several patents have appeared on the biochemical production of natural vitamin K_1 from callus tissue cultures (41). In addition, a patent has appeared which describes the concentration and purification of natural vitamin K_1 from deodorizer distillates (42). The biosynthesis of vitamin K_1 and vitamin K_2 has been reviewed (43).

VITAMIN K₂

As compared to vitamin K_1, vitamin K_2 is relatively unimportant industrially with only a few producers, such as Teikoku Kagaku Sangyo and Eisai, and is dominated by the manufacture of vitamin $K_{2(20)}$. The industrial synthesis parallels that of vitamin K_1 and involves as a key step alkylation of monosubstituted menadione with an appropriate (all-E) reagent (44,45). Several academic syntheses have been described (46–49).

In contrast to vitamin K_1, there has been considerably more activity on fermentative approaches to vitamin K_2 (50). The biosynthetic pathway to vitamin K_2 is analogous to that of vitamin K_1 except that poly(prenylpyrophosphates) are the reactive alkylating agent (51,52). Menaquinones of varying chain lengths from C_5 to C_{65} have been isolated from bacteria. The most common forms are vitamin $K_{2(35)}$, $K_{2(40)}$, and $K_{2(45)}$. A significant amount of $K_{2(20)}$ was observed in a *Flavobacterium* fermentation and was increased by the use of a mutant resistant to usnic acid (53). In other work, production of vitamin K_2 was enhanced by the use of 1-hydroxy-2-naphthoate (54). Many Japanese patents have appeared that describe production of menaquinones from bacteria (55).

VITAMIN K₃

Industrially, vitamin K_3 is prepared from the chromic acid oxidation of 2-methylnaphthalene (56). Although the yields are low, the process is economical owing to the low cost and availability of the starting material and the oxidizing agent. However, the process is complicated by the formation of isomeric 6-methyl-1,4-naphthoquinone. As a result, efforts have been directed to develop process technology to facilitate the separation of the isomeric naphthoquinone and to improve selectivity of the oxidation.

A process has been disclosed in which the mixture of naphthoquinones is reacted with a diene such as butadiene. Owing to the fact that the undesired product is an unsubstituted naphthoquinone, this dieneophile readily reacts to form a Diels-Alder adduct. By appropriate control of reaction parameters, little reaction is observed with the substituted naphthoquinone. Differential solubility of the adduct and vitamin K_3 allows for a facile separation (57,58).

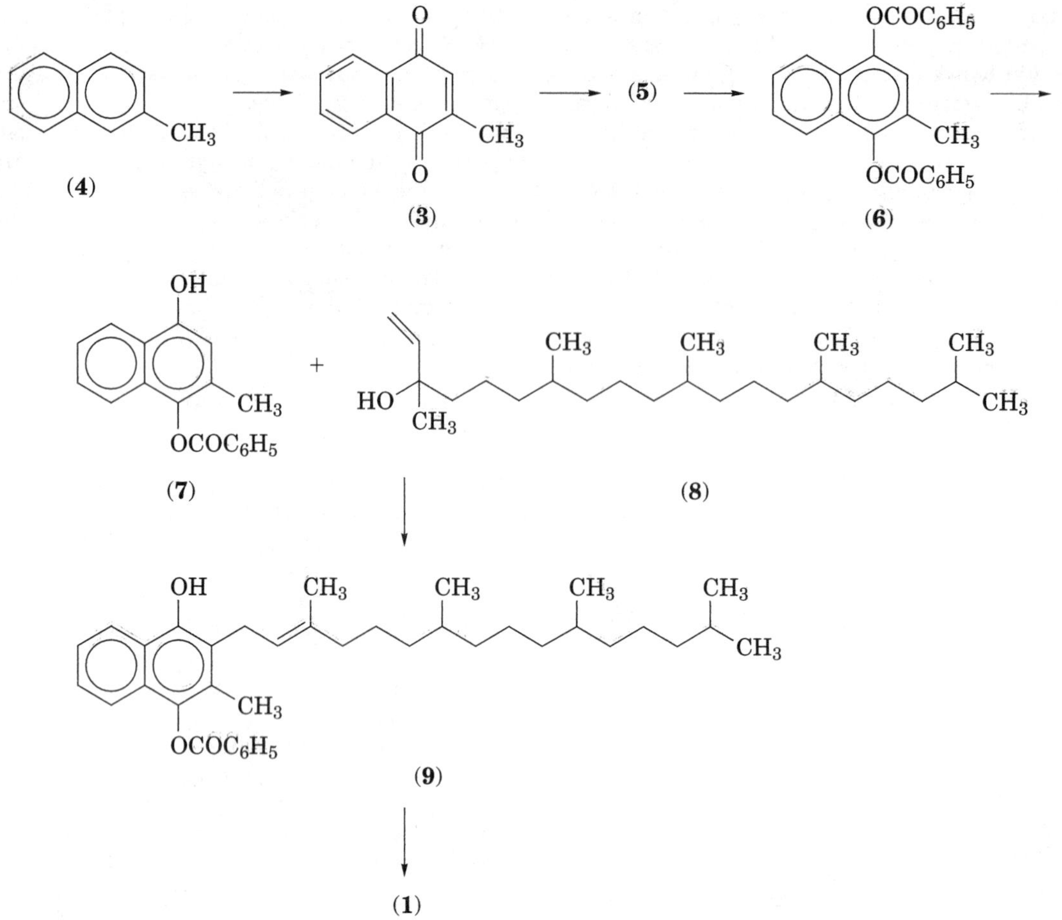

Figure 1. Synthesis of vitamin K₁.

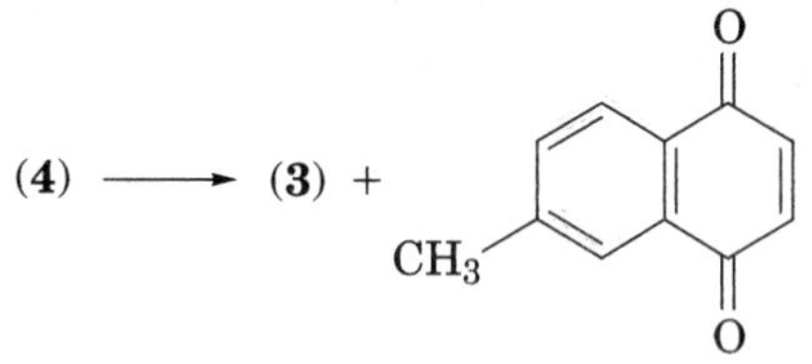

In an alternative approach, the desired 2-methyl-1,4-naphthoquinone is derivatized to form a water-soluble and commercially important sodium bisulfite adduct. Extraction of the organic phase with water separates the isomeric quinones (59).

In order to circumvent this problem, there has been significant activity directed toward the search for a less environmentally toxic and more selective oxidizing agent than chromium. For example, Hoechst has patented a process which uses organorhenium compounds. At a 75% conversion, a mixture of 86% of 2-methyl-1,4-naphthoquinone and 14% 6-methyl-1,4-naphthoquinone was obtained (60). Ceric sulfate (61) and electrochemistry (62,63) have also been used.

In a biotechnology-based approach, Japanese workers have reported on the microbial conversion of 2-methylnaphthalene to both 2-methyl-1-naphthol and menadione by *Rhodococcus* (64). The intermediate 2-methyl-1-naphthol can readily be converted to menadione by a variety of oxidizing agents such as heteropoly acids (65) and copper chloride (66). A review of reagents for oxidizing 2-methylnaphthalene and naphthol is available (67).

In addition to its industrial importance as an intermediate in the synthesis of vitamin K₁, menadione, or more specifically, salts of its bisulfite adduct, are important commodities in the feed industry and are used as stabilized forms in this application. Commercially significant forms are menadione dimethyl pyrimidinol (MPB) (**10**) and menadione sodium bisulfite (MSB) (**11**). MSB is sold primarily as its sodium bisulfite complex. The influence of feed processing, ie, pelleting, on the stability of these forms has been investigated (68). The biological availabilities and stability of these commercial sources has been determined (69,70).

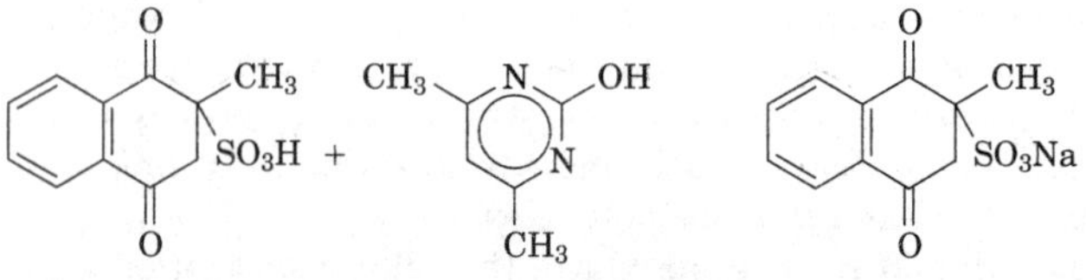

BIOSYNTHESIS

Animals cannot synthesize the naphthoquinone ring of vitamin K, but necessary quantities are obtained by ingestion and from manufacture by intestinal flora. In plants and bacteria, the desired naphthoquinone ring is synthesized from 2-oxoglutaric acid (12) and shikimic acid (13) (71,72). Chorismic acid (14) reacts with a putative succinic semialdehyde TPP anion to form o-succinyl benzoic acid (73,74). In a second step, ortho-succinyl benzoic acid is converted to the key intermediate, 1,4-dihydroxy-2-naphthoic acid. Prenylation with phytyl pyrophosphate is followed by decarboxylation and methylation to complete the biosynthesis (75).

ECONOMIC ASPECTS

The total market for vitamin K_1 is relatively small and the principal producer of vitamin K_1 is Hoffmann-La Roche. Nisshin Flour Milling Company is the predominant manufacturer for the optically active form of vitamin K_1. Total world market for vitamin K_3 is 1500 t with Vanetta Company as the dominant manufacturer. The list price for vitamin K USP grade is \$3.75–\$4.05/g. For the 1% spray dried formulated product, the price ranges from \$76–\$79/kg. The majority of vitamin K_1 is sold to the pharmaceutical industry, whereas vitamin K_3 is consumed primarily by the feed industry.

REQUIREMENTS

Owing to the ubiquitous natural occurrence of vitamin K and its production by intestinal bacteria, vitamin K defi-

ciencies are rare. However, they can be caused by certain antibiotics (qv) coupled with a reduced dietary intake. Newborn infants who do not possess the necessary intestinal bacterial population are at danger for vitamin K deficiency. As a result, vitamin K injections are routinely given to the newborn.

Table 2 lists the Recommended Dietary Allowances (RDA) for vitamin K. Although manufacture by intestinal bacteria represents a significant source of plasma menaquinone concentrations, reliance on this source alone is not sufficient to maintain healthy concentrations of menaquinone. Consequently, dietary supplementation is necessary (76).

BIOCHEMISTRY

As a result of empirical findings that alfalfa and other plants can alleviate a hemorrhagic situation in chicks, work was initiated to identify and isolate the causative agent. Concurrent to these activities was the observation that prothrombin levels in chicks given certain leafy green vegetables were enhanced. Studies on the role of vitamin K in blood clotting and more specifically, in the conversion of fibrinogen to fibrin, have led to the conclusion that the important function of vitamin K in the clotting mechanism is to act as a cofactor in the biosynthesis of the active form of prothrombin (11).

Work in the mid-1970s demonstrated that the vitamin K-dependent step in prothrombin synthesis was the conversion of glutamyl residues to γ-carboxyglutamyl residues. Subsequent studies more clearly defined the role of vitamin K in this conversion and have led to the current theory that the vitamin K-dependent carboxylation reaction is essentially a two-step process which first involves generation of a carbanion at the γ-position of the glutamyl (Gla) residue. This event is coupled with the epoxidation of the reduced form of vitamin K and in a subsequent step, the carbanion is carboxylated (77–80). Studies have provided thermochemical confirmation for the mechanism of vitamin K and have shown the oxidation of vitamin KH_2

Table 2. RDA Requirements for Vitamin K

Category	Age	Vitamin K (μg)
Infants	0.0–0.5	5
	0.5–1.0	10
Children	1–3	15
	4–6	20
	7–10	30
Males	11–14	45
	15–18	65
	19–24	70
	25–50	80
	>51	80
Females	11–14	45
	15–18	55
	19–24	60
	25–50	60
	>51	60
U.S. RDA	None established	

(15) can produce a base of sufficient strength to deprotonate the γ-position of the glutamate (81–83).

Although the role of vitamin K in blood clotting has clearly been demonstrated and involves carboxylation of multiple proteins in addition to prothrombin and includes factor VII, factor IX, and factor X, proteins containing Gla residues have been found in other tissues. For example, in 1975, Hauschka isolated an EDTA-soluble protein fraction of chick bones and identified the presence of Gla (84). Additional work sequenced the protein which was called bone Gla protein or osteocalcin (85). The properties of the protein as well as its function in bone mineralization have been extensively studied (86,87). However, its specific function is not completely understood. In addition, vitamin K-dependent carboxylase activity has been observed in cell cultures. For example, a vitamin K-dependent protein has been identified in a screen for growth arrest specific gene products. This protein, Gas6, has been identified as a ligand for tyrosine kinase (88). This observation has suggested that vitamin K may have a more general metabolic role (89–94).

CONCLUSIONS

Despite the fact that commercial synthesis in the 1990s of vitamin K has largely remained unchanged from its early roots, there has been significant activity in the area of process improvements to the basic approach. Also, several biotechnological systems have been described and this may be a future research area. A thorough understanding of the role of vitamin K in biological systems will continue as an active research area.

BIBLIOGRAPHY

1. H. Dam, *Biochem. Z.* **215**, 475 (1929).
2. H. Dam, *Biochem. J.* **29**, 1273 (1935).
3. H. Dam, *Nature,* **135**, 652 (1935).
4. H. J. Almquist, *J. Biol. Chem.* **114**, 241 (1936).
5. H. J. Almquist, *J. Biol. Chem.* **115**, 587 (1936).
6. H. Dam and F. Schonheyder, *Biochem. J.* **30**, 897 (1936).
7. S. A. Thayer, D. W. MacCorquodale, S. B. Binkley, and E. A. Doisy, *Science* **88**, 243 (1938).
8. D. W. MacCorquodale, L. C. Cheney, S. B. Binkley, W. F. Holcomb, R. W. McKee, S. A. Thayer, and E. A. Doisy, *J. Biol. Chem.* **131**, 357 (1939).
9. D. W. MacCorquodale, S. B. Binkley, R. W. McKee, S. A. Thayer, and E. A. Doisy, *Proc. Soc. Exptl. Biol. Med.* **40**, 482 (1939).
10. H. Mayer, U. Gloor, O. Isler, R. Ruegg, and O. Wiss, *Helv. Chim. Acta* **47**, 221 (1964).
11. J. W. Suttie, in L. J. Machlin, ed., *The Handbook of Vitamins*, Marcel Dekker, Inc., New York, 1991, pp. 145–169.
12. *Merck Index*, 11th ed., Merck & Co., Rahway, N.J., 1989, pp. 1580–1581.
13. R. W. McKee, S. B. Binkley, D. W. MacCorquodale, S. A. Thayer, and E. A. Doisy, *J. Am. Chem. Soc.* **61**, 1295 (1939).
14. S. B. Binkley, D. W. MacCorquodale, L. C. Cheney, S. A. Thayer, R. W. McKee, and E. A. Doisy, *J. Am. Chem. Soc.* **61**, 1612 (1939).
15. D. W. MacCorquodale, S. B. Binkley, S. A. Thayer, and E. A. Doisy, *J. Am. Chem. Soc.* **61**, 1928 (1939).
16. S. B. Binkley, L. C. Cheney, W. F. Holcomb, R. W. McKee, S. A. Thayer, D. W. MacCorquodale, and E. A. Doisy, *J. Am. Chem. Soc.* **61**, 2558 (1939).
17. H. Dam and E. Sondegarrd, in P. György and W. N. Pearson, eds., *The Vitamins*, 2nd ed., Vol. VI, Academic Press, Inc., New York, 1967, pp. 245–260.
18. J. T. Matshirer and W. V. Taggart, *J. Nutr.* **94**, 52 (1968).
19. M. M. A. Hussan, J. S. Mossa, and A. H. U. K. Taragan, *Anal. Profiles Drug Subst.* **17**, 449–531 (1988).
20. W. E. Lambert and A. P. De Leenheer, in A. P. De Leenheer, W. F. Lambert, and H. J. Nelis, eds., *Modern Chromatographic Analysis of the Vitamins*, Marcel Dekker, New York, 1992, pp. 197–233.
21. *The United States Pharmacopeia XXII* (USP XXII-NF XVII), United States Pharmacopeial Convention, Inc., Rockville, Md., 1990, pp. 1081–1082.
22. A. Ruettimann, *Chimia* **40**, 290 (1986); S. Yamada, T. Takeshita, and J. Tanaka, *Yuki Gosei Kagaku Kyokai Shi* **40**, 268 (1982).
23. H. J. Almquist and A. A. Klose, *J. Am. Chem. Soc.* **61**, 2557 (1939); H. J. Almquist and A. A. Klose, *J. Biol. Chem.* **132**, 469 (1940).
24. S. B. Binkley, L. C. Cheney, W. F. Holocomb, M. W. McKee, S. A. Thayer, D. W. MacCorquodale, and E. A. Doisy, *J. Am. Chem. Soc.* **61**, 2553 (1939).
25. L. F. Fieser, *J. Am. Chem. Soc.* **61**, 3467 (1939).
26. P. Karrer, H. Simon, and E. Zbinden, *Helv. Chem. Acta* **27**, 317 (1944).
27. O. Isler and K. Doebel, *Helv. Chim. Acta* **37**, 225 (1954).
28. R. Hirschmann, R. Miller, and N. L. Wendler, *J. Am. Chem. Soc.* **76**, 4592 (1954).
29. D. Bhatia, in *Encyclopedia of Food Science and Technology*, John Wiley & Sons, Inc., New York, 1991, pp. 2727–2732.
30. Jpn. Pat. 76-154299 (1979), Y. Tachibana (to Nisshin Flour Milling Co.).
31. Jpn. Pat. 661212 (1971), S. Kitamura, S. Matsuyama, and M. Morioka (to Teikoku Chemical Industry).
32. Jpn. Pat. 670414 (1970), S. Kitamura, Y. Fujita, and M. Morioka (to Teikoku Chemical Industry).
33. Jpn. Pat. 95-218011 (1989), A. Shinno, I. Teraro, T. Matsuda, and I. Tadashi (to Mitsubishi Gas Chemical Co., Inc.).

34. Jpn. Pat. 77-69415 (1979), S. Himoto, K. Miyahara, and Y. Yoshino (to Nisshin Flour Milling Co., Ltd.).

35. Jpn. Pat. 680309 (1971), S. Kitamura, Y. Fujita, and M. Morioka (to Teikoku Chemical Industry, Inc.).

36. Eur. Pat. 84-108332 (1985), A. Ruettimann and G. Buechi (to Hoffmann-La Roche Inc.).

37. Eur. Pat. 94-102979 (1994), K. Hamamura, T. IIwana, C. Seki, and M. Konishi (to Eisai Chemical Co., Ltd.).

38. S. Araki, T. Sato, H. Miyagawa, and Y. Bukugan, *Bull. Chem. Soc. Jpn.* **57**, 3523 (1984); Y. Naruta, *J. Org. Chem.* **45**, 4097 (1980); B. L. Chenard, M. J. Manning, P. W. Reynolds, and J. W. Swenton, *J. Org. Chem.* **45**, 378 (1980); K. H. Doetz and I. Pruskil, *J. Organomet. Chem.* **209**, C4 (1981); Ger. Pat. 2909091 (1980), K. H. Doetz (to Hoffmann-La Roche); Y. Tachibana, *Chem. Lett.*, 901 (1977); P. W. Raynolds, M. J. Manning, and J. S. Swenton, *J. Chem. Soc. Chem. Comm.*, 499 (1977); C. D. Synder and H. Rapport, *J. Am. Chem. Soc.* **96**, 8046 (1974).

39. H. Tso and Y. J. Chen, *J. Chem. Res. (S)*, **3**, 104 (1995).

40. R. Schmid, S. Antoulus, A. Rüttimann, M. Schmid, M. Vecchi, and H. Weiser, *Helv. Chim. Acta* **73**, 1276 (1990).

41. Jpn. Pat. 87-287021 (1990), T. Fujita and S. Takato (to Snow Brand Milk Products Co., Ltd.); Jpn. Pat. 87-287022 (1990), T. Fujita, H. Hatamoto, and S. Takato (to Snow Brand Milk Products Co., Ltd.).

42. Jpn. Pat. 85-90577 (1987), K. Kuihara and Y. Takagi (to Nisshin Oil Mills, Ltd.).

43. E. Leistner, *Recent Adv. Phytochem.* **20**, 243 (1986).

44. Jpn. Pat. 67-67366 (1973), S. Kitamura, M. Morioka, H. Tono, and Y. Fujita (to Teikoku Chemical Industry Co., Ltd.).

45. Jpn. Pat. 86-308171 (1989), T. Ichino, K. Nakano, T. Udagawa, Y. Kusaba, T. Muramatsu, and A. Katayama (to Eisai Chemical Co., Ltd.).

46. B. Liu, L. Gu, and J. Zhang, *Recl. Trav. Chim. Pays-Bas* **110**, 99 (1991).

47. M. Inoue, T. Uragaki, and S. Enomoto, *Chem. Lett.* **12**, 2075 (1986).

48. Y. Naruta, *J. Am. Chem. Soc.* **102**, 3774 (1980).

49. K. Sato, S. Inoue, and K. Saito, *J. Chem. Soc., Perkin Trans. I* **20**, 2289 (1973).

50. D. H. L. Bishop and H. H. King, *Biochem. J.* **85**, 550 (1962); D. H. L. Bishop, K. P. Pandya, and H. K. King, *Biochem. J.* **83**, 606 (1962); D. R. Threfall, *Vitam. Horm. (NY)*, **29**, 155 (1971).

51. H. Taber, "Vitamin K Metab.—Vitamin K-Dependent Proteins," in J. W. Suttie, ed., *Proc. Steenbock Symp. 8th*, University Park Press, Baltimore, Md., 1980, for a review of function of Vitamin K_2.

52. H. Taguchi and Y. Tani, *Microb. Util. Renewable Res.* **8**, 350 (1993).

53. Y. Tani, S. Asahi, and H. Yamada, *J. Nutr. Sci. Vitaminol.* **32**, 137 (1986).

54. Y. Tani, S. Asahi, and H. Yamada, *J. Ferment. Technol.* **62**, 321 (1984).

55. Jpn. Pat. 87-29722 (1988), Y. Tani (to Mitsubishi Gas Chemical Co., Inc.); Jpn. Pat. 86-237128 (1988), S. Uragami, Y. Tani, and Y. Tokunaga (to Mitsubishi Gas Chemical Co., Inc.); Jpn. Pat. 86-41840 (1988), Y. Tani (to Mitsubishi Gas Chemical Co., Inc.); Jpn. Pat. 85-55458 (1987), Y. Tani (to Mitsubishi Gas Chemical Co., Inc.).

56. Jpn. Pat. 92-72044 (1995), S. Nakoyooji (Daiwa Kasei KK).

57. U.S. Pat. 5,329,026 (July 12, 1994), N. Sugishiwa, N. Ikeda, and Y. Fujii (to Nippon Shokubui Kajaka Kogyo Co., Ltd.).

58. Eur. Pat. 490-122643 (1991), N. Sugishiwa, N. Ikeda, Y. Fujii, R. Aoki, and Y. Hatta (to Nippon Shokuba).

59. Jpn. Pat. 85-252445 (1986), T. Komatsui and T. Sumino (to Kawasaki Kasei Kogyo).

60. Eur. Pat. 665209 A1 950802 (1995), W. A. Hermann, J. D. Galamba Correia, R. W. Fischer, W. Adam, J. Lin, R. C. Saha-Moeller, and M. Shimizu (to Hoechst AG).

61. World Pat. 91-EP1958, H. M. Becher (1992) (to Shell International Research).

62. World Pat. 91-US3623 (1991), J. E. Toomey (Reilly Industries, Inc.).

63. S. Ito, Y. Karob, M. Iwata, K. Susuki, and A. Okika, *J. Appl. Electrochem.* **23**, 671 (1993).

64. Jpn. Pat. 92-178171 (1994), S. Tanyoshi (Kawasaki Steel Co.).

65. K. I. Matveev, E. G. Zhizhina, and V. F. Odyakov, *React. Kinet. Catal. Lett.* **55**, 47 (1995); R. Neumann and M. de la Vega, *J. Mol. Catal.* **84**, 93 (1993); K. I. Matveev, E. G. Zhizhina, V. F. Odyakov, and V. N. Parmon, *Izv. Akad. Nauk. Ser. Khim.* **7**, 1208 (1994).

66. Jpn. Pat. 92-352595 (1994), K. Nakao and H. Tamui (Kurary Co.); Jpn. Pat. 87-247363 (1989), N. Hirose and S. Kijima (Eisai Co., Ltd.).

67. U. M. Azizov and L. I. Leonteva, *Zhim. Farm. Zh.* **23**, 1488 (1989).

68. I. Nir, I. Kafri, and R. Cohen, *Poultry Sci.* **57**, 206 (1978).

69. P. W. Waldroup, W. D. Bussel, A. B. Burke, and Z. B. Johnson, *Arkansas Farm Res.* **26**, 11 (1977).

70. P. W. Waldroup, W. D. Bussell, A. B. Burke, Z. B. Johnson, *Animal Feed Sci. Technol.* **3**, 329 (1978).

71. R. Bentley, *Pure Appl. Chem.* **41**, 47 (1975).

72. R. Bentley and R. Megananthan, *Microbial. Rev.* **46**, 241 (1978).

73. I. M. Campbell, *Tetrahedron Lett.* **10**, 4777 (1969).

74. J. Soll and G. Schultz, *Biochem. Biophys. Res. Comm.* **99**, 907 (1981).

75. O. Samuel and R. Azerad, *FEBS Lett.* **2**, 336 (1969).

76. Committee on Dietary Allowances, Food and Nutrition Board, *Recommended Dietary Allowances*, 10th ed., National Academy Press, Washington, D.C., 1989.

77. J. Stenfo, *J. Biol. Chem.* **249**, 5527 (1974).

78. J. Stenflo, P. Fernlund, W. Egan, and P. Roepstorff, *Proc. Natl. Acad. Sci. USA* **71**, 2730 (1974).

79. G. L. Nelsestuen, T. H. Zytkovicz, and J. B. Howard, *J. Biol. Chem.* **249**, 6347 (1974).

80. S. Magnusson, L. Sottrup-Jensen, T. E. Petersen, H. R. Morris, and A. Dell, *FEBS Lett.* **44**, 189 (1974).

81. P. Dowd, S. W. Ham, and S. J. Geib, *J. Am. Chem. Soc.* **113**, 7734 (1991).

82. R. A. Flowers, II., S. Naganmathan, P. Dowd, E. M. Arnett, and S. W. Ham, *J. Am. Chem. Soc.* **115**, 9409 (1993).

83. E. A. Arnett, P. Dowd, R. A. Flowers, II., S. W. Ham, S. Naganmathan, *J. Am. Chem. Soc.* **114**, 9209 (1992).

84. P. V. Hauscka, J. B. Lian, and P. M. Gallop, *Proc. Natl. Acad. Sci. USA* **72**, 3925 (1975).

85. P. A. Price, A. S. Ofsuka, J. W. Poser, J. Kiristaposis, and N. Raman, *Proc. Natl. Acad. Sci. USA* **73**, 1447 (1976).

86. P. A. Price, J. W. Poser, and N. Raman, *Proc. Natl. Acad. Sci. USA* **73**, 3374 (1976).

87. J. B. Lian, P. V. Hauschksa, and P. M. Gallop, *Fed. Proc.* **37**, 2615 (1978).

88. T. N. Stitt et al., *Cell* **80**, 661 (1995).

89. J. B. Lianm, R. J. Levy, J. T. Levy, and P. A. Friedmann, in F. L. Siegel, E. Carafoli, R. H. Kretsinger, D. H. MacLennan, and R. H. Wassermann, eds., *Calcium-Binding Proteins: Structure and Function*, Elsevier North-Holland, New York, 1980, p. 449.

90. R. J. Lecvy, C. Gundberg, and R. Scheinmann, *Artherosclerosis* **46**, 49 (1983).

91. J. A. Helpern, S. J. McGee, and J. M. Riddle, *Henry Ford Hosp. Med. J.* **30**, 152 (1982).

92. J. W. Suttie, in A. T. Diplock, ed., *The Fat Soluble Vitamins*, William Heienemann, Ltd., London, 1985, p. 225.

93. C. Vermeer, *Haemostatsis* **16**, 239 (1986).

94. M. C. Roncaglioni, A. P. B. Dalessandro, B. Casali, C. Vermeer, and M. B. Donati, *Haemostatis* **16**, 295 (1986).

GENERAL REFERENCES

O. Isler and G. Brubacher, *Vitamine I*, Georg Thieme Verlag, New York, 1982, pp. 152–176.

W. E. Lambert and A. P. De Leenheer, in A. P. De Leenheer, W. F. Lambert, and H. J. Nelis, eds., *Modern Chromatographic Analysis of the Vitamins*, Marcel Dekker, New York, 1992, pp. 197–233.

S. Patai, ed., *The Chemistry of Quinonoid Compounds*, Parts 1 and 2, John Wiley & Sons, Inc., New York, 1974.

S. Patai and Z. Rappoport, eds., *The Chemistry of Quinonoid Compounds*, Vol. 2, Parts 1 and 2, John Wiley and Sons, Ltd., Chichester, U.K., 1988.

J. W. Suttie, in L. J. Machlin, ed., *The Handbook of Vitamins*, Marcel Dekker, Inc., New York, 1991.

Susan D. Van Arnum
Westfield, New Jersey

WASTE MANAGEMENT: ANIMAL PROCESSING

This section examines waste management from the standpoint of three animal-based food categories: dairy processing, egg processing, and meat and poultry processing. Waste management for an animal-based processing facility is essential because waste streams impact a plant's ability to operate within the state or local community. Permits to discharge wastewater under the National Pollution Discharge Elimination System (NPDES) (1) are restrictive for parameters such as biological oxygen demand (BOD), suspended solids, fats, oils, and greases (FOG), and ammonia. If compliance within permit limitations is not achieved, fines and civil penalties are triggered. Similar penalties can occur when biosolids are taken to landfills that are regulated under the Resource Conservation and Recovery Act (RCRA) (2). Restrictions on material composition determine the method of disposal, site, and quantity of biosolids that can be applied to land or buried in public landfills.

DAIRY-PROCESSING-RELATED WASTE MANAGEMENT

Figures 1 to 4 present utility input and waste output process schemes for fluid milk, cottage cheese, other cultured dairy products, and ice cream and frozen desserts (3–5).

Overview of Practices Contributing to Waste Generation

Table 1 presents a comparison of waste stream characteristics and coefficients for a 1000-kg milk throughput. Depending on the product being processed, the waste stream characteristics vary widely. Viscosity, product solute content, and time and temperature factors influence the waste stream's organic strength (3,4). Cleaning and sanitizing

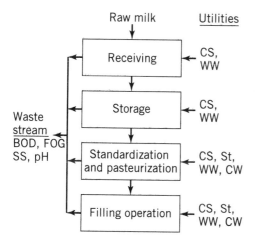

Figure 1. Fluid milk processing scheme, utility inputs and waste outputs. CS, cleaning and sanitizing; St, steam; WW, wash water; CW, cooling water; SS, suspended solids; BOD, biological oxygen demand; FOG, fats, oils, and greases.

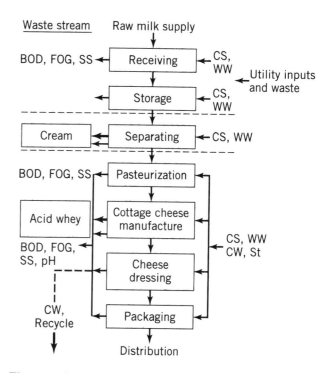

Figure 2. Cottage cheese process scheme, utility inputs and outputs. Abbreviations defined in Figure 1.

processes as well as product and water losses influence the waste stream's volume. The values presented reflect the milk handling and process practices that influence the final waste stream's organic composition and water volume.

In addition to the process practices adopted by the manufacturing operation, there are operational factors that contribute to waste generation (4,6,7). These include absence of salvage or recovery protocols when storage tanks, pipelines, and equipment are cleaned; loss of sealing integrity of pipelines, homogenizers, and valves as a result of poor maintenance; malfunction of equipment; leakage of product from damaged containers; failure to properly handle returned products; failure to isolate whey and curd wash water; inefficient use of water; and absence of recovery protocols for milk solids derived from rinses or products lost from pasteurization start-up and product changeover. Consistent day-to-day management of these areas is essential for minimizing pollutant releases and excessive hydraulic loads to the waste stream and the environment. The implementation of a sound management program can achieve up to a 50% reduction in the original waste load (8). The values given in Table 1 are approximated from both literature and field experience data and should be used only for estimating volume and waste parameters (3,5,6). These values should not be used for the design of a waste pretreatment or treatment system. Actual field measurements by qualified environmental design engineers should be used for such applications.

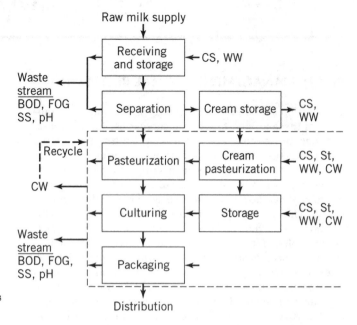

Figure 3. Dairy cultured product process scheme, utility inputs and waste outputs. Abbreviations defined in Figure 1.

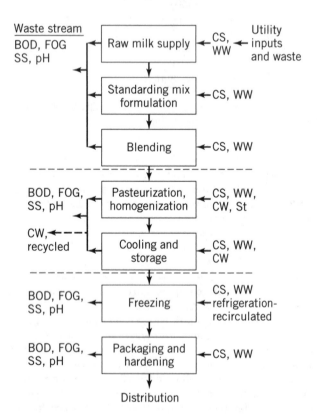

Figure 4. Ice cream and frozen dessert process scheme, utility inputs and waste outputs. Abbreviations defined in Figure 1.

Point Sources for Waste Generation

As fluid milk is handled and processed, the major point source for organic loading and water usage is pasteurization. Pasteurization includes three steps: heat exchange, separation, and homogenization. Approximately 30% of the total biological oxygen demand (BOD) and 50% of the

waste stream volume can be attributed to this operation. Milk receiving and storage account for approximately another 30% of the BOD and waste stream volume (3,5–7). Figure 1 identifies the key point sources for the fluid milk process.

The process scheme for cottage cheese is presented in Figure 2. The most important management consideration is the handling and disposal of acid whey (9,10). For purposes of this overview, it is assumed that the initial whey discharge is collected and not discharged to the dairy waste stream. The major point source for BOD is the curd washing step, which usually requires three wash cycles in which the rinses are discharged to the drain (3,5–7). Not surprising, this step accounts for 53% and 86% of the total BOD and water volume, respectively, released to the waste stream. Pasteurization contributes 25% of the total BOD and 4% of the water volume to the waste stream.

Figure 3 presents the flow scheme for other cultured dairy products. During the manufacture of cultured dairy products such as buttermilk, yogurt, and sour cream, the respective product's viscosity becomes a contributing factor to the quantity of BOD discharged to the waste stream. In most cases, milk receiving, storage, separation, blending, and pasteurization yield about the same BOD and wastewater volume values for a defined quantity of milk throughput (3,5–7). Collectively, these operations contribute 59% of the total BOD and 65% of the wastewater to the waste stream. However, 36% of the BOD and 43% of the wastewater come specifically from the pasteurization process. In contrast, viscosity and organic composition of the respective cultured products become evident as the *culturing phase* of manufacture is compared. Sour cream has the higher contribution, representing 18% of the total BOD generated, and buttermilk and yogurt contribute 14% and 16%, respectively.

Figure 4 presents the process flow scheme for ice cream and frozen desserts. It is important to remove product residue from pipelines and freezing equipment before clean-

Table 1. Waste Load Coefficients Based on 1000-kg Milk Throughput

Waste stream characteristics	Fluid milk processing	Cultured products processing	Cottage cheese	Ice cream
Water Usage (kL)	0.10–5.40	Approx. 1.16	0.80–12.40	0.80–5.60
Average (1000 L)	3.25	—	6.00	2.80
BOD (kg)	—	Buttermilk, 2.15	1.30–71.20, whey included	1.90–20.40
	0.20–7.80	Yogurt, 2.20	23.0, whey excluded	
		Sour cream, 2.25	23.0, whey excluded	
Average kg SS	4.20	—		5.76
approx. kg as % of BOD value	50%	35%	35%	25%
pH	7.6–11.5	4.1	4.4	5.8–6.8
Viscosity (cp)	2.2	Buttermilk, 500	Whey, 1.3	Mix, 35
		Yogurt, 9000		Frozen, 121
		Sour cream, 10,000		

Source: Refs. 3, 5, and 6.

up to minimize the discharge of this material to the waste stream (4). The combination of high viscosity and low temperature has a major influence on the amount of product residual in the equipment (6). The two major point source contributors to the waste stream are pasteurization and freezing (3–5,7). Pasteurization accounts for 27.5% and 43.5% of the total BOD and wastewater volume, respectively. The freezing step contributes 52% of the total BOD and 30% of the waste stream hydraulic load.

Waste Recovery, Disposal, and Treatment

The ideal situation is to reuse any milk, cream, or product recovered from the process equipment (4). This requires strict sanitary practices to avoid environmental contamination. In all cases, the recovered material must pass through the pasteurizer before it enters the final product package. Recovery practices must be approved by the state agency responsible for enforcement of dairy-related regulations.

In the cheese manufacturing sector, sweet whey is recovered and used to make whey powder, whey proteins, lactose, and delactosed whey powder (9,10). Membrane technologies are applied extensively for whey protein recovery and whey demineralization. Whey is also used extensively as a milk replacer for veal calf operations.

The more common practice throughout the dairy industry is to salvage, pool, and isolate recovered whey and dairy products for use as animal feed (9,11). Costs for handling, storing, and transporting these products to a pet food factory or to a farm are shared by the milk processor and the factory or farmer.

Land application is usually the last option for disposal of salvaged whey and dairy products (3,4,10,12). Application rates per acre and fertilizer value such as nitrogen and phosphates are important considerations when this method is used. Whey, dairy products with milkfat, and curd wash water that is high in sodium chloride present specific challenges because of the influence on the soil's pH, percolation, and flocculation properties (12). When land application is used, the salvaged materials must be "soil injected" to prevent pollutant runoff and odor releases into the atmosphere.

Figure 5 presents the possible treatment steps for dairy wastewater (6,11–13). The primary treatment may exist either as a pretreatment step or as the first phase of a waste treatment system. Pretreatment addresses a number of potential biological problems, such as sudden releases of extreme hydraulic volumes, excessive organic loads, or dramatic shifts in alkalinity or acidity. In cases where high concentrations of fats and proteins are present, physical removal is necessary in order to reduce the BOD strength of the waste stream. Pretreatment is a necessary management intervention to adjust the waste stream's characteristics to domestic strength so it is amenable to the subsequent treatment process. Generally, sewer use ordinances require industrial discharges to be at domestic wastewater strengths, or the industry is levied a costly surcharge.

The waste treatment process system is designed to achieve performance standards that meet limitations defined by the discharge permit. The waste treatment system may incorporate any of the unit operations identified in Figure 5. Generally, the process system selected is based mainly on organic assimilation by an active biological mass. Chemically oriented processes also can be incorporated into the waste treatment scheme. Waste treatment systems usually have four functional units of operation: (*1*) settling whereby fats and heavy particles are removed from the waste stream; (*2*) biological assimilation; (*3*) clarification whereby biologically active solids are recovered and recycled, and the treated water is discharged to a final treatment phase; and (*4*) polishing, which could involve a filtration process and disinfecting (13). Discharge standards to a receiving stream have a limit of 10 mg/L BOD, 10 mg/L suspended solids, 5 mg/L nitrogen as NH_3, and a pH value within a 6–9 range.

EGG-PROCESSING-RELATED WASTE MANAGEMENT

Figure 6 depicts an overview of an egg production operation (14). This section discusses the management of the processing plant wastes for shell eggs and liquid eggs.

Overview of Practices Contributing to Waste Generation

Major factors that influence the level of egg loss to the processing waste stream are strength of eggshell, egg-handling practices, and operation of washer equipment

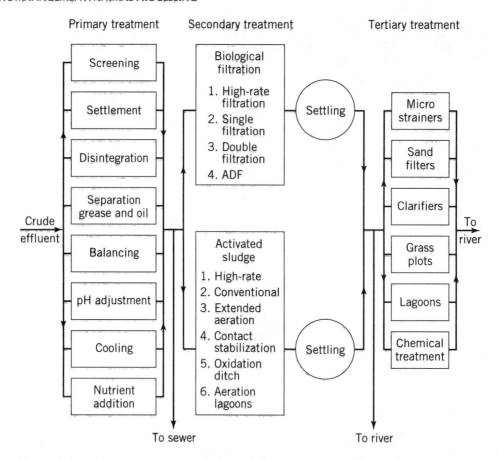

Figure 5. Possible treatment steps for dairy-food plant wastewaters. ADF, air dissolved flotation. *Source:* Ref. 12.

(14–16). Regarding the shell issue, two notable factors are the age of the laying hen and the extreme temperatures under which eggs are produced. The major egg-handling practices that generate waste are abusive handling during transportation from farm to processing facility and rough handling of the filler flats while the flats are being loaded to the egg washer conveyor. Equipment operational errors contribute to waste generation if wash water temperatures exceed 10°C above egg temperatures; wash water is discharged directly to the waste stream instead of being recycled; eggs are dropped when there is poor contact of suction cups with the eggshell surface during loading; or scrubber brushes exert too much pressure on the shell surface during the cleaning process.

Point Sources for Waste Generation

During shell egg washing, candling, and sizing, the washing and clean-up operations are the major point sources for waste generation. Incidental waste is generated from broken eggs that drop to the floor during handling and inspection. About 5% of the total shell egg production is lost either when it becomes inedible or when it enters the waste stream. Approximately 76% of the U.S. shell egg production goes directly to the retail marketplace (17). The remaining percentage goes to egg-breaking operations for processing as liquid egg-derived products. An estimated 7% of liquid products either becomes inedible or is lost to

the waste stream (18). In both shell egg and egg-breaking plants, the quantity of waste produced is *directly* proportional to the volume handled and the handling procedures practiced. Industrial plant waste surveys demonstrate that a well-managed liquid egg processing facility generates about 0.018 kg of BOD per kg of liquid egg throughput (15,19,20). Regarding water usage, each dozen of shell eggs washed generates approximately 2.5 L of wastewater (15,19,20).

For egg-breaking operations, the following management interventions can help to reduce product loss and waste (15,19,20): implement an efficient collection protocol for discarded eggs and shells with attached albumen; recover egg solids from storage units before rinsing and before the initial flush from blend tanks and the pasteurizer; eliminate storage vat spillovers; reduce the length of product lines; reduce water usage during plant clean-up; and dispose of on land the overflow and sump discharges from the egg-washing operation.

Waste Recovery, Disposal, and Treatment

Waste recovery usually involves the collection of cracked, broken, or leaking eggs during candling and the collection of broken shell residuals from liquid egg-derived products (21). The use of an apron underneath conveyors and egg-breaking equipment can be another means of salvaging egg solids. Egg materials collected can be centrifuged to

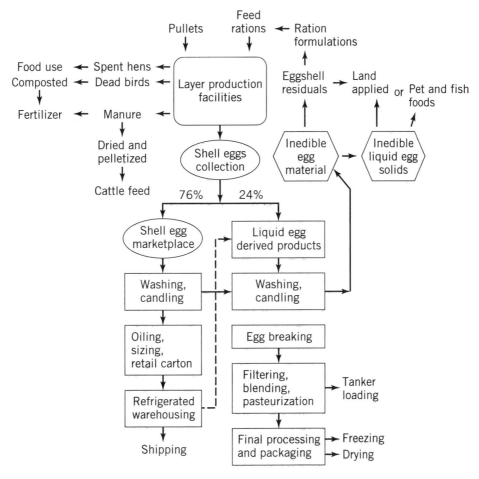

Figure 6. Material flow for an egg production breaking operation.

separate and collect the egg solids away from the shell material. Egg solids can be used in pet food formulations or as fish food. The shell portion of the egg can be used as a calcium-rich source for animal rations (22). Egg solids or shells require sterilization when used in feed applications. The shell material also has been used for industrial applications (23,24).

For egg-breaking facilities located in rural settings, most waste streams are applied to the land as fertilizer (15,25,26). On the other hand, many facilities in rural and urban locations discharge to a publicly owned treatment works (POTW) and must comply with limitations set forth by permit. A common permit issue is that egg-breaking facilities can be major contributors (over 25%) to the POTW's BOD and hydraulic inputs. In small waste treatment operations where a trickling filter system is used, it is not uncommon to encounter poor BOD removal performance and odor problems. This is because of the proteinaceous nature of the egg solids, which apply a high BOD loading rate to the system. Additionally, egg solids contain high quantities of sulfur-containing amino acids that ultimately develop into nuisance odors created by biomass activity and oxygen-deficient conditions.

To address permit limitations and to lower sewer surcharges, a pretreatment system could be installed at the egg-handling facility (12,13,27). This system would be designed to first average out the hydraulic flow to a constant flow rate. The second design feature would be to settle and float out suspended particles that are collected and disposed of to a POTW biosolids digester or on the land. A third design element would be to incorporate a dissolved air floatation (DAF) unit into the system to remove the colloidal materials that again are disposed of to a biosolids digester system or land. The fourth option would be to operate an anaerobic process to reduce the suspended solids and to have a subsequent discharge to the POTW. Depending on the overall influence of the egg solids and quantity of cleaning agents used on the pH of the waste stream, it may be necessary to provide an acid addition to maintain the pH value between 6 and 9.

Egg-handling and processing plants in rural locations generally are required to treat wastewater before disposal (12,13,19,20,28,29). Most plants adopt an aeration lagoon-polishing pond system and discharge to a receiving stream. This system is economical to maintain and does an adequate job of removing egg wastes from the aquatic environment. However, it has its share of operational problems, such as inadequate mixing and oxygen transfer, nuisance odors caused by biosolids accumulation that lead to an anaerobic condition, and increased biological activity due to atmospheric temperature changes, specifically during the summer months. Incorporating a pretreatment–treatment system is the best strategy. In the pretreatment phase, emphasis should be placed on suspended solids removal using either a DAF unit or an anaerobic process. For the aeration lagoon-polishing pond treatment, a supplemental filtering unit should be employed and a disinfectant added to meet the low discharge standards. Nitrogen content in the discharge stream also can become a regulatory issue if adequate denitrification is not achieved in the polishing pond.

MEAT- AND POULTRY-PROCESSING-RELATED WASTE MANAGEMENT

Figures 7 and 8 depict overviews of the slaughtering steps for beef and poultry, respectively (11). Meat is processed into human foods (edible meat) and pet foods and animal feed (inedible meat) (30). However, when the United Kingdom and Europe experienced an outbreak of bovine spongiform encephalopathy (BSE) in the late 1990s, strict regulations were adopted to disrupt possible BSE transmittal pathways. For example, materials containing nerve tissues are prohibited for food use. Bovine meat cannot be used in beef livestock feed, and inedible meat from sheep has been excluded as a source for animal feed.

Overview of Practices Contributing to Waste Generation

Factors that influence the ultimate composition of processing wastes are sizes and species of animals; quantity of daily throughput; water usage; sanitation and housekeeping; and method of product conveyance. Typical waste stream characteristics for beef and broiler processing activities are given in Table 2.

Point Sources for Waste Generation

The largest amount of waste in meat and poultry processing is generated at the abattoir or slaughter phase (Figs. 7 and 8). Common forms of waste are manure, paunch, hair, feathers, grit, blood, and body fluids.

In the case of beef livestock, approximately 60% of the animal's live weight either is rendered or enters the waste stream (33). Paunch contributes to the waste stream up to 5.5 kg of BOD/1000 kg live body weight at kill (LWK). Blood accounts for 33 kg of BOD/1000 kg LWK in the absence of a blood collection protocol versus the more common loss of 6.6 kg, with blood collection. Washing the carcass contributes the highest water usage rate at 350 gal/min.

For poultry, the following BOD waste loads are generated, expressed as kilogram units per 1000 birds processed (34): bleeding, 2.35; scalder, 0.98; feather, 0.94; eviscer, 1.68; offal, 1.16; washing, 1.06; prechiller, 0.96; and chiller, 0.76. The feather operation utilizes the largest amount of water at 21.45 gal/min.

Beef wholesale cutting operations contribute the least to the waste stream. However, waste materials common to this operation are meat and fat trimmings and bone dust. These wastes add another 0.3 to 3.3 kg of BOD to the waste stream per 1000 kg LWK processed.

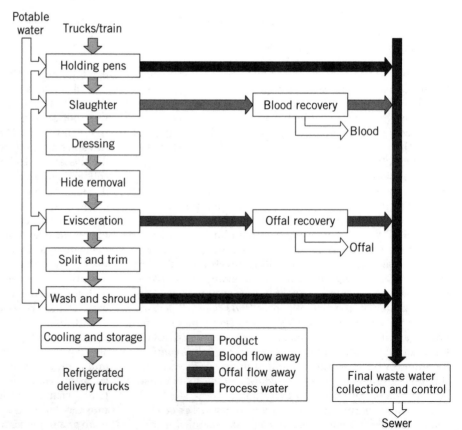

Figure 7. Process flow diagram for beef slaughtering operation.

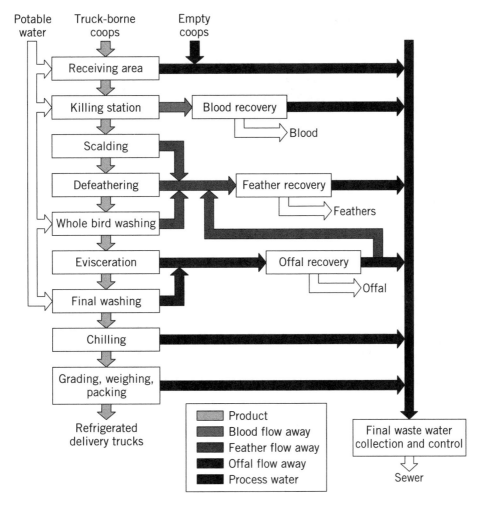

Potable water   Truck-borne coops   Empty coops

Receiving area

Killing station → Blood recovery → Blood

Scalding

Defeathering → Feather recovery → Feathers

Whole bird washing

Evisceration → Offal recovery → Offal

Final washing

Chilling

Grading, weighing, packing

Refrigerated delivery trucks

Product
Blood flow away
Feather flow away
Offal flow away
Process water

Final waste water collection and control

Sewer

Figure 8. Process flow diagram for poultry processing operation.

Table 2. Waste Characteristics for a Beef and Poultry Processing Operation

Animal type	Waste parameter contributed (lb/1000 animals)			Water (gal/animal)
	BOD	TSS	FOG	
Beef (31)	6710	6860	440	350
Poultry (32)	49	57	8	5.8

Waste Recovery, Disposal, and Treatment

Common by-products recovered from processing operations are blood, hide, bone, offal, and feathers. For example, blood is used commercially for microbial growth media, extraction of animal albumin, animal feed protein, and iron and nutrients for specific plants such as roses. Inedible meat materials are salvaged for use in pet and animal foods or sold for rendering purposes (30).

In addition to recovery of animal by-products, efforts are made to remove meat cutting and processing debris from the waste stream. For example, it is a common practice to remove debris by placing screens into the drain wells that collect wastewater from conveyance equipment,

clean-up, and sanitation activities. This collected material usually is disposed of to a landfill operation.

Further removal of suspended solids takes place in the pretreatment phase of the wastewater disposal operation. However, this presents a major challenge to the meat and poultry processing industries (11,35). Inadequately designed facilities, frequently with an under-capacity throughput, are common problems for these industries. Screens, sedimentation, floatation, and physicochemical methods are used to reduce the colloidal content in the discharged waste stream (12,13). However, in spite of the removal of suspended solids, considerable soluble BOD remains in the waste stream and must be treated further to remove the organic pollutants (2,11,33,34,36,37).

For ammonia control, blood and paunch collection practices have the greatest influence on nitrogen content in the waste stream. To address ammonia control, processing plants have adopted practices such as ammonia stripping (38), land application (11–13,33,34,36,39,40), deep-ditch aeration, and breakpoint chlorination (41).

Because of the high protein and fat content in the meat and poultry waste streams, it has been found that the use of an anaerobic-aerobic lagoon-polishing pond treatment sequence provides an effective means to remove organic pollutants (11–13,33,34,42,43). Attention must be given, however, to a routine biosolids removal protocol from the aerobic lagoon and polishing pond since solids accumulation can lead to odor and operational problems for these systems. For waste streams discharged to the POTW, the waste treatment process of choice is the activated sludge system.

BIBLIOGRAPHY

1. 40 *CFR* Parts 122, 405 and 432, U.S. Environmental Protection Agency, Washington, D.C., 1998.

2. 40 *CFR* Parts 243, 257, 258, and 503, U.S. Environmental Protection Agency, Washington, D.C., 1998.

3. R. E. Carawan, J. V. Chambers, and R. R. Zall, *Water and Wastewater Management in Food Processing, Dairy Spin Off Manual*, Extension Special Report No. AM-18B, North Carolina Agricultural Extension Service, Raleigh, N. Car., 1979.

4. W. J. Harper, J. L. Blaisdell, and J. Grosshopf, *Dairy Food Plant Wastes and Waste Treatment Practices*, U.S. Environmental Protection Agency, Washington, D.C. 1971.

5. *Development document for effluent limitations guidelines and new source performance standards for the dairy product processing point source category*, EPA-440/1-74-02/2, U.S. Environmental Protection Agency, Washington, D.C., 1974.

6. W. J. Harper, and J. L. Blaisdell, "State-of-the art of dairy food plant wastes and waste Treatment," *Proc. 2nd Food Wastes Symp.*, Denver, Colo., 1971, pp. 509–545.

7. *An Industrial Waste Guide to the Milk Processing Industry*, Public Health Services Publication No. 298, U.S. Department of Health, Education, and Welfare, Washington, D.C., 1963.

8. W. J. Harper, R. E. Carawan, and M. E. Parkin, *Wastewater Management Control Handbook for the Dairy Industry*, U.S. Environmental Protection Agency, Washington, D.C., 1984.

9. H. W. Moller, "Whey Utilization Systems," paper presented at the American Cultured Dairy Products Institute, San Antonio, Texas, 1981.

10. R. R. Zall, "Cost Effective Disposal of Whey," *Dairy Ind. Intl.* 30–39 (1980).

11. *Waste Management and Utilization in Food Production and Processing*, Task Force Report No. 124, Council for Agricultural Science and Technology, Ames, Iowa, 1995.

12. R. E. Carawan, J. V. Chambers, and R. R. Zall, *Water and Wastewater Management in Food Processing, Wastewater Treatment Spin Off Manual*, Extension Special Report No. AM-18J, North Carolina Agricultural Extension Service, Raleigh, N. Car., 1979.

13. Metcalf and Eddy, *Wastewater Engineering: Treatment, Disposal, Reuse*, 2nd ed., McGraw-Hill, New York, 1979.

14. O. J. Cotterill, "Egg-Products Industry," in W. J. Stadelman and O. J. Cotterill, eds., *Egg Science and Technology*, 4th ed., Harworth Press, Inc., New York, 1995, pp. 221–230.

15. R. E. Carawan, J. V. Chambers, and R. R. Zall, *Water and Wastewater Management in Food Processing, Egg Spin Off Manual*, Extension Special Report No. AM-18H, North Carolina Agricultural Extension Service, Raleigh, N. Car., 1979.

16. W. J. Stadelman, "The preservation of quality in shell eggs," in W. J. Stadelman and D. J. Cotterill, eds., *Egg Science and Technology*, 4th ed., Harworth Press, Inc., New York, 1995, pp. 67–80.

17. *Annual Agricultural Statistics*, National Agricultural Statistics Service, U.S. Department of Agriculture, Washington, D.C., 1997.

18. O. J. Cotterill and L. E. McBee, "Egg-Breaking," in W. J. Stadelman and O. J. Cotterill, eds., *Egg Science and Technology*, 4th ed., Harworth Press, Inc., New York, 1995, pp. 231–264.

19. W. J. Jewell et al., "Egg Breaking and Processing Waste Control and Treatment," EPA-660/2-75-019, U.S. Environmental Protection Agency, Corvallis, Oregon, 1975.

20. W. J. Jewell et al., "Egg Breaking and Processing Waste Control and Treatment," *Proc. 6th Natl. Symp. Food Processing Wastes*, EPA-600/2-76-224, December, 1976.

21. C. L. Barth, J. Brock, and R. Stovall, "Waste and Resource Management for a 720,000-Layer Complex," *Proc. 6th Intl. Symp. Agricultural and Food Processing Wastes*, Chicago, Ill., December 17–18, 1990, pp. 33–40.

22. C. Tadtiyanant, J. J. Lyons, and J. M. Vandepopuliere, "Extrusion Processing used to Convert Dead Poultry, Feathers, Egg-Shells, Hatchery Waste, and Mechanically Deboned Residue into Feedstuffs for Poultry," *Poultry Sci.* 72, 1515–1527 (1993).

23. J. M. Vandepopuliere et al., "Elimination of Pollutants by Utilization of Egg Breaking Plant Shell-Waste," EPA-600/2-78-044 U.S. Environmental Protection Agency, 1978.

24. G. W. Fronning and D. Bergquist, "Utilization of Inedible Egg-Shells and Technical Egg White Using Extrusion Technology," *Poultry Sci.* **69**, 2051–2053 (1990).

25. L. E. Carr, "Utilization of Poultry Solid Waste on Crops," *Proc. Natl. Poultry Waste Management Symp.*, Columbus, Ohio, April 18–19, 1988, pp. 133–140.

26. B. W. Jones, and G. Petitout, "In-House Sludge Management for Poultry Waste Utilizing Land Application: A Case Study," *Proc. 1989 Food Processing Waste Conf.*, Georgia Technology Research Institute, Atlanta, Ga., 1989.

27. D. E. Totzke, "Anaerobic Treatment Technology of Food Processing Wastes: A Review of Operating Systems," *Proc. 1987 Food Processing Waste Conf.*, Georgia Technology Research Institute, Atlanta, Ga., 1987.

28. J. C. Barker, P. W. Westerman, and L. M. Safley, "Lagoon Design and Management for Layer Waste Treatment and Storage," *Proc. Natl Poultry Waste Management Symp.*, Columbus, Ohio, April 18–19, 1988, pp. 26–33.

29. J. C. H. Shih, "Anaerobic Digestion of Poultry Waste and Byproduct Utilization," *Proc. Natl. Poultry Waste Management Symp.*, Columbus, Ohio, April 18–19, 1988, pp. 45–50.

30. A. N. Battacharya and J. C. Taylor, "Recycling Animal Waste as a Feedstuff: A Review," *J. Animal. Sci.* **41**, 1438–1457 (1975).

31. T. W. Stebor et al., "Case history: Anaerobic Treatment at Packerland Packing," *Proc. 1989 Food Processing Waste Conf.*, Georgia Technology Research Institute, Atlanta, Ga., 1989.

32. G. E. Valentine, C. C. Ross, and S. R. Harper, *High Rate Anaerobic Fixed-Film Reactor for the Food Processing Industry*, Final Report N 400-960 and A-4817, Georgia Technology Research Institute, Atlanta, Ga., 1988.

33. R. E. Carawan, J. V. Chambers, and R. R. Zall, *Water and Wastewater Management in Food Processing, Meat Spin Off Manual*, Extension Special Report No. AM-18C, North Carolina Agricultural Extension Service, Raleigh, N. Car., 1979, pp. 91–98.

34. R. E. Carawan, J. V. Chambers, and R. R. Zall, *Water and Wastewater Management in Food Processing, Poultry Spin Off Manual*, Extension Special Report No. AM-18D, North Carolina Agricultural Extension Service, Raleigh, N. Car., 1979.

35. M. C. Rooney and M. H. Wu, "Joint Treatment of Meat-Packing and Municipal Wastewater by Full-Scale AWT Facilities," *Proc. 36th Industrial Waste Conf.*, Ann Arbor Science Publishers, Ann Arbor, Mich., 1981, pp. 301–310.

36. R. E. Carawan, "Water Conservation and Waste Reduction for the Poultry Processing Industries," *Proc. Clean Water Act and the Poultry Industries*, Washington, D.C., 1989.

37. F. C. Blanc, J. C. O'Shaughnessy, and S. H. Corr, "Treatment of Beef Slaughtering and Processing Wastewaters Using Rotating Biological Contractors," *Proc. 38th Industrial Waste Conf.*, Ann Arbor, Mich., May 5, 1983, pp. 1133–1140.

38. Griggs, W. E., Damned if you do and damned if you don't, an ammonia discharge at a rendering plant. *Proceedings, 1989 Food Processing Waste Conference*, Georgia Technology Research Institute, Atlanta, 1989.

39. L. E. Carr et al., "Land Disposal of Dissolved Air Flotation Sludge from Poultry Processing," *Trans. Am. Soc. Agric. Eng.* **31**, 463–469 (1988).

40. O. C. Pancorbo and H. M. Barnhart, "Microbiological Characterization of Raw and Treated Poultry Processing Wastewater and Sludge: Preliminary Assessment of Suitability for Land Application," *Proc. 1989 Food Processing Waste Conf.*, Georgia Technology Research Institute, Atlanta, Ga., 1989.

41. Harper, S. R., R. Wallace, T. D. Shadburn, T. L. King, J. N. Mullis and J. T. Cook. Ammonia stripping and chlorination for nitrogen removal from poultry processing wastewater. Final Report. Project N 400-950. Environmental Sciences and Technology Division, Georgia Technology Research Institute, Atlanta, 1989.

42. *"Design of Anaerobic Lagoons for Animal Waste Management,"* Standards, EP403.2, American Society of Agricultural Engineers, St. Joseph, Mich., 1993. pp. 543–557.

43. E. R. Collins, "Lagoon Treatment of Poultry Wastes," *Proc. National Poultry Waste Management Symp.*, Raleigh, North Carolina, October 3–4, 1990, pp. 266–277.

JAMES V. CHAMBERS
Purdue University
West Lafayette, Indiana

WASTE MANAGEMENT: FRUITS AND VEGETABLES

Large quantities of liquid and solid wastes are produced annually in the processing of many different types of fruits and vegetables. According to Rose and coworkers (1), over 80 billion gallons of wastewater and nearly 9 million tons of solid residues are generated annually by the fruit and vegetable-processing industry in the United States (Table 1). Of all the residues generated by the food-processing industry, fruits and vegetables together account for 93% (2). The four crops processed in the largest quantities are citrus, corn, tomatoes, and white potatoes. These crops account for 67% of raw tonnage, 61% of wastewater, 45% of organic load expressed as the biochemical oxygen demand (BOD), 61% of suspended solids, and 71% of solid residues (1). The BOD value of a typical domestic sewage is usually about 250 mg/L. In general, the fruit and vegetable-processing industry produces a liquid waste approximately 10 times the BOD value of typical domestic sewage (3). The wastewater must be properly treated before it can be discharged to a stream or other water body. According to Hudson's survey results (2), nearly all the liquid waste is disposed of in water (stream, lake, bay, or ocean) and in public waste-treatment systems. Approximately 79% of the fruit

and vegetable solid residues is used for animal feed and the remaining 21% is handled as waste (2).

For the fruit and vegetable-processing industry to remain competitive, it is important that both liquid and solid wastes be properly managed. The objective of this article is to discuss the sources and characteristics of fruit and vegetable-processing wastes, to describe the in-plant waste-reduction practices, and to review the current developments in the processes for waste treatment and for recovery of useful by-products from the wastes.

SOURCES AND CHARACTERISTICS OF FRUIT AND VEGETABLE WASTES

The major contributing point sources for fruit-processing wastewater are washing, peeling, sorting and slicing, cooking, can cooling, and plant cleanup. Table 2 shows the estimated average volumes of wastewater if only fresh water is used in all the processing steps in the fruit-processing plant. If efforts are made to reuse water, the volumes of wastewater can be significantly reduced. For example, the cooling water can be reused for the can-cooling operation or washing the raw product. Reductions in the volume of fresh water consumed will result in corresponding reductions in the volume of wastewater to be treated (4). In vegetable processing, the major sources of wastewater include washing, grading (sorting), fluming, peeling, blanching, can cooling, and plant cleanup. The volumes of wastewater generated in the processing of vegetables can be significantly reduced if the water is reused in the plant according to the counterflow water reuse system (4). In this system, fresh water is used in the final step, collected and reused in a previous step, and recollected and reused in this manner one or several times. The water always flows countercurrent to the product. However, sufficient chlorine must be added to reduce bacterial contamination of the reused water used in each step. Most organic loads in the wastewater originate in the peeling and blanching operations at the fruit and vegetable-processing plants. For example, peeling and blanching contribute more than 80% of the BOD discharged from a beet-processing plant. Surplus brines generated in sauerkraut fermentation represent nearly 20% of the raw cabbage used (Table 3), and they have a BOD value of as much as 65,000 mg/L (Table 4). The soaking operations used in the processing of sour cherries and baked beans are a significant source of high BOD wastewater (7). The major sources of solid wastes from fruit and vegetable-processing plants are sorting, cutting, slicing, peeling, pulping, and pressing. In sauerkraut production, for example, trim losses represent over 35 kg per 100 kg of raw cabbage processed (Table 3). Fruit pomace is a waste product of the fruit juice industry, representing 10 to 20% of the fresh weight of the raw material. The strength and pH of fruit and vegetable-processing wastewater vary widely with the type of raw product being processed (Table 4). Some of the variations in strength and pH reflect differences in the processing methods used in individual plants. For example, the waste generated in the fermentation of cabbage to sauerkraut has a pH value of as low as 3.5 and its BOD value is as high as 65,000 mg/L (6).

Table 1. Wastes from Canned and Frozen Fruits and Vegetables

	Raw tons (1,000 tons)	Wastewater (million gal)	BOD (million lb)	Suspended solids (million lb)	Solids residuals (1000 tons)
Apple	1,000	5,000	40	5	320
Apricot	120	600	7	1	21
Cherry	190	400	4	1	27
Citrus	7,800	23,000	31	55	3,390
Peach	1,100	4,400	66	11	270
Pear	400	1,600	28	8	120
Pineapple	1,000	500	20	8	450
Other fruit	400	3,200	8	4	80
Asparagus	120	1,200	1	1	45
Bean, lima	120	1,100	3	10	10
Bean, snap	630	2,800	19	3	130
Beet	270	1,100	40	14	110
Carrot	280	1,100	15	11	150
Corn	2,500	4,500	62	25	1,660
Pea	580	2,900	29	6	78
Pumpkin, squash	220	700	18	3	150
Sauerkraut	230	100	3	1	75
Spinach, gr.	240	2,200	6	2	37
Sweet potato	150	1,000	30	12	(in other vegetable)
Tomato	5,000	10,000	60	20	400
Other vegetables	1,300	5,200	80	40	500
White potato	2,400	9,600	190	120	910
Total	*26,400*	*82,000*	*760*	*360*	*8,940*

Source: Ref. 1.

Table 2. Sources and Volumes of Wastewater from Fruit-Processing Steps

	Wastewater		Percentage of total flow
Operation	Gal/h	Gal/ton	
Peeling	1,200	48	2
Spray washing	11,000	385	17
Sorting, slicing, etc	3,000	120	5
Exhausting of cans	1,200	48	2
Processing	600	24	1
Cooling cans	24,000	945	37
Plant cleaning	21,000	840	33
Box washing	1,900	70	3

Source: Ref. 3.

Table 3. Material Balance of Sauerkraut Production

	Tons
Raw cabbage	100.00
Solid wastes (trim loss)	35.30
Shredded cabbage in vat	64.70
Salt added	1.70
Liquid wastes	
Early brine	11.00
Late brine	8.50
Total	*19.50*
Yield of finished product	46.90

Source: Ref. 5.

Table 4. Strength and pH of Fruit and Vegetable–Processing Wastes

Product	COD (mg/L)	BOD (mg/L)	Mean (BOD/COD)	pH
Apple	395–37,000	240–19,000	0.55	4.1–8.2
Beets	445–13,240	530–6,400	0.57	5.6–11.9
Carrots	1,750–2,910	817–1,927	0.52	7.4–10.6
Cherries	1,200–3,795	660–1,900	0.53	5.0–7.9
Corn	3,400–10,100	1,587–5,341	0.50	4.8–7.6
Green beans	78–2,200	43–1,400	0.55	6.3–8.3
Peas	723–2,284	337–1,350	0.61	4.9–9.2
Sauerkraut	470–65,000	300–41,000	0.66	3.6–6.8
Tomatoes	652–2,305	454–1,575	0.72	5.6–10.8
Wax beans	193–597	55–323	0.58	6.5–8.2
Wine	495–12,200	363–7,645	0.60	3.1–9.2

Source: Ref. 6.

At the other extreme, certain waste effluents are strongly alkaline because of the presence of lye-peeling liquors or alkaline cleaning compounds. Waste effluents with high and low pH values present problems for waste treatment. For optimal biological treatment, the pH of wastewater must be adjusted to between 6.0 and 9.0.

Waste Reduction

The pollution loads at the fruit and vegetable–processing plants can be reduced through the development of improved processing methods or through in-plant treatment and reuse.

Process Modification. Most pollution loads are generated in the wet peeling of fruits and vegetables. The conventional wet peeling process, for example, produces 10,200 lb of total solids/day, whereas the dry-caustic peeling process produces only about 1,050 lb of total solids/day (Table 5). The dry-peeling process also reduces the BOD load from 2,670 to 190 lb/day, representing a reduction of more than 90% (4). The conventional water- or steam-blanching process is also a major source of pollution loads. It inactivates enzymes and removes gases from the plant tissues but also leaches out organic matter. To minimize both the hydraulic and organic loads, a hot gas–blanching method has been developed for vegetable processing. Hot gas blanching has been reported to reduce the volume of vegetable processing wastewater, BOD, and suspended solids (SS), by as much as 99.9% (4).

Other water-conservation and waste-reduction techniques are available. These include the use of air cooling after blanching, the installation of low-volume, high-pressure cleaning systems, the substitution of mechanical conveyors for flumes, the use of automatic shutoff valves on water hoses, the separation of can cooling water from composite waste flow, the countercurrent reuse of wash/flume/cooling water, the separation of low- and high-strength waste streams, and use of air flotation units to remove suspended debris from raw crop materials (8).

In-Plant Treatment and Reuse. An important treatment system has been developed for reduction of the volume of water consumption at the fruit and vegetable–processing plants (4). This in-plant system consists of coagulation, sedimentation, and filtration. Chemicals such as alum and polymers are added to the screened wastewater to facilitate coagulation. Coagulated solids or sludge is discharged to a solid waste–disposal area and the clarified water is passed through activated carbon filters. The filtered effluent is chlorinated and reused for initial product washing, for product washing after blanching, and for carrying out solid wastes to the screening area. This system can remove all the suspended solids, including bacteria (4).

Waste Utilization

Fruit and vegetable–processing residues are rich in biodegradable organic matter, and disposal of these waste materials may pose many environmental pollution problems. Fortunately many opportunities exist for converting them into a wide variety of valuable by-products. Some examples follow below.

Animal Feed. Most fruit and vegetable–processing solid residues are presently used for animal feed (2). They may be fed fresh, as silage, or as dried waste solids. Citrus-, pineapple-, and potato-processing residues are almost entirely fed to animals. Other fruit and vegetable residues, however, are not widely used in animal feed rations because they are available only during the short processing season.

Natural Food Ingredients. Natural food ingredients can be isolated from fruit and vegetable–processing residues. Anthocyanins, tartrates, and grape-seed oil have been directly extracted from grape pomace (9). Carotenoids can be produced from sauerkraut brine by fermentation with the yeast *Rhodotortula rubra* (10). Apple- and citrus-processing solid wastes can serve as a good source of raw material for commercial production of pectins. Attempts are now being made to isolate health-promoting dietary fibers and other phytochemicals from fruit and vegetable residues. For example, a product with a dietary fiber content of over 56% has been produced from apple residues by Three Top Inc., Selah, Washington (7).

Biofuels. Biofuels have been generated from fruit and vegetable–processing residues by fermentation. Methane production is a two-stage process. The substrate is first converted to organic acids by acid-producing bacteria. The organic acids are then fermented anaerobically by methane-producing bacteria to methane. Capital and operating costs of methane production are dependent on the digester size, raw material, and operating conditions. The fermentation efficiency increases with decreased substrate particle size. Nearly 80% of the organic matter in apple pomace, for example, can be readily converted to methane (7). Ethanol can be directly produced from fruit-processing residues by yeast fermentation (7,9). Starch and cellulosic waste materials, however, must be hydrolyzed first to glucose before ethanol fermentation. For example, the ethanol yields from different grape pomace samples under solid-state fermentation conditions range from 68 to 86% (Table 6).

Citric Acid. Citric acid is a commercially important organic acid that has been produced from fruit and vegetable–processing residues by *Aspergillus niger* (7,9).

Table 5. Water Usage and Properties of Wastewater in Two Different Peeling Operations of Beets

Measurement	Conventional peeling	Dry-caustic peeling
Raw beets, input, ton/day	80	80
Water flow rate, gal/day	48,000	12,000
Total solids, lb/day	10,200	1,050
SS, lb/day	340	40
BOD, lb/day	2,670	190

Source: Ref. 4.

Table 6. Ethanol Production from Grape Pomace under Solid-State Fermentation Conditions

Variety	pH	Moisture, %	Yeast count, CFU/g		Sugar content g/kg		Ethanol content, g/kg		Yield[a] σ
			0 h	96 h	0 h	96 h	0 h	96 h	
Vinifera									
Chardonnay	3.7	74.5	5×10^3	3×10^7	125.0	32.5	0.4	40.8	86
Riesling	3.6	65.5	1×10^5	7×10^7	137.5	7.5	0.2	54.2	82
Geneva hybrid									
Cayuga white	3.6	64.3	5×10^6	2×10^7	97.5	9.4	2.9	40.6	86
French hybrid									
Aurora	3.6	64.5	3×10^5	2×10^7	70.0	15.6	2.5	24.6	80
Baco noir	3.3	57.0	7×10^4	1×10^5	121.2	12.5	0.4	42.7	76
Ravat	3.6	62.0	3×10^6	8×10^4	156.2	41.2	2.6	49.8	80
Red hybrid[b]	3.3	57.0	3×10^6	8×10^7	83.8	11.6	0.9	31.4	82
Seyval	3.8	68.4	9×10^5	4×10^7	137.5	13.0	3.8	58.0	86
Ventura	3.5	64.3	4×10^5	1×10^7	115.0	15.3	0.4	43.1	84
Verdelet	3.6	70.6	1×10^7	8×10^6	106.2	9.4	7.2	48.0	82
Labrusca									
Cascade	3.7	60.9	7×10^6	7×10^7	103.8	18.1	0.4	38.4	86
Concord									
Cold-pressed	3.3	66.7	2×10^4	3×10^7	58.1	10.5	0.9	20.8	82
Hot-pressed	3.3	54.8	1×10^5	8×10^7	103.8	20.9	0.6	32.0	74
Delaware	3.7	57.9	4×10^5	5×10^7	118.8	17.2	3.3	44.7	80
Diamond	3.5	68.1	5×10^4	2×10^7	76.2	38.1	15.0	28.4	68
Dutchess	3.6	64.9	5×10^5	3×10^7	130.0	7.2	0.4	46.3	72
Ives	3.5	56.7	5×10^4	1×10^5	87.5	12.5	0.8	32.2	82
Niagara	3.5	65.9	9×10^3	5×10^5	85.0	24.4	0.4	26.8	86

[a]Yield, % = Actual ethanol produced/Theoretical ethanol from sugar consumed $\times$ 100.
[b]Includes deChaunac, Rougeon, Chelois, Rosette.

The yields of citric acid from apple pomace, corncobs, kiwi peel, and grape pomace are about 320, 250, 230, and 270 g/kg dry matter, respectively.

Single-Cell Protein. Single-cell protein can be produced from fruit and vegetable–processing residues by fermentation. Hang (11) has described the processes for single-cell protein production from the waste effluents generated in the processing of cabbage, corn, baked beans, and potatoes. The processes not only produce single-cell protein but also reduce the BOD value by as much as 90%.

Enzymes. Enzymes of commercial importance can be directly extracted from fruit and vegetable–processing residues. Bromalain is a protease, for example, that has been commercially produced from pineapple stem (Sigma Chemical Company, St. Louis, Mo.). The enzyme has a variety of food-processing applications. Attempts have been made to utilize the fruit and vegetable–processing residues as a substrate for fermentative production of industrial enzymes by food-grade microorganisms. Sauerkraut waste, for example, has been found to be a favorable substrate for production of yeast enzymes such as polygalacturonase, invertase, diacetyl reductase, and beta-glucosidase (12).

Waste Treatment and Disposal. Liquid wastes generated in the processing of fruit and vegetables contain princi-

pally biodegradable organic matter in soluble and suspended form. The waste effluents are generally deficient in both nitrogen and phosphorus. For optimum biological treatment, it is essential to add inorganic nitrogen and phosphorus to give a BOD-to-nitrogen-to-phosphorus ratio of 100:5:1. Several waste-treatment processes have been used to make fruit and vegetable–waste effluents suitable for discharge. Aerated lagoons are the most widely used to treat the wastewater (4). In aerated lagoons, oxygen is added by either mechanical agitation or compressed-air diffusion. For solid separation, lagoons in series are used. The activated sludge process consists of a bioreactor that provides an environment for converting the BOD to microbial cells under aerobic conditions and a clarifier where microbial cells or sludge is settled. The settled sludge may be returned to the reactor or wasted. This system has been used to remove 80 to 99% of BOD depending on the waste composition, loading rate, and other operating conditions. The wastewater can also be treated in an anaerobic digester to produce methane. In pray irrigation, the wastewater is screened and sprinkled on land to simulate rainfall. Parameters that affect the system are the waste characteristics (solids, BOD, pH, salts, etc), site, and discharge limits. The organic loading rates range from 36 to 108 lb of BOD/acre/day and are dependent on the filtration and percolation capacity of the soil at the treatment site (4).

Most solid wastes are presently disposed of on land either by filling or by spreading (4). In land filling, solid wastes are dumped into an excavated pit and covered with soil. A crop is maintained to use the waste solids, minimize erosion, and enhance the appearance of the landfill site. The loading rates are about 5 to 50 tons of dry solids per acre per year (4). The leachates from the landfill sites are odorous and rich in organic matter. They must be properly treated before discharge to avoid environmental and public-health problems. In land speading, solid wastes are spread on land in thin layers and allowed to remain on the surfaces long enough to dry before they are tilled into soils. The loading rates by this method are about 3 to 5 tons of dry solids/acre/year (4).

Biotechnology and Waste Management. Research is needed to reduce the pollution loads generated in the processing of fruits and vegetables through genetic engineering of raw products and through development of new or modified processing methods. Fruits and vegetables can be genetically modified to have desirable processing characteristics such as increased solids content, uniform ripening, and size. For example, new hybrids of cabbage with a higher solids content have been developed for sauerkraut production. The genetically modified cabbage can increase the product yield with an accompanying decrease in the production of brine (13). The surplus brine generated in sauerkraut fermentation has been a serious pollution problem for kraut processors (5). To reduce the pollution loads at the processing plant, it is necessary to develop new processing techniques for processing raw materials at the harvest location. For the unavoidable processing wastes, new biotechnological processes should be developed to convert them into protein, fuels, biochemicals, or other valuable products with an accompanying decrease in the pollution loads. For example, to maximize the production of value-added products from grape pomace and to minimize the waste disposal problems, it is necessary to use several technologies in the following sequences (Fig. 1):

- Removal of seeds from the pomace for oil recovery
- Fermentation of the pomace minus seeds to ethanol or citric acid
- Extraction of anthocyanins and tartrates from spent pomace
- Anaerobic decomposition of the remainder for methane geneation (9)

CONCLUSION

Large quantities of both liquid and solid wastes are produced annually by the fruit and vegetable–processing industry. These waste materials contain principally biodegradable organic matter, and disposal of them can pose many environmental-pollution problems. Factors affecting the costs of waste treatment and disposal are the volume or hydraulic load and the strength or organic load. The waste load at the processing plant can be reduced significantly through the use of modified processing methods and through in-plant treatment and reuse. A number of waste-

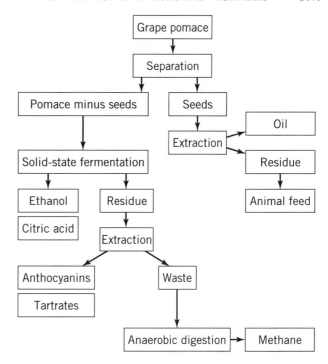

Figure 1. Resource recovery from grape pomace flow diagram. *Source:* Ref. 9.

treatment processes are available to remove the BOD and SS from the processing waste effluents. Biological treatment systems are widely used to make the wastewater suitable for discharge. Land irrigation of wastewater is used in rural areas. Most solid wastes are disposed of by returning them to the harvest fields for land treatment. The key to minimizing the disposal cost is to remove excessive moisture from the wastes. Many opportunities exist for better utilization of fruit and vegetable–processing wastes. A variety of processes have been developed for converting the waste materials into natural food ingredients, enzymes, fuels, and other valuable by-products.

BIBLIOGRAPHY

1. W. W. Rose et al., "Production and Disposal Practices for Liquid Wastes from Canning and Freezing Fruits and Vegetables," in *Proc. 2nd Natl. Symp. Food Processing Wastes*, Oregon State University, Corvallis, Or., 1971, pp. 109–127.

2. H. T. Hudson, "Solid Waste Management in the Food Processing Industry," in *Proc. 2nd Natl. Symp. Food Processing Wastes*, Oregon State University Corvallis, Or., 1971, pp. 637–654.

3. *Fruit Processing* Industry, U.S. Department of Health, Education, and Welfare, PHS Pub. No. 952, Washington, D.C., 1962.

4. U.S. Environmental Protection Agency, *Pollution Abatement in the Fruit and Vegetable Industry*, EPA-625/3-77-0007, Washington, D.C. 1977.

5. Y. D. Hang et al., "Wastes Generated in the Manufacture of Sauerkraut," *J. Milk Food Technol.* **35**, 432–435 (1972).

6. D. F. Spittstoesser and D. L. Downing, "Analysis of Effluents from Fruit and Vegetable Processing Factories," *New York State Agric. Exp. Sta. Res. Circ.* **17** (1969).

7. Y. D. Hang and R. H. Walter, "Treatment and Utilization of Apple Processing Wastes," in D. L. Downing, ed., *Processed Apple Products*, AVI Publishing Co., Westport, Conn., 1989, pp. 365–377.

8. *Waste Management and Utilization in Food Production and Processing*, Task Force Report No. 124, Council for Agricultural Science and Technology, Ames, Iowa, 1995.

9. Y. D. Hang, "Recovery of Food Ingredients from Grape Pomace," *Process Biochem.* **23**, 2–4 (1988).

10. C. T. Shih and Y. D. Hang, "Production of Carotenoids by *Rhodotorula rubra* from Sauerkraut Brine," *Food Sci. Technol.* **29**, 570–572 (1996).

11. Y. D. Hang, "Production of Single Cell Protein from Food Processing Wastes" in J. H. Green and A. Kramer, eds., Food Processing waste management, AVI Publishing Co., Westport, Conn., 1979, pp. 442–455.

12. S. L. Sim and Y. D. Hang, "Sauerkraut Brine: A Potential Substrate for Production of Yeast β-Glucosidase," *Food Sci. Technol.* **29**, 365–367 (1966).

13. J. R. Stamer and M. H. Dickson, "High Solids Cabbage for Sauerkraut Production: Its Effect Upon Brine Reduction and Yield of Product," *J. Milk Food Technol.* **38**, 688–690 (1975).

YONG D. HANG
Cornell University
Geneva, New York

WATER ACTIVITY: FOOD TEXTURE

FOOD TEXTURE

The textural properties of a food may be defined as that group of physical characteristics that arise from the structural elements of a food, are sensed by the feeling of touch, are related to the deformation, disintegration, and the flow of the food under a force and are measured objectively by functions of mass, time, and distance (1). The term textural properties is preferred over the word texture because texture implies a one dimensional property (eg, pH) while texture is multi-dimensional in nature and cannot be completely described by a single parameter (2). Because of the long time required to measure all the textural properties of a food the most common practice is to measure only one or a few of its most dominant textural properties.

A number of test principles are used for instrumental measurement of textural properties including force, time, distance, work, and rate of flow. Force measurements are the most common and can be classified according to the geometry of the test apparatus as puncture, extrusion, cutting shear, compression (single and multiple), bending—snapping, and torque (1).

Since texture is primarily a sensory perception (3), instrumental measurements need to be calibrated against the human palate. Sensory testing ranges from describing a single texture note up to a complete texture profile analysis that generally describes 15–30 texture notes depending on the food (4,5). Some of the most widely used sensory texture descriptors are firm, crisp, chewy, creamy, sticky, gummy, viscous, fibrous, granular, flakey, pulpy, moist, and greasy (2).

A promising hybrid sensory-instrumental technique is the use of a multiple-point sheet sensor (6–8). This comprises a thin flexible plastic film about 5–6 cm square on which are printed two series of stripes perpendicular to each other using electrically conductive ink. The resistivity of the ink is directly proportional to the force applied to it. When placed in the mouth, the pressure can be measured at many points over time, showing the distribution of forces rather than an average force. Although in an early stage of development, the technique shows great promise. For example, in testing crackers equilibrated to A_w levels of 0.14, 0.36, 0.63, and 0.80, the force-deformation curves detected by the multiple-point sheet sensor in the mouth closely matched force-deformation curves obtained with a compression test conducted in a universal testing machine (6).

WATER ACTIVITY

Water activity is defined as the ratio of the vapor pressure of water in a food to the vapor pressure of pure water at the same temperature:

$$A_w = \frac{P}{P_0}$$

where A_w is water activity, P is vapor pressure of water in the food, and P_0 is vapor pressure of pure water at the same temperature. Since solutes in food depress the water vapor pressure, P will always be less than P_0. Therefore A_w always has a value between 0 and 1.0, where 0 represents the total absence of water and 1.0 represents pure water. The effects of chemical composition on A_w and more rigorous definitions of A_w involving thermodynamic concepts are given elsewhere in this volume. However, the definition given above is adequate for the purposes of this article.

The older literature uses the term equilibrium relative humidity (ERH), which is the relative humidity of the air surrounding a food when the water vapor pressure in the air is the same as in the food (equilibrium state). Since relative humidity is expressed on a percentage basis, the old literature values for ERH can be easily transformed into A_w values by dividing by 100: $A_w = $ ERH/100. For example, if the ERH = 65%, then the $A_w = 65/100 = 0.65$.

The concept of water activity in food has many useful applications. A major application is to determine whether a food will absorb or lose water in a given environment. It is well known that liquid water always flows spontaneously from a high-pressure area to a low-pressure area (ie, downhill). Although invisible macroscopically, water vapor also flows spontaneously from high-pressure regions (high A_w) to low-pressure regions (low A_w). A food with $A_w = 0.60$ placed in air at 40% relative humidity (RH) (ie, $A_w = 0.40$) will give up water vapor to the air until the water activity in the food equals that in the air. In moving air at constant 40% RH the food will eventually reach $A_w = 0.40$; in a confined space with no change in the air the equilibrium A_w will be somewhere between 0.40 and 0.60. If the food at $A_w = 0.60$ is placed in air at 80% RH ($A_w = 0.80$) it will absorb water vapor from the air until equilibrium is reached at a level higher than A_w 0.60.

Figure 1 shows schematically the similarity between the flow of liquid water and the flow of water vapor. In both cases the spontaneous flow is always from the high-pressure region to the low pressure region. While most graphs plot water activity across the abscissa (horizontal axis), it conceptually helps to understand water activity if one imagines it on a vertical axis as shown in Figure 1. This rendition makes it easy to understand why the flow of water vapor is always from high A_w to low A_w. When two foods with different water activities are enclosed in the same airtight container but not in contact, an invisible transfer of water in the form of vapor will occur. Water vapor will migrate from the food with a high A_w into the air and then be absorbed from the air by the food with low A_w. Both foods and the air will eventually equilibrate at equal A_w levels. The only way to prevent this spontaneous transfer of water vapor is to place a barrier around the food that retards the rate of transfer of the vapor. The degree of resistance to the passage of water vapor is an important criterion in the selection of packaging materials for those foods whose quality will deteriorate if they pick up or lose water to the atmosphere during storage.

Fresh fruits and vegetables have a high water content, usually in the 80–90% range, and a water activity of about 0.99. They would quickly lose moisture to the air and become wilted, limp, and dry and lose their crispness if it were not for the skin, which acts as a barrier to impede the loss of water vapor. Many fruits and vegetables secrete a wax on the skin (cuticular wax), which makes the skin a more effective barrier to dehydration. Some commodities such as apples, oranges, cucumbers, winter tomatoes, and rutabagas have additional wax added to the skin to further impede loss of water vapor and extend their storage life. The waxing process also increases the gloss of the skin imparting a shiny appearance that makes it look more at-

tractive. Fruits and vegetables lose water rapidly in the air when the skin is punctured or broken. For this reason, care must be taken not to break the skin of those items that will be stored for any length of time.

GENERAL CONSIDERATIONS

It has been known for many years that foods may pick up or lose moisture from the air during storage and that these changes can affect the texture. For example, in 1945 Woodroofe and co-workers (9) demonstrated that raw unshelled peanuts stored at 80% RH maintained a good texture for 60 days; they became clammy after 90 days and soggy after 360 days. Peanuts stored at 65% RH maintained a good texture for 360 days. At 50% RH they maintained a good texture for 120 days, became dry after 180 days, and become slightly hard after 240 days. Shelled raw peanuts stored under the same conditions exhibited the same texture changes as did the unshelled ones when stored at 80 and 65% RH. Although these early results were more descriptive than quantitative, they clearly showed that losing moisture and gaining moisture can cause loss of textural quality. When moisture is lost, the peanuts become brittle or dry. When moisture is gained, they become soggy or clammy.

SORPTION ISOTHERMS

The relationship between percentage water content and water activity is complex. The A_w level almost always increases with increasing water content over the full A_w range, but in a nonlinear fashion. A plot of water content versus A_w usually takes the form of an S-shaped curve (see Fig. 2). These curves are determined experimentally. A food product is equilibrated at various A_w levels by exposing it to a series of atmospheres of constant RH until it reaches constant weight, and then the water content is measured by chemical analysis. Plotting these data points on a graph and connecting the points gives rise to curves

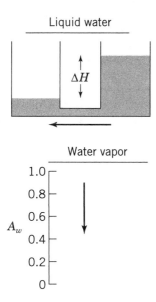

Figure 1. Top: The flow of liquid water is always from high-pressure to low-pressure regions at a rate proportional to the pressure difference ΔH. Bottom: Similarly the flow of water vapor is always from the high-A_w to the low-A_w region.

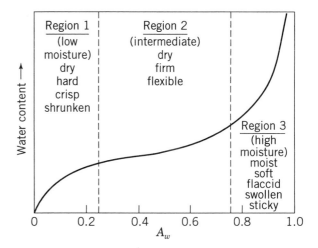

Figure 2. Food textures as a function of localized water sorption isotherms. *Source:* Ref. 10.

similar to those in Figure 2. Some researchers do not explore the full A_w range but present data over a limited range of A_w. In these cases the typical S-shaped curve may not be evident. When these experiments are performed at constant temperature, the graphs are called water sorption isotherms, when the food is gaining water and water desorption isotherms, when it is losing water. An excellent compilation of water sorption isotherms for many foods has been published (11). Note that in Figure 2 the slope of the water sorption isotherm is always positive, ie, an increase in water activity is always accompanied by an increase in percentage water content, although the rate of increase (ie, slope) varies across the A_w spectrum. There are occasional exceptions to this rule that water content always increases with increasing A_w. Some of these will be discussed under the section on sugars below.

A typical water sorption isotherm may be divided into three regions, each with characteristic textures (12). Region 1 is the lowest moisture level. Here the moisture is bound tightly to the hydrophilic constituents in the food. It is present at first in a monomolecular layer and then in multiple molecular layers. Typical textures in this region are hard, dry, crisp, brittle, and shrunken. In the intermediate moisture region 2 the moisture level is less tightly bound as additional molecular layers of water are absorbed into the food. This region is characterized by a shallower slope than regions 1 and 3. Dry, firm, flexible, and leathery are typical textures found in this region. Region 3 is the high-moisture region. A large increase in water content causes a relatively small increase in A_w. The water is probably held by the weak forces of capillary absorption. Typical textures in region 3 are moist, soft, flaccid, limp, tender, swollen, and sticky. A number of mathematical models have been developed that attempt to describe moisture sorption isotherms. A recent publication provides an up-to-date introduction to this literature (13).

Most foods are consumed at high A_w levels because people like food to have a moist and tender texture. Ready-to-eat foods with a low A_w are generally snack foods such as potato chips and crackers. These products must absorb saliva and soften rapidly during mastication to be acceptable. Foods that remain dry during mastication are not relished. This preference for foods with a moist and tender texture conflicts with their keeping properties. Molds will grow on the food with A_w above about 0.7, yeast grows at A_w above about 0.8, and bacteria above about 0.9. Although refrigeration inhibits the rate of growth of spoilage microorganisms and freezing, thermal sterilization or preservatives prevent microbial growth, the major volume of the present food supply is kept at $A_w < 0.7$ for long-term storage. Cereal grains, seeds such as beans and nuts, honey, dried fish, fruit, milk, and vegetables must be brought to $A_w < 0.7$ to prevent microbial growth. A large part of food processing and preparation is designed to convert a microbiologically stable low A_w food with unacceptable texture into another form that is texturally desirable, but renders the food subject to microbial spoilage. Spaghetti and other pasta shapes, rice, and bean seeds are stored at ambient temperature in the home and are moistened and softened during preparation in the kitchen. After preparation they must be promptly consumed or refrigerated. Grains of

wheat are converted into bread, cake, and similar tender-textured foods in readiness for consumption. Grains of barley are converted into beer. At the present time few foods are known to possess a tender texture at A_w low enough to prevent microbial growth. The goal of producing intermediate-moisture foods with a tender moist texture, acceptable flavor, and long shelf life has proved difficult, as many research laboratories will attest.

The relationship between water activity and texture is specific to each kind of food. These will now be discussed on an item-by-item basis.

APPLES

Bourne measured texture profile parameters of fresh apple and dehydrated apple equilibrated to A_w levels of 0.85, 0.75, 0.65, 0.55, 0.44, 0.33, 0.23, 0.12, and 0.01 (14). Cubes 10 mm on each side were cut from large pieces of apple after equilibration to the appropriate water activity level and subjected to 90% compression two times in a universal testing machine at a speed of 50 mm/min. The detailed procedure for instrumental texture profile analysis has been given (15,16). Six representative force—time curves generated by the Instron are shown in Figure 3. In each graph the left-hand curve represents the first compression and the right-hand curve, the second compression. In no case did the force—time plot fall below zero to a negative value which indicates zero adhesiveness at all A_w levels. The changes in textural properties of apple will be first described in qualitative terms referring to Figure 3 and then in quantitative terms referring to Figure 4. The curve for fresh apple with a calculated $A_w = 0.99$ is shown in the top right-hand side of Figure 3. The initial steep linear slope indicates that the apple is firm with low deformability. The sharp reduction in force that follows indicates sudden failure and that the apple is fracturable. The force continues to rise with continued compression reaching a maximum at the end of the compression; this force peak is defined as hardness. The force quickly falls to zero during decompression. In the second compression cycle the force rises smoothly to a peak with no sudden drops in force during the compression. The area under the compression curves is a measure of the work required to compress the apple. The second compression requires much less work than the first compression. The ratio of area under second compression curve to area under first compression curve is defined as cohesiveness. It is low for fresh apple. The horizontal distance between a perpendicular dropped from the force peak on the second compression curve and the point where the force curve moves off the baseline represents the height the apple recovered between the first and second compressions and is defined as springiness. It is low for fresh apple.

The curve for apple at $A_w = 0.85$ is shown in the top left-hand side of Figure 3. The shape of the first compression curve is greatly different from that for fresh apple. There is no fracturability peak. The initial slope is low, indicating that the rigidity has been lost and the deformability has become high. The maximum force (hardness) is about the same as for fresh apple at about 100 N. The area

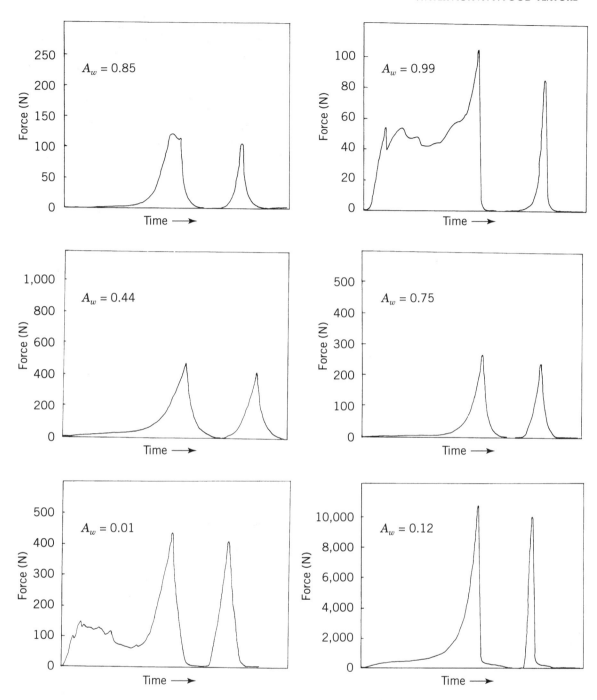

Figure 3. Typical force–time curves traced from Instron charts of texture profile analyses (two 90% compressions) on apple tissue dried to various water activity levels; note the differences in the force scales on the ordinates. *Source:* Ref. 14.

under the force curve in the first compression is much smaller than for the fresh apple, while the area under the second compression curve is about the same. This leads to an increased value of cohesiveness. There is little change in springiness. The Texture Profile Analysis (TPA) curve for apple at A_w levels 0.75, 0.65, 0.55, 0.44. 0.33, and 0.23 are qualitatively similar and are represented by the two curves with A_w = 0.75 and 0.44 in the center of Figure 3. They are characterized by no fracturability, high deform-

ability (low initial slope), and high cohesiveness. The hardness increases as the A_w is lowered. Note that the force scales in Figure 3 are increasing. The shape of these TPA curves resembles those for meat. Sensorial evaluation showed that the dominant textural characteristics are chewiness, toughness, and sponginess somewhat like that of meat. The texture profile at A_w = 0.12 is similar to those for higher A_w levels except that cohesiveness and springiness are lower and hardness are extraordinarily high. It is

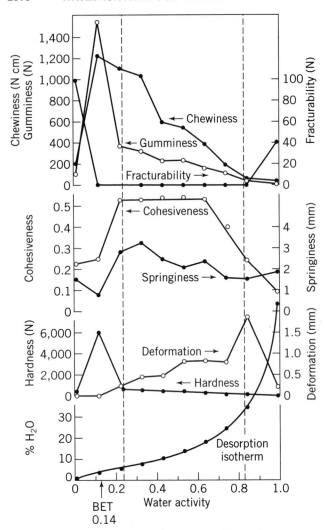

Figure 4. Moisture sorption isotherm and texture profile parameters of apple flesh equilibrated to various A_w levels. *Source:* Ref. 10.

remarkable that a little 1-cm cube of apple requires a force as high as 10,000 N (~1/2 ton!) to compress it by 90%.

There is an abrupt change in texture profile as A_w is reduced below 0.12 as shown in the lower left-hand graph in Figure 3. The fracturability peak reappears, the hardness drops and the shape of the curve resembles that of fresh apple but with higher fracturability and hardness.

The changes in texture profile parameters with decreasing moisture are consolidated in Figure 4. The desorption isotherm for apple as shown at the bottom of Figure 4 is consistent with desorption isotherms for apples and apple products that have been reported by others (11). The texture—A_w curves can be divided into three distinct stages during desorption of water. In the first stage the moisture content drops from 87 to 35% and A_w from 0.99 to 0.85. This causes total loss of the turgor pressure and fracturability of the fresh apple, a great increase in deformability and cohesiveness and minor changes in hardness, springiness, gumminess, and chewiness.

In the second stage the moisture content drops from 35 to 5% and the A_w from 0.85 to 0.23. This causes a progres-

sive increase in hardness, springiness, gumminess, and chewiness and a decrease in deformation. Cohesiveness increases from $A_w = 0.85$ to $A_w = 0.65$ and then remains at a high level from $A_w = 0.65$ to 0.23. This increase in hardness as apple is dried was confirmed by a later study reporting the cutting-shear force of dehydrated apple over the range A_w 0.77 to 0.29 (17).

The third stage is at a water content below 5% and $A_w < 0.23$ and is characterized by rapid seesaw changes in textural properties. Hardness and gumminess increase rapidly from $A_w = 0.23$–0.12 and decline just as sharply between $A_w = 0.12$ and 0.01. Chewiness falls rapidly between $A_w = 0.12$ and 0.01. Springiness and cohesiveness decrease sharply between $A_w = 0.23$ and 0.12. Fracturability is zero at $A_w = 0.23$ and 0.12 and reappears at $A_w = 0.01$ with a value of about 100 N, which is more than twice the fracturability value of 40 N for fresh apple. At the lowest A_w level (0.01) the textural parameters resemble that of fresh apple but with much higher hardness and fracturability. At all A_w levels between 0.12 and 0.99 the texture is tough, leathery, deformable, and chewy with zero fracturability and crispness.

The Brunauer–Emett–Teller (BET) value is the A_w level at which one complete monolayer of water molecules covers all the hydrophilic sites in the food. For apple the BET was calculated to be at $A_w = 0.14$, which is about the middle of the third stage. The transition from the multilayer to a monolayer of absorbed water and from a monolayer to almost zero water content is associated with major textural changes in apple flesh.

BEEF

Kapsalis, from the U.S. Army Natick Laboratories in Massachusetts, published some of the earliest and most comprehensive studies on texture and water activity in beef (18–20). He cooked beef semimembranous muscle to 63°C and, when cooled, cut it into 12-mm-thick slices which were frozen, freeze-dessicated over silica gel at 0.3 mmHg pressure, and then remoistened to a series of higher moisture levels. After equilibration the texture was measured in the standard 10-blade test cell of the Food Technology Corp. texture press (Kramer shear press). The cutting-shear force increased as the A_w was raised from 0 up to 0.8 and then decreased sharply at A_w levels above 0.8 (Fig. 5). There was considerable variation in the absolute values of the texture press force from one animal to another as demonstrated by the two curves in Figure 5, but the peak force was always around $A_w = 0.8$.

In another experiment beef semimembranosus muscle was cooked to 63°C and when cold cut into 12-mm-thick slices, freeze-dried, and then remoistened to a series of moisture levels. 38-mm-diameter disks were cut out and compressed 25% two times between extensive flat surfaces in a universal testing machine. This is less destructive than the texture press described above. The following mechanical properties were measured from this test:

Secant modulus—the ratio of stress to strain at 3.5% compression. This parameter is now called modulus of deformability.

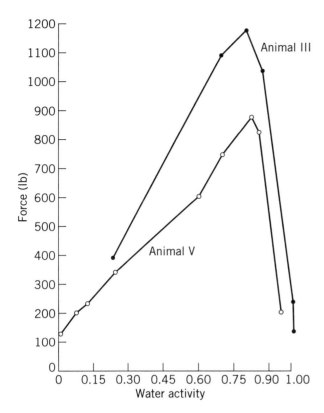

Figure 5. Effect of A_w on maximum cutting–extrusion force of precooked, freeze-desiccated beef that has been remoistened. *Source:* Ref. 19.

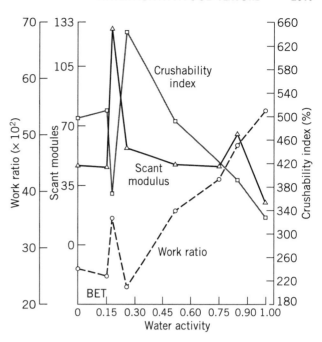

Figure 6. Effect of A_w on mechanical properties of precooked, freeze-dried beef that has been remoistened. *Source:* Ref. 19.

Degree of elasticity—the ratio of the elastic deformation to the sum of the elastic and plastic deformations derived from the load–unload curve of the first compression cycle.

Toughness—work per unit volume to compress the sample 25%. It is derived from the area under the curve up to 25% compression.

Crushability index—The ratio of nonrecoverable to recoverable work in the first compression. It is defined as (area under compression portion minus area under decompression portion of a curve)/(area under decompression portion of the curve).

Work ratio—the ratio of the area under the curve of the second compression to the area under the curve of the first compression. This value is similar to the cohesiveness parameter of the instrumental texture profile analysis.

The changes in work ratio, secant modulus, and crushability index are shown in Figure 6. There was little change in these three properties as the freeze-dried beef absorbed moisture from $A_w = 0$ to 0.15. Rapid seesaw changes occurred between $A_w = 0.15$ and 0.30, which spans the BET monolayer $A_w = 0.16$. Work ratio and secant modulus rapidly increased and just as rapidly decreased while the crushability index rapidly decreased and just as rapidly increased again. The crushability index steadily decreased and the work ratio increased as the A_w

was raised above 0.3 while the secant modulus showed a downward trend. The changes in secant modulus and work ratio are plotted again in Figure 7 but with both moisture content and A_w shown on the abscissa, and the A_w scale is nonlinear. Degree of elasticity and toughness are also shown in Figure 7. Elasticity follows a similar trend to work ratio and toughness is similar to secant modulus. Both of these properties change rapidly near the BET level of A_w. The texture press force shown in Figure 5 is often considered as an index of toughness of meat, and it peaks at $A_w = 0.8$ and shows no rapid changes near the BET A_w level. In contrast, the toughness curve shown in Figure 7 peaks at $A_w \approx 0.2$. The difference between texture press force and toughness is probably a result of the texture press being a highly destructive test while the toughness value is obtained by a different test that is approaching a nondestructive mode.

Figure 7 also shows the differential entropy change $\overline{-S°}$ with changing moisture content. It was pointed out that the sorption of water could cause two opposing trends: (a) $\overline{-S°}$ increases due to an increase in crystallization or similar effect or (b) $\overline{-S°}$ decreases due to increased movement of the polymer network chains, unfolding of peptide chains or local solubilization (19). The $\overline{-S°}$ curve in Figure 7 first decreases sharply, then increases sharply, and then alternately decreases and increases to a lesser extent with increasing A_w. Complex changes must be occurring in this system to account for this unusual $\overline{-S°}$ curve.

Other workers rapidly equilibrated cubes of freeze-dried beef at 0, 20, 40, 60, 80, and 100% RH and measured hardness and chewiness by a 35–40% compression of the cubes in a universal testing machine (21). Although plagued by variations in the operating conditions of the freeze-drier a statistical fit of the pooled data showed that

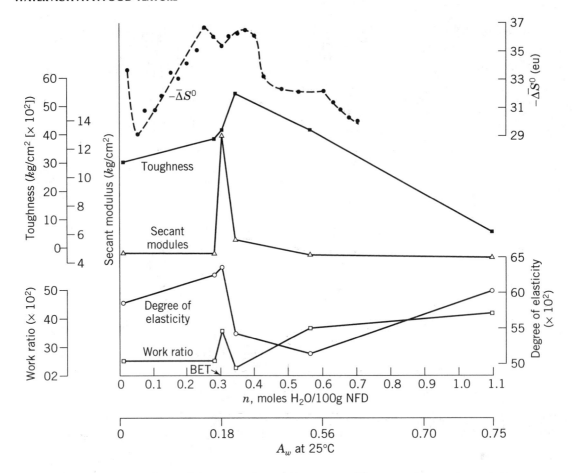

Figure 7. Relationships between textural properties measured by compression testing and standard differential entropy of water vapor sorption in precooked, freeze-dried beef as it is moistened (NFD = nonfat dry solids). *Source:* Ref. 19.

both hardness and chewiness increased from $A_w = 0$ to 0.4 and decreased from A_w 0.4 to 1.0. In a subsequent study (22) the same group reported that after 6-month storage both hardness and chewiness increased over the full A_w range (Fig. 8) with no maximum at $A_w = 0.4$.

Another group in England prepared intermediate moisture beef by immersing slices cut from longissimus dorsi muscle of one steer in a solution of 9.5% NaCl and 39.4% glycerol, cooking, and then storing at 38°C for up to 12 weeks under aerobic and anaerobic conditions (23). The samples equilibrated at $A_w = 0.83$ and 42% H_2O. The Warner–Bratzler shear values declined during storage. (This test is widely used as an index of toughness of meat.) An instrumental texture profile analysis was used to obtain four additional texture parameters: (*1*) hardness, which declined; (*2*) elasticity, which declined after 6 weeks and then increased up to 12 weeks; (*3*) cohesiveness, which followed the same down–up trend as elasticity and (*4*) adhesion, which decreased. These workers attributed the discrepancies between their results and the results discussed in the previous paragraph to traces of impurity in the glycerol reacting with the proteins in the beef.

It is evident from the differing reports between different researchers on beef texture that the relationships between

A_w and texture of beef is not clear. It is well known that a number of chemical and biochemical changes can occur in stored foods at all levels of A_w. Perhaps the time in storage has as great an effect on textural properties as the actual A_w level. It is noteworthy that one group of investigators working with beef (18–20) and another group working with apples (14) both found dramatic changes in textural properties around BET A_w level.

CEREAL GRAINS

Rice

Single grains of rice of the cultivar IR8 fashioned into little cylinders and equilibrated to various moisture levels were compressed to failure in a universal testing machine (24). Three compression speeds (2.54, 0.508, 0.127 mm/min) and five test temperatures (25, 36, 47, 58, 69°C) were used. Although the absolute values of failure stress changed with different compression speeds and temperatures, the same trend of decreasing strength with increasing moisture content was shown for all experimental conditions. The data for 36°C and 2.54 mm/min compression speed (plotted in Fig. 9) are typical of the trends for all the other

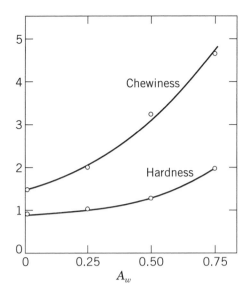

Figure 8. Effect of A_w on hardness and chewiness of freeze-dried beef after 6-month storage at 38°C. *Source:* Plotted from data in Ref. 21.

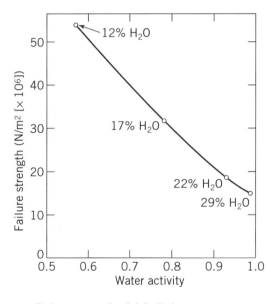

Figure 9. Failure strength of dehulled rice grains compressed 0.254 cm/min in a universal testing machine at 36°C. *Source:* Plotted from data in Ref. 24.

data. These researchers expressed moisture content on a percentage dry-matter basis. A water sorption isotherm for rice (25) was used to convert these data to an A_w basis for illustration in Figure 9. There is a fourfold decrease in failure strength as the moisture content of the whole rice grain increases from 12% H_2O (0.57 A_w) to 29% H_2O (0.93 A_w). In a later report, the same authors (26) showed that the uniaxial compression modulus and relaxation modulus, Young's modulus (E), shear modulus (G), and bulk modulus (K) of rice kernels all decreased with increasing moisture content. However, Poisson's ratio (μ) was practically unchanged.

Wheat

The apparent modulus of elasticity of whole grains of wheat of the Talent variety over a wide range of moisture has been measured (27). One interesting feature of this work was that the modulus of elasticity was measured after both sorption and desorption to different water activity levels to determine whether any hysteresis effects were present. One-half of the grain at 16% H_2O (dry-matter basis) was dried under vacuum with no heating to 5% H_2O. The other half was mixed with sufficient water containing dye to raise the moisture level to 33%. Blends of the dried grain and moistened grain were made, stored in closed containers, and slowly and continuously mixed for 3 weeks until the moisture in the grains equilibrated.

Whole grains of wheat were compressed at 0.5 mm/min in a universal testing machine, and the apparent modulus of elasticity was calculated from the slope of the initial linear section of the force deformation curve. The Hertz equation was used to correct for differences in dimensions between individual kernels. The uncolored grains were known to have absorbed water to reach their equilibrium water content, and the colored grains were known to have desorbed water to reach the same equilibrium water content (27). Two plots of percentage water (dry-matter basis) versus water activity were obtained: an absorption curve and a desorption curve. The absorption curve lay below the desorption curve.

Figure 10 shows that for both absorption and desorption the elastic modulus decreases rapidly with increasing A_w and moisture content. There is about a sevenfold decrease in the modulus over the moisture level range studied. In

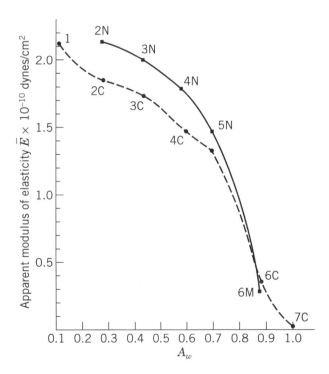

Figure 10. Apparent modulus of elasticity ($E \times 10^{-10}$ dyn/cm²) of wheat grains versus A_w; (—) absorption curve; (---) desorption curve. *Source:* Ref. 27.

the 0.2–0.6 A_w range the apparent modulus of elasticity for desorbing grain was about 13% less than that for absorbing grains at the same A_w level. The steep decrease in modulus of wheat grains with increasing moisture level is similar to the decrease in failure strength of rice grains discussed above (24) and in Figure 9.

A Canadian group measured physical properties of wheat grains equilibrated to 9.5, 11.0, 12.5, 14.0, and 15.5% H_2O, which corresponds to a range of 0.44–0.75 A_w (28). Three kinds of wheat—hard red spring, soft white, and durum—were studied and six different parameters were measured:

1. Wheat hardness index (WHI) which is the maximum torque produced by grinding in a two step Brabender hardness tester (BHT) divided by the percentage yield of flour.
2. Wheat hardness index measured in a one-step BHT.
3. Time (in seconds) to grind a standard sample of wheat into flour in a two-step BHT.
4. Pearling resistance index (PRI), which is the weight in grams after pearling 20 g of wheat grains for 20 s in a Strong Scott barley pearler.
5. Torque needed to grind grain in a two-step BHT.
6. Particle size index (PSI), which is the percent of flour passing through a 177-μm sieve after grinding in a two-step BHT.

The results are shown in Figure 11. The torque and both WHIs decreased with increasing moisture content (A,B,E). The time to grind (C) and the pearling resistance index (D) increased. Particle size (F) is practically constant, indicating that the locus of the failure planes in the wheat kernels is unaffected over this moisture level range.

Corn

The stress-relaxation properties of yellow dent corn were measured over the A_w 0.12–0.86 (H_2O 9.7–26% d.b.) (29) and temperature range 25–100°C. The relaxation modulus decreased with increasing A_w. Although the decrease was unidirectional, there was a transition between A_w 0.55 and 0.76; the corn kernels were brittle below A_w 0.55 and ductile above A_w 0.76. The same group found a similar brittle-to-ductile transition in durum semolina at about the same moisture content.

CEREAL-BASED FOODS

Changes in textural quality of a limited number of processed cereal foods have been measured over a range of water activities. Sensory and physical textural parameters of three commercially available snack foods pre-equilibrated to water activity levels A_w = 0, 0.549 and 0.653 have been measured (30). These data are summarized in Table 1. In every case sensory crispness and sensory crunchiness declined with increasing A_w levels while compression force and compression increased. The same trends were noted for saltine crackers equilibrated at A_w = 0, 0.44 and 0.653 (30).

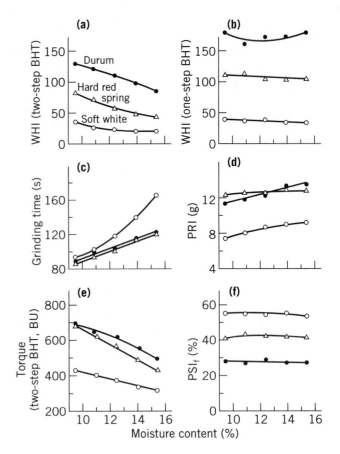

Figure 11. Hardness indices for three classes of wheat as a function of moisture content equivalent to A_w = 0.44–0.75; details of the six tests are given in the text. *Source:* Ref. 28.

Another group measured the sensory crispness of four commercial snack foods at A_w = 0–0.75 (30). Saltines, potato chips, puffed corn curls, and popcorn each lost their sensory crispness approximately linearly with increasing A_w. The puffed corn curl had almost zero crispness at A_w = 0.52, saltines and potato chips had zero crispness at A_w = 0.75, and popcorn was still slightly crisp at A_w = 0.75. Figure 12 shows the crispness versus water activity curve for popcorn. Hedonic scores declined in a similar fashion to crispness scores. These workers also reported that the initial slope of the force–distance compression curve for saltine crackers decreased with increasing water activity.

A group in France measured sensory crispness and crushing strength of breakfast cereals over the range A_w = 0.84 (32). The sensory crispness of corn flakes and rice crispies declined slowly from A_w = 0.53 then rapidly from A_w = 0.53 to 0.71. Crispness was very low at A_w > 0.71. The crispness curve for corn flakes is shown in the right-hand side of Figure 13. The crushing force for these breakfast cereals showed a remarkably similar pattern to the crispness curve (left-hand side of Fig. 13). It decreased slowly from A_w = 0 to 0.53 and then rapidly above A_w = 0.53.

Another group in Massachusetts (33) observed the characteristic sigmoid shape of the above data from France and

Table 1. Sensory and Physical Properties of Three Snack Foods

Property	Crunch twist[a]			Potato chips[b]			Rippled potato chips[c]		
	$A_w = 0$	$A_w = 0.549$	$A_w = 0.653$	$A_w = 0$	$A_w = 0.549$	$A_w = 0.653$	$A_w = 0$	$A_w = 0.549$	$A_w = 0.653$
Sensory crispness[d]	10.7	7.7	6.7	10.7	9.0	6.0	10.0	6.7	5.0
Sensory crunchiness[d]	9.7	6.7	5.7	11.3	8.7	6.3	10.0	7.0	5.0
Compression force, N	92	326	537	846	1042	1224	415	592	754
Compression work,[a] mJ	345	1057	1598	2630	2972	3537	566	796	1045

[a]Keebler and Co., Chicago.
[b]O'Grady's, Frito-Lay, Dallas.
[c]Pringles, Procter and Gamble, Cincinnati.
[d]Trained panel using a 0–13-point scale.
[e]Units of product compressed under a standard Food Technology texture press 10-blade plunger mounted in a universal testing machine (selected data from Ref. 30).

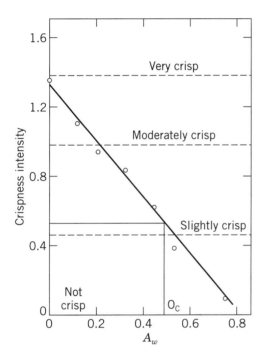
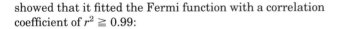

Figure 12. Sensory crispness intensity of popcorn vs A_w. *Source:* Ref. 31.

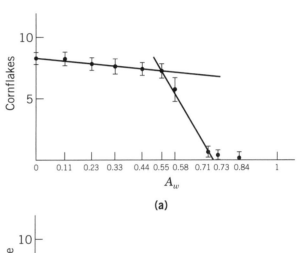

(a)

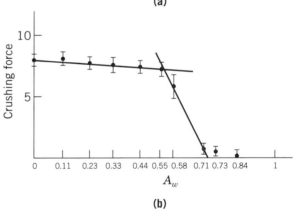

(b)

Figure 13. (a) Sensory crispness of cornflakes vs A_w; (b) crushing force of cornflakes vs A_w. *Source:* Ref. 32.

showed that it fitted the Fermi function with a correlation coefficient of $r^2 \geqq 0.99$:

$$Y(A_w) = Y_0/\{1 + \exp[(A_w - A_{wc})/b]\}$$

where $Y(A_w)$ is the sensory score at a given A_w level and Y_0 the sensory score of the cereal in the dry glassy state, A_w is the water activity of the sample, A_{wc} the critical water activity representing the inflection point in $Y(A_w)$, and b a constant representing the steepness of $Y(A_w)$ in the water-activity region where much of the plasticization occurs.

The same group (34) showed that the above equation fitted the force to compress other brittle cellular cereal foods with correlation coefficients $r^2 \geqq 0.955$ for cheese

balls, cheese puffs, French bread croutons, and pumpernickel croutons and r^2 values from 0.79 to 1.00 for the acoustic and mechanical signatures of the same products (35). They also showed (36) r^2 values of the Fermi equation of $r^2 \geqq 0.988$ for almonds and hazelnuts and $r^2 \geqq 0.99$ for kidney beans and chick peas.

The effect of moisture content on mechanical and textural properties of two cornmeal extrudates has been measured (36). Yellow cornmeal was passed through a Brabender one inch diameter single screw extruder fitted with a 5-mm-diameter die. Sample A was prepared with a feed moisture content of 25%, barrel temperature 60°C in zone I and 80°C in zone 2 and a screw speed of 120 rpm. The extrudate had a solid spaghetti like structure, a bulk density of 1.25 g/mL, and a mean diameter of 7.4 mm. Sample B was prepared with a feed moisture content of 50%, barrel temperatures 120°C in zone 1 and 180°C in zone 2, and a screw speed of 180 rpm. There were many tiny pores in the extrudate due to the violent puffing. The bulk density was 0.914 g/mL and mean diameter, 6.3 mm. After extrusion the samples were equilibrated at moisture contents ranging from 3.4 to 31.3% and compressed 30% two times in a universal testing machine. Figure 14 shows the compressive strength (maximum force per initial unit area) as a function of moisture content. For the high-density extrudate A, the strength increases from 3.9 to 8.9% moisture, remains level from 8.9 to 15.3% moisture and declines again at moisture levels above 15.3%. The low-density extrudate B shows a similar pattern except that the strength is about half that of sample A. The values for stiffness and fracturability show an inverted U shape similar to compressive strength. The values for springiness increase and for relaxation modulus decrease monotonically with increasing moisture content. The relaxation time for both samples declines as the moisture rises from 4.7 to 15.7% and increases at moisture levels above 15.7%.

The breaking strength of freshly baked sugar snap cookies after equilibration for 5 days in atmospheres ranging from 11 to 93% RH was measured with a single-blade compression bar mounted in a Food Technology Corp. texture press (38). The results are shown in Figure 15. The breaking strength of cookies prepared from hard wheat flour (low quality) was always less than for cookies made from soft wheat flour (high quality). The breaking strength decreased with increasing A_w with the fastest rate of increase occurring between A_w = 0.11 and 0.33. A parallel trend was found for tenderness of these cookies (38).

The texture of puffed whole-grain brown rice cakes equilibrated to A_w levels from A_w = 0.23–0.84 was measured by means of a punch-and-die compression test performed in an Instron (39). Crispness decreased with in-

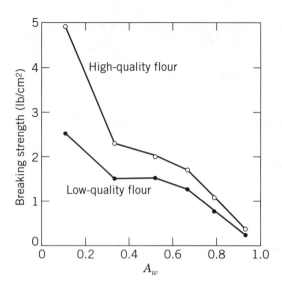

Figure 15. Breaking strength of sugar-snap cookies made from high-quality cookie flour (soft red winter wheat) and a low-quality cookie flour (hard red winter wheat). *Source:* Plotted from data in Ref. 38.

creasing A_w reaching zero at A_w = 0.57. Hardness increased from A_w = 0.23–0.65 and decreased from 0.65 to 0.84. Rice cakes with A_w between 0.23 and 0.44 were crisp, low in hardness, and required less work to snap.

The texture of a breakfast cereal made from barley flakes was studied over the A_w range 0–0.96 (40) using three-point bending and punch-and-die compression tests. In the bend test, the fracture force and maximum fiber stress increased more than threefold as A_w increased from 0 to 0.67 and then decreased above A_w = 0.67, reaching a low figure at A_w = 0.87. In the punch-and-die test, deformation increased from A_w = 0–0.32, remained reasonably constant from A_w = 0.34–0.75, and increased rapidly from A_w = 0.75–0.87; the modulus of elasticity declined slowly from A_w = 0–0.75 and then declined rapidly above A_w = 0.75, giving a curve similar to that shown for breakfast cereal flakes made from corn and rice in (32).

POWDERS

It has been observed that the cohesiveness of finely powdered foods increases with increasing moisture content (41). Food powders with a higher moisture content had a lower bulk density, lower relaxation constant, and higher compressibility than did low-moisture powders (Table 2). When a cohesive powder is poured into a container, it forms an open structure supported by interparticle forces and contains many voids, imparting a low bulk density, highly compressible, mechanically weak structure. A noncohesive power contains smaller voids created by random orientation of the particles. After compression the cohesive power has a higher tensile strength than the low-moisture powder that results from the greater interparticle attractive forces and a greater number and area of contact points between its particles (42).

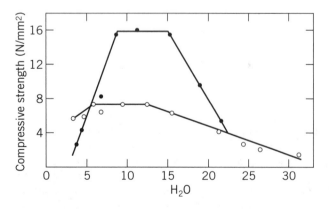

Figure 14. Compressive strength of two cornmeal extrudates as a function of water content. ● high-density sample; ○, low-density sample. *Source:* Plotted from data in Ref. 37.

Table 2. Some Physical Characteristics of Food Powders

Materials[a]	Moisture (%)	Bulk density (g/mL)	Compressibility	Relaxation constant
Baby formula	dry	0.49	6.4	2.6
Baby formula	1.0	0.46	8.1	1.5
Bran	2.8	0.55	5.0	2.0
Bran	5.4	0.42	6.7	1.4
Onion powder	0.7	0.66	8.6	2.0
Onion powder	1.2	0.55	9.5	1.4
Onion powder	2.3	0.50	8.1	1.0
Sucrose	dry	0.77	4.5	4.0
Sucrose	0.17	0.71	6.0	2.5

Source: Ref. 41.
[a]All material had a particle size − 100 + 200 mesh.

It has also been reported that the rate of caking of onion powder with a moisture content of 4–5% (which is equivalent to $A_w = 0.30$–0.35) increased rapidly with increasing temperature (43). At 15°C there was no tendency for the onion powder to cake after 6 months of storage. At 25°C agglomerates formed in 7–10 days, at 30°C caking occurred in 7–10 days, and at 35°C caking occurred in 3 days. With onion powder of 7% moisture no anticaking agents were successful while at 3% moisture the powder remained free-flowing even without caking agents after 30 days at 35°C. Since increasing the temperature increases the A_w of foods held at constant moisture content it seems likely that the caking of onion powder is directly related to its water activity.

A group on Brazil studied caking phenomena in tropical fruit powders (44). They found that the degree of caking of freeze-dried guava powder increased steadily but slowly from about A_w 0.3 to 0.6. From A_w 0.6 to 0.85 the degree of caking increased very rapidly with or without anticaking agents present. These authors also showed that at high temperatures (above about 40°C) caking always occurred regardless of the water activity in the powder and they point out that other workers have also had similar results at high temperatures.

SUGARS

Dry sugar can exist in two forms. One is as a glassy amorphous solid. Fine powders of glassy sugars are usually free-flowing and not sticky. The other form is crystalline, which is thermodynamically more stable than the glassy state but is often sticky. Under suitable conditions the glassy form spontaneously changes to the crystalline form. Each form has its own characteristic moisture sorption isotherm with the curve for the crystalline sugar lying well below that for the glassy sugar. When molten sugars are rapidly cooled, or sugar solutions are sprayed-dried or freeze-dried, they solidify in the glassy form, which is a metastable state. When this dry glassy sugar is exposed to increasing relative humidity atmospheres, it absorbs moistures and begins to form a conventional S-shaped water sorption isotherm. However, when the A_w rises sufficiently, the glassy sugar spontaneously changes into the crystalline form, which has an entirely different water sorption isotherm and a sharp negative slope appears in the water

sorption isotherm (Fig. 16 and Ref. 41). This is an example of an exception to the rule discussed earlier (beginning of section on sorption isotherms) that A_w always increases with increasing moisture content. It is caused by change in state of the sugar.

Moisture content of powdered sugars exposed to various relative humidities at 25°C for up to 3 years has been measured (46). The data for sucrose are plotted in Figure 17 using a logarithmic time scale. At 4.6 and 11.8% RH the powders absorbed a little moisture as they reached equilibrium and then the moisture level remained constant for 1,000 days. At 24% RH the moisture content was level for about 80 days and then declined because the sucrose began to crystallize. At 28.2% RH the moisture level began to fall after about 10 days. At 33.6% RH it began to fall immediately, indicating that conversion of the sugar into the crystallized form began immediately. The crystalline sugar caked badly even though it had a much lower moisture content than the glassy sugar. Those authors obtained similar results with amorphous glucose powder (42). This result has been confirmed by others (47) and also has been shown to occur with lactose (45). The addition of gums and stabilizers to powdered sucrose delayed the crystallization process and loss of moisture but had a negligible effect on the moisture content in the sugar after crystallization was complete (48).

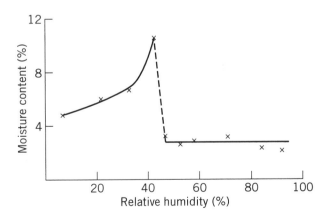

Figure 16. Absorption moisture isotherm for lactose at 25°C; on left-hand side, lactose is in glassy state; on right-hand side, it is in a crystalline state. *Source:* Ref. 45.

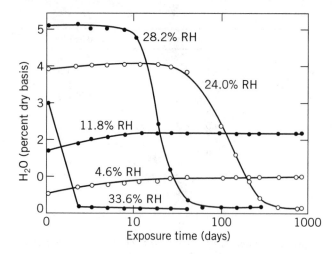

Figure 17. Changes in moisture content of amorphous sucrose powder stored in atmospheres of various relative humidities (RH) at 25°C. *Source:* Ref. 10.

The fact that crystallized sugars equilibrate at a lower moisture content than glassy sugars or sugars in solution is the basis of a clever process to impart shelf-stable contrasting textures to cookies (49). Fresh home-baked cookies are prepared with sucrose and are characterized by a crisp, firm texture on the outside and a soft, moist, chewy texture on the inside because the outside of the cookie is dry while the inside of the cookie is still moist when the cookies are removed from the oven. The moisture begins to migrate from the moist interior to the dry exterior of the cookie as soon as it is taken from the oven until there is uniform A_w throughout the cookie. The sucrose crystallizes, the cookie becomes uniformly dry and hard throughout and the highly relished contrasting textures of the freshly baked cookie are rapidly lost. Home-baked cookies lose their desirable dichotomous texture within a day or two. Traditional store-bought cookies are either hard and dry throughout, or soft and moist throughout because the A_w of all parts of the cookie equilibrates during distribution.

The new process uses two doughs, an outside dough that is extruded around an inside dough with different formulation (49). The inside dough contains a high proportion of crystallization-resistant sugars such as corn syrup or high-fructose corn syrup. During baking and storage these sugars do not crystallize. The outside dough has about the same proportions of flour, shortening, and sugar as the inside dough, but the sugar is mostly sucrose, which crystallizes easily. During the first few days after baking the A_w equilibrates throughout the cookie, the sucrose in the outer part crystallizes and releases most of the water it has absorbed. The center part of the cookie absorbs this moisture because its sugars do not crystallize. Moisture equilibrium is reached throughout the cookie after about 2 weeks of storage at $A_w \approx 0.45$–0.55 but the center of the cookie has a moisture content 2–4% higher than the outside. The result is a cookie that is dry and crisp on the outside and moist and soft on the inside similar to fresh home-baked cookies. The texture and moisture differential

is shelf stable and is maintained for many months, unlike fresh home-baked cookies, which lose their dichotomous texture within a few days.

MISCELLANEOUS FOODS

Some reports relating changes in texture with changes in water activity use such a narrow A_w range or so few A_w levels that one cannot confidently predict the textural properties over a wide range of A_w levels. Such reports will not be discussed here. Other researchers vary the method of processing or the formulation in so many ways (eg, water content, fat content, protein content, and salt content of sausages) that one cannot find any meaningful relationship between textural properties and water activity. Reports of this kind will not be discussed here because they throw no useful insights into texture—A_w relationships.

Walnuts

Some physical properties of shelled walnut kernels have been studied over the range $A_w = 0.1$–0.85 (50) using three test procedures:

1. Breakage. Cans of unbroken kernels were dropped 915 mm two times onto a concrete floor. The broken kernels were separated and weighed. The percentage of broken kernels decreased continuously with increasing A_w (see Fig. 18), indicating that the walnut meats became less brittle and more difficult to fracture as the moisture content increased.
2. Abrasion. 250-g lots of kernels were placed in 1-L tin cans and vibrated at 7 Hz and 5-mm amplitude for 15 min, and then the kernels were emptied into a tray and the chaff brushed off and weighed. The amount of chaff lost decreased rapidly from $A_w = 0.1$ to 0.55 and very rapidly from $A_w = 0.55$ to 0.75.
3. Pellicle content (skin). An estimate of the amount of walnut meat in the chaff was made on the basis of its fat content and subtracted from the value for the chaff to give the pellicle content. It followed a similar decreasing pattern as the chaff (Fig. 18).

Freeze-Dried Foods

Freeze-dried foods constitute a large portion of the food supply of astronauts because this reduces the weight of food that must be lifted into orbit. A modified texture profile analysis on bite-sized freeze-dried beef, chicken and cheese sandwiches, and chicken bites has been reported (18). Hardness (maximum force on the first bite) and cohesiveness (ratio of work done in second compression/work done in first compression) increased with increasing A_w for all four items (see Fig. 19). The crushability index (which is explained in the section on beef above) decreased for the beef and the chicken sandwiches and the chicken bites. For the cheese sandwich it increased from $A_w = 0$ to 0.5 and then decreased (Fig. 19).

Xixona Turrón

The effect of water activity on the textural properties of *xixona turrón*, a typical Spanish confectionery product

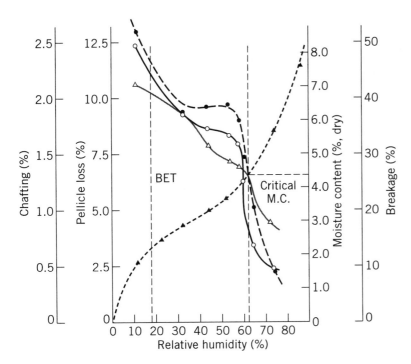

Figure 18. Moisture sorption isotherm and effect of A_w of walnut kernels on pellicle loss, chaffing, and kernel breakage. *Source:* Ref. 50.

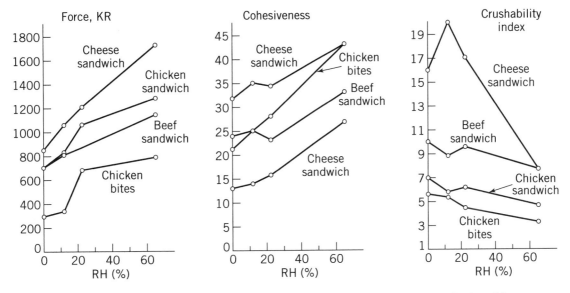

Figure 19. Hardness, cohesiveness, and crushability index vs relative humidity for four dehydrated space foods. *Source:* Ref. 15.

made from toasted almonds, sugar, inverted sugar, honey, and a little egg white is reported (51). The texture profile analysis (TPA) was measured at ~25°C. Hardness and chewiness increased sharply from $A_w = 0.11$ to $A_w = 0.22$, decreased sharply from $A_w = 0.22$ to $A_w = 0.33$, and then continued to decrease at a slower rate from $A_w = 0.33$ to $A_w = 0.75$ while springiness (elasticity) and cohesiveness increased slowly from $A_w = 0.11$ to $A_w = 0.43$ and steeply from $A_w = 0.43$ to $A_w = 0.75$. The glass-transition temperature (T_g) was 24°C at $A_w = 0.11$ and dropped steadily to -46°C at $A_w = 0.75$. Fracturability was only seen in samples with a moisture content lower than the critical

water content (CWC), the level at which the product changes from the rubbery state to the glassy state.

Although these authors did not elaborate the point, their data suggest that the textural properties of a food with moisture content near the CWC can change rapidly when the temperature is changed a few degrees to take it through the T_g region. Such a change could be expected to make the food brittle, fracturable, and crisp when cooled below the T_g, changing to leathery and chewy when warmed above the T_g. This data indicates the promising potential for studies involving T_g, A_w, and texture near ambient temperatures.

Fruit Bar

A Canadian group studied the change in cutting force of restructured fruit bars manufactured from apple puree, apple concentrate, water, maltodextrin, low-methoxyl pectin, citric acid, sodium citrate, and $CaHPO_4$ (52). The cutting force using a single blade with a 3×70-mm face increased 100-fold from $A_w = 0.86$ to $A_w = 0.33$ and then fell sharply from $A_w = 0.33$ down to $A_w = 0.30$ due to the change from cutting failure to brittle fracture. The general pattern of hardness change was similar to that found with dried apple discussed earlier (14), but the 100-fold increase in the formulated bar was much greater than the 6-fold increase found in dehydrated apple. The authors attribute the difference to the fruit bar being subjected to a cutting type of test and the apple to a uniaxial compression test. Another notable difference between these reports is that the apple test used 10-mm cubes for all A_w levels, whereas the fruit bars had a thickness of 5–6 mm at $A_w = 0.87$, which decreased with decreasing A_w until a thickness of only 1.5 mm was reached at $A_w = 0.3$. Another difference is that the fruit bars appeared to be case hardened at the lowest A_w levels.

Lempuk

This is a traditional semimoist product in Malaysia and nearby countries. It is made by cooking the flesh of durian fruit with sugar until the desired consistency is attained (53). TPA tests on lempuk over the range $A_w = 0.57$–0.84 (26–50% H_2O d.b) showed no fracturability and no change in cohesiveness and springiness over this range of water activity. Hardness and chewiness decreased and adhesiveness increased as A_w increased. There was a sharp change in slope, but not direction, for hardness, chewiness, and cohesiveness at $A_w = 0.75$ (33.6% H_2O), which was attributed to the glass transition for this product at about 33% H_2O and 25°C. However, the T_g for this product was not reported.

Agricultural Commodities

The physical properties of nonfood agricultural commodities such as seeds, hay, and tobacco leaf are often affected by water activity. For example, the breaking force in compression and deformation-to-break of bean seeds both increased as A_w increased from 0.35 to 0.65 (54). With soybean seeds, fewer split seeds and a higher germination was found as the moisture content was increased from 7.2 to 16.2% when the seeds were impacted against a steel plate, but the reverse effect was found when the soybean seeds were impacted against a hard polyurethane plate (55). See also Refs. 6 and 29.

This subject will not be pursued further because these are not foods. The four examples cited above should be sufficient to acquaint the reader with the fact that changing the A_w often has a profound effect on physical properties of nonfood agricultural commodities.

SUMMARY

The results discussed above do not paint a clear unequivocal of the relationships between A_w and texture. There are a number of reasons for this complex behavior.

1. Complex Texture. Every food possesses diverse textural properties that act independently with changing A_w. Over a given A_w range different textural properties in the same food may increase, decrease, or remain steady (Figs. 4, 6, 11, 19).

2. Water Activity Range. Some textural properties change in slope but not in direction with changing A_w (Figs. 8–10, 12, 13, 18). Very rapid changes in textural properties can occur near the BET monolayer (Figs. 4, 6, 7).

3. Previous history of sample treatment. A food that has desorbed water to reach a given A_w may be quantitatively different in texture than when it has absorbed moisture to reach the same A_w (Fig. 10). The time a food is stored can affect its textural properties because of chemical reactions that are occurring in it (22). Polymers such as proteins and complex carbohydrates can undergo complex molecular changes such as crystallization, cross-linking, folding, and unfolding, which can affect the texture.

4. Quality of ingredients. Different quality flours produce cookies with different breaking strengths (Fig. 15). Impurities in ingredients can cause anomalous trends in texture of beef (22).

5. Phase changes. When sugars change from the amorphous into the crystalline form, there are changes in the texture of the pure sugars (16, 17) and in products with a high sugar content (49).

6. Processing conditions. Extruded corn meals at the same A_w manifested substantial differences in textural quality because the different processing conditions in the extruder caused wide variations in structure (14 and Ref. 32).

7. Structure. Native foods such as fish, fruit, meat, and vegetables have a nonhomogeneous cellular structure that is substantially retained during preservation and storage. Some cellular regions may be affected by changes in moisture level more than other regions, eg, fiber bundles vs parenchyma cells in vegetables. The texture of fabricated foods such as cookies, crackers, and sausages can be affected by changes in formulation. For example, changing the fat content (which is hydrophobic and essentially inert to changes in moisture content) can affect the structure and the manner in which overall texture responds to changes in A_w.

8. Fundamentals. Theories such as water acting as a plasticizer in food and of glass transitions in food polymers are being developed that may eventually explain fundamental relationships between A_w and textural properties of foods (26,35,39,56). In the meantime we have to continue to gather empirical data for each food for each A_w level. This information will support or discredit the fundamental theories that are presently under development.

BIBLIOGRAPHY

1. M. C. Bourne, *Food Texture and Viscosity*, Academic Press, New York, 1982.

2. A. S. Szczesniak, "Classification of Textural Characteristics," *Journal of Food Science* **28**, 385–389 (1963).

3. A. S. Szczesniak, "Consumer Awareness of Texture and of Other Food Attributes," *Journal of Texture Studies* **2**, 196–200 (1971).

4. A. S. Szczesniak, "Correlating Sensory with Instrumental Texture Measurements—an Overview of Recent Developments," *Journal of Texture Studies* **18**, 1–15 (1987).

5. M. C. Bourne, A. M. R. Sandoval, M. Villalobos, and T. S. Buckle, "Training a Sensory Texture Profile Panel and Development of Standard Rating Scales in Colombia," *Journal of Texture Studies* **6**, 43–52 (1975).

6. K. Kohyama, M. Nishi, and T. Suzuki, "Measuring Texture of Crackers with a Multiple Point Sheet Sensor," *Journal of Food Science* **62**, 922–925 (1997).

7. K. Kohyama et al., "Evaluation of Texture of Cooked Rice Using Electromyography," *Journal of Texture Studies* **29**, 101–113 (1998).

8. W. E. Brown et al., "Use of Combined Electromyography and Kinesthesiology during Mastication to Chart the Oral Breakdown of Foodstuffs, Relevance to Measurement of Food Texture," *Journal of Texture Studies* **29**, 145–167 (1998).

9. J. G. Woodroofe, S. R. Cecil, and H. H. Thompson, *The Effects of Moisture on Peanuts and Peanut Products*, Georgia Experiment Station, Bulletin 241, May 1945.

10. M. C. Bourne, "Effects of Water Activity on Textural Properties of Food," in L. B. Rockland and L. A. Beuchat, eds., *Water Activity: Theory and Applications to Food*, Marcel Dekker, New York, 1987, pp. 75–99.

11. H. A. Iglesias and J. Chirife, *Handbook of Food Isotherms: Water Sorption Parameters for Food and Food Components*, Academic Press, New York, 1982.

12. L. B. Rockland, "Water Activity and Storage Stability," *Food Technology* **23**, 11–21 (1969).

13. P. P. Lewicki, "A Three Parameter Equation for Food Moisture Sorption Isotherms," *Journal of Food Process Engineering* **21**, 127–144 (1998).

14. M. C. Bourne, "Effect of Water Activity on Texture Profile Parameters of Apple Flesh," *Journal of Texture Studies* **17**, 233–260 (1986).

15. M. C. Bourne, *Food Texture and Viscosity*, Academic Press, New York, 1982, pp. 114–117.

16. M. C. Bourne, "Texture Profile Analysis," *Food Technology* **32**(7), 62–66, 72 (1978).

17. T. Beveridge and S. E. Weintraub, "Effect of Blanching Pretreatment on Color and Texture of Apple Slices at Various Water Activities," *Food Research International* **28**, 83–86 (1995).

18. J. G. Kapsalis, B. Drake, and B. Johansson, "Textural Properties of Dehydrated Foods. Relationships with the Thermodynamics of Water Vapor Sorption," *Journal of Texture Studies* **1**, 285–308 (1970).

19. J. G. Kapsalis, J. E. Walker, and M. Wolf, "A Physicochemical Study of the Mechanical Properties of Low and Intermediate Moisture Foods," *Journal of Texture Studies* **1**, 464–483 (1970).

20. J. G. Kapsalis, "The Influence of Water on Textural Parameters in Foods at Intermediate Moisture Levels," R. B. Duckworth, ed., *Water Relations in Foods*, Academic Press, New York, 1975, pp. 627–637.

21. G. A. Reidy and D. R. Heldman, "Measurement of Texture Parameters of Freeze-Dried Beef," *Journal of Texture Studies* **3**, 218–226 (1972).

22. D. R. Heldman, G. A. Reidy, and M. P. Palnitkar, "Texture Stability during Storage of Freeze-Dried Beef at Low and Intermediate Moisture Contents," *Journal of Food Science* **38**, 282 (1973).

23. D. A. Ledward, S. K. Lynn, and J. R. Mitchell, "Textural Changes in Intermediate Moisture Meat during Storage at 38°C," *Journal of Texture Studies* **12**, 173–184 (1981).

24. P. K. Chattopadhyay, J. R. Hammerle, and D. D. Hamann, "Time, Temperature and Moisture Effects on the Failure Strength of Rice," *Cereal Foods World* **24**, 514–516 (1979).

25. J. T. Hogan and M. L. Karon, "Hygroscopic Equilibria of Rough Rice at Elevated Temperatures," *Journal of Agricultural and Food Chemistry* **3**, 385 (1955).

26. P. K. Chattopadhyay and D. D. Haman, "The Rheological Properties of Rice Grain," *Journal of Food Processing Engineering* **17**, 1–17 (1994).

27. J. L. Multon, H. Bizot, J. L. Doublier, J. Lefebre, and D. C. Abbott, "Effects of Water Activity and Sorption Hysteresis on Rheological Behavior of wheat Kernels," in L. B. Rockland and G. F. Stewart, eds., *Water Activity: Influences on Food Quality*, Academic Press, New York, 1981, pp. 180–197.

28. W. Obuchowski and N. Boshuk, "Wheat Hardness: Comparison of Methods of Evaluation," *Cereal Chemistry* **57**, 421–425 (1980).

29. K. K. Waanen and M. R. Okos, "Stress Relaxation Properties of Yellow-Dent Corn Under Uniaxial Loading," *Transactions of the American Society of Agricultural Engineers* **35**, 1249–1258 (1992).

30. S. K. Seymour and D. D. Hamann, "Crispness and Crunchiness in Low Moisture Foods," Ph.D. dissertation, North Carolina State University at Raleigh, 1985.

31. E. E. Katz and T. P. Labuza, "Effect of Water Activity on the Sensory Crispness and Mechanical Deformation of Snack Food Products," *Journal of Food Science* **46**, 403–409 (1981).

32. F. Sauvageot and G. Blond, "Effect of Water Activity on Crispness of Breakfast Cereals," *Journal of Texture Studies* **22**, 423–442 (1991).

33. M. Peleg, "A Mathematical Model of Crunchiness/Crispiness Loss in Breakfast Cereals," *Journal of Texture Studies* **25**, 403–410 (1994).

34. M. Harris and M. Peleg, "Patterns of Changes in Brittle Cellular Cereal Foods Caused by Moisture Sorption," *Cereal Chemistry* **73**, 225–231 (1996).

35. R. Tesch, M. D. Normand, and M. Peleg, "Comparison of the Acoustic and Mechanical Signature of Two Cellular Crunchy Cereal Foods at Various Water Activity Levels," *Journal of Science of Food and Agriculture* **70**, 347–354 (1996).

36. A. Borges and M. Peleg, "Effect of Water Activity on the Mechanical Properties of Selected Legumes and Nuts," *Journal of Science of Food and Agriculture* **75**, 463–471 (1997).

37. G. W. Halek, S. W. Paik, and K. C. B. Chang, "The Effect of Moisture Content on Mechanical Properties and Texture Profile Parameters of Corn Meal Extrudates," *Journal of Texture Studies* **20**, 43–55 (1989).

38. M. E. Zabik, S. G. Fierke, and D. K. Bristol, "Humidity Effects on Textural Characteristics of Sugar-Snap Cookies," *Cereal Chemistry* **56**, 29–32 (1979).

39. F. Hsieh et al., "Effects of Water Activity on Textural Characteristics of Puffed Rice Cake," *Lebensmittel-Wissenschaft und Technologie* **23**, 471–473 (1990).

40. C. K. Mok et al., "Effects of Water Activity on Crispness and Brittleness, and Determination of Shelf-life of Barley Flake," *Korean Journal of Food Science and Technology* **13**, 289–298 (1981).

41. R. Moyreyra and M. Peleg, "Compressive Deformation Patterns of Selected Food Powders," *Journal of Food Science* **45**, 864–868 (1980).

42. M. Peleg, "The Role of Water in the Rheology of Hygroscopic Food Powders," in D. Simatos and J. L. Multon, eds., *Properties of Water in Foods*, Martinus Nijoff Publishers Dordrecht, 1985, pp. 393–404.

43. Y. Peleg and C. H. Mannheim, "Caking of Onion Powder," *Journal of Food Technology* **4**, 157–160 (1969).

44. J. Cal-Vidal, R. F. DeCarvalho, and S. C. S. Santos, "Caking Phenomena in Tropical Fruit Powders," in M. La Maguer and P. Jelen, eds., *Food Engineering and Process Calculations*, Vol. 1, Elsevier Applied Science Publishers, London, 1986, pp. 483–497.

45. S. Warburton and S. W. Pixton, "The Moisture Relations of Spray Dried Skimmed Milk," *Journal of Stored Products Research* **14**, 143–158 (1978).

46. B. Makower and W. B. Dye, "Equilibrium Moisture Content and Crystallization of Amorphous Sucrose and Glucose," *Journal of Agricultural and Food Chemistry* **4**, 72–77 (1956).

47. H. A. Iglesias, J. Chirife, and J. L. Lombardi, "Comparison of Water Vapour Sorption by Sugar Beet Root Components," *Journal of Food Technology* **10**, 385–391 (1975).

48. H. A. Iglesias and J. Chirife, "Delayed Crystallization of Amorphous Sucrose in Humidified Freeze-Dried Model Systems," *Journal of Food Technology* **13**, 137–144 (1978).

49. C. A. Hong and W. J. Brabbs, "Doughs and Cookies Providing Storage-Stable Texture Variability," U.S. Patent 4,455,333 issued June 1984 (to The Procter and Gamble Co.).

50. P. Veeraju, J. Hemuvathy, and J. V. Prabbhakar, "Influence of Water Activity on Pellicle Chaffing, Color and Breakage of Walnut (*Juglans regia*) Kernels," *Journal of Food Processing Preservation* **2**, 21–31 (1978).

51. N. Martinez and A. Chiralt, "Glass Transition and Texture in a Typical Spanish Confectionary Product Xixona Turrón," *Journal of Texture Studies* **26**, 653–663 (1995).

52. S. R. Owen, M. A. Tung, and T. D. Durance, "Cutting Resistance of a Restructured Fruit Bar as Influenced by Water Activity," *Journal of Texture Studies* **22**, 191–199 (1991).

53. C. C. Seow, "Lempuk, A Traditional Malaysian Intermediate Durian Product," *ASEAN Food Journal* **9**, 127–131 (1994).

54. A. P. M. Bay, M. C. Bourne, and A. G. Taylor, "Effect of Moisture Content on Compressive Strength of Whole Snap Bean (*Phaseolus vulgaris* L.) seeds and Separated Cotyledons," *International Journal of Food Science and Technology* **31**, 327–331 (1996).

55. M. D. Evans, R. G. Holmes, and M. B. McDonald, "Impact Damage to Soybean Seed as Affected by Surface Hardness and Seed Orientation," *Transactions of American Society of Agricultural Engineers* **33**, 234–239 (1990).

56. G. Roudat C. Dacremont, and M. LeMeste, "Influence of Water on the Crispness of Cereal-Based Foods: Acoustic, Mechanical and Sensory Studies," *Journal of Texture Studies* **29**, 199–213 (1998).

GENERAL REFERENCES

J. M. V. Blanshard and P. Lillford, eds., *Food Structure and Behavior*, Academic Press, New York, 1987.

J. M. V. Blanshard and J. R. Mitchell, eds., *Food Structure—Its Creation and Evaluation*, Butterworths, London, 1988.

M. C. Bourne, *Food Texture and Viscosity, Concept and Measurement*, Academic Press, New York, 1982 (reprinted 1994).

D. Holcomb and M. Kalab, eds., *Studies of Food Microstructure*, Scanning Electron Microscopy Inc., AMF O'Hare, Ill., 1981.

M. Peleg and E. B. Bagley, eds., *Physical Properties of Foods*, AVI Publishing Co., Westport, Conn., 1983.

M. A. Rao and S. S. H. Rizvi, eds., *Engineering Properties of Foods*, Marcel Dekker, New York, 1986.

L. B. Rockland and L. R. Beuchat, eds., *Water Activity: Theory and Applications to Food*, Marcel Dekker, New York, 1987.

L. B. Rockland and G. F. Stewart, eds., *Water Activity. Influences on Food Quality*, Academic Press, New York, 1981.

A. J. Rosenthal, ed., *Food Texture: Measurement and Perception*, Aspen Publishers, Gaithersburg, Md., 1999.

C. C. Seow, ed., *Food Preservation by Moisture Control*, Elsevier Science Publishing Co., New York, 1988.

P. Sherman, ed., *Food Texture and Rheology*, Academic Press, London, 1979.

D. Simatos and J. L. Multon, eds., *Properties of Water in Foods*, Martinus Nijhoff, Dordrecht, 1985.

J. A. Troller and J. H. B. Christian, *Water Activity and Food*, Academic Press, New York, 1978.

MALCOLM C. BOURNE
Cornell University
Geneva, New York

WATER ACTIVITY: GOOD MANUFACTURING PRACTICE

Water activity (a_w) is a measure of the free moisture in a product. It is defined as the quotient of the water vapor pressure of the substance (p) divided by the vapor pressure of pure water (p_0) at the same temperature; $a_w = p/p_0$. The amount of free moisture is of critical importance in supporting the growth of microorganisms. The primary public health concern associated with improperly processed low-acid foods is the possibility of a foodborne intoxication due to botulinum toxin.

The current good manufacturing practice (GMP) regulation for thermally processed low-acid foods packaged in hermetically sealed containers, Title 21 Code of Federal Regulations, Part 113 (21 CFR 113), establishes the specific level of a_w and pH to define the low-acid foods. Low-acid foods means any foods, other than alcoholic beverages, with a finished equilibrium pH >4.6 and an a_w >0.85. Commonly recognized low-acid canned foods (LACFs) with water activity level near 1.0 depend on thermal destruction of spores of pathogenic bacteria for safety. Water-activity-controlled LACFs are protected by achieving a safe a_w level, which is unfavorable for the germination and outgrowth of spores having public health significance. The application of mild heat treatment is intended to eliminate vegetative forms of microorganisms.

Control of water activity may be achieved through the proper design of product formulation. The rate of destruction of microorganisms when subject to heat is believed to be logarithmic in order. Knowledge of a_w control, thermal processing, and GMP requirements enable us to study this unique method of food preservation.

WATER ACTIVITY AND CONTROL OF MICROBIAL GROWTH

The purposes of food preservation are to extend shelf life, prevent microbial spoilage, and destroy potentially harmful microorganisms. The growth of microorganisms is inhibited at lower water activity levels. In the a_w range below 1.0 to 0.93, all known foodborne bacterial pathogens can grow in the upper part of this range. At a_w, 0.93 to 0.86, the only foodborne bacterial pathogen that can grow and cause an outbreak is *Staphylococcus aureus*. Under anaerobic conditions this nonsporulating bacteria is inhibited at an a_w level of 0.91, but aerobically it is inhibited at 0.86. Scientific evidence showed that the decimal reduction time D for microorganisms is greater at lower water activity levels. In one investigation, the minimum water activity for the growth of spores and toxin production was found to be 0.95 for *Clostridium botulinum* types A and B, and 0.97 for type E, at the optimum growth temperature (30–40°C) and pH (7.0). Another work showed that the minimum a_w for supporting the growth of botulinum types A and B was at 0.94 (1).

Thermal processing of commonly canned low-acid foods resembles wet heat sterilization (a_w = 1.0), whereas water-activity-controlled low-acid canned foods resemble dry-heat sterilization in that the water activity level inside the sealed container will be less than 1.0 during and after sterilization. Scientific evidence reveals that greater thermal resistance of microorganisms occurs when a reduced water activity level is encountered. In other words, the heating rate of an a_w-controlled low-acid canned food is lower than that of a wet low-acid food (a_w = 1.0).

A published paper showed that when a_w decreases from 1.0 to 0.90, and 0.80, resembling dry-heat sterilization, the decimal reduction time D increases from 7-fold and 33-fold, respectively. The resistance of bacterial spores to thermal kill increases with the decrease in a_w levels consistent with those for most a_w-controlled low-acid foods. The D value of *C. botulinum* spores increases with an increase in pH; the effect is greatest at lower processing temperatures (2). The synergistic effects of low a_w, pH, F^Z_T, and preservatives have clearly been proved to be effective against the growth of microorganisms. Generally, a_w-controlled foods contain large amounts of solutes and their a_w levels are between 0.8 and 0.95. Intermediate-moisture (IM) foods with a_w = <0.85 are not covered by 21 CFR 113. The microbial resistance to thermal kill will be different from that of wet-heat sterilization. For example, the $D_{65°C}$ values of *S. aureus* in skim milk, with 10 and 57% sugar, are 4.88 and 22.35, respectively. Although microorganisms show a greater resistance to thermal kill when at lower a_w levels, the addition of humectant(s) such as a high concentration of sugar, salt, or a combination of the two, for example, have proved to be effective in the preservation of foods. Thus, the growth of microorganisms in foods may be prevented by a_w control.

USDA's Agricultural Research Service developed a pathogen-modeling program (PMP) to predict the growth characteristics of various bacteria in food systems. This PMP is based on the equations derived by response surface analysis of growth data fitted using the Gompertz function in conjunction with nonlinear regression analysis. The program considers effects and interactions of sodium chloride (or a_w), acidity (or pH), temperature, under aerobic and anaerobic conditions, and so on. Evaluation of the models has indicated that they provide useful estimates of the microorganisms' growth characteristics in food systems.

ESTABLISHING SCHEDULED PROCESSES

Many factors influence the intrinsic characteristics and the microbial stability of foods. The five most important factors governing the growth of microorganisms in processed foods are a_w, pH, container seal integrity, temperature of storage and distribution, and thermal processing or equivalent treatment (F_0).

The thermal death time of 2.45 min with a z value of 17.6 for the 60 billion *C. botulinum* spores in pH 7.0 phosphate buffer when heated at 250°F (3) is generally accepted as a guide in process computations. The correlations between the spore resistance in phosphate and in a canned food may be expressed by phosphate factor (= resistance in food/resistance in standard phosphate buffer). The use of the food phosphate factor is a mechanism for adjusting the minimum public health sterilization value for a specific product. Perkins (4) indicated that many commercial processes are still computed on the basis of Esty and Meyer's classic *C. botulinum* resistance values, but there are many occasions when an F_0 of 5 to 7 min and even higher are used to achieve the commercial sterility cook for low-acid canned foods.

Alstrand and Ecklund proposed a guide for the selection of F_0 in relation to pH and carbohydrate content of canned foods (5). Many calculated F_0 values of successful commercial processes were presented, and most of those were greater than 3.0 rain. Pflug and coworkers (6) developed a model showing that at pH 6.0, 5.5, 5.0, and 4.6, the calculated F_0 value for public health processes were 3.0, 2.3, 1.6, and 1.2, respectively. Pflug (7) pointed out that when designing a low-acid canned food–sterilization process, three preservation conditions must be considered: (1) public health (F_0 = 3.0), (2) nonpathogenic mesophiles (F_0 = 5–7), and (3) thermophiles (F_0 = 5–7). However, if the product is expected to be stored and distributed at elevated temperatures, then F_0 should be 15 to 21. Low-acid foods packed in hermetically sealed containers are commonly processed in a retort to render the foods commercially sterile. As pointed out earlier, a minimum public health process of F_0 = 3.0 min or greater is usually called for. However, commercial sterility may also be achieved through inhibition instead of destruction of spores. To achieve this goal, factors including the intended level of preservation (or length of shelf life), the processing system, the characteristic of the food, the container type and size to be processed, and the expected commercial production conditions are to be considered.

The severe retorting process may be detrimental to the quality of certain foods heated by conduction. Reduced thermal processing requirements may be possible in that a scheduled process may include a proper combination of safe a_w level, a pasteurization process, and other treatment(s). The proper combination treatments would render the food commercially sterile. One investigator (8) pointed out that the canned cured meats with suitable amount of salt and nitrite are processed to achieve an F_0 value generally in the order of 0.05 to 0.6, but it may vary from 0 to 1.5. This reduced thermal treatment would also provide an immediate benefit of energy savings.

Safe a_w level is a level of water activity low enough to prevent the growth of undesirable microorganisms in the finished product under the intended conditions of manufacturing, storage, and distribution. The maximum safe a_w level is an a_w that will be considered safe for a food if adequate data are available that demonstrate that the food at or below the given a_w will not support the growth of undesirable microorganisms.

FDA Title 21 CFR 113.83 specifies that scheduled processes for low-acid foods shall be established by qualified persons having expert knowledge of thermal processing requirements for low-acid foods in hermetically sealed containers and having adequate facilities for making such determinations (9).

FDA GMP REGULATIONS GOVERNING THE PROCESSING REQUIREMENT

Canning is the most economical and widely used method of food preservation. Commercial sterility must be achieved for all thermally processed low-acid food packaged in hermetically sealed containers, including those low-acid intermediate-moisture foods. As defined by FDA's GMP regulation 21 CFR 113.3(e), commercial sterility of thermally processed food means the condition achieved (9):

1. By the application of heat which renders the food free of
   a. Microorganisms capable of reproducing in the food under normal nonrefrigerated conditions of storage and distribution; and
   b. Viable microorganisms (including spores) of public health significance; or
2. By the control of water activity and the application of heat, which renders the food free of microorganisms capable of reproducing in the food under normal nonrefrigerated conditions of storage and distribution.

Water activity is one of the two critical parameters used to determine the applicability of FDA's GMP regulations, 21 CFR 108, 110, 113, and 114. Figure 1 shows the type of foods and the applicable GMP regulations based on product a_w and pH levels.

Title 21 CFR 108.35 specifies that processors must register and file their scheduled processes with the FDA. The scheduled processes are reviewed for their adequacy for public health protection. When a process is considered potentially inadequate, follow-up action is generally required. Request for process establishment information, including the microbiological basis of a scheduled process,

has been the course of action most often recommended. Of the possible concerns are processes based on the lowest product sterilization values (<3 min) (10). The type of information needed by the FDA to evaluate the adequacy of a process for a water-activity-controlled low-acid food is specific under the following three conditions:

1. When the filed $a_w <= 0.85$ and the data appear reliable, the product in question is not covered by the LACF (low-acid food) regulations.
2. When $a_w <= 0.85$ to 0.90, the production in question is covered by the regulations 21 CFR 108.35 and 113, and it is necessary for the processor to register and file the scheduled process. The filing should include the critical maximum a_w necessary to control germination and outgrowth of *C. botulinum* spores and the critical thermal treatment necessary to destroy vegetative cells of pathogenic bacteria as well as prevent spoilage.
3. When the $a_w <= 0.90$, the product is covered by the regulations 21 CFR 108.35
4. and 113. The processor must register and file the scheduled process and will be requested to provide appropriate microbiological supporting data demonstrating process adequacy, in addition to the critical factors already mentioned for a_w-controlled products. The process-supporting data are to be obtained and reviewed and must be acceptable before the process filing forms are accepted for filing.

Title 21 CFR 113.40(i) requires that critical factors specified in the scheduled process, such as the a_w used in conjunction with thermal processing, shall be measured with instruments having the accuracy and dependability adequate to ensure that the requirements of the scheduled process are met. Official method of analysis for a_w has been adopted by the Association of Official Agricultural Chemists (AOAC) as the official final action method (Ref. 11).

Food preservation based on the principle of hurdle technology is recognized. 21 CFR 113.81(f) states that "when a low-acid food requires different solute to permit safe processing at low temperatures, such as in boiling water, there shall be careful supervision to ensure that the equilibrium water activity of the finished product meets that of the scheduled process." Specific parts and paragraphs of applicable GMP regulations from Title 21 of the Code of Federal Regulations delineating a_w in relation to control measures and safety requirements are:

1. 21 CFR 110: Current GMP in manufacturing, packing, or holding human food; 21 CFR 110.3(n), 110.3(r), 110.40(f), 110.80(b)(2), 110.80(b)(4), and 110.80(b)(14)
2. 21 CFR 113: Thermally processed low-acid foods packaged in hermetically sealed containers; 21 CFR 113.3(e)(ii), 113.3(n), 113.3(w), 113.10, 113.40(i), 113.81(f), 113.83, 113.89, 113.100(a)(6)

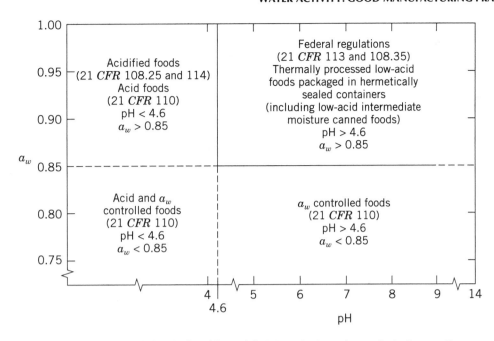

Figure 1. Applicability of the Food and Drug Administration's good manufacturing practice regulations.

3. 21 CFR 114: Acidified foods; 21 CFR 114.3(b), 114.3(d), and 114.3(h).

HUMECTANTS AND THEIR REGULATORY USE

Manipulation of moisture as a means of extending shelf life in perishable foods can be accomplished through either dehydration or a_w control. Dehydration and a_w control may be viewed as two different technologies. Dehydration is essentially direct removal of moisture, whereas a_w control is the addition of humectant(s) to bind or reduce free moisture, and not necessarily for the removal of free moisture from foods. Foods with the same amount of moisture could exhibit different levels of a_w. Foods with a higher amount of moisture content could be adjusted to exhibit a lower level of a_w than those with less moisture content. Formulated foods and foods preserved by the addition of salt, sugar, and glycerol are typical examples. According to a summation of experimental work over many years, it is generally conceded that 50% sugar or 10% salt is completely inhibitory to outgrowth of the *C. botulinum* spores of the proteolytic types A and B (12). These concentrations of sugar and salt correspond to the same a_w of 0.935. Through the use of salt, sugar, glycerol, and other humectants to control a_w levels, many low-acid foods are now processed at low temperatures. Most intermediate-moisture low-acid foods contain a large amount of solutes or humectants. Typical intermediate-moisture low-acid foods include salt-cured meat and meat products, salted fish and fish products, soup bases, specialty sauces, syrups, some preserves, puddings, some preserved vegetables, processed cheese products, and some snack foods.

Some humectants and their CFR-specified uses are summarized as follows:

1. Sucrose, 21 CFR 184.1854 and 582.1
2. Salt, 21 CFR 182.1 and 582.1
3. Glycerin, 21 CFR 182.1320 and 582.1320
4. Sodium nitrite, 21 CFR 172.175, 181.34, and 573.700
5. Potassium sorbate, 21 CFR 182.3640
6. Sorbitol, 21 CFR 184.1835
7. Potassium chloride, 21 CFR 184.1622
8. Propylene glycol, 21 CFR 184.1666 and 582.1666
9. Propylparaben, 21 CFR 184.1670
10. Lactic acid, 21 CFR 184.1061 and 582.1061
11. Invert sugar, 21 CFR 184.1859
12. High-fructose corn syrup, 21 CFR 184.1866
13. Modified food starch, 21 CFR 172.892

WATER ACTIVITY DETERMINATIONS

There are many ways to estimate the water activity of foods. The principles governing water activity determinations may be illustrated as follows:

$$a_w = p/p_o \qquad \text{(by definition)} \qquad (1)$$
$$= n_o/(n_o + n_s) \qquad \text{(Raoult's law)} \qquad (2)$$
$$= (a_w)_1\,(a_w)_2\,(a_w)_3 \cdots \quad \text{Gibbs-Duhem} \qquad (3)$$
$$= \text{ERH}/100 \qquad \text{by thermodynamic principle} \qquad (4)$$

where ERH is the equilibrium relative humidity. Equation 1 is the basis of the vapor pressure manometric method, in which the vapor pressure in the evacuated headspace above the food sample is determined. This provides a direct measurement of the vapor pressure.

Equation 2 is often used in work in formulating the intermediate-moisture (IM) foods and in predicting their a_w. The biochemical potential of the idealized solutions obeys Raoult's law. Most nonelectrolytes are also adequately represented by Raoult's law up to several molal concentrations. A modified equation, $a_w = n_o/(n_o + 1.5n_s)$ was found to fit experimental values up to high concentrations of sugars.

Equation 3 is commonly used in formulating the IM foods and in predicting their a_w. This equation (13) is very useful for aqueous systems involving more than two solutes. It is assumed that in a food system each a_w-lowering component behaves independently. The final a_w is a product of each component $(a_w)_n$, based on each solute being dissolved in all the water in the system.

Equation 4 is the basis for water activity measurement by hygrometers and isopiestic equilibrium methods. According to the thermodynamic principle, when at equilibrium, the activity of a water vapor equals the activity of the corresponding aqueous phase in a closed system. Therefore, the measurement of the ERH of a food is a measure of the water activity of that food, and a_w is numerically equal to the ERH of that food expressed as a decimal fraction; $a_w = \text{ERH}/100$.

DISCLOSURE STATEMENT

Instruments and methods for water activity determinations are detailed in the 1995 (16th) edition of AOAC Official Methods of Analysis, Method 978.18.

ACKNOWLEDGMENTS

This article was written by the author in his private capacity. No official support or endorsement by the U.S. Food and Drug Administration is intended or should be inferred.

BIBLIOGRAPHY

1. D. A. Kautter et al., "The Detection, Identification, and Isolation of *C. botulinum*," in M. Herzberg, ed., *Toxic Microorganisms*, Government Printing Office, Washington, D.C., 1970.

2. H. Xezones and I. J. Hutchings, "Thermal Resistance of *C. botulinum* (62A) Spores as Effected by Fundamental Food Constituents," *Food Technol.* **19**, 113–115 (1965).

3. I. J. Esty and K. F. Meyer, "The Heat Resistance of the Spores of *C. botulinum* and Allied Anaerobes, XI," *J. Infectious Dis.* **31**, 650 (1922).

4. W. E. Perkins, "Prevention of Botulism by Thermal Processing," in *Book Botulism*, U.S. Department of Health, Education, and Welfare, Public Health Service Publication No. 999-FP-1.

5. D. V. Alstrand and O. F. Ecklund, "The Mechanics and Interpretation of Heat Penetration Tests in Canned Foods," *Food Technol.* **6**, 185 (1952).

6. I. J. Pflug, T. E. Odlaug, and R. Christensen, "Computing a Minimum Public Health Sterilizing Value for Food with pH Values from 4.6 to 6.0," *J. Food Protection* **48**, 848–850 (1985).

7. I. J. Pflug, "Factors Important in Determining the Heat Process Value, F_t, for Low-Acid Canned Foods," *J. Food Protection* **50**, 528–530 (1987).

8. A. H. W. Hauschild, "*Clostridium botulinum*," in M. P. Doyle, ed., *Foodborne Bacterial Pathogens*, Marcel Dekker, New York, 1989.

9. U.S. Food and Drug Administration, Title 21 *Code of Federal Regulations*, Parts 108, 110, 113, 114, 121, 172, 175, 182, and 184, U.S. Government Printing Office, Washington, D.C., 1990.

10. R. T. Mulvaney, "Regulatory Review of Scheduled Thermal Process," *Food Technol.* **78**, 73–75 (1978).

11. Official Methods of Analysis, 15th ed., Method 978.18, Association of Official Analytical Chemists, Arlington, Va., 1990.

12. C. F. Schmidt, "Spores of *C. botulinum*: Formation, Resistance, Germination," in *Book Botulism*, U.S. Department of Health, Education, and Welfare, Public Health Service Publication No. 999-FP-1.

13. K. Ross, "Estimation of Water Activity in Intermediate Moisture Foods," *Food Technol.* **29**, 26–28 (1975).

GENERAL REFERENCES

A. C. Baird-Parker and B. Freame, "Combined Effect of Water Activity, pH and Temperature on the Growth of *Clostridium botulinum* from Spore and Vegetative Cell Inocula," *J. Appl. Bacteriol.* **30**, 420–429 (1967).

M. R. Johnston and R. C. Lin, "FDA Good Manufacturing Practice Regulations," Food Quality **3**, 109–118 (1980).

M. R. Johnston and R. C. Lin, "FDA Views on the Importance of a_w in Good Manufacturing Practice," in L. B. Rockland and L. R. Beuchat, eds., *Water Activity: Theory and Applications to Food*, Marcel Dekker, New York, 1987.

M. R. Johnston, R. C. Lin, and F. A. Philips, "FDA Views on the Importance of pH in Good Manufacturing Practice," *Food Product Devel.* **11**, 75–77 (1977).

R. C. Lin, P. H. King, and M. R. Johnston, "Examination of Containers for Integrity," *FDA Bacteriological Analytical Manual*, 6th ed., Association of Official Analytical Chemists, Arlington, Va., 1985.

W. G. Murrel and M. J. Scott, "The Heat Resistance in Bacteria Spores at Various Water Activities," *J. General Microbiol.* **43**, 411–425 (1966).

I. J. Pflug and T. E. Odlaug, "Review of z and F values Used to Ensure the Safety of Low-Acid Canned Food," *Food Technol.* **32**, 63–70 (1978).

C. R. Stumbo, K. S. Purohit, and T. V. Ramakrishnan, "Thermal Process Lethality Guide for Low-Acid Foods in Metal Containers," *J. Food Sci.* **40** (1975).

J. A. Troller, "Effect of Water Activity on Enterotoxin B Production and Growth of *Staphylococcus aureus*," *Appl. Microbiol.* **21**, 435–439 (1971).

Rong C. Lin
U.S. Food and Drug Administration
Washington, D.C.

WATER ACTIVITY: MICROBIOLOGY

The influence of water activity (a_w) on food product quality and stability began to receive considerable attention during the early 1950s. Such interest was promoted by inconsistent empirical observations of total moisture content and product stability. Microbiologists were among the first to recognize that a_w rather than total moisture content controlled microbiological growth, death, survival, sporulation, and toxin production by diverse microorganisms. Between the 1970s and 1980s, this group of scientists focused their interest on the influence of a_w by studying how microorganisms respond under different conditions of temperature, pH, and a_w (1).

DEFINITION OF WATER ACTIVITY

Water activity as a broad thermodynamic concept is defined as a relation of fugacities between liquid and vapor phases of a confined system while it is in equilibrium. A review of basic and applied thermodynamics is required to understand the a_w concept (2). Assuming that fugacity is a function of pressure and that the correction factor for vapor pressure is the same for an aqueous solution and pure water, then the fugacity ratio can be replaced by the pressure ratio (3):

$$a_w = P_w/P_w^\circ = \text{RHE}/100$$

where P_w is the vapor pressure of water in equilibrium with a food, P_w° is the vapor pressure of pure water at the same temperature of the food, and RHE is the relative humidity in equilibrium. Therefore, the vapor pressure of water in equilibrium with a food can be measured or related with the RHE. However, it should be taken into account that a_w is defined only for a system that presents a thermodynamic equilibrium, and its value is applicable only for specific conditions of temperature and pressure.

Many multicomponent food systems constituted by two or more phases (solid, liquid, aqueous, and oil) do not present equilibrium from one phase to the other (4). For this reason, a_w may not be an adequate thermodynamic parameter. Franks (5) proposed that the presence of hysteresis in food sorption isotherms (Fig. 1) may imply the presence of nonreversible equilibrium. These food systems are not really stable, but are in a pseudoequilibrium or metastable state. Thus, in these cases, the term water activity is not applied; instead, relative vapor pressure or relative humidity is used. However, when hysteresis is not shown, vapor pressure can be a measure of a_w.

WATER ACTIVITY AS AN INDICATOR OF MICROBIAL SAFETY AND SHELF LIFE

The influence of food water content on perishability has been known since ancient times. Almost every primitive culture found a convenient way to reduce food water content to a level that would protect against microbial spoilage (eg, drying, salting, adding sugars, concentrating). However, the occasional failure of these preservation techniques produced predictable results: salted products turned red because of the development of halophilic bacteria, sugar-preserved foods were fermented by osmophilic yeasts, and dried products were occasionally attacked by xerophilic molds (6). Also discovered was that food water content did not fully define microbial stability, because dried milk would rapidly spoil if it contained about 12% water, whereas cereals would remain fully sound at about 14% and dried fruits even at 18%. A better quantitative approach to define the influence of water on microbial response in foods was introduced by Scott (7) with the concept of a_w.

Considering a_w in relation to microbial stability, the minimum a_w values that permit microbial growth for different types of microorganisms are of considerable concern. Table 1 presents the a_w and microbial spoilage of foods with a classification of microorganisms as osmosensitive or osmotolerant. Corry (8), Beuchat (9,10), and Gould (11) generated extensive tables with minimum a_w values for the growth and toxin production of several pathogenic and spoilage microorganisms. Table 2 shows minimal a_w for the growth of foodborne bacterial pathogens at their optimum pH and temperature. With the exception of *Staphylococcus aureus*, for which a minimum a_w of 0.86 is needed to avoid growth under aerobic conditions, the growth of all other pathogens can be inhibited by a reduction in a_w to approximately 0.92. The major findings of Corry (8), Beuchat (9,10), and Gould (11) can be summarized as follows:

- At $a_w < 0.90$ to 0.92, most pathogenic bacteria are usually inhibited, except *Staphylococcus aureus*, which can grow at a_w 0.86.
- Minimum a_w for growth is always equal or lower than minimum a_w for toxin production.
- The minimum a_w for growth depends on the solute used to control a_w, the so-called solute effect.

Water activity is a major factor governing the microbial responses in foods. The concept of a_w has been successfully applied to the water relations in microorganisms because measured values of the food media in which they are immersed generally correlate well with the potential for growth and metabolic activity. Ecological factors and their interactions control the microbial responses in foods that determine what growth may occur. The combination of the

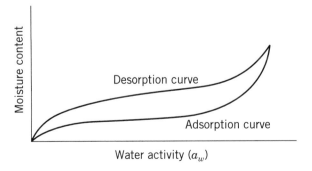

Figure 1. Typical sorption isotherm showing hysteresis.

Table 1. Wateer Activity and Food Microbial Spoilage

Range of a_w	Microorganisms inhibited	Examples of foods
1.00–0.95	Some yeasts, Gram-negative rods, bacterial spores	Foods containing 40% sucrose or 7% salt
0.95–0.91	Most cocci, lactobacilli, vegetative cells of bacilli, some molds	Foods containing 50% sucrose or 12% salt
0.94	Growth and toxin production by all types of *Clostridium botulinum*	Meats packaged under anaerobic conditions
0.91–0.87	Most yeasts	Foods containing 65% sucrose or 15% salt
0.8–0.80	*Staphylococcus aureus*, most molds	Flour, rice, etc. containing 15–17% water
0.86	Aerobic growth of *Staphylococcus aureus*	Mushrooms
0.80–0.75	Most halophilic bacteria	Foods with 26% salt
0.80	Production of micotoxins	
0.75–0.65	Xerophilic molds	Foods containing less than 10% water
0.68	Practical limit for fungi	
0.65–0.60	Osmophilic yeasts	Confectionery products, fruits containing 15–20% water

food a_w and pH frequently determine whether bacterial or fungal growth may take place, and the type of these species that will develop is often determined by temperature, gas atmosphere, redox potential, nutritional status, physical states, antimicrobial substances, and chemical preservatives (12).

The predominant microbial flora of a particular food (Figs. 2 and 3) can be determined by the interaction between its a_w and pH. Although bacteria are the major spoilage organisms in high pH (>4.6) and a_w (0.980–0.999) foods, yeasts and molds will dominate when the pH is below 4.0. In high a_w foods with a pH below 4.0, such as fruit juices and yogurt, yeasts as well as some lactic acid bacteria are most likely to be dominant. Some filamentous fungi also compete well in high a_w–low pH environments. The majority of yeasts and molds that cause spoilage are nonxerophilic, growing well at high a_w (>0.99) but also at a_w values below 0.90. In sugar-rich environments where bacteria do not seem to adapt, yeasts and molds take over as the major spoilage organisms. However, in foods such as meat with neutral pH and high a_w, bacteria dominate even though they may not be capable of growing in the food

at a_w values below 0.95. Bacteria rarely spoil foods with a_w values below 0.85, but brines and salted foods may be spoiled by moderate and extreme halophilic bacteria (12).

The a_w concept has assisted food scientists and microbiologists in their efforts to predict the onset of food spoil-

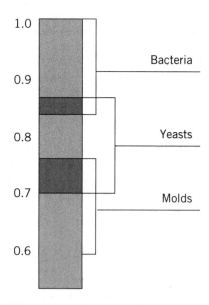

Figure 2. Effects of a_w on the growth of bacteria yeasts and mold microorganisms.

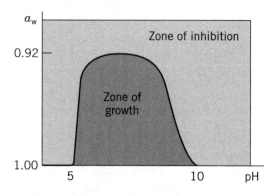

Figure 3. Interacting effects of pH and a_w on the growth of selected microorganisms.

Table 2. Minimal a_w for the Growth of Foodborne Bacterial Pathogens Under Optimum pH and Temperature Conditions

Bacteria	a_w
Campylobacter jejuni	0.990
Aeromonas hydrophila	0.970
Clostridium botulinum E	0.965
Clostridium botulinum G	0.965
Shigella spp.	0.960
Yersinia enterocolitica	0.960
Clostridium perfringens	0.945
Clostridium botulinum A & B	0.940
Salmonella spp.	0.940
Escherichia coli	0.935
Vibrio parahaemolyticus	0.932
Bacillus cereus	0.930
Listeria monocytogenes	0.920
Staphylococcus aureus (anaerobic)	0.910
Staphylococcus aureus (aerobic)	0.860

age and identify and control foodborne disease hazards that may exist in various food products (13). However, the a_w of a medium is not the only determining factor regulating biological response because the nature of the a_w controlling solute in the so-called solute effect occasionally plays a role as well (14). Consequently, the concept of a_w in foods as a determinant of microbial growth has been challenged (13). The occurrence of specific effects from different solutes already posed questions about the role of a_w as a credible measure of physiological viability because solute effects can mislead the prediction of microbial growth based on measured levels of a_w (9,13). During the 1980s, an approach that focused attention on the dynamic aspects of food systems of the low to intermediate moisture type that are nonequilibrium systems also received considerable attention (13,15). This perspective, which addresses the importance of maintaining foods in kinetically metastable, dynamically constrained glassy states, has incorporated the theoretical bases of fundamental structure and property principles from the field of synthetic polymer science, which includes the innovative concepts of water dynamics and glass dynamics in what has been called food polymer science (15). In many foods and biological materials, the solids are in an amorphous metastable state that is very sensitive to changes in moisture content and temperature (15). This amorphous matrix may exist either as a very viscous glass or a more liquidlike rubbery structure. The characteristic temperature (T_g) at which the glass–rubber transition occurs has been proposed as a physicochemical parameter that can determine product properties such as stability and safety better than a_w (15). However, the food science polymer approach may not constitute a better alternative to the concept of a_w as a predictor of microbial growth in foods despite the limitations of the a_w concept and possible influences of nonequilibrium situations (13).

OSMOTIC STRESS AND OSMOREGULATION IN MICROORGANISMS

Changes in medium osmolality (decreasing a_w) lead to a passive flow of water across microbial cell membranes, which results in a loss of turgor. The universal cell mechanism to overcome the initial loss following a hyperosmotic shock is the accumulation of cytoplasmatic solute(s) that increase the internal osmotic pressure, which in turn can restore turgor (11,16). There is great variation in the type of solutes accumulated by various microorganisms (eg, cations, amino and imino acids, amines, quaternary amines, polyols, sugars), but all exhibit a number of common properties, such as allowance of continued activity of cytoplasmic enzymes at lower a_w than the external solutes commonly present in foods and solubility to high concentrations.

Control of microbial growth in foods by a_w modifications can be achieved by two different strategies:

1. Using a_w as the only stress factor and therefore reducing it to very low levels (ie, fully dehydrated foods of $a_w < 0.6$).

2. Using other stress factors in combination with the low a_w (hurdle approach) that reduce the amount of energy available for osmoregulation, or impairing some other stresses that increase energy consumption (ie, intermediate moisture or high a_w foods).

WATER ACTIVITY IN DETECTION AND ENUMERATION OF MICROORGANISMS

All the ecological factors that are determinants of microbial response cannot always be considered in the microbial examination of foods. As discussed earlier, both the a_w and pH of a food substrate have a major influence on the type of microflora capable of colonizing and causing spoilage or foodborne diseases. These are facts that need consideration for reliable isolation and enumeration of significant microorganisms. The majority of nonxerophilic fungi that cause spoilage in moisture-rich foods can be detected, without special considerations, on general-purpose enumeration media such as acidified potato dextrose agar, malt extract agar, or dichloran rose bengal chloramphenicol (DRBC) agar (17). Conversely, the consideration of product a_w in intermediate and low moisture foods, such as syrups, confections, fruit concentrates, honey, dried fruits, salted dry fish and meat, grains, legumes, nuts, or spices, is of major relevance. Such products have characteristic mycoflora that are determined by factors such as a_w, type of solutes present (sugar or salt), and storage conditions, such as temperature, CO_2 concentration, and available oxygen (12). Because many fungi species require reduced a_w media; they are not capable of growth at a_w 0.99 and may be underestimated or undetected when traditional methods and media are used (ie, general plate enumeration in acidified potato dextrose agar, malt extract agar, or DRBC agar) (18).

Xerophilic fungi have been described as capable of growth below a_w 0.85. Under at least one set of environmental conditions they include osmophilic yeasts and xerophilic filamentous fungi: *Debaryomyces hansenii*, *Zygosaccharomyces bailii*, and *Z. rouxii* among the yeasts and several fungi species of *Aspergillus* (moderate xerophilic: *A. restrictus* series and *A. glaucus* series or *Eurotium*) and *Penicillium* as well as *Chrysosporium*, *Eremascus*, *Paecelomyces*, *Wallemia*, *Xeromyces*, *Basipetospora*, *Polypaecylum*, *Geomyces* and *Monascus*. Among the fastidious extreme xerophiles that require reduced a_w to grow are *Chrysosporium farinicola*, *C. fastidium*, *C. inops*, *C. xerophilium*, *Eremascus albus*, *E. fertilis* and *Xeromyces bisporus*; *Basipetospora halophila* and *Polypaecilum pisce* are halophilic molds. The inadequacy of media with high a_w to support the growth of xerophilic fungi is magnified when foods under examination contain inhibitory chemicals or have additional stress factors of a physical nature that affect the ability of cells to grow on enumeration media (18).

Among the common media for detection or enumeration of fungi in intermediate and low moisture foods is dichloran 18% glycerol (DG18) agar (a_w 0.95, pH 6.5), developed by Hocking and Pitt (17) for enumeration of moderately xerophilic molds in commodities such as grains, flours, nuts, and spices. The presence of dichloran in this media restricts the growth of rapidly growing mucoraceous species such as *Eurotium* spp. that tend to spread over the plates and obscure the growth, detection, and recovery of slower growing xerophiles, although this recovery is present in all but fastidious xerophiles. To overcome this problem, the addition of nontoxic chemicals (Triton X-301 or iprodione) into DG18 to enhance the case of counting colonies formed by the maximum number of viable propagate of moderately xerophilic molds present in intermediate moisture and dehydrated foods has been investigated. DRBC and DG18 are presently recommended as general purpose isolation and enumeration media that discriminate among foods depending on their a_w values (>0.90); DG18 for instance, is less suitable for fresh fruits and vegetables than DRBC.

To analyze foods with lower a_w, the major selective principles should be high concentrations of carbohydrates or sodium chloride. Malt salt agar (MSA) has long been used for enumerating molds in flour, but has recently been replaced by the more efficient glycerol-containing medium DG18. Other media are glucose citric acid tryptone (50 GCT) agar (pH 4.0) for enumeration of sugar-tolerant yeast in concentrated orange juice; Scarr's osmophilic wort (SOW) agar for xerophilic yeasts in high sugar products; a series of malt extract glucose (MYG) agar up to 60% glucose (a_w 0.85); malt extract yeast 70% glucose fructose (MY70GF) (a_w 0.76) for isolation of extreme xerophiles that may be accompanied by *Eurotium* spp.; and malt extract-yeast extract 5% salt 12% glucose agar (a_w 0.96) (18). In bacteriology, there are relatively few uses for salt-based media of reduced a_w, but because salt is an important component of selective enrichment broths and plating identification media, these are used to detect and enumerate salt-requiring or salt-tolerant bacterium such as many *Vibrio* species that have a physiological requirement for NaCl.

It is important to consider the composition and process history of reduced a_w foods when selecting methods for mycological analysis (18), cell adaptation to low a_w can occur because tolerance to low a_w can be gained or lost in the presence of various concentrations of solute. Therefore, the state of microbial cell adaptation will influence its ability to develop colonies on various enumeration media. As for diluents, peptone (0.1%) water and 0.05 to 0.1 M potassium phosphate buffers are common for general yeast and mold enumeration. Diluents containing a solute are recommended for analyzing intermediate and low-moisture foods for xerophilic fungi in order to minimize osmotic shock to fungal cells when making serial dilutions before plating. However, if cells are stressed, the use of diluents with reduced a_w can be critical. Rehydration of low a_w foods before homogenizing in diluent and plating is expected to enhance the resuscitation of injured cells (18).

The choice of solutes used to achieve the desired a_w value in diluents and detection or enumeration media depends on the major solute present in the food to be analyzed. As such, there continues to be a need for improved and new media that will enable a more rapid and accurate assessment of foods in which a_w plays a significant role as a determinant of microbial growth.

INFLUENCE OF WATER ACTIVITY ON MICROBIAL DEATH AND SURVIVAL

When a microorganism is transferred to a new environment, it either survives or dies depending on its ability to adapt. A generalized growth response of a microbial vegetative cell subjected to a reduced a_w is presented in Figure 4. Other than the presence of nutrients, the most important factors for growth and toxin production are temperature, a_w, pH, and oxygen. The basis for survival and death of microorganisms as influenced by a_w is complex. Several intrinsic and extrinsic factors differing within food types and processes and among types of microorganisms may affect this relation. Temperature, oxygen, and chemical and other physical treatments are some extrinsic factors that influence microbial spoilage in foods and the a_w-microorganism response. The use of combinations of extrinsic and intrinsic factors together with lowered a_w levels are common in the food industry. Generally, as the minimal a_w for growth of a microorganism is approached, changes in other environmental factors will have a greater impact on death or survival.

The effects of temperature on the survival of microorganisms are widely documented; the heat resistance of vegetative cells and spores as influenced by a_w is probably the most extensively studied area in terms of microbial inactivation. In general, vegetative cells and fungi spores are more resistant as the a_w of a heating menstruum is reduced. The type of solute used to adjust a_w to some value may result in significant differences in the heat resistance of a given microorganism. Small molecular weight compounds such as sodium chloride and glycerol have been shown to decrease heat resistance in *Salmonella* strains, whereas larger molecular weight solutes such as sucrose exerted a more protective effect against heat inactivation.

WATER ACTIVITY MEASUREMENT

The demand for quality control in the food industry necessitates accurate, fast, and convenient methods of measuring a_w. This is particularly true for food in which a_w is a controlling factor of microbial growth. This section pres-

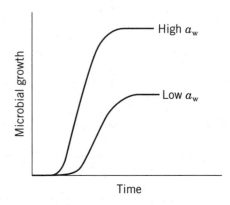

Figure 4. Generalized microbial growth response as affected by a_w.

Table 3. Relative Humidities Generated by the Saturated Solutions of Different Salts

Salt	10°C	15°C	20°C	25°C	30°C
Lithium bromide	7.1	6.9	6.6	6.4	6.2
Sodium hydroxide	—	9.6	8.9	8.2	7.6
Lithium chloride	11.3	11.3	11.3	11.3	11.3
Potassium acetate	23.5	23.5	23.0	22.5	22.0
Magnesium chloride	33.5	33.0	33.0	33.0	32.5
Potassium carbonate	44.0	43.5	43.0	43.0	43.0
Sodium bromide	60.0	59.0	58.0	57.5	56.5
Copper chloride	68.0	68.0	68.0	67.5	67.0
Potassium iodide	72.0	71.0	70.0	69.0	68.0
Sodium chloride	76.0	75.5	75.5	75.5	75.0
Ammonia sulfate	81.0	80.5	80.5	80.0	80.0
Potassium chloride	87.0	86.0	85.0	84.5	84.0
Sodium benzoate	88.0	88.0	88.0	88.0	88.0
Potassium nitrate	95.5	95.0	94.0	93.0	92.0
Potassium sulfate	98.0	98.0	97.5	97.0	97.0

Source: Ref. 21.

ents some of the available methods used to measure a_w in foods (19–21).

Vapor Pressure Measurement

Based on the definition of a_w, food is placed under vacuum conditions until it reaches equilibrium (at controlled temperature) with the surrounding atmosphere of known and constant relative humidity (Table 3), and then the vapor pressure of the atmosphere that is in equilibrium with the sample is measured with a manometer or pressure transducer. Therefore, the vapor pressure could be a direct measure of the a_w. The disadvantage is that the time needed to reach equilibrium is long, making this method inadequate for quick routine analysis.

Depression of the Freezing Point

This colligative property can be expressed by the Robinson and Stokes equation (3), which is valid for liquid foods of $a_w > 0.97$, although it has been acceptable for values as low as 0.80.

Dew Point Hygrometer

This method, which can measure a wide range of a_w at different temperatures, is based on the condensation of water vapor on the surface of a mirror that is cooled to a photoelectrically detected dew temperature of the atmosphere generated by the studied sample. Until recently, a_w meters based on the chilled mirror technique (dew point) were the only ones to feature measuring times of only a few minutes per product sample.

Thermocouple Psychrometer

This apparatus measures a_w by taking into account a wet bulb temperature, which is the temperature an air volume would have if it were cooled to saturation at constant pressure by evaporating water into it. The a_w measurement is based on relative humidity values that are a measure of the difference between dry and wet bulb temperatures.

Electric Hygrometers

The performance of this hygrometer is based on the use of hygrosensors formed by an electrical wire covered by a high hygroscopic salt such as lithium chloride. When the salt absorbs the water vapor released by the sample, a change on the conductance of the wire is provoked.

Filament Hygrometer

This low-price instrument uses a synthetic fiber that shrinks when exposed to high relative humidity. The dimensional modification is recorded and related to the sample a_w. This hygrometer is affected in an important way by temperature changes and the presence of volatiles because it has a low sensitivity, although it is functional within the 0.70–0.95 a_w range.

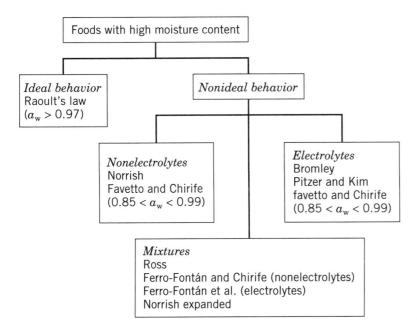

Figure 5. van den Berg's scheme to predict a_w.

PREDICTION OF WATER ACTIVITY

van den Berg and Bruin (4), Chirife (21), and Roa and Tapia (22) have presented a series of comprehensive analyses of the procedures traditionally used to calculate or predict a_w. Each discusses the applicability of diverse theoretical and empirical models on foods, including descriptive examples. Figure 5 summarizes several of these methods.

FINAL REMARKS

Even though there are restrictions and problems associated with use of the a_w term in some food products, its measurement or prediction continues to be useful in process design, food formulation, and the selection of storage conditions. Water activity is an important factor in food microbial responses because their relation to a_w has significant impact and practical implications in the food industry. Some of the areas where a_w has practical applications are in the storage of raw materials, quality of foods, and shelf life of foods stored under various conditions. It is clear that a_w plays a significant role in the survival and death of microorganisms in foods and is also an important influence on microbial response when combined with other intrinsic and extrinsic preservation factors.

BIBLIOGRAPHY

1. L. M. Lenovich, "Survival and Death of Microorganisms as Influenced by Water Activity," in L. B. Rockland and L. R. Beuchat, eds., *Water Activity: Theory and Applications to Food*, Marcel Dekker, New York, 1987, pp. 119–136.

2. J. Welti and F. T. Vergara, "Actividad de Agua: Concepto y Aplicación en Alimentos con Alto Contenido de Humedad," in J. M. Aguilera, ed., *Temas en Tecnologia de Alimentos*, CYTED-Instituto Politécnico Nacional, México, DF, 1997.

3. R. A. Robinson and R. M. Stokes, *Electrolyte Solutions*, 2nd ed., Butterworths Scientific Publishers, London, 1965.

4. C. van den Berg and S. Bruin, "Water Activity and its Estimation in Food Systems: Theoretical Aspects," in L. B. Rockland and G. F. Stewart, eds., *Water Activity: Influences on Food Quality*, Academic Press, New York, 1981.

5. F. Franks, "Water Activity: A Credible Measure of Food Safety and Quality?," *Trends in Food Sci.* **2**, 68–72 (1991).

6. D. A. A. Mossel, "Water and Microorganisms in Foods—A Synthesis," in R. B. Duckworth, ed., *Water Relations in Foods*, Academic Press, New York, 1975.

7. W. J. Scott, "Water Relations of Food Spoilage Microorganisms," *Advances in Food Res.* **7**, 83–127 (1957).

8. J. E. L. Corry, "The Water Relations and Heat Resistance of Microorganisms," *Progress Industrial Microbiol.* **12**, 73–80 (1973).

9. L. R. Beuchat, "Influence of Water Activity on Growth, Metabolic Activities and Survival of Yeasts and Molds," *J. Food Protection* **46**, 135–140 (1983).

10. L. R. Beuchat, "Influence of Water Activity on Sporulation, Germination, Outgrowth, and Toxin Production," in L. B. Rockland and L. R. Beuchat, eds., *Water Activity: Theory and Applications to Food*, L. B. Marcel Dekker, New York, 1987.

11. G. W. Gould, "Drying, Raised Osmotic Pressure and Low Water Activity," in G. W. Gould, ed., *Mechanisms of Action of Food*

Preservation Procedures, Elsevier Applied Science, New York, 1989.

12. A. D. Hocking and J. H. B. Christian, "Microbial Ecology Interactions in the Processing of Foods," in G. V. Barbosa-Cánovas and J. Welti-Chanes, eds., *Food Preservation by Moisture Control, Fundamentals and Applications*, Technomic Publishing, Lancaster, Pa., 1996.

13. J. Chirife and M. P. Buera, "Water Activity and Glass Dynamics, and the Control of Microbiological Growth in Foods," *Crit. Rev. Food Sci. Nutrition* **36**, 465–512 (1996).

14. J. H. B. Christian, "Specific Solute Effects on Microbial Water Relations," in L. B. Rockland and J. F. Stewart, eds., *Water Activity: Influences on Food Quality*, Academic Press, New York 1981.

15. L. Slade and H. Levine, "Beyond Water Activity: Recent Advances Based on an Alternative Approach to the Assessment of Food Quality," *Crit. Rev. Food Sci. Nutrition* **30**, 115–360 (1991).

16. C. Gutiérrez, T. Abee, and I. R. Booth, "Physiology of the Osmotic Stress Response in Microorganisms," *Int. J. Food Microbiol.* **28**, 233–244 (1996).

17. A. D. Hocking and J. I. Pitt. "Media and Methods for Enumeration of Microorganisms with Consideration of Water Activity," in L. B. Rockland and L. R. Beuchat, eds., *Water Activity: Theory and Applications to Food*, Marcel Dekker, New York, 1987.

18. L. R. Beuchat, "Detection and Enumeration of Microorganisms in Hurdle Technology Foods, with Particular Consideration of Foods with Reduced Water Activity," in G. V. Barbosa-Cánovas and J. Welti-Chanes, eds., *Food Preservation by Moisture Control, Fundamentals and Applications*, Technomic Publishing, Lancaster, Pa., 1996, pp. 603–612.

19. J. A. Troller and J. H. B. Christian, *Water Activity and Food*, Academic Press, New York, 1978.

20. L. B. Rockland and S. K. Nishi, "Influence of Water Activity on Food Product Quality and Stability," *Food Technol.* **34**(4), 42–51, 59 (1980).

21. J. Chirife, "An Update on Water Activity Measurements and Prediction in Intermediate and High Moisture Foods: The Role of Some Nonequilibrium Situations," in G. V. Barbosa-Cánovas and J. Welti-Chanes, eds., *Food Preservation by Moisture Control. Fundamentals and Applications*, Technomic Publishing, Lancaster, Pa., 1996, pp. 169–189.

22. V. Roa and M. S. Tapia, "Estimating Water Activity in Systems Containing Multiple Solutes Based on Solute Properties," *J. Food Sci.* **63**, 559–564 (1998).

A. LÓPEZ-MALO
University of las Américas
Puebla, México

M. S. TAPIA
Central University of Venezuela
Caracas, Venezuela

S. M. ALZAMORA
University of Buenos Aires
Buenos Aires, Argentina

J. WELTI-CHANES
University of las Américas
Puebla, México

M. M. GÓNGORA-NIETO
Washington State University
Pullman, Washington

G. V. BARBOSA-CÁNOVAS
Washington State University
Pullman, Washington

WHEAT SCIENCE AND TECHNOLOGY

References to wheat are worldwide, from the beginnings of recorded time (1–14). Ancient Chinese writings describe the growing of wheat 2,700 years before Christ, and even today wheat is considered a sacred crop in some parts of China.

Theophrastos, a Greek, wrote in 300 B.C. of the many different kinds of wheat grown along the Mediterranean Sea. Written records, works of art, and the excavation of ancient cities show the progressive advancement of the art of milling and baking in Greece and Rome and through the Middle Ages.

No one knows where the wheat plant originated, although it was cultivated where modern humans supposedly first appeared—in southwestern Asia. As early as 10,000–15,000 B.C. humans probably used wheat as food. In 1948 archaeologists from the University of Chicago uncovered an ancient village in Iraq, established 6,700 years before. In the ruins they found two different kinds of wheat similar to those grown today.

Most wheaten foods are made from flour and most of our flour is used by commercial bakers. Over 40 million loaves of bread are sold daily in the United States, and some bakers offer as many as 200 different products—loaves of bread of various shape, crust and texture, or rolls, buns, cookies, crackers, cakes, and pastries.

White flour is also found in the kitchens of homes and restaurants. It is used to prepare home-baked products; as a thickening agent in gravies, sauces, soups, puddings; and in the filling of pies and cakes.

Among the most popular forms of wheat are the macaroni foods—spaghetti, macaroni, and noodles. The basic ingredient of the best macaroni foods is obtained from a hard, amber-colored wheat called durum, grown especially for the macaroni market. The durum wheat is milled with special equipment into a golden-toned, coarse product called semolina, or into granulars that contain a higher percentage of flour, or into flour itself (15).

Macaroni foods made from durum wheat are superior because of their desirable yellow-amber color and nutty flavor, and because they hold their shape and firm texture when cooked.

Many canned, refrigerated, frozen, dried, or ready-to-eat packaged foods contain some wheat or are flour-based: crackers, cookies, snack food items, complete meals; frozen casseroles, soups, prepared sauces, dressings, and gravies.

Breakfast cereals include puffed wheat, wheat flakes, shredded wheat, and bran flakes in sugar-coated, flavored, or plain ready-to-eat form; other cereals are wheat meal, malted breakfast food, and farina—to be cooked and served hot.

By-products of wheat milling are used in rations for cattle, poultry, and other animals. Wheat starches are used industrially, and when priced advantageously, may be substituted for other starches, as in paper sizing, laundry starch, marshmallows, or other products (16–18).

WHEAT

Botanically, wheat belongs to the grass family Gramineae and the genus Triticum. The 14 species of wheat are (1) *Triticum aegilopoides* (wild einkorn), (2) *T. monococcum* (einkorn), (3) *T. dicoccoides* (wild emmer), (4) *T. dicoccum* (emmer), (5) *T. durum* (macaroni wheat), (6) *T. persicum* (Persian wheat), (7) *T. turgidum* (rivet wheat), (8) *T. polonicum* (Polish wheat), (9) *T. timopheevi* (which has no common name), (10) *T. aestivum* (common wheat), (11) *T. sphaerococcum* (shot wheat), (12) *T. compactum* (club wheat), (13) *T. spelta* (spelt), and (14) *T. macha* (macha wheat). The first two species have 7 chromosomes; the following seven species, 14 chromosomes; the last five species, 21 chromosomes (1).

All common wheat such as hard, red winter and spring, soft, red and white wheats, and the club wheats are included in the *Triticum aestivum* classification. Wheat is also classified in several other ways. Some of the factors involved are winter or spring habit of growth, time needed to mature, height, stem leaf, spike, glume, awn or beard, kernel characteristics, and yield.

The common ancestor of all wheats is believed to be wild einkorn (one-seed), which cross-bred with an unknown grass to produce our 7-chromosome wheat, einkorn. Kernels of both the wild and cultivated einkorn were found in the excavated ruins of a village known as Jarmo, dating from 6700 B.C. It is situated in the upper reaches of the Tigris-Euphrates basin in southwestern Asia, called the Fertile Crescent, the presumed birthplace of modern civilization. Einkorn wheats still grow in the Middle East.

The 14- and 21-chromosome species of wheats were believed to have developed as natural hybrids of the original einkorn and possibly emmer. Of the 14 species of wheat, only three—common, club, and durum—account for 90% of all wheat grown in the world today (1).

At the turn of the century, plant scientists began to produce new varieties of wheats artificially from parents selected for desirable traits: greater yield, resistance to diseases and insect infestation, a shorter growing season, better milling and baking qualities, and shorter straw to reduce the chance that wind and weather might flatten the wheat and make harvesting more difficult. Most of the varieties of wheat now grown commercially in the United States were unknown even 10 years ago. While about 30,000 wheat varieties belonging to 14 species are grown throughout the world, only about 1,000 varieties are of commercial significance (1).

PRODUCTION, TRADE, AND USES

Wheat is cultivated in most countries on all continents. World wheat production, consumption, and net exports (imports) are summarized in Table 1.

The top five wheat producing countries are the former Soviet Union, the People's Republic of China, the United States, India, and Canada. Of these five countries, only the United States and Canada grow more wheat than they use and export to other countries. The other three nations are large wheat importers. Some of the top wheat customers of the United States have been China, India, the former Soviet Union, Japan, and Brazil.

A nation of one billion people, China is traditionally regarded as a rice-eating nation. But China grows almost as

Table 1. Wheat: World Production, Consumption, and Net Exports, 1985/1986 (million metric tons)[a]

Country	Production	Consumption	Net exports
Major exporters			
United States	64.68	30.21	32.53
Canada	23.50	5.50	17.50
Australia	17.00	3.10	15.20
EC-10	70.07	53.07	14.30
Argentina	11.50	4.85	6.70
Turkey	13.00	13.70	−0.15
Major importers			
USSR	83.00	100.00	−19.00
China	87.00	94.00	−7.00
Eastern Europe	38.98	38.97	−0.21
Other Western Europe	9.57	9.68	−0.29
Brazil	2.20	6.40	−4.40
Mexico	4.40	4.50	−0.30
Other Latin America	.05	2.72	2.69
Japan	.79	6.35	−5.50
India	45.00	43.00	1.40
South Korea	.02	2.37	−2.30
Indonesia	0	1.37	−1.15
Other Asia	17.56	25.21	−8.14
Egypt	1.85	8.45	−6.70
Morocco	1.83	4.14	−2.30
Other northern Africa–Mideast	11.83	25.47	−14.23
Other Africa	3.58	9.48	−5.85
Residual	2.43	10.34	−3.92
Total world	*509.84*	*502.88*	

Source: Grain and Food Division, USDA.
[a]Trade on July–June years.

much wheat as the United States and buys and uses more wheat than any other country in the world. Each person in China on the average consumes 180 lb of wheat every year, mostly in the form of noodles. The average American eats only about 116 lb of wheat flour per year in all types of wheat-based products. Some nations have much higher per capita consumption, up to 300 lb of wheat per year per person (1,19,20).

Wheat is grown in most of the 50 states of the United States. The kind of wheat grown and quantity vary widely from one region to another. In total over 200 varieties are grown annually. Winter wheats are planted in the fall. After the grasslike seedlings emerge from the ground, they lie dormant during the winter. They come up again in the spring, ripen, and are harvested in early summer. Spring wheats are planted in the spring and harvested in late summer. Spring wheats grow best in the northern areas of the United States where the summers are not too hot for the young plants. Winter wheats grow best in areas of the country where the winters are not too harsh for the young plants.

The many varieties of winter and spring wheat are grouped into five official classes. The class a variety fits into is determined by the hardness, the color of its kernels, and its planting time. Each class of wheat has its own rela-

tively uniform characteristics, including those related to milling, baking, or other food uses (1,19,20). Protein range and flour use of major wheat classes are listed in Figure 1 (21).

Hard red winter (HRW) is an important bread wheat that accounts for more than 40% of the United States' wheat crop and wheat exports. This fall-seeded wheat is produced in the Great Plains, which extend from the Mississippi River west to the Rocky Mountains, and from the Dakotas and Montana south to Texas. Significant quantities are also produced in California. HRW has moderately high protein content, usually averaging 11–12%, and good milling and baking characteristics.

Hard red spring (HRS), another important bread wheat, has the highest protein content, usually 13–14%, in addition to good milling and baking characteristics. This spring-seeded wheat is primarily grown in the north central United States—North Dakota, South Dakota, Minnesota, and Montana. HRS constitutes about 15% of U.S. wheat exports. Subclasses based on the dark, hard, and vitreous (DHV) content, include Dark Northern Spring, Northern Spring, and Red Spring.

White wheat (WW) is a preferred wheat for noodles, flat breads, and bakery products other than loaf bread. WW, which includes both fall and spring-seeded varieties, is grown mainly in the Pacific Northwest. This low-protein wheat, usually about 10%, comprises about 15% of U.S. wheat exports, destined primarily for East Asia and the Middle East. Subclasses include hard white, soft white, western white, and white club.

Soft red winter (SRW), which is grown in the eastern third of the United States, is a high-yielding wheat, but relatively low in protein, usually about 10%. SRW best provides flour for cakes, pastries, quick breads, crackers, and snack foods. This fall-seeded wheat constitutes about one-quarter of U.S. wheat exports.

Durum, the hardest of all U.S. wheats, provides semolina for spaghetti, macaroni, and other pasta products. This spring-seeded wheat is grown primarily in the same northern areas as hard red spring, but small winter sown quantities are also grown in Arizona and California. Durum represents about 5% percent of total U.S. wheat exports. Subclasses are hard amber durum, amber durum, and durum.

Grades and grade requirements for all U.S. wheat classes are given in Table 2. Table 3 compares some important quality characteristics of the five classes. Soft white samples are listed separately for the East (SWE) and West (SWW) and club wheats. Values of the total domestic consumption of wheat products as foods in the United States are listed in Table 4.

GROSS COMPOSITION

The chemical composition of wheat, in common with that of other cereal grains, varies widely, for it is influenced by genetic, soil, and climatic factors. Variations are encountered in the relative amounts of proteins, lipids, carbohydrates, pigments, vitamins, ash, and mineral elements.

Wheat is characterized by relatively low protein and high carbohydrate levels; the carbohydrates consist essen-

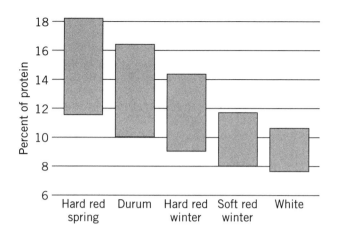

Floor uses:
• Used to blend with weaker wheats for bread flour

• Whole wheat bread, hearth breads

• Egg noodles (U.S.), macaroni and other alimentary pastes

• White bakers' bread, bakers' rolls

• Waffles, muffins, quick yeast breads, all-purpose flour

• Noodles (Oriental) kitchen cakes and crackers, pie crust, doughnuts, and cookies, foam cakes, very rich layer cakes

Figure 1. Protein range and flour uses of major wheat classes. Flour uses are listed according to the approximate level of protein required for specified wheat products. Durum is not traded on the basis of protein content. *Source:* Ref. 21.

Table 2. Official U.S. Standards for Grain (USDA)

	Minimum limits of		Maximum limits of						
	Test weight per bushel (lb)							Wheat of other classes[a]	
	Hard red spring wheat or white club wheat[b]	All other classes and subclasses	Heat-damaged kernels (%)	Damaged kernels, total (%)[c]	Foreign material	Shrunken and broken kernels (%)	Defects, total (%)[d]	Contrasting classes (%)	Wheat of other classes (%)[e]
U.S. no. 1	58.0	60.0	0.2	2.0	0.5	3.0	3.0	1.0	3.0
U.S. no. 2	57.0	58.0	0.2	4.0	1.0	5.0	5.0	2.0	5.0
U.S. no. 3	55.0	56.0	0.5	7.0	2.0	8.0	8.0	3.0	10.0
U.S. no. 4	53.0	54.0	1.0	10.0	3.0	12.0	12.0	10.0	10.0
U.S. no. 5	50.0	51.0	3.0	15.0	5.0	20.0	20.0	10.0	10.0
U.S. sample grade[f]									

[a]Unclassed wheat of any grade may contain not more than 10% of wheat of other classes.

[b]These requirements also apply when Hard Red Spring wheat or White Club wheat predominate in a sample of mixed wheat.

[c]Includes heat-damaged kernels.

[d]Defects (total) include damaged kernels (total), foreign material, and shrunken and broken kernels. The sum of these three factors may not exceed the limit for defects.

[e]Includes contrasting classes.

[f]U.S. sample grade shall be wheat that (a) does not meet the requirements for U.S. grades 1, 2, 3, 4, or 5; or (b) contains 8 or more stones, 2 or more pieces of glass, 3 or more crotalaria seeds (*Crotalaria* spp.), 2 or more castor beans (*Ricinus communis*), 4 or more particles of an unknown foreign substance(s) or a commonly recognized harmful or toxic substance(s), or 2 or more rodent pellets, bird droppings, or equivalent quantity of other animal filth per 1,000 g of wheat; or (c) has a musty, sour, or commercially objectionable foreign odor (except smut or galic odor); or (d) is heating or otherwise of distinctly low quality.

tially of starch (≥90%), dextrins, pentosans, and sugars. The various components are not uniformly distributed in the different kernel structures. Composition of different parts of the wheat kernel is shown in Table 5 (2). The hulls and bran are high in cellulose, pentosans, and ash; the germ is high in lipid, protein, sugar, and ash. The endosperm consists primarily of starch and is lower in protein content than the germ and, in some cereals, than the bran. It is also low in crude fat and ash. In milling, the hulls, germ, and bran are removed. The composition of the main anatomical parts of wheat and of flours that differ in extraction rate are compared in Table 6. The values in the table are representative averages. Analyses of ash, protein, crude fat, crude fiber, and starch of whole wheat and mill streams are summarized in Table 7.

Starch

The starch content of wheat flour is, in general, inversely related to protein content. In flours below 80% extraction, the starch content ranges from about 65–70% (on an as-is moisture basis) and the proportion of linear (amylose) to branched (amylopectin) fractions in starch is about 1:3.

Proteins

The protein content of wheat is an important index of its quality for the manufacture of different food products and is influenced by climate, weather, soil, and the variety of grain (25,26). The breadmaking potential of bread-wheat is largely associated with the quantity and quality of its protein. The quantity of protein is influenced to a large

Table 3. Mean Values of Studied Parameters in Wheats

Parameter	HRW	HRS	SRW	SWE	SWW	CLB	DUR
Test weight, lb/bu	60.9	61.4	61.9	61.8	61.0	60.7	61.8
1,000-kernel weight, g	28.0	32.1	33.3	37.8	36.0	29.1	39.3
Wheat							
Density, g/mL	1.4594	1.4605	1.4252	1.4224	1.4417	1.4451	1.4583
Ash, %	1.57	1.56	1.48	1.49	1.46	1.34	1.59
Protein, %	11.7	13.7	9.7	9.3	10.0	10.5	13.4
Flour yield, %	72.9	75.2	71.8	69.6	72.1	72.7	70.4
Mill rating (hardness)	6.1	7.3	3.3	1.9	3.2	3.4	9.5
Starch damage, %	7.0	7.2	4.4	4.2	5.0	4.4	13.8
Particle size index, %	28.0	26.5	41.7	41.3	36.6	38.4	15.7
Flour protein, %[a]	10.7	13.1	8.6	8.2	8.9	9.4	12.6
Bake absorption, %	63.9	67.5					
Mixograph development time, min	3.7	3.6	3.3	3.0	2.2	2.0	2.8
Loaf volume, mL	879	985					
Crumb grain[b]	3.7	2.5					
Specific loaf volume, mL/1% protein	58.7	56.0					
Cookie diameter, cm			9.5	9.5	9.2	9.4	

Source: Ref. 22.

[a]14% moisture basis, $N \times 5.7$.

[b]The higher the number, the harder the wheat. Average = normal hard-wheat hardness, 6–8; S/A = softer than average, 5; Q-S = too soft to be milled as hard wheat, 3, 4; U-S = satisfactory for use with hard-wheat blends, 1, 2; Q-U = harder than normal hard wheat, 9,10. The higher the score, the poorer the crumb grain. On a scale of 1–10, 1 = satisfactory, 10 = very unsatisfactory.

Table 4. Value of Total Domestic Consumption of Grain Products, 1983[a]

Product	Value of consumption (million U.S. $)
Baked goods[b]	26,052.9
Bread and rolls	9,354.1
White bread	6,494.7
Other bread	1,595.7
Rolls	1,263.8
Crackers, cookies[c]	10,498.1
Sweet goods[b]	6,200.6
Frozen and refrigerated baked goods	1,868.6
Cereal, flour, pasta[c]	8,390.5
Cereals[c]	4,170.6
Flour[g]	2,141.3
Pasta[h]	1,526.4
Macaroni	373.9
Noodles	273.9
Spaghetti	480.4
Total	*36,312.0*

Source: Ref. 23.

[a]All consumption, whether at home, in restaurants, or institutions, in terms of retail store valuation.

[b]Total minus cereal, flour, pasta, frozen and refrigerated baked goods.

[c]Includes snacks (chips, pretzels, popcorn).

[d]Cakes, pastries, doughnuts, pies, and in-store bakery.

[e]Includes rice.

[f]Cold and hot cereals, hominy grits, infant cereals.

[g]Cake flour, corn meal, family flour, pancake and waffle mixes, and prepared mixes, including cakes and frosting.

[h]Includes canned pasta.

extent by environmental factors; the quality of protein is heritable. In wheat varieties grown under comparable environmental conditions, a high-quality wheat produces good bread over a fairly wide range of protein levels, whereas a low-quality wheat produces relatively poor-quality bread even when its protein content is high. For many years it has been believed that differences in protein content among varieties of wheat grown under comparable conditions are small compared with differences due to environment. In recent years, however, it has been shown that the protein content of wheat can be greatly increased by selective breeding.

The proteins in wheat (Fig. 2) include water-soluble proteins (albumins), salt-soluble proteins (globulins), alcohol-soluble proteins (prolamins or gliadins), and acid-and alkali-soluble proteins (glutelins). When water is added, the wheat endosperm proteins (gliadin and glutenin) form a tenacious colloidal complex known as gluten, which is responsible for the superiority of wheat over the other cereals for the manufacture of leavened products. Gluten is responsible for the retention of carbon dioxide produced by yeast or chemical leavening agents.

As a class, cereal proteins are not as high in biological value as are the proteins from certain legumes, nuts, or animal products (28). The limiting amino acid in wheat endosperm proteins is lysine. Although biological values of the proteins of entire cereal grains are greater than those of the refined mill products, which consist chiefly of the endosperm, many diets normally include animal products as well as cereal products. With mixed diets, different proteins tend to supplement each other, and thus cereals can be important and valuable sources of amino acids for the synthesis of body proteins.

Lipids

Total wheat flour lipids (ca 1.4–2.0%) contain about equal amounts of nonpolar and polar components. Triglycerides are a major component of nonpolar lipids; digalactosyldiglycerides, of glycolipids; and lysophosphatidylcholines and phosphatidylcholines, of phospholipids. Wheat germ is an abundant source of α-tocopherol, as vitamin E (28).

Table 5. Composition of Different Parts of the Wheat Kernel

Kernel tissue[a]	Crude protein (%)	Lipid (%)	Starch (%)	Reducing sugars (%)	Pentosans (%)	Cellulose (%)	Ash (%)
Whole kernel	12.0	1.8	58.5	2.0	6.6	2.3	1.8
Pericarp	7.5	0.0	0.0	0.0	34.5	38.0	5.0
Testa and hyaline	15.5	0.0	0.0	0.0	50.5	11.0	8.0
Aleurone layer	24.0	8.0	0.0	0.0	38.5	3.5	11.0
Outer endosperm	16.0	2.2	62.7	1.6	1.4	0.3	0.8
Inner endosperm	7.9	1.6	71.7	1.6	1.4	0.3	0.5
Embryo and scutellum	26.0	10.0	0.0	26.0	6.5	2.0	4.5

Source: Ref. 2.
[a]Protein content calculated by nitrogen × 5.83; moisture content, 15%.

Table 6. Weight, Ash, Protein, Lipid, and Crude Fiber Contents of Main Anatomical Parts of the Wheat Kernel and of Flours Varying in Milling Extraction Rate[a]

Parameter	Wheat kernel fractions (%)				Milling extraction		
	Pericarp	Aleurone layer	Starchy endosperm	Germ			
Weight	9	8	80	3	75	85	100
Ash	3	16	0.5	5	0.5	1	1.5
Protein	5	18	10	26	11	12	12
Lipid	1	9	1	10	1	1.5	2
Crude fiber	21	7	>0.5	3	>0.5	0.5	2

Source: Ref. 24.
[a]Expressed on a 14% moisture basis.

Table 7. Gross Chemical Composition of Wheat and Mill Streams[a]

	Yield (%)	Ash (%)	Protein, $N \times 5.7$ (%)	Crude fat (%)	Crude fiber (%)	Starch (%)
Wheat	100	1.2–1.7	9.2–13.8	1.1–1.9	1.7–2.6	54.1–61.8
Flour	72.6–77.0	0.35–0.42	8.6–13.0	0.8–1.0	—	64.3–73.7
Germ	0.6–1.1	3.5–4.3	21.7–24.5	6.3–10.6	2.8–4.0	14.0–23.9
Red dog	1.4–4.7	1.5–2.7	12.7–15.2	2.3–4.7	1.2–3.2	37.0–47.8
Shorts	6.6–8.9	3.1–4.3	13.8–16.5	3.7–6.3	5.6–7.2	15.9–21.7
Bran	12.5–16.9	4.7–7.1	12.1–15.4	3.0–4.2	9.2–11.6	4.6–7.2

Source: Ref. 24.
[a]Expressed on a 14% moisture basis.

Vitamins

Wheat has a relatively high content of thiamine and niacin compared to other cereal grains. It is low in riboflavin and contains no vitamin A or C (28).

MILLING

Conventional, Modern Milling

Before wheat can be used in the production of most foods, it must undergo several mechanical and chemical changes. The first change involves milling of wheat into flour (1,25). The steps involved in wheat-to-flour production are wheat selection, blending, cleaning, conditioning, milling, and maturing. The endosperm, which forms about 83% of the kernel, is the source of white flour and contains 70–75% of the kernel's protein. The bran, forming about 14% of the kernel, is included in whole wheat flour, but is more often removed and used in animal or poultry feed. Because the cellulosic material of the bran cannot be digested and tends

to accelerate the passage of food through the human digestive tract, the total nutritive contribution of whole wheat flour is less than that found in enriched white flour products. The germ, forming about 3% of the kernel, is the embryo or sprouting tissue of the seed. It is usually separated out because it contains oil, which limits the keeping quality of flours. Although the germ is available as human food, it is usually added to animal or poultry feed. The modern milling process is depicted in a diagrammatic form in Figure 3 (1).

The mill flow diagram begins with a separator, where the wheat first passes through a vibrating screen that removes straw and other coarse materials, and then over a second screen through which drop small foreign materials like seeds. An aspirator lifts off lighter impurities in the wheat. After the aspirator, wheat moves into a disk separator, consisting of disks revolving on a horizontal axis. The disk surfaces are indented to catch individual grains of wheat but reject larger or smaller material. The blades push the wheat from one end of the machine to the other. The revolving disks discharge the wheat into a hopper or

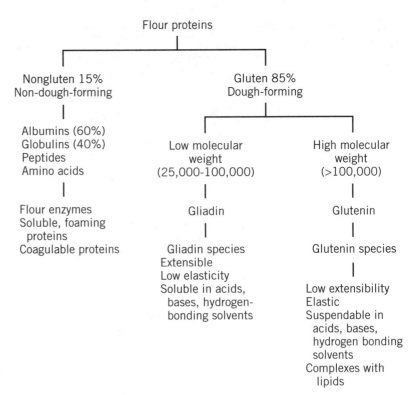

Figure 2. Schematic presentation of the main protein fractions of wheat flour. *Source:* Ref. 27.

into the continuing stream. The wheat then moves into a scourer—a machine in which beaters attached to a central shaft throw the wheat against a surrounding drum. Scourers may be either horizontal or upright, with or without brushes, and adjusted for mild, medium, or hard scouring. Air currents carry off the dust and loosened particles of bran coating.

The stream of wheat next passes over a magnetic separator that pulls out iron and steel particles. A washer–stoner may be the next piece of equipment. High-speed rotors spin the wheat in a water bath. Excess water is thrown out by centrifugal force. Stones drop to the bottom and are removed. Lighter materials float off, leaving only the clean wheat. In the production lines of some mills, a dry stoner is also used. The wheat passes over an inclined, vibrating table that pushes stones and heavy material away from the lighter wheat, discharged separately.

The clean wheat is then tempered before the start of grinding; in the process moisture is added. Tempering aids the separation of the bran from the endosperm and helps provide a constant, controlled amount of moisture and temperature throughout the milling process. The percentage of moisture, length of conditioning, and temperature are the three important factors in tempering, with different requirements in soft, medium, and hard wheats (1). Some water for tempering may be added, dampening the wheat; the wheat may be washed until a certain amount of moisture is absorbed; or the wheat may be subjected to steam under low pressure. The dampened wheat is held in a bin for a prescribed period—usually 8–24 h, depending on the type of wheat. The outer layers of the wheat berry tend to be brittle, and tempering toughens the bran coat for better separation of the endosperm. Within the kernel,

tempering also mellows the endosperm so that the floury particles break more freely in milling. When the moisture content is properly dispersed in the wheat for efficient milling, up to approximately 16%, the grain is passed through an Entoleter scourer–aspirator as a final step in cleaning. Disks revolving at high speed in the scourer–aspirator throw the wheat against fingerlike pins. The impact cracks unsound kernels (including insect-damaged), which are rejected. The sound wheat flows to a grinding bin or hopper from which it is fed in a continuous metered stream into the mill itself (1).

The first-break rolls of a mill are corrugated rather than smooth like the reduction rolls that reduce the particles of endosperm further along in the process. The rollers are paired and rotate inward against each other and at different speeds. The clearance between rollers and the pressure as well as the speed of each separate roller, can be adjusted. At each breaking step, the miller selects the milling surface and the corrugations; the speed of and interrelation between the rollers depend on the type and condition of the wheat.

The next major step introduces the broken-ground particles of wheat and bran into a sifter where they are shaken through a series of bolting cloths or screens to separate the larger from the smaller particles. The fractions, and particles of endosperm graded by size are carried to purifiers. In a purifier, a controlled flow of air lifts off bran particles while bolting cloth at the same time separates and grades coarser fractions by size and quality. Four or five additional break rolls, with successively finer corrugations and each followed by a sifter, are usually used to rework the coarse stocks from the sifters and reduce the wheat particles to granular middlings as free from bran as possible. Germ

How flour is milled
(A simplified diagram)

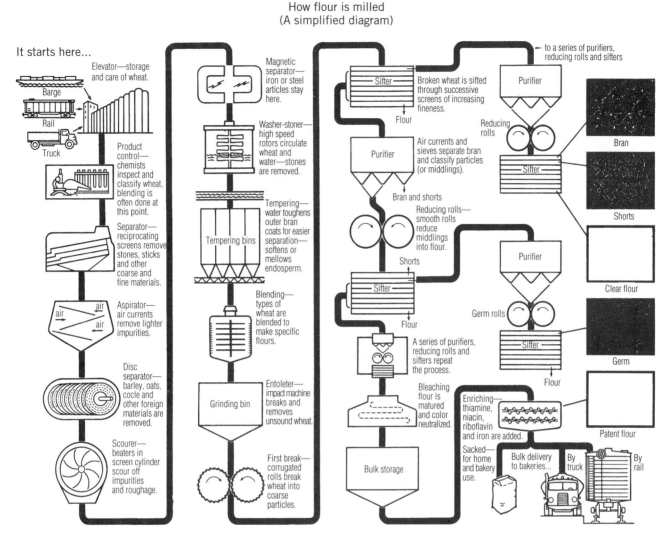

Note: This chart is greatly simplified. The sequence, number and complexity of different operations vary in different mills.

Figure 3. Milling process of flour. *Source:* Wheat Flour Institute, Chicago, Ill.

particles, being somewhat plastic, are flattened by passage through the smooth reduction rolls and can be separated. The reduction rolls reduce the purified, granular middlings or farina to flour.

The process is repeated until the maximum amount of flour is separated, consisting of at least 72% of the wheat. This flour is made up of the various grades shown in Figure 4 (29). The remaining percentage of the wheat berry is classified as millfeed. The flour can be classified in several ways. Straight flour is all the flour produced, with various streams of flour mixed into one. Flour emerges at a number of points in the milling process, with the purified middlings yielding the extrashort or patent or bread flour. In hard wheat mills, as much as 75–80% may be run together as first clear or split into fancy clear and second clear.

In a soft wheat mill, 40–60% of the fancy patent may be taken off separately, leaving about 55% of the remaining flour to be classified as fancy clear. The chart (Fig. 4) is a generalization rather than an exact description of the yield

of and single mill from a particular kind of wheat. It shows how the various streams of flour may be classified, starting with fancy patent, through short, medium, and long patent flours, leaving less and less to be classified as clear flour. The extrashort or fancy patent is the finest, with grades dropping down the scale to clears (30,31).

Toward the end of the millstream, the finished flour flows through a device that releases a bleaching–maturing agent in measured amounts. For bakery customers, the finished flour flows into hoppers for bulk storage, since most bakers add their own form of the enrichment formula to dough—a combination of thiamine, niacin, riboflavin, and iron. For packing as family flour, the enrichment ingredients are added in another mixing machine as the flour flows to the packing room. If the flour is self-rising, a leavening agent and salt are also added.

In milling of durum wheat for the macaroni trade, special equipment is required, especially additional purifiers to separate the bran from the semolina, a coarse granula-

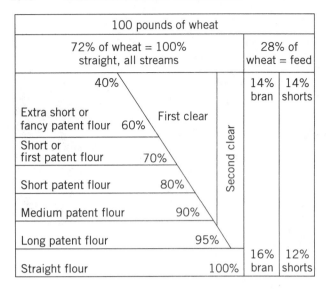

Figure 4. Grades of flour. *Source:* Ref. 29.

tion of the endosperm (15). By federal definition, semolina is prepared by grinding and bolting durum wheat, separating the bran and germ, to produce a granular product with no more than 3% flour. Durum millers also make granulars, or a coarse product with greater amounts of flour; or they grind the wheat into flour for special use in macaroni products, particularly in noodles.

New Milling Developments

In the early 1960s it was found that wheat could be ground to a very fine flour by impact milling and that the product could be further separated into products varying widely in protein. Finished flour from a conventional roller mill is further reduced in particle size in a highspeed grinder, an impact or pin mill. Disintegration of the flour particles takes place as they strike one another, the surface of the rotors, and the pins. The flour is fractured in granular form rather than pressed and broken as in a roller system. The reground flour contains a mixture of relatively coarse endosperm chunks, intermediate fragments, and fines.

The reground flour is channeled to a classifier, where swirling air funnels the larger particles down and away while the smaller fines are lifted up and separated. Repeating the process as shown in Figure 5 (1), 20–30% of the flour is separated into a low-protein product suitable for cakes and pastries; 5–15% makes up the fine fraction, containing 15–22% protein.

The high-protein flour may be used to fortify or blend with other flours. Recombining some of the high-protein fraction with the coarse portions permits a miller to tailor a flour of protein value to a buyer's specifications.

Fine grinding and air classification make possible the production of some cake flour from hard wheat and some bread flour or high-protein fractions from soft wheat. Application of the process theoretically frees the miller from dependence on different wheats, either hard or soft, that change each crop year. The problem is how to market the larger volume of low-protein or starch fractions at prices

adequate to justify the installation and operation of the special equipment (1).

Bleaching and Maturing Agents

Flour is often bleached, both for use by bakers and for the family flour trade. Bleaching improves the color of the flour, and some bleaching agents mature the flour and condition the gluten, improving baking quality. Flour fresh from the mill may be aged in storage or treated with a bleaching-maturing agent to improve baking quality. In aged, unbleached flour or in treated flour, the proteins are slightly oxidized by oxygen from air. The oxidation of flour makes gluten stronger or more elastic and produces better baking results. Cookie, pie crust, and cracker flours usually perform better when no bleach is used.

Dough handling properties can be modified beneficially, or adversely, by the addition of minute amounts of reducing agents (such as cysteine). In addition, the performance of a flour can be improved significantly by the addition of certain oxidizing agents. A list of oxidizing and reducing agents used to improve the baking quality of wheat flour is given in Table 8.

FLOUR TYPES

Hard-wheat flours are usually higher in protein than are soft-wheat flours. They may be milled from either winter or spring wheat varieties. Those with highest protein content, characterized by their capacity to develop the strongest gluten, are used in commercial bread production where doughs must withstand the rigors of machine handling. Other hard-wheat flours with more mellow gluten, easier to develop in kneading by hand, are packed as family flour, all-purpose flour, and self-rising flour. The protein in hard spring wheat flour runs from 11 to 16%; in hard winter wheat flour from 10–14% (19,20).

Soft-wheat flours are sold for general family use, as biscuit or cake flours, and for the commercial production of crackers, pretzels, cakes, cookies, and pastry. The protein in soft wheat flour runs from 7 to 10%. There are differences in appearance, texture, and absorption capacity between hard- and soft-wheat flour subjected to the same milling procedures. Hard-wheat flour falls into separate particles if shaken in the hand whereas, soft-wheat flour tends to clump and hold its shape if pressed together. Hard-wheat flour feels slightly coarse and granular when rubbed between the fingers; soft-wheat flour feels soft and smooth. Hard-wheat flour absorbs more liquid than does soft-wheat flour. Consequently, many recipes recommend a variable measure of either flour or liquid to achieve a desired consistency.

Whole wheat flour, according to FDA specifications, is a coarse-textured flour ground from the entire wheat kernel. It contains the bran, germ, and endosperm. The presence of bran reduces gluten development. Baked products made from whole wheat flour tend to be heavier and denser than those made from white flour. Whole wheat flour is rich in B-complex vitamins, vitamin E, fat, protein, and contains more trace minerals and dietary fiber than does white flour. In most recipes, whole wheat flour can be mixed half

How flour is fractioned

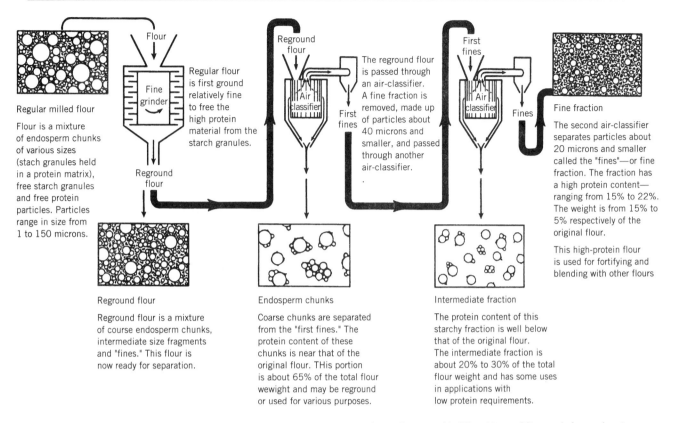

Notes: • Schematic drawings of flour are approximately 100 times actual size   • White areas in circles indicate starch granules
       • Flour particles measured in microns–1 micrin = 1/25,000 of an inch   • Black portions in schematics indicate protein

Regular milled flour

Flour is a mixture of endosperm chunks of various sizes (stach granules held in a protein matrix), free starch granules and free protein particles. Particles range in size from 1 to 150 microns.

Regular flour is first ground relatively fine to free the high protein material from the starch granules.

The reground flour is passed through an air-classifier. A fine fraction is removed, made up of particles about 40 microns and smaller, and passed through another air-classifier.

The second air-classifier separates particles about 20 microns and smaller called the "fines"—or fine fraction. The fraction has a high protein content—ranging from 15% to 22%. The weight is from 15% to 5% respectively of the original flour.

This high-protein flour is used for fortifying and blending with other flours

Reground flour

Reground flour is a mixture of course endosperm chunks, intermediate size fragments and "fines." This flour is now ready for separation.

Endosperm chunks

Coarse chunks are separated from the "first fines." The protein content of these chunks is near that of the original flour. THis portion is about 65% of the total flour wewight and may be reground or used for various purposes.

Intermediate fraction

The protein content of this starchy fraction is well below that of the original flour. The intermediate fraction is about 20% to 30% of the total flour weight and has some uses in applications with low protein requirements.

Note: This is a generalization of the fractionation process. The number and sequences of operations vary with different types of flour made from various types of wheat and for the specific uses of the flour.

Figure 5. How flour is fractionated. *Source:* Wheat Flour Institute, Chicago, Ill.

and half with white flour for a satisfactory product. Graham flour is synonymous with whole wheat flour, named after a physician, Dr. Sylvester Graham, who advocated the use of whole wheat flour in the first half of the nineteenth century (19,20).

According to FDA standards, wheat flour must be milled from cleaned wheat, essentially free of bran and germ, and be ground to such a degree that the product will pass through a sieve having 65 openings per lineal inch (or 4,225 openings per square inch). The moisture content must not exceed 15%.

All-purpose flours are designed for home baking of a wide range of products—yeast breads, quick breads, cakes, cookies, and pastries. Such flours may be made from low-protein hard wheat, from soft or intermediate wheats, or from blends of hard and soft wheat designed to achieve mellow gluten a homemaker can manipulate, and tenderness of structure in the final baked product. All-purpose flour, sometimes called general-purpose or family flour, is most commonly used for home baking. Most modern recipes for home baking are designed for use with all-purpose flour.

Self-rising flours are all-purpose flours to which leavening agents and salt have been added. The leavening agents used are sodium bicarbonate and an acid-reacting substance, monocalcium phosphate, sodium acid pyrophosphate, sodium aluminum phosphate, or a combination of these acids. The sodium bicarbonate and the acid ingredient react in the presence of liquid to release carbon dioxide. The leavening agent and salt are added in amounts sufficient to produce not less than 0.5% of their combined weight of carbon dioxide. Their combined weight must not exceed 4.5 parts per hundred parts of flour. Phosphated flour is all-purpose flour to which the acid-reacting ingredient, monocalcium phosphate, has been added in a quantity of not more than 0.75% of the weight of the finished phosphated flour. It assists in stabilizing gluten and helps nourish yeast. Bromated flour contains potassium bromate in a quantity not exceeding 50 ppm. The bromate has an oxidizing effect that improves the baking qualities of the flour. Such flour must be labeled as bromated.

Cake flours are milled from low-protein soft wheat especially suitable for baking cakes and pastries or from low-protein fractions derived in the milling process. Cake

Table 8. Flour Treatment Agents[a]

	Bromate	Iodate	Azodicarbonamide	Chlorine dioxide	Chlorine	Benzoyl peroxide	L-Ascorbic acid	L-Cysteine
Structure	$KBrO_2$	KIO_3	$H_2N-C(=O)-N=N-C(=O)-NH_2$	ClO_2	Cl_2	(structure)	(structure)	$HS-CH_2-CH(COOH)(NH_2)$
Form	Powder	Powder	Powder	Gas	Gas	Powder	Powder	Powder
Introduced	1916	1916	1962 patent	1948	1912	1921	1935	1962 patent
Regulations Flour	50 ppm max. U.S. whole wheat, 75 ppm max.	Not allowed	45 ppm max.	Sufficient for bleaching	Sufficient for bleaching	U.S.—sufficient for bleaching; Canada—150 ppm max.	200 ppm max.	U.S. pot allowed; Canada—90 ppm max.
Bread	U.S.—75 ppm max. (see below) Can.—100 ppm max. (see below)	45 ppm max. (see below)	U.S. and Canada, 45 ppm max.				U.S.—no limit; Canada—200 ppm max.	U.S. and Canada—90 ppm max.
Labeling	Potassium bromate or bromated flour	Potassium iodate	Bleached flour	Bleached flour	Bleached flour	Bleached flour	Ascorbic acid added as a dough conditioner	L-Cysteine
Used in	Bread flour and dough	Bread dough	Bread flour and dough	Bread flour	Cake flour	Bread and cake flour	Bread flour and dough	Flour (Canada only), miscellaneous doughs
Action	Oxidant—improver	Oxidant—improver, mix reducer	Oxidant—improver, maturing	Oxidant—maturing, bleaching	Oxidant—maturing, bleaching	Oxidant—bleaching	Reductant—oxidant—mix reducer, improver	Reductant—mix reducer, relaxer
Reaction rate	Slow	Fast	Fast	Fast	Fast	1–3 days	Medium	Fast
Reaction product	Bromides	Iodides	Biurea	Chlorides	Chlorides	Benzoic acid		
Usage rates, ppm flour, normal	15	0	5	15	120	50	70	30
total range	10–75	10–20	2–20		25–500	30–100	50–100	15–70
Overtreatment effects	Open grain holes Keyholling Incorrect volume	Low volume Weak, sticky doughs	Dry, bucky dough; low volume; poor grain; crust checking	Poor volume; poor grain	Low volume, dense crumb	None	None	Weak, sticky doughs; poor grain and texture
Testing Flour, spot	AACC 48-02	Yes—difficult	No	No	No	AACC 48-06B	Yes—good	Yes—fair
Flour, quantitative	AACC 48-42		No	pH used	Flour color	Yes	No	
Bread	Very difficult	Difficult	No	No	No	Benzoic acid—hard	No	No
Pennwalt products	Bromolux No—D-Lay		Maturox Dependox	Dyox	Chlorination systems	Novadelox Zerolux	No—D-Lay AA-25	No—D-Lay Assist

Source: Courtesy: Pennwalt, Buffalo, NY.

[a] These oxidating–reducing agents are used to improve the baking quality of wheat flour. They are added at the flour mill, the bakery, or both. Having the proper oxidation treatment is critical; it ranks second only to protein quality and quantity in determining flour quality.
Government regulations on bakery use are as follows: United States—potassium bromate, calcium bromate, potassium iodate, calcium iodate, and calcium peroxide, or any combination of these, cannot exceed 75 ppm. This includes any bromate added to the flour. Canada—potassium bromate, calcium peroxide, ammonium persulfate and potassium persulfate, or any combination of these, cannot exceed 100 ppm. This includes any bromate added to the flour. Calcium iodate, potassium iodate or any combination of iodates cannot exceed 45 ppm.

flours are usually not enriched, but are bleached. The sale of cake flours declined during the 1950s as packaged cake mixes became more popular. Instantized, instant blending, or quick-mixing flour is a granular or more dispersible type of product for home use. It is free-pouring like salt, and dust-free compared to regular flour. It eliminates the need for sifting, since it does not pack down in the package and since it pours right through a screen or sieve. Granular flour instantly disperses in cold liquid rather than balling or lumping as does regular flour. The granular flour is produced by special processes of grinding and bolting, or from regular flour subjected to a controlled amount of finely dispersed moisture that causes the flour to clump or agglomerate. It is then dried to a normal moisture level (19,20). Gluten flour is milled to have a high wheat gluten and a low starch content and is used primarily by bakers for dietetic breads, or mixing with other flours of a lower protein content.

Semolina is the coarsely ground endosperm of durum wheat. High in protein, it is used by American and Italian manufacturers for high-quality pasta products such as macaroni and spaghetti. In Africa and Latin America it is also used for a dish called couscous. Durum flour, a by-product in the production of semolina, is used to make commercial American noodles. Farina is the coarsely ground endosperm of hard wheats. It is the prime ingredient in many American breakfast cereals. It is also used by manufacturers for inexpensive pasta.

The following is a list of additional basic wheat products (20):

1. The *wheat berry* is another name for the wheat kernel. Wheat berries can be used to grow sprouts which are then added to salads or baked goods. The cooked whole kernel can be used as a meat extender, breakfast cereal, or substitute for beans in chili, salads, and baked dishes.

2. *Bulgur* is processed by soaking and cooking the whole wheat kernel, drying it, and then removing 5% of the bran and cracking the remaining kernel into small pieces. Bulgur absorbs twice its volume in water and can be used in place of rice in some recipes.

3. *Cracked wheat* is similar in nutrition and texture to bulgur. It is the whole kernel broken into small pieces, but has not been precooked. Cracked wheat can be added to baked goods for a nutty flavor and crunchy texture or cooked as hot cereal.

4. *Wheat germ* is often added to baked goods, casseroles, and even beverages to improve the nutritional value and give a nutty, crunchy texture.

5. *Bran* makes a nonnutritive addition to baked goods for additional dietary fiber.

6. *Commercial cereals* can be prepared from wheat and eaten as snacks, breakfast cereals, or added to baked products. Ready-to-eat cereals can also be used as coatings for meats, toppings for casseroles, and extenders for meat loaves and casseroles. A variety of ready-to-eat wheat cereals are available. The wheat may be shredded, puffed, flaked, or rolled. The bran may be in the form of flakes or granules. Bran cereals are a nutritious addition for muffins, breads, biscuits, and other baked goods providing an excellent source of fiber in the diet (19,20).

BIBLIOGRAPHY

1. *From Wheat to Flour*, Wheat Flour Institute, Chicago, Ill., 1965.

2. W. R. Akroyd and J. Doughty, *Wheat in Human Nutrition*, FAO Nutritional Studies No. 23, FAO, Rome, 1970.

3. H. Hanson, N. E. Borlaug, and R. G. Anderson, *Wheat in the Third World*, Westview Press, Boulder, Colo., 1982.

4. D. W. Kent-Jones and A. J. Amos, *Modern Cereal Chemistry*, 6th ed., Food Trade Press, London, 1967.

5. Y. Pomeranz, *Modern Cereal Science and Technology*, VCH Publishers, New York, 1987.

6. Y. Pomeranz, ed., *Wheat Chemistry and Technology*, 3rd ed., Vol. 2, American Association of Cereal Chemists, St. Paul, Minn., 1988.

7. Y. Pomeranz, ed., *Advances in Cereal Science and Technology*, Vols. I–IX, American Association of Cereal Chemists, St. Paul, Minn., 1976–1988.

8. E. G. Heyne, ed., *Wheat and Wheat Improvement*, 2nd ed., American Society of Agronomy, Madison, Wisc., 1987.

9. G. E. Inglett, ed., *Wheat: Production and Utilization*, AVI Publishing Co., Westport, Conn., 1973.

10. I. D. Morton, ed., *Cereals in a European Context*, VCH Publishers, New York, 1987.

11. R. C. Hoseney, *Principles of Cereal Science and Technology*, American Association of Cereal Chemists, St. Paul, Minn., 1986.

12. N. L. Kent, *Technology of Cereals with Special Reference to Wheat*, Pergamon Press, Oxford, 1960.

13. Y. Pomeranz, ed., *Wheat is Unique*, American Association of Cereal Chemists, St. Paul, Minn., 1989.

14. J. Storck and W. D. Teague, *A History of Milling, Flour for Man's Bread*, University of Minnesota Press, Minneapolis, 1952.

15. G. Fabriani and C. Lintas, eds., *Durum Wheat, Chemistry and Technology*, American Association of Cereal Chemists, St. Paul, Minn., 1988.

16. Y. Pomeranz, ed., *Industrial Uses of Cereals*, American Association of Cereal Chemists, St. Paul, Minn., 1973.

17. Y. Pomeranz, *Functional Properties of Food Components*, Academic Press, Orlando, Fla., 1985.

18. Y. Pomeranz and L. Munck, eds., *Cereals: A Renewable Resource, Theory and Practice*, American Association of Cereal Chemists, St. Paul, Minn., 1981.

19. *From Flour to Bread*, Wheat Flour Institute, Chicago, Ill., 1971.

20. *Nature's Best Wheat Foods*, Wheat Foods Council, Washington, D.C., 1984.

21. W. G. Heid, *U.S. Wheat Industry*, Agricultural Economics Report No. 432, U.S. Department of Agricultural Economics, Statistics and Cooperative Service, Washington, D.C., 1979.

22. Y. Pomeranz and co-workers, "Hardness and Functional (Bread and Cookie-making) Properties of U.S. Wheats," *Cereal Foods World* **33**, 297–304 (1988).

23. National Food Review, "1985 Wheat Facts," *The Wheat Grower* **8**, 28 (1985).

24. Y. Pomeranz and M. M. MacMasters, "Structure and Composition of the Wheat Kernel," *Baker's Digest* **42**, 24–26, 28, 29, 32 (1968).

25. Y. Pomeranz, "From Wheat to Bread, a Biochemical Study," *American Scientist* **61**, 683–691 (1973).

26. J. Holas and J. Kratochvil, eds., *Progress in Cereal Chemistry and Technology*, 2 vols., Elsevier Science Publishers, Amsterdam, 1983.

27. J. Holme, "A Review of Wheat Flour Proteins, and Their Functional Properties," *Bakers' Digest* **40**, 38 (1966).

28. Y. Pomeranz, ed., *Cereals' 78: Better Nutrition for the Worlds' Millions*, American Association of Cereal Chemists, St. Paul, Minn., 1978.

29. C. O. Swanson, *Wheat and Flour Quality*, Burgess Publishing Co., Minneapolis, Minn., 1938.

30. W. T. Yamazaki and C. T. Greenwood, eds., *Soft Wheat: Production, Milling, Uses*, American Association of Cereal Chemists, St. Paul, Minn., 1981.

31. J. H. Scott, *Flour Milling Process*, 2nd ed., Chapman and Hall, London, 1951.

Y. POMERANZ
Washington State University
Pullman, Washington

See also CEREALS SCIENCE AND TECHNOLOGY.

WHEY: COMPOSITION, PROPERTIES, PROCESSING, AND USES

In the dairy industry, the term whey denotes the greenish translucent liquid that separates from clotted milk during manufacture of cheese or industrial casein. More generally, residual streams from liquid fractionation processes of other food systems (especially of other protein solutions) have been referred to as whey on occasion, as in oilseed protein recovery processes (1) or in alkali extraction of protein from meat boning waste (2). This article is concerned solely with the traditional dairy whey and similar whey-like by-products of the modern dairy processing industry, using cow's milk as the principal raw material. With certain minor allowances for compositional differences, much of the information presented here is also relevant for whey resulting from processing of milk of other milking animals such as goats, buffaloes, and sheep.

TYPES AND FORMS OF WHEY

The traditional whey is produced as a result of processes aimed at recovering casein, the principal protein of milk. Separation of casein from the rest of the milk (as in cheese making or production of industrial casein and caseinates) is usually accomplished by acidification to pH 4.5–4.8 or through the action of rennet, a casein-coagulating enzyme preparation. In acid coagulation, the pH is lowered either by microbial fermentation of the milk sugar lactose into lactic acid or by direct addition of a mineral (phosphoric, hydrochloric, sulfuric, etc) or an organic (lactic, citric) acids. The fermentation route is most often used in the production of cottage cheese, quarg, and other fresh cheeses, and the direct acidification is typical for production of industrial casein and caseinate products; in both cases, the resulting whey is referred to as acid whey. In contrast, sweet wheys are obtained in the manufacture of most hard and semihard cheeses, including cheddar, Swiss, Gouda, and mozzarella, for which the rennet coagulation principle is employed as well as in the production of industrial rennet casein. Because enzymatic clotting of milk by rennet occurs at pH 6.0 or higher, the lactic acid content of freshly obtained sweet whey is low but may increase quickly if subsequent bacterial fermentation is not controlled by rapid pasteurization and/or by deep cooling. Approximately 9 L of whey will be obtained from 10 L of milk for every kg of cheese produced; this general ratio will vary somewhat depending on the type of cheese, fat content of the raw milk used, and other factors. In high-moisture fresh cheeses such as cottage cheese (where a portion of the original raw milk is returned to the cheese as cream dressing), the ratio may be as low as 6:1.

Either sweet or acid wheys can exist in various forms, depending primarily on whether any moisture has been removed in subsequent processing. Because whey contains on the average about 93–94% water, it is often desirable to remove some or most of its water for further handling. Application of common food processing operations such as evaporation or drying results in concentrated, semisolid, or dried forms of whey, the principal difference being the residual moisture content (Table 1). Liquid whey forms may be further processed into many additional modified whey products useful as ingredients for other food processors by reverse osmosis (RO), ultrafiltration (UF), electrodialysis (ED), ion exchange (IE), lactose hydrolysis (LH), and other modern unit operations now available to the whey processing industry.

OTHER BY-PRODUCTS OF MILK AND WHEY PROCESSING

The use of UF for cheese making and for further whey processing results in a wheylike liquid waste stream referred to as UF permeate. Although in principle UF permeate is similar to whey in terms of its physical state and volume, the main compositional difference between the permeate and the traditional whey is due to the almost complete protein recovery in the UF retentate that leaves the permeate virtually protein-free (Table 1). Minor compositional differences are normally observed between permeates from UF of milk and of the two types of whey, primarily because of the variations in the mineral fraction.

COMPOSITION AND PROPERTIES

Composition of Unprocessed Whey

Whey is a multicomponent solution of various water-soluble milk constituents in water; the dry matter of whey consists primarily of carbohydrate (lactose), protein (several chemically different whey proteins), and various minerals. Fat content of the freshly separated liquid whey may

Table 1. Proximate Composition Ranges for Major Whey Forms

Whey form	Total solids (%)	Total protein (%) (N × 6.38)	Lactose (%)	Minerals (% ash)
Liquid, unprocessed	6.0–7.0	0.4–0.8	4.4–4.9	0.5–0.8
Concentrated (RO)	20–25	1.2–3.0	14–20	1.0–3.0
Concentrated (evaporator)	40–60	4.0–8.0	26–48	3–8
Dried	96–97	10–13	70–75	7–12
Liquid UF Permeate	4.0–5.5	0.1–0.3	3.9–4.8	0.3–0.8

Note: After fat recovery by centrifugation, common for many types of cheese whey (eg, cheddar). Fat content of cottage cheese whey is negligible due to the use of skim milk. RO, reverse osmosis, UF, ultrafiltration.
Source: Orientation data based on various sources, including Refs. 3 and 4.

be up to 0.5–1% depending on the type of milk used and the efficiency of the cheese-making operation. However, the valuable butterfat is usually recovered by whey centrifugation and returned to the cheese or processed into whey butter. Thus, the fat content usually reported for the various types of whey is typically below 0.1%. Fat content of most acid wheys is negligible because skim milk is used in the manufacture of these cheeses. The proximate composition of various wheys may show significant variations due to the pretreatment of the cheesemilk (heating, centrifugation, homogenization); the processing differences characteristic for the various cheeses (cultures used, mechanical handling, use of processing aids such as the yellow color, use of membrane processes); and the whey handling and pretreatment processes (pasteurization, preconcentration, recovery of casein fines). Average compositional data for cheese wheys of the sweet and acid types are shown in Table 2.

Lactose, the principal component of the whey apart from water, constitutes about 4.4–4.9% of the whey "as is" depending on the whey type. Lower lactose content is usually found in the acid wheys due to the preceding fermentation process in which some of the lactose is converted to lactic acid. The most valuable whey component is the whey protein, constituting approximately 0.7% of the whey (about 9–11% of the dry matter). In addition, whey may contain about 0.2–0.3% of nitrogenous matter denoted as protein and reported as total whey protein (N × 6.38).

Table 2. Average Composition Data for Sweet and Acid Types of Whey

Component	Sweet whey (pH 5.9–6.4)	g/L	Acid whey (pH 4.6–4.8)
Total solids		63.0–70.0	
Protein (N × 6.38)	6.0–8.0		6.0–7.0
Lactose	46.0–52.0		44.0–46.0
Fat	0.2–1.0		0.1–0.5
Calcium	0.4–0.6		1.2–1.6
Magnesium	0.08		0.11
Phosphate	1.0–3.0		2.0–4.5
Citrate	1.2–1.7		0.2–1.0
Lactate[a]	2.0		6.4
Sodium		0.4–0.5	
Potassium		1.4–1.6	
Chloride		1.0–1.2	

[a]No lactate in uncultured rennet or mineral acid whey.
Source: From various sources including Refs. 5 and 6.

Whey protein is neither acid nor rennet coagulable and thus, when the classical cheese-making processes are used, the whey protein will pass from milk into the whey of either type. Finding suitable technological alternatives allowing recovery of whey protein in the cheese is one of the major concerns of current cheese research. Because about 50–60% of the whey proteins can be coagulated by heat, high heating of milk is now practiced for some types of fresh, acid coagulated cheese such as quarg (7) or cottage cheese where the quality is not greatly compromised and the diminished rennet clottability is not a major problem. As a result, whey obtained from such cheese-making operations will contain mainly the heat noncoagulable whey proteins, and its total protein content will be substantially lower, often around 0.4% (8). Table 3 shows the main components of the whey protein fraction of cow's milk, together with their relative content and heat stability characteristics.

After lactose and whey protein, minerals constitute the third major component group of whey dry matter. The mineral composition shows the greatest variations between the two types of whey, together with pH and the lactic acid content. Although the overall compositional values for each of the two types are dependent on many factors related to the cheese-making process, the main differences between the two types of wheys reflect the different casein coagulation pathways. In addition to lower pH and higher lactic acid (and correspondingly lower lactose) content, the acid wheys show substantially higher calcium and phosphorus contents caused by the solubilization of the calcium–phosphate complex of the casein micelle at the acid pH range, used for the acid coagulation of the casein. In contrast, the calcium removal from the casein micelle does not occur as a result of the rennet clotting at pH 6.0 or higher; thus, much of the milk calcium is retained in the cheese rather than being lost in the sweet whey. The differences in acidity as well as the higher calcium content of the acid whey appear to be the main reasons for variations in physicochemical properties of the two whey types, including the substantially lower heat stabilities of the dairy systems containing added UF retentates from acid wheys in comparison to sweet wheys (11); on the contrary, in the acidic pH range below 3.9, high-calcium containing acid whey appears to be more resistant against heat-induced instability (12).

Composition of Modified Whey Products

Using some of the typical whey processing unit operations such as UF, ED, or IE, numerous modified whey products

Table 3. Composition and Thermal Characteristics of the Whey Protein Fraction

Protein	Approx. content (g/L whey)	Total whey protein (%)	Denaturation temperature (°C)	Heat precipitation stability
α-Lactalbumin	0.6–1.7	20	61	Quite stable[a]
β-Lactoglobulin	2.0–4.0	55	82	Labile
Serum albumin	0.2–04	5	66	Very labile
Immunoglobulins	0.5–1.0	8	72	Extremely labile
Proteose-peptones	0.2–0.4	12		Heat stable
Other (caseins glycoproteins)	0.1	1		

[a]Due to its renaturation characteristics and despite its low denaturation temperature.
Source: Adapted from various sources, including Refs. 5, 6, 9, and 10.

Table 4. Characteristic Composition of Major Types of Modified Whey Products

Product	Total protein (%) (N × 6.38)	Lactose (%)	Minerals (%)	Fat (%)
Liquid WPC (10% TS)	3.3	5.5	0.8	0.3
Dried WPC (35% protein)	35.0	50.0	7.2	2.1
Dried WPC (80% protein)	81.0	3.5	3.1	7.2
Dried WPI (90% protein)	90.0	2.0	4.0	1.0
Lactose, edible grade	0.1	99.0	0.2	0.1
Whey powder				
Demineralized (70%)	13.7	75.7	3.5	0.8
(90%)	15.0	83.0	1.0	0.9
Lactose-reduced demin	30.5	51.5	7.8	2.0
Regular	12.5	73.5	8.5	0.8

Note: WPC, whey protein concentrate; WPI, whey protein isolate; TS, total solids.
Source: Refs. 3, 6, 13, and product specifications of various manufacturers.

may be obtained. Table 4 lists some of these products and gives their basic composition. Because most of these whey modification processes are capable of handling both sweet and acid wheys, the compositional differences of the final products often reflect the processing techniques used rather than the type of whey. Most of the modified whey products available on the market today are manufactured by various fractionation processes aimed at separation of the undesirable components and thereby enrichment of the target substance. The main exception is the enzymatically modified whey syrup where the constituent disaccharide lactose is converted to two monosaccharides, glucose and galactose, thus leaving the proximate composition virtually unchanged. Similar modifications involving enzymatic breakdown of the whey proteins are also of interest (14), one of the main industrial applications being the production of hypoallergenic baby formula (15).

Properties of Whey and Whey Components

Although whey is a rather dilute aqueous solution of lactose, whey protein, and some minerals, some physicochemical properties of unmodified wheys as determined by the individual system components are similar to those of water, but others differ rather substantially. As seen in Table 5, some of the physical properties (such as viscosity or surface tension) are very similar to skim milk, which can be expected from the compositional similarities. Freezing points of wheys and UF permeates can vary rather substantially depending on the type (19). The main difference between whey and skim milk is in the protein fraction, which accounts for a major difference in heat stability of

Table 5. Some Physicochemical Properties of Whey in Comparison to Skim Milk (Orientation Data)

Property	Whey	Skim milk	Ref.
Viscosity (mPa s)	1.2	1.5	16
Surface tension (dyn/cm)	42	48	16,17
Freezing point (°C)	−0.500/−0.600[a]	−0.520	16,18,19
Stability against heat coagulation (standard pH)	Unstable	Stable	16
Stability against acid coagulation (no heat)	Stable	Unstable	16

[a]Depending on type of whey.

the two systems. In the absence of casein, whey protein will coagulate and precipitate from the solution upon heating above the whey protein denaturation temperature, which varies depending on the type of protein (Table 3) and the heating time. As a general guideline, heating to more than 70°C will cause precipitation in whey systems above pH 3.9, the severity of the phenomenon being dependent on the heating time, temperature, and calcium content. Complete precipitation of the heat denaturable whey protein will occur upon heating to at least 90°C with holding for several minutes. Whey systems containing little calcium (such as rennet whey, decalcified acid whey, or solutions of whey protein isolates) will show increased turbidity but little or no precipitation upon heating (20). At pH lower than 3.7–3.9 depending on the calcium content, whey proteins become highly resistant to heat-induced precipitation; even severe heating above 90°C for 20 min or more will not result in protein precipitation or turbidity increase (12,21). Heating of sweet or acid whey systems will enhance the already high foamability of whey (22), probably due to the effect of heat on some of the heat-noncoagulable whey protein fractions.

Limited solubility and low sweetness are the most important properties of the main whey component, lactose, influencing many whey processing operations. Because the solubility of lactose is only about 20 g/100 g water at room temperature and 60 g/100 g water at 60°C, as compared with about 188 g/100 g and 235 g/100 g water for sucrose at the same conditions (23), whey concentration to more than about 36–38% total solids (TS) will result in formation of crystalline lactose. This is the principle of the lactose manufacturing process as well as the reason for specific processing steps aimed at avoiding formation of large lactose crystals in manufacture of dried whey, the Norwegian whey cheese mysost, table spreads based on whey (3,24,25), and other products containing high concentrations of lactose. To avoid the lactose crystal formation and/or to increase the sweetness of lactose-containing foods if desired, the lactose may be hydrolyzed to its two constituent monosaccharides, glucose and galactose. The hydrolysis may be accomplished enzymatically using either free or immobilized lactase (β-galactosidase, EC.3.2.1.23), by chemical hydrolysis through high temperature-low pH processes, or by sonicated or otherwise disrupted bacteria (26).

The main differences in physicochemical properties of sweet and acid wheys are due to the varying content of minerals and acids. Calcium appears to be particularly important in terms of whey protein heat stability, and sodium and potassium are primarily responsible for the saltiness and, together with lactose, for the osmolality of the whey (18). The mineral content of whey can be modified by ion exchange, electrodialysis, or certain specialized membrane processes (eg, nanofiltration).

Optical properties of the whey and wheylike systems are influenced by the content and type of protein present and by the most abundant whey vitamin, the riboflavin. The greenish color of most traditional whey systems is caused by the water-soluble and heat stable riboflavin. However, riboflavin is sensitive to light as well as to ionizing radiation treatments, and whey systems exposed to these conditions will show fading of the green color (27). As a small molecule, riboflavin will be retained by RO but not by most UF membranes, so both wheys as well as UF permeates of milk or whey processing have the same color character. Processing of whey on a metallic UF membrane cast within a porous carbon tube resulted in adsorption of riboflavin on the carbon support (28); the affinity of riboflavin for various adsorbents as crystalline lactose or carbon has been known for some time (29). The turbidity of most industrially produced wheys is due to the casein fines (30). Properly clarified wheys are virtually transparent and, in visual appearance, not substantially different from good quality UF permeates. Turbidity measurements can be used as indication of the heat effects in various whey systems; heating will cause aggregation of the whey protein into larger clusters that will change the optical character of the heated whey system even if no precipitation will result. Yellow discoloration of whey may occur after prolonged heating, especially in concentrated whey systems, because of the Maillard nonenzymatic browning reaction between the lactose and whey proteins. Yellowing may be also caused by the use of the water soluble annato color in the cheese-making process.

INDUSTRIAL PROCESSING OF WHEY

Traditionally, whey has been known as a troublesome by-product of cheese manufacturing and, because of its very high biological oxygen demand (BOD) (about 40,000 ppm), as one of the strongest industrial wastewater pollutants of any kind (31). The opportunities for using whey as a valuable resource have been recognized for some time, but so far, very few countries (eg, the Netherlands) have established a record of nearly complete utilization of all available whey. However, the question of whey utilization as opposed to whey disposal is becoming increasingly important in view of more stringent ecological constraints worldwide. Intensified research activities have resulted in many new approaches to industrial whey processing, and the value of cheese whey as a raw material for various consumer products or industrial food ingredients is now being recognized. Some of the largest whey processing facilities in the world are capable of handling in excess of 10^7 L of whey daily and may rely on supplies of raw or RO preconcentrated whey from cheese manufacturers several hundred kilometers away. The main whey disposal problem continues to be the acid whey from cottage cheese and other fresh cheese products, because the resulting amount of whey is not large enough to warrant separate whey processing facility. One of the possible solutions for these processors could be the production of high-value consumer products based on whey, such as whey beverages and whey cheeses.

Consumer Whey Products

Whey cheeses and whey drinks are traditional foods. The use of whey as a beverage has been recorded in history since the time of Hippocrates (32), and traditional whey cheeses such as ricotta and mysost (24,33–35) have been popular in various parts of the world for centuries. These products usually require simple processing facilities and thus, in their traditional versions, may be suitable for small whey processors supplying primarily local markets. However, neither whey cheeses nor whey beverages (with a few exceptions) have been particularly successful on large scale, possibly because of the limited consumer appeal outside of the traditional producing countries.

Whey Cheeses. There are two types of whey cheese: the Norwegian whey cheese mysost (also known as gjetost, primost, brunost, or Gudbrandsdalsost) and the Italian whey cheese ricotta. Figures 1 and 2 show schematically the basic processing steps used in production of these cheeses. The technology for mysost production is based on evaporation of water from the whole whey, followed by solidification of the evaporated mass under carefully controlled

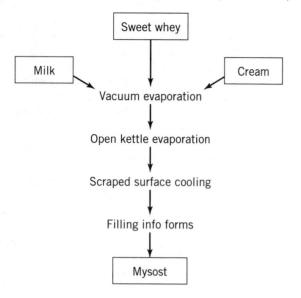

Figure 1. Main technological steps in the production of mysost whey cheese.

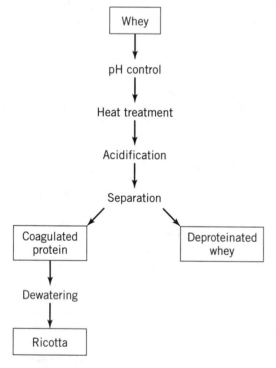

Figure 2. Main technological steps in the production of ricotta whey cheese.

conditions to avoid growth of lactose crystals beyond sizes that would be recognized as grittiness upon product consumption (24). In contrast, the ricotta-type whey cheese is based on the recovery of the whey protein coagulum obtained by high heating of the whey raw material, often containing additional skim milk. As a result, the composition of the two products differs rather substantially, as shown in Table 6. Because all traditional milk-based

Table 6. Illustrative Compositional Data for Ricotta- and Mysost-Type Whey Cheeses

Component	Ricotta type		Mysost type	
	Fresh	Pressed	Regular	Spread
Total solids (%)	25	40	80	66
Protein (%)	12	19	11.5	7.5
Carbohydrate (%)[a]	4	5	36	46[b]
Fat (%)	7	10	30	7.5

[a]Lactose only.
[b]May include additional carbohydrate sweeteners.
Source: Refs. 24, and 34–36.

cheeses are manufactured by retaining casein and removing whey proteins and lactose, describing mysost or ricotta as a cheese may be a misnomer, although allowed for in the modified FAO definition of cheese (36). The character of the protein-based ricotta is closer to the traditional cheese than the mysost, in which lactose is the most abundant component. From the standpoint of whey disposal, the mysost technology is much more suitable because it leaves no residue, whereas heat precipitation of protein from whey still results in a partially deproteinated whey stream almost as high in the BOD as the original whey. The BOD content of the deproteinated whey is dependent primarily on lactose, which is not recovered by the ricotta process. More details about the whey cheese manufacture, properties, and uses may be found elsewhere (33,35,36).

Whey Beverages. Incorporation of whey into various beverages and developments of whey drinks using numerous technological approaches have been attempted by large number of researchers. Extensive literature reviews exist (32,33,37) summarizing many of these attempts. However, there are still only a few truly successful whey beverages that can be found on market shelves in various countries. Virtually no products have established any major international reputation. Table 7 gives examples of whey beverage products known to have enjoyed some commercial success, as an illustration of the types of beverages that may be produced from whey. Figure 3 shows diagrammatically the main technological pathways available for

Table 7. Examples of Commercially Available Whey Beverage Products, 1983–1988

Product type	Commercial name	Country
Whey with fruit juices	Djoez	Netherlands
	Taksi	Netherlands
	Big M	Germany
	Mango-Molke mix	Germany
	Latella	Australia
	Hedelmatarha	Finland
	Nature's Wonder	Sweden, Canada, Germany
Drinkable yogurt with whey	Yor	Netherlands
	Interlac	Belgium
Soft drink (carbonated)	Rivella	Switzerland, Japan

Source: Ref. 33.

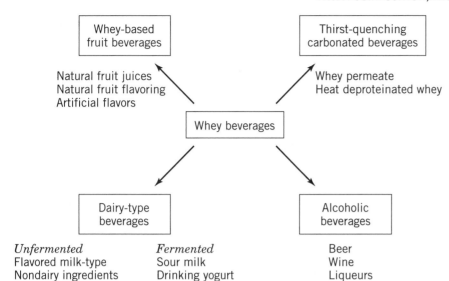

Figure 3. Alternative technological pathways for production of whey-based beverages.

production of a whey beverage. The most often used approach appears to be mixing whey with fruit juices, thereby combining the nutritional benefits of high-quality whey protein, calcium, and riboflavin from whey with the traditionally recognized source of various vitamins, the fruit. Nationally successful products of this type include the Taksi and Djoez drinks marketed in the Netherlands, the Austrian Latella, or the Finnish Hedelmatarha. However, frequent market failures of the less successful products in this category indicate that the seemingly simple technology of mixing cannot be applied without careful product and market development research. A typical quality problem found often in these products is a sediment formation, usually caused by the heat-labile whey proteins and their interactions with some of the fruit components (as pectins) even below the pH range of 3.9–3.7 where whey protein should be heat-stable (38). Similarly, drinkable-yogurt type products have been developed and, in view of the currently increasing popularity of yogurt-based drinks, should represent another avenue of success for further product development. Contrary to the precipitation in the fruit-juice-based products, the pectin can be used to stabilize these casein-containing beverages as in other fermented dairy products (39). A third type of whey beverage that has been successfully marketed is the clear product based on whey UF permeate or heat-deproteinated whey. In fact, the main representative of this approach, the Swiss product Rivella, is the longest existing, and the most successful, of all whey beverage products. Its popularity on the Swiss beverage market rivals that of Coca-Cola. This carbonated product, made by mixing the clear deproteinated whey serum with a flavor extract of various herbs, represents a general thirst-quenching beverage rather than a nutritious, dairy-based product such as the fruit-juice-based or the yogurt-based drinks. In this sense, the use of whey for Rivella production is similar to other industrial uses of whey, since the Rivella is manufactured by a specialized beverage company rather than a dairy processor.

Industrial Whey Products

Processing of whey into various industrial products has become the most attractive alternative to whey disposal. This includes the drying of whey, which may have been viewed earlier as a process merely reducing the disposal problem, since utilization of the dry whey often has been unprofitable. Significant research efforts worldwide have resulted in many new alternative technological approaches and uses for the industrial whey products, and the whey processing industry is today one of the most rapidly developing segments within the dairy field. In addition to the traditional dairy products made from whey (concentrated or dried whey, lactose, whey butter, partially delactosed whey, and heat-precipitated whey protein), many new whey-based products are now available as ingredients for the food industry; these include whey protein concentrates (WPC), isolates (WPI), or fractions (WPF); blends of whey or whey protein with other nondairy sources; demineralized whey; lactose hydrolyzed whey products; fractionated and enzymatically modified whey protein products for specialized food, pharmaceutical, and other nonfood uses; and various products of the whey fermentation processes, including biomass, food-grade, or fuel alcohol and methane. Some of the representative industrial whey products are listed in Table 4.

Whey Protein Products. The most valuable component of whey is the whey protein. The oldest process for production of a WPC consists of heating of whey and recovering the heat-precipitable whey proteins as an insoluble high-protein product (usually referred to as traditional lactalbumin). In the modern whey processing industry, various membrane-based separation technologies are being used to produce WPCs and WPIs with much improved functional properties, in particular regarding good water solubility. The predominant technology for production of modern WPCs is UF, which is often combined with diafiltration to remove additional lactose and ash and thereby increase the protein content of the final product. The typical protein content of the ordinary WPC, produced by simple UF as a skim milk replacement, is about 35%, but products with up to 80% protein are now being produced for many food ingredient uses. WPIs with protein content over 90% are also available for specialized applications. The modern WPCs and WPIs have some important functional properties differentiating them from other food protein ingredients, such as high solubility (including solubility in acidic systems), foaming properties, and heat coagulability combined with high water binding in the heat-denatured state. Because whey protein from cow's milk is similar to human whey protein, one of the main applications of the WPCs and WPIs is in the production of baby formulas. Enzymatic modification and/or fractionation techniques to separate the individual whey proteins are being developed to counteract the possible allergenicity of the most abundant whey protein, the β-lactoglobulin (15), and to maximize the functionality of the individual whey protein fractions (10,40). Other technological alternatives that have been suggested for whey protein recovery include the use of chemical precipitants, especially various biopolymers as pectin, carrageenan, or carboxymethylcellulose, as well as ion exchange resins or gel filtration (4,10). In general, whey protein is considered to be nutritionally one of the best proteins for human food use, and its versatile functional properties make it a valuable ingredient for the modern food industry.

Lactose Products.   Because lactose is the predominant whey component, any protein recovery operation not accompanied by some lactose processing alternative will result in a large lactose disposal problem. Traditionally, lactose has been recovered as an article of commerce by crystallization from highly concentrated whey (typically to 65–70% TS), with or without prior whey protein removal. Crystallizing lactose from unmodified whey will result in partially delactosed whey residue that can be dried to obtain a product with sufficiently high protein content to be valuable as a skim milk replacement in certain applications. When the whey protein is removed before the crystallization process, either by heat precipitation or by UF, yield of both products can be maximized and the rate of lactose crystallization optimized (23), but the residual mother liquor from the lactose crystallization process could constitute a disposal problem. However, lactose-producing plants are increasingly using UF permeate as the raw material for lactose manufacture (41). Lactose of edible grade or better can be produced also by spray drying of highly purified deproteinated and demineralized whey. Because the world market for lactose is rather static, new opportunities for lactose utilization would have to be found if lactose production were to increase substantially.

Hydrolysis of lactose into its two monosaccharide constituents, glucose and galactose, will result in increased sweetness of the mixture as well as reduced lactose crystallization problems in concentrated food systems (as ice cream). Sweetening syrups can be produced by lactose hydrolysis in purified lactose streams for use in canning, as brewing adjuncts, or for other applications. Lactose hydrolysis in whey or whey UF permeate for use in manufacture of whey drinks will decrease the caloric content of the final product as well as the required sweetener addition when

using some of the high potency noncaloric sweeteners (42). Lactose can be further modified chemically to obtain useful food or industrial ingredients such as lactitol, lactulose, or lactosyl urea (43); however, the raw material for these lactose derivatives is usually the isolated lactose rather than the whole whey.

Whole Whey Products.   The most common industrial whey product is spray-dried whey and its numerous variants. Whey drying is not an easy process because of the high lactose content and, in the case of acid whey, the high content of the nonvolatile lactic acid. For a typical spray-drying operation, the whey is first concentrated by multistage vacuum evaporation to more than 50% TS. Preconcentration by RO is often practiced to increase the capacity of the evaporator and/or to bring the raw material from a distant supplier to the central drying facility. To produce a nonhygroscopic whey powder, the lactose should be left to crystallize for at least 24 h before the spray-drying operation, otherwise noncrystalline, unstable, and highly hygroscopic lactose glass will be formed in the drying process. Sometimes, the whey proteins are also purposely heat denatured before the spray drying to minimize a potential for protein deposition on the heat-transfer surfaces, resulting in processing difficulties. As a result, the hygroscopicity, solubility, and other functional properties of whey powders may differ substantially depending on the processing conditions. Other drying procedures traditional for the dairy industry, such as drum drying, are seldom used because of the great operational difficulties and the inferior quality of the final product.

In addition to regular whey powders, products with modified lactose, protein, or mineral content can be produced using lactose crystallization, membrane processing, demineralization by IE or ED, or blending the liquid whey with other nondairy protein sources, including soy, pea, and other vegetable proteins as well as bran or other food materials. A U.S. patent (44) claims that wet grinding of dry cereal or oilseed protein sources (soy, rapeseed, oats) in whey or whey UF permeate followed by spray drying of the blends will result in improved functional properties, especially higher resistance toward lipid rancidity development in the dry product. Other whole whey products include sweetened and/or lactose-hydrolyzed concentrated whey as an in-house ingredient in ice cream and other dairy products and highly concentrated whey in the form of semisolid lick-blocks for cattle feeding purposes.

Fermentation Products from Whey.   Because lactose is fermentable by certain microorganisms, whey has been a traditional substrate for fermentation processes, including the production of antibiotics, lactic acid, or biomass. However, these established processes of the past have been replaced to a large degree by more economical alternatives, such as direct chemical synthesis. Recent developments of modern processing approaches as well as the preference of natural over synthetic ingredients may revitalize the interest in various fermentation processes based on whey. Production of lactic acid is one of the examples where modern technology (45) and the back-to-nature philosophy may make the traditional fermentation route successful again.

Likewise, production of food-grade alcohol from whey for use in dairy-based liqueurs have been successfully developed in Ireland (46). In New Zealand and other countries, whey fermentation into industrial alcohol in large commercial installations appears to be profitable (47,48). Use of liquid whey medium for bulk propagation of lactic starter cultures for in-house use is well established in many dairy processing plants. Further fermentation of whey for various whey-based beverages is a natural possibility, including production of whey wine or whey beer, even though lactose is not fermentable by many common yeasts without special pretreatments. Alcoholic whey-based beverages have been described in literature (32,37), but their commercial success is not well documented apart from the manufacture of whey beer in Germany during the period of scarcity in the World War II. In the nonfood area, whey has been used as a substrate for biomass production by yeast fermentation (the wheast process), for bacterial fermentation into lactic acid with continuous neutralization to produce fermented ammoniated concentrated whey (FACW) for ruminant feeding (49), or as a source of biogas generated in anaerobic waste treatment installations (50). However, these processes may be considered more as alternatives for whey disposal rather than profitable opportunities for manufacture of industrial whey products.

Whey Disposal and On-Farm Utilization

Although whey has become much too valuable to be disposed of on a large scale, there are still many smaller and isolated cheese manufacturing operations where whey disposal can be a major economical and/or ecological burden. Where industrial whey processing or manufacture of high-value consumer products is not feasible, other methods of whey disposal must be selected, based mainly on the associated cost. The down-the-drain approach is certainly the most wasteful although not always the most expensive alternative, especially when municipal or existing industrial waste treatment facilities are located nearby and the volume of whey to be discarded is not great. Spraying the whey onto fields is still being practiced by many isolated cheesemakers in rural areas, sometimes relying on sophisticated approaches with time-controlled application schedules to maximize the potential fertilizer value and to minimize soil salinity and acidity problems inherent in these techniques (51). Direct feeding of whey to agricultural animals, especially pigs, appears to be one of the most profitable on-farm uses of whey (52). When the whey feeding is combined with methane production from the animal excreta for use as a fuel in a cheese factory located nearby, a truly ecologically enclosed industrial system can be developed, as evidenced by several commercial installations in Switzerland.

NUTRITIVE VALUE AND USES OF WHEY

As a by-product of dairy processing, cheese whey contains about 50% of the nutrients found in milk. The most valuable whey components from a nutritional standpoint are the whey proteins, the water-soluble vitamins (especially riboflavin), and some of the nutritionally important minerals. Acid whey is a particularly good source of calcium; its content approaches that of regular cow's milk. Whey protein from either sweet or acid whey is one of the best proteins for human food use because of its balanced amino acid profile and superior digestibility (53). Dried cheese whey composition can be compared to that of regular wheat flour (31) in terms of the proximate composition and some of the micronutrients. In addition to the unquestionable value of the whey protein, calcium, and riboflavin, some sources also emphasize the beneficial effects of the high (L+) lactic acid content of whey produced with certain dairy cultures (53).

The use of many industrial whey products as ingredients in various foods is based not only on the nutritional qualities but also on the desirable functionality of some of the products. With the high lactose and protein content, dried whey is an ideal substrate for the nonenzymatic Maillard browning reaction in foods where such color development is desired, as in breaded meat or fish products or in toasted bread. Lack of sweetness in lactose is advantageous in many applications of whole or modified whey ingredients, especially when used in meat products, processed cheese foods, gravies, soups, sauces, salad dressings, etc. Whey protein is ideally suited for fortification of various foods both because of its nutritional quality and its functional attributes, including bland taste compatible with most food flavors and solubility in aqueous food systems (54). Many additional uses of whole whey and whey-based traditional or modern food ingredients have been suggested in various reviews (4,55–57). Inclusion of whey protein in cheese, either by varying the cheese processing technology or by reincorporation of the protein recovered from the whey, is one of the important research subjects concerning whey at the present time. Several alternative technological pathways can be used for incorporation of the heated whey protein precipitate into fresh cheeses such as quarg and quarg-based food products (7), in which the cooked flavor and quality impairment, often encountered in traditional cheeses, need not be a problem. This option is especially attractive for fresh cheese plants where whey from production of other cheese varieties is available.

Although food use of whey and whey-based products is the most desirable whey disposal alternative, various nonfood uses of whey also have been proposed, including incorporation into foamed insulation materials for the construction industry or use as a binder for iron ore pellets (58). Enzymatic modifications of whey proteins for use in the pharmaceutical industry may follow the successes predicted for the casein (14). Rapid advances in biotechnology will likely discover new opportunities for conversion of whey into valuable industrial food or nonfood materials such as biosorbents, biopolymers, stabilizer gums, flavoring or coloring compounds, bacteriocins or fat replacers such as the Simplesse developed from whey protein and egg white (59). With its high lactose content, surplus dried whey could be used as a quick source of energy in international hunger-relief programs; however, the prevalence of lactose intolerance in most third world countries makes such use difficult without careful development of foods that can be tolerated by the lactose malabsorbers. Incorporation of whey into high-solid indigenous foods should be considered because these foods appear to be better tolerated by lactose malabsorbers than more liquid products (60,61).

The current interest in functional foods and nutraceutical products (62) has opened yet another major avenue for utilization of whey as a raw material in production of special dietary aids and nutritional supplements. Whey protein isolates have become the most valued protein source for body-building formulas and similar high protein preparations (63) used by active athletes following intensive training regimes. Some of the minor whey proteins, such as lactoferrin, lactoperoxidase, or the various immunoglobulins, are rapidly becoming prized commodities for both food and nonfood uses, including personal hygiene products such as mouthwashes or skin creams (64); even the whey minerals are being marketed as natural food ingredients (62). A whey-protein-based dietary supplement has been patented (65) and is being promoted under the trade name Immunocal on the basis of its alleged stimulatory effect on immune functions through increased production of intracellular glutathione (66). The ultimate success of these and other proposed novel whey protein nutraceutical applications may depend on the proved physiological functionality, yet to be documented in humans in many cases, as well as on economically feasible utilization or disposal of the remaining whey constituents, especially lactose. In this regard, the rapidly increasing commercial interest in prebiotics of dairy origin, such as galactooligosaccharides, lactosucrose, and lactulose, for use in bifidobacteria-containing dairy foods may hold promise. Additional physiologically functional effects of some of these oligosaccharides, currently being investigated with animal models, include calcium absorption and reduction of the risk of colon cancer (67).

The possibilities for treating whey as an opportunity rather than a problem have increased dramatically in the recent years. In some dairy plants recently installed, all whey produced in cheesemaking is completely used on site in a variety of integrated processes, leaving no residue to dispose of. With continued technological developments, the emerging scarcity of inexpensive dairy ingredient sources, and the increasing emphasis on ecologically responsible industrial processes, the image of whey is rapidly changing from that of a bothersome by-product to a highly prized resource.

BIBLIOGRAPHY

1. J. T. Lawhon, L. J. Manak, and E. W. Lusas, "An Improved Process for Isolation of Glandless Cottonseed Protein Using Industrial Membrane Systems," *J. Food Sci.* **45**, 197 (1980).

2. P. Jelen, M. Earle, and W. Edwardson, "Recovery of Meat Protein from Alkaline Extracts of Beef Bones," *J. Food Sci.* **44**, 327 (1979).

3. G. Kjaergaard-Jensen and J. K. Oxlund, "Concentration and Drying of Whey and Permeates," in *Trends in Utilization of Whey and Whey Derivatives*, International Dairy Federation, Bull. 233, Brussels, Belgium, 1988, p. 21.

4. D. M. Irvine and A. R. Hill, "Cheese Technology," in M. Moo Young, ed., *Comprehensive Biotechnology*, Vol. 3, Pergamon Press, Toronto, Canada, 1985, p. 523.

5. K. R. Marshall, "Industrial Isolation of Milk Proteins: Whey Proteins," in P. F. Fox, ed., *Developments in Dairy Chemistry 1*, Elsevier, London, 1982, p. 340.

6. K. R. Marshall and W. J. Harper, "Whey Protein Concentrates," in *Trends in Utilization of Whey and Whey Derivatives*, IDF Bull. 233, Brussels, Belgium, 1988, p. 21.

7. P. Jelen and A. Renz-Schauen, "Quarg Manufacturing Innovations and Their Effect on Quality, Nutritive Value and Consumer Acceptance," *Food Technol.* **43**, 74 (1989).

8. H. Sheth et al., "Yield, Sensory Properties and Nutritive Qualities of Quarg Produced from Lactose—Hydrolyzed and High Heated Milk," *J. Dairy Sci.* **71**, 2891 (1988).

9. W. N. Eigel et al., "Nomenclature of Proteins of Cow's Milk: Fifth Revision," *J. Dairy Sci.* **67**, 1599 (1984).

10. J. L. Maubois et al., "Industrial Fractionation of Main Whey Proteins," in *Trends in Whey Utilization*, IDF Bull. 212, Brussels, Belgium, 1988, p. 150.

11. P. Jelen, W. Buchheim, and K.-H. Peters, "Heat Stability and Use of Milk with Modified Casein: Whey Protein Content in Yogurt and Cultured Milk Products," *Milchwissenschaft* **42**, 418 (1987).

12. J. Patocka, A. Renz-Schauen, and P. Jelen, "Protein Coagulation in Sweet and Acid Whey Upon Heating in Highly Acidic Conditions," *Milchwissenschaft* **41**, 490 (1986).

13. F. M. Driessen and M. G. van den Berg, "New Developments in Whey Drinks," in *Special Addresses at IDF Annual Sessions*, International Dairy Federation, Bull. 250, Brussels, Belgium, 1990, p. 11.

14. J. L. Maubois and J. Leonil, "Peptides du fait a activite biologique," *Le Lait* **69**, 245 (1989).

15. R. Jost, J. C. Monti, and J. J. Pahud, "Whey Protein Allergenicity and its Reduction by Technological Means," *Food Technol.* **41**, 118 (1987).

16. J. Sherbon, "Physical Properties of Milk," in N. P. Wong et al., eds., *Fundamentals of Dairy Chemistry*, 3rd ed., Van Nostrand Reinhold, New York, 1988, p. 426.

17. D. Roehl and P. Jelen, "Surface Tension of Whey and Whey Derivatives," *J. Dairy Sci.* **71**, 3167 (1988).

18. P. Jelen, "Physico-Chemical Properties of Milk and Whey in Membrane Processing," *J. Dairy Sci.* **62**, 1343 (1979).

19. W. Rattray and P. Jelen, "Protein Standardization of Milk and Dairy Products," *Trends in Food Sci. Technol.* **7**, 227–233 (1996).

20. M. Britten et al., "Composite Blends from Heat-Denatured and Undenatured Whey Protein-Emulsifying Properties," *Int. Dairy J.* **4**, 25–36 (1994).

21. P. Jelen and W. Buchheim, "Stability of Whey Protein Upon Heating in Highly Acidic Conditions," *Milchwissenschaft* **39**, 215 (1984).

22. P. Jelen, "Whipping Studies with Partially Delactosed Cheese Whey," *J. Dairy Sci.* **56**, 1505 (1973).

23. P. Jelen and S. T. Coulter, "Effects of Supersaturation and Temperature on the Growth of Lactose Crystals," *J. Food Sci.* **38**, 1182 (1973).

24. P. Jelen and W. Buchheim, "Norwegian Whey Cheese," *Food Technol.* **30**, 62 (1976).

25. P. Jelen, "Reprocessing of Whey and Other Dairy Wastes for Use as Food Ingredients," *Food Technol.* **37**, 81 (1983).

26. P. Jelen, "Lactose Hydrolysis Using Sonicated Dairy Cultures," in *Lactose Hydrolysis*, International Dairy Federation, Bull. 289, Brussels, Belgium, 1993, pp. 54–56.

27. T. H. Jones and P. Jelen, "Low Dose Gamma-Irradiation of Camembert, Cottage Cheese and Cottage Cheese Whey," *Milchwissenschaft* **43**, 233 (1989).

28. N. Dunford and P. Jelen, "Adsorption of Riboflavin by the CARBOSEP Ultrafiltration System," *8th World Congress of Food Sci. Technol.*, Toronto, Canada, September 29–October 4, 1991.

29. A. Leviton, "Adsorption of Riboflavin by Lactose," *Ind. Eng. Chem.* **36**, 744 (1944).

30. J. Patocka and P. Jelen, "Rapid Clarification of Cottage Cheese Whey by Centrifugation and Its Control by Measurement of Absorbance," *Milchwissenschaft* **44**, 501 (1989).

31. P. Jelen, "Industrial Whey Processing Technology: An Overview," *Agric. Food Chem.* **27**, 658 (1979).

32. V. H. Holsinger, L. P. Posati, and E. D. De Vilbiss, "Whey Beverages: A Review," *J. Dairy Sci.* **57**, 849 (1974).

33. P. Jelen, "Whey Cheeses and Beverages," in J. G. Zadow, ed., *Whey and Lactose Processing*, Elsevier, London, 1991, pp. 147–194.

34. H. W. Modler, "Development of a Continuous Process for the Production of Ricotta Cheese," *J. Dairy Sci.* **71**, 2003 (1988).

35. F. V. Kosikowski and V. V. Mistry, *Cheese and Fermented Milk Foods*, F. V. Kosikowski L.L.C., Westport, Conn., 1997.

36. R. Scott, *Cheesemaking Practice*, Elsevier, London, 1986, pp. 24, 316, 382.

37. E. F. Kravchenko, "Whey Beverages," in *Trends in Utilization of Whey and Whey Derivatives*, International Dairy Federation, Bull. 233, Brussels, Belgium, 1988, p. 61.

38. K. Devkota and P. Jelen, "Pectin–Whey Protein Interactions in Cottage Cheese Whey," *Posters and Brief Communications of the 23rd Int. Dairy Congress*, Montreal, Canada, October 8–12, 1990.

39. H. Trapp, *Technological Approaches in the Development of a Whey-Based Yogurt Beverage*, M.S. Thesis, University of Alberta, Edmonton, Canada, 1990.

40. R. J. Pearce, "Fractionation of Whey Proteins," in *Trends in Whey Utilization*, International Dairy Federation, Bull 212, Brussels, Belgium, 1987, p. 150.

41. A. K. Keller, "Permeate Utilization," in *Proc. Dairy Products Technical Conf.*, Madison, Wisc., April 25–26, 1990, p. 101.

42. C. Beukema and P. Jelen, "High Potency Sweeteners in Formation of Whey Beverages," *Milchwissenschaft* **45**, 576 (1990).

43. R. A. Visser, M. J. van den Bos, and W. P. Ferguson, "Lactose and Its Chemical Derivatives," in *Trends in Utilization of Whey and Whey Derivatives*, International Dairy Federation, Bull. 233, Brussels, Belgium, 1988, p. 33.

44. U.S. Pat. 4,968,521 (1990), P. Melnychyn.

45. J. R. Rosenau, G. Colon, and T. C. Wang, "Lactic Acid Production from Whey via Electrodialysis," in M. LeMaguer and P. Jelen, eds., *Food Engineering and Process Applications*, Vol. 2, Elsevier, London, 1986, p. 245.

46. J. A. Barry, "Alcohol Production from Cheese Whey," *Dairy Ind. Int.*, 19–21 (1982).

47. J. G. Zadow, "Fermentation of Whey and Permeate," in *Trends in Utilization of Whey and Whey Derivatives*, International Dairy Federation, Bull. 233, Brussels, Belgium, 1988, p. 53.

48. A. J. Mawson, "Bioconversions for Whey Utilization and Waste Abatement," *Bioresource Technol.* **47**, 195–203 (1994).

49. F. W. Juengst, "Use of Total Whey Constituents—Animal Feed," *J. Dairy Sci.* **62**, 106 (1979).

50. "South Caernarvon Creamery Converts Whey into Energy," *Dairy Ind. Int.* **49**, 16 (1984).

51. J. B. Bradford, D. B. Galpin, and M. F. Parkin, "Utilization of Whey as a Fertilizer Replacement for Dairy Pasture," *New Zealand J. Dairy Sci. Technol.* **21**, 65 (1986).

52. H. W. Modler et al., "Economic and Technical Aspects of Feeding Whey to Livestock," *J. Dairy Sci.* **63**, 838 (1980).

53. E. Renner, *Milk and Dairy Products in Human Nutrition*, Volkswirtschaftlicher Verlag, Munich, Germany, 1983.

54. A. Renz-Schauen and E. Renner, "Fortification of Nondairy Foods with Dairy Ingredients," *Food Technol.* **41**, 122 (1987).

55. T. Sienkiewicz and C. L. Riedel, *Whey and Whey Utilization*, Verlag T. Mann, Gelsenkirchen-Buer, Germany, 1990.

56. F. V. Kosikowski, "Whey Utilization and Whey Products," *J. Dairy Sci.* **62**, 1149 (1979).

57. B. P. Robinson, J. L. Short, and K. R. Marshall, "Traditional Lactalbumin—Manufacture, Properties and Uses," *New Zealand J. Dairy Sci. Technol.* **11**, 114 (1976).

58. J. V. Chambers and A. Ferretti, "Industrial Applications of Whey/Lactose," *J. Dairy Sci.* **62**, 112 (1979).

59. N. S. Singer, "Simplesse—All Natural Fat Substitute and the Dairy Industry," in *Proc. Dairy Products Technical Conf.*, Madison, Wisc., April 25–26, 1990 p. 85.

60. N. Shah and P. Jelen, "Lactose Absorption by Post-Weaning Rats from Yogurt, Quarg and Guarg Whey," *J. Dairy Sci.* **74**, 1512 (1991).

61. N. P. Shah, R. N. Fedorak, and P. Jelen, "Food Consistency Effects of Quarg in Lactose Malabsorption," *Int. Dairy J.* **2**, 257–269 (1992).

62. P. Jelen and S. Lutz, "Functional Milk and Dairy Products," in G. Mazza, ed., *Functional Foods—Biochemical and Processing Aspects*, Technomics Publishing Co., Lancaster, Pa., 1998, pp. 357–380.

63. F. Nettl, "The Shocking Truth About Whey Protein," *Muscle Media 2000*, 72–77 (1995).

64. G. O. Regester et al., "Bioactive Factors in Milk—Natural and Induced," in R. A. S. Welch et al., eds., *Milk Composition, Production and Biotechnology*, CAB International, New York, 1997, pp. 119–132.

65. Can. Pat. 1,338,682 (October 29, 1996), G. Bounous, P. Gold, and P. L. Kongshavn.

66. G. Bounous and P. Gold, "The Biological Activity of Undenatured Dietary Whey Proteins—Role of Glutathione," *Clin. Invest. Med.* **14**, 296–309 (1991).

67. R. Tanaka and K. Matsumoto, "Recent Progress on Prebiotics in Japan, Including Galaactooligosaccharides," Int. Dairy Fed. Bulletin 336, Brussels, Belgium, 1998, pp. 21–27.

P. Jelen
University of Alberta
Edmonton, Alberta
Canada

See also Dairy ingredients for foods.

WINE

INTRODUCTION

The word wine was first used for the fermented product of grapes, although fermented juices of many fruits (apples, berries, peaches, and even herbs) are now called wine (usually peach wine, etc). Wine has a long association with human artistic, cultural, and religious activities. The qualities of wine were praised in pre-Christian as well as post-Christian times. Addition of herbs and spices was common, to provide distinct flavors (and mask off-flavors) and to induce supposedly medicinal and aphrodisiac properties. The religious and allegorical significance of wine developed very early and has been utilized by many religions, including Christianity.

Grape cultivation first began around 6000 years ago. Although many of the wines were poor by modern standards, a developing devotion and taste for wine (beginning with the Roman era) encouraged improvement in grape and wine production to the point where the regional quality of special wines became recognized. In contrast to the invading Romans, who brought grapevines with them, the Moslems, in later years, virtually destroyed the wine industry of the countries they conquered. From then and throughout the Middle Ages, in central Europe monasteries maintained vineyards as a source of wine for religious ceremonies. By the seventeenth century, the coopers built more and better casks and barrels so that wines could be safely aged for years. At the same time, glass was strengthened when coal (instead of charcoal) was used in its manufacture. This permitted bottling of sparkling wines. The introduction of cork extended the time for which bottled wines could be stored. By the nineteenth century, the scientific revolution, as exemplified by the work of Pasteur and others, revolutionized the wine industry: the role of yeasts and bacteria and methods for their control changed the wine industry from a cottage industry of chance to a controlled business. Mechanization of operations in the vineyard and winery followed. Although producing quality wines is still something of an art, today's wine industry makes some of the best wine ever produced, with minimum spoilage and with much less human labor.

REGULATIONS AND APPELLATIONS

Wine is the fermented product of the fruit of several species of *Vitis*, mainly of cultivars of *Vitis vinifera*. The United States's definition of wine is contained in the U.S. Internal Revenue Code. Table and sparkling wines (<14.1% alcohol) are distinguished from dessert and flavored (vermouth, etc) wines (>14.0% alcohol) and from wine coolers (<7% alcohol). Each country has its own definitions, and the European Community (EC) has regulations for trade among its members and for imports from nonmember countries.

In Germany, the eastern United States, and parts of France, sugar, grape concentrate, or reduced musts (crushed grapes) may be added when the grapes do not ripen sufficiently *and* when local laws approve it. Gener-

ally addition of water and sugar is forbidden. Limits on chlorides, sulfates, lead, arsenic, volatile acidity (mainly acetic acid), and sulfur dioxide have been established by federal regulations (such as national public-health authorities). Because a small percentage of asthmatics are extremely sensitive to sulfur dioxide, U.S. government regulations require a warning about sulfites on wine labels.

Appellations of origin are used in many countries. In general, these delimit the region, the varieties of grapes used, maximum yield per hectare, and often the wine practices and in some cases (eg, Portugal [for Port], France, Germany, and South Africa) require sensory evaluation of each wine to qualify for a regional designation. Typically, use of the name of a specific vineyard on a wine label implies that the wine is of higher quality than those labeled for larger districts. In Germany, quality is also related to ripeness. The higher-quality wines are made with no sugar addition. Whereas France, Germany, Spain, and Italy have had protected geographical appellations for many years, it was not until 1978 that appellation procedures were instituted in the United States. In contrast to European laws, U.S. regulations define the boundaries of the delimited viticultural area and limit neither the variety of grapes grown nor the types of wine produced. In most countries (including the United States), where the appellation does not regulate which varieties may be planted, wines are labeled with the name of the grape variety from which it is made, hence the term varietal wines.

VITICULTURE (THE SCIENCE OF GRAPE GROWING)

V. vinifera, the main genus and species of grapes used for wine, produces the best grapes in temperate climates with mild winters and dry summers; vines are killed by very cold winters and are susceptible to mold, mildew, and diseases in humid summer regions. *V. vinifera* vines have no resistance to the root louse, *Phylloxera*, which virtually destroyed the wine-making industry in France and California in the late 1800s. Other species, such as *V. labrusca* (Concord grape), are resistant to the pest. Today *V. vinifera* vines in France and California are usually grafted to *Phylloxera*-resistant rootstocks.

Grape leaves appear in early spring, with grape flowering occurring about 45 days later. The small, hard berries continue to expand for 2 months until ripening is initiated (known as *veraison*). From this time on, the grape begins to color and soften and the sugar (glucose and fructose) rapidly increases. The concentrations of tartaric and malic acids, the major organic acids, peak at veraison. With the onset of ripening the acid level decreases as the berry expands, and in hot climates, the malic acid can decrease rapidly.

The maximum sugar accumulated by the grape during ripening depends on the coolness or warmth of the region where they are grown, on the variety of grape, and on other viticultural factors such as the crop level (amount of fruit per vine). For most varieties, 15 to 25% sugar is reached, but late-harvest fruit (such as those used for sauternes from France or *Beerenauslese* or *Trockenbeerenauslese* from Germany) may have 30 to 40% sugar. (High humidity during the final ripening of the grapes may result in infection with the mold *Botrytis cinerea*. This fungus loosens the skin and permits rapid loss of moisture and shriveling of the berry, resulting in a high sugar content and producing a honeylike flavor).

The pH and the concentration of acids, measured as titratable acidity, are further indicators of suitability of grapes for harvesting. During ripening, if malic acid decreases greatly, the grapes may have a high pH (pH ≥ 4). Addition of tartaric acid to crushed grapes is sometimes necessary to lower the pH in order to reduce susceptibility to microbial spoilage. In contrast, in cool-climate regions, the acidity may be so high at harvest that processing steps may be needed to reduce the acidity. Normal grape juice has pH values from 3.0 to 3.6 and with titratable acidities of 0.6 to 1.5% (calculated as tartaric acid). Under these conditions undesirable microorganisms grow slowly, if at all, whereas desirable yeasts grow well.

Although only a small amount of nitrogenous materials (0.3–1.0%) are found in mature grapes, they are significant for yeast and malolactic bacteria nutrition as well as for flavor development. With very low levels of nitrogen, the fermentation may proceed very slowly and may stop before the sugar is completely fermented, producing a "stuck" fermentation. However, nitrogen levels in the grape can be controlled by appropriate viticultural practices, such as fertilization or reduction in crop level.

Anthocyanins, the colored pigments of red grapes, are located in the epidermal (outer) cells of the skin, although a few varieties of grapes have pigments in the pulp as well. Tannins (phenolic compounds) are located primarily in grape skins and seeds. The tannins, which are found in higher concentrations in red wines, have some positive health benefits as antioxidants. Recently, evidence has been presented that drinking moderate amounts of red wine lowers susceptibility to heart disease and exerts anticarcinogenic effects.

MICROORGANISMS AND FERMENTATION

Wines are normally fermented with the yeast *Saccharomyces cerevisiae*, although *S. bayanus* and *S. oviformis* are sometimes used. The primary or alcoholic fermentation is the process of conversion of the grape sugars, glucose and fructose, to ethanol and carbon dioxide. In addition to *Saccharomyces*, other wild yeasts, such as *Kloeckera apiculata* and *Metschnikowia pulcherima*, can be found on the fruit and participate in the early stages of fermentation. However, these latter yeasts can impart compounds that alter the aroma of the wine.

Yeast is present at high enough concentrations on the grape skins and in the winery to initiate fermentation. However, pure strains of yeast cultures are often added to assure the completion of fermentation by *Saccharomyces*. To inhibit growth of the wild yeasts and other undesirable microorganisms, 25 to 75 mg/L sulfur dioxide (SO_2) is added in most white wine fermentations and frequently in reds.

In addition to yeast, bacteria are also associated with the wine-making process. Although lactic acid bacteria are found on the fruit and are present in the winery, a bacteria inoculum is usually added at the later stages of the alcoholic fermentation to ensure completion of the secondary or malolactic fermentation. During this secondary fermentation, malic acid from the grape is converted to lactic acid, carbon dioxide, and other compounds. These malolactic (ML) bacteria also produce diacetyl, which imparts the characteristic buttery aroma found in many wines. ML bacteria, of which the most frequently found is *Oenococcus oenii*, are very sensitive to SO_2; consequently, lower levels of SO_2 are used when the ML fermentation is desired.

Unlike many other products, no harmful bacteria or molds grow on grapes or in wine. However, spoilage organisms can be found in fermentations and in finished wine. The acetic acid bacteria that are inhibited by SO_2 can produce high levels of acetic acid in the presence of air, which is desirable only in production of wine vinegar. Once acetic acid is formed, the yeast and other microorganisms can convert it to ethyl acetate, which contributes a solvent odor similar to that of nail polish remover. The yeast *Brettanomyces* can contribute distinctive aromas ranging from barnyard, horsey, leathery, phenolic, and fecal to mousy. Although these "spoilage" organisms may enhance the complexity of wine aroma at low concentrations, they often result in "off-odors" at higher concentrations.

WINE MAKING

Still Wines

The basic steps used in making wine are very similar for small producers of the highest-quality wines, large producers of inexpensive wines, and home wine makers, although the size and sophistication of the equipment varies enormously. Grapes are first crushed in a stainless-steel crusher-destemmer. These operate at a high speed, separating the stems by centrifugal force and crushing the grapes with revolving paddles. The crushed grapes (called must) fall through holes in the cylinder as the stems are ejected.

Must pumps transfer the crushed white grapes to presses and the crushed red grapes to fermentors, with some exceptions. White wines are made from white grapes or by pressing crushed red grapes immediately after crushing. If the skins are rapidly separated from the juice, light or pink-colored musts can be produced from red grapes. This process is used to make the so-called blush wines such as White Zinfandel that comes from the red Zinfandel grapes; similarly, blanc de noir sparkling wines are made from the red Pinot noir grape by pressing rapidly so that minimum pigment is extracted.

The old-fashioned basket press is still found, but the new and larger wineries often use the Willmes press, which has an inflatable bag inside a revolving tank. As the bag is inflated, the musts are pressed against a perforated cylinder and the juice flows out to the bottom of the press. Some hydraulic presses are horizontally operated, with pressure exerted from both ends. The fermentors are usually tanks made of stainless steel that has the advantage of being chemically inert, thermodynamically uniform, and easy to clean. The fermentors are often jacketed to permit computerized control of temperature by the circulation of coolants. The size of the fermenting tanks varies with the winery from 500 gal to over 100,000 gal.

Sulfur dioxide (50–75 mg/L) may be added to white musts, which are then often centrifuged or settled overnight to clarify the juice before transfer to the fermentor, at which point a yeast inoculum is added to the white musts. In the case of red grapes, yeast and sulfur dioxide are added to the crushed grapes after they are pumped into the fermentation tank.

Fermenting tanks of white wines are usually maintained at 10 to 15°C, with the fermentation taking from 3 to 4 weeks. Red musts are fermented warmer than white, typically at temperatures between 18 and 27°C, so that fermentation is completed in 3 to 10 days.

For red musts, the skins rise to the surface, forming a cap during fermentation due to the evolution of carbon dioxide gas. To help extract the pigments from the skins, the fermenting must at the bottom of the tank is pumped over and sprayed onto the floating cap. Depending on the style of red wine being made, red musts can undergo an extended fermentation on the skins up to 4 weeks before they are pressed and then fermented to dryness (a level of sugar below 0.5%). Longer contact with the skins provides greater extraction of anthocyanin pigments and phenolic material, which have a direct effect on the sensory properties of wine. Accordingly, these wines are intended to be aged before they reach their maximum quality. The wine made with shorter skin-contact time is lighter in color and less bitter and astringent so that it can be consumed without aging.

When the fermentations have reached dryness, the wine is transferred from the fermentor to clean stainless-steel tanks for clarification or to wood barrels for aging. Wines that will not be exposed to oak cooperage are clarified by settling and transferring (racking) the clear wine to other tanks. To prevent formation of crystals of potassium acid tartrate (cream of tartar) in the consumer's refrigerator, wines are cold stabilized. The wine is stored at −4°C and the precipitated tartrate crystals are removed by filtration. Before bottling, the wine is fined by addition of an agent to clarify the wine by adsorbing and removing suspended or dissolved material. The most common fining agent for white wines is bentonite (a clay), whereas reds are fined with a variety of agents, including egg whites and gelatin. Other fining materials include isinglass, casein, polyvinyl-polypyrrolidone (PVPP), and silica suspensions.

Most red and some white wines are stored in oak barrels to extract flavor components from the oak. Wines are transferred to 50-gal oak barrels where they are held for periods from 3 months to 2 years. During this aging period, the level of wine in the barrels is frequently topped off to prevent exposure to air. After oak storage, the wines are fined and cold stabilized.

Before bottling, wine is usually filtered to be brilliantly clear. If the wine contains more than 0.7% sugar, it is microbially unstable. In this case a fine-mesh or sterile filtration is also used to remove microorganisms and prevent spoilage in the bottle. On automated bottling lines, the filtered wine is filled into bottles, corks are inserted, labels are attached, and the bottles are put in cases.

Sparkling Wines

Wines with a visible excess of carbon dioxide (CO_2) are called sparkling wines. In the United States a sparkling wine is one that contains a barely perceptible level of CO_2 (above 3.9 g/L at 15.7°C), although the typical level of carbon dioxide is much higher: 4 to 6 atm pressure. The most famous sparkling wine is Champagne, produced in the Champagne region of France, where it was first commercialized. Depending on local regulations, sparkling wines are also labeled *Sekt* (Germany), *cava* (Spain), or just sparkling. The two primary methods of production are bottle fermentation (*méthode champenoise*) and closed-tank fermentation (*Charmat* or *Tank*). Chardonnay, Pinot noir and Meunier wines are used for the premium sparkling wines, but other varieties, including Muscat and Riesling, are also used. The base wines are blended and yeast and sugar are added (24 g sugar per liter will produce a pressure of 6 atm). The resulting mixture of wine, sugar, and yeast is transferred to bottles (750 mL) or to stainless-steel pressure tanks for the secondary (alcoholic) fermentation.

The tank process is much cheaper to use, and fermentation is generally completed in less than 1 month. At that time the wine is filtered under pressure, sugar is added to achieve a desired level of sweetness, and the wine is filled into bottles. In contrast, although the *méthode champenoise* fermentation is completed in the filled bottles between 3 and 6 months, the wines are typically held in contact with the yeast for a minimum of 3 years. The bottles are intermittently jostled to force the yeast deposit to settle in the neck of the bottle. To remove the yeast deposit from the bottle, the neck of the bottle is frozen. When the cap is removed, the pressure in the bottle pushes out the frozen yeast deposit. The bottle is immediately filled with more of the same wine, and the dosage (sugar in wine or brandy) is added to achieve the desired sweetness before the bottle is closed with a cork.

SENSORY PROPERTIES

The appeal of wines is due to the complex flavors that are often difficult to describe without resorting to poetic images. Wine flavor is an overall perception, resulting from the integration of appearance, aroma, taste, and mouthfeel. Although wine aromas are often very complex, with little effort they can be described in terms that are precise and can be understood by others. For example, the wine aroma wheel (WAW) in Figure 1 lists terms that describe the aroma notes most often perceived in table wines. The flavor perceived in the mouth is primarily aroma, plus sweet, tart, or bitter tastes and an astringent mouth-feel. Flavors can be informally described using the WAW terminology or more formally profiled by trained judges using descriptive analysis to provide a "photograph of wine flavor."

The purpose of the WAW is to facilitate communication about wine flavor by providing a standard terminology. These terms are precise and not hedonic or judgmental. "Floral" is a general but analytical descriptive term, whereas "fragrant," "elegant," or "harmonious" are either

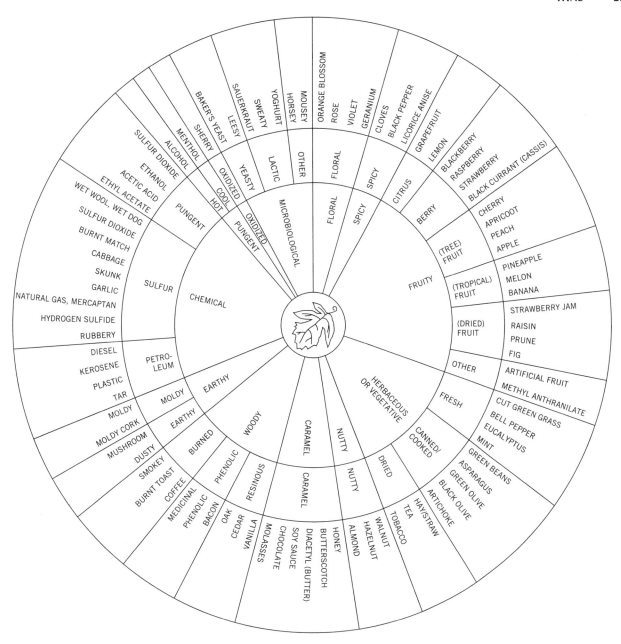

Figure 1. The wine aroma wheel presents descriptive terms for aroma notes that are often perceived in the aromas of table wines. General terms are located on the inside tier, with the most specific terms located on the outside tier. *Source:* Copyright 1990, A. C. Noble.

imprecise and vague (fragrant) or hedonic and judgmental (elegant and harmonious).

The center of the wheel has very general terms, whereas more specific terms are in the outer tiers. These terms are not the only ones that can be used to describe wines, but they represent aroma notes that are often encountered. The aromas for these descriptive terms can be defined by physical standards. For example, a small piece of bell pepper, a grapefruit section, or a few drops of vanilla extract can be presented to define "bell pepper," "grapefruit," or "vanilla," respectively.

The flavor of wine is influenced by the grape variety, the climate and location in which the grapes are grown, viti-

cultural practices, yeast and malolactic fermentations, skin-contact time, storage in oak cooperage, and aging. The characteristic aromas of many of the major *V. vinifera* cultivars used for making premium wines throughout the world arise from the grape. Zinfandel grapes and wines typically have black pepper and berry flavors, whereas red wines from Burgundy (made from Pinot noir grapes) and Pinot noir varietal wines from other locations are distinctive for their (straw)berry or cherry aromas. Red wines from Bordeaux and unblended red wines made from the same varieties (Cabernet Sauvignon, Merlot, and Cabernet Franc) typically have berry, black pepper, and herbaceous/vegetative aromas such as that of a green bell pepper or asparagus. The white varietal, Sauvignon blanc, is similarly characterized by the bell pepper note but also by many fruity attributes such as pineapple, peach, and apricot, which are found in most white wine varietals. Gewürztraminer, Riesling, and Muscat grapes and their respective wines are characterized by floral, citrus, and apricot notes.

Chardonnay is a white varietal that is noted for its citrus, pineapple, peach, and apricot aromas. However, it is more typically identified with aromas arising from winemaking practices. Chardonnay is sometimes fermented in oak and usually aged in oak barrels. Exposure to oak imparts spicy, clove, and vanilla aromas. If the barrels were heavily toasted or charred when coopered, then toasted/smoky aromas may also be perceived. Malolactic fermentation is encouraged in most Chardonnays, thereby infusing a distinct buttery aroma. Most red varietal wines, including reds from Bordeaux and Burgundy, undergo malolactic fermentation and aging in oak. Thus buttery, spicy, and vanilla notes are perceived in addition to the aromas coming from the grape.

Immediately after fermentation, wines tend to be very fruity, because high concentrations of fruity esters are formed in both red and white wines by yeast during the primary fermentation. However, these fruity notes diminish fairly rapidly and other very complex flavors develop during the aging process. Slow aging in oak cooperage or in the bottle results in the development of very complex aromas in red wines in which leather, tobacco, soy, and other related characteristics can be perceived. White wines are usually not aged as long as reds but can also develop complex aromas when stored at cool temperatures.

In sparkling wine, flavor is influenced most strongly by the time in which the wine is held in contact with the yeast. In the tank process, the wine is in contact with the yeast for only a few weeks, and the resulting sparkling wine is very fruity. In the bottle-fermented sparkling wines, the wines are usually not disgorged until after 3 years. The best sparkling wines are often disgorged after 10 years on the yeast. The extended contact with the yeast produces distinctive flavors through the slow breakdown (autolysis) of the yeast. Thus the *méthodé champenoise* wines are less fruity and have very complex flavors, such as caramel, malted, and nutty.

Although this objective approach describes wines well, it does not provide an overall preference or quality rating. Quality is a subjective and individual response, varying from person to person because of different experiences, expectations, and preferences. As a consequence, even wine experts will have differences of opinion about wine quality. Wine consumers also vary considerably in their preference ratings in that many factors other than wine flavor influence their purchasing behavior and quality judgments. Despite this, sales of many wines are influenced by the medals awarded at wine shows or by comments from wine critics.

Quality of wine is not directly related to price, which is influenced primarily by the cost of grapes and winemaking practices. The highest-quality grapes are grown in cooler climates at lower crop yields on very expensive land. Oak barrels are expensive and their maintenance is labor-

Figure 2. German white wine (Riesling) label. *Source:* SLFA Neustadt, Neustadt Germany, 1998.

Figure 3. A red wine (Pinot noir) label from the Burgundy region of France. *Source:* University of California, Davis, Department of Viticulture and Enology Archives, 1998.

intensive. Aging wines in the barrel or bottle represents a capital investment in time and assets.

With each wine-tasting experience, consumers learn more about wines and develop their own preferences and expectations, which may change over time. In the end, each person is the ultimate judge of his or her own preference and wine quality.

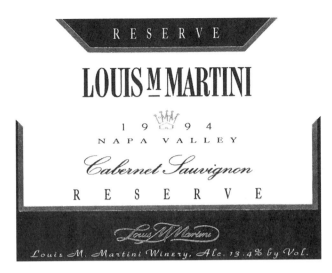

Figure 4. Cabernet Sauvignon wine label from Napa Valley, California. *Source:* Louis M. Martini Winery, St. Helena, California, 1998.

Figure 5. Cabernet Sauvignon wine label from Napa Valley, California, indicating grapes were harvested from the North Coast American Viticultural Area (AVA). *Source:* Louis M. Martini Winery, St. Helena, California, 1998.

READING WINE LABELS

Wine labels from different countries contain different information because of variation in legal regulations and appellations of origin (France, *Appellation d'Origine Contrôlée*; Spain, *Denominacion de Origen*; Italy, *Denominazione di Origine*). The first label (Fig. 2) is typical of those from Germany. *PFALZ* refers to the region, the Palatinate or Rheinpfalz, which is the second-largest wine region in Germany. The grapes for this wine were harvested in 1997 from the Johannitergarten vineyard in the town of Mussbach. The winery is a state winery (Staatsweingut). At least 85% of the grapes are Riesling. *Qualitätswein mit Prädikat* (QmP) means that it is a "quality wine with spe-

Figure 6. Cabernet Sauvignon wine label from Napa Valley, California, indicating a special vineyard as the source of the grapes. *Source:* Louis M. Martini Winery, St. Helena, California, 1998.

Table 1. Per Capita Wine Consumption by Country, 1970, 1980, and 1996

1996 rank per capita	L/gal		
	1996	1980	1970
1. Italy	61.1/16.2	79.8/21.2	110.8/29.3
2. France	59.9/15.8	90.9/24.0	108.2/28.6
3. Portugal	54.9/14.5	28.8/7.6	23.1/6.1
4. Luxembourg	53.6/14.2	N.A.	N.A.
5. Switzerland	41.1/10.9	47.0/12.4	38.9/10.3
6. Argentina	39.5/10.4	76.2/20.1	91.7/24.2
7. Slovenia	38.8/10.3	N.A.	N.A.
8. Spain	37.4/9.9	59.9/15.8	61.4/16.2
9. Austria	33.1/8.7	35.5/9.4	37.8/10
10. Romania	31.3/8.3	N.A.	N.A.
11. Greece	30.1/17.9	44.9/11.9	39.9/10.6
⋮			
15. Germany	22.9/6.1	25.5/6.7	16.2/4.3
16. Belgium	20.0/5.3	N.A.	N.A.
17. Australia	18.0/4.8	17.4/4.6	8.5/2.3
18. Chile	16.2/4.3	50.2/13.3	43.9/11.6
⋮			
25. New Zealand	10.5/2.8	4.38/1.2	2.34/0.62
26. U.S.A.	9.03/2.4	8.0/2.1	4.9/1.3
27. South Africa	8.99/2.37	9.1/2.4	11.2/3.0

Source: Ref. 1, provided by Wine Institute, San Francisco, 1998.

cial distinction" and did not have any sugar added. *Kabinett* means the grapes achieved a minimum required level of sugar. (If the grapes had been harvested at least 7 days later and were riper, the wine could have been labeled *Spätlese*). *TROCKEN* means the wine is dry (<9.0 g per liter sugar). The *A.P. Nr* . . . is the *Amtliche Prüfungsnummer* (State Institute quality control number), which identifies the details of the testing for chemical and sensory defects. The wine is 11% alcohol by volume and the bottle contains 750 mL (or 0.75 liter).

The second label (Fig. 3) is from a wine from the Burgundy region in France. The wine is a *Grand Cru* or "great growth," which is the highest-quality classification in the

Table 2. Leading Wine Producers by Country, 1996

	Hectoliters (in millions)	Gallons (in millions)
1. France	59.65	1576
2. Italy	59	1559
3. Spain	32.68	863
4. U.S.A.	21.95	580
5. Argentina	12.68	335
6. S. Africa	10	264
7. Portugal	9.53	252
8. Germany	8.3	219
9. Romania	7.66	202
10. Australia	6.78	179
11. China	4.3	114
12. Hungary	4.19	111
13. Greece	4.11	109
14. Chile	3.82	101
15. Yugoslavia	3.49	92

Source: Ref. 1, Wine Institute, San Francisco, 1998.

Table 3. Wine Consumption in the United States, Selected Years

	Total wine per capita[a] (gal)	Total wine (in million gals)	Total table wine (in million gals[b])
1997	1.95	523	462
1996	1.9	505	443
1995	1.79	469	408
1994	1.77	459	395
1993	1.74	449	381
1992	1.87	476	405
1991	1.85	466	394
1990	2.05	509	423
1989	2.11	524	432
1988	2.24	551	457
1987	2.39	581	481
1986	2.43	587	487
1985	2.43	580	378
1984	2.34	555	401
1983	2.25	528	402
1982	2.22	514	397
1981	2.2	506	387
1980	2.11	480	360
1970	1.31	267	133
1960	0.91	163	53
1950	0.93	140	36
1940	0.68	90	27

[a]All wine types, including sparkling wine, dessert wine, vermouth, other special natural wine, and table wine. Based on resident population. Per capita consumption will be higher if based on legal drinking age population.
[b]Because of changes in reporting, these numbers include all still wines not over 14% alcohol.
Note: To convert gallons to liters multiply by 3.785.
Source: Wine Institute, San Francisco, 1998.

Table 4. Wine Production by State, 1995–1997 (in Thousands of Gallons)

State	1995	1996	1997[a]	Rank 1997	% total 1997
AZ	N.A.	50	N.A.	N.A.	N.A.
AR	147	N.A.	N.A.	N.A.	N.A.
CA	397,042	418,376	422,560	1	90.57
CO	N.A.	147	110	17	0.02
CT	N.A.	64	50	20	0.01
FL	1,026	612	640	11	0.14
GA	1,241	1,163	1,120	7	0.24
ID	363	301	N.A.	N.A.	N.A.
KY	201	697	820	10	0.18
MD	N.A.	57	70	19	0.02
MA	50	68	90	18	0.02
MI	273	N.A.	N.A.	N.A.	N.A.
NC	176	190	210	14	0.05
NJ	742	843	840	9	0.18
NM	85	86	140	16	0.03
NY	21,096	25,157	23,180	2	4.97
OH	1,339	843	1,130	6	0.24
OR	1,586	1,544	1,860	4	0.4
PA	317	402	410	12	0.09
TN	132	130	150	15	0.03
TX	763	902	980	8	0.21
VT	N.A.	1,384	1,550	5	0.33
VA	391	292	320	13	0.07
WA	8,660	5,596	5,200	3	1.11
WI	119	N.A.	10	21	0
Other	1,292	1,178	5,100	N.A.	1.09
Total	435,749	458,904	461,440		100

[a]May through December are estimated.
Note: Based on volumes removed from fermenters.
Source: Ref. 2.

Gevrey–Chambertin (in the Côte d'Or region). The producer's name appears above the village name.

The third label (Fig. 4) comes from the Louis M. Martini Winery in St. Helena, California. The U.S. wine appellation of origin defines the geographical boundaries of American Viticultural Areas (AVA), which, unlike in the European system, are not an indication of the wine quality. By California law, 100% of the grapes must be grown in California if any California place name appears on the label. A minimum of 85% of the grapes used in making this wine must come from the AVA, Napa Valley. To be bottled as a cabernet sauvignon varietal wine, a minimum of 75% cabernet sauvignon grapes must have been used. To be vintage labeled, 95% of the wine must have come from grapes harvested in the vintage year, 1994. The word "Reserve" has no legal meaning in the United States but implies that the wine maker has used his best grapes and aged the wines longer before release.

The fourth label (Fig. 5) comes from the Louis M. Martini Winery in St. Helena, California. A minimum of 85% of the grapes used in making this wine must come from the AVA, North Coast.

The fifth label (Fig. 6) comes from the Louis M. Martini Winery in St. Helena, California. The vineyard designation here reflects that the winery chose to bottle the wine made from this particular vineyard separately and does not imply that this vineyard is an AVA on its own.

Burgundy area. Grand Cru wines can be named using only the name of the winery on the label, here *Charmes—Chambertin*. The regulations that determine the variety of grape (Pinot noir) and regulate specific viticultural and enological practices are defined by this appellation of origin (*Appellation Contrôlée*). The winery is located in the village

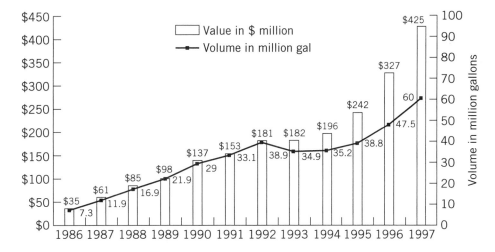

Figure 7. Dollar value (in millions US$) and volume (in millions of gallons) of U.S. exports from 1986 to 1997. In 1992, there was a decrease in the volume of U.S. wine exports. The price per gallon held steady at about $5/gal from 1986 to 1992 and then began a steady increase to approximately $7/gal in 1997.

Table 5. Commercially Produced Wines Entering U.S. Distribution Channels, Calendar Year Average (in Thousands of Gallons)

Type	Domestic Production 1985–1986	1995–1996	Imports 1985–1986	1995–1996	Total 1985–1986	1995–1996
Table	267,088	317,277	97,765	68,267	364,853	385,544
Sparkling	30,171	22,247	15,029	8,038	45,200	30,285
Dessert	31,259	18,034	3,380	2,473	34,639	20,507
Vermouth	3,804	2,411	2,845	1,823	6,649	4,234
Other spec (not over 14%)	12,793	17,431	2,603	491	15,396	17,922
Total	*345,115*	*377,400*	*121,622*	*81,092*	*466,737*	*458,492*

Source: Gomberg, Fredrikson & Associates from reports of the California State Board of Equalization, Bureau of Alcohol, Tobacco and Firearms, U.S. Treasury Dept., Bureau of Census, U.S. Dept. of Commerce. Data provided by Wine Institute, San Francisco, 1998.

PRODUCTION AND CONSUMPTION STATISTICS

Per capita consumption of wine in the United States is low (Table 1) compared to the other major wine-producing nations. The United States, in 1996, was fourth in world production (Table 2) but 26th in per capita consumption. In addition to the top ten consuming countries, Table 2 lists other countries that are both major wine producers and low per capita consumers. Overall, the global per capita consumption has decreased sharply since the 1980s, especially in France and Italy, who still, nonetheless, retain their prominent positions as the leading consumers and producers of wine. U.S. consumption declined by 20% from the mid-1980s high of 2.43 gallons per person annually to 1.95 gallons per capita in 1997 (Table 3).

Ninety one percent (410 million gallons) of the wine produced in the United States comes from California, which accounts for 71% of U.S. retail sales ($11.4 billion). Considerable development of the wine industry in other states occurred during the decade from 1980 to 1990. Consequently, increased plantings outside California, as well as shipment of California grapes and grape concentrate to other states for the purposes of wine production, catalyzed industry expansion into approximately 40 states. Table 4 lists the top 25 states in wine production from 1995 to 1997.

As one of the world's leading producers, the United States has witnessed a steady increase in the export of its wines (Figure 7). Despite the millions of gallons produced for both internal consumption and exportation, approximately 18% of the wine on the American market is imported (Table 5). However, the amount of imported wines in all categories (table, sparkling, dessert, etc) has declined by 33%, whereas domestic production of table and other special wines has increased by 20% and the domestic sparkling, dessert, and vermouth production has decreased by 23%.

BIBLIOGRAPHY

1. *International Review*, Vol. 71, Office International de la Vigne et Vin, Paris, 1998.

2. *U.S. Wine Stats* 2–16, Bureau of Alcohol, Tobacco and Firearms, Washington, D.C., 1997.

GENERAL REFERENCES

R. B. Boulton et al., *Principles and Practices of Winemaking*, Chapman and Hall, New York, 1996.

B. G. Coombe and P. R. Dry, eds., *Viticulture; Vol. 1 Resources; Vol. 2 Practices*, Winetitles, Adelaide, Australia, 1988.

K. C. Fugelsang, *Wine Microbiology*, Chapman and Hall, New York, 1997.

K. C. Fugelsang, B. H. Gump, and F. S. Nury, *Wine Analysis and Production*, Chapman and Hall, New York, 1996.

R. Jackson, *Wine Science: Principles and Applications*, Academic Press, New York, 1994.

H. Johnson, *The World Atlas of Wine*, 4th ed., Simon & Schuster, New York, 1994.

H. Johnson, *Hugh Johnson's Modern Encyclopedia of Wine*, 4th ed., Simon & Schuster, New York, 1998.

S. Koplan, B. H. Smith, and M. A. Weiss, *Exploring Wine. The Culinary Institute of America's Complete Guide to Wines of the World*, Van Nostrand Reinhold, New York, 1996.

C. S. Ough, *Winemaking Basics*, Food Products Press, New York, 1991.

C. S. Ough and M. A. Amerine, *Methods for Analysis of Musts and Wines*, 2nd ed., John Wiley and Sons, New York, 1988.

C. A. Rice Evans and L. Packer, eds., *Flavonoids in Health and Disease*, Marcel Dekker, New York, 1998.

J. Robinson, ed., *The Oxford Companion to Wine*, Oxford University Press, New York, 1994.

A. C. NOBLE
J. F. PFEIFFER
University of California
Davis, California

SUPPLEMENT

GENETIC ENGINEERING: ANIMALS

Animals in many capacities have been important components of human enterprise since prehistory, as food, shelter, transport, work, companionship, and, in more recent times, to determine safety and efficacy of therapeutics and as models to study disease. The development of the capacity to modify animals at the molecular level has expanded their roles, especially in the latter areas, and has added a new dimension to this compendium, "molecular pharming," the production of valuable products in milk.

For most of our long intertwined history, animals' most consistent contribution has been within the agricultural arena. With increased social awareness, we have begun to question and debate their role in this capacity, and with the advent of genetic engineering this has taken on a new level of complexity. This paper will provide a reasonable overview of the development and use of transgenic animals and will touch on the ethical and societal implications but will leave a more expanded analysis of these issues to more qualified reviewers.

New technologies have been applied in agriculture and food production as they evolved. Genetic engineering through the application of recombinant DNA methods and the use of DNA typing to aid traditional breeding are the new technologies currently having the greatest impact. In the latter half of the twentieth century, major improvements in agricultural productivity have been largely based on selective breeding programs for plants and animals; intensive use of chemical fertilizers, pesticides, and herbicides; advanced equipment developments; and widespread irrigation programs. This has been a very successful model at raising productivity, yet these improvements have brought corresponding problems of increasing uniformity in the genetic base of crop plants and domestic animals, pests resistant to chemical pesticides, adverse impacts on environmental quality, and capital-intensive production. In addition, from a global perspective, these advances have been the prerogative of more affluent regions. Farmers in developing countries have not had access to many of these technologies and capital-intensive methods of production. The emerging biotechnology revolution is stimulating hope that it will provide the basis for more sustainable agriculture.

The new aspects of biotechnology differ from previous agricultural technologies in a number of ways that are likely to be of significance. First, biotechnology can enhance productivity and product quality with minimal increase in production costs by providing tools to more effectively develop and incorporate desired characteristics of plants and animals. Second, biotechnology can provide more effective mechanisms for improving animal health. Third, biotechnology has the potential to conserve natural resources and improve environmental quality by reducing the dependency of chemical inputs through the development of natural fertilizers and of pest-resistant plants and

by contributing to the functioning of biocontrol systems. Finally, biotechnology can provide modified organisms to degrade wastes and pollutants and for biomass conversion to reduce dependency on nonrenewable resources.

In agriculture, biotechnology in the form of recombinant DNA technology is a powerful assistant in traditional plant and animal breeding. Traditional breeding programs are time-consuming, labor-intensive, and limited to transfers of genes between closely related species. In addition, because the breeder has no control at the genome level, many undesirable traits can also be incorporated, such as increased disease susceptibility and slower growth. Recombinant DNA technology permits the precise and predictable manipulation of genes. Single traits can be modified much more quickly than was possible using traditional selection and breeding methods alone. Because of the capability of moving genes between species, desirable traits from one organism can be transferred to another.

There is no evidence that rDNA techniques or rDNA-modified organisms pose any unique or unforeseen environmental or health hazards. In fact, a National Research Council study found that "as the molecular methods are more specific, users of these methods will be more certain about the traits they introduce into plants." Greater certainty means greater precision in safety assessments.

In a world whose population is increasing at a rate that threatens to confirm Malthus's worst predictions, it is hard to envisage feeding and sustaining these numbers in a livable environment without the use of biotechnology. It is difficult to imagine a "promising alternative" to biotechnology and industrial agriculture that will sustain such numbers without catastrophic consequences.

The application of recombinant DNA technology to the modification of organisms was first demonstrated in 1972 when Paul Berg of Stanford used enzymes to paste two DNA strands together to form a hybrid circular molecule. This was the first recombinant DNA molecule. The following year Stanley Cohen and Annie Chang of Stanford University and Herbert Boyer of UCSF "spliced" sections of two different plasmids (an extra-chromosomal piece of DNA) cut with the same restriction enzyme, creating a plasmid with dual antibiotic resistance. They then transformed bacteria with the plasmid, thereby producing the first recombinant DNA organism.

Since that time, rapid advances have been made in the application of rDNA techniques to increasingly complex organisms, from bacteria and yeasts to plants and mammalian species. Transgenic animals have tremendous potential to act as valuable research tools in the agricultural and biological sciences. They can be modified specifically to address scientific questions that were previously difficult if not impossible to determine.

Though the first directed "engineering" of animals was selection of animals with desirable traits for breeding purposes, there is no doubt that the first scientific contribution to reproductive physiology in animals was the successful attempt to culture and transfer embryos in 1891. The development of artificial insemination helped with the costs and control of breeding, but the first technological shift came with Gurdon's 1970 transfer of a nucleus of a somatic adult frog cell into an enucleated frog ovum and the birth of viable tadpoles. This experiment was of limited success in that none of the tadpoles developed into adult frogs. In 1977 Gurdon expanded the field further through the transfer of messenger RNA (mRNA) and DNA into toad (*Xenopus*) embryos where he observed that the transferred nucleic acids were expressed (turned on) (1). Also in the 1970s, Ralph Brinster developed a now-common technique used to inject stem cells into embryos. When these embryos became adults, they produced offspring carrying the genes of the original cells. In 1982, Brinster with his colleagues gained further renown by transferring genes for rat growth hormone into mice under the control of a mouse liver–specific promoter and producing mice that grew into "supermice"—twice their normal size (2).

During the two years 1980 and 1981, there were several reported successes at gene transfer and the development of transgenic mice. Gordon and Ruddle first coined the term transgenic to describe animals carrying exogenous genes integrated into their genome (3). Since that time this definition has been extended to include animals that result from the molecular manipulation of endogenous genomic DNA, including all techniques from DNA microinjection to embryonic stem (ES) cell transfer and "knock-out" mouse production.

Notwithstanding the advent of successful nuclear transfer technology with the dawn of Dolly (the cloned sheep), the most widely used technique for the production of transgenic animals including mice is by microinjection of DNA into the pronucleus of a recently fertilized egg. Using various transgenic tools such as antisense technology (putting a reverse copy to switch off expression), it is now possible to add a new gene to the genome, increase the level of expression or change the tissue specificity of expression of a gene, and decrease the level of synthesis of a specific protein. An additional factor added by the new nuclear transfer technology is the capability of removing or altering an existing gene via homologous recombination. See Table 1 for different mechanisms of creating transgenic animals.

Since Palmiter's mouse, transgenic technology has been applied to several species, including agricultural species such as sheep, cattle, goats, pigs, poultry, and fish (Table 2). The applications for transgenic animal research fall broadly into two distinct areas, namely, medical and agri-

Table 1. Transgenic Technologies

Integration of retroviral vectors into an early embryo
Injection of DNA into the pronucleus of a newly fertilized egg
Incorporation of genetically manipulated embryonic stem cells
  into an early embryo
Incorporation of genetically manipulated primordial germ cells
  into an early embryo
Sperm delivery
Nuclear transplantation
Microprojectile injection

Table 2. Examples of Species to Which Transgenic Technology Has Been Applied

Mammals	Birds	Fish
Mice	Chickens	Salmon
Rats	Japanese quail	Trout
Rabbits		Talapia
Cattle		Carp
Pigs		Channel catfish
Sheep		Medaka
Goats		Zebrafish
		Loach
		Goldfish
		Northern Pike

Source: Ref. 4.

cultural applications. The recent focus on developing animals as bioreactors to produce valuable proteins in their milk can be cataloged under both areas. Underlying each of these, of course, is a more fundamental application, that is, the use of those techniques as tools to ascertain the molecular and physiological bases of gene expression and animal development. This understanding can then lead to the creation of techniques to modify development pathways.

As noted by Pinkert and Murray, there are still fundamental limitations to the widespread use of transgenic technology in all animals except the mouse. Limitations include (1) lack of knowledge concerning the genetic basis of factors limiting production traits, (2) identification of tissue and developmentally specific regulatory sequences for use in developing gene constructs and expression vectors and in gene targeting, and (3) establishment of novel methods to increase efficiency of transgenic animal production. In medicine no doubt the most versatile animal model has been the mouse, and this especially applies because the technology has been perfected to customize the engineering of mice. Up until recently the principal advantage that the mouse held over all other species was the ability to isolate with relative ease ES cells, which remain pluripotent and amenable to engineering. ES cells are derived from the inner cell mass of the blastocyst formed during early embryogenesis. Distinguished from all other stem cells, they are pluripotent, able to develop into virtually any and all cells and tissues in the body; and, consistent with their expression of telomerase, self-renewing, a potentially limitless source of cells. One of the areas in which the mouse has been supreme is the ability, using these ES cells, to target with great specificity regions within chromosomes via what is termed homologous recombination. Using this method, researchers can (1) incorporate a novel foreign gene into a mouse genome, (2) modify an endogenous gene, or (3) delete a portion of a specific endogenous gene, creating a "loss-of-function" mutant, termed a "knock-out" mouse, from which to study phenotypic effects of inactivating genes. This flexibility makes the knock-out mouse a really good model from which to study disease and development in humans.

Up until November 1998, isolating ES cells in other mammals proved elusive, but in a milestone paper in the November 5, 1998, issue of *Science*, James A. Thomson, a developmental biologist at the University of Wisconsin–Madison, reported the first successful isolation, derivation, and maintenance of a culture of human ES cells (hES cells) (5). Obviously this is not the focus of this chapter, but it is interesting to note as an aside the leap made from mouse to man. As Thomson himself put it, these cells are different from all other human stem cells isolated to date, and as the source of all cell types, they hold great promise for use in transplantation medicine, drug discovery and development, and the study of human developmental biology.

The other great advance in the field of reproductive biology and methods for genetic engineering was Ian Wilmut's landmark work using nuclear transfer technology to generate the lambs Morag and Megan, reported in 1996 (from an embryonic cell nuclei), and the truly groundbreaking work of creating Dolly from an adult cell nucleus, reported in February 1997 (6). With the birth of Dolly, the sheep, Wilmut and his colleagues at the Roslin Institute demonstrated for the first time that the nucleus of an adult cell can be transferred to an enucleated egg to create cloned offspring. This technology supports the production of genetically identical and genetically modified animals. Thus, the successful "cloning" of Dolly has captured the imagination of researchers around the world. This technological breakthrough should play a significant role in the development of new procedures for genetic engineering in a number of mammalian species. It should be noted that nuclear cloning, with nuclei obtained from either mammalian stem cells or differentiated "adult" cells, is an especially important development for transgenic animal research. Since that time the clones have been arriving rapidly, with specific advances made by a Japanese group who used cumulus cells rather than fibroblasts to clone calves. They found that the percentage of cultured, reconstructed eggs that developed into blastocysts was 49% for cumulus cells and 23% for oviductal cells. These rates are higher than the 12% previously reported for transfer of nuclei from bovine fetal fibroblasts. Following on the heels of Dolly, Polly and Molly became the first genetically engineered transgenic sheep produced through nuclear transfer technology. Polly and Molly were engineered to produce human factor IX (for hemophiliacs) by transfer of nuclei from transfected fetal fibroblasts. Until then germline-competent transgenics had been produced in mammalian species, other than mice, only using DNA microinjection.

DNA microinjection has not been a very efficient mechanism to produce transgenic mammals. However, in November 1998, a team of Wisconsin researchers reported a nearly 100% efficient method for generating transgenic cattle. The established method of cattle transgenes involves injecting DNA into the pronuclei of a fertilized egg or zygote. In contrast, the Wisconsin team injected a replication-defective retroviral vector into the perivitelline space of an unfertilized oocyte. The perivitelline space is the region between the oocyte membrane and the protective coating surrounding the oocyte known as the zona pellucida.

In addition to ES cells, other sources of donor nuclei for nuclear transfer might be used such as embryonic cell lines, primordial germ cells, or spermatogonia to produce offspring. The utility of ES cells or related methodologies to provide efficient and targeted *in vivo* genetic manipulations offer the prospects of profoundly useful animal models for biomedical, biological, and agricultural applications. The road to such success has been most challenging, but recent developments in this field are extremely encouraging.

From Table 3 it can be seen that the development of transgenic animals has tremendous potential in medical research and therapeutic applications, including the creation of models to study disease, early development, aging, and specific gene function and the production of valuable proteins in milk (Tables 4 and 5). For example, pigs have been engineered as large-animal models for studying cone photoreceptor survival and degeneration in retinitis pigmentosa. A novel use also includes the creation of engineered animals where the surface antigens of the organs, such as the heart, have been altered so that they can be used for transplantation without rejection by the recipient because the latter's immune system will not see the transplants as foreign.

AGRICULTURAL APPLICATIONS

Since 1985, transgenic farm animals harboring growth-related gene constructs have been created, although ideal growth phenotypes were not achieved because of an inability to coordinately regulate either gene expression or the ensuing cascade of endocrine events.

Using DNA microinjection, the types of genes and regulatory sequences introduced into livestock species become important considerations. To date, for agricultural species, the types of transgenes used fall into two main types, those encoding growth factors and those encoding proteins for expression in the mammary gland.

The work with growth factors was carried out in an attempt to alter the efficiency of meat production and alter the partitioning of nutrient resources toward increased lean production. In addition to the work with livestock transgenic for growth factor, considerable effort has been directed toward increasing the efficiency of wool growth in Australian sheep by insertion of the two bacterial or yeast genes required for sheep to synthesize *de novo* the sulfur amino acid cysteine.

Milk is one of the principal targets for engineering productivity traits. This includes altering the properties or proportions of caseins, lactose, or butterfat in milk of transgenic cattle and goats. Another focus is engineering

Table 3. Use of Transgenic Animals in Medicine

Animal models to study

Disease—cancer, cystic fibrosis, bovine spongiform encephalopathy (BSE), sickle cell anemia

Development—embryology, aging

Epigenetic changes—imprinting and telomere shortening

Gene targeting—gene function, gene therapy

"Pharming"—producing valuable proteins in milk

Xenotransplants—immunocompatible organs for transplantation from animals

Table 4. Human Proteins in Animals

Protein	Use	Animal(s)
α-1-anti-protease inhibitor	α-1-antitrypsin deeficiency	Goat
α-1-antitrypsin	Anti-inflammatory	Goat, sheep
Antithrombin III	Sepsis and disseminated intravascular coagulation (DIC)	Goat
Collagen	Burns, bone fracture, incontinence	Cow
Factor VII and IX	Hemophilia	Sheep, pig
Fibrinogen	"Fibrin glue," burns, surgery localized chemotherapeutic drug delivery	Pig, sheep
Human fertility hormones	Infertility, contraceptive vaccines	Goat, cow
Human hemoglobin	Blood replacement for transfusion	Pig
Human serum albumin	Burns, shock, trauma, surgery	Goat, cow
Lactoferrin	Bacterial gastrointestinal infection	Cow
LAtPA	Venous stasis ulcers	Goat
Monoclonal antibodies	Colon cancer	Goat
Tissue plasminogen activator	Heart attacks, deep vein thrombosis, pulomary embolism	Goat

Source: Table modified from GEN.

Table 5. Transgenic Animals and Protein Yields, Estimated Herd Size for Selected Proteins Produced in the Milk of Transgenic Animals

Protein	Estimated annual need (kg)	Estimated herd size	Animal
α-1-antitrypsin	5,000	33,000	Sheep
		42,000	Goat
Factor IX	2	13	Sheep
		22	Pig
Human serum albumin	1,000	300	Cow
		8,300	Goat

Note: The number of lactating animals may be calculated as (expression in g/L) × (annual milk production in L) × (purification efficiency)

Table 6. Use of Transgenic Animals in Agriculture

To create tissue-specific gene expression
To study development
To improve productivity
To increase disease resistance
To improve meat and milk quality
To produce valuable proteins in milk, blood, or urine
To preserve rare or threatened animals
To cross species barriers

for enhanced resistance to viral and bacterial diseases, including development of "constitutive immunity," or germ-line transmission of specific, modified antibody genes. Castilla et al. reported in the April 1998 issue of *Nature Biotechnology* the generation of transgenic mice that secrete virus-neutralizing antibodies in their milk (7). These antibodies were directed against transmissible gastroenteritis coronavirus (TGEV). TGEV infection is an important disease in swine that causes a mortality close to 100% in 3-week-old piglets and severe diarrhea in young pigs. Therefore, the development of transgenic sows that synthesize virus-neutralizing antibodies in their milk could reduce the serious effects resulting from TGEV infection in newborn swine.

Work on the directed expression of new proteins with pharmaceutical value to the mammary gland of cattle, goats, pigs, and sheep has been more successful. A number of pharmaceutically important proteins have been expressed in the mammary gland (see Table 4).

The production of transgenic farm animals is costly and time-consuming (Table 6). In mouse experiments, less than 2 months is required from the time the construct is ready for microinjection through weaning of baby mice. In contrast, for pig experiments, 1 month to a year is required for a sufficient number of DNA injections and recipient transfers to ensure the likelihood of success. In addition,

the time frame from birth of a founder transgenic animal to the establishment of lines can be 1 to 2 years for pigs, sheep, and goats to 4 to 5 years for cattle. Neal First, UW-Madison, has come up with an ingenious mechanism for testing the feasibility of developing specific transgenics before heading down the expensive road of genetically engineering an entire animal. He first does a test run by using replication-defective retrovirus vectors to transiently express gene constructs in the mammary gland of animals. He has used an alpha virus containing a human growth hormone gene to produce human growth hormone in the milk of a goat. However, the transformed cells are sloughed off over time and the process needs to be repeated, but it does establish whether or not the construct will work.

Transgenic techniques have been developed for other vertebrate species in addition to mammals, including poultry and fish. In both instances, genetic selection is an exceedingly slow process. Because DNA microinjection into pronuclei of embryonic cells in poultry is not feasible, transfection using similar retroviruses to that used by Neal First has prevailed. As described, methods have included transfection of genes into cells of embryonic blastoderm, insertion of genes using replication-competent retroviruses, the use of replication-defective retroviruses, and sperm-mediated gene transfer. Although the last method has come under critical dispute, the other methods have led to the development of experimental models.

In contrast to poultry studies, work with fish has moved ahead with far greater speed. The principal area of research has focused on growth performance, and initial transgenic growth hormone (GH) fish models have demonstrated accelerated and beneficial phenotypes. DNA microinjection methods have propelled the many studies reported and have been most effective due to the relative ease of working with fish embryos. Bob Devlin's group in Vancouver has demonstrated extraordinary growth rate in coho salmon that were transformed with a growth hormone from sockeye salmon. The transgenics achieve up to 11 times the size of their littermates within 6 months, reaching maturity in about half the time. Interestingly, this dramatic effect is observed only in feeding pens where the transgenics' ferocious appetites demand constant feeding. If the fish are left to their own devices and must forage for themselves, they appear to be outcompeted by their smarter siblings.

DOMESTIC ANIMALS AS BIOREACTORS

The second general area of interest has been the development of lines of transgenic domestic animals for use as bioreactors. One of the main targets of these so-called "gene pharming" efforts has involved attempts to direct expression of transgenes encoding biologically active human proteins. In such a strategy, the goal is to recover large quantities of functional proteins that have therapeutic value from serum, urine, or the milk of lactating females. To date, expression of foreign genes encoding lysozyme, α-1-antitrypsin, tissue plaminogen activator, clotting factor IX, and protein C were successfully targeted to the mammary glands of goats, sheep, cattle, or a combination of these swine, (Tables 4 and 5). Similarly, lines of transgenic pigs and mice have been created that produce human hemoglobin or specific circulating immunoglobulins. The ultimate goal of these efforts is to harvest proteins from the serum of transgenic animals for use as important constituents of blood transfusion substitutes, or for use in diagnostic testing.

THE FUTURE

There is no question that the use of transgenic animals in medical research has already made contributions in helping us understand physiological processes. It shows much promise for the future in providing tools to elucidate the mechanisms of development, disease, and aging; to develop effective therapeutics, including gene therapy, and effective approaches to disease prevention and health maintenance; and to generate animals for the production of valuable proteins and modified organs for xenotransplantation. The recent acquisition of Roslin Bio-Med by Geron juxtaposes three of the major advances in this field in the last 2 years, namely, nuclear transfer technology; pluripotent stem cell, and telomerase, which allows cells to remain viable. The combined technologies are expected to enhance and accelerate the development of new transplantation therapies for numerous degenerative diseases such as diabetes, Parkinson's disease, cancer, and heart disease.

On the agriculture end, however, while transgenic animal technology continues to open new and unexplored frontiers, it also raises questions concerning regulatory and commercialization issues as demonstrated by molecular farming efforts. A number of major regulatory and public perception hurdles exist that may affect the time to commercialization of transgenic animals. These include safety issues related to food produced from genetically engineered animals and ethical considerations such as animal welfare. So the commercialization of genetically engineered animals in agriculture will have several hurdles to overcome on the way to market, including technical, regulatory, and consumer perception. Considering the long and winding path that genetically modified crops took on the road to acceptance, it is expected that for animals the path will be even more convoluted and fraught with obstacles. The potential is so great, however, that I remain optimistic that the future looks bright.

BIBLIOGRAPHY

1. J. B. Gurdon, "Nuclear Transplantation and Gene Injection in Amphibia," *Brookhave Symposia in Biology* **29**, 106–115 (1977).

2. R. D. Palmiter et al., "Dramatic Growth of Mice that Develop From Eggs Microinjected with Metallothionein-Growth Hormone Fusion Genes," *Nature* **300**, 611–615 (1982).

3. J. W. Gordon and F. H. Ruddle, "Integration and Stable Germline Transmission of Genes Injected into Mouse Pronuclei," *Science* **214**, 1244–1246 (1981).

4. C. A. Pinkert and J. D. Murray, "Transgenic Farm Animals," in J. D. Murray et al., eds., *Transgenic Animals in Agriculture*, CABI, London, 1999, pp. 1–18.

5. J. A. Thomson et al., "Embryonic Stem Cell Lines Derived from Human Blastocysts," *Science* **282**, 1145–1147 (1998).

6. I. Wilmut et al., "Viable Offspring Derived from Fetal and Adult Mammalian Cells," *Nature* **385**, 810–813 (1997).

7. J. Castilla et al., "Engineering Passive Immunity in Transgenic Mice Secreting Virus-Neutralizing Antibodies in Milk," *Nature Biotech.* **16**, 349–354 (1998).

GENERAL REFERENCES

R. L. Brinster, "Stem Cells and Transgenic Mice in the Study of Development," *Int. J. Dev. Biol.* **37**, 89–99 (1993).

M. R. Cappechi, "Altering the Genome by Homologous Recombination," *Science* **244**, 1288–1292 (1989).

A. W. S. Chan et al., "Transgenic Cattle Produced by Reverse-Transcribed Gene Transfer in Oocytes," *Proc. Natl. Acad. Sci.* **95**, 14028–14033 (1998).

R. A. Dunham, and R. H. Devlin, "Comparison of Traditional Breeding and Transgenesis in Farmed Fish with Implications for Growth Enhancement and Fitness," in J. D. Murray et al., eds., *Transgenic Animals in Agriculture*, CABI, London, 1999, pp. 209–230.

R. E. Hammer et al., "Production of Transgenic Rabbits, Sheep and Pigs by Microinjection," *Nature* **315**, 680–683 (1985).

Y. Kato et al., "Eight Calves Cloned from Somatic Cells of a Single Adult," *Science* **282**, 2095–2098 (1998).

E. A. Maga and J. D. Murray, "Mammary Gland Expression of Transgenes and the Potential for Altering the Properties of Milk," *Bio / Technol.* **13**, 1452–1457 (1995).

M. M. Perry and H. M. Sang, "Transgenesis in Chickens," *Transgenic Res.* **2**, 125–133 (1993).

R. M. Petters et al., "Genetically Engineered Large Animal Model for Studying Cone Photoreceptor Survival and Degeneration in Retinitis Pigmentosa," *Nature Biotechnol.* **15**, 965–970 (1997).

C. A. Pinkert, "Transgenic Pig Models for Xenotransplantation," *Xeno* **2**, 10–15 (1994).

C. A. Pinkert, "The History and Theory of Transgenic Animals," *Lab Anim.* **26**, 29–34 (1997).

C. A. Pinkert, M. H. Irwin, and R. J. Moffatt, "Transgenic Animal Modeling," in R. A. Meyers, ed., *Encyclopedia of Molecular Biology and Molecular Medicine*, Vol. 6, VCH Publishers, New York, 1997, pp. 63–74.

V. G. Pursel and C. E. Rexroad, Jr., "Status of Research with Transgenic Farm Animals," *J. Anim. Sci.* **71** (Suppl. 3), 10–19 (1993).

V. G. Pursel et al., "Genetic Engineering of Livestock," *Science* **244**, 1281–1288 (1989).

G. E. Rogers, "Improvement of Wool Production through Genetic Engineering," *Trends Biotechnol.* **8**, 6–11 (1990).

A. E. Schnieke et al., "Human Factor IX Transgenic Sheep Produced by Transfer of Nuclei from Transfected Fetal Fibroblasts," *Science* **278**, 2130–2133 (1998).

R. M. Shuman, "Production of Transgenic Birds," *Experientia* **47**, 897–905 (1991).

K. R. Thomas and M. R. Capecchi, "Site-Directed Mutagenesis by Gene-Targeting in Mouse Embryo-Derived Stem Cells," *Cell* **51**, 503–512 (1987).

R. J. Wall, H. W. Hawk, and N. Nel, "Making Transgenic Livestock: Genetic Engineering on a Large Scale," *J. Cellular Biochem.* **49**, 113–120 (1992).

S. A. Wood et al., "Simple and Efficient Production of Embryonic Stem Cell–Embryo Chimeras by Coculture," *Proc. Natl. Acad. Sci.* **90**, 4582–4585 (1993).

MARTINA MCGLOUGHLIN
University of California
Davis, California

NUTS

Since ancient times, nuts and seeds have been a staple in many healthy diets. Worldwide, these foods have been consumed in main dishes, in desserts, and alone as snacks. Nuts contain a variety of nutrients and are important in a healthy diet, especially for those who are consuming mainly plant-based diets.

Nuts are classified as either tree nuts or peanuts. Tree nuts are actually a one-seeded fruit in a hard shell. Tree nuts include cashews, almonds, hazelnuts, Brazil nuts, macadamia nuts, pecans, pine nuts, walnuts, and pistachio nuts.

WHERE THE SCIENCE BEGAN

Many large population studies show that the evidence linking frequent peanut and nut consumption to a 25 to 50% reduced risk of heart disease is strong. A landmark study in 1992 at Loma Linda University examined the diets of approximately 27,000 Seventh-Day Adventists. This study was one of the first to suggest that nut eaters are a healthy group. Researchers looked at the relationship between 65 different foods and coronary heart disease. Out

of all the foods studied, nuts by far had the strongest protective effect on the risk of having a heart attack or dying of heart disease. People eating nuts more than 5 times a week decreased their risk of heart disease by over 50%; those eating nuts 1 to 4 times a week decreased their risk by 27% (1). A further diet analysis showed that 32% of the nuts consumed by study participants were peanuts.

Participants enjoyed a wide variety of nuts, including peanuts, almonds, and walnuts. They ate them out of hand as a snack, in fruit shakes, and cooked in food. Interestingly, the participants who consumed nuts were significantly thinner than the rest of the group. Though this does not prove that eating nuts makes us thinner, it shows that it does not necessarily make us obese.

This study was validated by the Iowa Women's Study, which looked at the eating habits of 34,000 women. Over a 5-year period, the researchers found an inverse relationship between nut consumption and death from heart disease. The scientists determined that women who ate nuts were only 40% as likely to die from heart disease as those who never ate nuts (2).

The DASH diet, an eating plan that has been clinically proven to lower blood pressure in hypertensive men and women, may also be effective in preventing hypertension, according to a study in the *New England Journal of Medicine* (3). Although the DASH diet is an entire eating plan, a significant element is that it recommends eating 4 to 5 servings from the nuts, seeds, and legumes group each week.

Another study, in Australia, found that people who consumed diets rich in monounsaturated fats, mainly from peanuts, experienced a larger decline in LDL cholesterol as compared to the low-fat diet. Instead of raising triglyceride levels as on the low-fat diet, the peanut-enriched-diet eaters lowered their triglyceride levels (4).

Researchers note the favorable fatty acid profile of peanuts and nuts—low in saturated fat and high in monounsaturated fats (MUFA) and polyunsaturated fats—as one possible explanation for their protection against heart disease.

More recent research has further confirmed the relationship between consuming nuts and preventing heart disease. In the Nurses' Health Study published in the November 1998 *British Medical Journal*, researchers at Harvard School of Public Health found that "frequent consumption of peanuts and nuts was associated with a lowered risk of coronary heart disease (CHD)" (5). This is one of the most prestigious research studies, looking at the dietary habits of more than 86,000 female nurses. Those who ate more than 5 servings of nuts, peanuts, and peanut butter each week decreased their risk of heart disease by about one-third, compared to women who rarely or never ate nuts.

Preliminary results of the Physicians' Health Study, presented at the American Heart Association 1998 Scientific Sessions, found similar results. In a group of 22,000 male physicians, the risk of cardiac death and sudden death decreased as nut consumption increased (6).

Different types of clinical studies using a variety of nuts, such as peanuts, macadamia nuts, and walnuts, demonstrate that consumption of nuts can lower LDL cholesterol levels and possibly reduce the risks contributing to both fatal and nonfatal coronary heart disease (2,7–10).

PEANUTS IN A WEIGHT-LOSS DIET

Penny Kris-Etherton, professor of nutrition at Pennsylvania State University, found that a weight-loss diet high in MUFA provided by peanuts and peanut butter not only lowered cholesterol but also helped subjects lose weight. Participants on both low-fat (less than 20% calories from fat) and higher-fat (35% of calories from fat, which was mainly monounsaturated) diets lost an average of 2 lb per week (11). Both diets lowered total and LDL cholesterol, known risk factors for cardiovascular disease.

In this study, the healthy MUFA was added to the diets by using peanut butter instead of butter and jam on bagels and toast; snacking on peanuts instead of pretzels, cookies, or crackers, using peanut butter in sandwiches instead of lean luncheon meats; and using peanut-based dressing with salads and vegetables.

According to Kris-Etherton, "Consumers should understand that 'fat-free' does not always translate into weight loss and that healthy diets can include favorite foods, such as peanuts and peanut butter, while promoting weight loss and weight maintenance."

NUTS HELP PEOPLE STICK TO WEIGHT-LOSS DIETS

A real-life example of the benefits of adding peanuts and peanut butter to a healthy diet is a study conducted by Frank Sacks and Kathy McManus at Harvard Medical School and Brigham and Women's Hospital in Boston. Almost 3 times as many people were able to stick to a higher-fat diet that included peanuts and peanut butter during an 18-month free-living weight-loss study (12). One hundred one overweight men and women were assigned to either of two weight-loss diets: (1) a low-fat diet and (2) a higher MUFA "Mediterranean-style" diet.

After 18 months, the researchers found that subjects on the Mediterranean-style diet fared better throughout the study period. Those on the higher-fat Mediterranean-style diet were able to lose an average of 11 lb each and maintain their weight loss, whereas those following the low-fat diet gained almost half of the lost weight back, resulting in only a 6-lb weight loss. In addition, more than 80% of the participants on the low-fat diet dropped out of the study compared to fewer than half (46%) of the subjects on the higher-fat diet (12).

The Mediterranean-style diet included foods high in heart-healthy MUFA such as peanuts, peanut butter, peanut oil, nuts, avocados, and olive oil on a daily basis. When allowed to choose their own MUFA-rich foods, subjects consistently chose peanuts, peanut butter, and mixed nuts, favorite foods that used to be on a dieter's "forbidden" list.

Interestingly, those on the higher-fat diet ate more vegetables, fiber, and protein. This may be due to the fact that Mediterranean dishes typically consist of many vegetables and whole grains, sprinkled with peanuts and nuts or drizzled with peanut or olive oil to emphasize palatability.

McManus, manager of clinical nutrition at Brigham and Women's Hospital, adds, "As obesity becomes an epidemic in America, it is more important than ever to identify eating patterns that can not only promote weight loss, but help sustain that weight loss for a lifetime. Patients were delighted to be able to eat peanuts and peanut butter again" (12).

PEANUTS AND NUTS SATISFY HUNGER

Rick Mattes, a satiety expert at Purdue University, examined why peanuts and peanut butter may help people stick to weight-loss programs. Mattes and collaborator Corinna Lermer found that when 500 cal of peanuts were added to participants' regular diets, substituted in the diet for other fat, or eaten freely, the results were the same—the men and women automatically compensated for most of the additional calories, and they spontaneously commented on the high satiety of the peanuts (13). Those who either added peanuts to their regular diets or substituted peanuts for other fats had the added benefit of significantly lowering their triglyceride levels (TG), a known risk factor for cardiovascular disease (CVD).

Mattes notes, "Regular consumption of small amounts of peanuts does not necessarily lead to weight gain and may contribute to a healthy, more satisfying diet" (13).

NUT AND PEANUT EATERS HAVE BETTER BODY PROFILES

A study conducted by Brenda Eissenstat at Penn State found that the body mass index (BMI), a measure health professionals use to evaluate body weight, of peanut eaters was more favorable than that of non–peanut eaters (14). In addition, peanut eaters consistently tend to have higher levels of key nutrients and overall healthier diets than their non–peanut eating counterparts.

Additionally, a number of studies in the United States, Israel, and Australia have demonstrated the positive effects of almonds and other nuts on heart disease and cardiovascular disease risk factors. When suggesting that nuts can be a useful food for the population in terms of cardiovascular health, a common question is whether they will make people fat. In a growing body of evidence on the subject, it is interesting to note that in the large population studies, frequent consumers of nuts were more successful at maintaining their body weight than those who never ate nuts. Also, in studies investigating almonds' effects on blood cholesterol, participants consumed up to 3 oz of almonds a day, yet no one gained weight, despite the higher fat intake (15).

PEANUTS AND NUTS DECREASE THE RISK OF A SECOND HEART ATTACK

Lisa Brown at Harvard School of Public Health examined the eating patterns of participants in the Cholesterol and Recurrent Events (CARE) trial. People who ate 2 or more servings of peanuts or nuts a week decreased their risk of having another heart attack by 25% (16). Brown found that a healthy diet that includes peanuts and nuts can have benefits above and beyond those provided by drugs alone. This study gives further credibility to the diet and disease prevention link.

NUTRIENTS RECENTLY FOUND IN PEANUTS AND NUTS

Researchers at the United States Department of Agriculture (USDA) Agricultural Research Service at North Carolina State University have identified resveratrol, the phytochemical in red wine, in peanuts (17). The "French paradox" associates resveratrol's presence in red wine with reduced cardiovascular disease and refers to the fact that, despite consuming a relatively high-fat diet, the French have surprisingly low rates of heart disease.

Peanuts also contain other plant chemicals, such as phytosterols and isoflavones. A study at the State University of New York at Buffalo identified an important phytochemical, beta-sitosterol (SIT), in peanuts and peanut products. SIT has previously been shown to inhibit cancer growth, especially prostate cancer. Atif Awad examined the SIT content of peanuts, peanut butter, peanut flour, and peanut oil and found that the content varied from 44 mg SIT/100 g (peanut flour) to 191 mg SIT/100 g (unrefined peanut oil). Snack peanuts contain 160 mg SIT/100 g and peanut butter contains approximately 120 mg SIT/100 g (18). The amount of protective SIT in unrefined peanut oil is comparable to that of soybean oil (183 mg SIT/100 g).

In addition to phytochemicals, peanuts contain significant amounts of heart-healthy MUFA, plant protein, fiber, magnesium, and vitamin E, all of which may contribute to their healthfulness. Many hard-to-get nutrients such as copper, phosphorus, potassium, and zinc are also found in peanuts and peanut butter. All of these, plus the many plant sterols and phytochemicals that are still being discovered in peanuts, make them a complex food with many health benefits.

PEANUTS

Though many people think of peanuts as nuts, they are actually legumes and belong to the *Leguminosae* family. They are used in diets and cuisines as nuts, but their physical structure and nutritional benefits more closely resemble those of other legumes, such as beans.

History

The peanut, though grown in tropical and subtropical regions throughout the world, is native to the Western Hemisphere. It probably originated in South America and spread throughout the New World as Spanish explorers discovered the peanut's versatility. When the Spaniards returned to Europe, peanuts went with them. Later, traders were responsible for spreading peanuts to Asia and Africa. The peanut made its way back to North America during the slave trading period. Although there were some commercial peanut farms in the United States during the 1700s and 1800s, peanuts were not extensively grown. This lack of interest in peanut farming is attributed to the

fact that the peanut was regarded as food for the poor and to the slow, difficult growing and harvesting techniques. Until the Civil War, the peanut remained a regional food associated with the southern United States.

After the Civil War, the demand for peanuts increased rapidly. By the end of the nineteenth century, the development of equipment for production, harvesting, and shelling peanuts, as well as processing techniques, contributed to the expansion of the peanut industry. The new twentieth-century labor-saving equipment resulted in a rapid demand for peanut oil, roasted and salted peanuts, peanut butter, and confections.

Also associated with the expansion of the peanut industry is the research of George Washington Carver at Tuskegee Institute in Alabama at the turn of the century. The talented botanist recognized the intrinsic value of the peanut as a cash crop. Carver proposed that peanuts be planted as a rotation crop in the southeastern cotton-growing areas where the boll weevil threatened the region's agricultural base. Not only did Carver contribute to changing the face of southern farming, he also developed more than 300 uses for peanuts, mostly for industrial purposes.

Early Legislative Action

As peanut consumption continued to rise, the U.S. government instituted programs in 1934 to regulate the acreage, production, and price of this food item. Federal government production controls were lifted during World War II to meet the heavy demand for fats and oils required for the U.S. war effort. Controls were reestablished in 1949, and in 1977 a two-tier price-support system was initiated. This system has subsequently been revised, most recently by the 1995 Farm Bill.

Peanut Types and Production

Seven states account for approximately 98% of all peanuts grown in the United States: Georgia (37.7%) grows the major proportion of all peanuts, followed by Texas (23.2%), Alabama (10.5%), North Carolina (9.3%), Florida (6.4%), Virginia (5.4%), and Oklahoma (5.2%). There are approximately 40,000 peanut farms in the major producing regions. (Percentages are based on 1997 production of quota and nonquota peanuts.) The peanut-growing regions of the United States have direct access to port facilities of the Atlantic Ocean and the Gulf of Mexico. The United States produces four basic varieties of peanuts: runner, Virginia, Spanish, and Valencia. Each type is distinctive in size and flavor.

Runners

Runners have become the dominant type due to the introduction in the early 1970s of a new runner variety, the Florunner, which was responsible for a spectacular increase in peanut yields. Runners have rapidly gained wide acceptance because of their attractive kernel size range; a high proportion of runners are used for peanut butter. Runners, grown mainly in Georgia, Alabama, Florida, Texas, and Oklahoma, account for 73% of total U.S. production.

Virginia

Virginia have the largest kernels and account for most of the peanuts roasted and eaten as in-shells. When shelled, the kernels are sold as salted peanuts. Virginias are grown mainly in southeastern Virginia, northeastern North Carolina, and western Texas. Virginia-type peanuts account for about 22% of total U.S. production.

Spanish

Spanish-type peanuts have smaller kernels covered with a reddish-brown skin. They are used predominantly in peanut candy, although significant quantities are also used for salted nuts and peanut butter. They have a higher oil content than the other types of peanuts, which is advantageous when crushing for oil. Grown primarily in Oklahoma and Texas, Spanish-type peanuts account for 4% of U.S. production.

Valencia

Valencias usually have three or more small kernels to a pod. They are very sweet peanuts and are usually roasted and sold in the shell; they are also excellent for fresh use as boiled peanuts. Because of the greater demand for other varieties, Valencias account for less than 1% of U.S. production and are grown mainly in New Mexico.

Growing

Peanuts are the seeds of an annual legume that grows close to the ground and produces its fruit below the soil surface. U.S. peanuts are planted after the last frost in April or May when soil temperatures reach 65 to 70°F (20°C). Preplanting tillage ensures a rich, well-prepared seedbed. Seeds are planted about 2 in. (5 cm) deep, one every 2 to 4 in. (5–10 cm) in the Southeast, and 4 to 6 in. (10–15 cm) in the Virginia–Carolina area, in rows about 3 ft (1 m) apart. The row spacing is determined to a large extent by the type of planting and harvesting equipment utilized.

Cultivating

Peanuts may be cultivated up to three times, depending on the region, to control weeds and grasses. A climate with approximately 200 frost-free days (175 for Spanish peanuts) is ideal for a good crop. Warm weather conditions, coupled with rich, sandy soil, will result in the appearance of peanut leaves 10 to 14 days after the first planting. Farmers generally follow a 3-yr rotation pattern with cotton, corn, or small grains planted on the same acreage in intervening years to prevent disease. In addition, many farmers are using irrigation in an effort to reduce crop stress and thereby enhance opportunities for the production of high-quality peanuts.

Harvesting

The peanut-harvesting process occurs in two stages. Digging, the first stage, begins when about 70% of the pods have reached maturity. At optimum soil moisture, a digger proceeds along the rows of peanut plants, driving a hori-zontal blade 4 to 6 in. (10–15 cm) under the soil. The digger loosens the plant and cuts the taproot. A shaker lifts the plant from the soil, gently shakes the soil from the peanuts, and inverts the plant, exposing the pods to the sun in a windrow. The peanuts are now ready for the second phase of the harvest—curing. After they cure in the field for 2 or 3 days, a combine separates the pods from the vines, placing the peanut pods into a hopper on the top of the machine. The vine is returned to the field to improve the soil fertility or baled into hay for livestock feed. Freshly dug peanuts are then placed into drying wagons for further curing, with forced hot air slowly circulating through the wagons. In this final stage of the curing process, moisture content is reduced to 8 to 10% for safe storage.

Shelling and Grading

After proper curing, farmers' stock peanuts (harvested peanuts that have not been shelled, cleaned, or crushed) are inspected and graded to establish the quality and value of the product. The inspection process determines the overall quality and on-farm value of the shelled product for commercial sales and price-support loans.

The inspection and grading of peanuts by the Agricultural Marketing Service (AMS) of the USDA occur at buying stations or shelling plants usually located within a few miles of where the peanuts have been harvested. A pneumatic sampler withdraws a representative quantity of peanuts from the drying wagon, and from this sample the USDA inspector determines the meat content, size of pods, damaged kernels, foreign material, and kernel moisture content. Once the grade is established, the loan value is determined from USDA price-support schedules.

Peanuts are separated into three classifications at this farmers' stock marketing and grading stage: segregation I, segregation II, or segregation III. These classifications based on USDA grades are mainly concerned with the amount and type of damage in the kernels. Peanut shelters can buy only segregation I for use in edible products. Those peanuts not classified as segregation I are crushed for oil.

Segregation 1 peanuts move on to the shelling process, where they are first cleaned; stones, soil, bits of vines, and other foreign material are removed. The cleaned peanuts move by conveyor belt through shelling machines in which the peanuts are forced through perforated grates that separate the kernels from the hulls. Shakers separate the kernels and the pods. The kernels are then passed over the various screens where they are sorted by size into market grades. The edible nuts are individually inspected with electronic eyes that eliminate discolored or defective nuts as well as any remaining foreign material.

In-shell peanuts are usually produced from large Virginia-type peanuts or Valencias that have been grown in light-colored soil. Very immature and lightweight pods are removed by vacuum. The largest remaining pods are separated into size categories by screens. Stems are removed and any remaining immature pods are removed by specific gravity. Dark or damaged pods are then removed by electronic eyes so that only the most mature, brightest pods remain.

Manufacturing

Unlike in other countries where the end products are peanut oil, cake, and meal, the prime market for U.S. peanuts is edible consumption. Only 15% of U.S. production is normally crushed for oil. U.S. per capita consumption is approximately 6.5 lb per person (excluding crushing).

Peanut Butter

Peanut butter is one of America's favorite foods. Found in about 75% of American homes, peanut butter is considered by many to be a staple like bread and milk. By law, in the U.S., any product labeled "peanut butter" must be at least 90% peanuts. The remaining 10% may include salt, sweetener, and stabilizers that prevent the peanut oil from separating and rising to the top. About 40% of the U.S. peanut crop is used to make peanut butter. The peanuts are roasted and cooled, then the skins are removed. The kernels are inspected again and then ground, usually through two grinding stages, to produce a smooth, even-textured butter. The peanut butter is heated to about 170°F during grinding. After the stabilizer is added, the peanut butter is rapidly cooled to 120°F or below. This crystallizes the stabilizer, thus trapping the peanut oil that was released by the grinding. Peanut granules can be added at this point to make crunchy peanut butter.

Snack Peanuts

Americans consume more than 300 million lb of snack peanuts each year. Most snacking peanuts are shelled and blanched (to remove the skins) before roasting. Peanuts can be oil-roasted in continuous cookers that take a steady stream of peanuts through hot oil for about 5 min or dry-roasted in a large oven with dry, hot forced air. About 10% of the U.S. crop is sold as in-shell peanuts—usually the large Virginia and long Valencia types. The peanuts are conveyed over sizing screens that let the smaller pods fall through for shelling. In this way, only the largest pods are sold as in-shell peanuts. The peanuts are then roasted and packaged for sale. In-shell peanuts often are roasted with salt and occasionally with spicy seasonings.

Confectionery

The confectionery industry uses about 25% of the U.S. crop to make candy. Five of the eight top-selling chocolate confections contain peanuts or peanut butter. Peanut butter and chocolate seem to be one of America's favorite pairs.

Peanut Oil

Peanut kernels range in oil content from about 43 to 54%, depending on the variety of peanut and seasonal growing conditions. Peanuts supply one-sixth of the world's vegetable oil. Oil is extracted from shelled and crushed peanuts by one or a combination of the following methods: hydraulic pressing, expeller pressing, and solvent extraction. Peanut oil is an excellent-quality cooking oil with a high smoke point (440°F) and neutral flavor and odor. It allows food to cook very quickly with a crisp coating and little oil absorption. Peanut oil is liquid at room temperature. Highly aromatic 100% peanut oil and peanut extract are high-value products with a strong roasted peanut flavor and nut aroma. These products have applications in flavor compounds, confections, sauces, and baked goods.

Peanut Flour

Roasted and naturally processed to obtain a strong roasted peanut flavor, partially defatted peanut flour works well as a fat binder in confectionery products or to add flavor and extend shelf life. Peanut flour is also high in protein (40–45%).

Export Markets

World peanut exports are approximately 1.3 million t (shelled basis). The major suppliers to the export market are the United States, China, and Argentina. Although U.S. peanuts represent approximately 10% of world peanut production, the United States has become one of the leading world exporters, accounting for about one-fourth of world peanut trade. Other producers such as India, Vietnam, and several African countries, periodically enter the export market, depending on their crop quality and world market demand.

Sixty percent of U.S. raw peanut exports are destined for the European Union (EU). The major markets for peanuts within the EU are the United Kingdom, the Netherlands (which serves as the primary port of entry for peanuts), and Germany.

Demand in Europe for peanuts has been steady, although competition within a dynamic snack market has put considerable pressure on peanuts to compete with a growing range of products (potato chips, extruded products, pretzel sticks, etc). In addition, quality standards and import requirements continue to tighten, requiring the implementation of improved monitoring and quality-control standards at origin.

Exports of processed peanuts and specialty peanut products have steadily increased, representing approximately 25% of total U.S. peanut exports by value. The largest U.S. export market for processed peanut butter is Saudi Arabia, followed by Canada, Japan, Germany, and Korea. Major snack nut markets are the Netherlands, Spain, the United Kingdom, France, and Germany.

U.S. Quality Control and Research

Consumers throughout the world are concerned about consistently obtaining flavorful, wholesome peanuts that are uniform in size and free from foreign material and contamination. The U.S. peanut industry continues to invest heavily in plant modernization and the latest designs in automated cleaning and sorting equipment to ensure that all buyers receive the best possible product. U.S. government inspectors monitor processing at each stage of the peanut's journey from the farm to the manufacturer—and to the grocery shelf in the case of domestic production.

Marketing Agreement

In 1965, the AMS/USDA, the Food and Drug Administration, and the U.S. peanut industry initiated a cooperative quality-control program to prevent aflatoxin-contaminated peanuts from entering food channels. Under the provisions of the agreement, peanuts are subject to strict quality standards that are enforced by USDA and federal and state inspectors. These inspectors supervise, inspect, and grade peanuts from delivery at buying points to shipment from shelling plants. Strictly enforced government regulations ensure that U.S. peanuts are of consistently high quality. In addition, comprehensive lot identification systems enable peanuts to be tracked throughout their various stages of processing until final delivery to a domestic processor or upon export.

Industry Research

Due to the emphasis on production of edible peanuts, both the government and the U.S. industry allocate a considerable amount of time and money to peanut research to produce a high-quality food item. Government- and industry-funded peanut research provides farmers with the latest information on improving quality and yields. Since the late 1950s, peanut yields have tripled as a result of new varieties, advanced agronomic technology, irrigation, and totally mechanized operations.

CASHEWS

The cashew fruit consists of two distinct parts: a fleshy stalk in the form of a pear—also called the cashew apple—with a brilliant yellow or red skin that can measure from 5 to 10 cm; and a gray-brown-colored nut (the cashew) in the shape of a kidney, which hangs from the lower end of the stalk, or apple. The nut section of the cashew fruit is very rich in carbohydrates and vitamin A.

Cashews are believed to have originated in the northeast of Brazil, near the equator. It is likely that Spanish sailors first introduced the cashew to Central America in the sixteenth century. Later, Portuguese colonists brought cashews to territories in East Africa (Mozambique) and India (Goa), where its cultivation extended to Indonesia and the Philippines. Today the principal cashew-producing countries are Brazil, India, and Mozambique. Juices, syrups, preserves, wine, and liquors are obtained from the stalk or apple. However, the main commercial use is the cashew nut itself. Cashews are marketed in the shelled, roasted, and salted forms for use as a snack and as an ingredient (delicacies, chocolate, etc).

ALMONDS

Unlike other flowering fruit trees that bear edible fruit, the almond tree's "pearl" is the delicious nut found inside the fruit.

History

The exact ancestry of almonds is unknown, but they are thought to have originated in China and central Asia. Explorers ate almonds while traveling the Silk Road between Asia and the Mediterranean. Before long, almond trees flourished in the Mediterranean, especially in Spain and Italy. The almond tree was brought to California from Spain in the mid-1700s by the Franciscan padres. The moist, cool weather of the coastal missions, however, did not provide optimum growing conditions. It wasn't until the following century that trees were successfully planted inland. By the 1870s, research and crossbreeding had developed several of today's prominent almond varieties. By the turn of the twentieth century, the almond industry was firmly established in the Sacramento and San Joaquin areas of California's great Central Valley.

Throughout history, almonds have maintained religious, ethnic, and social significance. The Bible's Book of Numbers tells the story of Aaron's rod that blossomed and bore almonds, giving the almond the symbolism of divine approval. They were a prized ingredient in breads served to Egypt's pharaohs. The Romans showered newlyweds with almonds as a fertility charm. Today, Americans give guests at weddings a bag of sugared almonds, representing children, happiness, romance, good health, and fortune. In Sweden, cinnamon-flavored rice pudding with an almond hidden inside is a Christmas custom. Find it, and good fortune is yours for a year.

Marketing

The principal almond-producing countries are the United States, Spain, Italy, Portugal, Morocco, Tunisia, and Turkey. Almonds are marketed in various forms: in-shell, shelled, blanched, slivered, chopped, flour, roasted, sweetened, and salted. They are used as a snack, in confectionery (marzipan, *turron*, nougat, etc), in food products (almond milk, ice cream, chocolate), in culinary recipes, and also as a cosmetic base.

Growing

In the fall, flower parts begin to develop on the edges of the growing bud. By mid-December, pollen grains are present. The tiny bud remains dormant until early January, when it grows rapidly. A good chill during November and December followed by a warmer January and February coax the first almond tree blossoms from their buds. Because the almond tree is not self-pollinating, at least two different varieties of trees are necessary for a productive orchard. Bees pollinate alternating rows of almonds varieties. From February onward, orchards should be frost-free, having mild temperatures (55–60°F) and minimal rain so blossoms can flourish and bees can do their job.

After the petals drop and the trees have leafed out, the first signs of the fuzzy gray-green fruit appear. The hull continues to harden and mature, and in July it begins to split open. Between mid-August and late October, the split widens, exposing the shell, which allows the kernel (nut) to dry. The whole nut and stem finally separate and, shortly before harvest, the hull opens completely.

Harvest

State-of-the-art technology is used to ensure the highest-quality almonds. California's growing and sanitary standards lead the world, both in the field and in the almond-processing plant. To prepare for harvest, orchard floors are swept and cleared. Mechanical tree shakers knock unshelled nuts to the ground, where they are allowed to dry before they are swept into rows and picked up by machine. Finally, they are transported to carts and towed to the huller.

Packaging

At the processing plant, a random sample of almond shells are cracked open and the nuts inside are graded according to size and quality. Almonds are inspected to make sure they are whole, clean; well dried; and virtually free from decay, rancidity, insects, foreign matter, mold, and any kind of breakage or blemish. Almonds are then processed and packed to specification in an assortment of sizes and shapes. Stored properly at 40°F with low humidity, almonds have a shelf life of up to 3 years.

RESEARCH

Vitamin E (technically called tocopherol) is a strong antioxidant that has been associated with reduced risks of developing prostate cancer and coronary heart disease. The term vitamin E, however, does not refer to a single compound. Rather it refers to a category of compounds with similar, but not identical, biological functions. Stone performed a chemical analysis of vitamin E as it appears in almonds. He found that almonds are an excellent source of natural alpha-tocopherol, the most biologically active form of the various vitamin E compounds. A single 1-oz serving of almonds supplies approximately 10 IU (66%) of the recommended dietary allowance (RDA) for men and 83% of the RDA for women. Almonds also contain gamma-tocopherol, which may participate in neutralizing destructive chemicals purported to be involved in the development of degenerative diseases.

BRAZIL NUTS

Brazil nuts are the large seeds of an enormous evergreen tree of the Amazon district, *Bertholletia excelsa*. This tree can reach a height of more than 40 m, with a trunk diameter of nearly 2.5 m. Its round fruits, with a dark brown color, can weigh more than 1.5 kg each and resemble large coconuts with a woody shell.

The first historical reference to Brazil nuts dates from 1569, when a Spanish colonial official collected thousands of these nuts for his tired and hungry troops. Understandably, the soldiers recovered quickly, as Brazil nuts contain a high energy level, with high contents of digestible fats, calcium, phosporous, potassium, and vitamin B.

Growing Regions

Brazil, Peru, Bolivia, Colombia, Venezuela, and the Guianas are presently the principal Brazil nut–producing countries.

Marketing

Brazil nuts are marketed in in-shell and shelled form and are eaten raw, roasted, and salted as well as in ice creams, chocolate, and bakery and confectionery products.

Health

Brazil nuts contain a notably high amount of bioavailable selenium. Earlier research has shown that selenium supplementation can suppress some forms of cancer growth in different animal tumor models (19).

HAZELNUTS

Hazelnuts grow in clusters on the hazel tree in temperate zones around the world. The fuzzy outer husk opens as the nut ripens, revealing a hard, smooth shell.

History

One of the oldest agricultural crops, hazelnuts are believed to have originated in Asia, then extended into Europe. Today, the principal hazelnut-producing countries are Turkey, Spain, Italy, and the United States.

Since prehistoric times, proven by fossils found in Mesolithic and Neolithic sites in Sweden, Denmark, and Germany, the hazelnut formed part of primitive human diet. The origin of the hazelnut seems to be in Asia; from there it extended to Europe, where it constitutes one of the oldest agricultural cultivations. Several hazelnut varieties exist in the world, the most important being the European variety, *Corylus avellana L*. The hazelnut was present in Greek and Roman mythology and in the Bible, mentioned for its extraordinary nutritional and healing values and even as a tool for finding buried treasures and subterranean streams of water.

Marketing

Also known as filberts and cobnuts, these sweet, rich, grape-size nuts are marketed in in-shell and shelled forms, roasted or natural, and are used mostly in the baking and confectionery industries and for the preparation of food products such as chocolate, ice cream, and nougat. However, they can also be used in salads and main dishes or as snacks. Hazelnuts have been around for over 4,000 years. The hazelnut today is recognized by professional chefs, knowledgeable home cooks, and leading food processors around the world for its unique nutty-sweet flavor, rich, "creamy" texture; versatility; and consistent high quality. Hazelnuts are grown and processed using a combination of ancient artistry and the latest technology, then made available in either a natural or a dry-roasted state in a variety of forms for consumer, food service, or food-product manufacturing needs.

Growing Regions

Presently the principal hazelnut-producing countries are Turkey, Spain, Italy, France, and the United States. Hazelnuts are grown in the world's temperate zones. Two regions in particular stand out. Turkey is by far the world's largest hazelnut producer, accounting for more than three-quarters of annual worldwide production. The hazelnut has been an agricultural mainstay along the rugged and beautiful Black Sea coast for more than 2,000 years. Turkey's ancient hazelnut tradition today combines a rich heritage with twenty-first-century technology. Oregon is America's hazelnut capital, where rich valley soil, cool ocean breezes, and gentle Pacific Northwest winters offer an ideal climate for the hazel tree.

Harvest

Although hazel trees bloom in the middle of winter, their nuts are typically harvested between late summer and early autumn, when the mature nut turns from bright green to shades of hazel. Then the nuts are cracked and processed in either natural or dry-roasted form. The process of dry roasting gives hazelnuts a crunchier texture and a richer, more intense flavor.

Hazelnut Products

Hazelnuts can be processed into fine particles and are useful in adding texture and flavor to a dish. They can be used to replace added fat; as a binding agent for tempuras, to encrust fish, to thicken pies, or to dredge foods for sautéing or baking. They can also be used in cake batters, breads, fillings, and extruded snacks. Hazelnut butter is made of hazelnuts processed to a rich, buttery consistency. It can stand alone in a recipe or can be used with additional forms of hazelnuts for maximum flavor impact. Hazelnut paste is essentially hazelnut butter with added natural sweeteners. Hazelnut oil is a fragrant, full-flavored oil pressed from hazelnuts and can be used in salad dressings and sautéed foods.

Health

Ancient lore has it that hazelnuts held the cure for everything from baldness to stomachaches. Modern science has shown that although those claims may be somewhat overstated, hazelnuts are a good source of vital nutrients. High in dietary fiber, calcium, magnesium, potassium, and vitamin E, and one of the best nut sources of MUFA, hazelnuts are also an excellent source of protein. What's more, hazelnuts, like other plant foods, are naturally cholesterol-free. Mounting scientific evidence suggests that MUFA may work to lower low-density lipoprotein (LDL) cholesterol levels (the so-called "bad" cholesterol), thereby reducing the risk of coronary heart disease (CHD) (20). New research shows that the antioxidant vitamin E could play a role in preventing certain kinds of cancer and CHD (21–23).

Storage and Handling

Store hazelnuts at no more than 40°F for up to one full year. Allow the hazelnuts to warm to room temperature in an unopened container before using. Hazelnuts can also be stored frozen at 27°F or lower for up to 2 yr. Again, for optimum product performance, allow hazelnuts to warm to room temperature in an unopened container, in a well-ventilated area away from odor-producing substances. Hazelnut oil is best stored under refrigeration to prevent rancidity.

Quality and Sanitation Standards

Hazelnuts are grown and processed under exacting sanitary conditions. They meet or exceed quality and sanitation standards set by USDA and the Canadian Food Inspection Agency. Hazelnut equivalencies are as follows (24):

> 1 lb in-shell hazelnuts = 1 1/2 cups hazelnut kernels
> 1 lb (shelled) hazelnut kernels = 3 1/4 cups
> 1 cup hazelnut kernels = 1 1/8 cups large diced
> 1 cup hazelnut kernels = 1 1/4 cups small diced
> 1 cup hazelnut kernels = 1 1/3 cups meal

MACADAMIA NUTS

The macadamia is one of the youngest of the edible tree nuts; its commercial cultivation dates from 1858. The fruit consists of a fleshy husk that covers a spherical seed protected by a hard and durable shell. Inside this shell is the macadamia nut, with an exquisite flavor that is sometimes compared with that of a superfine hazelnut. In 1882 the macadamia was introduced to Hawaii, one of the principal production areas along with Australia and New Zealand. Although still little known to the consumer, the macadamia is, together with the pistachio and the pine kernels, one of the world's most expensive edible nuts. Very rich in calories and proteins, the macadamia is always marketed in shelled form and is used mostly as a snack, in confectionery, and for the elaboration of chocolate.

Though most people associate the macadamia with Hawaii, the nut is native to Australia. (In fact, it is the only commercial food crop that is native to Australia.) The colonization of Australia by the British began in 1788 with the establishment of a penal colony at Botany Bay. But it was not until 1875 that the recorded history of the macadamia began. Ferdinand Von Muller, Royal Botanist at Melbourne, and Walter Hill, director of the Botany Garden at Brisbane, were botanizing in the forest along the Pine River in the Moreton Bay district of Queensland. They discovered a species of tree in the family *Proteaceae* previously unknown to European and American botanists. This species did not fit into any previously established genera in that family, so in 1858 Muller established a new genus, Macadamia, named in honor of his friend John Macadam, secretary of the Philosophical Institute of Victoria.

Of course, the British were not the first inhabitants of Australia. At the time of their arrival Australia was inhabited by a primitive people generally referred to as aborigines whose population numbered around 300,000. Their food consisted mainly of fish, shellfish, turtle eggs, grubs of certain tree bark insects, kangaroo, koala, wombat, bandicoot, and other small animals and birds plus yarns and grass seeds. However, during the months of fall and winter (March to June), they would come from far and near to congregate on the eastern slopes of the Great Divide Range. Here they would feast on the seeds of two kinds of trees that were abundant in the area. One tree they called *BunyaBunya*, and the other they called *kindalKindal*. The former we now know botanically as *Araucaria bidwillii*, the latter as the macadamia.

The *Macadamia* genus consists of at least 10 species, but only two of those produce edible nuts, *integrifolia* and *tetraphylla*. The macadamia was introduced into California in late 1877 by C. H. Dwinelle of the University of California at Berkeley. He obtained seeds from Australia and planted several seedlings along Strawberry Creek on the Berkeley campus, and at least one of those trees is still growing in its original site. The next year the macadamia was introduced into Hawaii by Walter H. Perves, manager of the Honokoa Sugar Company at Haina, Hawaii. As of 1995 Hawaii had 22,000 acres planted in macadamias.

The first large planting of macadamias occurred in 1890 on the Frederickson estate at Rouse Mills, New South Wales. They planted around 250 trees as a source of nuts for the family. Many of those trees still exist and are still producing a good crop of nuts. Around 1910, two nurserymen, Ernest Braunton and Charles Knowlton, started selling seedling trees in southern California. But not until 1946 did a commercial planting finally occur in California. Robert W. Todd planted 2 acres of seedling trees on his property on Grandview Street in Oceanside. Several years later those trees were grafted, and today, more than 100 of them are still there and still producing. Macadamias are grown commercially in Hawaii, Australia, Malawi, Kenya, South Africa, Israel, Costa Rica, Guatemala, Mexico, Brazil, and many other tropical and subtropical regions, including Florida. The macadamia is grown in California from San Luis Obispo south to the Mexican border, west of the mountains.

The primary limiting factor in growing macadamias is temperature. Macadamia trees can take temperatures as low as 28°F for up to about 4 hours, and they start to get stressed around 105°F. Southern California has about 2,500 acres planted to the macadamia. Commercially, only *M. integrifolia* and *M. tetraphylla* and their hybrids are important. They are very similar to each other and botanically very closely related to a third species, *M. ternifolia F. Muell*, which produces a small, bitter kernel unsuitable as a table nut.

M. integrifolia is commonly referred to as the smooth-shell species. The fruit consists of a white kernel, high in oil content (72% oil and 4% sugar when dry), very uniform, and of excellent quality. It is enclosed in a round, hard shell about 1 in. (2.5 cm) in diameter surrounded by a smooth, bright green pericarp (husk). The leaves are stiff, oblong to lanceolate, 4 to 10 in. (10–25 cm) long in nodal whorls of three (rarely four), and either light green or bronze when young. The small, perfect white flowers are borne in racemes 4 to 8 in. (10–20 cm) long. Only a few flowers in a raceme will set fruit.

M. tetraphylla is called the rough-shelled species because of the pebbliness of the shell's surface. The husk is somewhat spindle shaped; grayish green; and covered with a dense, white pubescence. Kernels have a grayish base and are darker in color and more variable in quality than those of *M. integrifolia*. The oil content averages 67% in the dry nut, with 6 to 8% sugar. The leaves are characteristically sessile and serrated along the margins, in whorls of four at the nodes. The flowers are pink and in racemes 6 to 18 in. (15–45 cm) long. Cultivars that are hybrids of the two species possess characteristics of both, and the quality of their nuts compares favorably with that of *M. integrifolia*. The trees of both species are tall and spreading, reaching 60 ft (20 m) or more in height. The wood is hard and brittle; the exposed bark sunscalds very easily.

Cultivation and Production

There are approximately 40 described cultivars, most of them in Australia. The following varieties have been used for commercial plantings in Hawaii: Keauhou, Ikaika, Kakea, and Keaau. Kakea, however, is no longer propagated for commercial use. In California, Elimba, a *M. tetraphylla* cultivar, is also considered of commercial value. Beaumont, a productive hybrid cultivar, is also recommended for home plantings. With good care, cultivars of *M. integrifolia* begin producing 5 yr after planting, but appreciable yields are obtained only after the 8th year. A productive variety will bear as many as 150 lb of in-shell nuts per tree.

Before large-scale plantings are attempted, cultivars should be tested locally. In Hawaii, *M. integrifolia* cultivars are more productive on *M. tetraphylla* rootstocks in soils where they are not likely to suffer from iron deficiency. Where this deficiency may be a problem, *M. integrifolia* seedlings should be used as a rootstock. The season of production in Florida for *M. integrifolia* runs from July through November. Very few commercial cultivars have been tested in this state, and more information is needed to make specific recommendations.

Climate

Macadamias are well adapted to warm, subtropical conditions. Mature trees can withstand winter temperatures of as low as 25 to 26°F (3–5°C) for short periods with minor damage to the foliage. However, young trees and foliage are very tender and are killed by temperatures very near freezing. Temperatures below 28°F (−2°C) cause damage to flowers and young fruit and reduce production. In the tropics, macadamias are better adapted to medium elevations of 2100 to 3600 ft (700–1200 m), but in Hawaii, macadamias are not planted commercially above 2500 ft (800 m). Although the plant is quite resistant to drought, supplemental irrigation is very important, particularly during the flowering and fruit-setting season. Severe moisture stress results in considerable drop of young fruit.

Soils and Fertilizers

Macadamias are not demanding for soil fertility, but they do require good drainage. They also need relatively higher amounts of phosphorous in the fertilizer than other fruit crops, particularly when the trees are young.

Pests and Diseases

Leaves are occasionally infested by thrips and mites, which may become serious in large plantings. Green stinkbugs cause considerable damage at times by injuring very young fruit when the shell is still soft. Rats, squirrels, and nut borers also cause substantial losses if unchecked. Anthracnose (*Colletotrichum* spp.) attacks leaves and the husks of immature nuts. Diseased nuts do not drop when mature and usually spoil while still attached. *Phytophthora cinnamoni*, which causes root rot in avocados, produces a trunk canker in macadamia that may kill young seedlings. Fortunately, the tree is quite resistant to root rot caused by this fungus.

Harvesting and Processing

Mature nuts fall to the ground and have to be gathered manually every week to prevent spoilage, particularly from molds. Husking is done soon after harvesting and before the nuts are mechanically cracked. They are air dried at temperatures not higher than 110°F. The moisture content of the kernel is reduced to less than 1.5% to prevent development of off-flavors after roasting (either dry roasting or in a refined coconut oil at 275°F for 12–15 min).

Uses

Macadamias at present command premium prices because the demand is greater than current production. They are ordinarily offered on the gourmet shelves of supermarkets as salted nuts packed in glass jars. The largest use, however, is in confections. Only whole nuts are packed in jars.

PECANS

This native American tree nut is a member of the hickory family. Long before the arrival of the Europeans to the New World, pecans were part of the diet of the Indian tribes of the central and southern regions of the United States. The pecan slightly resembles the walnut but is longer in shape and has a smoother shell and a higher proportion of kernel in its shell (40–60%). The pecan nut is extraordinarily rich in fats and calories. The principal pecan-producing countries are the United States, Mexico, Australia, and Israel. Pecans are marketed in in-shell and shelled forms and can be eaten raw or roasted. They are used in the bakery, confectionery, and dairy industries and in chocolate and ice creams. Pecans are also added to cereals, breads, pastries and cookies, salads, and main dishes and are eaten as toppings on desserts and as a snack. The wood of the pecan tree is highly appreciated for its timber and is often used as decorative paneling.

History

Pecans can be traced back to the sixteenth century. The only major tree nut that grows naturally in North America, the pecan is considered the most valuable North American nut species. The name pecan is a Native American word of Algonquian origin that was used to describe all nuts requiring a stone to crack and a Natchez Indian word for the pecan plant. Originating in central and eastern North America and the river valleys of Mexico, pecans were widely used by pre-Colonial residents. Pecans were favored because they were accessible to waterways, easier to shell than other North American nut species, and great tasting.

The pecan's normal environment is the floodplains along the Mississippi, Ohio, Missouri, and Red rivers and along many of the large rivers in central Texas and northeastern Mexico. Due to its ready availability, many Native American tribes in the United States and Mexico used the wild pecan as a major food source during autumn. The kernel was used to produce a creamy fluid for cooking and drinking. It is speculated that the nut was used to produce a fermented intoxicating drink called *Powcohicora* (this was also the source of the name hickory). Because the pecan was a major food and trade item, it is said that the pecan tree was cultivated by Native Americans. It is reported that the Creek tribe possessed an "ancient cultivated field" of pecans.

Planting

One of the first known cultivated pecan tree plantings appears to have been accomplished in the late 1600s or early 1700s by Spanish colonists and Franciscans in northern Mexico. These plantings are documented to around 1711, about 70 yr before the first recorded planting by U.S. colonists. The first pecan planting in the United States was in Long Island, New York, in 1772. By the late 1700s nuts from the northern range reached the English portion of the Atlantic seaboard and were planted in the gardens of easterners such as George Washington (1775) and Thomas Jefferson (1779). Settlers were also planting pecans in community gardens along the Gulf Coast at this time.

Economics

In the late 1770s the French and Spanish colonists settling along the Gulf of Mexico realized the economic potential of pecans. By 1802 the French were exporting pecans to the West Indies. It is speculated that the nuts were exported to the West Indies and Spain earlier by Spanish colonists in northern Mexico.

Marketing

In 1805 it was advertised in London that the pecan was "a tree meriting attention as a cultivated crop." Because of its location near the mouth of the Mississippi, New Orleans became of great importance in the marketing of pecans. The city had a natural market and avenue for redistribution of nuts to other parts of the United States and the world. This market stimulated local interest in planting orchards, which led to the adaptation of vegetative propagation techniques and demand for trees that produced superior nuts. At least 15 commercial cultivars had developed in Louisiana by the end of the nineteenth century. In San Antonio, Texas, the wild pecan harvest was more valuable than row crops such as cotton. During the 1700s and the early 1800s the pecan nut had become an item of commerce for the American colonists, and the pecan industry had begun.

Because pecan groves (trees established by natural forces) and orchards (planted by farmers) consisted of diverse nuts of various sizes, shapes, shell characteristics, flavor, fruiting age, and ripening dates, a premium market for large, thin-shelled pecans was created. With such a variability in trees, there was the occasional discovery of a wild tree with unusually large, thin-shelled nuts, which were in high demand by customers. In 1822 Abner Landrum of South Carolina discovered a pecan-budding technique. This system provided a way to graft (unite with a growing plant by placing in close contact) plants derived from superior wild selections. However, this invention was lost or overlooked until the 1880s.

Grafting Pecans

In Louisiana in 1846 an African-American slave gardener, Antoine, successfully propagated pecan by grafting a superior wild pecan to seedling pecan stocks. This clone was named "Centennial" because it won the prize as best pecan exhibited at the Philadelphia Centennial Exposition in 1876. This planting, which became 126 "Centennial" trees, was the first planting of improved pecans. The successful use of grafting techniques led to grafted orchards of superior genotypes and was a milestone for the pecan industry. The adoption of these techniques was slow and had little commercial impact until the 1880s, when Louisiana and Texas nurserymen learned of pecan grafting and began propagation on a commercial level. This was the start of a booming pecan growing and shelling industry.

Pecan Timeline (25,26)

- American Indians utilized and cultivated wild pecans—1500s
- Spanish colonists cultivated orchards—late 1600s– early 1700s
- English settlers planted pecan trees—1700s
- George Washington planted pecan trees—1775
- Thomas Jefferson planted pecan trees—1779
- Economic potential for pecans realized—1700s
- Pecans exported by the French to the West Indies— 1802
- Pecan budding technique discovered—1822
- Successful grafting of the pecan tree—1846
- First planting of improved pecans—1876
- Commercial propagation of pecans begins—1880s

Pecan production has steadily increased in the United States since 1925, rising from 2.2 million lb in 1920 to 338.1 million lb in 1998. Before 1920 the few pecans that were consumed were hand shelled by consumers.

With the development of commercial shelling equipment, the pecan industry began to grow. In the early 1920s pecan processing was developed along with equipment used for sizing, separation of faulty nut meats and shells, cracking, grading of meats, drying, and packaging. The use of pecans was increased by the improvement of storage life through controlling temperature, relative humidity, air circulation, and storage atmospheres, as well as improved packaging. Since 1948 more than 80% of the pecans sold have been shelled before being marketed. Some shelling plants operate year-round, whereas others are seasonal (fall months). Many pecan plants have a capacity of 1 to 30 tons per day. Large plants can have as many as 14 cracking machines with a capacity of 150,000 lb a day and 30 million lb seasonally. Plants usually have a cold-storage facility, some capable of holding several million pounds of unshelled and shelled nuts.

Growing Regions

Not surprisingly, most shelling plants are located in the southeastern and southwestern United States, where the majority of pecans are grown. The leading production states (alphabetically) are Alabama, Arizona, Arkansas, California, Florida, Georgia, Kansas, Louisiana, Mississippi, New Mexico, North Carolina, Oklahoma, South Carolina, and Texas. Significant quantities of pecans also are produced in Mexico's states of Chihuahua, Coahuila, Durango, Nuevo Leon, and Sonora. Modest quantities are produced in Australia, Israel, Peru, and South Africa.

Technology

Machinery used in shelling plants includes (1) grading and sizing machinery equipped with screens and blowers for removal of faulty nuts and foreign materials and segregating nuts into sizes, (2) machines with vats or tanks for sterilizing and conditioning pecans for shelling, (3) cracking machines (called crackers) for cracking the shell, (4) shellers for separating shells from the meat, (5) screens for separation of halves and broken pieces, (6) dryers to remove excess moisture from meats, (7) grading belts or tables for handpicking of dried meats, and (8) electric eyes for sorting shell and foreign matter from the meats.

Shelling pecans reduces their weight from 50% to 65% and their volume approximately by half. To prevent the shattering of pecan meat, the nuts are moistened, or conditioned, before cracking. Two methods are used to accomplish this. The cold-water method soaks the pecans for as long as 8 h in water with chlorine. The pecans are then drained for 16 to 24 h in sacks, vats, or barrels for cracking within the next 24 h. Using the steam-pressure method, pecans are exposed to hot water or steam pressure for 6 to 8 min, then cooled and held for 30 to 60 min. This method is faster but can cause discoloration of the kernels.

The nuts are cracked by a machine applying force to the ends of the nuts. The cracked nuts are then placed on a conveyer that moves them to the sheller machines. Meats and shells are separated by a series of shaker screens. Shelled pecans are then separated into grades by the halve-size or pieces. The sizing is done as meats pass over rapidly vibrating machines with holes of different sizes (halves come in 8 sizes, based on the number per pound). Once shelled, the pecan meats contain from 7 to 9% moisture. For quality the moisture is reduced to 4% or lower. This is done by rapidly circulating dry air through the pecan meats. The meats are then ready for storage.

Storage

In-shell pecans are put in refrigerated storage as soon as possible after harvest. Refrigerated storage is necessary not only to maintain freshness and shelf-life, but also to avoid insect infestation during the warm-weather months. Freezing pecans from season to season is an excellent method for storage. Even at temperatures as low as −170°F (solid carbon dioxide), there is no seepage of oil or decrease in the nut quality.

Shelled pecans are packed in vacuum-packed cans, jars, glass, cellophane bags, and poly-lined boxes to protect against high humidities, foreign flavors, oxidation, insects, rodents, and light. After being packaged, the nuts are stored or shipped to retail, food-service, or commercial markets. There are five U.S. grades for shelled pecans; U.S. No. 1 halves, U.S. Commercial halves, U.S. No. 1 pieces, U.S. Commercial pieces, and Unclassified.

PINE NUTS

The pine tree is one of the most familiar trees in both Europe and North America, but it is mainly in the Mediterranean area that it obtains its highest importance with regard to production and consumption. The nuts are found inside the pinecone.

The pine kernel, or pine nut, is an edible nut with an exquisite flavor and high protein content. The Roman legions carried pine kernels as provisions, and all over Europe it is used as a culinary ingredient in the preparation of meat, fish, and vegetable dishes, as well as in the confectionery industry in chocolates and other delicacies.

Growing Regions

Spain, Italy, China, Portugal, and Turkey are the principal producing countries of pine kernels. Also called Indian nut, piñon, pinoli, and pignolia, the pine nut is marketed in shelled form.

Varieties

There are two main varieties. The Mediterranean or Italian pine nut is from the stone pine. It is torpedo shaped, has a light delicate flavor, and is the more expensive of the two. The stronger-flavored Chinese pine nut is shaped like a squat triangle. Its pungent flavor can easily overpower some foods.

PISTACHIO

With an antiquity of more than 10,000 years, the pistachio is one of the oldest edible nuts on earth. Originating from Southeast Asia, Asia Minor, Pakistan, and India, the pistachio belongs, like the cashew, to the family of Anacardiaceae. The fruit of the pistachio differs from all the other nuts because of its characteristic green color and the semi-opening of the shell, which in Iran is called the "smiling pistachio." This singular morphology makes the pistachio the only edible nut that does not need to be shelled for roasting and salting.

The pistachio has a delicious flavor and a high nutritional value, being very rich in proteins and vitamins. The principal pistachio-producing countries are Iran, Turkey, the United States, Greece, and Italy. Pistachios are always marketed in their in-shell form, roasted and salted, but they can be purchased shelled. They are mainly used as snack and confectionery and used in ice creams.

History

The history of pistachios includes royalty, perseverance, and pride. Pistachios date back to the Holy Lands of the Middle East, where they grew wild in high desert regions. Legend has it that lovers met beneath the trees to hear the pistachios crack open on moonlit nights for the promise of good fortune. A rare delicacy, pistachios were a favorite of the Queen of Sheba, who hoarded the entire Assyrian supply for herself and her court. The royal nut was imported by American traders in the 1880s, primarily for U.S. citizens of Middle Eastern descent. Some 50 years later, pistachios became a popular snack food, introduced in vending machines.

Storage

Pistachios are excellent keepers when stored properly. To keep them at their freshest, store pistachios in a refrigerated, airtight container, or keep them in the freezer for long-term storage. (A temperature of 32–40°F should be maintained with a relative humidity of 70%.) To restore crispness to pistachios that have lost their crunch, toast the nut meats at 200°F for 10 to 15 min. Any pistachios stored in bulk containers should be rotated frequently to maintain freshness. Kept in the refrigerator or freezer, pistachios can be stored for as long as a year.

Shelling

To remove the skin from nut meats by blanching, drop shelled nuts into boiling water and let them soak off the heat for about 1 min. Drain and rub the pistachios with a clean towel. To dry, spread on a large baking sheet in an oven preheated to 300°F for 10 to 15 min. The nut meat can also be removed by toasting. Spread the shelled nuts in one layer on a baking sheet. Bake in an oven preheated to 400°F for 4 to 5 min, watching carefully that they do not burn. Remove and let cool, then rub off the skins.

Red Color

The first pistachios available to consumers were imported from the Middle East. Importers dyed the shells red, both to disguise staining from antiquated harvesting methods and to make pistachios stand out among other nuts in vending machines.

Pistachios in California

Until the 1970s, there was no domestic pistachio industry in the United States. California harvested its first commercial crop in 1976. The entry of California pistachios into the marketplace made available nuts with clean, naturally tan shells. California's Kerman variety is also larger in size and a more vibrant green color.

Availability

Pistachios are harvested in September, but sufficient supply and state-of-the-art storage systems allow the industry to provide pistachios throughout the year.

Exports

Through the U.S. Foreign Agricultural Service, California pistachios are exported to the following markets: Japan,

Korea, China, Malaysia, the Philippines, Thailand, Indonesia, Canada, and the United Kingdom.

California pistachio exports exceeded 41 million lb during the 1996–1997 crop year and were expected to exceed 60 million lb during the 1997–1998 crop year. The 1997 harvest produced a record 179.5 million lb. Monthly shipment and inventory records as reported by industry processors are available.

WALNUTS

Consumed since prehistoric times, the walnut has various origins, in East Asia, southeastern Europe, and North and South America. There exist more than 15 varieties of walnuts, but the two most popular are the English (also called Persian) walnut and the black walnut. A close relative is the butternut, also referred to as the white walnut. By far the most delicious and commercially important variety is the English walnut. The Greeks gave it the name *kara* (head), because of its resemblance to the human skull and brain. This similarity is the source of medieval beliefs of the walnut's healing properties for any kind of headache.

Production

Today, the world production of walnuts is spread over the United States, France, China, India, Turkey, Italy, and Chile. Walnuts are marketed in shelled and in-shell forms and are primarily used as a snack, in desserts and confectionery, and for the preparation of several food products. Consumers and professionals enjoy adding walnuts to various culinary recipes including salads, pasta, main dishes, and baked goods.

History

Walnuts have been recognized as one of the oldest tree foods known to humankind, dating back to about 7000 B.C. Considered food for the gods in the early days of Rome, walnuts were named *Juglans regia* in honor of Jupiter. Today they are commonly called English walnuts, in reference to the English merchant marines whose ships once transported the product for trade to ports around the world. Historians prefer the name Persian walnuts, referring to Persia, the birthplace of walnuts. With its mild climate and deep, fertile soils, California provides ideal growing conditions for the California walnut, producing 98% of the total U.S. commercial crop and accounting for two-thirds of the world's trade.

The Franciscans are credited with bringing walnuts to California from Spain or Mexico. The first commercial planting began in 1867 when Joseph Sexton, an orchardist and nurseryman in the Santa Barbara County town of Goleta, planted English walnuts. For several years, walnuts were predominantly planted in the southern areas of California, accounting for 65% of all bearing acreage. Some 70 years after Sexton's first planting, the center of California walnut production moved north to the Stockton area. Better growing areas, improved irrigation, and better pest-control methods in the north resulted in greater yields, which gradually increase each year.

Coloring

California walnuts come in a spectrum of color shades ranging from extra light to amber. Color is caused naturally by a number of factors including the time of harvest, location of the tree in the orchard, variety, and probably factors we do not yet understand. Darkness of walnuts is not a defect. In fact, in parts of the world and in many industries, darker walnuts are prized because they normally have a stronger walnut taste. At the same time, lighter walnuts are always prized by bakers around the world as decoration for cakes and pastries.

NUTS IN DIET PATTERNS

Nuts are a good fit in a healthy diet. They help achieve optimal levels of protein, vitamin E, fiber, zinc, copper, and iron in the diet. Substituting nuts in place of saturated fat in any diet can also improve the ratio of unsaturated fats to saturated fats.

Because nuts are consumed in a variety of ways—in confections and baked goods, on salads and vegetables, in main dishes, or simply as a snack food—it is easy to incorporate them into the diet every day. In the USDA Food Guide Pyramid, nuts are found in the meat and legume category because of their high protein content. Nuts are a great source of protein for many people, especially those who choose not to eat animal meat.

MEDITERRANEAN DIET PYRAMID

The Mediterranean-style diet is based on the dietary traditions of the people of the many regions bordering the Mediterranean Sea, including Greece, southern Italy, Spain, southern France, Tunisia, Lebanon, and Morocco. People in this region eat an abundance of fruits, vegetables, nuts and legumes, and whole grains; some dairy products, fish, and poultry; and very small amounts of meats. The main source of fat, which usually constitutes 35 to 40% of calories, is MUFA from olive oil and nuts and omega-3 fat from fish (26). In addition, a Mediterranean-style diet includes moderate amounts of alcohol, particularly wine, which has been associated with a reduced risk of cardiovascular disease and seems to be compatible with a healthy lifestyle (27).

Mediterranean diets have long been known for their delicious tastes and flavors. Currently, the many health benefits of Mediterranean-style eating are being realized. In the Mediterranean Diet Pyramid, nuts are on the same level as fruits, vegetables, beans, and legumes. They fall closer to the bottom of the pyramid, meaning they should be eaten more often.

PEANUTS AND NUTS IN WORLD CUISINES

Peanuts have been used in healthy traditional cuisines around the world. Asian, Mediterranean, African, and Latin American people have included peanuts in their cooking for centuries. Stir-fries with peanuts, soups and satays, salads and vegetable dishes, sweets and confections—the list goes on. As nutrition scientists are unlocking more of the health benefits of peanuts, chefs are finding more and more ways to include them in healthy cuisines.

ALLERGIES

Although milk and eggs are the most common food allergens, a peanut or tree nut allergy can be among the most severe. A food allergy is an immunological response in which the protein from a specific food triggers a reaction that results in the release of massive amounts of histamines and other mediators. Symptoms that can occur from this reaction include rash, cough, hives, asthma, swelling, difficulty in breathing, cramps, vomiting, and a fall in blood pressure. The most severe reaction is called anaphylaxis and can result in shock or death. Some individuals are so allergic that a very small amount of the allergen can trigger a severe response. The best protection is avoidance. In addition, the allergic individual should carry an epi-pen, which contains medicine that can relieve even the most severe symptoms until the individual can reach a hospital.

The exact number of food allergy sufferers is unknown, but it is estimated that 1 to 2% of the population is affected by some food allergy (28). Though individuals who experience food allergies as children often outgrow them, peanut, tree nut, and shellfish allergies tend to be lifelong. Researchers, with the help of industry, are working toward treatments to prevent or decrease the symptoms of serious food allergies.

BIBLIOGRAPHY

1. G. Fraser et al., "A Possible Effect of Nut Consumption on Risk of Coronary Heart Disease," *Arch. Internal Med.* **152**, 1416–1424 (1992).

2. R. J. Prineas et al., "Letter to the Editor," *New Eng. J. Med.* **329**, 359 (1993).

3. L. J. Appel et al., "A Clinical Trial of the Effects of Dietary Patterns on Blood Pressure," *New Eng. J. Med.* **336**, 1117–1123 (1997).

4. D. M. Colquhoun et al., "Comparison of the Effects of a High-Fat Diet Enriched with Peanuts and a Low-Fat Diet on Blood Lipid Profiles," American Heart Association, 1996, p. 733 (abstract).

5. F. B. Hu et al., "Frequent Nut Consumption and Risk of Coronary Heart Disease in Women: Prospective Cohort Study," *Br. Medical J.* **317**, 1341–1345 (1998).

6. C. M. Albert et al., "Nut Consumption and the Risk of Sudden and Total Cardiac Death in the Physicians Health Study," American Heart Association, 1998 (abstract).

7. M. Abbey et al., "Partial Replacement of Saturated Fatty Acids with Almonds or Walnuts Lowers Total Plasma Cholesterol and Low-Density-Lipoprotein Cholesterol," *Amer. J. Clin. Nutrition* **59**, 995–999 (1994).

8. J. D. Curb et al., "Comparison of Lipid Levels in Humans on a Macadamia Nut Based High Monounsaturated Fat Diet and a High Fat Typical American Diet," *American Heart Association Scientific Conference on Efficacy of Hypocholesterolemic Dietary Interventions*, 1995, pp. 3–5.

9. J. Sabate et al., "Effect of Walnuts on Serum Lipid Levels and Blood Pressure in Normal Men," *New Eng. J. Med.* **328**, 603–607 (1993).

10. M. L. Dreher, C. Maher, and P. Kearney, "The Traditional and Emerging Role of Nuts in Healthful Diets," *Nutrition Rev.* **54**, 241–245 (1996).

11. T. A. Pearson et al., "Weight Loss and Weight Maintenance: Effects of High MUFA vs Low Fat Diets on Plasma Lipids and Lipoproteins," *Experimental Biology '99*, April 19, 1999.

12. K. McManus, L. Antinoro, and F. M. Sacks, "Weight Reduction: A Comparison of a High Unsaturated Fat Diet with Nuts Versus a Low-Fat Diet," *Experimental Biology '99*, April 19, 1999.

13. C. M. Lermer and R. M. Mattes, "Effects of Chronic Peanut Consumption on Body Weight and Serum Lipid Levels in Humans," *Experimental Biology '99*, April 19, 1999.

14. B. Eissenstat et al., "Impact of Consuming Peanuts and Peanuts Products on Energy and Nutrient Intakes of American Adults," *Experimental Biology '99*, April 19, 1999.

15. G. A. Spiller et al., "Nuts and Plasma Lipids: An Almond-based Diet Lowers LDL-C While Preserving HDL-C," *J. Amer. College of Nutrition* **17**, 285–290 (1998).

16. L. Brown et al., "Nut Consumption and Risk of Recurrent Coronary Heart Disease," *Experimental Biology '99*, April 19, 1999.

17. T. H. Sanders and R. W. McMichael, "Occurrence of Resveratrol in Edible Peanuts," *American Oil Chemists Society Meeting*, Las Vegas, Nev., 1998.

18. A. B. Awad et al., "Anticancer Properties," *Experimental Biology '99*, April 19, 1999.

19. C. Ip and D. J. Lisk, "Characterization of Tissue Selenium Profiles and Anticarcinogenic Responses in Rats Fed Natural Sources of Selenium-Rich Products," *Carcinogenesis* **15**, 573–576 (1994).

20. F. B. Hu et al., "Dietary Fat Intake and the Risk of Coronary Heart Disease in Women," *New Eng. J. Med.* **337**, 1491–1499 (1997).

21. L. H. Kushi et al., "Dietary Antioxidant Vitamins and Death from Coronary Heart Disease in Post-Menopausal Women," *New Eng. J. Med.* **334**, 1156–1162 (1996).

22. E. Rimm et al., "Vitamin E Consumption and the Risk of Coronary Heart Disease in Men," *New Eng. J. Med.* **328**, 1450–1456 (1993).

23. M. J. Stampfer et al., "Vitamin E Consumption and the Risk of Coronary Heart Disease in Women," *New Eng. J. Med.* **328**, 1444–1449 (1993).

24. The Hazelnut Council Web site, URL: *http://www.hazelnutcouncil.org*.

25. National Pecan Sheller's Association Web site, URL: *http://www.ilovepecans.org*.

26. A. Keys, *Seven Countries: A Multivariate Analysis of Death and Coronary Heart Disease*, Harvard University Press, Cambridge, Mass., 1980.

27. E. B. Rimm and R. C. Ellison, "Alcohol in the Mediterranean Diet," *Amer. J. Clinical Nutrition* **61**, 1378S–1382S (1995).

28. Food Allergy Network Web site, URL: *http://www.foodallergy.org*.

GENERAL REFERENCES

American Peanut Council Web site, URL: *http://www.peanutsusa.org*.

International Tree Nut Council Web site, URL: *http://www.nuthealth.org.*
Macadamia Society of Hawaii Web site, URL: *http://www.hmna.org.*

PATRICIA KEARNEY
PMK Associates
Arlington, Virginia

POSTHARVEST INTEGRATED PEST MANAGEMENT

Effective management of pests within systems where food is produced, processed, and distributed is critical for maintaining an abundant, affordable, and safe food supply. More than 100,000 pest species are known to cause losses in crop and livestock production and in the systems designed for processing and distributing food commodities (1). It has been estimated that at least one-third of the food supply potentially available to the population of the United States is lost on an annual basis as a result of pest infestations during production and after harvest. In addition, nearly $8.5 billion is spent annually on chemical pesticides applied in agriculture and industry as farmers and processors attempt to reduce losses (2). In food production and processing systems, the primary approach to pest control has been one of eradication, with the objective of total elimination of pest populations (3). This approach has resulted from concerns, arising foremost in production of fresh fruit and vegetables, that any pest injury would cause these commodities to become aesthetically unacceptable to consumers (4). Also, these commodities have such high value that it is considered unreasonable to risk the possibility of reduced yields or grades of produce (5). Demands for eradication of pests have resulted in excessive reliance on chemical pesticides to the extent that applications often been made according to schedules without assessment of pest infestation levels.

Gradual changes in approaches to pest control are now occurring in food production, processing, and distribution systems. Whereas once little consideration was given to alternatives to chemical pesticides, greater interest now exists for implementation of controls that are less threatening to the environment and nontarget species (6,7). Integrated pest management (IPM) concepts are gradually being incorporated into the systems that supply food for the world's population. With acceptance of principles of IPM has come willingness on the part of farmers and processors to implement control programs that are aimed not at eradication of pests, but at reduction of infestations and close monitoring of the benefits vs costs of control measures.

The goal of this article is to review general concepts that are basic to IPM as developed in the production agriculture venue and to provide an overview of the current issues, methodologies, and practical concerns relevant to pest management in the postharvest food industry. The topics included pertain to management of insects or other arthropod pests that infest durable commodities such as grains, nuts and dried foods, and all the processed food products derived from these commodities. Pest issues related to postharvest handling of fresh commodities are discussed briefly. Canned or otherwise processed fruits and vegetables and prevention of microbial contamination are discussed in other articles.

WHAT IS IPM?

IPM is a systematic approach to pest regulation that emphasizes increased sampling to assess pest infestation levels and promote improved decision making so that control costs can be reduced and social, economic, and environmental benefits can be maximized (8). From the standpoint of benefit vs cost in IPM, the basic goal of control programs is defined by the concept of economic injury levels. The economic injury level (EIL) is defined as the pest infestation level at which the loss due to pest is equal to the cost of available control measures. The EIL concept is used as the basis for determining economic thresholds (often called action thresholds), defined as the level of infestation at which the *potential* loss due to a pest infestation *exceeds* the cost of an available control measure. In decision making regarding chemical pesticide applications, the economic return from use of a pesticide is maximized when the application is made at the time the pest population reaches the economic threshold level. Implementation of IPM has typically resulted in reduction of purchased inputs, such as chemical pesticides, because of more effective assessment of pest infestation levels and well-defined criteria for determining when controls are warranted. In crop production systems, programs involving entomologists, plant pathologists, weed scientists, and economists have yielded cost-efficient tactics for suppressing pests while limiting contamination of the environment and the harvested food commodities. IPM concepts are now emphasized for postharvest systems using benefit-vs-cost principles to guide pest control decisions in storage and distribution systems, just as has occurred in field settings. Considering ecological, as well as economic concerns, IPM is regarded as an essential approach to preserving a safe food supply and healthy environment, while keeping U.S. agriculture competitive and profitable.

Explaining the context in which the terms *pest* and *management* are used can be quite helpful to understanding concepts of IPM. The term pest has historically referred to insects, weeds, plant pathogens, and rodents that compete for resources valued by humans in production, processing, and distribution systems for agricultural commodities. Thus, species attain pest status in the context of their association with plant and animal species used by humans as sources of food and fiber. The abundance of these species, and hence their importance as pests, is often enhanced in modern production and distribution systems for agricultural commodities because these systems result in abundant habitats and vast supplies of resources unlike anything that occurs in natural ecosystems.

IPM makes use of decision-making processes needed to produce commodities in carefully planned systems intended to keep pest populations from reaching economically damaging levels, while maintaining profitability of the enterprise with limited adverse environmental and social effects. The process of managing pests has significantly broadened in definition and scope during the past 30 years from unilateral controls applied against single species; to integrated controls employing multiple tactics against single species; to IPM programs designed to regulate insect pests, pathogens, and weeds in production systems; and to quite broad concepts of biointensive IPM (1) and integrated resource management (9). There is a substantial need for increased research and implementation efforts to keep pace with the theoretical development of management concepts. Practical aspects of implementation have tended to lag far behind IPM theory (1).

DEVELOPMENT OF IPM PROGRAMS

Basic changes in decision-making processes are important to development of effective IPM programs. In the past, eradication of pests was often the primary objective; now programs must address ecological, economic, and health-related concerns in conjunction with acceptable levels of pest regulation as defined by the EIL concept. IPM programs require time and expertise devoted to assessment of problems and decision making to gain maximum returns on inputs such as chemical pesticides. Significant resources must be committed to training of personnel or hiring consultants who have the necessary expertise to accurately monitor production fields and postharvest facilities for the presence of pest infestations and decide on appropriate management options. Before IPM programs were developed, applications of pesticides were routinely made according to schedules as insurance or prophylactic treatments. However, to address current economic, environmental, and social concerns, IPM programs require a greater time commitment for comprehensive assessment of pest infestation levels to ensure that control options are employed in a judicious, as opposed to routine, manner.

Food processors and distributors must realize that most pest management decisions have consequences far beyond the time and location that pests must be controlled, a lesson learned in field settings during the past 30 years. Comprehensive plans for pest management greatly improve effectiveness, profitability, and safety of pest control efforts over what can be achieved with piecemeal approaches. Important keys to effective, profitable, and environmentally safe pest management are careful monitoring of pest populations, complete records of infestation levels and controls applied, and comprehensive benefit-vs-cost analyses. To provide necessary training and technology, IPM programs must be supported by multidisciplinary research and extension efforts through the USDA and land grant universities and working with the food industry. A broad range of control options available for incorporation into IPM programs now exists.

1. *Biological controls.* Natural enemies such as parasites, predators, pathogens, and competitors of pest species in applied control programs.
2. *Cultural controls.* Methods such as tillage, host plant resistance, and crop rotation to reduce pest infestations in production systems.
3. *Chemical controls.* Chemical toxicants, repellents, protectants, growth regulators, germination inhibitors that are used to regulate populations of insect pests, plant pathogens, and weeds.
4. *Mechanical and physical controls.* Approaches that have greatest application in postharvest pest management include heat and cold treatments, sanitation, and protective packaging.
5. *Legislative controls.* Imposition of inspection and quarantine regulations to prevent importation and spread of introduced pests in living plants and animals or in harvested commodities.

IPM programs combine these control options to achieve the safest and most cost-effective regulation of all types of pests in food production, processing, and distribution systems.

SETTING THE STAGE FOR IPM: THE PESTICIDE ERA

Throughout most of the history of agriculture, a lack of highly effective, unilateral control measures (such as modern-day chemical pesticides) resulted in the use of various combinations of controls (eg, integrated control) to reduce losses caused by pest species. Controls included a variety of cultural measures such as cultivation and crop rotation; removal of insects and weeds from crops by hand; and application of inorganic pesticides containing active ingredients such as sulfur, lime, and arsenic. It was not until the modern insecticide era after World War II that use of highly effective compounds such as DDT and BHC (benzene hexachloride) introduced a new philosophy of pest control. Within a few years of their first applications, these organochlorine insecticides were used extensively to eradicate pest populations. Little concern was given to the potential for deleterious consequences to nontarget species, hazards to farm workers, or harmful residues in human food. The ready availability and high degree of effectiveness of these pesticides resulted in reliance on them as a unilateral control. By 1966, nearly 150 million pounds of insecticides were being applied per year in the United States, most of which was used in crop production (10). Although insecticides were the first of the chemical pesticides to be used in such vast quantities, research efforts led to the discovery of phytotoxic chemicals that were used as herbicides. Herbicide use peaked at more than 450 million pounds per year in 1982 (10). The increase in the quantity of fungicides applied has been much more modest, from 21 million pounds in 1966 to 37 million pounds in 1992 (2).

The availability of highly effective chemical insecticides has certainly been an important factor contributing to the significant reduction in incidence of arthropod-borne diseases in humans and domesticated animals. Chemical pesticides contributed to the 230% increase in agriculture productivity in the United States from 1947 to 1986 (11). Ease of application and the relatively low cost of these broad-spectrum pesticides have resulted in their tremendously high levels of use in food production and processing systems. By 1970, insecticide use in cotton in the United States had exceeded 70 million pounds per year. By 1985, herbicide use in corn had reached 250 million pounds per year (10).

The extensive reliance on chemical pesticides has led to serious problems. Questions relating to continued use of these compounds continue to be asked and have resulted in the cancellation of registrations for many uses of these products. In extreme instances, as many as 40 to 60 applications of pesticides have been made per year in cotton (12), with the result being serious issues of environmental pollution and mortality of nontarget species. The following are among the problems resulting from excessive use of chemical pesticides.

- Environmental contamination where pesticides in soils have entered surface and groundwater, threatening the safety of water supplies for human and animal consumption (2,10).

- Increasing human health risks resulting from exposure to residues in food and water supplies, particularly relating to the potential of some pesticides to cause cancer, disrupt the endocrine system, and interrupt normal development of the central nervous system in children (2). Despite cancellation of registrations for nearly all organochlorine insecticides in the U.S., residues of these compounds are still commonly detected in food products because of their use in other countries and long life in the environment.

- Threats to nontarget organisms such as wildlife species in both terrestrial and aquatic environments (2).

- Mortality of beneficial parasitic, predatory, and pathogenic species that often serve important roles in regulating pest populations (2,13). Destruction of beneficial organisms contributes to problems of pest resurgence and outbreaks of secondary pest species.

- Development of resistance in pest species to chemical pesticides. This is a growing problem for all types of pests worldwide, with more than 600 arthropod species, more than 100 species of plant pathogens, and more than 100 species of weeds exhibiting resistance to insecticides, fungicides, and herbicides, respectively, by 1996 (14).

- Increasing costs associated with intensive usage of chemical pesticides in crop production (10).

The increased urgency to find solutions to these problems has resulted in the passage of several major pieces of legislation by the U.S. government. The Federal Environmental Pesticide Control Act of 1972, comprised of extensive amendments to the Federal Insecticide, Fungicide, and Rodenticide Act of 1947, and the Food Quality Protection Act of 1996 will continue to have great impacts or availability and use patterns for chemical pesticides in the future. These problems have also generated great emphases for research and extension efforts to develop and employ alternative controls.

SETTING THE STAGE FOR IPM: ENTERING THE IPM ERA

Although environmental contamination and food safety concerns have resulted in some increased interest in de-velopment of integrated control programs to reduce reliance on chemical pesticides, the major impetus for reduction in their usage has resulted from political activism. Rachel Carson's *Silent Spring* (15) was a catalyst that resulted in a strong negative reaction by the general public to the extensive use of chemical pesticides, especially the organochlorine insecticides such as DDT. The public reaction to threats of environmental pollution and contamination of the food supply by pesticide residues (16) resulted in cancellation of all registrations for most organochlorine insecticides. Although insecticides belonging to the organophosphate and carbamate classes replaced the organochlorine compounds for most uses, growing support among farmers and consumers for integrated control programs has resulted in reduced reliance on these chemical insecticides and a movement from an industrial to an ecological model of pest management (17).

Public support for the IPM concept has led to gradual development of a philosophy that integrates pest control tactics and crop management from both agronomic and economic standpoints (18). From an original, relatively narrow focus on field-crop production (19), application of principles of IPM has now attained a much broader focus to include processing and distribution systems for food commodities. In relation to crop production, application of IPM provides a basis for continued profitable production of food and fiber commodities with much less threat of environmental degradation or harmful residues in food. Employing principles of IPM in processing and distribution of commodities is an additional, important step in maintaining a safe and wholesome food supply.

FOOD SAFETY AND IPM

Many countries maintain strict guidelines for the wholesomeness and safety of food products sold to end-user consumers. Food products must be free of contamination by harmful chemical residues and biological organisms, such as insects or microorganisms. They must also comply with standards of purity, composition, and preparation that are specified by regulatory agencies (see the article FOOD REGULATIONS: INTERNATIONAL, CODEX ALIMENTARIUS). The food safety challenge for industry requires delivery of products that are free of pest contamination while also containing pesticide residues that are below legal minimum levels.

Pests, Pesticides, and Regulations

Hundreds of species of insects and mites are known to infest grain and other raw commodities in storage, and many of these organisms infest grain-based food products after processing and packaging. Pest infestation in raw grain can begin a cascade of problems with serious economic and food safety implications. Infestation of bulk-stored grains by insects can result in weight loss of the grain, as well as, increased heating and moisture levels that facilitate growth of molds and spread of mycotoxins. The end result of these combined infestations by insects and molds is that grain may be rendered unusable (20).

Most insects infesting grain being prepared for milling can be removed with grain-cleaning equipment before grinding into flour; however, eggs of some species may survive cleaning and contribute to subsequent infestations in processed foods. Additionally, insect larvae and pupae inside grain kernels cannot be removed, but are milled with the flour and remain as fragmentary contaminants. Insect fragments, rodent hair, feces, and other foreign material are classified as filth that must be kept below legal threshold levels. Though not toxic, insect fragments can have direct human health effects as allergens for some persons (21). Contamination by insect-related filth signals an overall low level of quality and wholesomeness of food products.

Chemical insecticides are used in two principle ways to limit infestation of stored grains by insects. Insecticides in the form of residual protectants are used as grain enters storage facilities to prevent infestation during storage (see the article GRAINS AND PROTECTANTS). Fumigant insecticides are used to eliminate insect infestations that have developed after commodities are placed in storage. Fumigants such as phosphine and methyl bromide usually leave no detectable residue that may contaminate food. For this reason, fumigants are not considered significant threats to food safety; however, the lack of residual effects with these chemicals means that they do not protect grain from reinfestation after treatment. By contrast, residual protectant insecticides, which are used to treat raw grain and are applied throughout the food system to treat structures and equipment where foods are processed and packaged foods are stored, may pose problems through contamination by residues. Insecticide residues can remain in stored grain for months or years; chemical degradation is slow in the storage environment because of low light or darkness, stable or gradually changing temperatures, and relatively low moisture levels. Milling and, to a greater extent, baking or cooking cause substantial degradation of residues. However, some residues may still remain as the original chemical used in treatment or in the form of degradation products (22).

Pesticide residues in food pose both proved and suspected health risks to humans, and they generate much public concern and governmental regulatory action throughout the world (see the article PESTICIDE RESIDUES IN FOOD). The commonly used residual insecticides applied in grain storage facilities belong to the organophosphate and pyrethroid chemical classes. In the United States, no pyrethroids and relatively few organophosphate insecticides are registered for application directly to grain. However, additional organophosphates and certain pyrethroids can be used in structural surface treatments (see the article GRAINS AND PROTECTANTS). The Codex Alimentarius Commission has established international guidelines for permissible residues of organophosphates and pyrethroids in both raw grain and processed food products. Individual governmental regulatory agencies in most countries have generally concurred with these guidelines by passing appropriate regulations for enforcement.

Recent surveys of U.S. stored grain for detection of pesticide residues have not revealed levels above established tolerances, but the majority of samples contained detectable residues of organophosphates (Table 1). In spite of evidence for compliance by the food industry with established tolerance levels for pesticide residues, consumer advocacy

Table 1. Detection of Insecticide Residues in Wheat Samples from Grain Elevators in Major Wheat-Producing Regions of the United States

Insecticide	Percentage of samples with detectable levels in a given year[a]		
	1995 (600)	1996 (340)	1997 (623)
Carbaryl	0.5	0.3	0.3
Chlorpyrifos	19.5	14.4	6.4
Chlorpyrifos methyl[b]	54.2	73.2	55.6
Dichlorvos	0	0	0
Malathion[b]	71.0	70.3	68.2
Methoxyhlor	1.0	4.7	5.2
Methyl parathion	0	0.3	0.2

Source: Data compiled from Refs. 23–25.
[a]Number in parentheses refers to the number of samples analyzed in a given year.
[b]Registered by U.S. Environmental Protection Agency for direct admixture to grain as a protectant; other insecticides residues presumably resulted from other applications.

organizations have called for stricter tolerance guidelines. The Food Quality Protection Act (FQPA) passed by the U.S. Congress in 1996 mandates a reassessment of all existing pesticide registrations with strict attention to human safety. The FQPA emphasizes accumulated risk resulting from exposure to classes of chemical pesticides, not just to individual active ingredients. For example, all organophosphates used in a variety of applications to food commodities, inside homes, on lawns, and in public areas will be considered together in estimating accumulated exposure. Risks will be assessed based on the most sensitive members of the population, among those being infants and children. Residues in food appear to constitute the highest exposure risks because they result in direct ingestion of pesticides.

Organophosphate insecticides have been targeted as a class for initial review and potential elimination by the U.S. Environmental Protection Agency because of their activity as neurotoxins and their potential as endocrine disruptors (26). Cancellation of registrations for organophosphates currently applied to grain and used throughout the food industry would create a great need for effective controls as replacements.

The fumigant methyl bromide is a fast-acting, residue-free chemical that is used routinely in flour mills, food processing facilities, and warehouses to control insect infestations. It has perhaps been the most effective insecticide used in the value-added food industry, but registrations for this product are slated for cancellation. Because methyl bromide is an ozone-depleting substance and may pose a threat to the protective atmospheric ozone layer around the earth, its production and application are scheduled to cease in the United States by 2005 as mandated by the Clean Air Act. Other industrialized and developing countries worldwide are eliminating or severely reducing the use of methyl bromide over the next several decades in compliance with an international agreement known as the Montreal Protocol (27). In this case, an environmental rather than a food safety concern is fueling the campaign for cancellation of an effective and residue-free insecticide for the food industry. This cancellation could possibly lead to increased use of residual insecticides. Ideal substitutes for methyl bromide have not been found, but much research and discussion are aimed at developing safe alternatives for the food industry. One alternative that may become quite important is IPM, an approach that could keep chemical insecticide usage at a minimum while utilizing a variety of alternative controls to limit pest populations.

The Case for IPM in the Food Industry

It seems apparent that the impact of more stringent regulations on pesticide use in the food industry, in combination with growing consumer demands for more wholesome and pesticide-free foods, must result in more extensive adoption of IPM practices. Although purchase contract specifications for grain commonly stipulate that there can be no detectable pesticide residues, they also stipulate that the grain be of relatively high quality with little insect damage (28). In addition, the tolerance levels for insect contaminants in value-added, finished food products are even lower than those for raw commodities, essentially a zero tolerance. For the future, it is clear that raw commodities destined for use in food products must be kept in good condition by combined action of several preventative measures employed through an IPM program. Similarly, pest-free, finished products in warehouses and retail outlets that are at risk for reinfestation must be protected until purchased by the consumer.

Managers at all levels of the food processing and marketing continuum, from bulk storage of commodities to retail outlets of high-value, finished products, must adopt IPM strategies that include sanitation, prevention, effective monitoring, and informed decision making regarding pest suppression tactics. Informed decision making based on cost-vs-benefit analyses is the cornerstone of IPM and should guide pest control decisions at all levels of the food industry. The EIL concept explained earlier in this article must become the basis for these decisions. Because the relative value of products varies greatly through the different levels of the food industry, so also will the economic thresholds used to guide control decisions.

IPM FOR FRESH COMMODITIES WITH ZERO PEST TOLERANCE

Fresh fruits and vegetables are sold as soon after harvest as possible, and there is typically little time or need for postharvest IPM. However, IPM is practiced during production of these commodities to prevent damage, and those involved with postharvest handling must be concerned with quality preservation. Only undamaged, blemish-free fruits and vegetables are selected for the fresh table market. Producers must avoid damage that is evident in an exterior examination and also eliminate insects confined within tissues of harvested produce. These pests are not only serious problems for the domestic market, but in the export market they are often the targets of legal quarantines enforced to prevent the spread of pest species.

Fumigants such as methyl bromide and ethylene dibromide were once commonly used to ensure that fresh fruits and vegetables for export were pest-free (29), but these chemicals are now being replaced by nonchemical methods. Safe and effective physical controls are being used commercially, such as cold treatment to eliminate larvae of codling moths from apples and vapor heat treatment to kill fruit flies infesting papayas (30). Ionizing radiation is an effective treatment for killing immature insect life stages inside fruits and vegetables while maintaining product quality (31). However, radiation has only limited commercial adoption for use with fresh produce because of serious opposition in some locations by consumers who reject irradiated food. Research on controlled atmospheres using high carbon dioxide concentrations or low oxygen levels shows promise for controlling postharvest pests of fresh commodities, but practical limitations in maintaining proper treatment conditions has hindered commercial adoption of these methods (32).

POSTHARVEST IPM FOR RAW, DURABLE COMMODITIES

Although IPM methods discussed here address stored grain almost exclusively, the same concepts and methods apply to other durable commodities such as legumes, nuts, dried fruits, and other dried commodities that can be stored in bulk. Cereal and feed grains comprise the most important basic inputs to the world food supply, and because of their seasonal growth patterns, they must be safely stored and maintained for year-round consumption.

Grain Storage: Pest Prevention and Sanitation

Insect pest problems in stored grain can be prevented or delayed by limiting access to the storage facility. With exception of a few species that may infest ripening grain to a limited extent before harvest, stored grain insects do not infest crops in the field and normally do not enter storage facilities with the new harvest (33). Insects that become pests in storage facilities must have dried grains or other appropriate food on which to feed and reproduce. Typically, the source of insects infesting a newly stored commodity is from populations already infesting some previously stored material. A common source of infestation is from residues of grain in empty storage structures (bins or silos), harvesting and conveying equipment, and transportation vehicles. Grain storage structures and grain-handling machines must be thoroughly cleaned of residual grain before a subsequent crop is stored. If grain is stored in an insect-free structure then the only source of insects to infest grain is by immigration into the structure. Every effort must be made to repair leaks and close openings in grain bins and silos to prevent insect entry. Complete exclusion of insects from grain storages is extremely difficult, but minimizing entry points does reduce potential routes of infestation. Once storage structures are cleaned and sealed, surfaces should be treated with a residual spray of a registered insecticide to provide protection against insects that may remain hidden within the structure or that may immigrate into the bin soon after grain is stored.

Monitoring Storage Conditions

Insect population growth in grain is dependent primarily on temperature of the air and/or grain, moisture content of the grain, and the quality of the grain as a food source (33). Because most grains and grain products are nutritionally adequate, unless they have been severely degraded by microbial agents, food quality is rarely limiting for insects infesting stored grain. Temperature and grain moisture may vary greatly among storage situations and are important for insect survival and population growth. Biological factors, such as development time from egg to adult, survival rates, and the number of eggs produced by females all attain maximum levels within optimal ranges of temperatures and grain moisture. Typically, these life history variables decrease at temperatures and moisture contents either above or below the optimal zone. Insects that feed as larvae inside grain kernels are more sensitive to grain moisture content than are the external feeders. Optimal grain moisture levels for growth and survival of internal grain feeders ranges from 12 to 15%. Optimal temperatures for growth and development of stored grain insects are between 25 and 32°C. Most species experience mortality or reduced population growth at prolonged periods above 40°C and cease development below 15°C (34). It is clear, therefore, that insect population growth can be slowed by maintaining grain in a cool, dry condition.

For effective stored-grain IPM, it is essential to have the best information possible about the current status of the commodity being managed and the potential pest populations at any time. As grain enters the storage facility, it should be characterized for quality factors such as bulk density, percent composition comprised by broken kernels and foreign material, and moisture content. Information on initial moisture content is critical because high-moisture grain may need to be stored separately, mechanically dried, or later blended with drier grain to attain a composite grain mass that is safe for storage.

The storage structure should be equipped with thermocouple cables or similar devices to monitor grain temperature at various locations throughout the grain mass during the storage period. Temperature cables are typically hung from the roof of a storage structure and extend vertically through the grain mass. Thermocouples are positioned on cables at a spacing of 2 to 3 m from bottom to top of the mass, and a series of cables should be deployed at 5 to 10 m intervals so that the grain mass is evenly monitored. Temperatures from each monitoring point should be recorded on a regular basis, ideally weekly or biweekly, and any abrupt increases in temperature over a short period of time should be investigated by sampling the grain mass. Temperature cables are useful in detecting hot spots where moisture, mold growth, and insect infestation may have begun. Temperature monitoring is also important to check the progress of grain cooling during aeration.

The species and numbers of insects in stored grain can best be monitored by direct and systematic sampling of the grain and by use of traps. Direct grain sampling is the best monitoring approach, but it is labor intensive and requires a certain level of skill. Managers of grain storage facilities can easily learn the required methods and implement grain sampling for insects as part of their regular commodity management routine.

Samples can be obtained to a depth of 1 to 2 m by using a spear-shaped sampler to withdraw grain from the mass, or sampling may be done as grain is moved from bin to bin within a facility using a cup-type (pelican) sampler (Fig. 1). Increasing the number of samples enhances the level of accuracy achieved in estimating insect population density (35). For example, five 1-kg samples must be taken from each 27 metric tons (1000 bushels) of grain to determine with 90% accuracy if a grain mass contains as many as two insects per kg, which is a common economic threshold for control decisions. Grain samples must be processed with an appropriate sieve or sifting device to remove insects for identification and counting. Grain probe traps are also available that capture insects as they move through a grain mass (Fig. 1). These traps are effective for detection of insects at low population densities (36), but the data obtained are not easily converted to numbers of insects per unit of grain volume.

Decision Making for Insect Control

Correct interpretation of insect sampling data is important for making control decisions in stored grain IPM. As indicated earlier, decisions should be based on cost-vs-benefit analyses that incorporate assessments of the quality parameters. Raw grain is marketed based on quality, which is quantified by a numerical grade in the United States. The highest quality grain, in addition to having a high bulk density (test weight) and little or no foreign material, contains no live insects and has no measurable damage from insects. Damaged kernels result from feeding within by species such as weevils in the genus *Sitophilus* and the lesser grain borer, *Rhyzopertha dominica*, that chew holes through kernels. If one or more individuals of these species are found in a grain sample or probe trap sample, then more extensive sampling should be conducted. Discovery of damage by internal feeders may warrant a fumigation treatment to eliminate the infestation and prevent additional damage.

The presence of large numbers of external feeding insects, which cause no direct grain damage, may be revealed in grain or trap samples, but they can be tolerated until the time of sale as long as grain quality, temperature, and moisture remain acceptable. Live external feeding insects can be eliminated with an effective fumigation before grain is transported. It should be noted that many international markets will not tolerate even dead insects in grain, so treatment decisions must be adjusted accordingly.

Nonchemical Controls for Stored Grain Pests

Effective IPM programs for stored grain involve a variety of approaches intended to prevent the occurrence of damaging pest infestations. Sanitation was discussed earlier as a preventative measure used before grain enters the storage facility. Among the most effective nonchemical means for pest prevention or suppression after grain is in storage is use of ambient air cooling with aeration. This is achieved with air flow from fans mounted at the base of the storage structure that either blow air into the bottom of the mass and force it upward or draw air through roof vents downward through the grain mass. Effective aeration cooling occurs only when the outside air temperature is below the temperature of the grain. Thermostatically

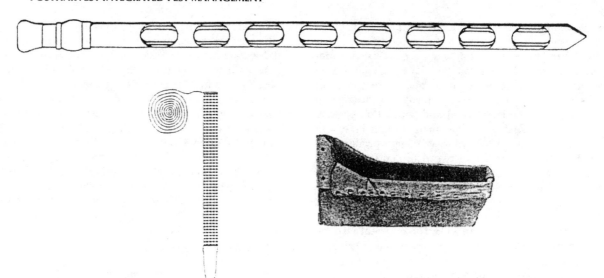

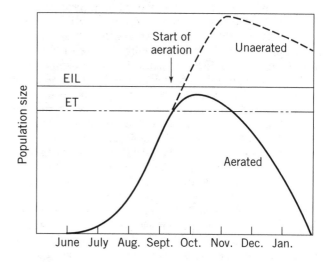

Figure 1. Devices used to sample grain for insect population monitoring. (**a**) A spear-shaped grain trier to extract a sample of grain from bulk storage. (**b**) An insect probe trap that is inserted into a grain mass; insects moving through the grain pass through the holes in the trap and fall into the collection tip. (**c**) A pelican sampler used to collect a sample of flowing grain as it falls vertically.

controlled fans can be used to ensure that they operate at the proper ambient temperatures (37). Because insect developmental rates are directly influenced by temperature, lowering the temperature of the grain mass below the optimum level for insect growth (below 15°C for most species) can significantly reduce population increase and the potential for damage. Low grain temperatures are achieved most efficiently in temperate regions when seasonal changes from summer to fall result in lower ambient air temperatures. However, even if grain cannot be cooled below 15°C immediately, any level of cooling can be effective in reducing rates of insect population increase (33).

Simulation models of temperature-dependent insect population growth rates have been developed for the most damaging stored grain pests. These models verify the benefits of aeration for suppression of pest populations (Fig. 2). Although costs of aeration depend on local electricity costs, in most cases expenditures are justified based on reductions in insect infestations that can be achieved (38).

Biological controls may provide additional nonchemical methods for pest control in stored grain. The community of insect species that infests stored grain includes those that consume the grain itself, those that feed on molds that infect the grain, and parasitoids and predators that feed on other insects. Studies have been published in which experimental populations of grain pests have been reduced by addition of natural enemies to storage containers (39). A recent study conducted in grain bins has validated the efficacy of introducing parasitic wasps to suppress populations of the lesser grain borer (40). The objective of these studies has been to identify beneficial species that can maintain pest populations below the EIL. Although some companies mass produce and sell parasitoids and predators of stored grain pests, the use of beneficial insects in grain storage facilities has not yet been widely adopted.

Figure 2. Conceptual plot of change in the size of a grain insect population, with and without aeration to cool grain. ET, economic threshold; EIL, economic injury level.

Chemical Insecticides for Stored Grain Pests

When alternative nonchemical controls fail to maintain insect pest populations below EILs, chemical insecticides remain the most effective tools for pest suppression in stored grain. Fumigant insecticides are ideal for bulk-stored commodities because they are released as gases that penetrate the commodity in place. Phosphine gas, or hydrogen phosphide, is the most commonly used fumigant for stored grain worldwide (41). Phosphine formulated as pellets or tablets remains a relatively inexpensive control costing a few cents per bushel (38). Phosphine is liberated from pellets or tablets of aluminum or magnesium phosphide upon reaction with water in the air. Because phosphine has a density close to that of air, it penetrates bulk commodities well. It must be stated, however, that this same low-density property can be a detriment because it allows phosphine to dissipate easily if structures are not sealed well. Aluminum phosphide pellets are typically added to grain that is being transferred from one storage structure to another or to a transport vehicle, allowing even distribution throughout the grain mass.

Recirculation methods have been developed that allow adequate penetration by the gas even when pellets are applied just to the top of a large grain mass and/or to a bottom access via the aeration plenum. Phosphine accumulated in the headspace above the grain is drawn through a pipe in the roof of the structure and reintroduced at the bottom of the mass, where it travels upward through the mass and attains even distribution (42). Phosphine is relatively slow acting, and several days exposure at optimum temperatures may be required to produce insect mortality. Long exposure time for bulk grain is usually not a problem because considerable time is available during storage or transport. Treatment times may be reduced by use of phosphine gas dissolved in carbon dioxide and contained in gas cylinders now becoming available in some countries. This formulation also provides a means for rapid, even distribution of gas throughout a storage structure.

Grain protectants are among the most commonly used controls for insects in stored grain. These residual insecticides are applied to grain at the time it initially goes into storage to provide protection against infestations for several months. Although more expensive than phosphine on a per bushel basis, protectants are the preferred control if a high risk of infestation exists. Among the disadvantages of grain protectants, as mentioned earlier, is the potential for residues to remain in the grain at the time of sale. Also, as the time grain is kept in storage increases, the possibility exists that the insecticide may lose effectiveness. There is also evidence that some species are not effectively controlled because of the development of resistance to the insecticides (43). Malathion, for example, remains a commonly used grain protectant in the United States, though it is well established that many populations of the common stored grain insects have evolved resistance to this compound (43). Because of these problems and the growing expense of protectant applications, the future of this approach for control of stored grain insects is in doubt. In addition, the use of protectants is not consistent with principles of IPM because they are often applied with little assessment of potential pest infestation levels. It is possible that use of insecticides as preventative measures will be restricted to applications as residual sprays in empty structures as part of a comprehensive sanitation program.

Diatomaceous earth is a naturally occurring absorbent dust that poses no chemical residue or food safety problems. Insects are naturally protected from dehydration by waxes deposited on the surface of the exoskeleton. Diatomaceous earth kills insects that contact it by absorbing and removing waxes from the exoskeleton, and death occurs due to desiccation (44). However, diatomaceous earth is as expensive to use as synthetic chemical insecticides, and it reduces the bulk density and the value of grain with which it is thoroughly admixed. Additionally, because its effectiveness is yet to be well established, the future contribution of diatomaceous earth to IPM in stored products is not clear.

IPM FOR FOOD PROCESSING

The information in this section relates to any facility in which a raw commodity such as grain is processed in any way to make a more valuable product; such a product may be ready for the consumer market or it may be an intermediate product that is processed further at another facility. Thus, food processing here refers to activities of flour mills, bakeries, pet food plants, breakfast cereal manufacturers, snack food plants, chocolate factories, and confectioneries.

Stored-product insects that infest food processing facilities are essentially the same species that infest raw grain, but the species composition and their relative importance is different. Internal grain feeding insects such as weevils and borers are not as important in processing environments because the concern for grain damage is no longer an issue. Species considered secondary in importance in raw grain because they feed only on broken kernels, such as flour beetles, sawtoothed grain beetles, and the Indianmeal moth, are considered of primary importance in food processing because they can easily infest milled products and contribute to problems with contaminants. For example, the Association of Operative Millers has listed the red flour beetle as the most important insect pest of flour mills because of its ability to survive and reproduce in flour. This species is also difficult to control in flour mills (45). By comparison, the red flour beetle, though common in stored grain, is rarely considered more than a nuisance. Also important is the consideration that once grain is milled into flour, its value increases greatly, and EILs for pests decrease accordingly.

Sanitation and Prevention

Food processing facilities must devote personnel and other resources to sanitation, pest monitoring, and pest control if they are to maintain a level of cleanliness and product quality that is demanded by consumers and mandated by governmental regulations. As with stored grain, preventive measures are basic to pest management programs. For flour mills and other processing plants, the sanitation routine should include daily or twice daily cleaning of equipment and floors where flour dust can accumulate (46). Use of compressed air to blow dust from machines and floors is a popular cleaning tool, but it should used sparingly so that dust is not simply blown into hard-to-clean areas. Food debris that accumulates in these areas can support residual populations of pests that constantly contribute to infestation of the product. Sweeping is effective when combined with vacuum cleaning. Problem areas for dust accumulation should be identified and regularly targeted with vacuum cleaning. Spills need to be cleaned immediately and broken packages discarded. Broken windows, window screens, and other structural components should be repaired promptly to limit entry of food pests into a facility. Rotation of raw and finished stock to utilize older products first represents a simple preventive measure to avoid the opportunity for growth of pest populations. Common-sense approaches to sanitation and prevention need to be adopted, but in many facilities these simple measures are often overlooked in favor of increased production.

Monitoring for Pest Infestations

Active efforts at pest detection and monitoring should be incorporated into a regular routine with sanitation and prevention efforts of food plants. Direct sampling of products and residues from the production line can alert managers to the presence of incipient pest problems (46). As dust and siftings from different points in the milling process are cleaned up, these materials should be carefully inspected daily for the presence of insects. Similarly, floor sweepings and the residue from machinery clean-outs must be inspected. Boots of elevator legs and the bottom trays of vertical conveyor systems contain spaces where food and pests can accumulate, and they should be cleaned and inspected regularly. Insect traps baited with attractants for certain species, or designed to capture any insects present, can be used throughout a facility to determine the presence, relative abundance, and location of pests (47). Traps baited with pheromones are highly sensitive to the presence of the target species for the attractant. By deploying traps in a regular gridlike arrangement throughout facilities, managers can identify areas with the highest likelihood of insect infestation and concentrate monitoring and control efforts in these areas. Monitoring data from traps or product and residue samples must be documented and the records maintained over time, so that trends in pest numbers can be observed and used in IPM decisions.

Controlling Pests Without Methyl Bromide

Methyl bromide has been used effectively for many years to suppress pest populations in structures such as flour mills and food processing plants (48). Grain millers and other processors consider this chemical to be an essential component of their pest control and sanitation programs. Methyl bromide will be unavailable for most uses in the United States and in most other industrialized countries beginning in 2005. No fumigant gas is available that can directly replace methyl bromide for rapid, effective insect control without the potential harmful residues in food products. Phosphine is registered for use in structures containing food and leaves no harmful residues on food. However, phosphine often requires several days to achieve effective insect control, compared to several hours required with methyl bromide. Large flour mills and similar plants continue production 7 days weekly and cannot operate profitably with the long shutdowns that would be required for phosphine fumigation. Another concern of many food processors when considering fumigation with phosphine is that copper and other metal components of machinery can be corroded by phosphine at high humidities and warm temperatures (49). Electrical wiring, computers, microprocessors, and other sensitive electronic equipment may be put at risk when phosphine is used in flour mills or food processing plants. One innovative application method for phosphine as a substitute for methyl bromide is use of relatively low doses with elevated temperatures and the addition of carbon dioxide. This method may increase activity of the insecticide and reduce potential damage to machines by shorter exposure times, especially in areas where sensitive electronic equipment is housed (50,51). The presence of carbon dioxide and higher temperatures apparently increases respiration rates of insects and enhances mortality.

Various other physical and chemical methods have been proposed as alternatives to use of methyl bromide for pest control in food processing; however, none can be considered an effective replacement in terms of efficiency and effectiveness (52). Controlled atmospheres with high carbon dioxide or low oxygen concentrations have been proposed but are not practical or cost effective for food processing facilities because of the need for enormous amounts of gases released in structures that cannot be sealed tightly. Heat treatment, or heat sterilization, of entire facilities has been attempted with some success; however, for effective insect control it is necessary to maintain the temperature at approximately 50°C for 24 to 36 h (53). A major problem with heat treatment is the difficulty in attaining temperatures lethal to insects inside walls and machines and close to floors in basements.

Numerous residual insecticides have been proposed as replacements for methyl bromide. Their effectiveness is limited, however, because they often cannot be applied in all areas where control is needed because of risks of contamination of food products by chemical residues. Consequently, control must be attained through residual activity on treated surfaces where insecticide residues are not likely to contaminate food products with the hope that insects may be controlled through contact with treated surfaces (43). Also being evaluated in flour mills is a combination of elevated temperatures with a thorough dusting of diatomaceous earth (54). This method also has limitations for control of insects harbored deep within structures. It seems apparent that when registrations for methyl bromide are finally cancelled, food processing facilities will be forced to develop more intensified efforts at prevention, vigilant monitoring, and specialized mitigation measures to replace the current unilateral applications of methyl bromide for pest control.

IPM FOR VALUE-ADDED FOOD PRODUCTS

At the end of the food processing continuum are the high-value, packaged products ready for retail marketing to consumers. After leaving a processing facility, these products are typically transported to regional wholesale distribution centers where they may be stored for periods ranging from a few days to many months. Although ownership of these products may reside with retail marketers, problems with pest infestations during storage and even at retail outlets are often blamed on the food processors. Regardless of who takes the responsibility for protection of these products, substantial investments in preventive measures must be made to protect finished products from infestation.

Internal corporate regulation of finished product quality must be maintained through a system of regular inspections and business-partner accountability. As a first step, conscientious food companies should set rigorous, internal standards and conduct sanitation audits on their own facilities and those of suppliers to ensure that their standards and those imposed through governmental regulations are met. Sanitation standards or specific good manufacturing practices (GMPs) are sometimes outlined in supplier contracts, placing suppliers at risk of losing business should their facilities fail to pass sanitation audits (55). All facilities should be inspected through the auditing system, including mills, processing plants, warehouses, and retail outlets.

Companies that produce value-added products can take several measures to prevent pest infestations through protective food packaging and even by engineering modifications to buildings. Many types of food packaging are at risk for penetration by insects; those comprised of paper and cellophane are least resistant. Polypropylene and similar polymer films provide some additional protection against boring and chewing insects, and their resistance to penetration improves with film thickness (56). The greatest protection is afforded by glass and metal containers. Certain food-safe, natural insect repellents have recently been approved by EPA for use in food packages to deter insect infestation, but the effectiveness of these agents has not been proved.

Structural design features for production facilities, warehouses, and retail outlets can be improved to enhance sanitation and pest control (57). Cracks, crevices, and other unnecessary openings in walls and floors should be sealed. Wall-to-floor junctions should be protected by concave molded tile to prevent harborage of pests, and wood-to-wood joints should be avoided or appropriately sealed. Voids in double-wall construction should be eliminated in new structures because they may allow accumulation of food debris and provide excellent harborage for pests. Physical separation of raw commodity storage areas, processing areas, and warehouses will help to prevent movement of pests from raw grain storage to high-value processed foods. Building exteriors should be free from vegetation, soil, and product spills. Exterior lights should not be mounted over doors or windows to avoid attracting insects to these openings at night. Rather, lights should be positioned away from the building and illuminate the structure from a distance.

ADOPTING IPM: CHALLENGE AND NECESSITY

The postharvest food industry faces serious challenges in pest management as a result of the attractiveness of food products to pests; greater public demands for safe, high-quality food products; and increasing regulatory constraints on use of chemical insecticides. It is apparent that the industry will no longer be able to rely on the quick fix provided by chemical pesticides such as methyl bromide. The industry must develop and adopt IPM programs to meet consumer demands and comply with governmental regulations. Fundamental components of the IPM programs for the food industry must include preventing infestations wherever possible through sanitation and pest exclusion by sealing of structures. Also, consistent moni-

toring programs and judicious decision making to guide the use of chemical pesticides will be critical. Alternatives to chemical pesticides, such as physical controls and biological control agents, must be used when possible. Thus, IPM in the food industry will be both preventive and responsive (58).

Decision-making parameters are not as well defined in the postharvest food industry as in production agriculture, but EILs and economic thresholds must be established for all key pests. Adoption of IPM practices is being encouraged through incentives at various levels in the industry. Some retailers are putting IPM labeling on products and report a favorable public response to these products. A positive public perception of IPM coupled with a desire to reduce pesticide inputs is giving momentum to the drive for adoption of IPM. Despite adoption of IPM, there are many unknowns regarding the safety and quality of the food supply with fewer pesticide inputs. Increased adoption of hazard analysis critical control points (HACCP) and related protocols will facilitate co-adoption of IPM for insect pests and should help ensure safety and quality. HACCP programs are customized for each food plant and are typically targeted at reducing risk from microbial contamination by detecting and identifying the potential for quality problems in a process or location, the so-called critical control points (see the article HAZARD ANALYSIS AND CRITICAL CONTROL POINTS [HACCP]). Flour mills also have HACCP plans centered around contamination (46), even though these facilities have fewer microbial problems compared to those with wet processes such as meat and dairy plants. Increased adoption of HACCP by the food industry has fostered a heightened sense of awareness for contamination and pest problems by facility managers. It should be possible for raw commodity and processed food managers to develop IPM programs modeled after a HACCP protocol. Such HACCP-based IPM would be custom designed for each facility and would identify key locations or activities that need to be addressed on a routine basis in order to prevent or mitigate pest problems.

BIBLIOGRAPHY

1. C. O. Knowles, "The Pesticide Revolution," *Proceedings of Pesticides: Risks, Management, Alternatives*, Virginia Polytechnic and State University, Blacksburg, Va., June 8–10, 1988.

2. C. M. Benbrook, *Pest Management at the Crossroads*, Consumers Union, Yonkers, N.Y., 1996.

3. D. J. Sissons and G. M. Teiling, "Agricultural Chemicals," in S. R. Tannenbaum, ed., *Nutritional and Safety Aspects of Food Processing*, Marcel Dekker, New York, 1979, pp. 295–367.

4. G. F. Stewart and M. A. Amerine, *Introduction to Food Science and Technology*, Academic Press, Orlando, Fla., 1973.

5. G. W. Ware and J. P. McMullum, *Producing Vegetable Crops*, Interstate Printers and Publishers, Danville, Ill., 1980.

6. J. M. Luna and G. J. House, "Pest Management in Sustainable Agricultural Systems," in C. A. Edwards et al., eds., *Sustainable Agricultural Systems*, Soil and Water Conservation Society, Ankeny, Iowa, 1990, pp. 157–173.

7. R. Frisbie and J. Magaro, "IPM in the Southwest," *Proc. Amer. Chem. Soc. Meet.*, Point Clear, Ga., August, 1989.

8. E. G. Rajotte et al., *The National Evaluation of Extension's Integrated Pest Management (IPM) Programs*, Publication No. 491-010, Virginia Cooperative Extension Service, Blacksburg, Virginia, 1987.

9. C. J. Scifres, "Decision-Analysis Approach to Brush Management Planning for Integrated Range Resources Management," *Journal of Range Management* **40**, 482–490 (1988).

10. National Research Council, *Alternative Agriculture*, National Academy Press, Washington, D.C., 1989.

11. P. I. Szmedra, "Pesticide Use in Agriculture," in D. Pimentel ed., *Handbook of Pest Management in Agriculture*, Vol. 1, CRC Press, Boca Raton, Fla., 1991, pp. 649–677.

12. W. H. Luckmann and R. L. Metcalf, "The Pest Management Concept," in R. L. Metcalf and W. H. Luckmann, eds., *Introduction to Insect Pest Management*, Wiley, New York, 1982, pp. 1–31.

13. B. Croft, "Pesticide Effects on Arthropod Natural Enemies: A Database Summary," in *Arthropod Biological Control Agents and Pesticides*, Wiley, New York, 1990, pp. 17–46.

14. T. M. Brown, "Applications of Molecular Genetics in Combatting Pesticide Resistance, An Overview," in T. M. Brown, ed., *Molecular Genetics and Evolution of Pesticide Resistance*, American Chemical Society, Washington, D.C., 1996, pp. 1–8.

15. R. Carson, *Silent Spring*, Fawcett, Greenwich, Conn., 1962.

16. Food and Drug Administration, *Residues in Food; 1988*, Washington, D.C., 1988.

17. R. Levins, "Perspectives in Integrated Pest Management: From an Industrial to an Ecological Model of Pest Management," in M. Kogan, ed., *Ecological Theory and Integrated Pest Management Practice*, Wiley, New York, 1986, pp. 1–17.

18. V. M. Stern et al., "The Integrated Control Concept," *Hilgardia* **29**, 81–101 (1959).

19. B. A. Croft, "Integrated Pest Management: The Agricultural Rationale," in R. E. Frisbee and P. L. Adkisson, eds., *Integrated Pest Management on Major Agricultural Systems*, Texas A&M University, College Station, Tex., 1985, pp. 712–728.

20. D. S. Jayas, N. D. G. White and W. Muir, eds., *Stored Grain Ecosystems*, Marcel Dekker, New York, 1995.

21. J. K. Phillips and W. E. Burkholder, "Health Hazards of Insects and Mites in Food," in F. J. Bauer, ed., *Insect Management for Food Storage and Processing*, American Association of Cereal Chemists, St. Paul, Minn., 1984, pp. 279–290.

22. J. T. Snelson, *Grain Protectants*, Australian Center for International Agricultural Research, Canberra, Australia, 1987.

23. U.S. Department of Agriculture, *Pesticide Data Program Annual Summary Calendar Year 1995*, Agricultural Marketing Service, U.S. Department of Agriculture, Washington, D.C., 1997.

24. U.S. Department of Agriculture *Pesticide Data Program Annual Summary Calendar Year 1996*, Agricultural Marketing Service, U.S. Department of Agriculture, Washington, D.C., 1998.

25. U.S. Department of Agriculture, *Pesticide Data Program Annual Summary Calendar Year 1997*, Agricultural Marketing Service, U.S. Department of Agriculture, Washington, D.C., 1998.

26. L. Keith, *Environmental Endocrine Disruptors: A Handbook of Property Data*, Wiley, New York, 1997.

27. N. Price, "Methyl Bromide in Perspective," in C. H. Bell, N. Price, and B. Chakrabarti, eds., *The Methyl Bromide Issue*, Wiley, Chichester, U.K., 1996, pp. 1–26.

28. P. Kenkel et al., "Biological Preservation of Grain Quality: Losses in Storage and Handling," in J. L. Steele and O. K. Chung, eds., *Proceedings of the International Wheat Quality Conference*, Grain Industry Alliance, Manhattan, Kans., 1997, pp. 385–389.

29. J. D. Stark, "Chemical Fumigants," in R. E. Paull and J. W. Armstrong, eds., *Insect Pests and Fresh Horticultural Products: Treatments and Responses*, CAB International, Wallingford, U.K., 1994, pp. 69–84.

30. J. W. Armstrong, "Heat and Cold Treatments," in R. E. Paull and J. W. Armstrong, eds., *Insect Pests and Fresh Horticultural Products: Treatments and Responses*, CAB International, Wallingford, U.K., 1994, pp. 103–119.

31. J. L. Nation and A. K. Burditt, Jr., "Irradiation," in R. E. Paull and J. W. Armstrong, eds., *Insect Pests and Fresh Horticultural Products: Treatments and Responses*, CAB International, Wallingford, U.K., 1994, pp. 85–102.

32. G. J. Hallman, "Controlled Atmospheres," in R. E. Paull and J. W. Armstrong, eds., *Insect Pests and Fresh Horticultural Products: Treatments and Responses*, CAB International, Wallingford, U.K., 1994, pp. 121–136.

33. D. W. Hagstrum, P. W. Flinn, and R. W. Howard, "Ecology," in B. Subramanyam and D. W. Hagstrum, eds., *Integrated Management of Insects in Stored Products*, Marcel Dekker, New York, 1996, pp. 71–134.

34. P. G. Fields, "The Control of Stored-Product Insects with Extreme Temperatures," *Journal of Stored Product Research* **28**, 89–118 (1992).

35. G. W. Cuperus and R. C. Berberet, "Training Specialists in Sampling Procedures," in L. P. Pedigo and D. G. Buntin, eds., *Handbook of Sampling Methods for Arthropods in Agriculture*, CRC Press, Boca Raton, Fla., 1994, pp. 669–681.

36. B. Subramanyam and D. W. Hagstrum, "Sampling," in B. Subramanyam and D. W. Hagstrum, eds., *Integrated Management of Insects in Stored Products*, Marcel Dekker, New York, 1996, pp. 135–193.

37. R. T. Noyes, G. W. Cuperus, and P. Kenkel, "Using Controlled Aeration for Insect and Mould Management in the South-Western United States," in E. Highley et al., eds., *Proceedings of the Sixth International Working Conference on Stored-Product Protection*, CABI Press, Canberra, Australia, 1994, pp. 323–334.

38. P. Kenkel et al., *Current Management Practices and Impact of Pesticide Loss in the Hard Red Wheat Post-Harvest System*, Circular E-930, Oklahoma Cooperative Extension Service, Stillwater, Okla., 1993.

39. J. H. Brower et al., "Biological Control," in B. Subramanyam and D. W. Hagstrum, eds., *Integrated Management of Insects in Stored Products*, Marcel Dekker, New York, 1996, pp. 223–286.

40. P. W. Flinn and D. W. Hagstrum, "Suppression of Beetles in Stored Wheat by Augmentative Releases of Parasitic Wasps," *Environmental Entomology* **25**, 505–511 (1996).

41. V. E. Walter, "Fumigation in the Food Industry," in J. R. Gorham, ed., *Ecology and Management of Food Industry Pests*, FDA Technical Bulletin No. 4, Association of Official Analytical Chemists Arlington, Va. 1991, pp. 441–457.

42. R. T. Noyes and P. Kenkel, "Closed-Loop Fumigation Systems in the South-Western United States," in E. Highley et al., eds., *Proceedings of the Sixth International Working Conference on Stored-Product Protection*, CABI Press, Canberra, Australia 1994, pp. 335–341.

43. F. H. Arthur, "Grain Protectants: Current Status and Prospects for the Future," *Journal of Stored Products Research* **32**, 293–302 (1996).

44. Z. Korunic, "Diatomaceous Earths, A Group of Natural Insecticides," *Journal of Stored Products Research* **34**, 87–97 (1998).

45. D. L. Faustini, "How to Use *Tribolium* Pheromone Traps," *Association of Operative Millers Bulletin* (Leawood, Kansas), 5715–5717 (1990).

46. R. Mills and J. Pederson, *A Flour Mill Sanitation Manual*, Eagan Press., St. Paul, Minn., 1990.

47. T. W. Phillips, "Semiochemicals of Stored-Product Insects: Research and Applications," *Journal of Stored Product Research* **33**, 17–30 (1997).

48. B. Chakrabarti, "Methyl Bromide in Storage Practice and Quarantine," in C. H. Bell, N. Price, and B. Chakrabarti, eds., *The Methyl Bromide Issue*, Wiley, Chichester, U.K., 1996, pp. 237–274.

49. E. J. Bond, T. Dumas, and S. Hobbs "Corrosion of Metals by the Fumigant Phosphine," *Journal of Stored Product Research* **20**, 57–63 (1984).

50. D. K. Mueller, "A New Method of Using Low Levels of Phosphine in Combination with Heat and Carbon Dioxide," in E. Highley et al., eds., *Proceedings of the Sixth International Working Conference on Stored-Product Protection*, CABI Press, Canberra, Australia 1994, pp. 123–125.

51. U.S. Pat. 5,403,597 (April 4, 1995), D. K. Mueller.

52. P. C. Annis and C. J. Waterford, "Alternatives-Chemicals," in C. H. Bell, N. Price, and B. Chakrabarti, eds., *The Methyl Bromide Issue*, Wiley, Chichester, U.K., 1996, pp. 275–321.

53. K. O. Sheppard, "Heat Sterilization (Superheating) as a Control for Stored-Grain Pests in a Food Plant," in F. J. Bauer, ed., *Insect Management for Food Storage and Processing*, American Association of Cereal Chemists, St. Paul, Minn., 1984, pp. 193–200.

54. P. G. Fields, A. Dowdy, and M. Marcotte, *Structural Pest Control: The Use of an Enhanced Diatomaceous Earth Product Combined with Heat Treatment for the Control of Insect Pests in Food Processing Facilities*, Agriculture Canada Publication, Ottawa, Ontario, 1997.

55. R. L. Hohman, "Food Industry Self-Inspection," in J. R. Gorham, ed., *Ecology and Management of Food Industry Pests*, FDA Technical Bulletin No. 4, Association of Official Analytical Chemists, Arlington, Va., 1991, pp. 519–527.

56. H. A. Highland, "Protecting Packages against Insects," in J. R. Gorham, ed., *Ecology and Management of Food Industry Pests*, FDA Technical Bulletin No. 4, Association of Official Analytical Chemists, Arlington, Va., 1991, pp. 345–350.

57. H. G. Scott, "Design and Construction: Building Out Pests," in J. R. Gorham, ed., *Ecology and Management of Food Industry Pests*, FDA Technical Bulletin No. 4, Association of Official Analytical Chemists, Arlington, Va., 1991, pp. 331–343.

58. D. W. Hagstrum and P. W. Flinn, "Integrated Pest Management," in B. Subramanyam and D. W. Hagstrum, eds., *Integrated Management of Insects in Stored Products*, Marcel Dekker, New York, 1996, pp. 399–408.

THOMAS W. PHILLIPS
RICHARD C. BERBERET
GERRIT W. CUPERUS
Oklahoma State University
Stillwater, Oklahoma

UNITED STATES FOOD MARKETING SYSTEM

The United States Food Marketing System connects roughly two and half million ranchers, farmers, and fishers to more than 270 million domestic consumers and a much larger and growing number of consumers worldwide. The primary role of the marketing system is to coordinate the vast array of economic activity involved in transporting and transforming producers' products to consumers—when, where, and how they prefer them. This coordination relies on economic markets, legal contracts, and direct ownership of various operations within the vertical marketing system to give order to the vertical flow of food products. Although the exact marketing channels used vary by product, the general vertical flow is depicted in Figure 1.

Consumers are the focal point of the marketing system because their demands are what everyone else attempts to profit from. It is this interplay of what consumers are willing (based on their needs, wants, and whims) and able (based on their incomes and prices) to buy and what sellers are willing and able to supply at a profit to themselves that makes the U.S. food marketing system an economic marvel. Americans today purchase essentially all their food from others—expending $561 million in 1997, a record amount for domestically produced farm foods (Fig. 2). Nearly 80% of that amount goes to pay the marketing bill to cover the costs of all the activities that lie between the primary producers and the ultimate consumers in the vertical system. The farm value share of food expenditures fell to a record low of 21% in 1997 (Fig. 3). In 1952, U.S. consumers spent $51 million and farmers received 40% of those expenditures. Over time, farmers have dramatically increased productivity with improved technology and spe-

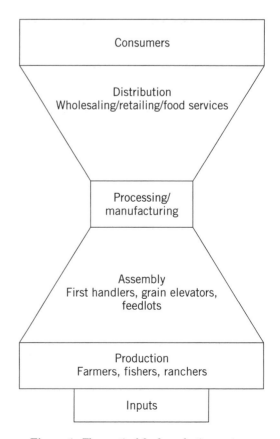

Figure 1. The vertical food marketing system.

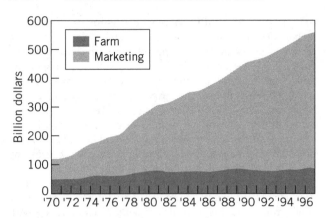

Figure 2. The distribution of food expenditures between farming and marketing, 1970–1997. The marketing bill is 79% of 1997 food expenditures. *Source:* Ref. 1.

cialization and have seen their share of the consumers' food dollar fall. Others, the so-called middlemen, have not only taken on the tasks abandoned by the producers but have specialized in providing all the marketing utilities to the eventual consumer by getting the right product (form utility), to the right place (space utility), at the right time (time utility), and at the right price (possession utility). It is these utilities that have increased the price of marketing bill over time as consumers demand more and more of these services.

AN OVERVIEW OF TRENDS

An overview should begin with consumers, the ultimate and somewhat elusive target of the entire marketing system. Then the focus shifts to the starting point of the vertical system to examine primary agriculture (and to a certain extent the input sector), and then in turn food processing, food wholesaling and retailing, and food service.

Consumer demands do not stand still, but change along with their preferences and incomes. Consumption patterns have seen some dramatic shifts over time, as shown in Figure 4. In beverages alone huge changes have occurred since 1970. Coffee consumption has fallen 32% despite the emergence of gourmet coffee houses. The only growth segments are in the smaller specialty-coffee beverages that include flavorings. Milk consumption is also down 23%, and only the no-fat and lowfat segments show increases. Calcium-fortified products have eliminated one of milk's traditional advantages; even orange juice companies suggest consumers substitute orange juice for milk. The big winner in the beverage aisle is soft drinks, up about 118%. The new bottled water market, flavored and unflavored, continues to grow as consumers consider tap water inferior (Table 1).

Consumption of eggs and red meat has also declined, 23% and 16% respectively, whereas poultry has increased dramatically and rivals beef as the leading meat consumed in the U.S. on a trimmed and boneless basis (3). The success of the broiler industry accounts for the majority of the poultry consumed. Fish consumption is up 24% since 1974, but remains a small part of the consumers' meat diet despite its known health benefits. Cheese consumption had the largest increase, up 146% since 1970. Americans have increased their consumption of fruits and vegetables 24%, but also had similar increases in fats and oils and caloric sweeteners. In addition, candy consumption reached a record 25 pounds per capita in 1997.

Much public and private effort has been directed at improving the diet of Americans. The Federal Food Guide

Figure 3. What a U.S. dollar spent for food paid in 1997. *Source:* Ref. 1.

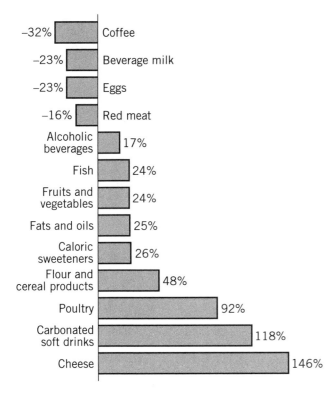

Figure 4. Changes in U.S. per capita food consumption, 1970–1997. *Source:* Ref. 2.

Table 1. Major Foods: U.S. Per Capita Food Supply

Food	1970	1980	1997
Pounds			
Total meats[a]	177.3	179.6	192.0
Beef	79.6	72.1	63.8
Pork	48.0	52.1	45.6
Veal	2.0	1.3	.9
Lamb and mutton	2.1	1.0	.8
Chicken	27.4	32.7	51.2
Turkey	6.4	8.1	14.8
Fish and shellfish	11.7	12.4	14.7
Eggs (number)	308.9	271.1	238.4
Cheese[b]	11.4	17.5	28.0
Ice cream	17.8	17.5	16.2
Lowfat ice cream	7.7	7.1	7.9
Fluid cream products	5.2	5.6	9.1
All dairy products[c]	563.8	543.2	579.8
Added fats and oils	52.6	57.2	65.6
Peanuts and tree nuts[d]	7.2	6.6	8.0
Fruit and vegetables[e]	573.2	598.8	710.8
Fruit	237.7	262.4	294.7
Vegetables	335.4	336.5	416.0
Caloric sweeteners[f]	122.3	123.0	154.1
Sucrose	101.8	83.6	66.5
Corn sweeteners	19.1	38.2	86.2
Grain products[a]	135.6	144.7	200.1
Wheat flour	110.9	116.9	149.7
Rice	6.7	9.4	19.5
Corn products	11.1	12.9	23.1
Gallons			
Beverage milks	31.3	27.6	24.0
Whole	24.8	16.5	8.2
Lower fat and skim	5.8	10.5	15.5
Coffee	33.4	26.7	23.5
Tea	6.8	7.3	7.4
Carbonated soft drinks	24.3	35.1	53.0
Fruit juices	5.7	7.4	9.2
Bottled water	N.A.	2.4	13.1
Beer	18.5	24.3	23.9
Wine	1.3	2.0	2.0
Distilled spirits	1.8	2.0	1.2

Source: Ref. 2.
Note: N.A., not available
[a]Boneless, trimmed weight.
[b]Excludes full-skim American, cottage, pot, and baker's cheese.
[c]Milk equivalent, milkfat basis.
[d]Shelled basis.
[e]Farm-weight, excludes wine grapes.
[f]Dry basis. Includes honey and edible syrups.
[g]Consumption of items at the processing level (excludes quantities used in alcoholic beverages and corn sweeteners).

Pyramid was developed as a simple guide to assist consumers in improving their nutrition (Fig. 5). The pyramid was controversial because industry groups positioned themselves to maximize benefit or minimize damage from the new recommendations. A study of American diets found some improvements, but some concerns as well. For example, consumption of fruits and vegetables did reach a record amount in 1996, but the use of caloric sweeteners also was a record because "on average, people consume too many servings of added fats and sugars and too few servings of fruits, vegetables, dairy products, lean meats, and foods made from whole grains" (5). The increased popularity of dining out has also contributed to nutrition concerns because consumers tend to eat more healthful foods at home than away from home (6).

Reasons for the changes in consumption can be grouped into three general areas: demographics, economics, and an explosion in available information (7). American households have changed tremendously since 1950. Total food consumption in an affluent society is linked to population because the amount of food consumed per person has not risen much, although more calories are being consumed. The U.S. population does continue to grow, roughly at 0.7% per year, driven by increased life expectancies, immigration, and fertility. In 1999, there were about 270 million Americans, but the census projects a population of 400 million by 2050. The U.S. population is aging; in 1996, 12.8% of the population was 65 and over, and it is projected to reach 20% in 2050 (8). Older people have different food interests and needs, and the marketing system will re-

spond to this major change. Although the average age is increasing, the proportion of young people is also increasing. In addition, the population is also becoming more ethnically diverse; by 2050 non-Hispanic whites will make up 53% of the population, down from 74% in 1996. African-Americans will increase only slightly from 11.5 to 13.6%, and Asians will increase from 1.5 to 9.6%. The majority of the population will be over 45 or under 18 by 2010, with nearly 50% of children belonging to an ethnic minority (9). The increased diversity should be reflected in the food sys-

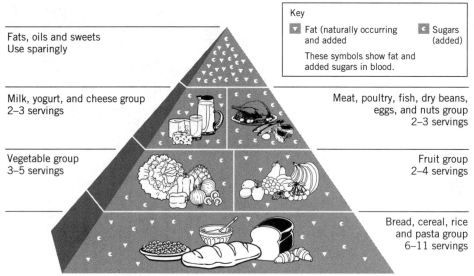

Source: U.S. Department of Agriculture/U.S. Department of Health and Human Services

Figure 5. Food guide pyramid. *Source:* Ref. 4.

tem. People's food choices are related to their cultural backgrounds, including religious beliefs, and as the population continues to diversify the food brought to the American table will reflect this diversity. Additionally, most Americans, especially younger people, will begin to experiment with the new ethnic foods.

The typical U.S. household has become smaller, older, and more likely to be a single-parent home. In 1955, 23% of households were a married couple, with two or more children under the age of 18, and the wife not in the labor force (7). By 1992, only 7% of American households fit that description. In 1960, 9.8% of households were single-parent households, and in 1995, the number increased to 27% (10). In 1960, 18.6% of married women with children under the age of 6 years old were in the labor force. By 1997, the percent was up to 63.6% (11). By 1996, nearly 60% of households with children under the age of 7 had personal computers (10). Such households have become very busy places, and the eating habits have changed accordingly. Children often prepare their own meals, usually by microwaving a ready-to-heat-and-eat meal.

Household economics influence consumer demands as relative prices change and as incomes rise allowing greater spending and increasing the value of leisure time. Rising incomes allow consumers to indulge more of their wants. Americans, on average, now spend less of their disposable income on food than any other country in the world, and the average percentage fell to a record low of 10.7% in 1997 (Table 2). This figure is both a statement about America's wealth and its efficient food system.

Averages, however, conceal the diversity that exists among U.S. households. Households with after-tax income of $5,000 to under $10,000, spend 34.2% of their after-tax income on food, whereas the food share for households with $70,000 and higher after-tax income was only 8.7% (2). The average U.S. household income has been flat for 15 years, resulting in a reduction in the number of middle-class

households and an increased number of affluent and low-income households (9). Some fear a permanent underclass, left behind by a modern information driven economy, and a growing affluent segment whose wealth allows dramatic lifestyle changes.

Whereas much of modern business marketing focused on the mass market and the efficiencies from the one-size-fits-all approach, today mass markets are segmenting, allowing customized marketing to cater to the needs of the various segments—from those with higher incomes to those with lower incomes; from those who stress health issues to those who prefer taste and convenience. Such segments are still large, and combined with new technologies and increased incomes, firms can pursue a strategy of mass customization where they attempt to retain the efficiencies of mass marketing while providing consumers with products better suited to their preferences. Dual-earning households have more money than time, and such consumers demand more convenient foods, higher quality foods, and other timesaving services. Affluent households now hire people to shop, cook, and clean for them. Internet shopping has exploded, yet firms find profits elusive to date. Firms seek to profit from saving consumers time by dramatically altering the way they shop for food. As one founder put it, "There is money to be made simplifying people's hectic lives" (12). Also, with or without Internet shopping but certainly more so with it, home delivery has returned.

For the middle classes, income growth has stalled, and the number of hours worked per week now exceeds 40 hours. Leisure time is short until retirement, but people are retiring earlier and living longer, which greatly increases the leisure time of retirees. Women have less leisure time than men because they still do 86% of the cooking and 91% of the shopping (7). Such time pressure adds stress to busy lives and has consumers looking for relief from timesaving products.

Table 2. Food Expenditurs As a Share of Disposable Personal Income, 1967–1997

Year	Disposal personal income Billions of dollars	Food at home Billions of dollars	Percent	Food away from home Billions of dollars	Percent	Total Billions of dollars	Percent
1967	562.1	60.3	10.7	19.8	3.5	80.1	14.2
1972	855.3	84.4	9.9	31.3	3.7	115.8	13.5
1977	1,401.4	131.8	9.4	58.5	4.2	190.3	13.6
1982	2,347.3	198.4	8.5	104.5	4.5	302.9	12.9
1987	3,363.1	249.0	7.4	146.3	4.3	395.3	11.8
1992	4,626.7	321.6	7.0	192.0	4.2	513.6	11.1
1997	5,885.2	390.3	6.6	239.1	4.1	629.4	10.7

Source: Ref. 2.

In single-parent households, there is not the benefit of two wage earners, but the time pressure can be even more severe. Adults increasingly view dinner as an impulse purchase. One study found 71% of adults do not plan their weekday dinner before 4 p.m., regardless of income level, marital status, or education (13). A Nabisco study put the figure at 61%, but found working parents are denied the fulfillment factor of doing something for their families (14). Those in single-parent households seldom have the luxury of using increased incomes to buy services to lessen the strain of their busy schedules, so they turn to convenient, yet inexpensive foods.

The explosion of studies relating health to diet have affected consumer demands. Although many such studies are controversial and are often sensationalized, the media cover them because there is great interest by consumers. Everyone wants to feel and look better, to live longer. Although some consumers have become jaded to the conflicting news, the information still moves markets. The markets for healthful products, including the explosion of nutraceuticals and functional foods, cater to these consumer desires. Consumers still do not want to exert too much effort to achieve dramatic results. The long list of diet and exercise books and the increasing proportion of Americans who are overweight (now a majority) suggest that there is a desire but also a lack of will.

As American households embraced the economic tenets of specialization and gains from trade, there has been some rethinking of the resulting lifestyle. The additional family income gained from dual wage earners is used to buy time-saving items and prepared meals. Dining out is no longer a luxury to many Americans but is done to save the meal preparation time. The dining experience itself must not take too long. Drive-up, fast-food windows proliferate because consumers cannot afford the time to have a full-service dinner experience, and they certainly lack the time to make traditional meals at home. The division of labor and both sexes in the paid labor force can lead to frustration, guilt, and even boredom. People long to do the things that their jobs prevent them from doing. Gardening and cooking, once dreaded chores, are considered enjoyable, but constrained by the lack of time. A recent study found only 13% of adults dislike cooking; 37% actively enjoy it and 50% do not mind it (15). Cooking is second only to pet care as the most enjoyable household task. Cooking has regained some popularity; 43% of adults indicated they often cook in their leisure time, which is up from 36% in 1991. Interest in food is up; the number of food magazines and cooking shows on cable TV is at an all-time high. Consumers, however, are left unable to find the time and energy to try their new ideas. The same study found 30% of adults always willing to try new or unusual flavors, with 21% having a strong interest in foods from other countries. Younger consumers were more likely to try something new.

The implications for the food marketing system are clear. As we move from a mass market where everyone eats similar foods to segmented markets with greater choices, firms will face new opportunities and challenges. Some consumers will avoid bioengineered foods; others will embrace it as science solving important problems. The controversies will be hotly debated, and food choice will contain plenty of political overtones. This is nothing new: the Pilgrims clashed over whether the making of a minced pie was a violation of their prohibitions against Catholic-style symbolism (16). Marketing firms strive to give consumers what they want, but firms often avoid full disclosure. "A 1997 survey of 1,000 U.S. consumers found that 93% wanted bio-engineered food to be labeled—presumably because many would avoid it" (17). Firms, however, do not want to label such foods, because they fear consumers are either ill-informed or merely paranoid about such tampering with foods. Instead, firms stress the advantages bioengineered food can offer—characteristics such as improved taste, reduced spoilage, and enhanced visual appeal, all factors consumers prefer. As technology advances, consumer sovereignty in the marketplace will be tested.

There is increasing demand for value-added products. Rising household incomes allow greater purchasing of value-added products, where the value can be in convenience, healthfulness, and quality. The aging baby boomers and other health-conscious consumers will seek more functional foods and nutraceuticals. Time-pressed consumers of all income levels will be searching for quick meals, and families will seek meals that children enjoy, thus lessening the stress involved in family dining. Taste remains paramount, and most foods must pass the taste test, even if they offer nutrition and convenience, especially in the away-from-home segment. McDonald's reduced-fat burger did not survive the consumer taste test.

Consumers will also respond to environmental concerns, so-called green marketing. However, again a discrepancy exists between one's desires and one's actual behavior. One study that tracks the attitudes of college freshmen found less interest in programs to clean up the environment in 1997 than in the early 1970s (18). Fast-paced lifestyles and demands for fresh produce year round lead to an increased use of transportation and packaging. Consumers face numerous conflicting decisions in balancing their views with their purchases. In addition to green marketing, consumers will respond to products where the added value is not product specific but combines nonfood attributes with the food item, such as products with "save the rainforest" and "save the family farm" themes. Consumers now realize that voting with their pocketbooks can also signal their views on the world in which they wish to live.

The U.S. food marketing system will continue its growing involvement in international trade. Immigration and increased interest in new foods will bring more global trade. Also, the demand for fresh produce year round will increase shipments from the southern hemisphere. Because industry growth is largely constrained by population, U.S. firms will continue to search for growth in overseas markets.

PRODUCTION

Farming still has the greatest number of businesses in any of the vertical stages in the food system, but consolidation continues. Farming once employed the largest number of people in the food system, but machines have replaced people, and it is now the food service sector that employs the most people (Table 3). The number of farms has fallen in every census year, the latest 1997 census shows 1.9 million farms, accounting for under 2% of the U.S. population, a record low number (Table 4). The other major trend is toward greater dispersion between the very small farm and

the very large. In 1992, nearly half the farms had sales of under $10,000 but accounted for less than 2% of total farm sales, whereas just over 2% of the farms had sales over $500,000 and did nearly 50% of total sales. There is also growing diversity among regions of the country and by crop speciality. California is the largest agricultural state in terms of dollar value, and the top 10 states accounted for nearly 53% of farm value in 1997 (Table 5). California is an amazing state because it is more commonly associated with nonagricultural images, such as Hollywood, yet it has almost twice the farm value of Texas, the second largest agricultural state. The other leading farm states, like Iowa and Nebraska, are often associated with agriculture and rural communities.

Increasingly, producers must align their production choices with consumer demands, rather than rely on others in the marketing system, mainly processors, to alter the primary foodstuffs to match consumer demands. Processors aware of what their consumers are willing and able to buy now ask producers to deliver raw products best suited to these demands. For example, a potato chip company that wants potatoes specially tailored for the chip making process will invest in research and technology to develop the best chipping potato and seek growers who will grow these potatoes under a legal contract. Even food service companies need to align their needs with producers to ensure a good match between consumer demand and product supply. For example, KFC (formerly Kentucky Fried Chicken) contracts with processors to supply the birds best suited to their cooking procedure.

This increased vertical coordination between farmers and processors has become known as the industrialization of agriculture (26). Farmers continue to specialize in a narrower slice of the food marketing system. Great efficiencies have been gained from advanced technology and special-

Table 3. Food System Employment, 1950–1997 (thousands)

Year	Agriculture	Manufacturing[a]	Wholesaling	Retailing	Food Service
1950	7160	1626	N.A.	1231	N.A.
1960	5458	1571	N.A.	1649	N.A.
1982	3364	1676	674	2348	4666
1992	3295	1541	812	2969	6548
1997	3399	1602	N.A.	3147	7636

[a]Includes tobacco manufacturing.
Note: N.A., not available.
Source: Refs. 19–24.

Table 4. Number of Food-System Business Establishments, 1972–1997

Year	Number of farms	Number of manufacturing establishments	Number of wholesale establishments	Number of retail establishments	Number of food-service establishments
1972	2,869,710	28,103	60,363	173,084	287,250
1977	2,455,830	26,656	59,473	171,592	308,614
1982	2,400,550	22,130	58,766	176,219	319,873
1987	2,176,110	20,201	42,075	190,706	391,303
1992	1,925,300	20,798	42,874	180,568	433,608
1997	1,911,859	20,913	N.A.	174,284	N.A.

N.A., not available.
Source: Refs. 19–22.

ized equipment. Inputs previously supplied on the farm are now purchased from the input sector. Dramatic growth has occurred in the chemical businesses that supply fertilizer, pesticides, and insecticides to farmers. Private seed companies have largely replaced public research, and under protection provided by the 1972 Plant Varieties Protection Act, they have invested heavily in patentable seeds. Given the protection to intellectual property rights, these companies now sell designer seeds with favorable characteristics. Others view this as dangerous tampering with the public's genetic seed stock that leaves farmers dependent on the seed companies that hold the patents to seeds engineered unable to reproduce.

The seed companies continue to consolidate into ever larger, worldwide firms. DuPont's announcement to buy the remaining 80% of Pioneer Hi-Bred International makes this a dominant two-firm industry, with DuPont and Monsanto controlling half of the U.S. soybean seed market and over half of the corn seed market—the two biggest crops (27). In the smaller produce seed markets, such as lettuce, tomatoes, and cucumbers, similar consolidation has occurred. A Mexican company has quietly amassed a dominant share in many of these markets; for example, it holds a 55% share of the U.S. lettuce seeds used by commercial farmers (17). It has linked with Monsanto to alter its lettuce seeds, making them immune to Monsanto's Roundup herbicide used by farmers to kill unwanted vegetation. Many consumers would love non-browning lettuce, attractively priced, but others view this technology as dangerous, sending them to the organic lettuce aisle.

PROCESSING AND MANUFACTURING

The processing stage has the fewest number of establishments in the vertical food system, but the processor/food manufacturer is often consider the most powerful, influential firm in the system—the marketing channel leader. These are the food firms the world knows by name: Philip Morris, Coca-Cola, Cargill, Kellogg's, and so on (Table 6). About 80% of all raw domestic food products pass through this stage; only produce and eggs avoid processing because they only require minimal market preparation services such as cleaning, sorting, and packaging (29). Processors and manufacturers, hereafter referred to as processors, add the form utility to the raw agricultural products and have invested heavily in market research to understand consumer demands. They buy or contract from farmers who have been advised (including through price signals) or legally bound to supply raw foodstuffs with desired characteristics for transforming into the products consumers eventually buy. In terms of value added, the economist's preferred measure of economic size, processing once contributed a smaller portion of value added to the GDP than farming, but since the 1950s, it has exceeded farming's value added and remained behind retailing and wholesaling and food service (Table 7).

The location decisions made by food processors involve a calculated tradeoff between processing costs, including input costs, and the costs of delivering their finished prod-

ucts to consumers. Since most of the country's consumers live near the coasts and most of the raw agricultural foodstuffs come from the middle of the country, the location decision is not always obvious. Over time, with modern transportation and refrigeration technologies, the balance has shifted to locating where the inputs are produced rather than where the people live. California is in the unique situation of being both the number 1 farm state and the number 1 food processing state by far (Table 5). It has both the agricultural commodities and the population. States such as Nebraska and Kansas, 4th and 5th in farm value, rank 24th and 27th, respectively, in processing. Overall, there is a strong association between farm value rank and food processing rank, with a simple correlation coefficient of 0.75. In certain crops it is even more pronounced, such as wine or broilers. Broiler processors prefer to locate within a 25-mile radius of where their chickens are raised to market weight, and the leading states in both production and processing closely follow a geographical pattern known as the broiler belt (3).

The largest food processors among the roughly 16,000 companies involved in food processing are huge, both in absolute terms and relative to the others. The largest 100 food and tobacco processors accounted for nearly 80% of the value added in 1995 (Fig. 6), almost doubling their share since 1954. The top 100 is itself skewed toward the very large, with the top 20 firms accounting for over 50% of total value added in 1995, more than doubling its 1967 share (Fig. 7). The remaining 80 firms among the top 100 firms actually lost share over the last 30 years. The sector is best described by a big–small model, where extremely large firms control leading positions in most markets and smaller companies, including start-ups, operate in a competitive fringe trying to serve a particular market niche or develop a new idea. The large companies know that if a new idea turns promising they can buy the entire company after the start-up has borne much of the risk.

The previous figures refer to overall size, or what economists call aggregate concentration, but market performance hinges on market concentration—the extent of market power held by leading firms in a well-defined economic market. Market power is what enables a firm to enhance prices to buyers, extract price reductions from its product suppliers, and subdue rivals. Although market definition is a complex task, it can be approximated by the census four-digit industry group, the four-digit SIC. The food and tobacco processing sector had 52 such industries in 1992, most of which remain too broadly defined, certainly so on the input side because substitution opportunities are much greater in consumption than in production (33). Although there are no monopolies and several industries are what economists call workably competitive (where the four largest firms have a combined market share of 40% or less), most have become oligopolies (Table 8). In oligopolies, firms get some of the advantages of market power without government regulation that would come if they were monopolies (34). Over time, most of these four-digit industries have lost companies, averaging a 25.5% reduction in company counts, and have increased in concentration as measured by the four-firm concentration ratio, CR4, which in-

Table 5. Ranking of States by Value of Agricultural Products Sold, (1997) and Value-Added in Food Manufacturing, (1996)

State	Rank		1997 value of agricultural products sold			Food manufacturing ($ million)	Percent of U.S. total	1996 value added in manufacturing	
	Agriculture	Food manufacturing	$ million	Percent of U.S. total	Cumulative percent			Total manufacturing ($ million)	Food manufacturing as percent of state's total
California	1	1	23,032	11.7	11.7	20,265	11.3	188,805	10.7
Texas	2	3	13,767	7.0	18.7	11,778	6.6	116,631	10.1
Iowa	3	10	11,948	6.1	24.8	6,367	3.6	27,021	23.6
Nebraska	4	24	9,832	5.0	29.8	2,549	1.4	9,218	27.6
Kansas	5	27	9,207	4.7	34.4	2,136	1.2	18,820	11.4
Illinois	6	2	8,556	4.3	38.8	12,602	7.1	92,011	13.7
Minnesota	7	15	8,290	4.2	43.0	5,023	2.8	34,716	14.5
North Carolina	8	12	7,677	3.9	46.9	5,144	2.9	76,475	6.7
Florida	9	14	6,005	3.1	49.9	5,071	2.8	38,621	13.1
Wisconsin	10	8	5,580	2.8	52.8	6,379	3.6	53,619	11.9
Arkansas	11	20	5,480	2.8	55.6	3,542	2.0	18,512	19.1
Missouri	12	9	5,368	2.7	58.3	6,377	3.6	40,208	15.9
Indiana	13	17	5,230	2.7	60.9	4,304	2.4	61,896	7.0
Georgia	14	6	4,993	2.5	63.5	6,796	3.8	51,753	13.1
Washington	15	19	4,768	2.4	65.9	3,582	2.0	31,929	11.2
Ohio	16	4	4,684	2.4	68.3	9,786	5.5	105,497	9.3
Colorado	17	21	4,534	2.3	70.6	2,919	1.6	19,215	15.2
Oklahoma	18	32	4,146	2.1	72.7	1,396	0.8	15,875	8.8
Pennsylvania	19	5	3,998	2.0	74.7	9,386	5.3	82,922	11.3
South Dakota	20	39	3,570	1.8	76.5	674	0.4	3,974	17.0
Michigan	21	11	3,568	1.8	78.3	5,149	2.9	85,688	6.0
Idaho	22	31	3,346	1.7	80.0	1,467	0.8	7,977	18.4
Mississippi	23	29	3,127	1.6	81.6	1,823	1.0	17,295	10.5
Alabama	24	30	3,099	1.6	83.2	1,551	0.9	27,451	5.7

25	Kentucky	23	3,064	1.6	84.8	2,685	1.5	35,040	7.7
26	Oregon	28	2,969	1.5	86.3	2,064	1.2	21,838	9.5
27	North Dakota	41	2,869	1.5	87.7	537	0.3	1,808	29.7
28	New York	7	2,835	1.4	89.2	6,419	3.6	90,665	7.1
29	Virginia	13	2,344	1.2	90.4	5,090	2.8	42,519	12.0
30	Tennessee	16	2,178	1.1	91.5	4,632	2.6	42,288	11.0
31	Louisiana	25	2,031	1.0	92.5	2,435	1.4	25,125	9.7
32	Arizona	33	1,903	1.0	93.5	1,365	0.8	22,850	6.0
33	Montana	49	1,871	1.0	94.4	165	0.1	1,707	9.7
34	New Mexico	45	1,618	0.8	95.2	333	0.2	11,745	2.8
35	South Carolina	34	1,588	0.8	96.0	1,312	0.7	30,769	4.3
36	Maryland	22	1,312	0.7	96.7	2,729	1.5	17,455	15.6
37	Wyoming	50	899	0.5	97.2	148	0.1	999	14.8
38	Utah	35	877	0.4	97.6	1,032	0.6	11,239	9.2
39	New Jersey	18	697	0.4	98.0	4,136	2.3	49,995	8.3
40	Delaware	37	691	0.4	98.3	920	0.5	5,791	15.9
41	Hawaii	42	497	0.3	98.6	533	0.3	1,609	33.2
42	Vermont	46	476	0.2	98.8	320	0.2	3,986	8.0
43	Massachusetts	26	454	0.2	99.0	2,272	1.3	44,047	5.2
44	West Virginia	47	447	0.2	99.3	257	0.1	8,965	2.9
45	Maine	40	439	0.2	99.5	541	0.3	6,675	8.1
46	Connecticut	36	422	0.2	99.7	947	0.5	24,772	3.8
47	Nevada	44	357	0.2	99.9	397	0.2	3,275	12.1
48	New Hampshire	43	149	0.1	100.0	478	0.3	10,815	4.4
49	Rhode Island	48	48	0.0	100.0	249	0.1	5,407	4.6
50	Alaska	38	25	0.0	100.0	683	0.4	1,470	46.5
	Total		*196,865*	*100.0*		*178,742*	*100.0*	*1,748,981*	*10.2*

Source: Refs. 24 and 25.

Table 6. The Top 25 Food Processing Companies, 1998

Rank	Company	Millions $ Food sales	Millions $ Total sales	Percent food
1	Philip Morris	31,527	71,592	44
2	Conagra	28,840	28,840	100
3	Cargill	21,400	51,000	42
4	Pepsico	20,917	20,917	100
5	Coca-Cola	18,800	18,868	100
6	Archer Daniels Midland	16,109	16,109	100
7	Mars	14,000	14,000	100
8	IBP	13,259	13,259	100
9	Anheurser-Busch	12,832	12,832	100
10	Sara Lee	10,800	20,000	54
11	H.J. Heinz	9,209	9,209	100
12	Nabisco	8,734	8,734	100
13	Bestfoods	8,400	8,400	100
14	Nestle USA	7,800	7,800	100
15	Dairy Farmers of America	7,000	7,000	100
16	Kellogg	6,830	6,830	100
17	Campbell Soup	6,696	6,696	100
18	Pillsbury	6,500	6,500	100
19	Tyson Foods	6,356	6,356	100
20	General Mills	6,033	6,033	100
21	Quaker Oats	5,010	5,010	100
22	Proctor and Gamble	4,376	37,154	12
23	Dole Food	4,336	4,336	100
24	Hershey Foods	4,300	4,300	100
25	Land O'Lakes	4,195	4,195	100

Source: Ref. 28.

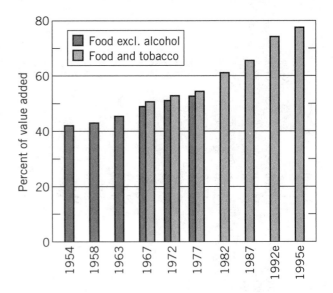

Figure 6. Largest 100 food manufacturing companies, census years 1954–1995. e, estimated. *Source:* Ref. 30; 1992 and 1995 were estimated from trade sources.

creased on average from 43.9 in 1967 to 53.3 in 1992, the last year data are available.

The immense size of processors has always concerned farmers, who fear the processors would exploit their bargaining power and pay farmers less than fair market value for their crops. Such fears led to agricultural cooperatives and the Capper-Volstead Act of 1922. Both economic theory and empirical studies (35) conclude that open membership cooperatives can negate market power imperfections and hence benefit both farmers and consumers. In Table 8, the share of an industry's shipments controlled for by the 100 largest cooperatives ranges from a high of 63% in the butter industry to several industries without any coopera-

tives, and averages 5.4% for all of food and tobacco processing (36).

Food processors have long used branded products and a pull marketing strategy where they create consumer demand for their products, so retailers are obliged to carry the products or lose sales. New products with strong media advertising support, especially television, are central to this strategy. Other marketing strategies (eg, coupons) are often correlated with new products and advertising efforts (29). New product introductions rose from 4540 items in 1983 to a high of nearly 17,000 in 1995 and fell back to 12,400 items in 1997 (Table 9). Most, by far the majority, do not represent truly new products but variations on existing products. Nevertheless, most new products fail in the marketplace, underscoring both the difficulty of knowing what the consumer wants and the wastefulness of new product launches. Even Coca-Cola, with its huge marketing muscle, misfired with New Coke. Consumers loudly voiced their preference for the original; hence, we now have Coke Classic and New Coke, with the former far outselling the newer product.

Table 7. Value Added by Industry Sector, Selected Years (in Billions of Dollars)

Industry sector	1947	1972	1982	1987	1996
Farming	19.5	22.0	48.1	51.0	95.0
Processing and manufacturing	11.3	30.6	65.0	85.7	131.0
Retailing and wholesaling	30.0	37.9	87.0	108.3	148.0
Transportation	15.3	6.5	16.9	20.5	24.0
Eating and drinking places	6.0	18.5	46.8	63.5	169.0
Other supporting sectors[a]	—	49.9	121.6	165.3	226.0
Total sector value added	82.1	165.4	385.4	494.3	793.0
U.S. gross national product	252.0	1,212.8	3,166.0	4,526.7	7,636.0
Total sector divided by GNP	32.6%	13.6%	12.2%	10.9%	10.4%

Source: Refs 11, 29, 31, and 32.
[a]Includes auxiliary activities such as packaging.

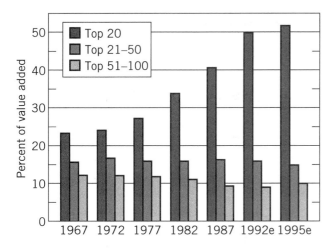

Figure 7. Increasing dominance by the top 20 among the largest 100 food and tobacco manufacturing companies, census years 1967–1995. e, estimated. *Source:* Ref. 30; 1992 and 1995 were estimated from trade sources.

Food processors outspend all other stages of the vertical marketing system advertising their products. In 1997, they accounted for over 65% of all media advertising in the food system (Table 10). Only restaurants, especially fast food, also spend large amounts on advertising, with a 27% share of the total. The bulk of the $55 million spent in farm-related expenditures was for farm chemicals and pest controls that were advertised by the large chemical companies targeting farmers as consumers of their chemical products. Media advertising accounts for somewhere between a fourth and a third of marketing dollars spent by processors. Within the food system, advertising created and maintained for product differentiation provides protection from new entrants and inroads from smaller rivals. Oligopolists often prefer to compete with their rivals in nonprice ways. Advertising and new product rivalry are perfectly suited to this strategy because they allow competition among the few in a manner that collectively erects barriers to entry to others not involved in the marketing fray. The cola wars are often mentioned as an example of the intense competition in the soft drink industry, but a former president of Pepsi once commented that such struggles did not involve "some gladiatorial contest where one of us has to leave on a stretcher. We're both winning" (28, p. 112).

Media advertising is not for the cash-starved start-up company, but it is standard operating procedure among the largest food and tobacco processors. Philip Morris, the number one food and tobacco advertiser, spent over a billion dollars promoting its brands, which include Miller beer, Marlboro cigarettes, Kraft cheese, General Foods Post cereals, Oscar Mayer hot dogs, and several other major brands. Of the 16,000 food and tobacco processing firms, the top 100 advertisers accounted for 96.4% of media advertising, and the top 8 alone accounted for over 50% in 1992 (Table 11). The vast majority of the remaining 15,900 food firms use media advertising minimally or resort to other marketing tactics. Because advertising-created-and-

maintained product differentiation is the major advantage in the food system, these firms are at disadvantage and are left to compete primarily on price and efficiency and the rare, truly new product that captures consumer interest. The concentration of advertising expenditures has risen sharply from 1967 to 1992, the last year the data were analyzed, with the four largest advertisers increasing their share from 19.4% in 1967 to 36.9% in 1992.

FOOD WHOLESALING, RETAILING, AND SERVICE

Wholesalers distribute the food products from points of production and processing to the retail stores and food service establishments located in every local community. Food wholesalers are classified into three general categories: (*1*) merchant wholesalers who take title of the products they distribute, (*2*) manufacturers' sales branches and offices that are the wholesaling arm of food processors but report their activities as wholesaling as long as the facility is located at a different address than the processor's location, and (*3*) agents, brokers, and commission merchants who do not take title of the products they distribute but work on commissions from the manufacturers whose products they represent. Merchant wholesalers account for the largest slice of the wholesaling function, with over 77% of the wholesale establishments and roughly 55% of total wholesaling sales in 1992 (39). The other two categories share the remainder, with manufacturers' sales branches and offices being slightly larger than the agents, brokers, and commission merchants.

In general, wholesalers are challenged by the processors' own sales forces on the one side and by retailers who self-supply on the other side. Nearly half of all retail food sales were from retailers who have integrated backward and provide their own wholesaling functions. Major wholesalers have integrated forward and own retail stores, but most remain as merely affiliated with retailers, either in a voluntary or a cooperative format.

Wholesalers must convince processors that they offer a more profitable method for distributing the processors' products than could be done through their own sales force. In addition, they must convince retailers that self-supply is less profitable than specializing in retailing and using a wholesaler to supply the products. Wholesalers work closely with both processors and retailers to coordinate product flows to benefit both parties while profiting from their activities and market expertise as well. They understand the local retail market and can use their personnel efficiently to represent processors' products and still provide retailer with supply savings, management, and merchandising services that only the largest retailers would find profitable to provide for themselves. One major service merchant wholesalers provide retailers is a line of private label products—products that they arrange for processors to pack and label with the wholesalers' label—that often sell at prices beneath the national brands and still offer a higher margin to the retailer (eg, IGA, President's Choice, and World Classics).

The wholesaling sector continues to consolidate, with larger firms gaining a larger share of the overall business.

Table 8. Concentration in Food and Tobacco Processing Industries, 1967 to 1992

SIC	Name	Concentration-CR4 1967	1987	1992	Change 1967–1987	Change 1987–1998	Number of companies 1967	1987	1992	% Change 1967–1987	% Change 1987–1998	VA/VS	Ag input share	Co-op VS share
20+21	All food and tobacco products[a]	51	66	75	15	9	26958	15790	16151	-41.4	2.3	38.8	—	5.4
2011	Meat packing plant products	26	32	50	6	18	2529	1328	1296	-47.5	-2.4	11.6	75.9	0.1
2013	Sausages and prepared meats	15	26	25	11	-1	1294	1207	1128	-6.7	-6.5	26.9	0.0	0.1
2015	Poultry and egg processing	15	28	34	13	6	709	284	373	-59.9	31.3	27.6	68.5	5.0
2021	Butter	15	40	49	25	9	510	44	31	-91.4	-29.5	9.4	19.1	62.8
2022	Cheese, natural and processed	44	43	42	-1	-1	891	508	418	-43.0	-17.7	20.2	47.2	23.4
2023	Condensed and evaporated milk	41	45	43	4	-2	179	124	153	-30.7	23.4	40.8	36.1	27.1
2024	Ice cream and ices	33	25	24	-8	-1	713	469	411	-34.2	-12.4	32.4	7.2	6.0
2026	Fluid milk	22	21	22	-1	1	2988	652	525	-78.2	-19.5	26.4	56.4	17.2
2032	Canned specialities	69	59	69	-10	10	150	183	200	22.0	9.3	49.6	5.7	0.5
2033	Canned fruits and vegetables	22	29	27	7	-2	930	462	502	-50.3	8.7	45.8	28.0	13.7
2034	Dehydrated fruits, vegetables, soups	32	39	39	7	0	134	107	124	-20.1	15.9	51.2	15.0	14.2
2035	Pickles, sauces, salad dressings	33	43	41	10	-2	479	344	332	-28.2	-3.5	50.4	10.3	1.8
2037	Frozen fruits and vegetables[b]		31	28	2	-3		194	182	42.6	-6.2	45.2	46.2	8.4
2038	Frozen specialities[b]		43	40	5	-3		244	308	-37.1	26.2	49.9	5.9	0.2
2041	Flour and other grain mill products	30	44	56	14	12	438	237	230	-45.9	-3.0	26.8	70.1	1.0
2043	Cereal breakfast foods	88	87	85	-1	-2	30	33	42	10.0	27.3	74.7	8.7	0.0
2044	Milled rice and by-products	45	56	50	11	-6	54	48	44	-11.1	-8.3	38.0	88.2	44.5
2045	Prepared flour mixes and refrigerated doughs	63	43	39	-20	-4	126	120	156	-4.8	30.0	48.7	0.0	0.0
2046	Wet corn milling	68	74	73	6	-1	32	31	28	-3.1	-9.7	43.3	53.3	0.0
2047	Dog, cat, and other pet food	46	61	58	15	-3		130	102	-11.6	-21.5	54.1	7.0	0.2
2048	Prepared feeds, n.e.c.[be]	22	20	23	-2	3		1182	1161	-25.1	-1.8	22.7	16.0	4.2
2051	Bread, cake, and related products	26	34	34	8	0	3445	1948	2180	-43.5	11.9	64.9	0.0	0.2
2052	Cookies and crackers	59	58	56	-1	-2	286	316	374	10.5	18.4	65.0	0.0	0.0
2053	Frozen bakery products		59	45		-14		103	160		55.3	51.5	0.0	0.0
2061	Sugar cane mill products	43	48	52	5	4	61	31	37	-49.2	19.4	40.7	81.3	10.7
2062	Refined cane sugar	59	87	85	28	-2	22	14	12	-36.4	-14.3	18.1	0.0	15.5
2063	Refined beet sugar	66	72	71	6	-1	15	14	13	-6.7	-7.1	33.5	75.3	29.3
2064	Candy and confectionary[c]		45	45		0			705			55.0	1.9	0.7

2066	Chocolate and cocoa products		69	75		6		173	146		−15.6	46.6	0.3 d	0.0
2067	Chewing gum[c]	86	96	96	10	0	19	8	8	−57.9	0.0	68.8	0.0	0.0
2068	Nuts and seeds		43	42		−1		79	102		29.1	39.8	35.5	25.9
2074	Cottonseed oil mill products	42	43	62	1	19	91	31	22	−65.9	−29.0	22.7	67.6	16.0
2075	Soybean oil mill products	55	71	71	16	0	60	47	42	−21.7	−10.6	11.1	76.0	16.8
2076	Vegetable oil mill products, n.e.c.	56	74	89	18	15	34	20	18	−41.2	−10.0	19.2	70.8	4.3
2077	Animal and marine fats and oils	28	35	37	7	2	477	194	159	−59.3	−18.0	42.7	0.0	1.5
2079	Shortening and cooking oils	43	45	35	2	−10	63	67	72	6.3	7.5	30.4	0.0	4.3
2082	Malt beverages	40	87	90	47	3	125	101	160	−19.2	58.4	53.5	1.9	0.0
2083	Malt and malt byproducts	39	64	65	25	1	32	15	16	−53.1	6.7	28.9	85.4	0.0
2084	Wines, brandy, and brandy spirits	48	37	54	−11	17	175	469	514	168.0	9.6	43.0	27.0	2.5
2085	Distilled liquor, except brandy	54	53	62	−1	9	70	48	43	−31.4	−10.4	59.5	2.0	0.1
2086	Bottled and canned soft drinks	13	30	37	17	7	3057	846	637	−72.3	−24.7	38.5	0.0	4.1
2087	Flavoring extracts and syrups n.e.c.	67	65	69	−2	4	401	245	264	−38.9	7.8	70.6	0.0	0.3
2091	Canned and cured seafood inc soup	44	26	29	−18	3	268	153	144	−42.9	−5.9	36.9	0.0 d	0.0
2092	Fresh or frozen packaged fish	26	18	19	−8	1		579	600		3.6	26.8	0.0 d	0.0
2095	Roasted coffee	53	66	66	13	0	206	110	134	−46.6	21.8	40.5	0.0 d	0.6
2096	Potato chips and similar products	41	62	70	21	8		277	333		20.2	65.5	19.4	0.0
2097	Manufactured ice	33	19	24	−14	5	688	503	513	−26.9	2.0	70.0	0.0	0.0
2098	Macaroni and spaghetti	34	73	78	39	5	190	196	182	3.2	−7.1	58.6	0.0	0.0
2099	Food preparations, n.e.c.	23	26	22	3	−4	1824	1510	1644	−17.2	8.9	52.4	8.3	0.6
2111	Cigarettes	81	92	93	11	1	8	9	8	12.5	−11.1	74.7	2.3	0.0
2121	Cigars	59	73	74	14	1	126	16	25	−87.3	56.3	55.5	4.7	0.0
2131	Chewing, smoking tobacco, snuff	51	85	87	34	2	41	23	23	−43.9	0.0	71.1	4.2	0.0
2141	Tobacco stemming and redrying	63	66	72	3	6	54	62	32	14.8	−48.4	19.0	49.5	0.0
Means for SIC 20–21		43.9	51.1	53.3	7.5	2.1				−25.5	3.0	42.8	24.1	6.9

Source: Ref. 30.

Note: CR4s aree from four-digit industry data, where available, or else 4-digit product class data from Rogers. VA/VS is the ratio of value added to the value-of-shipments, percent, 1987 data. Ag input share is the percentage of total cost of materials accounted for by agricultural inputs, 1987 data. Co-op VS share is the 1987 estimated percent of value-of-shipments accounted for by the 100 largest agricultural marketing cooperatives.

[a]For SIC 20 + 21, the concentration data are the percent of the sector's value added held by the top 100 food and tobacco companies.

[b]The changes are from 1972, not 1967.

[c]In 1992, SIC 2067, chewing gum, was combined with SIC 2064. The 1992 data for SIC 2067 are estimated by Rogers.

[d]Cocoa, coffee, and fish inputs were ignored.

[e]1967 CR4 is estimated.

Table 9. New Food Product Introductions, 1983–1997

Food category	1983	1985	1987	1989	1991	1993	1995	1997
Baby foods	24	14	10	53	95	7	61	53
Bakery foods	515	553	931	1,155	1,631	1,420	1,855	1,200
Baking ingredients	134	142	157	233	335	383	577	422
Beverages	506	625	832	913	1,367	1,842	2,854	1,606
Breakfast cereals	34	56	92	118	104	99	128	83
Condiments	906	1,146	1,367	1,355	1,885	2,043	2,462	2,505
Candy, gum, and snacks	775	904	1,145	1,701	2,787	3,147	3,698	2,631
Dairy products	486	671	1,132	1,348	1,111	1,099	1,614	862
Desserts	37	62	56	69	124	158	125	109
Entrées	319	409	391	694	808	631	748	629
Fruits and vegetables	126	195	185	214	356	407	545	405
Pet foods	62	103	82	126	202	276	174	251
Processed meats	348	383	581	509	798	453	790	672
Side dishes	133	187	435	489	530	680	940	678
Soups	135	167	170	215	265	248	292	292
Total food	*4,540*	*5,617*	*7,866*	*9,192*	*12,398*	*12,893*	*16,863*	*12,398*

Source: Refs. 9 and 37.

Table 10. Total Measured Media U.S. Advertising Spending by Category, 1997 and 1996

	Millions $		%
Category	1997	1996	Change
Total farm-related advertising	55.4	49.2	12.6
Food and food products	3,361.6	3,209.6	4.7
Beverages	1,320.7	1,324.5	−0.3
Beer, wine, and liquor	1,089.2	1,019.5	6.8
Candy and snacks	1,094.4	965.7	13.3
Pet foods and supplies	360.1	317.1	13.6
Cigarettes, cigars, and tobacco	455.1	488.3	−6.8
Total food and tobacco processing	*7,681.1*	*7,324.7*	*4.9*
Restaurants and fast food	3,147.1	2,960.8	6.3
Food and liquor retail	859.6	804.1	6.9
Food and liquor-direct mail	19.8	14.2	39.4
Total food and tobacco related	*11,763.0*	*11,153.0*	*5.5*
Total—U.S. media advertising	*73,214.7*	*66,711.0*	*9.7*

Source: Ref. 38.

Table 11. Concentration of Advertising Expenditures in Food and Tobacco Processing, 1967–1992

	1967 (%)	1982 (%)	1987 (%)	1992 (%)
Top 4	19.4	26.8	32.8	36.9
Top 8	29.9	39.3	47.3	51.0
Top 20	53.4	65.7	72.1	75.3
Top 50	78.1	88.7	90.6	91.1
Top 100	90.5	95.6	96.2	96.4

Note: Excludes advertising by associations, boards, and governments.
Source: Ref. 30.

Also, retailers continue to explore backward integration, and the larger processors often prefer to have their own sales force manage the distribution of their products in all markets where they have a sufficient volume to justify the expense. By 1992, the four largest wholesalers controlled 11.2% of wholesaling sales, up from 6.5% in 1982, and the 20 largest held a 25.2% share in 1992, up from 16.9% in 1982 (Table 12). The two largest wholesalers, Supervalu and Flemming Cos., are nearly of equal size and much larger than the other wholesalers. Both have invested in retail stores, and although wholesaling accounted for about three-fourths of their total sales, it has declined and their retailing sales have increased. The five largest wholesalers are listed in Table 13.

FOOD RETAILING AND FOOD SERVICE

Americans spent over $700 billion on food in 1997, including imports and fish, with food at home accounting for 55% of the total and the away-from-home market holding a 45%

Table 12. Aggregate Concentration of Largest Wholesale Firms, Grocery Stores, and Restaurants, 1982–1992

	Percent of total sales		
	1982	1987	1992
Wholesale firms			
4 largest	6.5	7.9	11.2
8 largest	10.1	12.2	16.9
20 largest	16.9	20.9	25.2
50 largest	27.0	32.5	35.2
Grocery stores			
4 largest	16.4	17.4	16.1
8 largest	24.1	26.5	25.3
20 largest	35.6	37.2	37.6
50 largest	44.7	47.7	49.9
Restaurants			
4 largest	5.6	6.9	7.8
8 largest	9.0	10.3	10.6
20 largest	12.6	14.4	15.2
50 largest	16.1	18.1	18.9

Source: Refs. 20 and 21.

Table 13. Top Five Food Wholesalers, 1998

Wholesaler	Total sales (millions $)	No. of shores	Private label share (%)
Supervalu	16,552	4,400	10
Fleming Cos.	16,487	3,500+	15
Nash Finch Co.	4,451	12,000	12
Wakefern Food Corp.	4,300	185	30
C&S Wholesale Grocers	3,400	1,250	10

Source: Ref. 40.

share (Table 14). This 55/45 percentage split of total food expenditures has remained roughly constant since 1988, but it is up substantially from the 75/25 percentage split in 1954 (42). Retail food stores have realized that they must compete with the away-from-home firms if they are to continue to grow as a food supplier. They have countered this trend toward away-from-home dining by offering ever more prepared foods from in-store delis to entire ready-to-serve meals prepared in their stores. No one has solved this marketing puzzle completely, but everyone is trying as they respond to time-pressed but more affluent shoppers seeking a better solution to their food needs and desires.

Food retailing is constantly changing. Store formats come and go as retailers strive to achieve greater efficiencies and a competitive advantage. A store's meat section was once the featured department to differentiate a store, but today a well-stocked, attractively displayed produce section is what consumers value most. Prices are always among the leading factors consumers mention when asked about why they shop at particular stores, but convenience and quality are also primary considerations. Affluent shoppers are willing and able to pay more for time saved. Home delivery, a retail method long thought dead, has returned, and in 1997 consumers spent $10.3 billion for food home delivery and mail order, up over three times in 10 years (43). It now accounts for 2.6% of food-at-home expenditures and 1.5% of total food expenditures. The proliferation of Internet shopping sites adds interest in this format; the best known Internet retailer, Amazon.com, has invested in an Internet grocery venture (44).

Retail food stores can be arrayed along two dimensions, variety and price, to capture most of the variation found in store formats today (Fig. 8). In the middle at the origin

Table 14. Expenditures for Food, 1977–1997

Year	Millions $ Food at home[a]	Food away from home[b]	Total (millions $)	Percent of total Home	Away
1977	137,332	84,835	222,167	61.81	38.19
1982	205,583	139,776	345,359	59.53	40.47
1987	254,629	198,926	453,555	56.14	43.86
1992	326,456	264,908	591,364	55.20	44.80
1997	394,593	320,275	714,868	55.20	44.80

Source: Ref. 41.
[a]Food for off-premise use.
[b]Meals and snacks.

is the conventional supermarket, with a full line of groceries, meat, produce, service deli and bakery and with general merchandise and health and beauty aids, which account for 6% to 8% of store sales. Alternative formats move out from the origin by adding or subtracting the number of products carried and by featuring lower or higher prices. The traditional mom-and-pop corner store continues to lose share but has not been eliminated. Nontraditional formats, such as wholesale clubs (eg, Costco and Sam's) and Supercenters (eg, WalMart, Kmart, Fred Myer) have increased their share of grocery volume (defined as items commonly found in a traditional channel) to roughly 10% (Table 15) (39). Newer formats that operate in the traditional channel, such as Superstores and Food Drug Combos, account for about two-thirds of store sales. The conventional supermarket's share of total sales peaked at 70% in 1965 and has fallen to just under 25% in 1995.

Overall or aggregate concentration in food retailing has not shown as much consolidation as found in the other sectors (Table 12) until recently. By 1992, the four largest food grocery retailers held 16.1% of total sales, down a point from 1987 and essentially unchanged since 1982. Among the top 20 (and 50), concentration did increase, but not dramatically, from 35.6% (44.7%) in 1982 to 37.6% (49.9%) in 1992 for the top 20 (top 50). This stability was sharply broken in the 1990s as the 10 largest chains increased their share of sales to 60% from 35% in 1995 (46). The five leading food retailers doubled their national share to 40% in 1998 from 20% in 1993 (Table 16).

Overall concentration is less interesting to economists than market concentration, and in food retailing, the geographic size of the market is local rather than national. Most economists use the Metropolitan Statistical Area (MSA) as the more appropriate geographic scope, but unfortunately the U.S. government does not publish concentration data at this level unless requested to perform a special tabulation, for which they charge. Such tabulations have been done in past census years. Alternatively, a trade publication does collect and publish data at the MSA level. When examined at the MSA level, food retailing markets are commonly oligopolies, with a CR4 of 50% or more (39). These data are consistent with a more thorough analysis of concentration in U.S. grocery MSAs done with special tabulations of the census data (48). Franklin and Cotterill found average market concentration did increase in the 1980s, with smaller markets more concentrated than larger markets on average.

The only challenger to the processor for control of the vertical marketing channel is the modern retailer. Retailers face consumers daily and have begun to harness the power of information generated by their checkout scanners. No longer does the retailer turn to the processor's personnel to explain what consumers are buying because they now control the most valuable asset in market research—data on consumers and their purchases. With the use of frequent shopper cards and other incentives, retailers can link demographic data to food sales in ways only dreamed about a decade ago. They are still learning to exploit this valuable information, which processors now must buy from retailers. No longer are they unsure of what sells and which products generate the highest profits. In Eu-

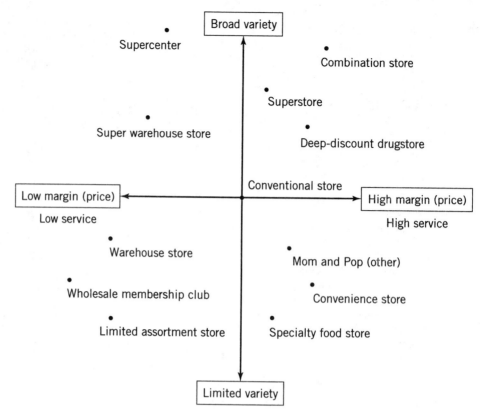

Figure 8. Food retail formats.
Source: Refs. 39 and 45.

Table 15. Store Format and All Commodity Volume (ACV) Share Trends

	1993		1994		1995	
	% of stores	% of ACV	% of stores	% of ACV	% of stores	% of ACV
Traditional channel[a]						
Conventional	9.3	26.1	9.5	25.6	9.5	24.3
Superstore	3.8	22.4	4.1	23.3	4.3	24.2
Food drug combo	1.3	10.2	1.5	11.2	1.7	12.3
Warehouse store	1.5	6.5	1.3	5.4	1.1	4.8
Super warehouse	0.3	3.4	0.3	3.5	0.3	3.7
Limited assortment	0.4	0.6	0.6	0.6	0.6	0.6
Convenience (trad)[a]	30.2	6.6	30.3	6.4	29.6	5.9
Convenience (petro)[a]	20.8	3.6	22.2	3.7	22.9	3.8
Other	31.3	11.8	29.1	11.4	28.6	12.1
Subtotal	98.9	91.2	98.8	91.1	98.8	91.6
Nontraditional channel[b]						
Hypermarket	0.0	0.2	0.0	0.1	0.0	0.1
Wholesale club	0.4	5.6	0.4	4.8	0.4	4.7
Mini club	0.1	0.3	0.1	0.3	0.1	0.3
Supercenter	0.2	1.5	0.2	2.0	0.3	2.5
Deep discounter	0.4	1.2	0.4	1.7	0.3	0.8
Subtotal	1.1	8.8	1.2	8.9	1.2	8.4
Total	*100.0*	*100.0*	*100.0*	*100.0*	*100.0*	*100.0*

Source: Ref. 39.
[a]Merchandise sales only (non-gas).
[b]Reflects share adjustments for items not commonly found in traditional grocery channel.

Table 16. Top Five Food Retailers, 1998

Rank	Company	Sales in billions $[a]	Market share (%)
1	Kroger	43.0	10
2	Wal-Mart[b]	39.0	9
3	Albertson's	36.0	9
4	Safeway	27.1	7
5	Ahold USA	22.2	5

Source: Ref. 47.

[a]Sales figures are based on fiscal year, and include mergers through 2/28/99.

[b]Includes grocery sales by Wal-Mart Supercenters and Sam's Clubs.

rope, the food retailer has long been the channel leader, and experts are predicting it to emerge as the leader in America as well. In terms of profitability, food retailers average only a little over 1% after-tax profit as percent of sales, whereas food manufacturers average at least four times that (Table 17). To compare these two vertical stages based on sales is misleading, and a more accurate comparison uses after-tax returns to stockholder equity. By this measure, manufactures and retailers both have averaged just over a 16% rate of return since 1980.

Today's consumer is shopping more for complete meals rather than meal ingredients to be prepared at home, even if such preparation is minimal. Food service outlets have long catered to this demand. Food service includes both commercial (eg, restaurants) and noncommercial (eg, school cafeterias, vending machines, hospitals) outlets, with the commercial outlets accounting for about 80% of the total sales volume. Consumers believe they have greater choice in selecting their food in the commercial sector. Currently, many noncommercial food service outlets are being replaced by commercial vendors.

Although food service has not reached an even split of the consumers' food dollar with food at home, it has been the growth segment of the retail food business for at least 40 years. Rising incomes, smaller households, younger ages, and time-pressed shoppers all lead to greater interest in dining out. The time pressure and younger ages along with a car culture drove the fast-food segment from a 4%

Table 17. Profit Margins of Food Manufacturers and Retail Food Chains, Industry Averages, 1981–1997

	After-tax profits as a percentage of			
	Food manufacturers		Retail food chains	
Year and quarter	Sales	Stockholder equity	Sales	Stockholder equity
1981	3.1	13.6	1.0	13.9
1983	3.3	13.3	1.1	13.6
1987	4.6	17.5	0.9	12.8
1989	4.2	17.1	0.8	20.7
1991	4.8	17.5	1.1	18.8
1993	3.7	13.5	0.8	11.7
1995	5.5	18.5	1.5	21.3
1997	5.6	19.8	1.6	17.4

Source: Ref. 49.

share of the away-from-home market in 1954 to 33% in 1996 (42). The rise of McDonald's alone charts the success of this segment.

Overall concentration is not high in food service, but like food retailing it is a local market rather than a national one. If all restaurants are combined, the largest four held only an 8% share and the largest 20 only a 15% share in 1992 (Table 12). Within local markets, the shares would be much higher, but no data are available, and most markets have several competing outlets. The overall concentration is highest in the fast-food segment, with fast-food burger restaurants having the greatest concentration. McDonald's held a 42% national share in 1997 (Table 18), and the top 5 chains held a 87.6% share. The pizza segment is not as highly concentrated, with the largest firm, Pizza Hut, holding a 22.6% national share in 1997 (Table 19). The largest chains control a larger share of the advertising expenditures in food service than they do in sales, especially so in the pizza segment.

Food retailers have responded to the loss of sales to the away from home market and now offer complete meals, called home replacement meals (HMR), and consumers show increased interest in these products. There has been some blurring of the lines between supermarkets and food service, fast-food restaurants now reside within retail stores and food retailers offer take-home meals such as fully cooked pizza and complete meals prepared while con-

Table 18. Top 10 Fast-Food Burger Restaurants, 1997

Rank	Brand	Share of market (%)	Share of media advertising (%)
1	McDonald's	42.2	41.8
2	Burger King	21.4	30.6
3	Wendy's	11.9	12.4
4	Hardee's	8.8	5.8
5	Jack in the Box	3.3	3.7
Total top 5 share		*87.6*	*94.4*
Total market		*$40,600.0*	*$1,380.9*

Source: Ref. 50.

Note: Dollars are in millions. Measured media from Competitive Media Reporting. Sales from Technomic. BK sales include major franchisors, Carrols, and AmeriKing, Sydran and Quality. Wendy's includes DavCo. Hardee's includes Boddie-Noell and the Advantica Hardee's franchises acquired in April. Hardee's media estimate.

Table 19. Top 5 Pizza Chains, 1997

Rank	Brand	Share of market (%)	Share of media advertising (%)
1	Pizza Hut	22.6	42.2
2	Domino's Pizza	11.4	35.1
3	Little Caesars	8.2	10.4
4	Papa John's	4.0	4.8
5	Chuck E. Cheese's	2.0	4.8
Total top 5 share		*48.2*	*97.3*
Total market in dollars (millions)		*$21,700.0*	*$349.2*

Source: Ref. 50.

Note: Measured media from Competitive Media Reporting. Sales from Technomic. Pizza Hut also includes franchisor NPC.

sumers shop for other groceries. Fast-food outlets are still the primary source for takeout food, with a 41% share in 1997 (Table 20). Other restaurants hold a 21% share, but supermarkets have increased their share to 22% from 12% a year earlier. If consumers remain satisfied with these meals and retailers can eventually profit from supplying them, then supermarkets will share in the growth of the takeout market. Busy people have less interest in dining out because it adds rather than detracts from daily stress. Such consumers are eager to take fully prepared meals home to be eaten in the comfort and privacy of their own homes.

FUTURE FORCES FOR CONTINUAL CHANGE

General

The U.S. food marketing system is a mature sector, but one that is constantly changing. To predict the characteristics of the food industry in the twenty-first century requires both the cautious extrapolation of the economic trends and insightful consideration of the potential impact from the most significant forces exerting pressure on the U.S. food marketing system.

Powerful forces—new technologies, industrialization and consolidation, globalization, and the changing consumer—will alter the future food system. The changing U.S. consumer was discussed earlier along with the implications for the food marketing system. Forecasting is always uncertain, and it is difficult to project quantitatively the impact these areas of change might have on the food system. It is impossible to predict the yet-to-be discovered information or technology, for example, that might alter the course of events in a monumental fashion, yet it must be expected that such discoveries can occur. Each individual firm's success, if not survival, and the development of the industry will be shaped by the industry's resilience and willingness to accept and adapt—indeed, to take advantage—of changes.

Technological Change

Much of the early development of the U.S. food system was catalyzed by the development and introduction of new technologies, especially in production and processing. Significant food science innovations from 1939 to 1989 include aseptic processing and packaging, minimum safe canning processes for vegetables, the microwave oven, frozen con-

centrated citrus juices, controlled atmosphere packaging for fruits and vegetables, freeze drying, frozen meals, the concept of water activity, food fortification, and ultrahigh temperature processing of milk and other products (51). Entire businesses have been developed on the basis of such technologies.

In the latter half of the twentieth century, food companies, in particular processors, concentrated most of their research efforts on the development and proliferation of new products rather than development of technology. Most of these new products do not represent great technological change. In fact, by most measures, the food industry has not positioned itself for significant internal technology development. Research and development investments are low compared to other industries. Public investment in food research for postproduction industries is low, and food-industry employment of engineers and scientists is considerably lower than other industries (52). Still, productivity in the food industry has grown at an annual rate of 3% since 1963, largely due to borrowing technologies from other industries.

A more complicated research effort, but still related to new product development, has been the search for fat alternatives and to offer fat-free versions of popular products. However, because animal and plant fats are difficult to duplicate in the lab, they have had limited success. Procter & Gamble's Olestra is best known and is being used commercially by companies, such as Pepsi Co's Frito Lay in their reduced-fat chips. Consumer acceptance is still unclear despite heavy promotional spending. Some consumer groups are strongly opposed to products using such alternative ingredients and have sought increased labeling of possible harmful effects and even the outright banning of such products. In addition, consumers show signs that their fat-avoidance days are waning.

Not all the technological effort in the last half of the century has been directed to minor tinkering with consumer products, and several new technologies have brought, or promise to bring, developments to rival any in our past history. Separation techniques allow basic commodities to be fragmented efficiently into components for restructuring or incorporation into other foods or for removal of undesirable compounds. Corn wet milling, introduced in the United States in the mid-twentieth century, is an example. The new separation technologies include passive membrane filtration (reverse osmosis, ultrafiltration, and microfiltration) and supercritical fluid extraction. These technologies have already achieved commercial success in the food industry.

Information processing and automation have had dramatic impacts on the entire food system as more efficient, productive methods replace outdated methods. Farmers can match fertilizer and pesticide applications to each acre's needs guided by on-board computers running advanced geographic information systems. Farm managers can harvest vast amounts of information and use it to improve operations from record keeping to computer-controlled nurseries for plants or animals. At the processing level, computer-aided measurement devices can efficiently evaluate product quality and pay producers accordingly. Computer-integrated manufacturing, which in-

Table 20. Shoppers' Primary Source of Takeout Food (%)

	1997	1996	Point change
Fast-food restaurant	41	48	−7
Restaurant	21	25	−4
Supermarket	22	12	10
Deli/pizza, bagel, coffee shop	5	4	1
Gourmet/specialty store	7	3	4
Convenience store	1	1	0
Other/don't know	3	6	−3

Source: Ref. 40.

tegrates order entry, scheduling, operations, inventory control, and other operations management activities, has been implemented in many facilities. At the retail level, information processing, including the development of direct product profitability strategies, has helped shift the balance of power in the food system from processors to retailers. New information processing hardware and software, coupled to on-line sensing and control devices, have encouraged increased vertical coordination in the industry and strategic alliances by linking retail activity with distribution, processing, and, perhaps, production. Retailers' scanning information has altered how consumer research is done, giving the retailer control of the best marketing information ever assembled.

The growth of the Internet and the growing comfort level with on-line shopping has led many visionaries to invest in Internet grocery businesses, some with home delivery. No company has managed to profit from these investments to date, but the appeal brings new investors and firms on a continual basis. Far more impressive results have been with company-to-company Internet sales, estimated to be five times the consumer retail total. Forrester Research, a market research firm, predicts on-line business-to-business sales will grow from $43 billion in 1998 to $1.3 trillion, or 9.4% of such sales within four years (53). The Internet will continue as a major force in altering businesses, especially in retailing.

Biotechnology and the related technologies, often called bioengineering or genetic engineering, have brought controversy and promise but few profits to date. Chemical companies and other input supply firms view biotechnology as the future and have several products now in wide use. Monsanto has been a pioneer in these developments. Production agriculture has seen the development of crops with unique characteristics, including disease and pest resistance and improved yield and quality. A gene, patented in 1998 and now purchased by Monsanto, causes genetically engineered crops to produce sterile seeds (54). Animal agriculture has adopted growth regulators, such as bovine growth hormone. Genetic manipulation of animals, including cloning, seeks commercial application. The lines between food firms and pharmaceutical companies continue to blur as these new products emerge. These new advances have raised both hopes of a better world and concerns that science is out of control and requires regulation before ethical lapses or mistakes result in unforeseen disasters.

The new processing technologies have provided more efficient means of converting commodities into high-quality foods and also have enhanced the consolidation of the entire food system. These new larger operations, from the biotech firms that supply the growth hormones to the mega-hog farms to the slaughtering firms that transform the animals into value-added products ready for the retailer's meat case, have their critics who charge that any benefits from these changes have been at an unacceptable cost in terms of environmental hazards, untold future dangers to health, and a decline in food safety. Massive meat recalls in the late 1990s and other food safety problems from salad bars to organic fruit drinks have damaged consumers' confidence in the safety of their food supply. Business and government have responded with programs such

as hazard analysis critical control points (HACCP) and by allowing greater use of food irradiation to kill harmful bacteria (55). The use of irradiation in the United States has been limited by consumer resistance over its safety, but given the routine occurrence of Listeria, Salmonella, and other bacteria outbreaks, consumers may accept this technology, which has been used in other countries (56). Others argue that the problems were created by technology in the first place and that a return to more traditional agricultural methods will eliminate the problems without resulting to technological fixes that may entail unknown repercussions.

History teaches that attempts to stifle technology are largely unsuccessful. Consumers must recognize that technological change itself will be shaped by the marketplace. Technologies will be developed and successfully introduced only when an economic advantage is afforded or a need in the marketplace is met and when consumers are willing to accept them or are passive and apathetic regarding their use. If consumers refuse to accept certain technologies, then businesses will abandon them or public officials will ban or regulate them. The Europeans have held firm to their disapproval of genetically-altered foods and their refusal to accept U.S. hormone-enhanced beef, but Americans appear more diverse in their views. Such consumer diversity leads to a variety of market niches from small organic farms to factory farming enhanced by biotechnology. Perhaps a huge event will grab attention and galvanize consumers against bioengineered foods. Alternatively, consumers may find their fears will fade as the products deliver promised benefits without side effects. The concern over bovine growth hormone in the milk supply did not result in any decline in consumption (57), yet an organic milk market has evolved during the same period (58).

Globalization

The global economy has had a huge impact on countries around the world. Trade has been embraced by even the most protectionist countries, and trading groups continue to form, from NAFTA to the European Union. The United States, with its abundant resources and huge size, began its international participation as primarily an exporter of surplus agricultural products. In the 1990s, however, the U.S. trade in high-value products, such as speciality products (fruit, nuts, breeder livestock, and processed foods) and dairy products, beverages, and prepared foods, overtook exports of bulk agricultural commodities, exceeding $30 billion in 1996 (Fig. 9). Additionally, Americans became major importers of agricultural products, going well beyond those products that cannot be grown profitably in the United States, such as cocoa, coffee, bananas, and palm oil. In 1996, the United States imported $33.6 billion, up from only $4 billion in 1959, on agricultural goods, with the vast majority being products that are grown in the U.S., such as meat, dairy, fruits, and vegetables (42). Some of this growth stems from foreign suppliers catering to U.S. consumer demand for fresh fruit and vegetables regardless of the season, but other imports represent lower priced foreign goods competing with domestic supplies. Overall, the United States maintains a trade surplus in agricultural products, some $22 billion in 1997 (11).

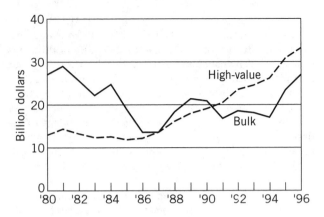

Figure 9. U.S. high-value and bulk agricultural exports, 1980–1996. *Source:* Ref. 42.

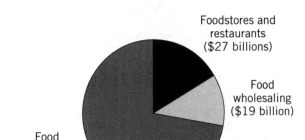

Figure 11. Estimated sales of foreign affiliates of U.S. food marketing firms, 1996. Largest share of sales came from food processing. *Source:* Ref. 32.

The United States is the world's leading importer and exporter of processed foods. Some of our leading domestic brands are also leading brands around the world, such as Coca-Cola soft drinks, Marlboro cigarettes, and Kellogg's cereals. Likewise, many popular brands are imported, such as Heineken beer and Perrier water. In 1996, the U.S. exported $30.1 billion of processed foods and imported $27.8 billion, leaving a surplus of $2.3 billion. Except in 1990, the United States has had a trade surplus in processed foods in the 1990s.

Actual international trade data fail to capture the full globalization of U.S. food marketing firms. U.S. food processing firms often prefer to invest in local facilities or in licensing agreements in a host country rather than export finished products from the U.S. Such foreign direct investment accounts for almost four times the value of sales than from exporting (Fig. 10). In 1996, U.S. food processing firms had estimated sales of $116 billion from their foreign affiliates (Fig. 11). In food retailing and food service, traditional exports are not possible, yet foreign direct investment allows transporting a firm's system and reputation to a foreign country. McDonald's operates in over 110 countries. The estimated sales of foreign affiliates of U.S. food

stores and restaurants reached $27 billion in 1996. Foreign direct investment offers firms several advantages over exporting from lower transportation costs to greater understanding of the consumers and familiarity with local government officials. It also avoids many tariffs and other barriers to trade, while retaining the brand name of the U.S. parent company.

Foreign direct investment also flows from abroad to the United States, with several leading firms, such as Nestle in food processing, Loblaw in food wholesaling and retailing, Ahold (owner of several food retailers including Stop & Shop and Giant Food Stores), and Diageo PLC (owner of Burger King) in food service. In 1996, the estimated sales of U.S. affiliates of foreign food marketing firms was $152 billion (Fig. 12), with roughly 77% of that coming form an even split between food processing ($59 billion) and food retailing ($58 billion).

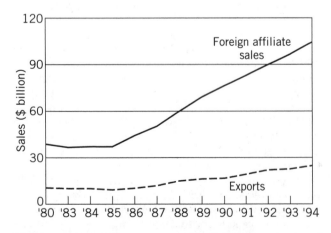

Figure 10. United States processed food exports and sales of U.S.-owned foreign affiliates, 1982–1994. *Source:* Ref. 42.

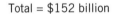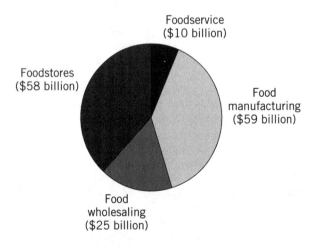

Figure 12. Estimated sales of U.S. affiliates of foreign marketing firms, 1996. Largest share of sales came from foodstores and processing. *Source:* Ref. 32.

The international flow of products and the mixing of cultures will continue as transportation improves and communication links increase around the globe. Jumbo jets now fly frozen fish from Taiwan to New York and fresh fruits and vegetables from Chile to the United States. Increased consumer diversity and rising incomes will encourage these flows. International trade agreements, such as GATT, also reduce the barriers to trade. The World Trade Organization (WTO) seeks to resolve trade disputes before they erupt into trade wars. Nevertheless, strategic moves and counter moves by countries and powerful firms will continue to seek an advantage. The current U.S.-Europe trade dispute over bananas and hormone-treated meat are suggestive of future squabbles, yet pose little threat to reversing the increased flow of goods and services across national borders. Many of the same issues exist in domestic trade as well because consumers vary on their acceptance of biotechnology and other practices that affect the food supply, such as pesticide residues.

Industrialization and Consolidation

American agriculture continues to consolidate and industrialize as consumers splinter into more segments and technology allows catering to the diversity of demands. Economic markets can be incredibly efficient in making sense out of the economic chaos involved in moving products from production to consumption because they summarize the information contained in buyers' demands and sellers' supplies. In the real world, however, traditional agricultural markets are not as perfect as the economist's model suggests; they miss opportunities to link producers and processors in more profitable arrangements. For example, major chicken processors have been fully integrated by ownership from the hatchery to the processing plant for decades. They merely hire contract growers to raise the birds to market weight without ever transferring ownership. They even supply the feed and other inputs required for the growout operation. Other industries have turned to legal contracts to secure input supplies tailored to their operations (eg, vegetables for processing) rather than using markets or ownership of the farms (59). Farmers benefit in lowering their risks and processors are assured of supplies with appropriate features. Several major processors have entered strategic alliances with growers where they contract for character-specific raw products in a relationship that all parties expect to be ongoing (60).

Economists understand the benefits of these nonmarket transactions, but also the costs. As more product volume moves through nonmarket methods, the less is known about true product values because key economic information summarized by price becomes more difficult to discover. To date, most of these nonmarket arrangements involve linking the processing and production stages of the marketing system. However, other stages have established nonmarket coordination in what has been termed supply chain management. Large retailers now contract for much of the produce they sell rather than buy their produce at the various regional markets. Much of the industrialization has featured improved information, tailored inputs, and reduced cost of production and processing. Consumer concerns arise from whether there will be sufficient competition to force such efficiencies to be passed on as lower prices and rural communities wrestle with major issues resulting from factory farms that reduce the number of family farms and add to environmental concerns. Even producers who entered these contracts worry whether they will receive fair prices for their products once the marketplace is removed or diminished.

All stages of the vertical system are becoming more concentrated as larger operations increase their size. At the same time, there is increased diversity as the larger firms get larger and the number of smaller firms increases. It is the middle-sized firm that is most endangered by the consolidation movement. Whether in farming or retailing, as the largest firms increase their share of the sector's output, a growing number of smaller firms emerge in the cracks and eddies left behind by the larger firms.

Food and tobacco processing has seen the most dramatic consolidation in this century. Merger patterns have followed the four great merger movements of the general economy. The first major merger wave occurred around the turn of the century and created some of the famous trusts that antitrust legislation was suppose to prevent (61). For example, American Tobacco and General Mills were formed during this merger wave. The next wave came during the 1920s when companies such as General Foods were being formed through merger. The third merger wave was in the 1960s and was characterized by amazing conglomerates being formed as unrelated firms sought management synergies, such as ITT buying Continental Baking. The fourth merger movement came in the late 1970s and 1980s and was a wild period of leveraged buyouts and hostile takeovers funded with questionable, often illegal, financial instruments. Food companies were at the forefront of these mergers with record-setting deals, such as the $25 million leveraged buyout of RJR Nabisco. The largest food and tobacco processor, Philip Morris, is essentially a case history in a merger-built business. Starting from its dominant position in cigarettes, Philip Morris purchased such already huge companies as Miller Brewing, General Foods (who already had bought Oscar Mayer), and Kraft Foods. Few American shoppers now know the parent company of the branded goods they bring home from the supermarket.

It appears that we are in the midst of a fifth merger wave, and again the food businesses are major players (Fig. 13). Most of the food-related mergers involve food processing and retailing firms, including some the largest mergers in history, but increasingly mergers in food service and wholesaling are commonplace. Wholesalers are increasing their ownership of retailers as they seek to survive in the modern food system. Some of the failures of the previous merger wave are being undone as firms now seek brands from other firms as they selectively add and subtract from their portfolio of brands. Others merely purchase firms whose brands fit well with their current offerings. This current merger wave is more horizontal in nature as processing firms seek merger partners among current rivals. Gone are the wild conglomerate mergers as firms now seek to consolidate their hold on leading positions in markets where they currently hold a strong position. Some economists have become concerned about the

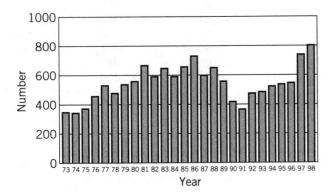

Figure 13. Food mergers and acquisitions, 1973–1998. *Source:* Ref. 62.

Table 21. Average Hourly Earnings of Production and Nonsupervisory Employees of Food Industries, 1977–1997 (Dollars Per Hour)

Year	Manufacturing food and kindred products	Wholesale trade groceries, and related products	Food stores	Eating and drinking places
1977	5.37	5.43	4.77	2.93
1982	7.92	8.25	7.22	4.09
1987	8.93	9.53	6.95	4.42
1992	10.20	11.09	7.56	5.29
1997	11.49	12.87	8.59	6.05

Source: Ref. 63.

growing concentration and march toward oligopoly in almost every market. There is little evidence of any positive benefits from such mergers outside of the stock market evaluation of these firms. The stock market rewards downsizing as a cost efficiency, and increased market share enhances profitability potential through uncontested price increases. The firm defenses of increased efficiencies and productivity gains have been overshadowed by a reduction in research and development and diminished employee morale. Professional organizations and associations, such as the Institute of Food Technologists and agricultural associations, have called the U.S. government to increase their oversight of this consolidation. Without public policy changes these consolidations will continue for several reasons: (*1*) the food industry is more predictable and less cyclical than most industries; (*2*) food companies generally generate high cash flows; (*3*) purchase of brands, often available through food company buyouts, is particularly attractive; and (*4*) international exchange rates and the stable political environment make the assets of U.S. food companies especially attractive to foreign companies looking to increase penetration into U.S. markets.

The Changing Consumer

It is appropriate to end this overview of the U.S. food system where we began—with the consumer—and to examine the changes that will challenge current firms and provide opportunities to new businesses. Consumer demand, in fact, will be the principal source of growth and development in the food industry in the future. Economists do not forecast significant changes in total caloric intake, although it has drifted upward slightly over time, but they do anticipate consumer demand to be affected by increases and demographic changes in the U.S. population, shifts in income distribution, and changes in consumer preferences. Increases in total consumer demand can be forecasted fairly accurately based on Census Bureau projections of the U.S. populations. These projections have been made, but generally they are based on the premise that consumers of certain age and income groups in the future will exhibit the same preference as persons of similar age and income groups in today's population. This premise is not universally accepted by marketers. In fact, changes in pref-

erence are almost certain to occur. Some preference changes will arise from continued immigration and assimilation of new cultural values. Others will arise from changes in household patterns. Most, however, will arise from poorly understood social phenomena. The interplay between technology and consumers will be critical to the future food system.

The workforce experiences the same changes, especially with respect to declining numbers of potential workers in the age groups normally associated with entry-level jobs. Wage rates in many food-related industries are low (Table 21), and firms have difficulty attracting workers. Some agricultural production will be forced offshore because of shortages of people willing and able to do manual labor for low wages. Another shift has been and will be the increasing assimilation of women, minorities, and immigrants into the workforce, especially in upper management positions. The changing workforce composition will place special demands on human resource management and development in the coming years, especially in the food industry, where a high percentage of entry-level workers are employed.

The increased diversity of the American consumer along with the diversity tapped through globalization provides many opportunities for businesses. Marketers no longer have to use the mass marketing methods that view all consumers as similar. They can target those consumers that they can best serve—from vegetarians who prefer organic foods to cigarette smokers who prefer red meat meals without the vegetables. The world is a fascinating place with a wide array of consumer preferences, and the food system has always reflected the entire distribution of humankind, making it an exciting place for career opportunities.

BIBLIOGRAPHY

1. U.S. Department of Agriculture, *1998 Agricultural Fact Book*, URL: *http://www.usda.gov* (last accessed Sept. 19, 1999).

2. J. J. Putnam and J. E. Allshouse, *Food Consumption, Prices, and Expenditures, 1970–97*, Statistical Bulletin No. 965, Food and Rural Economics Division, Economic Research Service, U.S. Department of Agriculture, Washington, D.C., 1999.

3. R. T. Rogers, "Broilers: Differentiating a Commodity," in L. L. Duetsch, ed., *Industry Studies*, 2nd ed., M. E. Sharp, Armonk, N.Y., 1998, pp. 65–100.

4. *Food Guide Pyramid Information*, U.S. Department of Agriculture and U.S. Department of Health and Human Services, URL: *http://www.nal.usda.gov/fnic/Fpyr/pyramid.html* (last accessed Sept. 19, 1999).

5. L. S. Kantor, *A Dietary Assessment of the U.S. Food Supply: Comparing Per Capita Food Consumption with Food Guide Pyramid Serving Recommendations*, Agricultural Economic Report No. 772, Food and Rural Economics Division, Economic Research Service, U.S. Department of Agriculture, Washington, D.C., 1998.

6. B.-H. Lin, E. Frazao, and J. Guthrie, *Away-from-Home Foods Increasingly Important to Quality of American Diet*, Agriculture Information Bulletin No. 749, U.S. Department of Agriculture, Washington, D.C., 1999.

7. J. Kinsey, "Changes in Food Consumption from Mass Market to Niche Markets," in Lyle P. Schertz and Lynn M. Daft, eds., *Food and Agricultural Markets: The Quiet Revolution*, NPA Report No. 270, Economic Research Service, U.S. Department of Agriculture, Washington, D.C., 1994, pp. 19–43.

8. *Population Trends 1996–2050*, The Food Institute, Fair Lawn, N.J., 1996.

9. "Publisher Categorizes Shoppers Six Ways," *Food Institute Report*, The Food Institute, Fair Lawn, N.J., February 15, 1999, p. 3.

10. M. Beck, "Next Population Bulge Shows Its Muscle," *Wall Street Journal*, February 3, 1997.

11. "Labor Force, Employment, and Earnings," *Statistical Abstract of the United States: 1998*, U.S. Department of Commerce, Bureau of the Census, Washington, D.C., September 25, 1998.

12. "These Grocery-Delivery Services Might Even Do Windows," *Wall Street Journal*, February 12, 1998.

13. "What's For Dinner? Ask Me Later," *Food Institute Report*, The Food Institute, Fair Lawn, N.J., March 9, 1998, p. 1.

14. "Nabisco Dumping Grey Poupon 'Pardon Me' Ads," *Wall Street Journal*, April 6, 1998.

15. "Americans Do Cook," *Food Institute Report*, The Food Institute, Fair Lawn, N.J., February 15, 1999, p. 2.

16. J. Krasner, "Pilgrim Dessert Returns, but Who Will Give Thanks?" *Wall Street Journal*, March 17, 1999.

17. J. Friedland and S. Kilman, "As Geneticists Develop an Appetite for Greens, Mr. Romo Flourishes," *Wall Street Journal*, January 28, 1999.

18. "Greed Is Good," *The Economist*, January 17, 1998, p. 26.

19. U.S. Department of Commerce, Bureau of the Census, *Statistical Abstract of the United States*, U.S. Government Printing Office, Washington, D.C., 1952–1998.

20. U.S. Department of Commerce, Bureau of the Census, *Census of Retail Trade*, U.S. Government Printing Office, Washington, D.C., 1972, 1977, 1982, 1987, 1992.

21. U.S. Department of Commerce, Bureau of the Census, *Census of Wholesale Trade*, U.S. Government Printing Office, Washington, D.C., 1972, 1977, 1982, 1987, 1992.

22. U.S. Department of Commerce, Bureau of the Census, *Census of Manufacturers*, U.S. Government Printing Office, Washington, D.C., 1972, 1977, 1982, 1987, 1992.

23. U.S. Department of Commerce, Bureau of the Census, *Economic Census*, U.S. Government Printing Office, Washington, D.C., 1972, 1977, 1982, 1987, 1992.

24. U.S. Department of Agriculture, *1997 Census of Agriculture*, U.S. Government Printing Office, Washington, D.C., 1997.

25. U.S. Department of Commerce, Bureau of the Census, *1996 Annual Survey of Manufacturers*, U.S. Government Printing Office, Washington, D.C., 1996.

26. J. S. Royer and R. T. Rogers, eds., *The Industrialization of Agriculture: Vertical Coordination in the U.S. Food System*, Ashgate, Aldershot, United Kingdom, 1998.

27. S. Warren and S. Kilman, "DuPont Co. Lands Huge Biotech Prize," *Wall Street Journal*, March 16, 1999.

28. M. E. Kuhn, "The 1998 Top 100 Food Companies," *Food Processing*, December 1998, pp. 19–44.

29. J. M. Connor et al., *The Food Manufacturing Industries*, Lexington Books, Lexington, Mass., 1985.

30. R. T. Rogers, "The Role of Cooperatives in Increasingly Concentrated Markets: Reaction," in M. Cook et al., eds., *Cooperatives: Their Importance in the Future Food and Agricultural System*, National Council of Farmer Cooperatives and The Food and Agricultural Marketing Consortium, Washington, D.C., 1997, pp. 49–65.

31. U.S. Department of Agriculture, *Food Marketing Review*, Commodity Economics Division, Economic Research Division, Agricultural Economics Report No. 614, 1989.

32. A. E. Gallo, *The Food Marketing System in 1996*, Agricultural Information Bulletin No. 743, U.S. Department of Agriculture, Washington, D.C., 1998.

33. R. T. Rogers and R. J. Sexton, "Assessing the Importance of Oligopsony Power in Agricultural Markets," *American Journal of Agricultural Economics* **76**, 1143–1150 (1994).

34. G. P. Zachary, "Many Industries Are Congealing into Lineup of Few Dominant Giants," *Wall Street Journal*, March 8, 1999.

35. R. T. Rogers and L. M. Petraglia, "Agricultural Cooperatives in Food Manufacturing: Implications for Market Performance," *Journal of Agricultural Cooperation* **9**, 1994, 1–12 (1994).

36. R. J. Sexton, "The Role of Cooperatives in Increasingly Concentrated Agricultural Markets," in Michael Cook et al., eds., *Cooperatives: Their Importance in the Future Food and Agricultural System*, National Council of Farmers Cooperatives, Washington, D.C., 1997, pp. 31–47.

37. "Gorman's New Product News," *Food Institute Report*, The Food Institute, Fair Lawn, N.J., January 21, 1989, p. 7.

38. Advertising Age Dataplace, URL: *http://adage.com/dataplace/archives/dp265.html* (last accessed March 25, 1999).

39. J. J. Belonax, Jr., *Food Marketing*, Simon and Schuster Custom Publishing, Needham Heights, Mass., 1999.

40. *Food Industry Review, 1998*, The Food Institute, Fair Lawn, N.J., 1998, p. 116.

41. A. E. Gallo, *Food Industry Review, 1999*, Economic Research Center, U.S. Department of Agriculture, Washington, D.C., 1999.

42. K. L. Lipton, W. Edmondson, and A. Manchester, *The Food and Fiber System: Contributing to the U.S. and World Economies*, Agriculture Information Bulletin No. 742, USDA Economic Research Service, Washington, D.C., 1998.

43. "Food Sales Top $10 Billion Mark," *Food Institute Report*, The Food Institute, Fair Lawn, N.J., February 8, 1999, p. 1.

44. G. Anders, "Amazon.com Buys 35% Stake of Seattle Online Grocery Firm," *Wall Street Journal*, May 18, 1999.

45. B. W. Marion, *The Organization of the U.S. Food System*, Lexington Books, D.C. Heath, Lexington, Mass., 1986.

46. "Nash Finch Does More Than Gnash Its Teeth," *Food Institute Report*, The Food Institute, Fair Lawn, N.J., April 5, 1999, p. 2.

47. "Five Retailers Now Account for 40% of Grocery Sales," *Food Institute Report*, The Food Institute, Fair Lawn, N.J., August 9, 1999, p. 1.

48. A. W. Franklin and R. W. Cotterill, *An Analysis of Local Market Concentration Levels and Trends in the U.S. Grocery Retailing Industry*, Food Marketing Policy Center, Research Report No. 19, May 1993.

49. U.S. Department of Commerce, "Profit Margins of Food Manufactures and Retailers," URL: *http://www.econ.ag.gov/briefing/foodmark/cost/data/profit.htm* (last accessed March 15, 1999).

50. Advertising Age, URL: *http://adage.com/dataplace/archives/dp240.html* (last accessed March 25, 1999).

51. Institute of Food Technologists, "Top Ten Food Science Innovations, 1939–1989," *Food Technol.* **43**(9), 308 (1989).

52. Research Committee, Institute of Food Technologists, "The Growth and Impact of the Food Processing Industry: "A Summary Report," *Food Technol.* **42**(5), 95–110 (1988).

53. "B-to-B On-Line Sales Dwarf Retail," *Food Institute Report*, April 26, 1999, p. 1.

54. *Food Regulation Weekly*, April 26, 1999, p. 16.

55. F. M. Biddle, "Titan Adapts Defense Technology to Food Irradiation," *Wall Street Journal*, May 13, 1999.

56. C. M. Bruhn and H. G. Shultz, "Consumer Awareness and Outlook for Acceptance of Food Irradiation," *Food Technol.* **43**(7), 93–94 (1988).

57. L. Aldrich and N. Blisard, *Consumer Acceptance of Biotechnology: Lessons from the rbST Experience*, Agriculture Information Bulletin No. 747-01, USDA, Economic Research Service, 1998.

58. "Organic Milk Making a Splash," *Food Institute Report*, March 22, 1999, p. 2.

59. M. Drabenstott, "Industrialization: Steady Current of Tidal Wave?," *Choices*, Fourth Quarter, 4–8 (1994).

60. E. van Duren, W. Howard, and H. McKay, "Forging Vertical Strategic Alliances," *Choices*, Second Quarter, 81–83 (1995).

61. J. M. Connor and F. E. Geithman, "Mergers in the Food Industries," *Choices*, Second Quarter, 1988.

62. *Food Business Mergers and Acquisitions*, The Food Institute, Fair Lawn, N.J., 1982–1998.

63. U.S. Department of Labor, "Employment and Earnings, March 1997," URL: *http://www.econ.ag.gov/briefing/foodmark/cost/data/bill/wages.htm* (last accessed March 15, 1999).

RICHARD T. ROGERS
University of Massachusetts
Amherst, Massachusetts

INDEX

Abomasum, 47
Absorption, 509–510
Absorption refrigeration cycle, 927–929
Absorption spectrophotometry, 868
Absorptive phase, of metabolism, 88
Acacia gum, 1222
Acceptable daily intake (ADI), 1212
Acerola, 1154
Acesulfame-K, 859, 2245, 2248, 2252–2253, 2255
Acetates, as antimicrobial in foods, 64–65
Acetic acid, as antimicrobial in foods, 64–65
Acetobacter, 1632
Acetylated glycerides, films of, 581–582
Acetylgeniposide, 397
Acid cheeses, 299
Acidic detergents, corrosion by, 432
Acidification, 1505–1506, 1615–1616
Acidified foods, 1501–1509
Acid-modified starches, 2207–2209
Acidophilus milk, 916
Acid phosphatase, in marine animals, 1528
Acids, food surface sanitation, 1039
Acidulants, 1–5, 134, 172, 624–625
Actellic, 1212
Actinidine, 30
Active modified atmosphere packaging, 1140–1141, 1563
Additives. *See* Food additives
Adenosine triphosphate (ATP), carbohydrate metabolism, 1059, 1062
Adenosylcobalamin, 2553
Adhumulone, 72, 158
Adlupulone, 72
Adsorption, 509
Advanced Systems Concepts Directorate (ASCD), 2364
Adverse reaction monitoring system (ARMS), 1053
Advertising, 937
Aeration, aquaculture, 116
Aerobic plate count, 1624
Aeromonas, 1958
Aeromonas hydrophia, 1087
Aerosol containers, 969
Affinity column, 1348
Affluence, nutrition-related diseases, 6–11
Aflatoxins, 1080–1081, 1353, 1354, 1700–1701, 2344–2347
Africanized honeybees (AHBs), 151

African saffron, 390
Agar, 1218, 1219
Agaricus mushrooms, 2338
Agglomerated cereals, 205
Agglomerated powders, 16–18
Agglomeration, 13–18
  caking, 13–14
  coffee, 360
  equipment, 569
  fluid-bed agglomeration, 567–571
  mechanism, 568–569
  methods, 14–16
  spray-bed dryer agglomeration, 571
  spray dryers, 567–571
  undesired, 13–14
Agglutination assay, 1347, 1357
Aging, meat, 1589–1590
Agitated batch (in-container) heating, 2324–2327
Agitated thin film evaporator, 678
Agitation, hydrogenation and, 1314
Aglycones, 375
AgResearch (New Zealand Pastoral Agriculture Research Institute Ltd.), 1404
Agriculture, 902. *See also* Crops; Food crops
  fumigants, 1172–1176
  functional foods, 1179
  Germany, 1394
  grain protectants, 1212–1215
  grapes, 2662–2663
  integrated pest management (IPM), 2690–2699
  organic farming, 2425
  peanuts, 2679
  radiation technology, 1428–1429
  South Africa, 1407
  Spain, 1410
  tea, 2292
  ultrafiltration and reverse osmosis, 2361
  U.S. Department of Agriculture, 2365–2381
  vegetables, 2421–2427
Agriculture and AgriFood Canada (AAFC), 1386
Agroclavine, 29
Agronomic crops, 902
Air, as leavening agent, 133
Air-blast freezers, 1118, 1119
Air-oxygen decarburization (AOD), 424
Air suspension coating, 622
Air-to-air heat recuperators, 565–566

Air-to-liquid heat recuperators, 566–567
Ajmaline, 29
Alaskan snow crab, 439
Albacore tuna, 831
Albumen, films from, 580
Alcohol
  absorption, 21–22
  fetal alcohol syndrome, 25
  hangover, 24–25
  as medicine, 24
  metabolism, 19–20
  Muslim dietary laws, 1683
  physiological actions of, 22–25
  sexual function and, 81
Alcohol dehydrogenase (ADH), 19
Alcoholic beverages, 7, 19. *See also* Beer
  alcohol absorption, 21–22
  alcoholism, 25
  alcohol metabolism, 19, 20
  Australia, 1381
  beer. *See* Beer
  champagne, 919–920
  distilled beverage spirits. *See* Distilled beverage spirits
  enzymes and, 654–655
  fermentation, 918–920
  flavor, 837
  hangover, 24–25
  history, 19
  nitrosamines in, 1713–1714
  physiological effects of alcohol, 22–25
  sake, 920
  Taiwan, 1419
  types of, 19–21
  wine. *See* Wine
Alcoholism, 25
Aldehyde dehydrogenase (ALDH), 19
Ale, 154
Algae, 403
Alginates, 580, 1217
Alginic acid, 1217
Algins, 1217, 1218
Alimentary system, food utilization, 1058–1059
Alimentary toxic aleukia, 2347
Aliphatic alcohols, in olive oil, 1765
Alitame, 860, 2245, 2248, 2254, 2255
Alkaline detergents, corrosion by, 432
Alkaline proteases, in marine animals, 1528
Alkaloids, 26–34
  in chocolate, 350
  classification, 26–30, 33–34

Monascorubramine, 398
Monascus, 398–399
Monellin, 884
Monitoring
    chilled foods, 343
    pesticide residues, 1870–1871,
        2698
Monoammonium glycyrrhizinate
    (MAG), 2256
Monoclonal antibodies, 1346
Monoglycerides, 614, 1523
Monosaccharides, 262–265, 1059
Monosodium glutamate (MSG), 1370,
    1864, 2080–2083
Monoterpenoid alkaloids, 30
Monounsaturated fatty acids, 9, 737,
    744, 1481
Monster plant, 1164
Monterey Jack cheese, 299, 308, 309
Moroccan olives, 1759
Morphine, 28
Most probable numbers (MPN), 1090
Mozzarella cheese, 299, 300, 301,
    303, 306, 308, 309, 332, 333
Muenster cheese, 301, 302, 306, 307,
    309, 332, 333
Mulberries, 1160
Mullet, 823–825, 1537
Multidimensional scaling (MDS),
    sensory analysis, 2138
Multiple regression analysis (MRA),
    sensory analysis, 2141–2142
Multiple-stage dryers, 561–563
Multistage band dryer, 552
Multitier freezer, 1122
Multivariate analysis (MVA), 2133
Multivariate analysis of variance
    (MANOVA), sensory analysis,
    2143–2144
Murexine, 32
Muscaridine, 32
Muscarine, 32
Muscazone, 34
Muscimol, 34
Muscle, 1584
    biochemical character of, 1724
    cell structure, 1585–1586
    energy supply, 1587
    fish, 1724–1725, 2078
    glycolysis in, 1588
    meat, 1585–1588, 1671–1672, 1725
    modified atmosphere packaging
        (MAP), 1564–1566
    scavenging of oxygen, 1562
Mushrooms
    cultivation, 1673–1677
    dehydration, 2404
    nucleotides in, 1725, 1726
    processing, 1678–1681
    solid-state fermentation and, 2169
    toxicants in, 2338

Muslim dietary laws, 1682–1683
Mussels, foodborne disease, 1534
Must, 2663
Mustard oil, 723
Mustard seed, 2200
Mycotoxins, 887, 1028, 1051, 1352,
    1356, 1698–1703, 2343–2349
    analysis, 1684–1693
    in food supply, 1354–1355
    natural, 1352–1355
    regulatory control, 1355
Myoglobin, 883, 1616
Myrcene, 158, 159
Myristicin, 2338–2339
Mysost, 2655

Nanofiltration (NF), 1627, 1628,
    1656
Narciclasine, 28
Narciprimine, 28
Naringin, 836
Naseberry, 1167
Nasogastric tube feeding, enteral
    formulas and feeding systems,
    633–640
Natamycin, 69
National Research Council (NRC),
    1010
National Shellfish Sanitation
    Program (NSSP), 797–798
National Soft Drink Association
    (NSDA), 175
Native Americans, ethnic foods of,
    443
Native flat breads, 139
Natto, 922, 2175, 2181–2182
Natural casings, 579
Natural circulation evaporators,
    666–668
Natural contaminants,
    immunological analysis,
    1352–1356
Natural extractives, spices,
    2191–2194
Natural foods, 2439
Natural language processing, 120
Natural sweetener, 859
Natural toxicants, 885–886,
    1051–1052, 1352–1356,
    2336–2351. See also Mycotoxins
Nectarines, dehydration, 1147
Nematodes, 1088
Neobetanidin, 381
Neohesperidin dihydrochalcone
    (DHC), 860, 2245, 2254–2255,
    2256
Neoxanthin, 386
Nervous system, of fish, 806–808
Net-pens, aquaculture, 102
Neural networks, 120, 121, 2144
Neurosporene, 274, 276

Neurotoxic shellfish poisoning,
    1534–1536
Neurotransmitters, eating and, 90
New Age products, 178
New England boiled dinner, 179
New Zealand
    electrical stimulation of meat,
        1550
    food industry, 1402–1406
New Zealand Dairy Research
    Institute (NZDRI), 1403
New Zealand Institute for Crop and
    Food Research Ltd. (Crop &
    Food), 1404
New Zealand Pastoral Agriculture
    Research Institute Ltd.
    (AgResearch), 1404
Niacin, 2442, 2444–2446, 2487–2495
Niacin deficiency, 1301, 2493
Nicholas Appert Award (IFT), 1372
Nickel alloys, 414, 416
Nickel catalysts, hydrogenation,
    1318
Nickel sulfur catalysts,
    hydrogenation, 1318
Nicotinamide, 2487–2495
Nicotine, 27, 32
Nicotinic acid, 2487–2495
Nicotinyl alcohol, 2494
Nisin, 68
Nitidine, 28
Nitrites, as antimicrobial in foods, 66
Nitrogen, as fumigant, 1175
Nitrosamines, 887, 1619, 1707–1715
Nitrosation, 1707–1708
NIZO method, 224
Nonagitated batch (in-container)
    heating, 2327–2329
Noncarbonated beverages, 177–179
Noncorrosive foodstuffs, 428–429
Nondairy frozen desserts, 1341, 1658
Nondestructive quality evaluation,
    food crops, 897–900
Nonfat ice cream, 1333
Non-insulin-dependent diabetes
    (NIDDM), 8
Nonionic emulsifiers, 614–617
Nonmeat protein additives,
    immunological analysis,
    1348–1351
Nonnutritive sweeteners, 848, 859,
    2245–2262
    acesulfame-K, 859, 2245, 2248,
        2252–2253, 2255
    alitame, 860, 2245, 2248, 2254,
        2255
    aspartame, 836, 850, 860, 1344,
        2245, 2248, 2251–2252, 2255
    cost, 2247–2248
    cyclamates, 860, 2245, 2248,
        2250–2251, 2255

Caspian Sea

Y0-AGT-281

ESOPOTAMIA

●Nineveh

●Ashur

ASSYRIA

Euphrates R.

Tigris R.

PERSIA

Zagros Mts.

BABYLONIA

CHALDEA

Haran

●Babylon

●Susa

●Erech

●Ur

IBLE LANDS
OF THE
TESTAMENT

SCALE OF MILES
100   200   300

Persian

Gulf

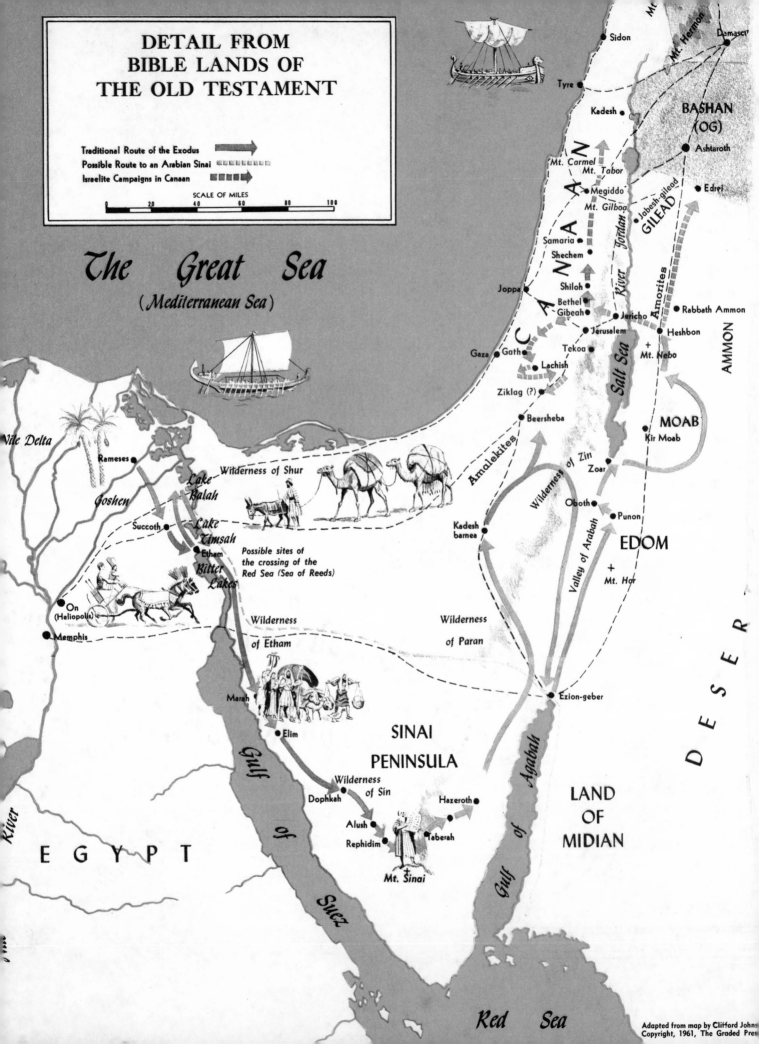

DETAIL FROM BIBLE LANDS OF THE OLD TESTAMENT

Traditional Route of the Exodus
Possible Route to an Arabian Sinai
Israelite Campaigns in Canaan

SCALE OF MILES

20   40   60   80   100

The Great Sea
(Mediterranean Sea)

Sidon
Tyre
Kadesh
Mt. Carmel
Mt. Tabor
Megiddo
Mt. Gilboa
Samaria
Shechem
Joppa
Shiloh
Bethel
Gibeah
Jericho
Jerusalem
Tekoa
Gaza
Gath
Lachish
Ziklag (?)
Beersheba

Mt. Hermon
Damascus
BASHAN (OG)
Ashtaroth
Edrei
Jabesh-gilead
GILEAD
River Jordan
Amorites
Rabbath Ammon
Heshbon
Mt. Nebo
AMMON
MOAB
Kir Moab
Salt Sea

Nile Delta
Rameses
Goshen
Succoth
Lake Balah
Lake Timsah
Etham
Bitter Lakes
On (Heliopolis)
Memphis

Wilderness of Shur

Possible sites of the crossing of the Red Sea (Sea of Reeds)

Wilderness of Etham

Marah
Elim

SINAI PENINSULA

Wilderness of Sin
Dophkah
Alush
Rephidim
Taberah
Hazeroth
Mt. Sinai

Amalekites
Kadesh barnea
Wilderness of Zin
Zoar
Oboth
Punon
EDOM
Mt. Hor
Wilderness of Paran
Valley of Arabah
Ezion-geber

Gulf of Suez
Gulf of Aqabah

EGYPT

River

LAND OF MIDIAN

DESERT

Red Sea

Adapted from map by Clifford Johns
Copyright, 1961, The Graded Press

YOUNG READERS BOOK
OF BIBLE STORIES

Helen Doss

illustrated by Tom Armstrong

Nashville ABINGDON PRESS New York

Scriptural quotations are based largely on the Revised Standard Version
of the Bible, copyrighted 1946 and 1952 by the Division of Christian
Education, National Council of Churches.

**For
George Wesley Anthony
who helped more than he knew**

TO THE READER:

The Bible stories in this book are told in chronological order and set in their historical context. The stories are based on the Revised Standard Version of the Bible. Other translations occasionally have provided the storyteller with a felicitous word or phrase or have thrown light upon a difficult passage.

Research on the history and cultures of the ancient Near-Eastern world has rounded out the basic story given in the Bible. Modern archaeology illustrates graphically the way of life at different periods and also provides substantiation of many historical events related in the Bible.

Biblical references included at the end of each story will help you to compare the stories in this book with corresponding sections in the Bible. A Revised Standard Version edition especially compatible with this volume of Bible stories is the YOUNG READERS BIBLE (Abingdon Press). A good dictionary is YOUNG READERS DICTIONARY OF THE BIBLE (Abingdon Press), and a good atlas with many large maps is the WESTMINSTER HISTORICAL ATLAS TO THE BIBLE by Wright and Filson (Westminster Press). There are fascinating articles on archaeology and modern biblical research, and a brief history of Bible times, in all of these.

Those who are intrigued with the finds of archaeology would enjoy DAILY LIFE IN BIBLE TIMES by Albert E. Bailey (Charles Scribner's Sons) and the middle sections on Babylonia and Assyria in GODS, GRAVES, AND SCHOLARS by C. W. Ceram (Alfred A. Knopf) There are also excellent articles and superb color photographs and drawings in the National Geographic Society's EVERY-DAY LIFE IN BIBLE TIMES. Another book rich in beautiful color photographs, as well as many in black and white, is ADAM TO DANIEL (The Macmillan Company) ; edited and written by Bible scholars and archaeologists in Israel, it is an illustrated guide to the Old Testament and its background.

These are only a few of the many interesting books which can be found to help make the Bible come alive for you.

The author hopes that the YOUNG READERS BOOK OF BIBLE STORIES will be a happy beginning.

—*Helen Doss*

CONTENTS

8

9

The Old Testament

PART I
IN THE BEGINNING

The ancestors of the Hebrew people were nomads and semi-nomads, moving from place to place on the edges of the desert. Their continual search was for life-giving springs of water where there would be grass enough for their precious sheep and goats.

Life was hard and uncertain for these desert people. Food was always scarce, water hard to find. Even children had work to do: churning butter in a goatskin bag, fetching desert thorns and brush for the fire, helping to patch the tents of black goathair, watching to see that the sheep did not stray.

When dusk came, everyone stopped for the evening meal, often the only meal of the day. It might be a bean or lentil stew—or flat, hard cakes of bread eaten with curds of sour milk. Boiled or roasted meat could be afforded only for a very special occasion. The animals were too valuable to kill for daily meat.

After supper the families would gather around the campfire as the evening stars began to brighten. Storytelling was the favorite pastime of these hours. Everyone would listen as a grandfather or one of the elders of the clan retold some treasured tale that had been passed along from one storyteller to another.

Many of these stories contained deep and lasting answers to such age-old questions as: Who created the world, the sun, the moon, and the stars? Why must men

12 eventually die? Why are men forced to labor so hard to earn a living? Were things once better? Why are there different languages?

It was much, much later when the finest of these stories came to be written down to make up what is now the book of Genesis.

It is really not surprising that the songs and stories of other peoples who lived in the ancient Near East are often somewhat like the stories in Genesis. Yet the tales told by the ancestors of the Hebrews had something special. There was a nobility and also a simplicity not found in the stories of neighboring peoples. These Hebrews had a growing awareness of a single, important God. He was the creator and ruler of all the world. He had a holy will which men were to obey, and he had a loving concern for all created things, and especially for man.

Creation

When God began to create the heavens and the earth, there was nothing but darkness. The darkness was empty and without form.

So God said, "Let there be light." And there was light, radiant and glowing. God called the light Day* and the darkness Night.

On the second day God said, "Let there be a sky above the waters." So there arched a great sky, a huge blue firmament above the rolling waters. And God named the sky Heaven.

On the third day God said, "Let the dry land rise out of the waters. Let the earth put forth living plants." God called the dry land Earth and the water Seas. Gentle flowers and herbs sprang up from the ground. Trees grew and spread their leaves. Grass and grains, vines and shrubs grew everywhere. God was pleased, for it was good.

On the fourth day God said, "Let there be lights to shine in the heavens." So the night sky glistened with stars. The blazing sun was made to rule the day, and the silver moon to rule the night. These heavenly lights would mark the passing of the days and months, the seasons and years.** And God saw that this was good.

On the fifth day God said, "Let the waters swarm with living things." So the sea was filled with whales and fish of every size and color. God made countless curious creatures of the ocean deep. And God said, "Let there be birds with feathered wings. Let them sing and fly above the earth." God made these creatures and blessed them saying, "Be fruitful and multiply and fill the waters in the seas, and let birds multiply on the earth."

On the sixth day God said, "Let animals live on the earth." So God made cattle, creeping things, and beasts of every kind. And then, as the final crowning of his creation, God made man to rule over the fish of the sea, the birds of the air, and the animals of the earth. God created man

14 in his own image, to be like him. God made both male and female, and he blessed them, saying, "Be fruitful and multiply and fill the earth and govern everything that I have made." God was pleased for it was very good.

On the seventh day God rested from his work. He blessed this special day of rest, and ever after that the seventh day was holy, the sabbath day.***

Genesis 1:1 through 2:3

* The light, which is called by God "Day," is created before the sun, so in this poem of the creation "Day" has a much larger meaning than a period of twenty-four hours.

** Most ancient peoples regarded the sun, moon, and stars as divine beings, and the heavenly bodies were often the objects of worship. Not so the Hebrews, for they clearly understood that God is the creator of these wonders.

*** This account of the creation emphasizes the sabbath, and many scholars think that it is the work of a writer whose viewpoint was like that of the priests. He was anxious to instruct his people, perhaps at a time when they had grown careless in keeping the sabbath day. The Hebrews made creative use of the seventh day, setting it aside for worship. For them the day was holy. Many ancient peoples held the seventh day as special, but for a strange reason. They regarded that day as belonging to the evil spirits. Nothing was done on that day because these evil spirits might interfere.

The First Man

God had created the earth and all the heavens above. But there was no life on the earth, for there had been no rain. However, a heavy mist rose up from the earth and watered it so that trees and flowers, grass and bushes began to grow.

Then God formed the first man out of the dust of the ground, breathing the breath of life into his nostrils. Thus it was that Adam* came to life and looked around him, and there was wonderment in his eyes. He discovered that there was a beautiful garden all around him. It was the garden of Eden, green with grass, gay with flowers, shaded with trees.

Adam's job was to take care of the garden which had been planted for his use. But he was lonely all by himself.

"It is not good that the man should be alone," God said. "I will make for him a worthy helper."

So God put Adam into a deep sleep and slipped out one of Adam's ribs. From this rib God created a woman.

When the man awoke, he sensed that something wonderful had happened. He leaped up and saw standing beside him a wonderful new creature, smaller than he, so finely shaped, with long hair tumbling down over the shoulders. He was delighted.

"You are bone of my bones, and flesh of my flesh," he cried out with great joy.

The woman smiled shyly and held out her soft, warm hand.

Hand in hand, naked and needing no clothes, innocent and without shame, the man and woman explored the garden. They nibbled on one delicious fruit and then another. There was more food about them than they could ever eat! When they had reached the center of the garden, Adam pointed to a particular tree, but he was careful not to touch it. God had given a special command about this tree. It was the tree of the knowledge of good and evil. They were never to eat its fruit.

Now God had also created all kinds of animals to live peacefully in this park. And he had allowed Adam the privilege of giving them each a name. To his wife Adam had given the name Eve.

Of all the animals in the garden the most deceitful was the serpent. One day this spectacular creature approached the woman and raised himself imposingly before her. His wily black eyes stared so hypnotically into hers that Eve was spellbound.

The creature spoke to Eve in a softly hissing tone, "Does God say, 'You shall not eat of any trees in the garden'?"

The woman shook her head. "No, we may eat all the fruit we want." Then she stopped short and frowned. She pointed to the forbidden tree at the center of the garden. "We must never touch that tree," she sighed, "or we'll die!"

The serpent swayed and hissed seductively. "Oh, pooh, you wouldn't die! It's just that God is afraid you'll eat and then your eyes will be opened. You will see things you never saw before and know

things you never knew before. Why, you will be like God, knowing good and evil."

Eve was astonished. She ran to the middle of the garden and looked closely at the tree.

"It looks good for food," she mused. "And isn't it good to be wise? It is a beautiful tree."

Eve stretched out her hand and took hold of a fruit. She plucked it and bit into it. Just then Adam came up to his wife. Eve stood there with the half-eaten fruit. She handed it to Adam. Without speaking he also took a bite.

But the taste of the fruit was not sweet. It was bitter. Adam and Eve found their eyes were opened. They did see things they had not seen before, but the sight was not beautiful. They looked at each other and were suddenly ashamed and uncomfortable.

They hurried to weave together fig leaves to cover themselves.

In the cool at the end of each day, God came walking in the garden. But now for the first time Adam and his wife knew the shivering, quaking sense of fear. They hid among the trees, but God called them out.

"I heard you coming," Adam confessed, "and I was afraid because I was naked. So I hid myself."

"Who told you that you were naked?" God demanded. "Have you eaten from the forbidden tree?"

Adam pointed at his wife. "This woman —the very one *you* gave to be with me, she gave me fruit of the tree." Then he mumbled, almost inaudibly, "so I ate."

The tears began to run down Eve's cheeks. She pointed to the serpent nearby. "This creature was cunning. It charmed and deceived me." Now she was sobbing aloud. "And so I ate."

God said to the serpent, "Because you have done this, you shall go on your belly in the dust. The woman and all her children will be your enemies. You shall strike at them, and they shall strike back at you."

Then God faced the weeping woman. "From now on your husband shall be master over you, and even your childbearing shall be in pain and worry."

And to Adam, God said, "Because you have listened to your wife and not to me, because you have eaten the fruit which I told you not to eat, you will have to work and sweat for your living. In the end you will die. You are made of dust, and to dust you shall return."

And so the first man and the first woman had to leave the garden and make their way in the world as best they could. The cherubim** were placed on guard with flaming swords that flashed like lightning in all directions so that man could never again return to the garden of Eden.

* In Hebrew there is an interesting wordplay on the word "Adam." It means "man" and is very similar to the word for "ground," the substance from which Adam was made. This account of the garden of Eden is probably the older of the two creation stories in Genesis.

** The cherubim (CHER-eh-bim) were familiar figures in the mythology of the ancient lands of the Near East. They were often represented as winged creatures, sometimes with two faces, one of a man and one of a lion. This figure is related to the sphinx in Egypt and the carved figures decorating gates and palaces of Assyrian ruins.

The Jealous Brother

Adam and Eve had many daughters and sons, grandchildren, and great-grandchildren. Among their descendants were two brothers, Cain and Abel.

Abel grew up to be a keeper of sheep. Cain became a farmer. He plowed and seeded the ground and raised grain.

It was the custom for men to offer sacrifices to God. Upon the altar of stones Abel laid the first lamb from his flock, and he knew that God was pleased. Instead of an animal, Cain offered his first sheaf of grain, but he seemed to feel a silence from God. This made Cain angry. He stood at the altar with a sullen face.

"Why are you so angry?" God asked. "If you do well, will you not be accepted?"

But Cain brooded, not listening. Was his brother more favored in God's eyes? Cain simmered with envy and jealousy until his resentment boiled out of control.

Cain pretended to be friendly and said to Abel, "Let's go out to the field together." When they were out of the sight and hearing of others in the distant field, Cain struck Abel. Abel lay on the ground, not moving. He was dead.

Cain was frightened at what he had done. He then heard God's voice accusing him, "Where is Abel, your brother?"

Cain's guilty heart was thumping wildly in his chest. "How should I know where my brother is?" he cried. "Am I my brother's keeper?"

"You have cursed the ground with your brother's blood," God thundered at him. "Abel was killed by your own hand! From now on whenever you plow and seed the earth, it shall no longer yield crops to you. You shall be a runaway, a fugitive, a lonely wanderer on the earth."

Cain hung his head, and there were tears in his eyes. "My punishment is greater than I can stand," he pleaded. "You drive me from your presence and away from the ground I love. Whoever finds me will slay me."

Then God answered, "Not so. Anyone who slays Cain will pay for it seven times over."

God put a protective mark on Cain so that he would be recognized and not killed. Cain went away from the presence of God and lived out the rest of his life in the land of Nod, which means "wandering."

Genesis 4:1-16

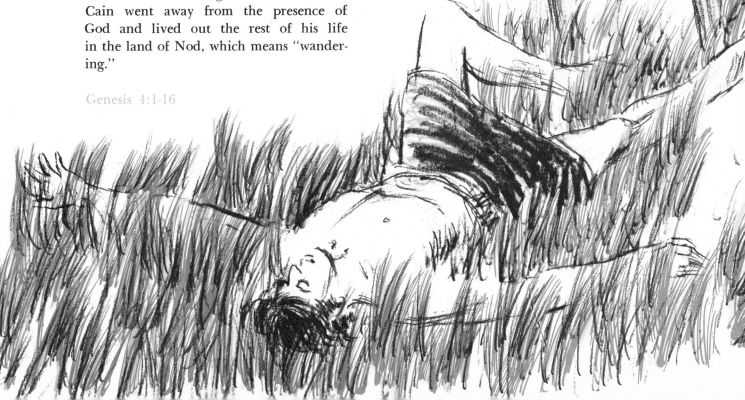

18 The Great Flood

As the people multiplied on the earth, their wickedness and wrongdoing multiplied even more. People were corrupt, cheating and stealing. Everyone could see drunkenness, murder, and rioting—evil piled on evil until it was hardly safe to be alive.

So God said, "I will destroy man whom I have created. I will blot out man from the face of the earth, for I am sorry that I made him."

Yet there was one good and righteous man who tried to know the will of God. His name was Noah.

God said to Noah, "Build yourself an ark with many rooms. Then cover it inside and out with pitch and asphalt to make it watertight."

God told Noah how to build this boat with three huge decks and a roof over the top, so it would be exactly the right size.

"Soon I will flood the earth to destroy every living thing," God said. "But you will ride safely in your ark along with your three sons, your wife, and your sons' wives. Now you shall bring two of every living creature into the ark: a bull and a cow, a ram and a ewe, an eagle and his mate. There shall be a male and a female of every kind of bird and animal. Also, store up all kinds of food."

The people around Noah went on with their wild and heedless ways. Many laughed and made fun of the family of Noah. "They must be crazy," the neighbors jeered. "Look at them, building a giant boat right in the middle of a dry field!"

At last the ark was finished. God said to Noah, "Start loading the ark. In seven days rain will begin. It will rain for forty days and forty nights."

In seven days the waters came. The fountains of the great deep burst forth and the windows of heaven were opened. It rained and rained and rained. The waters rose, bearing up the ark.

And still it continued to rain. The waters rose so high that even the tallest mountains were covered. The thundering downpour continued for forty days and forty nights.* Everything that had lived and breathed on dry land was now blotted out. Only Noah and his family were left alive, and all the creatures floating with them in the sturdy boat.

God did not forget Noah and his family and the animals in the ark. After a long time the waters began to go down.

At last the ark came to rest on a mountain. Noah opened a tiny window. Around him spread the endless lapping water.

At the end of forty days Noah sent forth a raven, but it could only fly back and forth, to and fro. Then Noah sent forth a dove to see if the waters had gone down anywhere. But the dove found no place to light and returned.

Noah waited another seven days. In the rosy dawn he sent forth the dove. In the evening the dove came back with a freshly plucked olive leaf in her mouth. So Noah knew the waters had receded at last.

God said to Noah, "Go forth from the ark, you and your family and the creatures with you. And now may you and your sons and all living things be fruitful and multiply upon the earth."

Then God said, "I will never again curse the ground because of man. Neither will I ever again destroy every living creature as I have done."

From this time on mankind would not find God only in the terrible acts of nature. Though he would be remembered in the raging flood, the thunderstorm, the cyclone, the hurricane, the erupting volcano, and in the earthquake, he would also be found in the quiet orderliness of the seasons, the daily rise and set of the sun, the coldness of winter and the warmth of summer. For God said to Noah, "While the earth remains—seedtime and harvest, cold and heat, summer and winter, day and night, shall not cease."

So God set a rainbow in the clouds as a sign of the agreement he made never again to destroy all life on earth.

"When I bring clouds over the earth," God promised, "and the rainbow is seen in the clouds, that will be forever the sign of my covenant."

Genesis 6:5 through 9:17

* Many accounts of great floods crop up in folklore throughout the world. They are so universal and so similar that most scholars believe they refer to actual disasters in nature. The stories almost always show the floods in terms of the dealings of the divine powers with man.

The Tower of Babel

Noah and his sons and their sons were fruitful, and the multitude of their descendants spread through the land. Now the whole earth had one language.

As men moved westward they found a fertile plain in a new land and settled there. There was no stone for building in this flat country, so they said to one another, "We will have to make bricks."

When they had great piles of bricks they said, "Come, let us build ourselves a mighty city with a tower so high we can reach the heavens! In this way we may make a name for ourselves and be as great as God himself!"

God came down to see the city and the tower which the arrogant men were building. "See what they have begun to do," God said. "And this is only the beginning! Nothing that they propose will now be impossible for them."

God decided to mix their language until they could no longer understand one another. Then they could not plan together or try to build a tower up to the heavens.

So God confused their speech, and everyone was babbling and jabbering. No man could find another able to understand him. They could no longer work together or cooperate with one another. So they left off building the tower and the city.

The abandoned city was called Babel. From here God scattered the people, with their many languages, across the face of the earth.*

Genesis 11:1-9

* Scholars cannot guess where this story of the origin of languages came from. The people of ancient Babylonia built towering terraced pyramids called ziggurats (ZIG-eh-rats) of sundried or oven-baked bricks. Since *Babel* is the Hebrew form of "Babylon," some connection is probable. Perhaps when early Hebrew nomads heard the tale, they recognized it as a parable of God's punishment for human arrogance and overreaching ambition. We may never know. But "tower of Babel" has become part of our way of speaking, a symbol for the hopeless, babbling confusion of human voices.

PART II
THE FIRST STORIES OF THE HEBREW PEOPLE

The history of the Hebrews begins with Abraham. He is the father of his people. Some of the great patriarchs seem almost legendary, but Abraham, Isaac, Jacob, and the Joseph who went to Egypt were not imaginary.

The stories grew up around real, if dimly remembered, heroes. The people told and retold tales of the adventures and trials of these great ones. The storytellers may well have arranged or added details to make a good story. But they always showed a special interest in the encounters of the heroes with God.

It was not until much later, after the Hebrews had settled in the land of Canaan, that the stories were written. When they had been collected, they were called the Books of Moses. This does not mean that they were written by him, but they do contain the things that Moses taught his people, and record their remembered history until the time of his death. They are found as the first five books of our Old Testament.

One important theme runs through all these books. Abraham and his descendants were considered leaders of a new nation, one chosen by God. A great deal of space is given over to the Law of God and to the rules by which the people were to live in order to please him and be his special people. These stories also show that

Abraham and his descendants did not always keep the Law. They were guilty of hating, cheating, lusting, lying. Yet none of these sins separated them completely from God, and in the stories of their lives their good qualities shine through.

Abraham and Lot

Abraham's ancestors had lived in Ur, an ancient city in the large fertile plain of the Tigris and Euphrates Rivers, in ancient Babylonia. The people of Ur worshiped many gods, but they were especially devoted to moon worship. Abraham's people migrated northward from here along the Euphrates and settled in the far northern country of Haran. It was in Haran that Abraham lived with his wife.

Abraham and his wife Sarah had one great disappointment. They were childless.

One day Abraham heard God's voice saying, "Go from this country and your father's house to a distant land. I will show you the way. There you will become the father of a great nation."

So Abraham packed up his belongings and took his wife Sarah with him. He also invited his nephew Lot and Lot's family. With their servants and their flocks of goats and sheep, they started out.

On foot Abraham's clan traveled slowly westward, over the desert, always on the lookout for a spring of water. At night they slept in their goathair tents. At last these weary wanderers came near the Great Sea and into the land of the Canaanites.

Here Abraham again heard God's voice saying, "To your descendants I will give this land." So Abraham built an altar beside a great oak tree which grew near the Canaanite town of Shechem, worshiping the Lord with the sacrifice of a burnt animal.

Abraham and Lot prospered in the land of Canaan. Their herds of cattle, sheep, and goats multiplied. They earned gold and silver by selling to the local people.

Finally their goods increased until the many servants and flocks and herds and tents began to get in the way of one another. Often Abraham's herdsmen and Lot's herdsmen fought for the use of the sparse pasturage and for the water supply.

The situation continually grew worse, and so Abraham said to his nephew, "Do you and I want to fight? Do we want our herdsmen fighting? Are we not kinsmen?"

Lot agreed that the problem must be solved.

"Let us go our separate ways," Abraham said. "You choose that part of the land you want."

The two men stood on a mountain ridge northeast of the ancient Canaanite town of Hebron. Below them was the deep, wide valley of the Jordan River, where the river rolled into the still water of the Dead Sea.

"I shall settle down there," Lot told his uncle Abraham, pointing to the valley and the great blue lake. The soil looked fertile and rich in the valley, and there was plenty of water.

So it was that they separated from each other. Abraham and Sarah continued to pitch their tents in the high, dry hills near Hebron.

Lot took his family, his flocks, and his servants to live in the lush green valley. After a time he moved his tents as far as the great cities of Sodom and Gomorrah, near the eastern shore of the bitter, salty lake.

At last Lot gave up his tents and became a city sojourner. He bought himself a house in Sodom. It was a wild and wicked city, as was the nearby city of Gomorrah. But Lot found that he could live in luxury here. It was quite different from the harsh, austere life he had shared with his uncle Abraham.

One evening Lot was sitting on the bench by the city gate. He saw two men coming into Sodom, but he did not know that these strangers were angels of God, disguised as men.

Lot rose to meet them and, bowing, gave the customary greeting to the strangers, "Let me be your servant! Come and spend the night at my house. Then you may rise early and go on your way."

They hesitated, as was polite. And Lot was equally polite, urging them more strongly. At last they followed him to his house. Lot ordered his servants to prepare a feast. So they ate boiled lamb, some greens, and freshly baked unleavened bread.

But before they lay down for the night, the men of the city, the wicked men of Sodom, gathered in front of Lot's house. They were half drunk and looking for trouble. "We know you are sheltering two strangers!" they shouted, banging on Lot's door, so that the whole house shook. "Send them out! Give them over to us, so we can do whatever we want to them!"

Lot went out into the noisy street, shutting his front door behind him. He tried to reason with the evil men of Sodom.

"I beg you," Lot pleaded, "do not be so vicious! These men are my guests—and the guests of your city."

But the mob began to mutter among themselves. "Look, this fellow came to live only awhile in our town. And now he plays the judge!" The mood became even uglier. They shouted and cursed at Lot, "Just wait, we'll treat you even worse than your visitors." They surged about him, ripping at his robe. Some of the mob pressed toward the house, ready to break down the door.

Just then Lot's front door flew open, and the two strangers snatched Lot inside to safety. At the same moment the men outside were struck with blindness. The men of Sodom were as helpless and confused as if they groped in a heavy fog. They could not even find Lot's house.

The angels warned Lot to leave the city at once. "Hurry, hurry! Run, with your wife and daughters, or you may be burned to death. This city will be punished!"

Lot stood there in his torn robes, looking about at his many possessions, uncertain and dazed. So the men seized Lot and his wife and his two daughters by the hand. They all escaped and made their way outside the city.

Here the strangers warned, "Flee for your life; do not look back or stop anywhere in the valley. Escape to the hills—or you will be burned alive!"

Lot and his family ran toward the high cliffs rising behind the town.

Then blazing, roaring fire and brimstone rained on the wicked cities of Sodom and Gomorrah. The fires raged. The cities were completely destroyed. Nothing was left but fires, smoke, and ashes.

Lot and his daughters made their way to safety. But Lot's wife disobeyed. She looked back. And immediately she became a pillar of salt.*

At dawn the next morning Abraham arose from his pallet on the ground. He walked from his tent out to an eastern slope, high above the Jordan valley and the huge bitter lake. Here he came to the place where he usually stood for his worship of God.

Abraham looked down toward Sodom and Gomorrah on the far shores of the still lake. This morning, where the two great cities had been, he could see nothing but blackened ruins, and the smoke of the land went up like the smoke of a furnace.**

Genesis 11:32 through 12:7; 13:2-18; 19:1-28

* To this day, pillars of salt and gypsum are left standing along the cliffs overlooking the Dead Sea after the rains have eroded the surrounding minerals. Perhaps the legend of Lot's wife originated around one of these curiously shaped salt formations. However it began, it has become a parable that warns the hesitant and the fearful not to look back.

** The story of the destruction of Sodom and Gomorrah may have been based on a remote prehistorical event. Earthquakes or other destructive forces seem to have wiped out several cities that once existed near the south and west shores of the Dead Sea. There are sulfurous gases and flammable pools of asphalt and petroleum products in this area, which might have produced explosions and fire along with a disastrous earthquake.

Abraham's Sons

Abraham grew tired of wandering from place to place. He came back to the southern part of Canaan to live and pitched his tents beneath some large oak trees near the town of Hebron.

Abraham was troubled. God had promised him and his descendants this land of Canaan. Yet it was already peopled by the Canaanites, who farmed the land, lived in strong, fortified cities, and worshiped strange gods in strange ways.

And what about his own descendants? God had said, "Abraham, look toward the heavens. Count the stars, if you can. Just that numerous will be your descendants!" Yet he and Sarah were growing old, and they had not a single child.

Sarah was touched by her husband's sorrow. "I have borne you no children, and I am growing old," she said. "Take my young maid as your wife."*

Abraham listened and agreed. So Sarah gave Hagar, her Egyptian maid, to her husband as a wife.

But the meek, quiet little desert girl turned into a different woman after she was the second wife of Abraham. She became haughty with her mistress. When Hagar found out she was going to bear a child for Abraham, she became even more arrogant.

At last Sarah could stand it no longer. "I gave you Hagar as a wife, and now she scorns me," she wept before Abraham. "And ever since she found that she is going to have your child, she shows me even more contempt!"

Abraham refused to take sides. "According to our Law," he told Sarah, "she is still in your power. So do with her as you please."

So Sarah became harsh and severe with the young woman. And at last Hagar ran away.

An angel from God found Hagar out in the wilderness, weeping by a spring of water. "Hagar," the angel said, "where have you come from and where are you going?"

"I am running away from my mistress," Hagar sobbed.

The angel said, "Listen to me, you are going to have a son. His name will be Ishmael. He shall be like a wild burro of the desert,** his hand against every man, and every man's hand against him. And I will greatly multiply your descendants!"

So Hagar, her head hanging, the stain of tears still upon her cheeks, returned to the tents of Abraham. There she bore a son and the baby was named Ishmael.

The years went by, and Ishmael grew into an active little boy. Abraham was fond of his son. He thought that Ishmael would be his only child.

One hot day Abraham was sitting under the oak tree which spread its great branches over his tent. He lifted up his eyes and stared straight at three strangers who had suddenly appeared. He did not know they had come, in disguise, directly from God.

Abraham rose to meet them, bowing low. He greeted the strangers with traditional nomadic hospitality: "Do not pass by. I will have water brought to wash your feet. Rest under this tree while I bring a little bread to refresh you."

Abraham hurried into his tent and whispered to Sarah, "Quickly measure out plenty of good flour and bake some bread."

Then Abraham selected a tender young calf from his herd. "Prepare this for my guests," he told one of his servants.

To this meal Abraham added a bowl of fresh curds. He stood under the tree while the three strangers ate.

The men looked about. "Where is Sarah, your wife?"

"She is in the tent," Abraham said.

"I will surely return to you in the springtime," one of the strangers said prophetically. "And at that time Sarah your wife shall have a son."

26

Sarah was listening from inside the tent. Her hair had turned gray, and she had given up all hopes of bearing a child. So she laughed to herself thinking, "After my husband and I have grown old, shall I now have the pleasure of conceiving a child?"

The stranger heard Abraham's wife laughing to herself in the tent. "Why did Sarah laugh and say, 'Shall I indeed bear a child, now that I am old?' " he asked. "Is anything too hard for God?"

True enough, God kept his promise to Abraham and Sarah. Although they were both old, Sarah did conceive. About the time the grain was ripening in the rock-bordered fields around Hebron, in the spring of the year, Sarah gave birth to a son. Abraham and Sarah were grateful to God and filled with joy. There never were more happy and grateful parents! They called their new son Isaac.

The baby grew and was weaned. Abraham made a great feast to celebrate.

But Sarah's happiness was spoiled by jealousy. Whenever she looked out from her tent and saw little Isaac playing with his older half-brother Ishmael, she fretted. Would Ishmael endanger the inheritance of Isaac? Sarah could no longer bear to see Hagar and her son.

So one day Sarah said to Abraham, "Cast out this slave woman with her son, for the son of this slave woman shall not be heir with my son Isaac!"

The whole affair was very distressing to Abraham. He adored Isaac, but he loved his son Ishmael too. But God comforted Abraham. "Do not grieve," he told the old patriarch. "Whatever Sarah says to you, do it. It is through Isaac that your descendants will be listed." Abraham listened, but his heart was still heavy for the sake of Ishmael. Then God promised, "I will make a nation of the son of the slave woman also, because he is your offspring."

So Abraham rose early in the morning. He gave Hagar some fresh bread and hung a skin-bag of water over her shoulder. He put their son's hand into hers. Then sadly he sent Hagar and Ishmael away.

So Hagar left the tents of Abraham. She trudged south, down from the hills of Hebron, out into the desert wilderness.

After several days there was not a drop left in the water bag. The desert sun blazed down on Hagar and her boy. Their mouths were dry; they were dying of thirst.

"Lie under the shade of that bush," Hagar said to Ishmael. "Lie there, my son, and rest awhile." Then she went away, at about the distance an archer could shoot an arrow, and sank with exhaustion upon the dry, stony ground. "I cannot look upon the death of my son," she wept. She could hear her son crying out, too.

Then God heard the boy's tearful calling. He said to the weeping mother, "Fear not, Hagar, for God has heard the voice of the boy. Go to your son and hold him by his hand, for I will make him the father of a great nation."

Then Hagar looked up, and suddenly she saw before her a well of cool water. She ran to fill her skin bottle with water and then hurried to give her son the first drink.

And so God was with the lad. Ishmael grew up, living in the desert wilderness. He hunted for his living and became an expert with the bow. When the time came for marriage, Hagar helped him find a wife in the land of Egypt, a girl who would become the mother of his many children.***

Genesis 15:1-6; 16:1-15; 18:1-15; 21:1-21

* This was a common custom in the ancient Near East. The children born to a slave woman were usually adopted by the first wife and treated as her own. Such a son might become a co-heir.

** When the angel said that Ishmael would become like a wild burro of the desert (literally "a wild ass of a man"), the description was not at all insulting. The wild ass was a noble animal, free and unfettered, the aristocrat of animal life in the desert.

*** The Ishmaelites, who traced their ancestry to Ishmael, were a nomadic people who engaged in caravan trading and camel herding. They lived for the most part east of Palestine in the Syrian desert.

The Sacrifice

A time came when God decided to test Abraham's faith. "Abraham!" he called.

The old man answered, "Here I am."

Then came God's voice again. "I want you to go on a journey with your only son Isaac, whom you love. At a place where I shall tell you, there you will offer Isaac as a burnt offering."

After a grief-filled, sleepless night Abraham rose at dawn and saddled his donkey. He felt numb and bewildered as he wrapped a small bundle of wood for the burning and tied it on the donkey's back. All his life he had waited for this boy. Had God granted him a son at last—only to demand him back? Would he and Sarah have no great line of descendants after all?

Gently Abraham awoke the sleeping boy. "Come, my son," the old man said. "We are going on a journey, you and I."

Isaac put his small hand trustingly into the work-gnarled hand of his father. Two young men, Abraham's servants, were chosen to go with them. One of the servants carried a small clay pot filled with glowing charcoal to start the altar fire.

On the third day of their journey Abraham knew that the place was not far off.

He tried to hide his heavy heart from his servants. "You men wait with my donkey," he said. "The lad and I are going a short distance beyond to worship."

Abraham untied the small bundle of wood and tied it on Isaac's strong young back. Into his belt Abraham tucked the bronze sacrificial knife. He tried to calm his shaking hands so that he could carry the clay pot with the glowing coals. He and Isaac started walking toward a grove of giant oak trees on the rise of a hill ahead.

Isaac turned up questioning eyes. "My father!" he said.

Abraham kept walking, looking down at the boy with great tenderness and love. "Here I am, my son."

The boy said, "I do not understand. You are carrying a knife and the fire. And I carry the wood for a sacrifice. But where is the lamb for a burnt offering?"

His father groaned softly, every breath aching. He raised his tear-laden eyes to the heavens. "God will provide the lamb for a burnt offering, my son."

When they reached the top of the hill, Abraham took stones and built an altar there. Isaac, his face puzzled, helped his father.

Abraham arranged the bundle of wood on the completed altar. Isaac's eyes grew large and round and frightened when his father took the rope that had bound the wood, tied it firmly all around him, and then laid him on the altar. Tears rolled down from his father's eyes but Isaac did not cry out.

Then Abraham slowly slipped the knife from his belt and raised it above his son.

Suddenly there was a voice from heaven. "Abraham, Abraham!"

Abraham looked up, and his voice trembled. "Here I am."

The voice warned, "Do not lay your hand on the lad or do anything to him; for now

I know that you fear God, since you have not withheld your son, your only son, from me."

Tears of thanksgiving wet Abraham's wrinkled, weatherbeaten cheeks. He heard a noise in the thicket behind him. There, caught by his horns in a large shrub, was a fine young ram.

Abraham's shaking fingers untied his beloved son. Tenderly, joyously, the old man helped Isaac down from the altar. He took the rope, bound up the ram, and offered the animal as a burnt offering—instead of his son.

Then Abraham took young Isaac by the hand, and they returned to the place where the two servants waited. After eating and resting, they all returned home.*

Genesis 22:1-19

* This story, in its original form, may have been a recital, used in a shrine, or other sacred place of worship, to show that sacrifice in Israel was different. They were to offer animal substitutes upon their altars—instead of the firstborn child. The offering of the firstborn of the flock, sometimes even the first child, was common in many Near East cultures. Long after Abraham's day there were warnings against child sacrifice in the Old Testament. Deuteronomy 18:10 says, "There shall not be found among you any one who burns his son or his daughter as an offering." Yet, especially in times of fear and superstition and pressure, this abominable practice sprang up again. Even two kings of later times, Ahaz and Manasseh, burned their sons as offerings. This short story states quite openly that human sacrifice has no place in the worship of the God of the Hebrews.

A Bride for Isaac

Isaac grew to manhood. His mother Sarah had died, and her body was laid in a burial cave near Hebron.

Abraham was well advanced in years. An illness left him weak and spent, confined to his bed. He was too old, too ill, too feeble to travel. He had one great concern, for Isaac was of marriageable age. Where could a suitable bride be found for his beloved son? She must not be a local Canaanite girl. The Canaanites worshiped the fertility god Baal. Isaac's bride must give her complete allegiance to the one God.

Abraham knew it was too late to make the search himself. The duty would fall to his faithful servant, who was in charge of all Abraham had.

The trusted servant, summoned to his master's bedside, came and bowed down before the feeble, white-bearded old man. There were tears in the servant's eyes. He knew his master did not have long to live.

"Put out your hand," Abraham whispered. "You must find a wife for Isaac. Swear that you will not choose a wife from the daughters of these Canaanites."

"I swear," the servant said, giving Abraham his hand.

"Promise that you will go to my country, back in Haran," Abraham insisted. "Seek among my relatives there for the right wife."

"Perhaps the woman may not be willing to return with me to this land," the servant said. "Must I then take Isaac back there to live in the land from which you came?"

Abraham struggled to raise himself to a sitting position. "See to it that you do not take my son back there!" He fell back on the pillows, exhausted. His voice dropped to a whisper. "God, who took me from my father's house and from the land of my birth, promised to give this very land to me, and to my descendants."

The old servant nodded, understanding.

"God will send his angel ahead of you," Abraham said. "You will have help in choosing Isaac's wife." He sighed. "But if the woman is not willing to follow you, then you will be free from your promise."

Careful preparations were made for the trip. Camels were loaded with food and water. Gold and silver jewelry, finely woven woolen garments, and other choice gifts were packed. With the laden animals the servant forded the Jordan River and began the long trek eastward over the vast stretches of burning desert.

It was many days before he arrived in Haran. From there it was only a short distance to Nahor, the city where Abraham had married Sarah.

Outside the city walls of Nahor was a deep, wide well with circular steps leading down to the pool of springwater at the bottom. At the top, near the edge of the well, was a water trough for animals.

It was late in the day, and the broiling sun was going down. The women of the town were coming to the well to fill their clay water jars, as they did every morning and evening.

The servant lifted his eyes and prayed silently. "God of my master Abraham, grant me success today! Here I stand by the spring, and the young maidens come to draw water. If I ask a maiden for a drink and she gives it to me and also offers to water my camels, then may she be the maiden you would choose for Isaac."

Just then a beautiful young woman, a veil over her long black hair, came up with a water jar on her shoulder. She padded, barefoot, down the cool, damp steps to the bottom of the well, filled her jar, and came up.

Abraham's servant met her and said, "May I have a drink from your jar?"

She said, "Drink, my lord."

Abraham's steward drank thirstily.

The young girl looked with compassion

on the huge beasts standing nearby. "I will draw water for your camels also." Her smile was friendly, but shy.

She poured the rest of the water from her jar into the drinking trough. Abraham's servant made the camels kneel down, which they did with their usual groaning and spitting. They began to drink noisily.

The girl ran back down the circular steps with her empty jar. When she returned it was brimming with cool water. Back and forth she went until the thirsty camels had drunk their fill. The servant watched all that she did; his heart quickened with hope. Would this be the right girl?

The servant took two gold bracelets and a gold ring from a camel's pack. He gave them to her, asking, "Whose daughter are you?"

The girl's eyes grew wide at the sight of the gifts. When she stammered out the name of her father, the old servant could hardly believe his ears. This girl belonged to the same tribe as Abraham!

"Is there room in your father's house," he asked politely, "where I might lodge?"

The maiden was almost speechless, but she managed to say, "My father's house has plenty of room." She added breathlessly, "And there will be food for you and straw for your animals. Follow me."

The girl ran on ahead through the city gates, down a narrow lane, to her house. She told her parents and her brother Laban about the stranger and excitedly showed them her new gold ring and bracelets.

Her brother Laban stared at the expensive gifts. Then he hurried out to greet the stranger, who was waiting in the brick-walled courtyard. When Laban saw the man he cried out, "Come in, blessed man of God! Why do you stand outside? I have ordered a feast and a place for your camels!"

So the old traveler came into the house of the young woman. Her brother Laban brought water to wash his hot dusty feet, and then a quickly prepared feast was set before him.

But the old man folded his hands and said, "I will not eat until I have told my errand."

"Speak on," the girl's brother Laban said.

The man explained that he was the steward of Abraham. "God has greatly blessed my master," he said. "He has gained flocks and herds, silver and gold, and many servants. His wife Sarah bore him one son when she was old. To the son, Isaac, my master has given all that he has."

The servant told of his promise to Abraham and the long trip across the scorching desert to seek a proper wife for Isaac. Isaac must not wed a Canaanite girl who worshiped strange gods. He must have a wife from among his own kinsmen, one who might know and worship the one true God.

The old man paused and looked with appreciation at the kind and beautiful young maiden, Rebekah, who stood in the doorway of her father's house. He turned to the girl's parents and to Laban her brother and told them of his prayer at the well and how the maiden seemed to be God's answer.

Then the girl's family withdrew to a back room to confer together.

At last Laban returned and said, "My father says this surely comes from God! Rebekah is before you, take her and go, and let her be the wife of your master's son."

When Abraham's servant heard their words, he brought out gifts for Rebekah, more fine jewelry and robes of wool and linen in soft colors. He also gave expensive gifts to her brother Laban and to her parents.

Then feasting together, they celebrated the betrothal of Rebekah to Isaac. Finally the old servant of Abraham went to rest for the night.

Early in the morning he said, "I must take the maiden and leave today to return to my master's household."

"Let her stay with us awhile," Rebekah's mother pleaded:

"At least ten days," protested Laban. "After that, she may go."

The old servant was worried. Abraham had not long to live, and the old father would be anxious to hear the good news. Besides, what if the girl or her family changed their minds? He dared not linger. "Please no, no," he said. "Let the maiden go with me now!"

Rebekah's family said, "We will call her and ask her."

So Rebekah was called into the room. Her parents and Laban said to her, "Will you go with this man?"

Abraham's servant held his breath. The girl looked tenderly at each member of her beloved family. Then she turned to Abraham's servant, and there was trust and expectation in her large, round eyes.

She said, "I will go."

So Rebekah's family prepared her for the trip; the girl, along with her nurse and her maidservant, mounted Abraham's camels. Rebekah's parents hugged her, tears in their eyes, and blessed her. Laban kissed her and said, "Dear sister, be the mother of thousands!"

The big beasts lunged to their feet, and the company rode away. Back they went across the hot, trackless desert, to Canaan.

One evening when Isaac had gone out to his field to think and to meditate, he looked up and saw camels coming from the East.

Rebekah, perched high on her swaying camel, had already noticed the man. "Who is that," she asked, "walking across the field to meet us?"

The servant saw Isaac and answered, "It is my master."

Rebekah, with shy and proper manners, pulled her veil down to cover her face. But her heart was beating faster because the sight of Isaac had made her happy.

It was with great joy that Isaac took Rebekah and brought her into his tent. She became his wife, and he loved her.

Genesis 24

Twin Sons

Abraham was an old man and full of years when he breathed his last. Sadly Isaac buried his father next to his mother Sarah in the cave near Hebron.

Isaac loved Rebekah, his wife. Yet they remained childless for many years. Isaac found himself yearning for a son to inherit the land of Canaan as promised by God. At last Isaac prayed that Rebekah might bear a child. God listened to Isaac. And Rebekah conceived.

When Rebekah's time came to give birth, Isaac saw that his prayer had been answered doubly! Rebekah bore him twins. The firstborn, a red-haired boy with a hairy body, was named Esau. Jacob, the twin brother, only a few minutes younger, was dark haired and smooth skinned.

As the boys grew up they proved to be as different in their natures as in their looks. The big, red-headed Esau became a skillful hunter with his bow and arrow, an outdoorsman who joyously roamed the fields and thickets in search of quail, gazelles, and other wild game.

Jacob was a quiet young man who stayed in the tents within the boundaries of his father's encampment.

Isaac was partial to Esau the hunter. He enjoyed sharing in his son's roasted or broiled game. Jacob, who stayed close to the tents, was Rebekah's favorite.

One day Jacob was boiling a savory stew of lentils and beans. Esau came in from the field, hot, tired, and famished for food. The delicious aroma of Jacob's stew made Esau's mouth water.

"I'm starving!" Esau said. "Let me eat some of your pottage."

Jacob seized this opportunity and said craftily, "First sell me your birthright! You are the oldest, and the blessing of the first-born will go to you. I want that blessing and the inheritance of Canaan that goes with it!"

Esau groaned. "I'm about to die of hunger. So what use is a birthright to me?"

Jacob insisted, "Swear to me first."

So Esau swore to him and sold his birth-right to Jacob for a mess of pottage, a mere bowl of bean stew.

Esau ate and drank, then rose and went his way. "After all," the rugged hunter thought to himself, "of what importance is a birthright when my father is still alive and healthy?"

The young men never told anyone what happened. And so Isaac never knew that his elder son had sold his birthright so cheaply. But Isaac and Rebekah were deeply hurt by something else that Esau did. He married two Canaanite women. Esau's wives, worshiping their strange gods, made life bitter for Isaac and Rebekah.

At last Isaac grew old. His eyes were dim so that he could not see. As he lay on the mat in his tent he called Esau to him. "My son, I am old, and I do not know the day of my death."

Esau knelt down on the ground beside his father's bed, and he tenderly held his father's frail hand in his own calloused, hairy one.

"Take your quiver and bow," the old man whispered. "Go hunt some game and prepare the savory food I love and bring it here. I wish to eat with you once more and bless you before I die."

Now Rebekah was listening behind the partition in the tent. So when Esau went out to hunt game, she ran to Jacob. "I heard your father ask your brother Esau to hunt game and prepare him meat. He wants to eat and then give Esau the blessing for the oldest son."

Jacob listened eagerly. Rebekah obviously had a plan. They both wanted him to have that blessing, and the inheritance of Canaan that went with it.

"Now go to our flock of goats," the mother said, smoothing her son's dark hair. "Kill and dress out two fat young kids from the flock, and I will prepare a savory stew. I will fix it just as Esau does. You will take it in to your father and receive the blessing, and he will never know the difference!"

So Jacob brought two fattened kids. Rebekah prepared savory food from the meat and baked fresh bread.

"What about my smooth skin?" Jacob worried. "My brother is a hairy man."

Rebekah took pieces of hairy skin from the kids and fastened them on her younger son's hands. She took another piece to cover his neck. Then she handed Jacob a sweat-stained robe of Esau.

"Put this on," she commanded him, "and take this savory meat and bread to your father."

So Jacob took it to the doorway of his father's tent. "My father," Jacob called, imitating the voice of his older brother.

"Here I am," came the faint voice of Isaac in the dimness of the tent. The old man's hearing was almost gone, and he could hardly distinguish the voice. "Who are you, my son?"

"I am Esau, your firstborn," Jacob lied, shouting loudly so that his father could hear. He laid the dish of succulent meat and the fragrant fresh bread beside the bed of his father. "I have done as you said. Now sit up and eat of my game, that you may bless me."

Isaac struggled to a sitting position. Jacob piled the pillows so that the old man could rest on them.

"But my son," Isaac asked, squinting his almost sightless eyes, "how is it you have found game so quickly?"

Jacob lied boldly, "Because your God granted me success!"

Then Isaac put out his thin, blue-veined hand. "Come closer, so I may feel you, my son. I must know whether you are really my son Esau or not."

So Jacob drew closer to Isaac. His father felt him, drawing his hands across the young man's neck, over the young man's hands.

The wrinkles on Isaac's face were deepened by a perplexed frown. He shook his head. "The voice is more like the voice of Jacob." Again he felt the hands of Jacob that were covered with hairy kidskin.

Jacob fearfully held his breath as his father felt him.

"But the hands!" his father sighed. "Yes, the hands must be Esau's hands." The unseeing face turned beseechingly toward Jacob. "Are you really my son Esau?"

Jacob lied yet again, his heart thumping. "I am," he said.

Isaac sagged back on the pillows of his bed. "Then give me the savory food," the old man said. "I want to eat of your game and bless you."

So Isaac ate, thinking it was the fresh game of Esau. When he was finished, he said, "Come near and kiss me, my son."

Jacob leaned over and kissed his father. His father smelled the familiar odor of Esau's sweat-stained robe, and he was completely deceived. So he gave Jacob the blessing for the firstborn, saying:

See, the smell of my son
  is like the smell of a field which the Lord
    has blessed!
May God give you the dew and rain from
    heaven,
  and plenty of fruit and grain.

Let peoples serve you,
  and nations bow down to you.
Be lord over your brothers
  and may your mother's sons bow down to
    you."

As soon as Isaac finished Jacob left the tent. At that very moment Esau came in. He had returned from his hunting and had prepared savory game for his father. He came hurrying into the tent, saying, "Let my father sit up and eat of this game and bless me!"

His father squinted, trying in vain to see his visitor. "Who are you?" he quavered.

The hunter answered, "I am your son, your firstborn, Esau."

The old man began to tremble so violently he could hardly speak. "Who was it then," old Isaac asked, "that hunted game and brought it to me, and I ate it all before you came? For I have blessed him!"

When Esau heard the words of his father, he cried out bitterly, "O my father, bless me, even me also."

Isaac's frail shoulders sagged helplessly. "Your brother has lied and tricked me. You know that a promise made before God cannot be taken back!"

Esau bowed his head. He beat his strong,

hairy fists on the ground. "My younger brother bought away my birthright, and now he has cheated me of my father's blessing!"

Then Isaac comforted his older son with the only blessing he had left to give:

Away from bountiful crops of fruit and grain
  shall you live,
  away from the refreshing dew and the rain.

You shall live by your sword and your bow,
  and you shall serve your brother;
But when you break loose,
  You shall break his yoke from your neck.

Esau left his father's tent, burning with anger and with hatred for his twin brother. "Soon my father will die. When the mourning is over, I shall kill my brother Jacob!"

Rebekah heard that Esau had sworn to kill Jacob. She called her younger son to her. "You must obey me again!" she whispered to him. "Flee to my brother Laban, across the desert in Haran. Stay with your uncle until your brother's anger dies away. Then I will send for you to come back!"

Needing an excuse for Jacob to leave, Rebekah began to complain to her dying husband of the idol-worshiping wives of Esau. "Jacob should marry one of our own kinsmen as you did," Rebekah argued. Then she began to weep. "If your youngest son also takes wives from here, what good will my life be?"

So the old man sent for his son Jacob. "You must not marry one of these Canaanite women," he ordered. "Arise, go to the house of your uncle Laban. Take a wife from the daughters of Laban's house."

And so it was that Jacob left the land of Canaan. He would never see his mother or his father again.

Genesis 25:8-34; 26:34 through 28:5

Jacob Earns a Wife

Jacob fled alone from his home. He came to a certain place and stayed there that night because the sun had set. For his supper, he munched a little bread from his packsack. Then he wrapped his cloak around him and lay down on the hard ground with a stone for his pillow. He fell into a restless sleep and dreamed.

In Jacob's dream there was a ladder set up on the earth, going all the way up to heaven, and there were angels going up and down the ladder. At the top was the radiance of God, and a voice said to Jacob, "I am the God of Abraham and Isaac. Your descendants shall spread over this whole land. I am with you and will watch over you wherever you go. And someday I will bring you back safely to this land."

So Jacob felt reassured as he made the long, tiresome journey across the desert. At last he came to Haran, the land of the people of the East. There were fields and pasturelands all about the city where his mother Rebekah once lived. In the middle of the nearest outlying field was a small well with a flat rock covering its mouth. The rock was so heavy, it took all the shepherds of the surrounding area to lift it. This guaranteed that one man would not steal more than his share of the water.

Jacob noticed that three separate flocks of sheep were grazing or lying near the well, and three shepherds sat idly talking together. Jacob addressed them politely, "My brothers, where do you come from?"

They said, "We are from Haran."

"I seek a man named Laban," Jacob said. "Do you know him?"

One of the shepherds chewed on a piece of grass and spat it out. "We know him," he said. He pointed to the far end of the meadow. "See that maiden coming? That is Rachel, Laban's daughter."

Jacob saw the slim figure approaching, and he was instantly attracted to her. She walked with such liveliness, and he liked the graceful way she tossed back her long black hair. He was eager to speak with her about his uncle. He looked impatiently at the shepherds who loitered at the well.

"Why don't you water your sheep," Jacob asked, "and take them to pasture?"

"We can't," the shepherds explained, shrugging their shoulders indifferently. "Not until the other shepherds come, and together we can roll the stone from the well."

As they spoke Rachel arrived at the well, her sheep milling about her. Jacob went up, took hold of the stone, and with a mighty heave he rolled it from the mouth of the well. The shepherds looked at Jacob, openmouthed. The girl's large black eyes rounded with amazement. She watched silently and with admiration as the strong, smooth-skinned stranger watered her sheep.

Then Jacob took the girl's hands and said, "Rachel, did you know that you are my cousin?" He kissed her. "Run quickly, and tell your father that Jacob the son of Rebekah is here!"

The girl ran and told her father. When Laban heard that the son of his beloved sister was outside the city, he hurried to meet Jacob. Laban hugged his nephew and kissed him and brought him into his house.

Jacob stayed on with his uncle and worked hard to care for his uncle's flocks and herds. Finally Laban said to Jacob, "Should you serve me for nothing, just because you are my kinsman? Tell me, what shall your wages be?"

Now Laban had two daughters. Leah, the older one, was unattractive and had weak eyes. But Rachel, the younger one, was lovely and enchanting. It was Rachel that Jacob loved with all his heart.

"I will serve you seven years for your younger daughter Rachel," Jacob proposed.

Laban said, "It is better that I give her to you than to any other. So stay with me!"

So it was that Jacob worked for Laban seven long years. Yet they seemed but a few days to Jacob because he had such great love for Rachel.

At last Jacob came joyfully to Laban and said, "My time is completed. Let me have my wife!"

So Laban invited everyone for a wedding feast. In the evening when the sun had set, the bride was brought out draped in a long, heavy veil; and Jacob was married to her.

In the morning light Jacob eagerly looked at his bride. His joy turned to anger. He was married to the homely Leah!

Jacob threw on his robes and stalked to the house of Laban. "What have you done to me?" he shouted angrily. "Did I not serve you for Rachel? Why then have you deceived me?"

Laban said smoothly, "In our country it is not the custom to give the younger girl in marriage ahead of the firstborn. Complete the marriage festivities this week with Leah. Then you may also marry the younger—in return for serving me another seven years."

So the marriage festivities went on for the rest of the week, to celebrate Jacob's union with Leah. Afterward the younger Rachel was given to Jacob as his second wife.* There was more feasting and merry-making.

Jacob was happy now that his beloved Rachel had become his wife. But he had to keep his promise to work for Laban for another seven years.

36   Both sisters were unhappy during these years. Leah gave birth to four sons, one after another, but she was miserable because she knew her husband did not love her. Rachel knew that she was dearly loved by Jacob, but she was envious and jealous of her older sister.

Finally Rachel turned on her loving husband. "Give me children, or I shall die!" she demanded.

Jacob replied angrily, "Am I in the place of God?"

So Rachel did with Jacob as Sarah had done with Abraham two generations before. "Here is my slave maiden," Rachel told Jacob. "Take her as your wife, so that she can bear children for me!"

In due time Rachel's maid bore a son. Rachel took the baby boy as her own son. "God has heard my voice and given me a son!" Rachel exulted. And the next year her maid bore a second son for her mistress.

Not to be outdone by her younger sister, Leah gave her slave maiden to Jacob for another wife. Leah's maid bore for her mistress two sons to be counted among the many sons of Leah. Then Leah herself bore two more sons.

At last it happened that Rachel found her joy complete. She bore a son of her own and named him Jospeh.

By this time Jacob had a large family: four wives and eleven sons. He had worked for his uncle Laban for fourteen long years. Now he wanted to return to the land of Canaan and claim that land as his inheritance.

"Let me take my wives and my children," Jacob said to his father-in-law. "I want to return to my own home and country."

But Laban protested. He had grown wealthy with the clever Jacob watching over his herds and his flocks. "Stay, and name your wages," Laban argued.

"No," Jacob said. "Now I must provide for my own household."

"Stay!" Laban begged. "Just tell me what I should give you."

Jacob thought how he might turn the tables on his uncle. "I will stay long enough to build a flock for myself," he said. "You need not pay me anything if you will let me keep all the speckled and spotted goats that are bred in your herd, and every black lamb that is born among your sheep."

Laban thought about this proposal. After all, were not most of his goats black, and not speckled or spotted? Also, a black lamb among all the white sheep was a rarity. "Good," Laban agreed. "Let it be as you have said."

Just to make doubly sure that he would lose nothing, Laban and his sons went among the sheep and took out the black rams and ewes. These odd-colored animals were herded off to a distant pasture, three days' journey away, and kept under a close watch by Laban's sons.

But Jacob was more wily than his uncle. He knew a trick that was part of every shepherd's lore. He took fresh branches and peeled white streaks on them and laid them before the strongest animals in Laban's flock during the mating season. The suggestive magic worked, curiously enough! All the strongest she-goats bore speckled and spotted kids, which Jacob set aside for breeding in a herd of his own. And the weaker she-goats bore black kids, to remain in Laban's herd. There were many black lambs, too, for Jacob's flock.

So it was that Jacob grew very rich in the next six years. He hired servants of his own and bought some camels and asses. Laban and his sons became fiercely jealous of Jacob's success. This made Jacob more eager than ever to leave and return to the land of his birth.

Rachel and Leah were equally eager to leave. "Has not our father sold us?" they said. "Has he not been using up the dowry given for us?"

One day when Laban was busy at sheep-shearing, Jacob mounted his wives and his sons on camels. "Come," he said to his household, "we will take our possessions and go! Let us hurry while Laban is away!"

With his servants driving his flocks and herds, they set their faces toward the west and started the long journey back to Canaan. Jacob's wives would never again return to the land where they were born. But Jacob's many sons would now grow up in the land of their inheritance.

Genesis 28:1 through 31:21

* It was not unusual in early Old Testament times for a man to have more than one wife at a time. The most important reason for this was the strong desire for sons who would secure an inheritance and provide manpower. After the Hebrews settled in the land of Canaan, the practice began to disappear, and monogamy, or marriage with only one wife at a time, became the custom for the ordinary man. Wealthy men and kings often continued to add wives to their households for reasons of prestige or as part of a common pattern of international diplomacy.

Jacob's Homecoming

Every night Jacob's servants pitched tents for Jacob's family and themselves. They were encamped in Gilead, the high and wooded hill country which rises east of the Jordan River, when the outraged Laban caught up with them.

Laban confronted his nephew Jacob and accused him. "Why did you flee secretly and cheat me? Why have you carried my daughters away like captives?"

Jacob still burned with resentment because his uncle had tricked him into working fourteen years to earn his two wives, plus the extra six years to care for his uncle's flocks. Yet Jacob always had been a faithful and honest shepherd for Laban.

"Because I was afraid," Jacob shouted back at his uncle. "I thought you might try to keep your daughters there by force!"

Laban threw up his hands to the sky. "You should have told me!" he lamented piously. "Then I might have sent you away with laughter and songs, with sweet music of singing and tambourines and with the strumming of lyres! And why did you not permit me to kiss my daughters and my grandsons farewell?"

Jacob retorted, "These twenty years I have been with you, your herds and your flocks have multiplied! If a sheep or a goat was torn by wild beasts, or stolen, I bore the loss of it myself. I lived the hard outdoor life for you, sweltering by day, and shivering with cold at night, and often sleep fled from my eyes! I served you fourteen years for your two daughters, and six years for my flock, and you have changed my wages ten times." Jacob put his work-worn hands on his hips and glared at his uncle. "What is my offense? What is my sin that you have hotly pursued me?"

Laban shouted at his nephew, "Why did you steal my household gods?"

Now Jacob did not know that Rachel had stolen the small carved idols belonging to her father.

"In the presence of our kinsmen," Jacob declared, with a sweep of his arm that took in the whole encampment, "point out what I have that is yours and take it."

So Laban went into Jacob's tent, but there were no idols there. Next he searched Leah's tent. Nothing. Then Laban went into the tent of his younger daughter Rachel. She sat demurely on a camel saddle in which the idols were hidden. Her father searched angrily. At last Laban gave up.

Before they parted, Laban and Jacob agreed to a covenant of friendship. They erected a heap of stones at that place as a memorial of their agreement. The stone pillar was named Mizpah. As they stood there by the pillar, Laban charged Jacob to take good care of his daughters. And he said, "The Lord watch between you and me when we are absent one from the other."

Jacob offered a sacrifice, and everyone gathered together to eat bread. Early the next morning Laban arose. He kissed his daughters and his grandchildren and blessed them. Then he returned home.

As he drew closer to Canaan, Jacob began to dread his first meeting with his brother Esau. Would Esau still have murderous feelings toward the twin brother who had cheated him of his birthright and his father's blessing?

Jacob sent messengers before him to Esau in the desert wilderness country of Edom, which was south of Canaan. "Treat my brother as a lord," Jacob instructed his messengers. "Tell him I have been twenty years with our uncle Laban and that I now have oxen, asses, flocks, and servants. Tell him that I hope to find favor in his sight!"

Soon the messengers returned. "Your brother Esau is coming," they told Jacob, "and four hundred men with him!"

This news filled Jacob with fear and worry. He turned to God for help. "I am not worthy of the steadfast love you have shown me," he prayed. "Deliver me from the hand of my brother Esau! I fear he may come and slay us all, the mothers with the children!"

The next morning he sent his messengers back to Esau, bearing a multitude of presents. He also sent goats, sheep, cows, and asses. "I may be able to appease him," Jacob said to himself. "Perhaps he will accept me!"

Jacob kept waiting, tense, anxious, and afraid. At last he lifted up his eyes and looked. There was his brother coming, and four hundred men with him. Jacob went out in front of his encampment, bowing himself humbly to the ground seven times.

But Esau ran to meet him and threw his arms around his brother's neck and kissed him. They were both grown men, but they wept.

After they visited, each went his own way. Esau and his men went back to the southern desert wilderness of Edom. And Jacob crossed the Jordan River, back into Canaan, and pitched his tents in the land where he was born.

After a time Jacob had another dream. In it God said to him, "You will now have another name: Israel. The multitude of your descendants shall be a great nation. They shall be called the children of Israel."

Genesis 31:17 through 33:17; 35:9-15

Joseph and His Brothers

Jacob lived on in the land where his father had lived. His family was very large, but Rachel had borne Jacob only one child, an adored son named Joseph.

At last, in the land of Canaan, Rachel gave birth to a long-desired second son. Then the beloved wife died.

Jacob buried Rachel, and with her he buried his greatest joy in life. He was left with twelve sons and three wives, but he had a special tenderness for the two sons of Rachel. He grew to love young Joseph and the baby Benjamin almost as much as he had loved their mother.

When Joseph was seventeen he began to help tend sheep and do work around the encampment with the brothers who were his own age. But he was the aging Jacob's favorite; Joseph's brothers, Judah, Simeon, Reuben, and the other sons of Leah and the sons of the two former maid-servants, tended the distant herds.

The older brothers wore short work-tunics of rough wool. When tending sheep on wintry nights, each man would girdle himself with a tanned sheepskin and try to keep warm by a flickering fire.

Joseph slept comfortably in his tent in the encampment. His favorite garment was a gift from his doting father, a full-length robe with long sleeves. His brothers resented the way Joseph strutted about in his beautiful robe. Joseph's cumbersome and impractical long sleeves were a symbol of luxury and leisure.

It was all too plain to the older brothers; their faither loved Joseph more than the rest of them. So they hated Joseph and could not speak peaceably to him.

One night Joseph had a dream. The next morning he took the first opportunity to tell it to his brothers. "Hear this dream which I have dreamed! It was harvesttime. We were all binding sheaves of wheat in the field. My sheaf arose, standing upright in the center. And your sheaves gathered around and bowed down to mine."

His brothers didn't like Joseph, and they didn't like his dream.

"Oh, are you indeed to reign over us?" Reuben jeered.

"Do you think you will someday rule us?" Judah asked sarcastically. "Will you, our younger brother, have dominion over us?"

The next night Joseph dreamed another dream. So the next morning he told it to his brothers. "I've dreamed another dream!" he said. "In it, the sun, the moon, and eleven stars were bowing down to me."

The older brothers were already tired of Joseph and these dreams were the last straw! So they complained to their father.

The dreams were too much even for the doting Jacob. He rebuked young Joseph. "What is this dream that you have dreamed? So the sun, the moon, and eleven stars were bowing down to you! Shall I, and the memory of your poor dead mother, and your eleven brothers indeed bow our-selves to the ground before you?"

But the older brothers were not satis-fied with their father's mild rebuke. They were so jealous of Joseph that they agreed to take the first opportunity to get even with him.

The brothers had been sent to watch the sheep in a distant pasture, in a lonely place among the hills and ravines and pits of the wilderness. Sometime later Jacob sent Joseph to see how the brothers fared.

The brothers could see Joseph coming from afar off. Simeon said bitterly, "Here comes this dreamer! Come, now, let us kill him!" "Yes!" another brother agreed. "And we'll throw his body into one of the pits."

"We could say that a wild beast has devoured him," suggested a third.

Reuben, the eldest, spoke up. "Let us not take his life." Reuben didn't like Jo-seph any more than the rest, but he knew how much Joseph meant to their father. "Shed no blood. Cast him into this pit

here in the wilderness, but lay no hand upon him." Reuben thought this might give the lad a needed scare. Later, he could be quietly rescued and restored to his father. Judah, another brother, nodded in agreement.

Joseph strolled up to his brothers. But before he could speak, they seized him and stripped him of his long-sleeved robe. They took him, clad only in his knee-length undershirt, and cast him into a deep pit. Then they went some distance away, behind a hill where they could not hear Joseph's screams and weeping. They kindled a fire, prepared bread and a stew, and sat down to eat.

Judah looked around at his brothers. "What profit is it if we slay Joseph and hide his spilled blood?" he said, dipping his bread into the stew and eating noisily. "Come, let us just sell him to the desert Ishmaelites."

The brothers nodded. Despite their jealous hatred of Joseph, they could not quite bring themselves to take his life.

As they talked and ate, some traders passed by and heard Joseph's cries. They threw down a rope and drew the young man up. Joseph's great joy at being freed was short-lived. The traders quickly bound him and sold him to a caravan of desert Ishmaelites for twenty pieces of silver. The Ismaelites took him to a marketplace in Egypt and sold him as a slave.

When Reuben returned to the pit and saw that Joseph was no longer there, he was alarmed. He hurried back to the brothers.

"The lad is gone," Reuben gasped hoarsely. His face was white and anxious. As the oldest son, he knew his father would hold him responsible for Joseph's safety. The magnitude of what they had done suddenly overwhelmed him.

All the brothers started talking at once.

They dared not tell their father the truth. Finally several brothers took Joseph's torn robe and dipped it into the blood of a goat. They took the bloodstained garment back to old Jacob's encampment.

"This we have found," they said. "See now whether it is your son's robe or not."

Jacob held the stained and torn garment in his hands, and he knew it was Joseph's. "It is my son's robe," he wept. "A wild beast has devoured him. Without doubt, Joseph is torn to pieces!"

Then Jacob mourned according to the custom of the times, tearing his clothes and wrapping himself in a rough piece of sackcloth.

His remaining children rose up to comfort him, but he refused to be comforted. He wept and wept for his lost son.

Genesis 35:16-20; 37:1-36

In Pharaoh's Prison

Joseph was sold as a slave in Egypt. He was bought by Potiphar, a captain of the guard in the royal court.

God was with Joseph, and he became a successful man. In the house of his Egyptian master all that he did prospered. The gifted Joseph found favor in his master's sight.

"You shall be overseer of my house," Potiphar told the young Hebrew. "I shall put you in charge of all that I have in my house and in my fields."

Joseph had become a very handsome man. After a time Potiphar's wife eyed

Joseph with great interest. She decided she wanted him to be her lover.

But Joseph refused. "My master has put everything that he has in my hand," Joseph said. "He has denied me nothing—except yourself because you are his wife. How can I be so wicked against my master and against God?"

Yet, though she followed him around day after day, begging him, teasing him, provoking him, Joseph would pay no attention to her. Then one day when she found him alone in the house doing his work, she grabbed the sleeve of his coat.

"Love me," she demanded.

Joseph's only thought was to get out of the house. He turned to flee from the room, but Popiphar's wife had fast hold of his coat sleeve. There was a rip, a tearing sound, and Joseph escaped from the room and from the house. But his torn coat remained behind.

So Potiphar's wife burning with vindictive anger, began to scream. Others in the household hurried into the room, and she pretended to weep. "See, my husband has brought among us a Hebrew to insult us," she cried out, showing everyone Joseph's torn coat. "He struggled with me, but when I cried out, he fled from the house leaving his garment behind!"

When Potiphar came home, his wife put on another scene, telling the same lies and showing the torn coat. Potiphar's anger was kindled. He ordered Joseph thrown into the prison of Pharaoh, king of Egypt.*

But even in prison, God was with Joseph. After a time Joseph found favor with the prison keeper. Joseph became his trusted helper.

Some time after this the chief butler and the chief baker of the king of Egypt angered their master. So Pharaoh jailed them in the same prison where Joseph was languishing.

One night they both dreamed, Pharaoh's chief butler and Pharaoh's chief baker. In the morning each was troubled. They wanted someone to interpret their dreams.

And Joseph said to them, "Do not interpretations belong to God? Tell me your dreams."

So the chief butler began. "There was a vine with three branches in my dream. It budded, and almost at once its blossoms shot forth, and then the clusters ripened into grapes. Pharaoh's cup was in my hand. I took the grapes and pressed them into Pharaoh's cup and placed the cup in Pharaoh's hand."

Then Joseph said to the butler, "This is the interpretation: The three branches are three days; within three days Pharaoh will lift your head and restore you to your office! And you shall place Pharaoh's cup in his hand as before, when you were his butler."

The butler was very happy with this interpretation and began to thank Joseph profusely.

"Just do me the kindness to remember me to Pharaoh, when all is well with you," Joseph said, "and so get me out of this place! For I was stolen from the land of the Hebrews, and I have done nothing that I should be put into a dungeon and forgotten."

The chief baker was now eager for his turn. "I also had a dream," he told Joseph. "There were three cake baskets on my head. In the uppermost basket were many kinds of pastries for Pharaoh. But birds were eating out of the basket on my head!"

Joseph answered, "This is its interpretation: The three baskets are three days; within three days Pharaoh will lift up your head—from you—and hang you on a tree, and the birds will eat the flesh from you."

Three days later it was Pharaoh's birthday. The king of Egypt made a feast for all the palace and for his servants. His chief butler was released from prison and restored to his old job. Indeed, the chief butler again placed the cup in Pharaoh's hand.

But the king of Egypt was still angry with the chief baker and ordered him hanged. It was just as Joseph had interpreted.

Yet the chief butler did not remember Joseph, but forgot him. And so Joseph remained in prison.

Genesis 39 through 40

* "Pharaoh" was the title of Egyptian kings. In the Bible it is often used as a proper name in place of the king's own name.

Joseph an Egyptian Ruler

Joseph was kept in the prison of Pharaoh for two long years.

Then Pharaoh himself had a disturbing dream. The Egyptian king dreamed that he was standing by the mighty Nile River. Seven fat cows came out of the water and fed in the reed grass. Then seven other cows, skinny and starving, came from the river and gobbled up the seven fat cows.

Pharaoh awoke feeling uneasy. When he went to sleep again, he dreamed a second time. Seven fat ears of grain were growing on one stalk. Then seven withered ears, blasted and blighted by the hot east wind, swallowed up the seven fat ones.

In the morning Pharaoh was haunted by his two dreams. He called for the magicians and wise men of Egypt to interpret the meaning, but no one could.

The chief butler belatedly remembered Joseph. "When your chief baker and I were in prison," he told Pharaoh, "we each had a dream. A young Hebrew was in the prison with us. He interpreted our dreams, and it all happened exactly as he said!"

Pharaoh sent for the Hebrew slave. When Joseph stood before him, the king challenged, "I have heard it said of you that when you hear a dream you can interpret it!"

Joseph bowed himself before the king. "It is not in me," he said. "God will give Pharaoh a favorable answer."

The king told Joseph his dreams about the seven fat cows eaten by seven skinny cows, and the seven fat ears of grain swallowed up by seven blighted ears.

Then Joseph said, "Both dreams have the same meaning, and they reveal what God is about to do. The seven good cows and the seven good ears of grain are seven good years that will come. But they will be followed by seven years of famine!"

"And what does your God say I should do?" Pharaoh asked.

Joseph replied, "Let Pharaoh select a man discreet and wise and set him over the land of Egypt. Then appoint overseers under him who will set aside one fifth of all the produce of the land during the seven plentiful years. Store this grain, and Egypt will have a great reserve during the seven years of famine that will follow."

Pharaoh said, "Since God has shown you all this, there is none so discreet and wise as you are. I shall set you over the land of Egypt, and all my people shall follow your command."

The king took his own signet ring and put it on Joseph's finger as a sign

44 of his new authority. Then Joseph was arrayed in garments of fine linen and given the second finest chariot in the land. And the daughter of a priest was presented to Joseph for his wife.* At this time Joseph was thirty years old.

Joseph's predictions came true. During seven plentiful years Joseph saw that great quantities of grain were heaped in the royal storehouses in every city. Seven years of famine followed. The shortage of food was not only in Egypt but in all the lands around. Everywhere people were hungry for bread. Joseph opened the storehouses and sold grain to the Egyptians. Starving people from other lands came by donkey and by camel train to buy the precious stored grain.

Back in the land of Canaan, the family of Jacob also suffered. Their crops were blighted and blasted. Old Jacob called his sons together and said, "I have heard there is grain in Egypt. Go down and buy wheat and barley for us there, or we will die of starvation."

So Joseph's ten brothers went down to Egypt. But Jacob did not send Benjamin, Joseph's younger brother. The old man was afraid some harm might come to the lad.

Thus it was that Joseph's brothers came to the Egyptian court. They bowed themselves to the ground before the great overseer, who was dressed in royal Egyptian garments.

Joseph recognized his brothers, and he remembered how in his dreams they had bowed down to him. He knew they did not recognize him, so he pretended they were strangers. "Where do you come from?" he asked in Egyptian, his voice brusque. A court interpreter translated to the sons of Jacob.

"From the land of Canaan," the men told Joseph through the interpreter. "We came to buy food."

"I don't believe that," Joseph accused. "I think you have come to spy out the weakness of the land!"

"No, my lord, only to buy food!" they protested. "Look, we are twelve brothers, the sons of one man, Jacob, called Israel. This very day the youngest is with our father." They hung their heads as one of them mumbled, "But one of us—he is no more!"

Joseph said, "I shall see whether or not you are a band of spies! One of you must stay here in prison while the rest go back to Canaan. You must bring back to me the youngest brother you speak about as proof of your story."

The brothers were deeply troubled. They were not willing that one of them should stay behind in a strange prison. They might never see him again! They gathered for a family council, standing in a pitiful circle there in the great hall in Egypt.

Reuben said, "Remember what we did to our brother Joseph? Now we suffer because we were the cause of his death!"

As they spoke among themselves they did not know that the great overseer of Egypt understood them. The tears came to Joseph's eyes as he looked at his brothers and listened to them, so that he had to turn away.

One of the brothers, Simeon, was chosen to stay in Egypt as a hostage in prison. Then Joseph told his servants to fill the bags of the remaining brothers with grain. The money each man paid for his grain was replaced in his own sack. And they were given provisions for their long journey home.

When the nine remaining brothers came to the land of Canaan, they told Jacob everything that had happened. When they opened their sacks and found the hidden money they were frightened.

The old man slumped down with sorrow. "You have bereaved me of my children: Joseph is no more, and Simeon is no more, and all this has come upon me. Now you would take Benjamin. No! My last son shall not go!"

But the famine was severe. Even though they ate only enough barely to stay alive, the wheat from Egypt did not last. They must find more food, and there was only one place to buy it—Egypt. But the brothers dared not return to Egypt without Benjamin. And besides, how else could they redeem their imprisoned brother Simeon?

"If it must be so," old Jacob said sadly, "take with you a present and your brother Benjamin and go back to this man."

So the brothers took a present, a double amount of money, and Benjamin, and set out on the long road back to Egypt. At last they stood in the Egyptian court before Joseph.

When Joseph saw them, he said to a palace steward, "Slaughter an animal and make a feast ready at my house. Those men are to dine with me at noon."

When the brothers were brought to Joseph's house, they were afraid. "It is because of the money in our sacks," they said to one another. "Now he has an excuse to make slaves of us!"

So they rushed up to the steward of Joseph's house and started explaining at once about the money in the sacks.

"Oh, no!" the steward said, pretending ignorance. "I received your money. If you found money, perhaps your God put it there?" Then he brought Simeon out, finely dressed and well cared for.

The brothers embraced one another. Just then Joseph came striding in, resplendent in his official robes. They all bowed low not knowing what to expect.

"Is your father well?" Joseph asked. The men nodded while Joseph stared with wonder and joy at the tall young boy with them. "Is this your youngest brother of whom you spoke?" The men nodded again, scarcely lifting their heads from the floor.

Joseph touched the dark hair of Benjamin, his own brother, his mother's second and last son. "God be good to you," he said. And then he could not hold back the tears which were coming into his eyes. Joseph wanted to be alone where he could weep, for he was not yet ready to reveal

46 himself to his brothers. He wheeled and went to his own bedchamber.

Later Joseph returned and ordered, "Let food be served!"

There was a banquet. Then the brothers' sacks were filled with grain and their money again secretly replaced in the sacks.

"And in the sack of the youngest," Joseph whispered to his steward, "place my large silver cup, along with his money."

At dawn the brothers started back toward Canaan with their laden donkeys. Then Joseph told his steward, "Catch up to the men just outside the city. Accuse them, and ask, 'Why have you returned evil for good? Why have you stolen the cup from which my lord drinks?'"

Joseph's steward took a squad of Egyptian guards and overtook the eleven brothers outside the city. The sacks were searched, and the cup was found in the sack of Benjamin. The brothers were forced to return to the city.

Judah, Simeon, Reuben, and the other brothers came to Joseph's house. They fell on the ground before him.

"What shall we say to you, our lord?" Judah said. "God is indeed punishing us for our long-time guilt!" He bowed his head. "We shall be your slaves."

"No," Joseph said, pretending to look severe. "Only the man in whose sack my cup was found shall be my slave. The rest of you shall return in peace to your father."

Then Judah wept and told how their aged father cherished Benjamin, the only remaining son of Rachel. "Let me stay in his place," Judah begged. "For how could I go back to our father Jacob if the lad is not with me? I fear that my father would die!"

Joseph could no longer control himself before all those in the room. He sent out everyone, except the sons of Jacob. Then Joseph wept and said, "Look—I am your brother, your long lost brother–Joseph!"

His brothers were too astonished to speak. Their faces were a study in amazement, guilt, and fear.

"Do not be distressed or angry with yourselves," Joseph consoled them. "As a result of what you did, I was sold into Egypt, but God sent me here to save your lives! For the famine has wasted the land for two years, and there will be yet five years more with no harvest. Now hurry to my father and tell him that Joseph is alive, and a lord over Egypt. All of you shall live here, you and your children and your children's children, and your flocks and your herds and all that you have. And I shall provide for you."

The brothers were so relieved they laughed and cried at the same time. It seemed too good to be true.

"Make haste, now," Joseph urged, "and bring my father down here."

Then the brothers returned to Canaan and told their father of all they had seen and that Joseph was still alive, a great lord in the land of Egypt.

And so it was that the whole household of Jacob and his sons moved down to live in the land of the Egyptians. Old Jacob was reunited with his son Joseph, with tears and rejoicing. After that they were settled by Pharaoh in the land of Goshen, in Egypt. And there they lived out their lives.

Genesis 41 through 47

* Joseph's position in Egypt was very like that of a prime minister. He was invested with this authority at a formal ceremony, and it was at this time that he was given an Egyptian name and also an Egyptian wife. The signet ring bore the seal of the pharaoh, and Joseph used it when he acted as the king's representative.

PART III
HEBREWS WITHOUT A HOMELAND

The Egyptian background for the Joseph stories seems quite authentic, with many details matching those found in Egyptian paintings and writings for the period.

This is also true for the Moses stories which follow. It is not unusual to find in the folktales of the world heroes who are marvelously protected in infancy so that they may grow to fulfill their heroic roles. And Moses is not the only baby said to have been found floating in a basket on the water. However, the story of Moses is unique in the amount of detail that is given about the man, his personality, his work, and his relation with God. The place he occupies in the history of the Hebrew people, as well as in the New Testament interpretation of that history, sets him apart.

We cannot be sure just how much Moses knew of his Hebrew background. There is good reason for thinking that some contact with his own family was kept up even after he was adopted by the Egyptian princess. Unknowingly, his adopted mother hired his real mother to be his nurse, and the child was sent into the home of his parents for his very earliest training. His sister may have served later as a palace maid, and there may have been visits and messages through the years from his old "nurse."

However it was, when he had grown to be a man, Moses was ready to identify himself with the downtrodden Hebrew slaves. He was willing to renounce the privileges and advantages of royalty which had become his by adoption.

When Moses killed an Egyptian and was forced to flee from the land, he found shelter among the Midianites. These were a semi-nomadic people, as the Israelites had been before they settled in Egypt. The two tribes were distantly related. Perhaps at the time of Moses some of the Midianites were living in the desert of the Sinai peninsula.

It was here among the Midianites that Moses had his famous encounter with God. This event marked the beginning of Moses' mission to lead the Hebrew people from slavery in Egypt.

When Moses and Aaron went to challenge Pharaoh, the snake episode was more than a magician's trick. It was a symbolic description of God's power. The serpent of Moses and Aaron devoured the serpents of the Egyptian wise men just as God would conquer all foreign powers and gods.

Later we are told of a series of miraculous feats performed by Moses and his brother before the Hebrew people escaped from Egypt. It is interesting that many of these happenings may occur naturally in that area. The waters of the Nile rise annually and become tinged with the red soil of the Ethiopian mountains. The river also is tinted red by algae during the summer months. Often there are plagues of frogs and of insects. Epidemics may rage among the cattle and the people. The significant thing is that the succession of wonders frightened and disrupted the Egyptians and their ruler until they let the Hebrew people get away. The same events inspired and encouraged the Hebrews themselves so much that this became the most important part of their history—their salvation history—which assured them that they were God's chosen people, that he had delivered them by these mighty acts, and that they had some great part to play in the world.

The Baby Moses

Old Jacob, whose God-given name was Israel, died at a good old age. And the time finally came when Joseph and all his brothers had died. But Jacob's descendants, the children of Israel, multiplied and grew exceedingly strong so that there were many of them in the land of Egypt.

Now there arose a new king over Egypt who had not known Joseph. And the new Pharaoh said to his people, "Look, the Israelites are too many and too mighty for us. Come, let us deal shrewdly with them, lest they multiply more. If there should be a war, they might join our enemies."

Besides, Pharaoh needed laborers to build new Egyptian cities. So he made slaves of the Israelite men and set taskmasters over them. But the more the Israelites were oppressed, the more they multiplied.

Now the Egyptians were more in awe of the power of the Israelites than before. Tasks were made harder. The Israelites were forced to work very long hours, making bricks and carrying heavy loads. Still they multiplied.

At last in desperation Pharaoh ordered, "Kill all the newborn boys!" The Israelites, of course, did what they could to hide their sons. In one Israelite family a strong and healthy baby boy was born, and his mother was determined to save his life. For three months she hid him. When she could hide him no longer, she took a basket woven of rushes, waterproofed it with bitumen and pitch, and lined it with soft linen. Then she laid her baby in the basket and placed it in the tall green rushes and reeds growing along the slow-moving river.

"Stay hidden nearby," the mother told Miriam, the baby's older sister. "Watch to see what happens."

Now Pharaoh had a grown daughter who regularly came to the river's edge to bathe. Her maidens-in-waiting walked along the shore while the princess waded into the warm, calm water.

On one of her trips to the river she saw the basket floating among the reeds. The princess called to one of her maids. "See that basket? Go fetch it!"

The basket was brought up on the shore. There lay the baby. He began to cry. Pharaoh's daughter took pity on the infant. She took him up and cradled him in her arms. "This is a child of the Hebrews," she said.

Then the baby's sister crept out of her hiding place and stood in front of the princess. "Shall I find you a nurse from among the Hebrew women?" Miriam asked. Her face reflected eagerness and worry. "Someone to nurse the child for you?"

Pharaoh's daughter smiled at the young girl. "Go," she said.

So the girl ran home to tell her mother the good news. "Our baby is safe!" she

50 gasped. "Come quickly, the daughter of Pharaoh wants a nurse for him!"

They hurried back to the bank of the river. The princess said to the mother, "Take this child home and nurse him for me, and I will give you wages."

Pharaoh's daughter adopted the boy and brought him up as her own son. She named him Moses. And Moses was instructed in all the wisdom of Egypt.

Exodus 1:1 through 2:10

Moses in Midian

When Moses was a grown man, he went out among his people and saw how they suffered under the burden of slavery.

One day Moses saw an Egyptian beating a Hebrew, one of his own people. Moses looked this way and that, and seeing no one, he killed the Egyptian and hid him in the sand.

The next day Moses saw two Hebrews struggling together. One man was trying to pick a fight, and so Moses said to him, "Why do you strike your fellow?"

The belligerent man growled back, "Who made you a prince and a judge over us?" Then he taunted, "Do you mean to kill me—as you killed the Egyptian?"

This frightened Moses, and he thought, "What I did yesterday is known!"

His fears were justified. People had seen and talked about the incident. Pharaoh heard of it and sent soldiers to kill Moses. But Moses fled eastward from Egypt. He came to the arid desert land of Sinai, where the wandering Midianites often pitched their tents. Tired, weary, and thirsty, he came to a well and sat down to rest.

There was a priest of Midian named Jethro who had seven daughters. The girls came to the well and drew up skin buckets of water and filled up the troughs to water their father's flock. But before the sheep belonging to Jethro could drink, some rough and surly shepherds came up and started to drive them away. They intended to use the water in the trough for their own sheep.

But Moses came to the rescue of the seven sisters. His passionate sense of justice was outraged, just as it had been when he had seen a slave beaten by an Egyptian overseer. He chased the rude and lazy shepherds away and helped the sisters water their sheep.

When the young women returned home with their sheep they told their father Jethro what had happened.

"Where is the man?" the Midianite priest asked his daughters. "Why have you left him? Call him, that he may eat bread with us."

So Moses was made welcome by Jethro's family. He stayed on to live with them for many years. He married one of the daughters, Zipporah. And Zipporah bore Moses a son.

Moses became the chief keeper of Jethro's flock. In the peace and quiet of the Sinai wilderness Moses had much time for thinking and meditation.

One day as he walked slowly with the grazing flock he came to a place where great mountains soared in jagged peaks toward the sky. As he neared Mount Horeb,* he saw a bush aflame with fire. It was a startling sight, for the bush was burning, yet it was not consumed. It flamed, yet did not burn up!

Moses stood there amazed. Then he heard a voice calling, "Moses, Moses!"

Moses looked about him. "Here I am," he answered. He and the sheep were alone

in the lonely wilderness at the foot of that high, forbidding mountain.

"Do not come near," came the awesome voice. "Put off your shoes from your feet, for the place on which you are standing is holy ground."

Moses quickly undid his sandals.

"I am the God of your fathers, of Abraham, of Isaac, and of Jacob," continued the majestic voice. "I have seen the suffering of my people who are in Egypt. I will rescue them. I will bring them up out of the land of their slavery to a good and broad land, a land flowing with milk and honey."

Moses felt a great joy in his heart—until he heard God's next words.

"You," the voice went on, "I will send to Pharaoh. You are to bring my people out of Egypt!"

Moses was deeply frightened. He had never had experience as a leader. "Who am I that I should do all this?" he asked. "And how will the Israelites believe me?"

God said, "Tell your people that the God of Abraham and Isaac and Jacob sent you. Tell them I will deliver them from Egypt and lead them to a land of milk and honey."

"I don't think the Israelites will believe me," Moses objected.

"Moses! What do you have in your hand?"

Moses held it up. "My shepherd's stick."

"Cast it down on the ground," God ordered.

Moses threw it down. It writhed and thrashed and turned into a serpent! Moses started to run from it.

"Wait," said the Lord. "Put out your hand and take it by the tail."

His heart thumping, fearful of being bitten, Moses grabbed the tail. It became a shepherd's stick again, wooden and lifeless.

"Do that for the people," the Lord said, "and they will know that the God of your fathers has spoken with you. And Pharaoh will know too."

But Moses was still full of excuses. "Oh, Lord, I am not eloquent. I am slow of tongue. I cannot speak well enough to Pharaoh. Send someone else."

The Lord's voice sounded angry. "Is there not Aaron, your brother? He can speak well! Very soon he will come to meet you. You shall speak to him and put the words in his mouth. And I will be your mouth and his mouth. I will teach you what you shall do. Now go back to Egypt for the men who once sought your life are dead. With your shepherd's rod you shall perform signs."

Moses went back to Jethro, his father-in-law. "Let me leave Sinai. Let me take my family and go back to my kinsmen in Egypt."

The old Midianite priest said, "Go in peace."

So Moses took his wife and his sons and set them on a donkey and went back to the land of Egypt, and in his hand Moses took the rod of God.

Exodus 2:11 through 4:20

* Mount Horeb is known also as Mount Sinai and as the mountain of God.

Moses and Pharaoh

Aaron met his brother Moses in the wilderness and greeted him with a kiss. Moses told Aaron of all he had heard from

God and of the plans to lead the Israelites into a new land.

Then Moses and his brother Aaron returned to Egypt and gathered together the elders of the people of Israel. Aaron delivered all the words which God had spoken to Moses. The people believed. They bowed their heads and worshiped God.

Afterward Moses and Aaron went to the court of the new king of Egypt. They said to Pharaoh, "Thus says the God of Israel, 'Let my people go, that they may hold a feast to me in the wilderness.'"

But Pharaoh would not listen to such a request. "Who is your God, that I should heed his voice and let Israel go worship him? I do not know your God, and I will not let Israel go!"

The two brothers pleaded, but the king had set his heart against them. "Why do you take the people away from their work?" he shouted at them. "Get back to your burdens!"

That same day Pharaoh sent instructions to the foremen of the Hebrew slaves, "You shall no longer provide the straw the people need to hold their bricks together. From now on they must gather their own straw. But they must still make the same number of bricks every day." Then he added to his courtiers, "Perhaps if they work harder, these Israelites will forget about making trips to sacrifice to their God!"

So the Israelites had to labor twice as hard as before. From the first faint light of dawn until it was too dark to see, they gathered stubble to use for straw. Yet they were required to make as many clay bricks as before. They were sick with their hard labor and came complaining to Moses.

Moses, in turn, complained to God.

God said, "I am the LORD, and you shall see what I will do to Pharaoh now! Go to the king and prove yourselves by working a miracle with the rod."

So Aaron and Moses went to Pharaoh in the great throne room. Moses handed Aaron his shepherd's rod. Aaron raised the rod high above his head, twirled it dramatically, and cast it down on the marble floor.

Everyone in the court gasped. For the rod writhed and thrashed and turned into a snake. Pharaoh stared openmouthed. Then he snapped his fingers angrily. He summoned his wise men, sorcerers, and magicians of Egypt to come forward and show up these two Hebrews.

The wise men, sorcerers, and magicians came. They knew secret arts and magician's tricks. They threw down their rods, and theirs became snakes too. Then a strange thing happened. All the magician's rods were swallowed up by the rod of Aaron.

"Let my people go!" cried Aaron, speaking for Moses.

But Pharaoh hardened his heart. His face was angry. He would not listen to Aaron and Moses, so they had to return to their people in sadness.

Then God said to Moses, "Go to Pharaoh in the morning as he is going out to the water. Ask him to let your people go. If he will not, strike the water with your rod."

Moses and Aaron did as they were commanded. They met the hard-hearted king of Egypt on the brink of the river Nile. He would not listen to them, so Moses said to Aaron, "Strike the water with the rod."

Suddenly the water turned blood-red. The fish in the Nile began to die. The river water had a foul taste so wells had to be dug to provide water to drink.

Yet Pharaoh returned to his house without relenting. So Moses went back to God for instructions.

Seven days later Moses and Aaron caused great numbers of frogs to plague the cities of Egypt. They came hopping, slimy and wet, out of the rivers, out of the canals and pools, into the cities, along the streets, and into the houses. They even swarmed over the shining marble floors of the throne room and palace quarters of Pharaoh.

This was too much for the king. He sent for Moses and Aaron. "Ask your God to take away the frogs, and I will let your people go out to the wilderness for their sacrifices!"

"When?" asked Moses.

"Tomorrow," Pharaoh snapped.

So Moses asked God to end the plague of frogs. They began dying in the houses and the courtyards and out in the fields. The dead frogs stank and were gathered together in heaps for disposal.

Now that there was relief from the frogs, Pharaoh changed his mind. He would not let the Israelite people go after all.

Next the land was filled with swarms of gnats and biting, stinging sand flies. They bit both man and beast. Pharaoh, slapping at the pesty insects, shouted for Moses and Aaron to appear before him.

"Go!" Pharaoh shouted. "Sacrifice to your God!" He shook his fist at the two brothers. "But have your ceremonies inside the borders of Egypt."

"Oh, no," Moses said. "We cannot do this within your borders. Our altar sacrifices offend the Egyptians, and they might stone us."

"Go out into the wilderness, then," Pharaoh huffed. "But see that you do not go very far away!"

So Moses asked God to remove the swarms of insects. But once they were gone, Pharaoh changed his mind again and would not let the people go.

Moses and Aaron went back again to Pharaoh and said, "How long will you refuse to humble yourself before our God? Let my people make their sacrifice in the wilderness. If you will not, then a plague of locusts will devastate Egypt."

Pharaoh groaned and said, "Who is to go with you for your sacrifices?"

"Everybody," the brothers told him. "All our people, including our wives and our children and our flocks and herds."

"You have some evil purpose in mind!" the exasperated Pharaoh shouted. "No! The men among you may go, but not your wives or your little ones."

So Moses waved his rod. A hot, dry wind came from the great sandy stretches of desert to the east. It blew across the land all day and all night. When it was morning, the east wind had brought the locusts.

The whirring, flying insects settled over the whole land of Egypt. Such dense swarms of these grasshoppers had never been seen before, nor since. They blotted out the sun and darkened the sky as they flew overhead. They crawled like a moving blanket of rasping legs and wings over the ground. They covered every living plant and tree, eating the grain, the fruit, and the leaves.

Pharaoh sent for Moses and Aaron in haste. "I have sinned against your God, and against you," he pleaded. "Please forgive my sins and beg your God to remove this curse from our land."

Moses prayed, and a strong west wind came up and blew the locusts back into the Red Sea so that they perished. Then Moses and Aaron appeared before Pharaoh in his great throne room.

"Go, serve your God," the king said, throwing up his hands. "And your children may go with you. But not your flocks and herds!"

Moses said calmly, "Our cattle and our sheep must go with us. Everything we own belongs to God, and we do not know what we may need until we get there. Not only that, but I am also requesting that you provide us with sacrifices and burnt offerings."

Pharaoh grew red in the face with anger. He rose from his throne menacingly. "Get away from my sight!" he roared. "You shall not go! Never see my face again, or you shall die!"

Moses left with his brother. Then he talked with God again.

"One more plague* will come upon Pharaoh," God said, "a plague of death. Afterwards he will let you go, and he will let you go completely." Then God gave Moses the instructions for a Passover feast. It was to be celebrated immediately, and it would be remembered and celebrated again down through the years. It would be an annual memorial of the day when the Israelites were freed from bondage in Egypt. The people would always recall the time when the plague of death passed over their houses and settled in the houses of Pharaoh and his people. They would also remember the night when they left their homes to "pass over" the borders of Egypt on their way to the Promised Land.

So on the tenth day of that month each Israelite family chose a perfect lamb from their flocks. On the fourteenth day the lamb was killed and roasted. At their Passover feast the lamb was eaten with lettuce, other bitter herbs, and flat loaves of unleavened bread.** The head of each Israelite household sprinkled some of the lamb's blood on the doorframe of his house to protect his family from the plague.

At midnight, after the Passover feast, the plague of death came upon Egypt. Every firstborn in the land died, from the firstborn of Pharaoh, who sat on the throne, to the firstborn of the captive who was in the dungeon, and all the firstborn of the cattle. But the destroyer "passed over" the houses of the Israelites, and none of them died.

There was weeping in every Egyptian house that night. Pharaoh rose up from his

56 bed, and he and the people of his court gathered together. The king summoned Moses and Aaron to come to him at once.

When they came, he said, "Rise up and go, you and the people of Israel. Take your flocks and herds and be gone!"

So the women of the Hebrews gathered their children. They took their unbaked bread, binding the dough-filled kneading bowls into their mantles. Some of the frightened Egyptians offered them parting gifts of clothing and silver and gold jewelry, hoping that the Israelite God would look favorably upon Egypt. The Israelite men gathered their herds and their flocks and drove them alongside the journeying multitude.

And God went before them by day in a pillar of cloud to lead them along the way, and by night in a pillar of fire to give them light. In this way they traveled, by day and by night.

Exodus 4:27 through 13:22

* Only five of the ten plagues are woven into the story here. The biblical account gives as the others: a sickness of the cattle, boils, hail, storms, and thick darkness.

** Unleavened bread was a flat, crisp bread made without the addition of yeast or anything to make it rise. Only pure unleavened bread could be offered to the Lord.

Israel's Escape

When the king of Egypt was told that the Israelites had fled, he said, "What is this we have done, that we have let Israel go from serving us?"

So Pharaoh made ready his chariot army.

There were more than six hundred chariots, with a fighting officer and a skilled charioteer in each one. "Let us force our valuable slaves to come back," they all agreed. The Egyptians thundered out across the plains in hot pursuit.

At this time the Israelites were encamped by the reed marshes. When Pharaoh's army drew near, the people of Israel realized they were being overtaken. Great dust clouds swirled on the horizon, raised by the pounding hooves of the horses and the spinning wheels of many chariots.

Gripped by sudden fear, the Israelites panicked. Some of them cried out to Moses, "Have you taken us away to die in the wilderness?" And others wailed, "It would have been better for us to serve the Egyptians than to die here!"

Moses alone had faith that God could still save them. "Fear not," he shouted. "Stand firm and see how God will work for you today."

Then God told Moses, "Lift up your rod and stretch out your hand toward the sea and divide it, that the people of Israel may go on dry ground to cross the water."

Then the pillar of cloud moved behind the Israelites, hiding them from the oncoming chariot army. Moses stretched his hand over the marshy lake. The Israelites gaped in wonder as a strong east wind came blowing across the water, making the ground dry to walk across. So the people crossed on dry ground. When they were all safely on the opposite side, the Egyptians reached the shore. Into the marsh the Egyptians plunged, recklessly pursuing the Hebrews.

But the heavy chariots bogged down in the mud, and the mud clogged their wheels so they would barely turn. Some of the charioteers said, "Let us flee and give up this chase. Their God fights for them against the Egyptians!"

Then God spoke to Moses who was on

the other side, "Stretch out your hand so that the water may come back on the Egyptians and their chariots."

So Moses stretched forth his hand over the marshland, and the waters returned and covered the chariots and the horsemen and the officers of Pharaoh that had followed them into the water. Not one of them remained!

Thus the Lord saved Israel that day from the hand of the Egyptians. And Israel saw the great work which the Lord did against the Egyptians, and the people feared the Lord; and they believed in the Lord and in his servant Moses.

Miriam, a prophetess and a sister of Moses and Aaron, took a tambourine in her hand. She shook the small, shallow drum so that the bronze pieces around the edges jingled, and she beat it rhythmically with her hand. All the women followed after her, tapping and jangling their tambourines and dancing. Miriam led them in singing:

"I will sing to the Lord, for he has triumphed
   gloriously;
    the horse and his rider he has thrown into
     the sea.
The Lord is my strength and my song,
   and he has become my salvation;
this is my God, and I will praise him,
   my father's God, and I will exalt him. . . .

"Pharaoh's chariots and his host he cast into
   the sea;
    and his picked officers are sunk in the Red
     Sea.* . . .
The enemy said, 'I will pursue, I will overtake, . . .
    I will draw my sword, my hand shall destroy
     them.'
Thou didst blow with thy wind, the sea
   covered them;
    they sank as lead in the mighty waters.
"Who is like thee, O Lord among the gods?

Who is like thee, majestic in holiness,
   terrible in glorious deeds, doing wonders? . . .

The Lord will reign for ever and ever." **
            (based on Exodus 15:1-18)

Exodus 14:5 through 15:21

  * The sea that was crossed is popularly translated as the Red Sea, but a more literal translation would be "Sea of Reeds" or "Lake of Rushes." Even today the phenomenon of an eastern desert wind blowing a dry place across a shallow lake or marshland is not unknown. A sandbar may become exposed and dry enough for crossing. Other scientific explanations have been proposed which rationally explain this miracle, but to the God-trusting Hebrews, no extra explanation was necessary.

  ** The first two lines of the song are known as the "Song of Miriam." It is one of the oldest fragments in the Bible, probably composed for the actual occasion and passed along by singing and repeating it for generations until it was finally written down. The other lines come from the "Song of Moses." This was probably composed later to celebrate the memories of this victorious event.

In the Desert

Moses led the Israelites onward from the reed marshes, and they went into the Sinai wilderness. For three days they journeyed without finding water. Finally they came to a pool, but the water was bitter, brackish, unwholesome. They could not drink it. The Israelites began to complain, and Moses prayed for help.

Moses was guided toward a scrubby tree with bitter wood. He broke off some of the branches and threw them into the pool.

58 When the people tasted the water, it had become sweet and crystal clear, good to drink.

As they traveled, the Israelites stopped whenever they came to an oasis with the welcome shade of palm trees and springs of fresh water. They would encamp and rest. After awhile the black goathair tents would be folded up and loaded on the patient donkeys. The kettles and bedding would be packed, and the migrating Israelites would slowly move out once more into the wastelands of the desert.

Again and again the people ran out of water. Once when they had used all the water in their waterskins and the last oasis was far behind them, they could not see a watering place in the lonely stretches of desert ahead of them. Their mouths were parched with thirst. The hot, sandy wind swirled around them, and the scorching sun shone down from a brassy sky.

"Give us water to drink!" the people cried out to Moses. "Water, water!" And again they began to complain, saying, "Why did you bring us up out of Egypt to kill us and our children and our cattle with thirst?" Those who passed by the tent of Moses would hold up their dried-out, empty waterskins. They hissed and threw stones at his tent.

"What shall I do with this people?" Moses cried out to God. "They are almost ready to stone me!"

God said, "Go ahead of the people. Take with you some of your leaders and also take your shepherd's rod. You shall see a great rock. Strike it with your rod. Water shall pour from that rock, and the people may drink."

Moses went on ahead, taking an eager young follower named Joshua and some of the respected older men of the tribes. Before long they came upon a large rock which stood out and caught their attention.

Then Moses took his rod and struck the porous limestone boulder, and water gushed out from a hidden underground spring. Everyone took turns coming up for long, cool drinks and for filling their empty waterskins. There was plenty for the herds and flocks and donkeys too.

After this the company traveled on. But the wilderness offered no new oasis. Finally they were forced to settle down for an encampment, but there was no food left for the empty stomachs.

The whole congregation of the Israelites murmured against Moses, grumbling and complaining and whining about their hardships. They forgot what they had suffered when they were oppressed slaves in Egypt. They remembered only the good things there. In fact, they remembered life in Egypt as having been far better than it ever was.

"We should have died in Egypt!" they moaned. "There we sat around kettles full of savory meat stew. We had bread to eat until we were full." They spoke accusingly to Moses, their leader. "You!" they shouted. "You brought us out into this barren wilderness. You will kill this whole assembly with hunger!"

Then God said to Moses, "Listen to me, I will rain food from heaven for you. Tell the people of Israel that they shall eat meat at twilight; and in the morning I shall provide a special, delicious food."

So Moses told the people, and some of them could not believe it. They looked about them. Barren, rocky, mighty mountains were on the horizon. The ground was sandy with nothing but boulders and dry, desert plants. Here and there were tamarisk trees and shrubs with tiny scalelike leaves on the odd, feathery branches; but they bore no fruit. It was nothing but a vast wasteland.

Yet at dusk that evening a wind blew across the wilderness from the distant sea. And the wind blew in a great cloud of migrating quail, off their course and weak from much flying. Each Israelite family caught enough of the birds to cook a pot of stew. Everyone ate and ate until he was full!

At dawn the next morning some began to wake up and ask, "What about that special food that was supposed to rain down in the morning?"

They looked around, and under every tamarisk tree were white, sugary, flakelike things scattered like frost on the ground.

"What is that?" they asked.

Moses said, "It is the special bread God promised you. Everyone may gather as much as he can eat."

They tasted it, and it was sweet with a flavor like wafers of honey. And there was enough for everybody.

The Israelites had no name for this strange new food, so they called it manna.* They were thankful that God had looked after them, sending them quail and also manna from heaven.

The flight of quail proved to be an exceptional treat. But they found the sticky-sweet manna under the tamarisk trees every morning during their long years of wandering.

Exodus 15:22 through 17:7

* Certain areas of the Sinai peninsula are still well known for the production of manna (MAN-eh). This is a granular sweet, exuded by scale insects living on tamarisk trees and shrubs. In order for the insects to get needed nitrogen from the plant sap they must consume great quantities of the sap and exude the excess sugars. The liquid honeydew excretions quickly solidify into sticky and often granular masses, which can be collected.

The Covenant

The fleeing Israelites traveled on and on, making their way across the barren wilderness. On the third new moon after leaving Egypt, they came to some mountains in Sinai that were familiar to Moses. Here were those high, jagged peaks, one higher and more awesome than the rest, which rose forbiddingly over the desert valley. The sun blazed in a hot, brassy sky. There was almost no vegetation, only an acacia tree or two and a few dry, thorny shrubs dotted here and there.

Moses recognized this as the very place where he had seen the burning bush and heard God's voice. It was here that God had commissioned him to go back to Egypt to free the people from their slavery under Pharaoh.

Moses sent out the order for the twelve tribes of Israelites to settle here for an encampment. The tents were erected, each family gathered bits of thornbush for the evening fires, and the smell of smoke and supper cooking filled the air.

Just before dawn the next morning, Moses climbed alone the rocky slope of the highest mountain. Behind him, scattered across the desert stretches of the valley, the Israelites began to awake. There was laughter as children dodged about between the tents. Boys and men began calling to their sheep, goats, and cattle, to take them out in search of a patch of wild grass, or a few edible shrubs. Women and girls took handfuls of their hoarded wheat from the grain sacks and began pounding it into flour on flat stones for the day's bread. The sun rose higher, burning away the chill of the night.

But on the mountain it was hushed and quiet. There was not even the chirp of a cricket or the song of a bird. In the stillness God spoke to Moses.

When Moses returned to the desert encampment, the sun was going down, making long shadows across the valley. The people dropped what they were doing and gathered in a great silent crowd to listen to their leader.

"God has spoken to me on this mountain!" Moses shouted so that all could hear. "God told me, 'You have seen what I did to the Egyptians, and how I bore you on eagles' wings and brought you to myself. Now therefore, if you will obey my voice and keep my covenant,* you shall be my own possession among all peoples. You shall be to me a kingdom of priests and a holy nation.'"

"Are you willing to make such an agreement with God?" Moses asked the people. "You will be his chosen people. In return, you must do as he asks."

And all the people answered together, "We will do whatever God asks."

So there was a three-day ceremony of preparation before learning what God's commandments would be. The people washed themselves and their clothes to be clean and pure for the holy occasion.

On the morning of the third day it was dark. A thick cloud hid the jagged peak of the great mountain. Lightning flashed from the clouds, and thunder rumbled so that the very air seemed to shake. The people followed their leader to the foot of the mountain. They were awed and quiet. This was the day on which Moses would learn what the people must do to keep a covenant with God.

As a ram-horn trumpet sounded, spears of lightning suddenly split the air, and thunder rolled and crashed. It sounded as if the whole mountain were quaking. The people trembled. They knew that God's voice was in the thunder and that he was calling Moses to go up into the mountain.

Moses left his people and started the climb up the steep sides of the mountain. He disappeared into the darkness of the clouds.

The people were afraid, and called after Moses, "You tell us what God says, but don't let God speak to us or we might die!"

Moses' voice, getting fainter and farther away, called back, "Do not fear."

That night when Moses came back, the people gathered round him again. Moses told of the laws God had given him, the Ten Commandments for the Israelites to keep. The people were not to worship other gods or make idols; they were not to make light of God's name or profane it; they were to keep the sabbath, honor their fathers and mothers, respect the unity of marriage and the family. They were not to commit murder, steal, lie against a neighbor, or jealously desire the possessions of others.

"God also laid down many other laws for us to follow," Moses went on. "He has told me how we should treat our families, strangers, and slaves. He has given the penalties for stealing and murder, as well as for other crimes. If you agree to these laws, I will write them down as part of our covenant with God."

As before, the people answered in one voice, "All the words which God has spoken we will do."

Then there was a sacrifice of animals upon an altar to celebrate Israel's agreement with God. It was followed by a feast.

The mountains quaked again, the thick clouds swirled around, and there was more awesome lightning and thunder. Moses said to the people, "I must return to the top of the mountain. God has promised to give me tablets of stone, on which will be written the commandments."

So Moses left the camp in the charge of Aaron, his brother, and started up the mountain with Joshua, his right-hand man.

Joshua waited at the bottom of the mountain while Moses climbed upward and disappeared into the clouds. Moses was alone on the mountain for forty days and forty nights.

As the days went on and Moses did not come back down, the people became restless and frightened. "We do not know what has become of Moses," they clamored before Aaron. "Make us gods to lead us. Make us a god we can see."

So Aaron sighed and said, "Bring your gold earrings and bracelets, and gold jewelry you have among you." The gold was melted and fashioned into a golden calf. The idol was mounted beside an altar, and the people bowed down to it, sacrificed to it, and worshiped it. Then they ate and drank and celebrated.

When Moses came down from the mountain, he met Joshua, who had a worried look on his face. "There is a noise of war in the camp," said the young man.

Moses listened. "No, it is singing."

As they went down to the camp, Moses was carrying two flat tablets of stone with the Ten Commandments chiseled on their surfaces. Suddenly Moses stopped and stood staring with horror at the people below. They were eating and drinking, singing and dancing before a golden idol. They were calling the calf their god! While he had been in the mountains receiving the tablets of stone and learning God's instructions for worship, the Israelites had already forgotten their own God!

Moses was so hot with anger he threw the stone tablets to the ground and broke them. He smashed the golden idol. Then he built a huge bonfire and cast the pieces of the molten calf into the fire.

Moses shouted to Aaron his brother, "What did the people do to you that you have brought a great sin upon them?"

Aaron shuffled uneasily. "Well, you know how the people are set on doing evil." He was beginning to look ashamed.

Some of the people began to be ashamed of themselves, too. They had already broken the first two of God's commandments: you shall worship no other gods, and you shall not make idols.

The next day Moses gathered the people together and said, "Now I will go up again to speak to God. I shall see if I can make up for this evil you have done. I will ask God to forgive you. And if God cannot, I will ask him to blot me out of his book. Perhaps I can make atonement for Israel so God may still consider us his people."

Again Moses climbed out of sight to the heights of the great mountain. Again he was gone for forty days and forty nights.

When he returned, he was carrying in his arms two tablets of stone. Upon them he had chiseled the Ten Commandments, which he had heard again from God. As Moses came down the mountain with the stone tablets in his hands, the people were waiting for him. Moses did not realize that his face was shining. It shone as if by an inner light—because he had been talking to God.

Exodus 19 through 34

* A covenant is a solemn agreement made between persons. The making of the covenant between God and Israel at Mount Sinai was considered by the Hebrews to be the most important event in their history.

The Land
of Plenty

The people were in no hurry to leave so special and holy a place as the foot of Mount Sinai. But at last the time came when there were no more patches of grass to be found in the area for pasturing hungry sheep, goats, and cattle. There were no edible shrubs left for miles around. The hungry goats had even eaten all the dry, thorny desert plants. So the Israelites had to pack up their tents and their few belongings and move on. Besides, Canaan was ahead, the Promised Land of milk and honey.

At first some of the Israelites were afraid to leave these awesome mountains. Some feared they would be leaving the place where God lived. But Moses told the people that God had given instructions for building an ark of the covenant to carry with them. It would be a very special box made of the desert acacia wood, just so in size, with poles held by gold rings along the sides so that it could be carried by four men. This ark would be a portable shrine. Inside, the tables of law could be safely kept. On top of the ark was to be the "mercy seat"—it would be a visible throne for the invisible God.

A very special tent was prepared to shelter the ark in the camp. People gave goathair cloth, animal skins, and beautifully colored linens for draperies. It was called the tent of meeting, for it was a place to meet God. It was also called the tabernacle.

Someday, when they reached the Promised Land, the Israelites said, they would replace the tabernacle with a great temple, permanent and magnificent.

But for the days of wandering in the wilderness, the ark of God would be kept in the most splendid tent that they could manage. It would stand in the sight of the people, to remind them that God had promised always to be with them.

Whenever the people had camped for too long at one place, whenever the forage for the animals gave out and it was time to move on, the ark was carried on ahead. Then Moses would say, "Arise, O Lord, and let your enemies be scattered!" When the ark rested on the site of the next encampment, Moses said, "Return, O Lord, to the thousands of Israel."

Sometimes an oasis would provide much grass and rich forage for the animals; new lambs, goats, and calves would be plentiful. Then there would be meat bubbling fragrantly in the stew kettles every evening. Yet many of the people still found things to complain about.

"Remember the variety of food we had in Egypt?" they would say around their campfires. "Fish! All the fish we wanted, there for the catching. Fish to roast on sticks over the hot coals of a fire. And the seasonings for our meat stews! Oh, the onions and the leeks and the garlic we had in Egypt! And the cucumbers, and the lush, juicy melons!"

And so the months went on and the years. There was slow progress across the barren wilderness and always much complaining. There were battles with roving desert tribes. Often their route looped backwards or zigzagged far off the course because they could only go where there was enough pasturage and water.

At one time toward the end of summer, they encamped close to the southern border of Canaan, and Moses sent men to spy out the land. The young man Joshua was in charge of an expedition of eleven other men.

"Go and find out what our Promised Land is like," Moses told them.

So Joshua and the other spies left on their mission. When they returned, they were carrying thick clusters of grapes and a large honeycomb dripping with honey. They told of seeing fat milk cows grazing in green pastures.

"Yes, Canaan is a good land," they reported. "Look at its fruit."

The spies also told of great fortified cities, and it became clear that the Israelites did not have the faith or the courage to go in and claim Canaan as their Promised Land.

"We cannot successfully fight those people," most of the spies said, "for they are larger and stronger than we are, and they are excellent warriors."

Now the discontent among the people swelled into an angry tide of rebellion. "We wish we had died in Egypt—or even in the wilderness," the people shouted at Moses. "Does God lead us to this land only to have us slain by a sword?"

The rebellion grew. They looked for a new leader to replace Moses, someone who would take them back to Egypt, back to slavery.

God was angry. He said to Moses and Aaron, "How long shall this wicked congregation murmur against me?" Then God gave them a message for the people: none of the older generation would live to enter the Promised Land! Only the children would grow up to inherit Canaan, the land that was promised to the descendants of Abraham. And so it was.

Moses remained the leader, and the people became wandering shepherds and nomads in the wilderness.

At last Moses' sister Miriam died and was buried. Moses wept for her, the one who had watched over him when he was a baby in a basket, floating among the rushes and reeds of the Nile.

Later Aaron the brother of Moses died. By God's wishes he had become the chief priest of Israel. Upon Aaron's death, his son became the new chief priest.

Moses was getting very old when the Israelites finally reached the mountains of Moab on the eastern side of the Jordan River. Moses and Joshua looked westward. Below them was the great Jordan River, winding its way into the deep, dead-end salty sea.

Between the Jordan and the Great Sea,* off in the dim western distance, there—as far as the eye could see—was the Promised Land. This was the land of Canaan, where Abraham and Isaac and Jacob once had pitched their tents. It was rocky and hilly with a rugged backbone of mountains. But there were green valleys and highlands too.

Tears welled in Moses' eyes as he stood on Mount Nebo and looked out over this land, for he knew that he would never enter it. He was an old, old man. Already he could feel the chill of death in his bones. But the hardy, skilled warrior beside him would go! Yes, Joshua would lead the Israelites into the Promised Land.

So it was that Moses died there in the land of Moab, on the eastern cliffs of the Jordan. There he was buried, but no man knows the place of his burial to this day. And there has not arisen since a prophet in Israel like Moses, whom God knew face to face.

Exodus 25:1-22
Numbers 10:33-36; 11:4-6; 13:1-33; 20:1-29
Deuteronomy 32:48-50; 34:1-12

* The Great Sea was the name given by the ancients to the Mediterranean Sea.

PART IV
BATTLES FOR A HOMELAND

The religious history of the Israelites after the death of Moses continues in the books of Joshua and Judges. These books tell stories of the time when the Hebrews were conquering and settling the land of Canaan. This was a heroic time in the history of Israel, rather like our own pioneer period. Leaders and warriors came forward to help win possession of the Promised Land.

Joshua, who had served as Moses' lieutenant in the days of the exodus from Egypt, was an able leader and military commander. When he and his forces invaded Canaan, they found the people dwelling in strongly fortified cities. The Hebrews planned and carried out attacks against the walled cities which lay in their path—beginning with the city of Jericho in the Jordan valley—and they wreaked havoc among the Canaanites.

Archaeology has confirmed that there was an invasion of Canaan during the thirteenth century B.C. Ruins of many towns mentioned in Joshua (Bethel-Ai, Lachish, Debir) have been found. Jericho is one of the most ancient cities in the world. Excavations have uncovered many cities in that location, each built on an earlier one. The last level occupied by the Canaanites actually did fall down flat. Blackened bricks show that a fire ravaged the city. This disaster in Jericho seems to have been at

66 an earlier time than the sieges of the other Canaanite cities, however, according to the archaeological evidence.

The conquest was not as easy as some of the stories might suggest, and it took quite a long time to accomplish. Some indication of this occurs in Joshua 13:1-6 where God says to Joshua, "You are old and advanced in years, and there remains yet very much land to be possessed." The passage goes on to list vast areas of Canaan which remained unconquered. It is here that we first learn of the coastal rivals of the Hebrews, the Philistines.

After the death of Joshua, during the twelfth to the tenth centuries B.C., the twelve tribes (named after the sons and grandsons of Jacob) were scattered over the land of Canaan. They settled down to an uneasy and uncertain existence among the Canaanite walled cities. Each Israelite tribe was allotted a territory of its own within which it was ruled by its elders. Whenever a territory was threatened, a judge—who was a combination of wise man, leader, and military chief—might rally about him as many fighting men as possible to protect and defend his people.

In the book of Judges we find the stories of such leaders, or judges, as Deborah and Barak, Gideon and Jephthah, and others. We also find Samson, a typical folk hero, pitted alone against the well-organized Philistines dwelling on the coast.

During the time of the judges the Israelites made some adjustment to living with their old enemies, the Canaanites. Despite the serious objections of the leaders of Israel, there was much borrowing of cultural ideas—some imaginatively rich and valuable, and some barbarous and demoralizing. It was feared this mixing would weaken the Hebrews' worship of and obedience to God.

There were still dangerous enemies who pushed in from all sides. Eastward across the Jordan on the high tablelands of Gilead, two distantly related Semitic groups, the Ammonites and the Moabites, caused much trouble.

To the west were the Philistines, a foreign people who had sailed in to establish settlements along the shores of the Great Sea. According to Egyptian wall paintings of that time, they had earlier attempted to invade the land of the Nile. When they were repulsed, they sailed northward and gained a toehold along the coast of Palestine. (Oddly enough, the word "Palestine" is derived from "Philistine.") Their five strong city-states were established at Gaza, Ashdod, Ashkelon, Gath, and Ekron. And the Philistines became an increasing menace to the Israelites.

There were also internal frictions among the tribes. Jealous differences sometimes provoked wars between one Israelite tribe and another. Their religion was virile and tremendously dynamic. Often it led to violent and bloody action. As the last verse in Judges says, "In those days there was no king in Israel; every man did what was right in his own eyes."

The book of Ruth also describes the period of the judges in Israel. It is quite a contrast to the accounts of the battles and conflicts contained in Joshua and Judges. It presents an idyllic, pastoral tale of life at that time, though it may have been written much later.

Joshua and the Battle of Jericho

After the death of Moses, Joshua became the leader of the Israelites.

The tribes of Reuben and Gad, along with part of the Manasseh tribe, made permanent settlements on the high table-lands east of the Jordan River. This land came to be called Gilead.

The rest of the Israelites intended to cross over the Jordan. There, to the west, they would try to conquer the land of the Canaanites between the Jordan and the Great Sea.

Just ahead, across the winding, shallow Jordan, was the city of Jericho. The mountains of the land of Canaan were cold in fall and winter, and so were the highlands of Moab and Gilead. But deep in the wide rift of valley, the weather was warm and semitropical. Hot winds blew across the thriving town of Jericho, waving the fronds of her palm trees.

This was the first Canaanite city that had to be conquered, and it would not be easy. Jericho was securely protected behind the high walls of mud bricks that encircled the whole city.

All the fighting men among the Israelites were mobilized. The priests, carrying the ark of God on poles, set out ahead; the fighters followed. When they came to the brink of the river Jordan, there was an ominous rumble, like an earthquake. A part of the high clay bluffs upstream tumbled into the river, damming it. Although the water usually overflowed its banks in the cold and rainy season of the year, now all the water was cut off. The whole army of Israel passed over on dry ground and thanked God for the miracle.

Now the people of Jericho shut their city gates, bolting them from within. But the people also were shut in from the outside because the Israelites were encamped just beyond the city. No one went into Jericho, and none went out.

Joshua told his army that God had given special instructions that this city should be besieged in a certain way. So the Israelites followed his commands.

On the first day seven priests blowing ram-horn trumpets were to lead a procession around the city walls. Other priests would follow bearing the ark, and then the whole army would bring up the rear in silence.

The warriors of Jericho, stationed on the walls to protect their city, watched in amazement. What was happening? No shouting, no hail of arrows, rocks, or spears aimed upon the city wall! No fighting.

The people of Jericho were confused. They did not know what to think.

The next day was the same with seven priests leading the strange procession, blowing those eerie blasts on their rams' horns. That gold-decorated wooden box, so reverently carried—did it hold some kind of magic, some evil spirit that would work against the inhabitants? And finally the whole army of Israelites, as quiet as if they were mutes, marching, marching—and never saying a word!

It happened again on the third day, and on the fourth, the fifth, and the sixth. Once around the city, then the strange, nerve-racking procession went silently back to camp.

What was the mysterious purpose of all this? The seasoned warriors of Jericho knew how to fight back in an ordinary siege. But this was ominous and scary. The inhabitants of Jericho were jittery and afraid.

On the seventh day it happened. Joshua gave the orders to his whole army, and they

marched out as before. This day they marched around the city seven times in total silence, silence broken only by the mournful, echoing blasts of the rams' horns.

Then Joshua gave the signal. The Israelites raised a great shout, a deafening roar. As their voices swelled, the earth under their feet began to rumble as if in an earthquake. The great walls trembled; they shook; they fell down flat!

And so it was that Joshua and the Israelites took the city.

Joshua 1:1 through 6:21

The Song of Deborah

Deborah was a prophetess of God. Her tribe of Ephraim looked up to her as their judge. She sat under "the tree of Deborah" near the town of Bethel, and her people often came to ask her advice.

The Ephraimites were doing well, as they became entrenched in the hill country of central Canaan. Far to the north, beyond a barrier-line of mountains, and beyond the great Kishon-watered Valley of Jezreel, two smaller Israelite tribes, Zebulun and Naphtali, were not doing well at all. They had tried to settle in the lovely northern hills above the Jezreel valley.* But there were powerful walled Canaanite cities there, and the new settlers found themselves enslaved and cruelly exploited. Jabin, king of the strong city of Hazor, was an especially harsh oppressor. He used his large chariot army to terrorize the tent-dwelling newcomers.

At last the people of Zebulun and Naphtali cried to God for help. Their pleas for aid became known among all the other Israelites.

When Deborah heard of the sad plight of the small northern tribes, she sent for Barak, leader of the tribe of Naphtali. "I have learned," the prophetess said to Barak, "that God has asked you to gather together an army of ten thousand men."

Barak was surprised that Deborah should know this. "Yes," he admitted, "I have heard God's command to free my people."

"Ten thousand men from your tribe and the tribe of Zebulun," Deborah said. "And God asked you to draw out the army of Jabin so that he will meet you by the river Kishon with his chariots and his troops."

"I know," Barak said. He looked uncomfortable standing before this stern woman. "But Jabin's army has chariots, and we do not. They have weapons of iron, while we have only wooden bows and arrows. Even our spears are made of wood. And the fearsome Sisera is the general of their army!"

"God said," Deborah challenged Barak with fire in her eyes, "that he will give Sisera into your hand!"

Barak was filled with admiration for her fighting ardor. "If you will go with me, I will go," he said.

Deborah rose from her seat beneath the tree. "I will go with you." She held up a cautioning hand. "But you must know that the road on which you are going will not lead to your own glory. For God will give Sisera into the hand of a woman."

So Deborah went with Barak through the pass in the mountain barrier, across the great Jezreel valley. Together they climbed to the northern hills where the tribes of Naphtali and Zebulun lived.

Ten thousand men of Naphtali and Zebulun now enlisted in Barak's army. And

70 Deborah went with them as they marched south. When they came to the head of the valley, they climbed Mount Tabor and encamped there.

King Jabin of Hazor heard that Barak and Deborah had mobilized an army and were encamped on Mount Tabor. He gave orders to Sisera to charge into battle and wipe out the Israelites.

The Canaanite general called out his chariots, nine hundred chariots of iron, and a large army of fighting men. Sisera's army came rolling into the valley of the river Kishon.

Then Deborah said to Barak, "Up! for this is the day on which God will give Sisera into your hand."

So Barak charged down from Mount Tabor into the Valley of Jezreel with his ten thousand men following him.

It was the time of the rains. Suddenly the clouds opened and poured down torrents of water. The river Kishon overflowed. The weighty Canaanite chariots mired down in the mud. There was a furious battle, and most of the army of Sisera perished.

When Sisera saw that his army was losing he leaped from his chariot and fled on foot. He escaped up into the hills, away from the rain and mud and the terrible slaughter.

After he had gone a good distance, the Canaanite general was so numb with weariness he craved rest more than anything else. He saw ahead a tent. It belonged to a Kenite, and these people were at peace with the Canaanites. Sisera thought that perhaps he could rest and get a drink of cool water there.

The Kenite's wife Jael was home alone. She welcomed the weary fugitive general to her humble tent. But Sisera was too tired to notice the gleam of hatred in her eyes.

Jael spoke reassuringly to him. "Have no fear, my lord."

The defeated general fell, exhausted, on the dirt floor of the tent. Jael covered him with a black goatskin rug.

Sisera raised up his weary head. "Pray give me a little water to drink," he begged, "for I am thirsty."

Jael opened a skin of clabbered milk. She poured some into her best bowl and gave it to him.

Sisera hungrily gulped down the refreshing milk. Then he wiped his mouth with the back of his hand. "Stand at the door of the tent," he ordered, "and if any man comes and asks you, 'Is anyone here?' say, 'No.'"

Then Sisera fell into a heavy sleep. As soon as he had begun to snore regularly, Jael took a tent peg in one hand and a hammer in the other. She tiptoed softly to where the general lay. With all her strength, she drove the peg into his temple till it went down into the ground.

The bloody deed over, Jael stepped outside her tent. At this moment Barak appeared in hot pursuit of Sisera. Jael went out to meet the Israelite leader. She recognized him, and she knew why he came. "Come," she said, "I will show you the man you seek."

Barak entered her tent, and there lay Sisera dead with the tent peg in his temple. It was just as Deborah had said, for Sisera's life had been taken by the hand of a woman.

Judges 4:1 through 5:31

* These northern hills—by a later corruption of the tribal name of Naphtali—were called "Galilee" in New Testament days.

Sometimes the drama of a memorable event was put into poetic form so that it could be chanted or sung and more easily remembered. Usually it was also told and retold as a prose story, embellished or elaborated upon by the creative imaginations of the storytellers. In this section of Judges the prose story of Deborah, Barak, Jael, and Sisera is preserved, as well as the small gem of poetry which follows it. The poem (Judges 5) is probably very close to the original version of the story, since the meter and rhythm of a poem help it to be handed down with few changes. The "Song of Deborah" and the "Song of Miriam" are considered by many to be among the oldest original compositions in our Bible.

Gideon the Captain

In the days of Moses the nomadic desert tribes of the Midianites were friendly to the Israelites. Now, in the days of the judges, things had changed. The Israelites had settled down in Canaan to become farmers wherever they could find a bit of land to plant and to plow. They put their sheep and goats, donkeys and cattle out to pasture. They built houses for themselves of stone or clay bricks.

The roving tribes of Midianites were no longer friendly. They would sweep in from the desert as numerous and unpredictable as a plague of locusts. Like ravaging insects, they swooped down upon the produce of the land. When they left they took every bushel of stored grain and all the fruits of the vineyards. They also drove off with them the Israelite flocks and herds.

The raids were especially frightening because the Midianites now rode camels. Many of the Israelites had never seen these animals before, much less an attacking army mounted on camels. Suddenly, out of the east, fording the Jordan, would come this horde of huge, terrifying beasts. Their

71

72

tremendously long, knobby legs carried the riders so fast, no village could be safe from them.

Whenever harvesttime was near, the Israelites kept looking nervously toward the east. When would the next raid come? Today? Tomorrow?

Everyone was fearful. Some abandoned their homes and moved into the mountains to hide in caves. Instead of threshing their wheat conveniently on a flattened hilltop, where a stiff breeze could blow away the chaff, many tried to thresh wheat secretly, down in their wine vats. Anything, so long as the harvest could be gathered, stored quickly, and hidden before the next raid.

But every raid left the Israelites poorer and more helpless. Many of them were killed or captured. So the people prayed for deliverance.

One day near the end of the harvest season an Israelite named Gideon was down in the stone chamber of his wine press, beating the chaff from his wheat and hoping to hide his work from the marauders. Suddenly he looked up and saw an angel of God standing there in the shape of a man. "God is with you, mighty warrior," the angel said.

"If God is with us," Gideon complained, "then why has all this happened?" He whacked angrily at the pile of wheat with the stick which was his threshing flail. "I think God has cast us off and given us to the Midianites!"

But Gideon was given a command. "You are a strong man," the angel said. "Deliver Israel from the Midianites."

Gideon was astonished. "Why me? My clan is the weakest of all the tribe of Manasseh." He kicked at the fluffy pile of chaff at his feet and said nervously, "And I am the least in my family."

"But God will be with you," the angel assured him. "You shall indeed smite the Midianites!"

Even while Gideon was thinking about his commission to lead the Israelites, bad news came. The Midianites had crossed the Jordan and had swept up the ravines to the Valley of Jezreel.

They were there now, pitching their tents in the same valley where Deborah and Barak once had fought Sisera and the Canaanites. Would Israel ever be free of enemies here in their Promised Land?

Gideon threw down his threshing flail, climbed the hill near his home, and blew a blast on his ram's horn. Then he sent out a summons for the men of Manasseh to rally for battle. Messengers were also sent to the tribes of Zebulun, Naphtali, and Asher, urging all available fighting men to join Gideon's forces.

As soon as his army was assembled, Gideon led his men northward. They encamped by Mount Gilboa. Here Gideon received a surprising message from God. There were too many men in the Israelite army; the campaign might be bungled. Besides, if the army were large, the Israelites might think credit for the victory belonged to them. On the other hand, if victory were achieved with a very small and select fighting force, then everyone would know the victory was God's doing.

So Gideon sent out an order: "Those who are afraid may go home."

Twenty-two thousand returned to their homes. Ten thousand Israelites remained.

There were still too many. So Gideon led them to a nearby spring of water and put the men to a test. They were told to file by one at a time and stop for a drink.

"If a man kneels down and drinks directly from the spring," Gideon told his officers, "he is to be sent home."

Then Gideon pointed to a man who warily leaned over, looking all about him, cupped his hands and brought water up to his mouth. "As our hand-picked men,"

Gideon went on, "we'll keep every fighter who cups his hands to bring the water up to his mouth and drink. A fighter like that will not be caught unaware."

Three hundred men passed the test. So Gideon said to them, "God has promised that you shall rescue Israel from the Midianites."

That night Gideon took his lieutenant and climbed over the crest of Mount Gilboa. There below them in the valley was the Midianite encampment. Torches and campfires burned in the nomads' camp. The warriors and their tethered camels filled the end of the valley.

Gideon and his man crept up to the outpost of the enemy camp and listened to two sentries talking. "I had a dream," one sentry said, "that a loaf of barley bread tumbled into our camp. It came to the tent and struck it. The tent fell, turning upside down, flattened. What do you suppose it meant?"

"That dream predicts defeat for us," the other Midianite whispered, fear in his voice. "That barley bread was a symbol for the sword of Gideon, a man of Israel. Their God has given Gideon a victory over us."

When Gideon heard this, his spirits soared. In the blackness of the night he and his lieutenant hurried back to their own camp.

"Arise!" Gideon called out to his three hundred picked men. "God has given us victory over the Midianites."

So Gideon divided his men into three companies. To each man he gave a torch stuck into a clay jar to hide its light. This was to be carried in the left hand.

"In your right hand you will each carry a ram-horn trumpet," Gideon ordered. "Follow me and do as I do. When I come to the outskirts of the enemy camp, I will blow my trumpet. Immediately you will all blow your horns, smash the jars, and show your torches. Everyone will shout, 'For God and Gideon!' "

So Gideon and the hundred men of the first company climbed over Mount Gilboa. They slipped down into the valley to the very edge of the Midianite camp. It was the beginning of the middle night watch.

The second company of one hundred circled around the mountain one way, and the third company went the opposite way, until the Midianite camp was surrounded by Israelites hiding in the dark.

Suddenly Gideon blew on his trumpet, smashed his earthen jar, and waved his blazing torch aloft. Immediately the other Israelites blasted their trumpets, making a strange and terrifying sound. At the same time jars were broken, the flaming torches were waved, and a mighty shout went up round the Midianite camp: "For God and for Gideon!"

The Midianite camp erupted into confusion. In panic they grabbed up their weapons and, thinking their camp was infiltrated by hordes of their enemies, began hacking with their swords at one another. Through the darkness the surviving Midianites fled toward the fords of the Jordan River, the frightened camels bucking and kicking at the running warriors.

Word was sent out summoning the rest of the fighting men to return. Gideon wanted the entire army to join in pursuing the Midianites back to their desert haunts.

The leaders of the Midianites were captured and slain. Their survivors were too badly beaten to challenge the Israelites again.

And Gideon was now the strongest man in Israel. Throughout his lifetime he kept peace and ruled Israel in the name of God. But after his death there was fighting again for Israel had many enemies.

Judges 6:1 through 8:28

Samson the Strong Man

To the west of Canaan, along the coast of the Great Sea, were another people who were becoming a threat to the Israelites. The people of the sea, the warrior Philistines, enlarged their coastal cities and grew more powerful.

Near the border between the territory of the Israelites and the land of the Philistines, there lived a young man named Samson.* He was handsome and strong, but he was not always a sensible fellow.

From birth Samson had been dedicated as a member of the Nazirites, a group of men whose lives were devoted to God. They vowed to drink no wine or strong drink, to eat no pork or other unclean meat.** No razor was to touch their heads. The hair of a Nazirite went uncut as a mark of his vows.

It happened that Samson fell in love with a young maiden of fully ripened beauty who lived in a nearby Philistine village. Samson decided to marry her despite the fact that she was not an Israelite. So he went down from his highland home along the road that led to the coastal plain. He was on his way to her village to ask her hand in marriage.

As he passed through a thicket near the bottom of the hill, a young lion sprang out at him roaring. Samson was not afraid. He leaped to one side, caught the half-grown animal by its mane, and wrestled it to the ground. With his bare hands he killed it. Then he threw the carcass into the bushes by the side of the road. Samson washed himself in a nearby stream and continued on his way.

In the Philistine village he called on the

plump and dimpled maiden. The girl seemed pleased to be courted by the strong and handsome Israelite. Her father agreed to the marriage.

Some months passed before arrangements for the wedding were complete, but the time finally arrived. On his way down the hillside road, Samson came to the thicket where he had battled the young lion. He looked on the roadside and saw the sun-bleached bones of the lion, right where he had tossed them. Bees had built a honeycomb within the white bones of the ribcage. Samson broke off a handful of the honeycomb and ate it.

In the marketplace of the Philistine village Samson took pieces of silver from the girdle around his waist. He bought a sheep to be roasted and some other banquet food. He ordered it sent to the house of his bride for the wedding feast.

All the maiden's relatives had been invited to her father's house. It was customary for the bridegroom to have thirty of his own companions, but Samson was alone. So the girl's father gathered together thirty young men of the village for the purpose. He warned the young men to keep a close eye on Samson. The people of the village were wary of him.

Samson had mischief in his eyes when he said to the thirty young men, "Let me put a riddle to you. If you can answer it by the end of the wedding feast, I will give to the lot of you thirty linen shirts and thirty fine robes. But if you cannot guess my riddle, you must give me thirty linen shirts and thirty festive robes."

The men talked this over. "All right," they agreed. "Now tell us your riddle."

Samson said,

"Out of the eater came something to eat.

Out of the strong came something sweet."

But the men could not make heads or tails out of this. They tried and tried, but they could not guess the riddle.

The week of feasting and merriment began. The actual wedding ceremony would come at the end.

About the middle of the week, a delegation from the young men came secretly to the bride. "Entice your bridegroom to tell you what the riddle is," they begged. "Have you invited us here to make us poor? We want the rich reward he offers."

The bride looked frightened. "No, no, I cannot," she said.

"Beg him, trick him, anything!" they threatened her. "And if you do not find out for us, we'll burn your father's house down."

So the girl waited until she could see Samson alone. "Tell me the riddle," she teased, wrapping her arms around him.

"No," Samson said. "If they cannot guess, they shall not gain my prizes." He laughed uproariously. "And the riddle is too hard. They will never guess."

"Please, tell me, please," she pleaded, running her soft hands through Samson's long hair. Samson refused again and again. But finally the Philistine maiden got the secret out of him. And she immediately sent word to the groomsmen.

Next day the thirty young Philistines swaggered up to Samson. Their leader said triumphantly:

"A lion is strong—is he not an eater?

Honey is sweet—what could be sweeter?"

Samson turned red with rage. He knew at once how they had learned his riddle, but a promise was a promise. So Samson charged off to Ashkelon, the nearest of the Philistine cities. Swiftly he killed thirty well-dressed men, stripped them of their fine clothes, and sped out of town. He returned to the village and hurled the pile of stolen garments at the feet of the amazed young men. He refused to look at the weeping girl who was to have been his bride. He no longer trusted her, and he wanted nothing to do with her or the wedding.

In hot anger he left the village and took the hill road leading back to the home of his father. And Samson's intended bride was given to the best man as his wife.

After awhile, by the time of wheat harvest, Samson's anger toward the Philistine girl who had betrayed him had cooled. He remembered only how pretty she was. After all, had she not been promised as his bride?

Samson took a young goat from his father's flock and went with his peace offering to the Philistine village. But the father of the girl would not let Samson see her. "I thought that you utterly hated her," the father protested. "Everyone was here for the festivities, and then you went off and left my poor daughter. So I wed her to the leader of the groomsmen." The father smiled uneasily. "Her younger sister is even prettier than she is. Take her instead."

Samson was too angry to listen. He rushed out and set fire to the ripened grainfields around the village. The blaze burned the shocks and the standing grain, as well as the olive orchards. Even foxes fled from the smoldering fields with their tails afire.

Then Samson returned to his home in the adjoining hills.

The Philistine villagers banded together and raided the nearby highland towns of the Judeans. The men of Judah had been trying to live peaceably with their Philistine neighbors. "Why have you come up against us?" they cried out.

"We have come to find Samson and bind him up," the raiders told the men of Judah. "We want to get even with him for the wrongs he did to us. Did you know that he set fire to foxes' tails and let them loose in our standing grain so that all was burnt up?"

So a delegation of men from Judah sought out Samson. "Why do you trouble the Philistines who are rulers over us?"

Samson shrugged. "They hurt me so I hurt them back."

The men of Judah felt this was a poor excuse. "We have come to bind you," they said, "so we may give you to the Philistines."

Samson allowed them to bind him with two new ropes. But when he was turned over to the Philistines, Samson flexed his powerful muscles and the ropes snapped.

Samson saw the jawbone of an ass lying on the ground near him. He seized it and began swinging, killing the Philistines around him with murderous blows. And so the strong man escaped.

Samson seemed unable to resist a pretty face. He fell in love with another Philistine woman whose name was Delilah.

The lords of the five great Philistine city-states saw an opportunity to deal with Samson. They came to Delilah and said to her, "Entice him, and see if you can discover the secret of his great strength. If you can find out how we can overpower our enemy, we will each give you eleven hundred pieces of silver."

This was a fortune! Delilah gave no thought to the fact that Samson loved her. She only plotted how she might win that fabulous reward. She could have a great house, and beautiful clothes and . . .

When Samson next came to see her she whispered in her most flattering manner, feeling the bulging muscles of his arm, "Samson, my love, you are so strong! Is there no way that someone could bind and subdue you?"

Thinking it was a game, Samson decided

to tease her. He told her that if he were tied with seven fresh bowstrings, he would be as weak as any man.

When Samson was asleep, Delilah tied him in just that way. Then she shook him and said, "The Philistines are upon you, Samson!" The strong man awoke and snapped the bowstrings as if they were thread. Then they laughed together, as if it were a great joke.

But inwardly Delilah was furious. "You have mocked me and told me a lie!" she pouted, refusing to have anything to do with him. Then she changed her tactics and became seductive and beguiling. "Tell me, my beloved, how you might be bound."

Samson couldn't help teasing her again. "If you weave the locks of my hair into the web of your loom," he smiled, running his fingers through his long locks, "then I shall be weak."

When he slept again, Delilah securely wove the locks of his hair in and out through the warp strands of her loom. She pinned them in tightly. "The Philistines are upon you, Samson!" she whispered.

Samson leaped to his feet, splintering her heavy wooden loom. He broke out the web of weaving and the pin, and shook his long mane of hair free.

Delilah burst into tears, but from angry frustration. "See, you smashed my loom, and I was only playing games." She wept and hoped he would think she wept for her broken loom. She saw her chance for great wealth slipping from her greedy fingers. "You keep mocking me and telling me lies. How can you say, 'I love you,' when your heart is not with me?"

Delilah became aloof then and told him he couldn't come near her—not until he told her the truth. She sniffled and wept, she teased and she begged, until he was vexed to death with her. In exasperation he finally told her the truth.

"A razor has never touched my head," Samson said, "for I was pledged to be a Nazirite. If I were shaved, I would have no more strength than an ordinary man."

So Delilah sent word to the lords of the Philistines. "Come up at once, for Samson has told me what you want to know."

The Philistine lords arrived with money in their hands for Delilah. While Samson lay asleep, the locks were shaved from his head. "The Philistines are upon you, Samson!" Delilah hissed in his ear.

Samson leaped up, but this time he had no strength. The Philistines seized and bound him, gouged out his eyes, and brought him captive to Gaza. There they shackled Samson to a great millstone in the prison, yoked him like an ox, and forced him to drag the millstone round and round. Shamed and eyeless, he ground the grain for the Philistine prison. But the hair of his head began to grow again.

Now the five lords of the Philistines gathered to offer a great sacrifice to Dagon, their god. They said, "Has not our god

given Samson our enemy into our hands?"

The hearts of the Philistines were merry. "Praise be to Dagon," they shouted. "We have captured the ravager of our country, the slayer of our men. Let us call him into the temple so we can make fun of him."

Samson was brought from the prison in chains to stand between the two middle pillars in the temple. And all the people hooted and hissed and whistled to see their enemy humiliated and blind.

Samson said to the lad who held him by the hand, "Let me feel the pillars on which this house rests, that I may lean against them."

The temple was packed with Philistines. More were on the roof. All were merry with wine and celebration.

Samson felt his strength returning, and he prayed to God that he would be strong enough to avenge himself once more upon the enemies of Israel. Then he grasped the two middle pillars which supported the temple.

"Let me die with these Philistines," Samson said. His muscles bulged, and he shoved with all his might. The roof began to crumble, the pillars cracked and fell, and the whole temple shattered as if torn by a mighty earthquake. The temple of Dagon fell on the lords and all the people who were in it.

So those whom Samson slew at his own death were more than those he had slain during his life. And his brothers came down and found his body. They sorrowfully buried him in the family tomb, back in the hills of Israel.

Judges 13 through 16

* The almost legendary stories of Samson's exploits and adventures caught the fancy of the people of Israel for several reasons. First, he was a man of unique strength and courage. Also, his strength had been dedicated to God. And, most important, he lived at a time when Israel was being oppressed by the Philistines. His own people were delighted to hear of this fine specimen of a man who was able to get the best of their powerful enemies in so many ways. They loved and admired him even with all his faults.

** As used in the Bible, "unclean" has a special meaning indicating that certain foods, objects, or actions made a person unfit to participate in the spiritual life of the community of Israel.

Ruth the Moabitess

It happened that at a time while the judges ruled Israel there was a famine in the land. A certain Israelite of Bethlehem, in the area claimed by the tribe of Judah, could not bear to see his wife and two sons go hungry. He packed up their few belongings. On foot they traveled eastward through the Judean wilderness, crossed the river Jordan, and climbed up the high hills overlooking the Dead Sea on the other side. There in the foreign land of Moab they found a place to stay. There was work for the husband and food to eat.

There the man died, but his wife, Naomi, stayed on in Moab because her sons were now grown and liked it there. Besides, both were in love with Moabite girls and had plans to marry soon.

Ten years passed and Naomi's sons also died, first one and then the other. The grieving mother was left with only her two Moabite daughters-in-law, Orpah and Ruth.

"I shall return now to my own country," Naomi said at last, wiping her tears. "I have heard that the crops have been bountiful in Bethlehem."

Her daughters-in-law, who had shared the same house with Naomi these last ten years, prepared to return with her. But Naomi hugged Orpah and Ruth. "No," she said regretfully, "go, return each of you to your mother's house. May the Lord be as kind to you as you have been to my loved ones and to me." Then Naomi kissed the two young women.

Ruth and Orpah wept. "No," they both agreed. "We have come to love you, and we will return with you to your people."

Naomi reasoned with them. "I am too old to marry again and bear other sons to be husbands to you. And even if I could, should you have to wait until they are grown?" She smiled at them. "No, stay here and marry again." Then she turned her head so they would not see her tears.

Orpah wept, and kissed her mother-in-law good-bye. But Ruth clung to Naomi.

"See," said Naomi gently, "your sister-in-law has gone back to her people and to her gods. You must return home with her."

Ruth's large brown eyes filled with tears. "Entreat me not to leave you or to return from following you; for where you go I will go, and where you lodge I will lodge; your people shall be my people, and your God my God; where you die I will die, and there will I be buried."

When Naomi saw that Ruth was determined to go with her, she said no more. So the two women made the long journey together to the little town of Bethlehem.

Bethlehem rested like a crown on a small hill, surrounded by vineyards and olive orchards and small rock-bordered wheat and barley fields. The small stone house of Naomi was still there in the village, dusty and unused. Naomi and Ruth set to work with straw brooms to sweep out the dirt and debris and cobwebs. A neighbor made them a new wooden lattice for the one small window. Another neighbor supplied them with a large clay water jar. Someone else brought over a stone bowl and pounder to grind their grain. But the two women had no grain to grind except for the remains of a small sack they had brought from Moab.

Now it was the beginning of the barley harvest, and there was a wealthy relative of Naomi's husband who owned a large barley field outside the town. His name was Boaz.

Ruth said to her mother-in-law. "Let me go into that field and glean among the ears of barley, gathering what falls to the ground from the hands of the reapers." * She smiled and smoothed back her lustrous, long black hair. "I may find favor in the eyes of your husband's kinsman."

Naomi kissed Ruth and said, "Go, my daughter."

So Ruth went out to the barley field of Boaz at dawn. She followed behind the men and women who were reaping and gleaned the grain that was left in the field.

Before the sun reached its noonday height, Boaz came from Bethlehem to see how the work progressed.

"God be with you," he said to his reapers.

"God bless you," the reapers answered.

Boaz saw Ruth out in the field picking up stray grains from the ground. "Who is this young woman?"

The head reaper answered, "She is the Moabitess who came back with Naomi. She asked permission to glean, and she has been here since early morning without resting."

Boaz walked between the stubbled rows and stood there in his fine, expensive robes, looking at the young woman.

Ruth glanced up. Her face was flushed and damp from the heat. Her scarf was tied about her waist to form a large pouch for the grain. She thought Boaz a handsome and attractive man, and she lowered her head shyly.

80

"Greetings, my kinswoman," Boaz said to her. "Don't go to glean in another field or leave this one, but keep close to my young women. I have charged my young men not to bother you. And when you are thirsty, go to the vessels and drink what the young men have drawn."

Ruth fell on her knees and bowed her face to the ground in a rush of gratitude. She looked up, her lovely face incredulous. "Why have I found favor in your eyes," she asked breathlessly, "that you should notice me?"

Boaz answered her, "All that you have done for your mother-in-law since the death of her husband has been fully told me, and how you left your father and mother and your native land and came to a people that you did not know before. May the God of Israel be good to you!"

When time for the noonday meal came, Boaz called to the young woman. "Come here, and eat some bread with us, and quench your thirst," he said. So Ruth had all she could eat and some left over.

When the lunch was finished, she went back to her gleaning. Boaz told his young men, "Let her glean even among the sheaves, and also pull out some bundles for her to pick up. And say nothing to embarrass her."

So Ruth collected the spilled heads of barley until it was almost dark. Then she beat out what she had gleaned and found she had a scarf full. She carried the grain back to the village and joyfully showed it to her mother-in-law. Ruth also gave Naomi the leftovers she had saved from her noonday meal.

Naomi ate the food with relish as she listened to Ruth's story of the day's happenings. "It is well, my daughter," Naomi said, brushing the crumbs from her mouth. "You should continue to go out with Boaz' gleaners. In another field you might not be safe."

So Ruth went out every day at dawn and gleaned after the reapers of Boaz in the barley field until evening. When the barley harvest was finished, she had brought home much grain to Naomi for their bread.

Then Naomi said to Ruth, "My daughter, it is time for me to seek a home for you—and I have a plan!"

Ruth sat down beside her mother-in-law and listened obediently.

"Tonight," Naomi said, "Boaz and his servants will be winnowing the chaff from his barley harvest. So wash yourself and make your skin smooth with olive oil and fragrant with perfume. Put on your best clothes and go down to his threshing floor. But stay hidden in the darkness. Don't let Boaz see you until the work is finished. Then Boaz will eat and drink and lie down by the pile of threshed grain, to guard it until the morning. When he is asleep, go and uncover his feet and lie down beside him."

"All that you say I will do," Naomi said, jumping up to get herself ready.

That night she went down to the threshing floor and did just as her mother-in-law had told her. And when Boaz had eaten and drunk and his heart was merry, he went to lie down by the heap of grain.

Ruth waited until she could hear his quiet breathing. She came softly and lay down beside him.

At midnight, the man turned over and was startled to find a young woman there. "Who are you?" he asked sleepily.

"I am Ruth," she answered softly. "I am your next of kin."

Boaz knew that according to the laws of Moses the nearest relative to a dead husband had the first right and duty to marry the widow.**

"You are a woman of worth," Boaz said, with admiration in his voice. "You have not chased after the young men in Bethlehem,

poor or rich. It is true I am a near kinsman. Yet there is a kinsman nearer than I, and he has the first right to marry you. If he will do the part of the next of kin for you, let him do it. If he is not willing, as God lives I will do it. Now lie down until morning."

So she lay at his feet until morning. At the first glimmer of dawn she arose and returned to Naomi, carrying a gift of barley.

"These measures of barley he gave me," Ruth said. "For he told me, 'You must not go back empty-handed to your mother-in-law.'"

Naomi listened to the story of all that had happened and nodded with satisfaction. "Wait, my daughter," she said, "until you learn how the matter turns out. For the man will not rest, but will settle it today."

That same morning, Boaz put on his best robes, carefully combed his hair and his beard, and walked to the gates of the town. There were benches in the wide place by the gates. Here, as in every Israelite town, the elders sat, giving judgment on the problems and disputes of the people of the city. Here legal matters were settled.

Boaz sat on one of the benches, and before long the kinsman he was watching for came by.

"Turn aside, friend," Boaz called out, indicating the bench beside him. "Sit down here."

The man sat down. Then Boaz beckoned to ten of the elders of the town. "I want you to be witnesses in this case," he said. He turned to the man. "As you know, Naomi has returned from the country of Moab. Now she wishes to sell the parcel of land which belonged to her husband—our kinsman. As the nearest relative, you have first right to buy it. But if you do not, my turn comes after you."

"I will buy it," the relative said, without hesitation. He began to reach into the folds of the cloth girdle, which belted his waist to find his money.

Boaz held up his hand. "The day you buy the field from the hand of Naomi, you must claim as your bride Ruth, her daughter-in-law from Moab. Only in this way can the name of the dead be carried on."

The next of kin changed his mind. "I cannot buy Naomi's land under those circumstances. Take the right yourself."

Then Boaz said to the elders and the people who stood around, "You are witnesses this day that I have bought from the hand of Naomi the land that belonged to her dead husband and sons, who were my kinsmen. I have also bought the right to take Ruth the Moabite for my wife so that through her there will be sons to remember the dead."

So Boaz took Ruth, and she became his wife and bore him a son.

"Blessed be God," the women of Bethlehem said to Naomi, the happy grandmother, as she cradled Ruth's baby in her arms. "This child shall be to you a restorer of life and a nourisher of your old age. For your daughter-in-law who loves you, who is more to you than seven sons, has borne him."

And the baby of Boaz and Ruth grew up to become the father of Jesse, who was the father of the great King David.

Ruth 1 through 4

* There was a provision in Hebrew law which allowed poor people to glean in the fields after the reapers had passed. This meant that they could pick up the grain which the reapers had missed. See Leviticus 19:9-10.

** Under Mosaic law the brother or other male kinsman of a man who had died was obligated to take his wife and have children by her so that the name of the dead man would not disappear in Israel. See Deuteronomy 25:5-10.

PART V
THE GOLDEN AGE

The books of Samuel (originally compiled as a single book) tell the story of the rise of a monarchy in Israel. We see a kingdom developing under Saul with a loose organization of the twelve independent tribes. Then the great King David welds the kingdom into a single powerful nation.

The opening chapters of the first book of Kings tell of the splendor of this unified monarchy under Solomon and its collapse under Solomon's arrogant son.

These stories will be told in Part V.

In the earliest Genesis tales of creation and the flood, we have myths which often were frankly adapted from the myths of even older cultures. Yet these stories embody the noble conceptions of God held by the early Hebrews.

The first stories of the Hebrew people, of Abraham, Isaac, Jacob, and Joseph, have in them more legend than fact. They recall the ancient memories of the nomadic days and show how the first ancestors of the Hebrews trusted in God to direct their affairs.

The later Moses stories and those of the book of Judges were closer in time to the people who wrote them down from the varying, lively oral versions. They are a fabric interwoven of history and legend, often lovingly embroidered with dramatic details. But these stories seem to have grown up around actual and outstanding persons.

In the books of Samuel, however, we come for the first

84 time to sections of biography and court history which were recorded soon after the happenings. Many of the principal figures and those who knew them were still alive; the writer himself may have been either a participant or an eyewitness.

In fact, no other culture of the time—or for centuries afterward—had such vivid, honest, realistic biographical reporting as we have in the original stories about King Saul and King David. They are fresh and unique.

King Saul and King David reigned about 1000 B.C. The story of the prophet-priest Samuel will begin a few years before this time.

Samuel, Prophet and Judge

In the days before David became king, Jerusalem was a walled city occupied by a Canaanite people called the Jebusites. There was a road which wound north from Jerusalem, along the mountain ridges, to the valley where the village of Shiloh nestled. Here it was that Joshua had pitched a tent for the ark of the covenant, the sacred box the people had carried with them through all their wandering.

Near the end of the time of the judges, the tabernacle, with the ark inside, was still there. Additions had been made, and there was now space to house the priest and his family who lived there.

It became the custom for the people round about to make a pilgrimage to this temple at Shiloh each year to worship and to sacrifice to God. Eli was now the priest, and his two sons assisted him in serving.

There was a certain man in the hill country of Ephraim who had two wives. The younger wife had children, but Hannah, the older wife, remained childless.

Now this Ephraimite used to take his two wives and his children, year by year, and travel to Shiloh to worship in a family festival. He brought with him a lamb or a goat for his sacrifice.

On the day of the sacrifice a fire would be built on the great outdoor altar. The priest would kill the animal, and the fat and certain other portions would be burned as an offering to God. The rest of the animal was boiled or roasted, and this provided meat for the family feast, except for a joint which was presented to the priest.

The Ephraimite gave the best and most generous portions of food to the wife who had borne him so many children. Although he loved Hannah, she received a smaller portion. Her rival jeered at Hannah's childlessness. Hannah wept and would not eat.

Her husband said, "Hannah, why do you weep and why do you not eat? Why is your heart sad? Am I not more to you than ten sons?"

But Hannah could only weep. Later she slipped alone inside the temple, dropped to her knees, and prayed. "Oh, God, if you will give me a son to stop my grief, I will dedicate him to your service for his whole life!"

Eli the priest heard her prayer. "Go in peace," he told Hannah. "May the God of Israel grant your petition."

Hannah rose and wiped the tears from her eyes. Already she felt better, and the terrible sadness was lifting from her heart.

In due time Hannah gave birth to a son. She was radiant. The child was named Samuel. The delighted mother nursed her son for the customary three years, and then brought him to Eli. Samuel was to grow up serving in the temple at Shiloh.

When Samuel was about twelve years old, he had a strange experience in the middle of the night. While asleep in the temple, he heard a call, "Samuel, Samuel!"

He opened his eyes and ran to the room where Eli slept. "Here I am," he said sleepily. "You called me?"

"No, I did not call," the old priest said. "Lie down again."

Samuel went back to his mat on the ground by the ark and lay down. Then he heard the same voice, "Samuel!" He rose and ran back to the priest, but Eli only said, "I did not call. Lie down again."

Again Samuel fell into uneasy sleep. Then he heard the voice a third time. Puzzled, the boy ran back to wake Eli. "I know you called me," he said.

"Go lie down," Eli said gently to the boy. "If you hear the voice again, answer, for it must be the voice of God."

When the boy did hear the voice again, he knew it was God's voice. And he listened.

This was the beginning of a closeness between Samuel and God which lasted all his life.

And it came to pass in those days that the Philistines gathered for war against Israel, and Israel went out against them to battle.

In the first skirmish the Philistines drew up in a line against Israel. When the battle spread, the well-armed Philistines made great slaughter among the Israelites.

As the weary Israelite troops straggled back to their camp, wounded and defeated, the elders said, "Why has God let us down today? Let us bring the ark of the covenant here from the temple of Shiloh. If we carry the ark into battle with us, perhaps God will save us from our enemies."

The two sons of Eli brought the ark from Shiloh.

The next day the Israelites charged again into battle with the Philistines, with the ark in their midst. But the Philistines outnumbered the Israelites and were superior in training and weapons. The Philistines fought with iron chariots and strong weapons of iron, and were skilled in strategy. The Israelites had only bows and arrows, clubs and spears of wood, and a few bronze swords. And they were poorly organized.

There was great slaughter. The Israelites were defeated, and they fled, every man to his house. The ark of God was captured by the enemy. And the two sons of old Eli lay dead on the battlefield.

When a runner returned to Shiloh with the news, he found Eli sitting on a stone bench by the road.

"What is the uproar?" the old priest called out, for he was blind in his old age and could not see.

"There has been a great slaughter, and Israel has fled from the Philistines," the runner panted. "Your two sons are dead, and the ark of God is captured!"

When he heard this, old Eli began to clutch at his chest and gasp for breath. He fell backward off the bench, and in the fall his neck was broken, and he died.

The Philistines did not keep the ark for

long. Soon afterward, mice began to over-run their cities; next a plague with boils broke out among their people. The Philistines decided that the ark must be bringing them this bad luck. So the ark was returned to the border on an oxcart. There the ark remained, in a small house on a hill, almost forgotten for the next twenty years.

When Samuel became a full-grown man, he was the most important religious leader in Israel. After a time the people came to Samuel, their priest, and complained, "The Philistines continue to fight us, and we are their slaves. We need a king to gather our tribes into one strong nation! Appoint for us a king, who will lead us against our enemies."

At first Samuel refused. One of the special things about Israel was the fact that God himself was their king. They did not need a human one as other peoples did. The people continued to beg for a king, however, so at last Samuel agreed. "I shall find you a king," he said. "Go, every man to his own city—and wait."

1 Samuel 1 through 8

Saul the First King

In the tribe of Benjamin was a handsome man named Saul. He was so tall, he stood head and shoulders above most men. Saul and his family lived in the house of his father, a wealthy farmer, in the hilltop village of Gibeah.

One day Saul's father found that some of his donkeys were missing. "Take a servant with you," he instructed his son, "and find where they have strayed."

For several days Saul and the servant searched through the countryside. But they couldn't find the lost animals.

"Come, let's go back," Saul said at last, "before my father stops worrying about the donkeys and becomes anxious about us."

"Wait," his servant suggested. "We are close to the town where the wise man Samuel lives. He knows many things, and all that he prophesies comes true. Let's speak with him."

Now the day before, God had spoken to Samuel in a dream, "Tomorrow I will send you a man from the tribe of Benjamin. You shall anoint him king over my people to save them from the Philistines."

So when Samuel saw this fine specimen of a man approaching his house with a servant, he knew that this was the future king of Israel. Samuel invited Saul and his servant to have a meal with him and to rest for the night. At daybreak Samuel took oil and secretly anointed Saul. "You shall be the king, and reign over Israel," the priest said solemnly. "You shall save Israel from our enemies round about."

Then Samuel issued a call for the elders and representatives of the tribes to meet together to determine by drawing lots who would be the king. The first lots showed that the king would come from the tribe of Benjamin. The second drawing fell on Saul, and Samuel presented the young man to the people.

There were a few among the people who complained because they felt Saul was inexperienced in leadership and the art of war. There was some murmuring. However, Samuel sent the people away, and Saul returned to his home. Everyone waited to see what would happen.

Then a challenge came that gave Saul a

chance to show his ability as a leader in battle.

The Ammonites, far to the east, were restlessly pushing in on the Israelites across the Jordan. Now the Ammonite king brought his army and surrounded the walled town of Jabesh in Gilead.

The men of Jabesh saw that they could not hold out long against the great army of the Ammonites. They stood on the city wall and called down to the besieging enemies, "Make a treaty of peace with us; we will serve you."

But the Ammonite king came up to the front ranks of his army and shouted back, "This is the condition on which I will make a treaty with you—that I gouge out your right eyes. Thus will I disgrace all Israel."

At this news the elders of Jabesh consulted together for a little time. Then they called down, "Give us seven days' respite. Give us a week's time to send messengers through all the territory of Israel. If there is no one to save us, we will give ourselves up."

The Ammonites agreed. What did they have to lose? The twelve Israelite tribes often fought among themselves. A single tribe would be no threat to them. And the Israelites had no leader who could command the allegiance and loyalty of all the tribes together. The Ammonites were sure such a call for help would go unanswered. And so the messengers were allowed to run to the tribes of Israel.

When the messengers came to Gibeah, in the territory of Benjamin, the townspeople wept to hear the news. Many of them had relatives in the besieged city of Jabesh. But what could the Benjaminites do? They were the smallest of the twelve tribes.

Now Saul was coming from plowing his field with oxen. Saul heard the weeping and asked, "What ails the people?"

When he was told about the message from Jabesh-gilead, Saul's face grew red with anger. He knew that the priest Samuel would want him to assert his kingship now.

With sudden decision Saul took his two oxen, killed them, and cut them into pieces. He sent messengers bearing the bloody pieces to the tribes and cities of Israel, saying, "Whoever does not come out after Saul and Samuel, so shall it be done to his oxen!"

The people were filled with dread. The men of Israel promptly mustered into Saul's army. They marched eastward to the Jordan, forded the river, and climbed the hills into Gilead. During the night Saul divided his army into three companies. In the morning watch they swooped down upon the surprised Ammonites. There was great slaughter of the enemy, and the survivors retreated back to their own country.

This was only the beginning of many successful battles fought by Saul and his Israelite army against their enemies. The twelve tribes provided men for the army, and Saul was recognized as their first king.

Saul built a fortress of rough-hewn stones in his native town of Gibeah. There was a large dirt-floored common room, with a fireplace for cold days and nights, where Saul could hold court in winter. In summer he usually held court under a tamarisk tree on the hill.

Upstairs there were sleeping chambers. There were also storage rooms for shields, bows and arrows, and spears. The Philistines had a monopoly on weapons of iron; the men of Israel had to make do with mostly wooden weapons.

The fortress was too crude and rustic to be called a palace, but it served as King Saul's headquarters.

Saul had several sons. The oldest was Prince Jonathan, who was as brave and fearless a fighter as his father. They fought side by side against the oppressing Philistines. Sometimes King Saul divided the

Israelite army and put one flank under the leadership of Prince Jonathan.

Once, accompanied only by his armor-bearer, Jonathan captured a Philistine garrison. He climbed down a steep ravine near the Israelite encampment; with only jagged rocks for footholds, he climbed up the sheer cliff on the other side. He and his armor-bearer fell upon the sentries and the guards with such sudden ferocity the few survivors panicked and fled. They thought that these fierce fighters were the vanguard and that the whole Israelite army must be right behind!

There continued to be hard fighting against the Philistines all the days of Saul. Yet their seacoast enemy was never completely conquered. The Philistines would be repulsed here, make some gains in another place, then be temporarily pushed back somewhere else.

The Philistines were not the sole enemies. There were difficult wars to be fought on the eastern side of the Jordan against the Ammonites and the Moabites. And there were always nomadic raiders from the deserts beyond—especially a tribe of fierce fighters called the Amalekites.

The old priest Samuel said to Saul, "God sent me to anoint you king over Israel, so you must listen to me. God has spoken to me about our enemy, the Amalekites. You must go to war against them, and utterly destroy them, and all that they have. You must not spare a single Amalekite! Kill every man, woman, and child, and all of their animals!"

So Saul marched south with his army to battle. He defeated the army and captured their king. Every man, woman, and child was put to the sword. But Saul admired the prowess of the Amalekite king and spared his life. Also, Saul did not see the sense of destroying fine animals. So he saved the best of the sheep, oxen, and fat young cattle.

When Samuel heard this, he was furious. King Saul had disobeyed!

When they met, King Saul bowed to the priest. "Blessed be you," Saul said, then added proudly, "I have performed my duty."

Samuel's face was hard. "Then what is this bleating of sheep in my ears, and the lowing of oxen which I hear?"

Saul said, "Oh, we spared the best of the animals, to sacrifice before God. And I brought back Agag, the king of the Amalekites, but the rest are utterly destroyed."

"Stop!" Samuel said angrily to Saul. "Why did you not obey the commands of God?"

So Samuel sent for Agag, the Amalekite king, and hacked him to pieces before the altar, saying, "As you dealt with the women and children of Israel, so do I deal with you."

As Samuel turned to go away, King Saul was distressed and grabbed at the old priest's robe to detain him. The robe tore.

Samuel said to him, "So God will tear the kingdom of Israel from you and shall give it to a better man." And Samuel returned to his home, and did not see King Saul again during his life.

After this King Saul became depressed and moody. He no longer believed that God was with him. His heart was not in his battles. He had lost his fire and his enthusiasm.

The men in Saul's court were worried about their brooding king.

"Look, an evil spirit is tormenting you," said Abner, who was a cousin of Saul, and also the general of Saul's army. "Let us seek out a man who is skillful in playing the lyre. Perhaps the gentle music may be soothing."

So Saul, in his misery, agreed. "Find a man who can play well and bring him to me."

A young soldier from the tribe of Judah

was in the court. "In my hometown of Bethlehem," he suggested, "I know a young man who can play well upon the lyre. He also sings songs of his own making."

"Who is this man?" Saul moaned, holding his fists to his throbbing temples.

"He is David, the son of Jesse," the young soldier said. "And he is prudent in speech, a handsome man of good presence, a man of valor and a man of war!"

Saul's head hurt, his eyes burned, and he felt sick and miserable. "Send for him," he ordered.

So messengers were sent to Bethlehem, in Judah. They found David in the hills, watching over his father's sheep. David left, with his father's blessings, to enter into service at Saul's court.

David was a handsome young man, ruddy-faced from his outdoor life on the hills, with flashing and intelligent eyes and a reddish glint to his dark hair. He was so openhearted, thoughtful, and full of good humor that everyone who knew him was drawn to him.

Saul loved this charming young man greatly; he made David his court musician and his armor-bearer. So the king sent word to Jesse of Bethlehem, saying, "Let your son David remain in my service, for he has found favor in my sight."

After this whenever the melancholy and sickness of spirit was upon Saul, David took his lyre and played it so that Saul would feel temporarily refreshed and relieved.

King Saul never saw the old priest Samuel anymore. Saul did not know what old Samuel knew in his heart: that this David of Bethlehem would become the next king of Israel.

1 Samuel 9 through 16

David and Goliath

Now the Philistines gathered their armies yet again for battle. They marched up into the southern part of Judah in the hill country. They camped on a low mountain overlooking a valley.

King Saul called up his army and marched south. The Israelites set up their camp on the oak-covered hills on the opposite side of the valley.

Out from the camp of the Philistines swaggered a champion warrior, Goliath by name. He was a giant of a fellow who towered above the other Philistines. He had a huge bronze helmet on his head. He was protected with a mighty coat of mail. And he carried an enormous iron-tipped spear and a sharp sword.

This giant came to the bottom of the valley, and shouted up to the ranks of Israel, "Why are you getting lined up for battle? Just choose a man for yourselves, and let him come down to me. If he can kill me, then we will be your slaves. But if I win, if I kill your champion, then you shall be our slaves!"

When King Saul and the Israelite soldiers heard this, they did not know what to do. How could any Israelite beat this huge armored warrior?

But David came up to King Saul and said, "I will fight this boastful Philistine."

The king looked at the young man, astounded. "How can you?" Saul asked. "You know nothing about fighting. And that Philistine giant is a seasoned warrior!"

David answered, "I used to keep sheep for my father. When a bear or a mountain lion came to seize a lamb from the flock, I went after him and killed him. If God has saved me from the teeth of a lion and the mighty claws of a bear, he can surely save me now."

"Go, then," the king said, "and God be with you." Then he gave David his own armor to wear, for David had none of his own. David strapped on the king's heavy breastplate and the bronze-and-leather leg protectors. Someone set the king's large bronze helmet on David's head and handed him the king's great sword.

David tried to walk, but he staggered under the unaccustomed weight. Also, Saul was an unusually big man, standing head and shoulders higher than David.

"I cannot go in the king's armor," David said. "I must do this my own way."

He took off all the armor. Wearing only his short tunic girded up at the waist, he marched down the hill with his wooden staff in one hand, and carrying his woolen shepherd's sling in the other.

When he reached the brook in the valley, he stopped to choose five smooth stones and put them into the leather shepherd's bag he wore by his side.

Goliath saw him and came closer, clanking his heavy armor.

"Am I a dog," the giant bellowed, "that you come at me with sticks? Come, I will give your flesh to the vultures and the jackals!"

David took a stone from his leather bag, fitted it into the wide place in his woolen sling, swung it round and round his head, and then let go of one end. The stone spun out with deadly accuracy. It struck the Philistine on his forehead. He stumbled and fell to the ground, stunned.

David ran up and grabbed the Philistine's sword, stabbed him, and cut off his head.

When the Philistines saw that their champion was dead and that the Israelites were charging down into the valley with a new excitement and fervor, they panicked and began to retreat.

David took out after them, followed by King Saul, Prince Jonathan, and the Israelite army. Many Philistines were killed, and the rest retreated temporarily back to their coastal cities.

1 Samuel 17

David the Outlaw

Saul's army marched homeward from their victorious battle with the Philistines. Word of the victory had quickly spread throughout the land. Everyone wanted to welcome the returning heroes. As the army passed through the Israelite cities, the women came out tapping and shaking their tambourines. "Saul has slain his thousands," they sang, "and David his ten thousands!"

This made King Saul very angry. He looked behind him, where David and Prince Jonathan marched side by side. Quite obviously the two young men had become the closest of friends.

"They have credited David with the slaying of ten times what I have done," Saul complained to Abner, his cousin and general of his army. Saul spat into the dust with contempt. "What more can David have— but the kingdom?"

And so Saul's jealousy of David began. It grew and grew until it became an obsession.

The next day they reached Saul's fortress home in Gibeah. It soon became apparent that the black mood was again upon the king. Saul strode back and forth in the large common room, his spear in his hand, complaining of a headache and raving like a madman. David sat on a bearskin rug on the floor, his lyre cradled on his knees, softly strumming a gentle melody to soothe the tortured king. Suddenly the king

92

whirled and threw the spear at David with all his might. It narrowly missed him. David paled, but he went on playing.

The next day, when the strange mood and the illness had temporarily eased, the king acted as if nothing had happened. Everyone thought it was a fit caused by an evil spirit and that Saul in his right mind would never want to hurt the young hero.

But Saul was insanely jealous of David and morbidly fearful of the young man's popularity. It enraged him that Prince Jonathan and the Princess Michal seemed to be so fond of the handsome young hero.

One day, when David's place was empty at the king's table, Saul roared, "Why has the son of Jesse not come to the meal?"

Jonathan explained that David had asked leave to attend his family's annual sacrifice in Bethlehem.

Saul turned his anger toward Jonathan. "You are a perverse, rebellious son! You have chosen the son of Jesse and turned away from me, to your own shame. Do you not know that as long as David lives upon the earth, neither you, nor your kingdom, shall be established? Now hear this: the young upstart shall surely die!"

"I have not rejected you—so why should my best friend be put to death?" Prince Jonathan angrily challenged his father. "What has David done? Is he not our country's hero? Should he not be given a worthy position, befitting a hero?"

So Saul begrudgingly commissioned David as the commander of a company of a thousand men and sent him out to do battle. Anything to keep him away from the royal family! David might kill some more Philistines for Israel. And there was also the chance that he might be killed and thus be out of the way.

But David was successful in every campaign. He became known and admired throughout the land.

Saul's daughter Michal also loved David, and David returned her love. Finally Saul realized that he could not stand in the way of their marriage. He gave his consent.

Yet if David became a member of the royal family, he would be in a better position to claim the throne. The king made up his mind to do away with him. King Saul gave a secret order for his guards to kill David in the morning when he came out of his house.

But Princess Michal heard of the plot. She warned her husband, "If you do not escape tonight, tomorrow you will be murdered!"

So Michal let David down through a back window, and David escaped in the middle of the night. He had no difficulty rounding up a band of loyal soldiers who had served under him. Together they escaped from Gibeah.

Nearby on a hilltop was the sanctuary of Nob where the priests served the people of the court. The soldiers hid in a ravine while David knocked on the door of the temple. A young lad came to the door and recognized David.

"My name is Abiathar," the boy said in an awed voice. "How may I serve my lord?"

"Who is in charge here?" David asked.

"My father is the priest here," the boy said. "And my uncles and older brothers and cousins are priests also." Then he ran to rouse up his father.

The priest trembled with an unknown fear when he saw David. "Why are you alone, and no one with you?"

"I am not alone," David said. "Some soldiers are with me, and we are on a secret mission for the king. But we need provisions. And I left in such haste, I did not bring a spear or a sword."

The boy Abiathar helped his father bring holy bread from the altar. Then the old priest brought a sword wrapped in a cloth.

"It is the sword of Goliath the Philistine, whom you killed. There is no other here."

"There is none like that," David said. "Give it to me."

So David took the supplies and rejoined his men. They fled to the caves and ravines of the wilderness.

When King Saul heard that the priests of Nob had helped David, he did not wait to find out whether they had done so innocently or not. In a rage he ordered all the priests of Nob massacred. Only the boy Abiathar escaped.

Abiathar searched until he found David in his wilderness stronghold. He told of Saul's cruel and angry order and the terrible, bloody massacre of his relatives.

David put his arm around the shoulders of the frightened lad. "I have been the cause of the death of those in your father's house. Stay with me and fear not. He that seeks your life seeks my life also. With me you shall be safe."

So Abiathar stayed with David. The boy was young, but he had been trained for the priesthood. And so Abiathar became David's priest, and also priest to the growing outlaw band.

More and more men came to join up with David and escape the unpredictable and often dangerous king. Some of those who came were restless, others fled from debts, and a few were fugitives from the law. David ruled them all with supreme tact and an iron discipline. They were fiercely loyal to him, to a man.

Life in the wilderness was hard for David and his men. It was difficult to get food for so many. They were forced to move their camp often, and always they must be alert, lest their hideout be discovered.

One day several men of a tiny village in the hills of southern Judah hiked north to the headquarters of King Saul. "We know where the outlaw army of David is hiding in the wilderness," they said, hoping for a fat reward.

So Saul called to his cousin Abner, commander of the army. "Come, muster up our army," the king ordered. "This time we shall surely capture and kill that upstart."

Saul and Abner marched to the wilderness with three thousand men and camped by a deep ravine. They did not know that David and his men were hidden in the mountains on the other side.

"When it is the middle of the night, I will cross the ravine and go into the camp of Saul," David told his men, "but only two of you may go with me."

That night, with only the faint glow of starlight to guide them, David and two of his men climbed as quietly as they could down into the steep gorge and made their way up the sheer cliffs on the other side. Here they tiptoed through the sleeping encampment, past sentries dozing by the ruddy embers of dying campfires. They found Saul on the ground, rolled in his cloak like the rest of his soldiers. His spear was stuck in the ground at his head, and his cousin Abner lay snoring beside him.

One of those with David whispered, "God has given your enemy into your hand this day! Let me pin him to the earth with one stroke of the spear, and I will not have to strike him twice."

But David looked with sorrow and pity at the sleeping king. In spite of all Saul had done, David could not find it in his heart to hate this man who once had been kind to a shepherd youth. Besides, Saul was his king and the father of Princess Michal and the devoted Prince Jonathan.

"Do not destroy him," David sighed. "The day will come when he shall fall in battle, and die."

David took only the spear and the water jar that were by Saul's head, and he and his men tiptoed back through the dark and quiet camp. No man saw or knew, for they were all asleep.

It was close to dawn when David and his lieutenants, having climbed down through the ravine, finally reached the top on the opposite side. Then David cupped his hands to his mouth and lifted his powerful voice over the great gap between them and the Israelite encampment.

"Abner!" he called. "Will you not answer, Abner?"

Against the paling early-morning sky David could make out the small silhouettes of men rising from the ground on the top of the other hill. Then one tall, sturdy figure strode to the edge. David recognized the husky voice of Abner.

"Who are you," Abner asked, "that calls to me?"

"Are you not a man?" David taunted the commander. "Who is like you in Israel? So why have you not kept watch over the king, your lord and cousin? For someone came in to destroy King Saul! As God lives, you deserve to die because you failed to keep watch! And now look, see where the king's spear is and the jar that was at his head."

There was amazement in the encampment on the other side. More men were rising and running about. Then the king, head and shoulders above the rest, strode to the rim of the wide ravine.

"Is that your voice, my son David?" Saul called.

David answered, "It is my voice, my lord."

They both stood there, straining to see in the early-morning light. The hush was broken by the stirring and twitter of hidden desert birds.

Then David shouted, "Why does my lord pursue me? What have I done? What guilt is on my hands? Why does the king of Israel come to seek my life, as a hunter seeks a partridge in the mountains?"

There was a pause, and the great shoulders of the figure across the ravine seemed to sag. "I have done wrong," Saul at last called back, his voice breaking. "Return, my son David! I will no more do you harm because my life was precious in your eyes this day. I have played the fool."

David held up Saul's spear. "Here is your spear, O king! Send one of your young men to fetch it. May God watch over my life as I have watched over yours this day!"

"Blessed be you, my son David," Saul called back, raising his hands as if in benediction. "You will do many things and will succeed in them."

So David and his lieutenants went back to their own encampment, and Saul's army packed up their gear and returned north to Gibeah.

But David said in his heart, "I will perish one day by the hand of Saul. He is possessed by an evil spirit. How can I trust him? It is better that I escape to the land of the Philistines. Then perhaps Saul will give up trying to hunt me down, and I shall escape out of his hand."

So David arose, with the six hundred men in his band, and with the two wives he had married during his years in the wilderness. They all traveled westward across the Judean border into the land of the Philistines.

And when it was told Saul that David had left the wilderness, had fled outside the borders of Israel, the king sought for him no more.

1 Samuel 18:1 through 27:4

David and the Philistines

David went to the great walled city of Gath. His army camped outside the city while David sought an audience with Achish, the Philistine lord who ruled Gath.

David gave Achish presents of sheep and fine garments. These had been stolen from other Philistine villages, but, of course, he did not mention how he had obtained the gifts.

"If I have found favor in your eyes," David said to the Philistine lord, "let a place be given me and my men in one of your border towns. The king of Israel is our enemy too, for he seeks to kill me. Perhaps we can be useful to you."

Achish thought this over. He knew it was true that King Saul regarded David as an outlaw and wanted him dead. That should cancel out David's loyalty to the Israelites. Also, after the hard life in the wilderness, David's men had become seasoned fighters. They could help protect the Philistine border against raids from the Israelites, to say nothing of the Amalekites and other desert raiders from the south.

Achish was also drawn to David because of the warmth and friendliness of his personality. It was difficult for anyone not to like David.

"I will give into your hands the border town of Ziklag," Achish decided. "You and your men and your families shall have a secure place where you can put down roots and become one with us." The Philistine lord raised his hand. "And—you shall bring me tribute from any raids you make against our common enemy, the Judeans, and the other Israelites."

So David and his men and all their families settled down in the Philistine border

town of Ziklag. They had no intention of becoming idol-worshipers like the Philistines. Secretly they yearned after the God of Israel. They looked forward to the time when they would be free to worship him on their own soil.

In the meantime David made the best of his situation in Ziklag. He wanted to keep in the good graces of his own countrymen, especially the people of Bethlehem, Hebron, and the other towns of Judah.

So David and his men made many daring raids on the Amalekites and other tribes in the southern deserts of Edom. These tribes had long been a scourge and a threat to the northern tribes of Israel, but even more to the people of Judah.

In these raids David and his men would gather together the sheep, oxen, donkeys, and camels, the garments and other booty and distribute most of it among the needy in the towns of Judah. A small portion of the booty he would take back to Achish, in Gath.

When Achish asked, "Against whom have you made a raid today?" David would say, "Against such-and-such a town in Judah."

Achish trusted David, thinking, "He has made himself utterly hated by his own people. Therefore he shall be my servant and my vassal always!"

But David was making himself loved in his own tribe of Judah.

Several years after this the Philistines began to gather their forces together for an all-out war on the Israelites. Achish called David into his audience hall and said to him, "You and your men will go out with me to war against Israel."

David made a guarded reply. "Very well, you know what your servant can do," he said.

"Good," Achish nodded, completely taken in by David's charm. "I will make you my bodyguard for life."

David was thinking to himself, "I can never fight against my own people. But I shall have to plan how to get out of this one step at a time."

Through his spies among the Philistines, David found out that Saul and Abner knew of the Philistine war preparations. The entire Israelite army had been mustered and was encamped by Mount Gilboa in the Valley of Jezreel. They were massing for a last-ditch battle to defend their homeland.

Then an order came to Ziklag, from Achish of Gath: "Come and join the men of Gath. The Philistines are ready to march."

David's six hundred men left their families in Ziklag and followed David to Gath. Many of the men muttered rebelliously. "Is David mad? Does he expect us to follow him to battle against our own countrymen?" Others said, "Wait and see. Perhaps David will find a way to turn this to our own good." But David was wondering desperately what to do.

At Gath, Achish welcomed David warmly. "We march north at dawn. You and your men will act as the rear guard for the division from Gath. We will meet in the Valley of Jezreel with the divisions from our other cities."

When the army of Achish and David reached the great Valley of Jezreel, Achish joined the other lords of the Philistines. They stood on a small hill with the commanders of their combined armies and reviewed their troops. When the men from Gath passed by, the commanders saw David and his men bringing up the rear.

"What are those Hebrews doing here?" one of the commanders yelled, his face dark with anger. "Are we not here to fight the Hebrews?"

Achish said, "Do not fear. This is the Israelite David, who has been with me now for days and years." He smiled confidently,

"And since he deserted to me, I have never found fault with him!"

But the commanders of the Philistine armies were angry with Achish. One said, "Send the man back! Let him and his men return to the place where you have assigned them." Another said, "Shall we have Hebrews in our midst to turn against us and knife us in the back while we are fighting?"

Another commander added, "How can you trust him? Is this not the David of whom they sing in Israel:

'Saul has slain his thousands,
And David his ten thousands'?

Do you not realize they sing of slain Philistines?"

The other lords agreed, and so the order was given. Achish went to David and said apologetically, "You have been honest, and I still trust you. To me it seems right that you should march out with us in this campaign. Nevertheless, the lords and commanders do not approve. So go back now. And go peaceably so you will not displease the lords of the Philistines."

David felt a great relief. He was saved from a painfully difficult situation, and he still had the trust of Achish. But David pretended anger. "But what have I done?" he blustered. "What have you found in your servant from the day I entered your service until now that I may not go and fight against the enemies of my lord?"

Achish was taken in by David's pretensions of anger. He tried to pacify the man he thought was his loyal friend. "I know you are blameless, my friend," Achish said, laying his hand on David's shoulder. "Nevertheless, these commanders have issued the order: 'He shall not go with us to battle.'" Then, not wanting David around to cause trouble, Achish urged, "You and your men rise up early in the morning and depart for Ziklag as soon as you have light. Remember, your wives and your little ones will be waiting to see you all safely back."

So David set out with his men early in the morning, to return to the land of the Philistines.

On the third day David and his marching men came over a rise and looked ahead expecting to see the familiar walls of Ziklag. What they saw made them gasp with surprise and horror. The walls were blackened and smoke rose from the city. As they ran forward, they met an aged man, one of the few survivors.

"The Amalekites!" the old man quavered, pointing south to the desert stretches. "The nomads came riding on their camels out of the desert. The women, the children, all taken captive!"

David called to Abiathar. "Shall I pursue this band?" he asked his young priest. "Shall I overtake them?"

Abiathar cast the holy stones* which he had brought with him when he fled from the massacre at the temple of Nob. Abiathar looked at the stones and nodded. "Pursue," he advised. "You shall surely overtake, and you shall surely rescue."

So David and his six hundred men set out across the desert wilderness without stopping for a rest. On the way they found an Egyptian slave lying half dead in the heat of the sun. David's men gave him water to drink, then fed him with bread, figs, and raisins. After he had eaten, the starving lad was able to talk and to tell them how he had been enslaved by the Amalekites. Then he guided them to a great oasis deep in the wilderness. There were the Amalekites, eating, drinking, and dancing because of the great spoil they had taken from the land of the Philistines and from Judah.

David and his men took the enemy by

98 surprise, slaughtering every Amalekite within reach of a slashing sword or a spear. Four hundred men escaped on camels.

The Hebrew wives and children were recovered, safe and unharmed, along with the possessions which the Amalekites had taken. The flocks and herds of the raiders were driven back to Ziklag, and some of the booty was secretly sent to the cities of Judah.

On the third day of their return to Ziklag a messenger came. The man had the swarthy complexion of a desert nomad, but he wore the clothes of an Israelite. When he came to David, he bowed down to the ground.

"Where do you come from?" David asked.

"I was with the Israelites—in the Valley of Jezreel—when the Philistines attacked," the man said, still breathing hard from his running.

David pulled the man to his feet. "How did it go?" he asked impatiently. "Tell me."

"The Philistines were victorious," the man gasped. "The Israelite army has fled from the battle. Many are dead. Saul and his son Jonathan are also dead!"

David began to tremble. He gripped the man so tightly that he winced with pain. "Tell me how they died," he said.

"By chance I happened to be on Mount Gilboa," the man continued hurriedly. "King Saul was badly wounded and was leaning on the point of his spear. The Philistines were close behind. When he saw me, the king begged me to slay him. So I stabbed him with his spear, for I saw that he could not live anyway."

As he said this, the man ventured a faint, knowing smile. He was expecting a reward for this news that would make David king. For he had heard the whispers all over Judah, and even in some parts of Israel, that David should be the next king.

"And I took the crown which was on his head and the armlet which he wore on his arm, and I have brought them here to my lord David."

David stood there, not seeing the crown and the armlet, ignoring the groveling messenger. Suddenly David turned with fury upon the fellow and cried, "How is it you dared to put forth your hand and destroy the king who was anointed of God?" Then he ordered the man to be put to death on the spot.

David and his men wept and fasted until evening in mourning for Saul and for Jonathan his son, and for all the fallen heroes of Israel. David composed a moving poem of lamentation for the fallen king who had been both loved and feared and for the princely son. "Thy glory, O Israel, is slain upon thy high places!" it began. David poured out his own grief for "Saul and Jonathan, beloved and lovely," and in doing it expressed the best sentiments of all Israel.

1 Samuel 27:5 through 2 Samuel 1:27

* The holy stones were called the Urim and Thummim (UHR-im, THUHM-im). They were objects used in Old Testament times to determine the will of God. They may have been stones or pieces of metal. They were cast by the priest rather as dice are thrown.

David the King

When Saul and Jonathan were killed by the Philistines on Mount Gilboa, the young nation was left without a ruler.

The northern tribes of Israel decided to continue together as one nation, although their land was overrun by Philistines. Abner, cousin of Saul and commander of Saul's army, escaped with a remnant of his forces. Eastward they fled, crossing the Jordan into Gilead.

There, in the highland town of Mahanaim, Abner established himself in a new capital city. Ishbosheth, the son of Saul, was set up as king, but Abner became the real authority.

However, Judah, the large and fiercely independent tribe in the south, wanted none of this kingship. They wanted David, who had been born and raised in Bethlehem, to be acclaimed their king. He was a member of the royal household through his marriage to Princess Michal. And none was more popular with the people or more skillful in battle.

So David and his men and their families turned their backs on the Philistines and left forever the blackened ruins of Ziklag. They moved to the ancient city of Hebron, the most important town in Judah.

The men of Judah came there and anointed David their king. Abiathar, David's long-time friend, became the priest of David's court. And Joab, the son of David's older sister, became the commander of the Judean army. Joab was wily, a master of battle strategy, tough, impulsive, and fearless.

There were many skirmishes between the house of Saul and the house of David, between squadrons from Abner's army and squadrons from Joab's army. But the son of Saul found his position weaker and weaker, while David grew stronger and stronger.

Joab was impatient with the civil war. He felt that Israel should reject the weakling son of Saul and give allegiance to David. Joab wanted David to be king over a united Judah and Israel, and the sooner the better.

The time came when Joab had his chance. King Ishbosheth had a quarrel with Abner. So Abner made up his mind to give the kingdom into the hand of David. He sent messengers to David at Hebron, saying, "Make your agreement with me, and my hand shall be with you to bring over all Israel to you."

David sent back saying, "Come to Hebron and confer with me. But you will not see my face unless you bring with you the wife I left behind—the Princess Michal!"

Abner came for his conference with David, bringing Michal. The burly commander reported that he had talked with the elders of the tribes of Israel. All agreed that David would be a better king than the son of Saul.

Afterward David entertained his guest with a feast. Then Abner left to gather Israel under the banner of David.

It was at this moment that Joab returned with his soldiers from a raid, bringing much plunder. Someone told Joab of Abner's visit.

Joab went storming in to see David. "What have you done?" he shouted. "If Abner came here, why have you let him go away in peace? Has he not been our enemy since the days we hid as outlaws?"

"Do not shout," David said. "Abner came to give all Israel into our hands."

Joab continued to sputter in anger. "You know that Abner came to deceive you! How can you trust him? He only wants to snoop into all you are doing!"

Joab wheeled and left the king's room. Being careful to keep his plans secret from the king, Joab dispatched messengers to run after the departing Israelite and ask him to return to the city gate.

Unsuspecting, Abner returned. When he reached the city gate, there was Joab waiting for him. David's commander drew out his sword and killed Abner on the spot.

When David heard of it, he cursed Joab. "May the blood of Abner be on your own head! He was a brave man. Now it is fitting that you weep and mourn with me and all the people for the fallen Abner."

When the news reached Mahanaim in Gilead, Ishbosheth trembled for his own life. Soon after this while he was taking his noonday rest, two rebels stole into his chamber and killed him. Now the ten northern tribes of Israel were left without a king.

The elders of Israel came to David at Hebron. King David made an agreement with them before God and was anointed king over Israel. David was thirty years old when he began to rule over the reunited kingdoms of Israel and Judah.

Now the king had to choose a new capital. Hebron was fiercely Judean in its loyalties and not acceptable to the Israelites in the north. Gibeah held too many memories of Saul and would not be acceptable to the Judeans. Mahanaim, Israel's temporary capital far across the Jordan, was not centrally located.

Jerusalem seemed to be the ideal location. An ancient city in a neutral area between Judah and Israel, it was well situated high on a hill and could hardly be conquered by attack. There was only one big problem—it had been occupied for centuries by a strong Canaanite tribe, the Jebusites.

David had a plan for capturing the city. At the head of the army of the newly united kingdom, David led his men toward Jerusalem. While the larger part of the army, under David's command, surrounded the city under the cover of night, Joab took a small band of handpicked men. He located the water shaft through which the city got its water supply in time of siege.* Joab crept up the shaft and into the city. There his band surprised and killed the Jebusite

sentries. The gates of the city were opened for David and his men. The city of Jerusalem was taken by surprise and with very little bloodshed.

And so it was that Jerusalem became the city of David, the capital of Judah and Israel. The king began at once to improve the city with a building program. David built a great palace for himself and his wives and children. He began to live like a real king.

The Philistines had been tolerant of David during the seven years that he reigned in Judah. But it was different when Israel made peace with David and crowned him their king. Now David was building a powerful kingdom with a great capital in Jerusalem. He could no longer be ignored.

The Philistine army marched inland, prepared to subdue David's kingdom.

David and Joab called out every fighting man in Judah and Israel. They attacked the Philistines here, outflanked them there, made furious assaults on one outpost after another. Joab's planning proved to be superb, and David was an inspired leader. All they had learned about Philistine military tactics was now used against their enemy.

At last the Philistines were forced to retreat. Finally they became a subject people to the Israelites.

Now David was ready to locate the long-neglected ark of God and set it up in his new capital city. He had in his heart plans for a splendid temple to be built for it, a place where all Israel would come to worship. But for now he would be content to put the ancient ark in the keeping of his priests on the highest spot in Jerusalem.

So David declared a festive holiday. With a company of priests and soldiers he went to fetch the ark. They brought it up the hill to Jerusalem with a great parade. The people ran along beside, shouting. Women

102 tapped their tambourines, priests blew on shophars of ram's horn. The symbol of God's presence was among them once more. And David the king, wearing a linen ephod,** leaped and danced at the head of the procession.

The ark was brought to a fine tent of goats' hair that David had ordered pitched on the high place. The tent was gaily decorated with wool and linen hangings. An altar was built there for burning sacrifices to God.

But the Princess Michal, watching the procession from her window, was angry with her husband. It annoyed her to see the king take such an enthusiastic part in the celebration.

"How the king of Israel honored himself today," she said bitterly, "stripping himself of his royal robes and dancing in the streets like any vulgar person."

David told her, "I will dance before the Lord, who chose me above your father, and above all his house, to appoint me as king over Judah and Israel. And so I will be humble, and I shall make merry before the Lord."

David and Michal were not close from that day. But the people of Israel loved David more than ever.

2 Samuel 2 through 6

* The water shaft may have been a passageway cut down through the rock as a shortcut to the valley spring. This was especially valuable during times of war, because inhabitants could use the shaft to get water without having to go outside the city walls.

** It is not certain just what the ephod (EF-od) was, though probably it was a garment of some kind worn by the priests. It possibly was sleeveless and may have been close-fitting. Some scholars think it may have covered only the front part of the body.

David and Bathsheba

Across the Jordan the Ammonites were becoming a problem again. Together with Syria, a powerful country farther north, they were massing thousands of foot soldiers on the borders of Gilead.

When David heard of this, he went into conference with his nephew, commander of his army. "Take your brother with you," David told Joab. "Divide the army in two parts so you may come between the Ammonites and their Syrian allies."

So Joab and his brother marched their army down the Jericho road, across the Jordan, and up into Gilead. Joab with some picked divisions charged the Syrians, while his brother thundered after the Ammonites with the balance of the army.

The Syrians fled north to their own country. When the Ammonites saw that their allies had deserted, they retreated back to their fortress city and barred their massive wooden gates.

The heavy winter rains were just beginning. Joab and the fighting men of Israel slogged their way back to Jerusalem. There could be no more fighting until winter was past and the rains were over and gone.

In the spring of the year, the time when kings go forth to battle, David sent Joab and the army back across the Jordan to finish the war against the Ammonites. King David remained at Jerusalem.

It happened that late one afternoon when David arose from his couch and was walking on the roof of the king's house, he saw below a woman bathing. She was very beautiful.

David asked his steward about her.

"She is the wife of Uriah the Hittite,

my lord, your army captain, now fighting the Ammonites with Joab."

"Bring her to me," David said, "for she pleases me."

So Bathsheba was brought to David, and the king made love to her. After she had stayed for some time with David, she returned to her home.

Messengers brought back word from across the Jordan that the siege was going well against the enemy. The Ammonites were bottled up inside their capital. They were unable to come and go. Their spring and waterworks, just outside the walls, were heavily fortified so they had water. But they could not reach their fields or their orchards, their flocks and herds. Eventually they would run out of food.

After a time Bathsheba sent word to David that she was to bear a child. David knew there might be serious trouble for Bathsheba if the truth came out. So David sent word to Joab, "Send me Uriah the Hittite."

When Uriah arrived, David showed great interest in the siege. "Tell me, brave captain, how is Joab doing? Does the war prosper?"

After Uriah had given his report, David said, "Now you deserve a short vacation. Go down to your house and wash your feet. Purify yourself from the grime of the battle and the dust of your journey. Rest awhile."

But Uriah did not go home to his own house. He slept on a mat in the forecourt of the palace, along with David's doorkeeper and some of the palace guard.

The next morning one of the guardsmen told David, "Uriah did not go to his own house last night. He slept here with us, by your own door."

David was vexed. He sent for Uriah. "Have you not come from a journey?" he asked. "Why did you not go to your house?"

Uriah bowed before the king and said, "My commander Joab and all the fighting men of Judah and Israel are camping in an open field. Shall I then go to my house, to eat and to drink and be with my wife? As your soul lives, I will not do it!"

David was now deeply perplexed. His plan to save Bathsheba from shame was not working out at all. He turned to Uriah and said, "Remain here today so that I may inquire after further news of our army. Tomorrow I will let you depart."

That evening David hosted a great feast in the palace, and Uriah ate at the king's table. The king instructed the wine steward to keep Uriah's cup full. He saw to it that many toasts were proposed and hailed with long drinks, until at last Uriah became drunk.

But Uriah still did not go to his house. Instead, he staggered over to where the palace guard slept and dropped onto a mat there.

In the morning David took a clay tablet and wrote a letter to Joab, "Set Uriah in the forefront of the hardest fighting, and then draw back from him that he may be struck down and die." He handed the message to Uriah, instructing him to give it to Joab.

As Joab was besieging the city to Rabbah he did as David had commanded him. The commander assigned Uriah to a place where there would be fierce fighting and much danger.

The Israelites charged the massive wall. The Ammonite bowmen sent clouds of deadly arrows flying toward the advancing Israelites. The soldiers and people of the city threw down spears, rocks, burning torches—even their grinding stones—on the heads of the men of Israel. At the end of the battle the field was covered with dead bodies. And among the slain was Uriah.

104

A messenger brought Joab's report of the battle to David: "The archers shot at your servants from the wall; some of the king's servants are dead; and your servant Uriah the Hittite is dead also."

David listened to the message and then sent word to Joab, "Do not let this matter trouble you, for the sword devours now one and now another; strengthen your attack upon the city, and overthrow it."

David knew that his kinsman Joab would be able to accomplish this. Soon the siege would be successful, and the Ammonites would threaten the Israelites no more.

When Bathsheba heard that Uriah was dead, she mourned for her husband. When the period for sorrow was over, David sent for her. She became David's wife and bore him a son.

There was a prophet in David's court named Nathan. He came to David and told him a story. "There were two neighbors, one rich and the other poor. The rich man had many flocks and herds, while the poor man had only one lamb. The poor man raised his lamb like a pet, sharing with it bits of his own bread and curds of milk. The lamb even curled up and slept on the same mat with him. Now the rich man had a visitor, but he didn't touch any of his own flock to provide food for his guest. Instead, he took the poor man's lamb and butchered it."

When David heard this story, he cried out angrily, "That rich man deserves to die!"

Nathan said to David, "You are the man! God has anointed you king over Judah and Israel. He gave you wives, and he has given you princes to carry on your name. Why have you ignored God? You have slain Uriah the Hittite with the sword of the Ammonites. You have taken his one and only wife, when you already had wives

of your own. And now, because you have turned your back on God, you will find evil raised up against you in your own house."

David hung his head. "I have sinned against the Lord," he admitted.

"Nevertheless," Nathan continued, "the child that is born to you shall die."

The child of David and Bathsheba did grow weaker and weaker, just as Nathan had said it would.

David the king humbled himself. He refused all food. While the child lay ill, he spent the days and nights begging God to spare its life. The king responded to nothing and to no one. He lay on the ground weeping and praying.

After seven days the baby died.

The servants of David were afraid to tell him that the child was dead, for they said, "While the child was yet alive, we spoke to him, and he did not listen to us. How then can we say to him that the child is dead? He may do himself some harm."

But when David saw his servants whispering together, he knew what had happened. "Is the child dead?" he asked.

They nodded, fear showing on their faces. "He is dead."

Much to their surprise David rose up, bathed, and put on fresh robes. Then he went into the house of God and worshiped. Afterward he returned to the palace and ate. His palace officials asked, "What is this thing that you have done? You fasted and wept for the child while it was alive. But when the child died, you arose and ate food."

David replied, "While the child was still alive, I thought that he might be spared. Now that he is dead, why should I fast? Can I bring him back again? I shall go to him, but he will not return to me."

Then David comforted his grieving wife Bathsheba, and after a time, she bore him another son. And David called his name Solomon. Nathan became quite fond of this young prince and called the boy "Beloved of the Lord."

2 Samuel 10 through 12

Absalom the Rebel

Amnon was the oldest of David's sons. But a younger son, Absalom, was the favorite of his father.

Amnon fell madly in love with Absalom's sister Tamar. Amnon was so tormented by his feelings that he plotted to take advantage of the beautiful girl. After he had done this, he was disgusted. He humiliated Tamar so that she fled weeping to Absalom.

Such treatment of his sister enraged Absalom, but he did not show his feelings right away. He nursed his anger and waited for a chance to get even with Amnon.

It was two years later when Absalom was ready to act. He held a feast to celebrate the end of his sheepshearing and invited all his brothers. Before the guests arrived, Absalom instructed his servants, "Wait until Amnon's heart is merry with wine. When I give the signal, kill him!"

The servants looked distressed. "But Amnon is a prince and beloved of David."

Absalom shrugged. "Am I not even more beloved of David?" He gave his servants sharp daggers, small enough to hide among their robes. "Just do as I have commanded you."

They did as they were told. At the height of the banquet Amnon was slain. Instantly there was confusion and panic. Who else might be killed? Did Absalom intend to eliminate all other rivals for the throne? The other princes fled for their lives.

Absalom took his servants and escaped. He found sanctuary in the northern country where his maternal grandfather was king. For three years he remained in exile.

But even though Absalom was guilty of the murder of his half-brother, David mourned the young man's absence. Finally Joab interceded so Absalom might return.

At last, after another two years, Absalom was summoned to the court of his father. The handsome prince entered the audience chamber and bowed with his face to the ground before the king. "If there is guilt in me," he cried out, "then kill me."

David looked tenderly on the beloved son who had been exiled from his sight these many years. He felt the tears on his cheeks as he stepped down from his throne, lifted Absalom up, and kissed him.

Soon after this reunion, Absalom set out on a plan to win the affection of the people away from the king. Now Absalom was an exceedingly attractive young man. His bearing was regal, and he had a wealth of dark curling hair. Each day he dressed himself in a splendid robe and went out among the people in a chariot drawn by horses, with fifty men running ahead of him.

He made frequent trips back to Hebron. There he gathered to his cause loyal Judeans who felt Judah was taking second place to the ten tribes of Israel. Many were still resentful that David had moved his capital from Hebron to Jerusalem.

"My father plays politics with the north," Absalom told them. "I would not do that. I would support Judah above all."

Yet when he traveled about in the north, he told the people of Israel the opposite. "My father is a Judean," he said. "He does not understand or care about the problems of northern Israel."

In Jerusalem he arose early and stood by the gates. When a man entered the city with a problem or a case for judgment, Absalom lamented loudly that his father was too busy to see everyone. "Your claims are good and right," he would say, "but see, there is no man appointed by the king to hear you. Oh, that I were a judge and a leader in this land! Then every man might come to me, and I would give him justice!"

Thus Absalom stole the hearts of the people.

At the end of four years Absalom said to his father, "Please let me go and offer worship in Hebron as I have vowed."

"Go in peace," the king said to his son.

So Absalom went to Hebron, and from there he sent secret messengers throughout the land, saying, "As soon as you hear the sound of the trumpet, cry out: 'Absalom is king at Hebron!'"

Helping to guide Absalom was Ahithophel, one of David's court advisers.

The conspiracy grew strong, and the people with Absalom kept increasing. At last a messenger came to David, saying, "The hearts of the men of Israel have gone after your son Absalom. Even your trusted adviser Ahithophel! Already Absalom is massing an army for a march on Jerusalem."

David was stunned, caught unaware by this subversion. He had no standing army. His fighting men had returned to their homes. David held counsel with Joab and his priest Abiathar and another of his advisers, Hushai. At last David said, "At the moment there is nothing we can do except to arise and flee. Otherwise Absalom will smite the city with a sword."

The friends of David made ready to flee immediately. "But not you, Abiathar, my old friend," David said to the priest. "You stay here in the city with the ark. You can keep me in touch with what is happening in Jerusalem by sending your son as a messenger."

Then David told Hushai, his trusted adviser, "You are a loyal friend and can be of much value to me if you stay in the city and say to Absalom, 'I will serve you, O king.' Then you may be able to give Absalom false counsel and defeat the advice of Ahithophel. Use Abiathar's son as a messenger and send news of anything you hear."

So the royal household left in haste, the king and his wives and children, along with the men and families attached to the court and the loyal palace guard. They traveled as fast as they could through the barren hills, along the hot and dusty road to Jericho. They came to the Jordan where they rested briefly and refreshed themselves. Then they crossed over and climbed up into the land of Gilead. They came to Mahanaim, the former capital of the son of Saul when he ruled with Abner as his regent.

The people of Mahanaim had grown to love David. They had not joined with Absalom. They welcomed the king and brought beds, basins, and earthen vessels, wheat, barley, beans and lentils, great combs of honey, skins of clabbered milk, cheese, a herd of sheep—everything to make him comfortable.

Even the Ammonites brought provisions. Since their defeat and the signing of a treaty, the Ammonites had been living in peace with the Israelites.

Word of David's plight also traveled throughout the land of Israel. More and more volunteers crossed the Jordan to join the army of their king.

Back in Jerusalem, Absalom had marched in and taken over the city. Immediately the white-haired Ahithophel said to Absalom, "Let me choose twelve thousand men, and I will set out and pursue David tonight. I will come upon him while he is weary and discouraged and before too many men have rallied to his aid. I will throw him into a panic, and his supporters will flee. I will strike down only the king. The rest of the people will return unharmed. Then our people will be at peace."

Ahithophel's advice pleased Absalom and the elders who were with him. If Absalom had followed it, everything might have happened as old Ahithophel predicted.

But Absalom turned to Hushai, who was noted for his wise counsel. "What do you say?"

Hushai appeared to be considering the matter from every angle, his face grave. "This time Ahithophel has not given good counsel," he said at last. "Now you know that your father and his men are mighty men. They are enraged like a bear robbed of her cubs. Besides, your father is expert in war. He will not be easily found or caught. Better to take time to muster up a large army, as numerous as the sands of the sea. You should lead your army in person and light upon David's smaller force as the dew falls upon the ground. There will not be one left, of him or his men!"

Absalom and the elders listened with admiration to the words of Hushai. "The counsel of Hushai is better than the counsel of Ahithophel," Absalom said.

When Ahithophel saw that his advice was not being followed, he knew that he was doomed. Now David would have time to organize his forces, and he and Joab were past masters at battle strategy. David might forgive the folly of his son, but he was not likely to forgive the treachery of a trusted adviser. Ahithophel saddled his donkey, went home to his own city to set his affairs in order, and then hanged himself.

While Absalom was preparing a large army to cross over the Jordan, David was organizing his own loyal army. He divided his troops into three sections. Joab was in charge of one third, Joab's brother was in charge of another third, and another seasoned commander was in charge of the last third. One section was to charge directly at Absalom's army, and the other two were to close in from either side.

When the word came that Absalom's army was crossing the Jordan, the king stood by the city gates of Mahanaim and watched his own forces march out. His advisers insisted that he stay within the safety of the town walls.

To Joab and the commanders David said, "Remember only this—deal gently with the young man Absalom."

So the army of David went out against the army of his son, and the battle was fought in the forest of Ephraim. Absalom's army was defeated there by Joab and the loyal soldiers of David. And the army of Absalom fell apart, every man fleeing for his life to recross the Jordan River.

Absalom found himself deserted by his troops. He jumped on his white mule, mercilessly spurring the animal as he tried to catch his fleeing army. Absalom twisted himself around to watch for pursuers just as the mule raced under the low branches of a great oak. The young man's head struck a heavy limb with such force that it was wedged into the fork of the branch. There he hung, feet dangling, as the mule galloped on without its rider. The more Absalom twisted to get free, the more tightly his head became wedged, and his heavy hair was entangled in the branches.

One of David's soldiers saw him and reported to Joab. "What!" Joab cried. "You saw the traitor? Why did you not strike him to the ground? I would have given you ten pieces of silver."

But the soldier said, "Even if I felt in my hand the weight of a thousand pieces of silver, I would not put forth my hand against the king's son. For in our hearing the king commanded you to deal gently with Absalom."

Joab spat out, "I will not waste time like this with you." He hurried to the spot described by the soldier and found Absalom caught in the tree. Joab thrust three darts into the heart of the prince. His soldiers surrounded Absalom and finished the deed.

Joab blew the trumpet. His troops came back from pursuing the rebel army. They threw the body of Absalom into a pit in the forest and heaped his grave with stones. And Joab dispatched a messenger to the king with the news.

Now David was waiting anxiously at the city gates. When the messenger arrived and told of the victory, the king had only one question. "Is it well with the young man Absalom?"

The messenger said, "May all your enemies be as dead as he is!"

The king turned his face away and climbed to a little room in the gate tower, where he broke down in racking sobs. Remembering the dark-haired little boy who used to twine his arms around his father's neck, the handsome young man so recently alive and full of charm, David wept. "O my son Absalom, my son, my son. Would I had died instead of you, O Absalom, my son!"

When Joab returned with his men from their victory, one of the guardsmen at the gate said, "Look up there in the tower. The king is weeping and mourning for Absalom."

So the general feeling of jubilation began to fade. People whispered, "The king is grieving for his son!" They were bewildered, because the king, instead of rejoicing in his victory, grieved for his lost son.

Once again Joab took matters into his

own hands. He marched up into the tower, flung open the door, and glared at his kinsman the king.

"A fine thing!" he shouted angrily. "Today your loyal soldiers risked their lives—to save your life, and the lives of your sons and daughters and wives. But you are completely mixed up. You prefer to love those who hate you! And when you ignore the people in this day of victory, it is as if you said you hated all these who so dearly love you!"

The king looked up, his face wet with his tears.

"Yes, it is true," Joab went on relentlessly. "You have made it clear today, when you did not come out to greet us. Your commanders and your archers and your footsoldiers are nothing to you. I think that if Absalom were alive and the rest of us were dead today, you would actually be pleased!"

David stirred, for the words of Joab had knifed through the fog of his grief.

"So arise, now," Joab commanded gruffly. "Wipe the tears from your face, and go out and speak kindly to your people. If you do not, I swear they will desert you. And I would not blame them!"

So the king arose, wiped the stain of tears from his face, and took his seat in the gate. His people came before him. He thanked them for their courage and blessed them.

With Prince Absalom dead and his battle lost, the rebellion died. David was invited to return to Jerusalem. Again he reigned over the united kingdoms of Judah and Israel.

2 Samuel 13:1 through 19:15

A Successor to David

Now that both Prince Amnon and Prince Absalom were dead, the oldest surviving son of King David was Adonijah.

Adonijah was a handsome young man with a burning desire to follow his father on the throne. Since King David had always been indulgent with him, he was used to having his own way. As Absalom had done, Adonijah provided himself with a chariot and charged through Jerusalem and the countryside, runners sprinting ahead to shout, "Make way for the royal prince!"

Joab liked Adonijah and felt that the oldest living son of David should be next on the throne. The old priest Abiathar agreed.

But there was dissension and bitter rivalry within the palace. Nathan, the fierce old prophet, was devoted to young Prince Solomon. A powerful priest named Zadok also preferred to have Solomon succeed his father. And so did the captain of David's palace guardsmen.

While the weary old king lay sick and ailing in his palace chamber, Adonijah worked out a plan to have himself anointed king. He held a feast at a spring just south of Jerusalem, but Solomon and his supporters were not invited. The priest Abiathar was to anoint the ambitious prince with holy oil. And then Adonijah would be the next king!

But word of the plot reached Nathan. The prophet took Bathsheba aside and whispered, "Did you know that Adonijah is to be anointed king at this moment and David does not know of it? If you would save your own life and that of your son Solomon, you must go immediately to King David's bedside. Remind him of his promise that the son you bore him would follow him on the throne. Then ask why Adonijah sets himself up to be king! While you are still there, I will come in and confirm your words."

So Bathsheba did as the prophet told her. "My lord," she implored her husband, "you swore to me that Solomon our son would reign after you. And now look, Adonijah is making himself king."

The old king tried to raise his head from the pillows. "Adonijah—king?"

Bathsheba nodded, weeping. "He has made a great feast below the city, and he's invited all the younger sons of the king except Solomon. Abiathar is there to anoint Adonijah, but neither Zadok nor Nathan is there."

While she was still speaking, Nathan the prophet came in. "Did you know that the people are eating and drinking with your son Adonijah, and they are shouting, 'Long live King Adonijah'? Is this your doing, and you have not told us?"

David shook his head. He took the hand of Bathsheba and whispered, "As God lives, I did swear that Solomon would reign after me. Call to me Zadok the priest and the captain of my palace guard."

The priest and the burly captain were quickly summoned. David said to them, "Put Solomon my son on my royal mule. Ride after him in a parade through the city. Let Zadok the priest and Nathan the prophet anointed him with holy oil. Then blow the trumpets, and let everyone shout, 'Long live King Solomon!' Then he shall come and sit on my throne, for I am old and about to die. He shall be ruler over Israel and over Judah."

It was done as King David had said. After the anointing, the trumpets sounded, flutes piped, and the people of Jerusalem shouted, "Long live King Solomon!"

Meanwhile, at the spring below the city, Adonijah and his guests heard the music and the shouting.

When Joab heard the blast of trumpets, he asked, "What does this uproar in the city mean?"

While he was still speaking, a breathless runner came from the city. "King David has made Solomon king," the messenger gasped. "Solomon has been anointed by the priest Zadok and the prophet Nathan, and the city rejoices."

Then the guests of Adonijah trembled with fear for themselves, for had they not been part of the plot to keep the throne from Solomon? Each hurried for safety to his own house, afraid for his life.

Soon after that the great King David died. Joab and Adonijah were put to the sword. But the old priest Abiathar was banished to the small town of Anathoth nearby, where he lived in retirement until his death. So it was that Solomon remained king without opposition.

1 Kings 1 through 2

Solomon's Reign

At first, when Solomon began his reign, he had a humble attitude toward his job as king. He worshiped the God of his fathers. He tried to know God's will and follow it.

One day he went to the nearby town of Gibeon, and there he worshiped at the ancient altar.

That night Solomon had a dream, and he saw God in his dream. "Ask what I shall give you," God said in the dream. And Solomon replied, "Give me an understanding mind and a heart that listens so I may wisely govern your people, and so I may know good from evil."

Then God answered, "I am pleased that you have asked for wisdom, rather than for long life or riches. Therefore I will give you wisdom so that you shall be wise above all men. I give you also what you have not asked, riches and honor. And if you walk in my ways, as your father David walked, keeping my laws and my commandments, then I will increase the days of your life."

Solomon awoke and knew it was a dream. But he felt he had come close to God and his will. For a long time he really strove to govern wisely, to have an understanding heart and an attentive mind.

One of the most important duties of the king in those days was hearing the complaints of people who had been wronged, and making decisions on disputes between his subjects.

One day two women came before the king quarreling loudly. One woman, weeping, clutched a baby. "It's my baby," she sobbed.

"My lord Solomon," the second woman complained, "this woman and I live in the same house. We both gave birth to sons. We were alone, and there was no one else in the house. This woman's baby died in the night. And she arose in the dark, while I slept, and took my son and then laid her dead baby in my arms. In the morning I wept to see that my child was dead. Then I looked closely, and I saw that it was not the same child I gave birth to."

The first woman clutched the baby more tightly. "No!" she interrupted. "The living child is mine, and the dead child is yours!"

The women tugged at the child and argued bitterly. Everyone in the court looked at one another. How could Solomon pass judgment in such a case when there had been no witnesses?

But Solomon commanded, "Bring me a sword!" It was brought to him, and then the king ordered, "Divide the living child in two. Give half to one and half to the other."

Then the woman who held the baby broke into tears and laid the infant at the feet of the king. "Oh, my lord," she sobbed, "give her the living child, and by no means harm him in any way!"

But the other woman folded her arms and said, "It shall be neither mine nor yours. Divide it!"

Then Solomon indicated the sobbing woman, "Give the child back to this woman and by no means slay it. She is its mother."

And all Judah and Israel heard of the judgment of Solomon, and they stood in awe of the wisdom of their king. Because of such wise decisions, Solomon's fame spread to the nations round about. His wise proverbs were repeated over and over again, written down, and referred to. Like his father before him, he wrote songs, and many of them were sung in the temple. So admired and popular was he that if anyone heard a wise saying or a beautiful hymn he said, "It must have been written by Solomon."

Men came from many lands to hear the wisdom of Solomon. Even kings and queens sent to know what wise judgments Solomon had made.

This brilliant son of David began to give his serious attention to building an empire. He sought to increase his wealth and influence through foreign alliances.

First he married the daughter of the Pharaoh of Egypt. Then he enlarged Jerusalem to the north. He built an expensive and beautiful new palace for himself, a great audience hall, and a special house just for the princess from Egypt. And he built a temple to house the ark of God, a temple far more splendid than David ever dreamed of.

For all these magnificent building projects, Solomon made contracts with Hiram, the king of the great Phoenician coastal city of Tyre. From Hiram, Solomon bought cedar and cypress wood, sending Hiram wheat and jars of olive oil in trade.

Solomon developed his own copper mines in the great deserts to the south. The copper was smelted, for use in building and decoration, in the new port city of Ezion-geber at the northern tip of the Red Sea. Gold and ivory, also for decoration, were purchased even farther south and brought to Jerusalem in camel caravans.

To do all the vast building and pay for it, the men of Israel were drafted to serve a term at forced labor. At the king's command they quarried out from the hills great costly stones. These were chiseled and dressed out on the site into white limestone blocks, then dragged to the hill above Jerusalem to be used in the building of foundations. Skilled workmen and architects from Phoenicia were hired to come down and take charge of the building.

The temple was made with an inner sanctuary where the ark of God could rest in perfect darkness. There was a main room outside this, and decorations of carved work and gold were everywhere. At the front was a porch with two great bronze pillars.

In the open-air court which surrounded the temple was a great altar for burnt offerings. There was also the molten sea, a huge bronze tank of water, supported by twelve bronze bulls.

At the dedication of the temple Solomon said, "But will God indeed dwell on the earth? Behold, the highest heaven cannot contain thee. Yet regard the prayer of your servant, O Lord, this day, that your eyes may be open night and day toward this house."

Solomon also built a fleet of ships and developed a harbor at Ezion-geber on the Red Sea, near his copper-smelting settlement. He hired skilled seamen from Phoenicia, for they were expert seafarers. Solomon's ships traded with other nations and brought gold from a mysterious and distant land known as Ophir.

Now the queen of Sheba heard of the fame of Solomon and decided to make a visit to his court. Up from her kingdom she came, she and her party riding in a great camel caravan. Over the vast Arabian deserts they rode, along the shores of the Red Sea, past Solomon's port of Ezion-geber, past his copper refinery, up through the desolate wastelands of Edom.

At last the queen of Sheba and her party reached Jerusalem. She was received with great ceremony at the court of Solomon. There she made a lavish display of the kinds of goods in which her country traded: spices, gold, and precious stones.

Solomon took the queen on a tour of his capital, showing her the dazzling new buildings in the northward extension of Jerusalem. The queen looked about in amazement, impressed by this magnificence.

The queen was also dazzled by Solomon. The king was richly dressed, and even his servants wore beautiful clothes. The food served at Solomon's table was bountiful, and there were delicious imported dainties of all kinds.

"The report was true," the queen said, "which I heard in my own land about your affairs and your wisdom. But I did not believe the reports until I came and my own eyes had seen it! And the half of it was not told me! Your wisdom and your prosperity are even greater than I heard. How happy your wives must be!"

Then she gave King Solomon a hundred and twenty pieces of gold and a very great quantity of spices and precious stones. And King Solomon gave the queen of Sheba gifts of wheat and oil from Judah, and anything else that she asked. Then she returned to her own land.

Solomon's kingdom of Judah and Israel was now even larger than it had been under King David. He had trade agreements and treaties with the great nations around him. But beneath the building and the magnificence and the splendor, the foundations of the country were beginning to totter.

King Solomon began to lose his sense of closeness with God because he loved and married many foreign women. The princesses and other wives brought their idols with them. Small temples appeared for the worship of Egyptian, Moabite, and Phoenician gods. As Solomon grew older, his wives encouraged him to worship their pagan gods, and many of the people in Jerusalem worshiped them too.

Solomon's government began to stagger under a growing mountain of public debt. Solomon had to turn over to Hiram of Tyre two Israelite cities in Galilee because there was no money for what the king owed him.

Edom, the nomadic nation just to the south, had been subdued by the powerful King David. Now, under Solomon, Edom revolted and formed a powerful alliance with Egypt. There was also another revolt in the north in the country of Syria. A strong rival kingdom was being established there, with the ancient city of Damascus as its capital.

So Solomon's territory of influence began to shrink. Within Judah, and even more throughout the areas of the northern ten tribes of Israel, there was increasing unrest.

"What do we get from Solomon's lavish spending?" many people asked. "He lives in luxury while the public debt grows. We become poor with taxes, and our men groan under his forced labor."

One of those who openly began to sympathize with the people was an able young man named Jeroboam. Solomon had put him in charge of the forced labor gangs from the northern tribes of Israel. When the Israelites complained to Jeroboam, the labor overseer readily admitted that their complaints were justified.

One day a prophet met Jeroboam on the road. The prophet took off his robe and tore it. "Listen to me, I know that God is going to rip apart this kingdom," he said to Jeroboam, "just as I have torn my robe. One part will remain for the house of David and Solomon. But the rest of the kingdom will go to you."

When Solomon heard of this, he ordered Jeroboam killed. But Jeroboam escaped into Egypt, and he stayed there until the death of Solomon.

1 Kings 3 through 11

The Kingdom Divided

Rehoboam, the son of Solomon, was spoiled and haughty. After his father died, he became even more headstrong. He rarely took advice from his elders.

Soon after Rehoboam was crowned in Jerusalem, he was told that the northern tribes of Israel were meeting in the mountain town of Shechem. If he wanted also to be crowned king of Israel, he must journey up there for the ceremony.

When Jeroboam, the former overseer for Solomon, received word of the planned

114 meeting, he quickly left Egypt and made the journey home.

So Jeroboam was standing among the elders of the ten northern tribes of Israel when King Rehoboam arrived from Jerusalem.

But before there would be any crowning of Solomon's son, the men of Israel wanted to get things straight with him. "Your father made our yoke heavy," they told him. "If we are to crown you our king, then you must promise to lighten this burden of taxes and hard labor. If you will do this, then we will serve you."

The king waved his hand to dismiss them. "Come back in three days and I'll give you my answer."

The wise old men from the Jerusalem court advised Rehoboam, "Speak good words to them, and they will be your servants forever."

But Rehoboam turned to his friends, fellows his own age. "What do you think?"

The young men said, "Tell those Israelites that you are mightier than your father and that you will add to their heavy yoke!"

So when Jeroboam and the people of Israel returned on the third day, King Rehoboam answered them harshly. "My father made your yoke heavy, but I will add to your yoke!" He laughed uproariously when he saw their stunned and angry faces. Then he lashed out at them, "My father only chastised you with whips—but I will chastise you with spiked scourges!"

And when the elders of Israel saw that the king was headstrong and harsh, a man who would not listen to them, they said angrily, "What share do we have in the house of David? We have no inheritance in the house of David and Solomon!" Some began to pelt the king's party with stones.

King Rehoboam, in fear for his own life, made haste to mount his chariot and flee back to Jerusalem.

So it came about that King Rehoboam was left with only his capital at Jerusalem and the small kingdom of Judah.

The people of the ten tribes of Israel established their own nation to the north, with their capital at the ancient city of Shechem. Jeroboam was chosen as their king and crowned there.

And from that time on, the kingdom was no longer one kingdom, a united kingdom, as it had been during the reigns of Saul, David, and Solomon. Now there was Judah, and there was Israel. And divided, they would fall.

1 Kings 12 through 14

PART VI
THE LAST YEARS
OF ISRAEL

There was a span of about two hundred years from the time of Solomon's son, when the United Kingdom was divided into separate nations of Judah and Israel (about 931 B.C.), until the fall and capture of Israel by the fearsome nation of Assyria (about 722 B.C.).

During this time there were off-and-on wars between Israel and her sister state to the south. There was trouble in the north also. Syria, with its capital at Damascus, continued to grow stronger, often warring upon Israelite cities, especially those east of the river Jordan.

During periods of peace and in spite of the wars, Israel grew in prosperity. The land was a natural and lucrative crossing point for trade routes going south to the Philistine cities and to Egypt, and north to Phoenicia and Syria. Wide, fertile valleys made agriculture very profitable, and the excess wheat, barley, wool, and olive oil were sold to foreign traders.

The native Canaanites were no problem now. The people, and many of their ways, had been assimilated by the Israelites. This was partly for the good. The Canaanite culture was more sophisticated than that of the Hebrews, especially in the fields of the arts, architecture, and city government.

But the religion of the Israelites suffered with this

infusion. Under Moses there had been a strong feeling of solidarity based on their feeling of oneness as God's chosen people. There was also a sense that the large group was responsible for each man and that each man owed complete allegiance to the whole community. This led to a fairly high-minded, responsible morality.

But when the Israelites moved into the Canaanite cities and farmlands, they settled among people who believed they were dominated by local gods. These Baal-gods were thought to be responsible for the productivity and fertility of the land. Many Israelites saw no harm in seeking the aid of the Baals in such matters as the increase of crops and herds. Their neighbors, the Canaanites, seemed to be successful in courting the help of their local Baal. Each Baal was thought to control the weather in his small area and to cause crops and animals to multiply. Soon many Israelites joined in the wild, idolatrous worship at the local Canaanite shrines, often while still professing a general belief in the God of their ancestors.

When the princess Jezebel came to Israel in the ninth century B.C. as a bride for King Ahab, she brought the worship of her own Baal with her, and a large number of priests. The prophets Elijah and Elisha fought her attempts to promote Baal worship in Israel.

In the eighth century came the first prophets to write down their messages. In Israel both Amos and Hosea attacked the demoralizing influence of the Baal-cult. They criticized the people for letting their worship of God become empty and meaningless.

In its last years the nation of Israel had to contend with an even deadlier foe than Baalism or Syria or civil war with Judah. Far to the east, in the fertile valleys of the Tigris and Euphrates Rivers, the vast and powerful nation of Assyria was growing from the remnants of the more ancient cultures of that area. A mighty king built up a huge capital city at Nineveh. Eventually the mighty army of Assyria would steamroller its way westward laying siege to cities, taking captives, and dominating the small nations in its path.

Assyria developed a way of keeping vassal nations under control. She simply intermingled the population of one with another. The leading citizens of a conquered nation were deported and scattered around in other conquered lands. Only a docile peasant class was left, and they soon had to absorb a polyglot assortment of foreigners from other vassal nations.

When Israel's capital of Samaria was finally captured by the Assyrians, the people of the land were deported. Later literature sometimes referred to the "ten lost tribes of Israel." However, it was only the leaders of these tribes who were scattered and whose identity as Israelites was lost. At the same time, the population in the area of Samaria became intermixed with assorted refugees from other lands. So much intermarriage resulted that the Judeans (later called "Jews") felt that the people of Samaria had lost their racial "purity." It was partly because of this racial mixing that the Samaritans were looked down on and considered to be outcasts in the time of Jesus.

Ahab and Jezebel

Several kings of the nation Israel lived, ruled, and died after the time of Jeroboam.

Then King Ahab came to the throne. His father Omri had been a strong king who had taken the throne by force and established Samaria as his new capital. Before he died, King Omri arranged a royal marriage for Ahab, in diplomatic alliance with the powerful land of Phoenicia. The bride-to-be was Jezebel, a princess from the great coastal city of Tyre.

After Ahab had been anointed king of Israel, Jezebel came down in a regal procession from Tyre with her vast bridal party. She brought not only her own slaves and advisers, but also a great company of priests trained in her foreign religious cult.*

Though she was now a queen in Israel, Jezebel continued to worship as she had been taught by her priest-king father. She persuaded Ahab to build in Samaria a great temple to her Baal and to join in her worship there. Jezebel began to make life very hard, even dangerous, for the priests of God. Many Israelite priests were slain on the queen's orders. Many fled the country, or were forced to hide out in caves.

So it was that Queen Jezebel began to have a bad effect on her husband, and on all of Israel.

Ahab's winter palace, decorated with cedar wood and inlaid with ivory, was in his capital city on the hill of Samaria. Farther north in Jezreel, at the eastern end of the great valley, the royal couple maintained a two-story summer palace. This residence was also richly decorated and had great latticed windows and balconies.

Next door to this summer palace was a vineyard that belonged to a man named Naboth. Ahab looked enviously at this patch of land. He said to Naboth, "Give me your vineyard. I want it for a vegetable garden, because it is near my house. Come, I will give you a better vineyard for it! Or, if it seems good to you, I will give you its value in money."

This seemed fair enough to King Ahab. But the land had belonged to Naboth's family for generations, and he didn't want to sell it. It belonged not only to him, but to the future generations of his family. So Naboth refused the king's offer. "God forbid that I should give you the inheritance of my fathers," he said with finality.

The king reacted in a childish way. He went back to his palace annoyed and sullen. He lay down on his bed, turned his face to the wall, and refused to eat.

Then Jezebel his wife came to him and asked, "Why are you so irritated that you eat no food?"

Ahab lay on his bed, his face still to the wall. "Because I spoke to Naboth, next door," Ahab pouted. "I said, 'Please give me your vineyard for money. Or else, if it please you, I will give you another vineyard for it,' and he told me he would not."

Jezebel stamped her pretty foot. She had learned from her tyrant father that no one stood in the way of the king. A king could do anything he wanted.

"Do you not govern Israel?" Jezebel scolded. She tugged at Ahab's garments. "Arise, my love, and let your heart be cheerful. Look, I will give you the vineyard of Naboth."

So Jezebel conceived a plot. She wrote letters in Ahab's name and sealed them with his seal. Jezebel wrote to the elders in Jezreel: "Proclaim a fast and set Naboth on high among the people. Then find two good-for-nothings who can be bribed and set them opposite Naboth. Prompt these two fellows to bring a charge against

Naboth, saying, 'You have cursed God—and the king!' Then, take Naboth out and stone him to death!"

The men of the city were afraid to go against the king's instructions. They did not want the same thing to happen to them so they followed the orders.

Naboth was set up before the people. The false accusation was made. The people of the city were shocked, for they believed that the uttering of a curse could result in real danger; a curse might be far worse than physical violence! And so Naboth was found guilty. He was hustled outside the city and stoned.

Then the elders sent a message to the palace saying, "Naboth has been found guilty of cursing our God and our king. He has been stoned, and now is dead."

Jezebel smiled at her husband. "See, now there is nothing between you and the kitchen garden you coveted. Arise, and take possession of the vineyard of Naboth of Jezreel."

Ahab arose from his couch, happily decked himself in his kingly robes, and walked next door to claim the vineyard of his dead neighbor.

But there was a rugged prophet from Gilead who lived in Israel at this time. His name was Elijah. His was almost the only voice in the land which dared to speak out against Ahab and Jezebel. So when Ahab walked into the vineyard that had belonged to Naboth, there was the fearless prophet Elijah, ready to confront his king. When Ahab saw Elijah, he was distressed. "Have you found me, O my enemy?" the king asked.

Elijah, his eyes flashing fire, said, "I have found you, because you have been part of an evil act in the sight of God! And have you not also built a house of pagan worship for Baal?" Elijah's voice trembled with passionate anger. "God will punish you,

King Ahab, and take the throne away from your descendants! For you have led Israel to sin!"

Ahab turned pale. He feared the fierce prophet and hated his words. But he believed him.

Then Elijah turned to go back into the quiet of the distant wilderness. But he paused for one last gruesome prediction. "And wild dogs shall someday devour Jezebel, within this city of Jezreel!"

King Ahab rushed back to his palace. Already the gaining of the vineyard of Naboth was as bitter ashes in his mouth. He tore his robes, put on rough sackcloth, and fasted. He was sorrowful and repentant.

But Jezebel tossed her imperious head and was not impressed. Was she not a princess from Tyre, a priestess of Baal, and the queen of Israel?

1 Kings 16:23 through 21:29

* Many scholars believe that Psalm 45 was a poem addressed to King Ahab and to his bride from Tyre. Jezebel worshiped Baal-Melkart, the ancient god of Tyre, who was somewhat similar to the local Baal-gods of the Canaanites.

Elijah Against Baal

For two long years the kingdom of Ahab suffered the most severe drought anyone could remember. Almost no rain fell in Israel.

The crops withered and died without maturing. There was not enough wheat and barley to feed the people. Even the sheep and cattle began to sicken and die for lack of food and water. The leaves curled on the fruit trees, and yellowed and fell. There were no figs and persimmons those years, no apricots and pomegranates. The olive trees sickened and dropped their withered fruit. The vineyards were parched and blasted.

In the third year of the drought Elijah left the peace and stillness of his wilderness retreat. The gaunt, weather-beaten prophet arranged for a meeting with King Ahab. Dressed in his rough haircloth garment cinched at his waist with a strip of leather, Elijah looked out of place next to the richly robed king.

When King Ahab saw Elijah, Ahab said to him, "Is it you, you troubler of Israel?"

"I have not troubled Israel," Elijah shot back, "but you have! You have followed the unholy ways of your wife Jezebel. You have forsaken the laws and the priests of God, and you turn to Baal." Ahab frowned.

"Now," Elijah ordered, "gather all Israel at Mount Carmel. And tell the four hundred and fifty prophets of Baal, who eat at the table of Jezebel, to be there, too!"

Mount Carmel was part of that great ridge of mountains stretching between Samaria and the Valley of Jezreel, a backbone of rocky heights that rises near Mount Gilboa, forms the southern barrier of the valley, and at last plunges into the Great Sea.

Up on the mountain one could look westward across the blue-green waters to the distant horizon, where the blue bowl of the sky touched the seemingly infinite expanse of the sea. Northward the eye could follow the rugged seacoast which led into Phoenicia, the homeland of Queen Jezebel. To the east the Jezreel valley spread for many miles, the Kishon River meandering its whole length and spilling into the sea below. Here the Midianites had been defeated by Gideon. Here the chariots of Sisera bogged down in the mud of the river Kishon. Here the Philistines gathered for their assault on King Saul.

Mount Carmel commanded a view that was rich in memories of the history of Israel, memories filled with a sense of God and his part in that history. Perhaps Elijah had chosen this spot for just this reason.

But the people of Israel seemed to be forgetting that history. They were now accustomed to filling Jezebel's great Baal temple in Samaria. Regularly they visited the pagan shrines to pray to the local Baal for the fertility of their crops and herds.

When Elijah climbed to the top of Mount Carmel and saw King Ahab and the Israelites assembled there, angrily he scolded them for their unfaithfulness.

"How long will you hop back and forth between two different opinions?" Elijah shouted. "If the God of Israel is indeed the only God, then follow him! You cannot worship God and Baal!"

The people did not answer.

"I am a single prophet of God," Elijah told them. He waved his hand contemptuously. "But see, here come Baal's prophets, four hundred and fifty strong!"

When the Baal prophets arrived, Elijah laid down the conditions for the contest. "Let two bulls be given to us. Let them choose one bull for themselves. They shall cut it in pieces and lay it on the wood of their altar, but they shall put no fire to it. I will prepare the other bull and lay it on the wood on my altar, but I will put no fire to it."

The people listened intently. Then Elijah faced the four hundred and fifty prophets of Baal. "You will call on the name of your god," Elijah told them, "and I will call on the name of the God of Israel. Whichever answers by fire, he is God!"

And the people answered, "It is well spoken."

120 So the prophets of Baal prepared a bull for sacrifice, and laid it on the pile of wood on their rough stone altar.

"Baal, Baal, send down fire!" the four hundred and fifty prophets chanted over and over. Nothing happened. After a while, it was "Baal, Baal, please send down fire." But still there was no answer, though they cried out steadily all morning long, until they were hoarse.

Finally the prophets were croaking desperately, "O Baal, answer us!"

There was no voice; no one answered.

At noon Elijah mocked them, saying, "Cry louder for he is a god. Either he is musing, or he has gone aside, or he is on a journey. Or perhaps he is asleep—and must be awakened?"

The prophets increased their hopping, limping dance around the altar. Soon they were gyrating madly, shrieking and screaming, gashing themselves, after their custom, until the blood ran.

As midday passed, the prophets of Baal raved on. But there was no answer. The blood congealed on the cut pieces of their bull, and the wood under it was not lighted. The prophets were dropping to the ground, exhausted.

Then Elijah said to the people of Israel, "Come closer."

The crowd moved closer to Elijah and watched him. The old prophet put together a simple altar of twelve stones, representing the twelve tribes of Israel. He dug a trench around it. On the altar he laid the wood, then the pieces of the slaughtered bull.

"Fill four jars with water," Elijah said to the elders nearest him. "Pour it over the bull and on the wood."

The elders followed his instructions. Then Elijah ordered them to do it a second time, and again a third time. It was done, and the water ran round about the altar, and filled the trench also.

Elijah stood there in simple dignity and said quietly, "O God of Abraham, Isaac, and Israel, let it be known this day that you are truly God in Israel, and that I am your servant, following your word."

Suddenly a bolt of flame struck at the altar of Elijah. Fire danced over the burnt offering, the wood began to smoke and burn, and flames even licked up the water that was in the trench.

When the people saw it, they fell on their faces, crying out, "The God of Israel! This is our God!"

Elijah shouted, "Seize the prophets of Baal! Let not one of them escape."

And so the four hundred and fifty prophets of Jezebel, who had led the Israelites astray, were slain upon Mount Carmel.

Now a small dark cloud appeared on the horizon and swirled closer and closer.

"I hear the sound and the roar of a downpour coming," Elijah said to King Ahab. "Get in your chariot and start back to Jezreel, before the rains come and the river Kishon overflows."

In a little while the heavens grew black with clouds and wind, and there was a great rain. The drought was broken in Israel.

But when King Ahab told Jezebel all that Elijah had done, and how her prophets were slain, Jezebel vowed to have Elijah killed. So the prophet fled to the wilderness.

1 Kings 18:1 through 19:4

The Honest Prophet

About this time there was a three-year pause in the warfare between Israel and

122 Syria. King Ahab had won a peace treaty after a spectacular victory over the Syrians.

According to the treaty, Syria had pledged to return the captured Israelite city of Ramoth. But the town was still in Syrian hands. This angered King Ahab. Finally he became so angry that he decided he must take action. So he sent a message down to King Jehoshaphat of Judah asking him to come at once to Samaria for a conference.

Under King Ahab's father, peace had been established between Judah and Israel. These friendly relations had been further cemented by a marriage between children of the two royal families.

King Jehoshaphat arrived at the gates of Samaria, wondering what Ahab could want so urgently. At a wide place by the gates, King Ahab had set his throne. Another throne had been placed beside it for the king of Judah.

When both kings were seated, King Ahab looked at the Israelites who thronged about. "Do you know that the town of Ramoth in Gilead belongs to us?" Ahab shouted to his subjects. "Why do we keep quiet, and make no move to take it out of the hand of the king of Syria?"

"Hear, hear!" the people shouted.

King Ahab turned to the king of Judah. "Will you go with me to battle at Ramoth-gilead?"

King Jehosphaphat was not enthusiastic. He did not want to get involved, but he didn't know exactly how he could get out of it. After all, Israel was the stronger power, and Judah was really just a vassal state. Jehoshaphat bowed his head, and said with diplomatic silkiness, "I am as you are, my people as your people, my horses as your horses."

Ahab nodded with great satisfaction. He would need all the help he could get.

Suddenly King Jehoshaphat thought of a possible way out. Turning to Ahab he said, "But first we must inquire for the word of God!"

Ahab was prepared for that. He beckoned to a court official, and four hundred men were summoned. They were dressed alike in short white linen tunics. "These are the official seers and soothsayers of my court," Ahab explained. "They can prophesy and tell me what I want to know."

Then Ahab turned to the crowd of advisers. "Tell me, shall I go to battle against Ramoth-gilead, or shall I not?"

The professional wise men looked at one another. It was obvious that the king was itching for a military victory. "Go up!" their spokesman said obediently. "God will give Ramoth-gilead into the hand of the king."

"Go up, go up!" the four hundred chanted.

"There!" Ahab said triumphantly to Jehoshaphat. "Four hundred seers cannot be wrong. Now are you satisfied?"

"Not yet," Jehoshaphat complained. "Is there not another prophet of God?"

King Ahab threw up his hands. "There is yet one man by whom we may inquire of God," Ahab finally admitted. "His name is Micaiah." Ahab's face reddened with anger. "But I hate him! This prophet is a bird of ill omen. He never prophesies anything good for me. Only evil!"

Jehoshaphat was eager for a differing opinion. "Now, now," he pacified. "Let not the king say so."

So Ahab sent for Micaiah the prophet.

While they waited, the leader of the professional seers picked up two long, curved daggers. He held the sharp iron weapons against the sides of his head as if they were the horns of a fighting animal. He jumped up and down before the two kings and shouted to Ahab, "With horns of iron you shall push the Syrians until they are destroyed."

Behind him the chorus chanted, "Go up!

Go up to Ramoth-gilead and triumph!"

Meanwhile, the messenger found the prophet Micaiah. "Listen," he advised, "the words of the king's soothsayers with one accord are favorable to the king. If you value your life, let your own words be the same. Speak favorably!"

But Micaiah said, without fear or trembling, "Whatever God wants me to say, that I will say."

When Micaiah had come before the two kings, Ahab said to him, "Micaiah, shall we go to Ramoth-gilead to battle, or not?"

Micaiah said, "Oh, do go up, and triumph! Perhaps God will give the city into the hand of our king!" There was sarcasm in his voice.

Ahab sensed this and said, "How many times must I tell you to say nothing but the truth?"

Then Micaiah said to the king, "I saw all Israel scattered upon the mountains of Gilead, as sheep—sheep that have no shepherd! Then I heard God say, 'These sheep have no master. Let them return home in peace.'"

Ahab let out a bellow of rage. He turned to Jehoshaphat. "Did I not tell you," he roared, "that he would not prophesy anything good about me? Only evil!" The king signaled to his captain of the guards. "Put this fellow in prison, and feed him with short rations of bread and water, until I return in peace."

Micaiah said to Ahab, loud and clear so that the thronging people could hear, "If you return in peace, then God has not spoken by me."

So the king of Israel and the uneasy king of Judah ignored Micaiah. They prepared to go over to Ramoth-gilead and fight.

"You will, of course, wear your royal robes in the battle," the wily Ahab told King Jehoshaphat.

Then Ahab himself put on the disguise of a common soldier. With chariots and foot soldiers the army went eastward into Gilead, crossing the river Jordan at the fords.

Now the king of Syria had told the captains of his chariots, "Fight with neither the small nor the great, but only with the king of Israel." For the king of Syria knew that Ahab was a powerful and crafty warrior. If the king were killed, his troops would be without a leader, and the battle would be won.

So when the Syrian captains saw Jehoshaphat riding in a royal chariot, dressed in regal robes, they said, "Surely it is the king of Israel!" Immediately, they turned to mass their fight against him. The terrified Jehoshaphat cried out, "Spare me! I am not an Israelite! I am only a Judean!"

When the Syrian captains saw that he was not Ahab, they left him alone.

But as for Ahab, even though he was not recognized, a stray arrow struck him.

"Take me out of the battle," Ahab gasped to his chariot driver. "I am wounded."

The battle raged all day, as the Syrians sought to find and kill Ahab. But Ahab was behind his own lines, propped up in his chariot so that his army could see him and take heart. By evening King Ahab had bled to death and lay slumped on the bottom of his chariot.

About sunset, a cry went up, "Our king is dead! Every man to his city, and every man to his country!"

The army fled back across the Jordan, every man going back to his house. And King Jehoshaphat, with his own men, returned to Jerusalem in Judah.

The dead King Ahab was brought back to Samaria and buried. And so the words of Micaiah came true—Micaiah, the prophet who never had anything good to say about his king.

1 Kings 22

Jehu and Jezebel

After the death of King Ahab his oldest son ruled Israel, but his reign lasted only two years. The new king accidentally fell through the latticed window of an upstairs room in the palace and died from the injuries. His brother Joram, the next son of Ahab, now became the king of Israel and reigned at Samaria. The widowed queen mother Jezebel took over the summer palace in Jezreel.

Some years after this another king came to the throne in Judah, a young prince related by blood to the royal house of Israel. King Joram sent down word to his Judean kinsman, "Come and join forces with me. Let us finish the battle my father and your grandfather started with Syria. We still have not recaptured Ramoth in Gilead."

So off in their chariots went the two kings with their men of war. But they did not do well in their skirmish with the well-armed Syrians. Joram, king of Israel, was badly wounded. He left the combined armies of Israel and Judah in the able hands of Jehu his commander and rode back to the summer palace at Jezreel.

There his mother Jezebel cared for him while his wounds healed. And his kinsman, the king of Judah, went to Jezreel to cheer King Joram in his convalescence.

Meanwhile, there was in Israel a vigorous young prophet who was active in the political and religious affairs of his country. His name was Elisha.

Elisha had been a wealthy young farmer when he heard of the activities of the revered prophet Elijah. Elisha left his farming and became Elijah's most devoted disciple. He followed the prophet wherever he went, listening to his words and observing what he did.

It was to this young follower that the authority of Elijah passed when the old prophet's career was ended. Elisha had been witness to a remarkable event. He told how a "chariot of fire and horses of fire" descended, and "Elijah went up by a whirlwind into heaven." The mantle that the old prophet had worn was left behind. Elisha took it up and wore it for the rest of his life as a sign of his own authority as a prophet.

Elisha now became the most influential prophet in Israel, and he had disciples of his own.

One day Elisha called one of his disciples to him. "Gird up your loins," Elisha told the young man. "Tuck up your tunic beneath your belt so that you can run freely, for I have an errand for you." Then Elisha solemnly handed him a small stoppered bottle made of clay.

"Take this flask of oil," Elisha went on, "and look for the commander Jehu who is in charge of the military camp at Ramoth-gilead. Get him apart from his fellows and into a room where you can be alone. Then take this holy oil and pour it on his head, and say, 'Thus says our God: I anoint you king over Israel.' Then open the door and flee. Do not tarry."

So the young prophet did as he was told. When he arrived at the camp, he found the commanders of the army in council. He cleared his throat and called out, "I have an errand to you, O commander."

Jehu turned his head and looked at the breathless young man. "To which of us?" Jehu asked.

"To you, O commander," the young man said, pointing directly at Jehu.

So Jehu followed the young man into a small building nearby, where they could talk privately. The prophet took the flask from his belt, poured the drops of oil on Jehu's black hair, and said to him, "Thus says the God of Israel, I anoint you king over the people of Israel! And you shall strike down the sons of Ahab, your former master. You will avenge on Jezebel the blood she has shed. And the oracle of Elijah will come to pass, that the dogs shall eat Jezebel in the territory of Jezreel!"

Then, without another word, the young man opened the door and fled.

When Jehu returned to his conference, the other commanders who were under him asked, "Is all well? What news did this fellow bring to you?"

Jehu shrugged, trying not to show his growing excitement. "It was nothing," he said, making an effort to sound casual.

But they were not satisfied and said, "That is not true! Tell us. Tell us now!"

So at last Jehu told them everything, that the young man had come from the prophet Elisha, and all he had done and said.

The other commanders quickly caught the importance and excitement of what was happening. Pulling off their colorful robes, they threw them down on the steps of a building nearby. When Jehu had mounted this improvised throne, they blew on a ram-horn trumpet.

"Jehu is king!" they shouted. "Long live Jehu!"

Jehu held up his hand. "If this is your mind," he said, "then set guards around the camp. Let no one slip out of the city to go and tell the news ahead of me in Jezreel."

Then Jehu mounted his chariot, called a guard of men to accompany him, and rode furiously toward Jezreel.

Now the watchman who was standing on the tower by the city gates saw the dust boiling up on the horizon and sent word to the king.

Joram came painfully to the gates, bandages still over his seeping wounds, his royal kinsman from Judah assisting him. He

126

called up to the tower, "What do you see?"

The watchman answered, "I can barely make out a company of men riding. And out in front, it looks like a chariot."

Joram gave orders to the two guards at the gate. "One of you quickly mount a horse. Go meet this company, and say, 'Is it peace?'"

So the guardsman on horseback thundered out to meet the approaching company. When he reached Jehu and his party, he asked, "Is it peace?"

Jehu barked, "What have you to do with peace? Turn round and ride behind me."

The watchman reported to the anxious king, "Your messenger reached them, but he is not coming back."

So King Joram sent out the second guard. When this horseman reached Jehu, he called out, "Is it peace?"

And Jehu again answered, "What have you to do with peace? Turn round and ride behind me."

Back at the tower, the watchman called down, "He reached them, but he is not coming back." He peered again at the nearing cloud of dust. "And the driving of the chariot in front—it is like the driving of your commander Jehu. For he drives furiously!"

King Joram ordered, "Bring out our chariots!"

Then the king rode out to meet Jehu. Joram was accompanied by the visiting king of Judah in his own chariot.

When King Joram saw Jehu, he reined his chariot beside the chariot of the commander. "Is it peace, Jehu?" he asked.

Jehu demanded grimly, "What peace can there be, so long as the sins and sorceries of your mother Jezebel are so many?"

Then the king of Israel reined his chariot about to flee, shouting a warning to the king of Judah behind him. "Treachery, cousin, treachery!" he yelled.

But Jehu had drawn a bow with his full strength, and he shot Joram between the shoulders. The arrow pierced the heart of the king of Israel. He sank, dying, to the bottom of his chariot.

"Take up the body of Jezebel's son," Jehu ordered a man on horseback beside him. "Throw the body of the son of Ahab into the vineyard that his father stole from Naboth!"

The youthful king of Judah knew he was equally doomed. Was he not a blood relative of the hapless Joram? The hated Jezebel was his grandmother.

The king of Judah was urging the horses of his chariot at topmost speed, in a frenzy to escape. But Jehu and his men followed in hot pursuit, and this king also was mortally wounded by a swift arrow. He died before his chariot could carry him out of the valley.

When Jehu came riding back to Jezreel, the queen mother Jezebel heard the news, and she knew what it meant. Jezebel had always been a proud and beautiful woman, and she now determined to die looking her best. Her hand was almost steady as she held her polished bronze mirror and blackened the edges of her eyelids in the fashion of the Phoenicians and the Egyptians. When her eyes were painted, when her hair was arranged and her head adorned, she looked out of her upstairs balcony window.

As Jehu entered the palace courtyard, Jezebel called down from the window, her voice edged with sarcasm. "Is it peace, you king-murderer?" she asked.

Jehu shouted, "Who up there is on my side?" Several palace servants looked out at him. "Throw her down!" Jehu commanded. So they threw Jezebel down to her death.

Then Jehu went into the palace as the new master. He ordered a banquet for himself and his men. As he ate and drank, the thought came to him that, after all, Jezebel

128 had been born a princess and had been queen of Israel. He laid aside the joint of roast lamb he was eating long enough to instruct his orderlies. "See now to this cursed woman. Bury her, for she is a king's daughter."

But when they went to bury her, they found that half-wild scavenger dogs had devoured the body. The long-ago prediction of the prophet Elijah had come to pass.

Thus began the rule of King Jehu over Israel.* He continued his bloodbath against remaining members of the royal house of Ahab and Jezebel until all had been killed. He plotted the deaths of Jezebel's Baal priests.

Jehu ended his reign as a political failure. Syria waged a war for half a century that was disastrous for Israel, with Israel's land east of the Jordan being captured. Farther to the east, in the Tigris and Euphrates valleys, the growing and fearsome nation of Assyria was a threat to Israel, too. To keep the Assyrian armies from his borders Jehu had to pay staggering sums of tribute to their kings.** Under Jehu and his son, Israel became small, weak, and almost bankrupt.

So it was that the rise of Jehu was also the beginning of his fall.

2 Kings 1 through 2; 9 through 10

* During the reign of Jehu, the prophet Elisha continued to be active in politics and religion, following in the footsteps of the great Elijah. In fact, there were soon as many or more miracle tales and wonder stories growing up around the name of this disciple, than there had been around his master.

** Archaeologists have unearthed the famous Black Obelisk, a four-sided monument about as tall as a man, which was erected by the powerful Assyrian king, Shalmaneser III. On it are chiseled illustrations of some of his war records; the inscriptions below, in the curious wedge-shaped cuneiform script, tell the details. One of the panels shows Jehu of Israel kneeling and prostrating himself before the Assyrian king. On the bottom panel are seen thirteen Israelites from Jehu's court carrying tribute to the Assyrian king: "silver, gold, golden bowls, a golden case, golden tumblers, vessels, a block of tin, a royal staff for a king, and fruit piled in baskets."

Amos, Prophet of Justice

Several generations had passed since the time of Ahab and Jehu. Jeroboam II, the great-grandson of Jehu, now ruled Israel. Outwardly there was prosperity again in the northern kingdom. Jeroboam was a strong king, and his court at Samaria glittered with lavish splendor.

Yet there was misery and a decay of spirit underneath all that wealth, and one man saw it clearly. His name was Amos.

Amos was not an Israelite, however. He lived in the remote village of Tekoa in the mountains of Judah. Here was wilderness; nearby, cliffs rose above the dead silence of the bright blue Salt Sea. It was a place where one could feel close to the elements and to God.

Amos was only a rustic shepherd, and not a professionally educated man. Yet he had taught himself to read and write. He knew and loved the history of his people. And he could set his prophecies in the rhythm of magnificent poetry.*

Though he lived in the country, Amos had traveled. Every spring the wool sheared from his sheep had to be taken to market and sold.

It was only a morning's walk to the markets at his capital city Jerusalem. More often Amos went across the border and journeyed farther north to sell his wool in the more profitable markets in Israel, especially at Bethel and Samaria.

On these journeys into Israel, Amos had found himself shocked by what he saw there. The upper classes were very rich and lived in the midst of extravagant luxuries. Among the great masses of the people, life was hard and bitter. Amos became indignant about the injustices he saw. Often a poor man would be sold into slavery to pay a debt amounting to the price of a pair of sandals.

In the city marketplaces, merchants used dishonest weights and measures cheating everyone—the farmers bringing in produce for sale, and the customers from the street.

The worship of Baal, suppressed for a while by Jehu, was rampant again. Israelites forgot their own religion as they joined in the licentious Baal festivals.

"These people have gone back on their covenant with God," Amos often muttered angrily to himself. Amos knew the stories about God's giving this land to Israel, but not to use as they were using it. If Israel continued to break God's commandments and to oppress the poor, then surely God would act.

One spring day Amos sat watching his sheep. The rains had turned the Judean hillsides green with new grass. Suddenly, as in a vision, Amos saw a swarm of locusts devour the grass to the roots. A chill gripped him. He had been thinking of the northern state of Israel, so lush, so prosperous. Surely this was symbolic! Just so, hordes of enemies would someday sweep across Israel and devour the people, leaving the land barren and devastated, stripped as if by a locust plague.

In the middle of the summer Amos had his second vision. At this time of year Amos

worked down in the dry lower valleys where the sycamore figs grew.** This inferior kind of fig had to be pinched or slit to allow excess fluid to escape so they could ripen properly. The poor had to be content with these because they could not afford the delicious fresh or dried common figs in the markets.

It was a blistering hot day, as Amos walked through the stubbled fields on his way to where the fig trees grew. Shimmering heat waves rose from the sunburnt hills. Amos half-closed his eyes. It was as if a great fire were devouring the land, flames licking at the dried stubble. Amos felt that God was showing him, in this pictorial imagery, things that were to happen to Israel!

Summer ended. Autumn crispness was in the air. Amos found himself meditating as his eyes wandered idly over a basket of summer fruit. He walked closer. How ripe and soft the fruit was, overripe, really, on the point of decay.

As Amos thought, he realized that here was the most fitting symbol of all. Israel also looked good—from a distance. But if you examined her more closely, you could see the rapid decay!

Amos became restless. He was acutely aware that God had appointed him as a prophet to Israel. The knowledge he had gained from the visions were not for his own learning, but for Israel.

He made up his mind to go up to Bethel and Samaria. He would bring the word of God to the Israelites.

"But how can you prophesy?" his friends asked. They knew he was a quiet man, a man of few words.

"When a lion roars," Amos said, "does it not follow that whoever hears becomes afraid?"

His friends nodded.

"So also, if God speaks to a man," Amos explained. "It follows, he will prophesy!"

Amos began his prophetic teachings across the border in Israel at Bethel. It was a festival day. People were thronging into Bethel to worship at its famous temple, where a golden calf had been placed by the first King Jeroboam. That king had hoped to entice his people to worship there, so they would not go down to Jerusalem in Judah. And the succeeding kings of Israel had preserved the custom.

Chattering people jostled each other in the marketplace. It would not be easy to get the attention of that busy, noisy crowd.

Amos leaped on an empty cart. "You hate the Syrians at Damascus," he shouted. "Many are their crimes, and God will punish them!"

People stopped, and a few cheered, for the Syrians were their long-time enemies.

"You hate the Phoenicians, especially in the city of Tyre," Amos went on. "Many are their crimes, and God will punish them."

More people stopped and applauded. Was not Tyre notoriously wicked? And the hated Phoenicians often sold captured Israelites into slavery in Edom.

Amos continued to describe enemy nations of Israel and to predict their doom. The Ammonites, who had killed women and children in Gilead—just to enlarge their borders. The Moabites, also from the other side of the Salt Sea—had they not fought with the Israelites since the days of Moses? Amos promised that God would punish each of these nations.

The growing crowd was loud in its approval. They nudged, grinned, and cheered.

Having obtained their attention, Amos then shouted, "Listen to what God says: many are the crimes of Israel! And God will punish you—"

The crowd looked baffled. The cheers died away. People whispered uneasily.

Amos stamped noisily on the cart raising a whirl of dust and wheat chaff. "Because you trample the head of the poor into the dust of the earth," he accused them. "You sell honest folk into slavery for a small debt."

A man in the front row of the crowd ducked his head, as if the words of this rustic stranger were too much to hear. He fondled a small clay image in his hands.

Amos reached out and plucked the image from the man's hands. He held it up, contempt written upon his stern face. "You worship man-made idols with a lust that profanes the name of our God!" Amos hurled the pagan image to the ground so fiercely that it shattered.

A cold wind was blowing through the marketplace. Most of the people were well-to-do, and warmly wrapped in woolen cloaks and mantles. One poor man, who looked hungry and emaciated, shivered at the back of the crowd. He wore only a thin, knee-length shirt.

Amos pointed to the man. There were many like him in the towns of Israel. "The rich take a debtor's only coat as a pledge. They break the law of Moses, for they leave him with no garment to keep out the cold!" Now the prophet from Tekoa shook his fist in a gesture of warning. "Listen well, a day of judgment comes. And no one shall escape!"

Amos leaped down from the cart and disappeared in the crowd. He strode home.

Even in the remoteness of Tekoa, Amos heard the fearsome rumors about Assyria. That great nation to the east was increasing its military might, gobbling up new nations right and left. Would Assyria soon pounce on Israel? Amos asked himself. Would God use Assyria to punish Israel, and other nations, for their sins?

Amos felt a compelling urge to return to Israel and warn the people again.

This time the stern, rugged prophet hiked up to Israel's great walled capital, Samaria, set like a queen on a commanding hill.

As he stood up to speak by the gate, a group of overfed, plump, richly dressed matrons came by. They tittered together, a bit tipsy from drinking.

"Hear the word of God, you fat cows!"

The women turned, incredulous. Was this country bumpkin speaking thus to them?

"Yes, you, on the hill of Samaria," Amos said. "You, who oppress the poor, who crush the needy. You, who say to your husbands, 'Bring goblets, that we may drink!' "

Some of the women, red-faced with anger, shook their jeweled fists at him.

Amos went on, imperturbable. "Look, the days are coming when you shall be dragged from the city, as dead animals are dragged out."

The ladies of Samaria began to scream insults at this rude shepherd. They made such a noise a crowd gathered to see what was going on. Amos shouted fearlessly to the people.

"You think the Assyrians are cruel and wicked? That the Egyptians are pagan and sinful? They should come to Samaria and see the cruelty and wickedness here! You do not know how to live honestly. You cheat and steal, and you encourage violence! Do you know what God will do if you do not change your ways?"

The crowd hushed suddenly, to hear God's word from this strange man.

"Thus says the Lord God: 'An enemy shall surround your land. Your defenses shall be brought down, and Samaria shall be plundered.' "

The whispering began again. "Are we going to be attacked?" people were asking. "Who could do it—Assyria? Or maybe Egypt?"

"Woe to those who feel secure on the hill

of Samaria," Amos thundered at the well-dressed crowd. "Woe to those who stretch out on ivory-trimmed couches, feasting on roast lamb and tender calves, drinking wine by the bowlful. Woe to you who are not concerned—and heartsick—over the ruin and poverty of your fellowman! God says you shall be the first to go into exile."

Again there was quiet. There was something about the forceful voice, the stern and earnest face of the prophet that awed this crowd.

Then a man raised his voice, trembling with a sudden fear. "If all this happens as you say, will no one be rescued?"

Amos frowned. "As the shepherd rescues from the mouth of a lion two legs, or a piece of an ear, so much, and no more of Samaria will be rescued." Gloomily he shook his head. "Only a remnant will be left."

Amos turned from the people, who listened but did not repent of their ways. They felt safe behind the massive walls of their capital.

The prophet left Samaria, going south. He decided to preach again at the border city of Bethel.

It was the sabbath when Amos arrived. People were supposed to be at rest or at worship. In the Bethel marketplace the merchants already were setting out a display of their wares. They were impatient to begin selling.

"Hear this, you merchants who trample upon the needy," Amos shouted at them in his anger. "You have two sets of measures, one for buying and one for selling. You sell short weight and you cheat the poor, making them even poorer. Watch out, evil days are coming for you."

Amos walked down the crooked streets to the temple. Here people were converging with their animal sacrifices. The priests slew the animals and burned them on the altar. The priests took a choice part of the roasted meat as their portion, and a part was returned to the worshipers for their own feasting.

Amos watched the people. They seemed unaware that this had nothing to do with a real worship of the Lord. It was more like a carnival. Amos challenged them, "Do you know what God has said to me?"

The people looked at this stranger, and some stopped to listen.

"God says, 'I hate, I despise your feasts, and I take no delight in your solemn assemblies. Even though you offer me your burnt offerings, I will not accept them.' "

The people looked surprised and confused. They had become so sluggish in their prosperity that they had lost sight of the fact that there was more to worship than going through the motions of offering and sacrifice. With things going along so well they felt no need for a close relationship with God. They dismissed their religious duties with careless gestures in the temple, forgetting that these were meant to be signs of their inner devotion.

"What does God want of us," someone called out, "if he doesn't want our sacrifices?"

Amos explained that God wanted his people to be faithful in their worship and to show justice for the downtrodden and needy. He cried out,

"Let justice roll down like the waters,
    and righteousness like an ever-flowing
        stream."

Many of the people were uncomfortable with these words. One man cupped his hands to his mouth and yelled, "Go back to the country, you hayseed. You haven't anything but criticism."

"I know," Amos admitted. "People hate reproof. They turn away from him who speaks the truth."

"But we are God's chosen people," said

another man. "God will protect us against all other nations."

Amos shook his head. "Do not believe that. God also loves and guides the Ethiopians, the Philistines, the Syrians—"

"The eyes of God are upon us," someone protested loudly.

"Of course," Amos agreed. "The eyes of God are especially upon this kingdom! God knows how you keep breaking his covenant with you. So I warn you. Israel will be destroyed, down to the ground."

A soldier in the crowd boasted, "Even Assyria cannot harm us. We have a great army in Israel. And Jeroboam is a powerful king."

Amos said, "The sons of Jeroboam shall die by the sword. And Israel must go into exile, far away from his land."

"Treason!" several persons cried out.

Amaziah, the chief priest of the temple, ordered a messenger to hurry immediately to Samaria. "Take this message to Jeroboam our king," Amaziah said. "Tell him that Amos has conspired against the throne in the midst of Israel. The land is not able to bear all his words!"

Then Amaziah decided he could not wait to hear from the king. The chief priest called his armed guard and confronted the prophet.

"O seer," the priest roared at Amos, "go! Flee to the land of Judah and be a seer or a soothsayer there. But never again prophesy at Bethel!"

Then Amos answered Amaziah. "I am no seer, and I am no soothsayer. I am a herdsman and a dresser of sycamore fig trees. God took me from following the flock and told me to go and prophesy to his people. Now, remember all that I have said! For your land shall be parceled out to others. You yourself shall die in an unholy land. And Israel surely shall go into exile."

Amos went back to Judah, back to his sheep and his fig gathering at Tekoa. He wrote down the visions he had seen that had impelled him to be a prophet of doom to the heedless people of Israel.

Amos 1 through 9

* Amos lived in the middle of the eighth century B.C., the same century that produced Hosea, Isaiah of Jerusalem, and Micah. Amos was the first of these great writing prophets, a new breed of man who had the courage to speak the truth and the ability to write it down.

Amos was a man of his time. Like the prophets who followed him, he spoke to the people of his age. Yet the insights of Amos are so universal in scope that their basic elements can apply to almost any country in any time.

** Sycamore figs were the fruit of a tree belonging to the mulberry family. They were not related to true figs or true sycamores. It was the shape of the fruit which earned it this name. The term is sometimes spelled "sycomore figs."

Hosea, Prophet of Love

Most Israelites had not been impressed when Amos came up from Judah to warn them. Those who lived well continued to be callous and indifferent to the plight of the poor and the downtrodden. They continued to worship the Baal-gods. But the impassioned words of Amos did awaken a few responsive Israelites.

While Jeroboam II was still on Israel's throne a young man named Hosea became afraid for the future of his beloved country. Like Amos, Hosea was pained by the way people had turned away from God, and by the inhumanity which resulted.

134

This sensitive young Israelite also was disgusted with the widespread Baal worship in his country. But he had an extra and personal reason to hate the licentious cult. Hosea was married to Gomer, once a priestess in a Baal temple.

Hosea thought his young wife had turned her back on the seductive lure of the fertility cult. She seemed happy to be married. She shared Hosea's delight in their first child.

About this time Hosea felt the call to carry God's word to his people. Israel must be warned before it was too late, before God punished the nation with destruction and exile.

Hosea used the name of his child to dramatize his first message. To all who would listen he said, "God told me to name my son Jezreel. Why? Because it was in the Valley of Jezreel that Jehu cruelly spilled much blood. God will punish the wicked house of Jehu in that same valley and bring an end to his kingdom!"

Then Hosea would hold up young Jezreel. "And whenever you hear the name of my son, you will remember my prophecy."

Not long after this, King Jeroboam II, great-grandson of Jehu, died.

"See?" Hosea's neighbors whispered among themselves. "That young man's prophecy never came true. Jehu's line was not punished after all. Our king just died of old age."

But before six months had passed, the son of Jeroboam was struck down at the south entrance to the Valley of Jezreel. The assassin had himself crowned king.

One month later this murderous usurper was killed by Menahem, captain of the army. Menahem moved into the palace at Samaria and was crowned king. More bloodshed followed, as Menahem cruelly butchered any rebels.

"Who is really aware of God in this land?" Hosea asked his countrymen. "There is no loyalty or kindness here. There is swearing, lying, and stealing. Marriage vows are broken! There is killing, and murder follows murder!"

Hosea felt especially bitter toward the priests. Did they not have the opportunity to feel very close to God? Were not the priests supposed to guide the people toward the right? Yet the priests were as unfaithful as the people.

Hosea strode to the temple and called out to the chief priest. "My quarrel is with you, you so-called priest! Why have you forgotten the law of God?"

The overfed, richly dressed chief priest came out, frowning. He was surrounded by the many other expensively robed priests of the temple. A crowd began to gather.

Hosea went on. "The people's souls are being destroyed, because they are not sensitive to wrongs. They are not aware! They will die, because they do not know or understand!"

The crowd began to murmur. The priests looked uncomfortably at each other, wondering how to force this impudent young man to shut up.

Hosea pointed an accusing finger at the chief priest. "And because you are without insight, because you grow fat on the people's wrongdoing, I reject you from being a priest to me!"

Now Hosea pointed to a wooden pole set up in the middle of the temple court. On the top was a carved image. Hosea cried out in despair, "The people seek advice from a thing of wood! They bow down to a wooden pole and ask it to tell their future!"

Again the people stirred and murmured.

Hosea whirled and addressed them. "You are unfaithful to our God. You lust after Baal, and your daughters degrade themselves in the Baal shrines. You go outside your city and make pagan sacrifices on the hilltops." His face reflected his disgust.

"People who are not sensitive and understanding will come to ruin," he cried.

Hosea's concern over the immorality of the Baal-cults became a personal sorrow. Gomer was tempted back into the sensual ways she had followed, before her marriage. She began to slip out to meet her lovers and to join in the wild rituals of Baal worship.

Gomer gave birth to another baby, but Hosea could not be sure he was the father. When neighbors asked Hosea the name of his infant daughter, the prophet said, "Call her name *Unpitied*. For she is like Israel! Our people continue their deliberate wrongdoing. How can God still have pity on us?"

When a third child was born, Hosea said, "Call his name *No-kin-of-mine*. For he is like Israel. How can God consider that we belong to him, when we do not accept him as our God?'"

At last the fickle Gomer left Hosea completely, running away with one of her lovers.

Hosea found a neighbor woman who could care for his three motherless children. He was crushed with grief. He loved his wayward young wife dearly, even though she had sinned against him.

"And if I grieve for Gomer," Hosea realized, "how much more God must grieve for the wayward Israelites."

Hosea began to use the parable of marriage in talking to his countrymen.

"With the covenant in the desert, God took Israel to be his wife. God remains faithful. But see how Israel runs after her lovers, the Baals!" Hosea shook a warning forefinger at the people. "You think your wooden idols bring you grain and vineyards and olive oil. You thank the Baals for your silver and gold. You are wrong. Your gifts come from the only God in heaven. Now God will punish Israel for her unfaithfulness. He will put an end to her gaiety and laughter. Her fields, vines, and fig trees will be laid waste. Israel shall become a desolation!"

But the impassioned words of Hosea did not move the people to change their ways. Life went on as before. Menahem, the king who had gained his throne by slaughter, continued to rule from the luxurious palace in Samaria.

During Menahem's reign, the most important nation on the international scene was Assyria. Even the venerable might of Egypt was eclipsed by this rising military power. Assyria had great cities in the Tigris and Euphrates valleys. Ancient Babylonia also was under her domination.

Menahem recognized the superior power of this gigantic nation to the east, and made treaties of alliance. He taxed the people heavily to pay the exhorbitant tribute Assyria demanded.

The people groaned under their tax burden. A rival political party advocated breaking off diplomatic relations with Assyria. This opposition party, which was secretly supported by a captain of the king's guards named Pekah, wanted to appeal for help to Egypt.

After almost a decade of rule, King Menahem died. His son took over the throne.

The rebellious Pekah bided his time for two years. Then, with the help of fifty picked men, the wily captain murdered the young king and seized the throne.

Hosea pushed aside his private grief. He lashed out at the political chaos and the immorality. "Listen to what God has said: 'Israel has made kings, but not through me. They set up princes, but I never knew of it. And now they count for nothing among the nations!'"

The people gaped, as Hosea graphically flapped his hands this way and that.

"Israel is like a silly, senseless bird, running to Assyria, calling to Egypt," Hosea said scornfully. "Do you know what God says? 'I will spread my fowler's net over Israel. I will catch them like a bird in a trap. I will punish them for their wicked deeds.' That is the word of God!"

The people looked stunned. Hosea pointed at them. "You have sowed the wind. Now you shall reap the whirlwind!"

The country was in a political turmoil. It was so demoralized, Egypt could easily march northward and take over the nation. How ironical, if the ancient enemy dragged the Israelites back into another period of slavery by the Nile!

On the other hand, Hosea warned, it might be powerful Assyria which would lay waste to Israel and take the people away in chains. It would be God's punishment.

"You put your trust in your chariots and in the multitude of your warriors," Hosea accused the people. "But the tumult of war shall arise among you. Because of your wickedness, all your fortresses shall be destroyed. The sword shall rage against your cities and devour you. Nettles shall cover your precious things, and thistles will grow in your abandoned homes. In Assyria you will eat unclean food. And your useless, man-made golden calf will be carried off as spoil for the Assyrian king."

It was after he had made these prophecies that Hosea walked through the marketplace and saw a slave woman offered for sale. Hosea hardly recognized her. She had suffered, and her ravaged face and figure showed how much. This degraded slave was the once-beautiful Gomer, his wife.

The compassionate prophet bought her for fifteen pieces of silver and two measures of barley. He took the broken woman home and gave her a room to live in. Finally, with Gomer chastened and repentant, love overcame the barriers. Hosea and Gomer were reunited.

Then a realization, a great insight, came to Hosea. Perhaps his personal suffering had not been in vain. The gentle, loving prophet became convinced that God had allowed his marriage to an unfaithful wife, perhaps even willed it. Had this not helped him to interpret to the Israelites how God felt about their unfaithfulness?

In the past, God had been feared as a stern and just God. Amos had emphasized his justice.

Now Hosea realized that God must also be a kind and loving God! He is not a God of blood and vengeance. God is both loving and forgiving.

If he, Hosea, a mere man, could continue to love Gomer after all her unfaithfulness, if he could forgive her after she was chastised and repentant, would not the same be true of God? For God's qualities would be more noble than those of any man!

Now Hosea could hold out a distant, future hope to his fellow Israelites. Israel would go into captivity. Yet someday the heedless people would repent, and God would forgive them.

Hosea 1 through 14
2 Kings 15:8-28

Israel Falls

King Pekah ruled Israel from his blood-stained throne.

Now that his anti-Assyrian party was in power, Pekah ejected those who sympathized with Assyria from his government. He refused to send any tribute to Assyria.

"Now is the time to free ourselves from domination," Pekah told his cabinet. "But Israel is too small to fight Assyria alone. We must form an alliance with all the small nations here in the west."

Rezin, king of Syria, was eager for such an alliance. The Philistines also wanted to end payments of tribute to Assyria. Both nations agreed to put aside old quarrels and join forces with Israel.

Ahaz was now king in Judah. Ahaz was too much afraid of gigantic, menacing Assyria to join. He sent word: "No!"

King Pekah, pacing the palace floors in Samaria, was furious. He met with King Rezin at Damascus. Together they plotted to make war against Judah. They would force King Ahaz to join their alliance.

Rezin marched down from Damascus with his Syrian army. King Pekah called out his army of Israelites and crossed the southern border. The two armies closed in on Jerusalem, besieging the city.

At this King Ahaz panicked. He rushed messengers to the king of Assyria, saying,

"I am your humble servant! Come quickly and rescue me from the hand of the king of Syria and the king of Israel who are attacking me!" Along with the message Ahaz sent rich gifts of gold and silver, taken from Solomon's temple and from his own palace.

The king of Assyria was glad for this opportunity. He mobilized his great army and marched westward, toward the boundaries of Syria and Israel.*

When King Pekah and King Rezin heard this, they abandoned their siege of Jerusalem. They rushed their armies back to defend their own nations.

Onward came the Assyrians. They marched down the coast of the Great Sea and captured the ancient Philistine stronghold of Gaza. The Philistines became powerless to aid their allies.

The Assyrian king circled back and captured the northernmost areas of Israel. Next he devastated Syria and successfully besieged their capital of Damascus. Syria's King Rezin was killed. The leading citizens were chained and taken away as captives.

Hosea had prophesied, "Samaria's king shall perish, like a chip on the waters." The prophecy came true. Israel's allies were gone, the state was in confusion. King Pekah was murdered. The leader of the conspirators, with the help of Assyria, put himself on the throne.

Then the old king of Assyria died. The new monarch, Shalmaneser V, proved to be equally strong. The shattered nation of Israel was forced to pay Assyria great annual sums in tribute.

Time passed, and Shalmaneser found treachery in his vassal Israelite king. Spies reported that messengers had gone to Egypt for help, in a secret plot against Assyria.

Shalmaneser ordered the Israelite king to be chained and brought back to an Assyrian prison. Then the Assyrian monarch swept westward with his mighty army and laid siege to rebellious Samaria.

The policy of the Assyrians was to destroy every city and every kingdom that dared to resist or refused to pay tribute. The captured people were scattered in faraway places throughout the empire. And the deported people were replaced by assorted captives from many other places. This thorough mixing of captives was planned in order to discourage further uprisings.

The inhabitants of Samaria knew that if they were captured they were doomed. The destruction of their capital would be the end of their little nation. So the Israelites resisted bravely. They shut themselves within their hilltop city, behind its massive stone walls.

For three long years Samaria was besieged by King Shalmaneser. Food grew so scarce that citizens had to scrounge for refuse. Some tried to cut leather sandals into pieces and boil them for soup. Many people died of starvation. Most of the survivors were weak and sick.

At last the Assyrians captured Samaria. Some of the defenders were cruelly impaled on sharp stakes. The leading citizens were marched away, weeping, disconsolate captives. Some were picked to be slaves at the Assyrian palace. The rest were scattered and resettled in far eastern provinces of Assyria.

The fearsome Sargon was now king of Assyria. Sargon brought enslaved people from Babylon and other provinces to rebuild the war-ravaged homes of the people of Samaria. These assorted strangers, with their foreign ways and their foreign religions, took possession of Samaria and dwelt there.

These new people made gods of their own and put them in the local shrines and places of worship. Some of the newcomers also began to worship Israel's God, believing that he might be powerful in his own land.

But the independent kingdom of Israel was no more, and never again would be.

2 Kings 15:29 through 17:41

* Records, inscriptions, and monuments have been dug up at the site of the ancient capital of Nineveh. They tell of wars against the small western nations near the Great Sea.

Because of the relative accuracy of Assyrian records, the fall of Samaria can be dated as about 722 B.C. The Assyrian king Sargon reported that he deported 27,290 Israelites and took 200 chariots as booty. He also lists the provinces where the rebellious Israelites were resettled.

A governor was appointed by Assyria. He was to rule over the Israelites who were allowed to remain in their homes, as well as the captured foreigners who were resettled in Samaria. These Samaritans (as they came to be called) continued to pay tribute to Assyria.

The kingdom of Israel was never revived in spite of several desperate attempts by the survivors. Whenever there were uprisings, the Assyrians continued their policy of reshuffling their captive populations.

PART VII
THE LAST YEARS
OF JUDAH

Many kings of David's line had lived and ruled and
died in Judah since the kingdom was split off from
Israel by Solomon's son.

Some kings, like Jehoshaphat, had enjoyed relatively
long reigns, with no war against the brother country to
the north.

There was only one brief time when a king of the
royal house of David did not rule at Jerusalem. Jezebel's
daughter Athaliah, the widow of a Judean king, usurped
the throne and tried to kill off the eligible princes. Only
one small boy, Joash, survived the murderous rampage
of his Phoenician-Israelite grandmother. Young Joash
was crowned king when he was seven years old. His reign
was remembered because he restored the temple of Solomon.

A later Judean king, called Azariah but also known
as Uzziah, had an exceptionally long, prosperous reign
in the eighth century B.C. During that same time, Jeroboam II had an equally long and prosperous rule in
Israel. King Uzziah contracted leprosy in his old age
and his son reigned as regent until the king died.

Then, while the government of the northern kingdom
of Israel was falling apart with one bloody assassination
after another, Uzziah's grandson Ahaz came to the throne
of the kingdom of Judah.

In the stories told of the last years of Israel, King Ahaz played an important role. This was during the tense international situation when Israel's King Pekah joined with King Rezin of Syria, and tried to force Ahaz into their coalition against the mighty nation of Assyria.

The part which King Ahaz played in this crisis is retold—in greater detail—in the first story of the prophet Isaiah.

The City Prophet

In the days of Ahaz, king of Judah, there was in the royal court at Jerusalem a young man named Isaiah. He was from a family of some wealth and prominence.

One day as Isaiah stood alone in the vestibule of the temple, he had a vision. He was filled with the awesome sense of God's immediate presence there, and yet he was overwhelmed at the same time by a realization that God was lofty and holy, far beyond man's reach and understanding. As he stood entranced he heard a celestial choir singing:

Holy, holy, holy is the Lord of all,
And the whole earth is full of his glory!

Suddenly Isaiah realized that God was asking for a messenger to go to the people of Judah; to tell them that the Lord, who was so far beyond man, loved this nation enough to choose them for his own people. Isaiah said aloud, "Here I am. Send me."

After that the young man walked through the streets of Jerusalem, ready to speak the word of God whether the people would listen or not.

The people thronged to the outskirts of Jerusalem to worship at pagan altars. Many were caught up in the worship of the eastern cults which had been imported from Assyria.

"The ox knows its owner," Isaiah told the people, "and the ass knows its master's stable. But you do not know to whom you belong."

When he visited the Jerusalem temple Isaiah was deeply distressed. His countrymen committed shameless wickedness and idolatry. Then they blithely brought sacrifices to the temple as if expecting these to make up to God for their indifference to his worship and his law. Isaiah accused these people. "God is tired of the burnt rams you offer," he told them. "Who asks such things of you? How can you expect God to listen to you when you spread out your hands? They are full of blood."

Everywhere he went, Isaiah preached. He spoke out fearlessly on the street corners and in the marketplace. He even went into the court of King Ahaz.

One official in the court of the king spread his hands helplessly. "But what do you expect us to do?"

"Cease to do evil, learn to do good," Isaiah told him. "Seek justice, correct

oppression. Defend the fatherless. Champion those who are helpless!"

Most of the people liked and respected this noble young man. For a time King Ahaz and his courtiers seemed to be listening. But they did not want to bother with anything that might interfere with their carefree riotous living. Then, too, they did not wish to offend Assyria by giving up the worship of their idols.

Isaiah warned, "The Lord God says to us, Come now, let us reason together. If you are willing and obedient, you shall eat the good of the land. But if you refuse and keep rebelling, Judah shall be devoured by the sword!"

At this time the international situation was becoming very tense. Pekah the king of Israel and Rezin the king of Syria were in the midst of plotting against Assyria. They had sent ambassadors to see King Ahaz in Jerusalem. The ambassadors presented their case with syrupy politeness.

"Assyria has become too demanding for our western nations to bear," the Syrian ambassador told King Ahaz. "Our nation of Syria must pay tribute to Assyria." He gestured to another ambassador. "The nation of Israel is forced to pay tribute. The Philistines pay tribute. Phoenicia pays. Even you, O King Ahaz, must pay your annual tribute to Assyria."

"I know, I know," King Ahaz said irritably, drumming with nervous fingers on the carved lions which formed the sides of his great throne. "But by paying that tribute, we remain at peace."

"It is too great a price to pay for peace." The voice of the ambassador from Israel had an edge to it. "Now we all know that Assyria is too mighty for Israel alone. Or Syria alone."

"King Pekah of Israel has decided to join forces with us," the Syrian ambassador informed Ahaz with a flourish of triumphant gestures. "Already the Philistines have agreed to join.'"

"But we must have your help, King Ahaz," the Israelite ambassador said. "We cannot succeed unless we add the strength of Judah to our alliance."

King Ahaz was at first attracted by the idea. He talked it over with his military advisors and the princes in his court. But in the end, King Ahaz was afraid to anger the powerful nation of Assyria. So he refused to join with Israel and Syria.

The alliance would fall through without Judah's help. King Pekah assembled his forces at Samaria and marched south. King Rezin gathered his army at Damascus and also marched south. The two armies were heading for Jerusalem, to force Ahaz and Judah to join with them.

The news of the impending war with Israel and Syria swept over Judah. King Ahaz was frightened. He had not expected such a turn of events. His people were unprepared, and the news caused them to panic.

The prophet Isaiah understood their fear, but he had a reassuring message from God.

He took his young son by the hand and went to find the king.

Isaiah walked boldly up to King Ahaz and said, "I shall speak to you as God has instructed me. Listen, and do not fear! You must not be fainthearted because of the fierce anger of King Rezin and King Pekah. You think of them as dangerous blazing torches. But they are only two stumps of firebrands, already smoldering out."

"I have heard what they plan to do," King Ahaz said hoarsely, trembling as he spoke. "The Israelites and the Syrians intend to conquer Judah. Then Pekah and Rezin intend to put their own puppet king on my throne—"

"Their plan shall not succeed," Isaiah

interrupted. "Damascus, Syria, and their King Rezin are nothing! Samaria, Israel, and their King Pekah are nothing!"

King Ahaz turned to consult with his Assyrian astrologer.

"Listen to God's word," Isaiah interrupted. "If your belief is not steadfast, then you cannot stand fast."

But King Ahaz was too frightened. "There is only one thing we can do. We must immediately beg Assyria to rush an army here to rescue us."

So Ahaz sent messengers to Assyria, along with presents of silver and gold. The king of Assyria responded at once and marched west with his great army.

The rebelling Philistines were beaten into submission. The armies of King Pekah and King Rezin rushed back to protect their own borders. It was too late. The northernmost provinces of Israel were conquered by the relentless Assyrians. Panic reigned in Israel, and King Pekah was murdered. Then the Assyrians moved to devastate the land of Syria. They captured its greatest city, Damascus.

Now the king of Assyria sent an imperial summons to King Ahaz. "I have conquered Damascus. Come up for a meeting and bring a caravan loaded with tribute."

King Ahaz went more like an obedient servant or vassal than a king. To get enough treasure he had to strip the temple. When he arrived at the great fallen city, he saw a magnificent new altar in the midst of the ruins. The Assyrian priests had erected it there for the use of their king in his pagan worship.

Immediately King Ahaz sent back to Jerusalem a model of the Assyrian altar. With it went an order to the Jerusalem priests: "Build an altar like this in the court of our temple."

So Ahaz returned to his country more than ever a slave to Assyria and its ways.

Isaiah went to the king and scathingly said, "You have let loose a flood! Do you think Assyria will stop with conquering Damascus and northern Israel? Samaria also will fall. And someday the mighty army of Assyria, like a flooding river, will sweep on, drowning even Judah!"

Then Isaiah took a scroll and wrote down the things he had been preaching to the people. It was all there in forceful poetic lines. He wrote of the injustices in Judah, the rampant greed, the self-indulgence, and, above all, the unfaithfulness to God. Isaiah also wrote his prophecies. He warned again of the Assyrian threat that hung like a heavy sword over little Judah.

"This is my testimony," Isaiah told King Ahaz, and all the people. "In this scroll I have written my teaching. Tie it with cords, and put a seal upon it. Later the scroll may be opened. Then you shall see that my prophecies have come true!"

After that, Isaiah withdrew for a long time from public life.

2 Kings 16:5-20
Isaiah 1:1 through 8:16; 10:28-32; 17:1-7

Isaiah and King Hezekiah

King Ahaz of Judah died. Preparations began for the anointing of his son Hezekiah.

The coronation was a solemn ceremony in which the new king was designated as the especially adopted son of God and as God's special representative to the people.

Isaiah returned to public life at this time for he had a great hope in Hezekiah. He may have composed a poem* at this time.

The people who walked in darkness
  have seen a great light!
For to us a child is born,
  to us a son is given.

He will sit upon the throne of David,
  and reign over his kingdom.
The spirit of God shall rest upon him,
  the spirit of wisdom and understanding.

He will provide justice for the downtrodden,
  and make fair decisions for the poor.
The wolf shall dwell with the lamb;
  and a little child shall lead them.
                    (based on Isaiah 9 and 11)

A century earlier, Solomon's great temple had been repaired and restored by the boy-king Joash. Now the young king Hezekiah found that the temple was again in a sad state. Its treasurers had been stripped away by his father and turned over to Assyria.

Hezekiah ordered the temple cleansed, repaired, and sanctified. He levied a tax to provide the money needed. Then Hezekiah ordered the foreign idols and altars smashed and destroyed. He followed the advice of Isaiah in all these matters. For Isaiah had warned that the people must return to their faith in God, or their kingdom would end in ruin.

Hezekiah also tried to change Judah from a weak nation into a strong one. He repaired the great city walls of Jerusalem. And he solved a knotty and vital problem: how to get enough water into the city in case of enemy siege.

East of Jerusalem, down in the valley outside its walls, were the precious waters of the Gihon Spring. King Hezekiah's engineers planned a long rock tunnel which would burrow under the city walls, allowing the springwater to flow into a pool inside the city.

Workmen began with picks at the cavern by the spring. Another crew began chipping through the solid limestone from the site of the new pool. The great tunnel was completed under the city, and a continuous water supply was assured.**

About this time word came that Sargon of Assyria was dead. He was succeeded by King Sennacherib.

Immediately there was rebellion in Babylon. Restlessness spread among other small kingdoms in the vast Assyrian Empire.

About this time King Hezekiah became sick with a deeply infected boil. As the king tossed feverishly on his bed, Isaiah came to him and said, "Set your house in order, O king, for you shall die."

When he heard this, Hezekiah turned his face to the wall and prayed, saying, "Remember, O Lord, how I have walked before you in faithfulness. I have done what is good in your sight." And the king wept bitterly.

As Isaiah was leaving the outer court of the palace, God spoke to him, "Turn back, and tell Hezekiah I have heard his prayer. I will heal him."

When the king of Babylon heard that Hezekiah had been sick, he sent envoys with letters and a present for Hezekiah. Israel's king, now fully recovered, was delighted. He proudly took the envoys from Babylon on a tour of his treasure-house, his armory, his storehouses.

Isaiah rebuked his king for revealing the total wealth of Judah to these outsiders. "Hear the word of God! The days are coming when all that is valuable will be carried to Babylon! Even some of your descendants shall be taken away."

Despite this dire prediction, Hezekiah did not worry. He only cared that there be peace and security in his own days.

But the peace and security for which

Hezekiah hoped did not last. In Egypt there was a strong new pharaoh. This Egyptian ruler began to conspire against Assyria, and he urged all the small western nations to join him.

King Hezekiah saw this as his chance to pull Judah from under the crushing Assyrian thumb. "The city of Babylon dares to revolt against Assyria," the king said to his military men. "Babylon may help us if we revolt, too. With Egypt as our ally, perhaps we can be free."

Hezekiah sent envoys to Egypt to arrange a pact and to buy arms.

Isaiah went to warn the king, "Woe to the rebellious people who go down to Egypt, without asking for God's counsel, to take refuge in the protection of Pharaoh."

Hezekiah tried to defend his plans. "Can we afford to be Assyria's slave forever? Egypt calls herself a mighty dragon. She will be a loyal ally, and a strong one."

"Egypt's help is worthless and empty," Isaiah said with scorn. "A huff-and-puff dragon that sits on its tail."

"Go away, O prophet," some of the advisers of Hezekiah said. "You give us nothing but bad news."

"You are a rebellious people," Isaiah answered. "Lying sons, who shut their ears to God's warnings. You say to the prophets, 'Don't tell us what is right.' You just want to hear agreeable things—even if they are false."

"How can we be saved from destruction, unless we make this pact with Egypt?" Hezekiah asked.

"Return to God and his ways," Isaiah told them. "Let quietness and trust be your strength. But you will not! Instead you say, 'No, we will speed on war-horses.' Indeed you shall speed—to your defeat!"

In spite of Isaiah's warning, Judah continued her plotting against Assyria. In the fourteenth year of King Hezekiah, the new ruler of Assyria came to end the revolts in his empire. First Babylon was subdued, next the resisting Philistines. Then the Assyrian king Sennacherib marched on to Judah.

Sennacherib ravaged and pillaged the countryside. He methodically besieged the fortified cities of Judah and took them one by one, including the border fortress of Lachish. Only the great capital city of Jerusalem remained out of the hands of the enemy.

King Hezekiah realized that his recent conspiracy with Babylon and Egypt was known. Quickly he sent a messenger to the Assyrian monarch at Lachish, saying, "I have done wrong! Only withdraw from my country, and whatever penalty you impose, I will bear."

Sennacherib replied by demanding a large amount of silver and gold.

Hezekiah was forced to take all the silver and gold from the nation's treasury. He stripped the gold from the doors and posts in the temple. He rushed this treasure to the emperor.

This was not enough. Sennacherib sent his chief deputy and a large contingent of his army up to the gates of Jerusalem. The Assyrians camped in the valley beside the great city.

The chief deputy strode close to the city wall. Cupping his hands, he shouted up to the guards patrolling the top of the wall. "I wish to speak to your king!"

Hezekiah sent a delegation made up of the secretary of state, the royal treasurer, and the keeper of the records. They were to listen, but to make no promises.

The three Judean officials appeared on the wall and identified themselves as the representatives of their king. The Assyrian chief deputy spoke in perfect Hebrew with only a trace of an accent. "Ask Hezekiah," he taunted, "if he thinks he can rely on

146 Egypt? Egypt is a broken reed of a staff. It will pierce the hand of any man who leans on it. How can you hope to push back a mighty army like ours when you must rely upon Egypt for chariots and horsemen?"

Many curious townspeople had gathered on the top of the wall together with the soldiers who were defending the city. The three officials did not want the citizens of Jerusalem to hear.

"Please," they begged, "use Aramaic, the diplomatic language, for we understand it. Do not speak to us in the language of Judah in the hearing of the people who are on the wall."

But the Assyrian replied loudly, "Has Sennacherib my master sent me to speak these words to your master and to you, and not to the men sitting on the wall? They, along with you, are doomed to have disgusting filth for food. Sennacherib will set siege to your city unless you make your peace with me and come out. If you do, then every one of you will eat the fruit of his own vine and fig tree, until my master takes you away to a better land. There you may live and not die!"

There was a murmur along the walls. Some of the people were trying to reassure each other. "Surely God will save us."

Then the Assyrian envoy shouted up, "Hezekiah cannot save you. Neither can your God save you." He hurled the final insult. "Did your God save Samaria from us?"

Hezekiah's officers rushed back to the palace to tell the king what the envoy had said. When he heard the report, the king mourned. He dressed in rough sackcloth and went into the temple to pray for God's mercy.

He sent some of his officers and priests to consult with Isaiah. The prophet received them calmly. "Tell our king not to be afraid," Isaiah instructed the delegation.

"Sennacherib will not come into this city, or shoot an arrow here. He shall hear a rumor and return to his own land, and there he will fall by the sword."

The next morning the enemy camp was deserted. Sennacherib, king of Assyria, had gone back to Nineveh. For the time being, Jerusalem was saved.***

2 Kings 18:9 through 20:21
Isaiah 9:2 through 11:16; 21 through 32; 36 through 39

* This poetry has come to be associated with the Christmas season because it applies so well to the coming of Jesus the Christ. However, many scholars believe these verses were first written for the crowning of Hezekiah.

** The Siloam tunnel, built in Hezekiah's reign, still runs under the most ancient part of Jerusalem. It is cut through the solid limestone rock. An inscription has been found on the tunnel wall, rudely scratched out. Archaeologists have identified the writing as the Old Semitic script of the eighth century B.C. An on-the-scene account of the digging has been deciphered:

The piercing-through: Now this is the way the cutting through happened. While yet the stone-cutters were wielding the pick, each toward his fellow, and while there yet were three cubits to be cut through, there was heard the voice of someone calling to another, for there was a crack in the rock. Then on the day that the cutting through was completed, the stone-cutters struck, each to meet his fellow, pick against pick. So the water flowed from the spring into the pool.

*** Archaeologists have dug up at the ancient Assyrian capital of Nineveh other supporting evidence for this biblical story. Sennacherib had chiseled in stone a complete description of his military campaign, including many detailed pictures of his siege of the Judean border fortress of Lachish. Inscribed on a barrel-shaped clay prism are the "Annals of Sennacherib." In this, his own record of his victories, the Assyrian king boasts: "Hezekiah himself I shut up like a bird in a cage, in his royal city."

According to Assyrian history, there was a new revolt in Babylon about this time. Perhaps this was the rumor, prophesied by Isaiah, which took the invaders back to the land of the two rivers. Assyrian annals also confirm that at a later date Sennacherib was slain by one of his own sons, and was succeeded by another son, Esarhaddon. This seems to be the actual ending of this particular siege of Jerusalem, although the Bible also includes another story having legendary characteristics, about an avenging angel which slew the forces of Sennacherib.

Micah the Country Prophet

The tiny village of Tekoa, hometown of Amos the prophet, was perched on the mountain backbone of Judah south of Jerusalem. Amos may have been alive in the dark days of Assyrian oppression. He would have been a very old man, sitting, perhaps, in the doorway of his small house, remembering how little the people of Israel had heeded his warnings.

Westward from Tekoa, between the Judean mountains and the narrow Philistine coastal plain, was a fertile strip of rolling foothills. Here, in the village of Moresheth, close to the Philistine border, lived a sensitive country boy named Micah. It is possible that more than once young Micah went to Tekoa to listen to Amos. Winding roads ran inland to the mountaintops, up to the isolated hamlet where one could sit at the feet of the aged prophet.

It is even more likely that young Micah read, with great enthusiasm, the prophecies of Amos which had been written down.

Micah also hiked up to the capital. Perhaps he heard Isaiah preaching on the Jerusalem streets or in the marketplace.

The impassioned words of Amos and Isaiah surely would influence such a sensitive lad. At any rate, Micah's heart was directly touched by God. He felt the call to carry God's message to the people. He was gripped with a great sense of urgency. He must warn the heedless people of Judah, before they suffered the same terrible fate as the Israelites.

Micah was a man of the country, at home with the farmers and the unsophisticated artisans and craftsmen of the villages. He found the same corruption and lack of faith in the country that Amos and Isaiah had found in the large cities.

Micah rebuked the wealthy farmers in his neighborhood, "Woe to you who plan wrongdoing when you should be sleeping. As soon as it is morning, you rush to carry out your evil plans. You are so greedy to own more fields and more houses, you seize them illegally. Poor men lose their property, and their children are cheated of their rightful inheritance!"

When he did go to Jerusalem, the young man from the country rebuked the vice and corruption he found there.

"Where did Israel go wrong? Was it not in her capital of Samaria?" Micah asked. "And where does Judah go wrong? Is it not in her capital of Jerusalem?"

The people of Jerusalem did not like to be criticized in this way. They clung to the belief that Jerusalem was special in the eyes of God, and would be preserved by him.

Micah reminded his listeners that God had caused Samaria to become a heap of stones, with foundations laid bare. Why did they think Jerusalem was any better?

The most important leaders in Jerusalem had gathered for a meeting at the palace. Micah stormed in and fearlessly called for their attention. Using harsh words, he told them they were guilty of practically skinning their fellow citizens alive.

"Hear, you heads of businesses!" The young man glared at the royal court. "Hear, you rulers and government leaders! How can you be so ignorant of what justice means? You are all mixed up, for you hate what is good and you love what is evil!"

"How dare you?" one shouted. Others yelled, "Go away."

Micah ignored the heckling.

"You tear the skin from my people, and their flesh off their bones," Micah shouted, biting off the words. "You break their bones in pieces, and chop them like meat in a kettle. You are destroying this city with your bloody wrongs."

One of the nobles said haughtily, "Do you not realize that Jerusalem will always be protected by God? It is the apple of his eye."

"Because of you, Jerusalem shall become a heap of ruins," Micah predicted. "The temple will be nothing more than a hilltop with thickets growing on it."

The crowd became angry at the young prophet's words. Someone screamed, "Stop! Do not preach such things. It is blasphemy."

Micah answered with scorn, "If a man would be a soft-soaping windbag uttering lies, he would be just the preacher you want."

Micah turned on his heels and went back to his village of Moresheth. He continued his struggle to arouse the people to their danger.

One day Micah heard about a dramatic demonstration by Isaiah of Jerusalem. For three years that prophet, who was an aristocrat by birth, kept his hair and his beard shaved off. He walked the streets of the capital wearing only a loincloth, saying, "So will you be shaven and naked when you are dragged off into captivity for your sins."

Micah did the same. He walked barefoot, wearing only a loincloth, even in the chilly drizzle and frost of winter.

"Hear, all of you," he cried out. "Let the Lord God be a witness against you. Your rich men are full of violence. Your inhabitants speak lies. Therefore God has begun to smite you. For this I will lament and wail. I will go stripped and naked. For Jerusalem's wound is incurable. All Judah is deathly sick with evil."

When Sennacherib had marched westward with his Assyrian army, his hard-riding cavalry came first. They pounded down the road along the Great Sea.

From the hillside village where he lived, Micah was among the first in Judah to see the invaders. The young prophet watched helplessly as the hordes of Assyrians descended upon the land, pillaging, ravaging, conquering.

The Assyrian soldiers wore their black hair to their shoulders, their curly black beards trimmed squarely across the bottom. They wore short fringed kilts and high pointed helmets. Their chariots and horses were as the sands of the sea.

Outside Judah's fortress at Lachish, the Assyrians built a huge camp. First the small Judean villages surrendered, then the scattered towns. After that the conquerors prepared to take Lachish itself.

The city was built upon a large flattened hill, surrounded by a high thick wall of sun-baked bricks. Partway down the hill was a second stout wall, this one of stones.

Each morning at dawn, the Assyrians set up their battering rams against the walls. All day the pounding went on. Logs and firewood also were piled against the walls and set on fire. Then holes were dug through the burnt and crumbled places. Assyrians fighting close to the walls were protected by great shields, and by a rain of arrows from the archers stationed behind them. Mechanical torch-throwers, working like giant slingshots, hurled many

flaming brands over the walls and into the beleaguered fortress.

At last Lachish was forced to surrender. The victorious King Sennacherib sat on a resplendent portable throne, nodding with approval as captured slaves and rich booty were paraded before him.

Micah and the people of his village heard that Sennacherib had sent part of his army on to Jerusalem. They knew that he was making insolent demands that the city submit or be crushed like Lachish.

The frightened people of the countryside waited anxiously, almost numbly. Would Hezekiah capitulate? Would they be dragged off in exile, like their fellow Hebrews from Israel?

Then a message came that the miracle—the unexpected—had happened. The Assyrians were gone, as suddenly as they had come!

Yet hunger and suffering followed, because crops had been burned and the land despoiled. Perhaps Micah hoped the terrible aftermath of war would frighten the people into following God's commandments. If so, he was disappointed. They went on as before.

Great prophets had spoken to the Hebrews, and yet the people did not listen. Despairingly Micah cried out, in words that echoed the basic teachings of Amos, Hosea, and Isaiah:

"God has showed you, O man, what is good;
    and what does the Lord require of you
but to do justice, and to love kindness,
    and to walk humbly with your God?"

Then King Hezekiah died. Micah's hopes for his country sagged when Hezekiah's young son Manasseh came to the throne of Judah. For Manasseh soon reversed the liberal policies of his father. He returned to the pro-Assyrian programs of his grandfather Ahaz. The temple in Jerusalem was neglected. Assyrian altars were built. Manasseh turned so completely from the worship of God that he followed the pagan custom of offering up his son as a burnt sacrifice.

When Micah was an old man and full of years, he heard a terrible rumor from Jerusalem. It was whispered that the aged prophet Isaiah had been put to death by Manasseh, that he had been sawed in half.

Isaiah 20
Micah 1 through 7

Josiah the Reformer

At last after a long, long reign, the wicked King Manasseh died. His son ruled Judah for only two years before he was assassinated. Old Manasseh's eight-year-old grandson Josiah was crowned as Judah's new king.

During his lifetime Manasseh had remained strongly pro-Assyrian. He killed off any dissenters who tried to stem the pollution of Judah's religion and culture.

Yet brave priests and prophets, working underground, continued in the footsteps of Amos and Hosea, Isaiah and Micah. They refused to bow down on the rooftops to worship the Assyrian gods of the sun, the moon, and the stars. Some of these courageous religious men became the tutors and advisers of the boy-king Josiah.

By the time Josiah came of age, he was afire to purify his nation's religion. He also dreamed of restoring her lost national glory. He planned to enlarge Judah's borders until his kingdom was again as large as it had been under David and Solomon.

The new king of Assyria was not strong enough to control unrest and rebellion. The vast Assyrian Empire was crumbling.

Josiah stopped paying tribute to Assyria. He banished the Assyrian counselors from his court. Then King Josiah marched his army up to Samaria. Here lived some of the former Israelites, now intermarried with foreigners. Josiah took control of the land from the weakening reins of Assyria.

But now there were rumors of a new enemy. A horde of barbarians was sweeping down from the far north. They were the fearsome nomadic Scythians,* fast-riding, cruel. They looted and killed, not for conquest but for plunder.

There was a prophet in the land who was a cousin of King Josiah. Zephaniah was his name, and he proclaimed that God was using these ferocious barbarians to punish Judah. Zephaniah felt that the nation had become shameless and pagan. Even with the Assyrians banished from Josiah's court the people continued with their heathen ways. Would not God punish them?

The Scythians swept down the Mediterranean coast, stealing, murdering, pillaging. The people of Judah waited uneasily. The dreaded invaders reached the narrow Philistine plain, only a short distance from Judah's border. Then they turned back. They returned to the northern countries in search of richer plunder.

The people of Judah sighed with relief. The prophecy of Zephaniah had not immediately come true. This particular danger was past. As soon as the threat was over, the people lapsed into their old ways. The moral degradation continued. People returned to the worship of Assyrian gods and to the immoral rites connected with the old Canaanite Baals.

In the eighteenth year of his reign King Josiah decided to take action. "We must encourage our people to come back to the temple," he said to his counselors. "Only there can we discover the will of God."

The king's advisers pointed out, "Our temple has been neglected for many, many years. It is so shabby. It will take a great deal of money to repair it."

King Josiah called his secretary Shaphan to carry a message to the chief priest Hilkiah. "Tell him to take the money that has been collected from the people and hire carpenters, masons, and other workmen. We must restore the house of God."

And so the work was begun. It moved along well, and the people began to take an interest in the project.

Then one day the chief priest sent for Shaphan. "I have found this scroll," he said. "It was discovered by the workmen while cleaning out a room in the temple. It is a book of the law."**

Carefully Shaphan carried the scroll back to the palace and showed it to the king. Josiah was intrigued. "I wish to know what it says. Read it to me," the king ordered.

So Shaphan began to read the scroll. It was a list of laws. Many were repetitions of the old laws of Moses, known for so long and yet almost forgotten. Some were newer laws which breathed a spirit very like that of the prophets Amos, Hosea, Isaiah, and Micah. These laws stressed the importance of showing consideration for the poor and defenseless among the people.

The more Shaphan read, the more Josiah understood just how far the people of Judah had fallen away from God.

King Josiah listened intently as his secretary read. "Those are wise laws," he said. "The worship of God must be made pure," the king said. "We must abolish the bowing-down to foreign idols."

"Wait, there is more," Shaphan said. He read a list of the blessings that would enrich the nation if Judah kept its covenant with God. This was followed by a solemn recital of the disasters that would befall the people if the laws of God were flouted and broken.

When he heard these things the king began to tremble in distress. Shaphan went on reading, "The Lord will bring a nation against you from afar, a nation you have not known. They shall besiege you in your towns, until your high and fortified walls come down. And you shall begrudge food to your own kin, because there will be almost nothing to eat. And afflictions, sickness, and destruction will come to you, if you are not careful to do all the words of the law as written in this scroll."

King Josiah rose from his throne and tore his robes. He sent for Hilkiah, for the other priests and prophets, and for the leading men of Judah. Then he called a solemn assembly of the people of Jerusalem.

The king read to them the words of the book of the covenant. Then he said to them, "We must serve and follow God because we love him, because we are grateful to him for the many gifts he has given. Worship should not be out of fear, or to 'buy' God's favors.

"I hereby make a solemn promise to God to obey the words of this book. Will you join me in that promise?"

The people were touched when they heard the king read. They did agree with him and made a solemn vow, and so began the great reform of King Josiah.

Astrologers, soothsayers, and wizards of the black arts, which had flourished under Assyrian influence, were driven out of Judah. Idols and foreign altars were smashed. The local high places, where the country people of Judah worshiped, were abolished. It was decreed that everyone must come to the restored temple in Jerusalem and worship there.

During this time of spiritual vitality in Judah, the vast empire of Assyria slowly died. The kingdom of Chaldea, helped by several other strong kingdoms, administered the death blow to Nineveh. After a terrible battle the great capital of the Assyrians became a heap of ruins.***

The Assyrian king fled to Haran where he made an attempt to rally his forces for a comeback. He appealed to his old ally, Egypt.

Pharaoh Neco mobilized an army of horsemen and chariot divisions. The Egyptians charged northward, along the road by the Great Sea.

King Josiah took his army and marched

152 out to stop the Egyptians at the pass into the Jezreel valley and so prevent their joining the Assyrians.

The Egyptian army and the small army of Judah met head on at Megiddo. Pharaoh Neco killed King Josiah there. Josiah's servant carried his master from the battlefield. He put the body in the royal chariot and returned to Jerusalem. There the monarch was buried.

The people of Judah mourned for the good King Josiah, who had lived a righteous life and had been cruelly cut off in his prime.

2 Kings 21:1 through 23:30
2 Chronicles 33 through 35
Zephaniah 1:1-18

* The Scythians were a nomadic people who invaded the Near East at the beginning of the eighth century B.C. They came from what would now be southern Russia, riding on horseback through the Caucasus Mountains. They became allies of the Assyrians.

** The "book of the law" may have been an ancient book which had been lost over the years. Or it may have been written at a much later time by priests or prophets who were forced to go into hiding during the time of the evil King Manasseh. The book shows the influence of the great writing prophets on the laws of the Hebrews. It forms the core of the book of Deuteronomy (chapters 12 through 26).

*** The fall of Nineveh, by modern reckoning, came in 612 B.C. The short book of the prophet Nahum gives vivid expression to the relief—and even rejoicing—that the people of Judah felt at the fall of their cruel enemy. Nahum pointed out that the idolatry and cruelty of the Assyrians were the cause of their downfall. He did not give attention to the sins of Judah, which were leading her in the same direction.

Jeremiah Warns Judah

During the reign of King Josiah, an intense young man named Jeremiah was growing up in a family of priests at Anathoth. This small village was a short northeasterly walk from Jerusalem.

When King Josiah abolished the small, local places of worship scattered across Judah, people had been forced to journey to the temple in Jerusalem for their worship. Josiah's reform left Jeremiah's family at Anathoth jobless. Like other country priests, they now had no altar to serve.

So Jeremiah moved to Jerusalem. There in the holy city the word of God came to him. He wrote down the oracles:* "Before you were born, I consecrated you! I appointed you to be a prophet to the nations."

Just as Moses had once been reluctant to accept God's call, now Jeremiah found himself unwilling. "I don't know how to speak," he objected. "And I am too young."

Then Jeremiah heard God's answer. "Whatever I command you, you shall speak. You shall not be afraid, for I am with you. You shall be as my mouth."

Jeremiah had read the words of Amos and Hosea. These prophets had spoken in Israel. The people had ignored the warnings and gone to their destruction. Jeremiah knew also how Isaiah and Micah had reproved the people of Judah, and how their message had been ignored.

Now Jeremiah began to preach the same message. "God brought our nation up from the land of Egypt," he told the people of Jerusalem. "He brought us into this plentiful land. Why do you defile it? Why do you persist in running after the Baal-gods? Why do you continue to bow down—like heathens—under every green tree which is on a high hill? Why do you abandon your own God?"

Jeremiah gave his full support to the reforms that King Josiah began. The prophet also took every opportunity to speak to gatherings of the people in the marketplace, giving them oracles of God and warning them to change their ways.

When King Josiah was killed by the Egyptians at the battle of Megiddo, the Judeans were shocked, grieved, and frightened.

Eliakim, Josiah's eldest son, hoped to be crowned as the new king of Judah. Although the Egyptians had just killed his father, Eliakim openly admired that ancient nation.

Judah's court, grieving for the slain king, was not interested in this eldest son. Eliakim was passed over, and a younger half-brother named Shallum was chosen. The young monarch, who took the throne name of Jehoahaz, reigned only three months.

When the Egyptian army returned to Judah after an unsuccessful attempt to aid the Assyrians against the conquering Babylonians, Pharaoh Neco took charge of affairs. He ruthlessly deposed Shallum and took him captive to Egypt.

The Judeans were still stunned by the loss of their beloved King Josiah. No one seemed to comprehend that the young king had been carried away in chains to an Egyptian prison. People acted as if he were there on a visit and might return at any time.

Jeremiah cried out, "Weep not for the dead Josiah, but for him who is taken away. He shall not return to his native land!" Jeremiah's prophecy came true. The unfortunate Shallum died in captivity in Egypt.

Pharaoh Neco ordered Judah to crown Eliakim, the pro-Egyptian elder brother. He took the throne name Jehoiakim.

When Jehoiakim was crowned, people thronged to the temple in Jerusalem, hoping to catch a glimpse of the coronation. Jeremiah took advantage of this opportunity to speak to the people. He told them that they were making a mistake to feel that no harm could come to Jerusalem because the temple of the Lord was located there. The temple was indeed the holy place of God, but the Lord was not confined to any building. The wickedness of the people was leading to destruction, and this destruction would include Jerusalem and the temple as well.

His remarks caused an uproar. Angry people grabbed the young prophet. "You shall die. Why have you prophesied such terrible things against the holy city?" The priests and professional prophets added their own cries of outrage. "God would never let Jerusalem fall. Kill that man!"

But some of the royal officials were impressed with Jeremiah and his sincerity. They gave him a chance to defend himself.

"I only spoke what God has told me," Jeremiah said. "Look, I am in your hands. Do what seems right and good to you. Only know for certain that if you put me to death, you will shed innocent blood."

The king's officials said to the priests, "This man does not deserve the sentence of death." Certain of the elders of the land also pointed out that in the days of Hezekiah, the prophet Micah had prophesied, "Jerusalem shall become a heap of ruins, and the temple hill a height overgrown with thickets." And Hezekiah had not put Micah to death for saying this.

Jeremiah was allowed to go free in the end, but the priests hated him for his predictions. They were especially angry that he spoke of the temple as a "den of thieves." So the prophet was no longer allowed to speak within the temple grounds.

Jeremiah's preaching was so unpopular

that he came near death many times. Priests, including his own brothers and relatives in Anathoth, laid a trap for the unsuspecting prophet. They almost succeeded in murdering him.

"I was like a lamb led to the slaughter!" Jeremiah complained to God. "I never knew they plotted against me." Then, with a cry of despair, Jeremiah asked, "Why does the way of the wicked prosper? Why do those who are treacherous thrive?"

After long and agonizing meditation Jeremiah heard God counseling him to have greater faith and courage, for these were dark days. Jeremiah would have to steel himself to face evil men and even more evil times.

Four years later the Egyptians again charged northward to aid a regrouped army of Assyrians. This was to be a last-ditch fight against the Chaldeans to maintain control of the vast empire built up by the Assyrians.

The great armies met in a ferocious battle.** Nebuchadrezzar of Babylon, king of Chaldea, overpowered the combined army of Egypt and Assyria.

In the year of that important battle Jeremiah was commanded by God to put his oracles into one written document. Jeremiah called upon his closest disciple, Baruch. Jeremiah dictated, and Baruch wrote the oracles on a scroll.

"They will not let me go into the temple," Jeremiah told Baruch. "So I want you to go there on this day of fasting, when the people will be at the temple to hear. You shall read the scroll. It may be not too late for the people to repent of their wrongs."

So Baruch went to the temple and read Jeremiah's words.

When the officials in the palace were told what Baruch was doing, they sent for him. "Sit down and read the scroll to us," they invited. Many of the men were impressed by the things Jeremiah said. "The king must hear this!" they agreed. But they also knew that these oracles might arouse the king's wrath, so they warned Baruch to hide himself and Jeremiah. "Let no one know where you are."

Then the officials took the scroll to the king's chamber. It was a cold day in early spring. The weak afternoon sun streamed through the open lattices of the windows. On tripod legs, a large, shallow firepan of bronze stood heaped with glowing, burning charcoal.

King Jehoiakim sat close to the warmth of the brazier. He scowled when he heard that the scroll contained the oracles of Jeremiah.

"He's a prophet of ill omens," the king muttered.

As the officer read three or four columns from the scroll, the king would cut them off with a penknife and throw the pieces into the fire. Even though a few of the listeners protested, the king continued to burn the sections as fast as they were read.

At last it was all burned. Neither the king nor most of his courtiers were the least afraid of Jeremiah's terrible warnings for Jerusalem.

The king still scowled as he waved a jeweled hand. "Seize Jeremiah!" he ordered. "Also his secretary, who wrote this down."

But fortunately Baruch and Jeremiah were safe in hiding.

Once again Jeremiah dictated the oracles to Baruch. The words of the burned document were recorded, along with a new prophecy that Babylon would surely come and destroy Judah.

Many Judeans were afraid of this mighty new empire ruled from the ancient city of Babylon. The Chaldeans were greedy conquerors, rapidly growing even more powerful than the Assyrians had been.

But King Jehoiakim did not seem to be concerned about what might happen to his nation. For three years he paid tribute to Chaldea. And Judah groaned under the increased taxes.

To make matters worse, Jehoiakim had expensive personal tastes. He enlarged and rebuilt his royal palace, trying to make it even more luxurious than it had been in the splendid days of Solomon. He taxed the people, and he conscripted the free citizens of Jerusalem to provide the labor.

Jeremiah risked his life again by publicly criticizing his king, saying, "Woe to him who builds his palace on injustice, and his upper rooms on fraud, who forces his neighbors to serve him for nothing!"

Then Jeremiah compared King Jehoiakim unfavorably with his father.

"Did not your father Josiah do justice? And live righteously? He helped the poor and the needy, and it was well with him. But you have eyes and heart only for dishonest gain!" Then Jeremiah hurled a dire prediction at the king, "You shall come to a violent end!"

The fearless prophet was not through. Jeremiah took a pottery flask, and walking up to a group of citizens, smashed the flask on the ground. "God says as this pottery is shattered, so will this people and this city be broken," the prophet told them. "It can never be mended. He will make this city desolate. Everyone who passes by will be horrified because of its disasters."

"Is it too late?" someone asked. "Can't Judah change?"

"Can the Ethiopian change his skin?" Jeremiah answered. "Or the leopard his spots?"

Then Jeremiah walked boldly up to the temple and announced, "God says that because you are a stiff-necked people who refuse to hear his words, he is going to bring evil upon your country."

An officer of the temple rushed over and arrested Jeremiah. The prophet was beaten severely and locked in the stocks by the temple gate. Here people could jeer at him as they passed.

When Jeremiah finally was released, he was in despair. Aching from his wounds, Jeremiah moaned, "Cursed be the day on which I was born! Why did I come forth from the womb to see toil and sorrow, and spend my days in shame?

"I have become a laughingstock all the day. Whenever I speak, everyone mocks me. Always I hear ominous whisperings. Terror is all around me. 'Denounce him,' say all my familiar friends. They are watching for my fall!"

Yet in spite of his gloom Jeremiah kept his faith in God and in God's word.

Jeremiah had predicted terrible events for Judah. Soon they began to come true.

King Jehoiakim kept his strong ties with Egypt. They had put him on his throne so he intrigued with the pharaoh. When Egypt encouraged him to rebel against the power of Babylon, Jehoiakim obliged and stopped paying tribute to the Chaldeans.

First King Nebuchadrezzar sent marauding bands of Chaldeans, Syrians, and others. They were to devastate Judah and carry on guerrilla warfare. Then Nebuchadrezzar himself came charging out of Babylon at the head of his mighty army.

In the midst of this crisis, King Jehoiakim died suddenly. The heavy responsibilities dropped on the slender shoulders of his eighteen-year-old son Coniah. The prince was hurriedly crowned, and given the throne name of Jehoiachin.

It was at about this time that the forces of Nebuchadrezzar surrounded Jerusalem.

The siege was cruel and complete. The Chaldean army bombarded the walls; arrows and flaming torches were rained upon

156 the defenders. All food was cut off from the city.

After three months the young king surrendered to save his city from total destruction.

King Nebuchadrezzar took as booty the treasures of the country. This included the gold and silver vessels from the king's palace and from the temple.

The Chaldeans also took back to Babylon, as prisoners, the youthful king and his mother, the government officials and members of the royal court, the priests, the weavers, potters, ironsmiths and other craftsmen. But Jeremiah was left in Jerusalem, along with a remnant of the people.

The prophet mourned, "Who will have pity on you, O Jerusalem? You have rejected your God, and you did not turn from your ways. Now your widows are more in number than the sand of the sea. A destroyer came to you at noon, and your sun has set!"

2 Kings 23:29 through 24:16
Jeremiah 1 through 36

* Oracle is the special name given in Old Testament times to the divinely inspired messages of the prophets.

** Assyria received its deathblow in the battle of Carchemish in 605 B.C. Jeremiah wrote, "In the north by the river Euphrates they have stumbled and fallen." The conquering Chaldeans are often referred to as the Babylonians, though Babylon was just the capital city.

The Destruction of Jerusalem

The victorious Chaldeans went back to Babylon, taking the young king and many other prisoners.

The conquerors left one member of the royal family to rule over the devastated country. It was Mattaniah, a younger brother of the unfortunate Shallum, just a ten-year-old boy when their father Josiah died. The prince was twenty-one when the Babylonians crowned him. He was given the throne name Zedekiah.

Many of the people left in Jerusalem bitterly rejected Zedekiah as their king because he had been put on the throne by the conquerors. Coniah, they kept saying to each other, was their rightful king. Coniah, who had borne the throne name of King Jehoiachin, had been carried away to Babylon after his short three-month reign. Coniah, or one of his sons, should be back in Jerusalem, ruling the people.

"Is Coniah a despised, broken pot?" these people mourned. "Why are he and his children hurled and cast into a land they do not know?"

"O land, land, land, hear the word of the Lord," Jeremiah cried out. "Write this man down as if he were childless. For none of his offspring shall succeed in sitting on the throne of David. Coniah's sons will not rule in Judah."

Jeremiah also wrote to give advice to the restless exiles who had been taken away into captivity. They too must accept the situation in which they found themselves: "Plant gardens and eat their produce. Take wives and have sons and daughters. Multiply there and do not decrease. Seek the welfare of the city where you are exiled. Pray to God on its behalf, for in its welfare you will find your welfare."

King Zedekiah was required to send tribute from Judah to Nebuchadrezzar at Babylon. After all he owed his throne to the Chaldeans. His country was in their power, and he had taken an oath of allegiance to them.

The other small conquered kingdoms

around Judah decided to rebel. Their ambassadors came to Jerusalem, to persuade King Zedekiah to join them. "Where is your courage and determination?" they asked the young king. "What little wealth you have left is being drained off by the enemy!"

The king, with no experienced older officials left to give him good counsel, found himself tempted toward rebellion.

Jeremiah was against such a rash policy. To dramatize his opposition, the prophet made himself a heavy wooden yoke, such as oxen wore to plow the fields. He strapped it on his shoulders and walked with his burden into the palace. "We must submit to the yoke of Babylonia," Jeremiah proclaimed. "It is our fate."

The startled ambassadors looked at this strange sight. Jeremiah said to King Zedekiah, "Bring your necks willingly under the yoke of the king of Babylon. Serve him and his people, and live. Why will you choose to die by the sword, by famine, and by plague in another siege? I bring you the warning of God! All this will happen, if we do not continue to serve Babylon."

But the king's inexperienced advisers and false prophets urged him to go ahead with the planned revolt. "Don't worry," they promised, "God will be with you."

"They lie!" Jeremiah cried out. "Do not listen to them, or this city will become a desolation!"

At about this time, a false prophet named Hananiah announced in the temple that he had a hopeful message directly from God. "Within two years Coniah our exiled king shall return!" he shouted. "With him shall come all our exiled people. And our temple treasures shall be returned."

Jeremiah had continued to wear his yoke. Now Hananiah reached out, yanked it from his shoulders and smashed it on the stone floor. The false prophet cried out, "Thus says the Lord: 'So will I break the yoke of Nebuchadrezzar king of Babylon from the neck of all the nations within two years!' "

Jeremiah went his way sadly, because he wished that Hananiah's lies could be true.

Then God spoke again to Jeremiah. The prophet returned to the palace with a new message. He found Hananiah there, too.

"You have broken a wooden yoke," Jeremiah accused Hananiah, "but God will make instead an iron yoke!" Jeremiah bowed before his king. "For it is the will of your Lord that these wicked and unrepentant nations wear the iron yoke of servitude to Nebuchadrezzar king of Babylon."

King Zedekiah was swayed by the words of Jeremiah. He sent the ambassadors back to their own countries. He rejected the advice of the priests, the false prophets, and the others who wanted to revolt. This happened in the fourth year of young Zedekiah.

As time went on, however, more and more Judeans wanted to rebel. This faction became strong and persuasive. "Let us ally ourselves with the strength of Egypt!" they drummed continuously into the king's ears. "Together with other small countries we can break the galling yoke of these Chaldeans."

At last, in the ninth year of his reign, Zedekiah took the foolish step. He rebelled against the king of Chaldea.

Nebuchadrezzar charged out from Babylon with his mighty army. The Chaldeans built siege works around Jerusalem. They cut off food supplies into the city.

Then came hopeful news. The Egyptians were marching north to aid their beleaguered allies! When the Chaldean army heard of it, they withdrew their siege.

But Jeremiah did not join in the delirious celebration of the people. Somberly he told the king, "I have God's word that Pharaoh's army is about to turn around and flee back to Egypt. The Chaldeans will return. They shall take Jerusalem and burn our city with fire!"

King Zedekiah refused to listen to such gloomy predictions. He joined in the rejoicing of his people. Everyone counted happily on a coming deliverance.

During this military lull, Jeremiah decided to take care of some personal business in his hometown of Anathoth. As the prophet was walking out through the city gate, a sentry seized him. "You are deserting to the Chaldeans!" he accused.

Jeremiah tried to explain that he was only walking the short distance to Anathoth and back, but the sentry refused to listen. The prophet was dragged to the government house. There he was beaten and thrown into a dungeon.

For many days Jeremiah sat, bruised and aching, in his dank cell. He yearned for fresh air and sunlight. Then the king sent for him, secretly. "I'm worried about whether the Chaldeans are going to come back," the king admitted. "Is there any new word from the Lord?"

The prophet, wearing only his torn and bloody tunic, bowed low. "You shall be delivered into the hand of Nebuchadrezzar the Chaldean." Then he challenged his king. "Why should I languish in a foul dungeon when I speak only God's truth?"

"I cannot release you," Zedekiah said. "You have too many enemies in high places who would take your life. But at least I can see that you are removed from that dungeon." The king arranged for Jeremiah to be transferred to protective custody in the open-air court of the guard. He was given a small loaf of bread each day for his rations.

But things happened as Jeremiah had predicted. The Egyptian army retreated to their own land. The Chaldeans returned to the siege of Judah's capital. The shortage of food in Jerusalem became desperate. Many of the people were reduced to eating filth, anything to try to stop the pain of empty stomachs.

Jeremiah was able to talk to the townspeople from the courtyard where he was held. He counseled them to surrender. "Listen to God's warning," he begged. "Those who stay in this city will die by the sword. They will die by famine and by plague. But those who go out to the Chaldeans will live, even though they be taken as captives."

The enemies of Jeremiah complained to the king, "This man preaches treason and surrender. He should be put to death!"

Zedekiah felt helpless before their demands. "I can do nothing against you," he sighed.

So the men rushed into the court of the guards and bound Jeremiah and threw him into a cistern. Luckily for Jeremiah, most of the water in the cistern had been used in the siege. He did not drown, but sank to his waist in the muck. There, without food or water, he knew he would soon die.

Ebed-melech was an Ethiopian official in the palace. When he heard of Jeremiah's plight, he hurried to the king and asked permission to rescue the prophet from the cistern.

The king's private sympathies were for Jeremiah, but he did not have the courage to admit it openly. He looked about nervously to see if any of Jeremiah's enemies were within hearing. "Go," Zedekiah whispered. "Do it quickly, but be quiet about it."

Ropes were let down and the prophet, covered with mire and slime, was drawn out of his pit of death. Again Jeremiah was kept in custody in the court of the guards.

In the sixth month of siege the famine

became extremely severe. People were dying of starvation. The cries of hungry children filled the streets.

The Chaldeans pounded the wall daily with their powerful siege machines. They built huge fires against the wall, weakening and crumbling it. Their engineers tunneled underneath. Suddenly one evening about sunset, the Chaldeans broke through Jerusalem's wall.

King Zedekiah, with his young sons and a band of soldiers, fled that night. They sneaked through the gate leading to the king's garden, then picked their way eastward through the wilderness.

The Chaldean warriors overtook Zedekiah's party near the Jordan River, on the plains of Jericho. The captives were taken before King Nebuchadrezzar. The Chaldean king, wanting to set an example for these troublesome and rebellious people, passed a cruel sentence. The young sons of Zedekiah were slain before their father's eyes. This terrible sight was the last thing that poor Zedekiah ever was to see, for at that point they put out his eyes. The king was bound, dumped into a cart, and carried off to prison in Babylon.

This time King Nebuchadrezzar's army did not leave Jerusalem standing. The city was sacked and burned. The once beautiful palace was leveled. The temple built by Solomon, refurbished by many of the kings who followed him, was left as a burned-out shell. The great city walls were broken down into rubble, the gigantic wooden gates burned.

The survivors in the city, as well as the deserters who had surrendered, were carried off into exile in Babylon. Only a few of the poorest farmers were left, scattered here and there in the countryside, to plant and plow the land.

The Chaldean king decided that the level-headed old prophet was a good influence upon his people. He sent word that Jeremiah was free to stay in Judah.

There would be no king. Judah was to be only a small Chaldean province, and Gedaliah, the grandson of good King Josiah's secretary, was appointed as governor. Jerusalem was in complete ruin, so a small house was provided for him in the nearby village of Mizpah.

Jeremiah encouraged the remaining people to plow and plant and harvest. He supported the new governor. But there were Judeans left who turned against Gedaliah because he had accepted his appointment from Babylon. "That makes him a traitor!" they said.

A hot-headed group assassinated Gedaliah. Then they fled the country.

To add to the confusion and unrest another group of Jews, who were sympathetic toward Egypt, decided that they had better flee, too. They feared the Chaldeans would come back, full of revenge, and kill the remaining Judeans. They went to Jeremiah for his advice.

"Remain in this land," Jeremiah told them. "Plow and plant. God is here and will watch over you."

This advice made the men angry. "You lie to us," they yelled. "You want to deliver us into the hand of the Chaldeans so they may kill us! Or else take us into exile!"

The band of rebels hurriedly fled to Egypt. They dragged Jeremiah with them, a captive. In Egypt, far from his beloved and ravaged land, the aged Jeremiah at last met his death.

2 Kings 24:8 through 25:26
Jeremiah 21 through 43

Mournful poetry in the book of Lamentations provides a vivid picture of the distress and suffering in Jerusalem during this period of its history. A "minor" prophet, Habakkuk, adds his woes, and urges his people to have trust and faith.

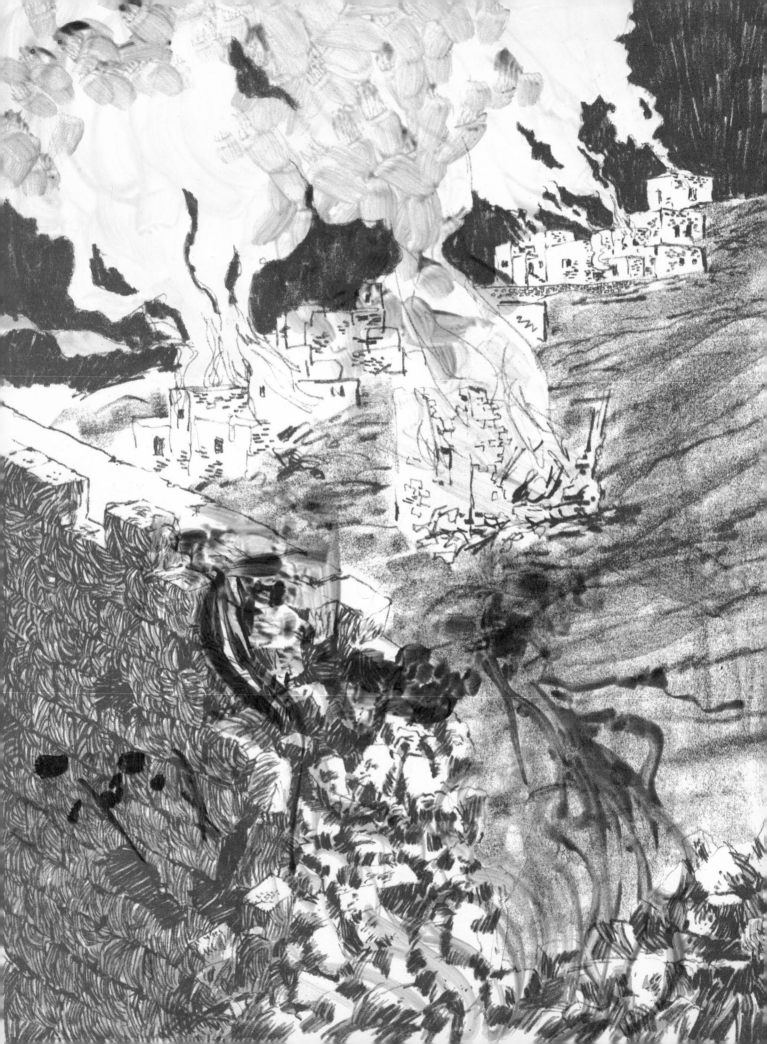

PART VIII
THE EXILE
AND RETURN

The first deportation of captives from Judah to Babylonia has been dated at about 597 B.C. The second, at the time of the sacking and burning of Jerusalem, was about 587 B.C.

The policy of the Chaldeans in dealing with their captives was a relatively lenient one. They did not scatter and forcibly mix their captive populations, as the Assyrians had done. Families were allowed to stay together. Often a whole community of Jews were given a piece of land where they might settle.

Those who wanted to put down enduring roots in this new land were encouraged to do so. Probably those who renounced their own religion and began to worship the Babylonian gods had better opportunities for advancement, especially in government positions.

Yet the Jews were allowed to meet in one another's homes for quiet worship. Since there was no temple or sacrifice, the people turned to the study of their religious laws. These were read every sabbath in the meetings, along with the cherished written oracles of the prophets. Perhaps the synagogue, so important in later Jewish life, developed from such sabbath gatherings.

New writers were at work, too.

At least one unknown poet, and perhaps more, wrote of the sadness of the Exile; later these poems were gathered together in the book of Lamentations, and credit for them was ascribed to Jeremiah. Ezekiel's teachings and writings were a great inspiration for a defeated people who felt shorn of hope. Then there was the unknown prophet, whose beautiful and profound writings were added to the scroll containing the written words of Isaiah of Jerusalem.

Also, since so much attention was being given to the Jewish laws, much editing and polishing was done during this time. It is probable that the books of Exodus, Leviticus, and Numbers were given substantially their present form.

The period known as the Exile dated from the deportation until the Jews received their freedom with the edict of Cyrus in 538 B.C. By this time Babylon and the Chaldean Empire had fallen to the conquering Persians. Cyrus the Persian emperor reversed the policies of the Chaldeans and allowed exiles to return to their homelands.

Few of the original captives were still alive when Cyrus freed the Jews. In the half century of exile most of the Jews had adapted to life in their new land. Many were prosperous there. They did not know or remember Judah, and had no wish to leave Babylonia.

Relatively few families went back to rebuild and colonize Jerusalem. Only the ones who had kept alive a devout religious loyalty along with a fierce patriotism returned. This group helped to complete the temple of Zerubbabel about 515 B.C.

After this, some of the Jews who had scattered to other lands also returned to Judah. These, who lived in the far-flung areas of the Mediterranean world, have been called the Jews of the Dispersion. For all the people of Judah had not been captured by the Chaldeans. Many had fled to safety in Egypt, joining small colonies of Jews already established there. Some escaped to other areas of Palestine. Some traveled farther to find new homes in Asia Minor, Greece, or North Africa.

Again, only a few of the Jews of the Dispersion returned when Jerusalem was rebuilt. These colonies of Jews—from Rome to Babylon, Corinth to Alexandria—were old establishments by the time of Jesus. Cut off from temple worship, they developed the synagogue as a place for study and prayer, the focal point of their religious lives.

Archaeology has provided us with a detailed picture of life during these times. The immense city of Babylon has yielded clay tablets written during the Chaldean era. On one Chaldean record King Jehoiachin of Jerusalem and five of his sons are listed among those who received government rations of food. Part of King Nebuchadrezzar's palace at Babylon has been unearthed, and the huge Ishtar Gate into the city. Precise Chaldean and Persian records have provided us with dates and a great deal of information to supplement the Bible story.

Unhappy Captives

The Jews of Judah who were deported to Babylon found it hard to settle down to a new life in a strange land.

Their last memories of Jerusalem were almost too sad to bear. When the captives were led away, they had looked back to see their beloved and sacred city in blackened, smoking ruins.

For many years they lamented the plight of the Jerusalem they had left behind them. An unknown Jewish poet used a sad allegory for his desolate city:

How lonely sits the city
that was full of people!
How like a widow she has become,
  she that was great among the nations!
She that was a princess among the cities
  has become a slave.
She weeps bitterly in the night.

The disgrace of captivity was almost too much for the proud Jews. The unknown poet mourned that Judah was in exile:

She dwells now among the nations,
  but finds no resting place.
When her people fell into the hands of the
    foe,
  and there was none to help her,
the foe gloated over her,
  mocking at her downfall.
The enemy has stretched out his hands
  over all her precious things.

Then the poet cried out:

  Is it nothing to you, all you who pass by?
    Look and see
  if there is any sorrow like my sorrow.

The Chaldean conquerors gave the Jews seed and also fertile soil so that they could be productive in their new land. They could even find places to serve in the Chaldean government. Brickmakers, woodworkers, stonecutters, and carpenters were given jobs on the many buildings going up in the immense city of Babylon. Potters, silversmiths, ironsmiths, copperworkers, weavers, and dyers were encouraged to work at their trades.

The Jews could put down roots here, if they wished. They were free to amass wealth and personal property. But the captives missed their native Judah. They missed their independence. They mourned the destruction of the temple that Solomon had built for them.

The exiled Jewish poet lamented:

  For these things I weep;
    my eyes flow with tears.
  There is no comforter near me,
    no one to revive my courage;
  my children are desolate,
    because the enemy has prevailed.

Babylonia was a great fertile land, crisscrossed by streams and canals, with willow and poplar trees along the banks. Here there were opportunities to find joy and merriment again. The Babylonians were not as cruel as the Assyrians had been, but many of the captives could not see any bright side to the picture.

"By the waters of Babylon, there we sat down and wept," a songwriter noted, "when we remembered Zion,* our Jerusalem."

The Chaldeans must have enjoyed the music of the Jews. They encouraged the exiles to play their instruments and to sing songs of their homeland. Sometimes the Jews were asked to provide music for entertainments. A plaintive psalm records the stubborn Jewish reaction to such requests. "On the willows there we hung up our lyres. For our captors wanted laughter and songs

164   from us, saying, 'Sing us one of the songs of Zion!' "

The unhappy exiles simply had no heart for singing. The psalmist wrote,

How shall we sing the Lord's song
   in a foreign land?
If I forget you, O Jerusalem,
   let my right hand shrivel up!
Let my tongue stick to the roof of my mouth,
   if I do not remember you!

The exiles were so wrapped up in their grief that they often burned with desire for revenge. They would lay bloodthirsty curses upon the inhabitants of Babylon. "Bring the day you have announced," many an unhappy exile prayed. "Let them be as I am! O Lord, be aware of their evildoing! Deal with them as you have dealt with me."

The exiles felt vengeful and angry not only toward Babylon, but also against the nation of Edom. The Edomites were descendants of Esau, as the Jews were the descendants of Jacob. And Jacob and Esau were twins, the exiled Jews muttered. Therefore the Edomites really were blood brothers of the men of Judah. Should not brothers stand together? But the Edomites had joined in with the enemy in the pillage and sacking of Jerusalem.

So the memory of Jerusalem's fall was doubly bitter. The exiles could not forgive the traitorous Edomites who had given help to the Babylonians. These tribal cousins from the desert had cast lots, along with the Babylonians, for the division of the spoil and plunder.

Even more maddening were the reports filtering back to Babylon that the Edomites were moving north into Judah, taking possession of Judean houses and fields.

An angered poet named Obadiah cried out in bitterness to the Edomites:

On the day that you stood as one of the enemy,
   on the day that strangers carried off his
      wealth,
and foreigners entered his gates,
   and cast lots for Jerusalem,
you were like one of them.
But you should not be gloating over the misfortunes
   of your brother!
You should not be rejoicing over the ruin
   of the people of Judah. You should not have
      entered the gate of my people.
You should not have looted his goods
   in the day of his calamity. You should not
      have blocked the ways,
   to cut off the fugitives. You should not have
      delivered up his survivors
   in the day of distress.

So Obadiah called down a curse upon the heads of the infiltrating Edomites. "As you have done, it shall be done to you. Your deeds shall return on your own head!" **

And so the exiled Jews continued to weep and to mourn for the lost glory of Judah and for the lost splendor of Jerusalem.

Psalm 137
Lamentations 1:1-22
Obadiah 1:1-15

* Zion was strictly the name of the hill on which the temple stood, but it is often used poetically to mean the entire city of Jerusalem.

** Obadiah's threat against the Edomites is a reflection of the prophetic idea of the Day of the Lord. This was viewed as a day in which God would judge the nations and reward or punish them according to their deserts. It is interesting to note that the words of Obadiah's threat were reversed and used centuries later by Jesus in the Golden Rule: "Whatever you wish that men would do to you, do so to them." (Matthew 7:12)

Ezekiel the Prophet

Ezekiel, son of a priest, was among the first group of captives marched off to Babylonia. After a tiring journey across the hot desert, they came to a lush expanse watered by the great Tigris and Euphrates Rivers. Young Ezekiel was with those Judeans who were settled at a place called Tel-abib, by the canal Chebar.

It was here that Ezekiel was called to be a prophet. He experienced a remarkable vision of something very like a storm. A strong wind whirled down out of the north. Flashes like lightning shot out from the clouds, and in the midst were winged cherubim with human heads and animal bodies. Then Ezekiel saw a constantly moving chariot of great splendor, with wheels within wheels turning in all directions at once. The chariot of God could move with tremendous speed, power, and unlimited freedom.

This was a revealing symbol. God could be everywhere, not just in Solomon's temple or within the boundaries of Judah. God was available and near, in every part of the earth. God was close to his people, even in a strange and faraway land.

Later, Ezekiel wrote how he had heard God's voice speaking to him, telling him to stand up and listen. And a scroll appeared, with words on it.

"Mortal man," the awesome voice said, "eat this scroll and go speak to the people!"

Ezekiel ate the scroll. It melted in his mouth, as sweet as honey, like the manna sent to Moses in the wilderness.

"You are not sent to a people of foreign speech and a hard language whose words you cannot understand," Ezekiel heard the voice go on. "Surely, if I sent you to such,

they would listen to you. But your own countrymen will not always listen to you, because they will not listen to me. They have stubborn hearts, and are blockheads! Now go, get you to the exiles, to your people and say to them, 'Thus says the Lord God!' Tell them my word, whether they hear or refuse to hear."

So it was that Ezekiel began to give the oracles of God to the exiles in Babylonia.

Here, in Tel-abib, Ezekiel's beloved wife took ill and died. Ezekiel grieved for his wife, but he felt that God wanted him to show no outward signs of mourning. When news came that Jerusalem had completely fallen, and that the rest of the Judean captives were on their way to this foreign land, Ezekiel felt he should be an example of quiet self-control to his people.

In their hearts the captives grieved deeply for the total loss of their nation. The Promised Land was in ruins. It was no longer their country.

Before this final and complete destruction of Jerusalem, Ezekiel preached words of warning and doom, just as Jeremiah and the earlier prophets had done. After Jerusalem was no more a city, Ezekiel's messages softened. He spoke to his fellow captives with understanding and affection.

But the exiles were in despair. They felt they had no future, nothing to hope for. They were defeated. Did this mean that God was also defeated? "God's land" was conquered, "God's people" were scattered. Some may even have thought, "God is dead!"

Ezekiel brought a vital message to his disheartened people: God was bigger than the small nations of Judah and Israel. God was everywhere, even in this faraway land of the enemy. God was always ready to join in the life of his people. A nation might reject him. God could withdraw temporarily and

168 let a nation destroy itself, but God was still there, and was still available!

This was hard for the discouraged exiles to believe. "What can we look forward to now?" they mourned. Ezekiel tried to tell them that out of the ashes of destruction, a new hope could rise. He proclaimed that God could make the difference between life and death. The presence of God would make the difference between hope and despair, even in Babylon.

Ezekiel wrote down for his people another vision: "The hand of the Lord was upon me," Ezekiel said, "and he took me to a valley." The prophet said it was as if a great army had fallen there at one time. Dry bones lay scattered about.

God spoke to Ezekiel, "Mortal man, can these bones live?"

Ezekiel looked across the broad valley. God spoke again, "Say to these bones, come together and live!"

So, in the vision, Ezekiel obeyed and spoke. There was a noise, a rattling. The bones came together, bone to bone. Then the muscles, the tendons, and flesh came upon the bones, and skin covered them.

So the slain men were whole again, but there was no breath in them. Then God told Ezekiel to speak to the four winds, and Ezekiel did. The winds of God blew upon the slain. Breath came into them; and they lived and stood upon their feet!

In his vision Ezekiel heard God say, "Mortal man, these bones are a parable for your nation. Your people say, 'Our bones are dried up, and our hope is lost. We are cut off from our land.' Now tell your people that their God will breathe his spirit into them. They shall live! And someday your people will return to their land."

For twenty-two years Ezekiel served the people, meeting with the elders in his house, preaching, counseling, and writing.

Because of Ezekiel the people clung to hopes for the future. They preserved a dream of the day when they would return to Jerusalem and build again their ruined temple.

Ezekiel 1:1 through 3:11; 24; 37

The book of Ezekiel developed a special kind of image using mysterious and fantastic symbols whose meaning is very difficult to understand. The same type of symbolic writing appears in the last half of Daniel and in the Revelation.

Chapters 40 through 48 give a very detailed description of Ezekiel's vision of the restored Jerusalem temple. While the long and involved descriptions seem tiresome to the modern reader, they were welcomed and treasured by the exiles who cherished every mention and recollection of their beloved temple. Many scholars conclude that the prophet's vision reflected his actual firsthand knowledge of the temple, and so this detailed information is very valuable in attempts to picture the temple of Solomon.

The Unknown Prophet

During the fifty years of captivity, a whole new generation of Jews had grown up in Babylon and other parts of the Chaldean Empire.

Most of the exiles made good use of the liberty given them by the Chaldeans. Many thrifty farmers now owned their own land. Craftsmen prospered. There were men of Jewish descent holding positions in the Chaldean government.

The Jews were allowed to worship as they wished. Though they had no temple for formal sacrifices and rites, they did gather together in homes every sabbath to study their laws and history and to read the cherished words of the great writing prophets.

These meetings were simple and quiet compared to the Babylonian religious festivals. The Babylonians worshiped their many gods at splendid temples. Their rituals blazed with color and pageantry. Every New Year's Day—the first day of spring—there was an extravagant parade. Ornate idols of iron or wood were carried upon the backs of gaudily decorated animals, and people lined the streets to watch.

As time went on, many of Jewish descent were seduced into the ways of the land. They began to worship Babylonian idols.

A sensitive and greathearted prophet, whose real name is now unknown, tried to reason with the Jews who were attracted to this idol worship.

"Whoever fashions an idol is making something which is both pointless and useless," he told them. "A blacksmith takes an iron in his tongs and hammers and shapes it over the burning coals to make an idol. Or a carpenter stretches a line on a block of wood and uses a compass, marking the shape with a pencil. He chisels and planes it. He fashions a figure to sit in a shrine."

It was hard for this prophet to see how intelligent men actually could worship a thing of wood made by human hands.

"Think!" he said with great scorn. "A man cuts down a cedar tree. Half of it he uses as firewood. By this fire he warms himself, saying, 'Aha, I am warm now.' With the heat from this wood he bakes his bread and roasts his meat. The other half of that same wood he turns into an idol. Then he bows down and worships it, saying, 'Save me, for you are my god.'"

This unknown "Prophet of the Exile" usually expressed himself in beautiful poetry which followed the tradition and style of the revered Isaiah of Jerusalem.* This Unknown Prophet also appreciated the holiness and wonder of God. He shared the same concept of God at work in history.

This might explain why his anonymous writings were attached to the end of the scroll containing the writings of Isaiah.

There are great differences in the messages of these two great prophets. Isaiah of Jerusalem lived long before the fall of his country. He scolded and condemned the people of Judah.

The Unknown Prophet of the Exile found a more pressing need. "Comfort, comfort my people," he wrote.

Speak tenderly to Jerusalem,
and tell her
that her war and servitude are over,
that her guilt is pardoned,
for she has been fully punished by God
for all her sins.

Even though living in exile, the Jewish people must not forget their God. He was the Lord of nature and of history. He was eternal. "The Lord is the everlasting God, the Creator of the ends of the earth," the prophet wrote. "The grass withers, the flower fades, but the word of our God will stand for ever."

This poet-in-exile had a large view of God. The Lord was more than the little tribal or national gods which were worshiped so fanatically by foreign peoples. The whole earth was God's domain.

"Who ever measured the waters in the hollow of his hand? And marked off the heavens with a measuring stick?" the poet asked. "Who has weighed the mountains in scales and the hills in a balance? Who else, but God?"

Many exiles were impressed by the might of Chaldea, and the kings who ruled from the great palace in Babylon. The prophet rebuked them, "Look, the nations are like a drop from a bucket, no more than specks of dust on the scales! It is God who brings the kings to nothing, and ruins the rulers of the earth."

When the people felt weary and leaderless in this strange land, the prophet spoke soothingly to them. He pictured God as a good shepherd, watching tenderly over his people, caring about their needs:

He will feed his flock like a shepherd,
  he will gather the lambs in his arms.

A few of the captive Jews never lost hope that circumstances would change so they could return to Judah. To wait patiently for an opportunity and to take advantage of any opportunity that came, the Jews would need energy, vigor, and endurance.

The prophet reminded them that their God never tired nor grew weary.

They who look to the Lord shall renew their
    strength,
  they shall mount up with wings like eagles,
they shall run and not be weary,
  they shall walk and not faint.

The Unknown Prophet told his people that they were called by God to be his servant. There was a purpose in their grief and suffering. Their job was not just to help restore their homeland. God wanted them to help redeem and save the whole world!

"It is too small a thing that you should be God's servant to raise up the tribes of Jacob, and to restore Israel," the prophet challenged them. "Your God says: 'I will give you as a light to the nations!' "

This prophet also spoke mysteriously of a single individual who would be the "suffering servant of God." ** "Despised and rejected by men, a man of sorrows, and acquainted with grief," was the prophet's description of this individual. "Surely he has borne our griefs and carried our sorrows. He was wounded for our transgressions, and with his bruises we are healed. All we like sheep have gone astray; and God has let him bear our sins. So he was cut off from the land of the living. He poured out his soul to death."

As the prophet was writing these words of hope and prophecy, the whole Near East bubbled with political unrest. The widespread power of Babylon was slowly but surely weakening. In the northeast a strong and ambitious man named Cyrus had made himself king of Persia. He began to subjugate nearby small nations.

The prophet was excited about the news of this new king. Word came to Babylon that Cyrus was marching relentlessly westward toward the Great Sea.

In the days before the Exile, Isaiah and Jeremiah had preached that God might use other nations to carry out his will. The Unknown Prophet believed this, too. To him, Cyrus was stirred up by God. He would be the conqueror of Babylon and the liberator of the Jewish people. "The coastlands have seen and are afraid. The ends of the earth tremble. Cyrus shall fulfill the purpose of God," he prophesied.

Cyrus marched westward and conquered the fabulously wealthy King Croesus of Lydia in Asia Minor. The rich booty helped the Persian king to finance further conquests.

Some of the Jews living in Babylon did not think this powerful new conqueror could batter through the huge walls and defenses of that city. "The Chaldean kings have made Babylon unconquerable," they moaned.

But when at last the triumphant army of Cyrus marched on Babylon, some rebels within the city opened the gates to him. Babylon surrendered without a fight.

Cyrus was now king of an immense new empire, which stretched from the Mediterranean Sea to the eastern lands beyond the Tigris and Euphrates.

Cyrus proved to be a benevolent ruler.

He allowed the captive peoples to return to their old homelands if they wished. Word went out among the Jewish exiles of the edict of Cyrus, which freed them. The Unknown Prophet sang of an exuberant return of the Jews to their beloved homeland:

For you shall go out in joy,
    and be led forth in peace;
the mountains and the hills before you
    shall break forth into singing,
        and all the trees of the field shall clap their
            hands.
Instead of the thorn shall come up the cypress;
    instead of the brier shall come up the
        myrtle;
and it shall be to the Lord for a memorial.

Isaiah 40 through 55

* The writings of Isaiah of Jerusalem are generally considered to fall within Isaiah chapters 1 through 39. Writings ascribed to the later Unknown Prophet fall within chapters 40 through 55. Although his name was not necessarily Isaiah, he often is referred to as Second Isaiah or Deutero-Isaiah.
The remaining passages of Isaiah, chapters 56 through 66, show that they were written at a time still later, possibly in Jerusalem when it was being rebuilt by the returning exiles. These chapters may have been written by disciples of the Unknown Prophet.

** Some of the most famous passages written by the Unknown Prophet are those about the "suffering servant." Writers in Christian times have seen in these verses striking parallels to the life of Jesus.

The Exiles Return

When Cyrus the Great became king of the vast Persian Empire, he freed all exiled peoples. But many exiles had no desire to leave the rich, fertile lands where they had put down roots. There were many who had never known Judah. Why should they make the long and arduous trek across burning deserts, they asked, to go live on the distant devastated mountains in what was now the Persian province of Judea?

Ezekiel and the Prophet of the Exile had stirred some of the people with their glorious visions of a rebuilt temple and a restored homeland. So there were some exiles who sold what land or houses or businesses they had acquired, packed as much as they could carry, and started back.

It was a weary, raggle-taggle band that completed the long, tiring journey westward. When they finally arrived at Jerusalem, the repatriates stared at the ruined city of their dreams. The huge city walls were in rubble. The burnt sides of the temple jutted upward in gaunt nakedness. Weeds and brambles choked the streets. Heaps of fire-blackened stones marked where houses once stood. After dark, bats flew in and out of the ruins. Owls hooted from the roofless walls of the temple.

In these disheartening surroundings the people settled down to the task of eking out a living. Gradually they built little one-room stone houses with earthen roofs, outside the devastated city. They cleared and plowed patches of the stony land; they planted and harvested meager crops of wheat and barley.

"We must begin to build a new temple," the leaders of the new settlers agreed.

Attempts were made to gather timber, but the native trees were not the best quality for building. They were not like those great cedar logs King Solomon had imported from Lebanon for his building.

Men took turns going down into the stone quarries to chisel out blocks of limestone for the new foundation. They were not skilled however, and the going was very slow.

The northern people of mixed nationalities, who now inhabited Samaria, heard of the temple project. They offered to help. "Look," they said, "we worship your God. Some of us are descendants of the original Israelites, and so we are your blood brothers. The rest of us have been sacrificing to the God of Abraham and Moses ever since we were settled in this land. Let us help."

But the returned exiles were in no mood to accept such an offer. A great pride in their history and national heritage had developed during the years of their separation from their homeland. Also during these years their faith in God had been the one thing that bound them together. They still possessed and shared their religion when there was nothing else they could call their own. They also remembered that the destruction and exile of Judah had come about because of foreign religious influences that the Jews seemed unable to resist.

This was the kind of thinking that caused the leaders of the Jews to insist upon exclusiveness in restoring the temple at Jerusalem. "We alone will build a house for our God," they declared.

After this rejection the Samaritans became angry and hostile. They did everything they could to frustrate the newcomers. They spread malicious rumors and turned the Persian overseers of the province against the Jews.

It was just too much for the small band of Jews. They could not cope with the mounting resentment of their neighbors, the lack of money, the absence of real enthusiasm. So the building of the temple was abandoned.

When Cyrus died there was a period of confusion at the Persian capital. After a time King Darius took the Persian throne.

A new governor was appointed then for Jerusalem. He was Zerubbabel, born in exile to a son of King Jehoiachin. This young prince was given the difficult job of governing this province. He brought to Judah a high priest, who would organize the religious life of the community.

A prophet named Haggai came forth to encourage Zerubbabel and the new high priest in the rebuilding of the temple.

"You all say," Haggai challenged the apathetic people, "that the time has not yet come to rebuild the temple." He pointed his finger at one farmer after another. "Is it time for you to build roofs for your own houses, when God's house is roofless and lies in ruins?"

These men stood around Haggai and listened. They were a glum lot, thin and haggard from their meager diet. They wore patched, worn clothes.

"Consider how you have fared," Haggai said. "You have sown much and harvested little. You eat, but you never have enough. You drink, but you never have your fill. You clothe yourselves, but no one is warm. Your money has no value. He who earns wages puts them in a bag with holes!"

"You are right," several Jews shouted. The men began to grumble, nodding their heads in agreement.

"Why do you think the land has withheld the dew and the rain?" Haggai needled them. "Why do you think the earth has withheld its bounties and its produce? It is because you neglect the house of God! You busy yourselves each with his own house."

The zeal of Haggai was contagious. The governor Zerubbabel and the high priest were eager to begin the new temple. The people were ready to help.

Rumors came of unrest back at the Persian capital. People wondered if Darius could dominate such a vast, rebellious empire. In this general political uncertainty Jerusalem's temple-building program flourished. Their governor was a man of royal blood, a descendant of the house of David. If the Persians were busy enough with other problems, Judea might be able to forge an independent kingdom once more.

The prophet Haggai encouraged his people to think of Zerubbabel as their anointed one, their messiah. Zerubbabel was the royal descendant of David who had come to reign, to bring back the lost Golden Age.

But the new Persian king proved to be a strong ruler. Darius soon restored peace and order in his empire. So the nationalistic vision of the Jews was pricked like a bubble. How could they rebel against such power? Jerusalem was an empty city with no walls. The Jews had no arms, no weapons. It was hopeless.

As their patriotic fervor dwindled, the work on the temple slowed. Finally all building stopped. Weeds and vines began to grow over the new foundations.

For a while it seemed the exhortations of the prophet Haggai had been in vain.

Then a man named Zechariah walked through the rubble and the abandoned foundations of the temple. He felt the call of God to stir up his people.

"Let your hands be strong, in continuing to build the temple," Zechariah cried out to the repatriated Jews. "Do you remember the days before you began to lay the foundations? In those days there was no wage for man or beast. There was no safety from our foes!"

174 Most of the Jews were still poor. There still was no prosperity in the land.

"But build the temple," Zechariah promised, "and the vine shall yield its fruit. The ground shall yield abundant crops. This remnant of the people shall possess all things! So let your hands be strong!"

The enthusiasm of the people returned. They worked hard on the rebuilding of the house of God once more. They quarried out large blocks of limestone and chiseled them to fit together evenly. They imported cedar logs for roof beams and paneling.

The prophet Zechariah, like Haggai, encouraged the people to renew their national hopes for Zerubbabel. Perhaps their royal governor could be enthroned someday as their king-messiah!

Tattenai, the governor of a neighboring province, heard what was being done in Jerusalem. He came to inspect the work. He scowled at the craftsmen working on the scaffolds. "Who gave you permission to build a temple in such a way it looks more like a fortress than a sanctuary?" The military appearance of this rebuilding made him worried and nervous. Tattenai plotted together with several other jealous and uneasy neighboring governors, and they drafted a letter, sending it to the king in the Persian capital.

It read, "To Darius the king, all peace. We went to the province of Judah. We found, in the city of Jerusalem, the Jews building a great house for their God. It is being built with polished and costly limestone. Timber is being laid upon the walls. We asked who gave them permission to build such a great structure. They said that in the first year after capturing Babylon, King Cyrus made a decree that this temple could be rebuilt. Therefore, if it seems good to the king, let a search be made in the royal archives there in Babylon to see whether such a decree was issued."

King Darius ordered a search made in the house of the archives, where official documents were stored. A dusty scroll was uncovered. It carried the written decree of Cyrus.

So Darius answered Tattenai and the other complaining governors. "Let the work on the temple continue without interruption."

However, from this time on, there is no more mention of the princely Zerubbabel in the Bible records. The eager Jewish governor in whom the people had such hopes simply dropped out of sight. Did word of the Jewish expectation of a messiah-king, to be realized in the royal person of Zerubbabel, filter back to the Persian court? Was Zerubbabel eliminated by murder? Or was he ordered to return to a restricted life in Babylon?

We may never know.

But the temple finally was completed under the leadership of Zerubbabel's high priest. There was a joyous dedication service with festival sacrifices.

The revered temple of Solomon had been destroyed and left in ruins. Now, in its place, stood the shining new temple of Zerubbabel.

Ezra 1:1-11; 2:68 through 4:5; 5 through 6
Haggai 1 through 2
Zechariah 8:9-13

Rebuilding Jerusalem

Almost a century after the first band of exiles had returned to the area around Jerusalem, the city itself still lay in ruins.

The rebuilt temple of Zerubbabel stood on the hill of Zion. But the rest of the city was devastated. At night jackals howled in the deserted, weedy streets. There was no wall around the city, only a circle of rubble.

The high priest had organized formal worship at the temple. But the priests were slovenly in their duties. The worship rituals were hit-or-miss, mechanical and meaningless. People showed contempt for God by bringing crippled or scrawny animals for sacrifice.

It was on this discouraging scene that the prophet Malachi looked. The spiritual ruin of the people appalled him even more than the material ruin of the city.

"Priests," he cried out, "you offer animals that are blind, sick, or lame! It would be better to shut the doors of the temple than to light fires on these polluted altars."

Malachi was also distressed to see how little understanding there was among these people who had suffered so much together. He asked why each man was trying to get the better of his neighbor. Why did the Jews of the community try to exploit each other?

"Have we not all one father?" Malachi demanded. "Has not one God created us? Why then are we faithless to one another?"

When Malachi saw how many of the Jews were intermarrying with their non-Jewish neighbors, he became very angry. He berated the older men who were divorcing their wives to marry young and attractive foreign wives.

"You have been faithless," he scolded. "The wife of your youth has been your companion all these years. She is your wife by covenant."

The land of Judah was as stony and unyielding as ever, and the people were still poor. Often there were long dry spells when no rain came to moisten the parched earth. Even periods of plentiful rain and good crops were apt to be turned to famine by a catastrophe such as a plague of locusts.* These grinding physical hardships did much to prevent the dreams for a new Jerusalem from materializing.

In the midst of their privations the people did not return to the life of faith. They sank deeper into discouragement and apathy. They felt depressed whenever they went into their ruined city, the empty city with stacks of rubble for houses, without walls or security.

At last, reports of the sad state of Jerusalem drifted back to the great eastern city of Susa, the winter residence of the Persian kings. In this palace was a young Jew named Nehemiah. He was a respected official, the cupbearer to King Artaxerxes.

When young Nehemiah heard of the deplorable conditions in the homeland of his ancestors, he wept. Then he prayed to God that he might do something to help.

Nehemiah always had been a happy youth, and King Artaxerxes had become quite fond of him. Now, as Nehemiah attended the Persian king at his meals and filled the royal wine cup, the king noticed a striking change in him. Nehemiah's sunny smile was gone.

"Why is your face sad?" the king asked.

"Let my king live forever," Nehemiah said dutifully. Then he cried out in the anguish of his heart, "Why should my face not be sad? The city of my ancestors lies like a wasteland. Its walls are down; and its gates have been destroyed by fire. And so my people are disheartened and discouraged."

King Artaxerxes raised his eyebrows. "So?"

Tears came into the young man's eyes. He stammered, "If only I—"

"Come now," the king said sympathetically. "What is your request?"

Nehemiah breathed a quick, silent prayer to God. Then he blurted out, "If it pleases the king, and if your servant has found favor in your sight, send me to Jerusalem. I yearn to rebuild the city of my forefathers."

The king hesitated. Nehemiah was a vigorous and capable young man. The whole court liked him for his pleasant manner.

"You may go for just long enough to do this task," King Artaxerxes finally decided. "And during this time, Nehemiah, you shall be my new governor in Judea!"

The king gave Nehemiah letters to the administrators of neighboring Persian provinces so he could pass through unchallenged. Along with Nehemiah the king sent officers and horsemen of the Persian army.

When Sanballat, the administrator at Samaria, and Tobiah, the part-Jewish administrator in Ammon, heard of Nehemiah's coming, they were jealous and uneasy about his mission. They began to plot against him.

After his arrival at Jerusalem, Nehemiah immediately inspected the broken-down walls. He looked with sorrow at the charred remnants of the Valley Gate, the Dung Gate, the Fountain Gate, and the others.

Then Nehemiah called together the priests, the nobles, and the common people. He showed them his credentials from the Persian king.

"You see the trouble we are in," the new young governor said. "Jerusalem lies in ruins with its gates burned. Come, let us build the wall of Jerusalem, that we may no longer suffer disgrace."

The people cheered their new governor. "Let us rise up and build," they cried out. The rebuilding of the city walls was begun.

When Sanballat heard this news up in Samaria, he was angry and enraged. He marched down with his provincial army and looked over the work.

"What are these feeble Jews doing?" he said loudly making fun of them. "Will they finish up in a day? Will they revive the stones out of heaps of rubbish—and burned ones, at that?"

Tobiah had come from his official residence, east of the river Jordan in Ammon, to confer with Sanballat. "Yes," he cried out, shouting with laughter, "if a fox were to climb on it, their stone wall would tumble down!"

Nehemiah and the Jews went grimly on with their work, ignoring the jeers and taunts and ridicule of their neighbors. For the people had a mind to work.

When the walls were joined together up to half their height, the leaders of the neighboring provinces became even more hostile. They plotted together to come and fight against the men of Jerusalem.

The enemies whispered, "They will not guess, until we come into the midst of them and kill them. That will stop the work!"

But word of the plotting reached Nehemiah. So, in the lowest parts of the space behind the wall, in the open places, Nehemiah stationed men with their weapons.

"Do not be afraid of our enemies," Nehemiah told the people. "Remember God. Fight for your families and your homes."

The enemies learned that their plans were known, that they could not go ahead with a surprise attack. From that day on, half the Jews worked on the construction of the walls, and the other half stood guard. Their spears and swords were ready, their bows and arrows at hand. Each of the stonemasons had his sword girded at his side while he built. And a trumpeter stood ready to blast a warning at any sign of danger.

So they labored from the break of dawn until the stars came out at dusk. At night they slept in the ruined city. None took off his clothes, and each slept with a weapon in his hand.

At last the wall was completed. Now Nehemiah and his men were ready to prepare the great timbers to be used in making new gates.

This progress was reported to Sanballat in Samaria and to Tobiah in Ammon. The two men met together and plotted. Then they sent an urgent message to Nehemiah, asking him to meet them at a certain place near the Mediterranean Sea.

Nehemiah guessed that his enemies intended to do him harm. But he did not let on about his suspicions. He simply sent messengers, saying, "I am doing a great work. I cannot leave it right now."

Four times Sanballat and Tobiah repeated their summons. Each time Nehemiah put them off by sending back a message that he was too busy. Then Sanballat sent one of his officers to Nehemiah. The Samaritan officer carried an open letter in his hand. In it was written, "It is reported among the nations that you and the Jews intend to rebel. That is why you are building the wall. You wish to become their king. Don't you know this will be reported to our Persian ruler? Come, meet me at the place I suggest. Let us take counsel together."

Nehemiah sent back an indignant reply: "No such things as you say have been done. You are inventing them out of your own mind."

Nehemiah was not going to be trapped into an ambush. And he knew that his enemies wanted to frighten the Jews so that the work on the gates would be dropped. The young governor prayed, "O God, strengthen my hands!"

Sanballat and Tobiah found they could not entice Nehemiah out of Jerusalem, where they might kill him. So they decided they would try to betray him through false friends. They would lay a well-planned trap, with armed men hidden inside the Jerusalem temple.

Nehemiah was summoned to meet a prominent citizen of the town. When Nehemiah arrived, the man pretended to be shaking with fear. "Let us hurry to the house of God and lock ourselves inside," he urged, grabbing the governor by his robe. "I just heard they are coming to kill you."

But again Nehemiah recognized the attempts of his enemies to snare him. He suspected that this fellow had been bribed by Tobiah and Sanballat. "Should such a man as I flee?" Nehemiah said angrily. Besides he knew that he was not consecrated or eligible to enter the holy interior of the temple. "I will not go in there."

Now the frustration and chagrin of Nehemiah's enemies exceeded all bounds. They composed a letter in Aramaic, the common language of the Empire, and sent it by fast riders to the Persian capital.

To Artaxerxes the king: your servants send greeting. Be it known to the king that the Jews are rebuilding their wicked and rebellious city. They are finishing the walls of Jerusalem. If this city is rebuilt and the walls finished, they will not pay tribute, custom, or toll, and the king will suffer losses! If you will search in the book of the records of your fathers, you will learn that this has always been a rebellious city, hurtful to kings and provinces. That is the very reason that this city was once laid waste.

King Artaxerxes sent back a message that the work of the Jews was to be halted until the matter was studied.

At last, however, all obstacles were overcome. Nehemiah convinced his king that there was no subversive activity in the building. Artaxerxes made a decree that the work could continue.

After much hard work, the governor and his people saw the stone walls of Jerusalem

standing proud and high around the old city. Each of the exits was fitted with a huge timbered gate.

Yet Jerusalem was still empty. A few people were camping in makeshift homes inside the ruined city. The walls defended nothing except Zerubbabel's temple and piles of rubble.

First, Nehemiah ordered a house to be built inside Jerusalem for the governor, and houses for the officials and leaders among the Jews. Then he cast lots among the scattered farmers who lived on the stony hills, and among the inhabitants of the small Judean villages. One in ten was picked to move into Jerusalem and build himself a home. Finally, the beloved city was on its way to restoration and repopulation.

Nehemiah called the people and their leaders together for a joyful dedication of the new walls. There was a triumphant parade led by the priests. They were followed by musicians. The air vibrated with the plucking of lyres and the banging of cymbals. The temple Levites sang:

> Walk about Zion, go round about her,
>   number her towers,
> consider well her ramparts,
>   go through her citadels;
>   tell the next generation
>   that this is God,
> our God for ever and ever.

Many animals were sacrificed at the temple. There followed a great feast of celebration, with roasted beef and lamb, fragrant loaves of wheat and barley bread, and fresh fruit.

His building task finished, Nehemiah returned to the Persian capital, to the services of his king.

After some time, disturbing news came from Jerusalem. It was said that Tobiah, the Ammonite, who had been so opposed to Nehemiah, was actually using a room in the holy temple to store his personal possessions. Jews were intermarrying at an alarming rate with other peoples in the land. And borrowers were losing their lands and their houses because of excessive interest rates from fellow Jews.

Nehemiah was grieved. He bowed before his master. Humbly he asked King Artaxerxes if he might be allowed to return to Jerusalem. A strong hand was needed to institute some vital reforms. Once again, the king allowed his official a vacation from his duties.

After a long trip over the interminable desert stretches, Nehemiah reached Jerusalem. He passed through the gates of the wall he had built. First he checked conditions at the temple. He noticed that the Levites, who were supposed to assist the priests, were not around.

"Why is the house of God forsaken?" Nehemiah asked sharply.

"The Levites have not been given their daily portions of food," a nervous priest admitted. "So they moved outside the city to raise their own food."

Nehemiah ordered the Levites to return to their work and appointed an official to see that they were paid regularly from the temple treasury.

Some of the temple storerooms opened out onto the surrounding court. Nehemiah discovered that one chamber was crowded with the possessions of Tobiah. The priests were apologetic. They explained that this Ammonite was influential in Jerusalem's affairs. He was married to the daughter of a Jewish nobleman, and his son was married to a Jewish girl from a prominent family.

Nehemiah was furious. He ordered the possessions of the intruder thrown out. The chamber was cleansed and returned to temple use. Nehemiah felt that this was the

180 final straw. Intermarriage with non-Jews must stop. As he toured the city, Nehemiah saw Jewish men going about with Philistine, Ammonite, and Moabite wives. Their children could not even speak the Jewish language clearly.

Nehemiah called the citizens together. He reminded them that Solomon had been led into sin on account of his marriages with foreign women. Nehemiah outlined a harsh and unyielding policy with regard to marriage with non-Jews. Some of the marriages already in existence were forcibly broken up, and those who would not obey these rules were sent away.

Next, Nehemiah turned his attention to other pressing problems. He listened to the great outcry of the people against the loan sharks in the land.

"We had a long dry spell and a famine," many inhabitants told him. "Now, to buy grain, we have had to borrow money and mortgage our fields and our homes. And many of us have lost everything because we could not repay the loans."

"We had to borrow to pay the king's taxes," others added. "Our moneylenders force us to sell our sons and daughters into slavery if we don't repay the money. What can we do? Our fields and vineyards already are mortgaged!"

Nehemiah became violently angry when he heard this. He called a great assembly and publicly rebuked the moneylenders. "The thing you are doing is not good. You must stop charging this excessive interest. Return to the people their fields, vineyards, olive orchards, and houses. Do it this very day, for the land cannot prosper while you put such burdens on the people.

The moneylenders agreed, taking a solemn oath in the presence of the priests.

Nehemiah felt that now he had relieved the worst ills of Jerusalem. He kept his promise to return to the Persian capital. There he continued to serve his kingly master.

Nehemiah 1 through 13
Ezra 4:7-23
Joel 1 through 3
Malachi 1 through 4
Psalm 48

* The book of Joel describes with poetic imagery not only the drought where "the seed shrivels under the clods," but also the devastation brought by a plague of locusts.

The actual order of events in this period of return and attempted restoration in Jerusalem can best be followed by reading the scriptural passages in the following order: Malachi 1:1–4:5; Joel 1:1–3:21; Nehemiah 1:1–2:20; 4:1-23; 6:1-16; Ezra 4:7-23; Nehemiah 7:1-4; 11:1-2; 12:27-43; Psalms 48:1-2, 12-14; Nehemiah 13:4-29; 5:1-19; 13:29-30.

PART IX
A TIME FOR WRITING AND EDITING

The period from Nehemiah to Jesus was about four centuries. The mighty Persian Empire flourished for a time, then was challenged by the growing power of the Greeks. In 331 B.C., Alexander the Great conquered the Persians. From that time on most of the Old Testament world was ruled by the Greeks—until the century before Jesus, when the Romans took over.

For the Jews this was a time for writing and for editing. The Pentateuch (the first five books of our Bible) was accepted by the Jews as their Torah. It became their authoritative religious guidebook.

Next, the historical books of Joshua, Judges, Samuel, and Kings were edited into a final form and accepted as holy scripture, along with the revered works of the great writing prophets.

When a book was accepted by the official body of religious leaders and scholars as a holy book, it became canonized. Other books, which were being composed at this time, were examined by this body. If they were accepted they also became part of the Canon.

After the Exile the priests occupied the central place in the life of the Jews. The high priest was not only the head of religious affairs but also of civic life. The men of the priesthood emphasized the ritual of the temple and ceremonial obedience to the Law. Their insistence on the outward form of worship may have led to the popular belief that this was the way to prosperity for the individual and for the nation.

During this period Chronicles I and II were written as a single book. It was a retelling of the history of the Hebrew people from priests' point of view. The writer is called the Chronicler, though the book may have been written by a group of priests. The memoirs of Nehemiah were taken over and adapted as part of this same work. Later the books of Chronicles I and II, Ezra, and Nehemiah were canonized.

From the days of Nehemiah, the Jews were sharply divided over how to treat non-Jewish people, the Gentiles of the world. Some Jews, like Nehemiah and the authors of Malachi and Joel, urged a rigid aloofness. This attitude, combined with an ardent patriotism, is expressed in the book of Esther.

But there were other Jews of opposite leanings and heart. Two unknown writers of this Greek-influenced period wrote two of the greatest short stories ever written, Ruth and Jonah. Each of these short books is an eloquent plea for tolerance and goodwill.

The book of Ecclesiastes, whose title means the "Preacher," was written during this Greek period. It forms part of the wisdom literature of the Jews, and is perhaps remembered by the opening and despairing thought, "Vanity of vanities, all is vanity." Also much quoted is the poem in the third chapter that begins, "For everything there is a season, and a time for every matter under heaven: a time to be born, and a time to die."

Other wisdom writings include the psalms and proverbs, which were brought into their final form at this time. Throughout the world of Old Testament times wisdom was respected and sought after. It was very often linked with magic; in Israel, however, there was a different emphasis. The wise men of the Hebrews were not magicians, for the Hebrews recognized God as the source of perfect wisdom. It was the duty of every man to acquire as much understanding as possible. The priests and those who had proved themselves to be possessors of true wisdom were the instructors of the people.

The psalms and proverbs had long circulated orally, and there were differing versions and various collections. The final books, as we have them now, seem to be collections of collections. The Song of Songs also may have been gathered together by editors, instead of being the work of a single man.

Perhaps the single greatest work by a wisdom writer is the main portion of the book of Job. The unknown author of Job may have borrowed a well-known folktale to provide the prologue and epilogue of his fine poetic drama. Within this setting the philosopher-poet has composed a series of poetic dialogues. First there are increasingly angry exchanges between Job and his three well-meaning friends. These fellows chide with platitudes and popular reasoning; they repeat the priestly idea that the good prosper while the wicked are overthrown by misfortune. Job poses the question of why the righteous man suffers. The answer is not easy, and it comes from God himself.

The Triumph of Job

There was a man in the land of Uz, whose name was Job. He was blameless and upright, a man who was devoted to God and who turned away from evil.

Job was also very rich.

Now there came a day when Satan met with God, in heaven.

"Where have you come from?" God asked.

"I've been going to and fro on the earth," Satan answered, "walking up and down upon it."

"Have you seen my servant Job?" God asked. "There is none like him on earth! He is blameless and upright, he fears and respects me, and he turns away from all evil."

"No wonder," Satan sneered. "Have you not blessed him with great possessions? Just take away all that he has. Then he will begin to curse you."

"He is in your power," God said. "Test him to see whether he worships me for my sake alone, or if he just does it for the rewards." Then he cautioned Satan, "Only do not lay hands on Job himself."

Soon after that, as Job sat at ease outside his home, a messenger arrived. Your oxen and asses have been stolen by desert raiders!" the man gasped out to Job. "And your herdsmen have been slain by the sword."

Just then another messenger arrived, his clothes in shreds and his face burnt by wind and hot sand. "Oh, master," he whispered hoarsely, "God sent the hot desert winds against your flocks, and your sheep and your shepherds have perished. I alone have escaped to tell you."

While he was still speaking, another ran up and said, "Some raiders came against us and have taken away your camels, and your camel drivers are slain! I alone have escaped to tell you—"

Still another messenger arrived breathless and frightened. "A great tornado swept across the wilderness where your sons and daughters were gathered together in one house. The house collapsed on them, and they are all dead!"

Job arose, tears streaming down his face. He tore his robe and shaved the hair from his head. Then he fell down on the ground, murmuring to God: "Naked I came from my mother's womb, and naked shall I return; the Lord gave, and the Lord has taken away; blessed be the name of the Lord!"

Satan and God heard this. Satan had to agree Job had not sinned and let his sorrow and misfortune separate him from God.

"But he still has his life and his health," Satan snapped. "Just touch his bone and his flesh, and you will hear him curse you!"

So it was agreed to further test Job's trust in God.

"He is in your power," God said. "Only spare his life."

Suddenly Job found himself afflicted with a loathsome skin disease. Sores and boils broke out all over his body, from the crown of his head to the soles of his feet.

Job was no longer a respected, wealthy citizen. He was an outcast, tormented by pain, apparently soon to die. He hobbled outside the town to the ash dump. There he sat, with only a broken piece of pottery to scrape the crusts from his sores.

His grieving wife followed him out to the ash heap. "How can you believe that God is righteous?" she asked. "If he is, why doesn't he recognize that you are not a sinner?" She shook her fists at the heavens, then looked, weeping, at her miserable husband. "Why not curse God, and die? At least that would end your suffering!"

But Job reprimanded her. "You speak foolish words. Shall we forget the good we

184 have received from God just because we also receive evil?"

Now Job had three friends named Eliphaz, Bildad, and Zophar. When they heard about Job's misfortunes they came to comfort him. When they saw him they wept and sat for a long time without speaking, for they saw his suffering was very great. Then at last they talked with Job.

Job: Let the day perish wherein I
     was born;
     Let that day be darkness!
     Why did I not die at birth?

Eliphaz: If one ventures a word with
     you, will you be offended?
     Behold, you have instructed
     many,
     Your words have upheld him
     who was stumbling.
     But now it has come to you,
     and you are impatient.
     Think now, who that was innocent ever perished?
     Behold, happy is the man
     whom God reproves;
     Therefore despise not the punishment of the Almighty.

Job: And what is my end, that I
     should be patient?
     Is my strength the strength of
     stones?
     Make me understand how I
     have erred.
     How forceful are honest words!
     But what does reproof from
     you reprove?
     Is there any wrong on my
     tongue?

Bildad: How long will you speak so,
     And the words of your mouth
     be a great wind?
     Does God pervert justice?

Job: God will not turn back his anger;
     Though I am innocent, I can
     not answer him.

There is no umpire between us
Who might lay his hand upon
us both.
I loathe my life;
I will speak in the bitterness of
my soul.

Zophar: Should a multitude of words
     go unanswered?
     Should your babble silence
     men?
     And when you mock, shall no
     one shame you?
     But the eyes of the wicked will
     fail;
     And their last hope is to
     breathe their last.

Job: What you know, I also know;
     I am not inferior to you.
     But I would speak to the Almighty;
     I desire to argue my case with
     God.
     As for you, you whitewash with
     lies!
     Will you speak falsely for God?
     Your maxims are proverbs of
     ashes.

Man that is born of a woman
Is of few days, and full of
trouble.
He comes forth like a flower,
and withers.
Man breathes his last, and
where is he?
He feels only the pain of his
own body,
And he mourns only for himself.

Eliphaz: Your own mouth condemns
     you, and not I.
     Why does your heart carry you
     away?
     And why do your eyes flash,
     That you turn your spirit
     against God,
     And let such words go out of
     your mouth?

Job: Miserable comforters are you all!
I also could speak as you do,
If you were in my place;
I could join words together against you.
Surely now God has worn me out;
He has torn me in his wrath, and hated me.
Men have gaped at me with their mouth,
They have struck me insolently on the cheek,
They mass themselves together against me.
My face is red with weeping,
And my eye pours out tears.
My spirit is broken, my days are extinct,
The grave is ready for me.

Bildad: Why are we stupid in your sight?
You who tear yourself in your anger,
Shall the earth be forsaken for you?
Yea, the light of the wicked is put out,
His strong steps are shortened
And his own schemes throw him down.

Job: How long will you torment me,
And break me in pieces with words?
Are you not ashamed to wrong me?
I am repulsive to my wife.
All my intimate friends abhor me.
My bones cleave to my skin and my flesh,
And I have escaped death by the skin of my teeth.
Yet I know that my Redeemer lives!

Zophar: Did you not know this from of old,

That the exulting of the wicked is short?
Though wickedness is sweet in his mouth,
It is the gall of asps within him.
He will get no enjoyment;
Terrors come upon him!

Job: Bear with me, and I will speak,
And after I have spoken, mock on.
Why do wicked men live,
Reach old age, and grow mighty in power?
They say to God, "Depart from us!
What is the Almighty, that we should serve him?
And what profit do we get if we pray to him?"

How often is it that the lamp of the wicked is put out?
That their calamity comes upon them?
How then will you comfort me with empty nothings?
There is nothing left of your answers but falsehood.

Eliphaz: Agree with God, and be at peace;
Thereby good will come to you.
If you return to the Almighty and humble yourself,
You will make your prayer to him,
And he will hear you.
For God abases the proud,
But he saves the lowly.

Job: Today my complaint is bitter.
Oh, that I knew where I might find him!
I would lay my case before him
And fill my mouth with arguments.
I would learn what he would answer me,
And understand what he would say to me.

Behold, I go forward, but he is not there;
And backward, but I cannot perceive him.
God has made my heart faint;
The Almighty has terrified me;
For I am hemmed in by darkness,
And thick darkness covers my face.

Voices of the Lord
out of the Whirlwind: Who is this that darkens counsel
By words without knowledge?
Gird up your loins like a man,
I will question you, and you shall answer me.

Where were you when I laid the foundation of the earth?
Who determined its measurements—surely you know!
Who laid its cornerstone,
When the morning stars sang together?
Have you entered into the springs of the sea,
Or walked in the recesses of the deep?
Have the gates of death been revealed to you?
Do you know when the mountain goats bring forth young?
Who has let the wild ass go free?
Is the wild ox willing to serve you?
Is it by your wisdom that the hawk soars?
Is it at your command that the eagle mounts up
And makes his nest on high?
Shall a faultfinder contend with the Almighty?
He who argues with God, let him answer it.

Job: Behold, I am of small account;
What shall I answer thee?
I lay my hand on my mouth.

The Lord: Will you even put me in the wrong?
Will you condemn me that you may be justified?
Have you an arm like God,
And can you thunder with a voice like his?

Job: I know that thou canst do all things,
And that no purpose of thine can be thwarted.
I had heard of thee by the hearing of the ear,
But now my eye sees thee;
Therefore I despise myself,
And repent in dust and ashes.

(based on Job 3:1—42:6)

Job 1 through 42

Epilogue

And so Job found himself at peace, though he was still a man who had lost everything, and was still afflicted with a loathsome disease.

After that God healed Job. The Almighty gave to Job twice as much as he ever had possessed before. God blessed these last years of Job more than the earlier years. Job had many sheep, camels, oxen, and asses. Seven sons and three daughters were born to him to bring him joy. And Job lived a long life, surrounded by his children and his many grandchildren.

Queen Esther

King Ahasuerus* sat on his royal throne in Susa and ruled his great empire. In those days Persia stretched from India to Ethiopia and on to the edges of Greece.

In the third year of his reign he gave a banquet for the officials and noblemen of his vast realm. The banquet lasted seven days. The guests reclined upon couches plated with gold and silver, in a splendid room with floors and pillars of marble, set in the midst of the king's garden. Drinks were served in golden goblets, and the royal wine flowed freely. By the end of the week, the king and his guests were more than filled with rich food and drink.

Usually the Persian women ate with their men. However, since there were so many guests, Queen Vashti gave a separate banquet for the women in another great room of the palace.

On the seventh day the drunken king sent a messenger to Queen Vashti. He ordered her to come before the men in the garden banquet hall and display her beauty. Queen Vashti modestly refused to come.

The king sputtered with rage. "According to the law," he shouted almost incoherently to his legal advisers, "what's to be done with Queen Vashti, because she will not obey me?"

His officials conferred as best they could. The wine had muddled their thinking. "She has not only wronged you," the leader told his king, "but all men who are your subjects. For when the women of the realm hear of this, they will not give due honor to their husbands. So let a royal order be made that Queen Vashti will be put away and divorced. Be sure to write it into the Persian laws, for what is written in the laws can never be changed."

And so it was decreed, and so it was written in the laws.

When the king had sobered up, he was no longer angry. He remembered the beautiful Vashti. But it was too late. The divorce was written in the laws, so King Ahasuerus had to find a new queen.

The officers of the palace sent out an order: "Let beautiful young maidens be sought out in all parts of the realm. And let the maiden who pleases the king be queen instead of Vashti."

This idea pleased the king.

Now there was a Jew named Mordecai who lived in Susa. He was a descendant of the Jews who had been carried into captivity by the Babylonians. After the Persians won control of the empire, Mordecai had come to live in this magnificent capital city. He was the foster father of a young orphaned relative named Esther. Esther was very beautiful, and she was among the maidens who were escorted to the king's palace.

In the women's quarters of the palace, Esther followed Mordecai's advice and told no one that she was a Jew. She was given creams to make her skin soft, and costly robes to wear. She waited anxiously until it was her turn to visit the king.

When King Ahasuerus saw Esther, he was charmed and delighted. He set a crown upon her head and proclaimed her his new queen. Esther kept her secret; she did not reveal to the king, or to anyone, that she was Jewish.

One day Mordecai was sitting at the gate by the king's palace. He overheard two men plotting to take the life of King Ahasuerus.

Mordecai sent word to Queen Esther. Esther quickly reported to the king, "A loyal servant of yours, named Mordecai, has overheard plans made to kill you!"

The assassination plot was foiled, and the two conspirators were hanged on the gallows.

About this same time King Ahasuerus had appointed an ambitious man named Haman as prime minister. Everyone except the king had to bow down to the powerful Haman.

Now there was enmity between Mordecai and Haman. Haman knew that Mordecai

187

was a Jew. And Mordecai knew Haman was a descendant of Agag the Amalekite king, a bitter enemy of the Hebrews during the time of King Saul. There had been anger between the Israelites and the Amalekites long ago, and there was the same anger between Mordecai and Haman now. So when Haman approached the king's gate and everyone bowed down, Mordecai did not bow.

The king's officers at the gate asked, "Why do you go against the king's command to bow down?"

Mordecai did not bother to answer. He continued to stand and sneer whenever the prime minister passed by, though everyone else bowed and showed great respect.

Haman was filled with fury. He resolved to destroy not only Mordecai but all the Jews in the Persian Empire.

On a spring day during the first month of the new year, Haman came before King Ahasuerus. "There are people in your kingdom with different laws of their own," he said slyly. "They do not keep the king's laws. It is not for the good of the king to put up with them!"

"So?" King Ahasuerus frowned.

"If it please the king," Haman smirked, rubbing his hands together, "I will pay ten thousand pieces of silver from my own estate into the treasury of the king—if the king will decree that these people be destroyed!"

The king said absently, "Do with these unruly people as it seems good to you." And he gave Haman his signet ring with his official seal on it.

Now Haman knew of the Persian custom of throwing dice to decide a matter; it was called casting lots or "purim." By this method he chose the time when he would kill the Jews. The lot fell on the twelfth month, the last month of the winter season.

On the thirteenth day of the first month Haman summoned the king's secretaries and dictated an edict. In the name of King Ahasuerus the edict ordered that all Jews were to be slain at the end of the year, on the thirteenth day of the twelfth month, the month of Adar. As a reward the slayers could take possession of all Jewish property. The document was stamped with the king's seal, and copies were sent to all parts of the Persian Empire.

When the people of Susa heard this news, the whole city was thrown into an uproar. Many goodhearted and peace-loving Gentiles agreed that the edict was cruel, outrageous, and arbitrary.

When Mordecai learned of the fate of his people, he tore his clothes, dressed himself in rough sackcloth, and put ashes on his head. He walked through the city, wailing with a loud and bitter cry in the traditional Jewish way of mourning.

Then Mordecai sent word to Queen Esther, telling of the amount of money Haman had promised to pay into the king's treasury for the destruction of the Jews. He also enclosed a copy of the decree issued in Susa. "I charge you to go to the king and beg mercy for our people," Mordecai added.

Esther felt her heart thumping with fright and horror at the message of her foster father. It was the first she had heard of the evil edict of Haman.

But how would she approach the capricious and undependable king? He had been moody lately, and she felt out of favor with him.

Esther sent this message back to Mordecai: "If anyone goes to the king's inner court without being called, it is the law that he be put to death—unless the king holds out the golden scepter to signify that he may live. And I have not been called before the king these thirty days!"

Mordecai returned his answer. "Think not that in the king's palace you will escape any more than the other Jews." Then he appealed to her sense of destiny: "And who knows whether you have not come to the kingdom for such a time as this?"

So Esther sent her decision back to Mordecai. "Gather the Jews in Susa for a three-day fast on my behalf. I and my maids will also go without eating. Then I will venture in to see the king. And if I perish, I perish."

On the third day Esther put on her royal robes and stood in the court outside the king's audience hall. Her mouth was dry with fear, and she was trembling inside. But she held herself steady.

The king, sitting in splendor on his throne, looked up and saw Queen Esther standing in the entrance. She looked so slender and lovely that the king's heart melted and he held out the golden scepter that was in his hand.

Esther approached timidly, and her fingertips touched the top of the scepter.

"What is it, Queen Esther?" King Ahasuerus said gently. "What is your request? It shall be given you, even to half of my kingdom."

Esther took a deep breath and tried to find her voice. "If it please the king," she murmured, "let the king and Haman come this day to a dinner I have prepared for the king."

The king beamed at her. "Bring Haman quickly," he roared to an officer, "that we may do as Queen Esther desires." So the king and Haman came to the apartment of Esther and dined with her. When they had finished eating, the king said to Esther, "Now what is your petition?"

But Esther stalled. "If it please the king," she pleaded, "let the king and Haman come once again to dinner here tomorrow, and I will tell him then of my petition." For she wanted yet more time in which to please the king, so that he would indeed grant her request.

Haman's pride was inflated because he was invited to share two successive banquets with the king and queen. He burst out of the palace like a conquering lion. There, sitting at the gate, was Mordecai. Everyone else who was there, or passing by, bowed to the ground and showed respect to the prime minister. Mordecai did not. He continued to sit. He completely ignored Haman.

The pompous prime minister was filled with rage against this obstinate Jew. He wanted to murder him on the spot. Nevertheless he restrained himself and hurried home. He called for his wife and sons and all his friends to gather around him. Haman recounted in great detail the splendor of his wealth and land, the promotions with which the king had honored him, and his advancement to the highest civil post in the land.

Then Haman boasted, "Even the queen let no one come to a banquet she had prepared for the king, none but myself!" He threw out his chest and looked around to take in the admiring glances. "And tomorrow also, I am invited to dine with her and the king again!"

Haman thought of the insolent Mordecai at the gate, and his wrath returned. The blood surged into his face and neck, and his eyes bulged with the force of his anger. "Yet all this does me no good, so long as I see Mordecai the Jew sitting at the king's gate!"

So his wife and friends said, "Why don't you erect a large gallows? You can ask the king to have Mordecai hanged on it. Then you can fully enjoy your dinner tomorrow."

This counsel pleased Haman. He ordered his servants to build a huge gallows there in his own courtyard.

On that very night, it happened the king tossed and turned, unable to sleep. When nothing else worked, he called his servants

to bring the book of memorable deeds, the chronicles of his realm. "Read them to me," he ordered. "Perhaps the reading will lull me to sleep."

Through the early hours of morning, the king's servants took turns reading while the king drowsed fitfully. The pale light of dawn was creeping through the window when an item in the book caught the monarch's attention. It was reported how Mordecai had foiled a plot on the king's life.

The king yawned and held up his hand. "What honor has been bestowed upon this man Mordecai for what he has done for me?"

The servants checked the records. "Nothing," they said.

"Well, now, this should not be—" the king began. He sat up and leaned against his pillows.

Just then footsteps were heard in the court outside. The king cocked his head, scratching the black ringlets of his long hair. "Who is there in my courtyard?" he called.

A servant poked his head in the door. "Haman is here, my lord, standing in the court."

The king combed his long, black curly beard with his fingers. "Let him come in."

Haman entered, bowing and smiling. "I'm glad you're here," the king said. "I have a question to ask you. Tell me, what shall be done for the man whom the king delights to honor?"

The conceited Haman thought to himself, "Whom would the king delight to honor more than me?" So he said out loud, "Let this man be dressed in royal robes and set him on your own horse. Have him conducted through the city. And appoint a servant to go before him to proclaim that you are honoring the man!"

The king nodded. "Then it is my wish that you arrange this for Mordecai. He is the Jew who sits at my gate." The king snapped his fingers at the stupefied prime minister. "Now make haste! And leave nothing out that you have mentioned."

So Haman, burning inside with fury, followed the king's orders. He took some of the king's robes and arrayed his enemy. Then he led Mordecai on one of the king's horses through the city square, proclaiming sullenly, "This is a man whom the king delights to honor!"

Now Haman hated Mordecai more than ever. But how could he arrange to have such an honored man swing from the gallows which stood waiting?

Haman was still brooding on this when he went to dinner at the queen's apartment. As they ate and drank together, the king said to Esther, "Will you tell me your petition today, Queen Esther? As I have said, it shall be granted to you, even to the half of my kingdom."

Then Queen Esther answered, "If I have found favor in your sight, O king, let my life be given to me—and the life of my people!" Her face was pale, and her large dark eyes beseeched her royal husband. "For we are sold, I and my people, to be destroyed! We are all to be slain and annihilated!"

"Who is he?" roared King Ahasuerus, "and where is he, that would dare to do this?"

"A foe and enemy," gasped Esther, trembling. She whirled and pointed a slender jeweled finger at her other dinner guest. "This wicked Haman!"

Haman's face was ashen. He began to shake.

The king rose from the feast, steaming with anger. He strode out into the garden, slapping one fist into the great palm of his other hand. He was muttering and cursing to himself, trying to decide on a fit punishment for his prime minister.

As soon as the king had left the room, Haman flung himself to the ground by Queen Esther. He clasped her sandaled feet, begging her to save his life.

Just then the king returned from the garden. "Will he even assault the queen in my presence," Ahasuerus shouted, "and in my own house? Seize him!"

One of the servants said to the king, "Did you know that Haman has prepared a gallows for Mordecai, this same Jew who once saved your life?"

The king said tersely, "Haman shall hang on that."

So they hanged Haman on the gallows in his own courtyard which he had prepared for Mordecai. On that day King Ahasuerus gave to Esther the house and possessions of the dead prime minister.

Then Mordecai came before the king, for Esther had revealed to King Ahasuerus that this Jew was really her foster father. The king took his signet ring, which he had retrieved from Haman, and gave it now to Mordecai. And Esther put Mordecai in charge of the house and possessions of Haman.

Because it was already written into the laws, the king could not change the evil edict of Haman. So how could King Ahasuerus keep his promise to Esther and save all the Jews from being slaughtered on the thirteenth day of the twelfth month? A new edict was written: The Jews throughout the land could gather together to defend their lives. It was proclaimed in Susa and sent out by fast courier to every province. So, on the thirteenth day of the twelfth month, a chill day at the end of winter, the Jews gathered together in every city. The only persons slain were the ones who tried to attack the Jews in hopes of acquiring the money and possessions of these exiles.

And now Mordecai was the new prime minister, elegant in his royal robes of blue and white with a great golden crown and a purple mantle thrown over his shoulders.

The city of Susa shouted and rejoiced. There was gladness and feasting among the Jews for their deliverance.

Mordecai recorded all these things and declared this to be an annual holiday for the Jews. Every year at this time there would be days of feasting and drinking and gladness, a time to send gifts to one another and to the poor. Since this time had been fixed by casting dice, the purim, it would be called the annual feast of Purim. These days of deliverance and joy would now be remembered and kept throughout every generation, in every family, province, and city.**

As for Queen Esther, she would be remembered for her courage and her beauty.

Esther 1 through 10

* In the King James and RSV translations of the Bible, the king of Persia is called Ahasuerus, the Hebrew version of the Greek name Xerxes. The historical Xerxes had a different queen and did not possess the weak character given to this fictional king.

** The book of Esther is part of Jewish literature, not Jewish history. It is religious and patriotic propaganda in the form of a short story, probably written just before or soon after the Persians were conquered by the Greeks. It came to be added to the Jewish holy books because it promoted a new yearly festival, the Feast of Purim, to celebrate the deliverance of the Jews from a plot against them in Persian times. Though the book of Esther is fictionalized the event may have some historical basis.

Though God is not mentioned in this book, to the Jews the story was a symbol of divine deliverance, of hope in the midst of suffering.

Reluctant Jonah

Back in the days of Jeroboam II, the Assyrians were the scourge of the land of the two rivers. When their great army came charging westward, the Assyrians burned, looted, killed, and took captives back to their own country. Judah, Israel, and Syria hated and feared the cruel and warlike Assyrians. They never knew when the Assyrians would return to conquer the small helpless nations near the Great Sea.

One night a young man named Jonah heard God calling him, "Arise, go to Nineveh, that great capital city of the Assyrians. Cry against it and warn the people, for their wickedness is known to me."

But Jonah tried to flee from God and from what he thought was God's unreasonable demand.

Nineveh was far to the east. "Then I will go west," Jonah decided, "as far as I can go."

Jonah hiked from his village home in Israel to the seacoast town of Joppa. He paid his fare on a ship sailing for Tarshish, a distant port somewhere on the western end of the Great Sea. Surely God could not find him there!

But God sent a great wind on the sea. A fierce rain lashed the deck of the ship. Rocked by overpowering waves, the small vessel seemed ready to break up.

The sailors were a sturdy lot and used to storms at sea. But this tempest was the worst they had ever seen. They began to cry out for help, each to his own god.

"Throw the cargo overboard," the captain shouted. "We must lighten our load."

In the hold of the ship, Jonah lay fast asleep. The captain found him there and shook him awake. "What do you mean, you sleeper? Arise, call on your god! Do you want us to perish in this storm?"

Jonah made his way to the deck of the heaving ship. He found the sailors casting lots. "Everyone must draw a stone from this bag," one burly sailor ordered. "A white stone will show innocence. The one who draws the black stone is to blame for this evil which the gods have brought upon us."

The lot fell on Jonah. The sailors hanging to the rail for their lives looked at Jonah holding the black stone. "Tell us, who are you?" they demanded. "What have you done that your god punishes you so?"

"I am a Hebrew," Jonah gasped out, "and I fear the Lord God of heaven, who made the sea and the dry land. I was commanded by the Lord to go to Nineveh. Instead, I took this ship to run away to Tarshish."

They asked, "What shall we do with you so this storm will end?"

"Throw me into the sea," Jonah said despondently. "I know God has caused this tempest because of me."

The frightened men began to call upon the God of Jonah. They were reluctant to do as Jonah had said. They even manned the oars and rowed with all their might in an effort to bring the ship safely into land. Yet it was no use. The huge waves tossed the creaking ship and threatened to sink it. The sailors had to give up. They begged the Lord not to hold them responsible for what they had to do to Jonah. Then they took the prophet and tossed him into the raging sea.

The sea calmed almost immediately. But Jonah did not drown. God had appointed a great fish to swallow him.

Jonah found himself deep inside the great fish; there he stayed for three days and three nights. Jonah cried out in his distress, and made up his mind to obey the command of God. Then God spoke to the fish, and it spat Jonah out on the dry land.

194 Again Jonah heard God's voice. "Arise, go to Nineveh, and give them my message."

Jonah girded up his loins and traveled eastward across the vast, sandy desert. At last he reached the green and pleasant land of the Assyrians, the fertile country between the two rivers.

Nineveh, the Assyrian capital, was a large city, much larger than any Jonah had ever seen. Within its towering walls were huge palace grounds, temples, and thousands of mud-brick houses. Nineveh was said to be so big that it would take three days to walk through it.

Jonah crossed the broad, floating bridge to the western gate and entered the bustling city. He began at once to preach to the people. "Nineveh is going to be destroyed! In a very short time this great city will be no more."

Much to Jonah's surprise, the people listened to him. They believed his message. A great fast was proclaimed, and the people dressed themselves in sackcloth and covered their heads with ashes.

The news reached the king in his magnificent palace. He rose from his throne, threw off his royal robes, and dressed in rough sackcloth. He went out and sat among the ashes to grieve for his sins and those of his people.

"I make a solemn decree," said the king. "Let every man and beast fast and mourn, and we will pray to God for forgiveness of our sins. Let every man avoid violence and evil ways. Then God may turn his anger away from us and not let us perish." This decree was published throughout Nineveh.

When God saw how the Assyrians had repented of their evil ways, he did not destroy Nineveh. Jonah was furious and disappointed that the doom he had predicted for Nineveh would not come about. He cried out to the Lord, "I tried to flee to

Tarshish, because I knew that you were a gracious God, merciful, slow to anger, loving and forgiving. I didn't want to warn these Assyrians because I was afraid you might forgive them. And now you have." He wept in his frustration. "Take my life away from me. I'd rather die than to see my enemies live."

God replied, "Do you do well to be so angry?"

But Jonah paid no attention. He hiked outside the city and climbed to a small hill overlooking Nineveh. Here he sat down to wait. An unreasonable hope grew in him that God might destroy the city after all.

It was a hot day. The sun beat down on his head so Jonah built a frame shelter from a few thin saplings. The latticed saplings and twigs did not provide enough shade to help much. But God appointed a fast-growing plant to shoot up, leaf out, grow, and cover the booth with a cooling shade. And Jonah was pleased.

But when dawn lightened the skies the next morning, God appointed a cutworm which attacked the stem of the plant. The plant withered and died. There was no shade left, only the few leafless branches.

A sultry east wind blew in from the desert, and the sun beat mercilessly down on the head of Jonah. Perspiration flowed down his reddened face. His clothes were sweat-soaked. He became dizzy and faint. His mouth was parched, his lips cracked and dry.

Then Jonah wept for the plant which had died and no longer gave him the shade he needed. "Let me die," Jonah moaned. "It is better for me to die than to live."

But God spoke to Jonah and said, "Is it right that you should be so angry because a plant dies?"

"It is right that I should weep for that plant," Jonah cried.

Then God replied, "You pity a mere plant for which you never labored. You did not make it grow. It was a passing thing that came into being in one night and perished in another. How much more should I pity Nineveh! In this great city are also my people, and there are more than a hundred twenty thousand children who are not old enough to know their right hand from their left."

And so it was that God taught Jonah a lesson. God did care for all the people in the world, no matter where they lived.

Jonah 1 through 4

The book of Jonah is a short story written from a point of view that shows concern for the welfare of those who were not Jews. It was probably composed between the fourth and the second centuries B.C. The background of Jonah is Israel of the eighth century B.C., during the time of Jeroboam II. The hero is a real prophet of that name mentioned in 2 Kings 14:23-25.

This allegorical story is not meant to be literal fact. It is more of a parable told to illustrate a spiritual truth. In the story Jonah is pictured as a narrow-minded, intolerant man who did not believe that God was concerned for people outside of Israel. His experiences reveal to him that God cares for all people, no matter where they live.

Psalms for Singing

The everyday life of a faithful Hebrew family was studded with small acts of prayer and praise. The Jewish confession of faith, called the Shema (Deuteronomy 6:4), was recited in the morning and evening. Each activity begun during the day was commended to God with a small ritual prayer, often using phrases from the scriptures.

There was also public worship which centered around the temple, especially at the great feasts. The book of Psalms contains some of the songs and prayers for these occasions. The psalms also played an important part in the devotional life of the people. A closer look at the activities of an imaginary Hebrew family would make this clear.

In the days after the return from Babylon, there lived in Jerusalem a man named Eliam. He had provided his family with a small, flat-roofed house set close to the city wall. His wife Dinah had borne him twins: Jeshua was now a young man, almost as tall as his father, and Abi, his sister, was already of marriageable age. There was also a younger son, Benjamin.

The summer had been long and hot. In the orchard surrounding the city, the olives had ripened from green to a blue black on the gnarled old trees. They had been beaten to the ground, gathered up, and pressed into golden oil. Some of the olives had been prepared for eating by a long soaking in dissolved wood ashes. When they had been rinsed over and over, they were preserved in brine.

When the grapes hung ripe in heavy bunches among their green leaves, many people went to the hillside vineyards to help with the harvesting. This was always a joyous occasion. Part of the fruit was laid in the sun to dry into clusters of sweet raisins. The rest was spread in vats hewn from the rocky hillside. Then the harvesters pressed the juice from the grapes by treading them with their feet. Much dancing and merriment went with this. The juice was stored for winemaking in great earthenware jars.

The ripe apricots and almonds had been picked from the orchards, and the pomegranates with their ruby-red, fleshy seeds. The summer fruits had been gathered in, and the year was ending.

The Feast of Booths at the gathering in

of the harvest was a most joyous festival. It was the only topic of conversation early one evening as the members of Jeshua's family gathered for their simple meal of bread with a stew of vegetables. The feast was only a day away, and every member of the family was full of holiday excitement.

Jeshua, his father, and his brother had spent three days cutting and gathering suitable branches from olive trees. They had also cut a number of fragrant myrtle boughs and limber willow. The stack of branches outside the door of their house had grown steadily. There were plenty now to make a booth large enough for all the family. They would all make their home in this leafy shelter for the seven days of the festival.

The following day, as soon as the morning psalm had been sung, the family set to work building the festival booth, piling the branches and twining them together with the willow switches. As they worked, the father told once more how their forefathers had lived in such shelters as these in the wilderness. He told again how God had supplied all the needs of the Israelites as he led them to the Promised Land.

Since Eliam's family lived inside the city, their booth could be built upon the roof of their house. But the pilgrims who came from a distance built theirs wherever they could find space in the courts and squares throughout the city.

Abi and her mother had been at work for days preparing the food for the feast. Many loaves of unleavened bread had been made. Some of them would be used as an offering to the Lord. When the booth was finished, Dinah and Abi set to work carrying vessels and baskets of fruit and olives, cheese and wine up to it. The bread wrapped in fresh linen napkins was also taken up.

As the sun set, the family completed their work. Then the family gathered around as Eliam raised his arms to the night sky. Together they chanted a familiar psalm of praise:

O Lord, our God,
What majesty is yours through all the earth!
When I look at the heavens, the work of your
    fingers,
    the moon and the stars which you have
    made;
what is mere man that you are even aware of
    him?
    What is insignificant man that you care for
    him?

(based on Psalm 8)

At the first light of dawn, parents and children arose and dressed in their best clothing. Then they hurried to join the procession of worshipers and pilgrims. The people fell in behind a line of priests marching to the Pool of Siloam. There the priests drew a vessel of water to be used in the morning ceremonies for the rest of the week.

As the procession made its way with the water from the pool to the courts of the temple, the people followed it singing the Hallel.* In their hands they carried fronds of palm or myrtle which they waved. The priests made their solemn way around the altar with the holy water, singing:

Praise the Lord!
Praise, O servants of the Lord,
    praise the name of the Lord! . . .
When Israel went forth from Egypt,
    the house of Jacob from a people of strange
    language,
Judah became his sanctuary,
    Israel his dominion.
O give thanks to the Lord, for he is good.

To this the people answered with a glad shout:
    For his steadfast love endures for ever!

Again the priests chanted:
    Let Israel say,

    "His steadfast love endures for ever."

198 The people answered. They sang back and forth in this manner until a great shout went up from everyone:

Save us, we beseech thee, O Lord!
  O Lord, we beseech thee, give us success!

This ceremony of circling the altar, singing, and offering the precious water was repeated each morning for the seven days of the feast.

There was also a great celebration on each of the seven nights. Four huge candlesticks, or lamps, were set up. The wicks for the olive-oil lamps were made from the old robes worn by the priests during the year just ended. After the lamps were lit, the Levites chanted a special set of psalms, repeating them on each of the steps leading from the court of the Israelites down to the court of the women. This singing was done by the trained temple choir. It sounded very beautiful echoing through the stone courtyards in the clear night air. Jeshua listened to catch the phrases:

Deliver me, O Lord,
  from lying lips,
  from a deceitful tongue.

I lift up my eyes to the hills.
  From whence does my help come?
My help comes from the Lord,
  who made heaven and earth.

Pray for the peace of Jerusalem!
  "May they prosper who love you!"

When the Lord restored the fortunes of Zion,
  we were like those who dream.
Then our mouth was filled with laughter
  and our tongue with shouts of joy.

Come,
bless the Lord, all you servants of the Lord,
  who stand by night in the house of the Lord!
Lift up your hands to the holy place,
  and bless the Lord!
May the Lord bless you from Zion,
  he who made heaven and earth! **

After this final blessing the grownups joined in a merry torch dance. And Jeshua, Abi, and Benjamin went reluctantly but wearily home to their leafy shelter.

Jeshua and Abi found Jerusalem an exciting place when it was filled with pilgrims. They met new friends, and there were travelers with tales to tell. Every day there were great sacrifices and feasting, and the solemn processions with singing.

But all good things must come to an end. And so on the eighth day there was a great assembly. Once again the Hallel was sung. When it was concluded there was a reading of the Law of Moses. At the end of the reading, the people made a renewal of their covenant with God before they went back to their everyday routine.

Exodus 23:16; 34:22
Leviticus 23:39-44
Nehemiah 8:13-18
Psalms 1 through 50

* The Hallel was the great song of praise composed of Psalms 113 through 118. This was also used at Passover. It is sometimes called the Egyptian Hallel. "Hallel" means praise in Hebrew; "Hallelujah" means to praise God.
** Each of these psalms (120 through 134) is called A Song of Ascents. This may refer to the pilgrims going up, or ascending, to Jerusalem— or to the Levites ascending the steps of the temple courts. Psalms 120 through 136 are known as the Great Hallel.

Words of Wisdom

Jeshua was now of an age to be instructed with the young men of the city at the nearby synagogue. This morning found him in a hurry to go to his class.

Abi looked up proudly at her twin and smiled. "How lucky you are to be going off to school."

"Yes, I am," Jeshua agreed. "But one needn't go to school to learn. Mother teaches you what you must know. How to be a good wife and mother!" Abi only nodded. "And that is important," continued her brother. "A good wife who can find? She is far more precious than jewels. The heart of her husband trusts in her."

"A true and wise proverb," said Eliam as he and young Benjamin walked up.

"What is a proverb?" the younger brother wanted to know.

"Well," Jeshua replied, pausing a moment, "it is a short statement of something wise and true. And it is put into words so neatly that it catches your attention. It is also easy to remember."

The father smiled at Benjamin. "You probably know some proverbs without realizing what they are," he said.

" 'Train up a child in the way he should go, and when he is old he will not depart from it.' That is a proverb," said the mother as she passed by.

"Of course one does not have to go to school to learn proverbs," Abi said. "I can think of several right now: 'Like cold water to a thirsty soul, so is good news from a far country,' or 'A wise son makes a glad father, but a foolish son is a sorrow to his mother.' "

Jeshua looked at his sister with a mischievous twinkle in his eye. "We learn more than proverbs at the school," he said. "We learn the situations in which to use them. 'Like a gold ring in a pig's snout is a beautiful woman without good sense.' "

Abi turned quickly and glared at him. "One does not have to go to school to learn that either," she snapped. " 'Do not descend to the level of a fool when you answer him, lest you be like a fool yourself.' "

To this Jeshua replied with a superior calm, " 'Puffed-up pride is followed by destruction, and a haughty spirit will lead to a fall.' "

At this point the father stepped between them. " 'A soft answer turns away wrath, but a harsh word stirs up anger,' " he said. "Both of you have a lot to learn before you can approach true wisdom."

The mother had been preparing food for the later morning meal. The smell of fresh baked bread wafted across the dooryard. She heard the bickering as she approached with some bread and cheese for Jeshua to take with him to his class, and for Eliam and the younger son to carry out to the fields. The family normally ate only two meals a day, a light one in mid-morning to break the night's fast and a larger dinner all together in the evening.

"Better is a dry crust of bread, with peace and quiet," Dinah said to her family, "than a great feast where there is contention and strife."

When the father and sons had left on their separate ways, the mother and Abi went on with their household tasks. The sleeping mats must be aired, the saucers of the lamps cleaned and filled with fresh oil and wicks, the floor of the house swept. There were also garments to mend.

The father and Benjamin took their snacks of bread and cheese and set off for the fields which Eliam owned outside the city. Their job for the day was to clean the new vineyard section of stones. As they walked along together Eliam said, "For the time being, my son, your school will be the field of your father. There is much to be learned here. Jeshua also attended this school." Eliam patted his son's shoulder. "The instructors from whom your brother now learns know the value of these early lessons at home. There is a proverb that says:

"Know well the condition of your flocks,
  and give attention to your herds; . . .
the lambs will provide your clothing,
  and the goats the price of a field;
there will be enough goats' milk for your
  food."
        (based on Proverbs 27:23, 26, 27)
  Meanwhile, Jeshua hurried to the synagogue where other young men were gathering. Their teacher was a Levite who had trained hard for his work. As he entered the room, he greeted his students. Then he began:
"Hear, O sons, a father's instruction,
  and be attentive, that you may gain insight;
for I give you good precepts:
  do not forsake my teaching. . . .
Do not forget, and do not turn away from
  the words of my mouth.
Put away from you crooked speech,
  and put devious talk far from you.
Do not enter the path of the wicked,
  and do not walk in the way of evil men.
The way of the wicked is like deep darkness;
  they do not know over what they stumble.
If you receive my words
  and treasure up my commandments with
    you,
then will you understand the fear of the
  Lord and find the knowledge of God."
        (based on Proverbs 2:1, 5; 4:1-27)
  "It is by gaining wisdom, my sons," the instructor continued, "and filling your mind with her precepts, that a man learns to deal with his family, his friends, his enemies, his business associates. A good life is impossible for the man who has no wisdom.
  "Now, think," he said, "and from your store of maxims see if you can tell me how a wise man would deal with his neighbor."
  A young man rose and began:
"Do not withhold good from those to whom
  it is due,

when it is in your power to do it.
Do not say to your neighbor, 'Go, and come
  again,
tomorrow I will give it'—when you have
  it with you.
Do not plan evil against your neighbor
  who dwells trustingly beside you.
Do not make trouble for no reason,
  when he has done you no harm."
        (based on Proverbs 3:27-30)
  "Well spoken," said the teacher. "What more?"
  A younger fellow with round, earnest eyes stood up: "Let your foot not be too often in your neighbor's house, lest he become weary of you and hate you."
  "Very true," said the teacher.
  " 'Argue your case with your neighbor himself; do not gossip about his affairs with the whole neighborhood,' " said a somewhat older student.
  During the afternoon the teacher held a discourse on laziness and sloth. In his talk he pointed out the waste and ruin that come to the man who does not do his work. There were a number of proverbs to be copied on this subject. Jeshua repeated one of them to himself as he walked toward his home when school was over:

Go to the ant, O sluggard;
  consider her ways, and be wise.
Without having any chief,
  officer, or ruler,
she prepares her food in summer,
  and gathers her sustenance in harvest.
        (Proverbs 6:6-8)

Proverbs 1 through 31

The proverbs are not statements of theology but are maxims for good living. This type of practical wisdom was greatly admired throughout the Near Eastern world of OT times.

Many of the psalms were ascribed to David, because he was the singer of sweet songs, even though he did not write them; in like manner

202 the proverbs were ascribed to Solomon, who was noted for his wisdom.

Most of the proverbs were the creation of ordinary people; parents and teachers repeated them for the instruction of their children. Because several collections eventually were joined together to form our present book of Proverbs, there are duplications and slightly varying statements of the same idea.

The majority of the proverbs were probably circulating among the people long before the Exile. They were compiled and edited into something like their final form sometime after 400 B.C.

Songs for a Wedding

Abi was very excited for today was her wedding day. Her mother and several of her friends had helped her dress in her wedding finery.

Abi fastened her new leather sandals. Then she stood up and wrapped her richly embroidered girdle about her slim waist, tucking in the ends. It looked so nice that she felt glad for spending all the long hours taking those tiny stitches.

Abi's mother brought out a new ivory comb from Egypt. It had been a gift from the bridegroom. Dinah combed and brushed her daughter's long dark hair until it shone in the morning sunlight. Then she fastened the comb high on the back of Abi's head.

"You look lovely, my dear daughter," said the mother. She took a tiny bag of myrrh, a fragrant and costly substance imported from far-off Arabia. She slipped the linen cord over her daughter's head so that the bag hung like a pendant from her neck. Abi sniffed the scented bag and then tucked it inside the neck of her bridal dress. She

gave her mother's hand a squeeze to let her know how delighted she was with this special gift.

Four of Abi's girl friends were serving as her attendants. They were also dressed in their festive best. Now they sat with Abi to wait for the bridegroom to arrive. As they waited, the bridesmaids helped Abi arrange her heavy veil over her face and hair.

Dinah said softly, "Your father has chosen a fine and fair young man for you. He seems honest and upright. He will provide a good home for you." She paused a moment and then continued, "I hope, my daughter, that you will come to love your husband as I have loved the man my father chose for me. I hope, too, that you will make him happy by bearing him many sons." A gentle smile curved the mother's lips as she added, "and daughters to bring you happiness."

They sat quietly listening for the approach of the bridegroom, but they could hear only the twitter of birds in the shrubs. Overhead the clouds drifted lazily in the blueness of the sky.

Abi's father and her two brothers came to sit with the party. As Eliam seated himself he said, "I have one word of advice for my daughter. Do not weary your husband by arguing and nagging. They say that a wife's quarreling is like continual dripping of rain."

"I know another such proverb," one of the maids-in-waiting said. "My mother likes to say, 'Better a serving of vegetables, and love, than a dinner of roast lamb served with hatred.'"

"My uncle warned me that it's better to live in a corner of the housetop, than in a house shared with a nagging woman," said another.

At this moment Benjamin, who had been watching on the street, called excitedly,

"They're coming, they're coming!"

The bridegroom's party could be heard before they could be seen. There was the sound of cymbals, horns, and flutes. Then the young men came into view. Some of those in front were leaping and dancing about.

The bridegroom had brought with him a sturdy brown donkey with a white mouth and white shanks, gaily bedecked with garlands of fresh flowers. He set the veiled bride upon the patient animal, and everybody followed the procession back to the bridegroom's home.

The groom's house was actually a large square room which had been added to the side of his father's house. It was built of blocks of stone with a flat roof of clay spread

204 over sturdy beams. The wooden door was swung open on its doorpost. Abi noted the high window covered with a fine lattice. Everything was decorated with flowers and boughs.

The bride glowed underneath the covering veil. Her handsome bridegroom was all that she had hoped for. He looked so lean and strong, yet so gentle. Abi stirred with excitement to know that this would be her new home, her lifetime home. Here she would bear and raise her children, and her sons would raise their children. As she sat trembling on the donkey, her father and the bridegroom signed the formal covenant of marriage. Several elders of the community and other witnesses looked on.

As soon as this was completed, the feasting and merriment began. The groom and his family had spread a great banquet: roast lamb, fresh loaves of bread, cheese, fresh fruit, raisin cakes, and wine.

As the wedding guests were eating, two young singers began to chant the verses of a marriage song. They took the parts of a bride and bridegroom singing to each other.

Bridegroom: Behold, you are beautiful, my love;
you are very beautiful.
Your eyes are like soft doves;
you are truly lovely.
As a flower amid the thorns,
So is my beloved among maidens.
Bride: The voice of my beloved!
Look, here he comes,
leaping upon the mountains,
bounding over the hills.
Bridegroom: Arise my love, my fair one,
and come away;
for lo, the winter is past,
the rain is over and gone.
The flowers appear on the earth,

the time of singing has come,
And the voice of the turtledove is heard in our land.
The figs are ripening on the trees,
the blossoming grapevines are fragrant.
Come, my dear, my beautiful one,
come away.
Bride: I went down to the almond grove,
to look at the flowers of the valley,
to see whether the grapevines were budding,
whether the coral-red pomegranates were blooming.
Before I knew it, I found myself
in a chariot beside my prince.
Bridegroom: Set me like a seal upon your heart,
like a ring upon your hand;
for love is as strong as death,
and jealousy cruel as the grave.
Many waters cannot quench love,
neither can floods drown it.
(based on Song of Songs 1; 2; 6; 8)

Song of Songs 1 through 8

The love songs sung at this typical wedding feast are excerpts from The Song of Songs, sometimes called the Song of Solomon. Selections from this collection of folk poetry, many scholars believe, were often used at such festive occasions.

According to ancient tradition, Solomon was the author of the poems. Other traditions say that the poems were dedicated to Solomon while the writer remains unknown.

Perhaps the book came to be accepted as religious literature because of its interpretation as an allegory in which God was the bridegroom and the Jewish nation the bride.

PART X
THE CLOSE OF THE OLD TESTAMENT PERIOD

After the death of Alexander the Great his empire was divided among three of his generals: Antipater, Seleucus, and Ptolemy.

Antipater took over the home country of Greece and Macedonia. Seleucus ruled most of what had been the Persian Empire; his successors, known as the Seleucid kings, reigned from their capital of Antioch in Syria. Egypt was ruled by a succession of Ptolemies from Alexandria, the capital.

Palestine suffered now from a prolonged tug-of-war between the Ptolemies and the Seleucids. For a century and a half, armies of the opposing forces tramped back and forth across Judea.

During this time the Greek influence was strong in the land. The old style of dress gave way to the softer and more graceful Greek robes, especially among the upper classes. Streets were broadened and paved, and new large buildings had Grecian architecture. The Greek language was gradually adopted.

At last, about 200 B.C., the Alexandrian rulers gave up the unprofitable struggle for Palestine. The Jews were left to the more severe control of the Seleucids in Antioch.

When King Antiochus (IV) Epiphanes, a particularly harsh ruler, came to power he made the mistake of trying to force the Jews to abandon their religion and to accept the gods of the Greeks.

Anyone caught reading Jewish scriptures was killed, and death was ordered even for those who were found with the Scriptures in their homes. In Jerusalem, sacrifices and feasts were discontinued. Instead, a Greek altar was erected in the temple, and swine's flesh (an abomination for the Jews) was sacrificed there.

It was in these times of terrorism and persecution that the book of Daniel was written. Its purpose was to buoy up the courage of the Jewish people and give them hope.

The book is in two parts. In the first six chapters the stories depict the time of the Babylonian Exile. Since the stories were actually written much later, there are understandable errors in the historical background.

This, however, does not detract from the aim of the unknown author who wanted to show the persecuted people that they could resist the tyrant Antiochus—just as Daniel and his friends resisted their overlords in Babylon. During times of exile the Jews could do little against the might of Gentile tyrants who mocked their religion and scoffed at their God. These stories revealed that God himself might humble the tyrants and that the strongest kingdoms come to an end.

Daniel and His Friends

In the days when Nebuchadrezzar* was king at Babylon, the Jews were there as captives. It was the custom of their captors, the Chaldeans, to choose from among their captives youths from noble families, who were handsome, intelligent, and skillful, to be educated for three years by the wise men of Chaldea. They were taught the language and the wisdom of the Babylonians. At the end of this time, they were brought to the king to serve as wise men, scribes, astrologers, and soothsayers.

The students were royally treated. Every day they were served portions of the rich food from the king's table. But there was a young Jew named Daniel, who did not want to eat the king's meat or drink his wine. Some dishes on the royal menu were forbidden to Jews. So Daniel and three of his friends asked permission of the chief steward to reject this rich food and drink.

The chief steward seemed worried at this request. "I don't know," he said. "I'm afraid my lord the king might soon find you in poor condition." He shook his head. "You would endanger my head with the king."

But Daniel was very confident. "Test us

for ten days," he said. "Let us be given vegetables to eat and water to drink. See if we have a poorer appearance than those who eat the king's rich food and drink his wine."

So the steward reluctantly agreed.

At the end of the ten days Daniel and his friends, Shadrach, Meshach, and Abednego, looked stronger and healthier than the other young men who had eaten from the king's table. And so these four were allowed to remain on their special diet.

When the time came for the students to be summoned before the king, Daniel and his friends were tested with the others. They proved to be the wisest of all.

Daniel 1

* The king's name is spelled "Nebuchadnezzar" in the book of Daniel.

Nebuchadrezzar's Dream

Not long after this Nebuchadrezzar found himself troubled with dreams. So the king summoned his magicians and wise men. "I have had a dream," he told them, "and my spirit is troubled to know the meaning of it."

"O king, live forever," said the magicians to their king. "Tell us the dream, and we will then interpret it."

But the king only scowled. "You must tell me the dream, as well as its interpretation. Otherwise, you will be torn limb from limb, and your houses laid in ruins."

The Chaldean wise men were now trembling with fear. "Let the king tell us the dream," they begged. "Then we will show its interpretation."

"You are only trying to gain time," the king roared at them. "Can none of you tell me my dream?"

His advisers shook more violently. "No man on earth can meet the king's demand," they implored. "Only the gods could do what you ask."

When they said this the king grew furious. He commanded that all the wise men of Babylon be destroyed.

The decree went out. The king's soldiers sought out young Daniel and his companions to slay them with the other counselors.

"Why is the king so angry?" asked Daniel. "And why is the decree so severe?" The captain told him how the wise men had failed to interpret the king's dream and how the king's anger had gone out against all the wise men of the empire.

When he heard this, Daniel sent word to the king asking for an appointment. Then Daniel returned to his house.

After he fell asleep that night, Daniel had a vision. In the vision the mystery of the king's dream was revealed. At the first light of dawn Daniel sent to the captain of the guard.

"Do not destroy the wise men of Babylon," he instructed. "Just take me before the king, and I will tell him what he wants to know."

Daniel was taken to the king in haste. The captain addressed the king breathlessly, "I have found among the exiles from Judah a man who can tell the king his dream—and the meaning of it."

The king looked at the young man. "Is this true?" he asked.

Daniel bowed low. "Your wise men, magicians, and astrologers cannot reveal such a mystery. But there is a God in heaven who reveals these things.

208 "You saw, O king, a great image," Daniel said. "This image, of tremendous brightness, stood before you. Its head was of gold; its breast and arms were silver; its belly and thighs were bronze, and the lower legs iron. Its feet were a mixture of clay and iron, which did not hold together well."

The king looked amazed. "Yes, yes, go on."

"A great stone knocked down the frightening image, smashing it to bits. The wind blew the fragments away, leaving no trace; the great stone became a mountain filling the earth."

"That was exactly my dream," the Chaldean king cried. "Now, the interpretation?"

"You are the head of gold," Daniel replied. "After you shall arise another kingdom of silver, inferior to you. Then a third kingdom of bronze shall rule over the earth. Then a fourth kingdom, strong as iron, shall arise. But iron breaks into pieces and shatters all things. The kingdom shall be divided and, like the feet of the idol, shall have some of the firmness of iron, but also the brittleness of clay. It will not hold together."

"And what is the great stone?" asked the king.

Daniel answered, "The God of Heaven will set up a kingdom which shall never be destroyed. It shall break in pieces these kingdoms and bring them to an end."

Then the king said to Daniel, "Truly, your God is the God of gods and the Lord of kings. He is a revealer of mysteries, for you have been able to reveal this mystery."

The king rewarded Daniel with high honors and many gifts. At Daniel's request his friends—Shadrach, Meshach, and Abednego—were given high posts in the provinces. But Daniel remained with honor at the king's court.

Daniel 2

The Fiery Furnace

King Nebuchadrezzar had made a huge image of gold and had it set up on a plain outside Babylon. Then the king sent word to the officials of his kingdom, in all the provinces, ordering them to come to the dedication of the image.

Everyone came—everyone except Shadrach, Meshach, and Abednego.

A herald proclaimed to the assembly, "When you hear the blast of music, you are to fall down and worship the golden image. And whoever does not shall immediately be cast into a burning furnace."

At that the music blared, and everyone standing about the image fell to the ground and worshiped the idol.

But certain jealous Chaldeans came forward and said to King Nebuchadrezzar, "O king, live for ever! You, O king, have made a decree that whoever does not fall down and worship the golden image shall be cast into the fiery furnace."

"Yes," the king snapped impatiently. "Go on."

"You have appointed three Jews as officials in this province," the spokesman said. "These men have not come. They pay no heed to you. They do not serve your gods or worship the golden image you have set up."

Nebuchadrezzar flew into a rage. He commanded that Shadrach, Meshach, and Abednego be found and brought to him.

When the three men were brought, King Nebuchadrezzar roared at them, "Is it true that you do not serve my gods or worship the golden image that I have set up? Don't you know that if you do not you shall be cast into a furnace? And who is the god that will deliver you out of my hands?"

Shadrach, Meshach, and Abednego said, "O King, we have no need to answer you in this matter. It may be that our God will deliver us out of your hand, O King, and out of the fiery furnace. Yet if not, we shall not serve your gods or worship the golden image, because we will serve God only."

At this, Nebuchadrezzar boiled with fury. He ordered the great furnace to be heated to its highest temperature. It was a giant kiln shaped like a beehive, with an opening on the top and a small door on one side. He ordered his soldiers to bind Shadrach, Meshach, and Abednego and to hurl them down into the fire. The furnace was so hot the soldiers themselves were fatally burned as they threw the three young men into the blazing fire.

Sometime later, when the flames were lower, King Nebuchadrezzar stooped to look through the low door on the side. In astonishment he said to his counselors, "Did we not cast three men, bound, into the fire?"

They answered, "True, O King."

Nebuchadrezzar began to stammer, "But —but—I see four men walking free and unhurt in the midst of the fire." The king's voice dropped to an awed whisper. "The appearance of the fourth is like a son of the gods!"

Then Nebuchadrezzar called out, "Shadrach, Meshach, and Abednego, servants of the most high God, come forth."

The young men walked out of the fire, unharmed. Not a hair on their heads was singed, nor a thread of their clothing scorched.

Nebuchadrezzar said, "Blessed be your God, who has sent his angel and delivered you. You trusted him at the risk of your lives. You yielded up your bodies rather than worship any god except your God."

Daniel 3

The Handwriting on the Wall

King Nebuchadrezzar and several other Chaldean kings were now dead. Young Belshazzar* ruled in Babylon.

When the Jews were taken into captivity by the Chaldeans, their temple was plundered of its treasures. Their holy vessels of gold and silver were put into the king's treasury in Babylon.

Young Belshazzar had no reverence for such things. He gave a great state banquet for the lords of his land and ordered that the wine be served in the vessels from the temple of Jerusalem. The king and his lords and ladies lifted the stolen goblets and drank to the idols made of wood and stone, bronze, silver, and gold that stood around the banquet hall.

Immediately the fingers of a mysterious hand appeared and wrote strange words on the white plaster wall opposite the king. The king's eyes bulged as he watched the hand. The flush from too much food and wine quickly drained away from his face, and he was as gray as cold ashes. He tried to stand up, but his legs gave way and his knees knocked together. His mind was in confusion.

Belshazzar cried aloud, "What thing is this?" When no one answered him, he shouted again, "Bring in the soothsayers and astrologers."

When the wise men were brought, Belshazzar pointed with a shaking hand to the words on the wall. "Who can tell me what this writing means?" the king quavered. "Speak out, and you shall be clothed with purple! I shall place a chain of gold around your neck, and you shall be given one of the highest offices in the kingdom."

The wise men studied the writing on the

wall. They talked together and shook their heads. They could make nothing of it.

Belshazzar became so impatient with the mumbling, stammering counselors that his face grew red with anger.

The aged queen mother heard the confusion in the banquet hall and came in to see what was the matter. Now she said to Belshazzar, "Let not your thoughts be unsettled. Do not be alarmed. There is in the kingdom a man named Daniel who has the knowledge and understanding to interpret dreams, explain riddles, and solve problems. Long before your time, when King Nebuchadrezzar was on the throne, Daniel did such things."

So Belshazzar sent for Daniel. Daniel was an old man now, but he hurried to see the king.

"If you can tell me the meaning of that writing," Belshazzar said to Daniel, "you shall be clothed in purple and have a chain of gold about your neck. I will give you a high office in the land."

Daniel answered calmly, "You may keep your gifts and rewards. Do you not recall that one of the kings of Babylon suffered madness? He roamed the fields like an animal eating grass, his body wet with the dew of heaven. He was in this state until he realized that the Most High God rules the kingdom of men and sets over it whom he will." **

"Yes, yes, I have heard that," Belshazzar answered impatiently. He waved his hand toward the wall. "Tell me what this means."

Daniel went on, "Though you knew all this, you have dared to profane the holy vessels from the temple of our God. You and your lords and your ladies have drunk wine from these holy cups. You have used them to toast your gods of metal, stone, and wood—gods which cannot see or hear or know. You have tried to lift yourself above

the Lord of heaven! So the hand was sent from God to write—"

"Yes, yes?" cried out Belshazzar hoarsely. "What do these words mean?"

"The words are *mene, mene, tekel* and *parsin,*" Daniel said. "*Mene* means that God has numbered the days of your kingdom. It has come to an end. *Tekel* means that you have been weighed in the balances and found wanting. *Parsin* means that your kingdom will be taken from you and given to the Persians." ***

That very night Belshazzar was slain in his palace, and the Persians took over the kingdom.

Daniel 5

* In actual history Belshazzar was only a prince, though he reigned with the powers of a king of Babylon. This was during the last years of Chaldean power, when his father, King Nabonidus, left Belshazzar in charge while he was securing the western part of the empire.

** Daniel 4 gives an account of the fit of madness suffered by Nebuchadrezzar.

*** Literally, the three words on the wall are related to designations of weights: the *mina,* the *shekel,* and the *parsin* (two half-minas). Before money was coined, a shekel meant a certain weight. Later on, a shekel came to be coin made of metal weighing that amount.

Although the actual words would have been familiar to Belshazzar and his court, their symbolic meaning was not intelligible to them.

Daniel in the Lions' Den

King Darius* changed the government of the huge territory now ruled by Persia. And Daniel was given one of the highest offices in the land.

But the favor of the king toward this Jewish exile only made the other court officials jealous of Daniel.

These men conspired against him. "He is too clever, and he never makes mistakes," they muttered to each other. "We shall never find any ground for complaint against him."

"Unless—" one official suggested, a sneer of contempt on his face, "unless it is in connection with the law of his God."

"Ah ha," another official said. "I think I see how we can lay a trap for him."

So after they schemed together, they went to the Persian king. Bowing low, they said, "O King, live for ever. We are all agreed that in your glorious power you should establish an ordinance. Decree that no man may make a petition to any god or man except to you, O King, for thirty days. Let those who break your ordinance be cast into a den of lions." So the king could not change his mind they craftily added, "And sign the document so that it cannot be changed or revoked, according to the law of the Persians." **

This pleased King Darius, and he signed the document.

The aging Daniel heard the public reading of the new law, but he could not let such a decree interfere with his private prayers and devotions. Three times a day he continued to go to his house. There he opened the windows of the upper room toward Jerusalem. Three times a day he dropped to his knees and prayed, giving thanks before God, just as he had always done.

Daniel's enemies watched him closely. Soon they were hurrying to the king. "O King, did you not sign a law saying that the man who petitions anyone except yourself within these thirty days shall be thrown into a den of lions?"

"So I did," the king agreed, "and the Persian law cannot be changed."

Then the officials said solemnly, trying to hide their pleasure, "Daniel the Jew pays no heed to your law. He prays to his God three times a day."

The king was much distressed to hear this because he had great affection and respect for this Jewish official. Until sundown the king paced the floor, trying to think of some way to rescue Daniel.

But the jealous officials came back to the palace, with insistent reminders. "Remember, O King, that you cannot change a Persian law which you have signed."

So Daniel was brought before the court. He stood quite calmly and showed no fear.

Sadly the king pronounced the sentence; then he added, "May your God, whom you serve continually, deliver you, Daniel."

Then Daniel was taken to the pit that was the den of lions. He was thrown inside, where the hungry animals paced and growled. A large stone was laid over the opening, and a hot wax seal was put on the edge of the stone. The king pressed his royal signet ring on the wax so that the stone could not be removed without breaking the seal.

Then the king returned to his palace. But he found no peace. He turned away from his dinner. All night he paced in the loneliness of his room, for he could not sleep.

At the break of day the king dressed and went in haste to the den of lions. As he came near, he cried out, "O Daniel, has your God been able to deliver you from the lions?"

From inside the pit came Daniel's voice, strong and calm. "God sent his angel and shut the mouths of the lions. They have not hurt me, because I was found blameless before my God, and also before you, O King."

214

The king was overcome with relief and gladness. The royal seal was broken as the stone was taken from the opening. A rope was lowered from the top, and Daniel was lifted out. There was no scratch or hurt on him.

Then the king commanded that the ringleaders of the plot against Daniel be thrown into the lions' den. Even before they had reached the floor of the den, the hungry lions were tearing at them.

As for King Darius, he acknowledged that the God of Daniel was the true and living God whose kingdom would endure forever.

Daniel 6

* This is not the actual Darius of Persian history whose reign comes after that of Cyrus the conqueror. Folktales may have confused these two names, but such details are not important. The writer of Daniel merely uses the old folktales as a framework for his message of encouragement.

** The supposed irreversibility and changelessness of a Persian law is an exaggerated literary device. Here, and in the story of Esther, it heightens the drama, but the fact that such devices were used in popular stories gives some insight into the absolute power of the ancient Near Eastern kings, who were often treated as gods or demigods.

The remaining chapters of Daniel (7 through 12) are called the Visions of Daniel. To understand these visions—full of fantastic animal symbols and other imagery—one needs to know some details of the history of that time. For the contemporary readers of Daniel who were familiar with current events these vivid images were clear references to their enemies, such as Antiochus, who is referred to as a "little horn" sprouting human eyes and a mouth. Also the "beasts" represent the empires of Babylonia, Persia, and Greece.

The vision of chapters 10 through 12 is given a setting during the third year of Cyrus. In it an angel reveals the future to Daniel. Most of the allegory can be identified with actual historical events; for this reason, many scholars conclude that the things spoken of had already happened at the time of writing and that the prophetic tone is a literary device.

The Persians did war with the Greeks. A mighty king (Alexander the Great) did arise, and his kingdom was later divided among princes not related to him. The strong kings of the south (Ptolemies of Egypt) did fight with the strong kings of the north (Seleucids in Antioch). A contemptible king (Antiochus Epiphanes) did succeed to the throne in the north. He did set himself to wipe out the religion of the Jews. His forces profaned the temple and offered abominable sacrifices. Those who resisted were persecuted and imprisoned or killed.

The author of the vision predicted that the tyrant would die, but that his death would be followed by a period of great trouble and hardship for the people. This would be followed by the resurrection of the dead.

This portion of the book of Daniel is a good illustration of what is known as apocalyptic literature. These are allegorical writings with hidden meanings concerning "things to come." The visions of Daniel served as disguised messages of hope to the Jews in a time of hardship. In the New Testament the book of Revelation contains the same type of writing.

The book of Daniel was the last book produced by the Jews of Palestine which found a place in their Sacred Scriptures. It is not the last book in our Old Testament, however, because the biblical books are not arranged in chronological order.

A brief bit of history will bridge the gap from the time of the writing of Daniel to the time of Jesus.

At first the resistance to the cruel policy of Antiochus was stubborn but passive. Then it burst into the hot flame of open rebellion.

The Maccabean War broke out when a Jewish priest refused to eat an unclean sacrifice. His five sons mustered a fighting force made up of loyal, outraged Jews. These fierce fighters attacked the Syrian armies.

Although the priest and his descendants belonged to the Hasmonean family, they came to be called the Maccabees because one of the sons, a masterful soldier, was named Judas Maccabeus, meaning Judas the "hammerer."

Under the inspired leadership of the fighting Maccabees, the city of Jerusalem was wrested from the power of the Syrians. The temple was cleansed and restored. The rededication of the temple was celebrated with a "feast of lights" called Hanukkah. This yearly festival has been celebrated by the Jews ever since.

The Maccabean War—sometimes called the War for Independence—lasted almost a quarter of a century. A succession of Maccabees became the undisputed leaders of the Jewish people. Each ruler became high priest rather than king, but his authority was political as well as religious.

During this period three new distinct power groups developed among the Jews. These parties—the Essenes, the Pharisees, and the Sadducees—were both political and religious in character. Each of these groups, as we shall see later, took a different attitude toward the spread of foreign culture, especially that of Greece.

Finally the intrigue and disorder became so bad that the Roman authorities moved in and took over in 63 B.C. During his reign, Julius Caesar restored full religious liberty to the Jews. When he was assassinated, however, the turbulent conditions that prevailed throughout the Roman Empire extended into Palestine.

In 40 B.C. the Romans appointed as king over Judea an Idumean (the Latin form of "Edomite") named Herod, whose parents had accepted the Jewish faith. This was the state of affairs at the close of the Old Testament period.

The New Testament

PART I
THE BIRTH AND CHILDHOOD OF JESUS

Palestine had won for herself almost a hundred years of independence. Yet her rulers, the Jewish priests and members of the aristocratic Maccabean family, had quarreled continuously.

In 63 B.C. the Romans moved into Palestine to enforce order. They were wise enough to see that the Jews would never become a docile province. For such trouble spots as this, where local or racial loyalties were strong, the Romans made it their policy to appoint a king who was a native of that region, but who ruled at their bidding.

So it was that Herod, an Idumean, was made puppet-king over much of Palestine, including Idumea. Even though his parents had practiced the Jewish faith, Herod's Edomite ancestry was held against him by many Jews.

Herod the Great ruled for forty-four years. He was an able administrator and managed to stay on excellent terms with the Roman authorities. There was relative peace and quiet, but the burden of taxation was heavy.

Much of the tax money went into an ambitious building program that Herod pursued to beautify and enlarge Jerusalem. The Jewish temple was included in this renewal; a splendid Roman structure was constructed to replace the temple built by Zerubbabel.

On the other hand, Herod's family life was tragic. His best-loved wife, Mariamne, was a Maccabean princess and the last of her line. Family intrigues and jealousies involving his ten wives seemed to hasten Herod's growing mental unbalance. In a fit of rage the king had Mariamne put to death. Almost insane after this tragedy, Herod turned more and more to murder to solve his problems, whether personal or political. So it is not surprising to read in the New Testament of his order to slaughter the innocent Jewish babies at the time of Jesus' birth.

Although there was a later error in computing time, it is likely that Jesus was born during the last years of King Herod, probably around 6 B.C.

Little is known of the birth, childhood, and early life of Jesus. When the writers of the gospels came to set down their material, they had no intention of producing work that resembled modern biography. For this reason there is a great lack of the kind of detail that we would now find most interesting.

Jesus never mentioned his own birth. His few references to his family would indicate that it was similar to any pious Jewish family of the time.

Peter and the other disciples who were closest to Jesus were not concerned about Jesus' beginnings. They seemed to feel that their master's teachings—and his mighty acts—were enough to mark him as an absolutely unique individual.

The later evangelists who wrote the gospels were believers in the divinity of Jesus the Christ. They were anxious to share with all mankind what they understood as the great act of God: sending Jesus to bring about a redeemed and Godlike world through his ministry, his death, and his resurrection. They recorded those things that they felt would surely convince others.

For years the deeds and sayings of Jesus had been passed on by word of mouth among the members of the Christian community. It was only when it was obvious that eyewitnesses would not be on the scene always that effort was made to get their knowledge into written form. The four short books that tell something about Jesus are called gospels—meaning literally "the good news."

Scholars generally agree that the earliest of the four gospels was Mark's. It seems to have been written shortly before the final rebellion of the Jews and the consequent destruction of Jerusalem by the Romans in A.D. 70. According to tradition Mark was a close friend and disciple of Peter, gaining most of his information from this important apostle.

It is generally believed that the gospels of Matthew and Luke were written not long afterward.

The gospel of Matthew is distinguished by its literary style and its unique emphasis on the church as the living body of believers in Jesus Christ. More than the other Gospels, Matthew tries to point out the ways in which the prophecies of the Old Testament are fulfilled in the life and work of Jesus. It adds something Mark did not include: a brief account of Jesus' birth including the story of the three wise men.

The "good news" of Luke, like the other Gospels, attempts to preserve for the early Christian community the accumulated knowledge and beliefs about Jesus. Luke the "beloved physician" is especially interested in human relations and healings of the sick. He has a gentle regard and respect for women and relates incidents about them. His addition of the story of Jesus being born in a manger has been a favorite down through the centuries.

The Gospel of John may have been written near the close of the first century, or earlier; there is more diversity of opinion among Bible scholars over the date and contents of this book than for any other Gospel. It is the author's deeply spiritual meditation upon the life, death, and resurrection of Jesus.

It is often pointed out that the Gospels are not in complete agreement about the details and events they report. This is true, but it should be remembered that from this variety of detail there has emerged a most vivid and lasting impression of Jesus.

Elizabeth and Mary

In the days of Herod, king of Judea, an elderly priest named Zechariah served in the Jerusalem temple. For many years he and his wife Elizabeth had prayed to God for a child. Now they were old and filled with sadness, for it seemed too late for their prayer to be granted.

One day when it was his turn to enter the inner sanctuary of the temple to burn holy incense, Zechariah reverently climbed the steps, opened the great door, and walked into the windowless sanctuary.

Outside in the open-air courtyard of the temple the congregation was at prayer. The old priest blinked his eyes to accustom himself to the dim interior. He saw the veil hanging before the small innermost room of the temple, the Holy of Holies; the golden altar, with its constant flame; the pale light flickering from the seven-branched candlestick.

With a reverent, slow sweep of his hand, the priest sprinkled holy incense into the flame on the golden altar. Sweet scent filled the air. Suddenly, Zechariah staggered backward, fear on his face. In the drifting haze of the incense the old priest saw the form of an angel. This celestial vision said softly, "Don't be afraid, Zechariah. God has heard your prayer. Elizabeth will bear you a son, and you shall name him John."

Zechariah, mouth agape, was unable to believe his eyes or his ears. The angel said, "You will have a son who will be filled with the Holy Spirit. He will drink no wine or strong drink. He will grow up to be like Elijah, a forerunner, to prepare the people."

When at last he could speak, Zechariah asked the angel, "How shall I know this? I am an old man. My wife is also well on in years."

"I am Gabriel," the angel said. "I was sent to bring you this good news. But now, because you did not believe my words you will be unable to speak. Until your son is born, you will be mute."

Outside, the people waited for the chosen priest to finish burning the incense. When Zechariah at last came stumbling into the brilliant sunlight of the temple court, he was blinking and waving his arms. He moved his lips, and his tongue wagged. But no words—not a sound—could be heard.

Through sign language the old priest managed to make the people understand that he had just had a vision and could not speak. So they let him go, wondering.

Zechariah's period of duty at the Jerusalem temple was over. He returned to his home nearby in the hill country of Judea.

Within a short time, Elizabeth found that she was pregnant. For the next five months she stayed in seclusion. She was filled with happiness and contentment because the prayers for a child had been answered.

In Elizabeth's sixth month of waiting, God sent the angel Gabriel to visit Elizabeth's young relative, Mary, in the northern village of Nazareth. Mary was a modest and pious young virgin engaged to marry an older man named Joseph.

The angel appeared to Mary when she was alone. "Hail, Mary," the angel greeted her. "The Lord God is with you!"

Mary was startled. The angel reassured her. "Do not be afraid, Mary, for you have found favor with God. You will conceive and bear a son. And you will call him Jesus."

Mary asked, "How can this be, when I am not yet married?"

"The Holy Spirit will come upon you,"

the angel said. "And the child to be born will be called holy, the Son of God."

Mary bowed down before the angel, wonder in her heart.

"Your cousin Elizabeth," the angel went on, "even at her advanced age, she has conceived a son. She is now in her sixth month. With God nothing is impossible."

Mary said, "I am the handmaid of my God. Let it be according to your word."

The angel left. And Mary dreamed of the new life that would begin to grow within her. She felt a great desire to visit her kinswoman Elizabeth.

A donkey was saddled, and Mary began the long two-day journey southward from Nazareth. She followed the road that climbed over the Galilean hills dotted with early spring flowers, went across the lush, broad valley of Jezreel and past Mount Gilboa, where King Saul had died in battle long ago. On and on the patient donkey plodded south, through territory of the Samaritans, through valleys green with new wheat, over hills terraced with vineyards.

In the late afternoon of the second day Mary reached the stony hills of Judea. At last she came to the village where Zechariah and Elizabeth lived.

Elizabeth was at work inside the house, cleaning the clay saucer-lamps, refilling them with olive oil and fresh wicks. She rose to meet her young relative as Mary came in the open door.

The two women embraced each other with tears in their eyes. It had been a long time since they had seen one another.

As Elizabeth looked at Mary she suddenly felt the child within her leap. Elizabeth understood that her young cousin was also to bear a child. "Blessed are you among women, and blessed is the fruit of your womb. Why should I be so fortunate, that the mother of my Lord should come to me?"

"My soul magnifies the Lord," Mary said, closing her eyes for a moment, "and my spirit rejoices that God considers me as his handmaiden." Now she looked, wide-eyed and wondering, at her cousin. She almost whispered. "He who is mighty has done great things for me."

Mary stayed with Elizabeth for about three months. Then she returned home to Nazareth.

Elizabeth's time came, and she gave birth to a son. Her relatives, friends, and neighbors came to congratulate her.

On the eighth day the child was taken to the temple in Jerusalem. Those who came along for the dedication of the child insisted, "You must name your son Zechariah after his father."

"Not so," Elizabeth said firmly. "He shall be called John."

"But none of your relatives are called by this name!" They hurried off to another part of the temple, and found Zechariah. "What are you going to call your son?"

Zechariah made signs to indicate that he wanted a writing tablet. They brought him a sheet of wood covered with a thin coat of wax. They handed him a sharp-pointed stylus for writing.

Zechariah took the sharpened stick. He wrote on the tablet in the common Aramaic language, "His name is John."

Everyone read what Zechariah had written, marveling among themselves. "Look, the old priest has chosen the same name that Elizabeth said!"

Immediately Zechariah's mouth opened, and he found that he could talk. So he bowed down and prayed, giving thanks to God. The people were filled with amazement and awe, wondering if the child of Zechariah and Elizabeth would grow up to be someone special in the sight of God.

Luke 1

The Amazing Birth

221

In those days a decree went out from Caesar Augustus to the whole Roman world: Everyone must report to the city of his ancestors to be taxed.

Joseph and his wife Mary lived in Nazareth, in the hills of Galilee. However, Joseph was a descendant of the great King David. So Joseph had to make plans for the long trip to Bethlehem, far to the south in Judea, to register for the census.

Mary wanted to go with her husband, even though she expected the birth of her child at any time. Joseph agreed, and provided her with a patient donkey to ride. Small loaves of bread were packed in the saddlebags, along with several cheeses and a skin bag containing thick, clabbered milk. This would provide food along the way.

They started out, Joseph walking and holding the donkey's reins. It was late at night on the second day of the long trip south when they arrived in Bethlehem in Judea.

The evening was bitter cold. The stars twinkled sharply in the black winter sky. Pilgrims with flaming torches made their way through the crowded streets, mantles wrapped tightly to keep out the chill wind.

Joseph knocked on the door of the nearest inn. "No rooms left," the porter said, and slammed the door. Joseph tried another hostel, and another, and another. The town was full of pilgrims coming to register.

Joseph knocked wearily at the last inn in the town. "We are full, completely full," the owner said. Then he saw Mary, her face pale and tired as she sat uncomfortably upon the donkey.

"Your wife is ill?" the innkeeper asked with sympathy.

"My wife is soon to deliver her child." Joseph implored, "We just have to find a place."

"Go behind the inn," the owner said. "There you will find a warm stable. Make beds for yourselves in the hay. Perhaps a room will be available in the morning, and I can help find her a midwife."

But the baby was born that night, and his mother made a cradle for him in the manger. She wrapped her firstborn son in swaddling clothes and laid him in the sweet-smelling hay. Close beside this makeshift bed the donkeys munched the barley straw, and the cattle began to stir and to low because of the activity that interrupted their rest.

On one of the hills outside of Bethlehem shepherds huddled by the embers of a small fire, keeping watch over their flocks by night. Without warning an angel appeared to them and the glory of the Lord shimmered and shone around them.

The shepherds clutched at each other, their eyes wide with fear. The angel said, "Don't be afraid. I bring you good news of a great joy which shall come to all people. This very night in the city of David, a Savior is born for all the people, who is Christ the Lord. Here is the sign by which you can find him: He will be wrapped in swaddling clothes and lying in a manger."

Suddenly there appeared a host of angels, singing,

Glory to God in the highest!
On earth peace,
goodwill toward all men.

Then they disappeared.
The men looked about them, open-mouthed with wonder. Their sheep still lay sleeping, crowded together for warmth.

"Let us hurry to Bethlehem," one shepherd urged, "and see this thing that has happened." The others agreed. "Let's find out what God is making known to us!"

The shepherds drew lots to decide which of them would stay and watch over the sheep. Then the rest hurried toward the town.

They searched until they found Mary and Joseph behind one of the inns, in the stable. And there was the baby fast asleep in the manger, wrapped in soft woolen cloths.

The roughly dressed sheep herders were simple lowly men. Yet they knew they had seen the wonder of something great beyond imagining.

In their joy the shepherds told Joseph and Mary about their experience with the angels. Then they went back out into the street, marveling aloud, telling their story to every passerby. They walked on out through the city gates, back to the hills and their flock. They were full of praise for God and his mystical, wonderful glory.

And Mary picked up her newborn son and held him close to her. Wistful and puzzled, she pondered all these things in her heart.

Luke 2:1-20

The Flight to Egypt

When Jesus was born in Bethlehem in the days of Herod the king, a bright new star blazed out in the sky.

Far to the east, across the great desert

224 in the land of ancient Babylonia and Persia, cults of astrologers and soothsayers flourished. These trained wise men made a constant study of the stars and their movements. When the astrologers saw this shining ball of light where there had been none before, they said to each other, "What is this bright new star? Let us discover what it means!"

So three of the wise men mounted camels. Keeping the new star in view, they followed it and came to the land of Judea. In the capital city of Jerusalem they asked, "Where is the one who has been born king of the Jews? We have followed his star from the East so that we might worship him."

Rumors of these inquiring astrologers reached Herod. He stalked nervously through the corridors of his palace. A new king of the Jews? Such rumors were a threat to his own uneasy position. He knew that the people resented him, that they prayed for a king of their own nationality.

Herod called together the chief priests and the scribes, who were experts in the Jewish laws. He asked, "Do your scriptures prophesy where your Messiah, your Christ, will be born?"

"In Bethlehem of Judea," they told him.

Herod fumed to himself. Was there among these Jews a baby whose birth was bringing wise men from afar? Did anyone dare to think that this unknown child was destined to take away Herod's throne? This would be investigated.

Herod sent messengers to find the wise men and bring them secretly to his palace. When they were brought before his throne, the wily King Herod smiled. "The infant king of the Jews is supposed to be born in Bethlehem, according to the prophecies."

The wise men bowed low and thanked the king for this information.

"Go now," Herod told them, "and search diligently for the child. When you have found him bring me word." His deceptive smile deepened. "I too wish to worship him."

The wise men took the road south, to Bethlehem. It was night when they arrived. The star now shone down over the place. The wise men entered the door and found Mary and her son. The scene was not what they expected, and yet they knew that this was the child they were seeking. The three strangers fell to their knees, bowing down to worship the little one. They opened the treasures they had brought, and offered them. A little casket, with coins of gold, a carved box containing fragrant powdered incense, a tiny jar of spicy, luxurious oil of myrrh, for anointing the skin.

The wise men, having left the child, held a conference. "Let us not return to Jerusalem," one said. "I don't trust King Herod."

"I agree," the second one said. "I had a bad dream about him."

"We shall return by another way," the third one said. And so they did.

After the wise men had gone, Joseph also had a dream. In it an angel appeared, telling him, "Rise, take the child and his mother, and flee to Egypt. Remain there until you are told to return. The wicked Herod wants to find the child, to destroy him."

Joseph awoke, his heart beating with fear. Without waiting a moment he wakened Mary and told her his dream.

Hastily they packed their few belongings and fastened them upon their donkey. Then Mary mounted the patient animal. In the middle of the night they set out, Mary anxiously clutching the warmly wrapped baby.

Under the quiet stars in the cold of the night they hurried to the city gate, where a sleepy watchman let them out. Then they started on the long trek southward, through the lonely wilderness to Egypt.

When days passed and the wise men did not return to Jerusalem, King Herod realized that he had been tricked. He flew into a rage, stomping and shouting through his palace chambers. "That baby shall not live to grow up and steal my throne," he swore.

The cruel king sent for a detachment of soldiers. His orders were brutal. "Go to Bethlehem. Kill every male Jewish child who is two years old or under."

By this time Joseph had escaped with Mary and the baby Jesus. They were well on their way to Egypt.

Later, after Herod died, Joseph had another dream. Again an angel appeared in his dream with a message. "Rise, take the child and his mother, and go back to the land of Israel. For the enemy of your child is now dead."

Joseph packed their possessions upon his donkey. Again the small family journeyed through arid deserts and desolate wilderness. But this time Joseph and Mary traveled with joy in their hearts. Their beloved child was safe.

Now they were going home, home to Nazareth in the secluded hills of Galilee. It would be a good place to raise a family.

Matthew 2

A Boy with Questions

In Nazareth the child Jesus grew and became strong.

Younger children were born into the family. The small house was filled with noise and laughter. There were sisters, who helped their mother Mary with the household work and brought the daily water in pottery jars from the village well. And there were four younger brothers of Jesus: James, Joses, Judas, and Simon.

Joseph, Jesus' father, was a carpenter. It was a good trade. Sturdy tables and benches were always in demand. The roughhewn pieces were sanded smooth and sold to innkeepers or to less well-to-do families for their homes. The fine pieces, crafted by a master carpenter, exquisitely finished, and decorated by delicate carving or insets of gleaming wood or ivory, could be sold for a good price to the wealthy citizens of Galilee.

The Romans were ambitious builders. The cities hummed with their grand construction programs. A village carpenter could always hire out his labor as a builder in one of the nearby cities. And he might take his growing sons along as his helpers. It was the custom, in those days, for sons to be taught the trade of the father.

A Jewish boy did not spend his whole day working, however, if his father was a pious Jew like Joseph of Nazareth. He would be enrolled in school for a Jewish education.

First the youngster would be sent to the home of a neighborhood rabbi, a teacher, for his primary education. With a class of other small boys he would learn to read from a "primer" written on a waxed wooden writing tablet. Then portions of the scriptures would be studied from small rolls. Among the first quotations he would read—and memorize—was the Shema from Deuteronomy: "Hear, O Israel: The Lord our God is one Lord; and you shall love the Lord your God with all your heart, and with all your soul, and with all your might." *

The small boy Jesus particularly stored up in his heart a quotation from Leviticus: "You shall not take revenge or bear any grudge against the sons of your own people; you shall love your neighbor as yourself." **

When the boys were ready for advanced study of the scriptures their rabbi took them to the synagogue, where copies of all the sacred scrolls were kept in a guarded chest.

The synagogue, or house of prayer, was the center of worship in towns and villages away from Jerusalem and the temple. There were even synagogues within Jerusalem where people could gather together in small groups for regular study and praise of God. These buildings provided a special gallery, or balcony, where the women and girls might listen and worship. On the main floor, stone benches lined three walls, providing seats for the elders. The boys sat with their fathers, listening intently as continuing sections of the Torah (the first five books of our Old Testament) were read aloud.

Next a portion from the prophets, chosen for this particular sabbath day, was read. A member of the congregation might be called upon to read yet another passage from the writings in the Holy Scriptures.

On weekdays advanced students, guided by their teachers, studied these same scrolls of the Scriptures. Even rabbis did not have complete copies of the Jewish sacred writings available in their houses—or anywhere else. It was considered a real privilege to be able to study the great laws and teachings here in the local synagogue.

Galilee was an interesting place for a boy to grow up. There were roads to explore and hills to climb. From a hill just south of Nazareth the young Jesus could look out across the broad Jezreel valley, the site of many events in his people's history.

Here the Hebrew warriors had bravely fought the Canaanites in the distant time of the judges. On the slopes of Mount Gilboa, King Saul and Jonathan were killed by the Philistines, and their army scattered and defeated.

On the ridge of Carmel near the Mediterranean Sea, the prophet Elijah challenged the prophets of Baal to a fire-making contest. Across the valley, the inland side, King Ahab came home from battle, bleeding to death in his chariot.

Through this Jezreel valley the armies of the cruel Assyrians came. Then the Persians conquered and took over, followed by the Greeks. Finally the Roman legions marched in. And now in the time of Jesus the whole Mediterranean world was under the military thumb of Rome.

Much of the history of the Jewish people was here for a child to learn and to think about. But Galilee was also a bustling part of the life of the present. Along with the Jews, great numbers of foreigners now lived in Galilee. Syrians, Phoenicians, Greeks, and Romans built their homes in Galilean cities and went into business here. This prosperous and fertile hill country was crisscrossed with roads bringing travelers from near and far in the great Roman Empire.

Surely a boy growing up in such a place would be impressed with the vastness of the world—and would be made aware of God's love for all people.

In the spring of the year when Jesus was twelve years old, he went with his family on their annual pilgrimage to Jerusalem for the Passover feast.

This two-day journey south to Judea was a tedious trip for many of the adults, and a tiring one for the old people; but for the children it was a joyous lark.

Joseph's family traveled with several other families from Nazareth. The road was soon crowded with Jerusalem-bound pilgrims from all the towns and villages.

When the Nazarenes arrived in the capital city, Joseph rented a room in which the band of relatives and friends could celebrate the Passover feast. Then he pushed through the crowded streets thronged with people, donkeys, and herded sheep to the marketplace. He purchased a perfect, unblemished lamb for the sacrifice. At the wine-sellers he haggled for a skin of Passover wine.

At the temple, all the various orders of the priesthood were on duty for this great spring festival. In the afternoon of the appointed day, the temple singers chanted as the priests slew the sacrificial animals brought by the people. Each band of worshipers took back its ritually dedicated and dressed lamb to the room which had been made ready for the occasion. The lamb was roasted in a portable clay oven.

The family and friends, dressed in festive white, drank the wine and ate all the lamb, with the accompanying bitter greens. The deliverance of Israel from the bondage of Egypt was recited in song and story.

When the Passover feast was ended the pilgrims slept until just before dawn. By the time the eastern sky was beginning to lighten they were joining the throngs streaming out through the city gates. Joseph and his friends were on their way home.

But the boy Jesus stayed behind in Jerusalem. His parents did not know it. Mary and Joseph supposed Jesus to be somewhere among their relatives and friends. Perhaps he was running ahead with the other boys, or lagging behind to wrestle or to throw stones to see whose went the farthest. After all, a boy of twelve was almost a man, able to take a great deal of responsibility.

At the end of the first day's journey the families gathered together for their evening meal. And Jesus was not there. When Mary and Joseph had made certain that Jesus was not among their relatives and friends, they decided to return to Jerusalem.

After three days the anxious parents found their son in a court of the temple. Young Jesus was sitting among the teachers, listening to them and asking them questions. The teachers, and the elders who sat about listening, were amazed at his understanding and his answers.

His parents were astonished to find him there. His mother sounded distressed. "Son, why have you treated us so? Don't you realize your father and I have been looking for you anxiously?"

Jesus' voice was assured and quiet. "Why did you search? Did you not know that I must be in my Father's house?"

His parents did not understand what he meant. They did not know that already Jesus felt his strongest kinship with his Father in heaven.

Jesus left Jerusalem in the company of his parents. They returned to their home and their daily life in Nazareth. But his mother kept his sayings in her heart.

And Jesus continued to grow in wisdom and in years. He gained favor with God and with his fellowmen.

Mark 6:3; Luke 2:40-52

* Deuteronomy 6:4-9 became the Jewish confession of faith. This prayer is called the Shema from the first word, "Hear."

** Leviticus 19:18.

PART II
THE EARLY WORK
OF JESUS

Herod the Great died while Jesus was a very young child. For all his faults, Herod had proved to be an able administrator, increasing the boundaries of his kingdom and bringing order. Upon his death Herod's authority was divided among three of his remaining sons.

The first son, Archelaus, was given rule over the central part of Palestine between the Jordan River and the Mediterranean Sea; this area included Judea, Samaria, and Idumea. Archelaus was so cruel in his dealings with the Jews and the Samaritans that he was brought to trial by the Romans and banished to what is now France. After his banishment the main area of Palestine was watched over by a Roman military officer, a kind of local governor known as a procurator. During the adult life of Jesus, this officer was Pontius Pilate.

A second son of King Herod the Great, named Herod Philip, inherited authority over a large area northeast of the Sea of Galilee. He was not made king, but was given a lesser title of tetrarch, which meant he was prince over only a part of the kingdom. Philip rebuilt Beth-saida on the north shore of Lake Galilee to be his capital city; Jesus later preached here.

The rule of the fertile, hilly country of Galilee was given to a third son, Herod Antipas. As a tetrarch he he also governed a strip of land along the east side of the Jordan. Jesus spoke of him as "that fox."

What sort of life did the people have in this Roman province of Palestine, during the adult years of Jesus?

On the face of it, things seemed orderly and controlled. There was the stability of universal Roman coinage. Roman law provided for a rather fair justice for all. The soldiers of the famous Roman legions kept the area, like the rest of the empire, in relative peace and security.

The citizens of Rome worshiped so many gods and in so casual a way that they were very tolerant toward the deities of minority groups within the empire. The Jews were left pretty much alone, to worship as they pleased, to administer their own laws, and to levy their own taxes.

Underneath this orderly surface, however, Palestine rumbled with unrest and rebellion. To the nationalistic Jews, Rome was the capital of a wicked empire. The occupation of their country by an army, even a token force, was to them an outrage.

The priests in Jerusalem, who continued to be the highest civil authorities among the Jews—except in dealings with Rome—added their own taxation to the financial burdens of the people. Making a living was very hard for most of the populace. Very few had much margin above the bare cost of living. By the time they paid their taxes to Rome and to the priest, and gave their tithes and offerings to the temple, the common people had very little left.

Those who were most dissatisfied with the political and economic conditions were the Zealots. Fanatics in religion and in loyalty to the Jewish nation, the Zealots were ready to light the fuse for a bloody revolt against everything foreign. They constantly looked for another mighty military leader, someone who could rescue the country as the patriotic Maccabees did during the Maccabean Wars.

The scribes and Pharisees were another influential group among the Jews. They were passive in resistance to the foreign element among them. Their response was one of rigid and formal piety that insisted on minute and detailed attention to the provisions of the law of Moses and to all the complicated interpretations that had grown up around it. They advocated keeping the Jewish people separate by accenting the peculiarities of Jewish faith and worship.

The Sadducees made a more congenial adjustment with the Roman authorities. Their readiness to participate in the political, social, and economic life of the empire, made for a free and easy relationship and often resulted in prosperity and high position for members of this group.

Still another type of reaction was demonstrated by a group called the Essenes. They established communities where they could withdraw completely from the everyday world, dedicating themselves wholly to meditation and prayer.

The majority of the common people, the plain, working people, yearned for the appearance of their Messiah, "the anointed one," who would be their own king to lead them out from under the Roman yoke and on to a great fulfillment. All the treasured scriptures of the Jewish people had led them to believe that they had a great purpose to accomplish in the world, a grand place to occupy among the nations.

Into this air of expectancy Jesus was born. When he began his ministry he came into contact with first one and then the other of these groups, these political-religious parties. He did not fit the pattern any one of them had pictured for the Messiah.

He was too peaceful a man to please the fanatic, revolutionary Zealots. He constantly offended the quibbling scribes and Pharisees by showing the shallowness of their insistence on the strict letter of the law rather than its deeper meaning.

Jesus was too plainspoken and open to have any appeal for the sophisticated, worldly Sadducees.

Jesus was too lively and practical to draw support from the ascetic, monastic Essenes.

But Jesus was exactly the man to inspire and comfort the plain people, the workers, the poor, and the downtrodden. They knew it, and they flocked to hear him.

John the Baptist

In the fifteenth year of the reign of the emperor Tiberius Caesar, Pontius Pilate was the Roman governor of Judea.

Prince Philip, one of the sons of the dead King Herod, ruled over a northern region of Palestine between Galilee and Syria. One of Philip's half-brothers was Herod Antipas, sometimes referred to in whispers as Herod the Fox; his territory included the hilly region of Galilee and the eastern side of the Jordan River.

It was not unusual in those parts to hear about one or another strange new prophet. So no one thought very much about the first reports of John the Baptist.

"He's preaching out in the Judean wilderness," a traveler in Nazareth said.

"He's the son of the priest Zechariah," a second traveler volunteered. "He baptizes people in the Jordan River."

"But only after a person truly repents of his sins," the first traveler interrupted. "He says baptism is a sign that their sins have been washed away."

"Could this prophet be the Messiah, the Anointed One?" a woman asked. She clasped her work-worn hands, and raised her eyes to the sky. "For so long we have waited for our king to appear and deliver us from our troubles."

A richly dressed merchant of Nazareth burst out, "This John is not the Messiah. He is no Christ. He's a crazy man—"

"John is not crazy!" the first traveler shouted. "He speaks God's truth. He prophesies that the Messiah is coming very soon. I tell you John is right. We must all repent of our sins so we will be ready when God sends Christ down to us."

"Ho!" the wealthy merchant spat out. "Who else but a madman would live out in the wilderness all by himself? I have heard he eats nothing but locusts and wild honey. He despises wine and drinks only water. He doesn't even have decent clothes to cover himself—just a haircloth shirt."

Someone laughed. An old woman murmured, "John the Baptizer sounds like the prophet Elijah to me." There was wonder in her voice.

The first traveler nodded vigorously. "He is! He's a pure man, eating only natural foods. And he's very close to God."

The second traveler agreed. "John says that God will soon judge us. And the wrath of God will destroy those who will not repent."

Two elderly rabbis stood to one side listening. Now they nodded to each other.

"Does this not fulfill the prediction of Malachi?" one of them asked, for all to hear. "That prophet said, 'I send my messenger to prepare the way before me.' * Perhaps John is God's messenger, to prepare the way for the glorious Messiah!"

The other rabbi grew excited as he answered, "And it is also written in Isaiah: 'A voice will be crying in the wilderness: Make straight in the desert a highway for our God. And the glory of the Lord shall be revealed.' " **

People of all sorts became curious to see and hear this desert prophet. They journeyed from Nazareth, and all the hill towns and lakeside cities of Galilee, going down to the river Jordan to hear the Baptizer preach. And Jesus of Nazareth, now a tall and earnest young man, also journeyed to listen to John.

Here on the bank of the Jordan people had gathered from Judea, Galilee, and all Palestine.

The serpentine river was muddy and shallow in this spot, but it rushed swiftly. Tangled thickets and patches of rank brush grew here and there along the winding banks. There were patches of barren poisonous earth, strewn with heaps of rocks. It was a bleak spot, wild and desolate. The hills rising on each side of the deep gorge that was the Jordan valley looked forbidding. A bright sun blazed above, and the people sweltered in the midday heat.

Down by the river, John stood on a rock, talking to his disciples. People pushed and crowded closer to hear him.

John was still a young man, but his rugged life in the wilderness had made him look much older. His skin was burned to a deep color, his face seamed. He wore a rough, brown shirt that reached to his knees, woven of coarse camel's hair and cinched at the waist with a wide strip of leather. Beads of perspiration plastered his long matted hair against his forehead and the sides of his face and dampened his long, untrimmed beard.

"Turn away from your sins," John shouted. "For the kingdom of Heaven is about to come upon you!"

A delegation of priests from Jerusalem came striding up to John. Their resplendent robes were quite a contrast to John's rough clothing. With them were some Levites who were workers and helpers in the temple.

"We have come out from Jerusalem to speak to you," the leader of the priests announced. "We were sent from the temple to ask who you are."

John knew what was hidden in their minds. He looked squarely at them. "I am not the Messiah," he said. "I am not the Christ."

"What, then?" the leader wanted to know. "Are you Elijah, come back?"

"No, I am not Elijah."

The priests and Levites muttered among themselves and threw up their hands. "Then who are you? What are you up to? By whose authority do you baptize? We must take an answer back to the Jerusalem temple. What do you say about yourself?"

John spread out his hands. "I am the voice of one crying in the wilderness. I have come to prepare the way—"

There were also some Pharisees who had come along. These fellows, who were very harsh and prudish in religious matters, now spoke sharply to John. "Then why are you baptizing, if you are neither the Christ nor Elijah?"

John answered quietly. "I baptize with water. But there is one who is coming after me, who is mightier than I. I am not worthy to stoop down and untie the strings of his sandals. He will baptize with the Holy Spirit of God."

Some wealthy Sadducees came up at this point, eager for John to baptize them. They wore Grecian-style robes or Roman togas. Their beards were shaved off, their hair meticulously cut and curled in the latest fashion. John understood the character of these men who were always ready to accept new notions and adjust to any change, especially if it offered them an opportunity to gain something in the way of wealth or position from it. He turned to them fearlessly. "You brood of vipers!" he said. "You are like snakes that flee before a brush fire in the desert. Who warned you to run from the wrath to come? If you honestly have repented, then prove it first by the lives you lead! Do you think you will be saved from God's judgment because you claim your descent from Abraham?"

An earnest onlooker asked, "What, then, should we do?"

John glanced at the well-dressed, well-fed Sadducees, Pharisees, and priests. "He who has two coats, let him share with him who has none," John said. "And he who has food, let him do likewise."

Two tax collectors had pushed up to the front. They were as richly robed as the Sadducees.

A peasant in rough country clothing recognized them. "Tax collectors! Robbers!" he hissed, and spat on the ground to show his contempt. He turned to the half-grown boy who stood beside him. "If you grow up and become a tax collector, my son, may God send a fire upon your bones."

"Why?" the boy asked.

"The idol-worshiping Romans levy taxes on us all," the father explained, hate in his voice. "Then they sell to the highest bidders the right to collect the taxes. And traitor Jews, like those two over there, go out and bleed their own people for all they can. And keep the difference."

The tax collectors ignored this remark.

"Teacher," one said respectfully to John, "what shall we do?"

John said firmly, "Collect no more than is appointed you."

Several Jewish soldiers now swaggered over. Their uniforms showed that they served under Herod the Fox.

"How about us?" one of the soldiers asked John. "What shall we do?"

The tall, gaunt preacher stared at them with keen, knowing eyes. Several began to hang their heads. Soldiers often took advantage of the common people, by intimidation or force, to gain extra money.

"Rob no one by violence or by false accusation," John admonished them. "And be content with your wages."

After he was through preaching, one of John's followers whispered to him. "Did you notice how many soldiers are here from the palace guard of Herod Antipas? I think he wants you watched."

"That fox knows that I watch him," John thundered. "What kind of a prince is Herod for our people? He has stolen the legal wife of his own half brother. What right has he to wed Herodias? I have twice sent word, condemning him."

"Yes, and now his new wife is furious because of your words." The disciple looked anxiously about. "And Herod is angry with you for more than that. He's afraid your prophecies about a coming Messiah will encourage the people to revolt."

Another disciple nodded. "You have dangerous enemies. You must be careful."

But John turned from their warnings to go with the crowd of people down the steep mud banks, to the edge of the fast-moving river. He walked waist deep into the shallow, silt-laden current and began to baptize those who repented of their sins and wanted to dedicate their lives to God and to holy living.

Jesus himself was among those who walked out into the waters of the Jordan to be baptized by John.

And when Jesus rose from the water, a dove came down from the heavens and fluttered for an instant above him. Then a voice was heard from heaven, saying, "This is my beloved Son, with whom I am well pleased."

After his baptism Jesus went into the wilderness to spend some time there alone with his thoughts, in deep and prayerful preparation for his ministry.

Matthew 3:1 through 4:1; Mark 1:4-12; Luke 3:1-22; John 1:6-34

* Malachi 3:1.

** Isaiah 40:3, 5a.

Temptation in the Wilderness

After Jesus was baptized he went out into the wilderness where he could be alone. It was a deserted place, with gnarled bare hills. No water flowed in the deep twisted creek beds, called wadies, except during spring rains. Almost nothing grew in that contorted, barren land, only occasional wizened shrubs and dried thornbushes.

In the daytime hot sand blew. A hawk swooped in the brassy sky, a dust-colored lizard skittered over the blistering rocks. They were the only living creatures to be seen.

At dusk bats circled out from the cliffs. During the night, mountain lions padded through the silent dark, looking for prey.

Occasionally jackals broke the stillness with their eerie yapping and howling.

For forty days and nights Jesus lived in the lonely wilderness. He fasted and prayed, as Moses had done upon the mountain in the Sinai desert.

Jesus had important decisions to make. He had heard God's voice calling him "my Son." Did this give him special privileges? And special responsibilities?

At last Jesus realized that he had become very hungry. At that moment it was as if temptation stood beside him, in the shape of the devil. Jesus was conscious of the devil's presence, a presence so real that the devil could be seen and heard. He taunted Jesus. "If you are indeed the Son of God, then command these stones to become bread!"

Jesus was ravenously hungry after the long days of fasting. He could hunt locusts, as John did, and roast them on a stick over a fire. He could find a hive of wild bees, where there would be sweet honey. But would it not be easier just to perform a miracle? It would prove his power as God's Son. If he performed such a miracle once, he could do it again and again. Imagine the crowds who would follow him if he offered them free bread!

Jesus struggled with the temptation and turned it down. No, that was not the way. People must be won by the message itself and not by an appeal to their stomachs.

So Jesus rebuked the devil with a quotation from Deuteronomy, which he knew by heart: "Man shall not live by bread alone. He shall live by the word of God." *

The young Galilean hiked to the top of the highest mountain and stood on the edge of a high, sheer cliff. Suddenly he sensed the devil beside him again, and was transported to the holy city of Jerusalem, and was standing on the highest pinnacle of the temple.

236 "If you are the Son of God, throw your-self down from here," the tempter whis-pered. "For it is written in the psalms: 'He will give his angels charge of you, to guard you. On their hands they will bear you up, lest you stub your foot against a stone.' " **

Jesus knew that the devil could quote scriptures to suit his own purposes. Jesus also knew that he must not push God for a special sign. He must face death when it was time, and leave his future in God's hands.

"It is also written," Jesus retorted, "you shall not tempt the Lord your God.' " ***

Then Jesus was offered an even greater temptation. "Beyond us are all the king-doms of the world," the devil wheedled. "All these I will give you. I only ask that you will respect and give honor to me."

Jesus knew he could win all the Jews if he were their worldly king with power over all the nations. But he also knew that political power was a passing thing.

"Begone, Satan!" Jesus cried out. "For it is written in the Law: 'You shall wor-ship the Lord your God and him only shall you serve.' " ****

The devil left him. Jesus had defeated the power of temptation.

Matthew 4:1-11; Mark 1:13; Luke 4:1-13

Jesus used quotations from the Old Testament to answer the temptations of Satan.

  * Deuteronomy 8:3.

  ** Psalm 91:11-12.

  *** Deuteronomy 6:16.

  **** Deuteronomy 6:13.

Calling the Disciples

When Jesus came out of the wilderness filled with the spirit of God, he heard some disturbing news. John the Baptist had been arrested.

Herod the Fox had thrown the outspoken preacher into jail. It was said that John languished in the dungeon of Herod's fortress-palace, on the cliffs overlooking the lonely Salt Sea. There was nothing John's friends could do to rescue him.

Jesus returned to Nazareth in Galilee. He went to the synagogue on the sabbath as was his custom. There in the synagogue Jesus stood up before the assembled people of his village. It was his turn to be the reader of the day. The scroll containing the writings of Isaiah was handed to him. He began to read:

The Spirit of the Lord is upon me,
  because he has anointed me;
He has sent me to preach
  the good news to the poor,
To proclaim release to the captives
  and recovering of sight to the blind,
To set at liberty those who are oppressed,
  to proclaim the year of the Lord.
    (based on Isaiah 61:1-2; 58:6)

Jesus rerolled the scroll. He gave it back to the attendant and sat down. The eyes of all the people were fixed on this intense young man. "Today," Jesus said solemnly, "the scripture that you have just heard has been fulfilled."

As Jesus continued to speak to them, some of the people realized what he meant. This son of Mary and Joseph was the "anointed one," the Messiah, the Christ. He was among them, preaching the good news about the coming of God's kingdom.

After the worship and study were over, the people gathered in clusters. Many spoke well of their young neighbor. "Is this not the son of Joseph the carpenter?" they marvelled.

Then Jesus went to other hill towns in Galilee, preaching the gospel of God, the good news.

"The time is fulfilled, and the kingdom of God is coming," Jesus told the people. "Repent, and believe this good news."

Jesus walked eastward to Capernaum, on the north shore of Lake Galilee. Here the crossing of trade routes had brought great prosperity. In this bustling lakeside city, Jesus taught in the synagogue. The people were astonished at his teaching, for he spoke with authority. Not only did Jesus teach in the synagogue, he also healed those who were ill, with mental and physical diseases and afflictions. Everywhere he went the people sought him out and followed him. When Jesus was ready to leave, the people begged, "Stay here with us."

But Jesus knew that he had much work to do. "I must preach the good news of the kingdom of God to the other cities, also," he told them. "For this is why I came."

As he was walking along the shore of Lake Galilee, Jesus saw two fishermen up to their knees in the water, casting a great circular net out into the lake. Jesus knew of these men. They were brothers named Simon and Andrew. Simon was a big man, impulsive, headstrong. He was also intelligent, with a great capacity for loyalty. Andrew was an earnest, conscientious Jew, one of those baptized by John.

Jesus called out to them. "Follow me, and I will make you fishers of men."

Immediately the brothers left their nets and followed him.

Jesus gave the first brother, Simon, a new name—Peter, which means "rock."

Going a little farther, Jesus saw two other brothers, James and John, sitting in a boat anchored just offshore. With them was Zebedee their father. They were mending their nets.

"James and John!" Jesus called. "Come, follow me and be my disciples, be my assistants in preaching the good news."

Immediately James and John put down their nets. They left their boat and their father and waded ashore to join Jesus, Peter, and Andrew.

Jesus and his disciples began to travel on foot throughout Galilee. Jesus taught in the synagogues of the towns and cities and healed the sick. As he went, Jesus found more disciples. There was Judas Iscariot, a patriotic Jew who seemed to be honest and careful with money. He became the treasurer for the band of Jesus' disciples.

A man called Matthew sat in his tollbooth collecting taxes on the goods passing along the trade routes that ran beside Lake Galilee. Tax collectors were looked on with contempt by most Jews, who resented every penny paid to the Roman authority. The collectors were considered as near traitors. Jesus did not despise them, however, or avoid their company. In fact, he saw great possibilities in Matthew.

Jesus spoke quietly to the tax collector. "Follow me."

Without hesitation, Matthew rose and left his tollbooth. Immediately he became a disciple, and joined the growing band.

At last Jesus had called to him twelve special disciples. He gave them authority to heal the sick and sent them out on their own. "Go, preach only to your fellow Jews," he charged. "We must reach the lost sheep of Israel."

"What shall we take with us?" asked Judas Iscariot.

"You received without pay, now you

must give without pay," Jesus said. "Take no gold, nor silver, nor copper in the purses in your belts. You will not need to take an extra tunic or sandals. You will earn your food as you go by preaching about the coming kingdom of heaven, and by healing the sick."

"Teacher," Andrew asked, "where shall we sleep?"

Jesus said, "Whatever town or village you enter, find out who is worthy in it. Stay with him until you depart."

One of the disciples named Thomas, who tended to be on the cautious side, asked, "But Master, what if people won't listen?"

"If anyone will not receive you or listen to your words, shake the dust from your feet as you leave that house, or that town. And on the day of judgment it shall be more tolerable for Sodom and Gomorrah than for that town!"

Then Jesus looked with a wistful smile at the trusting face of John, youngest of the disciples. "I send you out as sheep in the midst of wolves. So be as wise as serpents and as innocent as doves. Beware of men! For they will deliver you up to councils and flog you in their synagogues. You will be dragged before governors and rulers for my sake."

Some of the disciples moved restlessly, their faces betraying their nervousness. But Jesus continued. "Do not be anxious how you are to speak or what you are to say. Your Father in heaven will tell you what to say when the time comes."

"We go out to proclaim the good news and to heal the sick," one of the disciples sighed. "And yet, we will be badly treated."

Jesus nodded. His face was sad. "Yes, you will be hated, for my sake." Then his face was lighted with a radiant smile. "Yet he who endures will be saved. When they persecute you in one town, flee to the next."

Another disciple asked, "Yet is there no way we can avoid persecution?"

"A disciple cannot expect to fare better than his master," Jesus warned them. There was great compassion on his face. "Are not two sparrows sold for a penny? And not one of them falls to the ground without your Father's will. So do not be fearful. You are of more value than many sparrows. Just remember, to follow me will not be easy."

Jesus blessed the twelve men. "He who receives you, receives me," he told them. "And he who receives me, also receives him who sent me."

When Jesus had finished with these instructions to his twelve special disciples, called apostles, they went on their mission. And Jesus continued to teach and preach throughout Galilee.

Matthew 4:18-22; 9:9; 10:1-11; Mark 1:12-20; Luke 4:14-22, 31-43; John 1:42

A Wedding at Cana

Happy excitement ran through the little village of Cana in the Galilean hills. There was to be a wedding in town!

This meant that the groom's family would invite all their relatives and new in-laws, their friends and neighbors, to come for seven days of feasting and merriment.

A supply of wine was bought, enough to last for the whole week. Mountains of wheat were ground into fine flour for the quantity of bread required. Seven fat sheep were chosen from the family herd. One would be roasted on each of the days.

There was not room for all the guests to eat in the small main room of the house. So tables and benches were borrowed. They were set up under the leafy fig trees in the walled courtyard.

Several servants had been hired to prepare and serve the roast lamb, and to keep the wine pitchers full on the tables.

A close friend of the family was chosen as the "steward of the feast." His job was to see that there was food and drink for everyone. He would also check on each newly filled pitcher of wine brought to the table to be sure it was sweet and full-flavored, and had not turned to vinegar.

Now the guests began to arrive. Among them was Mary of Nazareth, a close friend of the groom's family.

Every day there was celebration, with the music of flute and tambourines. The men danced, leaping and singing, as the music grew faster and gayer. Then the women danced, their bare or sandaled feet beating a happy rhythm on the hard-packed dirt of the courtyard, their arms swaying in graceful motion.

Later in the week news came that Jesus of Nazareth was traveling toward Cana. The happy bride and groom sent a servant hurrying to the edge of town with a message for Jesus inviting him to join the wedding party, together with any friends who were with him. Jesus seemed happy to receive the invitation. But he motioned toward his twelve disciples. "I have many with me."

The servant bowed low. "My master begs all of you to come and be guests at his humble house."

So Jesus followed the servant along the narrow village street to the courtyard of the house where much laughter and music could be heard.

The groom came out and embraced Jesus. "Come in, my good friend! Bring your traveling companions, and find places at the table. There is food and wine for all."

Jesus saw his mother there, and greeted her with a kiss. He and his men were served. The happy laughter and toasting and eating continued. Jesus and his disciples were warmly welcomed by all the guests. They joined in with the merrymakers.

After a while Mary quietly plucked at her son's coat sleeve. She beckoned for him to follow her. Over by the side of the house, Mary pointed to some pottery wine jars. "Look, they are empty already," she whispered. "So many extra guests have come, and now the pitchers on the tables are but half full. Our hosts will not be able to refill the pitchers."

"Dear lady," Jesus said. "What has that to do with you and me?"

Mary looked flustered. "The mother of the groom is one of my best friends. This is such a proud and happy time for the whole family. They'll feel so ashamed and embarrassed if they discover they haven't ordered sufficient wine—"

Jesus smoothed the worried brow of his mother, then sighed. "My hour has not yet come."

Mary could not understand the meaning of her son's words. He often spoke strange things now that he was grown.

But as the mother looked up into the compassionate face, into those understanding eyes which seemed to know all, her anxiety ebbed away. She felt again the lighthearted spirit of the party. Above all, she felt a supreme confidence in her son.

Mary turned to a couple of servants who stood near, holding empty wine jugs in their hands. She gestured toward the husky, sun-bronzed young man. "Do whatever my son tells you." She smiled and hurried back to the tables under the spreading fig trees.

Jesus pointed to six stone jars that stood by the door of the house. Each held a number of gallons of water. Special water jars like these stood by the doorway of every good Jewish home, ready for the Jewish rites of purification as prescribed by Moses. Hands always were washed before eating so that the body would not be defiled. The water was used to purify members of the house when they returned from the marketplace, or wherever they might have become ritually contaminated. It also served for the traditional cleansing of cups, pots, and other vessels.

Each stone jar was now brimming with fresh water drawn from the village well. "Take your jugs," Jesus said, "and draw out some of the water."

The servants did so, looking at the young teacher with wonderment on their faces.

"Take it to the banquet tables and fill the wine pitchers there," Jesus said. "Then present the pitchers to the steward."

The servants, still wondering, carried their jugs and filled the half-empty pitchers on the table. When the steward of the feast tasted from the pitchers, he smacked his lips with delight. He called to the bridegroom.

"Every man serves the best wine first," he whispered to the happy bridegroom. "Then, after the guests have drunk freely, he brings out the poorest wine." The steward nudged his smiling host playfully. "But you, aha!" He tasted again from the pitcher. "You have kept the best wine until now!"

And so the wedding feast ended that night with everyone happy and gay.

Now this deed, reported at Cana in Galilee, was the first sign by which Jesus revealed his nature to the people. And his disciples, hearing what had happened, believed in him.

Mark 7:3-4; John 2:1-11

The Death of John

During the early months of Jesus' ministry, John languished in prison after his arrest by Herod. The disciples of John stayed together in the wilderness of Perea, on the other side of the Jordan. They continued to live the austere and ascetic life that John had taught them.

Some of the people criticized Jesus, saying, "The disciples of John fast often and live without comforts. But you and your disciples eat and drink wine."

Jesus chided his critics. He told them that they were like little children, sitting in two groups in the marketplace. One group of children wanted to play "wedding" with music and dancing, but the other group wouldn't play. The dissidents wanted to play "funerals," with weeping and wailing.

"We piped to you, and you did not dance," the first group of children accused.

"We wailed, and you did not weep," the funeral-minded cried back.

So in the end no one played.

Then Jesus said, "You are as obstinate as those children. John the Baptist has come eating no bread and drinking no wine. And you say, 'That man is crazy.' I myself have come eating and drinking, and you say, 'Look, a glutton and a drunkard, a friend of tax collectors and sinners.' Thereby you reject us both."

In his prison cell John heard of the work and the teachings of Jesus. He called two of his own disciples to come to the window of his cell.

"Go to Jesus," John said, "and ask him if he is the Messiah, the Christ, the Anointed One. Or shall we look for another?"

So the disciples of John came to Jesus one day as he was teaching and healing the people. They asked John's question.

Jesus answered, "Go and tell John what you hear and see. The blind receive their sight, the lame walk, the deaf hear. The lepers are cleansed, and the poor have the good news preached to them. And blessed is he who takes no offense at me."

When John's disciples had gone Jesus said, "John may be in prison, but he is no soft reed to be shaken in the wind. John is a prophet and a messenger who prepared the way."

At this time John the Baptist's days were numbered.

John had become exceptionally popular with the common Jewish people. Many had journeyed to hear him preach and to be baptized, until Herod began to grow nervous about him. The Jews of Galilee always seemed to be on the verge of a revolt, and the prince feared that some new agitation might sweep the area because of John. As long as John remained alive, even though he was imprisoned, his disciples would be active. Herod the Fox wanted an excuse to do away with this unsettling preacher—permanently.

Then in an unexpected way the ruler at last found his excuse. The palace had seethed with hatred for John because of the things which the prophet had said.

"It is not lawful for you to have Herodias for your wife," John had told the prince and all who would listen. "She is the legal wife of your half brother. Neither did you have the right to divorce the Arabian princess who was your own lawful wife—"

Herodias had smoldered with a fierce resentment. It was not enough for her that John languished in the palace dungeon. She wanted her critic dead.

Her chance came on Herod's birthday. There was a big dinner for the prince, with much drinking. Then beautiful Salome,

the daughter of Herodias, danced for her royal stepfather and his guests.

As the tale was later told Salome pleased the prince and his guests with her dancing so much that Herod cried out, "Salome, ask me for anything you wish, and I will grant it."

Salome stood there out of breath and not knowing what to say.

"Come, come," Herod urged. "I vow it, whatever you ask I will give you, even half of my kingdom."

Salome saw her mother beckoning to her and ran to the portico beside the banquet hall. "What shall I ask?" Salome inquired breathlessly.

Her mother's eyes narrowed. "Ask for the head of my enemy John the Baptist," she hissed.

Without hesitation Salome hurried back to Herod. "I want you to give me, at once, the head of John the Baptist," she cried, "on a platter!"

Herod was shocked. This request seemed too cold and vindictive even for him. He looked about uncomfortably. The guests all seemed to be waiting for the next amusement. Had he not made a vow? He could not break his word to his stepdaughter without embarrassment.

So it was that John the Baptist was beheaded that night in the prison. And the gruesome story was whispered throughout the countryside, how the head was carried on a platter into the banquet hall and presented to Salome, and how Salome in turn gave it to her mother.

When John's disciples heard that their master had been executed, they came to the prison and claimed his body. They laid it reverently in a tomb.

When Jesus heard the news, he was saddened. He wanted to withdraw to a lonely place where he could be alone with his thoughts, but the crowds would not let him. They followed him wherever he went.

So Jesus taught them, for he had compassion on the common people. "Are they not like sheep without a shepherd?" he said to his disciples, who wanted to shield him from the crowds.

At last Jesus became so well known in Galilee that Herod heard about him. Even the courtiers at his palace were discussing Jesus' teachings. There were some who insisted, "This Nazarene is really Elijah!"

And others said, "He is a prophet like the ones of old."

But Herod was deeply troubled. "What if John has been raised from the dead, and this is his ghost?"

When the fame of Jesus continued to spread, the prince was perplexed and curious. "I should like to see this man," Herod muttered.

Some of Jesus' friends at the court were concerned for him. They knew his life was in danger. "You must get away from here," they warned. "Herod most likely wants to kill you just as he killed John."

But Jesus told them, "Go tell that fox that I cast out demons and heal the sick. I have my work to do, and I shall do it until my time comes."

Matthew 11:2-19; 14:1-13; Mark 6:14-29; Luke 5:33-35; 7:18-35; 9:7-9; John 5:36-37

PART III
THE MINISTRY OF JESUS

Most modern biblical scholars agree that no accurate records exist for the birth and youth of Jesus, and that the stories we have came from oral tales circulated long afterward.

This is also true for the ministry and teachings of Jesus. The Teacher from Nazareth kept no journal, nor did his disciples. His most striking sayings were remembered, told, and retold to all who would listen. Decades later, Mark, Matthew, Luke, and John, each in turn, began to write down as many incidents as could be gathered and considered reliable. By this time the exact sequence of happenings had been forgotten; each of the gospel writers put incidents and sayings in a somewhat different order. In many cases, nobody remembered the place or even the circumstances in which a particular saying was uttered.

Yet biblical scholars, for all that they may disagree on various interpretations, unanimously agree that Jesus was a real, historical person. His ministry began sometime in the second quarter of the first century A.D., and lasted only a short time—perhaps little more than three years.

The similarity of all the accounts shows that his main teachings and experiences were preserved remarkably well. For instance, the New Testament gives us six variations of a story about the miraculous feeding of a multitude: two in Mark, two in Matthew, one in Luke, and one in John. To have made such an impression, it seems certain that some such event actually happened. It is interesting to note, by the way, that none of the versions specifically shows Jesus multiplying the loaves and fishes by other than some natural means; in fact, for Jesus to have done so would have been in direct contradiction to his resolve on the Mount of Temptation not to create bread by use of supernatural power.

There are other miracle stories in the gospels, and many of these are concerned with healings. Modern minds have produced scientific explanations; we do know that great faith, with a resultant change in attitude, can cure psychosomatic symptoms. These miracles were recorded by the gospel writers, however, not as legends, not as allegory, but as a marvelous proof of Jesus' unsurpassed greatness.

The bulk of the ministry of Jesus was in Galilee, a beautiful hilly area, easily walked across in a couple of days. It was quite unlike the dry mountain-ridge heights of Jerusalem and Judea, which swept down into adjacent parched wilderness and desert regions. Galilee's rich soil was well watered by springs and rivers fed by the rain and melting snows of the towering Lebanon Mountains to the north. Galilee was a land of terraced vineyards, abundant fruit orchards, and lush grainfields.

Galilee was crisscrossed by important trade routes. Roads ran northeast to Damascus, northwest to Tyre, east and west to connect the Mediterranean with the Lake of Galilee and the Jordan, south to Jerusalem and Egypt. These not only brought money to the area, but also cosmopolitan people and ideas. Roman legions marched on these roads, and Roman soldiers and officials were stationed in every city.

The village of Nazareth was almost a suburb of a bustling non-Jewish metropolis (Sepphoris) that served for many years as the capital of Galilee. Just before Jesus began his ministry, the new capital city of Tiberias was built by Herod on the western edge of Lake Galilee, by the forbidding black basalt cliffs.

Jesus apparently turned his back upon the large Gentile cities and spent most of his time in the vicinity of the predominately Jewish fishing towns of Capernaum and Bethsaida at the north end of the lake. He also made journeys through many of the villages nestled back in the hills of Galilee.

The Lake of Galilee was perhaps the most dominant feature of this area, a center of activity, surrounded by towns and fishing villages and boat-building docks. This lake, which is roughly the shape of a human heart, is not much over seven miles across at its widest part, and is only thirteen miles long. The lake basin is almost tropical in its warmth, set down in the great Jordan rift with the surface of its water at seven hundred feet below the level of the nearby Mediterranean. Little streams flow down into it from both sides. In the north the upper Jordan comes plunging down from greater heights, bringing fresh water and a continual new supply of edible fish. From the lower edge of the lake the Jordan tumbles southward through its deep valley until it reaches the bitter and salty Dead Sea from which there is no outlet.

During the lifetime of Jesus, Galilee was ruled by Herod Antipas, whom Jesus called "that fox." Although this ruler hated John the Baptist and had him put to death, it does not appear that he interfered with the preaching of Jesus. Neither was any opposition recorded on the part of Herod's half-brother Philip, when Jesus taught in the villages around Caesarea Philippi.

Twined through the area ruled by Philip, mostly east of the Jordan, were ten independent Greek-influenced cities, known as the Decapolis. They were interconnected with wide paved roads, making for easy travel and commerce between them. Jesus was reported to have made a teaching tour through this territory, also.

However we are most interested in Galilee itself. This is the district where Jesus grew up. It was in Galilee that Jesus concentrated most of his active ministry.

An Unforgettable Sermon

News about the new Galilean preacher spread rapidly. People were curious. They journeyed from all the villages of the area looking for Jesus, eager to hear him speak.

On one occasion when a crowd had gathered, Jesus went up on a nearby hillside. His disciples went with him, and all the people followed. Jesus sat down on a rock, his disciples at his feet. The people moved in close so they could hear whatever the Teacher said.

Jesus began teaching his disciples some beatitudes,* saying:

Happy are the lowly, unspoiled folk,
  for the kingdom of heaven is theirs.
Happy are those who can accept sorrow,
  for they shall be comforted.
Happy are the patient and gentle,
  for they shall inherit the earth.
Happy are those who hunger to do the right,
  for they shall be satisfied.
Happy are those who show mercy,
  for mercy shall be shown to them.
Happy are those who have pure hearts,
  for they shall see God.
Happy are the peacemakers,
  for God shall call them his sons.
Happy are those who are persecuted for a
    righteous cause,
  for theirs is the kingdom of heaven.

The people, crowding closer about the inner circle of the chosen disciples, cried out, "More, more! Tell us more."

So Jesus went on. "Do not judge others, and you will not be judged. For just as you pronounce judgment on others so will you yourself be judged. Whatever measure you deal out to others will be dealt to you. Tell me, why do you notice the speck of sawdust in your brother's eye and yet never notice the plank in your own? How can you say to a fellowman, 'Let me take that speck out of your eye,' when a whole log is in your own? Don't be a hypocrite. First take the beam out of your own eye. Then you will see more clearly to take the speck out of your brother's eye!"

A scribe, a lifetime scholar of the Hebrew Law, made his way to the front of the crowd. "Are the rules you have given us," he asked sharply, "to take the place of the Laws of Moses and the teachings of the prophets?"

Jesus frowned. "Now don't think that I've come to do away with the law and the prophets. I have come not to abolish the old laws, but to fulfill them. The new law is the completion of the old. It is not its destruction." He looked around at the tight-lipped scribes and Pharisees in his audience. "For I say to you that you must live even more uprightly than do the scribes and the Pharisees, who talk so much about the law."

"Master, give us some examples of what you mean," begged one of the disciples.

Jesus nodded. "You have heard that it was said to our forefathers, 'Do not commit murder. Whoever commits murder shall be brought to judgment.' But I tell you that every one who nurses a grudge against his fellowman shall be liable to judgment! Whoever speaks insulting words to his brother must be brought to judgment. So if you are offering your gift at the altar, and there remember that someone has a claim against you, leave your gift at the altar and go. First be reconciled. Then return and offer your gift."

Another disciple said, "Teacher, give us some more examples."

Jesus said, "You have always been told, 'Don't break faith with your marriage vows and make love to someone else.' But I tell

you this: if you harbor lust for another in your heart, already you have broken the faith of your marriage vows."

The people stirred and looked at each other. "This teacher goes farther than any of the prophets," a farmer whispered to his wife.

"Again, you have heard, 'An eye for an eye and a tooth for a tooth.'" Jesus continued, "But I say, do not resist one who is evil. If anyone strikes you on the right cheek, turn to him the other also. If anyone would sue you and take your shirt, let him have your coat also. And if anyone insists that you go one mile, go with him two miles. Give to him who begs from you. Do not refuse him who would borrow from you."

"Indeed he does go far," the farmer's wife answered softly. "These teachings are not easy ones, are they?"

"You have heard, 'You shall love your neighbor and hate your enemy.'" Jesus nodded. "Yet I say, love your enemies. Pray for those who persecute you. For if you love those who love you, what further reward can you expect?

"Do not even the tax collectors do this much? And if you bless only your relatives and friends, what more are you doing than others? Do not even the Gentiles, who do not know our Father—do they not do that much?"

Jesus looked around at all the listening faces and added gently, "You must be honest, sincere, dependable, without vengefulness, even as your heavenly Father is all these things."

When Jesus had finished these sayings, the crowds were astonished at his words. For he did not teach as the scribes did, quoting endless and dry traditions to prove each minute point.

This Jesus of Nazareth taught as one who had authority.

Matthew 5 through 7; Luke 6:17-45

* These sayings of Jesus are called beatitudes because most of them begin with the word "blessed" (translated "happy" in RSV).

The Man with Palsy

On the north shore of Lake Galilee was the bustling city of Capernaum. Many fishing boats docked at her wharves. Along the shore fishermen called, "Fish, fish, fresh fish! Come get your fish here!" Roads went through the city bringing trade from Damascus and Jerusalem and all the towns of Galilee. "Brassware from Damascus!" hawkers in the market called out. "Judean olives. Dates from Egypt."

There was a large synagogue here, built of stone. The Jews who came regularly for study and worship were friendly to Jesus. Had he not cured a demon-possessed man in their congregation? "Stay here in our city and teach us," many Jews in Capernaum begged Jesus.

Simon Peter and Andrew had lived in this city before they became disciples. Jesus had cured Peter's mother-in-law of a fever. "Yes, stay here for awhile," everyone pleaded. "You could reach many people here." They found a house in the city where Jesus could live while he taught in their city and in the countryside around.

One day Jesus had returned to the house after a trip to other villages in Galilee. Word spread through the city that the Teacher was back. People flocked to his house, crowding into the one large room.

250 On the other side of the city lived a young man who had not been able to walk for some time. He called four faithful friends and said, "Look, Jesus has returned to this city. Please take me to the house where he stays. Perhaps he will have mercy on me and heal me."

"A good idea," one said. "We'll make a stretcher out of the pallet on which you sleep."

They hastily improvised a stretcher and laid the paralytic on it. Quickly they jogged through the narrow, crowded streets, dodging walking people, a few camels, a small herd of sheep being driven to market, and men riding donkeys.

Finally they reached the house where Jesus was. But they could not get in. A great crowd was glued to the doorway, intent on the discussion within. Apparently every last place to sit or stand was taken.

"Excuse us," the stretcher-bearers said, tapping some of the outsiders. "In the name of God, will you let us through? We have a paralyzed man here."

The people in the doorway paid no attention. They were too busy trying to catch every word that was said inside the room.

One of the stretcher-bearers looked discouraged. "We can never get in this door."

One of the men looked up and saw a child on the roof. The child seemed to be peeking down through a hole.

"Let's go up and see if we can hear better from the roof," the man said.

The four friends climbed the outside stone steps with their burden. The roof was surrounded by a low stone parapet. The sunbaked clay top was crumbling in several places. It would need more coatings of wet clay, well rolled, to be ready to repel the first rains of autumn.

The boy was peeking down through a small hole in the middle of the roof. The faces of the four stretcher-bearers lighted up. They were jubilant for they all had the same idea at the same time.

"Indeed, you shall soon be at the feet of the Teacher himself," one friend chuckled. Then he turned to the small boy. "Do you think you can find us some strong rope?"

The boy pointed to the house next door. "My father is a fisherman. He has lots of ropes there." He ran off, padding down the stone stairs in his bare feet.

The four friends set to work, widening the hole in the roof. They reached a layer of twigs matted over stouter branches. By the time the boy came back with the ropes the friends had made a hole between the roof beams as long and as wide as the stretcher.

Jesus had stopped speaking. He and all in the room turned their faces toward the gaping hole in the roof, where the four faces peered at them.

Carefully the pallet holding the man was let down to the feet of Jesus. When Jesus saw the faith of the young cripple and his four friends, he said, "My son, your sins are forgiven."

The young man looked up with glowing eyes at Jesus.

"I say to you, arise!" Jesus said.

The young man rose, unsteadily. It was the first time he had stood on his legs since the paralysis began. There were tears of gratitude in his eyes.

"Now take up your bed," Jesus smiled, "and go home."

The young man stooped down, rolled up his pallet, and put it under his arm. There was a stir in the room as people moved back to make a pathway for him.

Everyone was amazed. They praised God and said, "We have never seen anything like this!"

Matthew 9:2-8; Mark 1:21-31; 2:1-12; Luke 5:17-26

Adventures in Galilee

Galilee was rolling country with scattered villages. Wheat fields flourished in fertile valleys; hills were terraced with vineyards. Jesus often took to the roads which led from one village to the next. Sometimes he traveled alone, sending his disciples in other directions.

Once Jesus returned to Nazareth—the village where he had grown up—and preached to the people in the synagogue there. "The kingdom of God is at hand," he told them. "Repent and believe the good news."

This had been the main message of John the Baptist, but when repeated by Jesus, the message had more depth and power. Many who heard him—some for the first time—were astonished.

One demanded, "Where did this man get this message?"

"How does he work all the miracles that are reported?" another wanted to know.

"Is he not the son of Mary and the carpenter?" an old man asked skeptically. "Is he not the brother of James and Joses, Judas and Simon, and are not his sisters here in this town with us?"

"We've heard what you did at Capernaum," someone shouted with derision. "Why don't you do that kind of magic healing here in your own hometown?"

Jesus was saddened by this surge of hostility. "A prophet is without honor in his own country," he said ruefully, "and among his own relatives."

Some fellows in the crowd were gesturing for Jesus to be quiet. They began to pull on his coat to make him come away. "This man is beside himself," the men

explained. Some of Jesus' kinsmen tried to get to him and hurry him away to a safe and quiet place until he should recover his senses, for they could not understand the work he was trying to do.

But a noisy group in the synagogue had become angry about the teachings of Jesus. They surged around him and began making threatening gestures. Then the mob hustled him out of the synagogue and out beyond the edge of the village.

"Let's bind him and stone him!" one hothead cried out.

"No, let's take him to the top of the hill and throw him over headlong," someone else ranted.

"Let's just beat him up as a lesson," another shrilled, but he was shouted down.

In all the pushing and bickering Jesus escaped the mob. He shook the dust of Nazareth from his feet. He was appalled by the lack of faith of these people, the very ones who had known him so well.

In another village of Galilee, Jesus came upon a man with the fearsome disease of leprosy. Ugly scabs crusted large areas of the suffering man's body. He could no longer live in his own house, for his illness was considered too contagious. He had to avoid his own family lest he make them unclean too. He slept near the city refuse dump. He poked through the trash to find scraps to eat or a discarded garment to wear when the weather turned cold.

When Jesus came by, the pitiable creature called out the usual warning, "Unclean, unclean! Have mercy upon a poor leper!"

Then the leper realized who this stranger was. He was the man from Nazareth, the one people were talking about, the Teacher who could cure all kinds of sickness.

"Unclean, unclean," the leper cried out in his agony of loneliness. He fell on his

252 face and begged, "Lord, if you will, you can make me clean."

Jesus turned to him and said, "I will." Fearlessly he stretched out his hand and touched the man, saying, "Be clean."

It was as if a great weight had rolled away from the man's tortured spirit. Already he was feeling better.

"Now tell no one," Jesus commanded him. "Just go and show yourself to the priest. Then make an offering for your cleansing as Moses prescribed." *

The man's heart sang with joy, and his step was light. He was filled with boundless gratitude, for he knew that he was healed. He could live once more like a human being among his own family and friends.

After that Jesus went to preach in a house in a nearby town. People had crowded into the single small room, sitting closely packed around Jesus as he talked. Several hunched on a bench along one wall of the room, their hands clasping their knees. One small fellow perched himself on the large clay water jar in one corner. Several lamps sputtered and smoked in their wall niches, adding to the shaft of light shining through the small, high window.

Suddenly, there was a disturbance among the cluster of people jammed into the doorway. Word was passed along inside to Jesus, "Your mother and your brothers are outside asking for you."

Jesus replied with a question. "Who are my mother and my brothers?" He looked around at those who sat about him, those who believed in him and had such deep interest in his teachings. He waved his arm to include all his eager listeners.

"Here are my mother and my brothers," Jesus said. "Whoever does the will of God is my brother, and sister, and mother."

Matthew 4:23-24; 8:1-4; 13:53-58; Mark 1:21-45; 2:1-5; 3:20-21, 31-35; 6:1-6; Luke 4:23-30; 5:12; 17:11-19

* This cleansing ritual is described in Leviticus 14. See especially verses 2-7, 10, 21-22.

Dark Days to Come

With his chosen disciples Jesus journeyed far to the north of Lake Galilee to the district of Caesarea Philippi. The city, enlarged and beautified by Herod's son Philip, was renamed after that prince and the reigning Caesar. It had many temples built for the worship of Greek and Roman gods.

Jesus did not preach in the city itself for mostly Gentiles lived here. Instead he took his message to Jews in the small villages in that district. He found them eager to listen.

One day the Twelve gathered around their master to take a rest under a spreading oak tree in a green valley. The everlastingly snow-crowned Mount Hermon rose high to the north of them against a blue sky. Near their feet a swift stream tumbled over rocks and boulders on its way to join the upper Jordan.

"Who do the people say that I am?" Jesus asked.

His disciples looked at one another. "Some say you are John the Baptist, come back to life, others say you are Elijah."

"And there are those who say you are one of the prophets of old, risen again—"

Jesus asked them, "But who do you say that I am?"

Peter answered quickly, "You are the Christ."

When Jesus saw that Peter and the others understood who he was, he went on, his face solemn and reflective. "I must suffer many things and be rejected by the elders, the chief priests, and the scribes. And I shall be killed—"

There was an immediate outcry from the disciples, and a look of horror on their faces.

Jesus held up his hand, quieting them.

"The Christ must be rejected and be killed," he went on calmly. "And on the third day be raised."

Again the disciples whispered among themselves. They could not understand.

"If any man would come after me," Jesus warned, "let him deny his own wants and wishes. Let him take up his cross daily and follow me. For whoever would try to save his life will lose it, and whoever loses his life for my sake, he will save it."

It was quiet in the small valley. There was no sound except for the humming of bees and the liquid splashing of water over the rocks.

"For what good does it do," Jesus asked, "if a man gains the wealth and power of the world and loses or forfeits his true self?"

"May we tell everyone that you are the Christ?" one of the disciples asked. "May we proclaim we have found the true Messiah?"

Jesus shook his head. "No, the people do not yet understand the kingdom of heaven. So I charge you all to tell no one that I am the Christ."

From that time, Jesus began to tell his disciples how he must someday go up to Jerusalem, and suffer many things from the elders and chief priests and scribes.

On an evening about a week later Jesus took Peter, and the brothers John and James, and led them up on a mountain to pray.

As Jesus prayed on the mountaintop, Peter, James, and John grew sleepy.

Suddenly they were wide awake for they saw that the face of Jesus shone like the sun. His robe seemed as dazzling white as light. He was transfigured, there on the mountaintop.

And then suddenly Moses appeared on one side of Jesus, and the prophet Elijah on the other. They were talking with him.

Peter was like a man in a trance, speaking slowly and strangely, not knowing what he said. "Master, it is well that we are here. If you wish, I will make three shelters here. One for you, one for Moses, and one for Elijah."

As he said this, a cloud came and overshadowed them. Peter felt his heart thumping in his fear. But a voice from the cloud said, "This is my beloved son, with whom I am well pleased. Listen to him!"

Peter, swept by a sense of awe at the vision, bowed his head and closed his eyes. Then he felt the touch of Jesus' hand on his shoulder.

"Rise," Jesus said, "and have no fear."

When Peter lifted up his eyes, he saw no one but Jesus before him. John and James were behind him, rubbing their eyes.

As they came down from the mountain, Jesus said to the three men, "Tell no one the vision until I am raised from the dead."

So the three disciples kept the matter to themselves, wondering what Jesus meant about being raised from the dead. Actually, they found it hard to think of a time to come when their Master might die. So they put it from their minds, and would not dwell on it.

Matthew 16:13-23; 17:1-9; Mark 8:27-31; 9:2-10; Luke 9:18-36

When Jesus returned to his ministry in Galilee, the people flocked to hear him. Those who had heard him before came seeking him out again, bringing with them their friends and neighbors.

Sometimes people walked miles to hear Jesus, and to ask him questions. They brought bread, in a leather shepherd's bag or tucked inside a sash, and sometimes cheese and olives to sustain them.

"What religious acts should we do if we want to be saved?" a vinedresser in rough clothing asked Jesus one day.

Jesus looked about at the crowd which had gathered. "Beware of practicing your piety before men in order to be seen by them," he warned. "For then you will have no reward from your Father who is in heaven. When you give money for the poor, don't blow your own horn to call attention to your good deed as the showoffs do in the synagogues and in the streets." A smile flickered on his lips. "They already have their reward."

Some of the people smiled at the subtle humor of their teacher.

"But when you give gifts for the needy," Jesus emphasized, "do not let your left hand know what your right hand is doing. Let your charity be done quietly and in secret, and your Father will reward you.

"And when you pray, you must not be like the hypocrites, putting on a show for the benefit of others. They love to stand and pray in public in the synagogues and on the street. They enjoy being seen."

Again his smile flickered. "Honestly, they too have already had their reward! When you pray, go into your room and shut the door. Or go to some quiet place. There you can talk to your Father alone."

Then one of the disciples asked, "But Master, what should we pray for? What shall we say?"

"Pray like this," Jesus said.

> Our Father who is in heaven,
> Holy be your name.
> May your kingdom come,
> And your will be done,
> On earth as it is in heaven.
> Give us today the bread we need;
> And forgive us our wrongdoing,
> As we forgive those who hurt us.
> And let us not fall into temptation,
> But deliver us from evil.

The disciples and the followers looked at one another and nodded with satisfaction. This prayer appealed to them with its simplicity and its completeness. They would learn it and use it.

"Remember that if you forgive others their injustices toward you," Jesus cautioned the people, "then your heavenly Father also will forgive you. But if you cannot find it in your heart to forgive the sins of others, then neither will your Father forgive your own sins."

Jesus continued looking first into one face, and then into another, in the crowd. "You have all seen those who make a public spectacle of their fasting. They go unwashed, with streaks of ashes on them, with torn mantles, and long sorrowful faces. Those who put on such a show have their reward when they are noticed and admired for their hardships.

"If you would keep a true fast, keep it secretly. Dress yourself as carefully as you usually do and wear a cheerful face. Then your fasting is in the heart, where it is seen and accepted by your heavenly Father."

Someone asked, "Will God give us anything we pray for?"

Jesus said, "If you live in me, and my words live in you, ask whatever you will. Under those conditions you will ask only what is right and possible; and it will be done for you. Yes, ask, and it will be given to you. Seek, and you will find. Knock, and it will be opened to you. For everyone who asks will receive just what he needs. He who seeks, will find. And to him who knocks, a door will be opened."

Jesus looked around the crowd at the fathers, holding their sons by the hand, or in their arms, or perched upon their shoulders. "What father among you, if his son asks for a loaf of bread, will instead give him a stone?" Jesus asked. "Or if he asks for a fish, will give him a serpent? Or if he asks for an egg, will hand him a biting scorpion? If you, then, who are less than perfect, know how to give good gifts to your children, how much more will the heavenly Father give good things to those who ask him."

It was late. Jesus had been teaching all day, and he was very tired. But someone called out, "Before you go, give us a final teaching that we can never forget."

Jesus looked at the people and said, "Whatever you wish that men would do to you, do so to them; for this is the teaching of the law and the prophets."

So the crowd broke up, and the people began going in different directions to their homes, thinking upon this golden rule.

Matthew 6:1-18; 7:7-12; Luke 6:31; 11:1-13; John 15:7

Along the Lakeshore

Andrew and John had been fishermen from Capernaum before Jesus called them to be his disciples. Two other brothers, James and John, also had been fishermen on Lake Galilee. They were all at home in rowboats or fishing boats. Traveling by water from one point on the lakeshore to another was often quicker and easier than hiking overland.

One day Jesus decided he wanted to go from Capernaum to Beth-saida. He chose one of his disciples, and they climbed into a rowboat and pushed out from the Capernaum wharf.

In tandem they rowed west, following the north shore of the lake. Fishermen were busily casting out their weighted nets. They rowed past the mouth of the upper Jordan River. There was a rippling current where the river flowed swiftly into the blue waters of the lake. This current brought more freshwater fish into the lake.

Just beyond the mouth of the river was the natural harbor of Bethsaida, crowded with wharves and small boats. On the large hill beyond the harbor, connected by a broad ribbon of paved Roman road and the leaping arches of an aqueduct, was the bustling city, the capital of Prince Philip's territory.

Jesus and his disciple rowed to a wharf and tied the boat. Here graybearded fishermen sat mending their nets. Along the shore fish were spread out to dry in the sun.

A man of the town said to his companion, "Is that not Jesus, the healer of Galilee?"

His friend hurried down the beach and tapped the shoulder of a blind man who

sat begging for coins. He whispered, and the beggar scrambled awkwardly to his feet, feeling for his cane.

"Wait, wait!" one of the blind man's friends called out to Jesus, who was making his way through the crowd. "Are you not the great healer Jesus?"

"We have heard of the wonders you did in Galilee," another burst out, pulling forward his blind friend. "We heard about the two blind men whose sight you restored by touching their eyes."

The first friend added, "They have spread your fame all over so the story was told even in Beth-saida. So now," he pleaded, "please will you touch our friend's eyes so he can see again?"

Jesus looked with pity on the blind man who stood there clasping his hands, his face strained with eagerness and hope. Jesus smiled and dismissed the two friends. "You two go on about your business. Let this blind man come with me." He took the

sightless man by the hand and led him out of the little fishing village.

When they were outside the village and alone, Jesus spat on his fingertips. He laid them on the man's eyes. "Do you see anything?"

The man looked up. He was staring in the direction of a nearby road, where travelers were going in and out of the village. He blinked. "I see light! I can tell light from darkness!" He squinted intently toward the road. "And I think I see the shapes of men coming and going over there." He shook his head. "But it is blurred. They look—well, like trees walking."

"Close your eyes," Jesus said. Again he laid his moist fingertips on the man's eyelids. "Now open them," he commanded.

The man obeyed. He looked all around, at everything, astonishment on his face. "I am restored!" he shouted. "Everything—I see everything clearly!"

"Do not announce this to the people," Jesus admonished, knowing that that would only bring unmanageable crowds of the curious. "Go directly to your own home without returning to the village."

But no matter how many times Jesus asked the cured ones to be quiet about their healing, the news continued to spread.

On another day Jesus was in a boat on the lake with his disciples. Word had spread around the countryside that Jesus was nearby. The people came out of the towns and followed his boat along the shore.

When Jesus went ashore he found a great throng awaiting him. "Teach us, Master," the people begged. Many sick persons pleaded with Jesus to cure them.

So Jesus had compassion on them. All day he taught the people and healed their sick. At sunset Jesus told his disciples to return to the boat and wait for him.

After he had sent the people home Jesus went up into the darkening hills by himself to pray.

Down by the lake a cool evening breeze was whipping up the waters. The twelve tired disciples nodded and dozed as they waited in the boat.

On and on into the night Jesus prayed. The disciples were fast asleep. Their boat, slapped and beaten by the steady waves, drifted out into the lake.

Late in the night, when the moon had risen, one of the disciples awoke. As he looked out across the moonlit waters he was startled by the sight of a vague white form approaching.

Quickly he awakened the other disciples. He pointed at the oncoming apparition.

"It is a ghost!" one of them cried out. They were terrified as the white figure came closer, walking on the night-black water.

"But—but he has the face and form of our M-Master," another disciple stuttered.

Then the unmistakable voice of Jesus came to them. "Take heart, it is I," he assured them. "Have no fear."

Peter stood up by the edge of the boat. "Lord, if it is you, then invite me to come to you on the water."

Jesus said, "Come."

Peter climbed out of the boat. He looked at Jesus, and was unafraid. He found himself walking across the water. Then he began to be afraid. As the fear gripped him, he started to sink.

"Lord, save me!" Peter cried out, stretching his hand toward Jesus.

Jesus immediately reached out his hand and caught Peter. "O man of little faith, why did you doubt?"

When they climbed into the boat, the wind ceased and the water grew very calm.

Then the disciples in the boat worshiped Jesus, saying, "Truly you are the Son of God."

Jesus was weary. He had been teaching since dawn.

All day long people had pushed and shoved, trying to get closer to the teacher. Finally Jesus climbed into a small boat where he sat and talked to them.

Now the end of the day was near, and the sun dipped behind the darkened hills of Galilee. Rosy beams rippled across the lake. The stark hills on the opposite side were suffused with pink.

But still the people would not leave, not so long as he sat there. It was getting dark. A few stars began to glimmer overhead, and a brisk evening wind ruffled the water.

Jesus whispered to his disciples. "Let's untie the boat and go across to the other side."

The disciples crowded into the fishing boat, and Peter pushed them out toward

258

deeper water, then vaulted in with the rest.

Jesus found a pillow in the stern. He curled up on the bottom of the boat, his head on the pillow. In a few minutes he was fast asleep.

Soon the wind began to increase. By the time they were in the middle of the lake dark clouds were scudding overhead and blotting out the stars.

A bit later the wind was howling and lashing at the sail. The boat rocked perilously on the waves. It began to rain in torrents. The boat began to fill with water as waves washed over the sides.

The disciples cowered in fear. "The boat will be swamped," they cried to each other. "We shall all be drowned."

Then they shook Jesus awake. "Master," they implored him. "Do you not care if we perish?"

Jesus rose and rebuked the wind. He said to the raging waves, "Quiet! Be still!"

The wind died down and ceased. There was a great calm.

"Why are you so afraid?" Jesus asked his disciples. "Have you no faith?"

The twelve men were awed by this experience. They asked each other, "Who can he be, that even the wind and the sea obey him?"

Matthew 8:23-27; 9:27-31; 14:22-23; Mark 4:35-41; 8:22-26; Luke 8:22-25

The Madman Healed

On the farther side of Lake Galilee, near the high cliffs along the eastern shore, was a city where mostly Gentiles lived. In the countryside around their city these non-Jews raised herds of swine. They enjoyed pork; pigs were not unclean animals to them as they were to the Jews.

Here lived a man who was so perplexed and burdened with worries, he lost his sanity. He couldn't talk sense. He even forgot his name and everything about himself. He became like a wild, untamed animal, uttering gibberish.

The people of his town whispered, "He is possessed by demons and evil spirits!"

They tried to drive the evil spirits out of the man by beating his body with sticks. They pelted him with stones. Yet the man remained as wild as ever.

"He must be plagued by a thousand devils," his friends said, awed by this frightening mystery. "He must have as many demons inside him as there are soldiers in a whole Roman legion."

His neighbors bound him with chains and fetters, but the tormented demoniac wrenched them apart. No one had the strength to subdue him. At last he ran from the city and found shelter in an empty cave by the cliffs overlooking the lake. There were many caves here used for burial of the dead.

The poor, afflicted man lived among these tombs, naked, often hungry, at the mercy of the summer heat and the winter cold. He often cried out and bruised himself with stones. The townspeople were afraid of him and kept their distance.

One day the pathetic creature looked out over the lake and saw a boat approaching. It pulled up to the narrow shore and landed. One man, who walked with grace and authority, led the way. A dozen other men climbed out of the boat and followed him.

Up the steep hill they came. Then the tortured man was face to face with the leader.

"Watch out, Master!" one of the disciples

called out, pointing to the naked, wild-eyed man.

Jesus seemed to be without fear. He put his hand on the man's shoulder and looked deeply into his eyes. The sufferer's aimless movement quieted. Docilely, like an eager child, he listened as the man spoke of his heavenly Father, of the God that cares, and the faith that cures.

As the teacher gave him new confidence and quieted his fears, a ray of clear thought flashed through his tumbling mind. He suddenly realized who this wonderful healer was.

"You are the teacher from Galilee I heard about before the demons possessed me," he cried out. "You are Jesus of Nazareth." He threw himself on the ground at the feet of Jesus. "Please, Jesus Son of God, deliver me from my demons."

"Come out of the man, you unclean spirits," Jesus commanded. The man fell down trembling. "What is your name?"

It was as if the demons were speaking for themselves. "My name is Legion, for we are many." And as if he were the mouthpiece for his demons, the man said, his face once more contorted and his eyes glassy, "Don't send us out of the country." As if in a trance, he pointed to a herd of pigs nearby. "Send us into the swine."

So Jesus gave the demons permission to enter the detested beasts. And the herd of about two thousand pigs rushed down the steep bank into the lake. All were drowned.

The herdsmen, who had been watching Jesus and the madman, became frightened. They fled back to the city and babbled the whole story.

The people came out of the town to see what had happened. They found Jesus with his disciples. And there, clothed and in his right mind, sat the former demoniac. They looked over the cliff and saw the floating carcasses of the dead swine.

They also grew fearful. "Is he a magician?" someone asked.

"He can drive out evil spirits," another worried. "His powers are too much for us." So they asked him to leave their neighborhood.

When Jesus and his disciples made their way down the steep bank going back to their boat, the cured man followed them.

"Let me go with you," he begged. "Let me also be a disciple."

Jesus said gently, "No. Go home to your family and your friends. Tell them how much God has done for you."

So the man went back to his own city. He told everyone about his experiences. He also traveled among the Greek-speaking towns of the Decapolis. In all these towns he repeated his story of what Jesus had done for him. And everyone marveled.

Later, when Jesus was traveling through the area of the Ten Towns, he found that people were eager to meet him.

"Even here we have heard how you heal people of their afflictions," they said to him.

A blank-faced man was brought to Jesus. His friends pleaded, "Will you not have pity? He is deaf and cannot speak clearly. Will you not heal him with your hands?"

A crowd was gathering. Jesus felt sorry for the man, but he did not want to work miracles just to satisfy the idle curiosity of a crowd. So he took the man away and found a place where they could be alone.

The afflicted man watched, wide-eyed and open-mouthed, as Jesus dampened his fingertips on his own tongue. Then the Healer reached out and touched the tongue of the deaf-mute.

The man took a deep breath, feeling new energy stirring within him.

Then Jesus put his fingers on the ears of the deaf man. Looking up to heaven, Jesus sighed, as if weighed down with the grief

of all the suffering he found in the world.

"Be opened," Jesus said.

The man could hear. He could hear Jesus talking and the babble of sellers and buyers in the nearby marketplace. He heard the tinkle of a camel bell, the baaing of sheep, the harsh bray of a donkey. He could even hear the buzz of the flies around him.

Now that he could hear once more, he could also speak intelligibly. He dropped to the ground at the feet of Jesus and poured out his thanks.

So the fame of Jesus continued to spread.

Matthew 8:28-34; Mark 5:1-20; 7:31-37; Luke 8:26-39

The Loaves and Fishes

It was a bright spring day. The usually bare hills were clothed in a fresh new coat of green. The blue water of Lake Galilee sparkled in the morning sun.

Jesus and his disciples headed their boat toward the northwestern shore. Everyone was tired, and Jesus felt the need to spend a day in a quiet place.

"Come away with me to a lonely place," Jesus said. "Let us rest awhile."

But when they had landed on the shore, people began to converge upon them from all directions. Word had been passed around, and the crowds had traveled on foot from all the towns nearby.

Jesus had compassion upon this great throng. "They are like sheep without a shepherd," he told his disciples.

So the people followed Jesus and his disciples up into the hills. All day Jesus taught them, telling them many things.

When the afternoon shadows grew long across the hill, Jesus called his disciples to him. "I feel sorry for all these people," he said. "All day they have been here with me, with nothing to eat."

"Surely they can find no food to eat up here in these hills," one of the disciples said.

"Send them away," another disciple urged. "Let them go home. Or they can go down to the villages round about and buy themselves something to eat."

Yet Jesus was not willing to abandon these eager people who had feasted on his words. "If I send them away hungry to their homes, some will faint on the way. And many have come a long way." He looked at his chosen twelve. "How much bread do we have?"

Andrew, Peter's brother, pushed to the front of the little circle. A young boy was with him. "There is a lad here who has five barley loaves and two fish."

The boy grinned up at the smiling teacher. He held up a sash that had been knotted to make a bag. In it were five small flat loaves and two tiny smoked fish. "My mother packed it for my supper," he said. "You can have it all."

Andrew looked back over his shoulder at the great, waiting crowd. He threw out his hands in a helpless gesture. "But what is this small lunch among so many?"

In his quiet voice, Jesus told his disciples, "Have the people sit down on the grass."

So the people sat down in groups and companies on the green grass. There were about five thousand there.

Jesus said to the disciples, "Each of you bring me your food basket."

The disciples did so, holding up their baskets before their master. Their rugged faces were puzzled. How could the Master hope to feed so many, with so little?

262

Jesus took the loaves of barley bread and gave thanks to God for them. He broke them into pieces, distributing them among the baskets which the disciples carried.

Then Jesus did the same with the two small fish, giving thanks and breaking them into pieces among the baskets.

"Now take the food," Jesus commanded, "and distribute it among the people."

So the Twelve circulated among the people, passing out food to any who were hungry.*

When the people had eaten their fill, Jesus told his disciples, "Gather up the fragments left over, that nothing may be lost or wasted." The disciples gathered up the remnants of the supper; and there were twelve baskets full of leftovers.

"This is indeed the prophet!" people began shouting. "Here is our Savior, who is come into the world!"

Jesus saw that the excited crowd was about to come and take him by force, to try to make him their king.

So Jesus and his disciples eluded the frenzied crowd. When it was dark they went down to their boat, to go back across the north end of the lake to the fishing village of Capernaum.

The next day, many of those in yesterday's crowd kept seeking Jesus. They found him in the synagogue in Capernaum. Here they flocked around him and questioned him.

"Rabbi, Teacher," they said breathlessly, "when did you get here?"

Jesus ignored this. "Honestly, I can tell you why you seek me," he chided. "Not because you saw a miracle, but because you ate your fill of bread."

They babbled their protests. But Jesus knew that these people were greedy for material things. They thought Jesus could provide them—the easy way.

Jesus knew that people were jumping to the wrong conclusions, after last night. Why couldn't they realize he was more interested in providing food for their souls than for their stomachs?

"Do not strain yourselves to get the food which perishes," Jesus told them. "Your first energy should go toward seeking the spiritual sustenance. For that will nourish you forever!"

One disappointed listener complained, "When our ancestors were fleeing from the angry Pharaoh in Egypt, Moses gave them manna." His expression was petulant, like that of a spoiled child. "Moses gave them delicious bread from heaven to eat."

"It was not Moses who gave the bread from heaven," Jesus corrected. "My heavenly Father supplies the bread which gives life to the world."

The people wondered what this young man was talking about. One asked, "Where can we find this special food from God?"

"I am the bread of life," Jesus said. "He who comes to me shall not hunger, and he who believes in me shall never thirst. I have come to do not my own will, but the will of my Father in heaven who sent me."

The Jews began to murmur among themselves and to say loud enough for Jesus to hear, "Is this not Jesus, the son of Joseph? What does he mean about his Father in heaven, and being sent here?"

"Do not murmur among yourselves," Jesus said. "I tell you, he who believes has eternal life. Your ancestors ate manna in the wilderness. Yet they died! But if you take me, and my teachings, inside of you, as part of you, you will not die."

The people could not see that he spoke in parables and allegories. They insisted on taking Jesus literally. "Does he want us to be cannibals, eating his very flesh and drinking his blood?" they scoffed.

"This is a hard saying," some of them muttered. "Who can listen to it?"

"Am I a stumbling block for you?" Jesus sighed. "Do you take offense at what I say? It is the spirit that gives life, not the actual flesh of your bodies. The words I have spoken to you are spirit—and life!"

This was so difficult to understand that many followers turned away from Jesus and left the synagogue. But the Twelve were more loyal than before.

Matthew 14:13-21; 15:32-39; Mark 6:35-44; 8:1-10; Luke 9:10-17; John 6

* We do not know how Jesus worked this miracle. Did many respond to the generosity of the little boy and to the compassion of Jesus and his disciples? Did Jesus work a miracle in the hearts of the people, changing their usual selfishness into generosity so that they pulled out hoarded picnic bits to add to the supply?

Raising the Dead

As Jesus continued his preaching mission throughout Galilee, great crowds kept coming to him. They brought with them the lame, the maimed, the blind, the dumb, and many others. They put them at his feet, and he healed them.

The people wondered, when they saw the dumb speaking, the blind seeing, and the lame walking. They were filled with awe that these miracles could be done in the name of ancient Israel's God.

Among the hill towns of Galilee was the town of Nain. Jesus came there with his disciples, and a crowd of people followed him.

As Jesus drew near to the gate of the town, a funeral procession was coming out. A young man had just died. It was a hot summer day, so arrangements had to be made to bury him immediately. He was now being carried on his funeral bier to the cemetery outside the town walls.

Among the large crowd of mourners one heavily veiled woman wept more than all the rest.

"She's the young man's mother," one of the mourners explained to Jesus, as the crowds made way for each other. "So sad! A widow, and that dead lad her only son. All the family she had in this world!"

Jesus felt great pity for the sorrowing widow. He walked to her side and said, "Do not weep." Then he touched the bier. The coffin-bearers stood still.

Jesus looked at the quiet body that lay on the stretcher, wrapped in burial clothes.

"Young man!" Jesus called out. "I say to you, arise!"

The eyelashes began to quiver. Then the eyes slowly opened. The young man who had appeared to be dead now raised his head. He sat up and began to speak.

"I give you back your son," Jesus said to the mother. Then he quietly went on his way into the town.

Fear and awe seized the people who saw this.

"A great prophet has arisen among us," some said.

Others exclaimed, "God has visited his people."

And this report concerning Jesus spread far and wide.

At another time when Jesus was returning to Capernaum from a trip, he was met by Jairus, one of the elected elders of the synagogue.

Jairus fell at the feet of the Teacher, and begged him, saying, "My little daughter is at the point of death! Come and lay your hands on her so that she may be made well, and live."

Jesus said, "Take me to your house."

Jairus led the way, and a great crowd followed and thronged about them. In the crowd was a suffering woman who had been ill for twelve years. She had been to many physicians, and spent all she had looking for a cure. Yet she was no better, but instead grew worse.

When she heard the reports about the healings of Jesus, she began to look for him. Now she saw him in the midst of this crowd.

Jesus wore the traditional tassels, tied by blue thread and fastened to each of the four corners of his outer cloak. This was as commanded by Moses, to remind a Jew of his obligations to his religious laws.*

The ailing woman came up behind Jesus in the crowd. "If only I can touch his garment," she said to herself, "I know I shall be made well!" With trembling fingers she reached out to touch the fringe of the tassel on his cloak. Immediately, for the first time in twelve years, she began to feel well again.

Jesus turned about in the crowd. "Someone touched me," he said. "For I feel that power has gone out from me."

In fear and trembling, the woman came forward and fell down before him. "I touched you, Teacher," she confessed. She blurted out the story of her long illness, and how she began to recover her health the very minute she touched his garment.

Jesus said kindly, "Daughter, your faith has made you well. Go in peace, and be completely healed of your disease."

Just then a breathless runner came from the Jewish elder's house and gasped, "Your daughter is dead!" There were tears in the messenger's eyes.

The elder began to weep. They were too late! But Jesus comforted Jairus. "Do not fear. Only believe."

Then Jesus told the rest of the disciples, "I am going on to this man's house with him. Let no one follow us, except for Peter, James, and John."

When Jesus and Jairus arrived at the official's home the place was in a tumult. Relatives and friends were wailing loudly, as was the custom.

Jesus held up his hand to silence them. "Why do you make such a racket, and weep?" he asked. "The child is not dead. She is only sleeping."

The people made rude and bitter remarks, because of their great sorrow.

Jesus sent them all outside the house to wait. Then he, and the mother and father, and the three disciples went inside where the twelve-year-old girl lay, as quiet as death.

Jesus walked to the bed and took the child's limp hand in his. "Little girl," he said with great tenderness, "I say to you, arise."

The girl's long dark eyelashes trembled; her large dark eyes opened. She saw her parents. "Mother!" she cried out. "Father!"

Jesus, still holding her hand, helped her to sit up. Then she swung her feet to the floor. She stood up, and walked.

The parents were filled with amazement and great joy.

"Now," Jesus said, placing the girl's hand in that of her mother, "give this girl something to eat. For she is hungry."

And the report of this went all through that district.

Matthew 9:18-26; 15:30-31; Mark 5:22-43; Luke 7:11-17; 8:41-55

* See Numbers 15:38-40.

The Boy with Epilepsy

Some of the disciples of Jesus had gone on ahead into the next Galilean town. They were to arrange for food and lodging for the night. Jesus and the remaining disciples walked leisurely along the country road.

It was fall. The skies were blue, the sunshine warm, but there was a cool tang in the air. Back during the time of the spring rains the rocky hills had been green with new grass, but the long months of summer heat had left them dry and brown.

Where the hills were terraced and edged with stones, the grapevines were heavy with clusters of ripe fruit. The wheat fields in the valleys were bare and stubbled. They waited for the first fall rains to moisten the parched earth so it could be plowed for winter planting.

The road was bordered with wild mustard plants. The plants were similar to the domesticated mustard grown in fields and gardens; the leaves provided greens for eating, and the seeds yielded oil and seasonings.

In the springtime Jesus and his disciples had come along this same country road. The mustard plants were pushing upward then, their leaves still bright green, young and tender. They had borne a riot of small blossoms, as yellow as fresh butter.

Now in autumn, the wild mustard had grown into large many-branched plants, with dark green, coarse and hairy leaves. The stems had become quite thick, hard, and brittle.

Small sparrows flew and twittered among the mustard plants. Some perched on the branches and picked at the contents of the dead seedpods with obvious delight.

"Teacher, tell us," one of the disciples asked, "what is the kingdom of God like?"

"To what shall I compare it?" Jesus mused. He ran his fingers over the coarse leaf of an especially large mustard plant. He plucked a dried pod and took out a single seed. It was very tiny, a mere speck in the middle of his palm.

"The kingdom of heaven is like a grain of mustard seed," Jesus said. "A small beginning, like this."

One of the disciples peered into his Master's palm. "That's a very small beginning, for anything. Much smaller than a grain of wheat or barley."

"Far smaller than a cucumber or melon seed," another disciple agreed.

"Yet a man can plant this tiny seed in his garden or his field," Jesus reminded them. "And when it is grown, it becomes one of the greatest of the herbs—" He waved his hand toward the roadside. "It becomes so great and strong, the birds of the air can alight on its branches."

The first disciple felt his question had been answered. The kingdom of God would have a very small beginning; but it would grow upon this earth into something vastly more great.

A crowd was waiting for Jesus and his disciples when they reached the town that was their destination. Word of the Teacher's coming had spread.

A man rushed out from the crowd, holding the hand of a small boy. He kneeled down before Jesus and cried out, "Teacher, I beg you to look upon my son, for he is my only child."

The boy, with dark hair and large dark eyes, stared up at Jesus. He smiled, flashing dimples in his round cheeks. He looked perfectly normal.

"You see," explained the unhappy father, "my son is an epileptic. How he suffers, when the demon seizes him and shakes him! Suddenly the boy will cry out and fall to the ground when the demon convulses him. He writhes until he is exhausted. He foams at his mouth and grinds his teeth. Then he grows rigid."

The father pointed to the disciples, who had come ahead of Jesus to the city.

"I brought him to your disciples," the man said, "but they could not heal him."

Jesus sighed. "Bring your son closer."

Just then the boy was seized with a convulsion. He fell to the ground and rolled about, jerking and writhing. His face contorted, and saliva foamed in his mouth.

"How long has he had this?" Jesus asked.

"Since childhood," the father wept. "Often he falls into the fire, or into the water. If you can do anything, have pity on us and help us."

Jesus said to him, "If? But all things are possible to him who believes!"

Immediately the father of the boy cried out, "I believe." Then he kneeled again at Jesus' feet. "Yet help me, whatever I lack in faith."

Jesus spoke to the demon that possessed the boy, and said, "Come out of him, and never enter or trouble him again."

The boy gave another convulsive shudder and lay quiet.

"He is dead," people whispered.

But Jesus took the boy by the hand and lifted him up. The boy rose unsteadily to his feet. He looked around wonderingly. Then he looked up at Jesus and smiled.

The grateful father scooped his son into his arms and took him home.

Later that night Jesus and his disciples were settled in the privacy of their rented quarters. Through a latticed window they could see the moon rising above a nearby mountain, just outside town.

The disciples asked Jesus, "Why were we not able to heal the boy?"

Jesus said to them, "Because of your little faith. I tell you, if you have faith as a grain of mustard seed, you could say to that mountain out there: 'Move from here to there,' and it would move. And nothing would be impossible to you."

And they knew that he was speaking to them in parables.

Matthew 13:31-32; 17:14-21; Mark 9:14-27; Luke 9:37-43; 13:19; 17:6

PART IV
STORIES JESUS TOLD

Jesus was known for his parables. He used them in his sermons and teachings.

A parable is a brief simple story that is told to teach a spiritual truth. It might be true, or invented, to illustrate a point.

In the Old Testament, when Nathan wanted to illustrate an injustice to David, he told the parable of the rich man and his poor neighbor. And the entire book of Jonah is a parable.

Some of the "parables" of Jesus are so short they are just figures of speech. "The kingdom of heaven is like a mustard seed," is a simile. So is "the kingdom of heaven is like a bit of leaven hidden in a great amount of dough." In the Gospel of John we find such metaphors as, "I am the good shepherd," "I am the door," and "I am the way."

Jesus, knowing the typically Oriental delight in exaggeration, often uses deliberate hyperbole. "If you have faith as a grain of mustard seed, you can move a mountain." His listeners would know this was meant figuratively and not literally; it was a picturesque way of saying that real faith could move a tremendous obstacle, although it might loom as big and as immovable as a mountain.

Jesus said, "It is easier for a camel to go through the

eye of a needle, than for a rich man to enter into heaven." Some Westerners have tried to explain that the small gate within the large city gate was called "the eye of the needle." If a traveler arrived late at night, after the great city gates were closed, he had to unload his camel and force the loudly protesting beast to scoot through the small door on his knees. Thus, it was difficult for a camel "to go through the eye of the needle." Yet it is obvious that Jesus spoke figuratively. He meant it would be difficult for a man who was obsessed and burdened with his riches to go to heaven, just as it would be difficult for a real

camel to go through a real needle. Jesus often exaggerated in order to emphasize his point.

The rich variety of the parables reflects the interest Jesus had in all the details of life around him: the sowing of a field, the building of a house, the search for a lost sheep or a lost coin, the relationships of family members or neighbors. With such down-to-earth illustrations, no wonder the crowds followed Jesus and listened with spellbound attention to his message.

The parables of Jesus are so beloved, and so often alluded to, it is worthwhile to gather them into a special section of their own.

The Lost Sheep, and the Lost Coin

The Jewish Pharisees tended to be smug and self-righteous. They considered themselves too pious to associate with ordinary people. The Pharisees continually criticized Jesus, not only because he associated with farmers and fishermen, tradesmen and housewives, but also because he ate and talked with sinners.

So Jesus told his critics this parable:

What if you were a shepherd, with a hundred sheep in your flock? You would find hills green with grass where your sheep could safely graze. You would lead the sheep to a spring of good water, or to the still waters of a stream so that they might quench their thirst.

Zealously you would watch over them, to protect them from the danger of bears, lions, or other wild animals. In the evening you would herd them into a cave or a rock-bordered sheepfold so they might sleep in safety through the night.

But what if one sheep wandered from the green pasture, out into the barren wilderness? Would you not leave the ninety-nine to graze safely on the hills and go look for the one that went astray? And when you had searched and searched, and at last had found the lost sheep, would you not rejoice? You would pick up the lamb and carry it back to be with the rest of the flock.

And when you returned home, you would call together your friends and neighbors, saying, "Rejoice with me! For I have found my sheep which was lost."

Then Jesus finished his parable, saying,

"Even so, I tell you, there will be more joy in heaven over one sinner who repents than over ninety-nine righteous persons who need no repentance."

At another time Jesus told the Pharisees a second parable, which made the same point:

Suppose a woman has only ten silver coins. If she loses just one coin somewhere in the dark corners of her house, does she ignore it? No, she lights a lamp and sweeps the house. She seeks diligently until at last she finds it.

And when she has found it, she calls together her friends and neighbors. "Come, rejoice with me. For I have found the coin which I lost."

Then Jesus turned to the Pharisees, "In the same way, I tell you, there is joy before the angels of God over one sinner who repents."

Matthew 18:12-14; Luke 15:1-10

The Loving Father

No matter what Jesus said, the Pharisees did not see how common and ordinary people, including all the sinners, could be spiritual brothers. How could God as a heavenly Father yearn for the return of a sinner?

"God should reject a sinner," the Pharisees fumed. They thought God should care far more for a Pharisee, a man who stayed close to the house of God, said his prayers often, and obeyed the smallest of God's laws.

So Jesus told a parable about the prodigal son.

There was a man who had two sons. He raised them with love, giving them all that they needed. There were servants to help with the work. There was plenty of meat and good food. The sons were supplied with good clothing and sandals for their feet.

But the younger son grew restless. He yearned to travel to far places, to see more of life, and perhaps to amass a fortune. "Father," he said one day, "may I have now my share of the property?"

It was customary to leave the younger son a bequest of one-third of the property, but the father agreed to turn it over to the young man immediately. So the younger son sold his share of the farm. He took the money and went on a journey into a far country.

At first the young man enjoyed his money and his newfound freedom. He began a prodigal life, squandering his fortune with extravagant, wasteful spending. He was sucked into a round of riotous and immoral pleasures. After a while his strength, his health, and his money were gone.

About this time a great famine arose in that country. The young man was ill, penniless, and friendless.

Because of the famine, good farm jobs were not to be had. The only work he could find was as a swineherd. How much lower could he fall? To a Jew, the pig was not only the most despised animal he could think of, but was religiously "unclean." Anyone who worked with pigs became contaminated, ritually unfit to worship or associate with other Jews. Pigs were eaten in the religious rites of some heathen cults.

The young man was starving so he took the job. He had nothing to eat, except for the long dark brown pods of the carob tree, the food of the swine.

He sat by the grunting, squealing pigs as they fed. Barefoot and in rags, he chewed

270 hungrily on the slightly sweet, seeded pods. The dry fibrous husks filled his mouth and his belly, but he did not feel nourished.

After almost swooning from hunger and illness, he finally came to himself and saw his position clearly.

"How many of my father's servants have bread enough—and to spare," he said aloud to himself. "But I am perishing here with hunger."

He threw down a dry, half-eaten pod. "I will arise and go to my father. I will say to him, 'Father, I have sinned against heaven and before you.' I will tell him I am no longer worthy to be called his son. I will beg to be taken as a hired servant."

The young man arose and began the long journey home, barefoot and in his thin rags.

Back at home the father had been running the farm with the help of the obedient elder brother. But the father missed his younger son. Every day he gazed down the road his son had taken when he left home.

One day the father looked down the road, and his heart quickened. How could a man not know his own son, even at a distance? Even if the young man limped, and walked with stooped shoulders—

The father had compassion. He ran and embraced his younger son, and kissed him.

The young man hung his head. "Father, I have sinned against heaven and before you. I am no longer worthy to be your son." Tears welled from his eyes. "All I ask is that you treat me as one of your servants."

But the father said to his servants, who had come running up, "Quick, bring the best robe, the one we keep for our most distinguished guest! Put it on him, and bring shoes for his feet. Then put a ring on his finger to show that he is still my son and heir. Go fetch the fatted calf and kill it. Let us eat and be merry. For this my son was dead, and is alive for us again. He was lost, and is found."

And so a happy celebration began.

The elder brother was in a far field, working. As he walked toward the house, he heard the laughter, the sounds of music and dancing.

He called a servant. "What is going on?"

"Your brother has come home," the servant told him, a big grin on his face. "Your father has killed the fatted calf."

The elder brother's face hardened. He looked toward the house with contempt. He refused to welcome his brother or join in the festivities.

The father saw his elder son stomping away from the house. He ran out and caught up with his son. "Come in, come in," he begged. "Help me welcome your brother home."

The elder son's face was bitter. "These many years I have stayed close to you and served you," he spat out. "I never disobeyed any command you gave. Yet you never offered to me a great roast of meat so I might have a celebration with my friends! Now this worthless son of yours has returned after wasting your money in loose living. And for him you kill the fatted calf!"

The father's face was sad. He put his arm around this self-righteous older son who had never suffered and been deprived in a lonely, sinful life.

"Son," he said, "you are always with me. And all that is mine is yours. But it was fitting to make merry and be glad, for your brother was dead to us, and now is alive. He was lost, and is found. So come, make your peace with him, and help me to welcome him home."

Luke 15:11-32

The Good Samaritan

A religious lawyer called a "scribe" decided to put Jesus to the test. "Teacher," he asked, "what shall I do to inherit eternal life?"

Jesus countered by asking, "What is written in the law?"

So the scribe, learned in all these matters, quoted quickly from Deuteronomy and from Leviticus. "You shall love the Lord your God with all your heart, and with all your soul, and with all your strength," * he chanted. "And your neighbor as yourself." **

"You have answered right," Jesus said. "Do this, and you will live."

But the lawyer wanted to justify his question so he asked arrogantly, "Who is my neighbor?"

Jesus looked at the man. Surely this expert in the law knew that in the eyes of God every man on earth was one's neighbor. The real question was: How should I treat my neighbor?

So Jesus replied by telling this parable:

A Jewish man was making the trip from Jerusalem to Jericho, down through the lonely and forbidding hills of the Judean wilderness. Robbers, hiding behind boulders, leaped out upon the traveler. They beat him until he was bloody and unconscious, stripped him of his clothes, and took his money belt. They went off, leaving him half dead.

A priest from the Jerusalem temple was going down that same Jericho road, perched comfortably on a large donkey. He saw the wounded man but did not want to get involved. Robbers might still lurk nearby. He prodded his beast to trot faster, passing by on the other side of the road.

Then a Levite came hiking up the steep road on his way back to Jerusalem and his duties as an assistant in the great temple. When he saw the injured man, unconscious and groaning, he quickly looked away and walked faster. He did not want to get involved either.

Then a Samaritan came along. People from his country, located midway between Judea and Galilee, were scorned by the Jews. The Samaritan had trading that took him to Jerusalem and to Jericho; he went about his business, not expecting to be treated well.

Riding upon his donkey, the Samaritan came around a bend in the bleak Jericho road. When he saw the bleeding victim of the robbers, his compassion was greater than religious or national differences. Immediately, the traveler leaped to the ground. He did not stop to consider his own safety, but he ran across the road to the injured man. He used a popular remedy for wounds, pouring oil and wine from small flasks he had with him. He bound up the wounds with strips torn from his own clothing.

The injured man opened his eyes, slowly coming back to consciousness. The Samaritan reassured him.

"You will be all right, now," this new friend said soothingly. "I will take care of you."

The Samaritan set the victim on his own beast. He brought him to an inn farther down the lonely road.

The next day the man's wounds were healing; but he needed more rest before he would be able to travel. The Samaritan gave two silver coins to the innkeeper, saying, "Take good care of him. Whatever more you spend, I will repay you on my return trip."

After telling this parable, Jesus looked

the questioning lawyer in the eye. "Which of these three," he asked, "do you think was a real neighbor to the man who fell among the robbers?"

The lawyer had to admit, "The one who showed mercy upon him."

Jesus said to this scribe, "Go and do likewise."

Luke 10:25-37

*Deuteronomy 6:4.

** Leviticus 19:18.

On Being Prepared

As Jesus was preaching one day he spoke of the patriarch Noah.

Noah was a man who listened for the word of God. He learned what God wanted him to do and did exactly that. And so he was prepared for the flood. In his ark Noah was able to ride out the storm and begin a new life.

"So it is today," Jesus told the people. "Great things are in store for those who are prepared. And those who are unprepared will be left behind, or left out. For the unprepared there will be only wailing and gnashing of teeth."

Then Jesus illustrated his point again with another parable.

A certain village was making excited preparations for a wedding. The festivities would begin when the bridegroom and his friends came for the bride at her father's house. There would be a gala trip to the home of the groom.

Maids-in-waiting would gather round the bride as she counted the hours and the minutes until her bridegroom came. The chosen girls were all aflutter. One of their duties was to carry lamps to light the way to the new house. They would be among the most honored guests at the wedding feast to follow. There would be gay music and dancing all night long.

On the eve of this particular wedding ten maidens waited with the bride, their festive lamps burning.

Five of the maids-in-waiting were wise, and five were foolish. When the foolish brought along their lamps, they brought no extra oil with them. The wise maidens were prepared for an emergency; they had brought extra flasks of oil.

The bridegroom was delayed. As the light from the ten clay saucer-lamps flickered lower and lower the girls' heads began to nod. The hour grew late; they all dozed and slept.

At midnight, there was a cry. "Look, the bridegroom comes! Let's all go out to meet him!"

The five unprepared girls pleaded with the wise ones, "Give us some of your oil! Our lamps are going out."

But the wise girls replied, "There is surely not enough to keep our lamps going, and yours as well. Go see if you can find an oil merchant, and waken him. Then you can buy a new supply for your lamps."

The foolish, unprepared girls rushed out into the darkened town, hoping to find the house of a seller of oil. While they were gone, the bridegroom arrived and claimed his bride.

The other five girls escorted the bride and groom to the marriage feast. Their freshly filled lamps lighted the way.

When the wedding party reached the groom's house, they hurried in, laughing and dancing. Because the hour was late, the heavy outside door was shut and bolted.

Later, the five foolish girls came running up, out of breath. They found the door shut and locked. They pounded and called, but nobody opened the door to them. They missed out on the joy and the feasting, the most exciting event in their village.

Jesus concluded his parable by saying, "Watch, therefore, and always be prepared. Find out what God wants you to do with your life, and do it. Then you will be prepared for whatever will come in God's kingdom."

Matthew 24:37 through 25:13

The Two Houses

Jesus knew about the work of a carpenter, how important it was to choose a good building site for a house and to lay a good foundation. As Jesus and his disciples hiked through the countryside, they often saw examples of bad house planning and poor construction.

During the winter and spring rainstorms most riverbeds ran full with water from the hills and mountains. During the long hot summer these twisting ravines, or wadies, were dry.

"See what that foolish man does?" one of the disciples said to Jesus, pointing. Down below, they saw a man building his house on the smooth dry sand of a broad wadi. He was using dried clay bricks.

"He thinks he'll do it the easy way, choosing that site," another disciple said scornfully. "It is summer now, but does he not know that a flood will likely roar down through here this winter? His house will surely wash away."

"Now that is how a wise man builds," Jesus said, pointing to another builder working on top of the rocky cliffs nearby. The man had laboriously hewn limestone blocks to be the basis of a sturdy foundation. He had dug down to the solid rock, and built his foundation upon that. It was far harder to build in that way; but he would have a house that could withstand the storms.

Not long after that, Jesus was preaching to a great crowd. Again he taught them from the store of his wisdom about the will of God for them.

All the people had bowed down to Jesus, giving him extravagant praise. Yet they showed no readiness to put his teachings into practice in their lives.

Jesus grew exasperated. "Why do you call me 'Lord, Lord,' and never do what I tell you?"

Then he told them a parable.

"Everyone who comes to me and hears my words and does them, I will show you what he is like," Jesus said. "He is like a wise man who built his house upon a rock. He dug deep and laid the foundation upon that rock. The rain fell, and the floods came, and the winds blew and beat upon that house. But the house did not fall, because it had been founded upon the rock."

Jesus looked searchingly at the listening people.

"But everyone who hears these words of mine and does not do them will be like a foolish man who built his house upon the sand. The rain fell, and the floods came, and the winds beat against that house. It fell, and great was the fall of it."

And the people knew what Jesus meant.

Matthew 7:24-27; Luke 6:47-49

The Buried Talent

It was evening. The Twelve and some of the other followers of Jesus were sitting by the shore of Lake Galilee. They had just dined on fresh fish caught in the lake by several of the men.

Now their campfire had flickered down to glowing red coals. A full moon rose above the hills on the eastern side of the lake, sending a dancing reflection across the dark waters. The men sat around their campfire listening to the steady lap of waves on the pebbled beach.

After a bit they began to discuss some of the stories Jesus had told in recent days about what the kingdom of God was like.

Jesus had been standing to one side looking out across the darkened lake meditating. Now he walked over to the group. And they were quiet for they knew he had heard their discussion.

Jesus smiled at them. "I will tell you a parable," he said. And this was the story he told:

A man who had a great deal of money was going on a journey. Before he left, he called his servants together.

"I am going on a trip and will not be seeing you for a long time," the master told them. "During the time that I am gone I am going to entrust my property to you."

Five talents* of gold, a small fortune in itself, went into the keeping of his first and most capable servant. The second was given two talents, and the third received one talent, each according to his ability.

Then the master went away.

As soon as they were on their own, each of the servants began to make plans for the use of the money in his care.

The first servant surveyed the business world and invested the five talents wisely. After a time he had doubled this amount.

The second servant did the same with his two talents, making a good profit on his investment.

But the third servant was lazy, overly cautious, and afraid to take a chance on using what he had. So he took his one piece of money, buried it in the ground, and left it there.

The day of reckoning came. The master returned and summoned his servants. "I am ready to settle my accounts with you. Come tell me what you have done with the talents."

The first servant brought his master ten talents. "Master, you gave into my keeping five talents," he said. "Here, I have added five more to them."

"Well done, good and faithful servant," his master said. "You have been faithful over a little so I will set you over much more. Share in my joy!"

The second servant brought forward four talents. "Master, you entrusted me with two talents. Here are the two, plus two more that I have gained for you."

"Well done, good and faithful servant," his master said. "You have been faithful over a little, I will set you over much more. Share in my joy!"

The third servant came forward saying, "Master, I knew you to be a hard man." He sighed, excusing himself. "I was afraid, and I hid the talent you gave me. I buried it in the ground." The man thrust the one talent forward, defiantly. "Here! Now you have what is yours."

His master bellowed at him, "You lazy and unprofitable servant! You knew that I was a hard taskmaster. You ought to have invested my money and put it to work. When I returned I should have received what was my own—with interest."

The master took the third servant's money and handed it to his first servant. "I will take the talent from you," he told the third man, "and give it to him who now has the ten."

When Jesus had finished telling this parable, he looked expressively at his disciples.

"For to every one who makes good use of what he has, more will be given; and he will have abundance," Jesus said. "But from him who has little and makes no use of it, all will be taken away."

The disciples remained quiet, thinking.

Matthew 25:14-30

* In the lands of Babylonia and Palestine a talent was a brick of gold or silver of a certain weight. One gold talent was worth roughly a thousand dollars of our money.

We have come to use the word talent to mean an aptitude as in "he has a talent for music," or for leadership, or "for making things with his hands." Our present use of the word was derived from this parable told by Jesus.

The Sower and the Seed

One cool winter day Jesus left his house in Capernaum. He walked through the crowded streets of the city and then westward along the shore of Lake Galilee, where a great plain pushed back the crowding hills.

Then Jesus sat down on a rock by the shore to meditate. The wind whipped his robe about him and ruffled the waters of the lake.

A farmer was sowing his field on the plain. He had a leather bag full of seed wheat slung over his shoulder. As he walked along, he reached for handfuls of grain and scattered it over the land.

Jesus turned and watched the farmer thoughtfully. Some of the farmer's seed fell on a hard-trodden path that crossed the field. Already birds were swooping down to peck at the fat wheat kernels that lay there. Some fell on the pasture where the soil looked rich and thick. But Jesus knew that in rocky Palestine even pastures like this had places where the rock lay just under the surface. The wheat could sprout and start to grow in such rock-underlaid places; but it could not send down roots deep enough to sustain life. And so it would wilt and die in the warmth of the spring.

People coming from nearby towns saw Jesus sitting by the shore. They crowded around him; more came, and they gave him no peace. Then the disciples arrived. They tried to push the people back, for by now Jesus was being crowded out, ankle deep, into the cold, shallow water.

Jesus, with great good nature, just climbed into a small rowboat that was anchored there by the shore. He sat in the boat and talked to the people who were gathered on the pebbled beach.

Two of the disciples waded out to the boat where Jesus sat. "Master, shall we tell these people to go back to their homes?" one asked. "Can't they see you are tired and hungry and cold?"

The other disciple looked discouraged. "A lot of these people just came out of idle curiosity. What you say will never take root in them! And see those Pharisees standing there? They are a hard and hostile lot. They'll reject anything you say." He threw out his hands. "So what's the use?"

Jesus smiled. He raised his voice and told the people a parable.

"Listen," he said. "A sower went out to sow his seed. As he sowed, some fell along the path, and was trodden underfoot, and the birds of the air devoured it. Some seeds fell on rocky ground, where the soil was thin. The seeds sprouted and grew. But when the sun waxed hot, the shoots withered away, because they had shallow roots.

"Some seeds fell among thorns. And the thorns grew up and choked them. So they yielded no grain. But other seeds fell on good soil and brought forth grain, growing up and increasing and yielding thirty times as much as the original seed! And some even yielded sixtyfold, and some a hundred-fold."

Matthew 13:1-9; Mark 4:1-9; Luke 8:4-8

The Rich Fool

After Jesus had preached to a crowd, a nervous, anxious little man pushed his way to the front.

"Teacher," he said. "You are learned in the laws, and a just man. "He glowered at a man nearby, and his arms waved belligerently. "There is my brother who keeps the lion's share of our father's inheritance for himself." The little man's face flamed with his anger. "Speak to my brother and tell him to divide the inheritance with me."

Jesus looked quizzically at the worried, anxious, and angry petitioner. He said, "Man, who made me a judge or a divider over you?"

Then Jesus said to the whole crowd, "Take heed, and beware of greed and covetousness. A man's life does not consist in the abundance of his possessions. His life is more than the owning of things."

Then Jesus told the people this parable:

There was a rich man who owned much fertile land. His crops were plentiful. The man thought to himself, "What shall I do? For I don't have the space to store all my crops."

He decided to pull down his old barns and build larger ones. "I will store all my grain and the fruits of my land," he thought smugly. "And I will say to my soul, Soul, you have ample goods laid up for many years, take your ease, eat, drink, and be merry."

But God said, "Fool! This very night your soul shall be required of you. You will die. But to whom will go all your possessions?"

And Jesus finished his parable by saying, "Any person is like this rich fool if he concentrates on gaining wealth for himself but does not have a rich relationship with God."

Then Jesus said to the distressed little man and to all the people, "Therefore I tell you, do not be excessively anxious about your life, what you shall eat or what you shall wear. For life is more than food, and the body more than clothing."

Jesus lifted a sunbrowned arm. He gestured toward the nearby field. "Look at the birds of the air," Jesus said. "They do not sow grain nor do they reap nor gather into barns. Yet your Heavenly Father feeds them. Aren't you worth more than the birds? And which of you by worrying can add extra hours to your life?"

Now Jesus waved toward the hills which soon would be alive with spring blossoms. "Consider the lilies of the field, how they grow. They neither toil nor spin. Yet I tell you, even Solomon in all his glory was not as splendidly clothed as one of these. If God dresses up the plants of the field like this, how much more will he clothe you? When you are so anxious and fretful about material things you show how little faith you have."

"What then can I do?" the little man asked despairingly.

"Just do not be anxious," Jesus said. "Seek God's kingdom, and you will find that the things you really need will come to you as well. It does no good to fret and worry, for each day has enough trouble of its own. Tomorrow will take care of itself when it comes."

Matthew 6:19-21, 24-33; Luke 12:13-34

The Rich Man and the Beggar

It was almost dusk. The afternoon light was fading fast; the birds were twittering as they settled down for the night. Jesus and his disciples walked to Capernaum on the path that followed the lakeshore. Two men of the town came with them, wanting to ask more questions of Jesus.

"I have listened to what you said about riches," a thin older man said as he jogged along beside Jesus. His clothes were ragged, his face was wistful. "But it is hard not to wish for more. Life can be so hard—"

Jesus looked at the man. "I will tell you a parable," he said. So he told this story.

There was a rich man who was clothed in expensive purple robes and undergarments of fine linen. Every day he feasted sumptuously at his table. There was also a poor man, thin and covered with sores, who lay at the gate of the rich man's house to beg for food.

The scraps and leftovers from the rich man's table were taken out by the servants and given to the unhappy beggar. And the dogs of the city came and licked his sores.

The poor man, who was called Lazarus, died, and was carried to the side of Abraham by the angels. The rich man also died, and was buried. He ended up in hell in torment. When he lifted up his eyes he saw Abraham in the distance and Lazarus beside him. The rich man called out, "Father Abraham, have mercy on me! Send Lazarus to dip the end of his finger in water and cool my tongue, for I am in anguish in this flame."

But Abraham said, "Son, remember that you in your lifetime received your good things. And Lazarus did not. But now he is comforted here, while you are in anguish. And besides, none may cross between here and there."

The agonized rich man cried out to Abraham, "Then please send Lazarus to my father's house, for I have five brothers. Let him warn them so that they will not make the same mistakes I did."

But Abraham replied, "Your five brothers already have the warnings of Moses and all the prophets. Let your brothers listen to their words in our scriptures."

The rich man protested, "Those words will not change them. But if someone goes to them from the dead, I know they will believe and repent."

Abraham answered, "If they will not listen to Moses and the prophets, neither will they be convinced if someone should rise from the dead."

When Jesus had finished his parable, he took off his warm coat and wrapped it around the shivering man. Then he turned and strode out again in the direction of Capernaum.

"He who has ears," he said, "let him hear."

Luke 16:19-31

The Unmerciful Servant

One day Peter came to Jesus. "Lord," he complained, exasperation in his voice, "how often must I forgive my brother when he sins against me? As many as seven times?"

Jesus said to him, "I do not say to you seven times, but seventy times seven." Then he told them this parable about forgiveness:

There was a king who wanted to settle accounts with the officers under him, his servants in the kingdom. When the reckoning was under way, one of his officers was brought into the court. He was a man trusted by the king to whom the king had loaned a great amount of money. It was figured that the man owed his king ten thousand talents—a fortune—and the officer was unable to pay it back.

The king said, "This debt has gone on long enough. Sell this servant into slavery. Sell also his wife and children and all that he has so that payment can be made."

The unhappy debtor fell on his knees and begged the king to have pity on him. "Lord," the officer implored, "have patience with me, and I will repay you everything."

The king felt sorry for the man. "Release this man," he ordered. "I forgive him his debt. Let it be cancelled."

Now this same officer, as he went out rejoicing, came on a friend who owed him a hundred denarii,* a fairly small sum. He grabbed the man by the throat and demanded, "Where is the money you owe me? Pay me at once!"

His friend fell down and begged for mercy. "Have patience with me," he pleaded, "and I will repay you."

But the officer of the king was unmerciful. He refused to wait. He called the palace guards and had his friend thrown into prison, to stay there until the debt was paid.

When the king heard what had happened he summoned the officer. Sternly he said to him, "You wicked servant! I forgave you your entire debt, because you begged me. Why didn't you have mercy on your fellow servant?"

So the king treated the unmerciful officer as that officer had treated his debtor. The king threw him into prison.

Jesus ended this parable by saying, "My heavenly Father will do the same to everyone who does not forgive his brother from his heart. As you judge others, so will you be judged."

Matthew 18:21-35

* The denarius (plural, denarii) was the most current silver coin in the first century of the Christian era. It was worth about forty cents, the daily wage of a common laborer, and was used to pay the Roman troops. It was also used to pay taxes to Rome.

PART V
FRIENDS AND
ENEMIES OF JESUS

Jesus had many friends: his twelve special disciples as well as many others who became disciples and followed him from place to place. There were fishermen who dropped their nets and went to hear Jesus talk whenever he was nearby, boatbuilders who left the docks on Lake Galilee, farmers who laid aside their plows, weavers who temporarily abandoned their looms, dyers who wiped their stained hands and left their dye vats.

The Bible also names many women among Jesus' followers. Some, like Joanna and Susanna, provided money and food from their own means. Mary, from the lakeshore town of Magdala, was healed by Jesus; gratefully she sold her home and became the devoted follower known as Mary (the) Magdalene. Others, like the sisters Mary and Martha, did not travel with Jesus and his company, but welcomed him with food and hospitality whenever he came south to Judea.

The power and force of Jesus' personality were so great he kept making new friends: men, women, children—even a Roman centurion and a woman of the hostile district of Samaria.

Not everyone was convinced by Jesus and his message, however. At one time Herod Antipas was reported to have made inquiries about the man from Nazareth who was teaching throughout the territory he ruled; yet apparently "the fox" did not interfere with Jesus.

The most bitter enemies of Jesus were the Pharisees, the scribes, and the Sadducees.

The religious-political party of the Pharisees has been mentioned before. The Pharisees' rigid adherence to every jot and tittle of their Jewish religious laws, their satisfaction with their own righteousness, made them very quick and harsh in judging others.

The scribes were a class of professional scholars who made a lifelong study of Jewish religious laws. These scribes might be called religious lawyers, or doctors of the law. They were only laymen, but they felt as qualified to elaborate on the laws as any priest.

The Sadducee party, on the other hand, was made up largely of priests and the wealthy landed nobility. They were on rather friendly terms with the Roman authorities and adjusted well to the Roman laws. However, they did not believe the laws of Moses should be added to or tampered with. They felt that Jesus was a heretic and a radical, someone who should be eliminated.

The Pharisees particularly resented Jesus' way of interpreting the old laws so that they had a new and deeper meaning. The Law, as Jesus preached it, required more of a man than certain gestures and rituals; the whole person had to be involved.

The Pharisees were also bitter because of Jesus' exposure of the shallowness and emptiness of their "righteousness" in which they took such pride.

Jesus and the Children

An argument arose among the disciples as to which one of them was the greatest, and would be the most honored in heaven. They came to Jesus to settle the matter. "Which of us is the greatest in the kingdom of heaven?" they asked, looking around at each other.

Jesus called a child to him. He put the child in the midst of them. "Unless you turn and become like children," he told them, "you will never enter the kingdom of heaven."

The men looked at the innocent face of the child and stopped their quarreling.

"Whoever keeps such simplicity and such humility," Jesus said, "is the greatest in the kingdom of heaven."

Jesus took the child into his arms. "Whoever cares for one such child for my sake cares for me," he said softly. "But whoever causes one of these little ones to stumble or wander into wrongdoing, it would be better for him to have a great millstone fastened around his neck and to be drowned in the depth of the sea."

The child looked up into the strong, kind face of the Teacher and smiled at him. "See that you do not despise one of these little ones," Jesus murmured as he stroked the child's hair. "In heaven their angels always behold the face of my Father."

On another day Jesus had been out in the countryside all day preaching to the people. A mother pressed forward with her little girl, asking the Master to touch her and bless her.

Another mother rushed up and begged, "O Master, bless my son." Her face was flushed with her earnestness.

Other women with infants in their arms pushed forward. All were clamoring, "Master, please touch my baby."

The disciples saw that Jesus was tired; they decided to protect him from requests which seemed trivial. "Please go away," they said, trying to push the mothers back. "Can't you see the Master is weary? Don't annoy him with your children."

When Jesus saw this he rebuked the disciples. "Let the children come to me, and do not stand in their way." Tired as he was, Jesus smiled at the mothers and the children. "Let the children come, for to such belongs the kingdom of God. Honestly I say to all of you, whoever does not receive the kingdom of God like a child shall not enter it."

Jesus sat down on a low, flat rock and welcomed the youngsters. He laid his hand on the heads of the older boys and girls and blessed them. And he took the babies into his arms and blessed them too.

The children loved this strange young teacher with the loving face, the warm smile, and the strangely sad eyes.

Matthew 18:1-6, 10; 19:13-15; Mark 10:13-15; Luke 9:46-48; 18:15-17

Mary and Martha

Two of Jesus' friends were sisters, Mary and Martha. They owned a small house in Bethany, a village only a brief walk east from Jerusalem.

Martha was a bustling, commonsense woman. She was very efficient herself, and was inclined to be impatient with anyone who was not. The younger sister Mary was quite different. Mary was quieter and more thoughtful. The two women had a younger brother, Lazarus, whom they both adored.

Now it happened that when spring came Jesus interrupted his preaching mission in Galilee. He began a trip south to Jerusalem so he might celebrate the annual Passover feast there. On his way Jesus passed through Bethany.

Word came to Martha that their friend was entering the village. Martha came running out of her house, down the narrow, twisting dirt street. She arrived at the edge of the village, panting for breath, just as Jesus came striding up.

"Jesus of Nazareth," she gasped out. "Peace be with you. And welcome to Bethany."

"Peace," said Jesus. "May God bless you and keep you." His eyes were warm with friendship as he looked at Martha.

"Lord, will you come to my house," Martha begged him, "and eat a bit of food that my sister Mary will help me prepare?"

Jesus went with Martha through the narrow crooked streets of Bethany to her house.

284 Martha put a cushion on the bench by the doorway, under the big-leaved fig tree that was already laden with new figs. A small pomegranate tree nearby was aflame with bright orange red blossoms.

"It is nicer outside, today, than in the house," Martha said. "Sit here and rest, Lord." She pushed some stray strands of dark hair from her flushed face. "I shall find Mary, and we shall soon have a bite for you to eat."

Martha hurried into the house. She found Mary strumming on a small harp-like zither and singing to a small bird that was making a nest just outside the high back window.

"Mary, Mary," Martha said excitedly. "The teacher from Galilee is here. Hurry now, go up to the roof and get the chicken from the coop. Clean it and dress it so we can make a savory stew."

Mary dropped her beloved stringed instrument with a clatter. "Jesus of Nazareth —here?" Mary rushed straightway to the small front courtyard, to welcome her friend.

Several times Martha called Mary. But Mary sat at the feet of Jesus, listening to all he had to say. Her face was rapt with wonder and joy. Her complete attention was upon Jesus. Martha decided that it was as if the girl had no ears at all.

The older sister sighed. She trudged up the outside stone steps to the flat rooftop and took the protesting hen from its wooden cage. She killed it in the ritual manner so that all of its lifeblood was drained from the meat. Deftly she stripped off the feathers and cleaned the bird. Soon the yeasty fragrance of baking bread was added to the delicious aroma of bubbling stew.

Martha kept bustling about, filling one small wooden bowl with dried raisins, another with salty olives. Next she made some honey cakes, her specialty.

At last her patience gave out. Warm and flushed from all her cooking and rushing about, she stalked into the courtyard and interrupted Jesus as he spoke to Mary.

"Lord," she complained, "do you not care that my sister has left me to serve alone?" She pushed back damp, stray strands of hair and sighed with exasperation. "She does not listen when I call her. Tell her to help me."

Jesus looked at the younger sister, who sat with eager, upturned face at his feet, learning all that she could while the opportunity was here. Then he turned his steady and compassionate gaze upon his distracted hostess, who was weary from her elaborate dinner preparations.

"Martha, Martha," Jesus chided her. "You are anxious and troubled about too many unimportant things. Mary has chosen to do what is most important. And it shall not be taken away from her."

So Martha learned from Jesus that day.

But Mary, who listened all afternoon at Jesus' feet, learned much more.

Luke 8:1-3; 10:38-42

The Believing Soldier

One day as Jesus entered Capernaum, a Roman centurion* rushed up to him.

The burly soldier confronted the teacher, his face grave with concern. "Lord, would you hear the petition of a centurion?" he burst out. He knew that being Roman he was hated and despised by most Jews of Palestine.

Jesus studied the man, and nodded.

"I have a servant who is as dear to me as a son," the Roman officer said. "At this moment he is at my home, lying on his bed with a fever, sick, and at the point of death." Tears welled up in the soldier's eyes. His voice seemed to catch in his throat. "He just lies there paralyzed, unable to move. And in such pain!"

One of the elders of the Jewish synagogue in Capernaum stepped forward. "This man is worthy to have you do this for him," the Jewish elder said. "He has grown to love the people of our nation. Did you know he contributed money, and the labor of his men, to build our synagogue?"

The centurion implored Jesus. "I have heard of the great miracles you have performed. I have heard how you heal the lame, the sick, the blind, and the paralyzed—"

"I will come to your house with you," Jesus interrupted. "I will come and heal him. Take me there now."

But the centurion protested. "Lord, I am not worthy to have you come under my roof." He gazed trustingly into the face of Jesus. "Only say the word, and my servant will be healed."

A murmur went around the crowd that had gathered by the city gate. It was indeed a miracle when this wonderful healer laid his hands upon a sick person and healed him. But how could any such thing be done at a distance?

"You see," the officer explained, "I am a man under authority with soldiers under me. I can tell them to go or to come, and they do as I say. So I understand how it is that you can give a command, and it will happen as you say."

Jesus listened, and his eyes, as he watched the centurion, were warm with appreciation. He turned toward the crowd.

"Honestly, I tell you," Jesus began, "not even among my fellow Jews have I found such faith." He turned to the waiting centurion. "Go, it shall be done for you as you have believed."

And when the Roman officer returned to his house he found his beloved slave already recovering, feeling healthy and well.

Matthew 8:5-13; Luke 7:1-10; John 4:46-53

* A centurion was the commanding officer of a hundred foot soldiers in the Roman army. They were actually the backbone of the army.

Jesus Among the Samaritans

The hostile country of the Samaritans lay between the Jewish areas of Galilee and Judea. When a Jew wished to travel from Galilee to Jerusalem, or to some other place in Judea, he often perferred to avoid Samaria, to ford the Jordan and go southward along the river road, recrossing the river opposite Jericho. He usually found this roundabout route safer and more pleasant.

Since the end of the Exile—when the Jews had rejected the help of the Samaritans in rebuilding Jerusalem and they, in turn, had done all they could to interfere—there had been unhealed and bitter feelings between them.

The Jews felt their bloodlines and religion had remained unpolluted by foreigners during their Babylonian captivity. To them the Samaritans were renegade, halfbreed Israelites.

The Samaritans, on the other hand, considered themselves the true remnant of Israel. They had build their own rival temple on Mount Gerizim.

On one of the trips Jesus made to Jerusalem he decided to take the direct route south, preaching as he went. His disciples accompanied him. On the first afternoon within Samaritan territory Jesus sent messengers ahead to make overnight arrangements at the nearest village.

But when the Samaritans heard that this Jewish teacher was on his way to Jerusalem, they would have nothing to do with him. They refused to deal with his messengers. The disciples returned to Jesus, their faces red with embarrassment and anger. When they told their story, the rest of the disciples began to mutter.

James and John shook their fists at the inhospitable village. "Lord, do you want us to call down fire from heaven to burn them up?" they shouted.

One cried, "Curse them like Elijah did."

But Jesus rebuked his angry followers. His face was troubled. Why were his disciples so out of tune with his feelings? "No," he said steadily, "we shall go farther and look for another village."

As they hiked along the road several new people joined the group. One young man said with great eagerness, "Master, I have heard you teach in Galilee. I want to follow you wherever you go."

Jesus warned the young man that life as a follower would not be easy. "Foxes have holes, and birds of the air have nests," he said. "But as for me, I have nowhere to lay my head."

Looking over the crowd Jesus recognized a big burly fellow as one who had listened eagerly to his teachings in several places. "Follow me," he invited.

But the man said, "First let me go and bury my father, who has just died."

"Leave those who are spiritually dead to bury the dead," Jesus commanded him. "But as for you, go now, and announce the kingdom of God."

Another said earnestly, "I will follow you, too, Lord." He hesitated, "But first I must say good-bye to those at my home."

Jesus said, "Take a lesson from the farmer, plowing with his ox. See how he grips the handles of his plow. He is intent, always looking ahead, careful lest a rock or clod should bounce the plow from his hands." He nodded gravely. "No one who puts his hand to the plow and then keeps looking back is fit for the kingdom of God."

When Jesus reached Jerusalem he launched into a preaching mission. His disciples baptized all those who repented and wanted to live the new life. It was then that the jealous Pharisees began to make trouble for Jesus. "Does not this Galilean make even more disciples than John did?" they asked. They persecuted and harassed Jesus and made it difficult for him to continue his preaching there.

Jesus left Jerusalem to return to Galilee. He decided to journey again through Samaria.

Hot, dusty, and tired, Jesus and his disciples arrived at the long-revered Jacob's well at noontime. This ancient well was in the center of Samaritan country with the twin heights of Mount Gerizim and Mount Ebal looming nearby.

Jesus sat near the well to rest while the disciples hiked the short distance to the town of Sychar to buy food. To the east, great fields of wheat and barley were ripening to a pale gold—almost white in the glare of the noonday sun. Here Jacob once gave a field to his beloved son Joseph. Jesus shaded his eyes with his hand as he looked about him.

A woman of Samaria came to the wide, deep well to draw water. Jesus looked at her with a gracious smile and asked for a drink of water. His accent betrayed that he was a Galilean Jew.

The Samaritan woman set her leather bucket with its long rope beside the well. She stood there, hands on her ample hips, a quizzical look on her face. "How is it that you, a Jew, ask a drink of me, a woman of Samaria?" Her eyes were faintly contemptuous. "For the Jews have no dealings with Samaritans."

Jesus ignored her taunt and said gently, "If only you knew the gift God can give, and who it is asking you for a drink, you would have asked for and received living water."

The woman laughed. "Sir, you have nothing to draw with, and the well is deep. Where would you get that living water?" She flounced her skirts and sat beside the well. "Are you greater than our ancestor Jacob, who gave us this well, and drank from it himself?"

"Everyone who drinks from this well will thirst again," Jesus said quietly. "But whoever drinks of the water that I shall give him will never thirst. For it will become a spring within him that wells up to eternal life."

The woman dropped her flippant manner. "Sir, give me this water, that I may not thirst, nor come here to draw again and again."

"Go," Jesus said. "Call your husband, and bring him here."

The woman hung her head. "I have no husband."

"You are right in saying that you have no husband," Jesus nodded. "Although you have had five husbands, the one you have now is not your husband. So you told me the truth."

The woman's jaw dropped. She looked with awe at this man. Finally she gulped and stammered, "Sir, I see that you are a prophet." She thought a moment, then pointed at Mount Gerizim. "Our ancestors have worshiped on this very mountain, and it is holy to us. But you Jews say that only in Jerusalem should men worship."

"Believe me," Jesus said to her, "the hour is coming when neither on this mountain nor in Jerusalem will you worship the heavenly Father. The time approaches—and now is—when the true worshipers will worship the Father in spirit and in truth. God is looking for such worshipers. Do you not know that God is 'Spirit'? Therefore those who worship him must do so in spirit and truth."

The woman said earnestly, "Everyone knows that the Messiah is coming, the Christ. When this anointed one comes, he will show us all things."

Jesus said, "I who speak to you am he."

"You are the Christ?" The woman clapped her hands to her mouth. Her eyes widened.

Just then the disciples came back from the town laden with the food they had bought in the marketplace.

"Look at the Master," one whispered. They were surprised to find Jesus talking with this Samaritan woman, but they did not question him.

The woman was so excited by the disclosure of Jesus she left her water bucket by the well and ran off to the town. There she cried out to the people, "Come, see a man who told me all I ever did. Can this be our Messiah?"

The Samaritans began to stream out of town, all coming to hear Jesus.

Meanwhile, the disciples laid out the bread and fruit, saying, "Teacher, come and eat."

But Jesus answered, "I have food to eat which you do not know about."

The disciples stared in bewilderment at each other. "Has anyone brought him food?"

Jesus said, "My food is to do the will of him who sent me, and to accomplish his work." He smiled at his puzzled disciples. "Do you not say in the winter, 'There are yet four months, and then comes the harvest'?" He waved toward the ripening fields of grain. "I tell you, lift up your eyes, and see how the fields are already white for the harvest."

The disciples realized what Jesus meant. Seeds had been sown among the Samaritans. Soon there would be a harvest of souls.

And indeed, many Samaritans from that city did come to hear the message of the coming of God's kingdom. They believed in Jesus because of the woman's testimony, and also because they heard the good news for themselves.

Luke 9:51-62; John 4:4-43

The Dinner Guests

One sabbath day Jesus was invited to dine at the house of an important Pharisee. This man was a little more tolerant and open-minded than most. He wanted to listen to Jesus, and he invited many of his friends.

In the middle of the meal a man appeared. It was obvious that he suffered from an illness. The guests around the table were horrified as the man knelt at the feet of Jesus begging to be healed.

Was Jesus going to work a miracle on the sabbath? Any good Jew knew that no work should be done on the day of rest, according to the laws of Moses. Even this banquet had been prepared the day before and kept warm so that no cooking would have to be done on the holy day.

Jesus spoke to the lawyers and Pharisees present, saying, "Is it lawful to heal on the sabbath or not?"

They sat stiffly silent, glaring at this presumptuous teacher who was the son of a lowly carpenter.

Then Jesus healed the man and let him go on his way.

"Which of you," Jesus asked the group, "having a son or a donkey or an ox that has fallen into a well will not immediately pull him out—even if it is a sabbath day?"

They could not reply to this.

At the head of the table the host appeared smug as he presided over the banquet. He smiled at his many pious friends and tried to change the subject. He said to Jesus, "Surely God is pleased with me because I invited godly men to my banquet."

Jesus smiled. "When you give a dinner and invite your friends or your relatives or rich neighbors or people you want to impress, then you have already given yourself your own reward. For you have enjoyed their company, and you know they will invite you in return."

The host looked astounded.

"But if you invite the poor, the maimed, the lame, and the blind," Jesus said earnestly, "then you will be blessed. They cannot repay you. Instead you will be repaid by God on the resurrection day."

One of the elderly Pharisees gave a pious sigh. He had been only half-listening. "Happy will be the man who is summoned to feast at the table of God." And all the dinner guests solemnly nodded.

Jesus knew that these lawyers and Pharisees easily could picture themselves as honored guests at a banquet in the kingdom of heaven. But they did not realize how many excuses and rationalizations they had which kept them from understanding God's message. This was why the invitation "to come to God's table" would have to be taken to the common people. Perhaps it must be taken even further—even outside Jewish territory?

Jesus tried to explain. He used this parable:

A man once gave a great banquet and invited many guests. At the time of the banquet he sent his servant, according to the custom, to say to those who had been invited, "Come, for all is now ready."

But they all began to make excuses. The first said, "I have just contracted to purchase a field. I must go and see it and attend to my business. Please excuse me from coming."

Another said, "I'm sorry, but I have just bought five yoke of magnificent oxen. I must go and examine them."

Another excused himself by saying, "I have just married a wife. She is waiting for me now."

So everyone pleaded one excuse or another. The host was angry and disappointed. He told his servant, "Go then to the streets and alleys of this city. Bring in the poor and the maimed, the blind and the lame."

The servant went out and did as he was commanded. Poor and disabled people quietly filed into the banquet hall. Yet there were still empty places. The master said to his servant, "Go beyond the city, out into the countryside. Find people by the highways and the hedges, and urge them to come in. I want my house to be filled."

Jesus looked at the feasting lawyers and Pharisees who kept rejecting him and the invitation he brought from his Father in heaven.

"I tell you," Jesus said as his parable was finished, "all who are invited to join with me and who turn down the invitation shall not taste of the banquet that I am serving."

But still the Pharisees did not seem to understand.

Luke 14:1-24

The Pharisees

The Pharisees were known as the most self-righteous of Jews, studying every detail of the laws given in the first five books of the Scriptures. They argued endlessly about each minute shade of meaning. Then they set themselves to keep this complicated version of the law down to the last crossing of a *t* or dotting of an *i*. Much of their time was spent going through rituals to cleanse and purify themselves after they had come in contact with anything impure and sinful.

The Pharisees carefully avoided any contact with murderers, thieves, and others who led immoral lives. They also shunned Romans, Greeks, Syrians, and other Gentiles, who were impure because they did not keep the Jewish laws.

Even the common ordinary Jews were looked upon as sinners, though they might go regularly to the temple and synagogue. They had not been carefully instructed in keeping the fine points of the religious law. So the Pharisees were convinced that none of the common people could be sinless and ritualistically pure.

The Pharisees had a special hatred for the tax collectors, officials not popular with any Jews. The Romans allowed men to pay for the right to become "publicans," and

292 then they were permitted to collect the tolls and custom duties in each area of Palestine. These chief tax collectors, some of them Jewish, then hired other Jews to be local collectors among their own people, gouging them for as much as they could in order to make a profit. Some of the money they handled stuck to their fingers, and many became quite wealthy.

"Publicans and tax collectors?" cried the people scornfully. "Might as well call them robbers."

"Tax collectors!" the Pharisees ranted. "To come near a tax collector would make us as unclean as to come near swine or a dead body." They were appalled when they discovered that one of the twelve apostles was a tax collector. They were even more outraged when they heard that Jesus sat down and ate with common people, and some who were known sinners.

The Pharisees took some of the disciples of Jesus to one side and upbraided them. "Why does your teacher eat with tax collectors and sinners? One should keep away from evil people."

There was a long pause, then one of the Pharisees said in an injured tone, "I invited this Jesus to my own house because I believed he might be a true prophet. An immoral woman of the town came in from the streets. She washed his feet with her tears and wiped them with her long hair, and then she anointed him with expensive ointment from a flask she carried. He didn't even push her away. He told that woman that her sins were forgiven because of her faith."

"Now there is nothing wrong with receiving the repentant sinner who has been properly purified," one of the Pharisees said. "But to seek out their company as your teacher does—"

The fuming disciples reported to Jesus all that the Pharisees said.

Jesus shrugged. "Those who are well have no need of a physician. It is those who are sick who need the doctor." He smiled, and his face was filled with a radiance that was beautiful to behold. "I did not come to call the righteous. I came to call the sinners to repent."

Matthew 9:10-13; 18:12; Luke 7:36-50; 18:9-14; 19:1-10; John 7:47-49

More Enemies

Often when Jesus healed the sick and suffering throughout the countryside, he would say, "Go your way, your sins are forgiven." This infuriated the Pharisees and the scribes.

"Just who are you?" they demanded of Jesus. "Do you think you are God himself that you can forgive sins?"

They stroked their beards and put their heads together. "This is blasphemy," they muttered to each other. The more the people listened to Jesus, the more jealous they became. "How can he forgive sins? It is an affront to God, a great indignity. What shall we do about this man?"

One sabbath day Jesus and his hungry disciples were walking through a field of grain. The disciples plucked some ears of wheat. They rubbed the ripe kernels between their hands, shucking the husks. They blew away this chaff and munched the chewy seeds which remained.

"Aha," said some passing Pharisees. "Do you see what these men are doing on the sabbath? They are threshing and winnowing grain."

They spoke severely to Jesus, "Why do you allow your men to do what is not lawful to do on the sabbath?"

Jesus answered, "Have you never read what David did when he and his men were hungry? David entered the temple at Nob and took the bread off the altar, which is forbidden to all except priests. Yet David ate of it, and also gave it to the men who were fleeing with him."

"That's right," one of the disciples said, spitting out a husk. "If David, the beloved of God, could do this for his hungry men, then is it not just as fitting for the disciples of Jesus to pluck a few ears of raw grain?"

Jesus said firmly, "The sabbath was made for man, not man for the sabbath."

On another sabbath Jesus entered the synagogue and taught. A man came up to him holding up a maimed right hand. "I am a stonemason, and I must make my living with my hands," he told Jesus. "Please restore me to health so that I don't have to beg in shame."

Some scribes and Pharisees stood nearby watching with narrowed eyes to see what Jesus would do. Jesus smiled at the man with the crippled hand. "Come and stand here," he said.

He turned his head and looked directly at his critics. "I ask you, is it lawful on the sabbath to do good—or to do harm?" Then Jesus looked at the man trembling before him. "Stretch out your hand." And he healed the hand.

The doctors of law and the Pharisees were filled with fury. Afterward, they discussed with one another what might be done to get rid of Jesus.

"He is an unlearned fellow. Surely we

can trip him up on the law. Let's ask him questions to test him."

They followed Jesus outside the city. They joined a crowd which had gathered to hear him.

"Teacher," one of the scribes said with mock deference, "which is the great commandment in the law?"

Jesus said, "You shall love the Lord your God with all your heart, and with all your soul, and with all your mind. This is the great and first commandment." He held up his hand. "And a second is like it: You shall love your neighbor as yourself. Upon these two commandments depend all the law and the prophets."

This silenced the scribes and the Pharisees. According to the laws which they treasured, Jesus had given the correct answer. Yet they still resented and hated this unorthodox country teacher.

"We have seen you and your disciples eat without the proper washing of hands," one of them accused Jesus. "Why do you not live according to all the traditions set down by our elders?"

Jesus threw up his hands and looked heavenward, his face a study in exasperation. "It is just as Isaiah said. The people try to honor God with their lips, when their hearts are far from him. They do not follow God's laws—but they follow slavishly the traditions of men."

One of the older Pharisees muttered, as if he had not heard the answer of Jesus, "They don't wash properly, that man and his followers. Not even their cups and their plates."

Jesus said with disgust, "You blind so-called guides! You try to strain out a gnat, and you swallow a camel. It is the inner purity of a man which is important to God. Don't you know that? Take care of inner purity first, and the proper outer cleanliness will follow."

Jesus raised his voice now, to reach the whole crowd. "Woe to you, scribes, Pharisees—hypocrites! For if your 'righteousness' is all on the outside, then you are like whitewashed tombs, full of dead bones and uncleanness within."

"If inner cleanness is so important," one white-bearded scribe, his face an angry red, accused, "then why do you sometimes eat unclean food—food that has not been purified by a proper religious ritual?"

Jesus sighed. The scribes and the Pharisees were so intent on the exact letter of the law they constantly missed the spirit behind it.

"Hear me, all of you, and understand," Jesus said. "Nothing outside a man, by going in to him, can defile him. Food enters a man's stomach, and is passed on through. It does not enter his heart! But what is harbored and nurtured in a man's heart defiles him: evil, hateful, envious, murderous thoughts. Puffed-up, self-centered pride! Lies and deception! Malicious gossip and cutting words! Such evil things, when they are encouraged within, can defile a man."

The common people heard the teachings of Jesus with a surge of hope, for they could understand this man. Jesus offered a way to cut through the fence of legal restrictions which the scribes and the Pharisees had built around the worship of God.

But the scribes and the Pharisees, and the elected rulers of the synagogues, were threatened by the way Jesus undermined their authority.

And they continued to plot how they might get rid of him.

Matthew 15:1-20; 22:34-40; 23; Mark 2:1-12; 2:23 through 3:6; 7:1-15; Luke 5:17-24; 6:1-12; 11:37-54; 14:1-6

PART VI
THE LAST DAYS
OF JESUS

The last days of Jesus were the best-remembered part of his life. Most of our information about them comes from the Gospel of Mark. It is probable that Mark has arranged this last period in six days—from the Palm Sunday entry into Jerusalem to the crucifixion on Friday—for the purpose of ritual readings among the early Christians.

Matthew's retelling is similar to Mark's with some elaborations. Luke adds some additional incidents he has heard. John follows the other three gospel writers more closely in recording these last events than he does in his version of the earlier part of the life of Jesus.

The story begins with the journey to Jerusalem. The entry of Jesus into the city was greeted with such enthusiasm that it is described as a triumph. Our Palm Sunday memorializes this event. It is not likely that the entire city turned out to do him homage. However, he did command enough interest and generate enough excitement to disturb the Jewish rulers and priests of the city.

The judicial body of the Jews was composed of priests, scribes, and elders (including both Pharisees and Sadducees); it was known as the Council, or the Sanhedrin. It was this powerful group that most feared the power and growing popularity of Jesus. It was the Sanhedrin that decided that Jesus must be killed, denouncing him as a seditious traitor and troublemaker to the Roman administrator, Pontius Pilate.

The Romans commonly used crucifixion as a punishment for rebels, slaves, and criminals from the lowest classes. It was a cruel and slow way to die. Death came from thirst, dehydration, exhaustion, and shock; or from the heart at last ceasing to work when blood stagnated in the veins.

There is little doubt that the record of the death of Jesus is reliable, for his followers would never have invented or imagined such a humiliating and tragic end for their beloved leader.

After Jesus had been crucified his followers were afraid for their own lives, and beside themselves in their grief. Everything seemed lost and hopeless. Like sheep without a shepherd they huddled together in forlorn little groups, or just wandered about Jerusalem.

Then came the unexpected news that Jesus had risen from the dead. The Gospels often spoke of portents, signs, and miracles, and now there were reports of the greatest miracle of all.

If it had not been for this realization that Jesus was alive, the disciples might have been too disheartened to go on. The impact of the Resurrection united the followers into a courageous band—led by the personal conviction of Peter and many others.

The Gospels report several appearances of the risen Lord. The details of these accounts differ, but all the Gospels and the whole New Testament breathe a sure joy that his resurrection was a reality. Jesus had died for the sins of all, and now he lived, and had sent his Spirit to lead his followers in the Way. The cross soon became a symbol of victory over the grave.

Jesus Leaves Galilee

As Jesus preached among the villages and lakeside cities of Galilee, many listened eagerly, but many more remained indifferent.

Galileans lived close to nature. They looked up and predicted rain when clouds blew in from the Mediterranean. When the wind swirled in from the deserts to the southwest, they scanned the sky and nodded sagely. "We're in for a couple of hot, dry days."

Jesus occasionally chided the people because they were so alert to weather signs, yet so unseeing and indifferent to the signs of more important things about to happen. Were the people blind to the economic inequalities? There was violence, injustice, suffering, and restlessness in their land. These were the signs of the times. How did the people interpret them?

The kingdom of God was breaking in upon history. God had sent his Son to bring a message to the people, but the people did not listen. Or they only half-listened, then went on living in the same ways.

Jesus rebuked these indifferent people. "When you see a cloud rising in the west, you say at once, 'A shower is coming.' And so it happens. And when you see the south wind blowing, you say, 'There will be scorching heat.' And it happens. You know how to interpret the appearance of earth and sky. Why can't you figure out what's going on in your time?"

After Jesus finished talking to a crowd, a well-dressed man came to him. His feet were shod in expensive tooled-leather sandals, and jeweled rings flashed on his fingers.

He said, "Teacher, what must I do to gain everlasting life?"

"You know the commandments," Jesus said. "Do not kill. Do not steal. Do not tell lies that will bring harm to another. Don't cheat others out of what belongs to them. Honor your father and mother. And love your neighbor as yourself."

"I have obeyed these commandments from my youth," the man said.

As he looked upon the wealthy man, so personable and eager, Jesus felt love and compassion for him. "If you would be perfect, there is one more thing. Go, sell what you have, and give it to the poor so you will have treasure in heaven. Then come and follow me."

The man looked sad. He turned and walked away. A friend of the rich man lingered a moment. "But his mammon." He used the Aramaic word for money and property. "My friend has such great riches —he owns much land. How can he bear to give it away?"

"No one can serve two masters," Jesus said. "Either he will hate the one and love the other, or he will be devoted to one and despise the other. You simply cannot serve both God and mammon."

The man hurried off to join his friend.

Jesus turned to his disciples "It is easier for a camel to go through the eye of a needle than for a rich man to enter the kingdom of God."

His disciples showed their surprise. Were there not many passages in the scriptures, telling how pious men were rewarded with wealth? Was not Job a rich man?

Jesus smiled at them, though his eyes watched sorrowfully the two men who were leaving the crowd. "Seek first the kingdom of God," he said. "After that, these minor things may be yours as well."

Those crowding around murmured among themselves. "Sell your extra possessions so you can help the poor," Jesus advised the people. "Provide yourselves with purses that do not grow old, with a treasure in the heavens that cannot fail." Then he summed it up so his listeners might memorize and remember it: "Do not lay up for yourselves treasures on earth, where moth and rust eat them away, where thieves break in and steal. But lay up for yourselves treasures in heaven, where neither moth nor rust can consume them, where thieves cannot break in and steal. For where your treasure is, there your heart will be also."

For three years he had hiked around the lakeshores and hills of Galilee, preaching and healing. Now Jesus knew the time had come for his Galilean ministry to end. So he set his face toward Jerusalem.

The disciples were amazed that Jesus would want to go there again. "Your life will be in danger," they argued. "The wealthy Sadducees fear you, and the priests in the temple hate you. The scribes and the Pharisees plot to have you killed."

"The Son of man must suffer many things," Jesus said. "Do you know what the prophets have written? Just so it will happen. For the Son of man will be delivered to the Gentiles to be mocked and shamefully treated and spit upon. They will scourge him and kill him, and on the third day he will rise."

The Twelve did not understand exactly what their Master meant. But they made preparations to accompany him on his trip to Jerusalem.

Matthew 6:19-21, 24, 33; 20:17-19; Mark 10:17-25, 32-34; Luke 12:54-56; 13:22-33; 18:31-33

Journey to Jerusalem

Jesus and his disciples forded the Jordan River just south of Lake Galilee. They joined the stream of travelers who were hiking south on the Jordan valley road on their way to Jerusalem for Passover. To their left towered the steep, gullied cliffs; to the right, usually hidden in the jungle growth of its banks, flowed the muddy river.

At last Jesus and his party came to a shallow place, opposite the city of Jericho. They went back across the Jordan, wading hip-deep in the swirling cold water. A hot breeze whipped sand into their eyes and filled their hair and clothes with grit.

Jesus decided to stop over in this ancient city, with its date-palm trees and gardens. He and his men needed rest, food, and fresh water. As they drew near the city gate, a curious crowd began to gather. Even this far from Galilee, Jesus was known and recognized by many.

A blind man was sitting by the roadside, begging. Someone tapped his shoulder and told him, "Bartimaeus, there goes Jesus of Nazareth into our city."

The afflicted man had heard of this miracle-worker. He cried out, "Jesus, Son of David, have mercy!"

The people around tried to hush him. "Quiet, you blind beggar." One of his friends whispered, "Don't bother the Teacher, Bartimaeus."

But Bartimaeus continued to shout until Jesus stopped and spoke to him. Bartimaeus sprang up and stumbled toward the voice, throwing off his cloak in his excitement.

Jesus spoke gently. "What do you want me to do for you?"

The blind man begged, "Master, let me receive my sight."

Jesus said with sympathy, "Go your way. Your faith has made you well."

Bartimaeus' hands flew to his face, and he rubbed his eyes. Then he began to look around, blinking. "I can see, I can see!" he shouted with joy. Tears began to roll down his sunburned face. He picked up his fallen cloak and joined those who were following Jesus.

By the time they entered Jericho the number gathering to see Jesus had grown. People lined the main street. Among them there was a wealthy tax collector named Zacchaeus. He had heard of the Nazarene preacher; like everyone else, he was hoping to catch a glimpse of him. But Zacchaeus was a small man, and he could not see over the heads of the crowd. Even on tiptoe, he could see nothing. Then he had an idea. He sprinted ahead and climbed into the branches of a sycamore-fig tree beside the road.

When Jesus came to the tree, he looked up at the small man, so richly dressed, wide-eyed, perched up there on an overhanging branch.

A local resident whispered, "Look at Zacchaeus. What a way for a wealthy man to behave!"

Someone else hissed, "I hope he falls and

breaks his perfidious neck. A tax collector!" And he spat with contempt upon the ground.

"Zacchaeus," Jesus called up to the little man. "Make haste and come down. I will stay at your home tonight."

The tax collector scrambled down from his perch, forgetting all dignity. He was so excited he couldn't talk. He just nodded and smiled, brushing the bark and leaves from his hair and his beard and his clothes. Then he snapped his fingers for his servant. "Hurry home," he whispered, "and see that a worthy banquet is prepared." And he led the way to his house.

The people in the crowd murmured, "The Galilean is the guest of a sinner."

Jesus turned and told them, "Today salvation comes to this man's house. Is he not as much a descendant of Abraham as any of you? You must know that I have come to seek and save the lost."

When Jesus and the disciples left Jericho they trudged upward along the lonely Jericho road through the wild and barren hills. And a great crowd of admirers from the city followed. When they reached the Mount of Olives they could look westward across the Kidron valley and see the holy city like a crown on its hill. Sunshine danced on the white limestone of the holy temple and its gold decorations. Nearby was the village of Bethany where Mary and Martha lived.

Jesus spoke to two of the disciples, "Go into that village, and you will find a young donkey. Untie it and bring it back. If anyone asks why, say, 'The Lord has need of it and will send it back.' "

The two disciples did as Jesus had told them, and the donkey, which had never been ridden, was brought to Jesus. Jesus mounted and began to ride down the valley road to Jerusalem.

An excited crowd heard of Jesus' coming. They streamed out from the city gates and welcomed him exuberantly. Many spread their cloaks across the road; others rushed to cut leafy branches to lay across the path.

"Hosanna,"* the welcomers cried. "Blessed be he who comes in the name of the Lord! Hosanna to the Son of David! Hosanna in the highest!"

As he rode into the great city more people lined the road and sang out, "Blessed be the King who comes in the name of the Lord! Glory in the highest!"

Despite this welcome the disciples were uneasy.

"The Master's life is in danger here," one whispered.

The disciple nearest him looked about warily. "The Pharisees and the Sadducees swarm in this city. They will set a trap for him."

A third disciple muttered, "I don't even trust these cheering people. Are they not like a donkey between two bundles of hay? The mood of a crowd is undependable."

"You are right," the first disciple agreed. "A crowd can be dangerously fickle."

And so their joy was mixed with foreboding.

Matthew 20:29 through 21:11; Mark 10:46 through 11:11; Luke 18:35 through 19:10, 28-40

* The literal meaning of hosanna (Hebrew *hoshiana*) is "oh save!" See Psalm 118:25-26.

Still More Enemies

Jesus and his disciples threaded their way through the crowded streets of Jerusalem. They headed for the temple area, the highest part of the city.

Jesus and all who followed him went through the temple gate into the outermost courtyard, which is called the Court of the Gentiles. Here the Teacher halted abruptly, staring at a scene of noisy confusion. His lips tightened; there was unaccustomed anger in his eyes.

Money-changers had set up a banking business at little tables. Incoming worshipers were negotiating with the bankers for correct change, to pay their temple taxes, or to provide themselves with small change for the poor box. Jews who had come from a distance were exchanging their foreign currency for local money. All seemed to be talking at the tops of their voices.

Also, many wanted to offer animal sacrifices at the altar in the innermost courtyard. If they were Jews with a farm, or from too great a distance to bring something, they expected to buy a sacrificial animal locally. Merchants had adopted these holy precincts as a marketplace to meet the need. Birds and animals were being sold here as if the temple grounds were a stockyard or a bazaar!

"Pigeons and doves," one man called in a shrill voice as he leaned against a stack of small slatted cages. "Buy the finest pigeons and doves right here."

In one corner a barricade hemmed in a flock of baaing sheep and goats. There were stalls of oxen champing on hay. Manure splotched the white limestone paving. Flies swarmed and buzzed. A stench from the animals hung in the air.

"Is it not written," Jesus cried out, his voice ringing above the babble, "that God's house is a house of prayer? But you have made it a den of thieves." He whirled upon a nearby pigeon-seller. "Take these things away!"

The seller stared at this rude stranger. "What are you saying?" He pointed at some of the sheep and young bulls. "Don't you know we rent this space from the high priests?"

"Out, out!" Jesus ordered. "You shall not make my Father's house a marketplace for trade."

Some sellers replied with insolent taunts. Others ignored this rustic stranger. All continued to hawk their wares.

Jesus boiled over with righteous indignation. He pushed over the stacked cages. Pigeons and doves fluttered everywhere. Then the fearless Nazarene smashed open the stalls and knocked away the barricades for the goats and the sheep. He knotted a bundle of cords to use as a drover's whip, herding all the animals out of the holy courtyard and into the street. And the sellers were driven out with them.

In the excitement the money tables were overturned. The bankers were down on their hands and knees, scrambling among the running hooves and feet for their coins, uttering terrible curses against Jesus.

The many followers of Jesus watched, openmouthed with wonder and admiration.

Then Jesus went into the middle courtyard, called the Court of the Women. Only Jews could go this far into the temple precincts; only the most pious and purified Jewish men could go into the innermost court of all, where the priests sacrificed animals upon the outdoor altar.

Here, in the middle courtyard, Jesus saw some wealthy Jews. Dressed in ornate Roman togas, they came sweeping up to the treasury box with purses in hand. Each

let his fingers rummage through his fat money bag, then dropped a gold coin—after looking around to make sure that such a generous gift had been noticed.

A poor widow hurried in. Her clothing was neatly patched and mended. She reached into her small purse and took out all that was there, two copper pennies. Quickly she dropped in her mite, and then she left.

Jesus turned to his followers. "Notice that the rich men contributed out of their abundance. But the widow, in spite of her poverty, put in all the money she had!"

In their own quarters the priests had hastily convened a council meeting with some of the leading Pharisees and Sadducees.

An irate priest shook his fist and shouted, "Did any of you see how that Nazarene imposter took the authority in his own hands? He drove out those merchants who were paying us rent money!"

"Does this country fellow dare to say how we should conduct our temple?" another priest asked, bristling with indignation.

A Pharisee held up an arm for attention. "We are dealing with a dangerous man. The people acclaim him as a miracle-worker, but perhaps he is some kind of a magician. The latest rumor is about a man of Bethany named Lazarus. He died and lay in his burial cave for four days. Then this Nazarene supposedly raised him from the dead."

"It's true," another Pharisee cried out. "Several of the villagers of Bethany came and told me about it. They were witnesses. Lazarus has two sisters in that village, of impeccable character, Mary and Martha. I know them myself. This Nazarene performs many signs, many miracles."

One of the Sadducees said soberly, "Let's be practical. If we let him go on, everyone will believe in him. There will be a revolt.

302 The Romans will come and destroy us and our temple."

The high priest agreed. "It is better that this one man should die, than to have a whole nation destroyed."

So from that day, they plotted how Jesus might be put to death. But they were afraid to arrest him, yet. He was so popular with the common people.

Matthew 21:12-13; Mark 11:15-19; Luke 19:45-46; 21:1-4; John 2:13-16; 11:1-54

The Last Teachings of Jesus

When evening came, Jesus went out of the crowded city and crossed the valley to the east. He had made arrangements to sleep in the village of Bethany.

The next day, and every day, regardless of the danger, he went back to teach in the temple courtyard.

"I am the light of the world," Jesus said to the people. "He who follows me will not walk in darkness, but will have the light of life."

The Pharisees in the crowd taunted him. "You are your own witness. Therefore, your testimony is not true!"

Jesus answered, "Yes, I bear witness to myself, and the Father who sent me bears witness to me." He smiled. "Your law says that the testimony of two witnesses is true."

One of the Pharisees sneered. "And where is this father of yours?"

"You know neither me nor my Father!" Jesus rebuked him. "If you knew me, you would know my Father also."

"I think he has a demon," someone in the crowd shouted. "That's way he says such crazy things."

"I do not have a demon." Jesus turned his clear, level gaze upon the heckler. Then he spoke to the Jews who were his followers. "If you continue to believe in my word and to live as I tell you, you are all truly my disciples. You will know the truth, and the truth will make you free."

Many of the people hung on the words of Jesus. But some of the crowd were so bothered by the things he said they picked up stones to throw at him. Jesus was forced to hide himself. Then he and his disciples slipped out another way, and went quietly out of the city.

At the end of each day Jesus liked to stop in the grove of olive trees on the side of the Mount of Olives, eastward across the valley from Jerusalem. Here in the quiet and darkness of the grove, which was called the Garden of Gethsemane, he could meditate and pray. After that, he might continue on to the nearby village of Bethany for the night.

This night, stars were twinkling out overhead. After Jesus had finished his prayer, he and his little company were trudging up the hillside road. They were tired after the long day at the temple.

"Let's return to Galilee," one of the disciples pleaded with Jesus. "You were almost killed when they threw rocks at you today. The Pharisees and scribes look on you as their enemy. The priests and the Sadducees plot to do away with you."

"Even the crowds are fickle, and cannot be trusted," another disciple warned.

But the disciples could not dissuade their Master from his plans. "There is a cup of suffering which I must drink," Jesus murmured. "My work here in Jerusalem is not yet finished—"

So Jesus returned again to teach in the courtyards and the porticoes surrounding the main temple structure. The blind, the lame, and the ailing came to him there, and he healed them.

One day, after teaching and healing in the outermost Court of the Gentiles, Jesus went walking through the colonnaded section known as Solomon's Porch. A delegation of chief priests, scribes, and elders cornered him.

"By what authority are you doing all these things?" they demanded officiously. "Who gave you any authority?"

Jesus looked thoughtful. "First I will ask you a question," he said softly. "Answer me, and I will tell you by what authority I do these things. Was the baptism of John from heaven or from man?"

The foes of Jesus drew back, gathering in a circle. They debated in whispers.

"If we say John's baptism was from heaven then he will ask us why we did not believe him," said one.

"But how can we say that John's baptism was only from men?"

"That's right," said another, "we can't do that either. We would rile up the anger of the people because they believe John the Baptist was a true prophet."

At last the priests, the learned and scholarly scribes, and the powerful elders faced Jesus again. "We don't know," they said. "We cannot say."

And Jesus said, "Then neither will I tell you by what authority I do these things."

The frustrated enemies of Jesus were tempted to arrest this presumptuous Galilean on the spot, but they feared the people.

"No, we'll have to trap him into saying something that will convict him," a cooler head counseled the others.

"Neither can we do it in broad daylight with crowds streaming in and out of the temple," another whispered to his confederates. "The crowds might riot against us. For many of them think this man, too, is a prophet."

So they left him and went their way grumbling.

The next day when Jesus was teaching before a crowd in the Court of the Gentiles a group of Pharisees came to spy on him. With them were some supporters of the Herod family. They hoped to get evidence to prove Jesus was a traitor so that they could deliver him up to the Roman authorities.

"Teacher," one of them spoke up, putting on a great pretense of sincerity. "You show no partiality because you honestly teach the way of God. Now tell us, is it lawful for us to pay tribute to Caesar or not?"

Jesus saw through their craftiness. Impassively he said, "Show me a coin."

The chief questioner dug into the purse he carried, in the girdle which belted his waist. He held out a denarius in his palm.

Jesus said, "Whose likeness is on the coin?"

The men in the group chorused, "Caesar's."

Jesus said to them, "Then give to Caesar the things that are Caesar's and to God the things that belong to God."

Then some Sadducees cornered Jesus. These conservatives, usually from noble priestly families, had gained lands and wealth under Greek and Roman overlords. They wanted to keep things as they were; but this man taught strange ideas which might rock the boat. They had little in common with the Pharisees, and fought them politically; but they were as anxious as were the Pharisees to be rid of this disturbing teacher.

The Sadducees rested all their beliefs in the first five books of Moses. Unlike their rivals, the Pharisees, they did not believe

304

in angels or spirits, or in the resurrection of the dead, since Moses did not mention them.

"Teacher," the Sadducees challenged, "Moses wrote for us that if a man's brother dies and leaves a wife, but no children, the man must marry the wife and raise up children for his brother."

Jesus nodded. All Jews knew this ancient Mosaic law.

"There were seven brothers," the Sadducean spokesman said, smoothing the folds of his Roman toga. "The first took a wife, and when he died, left no children. The second married her, and died, leaving no children. And the third likewise. In fact, not one of the seven had children by the first man's wife, who in turn became the wife of each of them. At last the woman died." He rolled his eyes solemnly heavenward; then he suddenly focused a steady gaze upon this Galilean in his coarse peasant robe. "You talk about the resurrection of the dead. If such a thing should come to pass, then whose wife will she be? For each of the seven had her as his wife."

Jesus said to them, "You show that you know neither the Scriptures nor the power of God! When persons rise from the dead, they do not marry. They become like the angels in heaven."

Someone began to object. Jesus raised his hand, to finish.

"As for the dead being raised," Jesus said, "have you not read in the book of Moses, in the passage about the burning bush, how God said, 'I am the God of Abraham and of Isaac and of Jacob?' The Heavenly Father is not the God of the dead, but of the living. You are quite wrong."

The Sadducees smoldered, hating this man all the more. For they could not answer him.

Scribes standing by listening reluctantly agreed among themselves that Jesus had correctly interpreted the Scriptures to the Sadducees. But still they resented Jesus because he had told the people, "Beware of the scribes, who like to go about in long robes and to be reverently addressed in the marketplaces and to have the best seats in the synagogues and the places of honor at feasts!"

Yet the common people continued to love Jesus. They liked his parables and stories. They hung on his words and were awed by the miracle of his healings.

In the outer temple court children cried out, "Hurray for the Son of David!"

Some chief priests and scribes immediately turned on Jesus, boiling with indignation. "Why do you let them call you that?" they accused Jesus. To call a man "the Son of David" was to imply that he was in the succession of Jewish kings, an "anointed one," a messiah. One irate priest shook Jesus by the shoulder. "Don't you hear what they are saying?"

An enigmatic half-smile crossed Jesus' face as he gently removed the clutching hand from his shoulder. "Have you never read, 'Out of the mouth of babes, God has brought perfect praise'?"

Leaving his critics, Jesus went out of the city to Bethany, stopping first on his way to meditate in the Garden of Gethsemane.

Matthew 21:14-17, 23-27; 22:15-33; Mark 11:27-33; 12:12, 38-39; 14:1; Luke 19: 47 through 20:8, 19-47; John 8:12-59

The Last Supper

It was now two days before the holy time of Passover and the Feast of Unleavened Bread.

Judas Iscariot had decided to betray Jesus. He separated from the other disciples and slipped quietly into the temple quarters to meet with the priests and the scribes.

Who knows what feelings were churning within him? Was he a singleminded patriotic Zealot, who had supported Jesus as a possible military leader, a new king of Israel? Had he been disillusioned by the peaceful, spiritual nature of Jesus? Or was this treasurer among the Twelve a Pharisee at heart? Whatever the reasoning, he no longer believed that Jesus was the Christ.

The priests and the scribes were startled when Judas was ushered into their private council chambers.

"You want to arrest this Jesus in a quiet place?" Judas asked. "If so, I can help."

Judas had arrived at a time when the enemies of Jesus were in the midst of their scheming. They were overjoyed at the offer of cooperation by Judas. They didn't question his motives. These Jewish elders and priests quickly agreed to pay Judas thirty silver coins for his help.

They began to plot with the traitor, planning how the thing might be done.

On the first feast day the disciples said to Jesus, "Where will you have us go and prepare for you to eat the Passover?"

Jesus appointed Peter and John to make the arrangements. "Go into the city," he told them. "A man carrying a jar of water will meet you. Follow him. When he goes into a house, ask the head of that household, 'Where is the guest room for the Teacher from Galilee, where he is to eat the Passover with his disciples?' Then he will show you a large upper room furnished and ready. There prepare for us."

Peter and John hiked down the road from Bethany past the feathery-leaved olive trees in the Garden of Gethsemane. They jogged down into the Kidron valley and up the road into Jerusalem. Everything happened as Jesus had told

them. Peter and John bought supplies in the marketplace, took them back to the upper room, and prepared for the Passover.

At twilight Jesus came with the other ten disciples and climbed the stairs to the upper room. It was comfortably furnished with rugs and cushions for sitting on the floor, and a low table for the food.

As they were eating at the table, Jesus sighed and looked around at these twelve who had been so close to him. "One of you who is eating with me will betray me."

The men looked startled. Even Judas. Each interrupted the others, asking, "Is it I?" One burst out incredulously, "Which one of us could do such a thing?"

They were clustered around the low table, dipping their bread in a common bowl of lamb broth. Jesus said, "It is one of the Twelve, who is dipping bread in the same dish with me." In a low voice he added, "It would have been better for that man if he had not been born."

Then Jesus took bread and said a blessing. He broke the bread into pieces, giving some to each. "Take, and eat this," he said. "This is my body."

Jesus reached his sunbrowned hand for a cup filled with wine. Again he said a blessing. He passed the cup around to each. "Drink of it," he said, "all of you. This is my blood of the new covenant, which is poured out for many. I tell you I shall not drink again of the fruit of the vine until that day when I drink it new with you in my Father's kingdom."

The disciples wondered what Jesus meant. Their minds puzzled over his strange words while they sang a hymn. When they were finished they made their way out of the crowded city to the Mount of Olives for their evening meditation.

Matthew 26:17-30; Mark 14:1-2, 10; Luke 22:1-23

A Tragic Night

The disciples followed Jesus down into the Kidron valley and up along the slopes of the Mount of Olives. It was quite dark now. The stars were out, and a full moon was rising behind the hill ahead of them.

"You will all fall away," Jesus murmured with deep sorrow in his voice. "When the shepherd is taken the sheep will be scattered."

Impetuously Peter burst out, "Even if everyone else falls away, I will not."

Jesus sighed. "This very night, Peter, before the morning cock crows, you will deny me three times."

"Even if I must die with you," the big fisherman vowed, "I will not deny you."

Ten more voices echoed the same thing. Nobody seemed to notice that there were only eleven disciples with Jesus now. They came to the grove of olive trees, the place called Gethsemane.

Jesus said to the rest, "Sit here while I pray."

Then Jesus took Peter and James and John, and they walked between the gnarled trunks of the trees, sinister shapes in the darkness. An owl hooted in the distance.

They stopped in a clearing. Jesus' face looked pale and anguished in the moonlight. He said to the three, "My heart is heavy with a deathly sorrow. Wait here, stay awake with me, and pray."

Jesus walked on about a stone's throw. In his agony he threw himself on the ground and prayed. "Father, all things are possible with you. Remove this bitter cup from me." Then his voice broke. He bowed his forehead against the cold earth. His voice came out in a hoarse whisper. "Yet, not what I will, but what you will."

He rose and picked his way back to the spot where Peter, James, and John had stretched out on the ground and were fast asleep.

Jesus touched Peter, "Are you asleep? Could you not keep awake one hour?" And the big fisherman roused and sat up, blinking.

"I know," Jesus said sympathetically. "The spirit is willing, but the flesh is weak. Now stay awake. Pray so you won't be tempted to fall asleep again."

Jesus went away, back into the pitch blackness of the night. Again he dropped to the ground and prayed, saying the same words.

When he returned he found Peter, James, and John still sleeping. Jesus called to them, and they awoke with shamed faces. They did not know what to say.

A third time Jesus plunged into the darkness of Gethsemane, alone and tortured in his prayer.

When he came back the third time, a commotion had begun over by the road. Torches flared, and the sound of tramping men came nearer.

Jesus threw out his hands, a tragic figure standing there in the eerie moonlight. "Are you still sleeping and taking your rest?" He nudged Peter. "Rise, let us be going. See, my betrayer is at hand."

Just then Judas Iscariot came striding up, carrying a flaming torch. With him was a band of soldiers from the temple guard and some officers from the chief priest and Pharisees, all with lanterns and torches.

Judas came up to Jesus at once, holding his lantern. "Master," he cried out and kissed Jesus on the cheek in the usual manner of a disciple greeting his teacher. The kiss was his signal to the arresting officer.

"Judas," Jesus rebuked him sorrowfully. "Would you betray me—with a kiss?"

A squad of soldiers marched up, laid

hands on Jesus, and seized him. Peter drew his sword and slashed at one of the soldiers, striking his ear so that the blood spurted and ran down the side of his face.

Jesus turned to Peter, "Put your sword in its sheath. I shall drink the cup which the Father has given me."

The band of soldiers quickly shackled the hands and arms of Jesus. Jesus looked at his bound hands, frowning. "Am I an armed bandit, that you come out with swords and clubs to capture me?" He looked scornfully at the officers from the temple. "Day after day I was with you teaching, and you did not seize me."

All the disciples turned and fled for their lives, and Jesus was alone with his enemies.

The armed guard led Jesus to the great house of the high priest, where the chief priests and the elders and scribes were already assembled. Expecting the capture of this troublesome Galilean, they were ready to hold a preliminary hearing immediately to find some basis for a death sentence.

As Jesus stood before them many witnesses came forward to present false evidence, but their testimony did not agree.

Then two men stood up. Their eyes darted furtively toward the prisoner. "We heard him say that he would destroy our temple," one testified. The other one added eagerly, "And then he claimed that in three days he would build another not made with hands." Yet when they were closely questioned, their testimony did not agree.

The high priest rose to his feet, his face cold and hard. He glared at the Nazarene who had so often challenged his authority. "Have you no answer to make?" he barked at Jesus.

Jesus, standing tall, made no answer.

Then the high priest tried to question Jesus about his disciples and his teaching. He tried to prove that this man from Galilee was the leader of a secret and traitorous movement trying to overthrow the government.

Jesus rejected that charge. "I have spoken openly to the world," he said. "I have always taught in synagogues and in the temple, where all Jews come together. Why do you keep asking me? Ask those who have heard me. Find out what I have told them."

When he said this, one of the officers standing by struck Jesus across the face with his hand. "Is that how you answer the high priest?" he shouted.

The high priest tried to trick Jesus with another question. "Are you the Messiah? Do you claim to be the Son of God?"

"You have said so," Jesus answered. He looked upward. "And you will see the Son of man seated at the right hand of God and coming on the clouds of heaven."

When he heard this, the high priest tore at his robes and cried, "Blasphemy. Have we any need for further witnesses now? You have all heard his blasphemy. What is your verdict?"

The angry-faced members of the Sanhedrin chorused, "He is guilty." All together the leading scribes and the Pharisees, the priests and the elders, were shouting, "Let him die."

Jesus was turned over to the guards to be held until daylight. In the guardroom the soldiers spat in their prisoner's face and hit him with their fists. Then they blindfolded Jesus and pushed him back and forth, mocking, "Prophesy, if you can. Go ahead, prophet. Who just struck you, great Messiah? Which of us is hitting you now?"

Peter had followed at a distance to the great house of the high priest. While Jesus was inside being questioned, Peter slipped in through the street gate. He mixed unobtrusively with the guards and slaves who milled about in the paved central courtyard.

It was a cold night. The shivering guards were huddled around an ironlegged brazier which glowed with embers of charcoal. Peter went in to stand near the door, hoping to overhear what went on inside. A serving maid came by and looked curiously at Peter's rustic peasant clothes. "You also were with Jesus the Galilean," she joshed, trying to be sociable. Some of the guards turned around, casually listening. Peter rebuked her coldly. "I do not know what you mean!" he said in an overly loud voice. He shuffled over to the fire and held out his hands to warm himself, trying to look detached and unconcerned. But his hands were trembling.

One of the guards gave him a half-friendly push. "Oh, come on, admit it! I can tell by the way you slur your words you are a Galilean."

Peter blustered wildly, "Man, I am not!"

"Any one could recognize that accent," another guard said. He came close and peered at the face of Peter. This soldier was a relative of the guard who had been injured by Peter in Gethsemane. Now his voice became hostile. "Didn't I see you in the garden with him?" Peter cursed, shouting, "I do not know the man!"

Immediately, the cock crowed. Peter remembered the saying of Jesus, "Before the cock crows, you will deny me three times." He went out into the night alone, and wept.

Matthew 26:30-75; Mark 14:26-72; Luke 22:31-65; John 18:1-27

The Death Sentence

When morning came the chief priests, the elders, and the scribes opened their official trial in the council chambers of the Sanhedrin. Jesus was brought before this official governing body of the Jews.

"If you are the Christ," the high priest said coldly, "tell us."

Jesus was battered and bruised from his treatment at the hands of the soldiers. Yet he stood tall and straight. "If I tell you, you will not believe. And if I ask you, you will not answer."

"Are you the Son of God, then?"

"You say that I am," Jesus answered.

"We have heard the blasphemy ourselves, from his own lips," the high priest shouted to the rest. "There is no need for further testimony."

So they put Jesus in chains and led him away to the palace of Pontius Pilate, the Roman administrator of Judea.

Jesus was turned over to the Roman officers of Pilate and taken inside the palace. His Jewish accusers would not enter the Roman ruler's house. If they did so, they would become unclean, and therefore unable to join in the Passover festivities.

Pilate went out and said to them, "You have brought me a prisoner. What accusation do you make against him?"

They answered piously, evading a specific reply, "Would we have handed this man over if he were not an evildoer?"

Pilate looked exasperated. It was impossible to preserve order among these contrary Jews. Who could understand them, their strange ways, and their complicated religious laws? "Take him yourselves," Pilate said, throwing out his hands, "and judge him by your own law." The accusing Jews looked heavenward. "According to our law he deserves to die, but we are not allowed to put anyone to death."

Pilate knew that they were forcing his hand. The Sanhedrin could pass the death sentence, but they could not actually carry out that sentence without the consent of Roman authorities.

Pilate wheeled and went in to confront his new prisoner. "Are you the King of the Jews?" he asked.

"Is this your own question," Jesus countered, "or did others put it in your mouth?"

Pilate looked pained. "Am I a Jew?" he asked. "Who but a Jew could ever understand you complicated people. Look," he said, "your own nation and the chief priests have handed you over to me. Now tell me, what have you done?"

Jesus answered quietly, "My kingship is not of this world." He opened his hands, palms up. "If it were, my followers would fight, so that I would not be handed over. No, my kingdom is not of the world."

Pilate's fingers fidgeted with the golden clasp on his toga. "So you are a king?"

Jesus answered, "You say that I am a king." He leaned forward, his deep-seeing eyes earnest. "I came into the world to bear witness to the truth. For this I was born. Anyone willing to hear the truth hears my voice."

Pilate's eyes wandered from the face of Jesus. He stared across the great ornate room. "And what is truth?" he murmured, half to himself. He turned, and went out to the Jewish leaders again. He told them brusquely, "I find no crime in him."

The Jewish leaders began to murmur among themselves. A great crowd of curious passersby had gathered, joining the angry priests, scribes, and elders.

Pilate stood on the steps of his palace and held up his hand to hush the crowd. "Wait! You Jews have a custom that I should release one man for you at the Passover, one prisoner to go free. Listen! Shall I release the King of the Jews, this man from Galilee? Or shall I release the prisoner Barabbas, who is a murderer?"

Pilate hoped the crowd would choose the mild-mannered Galilean. Barabbas was a known Zealot and insurrectionist, who had killed a man during a recent rebellion against Roman authority.

Also, while the Roman administrator was in the judgment hall with the Galilean teacher, his wife had sent word to him, "Pilate, have nothing to do with that righteous man. I was troubled with a vivid dream about him last night."

The priests and scribes and elders were shouting, "Barabbas! Release to us Barabbas!" They mingled with the motley crowd, inciting everyone to yell, "Barabbas! We want Barabbas!"

Pilate stood uncertainly looking over the noisy mob. He raised his hand for silence. "Then what shall I do with this Jesus who is called the Christ?"

"Let him be crucified!" the priests and elders and scribes shouted. The whole mob took up the cry, following the example of their leaders. "Crucify him!"

Pilate looked uneasy. The Roman emperor was severe with his administrators who practiced unnecessary brutality. There had to be a just reason for condemning a man to death. "Why?" Pilate asked. "What evil has this man done?"

They shouted all the more, "Let him be crucified!"

On the other hand, if anyone were suspected of treason, the emperor was ruthless. A Roman official who allowed treason or rioting would be quickly removed and punished.

"Isn't the Galilean your King?" Pilate asked. "Shall I crucify your King?"

The chief priests and the scribes were smooth and devious hypocrites. "We have no king but Caesar," they shouted.

Pilate saw that he was gaining nothing. At this rate he would have a full-scale riot on his hands. He called for a basin of water.

"I am innocent of this man's blood," Pilate announced curtly, "and I wash my hands of it." He glared at the chief priest

and his cohorts. "See to it yourselves."

So Barabbas the murderer was released. And Jesus was taken away by the Roman soldiers because his own leaders had insisted upon it.

In the Roman prison the soldiers stripped Jesus of his clothes and scourged him with knotted whips. Then, with mock deference, the soldiers threw one of their own cloaks over the shoulders of their prisoner. They wove a crown of thorn branches and pressed it cruelly upon his head, piercing him so that blood trickled down his face. They put a long, stout reed in one of his hands and knelt down laughing.

"Hail, King of the Jews," they taunted.

When they were through mocking him, the soldiers yanked off the red cloak. They gave Jesus back his own shirt and homespun robe. Then they led him away to crucify him.

When Judas found out what had happened, he hurried back to the temple. As he waited the priests and the scribes and the elders returned, the ones to whom he had betrayed Jesus.

The eyes of Judas were red with weeping. "I have sinned in betraying innocent blood," he cried out.

The members of the Sanhedrin shrugged when they saw him. "What is that to us?" they went on, ignoring him.

In an agony of self-condemnation the weeping Judas fumbled in his purse. He brought out the thirty pieces of silver. It was as if they burned his hand. He flung them at the feet of the elders, then stumbled out. Eyes glazed, walking as a man in a nightmare, Judas went and hanged himself.

Matthew 27:1-31; Mark 15:1-20; Luke 22:66 through 23:25; John 18:28 through 19:16

The Cross

The marketplaces of Jerusalem hummed with the beginning of the day's activities. People thronged the busy, narrow streets.

Four Roman soldiers emerged from the guardhouse at Pontius Pilate's Jerusalem headquarters. They prodded their prisoner, who staggered under the weight of a heavy beam. The prisoner looked haggard and bruised. Dried blood marked his brow.

"Another Roman crucifixion," a passing merchant muttered. His companion shrugged. It was a common sight to see some condemned wretch forced to carry his own crossbar outside the city, where he would be hanged until he died. Some of the more sympathetic or curious stopped to watch.

"Isn't that the preacher from Galilee?" a burly farmer asked.

"Poor man," his wife said, clasping her workworn hands. "The Romans probably thought he might lead a bloody revolt."

"Not him," a man with a Galilean accent said. "That Teacher was a man of peace. He told people to love each other."

"Well, I can't love the Romans," a fierce-eyed native of Jerusalem whispered. "Someday we Jews will find a new leader, someone like King David, or Judas Maccabeus—"

The hollow-eyed prisoner, sleepless and sore from beatings, stumbled under the weight of the beam he carried. He fell and lay sprawled on the paving stones.

"Here, you," the head centurion ordered. "I mean you." He grabbed the coat sleeve of a husky, darkskinned, well-dressed Jew who was passing by. "What's your name?"

"I'm Simon," the man said, taken by surprise and almost stuttering. "S-Simon of Cyrene. I just came to Jerusalem from the country this morning on business. Look, I haven't done anything."

"All right, Simon," the officer interrupted. "I've got a job for you." He jerked his thumb toward his prisoner. "Take his cross and carry it for him."

The passerby hesitated, then shrugged. It was better not to argue with Roman soldiers.

As Simon bent down to pick up the heavy beam his eyes met the suffering eyes of the prisoner. Suddenly there was recognition, and memory sprang into life. That business trip to Galilee a year ago—or was it two? A sunny hill covered with flowers and wheat ripening in the fields. A preacher with a radiant face talking to a crowd of spellbound people. . .

Happy are the pure in heart, for they shall see God. . .
Happy are you when men revile you and persecute you . . .
Let your light so shine before men . . .
Everyone who is angry with his brother shall be liable to God's judgment . . .
Do not resist one who is evil. . .

Many thoughts flashed through Simon's mind as he looked into those sorrowful, suffering eyes. The brief moment seemed far longer.

"You are Jesus," Simon whispered, "the Teacher from Galilee. I heard you speak once—"

"Come on, come on," the centurion barked. "We haven't got all day. Don't talk to the prisoner."

Simon helped Jesus to his feet. Then he reached down and lifted the heavy beam to his own shoulder.

Onward, through the narrow streets, the march continued in silence.

Some women, stifling their tears in their veils, followed at a distance, along with curious bystanders who had joined the procession. Out through the city gates they went, beyond the outer wall. They came to a hilltop known as Golgotha, which means the place of the skull.

Some permanent poles were set in the ground here and there for the execution of rebels, slaves, and criminals by crucifixion. Already another squad of soldiers was there. They were methodically preparing two condemned bandits for their death.

The two criminals had been stripped of their clothes. Each had his hands outstretched and tied to the crossbeam which he had carried out from the city. Two soldiers raised the crossbeam until the prisoner's feet were lifted from the ground. Then other soldiers fastened the beam onto the great upright stake. After this the dangling feet were lashed to the stake.

A centurion offered wine mixed with myrrh to numb some of the pain. The condemned men drank this eagerly, fear on their faces. They were left slumping on their crosses. Each had a sign hung around his neck which read "Bandit."

The soldiers with Jesus yanked off his robe, his sandals, and his long undershirt. Instead of tying his hands to the crossbar the soldiers hammered nails through his palms—a common alternative. Grunting, they hoisted him to his stake and tied him there between the crosses of the two criminals. They lashed his ankles to the post and nailed his feet into place.

"Here, drink this," a soldier said gruffly. He offered his prisoner the pain-easing drink of wine and myrrh. Jesus turned his head and would not take it.

Pilate had prepared a sign to be fastened on the cross over the head of Jesus. It said in Hebrew, Greek, and Latin, "King of the Jews."

"The leaders of the Jews won't like that sign," one of the soldiers snickered.

The centurion in charge laughed. "One of their priests has already objected. He

314 wanted Pilate to write 'This man claimed to be King of the Jews.' But Pilate just answered, 'I have written what I have written.' "

"Well, that shows what we Romans think of any provincial who sets himself up to be king," one of the men said. He kicked the cross of Jesus. "Hey, up there, how does it feel to be a king?" Then he doubled over laughing.

As Jesus hung there in pain the soldiers divided his clothing. Were not such items the legitimate spoils of the executioners?

The prisoner's robe was finely woven, without any seam. "Let's not tear it," one said. "Shall we cast lots for it?" So it was agreed. They sat on the hill, casting their dice and making rude remarks while Jesus suffered.

A curious, morbid crowd had gathered. Some had followed the soldiers and their prisoners out from the city. A mocker from the crowd called out to Jesus, "Aha! You who would destroy the temple and build it again in three days, now save yourself. Let's see you come down from the cross."

A priest from the temple stood nearby. "He saved others," he jeered, "but look at him now. He cannot save himself."

Someone else derided, wagging his head, "Let the Christ, the King of Israel, perform his magic now so we may see and believe."

Jesus murmured, "Father, forgive them; for they know not what they do."

Even the thieves hanging on each side of Jesus yelled at him between their groans. One of the criminals kept railing at him, saying, "Are you not supposed to be the Christ? Save yourself and us."

Finally the other one, who had grown quiet except for his low moaning, burst out, "Is it not enough? All three of us are dying. But you and I die as a just punishment. This man between us has done nothing wrong."

The sun was rising high in the brassy sky. The men had been hanging there since nine o'clock in tne morning. Already their mouths were dry.

"Jesus," said the second bandit in a hoarse voice, "remember me when you come in your kingly power."

Jesus whispered, "Honestly, I tell you, today you will be with me in Paradise."

At noon the sun clouded over. There was a thick darkness over the land for the next three hours.

Then in the middle of the afternoon the suffering Jesus cried out, "My God, my God, why have you forsaken me?" He winced, shuddering, and his expression showed unbearable pain. Then a look of peace came over his face, as if he sensed the crisis had come and his torture would soon be over. He uttered a great shout with the last muster of his dying strength. His voice trailed away to a whisper as he said, "Father, into your hands I commit my spirit." Having said this he breathed his last.

The centurion stood in silent contemplation of the man who had just died on the cross. With growing awe on his face he murmured, "Truly this man was the Son of God." And there, at this terrible place, he made up his mind to discover more about the teachings of Jesus of Galilee.

Some of those who knew Jesus, and several women who had followed him from Galilee, stood at a distance. They saw all that happened and how Jesus died. Their hearts were breaking with their loss and their grief.

Matthew 27:32-56; Mark 15:21-41; Luke 23:26-49; John 19:17-30

The Resurrection

The sun hung low over Golgotha, the hill where the body of Jesus drooped on the cross. All the land seemed gray and cold. The birds stopped singing.

The followers of Jesus were fearful of being seen by the priests or the Roman authorities. They were afraid they too might be jailed or even crucified. They returned sorrowfully to the city and kept to their quarters.

One secret disciple of Jesus took the risk. He was Joseph, a wealthy and respected member of the synagogue in his hometown of Arimathea. He was incensed and heartbroken at the action of the Sanhedrin.

Now he presented himself before Pilate. "I am Joseph of Arimathea," he courageously told the Roman administrator. "As you may know, it violates Jewish law to leave bodies of the dead hanging overnight, and most especially over the sabbath. Someone should bury the dead if there is no family nearby." He bowed low. "So I ask to bury Jesus the Galilean, who was crucified today."

Pilate looked surprised. "Is the man dead so soon?" He snapped his fingers and asked for the centurion in charge.

The soldier came in. "Yes," he told Pilate, "the man called 'King of the Jews' expired this afternoon about three o'clock."

Pilate muttered, "These things usually take twice this long." He turned back to Joseph. "Now what was your request?"

"I have a new tomb just outside Jerusalem," Joseph said. "I bought it for my own use. I would like to bury Jesus there."

Pilate shrugged, "I see no harm." He ordered the body released.

So Joseph purchased a new linen shroud and hurried back to Golgotha, with a friend. Reverently they took down the body of Jesus. Together they wrapped the body and carried it to the small cave-tomb nearby. When the body had been laid inside the tomb, Joseph and his friend rolled a large stone against the low opening.

Some of the sorrowing women who had been friends of Jesus saw where the body was laid.

In the evening when the sabbath was over, Mary Magdalene went down to the spice market. With her were two other women, mothers of disciples and also followers of Jesus. They bought the finest embalming spices they could afford for the proper anointing of the body of their beloved Master.

Very early the next morning, the first day of the week, Mary Magdalene and her friends went to the tomb. The sun was just rising over the hills to the east, spilling golden rays over Jerusalem. Fingers of light slid down through the gray green olive trees of Gethsemane as if seeking someone who was no longer there.

A low morning mist over the garden where the tombs were. Mary Magdalene, shivering, said to her companions, "Perhaps we should have brought Peter with us, or someone strong enough to roll away the stone."

"Do you suppose the three of us can push that stone from the door?" one of the other women asked. "It is very large."

Just then they reached the tomb. The three women drew together, puzzled. The stone was rolled back. The tomb was open.

Hesitantly they tiptoed to the low entrance and looked in. A young man sat by the long niche that was supposed to hold the body. The narrow ledge was empty. The young man's white robe gleamed in the dimness of the cave.

Mary Magdalene clapped a hand over her mouth to stifle a cry. The women were astonished, for they expected to find the linen-shrouded body only.

"Do not be amazed," the young man said. "You seek Jesus of Nazareth, who was crucified. He has risen, he is not here. See the place where they laid him—"

The women left quickly to take this astonishing news to the disciples. Mary Magdalene could think only that the body had been stolen. Sobbing, stumbling over stones and protruding roots in the garden, she made her way back to the city behind the others. She sought Simon Peter.

"They have taken the Lord out of the tomb," Mary sobbed. "Where have they laid him?"

"Come, we shall see," the big fisherman said.

John, the youngest of the Twelve and a favorite with Jesus, was with Peter. They hurried with Mary back to the garden of the tombs. The Beloved Disciple ran fastest, reached the open tomb, and peered in. Then Peter arrived and pushed his big frame through the small opening. He studied the linen clothes lying there and the long napkin that had been around the head. It still held some of its rounded shape.

The younger man had seen, and he believed instantly. "Jesus has risen from the dead," he whispered. Awed and thoughtful, the two men rushed back to Jerusalem. But Mary stood weeping outside the tomb. At last, when she crept back to the cave for a last look, she was startled to see two angels in white there.

"Woman," the angels asked, "why are you weeping?"

"They have taken away my Lord," she sobbed, "and I do not know where."

When she turned away to hide her tears she saw someone was standing beside her.

"Why are you weeping?" the man asked gently. "Whom do you seek?"

Mary's vision was blurred by her tears, and she did not recognize him. Thinking he was the gardener, she blurted out, "Sir, if you have carried him away, tell me where you have laid him."

The man said in his quiet, familiar voice, "Mary."

She knew then. "Teacher," she cried and started toward him, reaching out her hands to touch him, to grab hold of his hand or his clothing.

Jesus cautioned her, "Do not try to hold on to me, for I have not yet ascended to my Father. Go to my followers and tell them I am going to my Father and your Father, to my God and your God."

Mary hurried to tell the assembled disciples, "I have seen the Lord!"

On the same eventful day two disciples were hiking along the road which led from Jerusalem to the village of Emmaus, a journey of about seven miles. Another traveler joined them, and they sorrowfully told him about Jesus' crucifixion.

"We had hoped that he would be the one to save and redeem our nation," one disciple explained, his face heavy with sadness.

"This is the third day since it happened," the other sighed.

They told the stranger about the things the women had seen when they went to the tomb.

The stranger chided them, "O foolish men, and slow of heart to believe!" He did not seem to think the things they reported were strange at all. He began with the writings of Moses and went through the oracles of the other prophets pointing to the many passages that told about the Christ and his suffering and his triumph.

They walked on, and the man continued to speak with them. He was so interesting to listen to that when they arrived at the village of Emmaus the disciples asked him to stay.

"Come and have supper with us," they begged. "The day is almost gone."

They went into an inn there and sat down at a table. When the stranger blessed and broke bread for the meal, the eyes of the two were suddenly opened and they recognized their crucified Master, risen and alive again. At the same moment he vanished.

Without waiting a moment the two hurried back to Jerusalem and told their story to a group of the disciples. As they were talking, Jesus appeared among them.

Someone cried out fearfully, "It is a spirit."

Jesus calmed them saying, "Why are you troubled and puzzled? See my hands and feet. It is I myself. Handle me and see." However, some of them still could not believe that he was real.

"Have you anything here to eat?" Jesus asked.

They handed him a piece of broiled fish, and he ate with them.

The disciples crowded around Jesus, and he showed them the nail prints in his hands and the place where the soldier had pierced his body with a spear. Then Jesus breathed on them and said, "I am now sending you out just as my Father sent me. Receive the Holy Spirit."

One of the apostles, Thomas, was not there at the time. When the rest told him that they had seen the risen Christ, the skeptical Thomas shook his head. "Unless I place my finger in the print of the nails and place my hand in his pierced side, I will not believe."

Eight days later all the disciples were together again in the house. Suddenly Jesus was standing in the midst of them, just as before. He said to Thomas, "Put your finger here, and see my hands. And place your hand in my side. Do not be faithless, but believing."

All doubts left Thomas, and he dropped to his knees and cried out, "My Lord and my God!"

Then Jesus said to him, "Have you believed because you have seen me? Blessed are those who have not seen and yet believe."

Some days later several of the disciples had returned to Galilee and were fishing along the western shore of the lake. It was early. On the pebbled beach, standing obscured in the morning mists, a stranger called out to the men in the boat. "Have you caught any fish?"

"No!" they called back, disappointment in their voices.

"Cast your net on the right side of the boat," he told them.

The fishermen did, and almost at once the net grew so heavy with fish they could not pull it into the boat. They began to tow the bulging net back to land.

Suddenly, John recognized his Master on the shore. He shouted, "It is the Lord!"

All of the disciples cried out with joy. Peter was so excited he jumped into the water and swam ashore. The other disciples followed in the boat.

Jesus was standing there beside a glowing fire on which a fine breakfast of fish was broiling. "Come and eat," Jesus said. He broke bread and handed it around the group. They breakfasted on bread and fish while Jesus spoke with them about the work they must do for God's kingdom.

These appearances of the risen Lord, which revived and inspired the discouraged disciples, continued for forty days. During this time Jesus gave them instructions

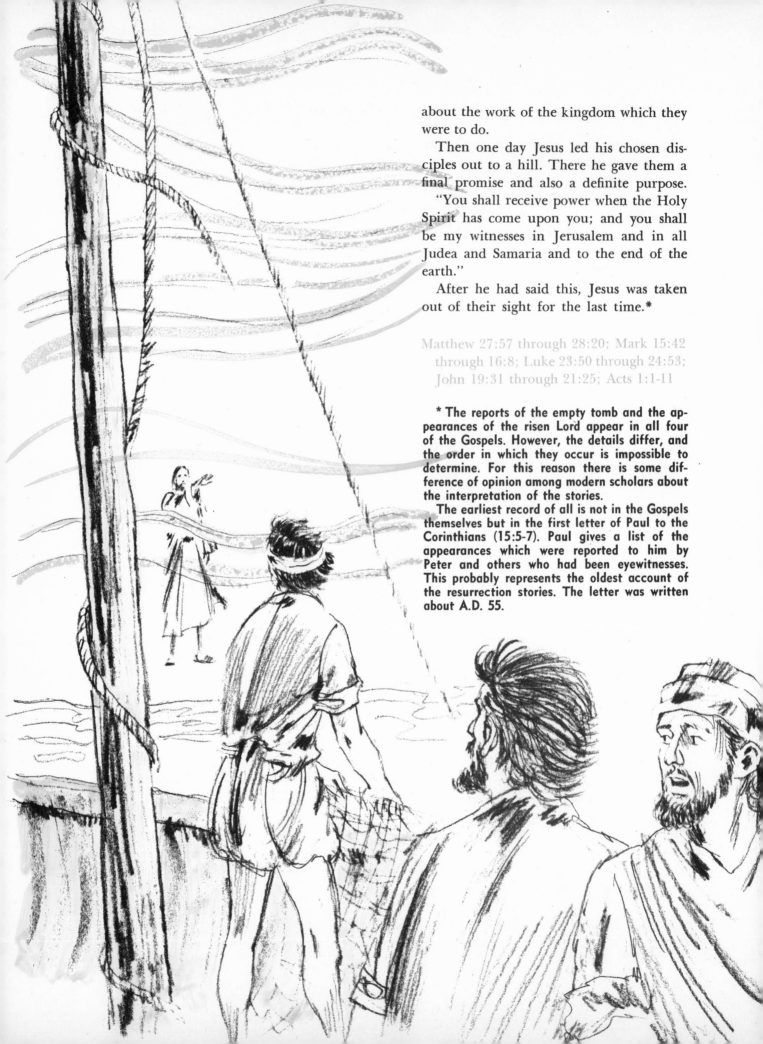

about the work of the kingdom which they were to do.

Then one day Jesus led his chosen disciples out to a hill. There he gave them a final promise and also a definite purpose.

"You shall receive power when the Holy Spirit has come upon you; and you shall be my witnesses in Jerusalem and in all Judea and Samaria and to the end of the earth."

After he had said this, Jesus was taken out of their sight for the last time.*

Matthew 27:57 through 28:20; Mark 15:42 through 16:8; Luke 23:50 through 24:53; John 19:31 through 21:25; Acts 1:1-11

* The reports of the empty tomb and the appearances of the risen Lord appear in all four of the Gospels. However, the details differ, and the order in which they occur is impossible to determine. For this reason there is some difference of opinion among modern scholars about the interpretation of the stories.

The earliest record of all is not in the Gospels themselves but in the first letter of Paul to the Corinthians (15:5-7). Paul gives a list of the appearances which were reported to him by Peter and others who had been eyewitnesses. This probably represents the oldest account of the resurrection stories. The letter was written about A.D. 55.

PART VII
THE BEGINNINGS
OF THE CHURCH

After the resurrection of Jesus the Christian church slowly came into being from a formless fellowship of believers. For a long time it was a movement within the Jewish community. As a Jew Jesus had preached largely to Jews. A separation did not occur to the first Christians. They worshiped daily at the temple, went to synagogue, obeyed thet Jewish laws; but they preached the good news of Jesus as they did these other things.

Most of our information about the beginnings of the Christian church comes from the book of Acts. Luke was a non-Jewish companion to Paul. He was a physician who lived in the second half of the first century. Scholars believe that he wrote not only the gospel bearing his name, but also the continuation of that book, which is called the Acts of the Apostles. Both are written in the same lively, descriptive style. His combined Luke-Acts is the longest contribution to the New Testament by one author. It was split apart when the New Testament was being assembled so the four gospels of the life of Jesus could stand together.

During Jesus' ministry his followers were called disciples. The special chosen ones were referred to as the Twelve. There were only eleven after the death of Judas. Later Matthias, a disciple who had been with Jesus throughout his ministry, was chosen to serve.

After the Master's death those who believed his teachings were still called disciples. Those who had been witnesses to the events in the life of Jesus became evangelists and missionaries, who went about preaching the gospel. The apostles were the leaders of this group.

The apostles and the disciples told and retold the stories of their personal experiences and memories. In time many of these were recorded in Acts.

It was inevitable that some stories would be colored and elaborated in the retelling, and others would take on elements of a legendary nature. Yet it must be remembered that these stories express the overwhelming joy the early followers felt in their very real sense of closeness to the risen Christ.

Peter's New Courage

Fifty days after Passover the Jews were celebrating the religious holiday of Pentecost. Again Jerusalem was crowded with incoming pilgrims.

The followers of Jesus met together to celebrate the feast day. Suddenly, they were filled with the Holy Spirit of God. They were afire with enthusiasm for the coming of God's kingdom. Their experience was so overwhelming there was no way to describe it except in poetic comparisons. "Suddenly a sound came from heaven like the rush of a mighty wind. . . . And there appeared to them tongues like fire, resting on each of them. . . . They began to speak in strange tongues."

Indeed, the followers were so intoxicated with the joy of feeling close to God, they did babble incoherently. Their thoughts were so exalted and overpowering, ordinary speech could not keep up. People passing by stopped and wondered, asking, "What does this mean?" Others, mocking, said, "They are drunk with new wine."

But Peter was bold. He seized the opportunity to stand up and speak to a gathering group.

"Men of Judea and all who dwell in Jerusalem," he shouted. "Listen carefully to what I say. These men are not drunk as you suppose. It is only nine o'clock in the morning."

There was a ripple of laughter through the crowd. More people paused to listen.

"You all know what the prophet Joel said," Peter reminded them. "In the last days your young men shall see visions, and your old men shall dream dreams. God will pour out his Spirit, and they shall prophesy." *

The people nodded gravely. A man who could quote the prophets was held in esteem.

"Listen," Peter commanded. "Jesus of Nazareth, who did mighty works and miracles among you, was crucified, and you consented to it. But God has raised him up. We are witnesses to that. Now, as he promised, God has poured out the Holy Spirit upon us. That is what you see and hear."

Peter, the gruff uneducated fisherman, speaking in Aramaic, was carried away with his message. He spoke rapidly and with assurance. There was so much excitement and fervor in his voice that even visiting Jews from other lands, who spoke mainly Greek or Latin, seemed to understand exactly what he was saying. Many were touched by the message, and cried out, "What should we do?"

"Repent," Peter said, with a new sense of authority. "Be baptized in the name of Jesus Christ for the forgiveness of your sins, and you shall receive the gift of the Holy Spirit. Save yourselves from this crooked generation."

Then Peter went on to tell them about the teachings of the Master. Many repented, and many were baptized as new followers of Jesus.

After that day of Pentecost, Peter's life was changed. The same fellow who was afraid to admit he knew Jesus now had the courage to speak everywhere of Jesus the Messiah. He even took his message to the very temple where Jesus had been sentenced to death by the Jewish council.

One afternoon Peter and John were going up to the temple for the three o'clock hour of prayer. A man lame from birth was being carried to his usual place at the courtyard entrance called Beautiful Gate. Here the cripple sat daily, begging from those who entered. As Peter and John came up, the beggar cried out, "Mercy, mercy," and held out his hands for money.

Peter commanded the beggar, "Look at us." The man was all attention, expecting money. Peter said, "I have no silver or gold, but I give you what I have. In the name of Jesus Christ of Nazareth, walk." Peter took him by the right hand and raised him up.

The man put his weight gingerly on his feet and ankles. He found them no longer so weak they could not hold him up. Leaping up, the man stood and walked. He ran into the temple courtyard with Peter and John. He walked and leaped and praised God in a voice that rang with joy.

When people saw the man walking and praising God, they recognized him as the beggar they had seen so long at the temple. "What do you suppose happened?" they asked each other. "Let's follow that man and find out."

The man stayed close to Peter and John as they walked in Solomon's Porch. "These men cured me of my lameness!" the man rejoiced before all the people.

Peter saw the crowd gathering around them and he saw an opportunity to preach again. "Men of Israel, why do you wonder at this?" he asked them. "Why do you stare at us as though by our own power or piety we made this man to walk? It is the God of our ancestors who glorified Jesus and gave us this power in his name."

"Jesus?" somebody asked uncertainly, trying to place the name.

"Yes, Jesus," Peter shouted at them, "whom you delivered up for a trial. You rejected him in the presence of Pilate, when Pilate was willing to release him. You repudiated Jesus, who was holy and righteous, and chose a murderer instead."

There was a murmur among the people. "But although you allowed Jesus to be killed," Peter continued, "God has raised him from the dead."

More murmurs came from the crowd. Their faces showed doubt and skepticism. "It's true!" Peter declared. "I myself saw

the risen Christ. And many others have been witnesses." He pointed to the former cripple, who was listening with open mouth. "It is the name of Jesus—faith in his name—that has made this man strong and healthy."

Some in the crowd seemed ready to believe. Others looked angry or disturbed. Peter held up his hand. "Now, my brothers, I know that when you let Christ be crucified you acted in ignorance. So did your rulers, who planned his death. Repent so your sins may be blotted out, and you will be ready for the coming of the kingdom of God."

As Peter was speaking, the priests and a group of Sadducees came up to the disciples. They were indignant because Peter and John were stirring up the people, getting them excited. Then too, the Sadducees did not believe in the resurrection of the dead. They felt that Peter and John were teaching false and foolish things.

"Arrest those men," an officer of the temple said to the temple guards. Peter and John were seized, bound, and thrown into prison.

The next day the council was gathered together in Jerusalem. The priests, the elders, and the Sadducees were there. Peter and John were brought into the middle of the courtroom. When the questions began, Peter answered boldly. "If you want to know how John and I were able to heal a cripple, I will tell you. It was done in the name of Jesus Christ. Yes, Jesus of Nazareth, the very one you crucified, whom God raised from the dead."

His accusers saw the boldness of Peter and the confident way that John stood up beside him. They saw the healed man standing beside these two uneducated peasants from Galilee. And the priests and elders didn't know how to argue with Peter and John. "Take the two prisoners from this room," the high priest ordered. "And keep them in custody outside."

Then the priests and elders conferred. "What shall we do with these men?" they said. "The word of what they have done will go all over Jerusalem. We cannot deny that it happened."

"It must spread no further!" the high priest said, agitated.

"Send for the two men," an elder urged. "Warn them to speak no more to anyone in the name of this crucified Jesus."

Peter and John were brought back and told they could speak no more in the name of Jesus.

Peter answered with unwavering courage, "Should we listen to you, or to God?" He glanced at John, and John nodded his agreement. Peter added firmly, "We must continue to speak of what we have seen and heard."

Outside in the temple courtyard the Jewish people were loudly praising God for the miracle of healing that had taken place there. The council of priests and elders knew this; they also knew the people might riot if the two apostles were punished right then. So the high priest made fearsome threats against Peter and John. Then he ordered them to be set free.

Peter and John, undaunted, returned to the company of their friends, the rest of the devoted followers of Jesus.

Acts 1:12 through 2:41; 3:1 through 4:31

* Joel 2:28.

The United Believers

At first the followers of Jesus stayed close to Jerusalem. They looked forward to a triumphal return of Jesus at any time. Any day, they thought. It could be any hour.

After this "second coming," God's new kingdom would be established on earth. And, since the present world was not expected to last very long, all worldly goods and pleasures had no appeal.

The followers banded together. They continued to go daily to the temple to pray, for they considered themselves good Jews. They ate together, in one home or another. And they offered all they owned for the use of the group. Those owning lands or houses sold them, bringing all proceeds to their society. These funds were used for distribution among the needy.

One loyal follower named Barnabas, a native of the island of Cyprus, was a Levite who assisted in the Jerusalem temple. He owned a fine field. "Why do I need a field, if our Lord will soon establish the new kingdom of God?" he said. He sold his field and brought the money to Peter and James.

But all followers could not be so completely dedicated. Much later, a story was whispered about a married couple who suffered a terrible fate when they were unable to forget their old, greedy ways.

A man named Ananias and his wife Sapphira dedicated a piece of property to the society. "We will sell our land," Ananias grandly told the others, "and bring in all the proceeds for the common use."

When the land was sold, Ananias and his wife counted the money.

"Our land brought more than I thought it would," Ananias whispered to his wife.

"We'll just keep back a part, in case we want to buy things for ourselves." The rest of the money he put into a leather bag.

At the next meeting he presented the money to Peter. The canny fisherman looked the giver in the eyes. "Ananias, why has Satan filled your heart so that you lie to the Holy Spirit? That is what you do, when you keep back part of the proceeds after you have dedicated all to God."

Ananias listened to the words of Peter with growing agitation. His heart pounded, and his face flushed red. His head hummed, he rocked on his heels, and he fell dead.

Some of the young men rose, wrapped him in his cloak, and took him out and buried him.

Later, his wife came to the meeting. She was nervous and edgy. Her eyes darted about the room.

Peter said to her, "Sapphira, tell me if you sold the land for—" And he named the amount her husband had brought.

Sapphira was afraid, for her conscience was bothering her. Yet she pretended to be calm. "Yes," she lied. "That was the exact amount."

Peter rebuked her. "Why is it you and your husband have conspired to put God to the test?" There was the noise of footsteps approaching. "Listen! The feet of those who buried your husband are at the door. And they will carry you out."

Sapphira clutched at her heart and swooned to the floor. The young men came in, found her dead, and carried her out to be buried beside her husband. When this story was whispered around, people were afraid to be liars or cheaters before God.

At first, Peter was the most important leader in Jerusalem, along with the brothers James and John, and the rest of the Twelve.

There was another James who was growing in importance. He was a brother of Jesus. He was pious, and well studied in the Jewish laws, tending to be somewhat conservative. Since his brother's resurrection, he was on fire with the new gospel.

So it was that the twelve apostles and James, brother of the Lord, assumed the leadership for the growing society. This suited the followers from Judea and Galilee.

But there were many new followers—Jews, who had lived abroad and who spoke in Greek and wore Greek clothes. These "Jews of the Dispersion" were known as Hellenists. "We are not fully represented in the leadership here," they complained. "Our widows are neglected in the daily distribution."

The complaint was considered just.

The Hellenists were told to choose seven good men from their number to represent their section of this new society. Seven were chosen: Stephen, Philip, and five other men who also bore Greek names. The Twelve laid their hands upon the newly elected seven, to consecrate them in the name of the Holy Spirit.*

It was a great honor for Stephen and Philip and the other five Greek-speaking Jews.

It was also an honor which could bring great danger.

Acts 2:42-47; 4:32 through 6:7

* These seven are often called the first deacons.

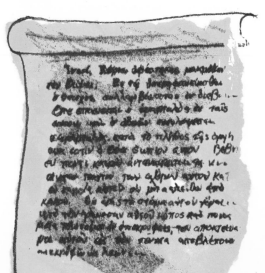

The Stoning of Stephen

325

Stephen set himself to preaching the new gospel of Jesus, in the Greek-speaking Jewish synagogues of Jerusalem. He spoke eloquently to these Hellenistic Jews. "Ever since the Exile, your ancestors have been scattered abroad," Stephen said. "Many of you grew up in Alexandria, or Greece, or the islands of the Mediterranean. But you have come back to Jerusalem to visit or to live because this place is our spiritual home."

"He is one of us," Jews whispered to each other in Greek. "He understands us."

But Stephen's criticisms could slash like a knife.

"You all know that our ancient ancestor Joseph was a great man in God's eyes, and full of promise. But what did his brothers do? They sold him into Egypt!"

Stephen's listeners moved uneasily on their benches in the synagogue. Why was Stephen pointing so accusingly at them? Did he class them with callous brothers who would plot to kill someone chosen by God?

"Yes," Stephen nodded, "and our revered Moses was also great in God's eyes. But when he tried to reconcile two quarreling Hebrews, they rejected him, saying, 'Who made you a ruler and a judge over us?' Moses had to flee into the desert. Yet Moses came back to save his exploited brothers from the Pharaoh. Was he appreciated? No! They thrust Moses aside and worshiped a golden calf. God was forced to punish this willful people, who could not recognize the beloved of God who were sent to save them."

The Hellenistic Jews prided themselves that they were more pious and God-fearing than the local Palestinian Jews. But

326 Stephen was making them feel guilty, and they did not like the feeling.

"We follow the prophets," someone in the congregation argued.

"Tell me, which prophets were not persecuted by your fathers?" Stephen challenged. "As your ancestors did, so do you. You are a stubborn, stiff-necked people, who will not bow your heads to the yoke of God's will." Stephen shook his fist at them. "Just a short time ago you consented to the betrayal and murder of Jesus of Nazareth, the Righteous One."

When they heard Stephen say these things the influential Jews in the synagogue were enraged. They seized Stephen and bound him. They dragged him to the temple, to appear before the council. They also brought along some witnesses, who gave false testimony against Stephen.

"He speaks against this holy place and the Law," one man lied. Another man stood up with more falsehood. "We have heard him say that Jesus of Nazareth will destroy this place and change the laws of Moses."

Stephen was asked what he had to say for himself. So the fearless young man preached to the whole Sanhedrin and to the other Jews crowded in the audience chamber. He was as outspoken as he had been in the synagogues. And when he spoke Stephen's face became like that of an angel.

Members of the council shouted, "Blasphemer!" The trial ended in a riot. Men rushed at Stephen, grabbed him, and hustled him out of the city.

But even as they were threatening him, Stephen looked up to heaven and said, "I see the heavens opened, and the Son of man standing on the right hand of God."

Then the accusers picked up rocks and began to pelt him.

A young Pharisee was there, who had come from Antioch. His name was Saul.* This young Pharisee watched over the pile of robes laid aside by the men doing the stoning. He did not participate, but his face showed that he thoroughly approved of the execution.

The bloodied Stephen knelt on the ground, as rocks continued to crash painfully against his body. "Lord Jesus, receive my spirit," Stephen prayed. Another large rock hit him on the forehead, and he cried out, "Lord, do not hold this sin against them."

When he had said this, he collapsed. Stephen, the first Christian martyr, was dead.

Devout men buried Stephen, and wept over his grave.

After this a great perecution arose in Jerusalem against the society of Jesus' followers. James, the brother of the Lord, and the twelve Galilean apostles went into hiding in Jerusalem.

The Greek-speaking disciples, who were most persecuted, fled farther, and were scattered throughout Judea and Samaria. Barnabas, who had given all his land to the followers, moved to the great Syrian city of Antioch to preach. With him went a group of converted friends, who also came from the island of Cyprus.

Philip, who had been ordained along with the martyred Stephen, fled to Samaria.

Acts 6:8 through 8:2

* Paul is the Romanized form of Saul.

Philip and the Ethiopian

After Stephen was killed, Philip found sanctuary in a city in Samaria. There he began to preach the good news of Jesus Christ.

People listened to Philip, for he was a persuasive man. They were amazed when they saw how he healed those possessed of unhealthy spirits. He also helped the paralyzed and the lame, just as Jesus had done.

Even a Samaritan named Simon, who practiced magic, was convinced by Philip's message. Up to this time the magician had gathered around himself a band of followers and had promoted himself as a great figure who had the power of God. After he listened to the good news about the kingdom of God, Simon believed. So Philip baptized Simon along with many other Samaritans.

When the apostles in Jerusalem heard how the gospel had been received in Samaria, they journeyed there and laid their hands on many of the people baptized by Philip. When Simon the Magician saw that these people received the Holy Spirit from the apostles, he offered Peter money to give him this power also. But Peter refused and spoke to him harshly for thinking that the gift of God could be bought.

After a time Philip felt called by God to go south on the road to Gaza to preach in the towns along the Mediterranean coast. On his way Philip stopped overnight in Jerusalem, staying secretly with other disciples. Before dawn he slipped quietly out of the holy city and continued his journey. He walked southward along the high backbone ridge of mountains that runs down through Judea. Philip took the western road to Gaza. That old Philistine city had been destroyed by Alexander the Great. There was nothing left but a desolate waste.

As Philip walked along the road through the deserted ruins he heard the noise of clopping horses' hooves and the rumbling of chariot wheels. He stepped off the road and looked behind.

The traveler was a dark-skinned Ethiopian whose trappings and dress befitted his position as treasurer to the Queen of Ethiopia. He had been on a pilgrimage to Jerusalem to worship and was returning now to his home in Africa. He sat in his chariot, the reins of the two high-stepping black horses looped over one arm, his dark hands holding a scroll.

As the Ethiopian passed, Philip could hear him reading the words of Isaiah. Philip ran alongside the chariot and called, "Do you understand what you are reading?"

The handsome black man smiled and answered in excellent Greek, "How can I, unless someone guides me?" He slowed his horses and smiled at Philip. "Come up into my chariot and sit with me."

Philip leaped into the chariot. The Ethiopian went over the passage in the scroll again:

> Like a lamb led to the slaughter,
>   or like a sheep before its
>   shearers is dumb,
>   so he opened not his mouth.*

"Is the prophet Isaiah talking about himself?" the Ethiopian asked. "Or someone else?" Then Philip talked about this scripture and went on to tell of the good news of Jesus.

As they jounced along the road they came to some water. "See, here is some water," the Ethiopian said, reining the horses to a stop in a cloud of dust. He

328 turned and faced the evangelist, and again the warm smile lighted up his dark face. "What is to prevent my being baptized right now?"

Hand in hand the Ethiopian and Philip walked into the water, and Philip baptized his new friend.

The Ethiopian official continued his way south rejoicing. But Philip turned north, preaching in all the towns of the seacoast until he came to his home in Caesarea, where his wife and four daughters awaited him. Here Philip continued to preach the gospel.

Acts 8:4-40

* Isaiah 53:7.

Herod's Persecution

About this time a turn in Roman politics placed a grandson of Herod the Great on a throne as ruler of most of Palestine. This Herod Agrippa was a middle-aged man who had grown bitter about life. He was now determined to be as powerful a king as his grandfather had been. How? Get the support of the powerful Jews who lived in his land. Please them!

Seeing that this new sect, the followers of Jesus, was a thorn in the flesh of the Jewish council, Herod Agrippa swung into action.

James, a brother of John and one of the original disciples of the Nazarene, was preaching on a street corner about the risen Christ. Herod sent a squad of soldiers to arrest James. He ordered the apostle put to death with the sword, and it was done.

So James, son of Zebedee, became the second Christian martyr. When King Herod heard that the powerful Jews in Jerusalem were pleased about James's death, he ordered the arrest of Peter also. Peter was seized and taken to prison.

The king intended to kill Peter as he had James, but first he wanted to display his prisoner to the Jewish people.

It was now Passover time again. "I will wait until the feast days are over," the king told the captain of the prison. "In the meantime keep a heavy guard on the prisoner."

On the night before Herod planned to bring Peter out for public trial the apostle was in his cell. He was bound with chains, sleeping between two soldiers. Sentries before the door were guarding the prison.

Suddenly Peter awoke. There was a bright, unearthly light. A figure that looked like an angel was speaking, "Get up quickly and dress."

As Peter stood up his chains dropped off, but the sleeping guards did not awake. Peter yanked on his robe, whipped his belt around his waist, and fastened his sandals.

The angel said, "Wrap your mantle around you and follow me."

Peter, thinking this was all a dream, followed the angel past the guards to the outside iron gate of the prison. Peter passed through the gate and found himself alone in the deserted street. A full moon shone down, a solitary dog picked at refuse in the narrow street. Peter shook his head, realizing suddenly that he was awake. He was free! He hurried along the street, thinking, "Now I am sure that the Lord has sent his angel to rescue me from the hand of Herod." He drew a deep breath of relief. "He has saved me from the hands of the rulers and the priests."

He headed straight for the house of Mary, the mother of an active young disciple named John Mark.* When he found the house he knocked at the courtyard gate. In the main room, he could see the flicker of lamplight.

A maid named Rhoda, carrying a small saucer lamp, answered his knock.

"Let me in," Peter whispered through the bars of the gate. "It is I, Peter."

When Rhoda heard his voice she knew it was Peter. In her excitement she flew back across the courtyard, forgetting to unlock the gate. She rushed into the crowded main room of the house, where many of Peter's friends had gathered to pray.

"Peter is no longer in prison!" the maid gasped. "Your prayers have been answered, for he is right now at the gate!"

"You are mad, Rhoda," they told the trembling, excited girl.

"I tell you, I recognized his voice,"

Rhoda insisted, "and he called out his name."

"It must be his ghost," an old woman said, rolling frightened eyes. "How could Peter get out of King Herod's prison?"

When they crossed the courtyard and opened the gate, they saw with amazement that Peter really was there. Some wept with joy to see him safe.

They gathered again in the big room, and Peter told them the story of his miraculous escape.

"It is getting too dangerous now for you to keep on preaching in Jerusalem," one of the older men said wisely. "You must move on to another place and continue preaching the gospel there."

"Yes, yes," the others agreed. "It is too dangerous, Peter. We cannot lose you, as we did the son of Zebedee."

Peter bowed his head, and they all prayed together. The brave apostle said sadly, "I will leave tonight and continue to preach the good news in other places. I leave you to the good leadership of James, the brother of our Lord."

Then Peter left Jerusalem.

Acts 12:1-24

* Though John Mark was not one of the twelve apostles, he was closely connected with the Twelve and perhaps had known Jesus himself. Many scholars think that this house may have been the same one in which the Last Supper was held. Mary, John Mark's mother, was also the sister of Barnabas, a helper of Paul.

Peter's Preaching

After Peter's escape from prison was discovered, King Herod's soldiers began to search Jerusalem for him. Peter was now a fugitive. With no fixed home, the apostle went here and there in Palestine to preach the good news of Jesus to the Jews. It did not occur to Peter to preach among non-Jews. The Gentiles did not recognize the Jewish religious laws, and so they were looked down upon, avoided, and despised.

So Peter worked hard among the God-fearing Jews to convert them into followers of Jesus.

Just as the Greek-speaking apostle Philip had preached in the towns along the coast of the Mediterranean, so now Peter went from town to town: Lydda, Joppa, and still farther north, the great seaport city of Caesarea.*

When he arrived in Lydda, Peter found a paralyzed man in need of help. His name was Aeneas, and he had been bedridden for eight years. Peter said to him, "Aeneas, Jesus Christ heals you. Rise and make up your bed." Immediately the man arose. After that many people in the town saw the healed man, and were converted to the new sect.

Among the followers of Jesus in that ancient seaport town of Joppa was a woman whose name was Tabitha.** Tabitha busied herself with good works and acts of charity. She sewed clothes for the poor, took them food, and nursed them when they were sick.

Then Tabitha became ill and suddenly died.

Her friends who were fellow Christians knew that Peter was nearby in Lydda. "Let's see if Peter will come," her grieving friends said.

Two men rushed south on the coastal road and found Peter at Lydda. "Tabitha has been deathly ill, and now she is being laid out for burial," the two men pleaded. "Please come without delay."

Peter rose and went with them. At the home of Tabitha in Joppa, the men took Peter to the upper room where Tabitha lay. Many of the poor of the town were there weeping. They showed Peter the garments she had made for them and described how Tabitha had taken care of them when they needed it. Peter nodded sympathetically and sent them all outside. He knelt down and prayed. Then he said, "Tabitha, rise."

The woman opened her eyes. When she saw Peter she sat up. He gave her his hand and helped her up. When he took her, alive and smiling, to the townspeople who streamed in, many more were convinced of the wonder-working power of Christ.

In fact, this town proved to be so responsive to the preaching of the gospel that Peter stayed on with another follower, a tanner of hides named Simon.

While residing in Joppa, Peter heard some good news. Farther north on the Mediterranean coast in the bustling seaport of Caesarea, King Herod Agrippa had died. The followers of Jesus were relieved and said, "An angel of God must have struck him down."

Caesarea was a showplace, rebuilt by the first King Herod. It was largely a pagan city for only a minority of Jews lived there in their own section of town. It was a beautiful city, with great Roman aqueducts carrying water, colonnaded paved streets, a Greek amphitheater for plays, a resplendent temple dedicated to Caesar, and a vast hippodrome for chariot racing and competitive games.

There had been no natural harbor on this coast. The resourceful engineers of Herod the Great had constructed a breakwater of huge stones out into the choppy blue waters of the Mediterranean so that ships could anchor at Caesarea.

Stationed in Caesarea was a centurion named Cornelius. A member of a company of soldiers sent from Italy, he had his own house in this city. Cornelius was a devout man drawn to the high ideals of the Jewish faith. He worshiped the Jewish God and gave to their poor.

At three in the afternoon this centurion knelt in the privacy of his room to pray, closing his eyes. Suddenly he saw in a vision an angel of God standing beside him saying, "Cornelius!"

The terrified soldier murmured, "What is it, Lord?"

"God knows of your prayers and your charity," the voice intoned. "Now send to Joppa for a man named Peter. He is lodging with Simon the Tanner, whose house is by the seaside." The vision faded.

Cornelius rose to his feet and rubbed his eyes. He looked dazed. He called an orderly and two servants and told them about his vision. "Go to Joppa and seek a man named Peter, lodging with Simon the Tanner in a house by the sea. Let us see what God wants of me."

The next morning the orderly and the two servants started south.

About noon Peter waited hungrily for the noon meal to be prepared. He walked up the outside stone steps to the roof of Simon's house. There was a low parapet around it, a bench and some cushions. As Peter stood looking out at the blue, sparkling Mediterranean and the white sails of boats in the shallow harbor he prayed and meditated.

After a time Peter grew sleepy in the hot noonday sun, and fell into a trance. He felt as if he were dreaming. The heavens opened. Something like a great white sheet or a huge square of sailcloth was being let down by its four corners upon the earth. In it were all kinds of animals, reptiles, and

332 birds of the air. A voice came. "Rise, Peter. Kill and eat."

Peter recoiled. He saw many creatures that Jews were forbidden to eat. "Oh no, Lord," he objected. "I have never eaten anything that is common and ritually unclean."

Sternly the voice came: "What God has cleansed, you must not call unclean!"

Three times this happened, and then Peter awoke.

While Peter was still on the rooftop puzzling over what this vision might mean, there came a loud knocking on the door below. Three men stood in Simon's courtyard asking for Peter.

The apostle hurried down the stairs to meet them. They were the messengers of Cornelius. They explained the centurion's vision, and Peter invited them into Simon's house to be his guests.

The next morning Peter accompanied the men to Caesarea. Six of Peter's fellow Christians from Joppa went with them.

When it was announced that the party of Jews had arrived, Cornelius ran out to meet them. Cornelius seemed to know that the big man was Peter. He threw himself down, in a Jewish bow of respect, at Peter's feet.

Peter did not take pride in this unusual scene of a Roman soldier at the feet of a lowly Jew. He held out his hand to Cornelius and helped him up.

"Stand up," Peter smiled. "I too am a man."

The centurion found his words rushing out, as he told Peter about his vision. "I have been expecting you. I've invited all my relatives and friends to come and hear what you have to say."

Peter and his party went inside. A crowd of Gentiles had gathered in the room. Peter began to speak. "You all know that it is not customary for a Jew to visit one of another belief. It just is not done." Then his big face broke into a broad grin. "But I too have had a vision. God has shown me that I should not call anything he has created common or unclean. So when I was sent for by Cornelius, I came without objection. I now see that God shows no partiality! All men are acceptable to God, if they believe in him and do what is right."

The Gentiles asked Peter to preach to them. With great eagerness, the big apostle told how Jesus was baptized, how he went about doing good and healing. He detailed the message of Jesus. "They put him to death by hanging him on a tree," Peter related sorrowfully. His rugged face lighted. "But God raised him on the third day, and many of us were witnesses. We were commanded to preach his gospel."

Even before Peter was through, the Gentiles began to believe. They even began to pray in strange tongues, because the gift of the Holy Spirit was poured out upon them.

Peter was amazed. "Can anyone forbid that we baptize these people in the name of Jesus Christ?" he asked his friends. "They have received God's Holy Spirit just as we have!"

The Jewish followers of Jesus were amazed, for they had taken for granted that their movement would grow only among their fellow Jews.

But now, Peter had opened the door to the Gentiles!

Acts 9:32 through 10:48

* Many scholars consider the stories of Peter's activities in these towns as legends which grew out of the enthusiasm and admiration of the people for this courageous and hardworking apostle.

** The Greek form of Tabitha is Dorcas which means "gazelle."

PART VIII
ADVENTURES
OF PAUL

Jesus lived, taught and preached in a very small area of Palestine; his ministry was among the Jews. Later, Peter and other disciples preached also to Jews, and only occasionally Gentiles.

The apostle Paul traveled through a much larger section of the Roman Empire. He was the first great Christian missionary to the Western world, opening the door for non-Jews to enter the faith.

What was the Roman world like in the time of Paul, twenty or thirty years after the death of Jesus?

The Romans were the real inheritors of the far-flung Macedonian-Greek Empire conquered by Alexander the Great. When this empire fell after the death of Alexander and before Christianity, Rome picked up the pieces.

Rome had a talent the Greeks did not have, the practical ability to organize and rule a patchwork empire made up of very differing peoples, religions, and cultures. Greece had never even been able to consolidate her own country politically. It had always been divided into warring city-states like Athens, Corinth, and Sparta.

The genius of Greece had been in other things: abstract thought, sports and the Olympic games, sculpture, architecture, literature, philosophy, and drama. Greek colonies had sprung up on most of the Mediterranean coasts and islands. There were ten Greek-influenced cities in eastern Palestine, connected by paved roads, called the Decapolis. Each was dominated by Greek architecture, Greek religion, and Greek ways of thinking. Long after the Romans had conquered the Mediterranean world, the Greek culture remained dominant. Greek was the most common language spoken. A Greek-speaking man like Paul could travel from one country to another, and be understood wherever he went.

The Romans provided a fairly just and stable government. Except when an occasional fanatic like Nero was in power, there was a great deal of freedom for the subject peoples, including freedom of religion.

There was slavery in the empire; prisoners of war were enslaved by the thousands. But slavery was taken for granted in the ancient world, even among the Jews. The great building enterprises of the Romans, including the paved roads and the sea lanes utilized by Paul, were made possible by slave labor.

As long as individuals or national groups lived peacefully under Roman rule, they were pretty much left alone. At the first sign of unrest or rebellion, however, the Roman fist smashed without mercy. In an uprising after the death of Herod the Great (when Jesus was an infant), two thousand Jews were crucified. In the ensuing years many more Jews, including Jesus, were executed on political charges of insurrection. The Jews at last erupted in a massive rebellion in A.D. 66. During the next four years, when Jerusalem was under siege, the Roman general Titus crucified as many as five hundred Jews in a single day outside the embattled walls.

Between the time of the death of Jesus and the destruction of Jerusalem by Titus, during the period when Paul lived and preached, Palestine and most of the Roman world were relatively calm—at least on the surface. Rome's power allowed a remarkable amount of freedom to travel safely from place to place, even among strange and inhospitable peoples. The sea lanes were kept free from pirates.

The apostle Paul was a great preacher. He was a devoted and untiring missionary. A creative religious thinker, Paul was most influential in organizing the teachings of Jesus into a creed and a philosophy. Detractors of Paul feel that his organization of the beliefs of Jesus into a "system" spoiled their simple freshness. Yet if it had not been for the efforts of Paul, Christianity might have died out as another small and unknown sect.

Almost all of what we know about Paul is from the biographical data in Acts, and in the letters (the epistles) he wrote to the young churches. When there is a conflict between information in Acts and in one of Paul's letters, biblical scholars prefer to believe what Paul himself wrote. The letters were a contemporary and firsthand source, though much of Acts was based on secondhand or even thirdhand information, written down almost a generation later.

Paul was born in Tarsus, the most important city of Cilicia. This area, bordering the northeastern end of Mediterranean, was a part of the Roman province of Syria.

Paul's parents were Jews, descended from the tribe of Benjamin. They named their son "Saul" after King Saul, one of the most illustrious of their tribe. The father also possessed Roman citizenship, which was passed on to his son. Perhaps the father early began to call his son by the Latin name "Paulus" (meaning small); or this preferred name may have been adopted when Paul began his work among the Gentiles. Paul always signed his own name thus.

Ancient Tarsus was a bustling and important city in the Roman Empire. Although located a short distance inland, Tarsus was bisected by a broad river that flowed into the Mediterranean. Merchant ships could sail up to the city's port to load and unload a dazzling array of wares upon the wharves.

Just north of Tarsus were the towering Taurus Mountains, separating the province of Cilicia from Galatia and the rest of the Asia Minor interior. A paved Roman road wound down through a deep and narrow pass, called the Cilician Gates. This trade route connected the Galatian towns and villages with the port of Tarsus and funneled more commerce and activity into Paul's home town.

So Paul grew up in a cosmopolitan city, among Jews, Cilicians, Romans, Greeks, Syrians, Egyptians—people of all races and nationalities and cultures. Among the Greek colonnades and in the busy marketplace he probably overheard discussions on Greek philosophy and other learned subjects.

With other boys Paul could go to the stadium and athletic field, to watch athletes practice for the Olympic Games and local competitions. These were exciting to watch: pole vaulting, discus throwing, broad jumps, the footraces. In the races it was customary to set the prize in a conspicuous place at the point where the race was to end. This urged the runners to "press on to the goal," forgetting all else.

This youthful exposure to the pagans, their ways of life, and their thinking, colored Paul's own beliefs and helped him to understand the Gentiles when he began his lifework among them. His childhood memories also provided him with colorful figures of speech.

Paul received a formal Jewish education, a grounding in religious history and literature. Like most Jewish boys, rich or poor, he also learned a trade. His apprenticeship was with a tentmaker.

As a young man, Paul journeyed to Jerusalem to continue his religious studies under the best expounders of Jewish law. He became a Pharisee; like them, he had a tremendous desire to attain righteousness by knowing and keeping the Jewish laws.

Paul was a unique person. He was proud, quick-tempered, sometimes sarcastic, and without the sense of humor possessed by Jesus. Wherever he went, Paul soon became the center of conflict. He himself admitted that he had a tendency to boast.

Yet Paul had a genius for making friends and inspiring loyalty. He is one of the most interesting and influential men in our Bible.

Saul of Tarsus

When the apostle Stephen was stoned to death outside Jerusalem, a young Pharisee guarded the coats of the persecutors. He was Saul, a student from the Asia Minor port of Tarsus.

Saul helped to capture followers of Jesus and drag them off to jail. As a result of this persecution, many others who believed in Jesus left Jerusalem and fled to Samaria, Galilee, or Damascus.

Saul, boiling over with hatred for the disciples of Jesus, went to the high priest in the Jerusalem temple. "I've heard that an active colony of these renegade Jews is growing in the Jewish section of Damascus," the young Pharisee said. "They are preaching their false doctrines there. If you will give me letters of introduction to our synagogues in Damascus, my friends and I will seek out the heretics. We'll bind them in chains and return them to Jerusalem for trial."

"Go," the high priest said. The Jewish senate, the Sanhedrin, had agreed that it was good politics to blot out all memories of the disturbing man they had sentenced to crucifixion. Was not Rome merciless when unrest or revolt threatened in her provinces? In the interests of self-preservation the Jewish authorities wanted to keep Palestine undisturbed and quiet.

So Saul and his friends took the Jericho road eastward from Jerusalem, through the barren wilderness. When they reached the Jordan they forded the river and hiked north. A couple of days later they skirted the pleasant shores of Lake Galilee.

"The Nazarene often preached here," one of the young Pharisees muttered.

"The accursed blasphemer," their leader snorted.

At last Saul and his companions drew near to Damascus. The snow-covered peak of Mount Hermon soared to their left. Ahead, mountain streams cascaded down to a fertile valley, feeding a network of sparkling streams and canals and relieving the dry and stark desert. The ancient city of Damascus flourished in the midst of her lush oasis. Wheat fields and orchards stretched beyond the city walls.

Suddenly, there on the road, it seemed to Saul that a light from heaven was flashing about him. His eyes were dazzled; he closed them and stumbled to the ground.

Then he heard an awesome voice. "Saul, Saul. Why do you persecute me?"

The young Pharisee's heart was pounding. His eyes, pained by the light, remained closed. "Who—who are you?" he stammered.

He heard the voice answer, "I am Jesus, whom you are persecuting. Get up and enter the city. You will be told what to do."

The companions did not know what was happening. They were frightened when he rose dizzily, opened his eyes and turned his head with a blank expression.

"I cannot see," he gasped. "I am blind."

His friends took him by the hand and led him into the city.

In the crowded Jewish section of Damascus a disciple of Jesus also had a vision. He heard his name called. "Ananias."

"Here I am, Lord," the disciple answered reverently, for he recognized the voice.

"Rise and go to the street called Straight," he was told. "Inquire at the house of Judas for a man who has just arrived there. He is Saul of Tarsus. You will find him at prayer, for he also has had a vision. Lay your hands upon him and restore his lost sight."

Ananias protested. "Lord, I've heard of this Saul of Tarsus. He has jailed and persecuted your disciples in Jerusalem. We hear that he comes to arrest your followers here."

"Go," the voice repeated. "He is chosen to carry my name to the Gentiles."

Ananias made his way through the crowded, twisting streets, through the age-old bazaar where merchants shouted their wares, until he came to the house where the blinded Pharisee was. He found him on his knees praying.

"Brother Saul," Ananias called out trying to forget the young man's past. He laid his hands on him saying, "The Lord Jesus has sent me that you may recover your sight and be filled with the Holy Spirit."

It was as if scales fell from Saul's eyes. He looked around. He could see.

Rejoicing, feeling like a brand new person, the young man leaped to his feet. His dazzling experience had converted him. "I want to be baptized, now," Saul said. And he was. Then he broke bread with his new friend.

After that Saul visited every synagogue in the Jewish quarters of Damascus. "Jesus is the Son of God!" he proclaimed.

The followers of Jesus were suspicious. Was this not the young persecutor who hated the Nazarene?

The unconverted Jews were equally puzzled. They knew Saul had arrived with the blessing of the high priest to arrest all heretics. Yet here he was speaking like a heretic himself.

So Saul found it difficult to make friends. Except for Ananias, no one would trust him or listen to him. The former Pharisee felt the need to retreat to the desert where he could meditate and make plans for the future. He hiked out into the wilderness of Arabia, an area which the Romans allowed King Aretas to rule. Whenever Saul wandered into an oasis for food and water he tried to tell the Arabs there about Jesus.

When Saul returned to Damascus he began to preach the gospel of Christ with great fervor. After some time he made new friends. They came to fear for his life.

"You must leave Damascus immediately," they warned him. "The unconverted Jews of this city are plotting to kill you. And King Aretas has instructed the governor to set guards at all the city gates. They watch day and night. They intend to catch you if you leave here."

So a plan was devised to help Saul escape. In the middle of the night when there was no moon, Saul and his friends went to a house built against the old city battlements. From an upstairs window they lowered him in a basket over the wall. Saul leaped out, wrapped his cloak tightly about him, and scurried silently away in the night.

Following the stars he made his way south in the darkness. He stumbled through fields of grain, sloshed through water, and finally came to the main road.

Saul of Tarsus was safely on his way back to Jerusalem. What would happen now?

Acts 7:58 through 8:4; 9:1-25

Paul and Barnabas

When Saul arrived in Jerusalem he tried to join the disciples there. But they were afraid of their former persecutor.

"How can we trust him?" they asked. "This may be a trick."

The apostle Barnabas stood up and pleaded Saul's case. "Do you know that he saw the Lord on the Damascus road?" Barnabas told them. "In Damascus he was converted, and he preached boldly there in the name of Jesus."

The Jerusalem Church still hesitated to accept Saul. Then his old friends, Pharisees and Hellenistic Jews, heard of Saul's change of heart—and that he was back in town. They were enraged because Saul was trying to convert Jews into followers of Jesus. They plotted how they might kill him.

Saul was warned, "You will lose your life if you stay in Jerusalem. Go back to your own city of Tarsus and preach there."

Saul left by night for the great Roman port of Caesarea. He bought passage on a ship which would put in at Tarsus. By the next afternoon he was on the choppy sea, sailing northward on the blue Mediterranean.

Some years went by. The followers of Jesus heard reports of a growing congregation of believers in Tarsus. This group was led by none other than Saul, but now he was increasingly known by his Roman name, Paul.

An even larger congregation had grown up in the huge city of Antioch, which had a large Jewish quarter. Antioch was strategically located at the northwestern end of the Mediterranean. It was an important and wealthy trade city, the third greatest city in the empire, surpassed only by Rome and Alexandria.

The disciples in Antioch were first dubbed "Christians," and the name stuck. After a while all the followers of Jesus Christ came to be called Christians. In time they accepted the name for themselves.

The apostle Barnabas, who grew up on the nearby island of Cyprus, joined this rapidly growing fellowship. After a while there were so many new Christians in Antioch that the congregation needed more apostles for the preaching.

"I can recommend a man who is full of zeal for the Lord," Barnabas said. "Paul, of Tarsus."

So Barnabas went to Tarsus. He found Paul and brought him back to Antioch as his coworker. For a year Paul and Barnabas preached and evangelized together.

Then word came of a terrible famine in Judea. People were starving. There was no grain or meat. Mothers had no milk, and infants and children were dying.

Quickly the Antioch church took up a collection to help. Syrian wheat and dried figs were bought and loaded on pack mules. Large numbers of sheep and goats were purchased to provide meat and milk.

Barnabas and Paul were in charge of delivering the relief shipments for distribution to Jewish Christians in Jerusalem.

Now for the first time Paul found that the Jerusalem church was willing to accept him. He had proved himself a true friend. While in Jerusalem the two apostles stayed at the home of Mary, the sister of Barnabas.

Mary's son, John Mark, was fascinated by his uncle's stories about the work he and Paul were doing in Antioch.

"Take me back with you," the young man begged his uncle.

Paul thought it might be a good idea. This nephew of Barnabas had a contagious eagerness. "Let the youth join us in our work," he said.

So Paul and Barnabas returned to Antioch taking John Mark with them.

Although Antioch was a few miles inland, its inhabitants were very conscious of the nearness of the island of Cyprus to them. There was much trade between Antioch and that large island. From Seleucia, the coastal port of Antioch, the mountains of Cyprus looked tantalizingly near on a clear day.

"We have grown big and prosperous," the people of the Antioch congregation said among themselves. "Let's send Paul and Barnabas over to Cyprus as missionaries. They can evangelize for Christ there."

Members of the congregation celebrated their decision with fasting and prayers. Then there was a laying on of hands, to bless their chosen apostles.

"Let me go too," the eager young John Mark begged. And so it was agreed.

Boats regularly plied the mighty Orontes River, taking paying passengers from Antioch down to its harbor of Seleucia. But Barnabas, Paul, and John Mark decided to walk to the seaport to save money.

The three crossed the Orontes on the Roman bridge and joined the many travelers coming and going on the paved road. Spring was beginning. The air was crisp and cool, and skies blue. Shrubs along the roadside wore new green leaves.

Often the sandaled or barefoot travelers were forced to the side of the road, as an iron-wheeled chariot thundered past or a horse-drawn loaded cart clattered by. Nearby, the river bore almost as much traffic as the road. There were boats loaded with grain, figs, woolens, and other wares from all parts of the Roman Empire.

At Seleucia the apostles bought passage on a ship crossing the channel to Cyprus.

The three men stood on deck as the ship made its way out of the harbor. The long breakwater piers were made of gigantic blocks of stone, fastened to each other with great iron clamps. Ahead lay the island's mountaintops, looming above the low-lying mist on the horizon.

A brisk wind bellied out the sails of the ship. In a few hours the three were disembarking at the large port city of Salamis on the eastern end of Cyprus. They went directly to the synagogues in the Jewish section of the town. Here Barnabas and Paul preached about Jesus, and John Mark assisted them.

"We've had a fruitful mission here," Barnabas said with great satisfaction after a month in Salamis. "Let's cross the island to the port of Paphos. It's not as large as Salamis, but it's important. The Roman governor of Cyprus has his palace there."

So the three climbed the road that traversed the mountains of Cyprus, joining the eastern end of the island to the western. At this altitude chilly winds whipped the cloaks about them; their lips and fingers were often blue and their feet icy cold in their open sandals.

At last the road wound down past a cavern-pocked mountainside to the capital city. Here Paphos nestled on a small plain along the western seashore.

Again their work bore good fruit. Even the Roman governor listened politely to their message.

"Let's not return the way we came," Paul suggested to his companions after a month in Paphos. "Why don't we take a ship north to the mainland? Perga is an important city, and I know we could reap a good harvest for the Lord there."

Barnabas and his nephew agreed. They

340

sailed northward into a deep bay and about seven miles farther upriver to Perga. This was a great mercantile harbor that seemed more Greek than Roman with its gracefully fluted Greek columns, many pagan temples, and a huge amphitheater. Behind the city rose the awesome Taurus Mountains.

Here as usual the three friends preached in a Jewish synagogue. But Paul was not content. Suddenly he was seized with a new ambition.

Looking up at the mountains, Paul said to Barnabas and John Mark, "Beyond these hills lies the great inland province of Galatia. Let us travel about there and evangelize it for Christ."

Barnabas was willing, but young John Mark was not. Had not everyone heard tales of the fierce marauding tribes who lived on those high plateaus of the interior? The influence of Greece and Rome had not changed many customs there. The coastal cities of Asia Minor were civilized, but who knew what unknown perils lurked in Galatia? Besides, the young man was homesick.

So John Mark left Paul and his Uncle Barnabas in Perga and went home.

Acts 9:26-31; 11:19-30; 12:25 through 13:5, 13

The Trip to Galatia

Paul and Barnabas trudged up the winding road leading northward from Perga, through craggy passes in the mountains. Oleander bloomed in a bright pink profusion along the lower watercourses. There was still snow on the highest cliffs above them.

Although this area was often infested with bandits the two apostles were lucky. They were not attacked on this journey.

At last they reached the heights of the vast treeless tableland of the interior. Once they hiked for hours past a dreary inland sea of salt. In places the barren ground was burnt and volcanic. Occasionally there were green highland meadows where sheep placidly grazed. Once there was a sweet-water lake, and more than once, blossoming orchards.

The towns were usually monotonous groups of small flat-roofed huts. Sometimes the travelers passed encampments of tents. There were horses tethered to stakes and browsing sheep and goats and grubby dark children watched the apostles pass by.

Not far from the northern shore of a beautiful blue lake, they came to a town that bore the familiar name of Antioch. This town was a Roman colony planted in the heart of Asia Minor by the same ruler who built and named the larger coastal city. This Antioch was about halfway between Tarsus and the great Grecian city of Ephesus on an important interior road.

This inland Antioch was not nearly so large and splendid as the Antioch of Syria. But there was a Jewish settlement in it, a good place to bring the Christian message.

On the sabbath Paul and Barnabas went to the Jewish synagogue to tell about Jesus. After the usual reading from the Holy Scriptures of Moses, Paul had his chance to rise and address the men of the synagogue. Behind the lattice of a small balcony, the women were also listening.

Paul reviewed the history of the Jews and the scriptural passages which mentioned the long-looked-for Messiah. Then he told how Jesus was indeed that very

Messiah, and how the leaders in Jerusalem had not recognized him and had caused Pilate to slay him.

"But God raised Jesus from the dead," Paul declared with emotion. "Many at Jerusalem saw our Christ again. Therefore you can believe in this man, my brothers, and through your belief in him, you can be redeemed from your sins."

After Paul's speech there was a benediction and a solemn "Amen." Then many of the people of Antioch crowded around Paul and Barnabas, wanting to know more. Some asked how they could be baptized.

"Come back next week and speak to us again," they begged Paul.

During the week Paul and Barnabas went around to private houses, speaking to the people. News of the two men and their message flew about the city. On the following sabbath the synagogue was packed to overflowing. Even curious Romans, Greeks, and other Gentiles turned out.

Again Paul spoke to the crowds. He made it clear that the gospel of Jesus Christ was open to all, Gentiles as well as Jews.

This was more than the Jewish members of the synagogue could bear. "Stone him!" some yelled. "Run these heretics out of town," others cried.

The meeting turned into a riot. Paul and Barnabas grabbed their cloaks and their books, beating a hurried retreat on the road that led eastward to Iconium. They shook the dust of Antioch from their feet.

As they hiked east on this bleak upland road, Barnabas sighed and quoted a saying of Jesus: "Blessed are you when men revile you and persecute you and say all manner of evil against you falsely, for my sake."

Paul nodded glumly. "Peter told me that Jesus also said, 'Rejoice and be very glad, for great is your reward in heaven. In the same way, the prophets before you were persecuted.'" So Paul and Barnabas looked for ways to rejoice.

After a tiring walk they came to Iconium, in the midst of a great plain. Again they went to the local synagogue.

Even the local Greeks came to find seats in the synagogue, to hear what these two newcomers had to say. The Greeks always were curious about any new religion.

Again there were those who opposed the teachings of the itinerant preachers. Some of the elders began to say, "These strangers should be stoned. They are stirring up unrest and division here."

Paul and Barnabas learned of the plot against them. They fled about twenty miles southward to the village of Lystra.

In Lystra, Paul noticed a crippled man listening to his sermon. He looked closely at the man and said, "I see that you have the faith to be made well. Stand up on your feet!" And the cripple sprang up and walked.

"Only gods have that kind of power," the awed crowds murmured. "Truly the gods have come down to us in the likeness of men."

Since Barnabas was tall and muscular, with the regal bearing of authority, they called him "Zeus," the name of the chief Olympian god. Paul was small, lithe, and quick, full of eager speech, so they called him "Hermes," after the herald and winged-sandaled messenger of the Greek gods. Even the priest of Zeus, who presided at the pagan temple there, came bringing garlands. He also prepared to sacrifice to the apostles.

When Barnabas and Paul heard of this, they tore at their robes in frustration. They rushed out among the crowd, crying, "Why are you doing this? We are not gods! We are men, like you, asking you to turn away from dead idols. We come bringing the good news of a living God."

Soon there was another interruption to their ministry at Lystra. Delegates arrived from the synagogues of the last two towns the apostles visited. They were boiling over with hate for such disturbing fellows.

"These strangers in your town are heretics and evil!" the delegates from Antioch and Iconium harangued the people of Lystra. "You say they cast out devils to heal a lame man? Of course, they are not gods! They are also devils, using diabolical magic. It's only so you will believe their heresy. Come, let us stone them to death before they do more harm!"

The aroused mob could not find Barnabas. But this time Paul did not escape. He was discovered preaching in a private home. The crowd pushed him into the village street and stoned him there. Then they dragged him outside the city and left him. He was covered with blood, and appeared to be dead.

Barnabas and some new Christian friends found the wounded Paul and took him back to the city.

One of the new converts was a young man named Timothy, whose mother was Jewish though his father was Greek. Timothy helped Barnabas to sponge clean Paul's wounds, and to bandage them. By dawn the next day, Paul could walk, but he was stiff and sore.

Paul said to Barnabas and to young Timothy, "Our Master has told us that when they persecute us in one city we should flee to another. I intend to continue on to Derbe."

"May I come with you as far as Derbe?" Timothy begged. "I want to hear more about the teachings of Jesus."

Paul and Barnabas agreed. The three walked across the plain the short distance to Derbe. There they preached and rested from their strenuous journeying. Young Timothy absorbed everything that the two apostles could tell him of this new gospel.

At last Paul and Barnabas knew they must report back to their home church in the Syrian capital of Antioch. Retracing their steps, they returned quietly to Lystra. There Paul and Barnabas met with their recent converts.

"We have come back to strengthen your souls," Barnabas told them. "We urge you all to continue in the faith."

Paul, still somewhat stiff and swollen from his stoning, agreed. "We must be prepared for persecution and suffering, if we wish to enter the kingdom of God," he reminded them.

Deacons and elders were appointed, as had been done back at Derbe. The new church needed organization, so it might continue to grow.

There were tearful farewells between the new friends. It was especially sad to part from Timothy. The young man stood at the edge of his village, watching until Paul and Barnabas became mere specks in the distance.

At Iconium there were greetings, instructions, prayers, the appointment of elders, and again fond farewells. And it was the same story at the interior town of Antioch.

Here Paul paused at the crossroads. He looked longingly at the highway which led eastward over the great interior plateau. Paul knew that this road led to the west coast port of Ephesus, the great and thriving Greek city.

"I wish we had time to go on to Ephesus now," Paul said wistfully to Barnabas. "Such opportunities for work there. And think how it would be to go on even to Rome itself! I dream of that."

But winter was soon coming. They must quickly go south and on through the mountains before the snows came.

At last the two weary travelers arrived again at the coastal city of Perga. They took passage on a ship and sailed back to their home church, in Antioch of Syria.

Thus ended the first great missionary journey of Paul and the genial Barnabas. They had covered the island of Cyprus and traveled to the cities of Galatia.

"Yes, Barnabas and I preached the gospel among the Jews," Paul told his assembled home church at Antioch. "And a few of them listened. But most of our converts were former pagans. God has opened a door of faith to the Gentiles."

Acts 13:14 through 14:28; 16:1

A Letter to the Galatians

Some time after his first missionary trip with Barnabas, Paul began to hear disturbing reports from some of his newly established Christian churches. There was news from the churches of Galatia which caused Paul to be concerned about arguments and quarreling in these congregations. The Gentiles among his converts were backsliding into the old ways of idol worship.

Also, other preachers had come along saying, "You must first become Jews, and then you can be Christians." This confused the new believers. "You must follow Moses and every Jewish law. Faith in Christ is not enough," the converts were warned.

This was the very opposite of what Paul had taught. So the apostle wrote a careful letter to the Galatians.*

Paul, an apostle, to the churches of Galatia:

Grace to you and peace from God the Father and our Lord Jesus Christ.

I am astonished that you are so quickly deserting me and turning to a different gospel—not that there is another gospel, but there are some who trouble you and want to pervert the gospel of Christ.

O foolish Galatians, who has bewitched you? Christ redeemed us from the curse of having to follow every letter of the laws of the Jews. Through faith in Christ Jesus the blessing of Abraham can come directly to the Gentiles.

Christ breaks down all barriers. There is neither Jew nor Greek, there is neither slave nor free, there is neither male nor female, for you are all one in Christ Jesus.

Sometimes I am afraid I have labored over you in vain.

You know of the bodily ailment I had when I first preached to you. Though my condition was a trial to you, yet you did not scorn or despise me. You received me as an angel of God, as Christ Jesus. What has become of the loyalty you felt? For I bear witness that, if possible, you would have plucked out your eyes and given them to me. Have I then become your enemy by telling you the truth?

I wish I could be present with you now, for I am perplexed about you.

You were called to freedom, my brothers. Use it to be loving servants of one another. For the whole law is fulfilled in this: "You shall love your neighbor as yourself." If you bite and devour

344

one another take heed that you are not consumed by one another.

Do not be deceived. God is not mocked, for whatever a man sows, that he will also reap. And let us not grow weary in well-doing, for in due season we shall reap, if we do not lose heart.

The grace of our Lord Jesus Christ be with your spirit, brothers.

(adapted)

The Letter of Paul to the Galatians

* This letter was probably written around A.D. 52. It is a very important writing because of Paul's explanation that faith in Jesus Christ rather than obedience to the Jewish law is the ground of salvation. This made Christianity attractive to people who were not Jews and gave the new faith universal appeal.

Some New Churches

Winter blew its frigid breath from the north. Storms whipped the Mediterranean, and ships stayed at anchor in the harbors. Ice and snow blocked the mountain passes.

When spring came to Antioch, Paul said to Barnabas, "Come, let's return and visit the brothers in Galatia."

"Fine," Barnabas agreed. "First let me send to Jerusalem so my nephew can join with us again."

Paul frowned and shook his head. "John Mark? No, he left us the last time."

Barnabas laughed. "He is young, and that was his first trip away from home." He flung his arm across Paul's shoulder.

"Surely we can give him another chance?"

Paul did not think this a very good idea, but Barnabas insisted. They had a heated argument about the matter. "Perhaps it would be better for you to go your way with John Mark," Paul told Barnabas. "I shall take Silas for my companion. He is one of our best workers in Jerusalem."

So Barnabas and John Mark set sail for their home island of Cyprus to settle down there and preach the gospel. And Paul took Silas, whom he usually called by his Greek name, Silvanus, and set out for Galatia.

Paul and Silas decided to go overland instead of by sea. They walked to Tarsus, then went up through the narrow gorge in the Taurus Mountains known as the Cilician Gates. They came to Derbe and Lystra in Galatia.

"Welcome, welcome back," said the citizens in both towns. They responded enthusiastically to Paul's preaching, and they listened to Silas also.

In Lystra, Timothy rushed joyfully to greet Paul. "Let me accompany you and Silas in your mission," he begged. "I want to be an assistant, and I'd like to help preach the gospel."

Paul was glad to add Timothy to his missionary team. Both he and Silas liked the young man's eagerness. The three went on together, revisiting the other new churches in Galatia.

From Antioch one road led northwest to a town named Troas near the ancient coastal town of Troy. Another road led straight west to Ephesus.

Paul gazed longingly down both roads stretching across the vastness of the great interior highlands. "I'd like to go on to Troas, and I also want to preach in Ephesus," he said to his companions. "But which shall it be?"

Then he made a quick decision suddenly

inspired. "First we'll take this road to Troas. Perhaps later we can work southward down the coast to Ephesus."

They started across the treeless plains. At last the tired missionaries reached Troas. There they met a warm and witty Greek physician, whose name was Luke. Luke listened to Paul's preaching and became an eager convert. Then he asked to join the missionary party.

Paul, Silas, and Timothy liked Luke. He was intelligent and charming, as skilled with the pen as he was in the use of bandages and medicinal herbs. They welcomed him to their team.

While still at Troas, Paul had a vision of a Macedonian who begged, "Come over to Macedonia and help us." So Paul and his three companions set sail across the blue Aegean for that area north of Greece, the home country of Alexander the Great.

Their ship docked at a small port city. This was familiar territory to Luke, so he led the way on the inland road. They came to a small Roman colony, Philippi, one of the most distinguished cities of Macedonia.

On the sabbath day Paul and his three companions went outside the city gate to the riverside. Here some Macedonians were gathered together for prayer. The missionaries joined the group, and each told of the wonderful gospel of Jesus Christ.

Lydia was a woman who listened most eagerly. She came from an Asia Minor town celebrated for its purple dyes. Here in Philippi, Lydia continued to work at her trade, dyeing and selling purple goods.

Lydia was convinced. This was the living faith she had been seeking all her life. "Will you baptize me?" she asked. "And my whole household with me?"

Lydia, her children, her slaves, and her hired workers were baptized there in the river.

"And now, if you count me as faithful to the Lord," Lydia said to the four missionaries, "you must accept the hospitality of my household."

So Paul, Silas, Timothy, and Luke were well cared for in this Roman colony.

As the evangelizing went on, an incoherent slave girl kept following Paul and his companions through the streets of Philippi. She muttered things like, "These men are the servants of the Most High God."

This mentally unbalanced girl was owned by several people; they had bought her, wanting to exploit her as a soothsayer who would predict the future for money.

Paul was exasperated by the girl's attention. He felt that the endorsement of a deranged person was of no value, even if the townspeople did regard her with superstitious awe. At last he turned to her. "In the name of Jesus Christ, you demon," Paul said, "come out of this girl." Immediately the girl became calm and rational.

Her owners were furious. "How can we make money from her now?" they cried out in anger. "She can no longer sell oracles for us." They called several burly servants and pointed out Paul and his companions. "Seize those men!"

Timothy and Luke ran and escaped into the crowd. Paul and Silas were seized and dragged before the Roman magistrates.

"These men are Jews, and they are disturbing our city," the irate slaveowners charged. One added cunningly, "They advocate strange customs that are not lawful for us as Romans to practice."

Many persons in the crowd were angry. They had been patrons of the slave girl. Because of Paul their source of oracles was gone. They began to riot. Surging forward, the crowd began to attack Paul and Silas.

The magistrates jumped in and ripped the cloaks and shirts from the missionaries. "Beat these two prisoners with rods," they ordered, trying to keep on top of the situation.

When the flesh of Paul and Silas was crisscrossed with bloody stripes the two men were dragged off to the city prison.

"Keep them safely," the jailer was told.

So Paul and Silas, in chains, were marched along moldy corridors, down torchlit steps, into the innermost dungeon. Their feet were clamped into stocks so they could not move. The iron door clanged shut, and they were left in the dark with no sound except for the scurrying of nearby rats.

Paul and Silas began to pray. Then they sang aloud. "The Lord is the stronghold of my life. Of whom shall I be afraid? . . . Be strong, let your heart take courage. . ."*

The other prisoners were listening. They were awed by this strange sound of triumphant singing from within a dark prison cell.

Suddenly, at midnight there was a great earthquake. The ground rumbled and shook. Stones tumbled from the walls. Great cracks appeared in the foundations of the prison. The doors swung open. Fetters were broken loose from the prisoners.

The trembling jailer ran from his adjoining house into the opened prison. He saw the cell doors were opened, and he trembled even more. Was he not responsible for the security of his prison? If his prisoners escaped, the Roman authorities would execute him.

"This is the end of my life," he cried out in despair. He pulled out his sword to kill himself.

"Do not harm yourself," Paul called in a loud voice, "for we are all here."

"Lights!" the jailer called out to his underlings. "Bring me torches." He rushed into the deepest part of the dungeon. There were Paul and Silas sitting calmly amidst the wreckage of their cell.

The jailer escorted them out, treating them with great awe. "Men, what must I do to be saved?"

Paul and Silas told him, "Believe in the Lord Jesus. Then you and your household will be saved."

So the story of Jesus and his gospel was preached to the jailer and his household, who gathered around these two extraordinary prisoners. Although it was still the middle of the night, they went to a place where there was water. First the jailer and his wife washed and bandaged the lash wounds of the missionaries. Then Paul and Silas baptized the whole family, along with their slaves and servants.

After that they all went back to the jailer's house. Food was brought out for Paul and Silas.

When it was daylight, the town magistrates sent some policemen with a message for the jailer: "Let those men go."

The jailer came to Paul and Silas, all smiles. "The magistrates have sent word that you may go. So come outside, and go in peace."

But Paul sat down and refused to budge. "They have beaten us publicly. Yet we were not even tried and condemned. Do you know we are Roman citizens? Roman citizens may not be beaten. And a citizen may appeal any judgment directly to the emperor!"

When this message was taken back to the magistrates they looked at each other fearfully. What if the emperor heard of this? The magistrates hurried to the jail in a body. Puffing from their haste, they bowed ceremoniously before Paul and Silas, "We apologize," they said, full of politeness. "Now, before there are any more complaints or riots, please leave our city."

Paul and Silas went from the prison to Lydia's house. Here they found Timothy and Luke, safe and unharmed.

Timothy and Luke were charged to stay on in Philippi, to help the new church get on its feet.

Paul and Silas traveled southward, then turned west to Thessalonica. This was the largest and most influential city in Macedonia, named for a sister of Alexander the Great. It sprawled on a plain overlooking a gulf of the sparkling blue Aegean Sea.

Paul and Silas found a place to stay in the Jewish section. They met many Jews in a synagogue there. For three weeks they argued with them, using the Jewish scriptures to prove their points.

"It was necessary for the Christ to suffer and to rise from the dead," Paul said. "And this Jesus, whom I tell you about, is the Christ."

A few of the Jews were persuaded, and joined Paul and Silas. Some of the Greeks and several of the leading women of the city were also won over to Christianity.

But the unconvinced Jews became very jealous. They gathered together a rabble-rousing crowd, incited this part of the city into an uproar, and attacked the house where Paul and Silas were staying.

This time Paul and Silas were able to avoid capture. But the new converts in Thessalonica were afraid for the lives of their two teachers.

"You must leave us," they told the two men sadly. Under cover of night the two missionaries were shepherded out of town. They took the road to the inland, mountainside town of Berea, about forty-five miles to the west.

At Berea, Paul and Silas went directly to the Jewish synagogue. Here they were joined by Timothy, who had been following in their footsteps.

These Jews in Berea were more open and tolerant than the ones back in the city of Thessalonica. They eagerly received the preaching of the three apostles. Every day they examined their scriptures to see if the things the missionaries told them were so. Many Jews were baptized as Christians, as well as a number of Greek women of high standing.

But when the angry Jews of Thessalonica heard how well the missionaries were doing in this small town, they journeyed here and began to stir up the people.

"These troublemakers from Thessalonica are primarily after you, Paul," some of the new converts told him. "You must move on. But please leave Silas and Timothy with us for a short while."

Paul agreed. His new friends took him down to the seaport. There they bought him passage on a ship sailing for Athens.

And Paul asked Silas and Timothy to join him as soon as they organized the new church in Berea.

Acts 15:36 through 17:15

* From Psalm 27.

In Greece

Paul welcomed the opportunity to preach in Athens. The most famous city in Greece with a glorious past, Athens was proud of being the cultural center of the Greek-speaking world.

Paul climbed the chiseled stone steps to the imposing Acropolis. This rocky hill rose like a resplendent crown over the city. On its summit were clustered temples and shrines dedicated to the numerous Greek gods. The greatest structure was the pillared Parthenon, dedicated to the mythical goddess Minerva.

Towering over the whole Acropolis—

even above the majestic Parthenon—and so high it could be seen far out to sea, was a statue of Minerva. This colossal bronze sculpture represented the helmeted goddess, armed with spear and shield, the "champion of Athens."

Standing on the Acropolis, Paul could look to the northwest and see the rise of a smaller hill, the Areopagus. This was popularly called "Mars Hill," because of a temple to the war god on it.

Paul climbed down from the Acropolis and walked through busy streets, past beautiful buildings with fluted columns. He came to the agora. This was a marketplace, not at all like the noisy, smelly bazaars that were common in Asia.

The vast open space of the agora was enclosed by stately buildings. It was the meeting place of philosophers, orators, statesmen, poets, artists, and well-informed citizens. Athenians loved to come here

daily, eager for the latest news. Opinions were exchanged on everything from art and science to politics.

As Paul walked alone through these arguing, gossiping crowds, his thoughts strayed back to the new congregations he had left at Philippi and Thessalonica. He felt alienated from these sophisticated Athenians, so well educated and yet so shallow.

Everywhere, everywhere were the carved idols, the pagan shrines and altars: Neptune, seated on horseback, hurling his trident toward the sea; Apollo, patron of the city; Bacchus, god of wine and merriment. The Athenians, wanting to be perfectly safe, and to cover everything, even had altars to fame, to pity, and to beauty.

Every day as Paul walked through this marketplace, he argued with any who would listen.

Some learned Epicurean and Stoic

philosophers gathered around Paul one day. One asked, "What is this babbler talking about?"

"He seems to be a preacher of foreign gods," someone replied. "He talks about a new god and goddess. A pair that seem to be named Jesus and the Resurrection."

The crowd thickened around Paul. A murmur arose, "Bring him up to the quiet of the Areopagus, so we may listen more carefully to this new teaching."

Paul was escorted down the street and up the carved stone steps of the nearby hill. On the summit was a platform with benches hewn from the rock where wise men of the city could listen to solemn debates. Official judges had sat on this same platform to pass sentence in cases of great crimes. Four and a half centuries ago, Socrates had been condemned here for not recognizing the city's mythical gods, and for corrupting the young people with his philosophical ideas.

This was a crowd of the curious, only wanting to hear the ideas of a stranger. Still, the setting gave a dignity to the event.

"You bring strange things to our ears," one Athenian told Paul. "We are curious to know what they mean."

Paul, standing in the middle of the Areopagus, began with a tinge of sarcasm in his voice. "Men of Athens, I see that you are very religious. For as I passed along and observed the many objects of your worship, I found also an altar with this inscription: 'To an unknown god.'" Then he began to speak with great earnestness. "What you worship as unknown, I shall now proclaim to you."

The apostle made a sweep with his arm, indicating all the shrines that crowded the hillside. "Idols are found in these shrines," he said. "But the living God who made the world and everything in it does not live in temples made by man."

Paul tried to challenge his audience. There were sophisticated Athenians, so exclusive in their relationships with outsiders. But there were also transplanted aliens from many places in the world. There were a number of curious travelers on their way through this free Greek city.

"God has made from one blood every nation of men to live on all the face of the earth," Paul told them passionately. "Being then God's offspring we ought not think that God is like an image made of gold or silver or stone."

Then Paul told them about Jesus Christ and his resurrection from the dead.

When the Athenians heard this some jeered and shouted insults at him. Others were dubious and said, "We will hear from you again about this story of a resurrection." A very few followed after Paul and became converts. But there was not enough interest shown to start a new church in Athens.

Discouraged, Paul traveled westward to Corinth. Corinth was strategically located on the western side of a narrow isthmus. If the isthmus had been under water, the whole southern bulge of Greece would have been a great island.

As a port city Corinth became one of the great trade centers of the Mediterranean world. To avoid the long voyage in rough seas around the southern cape, cargoes were hauled across the isthmus on the backs of slaves, and also by oxcart. Small boats were even put on rollers and hauled overland from one sea to another.

Athens and Corinth were as different as two cities could be. Athens was a free Greek city. Corinth was a Roman colony with Roman architecture and roads, using Roman money.

Athens was proud, intellectual, and yet provincial in its outlook. Corinth was a government and mercantile center, more cosmopolitan in population and outlook. It was crowded with Roman officials, businessmen, artisans, sailors, and traders.

Wealthy Corinth was also full of high living, immorality, and wild religious cults.

Paul did not intend to stay long in Corinth. By the time Timothy and Silas were able to rejoin him here, however, the mountain passes were closed with the first snowfall. All ships had anchored in the harbors for the winter.

"Anyway, you could not go back to Philippi or Thessalonica," Timothy and Silas warned Paul. "Your life would not be safe in either town."

Already Paul had converted a Corinthian man named Aquila and his wife Priscilla.* They were Jewish tentmakers, who recently had fled from Italy and settled here. Paul took his board and room at their house, joining them in the business of tentmaking.

"Then I will stay on with Aquila and Priscilla," Paul told Timothy and Silas, "and preach for a season in Corinth.

Silas and Timothy found lodging in Corinth, too. Every sabbath they went with Paul to preach and argue at the synagogue. Some of the Jews were persuaded and joined with Aquila and Priscilla as new Christians.

Other Jews opposed and reviled Paul. They would listen to no arguments about Jesus being the Christ, the Messiah. Finally Paul picked up his cloak and shook it out before them. "I shake the dust of your synagogue from myself," he told them indignantly. "From now on I will take the good news to the Gentiles."

He left the synagogue and went to the house next door. Here he converted a Gentile named Titus, who already had learned to worship the Jewish God. Not only that, but the ruler of the synagogue followed Paul to hear more, and he was converted. Corinthian Gentiles crowded to the house of Titus to hear Paul speak. Many believed and were baptized.

The missionary work was so successful, Paul stayed on in Corinth for eighteen months. Here Paul founded a church that was both his pride and his despair. Faith flourished here, but the congregation was always backsliding.

Paul continued to have trouble with the unconverted Jews in Corinth. Some of them came as an unruly mob before the Roman tribunal, dragging Paul with them.

"This man is persuading men to worship God contrary to the law," they complained to the judge.

Gallio, the Roman tribunal, interrupted before Paul could open his mouth to defend himself. "Contrary to what law?" he asked the accusers.

"Our Jewish law," they grumbled. "He says his Jesus is our Messiah, but we say—"

Gallio held up his hand for quiet. "If it were a matter of wrongdoing or vicious crime, I would bear with you." He stifled a yawn. "But since it is a matter of words and names and your own laws, see to it yourselves." He stood up, dismissing the court. "I refuse to be a judge of these things." He beckoned to the Roman guard and had them all driven out.

After this Paul stayed in Corinth for a while longer. It had been two years since he and Silas left on this second missionary journey from Antioch. He had preached again in Galatia; he had established thriving new churches in Philippi, Thessalonica, and Corinth. It was time to return to Antioch and report on his work.

Timothy remained in Corinth to work with the three new churches. Paul and Silas bought passage on a merchant ship going back to Caesarea and Antioch.

Aquila and Priscilla came to Paul with

a suggestion. "We have heard that your ship will stop over at Ephesus for a few days to load and unload cargo," Aquila said. "Let us accompany you as far as Ephesus. Perhaps we can begin the work there."

So it was agreed. In the spring of that year their ship sailed the busy sea lane across the Aegean. It took two weeks to reach the huge bustling port of Ephesus on the shore of Asia Minor.

Aquila and Priscilla found a house they could rent. Here they could go to work at once on their trade and begin the nucleus of a new Christian congregation.

While the ship was docked, Paul went to the synagogue and spoke with the Jews there. Some were intrigued by the gospel he brought. "Stay longer," they begged.

Paul explained to the Ephesians that he had yearned to come before, and visit them. Now he set his heart on going to Jerusalem, in time for the Passover; and he had a report to make to the Antioch church.

"But I will return to you," Paul promised the Ephesians, "if God wills." He and Silas rejoined their boat. They left Ephesus and set sail for Syria.

Acts 17:16 through 18:21

* Priscilla is called Prisca in 1 Corinthians 16:19.

A Letter to the Thessalonians

Before Paul rejoined his home congregation in Syria, he sent off a letter to the new church he had established at Thessalonica.

While in Greece, Paul had worried about his converts in this Macedonian city.

Persecution by jealous, unconverted Jews went on unceasingly. Christians were tempted to give up their new faith. Paul felt that his flock there needed his encouragement.

Also at this time the missionary clung to the expectation that they all would see the Christ return in their own lifetimes. He wanted to share this hope with his converts. So Paul dictated a letter, adding the names of his coworkers. The epistle read in part:

Paul, Silas, and Timothy,
To the church of the Thessalonians:
Grace to you and peace.

We give thanks to God always for you all, constantly mentioning you in our prayers. Like us and the Lord, you have endured much suffering so that you have become an example to all the believers in Macedonia and Greece.

What a welcome we had among you! You turned to God and away from idols. Now you serve a living and true God, and you wait for his Son from heaven, whom he raised from the dead.

I want you to know, my brothers, that our visit to you was not in vain. Though we had already suffered at Philippi, and were shamefully treated there—as you know—we had courage to declare to you the gospel of God in the face of great opposition. We were gentle among you, like a nurse caring for her children. We were ready to share with you not only the gospel of God but also our own selves, because you had become very dear to us.

You remember our labor and toil, brethren. We worked night

352 and day, so that we might not burden any of you while we preached the gospel.

I, Paul, again and again wanted to come to you to see you face to face. But now that Timothy has come to us from you and has brought the good news of your faith and love, we have been comforted.

Aspire to live quietly, to mind your own business, and to work with your hands as we charged you. Then you can command the respect of outsiders, and be dependent on nobody. Be at peace among yourselves. Scold the idle, encourage the fainthearted, help the weak, be patient with them all. See that none of you repays evil for evil, but always seek to do good to one another and to all.

Give thanks in all circumstances. Abstain from every form of evil.

Greet all the brethren with a holy kiss.

I ask that this letter be read to all the brethren.

(adapted)

The First Letter of Paul to the Thessalonians

* This is the earliest letter written by Paul that has been preserved. It was probably written about A.D. 50 from Corinth.

A Letter to the Corinthians

Paul spent some time at Antioch, his home church, reporting on the work of his second missionary journey.

Silas left Paul there. "It was good to work with you in the new churches," he told Paul. "Now I go back to Jerusalem. I have promised to work with my old friend Peter."

Paul began to feel restless at Antioch. He knew now that his great ministry was to be for the Gentiles. More Christian churches must be founded. The new churches should be revisited and encouraged.

Timothy rejoined Paul at Antioch. Paul asked, "Will you go with me on another missionary journey?"

Timothy was eager. Together they traveled overland through Syria and into the heart of Asia Minor. Here in the towns of Galatia, they taught, preached, baptized, and built up the local leadership.

From this inland area Paul and Timothy moved westward to the coast. "This time we'll go directly to Ephesus," Paul said.

As soon as they reached the edge of this great port city they began to preach the gospel in the streets. A dozen men surrounded Paul and claimed to be new disciples.

"Were you baptized in the name of Jesus Christ?" Paul asked them. He thought they might be converts of Priscilla and Aquila.

The Ephesians looked puzzled. "We have not heard of that baptism," one said. "We were baptized in the name of John the Baptist."

So Paul told them more of the gospel. Then he baptized them in the name of the Lord Jesus.

By the time dusk had darkened the city, Paul and Timothy came to the street where Aquila and Priscilla had rented a house. His friends were still there.

It was a joyful reunion, everyone talking at once. Paul told how he had baptized some men that day who knew only of baptism in the name of John.

"Aha," Aquila said. "They must have been baptized by Apollos, before we expounded the way of God to him more accurately."

Priscilla chuckled at the puzzlement on Paul's face. She nudged her husband. "Paul has never met Apollos," she reminded him.

So Priscilla and Aquila took turns telling Paul about this young Jew recently arrived from the cultured Egyptian metropolis of Alexandria.

"An eloquent and fervent man." Priscilla looked pleased. "Well versed in the scriptures. He spoke and taught accurately many things concerning Jesus. But he only knew about baptism as done by John."

"He spoke boldly in the synagogue here in Ephesus," Aquila said. He motioned to his wife. "When Priscilla and I heard him we took him under our wing and instructed him as you did us."

"Then he wanted to go and preach in Greece," Priscilla interrupted. "So we sent a letter to the church at Corinth to receive him. We hear that he's especially successful with the Greeks because of his learning and eloquence."

Soon afterwards Apollos returned from his mission to Corinth. He was happy to meet Paul. "I had great success in making converts among the Greeks," he told the older apostle. "But did Aquila and Priscilla tell you about the immorality in Corinth, and the backsliding? They all talk about you and want to see you. Until you can make a trip, why not a letter?"

So Paul dictated a brief note and sent it to the Corinthians. "It is not necessary for good Christians to cut themselves off from necessary business and social relations with pagans. You should also show friendliness and kindness to others, even sinners. Yet these close ties should not pull you into immoral situations or actions. Do not become thus yoked," he implored his congregation there.

The Corinthians did not take Paul's instructions to heart. They did not even save Paul's note for future rereading.*

After a time, however, the Corinthians decided that they wanted some advice on marriage and other problems. Also Paul had mentioned that he was collecting money for the poor in Jerusalem, and they wanted to know more about that.

The Corinthians wrote to Paul, "When will Apollos return to us?" they asked. "And when will you come?"

Paul sat down to dictate a long letter to the Corinthians. This is part of that epistle:

> Paul, called by the will of God to be an apostle of Jesus,
>
> To the church of God which is at Corinth:
>
> Grace to you and peace from God our Father and the Lord Jesus Christ.
>
> I appeal to you, my brothers, that there be no dissensions among you. It has been reported to me that there is quarreling among you.
>
> What I mean is that each one of you says, "I belong to Paul," or "I belong to Apollos," or "I belong to Peter." Is Christ divided? Was it Paul who was crucified for you?
>
> Who then is Apollos? What is

Paul? Servants through whom you believed, as the Lord assigned to each. I planted, Apollos watered, but God gave the growth!

Like a skilled master builder I laid a foundation, and another man has been trying to build upon it. Let each man take care how he builds upon it, for the foundation is Jesus Christ.

Do you not know that you are God's temple, and that God's spirit dwells in you?

I am sending to you Timothy, my beloved and faithful son in the Lord, to remind you of my ways in Christ. I myself will come to you soon, if the Lord wills.

It is reported that there is immorality among you. I wrote to you in my letter not to associate with immoral men, but I was not referring to outsiders. It is those inside your church who continue to live like unbelievers whom you are to judge? Leave it to God to judge those who are outside your church.

If any man in the church has a wife who is an unbeliever, and she consents to live with him, he should not divorce her. If any woman has a husband who is an unbeliever, and he consents to live with her, she should not divorce him. For the unbelieving one is consecrated by his partner, and the children are made holy. For God has called us to peace.   (*adapted*)

At this point Paul paused in his letter. He had dealt with everyday problems of a particular church at a particular time. Now his thoughts soared to more universal and lasting ideas. He went on to compose a beautiful passage on the nature of Christian love.

If I speak with the tongues of men and of angels, but have not love, I am like a noisy gong or a clanging cymbal.

If I have prophetic powers and understand all mysteries and all knowledge, and if I have enough faith to move mountains, but have not love, I am nothing.

If I give away all that I have to the poor, but have not love, I gain nothing.

Love is patient and kind. Love is not jealous or boastful; it is not arrogant or rude.

Love does not insist on its own way. It is not irritable or resentful. It does not rejoice at wrong, but rejoices in the right.

Love bears all things, believes all things, hopes all things, endures all things.

These three, faith, hope, and love, live on. But the greatest of these is love.   (*adapted*)

At the close of his letter to the Corinthians, Paul reminded the saints** of that church that on this trip he planned to collect money for the poor among the saints of the Jerusalem church:

Now concerning the contribution for the saints: As I directed the churches of Galatia, so you also are to do.

On the first day of every week, each of you is to put something aside and store it up, as he may prosper, so that contributions need not be made when I come.

I will visit you after passing through Macedonia. Perhaps I will even spend the winter. For I do not want to see you just in passing. I hope to spend some time with you, if the Lord permits. But I will stay in Ephesus until Pentecost. A wide door for effective work has opened to me here, yet there are many working against me.

When Timothy comes, put him at ease among you, for he does the work of the Lord.

As for our brother Apollos, I strongly urged him to visit you. But it was not God's will for him to go now.

Be watchful. Stand firm in your faith. Be courageous, be strong. Let all that you do be done in love.   (adapted)

Paul stopped dictating. He reached for the letter, which a scribe had neatly written. With a pen he added a postscript in his own large, bold handwriting:

The churches here in Asia send greetings, as do Aquila and Priscilla.

I, Paul, write this greeting with my own hand. My love be with you all in Christ Jesus. Amen.   (adapted)

The letter was sealed, and sent by messenger on the next ship sailing across the Aegean Sea to Corinth.

Acts 18:22 through 19:10; The First Letter of Paul to the Corinthians

* Many scholars think that this letter may be preserved in 2 Corinthians 6:14 through 7:1.

** The dedicated Christians of the early church were called saints.

The Riot at Ephesus

Ephesus was built on the hills and the plain by the mouth of a large river flowing into the Aegean Sea. Ships from all the ports of the Mediterranean docked at the fine harbor. Roads led inland to the great markets of the interior. It was the greatest metropolis of Asia Minor.

Near one hill, a stadium was the center for athletic contests. A huge semicircular outdoor theater hugged the concave side of another hill. Here were given Greek comedies and tragedies.

The temple of Artemis, the most magnificent edifice in town, was built of locally quarried marble, with fluted Ionian columns of green jasper. A spectacular statue of the goddess dominated the temple area. Romans and non-Greeks usually called this goddess by her Roman name Diana. Pilgrims came from all over the Mediterranean world to worship here.

In one of the shrines dedicated to Diana was a large meteorite that had fallen from the sky. It also was worshiped as one of the symbols of the goddess.*

Local silversmiths and other artisans grew rich by selling magic charms, small images of Diana, tiny models of their temple, and other souvenirs to the many visitors.

When Paul came to Ephesus and preached only among the Jews at their synagogue, the artisans and tradesmen of the metropolis were not affected. The Jews lived in their own section of the city and did not patronize the cults or the temples.

But three months of preaching in the synagogue produced no results, so Paul withdrew from the Jews. He began to preach daily to the Greeks, Romans, and Asiatics who lived in Ephesus. Now Paul began to have success. Gentiles were listening to his message. Many were converted. They stopped patronizing the temples. No longer did they buy silver, bronze, or pottery charms for themselves or as gifts for friends.

The Christian gospel began to reach some of the visitors, too. After Paul had been in Ephesus almost two and a half years, the sales of magic items had dropped so low the craftsmen held a meeting.

Demetrius, a silversmith, cried out, "Men, you know we get our wealth from this business. Now this Paul has persuaded and turned away a considerable company of people, saying that any gods made with hands are not really gods."

"That's right," another silversmith yelled, standing up. "Paul even converted my own wife into a Christian. Now she harps at me that I shouldn't even be in this business any more."

Angry murmurs swelled from the crowd. Demetrius held up his hand. "If this Paul is allowed to continue his nonsense," Demetrius shouted, "it is not only our trade which may come into disrepute. Do you realize that there is danger that the great goddess Artemis may count for nothing?" He raised his eyes solemnly to the sky, then glared back at the assembly. "Can we let her be deposed from her magnificence? Our great Diana of Ephesus is worshiped by the whole world!"

"Great is Diana of the Ephesians," the artisans and tradesmen cried out.

The angry craftsmen charged out into the marketplace and began to arouse and alarm the citizens. A growing, unruly crowd rushed into the great outdoor theater. A party of men came, dragging with them two of Paul's companions.

Paul was temporarily safe inside the nearby house of Priscilla and Aquila, and was as yet undiscovered by the mob. When he saw what was happening he started to bound out of the house to chase after the crowd. His fellow apostles pulled him back inside and slammed the door, bolting it.

"You cannot go out there," they told Paul. "They are like wild beasts. They would tear you limb from limb."

"But two of my friends are out there," Paul pleaded.

"To the mob, your companions are only accomplices," the other apostles argued. "Keep calm and pray. The mob may free the men, when they cannot find you."

At the theater the assembly was disorderly and in confusion. Most of the people had been swept along, and did not know why they had come together. The ringleaders began to chant, "Great is Artemis of the Ephesians!" Soon thousands in the theater were chanting together.

Alexander, a Jewish coppersmith, tried to get the attention of the assembly. Since they knew he was a Jew they refused to listen to him.

In the house of Priscilla and Aquila, Paul's friends waited. Their faces were drawn with anxiety. What if the mob found out where they were and charged the door, breaking it down? Paul paced up and down. His only concern was for his two missionary companions being held hostage.

The chanting continued at the theater. Paul's two friends, pinioned between two burly craftsmen on the stage of the theater, waited quietly. They were not sure when the crowd might take their lives.

358 At last, as the sun sank like a scarlet ball below the line of the darkening sea, the town clerk was able to quiet the crowd. "What man here does not know that our city of Ephesus is temple keeper for the great Diana?" He shouted, "And also that we are custodians of the sacred stone that fell from the sky?"

There were hoarse cheers of approval from the crowd. The town clerk waited for quiet. "Seeing that these things cannot be contradicted," he went on, "you ought to be quiet and do nothing rash. For you have brought these men here as prisoners, yet they are neither sacrilegious nor blasphemous of our goddess."

Both of Paul's companions opened their mouths as if they intended to object. The clerk angrily gestured them to be silent.

"If therefore," he continued sternly, "Demetrius and the craftsmen with him have a complaint against anyone, the courts are open. Let them bring orderly charges. It shall be settled in the regular assembly. Don't you realize we are in danger of being charged with rioting today? There is nothing to justify this commotion!"

The clerk told the crowd to leave. Paul's friends were released, and they fled.

After the uproar ceased, Paul sent for all the disciples and followers of Christ in Ephesus. "I have been advised that it is too dangerous for me to stay on here," Paul told them. "I must leave you now." They prayed together, and then Paul slipped out of town in the middle of the night. He made his way overland to Macedonia.

Acts 19:1 through 20:1

* This Artemis-Diana in the mythology of the Greeks and Romans was the virgin goddess of the hunt. She was believed to be a protector and patroness of women. But the Ephesians were half-Greek and half-Asiatic, and their worship had many elements of a primitive fertility cult.

More Letters from Paul

Sometime before the riot at Ephesus, Paul had resolved to visit the churches in Macedonia and Greece. Then he planned to go to Jerusalem. After that—on to Rome.

Now he was on his way to Macedonia, traveling with Timothy. He preached in every town and gathered money to be taken back to the poor in Jerusalem.

At one place many eager listeners were crowded together in an upstairs room of a large house. They broke bread and ate together. Then Paul began to talk, intending to depart in the morning. The longer he talked, the more subjects came to his mind that he wanted to expound upon. The full moon rose outside the window. The stars began to twinkle in the dark sky. On and on Paul talked. Midnight came, and his voice droned on.

A young man sitting on the sill of an open window began to yawn and nod. Suddenly he toppled and fell.

Paul rushed down the stairs and outside, the rest of the assembly at his heels. The young man lay sprawled on the ground, his body quiet. Some of the women in the group began to weep loudly.

Paul crouched by the victim and lifted his head. "Do not be afraid," he said. "The young man's life is in him."

The young man's eyes opened. He looked around, startled. He wiggled his fingers and tested his legs. Then he smiled at Paul. The apostle embraced the young man and helped him to his feet.

When the company realized that the young man was unhurt, everyone clamored for Paul to go on with his discussion. They all trooped back upstairs and hungrily finished the remains of a large basket of wheat loaves and cheese. Then Paul continued to talk to them until daybreak.

While on this trip Paul received news about the congregation in Corinth. It was brought by the Gentile convert and missionary, Titus. So Paul wrote his second long letter to the Corinthians.*

With the loyal Timothy acting as scribe, Paul dictated many admonitions about their reported behavior. He also said:

Our hope for you is unshaken; for we know that as you share in our sufferings, you will also share in our comfort.

We do not want you to be ignorant of the afflictions we experienced in Asia. We were so utterly, unbearably crushed we despaired of life itself. Why, we felt that we had received the sentence of death; yet that was to make us rely not on ourselves but on God who raises the dead. He delivered us from so deadly a peril.

Even when we came into Macedonia on this trip, our bodies had no rest. We were afflicted at every turn—fighting without and fear within.

But God, who comforts the downcast, comforted us by the coming of Titus; not only by his coming, but also by his telling us of your longing, your mourning, your zeal for me; so I rejoiced.

We want you to know, brothers, that the churches of Macedonia, out of their joy and in spite of their own poverty, have overflowed in a wealth of liberality. They gave to the collection, as I can testify, according to their means, and beyond their means, of their own free will. They considered it a favor to take part in the relief of the Jerusalem poor. We have urged Titus to complete this gracious work among you. Remember that God loves a cheerful giver.

On my missions I have undergone great labor, many imprisonments, countless beatings, and often been near death. Five times I have received lashes with whips. Once I was pelted with stones and left for dead. Three times I have been shipwrecked. A night and a day I have been adrift at sea.

On frequent journeys I've been in danger from rivers, danger from robbers, danger from my own people and from Gentiles. There has been danger in the city and in the wilderness.

I have been through toil and hardship, through many a sleepless night, in hunger and thirst, often without food, in cold and bitter exposure. And, apart from other things, there is the daily pressure on me of my anxiety for all the churches.

Also, to keep me from being too elated when I have an abundance of revelations, a

359

360

"thorn in the flesh" was given me. Three times I pleaded with God about this, that this recurring pain in my body should leave me. But I learned to be content with God's grace.

For the sake of Christ, then, I am content with weakness, insults, hardships, and calamities.

Examine yourselves to see if you are holding to your faith!

I write this while I am away from you, in order that when I come I may not have to be severe in my use of the authority which the Lord has given me.

Brothers, farewell. Mend your ways. Heed my appeal. Agree with one another. Live in peace, and the God of love and peace will be with you.

(adapted)

Paul had never been to Rome. He knew that a Christian congregation had been established there, and he had long wanted to visit them. The tentmakers, Priscilla and Aquila, no longer feeling safe at Ephesus, had recently returned to Italy and joined that Roman church.

Rome could not be included on this trip. First the collection must be finished, the money taken back to Jerusalem.

In the meantime Paul wanted to establish a bond with these more distant Christians in the empire's great capital. So he dictated a letter to the Romans:

Paul, a servant of Jesus Christ,
To all God's beloved in Rome, who are called to be saints:
Grace to you and peace from God our Father and the Lord Jesus Christ.
First I thank my God through Jesus for all of you, because your faith is proclaimed in all the world. As God is my witness, I mention you always in my prayers, asking that somehow by God's will I may now at last succeed in coming to visit you.

I long to see you, that I may give you some spiritual gift to strengthen you, and that we may be mutually encouraged by each other's faith.

I have often intended to come to you, but thus far have been prevented. I am eager to preach the gospel in Rome.

(adapted)

Paul wrote about the reports of idolatry and immorality in Rome. He spoke at length about the need for faith and the understanding of God's purposes. His letter continued:

I am sure that neither death, nor life, nor things present, nor things to come, nor height, nor depth, nor anything else in all creation will be able to separate us from the love of God in Christ.

I appeal to you, brethren, to present your bodies as a living sacrifice, holy and acceptable to God. Do not try to conform yourself to this age. Instead be transformed by the renewal of your mind, so you may prove what is the will of God, what is good and acceptable and perfect.

By the grace given to me I bid everyone among you not to think of himself more highly than he ought to think.

Having gifts that differ according to the grace given us, let us use them: if prophecy, in proportion to our faith; if service, in our serving; he who teaches, in his teaching; he who contributes, in liberality; he who gives aid, with zeal; he who does acts of mercy, with cheerfulness.

Bless those who persecute you; bless, and do not curse them. Rejoice with those who rejoice, weep with those who weep.

Live in harmony with one another. Do not be haughty, but associate with the lowly and give yourself to humble tasks. Never be conceited.

Repay no one evil for evil, but take thought for what is noble in the sight of all.

Beloved, never avenge yourselves, but leave it to the wrath of God. For it is written, "Vengeance is mine, I will repay, says the Lord."

Do not be overcome with evil, but overcome evil with good.

All the commandments are summed up in this sentence: "You shall love your neighbor as yourself." Love does no wrong to a neighbor.

I know and am persuaded in the Lord that nothing is unclean in itself. It is unclean for anyone who thinks it unclean.

Let us pursue what makes for peace and upbuilding.

Since I have longed for many years to come to you, I hope to see you in passing as I go to Spain, and to be sped on my journey there by you, once I have enjoyed your company for a little.

At the present, however, I am taking aid to Jerusalem. When I have delivered it I shall go to Spain, stopping to see you on my way.

Give my greeting to Priscilla and Aquila, who are my fellow workers in Christ. They risked their necks to save my life. I am grateful to them and so are all the Gentile congregations. Greet also the congregation at their house.

(adapted)

When the contributions had been gathered from his churches at Philippi, Thessalonica, and Corinth, Paul prepared to sail back to the Holy Land. In farewell to his friends the apostle said with prophetic sadness, "I go to Jerusalem, not knowing what will happen—except that the Holy Spirit tells me that imprisonment and afflictions await me." Then his weatherbeaten face lighted with a radiant smile. "But my life is of no value to myself. I only want to accomplish my ministry for the Lord Jesus."

Paul's friends wept and embraced him, feeling that they would never see him again.

Then they watched him board his ship.

Acts 19:21-22; 20:2 through 21:8
The Letter of Paul to the Romans
The Second Letter of Paul to the Corinthians

* Some biblical scholars think that 2 Corinthians is the combination of more than one letter, perhaps as many as three. The earliest short letter which the apostle wrote to the Corinthian church may be part of the book (see note, p. 355). Paul's second long letter is the book of 1 Corinthians. A third letter may be found in 2 Corinthians 10 through 13. A fourth letter seems to be in 2 Corinthians 1:1 through 6:13; 7:2 through 9:15.

Arrested by the Romans

When he landed at the port of Caesarea, Paul went directly to the house of Philip the evangelist for a brief rest.

Friends told Paul, "If you go to Jerusalem the Jews will deliver you into the hands of the Gentiles." With tears in their eyes, they begged Paul not to go.

Paul said, "What are you doing, weeping and breaking my heart? I am ready to be imprisoned, even to die for the name of the Lord Jesus."

Luke had made the trip back to Judea with Paul. He saw that Paul would not be persuaded. "The will of the Lord be done," the writer-physician sighed.

The Christians in Jerusalem received Paul gladly. They listened to the story of his perilous missions. They commended him for his work.

But when some zealous unconverted Jews saw Paul worshiping at their temple, they stirred up a crowd. Several angry men grabbed Paul and held him. "Men of Israel, help!" they shouted. "Here is the man who teaches men to forsake Moses and our laws."

"Not only that," a rabble-rouser cried out shrilly, "but once he actually brought a Gentile into the temple and defiled this holy place!"

This was a lie, but the excited people believed it. The crowd swelled. Tempers rose to the boiling point over Paul. The aging missionary was dragged out of the temple. The mob clearly intended to kill him.

Word of the riot came to a high-ranking Roman officer. He at once took soldiers and centurions and ran to the temple. When the crowd saw the Roman soldiers, they stopped beating Paul. They pushed their victim toward the soldiers, shouting, "Arrest this man! He is the one who started this riot!"

The Roman officer ordered Paul to be bound with chains. "Now who is this man?" he asked, trying to quiet the surging mob. "And what has he done?"

The whole crowd was shouting at once. Some were yelling one thing, others something different. The officer could not learn the facts because of the uproar.

"Away with him," the people yelled. "Away with the heretic, the traitor! He ought not to live." They jeered, waved their garments, and threw dust in the air.

The tribune* gave an order for Paul to be brought into the barracks for questioning. To get the truth the prisoner was tied up to be whipped. But when he had been bound with thongs Paul said to the centurion in charge, "Is it lawful for you to scourge a man who is a Roman citizen, one who has not had a fair trial nor been condemned?"

The centurion rubbed his smoothshaven chin in puzzlement. He decided to report this to his superior officer.

The tribune came to question the prisoner himself. "Are you really a Roman citizen?"

"Yes," Paul said. "I was born a citizen."

At this the officer was afraid that he had gone too far in punishing the man without a trial. He unbound Paul but kept him under guard in the barracks.

In the meantime the Jews planned to call a formal meeting of their council and to ask to have the prisoner brought before them. The real purpose was to ambush the party escorting Paul from the Roman prison and to kill the apostle. The forty or more conspirators in this scheme took a solemn oath not to eat or drink until Paul was dead.

Word of this plot was taken to the

tribune, and he immediately laid plans of his own to conduct Paul in safety to Caesarea. In the middle of the night the prisoner was led quietly from the barracks, mounted on a horse, and with an escort of two hundred soldiers, seventy horsemen, and two hundred spearmen, was taken to the provincial capital.

The centurion in charge carried a letter which read:

To his Excellency the governor Felix, greeting:

This man was seized by the Jews, and was about to be killed by them, when I came upon them with the soldiers and rescued him, having learned that he was a Roman citizen. I found that he was accused about questions of their law, but charged with nothing deserving death or imprisonment.

When Felix read this letter, he asked his new prisoner, "To what province do you belong?"

"Cilicia," Paul answered. "I was born there in the city of Tarsus."

"Keep this man in custody," Felix ordered his own soldiers. To Paul he said, "I will hear you when your accusers arrive."

Soon the high priest, along with a commission of elders, Pharisees, and Sadducees, came down to Caesarea from Jerusalem. They went to the great marble palace and laid before the governor their case against Paul.

"He is a pestilent fellow, and an agitator everywhere," they complained. "He's a ringleader of the sect of the Nazarenes. He even tried to profane the temple, but we seized him. Examine him yourself, and you will find this is true."

But Paul gave such an eloquent account of his work that Felix sent the accusers away. "I will await the arrival of the tribune before I decide the case," he told them.

In the meantime Felix kept Paul under light custody. Several times he visited Paul, listening to the arguments of the aging missionary.

There was also another reason why Felix put off passing judgment. He hoped that in time Paul's friends could raise enough money to offer a bribe for his freedom.

So Paul's custody dragged on for two years. Governor Felix was succeeded by another administrator named Festus.

When Festus made his first trip to Jerusalem the Jewish leaders from the temple hurried to petition that something be done about Paul.

"Send a delegation," Festus decided, wanting to be fair to all. "If there is anything wrong with the man, let him be accused face to face."

In Caesarea, Festus took the judge's seat and ordered Paul to be brought before the court. His accusers, all talking at once, began to make charges which they could not prove.

Paul replied, "Neither against the law of the Jews, nor against the temple, nor against Caesar have I offended at all."

Festus asked Paul, "Do you want to go to Jerusalem and be tried?"

Paul objected. He knew that he could not receive a fair trial in Jerusalem, where religious antagonisms were so strong. "I am standing before Caesar's tribunal, where, as a Roman citizen, I ought to be tried." He pointed accusingly at the Roman governor. "You know very well that I have done no wrong to the Jews. I do not beg off from dying. If there are still charges against me which you cannot dismiss, then I appeal to Caesar."

Festus said, "You have appealed to Caesar; to Caesar you shall go!"

But as Festus later explained to the tribune, Paul could have been freed quietly once the Jerusalem temple rulers had calmed down. If only the man had not appealed to the emperor.

Now the die was cast. Paul and several other prisoners were put in the keeping of a Roman centurion. They all boarded a mercantile ship, which would sail northward, then westward along the Asiatic coast.

Paul was finally on his way to Rome; but he went as a prisoner, shackled to a soldier. Would he be set free by the emperor? Or would he receive a death sentence?

Acts 21:8 through 27:2

* A tribune was the Roman military officer in charge of a cohort, about six or seven hundred men.

A Perilous Journey

Paul did not have to sail back to Rome without good friends near him. Since he was not on a military ship, Luke and another disciple were able to book passage also. They were determined to stay with their leader and to give him comfort on the way if they could.

Julius, the centurion in charge, treated his prisoners kindly. Paul was allowed the freedom of the little ship. He could talk to his two loyal companions.

Around the island of Cyprus they sailed, past the very familiar ports of Antioch and Tarsus. They passed the bay leading to Perga, where he and Barnabas once had landed; there they had parted with the homesick John Mark. Mark proved to be a good apostle for Christ after that incident. Now Paul thought warmly of the young man.

At last the ship docked at its destination, the port of Myra, farther west on the coast of Asia Minor. Here the centurion found a cargo ship from Egypt that was sailing for Italy. The centurion bought passage for himself, his prisoners, and his soldiers. Paul's two companions boarded this ship also.

The ship spread its canvas and put out to sea. Since the late autumn headwinds were increasing, it was not possible to cross the Aegean directly. Instead, the captain steered south and southwest, around the eastern cape of Crete. He coasted with difficulty along its southern shores, until he came to the harbor of Fair Havens.

Much time had been lost, due to the fierce headwinds. Also it was almost the end of the sailing season. From now on the voyage would be very dangerous. Only the foolhardy would sail the stormy Mediterranean during the winter.

The owner of the ship was on board. He agreed with the captain that it would be impossible to reach Italy until the sea calmed down in the early spring. Neither could they stay here. Fair Havens was a shallow harbor; in spite of its name, it did not offer enough storm protection.

"Farther up the coast of Crete is a good harbor at Phoenix," the captain told the owner. "We could winter there."

Paul was standing nearby. He interrupted. "I have a foreboding. There will be loss and injury if we set sail now!" He looked out at the choppy sea. "Not only of the cargo and the ship, but also of our lives."

The owner of the ship ignored Paul. He ordered the captain to make a run for the safe harbor.

A gentle south wind began to blow. The crewmen hurriedly let out the sails and weighed anchor. The ship sailed along the south coast, close to the protection of the shore. But soon a tempestuous wind struck. The ship was caught and driven off course.

The ship was so violently storm-tossed, many of the passengers became seasick. Paul and the other prisoners helped the crew to throw the hampers of wheat, bolts of linen cloth, and other cargo overboard.

Still the tempest continued. Great waves rocked the ship and slammed across its already soaked decks. The timbers creaked and groaned, as if the boat would be smashed to pieces at any moment.

On the third day the captain was so desperate that he ordered all the valuable tackle of the ship to be cast overboard.

Day after day fierce rain lashed the little ship. The waves swelled, rocking it like a toy. Neither sun nor stars had been seen for many a day. At last all hope of being saved was abandoned.

Paul found the captain, the owner of the ship, and the centurion clinging to the mast. They were soaked with icy sea-spray and shaking with cold. Their faces showed they were ready to give up and die.

"Men, you should have listened to me," Paul sighed. "You should never have set sail from Crete."

The three men raised their eyes to the stormy sky, their bearded faces bleak.

"But now I bid you to take heart," Paul added cheerfully, "for there will be no loss of life among you." He glanced sympathetically at the ship's owner. "There will only be the loss of your ship and its goods." He was full of confidence in his prediction. "I had this news in a dream last night, by an angel of the God to whom I belong and whom I worship. So take heart, men. I have faith that it will be exactly as I have been told. But—" Paul frowned. He tried to peer through the blackness of the pelting rain, beyond the heaving waves. "We shall have to run onto some island."

When the fourteenth night had come, the ship was drifting across the southern part of the Adriatic Sea. It was about midnight, when the sailors suspected that they were nearing land. They made a sounding.

"It's twenty fathoms here," they reported to the captain. A little farther on they sounded again. "Only fifteen fathoms now, sir."

"We dare not go on much farther," the captain decided. "We might run on the rocks. Let out four anchors from the stern."

The sky barely was beginning to lighten when Paul saw the crew seeking to escape from the ship. The sailors had lowered the small rowboat into the sea under the pretense of laying out extra anchors.

Paul called to the centurion and the soldiers, "Make those sailors stay on the ship, or you may not be saved!"

It was too late. The soldiers were so angry that the sailors would desert the ship, they slashed off the ropes of the rowboat, letting it go free. Then they shouted curses at the escaping crew.

Paul brought up some bread from the hold and broke it, passing pieces around to everyone. "Today is the fourteenth day that you have continued in fearful suspense, and without food. Take some bread and eat," he insisted. "It will give you strength to live." Paul gave thanks to God and began to eat. And the passengers took courage and managed to force down some food themselves.

When it was day, they did not recognize the nearby land. But there was a bay ahead, with a beach.

"Do you suppose we can sail this ship safely into that bay?" they asked each other. Everyone agreed to lend a hand. They cast off the anchors, leaving them sunk in the sea. They loosened the ropes that lashed down the rudders. Hoisting the foresail to the wind, they tried to make a run for the beach. The vessel struck a shoal and ran aground. The bow stuck and remained immovable. The stern, whipped by the sea, began to break up in the surf.

The soldiers had put their heads together, saying, "We must kill our prisoners. Otherwise some will swim away and escape." But the centurion, who liked Paul and wished to save him, ordered the soldiers not to carry out their plan.

"Throw yourselves overboard," the centurion ordered everyone. "Make for the land. If any cannot swim, lash yourselves to planks, or pieces of the ship." In spite of the rolling, pounding surf, all escaped to land.

The castaways learned that this island was called Malta. When some islanders appeared they showed kindness for the wet and shivering survivors. A fire was kindled, so they might huddle about it and find some comfort. Paul had gathered a bundle of sticks and put them on the fire. Suddenly, while everyone watched with horror, a snake lashed out from the edge of the fire, fastening its fangs on Paul's hand.

The superstitious natives, seeing a man with an ankle chain and the snake fastened on his hand, said to one another, "That man must be a murderer. He escaped from the sea, but justice will not allow him to live."

Paul coolly shook off the creature into the fire. There were ugly marks on his hand. The islanders waited, knowing no remedy for snakebite. They expected Paul's hand to begin to puff and swell up. They thought that very soon he would topple over dead.

But the next morning came, and the white-haired prisoner was just as spry and chipper as before. The natives were astounded. "Look, no misfortune has come to him!" they exclaimed among themselves, pointing to Paul. "Surely this man is a god!"

For three months the shipwrecked survivors lived on the island of Malta, and Paul found many opportunities to tell the good news about Jesus Christ there.

It was early in the spring when the almond trees were budding and the seas were more calm, before the travelers continued on their way. The centurion and the soldiers herded their manacled prisoners on board a grain ship which arrived from Alexandria. It was sailing directly to the west coast of Italy.

Paul's life was again in jeopardy for now he faced the law courts of Rome.

Acts 27:1 through 28:12

Paul's Last Journey

The Alexandrian grain ship made its way through the straits between the island of Sicily and the tip of Italy, and on up the west coast. At last the vessel sailed into a great bay, the Bay of Naples.

Paul, and the soldier to whom he was chained, stood near the prow of the boat. Julius the centurion began to point out the sights to Paul. "See that towering, cone-shaped mountain just back of the bay?" said the Roman officer, pointing. "That's Vesuvius. There on its slopes is the charming town of Pompeii. I was born there. Someday I want to retire to a farm there."

The ship headed for the northern part of the bay. "See that small bay, ahead, within this larger bay?" Julius went on. "It teems with oysters. They are taken to Rome and sold to the wealthy."

The wind was coming from the north. "What is that smell?" Paul asked.

Julius pointed. "Hot sulphur springs. Public baths there. Romans come down here for the healing waters or just for rest and relaxation. Over there is the port of Puteoli, very famous. See the people lined up along the wharves just beyond the lighthouse? Whenever word goes through Puteoli that a ship is arriving from Egypt, the curious come out to watch it sail in."

Word had also reached Puteoli that the apostle Paul was arriving on this ship. Among the spectators on the docks was a knot of Christians. "Stay with us for a week," they urged Paul, after making themselves known to him. "We have heard of the terrible trials of your trip." They turned to petition the centurion in charge of Paul.

The centurion could not let Paul stay that long. But he was a courteous and kind man, and he felt that Paul had helped them all escape death from the storm. So he allowed the shackled apostle to visit his new friends for a short while.

"Word has been sent to the brethren in Rome, too," these Christians of Puteoli told Paul. "They are sending a delegation to meet you."

Paul's party then went on, taking a crossroad to the famous Appian Way. Beside it were the soaring arches for the aqueduct that bore water into Rome.

At the Three Taverns inn outside Rome, a delegation of Christians from the Roman church had come out to meet Paul. Some were strangers, others were old friends who had recently moved to Rome from Asia Minor, Macedonia, or Greece.

When Paul saw them he thanked God, tears streaming down his weatherbeaten cheeks. He embraced them clumsily, for one hand was chained to a soldier. Paul's spirits rose, and he took courage.

So they came together upon the last ridge of hills, which overlooked Rome.

In the middle of a broad valley the great capital city sprawled thickly across seven low hills, almost obliterating them. Around the city spread the suburbs; beyond that were villages and the villas and gardens of the wealthy. From every direction, roads were converging upon Rome.

Paul and the other prisoners were marched into a military camp and turned over to the commander of the praetorian guard.* It was the job of this Roman official to keep in custody all accused persons who were to be tried before the emperor.

The letter concerning Paul's crime was read by the commander. "Is this man a dependable prisoner?" he asked Julius.

The centurion who had brought Paul to Rome answered, "He is a very honorable man, a model prisoner."

"Then he may live in his own rooms while awaiting trial," the commander decreed.

Paul rented a small house in the district beyond the river. He continued to wear a manacle on one wrist, with a light chain. This kept him always within sight and hearing of the guards. Yet Paul could invite in friends, and talk far into the night with them.

He preached to curious Jews who visited him. A few were converted. He bolstered the faith of the local Christians with his own boundless enthusiasm and confidence. They, in turn, saved him from loneliness.

He preached to Gentiles who heard about him and wandered in from the crowded, narrow streets. Many joined the growing Christian congregation of Rome.

For two years Paul lived there at his own expense, welcoming all who came to him. He preached the kingdom of God and taught about the Lord Jesus Christ quite openly and unhindered.

In all this time Nero the emperor did not grant Paul a hearing.

During these two years Paul's accusers in Jerusalem could have gathered much evidence, real or false, for a menacing case against him. What happened to the aging, zealous missionary? ** Was he brought to trial, condemned, and executed? Was he tried, and found not guilty of breaking Roman law? If so, was he able to complete his dreamed-of missionary journey to Spain? And did he then return to Rome, where he somehow angered the authorities and was again imprisoned?

Circumstantial evidence points to the possibility that Paul was imprisoned a second time. We do know that Nero's hatred of the Christians burst out in a violent persecution three years later, in the year 64. Like many other Christians, Paul may have been brought to trial on some charge, and condemned to death.

Acts 28:11-30

* The praetorian guard was composed of the soldiers in Rome who were the bodyguards of the emperor.

** It was at this point, probably about A.D. 61, that Luke ended his story in Acts, leaving his readers in suspense. Did the author intend to write more telling the rest of Paul's story?

Paul's Last Letters

While he was still alive and a prisoner at Rome, Paul continued to write letters.

In the heart of Asia Minor, about a hundred miles inland from Ephesus, was the town of Colossae. Paul did not report visiting this place; but a friend had evangelized there and started a Christian church.

Now this friend was disturbed about the Colossians. "They are superstitious," he told Paul, "and much influenced by pagan Asiatic cults. They believe in all kinds of demons and angels. They even think them to be as important as Jesus Christ! They know of you, and they would respect what you would say. Why not write them a letter?"

So Paul dictated an epistle to the Colossians. He emphasized the central and sole importance of Jesus Christ to Christianity. And he warned them:

See to it that no one makes a prey of you by philosophy and empty deceit. Let no one insist on worship of angels, taking his stand on visions, puffed up without reason by his sensuous mind.

Put away anger, malice, slander, and loud abuse from your mouth. Do not lie to each other. Let the peace of Christianity rule in your hearts.

Wives, be subject to your husbands. Husbands, love your wives. Children, obey your parents. Fathers, do not provoke your children, lest they become discouraged. Slaves, obey your masters. Masters, treat your slaves justly and fairly, knowing that you also have a Master in heaven.

If John Mark, the son of Barnabas' sister, comes to you, receive him. Mark has lately been a great comfort to me. Luke the beloved physician greets you.

(adapted)

When this had been dictated, Paul reached for a pen to add his usual postscript. The shackle chafed the indented place on his wrist, and the chain clanked as his hand moved across the page. "I, Paul, add this final greeting in my own hand. Remember my fetters. Grace be with you."

While under arrest, Paul wrote a short personal letter about a runaway slave. It was also an open letter to a congregation, which met in the slaveowner's house, asking for support in the matter.

The runaway slave was Onesimus, a young man who came in from the street to listen to Paul's teachings. Paul converted Onesimus to a faith in Christ, and he became attached to this young man who bore the mark of a slave, a hole in one ear.

Many a long night Paul and the youth discussed whether it was morally necessary for a slave to return to his master. As long as he was a runaway, his life was in danger. He would be executed by the Roman authorities if caught.

Why did Paul not denounce the whole institution of slavery?

In the first place, there was little market for free labor in those days. Most artisans and craftsmen were slaves, who sometimes could build up a sizable fortune for themselves while still receiving security and support from their masters. A gift of freedom was not always a blessing. Many freedmen became hangers-on of the wealthy, or eked out a precarious living on the charity of a public dole.

It was true that Paul and all leaders of Christianity gave slaves full rights in the church. Slaves worshiped on the same terms as their masters; they were considered equals in the sight of God.

Above all, Paul and the early Christians still looked forward to an early second coming of Christ. What need for long-term social reform, when everyone believed that the old order would soon be swept away? While waiting for the new kingdom of God to be established, Paul felt that the primary need was to live one's life with Christian virtue.

So Paul and Onesimus together decided that the young slave should return to his Christian master, Philemon. He should go back, even though the previous service was unsatisfactory to all concerned. Onesimus admitted that he had caused his master some loss.

Paul wished he could accompany the slave on his return. Of course this was impossible. So he sent along a messenger with this persuasive letter:

Paul, a prisoner for Christ Jesus,

To Philemon, our beloved worker, and the church in your house:

I remember you in my prayers, because I hear of your love, and the faith which you have toward the Lord Jesus.

Accordingly, though I am bold enough in Christ to command you to do what is required, yet for love's sake I prefer to appeal to you.

I, Paul, an old man, appeal to you for my child, Onesimus, whose father I have become in my imprisonment. (Formerly he was useless to you; but now he is indeed useful to you and to me.) I am sending him back to you, sending my very heart. I would have been glad to keep him with me, in order that he might serve me on your behalf during my imprisonment for the gospel; but I preferred to do nothing without your consent in order that your goodness might not be by compulsion, but of your own free will.

Perhaps this is why he was parted from you for a while, so you might have him back forever—no longer as a slave, but as a beloved brother.

If you consider me as your partner, receive him as you would receive me. If he has wronged you at all, or owes you anything, charge that to my account. I, Paul, write this with my own hand: I will repay it—to say nothing of your owing me even your own self.

Confident of your obedience, I write to you, knowing that you will do even more than I say.

(*adapted*)

It was possible, under Roman law, for an owner to beat his slave almost to death, especially for running away; it did not matter that the slave returned on his own.

However, since the letter was saved and cherished, there was probably a happy ending. Philemon no doubt responded with the generosity which Paul expected of him. We also have the letter of a later Christian writer, who mentions a Bishop Onesimus at the church of Ephesus. It seems quite likely that this was the same slave who was nurtured in the faith by Paul!

Not long before his death Paul wrote a

letter to his beloved Macedonian converts in the church at Philippi.

The Christians there had mourned Paul's imprisonment. They made personal sacrifices and collected money to help toward his expenses. They also sent one of their congregation to help him.

Through Paul's letter to the Philippians runs a song of joy, the philosophy of a mellowed old evangelist:

I thank my God for my memories of you. Every prayer of mine for you is joyful. I am thankful for your partnership in the gospel from the first day until now.

I want you to know, my brothers, that what has happened to me has really served to advance the gospel. It has become known throughout the whole praetorian guard, and to many more, that my imprisonment is for Christ. Now my fellow Christians are much more bold to speak the word of God without fear.

Let your manner of life be worthy of the gospel of Christ, so that whether I come and see you or am absent, I may hear that you stand firm in one spirit.

Do all things without grumbling or questioning. Be blameless and innocent in the midst of a crooked and perverse generation, among whom you shine as lights in the world. Then in the day of Christ, I can be proud that I did not run in vain or labor in vain.

Forgetting what lies behind and straining forward to what lies ahead, I press on toward the goal—for the prize of the upward call of God.

Rejoice in the Lord always. Again I will say, Rejoice!

Have no anxiety about anything. Let your thanksgiving and your requests be known to God. And the peace of God, which is beyond all understanding, will keep your hearts and minds in Christ Jesus.

Finally, brethren, whatever is true, whatever is honorable, whatever is just, whatever is pure, whatever is lovely, whatever is gracious, if there is any excellence, if there is anything worthy of praise, think on these things.

I rejoice that you were concerned for me. Not that I complain of want; for I have learned, in whatever state I am, to be content.

I hope to send Timothy to you soon, so that I may be cheered by news of you. I have no one like him, who will be genuinely anxious for your welfare. Timothy's worth you know; he has serve me as a son.

(adapted)

At last Paul's imprisonment ended in death. The young Pharisee from Tarsus, converted in a blinding experience on the Damascus road, was dead. Paul had not labored in vain. He had opened the doors of the Christian church to the whole world.

PART IX
THE REST OF THE NEW TESTAMENT

Paul's letters, whether to the churches or to individuals, were prompted by definite issues of a particular time and place.

Toward the end of the first century these letters began to be collected. Each church had made and passed around its own copies. Some of Paul's advice had such continuing and universal interest that the churches began to read aloud from Paul's letters.

Several versions of the Gospels were in circulation by now. Taken together, the stories about Jesus and the writing of Paul were valuable literature for the new church.

The collected scriptures of the Jews were also read in the Christian congregations and greatly respected. These Hebrew writings were God's Old Testament, his will for the Jews. The Letters of Paul and the Gospels formed the core of what came to be known as the New Testament. This was the record of God's will for mankind, Gentile as well Jew.

After Paul's death other disciples continued to produce essays and letters which the Christian church prized. In time, many of these would be added to Paul's Letters and the Gospels to form our completed New Testament.

Some Later Letters

The Letter to the Ephesians contains many ideas that Paul included in his letters to the other churches. The principles set out are very general. So the letter may well have been sent out to several young and struggling Christian congregations.* The writer emphasizes that Jesus Christ by his life and death for all mankind has broken down the walls that separate men. All people who believe in him become welcome members of the household of God.

The letters to Timothy and Titus are addressed to individuals rather than to a church. These three letters contain advice and counsel from a pastor to these young workers and so are called the Pastoral Letters.** They contain admonitions and warnings about heresies, and rules for church administration.

Paul's Second Letter to Timothy contains some interesting glimpses into Paul's last days of imprisonment in Rome:

> As I remember your tears, I long night and day to see you, so I may be filled with joy.
>
> I am already on the point of being sacrificed. I have fought the good fight, I have finished the race, I have kept the faith.
>
> Do your best to come to me before winter. Titus has gone to Dalmatia. Luke alone is with me. Get Mark and bring him with you. When you come, bring the cloak that I left, also the books, and above all the parchments.
>
> *(adapted)*

In the Letter to Titus, Paul deals with false teachers and heretics in the church. The epistle warns, "There are all too many men who are not willing to subordinate themselves to authority. They are given to wild talk, and they lead the true believers astray. They must be silenced, since they are upsetting whole families by teaching what they have no right to teach."

The title and the formal ending of the Letter to the Hebrews*** do not disguise the fact that this book is really an essay or a collection of sermons.

The writer is anxious to prove to the Jews that Jesus is the Messiah to which the Hebrew scriptures have pointed. And he uses examples and arguments that would be appreciated by the Jews. He writes of the great faith of the patriarchs: Abraham and Isaac, Jacob and Joseph, Moses and Joshua. It was faith in the still unseen Messiah that sustained these fathers of Israel:

> For time would fail me to tell of Gideon and Samson, of David, Samuel, and the prophets. Through faith they conquered kingdoms, enforced justice, stopped the mouths of lions, escaped the edge of the sword, won strength out of weakness. Some were tortured, others suffered mocking and scourging, even chains and imprisonment. They were stoned, they were sawn in two, they were killed with the sword.
>
> Since we are surrounded by so great a cloud of witnesses, let us also lay aside every weight, and let us run with perseverance the race that is set before us, looking to Jesus the pioneer and perfecter of our faith.

Lift your drooping hands, strengthen your weak knees, and make straight paths for your feet. Strive for peace with all men!

(adapted)

The Letter of Paul to the Ephesians
The First and Second Letters of Paul to
Timothy
The Letter of Paul to Titus
The Letter to the Hebrews

* Biblical scholars are not agreed whether the Letter to the Ephesians was actually written by Paul. Some believe that a circular letter was sent to many of the churches, and this is the copy addressed to the church at Ephesus. Other scholars think that it is a collection of Paul's sayings and ideas, put into the form of a letter.

** Traditionally these three were supposed to be from the hand of Paul. However, some scholars think that a later pastor, who admired Paul and who had taken over the apostle's task of guiding the growing Christian churches, may have written it in his name.

*** The Letter to the Hebrews was probably written about A.D. 85 to 95. Though it is written in the name of Paul it is not in his style, and most scholars agree that the author was someone else.

Letters from Other Apostles

When Paul was in military custody in Rome he never mentioned Peter's presence. Yet it is believed that Peter, too, died in Rome during Nero's persecutions.

Perhaps Peter arrived in Rome soon after Paul's death. Perhaps he, too, was tried and condemned. Tradition says that the stalwart old apostle was crucified upside down and was buried in Rome.

In the First Letter of Peter,* Christians are encouraged to be brave in dangerous times:

Rejoice, though now for a little while you may have to suffer various trials.

Gird up your minds, be sober, set your hope fully upon the grace that is coming to you at the revelation of Jesus Christ.

Above all hold unfailing your love for one another, since love covers a multitude of sins.

(adapted)

The Letter of James** provides a balance to the ideas of Paul. Paul emphasized that converts would be saved by their faith. Too many Christians took this to mean that one could ignore responsibility for righteous living. But James, with great moral earnestness, wrote:

Be doers of the word, and not hearers only. Be a doer that acts.

What good is it, my brothers, if a man says he has faith, but has done nothing to .prove it? Without works, can his faith save him?

If needy persons are ill-clad and in lack of daily food, and one of you says to them, "Go in peace, be warmed and filled," without giving them the things needed for the body, what good does it do?

So faith alone, without works or action, is dead.

(adapted)

Perhaps the most loved of the non-Pauline letters is the First Letter of John.*** It contains memorable words on

376

the meaning of love:

> We have passed out of death into life, because we love our brothers. Anyone who hates his brother is a murderer, and you know that no murderer has eternal life within him.
>
> Anyone who has the world's goods and sees his brother in need, yet closes his heart against him, how does God's love dwell in him?
>
> Dear friends, let us love one another; for love is of God, and he who loves is a son of God and knows God.
>
> He who does not love does not know God; for God is love. If anyone says, "I love God," and hates his brother, he is a liar. For he who does not love his brother whom he has seen, cannot love God whom he has not seen.

(adapted)

The Letter of James
The First Letter of Peter
The First Letter of John

* Many scholars believe, as did the early Christians, that the First Letter of Peter was written by the apostle himself. Though it is written in polished literary Greek, the unsophisticated Galilean could have used the services of a skilled Greek secretary and translator. Since Peter was probably martyred around A.D. 64-67, the book may have been written in Rome at about that time.

The Second Letter of Peter does not seem to be by the same author. He could have been a later person who admired the apostle and wished to offer instruction in his name. Much of 2 Peter repeats the ideas in Jude, the last epistle in the Bible; both are concerned about teachers of false Christian beliefs, in the second century.
** It was once thought that the James who wrote this letter was the brother of Jesus, but scholars now believe that it was written as late as A.D. 125 to 150, and by another James.

*** Traditional beliefs are that this first letter, the two short ones which follow, as well as the Revelation to John, all were written by John the gospeler. A number of modern scholars feel that the writing styles are different enough to question this. It is generally agreed that the letters of John were written about A.D. 100.

The Revelation to John

It was a bad time for Christians. They were being systematically persecuted and killed. In earlier times the Roman emperors had usually been tolerant of the many religions in their vast empire. As long as they obeyed the Roman laws Jews and Christians had freedom to worship.

During the reign of Nero in A.D. 64, a disastrous fire wiped out much of the city of Rome. Rumors began to spread that Nero himself had started the fire.

Some citizens who were irked by the growing Christian religion cried out that the Christians had started the fire. Nero took advantage of this second rumor to keep the accusations away from himself. He blamed the Christians, arresting, torturing, and killing many of them. Paul and Peter may have been martyred at this time. This persecution was confined to Rome. Christians in other parts of the empire continued to have freedom of worship.

Two years later the ever-restless Jews in Jerusalem finally rose up in a massive rebellion. Four years later they were crushed by a Roman general with the common Latin name of Titus. Proud Jerusalem was razed, and the temple was destroyed. Many Jews were killed or enslaved, and the rest were scattered.

Yet when this same Titus was later made emperor he turned out to be a benevolent ruler. During his reign there were three great disasters. The volcano Vesuvius suddenly erupted and buried Pompeii with all its inhabitants. Another great fire broke out and devastated Rome. Finally a dread pestilence stalked the stricken capital. Titus spent his own fortune to add to the public funds for relief.

Unfortunately for the Christians the successor to Titus was aggressive and ambitious. The emperor Domitian looked on Christian beliefs as a contagious superstition that must be wiped out. He ordered the persecution of Christians everywhere in the empire.

Centuries before, a poet and prophet had written the book of Daniel to encourage all Jews who were persecuted by the Syrian Greeks. Now John, imprisoned on the island of Patmos, wrote the Revelation* to encourage persecuted Christians throughout the Roman Empire.

The author of Daniel had used allegory and symbolism to hide his message from the persecutors. He knew that faithful readers would understand the real meaning and take heart. The author of Revelation used allegory and symbolism for the same purpose.

In the Revelation the city of Rome is the personification of evil. The doom of this world capital is foretold. It is another "Babylon" to the persecuted Christians.

Fallen, fallen is Babylon the great!
It has become a dwelling place of demons,
For all nations have fallen by the wine of her
    impure passion,
And merchants have grown rich with the
    wealth of her wantonness.

Come out of her, my people,
Lest you take part in her sins,
Lest you share in her plagues,
For her sins are heaped high as heaven.

She shall be burned with fire!
Alas, alas, for the great city
That was clothed in fine linen, in purple and
    scarlet,
Bedecked with gold, with jewels, and with
    pearls!

The themes of revenge and the visions of war sound strange to our ears, but the Christians were going through such terrible trials that these ideas were a part of the times.

And for all their suffering, the faithful faced it singing. From the beginning Christian congregations had been singing, and the Revelation has many fragments from early hymns:

Holy, holy, holy, is the Lord God Almighty,
Who was and is and is to come.

Worthy art thou, our Lord and our God,
To receive glory and honor and power,
For thou didst create all things;
By thy will they existed and were created.

The Revelation to John

* Little is known about this John. A poet and prophet, he seems to have been the pastor of seven churches in Asia Minor.

POSTSCRIPT TO THE NEW TESTAMENT

Sometime after the beginning of the second century, the four gospels were copied together as one unit. They were added to the letters of Paul which were already circulating. These became the official literature in the churches.

By the end of the second century the three Pastoral Epistles had been added to the previously accepted letters of Paul. The First and Second Letters of John and Jude were added. The Revelation was in many collections. The term "New Testament" was used to cover these generally accepted books.

In the third century Eusebius, a famous bishop of Caesarea in Palestine, made up a list of the twenty-seven books that we accept in our New Testament. He notes, however, that James, Jude, 2 Peter, and 2 and 3 John are still disputed; Revelation is to be included "if it seem proper."

Even in the fourth century there continued to be differences among Christian church leaders and scholars as to whether Hebrews, Revelation, and the non-Pauline letters should be in the canon.

At the opening of the fifth century, the head of a monastery in Bethlehem, Jerome, translated the Old and New Testaments into Latin. Since the Bible could now be read in the common "vulgar" tongue of the people, this Latin version was called the Vulgate.

During the barbarian invasion of Rome and the dark and troubled times that followed the collapse of the empire, the Bible was preserved in this Latin version. Most of the copies were in monasteries, where the monks carefully wrote out the sacred books and kept them safe from the turmoil that swept through the world.

By the time order began to emerge in Europe, about 800, the people were no longer able to read Greek and Latin and were dependent on the priests and teachers of the church to tell them what was in the Bible.

In the fourteenth century, John Wycliff and one of his pupils translated the Bible from the Vulgate into English. Each copy had to be printed by hand, taking ten months of hard labor. Very few people could afford a copy.

Then the printing press was invented.

Early in the sixteenth century William Tyndale took as his lifework the translation of the Bible from the early Greek directly into everyday English. He had finished the New Testament and was halfway through the Old, when he was imprisoned. He was condemned to die, and the executioner of King Henry VIII carried out the sentence by strangling him. Afterward his body was burned. Such was the opposition of many to making the Bible available to the individual.

Restrictions relaxed in time. King Henry finally approved a new version of the Bible in English. A number of translations followed. Best known is the one approved by King James in 1611.

In the late nineteenth century a group of scholars convened in England to bring the Bible up to date. Since the time of the King James Version many English words had become obsolete or had changed meanings. Also old manuscripts were continually being discovered that shed light on the original writings and suggested more accurate translations of words and phrases, and even whole ideas. When these men had finished with their labors they called their translation the Revised Standard Version.

The American scholars who assisted in this project published their own American Revised Standard Version in 1901. This RSV was again brought up to date in 1952. It is now commonly used in Protestant pulpits and church schools.

There are many other translations into English based on careful studies by individuals or groups of scholars. Some of the most popular of these are the Moffatt, the Smith and Goodspeed, and the J. B. Phillips translations. There is also the New English Bible.

The soaring literature of the Bible has shaped our language and inspired our cultural heritage. It is a treasury for those who hunger for spiritual riches.

PRONUNCIATION GUIDE

(containing significant proper names with selected
page references)

PHONETIC KEY

a	as in	**bad**
ay	as in	**say**
ah	as in	**father**
e	as in	**bet**
ee	as in	**me**
i	as in	**sit**
igh	as in	**high**
o	as in	**on**
oh	as in	**go**
oo	as in	**moon**
ow	as in	**cow**
u	as in	**use**
uh	as in	**hut**
eh	has the value of	
	a	in **alone**
	e	in **system**
	i	in **easily**
	o	in **gallop**
	u	in **circus**